MAGNETOHYDRODYNAMICS
of Laboratory and Astrophysical Plasmas

With 90% of visible matter in the Universe existing in the plasma state, an understanding of magnetohydrodynamics is essential for anyone looking to understand solar and astrophysical processes, from stars to accretion discs and galaxies; as well as laboratory applications focused on harnessing controlled fusion energy.

This introduction to magnetohydrodynamics brings together the theory of plasma behavior with advanced topics including the applications of plasma physics to thermonuclear fusion and plasma-astrophysics. Topics covered include streaming and toroidal plasmas, nonlinear dynamics, modern computational techniques, incompressible plasma turbulence and extreme transonic and relativistic plasma flows. The numerical techniques needed to apply magnetohydrodynamics are explained, allowing the reader to move from theory to application and exploit the latest algorithmic advances. Bringing together two previous volumes: *Principles of Magnetohydrodynamics* and *Advanced Magnetohydrodynamics*, and completely updated with new examples, insights and applications, this volume constitutes a comprehensive reference for students and researchers interested in plasma physics, astrophysics and thermonuclear fusion.

HANS GOEDBLOED is Advisor of the Dutch Institute for Fundamental Energy Research (DIFFER), and Professor Emeritus of Theoretical Plasma Physics at Utrecht University. He has been a visiting scientist at laboratories in the Soviet Union, the United States, Brazil and Europe. He has taught at Campinas, Rio de Janeiro, São Paulo, Massachusetts Institute of Technology, Katholieke Universiteit Leuven, Amsterdam Free University and Utrecht University. For many years he coordinated an interdisciplinary and large-scale computational effort with the Dutch Science Organisation on ‘Fast Changes in Complex Flows.’

RONY KEPPENS is Professor and Division Chair at the Centre for mathematical Plasma-Astrophysics, Katholieke Universiteit Leuven. He headed numerical plasma dynamics teams at FOM-Institute for Plasma Physics ‘Rijnhuizen’ (now DIFFER) and Leuven and frequently lectures on computational methods in astrophysics. His career started at the National Center for Atmospheric Research, Boulder, and the Kiepenheuer Institute for Solar Physics, Freiburg. He held a professorship at Utrecht University and a concurrent professorship at Nanjing University. His expertise ranges from solar to high-energy astrophysics and includes parallel computing and grid-adaptivity.

STEFAAN POEDTS is Professor and Chair of the Department of Mathematics at Katholieke Universiteit Leuven. He was a post-doctoral researcher at the Max Planck Institute for Plasma Physics, Garching, a senior researcher at the FOM-Institute for plasma physics ‘Rijnhuizen’, and research associate at the Centre for mathematical Plasma-Astrophysics, Katholieke Universiteit Leuven. His research interests include solar physics, space weather, thermonuclear fusion, MHD (in)stability, and multi-fluid modelling. He teaches basic math courses, and advanced courses on plasma physics of the Sun and numerical simulation.

MAGNETOHYDRODYNAMICS

of Laboratory and Astrophysical Plasmas

HANS GOEDBLOED
Dutch Institute for Fundamental Energy Research (DIFFER)

RONY KEPPENS
Centre for mathematical Plasma-Astrophysics,
Katholieke Universiteit Leuven

STEFAAN POEDTS
Centre for mathematical Plasma-Astrophysics,
Katholieke Universiteit Leuven

CAMBRIDGE
UNIVERSITY PRESS

CAMBRIDGE
UNIVERSITY PRESS

University Printing House, Cambridge CB2 8BS, United Kingdom

One Liberty Plaza, 20th Floor, New York, NY 10006, USA

477 Williamstown Road, Port Melbourne, VIC 3207, Australia

314–321, 3rd Floor, Plot 3, Splendor Forum, Jasola District Centre, New Delhi – 110025, India

79 Anson Road, #06–04/06, Singapore 079906

Cambridge University Press is part of the University of Cambridge.

It furthers the University's mission by disseminating knowledge in the pursuit of education, learning, and research at the highest international levels of excellence.

www.cambridge.org
Information on this title: www.cambridge.org/9781107123922
DOI: 10.1017/9781316403679

First published 2019

Printed in the United Kingdom by TJ International Ltd. Padstow Cornwall

A catalogue record for this publication is available from the British Library.

Library of Congress Cataloging-in-Publication Data
Names: Goedbloed, J. P., 1940– author. | Keppens, Rony, author. |
Poedts, Stefaan, 1962– author.
Title: Magnetohydrodynamics of laboratory and astrophysical plasmas / J.P. Goedbloed (Dutch Institute for Fundamental Energy Research (DIFFER)), Rony Keppens (Katholieke Universiteit Leuven, Belgium), Stefaan Poedts (Katholieke Universiteit Leuven, Belgium).
Description: Cambridge ; New York, NY : Cambridge University Press, 2019. |
Includes bibliographical references and index.
Identifiers: LCCN 2018029282 | ISBN 9781107123922
Subjects: LCSH: Magnetohydrodynamics. | Plasma astrophysics. | Astrophysics.
Classification: LCC QC718.5.M36 G63945 2019 | DDC 538/.6–dc23
LC record available at https://lccn.loc.gov/2018029282

ISBN 978-1-107-12392-2 Hardback

To Antonia, 陆蓉 (Rong Lu) and Małgosia

Contents

Preface

This book describes the two main applications of plasma physics, laboratory research on thermonuclear fusion energy and plasma-astrophysics of astronomical systems, from the single viewpoint of magnetohydrodynamics (MHD). This provides effective methods and insights for the interpretation of plasma phenomena on virtually all scales, ranging from the laboratory to the Universe. The key issue is understanding the complexities of plasma dynamics in extended magnetic structures. In the first half of the book, based on a revision of the previous volume *Principles of Magnetohydrodynamics* [1], the classical MHD model is developed in great detail without omitting steps in the derivations. This necessitated restriction to ideal dissipationless plasmas, in static equilibrium and inhomogeneous in one direction. In the second half of the book, based on a revision of the previous *Advanced Magnetohydrodynamics* [2], these restrictions are relaxed one by one: introducing stationary background flows, dissipation, two-dimensional toroidal geometry, linear and nonlinear computational techniques, turbulence, transonic flows and relativity. These topics transform the subject into a vital new area with numerous applications in laboratory, space and astrophysical plasmas. It is impossible to treat all topics that actually belong to the field of advanced MHD. Fortunately, books or chapters of books exist on some of those topics, like *dynamos* [444, 475, 174, 533], *solar magnetohydrodynamics* [510], *chaos* [649], *stellarators* [185], *spheromaks* [48] and *anomalous transport* [33, 662].

Inevitably, with the distinction between topics for Chapters 1–11 (mostly ideal linear phenomena described by self-adjoint linear operators) and Chapters 12–22 (mostly non-ideal and nonlinear phenomena), the difference between 'basic' and 'advanced' levels of magnetohydrodynamics could not strictly be maintained. The logical order required a quite advanced derivation of the MHD equations from kinetic theory (Chapter 3) at an early stage, different sections on advanced topics interspersed throughout the book, and rather complicated analyses of the initial value problem (Chapter 10) and flow in toroidal systems (Chapter 18). These parts are marked by a star (*) and can be skipped on a first study of the book. The same applies to text put in small print, in between triangles ($\triangleright \cdots \triangleleft$), usually containing tedious derivations or advanced material. The serious student is advised though not to skip the Exercises, which are also put in small print for typographical reasons. Frequent use of the vector expressions and tables of the Appendices is encouraged. Magnetohydrodynamics can only be mastered through intense practice.

An overview of the subject matter of the different parts and chapters of this book may help the reader to find his way.

Part I (Plasma physics preliminaries)

- Chapter 1 gives an introduction to laboratory fusion and astrophysical plasmas, and formulates provisional microscopic and macroscopic definitions of the plasma state.
- Chapter 2 discusses the three complementary points of view of single particle motion, kinetic theory and fluid description. The corresponding theoretical models provide the opportunity to introduce some of the basic concepts of plasma physics.
- Chapter 3 gives the 'derivation' of the macroscopic equations from the kinetic equations. Quotation marks because a fully satisfactory derivation can not be given at present in view of the largely unknown contribution of turbulent transport processes. The presentation provides some impression of the limitations of the macroscopic model.

Part II (Basic magnetohydrodynamics)

- Chapter 4 defines the MHD model and introduces the concept of scale independence. The central importance of the conservation laws is discussed at length. Based on this, the similarities and differences of laboratory and astrophysical plasmas are articulated in terms of a set of generic boundary value problems.
- Chapter 5 derives the basic MHD waves and describes their properties, with an eye on their role in spectral analysis and computational MHD. The theory of characteristics is introduced as a way to describe the propagation of nonlinear disturbances.
- Chapter 6 treats the subject of waves and instabilities from the unifying point of view of spectral theory. The force operator formulation and the energy principle are extensively discussed. The analogy with quantum mechanics is pointed out and exploited.

Part III (Standard model applications)

- Chapter 7 applies the spectral analysis of Chapter 6 to inhomogeneous plasmas in a plane slab. The wave equation for gravito-MHD waves is derived and solved in various limits. Here, all the intricacies of the subject enter: continuous spectra, damping of Alfvén waves, local instabilities, etc. The topic of MHD spectroscopy is shown to hold great promise for the diagnostics of plasma dynamics.
- Chapter 8 introduces the enormous variety of magnetic phenomena in astrophysics, in particular for the solar system (dynamo, solar wind, space weather, etc.), and provides basic examples of plasma dynamics worked out in later chapters.
- Chapter 9 is the cylindrical counterpart of Chapter 7, with a wave equation describing the various waves and instabilities. It presents the stability analysis of diffuse cylindrical plasmas (classical pinches and present tokamaks) from the spectral perspective.
- Chapter 10 solves the initial value problem for one-dimensional inhomogeneous MHD and the associated damping due to the continuous spectrum.
- Chapter 11 discusses resonant absorption and phase mixing in the context of heating mechanisms of laboratory plasmas, and solar or stellar coronae. Sunspot seismology is introduced as another example of MHD spectroscopy.

Part IV (Flow and dissipation)

- Chapter 12 initiates the most urgent extension of the theory of Chapters 1–10, viz. waves and instabilities in plasmas with stationary background flows, a theme of great interest for laboratory fusion and astrophysical plasma research. The old problem of how to find the complex eigenvalues of stationary plasmas is solved by means of the new method of constructing the Spectral Web in the complex plane.
- Chapter 13 applies the new theory of the Spectral Web to the two classical topics of shear flow in plane plasma slabs, including the Kelvin–Helmholtz instability, and rotation in cylindrical plasmas, including the magneto-rotational instability.
- Chapter 14 treats the considerable modification of plasma dynamics when resistivity is introduced in the MHD description, both in the linear domain of spectral theory and in the nonlinear domain of reconnection.
- Chapter 15 introduces the basic techniques of computational MHD, the discretization techniques, the methods of time stepping, etc. It thus provides the modern techniques needed to solve for the dynamics of plasmas in complicated magnetic geometries.

Part V (Toroidal geometry)

- Chapter 16 presents the classical theory of static equilibrium of toroidal plasmas, a topic of central interest in fusion research of tokamaks.
- Chapter 17 concerns the spectral theory of waves and instabilities in toroidal equilibria, again a central topic in tokamak research. Because of this important application, this part of MHD spectral theory is the most developed one, also with respect to comparison with experimental data. This activity is called MHD spectroscopy.
- Chapter 18 introduces the theory of transonic equilibria and spectral theory of toroidal equilibria rotating in both directions, a subject of great interest but still in its infancy.

Part VI (Nonlinear dynamics)

- Chapter 19 introduces the topic of 2D turbulence in magneto-fluids by deriving the scaling laws for MHD turbulence and presenting the high performance computing efforts needed to resolve the structures that occur.
- Chapter 20 presents the counterpart of Chapter 15 by introducing the numerical methods for nonlinear MHD, in particular for plasmas with large background flows, applied in the last two chapters of this book.
- Chapter 21 discusses the MHD shock conditions from a new perspective, scale independence leading to time reversal duality, and it introduces some of the important areas of application of nonlinear MHD, viz. astrophysical winds and transonic flows.
- Chapter 22 introduces special relativistic MHD, in particular the linear waves and nonlinear shocks that occur at relativistic speeds. The books ends with applications to astrophysical phenomena, like relativistic jets, and thus completes the panorama of the tremendously exciting field of magnetohydrodynamics dominated by flows.

The Appendices provide the essentials of the two indispensable tools for MHD calculations, vector relations and orders of magnitude of the plasma parameters.

We wish to acknowledge Guido Huysmans, Jelle Kaastra, Giovanni Lapenta, Sasha Lifschitz, Zakaria Meliani, Gábor Tóth, Ronald Van der Linden, Oliver Porth, Rob Rutten, Henk van der Vorst and Chun Xia for constructive comments on different chapters, Victor Land and Jan Willem Blokland for their contributions to the exercises, Paul Cally, Antoine Cerfon, Bassem Girgis, Pieter Groen, Ian Hutchinson, Ralph Kube and Ryan White for helpful criticism on the previous volumes, and Bram Achterberg, Hubert Baty, Sander Beliën, Nicolas Bessolaz, Tom Bogdan, Fabien Casse, Antoine Cerfon, Paul Charbonneau, Arpád Csík, Herman Deconinck, Peter Delmont, Hans De Sterck, Dan D'Ippolito, Jeff Freidberg, Ricardo Galvão, Marcel Goossens, Giel Halberstadt, Tony Hearn, Bart van der Holst, Hanno Holties, Alan Hood, Wolfgang Kerner, Rob Kleibergen, Max Kuperus, Boon Chye Low, Keith MacGregor, Daniel Mueller, Valery Nakariakov, Ronald Nijboer, Eric Priest, Jan Rem, Bernard Roberts, Andro Rogava, Ilia Roussev, Paulo Sakanaka, Brigitte Schmieder, Karel Schrijver and Lydia Van Driel-Gesztelyi for fruitful collaborations and discussions.

The first author is indebted to Jeff Freidberg for scientific stimulation and friendship for many years. He is also thankful to past and present management of DIFFER (formerly FOM-Institute for Plasma Physcis 'Rijnhuizen'), Marnix van der Wiel, Aart Kleyn, Niek Lopes Cardozo, Noud Oomens, Jan Kranenbarg, Richard van de Sanden, Tony Donné and Marco de Baar, for having provided optimum conditions to work on these books. We wish to thank Simon Capelin of Cambridge University Press for his support and patience over all the years of preparation of this book and the previous volumes. We are indebted to the content manager Roisin Munnelly for helpful assistance and to the copy editor Jon Billam for very careful editing of this book.

Part I

Plasma Physics Preliminaries

1
Introduction

1.1 Motivation

Under ordinary circumstances, matter on Earth occurs in the three phases of solid, liquid and gas. Here, 'ordinary' refers to the circumstances relevant for human life on this planet. This state of affairs does not extrapolate beyond earthly scales: astronomers agree that, ignoring the more speculative nature of dark matter, matter in the Universe consists of more than 90% plasma. Hence, *plasma is the ordinary state of matter in the Universe*. The consequences of this fact for our view of nature are not generally recognized yet (see Section 1.3.4). The reason may be that, since plasma is an exceptional state on Earth, the subject of plasma physics is a relative latecomer in physics.

For the time being, the following crude definition of plasma suffices. *Plasma is a completely ionized gas, consisting of freely moving positively charged ions, or nuclei, and negatively charged electrons.*[1] In the laboratory, this state of matter is obtained at high temperatures, in particular in thermonuclear fusion experiments ($T \sim 10^8$ K). In those experiments, the mobility of the plasma particles facilitates the induction of electric currents which, together with the internally or externally created magnetic fields, permits magnetic confinement of the hot plasma. In the Universe, plasmas and the associated large-scale interactions of currents and magnetic fields prevail under much wider conditions.

Hence, we will concentrate our analysis on the two mentioned broad areas of application of plasma physics, viz.

(a) *Magnetic plasma confinement for the purpose of future energy production by controlled thermonuclear reactions* (CTR); this includes the pinch experiments of the 1960s and early 1970s, and the tokamaks and alternatives (stellarator, spheromak, etc.) developed in the 1980s and 1990s and, at present, sufficiently matured to start designing prototypes of the fusion reactors themselves.

(b) *The dynamics of magnetized astrophysical plasmas*; this includes the ever growing research field of solar magnetic activity, planetary magnetospheres, stellar winds, interstellar medium, accretion discs of compact objects, pulsar magnetospheres, etc.

The common ground of these two areas is the subject of *plasma interacting with a magnetic field*. To appreciate the power of this viewpoint, we first discuss the conditions for laboratory fusion in Section 1.2, then switch to the emergence of the subject of plasma-astrophysics in Section 1.3, and

[1] In plasma physics, one can hardly avoid mentioning exceptions: in pulsar electron–positron magnetospheres, the role of positively charged particles is taken by positrons. In considerations of fusion reactions with exotic fuels like muonium, the role of negatively charged particles is taken by muons.

finally refine our definition(s) of plasma in Section 1.4. In the latter section, we also provisionally formulate the approach to plasmas by means of magnetohydrodynamics.

The theoretical models exploited lead to *nonlinear partial differential equations*, expressing *conservation laws*. The boundary conditions are imposed on an extended spatial domain, associated with the *complex magnetic plasma confinement geometry*, whereas the temporal dependence leads to *intricate nonlinear dynamics*. This gives theoretical plasma physics its particular, mathematical, flavour.

1.2 Thermonuclear fusion and plasma confinement

1.2.1 Fusion reactions

Both fission and fusion energy are due to nuclear processes and, ultimately, described by Einstein's celebrated formula $E = mc^2$. Hence, in nuclear reactions $\mathrm{A} + \mathrm{B} \rightarrow \mathrm{C} + \mathrm{D}$, net energy is released if there is a mass defect, i.e. if

$$(m_\mathrm{A} + m_\mathrm{B})\, c^2 > (m_\mathrm{C} + m_\mathrm{D})\, c^2 \,. \tag{1.1}$$

In laboratory fusion, reactions of hydrogen isotopes are considered, where the deuterium-tritium reaction (Fig. 1.1) is the most promising one for future reactors:

$$\mathrm{D}^2 + \mathrm{T}^3 \;\rightarrow\; \mathrm{He}^4\;(3.5\,\mathrm{MeV}) + \mathrm{n}\;(14.1\,\mathrm{MeV})\,. \tag{1.2}$$

This yields two kinds of products, viz. α particles (He^4), which are *charged* so that they can be captured by a confining magnetic field, and neutrons, which are *electrically neutral* so that they escape from the magnetic configuration. The former contribute to the heating of the plasma (so-called α particle heating) and the latter have to be captured in a surrounding $\mathrm{Li}^6/\,\mathrm{Li}^7$ blanket, which recovers the fusion energy and also breeds new T^3.

Deuterium abounds in the oceans: out of 6500 molecules of water one molecule contains a deuteron and a proton instead of two protons. Thus, in principle, 1 litre of sea water contains 10^{10} J of deuterium fusion energy. This is a factor of about 300 more than the combustion energy of 1 litre of gasoline, which yields 3×10^7 J.

A number of other reactions also occur, in particular reactions producing T^3 and He^3 which may be burned again. Complete burn of all available D^2 would involve the following reactions:

$$\begin{aligned}
\mathrm{D}^2 + \mathrm{D}^2 &\;\rightarrow\; \mathrm{He}^3\;(0.8\,\mathrm{MeV}) + \mathrm{n}\;(2.5\,\mathrm{MeV})\,,\\
\mathrm{D}^2 + \mathrm{D}^2 &\;\rightarrow\; \mathrm{T}^3\;(1.0\,\mathrm{MeV}) + \mathrm{p}\;(3.0\,\mathrm{MeV})\,,\\
\mathrm{D}^2 + \mathrm{T}^3 &\;\rightarrow\; \mathrm{He}^4\;(3.5\,\mathrm{MeV}) + \mathrm{n}\;(14.1\,\mathrm{MeV})\,,\\
\mathrm{D}^2 + \mathrm{He}^3 &\;\rightarrow\; \mathrm{He}^4\;(3.7\,\mathrm{MeV}) + \mathrm{p}\;(14.6\,\mathrm{MeV})\,,
\end{aligned} \tag{1.3}$$

so that in effect

$$6\mathrm{D}^2 \;\rightarrow\; 2\,\mathrm{He}^4 + 2\,\mathrm{p} + 2\,\mathrm{n} + 43.2\,\mathrm{MeV}\,. \tag{1.4}$$

In the liquid Li blanket, fast neutrons are moderated, so that their kinetic energy is converted into heat, and the following reactions occur:

$$\begin{aligned} \mathrm{n} + \mathrm{Li}^6 &\rightarrow \mathrm{T}^3\ (2.1\,\mathrm{MeV}) + \mathrm{He}^4\ (2.8\,\mathrm{MeV})\,, \\ \mathrm{n}\ (2.5\,\mathrm{MeV}) + \mathrm{Li}^7 &\rightarrow \mathrm{T}^3\ + \mathrm{He}^4 + \mathrm{n}\,. \end{aligned} \tag{1.5}$$

This provides the necessary tritium fuel for the main fusion reaction, described by the third item of Eq. (1.3) [442].

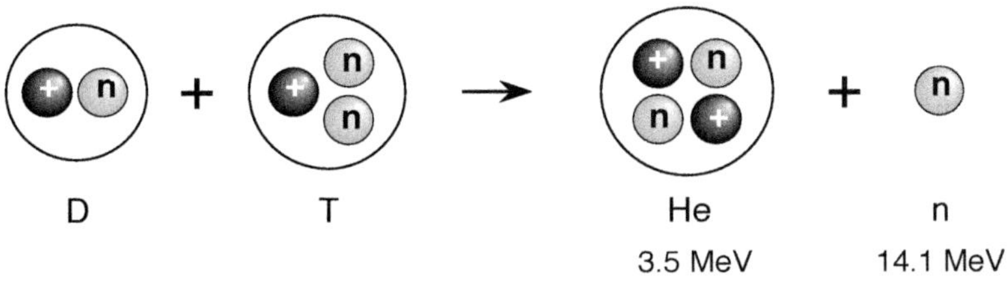

Fig. 1.1 Nuclear fusion reaction of the hydrogen isotopes deuterium and tritium.

Typical numbers associated with thermonuclear fusion reactors, as presently envisaged, are:

$$\begin{aligned} &\text{temperature} \quad T \sim 10^8\,\mathrm{K}\ (10\,\mathrm{keV})\,, \quad &&\text{power density} \ \sim 10\,\mathrm{MW\,m^{-3}}\,, \\ &\text{particle density} \ \ n \sim 10^{21}\,\mathrm{m^{-3}}\,, &&\text{time scale} \quad \tau \sim 100\,\mathrm{s}\,. \end{aligned} \tag{1.6}$$

It is often said that controlled thermonuclear fusion in the laboratory is an attempt to harness the power of the stars. This is actually a quite misleading statement since the fusion reactions which take place in, e.g., *the core of the Sun* are different reactions of hydrogen isotopes, viz.

$$\begin{aligned} \mathrm{p} + \mathrm{p} &\rightarrow \mathrm{D}^2\ + \mathrm{e}^+ + \nu_\mathrm{e}\ + 1.45\,\mathrm{MeV} \quad (2\times)\,, \\ \mathrm{p} + \mathrm{D}^2 &\rightarrow \mathrm{He}^3\ + \gamma + 5.5\,\mathrm{MeV} \qquad\quad (2\times)\,, \\ \mathrm{He}^3 + \mathrm{He}^3 &\rightarrow \mathrm{He}^4 + 2\,\mathrm{p} + 12.8\,\mathrm{MeV}\,, \end{aligned} \tag{1.7}$$

so that complete burn of all available hydrogen amounts to

$$4\,\mathrm{p} \rightarrow \mathrm{He}^4 + 2\,\mathrm{e}^+ + 2\,\nu_\mathrm{e}\ (0.5\,\mathrm{MeV}) + 2\,\gamma\ (26.2\,\mathrm{MeV})\,. \tag{1.8}$$

The positrons annihilate with electrons, the neutrinos escape, and the gammas (carrying the bulk of the thermonuclear energy) start on a long journey to the solar surface, where they arrive millions of years later (the mean free path of a photon in the interior of the Sun is only a few centimeters) [510]. In the many processes of absorption and re-emission the wavelength of the photons gradually shifts from that of gamma radiation to that of the visible and UV light escaping from the photosphere of the Sun, and producing one of the basic conditions for life on a planet situated at the safe distance of one astronomical unit (1.5×10^{11} m) from the Sun.

At higher temperatures another chain of reactions is effective, where carbon acts as a kind of catalyst. This so-called CNO cycle involves a chain of fusion reactions where C^{12} is successively converted into N^{13}, C^{13}, N^{14}, O^{15}, N^{15}, and back into C^{12} again. However, the net result of incoming and outgoing products is the same as that of the proton–proton chain, viz. Eq. (1.8).

Typical numbers associated with thermonuclear reactions in the stars, in particular the core of the Sun, are the following ones:

$$\text{temperature} \quad T \sim 1.5 \times 10^7\,\mathrm{K}\,, \qquad \text{power density} \ \sim 3.5\,\mathrm{W\,m^{-3}}\,,$$
$$\text{particle density} \quad n \sim 10^{32}\,\mathrm{m^{-3}}\,, \qquad \text{time scale} \quad \tau \sim 10^7\,\text{years}\,. \tag{1.9}$$

Very different from the numbers (1.6) for a prospective fusion reactor on Earth!

1.2.2 Conditions for fusion

Thermonuclear fusion happens when a gas of, e.g., deuterium and tritium atoms is sufficiently heated for the thermal motion of the nuclei to become so fast that they may overcome the repulsive Coulomb barrier (Fig. 1.2) and come close enough for the attractive nuclear forces to bring about the fusion reactions discussed above. This requires particle energies of $\sim 10\,\mathrm{keV}$, i.e. temperatures of about $10^8\,\mathrm{K}$. At these temperatures the electrons are completely stripped from the atoms (the ionization energy of hydrogen is $\sim 14\,\mathrm{eV}$) so that a plasma rather than a gas is obtained (cf. our crude definition of Section 1.1).

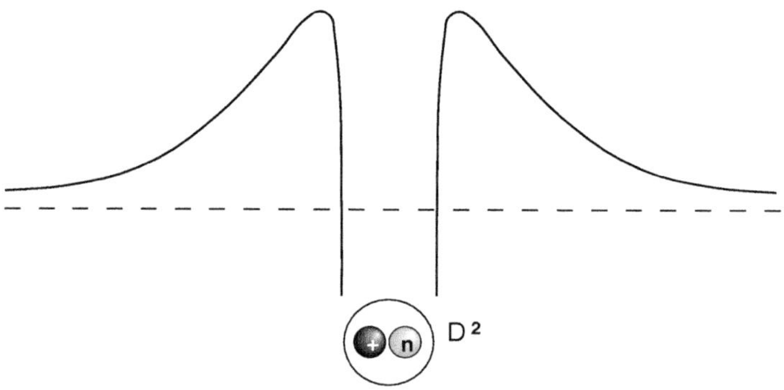

Fig. 1.2 Nuclear attraction and Coulomb barrier of a deuteron.

Because the charged particles (occurring in about equal numbers of opposite charge) are freely moving and rarely collide at these high temperatures, a plasma may be considered as a *perfectly conducting fluid* for many purposes. In such fluids, electric currents are easily induced and the associated magnetic fields in turn interact with the plasma to confine or to accelerate it. The appropriate theoretical description of this state of matter is called *magnetohydrodynamics* (MHD), i.e. the dynamics of magneto-fluids (Section 1.4.2).

Why are magnetic fields necessary? To understand this, we need to discuss the power requirements for fusion reactors (following Miyamoto [442] and Wesson [647]). This involves three contributions, viz.

(a) the thermonuclear output power per unit volume:

$$P_{\mathrm{T}} = n^2 f(\widetilde{T})\,, \qquad f(\widetilde{T}) \equiv \tfrac{1}{4}\langle\sigma v\rangle E_{\mathrm{T}}\,, \quad E_{\mathrm{T}} \approx 22.4\,\mathrm{MeV}\,, \tag{1.10}$$

where n is the particle density, σ is the cross-section of the D-T fusion reactions, v is the relative speed of the nuclei, $\langle\sigma v\rangle$ is the average nuclear reaction rate, which is a well-known function of temperature, and $E_{\rm T}$ is the average energy released in the fusion reactions (i.e. more than the 17.6 MeV of the D-T reaction (1.3)(c) but, of course, less than the 43.2 MeV released for the complete burn (1.4));

(b) the power loss by Bremsstrahlung, i.e. the radiation due to electron–ion collisions:

$$P_{\rm B} = \alpha n^2 \widetilde{T}^{1/2}\,, \qquad \alpha \approx 3.8 \times 10^{-29}\,{\rm J}^{1/2}\,{\rm m}^3\,{\rm s}^{-1}\,; \tag{1.11}$$

(c) the losses by heat transport through the plasma:

$$P_{\rm L} = \frac{3n\widetilde{T}}{\tau_{\rm E}}\,, \tag{1.12}$$

where $3n\widetilde{T}$ is the total plasma kinetic energy density (ions + electrons), and $\tau_{\rm E}$ is the energy confinement time (an empirical quantity). The latter estimates the usually anomalous (i.e. deviating from classical transport by Coulomb collisions between the charged particles) heat transport processes. Here, we have put a tilde on the temperature to indicate that energy units of keV are exploited:

$$\widetilde{T}\,({\rm keV}) = 8.62 \times 10^{-8} T\,({\rm K})\,,$$

since $\widetilde{T} = 1\,{\rm keV} = 1.60 \times 10^{-16}\,{\rm J}$ corresponds with $T = 1.16 \times 10^7\,{\rm K}$ (using Boltzmann's constant, see Appendix Table B.1).

If the three power contributions are considered to become externally available for conversion into electricity and back again into plasma heating, with efficiency η, *the Lawson criterion* [400],

$$P_{\rm B} + P_{\rm L} = \eta\,(P_{\rm T} + P_{\rm B} + P_{\rm L})\,, \tag{1.13}$$

tells us that there should be power balance between the losses from the plasma (LHS) and what is obtained from plasma heating (RHS). Typically, $\eta \approx 1/3$. Inserting the explicit expressions (1.10), (1.11) and (1.12) into Eq. (1.13) leads to a condition to be imposed on the product of the plasma density and the energy confinement time:

$$n\tau_{\rm E} = \frac{3\widetilde{T}}{[\eta/(1-\eta)]\, f(\widetilde{T}) - \alpha \widetilde{T}^{1/2}}\,. \tag{1.14}$$

This relationship is represented by the lower curve in Fig. 1.3. Since Bremsstrahlung losses dominate at low temperatures and transport losses dominate at high temperatures, there is a minimum in the curve at about

$$n\tau_{\rm E} = 0.6 \times 10^{20}\,{\rm m}^{-3}\,{\rm s}\,, \quad \text{for } \widetilde{T} = 25\,{\rm keV}\,. \tag{1.15}$$

This should be considered to be the threshold for a fusion reactor under the given conditions.

By a rather different, more recent, approach of fusion conditions, *ignition* occurs when the total amount of power losses is balanced by the total amount of heating power. The latter consists of α-particle heating P_α and additional heating power $P_{\rm H}$, e.g. by radio-frequency waves or neutral beam injection. The latter heating sources are only required to bring the plasma to the ignition point, when α-particle heating may take over. Hence, at ignition we may put $P_{\rm H} = 0$ and the power balance becomes

$$P_{\rm B} + P_{\rm L} = P_\alpha = \tfrac{1}{4}\langle\sigma v\rangle n^2 E_\alpha\,, \quad E_\alpha \approx 3.5\,{\rm MeV}\,. \tag{1.16}$$

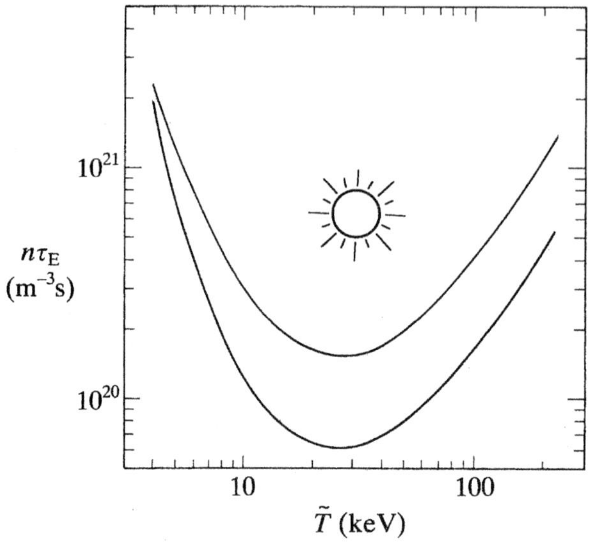

Fig. 1.3 Conditions for net fusion energy production according to the Lawson criterion (lower curve) and according to the view that power losses should be completely balanced by α-particle heating (upper curve). (Adapted from Wesson [647].)

Formally, this may be described by the same equation (1.14) taking now $\eta \approx 0.135$ so that a 2.5 times higher threshold for fusion is obtained:

$$n\tau_E = 1.5 \times 10^{20}\,\mathrm{m}^{-3}\,\mathrm{s}\,, \quad \text{for } \widetilde{T} = 30\,\mathrm{keV}\,. \tag{1.17}$$

This relationship is represented by the upper curve of Fig. 1.3.

Roughly speaking then, products of density and energy confinement time $n\tau_E \sim 10^{20}\,\mathrm{m}^{-3}\,\mathrm{s}$ and temperatures $\widetilde{T} \sim 25\,\mathrm{keV}$, or $T \sim 3 \times 10^8\,\mathrm{K}$, are required for controlled fusion reactions. As a figure of merit for fusion experiments one frequently constructs the product of these two quantities, which should approach

$$n\tau_E\widetilde{T} \sim 3 \times 10^{21}\,\mathrm{m}^{-3}\,\mathrm{s\,keV} \tag{1.18}$$

for a fusion reactor. To get rid of the radioactive tritium component, one might consider pure D-D reactions in a more distant future. This would require yet another increase of the temperature by a factor of 10. Considering the kind of progress obtained over the past 40 years though (see Fig. 1.1.1 of Wesson [647]: a steady increase of the product $n\tau_E T$ with a factor of 100 every decade!), one may hope that this difficulty eventually will turn out to be surmountable.

Returning to our question on the magnetic fields: no material containers can hold plasmas with densities of $10^{20}\,\mathrm{m}^{-3}$ and temperatures of 100–300 million K during times of the order of minutes, or at least seconds, without immediately extinguishing the 'fire'. One way to solve this problem is to make use of the *confining properties of magnetic fields*, which may be viewed from quite different angles:

(a) the charged particles of the plasma rapidly and tightly gyrate around the magnetic field lines (they 'stick' to the field lines, see Section 2.2);

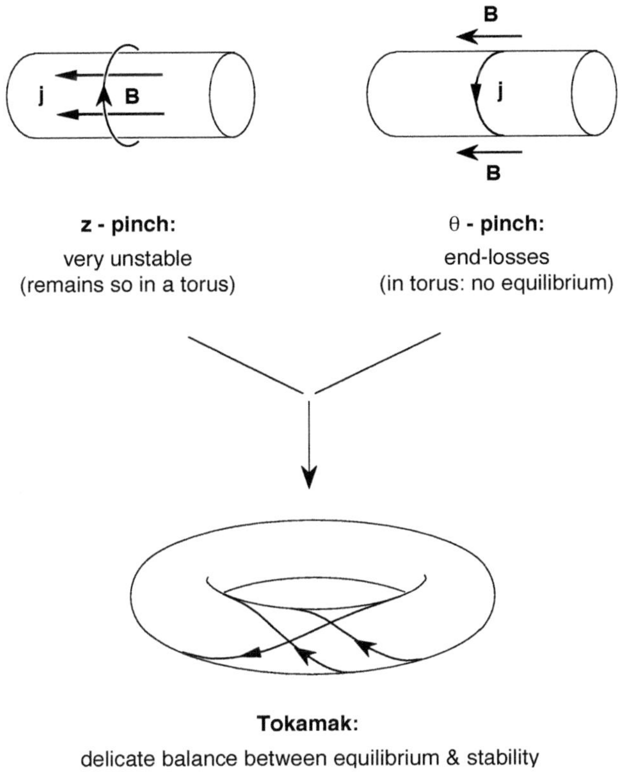

Fig. 1.4 Interaction of currents and magnetic fields: the history of plasma confinement experiments.

(b) fluid and magnetic field move together ('the magnetic field is frozen into the plasma', see Section 2.4), so that engineering of the geometry of the magnetic field configuration also establishes the geometry of the plasma;

(c) the thermal conductivity of plasmas is extremely anisotropic with respect to the magnetic field, $\kappa_\perp \ll \kappa_\parallel$ (Sections 2.3.1 and 3.3.2), so that heat is easily conducted along the field lines and the magnetic surfaces they map out, but not across.

Consequently, what one needs foremost is a *closed magnetic geometry* facilitating stable, static plasma equilibrium with roughly bell-shaped pressure and density profiles and nested magnetic surfaces. This is the subject of the next section.

1.2.3 Magnetic confinement and tokamaks

Controlled thermonuclear fusion research started in the 1950s in the weapons laboratories after the 'successful' development of the hydrogen bomb: fusion energy had been unleashed on our planet! The development of the peaceful, controlled, counterpart appeared to be a matter of a few years, as may become clear by considering the simplicity of early pinch experiments. The history of the subject is schematically illustrated in Fig. 1.4. In the upper part the two early attempts with the

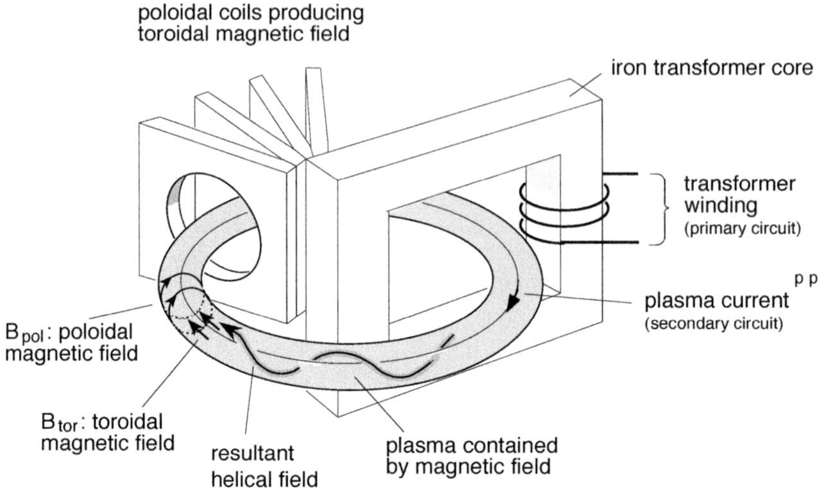

Fig. 1.5 Schematic presentation of magnetic confinement in a tokamak.

simple schemes of θ- and z-pinch are shown. Here, θ and z refer to the direction of the plasma current in terms of a cylindrical r, θ, z coordinate system. Since it is relatively straightforward to produce plasma by ionizing hydrogen gas in a tube, a very conductive fluid is obtained in which a strong current may be induced by discharging a capacitor bank over an external coil surrounding the gas tube. In a *z-pinch experiment*, this current is induced in the z-direction and it creates a transverse magnetic field B_θ, so that the resulting Lorentz force $(\mathbf{j} \times \mathbf{B})_r = -j_z B_\theta$ is pointing radially inward. In this manner, the confining force as well as near thermonuclear temperatures ($\sim 10^7$ K) are easily produced. There is only one problem: the curvature of the magnetic field B_θ causes the plasma to be extremely unstable, with growth rates in the order of microseconds. To avoid these instabilities, the orthogonal counterpart, the *θ-pinch experiment*, suggested itself. Here, current is induced in the θ-direction, it causes a radial decrease of the externally applied magnetic field B_z, so that the net Lorentz force $j_\theta \, \Delta B_z$ is again directed inward. In the θ-pinch, thermonuclear temperatures are also obtained, and the plasma is now macroscopically stable. However, pinching of the plasma column produces unbalanced longitudinal forces so that the plasma is squirted out of the ends, again terminating plasma confinement on the microsecond time scale. In conclusion, in pinch experiments the densities and temperatures needed for thermonuclear ignition are easily produced but the confinement times fall short by a factor of a million to a billion.

With these obstacles ahead, the nations involved with thermonuclear research decided it to be opportune to declassify the subject. This fortunate decision was landmarked by the Second International UN Conference on Peaceful Uses of Atomic Energy in Geneva in 1958, where all scientific results obtained so far were presented. Prospects then gradually became much brighter with the emergence of *the tokamak* alternative line (bottom part of Fig. 1.4 and Fig. 1.5) developed in the 1960s in the Soviet Union, and internationally accepted in the 1970s as the most promising scheme towards fusion. Crudely speaking, the tokamak configuration cures the main problems of the z-pinch (its instability due to the curvature of the transverse magnetic field) and of the θ-pinch (its end losses), both destroying the configuration on the microsecond time scale, by combining them into a single scheme. The vessel is now a torus rather than a straight tube and the magnetic field is helical, with a poloidal and a larger toroidal component. The latter component of the magnetic field provides the crucial longitudinal 'backbone' for stability. Whereas the toroidal geometry simply eliminates the end-loss problem of the θ-pinch, it is not quite true that the kink instability problem of the z-pinch is eliminated as well. Instead, the basic MHD problem of tokamak confinement turns out to be a delicate balance between equilibrium considerations, favouring a large toroidal current, and stability considerations, which favour a minimum current so as to eliminate the driving force of the kink instabilities. Thus, tokamak performance is an *intricate optimization problem* which makes it both interesting and impressive. Concerning the latter: to have improved upon a technological parameter by a factor of 10^6 in thirty years (from confinement times of microseconds in the sixties to seconds in the nineties) is a kind of progress which is only paralleled by developments in computer technology.

For more on the history of fusion research: see Braams and Stott [86].

1.3 Astrophysical plasmas

We have sketched the efforts in controlled thermonuclear confinement experiments, where the prospect of abundant energy has driven scientists to ever deeper exploration of the plasma state. At this point an entirely different line of research should enter the presentation. This is the rapidly growing field of plasma-astrophysics, which has much older credentials than laboratory plasma research. We will introduce this topic by means of the example of the solar system, where the usual gravitational picture completely masks the dynamics of the plasmas that are present. To understand how this picture has changed in recent times, we introduce some basic astrophysical notions and recall events in space research. We will also use the opportunity to introduce numerical values of certain quantities that may not be familiar to some readers.

1.3.1 Celestial mechanics

To set the stage, recall *the traditional picture of the solar system*: the Sun is the central massive object (a thousand times more massive than Jupiter) which keeps the nine planets orbiting around it by its gravitational attraction. (See Fig. 1.6 and the numerical values summarized in Table B.6.)

Recall that the planets move according to *Kepler's laws* (1610):

(a) *The planetary orbits are ellipses* lying in or close to the ecliptic (the orbital plane of the Earth) *with the Sun in one of the focal points*. The inclination of the orbit with respect to the ecliptic is modest ($< 4°$) for most of the planets, whereas the largest values occur for the innermost planet (Mercury: $7°$) and for the outermost 'planet' (Pluto: $17°$). The ellipses are characterized by the eccentricity parameter $e \equiv c/a = (1 - b^2/a^2)^{1/2}$, where c is the distance of the focal points to the origin and a and b are the lengths of the semi-axes of the ellipse. Again, the highest eccentricities occur for Mercury ($e = 0.206$) and Pluto ($e = 0.250$), whereas they are small for the other planets ($e < 0.1$). Incidentally, it is to be noted that the ellipticity as measured by the ratio of the semi-axes, $b/a = \sqrt{1 - e^2}$, is $\sqrt{0.96} \approx 0.98$ for Mercury and 0.97 for Pluto, i.e. just deviations of 2% and 3% from a circle, and much less for the other planets. The original approximation of circular motion by the ancients appears not all that stupid. The big effect is not the deviation from a circle though, but

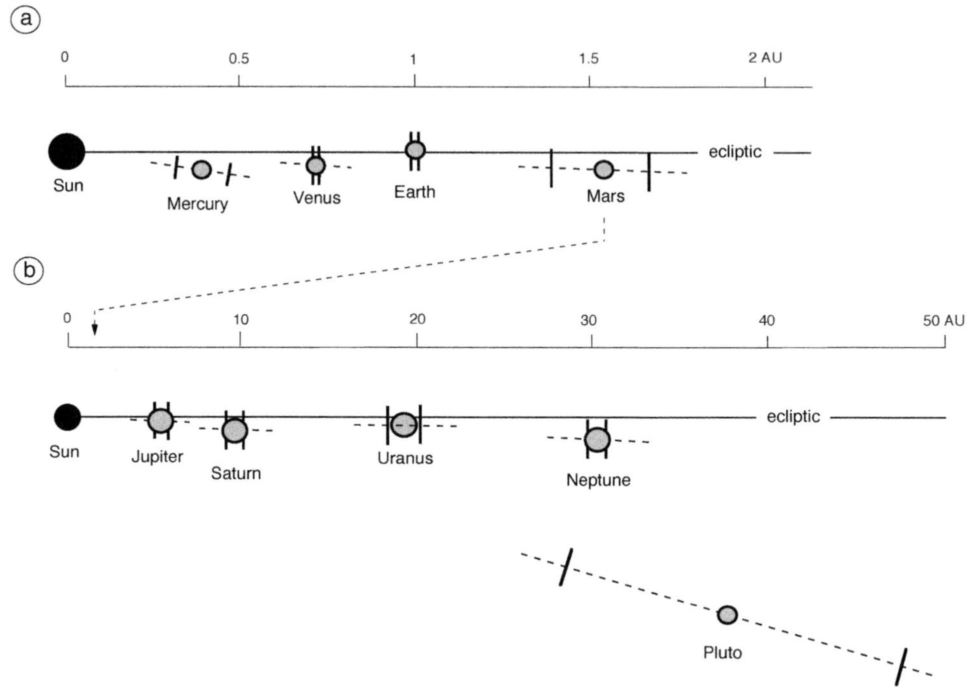

Fig. 1.6 Schematic representation of the size and inclination of the planetary orbits: (a) terrestrial planets, (b) giant planets and Pluto. The planets (not to scale) are drawn at a distance a from the Sun, where the vertical dashes indicate the extrema $a - c$ (perihelion) and $a + c$ (aphelion) of the orbital motion.

the eccentricity, i.e. *the shift* c of the near-circular orbit. This gives rise to significant variations in the distance to the Sun, as measured by the ratio $(a - c)/(a + c) = (1 - e)/(1 + e)$ which is 0.66 for Mercury and 0.60 for Pluto, as shown in Fig. 1.6.

(b) *The radius vector of the Sun to the planet sweeps out equal areas in equal times*. Hence, the orbital velocity is highest in the perihelion (the orbital point closest to the Sun) and smallest in the aphelion (the point farthest from the Sun). This *law of areas* is a consequence of conservation of angular momentum.

(c) *The harmonic law: The cubes of the semi major axis a of the orbits of the planets are proportional to the squares of the orbital period τ*,

$$a^3/\tau^2 = \text{const} \approx GM_\odot/4\pi^2 = 1\,(\text{AU})^3/\,\text{y}^2 = 3.38 \times 10^{18}\,\text{m}^3\,\text{s}^{-2}\,. \tag{1.19}$$

Here, G is the gravitational constant, $M_\odot$ is the mass of the Sun, $1\,\text{AU} = 1.5 \times 10^{11}\,\text{m}$ is the distance from the Earth to the Sun (the astronomical unit) and $1\,\text{y} = 3.16 \times 10^7\,\text{s}$ is, of course, the orbital period of the Earth.

▷ **Exercise** Use Table B.6 to check this number for the different planetary orbits. ◁

Next, Kepler's laws were then founded on the laws of mechanics, in particular *Newton's law of gravitational attraction* (1666):

$$F_{\rm gr} = G\,\frac{M_1 M_2}{r^2} = -\frac{dV_{\rm gr}}{dr}\,, \tag{1.20}$$

where $V_{\rm gr} = -G\,M_1 M_2/r$ is the gravitational potential energy. This law implies that the planets move as point particles in the gravitational field of the Sun whereas the whole solar system is kept together in dynamical equilibrium by gravity. All this belongs to the subject of *celestial mechanics* which is at the root of classical mechanics, which in turn constitutes the basis of physics. Thus, progress in understanding may schematically be depicted by the sequence Kepler (1609) → Newton (1687) → Lagrange (1782), Laplace (1799) → Hamilton (1845). After the work of these giants, the subject of classical mechanics (as, e.g., summarized by Goldstein [243]) has long been considered a closed subject. However, the sequence continues with the more recent names of Kolmogorov, Arnold and Moser (1964) associated with fundamental work on the stability of dynamical systems. At the present time, there is a resurgence of the subject of Hamiltonian mechanics through the development of the science of nonlinear dynamics.

So far, plasmas did not appear on the stage. Obviously, the gravitational attraction dominates everything. Gravitational and centrifugal acceleration balance perfectly in the leading order picture where the celestial bodies are treated as massive point particles. Since this is so, next order effects should be quite important (just like astronauts in an orbiting spacecraft may be accelerated by forces that are totally negligible as compared to gravity). Hence, when the internal structure of the stars (in this case, the Sun) and the planets is taken into account, the whole picture changes dramatically.

1.3.2 Astrophysics

In the nineteenth and twentieth century, there is a gradual shift away from exclusive interest in celestial mechanics towards the study of *the structure and evolution of stars and stellar systems*: the subject of astrophysics is born. Here, a basic postulate provides the guiding principle, viz. that *the laws of physics are valid throughout the Universe*. In historical perspective, the revolutionary character of this point of view can hardly be overestimated: Quintessence (according to *Webster's* Dictionary, 'the fifth and highest essence in ancient and medieval philosophy that permeates all nature and is the substance composing the heavenly bodies') is no longer essential, and 'heavenly' or 'celestial' are no longer descriptive adjectives for astronomical objects. A particularly relevant example is provided by the work of Kirchhoff and Bunsen (1859) who interpreted the observed dark lines in the spectrum of solar light, discovered by Fraunhofer (1814), as due to absorption by chemical elements in exactly the same way as spectra obtained in the laboratory. Consequently, most of our knowledge of the stars comes from *spectroscopy*, i.e. atomic physics applied to the photospheres of the stars where the spectra are determined by the temperature T of the surface and the different abundances of the chemical elements.

A measure for the relative brightness of a star is *the apparent magnitude* m,

$$m \equiv m_0 - 2.5 \times {}^{10}\!\log\,(l/l_0)\,, \tag{1.21}$$

where l is the flux, i.e. the amount of electromagnetic radiation energy passing per unit time through a unit area (taken at the position of our eye, or any other observing apparatus on Earth), and the subscript 0 refers to a reference star. The value of m_0 for the reference star is fixed by convention.

This definition bas been chosen to conform with the ancient classification based on what the human eye can distinguish, viz. five steps in a brightness scale ranging from $m = 0$ for the brightest star to $m = 5$ for the faintest one, corresponding to a decrease by a factor of $1/100$ in the flux.

Obviously, two stars of equal apparent magnitude may have a completely different value of the luminosity L, which is the total radiation energy output per unit time, since the flux l depends on the distance d from the star according to

$$l = L/(4\pi d^2)\,. \tag{1.22}$$

Hence, a quantity of more intrinsic physical interest is *the absolute magnitude* M, which is based on the flux $\hat{l}$ that would be produced at the position of the Earth (ignoring atmospheric extinction) if the star were moved from its actual distance d to a fictitious distance $\hat{d} = 10\,\text{pc}$ (parsec) from the Earth.[2] In other words, the absolute magnitude is defined as the apparent magnitude the star would have if positioned at $\hat{d}$, so that we obtain from Eqs. (1.21) and (1.22):

$$\begin{aligned} M &\equiv m_0 - 2.5 \times {}^{10}\log(\hat{l}/l_0) = m - 2.5 \times {}^{10}\log(\hat{l}/l) \\ &= m - 2.5 \times {}^{10}\log(d^2/100) = m + 5 - 5 \times {}^{10}\log d\,, \end{aligned} \tag{1.23}$$

where d is measured in pc. For the Sun, with $d_\odot = 1\,\text{AU} = 1.5 \times 10^8\,\text{km} = 5 \times 10^{-6}\,\text{pc}$, we get a huge difference between the absolute and the apparent magnitude: $M_\odot = m_\odot + 31.5$. (Note that we have used the same symbol $M_\odot$ already in Section 1.3.1 to indicate the solar mass; every now and then, we will not be able to avoid context-dependent notation.) The reason is clear: the apparent magnitude is based on night-time observation and, hence, totally out of range for the Sun. On the other hand, for the absolute magnitude of the Sun the very ordinary value $M_\odot = 4.7$ is obtained: apart from its proximity, the Sun is just an ordinary star.

A particularly effective way of representing the absolute magnitudes of a large number of stars is the celebrated *Hertzsprung–Russell diagram*, where the absolute magnitude M is plotted versus the effective surface temperature T_{eff}, or the associated spectral class indicated by the letters O, B, etc. (Fig. 1.7). A crude estimate of the curve for the main sequence stars may be obtained by using the Stefan–Boltzmann black-body radiation law for the luminosity,

$$L = 4\pi R^2 \sigma T_{\text{eff}}^4\,. \tag{1.24}$$

Here, R is the radius of the star, $\sigma = 5.67 \times 10^{-8}\,\text{W}\,\text{m}^{-2}\,\text{K}^{-4}$ is the constant of Stefan–Boltzmann and T_{eff} is the effective surface temperature of the star. For stars of equal size, we obtain from the first line of Eq. (1.23) the following difference in their absolute magnitudes:

$$\Delta M = -2.5 \times \Delta({}^{10}\log\,\hat{l}) = -2.5 \times \Delta({}^{10}\log\,L) = -10 \times \Delta({}^{10}\log\,T_{\text{eff}})\,. \tag{1.25}$$

This roughly checks with the overall slope of the Hertzsprung–Russell diagram.

For the Sun, $R_\odot = 700\,000\,\text{km}$ and $T_{\text{eff},\odot} = 5777\,\text{K}$ (i.e., spectral class G) so that $L_\odot =$

[2] Note on *distance scales*: a star at a distance $d = 1\,\text{pc}$ produces a parallax of $1'' = 4.85 \times 10^{-6}\,\text{rad}$, so that $1\,\text{pc} = (4.85 \times 10^{-6})^{-1}\,\text{AU} = 2.06 \times 10^5\,\text{AU} = 3.26\,\text{light-years} = 3.09 \times 10^{16}\,\text{m}$. The distance to the next nearest star, Alpha Centauri, is $1.3\,\text{pc}$. The size of our Galaxy (the Milky Way) is $50\,\text{kpc} = 1.6 \times 10^5$ light-years $= 1.3 \times 10^8 \times$ the size of the solar system (the diameter of the orbit of Pluto, i.e. $2 \times 40\,\text{AU}$).
Light also provides useful estimates for *time scales*: a photon would take $2\,\text{s}$ to travel from the centre of the Sun to the surface if the Sun were optically thin. (In reality, because of the innumerable absorptions and re-emissions it takes about 10^7 years, as we have already noted in Section 1.2.1.) It then takes $8.3\,\text{min}$ to reach Earth, 5.6 hours to reach Pluto, and 4.2 years to reach Alpha Centauri.

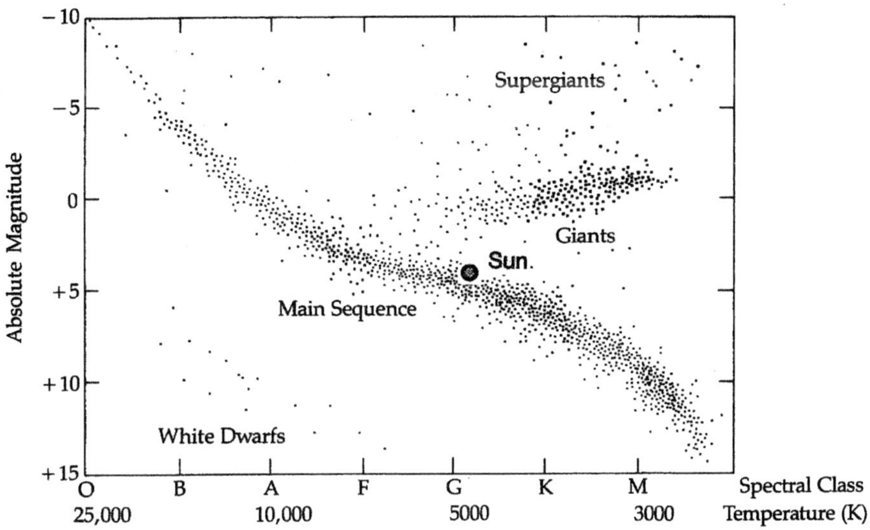

Fig. 1.7 Hertzsprung–Russell diagram: the Sun is an ordinary, main sequence, star. (Adapted from Zeilik and Smith [666].)

3.89×10^{26} W. Incidentally, the flux $l = L_\odot/(4\pi d^2)$ at the position of the Earth is called the solar constant. Since $d = 1$ AU, its value turns out be $l = 1.38\,\mathrm{kW/\,m^2}$: just the right value for human and other life. However, at this point of our exposition, we have turned away from Earth-centred considerations to the intrinsic properties of the stars. The central position of the point representing the Sun in the Hertzsprung–Russell diagram is then just another way of expressing that the Sun is but an ordinary main sequence star. Yet, as far as distance is concerned, we should consider ourselves lucky to have a typical star close enough to permit *spatially resolved* observations! This is crucial for our understanding of plasma dynamics in the Universe as a whole.

Not only is the solar system kept together by gravity, but the individual celestial bodies of the Sun and the planets are also kept together by gravity and, as a result, they *contract*. Stars with masses like that of the Sun (Jupiter is just too small to qualify as a star) contract so much that in the centre *densities* and *temperatures* are reached that are *high enough* for *thermonuclear burn by fusion* reactions of hydrogen, viz. $T_c = 1.5 \times 10^7$ K, and $\rho_c = 1.5 \times 10^5\,\mathrm{kg\,m^{-3}}$. We have encountered these fusion reactions and their conditions in Eqs. (1.8) and (1.9). Recall that, under these conditions, matter is ionized so that we encounter the plasma state again in the core of the Sun and the other stars. It appears that we have closed the circle and that the announcement of the theme for this book will simply be: laboratory fusion and astrophysical fusion reactions require the study of plasma physics. This is not the case. Reality is much more interesting (and subtle) than this.

1.3.3 Plasmas enter the stage

With the discovery of Bethe and von Weiszäcker in 1939 that thermonuclear fusion reactions take place in the centre of the Sun and the other stars, we know the ultimate source of the enormous amounts of power emitted in the form of visible and ultraviolet light. However, there is quite some distance in space and time between this source and the starlight it eventually produces. In the intermediate stages, this huge thermonuclear energy source indirectly excites a wide variety of additional, plasma dynamical, phenomena. Could we 'see' that? If only the blinding brightness of the solar disc were blocked for a few minutes we would be able to tell. Fortunately, provision has been made for that: the relative sizes of the Moon and the Sun, and their distances to the Earth, are precisely of the right magnitude to permit occultation of the Sun every now and then to exhibit an extremely beautiful phenomenon. At the moment of the eclipse, even the birds hold their breath, and a human being lucky enough to be at the right spot at the right moment can see a hot (millions of degrees) plasma with his own eyes: a diffuse light due to scattering of sunlight by the coronal plasma and stretching out over several solar diameters. Even the magnetic structures supporting it are visible to the (admittedly prejudiced) physicist in the form of streamers of plasma tracing out magnetic field lines and helmet structures associated with magnetic cusps (see Chapter 8). Hence, at a solar eclipse, one catches a wonderful glimpse of a huge magnetized plasma structure which engulfs the whole solar system.

This structure is the solar corona expanding into the *solar wind*, which forms magnetospheres when encountering the magnetic fields of the planets and which is a giant magnetosphere by itself, called the heliosphere, terminating only at distances beyond the solar system. The solar wind carries the wave-like signals of its creation, but it also carries the intermittent radiation and high-energy particle signatures of violent outbursts of magnetic energy releases by *flares* and *coronal mass ejections* (CMEs) at the solar surface (Chapter 8). This highly unsteady plasma dynamical state creates the critical conditions for magnetic storms in the magnetosphere. (*The* magnetosphere is always shorthand for the magnetosphere of the Earth, or *our* magnetosphere, like *the* Galaxy always stands for the Milky Way, or *our* Galaxy.) It forms a threat for safety of personnel and proper functioning of spacecrafts. This aspect of solar wind dynamics is called *space weather*.

Receding now to the interior of the star, closer to the thermonuclear energy source, we encounter the phenomena responsible for all this: radiation transport and convection which, together with the differential rotation of the star, create the conditions required for a dynamo. This dynamo produces magnetic fields that do not stay inside the star but are expelled, with the plasma, to form the extremely hot coronae and stellar winds that are the characteristics of X-ray emitting stars. Incidentally, the creation of magnetic fields in the interior of stars and the high temperatures of coronal plasmas are two plasma physical problems that are far from being solved at present. While we do not pretend to solve them here, we do believe that for progress one needs to delve deeply into basic magnetohydrodynamics, which is the subject of this book. Hence, the connection between laboratory and astrophysical plasmas is not the thermonuclear fusion reactions but their indirect result far away: *magnetized plasmas are present everywhere in the Universe!*

How do we know? High-resolution astronomical observations over the whole range of the electromagnetic frequency spectrum by means of 'telescopes', ground-based or from space vehicles, have produced irrefutable evidence for that. Whereas Sputnik (1957) and the Apollo flight to the Moon (1969) have spoken to the imagination of a large public, the less-known observations of the Sun and stars by means of X-ray telescopes on board rockets and the risky (manned) Skylab missions of 1973 and 1974 may have produced a more lasting revision of our scientific picture of the cosmos. It revealed the tremendously dynamic magnetic structure of the solar atmosphere and corona with myriads of closed magnetic flux tubes, containing hot plasma, bordering open magnetic regions, so-called coronal holes where the cooler plasma is associated with reduced X-ray emission. These early observations were finally superseded by the higher resolution images obtained from the Japanese satellite Yohkoh and the NASA-ESA Solar and Heliospheric Observatory SOHO, launched in 1992 and 1995, respectively. Meanwhile, the plasma physics picture of the solar system has been augmented considerably by planetary missions like Voyager 2 (launched 1989) travelling to the outer edges of the solar system and also measuring the magnetic fields of the giant planets (see Table B.6), or the flight of Ulysses (launched 1990), whereas Cluster II (launched in 2000) provides many more details of the three-dimensional structures of the magnetospheres. In the same period, the picture of the structure of the Galaxy and the Universe, essentially including galactic and cosmic magnetic fields, has evolved explosively due to the ever improved resolution of the traditional telescopes, the radio telescopes in large and very large arrays, and the numerous space missions, culminating in the launch of the Hubble Space Telescope in 1990. At the time of this writing, the Parker Solar Probe and the Solar Orbiter are about to be launched to reach positions extremely close to the Sun to finally, hopefully, disclose the mysteries of the coronal temperature and of the solar wind acceleration.

We summarize by making a few sweeping statements, obviously not meant to present final scientific truths:

– By means of X-ray observations, the few minutes of a solar eclipse have been extended almost indefinitely to provide a picture of the corona as a high-temperature plasma with *extremely complex dynamical magnetic structures* (Priest [510]).

– The interaction of the solar wind with the planetary magnetospheres is one of the most interesting plasma laboratories in space, offering the possibility of studying *spatially resolved plasma dynamics*.

– Finally, since the Sun is an ordinary star, what has been learned there may be extrapolated to other stars, of course with due modifications (Schrijver and Zwaan [544]). Going one step further, including neutron stars and pulsars (Mestel [438]), accretion disks about compact objects, etc.: what has been learned from magnetic plasma structures in the solar system may be extrapolated, again with due modifications, to the more exotic astrophysical objects that cannot be observed with spatial resolution but that do provide *intricate temporal signatures*.

Thus, a secondary layer (considering gravity and nuclear fusion as the primary layer) of phenomena has been revealed in the solar system that is present everywhere in the Universe. This brings us to our next subject.

1.3.4 The standard view of nature

Consider the standard view of nature, as developed in the twentieth century and widely held to provide the correct scientific representation of the Universe (Fig. 1.8). The four fundamental forces govern phenomena at immensely separate length scales, at least at times beyond 'The First Three Minutes' (Steven Weinberg, 1978) after the big bang. At the risk of caricaturing the wonderful achievements of elementary particle physics, on a scale of increasing dimensions, the weak and strong nuclear forces in the end just produce the different kinds of nuclei and electrons which constitute the main building blocks of matter. In a sense, these forces are exhausted beyond the length scale of 10^{-15} m. Since nuclei are positively charged and electrons negatively, the much longer

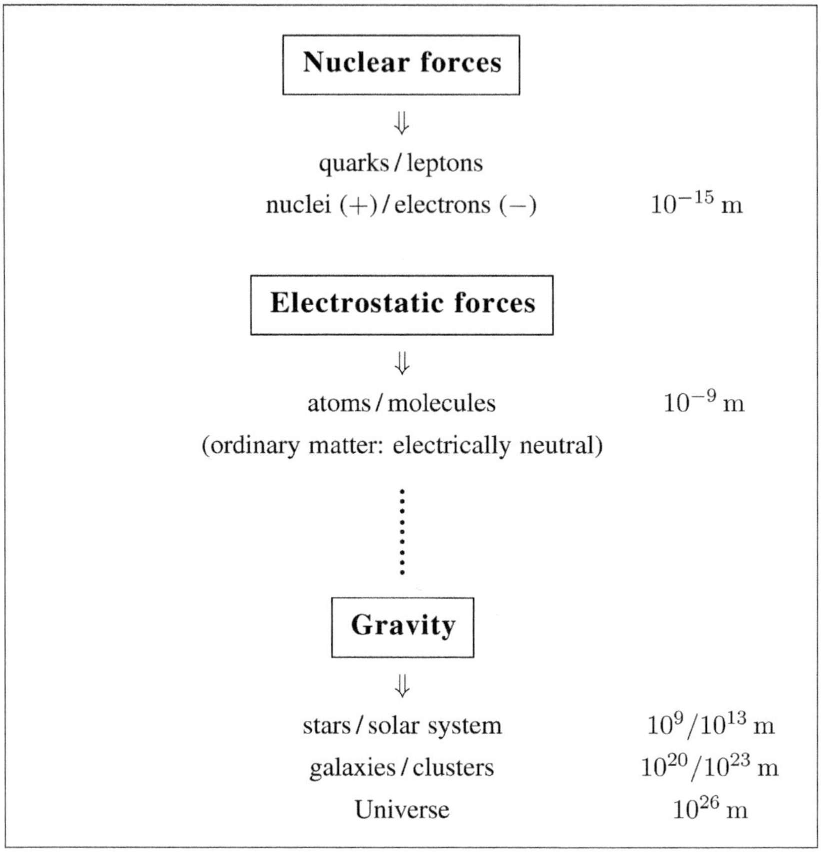

Fig. 1.8 The standard view of nature: not accounting for the abundance of plasma in the Universe.

range electric forces then take over, giving rise to the next stage of the hierarchy: 'ordinary' matter consisting of atoms and molecules with sizes of the order of 10^{-9} m. Since these particles are electrically neutral, all there appears to remain is the gravitational force which requires the collective effect of huge amounts of matter over length scales beyond 10^9 m in order to become sizeable. This gives rise to the different astronomical systems of stars, galaxies, clusters of galaxies, etc. Since the gravitational force is a long-range force which is solely attractive (there is no screening by repulsive negative mass particles), this force is only 'exhausted' at the scale of the Universe itself.

It will be noticed that the 'picture' of Fig. 1.8 jumps the eighteen orders of magnitude from atoms to stars (indicated by the dots) under the assumption that nothing of fundamental interest happens there. One could remark that we just happen to live on the least interesting level of the physical Universe, or one could dwell on the disproportion of man between the infinities of the small and the large (Pascal), or one could join the recent chorus of holistic criticism on the reductionism of physics. So much appears to be correct in the latter viewpoint that the given picture does not have any place for the complexities of solid state physics, fluid dynamics or biological systems, to name just a few. It should come as a big disappointment that nature would hang together from elementary particles to cosmology without really involving the intermediate stages.

For our subject, however, another misrepresentation is implicit in Fig. 1.8. We have started our discussion in Section 1.1 by noting that more than 90% of matter in the Universe is plasma so that *the Universe does not consist of ordinary matter (in the usual sense) but most of it is plasma*! It is true that plasma is usually also almost electrically neutral, like atoms and molecules, but the important difference is that the ions and electrons are not tied together in atoms but *move about freely as fluids*. The large-scale result of this dynamics is the formation of *magnetic fields* which in turn determine the plasma dynamics: a highly nonlinear situation. These magnetic fields not only bridge the gap between microscopic and macroscopic physics, but they also reach far into the astrophysical realm at all scales. Hence, the subject of plasma-astrophysics is of basic importance for understanding phenomena occurring everywhere in the Universe.

It will have been noted that we have ignored the unification of electric and magnetic forces brought about by Maxwell's theory of electromagnetism. There is a good reason for this since, in the domain of plasma dynamics, electric and magnetic forces are associated with quite different effects operating on immensely different length scales with the magnetic forces dominating on the longer length scales. Consequently, most of plasma dynamics is well described by exploiting the so-called pre-Maxwell equations, i.e. Maxwell's equations without the displacement term. We will see in later chapters that the dynamics of magnetic fields is so interwoven with the dynamics of the plasma itself that its proper description takes precedence over the one where electric and magnetic fields are treated on an equal footing.

The most important law for magnetic fields is $\nabla \cdot \mathbf{B} = 0$, which implies that there are no sources or sinks. This law is incompatible with spherical symmetry so that the simplest basic geometries of magnetised plasmas are completely different from the ones prevailing on the atomic and gravitational scales. In particular, large-scale tubular magnetic structures occur which move with the plasma so that magnetic forces are transmitted with the fluid. One could hardly imagine a bigger contrast with central electrostatic and gravitational forces decaying in vacuum with distance as r^{-2} ! Striking examples are solar flares, the X-ray emitting corona of the Sun, and coronal mass ejections (plasma expelled from the main body of the Sun against the gravitational pull), the interaction of the solar wind with the planetary magnetic fields, waves and flows in neutron star magnetospheres, extragalactic jets, spiral arm instabilities, etc.

In conclusion: the standard view of nature fails over a wide range of scales because it does not recognize the presence of magnetized plasmas all over the Universe. Magnetic fields are an important aspect of modern astrophysics. Hence, *the nonlinear interaction of plasma and complex magnetic structures* presents itself as an important common theme of laboratory and astrophysical plasma research.

1.4 Definitions of the plasma state

1.4.1 Microscopic definition of plasma

Turning now to the subject of plasmas proper, we need to refine the crude definition given in Section 1.1. This involves a closer study of the microscopic properties required for the plasma state. To that end, we follow the exposition given by F. F. Chen in the first chapter of his book on Plasma Physics [117].

First, we need to relax the condition of complete ionization given in our crude definition since plasma behaviour is already encountered when the ionization is only partial. A simpler definition of plasma would then be: *a plasma is an ionized gas*. However, how much ionization is required? An estimate may be obtained from the *Saha equation* which gives an expression for the amount of ionization of a gas in thermal equilibrium:

$$\frac{n_\mathrm{i}}{n_\mathrm{n}} = \left(\frac{2\pi m_\mathrm{e} k}{h^2}\right)^{3/2} \frac{T^{3/2}}{n_\mathrm{i}}\, \mathrm{e}^{-U_\mathrm{i}/kT} \approx 2.4 \times 10^{21}\, \frac{T^{3/2}}{n_\mathrm{i}}\, \mathrm{e}^{-15.8\times 10^4/T} . \qquad (1.26)$$

Here, n_i and n_n are the particle densities of ions and neutrals (in m^{-3}), U_i is the ionization energy (in J), T is the temperature (in K) and the other symbols have their usual meaning. The numbers on the RHS are obtained by exploiting Table B.1, $(2\pi m_\mathrm{e} k/h^2)^{3/2} = 2.4 \times 10^{21}\,\mathrm{K}^{-3/2}\,\mathrm{m}^{-3}$, and inserting the ionization energy of hydrogen, $U_\mathrm{i} = 13.6\,\mathrm{eV}$. (Ionization energies are usually given in eV, where $1\,\mathrm{eV} = 1.6 \times 10^{-19}\,\mathrm{J}$, which corresponds with $1.16 \times 10^4\,\mathrm{K}$ if one divides by the Boltzmann factor k.)

For *air* at room temperature, where $n_\mathrm{n} = 3 \times 10^{25}\,\mathrm{m}^{-3}$, $T = 300\,\mathrm{K}$, $U_\mathrm{i} = 14.5\,\mathrm{eV}$ (ionization potential for nitrogen), one finds a huge negative factor (-560) in the exponent of Eq. (1.26) so that the final ratio of the densities of ions and neutrals is extremely small: $n_\mathrm{i}/n_\mathrm{n} \approx 2 \times 10^{-122} \ll 1$. As expected, the degree of ionization is totally negligible: air is not a plasma. For hydrogen in *a tokamak machine* with $T = 10^8\,\mathrm{K}$ and $n \equiv n_\mathrm{e} = n_\mathrm{i} = 10^{20}\,\mathrm{m}^{-3}$, one finds that the expression in the exponent $U_\mathrm{i} \ll kT$ so that $\exp(U_\mathrm{i}/kT) \approx 1$ and $n_\mathrm{i}/n_\mathrm{n} \approx 2.4 \times 10^{13} \gg 1$: in such machines genuine plasmas are obtained. However, for *the core of the Sun* with $T = 1.6 \times 10^7\,\mathrm{K}$ and $n = 10^{32}\,\mathrm{m}^{-3}$, one finds that $n_\mathrm{i}/n_\mathrm{n} \approx 1.5$. Surprisingly, although thermonuclear reactions take place, ionization is not complete in the core of the Sun and plasma behaviour is not completely dominant! This is due to the extremely high densities there. On the other hand, *in the corona of the Sun*, with typical values of $T = 10^6\,\mathrm{K}$ (not thermonuclear, but anomalously high: a subject of concern in later chapters) and $n = 10^{12}\,\mathrm{m}^{-3}$, we have $n_\mathrm{i}/n_\mathrm{n} = 2.4 \times 10^{18}$: matter in the corona is an excellent plasma!

Even though we now have a measure for the degree of ionization required for plasmas, we still do not have a criterion for plasma behaviour. A much more precise definition, as given by Chen, reads: *a plasma is a quasi-neutral gas of charged and neutral particles which exhibits collective behaviour*.

In an ordinary gas neutral molecules move about freely (there are no net electromagnetic forces) until a collision occurs. This is a *short-range binary* event in which two particles hit each other. In a hard-sphere model of the molecules, the cross-section for such a collision is just the cross-section of the particles. In a plasma, on the other hand, the charged particles are subject to *long-range collective* Coulomb interactions with many distant encounters (so-called pitch angle scattering). Although the electrostatic force between two charged particles decays with the mutual distance ($\sim 1/r^2$), the combined effect of all charged particles may not even decay, since the interacting volume increases ($\sim r^3$). This is a typical collective effect, the result of the statistics of many particles, each moving in the average electrostatic field created by all the other particles.

We now discuss this electrostatic collective aspect quantitatively. For collective plasma behaviour, according to Chen [117], three conditions should be satisfied.

(a) The long-range Coulomb interaction between charged particles should dominate over the short-range binary collisions with neutrals. Indicating typical *time scales* of collective oscillatory motion by τ ($\sim 1/\omega$ when ω is the angular frequency of the oscillations), this implies that

$$\tau \ll \tau_{\rm n} \equiv \frac{1}{n_{\rm n}\sigma v_{\rm th}} \approx \frac{10^{17}}{n_{\rm n}\sqrt{T}}\,, \tag{1.27}$$

where $\tau_{\rm n}$ is the mean time between collisions of charged plasma particles with neutrals. The estimate on the RHS is obtained by writing $\tau_{\rm n} \approx \lambda_{\rm mfp}/v_{\rm th}$, where $\lambda_{\rm mfp}$ is the mean free path and $v_{\rm th}$ is the thermal speed of the particles. With $\lambda_{\rm mfp} = (n_{\rm n}\sigma)^{-1}$, where the cross-section $\sigma = \pi a^2 \approx 10^{-19}\,{\rm m}^2$ is obtained by taking the radius $a \approx 2 \times 10^{-10}\,{\rm m}$ of a neutral H atom, and $v_{\rm th} \approx (kT/m_p)^{1/2} \approx 100\sqrt{T}$, we obtain expression (1.27) for $\tau_{\rm n}$. Since we are interested in plasma conditions, we should convert this expression from neutral density $n_{\rm n}$ to ion density $n_{\rm i}$ by means of the Saha equation (1.26). For solar coronal plasma with $T = 10^6\,{\rm K}$ and $n_{\rm i} = 10^{12}\,{\rm m}^{-3}$, so that $n_{\rm n} = 4 \times 10^{-7}\,{\rm m}^{-3}$, this implies $\tau \ll \tau_{\rm n} \approx 2 \times 10^{20}\,{\rm s}$ (!). For tokamaks with $T = 10^8\,{\rm K}$ and $n_{\rm i} = 10^{20}\,{\rm m}^{-3}$, the condition becomes $\tau \ll \tau_{\rm n} \approx 2.4 \times 10^6\,{\rm s}$. Clearly, the condition (1.27) represents very mild restrictions on the time scales for plasma behaviour.

(b) The *length scale* of plasma dynamics should be much larger than the minimum size over which the condition of *quasi-neutrality* holds. Production of overall charge imbalance creates huge electric fields which in turn produce huge accelerations, so that such an imbalance is neutralized almost instantaneously and the plasma maintains charge neutrality to a high degree of accuracy. However, local charge imbalances may be produced by thermal fluctuations. To estimate their size, one should compare the thermal energy kT of the particles with their electrostatic energy $e\Phi$. The latter can be estimated through Poisson's law, $dE/dx = -d^2\Phi/dx^2 = -(1/\epsilon_0)\,en$, so that $kT \approx e\Phi \approx (1/\epsilon_0)\,e^2 n \lambda_{\rm D}^2$. Here, the gradient length has been equated to the *Debye length*, which is the typical size of a region over which charge imbalance due to thermal fluctuations may occur. Hence, length scales for a quasi-neutral plasma should satisfy

$$\lambda \gg \lambda_{\rm D} \equiv \sqrt{\frac{\epsilon_0 kT}{e^2 n}} \approx 70\sqrt{\frac{T}{n}}\,, \tag{1.28}$$

where $n \equiv n_{\rm e} \approx Z n_{\rm i}$ (with Z the ion charge number). Inserting the numbers for coronal plasma again, we find $\lambda_{\rm D} = 0.07\,{\rm m}$. For typical transverse length scales of coronal loops, $\lambda \sim 10000\,{\rm km} = 10^7\,{\rm m}$, the condition is easily satisfied.

▷ **Exercise** Exploit the tables of Appendix B to also find out what this condition means for other cases, like tokamak plasmas. ◁

Note that the concept of Debye length alleviates our original statement about long-range electrostatic forces considerably: sizeable regions with charge accumulation do not form through thermal fluctuations alone. A free charge, which in vacuum would have a potential $\Phi = q/r$, in a plasma is surrounded by a cloud of particles of opposite charge, which effectively shields the Coulomb potential for distances much larger than the Debye length: $\Phi_{\rm eff} = (q/r)\,\exp(-r/\lambda_{\rm D})$ (called *Debye shielding*). This just implies that $Z\,n_{\rm i} \approx n_{\rm e}$, i.e. *quasi* charge-neutrality holds. It does not mean that electric fields do not arise in plasmas. Actually, quite the opposite: electric fields arise almost automatically when plasmas move in a magnetic field. However, charge imbalances are extremely small when measured in terms of the total charge of the separate species:

$$|Zn_{\rm i} - n_{\rm e}|/n_{\rm e} \ll 1\,. \tag{1.29}$$

Hence $Zn_{\rm i} \approx n_{\rm e}$ holds to a high degree of accuracy in plasmas.

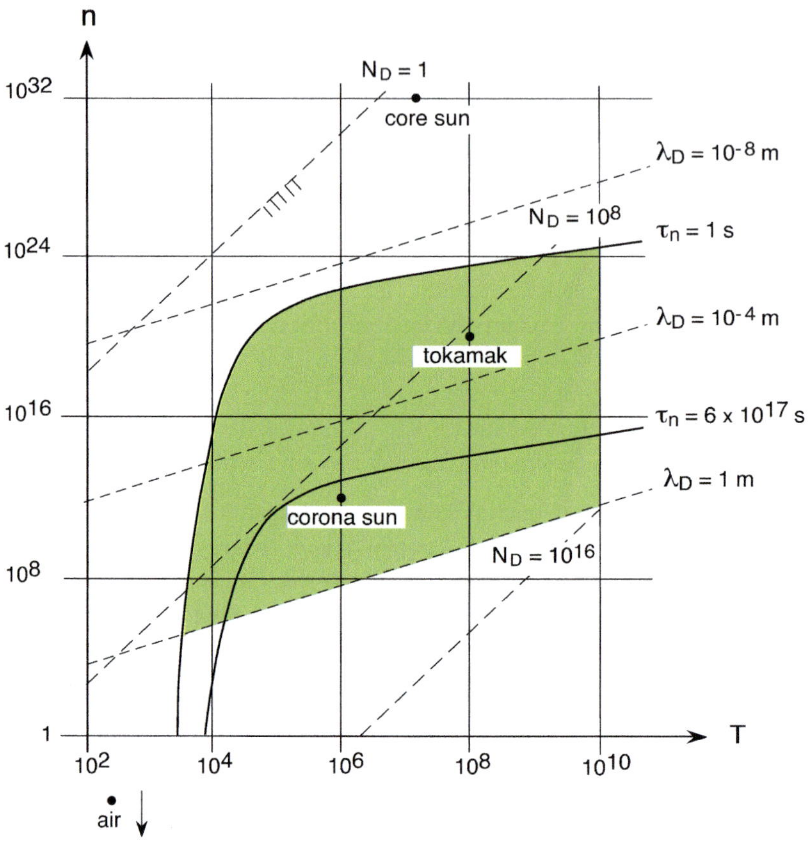

Fig. 1.9 Conditions for collective plasma behaviour, in terms of the density $n \equiv n_e \approx Zn_i$ and temperature $T \sim T_e \sim T_i$, are satisfied in the green area for time scales $\tau < \tau_n = 1\,\mathrm{s}$ and length scales $\lambda > \lambda_D = 1\,\mathrm{m}$, where $N_D \gg 1$. The restrictions on the upper time limit of low-density astrophysical plasmas quickly approach the age of the Universe, whereas the restrictions on the lower length limit for high density laboratory fusion experiments approach microscopic dimensions.

(c) Finally, in order for statistical considerations to be valid, sufficiently many particles should be present in a *Debye sphere*, i.e. a sphere of radius λ_D:

$$N_D \equiv \tfrac{4}{3}\pi\lambda_D^3\, n \approx 1.4 \times 10^6 \sqrt{T^3/n} \gg 1\,. \qquad (1.30)$$

For our example of a coronal plasma, this yields $N_D = 1.4 \times 10^9 \gg 1$, which is again easily satisfied. Note that both $\lambda_D \sim n^{-1/2}$ and $N_D \sim n^{-1/2}$ so that very high density plasmas are OK with respect to condition (1.28), but not with respect to condition (1.30). For example, for the core of the Sun, $\lambda_D = 3 \times 10^{-11}\,\mathrm{m}\,(!)$, but $N_D \approx 9$: not so good for the application of statistical mechanics.

In conclusion: collective plasma behaviour is encountered when the time scales are sufficiently short with respect to collision times with neutrals, $\tau \ll \tau_n$, the length scales are much larger than the Debye length, $\lambda \gg \lambda_D$ and there are many particles in a Debye sphere, $N_D \gg 1$. These conditions can be translated in terms of conditions on the density and the temperature, which are satisfied under a wide variety of conditions, as shown in Fig. 1.9. This picture confirms our statement of Section 1.1: plasma is a very normal state of matter in the Universe.

1.4.2 Macroscopic approach to plasma

So far, the most important physical variable in laboratory and astrophysical plasmas, viz. the magnetic field, has been conspicuously absent from our definition of the plasma state. The reason is that we have followed the traditional exposition of basic plasma theory, which starts with the microscopic point of view and stresses the collective phenomena involving electric fields. Whereas the length and time scales appropriate for these phenomena may be discussed in terms of the *local* values of the plasma density n and the temperature T, the magnetic field $\mathbf{B}$ brings in entirely different, *global*, considerations. (Incidentally, here one may detect one of the ways in which reductionism fails to recognize the emergence of new levels in the description of nature.) We have already observed the central importance of magnetic fields in confinement of fusion plasmas (Section 1.2) and in the dynamics of an enormous variety of astrophysical objects (Section 1.3), where we have stressed their basic non-locality. We now have to quantify these observations.

The macroscopic point of view does not set aside the microscopic conditions derived in Section 1.4.1 but it incorporates them as follows. A macroscopic description requires (1) frequent enough collisions between electrons and ions to establish fluid behaviour, (2) in addition to the microscopic conditions of length and time scales involving the density and temperature, global conditions on length and time scales involving the magnetic field. The latter quantities have to be large in order to permit *averaging over the microscopic dynamics*. To quantify this step requires the consideration of the cyclotron (or gyro) motion of the electrons and ions, which will only be discussed in the next chapter. Anticipating that discussion, the cyclotron radii $R_{e,i}$ and the inverse cyclotron frequencies $\Omega_{\rm e,i}^{-1}$ of the electrons and ions will be shown to be inversely proportional to the magnetic field strength, $R_{\rm e,i} \sim B^{-1}$ and $\Omega_{\rm e,i}^{-1} \sim B^{-1}$, where the ion expressions provide the most limiting conditions on macroscopic length and time scales. Consequently, 'large enough' means that macroscopic length and time scales should be much larger than $R_{\rm i}$ and $\Omega_{\rm i}^{-1}$, respectively. This is possible when the magnetic field is large enough for the plasma volume under consideration to contain many ion gyro radii and when the dynamic phenomena last many ion gyro periods.

Summarizing: *for a valid macroscopic model of a particular magnetized plasma dynamical configuration, size, duration, density and magnetic field strength should be large enough to establish fluid behaviour and to average out the microscopic phenomena (i.e. collective plasma oscillations and cyclotron motions).*

The distinguishing feature for macroscopic plasma dynamics is the interaction of plasma motion and magnetic field geometry. This fluid aspect of plasmas concerns the motion of the plasma as a whole, without considering the separate electrons and ions, under the influence of magnetic fields. These fields are, in turn, generated by the plasma motion itself: a highly nonlinear situation. *The* theoretical tool to describe this global interplay of plasma and magnetic field is called *MHD* $\equiv$ *magnetohydrodynamics*. The objective of this book is to demonstrate how this theory provides the common basis for the description of laboratory and astrophysical plasma dynamics.

The (surprisingly many) different aspects of the given definition of a macroscopic plasma model will be discussed one by one in the following chapters. In particular, Chapters 2 and 3 will provide the missing quantitative elements of microscopic plasma physics needed for the foundation of macroscopic plasma dynamics. The basic Chapter 2 may be skipped by readers that are already familiar with basic plasma physics. The advanced Chapter 3 may be skipped as well by readers that wish to start with magnetohydrodynamics proper as soon as possible.

One question, answered in detail in the next chapters, must be addressed at least provisionally here, viz.: why is the electric field not even mentioned in the above discussion of macroscopic plasma dynamics? The reason is that the electric field becomes, in fact, a secondary quantity in MHD. Large electrostatic fields due to charge imbalances only occur over Debye length scales, which are averaged out, and electromagnetic waves are absent in non-relativistic MHD since the displacement current is negligible. The electric field is then determined from the primary variables of the velocity $\mathbf{v}$ and the magnetic field $\mathbf{B}$ by means of 'Ohm's law' for a nearly perfectly conducting plasma: $\mathbf{E} + \mathbf{v} \times \mathbf{B} \approx 0$, i.e. the electric field in a frame moving with the plasma vanishes.

1.5 Literature and exercises

Notes on literature

Some general references for the whole book are given under the different headings below. The complete information on the references is given at the end of the book.

Introductory plasma physics

- Boyd and Sanderson, *Plasma Dynamics*, one of the older textbooks on plasma physics, has been revised completely in *The Physics of Plasmas* [84]. It may be recommended for complementary reading since it contains a thorough discussion of the various models used to describe plasma physics.
- Chen, *Introduction to Plasma Physics and Controlled Fusion* [117] is the most readable, and probably most widely used, basic textbook on plasma physics.
- Bittencourt, *Fundamentals of Plasma Physics* [70] is a basic theoretical course on plasma physics with detailed calculations.
- Sturrock, *Plasma Physics* [580] is a basic text on plasma physics written for graduate students from astrophysics, space science, physics and engineering departments.
- Goldston and Rutherford, *Introduction to Plasma Physics* [244] is a basic text on plasma physics based on teaching by two experts in tokamak physics.
- Hazeltine and Waelbroeck, *The Framework of Plasma Physics* [298] develops a basic understanding of plasma and the wide ranging applications of plasma behaviour in a thorough and systematic manner.
- Gurnett and Bhattacharjee, *Introduction to Plasma Physics* [272] is a comprehensive text on basic plasma theory with applications to space and laboratory plasmas.
- Fitzpatrick, *Plasma Physics, An Introduction* [178] provides the theoretical framework describing the most common plasmas of nature in a clear and concise way.

Topics in advanced plasma physics

- Leontovich (ed.), *Reviews of Plasma Physics*, Vols. 1–5 [402] contain the unsurpassed Russian expositions of the basics of plasma theory after the declassification of the 1958 Geneva Conference.
- Akhiezer, Akhiezer, Polovin, Sitenko and Stepanov, *Plasma Electrodynamics* [8] is another classic from one of the Soviet theory schools, systematically building up plasma physics by kinetic and hydrodynamic methods and progressing to the diverse linear and nonlinear manifestations of the plasma state.
- Dendy (ed.), *Plasma Physics: An Introductory Course* [150] contains the material taught at the yearly Culham summer schools on plasma physics.

- Donné, Rogister, Koch and Soltwisch (eds.), *Proc. Second Carolus Magnus Summer School on Plasma Physics* [162] contains the material taught at that summer /school held every other year.

Magnetohydrodynamics

- Freidberg, *Ideal MHD* [186] is a textbook on ideal MHD, based on lectures at MIT for graduate students and researchers, which puts perfect conductivity and the applications to fusion research centre stage.
- Lifschitz, *Magnetohydrodynamics and Spectral Theory* [410] is an advanced text on MHD stressing the unity of physics and mathematics through spectral theory. This kind of complex, yet faultless, calculations is a rare commodity in plasma physics.
- Davidson, *An Introduction to Magnetohydrodynamics* [140] is an introductory textbook on MHD for students in physics, applied mathematics and engineering, with stress on the fluid dynamics foundations and a wide variety of applications.
- Polovin and Demutskii, *Fundamentals of Magnetohydrodynamics* [502] is an introduction to MHD, written in the lucid style of the great Russian theoreticians.

Tokamaks

- Wesson, *Tokamaks* [647] is a veritable encyclopedia of the plasma physics involved in nuclear fusion research in tokamaks.
- White, *Theory of Toroidally Confined Plasmas* [649] contains the material of a graduate course at Princeton University on fundamental plasma theory of tokamaks.
- Hazeltine and Meiss, *Plasma Confinement* [296] provides the advanced theory of magnetic plasma confinement with stress on derivations from first principles.
- Braams and Stott, *Nuclear Fusion: Half a Century of Magnetic Confinement Fusion research* [86] gives the history of nuclear fusion research up to the present, leading up to the famous citation of Artsimovich "Fusion will be there when society needs it".

The Sun

- Priest, *Magnetohydrodynamics of the Sun* [510] is the classical introduction of magnetohydrodynamics of the Sun, in particular the solar corona.
- Stix, *The Sun* [569] is a textbook on the physics of the Sun with innumerable observational facts.
- Foukal, *Solar Astrophysics* [181] aims at making the advances in understanding of the Sun accessible to students and non-specialists by means of simple physical concepts and observations.

Space physics

- Hasegawa and Sato, *Space Plasma Physics* [293] is a monograph on the physics of stationary plasmas, small amplitude waves and the stationary magnetosphere.
- Kivelson and Russell (eds.), *Introduction to Space Physics* [368] is an introduction of all aspects of space and solar plasmas for senior undergraduate and graduate students, written by experts in the various fields.
- Baumjohann and Treumann, *Basic Space Plasma Physics*, and (same authors in reverse order) *Advanced Space Plasma Physics* [42] are the basic material presented in a space plasma physics course at the University of Munich, and the advanced nonlinear aspects of the various waves and instabilities.

Plasma astrophysics

- Battaner, *Astrophysical Fluid Dynamics* [40] is a systematic theoretical treatise of the dynamics of classical, relativistic, photon and plasma fluids, progressing from stars to the Universe at large.
- Choudhuri, *The Physics of Fluids and Plasmas* [122] is an introduction to fluid dynamics, plasma physics and stellar dynamics for graduate students of astrophysics.
- Mestel, *Stellar Magnetism* [438] is a monograph on MHD applied to the magnetism of stars, including stellar dynamos, star formation and pulsar electrodynamics.
- Kulsrud, *Plasma Physics for Astrophysics* [385] introduces plasma physics as a comprehensible field that can be grasped largely on the basis of physical intuition and qualitative reasoning.

Exercises

The exercises are meant to increase understanding of the principles of plasma dynamics. Estimating orders of magnitude is an essential part. Frequent use of the numerical appendices is recommended. Difficult problems are marked with a star.

[1.1] *Fusion reactions*

We know two methods of energy production by nuclear processes, namely nuclear fission and nuclear fusion. For both, the net energy released is described by the same formula.

- What is expressed by that formula? What is the major difference between fission and fusion? What is actually expressed by the following formula for the most likely reaction in future fusion reactors: $\mathrm{D}^2 + \mathrm{T}^3 \;\rightarrow\; \mathrm{He}^4\,(3.5\,\mathrm{MeV}) + \mathrm{n}\,(14.1\,\mathrm{MeV})$?

[1.2] *Fusion power*

If we want a tokamak reactor to produce energy and sustain itself, we need balance between thermonuclear power output and power losses. This leads to a condition on required particle density n, energy confinement time τ_{E}, and temperature $\widetilde{T}$.

- If the power output is given by $P_{\mathrm{T}} = n^2 f(\widetilde{T})$, where f is a known function of temperature, and the power losses consist of Bremsstrahlung, $P_{\mathrm{B}} = \alpha n^2 \widetilde{T}^{1/2}$, and heat transport, $P_{\mathrm{L}} = 3n\widetilde{T}/\tau_{\mathrm{E}}$, derive the criterion for fusion energy production. Assume that all power contributions can be converted in plasma heating with efficiency η.
- A more recent approach states that for ignition of a fusion reactor, α-particle heating of the plasma, P_α, should make up for the power losses. Express this criterion by a similar equation to that above. In that case, the product of the three mentioned quantities will have to be $n\tau_{\mathrm{E}}\widetilde{T} \approx 3 \times 10^{21}\,\mathrm{m}^{-3}\,\mathrm{s\,keV}$. Give some estimates for n, τ_{E} and $\widetilde{T}$. Using Table B.1, convert the temperature $\widetilde{T}$ in keV to T in degrees K.
- Why do magnetic fields play such an important role in thermonuclear fusion?

[1.3] *Solar plasmas*

All the light we receive is the result of specific nuclear reactions which release energy. For instance, nuclear reactions in the centre of the Sun are the ultimate cause of light escaping at positions where the Sun becomes optically thin for this radiation.

- Which nuclear reactions take place in the centre of the Sun? What kind of radiation is produced? Why doesn't this extremely energetic light escape right away? Why is the light we collect on Earth mostly in the visible range? Why is it the less energetic light that escapes from the Sun? Explain what this says about the density profile.
- Normally we cannot see the solar atmosphere further outwards since we are blinded by the escaping light. A beautiful exception occurs during a solar eclipse. Explain why we can observe it at all on Earth.
- The coronal structures visible during a solar eclipse are the footprints of a giant engine that produces the solar activity. What is the mechanism and how is that connected to the structures observed?

[1.4] *Plasma definitions and applications*

Putting plasmas in a wide perspective, discuss the following aspects.

- What is a plasma? How is it different from ordinary gases and fluids?
- Name some of the numerous applications of plasma physics.
- How can plasmas be confined?

[1.5] *Forces in nature*

Explain the major forces present in nature, together with their relative strength and decay distance and, thus, the scale at which they are dominant. Explain why gravity is such a special force. What forces dominate the plasma regime?

2
Elements of plasma physics

2.1 Theoretical models

Plasma processes are described by quite different theoretical models. Which one is to be chosen depends on the kind of phenomenon one is interested in. Broadly speaking, three kinds of theoretical description are used:

(a) the theory of the motion of *individual charged particles* in given magnetic and electric fields (Section 2.2);

(b) the *kinetic theory* of a collection of such particles, describing plasmas microscopically by means of particle distribution functions $f_{\mathrm{e,i}}(\mathbf{r}, \mathbf{v}, t)$ (Section 2.3);

(c) the *fluid theory* (magnetohydrodynamics), describing plasmas in terms of averaged macroscopic functions of $\mathbf{r}$ and t (Section 2.4).

By way of introduction, within each of these descriptions, we will give a simple example which illustrates the plasma property that is relevant for our subject, viz. plasma confinement by magnetic fields.

2.2 Single particle motion

2.2.1 Cyclotron motion

The motion of a charged, non-relativistic particle in an electric and magnetic field is described by the well-known equation of motion

$$m\frac{d\mathbf{v}}{dt} = q(\mathbf{E} + \mathbf{v} \times \mathbf{B}) \,, \tag{2.1}$$

where $\mathbf{E}(\mathbf{r}, t)$ and $\mathbf{B}(\mathbf{r}, t)$ are considered to be given (of course, in agreement with Maxwell's equations) and one has to solve for the particle velocity $\mathbf{v}(\mathbf{r}, t)$. For the moment, we do not specify the mass m and the charge q of the particles. They will be fixed later to correspond to either electrons ($m = m_{\mathrm{e}}$, $q = -e$) or ions with mass number A and charge number Z (i.e. multiples of the proton mass and charge: $m = m_{\mathrm{i}} = Am_{\mathrm{p}}$, $q = Ze$).

Consider a charged particle in a constant magnetic field, taken in the z-direction, in the absence of an electric field: $\mathbf{B} = B\mathbf{e}_z$, $\mathbf{E} = 0$. Performing two simple vector operations on Eq. (2.1) provides some preliminary insight: projecting on $\mathbf{B}$ (exploiting the vector identity (A.1) of Appendix A) gives

$$m\frac{dv_\parallel}{dt} = 0\,, \qquad \text{so that} \quad v_\parallel = \text{const}\,, \tag{2.2}$$

and projecting on **v** gives

$$\frac{d}{dt}(\tfrac{1}{2}mv^2) = 0 \Rightarrow \tfrac{1}{2}mv^2 = \text{const} \stackrel{(2.2)}{\longrightarrow} \tfrac{1}{2}mv_\perp^2 = \text{const} \Rightarrow v_\perp = \text{const}. \tag{2.3}$$

The expressions (2.2) and (2.3) already suggest the kind of orbits to be expected.

We now solve Eq. (2.1) systematically. With $\mathbf{v} = d\mathbf{r}/dt = (\dot{x}, \dot{y}, \dot{z})$ we obtain two coupled differential equations for the motion in the perpendicular plane:

$$\begin{aligned} \ddot{x} - (qB/m)\,\dot{y} &= 0\,, \\ \ddot{y} + (qB/m)\,\dot{x} &= 0\,. \end{aligned} \tag{2.4}$$

Defining *the gyro-frequency* or *cyclotron frequency*,

$$\Omega \equiv \frac{|q|B}{m}\,, \tag{2.5}$$

the solution to these equations represents a periodic circular motion about a point $x = x_c, y = y_c$ (*the guiding centre*):

$$\begin{aligned} x(t) &= x_c + (\dot{x}_0/\Omega)\sin\Omega t - \text{sgn}(q)(\dot{y}_0/\Omega)\cos\Omega t\,, \\ y(t) &= y_c + (\dot{y}_0/\Omega)\sin\Omega t + \text{sgn}(q)(\dot{x}_0/\Omega)\cos\Omega t\,. \end{aligned} \tag{2.6}$$

Here, $\text{sgn}(q)$ indicates the sign of the particle charge. Since the coupled differential equations constitute a fourth order system, there are four free constants. They are fixed by the choice of the initial positions x_0, y_0 and the initial velocities $\dot{x}_0$, $\dot{y}_0$, so that $x_c = x_0 + \text{sgn}(q)(\dot{y}_0/\Omega)$, $y_c = y_0 - \text{sgn}(q)(\dot{x}_0/\Omega)$. Check the property (2.3):

$$v_\perp = \sqrt{\dot{x}^2(t) + \dot{y}^2(t)} = \sqrt{\dot{x}_0^2 + \dot{y}_0^2} = \text{const}. \tag{2.7}$$

Also, $\sqrt{[x(t) - x_c]^2 + [y(t) - y_c]^2} = \sqrt{\dot{x}_0^2 + \dot{y}_0^2}\,/\Omega = v_\perp/\Omega = \text{const}$, so that *the gyro-radius* or *cyclotron radius*[1]

$$R \equiv \frac{v_\perp}{\Omega} = \text{const}\,. \tag{2.8}$$

Hence, the complete orbit consists of gyration $\perp$ **B** and inertial motion $\parallel$ **B**:

$$\begin{aligned} x(t) &= x_c + R\cos(\Omega t - \theta_0)\,, \\ y(t) &= y_c - \text{sgn}(q)R\sin(\Omega t - \theta_0)\,, \\ z(t) &= z_c(t) = z_0 + v_\parallel t\,, \end{aligned} \tag{2.9}$$

where θ_0 is the initial value of the polar angle in the perpendicular plane. This helical motion already

[1] The expressions (2.5) and (2.8) are usually called the Larmor frequency and the Larmor radius. Those are misnomers: the Larmor motion proper refers to the precession of a magnetic dipole in an applied magnetic field, as in the classical theory of nuclear magnetic resonance. That precession frequency is an entirely different physical effect, as evidenced by the fact that its value is half of that given by Eq. (2.5).

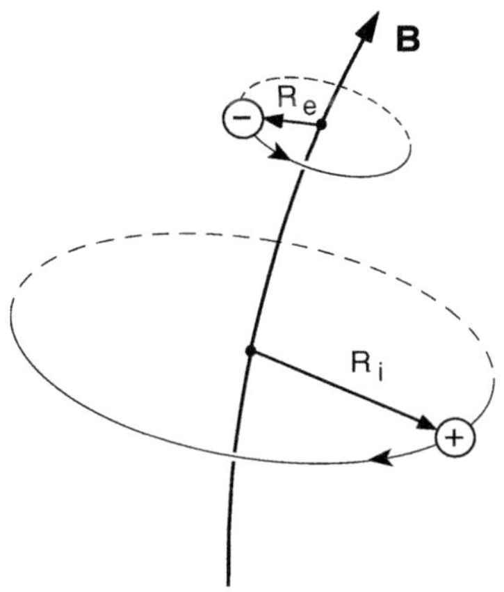

Fig. 2.1 Gyration of electrons and ions in a magnetic field: disparate length and time scales.

constitutes an important confining feature of a magnetic field: *charged particles stick to the field lines*. In other words: the magnetic field **B** determines *the geometry of the dynamics* of both kinds of particles and, hence, of the plasma.

Electrons and ions gyrate in opposite directions (Fig. 2.1) with quite different gyro-frequencies and gyro-radii because of the smallness of the mass ratio m_e/m_i:

$$\begin{aligned} \Omega_e &\equiv \frac{eB}{m_e} \quad \gg \quad \Omega_i \equiv \frac{ZeB}{m_i}\,, \\ R_e &\equiv \frac{v_{\perp,e}}{\Omega_e} \quad \ll \quad R_i \equiv \frac{v_{\perp,i}}{\Omega_i} \qquad \text{(assuming } T_e \sim T_i\text{)}\,. \end{aligned} \tag{2.10}$$

Inserting a typical value for the magnetic field in tokamaks (see Table B.3), viz. $B = 3\,\mathrm{T}$ (= 30 kgauss), and inserting the values for e, m_e and m_p of Table B.1, we find for the angular frequencies of protons and electrons

$$\begin{aligned} \Omega_e &= 5.3\times10^{11}\,\mathrm{rad\,s^{-1}} \quad \text{(i.e., a frequency of 84 GHz)}\,, \\ \Omega_i &= 2.9\times10^{8}\,\mathrm{rad\,s^{-1}} \quad \text{(i.e., a frequency of 46 MHz)}\,. \end{aligned} \tag{2.11}$$

To estimate the gyro-radii, consider particles with thermal speed[2] so that $v_\perp = v_{th} \equiv \sqrt{2kT/m}$. For electrons and protons at $\widetilde{T} = 10\,\mathrm{keV}$, i.e. $T_e = T_i = 1.16\times10^8\,\mathrm{K}$, this implies

$$\begin{aligned} v_{th,e} &= 5.5\times10^{7}\,\mathrm{m\,s^{-1}} \quad \Rightarrow \quad R_e = 1.1\times10^{-4}\,\mathrm{m} \approx 0.1\,\mathrm{mm}\,, \\ v_{th,i} &= 1.3\times10^{6}\,\mathrm{m\,s^{-1}} \quad \Rightarrow \quad R_i = 4.5\times10^{-3}\,\mathrm{m} \approx 5\,\mathrm{mm}\,. \end{aligned} \tag{2.12}$$

This gives an impression of the time and length scales for gyro-motion in a tokamak. Clearly, these time scales are very small compared to the typical time scales needed for thermonuclear fusion and the length scales are small compared to the macroscopic dimensions of a tokamak machine. This permits averaging over the gyro-motion in the macroscopic description of plasma dynamics.

[2] The convention for the definition of the thermal speed v_{th} differs for different authors: some have the factor $\sqrt{2}$, like we do, others do not have it.

▷ **Exercise** Consult Table B.3 for the values of the gyration parameters of a solar plasma. Also compute those values for solar wind, magnetosphere and your favourite plasma from Table B.5: it is important to familiarize yourself with the different orders of magnitude! ◁

One significant feature of the gyro-frequency is its dependence on the value of the magnetic field alone. Except for the fundamental constants of nature, no other quantities appear in the expressions (2.10)(a). (Here, and in the following, the letter in brackets will indicate the member of the set of equations.) Hence, detection of oscillatory motion at the cyclotron frequencies provides an important means for the determination of the value of the magnetic field in plasmas.

Since the equation of motion for the perpendicular motion is a fourth order differential equation (two coupled second order equations), quite complicated drift motions can occur if the electromagnetic fields $\mathbf{B}$ and $\mathbf{E}$ are not constant in time or if they are inhomogeneous in space. One then introduces expansions exploiting the smallness of the gyro-radius of the particles as compared to the length scale of the field inhomogeneities. This is the subject of individual *orbit theory*. For many applications, time and length scales of phenomena are large enough to permit averaging over the rapid gyration of the particles so that only the motion of the guiding centre needs to be considered. This is called the *guiding centre approximation*. We return to this subject in Section 2.2.3.

2.2.2 Excursion: Basic equations of electrodynamics and mechanics

We have treated the simple non-relativistic motion of a charged particle in a constant, prescribed, magnetic field. Before we proceed to motion in more complex fields, it is useful to recall some of the basic equations from electrodynamics and particle mechanics.

In plasma theory, where we consider the dynamics of a collection of charged particles in electromagnetic fields, the appropriate form of *Maxwell's equations* in mks units is given by:

$$\nabla \times \mathbf{E} = -\frac{\partial \mathbf{B}}{\partial t} \qquad \text{(Faraday)}\,, \tag{2.13}$$

$$\nabla \times \mathbf{B} = \mu_0 \mathbf{j} + \frac{1}{c^2}\frac{\partial \mathbf{E}}{\partial t} \qquad \text{('Ampère')}\,, \qquad c^2 \equiv (\epsilon_0 \mu_0)^{-1}\,, \tag{2.14}$$

$$\nabla \cdot \mathbf{E} = \frac{\tau}{\epsilon_0} \qquad \text{(Poisson)}\,, \tag{2.15}$$

$$\nabla \cdot \mathbf{B} = 0 \qquad \text{(no magnetic monopoles)}\,. \tag{2.16}$$

We have ignored polarization and magnetization effects, i.e. $\epsilon = \epsilon_0$ and $\mu = \mu_0$ so that $\mathbf{D} = \epsilon_0 \mathbf{E}$ and $\mathbf{H} = (\mu_0)^{-1}\mathbf{B}$, since these effects are absorbed in the definitions of charge and current density:

$$\tau = \sum_\alpha q_\alpha n_\alpha\,, \qquad \mathbf{j} = \sum_\alpha q_\alpha n_\alpha \mathbf{u}_\alpha \qquad (\alpha = \mathrm{e}, \mathrm{i})\,. \tag{2.17}$$

Here, n_α and $\mathbf{u}_\alpha$ are the particle density and the macroscopic velocity of particles of type α. This implies that *the plasma is viewed as existing of point charges moving in the electromagnetic fields, which they partially create themselves*. Obviously, adding an equation of motion of the form of Eq. (2.1) for every particle of the plasma would constitute a complete dynamical problem, but it would be foolish to proceed from that point of view since one would have to solve, say, 10^{20} equations of motion. In Section 2.3, we will see how this problem may be reformulated by means of a statistical approach.

In Section 2.4, we will introduce the next level of description, viz. the macroscopic approach to plasmas. For the majority of macroscopic plasma phenomena, the displacement current $\epsilon_0 \partial \mathbf{E}/\partial t$ and Poisson's equation are unimportant and may be dropped from Eqs. (2.14) and (2.15). In that case, Faraday's law expresses the dynamics of the magnetic field, 'Ampère's' law will become Ampère's law again, expressing the relation between the current and the magnetic field, and $\nabla\cdot\mathbf{B}=0$ is the usual restriction on the initial conditions for $\mathbf{B}$. In this description in terms of the so-called *pre-Maxwell's equations*, the magnetic field is a more basic quantity than the electric field.

Of course, to get *electromagnetic waves* the displacement current and Poisson's equation are essential. Keeping these terms, for a *vacuum* ($\tau = 0, \mathbf{j} = 0$) we get from the original set of Maxwell's equations (2.13)–(2.16) two identical wave equations for $\mathbf{E}$ and $\mathbf{B}$:

$$\nabla^2 \mathbf{E} - \frac{1}{c^2}\frac{\partial^2 \mathbf{E}}{\partial t^2} = 0 , \qquad \nabla^2 \mathbf{B} - \frac{1}{c^2}\frac{\partial^2 \mathbf{B}}{\partial t^2} = 0 , \tag{2.18}$$

where $c = (\epsilon_0 \mu_0)^{-\frac{1}{2}}$ is the velocity of light in vacuum. Considering plane wave solutions

$$\mathbf{E}(\mathbf{r}, t) = \hat{\mathbf{E}}\, \mathrm{e}^{\mathrm{i}(\mathbf{k}\cdot\mathbf{r} - \omega t)} , \qquad \mathbf{B}(\mathbf{r}, t) = \hat{\mathbf{B}}\, \mathrm{e}^{\mathrm{i}(\mathbf{k}\cdot\mathbf{r} - \omega t)} , \tag{2.19}$$

where $\hat{\mathbf{E}}$ and $\hat{\mathbf{B}}$ are complex (the physical quantities correspond to the real parts of $\mathbf{E}$ and $\mathbf{B}$), we just obtain the familiar relation between angular frequency ω and wave vector $\mathbf{k}$:

$$\omega^2 = k^2 c^2 . \tag{2.20}$$

Inserting these relations back into Eqs. (2.13)–(2.16) we find

$$\begin{aligned} &\omega\hat{\mathbf{B}} = \mathbf{k}\times\hat{\mathbf{E}} , \qquad && \omega\hat{\mathbf{E}} = -c^2\,\mathbf{k}\times\hat{\mathbf{B}} , \\ &\mathbf{k}\cdot\hat{\mathbf{B}} = 0 , && \mathbf{k}\cdot\hat{\mathbf{E}} = 0 , \end{aligned} \tag{2.21}$$

so that $|\hat{\mathbf{E}}| = c|\hat{\mathbf{B}}|$ and the vectors $\{\hat{\mathbf{E}}, \hat{\mathbf{B}}, \mathbf{k}\}$ form an orthogonal triad: electromagnetic waves are *transverse waves* where $\mathbf{E}$ and $\mathbf{B}$ oscillate in a plane perpendicular to the direction of propagation given by $\mathbf{k}$.

For the sake of reference, we present *the Lorentz transformation* of two inertial frames [331] moving with relative velocity $\mathbf{v}$:

$$\begin{aligned} \mathbf{x}' &= \mathbf{x} + \frac{\Gamma - 1}{v^2}\mathbf{v}\,\mathbf{v}\cdot\mathbf{x} - \Gamma\mathbf{v}\,t , \\ t' &= \Gamma\Big(t - \frac{1}{c^2}\mathbf{v}\cdot\mathbf{x}\Big) , \qquad \Gamma \equiv \frac{1}{\sqrt{1 - v^2/c^2}} . \end{aligned} \tag{2.22}$$

The electromagnetic fields transform according to

$$\begin{aligned} \mathbf{E}' &= \Gamma\Big(\mathbf{E} + \mathbf{v}\times\mathbf{B}\Big) - \frac{\Gamma^2}{(\Gamma+1)\,c^2}\mathbf{v}\,\mathbf{v}\cdot\mathbf{E} , \\ \mathbf{B}' &= \Gamma\Big(\mathbf{B} - \frac{1}{c^2}\mathbf{v}\times\mathbf{E}\Big) - \frac{\Gamma^2}{(\Gamma+1)\,c^2}\mathbf{v}\,\mathbf{v}\cdot\mathbf{B} . \end{aligned} \tag{2.23}$$

Notice that the symmetry between $\mathbf{E}$ and $\mathbf{B}$ is lost since we exploit mks units.

A consistent treatment of particles in electromagnetic waves would require the replacement of the classical equation of motion (2.1) by the relativistic one:

$$\frac{d\mathbf{p}}{dt} = q(\mathbf{E} + \mathbf{v}\times\mathbf{B}) , \tag{2.24}$$

where $\mathbf{p}$ is the relativistic momentum of the particle. *The relativistic expressions* for the energy E (not to be confused here with the electric field $\mathbf{E}$) and the momentum $\mathbf{p}$ of particles moving with the velocity $\mathbf{v}$ are given by:

$$\begin{aligned} E &= \Gamma mc^2 \quad (\approx mc^2 + \tfrac{1}{2}mv^2 \text{ for } v \ll c)\,, \\ \mathbf{p} &= \Gamma m\mathbf{v} \quad (\approx m\mathbf{v} \text{ for } v \ll c)\,. \end{aligned} \tag{2.25}$$

Hence, $E = \sqrt{p^2c^2 + m^2c^4}$, where E includes the rest mass energy mc^2. For photons, the rest mass $m \to 0$ and $v \to c$ so that the expressions (2.25) for E and $\mathbf{p}$ become undetermined but their relation is still given by $E = pc$.

Quantum mechanical expressions for the energy and momentum of particles, including photons, involve Planck's constant h:

$$E = h\nu\,, \qquad p = h/\lambda\,,$$

where ν and λ are the de Broglie frequency and wavelength of the wave functions associated with the particles. It is convenient to include the direction of the momentum vector $\mathbf{p}$ in these relations. This is done by introducing the wave vector $\mathbf{k}$, where $|\mathbf{k}| \equiv 2\pi/\lambda$, and defining the angular frequency $\omega \equiv 2\pi\nu$ and $\hbar \equiv h/2\pi$:

$$E = \hbar\omega\,, \qquad \mathbf{p} = \hbar\mathbf{k}\,. \tag{2.26}$$

These expressions are also valid for photons, for which frequency and wavelength are related by

$$\nu = c/\lambda\,, \qquad \text{or} \quad \omega = kc\,, \tag{2.27}$$

according to the dispersion equation (2.20), so that we recover the relation $E = pc$ again.

In this book, quantum mechanical effects will not play an important role (see Ref. [273] for that topic) and relativistic effects only enter in Chapter 22. It is of interest though to extend the analysis of Section 2.2.1 to the *cyclotron motion of relativistic particles* since high-energy non-thermal electrons frequently occur in plasmas (e.g., 60 keV electrons require a relativistic correction $v^2/c^2 \sim 0.2$). Consider again a charged particle moving in a constant magnetic field $\mathbf{B}$, but now exploit Eq. (2.24) rather than Eq. (2.1):

$$\frac{d\mathbf{p}}{dt} = \frac{q}{\Gamma m}\,\mathbf{p} \times \mathbf{B}\,. \tag{2.28}$$

Performing the operations of projection onto $\mathbf{B}$ and $\mathbf{p}$ again, we find that $p_\parallel = \text{const}$ and $|\mathbf{p}| = \text{const}$ so that, according to Eq. (2.25), $v = \text{const}$ and the relativistic factor $\Gamma = \text{const}$. Hence, particles turn out to gyrate around the magnetic field with the modified gyro-frequency

$$\Omega = \frac{|q|B}{\Gamma m}\,, \tag{2.29}$$

whereas the corresponding gyro-radius becomes

$$R = \frac{p_\perp}{|q|B} = \frac{v_\perp}{\Omega}\,. \tag{2.30}$$

The ratio $p_\perp/|q|$ $(= RB)$, which depends on the particle properties only, is a measure of how little a particle can be deflected by a magnetic field. It is called *the magnetic rigidity* of a particle. Energies of cosmic ray particles are often expressed in terms of this quantity.

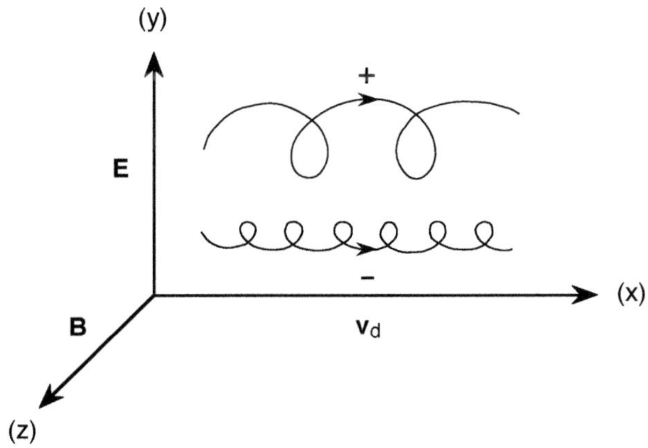

Fig. 2.2 Drift of ions and electrons in crossed electric and magnetic fields.

After this excursion, we *return to classical plasmas* in the double sense of non-relativistic (i.e. neglect of v/c terms and displacement current in 'Ampère's' equation) and non-quantum mechanical (i.e. neglect of the discreteness of the energy expressed by Eq. (2.26)).

2.2.3 Drifts, adiabatic invariants

In Section 2.2.1, we considered the gyro-motion of charged particles in a constant magnetic field $\mathbf{B}$. Let us now add a constant electric field $\mathbf{E}$ to the problem. Projecting Eq. (2.1) on $\mathbf{B}$ then gives

$$m\frac{dv_\parallel}{dt} = qE_\parallel \,, \tag{2.31}$$

which represents a constant acceleration along the magnetic field. In hot plasmas, such accelerations may lead to a high-energy tail of so-called *runaway electrons* when the electric field exceeds a certain critical value which depends on the electron–ion collision frequency.

For the present purpose, it is more relevant to consider the effect of a perpendicular electric field: $\mathbf{E}\ (= E\mathbf{e}_y) \perp \mathbf{B}\ (= B\mathbf{e}_z)$. The transverse motion is then described by the differential equations

$$\begin{aligned} \ddot{x} - \frac{qB}{m}\dot{y} &= 0\,, \\ \ddot{y} + \frac{qB}{m}(\dot{x} - E/B) &= 0\,, \end{aligned} \tag{2.32}$$

which just differ from the previous ones (2.4) by a transformation $\dot{x} \to \dot{x} - E/B$. Hence, the motion is virtually the same as before, except that it is superposed on a constant drift in the x-direction. This drift, which is called the $\mathbf{E} \times \mathbf{B}$ *drift*, may be written as

$$\mathbf{v}_\mathrm{d} = \frac{\mathbf{E} \times \mathbf{B}}{B^2}\,. \tag{2.33}$$

Notice that it is independent of the charge so that electrons and ions both drift in the same direction. The reason is the periodic increase and decrease of the perpendicular velocity due to the acceleration and deceleration of the charge as it moves in the electric field (Fig. 2.2). This results in a periodic

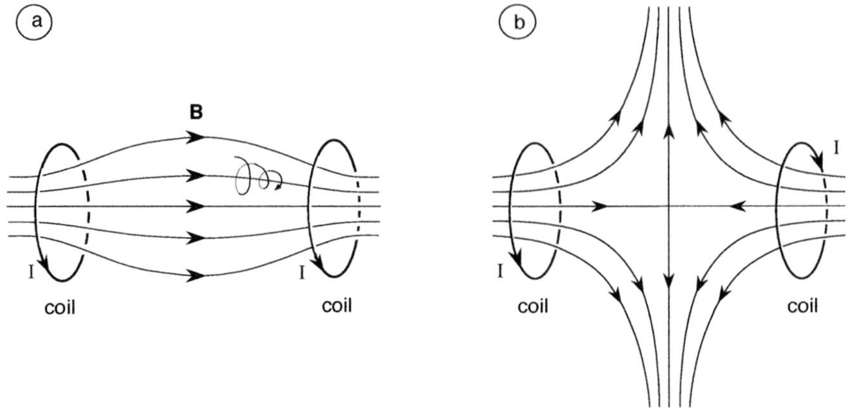

Fig. 2.3 (a) Mirror magnetic confinement and (b) cusp magnetic confinement schemes.

change of the size of the gyro-radius. Electrons are decelerated and ions are accelerated but, since the two orbits have an opposite sense of circulation, the net effect is a drift of ions and electrons in the same direction.

The $\mathbf{E} \times \mathbf{B}$ drift (2.33) permits a more general interpretation. Consider an inertial frame moving with a velocity $\mathbf{v}$ in the direction of $\mathbf{E} \times \mathbf{B}$ (i.e., in the x-direction). According to Eq. (2.23), since $\mathbf{E}, \mathbf{B}$ and $\mathbf{v}$ are mutually orthogonal, the fields in the moving frame are given by the Lorentz transformation $\mathbf{E}' = \Gamma(\mathbf{E} + \mathbf{v} \times \mathbf{B})$, $\mathbf{B}' = \Gamma(\mathbf{B} - c^{-2}\mathbf{v} \times \mathbf{E})$. Choosing $\mathbf{v} = \mathbf{v}_\mathrm{d}$ according to Eq. (2.33), we get $\mathbf{E} + \mathbf{v} \times \mathbf{B} = \mathbf{E} + (\mathbf{E} \times \mathbf{B}) \times \mathbf{B}/B^2 = 0$, so that $\mathbf{E}' = 0$: the particles move in such a way that the electric field in the moving frame vanishes! This is in precise agreement with the motion of a perfectly conducting plasma consisting of a huge number of oppositely charged particles: $\mathbf{E} + \mathbf{v} \times \mathbf{B} = 0$ is one of the fundamental equations for such a plasma (see Section 2.4.1). Note that the argument does not require relativistic velocities: whereas $\mathbf{B}' \approx \mathbf{B}$ for $v \ll c$, the electric field $\mathbf{E}'$ in the moving frame is certainly very different from $\mathbf{E}$ for $v \ll c$.

If we replace the electric force $q\mathbf{E}$ by some other force $\mathbf{F}$, e.g. the gravitational force $m\mathbf{g}$, we find a similar expression for the resulting drift:

$$\mathbf{v}_\mathrm{d} = \frac{\mathbf{F} \times \mathbf{B}}{qB^2} \,. \tag{2.34}$$

If the force $\mathbf{F}$ is charge-independent, like gravity, the drift itself becomes charge-dependent so that electrons and ions drift in opposite directions (neglecting interactions between the particles). This implies the flow of an electric current.

In inhomogeneous magnetic fields many more drifts occur, e.g. one due to the gradient of $\mathbf{B}$. This can again be understood, like the $\mathbf{E} \times \mathbf{B}$ drift, as caused by the periodic variation of the size of the gyro-radius. This so-called $\mathbf{B} \times \nabla\mathbf{B}$ drift is again charge-dependent so that it is associated with a current flow. Another drift is due to the curvature of the magnetic field lines, resulting in a centrifugal force for particles moving along the field lines. According to Eq. (2.34) this force gives rise to an additional drift velocity.

An important application of orbit theory is the *mirror effect*: particles entering the regions of higher magnetic field strength (created by so-called magnetic mirrors, see Fig. 2.3(a)) are reflected back into the region of smaller magnetic field strength where the gyro-radius is larger and the perpendicular velocity smaller. A proper treatment requires the consideration of the magnetic moment of the gyro-motion, which may be shown to be an adiabatic invariant (see below). Like many other confinement schemes considered in research on controlled thermonuclear reactions, the magnetic mirror (also called magnetic bottle) has been investigated extensively and subsequently abandoned as a candidate for fusion reactors. However, it remains an extremely important concept in plasma dynamics since all magnetic confinement schemes involve highly inhomogeneous magnetic fields with associated *trapping of particles* in between the maxima of the field. Also, the magnetic mirror concept plays a prominent role in space and plasma astrophysics. Examples are the Van Allen belts in the magnetosphere of the Earth and acceleration mechanisms for cosmic rays.

A definite disadvantage of the mirror confinement scheme is the curvature of the field lines, which is convex with respect to the confined plasma and, hence, subject to *interchange instability*. It was realized early in controlled fusion research that this instability is eliminated in the cusp confinement scheme (see Fig. 2.3(b)) where the magnetic field geometry consists of two mirrors connected with a cusped structure which is produced by simply reversing the direction of the current in one of the coils. The plasma is now stable up to very high values of $\beta \equiv 2\mu_0 p/B^2$ (an important parameter in plasma confinement, which we will encounter extensively later on) since the field lines are concave with respect to the plasma. However, confinement of particles has even more severe limitations than in the mirror scheme.

Periodic motion in inhomogeneous magnetic fields calls for a more systematic treatment exploiting the just-mentioned *adiabatic invariants*. This is based on the notion that there are distinct spatial scales in the problem, viz. one scale associated with the gyro-motion and another, much larger, scale associated with the field inhomogeneities. Action variables $J \equiv \oint P\,dQ$ (Goldstein [243]) are exploited, where P is the generalized momentum conjugate to a periodic coordinate Q. For non-relativistic particles in an electromagnetic field the generalized momentum is defined by $\mathbf{P} \equiv m\mathbf{v} + q\mathbf{A}$ (Jackson [331]), where $\mathbf{A}$ is the vector potential corresponding to the magnetic field $\mathbf{B} = \nabla \times \mathbf{A}$.

We now consider the gyro-motion in local cylindrical coordinates r, θ, z about the magnetic field, with $\mathbf{B} = B\,\mathbf{e}_z$ and $\mathbf{A} = \frac{1}{2}Br\,\mathbf{e}_\theta$, where the particle moves on an orbit $r = R = v_\perp/\Omega$ in the direction of *decreasing* angle θ (as the ion in Fig. 2.1). In the first adiabatic invariant, the transverse momentum corresponding to the gyro-motion enters:

$$J_1 \equiv \oint \mathbf{P}_\perp \cdot d\boldsymbol{l} = \oint (mv_\perp - \tfrac{1}{2}qBR)R\,d\theta = \frac{\pi m v_\perp^2}{\Omega}\,. \tag{2.35}$$

This invariant may be expressed in terms of *the magnetic moment* $\mu \equiv \pi R^2 I$ of the gyro-motion, where $I = (2\pi)^{-1}q\Omega$ is the current due to the circulating charge. Inserting the expressions (2.5) and (2.8) for Ω and R gives the required result:

$$\mu \equiv \pi R^2 I = \tfrac{1}{2}qR^2\Omega = \frac{\frac{1}{2}mv_\perp^2}{B} \quad \Rightarrow J_1 = \frac{2\pi m}{q}\,\mu\,. \tag{2.36}$$

According to convention, the corresponding vector of the magnetic moment, $\boldsymbol{\mu}$, is defined as pointing opposite to $\mathbf{B}$. The first adiabatic invariant may also be expressed in terms of the magnetic flux Ψ_R enclosed by the gyro-orbit: $\Psi_R = \pi R^2 B = (2\pi m/q^2)\,\mu$, so that $J_1 = q\Psi_R$. The adiabatic invariance implies that J_1 is constant when the external parameters only vary slowly. Consequently,

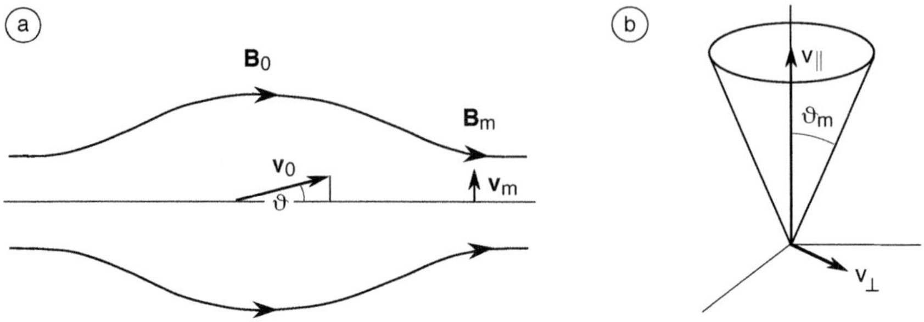

Fig. 2.4 (a) Magnetic mirror geometry and (b) Loss cone in velocity space.

for rapid gyro-motion in a slowly varying magnetic field, the magnetic moment μ and the contained magnetic flux Ψ_R are also constant.

The use of such adiabatic invariants can be demonstrated with the motion in a mirror field (Fig. 2.4). Because $\mu = \text{const}$, if the particle moves into the mirror, $v_\perp$ must increase since B increases. Because of energy conservation, this can happen only if $v_\parallel$ decreases. As a result, particles are reflected by the magnetic mirror. (Hence, its name.) Clearly, for particles with a very small value of $v_\perp/v_\parallel$, the mirror does not work since these particles are lost along the axis. This loss is determined by *the mirror ratio* B_m/B_0, where the subscript m refers to the mirror throat, where B has a maximum, and the subscript 0 refers to the mid-plane, where B has a minimum. One can express the pitch angle of *the loss cone* in velocity space (Fig. 2.4(b)) in terms of this mirror ratio by exploiting the constancy of μ,

$$v_{\perp,0}^2/B_0 = v_{\perp,\mathrm{m}}^2/B_\mathrm{m}\,, \tag{2.37}$$

and energy conservation for particles with velocities at the transition from trapped to untrapped ($v_{\parallel,\mathrm{m}} = 0$),

$$v_{\parallel,0}^2 + v_{\perp,0}^2 = v_{\perp,\mathrm{m}}^2\,, \tag{2.38}$$

so that

$$\left(1 + v_{\parallel,0}^2/v_{\perp,0}^2\right) B_0 = B_\mathrm{m}\,.$$

The pitch angle is defined by $\vartheta \equiv \arctan\left(v_\perp/v_\parallel\right)$. Hence, particles with large enough parallel velocity $v_\parallel$, such that

$$\vartheta < \vartheta_\mathrm{m} \equiv \arctan\sqrt{\frac{B_0}{B - B_0}}\,, \tag{2.39}$$

are lost.

The particles outside the loss cone are trapped. They bounce back and forth between the mirrors. With this motion a second adiabatic invariant is associated, viz. the *longitudinal invariant*

$$J_2 \equiv \oint \Gamma_\parallel\, dl \approx \oint m v_z\, dz = \oint m v_z^2\, dt = \frac{\pi m \hat{v}_z^2}{\omega_\mathrm{b}}\,, \tag{2.40}$$

where the motion along the symmetry axis of the configuration (the z-direction) has been assumed to be a harmonic oscillation with velocity $v_z = \hat{v}_z \cos\omega_\mathrm{b} t$, where ω_b is the bounce frequency.

A *third adiabatic invariant* is associated with the slow drift of the guiding centres of the particles across the field lines. This drift is caused by curvature and gradients of the magnetic field, with the associated variation of the size of the gyro-radius. For example, electrons gyrating about magnetic field lines of the Earth's dipole drift eastward and ions drift westward, creating a huge *ring current* system around the Earth. The guiding centres drift across the magnetic field lines while they stay on the 'magnetic surface' (called drift shell) mapped out by those field lines. Accordingly, the third adiabatic invariant is characterized by the value of the magnetic flux $\Psi_{\rm d}$ enclosed by the surface to which the drift is confined. This flux may be calculated by exploiting another cylindrical coordinate system r, ϕ, z with the magnetic field in the r, z-plane and ϕ the ignorable azimuthal coordinate in the direction of the drift (see Fig. 2.5). For the example of the Earth's dipole field, it is easiest to evaluate the flux at the equatorial plane. From $\mathbf{B} = \nabla \times \mathbf{A}$, we obtain $B_z = (1/r)\partial(rA_\phi)/\partial r$, so that the flux enclosed between the drift shell and a reference shell through $r = r_0, z = 0$, where we chose $A_\phi(r_0, 0) = 0$, may be written as

$$\Psi_{\rm d} = 2\pi \int_{r_0}^{r} B_z r\, dr = 2\pi r A_\phi(r, 0)\,.$$

Hence, the expression for the third adiabatic invariant becomes:

$$J_3 \equiv \oint P_\phi\, r d\phi = \oint (mv_{\rm d} + qA_\phi)\, r d\phi \approx 2\pi q\, r A_\phi = q\, \Psi_{\rm d}\,. \tag{2.41}$$

Here, the term with the drift velocity has been neglected since it is much smaller than the contribution of the vector potential A_ϕ (see Hasegawa and Sato [293], Chapter 1).

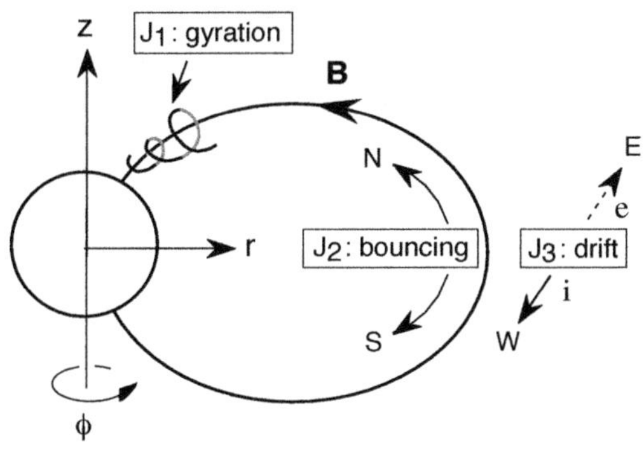

Fig. 2.5 Adiabatic invariants for particles in the magnetosphere.

Summarizing the use of adiabatic invariants describing the motion of charged particles in an inhomogeneous magnetic field, e.g. that of the Earth, associated with the Van Allen belts (Fig. 2.5):

(a) electrons and ions execute a fast gyration in opposite directions about the magnetic field lines conserving the first adiabatic invariant J_1, i.e. the magnetic moment of the guiding centres;

(b) they bounce back and forth between the mirrors on the northern and southern hemispheres on a slower time scale, conserving J_2;

(c) they drift on a slower time scale yet in opposite longitudinal directions conserving the third adiabatic invariant J_3, i.e. the magnetic flux inside the drift shell.

Obviously, the three adiabatic invariants are conserved in decreasing order of robustness. The fluctuating interaction of the solar wind with the magnetosphere will not invalidate the assumptions underlying the adiabatic invariance of J_1, since this invariant concerns very fast motion, but it may easily invalidate the invariance of the third adiabatic invariant J_3.

Thus, a very effective description of charged particle motion in inhomogeneous magnetic fields has been *sketched*. The precise formulation of the conditions of validity and *proofs* of adiabatic invariance would be another matter. This would lead from the early seminal work by Northrop [469] and by Kruskal to modern developments in Hamiltonian mechanics (see, e.g., Goldston and Rutherford [244], Chapter 4, and Balescu [32], Chapter 1).

2.3 Kinetic plasma theory

The motion of a *single* (non-relativistic) charged particle in electric and magnetic fields is described by the equation of motion (2.1). Single particle orbit theory is only valid when the density of charged particles is so low that the interactions between the particles can be ignored. However, a plasma consists of a very large number of *interacting* particles and, hence, it is appropriate to use a *statistical approach* for its analysis. It is the task of *kinetic plasma theory* to derive equations describing the *collective behaviour* of the many charged particles that constitute a plasma by applying the methods of statistical mechanics. One should be aware of the formidable amount of theoretical analysis involved in even a partial performance of this task. (See, e.g. the basic papers by Trubnikov [611] and Braginskii [92] in the first volume of the excellent series 'Reviews of Plasma Physics' for early contributions, and the more recent comprehensive treatise by Balescu [32].) Here, we exploit one of the end results of this program, viz. *the Boltzmann equation*, which may be derived by heuristic arguments as long as no specific expression for the collision term is needed. Single particle orbit theory (Section 2.2) ignores collective effects, and the fluid description of plasmas (Section 2.4) averages out microscopic fluctuations. Kinetic theory includes these important aspects of plasma dynamics and it is, therefore, more comprehensive than both orbit theory and the fluid description of plasmas, but it is also much more complicated.

First, we just give an introduction to some of the basic kinetic concepts (Section 2.3.1). A more detailed exposition is relegated to Chapter 3, which may be consulted for details on the derivations of the equations. Next, we consider a simple example of collective behaviour (Section 2.3.2), viz. electron plasma oscillations, associated with a fundamental plasma parameter: *the plasma frequency*. Section 2.3.3 deals with the damping of these oscillations through kinetic effects.

2.3.1 Boltzmann equation and moment reduction

Consider a plasma consisting of electrons and ions of one kind. In the statistical description, the information on the individuality of the particles is lost but the relevant physical information on the plasma as a whole is retained and expressed in terms of time-dependent *distribution functions* $f_\alpha(\mathbf{r}, \mathbf{v}, t)$ for the electrons and ions (α = e, i). These are defined as the density of the representative points of particles of type α in a six-dimensional *phase space* formed by the three position coordinates (x, y, z) and the three velocity coordinates (v_x, v_y, v_z) (see, e.g., Bittencourt [70]). The probable number of particles of type α in the six-dimensional volume element $d^3r\, d^3v$ centred at $(\mathbf{r}, \mathbf{v})$ is then given by $f_\alpha(\mathbf{r}, \mathbf{v}, t)\, d^3r\, d^3v$. The total number of particles, $N_\alpha \equiv \iint f_\alpha d^3r\, d^3v$, will be assumed constant. Clearly, to describe more general plasmas like thermonuclear ones, where fusion reactions create and annihilate particles, more than two distribution functions are needed and the respective total number of particles will not be constant.

The motion of the swarm of representative points in phase space is described by the total time derivative of the distribution function $f_\alpha(\mathbf{r}, \mathbf{v}, t)$:

$$\begin{aligned}\frac{\mathrm{d}f_\alpha}{\mathrm{d}t} &\equiv \frac{\partial f_\alpha}{\partial t} + \frac{\partial f_\alpha}{\partial \mathbf{r}} \cdot \frac{d\mathbf{r}}{dt} + \frac{\partial f_\alpha}{\partial \mathbf{v}} \cdot \frac{d\mathbf{v}}{dt} \\ &= \frac{\partial f_\alpha}{\partial t} + \mathbf{v} \cdot \frac{\partial f_\alpha}{\partial \mathbf{r}} + \frac{q_\alpha}{m_\alpha}(\mathbf{E} + \mathbf{v} \times \mathbf{B}) \cdot \frac{\partial f_\alpha}{\partial \mathbf{v}}, \end{aligned} \tag{2.42}$$

where the expression (2.1) for the acceleration $d\mathbf{v}/dt$ of the particles has been inserted in the second line. The notation of inner products involving derivatives with respect to the vectors $\mathbf{r}$ and $\mathbf{v}$ just indicates that the sum over the products of the three vector components is to be taken: $\mathbf{v} \cdot \partial/\partial\mathbf{r} \equiv v_x\, \partial/\partial x + v_y\, \partial/\partial y + v_z\, \partial/\partial z$, and similarly for the term with $\partial/\partial\mathbf{v}$. Also note a subtle, but important, difference in notation: $\mathrm{d}/\mathrm{d}t$ for the total time derivative and d/dt for ordinary time derivatives. In the absence of binary interactions between particles, the density of representative points in phase space remains constant in time so that $\mathrm{d}f_\alpha/\mathrm{d}t = 0$ (Liouville's theorem, Goldstein [243]).

Of course, the interesting part of kinetic theory comes with the introduction of interactions or, rather, collisions between the particles. The variation in time of the distribution functions of both electrons and ions is then found from a kinetic equation, known as *the Boltzmann equation*:

$$\frac{\partial f_\alpha}{\partial t} + \mathbf{v} \cdot \frac{\partial f_\alpha}{\partial \mathbf{r}} + \frac{q_\alpha}{m_\alpha}(\mathbf{E} + \mathbf{v} \times \mathbf{B}) \cdot \frac{\partial f_\alpha}{\partial \mathbf{v}} = C_\alpha \equiv \left(\frac{\partial f_\alpha}{\partial t}\right)_{\mathrm{coll}} . \tag{2.43}$$

Now, $\mathbf{E}(\mathbf{r}, t)$ and $\mathbf{B}(\mathbf{r}, t)$ consist of the contributions of the external fields *and* of the averaged internal fields originating from the long-range inter-particle interactions. The symbolic expression on the RHS of Eq. (2.43) represents the rate of change of the distribution function due to the short-range inter-particle interactions, which are somewhat arbitrarily called *collisions*. In a plasma, these may be considered as the cumulative effect of many small-angle velocity changes effectively resulting in large-angle scattering, described by a Fokker–Planck type collision operator. It is the first objective of kinetic theory to justify the distinction between long-range interactions and binary collisions, to determine the ranges of its validity, and to derive suitable expressions for the collision term. One such expression, discussed in Section 3.2, is the Landau collision integral (1936) [392]. On the other hand, neglect of the collisions leads to the *Vlasov equation* (1938) [634]:

$$\frac{\partial f_\alpha}{\partial t} + \mathbf{v} \cdot \frac{\partial f_\alpha}{\partial \mathbf{r}} + \frac{q_\alpha}{m_\alpha}(\mathbf{E} + \mathbf{v} \times \mathbf{B}) \cdot \frac{\partial f_\alpha}{\partial \mathbf{v}} = 0, \tag{2.44}$$

where it is to be realized that the particles still interact through the long-range interactions represented by the averaged internal parts of the $\mathbf{E}$ and $\mathbf{B}$ fields.

A closed system of equations is now obtained by combining either the Boltzmann equation (2.43) or the Vlasov equation (2.44), determining the distribution functions $f_\alpha(\mathbf{r}, \mathbf{v}, t)$, with Maxwell's equations (2.13)–(2.16), determining the electric and magnetic fields $\mathbf{E}(\mathbf{r}, t)$ and $\mathbf{B}(\mathbf{r}, t)$, and the expressions (2.17) for the charge and current density source terms $\tau(\mathbf{r}, t)$ and $\mathbf{j}(\mathbf{r}, t)$. The latter are related to the particle densities and the average velocities:

$$n_\alpha(\mathbf{r},t) \equiv \int f_\alpha(\mathbf{r},\mathbf{v},t)\, d^3v\,, \qquad \tau(\mathbf{r},t) \equiv \sum q_\alpha n_\alpha\,, \tag{2.45}$$

$$\mathbf{u}_\alpha(\mathbf{r},t) \equiv \frac{1}{n_\alpha(\mathbf{r},t)} \int \mathbf{v} f_\alpha(\mathbf{r},\mathbf{v},t)\, d^3v\,, \qquad \mathbf{j}(\mathbf{r},t) \equiv \sum q_\alpha n_\alpha \mathbf{u}_\alpha\,. \tag{2.46}$$

This completes the microscopic equations.

A systematic procedure to obtain macroscopic equations, not involving details of velocity space any more, is to expand in *a finite number of moments of the Boltzmann equation* (2.43), obtained by first multiplying the expressions with powers of $\mathbf{v}$ and then integrating over velocity space:

$$\int d^3v \cdots\,, \quad \int d^3v\, \mathbf{v} \cdots\,, \quad \int d^3v\, v^2 \cdots \Big|_{\text{truncate}}\,. \tag{2.47}$$

This, in turn, involves the moments of the distribution function itself, like the zeroth moment associated with the particle density $n_\alpha(\mathbf{r},t)$ and the first moment associated with the average velocity $\langle \mathbf{v} \rangle_\alpha \equiv \mathbf{u}_\alpha(\mathbf{r},t)$, just defined. In order for this expansion to be practical, it needs to be truncated at a very limited number of terms, like the five (one scalar + one vector + one scalar) indicated in Eq. (2.47). The justification of this truncation is part of transport theory, to be discussed below and more extensively in Chapter 3. In general, macroscopic variables $\langle g \rangle_\alpha(\mathbf{r},t)$ will appear as the average of some phase space function $g(\mathbf{r},\mathbf{v},t)$ over velocity space:

$$\langle g \rangle_\alpha(\mathbf{r},t) \equiv \frac{1}{n_\alpha(\mathbf{r},t)} \int g(\mathbf{r},\mathbf{v},t)\, f_\alpha(\mathbf{r},\mathbf{v},t)\, d^3v\,. \tag{2.48}$$

This definition obviously requires that the distribution functions f_α fall off rapidly enough for $v \to \infty$ to yield a finite answer.

The different moments of the collision term in the RHS of the Boltzmann equation should also be determined. Without specifying the particular form of the collision operator, important conclusions can be drawn from general physical principles. To that end, the collision term on the RHS of Eq. (2.43) is decomposed in contributions $C_{\alpha\beta}$ due to collisions of particles α (e.g. electrons) with particles β (i.e. electrons as well as ions):

$$C_\alpha = \sum_\beta C_{\alpha\beta}\,. \tag{2.49}$$

In the absence of fusion reactions, the total number of particles α at a certain position does not change by collisions with particles β, so that

$$\int C_{\alpha\beta}\, d^3v = 0\,. \tag{2.50}$$

Similarly, momentum and energy conservation lead to corresponding expressions. Details of these and other manipulations are worked out in Chapter 3. Here, we only present the derivation of the lowest moment equation (describing mass conservation) to give an impression of the procedure.

The *zeroth moment* of Eq. (2.43), obtained by integrating over velocity space, yields the following terms:

$$\int \frac{\partial f_\alpha}{\partial t}\, d^3v = \frac{\partial n_\alpha}{\partial t} \qquad \text{(definition (2.45))}\,,$$

$$\int \mathbf{v} \cdot \frac{\partial f_\alpha}{\partial \mathbf{r}} \, d^3v = \nabla \cdot (n_\alpha \mathbf{u}_\alpha) \quad \text{(definition (2.46))},$$

$$\int \frac{q_\alpha}{m_\alpha} (\mathbf{E} + \mathbf{v} \times \mathbf{B}) \cdot \frac{\partial f_\alpha}{\partial \mathbf{v}} \, d^3v = 0 \quad \text{(integrating by parts)},$$

$$\int C_\alpha \, d^3v = 0 \quad \text{(summing Eq. (2.50))}.$$

Adding these expressions gives *the continuity equation* for particles of species α:

$$\frac{\partial n_\alpha}{\partial t} + \nabla \cdot (n_\alpha \mathbf{u}_\alpha) = 0 . \tag{2.51}$$

In the same vein, the *first moment* of Eq. (2.43), obtained by multiplying with $m_\alpha \mathbf{v}$ and integrating over the velocities, yields *the momentum equation*:

$$\frac{\partial}{\partial t}(n_\alpha m_\alpha \mathbf{u}_\alpha) + \nabla \cdot \Big(n_\alpha m_\alpha \langle \mathbf{v}\mathbf{v} \rangle_\alpha \Big) - q_\alpha n_\alpha (\mathbf{E} + \mathbf{u}_\alpha \times \mathbf{B}) = \int C_{\alpha\beta} \, m_\alpha \mathbf{v} \, d^3v . \tag{2.52}$$

Finally, *the scalar second moment* of Eq. (2.43), obtained by multiplying with $\frac{1}{2} m_\alpha v^2$ and integrating over velocity space, yields *the energy equation*:

$$\frac{\partial}{\partial t}\Big(n_\alpha \tfrac{1}{2} m_\alpha \langle v^2 \rangle_\alpha \Big) + \nabla \cdot \Big(n_\alpha \tfrac{1}{2} m_\alpha \langle v^2 \mathbf{v} \rangle_\alpha \Big) - q_\alpha n_\alpha \mathbf{E} \cdot \mathbf{u}_\alpha = \int C_{\alpha\beta} \tfrac{1}{2} m_\alpha v^2 \, d^3v . \tag{2.53}$$

See Section 3.2.2 for the explicit steps in the derivation of these equations.

Clearly, this sequence can be continued indefinitely, but Eqs. (2.51)–(2.53) can be turned into a closed set by making additional assumptions. In broad outlines, the procedure is as follows:

(a) Split the particle velocity $\mathbf{v}$ into an average part $\mathbf{u}_\alpha$ and a random part $\tilde{\mathbf{v}}_\alpha$ defined as

$$\tilde{\mathbf{v}}_\alpha \equiv \mathbf{v} - \mathbf{u}_\alpha , \quad \text{where } \langle \tilde{\mathbf{v}}_\alpha \rangle = 0 . \tag{2.54}$$

This permits the definition of *thermal quantities*:

$$T_\alpha(\mathbf{r}, t) \equiv \frac{m_\alpha}{3k} \langle \tilde{v}_\alpha^2 \rangle \quad \textit{(temperature)}, \tag{2.55}$$

$$\mathsf{P}_\alpha(\mathbf{r}, t) \equiv n_\alpha m_\alpha \langle \tilde{\mathbf{v}}_\alpha \tilde{\mathbf{v}}_\alpha \rangle = p_\alpha \mathsf{I} + \boldsymbol{\pi}_\alpha , \quad p_\alpha \equiv n_\alpha k T_\alpha \quad \textit{(stress tensor)}, \tag{2.56}$$

$$\mathbf{h}_\alpha(\mathbf{r}, t) \equiv \tfrac{1}{2} n_\alpha m_\alpha \langle \tilde{v}_\alpha^2 \tilde{\mathbf{v}}_\alpha \rangle \quad \textit{(heat flow)}, \tag{2.57}$$

$$\mathbf{R}_\alpha(\mathbf{r}, t) \equiv m_\alpha \int C_{\alpha\beta} \, \tilde{\mathbf{v}}_\alpha \, d^3v \quad \textit{(momentum transfer)}, \tag{2.58}$$

$$Q_\alpha(\mathbf{r}, t) \equiv \tfrac{1}{2} m_\alpha \int C_{\alpha\beta} \, \tilde{v}_\alpha^2 \, d^3v \quad \textit{(heat transfer)}. \tag{2.59}$$

Here, I is the unit tensor, so that $\boldsymbol{\pi}_\alpha$ represents the off-diagonal terms of the pressure tensor P. A particular example, consistent with these definitions, is *the Maxwell distribution* for thermal equilibrium:

$$f_\alpha^0(\mathbf{r}, \mathbf{v}, t) = n_\alpha \left(\frac{m_\alpha}{2\pi k T_\alpha} \right)^{3/2} \exp\left(-\frac{m_\alpha \tilde{v}_\alpha^2}{2kT_\alpha} \right) . \tag{2.60}$$

For this distribution, the LHS of the Boltzmann equation (2.43) vanishes so that the collision term on the RHS should vanish as well, i.e. when the two distributions have equal average velocities ($\mathbf{u}_\mathrm{e} = \mathbf{u}_\mathrm{i}$) and temperatures ($T_\mathrm{e} = T_\mathrm{i}$). Plasma kinetic theory is concerned with deviations from this thermal equilibrium and the way in which collisions cause relaxation to thermal equilibrium in the course of time (Braginskii [92]).

(b) The equations of continuity, momentum and heat balance then take the form:

$$\frac{\partial n_\alpha}{\partial t} + \nabla\cdot(n_\alpha \mathbf{u}_\alpha) = 0\,, \tag{2.61}$$

$$n_\alpha m_\alpha \Big(\frac{\partial \mathbf{u}_\alpha}{\partial t} + \mathbf{u}_\alpha\cdot\nabla\mathbf{u}_\alpha\Big) + \nabla\cdot\mathbf{P}_\alpha - n_\alpha q_\alpha(\mathbf{E} + \mathbf{u}_\alpha\times\mathbf{B}) = \mathbf{R}_\alpha\,, \tag{2.62}$$

$$\tfrac{3}{2} n_\alpha k\Big(\frac{\partial T_\alpha}{\partial t} + \mathbf{u}_\alpha\cdot\nabla T_\alpha\Big) + \mathbf{P}_\alpha : \nabla\mathbf{u}_\alpha + \nabla\cdot\mathbf{h}_\alpha = Q_\alpha\,. \tag{2.63}$$

In Eq. (2.62), the divergence of the stress tensor may be decomposed into an isotropic part, involving the scalar pressure p_α, and an anisotropic part, involving the traceless pressure tensor $\boldsymbol{\pi}_\alpha$:

$$\nabla\cdot\mathbf{P}_\alpha = \nabla p_\alpha + \nabla\cdot\boldsymbol{\pi}_\alpha\,. \tag{2.64}$$

The double dot in Eq. (2.63) indicates that a double sum over the Cartesian components is to be taken, $\mathbf{P} : \nabla\mathbf{u} \equiv \sum_i \sum_j P_{ij}\partial u_i/\partial x_j$, so that decomposition into isotropic and anisotropic contributions gives

$$\mathbf{P}_\alpha : \nabla\mathbf{u}_\alpha = p_\alpha\nabla\cdot\mathbf{u}_\alpha + \boldsymbol{\pi}_\alpha : \nabla\mathbf{u}_\alpha\,. \tag{2.65}$$

The logical next step is to transform the temperature evolution equation (2.63) into a pressure evolution equation by exploiting Eqs. (2.56) and (2.61):

$$\frac{\partial p_\alpha}{\partial t} + \mathbf{u}_\alpha\cdot\nabla p_\alpha + \gamma p_\alpha\nabla\cdot\mathbf{u}_\alpha + (\gamma-1)(\boldsymbol{\pi}_\alpha : \nabla\mathbf{u}_\alpha + \nabla\cdot\mathbf{h}_\alpha) = (\gamma-1)Q_\alpha\,. \tag{2.66}$$

Here, we have introduced the ratio of specific heats, $\gamma \equiv C_p/C_v = 5/3$, to demonstrate the connection with gas dynamics. The equations (2.61) for n_α, (2.62) with (2.64) for $\mathbf{u}_\alpha$, and (2.66) for p_α now appear rather macroscopic, but they hide the unsolved kinetic dependence in the variables $\boldsymbol{\pi}_\alpha$ and $\mathbf{h}_\alpha$, which involve higher order moments, and the variables $\mathbf{R}_\alpha$ and Q_α, which involve the unspecified collision operator.

(c) The truncated set of moment equations is closed by exploiting the transport coefficients derived by *transport theory* (Braginskii [92], Balescu [32]). This theory concerns deviations from local thermodynamic equilibrium, expressed by Eq. (2.60), where the distribution functions are developed in powers of a small parameter measuring that deviation. This results in relationships, involving *transport coefficients*, between the thermal quantities defined in Eqs. (2.55)–(2.59) and the gradients of the macroscopic quantities. It is the second objective of kinetic theory (the first one being the derivation of the kinetic equation with a collision operator) to provide these coefficients: another formidable task. Just exploiting the final outcome, i.e. the explicit expressions of the transport coefficients, the closing relationships schematically take the form:

$$\boldsymbol{\pi}_\alpha \sim \mu_\alpha\nabla\mathbf{u}_\alpha \qquad \textit{(viscosity)},$$

$$\mathbf{h}_\alpha \sim -\kappa_\alpha\nabla(kT_\alpha) \qquad \textit{(heat conductivity)},$$

$$\mathbf{R}_\alpha \approx -q_\alpha n_\alpha \eta\,\mathbf{j}\,, \qquad \sum Q_\alpha \approx \eta\,|\mathbf{j}|^2 \qquad \textit{(resistivity)}. \tag{2.67}$$

Here, we have just indicated the form of the expressions, omitting many terms, suppressing the anisotropic tensor structure of the transport coefficients, and leaving their dependencies on the densities, temperatures and magnetic field unspecified. (See Section 3.3.2 for the explicit expressions.)

With respect to the anisotropy of the transport coefficients, an example is the huge difference between the electron heat conductivities parallel and perpendicular to the magnetic field: $\kappa_{e\perp}/\kappa_{e\parallel} \sim (\Omega_e \tau_e)^{-2} \ll 1$, where Ω_e is the electron gyro-frequency and τ_e is the electron collision time. This anisotropy is crucial for the possibility of magnetic confinement in fusion machines. On the other hand, the perpendicular resistivity only differs by a factor 2 from the classical value of the parallel resistivity, the so-called Spitzer resistivity:

$$\eta_\parallel \approx \frac{m_e}{2e^2 n_e \tau_e} = \frac{e^2 \sqrt{m_e} Z \ln \Lambda}{6\epsilon_0^2 (2\pi k T)^{3/2}} \approx 1.63 \times 10^{-9} \frac{Z \ln \Lambda}{\widetilde{T}^{3/2}} , \qquad \eta_\perp \approx 2\eta_\parallel , \tag{2.68}$$

where $\ln \Lambda$ (~ 20) is the Coulomb logarithm. (A crude isotropic resistivity model that is frequently exploited actually assumes $\eta = 2\eta_\parallel = \eta_\perp$; see Section 3.3.3.) These two loose remarks suffice to illustrate the intricacies of the subject of classical transport. All of this will be discussed more fully in Chapter 3. Here, we will not further dwell on this but just remark that in the derivation of the macroscopic equations most of the transport is neglected, i.e. assumed to operate on time scales which are much longer than those of interest for macroscopic dynamics. The moment equations (2.61)–(2.63), together with Maxwell's equations (2.13)–(2.16), then transform into the closed set of *two-fluid* and *one-fluid plasma equations*. This subject will be continued in Section 2.4.

We will now present a highly simplified application of the two-fluid description (Section 2.3.2). In that application, most of the complicated terms discussed do not occur. Nevertheless, it illustrates an important basic physical mechanism at work, viz. collective electrostatic oscillations. After that, we return to the kinetic description in terms of distribution functions and show how velocity space effects lead to the surprising kinetic phenomenon called Landau damping (Section 2.3.3).

2.3.2 Collective phenomena: plasma oscillations

We have encountered the concepts of quasi-neutrality and Debye length in Section 1.4. We extend these electric field concepts in two steps. First, we study perturbations of quasi-neutrality in a cold plasma by *plasma oscillations*, also called Langmuir waves (1929) after the name of the author who also introduced the term 'plasma' in 1923. Next, we study the thermal effects on these oscillations in terms of the Debye length.

Consider the highly simplified case of a cold plasma in the absence of a magnetic field ($\mathbf{B} = 0$). This implies that all thermal effects are neglected ($\mathbf{P}_\alpha$, $\mathbf{h}_\alpha$, $\mathbf{R}_\alpha$, and Q_α vanish), so that all complicated terms in the equations of motion (2.62) disappear and the energy equations (2.63) may be dropped. We then just need to exploit the continuity equations (2.61),

$$\frac{\partial n_\alpha}{\partial t} + \nabla \cdot (n_\alpha \mathbf{u}_\alpha) = 0 \qquad (\alpha = e, i) , \tag{2.69}$$

and the simplified momentum equations (2.62),

$$m_\alpha \Big(\frac{\partial \mathbf{u}_\alpha}{\partial t} + \mathbf{u}_\alpha \cdot \nabla \mathbf{u}_\alpha \Big) = q_\alpha \mathbf{E} \qquad (\alpha = e, i) . \tag{2.70}$$

The electric field can be determined self-consistently from the *Poisson equation* (2.15), where *the charge density* is obtained from Eq. (2.17)(a):

$$\nabla \cdot \mathbf{E} = \frac{\tau}{\epsilon_0} = \frac{e}{\epsilon_0}(Zn_\mathrm{i} - n_\mathrm{e})\,. \tag{2.71}$$

These equations constitute a complete set for the variables $n_{\mathrm{e,i}}(\mathbf{r},t)$, $\mathbf{u}_{\mathrm{e,i}}(\mathbf{r},t)$ and $\mathbf{E}(\mathbf{r},t)$ describing the problem of electrostatic oscillations.

We have already encountered one of the most fundamental properties of plasmas, viz. that *plasmas maintain approximate charge neutrality*. Indeed, charge imbalances on a macroscopic scale L would create huge electric fields ($E \sim \tau L/\epsilon_0$) which would neutralize these imbalances extremely fast by accelerating the electrons, so that the plasma maintains charge neutrality to a high degree of accuracy.

Considered on a finer time and length scale, however, charge imbalances do occur in the form of oscillations which are very typical for plasmas. For these oscillations, the heavy ions ($m_\mathrm{i} \gg m_\mathrm{e}$) may be considered as a fixed ($\mathbf{u}_\mathrm{i} = 0$) neutralizing background in which only the light electrons move ($\mathbf{u}_\mathrm{e} \neq 0$). Perturbing a small region inside the plasma by displacing the electrons there, charge neutrality will be disturbed ($n_\mathrm{e} \neq Zn_\mathrm{i}$). The electron variables then determine the problem:

$$\begin{aligned} n_\mathrm{e} &\approx n_0 + n_1(\mathbf{r},t)\,, \\ \mathbf{u}_\mathrm{e} &\approx \qquad\ \mathbf{u}_1(\mathbf{r},t)\,, \end{aligned} \tag{2.72}$$

whereas the ion variables simplify to

$$n_\mathrm{i} \approx n_0/Z = \mathrm{const}\,, \qquad \mathbf{u}_\mathrm{i} \approx 0\,. \tag{2.73}$$

Hence, the two ion equations (2.69) and (2.70) for $\alpha = i$ may be dropped.[3] The subscripts $_0$ and $_1$ refer to the constant background and the perturbations, respectively. The density perturbation $|n_1(\mathbf{r},t)| \ll n_0$ occurs in a small region of the plasma and is zero elsewhere so that *linearization* is appropriate, i.e. terms involving products of perturbations are neglected since they are small compared to linear terms. The electric field $\mathbf{E}_1$ that is created is then proportional to n_1. This small electric field creates a small electron flow velocity $\mathbf{u}_1$, which is also proportional to n_1.

The linearized electron density equation (2.69), the momentum equation (2.70) (both with $\alpha = \mathrm{e}$) and the Poisson equation (2.71) then yield a complete set of equations:

$$\frac{\partial n_1}{\partial t} + n_0 \nabla \cdot \mathbf{u}_1 = 0\,, \qquad m_\mathrm{e}\frac{\partial \mathbf{u}_1}{\partial t} = -e\mathbf{E}_1\,, \qquad \nabla \cdot \mathbf{E}_1 = \frac{\tau_1}{\epsilon_0} = -\frac{e}{\epsilon_0} n_1\,. \tag{2.74}$$

These equations may be reduced to a single wave equation for n_1:

$$\frac{\partial^2 n_1}{\partial t^2} = -n_0 \nabla \cdot \frac{\partial \mathbf{u}_1}{\partial t} = \frac{n_0 e}{m_\mathrm{e}} \nabla \cdot \mathbf{E}_1 = -\frac{n_0 e^2}{\epsilon_0 m_\mathrm{e}} n_1\,. \tag{2.75}$$

The solutions $n_1(\mathbf{r},t) = \hat{n}_1(\mathbf{r})\exp(-\mathrm{i}\omega t)$ represent electron density oscillations, called *plasma oscillations*, with a characteristic frequency, called *the electron plasma frequency*:

$$\omega = \pm\omega_\mathrm{pe}\,, \qquad \omega_\mathrm{pe} \equiv \sqrt{\frac{n_0 e^2}{\epsilon_0 m_\mathrm{e}}}\,. \tag{2.76}$$

[3] Heavy immobile ions imply taking the limits $m_\mathrm{i} \to \infty$, $\mathbf{u}_\mathrm{i} \to 0$ such that the LHS of Eq. (2.70) becomes undetermined. Such procedures always require justification in terms of a small parameter, in this case the mass ratio $m_\mathrm{e}/m_\mathrm{i} \ll 1$. The ion dynamics would enter as a higher order correction in terms of this parameter.

This frequency is one of the fundamental parameters of a plasma. Since it depends only on the plasma density, detection of plasma oscillations provides a diagnostic for the determination of the plasma density.

The plasma frequency is usually very high because m_e is very small. In tokamak plasmas, e.g., a typical density $n_0 = 10^{20}\,\mathrm{m}^{-3}$ gives

$$\omega_{pe} = 5.7 \times 10^{11}\,\mathrm{rad\,s^{-1}} \quad (\text{i.e. } 91\,\mathrm{GHz}).$$

Comparing this frequency with the gyro-frequencies of Section 2.2, we find that the electron plasma frequency is of the same order of magnitude as the electron cyclotron frequency for tokamaks with strong magnetic fields ($B \sim 3\,\mathrm{T}$).

▷ **Exercise** In the solar corona, the density is much lower, so that the electron plasma frequency is also much lower than for tokamaks. The representative numbers in Table B.3 have been chosen somewhat artificially such that the ratio ω_{pe}/Ω_e is the same for tokamaks and coronal loops. What is the plasma frequency for the latter if $n_0 = 10^{14}\,\mathrm{m}^{-3}$ is taken? How does this change the ratio ω_{pe}/Ω_e? ◁

Note that the spatial form of the amplitude $\hat{n}_1(\mathbf{r})$ of the plasma oscillations is not determined in cold plasma theory. This becomes different for 'warm' plasmas, where deviations from charge neutrality due to thermal fluctuations occur in small regions of a size of the order of the *Debye length*

$$\lambda_D \equiv \sqrt{\frac{\epsilon_0 k_B T_e}{n_0 e^2}} = \frac{v_{th,e}}{\sqrt{2}\,\omega_{pe}}. \tag{2.77}$$

We here indicate the Boltzmann constant with a subscript, k_B, to distinguish it from the wave number k of the waves that now enters the analysis. Inserting numbers again for thermonuclear plasmas, with $\widetilde{T} = 10\,\mathrm{keV}$, $v_{th,e} = 5.9 \times 10^7\,\mathrm{m\,s^{-1}}$, $\omega_{pe} = 5.7 \times 10^{11}\,\mathrm{rad\,s^{-1}}$ gives

$$\lambda_D = 7.4 \times 10^{-5}\,\mathrm{m} \approx 0.07\,\mathrm{mm},$$

i.e. of the order of the electron gyro-radius R_e.

▷ **Exercise** What is the Debye length in the solar corona for $\widetilde{T} = 100\,\mathrm{eV}$ and $n_0 = 10^{14}\,\mathrm{m}^{-3}$? For what value of the coronal magnetic field does this become of the order of the electron gyro-radius? ◁

Because of these thermal fluctuations, the frequency of the plasma oscillations becomes dependent on the wavelength. This part of the thermal contributions may be computed by means of the two-fluid equations (2.61)–(2.66) for an unmagnetized plasma ($\mathbf{B} = 0$), assuming an isotropic pressure and neglecting heat transport and collisions:

$$\frac{\partial n_\alpha}{\partial t} + \nabla \cdot (n_\alpha \mathbf{u}_\alpha) = 0, \tag{2.78}$$

$$n_\alpha m_\alpha \Big(\frac{\partial \mathbf{u}_\alpha}{\partial t} + \mathbf{u}_\alpha \cdot \nabla \mathbf{u}_\alpha \Big) + \nabla p_\alpha = n_\alpha q_\alpha \mathbf{E}, \tag{2.79}$$

$$\frac{\partial p_\alpha}{\partial t} + \mathbf{u}_\alpha \cdot \nabla p_\alpha + \gamma p_\alpha \nabla \cdot \mathbf{u}_\alpha = 0. \tag{2.80}$$

Again assuming immobile ions and linearizing the equations (2.78)–(2.80) for the electrons, we now get a modified eigenvalue problem where the pressure $p_0 = n_0 k_B T_0$, i.e. the temperature, of the background plasma enters:

$$\frac{\partial n_1}{\partial t} + n_0 \nabla \cdot \mathbf{u}_1 = 0, \tag{2.81}$$

$$n_0 m_e \frac{\partial \mathbf{u}_1}{\partial t} + \nabla p_1 = -e n_0 \mathbf{E}_1 \,, \tag{2.82}$$

$$\frac{\partial p_1}{\partial t} + \gamma p_0 \nabla \cdot \mathbf{u}_1 = 0 \,, \tag{2.83}$$

$$\nabla \cdot \mathbf{E}_1 = -\frac{e}{\epsilon_0} n_1 \,. \tag{2.84}$$

Assuming plane waves in the x-direction, and no spatial dependences in the y- and z-directions,

$$n_1(x,t) = \hat{n}_1 \mathrm{e}^{\mathrm{i}(kx-\omega t)} \tag{2.85}$$

(and similar expressions for $\mathbf{u}_1, p_1, \mathbf{E}_1$), the gradients $\nabla \to \mathrm{i}k\mathbf{e}_x$ and the time derivatives $\partial/\partial t \to -\mathrm{i}\omega$, so that Eqs. (2.81)–(2.84) become an algebraic system of equations for the amplitudes $\hat{n}_1$, $\hat{\mathbf{u}}_1$, $\hat{p}_1$, and $\hat{\mathbf{E}}_1$. The determinant provides the dispersion equation:

$$\omega^2 = \omega_{\mathrm{pe}}^2 (1 + \gamma k^2 \lambda_{\mathrm{D}}^2) \,. \tag{2.86}$$

Here, since the oscillations are one-dimensional, we should exploit the value $\gamma = 3$ (see Chen [117], Chapter 4). Note that the old result (2.76) is recovered for long wavelengths, where $k^2\lambda_{\mathrm{D}}^2 \ll 1$, but there is a large effect now on the oscillations for wavelengths of the order of or smaller than the Debye length. However, this effect is not quite correctly described since the fluid description actually breaks down because of a peculiar kinetic effect that will be discussed in the next section.

2.3.3 Landau damping

A more refined analysis of longitudinal plasma oscillations for 'warm' plasmas should take velocity space effects into account, exploiting the Vlasov, or *collisionless* Boltzmann, equation (2.44) for the perturbations $f_1(\mathbf{r}, \mathbf{v}, t)$ of the electron distribution function. Taking again plane wave solutions $\sim \exp \mathrm{i}(\mathbf{k}\cdot\mathbf{r} - \omega t)$, one immediately runs into a mathematical problem:

$$\frac{\partial f_1}{\partial t} + \mathbf{v} \cdot \frac{\partial f_1}{\partial \mathbf{r}} = -\mathrm{i}(\omega - \mathbf{k}\cdot\mathbf{v})\, f_1 = \frac{e}{m_e} \mathbf{E}_1 \cdot \frac{\partial f_0}{\partial \mathbf{v}} \,, \tag{2.87}$$

so that inversion of the operator $\partial/\partial t + \mathbf{v}\cdot\partial/\partial\mathbf{r}$, to express f_1 in tems of $\mathbf{E}_1$, leads to singularities when $\omega - \mathbf{k}\cdot\mathbf{v} = 0$. Incorporated in a proper treatment of the initial value problem, these singularities were shown by Landau (1946) [393] to give rise to damping of the plasma oscillations. This *Landau damping* is a surprising phenomenon since it occurs in a purely collisionless medium, i.e. there is no dissipation! Much later, experiments by Malmberg and Wharton [426] first verified the phenomenon of Landau damping (1966), and then also demonstrated that, in fact, the information contained in the initial signal is not irreversibly lost but that it may be recovered by means of plasma wave echos (1968).

A complementary approach to the electrostatic plasma oscillations by means of a normal mode analysis was given by Van Kampen (1955) [628, 629]. He showed that the singularities $\omega - \mathbf{k}\cdot\mathbf{v} = 0$ lead to a continuous spectrum of singular, δ-function type, modes (the Van Kampen modes), which constitute a complete set of 'improper' eigenmodes for this system. Damping occurs because a package of those modes rapidly loses its spatial phase coherence (*phase mixing*). The occurrence of a continuous spectrum is a very intriguing aspect of the analysis of plasma oscillations, which is also encountered in the fluid description of macroscopic waves (as we will see in Chapter 6).

Let us analyse the problem in some more detail for one space dimension x and one velocity space dimension v ($\equiv v_x$). Since Landau damping is due to velocity space effects we have to redo the problem of Section 2.3.2 in terms of distribution functions, leading to the so-called Vlasov–Poisson problem:

$$\frac{\partial f_1}{\partial t} + v\frac{\partial f_1}{\partial x} = \frac{e}{m_\mathrm{e}}\frac{\partial f_0}{\partial v}E_1\,,$$
$$\frac{\partial E_1}{\partial x} = -\frac{e}{\epsilon_0}n_1 = -\frac{e}{\epsilon_0}\int_{-\infty}^{\infty} f_1\,dv\,. \tag{2.88}$$

Inserting plane wave solutions

$$f_1(x,v,t) = \hat{f}_1(v)\,\mathrm{e}^{\mathrm{i}(kx-\omega t)}\,, \qquad E_1(x,t) = \hat{E}_1\,\mathrm{e}^{\mathrm{i}(kx-\omega t)}\,, \tag{2.89}$$

in these equations, i.e. making the replacements $\partial/\partial x \to \mathrm{i}k$, $\partial/\partial t \to -\mathrm{i}\omega$, they transform into

$$-\mathrm{i}(\omega - kv)\,\hat{f}_1 = \frac{e}{m_\mathrm{e}}\frac{\partial f_0}{\partial v}\hat{E}_1\,,$$
$$\mathrm{i}k\hat{E}_1 = -\frac{e}{\epsilon_0}\int_{-\infty}^{\infty}\hat{f}_1\,dv\,. \tag{2.90}$$

Just expressing $\hat{f}_1$ in terms of $\hat{E}_1$ by means of the first equation *(assuming $\omega \neq kv$!)*, and inserting $\hat{f}_1$ into the equation for $\hat{E}_1$, we obtain

$$\left[1 - \frac{e^2}{\epsilon_0 m_\mathrm{e} k^2}\int_{-\infty}^{\infty}\frac{1}{v-\omega/k}\frac{\partial f_0}{\partial v}\,dv\right]\hat{E}_1 = 0\,. \tag{2.91}$$

Hence, the expression inside the square brackets should vanish, providing the *dispersion equation*, i.e. the relation between ω and k, we are looking for:

$$D_\mathrm{V}(k,\omega) \equiv 1 - \frac{\omega_\mathrm{pe}^2}{k^2 n_0}\int_{-\infty}^{\infty}\frac{1}{v-\omega/k}\frac{\partial f_0}{\partial v}\,dv = 0\,. \tag{2.92}$$

Here, we have inserted the plasma frequency ω_pe defined in Eq. (2.76).

For definiteness, we now have to specify the distribution function f_0 of the background. It is logical to choose the one-dimensional form of the Maxwell equilibrium distribution introduced in Eq. (2.60):

$$f_0(v) = \frac{n_0}{\sqrt{\pi}v_\mathrm{th}}\mathrm{e}^{-v^2/v_\mathrm{th}^2}\,, \qquad v_\mathrm{th} \equiv \sqrt{\frac{2k_\mathrm{B}T_\mathrm{e}}{m_\mathrm{e}}}\,. \tag{2.93}$$

Since we wish to concentrate on velocity space effects, we assume spatial homogeneity of the background equilibrium so that the density n_0 and the thermal speed v_th (i.e. the electron temperature T_e) are constant. These apparently innocent assumptions imply that a whole 'zoo' of kinetic and macroscopic instabilities is eliminated at once.

Obviously, the assumption $\omega \neq kv$ cannot be justified if the frequency ω of the plane waves is real since the integration in Eq. (2.92) is then right across the singularity. This singularity occurs for particles with speeds that are resonant with the phase velocity of the waves: $v = v_\mathrm{ph} \equiv \omega/k$ (vertical line in Fig. 2.6). An *apparent* way out, proposed by Vlasov [634], is to exploit the principal

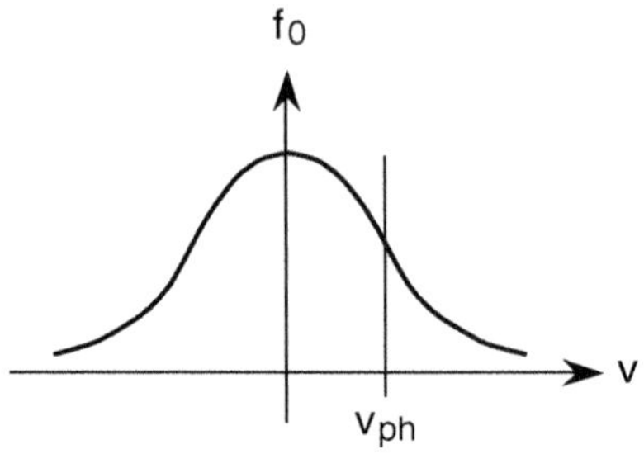

Fig. 2.6 Maxwell velocity distribution and a phase speed $v_{\rm ph} = \omega/k$ of plasma oscillations in the thermal region.

value of the integral for real ω, defined as:

$$\mathcal{P}\int_{-\infty}^{\infty} dv \cdots \equiv \lim_{\delta\to 0}\left[\int_{-\infty}^{v_{\rm ph}-\delta} dv \cdots + \int_{v_{\rm ph}+\delta}^{\infty} dv \cdots\right]. \tag{2.94}$$

One can crudely estimate this integral for long wavelengths ($k \to 0$), when $v_{\rm ph} \gg v_{\rm th}$, so that the Maxwell distribution f_0 becomes quite small at the singular point:

$$\mathcal{P}\int_{-\infty}^{\infty} \frac{1}{v-\omega/k}\frac{\partial f_0}{\partial v}\, dv = -\frac{2n_0}{\sqrt{\pi}v_{\rm th}^3}\,\mathcal{P}\int_{-\infty}^{\infty}\frac{v}{v-v_{\rm ph}}\,{\rm e}^{-v^2/v_{\rm th}^2}\, dv \approx \frac{n_0}{v_{\rm ph}^2}\left(1+\tfrac{3}{2}\frac{v_{\rm th}^2}{v_{\rm ph}^2}\right). \tag{2.95}$$

This gives the following approximate dispersion equation:

$$\omega^2 \approx \omega_{\rm pe}^2 + \tfrac{3}{2}k^2 v_{\rm th}^2 = \omega_{\rm pe}^2(1+3k^2\lambda_{\rm D}^2), \tag{2.96}$$

i.e. one obtains the thermal correction of the frequency (2.76) of the plasma oscillations in terms of the product of the wave vector of the oscillations and the Debye length. This correction agrees with Eq. (2.86) from the fluid approximation.

However, this procedure is much too cavalier, as pointed out by Landau [393] in his severe criticism of the work of Vlasov [634][4]: There is no justification for the use of the principal value integral, and a more careful analysis of the singularity reveals that there is an imaginary contribution (the Landau damping) to the frequency of the waves. The more careful analysis is somewhat beyond the level of this chapter, but too fundamental to be skipped altogether. Therefore, we put it in small print. (Recently, a nonlinear version of Landau damping was presented in Ref. [451].)

▷ **Landau's solution of the initial value problem** We return to the basic equations (2.88) of the Vlasov–Poisson problem. Instead of the Ansatz (2.89) of plane wave solutions, we now keep the $\exp(ikx)$ spatial dependence but treat the time dependence through the Laplace transform:

$$\tilde{f}_1(v;\omega) \equiv \int_0^{\infty} f_1(v;t){\rm e}^{{\rm i}\omega t}\, dt\,, \qquad \tilde{E}_1(\omega) \equiv \int_0^{\infty} E_1(t){\rm e}^{{\rm i}\omega t}\, dt\,. \tag{2.97}$$

(We exploit the variable ω instead of the standard Laplace variable $p \equiv -{\rm i}\omega$ so that convergence in the

[4] The cited two references of Vlasov both contain the important contribution of the collisionless Boltzmann equation (with the justification of the neglect of collisions for many plasma phenomena), but, unfortunately, also the incorrect wave analysis of the plasma oscillations.

right half p-plane is replaced by convergence in the upper half ω-plane.) The Laplace transform of the time derivative $\partial f_1/\partial t$ in Eq. (2.88) produces a similar expression $-\mathrm{i}\omega\tilde{f}_1$ as before,

$$\int_0^\infty \frac{\partial f_1}{\partial t} \mathrm{e}^{\mathrm{i}\omega t}\, dt = \left[f_1 \mathrm{e}^{\mathrm{i}\omega t}\right]_{t=0}^{t\to\infty} - \mathrm{i}\omega \int_0^\infty f_1 \mathrm{e}^{\mathrm{i}\omega t}\, dt = -g(v) - \mathrm{i}\omega \tilde{f}_1(v,\omega)\,, \tag{2.98}$$

but with an additional contribution of the initial value of the perturbation f_1 of the distribution function:

$$g(v) \equiv f_1(v, t=0)\,. \tag{2.99}$$

The contribution for $t \to \infty$ vanishes since we assume that $\operatorname{Im}\omega > 0$.

In effect, the Laplace transform of Eqs. (2.88) is virtually the same as Eqs. (2.90), except for the additional term $-g(v)$. We again express $\tilde{f}_1$ in terms of $\tilde{E}_1$,

$$\tilde{f}_1(v,\omega) = \frac{\mathrm{i}}{\omega - kv}\left[\frac{e}{m_\mathrm{e}}\frac{\partial f_0}{\partial v}\tilde{E}_1(\omega) + g(v)\right], \tag{2.100}$$

and insert $\tilde{f}_1$ into the equation for $\tilde{E}_1$:

$$\tilde{E}_1(\omega) = \frac{e}{\epsilon_0 k^2}\,\frac{\displaystyle\int \frac{1}{v-\omega/k}\, g\, dv}{\displaystyle 1 - \frac{\omega_\mathrm{pe}^2}{k^2 n_0}\int \frac{1}{v-\omega/k}\frac{\partial f_0}{\partial v}\, dv} \equiv \frac{e}{\epsilon_0 k^2}\,\frac{N(\omega)}{D(\omega)}\,. \tag{2.101}$$

Maybe not too surprising, the denominator *appears to be identical* to the expression D_V of the dispersion equation (2.92): the zeros of $D(\omega)$, i.e. the solutions of the dispersion equation $D(\omega) = 0$, become poles of the complex function $\tilde{E}_1(\omega)$. However, in contrast to Vlasov's approach, we now have a procedure to make sense of the singularities $v = \omega/k$ in the velocity integrals, viz. by completing the solution of the initial value problem by means of the inverse Laplace transform:

$$E_1(t) = \frac{1}{2\pi}\int_{\mathrm{i}\nu_C-\infty}^{\mathrm{i}\nu_C+\infty} \tilde{E}_1(\omega)\mathrm{e}^{-\mathrm{i}\omega t}\, d\omega\,, \quad \text{and similarly for } f_1(v,t) \quad (\operatorname{Im}\omega = \nu_C > 0)\,. \tag{2.102}$$

Formally, the problem is now solved since these integrals avoid the singularities by staying away from them on the path C_ω in the upper half of the ω-plane, indicated in Fig. 2.7(a).

To describe the collective plasma oscillations, the expressions (2.102) need to be evaluated with respect to their asymptotic time dependence for $t \to \infty$. To that end, it is expedient to deform the contour C_ω into the contour C'_ω in the lower half of the ω-plane. For the latter contour, the contribution of the straight pieces may be neglected ($\operatorname{Im}\omega \ll 0$) whereas the residue of the uppermost pole (indicated by ω_0) will survive the longest:

$$\tilde{E}_1(\omega) \approx A(\omega-\omega_0)^{-1} \quad \Rightarrow \quad E_1(t) \approx -\frac{A}{2\pi}\oint \frac{1}{\omega-\omega_0}\,\mathrm{e}^{-\mathrm{i}\omega t}\, d\omega = -\mathrm{i}A\mathrm{e}^{-\mathrm{i}\omega_0 t}\,. \tag{2.103}$$

Of course, the explicit value of ω_0 is to be computed yet. This requires knowledge of the complex function $\tilde{E}_1(\omega)$ for $\operatorname{Im}\omega \le 0$, whereas it was only defined for $\operatorname{Im}\omega > 0$. The canonical way to obtain that knowledge is by means of *analytic continuation* of the functions $N(\omega)$ and $D(\omega)$. If one assumes that $g(v)$ is an entire function (i.e. an analytic function which is regular for all finite values of v; see e.g. Nehari [459], Chapter 3), then the Cauchy contour integral along the closed path C_v ($= \mathrm{C}_{v1} + \mathrm{C}_{v2}$) will just pick up the residue at $v = \omega/k$ when $\operatorname{Im}\omega > 0$ (Fig. 2.7(b_1)). Whereas this implies that one could relate the value of the integral along the real axis C_{v1} (plus the residue) to the value of the integral along the semi-circular path C_{v2} with $R \to \infty$, this does not provide us with the usual simplification of the algebra, since the latter integral does not vanish. However, it does show what to do when $\operatorname{Im}\omega \le 0$ (when the singularity crosses the real axis): one should deform the contour C_{v1} to remain on the same side of the singularity, as indicated in Figs. 2.7(b_2) and 2.7(b_3). Then, the appropriate analytic continuation of $N(\omega)$ is obtained by integrating along the real v-axis

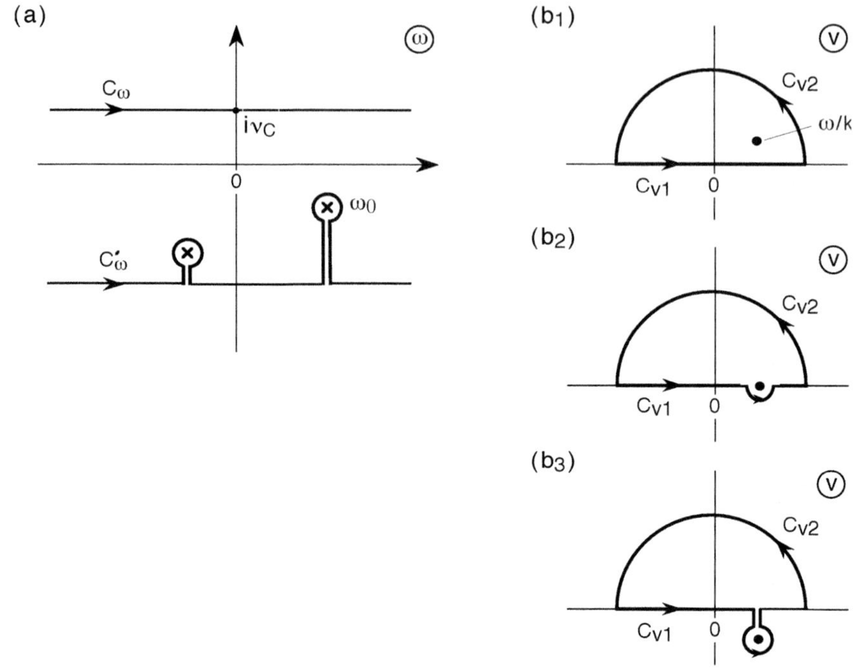

Fig. 2.7 Techniques used by Landau in his solution of the initial value problem:
(a) deforming the contour C_ω in the complex ω-plane of the inverse Laplace transform to a contour C'_ω above the solutions of the dispersion equation;
(b) analytic continuation of the velocity integrals, considered as functions of ω, by means of a Cauchy contour integral in the complex v-plane along a path $\mathrm{C_v}$ $(= \mathrm{C_{v1}} + \mathrm{C_{v2}})$ enclosing the singularity $v = \omega/k$ for values $(\mathrm{b_1})\ \mathrm{Im}\,\omega > 0$, $(\mathrm{b_2})\ \mathrm{Im}\,\omega = 0$ and $(\mathrm{b_3})\ \mathrm{Im}\,\omega < 0$.

and adding no residue, half a residue, or the full residue, depending on the value of $\mathrm{Im}\,\omega$:

$$\int_{\mathrm{C_{v1}}} \frac{1}{v-\omega/k}\, g(v)\, dv = \begin{cases} \displaystyle\int_{-\infty}^{\infty} \frac{1}{v-\omega/k}\, g(v)\, dv & (\mathrm{Im}\,\omega > 0) \\ \displaystyle\mathcal{P}\int_{-\infty}^{\infty} \frac{1}{v-\omega/k}\, g(v)\, dv + \mathrm{i}\,\pi\, g(\omega/k) & (\mathrm{Im}\,\omega = 0) \\ \displaystyle\int_{-\infty}^{\infty} \frac{1}{v-\omega/k}\, g(v)\, dv + \mathrm{i}\, 2\pi\, g(\omega/k) & (\mathrm{Im}\,\omega < 0) \end{cases} . \tag{2.104}$$

The integral in $D(\omega)$ may be analytically continued in a similar way using the fact that, for the Maxwell distribution (2.93), $\partial f_0/\partial v$ is also an entire function.

The three expressions (2.104) may be conveniently combined in a single one, viz. the middle expression (2.104)(b), which may be considered as *valid for all values of* ω if the principal value integral is interpreted as the average of the two integrals along paths just above ($\mathrm{C_a}$) and just below ($\mathrm{C_b}$) the singularity:

$$\mathcal{P}\int dv \cdots \equiv \tfrac{1}{2}\int_{\mathrm{C_a}} dv \cdots + \tfrac{1}{2}\int_{\mathrm{C_b}} dv \cdots . \tag{2.105}$$

For $\mathrm{Im}\,\omega = 0$, the addition of two semi-circles above and below the singularity does not change anything since those contributions cancel. However, for $\mathrm{Im}\,\omega > 0$, the singularity contributes $-\mathrm{i}\,\pi\, g(\omega/k)$ from the path C_a, canceling the second term of Eq. (2.104)(b) so that the expression (2.104)(a) is obtained. For $\mathrm{Im}\,\omega < 0$, the singularity contributes $+\mathrm{i}\pi\, g(\omega/k)$ from the path $\mathrm{C_b}$, doubling the second term so that the expression (2.104)(c) is obtained.

Hence, the correct dispersion equation, valid for all values of ω, may be written as

$$D(k,\omega) \equiv 1 - \frac{\omega_{\rm pe}^2}{k^2 n_0}\left[\mathcal{P}\int_{-\infty}^{\infty} \frac{1}{v-\omega/k}\,\frac{\partial f_0}{\partial v}\,dv + \mathrm{i}\pi\,\frac{\partial f_0}{\partial v}\bigg|_{\omega/k}\right]$$
$$= 1 + (k\lambda_{\rm D})^{-2}\left[1 + \frac{1}{\sqrt{\pi}}\,\zeta\,\mathcal{P}\int_{-\infty}^{\infty}\frac{1}{\bar v - \zeta}\,\mathrm{e}^{-\bar v^2}\,d\bar v + \mathrm{i}\sqrt{\pi}\,\zeta \mathrm{e}^{-\zeta^2}\right] = 0\,, \qquad (2.106)$$

where the Maxwell equilibrium distribution (2.93) with $\bar v \equiv v/v_{\rm th}$ and $\zeta \equiv \omega/(kv_{\rm th})$ have been inserted in the second line. Clearly, the interpretation of the integrals in Eq. (2.101) (where we purposely omitted the integration path since that was yet to be determined) is completely different from the one given by Vlasov in Eqs. (2.92) and (2.94): there is an imaginary part so that one does not obtain oscillations with a real frequency.

All ambiguity about the mathematical physics of the plasma oscillations has now been resolved. The rest is technical (though intricate) details about the evaluation of the velocity integrals for complex values of ω. That part of the problem is usually collected in the properties of the plasma dispersion function, which is a well-known tabulated complex function described in many textbooks on waves in plasmas (see, e.g., Stix [570] or Swanson [582]):

$$Z(\zeta) \equiv \frac{1}{\sqrt{\pi}}\int_{-\infty}^{\infty}\frac{1}{\bar v - \zeta}\,\mathrm{e}^{-\bar v^2}\,d\bar v \qquad (\mathrm{Im}\,\zeta > 0)\,, \qquad (2.107)$$

where the analytic continuation for $\mathrm{Im}\,\zeta \le 0$ is obtained as above. In terms of $Z(\zeta)$, the dispersion equation becomes

$$D(k,\omega) \equiv 1 + (k\lambda_{\rm D})^{-2}\Big[1 + \zeta Z(\zeta)\Big] = 0\,, \quad \zeta \equiv \frac{\omega}{kv_{\rm th}} \equiv \frac{\omega/\omega_{\rm pe}}{\sqrt{2}\,k\lambda_{\rm D}}\,. \qquad (2.108)$$

The power series of the plasma dispersion function $Z(\zeta)$ for $|\zeta| < 1$ reads

$$Z(\zeta) = \mathrm{i}\sqrt{\pi}\mathrm{e}^{-\zeta^2} - 2\zeta + \frac{4}{3}\zeta^3 - \frac{8}{15}\zeta^5 \cdots\,. \qquad (2.109)$$

A more useful expression for physical applications is the asymptotic expansion for $|\zeta| \gg 1$:

$$Z(\zeta) \approx -\frac{1}{\zeta} - \frac{1}{2\zeta^3} - \frac{3}{4\zeta^5}\cdots + \mathrm{i}s\sqrt{\pi}\mathrm{e}^{-\zeta^2}\,, \qquad s = \begin{cases} 0 & (\,\mathrm{Im}\,\zeta\, > 0) \\ 1 & (|\mathrm{Im}\,\zeta| < |\mathrm{Re}\,\zeta|^{-1} \ll 1) \\ 2 & (\,\mathrm{Im}\,\zeta\, < 0) \end{cases}\,, \qquad (2.110)$$

where the middle expression can be applied to a finite region about the real axis. (As usual in asymptotic expansions, the boundaries indicated in brackets do not imply that the expansion is valid up to that boundary but, rather, that the expression is no longer valid there because of the Stokes phenomenon; see Bender and Orszag [50], Chapter 3.) Iterating on the real and imaginary parts of the solution ζ of the dispersion equation for long wavelengths ($k\lambda_{\rm D} \ll 1$) yields the expression (2.111) below for the complex frequency of the plasma oscillations. ◁

In conclusion: Landau's study of the initial value problem of electrostatic plasma oscillations shows that there is an important contribution of the singularities $v = v_{\rm ph} \equiv \omega/k$ where the particles are in resonance with the phase velocity of the waves. For a Maxwell distribution, the solution of the dispersion equation (2.108) for long wavelengths ($k\lambda_{\rm D} \ll 1$) is given by

$$\omega \approx \omega_{\rm pe}\Big\{1 + \tfrac{3}{2}k^2\lambda_{\rm D}^2 - \mathrm{i}\sqrt{\tfrac{\pi}{8}}\,(k\lambda_{\rm D})^{-3}\exp\Big[-\tfrac{1}{2}(k\lambda_{\rm D})^{-2} - \tfrac{3}{2}\Big]\Big\}\,, \qquad (2.111)$$

where the imaginary part represents *damping of the waves*. For long wavelengths, this damping is exponentially small. For short wavelengths ($k\lambda_{\rm D} \sim 1$), the damping becomes very strong so that wave motion with wavelengths smaller than the Debye length becomes impossible.

2.4 Fluid description

Kinetic theory involves details of the distribution functions that evolve on very short length and time scales, like the Debye length λ_D and the plasma frequency ω_{pe}. Since the subject of this book is the macroscopic dynamics of magnetized plasmas, we now have to face the main difficulty, viz. how to bridge the enormous difference between these scales and the macroscopic ones. The development of the fluid picture of plasmas involves three major steps, illustrated in Fig. 2.8 and elaborated in the next chapter:

(a) Collisionality A first step has been taken in Section 2.3.1 with the formulation of the lowest moments (2.61)–(2.63) of the Boltzmann equation and the transport closure relations indicated in Eqs. (2.67). In this manner, a system of *two-fluid equations* is obtained describing the plasma dynamics in terms of the ten variables $n_{e,i}$, $\mathbf{u}_{e,i}$, $T_{e,i}$. To justify such a fluid description, the electrons and ions must undergo *frequent collisions* to establish the separate electron and ion fluids. Transport theory provides the quantitative criterion for the time scale τ_H on which the hydrodynamic description is valid, compared to the collisional relaxation times τ_e and τ_i of the electrons and ions:

$$\tau_H \gg \tau_i \left[\gg \tau_e \right]. \tag{2.112}$$

The explicit expressions for τ_e and τ_i may be found in Section 3.2.4 (Eqs. (3.50) and (3.53), which demonstrate the faster relaxation of the electrons due to the smallness of the mass ratio: $\tau_e/\tau_i \sim (m_e/m_i)^{1/2} \ll 1$). Once more: transport theory is an enormous field of research by itself, the needed results of which have been collected in Chapter 3. Here, we just indicate the main line of thought leading to the fluid description. In particular, from the explicit expressions for τ_e and τ_i, frequent collisions imply that the plasma densities $n_{e,i}$ should be high enough for given values of the temperatures $T_{e,i}$.

(b) Macroscopic scales The plasma dynamics described by the two-fluid equations still involves the small length and time scales of the fundamental phenomena we have encountered, viz. the plasma frequency ω_{pe}, the cyclotron frequencies $\Omega_{e,i}$, the Debye length λ_D and the cyclotron radii $R_{e,i}$ (and also a quantity not yet encountered, viz. the electron skin depth $\delta_e \equiv c/\omega_{pe}$; see Eq. (3.100)). Therefore, the essential second step towards the *magnetohydrodynamics* (MHD) description of plasmas is to consider *large length and time scales*:

$$\lambda_{MHD} \sim a \gg R_i, \qquad \tau_{MHD} \sim a/v_A \gg \Omega_i^{-1}. \tag{2.113}$$

Here, the magnetic field crucially enters: the larger the magnetic field strength, the more easy these conditions are satisfied. On these scales, the plasma is considered as a *single* conducting fluid without distinguishing its individual species: the MHD equations describe the behaviour of the plasma as a whole. Therefore, the dimension a of the plasma appears in the estimate of the length scale and the Alfvén velocity v_A (a new fundamental plasma quantity, to be introduced in Section 2.4.2) appears in the time scale. The derivation of the MHD equations from the two-fluid equations will be sketched in Section 2.4.1, and presented in full detail in Section 3.4.

(c) Ideal fluids A third step is to consider the plasma dynamics on time scales *faster* than the *slow dissipation* connected with the decay of the macroscopic variables, in particular the resistive decay

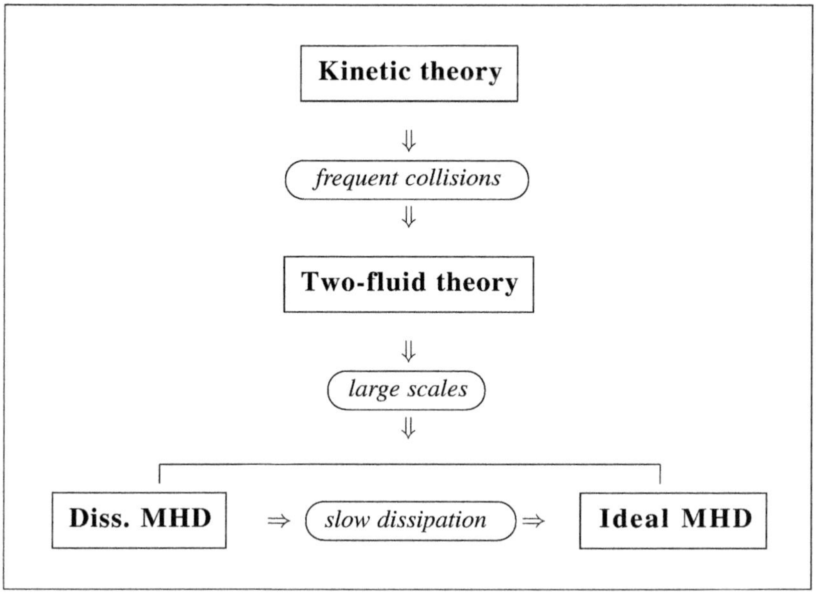

Fig. 2.8 The various theoretical models to describe plasma dynamics and their mutual connections.

of the magnetic field:

$$\tau_{\rm MHD} \ll \tau_{\rm R} \sim a^2/\eta \,. \tag{2.114}$$

Here, the actual value of the resistivity given by Eq. (2.68) is so small that the condition may be satisfied even for the relatively small sizes of fusion machines, and very easily for the huge sizes of astrophysical plasmas. (Actually, too easily: a major problem is to find dissipation mechanisms fast enough to explain the onset of frequently observed disruptions, like stellar flares.) This condition leads to the model of *ideal MHD*: the most robust macroscopic description of magnetized plasmas. An ideal two-fluid counterpart may also be formulated since the condition (2.114) is certainly satisfied on the two-fluid time scale when it is satisfied on the MHD time scale ($\tau_{\rm H} \ll \tau_{\rm MHD}$).

2.4.1 From the two-fluid to the MHD description of plasmas

To derive the MHD equations, we continue the exposition of Section 2.3.1. Again, the derivation will omit most of the intermediate steps, which may be found in Chapter 3. We now specify the plasma to consist of electrons, $q_{\rm e} = -e$, and ions of one kind, with $q_{\rm i} = Ze$. Eqs. (2.61)–(2.63) then give the double set of *two-fluid moment equations*, which is closed by the specification of the transport coefficients, like in Eqs. (2.67). Since macroscopic dynamics on the MHD time scale is generally much faster than changes due to dissipative transport, as indicated in Eq. (2.114), we neglect most of the dissipative terms:

$$\boldsymbol{\pi}_{\rm e,i} \to 0 \,, \quad \mathbf{h}_{\rm e,i} \to 0 \qquad \textit{(neglect of viscosity and heat flow)} \,, \tag{2.115}$$

However, we still keep the small terms due to momentum transfer and generated heat associated with resistivity, for reasons explained below, so that

$$\mathbf{R}_e = -\mathbf{R}_i \approx e n_e \eta \mathbf{j}\,, \tag{2.116}$$

$$Q_e + Q_i = -(\mathbf{u}_e - \mathbf{u}_i) \cdot \mathbf{R}_e \approx \eta |\mathbf{j}|^2 \qquad \textit{(scalar resistivity)}\,. \tag{2.117}$$

From Eqs. (2.61), (2.62) and (2.66), we then get the following set of *resistive two-fluid equations* (with $\alpha = \mathrm{e}$ for the electrons and $\alpha = \mathrm{i}$ for the ions):

$$\frac{\partial n_\alpha}{\partial t} + \nabla \cdot (n_\alpha \mathbf{u}_\alpha) = 0\,, \tag{2.118}$$

$$n_\alpha m_\alpha \Big(\frac{\partial \mathbf{u}_\alpha}{\partial t} + \mathbf{u}_\alpha \cdot \nabla \mathbf{u}_\alpha \Big) + \nabla p_\alpha - n_\alpha q_\alpha (\mathbf{E} + \mathbf{u}_\alpha \times \mathbf{B}) = \mathbf{R}_\alpha\,, \tag{2.119}$$

$$\frac{\partial p_\alpha}{\partial t} + \mathbf{u}_\alpha \cdot \nabla p_\alpha + \gamma p_\alpha \nabla \cdot \mathbf{u}_\alpha = (\gamma - 1) Q_\alpha\,. \tag{2.120}$$

This set is completed by adding Maxwell's equations (2.13)–(2.16).

Next, we combine the two-fluid equations such that a set of nearly equivalent one-fluid equations is obtained. This is done by defining macroscopic one-fluid variables that are linear combinations of the two-fluid variables:

$$\rho \equiv n_e m_e + n_i m_i \qquad \textit{(mass density)}\,, \tag{2.121}$$

$$\tau \equiv -e\,(n_e - Z n_i) \qquad \textit{(charge density)}\,, \tag{2.122}$$

$$\mathbf{v} \equiv (n_e m_e \mathbf{u}_e + n_i m_i \mathbf{u}_i)/\rho \qquad \textit{(centre of mass velocity)}\,, \tag{2.123}$$

$$\mathbf{j} \equiv -e\,(n_e \mathbf{u}_e - Z n_i \mathbf{u}_i) \qquad \textit{(current density)}\,, \tag{2.124}$$

$$p \equiv p_e + p_i \qquad \textit{(pressure)}\,. \tag{2.125}$$

(Notice the new meaning of $\mathbf{v}$, which can now be used without confusion with the particle velocities since distribution functions will not be used any more in this chapter.) This implies that no information of the mass and momentum equations is lost since the one-fluid variables ρ, τ, $\mathbf{v}$, $\mathbf{j}$ are just linear combinations of the two-fluid variables n_e, n_i, $\mathbf{u}_e$, $\mathbf{u}_i$. The essential assumption of one-fluid dynamics is that the temperature equilibration time between electrons and ions is short compared to other characteristic times, so that $T_e = T_i$. (This assumption presents a significant simplification of the model, but one has to keep in mind that some plasmas in nature (e.g. the solar wind) are not adequately described this way.) This implies that the information on the separate electron and ion temperatures is annihilated so that there is one variable fewer in one-fluid theory and, hence, one equation fewer to be solved.

The one-fluid equations are then obtained by operating on the two-fluid equations (2.118)–(2.120) by adding pairs, multiplied with mass and charge factors:

$$m_e\,(2.118)_e + m_i\,(2.118)_i \;\Rightarrow\; \partial\rho/\partial t\,,$$

$$-e\,(2.118)_e + Ze\,(2.118)_i \;\Rightarrow\; \partial\tau/\partial t\,,$$

$$(2.119)_e + (2.119)_i \;\Rightarrow\; \partial\mathbf{v}/\partial t\,,$$

$$-\frac{e}{m_e}\,(2.119)_e + \frac{Ze}{m_i}\,(2.119)_i \Rightarrow \partial\mathbf{j}/\partial t\,,$$

$$(2.120)_e + (2.120)_i \;\Rightarrow\; \partial p/\partial t\,.$$

In principle, this results in a set of one-fluid equations in terms of the variables $\rho, \tau, \mathbf{v}, \mathbf{j}$ and p alone. However, to remove all two-fluid variables from the equations one needs to exploit the inverses of the relations (2.121)–(2.124). These simplify significantly (see Eqs. (3.132) of Section 3.4.1) if one exploits the approximation

$$|n_e - Zn_i| \ll n_e \qquad \textit{(quasi charge-neutrality)}\,, \tag{2.126}$$

which is extremely well satisfied already at the microscopic level, as we have seen in Sections 1.4.1 and 2.3.2. The resulting one-fluid equations still contain the small length and time scale phenomena of the two-fluid equations (see Eqs. (3.135)–(3.139) of Section 3.4.1). Those are removed by the use of two additional approximations:

$$|\mathbf{u}_i - \mathbf{u}_e| \ll v \qquad \textit{(small relative velocity of ions and electrons)}\,, \tag{2.127}$$

which implies that the electron skin depth should be small, $\delta_e \ll (m_e/m_i)^{1/2}$, and

$$v \ll c \qquad \textit{(non-relativistic speeds)}\,. \tag{2.128}$$

As a result, the evolution expressions for the variables τ and $\mathbf{j}$ disappear.

Combining the one-fluid moment equations thus obtained with the pre-Maxwell equations (i.e., according to Section 2.2.2, dropping the displacement current and Poisson's equation), results in *the resistive MHD equations*:

$$\frac{\partial \rho}{\partial t} + \nabla\cdot(\rho\mathbf{v}) = 0 \qquad \textit{(continuity)}\,, \tag{2.129}$$

$$\rho\left(\frac{\partial \mathbf{v}}{\partial t} + \mathbf{v}\cdot\nabla\mathbf{v}\right) + \nabla p - \mathbf{j}\times\mathbf{B} = 0 \qquad \textit{(momentum)}\,, \tag{2.130}$$

$$\frac{\partial p}{\partial t} + \mathbf{v}\cdot\nabla p + \gamma p\nabla\cdot\mathbf{v} = (\gamma-1)\eta|\mathbf{j}|^2 \qquad \textit{(internal energy)}\,, \tag{2.131}$$

$$\frac{\partial \mathbf{B}}{\partial t} + \nabla\times\mathbf{E} = 0 \qquad \textit{(Faraday)}\,, \tag{2.132}$$

where

$$\mathbf{j} = \mu_0^{-1}\nabla\times\mathbf{B} \qquad \textit{(Ampère)}\,, \tag{2.133}$$

$$\mathbf{E}' \equiv \mathbf{E} + \mathbf{v}\times\mathbf{B} = \eta\mathbf{j} \qquad \textit{(Ohm)}\,, \tag{2.134}$$

whereas

$$\nabla\cdot\mathbf{B} = 0 \qquad \textit{(no magnetic monopoles)} \tag{2.135}$$

is just an initial condition on Faraday's law.

We will have plenty of opportunity to come to appreciate the power of this set of equations for the description of macroscopic plasma dynamics. For now, it suffices to make a few remarks.

(a) The momentum equation (2.130) represents balance between acceleration, on the one hand, and pressure gradient and Lorentz force (per unit volume), on the other. Additional forces, like the gravitational force $\mathbf{F}_{grav} = \rho\mathbf{g}$, may be put on the RHS of this equation. In laboratory plasmas, that force is completely negligible compared to the pressure gradient and the Lorentz force. Note that the Lorentz force is directed *perpendicular to* the magnetic field, so that acceleration *along* magnetic field lines must be produced by pressure gradients or gravity.

(b) The electric current density **j** and the electric field **E** have become secondary quantities in MHD, to be derived from Ampère's law and Ohm's law. Substituting them in Faraday's law (2.132) yields *the induction equation*:

$$\frac{\partial \mathbf{B}}{\partial t} = \nabla \times (\mathbf{v} \times \mathbf{B}) - \mu_0^{-1} \nabla \times (\eta \nabla \times \mathbf{B}) \,. \tag{2.136}$$

This equation couples the dynamics of the magnetic field to that of the plasma through the velocity term. When **v** was known, the induction equation could be used to determine **B**, where $\nabla \cdot \mathbf{B} = 0$ should be imposed as an initial condition.

(c) The algebraic relation (2.134) between the electric field **E** and the electric current density **j** is the generalization of Ohm's law for moving conducting media. According to this law, **j** is proportional to the electric field $\mathbf{E}'$ in a frame moving with the plasma. Many laboratory and astrophysical plasmas are nearly perfectly conducting, so that

$$\mathbf{E}' \equiv \mathbf{E} + \mathbf{v} \times \mathbf{B} = 0 \qquad \textit{(perfect conductivity: ideal MHD)} \tag{2.137}$$

almost everywhere. (Also recall from Section 2.2.3 that single particles drift in such a way that $\mathbf{E}' = 0$.) Therefore, the resistive form (2.134) of Ohm's law is only needed in regions of high current concentration, which are usually extremely thin. However, such current sheets do occur and play an important role in models for the disruptive phenomena mentioned above. That is why we have kept the term $\eta\mathbf{j}$ in Ohm's law. For the same reason, we have kept the term $(\gamma - 1)\eta|\mathbf{j}|^2$, representing heating due to Ohmic dissipation, in the energy equation (2.131). This term is usually very small compared to the terms on the left hand side.

(d) The slow time scale for *resistive diffusion* of the magnetic field, introduced in Eq. (2.114), can be estimated from the induction equation (2.136):

$$\tau_{\rm R} \sim \frac{\mu_0 l_0^2}{\eta} \,, \tag{2.138}$$

where $l_0 \sim a$ is the length scale of gradients. In terms of a representative plasma speed v_0 and length scale l_0, the ratio of the magnitudes of the convective and diffusive terms on the RHS of the induction equation (2.136) is a dimensionless number,

$$R_{\rm m} \equiv \frac{\mu_0 l_0 v_0}{\eta} \,, \tag{2.139}$$

called the *magnetic Reynolds number*, in analogy with the Reynolds number $R \equiv v_0 l_0 / \nu$ (where $\nu \equiv \mu/\rho$ is the kinematic viscosity), measuring the relative importance of inertial and viscous effects in ordinary hydrodynamics. The magnetic Reynolds number is a measure of the strength of the coupling between the flow and the magnetic field. Since η is extremely small for most plasmas of interest (e.g. tokamaks and the solar corona), $R_{\rm m} \gg 1$ so that this coupling is very strong and resistive decay is negligible on the time scales specified in Eqs. (2.114).

The neglect of resistivity, and the substitution of **j** and **E**, finally leads to *the ideal MHD equations* in their most compact form:

$$\frac{\partial \rho}{\partial t} + \nabla \cdot (\rho \mathbf{v}) = 0 \,, \tag{2.140}$$

$$\rho\left(\frac{\partial \mathbf{v}}{\partial t} + \mathbf{v}\cdot\nabla\mathbf{v}\right) + \nabla p - \mu_0^{-1}(\nabla\times\mathbf{B})\times\mathbf{B} = 0\,, \tag{2.141}$$

$$\frac{\partial p}{\partial t} + \mathbf{v}\cdot\nabla p + \gamma p\nabla\cdot\mathbf{v} = 0\,, \tag{2.142}$$

$$\frac{\partial \mathbf{B}}{\partial t} - \nabla\times(\mathbf{v}\times\mathbf{B}) = 0\,, \qquad \nabla\cdot\mathbf{B} = 0\,. \tag{2.143}$$

These *coupled nonlinear partial differential equations* govern the evolution of the density, the velocity, the pressure and the magnetic field. Their properties and solutions will occupy us for most of the remainder of this book.

2.4.2 Alfvén waves

Let us consider wave propagation in a *homogeneous* plasma occupying all space, with a constant magnetic field in the z-direction:

$$\rho_0 = \text{const}\,, \quad \mathbf{v}_0 = 0\,, \quad p_0 = \text{const}\,, \quad \mathbf{B}_0 = B_0\,\mathbf{e}_z \ \Rightarrow\ \mathbf{j}_0 = 0\,. \tag{2.144}$$

We perturb this plasma with small deviations $\rho_1, \mathbf{v}_1, p_1, \mathbf{B}_1$ from the background state. This permits us to *linearize* the equations (2.140)–(2.143):

$$\frac{\partial \rho_1}{\partial t} = -\,\rho_0\nabla\cdot\mathbf{v}_1\,, \tag{2.145}$$

$$\rho_0\frac{\partial \mathbf{v}_1}{\partial t} = -\,\nabla p_1 + \mu_0^{-1}(\nabla\times\mathbf{B}_1)\times\mathbf{B}_0\,, \tag{2.146}$$

$$\frac{\partial p_1}{\partial t} = -\,\gamma p_0\nabla\cdot\mathbf{v}_1\,, \tag{2.147}$$

$$\frac{\partial \mathbf{B}_1}{\partial t} = \nabla\times(\mathbf{v}_1\times\mathbf{B}_0)\,, \tag{2.148}$$

producing a complete set of equations for the unknowns $\rho_1, \mathbf{v}_1, p_1$ and $\mathbf{B}_1$.

Note that the density perturbation ρ_1 only appears in Eq. (2.145) so that it need not be determined to solve the other equations. Moreover, in order to concentrate on the magnetic effects, let us neglect the equilibrium pressure ($p_0 = 0$ or, in terms of the parameter β introduced in Section 2.2.3, $\beta = 0$) so that Eq. (2.147) implies that $p_1 = 0$ as well. We then obtain from the remaining Eqs. (2.146) and (2.148) a wave equation for the velocity $\mathbf{v}_1$:

$$\rho_0\frac{\partial^2 \mathbf{v}_1}{\partial t^2} = \mu_0^{-1}\left(\nabla\times\frac{\partial \mathbf{B}_1}{\partial t}\right)\times\mathbf{B}_0 = \mu_0^{-1}\mathbf{B}_0\times\Big(\nabla\times\big(\nabla\times(\mathbf{B}_0\times\mathbf{v}_1)\big)\Big)\,. \tag{2.149}$$

Inserting plane wave solutions of the form

$$\mathbf{v}_1(\mathbf{r},t) = \hat{\mathbf{v}}\,\mathrm{e}^{\mathrm{i}(\mathbf{k}\cdot\mathbf{r}-\omega t)}\,, \tag{2.150}$$

i.e. replacing $\partial/\partial t \to -\mathrm{i}\omega$ and $\nabla \to \mathrm{i}\mathbf{k}$, Eq. (2.149) transforms into an algebraic eigenvalue equation

$$-\rho_0\omega^2\,\hat{\mathbf{v}} = -\mu_0^{-1}B_0^2\,\mathbf{e}_z\times\Big(\mathbf{k}\times\big(\mathbf{k}\times(\mathbf{e}_z\times\hat{\mathbf{v}})\big)\Big)\,. \tag{2.151}$$

Because the expression on the RHS is a vector $\perp\ \mathbf{e}_z$ (i.e. $\perp\ \mathbf{B}_0$), the immediate result is that the parallel velocity vanishes: $\hat{v}_\parallel \equiv \mathbf{e}_z\cdot\hat{\mathbf{v}} = 0$. This leaves two components for $\hat{\mathbf{v}}_\perp$, which turn out to oscillate independently, each with its own characteristic frequency. The flow fields of the two modes differ with respect to their direction relative to the plane through $\mathbf{k}$ and $\mathbf{B}_0$. To distinguish them, it is convenient to fix the direction of the wave vector $\mathbf{k}$ to lie in the x-z plane, as indicated in Fig. 2.9: $\mathbf{k} = (k_\perp, 0, k_\parallel)$. Obviously, this can be done without loss of generality. Focusing on the most significant branch, viz. the one with a flow velocity perpendicular to both $\mathbf{k}$ and $\mathbf{B}_0$ (i.e. $\hat{\mathbf{v}} \sim \mathbf{k} \times \mathbf{B}_0$), and manipulating with the vector products (using Eq. (A.2) of Appendix A), the eigenvalue problem becomes

$$(\omega^2 - k_\parallel^2 v_\mathrm{A}^2)\, \hat{v}_y = 0\,, \tag{2.152}$$

where

$$v_\mathrm{A} \equiv \frac{B_0}{\sqrt{\mu_0 \rho_0}} \tag{2.153}$$

is the celebrated *Alfvén velocity*. Hence, one obtains the two *Alfvén waves*, discovered by Alfvén in 1942 [10], with frequency

$$\omega = \pm\omega_\mathrm{A}\,, \quad \text{where } \omega_\mathrm{A} \equiv k_\parallel v_\mathrm{A} = \mathbf{k}\cdot\mathbf{B}/\sqrt{\mu_0\rho_0}\,, \tag{2.154}$$

and phase velocity v_A, corresponding to waves that run along the magnetic field to the right $(+)$ or to the left $(-)$.

▷ **Magneto-sonic waves** For $\omega = 0$, an extra solution is obtained, viz. the remnant of the slow magneto-sonic mode, with $\hat{v}_\parallel \neq 0$. Another branch, $\omega^2\, \hat{v}_x = (k_\perp^2 + k_\parallel^2) v_\mathrm{A}^2\, \hat{v}_x$, represents degenerate fast magneto-sonic waves which, accidentally, also propagate with the Alfvén speed. These waves are not well described here since the pressure has been neglected. They will be discussed more extensively later (in Chapter 5). ◁

The Alfvén waves are caused by tension of the magnetic field lines, which tends to restore the initial shape (as schematically indicated in Fig. 2.9). Another important aspect of the macroscopic MHD waves is that their frequency depends on the value of the wave vector $\mathbf{k}$, i.e. on the wavelength, in contrast to the microscopic plasma oscillations with the plasma frequency (2.76), which is independent of the wavelength. This points to an extremely significant property of Alfvén waves, and of MHD perturbations in general: the slowest of them have the longest wavelengths and, hence, they sample the magnetic field in the large, i.e. they 'feel' the global magnetic geometry. Of course, infinite homogeneous plasmas do not have a particularly interesting magnetic geometry. However, when the magnetic field is bent into the relevant shape for plasma confinement, e.g. a torus, this property survives: *Alfvén waves carry the information of the overall magnetic geometry*.

Inserting numbers for a typical tokamak experiment, viz. $n = 10^{20}\,\mathrm{m}^{-3}$, $B = 3\,\mathrm{T}$, so that $\rho = n m_\mathrm{i} = 1.7 \times 10^{-7}\,\mathrm{kg\,m}^{-3}$, we obtain a phase velocity

$$v_\mathrm{A} \approx 6 \times 10^6\,\mathrm{m\,s}^{-1}$$

(rather fast, but still non-relativistic: only 2% of the velocity of light). For long wavelengths, e.g. $\lambda_\parallel = 20\,\mathrm{m}$ ($\approx 2\pi R$ for a torus with major radius $R \approx 3\,\mathrm{m}$), so that $k_\parallel \equiv 2\pi/\lambda_\parallel \approx 0.3\,\mathrm{m}^{-1}$, this gives an Alfvén frequency of

$$\omega_\mathrm{A} = k_\parallel v_\mathrm{A} \approx 1.8 \times 10^6\,\mathrm{s}^{-1}\,.$$

Hence, with wavelengths of the order of the size of the confinement experiment, we are now on the microsecond time scale: $\tau = 2\pi R/v_\mathrm{A} = 2\pi/\omega_\mathrm{A} \approx 3$ microsec.

▷ **Exercise** Consult Table B.4 for corresponding values of a solar plasma. What is the Alfvén transit time for

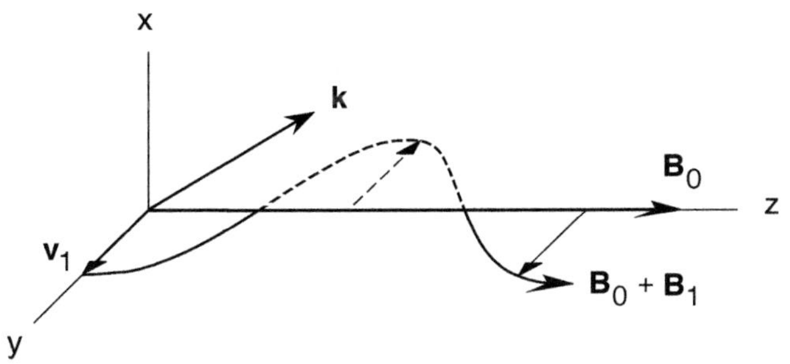

Fig. 2.9 Velocity and magnetic field perturbations of an Alfvén wave.

a solar coronal loop of length $L = 100\,000\,\mathrm{km}$, $n = 10^{14}\,\mathrm{m}^{-3}$, $B = 0.003\,\mathrm{T}$? How does this compare to the 'times of interest' indicated in this table? The latter refer to time scales needed for fusion in tokamaks or typical life times of coronal loops. How do tokamaks compare with coronal loops as far as the number of Alfvén transit times during that time span is concerned? Also complete the empty column of Table B.5. ◁

On the Alfvén transit time scale, the overall magnetic confinement geometry becomes 'known' to the plasma, which then will exhibit the dynamics corresponding to the intrinsic stability or instability of the configuration. These possibilities necessarily involve inhomogeneous plasmas with curved magnetic fields. This brings us to our next topic.

2.4.3 Equilibrium and stability

The basic theoretical approach to plasma confinement for a given magnetic geometry typically consists of, first, fixing an *equilibrium* state, next, determining the different kinds of *waves* produced by perturbing this state, and, finally, finding out whether amongst those perturbations there are *instabilities* that would lead to destruction of the configuration. We have just encountered an example of the second topic, viz. Alfvén waves. We will now treat simple examples of the first and last topic.

Consider an *inhomogeneous* plasma in *static* equilibrium, i.e. the fluid is at rest ($\mathbf{v} = 0$) and all time derivatives in the MHD equations (2.140)–(2.143) vanish. In contrast to ordinary hydrodynamics, even this highly idealized situation is non-trivial because of the possibility of balancing pressure gradients by the Lorentz force:

$$\nabla p = \mathbf{j} \times \mathbf{B}\,. \tag{2.155}$$

This is the basis of all *magnetic confinement* systems for controlled thermonuclear fusion experiments. (In laser fusion, or inertial confinement experiments, the inertial forces replace the magnetic ones.) The current, of course, has to satisfy Ampère's law

$$\mathbf{j} = \mu_0^{-1} \nabla \times \mathbf{B}\,, \tag{2.156}$$

and the magnetic field should satisfy

$$\nabla \cdot \mathbf{B} = 0\,. \tag{2.157}$$

The latter equation is no longer an initial condition now, but it remains a restriction on the kind of vectors that qualify as magnetic fields. These are the equations which the variables p, $\mathbf{j}$ and $\mathbf{B}$ have to satisfy in order to produce a *static equilibrium* state.

(a) Equilibrium of a z-pinch As an example of a simple equilibrium, consider a cylindrical plasma column with a strong electric current. This current creates a magnetic field, resulting in a force $\mathbf{j}\times\mathbf{B}$ directed radially inward: *the pinch effect*. On the basis of this effect, two interesting plasma confinement configurations can be distinguished, viz. the θ- and the z-pinch illustrated in Fig. 1.4. Here, the labels θ and z refer to the direction of the current in the cylindrical coordinate representation of the plasma column. As we have seen in Section 1.2.3, both types of configurations have been considered in the early days of nuclear fusion research, both without success, at least in their pure form, but the concepts remain important because they represent the basic forces in plasma confinement.

For *the z-pinch* in cylinder geometry (current in the z-direction, magnetic field in the θ-direction, gradients in the r-direction; see Fig. 2.10) the equilibrium equations (2.155)–(2.157) reduce to

$$\frac{dp}{dr} = -j_z B_\theta\,, \quad \text{and} \quad j_z = \frac{1}{\mu_0 r}\frac{d}{dr}(rB_\theta)\,. \tag{2.158}$$

This gives the following relationship between the pressure profile $p(r)$ and the magnetic field profile $B_\theta(r)$:

$$\frac{dp}{dr} = -\frac{B_\theta}{\mu_0 r}\frac{d}{dr}(rB_\theta)\,. \tag{2.159}$$

This is the only restriction on these profiles as far as the equilibrium is concerned. We are free to choose e.g., for simplicity, a constant current profile j_z = const. This choice determines the other profiles:

$$B_\theta = \tfrac{1}{2}\mu_0 r j_z\,, \quad \text{and} \quad p = p_c(1 - r^2/a^2)\,, \quad p_c \equiv \tfrac{1}{4}\mu_0 a^2 j_z^2\,, \tag{2.160}$$

where a is the radius of the plasma cylinder.

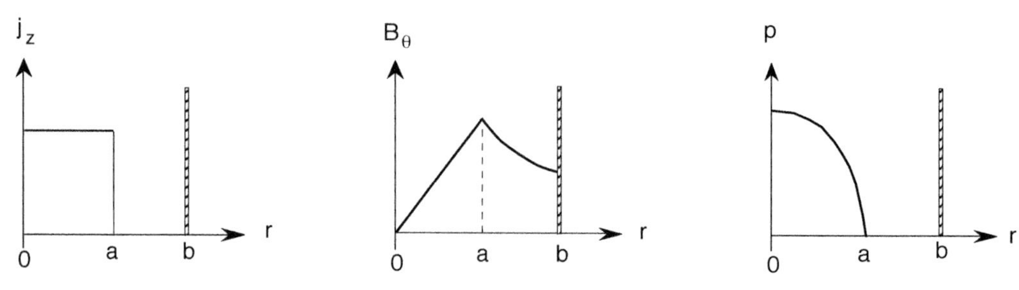

Fig. 2.10 Radial distributions of current density, magnetic field and pressure in a z-pinch.

The plasma cylinder may be surrounded by a vacuum with a magnetic field $\hat{\mathbf{B}}$ satisfying $\nabla\times\hat{\mathbf{B}} = 0$ and $\nabla\cdot\hat{\mathbf{B}} = 0$, so that $\hat{j}_z = 0$ and $\hat{B}_\theta(r) = B_\theta(a)\,a/r$ there. Obviously, this radially decaying magnetic field is produced by the total current I_z flowing within the plasma interval $0 \le r \le a$. Finally, the configuration may be closed off by putting a perfectly conducting wall at some radius $r = b$. The central pressure p_c is related to the total current I_z:

$$I_z = \pi a^2 j_z\,, \quad \text{so that} \quad p_c = \frac{\mu_0 I_z^2}{4\pi^2 a^2}\,. \tag{2.161}$$

Inserting typical numbers for the early laboratory pinch plasmas, viz. $n = 10^{22}\ \mathrm{m}^{-3}$ (i.e. two orders higher than present-day tokamaks), $T = 10^8\,\mathrm{K}, a = 0.1\,\mathrm{m}$, we find

$$p_c = nkT = 1.38 \times 10^7\,\mathrm{N\,m^{-2}}\ (= 136\,\mathrm{atm}!)\,,$$
$$I_z = 2\pi a(p_c/\mu_0)^{1/2} = 2.1 \times 10^6\,\mathrm{A}\,,$$
$$B_\theta = \mu_0 I_z/(2\pi a) = 4.2\,\mathrm{T}\ (= 42\,\mathrm{kgauss})\,.$$

Although these currents and fields are quite large and exert huge pressures, they were already well within reach of the technology of the 1950s: a thermonuclear reactor by just passing a current through a linear plasma column?

(b) Kink instabilities In Section 2.4.2, we found the frequency of Alfvén waves for a homogeneous plasma with straight magnetic field lines. However, for the z-pinch equilibrium, the field lines are curved. Consequently, the analysis of the dispersion equation is significantly more complicated. We will not enter that analysis here since it will be the subject of Chapter 9, but just indicate a result that is relevant to the present discussion. Amongst the normal mode solutions $\sim \exp[\mathrm{i}(m\theta + kz - \omega t)]$ of the cylindrical configuration there are some 'Alfvén' waves, with low mode number m (typically $m = 1$), that have a complex frequency ($\omega^2 < 0$) so that they are exponentially growing. These modes are called *kink instabilities* because of the associated helical deformation of the plasma column. For wavelengths $k_z^{-1} \gg a$, the expression for their growth rate demonstrates the cause of the instability, viz. the curved magnetic field component $B_\theta(a)$ at the plasma edge:

$$\omega^2 \approx -\frac{B_\theta^2(a)}{2\mu_0\rho_0 a^2}\,. \tag{2.162}$$

As illustrated by Fig. 2.9, the magnetic force $\mathbf{F} = \mathbf{j}_1 \times \mathbf{B}_0$ in the RHS of Eq. (2.146) is opposite to $\boldsymbol{\xi}$ (the displacement corresponding to $\mathbf{v}_1 = \partial\boldsymbol{\xi}/\partial t = -\mathrm{i}\omega\boldsymbol{\xi}$) in the case of genuine Alfvén waves. Hence, this force is restoring, i.e. stabilizing (the effect counteracts its cause). However, in the case of a z-pinch equilibrium there is an extra term $\mathbf{j}_0 \times \mathbf{B}_1$, due to the plasma current, which points in the same direction as $\boldsymbol{\xi}$ so that it is destabilizing (the effect enhances its cause). That term involves the detailed distribution of the magnetic field B_θ in the plasma ($r \leq a$) *and* the surrounding vacuum ($r > a$) shown in Fig. 2.10. For the latter reason, these modes are called *external* kink instabilities.

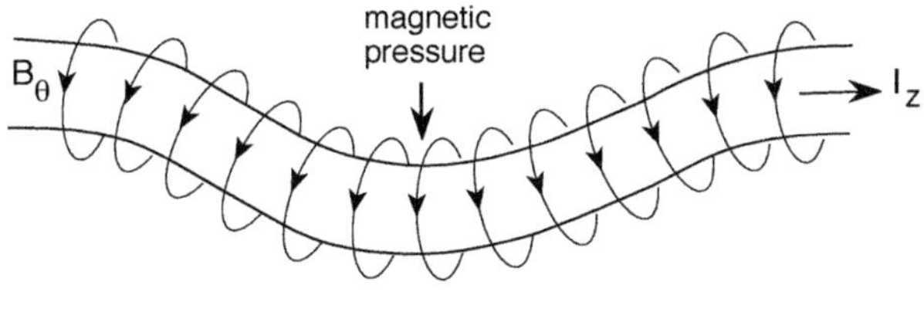

Fig. 2.11 External kink instability of a z-pinch.

The mechanism of the external kink instability can easily be illustrated from the similar situation of a current carrying wire. Here, a helical deformation of the wire tends to be magnified due to the compression of the field lines on the inside and expansion on the outside bends of the deformation. The net result is an unbalanced increase of the *magnetic pressure* $B^2/(2\mu_0)$ on the inside as compared to the outside bends (see Fig. 2.11). This is fatal for the z-pinch since time scales for the instability are of the same order of magnitude as the time scale of Alfvén waves, viz. microseconds, whereas at least seconds are needed for thermonuclear ignition. That answers the question posed above: no simple fusion reactor is obtained this way; no free rides in science and technology!

For thermonuclear confinement, this is the first reason that something more clever had to be done than just a straight z-pinch. Here, the long history of thermonuclear research starts: the next step was the linear θ-pinch, which was equally disastrous because of the open ends. Success gradually came when these two concepts were combined in the tokamak (illustrated in Fig. 1.4).

How is a violent instability like the external kink mode remedied in a tokamak? For a change, this turned out to be quite simple. The cylinder is replaced by a torus and, since kink modes are long-wavelength instabilities, the parameters may be chosen such that the range of unstable wavelengths simply does not fit in the torus. This leads to the famous *Kruskal–Shafranov condition* for external kink mode stability, which puts a limit on the total plasma current:

$$I_z(a) < \frac{2\pi a^2 B_z}{\mu_0 R_0} \,. \tag{2.163}$$

Notice that this requires a contribution of the θ-pinch concept: a stabilizing 'backbone' of longitudinal magnetic field B_z is necessary. As is evident from this expression, a z-pinch ($B_z = 0$) does not have a stable range of the current.

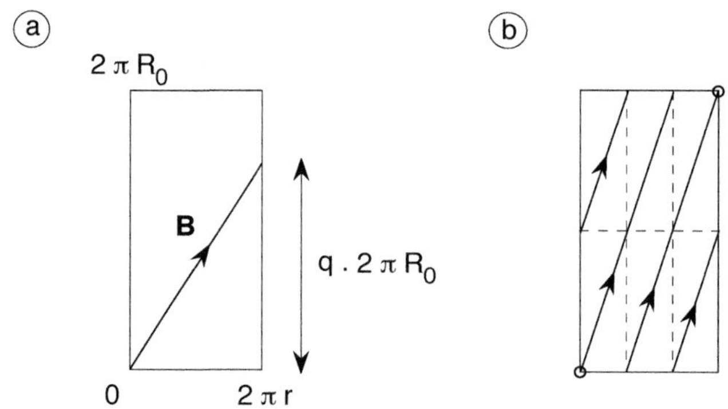

Fig. 2.12 'Safety factor' in periodic cylinder model of a toroidal plasma: (a) field line with $q < 1$; (b) rational field line with $q = 3/2$.

The Kruskal–Shafranov condition can also be written in terms of the *'safety factor'* $q(a)$. This concept is easily understood by considering a double periodic cylinder (periodic in θ with period 2π and in z with period $2\pi R_0$) as a mathematical model for a torus with an inverse aspect ratio $\epsilon \equiv a/R_0$. The physical justification of such a model is that it provides a first approximation to a genuine, but slender ($\epsilon \ll 1$), torus which neglects the toroidal curvature but accounts for the more important toroidal periodicity. The cylindrical helical magnetic field $\mathbf{B} = B_\theta(r)\mathbf{e}_\theta + B_z(r)\mathbf{e}_z$ then provides a model of nested magnetic surfaces (cylinders of radius $r \leq a$ and length $2\pi R_0$) with field lines having a pitch that is constant on each surface, but varies from surface to surface. This may be visualized by unrolling one of the cylinders (Fig. 2.12). The parameter $q(r)$ is defined as the pitch of the field lines with a normalization chosen such that $q = 1$ corresponds to a topology where the field lines close upon themselves after one revolution the short way and one revolution the long way around the torus:

$$q(r) \equiv \frac{rB_z(r)}{R_0 B_\theta(r)} = \frac{2\pi r^2 B_z(r)}{\mu_0 R_0 I_z(r)} \,. \tag{2.164}$$

Hence, the Kruskal–Sfafranov limit can be expressed as $q(a) > 1$. This *accidental agreement* with an integer value is rather unfortunate since it has led to many erroneous statements in the literature about this limit being connected with rational field lines 'biting' in their own tail. Actually, this has nothing to do with the external kink mode mechanism, which is driven by the total plasma current $I_z(a)$ creating a curved magnetic field $B_\theta(a)$ on the outside of the plasma column. In fact, in more general configurations, like a genuine torus with non-circular cross-section, the external kink mode is no longer associated with integer values of $q(a)$.

Curing the remaining instabilities (the internal ones, not perturbing the vacuum magnetic field) was not so simple. Here, rational values of the safety factor do play an important role since most internal instabilities are localized about rational field lines and surfaces where $q = -m/n$, when m and n indicate the mode numbers in the periodic directions (see Fig. 2.12(b)). This has led to a long period of steady experimental trial and error, eventually leading to confinement on the second and even minute time scales in the last part of the twentieth century.

2.5 In conclusion

In this chapter, we have introduced the three main theoretical approaches of plasmas, viz. the theory of single particle motion in prescribed electric and magnetic fields, the kinetic theory of collections of many such particles (with roughly equal numbers of the opposite charges) and the theory of magnetohydrodynamics pertaining to the global macroscopic aspects of the dynamics of plasmas in complex magnetic field geometries. We have also considered a sample of the many different dynamical phenomena associated with these theoretical models. Three effects were encountered which give plasmas the *coherence* that is necessary for thermonuclear confinement of laboratory plasmas and which is also characteristic for magnetized plasmas encountered in nature.

(a) Within the single particle picture, *particles of either charge stick to the magnetic field lines by their gyro-motion* restraining the perpendicular motion.

(b) Because of the large electric fields that occur when electrons and ions are separated, deviations from charge neutrality can occur only in very small regions (of the size of a Debye length). Over larger regions, *ions and electrons stay together to maintain approximate charge neutrality.*

(c) In the fluid picture, it was found that *currents in the plasma create their own magnetic field, pinching the plasma,* and that *Alfvén waves act as a restoring agent on magnetic field distortions.* However, we also encountered the first effect *destroying the coherence of plasmas, viz. the global external kink instability.*

In the next chapter, we delve more deeply into the kinetic theory aspects (loosely considering two-fluid theory also under this heading, although it is strictly a fluid theory, because of the microscopic scales involved). The exposition of MHD is continued in Chapter 4.

2.6 Literature and exercises

Notes on literature

The list below is a very limited choice from the numerous books and articles on the topics introduced in this chapter, which are discussed in all textbooks on plasma physics.

Basic texts on classical mechanics and electrodynamics

- Goldstein, *Classical Mechanics* [243].
- Jackson, *Classical Electrodynamics* [331].

Single particle motion

- Chen, *Introduction to Plasma Physics* [117], Chapter 2 on single-particle motions (elementary).
- Sturrock, *Plasma Physics* [580], Chapters 3–5 on orbit theory and adiabatic invariants (intermediate).
- Northrop, *The Adiabatic Motion of Charged Particles* [469]: a whole monograph on adiabatic motion of charged particles (advanced).
- Balescu, *Transport Processes in Plasmas* [32], Chapter 1 on modern developments in Hamiltonian mechanics of charged particles (advanced).

Kinetic theory and Landau damping

- Bittencourt, *Fundamentals of Plasma Physics* [70], Chapters 5–8 on kinetic theory and the derivation of macroscopic equations, Chapter 18 on waves in hot plasmas with Landau damping (elementary).
- Goldston and Rutherford, *Introduction to Plasma Physics* [244], Chapters 22–24 on kinetic theory and Landau damping (elementary).
- Hasegawa and Sato, *Space Plasma Physics* [293], Chapter 1 on the physics of stationary plasmas with motion of charged particles, adiabatic invariants, kinetic theory and MHD equations (intermediate).

Fluid theory and MHD

- Freidberg, *Ideal MHD* [186], Chapter 2 on the derivation of the MHD equations from the Boltzmann equation (intermediate).
- Roberts, *An Introduction to Magnetohydrodynamics* [522], Chapter 1 on the MHD equations as a part of continuum mechanics (intermediate).
- Alfvén's 1942 paper on the 'Existence of electromagnetic-hydrodynamic waves' [10] is just one column in Nature, with one paragraph on the discovery of Alfvén waves, that earned him the Nobel prize. The application at the end to sunspot migration is an unfortunate example (the waves he discovered propagate along the magnetic field, not across) at a time when solar coronal dynamics was still largely unknown.

Exercises

[2.1] *Cyclotron motion*

We start from the equation of motion for an electron or ion in an electromagnetic field, with $\mathbf{B}$ in the z-direction and $\mathbf{E}$ in the (y, z)-plane.

- Write down the component for the motion parallel to the magnetic field. What happens to electrons when collisions are negligible?
- Now, choose $\mathbf{E}$ perpendicular to $\mathbf{B}$ and solve the full equation of motion. What is different compared to a situation with $\mathbf{E} = 0$? Is there an extra velocity component and, if so, in which direction? Is there a difference between electrons and ions?
- Assume particles with thermal perpendicular speed, $v_{\rm th} = (2kT/m)^{1/2}$, at room temperature! Calculate the gyro-frequencies and radii for electrons and protons for $B = 1\,\mathrm{T}$. Which particles have the largest gyro-radius? How do they encircle the field lines?

- For a background density typical of tokamaks, $n = 10^{20}\,\mathrm{m}^{-3}$, calculate the plasma frequency and the Debye length. Compare this to the gyro-radius. What do you conclude?

[2.2] *Drift velocities*

Assume we have a cyclotron device with a magnetic field strength of 1 T in the z-direction. The radius of the device is 2 metres.

- On both sides we put a plate, with a potential difference of $+300$ V between the right and left one. Calculate the drift velocities for electrons and protons.
- We remove the plates but take gravity into account. Unfortunately, the apparatus is old and the magnetic field lines are not vertical any more but slightly tilted, 2° to be specific. What are the drift velocities now? In what direction are they pointing?

[2.3] *Momentum equation*

Derive the momentum equation from the Boltzmann equation. (If you want to check your answer, the complete derivation is given in Section 3.2.2.)

[2.4] *Plasma oscillations*

We displace electrons with respect to ions in a cold, unmagnetized, plasma and study the resulting oscillations from the two-fluid equations and Poisson's law.

- Give an argument why the ion equations will not play a significant role.
- Since we are left with the electron equations now, linearize them and Poisson's law to construct the wave equation for the perturbed density. Assuming harmonic time dependence, find the frequency of the oscillations. What does it depend on?
- Try to design a device that can measure the background density profile of a plasma.

[2.5] *Landau damping*

What is Landau damping? What are the implications for possible wave motions in a plasma?

[2.6] *Alfvén waves*

We start from the ideal MHD equations (2.140)–(2.143). Assuming a static homogeneous low β plasma with a constant magnetic field in the z-direction, perturb all quantities and linearize.

- Find the wave equation for the velocity. Insert plane wave solutions and construct the eigenvalue equation. What can you say at once about the velocity?
- Choose the $\mathbf{k}$ vector in the x-z plane and the velocity in the y-direction. Give the solutions to the new eigenvalue problem. How many are there?
- What does the Alfvén velocity depend on? Could that be useful for plasma diagnostics?

[2.7] *MHD time scales*

Estimate the longest time scale $\tau_{2\mathrm{F}}$ of the different high-frequency waves described by the two-fluid model. Also estimate the slow diffusion time scale τ_{R} of the resistive MHD model. Use a characteristic length scale l_0 for the gradients. Show that the window $\tau_{2\mathrm{F}} \ll \tau_{\mathrm{MHD}} \ll \tau_{\mathrm{R}}$ easily accommodates Alfvén wave dynamics. Insert numbers for the different plasmas of Table B.5.

[2.8] *Equilibrium of a plasma cylinder*

Consider a cylindrical plasma column in static equilibrium, with a diameter of 2 metres and carrying a current of 10^6 A in the z-direction.

- Write down the equations for the pressure, the current density and the magnetic field in cylindrical coordinates. Determine the number of free profiles and parameters.
- Assuming a constant current density profile, determine the pressure profile and calculate the pressure at the centre of this cylinder. How does it compare with atmospheric pressures?

[2.9] *Kink instabilities*

How is the violent kink instability mechanism of the z-pinch eliminated in a tokamak? Assuming a 'straight tokamak' model of a finite-length periodic cylinder, determine the limit on the total current from the order of magnitude estimate $q = 1$ for the 'safety factor'.

3

'Derivation' of the macroscopic equations*

3.1 Two approaches*

There are basically two ways of introducing the equations of magnetohydrodynamics:

(a) pose them as reasonable postulates for a hypothetical medium called 'plasma';

(b) derive them by appropriate averaging of kinetic equations.

Our approach, starting with Chapter 4, is mainly along the lines of the first method, pioneered by Grad [253, 72] in a series of lecture notes, using physical arguments and mathematical criteria to justify the results. In this chapter, the main steps of the second method will be discussed and shown to be somewhat unsatisfactory since they involve a number of approximations that are often difficult to justify. The reason for going through this analysis anyway is that it provides understanding of the domain of validity of the MHD description and that it indicates what kind of modifications are in order when this description fails.

Mathematically inclined readers may skip this digression, where many results from kinetic theory are not derived but simply stated, and continue reading with Chapter 4. Also, the serious student of magnetohydrodynamics is advised to turn to a detailed study of the present chapter only after a first reading, e.g. of the Chapters 4–11 on basic MHD and the applications to standard models, since the level of this chapter is essentially that of the advanced theory of the later chapters. It has been placed here because this is where it logically belongs.

We give a 'derivation' of the MHD equations by averaging the kinetic equations for plasmas. The quotation marks are meant to remind the reader of the embarrassing assumptions that have to be made when starting from kinetic equations and then averaging them to obtain the transport coefficients and the macroscopic equations (like in the classics by Chapman and Cowling [116] and Braginskii [92]). The point is not so much that classical transport theory is in a poor state (an excellent monograph by Balescu [32] exists), but the justification of local transport coefficients with respect to the observed dissipative processes. Nature just appears to mock classical transport. For example, the very existence of the solar cycle, a magnetohydrodynamic phenomenon with a period of about 22 years, requires a turbulent resistivity which is a factor of about 10^{10} larger than the classical Spitzer value! Clearly, such order of magnitude 'anomalies' cannot be maintained indefinitely in a scientific enterprise. This challenge is presently taken up by intensive efforts in plasma turbulence theory, see e.g. Refs. [33] and [662], in particular applied to magnetic confinement in nuclear fusion machines, but honesty demands to say that this field is far from closing the gap between observed and computed transport phenomena.

3.2 Kinetic equations★

3.2.1 Boltzmann equation★

Recall our introductory exposition in Section 2.3: The plasma is considered as a collection of charged particles moving in an electromagnetic field $\mathbf{E}, \mathbf{B}$. Different species of particles, specifically ions and electrons, are distinguished by a subscript α. The electrons and ions are described by time-dependent distribution functions in six-dimensional phase space:

$$f_\alpha = f_\alpha(\mathbf{r}, \mathbf{v}, t) \qquad (\alpha = \mathrm{e}, \mathrm{i})\,. \tag{3.1}$$

The probable number of particles of species α in the six-dimensional volume element $d^3r\, d^3v$ centred at $\mathbf{r}, \mathbf{v}$ will then be $f_\alpha(\mathbf{r}, \mathbf{v}, t)\, d^3r\, d^3v$. The variation in time of these distribution functions is determined by a *Boltzmann equation* for each particle species:

$$\frac{\mathrm{d}f_\alpha}{\mathrm{d}t} \equiv \frac{\partial f_\alpha}{\partial t} + \mathbf{v} \cdot \frac{\partial f_\alpha}{\partial \mathbf{r}} + \frac{q_\alpha}{m_\alpha}(\mathbf{E} + \mathbf{v} \times \mathbf{B}) \cdot \frac{\partial f_\alpha}{\partial \mathbf{v}} = C_\alpha\,. \tag{3.2}$$

Here, $\mathbf{E}$ and $\mathbf{B}$ consist of contributions of the external fields and of the averaged internal fields originating from the long-range inter-particle interactions. The RHS of Eq. (3.2) gives the rate of change of the distribution function due to short-range binary particle interactions called collisions. Neglect of these collisions leads to *the Vlasov equation*, also called the *collisionless Boltzmann equation* (punned the 'Boltzmannless' Boltzmann equation by Rosenbluth):

$$\frac{\partial f_\alpha}{\partial t} + \mathbf{v} \cdot \frac{\partial f_\alpha}{\partial \mathbf{r}} + \frac{q_\alpha}{m_\alpha}(\mathbf{E} + \mathbf{v} \times \mathbf{B}) \cdot \frac{\partial f_\alpha}{\partial \mathbf{v}} = 0\,. \tag{3.3}$$

A closed system of equations is obtained by adding Maxwell's equations to determine $\mathbf{E}(\mathbf{r}, t)$ and $\mathbf{B}(\mathbf{r}, t)$.

At this point, some remarks are in order.

(a) In this chapter, we assume that the plasma consists of electrons and ions of one kind so that we just have the two distribution functions f_e and f_i to worry about (more than enough!). The temperature is assumed to be high enough to have complete ionization, so that ionization, recombination and charge exchange processes no longer occur. We also assume that fusion reactions do not yet occur. There will be conservation of the total number of each of the two kinds of charged particles separately. These assumptions can be relaxed, but at the cost of enormously complicating the analysis by the appearance of even more unknown distribution functions. However, even with a large fraction of neutral particles present, one still can show dominance of magnetohydrodynamic processes.

(b) Following standard terminology in plasma physics, the basic kinetic equation (3.2) has been called the 'Boltzmann' equation but it should be stressed that the expression C_α for the rate of change of the distribution function by collisions is completely different from the one derived by Boltzmann himself. He was concerned with collisions of neutral particles, generally resulting in large-angle scattering, described by an integral of the product of distribution functions evaluated for the different relative velocities of the colliding particles. Whereas neutral particles only collide when they hit each other, charged particles interact through the long-range Coulomb force. In high-temperature plasmas, most of those scattering events are small-angle deflections. By judiciously exploiting this feature, Landau [392] (1936) was able to convert the Boltzmann collision integral into an expression valid for charged particles: Eq. (3.13) below. The proper way of handling the statistics of many small-angle deflections is by means of a Fokker–Planck equation. This was done by Rosenbluth et al. [530] (see also Trubnikov [611]), resulting in an expression that is equivalent to the Landau form (Eq. (3.16) below).

(c) A fundamental assumption for the validity of the Landau collision integral (as well as the Boltzmann integral) is the dependence on the so-called *one-particle* distribution functions $f_\alpha(\mathbf{r}, \mathbf{v}, t)$ alone. These functions express the probability of finding particles of type α at a certain time at the phase space position $\mathbf{r}, \mathbf{v}$. However, binary collisions essentially depend on the two-particle distribution function, i.e. on the probability of finding one particle at $\mathbf{r}_1, \mathbf{v}_1$, simultaneous with another particle at $\mathbf{r}_2, \mathbf{v}_2$, where $\mathbf{r}_1 \approx \mathbf{r}_2$. The evolution of the one-, two- and more-particle distribution functions leads to the BBGKY hierarchy of equations (see, e.g., [629], [323], [32]). To truncate this infinite set and to reduce the expressions to only involve the one-particle distribution function requires the formulation of a kinetic regime for weakly coupled plasmas, where the potential energy of the particles is very small compared to their mean kinetic energy [32]. For the validity of this regime, the plasma must be not too hot and not too dense (essentially the coloured region of Fig. 1.9). In conclusion: the three conditions (1.27), (1.28) and (1.30) of Section 1.4 for collective plasma behaviour should be satisfied to justify a description of plasmas by means of the Boltzmann equation (3.2) with a Landau-type collision integral (3.13).

In order to determine the charges and currents that occur in Maxwell's equations we take moments of the distribution functions. The zeroth moment gives the number of particles of species α per unit volume:

$$n_\alpha(\mathbf{r}, t) \equiv \int f_\alpha(\mathbf{r}, \mathbf{v}, t)\, d^3v \qquad \textit{(particle density)}. \tag{3.4}$$

Exploiting the definition (2.48) for averages, the first moment yields the average velocity:

$$\mathbf{u}_\alpha(\mathbf{r}, t) = \langle \mathbf{v} \rangle_\alpha \equiv \frac{1}{n_\alpha(\mathbf{r}, t)} \int \mathbf{v} f_\alpha(\mathbf{r}, \mathbf{v}, t)\, d^3v \quad \textit{(average velocity)}. \tag{3.5}$$

The charge and current density then follow by multiplying with q_α and summing over particle species:

$$\tau(\mathbf{r}, t) = \sum_\alpha q_\alpha n_\alpha(\mathbf{r}, t) \qquad \textit{(charge density)}, \tag{3.6}$$

$$\mathbf{j}(\mathbf{r}, t) = \sum_\alpha q_\alpha n_\alpha(\mathbf{r}, t)\, \mathbf{u}_\alpha(\mathbf{r}, t) \qquad \textit{(current density)}. \tag{3.7}$$

Since all charges and currents in the plasma are supposed to be free, polarization and magnetization effects are negligible so that Maxwell's equations only involve $\mathbf{E}$ and $\mathbf{B}$. In the rationalized mks system of units we then have:

$$\nabla \times \mathbf{E} = -\frac{\partial \mathbf{B}}{\partial t}, \tag{3.8}$$

$$\nabla \times \mathbf{B} = \mu_0 \mathbf{j} + \frac{1}{c^2}\frac{\partial \mathbf{E}}{\partial t}, \tag{3.9}$$

$$\nabla \cdot \mathbf{E} = \frac{\tau}{\epsilon_0}, \tag{3.10}$$

$$\nabla \cdot \mathbf{B} = 0, \tag{3.11}$$

where $c = (\epsilon_0\mu_0)^{-1/2}$. In the Vlasov theory of plasmas, Eqs. (3.3)–(3.11) constitute a complete set of equations for the variables $f_\alpha(\mathbf{r},\mathbf{v},t)$, $\mathbf{E}(\mathbf{r},t)$ and $\mathbf{B}(\mathbf{r},t)$.

With collisions, an explicit expression for the collision term on the RHS of Eq. (3.2) is needed. We decompose this term into contributions $C_{\alpha\beta}$ due to collisions of particles of species α with particles of species β:

$$C_\alpha = \sum_\beta C_{\alpha\beta}(f_\alpha, f_\beta)\,. \tag{3.12}$$

We consider two kinds of particles, viz. electrons and ions (subscripts e and i), so that the indices α and β just run over the two values e and i (giving collision terms $C_{\mathrm{ee}}, C_{\mathrm{ei}}$ for the electron Boltzmann equation, and $C_{\mathrm{ie}}, C_{\mathrm{ii}}$ for the ion Boltzmann equation). The dependence on f_α and f_β in brackets indicates that the collision operator is a quadratic form where f_β is integrated over velocity space.

We now give the explicit form of the collision operator in terms of the *Landau collision integral* (1936) [392]. We suppress the dependence on $\mathbf{r}$ and t in $f_\alpha(\mathbf{r},\mathbf{v},t)$ since only operations on the velocity variable occur. The effect on the distribution function $f_\alpha(\mathbf{v})$ of binary collisions with particles of type β is then expressed by the following integral over velocity space:

$$C_{\alpha\beta} = \frac{q_\alpha^2 q_\beta^2}{8\pi\epsilon_0^2}\ln\Lambda_{\alpha\beta}\int\frac{1}{m_\alpha}\frac{\partial}{\partial\mathbf{v}}\cdot\left[\mathsf{G}(\mathbf{v}-\mathbf{v}')\cdot\left(\frac{1}{m_\alpha}\frac{\partial}{\partial\mathbf{v}}-\frac{1}{m_\beta}\frac{\partial}{\partial\mathbf{v}'}\right)f_\alpha(\mathbf{v})f_\beta(\mathbf{v}')\right]d^3v', \tag{3.13}$$

where the prime indicates the integration variable. The integrand involves a double contraction (defined by $\mathbf{a}\cdot\mathbf{T}\cdot\mathbf{b} \equiv \sum_{\mathrm{i}}\sum_j a_{\mathrm{i}}T_{ij}b_j$ for Cartesian components) with the Landau tensor function of the relative velocity

$$\mathsf{G}(\mathbf{v}-\mathbf{v}') \equiv \frac{1}{|\mathbf{v}-\mathbf{v}'|}\left[\mathsf{I}-\frac{(\mathbf{v}-\mathbf{v}')(\mathbf{v}-\mathbf{v}')}{|\mathbf{v}-\mathbf{v}'|^2}\right]. \tag{3.14}$$

It satisfies the useful equality $(\mathbf{v}-\mathbf{v}')\cdot\mathsf{G}(\mathbf{v}-\mathbf{v}') = 0$. The Coulomb logarithm $\ln\Lambda_{\alpha\beta}$ represents the screening of the Coulomb potential in a plasma for volumes outside a Debye sphere. Since the logarithm strongly suppresses the numerical dependence on the physical quantities, so that $\ln\Lambda_{\mathrm{ee}} \sim \ln\Lambda_{\mathrm{ei}} \sim \ln\Lambda_{\mathrm{ii}}$, one may as well drop the subscripts and exploit the simple expression (see Bittencourt [70]) for a plasma with singly charged ions ($Z = 1$) and equal temperatures ($T_{\mathrm{e}} = T_{\mathrm{i}}$):

$$\ln\Lambda = \ln(9N_{\mathrm{D}}) = \ln\frac{12\pi(\epsilon_0 kT)^{3/2}}{e^3 {n_{\mathrm{e}}}^{1/2}} \approx \ln\left(4.9\times10^{17}\,\frac{\widetilde{T}^{3/2}}{{n_{\mathrm{e}}}^{1/2}}\right). \tag{3.15}$$

The value of $\ln\Lambda$ just ranges from 10 to 30 for plasmas of laboratory and astrophysical interest. For example, for $\widetilde{T}_{\mathrm{e}} = \widetilde{T}_{\mathrm{i}} = 1\,\mathrm{keV}$ and $n = 10^{20}\,\mathrm{m}^{-3}$, one gets $\ln\Lambda \approx 17.7$.

By some elementary algebra, the Landau collision integral (3.13) may be transformed into the following form:

$$C_{\alpha\beta} = \frac{q_\alpha^2 q_\beta^2}{4\pi\epsilon_0^2}\frac{\ln\Lambda_{\alpha\beta}}{m_\alpha^2}\left\{-\frac{\partial}{\partial\mathbf{v}}\cdot\left[\frac{\partial h_\beta(\mathbf{v})}{\partial\mathbf{v}}f_\alpha(\mathbf{v})\right]+\tfrac{1}{2}\frac{\partial^2}{\partial\mathbf{v}\partial\mathbf{v}}:\left[\frac{\partial^2 g_\beta(\mathbf{v})}{\partial\mathbf{v}\partial\mathbf{v}}f_\alpha(\mathbf{v})\right]\right\}, \tag{3.16}$$

where the Rosenbluth potentials g_β and h_β [530], [611] are defined as integrals of the distribution function f_β:

$$g_\beta \equiv \int|\mathbf{v}-\mathbf{v}'|f_\beta(\mathbf{v}')\,d^3v'\,,\qquad h_\beta(\mathbf{v}) \equiv (1+\mathrm{m}_\alpha/m_\beta)\int\frac{f_\beta(\mathbf{v}')}{|\mathbf{v}-\mathbf{v}'|}\,d^3v'\,. \tag{3.17}$$

The two terms of the Fokker–Planck expression (3.16) represent dynamical friction and diffusion in velocity space, respectively [244].

Except for its beauty, the symmetric form of the Landau collision operator also guarantees the satisfaction of the conservation properties (3.18)–(3.22) below, that are essential for the derivation of macroscopic equations. The generalization by Balescu [31] and Lenard [401] to include the collective effects of plasma oscillations leads to a similar form where the tensor **G** contains an additional summation over the wave vector **k** associated with the collective modes.

3.2.2 Moments of the Boltzmann equation★

The distribution function is a function of seven independent variables. This presents formidable complications in the analysis. Since we wish to study plasmas in complex magnetic geometries, we have to get rid of some of the independent variables in order to make progress. The most logical approach is to remove the velocity as an independent variable by taking *moments of the Boltzmann equation*. This approach will run into the problem of producing an infinite chain of equations which somehow has to be truncated in order to make sense. At that point assumptions need to be made that restrict the domain of validity of the theory.

The different moments of the Boltzmann equation are obtained by multiplying the Boltzmann equation (3.2) with powers of $\mathbf{v}$ and integrating over velocity space. In the derivations below, integration by parts will produce surface integrals over a surface at $|\mathbf{v}| = \infty$. It is assumed that the distribution functions $f_\alpha(\mathbf{v})$ tend to 0 sufficiently rapidly as $|\mathbf{v}|$ tends to ∞, in particular that $\lim_{|\mathbf{v}|\to\infty} [g(\mathbf{v})f_\alpha(\mathbf{v})] = 0$ for all functions $g(\mathbf{v})$ that appear, so that these surface integrals do not contribute.

For now, we do not need to go into the specific form of the collision term (3.12). It suffices to list a few general properties following from conservation principles. Since the total number of particles of species α at a certain position is not changed by collisions with particles of species β (only their velocities change), we have

$$\int C_{\alpha\beta}\, d^3v = 0 \qquad \text{(including } \beta = \alpha\text{)}\,. \tag{3.18}$$

Also, momentum and energy are conserved for collisions between like particles,

$$\int m_\alpha \mathbf{v}\, C_{\alpha\alpha}\, d^3v \;=\; 0\,, \tag{3.19}$$

$$\int \tfrac{1}{2} m_\alpha v^2 C_{\alpha\alpha}\, d^3v \;=\; 0\,, \tag{3.20}$$

whereas for collisions between unlike particles ($\beta \neq \alpha$):

$$\int m_\alpha \mathbf{v}\, C_{\alpha\beta}\, d^3v + \int m_\beta \mathbf{v}\, C_{\beta\alpha}\, d^3v \;=\; 0\,, \tag{3.21}$$

$$\int \tfrac{1}{2} m_\alpha v^2 C_{\alpha\beta}\, d^3v + \int \tfrac{1}{2} m_\beta v^2 C_{\beta\alpha}\, d^3v \;=\; 0\,. \tag{3.22}$$

Obviously, the Landau form (3.13) of the collision operator should satisfy these properties and it does (prove!). For a Maxwellian distribution function, as given by Eq. (2.60), the separate terms of Eqs. (3.21) and (3.22) also vanish. We will return to this fundamental fact in Section 3.2.3.

We derive the zeroth moment of the Boltzmann equation (3.2) by just integrating over velocity space. As we have already seen in Section 2.3.1, this results in *the continuity equation* for particles of species α:

$$\frac{\partial n_\alpha}{\partial t} + \nabla \cdot (n_\alpha \mathbf{u}_\alpha) = 0 \qquad \textit{(mass conservation)}. \tag{3.23}$$

This is the most robust macroscopic equation since its derivation does not require any restrictive assumptions on transport coefficients. In brackets we have indicated the conservation law that is expressed by this equation.

The first moment of the Boltzmann equation (3.2) is obtained by multiplying it with $\mathbf{v}$ and integrating over velocity space. This results in the terms

$$\begin{aligned}
\int \frac{\partial f_\alpha}{\partial t} \mathbf{v}\, d^3v &= \frac{\partial}{\partial t}(n_\alpha \mathbf{u}_\alpha), \\
\int \mathbf{v} \cdot \frac{\partial f_\alpha}{\partial \mathbf{r}} \mathbf{v}\, d^3v &= \nabla \cdot \int \mathbf{v}\mathbf{v} f_\alpha\, d^3v = \nabla \cdot \Big(n_\alpha \langle \mathbf{v}\mathbf{v} \rangle_\alpha \Big), \\
\int \frac{q_\alpha}{m_\alpha}(\mathbf{E} + \mathbf{v} \times \mathbf{B}) \cdot \frac{\partial f_\alpha}{\partial \mathbf{v}} \mathbf{v}\, d^3v &= -\frac{q_\alpha n_\alpha}{m_\alpha}(\mathbf{E} + \mathbf{u}_\alpha \times \mathbf{B}), \\
\int C_\alpha \mathbf{v}\, d^3v &= \int C_{\alpha\beta}\, \mathbf{v}\, d^3v \qquad (\beta \neq \alpha).
\end{aligned}$$

Adding them, we obtain *the momentum equation* for particles of species α:

$$\frac{\partial}{\partial t}(n_\alpha m_\alpha \mathbf{u}_\alpha) + \nabla \cdot \Big(n_\alpha m_\alpha \langle \mathbf{v}\mathbf{v} \rangle_\alpha \Big) - n_\alpha q_\alpha (\mathbf{E} + \mathbf{u}_\alpha \times \mathbf{B}) = \int C_{\alpha\beta}\, m_\alpha \mathbf{v}\, d^3v$$
$$\textit{(momentum conservation)}. \tag{3.24}$$

Here, new averaged quantities like $n_\alpha \langle \mathbf{v}\mathbf{v} \rangle_\alpha$ and the collision term appear that require further evaluation.

The final relevant equation is obtained from the scalar second moment equation resulting from multiplying Eq. (3.2) with v^2. The following terms result:

$$\begin{aligned}
\int \frac{\partial f_\alpha}{\partial t} v^2\, d^3v &= \frac{\partial}{\partial t}\Big(n_\alpha \langle v^2 \rangle_\alpha \Big), \\
\int \mathbf{v} \cdot \frac{\partial f_\alpha}{\partial \mathbf{r}} v^2\, d^3v &= \nabla \cdot \Big(n_\alpha \langle v^2 \mathbf{v} \rangle_\alpha \Big), \\
\int \frac{q_\alpha}{m_\alpha}(\mathbf{E} + \mathbf{v} \times \mathbf{B}) \cdot \frac{\partial f_\alpha}{\partial \mathbf{v}} v^2\, d^3v &= -2\frac{n_\alpha q_\alpha}{m_\alpha}\, \mathbf{E} \cdot \mathbf{u}_\alpha, \\
\int C_\alpha v^2\, d^3v &= \int C_{\alpha\beta}\, v^2\, d^3v \qquad (\beta \neq \alpha).
\end{aligned}$$

Multiplying these terms by $\frac{1}{2} m_\alpha$ and adding them gives *the energy equation*:

$$\frac{\partial}{\partial t}\Big(n_\alpha \tfrac{1}{2} m_\alpha \langle v^2 \rangle_\alpha \Big) + \nabla \cdot \Big(n_\alpha \tfrac{1}{2} m_\alpha \langle v^2 \mathbf{v} \rangle_\alpha \Big) - n_\alpha q_\alpha \mathbf{E} \cdot \mathbf{u}_\alpha = \int C_{\alpha\beta} \tfrac{1}{2} m_\alpha v^2\, d^3v$$
$$\textit{(energy conservation)}. \tag{3.25}$$

Again, averages and a collision term appear that require further reduction to really give a self-consistent macroscopic equation.

This chain of moment equations can be continued indefinitely. As a matter of fact, each moment introduces a new unknown whose temporal evolution is described by the next order moment of the Boltzmann equation. For example, the zeroth-order moment (3.23) is an evolution equation for the particle density n_α and it introduces the average velocity $\mathbf{u}_\alpha$ as a new unknown; the first-order moment (3.24) then yields an evolution equation for this $\mathbf{u}_\alpha$ but it contains the unknown $\langle \mathbf{vv} \rangle_\alpha$, etc. This infinite procedure will have to be truncated at some point to get a closed set of equations. The 'art' of obtaining macroscopic equations resides in an appropriate closure of the chain at a very limited number of moments. In the usual fluid theories this number is just five: the (scalar) continuity equation (3.23), the (vector) momentum equation (3.24) and the (scalar) energy equation (3.25).

3.2.3 Thermal fluctuations and transport★

The derived moment equations (3.23)–(3.25) are the only ones needed here. In order to turn them into a closed set, a number of assumptions have to be made. Before we do this it is useful to transform the momentum and energy equation into a form that has a more macroscopic appearance. To that end, as already discussed in Section 2.3.1, we separate the effects of thermal fluctuations from the macroscopic background, by defining the random velocity $\tilde{\mathbf{v}}_\alpha$ of the particles with respect to the average velocity $\mathbf{u}_\alpha$:

$$\tilde{\mathbf{v}}_\alpha \equiv \mathbf{v} - \mathbf{u}_\alpha \,, \quad \text{where } \langle \tilde{\mathbf{v}}_\alpha \rangle = 0 \,. \tag{3.26}$$

The random velocity part of the scalar $\langle v^2 \rangle_\alpha$ in the energy equation (3.25) then gives rise to a quantity measuring the mean kinetic energy of the particles in a frame moving with the velocity $\mathbf{u}_\alpha$, which is *the temperature* T_α:

$$T_\alpha(\mathbf{r}, t) \equiv \frac{m_\alpha}{3k} \langle \tilde{v}_\alpha^2 \rangle \,. \tag{3.27}$$

Likewise, the random velocity part of the term $\langle \mathbf{vv} \rangle_\alpha$ in the momentum equation (3.24) gives rise to *the stress tensor* P_α, defined as

$$\mathsf{P}_\alpha \equiv n_\alpha m_\alpha \langle \tilde{\mathbf{v}}_\alpha \tilde{\mathbf{v}}_\alpha \rangle = p_\alpha \mathsf{I} + \boldsymbol{\pi}_\alpha \,, \tag{3.28}$$

where the isotropic part is directly related to the temperature,

$$p_\alpha(\mathbf{r}, t) \equiv \tfrac{1}{3} \mathrm{Tr}\,(\mathsf{P}_\alpha) = \tfrac{1}{3} n_\alpha m_\alpha \langle \tilde{v}_\alpha^2 \rangle = n_\alpha k T_\alpha \,, \tag{3.29}$$

and the traceless tensor $\boldsymbol{\pi}_\alpha(r, t)$ is the contribution due to the anisotropy of the distribution function:

$$\boldsymbol{\pi}_\alpha(\mathbf{r}, t) \equiv n_\alpha m_\alpha \langle \tilde{\mathbf{v}}_\alpha \tilde{\mathbf{v}}_\alpha - \tfrac{1}{3} \tilde{v}_\alpha^2 \mathsf{I} \rangle \,. \tag{3.30}$$

Finally, the random velocity part of the vector $\langle v^2 \mathbf{v} \rangle_\alpha$ in the energy equation (3.25) gives rise to a quantity

$$\mathbf{h}_\alpha(\mathbf{r}, t) = \tfrac{1}{2} n_\alpha m_\alpha \langle \tilde{v}_\alpha^2 \tilde{\mathbf{v}}_\alpha \rangle \,, \tag{3.31}$$

which is *the heat flow* by random motion of the particles of species α.

The collision terms may also be simplified by transforming to a frame moving with the velocity

$\mathbf{u}_\alpha$. From Eq. (3.18) it follows that only the random part contributes to the RHS of the momentum equation (3.24):

$$\int C_{\alpha\beta}\, m_\alpha \mathbf{v}\, d^3v = \mathbf{R}_\alpha \equiv \int C_{\alpha\beta}\, m_\alpha \tilde{\mathbf{v}}_\alpha\, d^3v \qquad (\beta \neq \alpha)\,, \tag{3.32}$$

which is the friction force, i.e., *the mean momentum transfer* from particles β to particles α. Similarly, the RHS of the energy equation (3.25) may be written as

$$\int C_{\alpha\beta}\, \tfrac{1}{2} m_\alpha v^2\, d^3v = \int C_{\alpha\beta} \Big(m_\alpha \mathbf{u}_\alpha \cdot \tilde{\mathbf{v}}_\alpha + \tfrac{1}{2} m_\alpha \tilde{v}_\alpha^2 \Big) d^3v = \mathbf{u}_\alpha \cdot \mathbf{R}_\alpha + Q_\alpha, \tag{3.33}$$

$$\text{where} \quad Q_\alpha \equiv \int C_{\alpha\beta}\, \tfrac{1}{2} m_\alpha \tilde{v}_\alpha^2\, d^3v \qquad (\beta \neq \alpha) \tag{3.34}$$

is *the heat transferred* to the system of particles α due to collisions with the unlike particles β.

Whereas the mass conservation equation (3.23) already has the required macroscopic form, the momentum and energy conservation equations (3.24) and (3.25) still need further transformation. Substituting the definitions (3.27)–(3.34), they become

$$\frac{\partial}{\partial t}(n_\alpha m_\alpha \mathbf{u}_\alpha) + \nabla \cdot (n_\alpha m_\alpha \mathbf{u}_\alpha \mathbf{u}_\alpha) + \nabla \cdot \mathbf{P}_\alpha - n_\alpha q_\alpha (\mathbf{E} + \mathbf{u}_\alpha \times \mathbf{B}) = \mathbf{R}_\alpha, \tag{3.35}$$

$$\frac{\partial}{\partial t}\Big(\tfrac{1}{2} n_\alpha m_\alpha u_\alpha^2 + \tfrac{3}{2} n_\alpha k T_\alpha \Big) + \nabla \cdot \Big(\tfrac{1}{2} n_\alpha m_\alpha u_\alpha^2 \mathbf{u}_\alpha + \tfrac{3}{2} n_\alpha k T_\alpha \mathbf{u}_\alpha + \mathbf{u}_\alpha \cdot \mathbf{P}_\alpha + \mathbf{h}_\alpha \Big) - n_\alpha q_\alpha \mathbf{E} \cdot \mathbf{u}_\alpha = \mathbf{u}_\alpha \cdot \mathbf{R}_\alpha + Q_\alpha\,. \tag{3.36}$$

The momentum equation (3.35) may be simplified by using the continuity equation (3.23) to remove contributions $\partial n_\alpha / \partial t$, whereas the energy equation (3.36) may be simplified by removing the bulk kinetic energy part by means of Eqs. (3.23) and (3.35). The three lowest moments of the Boltzmann equation then take the compact form

$$\Big(\frac{\partial}{\partial t} + \mathbf{u}_\alpha \cdot \nabla \Big) n_\alpha + n_\alpha \nabla \cdot \mathbf{u}_\alpha = 0 \qquad \textit{(mass)}, \tag{3.37}$$

$$n_\alpha m_\alpha \Big(\frac{\partial}{\partial t} + \mathbf{u}_\alpha \cdot \nabla \Big) \mathbf{u}_\alpha + \nabla (n_\alpha k T_\alpha) - n_\alpha q_\alpha (\mathbf{E} + \mathbf{u}_\alpha \times \mathbf{B}) = -\nabla \cdot \boldsymbol{\pi}_\alpha + \mathbf{R}_\alpha \qquad \textit{(momentum)}, \tag{3.38}$$

$$\tfrac{3}{2} n_\alpha \Big(\frac{\partial}{\partial t} + \mathbf{u}_\alpha \cdot \nabla \Big) k T_\alpha + n_\alpha k T_\alpha \nabla \cdot \mathbf{u}_\alpha = -\boldsymbol{\pi}_\alpha : \nabla \mathbf{u}_\alpha - \nabla \cdot \mathbf{h}_\alpha + Q_\alpha \qquad \textit{(heat)}, \tag{3.39}$$

which we already encountered in Section 2.3.1. They constitute the equations of continuity, of motion and of heat balance for particles of species α. It will not have escaped the attentive reader that apparent progress has been made by just hiding the problems in simple looking variables that are abbreviations of intricate kinetic processes. Clearly, we need substantial additional information concerning the variables $\boldsymbol{\pi}_\alpha, \mathbf{h}_\alpha, \mathbf{R}_\alpha$ and Q_α to be able to express them in terms of the macroscopic variables $n_\alpha, \mathbf{u}_\alpha, T_\alpha$ and the electromagnetic fields $\mathbf{E}$ and $\mathbf{B}$, to really close the set of macroscopic equations (3.37)–(3.39) so that they become genuine hydrodynamic equations. Such information comes from transport theory.

An important equilibrium distribution function in kinetic theory, conforming to the temperature definition (3.27), is *the Maxwell distribution* already introduced in Eq. (2.60), which we here repeat for convenience:

$$f_\alpha^0(\mathbf{r},\mathbf{v},t) = n_\alpha\left(\frac{m_\alpha}{2\pi kT_\alpha}\right)^{3/2}\exp\left(-\frac{m_\alpha\tilde{v}_\alpha^2}{2kT_\alpha}\right), \qquad \tilde{v}_\alpha^2 \equiv |\mathbf{v}-\mathbf{u}_\alpha|^2 . \tag{3.40}$$

This function represents *local thermal equilibrium*: it just depends on the local values of the macroscopic variables $n_\alpha(\mathbf{r},t)$, $\mathbf{u}_\alpha(\mathbf{r},t)$, $T_\alpha(\mathbf{r},t)$ which are supposed to evolve according to the hydrodynamics equations, that are still to be derived by the appropriate closure of Eqs. (3.37)–(3.39). Substitution of f_α^0 in the definitions (3.30) and (3.31) for the anisotropic pressure tensor and the heat flow gives $\boldsymbol{\pi}_\alpha[f_\alpha^0] = 0$ and $\mathbf{h}_\alpha[f_\alpha^0] = 0$. Moreover, due to the equality below Eq. (3.14), the Landau collision integral vanishes for like-particle collisions of Maxwellian particles: $C_{\alpha\alpha}(f_\alpha^0, f_\alpha^0) = 0$. If the average velocities and the temperatures were equal, i.e. $u_\alpha = u_\beta$ (no electric current!) and $T_\alpha = T_\beta$, the collision integral for unlike-particle collisions would vanish as well, $C_{\alpha\beta}(f_\alpha^0, f_\beta^0) = 0$, so that also the momentum transfer and the heat transfer would vanish: $\mathbf{R}_\alpha[f_\alpha^0, f_\beta^0] = 0$ and $Q_\alpha[f_\alpha^0, f_\beta^0] = 0$. Under these circumstances, the two sets of hydrodynamical variables evolve according to the equations (3.37)–(3.39) with vanishing RHSs, whereas initial charge neutrality and the equality of average velocities and temperatures would even remove the separate identities of the electron and ion fluids. If this state were established on a collision time scale, thermonuclear fusion would be impossible and most of the interesting plasma-astrophysical phenomena would not occur. However, the global boundary conditions on the hydrodynamical variables constantly drive the system away from this state of equilibrium and collisions establish a state of quasi-equilibrium with non-vanishing dissipative quantities $\boldsymbol{\pi}_\alpha$, $\mathbf{h}_\alpha$, $\mathbf{R}_\alpha$ and Q_α that are due to the systematic deviations from Maxwellian distributions f_α^0 and f_β^0. *Transport theory* is concerned with the analysis of these deviations from the state of local thermodynamic equilibrium.

The fundamentals of classical transport theory are clearly explained in the seminal paper by Braginskii [92]. Here, we follow the theoretical framework expanded by Balescu [32] since it is more general and incorporates later developments in transport theory. While it is impossible to do justice to the diversity and beauty of this field, we just collect the results needed for our present purpose, in particular the relaxation processes (Section 3.2.4) and the transport coefficients (Section 3.3.2). For full details the reader is referred to the quoted references.

To analyse the consequences of deviations from local thermodynamic equilibrium, the distribution functions are developed in powers of some small parameter (to be specified in Section 3.2.4) measuring these deviations (called the Chapman–Enskog procedure [116]):

$$f_\alpha \approx f_\alpha^0 + f_\alpha^1 , \quad \text{where } |f_\alpha^1| \ll f_\alpha^0 . \tag{3.41}$$

It is expedient to extract a factor f_α^0 from f_α^1,

$$f_\alpha(\mathbf{v};\mathbf{r},t) \approx f_\alpha^0(\mathbf{v};\mathbf{r},t)\,[\,1 + \chi_\alpha(\mathbf{v};\mathbf{r},t)\,] , \tag{3.42}$$

so that the deviations from the local Maxwellians are now given by the functions χ_α. These functions can be normalized such that n_α, $\mathbf{u}_\alpha$, and T_α not only measure the average with respect to the Maxwellian f_α^0 but also with respect to the full distribution function f_α. In the representation of Balescu, χ_α is systematically expanded in irreducible tensorial Hermite polynomials $H^{(\cdots)}_{\cdots}(\mathbf{w}_\alpha)$ in the dimensionless fluctuating velocity variable $\mathbf{w}_\alpha \equiv (2kT_\alpha/m_\alpha)^{-1/2}\,\tilde{\mathbf{v}}_\alpha$:

$$\chi_\alpha(\mathbf{w}_\alpha;\mathbf{r},t) = \sum_{n=1}^{\infty}\Big[\bar{h}_\alpha^{(2n+2)}H^{(2n+2)} + \bar{\mathbf{h}}_\alpha^{(2n+1)}\cdot\mathbf{H}^{(2n+1)} + \bar{\mathsf{h}}_\alpha^{(2n)}:\mathsf{H}^{(2n)} + \cdots\Big]\,, \tag{3.43}$$

where the first three terms represent the scalar, vector and traceless second rank tensor contributions, respectively. The expansion coefficients $\bar{h}_{\alpha\ldots}^{(\cdots)}(\mathbf{r},t)$ are the *unknown* dimensionless Hermitian moments. The lowest order ones correspond to the anisotropic pressure $\boldsymbol{\pi}_\alpha \equiv 2^{1/2}\,n_\alpha kT_\alpha\bar{\mathsf{h}}_\alpha^{(2)}$ and the heat flow $\mathbf{h}_\alpha \equiv (5/2)^{1/2}n_\alpha m_\alpha(kT_\alpha)^{3/2}\bar{\mathbf{h}}_\alpha^{(3)}$. The program of transport theory now consists of substituting (3.43) in the Boltzmann equation (3.2) and solving for the Hermitian moments. This results in closure of the hydrodynamic equations.

The expansion (3.43) may be truncated at different levels, where the number of moments kept will determine the accuracy. For example, the thirteen moment (13M) approximation just keeps the five plasma dynamical moments n_α, $\mathbf{u}_\alpha$, T_α, the three components of the heat flow vector $\mathbf{h}_\alpha$, and the five independent components of the anisotropic pressure tensor $\boldsymbol{\pi}_\alpha$. The 21M approximation adds the three components of the fifth order Hermitian moment $\bar{\mathbf{h}}_\alpha^{(5)}$ and the five components of the fourth order Hermitian moment $\bar{\mathsf{h}}_\alpha^{(4)}$, etc. In this manner, a general framework of transport theory is constructed that may be, and has been, used to extend the classical transport theory to incorporate the effects of particle orbits in toroidal geometry (called the neo-classical transport theory, initiated by Galeev and Sagdeev [203]) and, currently, also turbulent effects (leading to anomalous transport). Balescu [32] demonstrates that the classical transport coefficients in the 21M approximation agree to within 1% with the ones obtained much earlier by Braginskii [92] by a very different representation, but it deviates in some important aspects of the interpretation (see Section 3.3.2).

3.2.4 Collisions and closure★

So far, the Hermitian moment expansion is just a formal representation. To actually calculate the moments, the specific properties of the collision operator need to be exploited. At this point, the nice symmetry of the Landau collision operator with respect to electrons and ions should be taken for granted and, instead, the consequences of the smallness of the mass ratio,

$$m_\mathrm{e}/m_\mathrm{i} \ll 1\,, \tag{3.44}$$

should be utilized. This condition implies that typical ion velocities are much smaller than typical electron velocities, so that the ion distribution $f_\mathrm{i}(\mathbf{v})$ is a much narrower function than the electron distribution $f_\mathrm{e}(\mathbf{v})$. Consequently, e.g. in the electron–ion collision integral $C_\mathrm{ei}(f_\mathrm{e}, f_\mathrm{i})$, there is only interaction over a limited range of velocities, so that the Landau tensor $\mathsf{G}(\mathbf{v}-\mathbf{v}')$ can be expanded in powers of $|\mathbf{v}'-\mathbf{u}_\mathrm{e}|/|\mathbf{v}-\mathbf{u}_\mathrm{e}|$, giving significant contributions only for the lowest order terms. Systematically exploiting such asymmetries between electron and ions, the expressions for the linearized collision integrals may be grouped as follows.

(a) The *electron–ion collision integral* splits into two contributions, where we only indicate the most important term representing pitch angle scattering of the electrons on the heavy, virtually immobile, ions in thermal equilibrium:

$$C_\mathrm{ei}(f_\mathrm{e}, f_\mathrm{i}) \approx C_\mathrm{ei}(f_\mathrm{e}^1, f_\mathrm{i}^0) + C_\mathrm{ei}(f_\mathrm{e}^0, f_\mathrm{i}^1) \approx \nu_\mathrm{ei}(w_\mathrm{e})\,\mathcal{L}_\mathrm{e}(f_\mathrm{e}^1) + \cdots\,. \tag{3.45}$$

The electron–ion collision frequency,

$$\nu_{\rm ei} \equiv \frac{3\sqrt{\pi}}{4w_{\rm e}^3}\,\tau_{\rm e}^{-1} \tag{3.46}$$

involves the relaxation time $\tau_{\rm e}$, defined below in Eq. (3.50), and the operator

$$\mathcal{L}_{\rm e} \equiv \tfrac{1}{2}\,\frac{\partial}{\partial \mathbf{w}_{\rm e}}\cdot\Big(w_{\rm e}^2\,\mathbf{I} - \mathbf{w}_{\rm e}\mathbf{w}_{\rm e}\Big)\cdot\frac{\partial}{\partial \mathbf{w}_{\rm e}}\,, \tag{3.47}$$

is due to the mentioned expansion of the Landau tensor.

Like-particle collisions require a different expansion. The dominant contribution of the resulting *electron–electron collision integral* reads:

$$C_{\rm ee}(f_{\rm e}, f_{\rm e}) \approx C_{\rm ee}(f_{\rm e}^1, f_{\rm e}^0) + C_{\rm ee}(f_{\rm e}^0, f_{\rm e}^1) \approx \nu_{\rm ee}(w_{\rm e})\,\mathcal{L}_{\rm e}(f_{\rm e}^1) + \cdots\,. \tag{3.48}$$

The electron–electron collision frequency,

$$\nu_{\rm ee} \equiv \frac{3\sqrt{\pi}}{4w_{\rm e}^3}\,H(w_{\rm e})\,(Z\tau_{\rm e})^{-1}\,, \tag{3.49}$$

involves the same relaxation time $\tau_{\rm e}$ as the electron–ion collision frequency, but it also depends on the Chandrasekhar function $H(w_{\rm e})$, which is a smoothly increasing function from $H(0) = 0$ to $H(\infty) = 1$. Hence, $\nu_{\rm ee}$ is of the same order as $\nu_{\rm ei}$, but always numerically smaller. Since $C_{\rm e} = C_{\rm ee} + C_{\rm ei}$, the expressions (3.45) and (3.48) should be added so that the two collision processes, represented by $\nu_{\rm ei}$ and $\nu_{\rm ee}$, both contribute to the *relaxation to an electron fluid* on the time scale

$$\tau_{\rm e} \equiv 6\pi\sqrt{2\pi}\epsilon_0^2\,\frac{m_{\rm e}^{1/2}(kT_{\rm e})^{3/2}}{\ln\Lambda Z^2 e^4 n_{\rm i}} \approx \frac{1.09\times 10^{16}}{\ln\Lambda}\,\frac{\widetilde{T}_{\rm e}^{3/2}}{Zn_{\rm e}}\,. \tag{3.50}$$

On this time scale, the electron fluid is established, i.e. the collection of electrons relaxes to its state of near local thermodynamic equilibrium. Recall that $\ln\Lambda \approx 17.7$ for laboratory plasmas with $\widetilde{T}_{\rm e} = 1\,$keV and $n = 10^{20}\,{\rm m}^{-3}$, so that a typical value of $\tau_{\rm e} \approx 6.1\times 10^{-6}\,$s is obtained for those plasmas.

(b) In the case of ions, $C_{\rm i} = C_{\rm ii} + C_{\rm ie}$, but the two collision processes do not operate on the same time scale. The effect of scattering of ions on electrons, represented by $C_{\rm ie}$, is negligible (an order $(m_{\rm e}/m_{\rm i})^{1/2}$ smaller) on the ion relaxation time scale. Hence, the dominant process is represented by the *ion–ion collision integral* only:

$$C_{\rm ii}(f_{\rm i}, f_{\rm i}) \approx C_{\rm ii}(f_{\rm i}^1, f_{\rm i}^0) + C_{\rm ii}(f_{\rm i}^0, f_{\rm i}^1) \approx \nu_{\rm ii}(w_{\rm i})\,\mathcal{L}_{\rm i}(f_{\rm i}^1) + \cdots\,, \tag{3.51}$$

where the operator $\mathcal{L}_{\rm i}$ is defined as in Eq. (3.47) with $w_{\rm e}$ replaced by $w_{\rm i}$. The ion–ion collision frequency,

$$\nu_{\rm ii} \equiv \frac{3\sqrt{\pi}}{4w_{\rm i}^3}\,H(w_{\rm i})\,\tau_{\rm i}^{-1}\,, \tag{3.52}$$

involves the ion relaxation time $\tau_{\rm i}$ and the Chandrasekhar function $H(w_{\rm i})$. Due to the ion–ion collisions, *relaxation to an ion fluid* proceeds on the time scale

$$\tau_{\rm i} \equiv 6\pi\sqrt{2\pi}\epsilon_0^2\,\frac{m_{\rm i}^{1/2}(kT_{\rm i})^{3/2}}{\ln\Lambda Z^4 e^4 n_{\rm i}} \approx \frac{4.66\times 10^{17}}{\ln\Lambda}\,\frac{A^{1/2}\widetilde{T}_{\rm i}^{3/2}}{Z^3 n_{\rm e}}\,. \tag{3.53}$$

On this time scale, the ion fluid is established, i.e. the collection of ions relaxes to its own state of near local thermodynamic equilibrium. For a laboratory hydrogen plasma with $\widetilde{T}_{\rm i} = 1\,{\rm keV}$ and $n = 10^{20}\,{\rm m}^{-3}$, a typical value of $\tau_{\rm i} \approx 2.6 \times 10^{-4}\,{\rm s}$ is obtained. If $T_{\rm i} \sim T_{\rm e}$, due to the smallness of the mass ratio, $\tau_{\rm e}/\tau_{\rm i} \sim (m_{\rm e}/m_{\rm i})^{1/2} \ll 1$, so that the ions relax much slower than the electrons.

(c) Finally, the most important part of the *ion–electron collision integral* is due to a non-vanishing contribution from the two Maxwellians $f_{\rm i}^0$ and $f_{\rm e}^0$:

$$C_{\rm ie}(f_{\rm i}^0, f_{\rm e}^0) \approx \tau_{\rm eq}^{-1}\Big(\tfrac{1}{2}Z\,\frac{\partial}{\partial \mathbf{w}_{\rm i}}\cdot \mathbf{w}_{\rm i} f_{\rm i}^0 + \tfrac{1}{4}\frac{m_{\rm e}}{m_{\rm i}}\frac{\partial}{\partial \mathbf{w}_{\rm e}}\cdot\frac{\partial}{\partial \mathbf{w}_{\rm e}} f_{\rm e}^0\Big) + \cdots . \tag{3.54}$$

This process proceeds on the temperature equilibration time scale, which is the longest relaxation time scale:

$$\tau_{\rm eq} \equiv \tfrac{1}{2}(m_{\rm i}/m_{\rm e})\,\tau_{\rm e} . \tag{3.55}$$

On this time scale a fluid *with equal electron and ion temperatures* is established. Again, for the mentioned laboratory plasma, a typical value of $\tau_{\rm eq} \approx 5.6 \times 10^{-3}\,{\rm s}$ is obtained.

Next, having obtained the collisional expressions, the two linearized Boltzmann equations for $f_{\rm e}^1$ and $f_{\rm i}^1$ are solved to produce the closure relations. Exploiting the Hermitian expansion (3.43), these relations take the form of evolution equations for the non-plasmadynamical moments $\bar{h}_{\alpha\ldots}^{(\cdots)}(\mathbf{r},t)$ connecting them to the hydrodynamical sources, i.e. the gradients of $n_{\rm e,i}$, $\mathbf{u}_{\rm e,i}$, $T_{\rm e,i}$ and the fields $\mathbf{E}$ and $\mathbf{B}$. The latter are obtained from the hydrodynamical equations (3.37)–(3.39), neglecting the RHSs, and Maxwell's equations (3.6)–(3.11). At this point, a crucial distinction is to be made between the relaxation time scales $\tau_{\rm e,i}$ of the non-plasmadynamical moments and the hydrodynamical evolution time scale $\tau_{\rm H}$ of the hydrodynamical variables. In order for this split to make sense at all, we should have $\tau_{\rm e,i} \ll \tau_{\rm H}$. Hence, the small parameter that was implicit in the approximations (3.42) for $\chi_{\rm e}$ and $\chi_{\rm i}$ can now be specified:

$$\epsilon_{\rm e} \equiv \tau_{\rm e}/\tau_{\rm H} \ll 1 , \qquad \epsilon_{\rm i} \equiv \tau_{\rm i}/\tau_{\rm H} \ll 1 . \tag{3.56}$$

Usually, $\tau_{\rm e} \ll \tau_{\rm i}$, so that the second restriction is the most severe one. This implies that ion relaxation should proceed faster than hydrodynamic evolution, whereas temperature equilibration might proceed on the hydrodynamical time scale, $\tau_{\rm eq} \sim \tau_{\rm H}$.

Balescu demonstrates that, under these circumstances, the initial value problem for the non-plasmadynamical moments $\bar{h}_{\alpha\ldots}^{(\cdots)}(\mathbf{r},t)$ has a very simple solution. Due to the frequent collisions, the system forgets about the initial data after a few relaxation times so that the $\partial/\partial t$ terms may be neglected. In effect, the system tends towards a state where the non-plasmadynamical moments are linearly related to the hydrodynamical variables: *closure!* The resulting relations are just four (two for the electrons and the ions times two for the vectors and the tensors) uncoupled sets of algebraic equations relating the thermodynamic fluxes to the sources. They are explicitly listed in Section 3.3.2 for the simplified case of singly charged ions ($Z = 1$) and large magnetic field,

$$\Omega_{\rm e}\tau_{\rm e} \gg 1 , \qquad \Omega_{\rm i}\tau_{\rm i} \gg 1 , \tag{3.57}$$

in the 21M approximation. The latter conditions are not required for the transport coefficients derived by Balescu, but they greatly simplify the expressions. Moreover, they are easily satisfied for cases of interest: $\Omega_{\rm e}\tau_{\rm e} \approx 3.2 \times 10^6$, $\Omega_{\rm i}\tau_{\rm i} \approx 7.5 \times 10^4$ for tokamak plasmas with $\widetilde{T}_{\rm i} = 1\,{\rm keV}$, and

$\Omega_e\tau_e \approx 1.0 \times 10^8$, $\Omega_i\tau_i \approx 2.4 \times 10^5$ for coronal loop plasmas with $\widetilde{T}_i = 0.1\,\text{keV}$ (exploiting the values of Table B.3).

Unfortunately, the numerical values just given also reveal the tremendous restrictions imposed by the conditions (3.56), in particular for laboratory fusion plasmas (see Freidberg [186]). Exploiting the Alfvén wave crossing time $\tau_A \equiv L/v_A$ along the plasma as a measure for τ_H, these conditions become

$$\epsilon_e \equiv \tau_e v_A/L \ll 1\,, \qquad \epsilon_i \equiv \tau_i v_A/L \ll 1\,. \tag{3.58}$$

Inserting the numerical values of Table B.4, and even lowering the temperature for tokamaks to $\widetilde{T} = 1\,\text{keV}$, we get $\epsilon_e = 2$ and $\epsilon_i = 87$; not small at all! On the other hand, for astrophysical plasmas, the occurrence of L in the denominator completely saves the day: $\epsilon_e = 1.3 \times 10^{-5}$ and $\epsilon_i = 5.3 \times 10^{-3}$ for coronal loops with $\widetilde{T} = 0.1\,\text{keV}$. Of course, the choice of large L is not an available option in fusion research since the size of plasma confinement devices is determined by economical considerations, not by theoretical convenience.

Some comments are in order. (1) MHD fusion theory is not concerned about Alfvén wave propagation on the microsec time scale, but with the much slower residual instabilities (that are left after considerable experimental effort to eliminate the fastest ones) growing on the msec time scale. The description of these phenomena brings in small geometrical factors increasing the effective value of τ_H. (2) Neoclassical transport theory, incorporating the effects of particle trapping in toroidal magnetic fields [203], brings in similar geometrical factors that reduce the effective value of the relaxation times τ_e and τ_i. (3) Usually, turbulent transport processes completely swamp the classical ones so that the concern is not so much satisfaction of the conditions (3.56), but finding regimes where anomalous transport is sufficiently reduced to make fusion possible. (4) Of course, the option always remains open to exploit one of the hybrid models, e.g. with a fluid description for the electrons but a kinetic description for the ions.

With these caveats, we just proceed on the basis of classical transport theory, noting that asymptotic results from a rigid theoretical framework frequently remain valid far outside the strict domain of validity.

3.3 Two-fluid equations★

3.3.1 Electron–ion plasma★

We now collect the two-fluid equations for a plasma consisting of electrons, $q_e = -e$, and ions of one kind with charge number Z, $q_i = Ze$. From the moment equations (3.37)–(3.39) a double set of equations for the electrons and ions is obtained, where the explicit expressions for the RHS quantities $\boldsymbol{\pi}_{e,i}$, $\mathbf{h}_{e,i}$, $\mathbf{R}_{e,i}$ and $Q_{e,i}$ in terms of the macroscopic variables are listed in Section 3.3.2. The latter two pairs of quantities are not independent because of the momentum and energy conservation properties (3.21) and (3.22) of unlike-particle collisions. From momentum conservation one derives

$$\mathbf{R}_e = -\mathbf{R}_i\,, \tag{3.59}$$

whereas energy conservation, by the use of Eq. (3.33), yields

$$Q_e = -(\mathbf{u}_e - \mathbf{u}_i)\cdot\mathbf{R}_e - Q_i\,. \tag{3.60}$$

Consequently, the set of dissipative two-fluid equations becomes

$$\begin{aligned}\Big(\frac{\partial}{\partial t}+\mathbf{u}_{\rm e}\cdot\nabla\Big)n_{\rm e}+n_{\rm e}\nabla\cdot\mathbf{u}_{\rm e} &= 0\,,\\ \Big(\frac{\partial}{\partial t}+\mathbf{u}_{\rm i}\cdot\nabla\Big)n_{\rm i}+n_{\rm i}\nabla\cdot\mathbf{u}_{\rm i} &= 0\,,\end{aligned}\qquad(3.61)$$

$$\begin{aligned}n_{\rm e}m_{\rm e}\Big(\frac{\partial}{\partial t}+\mathbf{u}_{\rm e}\cdot\nabla\Big)\mathbf{u}_{\rm e}+\nabla p_{\rm e}+en_{\rm e}(\mathbf{E}+\mathbf{u}_{\rm e}\times\mathbf{B}) &= -\nabla\cdot\boldsymbol{\pi}_{\rm e}+\mathbf{R}_{\rm e}\,,\\ n_{\rm i}m_{\rm i}\Big(\frac{\partial}{\partial t}+\mathbf{u}_{\rm i}\cdot\nabla\Big)\mathbf{u}_{\rm i}+\nabla p_{\rm i}-Zen_{\rm i}(\mathbf{E}+\mathbf{u}_{\rm i}\times\mathbf{B}) &= -\nabla\cdot\boldsymbol{\pi}_{\rm i}-\mathbf{R}_{\rm e}\,,\end{aligned}\qquad(3.62)$$

$$\begin{aligned}&\tfrac{3}{2}n_{\rm e}\Big(\frac{\partial}{\partial t}+\mathbf{u}_{\rm e}\cdot\nabla\Big)kT_{\rm e}+p_{\rm e}\nabla\cdot\mathbf{u}_{\rm e}=-\boldsymbol{\pi}_{\rm e}:\nabla\mathbf{u}_{\rm e}-\nabla\cdot\mathbf{h}_{\rm e}-(\mathbf{u}_{\rm e}-\mathbf{u}_{\rm i})\cdot\mathbf{R}_{\rm e}-Q_{\rm i}\,,\\ &\tfrac{3}{2}n_{\rm i}\Big(\frac{\partial}{\partial t}+\mathbf{u}_{\rm i}\cdot\nabla\Big)kT_{\rm i}+p_{\rm i}\nabla\cdot\mathbf{u}_{\rm i}=-\boldsymbol{\pi}_{\rm i}:\nabla\mathbf{u}_{\rm i}-\nabla\cdot\mathbf{h}_{\rm i}+Q_{\rm i}\,,\end{aligned}\qquad(3.63)$$

where $$p_{\rm e}=n_{\rm e}kT_{\rm e}\,,\qquad p_{\rm i}=n_{\rm i}kT_{\rm i}\,.\qquad(3.64)$$

Of course, this set only becomes complete when augmented with Maxwell's equations (3.8)–(3.11) with the charge and current density,

$$\tau\equiv-e\,(n_{\rm e}-Zn_{\rm i})\,,\qquad \mathbf{j}\equiv-e\,(n_{\rm e}\mathbf{u}_{\rm e}-Zn_{\rm i}\mathbf{u}_{\rm i})\,,\qquad(3.65)$$

acting as sources.

3.3.2 The classical transport coefficients★

Since the transport closure relations involve vector and tensor expressions with highly anisotropic coefficients (due to the presence of the magnetic field), it is convenient to introduce a notation that highlights these anisotropies. First, associated with any vector $\mathbf{V}$, we define three auxiliary vectors

$$\mathbf{V}_{\parallel}\equiv\mathbf{bb}\cdot\mathbf{V}\,,\quad \mathbf{V}_{\wedge}\equiv\mathbf{b}\times\mathbf{V}\,,\quad \mathbf{V}_{\perp}\equiv(\mathbf{b}\times\mathbf{V})\times\mathbf{b}\,,\qquad(3.66)$$

where $\mathbf{b}\equiv\mathbf{B}/B$ is the unit vector in the direction of the magnetic field. Hence, $\mathbf{V}\equiv\mathbf{V}_{\parallel}+\mathbf{V}_{\perp}$, and $\mathbf{V}_{\wedge}$ is a vector of length $V_{\perp}$ but orthogonal to both $\mathbf{V}$ and $\mathbf{B}$. In a projection with respect to a triad of orthogonal unit vectors $\mathbf{e}_1,\mathbf{e}_2,\mathbf{b}$, we have $\mathbf{V}=(V_1,V_2,V_{\parallel})^{\rm T}$, so that the auxiliary vectors are represented by

$$\mathbf{V}_{\parallel}=(0,0,V_{\parallel})^{\rm T}\,,\quad \mathbf{V}_{\wedge}=(-V_2,V_1,0)^{\rm T}\,,\quad \mathbf{V}_{\perp}=(V_1,V_2,0)^{\rm T}\,.\qquad(3.67)$$

Next, we consider a typical expression $\boldsymbol{\vartheta}\cdot\mathbf{V}$, where $\boldsymbol{\vartheta}$ is a second rank tensor of transport coefficients. Symmetry with respect to rotations about the magnetic field implies that $\boldsymbol{\vartheta}$ can only have three independent elements, which we denote by $\vartheta_{\parallel}$, $\vartheta_{\wedge}$ and $\vartheta_{\perp}$:

$$\boldsymbol{\vartheta}=\begin{pmatrix}\vartheta_{\perp} & -\vartheta_{\wedge} & 0\\ \vartheta_{\wedge} & \vartheta_{\perp} & 0\\ 0 & 0 & \vartheta_{\parallel}\end{pmatrix}\quad\Rightarrow\quad \boldsymbol{\vartheta}^{-1}=\begin{pmatrix}\vartheta_{\perp}/D & \vartheta_{\wedge}/D & 0\\ -\vartheta_{\wedge}/D & \vartheta_{\perp}/D & 0\\ 0 & 0 & 1/\vartheta_{\parallel}\end{pmatrix},\qquad(3.68)$$

where $D \equiv \vartheta_\perp^2 + \vartheta_\wedge^2$. (The expression for the inverse tensor $\boldsymbol{\vartheta}^{-1}$ will be needed below.) One then easily checks that multiplication of the left matrix (3.68) with the vector $\mathbf{V}$ is equivalent to the sum over the three auxiliary vectors (3.66) multiplied with the three transport coefficients:

$$\boldsymbol{\vartheta} \cdot \mathbf{V} = \sum_\lambda \vartheta_\lambda \mathbf{V}_\lambda \qquad (\lambda = \parallel, \wedge, \perp). \tag{3.69}$$

Thus, the tensor expressions are conveniently compressed by this notation.

We now list the electron vector transport coefficients under (a) and the electron tensor coefficients under (d). Recall that these coefficients are established on the time scale τ_e. The ion vector and tensor coefficients, established on the time scale τ_i, are listed under (b) and (d). The expression for the heat transfer from the electrons to the ions, established on the time scale τ_{eq}, is given under (c).

(a) Electron electrical and thermal coefficients The expressions for the electron electrical and thermal transport coefficients are the most interesting ones since they exhibit the characteristic Onsager symmetry of the transport matrix, relating thermodynamics fluxes and forces in non-equilibrium thermodynamics. In Balescu's treatment, the current density $\mathbf{j}$ and electron heat flow $\mathbf{h}_e$ are considered as the fluxes that are driven by the modified electric field, $\widehat{\mathbf{E}} \equiv \mathbf{E} + \mathbf{u}_i \times \mathbf{B} + (en_e)^{-1}\nabla(n_e kT_e)$, and the negative gradient of the electron temperature, $-\nabla(kT_e)$:

$$\mathbf{j} = \boldsymbol{\sigma} \cdot \widehat{\mathbf{E}} - \boldsymbol{\alpha} \cdot \nabla(kT_e), \tag{3.70}$$

$$\mathbf{h}_e = kT_e\, \boldsymbol{\alpha} \cdot \widehat{\mathbf{E}} - \boldsymbol{\kappa}_e \cdot \nabla(kT_e). \tag{3.71}$$

(**NB:** The appearance of the Boltzmann constant in these equations implies that our transport coefficients may differ by a factor k from those exploited by other authors.) The explicit expressions for the *electrical conductivity* σ_λ, the *thermo-electric coupling* α_λ and the *electron thermal conductivity* $\kappa_{e\lambda}$ for $Z = 1$ read:

$$\begin{aligned} &\sigma_\parallel \approx 1.95\,\hat{\sigma}, && \alpha_\parallel \approx -1.39\,\hat{\alpha}, && \kappa_{e\parallel} \approx 4.15\,\hat{\kappa}_e, \\ &\sigma_\wedge \approx 1.0\,\hat{\sigma}(\Omega_e\tau_e)^{-1}, && \alpha_\wedge \approx -6.60\,\hat{\alpha}(\Omega_e\tau_e)^{-3}, && \kappa_{e\wedge} \approx 2.5\,\hat{\kappa}_e(\Omega_e\tau_e)^{-1}, \\ &\sigma_\perp \approx 1.0\,\hat{\sigma}(\Omega_e\tau_e)^{-2}, && \alpha_\perp \approx 1.5\,\hat{\alpha}(\Omega_e\tau_e)^{-2}, && \kappa_{e\perp} \approx 4.66\,\hat{\kappa}_e(\Omega_e\tau_e)^{-2}, \end{aligned} \tag{3.72}$$

where the dimensional factors are given by

$$\hat{\sigma} \equiv \frac{e^2 n_e \tau_e}{m_e}, \qquad \hat{\alpha} \equiv \frac{e n_e \tau_e}{m_e}, \qquad \hat{\kappa}_e \equiv \frac{n_e k T_e \tau_e}{m_e}. \tag{3.73}$$

The inverse of the conductivity tensor is the resistivity tensor $\boldsymbol{\eta} \equiv \boldsymbol{\sigma}^{-1}$. Its coefficients are obtained by means of the right matrix (3.68):

$$\eta_\parallel = \frac{1}{\sigma_\parallel} \approx 0.51\,\hat{\sigma}^{-1}, \qquad \eta_\wedge \approx -\frac{1}{\sigma_\wedge} \approx -\hat{\sigma}^{-1}\,\Omega_e\tau_e, \qquad \eta_\perp \approx \frac{\sigma_\perp}{\sigma_\wedge^2} \approx \hat{\sigma}^{-1}. \tag{3.74}$$

As stressed by Balescu, this shows the following peculiarities: (1) the perpendicular resistivity is not the inverse of the perpendicular conductivity, $\eta_\perp \neq (\sigma_\perp)^{-1}$, and the simple relationship $\eta_\perp \approx 2\,\eta_\parallel$ of the Spitzer resistivity (introduced in Eq. (2.68)), with no dependence on the magnetic field to leading order, is essentially due to the contribution of the Hall conductivity $\sigma_\wedge$; (2) in sharp contrast to the corresponding component $\eta'_\wedge$ of the Braginskii resistivity tensor, discussed below, the component $\eta_\wedge$ becomes very large when the magnetic field is large ($\Omega_e\tau_e \gg 1$).

This does not yet provide the necessary expression for the friction force $\mathbf{R}_e$. At this point, the Hermitian moment expansion brings in an additional number of higher order moments, depending on the level of accuracy desired. In the 21M approximation,

$$\mathbf{R}_e = \frac{m_e}{e\tau_e}\Big[\mathbf{j} + 0.6\, e(kT_e)^{-1}\mathbf{h}_e - 0.896\, en_e(kT_e/m_e)^{1/2}\,\bar{\mathbf{h}}_e^{(5)}\Big], \tag{3.75}$$

where the heat flow $\mathbf{h}_e$ corresponds to the dimensionless third order Hermitian moment $\bar{\mathbf{h}}_e^{(3)}$, and $\bar{\mathbf{h}}_e^{(5)}$ is the dimensionless fifth order Hermitian moment. The latter is related to the driving thermodynamic forces by

$$\bar{\mathbf{h}}_e^{(5)} = \frac{1}{en_e}\sqrt{\frac{m_e}{kT_e}}\,\Big[\boldsymbol{\gamma}\cdot\widehat{\mathbf{E}} - \boldsymbol{\delta}\cdot\nabla(kT_e)\Big], \tag{3.76}$$

where $\boldsymbol{\gamma}$ and $\boldsymbol{\delta}$ are pseudo-transport coefficients:

$$\begin{aligned}
&\gamma_\| \approx 0.132\,\hat{\sigma}\,, && \delta_\| \approx 1.23\,\hat{\alpha}\,,\\
&\gamma_\wedge \approx 4.64\,\hat{\sigma}(\Omega_e\tau_e)^{-3}\,, && \delta_\wedge \approx 12.4\,\hat{\alpha}(\Omega_e\tau_e)^{-3}\,,\\
&\gamma_\perp \approx -0.896\,\hat{\sigma}(\Omega_e\tau_e)^{-2}\,, && \delta_\perp \approx -2.57\,\hat{\alpha}(\Omega_e\tau_e)^{-2}\,.
\end{aligned} \tag{3.77}$$

This completes the closure of the electron equations with respect to $\mathbf{R}_e$ and $\mathbf{h}_e$.

▷ **Braginskii's transport expressions** [92] originate from an entirely different approach, leading to the following relations for the momentum transfer $\mathbf{R}_e$ and the heat flow $\mathbf{h}_e$:

$$\mathbf{R}_e = en_e\,\boldsymbol{\eta}'\cdot\mathbf{j} - \boldsymbol{\beta}\cdot\nabla(kT_e)\,, \tag{3.78}$$

$$\mathbf{h}_e = -\frac{kT_e}{en_e}\,\boldsymbol{\beta}\cdot\mathbf{j} - \boldsymbol{\kappa}'_e\cdot\nabla(kT_e)\,. \tag{3.79}$$

Here, $\mathbf{j}$ and $-\nabla(kT_e)$ are considered as the driving thermodynamic forces and $-\mathbf{R}_e$ and $\mathbf{h}_e$ as the fluxes (a 'rather unnatural choice', according to Balescu, although Onsager symmetry is also obtained this way). The connection between Braginskii's and Balescu's expressions for the electron electrical and thermal coefficients is given by

$$\begin{aligned}
\boldsymbol{\eta}' &= \frac{m_e}{e^2 n_e\tau_e}\Big[\mathbf{I} + (0.6\, e\,\boldsymbol{\alpha} - 0.896\,\boldsymbol{\gamma})\cdot\boldsymbol{\sigma}^{-1}\Big]\\
&= \boldsymbol{\sigma}^{-1} + \frac{m_e}{e^2 n_e\tau_e}\Big[\boldsymbol{\alpha} + 0.6\, e(kT_e)^{-1}\,\boldsymbol{\kappa}_e - 0.896\,\boldsymbol{\delta}\Big]\cdot\boldsymbol{\alpha}^{-1}\,,
\end{aligned} \tag{3.80}$$

$$\boldsymbol{\beta} = -\,en_e\,\boldsymbol{\alpha}\cdot\boldsymbol{\sigma}^{-1}\,, \tag{3.81}$$

$$\boldsymbol{\kappa}'_e = \boldsymbol{\kappa}_e - kT_e\,\boldsymbol{\alpha}\cdot\boldsymbol{\sigma}^{-1}\cdot\boldsymbol{\alpha}\,, \tag{3.82}$$

where the resistivity $\boldsymbol{\eta}'$ now involves the pseudo-transport coefficients $\boldsymbol{\gamma}$ and $\boldsymbol{\delta}$. This yields the following (very different from Balescu's!) explicit expressions for the *resistivity* η'_λ, the *thermo-electric coupling* β_λ and the *electron thermal conductivity* $\kappa'_{e\lambda}$ for $Z = 1$:

$$\begin{aligned}
&\eta'_\| \approx 0.51\,\hat{\sigma}^{-1}\,, && \beta_\| \approx 0.70\,n_e\,, && \kappa'_{e\|} \approx 3.2\,\hat{\kappa}_e\,,\\
&\eta'_\wedge \approx -1.70\,(\hat{\sigma}\,\Omega_e\tau_e)^{-1}\,, && \beta_\wedge \approx 1.5\,n_e(\Omega_e\tau_e)^{-1}\,, && \kappa'_{e\wedge} \approx 2.5\,\hat{\kappa}_e(\Omega_e\tau_e)^{-1}\,,\\
&\eta'_\perp \approx 1.0\,\hat{\sigma}^{-1}\,, && \beta_\perp \approx 5.1\,n_e(\Omega_e\tau_e)^{-2}\,, && \kappa'_{e\perp} \approx 4.66\,\hat{\kappa}_e(\Omega_e\tau_e)^{-2}\,.
\end{aligned} \tag{3.83}$$

Note that Braginskii's resistivity tensor $\boldsymbol{\eta}'$ is not the inverse of the conductivity tensor $\boldsymbol{\sigma}$. Also, that there is significant difference between the two thermo-electric coefficients $\boldsymbol{\alpha}$ and $\boldsymbol{\beta}$, and between the two parallel components of the thermal conductivities $\boldsymbol{\kappa}_{\rm e}$ and $\boldsymbol{\kappa}'_{\rm e}$.

(The notations used by Balescu and Braginskii differ, where we mostly follow the first author with some exceptions dictated by consistency of the present text: $\boldsymbol{\eta} \equiv \boldsymbol{\sigma}^{-1}|_{\rm Bal}$, $\gamma \equiv \beta|_{\rm Bal}$, $\boldsymbol{\delta} \equiv \gamma|_{\rm Bal}$, $\boldsymbol{\eta}' \equiv \boldsymbol{\rho}|_{\rm Bal} \equiv (en_{\rm e})^{-1}\boldsymbol{\alpha}|_{\rm Brag}$, $\boldsymbol{\beta} \equiv -\mathbf{b}|_{\rm Bal} \equiv \boldsymbol{\beta}|_{\rm Brag}$, $\boldsymbol{\kappa}'_{\rm e} \equiv \boldsymbol{\kappa}'_{\rm e}|_{\rm Bal} \equiv \boldsymbol{\kappa}_{\rm e}|_{\rm Brag}$. Note that $\alpha_\wedge|_{\rm Brag}$ and $\kappa_{\rm i\wedge}|_{\rm Brag}$ have been defined with opposite sign. Also note that our sign of $\Omega_{\rm e}$ is positive.) ◁

(b) Ion thermal coefficients Since there is only ion–ion scattering on the time scale $\tau_{\rm i}$, the expressions for the ion vector moments are much simpler that those for the electrons. Only the expression for the ion heat flow appears here, viz.

$$\mathbf{h}_{\rm i} = -\boldsymbol{\kappa}_{\rm i} \cdot \nabla(kT_{\rm i}), \tag{3.84}$$

with the following *ion thermal conductivity* coefficients $\kappa_{i\lambda}$:

$$\begin{aligned}
&\kappa_{\rm i\parallel} \approx 5.52\,\hat{\kappa}_{\rm i}\,, \\
&\kappa_{\rm i\wedge} \approx -2.5\,\hat{\kappa}_{\rm i}(\Omega_{\rm i}\tau_{\rm i})^{-1}\,, \\
&\kappa_{\rm i\perp} \approx 1.41\,\hat{\kappa}_{\rm i}(\Omega_{\rm i}\tau_{\rm i})^{-2}\,, \qquad \text{with} \quad \hat{\kappa}_{\rm i} \equiv n_{\rm i}kT_{\rm i}\tau_{\rm i}/m_{\rm i}\,.
\end{aligned} \tag{3.85}$$

(c) Heat transfer Neglecting viscous heating processes, the heat transfer functions read:

$$Q_{\rm e} = \frac{1}{en_{\rm e}}\mathbf{R}_{\rm e} \cdot \mathbf{j} - Q_{\rm i}\,, \qquad Q_{\rm i} = \frac{3n_{\rm e}k(T_{\rm e} - T_{\rm i})}{2\tau_{\rm eq}}\,, \tag{3.86}$$

demonstrating that equilibration of the electron and ion temperatures proceeds on the time scale $\tau_{\rm eq}$ defined in Eq. (3.55).

(d) Electron and ion viscosities The stress tensors $\boldsymbol{\pi}_{\rm e}$ and $\boldsymbol{\pi}_{\rm i}$ are related to the traceless strain tensors $\mathbf{W}_{\rm e}(\mathbf{u}_{\rm e})$ and $\mathbf{W}_{\rm i}(\mathbf{u}_{\rm i})$, respectively, through fourth rank viscosity tensors which contain just five different elements $\mu_{{\rm e}\ell}$ and $\mu_{{\rm i}\ell}$ (now: $\ell = \parallel, 1, 2, 3, 4$) for each species. (Again, for consistency of the notation, we exploit the symbol μ for the viscosity with Balescu's numbering of the components: $(\mu_\parallel, \mu_1, \mu_2, \mu_3, \mu_4) \equiv (\eta_\parallel, \eta_1, \eta_2, \eta_3, \eta_4)|_{\rm Bal} \equiv (\eta_0, -\eta_4, \eta_2, -\eta_3, \eta_1)|_{\rm Brag}$.)

This is due to symmetry with respect to rotations about the magnetic field. Suppressing the indices e and i, these relationships are given by

$$\begin{aligned}
\pi_{11} &= -\tfrac{1}{2}\mu_\parallel(W_{11} + W_{22}) + \mu_3 W_{12} - \tfrac{1}{2}\mu_4(W_{11} - W_{22})\,, \\
\pi_{12} &= \pi_{21} = -\tfrac{1}{2}\mu_3(W_{11} - W_{22}) - \mu_4 W_{12}\,, \\
\pi_{13} &= \pi_{31} = \mu_1 W_{23} - \mu_2 W_{13}\,, \\
\pi_{22} &= -\tfrac{1}{2}\mu_\parallel(W_{11} + W_{22}) - \mu_3 W_{12} + \tfrac{1}{2}\mu_4(W_{11} - W_{22})\,, \\
\pi_{23} &= \pi_{32} = -\mu_1 W_{13} - \mu_2 W_{23}\,, \\
\pi_{33} &= -\mu_\parallel W_{33}\,,
\end{aligned} \tag{3.87}$$

where the strain tensors are defined by

$$W_{ij} \equiv \frac{\partial u_{\rm i}}{\partial x_j} + \frac{\partial u_j}{\partial x_{\rm i}} - \tfrac{2}{3}\delta_{ij}\nabla\cdot\mathbf{u}\,. \tag{3.88}$$

The *electron viscosity* coefficients $\mu_{{\rm e}\ell}$ for $Z = 1$ read:

$$\begin{aligned} &\mu_{\rm e\|} \approx 0.73\,\hat\mu_{\rm e}\,, \\ &\mu_{\rm e1} \approx 1.0\,\hat\mu_{\rm e}(\Omega_{\rm e}\tau_{\rm e})^{-1}\,, \qquad \mu_{\rm e3} \approx \tfrac{1}{2}\mu_{\rm e1}\,, \\ &\mu_{\rm e2} \approx 2.05\,\hat\mu_{\rm e}(\Omega_{\rm e}\tau_{\rm e})^{-2}\,, \qquad \mu_{\rm e4} \approx \tfrac{1}{4}\mu_{\rm e2}\,, \qquad \text{with}\quad \hat\mu_{\rm e} \equiv n_{\rm e}kT_{\rm e}\tau_{\rm e}\,. \end{aligned} \tag{3.89}$$

The *ion viscosity* coefficients $\mu_{{\rm i}\ell}$ read:

$$\begin{aligned} &\mu_{\rm i\|} \approx 1.36\,\hat\mu_{\rm i}\,, \\ &\mu_{\rm i1} \approx -1.0\,\hat\mu_{\rm i}(\Omega_{\rm i}\tau_{\rm i})^{-1}\,, \qquad \mu_{\rm i3} \approx \tfrac{1}{2}\mu_{\rm i1}\,, \\ &\mu_{i2} \approx 0.85\,\hat\mu_{\rm i}(\Omega_{\rm i}\tau_{\rm i})^{-2}\,, \qquad \mu_{\rm i4} \approx \tfrac{1}{4}\mu_{\rm i2}, \qquad \text{with}\quad \hat\mu_{\rm i} \equiv n_{\rm i}kT_{\rm i}\tau_{\rm i}\,. \end{aligned} \tag{3.90}$$

With the expressions listed under (a)–(d), the two-fluid equations (3.61)–(3.63) have become a closed set.

3.3.3 Dissipative versus ideal fluids*

We now come to a peculiar point in the exposition: in the end, the extensive discussion of the transport coefficients just serves to neglect most of them! This will be justified on the basis of the time scales $\tau_{\rm H}$ for the hydrodynamic and $\tau_{\rm D}$ for the dissipative phenomena. Paradoxically, whereas the relaxation times $\tau_{\rm e,i}$ measure the very *short* time scales needed to establish the electron and ion fluids, the associated dissipative diffusion (decay) of the macroscopic quantities takes place on the very *long* time scale $\tau_{\rm D}$. Hence, in the restricted range

$$\tau_{\rm e,i} \ll \tau_{\rm H} \ll \tau_{\rm D}\,, \tag{3.91}$$

the macroscopic fluid dynamics may be considered as dissipationless, or *ideal*. We will demonstrate that this restriction not only permits a very significant reduction of the number of variables, but also creates the essential window on the great majority of macroscopic plasma dynamical phenomena.

As an example, consider the transport of ion momentum and energy by viscosity and heat conduction. This is described by the momentum equation (3.62)(b), with the ion viscosities $\mu_{{\rm i}\ell}$ given in Eq. (3.90), and the energy equation (3.63)(b), with the ion heat conductivities $\kappa_{{\rm i}\lambda}$ given in Eq. (3.85). To estimate orders of magnitude, we just single out the two terms that give a diffusion equation:

$$n_{\rm i}m_{\rm i}\frac{\partial \mathbf{u}_{\rm i}}{\partial t} \approx -\nabla\cdot\boldsymbol{\pi}_{\rm i} \sim \mu_{{\rm i}\ell}\frac{\partial^2\mathbf{u}_{\rm i}}{\partial x_\ell^2}\,, \tag{3.92}$$

$$\tfrac{3}{2}n_{\rm i}\frac{\partial(kT_{\rm i})}{\partial t} \approx -\nabla\cdot\mathbf{h}_{\rm i} \sim \kappa_{{\rm i}\lambda}\frac{\partial^2(kT_{\rm i})}{\partial x_\lambda^2}\,. \tag{3.93}$$

For simplicity, we consider the indices ℓ and λ to just take the values $\|$ and $\perp$. Since both $\mu_{\rm i\|}$ and

$\kappa_{i\parallel} \sim kT_i\,\tau_i$, we obtain the following estimates for the parallel and perpendicular ion diffusion time scales:

$$\tau_{D,i\parallel} \sim \frac{L^2}{v_{th,i}^2\,\tau_i}\,, \qquad \text{and} \quad \tau_{D,i\perp} \sim \left(\frac{a}{L}\right)^2 (\Omega_i\tau_i)^2\,\tau_{D,i\parallel}\,. \tag{3.94}$$

This exhibits the mentioned paradox. Since the relaxation time scale τ_i is considered to be short, the parallel diffusion time $\tau_{D,i\parallel}$ will be long. Moreover, although the ratio a/L of perpendicular to parallel plasma dimensions is small, typically ~ 0.1, the factor $\Omega_i\tau_i$ is usually so large (see the numerical examples below Eq. (3.57)) that the ratio $[(a/L)\,\Omega_i\tau_i]^2$ between perpendicular and parallel diffusion times will be huge: classical perpendicular thermal isolation is nearly perfect! Of course, this is the very reason for using magnetic fields to confine thermonuclear plasmas in the laboratory. It is also the reason why large perpendicular temperature gradients are maintained so well in coronal magnetic flux tubes.

Let us again insert the numbers of our generic examples, discussed in Section 3.2.4 using the Tables B.3 and B.4. For a tokamak plasma with $\widetilde{T} = 1\,\mathrm{keV}$, $v_{th,i} = 0.4 \times 10^6\,\mathrm{m\,s^{-1}}$, $\tau_i = 2.6 \times 10^{-4}\,\mathrm{s}$, we obtain $\tau_{D,i\parallel} = 8 \times 10^{-6}\,\mathrm{s}$ and $\tau_{D,i\perp} = 4.5 \times 10^4\,\mathrm{s}$ ($\approx$ 12 hrs!). For a coronal loop with $\widetilde{T} = 0.1\,\mathrm{keV}$, $v_{th,i} = 1.4\times10^5\,\mathrm{m\,s^{-1}}$, $\tau_i = 8.6\times10^{-3}\,\mathrm{s}$, we obtain $\tau_{D,i\parallel} = 6.2\times10^6\,\mathrm{s}$ ($\approx$ 72 days!) and $\tau_{D,i\perp}$ is virtually infinite. This demonstrates, once more, the easy justification of hydrodynamic models for astrophysical plasmas, due to the large length scales L. It also justifies the use of *ideal* hydrodynamic models for those plasmas. On the other hand, satisfaction of the condition (3.91) for the parallel diffusion in thermonuclear laboratory plasmas is difficult, for the same reasons as discussed in Section 3.2.4 with respect to the conditions (3.58) for hydrodynamic behaviour. The comments raised there might serve to alleviate our concerns here as well. From a practical point of view, rapid parallel diffusion is actually desirable in fusion devices since it evens out temperature gradients in the magnetic surfaces.

Electron momentum and energy transport by viscosity and heat conduction lead to similar considerations. The estimation of the numerical magnitudes is left as an exercise for the reader. Consequently, under the restrictions (3.91), the electron and ion anisotropic pressure and heat flow terms may be neglected:

$$\boldsymbol{\pi}_{e,i} \to 0 \quad (\text{i.e. } \mu_{e,i\ell} \to 0)\,, \qquad \mathbf{h}_{e,i} \to 0 \quad (\text{i.e. } \kappa_{e,i\lambda} \to 0) \tag{3.95}$$

in the two-fluid equations (3.62)–(3.63).

The next step towards an ideal fluid description is the neglect of the ion–electron momentum transfer $\mathbf{R}_e$ in the two-fluid equations. This requires an entirely different type of argumentation since the pertinent transport coefficient is the electrical conductivity $\sigma_\parallel$ which, although proportional to τ_e like the other transport coefficients, should be large (nearly perfect conductivity) instead of small for the emergence of ideal fluid behaviour. We need to exploit some form of the generalized Ohm's law (see Sections 2.4.1 and 3.4.1), which is obtained by subtracting from the electron momentum equation (3.62)(a) the ion momentum equation (3.62)(b) multiplied with Zm_e/m_i and neglecting small terms in the mass ratio:

$$en_e\widehat{\mathbf{E}} = \mathbf{R}_e + \mathbf{j}\times\mathbf{B} \qquad \Rightarrow \qquad en_e(\mathbf{E} + \mathbf{u}_i \times \mathbf{B}) \approx \mathbf{R}_e\,. \tag{3.96}$$

Here, the left expression is the one exploited by Balescu [32] to obtain his transport coefficients keeping the appropriate number of Hermitian moments (typically 21M), whereas the right expression results from neglecting the contributions of the electron pressure gradient and the Hall term. The latter approximations are necessary to obtain a consistent representation in a low number of moments. As suggested by Balescu, the above neglect of the electron heat flow ($\boldsymbol{\kappa}_e \to 0$) implies that the thermo-electric coupling should be neglected as well ($\boldsymbol{\alpha} \to 0$) and, hence, that all higher moments in Eq. (3.75) disappear. Effectively, we are now down to a crude 5M approximation with a very simple relationship between $\mathbf{R}_e$ and $\mathbf{j}$:

$$\mathbf{R}_e = \frac{m_e}{e\tau_e}\mathbf{j} = en_e\eta_0\,\mathbf{j} \quad \Rightarrow \quad \eta_0 \equiv \frac{m_e}{e^2 n_e \tau_e} = 2\eta_\parallel = 2\sigma_\parallel^{-1}\,. \tag{3.97}$$

Consequently, we obtain an isotropic resistivity tensor $\eta_0\mathbf{I}$ which is off by a factor of 2 in the parallel direction and misses the off-diagonal elements. Nevertheless, the resulting resistive fluid model is the most widely used one in plasma physics.

We still have to demonstrate why and when the ion–electron momentum transfer, i.e. the resistivity, may be neglected. The crucial observation here is that resistivity is the cause of current decay, i.e. decay of magnetic field inhomogeneity. Obviously, in fusion devices, such dissipation should be small in order for the confinement configuration to be maintained. The relevant diffusion equation for magnetic field inhomogeneity is obtained by substituting in Faraday's law (3.8) the approximate electric field $\mathbf{E} \approx (\boldsymbol{\sigma})^{-1}\cdot\mathbf{j} \approx \eta_0\,\mathbf{j}$ (from Eq. (3.70), neglecting the ion flow and electron temperature gradient contributions) and exploiting Ampère's law $\mathbf{j} \approx \mu_0^{-1}\nabla\times\mathbf{B}$ (from Eq. (3.9), neglecting the displacement current):

$$\frac{\partial \mathbf{B}}{\partial t} \approx -\nabla\times\mathbf{E} \approx -\nabla\times\left(\frac{\eta_0}{\mu_0}\nabla\times\mathbf{B}\right) \sim -\frac{\eta_0}{\mu_0}\frac{\partial^2\mathbf{B}}{\partial x_\perp^2}\,. \tag{3.98}$$

This yields the following estimate of the resistive decay time:

$$\tau_R \sim \frac{\mu_0 a^2}{\eta_0} = \frac{\mu_0 e^2 n_e \tau_e a^2}{m_e} = a^2\left(\frac{\omega_{pe}}{c}\right)^2\tau_e \equiv \left(\frac{a}{\delta_e}\right)^2\tau_e\,. \tag{3.99}$$

Here, a new length scale enters the discussion, viz. the *electron skin depth*,

$$\delta_e \equiv c/\omega_{pe}\,, \tag{3.100}$$

which measures the thickness of the layer in which high frequency electromagnetic waves can penetrate a well-conducting plasma. Using the numbers of Tables B.3 and B.4 again, we find $\tau_R = 24\,\mathrm{s}$ for a 1 keV tokamak plasma and $\tau_R = 8\times10^{12}\,\mathrm{s}$ (!) for a coronal loop plasma; sufficiently long to assume $\tau_H \ll \tau_R$ and to put

$$\mathbf{R}_e \to 0 \quad (\text{i.e. } \eta_0 \to 0) \tag{3.101}$$

in the two-fluid equations (3.62)–(3.63).

Finally, we need to insert the expression (3.86) for the electron–ion energy transfer rate Q_i into the two-fluid energy equations (3.63). These will be simplified again by putting

$$Q_i \to 0\,. \tag{3.102}$$

In a two-fluid plasma model, this neglect of temperature equilibration may be justified for short time scales $\tau_H \ll \tau_{eq}$, when the two temperatures T_e and T_i still evolve by themselves. In the one-fluid model, considered in Section 3.4, the opposite will be assumed, viz. $\tau_{\rm H} \gg \tau_{\rm eq}$, when temperature equilibration has already taken place so that $T_{\rm e} = T_{\rm i}$. In both limits, the condition (3.102) is a valid assumption. Hence, the *ideal two-fluid equations* are obtained from Eqs. (3.61)–(3.64) in the limits (3.95), (3.101), (3.102), i.e. neglecting all the RHS terms.

Even though the condition $\tau_{\rm H} \ll \tau_{\rm R}$ appears to be well satisfied for plasmas of interest, there are important instances when resistive effects enter the hydrodynamic time scales after all. This happens, e.g., because certain resistive instabilities evolve on a time scale faster than $\tau_{\rm R}$, or turbulence creates current sheets with much smaller length scales than a. Effectively, internal resistive boundary layers with large gradients develop, very analogous to the boundary layers of ordinary fluids. A spectacular example is the solar flare with the release of huge amounts of magnetic energy, triggered by resistive phenomena on a very small length scale (totally unrelated to the astronomical scales L and a). In those cases, it is expedient to keep the resistive terms

$$\mathbf{R}_{\rm e} = e n_{\rm e} \eta_0 \, \mathbf{j} \,, \quad \text{and} \quad -(\mathbf{u}_{\rm e} - \mathbf{u}_{\rm i}) \cdot \mathbf{R}_{\rm e} = (e n_{\rm e})^{-1} \mathbf{j} \cdot \mathbf{R}_{\rm e} = \eta_0 |\mathbf{j}|^2 \tag{3.103}$$

in the momentum and energy equations, where the right expression represents Ohmic heating. Hence, the *resistive two-fluid equations* are obtained from Eqs. (3.61)–(3.64) in the limits (3.95) and (3.102), i.e. neglecting pressure anisotropies and heat flows, but keeping the resistive terms (3.103) in the RHSs. From now on, we will drop the subscript 0 on the resistivity η_0, which we recall is just a model representation of the actual anisotropic tensor.

In conclusion: the essential, but reduced, picture of ideal two-fluid plasma dynamics is valid in the wide range of time scales (3.91), intermediate between rapid kinetics and slow transport. These conditions clearly indicate how the theory is to be modified when the ideal description fails. When the left condition is violated, a kinetic description is in order, and when the right condition is not satisfied, the dissipative transport terms should be restored in the equations. A very relevant example of the latter procedure is the resistive two-fluid model, obtained after restoring only the resistive terms in the ideal two-fluid equations.

3.3.4 Excursion: waves in two-fluid plasmas★

An instructive example of dynamics in a two-fluid plasma is the enormous variety of waves it supports. We here derive the general *dispersion equation* for a homogeneous, resistive, two-fluid plasma, jumping details to keep this subsection within reasonable bounds. The dispersion equation will be solved explicitly for the ideal case, exposing the different length and time scales of the waves in order to facilitate the discussion of Section 3.4 where only the largest scales will survive.

Our starting point is the complete set of two-fluid equations (3.61)–(3.64), neglecting the RHSs except for the resistive terms (3.103), and Maxwell's equations (3.6)–(3.11). As in Section 2.3, we exploit the pressures $p_{\rm e,i}$ instead of the temperatures as variables. We assume a homogeneous background equilibrium with the electrons and ions at rest:

$$p_{\rm e0}\,, p_{\rm i0}\,, \mathbf{B}_0 \;\; \text{const}\,, \quad \mathbf{E}_0 = 0 \;\Rightarrow\; n_{\rm e0} = Z n_{\rm i0}\,, \quad \mathbf{u}_{\rm e0} = \mathbf{u}_{\rm i0} = 0 \;\Rightarrow\; \mathbf{j}_0 = 0\,. \tag{3.104}$$

We now perturb this equilibrium with small amplitude oscillations. Dropping the 0 on the equilibrium quantities and indicating perturbations with a tilde, the equations for the perturbations become:

$$\begin{aligned}
&\frac{\partial \tilde{n}_e}{\partial t} + n_e \nabla\cdot\tilde{\mathbf{u}}_e = 0\,,\\
&n_e m_e \frac{\partial \tilde{\mathbf{u}}_e}{\partial t} + \nabla\tilde{p}_e + e n_e(\tilde{\mathbf{E}} + \tilde{\mathbf{u}}_e\times\mathbf{B}) = \frac{m_e}{e}\nu\tilde{\mathbf{j}} = -n_e m_e \nu(\tilde{\mathbf{u}}_e - \tilde{\mathbf{u}}_i)\,,\\
&\frac{\partial \tilde{p}_e}{\partial t} + \gamma p_e \nabla\cdot\tilde{\mathbf{u}}_e = 0\,,
\end{aligned} \tag{3.105}$$

$$\begin{aligned}
&\frac{\partial \tilde{n}_i}{\partial t} + n_i \nabla\cdot\tilde{\mathbf{u}}_i = 0\,,\\
&n_i m_i \frac{\partial \tilde{\mathbf{u}}_i}{\partial t} + \nabla\tilde{p}_i - Zen_i(\tilde{\mathbf{E}} + \tilde{\mathbf{u}}_i\times\mathbf{B}) = -\frac{m_e}{e}\nu\tilde{\mathbf{j}} = n_e m_e \nu(\tilde{\mathbf{u}}_e - \tilde{\mathbf{u}}_i)\,,\\
&\frac{\partial \tilde{p}_i}{\partial t} + \gamma p_i \nabla\cdot\tilde{\mathbf{u}}_i = 0\,,
\end{aligned} \tag{3.106}$$

$$\begin{aligned}
&\nabla\times\tilde{\mathbf{E}} = -\frac{\partial\tilde{\mathbf{B}}}{\partial t}\,, \qquad \nabla\cdot\tilde{\mathbf{E}} = -\frac{e}{\epsilon_0}(\tilde{n}_e - Z\tilde{n}_i)\,,\\
&\nabla\times\tilde{\mathbf{B}} = \frac{1}{c^2}\frac{\partial\tilde{\mathbf{E}}}{\partial t} - \mu_0 e n_e(\tilde{\mathbf{u}}_e - \tilde{\mathbf{u}}_i)\,, \qquad \nabla\cdot\tilde{\mathbf{B}} = 0\,.
\end{aligned} \tag{3.107}$$

Here, $\gamma \equiv 5/3$ and we have introduced an effective electron–ion collision frequency,

$$\nu \equiv \tau_e^{-1} \quad \Rightarrow \quad \eta = m_e(e^2 n_e)^{-1}\nu\,, \tag{3.108}$$

which is a convenient way of bookkeeping the resistive damping of the waves.

Since all equilibrium quantities are constant in space and time, we assume plain wave solutions, $n_e(\mathbf{r},t) \sim \exp[\mathrm{i}(\mathbf{k}\cdot\mathbf{r} - \omega t)]$, etc., so that $\nabla \to \mathrm{i}\mathbf{k}$, $\partial/\partial t \to -\mathrm{i}\omega$. This turns the partial differential equations into a set of algebraic equations. The determinant of this set gives the dispersion equation $\omega = \omega(\mathbf{k})$ of the waves. The manner in which the time derivatives appear in these equations dictates the number of waves to be expected. Since there are five electron variables ($\tilde{n}_e$, $\tilde{\mathbf{u}}_e$, $\tilde{p}_e$), five ion variables ($\tilde{n}_i$, $\tilde{\mathbf{u}}_i$, $\tilde{p}_i$), and four EM variables (two independent components of $\tilde{\mathbf{E}}$ and two independent components of $\tilde{\mathbf{B}}$), we may expect 14 different types of waves!

To solve this system, we capitalize on all available geometrical and physical relationships between the variables. To that end, we project onto the three directions given by the unit vectors oriented with respect to $\mathbf{k}$ and $\mathbf{B}$, with $\mathbf{e}_3$ along $\mathbf{k}$:

$$\mathbf{e}_1 \equiv \mathbf{e}_2\times\mathbf{e}_3\,, \qquad \mathbf{e}_2 \equiv \mathbf{B}\times\mathbf{k}/|\mathbf{B}\times\mathbf{k}|\,, \qquad \mathbf{e}_3 \equiv \mathbf{k}/|\mathbf{k}|\,, \tag{3.109}$$

and indicate the direction of $\mathbf{k}$ by means of the angle ϑ between $\mathbf{k}$ and $\mathbf{B}$,

$$\lambda \equiv k_\parallel/k = \cos\vartheta\,, \qquad \tau \equiv k_\perp/k = \sin\vartheta\,. \tag{3.110}$$

In this projection, the components $\tilde{E}_3$ and $\tilde{B}_3$ $(= 0)$ may be eliminated and we obtain the mentioned system of 14 variables.

Next, we isolate a peculiar class of marginal 'waves' ($\omega = 0$) consisting of a two-parameter family of modes with $\tilde{n}_e \neq 0$ and $\tilde{n}_i \neq 0$ and associated pressure perturbations that are completely out of phase,

$$\tilde{p}_{\rm e} = -\tilde{p}_{\rm i} = {\rm i}\frac{en_{\rm e}}{k}\tilde{E}_3 = -\frac{e^2 n_{\rm e}}{\epsilon_0 k^2}\,(\tilde{n}_{\rm e} - Z\tilde{n}_{\rm i})\,, \tag{3.111}$$

whereas the remaining variables vanish. These waves do not move, they just sit there with the pressures kept in balance by the longitudinal electric field due to the space charge clouds. Even in the absence of the latter, there still is a subclass of charge-neutral modes with $\tilde{n}_{\rm e} = Z\tilde{n}_{\rm i} \neq 0$, and all of the remaining variables identically zero. In the ideal MHD limit, the two-parameter marginal modes transform into the one-parameter entropy modes. Those modes will be discussed in Section 5.2, and they will also be discarded there as being irrelevant for the dynamics.

For the remaining twelve solutions, which represent genuine waves ($\omega \neq 0$), we can freely divide by ω to express the density and pressure variables $\tilde{n}_{\rm e,i}$ and $\tilde{p}_{\rm e,i}$ in terms of $\tilde{u}_{\rm e,i3}$, and the magnetic field perturbations $\tilde{B}_{2,1}$ in terms of $\tilde{E}_{1,2}$. This yields a system of eight algebraic equations in the variables $\tilde{\mathbf{u}}_{\rm e,i}$ and $\tilde{E}_{1,2}$. At this point, it is expedient to account for the phase differences of the variables, expressed by the factors i, and to equalize the dimensions of the different terms. This is done by introducing the following variables for the perturbations:

$$\begin{aligned} \tilde{E}_1 &\equiv {\rm i}\sqrt{\epsilon_0}\,\tilde{\mathbf{E}}\cdot\mathbf{e}_1\,, & \tilde{E}_2 &\equiv \sqrt{\epsilon_0}\,\tilde{\mathbf{E}}\cdot\mathbf{e}_2\,, \\ \tilde{U}_{\rm e1,3} &\equiv \sqrt{n_{\rm e}m_{\rm e}}\,\tilde{\mathbf{u}}_{\rm e}\cdot\mathbf{e}_{1,3}\,, & \tilde{U}_{\rm e2} &\equiv {\rm i}\sqrt{n_{\rm e}m_{\rm e}}\,\tilde{\mathbf{u}}_{\rm e}\cdot\mathbf{e}_2\,, \\ \tilde{U}_{\rm i1,3} &\equiv \sqrt{n_{\rm i}m_{\rm i}}\,\tilde{\mathbf{u}}_{\rm i}\cdot\mathbf{e}_{1,3}\,, & \tilde{U}_{\rm i2} &\equiv {\rm i}\sqrt{n_{\rm i}m_{\rm i}}\,\tilde{\mathbf{u}}_{\rm i}\cdot\mathbf{e}_2\,, \end{aligned} \tag{3.112}$$

and equilibrium parameters characterizing the waves:

$$\begin{aligned} \omega_{\rm pe} &\equiv \sqrt{\frac{e^2 n_{\rm e}}{\epsilon_0 m_{\rm e}}}\,, & \Omega_{\rm e} &\equiv \frac{eB}{m_{\rm e}}\,, & v_{\rm e} &\equiv \sqrt{\frac{\gamma p_{\rm e}}{n_{\rm e}m_{\rm e}}}\,, \\ \omega_{\rm pi} &\equiv \sqrt{\frac{Z^2e^2 n_{\rm i}}{\epsilon_0 m_{\rm i}}} \equiv \sqrt{Z\mu}\,\omega_{\rm pe}\,, & \Omega_{\rm i} &\equiv \frac{ZeB}{m_{\rm i}} \equiv \mu\Omega_{\rm e}\,, & v_{\rm i} &\equiv \sqrt{\frac{\gamma p_{\rm i}}{n_{\rm i}m_{\rm i}}}\,. \end{aligned} \tag{3.113}$$

Here, $v_{\rm e}$ and $v_{\rm i}$ represent the electron and ion sound speeds (differing from the thermal speeds by a factor of $(\gamma/2)^{1/2} \approx 0.913$). The ratio of masses over charges,

$$\mu \equiv Zm_{\rm e}/m_{\rm i}\,, \tag{3.114}$$

is not yet assumed to be small in order to profit from the symmetry of electron and ion terms in the analysis. We will present the eigenvalue problem with and without dissipation.

(a) Eigenvalue problem for a resistive two-fluid plasma The above transformations yield the following eigenvalue problem:

$$\mathsf{A}\cdot\mathbf{X} = 0\,, \tag{3.115}$$

where

$$\mathbf{A} \equiv \begin{pmatrix} \omega^2 - k^2c^2 & 0 & \omega\omega_{pe} & 0 & 0 & -\omega\omega_{pi} & 0 & 0 \\ 0 & \omega^2 - k^2c^2 & 0 & -\omega\omega_{pe} & 0 & 0 & \omega\omega_{pi} & 0 \\ \omega_{pe} & 0 & \omega + i\nu & \lambda\Omega_e & 0 & -i\sqrt{\mu}\nu & 0 & 0 \\ 0 & -\omega_{pe} & \lambda\Omega_e & \omega + i\nu & \tau\Omega_e & 0 & -i\sqrt{\mu}\nu & 0 \\ 0 & 0 & 0 & \omega\tau\Omega_e & \begin{matrix}\omega^2 + i\nu\omega \\ -k^2 v_e^2 - \omega_{pe}^2\end{matrix} & 0 & 0 & \begin{matrix}\omega_{pe}\omega_{pi} \\ -i\sqrt{\mu}\nu\omega\end{matrix} \\ -\omega_{pi} & 0 & -i\sqrt{\mu}\nu & 0 & 0 & \omega + i\mu\nu & -\lambda\Omega_i & 0 \\ 0 & \omega_{pi} & 0 & -i\sqrt{\mu}\nu & 0 & -\lambda\Omega_i & \omega + i\mu\nu & -\tau\Omega_i \\ 0 & 0 & 0 & 0 & \begin{matrix}\omega_{pe}\omega_{pi} \\ -i\sqrt{\mu}\nu\omega\end{matrix} & 0 & -\omega\tau\Omega_i & \begin{matrix}\omega^2 + i\mu\nu\omega \\ -k^2 v_i^2 - \omega_{pi}^2\end{matrix} \end{pmatrix},$$

and

$$\mathbf{X} \equiv \big(\tilde{E}_1, \tilde{E}_2, \tilde{U}_{e1}, \tilde{U}_{e2}, \tilde{U}_{e3}, \tilde{U}_{i1}, \tilde{U}_{i2}, \tilde{U}_{i3}\big)^{\mathrm{T}}.$$

Because of the judicious choice of variables, most of the matrix elements are real now, except for the resistive terms ($\sim i\nu$). This implies that the waves will be damped: $\omega = \omega_0 + i\gamma_0$ with $\gamma_0 < 0$, or $k = k_0 + i\alpha_0$ with $\alpha_0 \neq 0$.

Such behaviour is easily checked for the special case of cold, unmagnetized electrons with heavy, immobile ions ($\omega_{pi} = 0, \Omega_e = \Omega_i = 0, v_e = v_i = 0$). The dispersion equation then renders two branches, one for the longitudinal plasma oscillations (Section 2.3.2),

$$\omega(\omega + i\nu) - \omega_{pe}^2 = 0 \quad \Rightarrow \quad \omega_0^2 \approx \omega_{pe}^2, \quad \gamma_0 \approx -\tfrac{1}{2}\nu, \tag{3.116}$$

and one for the transverse electromagnetic waves,

$$(\omega + i\nu)(\omega^2 - k^2c^2) - \omega\omega_{pe}^2 = 0 \quad \Rightarrow \quad \omega_0^2 \approx \omega_{pe}^2 + k^2c^2, \quad \gamma_0 \approx -\tfrac{1}{2}\nu\, \omega_{pe}^2/\omega_0^2, \tag{3.117}$$

where the damping is small when $\omega_{pe}\tau_e \gg 1$. For the transverse waves, we have assumed k to be real. On the other hand, if ω is imposed to be real, the wave number becomes complex, satisfying $k_0^2 - \alpha_0^2 \approx (\omega^2 - \omega_{pe}^2)/c^2$, with $\alpha_0 k_0 \approx \frac{1}{2}(\nu/\omega)(\omega_{pe}^2/c^2)$. Consequently, the electron skin depth $\delta_e \equiv c/\omega_{pe}$, as introduced in Eq. (3.100), naturally emerges as the spatial decay length of transverse EM waves.

(b) Eigenvalue problem for an ideal two-fluid plasma In the limit of vanishing resistivity ($\nu \to 0$), a significant simplification results from the elimination of the variables $\tilde{U}_{e2}$ and $\tilde{U}_{i2}$:

$$\mathbf{A}_0 \cdot \mathbf{X}_0 = 0, \tag{3.118}$$

where

$$
\mathbf{A}_0 \equiv \begin{pmatrix}
\omega^2 - k^2c^2 & 0 & \omega_{\rm pe}\omega & 0 & -\omega_{\rm pi}\omega & 0 \\
0 & \begin{matrix}\omega^2 - k^2c^2 \\ -\omega_{\rm pe}^2 - \omega_{\rm pi}^2\end{matrix} & \lambda\omega_{\rm pe}\Omega_{\rm e} & \tau\omega_{\rm pe}\Omega_{\rm e} & \lambda\omega_{\rm pi}\Omega_{\rm i} & \tau\omega_{\rm pi}\Omega_{\rm i} \\
\omega_{\rm pe}\omega & \lambda\omega_{\rm pe}\Omega_{\rm e} & \omega^2 - \lambda^2\Omega_{\rm e}^2 & -\lambda\tau\Omega_{\rm e}^2 & 0 & 0 \\
0 & \tau\omega_{\rm pe}\Omega_{\rm e} & -\lambda\tau\Omega_{\rm e}^2 & \begin{matrix}\omega^2 - k^2v_{\rm e}^2 \\ -\omega_{\rm pe}^2 - \tau^2\Omega_{\rm e}^2\end{matrix} & 0 & \omega_{\rm pe}\omega_{\rm pi} \\
-\omega_{\rm pi}\omega & \lambda\omega_{\rm pi}\Omega_{\rm i} & 0 & 0 & \omega^2 - \lambda^2\Omega_{\rm i}^2 & -\lambda\tau\Omega_{\rm i}^2 \\
0 & \tau\omega_{\rm pi}\Omega_{\rm i} & 0 & \omega_{\rm pe}\omega_{\rm pi} & -\lambda\tau\Omega_{\rm i}^2 & \begin{matrix}\omega^2 - k^2v_{\rm i}^2 \\ -\omega_{\rm pi}^2 - \tau^2\Omega_{\rm i}^2\end{matrix}
\end{pmatrix},
$$

and

$$
\mathbf{X}_0 \equiv \big(\tilde{E}_1, \tilde{E}_2, \tilde{U}_{\rm e1}, \tilde{U}_{\rm e3}, \tilde{U}_{\rm i1}, \tilde{U}_{\rm i3}\big)^{\rm T}.
$$

The twelfth order system is now represented by two EM variables, two electron variables and two ion variables. These yield a twofold degenerate dispersion equation since, due to the symmetry of the matrix, the eigenvalue parameter ω will only appear squared.

Since no assumption on the smallness of the mass ratio was made, the characteristic electron and ion frequencies may still be assumed to have equal orders of magnitude. Therefore, when frequencies are made dimensionless by means of the plasma frequency, the electron and ion contributions appear on an equal footing:

$$
\bar{\omega} \equiv \omega/\omega_{\rm p}\,, \quad \text{where } \omega_{\rm p} \equiv \sqrt{\omega_{\rm pe}^2 + \omega_{\rm pi}^2}\,. \tag{3.119}
$$

Similarly, wavelengths are made dimensionless by means of the combined skin depth for electromagnetic waves,

$$
\bar{k} \equiv k\delta\,, \quad \text{where } \delta \equiv c/\omega_{\rm p}\,, \tag{3.120}
$$

and velocities by means of the speed of light in vacuum, c. This yields six dimensionless parameters, which we indicate by one-letter symbols to facilitate the formidable algebra and numerical implementation:

$$
\begin{aligned}
e &\equiv \omega_{\rm pe}/\omega_{\rm p}\,, \qquad & E &\equiv \Omega_{\rm e}/\omega_{\rm p}\,, \qquad & v &\equiv v_{\rm e}/c\,, \\
i &\equiv \omega_{\rm pi}/\omega_{\rm p}\,, \qquad & I &\equiv \Omega_{\rm i}/\omega_{\rm p}\,, \qquad & w &\equiv v_{\rm i}/c\,.
\end{aligned} \tag{3.121}
$$

These parameters are not all independent. Charge neutrality, $n_{\rm e} = Zn_{\rm i}$, implies from the expressions (3.113) that e, i, and I may be eliminated in favour of the ratio of masses over charges:

$$
\mu = \frac{i^2}{e^2} = \frac{I}{E} \quad \Rightarrow \quad e^2 = \frac{1}{1+\mu}\,, \quad i^2 = 1 - e^2 = \frac{\mu}{1+\mu}\,, \quad I = \mu E\,. \tag{3.122}
$$

After removing dimensions with $\omega_{\rm p} \sim n_{\rm e}^{-1/2}$ and c, the dimensionless dispersion equation is determined by the direction λ of the wave vector and the four independent parameters μ, $E \sim B\, n_{\rm e}^{-1/2}$, $v \sim T_{\rm e}^{1/2}$ and $w \sim T_{\rm i}^{1/2}$, related to the physical variables of the background state.

The actual construction of the explicit form of the dispersion equation is one of those calculations which require, according to Stix [570], "in small proportion, insight, and in large proportion, stamina". Fortunately, the earlier enterprises of Braginskii [91], and in particular of Denisse and Delcroix [151], and Stringer [578], summarized by Swanson [582], permit conclusive checking of the correctness of the final expressions below. Once this is established, the short-hand notation (3.121) is of course replaced again by the physical expressions.

The *dispersion equation for the waves in an ideal two-fluid plasma*, finally obtained by brute force reduction of the determinant expressions, is a polynomial of the sixth degree in $\bar\omega^2$ and of the fourth degree in $\bar k^2$:

$$F(\bar k^2, \bar\omega^2) \equiv \sum_{m=0}^{6} \sum_{n=\max(0,3-m)}^{\min(4,6-m)} \alpha_{mn}\, \bar k^{2n}\, \bar\omega^{2m} = 0\,. \tag{3.123}$$

Arranging the terms of the dispersion equation (3.123) as shown in Table 3.1 facilitates the discussion of the different asymptotic limits for which analytic solutions may be obtained, as presented in Eqs. (3.124)–(3.129) below. The explicit expressions for the 19 coefficients $\alpha_{mn} = \alpha_{mn}(\lambda^2, \mu, E^2, v^2, w^2)$ are listed in Table 3.2. They just depend on the five background parameters indicated.

Table 3.1 *Dispersion equation for an ideal two-fluid plasma with asymptotic limits.*

cold (←)

$$\begin{aligned}
&\alpha_{60}\ \bar\omega^{12}\\
&+(\alpha_{50} + \alpha_{51}\bar k^2)\,\bar\omega^{10}\\
&+(\alpha_{40} + \alpha_{41}\bar k^2 + \alpha_{42}\bar k^4)\,\bar\omega^{8}\\
&+(\alpha_{30} + \alpha_{31}\bar k^2 + \alpha_{32}\bar k^4 + \alpha_{33}\bar k^6)\,\bar\omega^{6}\\
&\qquad+(\alpha_{21}\bar k^2 + \alpha_{22}\bar k^4 + \alpha_{23}\bar k^6 + \alpha_{24}\bar k^8)\,\bar\omega^{4}\\
&\qquad\qquad+(\alpha_{12}\bar k^4 + \alpha_{13}\bar k^6 + \alpha_{14}\bar k^8)\,\bar\omega^{2}\\
&\qquad\qquad\qquad+\alpha_{03}\bar k^6 + \alpha_{04}\bar k^8 \qquad = 0\,.
\end{aligned}$$

(1) cutoff

(2) resonance

(3) local, HF

(4) global, MHD

The numerical solution of this dispersion equation is shown in Fig. 3.1 for a representative choice of the parameters (the tokamak example of Table B.3). As expected, for each value of k^2, there are six twofold degenerate waves corresponding to propagation in two opposite directions (for $\omega < 0$

Table 3.2 *Coefficients of the dispersion equation for an ideal two-fluid plasma.*

$$
\begin{aligned}
\alpha_{60} &= 1\,,\\[4pt]
\alpha_{50} &= -(3+E^2+I^2)\,,\\
\alpha_{51} &= -(2+v^2+w^2)\,,\\[4pt]
\alpha_{40} &= 3+E^2+I^2+2EI+E^2I^2\,,\\
\alpha_{41} &= 4+2E^2+2I^2+(2+\lambda^2E^2+I^2)v^2+(2+E^2+\lambda^2I^2)w^2+i^2v^2+e^2w^2\,,\\
\alpha_{42} &= 1+2v^2+2w^2+v^2w^2\,,\\[4pt]
\alpha_{30} &= -(1+EI)^2\,,\\
\alpha_{31} &= -\Big\{2(1+EI)^2+(1+\lambda^2)(E^2+I^2-EI)\\
&\qquad+\big[1+I^2+\lambda^2(3+EI)EI\big]v^2+\big[1+E^2+\lambda^2(3+EI)EI\big]w^2\\
&\qquad+\big[2+(1-3\lambda^2)EI\big](i^2v^2+e^2w^2)\Big\}\,,\\
\alpha_{32} &= -\Big\{1+E^2+I^2+2(1+\lambda^2E^2+I^2)v^2+2(1+E^2+\lambda^2I^2)w^2\\
&\qquad+2(i^2v^2+e^2w^2)+\big[2+\lambda^2(E^2+I^2)\big]v^2w^2\Big\}\,,\\
\alpha_{33} &= -(v^2+w^2+2v^2w^2)\,,\\[4pt]
\alpha_{21} &= (1+EI)(1+\lambda^2)EI+(1+EI)(1+\lambda^2EI)(i^2v^2+e^2w^2)\,,\\
\alpha_{22} &= (1+EI)EI+\lambda^2(E^2+I^2-EI)\\
&\qquad+\big[(1+\lambda^2)I^2+2\lambda^2EI(2+EI)\big]v^2+\big[(1+\lambda^2)E^2+2\lambda^2EI(2+EI)\big]w^2\\
&\qquad+\big[2+(1-5\lambda^2)EI\big](i^2v^2+e^2w^2)+(1+\lambda^2EI)^2v^2w^2\,,\\
\alpha_{23} &= (I^2+\lambda^2E^2)v^2+(E^2+\lambda^2I^2)w^2+i^2v^2+e^2w^2+2\big[1+\lambda^2(E^2+I^2)\big]v^2w^2\,,\\
\alpha_{24} &= v^2w^2\,,\\[4pt]
\alpha_{12} &= -\lambda^2EI\Big\{EI+\big[2+(1+\lambda^2)EI\big](i^2v^2+e^2w^2)\Big\}\,,\\
\alpha_{13} &= -\lambda^2\Big\{E^2I^2v^2+E^2I^2w^2+(E^2+I^2)(i^2v^2+e^2w^2)+2EI(1+\lambda^2EI)v^2w^2\Big\}\,,\\
\alpha_{14} &= -\lambda^2(E^2+I^2)v^2w^2\,,\\[4pt]
\alpha_{03} &= \lambda^4E^2I^2(i^2v^2+e^2w^2)\,,\\
\alpha_{04} &= \lambda^4E^2I^2v^2w^2\,.
\end{aligned}
$$

and for $\omega > 0$). Other than this degeneracy, and that of the two EM waves at high frequencies (corresponding to two different states of polarization), *the six waves are non-degenerate for oblique propagation*. (To avoid confusion with dispersion diagrams shown in other textbooks: frequently,

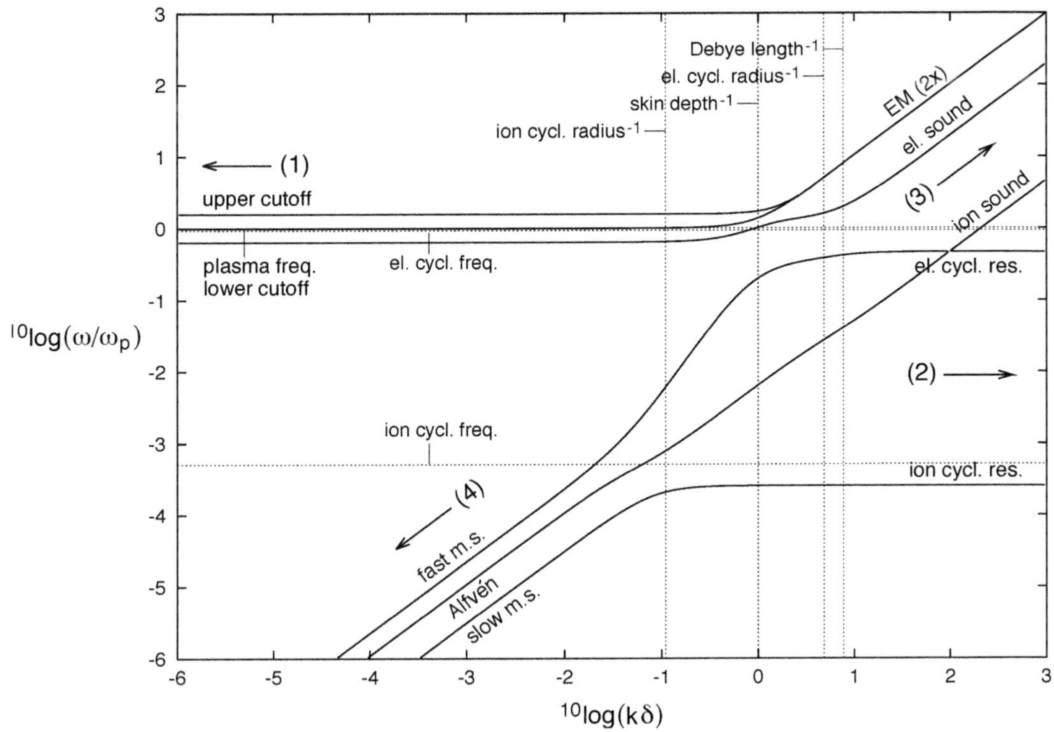

Fig. 3.1 Dispersion diagram for the oblique waves of an ideal two-fluid (hydrogen) plasma; $\lambda = 0.5$ ($\vartheta = 60^0$), $\mu = 5.4 \times 10^{-4}$, $E \equiv \Omega_e/\omega_p = 0.93$, $v \equiv v_e/c = 0.2$, $w \equiv v_i/c = 4.7 \times 10^{-3}$ ($\widetilde{T}_e = \widetilde{T}_i = 10\,\mathrm{keV}$). The arrows refer to the asymptotic limits (1)–(4) discussed in the text.

those are restricted to the degenerate case of parallel propagation.) This means that waves in two-fluid plasmas exhibit very intricate behaviour, where virtually all characteristic plasma frequency and length scales enter, as is indicated by the dotted lines in Fig. 3.1. It is beyond the scope of the present chapter to dwell on all of these aspects. The reader should consult any basic textbook on plasma physics, like Chen [117] and Goldston and Rutherford [244], or books entirely devoted to waves in plasmas, like Stix [570] and Swanson [582].

Our interest here is to demonstrate how the MHD phenomena emerge from the two-fluid ones. To that end, we first analyse how the six waves split apart for extreme values of k^2 and ω^2. By means of the asymptotic limits indicated in Table 3.1 and exploiting the explicit expressions for the coefficients α_{mn} listed in Table 3.2, one may easily determine those asymptotic limits of the dispersion equation.

(1) *Cutoff limits* ($k^2 \to 0$):

$$\omega^2 = \begin{cases} \omega_p^2 & \textit{(plasma frequency)} \\ \Big[\tfrac{1}{2}(\Omega_e - \Omega_i) \pm \sqrt{\tfrac{1}{4}(\Omega_e + \Omega_i)^2 + \omega_p^2}\,\Big]^2 & \textit{(upper and lower cutoff)}. \end{cases} \tag{3.124}$$

They represent the lower limits of the high-frequency waves. Excitation at frequencies below these limits results in wave motion that is spatially evanescent ($k^2 < 0$). The huge evanescent gap in Fig. 3.1 between these waves, where the electric field dominates, and the low-frequency MHD waves, where the magnetic field dominates, is the way in which the symmetry breaking of Maxwell's equations mentioned in Section 1.3.4 appears.

(2) *Resonance limits* ($k^2 \to \infty$):

$$\omega^2 = \begin{cases} \lambda^2 \, \Omega_\mathrm{e}^2 & \textit{(electron cyclotron resonance)} \\ \lambda^2 \, \Omega_\mathrm{i}^2 & \textit{(ion cyclotron resonance).} \end{cases} \tag{3.125}$$

These frequencies represent the asymptotic limits of spatially localized cyclotron waves. Different resonances (with hybrid contributions of ω_p, Ω_e and Ω_i) are obtained for cold plasmas ($v = w = 0$), as indicated by the dotted vertical line labelled 'cold' in Table 3.1 to the right of which all contributions to the dispersion equation vanish, whereas the remaining ones are also substantially simplified, as is evident from Table 3.2.

(3) *Local, high frequency limit* ($k^2 \to \infty \,, \omega^2 \to \infty$, but ω^2/k^2 finite):

$$\omega^2 = \begin{cases} k^2 c^2 & \textit{(two degenerate EM waves)} \\ k^2 v_\mathrm{e}^2 & \textit{(electron sound wave)} \\ k^2 v_\mathrm{i}^2 & \textit{(ion sound wave).} \end{cases} \tag{3.126}$$

In this limit, corresponding to localization in both space and time, the electromagnetic, the electron thermal and the ion thermal contributions split apart.

(4) *Global, low-frequency MHD limit* ($k^2 \to 0 \,, \omega^2 \to 0$, but ω^2/k^2 finite):

$$\begin{aligned} &(1 + v_\mathrm{A}^2/c^2)^2 \Big(\frac{\omega}{k}\Big)^6 - (1 + v_\mathrm{A}^2/c^2)\Big[(1+\lambda^2)v_\mathrm{A}^2 + v_\mathrm{s}^2 + \lambda^2 v_\mathrm{s}^2 v_\mathrm{A}^2/c^2\Big]\Big(\frac{\omega}{k}\Big)^4 \\ &\quad + \lambda^2 v_\mathrm{A}^2 \Big[v_\mathrm{A}^2 + 2v_\mathrm{s}^2 + (1+\lambda^2) v_\mathrm{s}^2 v_\mathrm{A}^2/c^2\Big]\Big(\frac{\omega}{k}\Big)^2 - \lambda^4 v_\mathrm{A}^4 v_\mathrm{s}^2 = 0 \,. \end{aligned} \tag{3.127}$$

In this opposite limit, where the waves are both spatially and temporally global, all characteristic plasma parameters ω_p, Ω_e, Ω_i, v_e and v_i contribute to produce the following *exact* expressions for the two characteristic MHD velocities:

$$\begin{aligned} EI &\equiv \frac{\Omega_\mathrm{e}\Omega_\mathrm{i}}{\omega_\mathrm{p}^2} \equiv \frac{B^2}{\mu_0(n_\mathrm{e}m_\mathrm{e} + n_\mathrm{i}m_\mathrm{i})c^2} \equiv \frac{v_\mathrm{A}^2}{c^2} && \textit{(Alfvén speed)}, \\ i^2v^2 + e^2w^2 &\equiv \frac{\omega_\mathrm{pi}^2 v_\mathrm{e}^2 + \omega_\mathrm{pe}^2 v_\mathrm{i}^2}{\omega_\mathrm{p}^2 c^2} \equiv \frac{\gamma(p_\mathrm{e} + p_\mathrm{i})}{(n_\mathrm{e}m_\mathrm{e} + n_\mathrm{i}m_\mathrm{i})c^2} \equiv \frac{v_\mathrm{s}^2}{c^2} && \textit{(sound speed)}. \end{aligned} \tag{3.128}$$

In other words: in this limit, the two-fluid plasma waves become entangled to produce the three MHD waves that are central to macroscopic plasma dynamics, as will be discussed extensively in Chapter 5. Consistent with the non-relativistic form of the two-fluid equations employed, one should drop the four terms with v_A^2/c^2 ($\ll 1$) in the asymptotic dispersion equation (3.127). With this single approximation, the expressions for the three MHD waves become:

$$\omega^2 = \begin{cases} \lambda^2 k^2 v_A^2 & \textit{(Alfvén wave)} \\ \tfrac{1}{2}k^2\left[v_A^2 + v_s^2 \pm \sqrt{(v_A^2+v_s^2)^2 - 4\lambda^2 v_A^2 v_s^2}\right] & \textit{(fast \& slow magneto-sonic wave)}. \end{cases} \tag{3.129}$$

They manifest the magnetic dominance of macroscopic plasma dynamics.

The two-fluid theory represents only a very partial picture of plasma wave dynamics in the high-frequency domain. This is evident from the fact that usually $\omega\tau_e \gg 1$, in violation of the left part of the condition (3.91) for fluid behaviour. In fact, in RF (radio-frequency) wave diagnostics and heating, velocity space effects determining Landau damping and the opposite phenomenon of microscopic instabilities (driven by non-monotonicities of the distribution functions) become important. Accordingly, the greater part of the mentioned books on plasma waves [570, 582] is devoted to the kinetic picture.

On the other hand, in the MHD limit of *large scales*, in space and time, the use of fluid theory is well justified. From Fig. 3.1 it is obvious how this condition is to be quantified with respect to the different inverse time scales Ω_i, Ω_e, ω_p, and inverse length scales R_i^{-1}, δ^{-1}, R_e^{-1}, λ_D^{-1} of the two-fluid model. Whereas we did not use the smallness of the mass ratio μ so far, it is expedient to exploit it now to find out that the ion scales determine the limits for the validity of the MHD model:

$$\omega \ll \Omega_i\,, \qquad k \ll R_i^{-1}\,, \tag{3.130}$$

i.e. the lower left quadrant of Fig. 3.1. Going down to larger and larger scales in this diagram, e.g. to study the transition to instability ($\omega^2 < 0$), we automatically encounter the finite size of the plasma so that we will have to drop the simplifying assumption of plasma homogeneity that was made in this section. Hence, the counterpart of kinetic theory, paying attention to inhomogeneities in velocity space, is the large-scale magnetohydrodynamic theory, paying attention to inhomogeneities in ordinary space. This will occupy us for most of the rest of this book.

3.4 One-fluid equations★

3.4.1 Maximal ordering for MHD★

We now derive the one-fluid MHD equations from the two-fluid equations (3.61)–(3.63). They are obtained as linear combinations of the pairs of the mass conservation equations for n_e and n_i, and of the momentum conservation equations for $\mathbf{u}_e$ and $\mathbf{u}_i$, and as the sum of the energy conservation equations for T_e and T_i (or, rather, p_e and p_i). At this point, we annihilate the information on the temperature difference between electrons and ions by imposing the time scale restriction $\tau_H \gg \tau_{eq}$ (temperature equilibration is assumed to have taken place already) associated with the approximation (3.102), i.e. the electron–ion energy transfer rate $Q_i \to 0$. Consequently, the number of moment equations is reduced by one. This permits the definition of the following one-fluid variables:

$$\rho \equiv n_e m_e + n_i m_i \qquad \textit{(total mass density)},$$

$$\tau \equiv -e\,(n_e - Z n_i) \qquad \textit{(charge density)},$$

$$\rho\mathbf{v} \equiv n_{\rm e}m_{\rm e}\mathbf{u}_{\rm e} + n_{\rm i}m_{\rm i}\mathbf{u}_{\rm i} \qquad \textit{(momentum density)}\,, \tag{3.131}$$
$$\mathbf{j} \equiv -e\,(n_{\rm e}\mathbf{u}_{\rm e} - Zn_{\rm i}\mathbf{u}_{\rm i}) \qquad \textit{(current density)}\,,$$
$$p \equiv p_{\rm e} + p_{\rm i} = (n_{\rm e} + n_{\rm i})\,kT \qquad \textit{(pressure)}\,,$$

as we have already seen in Section 2.4.1. Except for the temperature difference, the full information contained in the two-fluid equations is retained by the mentioned linear combinations, when the inverses of the one-fluid variables are exploited:

$$n_{\rm e} = \frac{Z\,[\,\rho - (m_{\rm i}/Ze)\tau\,]}{m_{\rm i}(1+\mu)} \approx \frac{Z}{m_{\rm i}(1+\mu)}\,\rho\,, \qquad n_{\rm i} = \frac{\rho + \mu(m_{\rm i}/Ze)\tau}{m_{\rm i}(1+\mu)} \approx \frac{1}{m_{\rm i}(1+\mu)}\,\rho\,,$$
$$\mathbf{u}_{\rm e} = \frac{\rho\,\mathbf{v} - (m_{\rm i}/Ze)\,\mathbf{j}}{\rho - (m_{\rm i}/Ze)\tau} \approx \mathbf{v} - \frac{m_{\rm i}}{Ze}\frac{\mathbf{j}}{\rho}\,, \qquad \mathbf{u}_{\rm i} = \frac{\rho\,\mathbf{v} + \mu(m_{\rm i}/Ze)\,\mathbf{j}}{\rho + \mu(m_{\rm i}/Ze)\tau} \approx \mathbf{v} + \mu\frac{m_{\rm i}}{Ze}\frac{\mathbf{j}}{\rho}\,,$$
$$p_{\rm e} = n_{\rm e}kT = \frac{n_{\rm e}}{n_{\rm e}+n_{\rm i}}\,p \approx \frac{Z}{1+Z}\,p\,, \qquad p_{\rm i} = n_{\rm i}kT = \frac{n_{\rm i}}{n_{\rm e}+n_{\rm i}}\,p \approx \frac{1}{1+Z}\,p\,. \tag{3.132}$$

Here, the ratio μ of masses over charges has been defined in Eq. (3.114), and the approximations on the RHS are due to the assumption of *quasi charge-neutrality:*

$$|n_{\rm e} - Zn_{\rm i}| \ll n_{\rm e}\,, \quad \text{or} \quad \frac{m_{\rm i}}{Ze}\,|\tau| \ll \rho\,. \tag{3.133}$$

As we have seen in Sections 1.4.1 and 2.3.2, this approximation is extremely well satisfied for plasma phenomena with a macroscopic (hydrodynamic) length scale

$$\lambda_{\rm H} \gg \lambda_{\rm D} \equiv v_{\rm th,e}/(\sqrt{2}\,\omega_{\rm pe})\,. \tag{3.134}$$

(For the tokamak and coronal loop examples of Tables B.3 and B.4, the values of the relevant small parameter are $\lambda_{\rm D}/a = 7\times10^{-5}$ and 7×10^{-11}, resp.) This does not imply that space charges do not occur, but just that they do not involve a sizeable fraction of the available free charges. Hence, for the time being (i.e. until we estimate the different terms in Maxwell's equations) we will keep the electrostatic contributions when they occur by themselves (like $\tau\mathbf{E}$ in the momentum equation), but drop them in the inverse expressions (3.132).

Multiplying the pair of equations (3.61) by the masses and adding them gives the equation of *mass conservation:*

$$\frac{\partial\rho}{\partial t} + \nabla\cdot(\rho\mathbf{v}) = 0\,, \tag{3.135}$$

whereas multiplication by the charges and subtraction results in the equation of *charge conservation*:

$$\frac{\partial\tau}{\partial t} + \nabla\cdot\mathbf{j} = 0\,. \tag{3.136}$$

Likewise, adding the pair of equations (3.62), while using the approximations on the RHS of Eqs. (3.132), results in *the equation of motion:*

$$\rho\frac{\partial\mathbf{v}}{\partial t} + \rho\mathbf{v}\cdot\nabla\mathbf{v} + \mu\Big(\frac{m_{\rm i}}{Ze}\Big)^2\nabla\cdot\Big(\frac{1}{\rho}\mathbf{j}\mathbf{j}\Big) + \nabla p - \tau\mathbf{E} - \mathbf{j}\times\mathbf{B} = -\nabla\cdot(\boldsymbol{\pi}_{\rm e} + \boldsymbol{\pi}_{\rm i})\,. \tag{3.137}$$

(This equation reduces to the Navier–Stokes equation when electric and magnetic effects are absent and the stress tensor is replaced by the hydrodynamic expression.) Multiplying the pair of equations

(3.62) by the charge over mass quotients and adding them results in an equation for the rate of change of the current density, which is known under the name *generalized Ohm's law:*

$$\frac{\partial \mathbf{j}}{\partial t} + \nabla \cdot \Big[\mathbf{jv} + \mathbf{vj} - \frac{m_\mathrm{i}}{Ze}(1-\mu)\frac{1}{\rho}\mathbf{jj} \Big]$$
$$+ \frac{1}{\mu}\frac{Ze}{m_\mathrm{i}}\Big[(1-\mu)\,\mathbf{j}\times\mathbf{B} - \frac{Z-\mu}{Z+1}\nabla p \Big] - \frac{1}{\mu}\Big(\frac{Ze}{m_\mathrm{i}}\Big)^2 \rho(\mathbf{E} + \mathbf{v}\times\mathbf{B})$$
$$= \frac{1}{\mu}\frac{Ze}{m_\mathrm{i}}\Big[\nabla\cdot(\boldsymbol{\pi}_\mathrm{e} - \mu\boldsymbol{\pi}_\mathrm{i}) - (1+\mu)\mathbf{R}_\mathrm{e} \Big]. \tag{3.138}$$

Finally, addition of the equations (3.63) results in *the heat balance equation:*

$$\frac{\partial p}{\partial t} + \mathbf{v}\cdot\nabla p + \gamma p \nabla\cdot\mathbf{v} + \frac{Z-\mu}{Z+1}\frac{m_\mathrm{i}}{Ze}\Big[(\gamma-1)\frac{1}{\rho}\mathbf{j}\cdot\nabla p - \gamma\mathbf{j}\cdot\nabla\Big(\frac{p}{\rho}\Big) \Big]$$
$$= -(\gamma-1)\Big[\boldsymbol{\pi}_\mathrm{e} : \nabla\mathbf{u}_\mathrm{e} + \boldsymbol{\pi}_\mathrm{i} : \nabla\mathbf{u}_\mathrm{i} + \nabla\cdot(\mathbf{h}_\mathrm{e} + \mathbf{h}_\mathrm{i}) - (1+\mu)\frac{m_\mathrm{i}}{Ze}\frac{1}{\rho}\mathbf{j}\cdot\mathbf{R}_\mathrm{e} \Big], \tag{3.139}$$

where the ratio $\gamma = 5/3$ of specific heats is introduced again.

To reduce the dissipative expressions on the RHSs of Eqs. (3.137)–(3.139), we again apply the time scale ordering (3.91) of Section 3.3.3, leading to the neglect of the electron and ion viscosities and heat conductivities:

$$\boldsymbol{\pi}_\mathrm{e,i} \to 0\,, \qquad \mathbf{h}_\mathrm{e,i} \to 0\,, \tag{3.140}$$

and to the reduction of the ion–electron momentum transfer $\mathbf{R}_\mathrm{e}$ in terms of the current density:

$$\mathbf{R}_\mathrm{e} = \eta e n_\mathrm{e}\mathbf{j} \approx \eta\frac{1}{1+\mu}\frac{Ze}{m_\mathrm{i}}\rho\mathbf{j}\,. \tag{3.141}$$

Here, the factor of proportionality, *the resistivity* η, is assumed to be a scalar.

To consistently derive the *large-scale* dynamics associated with the MHD description, we now apply a *maximal ordering*[1] of the one-fluid variables, starting from the expressions (3.128) for the Alfvén and sound speed:

$$v_\mathrm{A}^2 \equiv \frac{B^2}{\mu_0\rho} \sim v_\mathrm{s}^2 \equiv \frac{\gamma p}{\rho} \sim v^2\,. \tag{3.142}$$

(This implies that we admit arbitrary values of the ratio between kinetic and magnetic pressures, $\beta \equiv 2\mu_0 p/B^2 \sim 1$, and of the Mach number of the flow, $M \equiv v/v_\mathrm{s} \sim 1$.) The hydrodynamic length and time scales will now (in contrast to Section 3.3.3) be chosen to correspond to the size of the plasma:

$$\begin{aligned} \lambda_\mathrm{MHD} &\equiv |\nabla|^{-1} \sim a \gg R_\mathrm{i} \Big[\gg \delta \gg R_\mathrm{e} \gg \lambda_\mathrm{D} \Big], \\ \tau_\mathrm{MHD} &\equiv |\partial/\partial t|^{-1} \sim a/v_\mathrm{A} \gg \Omega_\mathrm{i}^{-1} \Big[\gg \omega_\mathrm{p}^{-1} \sim \Omega_\mathrm{e}^{-1} \Big]. \end{aligned} \tag{3.143}$$

(Again, admitting arbitrary values of the ratio between the transverse and longitudinal dimensions

[1] In the sense of the principles of asymptotology, i.e. "the art of dealing with applied mathematical systems in limiting cases", formulated by Kruskal [379].

of the plasma geometry, $a/L \sim 1$.) The magnitudes of the remaining, electrodynamic, variables $\mathbf{j}$, $\mathbf{E}$ and τ are chosen to be set by Ampère's law, Ohm's law and Poisson's law, resp.:

$$j \sim \frac{B}{\mu_0 a}\,, \quad E \sim vB\,, \quad \tau \sim \frac{\epsilon_0 E}{a} \sim \frac{\epsilon_0 vB}{a}\,. \tag{3.144}$$

This suffices to bring the large-scale MHD phenomena to the fore, as we will see.

Let us now get rid of the remaining local two-fluid effects in Eqs. (3.137)–(3.139). To that end, we first check the assumption (3.133) of quasi-neutrality. In terms of the maximal ordering (3.142)–(3.144), the relevant small parameter becomes

$$\frac{\mathrm{m_i}}{Ze}\frac{|\tau|}{\rho} \sim \frac{v_{\mathrm{th,i}}}{c}\frac{\lambda_{\mathrm{D}}}{a}\left[\sim \left(\frac{v_{\mathrm{th,i}}}{c}\right)^2 \frac{1}{\sqrt{\mu}}\frac{\delta_{\mathrm{e}}}{a}\right] \ll 1\,. \tag{3.145}$$

In fact, with the additional small factor $v_{\mathrm{th,i}}/c\,(\sim 5\times 10^{-3}$ for the tokamak example), this becomes an excellent approximation. Next, we assume the relative velocity of the ions and electrons to be small compared to the centre of mass velocity:

$$|\mathbf{u}_{\mathrm{i}} - \mathbf{u}_{\mathrm{e}}| \ll |\mathbf{v}|\,, \quad \text{or} \quad \frac{\mathrm{m_i}}{Ze}|\mathbf{j}| \ll \rho|\mathbf{v}|\,. \tag{3.146}$$

The latter restriction on the magnitude of the current density is to be understood in the sense of the maximal ordering, so that there is an upper limit to the current density also for static plasmas ($\mathbf{v} = 0$), where $|\mathbf{v}|$ should be replaced by v_{s} or v_{A} in the inequality. The magnitude of the associated small parameter is

$$\frac{\mathrm{m_i}}{Ze}\frac{|\mathbf{j}|}{\rho|\mathbf{v}|} \sim \frac{1}{\sqrt{\mu}}\frac{\delta_{\mathrm{e}}}{a} \ll 1\,, \tag{3.147}$$

i.e. much less extreme than the condition (3.145) for quasi-neutrality (note the expression in square brackets there), but still easily satisfied for plasmas of interest (and extremely easily for astrophysical plasmas due to their huge length scales).

With this ordering, and the condition (3.146) on the current density, the evolution equations for $\mathbf{v}, \mathbf{j}$, and p simplify to

$$\rho\frac{\partial \mathbf{v}}{\partial t} + \rho\mathbf{v}\cdot\nabla\mathbf{v} + \nabla p - \tau\mathbf{E} - \mathbf{j}\times\mathbf{B} = 0\,, \tag{3.148}$$

$$-\mu\Big(\frac{m_{\mathrm{i}}}{Ze}\Big)^2\frac{1}{\rho}\Big[\frac{\partial \mathbf{j}}{\partial t} + \nabla\cdot(\mathbf{jv}+\mathbf{vj})\Big] - \frac{m_{\mathrm{i}}}{Ze}\frac{1}{\rho}\Big[(1-\mu)\,\mathbf{j}\times\mathbf{B} - \frac{Z-\mu}{Z+1}\nabla p\Big] + \mathbf{E} + \mathbf{v}\times\mathbf{B} = \eta\mathbf{j}\,, \tag{3.149}$$

$$\frac{\partial p}{\partial t} + \mathbf{v}\cdot\nabla p + \gamma p\nabla\cdot\mathbf{v} = (\gamma-1)\,\eta|\mathbf{j}|^2\,. \tag{3.150}$$

Here, we have momentarily kept an intermediate form of the generalized Ohm's law with the time derivative term, which is small to second order in the small parameter (3.147), and the Hall current term, which is small to first order. Keeping those terms would conserve the evolutionary form of Ohm's law, but numerical computation of the current density this way would be extremely inaccurate since it would force the computation to proceed on the short time scale of the two-fluid model instead of on the much longer time scale of the MHD model. Obviously, for consistency of the

MHD description, we have to drop those terms so that the usual *Ohm's law* of the second line of Eq. (3.149) remains. However, this law gives the current density $\mathbf{j} = \sigma \mathbf{E}'$ with a large (and inaccurately known) parameter $\sigma \equiv \eta^{-1}$ multiplying the electric field in the moving frame: not very accurate either! Moreover, in the ideal MHD case ($\eta = 0$), Ohm's law $\mathbf{E} + \mathbf{v} \times \mathbf{B} = 0$ completely changes character from an equation that determines $\mathbf{j}$ into one that expresses $\mathbf{E}$ in terms of $\mathbf{v}$ and $\mathbf{B}$. Apparently, something else is required to restore the peace in the system. This will be the subject of the next section.

Concluding this section, it will have been noticed that we did not exploit the obvious further reduction of the expressions (3.132) for small values of the ratio μ of masses over charges. Although $\mu = 5.4 \times 10^{-4}$ for hydrogen plasmas, we have seen in Section 3.3.4 that MHD phenomena do not really depend on the assumption $\mu \ll 1$. This implies that the MHD equations will also be valid for other plasmas, like electron–positron plasmas in the classical, and non-relativistic, limit. However, this would require adaptation of the transport coefficients since the expressions of Section 3.3.2 essentially depend on the assumption $\mu \ll 1$. Also, the sequences of inequalities (3.143) in the square brackets will no longer apply since the quantities involved become of the same order in μ for those plasmas. This is not a serious problem though, since the hydrodynamic scales will then be defined by the largest one of them.

3.4.2 Resistive and ideal MHD equations★

We still need to subject Maxwell's equations to the maximal ordering (3.142)–(3.144). Before we do that, it is expedient to collect all equations derived so far, even though they are not quite consistent in this respect:

$$\frac{\partial \rho}{\partial t} + \nabla \cdot (\rho \mathbf{v}) = 0 \qquad \textit{(continuity)}, \tag{3.151}$$

$$\left[\frac{\partial \tau}{\partial t} + \nabla \cdot \mathbf{j} = 0 \qquad \textit{(charge)}, \right] \tag{3.152}$$

$$\rho \frac{\partial \mathbf{v}}{\partial t} + \rho \mathbf{v} \cdot \nabla \mathbf{v} + \nabla p \left[- \tau \mathbf{E} \right] - \mathbf{j} \times \mathbf{B} = 0 \qquad \textit{(momentum)}, \tag{3.153}$$

$$\frac{\partial p}{\partial t} + \mathbf{v} \cdot \nabla p + \gamma p \nabla \cdot \mathbf{v} = (\gamma - 1) \eta j^2 \qquad \textit{(internal energy)}, \tag{3.154}$$

$$\frac{\partial \mathbf{B}}{\partial t} + \nabla \times \mathbf{E} = 0 \qquad \textit{(Faraday)}, \tag{3.155}$$

$$\left[\frac{1}{c^2} \frac{\partial \mathbf{E}}{\partial t} + \right] \mu_0 \mathbf{j} - \nabla \times \mathbf{B} = 0 \qquad \textit{('Ampère')}, \tag{3.156}$$

where

$$\eta \mathbf{j} = \mathbf{E} + \mathbf{v} \times \mathbf{B} \qquad \textit{(Ohm)}, \tag{3.157}$$

and initially the following conditions need to be satisfied:

$$\left[\nabla \cdot \mathbf{E} = \tau / \epsilon_0 \qquad \textit{(Poisson)}, \right] \tag{3.158}$$

$$\nabla \cdot \mathbf{B} = 0 \qquad \textit{(no magnetic monopoles)}. \tag{3.159}$$

These equations would represent evolution equations for the variables ρ, τ, $\mathbf{v}$, p, $\mathbf{B}$ and $\mathbf{E}$, with Ohm's law determining $\mathbf{j}$ and the last two equations to be considered as initial conditions on the differential equations for $\mathbf{E}$ and $\mathbf{B}$.

First notice that the charge conservation equation (3.152) is really redundant since it follows from Eqs. (3.156) and (3.158). Therefore, we should drop it and determine τ from Poisson's law (3.158). Consequently, Poisson's law can no longer be considered as an initial condition on 'Ampère's' law, indicating that something should be changed there as well. This becomes immediately clear by estimating the displacement current in the ordering (3.142)–(3.144):

$$\frac{1}{c^2}\left|\frac{\partial \mathbf{E}}{\partial t}\right| \sim \frac{v^2}{c^2}\frac{B}{a} \ll |\nabla \times \mathbf{B}| \sim \frac{B}{a}\,. \tag{3.160}$$

Hence, for the non-relativistic flows we are considering, where

$$v^2/c^2 \ll 1\,, \tag{3.161}$$

the displacement current is negligible and we return to the *pre-Maxwell equations* [72]. These are characterized by the fact that Eq. (3.156) is replaced by Ampère's law proper (without the quotes, since this is the form in which Ampère posed it):

$$\mathbf{j} = \mu_0^{-1}\nabla \times \mathbf{B}\,. \tag{3.162}$$

We have now lost the evolution equation for $\mathbf{E}$ and obtained instead the simple expression for $\mathbf{j}$ as determined by the curl of $\mathbf{B}$. This appears to be consistent with the result of Section 3.4.1: Ohm's law should be read from right to left, i.e. it determines $\mathbf{E}$ rather than $\mathbf{j}$.

So far so good. But how about the charge conservation equation (3.152), that we have already dropped? Equation (3.162) implies that $\nabla \cdot \mathbf{j} = 0$ so that charge conservation now appears to tell us that $\partial\tau/\partial t = 0$. This would be in conflict with Eqs. (3.156) and (3.158), which imply that

$$\frac{\partial \tau}{\partial t} = \epsilon_0 \nabla \cdot \frac{\partial \mathbf{E}}{\partial t} = -\epsilon_0 \frac{\partial}{\partial t}\nabla \cdot (\mathbf{v} \times \mathbf{B}) \neq 0\,,$$

in general. The way out is to consistently apply an ordering in the small parameter v^2/c^2. One then finds that the electrostatic force $\tau\mathbf{E}$ in the momentum equation is one order smaller than the macroscopic Lorentz force $\mathbf{j} \times \mathbf{B}$, so that it should be dropped:

$$|\tau\mathbf{E}| \sim \frac{\epsilon_0 E^2}{a} \sim \frac{v^2}{c^2}\frac{B^2}{\mu_0 a} \ll |\mathbf{j} \times \mathbf{B}| \sim \frac{B^2}{\mu_0 a}\,. \tag{3.163}$$

After this, all equations are expressions of the same order, except for the charge conservation equation (3.152) which is one order in v^2/c^2 smaller. This justifies its elimination from the system. Poisson's equation may still be used to calculate the charge density, but, since τ does not occur in any of the other equations, that equation should be dropped as well. The resulting set of equations is a mathematically consistent set which, amongst the many other attractions exposed in this book, enjoys the property of being *Galilean invariant*.

Clearly, the approximations of the present section are of a different kind than those of the previous one. In the latter section, the conditions given are mandatory for the description of macroscopic dynamics, whereas the restriction (3.161) of the present section can be, and has been, lifted to construct a consistent relativistic MHD theory. This has been done for special as well as general relativity, see e.g. Lichnerowicz [409], Achterberg [4] and Anile [14]. Relativistic MHD is the subject of our final Chapter 22.

In conclusion, all expressions in square brackets in Eqs. (3.151)–(3.159) should be dropped and we obtain the following set of *resistive or ideal MHD equations* determining the evolution of the macroscopic variables $\rho, \mathbf{v}, p$, and $\mathbf{B}$ (already introduced in Section 2.4.1):

$$\frac{\partial \rho}{\partial t} + \nabla \cdot (\rho \mathbf{v}) = 0 \qquad \textit{(mass)}, \tag{3.164}$$

$$\rho \frac{\partial \mathbf{v}}{\partial t} + \rho \mathbf{v} \cdot \nabla \mathbf{v} + \nabla p - \mathbf{j} \times \mathbf{B} = 0 \qquad \textit{(momentum)}, \tag{3.165}$$

$$\frac{\partial p}{\partial t} + \mathbf{v} \cdot \nabla p + \gamma p \nabla \cdot \mathbf{v} = \begin{cases} (\gamma - 1)\, \eta j^2 & \text{(resistive)} \\ 0 & \text{(ideal)} \end{cases} \qquad \textit{(internal energy)}, \tag{3.166}$$

$$\frac{\partial \mathbf{B}}{\partial t} + \nabla \times \mathbf{E} = 0 \qquad \textit{(Faraday)}, \tag{3.167}$$

where the variables $\mathbf{j}$ and $\mathbf{E}$ may be eliminated by means of the algebraic equations

$$\mathbf{j} = \mu_0^{-1} \nabla \times \mathbf{B} \qquad \textit{(Ampère)}, \tag{3.168}$$

$$\mathbf{E} + \mathbf{v} \times \mathbf{B} = \begin{cases} \eta \mathbf{j} & \text{(resistive)} \\ 0 & \text{(ideal)} \end{cases} \qquad \textit{(Ohm)}, \tag{3.169}$$

and the magnetic field needs to satisfy the initial condition

$$\nabla \cdot \mathbf{B} = 0 \,. \tag{3.170}$$

This is how, in the end, the double set of five moment equations of kinetic theory, combined with Maxwell's equations, lead to a very powerful description of large-scale plasma dynamics, where collisions appear to have served no other purpose (i.e. in the ideal case) than to establish the coherence of the single fluid.

3.5 Literature and exercises★

Notes on literature

Kinetic theory and transport

- Braginskii's paper [92] on 'Transport processes in a plasma' in the first volume of the Russian *Reviews of Plasma Physics* is a masterpiece on the subject that has been used by most plasma physicists in one form or the other ever since its appearance.
- Balescu's two volumes [32] on *Transport Processes in Plasmas* are fruits of a life-long dedication to the science of statistical mechanics and non-equilibrium thermodynamics of charged particles. His love for the subject shines through every page of these books. The first volume is devoted to 'Classical Transport' and the second to 'Neo-classical transport'. The recent addition on 'Aspects of Anomalous Transport in Plasmas' [33] is a most welcome guide through the jungle of turbulent transport processes.
- Neo-classical transport in toroidal plasmas is discussed in the review paper by Galeev and Sagdeev [203].
- Chapman and Cowling, *The Mathematical Theory of Non-uniform Gases* [116] is the classical treatise of obtaining hydrodynamic equations by expansion of the distribution functions in a small parameter.

- The papers by Rosenbluth et al. [530] and Trubnikov [611] contain the derivation of the Fokker–Planck form of the collision operator for charged particles.
- Montgomery and Tidman, *Plasma Kinetic Theory* [446] is a critical exposition of the principles of kinetic theory of plasmas.

Derivation of fluid equations

- Grad and Blank et al., *Notes on Magnetohydrodynamics* [253, 72] present the pioneering lecture notes on the foundations of MHD.
- Most basic textbooks on plasma physics contain one or more chapters on the derivation and conditions of a macroscopic description, e.g.:
  Spitzer, *Physics of Fully Ionized Gases* [561], Chapter 2 and Appendix,
  Clemmov and Dougherty, *Electrodynamics of Particles and Plasmas* [124], Chapter 11,
  Schmidt, *Physics of High Temperature Plasmas* [542], Chapter 3,
  Boyd and Sanderson, *The Physics of Plasmas* [84], Chapter 3,
  Krall and Trivelpiece, *Principles of Plasma Physics* [378], Chapter 3,
  Akhiezer et al., *Plasma Electrodynamics* [8], Chapter 1.
- Hazeltine and Meiss, *Plasma Confinement* [296] contains a fundamental discussion of MHD and alternative closures of the moment equations.

Exercises

[3.1] *Collision integral*

Prove that the Landau collision operator (3.13) satisfies the conservation properties (3.18)–(3.22).

[3.2] *Transport coefficients*

Estimate the numerical magnitudes of the electron momentum and energy transport by viscosity and heat conduction for tokamak and coronal loop plasmas. What conclusions can be drawn from these numbers?

[3.3]★ *Waves in two-fluid plasmas*

Write a numerical program, in the computer language of your choice, for the numerical solution of the dispersion equation, derived in Section 3.3.4, for the waves in two-fluid plasmas. Carefully copy the explicit expressions of Table 3.2 in a separate subroutine! (As always in computations, you save time by spending enough time when coding up the algebra.) The graphical representation of the solution of the dispersion equation is most easily obtained by just contour plotting the implicit dispersion function $F(\bar{k}^2, \bar{\omega}^2) = 0$. For simplicity, you may restrict the analysis to cold plasmas by dropping the thermal terms. Now, enjoy the powerful tool you have obtained to study:

- Ordinary and extra-ordinary cutoffs, upper and lower hybrid resonances, whistler waves, etc. Compare the numerical results with the analytical asymptotic results.
- Compute the dispersion diagrams for perpendicular ($\lambda = 0$) and parallel ($\lambda = 1$) propagation and compare them with Fig. 3.1. Comment on the degeneracies obtained.
- Which of the MHD waves do you obtain in the large-scale limit?
- Suppose you re-derived the dispersion equation from scratch and kept the dimensional form of the parameters, would the resulting numerical program work?

[3.4] *Derivation of the one-fluid equations*

Complete the derivation of the one-fluid equations by constructing the sixth equation, for the temperature difference between electron and ions. Check for the consistency of omitting this equation under the conditions formulated.

Part II

Basic Magnetohydrodynamics

4
The MHD model

4.1 The ideal MHD equations

The dynamics of magnetically confined plasmas, as exploited in laboratory nuclear fusion research and observed in astrophysical systems, is essentially of a macroscopic nature so that it can be studied in the fluid (MHD) model introduced in Chapter 2. The 'derivation' of the MHD equations in Chapter 3 provided indications about the range of validity and the limitations of the equations. In the present chapter, we will develop the MHD model for the interaction of plasma and magnetic field in detail and, thus, obtain a powerful 'picture' for the dynamics of the mentioned plasmas.

Recall from the introduction of Chapter 3 that the equations of magnetohydrodynamics can be introduced either by just posing them as postulates for a hypothetical medium called 'plasma' or by the much more involved procedure of averaging the kinetic equations. Whereas Chapter 3 was mainly concerned with the second method, in the present chapter we exploit the first method: we simply pose the equations and use physical arguments and mathematical criteria to justify the result. We continue the exposition of Section 2.4.1, where we already encountered the relevant equations.

4.1.1 Postulating the basic equations

The ideal MHD equations describe *the motion of a perfectly conducting fluid interacting with a magnetic field*. Hence, we need to *combine Maxwell's equations with the equations of gas dynamics* and provide *equations describing the interaction*.

First, consider *Maxwell's equations*, already encountered in Chapters 2 and 3. They describe the evolution of the electric field $\mathbf{E}(\mathbf{r}, t)$ and the magnetic field $\mathbf{B}(\mathbf{r}, t)$ in response to the current density $\mathbf{j}(\mathbf{r}, t)$ and the space charge $\tau(\mathbf{r}, t)$:

$$\nabla \times \mathbf{E} = -\frac{\partial \mathbf{B}}{\partial t}, \tag{4.1}$$

$$\nabla \times \mathbf{B} = \mu_0 \mathbf{j} + \frac{1}{c^2}\frac{\partial \mathbf{E}}{\partial t}, \qquad c \equiv (\epsilon_0 \mu_0)^{-1/2}, \tag{4.2}$$

$$\nabla \cdot \mathbf{E} = \frac{\tau}{\epsilon_0}, \tag{4.3}$$

$$\nabla \cdot \mathbf{B} = 0. \tag{4.4}$$

Next, consider *the equations of gas dynamics* for the evolution of the density $\rho(\mathbf{r}, t)$ and the pressure

$p(\mathbf{r},t)$, encountered in Eqs. (2.140) and (2.142):

$$\frac{\mathrm{D}\rho}{\mathrm{D}t} + \rho\nabla\cdot\mathbf{v} \equiv \frac{\partial\rho}{\partial t} + \nabla\cdot(\rho\mathbf{v}) = 0\,, \tag{4.5}$$

$$\frac{\mathrm{D}p}{\mathrm{D}t} + \gamma p\nabla\cdot\mathbf{v} \equiv \frac{\partial p}{\partial t} + \mathbf{v}\cdot\nabla p + \gamma p\nabla\cdot\mathbf{v} = 0\,. \tag{4.6}$$

We will see later that these equations actually express mass conservation and conservation of entropy. Note that we have used the occasion to introduce the notation

$$\frac{\mathrm{D}}{\mathrm{D}t} \equiv \frac{\partial}{\partial t} + \mathbf{v}\cdot\nabla$$

for the *Lagrangian time-derivative*, evaluated while moving with the fluid, in contrast to the *Eulerian time-derivative* $\partial/\partial t$, which is evaluated at a fixed position.

So far, the two systems described by the variables $\mathbf{E}, \mathbf{B}$ and ρ, p do not appear to interact. Such interaction is introduced through the equations involving the velocity $\mathbf{v}(\mathbf{r},t)$ of the fluid. First, 'Newton's' *equation of motion for a fluid element*,

$$\rho\frac{\mathrm{D}\mathbf{v}}{\mathrm{D}t} = \mathbf{F} \equiv -\nabla p + \rho\mathbf{g} + \mathbf{j}\times\mathbf{B} + \tau\mathbf{E}\,, \tag{4.7}$$

expresses the acceleration of a fluid element (LHS) caused by the force $\mathbf{F}$ consisting of pressure gradient, gravity and electromagnetic contributions. (Actually, these expressions have the dimension of *force density*.) Next, one of the most characteristic equations describing the plasma state is the equation for the electric field in a perfectly conducting moving fluid,

$$\mathbf{E}' \equiv \mathbf{E} + \mathbf{v}\times\mathbf{B} = 0\,, \tag{4.8}$$

which expresses that the electric field $\mathbf{E}'$ in a co-moving frame should vanish.

The system of equations (4.1)–(4.8) is now complete, but not yet in a form suitable for self-consistent calculations for the majority of plasmas occurring in the laboratory and in nature. We need to make one additional assumption (repeating some of the discussion and expressions of Section 3.4.2). For most plasma phenomena it is sufficient to restrict the analysis to *non-relativistic velocities*:

$$v \ll c\,. \tag{4.9}$$

In that case, we can make the following estimates for the orders of magnitude of the different terms in Eq. (4.2):

$$\frac{1}{c^2}\left|\frac{\partial\mathbf{E}}{\partial t}\right| \sim \frac{v^2}{c^2}\frac{B}{l_0} \quad\ll\quad |\nabla\times\mathbf{B}| \sim \frac{B}{l_0} \qquad \text{(using Eq. (4.8))}\,,$$

where we have indicated the length scale of gradients by l_0 and the time scale by t_0, so that $v \sim l_0/t_0$. Hence, the displacement current (Maxwell's great contribution to electrodynamics) is small, $\mathcal{O}(v^2/c^2)$, and can be removed again from Eq. (4.2), so that the current $\mathbf{j}$ may be expressed directly in terms of $\mathbf{B}$ and the original form of Ampère's law is recovered:

$$\mathbf{j} = \frac{1}{\mu_0}\nabla\times\mathbf{B}\,. \tag{4.10}$$

Furthermore, the non-relativistic approximation implies a remarkable simplification of the equation of motion (4.7) as well since the electrostatic acceleration,

$$\tau|\mathbf{E}| \sim \frac{v^2}{c^2}\frac{B^2}{\mu_0 l_0} \quad \ll \quad |\mathbf{j}\times\mathbf{B}| \sim \frac{B^2}{\mu_0 l_0} \qquad \text{(using Eqs. (4.3), (4.8) and (4.10))},$$

is also $\mathcal{O}(v^2/c^2)$. Consequently, space charge effects can be neglected, and Poisson's law (4.3) may be dropped since it is no longer needed. The electric field then becomes a secondary quantity, to be determined from Eq. (4.8):

$$\mathbf{E} = -\mathbf{v}\times\mathbf{B}\,. \tag{4.11}$$

This shows that, for non-relativistic MHD motions, the order of magnitude of the electric field as compared to the magnetic field is given by $|\mathbf{E}| \sim |\mathbf{v}||\mathbf{B}|$, i.e. an order $\mathcal{O}(v/c)$ smaller than for electromagnetic waves, where $|\mathbf{E}| \sim c|\mathbf{B}|$,

Exploiting the mentioned approximations and eliminating $\mathbf{E}$ and $\mathbf{j}$ from the equations by means of Eqs. (4.10) and (4.11), *the basic equations of ideal MHD* of Sections 2.4.1 and 3.4.2 are recovered:

$$\frac{\partial\rho}{\partial t} + \nabla\cdot(\rho\mathbf{v}) = 0\,, \tag{4.12}$$

$$\rho\Big(\frac{\partial\mathbf{v}}{\partial t} + \mathbf{v}\cdot\nabla\mathbf{v}\Big) + \nabla p - \rho\mathbf{g} - \frac{1}{\mu_0}(\nabla\times\mathbf{B})\times\mathbf{B} = 0\,, \tag{4.13}$$

$$\frac{\partial p}{\partial t} + \mathbf{v}\cdot\nabla p + \gamma p\nabla\cdot\mathbf{v} = 0\,, \tag{4.14}$$

$$\frac{\partial\mathbf{B}}{\partial t} - \nabla\times(\mathbf{v}\times\mathbf{B}) = 0\,, \qquad \nabla\cdot\mathbf{B} = 0\,. \tag{4.15}$$

This is a set of eight nonlinear partial differential equations for the eight variables $\rho(\mathbf{r},t)$, $\mathbf{v}(\mathbf{r},t)$, $p(\mathbf{r},t)$ and $\mathbf{B}(\mathbf{r},t)$. The magnetic field equation (4.15)(b) is to be considered as a condition: once satisfied, it remains satisfied for all later times by virtue of Eq. (4.15)(a). Depending on the physical problem considered, the value of the gravitational acceleration $\mathbf{g}(\mathbf{r},t)$ is either an externally fixed quantity or partially determined by the plasma itself. This aspect is discussed below.

For *static equilibria* ($\partial/\partial t = 0$ and $\mathbf{v} = 0$) in the absence of gravity ($\mathbf{g} = 0$), the MHD equations reduce to

$$\nabla p = \frac{1}{\mu_0}(\nabla\times\mathbf{B})\times\mathbf{B}\,, \qquad \nabla\cdot\mathbf{B} = 0\,. \tag{4.16}$$

Now, the magnetic field equation (4.16)(b) fully counts, in contrast to the dynamical problem, where it is just an initial condition. Thus, we appear to obtain four equations for the determination of $p(\mathbf{r})$ and the three components of $\mathbf{B}(\mathbf{r})$. However, we have already seen in Section 2.4.3, for the specific example of a z-pinch, that the equilibrium equations leave much more freedom in the determination of these quantities than this remark suggests. This is due to the symmetry that is usually assumed for equilibria.

(a) Thermodynamic variables The above formulation of the ideal MHD equations exploits ρ, $\mathbf{v}$, p, $\mathbf{B}$ as the basic variables. It is of interest to also work out the evolution equations for the *other thermodynamical variables*, which could replace ρ and p, viz.: e – the internal energy per unit mass

(which is equivalent to T – the temperature) and s – the entropy per unit mass. These are defined by the ideal gas relations, with $p = (n_\mathrm{e} + n_\mathrm{i})kT$:

$$\begin{aligned} e &\equiv \frac{1}{\gamma - 1}\frac{p}{\rho} \approx C_v T\,, \qquad C_v \approx \frac{(1+Z)k}{(\gamma-1)m_\mathrm{i}}\,, \\ s &\equiv C_v \ln S + \mathrm{const}\,, \qquad S \equiv p\rho^{-\gamma}\,, \end{aligned} \tag{4.17}$$

where m_i is the mass of the ions, k is the Boltzmann constant and $\gamma \equiv C_p/C_v$ is the ratio of specific heats at constant pressure and volume, respectively. The variable $S \equiv \rho^{-\gamma}p$ is a slightly more convenient measure for entropy than the variable s itself because it does not contain awkward constants any more.

Neglecting thermal conduction and heat flow, i.e. considering adiabatic processes, *the entropy convected by the fluid is constant*:

$$\frac{\mathrm{D}s}{\mathrm{D}t} = 0\,, \quad \text{or} \quad \frac{\mathrm{D}S}{\mathrm{D}t} \equiv \frac{\mathrm{D}}{\mathrm{D}t}(p\rho^{-\gamma}) = 0\,. \tag{4.18}$$

Eliminating the density ρ from the latter equation by means of Eq. (4.12) yields again the pressure evolution equation (4.14). This substantiates our claim that Eq. (4.14) actually expresses entropy conservation. On the other hand, we could also exploit the internal energy e (or T) as a basic variable. From Eqs. (4.12) and (4.14) we then have:

$$\frac{\mathrm{D}e}{\mathrm{D}t} + (\gamma - 1)\, e \nabla \cdot \mathbf{v} = 0\,. \tag{4.19}$$

Which pair of the four state variables ρ, p, s, e one chooses to complement the description by means of $\mathbf{v}$ and $\mathbf{B}$ is usually a matter of convenience. However, for the expression of conservation laws and certain symmetry properties this choice matters, as we will see later (Sections 4.3.1 and 5.2.1).

▷ **Exercise** Anticipating the discussion of the conservation properties in Section 4.3.1: why is the internal energy convected by the fluid not constant? ◁

(b) Gravitation In the equation of motion (4.13), the gravitational acceleration $\mathbf{g}$ due to the masses of the plasma within the region under consideration may be derived from an internal gravitational potential $\Phi_\mathrm{gr}^\mathrm{in}$. It satisfies the Poisson equation

$$\nabla^2 \Phi_\mathrm{gr}^\mathrm{in} = 4\pi G \rho(\mathbf{r})\,, \tag{4.20}$$

having the solution

$$\Phi_\mathrm{gr}^\mathrm{in}(\mathbf{r}) = -G \int \frac{\rho(\mathbf{r}')}{|\mathbf{r} - \mathbf{r}'|}\, d^3 r'\,, \tag{4.21}$$

so that

$$\mathbf{g}^\mathrm{in}(\mathbf{r}) = -\nabla \Phi_\mathrm{gr}^\mathrm{in}(\mathbf{r}) = -G \int \rho(\mathbf{r}') \frac{\mathbf{r} - \mathbf{r}'}{|\mathbf{r} - \mathbf{r}'|^3}\, d^3 r'\,. \tag{4.22}$$

To check the solution (4.21), use the property of the Dirac delta function in three dimensions,

$$\nabla^2 \left(\frac{1}{|\mathbf{r} - \mathbf{r}'|} \right) = -4\pi \delta(\mathbf{r} - \mathbf{r}')\,, \tag{4.23}$$

which is also exploited in the problem of electrostatics (see Jackson [331]).

In many astrophysical systems, the internal gravitational force $\mathbf{F}_g^{\rm in} \equiv \rho \mathbf{g}^{\rm in}$ is completely negligible compared to the Lorentz force $\mathbf{F}_B \equiv \mathbf{j} \times \mathbf{B}$, but also compared to the gravitational force $\mathbf{F}_g^{\rm ex} \equiv \rho \mathbf{g}^{\rm ex}$ due to an external compact object. We will represent such an external gravitational field by a point mass M_* situated at a position $\mathbf{r} = \mathbf{r}_*$ far outside the plasma region. In that case, we have a Poisson equation

$$\nabla^2 \Phi_{\rm gr} = 4\pi G \Big[M_* \delta(\mathbf{r} - \mathbf{r}_*) + \rho(\mathbf{r}) \Big] \,, \tag{4.24}$$

having the solution

$$\Phi_{\rm gr}(\mathbf{r}) = -G \frac{M_*}{|\mathbf{r} - \mathbf{r}_*|} + \Phi_{\rm gr}^{\rm in}(\mathbf{r}) \,, \tag{4.25}$$

so that

$$\mathbf{g}(\mathbf{r}) = -\nabla \Phi_{\rm gr}(\mathbf{r}) = -G M_* \frac{\mathbf{r} - \mathbf{r}_*}{|\mathbf{r} - \mathbf{r}_*|^3} + \mathbf{g}^{\rm in}(\mathbf{r}) \,, \tag{4.26}$$

where the internal field is given by Eqs. (4.21) and (4.22).

Let us now estimate the relative magnitudes of the Lorentz and the gravitational forces. For a tokamak, with typical parameters as given in Table B.4, we get:

$$\begin{aligned} |\mathbf{F}_B| &\equiv |\mathbf{j} \times \mathbf{B}| \sim B^2/(\mu_0 a) &&= 7.2 \times 10^6 \,{\rm kg\,m^{-2}\,s^{-2}} \,, \\ |\mathbf{F}_g^{\rm ex}| &\equiv |\rho \mathbf{g}^{\rm ex}| \sim \rho G\, M_*/R_*^2 &&= 1.7 \times 10^{-6} \,{\rm kg\,m^{-2}\,s^{-2}} \,, \\ |\mathbf{F}_g^{\rm in}| &\equiv |\rho \mathbf{g}^{\rm in}| \sim \rho^2 G a &&= 1.9 \times 10^{-24} \,{\rm kg\,m^{-2}\,s^{-2}} \,, \end{aligned} \tag{4.27}$$

where a is the width of the plasma tube and M_* and R_* here refer to the Earth ($GM_{\rm E}/R_{\rm E}^2 = 10\,{\rm m\,s^{-2}}$). Clearly, the gravitational contributions are completely negligible for tokamak plasmas.

▷ **Exercise** Compute the corresponding numbers for solar coronal flux tubes from Table B.4 and also for 'your favourite plasma' (filling out the last column of Table B.5) by finding the relevant orders of magnitude for the different quantities from whatever source you can find. What do you conclude from these numbers? ◁

Let us also estimate the relative magnitudes of the forces for a typical astrophysical plasma, viz. an accretion disk surrounding a compact object (see Frank, King and Raine [183], and Balbus and Hawley [30]). A schematic picture of the system is shown in Fig. 4.1. The compact object is located in the origin and matter is accreting in a thin rotating disk of size $\sim R_{\rm d}$ and height $\sim H_{\rm d}$ (at a distance $\sim 0.1 R_{\rm d}$ from the centre). The compact object carries it own magnetic field, which is mainly dipolar and the disk features a protruding poloidal magnetic field $B_{\rm p}$. One important aspect is the frequent appearance of powerful jets ejected from the centre in the direction perpendicular to the disk (along the Z-axis). To estimate forces, we exploit the expressions (4.27) with $a = H_{\rm d}$ but replacing R_* by a reasonable distance from the central object, viz. $R = 0.1 R_{\rm d}$. This object may be a black hole, but our gravitational equations refer to a Newtonian potential, neglecting general relativity, so that we have to stay far outside the Schwarzschild radius, i.e.

$$R \gg R_{\rm Schw} = 2GM_*/c^2 \,. \tag{4.28}$$

This condition is easily met by our choice of the distance since $R_{\rm Schw}$ is just a multiple of 3 km for every solar mass unit in the black hole.

Figure 4.1 may be used to illustrate astrophysical plasmas on entirely different scales. Let us consider the following two examples.

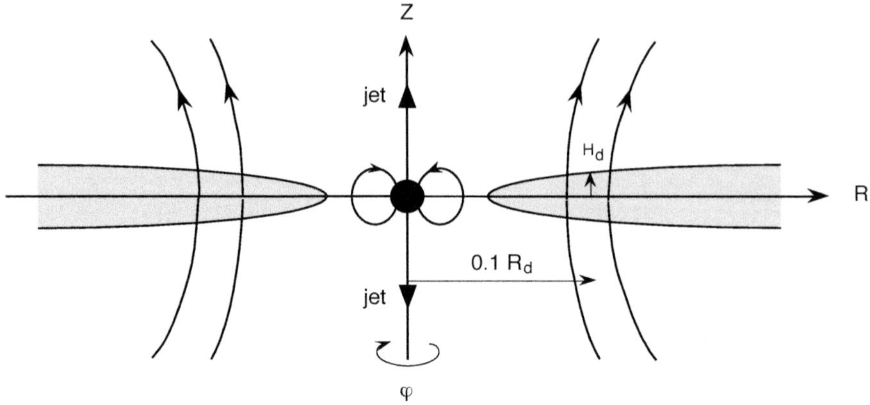

Fig. 4.1 Schematic representation of an accretion disc around a compact object with ejected jets.

(a) Accretion disc around a young stellar object (YSO), a protostellar cradle for stars and planets: $R_{\rm d} \sim 1\,{\rm AU}\ (= 1.5 \times 10^{11}\,{\rm m})$, $H_{\rm d} \sim 0.01\,{\rm AU}\ (= 1.5 \times 10^{9}\,{\rm m})$. The central object has a mass of the order of the solar mass, $M_* \sim 1 M_\odot\ (= 2.0 \times 10^{30}\,{\rm kg})$, and the accretion rate is of the order $\dot{M} \sim 10^{-7} M_\odot\,{\rm y}^{-1}\ (= 6 \times 10^{15}\,{\rm kg\,s^{-1}})$. Representative values of the other physical parameters are: $B = 10^{-4}\,{\rm T}\ (= 1\,{\rm G})$, $n = 10^{18}\,{\rm m}^{-3}$, $T = 10^4\,{\rm K}$, so that $\rho \approx 1.7 \times 10^{-9}\,{\rm kg\,m^{-3}}$ and $v_{\rm A} = 2.2 \times 10^3\,{\rm m\,s^{-1}}$. This gives the following estimate for the three forces:

$$|\mathbf{F}_B| = 5.3 \times 10^{-12}, \quad |\mathbf{F}_g^{\rm ex}| = 1.0 \times 10^{-9}, \quad |\mathbf{F}_g^{\rm in}| = 2.9 \times 10^{-19}\ [\,{\rm kg\,m^{-2}\,s^{-2}}]. \tag{4.29}$$

In this case, the contribution of the external gravitational field dominates.

(b) Accretion disc around an active galactic nucleus (AGN): $R_{\rm d} \sim 50\,{\rm kpc}\ (= 1.6 \times 10^5\,{\rm ly} = 1.5 \times 10^{21}\,{\rm m})$, $H_{\rm d} \sim 120\,{\rm pc}\ (= 3.3 \times 10^{18}\,{\rm m})$. The central object has a mass of the order of a hundred million solar masses, $M_* \sim 10^8 M_\odot\ (= 2.0 \times 10^{38}\,{\rm kg})$, and the accretion rate is of the order $\dot{M} \sim 0.1 M_\odot\,{\rm y}^{-1}\ (= 6 \times 10^{21}\,{\rm kg\,s^{-1}})$. Representative values of the other physical parameters are: $B = 10^{-4}\,{\rm T}\ (= 1\,{\rm G})$, $n = 10^{12}\,{\rm m}^{-3}$, $T = 10^8\,{\rm K}$, so that $\rho \approx 1.7 \times 10^{-15}\,{\rm kg\,m^{-3}}$ and $v_{\rm A} = 2.2 \times 10^6\,{\rm m\,s^{-1}}$. This gives:

$$|\mathbf{F}_B| = 2.2 \times 10^{-21}, \quad |\mathbf{F}_g^{\rm ex}| = 1.0 \times 10^{-27}, \quad |\mathbf{F}_g^{\rm in}| = 6.4 \times 10^{-22}\ [\,{\rm kg\,m^{-2}\,s^{-2}}]. \tag{4.30}$$

Now, even though there may be a huge black hole in the centre, the internal gravity field dominates over the external field *at the position chosen*.

▷ **Exercise** Complete the discussion of the different forces at work in the two types of accretion discs by also estimating the rotational and pressure terms. Do these estimates permit stationary equilibrium (i.e. solutions with $\partial/\partial t = 0$, but $\mathbf{v} \neq 0$)? What is the ratio of the plasma pressure p compared to the magnetic pressure $B^2/(2\mu_0)$? How is vertical equilibrium established? ◁

4.1.2 Scale independence

The MHD equations (4.12)–(4.15) can be made dimensionless by means of three quantities expressing a choice for the *units of length, mass and time*. For that purpose, a typical length scale

l_0 is chosen (e.g. the transverse length scale of the plasma), and values for the magnitude B_0 of the magnetic field and for the plasma density ρ_0 are chosen at some representative position (e.g. on the magnetic axis in a tokamak). The unit of time then follows by exploiting the basic speed of macroscopic plasma dynamics, viz. *the Alfvén speed* (Section 2.4.2):

$$v_0 \equiv v_{A,0} \equiv \frac{B_0}{\sqrt{\mu_0 \rho_0}} \quad \Rightarrow \quad t_0 \equiv \frac{l_0}{v_0} \,. \tag{4.31}$$

Thus, the relevant triplet of basic parameters becomes l_0, B_0, t_0.

By means of this triplet, and the derived quantities ρ_0 and v_0, we now create the dimensionless independent variables and their associated differential operators,

$$\bar{l} \equiv l/l_0 \,, \quad \bar{t} \equiv t/t_0 \quad \Rightarrow \quad \bar{\nabla} \equiv l_0 \nabla \,, \quad \partial/\partial\bar{t} \equiv t_0\, \partial/\partial t \,, \tag{4.32}$$

and the dimensionless dependent variables,

$$\bar{\rho} \equiv \rho/\rho_0 \,, \quad \bar{\mathbf{v}} \equiv \mathbf{v}/v_0 \,, \quad \bar{p} \equiv p/(\rho_0 v_0^2) \,, \quad \bar{\mathbf{B}} \equiv \mathbf{B}/B_0 \,, \quad \bar{\mathbf{g}} \equiv (l_0/v_0^2)\, \mathbf{g} \,. \tag{4.33}$$

With these transformations, the equations (4.12)–(4.15) remain unchanged, except that all quantities are equipped with a bar and that the awkward quantity μ_0 (associated with the mks system of units) disappears. Obviously, we will drop the bars again and just enjoy the fact that dimensions need no longer worry us.

Table 4.1 *Scales of different plasmas.*

	l_0 (m)	B_0 (T)	t_0 (s)
tokamak	20	3	3×10^{-6}
magnetosphere Earth	4×10^{7}	3×10^{-5}	6
solar coronal loop	10^{8}	3×10^{-2}	15
magnetosphere neutron star	10^{6}	10^{8} *	10^{-2}
accretion disc YSO	1.5×10^{9}	10^{-4}	7×10^{5}
accretion disc AGN	4×10^{18}	10^{-4}	2×10^{12}
galactic plasma	10^{21} ($= 10^{5}$ ly)	10^{-8}	10^{15} ($= 3 \times 10^{7}$ y)

* Some recently discovered pulsars, called magnetars, have record magnetic fields of 10^{11} T : the plasma Universe is ever expanding!

The important result of the exercise is the realization that the equations do not depend on the size of the plasma (l_0), on the magnitude of the magnetic field (B_0) and on the density (ρ_0), i.e. on the time scale (t_0): *the ideal MHD equations are scale independent* with respect to changes of these quantities. Notwithstanding the huge differences in magnitude of the parameters l_0, B_0 and t_0 encountered in nature and in the laboratory (see Table 4.1), yet the same equations of magnetohydrodynamics apply! This provides the basis for the description of macroscopic dynamics of

90% of matter in the Universe and, hence, for effective cross-fertilization between laboratory and astrophysical plasma physics (Section 1.1).

After the scaling with l_0, B_0, t_0, the value $\bar{p}_0$ of the dimensionless pressure at the reference point automatically becomes a quantity of intrinsic importance. It is directly related to the ratio of the plasma kinetic pressure to the magnetic pressure, commonly indicated by the symbol β:

$$\beta \equiv \frac{2\mu_0 p_0}{B_0^2} = 2\bar{p}_0 \,. \tag{4.34}$$

Since $\beta \ll 1$ for many plasmas of interest, pressure terms, labelled by the parameter β, are frequently neglected or only computed as a higher order correction.

▷ **Exercise** Compute the value of β for 'your favourite plasma' from the numbers you have collected in the last column of Table B.5 and compare it with the other plasmas. Can you now estimate the relative importance of the different terms in the momentum equation? ◁

4.1.3 A crucial question

Having obtained the complete set of partial differential equations (4.12)–(4.15) for the plasma variables, and realizing their huge potential from the previous section, we now ask a crucial question: do these equations provide us with *a complete model* for plasma dynamics?

This question is to be answered with an emphatic **no!** The reason is that the two most essential elements of a scientific model are still missing, viz.

(a) What is actually the physical problem we want to solve?

(b) How is this problem mathematically translated in conditions to be imposed on the solutions of the partial differential equations?

For example, one might want to study the stability of a laboratory fusion plasma or the evolution of an expanding flux tube in the solar corona. This brings in the specific space and time constraints embodied in the *boundary conditions* and *initial data*. The consideration of initial data will be our concern in the next chapter (Section 5.2). For now, this part is nearly trivial since it just amounts to prescribing arbitrary functions

$$\rho_i(\mathbf{r}) \equiv \rho(\mathbf{r}, t\!=\!0)\,, \qquad \mathbf{v}_i(\mathbf{r}) \equiv \mathbf{v}(\mathbf{r}, t\!=\!0)\,,$$
$$p_i(\mathbf{r}) \equiv p(\mathbf{r}, t\!=\!0)\,, \qquad \mathbf{B}_i(\mathbf{r}) \equiv \mathbf{B}(\mathbf{r}, t\!=\!0)\,, \tag{4.35}$$

on the domain of interest. There is no restriction on the initial data permitted other than $\nabla\cdot\mathbf{B}_i(\mathbf{r}) = 0$. However, the consideration of the appropriate boundary conditions is a much more involved one since it implies *the specification of a geometry* associated with a particular *magnetic confinement* scheme.

We will extensively dwell on the specific boundary conditions for the different magnetic confinement geometries encountered in the laboratory and nature in Section 4.6. Before we can effectively do that, we first have to introduce the central concepts of magnetohydrodynamics underlying it all, viz. magnetic flux tubes and flux conservation (Section 4.2), the general conservation properties of the MHD equations (Sections 4.3 and 4.4), and the discontinuities permitted (Section 4.5).

4.2 Magnetic flux

4.2.1 Flux tubes

Magnetic flux tubes occur in different kinds, e.g. closed onto themselves, like in thermonuclear *tokamak* confinement machines, or connecting onto a medium of so vastly different physical characteristics that one may consider the flux tube to be separated from the other medium by suitable jump conditions. The latter kind is the appropriate model for *coronal flux tubes*. These two generic plasma configurations are shown schematically in Fig. 4.2. The main point of this illustration is to show the global geometry, which is superficially similar but essentially different.

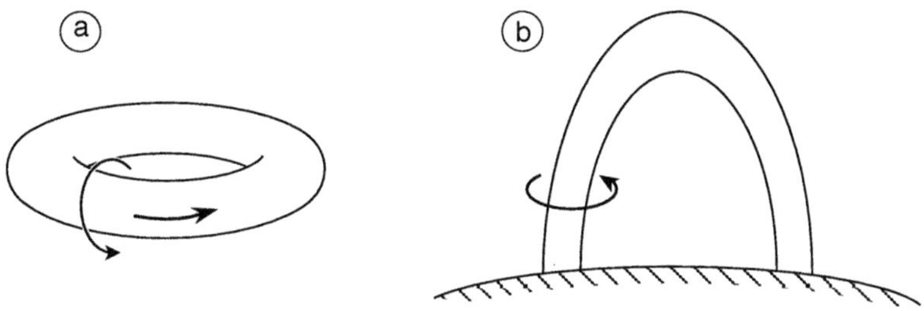

Fig. 4.2 Generic plasma confinement structures: (a) tokamak and (b) coronal magnetic loop (arrows indicate periodic directions).

▷ **Exercise** It is frequently said that solar coronal flux loops that are 'standing' for weeks on the Sun appear to mock the stability problems encountered in laboratory plasmas. Consult Table 4.1 and compare a tokamak with a plasma that is routinely confined for a minute (a typical operational feature at present) with a coronal loop that stands for a week on the Sun before erupting. Is the statement true or false? (Hint: stability is to be measured in terms of the number of Alfvén wave crossing times of the configuration.) ◁

The tokamak basically consists of closed and nested magnetic surfaces which have a toroidal shape, i.e. they are periodic in two directions (poloidal and toroidal). The typical solar coronal magnetic loop, on the other hand, is a finite flux tube bound by the photosphere, which is considered to be infinitely inert, so that there is no motion at the ends (so-called line-tying). Whereas the same MHD equations may be used to describe the macroscopic plasma dynamics in both configurations, *the boundary conditions* are quite different and, consequently, the exact form of the resulting dynamics will also be different. We will return to this point in Section 4.6.

The above examples demonstrate that magnetic fields confining plasmas, man made as well as naturally occurring, are basically *tubular structures*. This is the result of the magnetic field equation (4.15)(b), repeated here for convenience,

$$\nabla \cdot \mathbf{B} = 0 , \tag{4.36}$$

which does not permit, for example, spherically symmetric solutions. Instead, magnetic flux tubes become the essential constituents of a magnetically confined plasma configuration. Following Newcomb [460, 464], we consider a surface element $d\sigma_1$ and all magnetic field lines puncturing it (Fig. 4.3(a)). The magnetic flux through an arbitrary other surface element $d\sigma_2$ intersecting that

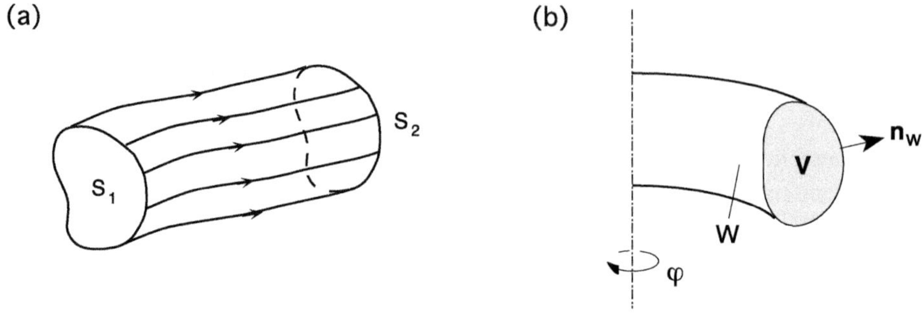

Fig. 4.3 (a) Magnetic flux tube: all field lines through S_1 also cross S_2. (b) Tokamak geometry.

field line bundle is the same. This follows from the application of Gauss' theorem (A.14):

$$\iiint_V \nabla\cdot\mathbf{B}\,d\tau = \oiint \mathbf{B}\cdot\mathbf{n}\,d\sigma = -\iint_{S_1}\mathbf{B}_1\cdot\mathbf{n}_1\,d\sigma_1 + \iint_{S_2}\mathbf{B}_2\cdot\mathbf{n}_2\,d\sigma_2 = 0\,,$$

where $\mathbf{n}_1$ and $\mathbf{n}_2$ are the unit normals to the surfaces S_1 and S_2. Hence, the magnetic flux

$$\Psi \equiv \iint_S \mathbf{B}\cdot\mathbf{n}\,d\sigma \tag{4.37}$$

through an arbitrary cross-section of the magnetic flux tube is a well-defined quantity, i.e. it does not depend on how S is taken. This holds for the flux tube as a whole (like the ones shown in Fig. 4.2), but also for the infinity of smaller flux tubes that one can imagine to be obtained by subdividing the cross-section S. Obviously, the important question to be asked next is: how do these flux tubes move when the plasma moves?

4.2.2 Global magnetic flux conservation

Whereas the concept of flux tube comes from the electromagnetic field equation (4.36), the dynamics of flux tubes requires the consideration of the two other electromagnetic field equations (4.1) and (4.8) absorbed in the induction equation (4.15)(a), which is repeated again for convenience:

$$\frac{\partial\mathbf{B}}{\partial t} = \nabla\times(\mathbf{v}\times\mathbf{B})\,. \tag{4.38}$$

Exploiting this equation, one should always remember that it is really a contraction of Faraday's law, $\partial\mathbf{B}/\partial t = -\nabla\times\mathbf{E}$, describing the time dependence of the magnetic field, and 'Ohm's' law with perfect conductivity, $\mathbf{E}+\mathbf{v}\times\mathbf{B} = 0$, relating the electric field to the plasma flow. These two aspects also guide the following analysis.

Consider the complete magnetic flux tube inside the toroidal vessel of a tokamak (Fig. 4.3(b)). The toroidal plasma volume is indicated by the letter V, the surrounding conducting wall by W and the normal vector on the wall by $\mathbf{n}_w$. We will investigate the implications of the electromagnetic field equations for the dynamics of that flux tube inside the vessel.

First of all, we need *boundary conditions (BCs) at the wall.* An obvious one is the requirement that the wall is 'perfect', i.e. it neither absorbs plasma nor emits gas (impurities). This translates into the boundary condition

$$\mathbf{n}_\mathrm{w} \cdot \mathbf{v} = 0 \qquad \text{(on W)}\,. \tag{4.39}$$

A second boundary condition is the requirement that the wall acts as a perfect conductor, i.e. it short-circuits tangential electric fields $\mathbf{E}_\mathrm{t}$. Interestingly enough, this condition is not a restriction on the kind of materials a wall could be made of, but rather on the resistivity of the plasma itself: if the wall is an isolator, a thin perfectly conducting plasma layer in front of the wall serves the same purpose of short-circuiting the electric field. Hence, 'Ohm's' law applied to this layer yields:

$$\mathbf{n}_\mathrm{w} \times \left[\, \mathbf{E} + (\mathbf{v} \times \mathbf{B}) \,\right] \stackrel{(A.2)}{=} \mathbf{n}_\mathrm{w} \times \mathbf{E}_\mathrm{t} + \mathbf{n}_\mathrm{w} \cdot \mathbf{B}\, \mathbf{v} - \mathbf{n}_\mathrm{w} \cdot \mathbf{v}\, \mathbf{B} = 0\,. \tag{4.40}$$

Since the tangential electric field vanishes, $\mathbf{E}_\mathrm{t} = 0$, and there is no flow across the wall, $\mathbf{n}_\mathrm{w} \cdot \mathbf{v} = 0$, the other contribution has to vanish as well:

$$\mathbf{n}_\mathrm{w} \cdot \mathbf{B} = 0 \qquad \text{(on W)}\,. \tag{4.41}$$

Hence, the magnetic field lines do not intersect the wall so that this boundary condition prevents plasma flowing along the field lines hitting the wall and being lost.

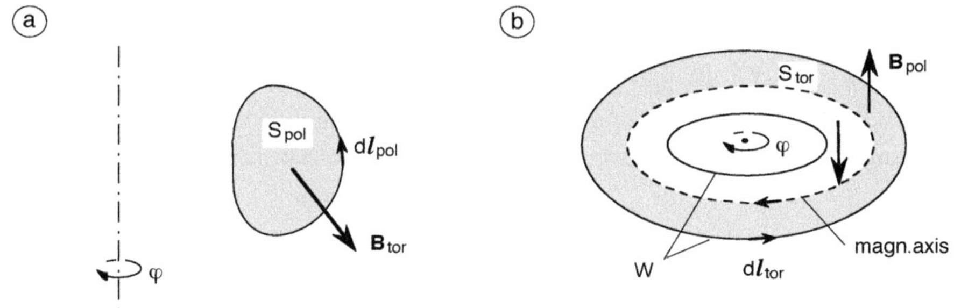

Fig. 4.4 Surfaces in tokamak geometry for: (a) toroidal magnetic flux, (b) poloidal magnetic flux.

Let us now study the consequences of the magnetic field evolution equation (4.38) for the total magnetic flux inside a tokamak (Fig. 4.4). Since there are two magnetic field components (poloidal and toroidal), there are also two magnetic fluxes, associated with the two different surfaces obtained by a cut along or across the torus. The toroidal component is the simplest. Starting from Eq. (4.37), with $\mathrm{S} \equiv \mathrm{S}_\mathrm{pol}$ indicating a poloidal cross-section of the torus, we find by the application of Stokes' law that the time derivative of the total toroidal flux vanishes:

$$\frac{\partial \Psi_\mathrm{tor}}{\partial t} \equiv \iint_{S_\mathrm{pol}} \frac{\partial \mathbf{B}_\mathrm{tor}}{\partial t} \cdot \mathbf{n}_\mathrm{tor}\, d\sigma = \iint \nabla \times (\mathbf{v} \times \mathbf{B}_\mathrm{tor}) \cdot \mathbf{n}_\mathrm{tor}\, d\sigma$$

$$\stackrel{(A.18)}{=} \oint \mathbf{v} \times \mathbf{B}_\mathrm{tor} \cdot d\boldsymbol{l}_\mathrm{pol} = 0\,, \tag{4.42}$$

because $\mathbf{v}$, $\mathbf{B}_\mathrm{tor}$, and $\boldsymbol{l}_\mathrm{pol}$ are tangential to the wall. In other words, the two boundary conditions (4.39) and (4.41) suffice to guarantee that the total toroidal magnetic flux inside the tokamak is conserved: $\Psi_\mathrm{tor} = \text{const.}$

A similar story holds for the poloidal flux:

$$\frac{\partial \Psi_{\rm pol}}{\partial t} \equiv \iint_{S_{\rm tor}} \frac{\partial \mathbf{B}_{\rm pol}}{\partial t} \cdot \mathbf{n}_{\rm pol}\, d\sigma = 0 \quad \Rightarrow \quad \Psi_{\rm pol} = {\rm const}\,, \tag{4.43}$$

where we exploit a surface $S_{\rm tor}$ bounded by a toroidal circle lying in the wall and by the magnetic axis. For the time being, this proof requires that the magnetic axis should be kept fixed. Relaxing the latter constraint requires the consideration of the local version of magnetic flux conservation. This is a slightly more complicated, but more powerful, property that will be dealt with in Section 4.3.3. Anticipating the outcome of that analysis, viz. that the magnetic flux through any surface moving with the plasma is constant, it becomes clear that *magnetic flux conservation is the central issue in magnetohydrodynamics*. To put this into proper perspective, we now turn to a systematic exposition of the conservation properties of the MHD equations.

4.3 Conservation laws

4.3.1 Conservation form of the MHD equations

We have already stated several times that the MHD equations express conservation of the main macroscopic quantities of a plasma, viz. mass, momentum, energy and magnetic flux. We will now substantiate this claim.

A system of quasi-linear partial differential equations is said to be in conservation form if all terms can be written as a generalized divergence (incorporating the time and space derivatives on an equal footing) of the dependent variables, or simple functions of them:

$$\frac{\partial}{\partial t}(\cdots) + \nabla \cdot (\cdots) = 0\,. \tag{4.44}$$

The use of such a form of the equations is that one can obtain *local and global conservation laws* and *jump conditions* from them. Moreover, powerful numerical algorithms exist for the solution of such equations. We will return to this point in Chapter 20 on computational nonlinear MHD.

Consider again the set of nonlinear ideal MHD equations discussed in Section 4.1, exploiting *the internal energy* e rather than the pressure p as a variable, according to Eq. (4.19), and dropping the constant μ_0 for convenience:

$$\frac{\partial \rho}{\partial t} + \nabla \cdot (\rho \mathbf{v}) = 0\,, \tag{4.45}$$

$$\rho \frac{\partial \mathbf{v}}{\partial t} + \rho \mathbf{v} \cdot \nabla \mathbf{v} + \nabla p - \mathbf{j} \times \mathbf{B} = -\rho \nabla \Phi_{\rm gr}\,, \quad p = (\gamma - 1)\rho e\,, \quad \mathbf{j} = \nabla \times \mathbf{B}\,, \tag{4.46}$$

$$\frac{\partial e}{\partial t} + \mathbf{v} \cdot \nabla e + (\gamma - 1) e\, \nabla \cdot \mathbf{v} = 0\,, \tag{4.47}$$

$$\frac{\partial \mathbf{B}}{\partial t} + \nabla \times \mathbf{E} = 0\,, \qquad \mathbf{E} = -\mathbf{v} \times \mathbf{B}\,, \qquad \nabla \cdot \mathbf{B} = 0\,. \tag{4.48}$$

(From now on, the constant μ_0 will be omitted consistently. To restore mks units one should replace: $\mathbf{B} \rightarrow \mathbf{B}/\sqrt{\mu_0}\,, \mathbf{E} \rightarrow \mathbf{E}/\sqrt{\mu_0}\,, \mathbf{j} \rightarrow \sqrt{\mu_0}\,\mathbf{j}$ in the formulas.) Evidently, only the mass conservation equation (4.45) and the magnetic field equation (4.48)(c) have the required conservation form. We only consider the effect of an *external* gravitational potential $\Phi_{\rm gr}$ in the momentum equation (4.46), where $\mathbf{g} = -\nabla\Phi_{\rm gr}$, since it demonstrates how such a field spoils the strict conservation property of the equations. This is why this term is put on the right hand side of the equality sign.

The following text, in small print, may be skipped because we will just exploit the end result, viz. the conservation form of the ideal MHD equations summarized in Eqs. (4.59)–(4.62).

▷ **Transformation to conservation form** The mass conservation form is repeated in Eq. (4.59) below. In order to bring the other equations into conservation form one makes use of the following vector identities (where equation numbers above the equal signs refer to auxiliary equations like those of Appendix A):

$$\nabla\cdot(\mathbf{ab}) \overset{(A.11)}{=} \mathbf{a}\cdot\nabla\mathbf{b} + \mathbf{b}\,\nabla\cdot\mathbf{a}\,, \tag{4.49}$$

$$\mathbf{a}\times(\nabla\times\mathbf{b}) \overset{(A.8),(A.11)}{=} (\nabla\mathbf{b})\cdot\mathbf{a} - \nabla\cdot(\mathbf{ab}) - \mathbf{b}\,\nabla\cdot\mathbf{a}\,, \tag{4.50}$$

$$\nabla\times(\mathbf{a}\times\mathbf{b}) \overset{(A.13)(a)}{=} \nabla\cdot(\mathbf{ba}-\mathbf{ab})\,, \tag{4.51}$$

$$\nabla(\mathbf{a}\cdot\mathbf{b}) \overset{(A.10)(a)}{=} (\nabla\mathbf{a})\cdot\mathbf{b} + (\nabla\mathbf{b})\cdot\mathbf{a}\,. \tag{4.52}$$

The first two terms of the momentum equation (4.46) may then be transformed to

$$\rho\frac{\partial\mathbf{v}}{\partial t} + \rho\mathbf{v}\cdot\nabla\mathbf{v} \overset{(4.45)}{=} \frac{\partial}{\partial t}(\rho\mathbf{v}) + \mathbf{v}\nabla\cdot(\rho\mathbf{v}) + \rho\mathbf{v}\cdot\nabla\mathbf{v} \overset{(4.49)}{=} \frac{\partial}{\partial t}(\rho\mathbf{v}) + \nabla\cdot(\rho\mathbf{vv})\,, \tag{4.53}$$

and the last term to

$$-\mathbf{j}\times\mathbf{B} = \mathbf{B}\times(\nabla\times\mathbf{B}) \overset{(4.50),(4.52)}{=} \nabla(\tfrac{1}{2}B^2) - \nabla\cdot(\mathbf{BB})\,, \tag{4.54}$$

so that we obtain the conservation form of the momentum equation, Eq. (4.60) below. Similarly, the second term of Faraday's law (4.48) becomes

$$\nabla\times\mathbf{E} = -\nabla\times(\mathbf{v}\times\mathbf{B}) \overset{(4.51)}{=} \nabla\cdot(\mathbf{vB}-\mathbf{Bv})\,, \tag{4.55}$$

so that we obtain the conservation form of the magnetic flux equation, Eq. (4.62) below.

Finally, the internal energy equation (4.47) cannot be brought into conservation form for the obvious reason that it contains only part of the energy, which can be converted into other forms of energy. We therefore need a conservation equation for the total energy density. This is obtained by adding the separate contributions of the kinetic, the internal and the magnetic energy:

$$\begin{aligned}
\mathbf{v}\cdot\text{Eq. (4.46)} \quad &\Rightarrow \quad \rho\mathbf{v}\cdot(\frac{\partial\mathbf{v}}{\partial t} + \mathbf{v}\cdot\nabla\mathbf{v}) + \mathbf{v}\cdot\nabla p - \mathbf{v}\cdot\mathbf{j}\times\mathbf{B} = -\rho\mathbf{v}\cdot\nabla\Phi_{\rm gr}\\
&\longrightarrow \quad \frac{\partial}{\partial t}(\tfrac{1}{2}\rho v^2) - \tfrac{1}{2}v^2\frac{\partial\rho}{\partial t} + \tfrac{1}{2}\rho\mathbf{v}\cdot\nabla v^2 + \mathbf{v}\cdot\nabla p - \mathbf{v}\cdot\mathbf{j}\times\mathbf{B} = -\rho\mathbf{v}\cdot\nabla\Phi_{\rm gr}\\
&\overset{(4.45)}{\longrightarrow} \quad \frac{\partial}{\partial t}(\tfrac{1}{2}\rho v^2) + \nabla\cdot(\tfrac{1}{2}\rho v^2\mathbf{v}) + \mathbf{v}\cdot\nabla p - \mathbf{v}\cdot\mathbf{j}\times\mathbf{B} = -\rho\mathbf{v}\cdot\nabla\Phi_{\rm gr}\,,
\end{aligned} \tag{4.56}$$

$$\begin{aligned}
\rho\;\text{Eq. (4.47)} \quad &\Rightarrow \quad \rho\frac{\partial e}{\partial t} + \rho\mathbf{v}\cdot\nabla e + (\gamma-1)\rho e\,\nabla\cdot\mathbf{v} = 0\\
&\longrightarrow \quad \frac{\partial}{\partial t}(\rho e) - e\frac{\partial\rho}{\partial t} + \rho\mathbf{v}\cdot\nabla e + p\nabla\cdot\mathbf{v} = 0\\
&\overset{(4.45)}{\longrightarrow} \quad \frac{\partial}{\partial t}(\rho e) + \nabla\cdot(\rho e\mathbf{v}) + p\nabla\cdot\mathbf{v} = 0\,,
\end{aligned} \tag{4.57}$$

$$\begin{aligned}
\mathbf{B}\cdot\text{Eq. (4.48)} \quad &\Rightarrow \quad \mathbf{B}\cdot\frac{\partial\mathbf{B}}{\partial t} - \mathbf{B}\cdot\nabla\times(\mathbf{v}\times\mathbf{B}) = 0\\
&\overset{(A.12)}{\longrightarrow} \quad \frac{\partial}{\partial t}(\tfrac{1}{2}B^2) + \nabla\cdot[\mathbf{B}\times(\mathbf{v}\times\mathbf{B})] - (\mathbf{v}\times\mathbf{B})\cdot\nabla\times\mathbf{B} = 0\\
&\overset{(A.2)}{\longrightarrow} \quad \frac{\partial}{\partial t}(\tfrac{1}{2}B^2) + \nabla\cdot[\mathbf{B}\cdot\mathbf{B}\,\mathbf{v} - \mathbf{v}\cdot\mathbf{B}\mathbf{B}] + \mathbf{v}\cdot\mathbf{j}\times\mathbf{B} = 0\,.
\end{aligned} \tag{4.58}$$

Eqs. (4.56) + (4.57) + (4.58) yield the energy conservation form (4.61) below. ◁

Recapitulating, *the conservation form of the ideal MHD equations* reads:

$$\frac{\partial \rho}{\partial t} + \nabla \cdot (\rho \mathbf{v}) = 0\,, \tag{4.59}$$

$$\frac{\partial}{\partial t}(\rho \mathbf{v}) + \nabla \cdot \Big[\rho \mathbf{v}\mathbf{v} + (p + \tfrac{1}{2}B^2)\,\mathbf{I} - \mathbf{B}\mathbf{B} \Big] = -\rho \nabla \Phi_{\rm gr}\,, \tag{4.60}$$

$$\frac{\partial}{\partial t}(\tfrac{1}{2}\rho v^2 + \rho e + \tfrac{1}{2}B^2) + \nabla \cdot \Big[(\tfrac{1}{2}\rho v^2 + \gamma \rho e + B^2)\mathbf{v} - \mathbf{v}\cdot\mathbf{B}\mathbf{B} \Big] = -\rho \mathbf{v} \cdot \nabla \Phi_{\rm gr}\,, \tag{4.61}$$

$$\frac{\partial \mathbf{B}}{\partial t} + \nabla \cdot (\mathbf{v}\mathbf{B} - \mathbf{B}\mathbf{v}) = 0\,, \qquad \nabla \cdot \mathbf{B} = 0\,, \tag{4.62}$$

where $p = (\gamma - 1)\rho e$. Note that the energy conservation form (4.61) for the total energy, which replaces the evolution equation (4.47) for the internal energy e, required the most extensive transformation since e is not conserved by itself.

From the preceding analysis we conclude that the best representation of the evolution equations is in terms of the variables ρ, $\mathbf{v}$, e, and $\mathbf{B}$, as expressed by the Eqs. (4.45)–(4.48). A peculiar additional variable is *the specific entropy* s (the entropy per unit mass), introduced in Section 4.1.1. For adiabatic processes of ideal gases, which is applicable here, we have

$$S \equiv p\rho^{-\gamma} = f(s)\,, \quad \text{or} \;\; s = C_v \ln (p\rho^{-\gamma}) + \text{const}\,. \tag{4.63}$$

Hence, from Eqs. (4.45) and (4.47),

$$\frac{\mathrm{D}S}{\mathrm{D}t} \equiv \frac{\partial S}{\partial t} + \mathbf{v} \cdot \nabla S = 0\,, \tag{4.64}$$

which is not in conservation form, but expresses the conservation of specific entropy co-moving with the fluid. A genuine conservation form is obtained by transforming to the variable ρS, associated with *the entropy per unit volume*. In that variable we get

$$\frac{\partial}{\partial t}(\rho S) + \nabla \cdot (\rho S \mathbf{v}) = 0\,, \tag{4.65}$$

which is in conservation form now so that one may derive the global conservation law for the entropy of the whole system from it.

4.3.2 Global conservation laws

To understand the physical meaning of the different terms in the conservation equations (4.59)–(4.62) of ideal MHD, we define the following quantities:

– *momentum density* $\quad \boldsymbol{\pi} \equiv \rho \mathbf{v}\,,$ (4.66)

– *stress tensor* $\quad \mathbf{T} \equiv \rho \mathbf{v}\mathbf{v} + (p + \tfrac{1}{2}B^2)\,\mathbf{I} - \mathbf{B}\mathbf{B}\,,$ (4.67)

– *total energy density* $\quad \mathcal{H} \equiv \tfrac{1}{2}\rho v^2 + \dfrac{p}{\gamma - 1} + \tfrac{1}{2}B^2\,,$ (4.68)

– *energy flow* $\quad \mathbf{U} \equiv \Big(\tfrac{1}{2}\rho v^2 + \dfrac{\gamma}{\gamma - 1} p\Big)\mathbf{v} + B^2\mathbf{v} - \mathbf{v}\cdot\mathbf{B}\,\mathbf{B}\,,$ (4.69)

– (no name) $\quad \mathbf{Y} \equiv \mathbf{v}\mathbf{B} - \mathbf{B}\mathbf{v}\,.$ (4.70)

On purely formal grounds, we have introduced the tensor $\mathbf{Y}$ without giving it a name because it appears to have no direct intuitive meaning. We exploit it in the flux equation (4.74) below since one obtains the jump conditions most easily from it (see Section 4.5).

Exploiting these quantities and neglecting gravity, the conservation equations (4.59)–(4.62) may be written as evolution equations for the variables $\rho, \boldsymbol{\pi}, \mathcal{H}$, and $\mathbf{B}$ in conservation form:

$$\frac{\partial \rho}{\partial t} + \nabla \cdot \boldsymbol{\pi} = 0 \qquad \textit{(conservation of mass)}, \tag{4.71}$$

$$\frac{\partial \boldsymbol{\pi}}{\partial t} + \nabla \cdot \mathbf{T} = 0 \qquad \textit{(conservation of momentum)}, \tag{4.72}$$

$$\frac{\partial \mathcal{H}}{\partial t} + \nabla \cdot \mathbf{U} = 0 \qquad \textit{(conservation of energy)}, \tag{4.73}$$

$$\frac{\partial \mathbf{B}}{\partial t} + \nabla \cdot \mathbf{Y} = 0 \qquad \textit{(conservation of magnetic flux)}. \tag{4.74}$$

Note that the quantities appearing in the divergence terms can all be expressed in terms of these four variables so that they constitute yet another basic set of variables to describe ideal MHD.

▷ **Computation of primitive variables from conserved ones** The conservation equations (4.71)–(4.74) are used in many numerical schemes to compute the time-advance of the conserved variables $\rho, \boldsymbol{\pi}, \mathcal{H}, \mathbf{B}$. The calculation of the primitive (original) variables $\rho, \mathbf{v}, e$ or $p, \mathbf{B}$ from them is straightforward. However, calculation of the pressure,

$$p = (\gamma - 1)\big[\mathcal{H} - \tfrac{1}{2}(\pi^2/\rho + B^2)\big] , \tag{4.75}$$

presents a particular numerical problem. It involves subtraction of the sum of kinetic energy $\frac{1}{2}\pi^2/\rho$ and magnetic energy $\frac{1}{2}B^2$ from the total energy $\mathcal{H}$, which may not result in a positive numerical value for p. This requires care in the construction of numerical schemes since $\beta \equiv 2p/B^2 \ll 1$ for many plasmas of interest. ◁

The stress tensor $\mathbf{T}$ consists of the *Reynolds stress tensor* $\rho\mathbf{vv}$, the *isotropic pressure* $p\,\mathbf{I}$ and the *magnetic part* $\frac{1}{2}B^2\,\mathbf{I} - \mathbf{BB}$ *of the Maxwell stress tensor*. Projecting onto the velocity $\mathbf{v}$, the only non-vanishing contribution to the Reynolds stress is a positive stress ('pressure') ρv^2 along $\mathbf{v}$. The remaining part is more clearly represented in a projection based on the magnetic field $\mathbf{B}$:

$$\begin{pmatrix} p + \frac{1}{2}B^2 & 0 & 0 \\ 0 & p + \frac{1}{2}B^2 & 0 \\ 0 & 0 & p - \frac{1}{2}B^2 \end{pmatrix} \quad \begin{matrix} \perp \\ \perp \\ \parallel . \end{matrix}$$

Hence, the magnetic field provides positive stress, *magnetic pressure*, in directions perpendicular to $\mathbf{B}$ and negative stress, *magnetic tension*, parallel to $\mathbf{B}$ (see Fig. 4.5).

The different terms of the total energy density $\mathcal{H}$ may be grouped into two parts:

$$\mathcal{H} = \mathcal{K} + \mathcal{W} , \tag{4.76}$$

where $\mathcal{K}$ is *the kinetic energy density*,

$$\mathcal{K} \equiv \tfrac{1}{2}\rho v^2 , \tag{4.77}$$

and $\mathcal{W}$ is *the potential energy density*,

$$\mathcal{W} \equiv \rho e + \tfrac{1}{2}B^2 = \frac{p}{\gamma - 1} + \tfrac{1}{2}B^2 . \tag{4.78}$$

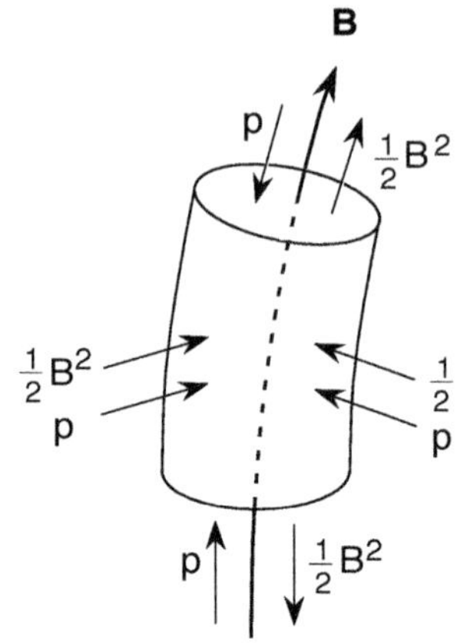

Fig. 4.5 Magnetic stress and tension in a flux tube. (Except for **B**, arrows indicate pressure increase/decrease when pointing towards/away from the tube tube.)

In many plasmas of interest, the magnetic energy density $\frac{1}{2}B^2$ represents a huge energy reservoir that may be released suddenly, e.g. in violent disruptions in tokamaks, in solar flares, in coronal mass ejections, and in many other explosive events in astrophysics. The energy flow vector **U** is again composed of a hydrodynamic part (the term with the bracket in Eq. (4.69)) and a magnetic part. The latter part may be transformed into the usual *Poynting vector*:

$$\mathbf{U}_{\rm magn} \equiv B^2\mathbf{v} - \mathbf{v}\cdot\mathbf{B}\,\mathbf{B} \overset{(A.2)}{=} -(\mathbf{v}\times\mathbf{B})\times\mathbf{B} \overset{(4.48)}{=} \mathbf{E}\times\mathbf{B} \equiv \mathbf{S}\,, \tag{4.79}$$

which represents the flow of electromagnetic energy.

Whereas global conservation laws for mass, momentum and energy are obtained by the application of *Gauss' theorem* on the equations for ρ, $\boldsymbol{\pi}$ and $\mathcal{H}$, to get a global conservation law for the magnetic flux one should apply *Stokes' theorem*. For that reason, the previously exploited form of Faraday's law (4.15), with the curl operator, is far to be preferred over that of Eq. (4.74), with the divergence.

Consider now a plasma surrounded by a perfectly conducting wall so that both $\mathbf{v}\cdot\mathbf{n} = 0$ and $\mathbf{n}\cdot\mathbf{B} = 0$ at the wall (which are the BCs (4.39) and (4.41) derived in Section 4.2.2). Define the following quantities:

– *total mass* $$M \equiv \int \rho\, d\tau\,, \tag{4.80}$$

– *total momentum* $$\boldsymbol{\Pi} \equiv \int \boldsymbol{\pi}\, d\tau\,, \tag{4.81}$$

– *total energy* $$H \equiv \int \mathcal{H}\, d\tau\,, \tag{4.82}$$

– *total magnetic flux* $$\Psi \equiv \int \mathbf{B}\cdot\tilde{\mathbf{n}}\, d\tilde{\sigma}\,, \tag{4.83}$$

where we now abbreviate the previously used triple and double integral signs for volume and surface integration by just a single integral. In the definitions of the total mass, momentum and energy, $\int d\tau$

is the total plasma volume enclosed by the surface $\oint d\sigma$ of the wall. However, in the definition (4.83) of the total magnetic flux, we have put a tilde on the surface element $d\tilde{\sigma}$ and the normal vector $\tilde{\mathbf{n}}$ to indicate that they refer to a cross-section of the plasma enclosed by a boundary curve $\oint d\boldsymbol{l}$ lying in the wall (like in Fig. 4.4 for the fluxes in a tokamak).

By applying Gauss' theorem (A.14) to the local mass conservation equation (4.71) we find the time derivative of the total mass,

$$\dot{M} = \int \dot{\rho}\, d\tau = -\int \nabla \cdot \boldsymbol{\pi}\, d\tau = -\oint \boldsymbol{\pi} \cdot \mathbf{n}\, d\sigma = 0\,, \tag{4.84}$$

which vanishes by virtue of the BC (4.39). Hence, *the total mass is conserved*. Applying Gauss' theorem, and the BCs (4.39) and (4.41), to the local momentum conservation equation (4.72) gives

$$\mathbf{F} = \dot{\mathbf{\Pi}} = \int \dot{\boldsymbol{\pi}}\, d\tau = -\int \nabla \cdot \mathbf{T}\, d\tau = -\oint (p + \tfrac{1}{2}B^2)\, \mathbf{n}\, d\sigma\,, \tag{4.85}$$

so that *the total momentum is conserved*, however, only *if the total force exerted by the wall vanishes*. If this were not the case, there would be an imbalance of the last term involving the total pressure. This is a logical proviso if the configuration is to remain in place (we exclude disasters like earthquakes or disruptions of the vessel). Applying Gauss' theorem, and the mentioned BCs, to the local energy conservation equation (4.73) gives

$$\dot{H} = \int \dot{\mathcal{H}}\, d\tau = -\int \nabla \cdot \mathbf{U}\, d\tau = -\oint \mathbf{U} \cdot \mathbf{n}\, d\sigma = 0\,, \tag{4.86}$$

which states that *the total energy is conserved*.

On the other hand, applying Stokes' theorem (A.18), and the BCs (4.39) and (4.41), to the induction equation (4.48), we obtain

$$\dot{\Psi} = \int \dot{\mathbf{B}} \cdot \tilde{\mathbf{n}}\, d\tilde{\sigma} = \int \nabla \times (\mathbf{v} \times \mathbf{B}) \cdot \tilde{\mathbf{n}}\, d\tilde{\sigma} = \oint \mathbf{v} \times \mathbf{B} \cdot d\boldsymbol{l} = 0\,, \tag{4.87}$$

since $\mathbf{v}$, $\mathbf{B}$ and $d\boldsymbol{l}$ are tangential to the wall (as discussed already in Section 4.2.2). Hence, *magnetic flux is conserved* as well: it cannot leave or enter the vessel.

Consequently, in a plasma enclosed by a rigid shell (this will be called model I in Section 4.6), *the boundary conditions* $\mathbf{v} \cdot \mathbf{n} = 0$ *and* $\mathbf{n} \cdot \mathbf{B} = 0$ *guarantee that all physical quantities of interest are conserved*, so that *the system is closed*. This remains true for a plasma surrounded by vacuum (model II), as will be proved in Section 4.6 by appropriate modifications of these boundary conditions. It is no longer true if the wall is replaced by a system of external coils with time-dependent currents (model III): magnetic flux and Poynting flux may be pumped into the system. Similarly, coronal plasmas bounded by an immobile photosphere (models V and VI of Section 4.6), can be extended to incorporate coronal changes of mass, momentum, energy and magnetic flux by photospheric boundary motions. We conclude that the conservation equations (4.71)–(4.74) are an extremely powerful consequence of the nonlinear macroscopic dynamics of plasmas.

4.3.3 Local conservation of magnetic flux

We now wish to describe the local conservation properties, in particular flux conservation, in terms of the dynamics of fluid elements. To that end, it is helpful to derive the *kinematic expressions* for the change in time of the geometric quantities.

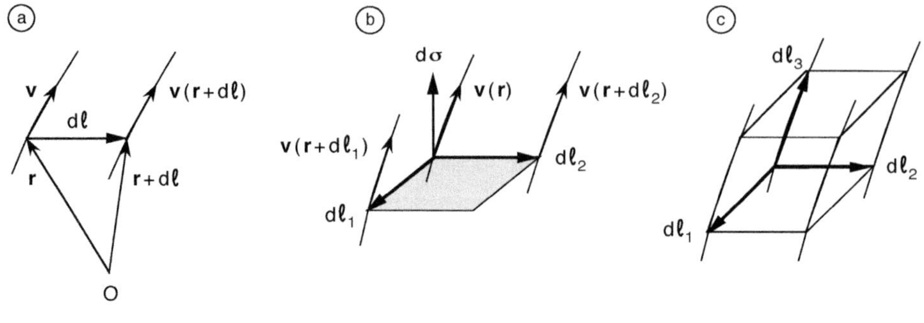

Fig. 4.6 Kinematics of (a) line element $d\boldsymbol{l}$, (b) surface element $d\boldsymbol{\sigma}$, (c) volume element $d\tau$.

First, consider the fluid flow $\mathbf{v}(\mathbf{r})$ at the positions $\mathbf{r}$ and $\mathbf{r} + d\boldsymbol{l}$ and let us find the equations describing the motion of the *line element* $d\boldsymbol{l}$ connecting those positions (Fig. 4.6(a)). This is given by the Lagrangian derivative of $d\boldsymbol{l}$,

$$\frac{\mathrm{D}}{\mathrm{D}t}(d\boldsymbol{l}) = \frac{\mathrm{D}(\mathbf{r} + d\boldsymbol{l})}{\mathrm{D}t} - \frac{\mathrm{D}\mathbf{r}}{\mathrm{D}t} = \mathbf{v}(\mathbf{r} + d\boldsymbol{l}) - \mathbf{v}(\mathbf{r}) = d\boldsymbol{l} \cdot (\nabla \mathbf{v}), \tag{4.88}$$

which describes the kinematics of a line element.

Next, we define the *surface element* by the vector product $d\boldsymbol{\sigma} \equiv d\boldsymbol{l}_1 \times d\boldsymbol{l}_2$ (Fig. 4.6(b)). Its Lagrangian derivative is derived by applying Eq. (4.88) twice:

$$\frac{\mathrm{D}}{\mathrm{D}t}(d\boldsymbol{\sigma}) = \frac{\mathrm{D}(d\boldsymbol{l}_1)}{\mathrm{D}t} \times d\boldsymbol{l}_2 + d\boldsymbol{l}_1 \times \frac{\mathrm{D}(d\boldsymbol{l}_2)}{\mathrm{D}t} = d\boldsymbol{l}_1 \cdot (\nabla \mathbf{v}) \times d\boldsymbol{l}_2 - d\boldsymbol{l}_2 \cdot (\nabla \mathbf{v}) \times d\boldsymbol{l}_1 .$$

Note the brackets around the expression $(\nabla \mathbf{v})$, introduced here to indicate that the gradient is meant to operate only on the quantity inside. To further reduce this kinematic relation, we need some vector identities that are rather involved because of the tensor $(\nabla \mathbf{v})$ and, therefore, put in small print.

▷ **Horrible derivation of a useful expression** Use vector identity (A.2) of Appendix A:

$$(\mathbf{a} \times \mathbf{b}) \times \mathbf{c} = \mathbf{a} \cdot \mathbf{c}\, \mathbf{b} - \mathbf{b} \cdot \mathbf{c}\, \mathbf{a} = \mathbf{b}\, \mathbf{a} \cdot \mathbf{c} - \mathbf{a}\, \mathbf{b} \cdot \mathbf{c},$$

where $\mathbf{c}$ has been moved to the utmost right since we will replace it by ∇:

$$(\mathbf{a} \times \mathbf{b}) \times \nabla = \mathbf{b}\, \mathbf{a} \cdot \nabla - \mathbf{a}\, \mathbf{b} \cdot \nabla .$$

Applying this operator on $\mathbf{v}$ gives

$$[\,(\mathbf{a} \times \mathbf{b}) \times \nabla\,] \times \mathbf{v} = [\mathbf{b}\mathbf{a} \cdot \nabla] \times \mathbf{v} - [\mathbf{a}\mathbf{b} \cdot \nabla] \times \mathbf{v} = -\mathbf{a} \cdot (\nabla \mathbf{v}) \times \mathbf{b} + \mathbf{b} \cdot (\nabla \mathbf{v}) \times \mathbf{a} .$$

Setting $\mathbf{a} \equiv d\boldsymbol{l}_1$ and $\mathbf{b} \equiv d\boldsymbol{l}_2$, so that $\mathbf{a} \times \mathbf{b} \equiv d\boldsymbol{\sigma}$, we obtain the required expression:

$$-d\boldsymbol{l}_1 \cdot (\nabla \mathbf{v}) \times d\boldsymbol{l}_2 + d\boldsymbol{l}_2 \cdot (\nabla \mathbf{v}) \times d\boldsymbol{l}_1 = (d\boldsymbol{\sigma} \times \nabla) \times \mathbf{v} = (\nabla \mathbf{v}) \cdot d\boldsymbol{\sigma} - \nabla \cdot \mathbf{v}\, d\boldsymbol{\sigma} ,$$

where Eq. (A.9) has been used in the last step. ◁

Hence,

$$\frac{\mathrm{D}}{\mathrm{D}t}(d\boldsymbol{\sigma}) = -(d\boldsymbol{\sigma} \times \nabla) \times \mathbf{v} = -(\nabla \mathbf{v}) \cdot d\boldsymbol{\sigma} + \nabla \cdot \mathbf{v}\, d\boldsymbol{\sigma} , \tag{4.89}$$

which is the kinematic expression for the motion of a surface element.

For the derivation of local magnetic flux conservation the expression (4.89) would be sufficient. However, to derive the local conservation laws for the other variables, we also need the kinematics of a *volume element*. Its definition, $d\tau \equiv d\boldsymbol{\sigma} \cdot d\mathbf{l}_3 = (d\mathbf{l}_1 \times d\mathbf{l}_2) \cdot d\mathbf{l}_3$ (Fig. 4.6(c)), gives by straightforward application of Eqs. (4.89) and (4.88):

$$\begin{aligned} \frac{\mathrm{D}}{\mathrm{D}t}(d\tau) &= \frac{\mathrm{D}(d\boldsymbol{\sigma})}{\mathrm{D}t} \cdot d\mathbf{l}_3 + d\boldsymbol{\sigma} \cdot \frac{\mathrm{D}(d\mathbf{l}_3)}{\mathrm{D}t} \\ &= -\, d\mathbf{l}_3 \cdot (\nabla \mathbf{v}) \cdot d\boldsymbol{\sigma} + \nabla \cdot \mathbf{v}\,(d\boldsymbol{\sigma} \cdot d\mathbf{l}_3) + d\mathbf{l}_3 \cdot (\nabla \mathbf{v}) \cdot d\boldsymbol{\sigma} = \nabla \cdot \mathbf{v}\, d\tau\,. \end{aligned} \tag{4.90}$$

This is the kinematic expression for the motion of a volume element.

Combining now the dynamic equation (4.71) with the kinematic relation (4.90) for the motion of *the mass of a fluid element*, $dM \equiv \rho d\tau$, gives the following expression for its rate of change:

$$\frac{\mathrm{D}}{\mathrm{D}t}(dM) = \frac{\mathrm{D}}{\mathrm{D}t}(\rho d\tau) = \frac{\mathrm{D}\rho}{\mathrm{D}t} d\tau + \rho \frac{\mathrm{D}}{\mathrm{D}t}(d\tau) = -\rho \nabla \cdot \mathbf{v}\, d\tau + \rho \nabla \cdot \mathbf{v}\, d\tau = 0\,. \tag{4.91}$$

Hence, the mass of a moving fluid element is constant.

Similarly, the rate of change of *the momentum of a fluid element*, $d\mathbf{\Pi} \equiv \boldsymbol{\pi} d\tau$, is found from Eqs. (4.72) and (4.90):

$$\begin{aligned} \frac{\mathrm{D}}{\mathrm{D}t}(d\mathbf{\Pi}) &= \frac{\mathrm{D}\boldsymbol{\pi}}{\mathrm{D}t} d\tau + \boldsymbol{\pi} \frac{\mathrm{D}}{\mathrm{D}t}(d\tau) = -\nabla \cdot \mathbf{T}\, d\tau + (\mathbf{v} \cdot \nabla \boldsymbol{\pi})\, d\tau + \boldsymbol{\pi} \nabla \cdot \mathbf{v}\, d\tau \\ &= -\nabla \cdot \Big[\rho \mathbf{v}\mathbf{v} + (p + \tfrac{1}{2}B^2)\,\mathbf{I} - \mathbf{B}\mathbf{B} - \mathbf{v}\boldsymbol{\pi}\Big]\, d\tau = (-\nabla p + \mathbf{j} \times \mathbf{B})\, d\tau \neq 0\,, \end{aligned} \tag{4.92}$$

where the equality (4.54) has been used in the last step. Hence, the momentum of a moving fluid element is *not* constant: it changes through the force density $-\nabla p + \mathbf{j} \times \mathbf{B}$ acting upon it.

For the change of *the energy of a fluid element*, $dH \equiv \mathcal{H}\, d\tau$, from Eqs. (4.73) and (4.90):

$$\begin{aligned} \frac{\mathrm{D}}{\mathrm{D}t}(dH) &= \frac{\mathrm{D}\mathcal{H}}{\mathrm{D}t} d\tau + \mathcal{H} \frac{\mathrm{D}}{\mathrm{D}t}(d\tau) = -\nabla \cdot (\mathbf{U} - \mathbf{v}\mathcal{H})\, d\tau \\ &= -\nabla \cdot \Big\{ \Big[(p + \tfrac{1}{2}B^2)\,\mathbf{I} - \mathbf{B}\mathbf{B}\Big] \cdot \mathbf{v} \Big\}\, d\tau \neq 0\,. \end{aligned} \tag{4.93}$$

Hence, the energy of a moving fluid element changes through the work performed by the total pressure (isotropic + anisotropic) on the element. So far, in the Lagrangian picture, we have only obtained the triviality of local mass conservation. Momentum and energy are not conserved.

There is one local Lagrangian conservation law though which is non-trivial and truly important. This does not follow from the motion of a volume element, but from that of a surface element. Consider *the magnetic flux through a surface element*, $d\Psi \equiv \mathbf{B} \cdot d\boldsymbol{\sigma}$. Departing now from the original evolution equation (4.48) for $\mathbf{B}$, rather than the artificial one (4.74), we have

$$\frac{\partial \mathbf{B}}{\partial t} = \nabla \times (\mathbf{v} \times \mathbf{B}) \overset{(\mathrm{A}.13)}{=} \mathbf{B} \cdot \nabla \mathbf{v} - \mathbf{B} \nabla \cdot \mathbf{v} - \mathbf{v} \cdot \nabla \mathbf{B} \quad , \tag{4.94}$$

so that

$$\frac{\mathrm{D}\mathbf{B}}{\mathrm{D}t} \equiv \frac{\partial \mathbf{B}}{\partial t} + \mathbf{v} \cdot \nabla \mathbf{B} = \mathbf{B} \cdot \nabla \mathbf{v} - \mathbf{B} \nabla \cdot \mathbf{v}\,. \tag{4.95}$$

From this dynamic equation for $\mathbf{B}$ and the kinematic relation (4.89) for the surface element, we obtain:

$$\begin{aligned} \frac{\mathrm{D}}{\mathrm{D}t}(d\Psi) &= \frac{\mathrm{D}}{\mathrm{D}t}(\mathbf{B} \cdot d\boldsymbol{\sigma}) = \frac{\mathrm{D}\mathbf{B}}{\mathrm{D}t} \cdot d\boldsymbol{\sigma} + \mathbf{B} \cdot \frac{\mathrm{D}}{\mathrm{D}t}(d\boldsymbol{\sigma}) \\ &= \big(\mathbf{B} \cdot \nabla \mathbf{v} - \mathbf{B} \nabla \cdot \mathbf{v}\big) \cdot d\boldsymbol{\sigma} + \mathbf{B} \cdot \big(-(\nabla \mathbf{v}) \cdot d\boldsymbol{\sigma} + \nabla \cdot \mathbf{v}\, d\boldsymbol{\sigma}\big) = 0\,. \end{aligned} \tag{4.96}$$

Hence, *the magnetic flux through a co-moving surface element is constant*. But, since this holds for any surface element, *the flux through any surface bounded by a contour C moving with the fluid is conserved*:

$$\Psi = \int_{\mathrm{c}} \mathbf{B}\cdot\mathbf{n}\,d\sigma = \text{const}\,. \tag{4.97}$$

This completes the exposition of Sections 4.2.1 and 4.2.2 centred on the magnetic field and induction equations (4.36) and (4.38). To summarize:

(1) the magnetic flux Ψ of an arbitrary flux tube is a well-defined quantity;

(2) this flux remains constant as the flux tube moves.

Since we may shrink the cross-section of the flux tube to an arbitrarily small size, the dynamics of single magnetic field lines now also comes into view. However, in this limit the magnetic flux vanishes so that we have to divide by another quantity, that also vanishes in this limit, to get a finite result. For this we may take the mass of a segment of the flux tube. To that end, combine the induction equation (4.95) in Lagrangian form with the mass conservation equation in Lagrangian form,

$$\frac{\mathrm{D}\rho}{\mathrm{D}t} = -\rho\nabla\cdot\mathbf{v} \quad\Rightarrow\quad \frac{\mathrm{D}}{\mathrm{D}t}\Big(\frac{\mathbf{B}}{\rho}\Big) = \frac{1}{\rho}(\mathbf{B}\cdot\nabla\mathbf{v} - \mathbf{B}\nabla\cdot\mathbf{v}) + \frac{\mathbf{B}}{\rho}\nabla\cdot\mathbf{v} = \Big(\frac{\mathbf{B}}{\rho}\Big)\cdot\nabla\mathbf{v}\,. \tag{4.98}$$

Compare this expression with the kinematic relation (4.88) and note that a line element $d\boldsymbol{l}$ parallel to the magnetic field $\mathbf{B}$ moves in exactly the same fashion as the quantity $\mathbf{B}/\rho$. Plasma on this line-element and magnetic field line move together. In the words of Alfvén (1950) [11], p. 82: *the lines of force are thus 'frozen' in the body*. Indeed, in ideal MHD (perfect conductivity!), the concept of magnetic field lines obtains more physical reality than it even had in Faraday's time.

Thus, field lines and magnetic flux moving with the plasma manifest the principal conservation property of plasmas, viz. *conservation of magnetic flux*.

4.3.4 Magnetic helicity

After the discussion of the local and global conservation properties of the basic MHD equations, the exposition of conservation laws appears to be complete. However, even though the four partial differential equations for ρ, $\mathbf{v}$, p and $\mathbf{B}$ are a complete set expressing these conservation properties, this is not so. Magnetohydrodynamics permits many different approaches which shed light on this fascinating subject, and which all start afresh from the basic equations. In particular, in studies of the dynamo mechanism for magnetic field generation, the development of MHD turbulence, and resistive reconnection of magnetic field lines, the discussion of the *magnetic topology* properties of global plasma configurations is greatly facilitated by a representation of the magnetic field in terms of the *vector potential*.

Let us, therefore, concentrate on the magnetic field equations, taking the determination of the density, velocity and pressure from the first three MHD equations for granted (assuming some magic black box taking care of those variables). We decompose the induction equation (4.15)(a) in Faraday's law and the ideal MHD version of Ohm's law and take into account that $\mathbf{B}$ should be divergence free,

$$\frac{\partial\mathbf{B}}{\partial t} = -\nabla\times\mathbf{E}\,, \qquad \mathbf{E} = -\mathbf{v}\times\mathbf{B}\,, \qquad \nabla\cdot\mathbf{B} = 0\,. \tag{4.99}$$

The latter constraint is also called the *solenoidal condition* for the magnetic field. It may be satisfied once for all by means of the vector potential $\mathbf{A}$:

$$\mathbf{B} = \nabla \times \mathbf{A} . \tag{4.100}$$

Ohm's law then becomes

$$\mathbf{E} = -\mathbf{v} \times (\nabla \times \mathbf{A}) , \tag{4.101}$$

and integration of Faraday's law provides the induction equation in terms of $\mathbf{A}$:

$$\frac{\partial \mathbf{A}}{\partial t} = \mathbf{v} \times (\nabla \times \mathbf{A}) - \nabla \Phi , \tag{4.102}$$

where Φ is a scalar potential. Since the equations (4.100)–(4.102) are invariant under a gauge transformation $\mathbf{A} \to \mathbf{A} + \nabla \chi$, $\Phi \to \Phi - \partial \chi / \partial t$, with arbitrary scalar function χ, there is no loss in generality if we choose $\Phi = 0$ as a gauge condition; see Jackson [331].

Next, we introduce the *magnetic helicity*

$$K(\mathrm{V}) \equiv \int_{\mathrm{V}} \mathbf{A} \cdot \mathbf{B} \, d\tau , \tag{4.103}$$

where integration is over the volume V of some flux tube, as in Section 4.2.1. The Lagrangian rate of change of this quantity is easily determined:

$$\begin{aligned} \frac{\mathrm{D}K}{\mathrm{D}t} &= \int \Big[\frac{\mathrm{D}\mathbf{A}}{\mathrm{D}t} \cdot \mathbf{B} + \mathbf{A} \cdot \frac{\mathrm{D}\mathbf{B}}{\mathrm{D}t} \Big] d\tau + \int \mathbf{A} \cdot \mathbf{B} \frac{\mathrm{D}}{\mathrm{D}t} (d\tau) \\ &\overset{(A.6)}{=} \int \Big[(\mathbf{v} \times \mathbf{B}) \cdot \mathbf{B} + \mathbf{A} \cdot \nabla \times (\mathbf{v} \times \mathbf{B}) + \nabla \cdot (\mathbf{A} \cdot \mathbf{B} \, \mathbf{v}) \Big] d\tau \\ &\overset{(A.12)}{=} \int \nabla \cdot \Big[(\mathbf{v} \times \mathbf{B}) \times \mathbf{A} + \mathbf{A} \cdot \mathbf{B} \, \mathbf{v} \Big] d\tau \\ &\overset{(A.14)}{=} \int \Big[\mathbf{A} \cdot \mathbf{v} \mathbf{B} - \mathbf{A} \cdot \mathbf{B} \, \mathbf{v} + \mathbf{A} \cdot \mathbf{B} \, \mathbf{v} \Big] \cdot \mathbf{n} \, d\sigma = 0 , \end{aligned} \tag{4.104}$$

because $\mathbf{B} \cdot \mathbf{n} = 0$ on the boundary of the flux tube. Hence, as first shown by Woltjer [653], *the magnetic helicity of any flux tube is conserved* in ideal MHD. Clearly, since the plasma can be decomposed in infinitely many different ways in flux tubes, from infinitesimally small to globally large, there is an infinity of different magnetic helicities which are all conserved.

To appreciate the subtleties of the magnetic helicity concept, let us compute it for some magnetic field distributions in an infinitely long plasma cylinder,

$$\mathbf{B} = B_\theta(r) \, \mathbf{e}_\theta + B_z(r) \, \mathbf{e}_z , \tag{4.105}$$

where r, θ, z are cylindrical coordinates. This helical magnetic field may be characterized by the distribution of the inverse pitch μ of the field lines,

$$\mu(r) \equiv B_\theta / (r B_z) . \tag{4.106}$$

(For the finite length $L = 2\pi R_0$ periodic cylinder model of a torus, this quantity is related to the safety factor $q(r) \equiv (\mu R_0)^{-1}$, defined in Eq. (2.164).) For the θ-pinch ($\mu = 0$) and z-pinch

($\mu \to \infty$) examples of Section 2.4.3,

$$B_\theta = 0 \;\Rightarrow\; A_z = -\int_0^r B_\theta \, dr = 0 \qquad (\theta\text{-pinch}),$$
$$B_z = 0 \;\Rightarrow\; A_\theta = \frac{1}{r}\int_0^r r B_z \, dr = 0 \qquad (z\text{-pinch}), \tag{4.107}$$

so that the integrand $I \equiv A_\theta B_\theta + A_z B_z$ of the helicity integral (4.103) vanishes, and, hence, the helicity vanishes as well. Apparently, the magnetic field should be at least helical to have $K \neq 0$. However, for a general helical magnetic field in cylindrical geometry, the expression for the integrand may be written as

$$I(r) = \frac{B_\theta}{r}\int_0^r r B_z \, dr - B_z \int_0^r B_\theta \, dr = \mu B_z \int_0^r \frac{B_\theta}{\mu} \, dr - B_z \int_0^r B_\theta \, dr . \tag{4.108}$$

For a constant pitch ($\mu = $ const) helical magnetic field in a cylinder, the helicity vanishes: $I(r) = 0 \Rightarrow K(r) = 0$. Hence, magnetic helicity is *not* the property that field lines are helical. In an infinite cylinder, the field lines should at least have different values of μ at different radii, i.e. they should have *magnetic shear*.

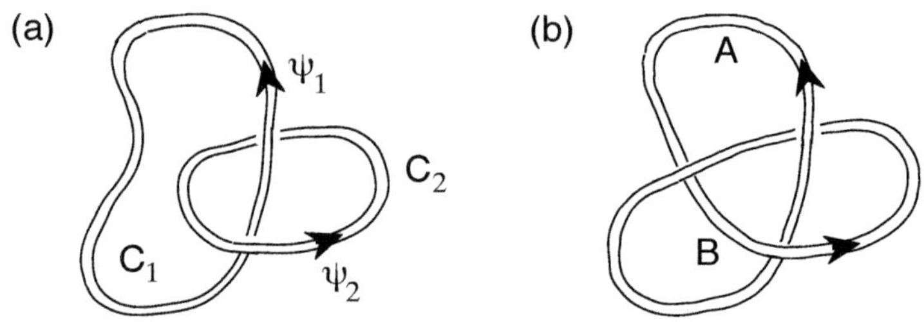

Fig. 4.7 Thin flux tube topologies: (a) two linked magnetic loops; (b) one knotted loop. (Adapted from Moffatt [444].)

More important than these local considerations are the global magnetic topology implications for the magnetic helicity. This may be demonstrated from the example of *linked magnetic loops* constructed by Moffatt [444] (see Fig. 4.7). Two infinitesimally thin flux tubes C_1 and C_2 with longitudinal magnetic fluxes Ψ_1 and Ψ_2 are linked as shown in Fig. 4.7(a). Outside these loops, $\mathbf{B} = 0$. The helicity of the first loop then becomes

$$K_1 = \int_{V_1} \mathbf{A}\cdot\mathbf{B}\, d\tau = \oint_{C_1} \mathbf{A}\cdot d\boldsymbol{l} \int_{S_1} \mathbf{B}\cdot\mathbf{n}\, d\sigma = \Psi_1 \oint_{C_1} \mathbf{A}\cdot d\boldsymbol{l} . \tag{4.109}$$

Since the magnetic field vanishes in the intermediate region, the contour C_1 of the latter line integral over $\mathbf{A}$ may be shrunk to a small contour bounding the cross-section of the second loop, so that

$$\Psi_2 = \int_{S_2} \mathbf{B}\cdot\mathbf{n}\, d\sigma = \int_{S_2} (\nabla\times\mathbf{A})\cdot\mathbf{n}\, d\sigma \overset{(A.18)}{=} \oint_{C_1} \mathbf{A}\cdot d\boldsymbol{l} . \tag{4.110}$$

A similar argument may be applied to the second loop, so that we obtain the following significant expression for the total helicity:

$$K_1 = K_2 = \Psi_1\Psi_2 \quad\Rightarrow\quad K = K_1 + K_2 = 2\Psi_1\Psi_2 . \tag{4.111}$$

When the tubes wind n times around each other, we get

$$K_1 = K_2 = \pm n\Psi_1\Psi_2 \quad \Rightarrow \quad K = K_1 + K_2 = \pm 2n\Psi_1\Psi_2 \,, \tag{4.112}$$

where the $+$ or $-$ refers to right- or left-handed orientation.

Magnetic flux tubes may also be *knotted*, with non-vanishing helicity in general. A single right-handed trefoil knot is shown in Fig. 4.7(b). Since the magnetic field has opposite directions at the points A and B, pinching the loop by moving these points towards each other produces a linking equivalent to Fig. 4.7(a). Hence, $\Psi_1 = \Psi_2 = \Psi$, so that $K = 2\Psi^2$ for this knot.

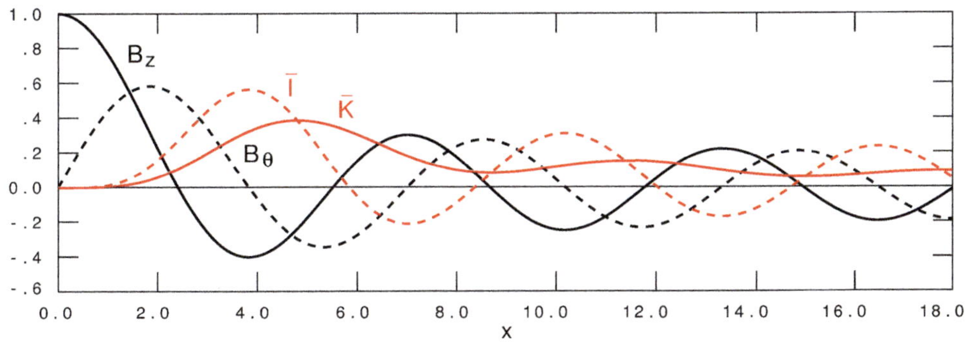

Fig. 4.8 Lundquist field: magnetic field components (black) and normalized helicities (red).

Returning to our infinite cylindrical flux tubes, an example of sufficient complexity in the radial direction is obtained for *force-free magnetic fields*. Such fields, for which the Lorentz force vanishes so that $\mathbf{j} \parallel \mathbf{B}$, obey the differential equation

$$\mathbf{j} = \nabla \times \mathbf{B} = \alpha \mathbf{B} \,, \tag{4.113}$$

where α is an arbitrary function of $\mathbf{r}$. A particularly useful example is the cylindrical solution for constant α constructed by Lundquist [420]:

$$B_z = C J_0(\alpha r) \,, \quad B_\theta = C J_1(\alpha r) \,, \tag{4.114}$$

where $C \equiv B_z(0)$ is the amplitude of the magnetic field on axis, and J_0 and J_1 are the zeroth and first order Bessel functions of real argument [3](a), Chapter 9. Because the Bessel functions oscillate with increasing values of r, the direction of this helical magnetic field constantly changes for the different annularly nested cylinders, which all may be considered as separate flux tubes (Fig. 4.8). Using the differential properties of the Bessel functions, $J_1 = -J_0{}'$, $xJ_0 = (xJ_1)'$, where $x \equiv \alpha r$, the expressions for the vector potential become

$$\begin{aligned} A_z &= -C \int_0^r J_1(\alpha r)\, dr = \frac{C}{\alpha} \Big[J_0(\alpha r) - 1 \Big] \,, \\ A_\theta &= \frac{C}{r} \int_0^r r J_0(\alpha r)\, dr = \frac{C}{\alpha} J_1(\alpha r) \,. \end{aligned} \tag{4.115}$$

Hence,

$$\bar{I}(r) \equiv \alpha I(r) \equiv \alpha \,(A_\theta B_\theta + A_z B_z) = C^2 \Big[J_0^2(\alpha r) + J_1^2(\alpha r) - J_0(\alpha r) \Big] \,, \tag{4.116}$$

and the renormalized helicity per volume of unit length L of the cylinder becomes

$$\bar{K}(r) \equiv \frac{\alpha}{\pi L r^2} K(r) = \frac{2}{r^2}\int_0^r \bar{I}(r)\, r\, dr = \frac{2C^2}{\alpha^2 r^2}\int_0^{\alpha r} \Big[J_0^2(x) + J_1^2(x) - J_0(x)\Big]\, x\, dx$$
$$= \frac{C^2}{\alpha^2 r^2}\Big\{\big[\alpha r J_0(\alpha r)\big]^2 + \big[\alpha r J_1(\alpha r) - 1\big]^2 - 1 + 4\sum_{n=1}^{\infty} 2n J_{2n}^2(\alpha r)\Big\}. \tag{4.117}$$

The integrals of the Bessel functions are obtained from standard formulas, see [3](a), Chapter 11. The essential point is that the integral $I(r)$ oscillates but the integral $K(r)$ remains positive definite (for $\alpha > 0$) or negative definite (for $\alpha < 0$). Hence, the total helicity of a force-free magnetic flux tube of constant α is non-zero, in contrast to the constant pitch magnetic field (which could be force-free as well) considered above. Moreover, since the integrand $I(r)$ oscillates, the helicities of the constituent annular flux tubes have consecutively opposite signs. All these flux tubes are linked in the sense of Fig. 4.7.

Consequently, the dynamics of force-free magnetic fields, conserving all these helicities, is an intricate subject. Considering the finite length periodic cylinder as a model for toroidal configurations, one should constrain the helicity to remain single-valued on the domain (which is multiply-connected in a torus). This may be done by means of a gauge-invariant generalization of the helicity; see Reiman [518]. Similarly, when the solar corona is considered as a dynamic magnetized plasma bounded by the photosphere, one should pay attention to the formulation of gauge-invariant boundary conditions at the photosphere; see Berger [52]. For further discussions of the helicity concept: see Moffatt [444] for applications to dynamo theory, see Biskamp [65] for the theory of Taylor [591] on the attainment of minimum energy states by self reversal of magnetic fields, and see Priest and Forbes [511] for applications to coronal magnetic fields.

4.4 Dissipative magnetohydrodynamics

4.4.1 Resistive MHD

Conservation of magnetic flux is directly connected with perfect conductivity, i.e. zero resistivity. Like in ordinary fluid mechanics, ideal fluids do not provide the full story but have to be complemented with a consideration of dissipative effects in *boundary layers* that generally occur, even though the dissipation coefficient may be extremely small; see e.g. Batchelor [38]. This happens because this coefficient multiplies the gradient of a physical quantity that is usually bounded but becomes very large in the boundary layer, producing a finite (and important) effect there. In contrast to ordinary fluid mechanics, in resistive MHD such 'boundary' layers are not associated with the physical boundaries of the system but with internal ones where gradients of the magnetic field (i.e. the current density) become very large.

The ideal MHD equations (4.12)–(4.15) for the variables ρ, $\mathbf{v}$, p and $\mathbf{B}$ permit a straightforward generalization to include one form of dissipation, viz. Ohmic dissipation through the plasma resistivity η. The resistive MHD equations were introduced in Section 2.4.1. Recall the main points: Ampère's law (4.10),

$$\mathbf{j} = \frac{1}{\mu_0}\nabla \times \mathbf{B}\,, \tag{4.118}$$

and Ohm's law (4.8) in the moving frame, extended with the resistive term,

$$\mathbf{E}' \equiv \mathbf{E} + \mathbf{v} \times \mathbf{B} = \eta \mathbf{j}, \tag{4.119}$$

are substituted in Faraday's equation (4.1) and combined with the classical fluid equations. This gives the set of evolution equations for resistive MHD expressing

– *conservation of mass:*

$$\frac{\partial \rho}{\partial t} = -\nabla \cdot (\rho \mathbf{v}), \tag{4.120}$$

– *conservation of momentum:*

$$\rho\Big(\frac{\partial}{\partial t} + \mathbf{v} \cdot \nabla\Big)\mathbf{v} = -\nabla p + \rho\, \mathbf{g} + \frac{1}{\mu_0}(\nabla \times \mathbf{B}) \times \mathbf{B}, \tag{4.121}$$

– *(near) conservation of entropy:*

$$\Big(\frac{\partial}{\partial t} + \mathbf{v} \cdot \nabla\Big)\, p = -\gamma p \nabla \cdot \mathbf{v} + (\gamma - 1)\, \frac{\eta}{\mu_0^2}\, (\nabla \times \mathbf{B})^2, \tag{4.122}$$

– *(near) conservation of magnetic flux:*

$$\frac{\partial \mathbf{B}}{\partial t} = \nabla \times (\mathbf{v} \times \mathbf{B}) - \frac{1}{\mu_0} \nabla \times (\eta\, \nabla \times \mathbf{B}), \qquad \nabla \cdot \mathbf{B} = 0. \tag{4.123}$$

Note that, in this section, we temporarily reintroduce the constant μ_0 to establish the dimensional expressions for the dissipative parameters. This requires the replacements $\mathbf{B} \to \mathbf{B}/\sqrt{\mu_0}$, $\mathbf{E} \to \mathbf{E}/\sqrt{\mu_0}$, $\mathbf{j} \to \sqrt{\mu_0}\,\mathbf{j}$ and $\eta \to \eta/\mu_0$ in the previously obtained dimensionless equations.

We have seen in Section 4.3.1, for the ideal MHD case, that the equations need substantial reworking to bring them into conservation form and to demonstrate the actual conservation of the indicated quantities. The resistive counterpart, i.e. demonstration of non-conservation (or, rather, conservation excepting small dissipative contributions) will be given in Section 4.4.2. The equations have been written as evolution equations, i.e. in a form that is suitable for numerical integration. In this context, the constraint $\nabla \cdot \mathbf{B} = 0$ on the magnetic field complicates the structure of the evolution problem (4.120)–(4.123) significantly. Since the dynamics of plasmas is associated with an evolving magnetic geometry, its satisfaction is an important issue in computational MHD.

(a) Dimensionless parameters The basic equations of ideal MHD, Eqs. (4.12)–(4.15), are obtained from Eqs. (4.120)–(4.123) by dropping the terms with η, i.e., the Joule heating term in Eq. (4.122) and the magnetic field dissipation term in Eq. (4.123), so that exact conservation is obtained. This is justified if the *magnetic Reynolds number*, introduced in Section 2.4.1, is large:

$$R_{\rm m} \equiv \mu_0 l_0 v_0/\eta \gg 1. \tag{4.124}$$

Here, l_0 and v_0 are characteristic length and velocity scales of the plasma flow. In Section 4.1.2, we have introduced another characteristic velocity, viz. the Alfvén speed $v_{\rm A}$. Using that velocity, we may define a dimensionless parameter that is large for small resistivity, viz. the *Lundquist number*:

$$L_{\rm u} \equiv \mu_0 l_0 v_{\rm A}/\eta \gg 1. \tag{4.125}$$

(In laboratory plasma literature, this quantity is usually indicated by the symbol S. Since we already have another use for this symbol, we prefer the symbol $L_{\rm u}$ here.) Which of the two dimensionless numbers is to be preferred depends on the resistive problem considered. In turbulence problems, where flow dominates, the magnetic Reynolds number is the more significant one. In resistive instabilities, also present in the absence of flow, the Lundquist number is the more relevant one (Biskamp [65]). Their ratio is determined by a third dimensionless parameter, characteristic for ideal MHD flow problems, viz. the *Alfvén Mach number*:

$$R_{\mathrm{m}}/L_{\mathrm{u}} = v_0/v_{\mathrm{A}} \equiv M_{\mathrm{A}} \,. \tag{4.126}$$

Note that we have not distinguished v_0 and v_{A} in the numerical tables of Appendix B. (In other words, $M_{\mathrm{A}} = 1$ has been assumed there.)

For fusion plasmas, the value of the resistivity is usually quite small, typically, $R_{\mathrm{m}} \sim 10^9$. For astrophysical plasmas, the resistivity may be somewhat larger, but then the length scales are very much larger too, so that huge values of the magnetic Reynolds number are obtained, e.g. $R_{\mathrm{m}} \sim 10^{13}$ for the solar corona. It would appear that the approximation (4.124) is an extremely good one so that there is no need for the consideration of dissipative effects. However, heating and reconnection (connected with the non-conservation of magnetic flux) are generally observed in plasmas, so that resistive processes must be operating. The crucial point is that η occurs in the equations in combination with gradients of the magnetic field. Hence, if very small-scale perturbations occur, e.g. due to specific resistive modes (like tearing modes, see below) or turbulence, the associated resistive terms may become sizeable. In addition, turbulence may increase the resistive coefficient itself, which then becomes an *anomalous* transport coefficient. For numerical calculations, all this implies extreme requirements on the spatial resolution.

(b) Tearing and reconnection of magnetic field lines To appreciate the global impact of small-scale resistive effects, consider the generic configuration illustrated in Fig. 4.9(a). A sheet of surface currents pointing into the plane of drawing creates a magnetic field with opposite directions in the upper and lower halves of the space. (The concept of surface current is precisely defined in Section 4.5.2; for now, it suffices to consider it as a thin plasma layer where the current density becomes very large.) Magnetic configurations of this kind frequently occur, e.g. on the day-side of the magnetosphere, when the solar wind impinges upon it with an embedded *IMF* (interplanetary magnetic field) that may have a direction opposite to that of the magnetosphere itself, and also on the night-side of the magnetosphere, where the solar wind drags the planetary magnetic field lines stretching them out to create the *magnetotail* over a distance of many planetary radii. These situations are inherently dynamic where the two magnetic fields may be pushed together by plasma motions. In ideal MHD perturbations, as shown in Fig. 4.9(b), such dynamics is essentially flux-conserving so that the magnetic topology cannot be changed. In resistive MHD, however, the two parts of the magnetic structure may be reconfigured to form the structure shown in Fig. 4.9(c): the magnetic field lines have been broken and rejoined to form an entirely new magnetic topology with an x-point separatrix (dotted line).

The driving force leading to reconnection is the fact that the magnetic configuration depicted in Fig. 4.9(c) represents a lower energy state than that of Fig. 4.9(a). The ideal MHD perturbation of Fig. 4.9(b) is not able to create that state because of flux conservation. Hence, the important remaining question to be answered is on what time scale the constraint of flux conservation can be broken so that reconnection can take place. In the absence of current concentration, this time scale is just determined by resistive diffusion, which is extremely slow, as we have seen in Sections 2.4

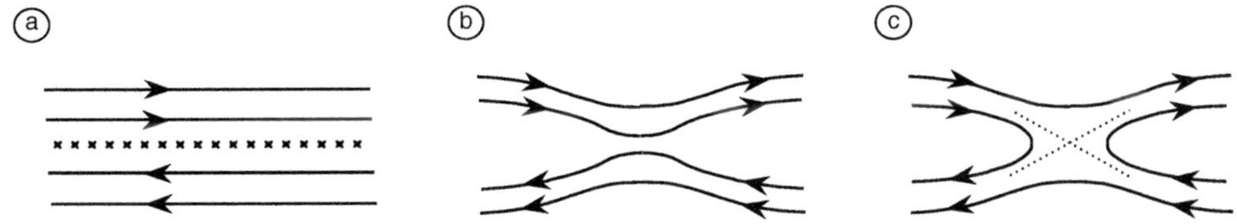

Fig. 4.9 Tearing and reconnection of magnetic field lines: (a) magnetic field of opposite directions created by sheet current (pointing into the plane); (b) ideal MHD perturbation; (c) reconnection by a resistive perturbation.

and 3.3.3. However, when external forcing creates current sheets this time scale apparently becomes short enough to permit all the violent plasma phenomena observed in astrophysical plasmas. On the other hand, in laboratory fusion research efforts have been successful enough to avoid such disruptions usually. We will return to resistive plasma dynamics in Chapter 14, when we have obtained sufficient preparation from ideal and one-dimensional analysis.

4.4.2 (Non-)conservation form of the dissipative equations*

It is instructive to consider how the introduction of resistivity and other dissipative effects spoils the conservation form of the MHD equations. This section is put in small print since it may be skipped on first reading of this chapter.

▷ **(a) Resistive effects** Introducing resistivity, the original equations (4.45) for ρ and (4.46) for $\mathbf{v}$ are unchanged, but Eq. (4.47) for the internal energy e is modified by the Ohmic dissipation term:

$$\frac{\partial e}{\partial t} + \mathbf{v}\cdot\nabla e + (\gamma-1)e\,\nabla\cdot\mathbf{v} = \frac{1}{\rho}\,\eta\, j^2\,, \qquad e \equiv \frac{1}{\gamma-1}\,\frac{p}{\rho}\,, \tag{4.127}$$

whereas Eq. (4.48) for the magnetic field $\mathbf{B}$ is changed by the modification of the electric field in Ohm's law:

$$\frac{\partial \mathbf{B}}{\partial t} + \nabla\times\mathbf{E} = 0\,, \qquad \mathbf{E} = -\mathbf{v}\times\mathbf{B} + \eta\,\mathbf{j}\,, \qquad \nabla\cdot\mathbf{B} = 0\,. \tag{4.128}$$

Consequently, the conservation equations (4.59) for ρ and (4.60) for $\rho\mathbf{v}$ remain unchanged, but the energy conservation (4.61) and the flux conservation equation (4.62) have to be modified.

In the derivation of the energy conservation equation, the contribution (4.56) remains the same, but the contributions (4.57) and (4.58) are to be replaced by

$$\frac{\partial}{\partial t}(\rho e) + \nabla\cdot(\rho e\mathbf{v}) + p\nabla\cdot\mathbf{v} = \eta\, j^2\,, \tag{4.129}$$

$$\frac{\partial}{\partial t}\Big(\frac{B^2}{2\mu_0}\Big) + \frac{1}{\mu_0}\nabla\cdot\Big[\mathbf{B}\times(\mathbf{v}\times\mathbf{B})\Big] + \mathbf{v}\cdot\mathbf{j}\times\mathbf{B} = -\frac{1}{\mu_0}\mathbf{B}\cdot\nabla\times(\eta\mathbf{j})$$

$$\stackrel{(A.12)}{=} -\frac{1}{\mu_0}\nabla\cdot(\eta\,\mathbf{j}\times\mathbf{B}) - \eta\, j^2\,. \tag{4.130}$$

Adding the contributions (4.56), (4.129), and (4.130) yields the resistive energy conservation equation:

$$\frac{\partial}{\partial t}\Big(\tfrac12\rho v^2 + \rho e + \frac{B^2}{2\mu_0}\Big) + \nabla\cdot\Big[\Big(\tfrac12\rho v^2 + \rho e + p\Big)\mathbf{v} + \frac{1}{\mu_0}\Big(-\mathbf{v}\times\mathbf{B} + \eta\mathbf{j}\Big)\times\mathbf{B}\Big] = -\rho\mathbf{v}\cdot\nabla\Phi_{\rm gr}\,. \tag{4.131}$$

This equation remains an energy conservation equation: magnetic energy may be converted into internal energy, but the sum is constant. Notice that resistivity here enters in the divergence term as $\mu_0^{-1}\mathbf{E}\times\mathbf{B}$, which is the

Poynting flux discussed in Section 4.3.2.

In contrast, the magnetic flux equation becomes essentially non-conservative, due to the magnetic diffusivity coefficient η/μ_0 (with dimension $[\eta/\mu_0] = \mathrm{m^2\,s^{-1}}$) on the RHS:

$$\frac{\partial \mathbf{B}}{\partial t} + \nabla\cdot(\mathbf{vB} - \mathbf{Bv}) = -\nabla\times\Big(\frac{\eta}{\mu_0}\nabla\times\mathbf{B}\Big) \overset{(A.7),(A.5)}{=} \frac{\eta}{\mu_0}\nabla^2\mathbf{B} + \mathbf{j}\times\nabla\eta\,, \tag{4.132}$$

where, as always $\nabla\cdot\mathbf{B} = 0$. Finally, the entropy conservation equations (4.64) and (4.65) change into equations that clearly exhibit non-conservation due to Ohmic dissipation:

$$\frac{\mathrm{D}S}{\mathrm{D}t} \equiv \frac{\partial S}{\partial t} + \mathbf{v}\cdot\nabla S = (\gamma-1)\rho^{-\gamma}\eta j^2\,, \qquad \frac{\partial}{\partial t}(\rho S) + \nabla\cdot(\rho S\mathbf{v}) = (\gamma-1)\rho^{-\gamma+1}\eta j^2\,. \tag{4.133}$$

The ideal equations (4.59) and (4.60) with the resistive equations (4.131) and (4.132) constitute the complete set of (non-)conservation equations for resistive MHD. The equations (4.133) are a mere consequence. ◁

▷ **(b) Other dissipative effects** It is instructive to generalize the equation of motion and the internal energy equation with contributions of viscosity, heat conduction, radiative losses, etc., to indicate how the MHD equations are related to the hydrodynamics equations (like the Navier–Stokes equation); see e.g. Roberts [522]. We here ignore the tremendous complexity of the plasma transport coefficients in the presence of a magnetic field, as summarized in Section 3.3.2, and just exploit scalar one-fluid transport coefficients.

Including viscous effects turns Eq. (4.121) into an equation expressing *(near) conservation of momentum*:

$$\rho\Big(\frac{\partial}{\partial t} + \mathbf{v}\cdot\nabla\Big)\mathbf{v} \;=\; -\nabla p + \rho\mathbf{g} + \frac{1}{\mu_0}(\nabla\times\mathbf{B})\times\mathbf{B} + \mathbf{F}_{\mathrm{visc}}\,, \tag{4.134}$$

$$\mathbf{F}_{\mathrm{visc}} \;\approx\; \rho\nu(\nabla^2\mathbf{v} + \tfrac{1}{3}\nabla\nabla\cdot\mathbf{v})\,, \tag{4.135}$$

where $\nu \equiv \mu/\rho$ is the kinematic viscosity coefficient (with dimension $[\nu] = \mathrm{m^2\,s^{-1}}$).

The internal energy equation (4.19) is modified by the effects of heat generation and heat flow, associated with the different dissipation mechanisms, as follows:

$$\rho\frac{\mathrm{D}e}{\mathrm{D}t} + (\gamma-1)\rho e\nabla\cdot\mathbf{v} = -\nabla\cdot\mathbf{h} + Q\,, \tag{4.136}$$

where $\mathbf{h}$ is the heat flow and Q is the generated heat per unit volume. The heat flow is given by the expression

$$\mathbf{h} \approx -\kappa\nabla(kT) = -\rho\lambda\nabla e\,, \tag{4.137}$$

where κ is the coefficient of thermal conductivity and $\lambda \equiv k\kappa/(C_v\rho)$ is the coefficient of thermal diffusivity (with dimension $[\lambda] = \mathrm{m^2\,s^{-1}}$). The total generated heat can be written as the difference between heating proper, H, due to resistivity, viscosity, thermonuclear fusion energy production, etc., and the thermal losses, L, e.g. due to radiation:

$$Q \equiv H - L\,, \quad H = H_{\mathrm{res}} + H_{\mathrm{visc}} + H_{\mathrm{fus}} + \cdots\,, \quad L = L_{\mathrm{rad}} + \cdots\,. \tag{4.138}$$

The resistive and viscous heating terms may be estimated as

$$H_{\mathrm{res}} \approx \eta j^2 = \eta\,\mu_0^{-2}\,|\nabla\times\mathbf{B}|^2\,, \quad H_{\mathrm{visc}} \sim \rho\nu|\nabla\mathbf{v}|^2\,, \tag{4.139}$$

where we stress once more that the anisotropies of the transport coefficients parallel and perpendicular to the magnetic field have been ignored.

This indicates how the kinematic diffusivity ν governs diffusion of the velocity, the thermal diffusivity λ governs diffusion of the internal energy, and the magnetic diffusivity η/μ_0 governs diffusion of the magnetic field. These effects are completely negligible when the MHD description of plasmas applies, except when small-scale structures are present. ◁

4.5 Discontinuities

4.5.1 Shocks and jump conditions

An important application of the MHD conservation equations (4.59)–(4.62), or (4.71)–(4.74), is the derivation of shock conditions. Here, it is crucial that the energy conservation equation (4.61), rather than the entropy conservation equation (4.65), is exploited because *a shock is an irreversible (entropy-increasing) transition* [98, 130, 394]. In ordinary gas dynamics, this transition is associated with supersonic flow upstream of the shock and subsonic flow downstream. Given the upstream state of the flow, the question there arises: what is the state on the downstream, subsonic, side of the shock? The Rankine–Hugoniot relations, i.e. the shock conditions, provide the answer to this question for ordinary gas dynamics. In that case, the characteristic speed of propagation of disturbances is the sound speed, c, and the Mach number, $M \equiv v/c$, determines whether the flow is supersonic ($M > 1$) or subsonic ($M < 1$). In MHD, there are three characteristic speeds (as we will see in Chapter 5). Consequently, the subject of MHD shocks is much richer than that of gas dynamic shocks. For two-fluid (or multi-fluid) plasmas, the consideration of the separate particle effects of the electrons and ions even leads to more characteristic speeds, corresponding to the different electron and ion wave motions. In the kinetic theory of plasma dynamics, the notion of collisionless shocks occurs, which is important, e.g., for the description of magnetospheric plasmas; see Burgess [98]. Here, we will restrict the analysis to MHD shocks.

At this point, our interest is even more restricted since we have not yet developed the MHD waves that are necessary to analyse the different shocks permitted. The discussion of genuine shocks is relegated to the later Chapter 21, devoted to transonic MHD flows and shocks. Here, we will just use the mechanism of shock formation to derive the appropriate *jump conditions for plasmas with an internal boundary*. Our immediate aim is to generalize the boundary conditions (4.39) and (4.41), which are valid for laboratory plasmas that are completely isolated from the outside world by a rigid wall, to boundary conditions describing more compound magnetic confinement structures.

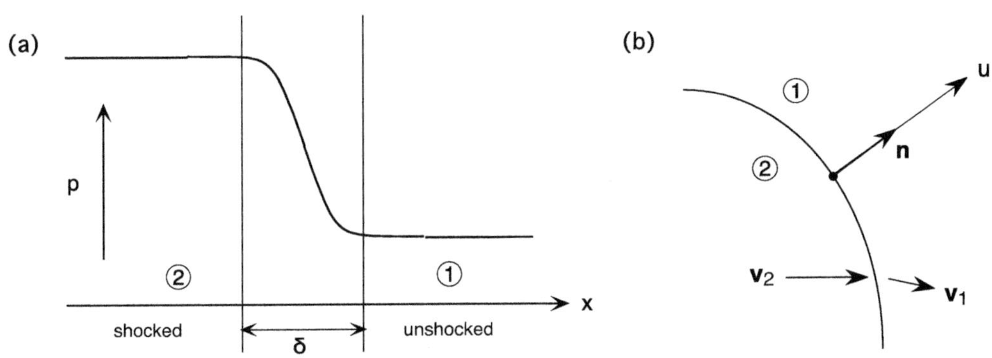

Fig. 4.10 (a) Shock formation: sudden increase of the pressure. (b) Directions in a shock front.

First, consider the one-dimensional flow of gas in which sound waves are excited. Local perturbations travel with the sound speed $c \equiv \sqrt{\gamma p/\rho}$. Their trajectories in the x–t plane, called *characteristics* (see Section 5.4), are two sets of parallel straight lines with derivatives $dx/dt = \pm c$. Suppose now that we suddenly increase the pressure, so that the sound speed increases (Fig. 4.10(a)). In

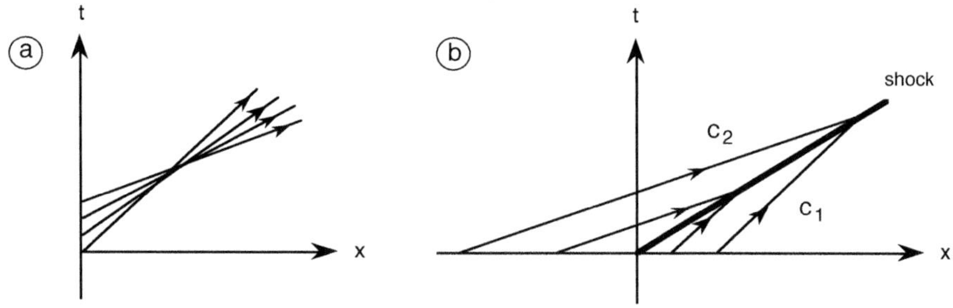

Fig. 4.11 (a) 'Crossing characteristics' due to a sudden change of the background variables. (b) In the ideal model, the characteristics meet at the shock discontinuity.

the x–t plane this means that the slopes of the characteristics (dt/dx) decrease. Therefore, the characteristics would cross (Fig. 4.11(a)), if some other mechanism did not interfere. The picture suggests what will happen: information originating from different space-time points accumulates. Consequently, gradients in the macroscopic variables build up until the point that the idealized model breaks down and dissipative effects due to the large gradients have to be taken into account. Eventually, a steady state will be reached where nonlinear and dissipative effects counterbalance: *a shock-wave* has been created. Neglecting the thickness δ of the shock, the steady state will consist of two regions with different sound speed, separated by the moving shock front. In the x–t plane, this front is located at the position where the forward characteristics (c_2) of the shocked part meet the forward characteristics (c_1) of the unshocked part (Fig. 4.11(b), also see LeVeque [404], p. 29).

Without specifying the kind of dissipation, one may arrive at the so-called *shock relations* that relate variables on the two sides of the propagating shock front. The idea is that the ideal model breaks down inside a layer of infinitesimal thickness δ (i.e., a thickness proportional to some power of the dissipation coefficient, which is assumed to be vanishingly small), but it holds on either side of the layer. In the limit $\delta \to 0$ the variables will jump across the layer, and the magnitude of the jumps is determined from the condition that mass, momentum, energy and magnetic flux should be conserved. Thus, one integrates the conservation equations (4.59)–(4.62) across the shock and keeps the leading order contributions arising from the gradients normal to the shock front only, since these gradients are infinitely large in the limit: $\partial f/\partial l \to \infty$, where f indicates any of the physical variables. Defining the jump in f by

$$[\![f]\!] \equiv f_1 - f_2 \,, \tag{4.140}$$

these contributions give

$$\lim_{\delta\to 0} \int_1^2 \nabla f \, dl = -\lim_{\delta\to 0} \mathbf{n} \int_1^2 \frac{\partial f}{\partial l} \, dl = \mathbf{n} \, [\![f]\!] \,, \tag{4.141}$$

where $\mathbf{n}$ is the normal to the shock front, chosen to point in the direction of the undisturbed fluid ahead of the shock (Fig. 4.10(b)). By convention, the integration across the shock, along l, is chosen in just the opposite direction, viz. from ① (undisturbed fluid) to ② (shocked part of the fluid). The time-derivatives $\partial f/\partial t$ also contribute to the shock conditions, as may be seen by transforming to a frame moving with the normal speed u of the shock-front, $(\mathrm{D}f/\mathrm{D}t)_{\rm shock} = \partial f/\partial t - u \, \partial f/\partial l$, where $(\mathrm{D}f/\mathrm{D}t)_{\rm shock}$ denotes the rate of change in a frame moving with the shock. Since this quantity remains finite and $\partial f/\partial l \to \infty$, for balance we must have $\partial f/\partial t \to \infty$ as well. Hence,

$$\lim_{\delta\to 0}\int_1^2 \frac{\partial f}{\partial t}\,dl = u\lim_{\delta\to 0}\int_1^2 \frac{\partial f}{\partial l}\,dl = -u\,[\![f]\!]\,. \tag{4.142}$$

In conclusion: from Eqs. (4.141) and (4.142) it follows that the shock relations are obtained by simply making the substitutions

$$\partial f/\partial t \to -u\,[\![f]\!]\,, \qquad \nabla f \to \mathbf{n}\,[\![f]\!]\,, \tag{4.143}$$

in the conservation equations.

For the ideal MHD conservation equations (4.59)–(4.62), the substitutions (4.143) result in the following general jump conditions:

$$-u\,[\![\rho]\!] + \mathbf{n}\cdot[\![\rho\mathbf{v}]\!] = 0\,, \tag{4.144}$$

$$-u\,[\![\rho\mathbf{v}]\!] + \mathbf{n}\cdot[\![\rho\mathbf{v}\mathbf{v} + (p+\tfrac{1}{2}B^2)\,\mathbf{I} - \mathbf{B}\mathbf{B}]\!] = 0\,, \tag{4.145}$$

$$-u\left[\!\!\left[\tfrac{1}{2}\rho v^2 + \frac{p}{\gamma-1} + \tfrac{1}{2}B^2\right]\!\!\right] + \mathbf{n}\cdot\left[\!\!\left[\left(\tfrac{1}{2}\rho v^2 + \frac{\gamma}{\gamma-1}p + B^2\right)\mathbf{v} - \mathbf{v}\cdot\mathbf{B}\mathbf{B}\right]\!\!\right] = 0\,, \tag{4.146}$$

$$-u\,[\![\mathbf{B}]\!] + \mathbf{n}\cdot[\![\mathbf{v}\mathbf{B} - \mathbf{B}\mathbf{v}]\!] = 0\,, \qquad \mathbf{n}\cdot[\![\mathbf{B}]\!] = 0\,, \tag{4.147}$$

where we have eliminated the internal energy variable e in favour of the pressure p. Recall that the original entropy conservation equation had to be replaced by the energy conservation equation since the latter remains valid in the presence of dissipation. In the limit $\delta \to 0$, the dissipative boundary layer contributions vanish and the variables $\rho, \mathbf{v}, p$ and $\mathbf{B}$ may become discontinuous, according to the jump conditions (4.144)–(4.147). However, the second law of thermodynamics demands that the entropy should have increased (or, rather, should not have decreased) when the shock has passed. Hence, Eqs. (4.144)–(4.147) have to be supplemented with the condition that *entropy increases across the shock*:

$$[\![s]\!] \le 0\,, \quad \text{or} \quad [\![S]\!] \equiv [\![\rho^{-\gamma}p]\!] \le 0 \qquad \textit{(entropy)}\,. \tag{4.148}$$

This miraculous condition is all that remains from the dissipative processes in the limit of infinitesimal thickness of the boundary layer.

For the steady shocks that we will now consider, it is convenient to transform to *the shock frame*, in which the shock is stationary, and the fluid velocities $\mathbf{v}' \equiv \mathbf{v} - u\mathbf{n}$ are evaluated with respect to this frame. The jump conditions (4.144)–(4.147) may then be written as

$$[\![\rho v_{\rm n}']\!] = 0 \qquad \textit{(mass)}\,, \tag{4.149}$$

$$[\![\rho v_{\rm n}'^{\,2} + p + \tfrac{1}{2}B_{\rm t}^2]\!] = 0 \qquad \textit{(normal momentum)}\,, \tag{4.150}$$

$$\rho v_{\rm n}'[\![\mathbf{v}_{\rm t}']\!] = B_{\rm n}[\![\mathbf{B}_{\rm t}]\!] \qquad \textit{(tangential momentum)}\,, \tag{4.151}$$

$$\rho v_{\rm n}'\left[\!\!\left[\tfrac{1}{2}\left(v_{\rm n}'^{\,2} + v_{\rm t}'^{2}\right) + \frac{1}{\rho}\left(\frac{\gamma}{\gamma-1}p + B_{\rm t}^2\right)\right]\!\!\right] = B_{\rm n}[\![\mathbf{v}_{\rm t}'\cdot\mathbf{B}_{\rm t}]\!] \qquad \textit{(energy)}\,, \tag{4.152}$$

$$[\![B_{\rm n}]\!] = 0 \qquad \textit{(normal flux)}\,, \tag{4.153}$$

$$\rho v_{\rm n}'\left[\!\!\left[\frac{\mathbf{B}_{\rm t}}{\rho}\right]\!\!\right] = B_{\rm n}[\![\mathbf{v}_{\rm t}']\!] \qquad \textit{(tangential flux)}\,, \tag{4.154}$$

where the momentum equation (4.145) and the magnetic flux equation (4.147) have been projected in the directions normal and tangential to the shock front. Of course, this representation of the shock conditions also has to be supplemented with the entropy condition (4.148).

We have now obtained six algebraic equations for the six jumps $[\![\rho]\!]$, $[\![v_{\rm n}]\!]$, $[\![\mathbf{v}_{\rm t}]\!]$, $[\![p]\!]$, $[\![B_{\rm n}]\!]$, $[\![\mathbf{B}_{\rm t}]\!]$, so that we may compute the values of all variables on the downstream side of the shock if their upstream values are known from the solution of the PDEs themselves. The values of the variables on the downstream boundary thus computed provide the *boundary conditions* to be imposed on the solution of the PDEs in the downstream region. The entropy condition just forbids solutions that do not correspond to an increase of the entropy on the shocked side.

4.5.2 Boundary conditions for plasmas with an interface

The jump conditions (4.148)–(4.154) provide the necessary tools to describe two quite different physical phenomena (see Landau and Lifschitz, *Electrodynamics of Continuous Media* [395]), viz.

(1) boundary conditions for *moving plasma–plasma interfaces*, where there is *no flow across the discontinuity* ($v'_{\rm n} = 0$);

(2) jump conditions for *genuine shocks,* where flow across the discontinuity ($v'_{\rm n} \neq 0$) is an essential feature.

In this chapter, we will not enter into the discussion of the second class of discontinuous phenomena, except to mention that our definition of shocks is chosen such that it contains rotational discontinuities (usually not considered to be proper shocks) as well as the regular magneto-sonic shocks. The full implications of these discontinuities for the analysis of transonic MHD flows and shocks, with the surprising variety of MHD shocks, will have to await the detailed exposition of Chapter 21. For now, we concentrate on the implications for the formulation of the different model configurations for confined plasmas (Section 4.6) that may be distinguished.

For the first class of discontinuous phenomena (co-moving interfaces, $v'_{\rm n} = 0$), the jump conditions reduce to:

$$[\![p + \tfrac{1}{2}B_{\rm t}^2]\!] = 0 \qquad \textit{(normal momentum)}, \qquad (4.155)$$

$$B_{\rm n}[\![\mathbf{B}_{\rm t}]\!] = 0 \qquad \textit{(tangential momentum)}, \qquad (4.156)$$

$$B_{\rm n}[\![\mathbf{v}'_{\rm t} \cdot \mathbf{B}_{\rm t}]\!] = 0 \qquad \textit{(energy)}, \qquad (4.157)$$

$$[\![B_{\rm n}]\!] = 0 \qquad \textit{(normal flux)}, \qquad (4.158)$$

$$B_{\rm n}[\![\mathbf{v}'_{\rm t}]\!] = 0 \qquad \textit{(tangential flux)}. \qquad (4.159)$$

This permits two distinct possibilities for jumps, viz.

(a) contact discontinuities, if the magnetic field intersects the interface ($B_{\rm n} \neq 0$), where the variables are alternatively

– jumping: $[\![\rho]\!] \neq 0\,,$

– continuous: $v'_{\rm n} = 0\,, \quad [\![\mathbf{v}'_{\rm t}]\!] = 0\,, \quad [\![p]\!] = 0\,, \quad [\![B_{\rm n}]\!] = 0\,, \quad [\![\mathbf{B}_{\rm t}]\!] = 0;$ (4.160)

(b) tangential discontinuities, if the magnetic field is parallel to the interface ($B_{\rm n} = 0$), where the variables are alternatively

$$- \text{jumping:} \qquad [\![\rho]\!] \neq 0\,, \quad [\![\mathbf{v}'_{\rm t}]\!] \neq 0\,, \quad [\![p]\!] \neq 0\,, \quad [\![\mathbf{B}_{\rm t}]\!] \neq 0\,,$$

$$- \text{continuous:} \qquad v'_{\rm n} = 0\,, \quad B_{\rm n} = 0\,, \quad [\![p + \tfrac{1}{2}B_{\rm t}^2]\!] = 0\,. \tag{4.161}$$

Note that the latter discontinuities are not a special case of the former ones.

On the basis of these two kinds of discontinuity, we may distinguish two types of magnetic configuration with an interface, viz. *astrophysical plasmas* where the magnetic fields typically originate in a planet, star, or other rotating object, with a dynamo operating inside, but intersect the surface (in the case of a star, the photosphere) where the plasma density may jump to the much lower values pertinent for a corona. Such jumps may be characterized as contact discontinuities. All variables should be continuous there, except for the density ρ (or the temperature T, or the entropy S). In *laboratory plasmas* aimed at thermonuclear energy production, on the other hand, the discontinuities of interest really serve to confine a high density plasma by a lower density one, that may even effectively qualify as a vacuum, in order to isolate it thermally from an outer wall. Such jumps are typically tangential discontinuities at a magnetic surface that is nested within the other ones. From Eq. (4.161) it is clear that these discontinuities permit much more freedom in the choice of the values of the variables. Except for the density, also the tangential velocity, the pressure and the tangential magnetic field may jump, as long as the magnetic field stays tangential and the total pressure remains balanced.

It should be stressed that this classification is not at all strictly one-to-one corresponding with that of astrophysical and laboratory plasmas. In many astrophysical plasmas of interest (e.g. at magnetospheric boundaries) tangential discontinuities occur, whereas in laboratory plasmas magnetic fields intersecting the boundary have become an important issue (e.g. in divertor plasmas). We will treat the two kinds of discontinuity, and their associated physical models, in reverse order since tangential discontinuities (laboratory plasmas) admit a much cleaner picture of confinement (Section 4.6.1). In this respect, the apparent simplicity of contact discontinuities is deceiving because these discontinuities separate regions with widely different physical properties so that the formulation of realistic boundary conditions usually becomes a much more involved problem. In Section 4.6.3, we will just sketch the kind of astrophysical models possible.

We finish our exposition of discontinuities with the formulation of the boundary conditions for plasma–plasma interfaces in the laboratory frame. In that frame, $v_{\rm n} = v'_{\rm n} + u$. Eliminating the normal speed u of the interface, the interface discontinuities are characterized by $[\![v_{\rm n}]\!] = 0$. The *boundary conditions for a tangential plasma–plasma interface* then become:

$$\mathbf{n} \cdot \mathbf{B} = 0 \qquad \textit{(at the interface)}\,, \tag{4.162}$$

$$\mathbf{n} \cdot [\![\mathbf{v}]\!] = 0 \qquad \textit{(at the interface)}\,, \tag{4.163}$$

$$[\![p + \tfrac{1}{2}B^2]\!] = 0 \qquad \textit{(at the interface)}\,. \tag{4.164}$$

The jump of the tangential magnetic field implies that there should be a *surface, or skin, current* flowing at the plasma–plasma interface. Such a current is obtained in the limit of a surface layer of thickness δ with large current density $\mathbf{j}$ when the limits $\delta \to 0$ and $|\mathbf{j}| \to \infty$ are taken in such a way that $\mathbf{j}^\star \equiv \lim_{\delta\to 0,\,|\mathbf{j}|\to\infty} (\delta\,\mathbf{j})$ remains finite. Note that the dimension of $\mathbf{j}^\star$ is that of current density times length, i.e. current per unit length. Application of our jump recipe (4.143) to Ampère's law, $\mathbf{j} = \nabla \times \mathbf{B}$, provides an expression for the magnitude of this surface current density:

$$\mathbf{j}^\star = \mathbf{n} \times [\![\mathbf{B}]\!] \,. \qquad (4.165)$$

When the plasma is rotating, so that there is a finite vorticity $\boldsymbol{\omega} \equiv \nabla \times \mathbf{v}$, there may be a *surface vorticity* as well:

$$\boldsymbol{\omega}^\star = \mathbf{n} \times [\![\mathbf{v}]\!] \,. \qquad (4.166)$$

For the present purpose, these expressions (representing singularities of the current density and the vorticity) are to be considered as mere consequences of the application of the interface BCs (4.162)–(4.164).

4.6 Model problems

We are now in a position to formulate the proper boundary conditions to distinguish two broad classes of magnetic confined configurations, of which the tokamak and the coronal loop shown in Fig. 4.2 are a particular example. We will identify six models, consisting of the MHD equations (4.12)–(4.15) + specification of a particular magnetic geometry with associated BCs. Models I–III refer to laboratory plasmas with tangential discontinuities, and models IV–VI refer to astrophysical plasmas with contact discontinuities.

4.6.1 Laboratory plasmas (models I–III)

The three models for confined laboratory plasmas are shown in Fig. 4.12. These configurations refer to toroidally symmetric tokamaks, i.e. the toroidal angle is an ignorable coordinate ($\partial/\partial\varphi = 0$) for all physical variables. Such problems involve the solution of partial differential equations (PDEs) in the two spatial coordinates associated with the poloidal cross-section (shaded in the upper part of Fig. 4.12). Therefore, they are called two-dimensional (2D) problems. The solution of 2D PDEs requires numerical analysis, which has been carried out extensively for tokamaks. One step more complicated is the stellarator, also a toroidal plasma confinement system but axi-symmetry is lost there by the introduction of asymmetric external coils and cross-sectional shaping in order to eliminate the toroidal plasma current (which is essential for tokamak confinement, but associated with dissipative decay and current-driven instabilities). Such configurations are called 3D. They require even more sophisticated numerical analysis.

However, in order to build up physical understanding and to develop mathematical techniques, the approximation of the toroidal geometry by that of an infinite or 'periodic' cylinder (bottom part of Fig. 4.12) is extremely useful. Since a circular cylinder has two directions of symmetry, the resulting equations will be non-trivial in only one of the three spatial coordinates (the radial one) and reduce to ordinary differential equations (ODEs) in that coordinate. For this reason, this geometry is considered to be one-dimensional (1D). The important point for the present discussion is that this reduction does not change the BCs discussed, but merely facilitates the solution of the differential equations. We will extensively exploit this simplification in the following chapters.

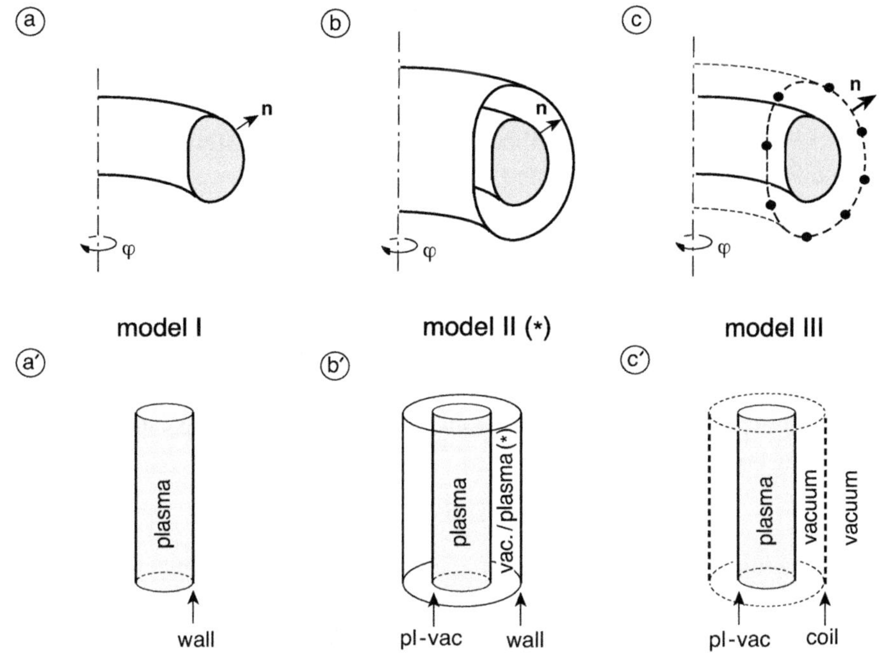

Fig. 4.12 Models for magnetic confinement of laboratory plasma (tokamak). (a) Model I: plasma surrounded by a wall; (b) model II (*): plasma isolated from the wall by a vacuum (*or another plasma); (c) model III: plasma excited by currents in external coils. Bottom part: cylindrical versions of the three models. Surfaces where boundary conditions have to be imposed are indicated by arrows.

(a) Model I: plasma confined inside rigid wall In this model, Fig. 4.12(a), the plasma is closed off from the outside world by a perfectly conducting wall. This is appropriate for the study of equilibrium, waves and instabilities of confined plasmas in closed vessels, as used in thermonuclear research. We already considered the appropriate boundary conditions for this model in Section 4.2.2. At the wall, both the normal magnetic field and the normal velocity have to vanish:

$$\mathbf{n} \cdot \mathbf{B} = 0 \qquad \text{(at the wall)}\,, \tag{4.167}$$

$$\mathbf{n} \cdot \mathbf{v} = 0 \qquad \text{(at the wall)}\,. \tag{4.168}$$

Recall from Section 4.3.2 that these BCs are sufficient to guarantee that the system conserves all important physical quantities (mass, momentum, energy, magnetic flux). Hence, this model is the simplest, and most relevant, one to describe confined plasmas. It will be used in Chapters 7 and 9 for the basic analysis of stability and waves in plasmas in plane slab and cylindrical geometries.

Amazingly, only two BCs need to be satisfied for eight variables. This is due to the intrinsic anisotropy introduced in the system by the magnetic field. Of course, the model is a restriction to facilitate analysis of confined plasmas. In reality, one may encounter $\mathbf{n} \cdot \mathbf{v} \neq 0$, e.g. when plasma is injected or ejected, but this is a complication to be introduced only (in Chapter 12 and following) when the more basic dynamics of this model is fully understood.

(b) Model II: plasma–vacuum system confined inside rigid wall In this model, Fig. 4.12(b), the plasma is confined inside a rigid wall and isolated from it by a region of low enough density to be treated as a 'vacuum'. This model again describes confined plasmas in a closed vessel, but separated from the wall by a vacuum region. The dynamics of the vacuum field variables $\hat{\mathbf{E}}$ and $\hat{\mathbf{B}}$ should correspond to the plasma dynamics described by the non-relativistic MHD equations (4.12)–(4.15), where time and length scales were assumed to satisfy $|\partial/\partial t|/|\nabla| \sim v \ll c$. This implies neglect of the displacement current also in Maxwell's equations for the vacuum:

$$\nabla \times \hat{\mathbf{B}} = 0\,, \qquad \nabla \cdot \hat{\mathbf{B}} = 0\,, \tag{4.169}$$

$$\nabla \times \hat{\mathbf{E}} = -\frac{\partial \hat{\mathbf{B}}}{\partial t}\,, \qquad \nabla \cdot \hat{\mathbf{E}} = 0\,. \tag{4.170}$$

Actually, this again degrades the electric field to a secondary variable that may be computed from the magnetic field $\hat{\mathbf{B}}$. Hence, in the vacuum only one basic variable is needed, viz. $\hat{\mathbf{B}}$, satisfying Eq. (4.169) and the boundary condition

$$\mathbf{n} \cdot \hat{\mathbf{B}} = 0 \qquad \text{(at the conducting wall)}\,. \tag{4.171}$$

This boundary condition is consistent with the assumption of vanishing tangential electric field,

$$\mathbf{n} \times \hat{\mathbf{E}} = 0 \qquad \text{(at the conducting wall)}\,, \tag{4.172}$$

since this also implies that $\mathbf{n} \cdot (\nabla \times \hat{\mathbf{E}}) = 0$ there, so that Eq. (4.170) then yields the boundary condition (4.171) on the normal magnetic field.

Of course, we also need BCs connecting the plasma variables with the vacuum magnetic field across the plasma–vacuum interface. In order to establish those, we consider the closely related *Model II*: plasma–plasma system confined inside rigid wall*. This model is also illustrated by Fig. 4.12(b) and indicated by the asterisk. Now, instead of a vacuum with $\hat{\mathbf{B}}$ satisfying Eq. (4.169), we have another plasma with variables that satisfy the ideal MHD equations. These variables are subject to the BCs (4.167) and (4.168) at the wall and the plasma–plasma interface BCs (4.162)–(4.164) derived in Section 4.5.2. Hence, this model is complete as well. Its significance is twofold: (1) it provides a useful alternative for the description of the outer region of tokamaks; (2) it is more widely applicable to astrophysical plasmas, e.g. a solar coronal magnetic loop where an external plasma may interact with the inner one to excite waves or to provide confinement. (In that case, confinement by the outer wall is dropped and replaced by a condition on behaviour at infinity.)

We now extrapolate model II* to obtain the BCs for model II. If the outer plasma is replaced by a vacuum, the BC (4.163) is no longer needed since the plasma velocity just determines the velocity of the plasma–vacuum interface and nothing else. Hence, the BCs for model II proper become:

$$\mathbf{n} \cdot \mathbf{B} = \mathbf{n} \cdot \hat{\mathbf{B}} = 0 \qquad \text{(at the plasma–vacuum interface)}\,, \tag{4.173}$$

$$[\![p + \tfrac{1}{2}B^2]\!] = 0 \qquad \text{(at the plasma–vacuum interface)}\,. \tag{4.174}$$

These boundary conditions are quite reasonable. If a normal magnetic field component were sticking through the plasma–vacuum boundary, i.e. if Eq. (4.173) did not hold, the vacuum region simply could not exist since plasma would freely flow along the magnetic field lines into the vacuum region. Next, if there were no balance of the total pressure, i.e. if Eq. (4.174) were not valid, the plasma–vacuum interface would simply be blown apart by the huge pressure imbalance. Finally, recall from the discussion in Section 4.5.2 that this pressure balance condition permits jumps in the pressure and the two tangential components of the magnetic field, associated with a surface current density

$$\mathbf{j}^\star = \mathbf{n} \times [\![\mathbf{B}]\!] \qquad \text{(at the plasma–vacuum interface)}\,. \tag{4.175}$$

This is not a separate BC but just a consequence of the BCs (4.173) and (4.174).

For later applications, it is of interest to also derive the boundary condition on the electric field at the plasma–vacuum interface. In Section 4.5, we derived the jump conditions starting from the MHD equations (4.12)–(4.15), where the electric field was already eliminated. Applying the substitutions (4.143) to Faraday's law (4.1) before this elimination, with the normal interface velocity determined by the plasma velocity, $u = \mathbf{n} \cdot \mathbf{v}$, we obtain the following jump condition for the electric field at model II interfaces: $\mathbf{n} \times [\![\mathbf{E}]\!] = \mathbf{n} \cdot \mathbf{v}\, [\![\mathbf{B}]\!]$. On the plasma side of the interface, Ohm's law (4.8) yields $\mathbf{n} \times \mathbf{E} = -\mathbf{n} \times (\mathbf{v} \times \mathbf{B}) = \mathbf{n} \cdot \mathbf{v}\, \mathbf{B}$. Hence, we obtain a basic relationship between the vacuum field variables $\hat{\mathbf{E}}$ and $\hat{\mathbf{B}}$ and the normal plasma velocity:

$$\mathbf{n} \times \hat{\mathbf{E}} = \mathbf{n} \cdot \mathbf{v}\, \hat{\mathbf{B}} \qquad \text{(at the plasma–vacuum interface)}\,. \tag{4.176}$$

This BC is actually redundant, just like the BC (4.172) at the conducting wall.

(c) Model III: plasma–vacuum system excited by external currents In this model, Fig. 4.12(c), instead of a wall we consider an open plasma–vacuum system excited by time-dependent magnetic fields $\hat{\mathbf{B}}(t)$ that are externally created. In laboratory plasmas, this external excitation may be caused by a system of coils. Such a system may be modelled by replacing the wall of model II by an auxiliary surface on which a time-dependent surface current $\mathbf{j}_c^\star(\mathbf{r}, t)$ forces oscillations onto the plasma–vacuum system. The effect is that the system is not isolated from the outside: energy flows into the system. The appropriate BCs at the coil surface are again obtained by applying our jump recipe (4.143) to the vacuum magnetic field equations $\nabla \cdot \hat{\mathbf{B}} = 0$ and $\nabla \times \hat{\mathbf{B}} = \hat{\mathbf{j}}$, giving

$$\mathbf{n} \cdot [\![\hat{\mathbf{B}}]\!] = 0 \qquad \text{(at the coil surface)}\,, \tag{4.177}$$

$$\mathbf{n} \times [\![\hat{\mathbf{B}}]\!] = \mathbf{j}_c^\star(\mathbf{r}, t) \qquad \text{(at the coil surface)}\,. \tag{4.178}$$

Note that the surface current $\mathbf{j}_c^\star(\mathbf{r}, t)$ is now cause, not effect as in model II, since it is prescribed. Also, the magnetic field outside the coils, in principle all the way up to ∞, is needed to solve this problem. Clearly, this case is of importance for laboratory plasma confinement because this always involves external magnetic fields that have to be created somehow. External excitation of MHD waves also gives rise to this time-dependent problem.

Model III can also be exploited for the analysis of waves in astrophysical plasmas, e.g. by mimicking the effects of excitation of MHD waves by an external plasma by means of a localized set of 'coils' when the response of the internal plasma is the main issue (e.g. in the problem of sunspot oscillations excited by sound waves in the photosphere; see Section 11.3.3.)

4.6.2 Energy conservation for interface plasmas

In Section 4.3.2 we proved energy conservation of the nonlinear system of ideal MHD equations for model I (plasma enclosed by a wall). In a later chapter (Section 6.6.3) we will need the law of conservation of the total energy for a plasma–vacuum system (model II). In such interface systems, the separate energies of plasma and vacuum are not conserved but the total energy is. Also, it is instructive to consider energy conservation for model III, where the outer boundary is replaced by a current-carrying coil which transfers energy to the system.

The generalization to *model II* is straightforward. The total energy for plasma and vacuum is

$$H = \int \mathcal{H}^{\rm p}\, d\tau^{\rm p} + \int \mathcal{H}^{\rm v}\, d\tau^{\rm v}\,, \qquad \mathcal{H}^{\rm p} \equiv \tfrac{1}{2}\rho v^2 + \frac{p}{\gamma - 1} + \tfrac{1}{2}B^2\,, \quad \mathcal{H}^{\rm v} \equiv \tfrac{1}{2}\hat{B}^2\,. \tag{4.179}$$

In the time dependence of these energies one needs to account for the rate of change of the volume elements as given by Eq. (4.90) of Section 4.3.3. Hence,

$$\frac{{\rm D}H^{\rm p}}{{\rm D}t} = \int \frac{{\rm D}\mathcal{H}^{\rm p}}{{\rm D}t}\, d\tau + \int \mathcal{H}^{\rm p}\, \frac{\rm D}{{\rm D}t}(d\tau) = \frac{\partial H^{\rm p}}{\partial t} + \int \mathcal{H}^{\rm p}\, \mathbf{v}\cdot\mathbf{n}\, d\sigma\,, \tag{4.180}$$

where Gauss' theorem (A.14) has been applied in the last step. Although $\mathbf{v}$ is only defined in the plasma, Eq. (4.180) obviously applies to the vacuum as well (with $\mathcal{H}^{\rm p}$ replaced by $\mathcal{H}^{\rm v}$) as it merely tells us that the rate of change of the energy is due to the changes of the energy density and of the total volume. According to the energy conservation equation (4.73), with the expression (4.69) for the energy flow inserted, we may integrate the plasma contribution by parts:

$$\frac{{\rm D}H^{\rm p}}{{\rm D}t} = -\int (p + \tfrac{1}{2}B^2)\, \mathbf{v}\cdot\mathbf{n}\, d\sigma\,. \tag{4.181}$$

For the vacuum contribution, we exploit Faraday's law (4.1) to introduce the electric field $\hat{\mathbf{E}}$ and the Poynting vector $\hat{\mathbf{S}} \equiv \hat{\mathbf{E}} \times \hat{\mathbf{B}}$:

$$\frac{{\rm D}H^{\rm v}}{{\rm D}t} = \int \frac{\partial \mathcal{H}^{\rm v}}{\partial t}\, d\tau^{\rm v} - \int \mathcal{H}^{\rm v}\, \mathbf{v}\cdot\mathbf{n}\, d\sigma \overset{(A.12)}{=} \int \hat{\mathbf{E}} \times \hat{\mathbf{B}}\cdot\mathbf{n}\, d\sigma - \int \tfrac{1}{2}\hat{B}^2\, \mathbf{v}\cdot\mathbf{n}\, d\sigma\,. \tag{4.182}$$

To remove the electric field from this expression again, we exploit the boundary condition (4.176) for model II interfaces derived above:

$$\frac{{\rm D}H^{\rm v}}{{\rm D}t} = \int \hat{B}^2\, \mathbf{v}\cdot\mathbf{n}\, d\sigma - \int \tfrac{1}{2}\hat{B}^2\, \mathbf{v}\cdot\mathbf{n}\, d\sigma = \int \tfrac{1}{2}\hat{B}^2\, \mathbf{v}\cdot\mathbf{n}\, d\sigma\,. \tag{4.183}$$

Adding Eqs. (4.181) and (4.183) for the energies, and applying the jump condition (4.174) for the total pressure, yields

$$\frac{{\rm D}H}{{\rm D}t} = \int [\![p + \tfrac{1}{2}B^2]\!]\, \mathbf{v}\cdot\mathbf{n}\, d\sigma = 0\,, \tag{4.184}$$

which is the desired result: *For model II, the total energy is conserved;* QED.

For *model III*, where the vacuum is enclosed by coils with surface currents $\mathbf{j}_{\rm c}^\star$, there is no conservation of energy for the interior region because the surface currents pump energy into the system. If we assume that these currents are arranged in such a way that no magnetic energy is lost external to these coils, i.e. $\hat{\mathbf{B}}^{\rm ext} = 0$, the rate of change of the energy is given by:

$$\frac{{\rm D}H^{\rm int}}{{\rm D}t} = \frac{\partial H^{\rm int}}{\partial t} = -\int \hat{\mathbf{S}}\cdot\mathbf{n}\, d\sigma_{\rm c} = \int \mathbf{n} \times \hat{\mathbf{B}}\cdot\hat{\mathbf{E}}\, d\sigma_{\rm c} = -\int \hat{\mathbf{E}}\cdot\mathbf{j}_{\rm c}^\star\, d\sigma_{\rm c}\,, \tag{4.185}$$

where we have exploited the jump condition (4.178) for the surface currents. (Note the change of sign of the Poynting flux and the magnetic field jump since $\mathbf{n}$ points into the vacuum for Eq. (4.182), but out of it here.) Hence, the rate of change of the energy internal to the coils is given by the Poynting flux across the coils, which equals the power transferred by the coils [331].

4.6.3 Astrophysical plasmas (models IV–VI)

The three models for astrophysical plasmas are shown in Fig. 4.13. These configurations refer, respectively, to closed and open coronal magnetic loops, and to solar or stellar wind outflow. All three examples are 2D exhibiting only azimuthal symmetry. This is indicated by the angle θ in the loop examples (described by cylindrical coordinates r, θ, z) and by the angle ϕ in the wind example (described by spherical coordinates r, θ, ϕ). The actual physical problems really involve genuine 3D geometries because, in general, the loops are not straight and the stellar wind outflows are not axi-symmetric. Nevertheless, it is useful to consider the simpler quasi-1D cylindrical versions of the loops (shown in the bottom part of Fig. 4.13) as well. It is to be noted though that this simplification does not lead to genuine 1D problems because of the photospheric BCs to be imposed.

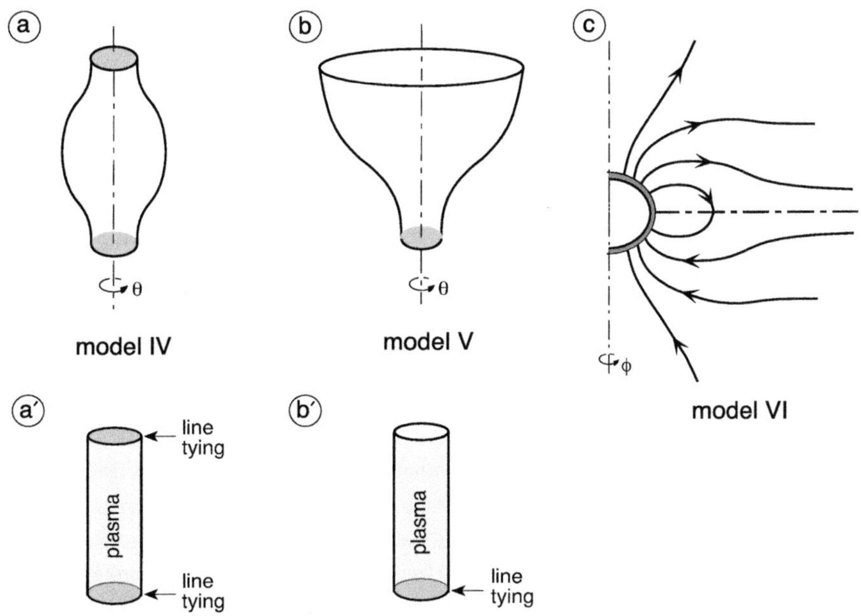

Fig. 4.13 Models for magnetic confinement in astrophysical plasmas. (a) Model IV: closed coronal magnetic loop, line-tied at both ends; (b) model V: open coronal loop, line-tied at one end and flaring at the other; (c) model VI: stellar wind outflow. Bottom part: quasi-1D (cylindrical) versions of two of the three models. Surfaces where boundary conditions have to be imposed are hatched.

(a) Model IV: 'closed' coronal magnetic loop In this model, Fig. 4.13(a), the magnetic field lines of a finite plasma column are line-tied on both sides to a plasma of so much higher density that it may be considered as immobile. Of course, the magnetic field lines do not end at the bounding planes, but the tubular domain is simply closed-off by the so-called *line-tying boundary conditions*

$$\mathbf{v} = 0 \qquad \text{(at the photospheric end planes)}\,. \tag{4.186}$$

These conditions model plasma–plasma interfaces with large density differences, where 'line-tying' refers to the fact that the magnetic field sticks through the interface and that it is effectively tied to the much heavier plasma below.

This model is appropriate for the study of waves in solar coronal flux tubes emanating from the photosphere because the density of the latter is about a factor of 10^9 higher than in the corona. In the consideration of the dynamics of a tenuous coronal flux tube, the back reaction on the photosphere may be neglected: the photosphere is too massive to be set in motion by the corona. The reverse problem is also meaningful and even quite important: the dynamics of the photosphere forces motion onto the coronal flux tubes. This problem may be represented by replacing the right hand side of the BC (4.186) by a prescribed velocity field at the photospheric boundary. In that case, Poynting flux enters the loop so that it is not closed at all then.

(b) Model V: open coronal magnetic loop In this model, Fig. 4.13(b), the magnetic field lines of a semi-infinite plasma column are line-tied on one side to a massive plasma. This model is appropriate for the description of open magnetic field lines, emanating from the coronal holes on the Sun, and associated with the solar wind escaping along them into interplanetary space. The boundary condition (4.186) now only applies to the end that is fixed into the photosphere under the coronal hole. One can consider either the passive problem of waves in open field lines by themselves, as represented by the BC (4.186), or the driven problem of wave generation by photospheric motion, which requires a non-zero right hand side of Eq. (4.186). Other applications of models IV and V are the dynamics of closed and open field lines in the magnetospheres of planets and pulsars.

(c) Model VI: stellar wind In this model, Fig. 4.13(c), a plasma is ejected from the photosphere of a star and accelerated along the open magnetic field lines into outer space. Clearly, this model is a composite of models IV and V, with the stress now on the outflow rather than on the waves.

This suffices for our purpose of introducing the self-consistent boundary value problems relevant for magnetically confined laboratory and astrophysical plasmas. The construction of the last two models V and VI has not been complete at all, since we have skipped over the most difficult problems there, viz. the appropriate conditions at infinity. It would be wrong to suggest that these problems have been solved as satisfactorily as those for the models I–III. It makes no sense though to discuss them separately from the dynamics. Hence, it is time to move onto the time domain. The next Chapter 5 is devoted to the analysis of the simplest of all geometries possible, viz. infinite homogeneous space, but replete with genuine plasma dynamics. In Chapter 6 we then merge the space and time domains and start the study the magnetohydrodynamics waves and instabilities of magnetically confined, i.e. inhomogeneous, plasmas.

4.7 Literature and exercises

Notes on literature

Flux tube dynamics

– The classical paper by Newcomb [460] on 'Motion of magnetic lines of force' is still very worth studying.

Ideal MHD model and conservation laws

– Goedbloed, *Lecture Notes on Ideal Magnetohydrodynamics* [218], containing the material of a course taught in Brazil 40 years ago, is the origin of this book that has been elaborated ever since. A chapter on the dynamics of the screw pinch [241], an illuminating illustration of the coupling of nonlinear plasma dynamics and external circuits, has been eliminated here since the experiment is outdated.

Magnetic helicity

- Moffatt, *Magnetic Field Generation in Electrically Conducting Fluids* [444], Chapter 2 on magnetokinematics on the topology implications of magnetic helicity.

Resistive MHD and reconnection

- Biskamp, *Nonlinear Magnetohydrodynamics* [65], Chapter 6 on the central importance of magnetic reconnection in nonlinear MHD.
- Priest and Forbes, *Magnetic Reconnection* [511], Chapter 4 on the solutions of steady reconnection.

Discontinuities and shocks

- Courant and Friedrichs, *Supersonic Flow and Shock Waves* [130], in particular Chapter IIIC on shocks in one-dimensional flows in ordinary fluids, with emphasis on irreversibility.
- Landau and Lifshitz, *Fluid Mechanics* [394], Chapter IX on shock waves in fluids.
- Landau and Lifshitz, *Electrodynamics of Continuous Media* [395], Chapter VIII on magnetic fluid dynamics with a section on MHD shock waves.

Exercises

[4.1] *Conservation laws*

Here comes the most important question on magnetohydrodynamics: what is the principal, i.e. most distinctive, conservation law of magnetized plasmas compared to ordinary fluids? (If you do not know, reread Chapter 4. If you still do not know, ask your professor. If he/she does not know, he/she should not be teaching this subject.)

[4.2] *Eulerian and Lagrangian time-derivatives*

Write down and explain these two ways of evaluating time derivatives in fluid dynamics.

[4.3] *Non-relativistic approximation*

Construct the MHD equations from Maxwell's equations and the equations for classical fluid dynamics. Show that, by assuming $v \ll c$, the displacement current and the electrostatic acceleration become negligible. Write down the resulting non-relativistic MHD equations in their most compact form.

[4.4] *Scale independence*

Why, and when, does it make sense to compare the plasma dynamics in small laboratory devices with the plasma dynamics in huge plasma-astrophysical systems?

[4.5]* *Conservation form*

The MHD equations can be brought into so-called 'conservation form'.

- What is the general structure of this form and can you give an interpretation of it?
- Try to construct the conservation form of the ideal MHD equations in terms of the variables ρ, $\mathbf{v}$, e and $\mathbf{B}$. If this is too much for you, try at least to indicate the basic steps involved in the construction.

[4.6] *Stress tensor*

In the conservation form of the MHD equations, one encounters the stress tensor defined by $\mathsf{T} \equiv \rho\mathbf{v}\mathbf{v} + (p + \frac{1}{2}B^2)\,\mathsf{I} - \mathbf{B}\mathbf{B}$, representing the Reynolds stress tensor, the isotropic pressure and the magnetic part of the Maxwell stress tensor.

- Project the Reynolds stress tensor on the velocity $\mathbf{v}$ and show that the only contribution is ρv^2 along $\mathbf{v}$. Decompose the remaining part of the stress tensor parallel and perpendicular to $\mathbf{B}$ and show how it is represented in matrix form.
- Draw a picture of a flux tube and comment on the different stresses acting on it.

[4.7] *Surface element kinematics and magnetic flux conservation*

We consider the magnetic flux $d\Psi \equiv \mathbf{B} \cdot d\boldsymbol{\sigma}$ through a surface element $d\boldsymbol{\sigma}$. When the latter element moves with the plasma, it changes in time as $\mathrm{D}(d\boldsymbol{\sigma})/\mathrm{D}t = -(\nabla\mathbf{v}) \cdot d\boldsymbol{\sigma} + \nabla \cdot \mathbf{v}\, d\boldsymbol{\sigma}$.

- Use the vector identities of Appendix A to show that the Lagrangian time derivative of the magnetic flux through that surface element vanishes: $\mathrm{D}(d\Psi)/\mathrm{D}t = 0$.
- What role does the assumption of perfect conductivity play here?

[4.8] *Magnetic helicity and force-free magnetic fields*

In Section 4.3.4, the magnetic helicity of a special class of cylindrical force-free magnetic fields of constant α has been computed. Let us extend these calculations in two directions.

- Calculate the components of a cylindrical force-free magnetic field with constant pitch μ. We already know that the helicity of this field vanishes. Now derive the expression for $\alpha(r)$ for this field and note that it tends to a constant value on the axis. How does that relate to the expression for the helicity of a constant α force-free field?
- Calculate the helicity for the compound configuration of a force-free magnetic field of constant α within a cylinder of radius $r = a$ surrounded by a vacuum magnetic field in the annular cylinder $a \leq r \leq b$. Does it make a difference if the latter is considered as a proper vacuum ($\eta = \infty$) or as a plasma ($\eta = 0$) without current?

[4.9] *Neglect of resistivity and dimensionless parameters*

We start with the resistive induction equation for the magnetic field $\mathbf{B}$ (Eq. (4.123)).

- Use a length scale l_0 for the gradients and velocity v_0 for the flow to construct the dimensionless parameter (R_m) to justify neglect of the resistive term when $R_\mathrm{m} \gg 1$.
- Another dimensionless parameter (L_u) may be constructed where v_0 is replaced by the characteristic velocity of Alfvén waves. When and why would that number be relevant?

[4.10] *Jump conditions*

Boundary conditions at plasma–plasma or plasma–vacuum interfaces may be derived from jump conditions obtained from the theory of supersonic flow and shocks in gas dynamics. Integrating the partial differential equation for a variable f across a shock results in jumps of ∇f and $\partial f/\partial t$ of magnitude $\mathbf{n}\, [\![f]\!]$ and $-u\, [\![f]\!]$, resp., where $\mathbf{n}$ is the normal to the shock front, the jump $[\![f]\!] \equiv f_1 - f_2$ and u is the normal velocity of the shock.

- Derive the ideal MHD jump conditions from the conservation laws.
- Transform these conditions to a frame moving with the shock, so that the fluid velocities relative to this frame may be written as $\mathbf{v}' \equiv \mathbf{v} - u\mathbf{n}$.
- Find the required interface conditions by assuming absence of flow across and a magnetic field tangential to the interface.

[4.11]* *Plasma–vacuum configuration between walls*

Consider a configuration of a plasma resting (against vertical gravity) on top of a vacuum with horizontal magnetic field confined between two horizontal, infinitely extended, conducting walls. (The stability of this configuration will be investigated in Section 6.6.4.)

- Derive all boundary conditions for the plasma variables at the top wall, the plasma–vacuum interface (assuming no flow across the boundary) and the vacuum variables at the bottom wall. Also give the equations for the magnetic field $\hat{\mathbf{B}}$ in the vacuum.

5

Waves and characteristics

5.1 Physics and accounting

5.1.1 Introduction

In the previous chapter, we concentrated on the *spatial* aspect of MHD by formulating the different boundary value problems associated with different plasma confinement geometries. In this chapter, we will concentrate on the *temporal* aspect by neglecting all effects of the geometry by considering, first, the linear waves of an infinite homogeneous plasma (Sections 5.1–5.3) and, next, the non-linear counterpart of these waves, viz. the characteristics and the associated initial value problem (Section 5.4). With these tools in hand, we can fruitfully study the effects of plasma geometry and inhomogeneity. They will be the subject of the following chapters, on spectral theory (Chapter 6) and plasma stability (Chapter 7).

The theory of wave propagation in plasmas involves a large number of algebraic manipulations to reduce a particular problem with many variables to one that can be solved explicitly. In this reduction, the bookkeeping should not obscure the physics. We will use the familiar example of sound waves to illustrate this.

5.1.2 Sound waves

As a preliminary to the study of linear MHD waves, we consider the simplest example of waves described by fluid equations, viz. sound waves. We start from the gas dynamic equations, contained in the MHD equations (4.12)–(4.14) as the special case of vanishing magnetic field and gravity ($\mathbf{B} = 0\,, \mathbf{g} = 0$). This yields the following set of partial differential equations for the density $\rho(\mathbf{r}, t)$, the velocity $\mathbf{v}(\mathbf{r}, t)$ and the pressure $p(\mathbf{r}, t)$:

$$\frac{\partial \rho}{\partial t} + \nabla \cdot (\rho \mathbf{v}) = 0\,, \tag{5.1}$$

$$\rho\Big(\frac{\partial \mathbf{v}}{\partial t} + \mathbf{v} \cdot \nabla \mathbf{v}\Big) + \nabla p = 0\,, \tag{5.2}$$

$$\frac{\partial p}{\partial t} + \mathbf{v} \cdot \nabla p + \gamma p \nabla \cdot \mathbf{v} = 0\,. \tag{5.3}$$

We now *linearize* these equations about a time-independent ($\partial/\partial t = 0$) infinite and homogeneous ($\nabla = 0$) background, characterized by constant values of ρ_0, $\mathbf{v}_0$ and p_0. The perturbations of this background may be written as

$$\begin{aligned} \rho(\mathbf{r},t) &= \rho_0 + \rho_1(\mathbf{r},t) && (\text{where } |\rho_1| \ll \rho_0 = \text{const})\,, \\ p(\mathbf{r},t) &= p_0 + p_1(\mathbf{r},t) && (\text{where } |p_1| \ll p_0 = \text{const})\,, \\ \mathbf{v}(\mathbf{r},t) &= \mathbf{v}_0 + \mathbf{v}_1(\mathbf{r},t) && (\text{where } |\mathbf{v}_1| \ll c)\,. \end{aligned} \tag{5.4}$$

Note that $|\mathbf{v}_1|$ is considered small, not with respect to $|\mathbf{v}_0|$ because we will consider a static ($\mathbf{v}_0 = 0$) background but with respect to another quantity that serves as a measure for velocities, viz. the sound speed c defined below.

Inserting these expressions in the differential equations and neglecting nonlinear coupling through quadratic and higher order terms, since the amplitudes of the waves are assumed small, we obtain the linearized equations of gas dynamics:

$$\Big(\frac{\partial}{\partial t} + \mathbf{v}_0 \cdot \nabla\Big)\rho_1 + \rho_0 \nabla \cdot \mathbf{v}_1 = 0\,, \tag{5.5}$$

$$\rho_0\Big(\frac{\partial}{\partial t} + \mathbf{v}_0 \cdot \nabla\Big)\mathbf{v}_1 + \nabla p_1 = 0\,, \tag{5.6}$$

$$\Big(\frac{\partial}{\partial t} + \mathbf{v}_0 \cdot \nabla\Big)p_1 + \gamma p_0 \nabla \cdot \mathbf{v}_1 = 0\,. \tag{5.7}$$

Note that Eq. (5.5) for ρ_1 does not couple to the other equations, so that it may be dropped or solved separately after the equations for $\mathbf{v}_1$ and p_1 have been solved. The latter two equations may be combined by applying the operator $\partial/\partial t + \mathbf{v}_0 \cdot \nabla$ to Eq. (5.6) for $\mathbf{v}_1$ and eliminating p_1 from it by means of Eq. (5.7). This yields *the wave equation for sound waves*:

$$\Big(\frac{\partial}{\partial t} + \mathbf{v}_0 \cdot \nabla\Big)^2 \mathbf{v}_1 - c^2\, \nabla\nabla \cdot \mathbf{v}_1 = 0\,, \qquad c \equiv \sqrt{\gamma p_0/\rho_0}\,, \tag{5.8}$$

where c is *the velocity of sound* of the background medium. (We may exploit the symbol c for this purpose since the displacement current has been removed from Maxwell's equations in Section 4.1.1, so that the plasmas will not support electromagnetic waves and the velocity of light does not appear any more.)

The wave equation has constant coefficients ($\mathbf{v}_0$ and c^2) so that the most general solution can be written as a superposition of *plane waves*:

$$\mathbf{v}_1(\mathbf{r},t) = \sum_{\mathbf{k}} \hat{\mathbf{v}}_{\mathbf{k}}\, \mathrm{e}^{\mathrm{i}(\mathbf{k}\cdot\mathbf{r} - \omega t)}\,, \tag{5.9}$$

for simplicity represented as a sum, corresponding to waves in a finite box. These plane waves do not couple, since the problem is linear and homogeneous, so that each harmonic by itself is a solution of the problem. Hence, we will consider them separately and drop the subscript $\mathbf{k}$. For such solutions, the differential operators of the PDEs turn into multiplications by algebraic factors:

$$\nabla \to \mathrm{i}\mathbf{k}\,, \qquad \partial/\partial t \to -\mathrm{i}\omega\,. \tag{5.10}$$

This transforms the wave equation (5.8) into an algebraic eigenvalue equation:

$$\Big[(\omega - \mathbf{k}\cdot\mathbf{v}_0)^2\,\mathbf{I} - c^2\,\mathbf{k}\mathbf{k}\Big] \cdot \hat{\mathbf{v}} = 0\,. \tag{5.11}$$

This eigenvalue problem for the frequency ω, measured in the laboratory frame, may also be considered as an eigenvalue problem for the *Doppler shifted frequency* ω', measured in the co-moving frame, $\omega' \equiv \omega - \mathbf{k} \cdot \mathbf{v}_0$. The difference between the two eigenvalue problems is trivial for homogeneous media ($\mathbf{v}_0 = \text{const}$) since the Doppler shift is constant. However, our main concern in the following chapters will be inhomogeneous media where the background flow significantly complicates the wave propagation problem. We will leave that subject for the later Chapter 12 and, for now, concentrate on wave propagation in *static* media ($\mathbf{v}_0 = 0$) where $\omega' = \omega$. For those problems, the eigenvalue ω only appears as a square.

Since there is no preferred direction in the system, we choose the waves to propagate in the z-direction, $\mathbf{k} = k\mathbf{e}_z$, so that the eigenvalue problem becomes

$$\omega^2 \hat{v}_x = 0 , \qquad \omega^2 \hat{v}_y = 0 , \qquad (\omega^2 - k^2c^2)\, \hat{v}_z = 0 . \tag{5.12}$$

The important solutions are given by

$$\omega = \pm k\, c , \qquad \hat{v}_x = \hat{v}_y = 0 , \qquad \hat{v}_z \text{ arbitrary} , \tag{5.13}$$

representing plane *sound waves* travelling to the right ($+$) or to the left ($-$). These waves are *compressible* ($\nabla \cdot \mathbf{v}_1 \neq 0$) and *longitudinal* ($\mathbf{v}_1 \parallel \mathbf{k}$). The other solutions,

$$\omega^2 = 0 , \qquad \hat{v}_x , \hat{v}_y \text{ arbitrary} , \qquad \hat{v}_z = 0 , \tag{5.14}$$

just correspond to time-independent incompressible transverse ($\mathbf{v}_1 \perp \mathbf{k}$) translations. They do not represent interesting physics, but simply establish the completeness of the velocity representation.

Recapitulating, we have transformed the original system (5.5)–(5.7) of first order PDEs for the primitive variables $\rho_1, \mathbf{v}_1, p_1$ into the second order wave equation (5.8) in terms of $\mathbf{v}_1$ alone. This transformation highlights the main physics of the problem, which is sound wave propagation, but it also yields some trivial $\omega = 0$ solutions that are not interesting from the physical point of view.

Now count: the first order system appears to have five degrees of freedom represented by the five primitive variables, whereas the second order system appears to have six degrees of freedom since there are three components of $\mathbf{v}_1$ and the eigenvalue is squared. One easily recognizes though that the second order system actually only has four degrees of freedom, since the quadratic dependence on ω does not double the actual number of translations (5.14). This spurious doubling of the eigenvalue $\omega = 0$ happened when we applied the operator $\partial/\partial t + \mathbf{v}_0 \cdot \nabla$ to Eq. (5.6) to eliminate p_1. One easily checks that the solutions (5.13) and (5.14), with $\omega = 0$, are solutions of the original system of equations (5.5)–(5.7). Hence, we have actually *lost one degree of freedom* in the reduction to the wave equation in terms of $\mathbf{v}_1$ only. This happened when we dropped Eq. (5.5) for ρ_1. By inserting $\mathbf{v}_1 = 0$ in the original system we find the signature of this lost mode:

$$\omega\hat{\rho} = 0 \quad \Rightarrow \quad \omega = 0 , \quad \hat{\rho} \text{ arbitrary} , \quad \text{but } \hat{\mathbf{v}} = 0 \ \& \ \hat{p} = 0 . \tag{5.15}$$

This mode is called an *entropy wave*: it represents a perturbation of the density and, hence, of the entropy function $S \equiv p\rho^{-\gamma}$, since the pressure is not perturbed. Like the translations (5.14), the entropy mode does not represent important physics in this case. It is needed to account for the degrees of freedom of the different representations. This is an important issue in computational studies. It will return in the MHD analysis below, where the system is much more involved, so that the distinction between genuine and spurious solutions is much less transparent.

5.2 MHD waves

5.2.1 Symmetric representation in primitive variables

We perform a similar analysis for the magnetohydrodynamic system. In Chapter 4, the basic MHD equations were presented in two forms, viz. Eqs. (4.12)–(4.15) for the variables ρ, $\mathbf{v}$, p, $\mathbf{B}$, and Eqs. (4.45)–(4.48) for ρ, $\mathbf{v}$, e, $\mathbf{B}$. Since the derivation of conservation laws was based on the latter set of equations, we will exploit that set here. We drop the gravitational term in the momentum equation, since it is incompatible with the assumption of a homogeneous plasma, and convert the two occurring cross-products by means of the vector identities of the Appendix:

$$-\mathbf{j}\times\mathbf{B} = -(\nabla\times\mathbf{B})\times\mathbf{B} \overset{(A.8)}{=} (\nabla\mathbf{B})\cdot\mathbf{B} - \mathbf{B}\cdot\nabla\mathbf{B}\,,$$

$$\nabla\times\mathbf{E} = -\nabla\times(\mathbf{v}\times\mathbf{B}) \overset{(A.13)}{=} -\mathbf{B}\cdot\nabla\mathbf{v} + \mathbf{B}\nabla\cdot\mathbf{v} + \mathbf{v}\cdot\nabla\mathbf{B}\,.$$

This yields the following set of equations:

$$\frac{\partial\rho}{\partial t} + \nabla\cdot(\rho\mathbf{v}) = 0\,, \tag{5.16}$$

$$\rho\frac{\partial\mathbf{v}}{\partial t} + \rho\mathbf{v}\cdot\nabla\mathbf{v} + (\gamma-1)\nabla(\rho e) + (\nabla\mathbf{B})\cdot\mathbf{B} - \mathbf{B}\cdot\nabla\mathbf{B} = 0\,, \tag{5.17}$$

$$\frac{\partial e}{\partial t} + \mathbf{v}\cdot\nabla e + (\gamma-1)e\nabla\cdot\mathbf{v} = 0\,, \tag{5.18}$$

$$\frac{\partial\mathbf{B}}{\partial t} + \mathbf{v}\cdot\nabla\mathbf{B} + \mathbf{B}\nabla\cdot\mathbf{v} - \mathbf{B}\cdot\nabla\mathbf{v} = 0\,, \qquad \nabla\cdot\mathbf{B} = 0\,, \tag{5.19}$$

which is yet another form of the basic nonlinear MHD equations.

We again choose an *infinite homogeneous plasma at rest* ($\mathbf{v}_0 = 0$) as the background state,

$$\rho = \rho_0\,, \qquad e = e_0 \equiv p_0/[(\gamma-1)\rho_0]\,, \qquad \mathbf{B} = \mathbf{B}_0\,, \tag{5.20}$$

where ρ_0, e_0 and B_0 are constants in space and time. Linearization proceeds as in Section 5.1. Since there are no gradients of the equilibrium quantities, terms like $\mathbf{v}_1\cdot\nabla\rho_0$ disappear. Again, we only keep linear terms but neglect nonlinear terms, like $\rho_0\mathbf{v}_1\cdot\nabla\mathbf{v}_1$. This results in the set of linearized MHD equations:

$$\frac{\partial\rho_1}{\partial t} + \rho_0\nabla\cdot\mathbf{v}_1 = 0\,, \tag{5.21}$$

$$\rho_0\frac{\partial\mathbf{v}_1}{\partial t} + (\gamma-1)(e_0\nabla\rho_1 + \rho_0\nabla e_1) + (\nabla\mathbf{B}_1)\cdot\mathbf{B}_0 - \mathbf{B}_0\cdot\nabla\mathbf{B}_1 = 0\,, \tag{5.22}$$

$$\frac{\partial e_1}{\partial t} + (\gamma-1)e_0\nabla\cdot\mathbf{v}_1 = 0\,, \tag{5.23}$$

$$\frac{\partial\mathbf{B}_1}{\partial t} + \mathbf{B}_0\nabla\cdot\mathbf{v}_1 - \mathbf{B}_0\cdot\nabla\mathbf{v}_1 = 0\,, \qquad \nabla\cdot\mathbf{B}_1 = 0\,. \tag{5.24}$$

Two characteristic speeds describe the background state, viz. *the sound speed* and *the vectorial Alfvén speed*,

$$c = \sqrt{\gamma p_0/\rho_0}\,, \qquad \mathbf{b} \equiv \mathbf{B}_0/\sqrt{\rho_0}\,, \tag{5.25}$$

where $|\mathbf{b}| \equiv v_A$ is the scalar Alfvén speed, defined in Eq. (2.153) of Section 2.4.2. Together, they govern the speed of propagation of magnetohydrodynamic waves.

We then construct *dimensionless variables* for the perturbations,

$$\bar{\rho} \equiv \rho_1/(\gamma\rho_0)\,, \quad \bar{\mathbf{v}} \equiv \mathbf{v}_1/c\,, \quad \bar{e} \equiv e_1/(\gamma e_0)\,, \quad \bar{\mathbf{B}} \equiv \mathbf{B}_1/(c\sqrt{\rho_0})\,, \tag{5.26}$$

so that the linearized MHD equations only involve the coefficients c and $\mathbf{b}$ (and γ), and we consider *plane wave solutions*,

$$\bar{\rho} = \bar{\rho}(\mathbf{r}, t) = \hat{\rho}\, \mathrm{e}^{\mathrm{i}(\mathbf{k}\cdot\mathbf{r}-\omega t)}\,, \quad \text{etc.} \tag{5.27}$$

This yields the following algebraic system of eigenvalue equations:

$$\begin{aligned} c\mathbf{k}\cdot\hat{\mathbf{v}} &= \gamma\omega\,\hat{\rho}\,, &(5.28)\\ \mathbf{k}c\hat{\rho} + \mathbf{k}c\hat{e} + (\mathbf{kb}\cdot - \mathbf{k}\cdot\mathbf{b})\,\hat{\mathbf{B}} &= \omega\,\hat{\mathbf{v}}\,, &(5.29)\\ c\mathbf{k}\cdot\hat{\mathbf{v}} &= \frac{\gamma}{\gamma-1}\,\omega\,\hat{e}\,, &(5.30)\\ (\mathbf{bk}\cdot - \mathbf{b}\cdot\mathbf{k})\,\hat{\mathbf{v}} &= \omega\,\hat{\mathbf{B}}\,, \qquad \mathbf{k}\cdot\hat{\mathbf{B}} = 0\,. &(5.31) \end{aligned}$$

Ignoring the constraint $\mathbf{k}\cdot\hat{\mathbf{B}} = 0$ for the moment, this system is solvable if the determinant of the 8×8 matrix vanishes. The medium is now anisotropic through the presence of the background magnetic field $\mathbf{B}_0$, but there is no loss of generality if we choose $\mathbf{b}$ along the z-axis and $\mathbf{k}$ in the x–z plane (like in Fig. 5.1(a)):

$$\mathbf{b} = (0, 0, b)\,, \qquad \mathbf{k} = (k_\perp, 0, k_\parallel)\,. \tag{5.32}$$

This leads to the following matrix representation of the eigenvalue problem:

$$\begin{pmatrix} 0 & k_\perp c & 0 & k_\parallel c & 0 & 0 & 0 & 0\\ k_\perp c & 0 & 0 & 0 & k_\perp c & -k_\parallel b & 0 & k_\perp b\\ 0 & 0 & 0 & 0 & 0 & 0 & -k_\parallel b & 0\\ k_\parallel c & 0 & 0 & 0 & k_\parallel c & 0 & 0 & 0\\ 0 & k_\perp c & 0 & k_\parallel c & 0 & 0 & 0 & 0\\ 0 & -k_\parallel b & 0 & 0 & 0 & 0 & 0 & 0\\ 0 & 0 & -k_\parallel b & 0 & 0 & 0 & 0 & 0\\ 0 & k_\perp b & 0 & 0 & 0 & 0 & 0 & 0 \end{pmatrix} \begin{pmatrix} \hat{\rho}\\ \hat{v}_x\\ \hat{v}_y\\ \hat{v}_z\\ \hat{e}\\ \hat{B}_x\\ \hat{B}_y\\ \hat{B}_z \end{pmatrix} = \omega \begin{pmatrix} \gamma\hat{\rho}\\ \hat{v}_x\\ \hat{v}_y\\ \hat{v}_z\\ \frac{\gamma}{\gamma-1}\,\hat{e}\\ \hat{B}_x\\ \hat{B}_y\\ \hat{B}_z \end{pmatrix}. \tag{5.33}$$

Except for a few shortcomings, to be addressed, this expression is satisfactory since it clearly identifies the two new features of MHD waves as compared to sound waves, viz. the occurrence of the Alfvén speed in addition to the sound speed and the anisotropy expressed by the appearance of two components of the wave vector. We could compute the dispersion equation from the determinant and study the associated waves, but we have good reasons to postpone this to Section 5.2.3.

In the formulation (5.33), one purpose has been realized, viz. the demonstration of *the symmetry of the operator describing the linearized system*, which is one of the most important properties of ideal magnetohydrodynamics. This has been demonstrated now for the simplest case, that of homogeneous plasmas. Its generalization to inhomogeneous plasmas will be an important issue in the spectral theory of MHD, to be discussed in Chapter 6. The symmetry of the linearized system is closely related to an analogous property of the original nonlinear equations, viz. the fact that *the nonlinear ideal MHD equations are symmetric hyperbolic partial differential equations*. This will be proved in Section 5.4.3. These symmetry properties were behind our choice for the representation in terms of the variables $\rho, \mathbf{v}, e, \mathbf{B}$, corresponding to the variables of the moment expansion of the kinetic equations (Chapter 3). This formulation lends itself most naturally for extensions, e.g. with dissipative effects, so that it is frequently exploited in large-scale numerical efforts for the solution of the MHD equations in practical applications.

▷ **Generalized eigenvalue problem** A minor shortcoming of the eigenvalue problem (5.33) is that it still contains the factors γ and $\gamma/(\gamma-1)$ in the vector on the RHS. Formally, this implies that we are dealing with a *generalized eigenvalue problem*:

$$\mathsf{A}\cdot\mathbf{x} = \lambda\,\mathsf{B}\cdot\mathbf{x}\,. \tag{5.34}$$

Here, A denotes the 8×8 matrix and $\mathbf{x}$ the 8-vector of unknowns on the LHS of Eq. (5.33), $\lambda \equiv \omega$ is the eigenvalue, and B is a diagonal matrix which deviates from the unit matrix by the two mentioned factors multiplying the variables $\hat{\rho}$ and $\hat{e}$. By the transformation $\mathsf{A}' \equiv \mathsf{B}^{-1/2}\cdot\mathsf{A}\cdot\mathsf{B}^{-1/2}\,,\ \mathbf{x}'\equiv\mathsf{B}^{1/2}\,\mathbf{x}$, an *ordinary* eigenvalue problem is obtained:

$$\mathsf{A}'\cdot\mathbf{x}' = \lambda\,\mathbf{x}'\,, \tag{5.35}$$

where

$$\mathsf{A}' = \begin{pmatrix} 0 & \frac{1}{\sqrt{\gamma}}\mathbf{k}^{\mathrm{T}}c & 0 & 0 \\ \frac{1}{\sqrt{\gamma}}\mathbf{k}\,c & 0 & \sqrt{\frac{\gamma-1}{\gamma}}\,\mathbf{k}\,c & \mathbf{k}\,\mathbf{b}^{\mathrm{T}}-\mathbf{k}\cdot\mathbf{b}\,\mathsf{I} \\ 0 & \sqrt{\frac{\gamma-1}{\gamma}}\,\mathbf{k}^{\mathrm{T}}c & 0 & 0 \\ 0 & \mathbf{b}\,\mathbf{k}^{\mathrm{T}}-\mathbf{k}\cdot\mathbf{b}\,\mathsf{I} & 0 & 0 \end{pmatrix}, \qquad \mathbf{x}' = \begin{pmatrix} \sqrt{\gamma}\,\hat{\rho} \\ \hat{\mathbf{v}} \\ \sqrt{\frac{\gamma}{\gamma-1}}\,\hat{e} \\ \hat{\mathbf{B}} \end{pmatrix}.$$

The superscript 'T' indicates row vectors. The 'cost' of this transformation is the appearance of odd square root factors in the matrix A' and the vector $\mathbf{x}'$. These factors are really spurious since they will not appear in the final dispersion equation, as one can directly check from the present form (but much easier from the one derived in Section 5.2.3). ◁

5.2.2 Entropy wave and magnetic field constraint

We will transform to the more transparent velocity representation in Section 5.2.3. But first, in analogy with our exposition of sound waves of Section 5.1.2, we have to account for two subtleties related to the presence of a genuine and a spurious solution $\omega = 0$ of the eigenvalue problem. Such solutions are called *marginal* since the transition from stability to instability occurs at that value of ω, as we will see in Chapter 6. The analysis of these modes is facilitated by means of the apparent detour of replacing the thermodynamic variables ρ and e by the entropy S and the pressure p. As a bonus, the redundant factors γ will disappear.

(a) Marginal entropy wave To that end, we linearize the dynamical equations for the entropy function, $S \equiv p\rho^{-\gamma}$, and the pressure, $p = (\gamma-1)\rho e$, which may replace Eqs. (5.28) for the density ρ and (5.30) for the internal energy e:

$$\frac{\partial S_1}{\partial t} = 0\,, \tag{5.36}$$

$$\frac{\partial p_1}{\partial t} + \gamma p_0 \nabla\cdot\mathbf{v}_1 = 0\,, \tag{5.37}$$

and construct the appropriate dimensionless variables,

$$\bar{S} \equiv S_1/(\gamma S_0) = -(\gamma-1)\bar{\rho} + \bar{e}\,, \qquad \bar{p} \equiv p_1/(\gamma p_0) = \bar{\rho} + \bar{e}\,. \tag{5.38}$$

Any two of the four variables $\bar{\rho}, \bar{e}, \bar{S}, \bar{p}$, together with $\bar{\mathbf{v}}$ and $\bar{\mathbf{B}}$, may be chosen to express the eigenvalue problem. However, not every choice leads to a symmetric eigenvalue problem! It turns out to be expedient to transform to the unusual $(\bar{S}, \bar{\mathbf{v}}, \bar{p}, \bar{\mathbf{B}})$ representation for the eigenvalue problem:

$$-\omega\,\hat{S} = 0\,, \tag{5.39}$$

$$-\omega\,\hat{\mathbf{v}} + \mathbf{k}\,c\,\hat{p} + (\mathbf{k}\mathbf{b}\cdot - \mathbf{k}\cdot\mathbf{b})\,\hat{\mathbf{B}} = 0\,, \tag{5.40}$$

$$-\omega\,\hat{p} + c\,\mathbf{k}\cdot\hat{\mathbf{v}} = 0\,, \tag{5.41}$$

$$-\omega\,\hat{\mathbf{B}} + (\mathbf{b}\mathbf{k}\cdot - \mathbf{b}\cdot\mathbf{k})\,\hat{\mathbf{v}} = 0\,, \qquad \mathbf{k}\cdot\hat{\mathbf{B}} = 0\,. \tag{5.42}$$

This representation is more compact than the previous one, fully symmetric, and free of redundant factors γ. More important for our present purpose is that it clearly exhibits the peculiar marginal *entropy wave*, for which the entropy and, hence, density and internal energy perturbations are finite, but all other variables (pressure, velocity and magnetic field perturbations) vanish:

$$\omega = 0\,, \qquad \hat{S} = -\gamma\hat{\rho} = \gamma\hat{e} \neq 0\,, \qquad \hat{p} = \hat{\rho}\hat{e} = 0\,. \tag{5.43}$$

Superposition of the different Fourier harmonics, like in Eq. (5.9) of Section 5.1.2, gives a mode with a completely arbitrary spatial distribution of the entropy. This may be realized, for example, by a local increase $\hat{e}$ of the internal energy (or the temperature) and a decrease $\hat{\rho}$ of the density such that there is no net pressure change $\hat{p}$. *This marginal mode is genuine but does not represent important physics*, at least not in the context of the ideal MHD model. Since it does not propagate and since the entropy does not couple to the other variables, not much is lost if this part of the problem is dropped altogether so that, actually, a 7×7 representation is obtained. This is effected by the transformation to the velocity representation, to be discussed next, where also a spurious solution is removed, so that, effectively, a 6×6 representation is obtained.

(b) Magnetic field constraint So far, we have ignored accounting for the constraint $\mathbf{k}\cdot\hat{\mathbf{B}} = 0$, originating from the general condition $\nabla\cdot\mathbf{B} = 0$ on magnetic fields. If one were to just solve the eigenvalue problem as it is, one would get eight solutions with *one spurious eigenvalue* $\omega = 0$ (to be distinguished from the genuine eigenvalue $\omega = 0$ of the entropy wave). This may be seen by operating with the projector $\mathbf{k}\cdot$ onto Eq. (5.31), or (5.42), which gives

$$\omega\,\mathbf{k}\cdot\hat{\mathbf{B}} = 0\,,$$

suggesting that $\omega = 0$ is an eigenvalue with eigenvectors satisfying $\mathbf{k}\cdot\hat{\mathbf{B}} \neq 0$, in glaring conflict with the constraint. The problem of spurious marginal eigenvalues is a more serious one than it appears at this stage. For example, in numerical MHD programs calculating eigenvalues to determine whether a certain plasma configuration is stable or not, a spurious eigenvalue $\omega = 0$ may not be distinguishable from a genuine transition from stability to instability.

To remove this spurious marginal eigenvalue, the constraint has to be incorporated in the eigenvalue problem by eliminating one of the magnetic field variables to obtain a 7×7 matrix representation. For example, one could eliminate $\hat{B}_z$ by simply substituting $\hat{B}_z = -(k_\perp/k_\parallel)\,\hat{B}_x$ in the second row and dropping the last row of Eq. (5.33). However, the symmetry of the matrix would then be lost. A more satisfactory approach to eliminate the spurious eigenvalue is obtained by exploiting the wave vector projection.

▷ **Wave vector projection** This approach exploits *new variables reflecting the physics of the problem* by projecting the Fourier components of the vectors $\hat{\mathbf{v}}$ and $\hat{\mathbf{B}}$ on the three directions associated with the wave vector $\mathbf{k}$. This amounts to exploiting the two components of *the vorticity* $\nabla\times\hat{\mathbf{v}}$ and *the compressibility* $\nabla\cdot\hat{\mathbf{v}}$ for the velocity, and restricting the magnetic field variables to two components of *the current density* $\nabla\times\hat{\mathbf{B}}$:

$$\hat{v}_{1,2}\equiv[(\mathbf{k}/k)\times\hat{\mathbf{v}}]_{x,y}\,,\qquad \hat{v}_3\equiv(\mathbf{k}/k)\cdot\hat{\mathbf{v}}\,,\qquad \hat{B}_{1,2}\equiv[(\mathbf{k}/k)\times\hat{\mathbf{B}}]_{x,y}\,. \tag{5.44}$$

Operating with the projectors $(\mathbf{k}/k)\cdot$ and $(\mathbf{k}/k)\times$ on Eqs. (5.40) and (5.42) yields the following 7×7 representation in terms of the variables $\hat{S}, \hat{v}_1, \hat{v}_2, \hat{v}_3, \hat{p}, \hat{B}_1$ and $\hat{B}_2$:

$$\begin{pmatrix} 0 & 0 & 0 & 0 & 0 & 0 & 0 \\ 0 & 0 & 0 & 0 & 0 & -k_\parallel b & 0 \\ 0 & 0 & 0 & 0 & 0 & 0 & -k_\parallel b \\ 0 & 0 & 0 & 0 & kc & -k_\perp b & 0 \\ 0 & 0 & 0 & kc & 0 & 0 & 0 \\ 0 & -k_\parallel b & 0 & -k_\perp b & 0 & 0 & 0 \\ 0 & 0 & -k_\parallel b & 0 & 0 & 0 & 0 \end{pmatrix} \begin{pmatrix} \hat{S} \\ \hat{v}_1 \\ \hat{v}_2 \\ \hat{v}_3 \\ \hat{p} \\ \hat{B}_1 \\ \hat{B}_2 \end{pmatrix} = \omega \begin{pmatrix} \hat{S} \\ \hat{v}_1 \\ \hat{v}_2 \\ \hat{v}_3 \\ \hat{p} \\ \hat{B}_1 \\ \hat{B}_2 \end{pmatrix}. \tag{5.45}$$

The representation is symmetric and even simpler than Eq. (5.33). Discarding the entropy wave by dropping the first row and column results in the 6×6 representation, equivalent to the velocity representation discussed in Section 5.2.3, for the main MHD waves. ◁

▷ **Numerical 8-wave scheme with a $\nabla\cdot\mathbf{B}$ wave** In numerical studies of the nonlinear MHD equations, the constraint $\nabla\cdot\mathbf{B}=0$ also poses a problem since it does not correspond to a proper evolution equation. This problem has been addressed by Powell [507] [508], following an older idea of Godunov [213], by artificially extending the MHD system with a '$\nabla\cdot\mathbf{B}$ wave'. To that end, the RHSs of the conservation equations (4.59)–(4.62), or rather (4.71)–(4.74) without gravity, are replaced by source terms proportional to $\nabla\cdot\mathbf{B}$:

$$\frac{\partial\rho}{\partial t}+\nabla\cdot\boldsymbol{\pi}=0\,, \tag{5.46}$$

$$\frac{\partial\boldsymbol{\pi}}{\partial t}+\nabla\cdot\Big[\boldsymbol{\pi}\mathbf{v}+(p+\tfrac{1}{2}B^2)\,\mathbf{I}-\mathbf{B}\mathbf{B}\Big]=-\mathbf{B}\,\nabla\cdot\mathbf{B}\,, \tag{5.47}$$

$$\frac{\partial\mathcal{H}}{\partial t}+\nabla\cdot\Big[(\mathcal{H}+p+\tfrac{1}{2}B^2)\mathbf{v}-\mathbf{v}\cdot\mathbf{B}\,\mathbf{B}\Big]=-\mathbf{v}\cdot\mathbf{B}\,\nabla\cdot\mathbf{B}\,, \tag{5.48}$$

$$\frac{\partial\mathbf{B}}{\partial t}+\nabla\cdot(\mathbf{v}\mathbf{B}-\mathbf{B}\mathbf{v})=-\mathbf{v}\,\nabla\cdot\mathbf{B}\,. \tag{5.49}$$

The source terms should become vanishingly small in the limit of vanishing step size. This provides a numerical scheme of (near) conservation equations, that has been employed, for example, for the construction of solutions of the interaction of the solar wind with the magnetosphere of the Earth.

Of course, the addition of $\nabla\cdot\mathbf{B}$ terms in these equations implies that unphysical, numerically created, magnetic monopoles are permitted [89]. As noticed by Janhunen [332], the problem of possible non-positive numerical values of the pressure, calculated by using Eq. (4.75), is enlarged this way since the above modified momentum and energy 'conservation' equations involve unbalanced forces on those monopoles. By means of the modified Maxwell's equations involving hypothetical magnetic monopoles, he then shows that a more consistent set of modified MHD 'conservation' equations is obtained by restoring the momentum and energy equations (5.47) and (5.48) to have a vanishing RHS, so that only the induction equation keeps the $\nabla\cdot\mathbf{B}$ source term. Shortly afterwards, it was shown by Dellar [147] that this set is easily extended to a modified set of relativistic, Lorentz invariant, MHD equations, which contain the earlier, Galilean invariant, set in the limit $v/c\to0$.

The modification of Faraday's law, Eq. (5.49), is the essential one. With some algebra one obtains an evolution equation for $\nabla\cdot\mathbf{B}$ from it,

$$\frac{\partial}{\partial t}(\nabla\cdot\mathbf{B})+\nabla\cdot(\mathbf{v}\,\nabla\cdot\mathbf{B})=0\,, \tag{5.50}$$

replacing the constraint $\nabla\cdot\mathbf{B}=0$. This equation is in genuine conservation form now: $\nabla\cdot\mathbf{B}$ is convected with the fluid in the same way as the density ρ. This implies that if $\nabla\cdot\mathbf{B}$ is small enough initially, and at the boundaries, it should remain small. Numerical errors $\nabla \cdot \mathbf{B} \neq 0$ are then convected with the flow and, hopefully, out of the computational domain. Since Maxwell's equations are modified here, the physicist cannot offer assistance to the accountant here. All that matters for him is whether the books finally balance, i.e. whether the numerical scheme is accurate enough to keep $\nabla \cdot \mathbf{B}$ small enough to qualify as vanishing. For more on the computational aspects of the implementation of the $\nabla \cdot \mathbf{B} = 0$ constraint: see Tóth [603]. ◁

5.2.3 Reduction to velocity representation: three waves

We now transform to the velocity representation since it is the more powerful one. It gets rid of the entropy wave and absorbs the constraint $\mathbf{k} \cdot \hat{\mathbf{B}} = 0$ so that the associated spurious eigenvalue $\omega = 0$ is eliminated as well. For the reduction to the velocity variable, the perturbations $\rho_1, e_1, \mathbf{B}_1$ are expressed in terms of $\mathbf{v}_1$ by means of Eqs. (5.21), (5.23) and (5.24), and substituted into the momentum equation (5.22). This yields *the MHD wave equation for a homogeneous medium*:

$$\frac{\partial^2 \mathbf{v}_1}{\partial t^2} - \Big[(\mathbf{b} \cdot \nabla)^2 \, \mathbf{I} + (b^2 + c^2) \, \nabla \nabla - \mathbf{b} \cdot \nabla \, (\nabla \mathbf{b} + \mathbf{b} \nabla) \Big] \cdot \mathbf{v}_1 = 0 \,. \tag{5.51}$$

Note that this equation contains the sound wave equation (5.8), for $\mathbf{v}_0 = 0$, as the special case $\mathbf{b} = 0$ (as it should). Inserting plane wave solutions gives the required eigenvalue equation:

$$\Big\{ \Big[\omega^2 - (\mathbf{k} \cdot \mathbf{b})^2 \Big] \, \mathbf{I} - (b^2 + c^2) \, \mathbf{k} \mathbf{k} + \mathbf{k} \cdot \mathbf{b} \, (\mathbf{k} \mathbf{b} + \mathbf{b} \mathbf{k}) \Big\} \cdot \hat{\mathbf{v}} = 0 \,, \tag{5.52}$$

or, in components:

$$\begin{pmatrix} -k_\perp^2 (b^2 + c^2) - k_\parallel^2 b^2 & 0 & -k_\perp k_\parallel c^2 \\ 0 & -k_\parallel^2 b^2 & 0 \\ -k_\perp k_\parallel c^2 & 0 & -k_\parallel^2 c^2 \end{pmatrix} \begin{pmatrix} \hat{v}_x \\ \hat{v}_y \\ \hat{v}_z \end{pmatrix} = -\omega^2 \begin{pmatrix} \hat{v}_x \\ \hat{v}_y \\ \hat{v}_z \end{pmatrix} . \tag{5.53}$$

Hence, a 3×3 symmetric matrix equation is obtained in terms of the variable $\hat{\mathbf{v}}$, with *quadratic eigenvalue* ω^2, corresponding to the original 6×6 representation where the eigenvalue ω appeared linearly.

The reduction to a description in terms of the velocity alone may be generalized to inhomogeneous plasmas, where it leads to the very powerful force operator formalism (Section 6.2). It should be kept in mind that this reduction is only possible in *ideal* MHD, i.e. in the absence of dissipative effects. With dissipation, the basic equations (5.16)–(5.19) have additional terms [see, e.g., Eq. (2.136)], spoiling the possibility of reduction in terms of the velocity alone. However, the analysis leading to Eqs. (5.39)–(5.42) may be extended easily. That is why we have presented both the representation in primitive variables and the velocity representation.

With the description in terms of the velocity we have lost the marginal entropy mode $\omega = 0$. For the sake of completeness, we will include this mode again. This is done by simply multiplying the determinant of Eq. (5.53) with the factor ω to yield the determinant of the original system (5.39)–(5.42), or rather (5.45):

$$\det = \omega \, (\omega^2 - k_\parallel^2 b^2) \, \Big[\omega^4 - k^2 (b^2 + c^2) \, \omega^2 + k_\parallel^2 k^2 b^2 c^2 \Big] = 0 \,. \tag{5.54}$$

This algebraic expression, with solutions $\omega = \omega_i(\mathbf{k})$ $(i = 1, \ldots 7)$, is called *the dispersion equation* for the MHD waves. Subsequently putting each of the three factors equal to zero gives the eigenfrequencies of the waves, and substituting those frequencies back into (5.39)–(5.42) or Eqs. (5.53) gives the relationships between the amplitudes $\hat{s}$, $\hat{\mathbf{v}}$, $\hat{p}$, $\hat{\mathbf{B}}$ which characterize the eigenfunctions. The dispersion equation (5.54) admits the following four kinds of solutions:

(a) Entropy 'waves' As we have already seen in Section 5.2.2(a), the eigenfrequency of these waves vanishes:

$$\omega = \omega_{\mathrm{E}} \equiv 0\,, \tag{5.55}$$

whereas the eigenfunctions just involve the entropy:

$$\hat{\mathbf{v}} = \hat{\mathbf{B}} = 0\,, \quad \hat{p} = 0\,, \quad \text{but} \quad \hat{S} \neq 0\,. \tag{5.56}$$

The entropy waves are quite degenerate: they do not propagate, they do not involve flow, magnetic field, or pressure perturbations. They just constitute a perturbation of the entropy (or density and internal energy) which would be carried with the flow if there were a background velocity field. Now, they just sit there: not very exciting. The use of such solutions is usually that they serve as a reminder for the possibility of new waves if additional physics is brought into the model. For the present discussion, they do not serve any purpose. That is why the formulation in terms of $\hat{\mathbf{v}}$ is so attractive: these marginal modes are automatically excluded.

(b) Alfvén waves The eigenfrequency of these waves is determined by the parallel wave vector and the Alfvén velocity:

$$\omega = \pm\omega_{\mathrm{A}} = \pm k_{\parallel} b\,, \qquad \omega_{\mathrm{A}}^2 \equiv k^2 b^2 \cos^2\vartheta\,, \tag{5.57}$$

where ϑ is the angle between $\mathbf{k}$ and $\mathbf{b}$, i.e. $\mathbf{B}_0$. There are two solutions, one for $\omega = \omega_{\mathrm{A}}$ corresponding to waves propagating in the direction of $\mathbf{B}_0$, and another one for $\omega = -\omega_{\mathrm{A}}$ corresponding to waves propagating in the opposite direction.

The eigenfunctions of the Alfvén waves just involve the perpendicular components of the velocity and the magnetic field (Fig. 5.1(a)), whereas all other components and the perturbations of all thermodynamic variables vanish:

$$\hat{B}_y = -\hat{v}_y \neq 0\,, \qquad \hat{v}_x = \hat{v}_z = \hat{B}_x = \hat{B}_z = \hat{S} = \hat{p} = 0\,. \tag{5.58}$$

These are the waves that were already introduced in Section 2.4.2. They are incompressible and purely transverse, as regards both $\hat{\mathbf{v}}$ and $\hat{\mathbf{B}}$. Alfvén waves are the most important MHD waves since they are a direct result of flux conservation and magnetic field lines frozen into the fluid: a perpendicular flow velocity $\hat{\mathbf{v}}$ results in a field perturbation $\hat{\mathbf{B}}$ of opposite sign which just causes the field lines to follow the flow (as illustrated in Fig. 2.9).

(c) Fast and slow magneto-acoustic waves The eigenfrequencies of these waves are obtained from the quartic factor of the dispersion equation (5.54):

$$\omega = \pm\omega_{\mathrm{s,f}}\,, \qquad \omega_{\mathrm{s,f}}^2 \equiv \tfrac{1}{2}k^2(b^2+c^2)\Big(1 \pm \sqrt{1-\sigma\cos^2\vartheta}\,\Big)\,, \tag{5.59}$$

where the first $\pm$ sign refers to wave propagation to the right (+) or to the left (−), and the second refers to *the fast* (+) or *slow* (−) *magneto-acoustic wave*, respectively. The auxiliary parameter

$$\sigma \equiv \frac{4b^2c^2}{(b^2+c^2)^2} = \Big[\tfrac{1}{2}(b/c + c/b)\Big]^{-2} \qquad (0 \le \sigma \le 1) \tag{5.60}$$

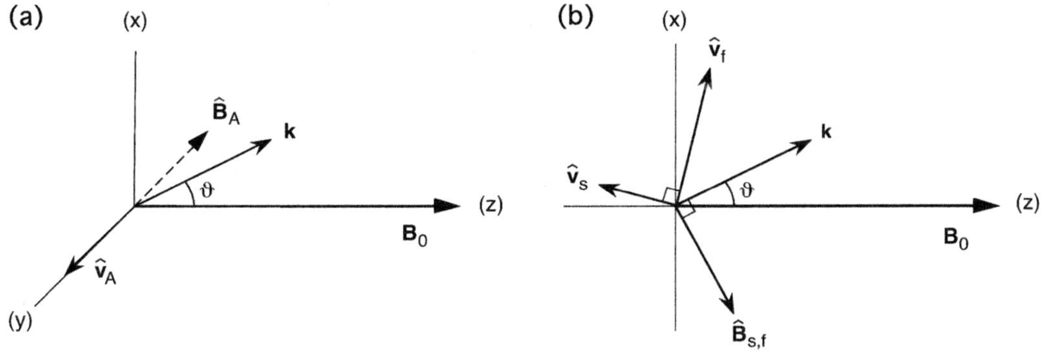

Fig. 5.1 Velocity and magnetic field perturbations for: (a) Alfvén waves, (b) magneto-sonic waves.

is a function of the ratio of the two characteristic speeds b and c in the problem, where $\sigma = 0$ when $b = 0$ or $c = 0$, and $\sigma = 1$ when $b = c$.

The eigenfunctions of the magneto-sonic waves exhibit a complicated dependence on almost all of the variables (Fig. 5.1(b)):

$$\begin{aligned} \hat{v}_z &= \alpha_{s,f}\,(k_\parallel/k_\perp)\,\hat{v}_x \neq 0\,, & \hat{v}_y &= 0\,, \\ \hat{B}_z &= -(k_\perp/k_\parallel)\,\hat{B}_x = (k_\perp b/\omega_{s,f})\,\hat{v}_x \neq 0\,, & \hat{B}_y &= 0\,, \\ \hat{p} &= \big[\alpha_{s,f}\,\omega_{s,f}/(k_\perp c)\big]\,\hat{v}_x \neq 0\,, & \hat{S} &= 0\,, \end{aligned} \tag{5.61}$$

where the factor $\alpha_{s,f}$ distinguishes between the fast and slow eigenfunctions:

$$\alpha_{s,f} \equiv 1 - k^2 b^2/\omega_{s,f}^2\,, \qquad \text{so that} \quad \alpha_s \leq 0 \quad \text{and} \quad \alpha_f \geq 0\,. \tag{5.62}$$

The magneto-acoustic waves are composed of both magnetic (b) and acoustic (c) constituents. The perturbations $\hat{\mathbf{v}}$ and $\hat{\mathbf{B}}$ are lying in the plane through $\mathbf{k}$ and $\mathbf{B}_0$, so that the total perturbed field $(\mathbf{B}_0 + \mathbf{B}_1)$ also lies in that plane. The fast and slow velocity fields are orthogonal to each other, whereas both magnetic field perturbations are orthogonal to $\mathbf{k}$ (since $\mathbf{k} \cdot \hat{\mathbf{B}} = 0$). Because the normalization of the waves is arbitrary, we have chosen $\hat{\mathbf{B}}_s = \hat{\mathbf{B}}_f$ in the picture for simplicity.

5.2.4 Dispersion diagrams

From now on, we will ignore the entropy waves. We then notice two quite general and important properties of the eigenfrequencies and eigenfunctions of the remaining three (slow, Alfvén and fast) MHD waves, viz.:

– *the eigenfrequencies are well ordered*,

$$0 \leq \omega_s^2 \leq \omega_A^2 \leq \omega_f^2 < \infty\,; \tag{5.63}$$

– *the eigenfunctions are mutually orthogonal*,

$$\hat{\mathbf{v}}_s \perp \hat{\mathbf{v}}_A \perp \hat{\mathbf{v}}_f\,. \tag{5.64}$$

The first property will turn out to play a crucial role in the spectral theory of MHD waves discussed in Chapter 6. The second property guarantees that an arbitrary velocity field may be decomposed

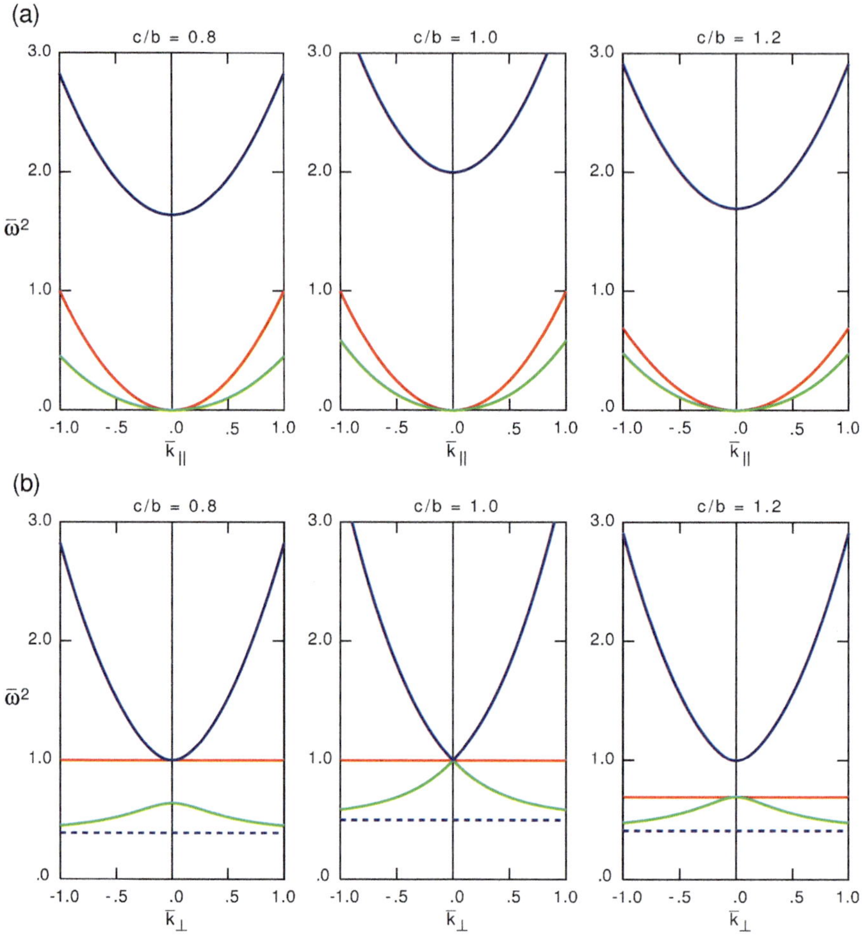

Fig. 5.2 Dispersion diagrams for the fast magneto-sonic (blue), the Alfvén (red), and the slow magneto-sonic (green) waves for three values of the ratio c/b of the sound speed to the Alfvén speed: (a) $\omega^2 = \omega^2(k_\parallel)$ with $\bar{k}_\perp = 1$, (b) $\omega^2 = \omega^2(k_\perp)$ with $\bar{k}_\parallel = 1$. Eigenvalues and wave vectors are normalized as $\bar{\omega}^2 \equiv \omega^2 \ell^2 / \max(b^2, c^2)$ and $\bar{k} \equiv k\ell$, with unit length ℓ.

at all times (e.g. at $t = 0$) in the three MHD waves. This implies that *the initial value problem is a well-posed problem.*

The three MHD waves exhibit a strong anisotropy depending on the direction of the wave vector $\mathbf{k}$ with respect to the magnetic field $\mathbf{B}_0$. This is expressed by the dispersion diagrams $\omega^2 = \omega^2(k_\parallel)$ and $\omega^2 = \omega^2(k_\perp)$ shown in Fig. 5.2. They are obtained from the dispersion equations (5.57) for the Alfvén waves and (5.59) for the magneto-sonic waves, where $k_\perp$ is kept fixed in the upper diagram and $k_\parallel$ in the lower. They clearly illustrate the ordering (5.63). The limits $k_\perp \to \infty$ of the functions $\omega^2 = \omega^2(k_\perp)$ provide the most distinctive property of the three dispersion curves. We return to this point in Section 5.3.3.

The behaviour of the magneto-sonic eigenfrequencies at small $k_\parallel$ or $k_\perp$ is obtained by expanding the square root factor in the dispersion expression (5.59): For *approximately perpendicular propa-*

gation ($\vartheta \approx \pi/2$, i.e. $k_\parallel \ll k$), the square root factor becomes $\sqrt{1-\sigma\cos^2\vartheta} \approx 1 - \frac{1}{2}\sigma\,(k_\parallel/k)^2$, so that

$$\omega_s^2 \approx k_\parallel^2 \frac{b^2c^2}{b^2+c^2} \quad \leq \quad \omega_A^2 = k_\parallel^2 b^2 \quad \ll \quad \omega_f^2 \approx k_\perp^2(b^2+c^2)\,. \tag{5.65}$$

Hence, the slow and the Alfvén frequencies approach each other and vanish for $k_\parallel \to 0$. This will turn out to be important in the stability studies of later chapters. For *purely parallel propagation* ($\vartheta = 0$, i.e. $k_\perp = 0$), the square root factor becomes $\sqrt{1-\sigma} = |b^2 - c^2|/(b^2+c^2)$, so that

$$\omega_s^2 = k_\parallel^2 \min(b^2, c^2) \;\leq\; \omega_A^2 = k_\parallel^2 b^2 \;\leq\; \omega_f^2 = k_\parallel^2 \max(b^2, c^2)\,. \tag{5.66}$$

Hence, either the slow frequency coincides with the Alfvén frequency (if $b^2 < c^2$) or the fast one coincides with ω_A^2 (if $b^2 > c^2$). Complete degeneracy of the three MHD waves obtains for $k_\perp = 0$ and $b = c$ (middle frame of Fig. 5.2(b)).

The dispersion diagrams shown in Fig. 5.2 depend qualitatively on the parameter c/b, measuring the relative contributions of hydrodynamic and magnetic effects. Notice that the parameter β, which measures the ratio of the plasma pressure to the magnetic pressure [defined in Eq. (4.34)], is related to the square of this ratio between the sound and the Alfvén frequency:

$$\beta \equiv 2\mu_0 p_0 / B_0^2 = (2/\gamma)\, c^2/b^2\,. \tag{5.67}$$

For $\gamma = 5/3$, this becomes $\beta = 1.2\, c^2/b^2$. The approximation $\beta \ll 1$, or $c^2 \ll b^2$, is frequently exploited since it applies to quite some relevant plasmas, like in tokamaks and the solar corona.

We will frequently return to the dispersion diagrams of Fig. 5.2 when discussing wave propagation in inhomogeneous media. But already we wish to draw the attention to some generic features.

(a) The Alfvén and slow frequencies ω_A and ω_s vanish for $k_\parallel = 0$. In that case, $\mathbf{k} \perp \mathbf{B}_0$, which implies that the waves do not bend the background magnetic field $\mathbf{B}_0$. This condition has important consequences for *stability*, as we will see.

(b) The eigenfrequencies ω^2 of the different waves depend monotonically on the value of either one of the two components of the wave vector $\mathbf{k}$. In particular, notice the unusual decreasing dependence $\omega^2 = \omega^2(k_\perp)$ for the slow magneto-sonic waves. This property determines the *general structure of the spectrum of MHD waves* in plasmas.

(c) For large values of the wave vector the eigenfrequencies asymptotically tend to either ∞ or some finite value, which turns out to be rather independent of the assumption of homogeneity of the plasma. This has extremely important consequences for *local wave propagation*.

In one form or another, all of these properties survive in the more general inhomogeneous plasmas which we will investigate in later chapters.

5.3 Phase and group diagrams

5.3.1 Basic concepts

We have discussed the peculiar properties of the MHD dispersion diagrams. We now analyse the general implications of a *dispersion equation* $\omega = \omega(\mathbf{k})$ of wave phenomena, relating the angular frequency ω to the wave vector $\mathbf{k}$. For the ideal MHD waves this is implicitly expressed by

Eq. (5.54). From this dispersion equation two important quantities may be derived. The first one is *the phase velocity,*

$$\mathbf{v}_{\mathrm{ph}} \equiv (\omega/k)\,\mathbf{n}\,, \qquad \mathbf{n} \equiv \mathbf{k}/k\,, \tag{5.68}$$

giving the speed of propagation of a single plane wave in the direction of $\mathbf{k}$. The magnitude of the phase velocity of the three MHD waves depends on the angle ϑ between $\mathbf{k}$ and the background magnetic field $\mathbf{B}_0$, so that $v_{\mathrm{ph}} \equiv \omega/k = f(\vartheta)$, but it does not depend on the magnitude k itself of the wave vector. Such waves are called *non-dispersive* because a wave packet, consisting of many components with different wave numbers, may propagate without distortion (at least in one direction). Such a packet propagates with *the group velocity,*

$$\mathbf{v}_{\mathrm{gr}} \equiv \frac{\partial\omega}{\partial\mathbf{k}} \quad \left[\equiv \frac{\partial\omega}{\partial k_x}\mathbf{e}_x + \frac{\partial\omega}{\partial k_y}\mathbf{e}_y + \frac{\partial\omega}{\partial k_z}\mathbf{e}_z \right], \tag{5.69}$$

which also gives the direction of the flow of the energy carried by the wave packet (see Braddick [90], Section 5.2).

The concept of group velocity deserves some further amplification (see Bittencourt [70], Section14.6). Consider a wave packet consisting of a superposition of plane waves obeying a dispersion equation $\omega = \omega(\mathbf{k})$:

$$\Psi_i(\mathbf{r},t) = \frac{1}{(2\pi)^{3/2}} \int_{-\infty}^{\infty} A_i(\mathbf{k})\, \mathrm{e}^{\mathrm{i}[\mathbf{k}\cdot\mathbf{r} - \omega(\mathbf{k})t]}\, d^3k\,. \tag{5.70}$$

This packet evolves from an initial shape $\Psi_i(\mathbf{r},0)$ related to the amplitudes $A_i(\mathbf{k})$ by Fourier synthesis, and vice versa by Fourier analysis:

$$\begin{aligned} \Psi_i(\mathbf{r},0) &= \frac{1}{(2\pi)^{3/2}} \int_{-\infty}^{\infty} A_i(\mathbf{k})\, \mathrm{e}^{\mathrm{i}\mathbf{k}\cdot\mathbf{r}}\, d^3k\,, \\ A_i(\mathbf{k}) &= \frac{1}{(2\pi)^{3/2}} \int_{-\infty}^{\infty} \Psi_i(\mathbf{r},0)\, \mathrm{e}^{-\mathrm{i}\mathbf{k}\cdot\mathbf{r}}\, d^3r\,. \end{aligned} \tag{5.71}$$

In our case, Ψ_i and A_i may represent the primitive variable representation (5.33) with the perturbations ρ_1, $\mathbf{v}_1$, e_1, $\mathbf{B}_1$ and their Fourier amplitudes, so that $i = 1, 2, \ldots 8$, or the wave vector representation (5.45), where $i = 1, 2, \ldots 7$, or the velocity representation (5.53), where $i = 1, 2, 3$. Furthermore, the pertinent dispersion equation $\omega = \omega_{\mathrm{A}}(\mathbf{k})$ of the Alfvén waves, given by Eq. (5.57), or $\omega = \omega_{\mathrm{s,f}}(\mathbf{k})$ of the magneto-sonic waves, given by Eq. (5.59), should be inserted. It is to be noted that only one of the amplitudes $A_i(\mathbf{k})$ can be chosen freely as a normalization, whereas the others should be chosen in agreement with the eigenfunction relations (5.58) for the Alfvén waves, or (5.61) for the magneto-sonic waves.

Let us assume that the wave packet consists of harmonics with wave vectors centred about some central value $\mathbf{k}_0$. A typical example is the Gaussian distribution,

$$A_i(\mathbf{k}) = \hat{A}_i\, \mathrm{e}^{-\frac{1}{2}|(\mathbf{k}-\mathbf{k}_0)\cdot\mathbf{a}|^2}\,, \tag{5.72}$$

where the components of the auxiliary vector $\mathbf{a}$ measure the width of the distribution in each of the three directions. This corresponds to an initial wave packet consisting of a main harmonic, with wave vector $\mathbf{k}_0$, and a modulated amplitude centred at the origin $\mathbf{r} = 0$:

$$\Psi_i(\mathbf{r},0) = \mathrm{e}^{\mathrm{i}\mathbf{k}_0\cdot\mathbf{r}}\, \frac{\hat{A}_i}{a_x a_y a_z}\, \mathrm{e}^{-\frac{1}{2}[(x/a_x)^2 + (y/a_y)^2 + (z/a_z)^2]}\,. \tag{5.73}$$

An example is the δ-function distribution, $A_i(\mathbf{k}) = (2\pi)^{3/2}\,\hat{A}_i\,\delta(\mathbf{k}-\mathbf{k}_0)$, corresponding to a plane wave $\Psi_i(\mathbf{r},0) = \hat{A}_i \exp(\mathrm{i}\mathbf{k}_0\cdot\mathbf{r})$ with wave vector $\mathbf{k}_0$.

For an arbitrary wave packet with a reasonably localized range of wave vectors (i.e. not infinitely narrow or infinitely wide), we may expand the dispersion equation about the central value $\mathbf{k}_0$,

$$\omega(\mathbf{k}) \approx \omega_0 + (\mathbf{k}-\mathbf{k}_0)\cdot\left(\frac{\partial\omega}{\partial\mathbf{k}}\right)_{\mathbf{k}_0}, \qquad \omega_0 \equiv \omega(\mathbf{k}_0). \tag{5.74}$$

Inserting this approximation into the expression (5.70) for the wave packet gives

$$\Psi_i(\mathbf{r},t) \approx \mathrm{e}^{\mathrm{i}(\mathbf{k}_0\cdot\mathbf{r}-\omega_0 t)}\,\frac{1}{(2\pi)^{3/2}}\int_{-\infty}^{\infty} A_i(\mathbf{k})\,\mathrm{e}^{\mathrm{i}(\mathbf{k}-\mathbf{k}_0)\cdot(\mathbf{r}-(\partial\omega/\partial\mathbf{k})_{\mathbf{k}_0}t)}\,d^3k\,, \tag{5.75}$$

representing a carrier wave $\exp\mathrm{i}(\mathbf{k}_0\cdot\mathbf{r}-\omega_0 t)$ with an amplitude-modulated envelope. Through constructive interference of the plane wave harmonics, the envelope maintains its shape during an extended interval of time, whereas the surfaces of constant phase of this envelope move precisely with the group velocity,

$$\mathbf{v}_{\rm gr} = \left(\frac{d\mathbf{r}}{dt}\right)_{\rm const.\,phase} = \left(\frac{\partial\omega}{\partial\mathbf{k}}\right)_{\mathbf{k}_0}, \tag{5.76}$$

in agreement with the definition (5.69).

5.3.2 Application to the MHD waves

The relation with the geometric construction of wave fronts by Huygens principle is well known. For MHD waves, this is most strikingly illustrated for the case of Alfvén waves. From Eq. (5.57), the phase velocity for Alfvén waves is given by

$$(\mathbf{v}_{\rm ph})_{\rm A} \equiv (v_{\rm ph})_{\rm A}\,\mathbf{n}\,, \qquad (v_{\rm ph})_{\rm A} = b\cos\vartheta\,, \tag{5.77}$$

so that the locus of the end points of the vector $(\mathbf{v}_{\rm ph})_{\rm A}$, i.e. the *phase diagram*, consists of two circles touching the origin (one of which is shown in Fig. 5.3(a)). The anisotropy of Alfvén wave propagation is even more strongly manifested by a localized wave packet resulting from a point perturbation in the origin at $t = 0$. This gives rise to the *group diagram*, which is the envelope at unit time of the wave fronts of a superposition of plane waves having passed in all directions through the origin at $t = 0$. Because the end points of the phase velocity vectors $(\mathbf{v}_{\rm ph})_{\rm A}$ of plane Alfvén waves lie on a circle, these wave fronts all go through a single point (Fig. 5.3(b)). Hence, the group diagram (or *caustic*) for Alfvén waves consists of the two points $\pm b$ along $\mathbf{B}_0$, so that the group velocity of Alfvén waves is given by

$$(\mathbf{v}_{\rm gr})_{\rm A} = \mathbf{b}\,. \tag{5.78}$$

This provides the most extreme example of anisotropy of plasma waves guided by a magnetic field: *Alfvén wave point disturbances, and their associated energy flow, just propagate along single magnetic field lines*. Of course, Eq. (5.78) is also obtained algebraically from the dispersion equation (5.57).

For the magneto-acoustic waves similar geometrical constructions can be made, but, in this case, it is easier to exploit the algebraic expressions for the phase and group velocities following from the

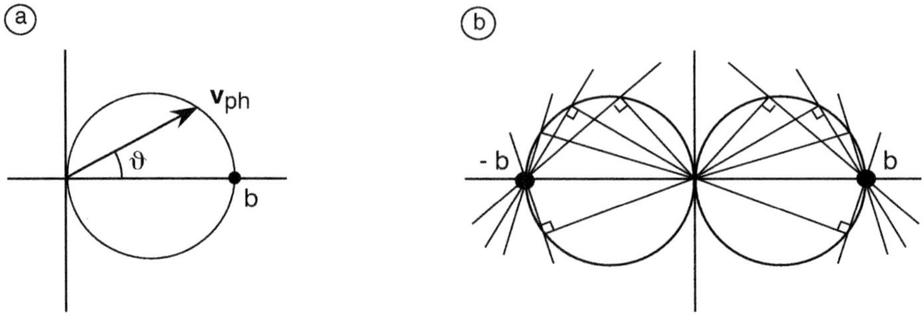

Fig. 5.3 Construction of (a) phase and (b) group diagram for Alfvén waves.

dispersion equation (5.59). It is expedient to introduce unit vectors in the x–z plane, i.e. the plane of $\mathbf{k}$ and $\mathbf{b}$:

$$\mathbf{n} \equiv \frac{\mathbf{k}}{k} = (\sin\vartheta, 0, \cos\vartheta)\,, \qquad \mathbf{t} \equiv \Big(\frac{\mathbf{b}}{b} \times \mathbf{n}\Big) \times \mathbf{n} = (\cos\vartheta, 0, -\sin\vartheta)\,. \tag{5.79}$$

The expression for the phase velocity of the magneto-acoustic waves then reads

$$(\mathbf{v}_{\rm ph})_{\rm s,f} \equiv (v_{\rm ph})_{\rm s,f}\,\mathbf{n}\,, \quad (v_{\rm ph})_{\rm s,f} = \sqrt{\tfrac{1}{2}(b^2+c^2)\Big[1 \pm \sqrt{1-\sigma\cos^2\vartheta}\,\Big]}\,, \tag{5.80}$$

whereas the expression for the group velocity becomes

$$(\mathbf{v}_{\rm gr})_{\rm s,f} = (v_{\rm ph})_{\rm s,f}\left[\mathbf{n} + \frac{\sigma\sin\vartheta\cos\vartheta}{2\big[1-\sigma\cos^2\vartheta \pm \sqrt{1-\sigma\cos^2\vartheta}\,\big]}\,\mathbf{t}\right]. \tag{5.81}$$

The derivation of the latter expression requires some straightforward algebra which we leave as an exercise.

▷ **Exercise** Carry this out. Hint: compute $(v_{\rm gr})_\perp = \partial\omega/\partial k_\perp = (\sin\vartheta/v_{\rm ph})\,\partial\omega^2/\partial k_\perp^2$, and similarly for $(v_{\rm gr})_\parallel$, and project on the unit vectors $\mathbf{n}$ and $\mathbf{t}$. ◁

All of the above expressions are brought together in the phase and group diagrams shown in Fig. 5.4 depicting $\mathbf{v}_{\rm ph}(\vartheta)$ and $\mathbf{v}_{\rm gr}(\vartheta)$ for the three MHD waves for three different values of c/b. Pictures of this kind were first constructed by K. O. Friedrichs [195] and, therefore, rightly named *Friedrichs diagrams*. The very strong anisotropy of the MHD waves is manifest. *The fast magneto-sonic waves may be considered as generalized sound waves with significant contributions of the magnetic pressure*. These contributions increase the perpendicular speed of propagation. In accordance with the orthogonality condition (5.64), expressing a fundamental duality of the waves, the slow waves behave in exactly the opposite manner. *The slow magneto-sonic waves may be considered as sound waves with strong magnetic guidance*. Focusing by the magnetic field produces wave packages that propagate predominantly along the magnetic field. This leads to the peculiar cusp-shaped caustics of the slow wave group diagrams in the bottom part of Fig. 5.4. Note that the degeneracy of the group velocities of the three MHD waves at $c/b = 1$ for $\vartheta = 0$, i.e. $\bar{v}_{\rm gr,s} = \bar{v}_{\rm gr,A} = \bar{v}_{\rm gr,f} = \mathbf{n}$, is resolved by any small value $\vartheta = \epsilon$: the group velocities of the magneto-sonic waves then jump to $\bar{v}_{\rm gr,s} = \mathbf{n} - \mathbf{t}$ and $\bar{v}_{\rm gr,f} = \mathbf{n} + \mathbf{t}$, as illustrated in the middle frame of Fig.5.4(b).

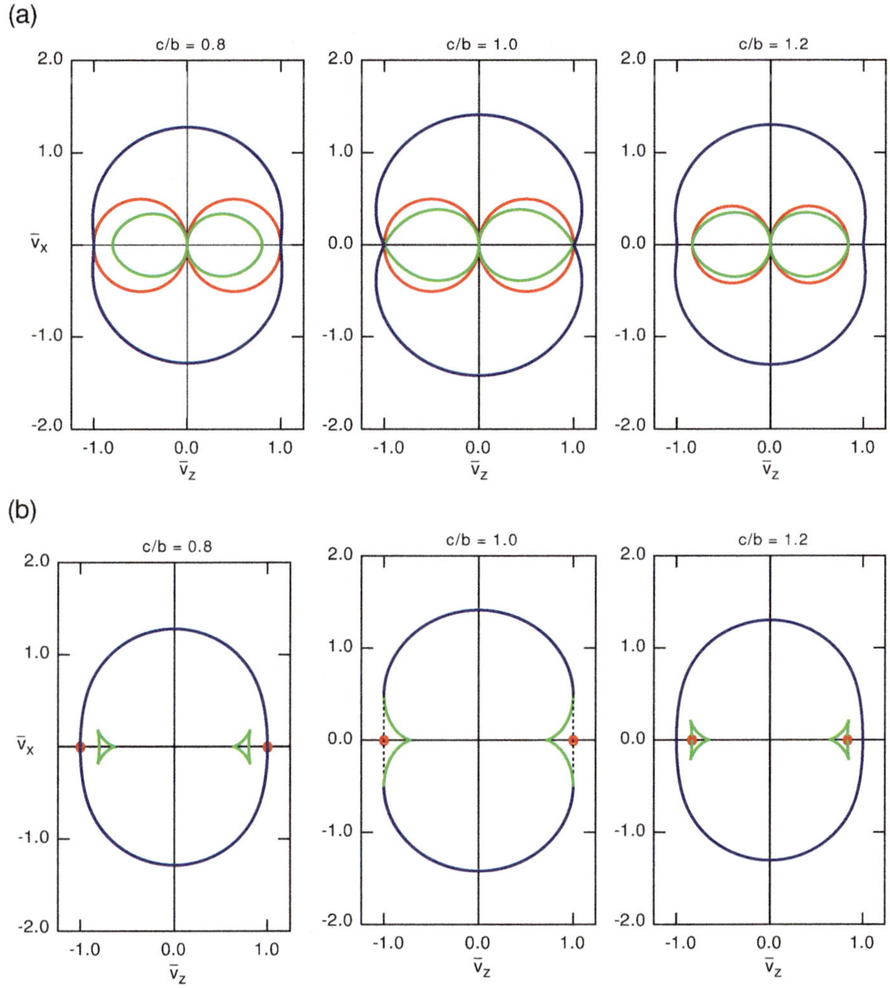

Fig. 5.4 (a) Phase diagrams and (b) group diagrams for the fast (blue), Alfvén (red) and slow (green) MHD waves for three values of the ratio c/b of the sound to the Alfvén speed. Phase and group velocities are normalized as $\bar{v} \equiv v/\max(b,c)$.

Clearly, all MHD wave properties depend strongly on the value of the parameter c/b, related to the usual parameter β through Eq. (5.67). In Fig. 5.4, we have depicted three curves for values of c/b deviating little from unity, to show the interesting degeneracy at $c = b$, but also because the slow wave curves are still visible for this choice. For the important case of low β plasmas (e.g. $c/b = 0.1$, so that $\beta = 0.012$ for $\gamma = 5/3$), the slow branches virtually disappear at the origin. This wide separation from the Alfvén and fast branches is an important property that is frequently exploited to simplify wave and stability studies, e.g. of tokamaks. The theoretical framework is called *the low-β expansion* scheme.

▷ **Exercise** What happens in the opposite limit ($c/b \gg 1$ or $\beta \gg 1$)? Also discuss the limiting case of pure gas dynamics ($b = 0$ or $\beta \to \infty$). ◁

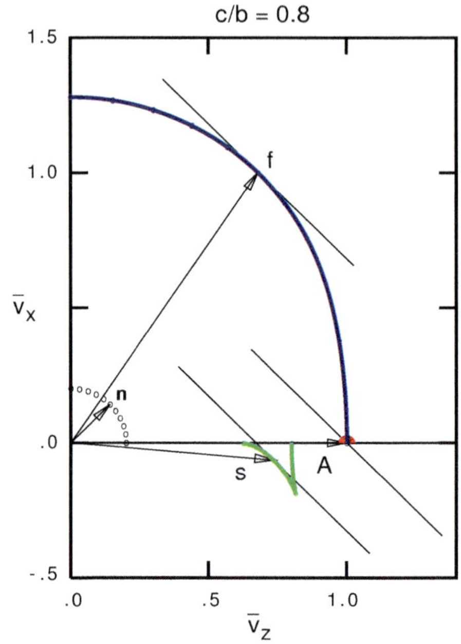

Fig. 5.5 Group diagrams of the three MHD waves in the first quadrant with the group velocity vectors relative to the normal $\mathbf{n}$.

Construction of the phase diagram is easier than that of the group diagram since $\mathbf{v}_{\rm ph}$ and $\mathbf{n}$ have the same direction. However, the group diagram has a wider applicability since it represents the response to local disturbances, which remains valid for inhomogeneous plasmas. The complicating factor, not visible in Fig. 5.4, is that the direction of the group velocity $\mathbf{v}_{\rm gr}(\vartheta)$ deviates significantly from the direction $\mathbf{n}(\vartheta)$ of the central wave vector, as is evident from Eqs. (5.78) and (5.81). The dispersion equations (5.57) and (5.59) yield the following limiting values of the phase and group velocities for parallel and propagation:

$$
\begin{aligned}
&\vartheta = 0: \\
&\quad (v_{\rm ph/gr})_{\rm s} = \min(b,c) \le (v_{\rm ph/gr})_{\rm A} = b \le (v_{\rm ph/gr})_{\rm f} = \max(b,c)\,, \\
&\vartheta = \pi/2: \\
&\quad (v_{\rm ph})_{\rm s} = 0 = (v_{\rm ph})_{\rm A} = 0 < (v_{\rm ph})_{\rm f} = \sqrt{b^2+c^2}\,, \\
&\quad (v_{\rm gr})_{\rm s\,(cusp)} = \frac{bc}{\sqrt{b^2+c^2}} \le (v_{\rm gr})_{\rm A} = b \le (v_{\rm gr})_{\rm f} = \sqrt{b^2+c^2}\,,
\end{aligned}
\tag{5.82}
$$

in agreement with the expressions (5.66) and (5.65), respectively.

Finally, we present the peculiar differences of the relationship between the directions of $\mathbf{n}(\vartheta)$ and $\mathbf{v}_{\rm gr}(\vartheta)$ for the three MHD waves. In Fig. 5.5 one of the group diagrams is depicted once more with the central wave vector $\mathbf{n}$ lying in the first quadrant and the three group velocities exhibiting their mutually exclusive directions of propagation: when the direction of $\mathbf{n}$ changes from $\vartheta = 0$ (parallel to $\mathbf{B}$) to $\vartheta = \pi/2$ (perpendicular to $\mathbf{B}$), the fast group velocity also changes from parallel to perpendicular (though it does not remain parallel to $\mathbf{n}$), the Alfvén group velocity just remains purely parallel, but the slow group velocity initially changes *clockwise* from parallel to some negative angle at $\vartheta = \vartheta_{\rm m}$ (computed below) and then back again (anti-clockwise) to purely parallel. For the latter direction, the value of the slow group velocity, called the *cusp velocity* (indicated by the letter c) is given on the last line of Eq. (5.82). It plays an important role in the analysis of local waves in inhomogeneous plasmas. Notice that slow wave packages propagate in the perpendicular direction opposite to the direction of $\mathbf{n}$! Somehow, magnetic focusing is perfect for Alfvén waves, but slightly overshoots for slow waves.

▷ **Computation of the return angle in the slow wave group diagram** The angle $\vartheta_{\rm m}$ between $\mathbf{k}$ and $\mathbf{B}_0$ where the slow wave group vector $(\mathbf{v}_{\rm gr})_{\rm s}$ returns may be computed from Eq. (5.81) by defining the square root expression $R(\vartheta) \equiv \sqrt{1 - \sigma \cos^2 \vartheta}$. This quantity increases monotonically with ϑ so that it can be used as a parameter measuring the angle instead of ϑ. In terms of R, the magnitude of the slow group velocity reads:

$$(v_{\rm gr}^2)_{\rm s} = \frac{1}{8}(b^2 + c^2)\,\frac{-3R^3 + 5R^2 - (1-\sigma)R - (1-\sigma)}{R^2}\,. \tag{5.83}$$

Its maximum is reached for $R_{\rm m} \equiv R(\vartheta_{\rm m})$ satisfying the cubic equation

$$3R_{\rm m}^3 - (1-\sigma)R_{\rm m} - 2(1-\sigma) = 0\,, \tag{5.84}$$

which has only one physically acceptable solution:

$$R_{\rm m} = 3\tau^{1/3}\left[(1+\sqrt{1-\tau})^{1/3} + (1-\sqrt{1-\tau})^{1/3}\right], \qquad \tau \equiv \frac{1}{81}(1-\sigma)\,, \tag{5.85}$$

from which the required angle $\vartheta_{\rm m} = \arccos\big(\sqrt{(1-R_{\rm m}^2)/\sigma}\big)$ is obtained. The corresponding magnitude of $(\mathbf{v}_{\rm gr})_{\rm s}$, and its angle with $\mathbf{B}_0$ (not to be confused with $\vartheta_{\rm m}$!), may be obtained from Eq. (5.81) by substituting $\vartheta_{\rm m}$. [Check that $(v_{\rm gr})_\perp < 0$!] ◁

5.3.3 Asymptotic properties

The three MHD waves exhibit a distinct difference with respect to their propagation properties in the different directions, as already demonstrated in Section 5.2.4 by the dispersion diagrams of Fig. 5.2. In Fig. 5.6, we show the dispersion diagrams once more, schematically now, in order to highlight their asymptotic properties for large values of $k_\parallel$ and $k_\perp$. In the left diagram, showing $\omega^2(k_\parallel)$ for fixed $k_\perp$, the fast branch starts at $\omega_1^2 \equiv k_\perp^2(b^2 + c^2)$ and tends to $\omega_2^2 \equiv k_\parallel^2 b^2 \equiv \omega_{\rm A}^2$ as $k_\parallel \to \infty$, whereas the slow branch starts at 0 and tends to $\omega_3^2 \equiv k_\parallel^2 c^2$. In the right diagram, showing $\omega^2(k_\perp)$ for fixed $k_\parallel$, the fast branch ranges now from $\omega_4^2 \equiv k_\parallel^2 b^2 \equiv \omega_{\rm A}^2$ at $k_\perp = 0$ to $\omega_7^2 \equiv k_\perp^2(b^2 + c^2) \to \infty$, and the slow one ranges from $\omega_5^2 \equiv k_\parallel^2 c^2$ to $\omega_6^2 \equiv k_\parallel^2 b^2 c^2/(b^2 + c^2)$. These asymptotic properties will return when we investigate the local waves in inhomogeneous equilibria since large wave numbers correspond to small wavelengths.

While the group velocity in the parallel direction is positive, $\partial\omega/\partial k_\parallel > 0$, for all three kinds of waves, the group velocity $\partial\omega/\partial k_\perp$ in the perpendicular direction and the asymptotic value of the frequency for $k_\perp \to \infty$ display a very characteristic difference for the three waves:

$$\begin{cases} \partial\omega/\partial k_\perp > 0\,, & \omega_{\rm f}^2 \to \infty \qquad \text{for the fast waves,} \\ \partial\omega/\partial k_\perp = 0\,, & \omega_{\rm A}^2 \to k_\parallel^2 b^2 \qquad \text{for the Alfvén waves,} \\ \partial\omega/\partial k_\perp < 0\,, & \omega_{\rm s}^2 \to k_\parallel^2 \dfrac{b^2 c^2}{b^2 + c^2} \qquad \text{for the slow waves.} \end{cases} \tag{5.86}$$

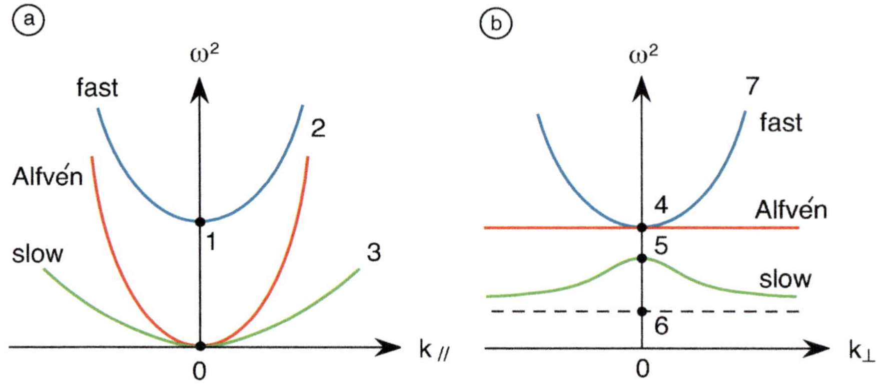

Fig. 5.6 Schematic dispersion diagrams and asymptotics for large wave numbers: (a) $\omega^2 = \omega^2(k_\parallel)$ for fixed $k_\perp$; (b) $\omega^2 = \omega^2(k_\perp)$ for fixed $k_\parallel$.

Note that this implies that the energy propagation of slow wave packets is opposite to that of fast wave packets in the perpendicular direction! This is caused by the peculiar behaviour of the slow group velocity depicted in Fig. 5.5.

5.3.4 *Self-gravity and contraction in homogeneous media*★

Since self-gravity is an important mechanism in many astrophysical contexts, we also discuss the slightly artificial case of an infinite uniform magnetized medium with constant density and pressure, where – to some extent – self-gravity is incorporated [346]. Following Strittmatter [579] and Chandrasekhar and Fermi [113], we adopt what is known as *the Jeans swindle,* originally used by Jeans [334] to describe the perturbations of a uniform, unmagnetized, infinite gas subject to self-gravity. Self-gravity entails adding a term $-\rho\nabla\Phi_{\rm gr}$ in the RHS of Eq. (5.17), and to supplement the system (5.16)–(5.19) with the Poisson law relating the gravitational potential $\Phi_{\rm gr}$ with density through

$$\nabla^2\Phi_{\rm gr} = 4\pi G\rho\,, \tag{5.87}$$

where G is the gravitational constant. Linearization of the MHD equations about a static medium still leads to Eqs. (5.21)–(5.24), but with terms $-\rho_1\nabla\Phi_{\rm gr,0} - \rho_0\nabla\Phi_{\rm gr,1}$ in the RHS of Eq. (5.22). The 'swindle' boils down to neglecting the term proportional to $\nabla\Phi_{\rm gr,0}$ (this is opposite to the so-called Cowling approximation, which instead only keeps this term when dealing with gravity). For plane wave solutions, one then finds from Eq. (5.87) the relation between the Fourier amplitudes,

$$\omega k^2\hat{\Phi}_{\rm gr} = -4\pi G\rho_0\,\mathbf{k}\cdot\hat{\mathbf{v}}\,. \tag{5.88}$$

By again manipulating Eqs. (5.21)–(5.24) to the velocity representation, we get the following modification of the eigenvalue equation (5.52):

$$\Big\{\Big[\omega^2 - (\mathbf{k}\cdot\mathbf{b})^2\Big]\,\mathsf{I} - (b^2+\widetilde{c^2})\,\mathbf{k}\,\mathbf{k} + \mathbf{k}\cdot\mathbf{b}\,(\mathbf{k}\,\mathbf{b}+\mathbf{b}\,\mathbf{k})\Big\}\cdot\hat{\mathbf{v}} = 0\,,$$
$$\text{where}\quad \widetilde{c^2} = \widetilde{c^2}(k^2) \equiv c^2 - 4\pi G\rho_0/k^2\,. \tag{5.89}$$

From here on, the analysis is analogous to that of Section 5.2 for a homogeneous plasma without self-gravitation, except that the square of the sound velocity, c^2, is now replaced by a related quantity, distinguished by the tilde, that depends on the wave vector, $\widetilde{c^2} = \widetilde{c^2}(k^2)$. Since *it may become negative,* it can no longer be considered as a squared wave speed. We will investigate the physical consequences.

The solutions of the modified dispersion equation (5.89) are, first, two unaffected Alfvén waves with squared frequency $\omega^2 = \omega_A^2$ and phase speeds $\pm b \cos\vartheta\, \mathbf{n}$. The group speeds $\pm\mathbf{b}$ keep the extreme anisotropy of Alfvén waves, just sampling individual magnetic field lines. The other four solutions yield slow/fast magneto-acoustic wave pairs, modified by gravity and satisfying

$$\omega^2 = \omega_{s,f}^2 \equiv k^2 v_{\text{ph s,f}}^2 = \tfrac{1}{2}k^2\Big(b^2 + \widetilde{c^2} \pm |b^2 + \widetilde{c^2}|\sqrt{1 - \widetilde{\sigma}\cos^2\vartheta}\Big)\,,$$

$$\text{where}\quad \widetilde{\sigma} = \widetilde{\sigma}(k^2) \equiv \frac{4b^2\widetilde{c^2}}{(b^2 + \widetilde{c^2})^2}\,. \qquad (5.90)$$

These expressions correspond to Eqs. (5.59) for $\omega_{s,f}^2$ and (5.60) for σ, with the crucial difference that $\widetilde{\sigma}(k^2)$ *becomes negative for sufficiently small* $|k|$, *and even* $b^2 + \widetilde{c^2}$ *may become negative.*

It is of interest to discuss the implications of these expressions. First, note that the case of vanishing magnetic field has both the Alfvén and one pair of the magneto-sonic waves collapse into marginal frequencies, while the other pair of magneto-sonic 'waves' shows the gravitational modification to sound waves as originally analysed by Jeans: $\omega^2 = k^2\widetilde{c^2} \equiv k^2c^2 - 4\pi G\rho_0$. For positive values, they represent sound-like waves, with smaller phase speed. When the RHS is negative, which happens for sufficiently small values of $|k|$ (i.e. large wavelengths), they represent instabilities growing exponentially as $\exp(\nu t)$, where $\nu \equiv \mathrm{Im}(\omega)$ is the growth rate: *the Jeans instability.* Apparently, when we consider a large enough volume of a homogeneous, compressible, unmagnetized plasma, gravitational contraction can overcome compressibility, giving a runaway condensation.

For a static self-gravitating, homogeneous, magnetized plasma state, described by $\mathbf{B}_0$, ρ_0, p_0, wave behaviour again depends on the absolute value of the wave number with respect to a critical wave number (or a critical wavelength) given by

$$k_{\text{crit}} \equiv \sqrt{4\pi G\rho_0}/c\,, \quad \text{or} \quad \lambda_{\text{crit}} \equiv c\sqrt{\pi/(G\rho_0)}\,. \qquad (5.91)$$

This now yields the following cases:

$$|k| \begin{cases} > k_{\text{crit}} & \Rightarrow \quad \omega_s^2 \geq 0 \ \text{(stable)}\,, \quad \omega_f^2 > 0 \ \text{(stable)} \\ = k_{\text{crit}} & \Rightarrow \quad \omega_s^2 = 0 \ \text{(marginal)}\,, \quad \omega_f^2 > 0 \ \text{(stable)} \\ < k_{\text{crit}} & \Rightarrow \quad \omega_s^2 \leq 0 \ \text{(unstable)}\,, \quad \omega_f^2 \geq 0 \ \text{(stable)} \end{cases} . \qquad (5.92)$$

Hence, for all wave numbers larger than k_{crit}, we find two pairs of stable magneto-sonic modes. At the critical wave number k_{crit}, where $\widetilde{c^2}(k_{\text{crit}}^2) = 0$, the slow pair becomes marginal and the fast modes become isotropic with $\omega_f^2 = k^2b^2$. For all wave numbers smaller than k_{crit}, when $\widetilde{c^2}(k^2) < 0$, the slow modes exhibit the MHD analogue of the Jeans instability and the fast modes that of the modified sound wave. In that range, for the special case $k^2 = k_0^2 \equiv 4\pi G\rho_0/(b^2 + c^2)$, so that $b^2 + \widetilde{c^2} = 0$, the solution pairs of Eq. (5.89) reduce to $\nu_s^2 = k_0^2b^2|\cos\vartheta|$ and $\omega_f^2 = k_0^2b^2|\cos\vartheta|$. Finally, in the limit $k \to 0$, the slow modes become purely gravitationally unstable, $\nu_s^2 \to 4\pi G\rho_0$, and the fast modes approaching the Alfvén frequency become marginal, $\omega_f^2 \approx k^2b^2\cos^2\vartheta \to 0$.

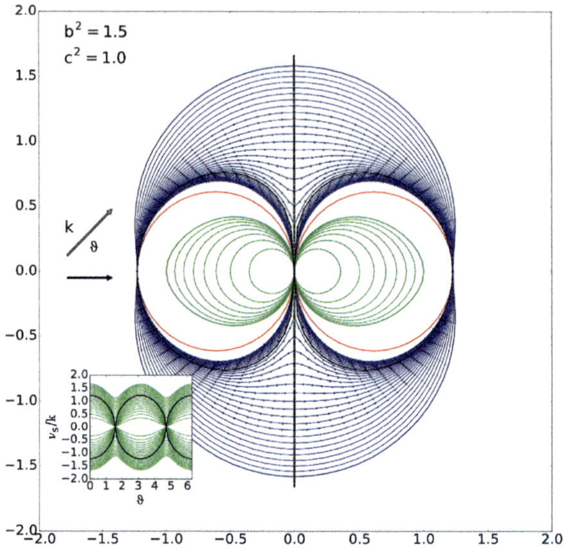

Fig. 5.7 A representative phase diagram, with and without self-gravity. Blue curves refer to the fast, green curves to the slow and red curves to the Alfvén waves (which are stable). Jeans unstable slow and stable fast branches appear. The inset shows the angular variation of the normalized growth rate of the unstable slow modes.

The phase diagram for a magnetized uniform plasma is shown in Fig. 5.7. This example has $b^2 = 1.5$ and is normalized by taking $c^2 = 1$. Blue curves refer to the fast wave family, red curves are for the Alfvén pair and green curves belong to the slow family. Each curve is obtained by varying the angle ϑ between the wave vector $\mathbf{k}$ and the horizontal magnetic field $\mathbf{b}$ in $[0, 2\pi]$. We also plotted the phase diagram for a case without gravity: the familiar triplet of fast (outermost curve), Alfvén (pair of red circles, unaffected by gravity) and slow modes (outermost green curves). We then added a representative set of phase speed diagrams for the case with self-gravity. This set was chosen with wave numbers k such that the important dimensionless combination takes on the values $4\pi G\rho_0/(k^2c^2) = j/j_{\rm m}$ with $j \in [1, 30]$ and $j_{\rm m} = 8$. The corresponding 30 phase diagrams start off as nested versions of the $G = 0$ case. The black curves indicate the slow and fast phase diagrams for the mentioned special case $|k| = k_0$. The unstable part of the slow wave diagram is plotted as if we display the location of the complex number in the complex plane, with the vertical axis as the imaginary axis (indicated by the symbols on it). The inset in the bottom left corner shows the variation of the growth rate $\nu_{\rm s}/k$ (which is normalized as a phase speed) with the angle ϑ for the slow modes.

In contrast to the gas dynamical Jeans instability, the growth rate $\nu_{\rm s}$ for the magnetohydrodynamical counterpart oscillates between maxima at $\vartheta = 0, \pi, 2\pi$ (where $\mathbf{k} \parallel \mathbf{B}$) and minima at $\vartheta = \frac{1}{2}\pi, \frac{3}{2}\pi$ (where $\mathbf{k} \perp \mathbf{B}$), whereas the maxima themselves are independent of the magnitude of the magnetic field:

$$\nu_{\rm s,max} = \sqrt{-k^2\widetilde{c^2}} \equiv \sqrt{4\pi G\rho_0 - k^2c^2}\,, \qquad (5.93)$$

Consequently, a crude estimate of the largest growth rate for a finite size but homogeneous (contradictio in terminis) astronomical object of size L along the magnetic field is obtained by the condition that half of the wavelength $\lambda \equiv 2\pi/k$ should fit into L, i.e. by substituting $k = k_{\rm min} = \pi/L$ in the expression for $\nu_{\rm s,max}$.

5.4 Characteristics★

The group diagram has a much wider applicability than just wave propagation in infinite homogeneous plasmas. This is so because the construction of a wave packet involves the contributions of large **k**-vectors (small wavelengths) so that *the concept of group velocity is essentially a local one*. This is one of the reasons why it returns in the context of nonlinear MHD of inhomogeneous plasmas, where the associated *characteristics* become the carriers of the information of disturbances of the plasma. This subject is discussed next.

5.4.1 The method of characteristics★

Consider the example of the linear advection equation in one spatial dimension,

$$\frac{\partial \Psi}{\partial t} + u \frac{\partial \Psi}{\partial x} = 0 \,, \tag{5.94}$$

where $\Psi(x,t)$ is the unknown and the advection velocity u is considered to be given. If $u =$ const, the solution is trivial:

$$\Psi = f(x - ut) \,, \quad \text{where} \quad f = \Psi_0 \equiv \Psi(x, t = 0) \,. \tag{5.95}$$

Hence, the initial data Ψ_0 are simply propagated along the set of parallel straight lines $dx/dt = u$, which are the *characteristics* for this case.

For our subject (the study of the MHD equations (5.16)–(5.19)), it is important that this procedure may be generalized to systems of first order partial differential equations (PDEs) in three spatial dimensions and time, so that Ψ becomes a vector and u a matrix. For example, the linearized equations for one-dimensional sound waves, following from Eqs. (5.5)–(5.7), read:

$$\frac{\partial}{\partial t}\begin{pmatrix} v_1 \\ p_1 \end{pmatrix} + \begin{pmatrix} 0 & 1/\rho_0 \\ \gamma p_0 & 0 \end{pmatrix} \frac{\partial}{\partial x}\begin{pmatrix} v_1 \\ p_1 \end{pmatrix} = 0 \,, \tag{5.96}$$

where we have omitted the redundant density equation. As we have seen in Section 5.1.2, for constant values of ρ_0 and p_0, the solutions (5.9) represent plane sound waves with frequencies $\omega = \pm kc$, where $c \equiv \sqrt{\gamma p_0/\rho_0}$ is the sound speed. Exploiting real notation, the general solution may be written as

$$v_1(x,t) = \sum_k \Big[\alpha_k \sin k(x \pm ct) + \beta_k \cos k(x \pm ct) \Big] \,, \tag{5.97}$$

and a similar expression for p_1. Here, the coefficients α_k and β_k follow from Fourier decomposition of the initial data,

$$v_1(x,0) = \sum_k (\alpha_k \sin kx + \beta_k \cos kx) \,, \tag{5.98}$$

demonstrating that the initial data do, in fact, propagate along the two sets of straight-line characteristics $dx/dt = \pm c$.

Returning to the linear advection equation (5.94), but now with u not constant, the characteristics become solutions of the ordinary differential equations (ODEs)

$$\frac{dx}{dt} = u(x,t)\,. \tag{5.99}$$

Along these curves, the solution $\Psi(x,t)$ is constant,

$$\frac{\mathrm{d}\Psi}{\mathrm{d}t} \equiv \frac{\partial \Psi}{\partial t} + \frac{\partial \Psi}{\partial x}\frac{dx}{dt} = 0\,. \tag{5.100}$$

For given initial data $\Psi_0(x)$, the solution can then be determined at any time $t_1 > 0$ by constructing the characteristics through a set of points $\{\ldots, x_i, x_{i+1}, \ldots\}$ such that $\Psi(x_i', t_1) = \Psi_0(x_i)$, where x_i' lies on the characteristic through $x = x_i$. This is illustrated in Fig. 5.8 for the case that a weak discontinuity (a 'tent' function with a discontinuous derivative) is applied at $t = 0$. The characteristic through $x = x_i$ propagates this discontinuity forward in space-time.

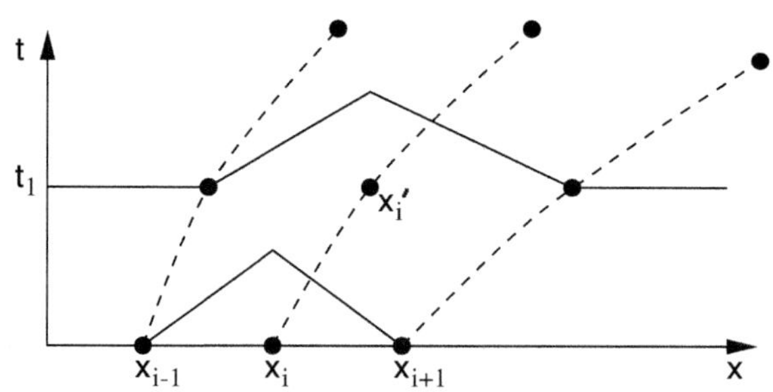

Fig. 5.8 Pointwise propagation of a solution along the characteristics.

Most important for the study of gas dynamics and magnetohydrodynamics, in particular for modern developments in computational fluid dynamics (CFD) and computational magneto-fluid dynamics (CMFD), is the fact that *the method of characteristics* generalizes to *nonlinear* partial differential equations. Since it lends itself naturally to numerical implementation, it is at the basis of many methods in CFD and CMFD. Following the clear exposition of LeVeque [404, 406] on this topic, this may be illustrated again with the advection equation which becomes *quasi-linear* when u is also a function of the unknown Ψ itself. A particularly relevant example is the case $u = \Psi$ which leads to Burgers' equation:

$$\frac{\partial \Psi}{\partial t} + \Psi\,\frac{\partial \Psi}{\partial x} = \nu\,\frac{\partial^2 \Psi}{\partial x^2}\,, \tag{5.101}$$

where a viscous term is added on the RHS to model the balance between nonlinear and dissipative processes when gradients build up in the solutions. At first neglecting this small term, the characteristics are the solutions of the ODE

$$\frac{dx}{dt} = \Psi\big(x(t), t\big)\,, \tag{5.102}$$

which are just a set of straight lines with slopes determined by the initial data, like in the first step of Fig. 5.8. Without dissipation, the characteristics would cross for large times, but the build-up of large gradients of Ψ is counteracted by smoothing through the dissipative term on the RHS of Burgers' equation. This will occur in a very narrow region, so that effectively a valid solution with a *shock* is obtained in the limit $\nu \to 0$, where the condition of increasing entropy across the shock is to be applied to the ideal model to eliminate unphysical solutions, as we have already seen in Section 4.5.1 and Fig. 4.11.

We will return to computational methods in the chapters 15 and 20 on this subject. For the present purpose, the essential issue of characteristics is the fact that the equations of MHD are actually *hyperbolic* partial differential equations. This means that they possess a complete set of real characteristics related to the eigenvalues of the linearized system, as will be demonstrated in Section 5.4.3.

5.4.2 Classification of partial differential equations★

Long before computational methods for solving partial differential equations became common tools, the theory of second order PDEs was already a central subject of mathematical physics, where the method of characteristics was applied extensively; see Courant and Hilbert [132](b), Garabedian [206], or Morse and Feshbach [450](a). It is useful to recall some of the concepts developed there before we return to the study of the MHD equations proper.

Consider the following second order PDE in two dimensions:

$$A\Phi_{xx} + 2B\Phi_{xy} + C\Phi_{yy} = D(\Phi_x, \Phi_y, x, y)\,, \tag{5.103}$$

where A, B and C are functions of only x and y for the time being. These independent variables may indicate two spatial dimensions as well as one space and one time coordinate, or linear combinations. Subscripts x and y indicate differentiation with respect to those variables: $\Phi_x \equiv \partial\Phi/\partial x$, etc. The *Cauchy problem* consists in finding the solution Φ away from a boundary C, when both Φ and its normal derivative $\Phi_{\rm n} \equiv \mathbf{n}\cdot\nabla\Phi$ are specified on it. Introducing new variables

$$\Psi_1 \equiv \Phi_x\,, \qquad \Psi_2 \equiv \Phi_y\,,$$

Eq. (5.103) is transformed into an equivalent system of first order equations

$$\begin{aligned} A\Psi_{1x} + B\Psi_{1y} + B\Psi_{2x} + C\Psi_{2y} &= D(\Psi_1, \Psi_2, x, y)\,, \\ \Psi_{1y} - \Psi_{2x} &= 0\,. \end{aligned} \tag{5.104}$$

The pertinent Cauchy problem is to determine Ψ_1 and Ψ_2 away from the boundary, when they are given on C.

To facilitate the solution of the Cauchy problem, let us replace the Cartesian coordinates x, y by boundary fitted coordinates ξ, η, where the boundary curve C is given by $\xi(x,y) = \xi_0$. These coordinates may be chosen orthogonal, but this is not essential for the discussion. The boundary data can then be expressed as

$$\Psi_1(\xi_0, \eta) = f_1(\eta)\,, \qquad \Psi_2(\xi_0, \eta) = f_2(\eta)\,. \tag{5.105}$$

We wish to investigate under which conditions $\Psi_1(\xi, \eta)$ and $\Psi_2(\xi, \eta)$ may be obtained by means of a power series solution about a particular point (ξ_0, η_0) on the boundary:

$$\Psi_1(\xi,\eta) = \Psi_1(\xi_0,\eta_0) + (\xi-\xi_0)\left(\frac{\partial\Psi_1}{\partial\xi}\right)_0 + (\eta-\eta_0)\left(\frac{\partial\Psi_1}{\partial\eta}\right)_0 + \cdots,$$
$$\Psi_2(\xi,\eta) = \Psi_2(\xi_0,\eta_0) + (\xi-\xi_0)\left(\frac{\partial\Psi_2}{\partial\xi}\right)_0 + (\eta-\eta_0)\left(\frac{\partial\Psi_2}{\partial\eta}\right)_0 + \cdots. \quad (5.106)$$

The expressions $\Psi_i(\xi_0,\eta_0) \equiv f_i(\eta_0)$ and $(\partial\Psi_i/\partial\eta)_0 \equiv df_i/d\eta\,(\eta_0)$ $(i = 1,2)$ are known from the boundary conditions (5.105), so that we need to investigate under which circumstances the remaining expressions $(\partial\Psi_i/\partial\xi)_0$ $(i = 1,2)$ can be calculated. Once the latter two derivatives are known, the higher derivatives in the expansion (5.106) may be found by successive differentiations of the original equations (5.104) so that the problem may be considered to be solved in the neighbourhood of $\xi = \xi_0$. The process is then repeated by moving C to $\xi = \xi_1$, etc., until the solution is known everywhere.

We transform the PDE (5.104) to ξ–η coordinates:

$$(A\xi_x + B\xi_y)\frac{\partial\Psi_1}{\partial\xi} + (B\xi_x + C\xi_y)\frac{\partial\Psi_2}{\partial\xi} = D - (A\eta_x + B\eta_y)\frac{\partial\Psi_1}{\partial\eta} - (B\eta_x + C\eta_y)\frac{\partial\Psi_2}{\partial\eta},$$
$$\xi_y\frac{\partial\Psi_1}{\partial\xi} - \xi_x\frac{\partial\Psi_2}{\partial\xi} = -\eta_y\frac{\partial\Psi_1}{\partial\eta} + \eta_x\frac{\partial\Psi_2}{\partial\eta}, \quad (5.107)$$

where the LHS contains the unknown derivatives and the RHS the known ones. The unknown derivatives $\partial\Psi_1/\partial\xi$ and $\partial\Psi_2/\partial\xi$ may then be determined from Eqs. (5.107) *if the determinant of the coefficients on the left hand side does not vanish*. On the other hand, the condition that the determinant vanishes,

$$-A\xi_x^2 - 2B\xi_x\xi_y - C\xi_y^2 = 0, \quad (5.108)$$

defines two directions in every point of the plane, *the characteristic directions*, along which posing Cauchy boundary conditions does not determine the solution. The *characteristics* are the curves in the x–y plane that are everywhere tangent to these characteristic directions. Since $\xi(x,y) = \xi_0 \Rightarrow d\xi = \xi_x dx + \xi_y dy = 0$ along C, those characteristic directions are given by

$$\left.\frac{dy}{dx}\right|_{\rm char} = -\frac{\xi_x}{\xi_y} = \frac{B \pm \sqrt{B^2 - AC}}{A}. \quad (5.109)$$

Three cases may be distinguished.

(a) $B^2 > AC$: the characteristics are real and the equation is called *hyperbolic* (example: the wave equation $\Phi_{xx} - (1/c^2)\Phi_{tt} = 0$);

(b) $B^2 = AC$: the characteristics are real but coincide and the equation is *parabolic* (example: the heat equation $\Phi_{xx} - (1/\lambda)\Phi_t = 0$);

(c) $B^2 < AC$: the characteristics are complex and the equation is *elliptic* (example: Laplace's equation $\Phi_{xx} + \Phi_{yy} = 0$).

We will be concerned mainly with hyperbolic equations. Cauchy *initial* conditions (the variable y then becomes t) may be considered appropriate if the boundary (in x–t now) is not a characteristic. For the example of the wave equation

$$\Phi_{xx} - \frac{1}{c^2}\Phi_{tt} = 0,$$

the characteristic directions are given by $dx/dt = \pm c$. The initial data propagate along those characteristics. For the Cauchy problem to be well-posed in spaces of higher dimension than 2, it is not sufficient that the boundaries do not coincide with a characteristic. One has to demand in addition that they are space-like (satisfying $0 \leq ct < |x|$). The reason is that the spatial part by itself is generally elliptic, so that Cauchy's problem is ill-posed if we consider time-independent solutions. In physical problems initial data are usually given along space-directions,[1] so that this does not really present a restriction.

Notice that the analysis above is unchanged when the coefficients A, B, and C depend on Ψ_1 and Ψ_2 as well, so that the method of characteristics also works for nonlinear equations, specifically *quasi-linear partial differential equations*.

5.4.3 Characteristics in ideal MHD★

We now generalize the discussion to PDEs in more than two independent variables and also more than two dependent variables, in particular the ideal MHD equations (5.16)–(5.19) for the variables ρ, $\mathbf{v}$, e, $\mathbf{B}$ as functions of $\mathbf{r}$ and t. Instead of a 2-vector (Ψ_1, Ψ_2), the unknowns will now be represented by an 8-vector Ψ_i $(i = 1, \ldots, 8)$. We will show that the MHD equations are *symmetric hyperbolic PDEs*, where the nonlinearity is only of a *quasi-linear* nature.

The equations of ideal MHD are PDEs with respect to the independent variables $\mathbf{r}, t$. Hence, the characteristics are three-dimensional manifolds

$$\xi(\mathbf{r}, t) = \xi_0, \tag{5.110}$$

in four-dimensional space-time $\mathbf{r}, t$. These manifolds may be visualized as being swept out by the motion of surfaces in ordinary three-dimensional space $(\mathbf{r})$ when the time t progresses. We apply the same techniques as in the previous section to determine when the manifold $\xi(\mathbf{r}, t) = \xi_0$ is a characteristic manifold.

Assume that boundary data for $\rho(\mathbf{r}, t), \mathbf{v}(\mathbf{r}, t), e(\mathbf{r}, t), \mathbf{B}(\mathbf{r}, t)$ are given on the manifold $\xi(\mathbf{r}, t) = \xi_0$. [Notice that *the initial value problem* corresponds to giving $\rho(\mathbf{r}, 0), \mathbf{v}(\mathbf{r}, 0), e(\mathbf{r}, 0), \mathbf{B}(\mathbf{r}, 0)$ on the domain of interest in ordinary three-space. In order for this problem to be well-posed, ordinary three-space should not be a characteristic. Here, we consider the opposite case that data are given on a characteristic, so that the Cauchy problem is not well-posed.] As in Section 5.4.2, we consider ξ as a coordinate and introduce additional coordinates η, ζ, τ, such that four-space $(\mathbf{r}, t)$ is covered by the coordinates ξ, η, ζ, τ. The boundary data may then be written as

$$\rho(\xi_0, \eta, \zeta, \tau) = \rho_0(\eta, \zeta, \tau), \quad \text{etc.}, \tag{5.111}$$

where η, ζ, τ parameterize the boundary $\xi(\mathbf{r}, t) = \xi_0$. Since $\rho_0(\eta, \zeta, \tau)$ is a known function, the derivatives $\partial\rho_0/\partial\eta, \partial\rho_0/\partial\zeta, \partial\rho_0/\partial\tau$ may also be considered to be known. Similarly, for the other variables $\mathbf{v}_0, e_0, \mathbf{B}_0$.

We want to find out under which conditions the solutions $\rho(\xi, \eta, \zeta, \tau), \mathbf{v}(\xi, \eta, \zeta, \tau), e(\xi, \eta, \zeta, \tau)$, $\mathbf{B}(\xi, \eta, \zeta, \tau)$ may be obtained away from the boundary $\xi = \xi_0$ or, rather, may not be obtained since

[1] An exception is the excitation of waves by time-dependent forcing terms at the boundary of the plasma. In that case data are given on time-like (satisfying $ct > |x|$) boundaries.

then $\xi = \xi_0$ is a characteristic. To that end, we write the variables in terms of a power series:

$$\rho(\xi, \eta, \zeta, \tau) = \rho_0(\eta_0, \zeta_0, \tau_0) + (\xi - \xi_0)\left(\frac{\partial \rho}{\partial \xi}\right)_0 + \cdots , \tag{5.112}$$

and likewise for $\mathbf{v}$, e and $\mathbf{B}$. As in the previous section, we may consider the problem to be solvable if $(\partial\rho/\partial\xi)_0$, $(\partial\mathbf{v}/\partial\xi)_0$, $(\partial e/\partial\xi)_0$ and $(\partial\mathbf{B}/\partial\xi)_0$ can be constructed, since the other first order derivatives follow from the boundary data (5.111), whereas the higher order ones may be obtained by subsequent differentiations of the original partial differential equations.

We denote the unknown derivatives with respect to ξ with a prime, so that the derivatives in the MHD equations (5.16)–(5.18) may be written as

$$\nabla\rho = \nabla\xi\,\rho' + \cdots , \qquad \frac{\mathrm{D}\rho}{\mathrm{D}t} = (\xi_t + \mathbf{v}\cdot\nabla\xi)\,\rho' + \cdots , \tag{5.113}$$

and similarly for the other variables. Hence, with respect to the primed variables, the coordinate transformation amounts to the replacements

$$\begin{aligned} \nabla f \quad &\to \quad \mathbf{n} f' + \cdots , \qquad \mathbf{n} \equiv \nabla\xi , \\ \frac{\mathrm{D}f}{\mathrm{D}t} \quad &\to \quad -\,u f' + \cdots , \qquad -u \equiv \xi_t + \mathbf{v}\cdot\nabla\xi . \end{aligned} \tag{5.114}$$

Here, $\mathbf{n}$ is the *normal to the characteristic* (where ξ has been chosen such that $|\nabla\xi| = 1$, so that $\mathbf{n}$ has unit length), and u is the *characteristic speed*, i.e. the normal velocity of the characteristic ξ measured with respect to the fluid velocity $\mathbf{v}$. [We just note the correspondence between these transformations for the characteristics (weak discontinuities) and the transformations (4.143) for the shocks (strong discontinuities). Further discussion of this intriguing correspondence has to await the analysis of transonic flows in Chapter 21.]

Inserting the expressions (5.113) into Eqs. (5.16)–(5.19), keeping the primed (unknown) variables on the left hand side and moving the known variables to the right hand side of the equation, we obtain the following set of algebraic equations:

$$\begin{aligned} -u\rho' + \rho\,\mathbf{n}\cdot\mathbf{v}' &= \cdots , \\ -\rho u\mathbf{v}' + (\gamma-1)\,\mathbf{n}\,(e\rho' + \rho e') + (\mathbf{n}\,\mathbf{B}\cdot - \mathbf{n}\cdot\mathbf{B})\,\mathbf{B}' &= \cdots , \\ -ue' + (\gamma-1)e\,\mathbf{n}\cdot\mathbf{v}' &= \cdots , \\ -u\mathbf{B}' + (\mathbf{B}\,\mathbf{n}\cdot - \mathbf{n}\cdot\mathbf{B})\,\mathbf{v}' &= \cdots , \qquad \mathbf{n}\cdot\mathbf{B}' = \cdots . \end{aligned} \tag{5.115}$$

The dots on the RHS indicate the known derivatives with respect to η, ζ and τ. In order to get equations of the same dimension we multiply the four lines of Eq. (5.115) by c/ρ, $1/\rho$, γ/c and $1/\sqrt{\rho}$, respectively, recalling that $c^2 = \gamma(\gamma-1)e$. By now, the reader will have noted that we have virtually reduced the algebra to that of Eqs. (5.28)–(5.31) of Section 5.2, where $\mathbf{n}$ has taken the place of the wave vector $\mathbf{k}$ and u that of the frequency ω (or, rather, the Doppler shifted frequency $\omega - \mathbf{k}\cdot\mathbf{v}$). We again choose $\mathbf{B}$ along the z-axis and $\mathbf{n}$ in the x–z plane,

$$\mathbf{B} = (0, 0, B) , \qquad \mathbf{n} = (n_x, 0, n_z) , \tag{5.116}$$

and introduce *the Alfvén speed* and *the sound speed*,

$$b \equiv B/\sqrt{\rho} , \qquad c \equiv \sqrt{\gamma p/\rho} . \tag{5.117}$$

The system of equations (5.115) then becomes

$$\begin{pmatrix} -\gamma u & n_x c & 0 & n_z c & 0 & 0 & 0 & 0 \\ n_x c & -u & 0 & 0 & n_x c & -n_z b & 0 & n_x b \\ 0 & 0 & -u & 0 & 0 & 0 & -n_z b & 0 \\ n_z c & 0 & 0 & -u & n_z c & 0 & 0 & 0 \\ 0 & n_x c & 0 & n_z c & -\frac{\gamma}{\gamma-1}u & 0 & 0 & 0 \\ 0 & -n_z b & 0 & 0 & 0 & -u & 0 & 0 \\ 0 & 0 & -n_z b & 0 & 0 & 0 & -u & 0 \\ 0 & n_x b & 0 & 0 & 0 & 0 & 0 & -u \end{pmatrix} \begin{pmatrix} c\,\rho'/(\gamma\rho) \\ v'_x \\ v'_y \\ v'_z \\ c\,e'/(\gamma e) \\ B'_x/\sqrt{\rho} \\ B'_y/\sqrt{\rho} \\ B'_z/\sqrt{\rho} \end{pmatrix} = \cdots\,, \tag{5.118}$$

where $n_z = B_{\rm n}/B$, $n_x = [1-(B_{\rm n}/B)^2]^{1/2}$, and the constraint on $\mathbf{n}\cdot\mathbf{B}'$ is ignored for the time being. Not surprising, the homogeneous part of this equation corresponds to the eigenvalue problem (5.33) for the MHD waves. If the algebra is the same, the physics probably is also the same. We will show this to be so.

The characteristics are obtained when the determinant of the LHS of Eq. (5.118) vanishes so that the full inhomogeneous problem cannot be solved. In that case, the solutions cannot be propagated away from the manifold $\xi = \xi_0$. This condition may be written as

$$\Delta = \frac{\gamma^2}{\gamma-1}u^2(u^2-b_{\rm n}^2)\left[\,u^4-(b^2+c^2)u^2+b_{\rm n}^2c^2\,\right]=0\,, \tag{5.119}$$

where $b_{\rm n}$ is the normal Alfvén speed, $b_{\rm n} \equiv \mathbf{n}\cdot\mathbf{B}/\sqrt{\rho}$. Hence, we have recovered the dispersion equation (5.54) for the linear MHD waves in the disguise of an equation for the characteristic speeds u. From Sections 5.2.1 and 5.2.2, we immediately conclude that a spurious root $u=0$ has been introduced from the condition $u\,\mathbf{n}\cdot\mathbf{B}'=0$, following from the homogeneous part of the first equation (5.115)(d), which should be eliminated on account of the second equation. Hence, *seven real characteristics* are obtained, corresponding to the eight variables minus the redundant one needed to describe the system. The matrix on the LHS of Eq. (5.118) is real, symmetric, and has only real eigenvalues. This implies that *the equations of ideal MHD are symmetric hyperbolic PDEs* and *the initial value problem*, where values are assigned to the variables $\mathbf{v}$, $\mathbf{B}$, e, and ρ in ordinary three-dimensional space at $t=0$, *is well-posed*. This important result is due to Friedrichs [195].

Disregarding the redundant root, Eq. (5.119) yields seven characteristic speeds:

$$u = u_{\rm E} \equiv 0\,, \tag{5.120}$$

$$u = u_{\rm A} \equiv \pm b_{\rm n}\,, \tag{5.121}$$

$$u = u_{\rm s} \equiv \pm\Big[\,\tfrac{1}{2}(b^2+c^2)-\tfrac{1}{2}\sqrt{(b^2+c^2)^2-4b_{\rm n}^2c^2}\,\Big]^{1/2}\,, \tag{5.122}$$

$$u = u_{\rm f} \equiv \pm\Big[\,\tfrac{1}{2}(b^2+c^2)+\tfrac{1}{2}\sqrt{(b^2+c^2)^2-4b_{\rm n}^2c^2}\,\Big]^{1/2}\,. \tag{5.123}$$

The solution (5.120) corresponds to *entropy disturbances* that are just convected by the flow. The pair of solutions (5.121) corresponds to *Alfvén disturbances* moving forward (+) or backward (−) with respect to the flow. The pair of solutions (5.122) are forward and backward *slow magneto-acoustic disturbances*, whereas the solutions (5.123) constitute forward and backward *fast magneto-*

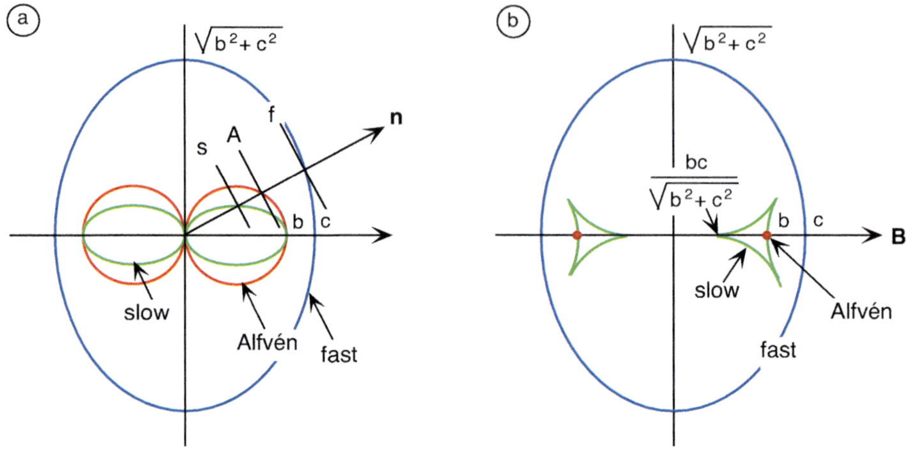

Fig. 5.9 Friedrichs diagrams: schematic representation of (a) reciprocal normal surfaces (or phase diagram) and (b) ray surfaces (or group diagram) of the fast (blue), Alfvén (red) and slow (green) MHD waves for $b < c$.

acoustic disturbances. The characteristic speeds are ordered according to the following sequence of inequalities:

$$0 = |u_{\rm E}| \le |u_{\rm s}| \le |u_{\rm A}| \le |u_{\rm f}| < \infty \,. \tag{5.124}$$

Degeneracies occur for $\mathbf{n} \parallel \mathbf{B}$:

$$|u_{\rm s}| = \min(b,c)\,, \qquad |u_{\rm A}| = b\,, \qquad |u_{\rm f}| = \max(b,c)\,, \tag{5.125}$$

and for $\mathbf{n} \perp \mathbf{B}$:

$$|u_{\rm s}| = |u_{\rm A}| = 0\,, \qquad |u_{\rm f}| = (b^2+c^2)^{1/2}\,. \tag{5.126}$$

The equations of gas dynamics are obtained in the limit $b \to 0$, when the slow and Alfvén waves disappear at the origin ($u \to 0$) and the fast magneto-acoustic waves degenerate into sound waves ($u_{\rm f} \to \pm c$). Another limit of interest is the case of incompressible plasma ($c \to \infty$), when the speed of the fast magneto-acoustic wave disappears at infinity (instantaneous propagation) whereas the slow magneto-acoustic speed and the Alfvén speed coincide.

Consider now the spatial part of a characteristic manifold at a certain time $t = t_0$. This is called the *ray surface*. It may be considered as a wave front, i.e. a surface across which weak discontinuities (as in Fig. 5.8) occur, emitted at $t = 0$ from the origin. For example, without a magnetic field, a characteristic manifold would be the spherical sound front $x^2 + y^2 + z^2 = c^2t^2$ so that the ray surface is the sphere with radius ct_0. Dropping the z-dependence, the characteristic in x, y, t space then becomes a cone through the point $x = y = t = 0$, whereas the circular intersection of this cone with the plane $t = t_0$ constitutes the ray surface.

To get the ray surface we first of all compute from Eqs. (5.120)–(5.123) the distance ut_0 which a plane wave front travels along $\mathbf{n}$ after having passed the origin at $t = 0$. The collection of these points gives Fig. 5.9(a) [which is a schematic rendering of the computed phase diagram of Fig. 5.4(a)]. However, this is *not* the ray surface, but the so-called *reciprocal normal surface*. To get the ray surface that is depicted in Fig. 5.9(b) [which is again a schematic rendering, but now of the group diagram of Fig. 5.4(b)], we have to take the envelope of the plane wave fronts since the ray surface corresponds to a wave front due to a point disturbance at the origin at $t = 0$. Taking the envelope of the fronts indicated by s, A and f in Fig. 5.9(a) results in the completely different and, in particular, more singular picture of Fig. 5.9(b). As we have seen in Section 5.3.2, the reciprocal normal surface for the Alfvén wave consists of two spheres touching the origin. Correspondingly, *the Alfvén ray surface* just consists of the two points $z = \pm b$: the Alfvén waves travel as point disturbances along the magnetic field. The ray surface for the slow magneto-acoustic wave also exhibits a quite anisotropic character: it consists of two cusped figures. The fast magneto-acoustic waves exhibit the least degree of anisotropy. They resemble ordinary sound waves most with respect to their propagation properties.

A simple representation of the MHD characteristics is obtained by dropping one more spatial dimension and exploiting the expressions (5.120)–(5.123) for the seven speeds to compute the x–t cross-sections of the characteristics. This yields the schematic illustration of the characteristics in Fig, 5.10, which will play an important role in our discussion of the Riemann problem in computational MHD (see Section 20.2.1). Since the characteristic speed u is measured with respect to the background velocity $\mathbf{v}$, $d\mathbf{r}/dt|_{\rm char} = u + \mathbf{v}\cdot\mathbf{n}$, the inclination of the entropy characteristic indicates that the plasma is assumed to flow to the right here.

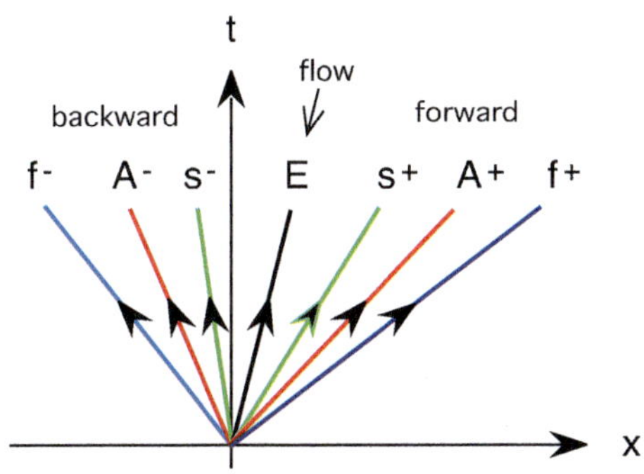

Fig. 5.10 The seven characteristic directions for MHD (when the x-axis is not perpendicular or parallel to $\mathbf{B}$).

On a characteristic manifold, where $\Delta = 0$, the homogeneous counterpart of Eq. (5.118) has a solution with fixed relations between the values ρ', $\mathbf{v}'$, e' and $\mathbf{B}'$. A physical interpretation is to consider those primed variables as discontinuities of the derivative across the characteristic, so that $\rho' \equiv \partial\rho/\partial\xi \gg \partial\rho/\partial\eta$, etc., justifying the neglect of the RHS of Eq. (5.118). We may consider these quantities as *weak discontinuities* of the flow that are propagated along with the characteristics. We then find the following relationships for the different characteristics which, of course, reproduce the relationships found for the waves in Section 5.2.3:

(a) Entropy characteristics ($u = 0$):

$$S'/S = -\gamma\rho'/\rho = \gamma e'/e \neq 0\,, \qquad \mathbf{v}' = 0\,, \qquad \mathbf{B}' = 0\,. \tag{5.127}$$

Only the thermodynamic variables are perturbed, in particular the entropy.

(b) Alfvén characteristics ($u = u_{\rm A}$):

$$B'_y = -\sqrt{\rho}\, v'_y \neq 0\,, \quad \rho' = 0\,, \quad e' = 0\,, \quad v'_x = v'_z = 0\,, \quad B'_x = B'_z = 0\,. \tag{5.128}$$

These are purely transverse disturbances where $\mathbf{v}'$ and $\mathbf{B}'$ are perpendicular to the plane through $\mathbf{n}$ and $\mathbf{B}$ (Fig. 5.1(a)). The thermodynamic variables are not perturbed.

(c) Magneto-acoustic characteristics ($u = u_{\rm s,f}$):

$$v'_x = \frac{n_x}{n_z}\frac{u^2}{u^2 - b^2}\, v'_z \neq 0\,, \qquad B'_z = -\frac{n_x}{n_z}\, B'_x = \frac{n_x b}{u}\sqrt{\rho}\, v'_x \neq 0\,,$$
$$\rho' = \frac{\gamma p}{c^2}\, e' = \frac{\rho u}{n_z c^2}\, v'_z \neq 0\,, \qquad v'_y = 0\,, \qquad B'_y = 0\,. \tag{5.129}$$

These disturbances are polarized in the plane $(\mathbf{n}, \mathbf{B})$, with the fast and slow polarizations perpendicular to each other (Fig. 5.1(b)). This difference arises through the factor $u^2 - b^2$ which is positive for the fast characteristic and negative for the slow one. The thermodynamic variables are perturbed, except for the entropy.

For characteristic directions $\mathbf{n} \cdot \mathbf{B} = 0$ (when the magnetic field is tangential to the space-part of the characteristic), the root $u = 0$ is fivefold degenerate. The homogeneous version of Eqs. (5.115) then reduces to the two conditions

$$\mathbf{n} \cdot \mathbf{v}' = 0\,, \tag{5.130}$$

$$p' + \mathbf{B} \cdot \mathbf{B}' = (p + \tfrac{1}{2}B^2)' = 0\,, \tag{5.131}$$

whereas we now also have to include the separate condition

$$\mathbf{n} \cdot \mathbf{B}' = 0\,. \tag{5.132}$$

All other components of the variables are arbitrary. These disturbances are called *tangential discontinuities*. An example would be an equilibrium of two adjacent plasmas with different pressure, density, tangential magnetic field, tangential velocity, but satisfying the relations (5.130)–(5.132). At the interface, a surface current $\mathbf{j}^{\star\prime} = \mathbf{n} \times \mathbf{B}'$ and a surface vorticity $\boldsymbol{\omega}^{\star\prime} = \mathbf{n} \times \mathbf{v}'$ produce disturbances of the tangential field and velocity. Thus, we recover the interface conditions of Section 4.5.2 and the circle is closed: boundary value problem and initial value problem are interwoven in MHD.

The fivefold degeneracy for vanishing characteristic speed for $\mathbf{n} \cdot \mathbf{B} = 0$ corresponds to the degeneracy of marginal entropy, Alfvén and slow magneto-sonic waves ($\omega = 0$ for $\mathbf{k} \cdot \mathbf{B} = 0$). In general, such degeneracies signal the need to extend the theory with more sophisticated assumptions. The present degeneracies originate from the assumption of *locality* on the group diagrams and characteristics, which neglects the global plasma properties associated with *inhomogeneity*. In order to make progress, we now have to enter the vast territory of waves and instabilities in inhomogeneous plasmas. This is the subject where MHD acquires its particular strength and beauty. It will occupy us for the rest of this book.

5.5 Literature and exercises

Notes on literature

Phase and group diagrams

- Bittencourt, *Fundamentals of Plasma Physics* [70], Section 14.6 and Chapter 15.

Characteristics

- Courant and Hilbert, *Methods of Mathematical Physics II* [132](b) Chapter VI.
- Garabedian, *Partial Differential Equations* [206], Chapters 2, 3, 4, 6, 14.

Computational fluid dynamics

- LeVeque, *Numerical Methods for Conservation Laws* [404], and his contributions to the Saas-Fe lecture Notes by LeVeque, Mihalas, Dorfi and Müller, *Computational Methods for Astrophysical Fluid Flow* [406].

Characteristics in MHD

- The demonstration that the MHD equations are symmetric hyperbolic equations is due to Friedrichs. It probably appeared for the first time in print in the Courant Institute lecture notes [195], which is still very worth reading.
- Akhiezer, Akhiezer, Polovin, Sitenko and Stepanov, *Plasma Electrodynamics*, Vol. 1 [8], Chapter 3.

Exercises

[5.1] *Sound waves*

We start with a plasma without magnetic field and neglect gravity.

- Which variables are involved? Write down the equations governing their evolution.
- We assume a time-independent, infinite and homogeneous background and perturb it. Derive the equations for the linear perturbations. Do all three variables still couple?
- Derive the dispersion equation for plane waves from the wave equation for the perturbed velocity, exploiting the sound speed c. Since there is no preferred direction in the plasma, choose a direction of wave propagation and solve the dispersion equation. How many solutions are there? What do they represent?

[5.2] *Towards MHD waves*

We now introduce a magnetic field into the problem and rewrite the vector products in terms of inner products by means of the vector identities of Appendix A.

- Express the MHD equations in terms of the variables $\rho, \mathbf{v}, e = p/[(\gamma - 1)\rho]$ and $\mathbf{B}$.
- Again perturb the quantities and derive the linearized MHD equations in terms of dimensionless variables $\bar{\rho}, \bar{\mathbf{v}}, \bar{e}, \bar{\mathbf{B}}$, using γ, c and the vectorial Alfvén speed $\mathbf{b}$.
- Is there a preferred direction in the plasma now? Choose general directions for $\mathbf{B}_0$ and $\mathbf{k}$ (can one choose them parallel?), and derive the matrix representation of the eigenvalue problem. Is the matrix symmetric, or should it be?

[5.3] *The marginal entropy wave*

To isolate this wave, we exploit the entropy $S \equiv p\rho^{-\gamma}$ and the pressure p as variables.

- Perturb these two quantities, construct dimensionless variables $\bar{S}$ and $\bar{p}$, and derive the evolution equations for them. Combine these equations with those of problem [5.2] to obtain the linearized MHD equations in terms of $\bar{S}, \bar{\mathbf{v}}, \bar{p}$ and $\bar{\mathbf{B}}$.
- Show that these equations have a special solution, called the marginal entropy wave, where $\omega = 0$ and most of the variables, except $\bar{S}$, vanish. What is the physical interpretation of this solution?

[5.4] *The velocity representation*

When the constraint $\nabla \cdot \bar{\mathbf{B}} = 0$ is not observed, a spurious marginal 'solution' is introduced. To get rid of it, we reduce the MHD equations to the velocity representation.

- From the linearized MHD equations for $\bar{\rho}$, $\bar{\mathbf{v}}$, $\bar{e}$ and $\bar{\mathbf{B}}$, derive the second order wave equation for the velocity $\bar{\mathbf{v}}$. Are the sound waves included in this equation?
- Again insert plane wave solutions and construct the matrix representation of the eigenvalue problem. What is the dimension of this matrix? Count the unknowns!
- What is the general condition for an equation like $\mathbf{A} \cdot \mathbf{x} = \lambda \mathbf{x}$ to have a solution where $\mathbf{x} \neq 0$? Give this equation for the matrix representation you derived.

[5.5] *The three MHD waves*

The dispersion equation has three genuine ($\omega \neq 0$) wave solutions: Alfvén waves (frequencies $\pm\omega_{\mathrm{A}}$), slow and fast magneto-sonic waves (frequencies $\pm\omega_{\mathrm{s,f}}$).

- Derive ω_{A} of the Alfvén waves. Are these waves longitudinal or transverse? In what direction do they propagate? Could they exist in a plasma without magnetic field?
- Derive the frequencies $\omega_{\mathrm{s,f}}$ of the magneto-sonic waves. Which of these waves survives in a plasma without a magnetic field? What are their frequencies in that case?
- Show that $0 \leq \omega_{\mathrm{s}}^2 \leq \omega_{\mathrm{A}}^2 \leq \omega_{\mathrm{f}}^2 < \infty$.

[5.6] *Phase and group velocities*

There are two important velocities in the propagation of wave trains. One is the phase velocity, $\mathbf{v}_{\mathrm{ph}} \equiv (\omega/k)\,\mathbf{k}/k$, and the other is the group velocity, $\mathbf{v}_{\mathrm{gr}} \equiv \partial\omega/\partial\mathbf{k}$.

- Derive the equations for these velocities for the three kinds of MHD waves. Give explicit expressions for propagation parallel and perpendicular to the magnetic field.
- Show that a wave packet with wave vectors localized about a central value $\mathbf{k}_0$ propagates with the group velocity $\mathbf{v}_{\mathrm{gr}}(\mathbf{k}_0)$.

[5.7]* *Characteristics in alternative representations.*

In the text, the characteristics are obtained from the matrix representation Eq. (5.118) in terms of the variables $\rho, \mathbf{v}, e, \mathbf{B}$. The same characteristics are obtained from the basic set $s, \mathbf{v}, p, \mathbf{B}$. Show that this is also true for the system ρ, $\mathbf{v}$, p, $\mathbf{B}$, or the one obtained from it for the variables ρ, $\mathbf{v}$, s, $\mathbf{B}$. Notice that in these representations the matrix is no longer symmetric, so that the first two should be considered as the more adequate ones.

6

Spectral theory

6.1 Stability: intuitive approach

6.1.1 Two viewpoints

How does one know whether a dynamical system is stable or not? Consider the well-known example of a ball at rest at the bottom of a trough or on the top of a hill (Fig. 6.1). There is a position (indicated by the full circle) where the potential energy W due to gravity has an extremum W_0. Displacing the ball slightly to a neighbouring position (at the open circle) results in either a higher or a lower potential energy W_1. This corresponds to a *stable* system in the first case ($W_1 > W_0$) and an *unstable* system in the second case ($W_1 < W_0$).

Already at this stage some important observations can be made, viz.:

(a) We have tacitly assumed that the constraining surface is curved, i.e. either convex or concave, so that there is a position of rest, which is called the *equilibrium* position. In this case, one may rescale the potential energy such that the equilibrium state corresponds to $W_0 = 0$, and W_1 becomes the potential energy of the displacement, which is called the *perturbation*.

(b) If the constraining surface is flat and inclined, the system is not in equilibrium and the ball simply rolls along the plane. This *lack of equilibrium*, when W has no extremum, should clearly be distinguished from *neutral* or *marginal stability*, when $W_1 = W_0$. The latter situation occurs when the surface is horizontal, so that the value $W = 0$ may be assigned to both W_0 and W_1.

This simple example illustrates the general theoretical approach to linear stability, where the study of the original nonlinear equations is simplified by means of a split in equilibrium and perturbations. First, a time-independent equilibrium state is to be found. This still involves the solution of the nonlinear equations, but with the simplifying condition that the time dependence of the variables vanishes and, usually, that the equilibrium is (translationally and/or rotationally) symmetric. An additional simplification, which is usually made for MHD and which is quite pertinent for fusion machines (not for astrophysical plasmas!), is the assumption of *static equilibrium*, not involving flow ($\mathbf{v} = 0$). Next, this equilibrium is subjected to small, i.e. linear, perturbations. This involves the study of the linearized, time-dependent equations.

Such a study may be conducted by means of two broad classes of methods, viz. by using variational principles involving *quadratic forms* (like the energy) or by solving *the (partial) differential equations* themselves. These methods are just a generalization of the two intuitive approaches illustrated in Fig. 6.1. The upper part illustrates the investigation of stability by itself by means of the so-called *energy principle*, i.e. a study of the sign of the potential energy W_1 of the perturbations ($W_1 > 0$: stable, $W_1 = 0$: marginally stable, $W_1 < 0$: unstable). The full dynamics of the system

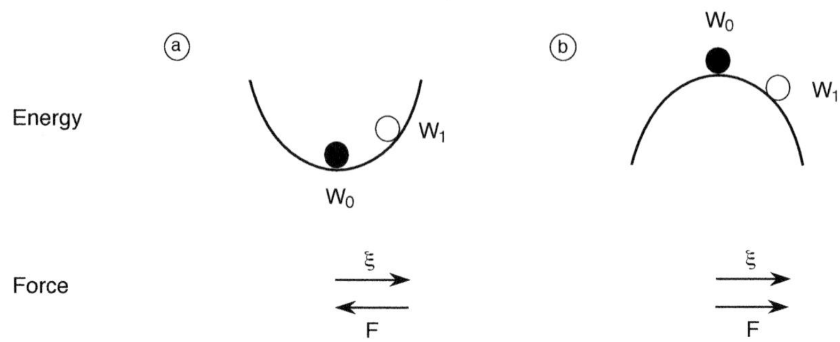

Fig. 6.1 Energy and force (a) in a linearly stable system, and (b) in a linearly unstable system.

may be obtained from a variational principle which not only involves the potential energy but also the kinetic energy of the perturbations. The more usual approach is the solution of differential equations, in particular the equation of motion, which involves a study of the *forces* acting on the system. With respect to stability, this method is illustrated in the bottom part of Fig. 6.1. If a displacement ξ creates a force F in the opposite direction, the state of equilibrium tends to be restored and the system is stable. On the other hand, if the resulting force is in the same direction as the displacement, the motion will be away from equilibrium and the system is unstable.

These intuitive notions on displacements, forces and energies may be generalized to continuous media, in particular magneto-fluids. This leads to the two alternative representations by means of *the equation of motion*, involving the plasma displacement vector field $\boldsymbol{\xi}(\mathbf{r}, t)$ and the linear force operator $\mathbf{F}(\boldsymbol{\xi})$ acting on that field, on the one hand (Section 6.2), and by means of *variational quadratic forms*, involving the potential energy functional $W[\boldsymbol{\xi}]$ and the kinetic energy functional $K[\dot{\boldsymbol{\xi}}]$, on the other hand (Section 6.4).

▷ **Question** When a glass of water is turned upside down the contents will drop out, as is generally known (first situation). However, if the glass is filled to the rim and covered by a piece of paper the water will not drop out when the glass is turned over (second situation). Discuss these facts in terms of equilibrium and stability properties of the configuration. Is the first situation lack of equilibrium or instability? How about the second one? What is the function of the piece of paper? ◁

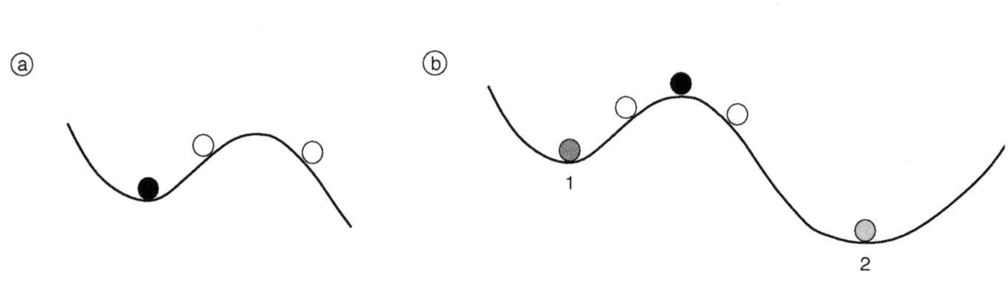

Fig. 6.2 (a) Nonlinear stability: Linearly stable, nonlinearly unstable system; (b) Linearly unstable, nonlinearly 'stable' system.

Nonlinear stability or instability concerns the behaviour of dynamical systems with respect to finite (non-infinitesimal) amplitudes of the displacements. Some examples are shown in Fig. 6.2. On the left a system is shown which is stable when subjected to small perturbations, but which becomes unstable when the amplitude of the perturbation is big enough. Vice versa, a system may be unstable with respect to small perturbations (Fig. 6.2(b)) but may possess neighbouring equilibrium states (indicated by the labels 1 and 2) which are stable. If those states are accessible, the system may turn out to be nonlinearly 'stable'. (Quotation marks since the original system is unstable but it evolves towards another state, which is stable.) This subject is quite relevant for the properties of confined plasmas, but too complicated to be covered in this chapter.

▷ **Subtlety** Consider the situation of Fig. 6.3. The potential energy of the ball has been lowered by taking it outside the cup. Is this situation unstable? Of course, this is cheating: we have tacitly agreed upon the *constraint* that the ball should stay on the surface of the cup. However, such subtleties are relevant for our study of plasma instabilities, as we will see later (Section 6.1.2) for the constraint of 'frozen' field lines. For the moment it suffices to state that it is important to make constraints explicit and to observe them. ◁

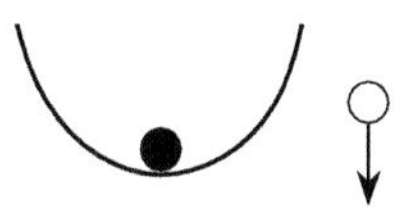

Fig. 6.3 Violating constraints.

6.1.2 Linearization and Lagrangian reduction

Let us now carry out the mentioned programme on force and energy for plasmas. The starting point is again the ideal MHD equations, which we write in the following particular order in terms of the variables $\mathbf{v}, p, \mathbf{B}$ and ρ:

$$\rho\Big(\frac{\partial\mathbf{v}}{\partial t} + \mathbf{v}\cdot\nabla\mathbf{v}\Big) = -\nabla p + \mathbf{j}\times\mathbf{B} - \rho\nabla\Phi_{\rm gr}\,, \qquad \mathbf{j} = \nabla\times\mathbf{B}\,, \tag{6.1}$$

$$\frac{\partial p}{\partial t} = -\mathbf{v}\cdot\nabla p - \gamma p\nabla\cdot\mathbf{v}\,, \tag{6.2}$$

$$\frac{\partial\mathbf{B}}{\partial t} = \nabla\times(\mathbf{v}\times\mathbf{B})\,, \qquad \nabla\cdot\mathbf{B} = 0\,, \tag{6.3}$$

$$\frac{\partial\rho}{\partial t} = -\nabla\cdot(\rho\mathbf{v})\,. \tag{6.4}$$

For the time being, we restrict the analysis to model I, i.e. application of the BCs

$$\mathbf{n}\cdot\mathbf{v} = 0 \qquad \text{(at the wall)}\,, \tag{6.5}$$

$$\mathbf{n}\cdot\mathbf{B} = 0 \qquad \text{(at the wall)}\,. \tag{6.6}$$

These BCs do not need to be linearized because they are already linear. (That is, with fixed $\mathbf{n}$; interface plasmas, where $\mathbf{n}$ is not fixed, will be discussed in Section 6.6.1).

Linearization proceeds by first defining a background equilibrium state about which the dynamics is supposed to take place. The simplest and most relevant choice is that of a *static equilibrium* ($\mathbf{v}_0 = 0$):

$$\mathbf{j}_0\times\mathbf{B}_0 = \nabla p_0 + \rho_0\nabla\Phi_{\rm gr}\,, \qquad \mathbf{j}_0 = \nabla\times\mathbf{B}_0\,, \qquad \nabla\cdot\mathbf{B}_0 = 0\,, \tag{6.7}$$

$$\text{subject to the BC} \quad \mathbf{n}\cdot\mathbf{B}_0 = 0 \qquad \text{(at the wall)}\,. \tag{6.8}$$

These equations only partly determine the equilibrium functions $\rho_0(\mathbf{r}), p_0(\mathbf{r}), \mathbf{j}_0(\mathbf{r})$ and $\mathbf{B}_0(\mathbf{r})$. A lot of freedom is left in the choice of the equilibrium profiles. In particular, in the absence of gravity, the density $\rho_0(\mathbf{r})$ is completely arbitrary.

▷ **Example: freedom in circular cylinder equilibria** We drop the gravity term and momentarily omit the subscript o on the equilibrium quantities. Exploiting cylinder coordinates r, θ, z, with $\partial/\partial\theta = 0$ and $\partial/\partial z = 0$, Eq. (6.7)(b) yields $\nabla \cdot \mathbf{j} = (1/r)\, d(rj_r)/dr = 0$, so that $j_r = 0$. Similarly, Eq. (6.7)(c) yields $B_r = 0$. Hence, the physical variables $p(r)$, $\mathbf{B} = B_\theta(r)\,\mathbf{e}_\theta + B_z(r)\,\mathbf{e}_z$, and $\mathbf{j} = j_\theta(r)\,\mathbf{e}_\theta + j_z(r)\,\mathbf{e}_z$ involve five scalar functions which depend only on r. In that case, the *magnetic surfaces*, which are tangent to $\mathbf{j}$ and $\mathbf{B}$ and orthogonal to ∇p, are nested circular cylinders. The five radial equilibrium functions are further restricted by the remaining relations of Eqs. (6.7), viz.

$$p' = j_\theta B_z - j_z B_\theta\,, \qquad j_\theta = -B_z'\,, \qquad j_z = \frac{1}{r}(rB_\theta)'\,, \tag{6.9}$$

$$\text{so that} \quad (p + \tfrac{1}{2}B^2)' = -B_\theta^2/r\,, \tag{6.10}$$

where the primes denote derivatives with respect to r. Consequently, Eq. (6.10) is the only restriction on the three basic functions $p(r), B_\theta(r), B_z(r)$, so that two of them may be chosen arbitrarily. Note that this freedom comes in addition to the arbitrariness of the density profile $\rho(r)$. For example, choosing $p = p_0(1 - r^2)$ and $B_z = B_0$, we get $B_\theta = \sqrt{p_0}r$, $j_\theta = 0$ and $j_z = 2\sqrt{p_0}$, which is a slight generalization (permitting a constant B_z field) of the z-pinch equilibrium considered in Section 2.4.3. ◁

We return to the general discussion of the split in equilibrium and dynamics. Time enters into the problem with the *perturbation* of the equilibrium:

$$\begin{aligned}
\mathbf{v}(\mathbf{r},t) &= \mathbf{v}_1(\mathbf{r},t)\,,\\
p(\mathbf{r},t) &= p_0(\mathbf{r}) + p_1(\mathbf{r},t)\,,\\
\mathbf{B}(\mathbf{r},t) &= \mathbf{B}_0(\mathbf{r}) + \mathbf{B}_1(\mathbf{r},t)\,,\\
\rho(\mathbf{r},t) &= \rho_0(\mathbf{r}) + \rho_1(\mathbf{r},t)\,,
\end{aligned} \tag{6.11}$$

where p_0, $\mathbf{B}_0$ and ρ_0 now correspond to an *inhomogeneous* equilibrium, satisfying Eqs. (6.7) and (6.8), and the perturbations should satisfy $|f_1(\mathbf{r},t)| \ll |f_0(\mathbf{r})|$ (except for the velocity $\mathbf{v}$). The resulting first order equations for the perturbations of Eqs. (6.1)–(6.4) then read:

$$\rho_0 \frac{\partial \mathbf{v}_1}{\partial t} = -\nabla p_1 + \mathbf{j}_1 \times \mathbf{B}_0 + \mathbf{j}_0 \times \mathbf{B}_1 - \rho_1 \nabla \Phi_{\rm gr}\,, \quad \mathbf{j}_1 = \nabla \times \mathbf{B}_1\,, \tag{6.12}$$

$$\frac{\partial p_1}{\partial t} = -\mathbf{v}_1 \cdot \nabla p_0 - \gamma p_0 \nabla \cdot \mathbf{v}_1\,, \tag{6.13}$$

$$\frac{\partial \mathbf{B}_1}{\partial t} = \nabla \times (\mathbf{v}_1 \times \mathbf{B}_0)\,, \qquad \nabla \cdot \mathbf{B}_1 = 0\,, \tag{6.14}$$

$$\frac{\partial \rho_1}{\partial t} = -\nabla \cdot (\rho_0 \mathbf{v}_1)\,, \tag{6.15}$$

whereas the BCs (6.5) and (6.6) for the perturbations require

$$\mathbf{n} \cdot \mathbf{v}_1 = 0\,, \quad \mathbf{n} \cdot \mathbf{B}_1 = 0 \qquad \text{(at the wall).} \tag{6.16}$$

At this stage, p_0, $\mathbf{j}_0$, $\mathbf{B}_0$ and ρ_0 are considered to be known from the zeroth order equilibrium equations, and $\mathbf{v}_1, p_1, \mathbf{B}_1$ and ρ_1 are unknown. Note that many terms have been eliminated by the assumption $\mathbf{v}_0 = 0$.

We are now ready for an extremely useful further reduction, due to Bernstein et al. (1958) [55]. We have promoted the momentum conservation equation (6.12) to the first place since it describes

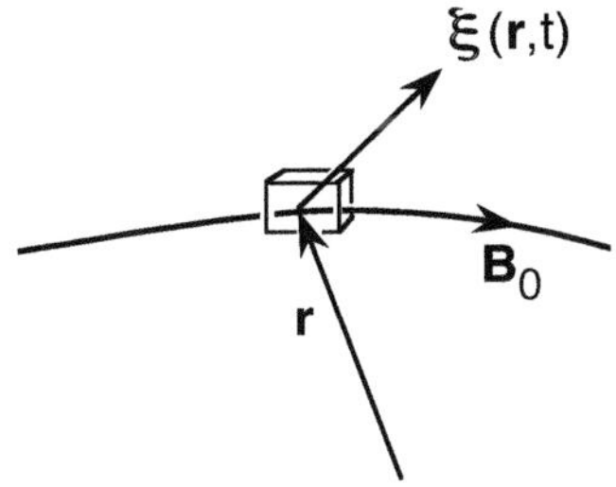

Fig. 6.4 Plasma displacement vector field.

the evolution of the velocity $\mathbf{v}_1$, whereas the perturbations p_1, $\mathbf{B}_1$ and ρ_1 may be considered as secondary, i.e. determined by $\mathbf{v}_1$ as expressed by the right hand sides of Eqs. (6.13)–(6.15). Elimination of the latter equations is straightforward if the momentum equation is differentiated with respect to time. However, the opposite procedure of integrating the equations (6.13)–(6.15) is more powerful. This is effected by means of a new variable, viz. *the Lagrangian displacement vector field* $\boldsymbol{\xi}(\mathbf{r}, t)$ of a plasma element away from the equilibrium state (see Fig. 6.4). The velocity is just the Lagrangian time derivative (i.e. the variation in time experienced in a local coordinate system co-moving with the fluid) of this variable:

$$\mathbf{v} = \frac{\mathrm{D}\boldsymbol{\xi}}{\mathrm{D}t} \equiv \frac{\partial \boldsymbol{\xi}}{\partial t} + \mathbf{v} \cdot \nabla \boldsymbol{\xi}\,, \tag{6.17}$$

which is an exact, but highly nonlinear, expression. However, for the linearized problem, we only need the first order, Eulerian, part of the time derivative:

$$\mathbf{v} \approx \mathbf{v}_1 = \frac{\partial \boldsymbol{\xi}}{\partial t}\,. \tag{6.18}$$

Inserting this expression, the equations (6.13)–(6.15) for p_1, $\mathbf{B}_1$ and ρ_1 can be integrated:

$$p_1 = -\,\boldsymbol{\xi} \cdot \nabla p_0 - \gamma p_0 \nabla \cdot \boldsymbol{\xi}\,, \tag{6.19}$$

$$\mathbf{B}_1 = \nabla \times (\boldsymbol{\xi} \times \mathbf{B}_0)\,, \tag{6.20}$$

$$\rho_1 = -\,\nabla \cdot (\rho_0 \boldsymbol{\xi})\,. \tag{6.21}$$

Note that the equation $\nabla \cdot \mathbf{B}_1 = 0$ can be dropped now since it is automatically satisfied by Eq. (6.20). Inserting these expressions into Eq. (6.12) leads to the desired result, viz. an equation of motion of the form

$$\rho_0 \frac{\partial^2 \boldsymbol{\xi}}{\partial t^2} = \mathbf{F}\Big(p_1(\boldsymbol{\xi}), \mathbf{B}_1(\boldsymbol{\xi}), \rho_1(\boldsymbol{\xi})\Big)\,, \tag{6.22}$$

where p_1, $\mathbf{B}_1$ and ρ_1 are determined by Eqs. (6.19)–(6.21). We will continue with this formulation in Section 6.2, where the explicit form of the operator $\mathbf{F}$ is given.

In the absence of gravity, the expression (6.21) for ρ_1 is actually not needed since it does not occur in the equation of motion. Hence, we could forget about this relation in that case. This changes when gravitational forces are taken into account, as needed in astrophysical plasmas. We will restrict the analysis to gravitational potentials that are external, i.e. not caused by self-gravitation of the plasma but by some external massive object.

▷ **Subtleties on the initial data** So far, we have not mentioned initial conditions. In particular, we have omitted explicit terms for $p_1(\mathbf{r},0)$, $\mathbf{B}_1(\mathbf{r},0)$ and $\rho_1(\mathbf{r},0)$ in Eqs. (6.19)–(6.21). This implies that we have tacitly assumed that finite initial values for these quantities only appear as a result of a finite initial displacement $\boldsymbol{\xi}(\mathbf{r},0)$. Since the ultimate equation (6.22) is of second order in time, we should consider both $\boldsymbol{\xi}(\mathbf{r},0)$ and $\dot{\boldsymbol{\xi}}(\mathbf{r},0)$ as independent initial data. The following possibilities arise: (1) $\boldsymbol{\xi}(\mathbf{r},0) = 0$, $\dot{\boldsymbol{\xi}}(\mathbf{r},0) \neq 0$; (2) $\boldsymbol{\xi}(\mathbf{r},0) \neq 0$, $\dot{\boldsymbol{\xi}}(\mathbf{r},0) = 0$; and (3) combinations of case (1) and (2).

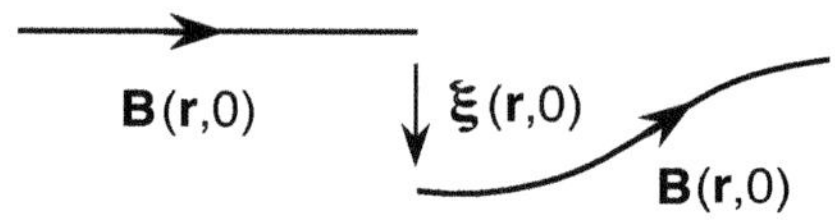

Fig. 6.5 Cheating on the initial data.

First, consider case (2). Should we consider $p_1(\mathbf{r},0)$, $\mathbf{B}_1(\mathbf{r},0)$, $\rho_1(\mathbf{r},0)$ as additional independent initial data, together with $\boldsymbol{\xi}(\mathbf{r},0)$? **No!**: If we start off with a finite displacement of the plasma, we better make sure that we got there by means of a motion consistent with the ideal MHD conservation laws. Otherwise, something could happen as illustrated in Fig. 6.5. At $t = 0$ the fluid has been displaced, but simultaneously a perturbation in $\mathbf{B}$ has been allowed that corresponds to broken field lines: that is not a meaningful operation in ideal MHD. Such a perturbation, *not satisfying the constraints*, should be considered as the analogue of taking the ball out of the cup illustrated in Fig. 6.3.

Similarly for case (1), we should assume that $p_1(\mathbf{r},0) = 0$, $\mathbf{B}_1(\mathbf{r},0) = 0$, and $\rho_1(\mathbf{r},0) = 0$ because $\boldsymbol{\xi}(\mathbf{r},0) = 0$. Hence, for physical reasons, we tacitly impose the additional condition that the initial data on p_1, $\mathbf{B}_1$, ρ_1 are consistent with ideal MHD. This implies that the equations (6.19)–(6.21) are considered to be valid at all times. Consequently, all perturbations are expressed in terms of $\boldsymbol{\xi}(\mathbf{r},t)$ alone and the possibility of Fig. 6.5 is eliminated. ◁

6.2 Force operator formalism

6.2.1 Equation of motion

We complete the discussion of the equation of motion (6.22), where the right hand side defines the *ideal MHD force operator* $\mathbf{F}(\boldsymbol{\xi})$. Inserting Eqs. (6.19) and (6.20) of Section 6.1.2, the explicit form of the linearized equation of motion becomes

$$\mathbf{F}(\boldsymbol{\xi}) \equiv -\nabla\pi - \mathbf{B}\times(\nabla\times\mathbf{Q}) + (\nabla\times\mathbf{B})\times\mathbf{Q} + (\nabla\Phi_{\rm gr})\,\nabla\cdot(\rho\boldsymbol{\xi}) = \rho\frac{\partial^2\boldsymbol{\xi}}{\partial t^2}\,, \qquad (6.23)$$

where

$$\pi \equiv p_1 = -\gamma p\nabla\cdot\boldsymbol{\xi} - \boldsymbol{\xi}\cdot\nabla p\,, \qquad (6.24)$$

$$\mathbf{Q} \equiv \mathbf{B}_1 = \nabla\times(\boldsymbol{\xi}\times\mathbf{B})\,. \qquad (6.25)$$

The symbols π (not to be confused with the earlier use of π in Chapter 3) and $\mathbf{Q}$ are introduced here for no other reason than to be able to drop the subscripts 0 and 1 on the equilibrium and perturbation variables. This we will do from now on. For model I (wall on the plasma), the only boundary condition needed to close the system is obtained from Eq. (6.16)(a) for $\mathbf{v}_1$:

$$\mathbf{n}\cdot\boldsymbol{\xi} = 0 \qquad \text{(at the wall)}\,. \qquad (6.26)$$

Once the geometry of the plasma has been specified (see Section 4.6), the equation of motion (6.23) and the boundary condition (6.26) completely determine the linearized dynamics of the plasma.

Consequently, introducing the plasma displacement vector $\boldsymbol{\xi}$ has enabled us to cast the linear equations of ideal MHD in the compact form (6.23), which may be considered to be *Newton's law for a plasma element*. This is the physical basis of the very powerful spectral methods that can be applied to these problems.

▷ **Magnetic field boundary condition** The boundary condition (6.16)(b) for $\mathbf{B}_1$, that is $\mathbf{n}\cdot\nabla\times(\boldsymbol{\xi}\times\mathbf{B})=0$, is automatically satisfied since $\mathbf{n}\cdot\boldsymbol{\xi}=0$ and $\mathbf{n}\cdot\mathbf{B}=0$. This is proved by noting that the latter two conditions imply that $\boldsymbol{\xi}\times\mathbf{B}=f\mathbf{n}$, with f some scalar function. Hence, $\mathbf{n}\cdot\nabla\times(\boldsymbol{\xi}\times\mathbf{B})=\mathbf{n}\cdot[(\nabla f)\times\mathbf{n}+f\nabla\times\mathbf{n}]=f\mathbf{n}\cdot\nabla\times\mathbf{n}=0$. The last equality is obtained from the expression (A.34) for the curl of a vector in curvilinear coordinates by choosing one coordinate along $\mathbf{n}$, so that $n_1=1$ and $n_2=n_3=0$. ◁

It is important to count unknowns and equations. In linearized ideal MHD only one vector field $\boldsymbol{\xi}(\mathbf{r},t)$ appears, whereas in the nonlinear theory the variables $\mathbf{v}$, $\mathbf{B}$, p and ρ are needed. This is a substantial simplification (in addition to the linearization). The order of the system should be independent of the choice of the variables exploited. Equation (6.23), with the definitions (6.24) and (6.25) substituted, constitutes a sixth order system[1] of three second order PDEs for the three components of $\boldsymbol{\xi}$. On the other hand, the equations (6.12)–(6.14) also constitute a sixth order system, viz. of six first order PDEs for the three components of $\mathbf{v}$, the scalar p, and two of the three components of $\mathbf{B}_1$. The third component of $\mathbf{B}_1$ is redundant because of the condition $\nabla\cdot\mathbf{B}_1=0$, which is an initial condition for Eq. (6.14), but automatically satisfied in Eq. (6.25). In the absence of gravity, the equation for ρ_1 is redundant in both representations since it does not couple to the rest of the equations. The presence of an external gravity field does not raise the order of the system since gradients of ρ_1 do not appear in the equations (6.12) and (6.15).

Recall the similar reduction from representation in primitive variables to velocity representation in Section 5.2.3, which we showed to be equivalent if the marginal entropy mode, for $\omega_{\rm E}S_1=0$ or $\omega_{\rm E}\rho_1=0$, was ignored. We conclude that the $\boldsymbol{\xi}$-representation, also called the *Lagrangian representation*, is only equivalent to the original *Eulerian representation* in primitive variables if the Eulerian entropy mode $\omega_{\rm E}=0$ is neglected. This is why the subscript E is put on this mode.

Since the equilibrium quantities $p(\mathbf{r})$, $\mathbf{B}(\mathbf{r})$ and $\rho(\mathbf{r})$ appearing in Eq. (6.23) do not depend on time we may consider solutions in the form of *normal modes*:

$$\boldsymbol{\xi}(\mathbf{r},t)=\hat{\boldsymbol{\xi}}(\mathbf{r})\mathrm{e}^{-\mathrm{i}\omega t}\,. \tag{6.27}$$

This transforms Eq. (6.23) into the eigenvalue problem

$$\mathbf{F}(\hat{\boldsymbol{\xi}})=-\rho\omega^2\hat{\boldsymbol{\xi}}\,, \tag{6.28}$$

involving the linear operator $\mathbf{F}$ (or rather $\rho^{-1}\mathbf{F}$) and its eigenvalues ω^2. Except for discrete eigenvalues, ideal MHD also allows for continuous (or 'improper') eigenvalues, as we will see in Chapter 7. The collection $\{\omega^2\}$ of these two kinds of eigenvalues is called *the spectrum of ideal MHD*. The central property, which puts this subject on an equal mathematical footing to quantum mechanics, is that *the operator* $\rho^{-1}\mathbf{F}$ *is self-adjoint* so that *the squared eigenvalues* ω^2 *are real*. This will be

[1] The attentive reader may be puzzled by the appearance of only one boundary condition, viz. Eq. (6.26). Actually, in order not to slow down the exposition by too much detail, we had to be a little cavalier here. This will be remedied in later calculations where additional regularity and periodicity conditions appear, dictated by the specific geometry which is chosen. Yet, it should be stressed at this point that the general statement about the sixth order character of the problem only holds with respect to two of the three spatial coordinates. Due to the extreme anisotropy of magnetically confined systems, the dependence on the coordinate normal to the magnetic surfaces turns out to reduce to a second order system in terms of $\mathbf{n}\cdot\boldsymbol{\xi}$ alone. Hence, regularity at the magnetic axis and the condition (6.26) suffice to fix the solutions.

proved in Section 6.2.3. An important physical consequence of the appearance of the *square* of the eigenvalues is that the eigenvalues ω themselves are either real or purely imaginary.

Writing $\omega = \sigma + \mathrm{i}\nu$, this implies that two quite different classes of solution occur, viz. *stable waves* for $\omega^2 > 0$, so that $\nu = 0$ and the temporal dependence is given by $\exp(\mp\mathrm{i}\sigma t)$, and *instabilities* for $\omega^2 < 0$, so that $\sigma = 0$ and the temporal dependence $\exp(\pm\nu t)$ represents exponential growth. For eigenmodes, we already see the connection with the bottom part of the pictures of Fig. 6.1 on the relationship between displacements and forces. According to Eq. (6.28), $\mathbf{F}(\hat{\boldsymbol{\xi}}) \sim -\hat{\boldsymbol{\xi}}$ for stable waves with $\omega^2 > 0$ and $\mathbf{F}(\hat{\boldsymbol{\xi}}) \sim \hat{\boldsymbol{\xi}}$ for instabilities with $\omega^2 < 0$. For general motions, consisting of a superposition of eigenmodes, such a simple relationship does not hold. In that case, the sign of the potential energy (corresponding to the upper parts of Fig. 6.1) will provide the test for stability (see Section 6.4.4).

In dissipative (e.g. resistive) MHD, different eigenmodes are possible. In particular, since ω^2 needs not be real then, complex values of ω may occur. This may give rise to stable, but damped, modes if $\mathrm{Im}(\omega) < 0$ and so-called 'overstable' modes if $\mathrm{Im}(\omega) > 0$. The term 'overstable' expresses the fact that the direction of the restoring force is opposite to the displacement, as in Fig. 6.1(a), but it is too big so that the resulting motion overshoots and the net result is again an instability. These additional possibilities are associated with the fact that the waves are non-conservative in dissipative MHD: energy may be dissipated or accumulated.

Clearly, ideal, conservative MHD of static equilibria presents a significant simplification for stability problems. Since ω^2 is real, the transition from stability to instability occurs only through the value $\omega^2 = 0$, or $\omega = 0$, i.e. through *marginal stability*. Consequently, to study the problem of stability one could study *the marginal equation of motion*

$$\mathbf{F}(\hat{\boldsymbol{\xi}}) = 0\,, \tag{6.29}$$

subject to the boundary condition (6.26). In general, this equation does not have a solution because it is obtained from the eigenvalue problem (6.28) and $\omega^2 = 0$ does not have to be an eigenvalue. In order to get genuine solutions one should arrange the equilibrium parameters such that $\omega^2 = 0$ becomes an eigenvalue. For example, a typical tokamak stability study would involve the variation of global equilibrium parameters like the value of $\beta \equiv 2\mu_0 p/B^2$ and the 'safety factor' at the plasma boundary, $q_1 \sim 1/I_\mathrm{p}$, while keeping other variables fixed. For a particular value of the plasma current I_p one would push the value of β until the marginal equation of motion Eq. (6.29) is satisfied, subject to the boundary condition (6.26). In this manner one would calculate one value β_crit where marginal stability is obtained. By varying the value of I_p one would trace out marginal stability curves in the β–q_1 diagram, as shown schematically in Fig. 6.6. Physical arguments usually indicate on which side of the curve the stable states are to be found. (For example, in Fig. 6.6, this would be on the low β side.) This is the most general, though not the easiest, method of studying stability problems.

Since $\mathbf{F}$ now contains all information on MHD waves and instabilities, let us rearrange the terms slightly to see what we have:

$$\mathbf{F}(\boldsymbol{\xi}) = \nabla(\gamma p \nabla\cdot\boldsymbol{\xi}) - \mathbf{B}\times(\nabla\times\mathbf{Q}) + \nabla(\boldsymbol{\xi}\cdot\nabla p) + \mathbf{j}\times\mathbf{Q} + (\nabla\Phi_\mathrm{gr})\,\nabla\cdot(\rho\boldsymbol{\xi})\,. \tag{6.30}$$

One recognizes, successively, an isotropic force due to plasma *compressibility*, associated with sound waves, and an anisotropic force ($\perp$ **B**) due to *bending of field lines*, which is responsible for Alfvén waves. These terms are always present, even in an infinitely homogeneous plasma ($\nabla p = 0$,

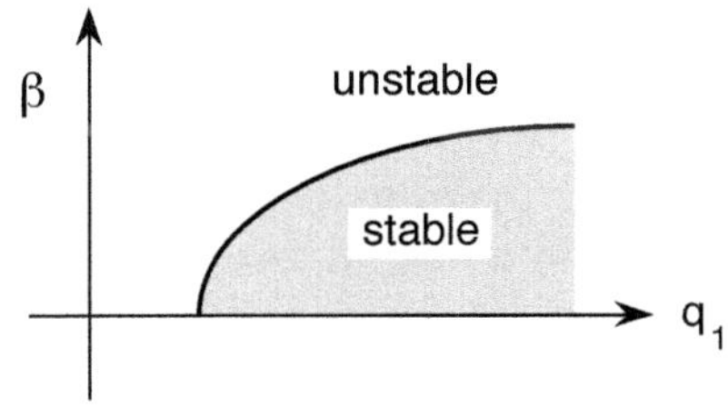

Fig. 6.6 Schematic stability diagram for tokamaks.

$\mathbf{j} = 0$), which has been shown (Chapter 5) to give rise to stable waves only. The third and fourth terms only occur in inhomogeneous plasmas, e.g. those encountered in confined plasmas for thermonuclear research or magnetic flux tubes in astrophysical plasmas. In those plasmas, the effect of equilibrium *pressure gradients* and *currents*, which are intrinsically associated with confinement, may give rise to instabilities. Of course, the art of designing thermonuclear confinement machines is to find regions in parameter space (as exemplified by Fig. 6.6) where these potentially unstable effects are counterbalanced by the stabilizing contributions. Finally, the gravitational term is also associated with inhomogeneity. It gives rise to confinement as well as gravitational instabilities in astrophysical plasmas.

For *homogeneous* plasmas, the spectral equation (6.28) may be written as

$$\rho^{-1}\mathbf{F}(\hat{\boldsymbol{\xi}}) = c^2\nabla\nabla\cdot\hat{\boldsymbol{\xi}} + \mathbf{b}\times(\nabla\times(\nabla\times(\mathbf{b}\times\hat{\boldsymbol{\xi}}))) = -\omega^2\hat{\boldsymbol{\xi}}\,, \tag{6.31}$$

where $c \equiv \sqrt{\gamma p/\rho}$ and $\mathbf{b} \equiv \mathbf{B}/\sqrt{\rho}$ are constants. In that case, plane wave solutions $\hat{\boldsymbol{\xi}} \sim \exp(\mathrm{i}\mathbf{k}\cdot\mathbf{r})$ give rise to the algebraic eigenvalue problem

$$\begin{aligned} \rho^{-1}\mathbf{F}(\hat{\boldsymbol{\xi}}) &= -c^2\mathbf{k}\mathbf{k}\cdot\hat{\boldsymbol{\xi}} - \mathbf{b}\times(\mathbf{k}\times(\mathbf{k}\times(\mathbf{b}\times\hat{\boldsymbol{\xi}}))) \\ &= \big[-(\mathbf{k}\cdot\mathbf{b})^2\,\mathbf{I} - (b^2+c^2)\,\mathbf{k}\mathbf{k} + \mathbf{k}\cdot\mathbf{b}\,(\mathbf{k}\mathbf{b}+\mathbf{b}\mathbf{k})\big]\cdot\hat{\boldsymbol{\xi}} = -\omega^2\hat{\boldsymbol{\xi}}\,. \end{aligned} \tag{6.32}$$

This equation for $\hat{\boldsymbol{\xi}}$ is the same as Eq. (5.52) for $\hat{\mathbf{v}}$, which produced the dispersion equation (5.54), so that the three stable MHD waves of Section 5.2.3 are recovered. Hence, analogous to Eq. (5.64), the slow, Alfvén and fast eigenvectors $\hat{\boldsymbol{\xi}}_{\mathrm{s}}$, $\hat{\boldsymbol{\xi}}_{\mathrm{A}}$ and $\hat{\boldsymbol{\xi}}_{\mathrm{f}}$ form an orthogonal triad. This guarantees that arbitrary vectors $\hat{\boldsymbol{\xi}}$ may be decomposed into the three MHD eigenvectors so that the eigenspaces of the force operator span the whole space. We will show that this space is a Hilbert space.

Finally, it is important to note that the eigenvalue problem (6.28), with $\mathbf{F}$ given by either Eq. (6.30) or (6.31), contains Alfvén waves as the dynamical centrepiece. Thus, Eq. (6.32), with $\mathbf{B}$ and $\mathbf{k}$ in the z-direction (see Fig. 5.1(a)) and transverse incompressible motion ($\mathbf{k}\cdot\boldsymbol{\xi} = 0$), yields the Alfvén wave equation

$$\rho^{-1}\hat{F}_y = b^2\frac{\partial^2\hat{\xi}_y}{\partial z^2} = -k_z^2 b^2\hat{\xi}_y = \frac{\partial^2\hat{\xi}_y}{\partial t^2} = -\omega^2\hat{\xi}_y\,, \tag{6.33}$$

with $\omega^2 = \omega_{\mathrm{A}}^2 \equiv k_z^2 b^2$ as eigenvalue. The formidable analysis that follows is, in a sense, just the machinery needed to trace this wave in inhomogeneous plasmas.

6.2.2 Hilbert space

We have cast the ideal MHD equations in a form which is general enough to benefit from comparison with another field of physics where *linear operators* play an important role, viz. quantum mechanics. Consider a plasma volume V enclosed by a wall W. Define two plasma displacement fields:

$$\begin{aligned} \boldsymbol{\xi} &= \boldsymbol{\xi}(\mathbf{r}, t) \quad (\text{on V}), \qquad \text{where} \quad \mathbf{n}\cdot\boldsymbol{\xi} = 0 \quad (\text{at W}), \\ \boldsymbol{\eta} &= \boldsymbol{\eta}(\mathbf{r}, t) \quad (\text{on V}), \qquad \text{where} \quad \mathbf{n}\cdot\boldsymbol{\eta} = 0 \quad (\text{at W}). \end{aligned} \tag{6.34}$$

An *inner product* may be defined with the plasma equilibrium density $\rho(\mathbf{r})$ acting as a weight function:

$$\langle \boldsymbol{\xi}, \boldsymbol{\eta} \rangle \equiv \tfrac{1}{2} \int \rho\, \boldsymbol{\xi}^* \cdot \boldsymbol{\eta}\, dV \,. \tag{6.35}$$

By means of this definition, one may also define *the norm* of the vector field $\boldsymbol{\xi}(\mathbf{r}, t)$:

$$\|\boldsymbol{\xi}\| \equiv \langle \boldsymbol{\xi}, \boldsymbol{\xi} \rangle^{1/2} . \tag{6.36}$$

For functions $\boldsymbol{\xi}$ with *a finite norm*, $\|\boldsymbol{\xi}\| < \infty$, a special linear function space is obtained, viz. a *Hilbert space*. At this stage, certain properties of this space with respect to completeness and separability are assumed. For a more complete discussion of these aspects of MHD spectral theory see Lifschitz [410].

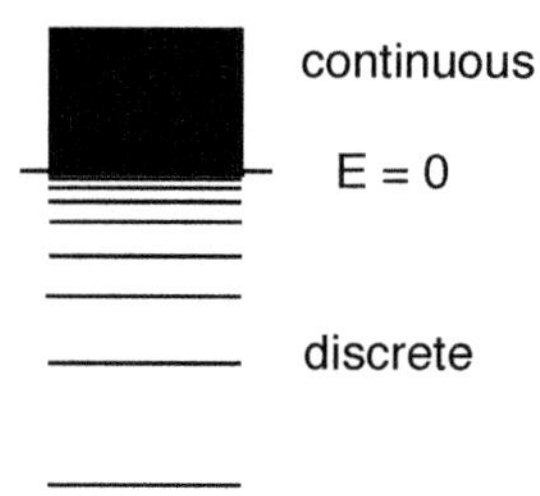

Fig. 6.7 Schematic spectrum of a quantum mechanical system.

The mathematical analogy with quantum mechanics is obvious. The Schrödinger equation for the wave function ψ,

$$H\psi = E\psi \,, \tag{6.37}$$

is an eigenvalue equation for the self-adjoint linear operator H (the Hamiltonian) with eigenvalues E (the energy levels). This eigenvalue problem leads to a spectrum of eigenvalues (Fig. 6.7), which may be discrete (for bound states with $E < 0$) or continuous (for free particles with $E > 0$). Ideal MHD also has a spectrum of eigenvalues ω^2, which may be discrete or continuous, with properties depending on the force operator $\mathbf{F}$. It is useful to learn the relevant methods from another field.

The inner product in quantum mechanics,

$$\langle \psi_1, \psi_2 \rangle = \tfrac{1}{2} \int \psi_1^* \psi_2 \, dV \,, \tag{6.38}$$

permits one to define the norm

$$\|\psi\| \equiv \langle \psi, \psi \rangle^{1/2} , \tag{6.39}$$

which defines the probability of finding the particle in the volume considered. In general, one chooses $\|\psi\| = 1$ to express the certainty that the particle will be located somewhere in the volume. Clearly, normalization and Hilbert space have a very clear-cut physical background in quantum mechanics.

What is the physical background for considering vector fields $\boldsymbol{\xi}(\mathbf{r}, t)$ with a finite norm $\|\boldsymbol{\xi}\|$ in ideal MHD? To answer that question, consider *the kinetic energy* of the plasma,

$$K \equiv \tfrac{1}{2}\int \rho \mathbf{v}^2\, dV \approx \tfrac{1}{2}\int \rho \dot{\boldsymbol{\xi}}^2\, dV = \langle \dot{\boldsymbol{\xi}}, \dot{\boldsymbol{\xi}} \rangle \equiv \|\dot{\boldsymbol{\xi}}\|^2 \,. \tag{6.40}$$

As far as the spatial dependence is concerned, $\dot{\boldsymbol{\xi}}(\mathbf{r}, t)$ belongs to the same class of functions as $\boldsymbol{\xi}(\mathbf{r}, t)$ (time t is simply a parameter). Hence, a bounded norm implies finite kinetic energy of the perturbations. Because the total energy is conserved, it also implies finite potential energy. This is a very reasonable assumption, which justifies the use of Hilbert space as the mathematical device for our investigations.

Hence, in quantum mechanics, we encounter linear operators acting on the wave function ψ, and in magnetohydrodynamics, we encounter linear operators acting on the displacement vector field $\boldsymbol{\xi}$. We also have defined an inner product and a norm for the MHD displacements. Finally, the analogy would be perfect if we could also show that there is a counterpart in MHD for the important role played by self-adjoint operators L, defined by

$$\langle \psi_1, L\psi_2 \rangle = \langle L\psi_1, \psi_2 \rangle \tag{6.41}$$

in quantum mechanics. In fact, one of the central results of linearized ideal MHD turns out to be that *the force operator* $\mathbf{F}$, or rather $\rho^{-1}\mathbf{F}$, *is a self-adjoint linear operator in the Hilbert space of plasma displacement vectors*:

$$\langle \boldsymbol{\eta}, \rho^{-1}\mathbf{F}(\boldsymbol{\xi}) \rangle \equiv \tfrac{1}{2}\int \boldsymbol{\eta}^* \cdot \mathbf{F}(\boldsymbol{\xi})\, dV = \tfrac{1}{2}\int \boldsymbol{\xi} \cdot \mathbf{F}(\boldsymbol{\eta}^*)\, dV \equiv \langle \rho^{-1}\mathbf{F}(\boldsymbol{\eta}), \boldsymbol{\xi} \rangle \,. \tag{6.42}$$

This provides linear MHD with a mathematical foundation that is the same as that of quantum mechanics, viz. linear operators in Hilbert space: a very solid ground indeed! On this basis, many analogies between MHD and quantum mechanical spectral theory should be expected.

6.2.3 Proof of self-adjointness of the force operator

Of course, after the excitement about a beautiful analogy, the dirty job of *proving the self-adjointness property* (6.42) remains; see Bernstein et al. [55](b). This consists of a lot of cumbersome vector manipulations with little beauty, but is does belong to the craft of MHD. Also, we need some of the intermediate steps in later sections. Hence, we just give the complete proof here, putting some of the more technical details in small print.

We start from the expression (6.30) for the force operator, where p and $\mathbf{B}$ should satisfy the equilibrium relations $\mathbf{j} \times \mathbf{B} = \nabla p + \rho \nabla \Phi_{\rm gr}$, $\mathbf{j} = \nabla \times \mathbf{B}$, and $\nabla \cdot \mathbf{B} = 0$. The first two terms of $\mathbf{F}(\boldsymbol{\xi})$ are already present in homogeneous plasmas, but the last three terms are associated with inhomogeneity. They will turn out to be a considerable complication. We will rework those three inhomogeneity terms to show that, in the absence of gravity, they give rise to *a force that is purely perpendicular to* $\mathbf{B}$ (like the second term of Eq. (6.30)).

▷ **Transformation of the inhomogeneity terms of the force operator** The transformation of the last three terms of the expression (6.30) for the force operator involves extensive use of the vector identities of Appendix A.1. They are indicated above the equal signs:

$$\begin{aligned}
\nabla(\boldsymbol{\xi}\cdot\nabla p) &\overset{(A.10)}{=} (\nabla\boldsymbol{\xi})\cdot\nabla p + \boldsymbol{\xi}\cdot\nabla\nabla p \\
&\overset{(A.9)}{=} (\nabla p\times\nabla)\times\boldsymbol{\xi} + (\nabla p)\nabla\cdot\boldsymbol{\xi} + \boldsymbol{\xi}\cdot\nabla\nabla p \\
&\overset{\text{equilibrium}}{=} ((\mathbf{j}\times\mathbf{B})\times\nabla)\times\boldsymbol{\xi} - \rho(\nabla\Phi_{\rm gr}\times\nabla)\times\boldsymbol{\xi} + (\nabla p)\nabla\cdot\boldsymbol{\xi} + \boldsymbol{\xi}\cdot\nabla\nabla p \\
&\overset{(A.2)\ \text{on}\ \mathbf{j},\mathbf{B},\nabla}{=} (\mathbf{B}\mathbf{j}\cdot\nabla - \mathbf{j}\mathbf{B}\cdot\nabla)\times\boldsymbol{\xi} - \rho(\nabla\Phi_{\rm gr}\times\nabla)\times\boldsymbol{\xi} + (\nabla p)\nabla\cdot\boldsymbol{\xi} + \boldsymbol{\xi}\cdot\nabla\nabla p \\
&\overset{(A.9)\ \text{on}\ \nabla\Phi_{\rm gr},\nabla,\boldsymbol{\xi}}{=} \mathbf{B}\times(\mathbf{j}\cdot\nabla\boldsymbol{\xi}) - \mathbf{j}\times(\mathbf{B}\cdot\nabla\boldsymbol{\xi}) - \rho(\nabla\boldsymbol{\xi})\cdot\nabla\Phi_{\rm gr} + \rho(\nabla\Phi_{\rm gr})\nabla\cdot\boldsymbol{\xi} \\
&\qquad + (\nabla p)\nabla\cdot\boldsymbol{\xi} + \boldsymbol{\xi}\cdot\nabla\nabla p\,,
\end{aligned}\tag{6.43}$$

$$\begin{aligned}
\mathbf{j}\times\mathbf{Q} &\overset{(A.13)}{=} \mathbf{j}\times(\mathbf{B}\cdot\nabla\boldsymbol{\xi} - \mathbf{B}\nabla\cdot\boldsymbol{\xi} - \boldsymbol{\xi}\cdot\nabla\mathbf{B}) \\
&= \mathbf{j}\times(\mathbf{B}\cdot\nabla\boldsymbol{\xi}) - \mathbf{j}\times\mathbf{B}\nabla\cdot\boldsymbol{\xi} - \boldsymbol{\xi}\cdot\nabla(\mathbf{j}\times\mathbf{B}) - \mathbf{B}\times(\boldsymbol{\xi}\cdot\nabla\mathbf{j})\,,
\end{aligned}\tag{6.44}$$

$$(\nabla\Phi_{\rm gr})\nabla\cdot(\rho\boldsymbol{\xi}) \overset{(A.6)}{=} \rho(\nabla\Phi_{\rm gr})\nabla\cdot\boldsymbol{\xi} + (\nabla\Phi_{\rm gr})(\nabla\rho)\cdot\boldsymbol{\xi}\,, \tag{6.45}$$

where the equality $\nabla(\mathbf{a}\times\mathbf{b}) = (\nabla\mathbf{a})\times\mathbf{b} - (\nabla\mathbf{b})\times\mathbf{a}$ has been used in the last step of the derivation of Eq. (6.44).[2] Hence, the three inhomogeneity terms may be written as

$$\begin{aligned}
&\nabla(\boldsymbol{\xi}\cdot\nabla p) + \mathbf{j}\times\mathbf{Q} + (\nabla\Phi_{\rm gr})\nabla\cdot(\rho\boldsymbol{\xi}) \\
&\quad= \mathbf{B}\times(\mathbf{j}\cdot\nabla\boldsymbol{\xi} - \boldsymbol{\xi}\cdot\nabla\mathbf{j}) + (\nabla p - \mathbf{j}\times\mathbf{B} + 2\rho\nabla\Phi_{\rm gr})\nabla\cdot\boldsymbol{\xi} \\
&\qquad - \rho(\nabla\boldsymbol{\xi})\cdot\nabla\Phi_{\rm gr} + \boldsymbol{\xi}\cdot\nabla\nabla p - \boldsymbol{\xi}\cdot\nabla(\mathbf{j}\times\mathbf{B}) + (\nabla\Phi_{\rm gr})\nabla\rho\cdot\boldsymbol{\xi} \\
&\overset{(A.13),\nabla\cdot\mathbf{j}=0}{=} -\mathbf{B}\times(\nabla\times(\mathbf{j}\times\boldsymbol{\xi})) - \mathbf{j}\times\mathbf{B}\nabla\cdot\boldsymbol{\xi} + (\nabla p - \mathbf{j}\times\mathbf{B} + 2\rho\nabla\Phi_{\rm gr})\nabla\cdot\boldsymbol{\xi} \\
&\qquad - \rho(\nabla\boldsymbol{\xi})\cdot\nabla\Phi_{\rm gr} + \boldsymbol{\xi}\cdot\nabla\nabla p - \boldsymbol{\xi}\cdot\nabla(\mathbf{j}\times\mathbf{B}) + (\nabla\Phi_{\rm gr})\nabla\rho\cdot\boldsymbol{\xi} \\
&\overset{\text{equil.},(A.10)}{=} -\mathbf{B}\times(\nabla\times(\mathbf{j}\times\boldsymbol{\xi})) - (\nabla p)\nabla\cdot\boldsymbol{\xi} - \rho\nabla(\boldsymbol{\xi}\cdot\nabla\Phi_{\rm gr})\,.
\end{aligned}\tag{6.46}$$

This proves our assertion: in the absence of gravity, when the last term vanishes and ∇p may be replaced by $\mathbf{j}\times\mathbf{B}$, the remaining two terms represent a force $\perp\mathbf{B}$. ◁

Using this expression, Eq. (6.30) may be rewritten in the form

$$\mathbf{F}(\boldsymbol{\xi}) = \nabla(\gamma p\nabla\cdot\boldsymbol{\xi}) - \mathbf{B}\times[\nabla\times\mathbf{Q} + \nabla\times(\mathbf{j}\times\boldsymbol{\xi})] - (\nabla p)\nabla\cdot\boldsymbol{\xi} - \rho\nabla(\boldsymbol{\xi}\cdot\nabla\Phi_{\rm gr})\,. \tag{6.47}$$

In the absence of gravity (using the equilibrium relation $\nabla p = \mathbf{j}\times\mathbf{B}$), the force operator reduces to

$$\mathbf{F}_{\rm g=0}(\boldsymbol{\xi}) = \nabla(\gamma p\nabla\cdot\boldsymbol{\xi}) - \mathbf{B}\times[\nabla\times\mathbf{Q} + \nabla\times(\mathbf{j}\times\boldsymbol{\xi}) - \mathbf{j}\nabla\cdot\boldsymbol{\xi}]\,, \tag{6.48}$$

and, in the absence of a magnetic field (using the equilibrium relation $\nabla p = \rho\mathbf{g}$), to

$$\mathbf{F}_{\rm B=0}(\boldsymbol{\xi}) = \nabla(\gamma p\nabla\cdot\boldsymbol{\xi}) - \rho\mathbf{g}\,\nabla\cdot\boldsymbol{\xi} + \rho\nabla(\mathbf{g}\cdot\boldsymbol{\xi})\,. \tag{6.49}$$

These very useful intermediate results in our derivation give an exposition, alternative to that of

[2] The brackets are used here to delimit the range of the gradient operator. For example, the expression $(\nabla\boldsymbol{\xi})\cdot\nabla p$ means that ∇ acts on $\boldsymbol{\xi}$ only, and contraction occurs between the components of $\boldsymbol{\xi}$ and ∇p, so that we get $\Sigma_j(\partial_i\xi_j)\,\partial_j p$ in Cartesian coordinates. Note that, in curvilinear coordinates, $\boldsymbol{\xi} = \Sigma_j\,\xi_j\mathbf{a}^j$ so that the derivatives of the basis vectors $\mathbf{a}^j$ also have to be taken into account.

Eq. (6.30), of the different forces at work in an inhomogeneous plasma. There is an isotropic (sound) contribution from the pressure and an anisotropic (magnetic) contribution orthogonal to the background equilibrium magnetic field $\mathbf{B}$. Gravity, $\mathbf{g} = -\nabla\Phi_{\rm gr}$, introduces a direction additional to the triad $\mathbf{B}, \mathbf{j}, \nabla p$. This contribution lifts the degeneracy implicit in the identification of ∇p and $\mathbf{j} \times \mathbf{B}$ (i.e., the identification of pressure and magnetic surfaces), which is at the heart of calculations on stability of thermonuclear plasmas. Hence, astrophysical plasmas exhibit a richer variety of waves and instabilities. It belongs to the intricacies of linear MHD theory that so many equivalent forms of the force operator (and the associated potential energy, see Section 6.4) may be given. This betrays the nonlinear origin which remains present in the equilibrium equations coupling the three inhomogeneity ingredients $\mathbf{j}$, ∇p and $\nabla\Phi_{\rm gr}$.

With respect to the proof of self-adjointness, we still have quite a way to go. We first establish two notational conveniences.

(1) Except for the two vector fields $\boldsymbol{\xi}$ and $\boldsymbol{\eta}$ defined in Eq. (6.34), we define their associated magnetic field perturbations:

$$\mathbf{Q}(\mathbf{r}) \equiv \nabla \times (\boldsymbol{\xi} \times \mathbf{B}), \qquad \mathbf{R}(\mathbf{r}) \equiv \nabla \times (\boldsymbol{\eta} \times \mathbf{B}) \qquad (\text{on V}). \tag{6.50}$$

(2) Concerning the integrand $\boldsymbol{\eta}^* \cdot \mathbf{F}(\boldsymbol{\xi})$ of the quadratic expression on the left hand side of Eq. (6.42), we will drop the asterisk on the variables and temporarily exploit *a real type inner product* so that we do not have to write "$\boldsymbol{\eta}^* \cdot \mathbf{F}(\boldsymbol{\xi})$ + complex conjugate" all the time. Of course, we will revert to the complex inner product again later on in the exposition.

The proof of self-adjointness of the force operator basically consists of rewriting the integrand $\boldsymbol{\eta} \cdot \mathbf{F}(\boldsymbol{\xi})$, with $\mathbf{F}$ given by the expression (6.47), into a symmetric part and a remainder which should be cast in the form of a divergence. Because of Gauss' theorem (A.14) and application of the BCs (6.34), the latter part then disappears upon integration over the volume. For example, by straightforward application of the vector identity (A.6) for the divergence, the first term of Eq. (6.47) produces the following expression:

$$\boldsymbol{\eta} \cdot \nabla(\gamma p \nabla \cdot \boldsymbol{\xi}) \stackrel{(A.6)}{=} -\gamma p (\nabla \cdot \boldsymbol{\xi}) \nabla \cdot \boldsymbol{\eta} + \nabla \cdot (\boldsymbol{\eta}\, \gamma p \nabla \cdot \boldsymbol{\xi}), \tag{6.51}$$

which obviously has the required form. Similarly, the second term may be reduced by means of the vector identities (A.1) and (A.12) and the definition (6.50) for the magnetic perturbation $\mathbf{R}$ associated with the displacement vector field $\boldsymbol{\eta}$:

$$\begin{aligned} -\boldsymbol{\eta} \cdot \mathbf{B} \times (\nabla \times \mathbf{Q}) &\stackrel{(A.1)}{=} -(\nabla \times \mathbf{Q}) \cdot (\boldsymbol{\eta} \times \mathbf{B}) \\ &\stackrel{(A.12)}{=} -\mathbf{Q} \cdot \nabla \times (\boldsymbol{\eta} \times \mathbf{B}) + \nabla \cdot [(\boldsymbol{\eta} \times \mathbf{B}) \times \mathbf{Q}] \\ &\stackrel{(6.50)}{=} -\mathbf{Q} \cdot \mathbf{R} + \nabla \cdot [(\boldsymbol{\eta} \times \mathbf{B}) \times \mathbf{Q}] \\ &\stackrel{(A.2)}{=} -\mathbf{Q} \cdot \mathbf{R} + \nabla \cdot [\mathbf{B}\, \boldsymbol{\eta} \cdot \mathbf{Q} - \boldsymbol{\eta}\, \mathbf{B} \cdot \mathbf{Q}]. \end{aligned} \tag{6.52}$$

Again, the required form of a symmetric expression and a divergence is obtained. The two expressions (6.51) and (6.52), corresponding to the homogeneous plasma part, are still relatively simple. The transformation of the next three terms of the integrand, which are due to plasma inhomogeneity, is much more complicated.

▷ **Transformation of the inhomogeneity terms of the integrand** We take the three inhomogeneity terms together, exploiting the expression (6.46):

$$
\begin{aligned}
&\boldsymbol{\eta}\cdot[\,\nabla(\boldsymbol{\xi}\cdot\nabla p)+\mathbf{j}\times\mathbf{Q}+(\nabla\Phi_{\rm gr})\nabla\cdot(\rho\boldsymbol{\xi})\,]\\
&\stackrel{(6.46)}{=}\quad \underbrace{-\boldsymbol{\eta}\cdot\mathbf{B}\times[\,\nabla\times(\mathbf{j}\times\boldsymbol{\xi})\,]}_{\text{reduce as (6.52) with } \mathbf{Q} \text{ replaced by } \mathbf{j}\times\boldsymbol{\xi}} -(\boldsymbol{\eta}\cdot\nabla p)\,\nabla\cdot\boldsymbol{\xi}-\rho\boldsymbol{\eta}\cdot\nabla(\boldsymbol{\xi}\cdot\nabla\Phi_{\rm gr})\\
&=\quad \overbrace{-\mathbf{j}\times\boldsymbol{\xi}\cdot\mathbf{R}+\nabla\cdot[\,(\boldsymbol{\eta}\times\mathbf{B})\times(\mathbf{j}\times\boldsymbol{\xi})\,]}\;-(\boldsymbol{\eta}\cdot\nabla p)\,\nabla\cdot\boldsymbol{\xi}-\rho\boldsymbol{\eta}\cdot\nabla(\boldsymbol{\xi}\cdot\nabla\Phi_{\rm gr})\\
&\stackrel{(A.2),(A.1)}{=}\quad \boldsymbol{\xi}\cdot\mathbf{j}\times\mathbf{R}+\nabla\cdot[\,\mathbf{j}\mathbf{B}\cdot(\boldsymbol{\xi}\times\boldsymbol{\eta})+\boldsymbol{\xi}\boldsymbol{\eta}\cdot(\mathbf{j}\times\mathbf{B})\,]\\
&\qquad\qquad -(\boldsymbol{\eta}\cdot\nabla p)\,\nabla\cdot\boldsymbol{\xi}-\rho\boldsymbol{\eta}\cdot\nabla(\boldsymbol{\xi}\cdot\nabla\Phi_{\rm gr})\\
&\stackrel{\text{equil.},(A.6)}{=}\quad \boldsymbol{\xi}\cdot\mathbf{j}\times\mathbf{R}+\nabla\cdot[\,\mathbf{j}\mathbf{B}\cdot(\boldsymbol{\xi}\times\boldsymbol{\eta})]+\boldsymbol{\xi}\cdot\nabla(\boldsymbol{\eta}\cdot\nabla p)\\
&\qquad\qquad +\underbrace{\nabla\cdot[\,\boldsymbol{\xi}\boldsymbol{\eta}\cdot\rho\nabla\Phi_{\rm gr}]-\rho\boldsymbol{\eta}\cdot\nabla(\boldsymbol{\xi}\cdot\nabla\Phi_{\rm gr})}_{\Downarrow\,(A.6)}\\
&\qquad\qquad \overbrace{\nabla\cdot[\,\rho(\nabla\Phi_{\rm gr})\cdot(\boldsymbol{\eta}\boldsymbol{\xi}-\boldsymbol{\xi}\boldsymbol{\eta})\,]+(\boldsymbol{\xi}\cdot\nabla\Phi_{\rm gr})\nabla\cdot(\rho\boldsymbol{\eta})}\\
&\stackrel{\text{rearrange}}{=}\quad \boldsymbol{\xi}\cdot[\,\nabla(\boldsymbol{\eta}\cdot\nabla p)+\mathbf{j}\times\mathbf{R}+(\nabla\Phi_{\rm gr})\nabla\cdot(\rho\boldsymbol{\eta})\,]\\
&\qquad\qquad +\nabla\cdot[\,\mathbf{j}\mathbf{B}\cdot(\boldsymbol{\xi}\times\boldsymbol{\eta})-\rho(\nabla\Phi_{\rm gr})\cdot(\boldsymbol{\xi}\boldsymbol{\eta}-\boldsymbol{\eta}\boldsymbol{\xi})\,]\,.
\end{aligned}
$$

Now, the first term on the right hand side is not symmetric in $\boldsymbol{\xi}$ and $\boldsymbol{\eta}$. However, this term is the exact mirror image of the left hand side. Hence, it is very easy to symmetrize the latter expression in the following way:

$$
\begin{aligned}
&\boldsymbol{\eta}\cdot[\,\nabla(\boldsymbol{\xi}\cdot\nabla p)+\mathbf{j}\times\mathbf{Q}+(\nabla\Phi_{\rm gr})\nabla\cdot(\rho\boldsymbol{\xi})\,]\\
&=\quad \tfrac12\boldsymbol{\eta}\cdot[\,\nabla(\boldsymbol{\xi}\cdot\nabla p)+\mathbf{j}\times\mathbf{Q}+(\nabla\Phi_{\rm gr})\nabla\cdot(\rho\boldsymbol{\xi})\,]\\
&\qquad +\tfrac12\boldsymbol{\xi}\cdot[\,\nabla(\boldsymbol{\eta}\cdot\nabla p)+\mathbf{j}\times\mathbf{R}+(\nabla\Phi_{\rm gr})\nabla\cdot(\rho\boldsymbol{\eta})\,]\\
&\qquad +\tfrac12\nabla\cdot[\,\mathbf{j}\mathbf{B}\cdot(\boldsymbol{\xi}\times\boldsymbol{\eta})-\rho(\nabla\Phi_{\rm gr})\cdot(\boldsymbol{\xi}\boldsymbol{\eta}-\boldsymbol{\eta}\boldsymbol{\xi})\,]\\
&\stackrel{(A.6),\text{equil.}}{=}\quad \tfrac12\nabla\cdot[(\nabla p)\cdot(\boldsymbol{\xi}\boldsymbol{\eta}+\boldsymbol{\eta}\boldsymbol{\xi})]-\tfrac12(\nabla p)\cdot(\boldsymbol{\xi}\nabla\cdot\boldsymbol{\eta}+\boldsymbol{\eta}\nabla\cdot\boldsymbol{\xi})\\
&\qquad -\tfrac12\nabla\cdot[(\mathbf{j}\times\mathbf{B}-\nabla p)\cdot(\boldsymbol{\xi}\boldsymbol{\eta}-\boldsymbol{\eta}\boldsymbol{\xi})]-\tfrac12\mathbf{j}\cdot(\boldsymbol{\eta}\times\mathbf{Q}+\boldsymbol{\xi}\times\mathbf{R})\\
&\qquad +\tfrac12(\nabla\Phi_{\rm gr})\cdot[\boldsymbol{\eta}\nabla\cdot(\rho\boldsymbol{\xi})+\boldsymbol{\xi}\nabla\cdot(\rho\boldsymbol{\eta})]+\nabla\cdot[\,\tfrac12\mathbf{j}\mathbf{B}\cdot(\boldsymbol{\xi}\times\boldsymbol{\eta})\,]\\
&=\quad -\tfrac12(\nabla p)\cdot(\boldsymbol{\xi}\nabla\cdot\boldsymbol{\eta}+\boldsymbol{\eta}\nabla\cdot\boldsymbol{\xi})-\tfrac12\mathbf{j}\cdot(\boldsymbol{\eta}\times\mathbf{Q}+\boldsymbol{\xi}\times\mathbf{R})\\
&\qquad +\tfrac12(\nabla\Phi_{\rm gr})\cdot[\,\boldsymbol{\eta}\nabla\cdot(\rho\boldsymbol{\xi})+\boldsymbol{\xi}\nabla\cdot(\rho\boldsymbol{\eta})\,]\\
&\qquad +\nabla\cdot[\,\boldsymbol{\eta}(\boldsymbol{\xi}\cdot\nabla p)-\tfrac12(\mathbf{j}\times\mathbf{B})\cdot(\boldsymbol{\xi}\boldsymbol{\eta}-\boldsymbol{\eta}\boldsymbol{\xi})+\tfrac12\mathbf{j}\mathbf{B}\cdot(\boldsymbol{\xi}\times\boldsymbol{\eta})\,]\,.
\end{aligned}
\tag{6.53}
$$

Again, a symmetric expression and a divergence is obtained! ◁

Finally, adding up the expressions (6.51), (6.52) and (6.53) gives the required symmetric form

$$
\begin{aligned}
\boldsymbol{\eta}\cdot\mathbf{F}(\boldsymbol{\xi}) \;=\;& -\gamma p(\nabla\cdot\boldsymbol{\xi})\nabla\cdot\boldsymbol{\eta}-\mathbf{Q}\cdot\mathbf{R}-\tfrac12(\nabla p)\cdot(\boldsymbol{\xi}\,\nabla\cdot\boldsymbol{\eta}+\boldsymbol{\eta}\,\nabla\cdot\boldsymbol{\xi})\\
&-\tfrac12\mathbf{j}\cdot(\boldsymbol{\eta}\times\mathbf{Q}+\boldsymbol{\xi}\times\mathbf{R})+\tfrac12(\nabla\Phi_{\rm gr})\cdot[\,\boldsymbol{\eta}\,\nabla\cdot(\rho\boldsymbol{\xi})+\boldsymbol{\xi}\,\nabla\cdot(\rho\boldsymbol{\eta})\,]\\
&+\nabla\cdot\Big[\boldsymbol{\eta}\,(\gamma p\,\nabla\cdot\boldsymbol{\xi}+\boldsymbol{\xi}\cdot\nabla p-\mathbf{B}\cdot\mathbf{Q})\\
&+\mathbf{B}\boldsymbol{\eta}\cdot\mathbf{Q}+\tfrac12\mathbf{j}\mathbf{B}\cdot(\boldsymbol{\xi}\times\boldsymbol{\eta})-\tfrac12(\mathbf{j}\times\mathbf{B})\cdot(\boldsymbol{\xi}\boldsymbol{\eta}-\boldsymbol{\eta}\boldsymbol{\xi})\Big]\,.
\end{aligned}
\tag{6.54}
$$

Integration over the plasma volume and application of Gauss' theorem (A.14) on the divergence term then yields the following very important result:

$$\begin{aligned}\int \boldsymbol{\eta}\cdot\mathbf{F}(\boldsymbol{\xi})\,dV \;=\; &-\int \Big\{ \gamma p(\nabla\cdot\boldsymbol{\xi})\nabla\cdot\boldsymbol{\eta} + \mathbf{Q}\cdot\mathbf{R} + \tfrac{1}{2}(\nabla p)\cdot(\boldsymbol{\xi}\nabla\cdot\boldsymbol{\eta} + \boldsymbol{\eta}\nabla\cdot\boldsymbol{\xi}) \\ &+ \tfrac{1}{2}\mathbf{j}\cdot(\boldsymbol{\eta}\times\mathbf{Q} + \boldsymbol{\xi}\times\mathbf{R}) - \tfrac{1}{2}(\nabla\Phi_{\rm gr})\cdot\big[\boldsymbol{\eta}\,\nabla\cdot(\rho\boldsymbol{\xi}) + \boldsymbol{\xi}\,\nabla\cdot(\rho\boldsymbol{\eta})\big]\Big\}\,dV \\ &+ \int \mathbf{n}\cdot\boldsymbol{\eta}\,\big[\gamma p\,\nabla\cdot\boldsymbol{\xi} + \boldsymbol{\xi}\cdot\nabla p - \mathbf{B}\cdot\mathbf{Q}\big]\,dS\,. \qquad (6.55)\end{aligned}$$

Here, the three last terms of Eq. (6.54) did not give a contribution to the surface integral since $\mathbf{n}\cdot\mathbf{B} = 0$ and $\mathbf{n}\cdot\mathbf{j} = 0$ on the plasma surface, whereas $\mathbf{j}\times\mathbf{B} \sim \mathbf{n}$ for the plasma-wall and plasma interface models I–III of Section 4.6. For the line-tying magnetic loop models IV and V, there is also a boundary intersecting the magnetic field and current lines so that the given argument does not apply. However, in that case, the line-tying boundary condition (4.186) translates into the conditions $\boldsymbol{\eta} = \boldsymbol{\xi} = 0$ at the photospheric boundary so that the three terms do not contribute there either. We conclude that the expression (6.55) is a very general one, applying to all of the laboratory and astrophysical problems I–V formulated in Section 4.6.

The surface integral in Eq. (6.55) is just the perturbation of the total pressure (apart from the sign). In plasmas with a free boundary (models II and III), this term couples to the external variables $\hat{\mathbf{Q}}$ and $\hat{\mathbf{R}}$ through the boundary conditions at the plasma interface so that a further analysis is needed to properly represent the contributions of the outer regions. This we leave for later (Section 6.6.2). For our present concern (wall on the plasma: model I), the term does not contribute. Consequently,

$$\begin{aligned}\int \Big\{\boldsymbol{\eta}\cdot\mathbf{F}(\boldsymbol{\xi}) - \boldsymbol{\xi}\cdot\mathbf{F}(\boldsymbol{\eta})\Big\}\,dV \;=\; &\int \Big\{ \mathbf{n}\cdot\boldsymbol{\eta}\,[\gamma p\,\nabla\cdot\boldsymbol{\xi} + \boldsymbol{\xi}\cdot\nabla p - \mathbf{B}\cdot\mathbf{Q}] \\ &- \mathbf{n}\cdot\boldsymbol{\xi}\,[\gamma p\,\nabla\cdot\boldsymbol{\eta} + \boldsymbol{\eta}\cdot\nabla p - \mathbf{B}\cdot\mathbf{R}]\Big\}\,dS = 0\,, \qquad (6.56)\end{aligned}$$

by virtue of the BCs $\mathbf{n}\cdot\boldsymbol{\eta} = \mathbf{n}\cdot\boldsymbol{\xi} = 0$ on the normal components of the displacement vector. $\Rightarrow$ $\mathbf{F}$ is a self-adjoint operator in model I; QED.

At this point we will revert to the original complex type scalar product (6.35). We will prove that the eigenvalues of the operator $\rho^{-1}\mathbf{F}$ are real. This follows directly from the self-adjointness property (6.56): Let $\boldsymbol{\xi}_n$ be the eigenfunction belonging to an eigenvalue $-\omega_n^2$,

$$\rho^{-1}\mathbf{F}(\boldsymbol{\xi}_n) = -\omega_n^2\,\boldsymbol{\xi}_n\,.$$

Then, the complex conjugate equation reads

$$\rho^{-1}\mathbf{F}^*(\boldsymbol{\xi}_n) = \rho^{-1}\mathbf{F}(\boldsymbol{\xi}_n^*) = -\omega_n^{2*}\,\boldsymbol{\xi}_n^*\,.$$

Multiplying the first equation by $\boldsymbol{\xi}_n^*$ and the second by $\boldsymbol{\xi}_n$, subtracting and integrating over the plasma volume yields

$$0 = (\omega_n^2 - \omega_n^{2*})\,\|\boldsymbol{\xi}\|^2\,,$$

by virtue of the self-adjointness property, so that

$$\omega_n^2 = \omega_n^{2*}\,. \qquad (6.57)$$

Hence, *the eigenvalues ω_n^2 of the operator $\rho^{-1}\mathbf{F}$ are real*, so that the spectrum is confined to the real and imaginary axes of the complex ω-plane; QED. Recall that this implies that the eigenvalues correspond to either waves ($\omega^2 > 0$) or exponential instabilities ($\omega^2 < 0$).

6.3 Spectral alternatives*

6.3.1 Mathematical intermezzo*

Recapitulating: we have seen in Section 6.2 that the spectral problem of linear ideal MHD arises from a study of the *dynamics*, described by the equation of motion

$$\rho^{-1}\mathbf{F}(\boldsymbol{\xi}) = \frac{\partial^2 \boldsymbol{\xi}}{\partial t^2}\,, \tag{6.58}$$

where normal mode solutions with an exponential time-dependence lead to the *spectral equation*

$$\rho^{-1}\mathbf{F}(\hat{\boldsymbol{\xi}}) = -\omega^2 \hat{\boldsymbol{\xi}}\,, \tag{6.59}$$

whereas the *stability problem* reduces to a study of the marginal equation

$$\rho^{-1}\mathbf{F}(\hat{\boldsymbol{\xi}}) = 0\,. \tag{6.60}$$

Before we proceed to analyse these three main physical problems from the second point of view, that of quadratic forms and variational principles (Section 6.4), it is useful to put them into a mathematical context since other alternative formulations are encountered there.

The spectral problem of partial differential equations like Eq. (6.59) is just a generalization of the methods used in linear algebra of finite-dimensional vector spaces. Three alternatives appear.

(a) *The eigenvalue problem* arises in the study of finite $N \times N$ matrices,

$$\sum_{j=1}^{N} L_{ij}\, x_j = \lambda\, x_i\,, \quad \text{or} \quad \mathbf{L}\cdot\mathbf{x} = \lambda\,\mathbf{x}\,, \tag{6.61}$$

where the eigenvalues are found from the condition

$$\det\left(L_{ij} - \lambda\delta_{ij}\right) = 0\,, \tag{6.62}$$

and substitution back into Eq. (6.61) yields the eigenvectors $\mathbf{x}_n$.

(b) Another formulation of this problem is obtained by constructing *the quadratic forms*, as e.g. entering the Rayleigh quotient

$$\Lambda \equiv \sum_{i=1}^{N}\sum_{j=1}^{N} x_i^* L_{ij} x_j \Big/ \sum_{i=1}^{N} x_i^2\,, \tag{6.63}$$

where the eigenvalues λ are the stationary values of Λ.

(c) A third formulation arises by considering *the inhomogeneous equation*

$$(\mathbf{L} - \lambda\,\mathbf{I})\cdot\mathbf{x} = \mathbf{a}\,, \tag{6.64}$$

where $\mathbf{a}$ is a known vector. Here, *the Fredholm alternative* states that either the homogeneous equation (6.61) has a solution, so that λ coincides with one of the eigenvalues, or the inhomogeneous equation (6.64) has a solution, so that λ is outside the spectrum of eigenvalues of the matrix $\mathbf{L}$.

Clearly, whereas our two physical viewpoints correspond to the first two spectral alternatives, the third alternative admits yet another class of physical applications. This is associated with the initial value problem (Section 6.3.2).

In the generalization of these ideas to the infinite-dimensional Hilbert space associated with the operator $\rho^{-1}\mathbf{F}$, two kinds of mathematical problem are encountered. The first one is the fact that the operator $\rho^{-1}\mathbf{F}$ is a differential operator and, therefore, *unbounded*. In contrast, bounded operators B have the property that

$$\|Bx\| \leq M\|x\| \quad \text{for all } x \in \mathcal{H}\,, \tag{6.65}$$

where $\mathcal{H}$ is the Hilbert space and M is some constant. Differential operators do not have this property. Operating on a bounded (square integrable) sequence of functions in Hilbert space they may produce a sequence that is unbounded and, therefore, leads outside Hilbert space. (For example, the differential operator d/dx transforms the bounded sequence $\sin{(n\pi x)}$ into the diverging sequence $n\pi\cos{(n\pi x)}$.) One usually tries to avoid this problem by transforming it into one that involves *completely continuous* or *compact operators*. These operators have the opposite property: they transform a sequence of bounded functions into one that converges in the mean. For these operators, the theory of infinite-dimensional Hilbert space is completely analogous to that of the finite-dimensional vector spaces of linear algebra. In the case of differential operators, this implies that one inverts them, which leads to the study of integral operators involving Green's functions. Those operators frequently do have the required property of compactness.

Another, more serious, problem is the existence of a third class of operators where the above trick does not work, viz. that of bounded operators that are not compact. (Example: the operator of multiplication by x.) Those operators may give rise to a *continuous spectrum*, which is roughly speaking the collection of 'improper eigenvalues', for which the eigenvalue equation is solved, but not by functions that belong to Hilbert space. In the mathematical description, one then may stick to the notion of Hilbert space by introducing the concept of *approximate spectrum*, where sequences are considered that do not converge (which is the approach of von Neumann [636] in his treatment of spectral theory for quantum mechanics), or one may consider wider classes of elements than those that belong to Hilbert space, viz. *distributions* (which is the approach of Dirac [161], perfected by Schwartz [545]). Loosely speaking, one could say that the diverging sequences of functions, that are considered in the first approach, converge to elements outside Hilbert space, which are the distributions considered in the second approach.

Using this terminology, the following generalization of the ideas of linear algebra expressed by the equations (6.61)–(6.64) may be formulated. The spectrum of a linear operator L is obtained from the study of the inhomogeneous equation

$$(L - \lambda)\, x = a\,, \tag{6.66}$$

where a is a given element in Hilbert space and we look for solutions

$$x = (L - \lambda)^{-1} a\,. \tag{6.67}$$

For complex values of λ, three possibilities arise, see Friedman [193]:

(1) $(L - \lambda)^{-1}$ does not exist because $(L - \lambda)\, x = 0$ has a solution

$\Rightarrow$ λ belongs to *the point* or *discrete spectrum* of L;

(2) $(L-\lambda)^{-1}$ exists but is unbounded

$\Rightarrow$ λ belongs to *the continuous spectrum* of L;

(3) $(L-\lambda)^{-1}$ exists and is bounded

$\Rightarrow$ λ belongs to *the resolvent set* of L.

Thus, a complex value of λ either belongs to the spectrum or to the resolvent set, so that one may say that the spectrum of L consists of the collection of λs where the so-called *resolvent operator* $R_\lambda \equiv (L-\lambda)^{-1}$ misbehaves.

Finally, it is useful to anticipate the exposition on computational MHD in Chapter 15 where the circle will be closed by the consideration of the discretized problem, which leads back to finite-dimensional vector spaces. It will be shown that extremely powerful numerical methods exist which provide the construction of solutions of the three basic linear problems of physical interest, viz.

– determining *the temporal evolution*,

$$\mathsf{L}\cdot\mathbf{x} = \frac{\partial \mathbf{x}}{\partial t}\,, \tag{6.68}$$

– determining *the spectrum*,

$$\mathsf{L}\cdot\mathbf{x} = \lambda\,\mathbf{x}\,, \tag{6.69}$$

– and determining *the stationary state of a driven plasma*,

$$(\mathsf{L} - \lambda_{\mathrm{d}}\,\mathsf{I})\cdot\mathbf{x} = \mathbf{f}\,, \tag{6.70}$$

where $\mathbf{f}$ represents the driving force with fixed frequency λ_{d}.

Clearly, the last problem corresponds again with the third spectral problem (6.64). It should be mentioned that the relevant computational methods are not restricted to ideal MHD, with the associated displacement vector description, but they may be generalized to dissipative MHD with an arbitrary number of unknowns.

6.3.2 Initial value problem in MHD*

To give the reader some feeling of what is yet in store, we digress on an advanced topic. According to the exposition given in the previous subsection, in particular in connection with Eqs. (6.66) and (6.67), the third, and most general, approach to the spectrum of a linear operator is to consider the inhomogeneous problem, i.e. the generalization of Eq. (6.64) to infinite-dimensional spaces. For the MHD operator $\rho^{-1}\mathbf{F}$, this approach leads to the following inhomogeneous problem:

$$(\rho^{-1}\mathsf{F} + \omega^2\mathsf{I})\cdot\hat{\boldsymbol{\xi}} = i\omega\mathbf{X}\,. \tag{6.71}$$

Here, we have introduced matrix notation for the force operator, $\mathsf{F}\cdot\hat{\boldsymbol{\xi}} \equiv \mathbf{F}(\hat{\boldsymbol{\xi}})$, and for the unit operator, $\mathsf{I}\cdot\hat{\boldsymbol{\xi}} \equiv \mathbf{I}(\hat{\boldsymbol{\xi}}) \equiv \hat{\boldsymbol{\xi}}$, and $\mathbf{X}$ is a known vector. Our task is then to construct the resolvent operator $(\rho^{-1}\mathsf{F} + \omega^2\mathsf{I})^{-1}$, and to study its behaviour for complex values of ω^2.

In order to see how this is connected with physics, we consider the initial value problem. (Also notice that Eq. (6.70), obtained from model III excitation, closely corresponds to this problem.) We

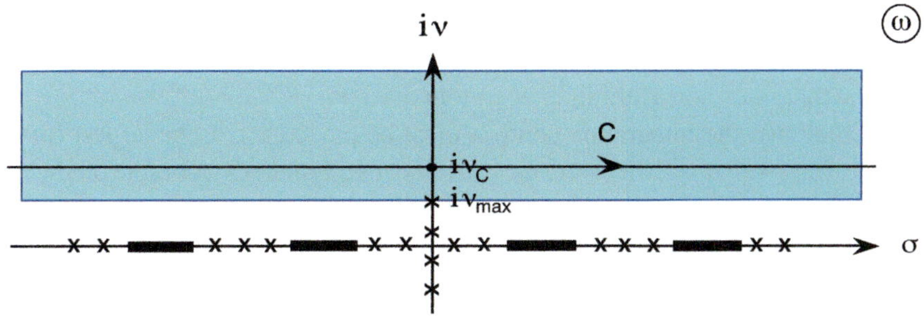

Fig. 6.8 Strip of convergence and contour C for the inverse Laplace transformation.

define *the Laplace transform* of $\boldsymbol{\xi}(\mathbf{r};t)$ in the complex ω-plane,

$$\hat{\boldsymbol{\xi}}(\mathbf{r};\omega) \equiv \int_0^\infty \boldsymbol{\xi}(\mathbf{r};t)\, \mathrm{e}^{\mathrm{i}\omega t}\, dt\,, \tag{6.72}$$

so that the equation of motion (6.58) takes the form

$$\rho^{-1}\mathbf{F}(\hat{\boldsymbol{\xi}}) = \int_0^\infty \frac{\partial^2 \boldsymbol{\xi}}{\partial t^2}\, \mathrm{e}^{\mathrm{i}\omega t}\, dt = -\omega^2 \hat{\boldsymbol{\xi}} + \left[\left(\frac{\partial \boldsymbol{\xi}}{\partial t} - \mathrm{i}\omega \boldsymbol{\xi} \right) \mathrm{e}^{\mathrm{i}\omega t} \right]_{t=0}^{t\to\infty}. \tag{6.73}$$

Writing $\omega = \sigma + \mathrm{i}\nu$, we then get for $\nu > 0$:

$$(\rho^{-1}\mathbf{F} + \omega^2 \mathbf{I}) \cdot \hat{\boldsymbol{\xi}}(\mathbf{r};\omega) = \mathrm{i}\omega \boldsymbol{\xi}_i(\mathbf{r}) - \dot{\boldsymbol{\xi}}_i(\mathbf{r}) \equiv \mathrm{i}\omega \mathbf{X}\,, \tag{6.74}$$

where the vector $\mathbf{X}$ of Eq. (6.71) turns out to be the function of initial displacement $\boldsymbol{\xi}_i(\mathbf{r})$ and initial velocity $\dot{\boldsymbol{\xi}}_i(\mathbf{r})$ defined in the RHS of Eq. (6.74). In order to find the response $\boldsymbol{\xi}(\mathbf{r};t)$ to a certain initial perturbation $\mathbf{X}$, one then first has to invert Eq. (6.74) to find the Laplace transformed variable $\hat{\boldsymbol{\xi}}$ in terms of $\mathbf{X}$,

$$\hat{\boldsymbol{\xi}}(\mathbf{r};\omega) = (\rho^{-1}\mathbf{F} + \omega^2 \mathbf{I})^{-1} \cdot \mathrm{i}\omega\, \mathbf{X}(\mathbf{r};\omega)\,, \tag{6.75}$$

and next perform *the inverse Laplace transformation*:

$$\begin{aligned} \boldsymbol{\xi}(\mathbf{r};t) &= \frac{1}{2\pi} \int_{\mathrm{i}\nu_C - \infty}^{\mathrm{i}\nu_C + \infty} \hat{\boldsymbol{\xi}}(\mathbf{r};\omega)\, \mathrm{e}^{-\mathrm{i}\omega t}\, d\omega \\ &= \frac{1}{2\pi} \int_C (\rho^{-1}\mathbf{F} + \omega^2 \mathbf{I})^{-1} \cdot \Big[\mathrm{i}\omega \boldsymbol{\xi}_i(\mathbf{r}) - \dot{\boldsymbol{\xi}}_i(\mathbf{r}) \Big]\, \mathrm{e}^{-\mathrm{i}\omega t}\, d\omega\,. \end{aligned} \tag{6.76}$$

Here, according to the theory of Laplace transforms, the contour C has to be placed in the strip of convergence (Fig. 6.8). The essential point is that, for the inverse transform, more is needed than just $\nu_C > 0$ because $\hat{\boldsymbol{\xi}}(\mathbf{r};\omega)$ may not exist for certain values of ω, or may be singular.

According to the discussion above, it is precisely when ω belongs to the spectrum of the operator $\rho^{-1}\mathbf{F}$ that we may expect trouble with Eq. (6.75). If ω is a point eigenvalue the operator $(\rho^{-1}\mathbf{F} + \omega^2\mathbf{I})^{-1}$ simply does not exist, whereas for improper eigenvalues (i.e. ω in the continuum) the operator $(\rho^{-1}\mathbf{F} + \omega^2\mathbf{I})^{-1}$ is unbounded. Before we know where to place the integration contour C for the inverse Laplace transform we, therefore, have to know the spectrum. Here, we get substantial help from the fact that $\rho^{-1}\mathbf{F}$ is Hermitian (self-adjoint) so that the eigenvalues (including the improper ones) have to be real, so that the spectrum is confined to the real and imaginary axes of the complex ω-plane. In fact, we would be in severe trouble if the operator $\mathbf{F}$ were not self-adjoint because a general theory of non-Hermitian operators does not exist.

We then conclude that the integration contour must be placed above the largest (most unstable) point eigenvalue $\nu_{\rm max}$ of $\rho^{-1}\mathbf{F}$, $\nu_{\rm C} > \nu_{\rm max}$. (Because Landau's solution of the initial value problem of the Vlasov equation was also obtained by means of Laplace transforms, the similarity with the discussion of Landau damping in Section 2.3.3, Fig. 2.7, is obvious.) In other words, the class of permissible functions $\boldsymbol{\xi}(\mathbf{r};t)$ is restricted to functions of exponential order $\exp(\nu_{\rm C}t)$ where $\nu_{\rm C}$ is larger than the largest growth rate of the system. In Fig. 6.8 we have schematically indicated our knowledge so far of the spectrum of ideal MHD of one-dimensional systems, which will be analysed in more detail in Chapters 7 and 9. One finds two pairs of continua on the real axis, whereas point eigenvalues can occur almost everywhere on the real σ-axis and also on a bounded part $-\nu_{\rm max} \le \nu \le \nu_{\rm max}$ of the imaginaryν-axis.

It is extremely difficult to obtain the explicit time-dependence of $\boldsymbol{\xi}(\mathbf{r};t)$ in situations of practical interest, so that one usually restricts the study to time-asymptotic solutions. For $t \to \infty$ one then wishes to deform the integration contour in the inverse Laplace transform to the lower half of the ω-plane to exploit the smallness of the exponential factor $\exp(-\mathrm{i}\omega t)$ in Eq. (6.76). For this advantage one must pay in the form of a study of the analytic continuation of $\hat{\boldsymbol{\xi}}$ about the occurring *poles* (point eigenvalues) and *branch points* of $\hat{\boldsymbol{\xi}}$ (associated with the continuous spectrum). The branch point singularities lead to different branches of the complex function $\hat{\boldsymbol{\xi}}(\mathbf{r};\omega)$ so that the inverse Laplace transform contour may be moved to another Riemann sheet, where it could pick up poles. Such poles could not correspond to point eigenvalues since these are confined to the real and imaginary axes of the principal branch of $\hat{\boldsymbol{\xi}}$, but they do have physical significance. We will continue the analysis of the initial value problem in Chapter 10, where we consider the explicit example of an inhomogeneous slab.

6.4 Quadratic forms and variational principles

6.4.1 Expressions for the potential energy

In Section 6.2, we have completed the description of linearized ideal MHD in terms of differential equations by deriving the equation of motion (6.23) with the force operator $\mathbf{F}(\boldsymbol{\xi})$, which we have proven to be self-adjoint. This expresses the basic fact of *energy conservation*: $H = W + K =$ const, where W is the potential energy and K is the kinetic energy of the perturbations (see Section 4.3.2). We now turn to the alternative description, viz. the one exploiting these quadratic forms. First of all, we need to derive expressions for these energies for the linearized dynamics, corresponding to the first order expression for $\mathbf{F}$. The direct derivation, discussed first, is put in small print since it turns out to be complicated, whereas the indirect one, discussed next, is much simpler.

▷ **Derivation from the nonlinear expression for the potential energy** The most obvious procedure is to start from the nonlinear expressions for W and K derived in Section 4.3.2, and to exploit the energy conservation law (4.86), where we neglect gravity now:

$$H = \int \Big(\underbrace{\tfrac{1}{2}\rho v^2}_{\mathcal{K}} + \underbrace{\frac{p}{\gamma-1} + \tfrac{1}{2}B^2}_{\mathcal{W}} \Big)\, dV = \text{const}\,. \tag{6.77}$$

We have already derived in Eq. (6.40) the linearized expression for the kinetic energy, corresponding to the first part of Eq. (6.77), which is a second order quantity in $\dot{\boldsymbol{\xi}}$:

$$K = \tfrac{1}{2} \int \rho \dot{\boldsymbol{\xi}} \cdot \dot{\boldsymbol{\xi}} \, dV + \text{third and higher order terms}\,. \tag{6.78}$$

Similarly, we could work out the potential energy W, starting from the second part of Eq. (6.77):

$$W = \int \Big[\underbrace{\frac{p}{\gamma - 1} + \tfrac{1}{2} B^2}_{\mathcal{W}_0} + \underbrace{\frac{\pi}{\gamma - 1} + \mathbf{B} \cdot \mathbf{Q}}_{\mathcal{W}_1} + \underbrace{\cdots\cdots}_{\mathcal{W}_2} \Big] \, dV\,. \tag{6.79}$$

Here, the zeroth order term corresponds to the energy of the background equilibrium, described by the pressure p and the magnetic field $\mathbf{B}$. (Note that the interpretation of p and $\mathbf{B}$ changes from Eq. (6.77), where they indicate the full nonlinear expressions, to Eq. (6.79), where they only indicate the equilibrium parts.) The first order term corresponds with the energy change produced by the plasma displacement $\boldsymbol{\xi}$, resulting in the pressure and magnetic field perturbations π and $\mathbf{Q}$ given by Eqs. (6.24) and (6.25). However, how could this term be balanced in the energy conservation law by the expression for the kinetic energy K, which is a second order quantity only? To see what goes on, work out the perturbations:

$$\begin{aligned}
\frac{\pi}{\gamma - 1} &\equiv \frac{1}{\gamma - 1}(-\boldsymbol{\xi} \cdot \nabla p - \gamma p \nabla \cdot \boldsymbol{\xi}) = \boldsymbol{\xi} \cdot \nabla p - \frac{\gamma}{\gamma - 1} \nabla \cdot (p \boldsymbol{\xi})\,, \\
\mathbf{B} \cdot \mathbf{Q} &\equiv \mathbf{B} \cdot \nabla \times (\boldsymbol{\xi} \times \mathbf{B}) \overset{(A.12)}{=} (\boldsymbol{\xi} \times \mathbf{B}) \cdot \nabla \times \mathbf{B} + \nabla \cdot [(\boldsymbol{\xi} \times \mathbf{B}) \times \mathbf{B}] \\
&\overset{(A.1),(A.2)}{=} -\boldsymbol{\xi} \cdot [\mathbf{j} \times \mathbf{B}] + \nabla \cdot [\boldsymbol{\xi} \cdot \mathbf{B}\mathbf{B} - B^2 \boldsymbol{\xi}]
\end{aligned}$$

$$\Rightarrow \quad W_1 = \int \boldsymbol{\xi} \cdot \underbrace{(\nabla p - \mathbf{j} \times \mathbf{B}) \, dV}_{=\,0 \text{ (equil.)}} + \underbrace{\int \nabla \cdot \Big[-\frac{\gamma}{\gamma - 1} p \boldsymbol{\xi} + \boldsymbol{\xi} \cdot \mathbf{B}\mathbf{B} - B^2 \boldsymbol{\xi} \Big] \, dV}_{=\,0 \text{ (Gauss + BCs)}}\,. \tag{6.80}$$

Hence, W_1 vanishes. Consequently, in order to get relevant, non-vanishing, expressions for the potential energy W one should compute the next order contribution W_2 (also indicated as δW in the literature, although $\delta^2 W$ would be a better notation):

$$W_2 = \int \Big(\frac{p_2}{\gamma - 1} + \tfrac{1}{2} Q^2 + \mathbf{B} \cdot \mathbf{B}_2 \Big) \, dV\,. \tag{6.81}$$

Hence, we should compute p and $\mathbf{B}$ to second order in $\boldsymbol{\xi}$. This can be done but it is complicated., We will not pursue this further since there is a much easier method to derive the potential energy quadratic. ◁

The easier method to derive the expression for the potential energy is to exploit Eq. (6.40) for the linearized kinetic energy and to construct the linearized potential energy from energy conservation. This is done by taking the inner product of $\dot{\boldsymbol{\xi}}^*$ with the equation of motion (6.23) and integrating over the plasma volume:

$$\int \dot{\boldsymbol{\xi}}^* \cdot \mathbf{F}(\boldsymbol{\xi}) \, dV = \int \rho \dot{\boldsymbol{\xi}}^* \cdot \ddot{\boldsymbol{\xi}} \, dV = \frac{d}{dt} \Big[\tfrac{1}{2} \int \rho |\dot{\boldsymbol{\xi}}|^2 \, dV \Big] = \frac{dK}{dt}\,. \tag{6.82}$$

From energy conservation and the self-adjointness of $\mathbf{F}$ one then easily derives that

$$\begin{aligned}
\frac{dW}{dt} &= -\frac{dK}{dt} = -\tfrac{1}{2} \int \dot{\boldsymbol{\xi}}^* \cdot \mathbf{F}(\boldsymbol{\xi}) \, dV - \tfrac{1}{2} \int \boldsymbol{\xi}^* \cdot \mathbf{F}(\dot{\boldsymbol{\xi}}) \, dV \\
&= \frac{d}{dt} \Big[-\tfrac{1}{2} \int \boldsymbol{\xi}^* \cdot \mathbf{F}(\boldsymbol{\xi}) \, dV \Big] \quad \Rightarrow \quad W = -\tfrac{1}{2} \int \boldsymbol{\xi}^* \cdot \mathbf{F}(\boldsymbol{\xi}) \, dV,
\end{aligned} \tag{6.83}$$

which is the expression for *the linearized potential energy* we were looking for. The intuitive meaning of Eq. (6.83) is clear. The increase of the potential energy due to the perturbation is just the work done against the force $\mathbf{F}$ to displace the plasma by an amount $\boldsymbol{\xi}$. The factor $\frac{1}{2}$ represents the averaging involved as the work builds up when the plasma is displaced from 0 to its actual value $\boldsymbol{\xi}$. (Note that here, and in the following, we drop the subscripts $_1$ and $_2$ on first and second order expressions since the interpretation is obvious for the context.)

We can now specify the loose remark made in Section 6.1.1 that a perturbation is stable ($W > 0$), or unstable ($W < 0$), if $\boldsymbol{\xi}$ and $\mathbf{F}$ are pointing in the opposite, or in the same direction. This has to be interpreted in the integrated sense given by the definition of the potential energy in Eq. (6.83). We will prove that W is a variational expression. This implies that a test function $\boldsymbol{\xi}$ can be substituted in W to check for the sign of that expression. If some trial function $\boldsymbol{\xi}$ is found such that $W[\boldsymbol{\xi}] < 0$, the system is certainly unstable. In many cases, this is a much quicker way to establish instability of a particular configuration than to study the marginal equation of motion $\mathbf{F}(\boldsymbol{\xi}) = 0$. We will return to this topic in Section 6.4.4.

A bonus of the awful proof in Section 6.2.3 of the self-adjointness of $\mathbf{F}$ is the intermediate expression (6.55) which, upon identification of $\boldsymbol{\eta}$ and $\boldsymbol{\xi}^*$, immediately yields a more useful form of W than that of Eq. (6.83):

$$W = \frac{1}{2}\int \Big[\gamma p |\nabla\cdot\boldsymbol{\xi}|^2 + |\mathbf{Q}|^2 + (\boldsymbol{\xi}\cdot\nabla p)\nabla\cdot\boldsymbol{\xi}^* + \mathbf{j}\cdot\boldsymbol{\xi}^*\times\mathbf{Q} - (\boldsymbol{\xi}^*\cdot\nabla\Phi_{\rm gr})\,\nabla\cdot(\rho\boldsymbol{\xi})\Big]\, dV\,. \qquad (6.84)$$

The five terms represent, successively, the acoustic and magnetic energy, which are positive definite so that *homogeneous plasmas are always stable*, and the additional energies due to the pressure gradient, current density and gravity, which may have either sign so that *inhomogeneous plasmas may be unstable*.

The expression (6.84) for W is only valid for *model I* (wall on the plasma), since we have discarded the divergence term appearing in Eq. (6.55) by applying the boundary condition

$$\mathbf{n}\cdot\boldsymbol{\xi} = 0 \qquad \text{(at the wall)}\,. \qquad (6.85)$$

The wall may be put at ∞ if one wishes to consider the limit of an infinitely extended plasma. If one wishes to incorporate the influence of an external vacuum, *model II* (plasma–vacuum system) should be studied and the divergence term in Eq. (6.55) then gives a further contribution to W. This contribution represents the energy of the moving boundary and of the external vacuum magnetic field region. The proof of self-adjointness and the derivation of the extended expression for W for that case is left for later (Section 6.6.2).

6.4.2 Hamilton's principle

We continue to derive the variational counterparts of the differential equations (6.23), (6.28) and (6.29), describing the dynamics, spectrum and stability. The most general way of describing the motion of a dynamical system is through Hamilton's principle, which may be generalized to continuous systems; see Goldstein [243]. The Lagrangian and Hamiltonian formulation of nonlinear MHD has been given in a classical paper by Newcomb [463], discussed in Chapter 12. In the same manner as above, we could linearize the expressions given by him to obtain the linearized Lagrangian. However, this labour is superfluous since we have the ingredients already: the linearized kinetic energy K is given by Eq. (6.40) and the linearized potential energy W by Eq. (6.83).

Hence, we may state *the linearized version of Hamilton's principle*:

The evolution of the system from time t_1 to time t_2 through the perturbation $\boldsymbol{\xi}(\mathbf{r}, t)$ is such that the variation of the integral of the Lagrangian vanishes,

$$\delta \int_{t_1}^{t_2} L\, dt = 0\,, \quad \textit{where } L \equiv K - W \tag{6.86}$$

$$\textit{with} \quad K = K[\dot{\boldsymbol{\xi}}] = \tfrac{1}{2} \int \rho \dot{\boldsymbol{\xi}}^* \cdot \dot{\boldsymbol{\xi}}\, dV\,, \quad \text{and} \quad W = W[\boldsymbol{\xi}] = -\tfrac{1}{2} \int \boldsymbol{\xi}^* \cdot \mathbf{F}(\boldsymbol{\xi})\, dV\,.$$

Carrying out the minimization of this expression (see Exercise [6.6]) directly leads to the following *Euler–Lagrange equation*:

$$\mathbf{F}(\boldsymbol{\xi}) = \rho \frac{\partial^2 \boldsymbol{\xi}}{\partial t^2}\,, \tag{6.87}$$

which reproduces the equation of motion (6.23). Clearly, the variational formulation for the dynamics is fully equivalent to the differential equation formulation.

6.4.3 Rayleigh–Ritz spectral variational principle

In the differential equations approach, we obtained the spectral eigenvalue equation (6.28) by considering the normal modes (6.27). Let us also consider the quadratic forms for normal modes by inserting (6.27) into the expressions for K and W. However, since these expressions have to be real whereas ω may be either real or imaginary, the result of this substitution is obtained much more easily by dropping the time-dependence and starting from the eigenvalue equation (6.28), $\mathbf{F}(\hat{\boldsymbol{\xi}}) = -\rho\omega^2 \hat{\boldsymbol{\xi}}$, with complex $\hat{\boldsymbol{\xi}}(\mathbf{r})$. The inner product of this equation with $\hat{\boldsymbol{\xi}}^*$ gives

$$W[\hat{\boldsymbol{\xi}}] \equiv -\tfrac{1}{2} \int \hat{\boldsymbol{\xi}}^* \cdot \mathbf{F}(\hat{\boldsymbol{\xi}})\, dV = \omega^2\, \tfrac{1}{2} \int \rho \hat{\boldsymbol{\xi}}^* \cdot \hat{\boldsymbol{\xi}}\, dV \equiv I[\hat{\boldsymbol{\xi}}]\,,$$

where $I[\hat{\boldsymbol{\xi}}] \equiv \|\hat{\boldsymbol{\xi}}\|^2$ is the square of the norm, defined in Eq. (6.36), so that

$$\omega^2 = W[\hat{\boldsymbol{\xi}}]/I[\hat{\boldsymbol{\xi}}] \qquad \text{for normal modes}\,. \tag{6.88}$$

This is a nice expression but rather useless, as it stands, since it is just a conclusion a posteriori, after the normal modes have been obtained. It does not provide a recipe for actually finding the eigenvalues ω^2 and eigenfunctions $\hat{\boldsymbol{\xi}}$. Such a recipe is obtained by considering the right hand side of Eq. (6.88) to be a variational expression for arbitrary trial functions $\boldsymbol{\xi}(\mathbf{r})$. Here, we drop the hat not only for simplicity of notation but also because these functions need not be eigenfunctions. Nevertheless, they do produce them, as we will prove now.

Spectral variational principle *The eigenfunctions $\boldsymbol{\xi}$ of the operator $\rho^{-1}\mathbf{F}$ make the Rayleigh quotient*

$$\Lambda[\boldsymbol{\xi}] \equiv W[\boldsymbol{\xi}]/I[\boldsymbol{\xi}] \tag{6.89}$$

stationary; the eigenvalues ω^2 are the stationary values of Λ. The quadratic forms

$$W[\boldsymbol{\xi}] \equiv -\langle \boldsymbol{\xi}, \rho^{-1}\mathbf{F}(\boldsymbol{\xi})\rangle \quad \text{and} \quad I[\boldsymbol{\xi}] \equiv \langle \boldsymbol{\xi}, \boldsymbol{\xi}\rangle \tag{6.90}$$

are the potential energy and the square of the norm, respectively.

▷ **Proof** Let ω^2 be a stationary value of $\Lambda[\boldsymbol{\xi}]$, i.e. $\delta\Lambda = 0$. Then

$$\delta\Lambda = \frac{\delta W}{I} - \frac{W}{I^2}\,\delta I = -2\,\frac{\langle \delta\boldsymbol{\xi}, \rho^{-1}\mathbf{F}(\boldsymbol{\xi}) + \omega^2\boldsymbol{\xi}\rangle}{\langle \boldsymbol{\xi}, \boldsymbol{\xi}\rangle} = 0\,,$$

where we used the self-adjointness (giving the factor 2) and substituted ω^2 for Λ. Since $\delta\boldsymbol{\xi}$ is an arbitrary variation of $\boldsymbol{\xi}$, it follows that $\rho^{-1}\mathbf{F}(\boldsymbol{\xi}) = -\omega^2\boldsymbol{\xi}$, which is the eigenvalue equation; QED. ◁

The Rayleigh–Ritz principle is extremely useful for the approximation of eigenvalues by means of finite-dimensional subspaces of the infinite-dimensional Hilbert space. Here, one selects a suitable class of functions $(\boldsymbol{\eta}_1, \boldsymbol{\eta}_2, \ldots, \boldsymbol{\eta}_N)$, with finite norm $\|\boldsymbol{\eta}_n\|$, which are used as *trial functions* in the Rayleigh quotient (6.89). The linear combination of these functions that minimizes the functional Λ constitutes an approximation for the lowest eigenvalue ω_0^2, where the minimum value of Λ is always larger than the actual value of ω_0^2. An approximation to the N lowest eigenvalues may be obtained by choosing the $\boldsymbol{\eta}_n$s to be orthonormal,

$$\langle \boldsymbol{\eta}_m, \boldsymbol{\eta}_n\rangle = \delta_{mn}\,. \tag{6.91}$$

Since these functions are supposed to be known, one may compute the matrix elements

$$W_{mn} = \langle \boldsymbol{\eta}_m, \rho^{-1}\mathbf{F}(\boldsymbol{\eta}_n)\rangle\,. \tag{6.92}$$

Writing

$$\boldsymbol{\eta} = \sum_{n=1}^{N} a_n \boldsymbol{\eta}_n\,, \tag{6.93}$$

one obtains the following approximation:

$$\Lambda[\boldsymbol{\xi}] \approx \Lambda[\boldsymbol{\eta}] = \frac{\sum_{m=1}^{N}\sum_{n=1}^{N} a_m^* W_{mn} a_n}{\sum_{n=1}^{N} |a_n|^2}\,. \tag{6.94}$$

Hence, the problem boils down to Eq. (6.63) of Section 6.3.1, i.e. the simultaneous diagonalization of the two finite-dimensional quadratic forms $W[\boldsymbol{\eta}]$ and $I[\boldsymbol{\eta}]$. In this case, since the $\boldsymbol{\eta}$s have been chosen to be orthonormal, only $W[\boldsymbol{\eta}]$ needs to be diagonalized. Consequently, the eigenvalues $\omega_{\boldsymbol{\eta}}^{2(i)}$ and eigenfunctions $\boldsymbol{\eta}^{(i)}$ of the matrix W_{mn} are approximations to the lowest N eigenvalues $\omega^{2(i)}$ and corresponding eigenfunctions $\boldsymbol{\xi}^{(i)}$ of the operator $\rho^{-1}\mathbf{F}$. Of course, the accuracy of this approximation depends on the choice of the basis functions $\{\boldsymbol{\eta}_n\}$.

6.4.4 Energy principle

The above proof demonstrates the equivalence of the Rayleigh–Ritz variational principle with the eigenvalue equation. It also provides a formulation for stability problems that is one step more useful than the force operator equation. Since $I[\boldsymbol{\xi}] \equiv \|\boldsymbol{\xi}\|^2 \geq 0$, one may insert *trial functions* in W. If $W[\boldsymbol{\xi}] > 0$ for all possible trial functions $\boldsymbol{\xi}$, one may conclude that eigenvalues $\omega^2 < 0$ do not exist and that the system is *stable*. On the other hand, if one can find a single $\boldsymbol{\xi}$ for which $W[\boldsymbol{\xi}] < 0$, at least one eigenvalue $\omega^2 < 0$ exists and the system is *unstable*. This is summarized in the following powerful statement, due to Bernstein, Frieman, Kruskal and Kulsrud [55] and, independently, Hain, Lüst and Schlüter [275].

Energy principle for stability *An equilibrium is stable if and only if*

$$W[\boldsymbol{\xi}] > 0 \tag{6.95}$$

for all displacements $\boldsymbol{\xi}(\mathbf{r})$ that are bound in norm and satisfy the boundary conditions.
$W[\boldsymbol{\xi}]$ is defined by Eq. (6.84) and the appropriate boundary condition is given in Eq. (6.85).

The variational approach offers three methods for determining the stability of a system:

(a) By physical intuition one may *guess* a trial function $\boldsymbol{\xi}(\mathbf{r})$ that picks up the unstable part of the potential energy, so that $W[\boldsymbol{\xi}] < 0$. This provides a direct demonstration of the instability of a certain system: *necessary stability (≡ sufficient instability) criteria* are obtained this way.

(b) More systematically, one may investigate the sign of W by exploiting *a complete set of trial functions that are normalized in any convenient way* (which need not be I): *necessary + sufficient criteria* for stability are obtained.

(c) Finally, by considering *a complete set of trial functions that are properly normalized with the correct physical norm $I[\boldsymbol{\xi}]$*, corresponding to the kinetic energy, *the complete spectrum* of eigenvalues $\{\omega^2\}$ is obtained.

Comparing the variational methods with the differential equations counterparts, notice that:

– Method (a) has no counterpart in the equation of motion approach. It is the most direct method to investigate stability problems. If one has a good physical intuition, one may be able to design a trial function that shows right away that a system is unstable by picking up the dominant part of the driving energy of the instability. For experimental fusion devices, this approach is of historical interest only because it is usually impossible to pursue it for configurations that have been optimized with respect to confinement. One may formalize it though by testing with a finite class of trial functions that may be considered as a subspace of the complete Hilbert space of the system.

– Method (b) is equivalent to solving the marginal equation of motion $\mathbf{F}(\boldsymbol{\xi}) = 0$, but much simpler to apply since one may use any normalization of the trial functions in the expression (6.95) for W to test for stability. For example, one could exploit a normalization involving only the component of the perturbation perpendicular to the magnetic surfaces, $\int \rho(\mathbf{n}\cdot\boldsymbol{\xi})^2\, dV$; see Bernstein et al. [55](b). The only limitation in the choice of the normalization of the trial functions is that the original norm $\|\boldsymbol{\xi}\|$ should remain finite. Of course, in the process of dropping the proper normalization of the Hilbert space, one loses the possibility of calculating the actual growth rates of the instabilities.

– Method (c) is equivalent to solving the spectral equation $\mathbf{F}(\boldsymbol{\xi}) = -\rho\omega^2\boldsymbol{\xi}$, i.e. to a full-blown normal mode analysis. This shows that variational method and differential equation are equivalent.

Although the above discussion of the energy principle appears to be rather solid, i.e. in no need of further proof, this turns out to be deceptive since the tacit assumption has been made that the spectrum consists of discrete eigenvalues only. In general, this is not the case in MHD, where the existence of a continuous spectrum is a rule rather than an exception. For stability analysis, this presents a substantial complication. This problem will be addressed in Section 6.5, where we first discuss stability analysis with normal modes only (Section 6.5.1), and next show that the energy principle can still be proved avoiding the subtleties of the continuous spectrum (Section 6.5.2). This will also facilitate a practical modification of the energy principle (Section 6.5.3). Having settled these more advanced issues, we will return to our starting point and put the spotlight on the structural beauty and applicability of MHD spectral theory (Section 6.5.4).

6.5 Further spectral issues

6.5.1 Normal modes and the energy principle★

Consider a pair of discrete normal modes $\exp(-\mathrm{i}\omega_n t)$ and $\exp(\mathrm{i}\omega_n t)$ belonging to the same eigenvalue $\omega^2 = \omega_n^2$. Neglecting all other modes, e.g. by preferentially exciting this one pair of modes, the solution (6.76) of the initial value problem (with the contour C as in Fig. 6.8) may be completed. Since

$$\rho^{-1}\mathbf{F}(\boldsymbol{\xi}) = -\omega_n^2\, \boldsymbol{\xi}\,, \tag{6.96}$$

the resolvent operator would simply be given by

$$(\rho^{-1}\mathbf{F} + \omega^2\mathbf{I})^{-1} = \big[(\omega^2 - \omega_n^2)\mathbf{I}\big]^{-1}\,. \tag{6.97}$$

Hence, the discrete eigenvalue ω_n^2 gives rise to two poles $\omega = \pm\omega_n$ which, by virtue of the reality condition (6.57), are situated on either the real or the imaginary axis of the complex ω-plane. Clearly, for $\omega^2 = \omega_n^2$ the resolvent operator does not exist, but everywhere else in the complex ω-plane it is now defined (ignoring the rest of the spectrum). We may now complete the integration (6.76) by deforming the contour around the two poles $\omega = \pm\omega_n$. Shifting the straight part of the contour to $\nu = -\infty$, so that $\exp(-\mathrm{i}\omega t)$ vanishes faster than everything else, the only contribution that survives for large t will be the two residues picked up at the poles. These may be computed one by one by means of Cauchy's integral formula

$$\oint \frac{f(z)}{z - z_0}\, dz = 2\pi\mathrm{i} f(z_0)\,, \tag{6.98}$$

where $f(z)$ should be analytic inside a closed contour encircling the point $z = z_0$. We then find for the asymptotic time-dependence of the normal modes:

$$\begin{aligned} \boldsymbol{\xi}(\mathbf{r};t) &= \frac{1}{2\pi}\int_{\mathrm{C}} \frac{\mathrm{i}\omega\boldsymbol{\xi}_i(\mathbf{r}) - \dot{\boldsymbol{\xi}}_i(\mathbf{r})}{(\omega+\omega_n)(\omega-\omega_n)}\, \mathrm{e}^{-\mathrm{i}\omega t}\, d\omega \\ &= \frac{[\mathrm{i}\omega_n\boldsymbol{\xi}_i(\mathbf{r}) + \dot{\boldsymbol{\xi}}_i(\mathbf{r})]\mathrm{e}^{\mathrm{i}\omega_n t} + [\mathrm{i}\omega_n\boldsymbol{\xi}_i(\mathbf{r}) - \dot{\boldsymbol{\xi}}_i(\mathbf{r})]\mathrm{e}^{-\mathrm{i}\omega_n t}}{2\mathrm{i}\omega_n}\,, \end{aligned} \tag{6.99}$$

where one should notice that the contour C deformed around a pole has just the opposite sense of a Cauchy contour.

Specifically, writing $\omega_n = \sigma_n + \mathrm{i}\nu_n$, we either have $\nu_n = 0$ or $\sigma_n = 0$. If $\nu_n = 0$, the poles are situated on the real axis (Fig. 6.9(a)) so that

$$\boldsymbol{\xi}(\mathbf{r};t) = \boldsymbol{\xi}_i(\mathbf{r})\cos\sigma_n t + \dot{\boldsymbol{\xi}}_i(\mathbf{r})\sigma_n^{-1}\sin\sigma_n t\,, \tag{6.100}$$

which is a stable undamped oscillation excited by an initial displacement $\boldsymbol{\xi}_i(\mathbf{r})$, or by an initial velocity $\dot{\boldsymbol{\xi}}_i(\mathbf{r})$, or by a combination of both. If $\sigma_n = 0$, the poles are situated on the imaginary axis (Fig. 6.9(b)) and we have

$$\boldsymbol{\xi}(\mathbf{r};t) = \boldsymbol{\xi}_i(\mathbf{r})\cosh\nu_n t + \dot{\boldsymbol{\xi}}_i(\mathbf{r})\nu_n^{-1}\sinh\nu_n t\,. \tag{6.101}$$

Since both $\cosh(\nu_n t)$ and $\sinh(\nu_n t)$ eventually grow as $\exp(\nu_n t)$ this is called an exponential instability, excited by initial displacements or velocities.

The important feature is that, in this case, normal modes, i.e. discrete eigenvalues, are either

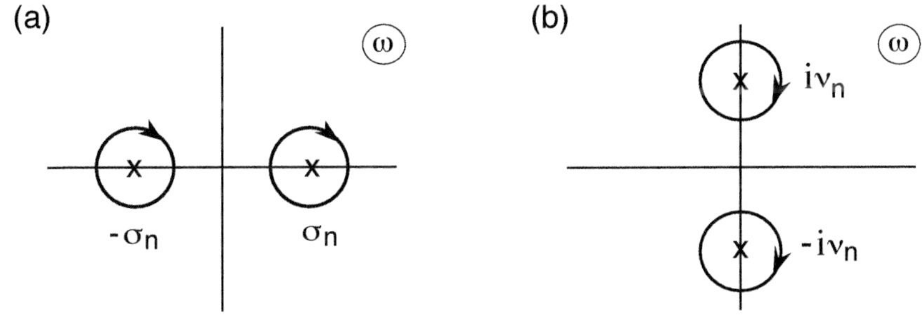

Fig. 6.9 Integration contours for (a) a stable oscillation, (b) an exponentially growing instability.

oscillatory or exponentially growing, but never damped. This is the real simplifying feature of *ideal* MHD perturbations of *static equilibria* expressed by the *self-adjointness* of the force operator. As a consequence, stability studies may be simplified considerably as compared to the analysis of stationary ($\mathbf{v}_0 \neq 0$) or dissipative systems, to be discussed in Chapters 12–14.

Intuitively clear as the energy principle may seem now, its proof is actually not quite straightforward, not even for the self-adjoint systems we are dealing with. If the force operator $\mathbf{F}$ would only allow for discrete eigenvalues satisfying

$$\rho^{-1}\mathbf{F}(\boldsymbol{\xi}_n) = -\omega_n^2\,\boldsymbol{\xi}_n\,, \tag{6.102}$$

it would be reasonable to assume that the set $\{\boldsymbol{\xi}_n\}$ constitutes a complete basis for the Hilbert space. The eigenfunctions $\boldsymbol{\xi}_n$ could then be chosen to be orthonormal,

$$\langle \boldsymbol{\xi}_m, \boldsymbol{\xi}_n \rangle = \delta_{mn}\,. \tag{6.103}$$

An arbitrary $\boldsymbol{\xi}$ could then be expanded in eigenfunctions:

$$\boldsymbol{\xi} = \sum_{n=1}^{\infty} a_n \boldsymbol{\xi}_n \quad \Rightarrow \quad W = -\langle \boldsymbol{\xi}, \rho^{-1}\mathbf{F}(\boldsymbol{\xi})\rangle = \sum_{n=1}^{\infty} a_n^2 \omega_n^2\,. \tag{6.104}$$

Hence, if one could find a $\boldsymbol{\xi}$ for which $W < 0$ at least one eigenvalue $\omega_n^2 < 0$ should exist. Such an eigenvalue would correspond with an exponential instability. This ‘proof’ was given in the original paper by Bernstein et al. [55], before it was known that ideal MHD systems usually have a *continuous spectrum* (see Chapter 7) *extending* $\omega^2 = 0$. This fact implies that more care is needed to establish necessity of the energy principle. This will be the subject of the next subsection.

6.5.2 Proof of the energy principle*

A proof of necessity and sufficiency of the energy principle without invoking the assumption of a complete basis of discrete eigenvalues, but also avoiding analysis of the continuous spectrum, has been given by Laval, Mercier and Pellat in 1965 [398]. That proof is based on energy conservation,

$$\dot H = 0\,, \qquad H \equiv K + W\,, \tag{6.105}$$

and the so-called virial equation

$$\ddot I \equiv \langle \boldsymbol{\xi}, \boldsymbol{\xi}\rangle^{\cdot\cdot} = 2\langle \dot{\boldsymbol{\xi}}, \dot{\boldsymbol{\xi}}\rangle + 2\langle \boldsymbol{\xi}, \ddot{\boldsymbol{\xi}}\rangle = 2K - 2W\,. \tag{6.106}$$

The proof of sufficiency is quite simple, but the proof of necessity is more involved.

Sufficiency of the energy principle If $W[\boldsymbol{\eta}] > 0$ for all trial functions $\boldsymbol{\eta}$, no motion $\boldsymbol{\xi}(t)$ exists such that the kinetic energy $K[\boldsymbol{\xi}(t)]$ grows without bound.

Proof Since $W \equiv H - K > 0$ and H is finite, unbounded growth for K would violate energy conservation; QED. (We exclude linearly growing instabilities, where $\boldsymbol{\xi} \sim t$ and $K = \text{const}$.)

Necessity of the energy principle If a function $\boldsymbol{\eta}$ exists such that $W[\boldsymbol{\eta}] < 0$, the system will exhibit an unbounded motion $\boldsymbol{\xi}(t)$.

First proof Start from $W[\boldsymbol{\eta}] < 0$ and choose initial data $\boldsymbol{\xi}(0) = \boldsymbol{\eta}$, $\dot{\boldsymbol{\xi}}(0) = 0$. From Eq. (6.105), $H(t) = H(0) = K(0) + W(0) = W[\boldsymbol{\eta}] < 0 \Rightarrow \ddot{I}(t) = 2K - 2W = 4K - 2H \geq -2H(t) > 0$. Hence, $\dot{I}$ grows without limit as $t \to \infty$ and I grows at least like $-Ht^2$. As a result, $\boldsymbol{\xi}$ grows at least linearly in t; QED.

▷ **Estimates of growth rates** *(Second proof)* Laval, Mercier and Pellat [398] also gave a sharper proof by estimating the growth rate. Again, start from $W[\boldsymbol{\eta}] < 0$ and define

$$\lambda \equiv -W[\boldsymbol{\eta}]/I[\boldsymbol{\eta}] > 0 . \tag{6.107}$$

We then prove that there exists a $\boldsymbol{\xi}(t)$ growing at least as $\exp(\sqrt{\lambda}\, t)$ (Fig. 6.10(a), lower curve). Choose as initial data $\boldsymbol{\xi}(0) = \boldsymbol{\eta}$, $\dot{\boldsymbol{\xi}}(0) = \sqrt{\lambda}\,\boldsymbol{\eta}$ (i.e., exciting a motion with the proper relationship between $\boldsymbol{\xi}$ and $\dot{\boldsymbol{\xi}}$ for an exponentially growing mode). Consequently, $H(t) = H(0) = K[\dot{\boldsymbol{\xi}}(0)] + W[\boldsymbol{\xi}(0)] = \lambda I[\boldsymbol{\eta}] + W[\boldsymbol{\eta}] = 0$. From Eq. (6.106),

$$\ddot{I}(t) = 2K - 2W = 4K - 2H = 4K(t) > 0 , \tag{6.108}$$

whereas the Schwartz inequality gives

$$\dot{I}^2(t) = 4\langle \boldsymbol{\xi}, \dot{\boldsymbol{\xi}} \rangle^2 \leq 4\langle \boldsymbol{\xi}, \boldsymbol{\xi} \rangle \langle \dot{\boldsymbol{\xi}}, \dot{\boldsymbol{\xi}} \rangle = 4I(t)K(t) = I(t)\ddot{I}(t) . \tag{6.109}$$

Since

$$\dot{I}(0) = 2\sqrt{\lambda}\,\langle \boldsymbol{\eta}, \boldsymbol{\eta} \rangle = 2\sqrt{\lambda}\, I(0) > 0 , \tag{6.110}$$

we have from Eq. (6.108) that $\dot{I}(t) > 0$ for $t > 0$, so that we may divide the inequality (6.109) by $\dot{I}(t)I(t)$, giving the following sequence of inequalities:

$$\frac{\dot{I}(t)}{I(t)} \leq \frac{\ddot{I}(t)}{\dot{I}(t)} \quad \Rightarrow \quad \ln\frac{I(t)}{I(0)} \leq \ln\frac{\dot{I}(t)}{\dot{I}(0)} = \ln\frac{\dot{I}(t)}{2\sqrt{\lambda}I(0)} \quad \Rightarrow \quad \frac{I(t)}{I(0)} \leq \frac{\dot{I}(t)}{2\sqrt{\lambda}I(0)}$$
$$\Rightarrow \quad \frac{\dot{I}(t)}{I(t)} \geq 2\sqrt{\lambda} \quad \Rightarrow \quad \ln\frac{I(t)}{I(0)} \geq 2\sqrt{\lambda}\, t \quad \Rightarrow \quad I(t) \geq I(0)\exp(2\sqrt{\lambda}\, t) .$$

Consequently, the perturbation $\boldsymbol{\xi}(t)$ grows at least as $\exp(\sqrt{\lambda}\, t)$; QED.

One may also prove the following theorem (pertaining to Fig. 6.10(a), upper curve):

Upper bound on the growth rate If the ratio $-W[\boldsymbol{\xi}]/I[\boldsymbol{\xi}]$ has a smallest upper bound

$$\Lambda \geq \lambda[\boldsymbol{\xi}] \equiv -W[\boldsymbol{\xi}]/I[\boldsymbol{\xi}] \quad \text{for all } \boldsymbol{\xi} ,$$

then $\boldsymbol{\xi}(t)$ cannot grow faster than $\exp(\sqrt{\Lambda}\, t)$.

Proof Start again from the virial expression (6.106):

$$\ddot{I}(t) = 2K(t) - 2W(t) = 2H(t) - 4W(t) \leq 2H(t) + 4\Lambda I(t) \quad \Rightarrow \quad \ddot{I}(t) - 4\Lambda I(t) \leq 2H(t) = 2H(0) .$$

Consequently, $I(t)$ grows at most like $\exp(2\sqrt{\Lambda}\, t)$, so that the perturbation $\boldsymbol{\xi}(t)$ cannot grow faster than $\exp(\sqrt{\Lambda}\, t)$; QED. ◁

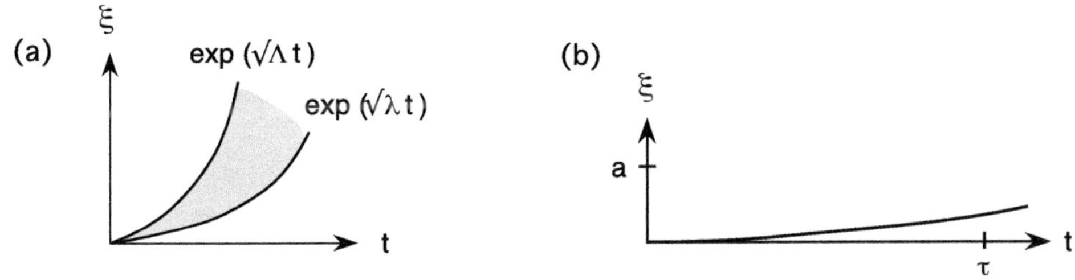

Fig. 6.10 (a) Bounds on exponential growth. (b) Tolerable exponential growth in σ-stability context.

6.5.3 σ-stability

For thermonuclear confinement of plasma, the stability concept used above may be relaxed. One is not really interested in whether the plasma is stable, one is interested in whether or not one can confine it long enough to obtain fusion. For example, if the worst instability of a particular configuration were to grow as illustrated in Fig. 6.10(b), where a is the radial dimension of the plasma vessel and τ is the characteristic confinement time needed for fusion, one would call this configuration stable for all practical purposes. One could also choose τ to be another time scale, e.g. the time scale for which one accepts the ideal MHD model as a valid description, or one may choose τ to be the time scale of an actual experiment determined by the decay of external currents, or τ could correspond to the limit posed by the accuracy of a certain numerical stability program. For all these purposes, one may allow perturbations that grow at most like $\exp(\sigma t)$, where $\sigma \equiv 1/\tau$. We will call equilibria *σ-stable* if they do not manifest growth faster than this. This extension of the stability concept was introduced by Goedbloed and Sakanaka in 1974 [240], and applied to the experimental devices in use at that time [536].

The concept of σ-stability is also useful for analysis. As will be shown in Chapter 7, the continuous spectrum nearly always reaches the origin $\omega = 0$ and it frequently carries with it infinitely many point eigenvalues that accumulate at the edge of the continuum. Hence, the marginal point $\omega = 0$ is a highly singular point in the spectrum (Fig. 6.11(a)) so that the supposed simplicity of a marginal stability analysis, as compared to calculating actual growth rates, often turns out to be illusory. In contrast, a σ-stability analysis avoids these difficulties by staying on the unstable side of the spectrum (Fig. 6.11(b)) which consists of point eigenvalues only. At least, that is a conjecture by Grad [254] to which no exceptions have been found yet. (One should actually exclude perturbations with infinitely large mode numbers since they may lead to dense sets of unstable point eigenvalues; the closure of those sets formally contains a continuous spectrum; see Spies [558], and Spies and Tataronis [560].) This is of particular importance for numerical stability studies where one wishes to avoid singularities as much as possible.

Since we are dealing now with point eigenvalues only, we may define an equilibrium to be σ-stable if no point eigenvalues $\omega^2 < -\sigma^2$ exist, and σ-unstable if such eigenvalues do exist. This problem may be studied by means of the *σ-marginal equation of motion*:

$$\mathbf{F}^\sigma(\boldsymbol{\xi}) \equiv \mathbf{F}(\boldsymbol{\xi}) - \rho\sigma^2\boldsymbol{\xi} = 0\,, \tag{6.111}$$

where the force $\mathbf{F}^\sigma$ available to drive a σ-instability is reduced by the amount $\rho\sigma^2\boldsymbol{\xi}$ with respect

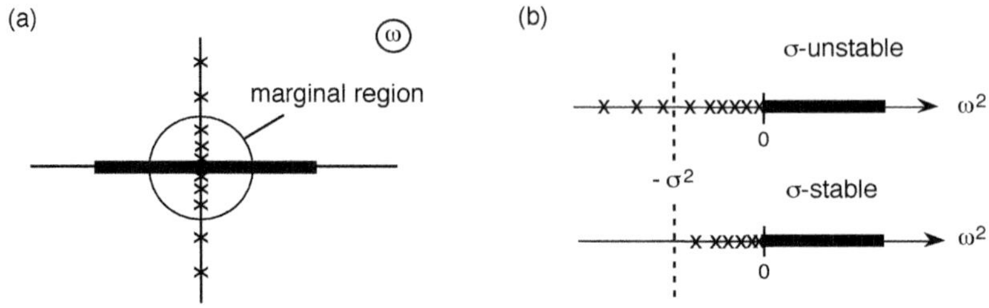

Fig. 6.11 (a) Marginal stability encounters singularities at the origin of the ω-plane; (b) σ-stability analysis avoids them by staying away from the origin.

to the force $\mathbf{F}$ to drive a regular ($\sigma = 0$) instability. The variational form of this problem is the following modification of the energy principle.

Modified energy principle for stability *An equilibrium is σ-stable if and only if*

$$W^\sigma[\boldsymbol{\xi}] \equiv W[\boldsymbol{\xi}] + \sigma^2 I[\boldsymbol{\xi}] > 0 \tag{6.112}$$

for all displacements $\boldsymbol{\xi}$ that are bound in norm and satisfy the boundary conditions. Hence, the amount of negative potential energy available for driving a σ-instability is reduced by the amount $\sigma^2 I[\boldsymbol{\xi}]$ as compared to that available for driving an instability under the original definition. This proof can be given in complete analogy with that of the ordinary energy principle given in the previous section.

Comparing Eq. (6.111) with the normal mode equation (6.28), and Eq. (6.112) with the energy principle (6.95), one observes that their formal structure is the same. One might even wonder whether the whole concept of σ-stability does not boil down to a normal-mode analysis. This is not the case, the crucial difference being that in a normal-mode analysis the eigenvalue ω has to be determined, whereas in a σ-stability analysis σ is a pre-fixed parameter. Hence, the problem is of the same nature as a stability analysis by means of the energy principle, although the equations are more complicated (they have more terms). The latter complication, which is unimportant for numerical applications anyway, is more than offset by the absence of continuum singularities at $\omega^2 = 0$.

6.5.4 Returning to the two viewpoints

We have seen how the two intuitive viewpoints of stability, illustrated in Fig. 6.1, lead to two alternative approaches of the linearized MHD equations, viz. one in terms of differential equations (the equation of motion) and another one in terms of the quadratic forms of the potential and kinetic energy. Since this duality is also present in the formalism of quantum mechanics, expressed in the language of linear operators in Hilbert space, it is expedient to highlight the analogy.

(a) Analogy with quantum mechanics As far as spectral theory is concerned, *the analogy between linearized MHD and quantum mechanics is complete*. In the terminology of Dirac [161], we have two 'pictures' of ideal MHD spectral theory (summarized in Fig. 6.12), viz. that of *the equation of motion* and that of *the variational principles* in terms of the potential and kinetic energies. They

Differential eqs. ('Schrödinger')	**Quadratic forms** ('Heisenberg')	
Equation of motion: $\mathbf{F}(\boldsymbol{\xi}) = \rho \dfrac{\partial^2 \boldsymbol{\xi}}{\partial t^2}$	*Hamilton's principle:* $\delta \int_{t_1}^{t_2} \Big(K[\dot{\boldsymbol{\xi}}] - W[\boldsymbol{\xi}] \Big)\, dt = 0$ $\Rightarrow$	Full dynamics: $\boldsymbol{\xi}(\mathbf{r}, t)$
Eigenvalue problem: $\mathbf{F}(\boldsymbol{\xi}) = -\rho\omega^2 \boldsymbol{\xi}$	*Rayleigh's principle:* $\delta \dfrac{W[\boldsymbol{\xi}]}{I[\boldsymbol{\xi}]} = 0$ $\Rightarrow$	Spectrum $\{\omega^2\}$ + eigenf. $\{\boldsymbol{\xi}(\mathbf{r})\}$
Marginal equation: $\mathbf{F}(\boldsymbol{\xi}) = 0$	*Energy principle:* $W[\boldsymbol{\xi}] \gtrless 0$ $\Rightarrow$	Stability $\binom{\mathrm{y}}{\mathrm{n}}$ + trial funct. $\boldsymbol{\xi}(\mathbf{r})$

Fig. 6.12 The two 'pictures' of ideal MHD spectral theory.

correspond to the 'Schrödinger picture' of wave mechanics (with a description in terms of the wave equation $H\psi = E\psi$), exploiting differential equations, and the 'Heisenberg picture' of matrix mechanics (with a description in terms of the representative matrix elements $\langle n|H|m\rangle$ of the Hamiltonian), exploiting quadratic forms.

Of course, this analogy is mathematical, not physical. The physical systems are totally different. For example, in the spectrum, the transition from bound to free states in quantum mechanics corresponds to the completely different physical problem of transition from stable to unstable modes in magnetohydrodynamics. More important, whereas the wave function ψ in quantum mechanics is a complex quantity that requires a physical interpretation to connect it to observable properties of the atomic system, the displacement vector $\boldsymbol{\xi}$ in MHD is a real quantity that refers directly to the observable macroscopic displacement of the classical plasma fluid. Using complex notation here is no more than a matter of convenience. (We do not pay much attention to the difference between a scalar ψ in quantum mechanics and a vector $\boldsymbol{\xi}$ in MHD since this appears to be of minor interest; quantum mechanical systems with wave functions with more components also occur, e.g. in quantum electrodynamics of $S = 1$ particles.)

An important reason to dwell on the analogy with quantum mechanical spectral theory is the need to disentangle useful mathematical techniques that can be transferred to another field from concepts that are essential to the physical formulation. In this respect, it should be clear (but frequently it is not for physics students because of their over-exposition to quantum mechanical problems as compared to classical ones) that linear operators in Hilbert space as such have nothing to do with quantum mechanics. In fact, the mathematical formulation by Hilbert in 1912 preceded the advent of quantum mechanics by more than a decade. Essentially, the two 'pictures' illustrated in Fig. 6.12 are just a translation to physics of the generalization of linear algebra concepts to infinite-dimensional vector spaces. (I am indebted to Jürgen Moser (1973) for enlightening discussions on this.)

There is yet another hurdle to overcome. Whereas quantum mechanics of atomic and sub-atomic particles applies to a large collection of relevant spherically symmetric systems, with the attraction of symmetry with respect to the rotation groups, MHD spectral theory refers to magnetically confined plasmas where the constraint of $\nabla \cdot \mathbf{B} = 0$ forbids spherical symmetry (as we saw in Section 1.3.4) and demands the consideration of extended magnetic structures with symmetries that are much less obvious than rotations. In fact, the application of symmetry groups to MHD spectral theory is really in its infancy, as should be evident from the more complex structure of the MHD equations compared to the quantum mechanical ones and the relatively insignificant effort in MHD compared to the impressive accomplishments of the great physicists of the twentieth century who created the quantum mechanical picture of the atomic world.

(b) Why does the water fall out of the glass? We now show how the machinery of spectral theory and energy principle works to solve the practical problem posed in Section 6.1.1. We consider a simple fluid (no magnetic field) with a varying density in an external gravitational field. For equilibrium, the pressure and density distributions should satisfy $\nabla p = -\rho \nabla \Phi_{\rm gr} = \rho \mathbf{g}$. The expression (6.84) for the energy W then simplifies to

$$\begin{aligned} W^{\rm f} &= \tfrac{1}{2} \int \big[\gamma p |\nabla \cdot \boldsymbol{\xi}|^2 + (\boldsymbol{\xi} \cdot \nabla p) \nabla \cdot \boldsymbol{\xi}^* - (\boldsymbol{\xi}^* \cdot \nabla \Phi_{\rm gr}) \, \nabla \cdot (\rho \boldsymbol{\xi}) \big] \, dV \\ &= \tfrac{1}{2} \int \big[\gamma p |\nabla \cdot \boldsymbol{\xi}|^2 + \rho \mathbf{g} \cdot (\boldsymbol{\xi} \, \nabla \cdot \boldsymbol{\xi}^* + \boldsymbol{\xi}^* \nabla \cdot \boldsymbol{\xi}) + \mathbf{g} \cdot \boldsymbol{\xi}^* (\nabla \rho) \cdot \boldsymbol{\xi} \big] \, dV. \end{aligned} \qquad (6.113)$$

Without gravity, the fluid is stable since only the positive definite first term, corresponding to compressive sound motions, remains. With gravity, the sign of W depends on the density gradient in a way that we now have to determine.

Let us specify to plane slab geometry so that pressure $p(x)$ and density $\rho(x)$ are functions of the vertical coordinate x alone, and gravity points in the negative x-direction, $\mathbf{g} = -g \mathbf{e}_x$. The equilibrium then becomes $p' = -\rho g$. The expression for $W^{\rm f}$ now simplifies to

$$W^{\rm f} = \tfrac{1}{2} \int \big[\gamma p \, |\nabla \cdot \boldsymbol{\xi}|^2 - \rho g (\xi_x \nabla \cdot \boldsymbol{\xi}^* + \xi_x^* \nabla \cdot \boldsymbol{\xi}) - \rho' g |\xi_x|^2 \big] \, dV \, . \qquad (6.114)$$

The energy principle according to method (a) of Section 6.4.4 is illustrated by the immediate guess suggested by this expression, viz. to exploit incompressible trial functions, $\nabla \cdot \boldsymbol{\xi} = 0$, so that

$$W^{\rm f} = -\tfrac{1}{2} \int \rho' g |\xi_x|^2 \, dV \geq 0 \quad \Rightarrow \quad \rho' g \leq 0 \quad \text{(everywhere)}, \qquad (6.115)$$

which is a *necessary criterion* for stability. This already more or less explains our glass of water experiment since this criterion clearly shows that lighter fluid should be on top of heavier fluid for stability. Actually, we here have derived the condition for stability with respect to *internal modes*, whereas the water–air system requires an extended form of the energy principle with a two-fluid interface (model II*), permitting *external modes*. This will be considered in Section 6.6.4. The physics will be the same though: the density gradient then becomes a jump that should be negative at the interface (light fluid above) for stability.

The expression (6.114) also permits the derivation of a much sharper stability condition, with respect to all possible perturbations (compressible as well as incompressible ones), according to method (b) of Section 6.4.4. In this case, it just implies rearranging terms such that two definite terms are obtained:

$$W^{\rm f} = \tfrac{1}{2} \int \Big[\gamma p \Big| \nabla \cdot \boldsymbol{\xi} - \frac{\rho g}{\gamma p} \xi_x \Big|^2 - \Big(\rho' g + \frac{\rho^2 g^2}{\gamma p} \Big) |\xi_x|^2 \Big] \, dV \,. \tag{6.116}$$

Since the components ξ_y and ξ_z only appear in the compressibility term $\nabla \cdot \boldsymbol{\xi}$, minimization with respect to those components is trivial: $\nabla \cdot \boldsymbol{\xi} = \big(\rho g/(\gamma p)\big)\, \xi_x$. Consequently,

$$\rho' g + \frac{\rho^2 g^2}{\gamma p} \le 0 \quad \text{(everywhere)} \tag{6.117}$$

is a *necessary and sufficient criterion* for stability.

Finally, questions like 'How long does it take for the instability to develop?' or 'What do the perturbations look like?' require a complete normal mode analysis of the spectrum and eigenfunctions, according to method (c) of Section 6.4.4. This will be one of the topics covered in Chapter 7.

6.6 Extension to interface plasmas

So far, we have been concerned with spectral theory of plasmas surrounded by a rigid wall (model I). For many applications, it is useful to be able to treat plasmas with an interface by the same techniques. For example, in laboratory fusion research it is appropriate to model the very low density region close to the wall (created by a limiter) as a vacuum so that effectively a *plasma–vacuum system* is obtained (model II). In astrophysical plasmas it is frequently expedient to model plasmas with a jump in the density (e.g. to a low-density force-free plasma) as a *plasma–plasma system* (model II*).

Recall the steps of the spectral analysis of model I plasmas. The nonlinear equations (6.1)–(6.4) for the plasma variables $\mathbf{v}$, p, $\mathbf{B}$, and ρ were linearized about a static equilibrium, prescribed by Eqs. (6.7)–(6.8), with perturbations satisfying the differential equations (6.12)–(6.15) and the boundary conditions (6.16). We then defined the plasma displacement vector $\boldsymbol{\xi}(\mathbf{r}, t)$, and cast the dynamical problem in the form of the equation of motion (6.23) involving the force operator $\mathbf{F}(\boldsymbol{\xi})$. Next, $\mathbf{F}$ was proved to be self-adjoint in Section 6.2.3, and the quadratic form (6.84) for the associated potential energy W was derived in Section 6.4.1. These steps will now be generalized to plasmas with an interface, following the original paper on the energy principle by Bernstein et al. [55] and the presentation by Kadomtsev [338].

The original nonlinear equations for model II and model II* interface plasmas of Section 4.6.1 need to be subjected to the same procedure. The vacuum magnetic field then splits into an equilibrium part $\hat{\mathbf{B}}$ (suppressing the subscript 0 again) and a perturbation $\hat{\mathbf{Q}}$. The *equilibrium* vacuum magnetic field $\hat{\mathbf{B}}$ satisfies

$$\nabla \times \hat{\mathbf{B}} = 0 \,, \qquad \nabla \cdot \hat{\mathbf{B}} = 0 \,, \tag{6.118}$$

subject to the boundary conditions

$$\mathbf{n} \cdot \mathbf{B} = \mathbf{n} \cdot \hat{\mathbf{B}} = 0 \,, \quad \Big[\!\Big[p + \tfrac{1}{2} B^2 \Big]\!\Big] = 0 \qquad \text{(at the interface S)} \,, \tag{6.119}$$

which imply surface currents $\mathbf{j}^\star = \mathbf{n} \times [\![\mathbf{B}]\!]$, and to the boundary condition

$$\mathbf{n} \cdot \hat{\mathbf{B}} = 0 \qquad \text{(at the conducting wall } \hat{\rm W}) \,. \tag{6.120}$$

For model II*, the outer region is a plasma so that the vacuum equations (6.118) should be replaced by plasma equations, like Eqs. (6.7), for $\hat{\mathbf{B}}$, $\hat{p}$ and $\hat{\rho}$, whereas the jump conditions and the outer BC remain the same.

The vacuum magnetic field *perturbations* $\hat{\mathbf{Q}}$ are described by

$$\nabla \times \hat{\mathbf{Q}} = 0\,, \qquad \nabla \cdot \hat{\mathbf{Q}} = 0\,, \tag{6.121}$$

subject to two entirely non-trivial boundary conditions (see below) connecting $\hat{\mathbf{Q}}$ to the plasma variable $\boldsymbol{\xi}$ at the plasma–vacuum interface, and one boundary condition

$$\mathbf{n} \cdot \hat{\mathbf{Q}} = 0 \qquad (\text{at } \hat{\mathrm{W}})\,. \tag{6.122}$$

Of course, for model II* plasmas, Eqs. (6.121) should be replaced by equations analogous to Eqs. (6.23)–(6.25) for the displacement vector $\hat{\boldsymbol{\xi}}$ of the outer plasma, whereas the outer BC (6.122) is to be replaced by

$$\mathbf{n} \cdot \hat{\boldsymbol{\xi}} = 0 \qquad (\text{at } \hat{\mathrm{W}})\,. \tag{6.123}$$

The two mentioned BCs *at the interface*, connecting $\boldsymbol{\xi}$ in the plasma to $\hat{\mathbf{Q}}$ in the vacuum (model II) or to $\hat{\boldsymbol{\xi}}$ in the outer plasma (model II*), are obtained after a rather laborious derivation, to be given in Section 6.6.1.

Note that, in model II*, the limit of a very tenuous plasma with $\hat{p} = 0\,, \hat{\rho} = 0\,, \hat{\mathbf{j}} = 0$ still implies very different dynamics from that of a vacuum because the MHD magnetic field equation (6.3) implies the picture of frozen field lines, whereas such a condition does not hold for a vacuum magnetic field.

▷ **Vector potential formulation for the vacuum field perturbations** For some of the derivations, it is useful to exploit the alternative representation of the vacuum perturbations in terms of the vector potential. Recall the pre-Maxwell equations (4.169) and (4.170) given in Section 4.6.1. They yield the following expressions for the vacuum field perturbations $\hat{\mathbf{Q}}$ and $\hat{\mathbf{E}}$ in terms of the vector potential $\hat{\mathbf{A}}$ and the scalar potential $\hat{\Phi}$:

$$\hat{\mathbf{Q}} = \nabla \times \hat{\mathbf{A}}\,, \qquad \nabla \times \nabla \times \hat{\mathbf{A}} = 0\,, \tag{6.124}$$

$$\hat{\mathbf{E}} = -\frac{\partial \hat{\mathbf{A}}}{\partial t} - \nabla \hat{\Phi}\,, \qquad \nabla^2 \hat{\Phi} = -\frac{\partial}{\partial t}(\nabla \cdot \hat{\mathbf{A}})\,. \tag{6.125}$$

With the choice of the Coulomb gauge condition $\hat{\Phi} = 0$, which implies that $\nabla \cdot \hat{\mathbf{A}} = 0$ as well, the vector potential $\hat{\mathbf{A}}$ becomes the only variable needed to describe the vacuum. The vacuum field variables $\hat{\mathbf{Q}}$ and $\hat{\mathbf{E}}$ are then obtained by

$$\hat{\mathbf{Q}} = \nabla \times \hat{\mathbf{A}}\,, \qquad \hat{\mathbf{E}} = -\frac{\partial \hat{\mathbf{A}}}{\partial t}\,, \qquad \text{where} \qquad \nabla^2 \hat{\mathbf{A}} = 0\,, \qquad \nabla \cdot \hat{\mathbf{A}} = 0\,. \tag{6.126}$$

The appropriate boundary condition at the perfectly conducting wall is obtained from the condition that the tangential electric field has to vanish there (cf. Eq. (4.172)):

$$\mathbf{n} \times \hat{\mathbf{A}} = 0 \qquad (\text{at } \hat{\mathrm{W}})\,. \tag{6.127}$$

One easily demonstrates that satisfaction of this boundary condition also implies satisfaction of the boundary condition $\mathbf{n} \cdot \hat{\mathbf{Q}} = \mathbf{n} \cdot (\nabla \times \hat{\mathbf{A}}) = 0$ at the conducting wall. ◁

6.6.1 Boundary conditions at the interface

The linearization of the two boundary conditions (4.173) and (4.174), for the normal magnetic field and the total pressure, provide the necessary connection between the plasma variable $\boldsymbol{\xi}$ and the vacuum variable $\hat{\mathbf{Q}}$ at the interface. The derivation is rather involved since we need to evaluate the physical variables *at the perturbed boundary* and we also need an expression for the *perturbation of the normal* to that boundary. Because of the importance of the interface boundary conditions for applications, we give their derivation in full and point out some pitfalls in passing. At this point, we temporarily revert to writing again the subscripts 0 and 1 for equilibrium and perturbations.

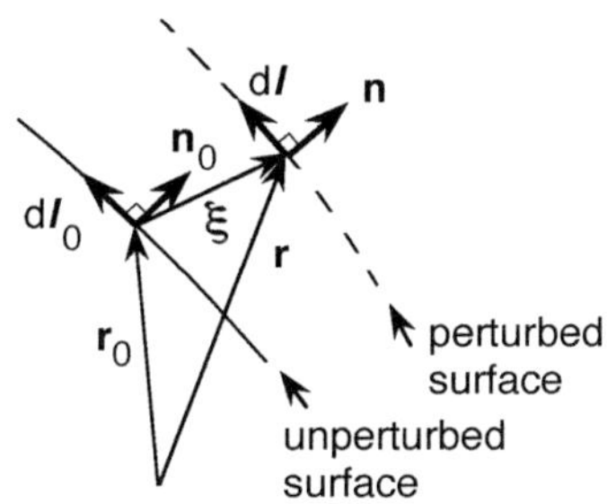

Fig. 6.13 Perturbation of the normal to the plasma boundary.

An expression for the perturbation of the normal is most easily obtained by integrating the Lagrangian time derivative of a line element (4.88), derived in Section 4.3.3, to give the perturbation of a line element moving with the fluid:

$$d\boldsymbol{l} \approx d\boldsymbol{l}_0 \cdot (\mathbf{I} + \nabla\boldsymbol{\xi}) \,. \tag{6.128}$$

This relation is correct to first order since the difference between Eulerian and Lagrangian time derivatives is of higher order. From this expression we obtain the following identity for a line element $d\boldsymbol{l}$ lying in the boundary surface (Fig. 6.13):

$$0 = \mathbf{n} \cdot d\boldsymbol{l} \approx d\boldsymbol{l}_0 \cdot (\mathbf{I} + \nabla\boldsymbol{\xi}) \cdot (\mathbf{n}_0 + \mathbf{n}_{1\mathrm{L}}) \approx d\boldsymbol{l}_0 \cdot [(\nabla\boldsymbol{\xi}) \cdot \mathbf{n}_0 + \mathbf{n}_{1\mathrm{L}}] \,,$$

where we have put an index L on $\mathbf{n}_{1\mathrm{L}}$ to indicate that this is a Lagrangian perturbation. Hence, $\mathbf{n}_{1\mathrm{L}} = -(\nabla\boldsymbol{\xi}) \cdot \mathbf{n}_0 + \boldsymbol{\lambda}$, where the vector $\boldsymbol{\lambda} \perp d\boldsymbol{l}_0$. But $d\boldsymbol{l}_0$ may have any direction in the unperturbed surface so that $\boldsymbol{\lambda}$ must be parallel to $\mathbf{n}_0$: $\boldsymbol{\lambda} = \mu\mathbf{n}_0$. Since $|\mathbf{n}| = |\mathbf{n}_0| = 1$, we have $\mathbf{n}_0 \cdot \mathbf{n}_{1\mathrm{L}} = 0$ so that $\mu = \mathbf{n}_0 \cdot (\nabla\boldsymbol{\xi}) \cdot \mathbf{n}_0$. This provides us with the required *Lagrangian perturbation of the normal:*

$$\mathbf{n}_{1\mathrm{L}} = -(\nabla\boldsymbol{\xi}) \cdot \mathbf{n}_0 + \mathbf{n}_0\, \mathbf{n}_0 \cdot (\nabla\boldsymbol{\xi}) \cdot \mathbf{n}_0 \overset{(\mathrm{A}.2)}{=} \mathbf{n}_0 \times \{\mathbf{n}_0 \times [(\nabla\boldsymbol{\xi}) \cdot \mathbf{n}_0]\} \,. \tag{6.129}$$

Note that the brackets in $(\nabla\boldsymbol{\xi})$ are essential since they indicate that the derivative is to be taken on $\boldsymbol{\xi}$ alone and not on any quantity appearing after this expression. In fact, in the first investigation of the Rayleigh–Taylor and kink instabilities of plasmas with a vacuum interface by Kruskal and Schwarzschild [381], the expression for the Lagrangian perturbation of the normal was incorrect in this respect. It was corrected later by Kruskal and Tuck [382] to the above form.

(a) Model II boundary conditions The evaluation of the boundary conditions (4.173) and (4.174) for the normal magnetic field and the total pressure requires the Lagrangian expressions for the perturbed magnetic field $\mathbf{B}$ and the pressure p at the perturbed position $\mathbf{r}$ of the boundary, evaluated to first order:

$$\begin{aligned} \mathbf{B}|_{\mathbf{r}} &\approx (\mathbf{B}_0 + \mathbf{Q} + \boldsymbol{\xi}\cdot\nabla\mathbf{B}_0)|_{\mathbf{r}_0}\,, \\ p|_{\mathbf{r}} &\approx (p_0 + \pi + \boldsymbol{\xi}\cdot\nabla p_0)|_{\mathbf{r}_0} = (p_0 - \gamma p_0 \nabla\cdot\boldsymbol{\xi})|_{\mathbf{r}_0}\,. \end{aligned} \tag{6.130}$$

Here, $\mathbf{Q}$ and π are the Eulerian perturbations, defined in Eqs. (6.25) and (6.24), and the terms $\boldsymbol{\xi}\cdot\nabla\mathbf{B}_0$ and $\boldsymbol{\xi}\cdot\nabla p_0$ are due to the shift of the boundary.

Inserting Eqs. (6.129) and (6.130) into the first part of the boundary condition (4.173) for the normal magnetic field gives

$$\begin{aligned} 0 &= \mathbf{n}\cdot\mathbf{B} \approx [\,\mathbf{n}_0 - (\nabla\boldsymbol{\xi})\cdot\mathbf{n}_0 + \mathbf{n}_0\,\mathbf{n}_0\cdot(\nabla\boldsymbol{\xi})\cdot\mathbf{n}_0\,]\cdot(\mathbf{B}_0 + \mathbf{Q} + \boldsymbol{\xi}\cdot\nabla\mathbf{B}_0) \\ &\approx -\mathbf{B}_0\cdot(\nabla\boldsymbol{\xi})\cdot\mathbf{n}_0 + \mathbf{n}_0\cdot\mathbf{Q} + \boldsymbol{\xi}\cdot(\nabla\mathbf{B}_0)\cdot\mathbf{n}_0 \\ &\overset{(A.13)}{=} -\mathbf{n}_0\cdot\nabla\times(\boldsymbol{\xi}\times\mathbf{B}_0) + \mathbf{n}_0\cdot\mathbf{Q}\,. \end{aligned}$$

This condition is automatically satisfied by virtue of the definition (6.25) for $\mathbf{Q}$. However, exactly the same derivation applies to the second part of the boundary condition (4.173) giving the required *first interface condition* relating $\boldsymbol{\xi}$ and $\hat{\mathbf{Q}}$:

$$\mathbf{n}\cdot\nabla\times(\boldsymbol{\xi}\times\hat{\mathbf{B}}) = \mathbf{n}\cdot\hat{\mathbf{Q}} \qquad \text{(at the plasma–vacuum interface S)}\,, \tag{6.131}$$

where we now definitively drop the 0s on the equilibrium quantities again.

▷ **Alternative expressions for the first interface condition** That this boundary condition only depends on the normal components of $\boldsymbol{\xi}$ and $\hat{\mathbf{Q}}$ may be shown by one of those tedious vector manipulations that abound in this field (see Freidberg and Haas [188]), giving:

$$\hat{\mathbf{B}}\cdot\nabla(\mathbf{n}\cdot\boldsymbol{\xi}) - \mathbf{n}\cdot(\nabla\hat{\mathbf{B}})\cdot\mathbf{n}\,\mathbf{n}\cdot\boldsymbol{\xi} = \mathbf{n}\cdot\hat{\mathbf{Q}} \qquad \text{(at S)}\,. \tag{6.132}$$

This form is actually to be preferred over (6.131) as it directly gives the relation between $\mathbf{n}\cdot\boldsymbol{\xi}$ and $\mathbf{n}\cdot\hat{\mathbf{Q}}$. Another alternative expression for the first interface condition is obtained by exploiting the vector potential $\hat{\mathbf{A}}$ instead of $\hat{\mathbf{Q}}$. Recall the exact model II boundary condition (4.176) in terms of the vacuum electric field $\hat{\mathbf{E}}$ and the plasma velocity $\mathbf{v}$ that was derived in Section 4.6.1. Expressing $\hat{\mathbf{E}}$ in terms of $\hat{\mathbf{A}}$ by means of Eq. (6.126), and $\mathbf{v}$ in terms of $\boldsymbol{\xi}$, this condition may be integrated to first order to yield the first interface condition in terms of the vector potential:

$$\mathbf{n}\cdot\boldsymbol{\xi}\,\hat{\mathbf{B}} = -\mathbf{n}\times\hat{\mathbf{A}} \qquad \text{(at S)}\,. \tag{6.133}$$

This form will be exploited in Section 6.6.2. ◁

Inserting Eqs. (6.130) into the boundary condition (4.174) for the total pressure, and exploiting the equilibrium equation (6.119) to remove the equilibrium boundary contributions, leads to the *second interface condition* relating $\boldsymbol{\xi}$ and $\hat{\mathbf{Q}}$:

$$-\gamma p\nabla\cdot\boldsymbol{\xi} + \mathbf{B}\cdot\mathbf{Q} + \boldsymbol{\xi}\cdot\nabla(\tfrac{1}{2}B^2) = \hat{\mathbf{B}}\cdot\hat{\mathbf{Q}} + \boldsymbol{\xi}\cdot\nabla(\tfrac{1}{2}\hat{B}^2) \qquad \text{(at S)}\,. \tag{6.134}$$

The left hand side is just the Lagrangian perturbation of the total plasma pressure.

For model II *plasma–vacuum interface systems*, the formulation of the spectral problem is now complete. The equation of motion (6.23) for $\boldsymbol{\xi}$, the Eqs. (6.121) and (6.122) for $\hat{\mathbf{Q}}$, and the boundary conditions (6.131) and (6.134) connecting $\boldsymbol{\xi}$ and $\hat{\mathbf{Q}}$ at the plasma–vacuum interface constitute a complete set of equations for the investigation of waves and stability properties of these systems.

(b) Model II boundary conditions* For *plasma–plasma interface systems*, extra care is needed in the derivation of the boundary conditions. In that case, $\hat{\boldsymbol{\xi}}$ is also defined for the outer plasma and $\hat{\mathbf{Q}} \equiv \nabla \times (\hat{\boldsymbol{\xi}} \times \hat{\mathbf{B}})$ there so that the condition (6.131) is replaced by the *first interface condition:*

$$\mathbf{n} \cdot \boldsymbol{\xi} = \mathbf{n} \cdot \hat{\boldsymbol{\xi}} \qquad \text{(at the plasma–plasma interface S)}\,, \tag{6.135}$$

which is obtained directly from linearization of Eq. (4.163).

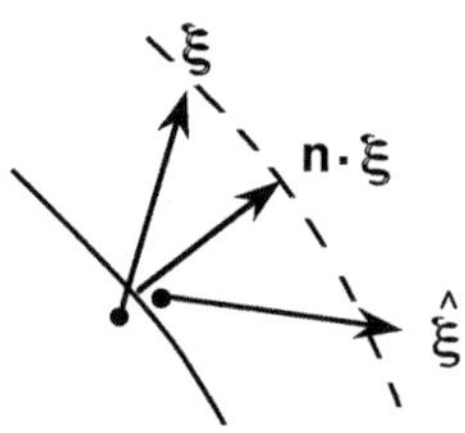

Fig. 6.14 Discontinuity of the tangential boundary displacement.

For the pressure balance equation one has to add the pressure terms of the exterior fluid to the boundary condition (6.134). One may then be tempted to infer from the continuity of the Lagrangian perturbation of the total pressure that the RHS of the boundary condition should be just the same expression as the LHS of Eq. (6.134) with $\boldsymbol{\xi}$, $\mathbf{Q}$, p and $\mathbf{B}$ replaced by $\hat{\boldsymbol{\xi}}$, $\hat{\mathbf{Q}}$, $\hat{p}$ and $\hat{\mathbf{B}}$. In fact, such a mistake has been made (Goedbloed [214](a), corrected in [214](d)). The point is that the two mentioned expressions do not refer to the same position on the interface since the tangential components of $\boldsymbol{\xi}$ are not continuous in general (Fig. 6.14). Thus, for the sake of symmetry between inner and exterior fluid, it is to be preferred to express the perturbation on the perturbed boundary at the position $\mathbf{r}_0 + (\mathbf{n} \cdot \boldsymbol{\xi})\mathbf{n}$, since the normal component of $\boldsymbol{\xi}$ is continuous. The expressions for the perturbations of the plasma pressure and the magnetic field pressure at that position read:

$$\begin{aligned} &\Delta p = \pi + \mathbf{n} \cdot \boldsymbol{\xi}\, \mathbf{n} \cdot \nabla p\,, \qquad \Delta(\tfrac{1}{2}B^2) = \mathbf{B} \cdot \mathbf{Q} + \mathbf{n} \cdot \boldsymbol{\xi}\, \mathbf{n} \cdot \nabla(\tfrac{1}{2}B^2) \\ &\quad \Rightarrow\ \Delta(p + \tfrac{1}{2}B^2) = -\gamma p \nabla \cdot \boldsymbol{\xi} - \boldsymbol{\xi} \cdot \nabla p + \mathbf{B} \cdot \mathbf{Q} + \mathbf{n} \cdot \boldsymbol{\xi}\, \mathbf{n} \cdot \nabla(p + \tfrac{1}{2}B^2)\,. \end{aligned} \tag{6.136}$$

The *second interface condition* on $\boldsymbol{\xi}$ and $\hat{\boldsymbol{\xi}}$ then becomes nicely symmetric:

$$\begin{aligned} &-\gamma p \nabla \cdot \boldsymbol{\xi} + \mathbf{B} \cdot \mathbf{Q} + \mathbf{n} \cdot \boldsymbol{\xi}\, \mathbf{n} \cdot \nabla(\tfrac{1}{2}B^2) \\ &\qquad = -\gamma \hat{p} \nabla \cdot \hat{\boldsymbol{\xi}} + \hat{\mathbf{B}} \cdot \hat{\mathbf{Q}} + \mathbf{n} \cdot \hat{\boldsymbol{\xi}}\, \mathbf{n} \cdot \nabla(\tfrac{1}{2}\hat{B}^2) \qquad \text{(at S)}\,. \end{aligned} \tag{6.137}$$

The difference between this correct BC and the incorrect one (assuming continuity of the Lagrangian total pressure perturbation) is a jump $-[\![\boldsymbol{\xi}_{\rm t} \cdot \nabla(p + \tfrac{1}{2}B^2)]\!]$, where $\boldsymbol{\xi}_{\rm t} \equiv \boldsymbol{\xi} - \mathbf{n} \cdot \boldsymbol{\xi}\, \mathbf{n}$. This term would vanish if $p + \tfrac{1}{2}B^2 = \text{const}$ on the interface, which is not the case for toroidal geometries.

6.6.2 Self-adjointness for interface plasmas

We now extend the proof of self-adjointness of the force operator $\mathbf{F}$ to plasmas with an interface. Recall the proof of Section 6.2.3 for model I plasmas, which resulted in the expression (6.55) with a boundary term that we purposely kept for the present reduction. That term will be transformed to a symmetric expression, both manifesting self-adjointness (this section) and resulting in the extensions of the energy expression (6.84) for model II interface plasmas (next section).

As in Section 6.2.2, consider two displacement vector fields $\boldsymbol{\xi}(\mathbf{r},t)$ and $\boldsymbol{\eta}(\mathbf{r},t)$ defined over the plasma volume V, not necessarily satisfying the ideal MHD equation of motion (6.23). These vector fields will be connected by means of the boundary conditions (6.131) and (6.134) to the associated magnetic perturbations $\hat{\mathbf{Q}}(\mathbf{r},t)$ and $\hat{\mathbf{R}}(\mathbf{r},t)$, defined over the vacuum volume $\hat{\mathrm{V}}$, that satisfy the vacuum equations (6.121) and the BC (6.122) on the wall W. Hence, the 'extensions' $\hat{\mathbf{Q}}$ and $\hat{\mathbf{R}}$ satisfy the following differential equations:

$$\nabla \times \hat{\mathbf{Q}} = 0\,, \quad \nabla \cdot \hat{\mathbf{Q}} = 0\,, \qquad \nabla \times \hat{\mathbf{R}} = 0\,, \quad \nabla \cdot \hat{\mathbf{R}} = 0 \qquad (\text{on } \hat{\mathrm{V}})\,, \tag{6.138}$$

subject to the boundary conditions

$$\mathbf{n} \cdot \nabla \times (\boldsymbol{\xi} \times \hat{\mathbf{B}}) = \mathbf{n} \cdot \hat{\mathbf{Q}}\,, \qquad \mathbf{n} \cdot \nabla \times (\boldsymbol{\eta} \times \hat{\mathbf{B}}) = \mathbf{n} \cdot \hat{\mathbf{R}} \qquad (\text{on S})\,, \tag{6.139}$$

$$-\gamma p \nabla \cdot \boldsymbol{\xi} + \mathbf{B} \cdot \mathbf{Q} + \boldsymbol{\xi} \cdot \nabla(\tfrac{1}{2}B^2) = \hat{\mathbf{B}} \cdot \hat{\mathbf{Q}} + \boldsymbol{\xi} \cdot \nabla(\tfrac{1}{2}\hat{B}^2)\,,$$

$$-\gamma p \nabla \cdot \boldsymbol{\eta} + \mathbf{B} \cdot \mathbf{R} + \boldsymbol{\eta} \cdot \nabla(\tfrac{1}{2}B^2) = \hat{\mathbf{B}} \cdot \hat{\mathbf{R}} + \boldsymbol{\eta} \cdot \nabla(\tfrac{1}{2}\hat{B}^2) \qquad (\text{on S}), \tag{6.140}$$

$$\mathbf{n} \cdot \hat{\mathbf{Q}} = 0\,, \qquad \mathbf{n} \cdot \hat{\mathbf{R}} = 0 \qquad (\text{on } \hat{\mathrm{W}})\,, \tag{6.141}$$

where $\mathbf{Q} \equiv \nabla \times (\boldsymbol{\xi} \times \mathbf{B})$ and $\mathbf{R} \equiv \nabla \times (\boldsymbol{\eta} \times \mathbf{B})$ in the LHSs of Eqs. (6.140).

The idea of the relations (6.138)–(6.141) is to 'extend' the function $\boldsymbol{\xi}$ into the vacuum by means of the magnetic field variable $\hat{\mathbf{Q}}$, and likewise to 'extend' $\boldsymbol{\eta}$ by means of $\hat{\mathbf{R}}$, by matching something like the function value and the normal derivative at the plasma–vacuum interface. It is a very remarkable property of ideal MHD that only two conditions need to be satisfied to connect two vector fields $\boldsymbol{\xi}$ and $\hat{\mathbf{Q}}$. Hence, it appears that we are dealing only with ordinary second order differential equations. The reason behind this is the extreme anisotropy of ideal MHD as regards motion inside and across the magnetic surfaces.

The quadratic from (6.55) was derived in Section 6.2.3 without invoking the solid wall BCs so that it remains valid for model II and model II* interfaces.

▷ **Transformation of the surface integral** By the application of the second interface condition (6.134), the surface integral may be transformed as follows:

$$\int \mathbf{n} \cdot \boldsymbol{\eta}\, (\gamma p \nabla \cdot \boldsymbol{\xi} + \boldsymbol{\xi} \cdot \nabla p - \mathbf{B} \cdot \mathbf{Q})\, dS$$
$$= - \int \mathbf{n} \cdot \boldsymbol{\eta}\, \mathbf{n} \cdot \boldsymbol{\xi}\, \mathbf{n} \cdot \left[\!\left[\nabla (p + \tfrac{1}{2}B^2) \right]\!\right] dS - \int \mathbf{n} \cdot \boldsymbol{\eta}\, \hat{\mathbf{B}} \cdot \hat{\mathbf{Q}}\, dS\,. \tag{6.142}$$

Here, we used the equilibrium jump condition $[\![p + \tfrac{1}{2}B^2]\!] = 0$, which implies that the tangential derivative of the jump vanishes as well: $\mathbf{t} \cdot [\![\nabla(p + \tfrac{1}{2}B^2)]\!] = 0$, where $\mathbf{t}$ is an arbitrary unit vector tangential to the interface. Next, we transform the last term. Introducing vector potentials in the vacuum, $\hat{\mathbf{Q}} \equiv \nabla \times \hat{\mathbf{A}}$ and $\hat{\mathbf{R}} \equiv \nabla \times \hat{\mathbf{C}}$, and exploiting the first interface condition (6.133) in terms of the vector potential $\hat{\mathbf{C}}$, we get:

$$\begin{aligned}
-\int \mathbf{n}\cdot\boldsymbol{\eta}\,\hat{\mathbf{B}}\cdot\hat{\mathbf{Q}}\,dS &= \int \mathbf{n}\times\hat{\mathbf{C}}\cdot\hat{\mathbf{Q}}\,dS = \int \mathbf{n}\times\hat{\mathbf{C}}\cdot\nabla\times\hat{\mathbf{A}}\,dS \\
&\overset{(A.1)}{=} -\int (\nabla\times\hat{\mathbf{A}})\times\hat{\mathbf{C}}\cdot\mathbf{n}\,dS \overset{(A.14)}{=} \int \nabla\cdot\Big[(\nabla\times\hat{\mathbf{A}})\times\hat{\mathbf{C}}\Big]\,d\hat{V} \\
&\overset{(A.12)}{=} \int \Big[\hat{\mathbf{C}}\cdot\nabla\times\nabla\times\hat{\mathbf{A}} - \nabla\times\hat{\mathbf{A}}\cdot\nabla\times\hat{\mathbf{C}}\Big]\,d\hat{V} \\
&\overset{(6.124)}{=} -\int \nabla\times\hat{\mathbf{A}}\cdot\nabla\times\hat{\mathbf{C}}\,d\hat{V} = -\int \hat{\mathbf{Q}}\cdot\hat{\mathbf{R}}\,d\hat{V}\,. \qquad (6.143)
\end{aligned}$$

A minus sign appears in the conversion to the volume term because $\hat{\text{V}}$ is outside S, and the contribution over $\hat{\text{W}}$ could be added since it vanishes because of the BC (6.127). ◁

Collecting terms yields

$$\begin{aligned}
\int \boldsymbol{\eta}\cdot\mathbf{F}(\boldsymbol{\xi})\,dV = -\int \Big\{ &\gamma p\,\nabla\cdot\boldsymbol{\xi}\,\nabla\cdot\boldsymbol{\eta} + \mathbf{Q}\cdot\mathbf{R} + \tfrac{1}{2}\nabla p\cdot(\boldsymbol{\xi}\,\nabla\cdot\boldsymbol{\eta} + \boldsymbol{\eta}\,\nabla\cdot\boldsymbol{\xi}) \\
&+ \tfrac{1}{2}\mathbf{j}\cdot(\boldsymbol{\eta}\times\mathbf{Q} + \boldsymbol{\xi}\times\mathbf{R}) - \tfrac{1}{2}\nabla\Phi_{\rm gr}\cdot\Big[\boldsymbol{\eta}\,\nabla\cdot(\rho\boldsymbol{\xi}) + \boldsymbol{\xi}\,\nabla\cdot(\rho\boldsymbol{\eta})\Big]\Big\}\,dV \\
&- \int \mathbf{n}\cdot\boldsymbol{\eta}\,\mathbf{n}\cdot\boldsymbol{\xi}\,\mathbf{n}\cdot[\![\nabla(p+\tfrac{1}{2}B^2)]\!]\,dS - \int \hat{\mathbf{Q}}\cdot\hat{\mathbf{R}}\,d\hat{V}\,, \qquad (6.144)
\end{aligned}$$

which is symmetric in the variables $\boldsymbol{\xi}$ and $\boldsymbol{\eta}$, and their 'extensions' $\hat{\mathbf{Q}}$ and $\hat{\mathbf{R}}$; QED.

6.6.3 Extended variational principles

Analogous to the procedure of Sections 6.2.3 and 6.4.1, we immediately obtain a meaningful expression for the potential energy W of interface plasmas by identifying $\boldsymbol{\eta}$ and $\boldsymbol{\xi}^*$, and $\hat{\mathbf{R}}$ and $\hat{\mathbf{Q}}^*$, in the quadratic form (6.144):

$$W[\boldsymbol{\xi},\hat{\mathbf{Q}}] = -\tfrac{1}{2}\int \boldsymbol{\xi}^*\cdot\mathbf{F}(\boldsymbol{\xi})\,dV = W^{\rm p}[\boldsymbol{\xi}] + W^{\rm s}[\xi_{\rm n}] + W^{\rm v}[\hat{\mathbf{Q}}]\,, \qquad (6.145)$$

where

$$\begin{aligned}
W^{\rm p}[\boldsymbol{\xi}] &= \tfrac{1}{2}\int \Big[\gamma p\,|\nabla\cdot\boldsymbol{\xi}|^2 + |\mathbf{Q}|^2 + (\boldsymbol{\xi}\cdot\nabla p)\,\nabla\cdot\boldsymbol{\xi}^* + \mathbf{j}\cdot\boldsymbol{\xi}^*\times\mathbf{Q} \\
&\qquad\qquad - (\boldsymbol{\xi}^*\cdot\nabla\Phi_{\rm gr})\,\nabla\cdot(\rho\boldsymbol{\xi})\Big]\,dV\,, \qquad (6.146)\\
W^{\rm s}[\xi_{\rm n}] &= \tfrac{1}{2}\int |\mathbf{n}\cdot\boldsymbol{\xi}|^2\,\mathbf{n}\cdot[\![\nabla(p+\tfrac{1}{2}B^2)]\!]\,dS\,, \qquad (6.147)\\
W^{\rm v}[\hat{\mathbf{Q}}] &= \tfrac{1}{2}\int |\hat{\mathbf{Q}}|^2\,d\hat{V}\,. \qquad (6.148)
\end{aligned}$$

Note that Eq. (6.146) is identical to the expression (6.84) for $W^{\rm p}$. This shows that the work done against the force $\mathbf{F}$ leads to an increase of the potential energy $W^{\rm p}$ of the plasma proper, of the potential energy $W^{\rm s}$ of the plasma–vacuum surface, and of the potential energy $W^{\rm v}$ of the vacuum: very plausible indeed.

However, how to exploit the expression $W[\boldsymbol{\xi},\hat{\mathbf{Q}}]$? If we were to minimize it by exploiting trial functions $\boldsymbol{\xi}$ with vacuum 'extensions' $\hat{\mathbf{Q}}$ that would have to satisfy the differential equations (6.121), the two interface conditions (6.131) and (6.134) on S, and the boundary condition (6.122) on $\hat{\mathrm{W}}$, we would have obtained a very awkward, asymmetrical, variational principle. Here is how symmetry in the use of the variables $\boldsymbol{\xi}$ and $\hat{\mathbf{Q}}$ is established.

Extended spectral variational principle Eigenfunctions $\boldsymbol{\xi}$ of the operator $\rho^{-1}\mathbf{F}$ with their vacuum 'extension' $\hat{\mathbf{Q}}$ make the Rayleigh quotient

$$\Lambda[\boldsymbol{\xi}, \hat{\mathbf{Q}}] \equiv \frac{W[\boldsymbol{\xi}, \hat{\mathbf{Q}}]}{I[\boldsymbol{\xi}]} \equiv \frac{W^{\mathrm{p}}[\boldsymbol{\xi}] + W^{\mathrm{s}}[\xi_{\mathrm{n}}] + W^{\mathrm{v}}[\hat{\mathbf{Q}}]}{\frac{1}{2}\int \rho|\boldsymbol{\xi}|^2\, dV} \tag{6.149}$$

stationary; the eigenvalues ω^2 are the stationary values of Λ.

Here, $\boldsymbol{\xi}$ and $\hat{\mathbf{Q}}$ have to satisfy the following *essential boundary conditions*:

$$(1)\ \ \boldsymbol{\xi} \text{ regular} \qquad\qquad (\text{on V}), \tag{6.150}$$

$$(2)\ \ \mathbf{n}\cdot\nabla\times(\boldsymbol{\xi}\times\mathbf{B}) = \mathbf{n}\cdot\hat{\mathbf{Q}} \qquad\qquad (\text{on S}), \tag{6.151}$$

$$(3)\ \ \mathbf{n}\cdot\hat{\mathbf{Q}} = 0 \qquad\qquad (\text{on } \hat{\mathrm{W}}). \tag{6.152}$$

In this formulation, $\hat{\mathbf{Q}}$ is a trial function, just like $\boldsymbol{\xi}$, that does *not* have to satisfy differential equations and it only has to satisfy the first interface boundary condition, not the second one! The reason is that they have been absorbed in the form of $W[\boldsymbol{\xi}, \hat{\mathbf{Q}}]$ and are automatically satisfied upon minimization. For that reason the second interface condition (6.136) is called a *natural boundary condition*. The distinction between essential and natural boundary conditions is of fundamental importance since it is connected with *counting* the number of equations and unknowns. It will return when we discuss numerical methods of solving the MHD equations by means of Galerkin methods (Chapter 15).

For the investigation of stability, the spectral variational principle can again be simplified considerably by dropping the requirement that the functions $\boldsymbol{\xi}$ are normalized. This yields the following formulation of the stability problem.

Extended energy principle for stability A plasma–vacuum interface equilibrium is stable if (sufficient) and only if (necessary)

$$W[\boldsymbol{\xi}, \hat{\mathbf{Q}}] > 0 \tag{6.153}$$

for all trial functions $\boldsymbol{\xi}(\mathbf{r})$ in the plasma, that are bound in norm, 'extended' with trial functions $\hat{\mathbf{Q}}(\mathbf{r})$ in the vacuum, satisfying the BCs (6.150)–(6.152).

The extended energy principle can be modified with the σ-stability contribution, as in Section 6.5.3: $W^{\sigma}[\boldsymbol{\xi}, \hat{\mathbf{Q}}] \equiv W[\boldsymbol{\xi}, \hat{\mathbf{Q}}] + \sigma^2 I[\boldsymbol{\xi}]$. This yields the *extended σ-stability principle*.

The three different methods of stability analysis discussed in Section 6.4.4 again apply with the extended expression (6.145) for $W[\boldsymbol{\xi}, \hat{\mathbf{Q}}]$. We will demonstrate them by completely solving a particular problem of gravitational stability in the next section. In order to simplify that analysis as much as possible, we will exploit an incompressible plasma model. This choice implies one more *counting* problem, associated with the incompressibility condition. We will settle this before becoming immersed in the detailed explicit calculations.

Incompressibility In the energy principle, one can always test for stability with respect to the restricted class of *incompressible displacements*, $\nabla \cdot \boldsymbol{\xi} = 0$. That is something else than considering an *incompressible plasma model*. Since such a plasma admits only incompressible displacements, one might be tempted to simply pose $\nabla \cdot \boldsymbol{\xi} = 0$ and to drop the contribution $-\gamma p \nabla \cdot \boldsymbol{\xi}$ from the pressure perturbation π, defined in Eq. (6.24), entering the equation of motion (6.23). However, that would lead to an overdetermined system of equations for the three components of $\boldsymbol{\xi}$! The problem is that the ratio of specific heats, γ, diverges for incompressible fluids. A consistent procedure to restore the required freedom in the dynamics of incompressible plasmas is, therefore, to apply the two limits $\gamma \to \infty$ and $\nabla \cdot \boldsymbol{\xi} \to 0$ simultaneously in such a way that the Lagrangian pressure perturbation $\pi_{\rm L} \equiv -\gamma p \nabla \cdot \boldsymbol{\xi}$ remains finite. Equation (6.24) for the Eulerian pressure perturbation π ($\equiv \pi_{\rm E}$) then has to be replaced by $\pi = \pi_{\rm L} - \boldsymbol{\xi} \cdot \nabla p$, so that an additional free variable $\pi_{\rm L}$ appears in the equation of motion. Hence, for incompressible plasmas, Eq. (6.23) is to be replaced by

$$\begin{aligned} \mathbf{F}_{\rm inc}(\boldsymbol{\xi}) \equiv \; & - \nabla \pi_{\rm L} + \nabla(\boldsymbol{\xi} \cdot \nabla p) - \mathbf{B} \times (\nabla \times \mathbf{Q}) \\ & + (\nabla \times \mathbf{B}) \times \mathbf{Q} + (\nabla \Phi_{\rm gr}) \nabla \rho \cdot \boldsymbol{\xi} = \rho \frac{\partial^2 \boldsymbol{\xi}}{\partial t^2}, \end{aligned} \tag{6.154}$$

where application of the condition

$$\nabla \cdot \boldsymbol{\xi} = 0 \tag{6.155}$$

implicitly determines the free variable $\pi_{\rm L}$. For incompressible interface plasmas, there is one additional adjustment, viz. that the second interface condition (6.134) is to be replaced by

$$\pi_{\rm L} + \mathbf{B} \cdot \mathbf{Q} + \boldsymbol{\xi} \cdot \nabla(\tfrac{1}{2} B^2) = \hat{\mathbf{B}} \cdot \hat{\mathbf{Q}} + \boldsymbol{\xi} \cdot \nabla(\tfrac{1}{2} \hat{B}^2) \qquad \text{(at S)}, \tag{6.156}$$

where the boundary value of $\pi_{\rm L}$ is determined by the solutions of the plasma equations (6.154) and (6.155).

The equivalent variational principle for incompressible interface plasmas departs from an expression for the Rayleigh quotient $\Lambda[\boldsymbol{\xi}, \hat{\mathbf{Q}}]$ as in Eq. (6.149), where the expression (6.146) for the plasma energy is to be replaced by

$$W^{\rm p}_{\rm inc}[\boldsymbol{\xi}] = \tfrac{1}{2} \int \Big[|\mathbf{Q}|^2 + \mathbf{j} \cdot \boldsymbol{\xi}^* \times \mathbf{Q} - (\boldsymbol{\xi}^* \cdot \nabla \Phi_{\rm gr}) \nabla \rho \cdot \boldsymbol{\xi} \Big] \, dV, \tag{6.157}$$

and minimization of Λ should be subject to the (unchanged) boundary conditions (6.150)–(6.152) plus the incompressibility constraint (6.155). The simplification due the incompressibility is evident. The Lagrangian pressure variable $\pi_{\rm L}$ enters when minimizing Λ subject to a constraint (6.155). This brings in a Lagrange multiplier which plays the role of $\pi_{\rm L}$, determined by the minimization. As before, it is not necessary to subject the trial functions to the second interface condition (6.156) but the functions satisfying it naturally emerge from the minimization, whereas the boundary value of $\pi_{\rm L}$ is found from the Lagrange multiplier.

6.6.4 Application to the Rayleigh–Taylor instability

We now apply the extended energy principle to the problem of the gravitational instability of a magnetized plasma that is supported from below by a vacuum magnetic field. Hence, we extend the problem of the ordinary gravitational stability of an inverted glass of water (Sections 6.1.1 and 6.5.4) to a genuine plasma with an interface. The gravitational instability for interface fluids is called the

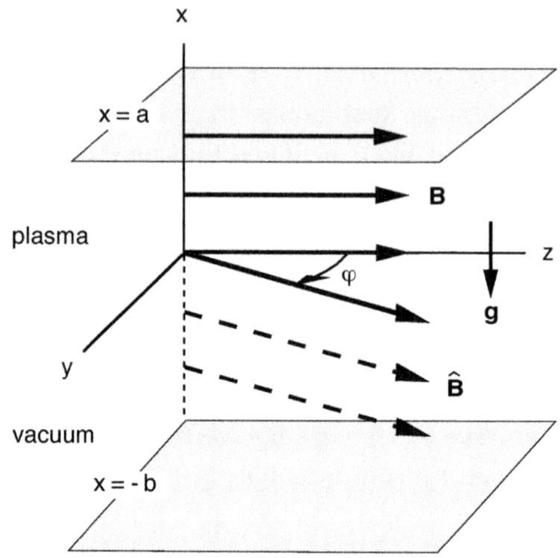

Fig. 6.15 Gravitating plasma slab supported from below by a vacuum magnetic field with conducting walls at $x = a$ and $x = -b$.

Rayleigh–Taylor instability. We will use the same term for the generalized instability of magnetized plasmas. This presents a model problem for plasma confinement with a clear separation of inner plasma and outer vacuum, where instabilities are preferentially localized at the interface (so-called free-boundary or surface instabilities). Of course, for laboratory plasmas, the driving force of the instability, gravity, is negligible, but very similar instabilities arise in a plasma with a curved magnetic field at the interface, as shown in the pioneering investigation by Kruskal and Schwarzschild [381]. Thus, the Rayleigh–Taylor instability of magnetized plasmas allows us to discuss a number of basic concepts, like *interchange instability*, stabilization by *magnetic shear*, and *wall stabilization*. On the other hand, gravitational instabilities do arise in a wide class of astrophysical situations, notably the *Parker instability* [482] in galactic plasmas (Section 7.3.3). Hence, it pays to carefully study the different steps in the analysis of its most simple representation.

In contrast to Section 6.5.4, we now assume constant density ρ_0 in the plasma and neglect compressibility, but allow for a constant uni-directional magnetic field $\mathbf{B}$ (no current), on top of a vacuum also with a constant uni-directional magnetic field $\hat{\mathbf{B}}$, but pointing in a different direction (Fig. 6.15). The plasma occupies a horizontal layer ($0 \leq x \leq a$) that is infinitely extended in the y and z directions, but vertically confined between a conducting wall at $x = a$ and a vacuum below. Likewise, the vacuum occupies a horizontal layer ($-b \leq x \leq 0$) confined between the plasma above and a conducting wall below at $x = -b$. The gravitational acceleration is constant and points downward in the vertical direction, $\mathbf{g} = -\nabla\Phi_{\rm gr} = -g\mathbf{e}_x$.

In the plasma volume V, $\rho = \rho_0$, $\mathbf{B} = B_0\mathbf{e}_z$, $\mathbf{j} = 0$, and gravitational equilibrium is due to balance by the pressure gradient:

$$\nabla p = -\rho\nabla\Phi_{\rm gr} \quad \Rightarrow \quad p' = -\rho_0 g \quad \Rightarrow \quad p = p_0 - \rho_0 g x \quad (p_0 \geq \rho_0 g a)\,. \tag{6.158}$$

Pressure balance at the plasma–vacuum interface S yields

$$p_0 + \tfrac{1}{2}B_0^2 = \tfrac{1}{2}\hat{B}_0^2\,; \tag{6.159}$$

and the magnetic field in the vacuum volume $\hat{\mathrm{V}}$ is described by

$$\hat{\mathbf{B}} = \hat{B}_0(\sin\varphi\, \mathbf{e}_y + \cos\varphi\, \mathbf{e}_z)\,, \tag{6.160}$$

where φ is the angle between $\hat{\mathbf{B}}$ and $\mathbf{B}$. The jump in the direction and magnitude of the magnetic field at the interface implies that a surface current flows there:

$$\mathbf{j}^\star = \mathbf{n} \times [\![\mathbf{B}]\!] = \mathbf{e}_x \times (\mathbf{B} - \hat{\mathbf{B}})\,, \tag{6.161}$$

where the unit normal $\mathbf{n}$ is defined to point into the vacuum, so that $\mathbf{n} = -\mathbf{e}_x$.

(a) Reduction to a one-dimensional problem The assumption of incompressibility permits the use of the simplified expression (6.157) for the plasma energy, derived in Section 6.6.3, and to illustrate how the incompressibility constraint is handled in practice. Some of the more subtle gravitational mechanisms involving compressibility have already been encountered in Section 6.5.4, and will return in Chapter 7. Assuming incompressibility is just a convenient way to defer the difficult issue of wave propagation and instabilities in inhomogeneous plasmas, with the associated singularities, to a later stage (Chapter 7). Since $p' \neq 0$ in the present equilibrium, the sound speed varies in the layer. However, the simplification of the incompressible model is really that the sound speed does not enter the analysis.

Applying the simplifications of the chosen equilibrium to the expressions $W^{\mathrm{p}}_{\mathrm{inc}}$, W^{s}, W^{v}, where the jump in the surface integral (6.147) exhibits the driving energy of the gravitational instability explicitly,

$$\mathbf{n}\cdot[\![\nabla(p + \tfrac{1}{2}B^2)]\!] = p' = -\rho_0 g\,, \tag{6.162}$$

we obtain the following potential energy contributions to $W[\boldsymbol{\xi}, \hat{\mathbf{Q}}]$:

$$W^{\mathrm{p}} = \tfrac{1}{2}\int |\mathbf{Q}|^2\, dV\,, \quad \mathbf{Q} \equiv \nabla\times(\boldsymbol{\xi}\times\mathbf{B})\,, \qquad \nabla\cdot\boldsymbol{\xi} = 0\,, \tag{6.163}$$

$$W^{\mathrm{s}} = -\tfrac{1}{2}\rho_0 g \int |\mathbf{n}\cdot\boldsymbol{\xi}|^2\, dS\,, \tag{6.164}$$

$$W^{\mathrm{v}} = \tfrac{1}{2}\int |\hat{\mathbf{Q}}|^2\, d\hat{V}\,, \qquad \nabla\cdot\hat{\mathbf{Q}} = 0\,. \tag{6.165}$$

Minimization of $\Lambda[\boldsymbol{\xi}, \hat{\mathbf{Q}}]$, or $W[\boldsymbol{\xi}, \hat{\mathbf{Q}}]$, is to be carried out for divergence-free trial functions $\boldsymbol{\xi}$ and $\hat{\mathbf{Q}}$ satisfying the essential boundary conditions (6.150)–(6.152).

Because the slab is translation symmetric in the y and z directions, we may represent the perturbations in terms of separate Fourier modes that do not couple:

$$\boldsymbol{\xi} = \big(\xi_x(x), \xi_y(x), \xi_z(x)\big)\, \mathrm{e}^{\mathrm{i}(k_y y + k_z z)}\,, \qquad \hat{\mathbf{Q}} = \big(\hat{Q}_x(x), \hat{Q}_y(x), \hat{Q}_z(x)\big)\, \mathrm{e}^{\mathrm{i}(k_y y + k_z z)}\,. \tag{6.166}$$

Hence, $\nabla \to (d/dx, 0, 0)$ for equilibrium quantities, but $\nabla \to (d/dx, \mathrm{i}k_y, \mathrm{i}k_z)$ for perturbations. In products like $\xi_x^*\,\xi_y$, which occur in the integrands of the quadratic forms, the exponential y and z dependencies of the complex conjugate Fourier factors cancel out so that we just obtain one-dimensional (1D) integrations of functions of x alone. For that reason, the indices k_y and k_z that should appear on the Fourier amplitude $\xi_{x;k_y,k_z}(x)$, etc., are simply suppressed in Eqs. (6.166). Also, the exponential factor is usually suppressed when working out the expressions, e.g. by writing $\nabla\cdot\boldsymbol{\xi} = \xi_x' + \mathrm{i}k_y\xi_y + \mathrm{i}k_z\xi_z = 0$. These are just shorthand notational conveniences. In case of doubt, one should restore the mentioned factors. Another matter of notation: since the slab is infinitely extended, so that the horizontal areas $A \equiv \iint dydz \to \infty$, the integrals will be renormalized to correspond to the finite contribution over a unit area, $\overline{W} \equiv W/A$. Below, the bars will be suppressed, leaving it understood that all expressions for the Ws and Is are renormalized.

Eliminating the contributions of ξ_z to $W^{\rm p}$, and of $\hat{Q}_z$ to $W^{\rm v}$, by exploiting the divergence conditions, we get the following 1D expressions:

$$\begin{aligned} W^{\rm p} &= \tfrac{1}{2}B_0^2 \int_0^a \Big[k_z^2(|\xi_x|^2 + |\xi_y|^2) + |\xi_x' + {\rm i}k_y\xi_y|^2 \Big]\, dx\,, && (6.167)\\ W^{\rm s} &= -\,\tfrac{1}{2}\rho_0 g|\xi_x(0)|^2\,, && (6.168)\\ W^{\rm v} &= \tfrac{1}{2}\int_{-b}^0 \Big[|\hat{Q}_x|^2 + |\hat{Q}_y|^2 + \frac{1}{k_z^2}|\hat{Q}_x' + {\rm i}k_y\hat{Q}_y|^2 \Big]\, dx\,. && (6.169) \end{aligned}$$

These are to be minimized subject to a normalization that may be chosen freely if we just investigate stability. For surface instabilities, the choice

$$\xi_x(0) = \text{const} \tag{6.170}$$

is quite effective since it stresses the contribution of the surface. (Under point (d), we will point out an important restriction though in the use of this normalization.) On the other hand, to obtain the growth rate of the instability, the full physical norm for incompressible plasmas should be exploited:

$$I = \tfrac{1}{2}\rho_0 \int_0^a \left[|\xi_x|^2 + |\xi_y|^2 + \frac{1}{k_z^2}|\xi_x' + {\rm i}k_y\xi_y|^2 \right] dx\,. \tag{6.171}$$

(For stability investigations, it would also be permissible to drop the second and third term and to keep a normalization involving the normal perturbation ξ_x only. However, in this particular case, that choice turns out not to be a clever one since it is much easier to keep the full norm, as we will see under point (c).) Finally, independent of whether the normalization (6.170) or (6.171) is used, the essential boundary conditions should be satisfied. They reduce to

$$\begin{aligned} \xi_x(a) &= 0\,, && (6.172)\\ \hat{Q}_x(0) &= {\rm i}\mathbf{k}_0\cdot\hat{\mathbf{B}}\,\xi_x(0)\,, && (6.173)\\ \hat{Q}_x(-b) &= 0\,, && (6.174) \end{aligned}$$

where $\mathbf{k}_0 \equiv (0, k_y, k_z)$ is the horizontal wave vector.

(b) Stability analysis We will investigate the stability by means of the normalization (6.170). Our first task is to get rid of the tangential components ξ_y and $\hat{Q}_y$. Since ξ_y does not occur in the reduced normalization, minimization with respect to ξ_y only involves minimization of $W^{\rm p}$. This is accomplished algebraically by splitting off a positive definite term which is then put equal to zero:

$$W^{\rm p} = \tfrac{1}{2}B_0^2 \int_0^a \left[\frac{k_z^2}{k_0^2}{\xi_x'}^2 + k_z^2\xi_x^2 + \left|\frac{k_y}{k_0}\xi_x' + {\rm i}k_0\xi_y\right|^2 \right] dx = \tfrac{1}{2}k_z^2B_0^2 \int_0^a \left(\frac{1}{k_0^2}{\xi_x'}^2 + \xi_x^2 \right) dx\,. \tag{6.175}$$

Similarly, minimization of $W^{\rm v}$ with respect to $\hat{Q}_y$ is trivial:

$$W^{\rm v} = \tfrac{1}{2}\int_{-b}^{0}\Big[\,|\hat{Q}_x|^2 + \frac{1}{k_0^2}|\hat{Q}_x'|^2 + \frac{1}{k_z^2}\Big|\frac{k_y}{k_0}\hat{Q}_x' + {\rm i}k_0\hat{Q}_y\Big|^2\Big]\,dx = \tfrac{1}{2}\int_{-b}^{0}\Big(\frac{1}{k_0^2}|\hat{Q}_x'|^2 + |\hat{Q}_x|^2\Big)\,dx\,. \tag{6.176}$$

Hence, just the two unknown functions ξ_x and $\hat{Q}_x$, joined by the first interface condition (6.173), remain to be determined, whereas the other components of the vectors follow from the relations that we have already found:

$$\begin{aligned} \xi_y &= {\rm i}(k_y/k_0^2)\,\xi_x'\,, \qquad & \xi_z &= {\rm i}(k_z/k_0^2)\,\xi_x'\,,\\ \hat{Q}_y &= {\rm i}(k_y/k_0^2)\hat{Q}_x'\,, \qquad & \hat{Q}_z &= {\rm i}(k_z/k_0^2)\,\hat{Q}_x'\,. \end{aligned} \tag{6.177}$$

Notice that there is no loss in generality if we assume ξ_x, $\hat{Q}_y$, $\hat{Q}_z$ to be real, and ξ_y, ξ_z, $\hat{Q}_x$ to be purely imaginary. This duality between the Fourier components in the directions of inhomogeneity (x) and homogeneity (y, z) will frequently be employed in the following chapters.

The stability problem has now been reduced to the minimization of

$$W = W^{\rm p}[\xi_x(x)] + W^{\rm s}[\xi_x(0)] + W^{\rm v}[\hat{Q}_x(x)]\,, \tag{6.178}$$

subject to the constraint (6.170) and the BCs (6.172)–(6.174), where $W^{\rm p}$ is given by (6.175), $W^{\rm s}$ by (6.168) and $W^{\rm v}$ by (6.176). This involves the standard variational problem of minimizing $W^{\rm p}$ with respect to real functions ξ_x on $0 \le x \le a$, and of $W^{\rm v}$ with respect to real functions ${\rm i}\hat{Q}_x$ on $-b \le x \le 0$.

To carry out these minimizations, recall some general results from *variational analysis*. Minimization of the quadratic form

$$W[\xi] = \tfrac{1}{2}\int_0^a (F\xi'^2 + G\xi^2)\,dx = \tfrac{1}{2}\Big[F\xi\xi'\Big]_0^a - \tfrac{1}{2}\int_0^a \Big[(F\xi')' - G\xi\Big]\xi\,dx\,, \tag{6.179}$$

where $F \ge 0$ but, otherwise, $F(x)$ and $G(x)$ are arbitrary functions on $[0, a]$, is effected by the variation $\delta\xi(x)$ of the unknown function $\xi(x)$:

$$\delta W = \int_0^a \Big(F\xi'\delta\xi' + G\xi\delta\xi\Big)\,dx = \Big[F\xi'\delta\xi\Big]_0^a - \int_0^a \Big[(F\xi')' - G\xi\Big]\delta\xi\,dx = 0\,. \tag{6.180}$$

Since $\xi(0)$ and $\xi(a)$ are prescribed, $\delta\xi = 0$ at the boundaries so that the solution of the *Euler–Lagrange equation* $(F\xi')' - G\xi = 0$ minimizes $W[\xi]$. For that minimizing solution, the value of W becomes

$$W_{\rm min} = \tfrac{1}{2}\Big[F\xi\xi'\Big]_0^a = -\tfrac{1}{2}[F\xi\xi'](x{=}0)\,, \tag{6.181}$$

where we imposed the boundary condition $\xi(a) = 0$.

Minimization of the integral (6.175) for $W^{\rm p}$ and of (6.176) for $W^{\rm v}$ yields the following Euler–Lagrange equations, with solutions satisfying the respective boundary conditions on the upper and lower walls:

$$\begin{aligned} \xi_x'' - k_0^2\xi_x = 0 \qquad &\Rightarrow \quad \xi_x = C\sinh\,[k_0(a - x)]\,,\\ \hat{Q}_x'' - k_0^2\hat{Q}_x = 0 \qquad &\Rightarrow \quad \hat{Q}_x = {\rm i}\hat{C}\sinh\,[k_0(x + b)]\,. \end{aligned} \tag{6.182}$$

Whereas the y- and z-dependencies of the Fourier modes (6.166) correspond to *wave-like* motions (with a real wave number k_0) in the horizontal direction, the exponential x-dependencies $\exp(\pm k_0 x)$ combine into cusp-shaped vertical disturbances about the interface: the waves are *evanescent* (with a purely imaginary wave number $\mathrm{i}k_0$) in the vertical direction. The constant C is related to the normalization (6.170) and $\hat{C}$ may be eliminated by means of the boundary condition (6.173):

$$\hat{C}\sinh(k_0 b) = \mathbf{k}_0\cdot\hat{\mathbf{B}}\,\xi_x(0) = C\mathbf{k}_0\cdot\hat{\mathbf{B}}\sinh(k_0 a)\,. \tag{6.183}$$

Inserting these solutions back into the integrals (6.178), using the relation (6.181), yields the final expression for the energy in terms of $\xi_x(0)$ alone:

$$W = \frac{\xi_x^2(0)}{2k_0\tanh(k_0 a)}\left[(\mathbf{k}_0\cdot\mathbf{B})^2 - \rho_0 k_0 g\tanh(k_0 a) + (\mathbf{k}_0\cdot\hat{\mathbf{B}})^2\frac{\tanh(k_0 a)}{\tanh(k_0 b)}\right]. \tag{6.184}$$

Here, the expression in square brackets has been arranged so as to correspond to the expression for the growth rate derived below.

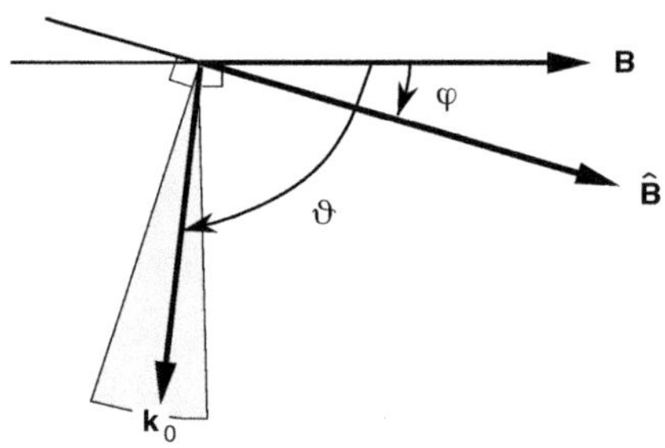

Fig. 6.16 Direction of the horizontal wave vector $\mathbf{k}_0$ for the least stable modes.

The perturbation energy W clearly exhibits the competition between the stabilizing field line bending energies, $\sim \frac{1}{2}(\mathbf{k}_0\cdot\mathbf{B})^2$ for the plasma and $\sim \frac{1}{2}(\mathbf{k}_0\cdot\hat{\mathbf{B}})^2$ for the vacuum, with the destabilizing gravitational energy, $\sim -\frac{1}{2}\rho_0 k_0 g\tanh(k_0 a)$, of the Rayleigh–Taylor instability due to the motion of the interface. Since the magnetic fields $\mathbf{B}$ and $\hat{\mathbf{B}}$ do not point in the same direction, representing *magnetic shear* at the plasma–vacuum interface, no choice of $\mathbf{k}_0$ exists for which the magnetic energies vanish. However, minimum stabilization is obtained for directions of $\mathbf{k}_0$ that are, on average, perpendicular to the field lines (the shaded area in Fig. 6.16). The Rayleigh–Taylor instability may then lead to *interchange instability*, where regions of high plasma pressure and vacuum magnetic field regions are interchanged.

To discuss the dependence on the magnitude of the wave vector $\mathbf{k}_0$, we exploit approximations of the hyperbolic tangent of some argument κ,

$$\tanh\kappa \equiv \frac{\mathrm{e}^{\kappa}-\mathrm{e}^{-\kappa}}{\mathrm{e}^{\kappa}+\mathrm{e}^{-\kappa}} \approx \begin{cases} 1 & (\kappa \gg 1:\text{ short wavelength}) \\ \kappa & (\kappa \ll 1:\text{ long wavelength}) \end{cases}. \tag{6.185}$$

For short wavelengths of the modes or far away walls ($k_0 a \gg 1$, $k_0 b \gg 1$), the magnetic terms ($\sim k_0^2$) dominate over the gravitational term ($\sim k_0$) and the system is stable. The more dangerous regime is for long wavelengths of the modes ($k_0 a \ll 1$), when the lower wall is far away or even removed ($k_0 b \to \infty$). (To keep p_0 finite in the present equilibrium, the upper wall cannot be moved to ∞.) In that case, the plasma magnetic energy and the gravitational energy compete ($\sim k_0^2$) but, since the vacuum magnetic energy is negligible ($\sim k_0^3$), the plasma becomes Rayleigh–Taylor unstable for interchange modes ($\mathbf{k}_0 \perp \mathbf{B}$) where the plasma magnetic energy vanishes. Only when $b/a \sim 1$ is there genuine competition between the three terms ($\sim k_0^2$) so that effective *wall stabilization* may be obtained.

(c) Growth rate analysis To get the growth rate of the instabilities, when W is negative, we need to minimize the full Rayleigh quotient Λ with the complete expression (6.171) for the norm I. Exceptionally, for the present case of incompressible surface instabilities, this turns out to be a minor extension since I and $W^{\rm p}$ have the same functional dependence:

$$(k_z^2 B_0^2/\rho_0)\, I = W^{\rm p} \,. \tag{6.186}$$

It should be pointed out that such blind luck is not at all typical in plasma stability calculations. Nevertheless, of course we will make use of it to get our answer quickly. Because of this relation, the expression for Λ simplifies to

$$\Lambda = k_z^2 B_0^2/\rho_0 + (W^{\rm s} + W^{\rm v})/I \,. \tag{6.187}$$

Minimization is now straightforward: $W^{\rm s}$ and $W^{\rm v}$ do not depend on the radial dependence of $\boldsymbol{\xi}$ so that the expressions (6.168) and (6.176) still hold, whereas, vice versa, minimization of I does not depend on the radial dependence of $\hat{\mathbf{Q}}$ so that it proceeds along the same line as minimization of $W^{\rm p}$. (Note that we obtain the largest growth rate by minimizing $W^{\rm s}$ and $W^{\rm v}$ (which are negative for instability) and also minimizing I (which is positive definite).) This yields

$$I = -\frac{\rho_0}{2k_0^2}\, \xi_x(0)\xi_x'(0) = \frac{\rho_0\, \xi_x^2(0)}{2k_0 \tanh(k_0 a)} \,, \tag{6.188}$$

so that we obtain the following *dispersion equation*:

$$\omega^2 = \frac{W}{I} = \frac{1}{\rho_0}\left[(\mathbf{k}_0 \cdot \mathbf{B})^2 - \rho_0 k_0 g \tanh(k_0 a) + (\mathbf{k}_0 \cdot \hat{\mathbf{B}})^2 \frac{\tanh(k_0 a)}{\tanh(k_0 b)} \right] . \tag{6.189}$$

The expression in square brackets is identical to that of W in Eq. (6.184). In hindsight, that justifies the discussion under point (b) of the wave number dependence: Whereas the energy expression W allows a discussion of the relative importance of the three contributions, there is really no invariant measure for their absolute magnitudes, like that of Eq. (6.189), in the context of the energy principle.

To construct a *scale independent expression for the growth rate* in terms of a minimum number of essential parameters (see Section 4.1.2), we exploit the trivial parameters a, ρ_0, and $\hat{B}_0$ to get rid of the dimensions. This yields the following six dimensionless parameters:

$$\bar{k}_0 \equiv k_0 a \,, \quad \beta \equiv 2p_0/\hat{B}_0^2 = 1 - B_0^2/\hat{B}_0^2 \,, \quad \bar{g} \equiv (a\rho_0/\hat{B}_0^2)\, g \,, \quad w \equiv b/a \,, \tag{6.190}$$

with the angle ϑ between $\mathbf{k}_0$ and $\mathbf{B}$, and the angle φ between $\hat{\mathbf{B}}$ and $\mathbf{B}$ (Fig. 6.16). Normalizing the squared eigenfrequency, $\bar{\omega}^2 \equiv (a^2\rho_0/\hat{B}_0^2)\,\omega^2$, then gives:

$$\bar{\omega}^2 = (1-\beta)\bar{k}_0^2 \cos^2\vartheta - \bar{k}_0 \bar{g} \tanh \bar{k}_0 + \bar{k}_0^2 \cos^2(\vartheta - \varphi)\, \frac{\tanh \bar{k}_0}{\tanh(\bar{k}_0 w)} \,. \tag{6.191}$$

Hence, $\bar{\omega}^2 = \bar{\omega}^2(\bar{k}_0, \vartheta; \beta, \bar{g}, \varphi, w)$, where parameters in front of the semicolon refer to the wave number of the perturbations and parameters behind it refer to the equilibrium. Their ranges are: $\bar{k}_0 \geq 0$, $0 \leq \vartheta \leq \pi$, $0 \leq \beta \leq 1$, $0 \leq \bar{g} \leq \frac{1}{2}\beta$, $0 \leq \varphi \leq \pi$, $w \geq 0$. Clearly, this is a final and complete answer that just remains to be studied in the relevant limits.

Approximating the hyperbolic tangents, we get the following expressions for the short and long wavelength limits:

$$\bar{\omega}^2 \approx \begin{cases} (1-\beta)\bar{k}_0^2\cos^2\vartheta - \bar{k}_0\bar{g} + \bar{k}_0^2\cos^2(\vartheta-\varphi) & (\bar{k}_0 \gg 1) \\ \bar{k}_0^2\Big[(1-\beta)\cos^2\vartheta - \bar{g} + w^{-1}\cos^2(\vartheta-\varphi)\Big] & (\bar{k}_0 \ll 1) \end{cases} . \tag{6.192}$$

For $\beta = 1$ (no magnetic field in the plasma) and $\vartheta - \varphi = \frac{1}{2}\pi$ (horizontal wave vector $\mathbf{k}_0 \perp \hat{\mathbf{B}}$), the Rayleigh–Taylor instability survives in its pure form:

$$\bar{\omega}^2 = -\bar{k}_0\,\bar{g}\tanh\bar{k}_0 \;\approx\; \begin{cases} -\bar{k}_0\,\bar{g} & (\bar{k}_0 \gg 1) \\ -\bar{k}_0^2\,\bar{g} & (\bar{k}_0 \ll 1) \end{cases} . \tag{6.193}$$

As we saw above, the Rayleigh–Taylor instability also survives for arbitrary β and long-wavelength interchange modes in the plasma, i.e. $\vartheta = \frac{1}{2}\pi$ ($\mathbf{k}_0 \perp \mathbf{B}$), when the lower wall is removed ($w \to \infty$). This implies that the vacuum wall is essential to obtain complete stability. Further study of the different limits, and their physical significance, is left to the reader (Exercise [6.8]).

(d) Concluding remarks We have derived the stability properties of a simple gravitating plasma–vacuum system bounded by conducting walls above and below. When this system is considered as an example of confined laboratory plasmas, gravity should be viewed as mimicking magnetic curvature effects. When the interest is in astrophysical applications, the walls should either be moved to infinity or considered as a device to limit the domain of the calculation. In computational MHD, such 'walls' are nearly always present for that purpose. One should check then that the solutions found are insensitive to the positions of these 'walls'.

The growth rates for the Rayleigh–Taylor instability depend on the wave number as $\bar{\omega}^2(\bar{k}_0^2, \vartheta)$, i.e. they just depend on the two horizontal components of $\bar{\mathbf{k}}_0$. Why no dependence on a vertical 'wave number'? Of course, strictly speaking, there is no wave number in that direction because the plasma is inhomogeneous. However, that is just a formal answer, since it is obvious that wave-like motion must exist, also when the spatial dependence is not precisely periodic. The point is that the incompressible plane plasma slab is a highly degenerate configuration, admitting oscillatory solutions of the Alfvén and slow magneto-sonic type, but these internal modes completely decouple from the external surface modes studied in this section. It is a useful exercise, left to the reader (Exercise [6.7]), to figure out from the equation of motion how this decoupling comes about.

The present analysis could be carried out completely in terms of elementary functions. The reason is that the inhomogeneities that occur are concentrated on a single surface ($x = 0$). This gives rise to surface waves with a cusped structure that is localized precisely there. For compressible plasmas with distributed inhomogeneities, the mentioned degeneracy is lifted so that all waves and instabilities are coupled and the solutions $\xi_x(x)$ are oscillatory, in general. In that case, the normalization (6.170) cannot be used since $\xi_x(0)$ may become zero. Such a zero marks a transition when the plasma becomes internally unstable. This demonstrates that the internal inhomogeneities, and their associated singularities, require a much more careful study than presented so far. This is the subject of Chapter 7.

6.7 Literature and exercises

Notes on literature

Stability

- Bernstein, Frieman, Kruskal and Kulsrud [55], 'An energy principle for hydromagnetic stability problems', is probably the most quoted paper in plasma physics. The Project Matterhorn report [55](a) is an appendix containing useful details on the derivations.
- Hain, Lüst and Schlüter [275], 'Zur Stabilität eines Plasmas', appeared simultaneously with [55] but is less known, probably because it is written in German and uses tensor notation.
- Kadomtsev [338] gives a clear exposition of the energy principle with interface extensions.
- Freidberg, *Ideal MHD* [186], Chapter 8, is an overview of the general theory of MHD stability.

Spectral theory

- Friedman, *Principles and Techniques of Applied Mathematics* [193] contains the elements of linear operator theory in a form that is well adapted to use by physicists.
- Lifschitz, *Magnetohydrodynamics and Spectral Theory* [410], Chapter 2, presents the mathematical preliminaries to linear MHD from the spectral theory of operators.

Proof of the energy principle and modifications

- Laval, Mercier and Pellat [398] presents the proofs of the 'Necessity of the energy principle for magnetostatic stability' for the first time.
- Goedbloed and Sakanaka [240] presents a 'New approach to magnetohydrodynamic stability' by introducing the σ-stability concept.

Rayleigh–Taylor instability

- Kruskal and Schwarzschild [381] gives the first analysis of the gravitational and kink instabilities of a plasma–vacuum interface system.
- Meyer [439] shows that the gravitational instability can be stabilized by magnetic fields with different directions in the plasma and the vacuum.
- Goedbloed [214](a) demonstrates stabilization over a wider parameter range by replacing the vacuum by a tenuous plasma with a force-free magnetic field.

Exercises

[6.1] *Stability*

Clarify, with a sketch, the meaning of: stable and unstable perturbations, marginal stability, nonlinear stability, and the difference between marginal stability and lack of equilibrium.

[6.2] *The force operator*

The plasma velocity is related to the displacement vector $\boldsymbol{\xi}$ by the Lagrangian time derivative $\mathbf{v} = \mathrm{D}\boldsymbol{\xi}/\mathrm{D}t$.

- Exploiting this definition, linearize the MHD equations for a plasma with a background flow. What is the difference between keeping and ignoring the flow?
- Ignore the background flow. Insert the expression for $\mathbf{v}$ in the linearized MHD equations and find the expressions for the perturbations p_1, $\mathbf{B}_1$ and ρ_1. Now, derive the equation of motion in terms of the force operator $\mathbf{F}$.

- Count the number of variables and equations in this formalism and compare it with the formalism in terms of primitive variables. Does it check? Comment on the difference.

[6.3] *Stable and unstable solutions*

Express the force operator equation of motion for normal modes with $\omega = \sigma + i\nu$.

- Discuss the different classes of solutions in terms of the signs of σ and ν. Which ones are stable and which ones are unstable?
- Show that for stable solutions the force of the perturbation is restoring, while for unstable solutions it is in the same direction as the plasma displacement.
- Show that, in ideal MHD, the transition from stable to unstable solutions has to go through the marginal state where the force operator $\mathbf{F}(\boldsymbol{\xi}) = 0$.
- What happens to the perturbations when dissipation cannot be neglected?

[6.4] *Force operator; meaning of the terms*

The force operator is given by the expression (6.30) of Section 6.2.1.

- Give the meaning of the different terms. Which one gives rise to the Alfvén waves?
- Simplify the force operator for infinite homogeneous plasmas and show that only stable waves can occur.
- Insert plane wave solutions in the spectral equation for a homogeneous plasma and derive the algebraic eigenvalue problem. Discuss the different solutions.

[6.5]* *The resolvent operator*

The spectrum of a linear operator L is obtained from the study of the inhomogeneous equation $(L - \lambda)x = a$, where a is a given element of the function space exploited, λ is one of the (complex) eigenvalues of the operator L, and x is the unknown solution.

- Invert this expression. Discuss the possibilities for solutions in dependence of λ.
- Convert the equation of motion, with the force operator $\mathbf{F}$, by means of the Laplace transform $\hat{\boldsymbol{\xi}} \equiv \int_0^\infty \boldsymbol{\xi} e^{i\omega t} dt$. Show that the resulting equation is of the above type.
- Invert this expression to find the solution for the Laplace transformed variable $\hat{\boldsymbol{\xi}}$.
- Find the formal solution for $\boldsymbol{\xi}$ by performing the inverse Laplace transformation.

[6.6] *The energy principle*

The energy for linear plasma perturbations is given by Eq. (6.84) of Section 6.4.1.

- What is the criterion for stability? Ignoring magnetic fields and gravity, is an ordinary fluid always stable?
- Specify to plane slab geometry, where the fluid is infinitely extended in the horizontal directions and the variables are functions of height only. With limited height of the fluid, and still ignoring magnetic fields but including gravity, impose the equilibrium equation and derive the expression for W. Construct a necessary stability criterion by means of incompressible trial functions and explain what it means.
- Rearrange the equation for W and obtain the Rayleigh–Taylor stability criterion by means of a complete minimization.

[6.7]* *Normal mode analysis for the Rayleigh–Taylor instability*

Work out the normal mode analysis for the Rayleigh–Taylor instability of incompressible interface plasmas by means of the equation of motion. Can you figure out where the Alfvén and slow magneto-sonic waves reside? Of course, the final expression for the growth rate should agree with the expression derived in Section 6.6.4 by means of the variational analysis. What is the role of the Lagrangian pressure perturbation?

[6.8]* *Stability criteria for plasma–vacuum configurations*

Derive stability criteria for general plasma–vacuum configurations with $\beta = 1$ and a curved interface. Discuss the qualitative difference between convex and concave curvature of the outer magnetic field. (Hint: consult Kadomtsev [338], p. 162.)

Part III

Standard Model Applications

7

Waves and instabilities of inhomogeneous plasmas

7.1 Hydrodynamics of the solar interior

We have studied the MHD waves for homogeneous plasmas in Chapter 5. This theory was transformed in Chapter 6 to the level of spectral theory in order to facilitate the much more complicated analysis of inhomogeneous plasmas, which we want to undertake in the present chapter. Plasma inhomogeneity is not just a complication in the analysis, but also provides qualitatively new physical phenomena like wave damping, wave transformation and, most important of all, a very wide class of global MHD instabilities of magnetically confined plasmas.

Explicit examples of inhomogeneous plasma dynamics abound in the solar system (see Chapter 8). For the Sun, a number of important phenomena may be described neglecting the magnetic field. Therefore, we will first simplify the MHD model to a purely hydrodynamic one and study the effects of sound and gravity separate from the three MHD waves. Since the hydrodynamic waves are clearly identified in solar observations, we will highlight the potential of observing MHD wave propagation for the investigation of astrophysical objects (Section 7.2.4).

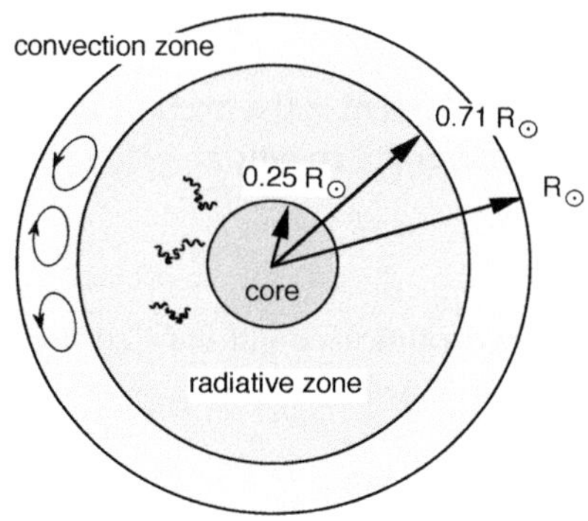

Fig. 7.1 Internal structure of the Sun: thermonuclear energy production in the core, radiation transport in the radiation zone and convection in the outermost layer.

We summarize some basic facts of *the standard solar model* (see Priest [510], Stix [569] or Foukal [181]). The Sun (Fig. 7.1) is a sphere of hot material, mainly plasma, of radius $R_\odot = 7.0 \times 10^8\,\mathrm{m}$ and mass $M_\odot = 2.0 \times 10^{30}\,\mathrm{kg}$. The total radiation power output of the Sun, *the solar luminosity*, is $L_\odot = 3.86 \times 10^{26}\,\mathrm{W}$. At the Earth, i.e. at a distance of $1\,\mathrm{AU}$ (= $1.5 \times 10^{11}\,\mathrm{m}$), this yields a heat flux of $L_\odot/[4\pi \times (1\,\mathrm{AU})^2] = 1.36\,\mathrm{kW\,m^{-2}}$: *the solar constant* (roughly corresponding to a crude estimate one might make based on sunbathing experience). Through the work of H. A. Bethe and C. F. von Weizsäcker (1939) we know that this enormous amount of energy is produced by *thermonuclear reactions*.

The thermonuclear energy source is located in *the core* ($r \leq 0.25R_\odot$) of the Sun, where the temperature $T_0 = 1.6 \times 10^7\,\mathrm{K}$ and the density $\rho_0 = 1.6 \times 10^5\,\mathrm{kg\,m^{-3}}$. This is sufficient for *the* p–p *fusion reaction chain* to occur, as described in Section 1.2.1. The energy produced in the form of gamma radiation is transported through *the radiative zone* ($0.25R_\odot \leq r \leq 0.713R_\odot$) to the outer layers. On average, this process takes millions of years per photon and, during this long trip, the wavelength gradually shifts to that of UV and visible light. The radiative transport is exceeded by convection in the outermost layer of the Sun, *the convection zone* ($0.713R_\odot \leq r \leq R_\odot$), as discussed in Section 7.1.2.

7.1.1 Radiative equilibrium model

The description of radiation transport of the thermonuclear power produced in the core of the Sun to the outer layers involves a *radiative equilibrium model* predicting the static equilibrium distributions of the internal variables of the Sun. We start from the *time-dependent* hydrodynamic equations, extending the MHD equations (4.12)–(4.14) with internal energy production and radiation transport, but neglecting the magnetic field. The equation of mass conservation (4.12) is unchanged:

$$\frac{\partial \rho}{\partial t} + \nabla \cdot (\rho \mathbf{v}) = 0\,. \tag{7.1}$$

In the momentum conservation equation (4.13), the gravitational attraction is specified to be spherically symmetric:

$$\rho \frac{\mathrm{D}\mathbf{v}}{\mathrm{D}t} + \nabla p - \rho \mathbf{g} = 0\,, \qquad \mathbf{g} = -g\mathbf{e}_r\,, \quad g = g(r) = G\,\frac{M(r)}{r^2}\,. \tag{7.2}$$

Here, G is the gravitational constant and $M(r)$ is the mass inside a sphere of radius r, related to the density ρ by means of the obvious differential equation

$$\frac{dM}{dr} = 4\pi r^2 \rho\,. \tag{7.3}$$

It is expedient to replace the pressure evolution equation (4.14) by the generalization of the internal energy equation (4.19) given by Eq. (4.136):

$$\rho \frac{\mathrm{D}e}{\mathrm{D}t} + p \nabla \cdot \mathbf{v} = \nabla \cdot [\lambda \nabla (kT)] + \rho \varepsilon\,, \qquad e \equiv \frac{1}{\gamma - 1}\,\frac{p}{\rho}\,. \tag{7.4}$$

The right hand side now contains two terms due to *radiative transport*, governed by *the thermal conduction coefficient*, $\lambda(r)$ (previously indicated by the symbol κ), and *the thermonuclear energy production per unit mass*, $\varepsilon(r)$. The latter is strongly dependent on the temperature and the concentration of hydrogen.

The radiative model actually requires more than just one fluid since thermonuclear energy production and thermal conduction depend on the concentration of the different particle species. This will be approximated by a modified equation of state taking the different masses of the particles into account. For a plasma consisting of electrons and one kind of ions with equal temperatures

($T \equiv T_e = T_i$), the expression for the pressure may be converted by means of the charge neutrality relation, $n_e = Z_c n_i$, and the definition for the density, $\rho \equiv n_e m_e + n_i m_i$:

$$p = (n_e + n_i)\,kT = \frac{1 + Z_c}{m_i(1 + Z_c m_e/m_i)}\,\rho kT \approx \frac{1 + Z_c}{A}\frac{\rho kT}{m_p}\,. \tag{7.5}$$

Here, A is the mass number (in multiples of the proton mass, $m_i = Am_p$) and Z_c is the charge number of the ions (in multiples of e). The latter is subscripted with c since the bare symbol Z is used for another quantity, introduced below.

So far, the equation of state (7.5) is just the usual one for a plasma consisting of electrons and one kind of ions, so that Z_c and A are constants. Now, we modify the equation of state by permitting different ion species that are radially distributed according to a function called the *mean molecular weight* μ, so that

$$p = \frac{1}{\mu}\frac{\rho kT}{m_p}\,, \quad \text{where } \mu \equiv \frac{A}{1 + Z_c} = \mu(r)\,. \tag{7.6}$$

The mean molecular weight μ measures the average weight of the constituent particles in units of m_p: $\mu = 1/2$ for ionized hydrogen ($Z_c = A = 1$), $\mu = 4/3$ for ionized helium ($Z_c = 2$, $A = 4$), and $\mu \approx 2$ for the completely ionized heavier atoms ($Z_c \gg 1$, $A \approx 2Z_c$). For a plasma with a mixture of these ions, $\mu \approx (2X + \frac{3}{4}Y + \frac{1}{2}Z)^{-1}$, where X, Y and Z indicate the mass fractions of hydrogen, helium, and the heavier elements, respectively. Hence, the radial function $\mu = \mu[\rho(r), T(r)]$ reflects the abundance of the various elements and their degrees of ionization, determined by Saha equations (Section 1.4.1). The dependence of p on μ implies that pressure decreases when hydrogen is converted into helium.

To construct a *static* equilibrium model of the Sun, set $\mathbf{v} = 0$ and $\partial/\partial t = 0$. The mass conservation equation (7.1) is then trivially satisfied, whereas Eqs. (7.2) and (7.3) provide the equations for *hydrostatic equilibrium:*

$$\frac{dp}{dr} = -\rho g = -G\frac{\rho M}{r^2} \quad \Rightarrow \quad \frac{1}{r^2}\frac{d}{dr}\left(\frac{r^2}{\rho}\frac{dp}{dr}\right) = -\frac{G}{r^2}\frac{dM}{dr} = -4\pi G\rho\,. \tag{7.7}$$

In static equilibrium, the LHS of the internal energy equation (7.4) vanishes so that

$$\frac{1}{r^2}\frac{d}{dr}\left[r^2\lambda\frac{d(kT)}{dr}\right] = -\rho\varepsilon\,, \tag{7.8}$$

expressing *radiative equilibrium*: the thermonuclear energy produced is transported outward by radiative transfer. The set of differential equations (7.7)–(7.8) is complete when supplemented with the equation of state (7.6) and equations determining the functions $\mu(r)$, $\lambda(r)$ and $\varepsilon(r)$.

The hydrostatic model is now complete, but it is useful to relate the differential equation (7.8) for the temperature with the fundamental quantities of radiative transport, viz. luminosity, radiation pressure and opacity. Radiative equilibrium implies that the local value of *the luminosity*, $L(r)$, which is the net radiation flux per unit of time through a spherical surface, is given by

$$L(r) = \int_0^r \varepsilon(r')\,dM = 4\pi\int_0^r \rho\,\varepsilon r'^2\,dr' = -\tfrac{16}{3}(\pi\sigma)\,\frac{r^2}{\kappa\rho}\frac{dT^4}{dr}\,. \tag{7.9}$$

The last equality is obtained by exploiting *the radiation pressure* $q = \frac{4}{3}(\sigma/c)T^4$, with the Stefan–Boltzmann constant $\sigma \equiv 2\pi^5 k^4/(15c^2h^3) = 5.67 \times 10^{-8}\,\mathrm{W\,m^{-2}\,K^{-4}}$, and the relation between

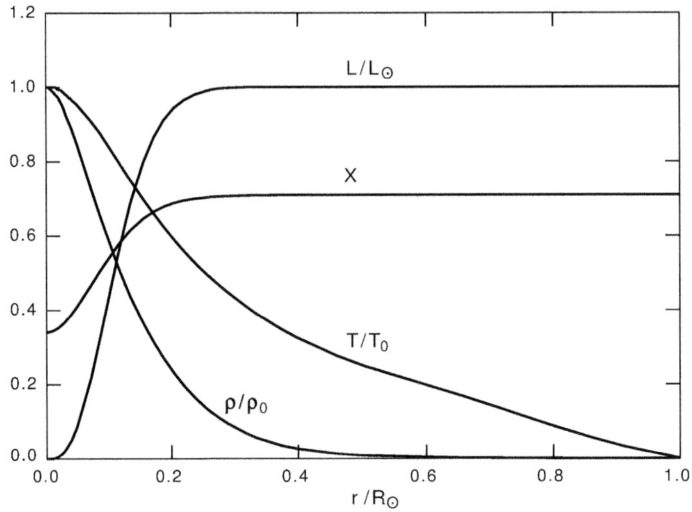

Fig. 7.2 Radial distributions of the physical quantities for a standard solar model. (After Foukal [181], with data from Bahcall and Ulrich, *Rev. Mod. Phys.* **60**, 297 (1988).)

the gradient of the radiation pressure and the absorbed radiation, $dq/dr = -\frac{1}{4}(\pi c)^{-1}(\kappa\rho L/r^2)$, where κ is *the opacity* (the absorption coefficient of the radiation per unit mass and per unit length). Of course, this differential equation for the temperature should be compatible with Eq. (7.8), so that the thermal conductivity λ is inversely proportional to the opacity κ:

$$\lambda = \tfrac{16}{3}\,(\sigma/k)\,\frac{T^3}{\kappa\rho}\,. \tag{7.10}$$

The opacity is a microscopic quantity determined by quite a number of different scattering processes between photons and atoms, ions or electrons, averaged over frequency. To compute radiative equilibrium, one considers $\kappa = \kappa(\rho, T)$ to be a known function, obtained in the form of a table from the extensive numerical calculations that have been developed for this purpose [569].

Finally, we need an explicit relation for the thermonuclear energy production rate by proton–proton reactions, as given, e.g., by Zirin [669]:

$$\varepsilon = \varepsilon(X, \rho, T) = 0.25\,\rho X^2 (10^6/T)^{2/3} \exp\Big[- 33.8\,(10^6/T)^{1/3} \Big]\,, \tag{7.11}$$

where the dimensions of the quantities are given by $[\rho] = \mathrm{kg\,m^{-3}}$, $[T] = \mathrm{K}$ and $[\varepsilon] = \mathrm{J\,kg^{-1}\,s^{-1}}$. It is clear that the function $\varepsilon(X, \rho, T)$ is strongly concentrated in the central regions of the Sun where the temperature is high: thermonuclear burn only occurs in *the core* of the Sun ($r \leq 0.25 R_\odot$).

The standard solar model consists of the two second order differential equations (7.7) and (7.8), supplemented with the equation of state (7.6) and numerically obtained values of the functions $\mu(\rho, T)$, $\kappa(\rho, T)$ and $\varepsilon(X(\rho, T), \rho, T)$. For the latter quantity one should exploit Eq. (7.11), which evidently spoils the possibility of obtaining explicit analytic solutions. Since two second order ODEs are equivalent to four first order ODEs, one may choose, for example, to solve for the following

ODEs for the four quantities $M(r), L(r), p(r)$ and $T(r)$:

$$\begin{aligned} \frac{dp}{dr} &= -G\,\frac{\rho M}{r^2}\,, \quad \rho = (m_{\rm p}/k)\,\frac{\mu p}{T}\,, & \frac{dM}{dr} &= 4\pi\, r^2 \rho \\ \frac{dT}{dr} &= -\tfrac{3}{64}\,(\pi\sigma)^{-1}\,\frac{\kappa\rho L}{r^2 T^3}\,, & \frac{dL}{dr} &= 4\pi\, r^2 \rho\varepsilon\,. \end{aligned} \tag{7.12}$$

Appropriate boundary conditions for these equations are

$$p(R_\odot) \approx 0\,, \quad M(0) = 0\,, \quad T(R_\odot) \approx 0\,, \quad L(0) = 0\,. \tag{7.13}$$

Of course, the solution obtained (see Fig. 7.2) should reproduce the known value of the solar mass, $M(R_\odot) = M_\odot$, and of the solar luminosity, $L(R_\odot) = L_\odot$, at the known position $r = R_\odot$ of the solar radius. Also, the values p_0 and T_0 should correspond to the numbers cited for the central density and temperature to yield thermonuclear burn. This procedure turns out to produce quite realistic solutions, except for the assumption of hydrostatic equilibrium *up to the solar surface*.

7.1.2 Convection zone

In the outer layers of the Sun, cooling is so strong that the absolute value of the temperature gradient exceeds a certain threshold given by the Schwarzschild criterion for convective stability. Since the temperature gradient is negative, that criterion is best discussed in terms of the quantity $-dT/dr$. Convective stability is achieved when $-dT/dr$ does not exceed a critical value, which is obtained as follows.

Introducing the mean particle mass, $m \equiv \mu m_{\rm p}$, the ratio of the Boltzmann constant k and m may be abbreviated as a kind of gas constant, $\mathcal{R} \equiv k/m$. Anticipating the outcome of the present section, $\mathcal{R}$ is actually constant in the convection zone because of the strong mixing that occurs, so that $\mu \approx \text{const}$ there. Consequently, the equation of state (7.6) may be written as

$$p = \mathcal{R}\rho T\,. \tag{7.14}$$

Hence, the temperature gradient is related to the density and pressure gradients:

$$-\frac{dT}{dr} = \frac{1}{\mathcal{R}}\left(\frac{p}{\rho^2}\frac{d\rho}{dr} - \frac{1}{\rho}\frac{dp}{dr}\right) \qquad \left[= \frac{1}{\mathcal{R}}\left(\frac{p}{\rho^2}\frac{d\rho}{dr} + g\right)\right], \tag{7.15}$$

where the equilibrium relation (7.7), $dp/dr = -\rho g$, has been inserted in the rightmost equality. Here, and below, we put square brackets around such equalities to indicate that this additional equilibrium information has been exploited.

The critical value for convective stability is obtained from the consideration that motions of the gas are neutrally stable only if the fluid is *isentropic*, i.e. if the value of the specific entropy is constant throughout the pertinent region of space:

$$S \equiv p\rho^{-\gamma} = \text{const} \quad \Rightarrow \quad \frac{d\rho}{dr} = \frac{\rho}{\gamma p}\frac{dp}{dr} \qquad \left[= -\frac{\rho^2 g}{\gamma p}\right], \tag{7.16}$$

so that

$$-\left(\frac{dT}{dr}\right)_{\rm isentr.} = -\frac{1}{\mathcal{R}}\,\frac{\gamma-1}{\gamma}\,\frac{1}{\rho}\frac{dp}{dr} \qquad \left[= \frac{1}{\mathcal{R}}\,\frac{\gamma-1}{\gamma}\,g\right]. \tag{7.17}$$

Convective instability occurs when the actual temperature gradient $-dT/dr$ exceeds this value, i.e. when *the Schwarzschild criterion for convective stability*,

$$-\frac{dT}{dr} \leq -\left(\frac{dT}{dr}\right)_{\rm isentr.} \quad \Rightarrow \quad \frac{d\rho}{dr} - \frac{\rho}{\gamma p}\frac{dp}{dr} \leq 0 \qquad \left[\Rightarrow \; \frac{d\rho}{dr} g + \frac{\rho^2 g^2}{\gamma p} \leq 0\right], \tag{7.18}$$

is violated. Interestingly enough, the criterion for gravitational stability (in square brackets) of Section 6.5.4 is recovered: convective and gravitational (Rayleigh–Taylor) instabilities amount to the same.

We here exploit the term 'isentropic' rather than the more usual term 'adiabatic' to stress that the background equilibrium state is not isentropic in general but that the assumption of adiabatic fluid *motion*, i.e. $DS/Dt = 0$, may still be made. Recall that this assumption is underlying the whole analysis of Chapter 6, including the derivation of the gravitational stability in Section 6.5.4. Hence, adiabatic motions are stable or unstable according to whether the condition (7.18) is satisfied or violated. Only at marginal stability do the two concepts coincide: the equilibrium is then characterized by constant entropy in space and marginal motion ($\omega = 0$) does not change this distribution.

The Schwarzschild criterion (7.18) is violated in the region $0.713R_\odot \leq r \leq R_\odot$ of the Sun, for that reason called *the convective region*. In this region turbulent mixing occurs, i.e. eddies of hot material are carried up and eddies of cool material are carried down so as to maintain the temperature gradient at the value given by the RHS of the first expression (7.18). Consequently, rather than applying the boundary conditions (7.13) at the solar surface ($r = R_\odot$), one obtains a more realistic solar model by separating the interior into the regions $r \leq 0.713R_\odot$ (core and radiative zone), bounded by a surface where the boundary condition

$$-\frac{dT}{dr} = -\left(\frac{dT}{dr}\right)_{\rm isentr.} \qquad (\text{at } r = 0.713R_\odot) \tag{7.19}$$

is imposed, and the convective region which is governed by a completely different set of equations. In particular, the assumption of static equilibrium is not appropriate for the convective region so that one should return to the time-dependent equations (7.1)–(7.6), where the flow also destroys the spherical symmetry of the problem. Unfortunately, there is little one can do about solving this problem without recourse to a large computer and, even then, there remains a large uncertainty about the nature of the solutions in the convection zone.

Ironically, there appears to be more certainty about the structure of the interior of the Sun, which is completely opaque (recall that the mean free path of photons is only a few centimeters there), than the outer regions. Also, the magnetic field, which we have ignored until now, turns out to play an important role in the outer layers of the Sun. In particular, the origin of the solar magnetic field, i.e. *the solar dynamo*, is thought to be situated at the bottom of the convection zone. The dynamo problem involves contributions of differential rotation, spherical geometry and magnetic fields. This significantly adds to the complexity of the problem since magnetic fields cannot be fitted in a spherically symmetric geometry. This fact is very evident from the huge number of complex magnetic structures that can be observed in the atmosphere of the Sun, see Chapter 8.

In conclusion, it appears hopeless to try to find solutions for the interior of the Sun, by means of the theoretical approach discussed so far, starting from such a complicated boundary. Fortunately, a powerful observational and analytical tool is available that can probe the deep structure of the Sun, viz. *helioseismology*, which will be treated in the following section.

7.2 Hydrodynamic waves and instabilities of a gravitating slab

The convective layer gets its name from the convective instability arising when the Schwarzschild criterion (7.18) is violated. Since this is associated with the transition from stability to instability of gravity waves, it is instructive to consider the more general subject of solar oscillations. Here, we will consider the Sun as a whole and, again, neglect the presence of magnetic fields. The symmetry of the problem would involve the treatment of wave propagation in spherically inhomogeneous systems (background quantities depending on the radial coordinate r), leading to the occurrence of spherical harmonics in the analysis. However, we will avoid these technical complications and just consider *the gravito-acoustic waves in a planar stratification* (background quantities depending on the vertical coordinate x). Obviously, the expressions derived require a generalization to spherical stratification if one wishes to compare calculated frequencies with observed ones. Our goal is to demonstrate how spectral analysis relates to practical applications, not to treat the subject of solar and stellar oscillations in detail. (The interested reader should consult the texts by Aerts et al. [6] or Unno et al. [616] on that.)

7.2.1 Hydrodynamic wave equation

Again, we exploit the equations of gas dynamics. We now keep the adiabatic law so that the relevant equations become

$$\frac{\mathrm{D}\rho}{\mathrm{D}t} + \rho\nabla\cdot\mathbf{v} = 0\,, \qquad \rho\frac{\mathrm{D}\mathbf{v}}{\mathrm{D}t} + \nabla p - \rho\mathbf{g} = 0\,, \qquad \frac{\mathrm{D}p}{\mathrm{D}t} + \gamma p\nabla\cdot\mathbf{v} = 0\,. \tag{7.20}$$

Note that this implies that, for the study of waves, we neglect the RHS of Eq. (7.4), i.e. *non-adiabatic effects* caused by thermal conduction and thermonuclear energy production. Also, for simplicity, we consider the gravitational acceleration as a given constant, i.e. we neglect perturbations of the gravitational potential, which is called the *Cowling approximation:*

$$\mathbf{g} = (-g, 0, 0)\,, \qquad g = \mathrm{const}\,.$$

We will again assume a static equilibrium, i.e. $\mathbf{v} = 0$ and $\partial/\partial t = 0$, so that

$$\nabla p_0 = \rho_0\mathbf{g} \quad \Rightarrow \quad p_0'(x) = -\rho_0(x) g\,, \tag{7.21}$$

where the prime indicates differentiation with respect to x.

Linearization leads to the following equations:

$$\frac{\partial\rho_1}{\partial t} + \mathbf{v}_1\cdot\nabla\rho_0 + \rho_0\nabla\cdot\mathbf{v}_1 = 0\,, \tag{7.22}$$

$$\rho_0\frac{\partial\mathbf{v}_1}{\partial t} + \nabla p_1 - \rho_1\mathbf{g} = 0\,, \tag{7.23}$$

$$\frac{\partial p_1}{\partial t} + \mathbf{v}_1\cdot\nabla p_0 + \gamma p_0\nabla\cdot\mathbf{v}_1 = 0\,. \tag{7.24}$$

These equations are straightforward generalizations of Eqs. (5.5)–(5.7) with additional terms due to gravity and the associated inhomogeneity. Hence, we could derive the wave equation by differentiating Eq. (7.23) with respect to t and inserting the expressions for $\partial\rho_1/\partial t$ and $\partial p_1/\partial t$ from

Eqs. (7.22) and (7.24). However, we rather integrate by means of the displacement vector $\boldsymbol{\xi}$ (Section 6.1.2), where $\mathbf{v}_1 = \partial\boldsymbol{\xi}/\partial t$, so that $\rho_1 = -\nabla\cdot(\rho_0\boldsymbol{\xi})$ and $p_1 = -\boldsymbol{\xi}\cdot\nabla p_0 - \gamma p_0\nabla\cdot\boldsymbol{\xi}$. Inserting these expressions into Eq. (7.23) and exploiting the equilibrium relation (7.21), yields *the wave equation for gravito-acoustic waves in a plane stratified medium:*

$$\rho\frac{\partial^2\boldsymbol{\xi}}{\partial t^2} - \nabla(\gamma p\nabla\cdot\boldsymbol{\xi}) - \rho\,\nabla(\mathbf{g}\cdot\boldsymbol{\xi}) + \rho\mathbf{g}\,\nabla\cdot\boldsymbol{\xi} = 0\,. \tag{7.25}$$

We have omitted the subscripts 0 and 1 for the equilibrium and perturbed quantities since there can be no confusion any more: perturbed quantities are expressed in $\boldsymbol{\xi}$, all other quantities refer to the background equilibrium.

Considering normal mode solutions of the form

$$\boldsymbol{\xi}(\mathbf{r},t) = \hat{\boldsymbol{\xi}}(x)\,e^{i(k_y y + k_z z - \omega t)}\,, \tag{7.26}$$

Eq. (7.25) may be written in matrix form as

$$\begin{pmatrix} \frac{d}{dx}\gamma p\frac{d}{dx} + \rho\omega^2 & ik_y\Big(\frac{d}{dx}\gamma p + \rho g\Big) & ik_z\Big(\frac{d}{dx}\gamma p + \rho g\Big) \\ ik_y\Big(\gamma p\frac{d}{dx} - \rho g\Big) & -k_y^2\gamma p + \rho\omega^2 & -k_y k_z\gamma p \\ ik_z\Big(\gamma p\frac{d}{dx} - \rho g\Big) & -k_y k_z\gamma p & -k_z^2\gamma p + \rho\omega^2 \end{pmatrix}\begin{pmatrix}\hat{\xi}_x\\ \hat{\xi}_y \\ \hat{\xi}_z\end{pmatrix} = 0\,. \tag{7.27}$$

Since there is no preferred horizontal direction in the system, there is no loss in generality if we rotate the coordinate system such that $k_z = 0$, so that $\rho\omega^2\hat{\xi}_z = 0$. Hence, ignoring marginal shifts ($\omega^2 = 0$, $\hat{\xi}_z \neq 0$), all terms with k_z and $\hat{\xi}_z$ may be neglected and $\hat{\xi}_y$ may be eliminated:

$$\hat{\xi}_y = -ik_0\frac{\gamma p\hat{\xi}_x' - \rho g\hat{\xi}_x}{\rho\omega^2 - k_0^2\gamma p}\,, \tag{7.28}$$

where the horizontal wave number $k_0 \equiv \sqrt{k_y^2 + k_z^2} = k_y$ now. Substitution into the first component of Eq. (7.27) gives a *second order differential equation* for $\hat{\xi}_x$:

$$\frac{d}{dx}\bigg(\frac{\gamma p\,\rho\omega^2}{\rho\omega^2 - k_0^2\gamma p}\,\frac{d\hat{\xi}_x}{dx}\bigg) + \bigg[\rho\omega^2 - \frac{k_0^2\rho^2 g^2}{\rho\omega^2 - k_0^2\gamma p} + \rho' g - \bigg(\rho g\frac{\rho\omega^2}{\rho\omega^2 - k_0^2\gamma p}\bigg)'\bigg]\hat{\xi}_x = 0\,. \tag{7.29}$$

To solve this differential equation one should specify the profiles $p(x)$ and $\rho(x)$, satisfying the equilibrium condition (7.21), and the boundary conditions.

Formulation of the appropriate boundary conditions obviously would involve the spherical geometry of the original problem. Since we have ignored that part of the problem, the next logical step is to consider modes that do not sensitively depend on the choice of the boundary. Such modes are obtained by the assumption that the boundaries are far away. In other words: we should study modes of finite extent in the y- and z-directions (either arbitrary wave numbers k_y and k_z, or quantized if one wishes to consider the y- and z-directions as periodic), but *sufficiently localized in the direction of inhomogeneity*. Hence, imposing rigid wall boundary conditions,

$$\hat{\xi}_x(x{=}0) = \hat{\xi}_x(x{=}a) = 0\,, \tag{7.30}$$

where $x{=}0$ corresponds to the centre and $x{=}a$ to the surface of the Sun, only makes sense if a is large compared to typical 'wavelengths' in the x-direction. For example, we might consider modes which rapidly oscillate in that direction:

$$\hat{\xi}_x(x) \sim \mathrm{e}^{\mathrm{i}qx} \quad \text{with} \quad qa \gg 1\,. \tag{7.31}$$

Of course, there are no walls on the Sun, and the boundary conditions (7.30) just serve to define a region of localization of the modes. In more general terms, one would consider so-called *cavity modes*, where the background pressure and density distributions create a kind of potential well in which the modes are confined.

The differential equations (7.29) with the boundary conditions (7.30) permit the study of the instabilities ($\omega^2 < 0$) and stable waves ($\omega^2 > 0$) which originate from the peculiar interaction of the sound and gravity terms. For the solar context, these two cases are exemplified by the convective instabilities of the convection zone and the solar acoustic-gravity waves (the p- and g-modes), respectively. We treat these two examples in the following Sections 7.2.2 and 7.2.3.

7.2.2 Convective instabilities

Since time scales of gravitational instabilities are much longer than time scales of acoustic oscillations, we may assume $\rho\,|\omega^2| \ll k_0^2\gamma p$ for the study of these instabilities. The wave equation (7.29) then simplifies to

$$\frac{d}{dx}\left(\frac{\rho\omega^2}{k_0^2}\frac{d\hat{\xi}_x}{dx}\right) - \rho\Big(\omega^2 - N_{\rm B}^2\Big)\,\hat{\xi}_x = 0\,, \tag{7.32}$$

where $N_{\rm B}$ is *the Brunt–Väisäläa frequency*:

$$N_{\rm B}^2 = N_{\rm B}^2(x) \equiv -g\left(\frac{1}{\rho}\frac{d\rho}{dx} - \frac{1}{\gamma p}\frac{dp}{dx}\right) \qquad \left[= -\frac{1}{\rho}\Big(\rho' g + \frac{\rho^2 g^2}{\gamma p}\Big)\right]. \tag{7.33}$$

Since ρ and $N_{\rm B}^2$ depend on x through $\rho(x)$ and $p(x)$, of which at least one function is completely arbitrary, in general, the solution of the differential equation (7.32) can only be obtained by numerical integration. However, assuming rapid spatially oscillatory modes of the kind expressed by Eq. (7.31) with $q \gg |\nabla| \sim 1/a$, where $|\nabla|$ represents a typical gradient scale length of the equilibrium, one obtains the following estimate for the eigenfrequencies of the local instabilities:

$$\omega^2 \approx \left[k_0^2/(k_0^2 + q^2)\right] N_{\rm B}^2(x)\,. \tag{7.34}$$

Hence, the system is locally unstable in the range of x where $N_{\rm B}^2 < 0$. Notice that the bracket in the definition (7.33) for $N_{\rm B}^2$ is just the expression entering the gravitational stability criterion (6.117) derived in Section 6.5.4 by applying the energy principle. We now also have information on the growth rates of this instability.

We have already seen in Section 7.1.2 that the Schwarzschild criterion (7.18) for convective stability and the gravitational stability criterion coincide when the equilibrium conditions are taken into account. This is now immediately evident from the connection with the Brunt–Väisäläa frequency:

$$\frac{dT}{dx} - \left(\frac{dT}{dx}\right)_{\rm isentr.} = \frac{T}{g}\,N_{\rm B}^2 \geq 0 \qquad \left[\Rightarrow\ -\frac{T}{\rho g}\left(\rho' g + \frac{\rho^2 g^2}{\gamma p}\right) \geq 0\right]. \tag{7.35}$$

Hence, $N_{\rm B}^2 \geq 0$ is nothing else than the Schwarzschild criterion.

If the Schwarzschild criterion is violated, i.e. $N_B^2 < 0$, Eq. (7.34) tells us that we should expect convective instabilities exponentially growing as $\exp(\sqrt{-\omega^2}\, t)$. Such modes rapidly enter the nonlinear domain, where the assumptions underlying the given derivation break down. This is associated with the formation of convective cells, which may be nonlinearly stable. However, the onset of this phenomenon is correctly predicted by the linear theory. On the other hand, if $N_B^2 > 0$, Eq. (7.34) provides us with an estimate of the frequency of slow gravitational modes (the g-modes). These will be treated in the next section, where the assumption of smallness of ω^2 made in this section will be dropped.

7.2.3 Gravito-acoustic waves

An instructive special case of gravito-acoustic waves in an inhomogeneous medium is obtained for an *exponentially stratified medium with constant sound speed*:

$$\rho = \rho_0 e^{-\alpha x}, \quad p = p_0 e^{-\alpha x}, \quad \text{so that} \quad c^2 = \frac{\gamma p}{\rho} = \frac{\gamma p_0}{\rho_0} = \text{const}. \tag{7.36}$$

From the equilibrium equation (7.21) this implies that $p' = -\alpha p = -\rho g$ (where we recall that g is also assumed constant), so that

$$\alpha = \frac{\rho g}{p} = \frac{\rho_0 g}{p_0} = \frac{\gamma g}{c^2} = \text{const}. \tag{7.37}$$

The expression (7.33) for the squared Brunt–Väisäläa frequency simplifies to

$$N_B^2 = \alpha g - g^2/c^2 = (\gamma - 1) g^2/c^2 > 0, \tag{7.38}$$

so that we will only find stable waves. Equation (7.29) then transforms to

$$\frac{d^2\hat{\xi}_x}{dx^2} - \alpha \frac{d\hat{\xi}_x}{dx} + \frac{\omega^4 - k_0^2 c^2 \omega^2 + k_0^2 c^2 N_B^2}{c^2 \omega^2} \hat{\xi}_x = 0. \tag{7.39}$$

This is a differential equation with constant coefficients: the solution is trivial. Of course, this is *not at all typical* for the problem of wave propagation in inhomogeneous media. One should be aware of the limitations intrinsic to the use of special cases that are analytically solvable. For example, here we have lost the important possibility of convective instability, discussed in the previous section, simply because $N_B^2 > 0$ due to the assumption of an exponential medium. In general, for arbitrary pressure and density profiles, the differential equation to be solved is just a general ODE, where numerical solution is the appropriate procedure.

The solutions of Eq. (7.39) are exponentials of the form $\hat{\xi}_x = C \exp[(\frac{1}{2}\alpha \pm iq)x]$, with

$$q \equiv \sqrt{-\tfrac{1}{4}\alpha^2 + \omega^2/c^2 - k_0^2 + k_0^2 N_B^2/\omega^2}\,. \tag{7.40}$$

The expression under the square root sign has to be positive in order to obtain spatially oscillatory solutions that can satisfy the boundary conditions (7.30). The latter conditions imply that q has to be quantized:

$$qa = n\pi \quad (n = 1, 2, \dots). \tag{7.41}$$

The *dispersion equation of the gravito-acoustic waves* is then obtained by inverting the expression (7.40) for q:

$$\omega^4 - (k_0^2 + q^2 + \tfrac{1}{4}\alpha^2) c^2 \omega^2 + k_0^2 c^2 N_B^2 = 0, \tag{7.42}$$

having the solutions

$$\omega_{\rm p,g}^2 = \tfrac{1}{2}k_{\rm eff}^2 c^2 \Big[1 \pm \sqrt{1 - 4k_0^2 N_{\rm B}^2/(k_{\rm eff}^4 c^2)}\,\Big]\,, \quad k_{\rm eff}^2 \equiv k_0^2 + q^2 + \tfrac{1}{4}\alpha^2\,, \tag{7.43}$$

where $k_{\rm eff}$ is the effective total 'wave number' (quotation marks since the vertical dependence of the perturbations does not correspond to a plane wave) and k_0 is the horizontal wave number. The branch with the $+$ sign refers to *the acoustic waves* or *p-modes* (pressure driven), and the branch with the $-$ sign refers to the *gravity waves* or *g-modes* (gravity driven).

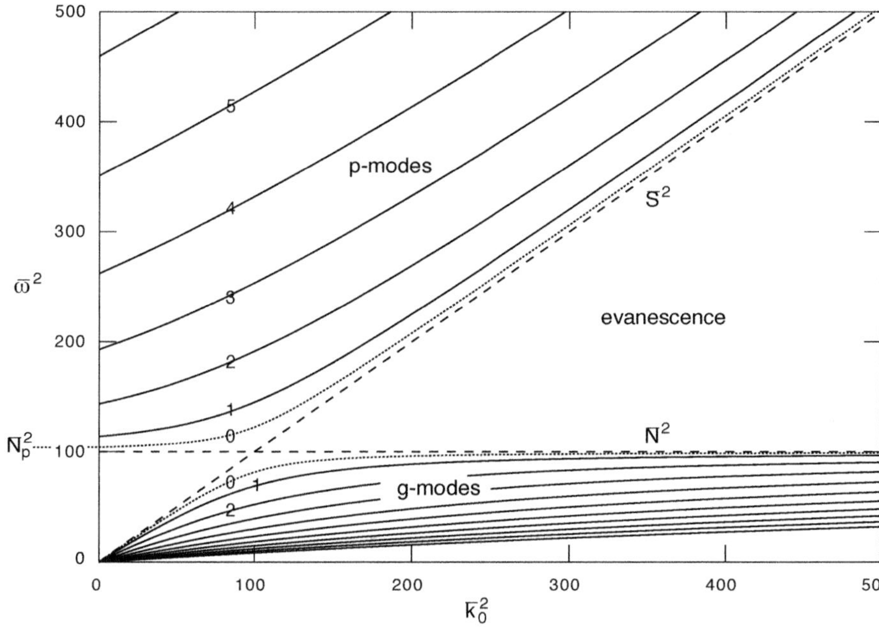

Fig. 7.3 Dispersion diagram for the p- and g-modes of a plane, exponential, atmosphere. The squared normalized frequencies $\bar\omega^2 = \bar\omega^2(\bar k_0^2, \bar q^2)$ are plotted for ten values of the vertical 'wave number' $\bar q = n\pi$ $(n = 1, 2, \ldots 10)$; $\bar N_{\rm B} = 10$, $\bar N_{\rm p} = \frac{1}{2}\bar\alpha = 10.21$. Dotted curves represent the turning point frequencies $(n = 0)$, dashed curves are the asymptotic values for $\bar k_0 \to \infty$.

In Fig. 7.3 we have drawn the curves for $\omega^2 = \omega^2(k_0^2)$ for different values of the vertical 'wave number' q. The curves for $q^2 = 0$ (dotted), which represent the so-called *turning point* frequencies $\omega_{\rm p0}^2$ and $\omega_{\rm g0}^2$, mark the boundaries between propagation (q real) and evanescence (q imaginary) of the waves. For increasing values of q^2, the eigenfrequencies ω^2 of the p-modes monotonically increase with an *accumulation* or *cluster point* at ∞: $\omega_{\rm p}^2 \to \omega_{\rm P}^2 = \infty$. This monotonicity of the eigenfrequency with the number of nodes of the eigenfunctions, so-called *Sturmian* behaviour, is a well-known property of sound waves and musical instruments. On the other hand, the eigenfrequencies of the g-modes monotonically decrease (*anti-Sturmian* behaviour) with an accumulation point at 0: $\omega_{\rm g}^2 \to \omega_{\rm G}^2 = 0$. The lowest branch $(q^2 = 0)$ of the p-modes crosses the ω^2 axis at the value $N_{\rm p}^2$, given by

$$N_{\rm p}^2 \equiv \omega_{\rm p,cut}^2 \equiv \omega_{p0}^2(k_0^2 = 0) = \tfrac{1}{4}\alpha^2 c^2 = \tfrac{1}{4}\gamma^2 g^2/c^2 \quad \Big(\geq N_{\rm B}^2\Big)\,, \tag{7.44}$$

which is called *the acoustic cutoff frequency*. For $k_0^2 \to \infty$, the acoustic branches asymptotically tend to the ordinary sound wave frequency, which is called the Lamb frequency in this context, $\omega_{\rm p}^2 \approx S^2 \equiv k_0^2 c^2$, whereas the g-modes asymptotically tend to the Brunt–Väisäläa frequency: $\omega_{\rm g}^2 \approx N_{\rm B}^2$ (dashed curves).

The split into Sturmian ($\partial\omega^2/\partial q^2 > 0$) and anti-Sturmian ($\partial\omega^2/\partial q^2 < 0$) discrete modes is a quite general one for wave propagation in inhomogeneous media. We will prove in Section 7.5.1, for the more general MHD case, that this difference is associated with the sign of the coefficient in front of the highest derivative of the differential equation, i.e. $\rho c^2\omega^2/(\omega^2 - k_0^2 c^2)$, for the gravito-acoustic waves described by Eq. (7.29). Hence, the marginal frequency ($\omega^2 = 0$) of the numerator and the Lamb frequency ($\omega^2 = k_0^2 c^2$) of the denominator delimit the frequency ranges where the modes are Sturmian or anti-Sturmian.

For the interpretation of the dispersion curves of Fig. 7.3, note that all frequencies and wave numbers have been made dimensionless by means of the sound speed c and the height a of the layer: $\bar\omega \equiv (a/c)\,\omega$, $\bar k_0 \equiv k_0 a$, $\bar q \equiv qa$, etc. To get estimates for the p-mode frequencies of the Sun, choose $a \equiv R_\odot = 7\times 10^8$ m and a value for c somewhere in the interior (using data from Unno et al. [616]):

$$r = 0.5 R_\odot\,, \quad c = 3\times 10^5\,{\rm m\,s^{-1}} \quad \Rightarrow \quad f_{\rm p} \equiv \frac{c}{2\pi a} = 6.8\times 10^{-5}\,{\rm Hz}\,. \tag{7.45}$$

With the factor $f_{\rm p}$ we may convert the normalized angular frequencies $\bar\omega$ into the actual frequencies ν of the modes, e.g.

$$\begin{aligned}
l &= 0\,, & n &= 0 & &\Rightarrow & \nu_{\rm p,cut} &\equiv \bar N_{\rm p} f_{\rm p} = 0.68\,{\rm mHz}\,,\\
l &= 0\,, & n &= 20 & &\Rightarrow & \nu_{\rm p} &\approx n\pi f_{\rm p} = 4.3\,{\rm mHz}\,,\\
l &= 100\,, & n &= 1 & &\Rightarrow & \nu_{\rm p} &\approx l f_{\rm p} = 6.8\,{\rm mHz}\,,
\end{aligned}$$

where $l \approx \bar k_0$ is the number of nodal lines on the spherical surface and $n = \bar q/\pi$ is the radial quantum number. These frequencies lie in the range of the observed *five minute oscillations* ($\nu_{\rm p} = 3.3$ mHz). To go beyond these order of magnitude estimates, one has to consider the actual geometry and inhomogeneities of the Sun.

▷ **Free-boundary modes** The addition of a separate low-density layer on top of the exponential one we considered so far would permit the investigation of surface modes, analogous to the analysis of the Rayleigh–Taylor instability of Section 6.6.4. In the solar context, these oscillations are called *f-modes*, which are free-boundary oscillations of the solar surface involving both the interior and the atmosphere of the Sun. Their frequency is intermediate between the lowest n branches of the p- and g-modes so that the f-mode could be considered as the common $n = 0$ branch of these modes. However, an estimate of the short-wavelength limit of the frequency of incompressible surface waves (which remains valid for compressible fluids) shows that the mechanism is really gravitational:

$$\omega_{\rm F}^2 \approx k_0 g\,. \tag{7.46}$$

In contrast to the problem of Section 6.6.4, gravity is now pointing into the heavier fluid so that the waves are stable. Inserting the value of the gravitational acceleration at the solar surface, we obtain the following approximation of the frequency of the f-modes:

$$\nu_{\rm f} \equiv \omega_{\rm F}/(2\pi) \approx \sqrt{l}\, f_{\rm f}\,, \qquad f_{\rm f} \equiv \sqrt{GM_\odot/R_\odot^3}/(2\pi) \approx 0.1\,{\rm mHz}\,. \tag{7.47}$$

Since helioseismology depends on surface oscillations for detection, the extension with free-boundary modes is an important step towards reality of the model. For further discussion of surface modes, and generalization to MHD, see Roberts [520]. ◁

7.2.4 Helioseismology and MHD spectroscopy

For educational purposes, we have treated the p- and g-modes in a highly idealized medium, characterized by constant sound speed and gravitational acceleration. Of course, in the Sun these conditions are not valid. Also, one should treat the problem in *spherical geometry*. This gives rise to a quantization condition for the horizontal wave number, $k_0 R_\odot = \sqrt{l(l+1)}$, where l is the degree of the mode, whereas the vertical wave number q and associated quantization condition (7.41) also require modification to account for the inhomogeneities of the interior of the Sun.

▷ **Spherical geometry** For spherical symmetry, the expansion (7.26) is to be replaced by

$$\begin{aligned} \xi_r(r,\theta,\phi,t) &= \hat{\xi}(r)\, Y_l^m(\theta,\phi)\, \mathrm{e}^{-\mathrm{i}\omega t}, \\ \xi_\theta(r,\theta,\phi,t) &= \Big[\hat{\eta}(r)\,\partial Y_l^m/\partial\theta + \hat{\zeta}(r)\,(\sin\theta)^{-1}\partial Y_l^m/\partial\phi\Big]\, \mathrm{e}^{-\mathrm{i}\omega t}, \\ \xi_\phi(r,\theta,\phi,t) &= \Big[\hat{\eta}(r)\,(\sin\theta)^{-1}\partial Y_l^m/\partial\phi - \hat{\zeta}(r)\,\partial Y_l^m/\partial\theta\Big]\, \mathrm{e}^{-\mathrm{i}\omega t}, \end{aligned} \tag{7.48}$$

where $Y_l^m(\theta,\phi)$ are spherical harmonics, well known from quantum mechanics of spherically symmetric systems. Like in quantum mechanics, the actual eigenfrequencies are determined from the radial wave equation which brings in the radial mode number n (number of nodes of the radial wave function). The degree l (total number of nodal circles on the spherical surface) takes the place of the horizontal wave number k_0, and the system is degenerate with respect to the longitudinal order m (number of nodal circles through the poles, $|m| \le l$). This degeneracy is lifted when differential rotation is taken into account.

In the Cowling approximation (neglecting perturbations of the gravitational potential), the original vector eigenvalue problem in terms of $\boldsymbol{\xi}$ may be reduced to the solution of just one ODE for $\hat{\xi}$, the radial wave equation. One of the three variables $\hat{\xi}$, $\hat{\eta}$, $\hat{\zeta}$ then vanishes,

$$\omega^2 \hat{\zeta} = 0 \quad \Rightarrow \quad \hat{\zeta} = 0 \qquad (\text{if } \omega^2 \neq 0), \tag{7.49}$$

when marginal toroidal shifts are ignored ($\omega^2 = 0$, $\hat{\zeta} \neq 0$). Eliminating

$$\hat{\eta} = -[(c^2/r^2)(r^2\hat{\xi})' - g\hat{\xi}]/[r^2(\omega^2 - S^2)], \tag{7.50}$$

yields the radial wave equation:

$$\frac{d}{dr}\left(\frac{\omega^2}{\omega^2 - S^2}\frac{\gamma p}{r^2}\frac{d}{dr}(r^2\hat{\xi})\right) + \left[\rho\omega^2 - \frac{\rho g^2}{r^2}\frac{l(l+1)}{\omega^2 - S^2} + \rho' g - r^2\left(\frac{\rho g}{r^2}\frac{\omega^2}{\omega^2 - S^2}\right)'\right]\hat{\xi} = 0, \tag{7.51}$$

where $S^2 \equiv l(l+1)c^2/r^2$ is the squared Lamb frequency. Note the similarity with the analogous equation for the H-atom in quantum mechanics, but also notice that there is considerably more structure to this problem (due to the presence of three background fields $\rho(r), p(r), g(r)$, and the reduction of a vector eigenvalue problem to a scalar wave equation, with a much more complicated dependence on the eigenvalue parameter). ◁

When the theory is modified in this manner and applied to the observations of the p-mode spectra, it provides a powerful tool for probing the interior of the Sun, called *helioseismology*; see Gough [252], Christensen-Dalsgaard [123] or Harvey [292]. Broadly speaking, this is done by 'inversion' of the observed spectra to obtain the equilibrium profiles that produced them. Figure 7.4 shows a typical example of a *classical* spectrum of solar oscillations observed by means of a measurement of the Doppler shifts of the *quantum mechanical* spectral lines of the sunlight. At this point, it is important to realize that the Doppler shifts themselves are the object of study here since they are directly related to the radial oscillations of the photosphere. With observed radial velocities $\tilde{v}_r \sim 1\,\mathrm{km/s}$, it is quite all right to exploit linear theory since the amplitudes reached in typical 5 min periods are not larger than $5 \times 10^{-4} R_\odot \ll R_\odot$.

Above, 'inversion' of the solar oscillation spectrum is put in quotation marks to indicate that it

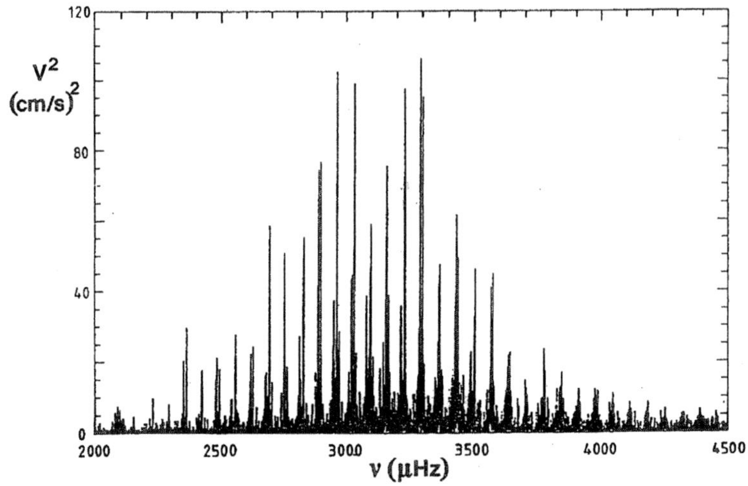

Fig. 7.4 Power spectrum of the solar oscillations from Doppler velocity measurements in light integrated over the solar disk. (From Christensen-Dalsgaard [123], citing A. Claverie, G. R. Isaak, C. P. McLeod, H. B. van der Raay, P. L. Palle, and T. Roca Cortes, *Mem. Soc. Astron. Ital.* **55**, 63 (1984).)

is not a mechanical procedure that can be applied blindly. Whereas computing a spectrum for a particular equilibrium can be carried out numerically (see Fig. 7.5), virtually to any degree of accuracy desired, the inverse problem requires a much larger effort. It implies computing the spectrum while iterating on infinitely many possible values of the equilibrium parameters until matching is obtained. This procedure is ill-posed, a mathematical way of expressing that it is really a kind of art. Also notice that a spectrum of solar oscillations, as shown in Fig. 7.4, does not exhibit the modes on an equal footing. First, out of the infinity of possible modes only a finite number are observed that have been excited by some mechanism, which itself is part of the puzzle. Second, the spectral peaks observed do not correspond to the energy contents of the modes but to the amplitudes of the eigenfunctions at the position where the Doppler shifts are produced, that is in the photosphere. Hence, there is strong biasing towards modes that have a sizeable amplitude there.

One finds that p-modes of low order l penetrate the deep interior of the Sun, whereas high l modes in general are more localized towards the outside. This high l localization is most pronounced for the f-modes, which are free-boundary gravitational oscillations of the solar surface. (Note that the frequency estimate given in Eq. (7.47) checks quite well with the numbers of Fig. 7.5.) In general, g-modes are *cavity modes* trapped in the region interior to the convection zone and, hence, quite difficult to observe. The spectrum of p-modes, however, has been determined with surprising accuracy. It has been compared with computed values based on theoretical solar models. It turns out that the frequencies deduced from the Doppler shifts of spectral lines agree with the calculated ones for the p-modes to within 0.1%! This impressive agreement serves as an example for what is possible in a purely classical (as opposed to quantum mechanical) kind of spectroscopy.

On the basis of this example of fluid dynamical spectroscopy, we have proposed a similar activity for laboratory fusion plasmas, called *MHD spectroscopy* [237]. In order to understand the common feature, the procedure of helioseismology is summarized in Fig. 7.6 to provide a kind of model example for other physical systems, like laboratory fusion machines. Notice that for the birth of

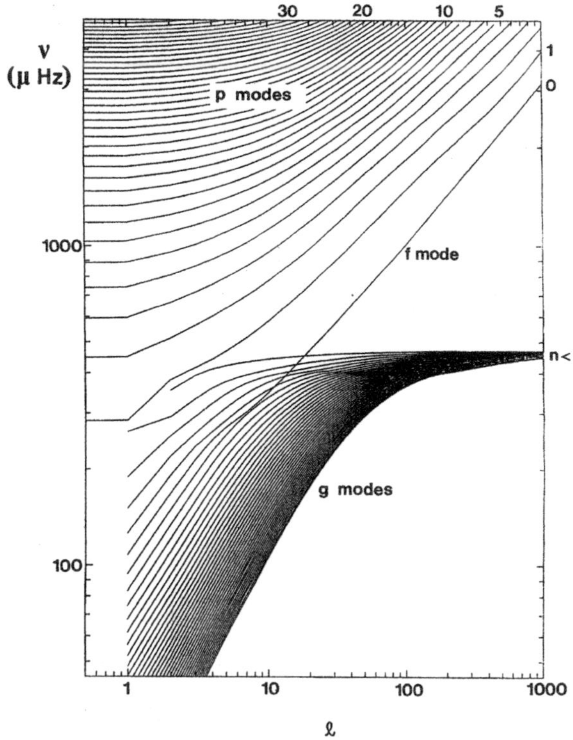

Fig. 7.5 Solar oscillations for a standard solar model: computed frequencies of the p- and g-modes are plotted as a function of the degree ℓ for different values of the radial mode number n. (From Christensen-Dalsgaard [123].)

helioseismology it was not sufficient to have collected a great number of observations of solar oscillations or to have calculated many spectra, but rather that the two activities should have matured to a sufficient degree to allow for a unique correlation between observed and calculated spectra. This is illustrated in the figure. The observations provide values for the collection $\{\omega_{l,n}\}$ of the frequencies of the different modes. Theory, consisting of a Solar Model and a Spectral Code, provides values for the same collection of frequencies. The discrepancies between the two are gradually eliminated by improvements of the Solar Model. To leading order, the model is a spherically symmetric equilibrium depending on the abundance of the different chemical elements H, He and the heavier ones (indicated by the symbols X, Y and Z), the temperature T, the density ρ, the luminosity L, etc. The perturbations of this equilibrium are studied by means of the Spectral Code. Improved models incorporate deviations from the spherical symmetry, due to differential rotation (Ω) and magnetic fields ($\mathbf{B}$), whereas the final interest is in stellar evolution, i.e. the very slow time-dependence of all the equilibrium quantities that are determined on a snapshot basis by helioseismology.

It is not difficult to recognize the counterparts of the three activities depicted in Fig. 7.6 for laboratory fusion research: diagnostics to determine the frequencies and other characteristics of the modes, spectral calculations of the associated fast MHD modes, and transport models to describe the resulting slow evolution of the background equilibrium. Clearly, the success of the similar activity

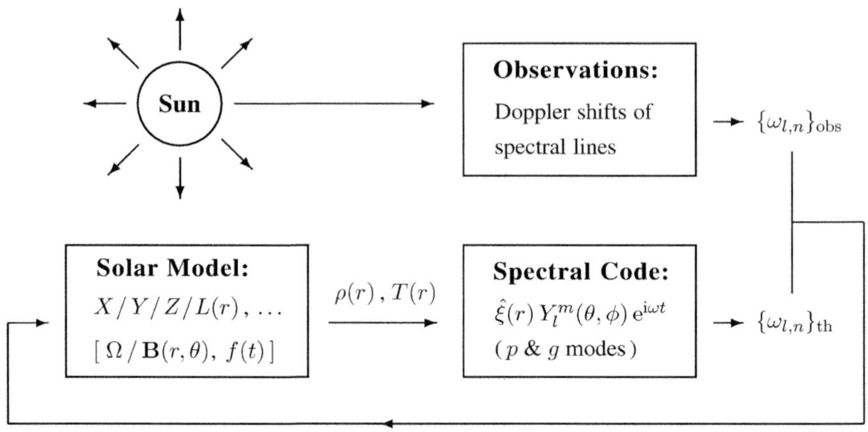

Fig. 7.6 Systematics of helioseismology: the comparison of observed and calculated spectra of solar oscillations is exploited to improve the solar evolution model.

of MHD spectroscopy hinges on improved diagnostics and spectral calculations. Helioseismology demonstrates that fluid dynamical states can be predicted with high accuracy. We will see in a later chapter (Section 17.3.4) that this should be realizable for tokamaks as well.

At this point, it might appear that extension of the 1D spherical HD model to MHD with magnetic effects is the next logical step. However, recall that spherically symmetric magnetic fields are incompatible with $\nabla\cdot\mathbf{B} = 0$ (Section 1.3.4). Hence, incorporation of magnetic fields in spherical geometry inevitably involves 2D magnetic configurations. In the following sections we will stay with 1D and introduce magnetic fields in the plane slab model. This will give rise to plenty of new phenomena associated with plasma inhomogeneity. The further complications of two-dimensionality will be the subject of the later Chapters 16–18.

7.3 MHD wave equation for a gravitating magnetized plasma slab

7.3.1 Preliminaries

With the instructive examples of hydrodynamic waves and instabilities in ordinary fluids of Section 7.2, we now turn to the influence of inhomogeneity on the spectrum of waves and instabilities in *magnetized* plasmas. Macroscopic plasma dynamics inevitably involves effects of the *finite* geometry of the magnetic confinement volume. This substantially complicates the analysis since the assumption (made in Chapter 5 for homogeneous plasmas) of plane waves represented by uncoupled Fourier harmonics breaks down. Waves with different wave numbers $\mathbf{k}$ couple through plasma inhomogeneity so that *wave transformation* takes place. It is perfectly well possible to have a wave that should be characterized as a fast magneto-sonic wave in one part of the plasma volume, but which exhibits Alfvén wave properties in another part.

More important yet, from a fundamental as well as from a practical point of view, is the occurrence of *instabilities* and *continuous spectra*. Both are essentially due to plasma inhomogeneity, but instabilities (as the hydrodynamic convective instability encountered in Section 7.2.2) obviously occur for $\omega^2 < 0$, whereas continua are mainly located on the stable side ($\omega^2 > 0$) of the spectrum in ideal MHD. These phenomena are extremely important for a wide class of laboratory and astrophysical plasmas. This will be extensively demonstrated in the later chapters, but first we have to get acquainted with the necessary analytical tools.

We will approach the problem stepwise and, at first, restrict the analysis of the systems to inhomogeneity in one direction only (called *one-dimensional* or 1D systems) leading to a description in terms of ordinary differential equations (ODEs). A generic example of such systems is the plane gravitating plasma slab with inhomogeneities $\rho(x), p(x)$ and $\mathbf{B}(x)$ of the density, pressure and magnetic field, caused by plasma currents and gravitational stratification. We have already encountered the hydrodynamic version of the basic ODE in the wave equation (7.29) of Section 7.2.1. The fundamental wave equation for the gravitating magnetized plasma slab is derived in Section 7.3.2. This equation describes a bewildering variety of waves, continua and instabilities, as will be shown in the Sections 7.4 and 7.5.

(a) Transformation of the homogeneous problem As a preliminary to the study of inhomogeneous plasmas, it is expedient to restate the *homogeneous* wave problem of Section 5.2 in a slightly different manner. Recall that the equilibrium background in that case was specified by

$$\mathbf{B} = B\mathbf{e}_z\,, \quad \text{with } \rho\,, p\,, B = \text{const}\,, \tag{7.52}$$

so that the sound and Alfvén speeds, $c \equiv \sqrt{\gamma p/\rho}$ and $\mathbf{b} \equiv \mathbf{B}/\sqrt{\rho}$, are constant. We have already transformed the eigenvalue problem (5.52) in terms of the force operator and associated displacement vector in the spectral equation (6.31) of Section 6.2.1. Since, for the time being, all equilibrium quantities are constant, we may write the normal mode amplitude $\boldsymbol{\xi}(\mathbf{r})$ as a Fourier integral (or a Fourier series if one considers a finite box) of plane wave solutions which do not couple:

$$\boldsymbol{\xi}(\mathbf{r}) = (2\pi)^{-3/2} \iiint \hat{\boldsymbol{\xi}}(\mathbf{k})\, \mathrm{e}^{\mathrm{i}\mathbf{k}\cdot\mathbf{r}}\, d^3k\,. \tag{7.53}$$

The normal modes $\hat{\boldsymbol{\xi}}(\mathbf{k};\omega)\exp \mathrm{i}(\mathbf{k}\cdot\mathbf{r} - \omega t)$ may be studied separately by substituting $\nabla \to \mathrm{i}\mathbf{k}$ in Eq. (6.31). This leads to the algebraic eigenvalue problem (6.32). The matrix representation of this equation is identical to Eq. (5.53) with $\hat{v}_{x,y,z}$ replaced by $\hat{\xi}_{x,y,z}$. However, we will now generalize the wave vector $\mathbf{k}$ to have a non-vanishing component k_y in order to be able to distinguish between the two perpendicular directions x and y. This will facilitate the transition to the analysis of the inhomogeneous problem, where x and y will no longer be equivalent since the eventual inhomogeneity will be in the x-direction.

The mentioned transformation amounts to a rotation of the coordinate system through an angle $-\chi \equiv -\arctan(k_y/k_x)$ about the magnetic field (i.e. about the z-axis). We indicate the old representation by primes, $\rho^{-1}\mathsf{F}'\cdot\boldsymbol{\xi}' = -\omega^2\boldsymbol{\xi}'$, where the sans serif bold symbol F' is used to denote the (old) matrix representation of the operator $\mathbf{F}$. The new representation then becomes

$$\rho^{-1}\mathsf{F}\cdot\boldsymbol{\xi} = -\omega^2\boldsymbol{\xi}\,, \quad \text{with} \quad \mathsf{F} = \mathsf{R}\cdot\mathsf{F}'\cdot\mathsf{R}^{-1} \text{ and } \boldsymbol{\xi} = \mathsf{R}\cdot\boldsymbol{\xi}'\,, \tag{7.54}$$

where the elements of the rotation matrix R are given by $R_{11} = R_{22} = \cos\chi$, $R_{21} = -R_{12} = \sin\chi$, $R_{33} = 1$, and the other ones vanish. This yields the following matrix representation of the spectral equation (6.32):

$$\begin{pmatrix} -k_x^2(b^2+c^2)-k_z^2b^2 & -k_xk_y(b^2+c^2) & -k_xk_zc^2 \\ -k_xk_y(b^2+c^2) & -k_y^2(b^2+c^2)-k_z^2b^2 & -k_yk_zc^2 \\ -k_xk_zc^2 & -k_yk_zc^2 & -k_z^2c^2 \end{pmatrix} \begin{pmatrix} \xi_x \\ \xi_y \\ \xi_z \end{pmatrix} = -\omega^2 \begin{pmatrix} \xi_x \\ \xi_y \\ \xi_z \end{pmatrix}, \qquad (7.55)$$

where we have dropped the hats on $\boldsymbol{\xi}$ for simplicity. Note that Eq. (5.53) is recovered in the limit $k_y \to 0$. Solutions are obtained by setting the determinant to zero. This results in the same dispersion equations (5.57) for the Alfvén eigenvalues ω_A^2 and (5.59) for the magneto-sonic eigenvalues $\omega_{s,f}^2$:

$$\omega_A^2 = k_\parallel^2 b^2, \qquad \omega_{s,f}^2 = \tfrac{1}{2}k^2(b^2+c^2)\left[1 \pm \left(1 - \frac{4k_\parallel^2 b^2 c^2}{k^2(b^2+c^2)^2}\right)^{1/2}\right], \qquad (7.56)$$

where now $k^2 = k_x^2 + k_y^2 + k_z^2$, $k_\perp = \sqrt{k_x^2 + k_y^2}$, $k_\parallel = k_z$. However, the expressions (5.58) and (5.61) for the eigenvectors have to be modified to account for the changed direction of the plane spanned by $\mathbf{k}$ and $\mathbf{B}$ (previously, the x'–z' plane) since the Alfvén eigenvectors $\boldsymbol{\xi}_A$ are perpendicular to this plane and the magneto-sonic eigenvectors $\boldsymbol{\xi}_{s,f}$ are lying in it.

(b) Essential spectrum The spectrum of MHD waves is modified significantly by plasma inhomogeneity. If the inhomogeneity is in the x-direction, the assumption of uncoupled plane waves breaks down with respect to the wave number k_x so that we have to replace k_x by $-i\partial/\partial x$ in Eq. (7.55). Hence, the dispersion equation $\omega^2 = \omega^2(k_x, k_y, k_z)$ loses its meaning with respect to the dependence on k_x. The consequences of this will be worked out in the next section. However, one essential feature of the spectrum remains intact, already alluded to in the discussion of the asymptotic properties associated with the group velocity (Section 5.3.3). This relates to *localized wave motion in the direction of inhomogeneity*, which can be understood already from the homogeneous theory by considering the limit $k_x \to \infty$. To that end, we plot ω^2 as a function of k_x, keeping k_y and k_z fixed (Fig. 7.7(a)). Note that this picture is very similar to Fig. 5.6(b), but the degeneracy of Alfvén and fast modes at $k_\perp = 0$ is lifted.

The effects of a confined inhomogeneous plasma can now be mimicked by considering the plasma as a slab of finite extension enclosed by plates at $x = 0$ and $x = a$. The wave number k_x will then be quantized: $k_x = n\pi/a$, where n is the *number of nodes* of the eigenfunction $\xi_x(x)$. Hence, the eigenvalues will be labelled by n, rather than by k_x, and this labelling still makes sense for an inhomogeneous medium varying in the x-direction. This quantization produces a discrete spectrum, with three branches that behave distinctly different (Fig. 7.7(b)). The essential features of these sub-spectra of point eigenvalues are as follows.

(1) The discrete eigenvalues of the fast sub-spectrum monotonically increase, so that

$$\omega_F^2 \equiv \lim_{k_x\to\infty} \omega_f^2 \approx \lim_{k_x\to\infty} k_x^2(b^2+c^2) = \infty \qquad (7.57)$$

is an *accumulation* or *cluster point* of the fast wave point spectrum.

(2) The eigenvalues ω_A^2 of the Alfvén sub-spectrum are *infinitely degenerate*, since

$$\omega_A^2 \equiv \lim_{k_x\to\infty} \omega_a^2 = k_\parallel^2 b^2, \qquad (7.58)$$

so that there is no distinction between the point eigenvalues ω_a^2 and the limit ω_A^2.

(3) The discrete eigenvalues of the slow sub-spectrum monotonically decrease, so that

$$\omega_S^2 \equiv \lim_{k_x \to \infty} \omega_s^2 = k_\parallel^2 \frac{b^2 c^2}{b^2 + c^2} \tag{7.59}$$

is a *cluster point* of the slow wave point spectrum.

Mathematically speaking, ω_F^2, ω_A^2 and ω_S^2 belong to what is called the *essential spectrum*, which is the manifestation of the continuous spectrum in this context. This turns out to be basic for the discussion of the inhomogeneous case, where the equilibrium varies with x. The infinite degeneracy of the Alfvén eigenvalues is then lifted by the appearance of a *continuous spectrum of improper Alfvén modes*, and the cluster point of the slow point spectrum is spread out in a *continuous spectrum of improper slow modes*, whereas the fast cluster point $\omega_F^2 = \infty$ is unaffected.

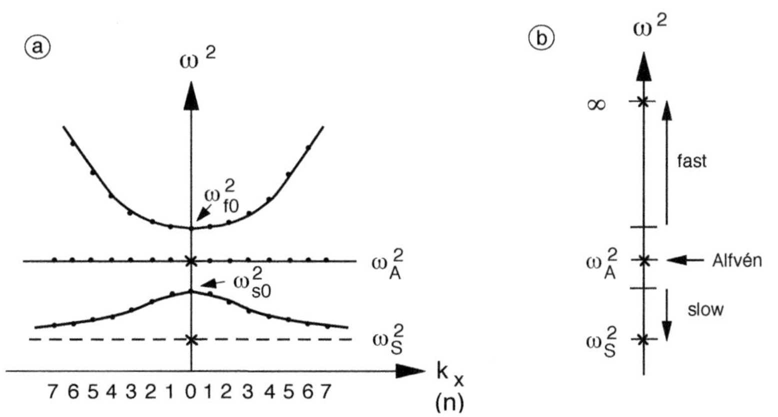

Fig. 7.7 (a) Dispersion diagram $\omega^2 = \omega^2(k_x)$ for k_y and k_z fixed; (b) corresponding structure of the spectrum with degeneracy of the Alfvén waves and clustering of the magneto-acoustic waves.

The two values of ω^2 denoted by ω_{s0}^2 and ω_{f0}^2, where the slow and the fast modes cross the vertical axis in Fig. 7.7(a), also turn out to play a special role in the spectral theory of inhomogeneous plasmas. Their values are given by

$$\omega_{s0,f0}^2 \equiv \omega_{s,f}^2(k_x = 0) , \tag{7.60}$$

where the expression for $\omega_{s,f}^2$ is given in Eq. (7.56). The function of these *turning point frequencies* ω_{s0}^2 and ω_{f0}^2 appears to be to separate the three branches of the spectrum, as evidenced by the following sequence of inequalities:

$$0 \le \omega_s^2 \le \omega_S^2 \le \omega_{s0}^2 \;\le\; \omega_A^2 \;\le\; \omega_{f0}^2 \le \omega_f^2 \le \omega_F^2 = \infty . \tag{7.61}$$

This clear separation of the three discrete sub-spectra for homogeneous media returns, in a modified form, when the plasma becomes inhomogeneous.

7.3.2 MHD wave equation for a gravitating slab

We now introduce genuine inhomogeneity in the model. Consider a plasma slab, infinite and homogeneous in the y- and z-directions, and contained between two plates at $x = x_1$ and $x = x_2$, see Fig. 7.8(a). The equilibrium is assumed to vary in the x-direction:

$$\mathbf{B} = B_y(x)\,\mathbf{e}_y + B_z(x)\,\mathbf{e}_z\,, \qquad \rho = \rho(x)\,, \qquad p = p(x)\,. \tag{7.62}$$

The magnetic field is confined to plane layers parallel to the y–z plane, but its direction varies with height. This is caused by current layers that are confined to the same planes (to be considered as the *magnetic surfaces* for this configuration):

$$\mathbf{j} = \nabla \times \mathbf{B} = -B_z'(x)\,\mathbf{e}_y + B_y'(x)\,\mathbf{e}_z\,, \tag{7.63}$$

where primes denote differentiation with respect to x . In general, $\mathbf{j}$ and $\mathbf{B}$ have different directions so that there is a Lorentz force $\mathbf{j} \times \mathbf{B} = (\nabla \times \mathbf{B}) \times \mathbf{B} = -\nabla(\frac{1}{2}B^2)$ in the x-direction. This force is to be balanced by the pressure gradient ∇p and the gravitational force $\rho(x)\mathbf{g}$, due to an external gravitational field $\Phi_{\rm gr}$ with constant acceleration,

$$\mathbf{g} = -\nabla\Phi_{\rm gr} = -g\mathbf{e}_x\,. \tag{7.64}$$

Consequently, the equilibrium requires satisfaction of just one ODE,

$$(p + \tfrac{1}{2}B^2)' = -\rho g\,, \tag{7.65}$$

which is the only restriction to be imposed on the possible choices of the functions $\rho(x)$, $p(x)$, $B_y(x)$, and $B_z(x)$: an enormous freedom of choice remains.

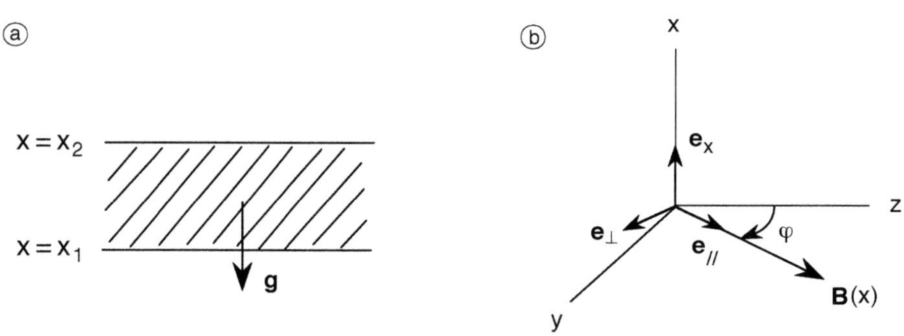

Fig. 7.8 (a) Plane gravitating plasma slab and (b) associated magnetic field line projection.

In this manner, a generic 1D inhomogeneous model for confined plasmas is obtained. The presence of the two bounding plates may appear to be a rather artificial constituent of the model, in particular for astrophysical plasmas. Their function is to simply model plasma confinement, either by the presence of a conducting shell (as in tokamaks), or by the presence of an immobile (very dense) neighbouring plasma. One could always try to minimize their influence by moving them out to infinity, while keeping their essential role of eliminating the flow of energy in or out of the system. Of course, the ultimate remedy is to extend the analysis to an appropriate choice of one of the models (II)–(VI) of Section 4.6.

The wave and spectral equations for the gravitating plasma slab are obtained from Eqs. (6.23) and (6.28), repeated here for convenience:

$$\mathbf{F}(\boldsymbol{\xi}) = \rho \frac{\partial^2 \boldsymbol{\xi}}{\partial t^2} = -\rho\omega^2 \boldsymbol{\xi} \,. \tag{7.66}$$

The most laborious part of the problem is the evaluation of the force operator

$$\mathbf{F}(\boldsymbol{\xi}) \equiv -\nabla\pi - \mathbf{B}\times(\nabla\times\mathbf{Q}) + (\nabla\times\mathbf{B})\times\mathbf{Q} + \nabla\Phi_{\rm gr}\,\nabla\cdot(\rho\boldsymbol{\xi})\,, \tag{7.67}$$

$$\text{with}\quad \pi = -\gamma p \nabla\cdot\boldsymbol{\xi} - \boldsymbol{\xi}\cdot\nabla p\,, \quad\text{and}\quad \mathbf{Q} = \nabla\times(\boldsymbol{\xi}\times\mathbf{B})\,. \tag{7.68}$$

This will be done in the following sequence of steps.

(a) Field line projection First, normal modes are studied satisfying the spectral equation (7.66), but now the 3D Fourier representation (7.53) is replaced by a *2D Fourier representation* for the two homogeneous directions only:

$$\boldsymbol{\xi}(\mathbf{r}) = \frac{1}{2\pi}\iint \hat{\boldsymbol{\xi}}(x;k_y,k_z)\,\mathrm{e}^{\mathrm{i}(k_y y + k_z z)}\,dk_y dk_z \,. \tag{7.69}$$

Our task is to find the x-dependence of the separate Fourier components, which simply will be denoted as $\boldsymbol{\xi}(x)\ \exp \mathrm{i}(k_y y + k_z z)$, dropping the hat. The horizontal part of the differential operator ∇ then produces the algebraic substitutions $\partial_y \to \mathrm{i}k_y$ and $\partial_z \to \mathrm{i}k_z$, whereas the component ∂_x in the direction of inhomogeneity will produce (ordinary) differential equations.

Next, all vectors are projected onto the three unit vectors $\mathbf{e}_x, \mathbf{e}_\perp, \mathbf{e}_\parallel$ (called *the field line triad*, see Fig. 7.8(b)), which represent the physically relevant directions:

$$\mathbf{e}_x \equiv \nabla x\,, \quad \mathbf{e}_\perp \equiv (\mathbf{B}/B)\times\mathbf{e}_x\,, \quad \mathbf{e}_\parallel \equiv \mathbf{B}/B\,. \tag{7.70}$$

This *orthogonal projection*[1] is based on the physical significance of *magnetic field lines and magnetic surfaces* for the description of perturbations. It may be generalized to more complicated equilibria, like tokamaks with nested magnetic surfaces where $\mathbf{e}_x$ represents the normal, and $\mathbf{e}_\perp$ and $\mathbf{e}_\parallel$ the two tangential directions with respect to these surfaces. Hence, we will call the three directions defined by Eq. (7.70) by generic names, viz. *normal* (with respect to the magnetic surfaces), *perpendicular* (with respect to the field lines, tangential to the magnetic surfaces) and *parallel* (with respect to the field lines). In this projection, the result of the gradient operator on the perturbation $\boldsymbol{\xi}(x)\ \exp \mathrm{i}(k_y y + k_z z)$ may be written as

$$\nabla = \mathbf{e}_x\partial_x + \mathrm{i}\,\mathbf{e}_\perp(x)\,k_\perp + \mathrm{i}\,\mathbf{e}_\parallel(x)\,k_\parallel\,, \tag{7.71}$$

where $\partial_x \equiv d/dx$ is the normal derivative, and $k_\perp$ and $k_\parallel$ represent the perpendicular and parallel

[1] One frequently encounters the statement in the literature that a local field line *coordinate* representation is exploited when, at best, a *projection* like the present one is meant. At worst, the existence of coordinates $u(x,y,z)$, $v(x,y,z)$, $w(x,y,z)$ is really assumed and one of them, say w, is supposed to measure distance along the magnetic field. The existence of such coordinates would imply $\mathbf{B} = f(u,v,w)\nabla w$, where f is some scalar function. In turn, this would imply $\mathbf{j} = \nabla\times\mathbf{B} = \nabla f\times\nabla w = (1/f)\nabla f\times\mathbf{B}$, so that $j_\parallel = 0$. Of course, such a severe equilibrium restriction is not justified in general. In plane slab geometry, it would imply that $\mathbf{B}$ is uni-directional. Only for such trivial fields could field line coordinates of that kind be constructed. Field line coordinates of an entirely different kind, so-called *Clebsch coordinates* α,β,γ, do exist though. For those, $\mathbf{B}$ is represented as the vector product of two gradients of them: $\mathbf{B} = \nabla\alpha\times\nabla\beta$. These coordinates are useful for the study of instabilities localized about magnetic field lines.

derivatives (resulting from the exponential Fourier factor):

$$\begin{aligned} k_\perp &\equiv k_\perp(x) \equiv -\mathrm{i}\mathbf{e}_\perp \cdot \nabla = G/B\,, \qquad G \equiv k_y B_z - k_z B_y\,, \\ k_\parallel &\equiv k_\parallel(x) \equiv -\mathrm{i}\mathbf{e}_\parallel \cdot \nabla = F/B\,, \qquad F \equiv k_y B_y + k_z B_z\,. \end{aligned} \tag{7.72}$$

The 'wave vectors' $k_\perp$ and $k_\parallel$ will be replaced by the symbols G and F in later chapters where division by B becomes awkward. The quantities $k_\perp$ and $k_\parallel$ may be considered as the wave vectors in the perpendicular and parallel directions, although they are functions of x in general. However, the magnitude of the resulting horizontal wave vector is constant, $k_0 \equiv (k_\perp^2 + k_\parallel^2)^{1/2} = (k_y^2 + k_z^2)^{1/2}$. We also project $\boldsymbol{\xi}$ on the field line triad:

$$\begin{aligned} \boldsymbol{\xi} &= \xi\,\mathbf{e}_x - \mathrm{i}\eta\,\mathbf{e}_\perp - \mathrm{i}\zeta\,\mathbf{e}_\parallel\,, \qquad \xi \equiv \mathbf{e}_x \cdot \boldsymbol{\xi} = \xi_x\,, \\ \eta &\equiv \mathrm{i}\mathbf{e}_\perp \cdot \boldsymbol{\xi} = \mathrm{i}(B_z\xi_y - B_y\xi_z)/B\,, \qquad \zeta \equiv \mathrm{i}\mathbf{e}_\parallel \cdot \boldsymbol{\xi} = \mathrm{i}(B_y\xi_y + B_z\xi_z)/B\,. \end{aligned} \tag{7.73}$$

The factors i have been inserted here since this turns out to lead to a representation where ξ, η and ζ may be assumed to be real.

When manipulating with vector equations, it is important to remember that the directions of the unit vectors $\mathbf{e}_\perp$ and $\mathbf{e}_\parallel$ vary with x when $\mathbf{B}$ is not unidirectional:

$$\partial_x \mathbf{e}_\perp = -\varphi' \mathbf{e}_\parallel\,, \quad \partial_x \mathbf{e}_\parallel = \varphi' \mathbf{e}_\perp\,. \tag{7.74}$$

Here, $\varphi \equiv \arctan(B_y/B_z)$ is the angle between $\mathbf{B}$ and the z-axis, see Fig. 7.8(b). It remains to evaluate the components of the force operator $\mathbf{F}(\boldsymbol{\xi})$ explicitly, i.e. to exploit the consequences of the equilibrium relation (7.65), of the 2D Fourier representation (7.69), and of the orthogonal projection (7.70)–(7.73). This involves straightforward but rather tedious analysis, which is put in small print.

▷ **Details of the construction of the wave equation matrix** In order to keep the analysis as mechanical as possible, we first derive the x, y, z components of the force operator $\mathbf{F}$ and only at the end we compose the perpendicular and parallel components. Writing the Fourier components of $\mathbf{F}$, we will drop the harmonic factor $\exp \mathrm{i}(k_y y + k_z z)$, but only after the gradient, divergence and curl operations have been applied. Straightforward application of the operations (7.71)–(7.73) then yields, for example, the important relation

$$\nabla \cdot \boldsymbol{\xi} = \xi' + k_\perp \eta + k_\parallel \zeta\,, \tag{7.75}$$

so that the fluid terms of $\mathbf{F}$ may be written as

$$\begin{aligned} -\nabla\pi &\equiv [p'\xi + \gamma p(\xi' + k_\perp\eta + k_\parallel\zeta)]'\,\mathbf{e}_x + \mathrm{i}k_y[p'\xi + \gamma p(\xi' + k_\perp\eta + k_\parallel\zeta)]\,\mathbf{e}_y \\ &\quad + \mathrm{i}k_z[p'\xi + \gamma p(\xi' + k_\perp\eta + k_\parallel\zeta)]\,\mathbf{e}_z\,, \\ \nabla\Phi_{\mathrm{gr}}\,\nabla \cdot (\rho\boldsymbol{\xi}) &\equiv [(\rho\xi)' + \rho k_\perp \eta + \rho k_\parallel \zeta]\,g\,\mathbf{e}_x\,. \end{aligned} \tag{7.76}$$

The evaluation of the magnetic terms requires some more diligence because of the repeated curl operations:

$$\begin{aligned} \boldsymbol{\xi} \times \mathbf{B} &= -\mathrm{i}B\eta\,\mathbf{e}_x - B_z\xi\,\mathbf{e}_y + B_y\xi\,\mathbf{e}_z\,, \\ \mathbf{Q} &= \mathrm{i}Bk_\parallel\xi\,\mathbf{e}_x - [(B_y\xi)' - k_z B\eta]\,\mathbf{e}_y - [(B_z\xi)' + k_y B\eta]\,\mathbf{e}_z\,, \\ \nabla \times \mathbf{Q} &= -\mathrm{i}[(Bk_\perp\xi)' + k_0^2 B\eta]\,\mathbf{e}_x + [(B_z\xi)'' - k_z Bk_\parallel\xi + k_y(B\eta)']\,\mathbf{e}_y \\ &\qquad [(B_y\xi)'' - k_y Bk_\parallel\xi - k_z(B\eta)']\,\mathbf{e}_z\,, \\ -\mathbf{B} \times (\nabla \times \mathbf{Q}) &= [B_y(B_y\xi)'' + B_z(B_z\xi)'' - B^2 k_\parallel^2 \xi + Bk_\perp(B\eta)']\,\mathbf{e}_x \\ &\qquad + \mathrm{i}B_z[(Bk_\perp\xi)' + k_0^2 B\eta]\,\mathbf{e}_y - \mathrm{i}B_y[(Bk_\perp\xi)' + k_0^2 B\eta]\,\mathbf{e}_z\,, \end{aligned}$$

$$\mathbf{j}\times\mathbf{Q} = [B_y'(B_y\xi)' + B_z'(B_z\xi)' + B(Bk_\perp)'\eta]\,\mathbf{e}_x + \mathrm{i}BB_y'k_\parallel\xi\,\mathbf{e}_y + \mathrm{i}BB_z'k_\parallel\xi\,\mathbf{e}_z\,. \tag{7.77}$$

Substitution of these expressions into the spectral equation (7.66), using the equilibrium equation (7.65), produces the three vector components of $\mathbf{F}$ in the x, y, z directions:

$$\begin{aligned}
-\rho\omega^2\xi_x &= [(\gamma p + B^2)\xi']' - B^2k_\parallel^2\xi + [k_\perp(\gamma p + B^2)\eta + k_\parallel\gamma p\zeta]' + \rho g(k_\perp\eta + k_\parallel\zeta)\,,\\
-\rho\omega^2\xi_y &= \mathrm{i}(k_\perp B_z B + k_y\gamma p)\xi' - \mathrm{i}k_y\rho g\xi + \mathrm{i}(k_0^2 B_z B + k_y k_\perp\gamma p)\,\eta + \mathrm{i}k_y k_\parallel\gamma p\zeta\,,\\
-\rho\omega^2\xi_z &= -\,\mathrm{i}(k_\perp B_y B - k_z\gamma p)\xi' - \mathrm{i}k_z\rho g\xi - \mathrm{i}(k_0^2 B_y B - k_z k_\perp\gamma p)\,\eta + ik_z k_\parallel\gamma p\zeta\,.
\end{aligned}$$

Projecting the last two components onto $\mathbf{e}_\perp$ and $\mathbf{e}_\parallel$ finally yields Eq. (7.78) below. ◁

(b) Wave equation in field line projection The matrix representation of the spectral equation for the plane gravitating slab then assumes the symmetric form

$$\mathbf{F}\cdot\mathbf{X} = -\rho\omega^2\mathbf{X}\,, \qquad \mathbf{X} \equiv (\xi, \eta, \zeta)^{\mathrm{T}}\,, \tag{7.78}$$

$$\mathbf{F} \equiv \begin{pmatrix}
\frac{d}{dx}(\gamma p + B^2)\frac{d}{dx} - k_\parallel^2 B^2 & \frac{d}{dx}k_\perp(\gamma p + B^2)g + k_\perp\rho g & \frac{d}{dx}k_\parallel\gamma p + k_\parallel\rho g\\
-\,k_\perp(\gamma p + B^2)\frac{d}{dx} + k_\perp\rho g & -k_\perp^2(\gamma p + B^2) - k_\parallel^2 B^2 & -k_\perp k_\parallel\gamma p\\
-\,k_\parallel\gamma p\frac{d}{dx} + k_\parallel\,\rho g & -k_\parallel\,k_\perp\,\gamma p & -k_\parallel^2\gamma p
\end{pmatrix}.$$

The important feature of this matrix representation of the force operator $\mathbf{F}$ is that it depends on x through $\rho(x), p(x), B^2(x), k_\parallel(x)$ and $k_\perp(x)$.

The matrix eigenvalue problem (7.78) for the three vector components only contains normal differential operators d/dx in the first row and column. Hence, it can be reduced to a single second order differential equation in ξ by eliminating η and ζ through the algebraic second and third row:

$$\begin{aligned}
\eta &= k_\perp\left\{[(b^2 + c^2)\omega^2 - k_\parallel^2 b^2 c^2]\,\xi' - g\,\omega^2\,\xi\right\}\big/D\,,\\
\zeta &= k_\parallel\left\{c^2(\omega^2 - k_\parallel^2 b^2)\,\xi' - g\,(\omega^2 - k_0^2 b^2)\,\xi\right\}\big/D\,,
\end{aligned} \tag{7.79}$$

where $b \equiv B/\sqrt{\rho}$, $c \equiv \sqrt{\gamma p/\rho}$, and D is the determinant of the lower right 2×2 sub-matrix. Substituting these expressions into the first component yields a *second order differential equation for* ξ, first derived by Goedbloed [214](a):

$$\frac{d}{dx}\frac{N}{D}\frac{d\xi}{dx} + \left[\rho(\omega^2 - k_\parallel^2 b^2) + \rho' g - k_0^2\rho g^2(\omega^2 - k_\parallel^2 b^2)/D - \left\{\rho g\omega^2(\omega^2 - k_\parallel^2 b^2)/D\right\}'\right]\xi = 0\,, \tag{7.80}$$

where
$$\begin{aligned}
N = N(x;\omega^2) &\equiv \rho(\omega^2 - k_\parallel^2 b^2)\left[(b^2 + c^2)\,\omega^2 - k_\parallel^2 b^2 c^2\right],\\
D = D(x;\omega^2) &\equiv \omega^4 - k_0^2(b^2 + c^2)\,\omega^2 + k_0^2 k_\parallel^2 b^2 c^2\,.
\end{aligned} \tag{7.81}$$

This equation has to be solved subject to the model I boundary conditions

$$\xi(x_1) = \xi(x_2) = 0\,. \tag{7.82}$$

This completes the formulation of this one-dimensional spectral problem.

Note that Eq. (7.80) contains the differential equation (7.29) for gravito-acoustic modes in the limit $b \to 0$. Also, the three MHD wave frequencies $\omega_S^2, \omega_A^2, \omega_F^2$ for homogeneous plasmas, given in Eq. (7.56), are recovered in the limit $\rho g \to 0$ and neglecting all x dependencies, so that $d/dx \to ik_x$, $k_\perp \to k_y$ and $k_\parallel \to k_z$.

It is of interest for later generalizations to observe the intriguing difference between the spectral problem posed by the '3D' vector eigenvalue equation (7.78) and the equivalent '1D' scalar eigenvalue equation (7.80), supplemented with the algebraic relations (7.79). Whereas the latter formulation is convenient for most of the analysis, and certainly for numerical integration, the eigenvalue character of the problem has been spoiled greatly because ω^2 is scattered throughout Eq. (7.80). To prove or disprove general properties (like monotonicity of the discrete spectrum) it may be advisable to return to the original formulation (7.78).

(c) Singular frequencies The differential equation (7.80), with the boundary conditions (7.82), describes all gravito-magnetohydrodynamic modes of a gravitating magnetized plasma slab with completely arbitrary equilibrium profiles. Before trying to solve it, one first needs to pay attention to the physical meaning of the different terms appearing in the equation and, in particular, to the different singularities.

The factor N/D in front of the highest derivative of the differential equation plays an important role. We may write it in terms of the four special frequencies that were already introduced in Eqs. (7.83)–(7.85) for a homogeneous plasma:

$$\frac{N}{D} = \rho(b^2 + c^2)\,\frac{[\,\omega^2 - \omega_A^2(x)\,]\,[\,\omega^2 - \omega_S^2(x)\,]}{[\,\omega^2 - \omega_{s0}^2(x)\,]\,[\,\omega^2 - \omega_{f0}^2(x)\,]}\,. \tag{7.83}$$

The numerator N involves the *Alfvén and slow magneto-sonic asymptotics*,

$$\omega_A^2(x) \equiv k_\parallel^2 b^2 \equiv F^2/\rho\,, \qquad \omega_S^2(x) \equiv k_\parallel^2 \frac{b^2 c^2}{b^2 + c^2} \equiv \frac{\gamma p}{\gamma p + B^2}\,\omega_A^2(x)\,, \tag{7.84}$$

whereas the denominator D involves the *slow and fast turning point frequencies*,

$$\omega_{s0,f0}^2(x) \equiv \tfrac{1}{2}\omega_m^2\Big[1 \pm \sqrt{1 - 4\omega_S^2/\omega_m^2}\,\Big]\,, \qquad \omega_m^2 \equiv k_0^2(b^2 + c^2)\,, \tag{7.85}$$

defined in Eq. (7.60). For inhomogeneous plasmas, all of these four special frequencies depend on x, through $k_\parallel^2(x)$, $b^2(x)$ and $c^2(x)$, so that their role is to be determined by a local analysis about the points where the different factors of N and D vanish. In particular, the ODE (7.80) becomes *singular* for $N \to 0$, when $\omega^2 \to \omega_A^2(x)$ or $\omega^2 \to \omega_S^2(x)$, i.e. when the local Alfvén or slow magneto-sonic frequencies are approached. These singularities deserve a separate treatment, given in Section 7.4, where it is shown that they give rise to non-square integrable solutions associated with *continuous spectra*. On the other hand, the ODE only develops *apparent singularities* for $D \to 0$, when $\omega^2 \to \omega_{s0}^2(x)$ or $\omega^2 \to \omega_{f0}^2(x)$, i.e. when the local magneto-sonic turning point frequencies are approached. An apparent singularity implies that there are cancellations in the series expansion of the solution, so that it remains finite in the end. This follows directly from the equivalent system of first order differential equations (next section), but it can also be demonstrated explicitly from the second order differential equation (Section 7.4.1). The history of this topic is summarized at the end of Section 7.4.2.

The function $F(x)$, that occurs in the frequencies $\omega_A(x)$ and $\omega_S(x)$ of the Alfvén and slow continua, deserves special attention since it represents the projection of the gradient operator parallel to the magnetic field:

$$F \equiv -\mathrm{i}\mathbf{B}\cdot\nabla = \mathbf{k}_0\cdot\mathbf{B}\,. \tag{7.86}$$

The locations in the plasma where $F = 0$ play an important role in the stability of plasmas since the stabilizing magnetic field line bending energy vanishes there, favouring interchange instability (see Section 7.5.3).

(d) Equivalent system of first order differential equations For numerical integration one usually converts a second order ODE into a pair of first order ODEs for the function and its first derivative. In this case, instead of ξ', it is expedient to exploit the (Eulerian) *perturbation of the total pressure* for that purpose,

$$\Pi \equiv \pi + \mathbf{B}\cdot\mathbf{Q} = -\gamma p\nabla\cdot\boldsymbol{\xi} - \boldsymbol{\xi}\cdot\nabla p + \mathbf{B}\cdot\mathbf{Q}\,. \tag{7.87}$$

Converting $\nabla\cdot\boldsymbol{\xi}$ by means of Eq. (7.75), $\mathbf{Q}$ with Eq. (7.77), and exploiting the equilibrium relation (7.65), this expression may be transformed into

$$\begin{aligned} \Pi &= -\rho(b^2+c^2)\xi' + \rho g\xi - \rho k_\perp(b^2+c^2)\eta - \rho k_\parallel c^2\zeta \\ &= -(N/D)\,\xi' + \rho g\left[\omega^2(\omega^2 - k_\parallel^2 b^2)/D\right]\xi\,, \end{aligned} \tag{7.88}$$

where η and ζ have been eliminated in the second line by means of Eq. (7.79). Hence, the peculiar derivative term in curly brackets in the second order ODE (7.80) comes from this expression for the total pressure perturbation Π.

By means of these expressions, we may convert the second order differential equation (7.80) into the following pair of first order differential equations:

$$\begin{pmatrix} \xi \\ \Pi \end{pmatrix}' = -\frac{1}{N}\begin{pmatrix} C & D \\ E & -C \end{pmatrix}\begin{pmatrix} \xi \\ \Pi \end{pmatrix}, \tag{7.89}$$

where

$$\begin{aligned} C &\equiv -\rho g\,\omega^2(\omega^2 - k_\parallel^2 b^2)\,, \\ E &\equiv -\Big[\rho(\omega^2 - k_\parallel^2 b^2) + \rho' g\Big]N - \rho^2 g^2(\omega^2 - k_\parallel^2 b^2)^2\,. \end{aligned} \tag{7.90}$$

The determinant of the matrix on the RHS of Eq. (7.89) is proportional to N:

$$DE + C^2 = -\Big\{\big[\rho(\omega^2 - k_\parallel^2 b^2) + \rho' g\big]D - k_0^2\rho g^2(\omega^2 - k_\parallel^2 b^2)\Big\}N\,. \tag{7.91}$$

This guarantees that the second singularity $N = 0$ that has been introduced in this formulation (additional to the first, genuine, one) is spurious. The vector eigenvalue equations (7.78) for ξ, η, ζ, the scalar eigenvalue problem (7.80) for ξ, and the two-variable eigenvalue problem (7.89) for ξ, Π are equivalent.

The formulation in terms of two first order ODEs was first exploited by Appert, Gruber and Vaclavik [18] for the analogous problem in cylindrical geometry. It shows right away that nothing blows up for $D = 0$ so that these singularities of Eq. (7.80) must be *apparent*, whereas the $N = 0$ singularities are genuine.

7.3.3 Gravito-MHD waves

Before we start the investigation of the general structure of the spectrum of MHD waves in a gravitating medium, where the mentioned singularities will be in the centre of our attention, it is useful to discuss a simple but non-trivial example of a complete spectrum where the singularities are still absent (or, rather, degenerate enough to permit a simple representation). This is obtained from the generalization of the problem considered in Section 7.2.3 to an *exponentially stratified atmosphere with constant sound and Alfvén speeds:*

$$\rho = \rho_0 \mathrm{e}^{-\alpha x}\,, \quad p = p_0 \mathrm{e}^{-\alpha x}\,, \quad \mathbf{B} = B_0 \mathrm{e}^{-\frac{1}{2}\alpha x}\mathbf{e}_z\,, \tag{7.92}$$

so that $c^2 = \gamma p/\rho = \gamma p_0/\rho_0$ and $b^2 = B^2/\rho = B_0^2/\rho_0$ are constant. It is useful to exploit the parameter β as a measure for the relative magnitude of the sound speed with respect to the Alfvén speed:

$$\beta \equiv 2p/B^2 = 2p_0/B_0^2 \quad \Rightarrow \quad c^2/b^2 = \tfrac{1}{2}\gamma\beta\,. \tag{7.93}$$

The equilibrium relation (7.65) then fixes the new relationship between the previously introduced parameter α and the gravitational acceleration g:

$$\alpha = \frac{\rho g}{p + \frac{1}{2}B^2} = \frac{\rho_0 g}{p_0 + \frac{1}{2}B_0^2} = \frac{2}{1+\beta}\,\frac{g}{b^2}\,. \tag{7.94}$$

Also, the wave numbers $k_\parallel$ $(= k_z)$ and $k_\perp$ $(= k_y)$ are constants now, so that the continuous spectra $\omega_S^2 \equiv k_\parallel^2 b^2 c^2/(b^2+c^2)$ and $\omega_A^2 \equiv k_\parallel^2 b^2$, defined in Eq. (7.84), just degenerate into two points on the ω^2-axis.

Through these simplifications, the spectral equation (7.80) transforms into a differential equation *with constant coefficients*:

$$\frac{N_0}{\rho_0 D_0}\left(\frac{d^2\xi}{dx^2} - \alpha\frac{d\xi}{dx}\right) + \left[\omega^2 - k_\parallel^2 b^2 - \alpha g + \frac{(\alpha g \omega^2 - k_0^2 g^2)(\omega^2 - k_\parallel^2 b^2)}{D_0}\right]\xi = 0\,, \tag{7.95}$$

where $N_0 \equiv N(0)$ and $D_0 \equiv D(0)$. This equation yields an exponential solution $\xi = C\exp[(\frac{1}{2}\alpha \pm \mathrm{i}q)x]$ with a vertical 'wave number'

$$q \equiv \Big\{ -\tfrac{1}{4}\alpha^2 + \rho_0\Big[(\omega^2 - k_\parallel^2 b^2)(D_0 + k_0^2 c^2 N_B^2) + \alpha g\, k_\perp^2 b^2 \omega^2\Big]\Big/N_0\Big\}^{1/2}, \tag{7.96}$$

that has to satisfy the quantization condition (7.41). Again, the Brunt–Väisäläa frequency N_B emerges:

$$N_B^2 \equiv -\frac{1}{\rho}\left(\rho' g + \frac{\rho^2 g^2}{\gamma p}\right) = \alpha g - \frac{g^2}{c^2} = \frac{(\gamma-1)\beta - 1}{1+\beta}\,\frac{g^2}{c^2}\,. \tag{7.97}$$

In the HD limit ($\beta \to \infty$), the expression (7.38) of Section 7.2.3 is recovered: $N_B^2 \to (\gamma-1)g^2/c^2$. Due to the finite β MHD modifications of the equilibrium, positivity of N_B^2 is no longer guaranteed but now depends on the value of β:

$$N_B^2 \geq 0 \quad \text{for } \beta \geq (\gamma-1)^{-1}\,, \qquad N_B^2 < 0 \quad \text{for } 0 \leq \beta < (\gamma-1)^{-1}\,. \tag{7.98}$$

This permits some analysis of the MHD version of the convective instability, but one should be aware of the intrinsic limitation of the stability results obtained on the assumption (7.92) of exponential profiles. Obviously, this assumption is not dictated by physical necessity but just made for analytical convenience.

(a) Dispersion equation and special solutions Inversion of Eq. (7.96) for q yields the *dispersion equation for the gravito-MHD waves*:

$$(\omega^2 - k_\parallel^2 b^2)\Big[\omega^4 - k_{\rm eff}^2(b^2+c^2)\omega^2 + k_{\rm eff}^2 k_\parallel^2 b^2 c^2 + k_0^2 c^2 N_{\rm B}^2\Big] + \alpha g\, k_\perp^2 b^2 \omega^2 = 0\,, \qquad (7.99)$$

where $k_\parallel = k_0 \cos\vartheta$ and $k_\perp = k_0 \sin\vartheta$ (with ϑ indicating the angle between $\mathbf{k}_0$ and $\mathbf{B}$) and we recall the definition of the effective total 'wave number',

$$k_{\rm eff}^2 \equiv k_0^2 + q^2 + \tfrac{1}{4}\alpha^2\,.$$

Since the dispersion equation (7.99) is a cubic equation in ω^2, with three independent solutions, we conclude that gravity does not increase the number of degrees of freedom of the MHD system. These waves transform into the ordinary MHD waves of Chapter 5 when $g \longrightarrow 0$ and into the gravito-acoustic waves of Section 7.2.3 when $b \rightarrow 0$ ($\beta \rightarrow \infty$). The latter limit deserves further analysis to clarify how the twofold HD spectral structure of Fig. 7.3 (with clustering at $\omega = \infty$ and $\omega = 0$) relates to the threefold MHD spectral structure (with clustering at $\omega = \infty, \omega = \omega_{\rm A}$ and $\omega = \omega_{\rm S}$). This will be discussed after we have obtained the MHD counterpart in Fig. 7.9.

Two misleadingly simple expressions are obtained from the dispersion equation for the two limiting cases of purely parallel and purely perpendicular propagation. For *parallel propagation* ($\mathbf{k}_0 \,\|\, \mathbf{B} \;\Rightarrow\; k_\parallel = k_0\,, k_\perp = 0$):

$$\omega_1^2 = k_0^2 b^2\,, \quad \omega_{2,3}^2 = \tfrac{1}{2}k_{\rm eff}^2(b^2+c^2)\left[1 \pm \sqrt{1 - \frac{4k_0^2 c^2 (k_{\rm eff}^2 b^2 + N_{\rm B}^2)}{k_{\rm eff}^4 (b^2+c^2)^2}}\,\right]. \qquad (7.100)$$

These *appear* to be the unaffected Alfvén waves and gravitational modifications of the fast and slow magneto-sonic waves (one of which giving rise to the Parker instability). For *perpendicular propagation* ($\mathbf{k}_0 \perp \mathbf{B} \;\Rightarrow\; k_\parallel = 0\,, k_\perp = k_0$):

$$\omega_1^2 = 0\,, \qquad \omega_{2,3}^2 = \tfrac{1}{2}k_{\rm eff}^2(b^2+c^2)\left[1 \pm \sqrt{1 - \frac{4k_0^2 N_{\rm m}^2}{k_{\rm eff}^4(b^2+c^2)}}\,\right], \qquad (7.101)$$

where another characteristic expression appears, viz.

$$N_{\rm m}^2 \equiv -\frac{1}{\rho}\left(\rho' g + \frac{\rho^2 g^2}{\gamma p + B^2}\right) = \frac{(\gamma-1)\beta + 1}{1+\beta}\,\frac{g^2}{b^2+c^2}\,, \qquad (7.102)$$

which is called the magnetically modified Brunt–Väisäläa frequency [510]. Note the different sign in the numerator compared to the expression (7.97) for $N_{\rm B}^2$, so that $N_{\rm m}^2 > 0$ for this class of equilibria, independent of the value of β. One of the solutions degenerates into a marginal mode and the two other solutions are magnetic compressional modifications of the p- and g-modes given by Eq. (7.43). The dispersion diagram of Fig. 7.3 remains valid for these solutions if we replace the sound speed by the magneto-sonic speed ($c^2 \rightarrow b^2 + c^2$), and adapt the definitions of the Lamb and Brunt–Väisäläa frequencies accordingly.

We have termed the expressions (7.100) and (7.101) misleadingly simple since they represent limiting cases of intricate mode couplings where the labels A, s and f may no longer be appropriate (which is why we have used the neutral labels 1, 2 and 3 instead) and they create a false impression on stability (with only the parallel expression admitting unstable solutions, whereas MHD instabilities in most laboratory plasmas operate in the perpendicular direction). To obtain the full spectral picture of the different HD and MHD effects operating in gravitating plasmas, and their relation to stability, we should study the cubic dispersion equation (7.99) for *oblique propagation*. This is done, most effectively, by means of a numerical representation (exploiting Cardano's explicit solutions) for a typical choice of the parameters. Since the latter part of the problem is less trivial than it may appear, we first pay attention to the construction of dimensionless parameters.

(b) Dimensionless scaling of the dispersion equation and its solutions It is always useful to convert the final result of an MHD calculation, like the dispersion equation (7.99), to a dimensionless form by exploiting the scale independence introduced in Section 4.1.2 to get rid of the trivial dimensional factors. To that end, we exploit the thickness a ($\equiv x_2 - x_1$) of the slab as a measure for length, and the density ρ_0 and the magnetic field strength B_0 at $x = 0$ as measures for the density and magnetic field determining the Alfvén speed $b \equiv B_0/\sqrt{\rho_0}$ (which is constant here). The relevant, dimensionless, parameters then become

$$\bar{\omega} \equiv (a/b)\,\omega\,, \quad \bar{k}_{\rm eff} \equiv k_{\rm eff}a\,, \quad \bar{k}_0 \equiv k_0 a\,, \quad \bar{q} \equiv qa\,, \quad \bar{\alpha} \equiv \alpha a\,, \tag{7.103}$$

so that the dimensionless form of the dispersion equation becomes

$$\begin{aligned}\bar{\omega}^6 &- \Big[(1+\tfrac{1}{2}\gamma\beta)\bar{k}_{\rm eff}^2 + \bar{k}_0^2\cos^2\vartheta\Big]\bar{\omega}^4 \\ &+ \Big[(1+\gamma\beta)\bar{k}_{\rm eff}^2\cos^2\vartheta + \tfrac{1}{4}(1+\beta)(\gamma\beta-\beta+1-2\cos^2\vartheta)\,\bar{\alpha}^2\Big]\,\bar{k}_0^2\,\bar{\omega}^2 \\ &- \Big[\tfrac{1}{2}\gamma\beta\bar{k}_{\rm eff}^2\cos^2\vartheta + \tfrac{1}{4}(1+\beta)(\gamma\beta-\beta-1)\,\bar{\alpha}^2\Big]\,\bar{k}_0^4\cos^2\vartheta = 0\,. \end{aligned} \tag{7.104}$$

The solution of this cubic may be written as $\bar{\omega} = \bar{\omega}(\bar{k}_0, \vartheta, \bar{q}; \bar{\alpha}, \beta, \gamma)$, where the parameters in front of the semicolon refer to the perturbation and the ones behind it refer to the equilibrium. For the numerical investigation of the stability of a particular equilibrium, one could fix the parameters $\bar{\alpha}, \beta, \gamma$ ($= 5/3$) of that equilibrium, but one should investigate the full range of the parameters $\bar{q}, \vartheta, \bar{k}_0$, since that is what nature does. Using the scale independence in this manner, we obtain certainty about the complete parameter space, avoiding useless scanning of redundant parameters.

In terms of these parameters, the 'singular' frequencies of Section 7.3.2 become

$$N_0 = 0 \;\Rightarrow\; \bar{\omega}_{\rm A}^2 = \bar{k}_0^2\cos^2\vartheta\,, \qquad \bar{\omega}_{\rm S}^2 = \frac{\tfrac{1}{2}\gamma\beta}{1+\tfrac{1}{2}\gamma\beta}\,\bar{k}_0^2\cos^2\vartheta\,, \tag{7.105}$$

$$D_0 = 0 \;\Rightarrow\; \bar{\omega}_{\rm s0,f0}^2 = \tfrac{1}{2}\bar{k}_0^2(1+\tfrac{1}{2}\gamma\beta)\left[1 \pm \sqrt{1 - \frac{2\gamma\beta}{(1+\tfrac{1}{2}\gamma\beta)^2}\cos^2\vartheta}\,\right]. \tag{7.106}$$

The dimensionless expressions of the two kinds of Brunt–Väisäläa frequencies, involving the gravitational parameter $\bar{\alpha}$, read:

$$\bar{\rm N}_B^2 = \tfrac{1}{4}\bar{\alpha}^2\,\frac{(1+\beta)\,[(\gamma-1)\beta-1]}{\tfrac{1}{2}\gamma\beta}\,, \qquad \bar{N}_{\rm m}^2 = \tfrac{1}{4}\bar{\alpha}^2\,\frac{(1+\beta)\,[(\gamma-1)\beta+1]}{1+\tfrac{1}{2}\gamma\beta}. \tag{7.107}$$

Whereas the 'singular' frequencies (7.105) and (7.106) do not involve $\bar{\alpha}$, they play a decisive role in the clarification of the MHD spectral structure.

In Fig. 7.9 we show plots of $\bar{\omega}^2(\bar{k}_0^2)$ for oblique propagation ($\vartheta = \pi/4$) for 30 values of the vertical mode number n. The value of $\bar{\alpha}$ ($= 20$) has been chosen approximately the same as for the HD p- and g-modes of Fig. 7.3. To demonstrate the connection with the latter diagram, a high

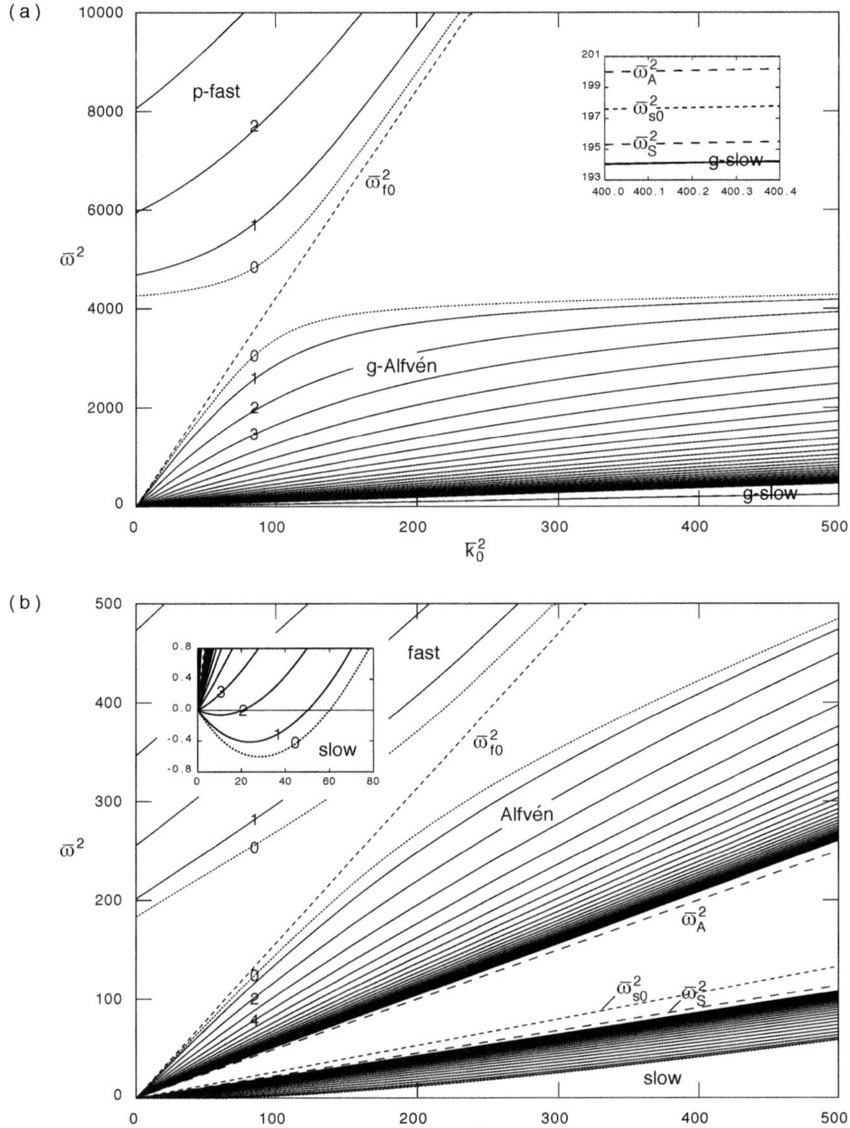

Fig. 7.9 Dispersion diagrams for the oblique gravito-MHD modes of an exponential plane plasma layer. The squared normalized frequencies $\bar{\omega}^2 = \bar{\omega}^2(\bar{k}_0^2, \bar{q}^2, \vartheta)$ are plotted for thirty values of the vertical 'wave number' $\bar{q} = n\pi$ $(n = 1, 2, \ldots 30)$; $\bar{\alpha} = 20$, $\vartheta = \pi/4$. (a) Near HD regime ($\beta = 50$), inset: blow-up of low-frequencies (slow waves not resolved); (b) Regular MHD regime ($\beta = 1$), inset: unstable slow modes.

value of β is taken in Fig. 7.9(a), whereas β is chosen to be unity in Fig. 7.9(b) to exhibit the more usual MHD spectral structure. The singular frequencies $\bar{\omega}_A^2$ and $\bar{\omega}_S^2$ are indicated by long dashes and the magneto-sonic turning point ones $\bar{\omega}^2_{s0,f0}$ by short dashes. The three dotted lines represent the zeroth ($n = 0$) 'member' of each family of solutions. Those are actually not solutions (they do not satisfy the boundary conditions), but they mark the boundaries of the most global modes. The spectral structure is determined by how these three families of solutions for $\bar{\omega}^2$ are distributed from $n = 1$ to $n \to \infty$, where the latter limit yields the essential spectrum of Section 7.3.1.

By means of Fig. 7.9, the following matters may be elucidated:

(1) The relationship between the spectral structures of HD and MHD The HD dispersion equation (7.42) is obtained from the MHD equation (7.104) by scaling with β, $\widetilde{\omega}^2 \equiv \frac{1}{2}\gamma\beta\,\bar{\omega}^2$ (so that the vertical sizes in Fig. 7.9(a) are a factor $\frac{1}{2}\gamma\beta$ larger than those in Fig. 7.3), and taking the limit $\beta \to \infty$:

$$\widetilde{\omega}^2\Big[\widetilde{\omega}^4 - \bar{k}_{\rm eff}^2\,\widetilde{\omega}^2 + \tfrac{1}{4}(\gamma-1)\bar{\alpha}^2\bar{k}_0^2\Big] = 0\,. \qquad (7.108)$$

This shows that, in addition to the p- and g-modes discussed in Section 7.2.3, the marginal horizontal shifts perpendicular to $\mathbf{k}_0$, that were discarded in Section 7.2.1, are needed to obtain a *threefold essential spectrum* ($\widetilde{\omega}_{1,2}^2 \equiv 0, \widetilde{\omega}_3^2 \equiv \infty$) that may be connected to the threefold essential spectrum of MHD. In fact, in the limit $\beta \to \infty$, the Alfvén and slow mode frequencies $\bar{\omega}_{\rm A,S}^2 \to 0$, where the vector $\boldsymbol{\xi}$ becomes dominantly horizontal, whereas $\bar{\omega}_{\rm F}^2 \equiv \infty$, corresponding to dominantly vertical displacements. Figure 7.9(a) shows that, for large but finite β, the p-modes transform into the fast MHD modes and the g-modes transform into the Alfvén modes. In addition, a new branch of gravitational slow MHD modes springs forth in the low-frequency domain. From the inset in Fig. 7.9(a) it is clear that for the value of the gravitational parameter chosen, $\bar{\alpha}^2 \sim \bar{k}_0^2$, the spectrum is strongly affected by gravity both for the Alfvén modes (infinitely degenerate in the absence of gravity) and for the slow modes (anti-Sturmian in the absence of gravity). Notice that there is still a lot of empty space left between the lowest Alfvén curve and $\bar{\omega}_{\rm A}^2$ and between the highest slow curve and $\bar{\omega}_{\rm S}^2$, which will be occupied by the curves for $n > 30$ (not shown) establishing the link with the essential spectrum.

(2) Influence of gravity on the approach to the essential spectrum Decreasing the value of β to unity (Fig. 7.9(b)), the three MHD sub-spectra become more clearly distinguished. The three sequences can be labelled as fast, Alfvén and slow by virtue of the monotonicity of $\bar{\omega}^2$ as a function of n (proved in general in Section 7.4.3), where $\bar{\omega}_{\rm F}^2$, $\bar{\omega}_{\rm A}^2$ and $\bar{\omega}_{\rm S}^2$ are obtained in the limit $n \to \infty$. By expanding the solutions of the dispersion equation (7.104) around these frequencies, one obtains the following expressions for the approach to the essential spectrum:

$$\begin{aligned}
\bar{\omega}_{\rm f}^2 &\approx (1 + \tfrac{1}{2}\gamma\beta)\,\bar{q}^2 && \to \quad \bar{\omega}_{\rm F}^2 \equiv \infty\,,\\
\bar{\omega}_{\rm a}^2 &\approx \bar{\omega}_{\rm A}^2 + \tfrac{1}{2}(1+\beta)\big(\bar{k}_0/\bar{q}\big)^2\bar{\alpha}^2\sin^2\vartheta && \to \quad \bar{\omega}_{\rm A}^2\,,\\
\bar{\omega}_{\rm s}^2 &\approx \bar{\omega}_{\rm S}^2 + \frac{\bar{\omega}_S^4 - \frac{1}{4}(1+\beta)(1+\beta-\gamma\beta\cos^2\vartheta)\,\bar{k}_0^2\bar{\alpha}^2}{(1+\frac{1}{2}\gamma\beta)\,\bar{q}^2} && \to \quad \bar{\omega}_{\rm S}^2\,.
\end{aligned} \qquad (7.109)$$

The deviation from the limiting frequencies is determined by the gravitational parameter $\bar{\alpha}$, but the essential spectrum itself is not affected by gravity: it is robust! These expressions show how the degeneracy of the Alfvén frequencies is lifted by gravity so that an anti-Sturmian sequence is obtained ($\bar{\omega}^2 \downarrow \bar{\omega}_{\rm A}^2$ as $n \to \infty$). The slow frequencies, which are anti-Sturmian ($\bar{\omega}^2 \downarrow \bar{\omega}_{\rm S}^2$) in the absence of gravity, either become Sturmian ($\bar{\omega}^2 \uparrow \bar{\omega}_{\rm S}^2$), as in the range of $\bar{k}_0^2$ plotted in Fig. 7.9(b), or remain anti-Sturmian when $\bar{k}_0^2$ is large enough (outside the range shown). For monotonicity of $\bar{\omega}^2(n)$ to remain valid, the transition from Sturmian to anti-Sturmian behaviour should occur at a point of *infinite degeneracy of the slow modes*. This exactly happens when the numerator of the expression (7.109)(c) vanishes.

We return to our reservations about the meaning of the expressions (7.100) for parallel propagation. It would appear logical to associate ω_1^2 with the Alfvén and ω_2^2 (with the minus sign) with the slow modes. However, as shown by Adam [5], when $N_B^2 c^2 > \frac{1}{4}\alpha^2 b^4$ (i.e. $\beta > 2$ for $\gamma = 5/3$), the curve for $\omega_2^2(q)$ crosses the value ω_1^2 so that some of the slow frequencies would exceed the Alfvén frequency, in conflict with the very name 'slow'. The paradox is eliminated by considering near parallel propagation in the limit $|\vartheta| \to 0$. For $\vartheta \neq 0$, the two solutions split apart into an anti-Sturmian Alfvén branch approaching ω_A^2 for $q \to \infty$ and an anti-Sturmian slow branch starting just below ω_A^2 $(= \omega_{s0}^2)$ at $q = 0$ and approaching ω_S^2 for $q \to \infty$. For purely parallel propagation $(\vartheta = 0)$, these curves develop a kink at the cross-over point $q = q_c$ so that we may still call the upper part (consisting of ω_2^2 for $0 \le q < q_c$ and of ω_1^2 for $q_c \le q < \infty$) 'Alfvén', and the lower part (consisting of ω_1^2 for $0 \le q < q_c$ and of ω_2^2 for $q_c \le q < \infty$) 'slow'.

(3) Convective Rayleigh–Taylor instability in the presence of a magnetic field The transition to Sturmian monotonicity of the slow modes, exhibited by both Figs. 7.9(a) and (b), is crucial for the well-posedness of the ideal MHD stability problem. It guarantees that, for $\bar{\omega}^2 < 0$, *the largest growth rates are obtained for the most global* $(n = 1)$ *instability*. For example, as shown by the inset of Fig. 7.9(b), the modes $n = 1$ and $n = 2$ are unstable but all the higher ones $(n \ge 3)$ are stable for this particular choice of the parameters.

To systematically investigate the unstable range of parameters, one needs a general stability criterion. For the present case of the exponential equilibrium, this is easily obtained from the dispersion equation (7.104). For unstable solutions, the last square bracket term should be negative:

$$\left(\tfrac{1}{2}\gamma\beta\bar{k}_{\rm eff}^2\cos^2\vartheta + \tfrac{1}{2}\gamma\beta\bar{N}_B^2\right)\cos^2\vartheta$$
$$\equiv \left\{\tfrac{1}{2}\gamma\beta(\bar{k}_0^2 + \bar{q}^2)\cos^2\vartheta + \tfrac{1}{4}\Big[(1+\beta)(\gamma\beta - \beta - 1) + \tfrac{1}{2}\gamma\beta\cos^2\vartheta\Big]\,\bar{\alpha}^2\right\}\cos^2\vartheta < 0\,. \quad (7.110)$$

This expression shows that gravitational instability is obtained when the first term in square brackets is not only negative $(\bar{N}_B^2 < 0)$ but also dominates over the remaining terms proportional to $\cos^2\vartheta$: the gravitational instability drive should dominate over the stabilizing field-line bending contribution of the slow mode perturbations. Hence, one expects the worst instability for perpendicular propagation $(\vartheta = \pi/2)$. On the other hand, recalling from Section 7.2.3 that the exponential atmosphere is stable in the HD limit $(\beta \to \infty)$, magnetic field perturbations are apparently essential to create gravitational instability in the MHD domain. Accordingly, the potentially negative expression in curly brackets is multiplied by $\cos^2\vartheta$ so that, at worst, marginal stability is obtained for $\vartheta = \pi/2$. *For this class of equilibria*, the MHD version of the Rayleigh–Taylor instability occurs in a range of near-perpendicular wave vectors $\mathbf{k}_0$, excluding the direction $\mathbf{k}_0 \cdot \mathbf{B} = 0$ itself. This range may even include parallel propagation $(\vartheta = 0)$ when $\bar{N}_B^2$ is negative enough, giving rise to the Parker instability (see below). The stabilizing slow mode contributions also vanish in the extreme limit $\beta \to 0$, opposite to HD, when the plasma becomes unstable for all $\vartheta \neq \pi/2$ and any value of $\bar{\alpha}$.

(c) Parker instability According to Parker [482], "the interstellar magnetic field in the general neighbourhood of the Sun is, on average, parallel to the plane of the Galaxy ... This galactic magnetic field must be confined by the weight of the gas (plasma) threaded by the field. This interstellar

gas-field system is subject to a universal Rayleigh–Taylor instability such that the interstellar gas tends to concentrate into pockets suspended in the field. The cause of the instability may be thought of as a hydromagnetic self-attraction in the interstellar gas, which may be ten times larger than the gravitational self-attraction of the gas. It is this hydromagnetic self-attraction which produces the observed tendency of the interstellar gas to be confined in discrete clouds."

The gas, which consists of electrons, ions and neutrals, is described by means of the ideal MHD equations, so that the equations of the present section apply. The instability is investigated for motions in the vertical $x-z$ plane only, i.e. longitudinal perturbations ($\vartheta = 0$). Parker also includes the effects of a relativistic cosmic ray particle population on the equilibrium pressure and the dynamics. Neglecting the latter, Eq. (7.110) yields the criterion for the Parker instability in our notation:

$$\bar{k}_{\rm eff}^2 + \bar{N}_{\rm B}^2 < 0 \quad \Rightarrow \quad \bar{k}_0^2 + \bar{q}^2 - \tfrac{1}{4}\bar{\alpha}^2 \cdot \frac{2(1+\beta)(1+\beta-\gamma\beta) - \gamma\beta}{\gamma\beta} < 0 \,. \tag{7.111}$$

The growth rate of the instabilities may be obtained from Eq. (7.100). This shows the connection with the slow magneto-sonic wave which is essential for the instability since it involves dominant perturbed flows parallel to the magnetic field. Hence, "the gas tends to drain downward along the magnetic lines of force into the lowest region along each line". Obviously, numerical values for the wave numbers k_0, q and for the equilibrium parameters α, β and γ have to be chosen to demonstrate the viability of the mechanism for the description of cloud formation along galactic spiral arms. We refer to the original literature [482] and the review paper by Mouschovias [452] for a discussion of all the astrophysical ramifications.

(d) Interchanges and quasi-interchanges We return to the full dispersion equation (7.104), for $\vartheta \neq 0$, since perpendicular and near-perpendicular propagation properties are of generic interest for the stability of confined plasmas in general. Expanding the dispersion equation for $\vartheta = \pi/2 \pm \epsilon$, with $\epsilon \ll 1$, we obtain two potentially unstable solutions:

$$\bar{\omega}^2 \approx \frac{\bar{k}_0^2}{\bar{k}_{\rm eff}^2}\,\bar{N}_{\rm m}^2 \qquad \text{(pure interchanges, } \vartheta = \pi/2)\,,$$
$$\bar{\omega}^2 \approx \frac{\bar{N}_{\rm B}^2}{\bar{N}_{\rm m}^2}\,\bar{\omega}_{\rm S}^2 = \frac{(\gamma-1)\beta - 1}{(\gamma-1)\beta + 1}\,\bar{k}_0^2\cos^2\!\vartheta \qquad \text{(quasi-interchanges, } \vartheta \neq \pi/2)\,. \tag{7.112}$$

This shows a severe limitation of the exponential equilibrium: pure interchanges, with a maximum growth rate at $\vartheta = \pi/2$, are not possible for these equilibria since they have $\bar{N}_{\rm m}^2 > 0$. However, the second branch, with maximum growth rate at $\vartheta \neq \pi/2$, may be unstable since it requires $\bar{N}_{\rm B}^2 < 0$. They have been called *gravitational quasi-interchanges* by Newcomb [462]. Both the regular and the modified Brunt–Väisäläa frequencies, $\bar{N}_{\rm B}$ and $\bar{N}_{\rm m}$, enter the expression for their growth rate, implying intricate stability transitions. In the stability analysis of general equilibria, the crossing of these two branches of the dispersion equation plays an important role in clarifying the apparent discrepancies that occur in the stability criteria for gravitational interchanges. We defer further discussion to Section 7.5.3, when we are no longer restricted to the choice of uni-directional magnetic field and exponential equilibrium profiles.

In Section 12.1.3, we will return to the analysis of the interchange instability (operating in tokamaks) and the Parker instability (operating in galaxies) to highlight the fundamental distinction between laboratory and astrophysical plasma dynamics with respect to the dependence on the parallel wave number (Fig. 12.2).

7.4 Continuous spectrum and spectral structure

"Singularity is almost invariably a clue."

(From Bender and Orszag [50], in their presentation of approximate solutions of linear differential equations, quoting Sherlock Holmes in *The Boscombe Valley Mystery* by Sir Arthur Conan Doyle.)

7.4.1 Singular differential equations

We have obtained important insight into the structure of the MHD spectrum for equilibria with exponential profiles, where the 'singular' frequencies determined by $N_0 = 0$ and $D_0 = 0$ are constant. However, we have evaded the most important issues of inhomogeneity in MHD, which are that, in general, these frequencies depend on the coordinate x of inhomogeneity and the magnetic field is not uni-directional: *continuous spectrum* and *magnetic shear* are essential for the MHD description of plasmas. We face these issues in this section and the next.

When a problem has been reduced to a *non-singular* ordinary second order differential equation it may be considered to have been solved, because one can always obtain the explicit answers numerically to any relevant degree of accuracy. A specific example has just been given in Section 7.3.3. Since this example actually hinges on simplification through the exponential factor, let us briefly consider how the numerical solution might proceed with a more general inhomogeneity. To that end, we write the basic differential equation (7.80) symbolically as

$$\frac{d}{dx}\left[P(x;\omega^2)\,\frac{d\xi}{dx}\right] - Q(x;\omega^2)\,\xi = 0\,, \tag{7.113}$$

where $P \equiv N/D$, and $Q \equiv -[\cdots]$ with the expression in square brackets as in Eq. (7.80). This equation is to be solved subject to the model I boundary conditions (7.82). Specifying the background equilibrium by a particular choice of the profiles $\rho(x)$, $p(x)$, $B_y(x)$ and $B_z(x)$, satisfying the equilibrium constraint (7.65) of Section 7.3.2, *and prescribing a particular value* $\omega^2 = \omega^{2(0)}$ for the unknown eigenfrequency, the functions $P(x;\omega^2)$ and $Q(x;\omega^2)$ are completely determined. One may then solve Eq. (7.113) by means of a standard numerical library routine, starting from the left boundary condition $\xi(x_1) = 0$ (where the derivative $\xi'(x_1)$ may be prescribed arbitrarily since the differential equation is homogeneous and linear), and stepping towards the right end of the interval $[x_1, x_2]$. Of course, arriving there, the right boundary condition $\xi(x_2) = 0$ will not be satisfied in general. This necessitates the construction of an algorithm for the choice of a new value $\omega^2 = \omega^{2(1)}$ that will bring the solution closer to satisfying that boundary condition in the next iteration. This part of the problem is important for the determination of the eigenvalue, but it does not present any difficulty. The point is that an extremely effective algorithm exists, based on shooting (Section 7.5.1) and the oscillation theorem (Section 7.4.3), which is valid *as long as the ODE is non-singular*. The essential problem left is, therefore, a proper treatment of the singularities of the basic differential equation. Hence, we now resume our discussion of the singularities that was started in Section 7.3.2(c).

In contrast to the non-singular integration procedure just sketched, where the boundary conditions determine the solution on the entire interval $[x_1, x_2]$, a singularity $\omega^2 = \omega_A^2(x_0)$ or $\omega^2 = \omega_S^2(x_0)$ at a point $x = x_0$ splits the interval into two parts that become *independent*, in a sense to be discussed. Hence, there is no way to produce the singular problem from the non-singular one by a small perturbation. There either is a singular point or there is not, and the presence of such a point changes the analysis completely. For example, in marginal stability theory (where $\omega^2 \equiv 0$, so that the mentioned singularities become zeros of the Alfvén frequency $\omega_A = F/\sqrt{\rho}$) there is always a distinct difference between configurations with magnetic shear ($F' \neq 0$) and without ($F' = 0$), where the limit $F' \to 0$ produces complicated boundary layers. We will analyse this in Section 7.5.

Before we delve into the singularity theory of Eq. (7.113), an example illustrates the different kinds of singularities encountered for inhomogeneous plasmas. For that purpose, we concentrate on the Alfvén singularity. In Fig. 7.10 three generic cases are depicted for the distribution of $\omega_A^2(x)$, where the slow singularities $\omega_S^2(x)$ (not shown) may be assumed to be well-separated from the Alfvén ones (cases (a) and (b)), but they necessarily coalesce at the marginal point (case (c)).

(a) If $\omega_A^2(x)$ is monotonic, any choice of ω^2 in the range $\omega_A^2(x_1) \leq \omega^2 \leq \omega_A^2(x_2)$ will lead to a singular point $x_1 \leq x_0 \leq x_2$ where the Alfvén factor of the function P may be expanded as follows:

$$\omega^2 - \omega_A^2 \approx -(\omega_A^{2\,\prime})_0\,(x - x_0) \quad \Rightarrow \quad P \sim x - x_0 \,. \tag{7.114}$$

This range will determine the *continuous spectrum* $\{\omega_A^2\}$ of Alfvén modes, and similarly for the frequency range $\{\omega_S^2\}$ of the slow modes.

(b) If $\omega_A^2(x)$ is not monotonic, the continuous spectrum develops a maximum or minimum at some point x_0 where it starts to fold over onto itself. Choosing ω^2 to correspond to that extremum permits expansion of the Alfvén factor as

$$\omega^2 - \omega_A^2 \approx -\tfrac{1}{2}(\omega_A^{2\,\prime\prime})_0\,(x - x_0)^2 \quad \Rightarrow \quad P \sim (x - x_0)^2 \,. \tag{7.115}$$

The value $\omega^2 = \omega_A^2(x_0)$ corresponds to an internal edge of the continuous spectrum which may become a *cluster point* of the discrete spectrum, depending on further conditions, as will be discussed in Section 7.4.4.

(c) For marginal stability ($\omega^2 = 0$), such candidate cluster points always occur in sheared magnetic fields at an *interchange point* $F^2(x_0) = 0$, where both the Alfvén and the slow continuum reach the origin: $\omega_S^2 \equiv \gamma p(\gamma p + B^2)^{-1}\omega_A^2 = 0$. Since, according to Eq. (7.85), also $\omega_{S0}^2 \approx \omega_S^2$ there, the factor ω_S^2 of the numerator $N(\omega^2 = 0)$ is cancelled by the leading order of the factor ω_{S0}^2 of the denominator $D(\omega^2 = 0)$ so that the Alfvén factor determines the behaviour at the singularity:

$$-\omega_A^2 \approx -\tfrac{1}{2}(\omega_A^{2\,\prime\prime})_0\,(x - x_0)^2 = -(\omega_A{}')_0^2\,(x - x_0)^2 \quad \Rightarrow \quad P \sim (x - x_0)^2 \,. \tag{7.116}$$

By means of this expansion, the conditions for *local interchange stability* may be derived, as will be demonstrated in Section 7.5.3.

We are now prepared to appreciate the physical significance of the different steps in the following mathematical analysis of Eq. (7.113).

Singularity of this differential equation is best discussed in terms of the standard form used in classical texts [325] (dating from the pre-computing era) as well as in modern ones [50] on linear

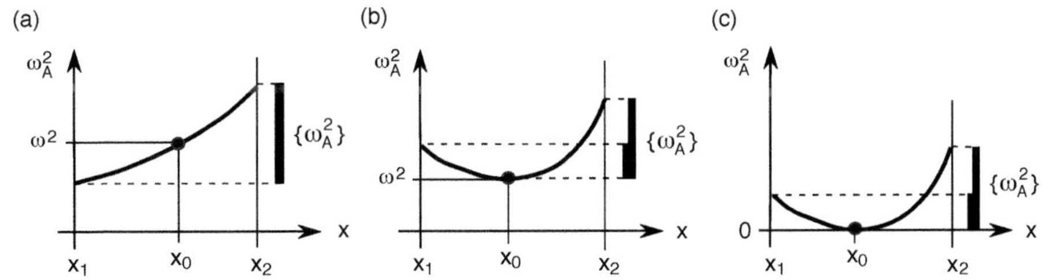

Fig. 7.10 The normal dependence of the squared Alfvén frequency $\omega_{\rm A}^2(x)$ determines: (a) a continuous spectrum singularity, (b) a potential cluster point for stable waves, (c) a potential cluster point at marginal stability (omitting the slow continua).

differential equations:

$$\xi'' + \frac{P'}{P}\,\xi' - \frac{Q}{P}\,\xi = 0\,. \tag{7.117}$$

This equation is non-singular if the two functions P'/P and Q/P are analytic on the whole domain of interest, i.e. in a finite region of the complex z plane containing the interval $x_1 \le x \le x_2$ of the real axis (where the prime is redefined as differentiation with respect to z). Theoretically, the solution may be carried from (ordinary) point to (ordinary) point by Taylor series expansion with a radius of convergence determined by the distance to the nearest singularity of the two mentioned functions. Hence, such singularities interfere with the numerical procedure described earlier. Obviously, the zeros of

$$N \equiv \rho(b^2 + c^2)\Big[\,\omega^2 - \omega_{\rm A}^2(x)\Big]\Big[\,\omega^2 - \omega_{\rm S}^2(x)\Big] \sim (x - x_0)^l \tag{7.118}$$

(where $l = 1$ or $l = 2$), produce singularities of the kind $P'/P \sim z^{-1}$ and $Q/P \sim z^{-l}$. Here, z is the complex continuation of the real variable $x - x_0$, which is the distance to a point x_0 where either $\omega_{\rm A}^2$ or $\omega_{\rm S}^2$ coincides with the eigenvalue parameter ω^2. We have seen above that it is sufficient for virtually all cases of interest to consider $l = 1$ and $l = 2$ only, i.e. *regular singularities*. In the neighbourhood of such singularities, the differential equation may be written as

$$\xi'' + \frac{1}{z}\,p(z)\,\xi' - \frac{1}{z^2}\,q(z)\,\xi = 0\,, \tag{7.119}$$

where $p(z)$ and $q(z)$ are analytic functions with the following Taylor expansion:

$$\begin{aligned} &p(z) \equiv z\,P'/P = p_0 + p_1 z + \cdots\,, \qquad q(z) \equiv z^2\,Q/P = q_0 + q_1 z + \cdots \\ &\text{with}\;\; p_0 = 1\,,\; q_0 = 0\;\; (l = 1)\,, \quad \text{or}\;\; p_0 = 2\,,\; q_0 \neq 0\;\; (l = 2)\,. \end{aligned} \tag{7.120}$$

The behaviour of the solutions in the neighbourhood of such points may be obtained by means of the *Frobenius method of expansion*.

The Frobenius expansion about a regular singularity assumes

$$\xi = z^\nu \sum_{n=0}^{\infty} a_n z^n\,, \tag{7.121}$$

where, in general, the index ν is not an integer and may even be complex. The values of ν and the coefficients a_n are obtained by substituting the expansion into the differential equation (7.119):

$$\sum_{n=0}^{\infty}(\nu+n)(\nu+n-1)a_n z^{\nu+n-2} + (p_0 + p_1 z + \cdots)\sum_{n=0}^{\infty}(\nu+n)a_n z^{\nu+n-2} \\ - (q_0 + q_1 z + \cdots)\sum_{n=0}^{\infty} a_n z^{\nu+n-2} = 0 . \qquad (7.122)$$

Balancing the different powers yields the following sequence of equalities:

$$z^{\nu-2}: \quad [\nu^2 + (p_0 - 1)\nu - q_0]\, a_0 = 0 ,$$

$$z^{\nu-1}: \quad [(\nu+1)\nu + p_0(\nu+1) - q_0]\, a_1 = (-\nu p_1 + q_1)\, a_0 , \quad \text{etc.} \qquad (7.123)$$

Hence, in general, the two independent solutions ξ_1 and ξ_2 start off with a term $a_0 z^{\nu_{1,2}}$, where a_0 is arbitrary and the indices $\nu_{1,2}$ are the solutions of the *indicial equation*, obtained by putting the factor in square brackets of (7.123)(a) equal to zero. The coefficients a_n ($n \geq 1$) follow by recursion from the relations (7.123)(b). The solutions of the indicial equation, and hence the solutions ξ_1 and ξ_2 themselves, are qualitatively different for the case (a) and the cases (b), (c) of Fig. 7.10:

$$l = 1 \ \text{(continuous spectrum)}: \qquad \nu_1 = \nu_2 = 0 , \qquad (7.124)$$

$$l = 2 \ \text{(cluster points)}: \qquad \nu_{1,2} = -\tfrac{1}{2} \pm \tfrac{1}{2}\sqrt{1 + 4q_0} . \qquad (7.125)$$

For the cluster point analysis ($l = 2$), this is sufficient information. It will be completed in Section 7.4.4 for the waves (case (b)) and in Section 7.5.3 for the marginal states (case (c)). For the analysis of the continuous spectrum ($l = 1$), we are not done yet since we have just obtained one value ($\nu = \nu_1 = \nu_2 = 0$) corresponding to an analytic function (called the *'small' solution*) which obviously does not exhibit singular behaviour. The *'large' solution*, revealing the character of the singularity, is to be obtained yet.

To complete the analysis for the case with equal indices ($\nu_1 = \nu_2 = 0$), the quoted literature (Ince [325], p. 397; Bender and Orszag [50], p. 73) shows how the 'large' solution is obtained. This involves the special trick of writing the 'small' solution obtained as a function of ν, not substituting the value 0 yet, differentiating it with respect to ν, and then inserting the value 0. This produces a second, independent, solution containing a logarithmic contribution, since

$$\frac{\partial}{\partial\nu}(z^\nu) = \frac{\partial}{\partial\nu}\left(\mathrm{e}^{\nu \ln z}\right) = z^\nu \ln z \;\rightarrow\; \ln z , \quad \text{for } \nu \rightarrow 0 . \qquad (7.126)$$

Hence, in the neighbourhood of the regular singular point $z = x - x_0 = 0$, two solutions are obtained:

$$\xi_1 = u(z) , \qquad u = u_0 + u_1 z + \cdots \qquad \text{('small' solution)} ,$$

$$\xi_2 = u(z)\,\ln z + v(z) , \qquad v = v_0 + v_1 z + \cdots \qquad \text{('large' solution)} , \qquad (7.127)$$

where the explicit form of the coefficients u_n and v_n is much less important than the appearance of the logarithmic factor. The latter introduces a branch point at $z = 0$ where a decision is to be taken on how to continue the solution past this point. This choice depends on the physical context in which the singularity is encountered.

In the solution of the initial value problem by means of the Laplace transform, as sketched in

Section 6.3.2 and elaborated in Chapter 10, the parameter ω is assumed to be complex and the continuation is obtained by deforming the Laplace contour around the singularity. This procedure is completely analogous to that exploited by Landau in his solution of the initial value problem for the Vlasov equation (Section 2.3.3), where the spatial coordinate x of the MHD problem corresponds to the velocity variable v of the Vlasov problem.

In the determination of the spectrum of ideal MHD, ω^2 remains real since there is no dissipation, so that we should consider the solutions (7.127) along the real axis, exploiting the principal value of $\ln z$, i.e. $\ln|x - x_0|$. The logarithmic singularity at $x = x_0$ then divides the real x-interval into two parts such that the solution ξ obtains just enough additional freedom (compared to that at an ordinary point) to satisfy the boundary conditions for any value of ω^2. This produces the continuous spectrum, as will be demonstrated in the next section.

▷ **Apparent singularities** The singularities $D = 0$ occur when either one of the factors $\omega^2 - \omega_{s0}^2$ or $\omega^2 - \omega_{f0}^2$ vanishes on the interval:

$$D \equiv \left[\omega^2 - \omega_{s0}^2(x)\right]\left[\omega^2 - \omega_{f0}^2(x)\right] \sim x - x_0 \quad \Rightarrow \quad P \sim (x - x_0)^{-1}\,, \quad Q \sim (x - x_0)^{-2}\,. \tag{7.128}$$

This produces singularities of the kind $P'/P \sim z^{-1}$ and $Q/P \sim z^{-1}$. The coefficients of the analytic functions $p(z)$ and $q(z)$ then yield solutions of the indicial equation with *an integer difference between the indices:*

$$p_0 = -1\,,\ q_0 = 0 \quad \Rightarrow \quad \nu_1 = 2\,,\ \nu_2 = 0\,. \tag{7.129}$$

This implies that only one 'small' solution $\sim z^2$ is obtained, whereas the other one would contain a logarithmic contribution, unless the coefficients of the expansion of $p(z)$ and $q(z)$ possess *the special property* [258]

$$(p_1 - q_1)q_1 - (p_0 - q_0)q_2 = 0\,. \tag{7.130}$$

In that case, none of the solutions is singular; see Ince [325], p. 404. The MHD wave equation (7.80) turns out to have precisely that property! This follows from how the expression D appears in the functions P and Q:

$$P \equiv N/D\,, \quad Q \equiv -A - B/D - \left(C/D\right)'\,, \tag{7.131}$$

where the explicit expressions for A, B and C can be read off from Eq. (7.80). They are related by

$$C^2 + NB = \rho^2 g^2 (\omega^2 - f^2 b^2)^2\, D \to 0 \quad \text{as } x \to x_0\,. \tag{7.132}$$

Translated in terms of the functions p and q, this produces the equality (7.130). Consequently, *the singularities $D = 0$ are apparent.* ◁

7.4.2 Alfvén and slow continua

For simplicity, we choose the equilibrium quantities such that the singular frequency functions $\omega_A^2(x)$ and $\omega_S^2(x)$ are well separated and monotonically increasing, as in Fig. 7.10(a). We will prove that the collection of frequencies $\omega^2 \in \{\omega_A^2(x)|\, x_1 \le x \le x_2\}$ and $\omega^2 \in \{\omega_S^2(x)|\, x_1 \le x \le x_2\}$ constitutes the continuous spectrum, i.e. the set of *improper eigenvalues* of the MHD force operator $\rho^{-1}\mathbf{F}$, with associated non-square integrable 'eigenfunctions'.

We concentrate again on the Alfvén singularities $\omega^2 \in \{\omega_A^2(x)\}$, noting that the analysis for the slow singularities $\omega^2 \in \{\omega_S^2(x)\}$, lying in a narrow band below the Alfvén ones, is completely analogous. For a given value of $\omega^2 = \omega_0^2$ from this set, the ODE (7.80) is singular ($N = 0$) at the position $x = x_0$ where $\omega_A^2(x_0) = \omega_0^2$, as shown in Fig. 7.10(a). The monotonically increasing function $\omega_A^2 = \omega_A^2(x)$ may be inverted to give a monotonically increasing function $x_A = x_A(\omega^2)$. In this manner, the position x of the singularity becomes a function of ω^2. As we have seen in Section 7.4.1, expansion about the singularity gives the dominant coefficient of the differential equation:

$$\omega^2 - \omega_A^2(x) \approx -(\omega_A^{2\,\prime})_0(x - x_0) = -(\omega_A^{2\,\prime})_0\, s\,, \qquad (7.133)$$

where, since everything is real now, the complex variable z is replaced by the real variable s, which is the distance to the singularity:

$$s \equiv x - x_0 = x - x_A(\omega^2)\,. \qquad (7.134)$$

Instead of the complex expressions (7.127), we now obtain real independent solutions of the form

$$\begin{aligned} \xi_1 &= u(s;\omega^2) && (\text{‘small’ solution})\,, \\ \xi_2 &= u(s;\omega^2)\ln|s| + v(s;\omega^2) && (\text{‘large’ solution})\,, \end{aligned} \qquad (7.135)$$

to the right ($s > 0$) as well as to the left ($s < 0$) of the singularity.

The most important consequence of the singularities $N = 0$ of the differential equation (7.80) is that certain jumps of the solutions are permitted, so that the boundary conditions (7.82) can always be satisfied for ω^2 in the singular domain. To prove this, we exploit the series expansions just derived. Because of the assumed monotonicity of $\omega_A^2(x)$, the interval (x_1, x_2) contains only one singular point for a fixed value of ω^2 so that the general solution may be written as

$$\xi = \big[A_1 u + C_1(u\ln|s| + v)\big]\, H(-s) + \big[A_2 u + C_2(u\ln|s| + v)\big]\, H(s)\,, \qquad (7.136)$$

where $H(s)$ is the Heaviside step function ($\equiv 0$ for $s < 0$ and $\equiv 1$ for $s > 0$), and we have to determine the values of A_1, C_1, A_2 and C_2. For a non-singular differential equation the solution should be continuous: $A_1 = A_2$ and $C_1 = C_2$. For a singular second order ODE we will prove that *the small solution may jump* whereas the large solution has to be continuous: $A_1 \neq A_2$, $C_1 = C_2$.

▷ **Proof that the small solution may jump** Write the differential equation (7.80) as

$$(P\xi')' - Q\xi = 0\,, \quad \text{where} \quad P \approx P_1 s + \cdots\,, \quad Q \approx Q_0 + \cdots\,. \qquad (7.137)$$

Substitution of a ‘small’ solution $\xi = uH(s)$ leads to the following equalities:

$$\begin{aligned} \xi' = u'H(s) + u\delta(s) \quad &\Rightarrow \quad P\xi' = Pu'H(s) + Pu\delta(s) = Pu'H(s)\,, \\ &\quad\; (P\xi')' = (Pu')'H(s) + Pu'\delta(s) = (Pu')'H(s)\,, \\ &\Rightarrow \quad (P\xi')' - Q\xi = \big[(Pu')' - Qu\big] H(s) = 0\,, \end{aligned}$$

because $u(s)$ is a solution of Eq. (7.137), and properties such as $H'(s) = \delta(s)$ and $s\delta(s) = 0$. Consequently, $A_1 uH(-s)$ is a solution of Eq. (7.137) but, likewise, $A_2 uH(s)$ is also a solution, so that the small solutions left and right are totally unrelated: $A_1 \neq A_2$. Performing a similar analysis for the large solution, one finds that the term $u\ln|s|H(s)$ produces a δ-function contribution that does not vanish, so that $C_1 = C_2$, QED. ◁

Consequently, the general solution to Eq. (7.137) may be written as

$$\xi = A_1 u\, H(-x + x_A) + A_2 u\, H(x - x_A) + C\,[u\ln|x - x_A| + v]\,. \qquad (7.138)$$

Due to the fact that we have now three (rather than two) constants available, the two boundary conditions $\xi(x_1) = 0$ and $\xi(x_2) = 0$ may always be satisfied for $\omega^2 \in \omega_A^2(x)$, i.e. when there is a singular point on the interval (x_1, x_2). By imposing these two boundary conditions, the constants

A_1 and A_2 may be eliminated so that *the improper eigenfunctions for an Alfvén continuum mode* become:

$$\xi_A(x;\omega^2) = C(\omega^2)\Big\{\Big[\ln\frac{x - x_A(\omega^2)}{x_1 - x_A(\omega^2)} - \frac{v_1(\omega^2)}{u_1(\omega^2)}\Big]\,u(x;\omega^2)H(-x + x_A(\omega^2))$$
$$+\Big[\ln\frac{x - x_A(\omega^2)}{x_2 - x_A(\omega^2)} - \frac{v_2(\omega^2)}{u_2(\omega^2)}\Big]\,u(x;\omega^2)H(x - x_A(\omega^2)) + v(x;\omega^2)\Big\}, \quad (7.139)$$

where $u(x;\omega^2) \equiv u(x - x_A(\omega^2))\,, \quad u_1(\omega^2) \equiv u(x_1;\omega^2)\,,$ etc.

The factor $C(\omega^2)$ may be fixed by 'normalizing' the eigenfunctions:

$$\langle\, \boldsymbol{\xi}_A(x;\omega^2), \boldsymbol{\xi}_A(x;\widetilde{\omega}^2)\,\rangle = \delta(\omega^2 - \widetilde{\omega}^2)\,. \quad (7.140)$$

Similar expressions may be derived for the improper eigenfunctions of the slow continuum modes. Therefore, 'solutions' have been obtained which satisfy the boundary conditions (Fig. 7.11(a)) for any $\omega^2 \in \{\omega_A^2(x)\}$ and $\omega^2 \in \{\omega_S^2(x)\}$, QED. The frequency ranges $[\omega^2_{A,\min}, \omega^2_{A,\max}]$ and $[\omega^2_{S,\min}, \omega^2_{S,\max}]$ are called the *Alfvén continuum* and the *slow magneto-sonic continuum*, respectively.

This appears to establish the existence of two continuous spectra, but the most characteristic part of the eigenfunctions is not yet discussed. Actually, if we restrict the analysis to the normal component of that function, we can not even prove that we have 'improper' eigenfunctions because the singularities $\ln|s|$ and $H(s)$ are square integrable. The dominant, non-square integrable, part of the 'improper' eigenfunction resides in the tangential components η and ζ, given by Eqs. (7.79), which involve the derivative $\xi' \gg \xi/a$. From Eq. (7.138), this derivative produces an $(x - x_A)^{-1}$ singularity and a δ-function:

$$\xi_A' = C\Big\{\Big[\mathcal{P}\frac{1}{x - x_A(\omega^2)} - \frac{A_1 - A_2}{C}\,\delta(x - x_A(\omega^2))\Big]\,u$$
$$+\Big[\frac{A_1}{C}H(-x + x_A(\omega^2)) + \frac{A_2}{C}H(x - x_A(\omega^2)) + \ln|x - x_A(\omega^2)|\Big]\,u' + v'\Big\}$$
$$\approx C(\omega^2)\Big\{\mathcal{P}\frac{1}{x - x_A(\omega^2)} + \mu(\omega^2)\,\delta(x - x_A(\omega^2))\Big\}\,, \quad (7.141)$$

where $\mathcal{P}$ indicates that the principal part is to be taken when integrating this function, and $\mu(\omega^2)$ is a function involving the boundary data of u and v:

$$\mu \equiv \frac{-A_1 + A_2}{C} = \ln\left|\frac{x_2 - x_A(\omega^2)}{x_1 - x_A(\omega^2)}\right| - \frac{v_1(\omega^2)}{u_1(\omega^2)} + \frac{v_2(\omega^2)}{u_2(\omega^2)}\,. \quad (7.142)$$

For the Alfvén continuum modes, the tangential component η perpendicular to $\mathbf{B}$ dominates (since ξ_A' is multiplied by $\omega^2 - \omega_A^2$ in the expression for ζ):

$$\eta_A \approx \frac{k_\perp(b^2 + c^2)(\omega_A^2 - \omega_S^2)}{D(\omega_A^2)}\,\xi_A' = -\frac{1}{k_\perp}\,\xi_A' \gg \xi_A \sim \zeta_A\,, \quad \text{for } \omega^2 \to \omega_A^2\,. \quad (7.143)$$

Similarly, for the slow continuum modes, the component ζ parallel to $\mathbf{B}$ dominates (since ξ_S' is multiplied by $\omega^2 - \omega_S^2$ in the expression for η):

$$\zeta_S \approx \frac{k_\parallel c^2(\omega_S^2 - \omega_A^2)}{D(\omega_S^2)}\,\xi_S' = -\frac{b^2+c^2}{k_\parallel c^2}\,\xi_S' \gg \xi_S \sim \eta_S\,, \quad \text{for } \omega^2 \to \omega_S^2\,, \tag{7.144}$$

where the formal expression ξ_S' is obtained from Eq. (7.141) by replacing the subscripts A by S. The tangential components of these 'improper' eigenfunctions are schematically shown in Fig. 7.11(b).

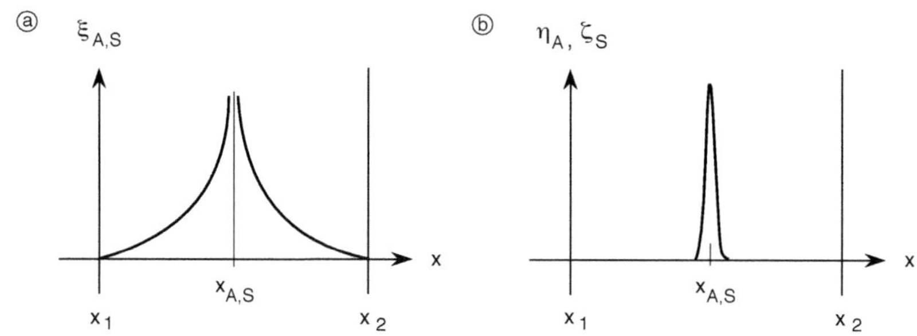

Fig. 7.11 Schematic representation of the (a) normal and (b) tangential components of the improper Alfvén and slow continuum 'eigenfunctions'.

In conclusion: the continuum modes are characterized by *a non-square integrable tangential component perpendicular to the magnetic field for the Alfvén modes* and *a non-square integrable parallel tangential component for the slow modes*. This shows the extreme anisotropy of ideal MHD waves as regards motion inside and across magnetic surfaces. This is a quite general property which remains true for other plasma geometries, like cylindrical and toroidal ones.

We have obtained the generalization of the essential spectrum, represented by the three cluster point singularities for the simpler inhomogeneous plasmas discussed in Section 7.3.3. What about the *fast magneto-sonic singularities* in the high frequency limit? In that limit, as we will show, ξ' also blows up but the normal component itself still dominates over the tangential components:

$$\xi_F' \gg \xi_F \gg \eta_F \sim \zeta_F\,, \quad \text{for} \quad \omega^2 \to \omega_F^2 \equiv \infty\,. \tag{7.145}$$

Hence, the three expressions (7.143)–(7.145) for the slow and Alfvén continuum modes and the fast cluster modes together establish the generalization for inhomogeneous plasmas of the orthogonality property (5.64) and of the asymptotic properties (5.86) for the three MHD waves that were introduced in Chapter 5.

▷ **Fast magneto-sonic cluster point singularities** For large ω^2, the derivatives with respect to x also become large, so that the differential equation for ξ reduces to

$$\xi'' + \frac{\omega^2}{b^2+c^2}\,\xi = 0\,. \tag{7.146}$$

The solution of this equation may be obtained by means of WKB analysis; see e.g. Bender and Orszag [50], p. 484. The leading order solution reads:

$$\xi \approx C(b^2+c^2)^{1/4}\sin\Big[\omega_n \int_{x_1}^{x} (b^2+c^2)^{-1/2}\,dx\Big]\,, \qquad \omega_n \approx n\pi \Big/ \int_{x_1}^{x_2} (b^2+c^2)^{-1/2}\,dx\,. \tag{7.147}$$

Although the sequence (7.147) contains functions of finite norm, it does not converge to an element of Hilbert space for $n \to \infty$ (see Section 6.3.1). Hence, the fast cluster point $\omega_F^2 \equiv \lim_{n\to\infty} \omega_n^2 = \infty$ belongs to the continuous or essential spectrum. From these expressions, the singular behaviour (7.145) of the fast magneto-sonic modes follows. ◁

▷ **Historical note** It is clear now that the proper description of the singularities of the MHD wave equation is crucial for the theory of inhomogeneous plasmas. The correct picture gradually evolved, mostly from the analysis of the analogous problem in cylinder geometry with various simplifying assumptions. The presence of a continuous spectrum of Alfvén waves in MHD was pointed out by Uberoi [614], who also drew attention to the analogy with the problem of electrostatic oscillations in inhomogeneous cold plasmas (another fluid model). The latter was analysed by Barston [36] by means of singular ('improper') normal modes, like the van Kampen modes [628] in the Vlasov description of plasmas, and by Sedláček [546] by means of the Laplace transform of the initial value problem, like in the celebrated Landau solution of the damping [393]. All this made it clear that these kinds of singularities are not restricted to the microscopic velocity space details of a kinetic description of plasmas, but also occur in fluid descriptions when the plasma is inhomogeneous in ordinary space. Accordingly, the possibility of dissipationless damping of Alfvén waves was pointed out by Tataronis and Grossmann [588], and the complementary theory of heating by Alfvén waves was developed by Chen and Hasegawa [118]. From there, applications to laboratory and astrophysical plasmas multiplied. These topics will be discussed in Chapter 10. In the meanwhile, Grad [254] had put the subject, including stability, in the context of MHD spectral theory, pointing out the presence of four types of singularities (for $N = 0$ and $D = 0$) and, unfortunately, associating a continuous spectrum with each of them. The mistake concerning the $D = 0$ singularities was pointed out immediately by Appert, Gruber and Vaclavik [18] (by means of the system of first order differential equations discussed in Section 7.3.2), but Grad maintained that this was insufficient proof since a full construction of the resolvent operator (appearing in the solution of the initial value problem; see Section 6.3) would be required. Such a proof was given by Goedbloed in a memorandum, that was only published 24 years later [221] when the mistake was repeated by other authors. Since the form of the resolvent operator is another display of the consistency of MHD spectral theory, it is reproduced in Section 10.2. ◁

7.4.3 Oscillation theorems

In Section 7.3.1, we have shown that the sequence of inequalities (7.61) already exhibits the central position of the essential spectrum for homogeneous plasmas, where it just consists of the single frequencies ω_S^2, ω_A^2, and ω_F^2. In the MHD wave equation (7.80) for inhomogeneous plasmas, these frequencies returned in the form of the *genuine singularities* $\omega_S^2(x)$ and $\omega_A^2(x)$, which give rise to continuous spectra consisting of their values on the entire plasma interval $x_1 \leq x \leq x_2$. In addition, certain *apparent singularities* $\omega_{s0}^2(x)$ and $\omega_{f0}^2(x)$ turned up, which could be proved not to give rise to logarithmic contributions in the normal component, nor to any non-square integrable contributions in the tangential components, so that they do not represent continuous spectra. In this section, we will get to appreciate the constructive role they play in determining the spectral structure. For inhomogeneous plasmas, similar to the inequalities (7.61), one easily demonstrates that the mentioned frequencies are well ordered at each position x of the plasma slab:

$$0 \leq \omega_S^2(x) \leq \omega_{s0}^2(x) \leq \omega_A^2(x) \leq \omega_{f0}^2(x) \leq \omega_F^2 = \infty\,. \qquad (7.148)$$

Hence, for a weakly inhomogeneous plasma, or for a thin slice of the plasma not containing points $F = 0$ (since ω_S, ω_{s0} and ω_A coalesce there), the genuine and apparent singularities (7.148) on the interval $x_1 \leq x \leq x_2$ might be distributed along the real ω^2-axis as schematically illustrated in Fig. 7.14 further in this section. In this section, we will demonstrate how the rest of the spectrum shown is obtained.

(a) Sturm's theorems To do that, we need to study the qualitative behaviour of the eigenfunctions ξ of the differential equation (7.80), subject to the boundary conditions (7.82), as a function of the eigenvalue parameter ω^2. The kind of qualitative behaviour we envision is exemplified by the classical *Sturm–Liouville system* (see e.g. [450](a)), which is described by the non-singular second order differential equation with eigenvalue parameter λ,

$$(P\xi')' - (Q - \lambda R)\,\xi = 0\,, \tag{7.149}$$

where P, Q and R are functions of x, with $P > 0$ and $R > 0$. Let $\xi^{(1)}$ and $\xi^{(2)}$ be linearly independent solutions of Eq. (7.149) for a fixed value of λ. Denote the two linear combinations of these solutions by

$$\xi_{\rm a} = a_1\xi^{(1)} + a_2\xi^{(2)}\,, \qquad \xi_{\rm b} = b_1\xi^{(1)} + b_2\xi^{(2)}\,. \tag{7.150}$$

(See Fig. 7.12(a), where we temporarily use x_1 and x_2 to indicate a sub-interval.) If $b_2/b_1 \neq a_2/a_1$, these solutions are linearly independent, i.e. the Wronski determinant $\xi_{\rm a}\xi_{\rm b}' - \xi_{\rm a}'\xi_{\rm b} \neq 0$ on the interval considered. *Sturm's separation theorem* [325] states that the zeros of these solutions separate each other: *If x_1 and x_2 are consecutive zeros of $\xi_{\rm a}$, then $\xi_{\rm b}$ vanishes once in the open interval (x_1, x_2).*

Proof Suppose $\xi_{\rm b}$ does not vanish on (x_1, x_2). Then, x_1 and x_2 are consecutive zeros of the finite function $\xi_{\rm a}/\xi_{\rm b}$. Hence, $d(\xi_{\rm a}/\xi_{\rm b})/dx$ must vanish at least once on the interval. However,

$$d(\xi_{\rm a}/\xi_{\rm b})/dx = (\xi_{\rm a}'\xi_{\rm b} - \xi_{\rm a}\xi_{\rm b}')/\xi_{\rm b}^2 \tag{7.151}$$

cannot vanish because that would imply that the Wronski determinant vanishes somewhere. This contradiction proves that $\xi_{\rm b}$ must vanish at least once. It cannot vanish more than once since then we could interchange the roles of $\xi_{\rm a}$ and $\xi_{\rm b}$ and again get a contradiction. Hence, $\xi_{\rm b}$ vanishes once, and only once; QED.

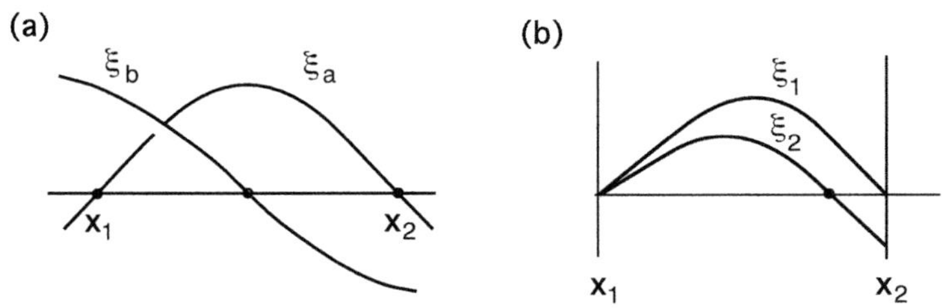

Fig. 7.12 Sturm's theorems: (a) separation theorem, and (b) oscillation theorem.

As far as the oscillatory properties of Eq. (7.149) are concerned, one could say that all solutions oscillate equally fast if λ is kept fixed. Considering solutions of Eq. (7.149) for different values of λ, one may compare their oscillatory behaviour by means of *Sturm's fundamental oscillation theorem* [325] stating the following: *If x_1 and x_2 are two consecutive zeros of the function ξ_1 satisfying Eq. (7.150) for $\lambda = \lambda_1$, then the solutions ξ_2 of Eq. (7.150) for $\lambda = \lambda_2$ oscillate faster than ξ_1 if $\lambda_2 > \lambda_1$*; see Fig. 7.12(b). Here, 'faster oscillating' means that the solution ξ_2 that vanishes at the left end point $x = x_1$ vanishes at least once on (x_1, x_2).

Proof Multiply the Sturm–Liouville equation (7.149) for $\lambda = \lambda_1$ with ξ_2 and also multiply the equation for $\lambda = \lambda_2$ with ξ_1, integrate over (x_1, x_2), and subtract:

$$\int_{x_1}^{x_2} \Big[\xi_2(P\xi_1')' - \xi_1(P\xi_2')'\Big]\, dx = \Big[\xi_2 P\xi_1' - \xi_1 P\xi_2'\Big]_{x_1}^{x_2} = (\xi_2 P\xi_1')\Big|_{x=x_2}$$
$$= (\lambda_2 - \lambda_1) \int_{x_1}^{x_2} R\,\xi_1\xi_2\, dx\,. \qquad (7.152)$$

Suppose that the solution ξ_2, which vanishes at $x = x_1$, does not vanish in the open interval (x_1, x_2). Then, the LHS of Eq. (7.152) is negative, whereas the RHS is positive. This contradiction proves that ξ_2 has to vanish at least once on the open interval (x_1, x_2); QED.

Sturm's oscillation theorem provides the qualitative behaviour of the solutions of the differential equation (7.149) on any sub-interval of the interval (x_1, x_2). Such an equation, with the property that the solutions oscillate faster upon increasing the eigenvalue parameter λ, is called *Sturmian*. We will call differential equations that have the opposite property (e.g. Eq. (7.149) when the sign of λ is reversed) *anti-Sturmian*. An immediate consequence of these properties is that one can label different discrete modes by just counting the number of nodes on the interval (x_1, x_2). If the system is Sturmian the eigenvalue λ is an increasing function of the number of nodes n, whereas for anti-Sturmian systems λ decreases as a function of n (Fig. 7.13). The classical example of the first kind of behaviour is the vibrating string. Examples of the second kind are much less familiar. However, in Sections 7.2.3 and 7.3.3, we have encountered some for the HD and MHD waves in a gravitating atmosphere. We will now generalize these properties to arbitrary inhomogeneity.

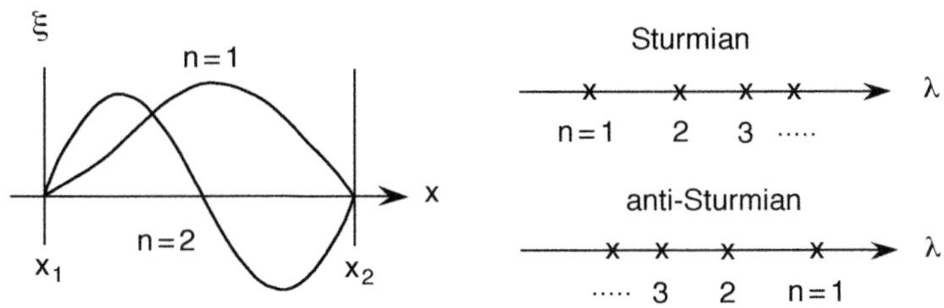

Fig. 7.13 Sturmian and anti-Sturmian dependence of the eigenvalue λ on the number of nodes.

(b) Oscillation theorem for the MHD wave equation Turning to the MHD wave equation (7.80), it is immediately clear that it is not an equation of the simple Sturm–Liouville kind (7.149). Nevertheless, we may ask whether it still has the Sturmian, or anti-Sturmian, property. It is clear that in order to prove such a property we have to exclude regions of ω^2 corresponding to the continuous spectrum, where the differential equation becomes singular, i.e. where the coefficient N/D in front of the highest derivative develops zeros. Moreover, it turns out that we also will have to exclude the regions of ω^2 corresponding to the apparent singularities, where N/D becomes infinite. Let us then study the monotonicity properties, if any, of the discrete spectrum of Eq. (7.80) for values of ω^2 outside the continua $\{\omega_A^2\}$ and $\{\omega_S^2\}$ and also outside the ranges $\{\omega_{s0}^2\}$ and $\{\omega_{f0}^2\}$ shown in Fig. 7.14 below. For those values of ω^2 the wave equation is free of both genuine and apparent singularities, but the way in which ω^2 appears in the equation makes it virtually impossible to prove anything directly from it. The reason is that the wave equation (7.80) has been derived by reducing

the original vector eigenvalue problem (with linear dependence on ω^2) to a scalar one. The only hope to prove general properties about the spectrum is to go back to first principles and to exploit the self-adjointness of the original operator $\rho^{-1}\mathbf{F}$.

We prove the *MHD spectral oscillation theorem*, due to Goedbloed and Sakanaka [240]:
If x_1 and x_2 are two consecutive zeros of the function ξ_1 satisfying the MHD wave equation (7.80) for $\omega^2 = \omega_1^2$, then the solutions ξ_2 of the MHD wave equation for $\omega^2 = \omega_2^2$ oscillate faster than ξ_1 if $\omega_2^2 > \omega_1^2$ and $N/D > 0$ (Sturmian), and slower if $N/D < 0$ (anti-Sturmian).

Proof Recall the proof of self-adjointness of the operator $\rho^{-1}\mathbf{F}$ of Section 6.2.3. In particular, let us start from the expression (6.56) for the inner product involving the two displacement vectors $\boldsymbol{\xi}$ and $\boldsymbol{\eta}$, which we now replace by $\boldsymbol{\xi}_1$ and $\boldsymbol{\xi}_2$:

$$\begin{aligned}\langle \boldsymbol{\xi}_2, \rho^{-1}\mathbf{F}(\boldsymbol{\xi}_1)\rangle - \langle \boldsymbol{\xi}_1, \rho^{-1}\mathbf{F}(\boldsymbol{\xi}_2)\rangle &= \tfrac{1}{2}\int \mathbf{n}\cdot\boldsymbol{\xi}_2\,(\gamma p \nabla\cdot\boldsymbol{\xi}_1 + \boldsymbol{\xi}_1\cdot\nabla p - \mathbf{B}\cdot\mathbf{Q}_1)\,dS \\ &\quad - \tfrac{1}{2}\int \mathbf{n}\cdot\boldsymbol{\xi}_1\,(\gamma p \nabla\cdot\boldsymbol{\xi}_2 + \boldsymbol{\xi}_2\cdot\nabla p - \mathbf{B}\cdot\mathbf{Q}_2)\,dS. \qquad (7.153)\end{aligned}$$

For simplicity, we here exploit the inner product definition (6.35) but assume the vectors to be real. The surface S bounding the plasma actually consists of two surfaces, viz. the planes $x = x_1$ and $x = x_2$. Note that the expressions in the brackets on the RHS correspond to the Eulerian perturbation Π of the total pressure defined in Eq. (7.87), and reduced in Eq. (7.88). Therefore, we may write

$$\begin{aligned}&\langle \boldsymbol{\xi}_2, \rho^{-1}\mathbf{F}(\boldsymbol{\xi}_1)\rangle - \langle \boldsymbol{\xi}_1, \rho^{-1}\mathbf{F}(\boldsymbol{\xi}_2)\rangle - \tfrac{1}{2}\int \mathbf{n}\cdot\boldsymbol{\xi}_2\,\Pi(\boldsymbol{\xi}_1)\,dS + \tfrac{1}{2}\int \mathbf{n}\cdot\boldsymbol{\xi}_1\,\Pi(\boldsymbol{\xi}_2)\,dS \\ &\quad = -\tfrac{1}{2}\Big[\xi_2\,\Pi(\boldsymbol{\xi}_1) - \xi_1\,\Pi(\boldsymbol{\xi}_2)\Big]_{x_1}^{x_2} = \tfrac{1}{2}\bigg[\xi_2\,\frac{N}{D} - \xi_1\,\frac{N}{D}\bigg|_{\omega_2^2}\xi_2'\bigg]_{x_1}^{x_2}, \qquad (7.154)\end{aligned}$$

where we have suppressed the (infinite) area $A \equiv \int dS$ of the y–z boundaries. Consider two solutions $\boldsymbol{\xi}_1$ and $\boldsymbol{\xi}_2$ of the MHD normal mode equation corresponding to different values ω_1^2 and ω_2^2 of the eigenvalue parameter,

$$\mathbf{F}(\boldsymbol{\xi}_1) = -\rho\omega_1^2\,\boldsymbol{\xi}_1, \qquad \mathbf{F}(\boldsymbol{\xi}_2) = -\rho\omega_2^2\,\boldsymbol{\xi}_2, \qquad (7.155)$$

but not necessarily satisfying the boundary conditions (7.82). Then, the LHS of Eq. (7.154) transforms to $(\omega_2^2 - \omega_1^2)\,\langle\boldsymbol{\xi}_1, \boldsymbol{\xi}_2\rangle$. Consider a sub-interval (x_1, x_2) of the complete physical interval, exploiting x_1 and x_2 again to indicate two consecutive zeros of the normal component ξ_1 of $\boldsymbol{\xi}_1$. Let ω_2^2 be close to ω_1^2, so that $\boldsymbol{\xi}_2$ is close to $\boldsymbol{\xi}_1$ and, hence, $\langle\boldsymbol{\xi}_1, \boldsymbol{\xi}_2\rangle > 0$. We will also choose $\boldsymbol{\xi}_2$ such that ξ_2 vanishes at $x = x_1$. We wish to find out whether or not ξ_2 has another zero on (x_1, x_2), i.e., whether ξ_2 oscillates faster or slower than ξ_1 for a given difference of ω_1^2 and ω_2^2. Under the mentioned conditions all that remains of Eq. (7.154) is

$$(\omega_2^2 - \omega_1^2)\,\langle\boldsymbol{\xi}_1, \boldsymbol{\xi}_2\rangle = \tfrac{1}{2}\bigg(\xi_2\frac{N}{D}\bigg|_{\omega_1^2}\xi_1'\bigg)\bigg|_{x=x_2}. \qquad (7.156)$$

Let $\xi_1 > 0$ on the open interval (x_1, x_2), so that $\xi_1'(x = x_2) < 0$. If $N/D > 0$ and $\omega_2^2 - \omega_1^2 > 0$, this implies that $\xi_2(x = x_2) < 0$ so that ξ_2 oscillates faster than ξ_1 (Sturmian behaviour). If, on the other hand, $N/D < 0$, ξ_2 oscillates slower than ξ_1 (anti-Sturmian behaviour); QED.

An important property, directly following from Eq. (7.156), concerns the *orthogonality* of the eigenfunctions of the discrete spectrum. Returning again to using x_1 and x_2 for the complete physical interval, if ξ_1 and ξ_2 both satisfy the left and the right hand boundary conditions (7.82), the RHS of Eq. (7.156) vanishes, so that

$$\langle \boldsymbol{\xi}_1, \boldsymbol{\xi}_2 \rangle = 0 \qquad \text{for} \quad \omega_1^2 \neq \omega_2^2 \,. \tag{7.157}$$

Hence, the discrete eigenfunctions form an orthogonal set, which may also be normalized to obtain an orthonormal set.

(c) Spectral structure From the oscillation theorem it follows that the discrete spectrum outside the ranges $\{\omega_{\rm A}^2\}$, $\{\omega_{\rm S}^2\}$, $\{\omega_{\rm s0}^2\}$, $\{\omega_{\rm f0}^2\}$ is Sturmian for $N/D > 0$, so that the eigenvalue ω^2 increases with the number of nodes of the eigenfunction, and anti-Sturmian for $N/D < 0$, with opposite behaviour of the eigenfunctions. Consequently, the discrete spectrum changes from Sturmian to anti-Sturmian, and vice versa, every time ω^2 crosses one of those four special frequency regions. Thus, the frequency regions $\{\omega_{\rm s0}^2\}$ and $\{\omega_{\rm f0}^2\}$ are 'separators' of the discrete sub-spectra where non-monotonicity may occur. Thus, the spectrum of an inhomogeneous plasma slab schematically may look as shown in Fig. 7.14.

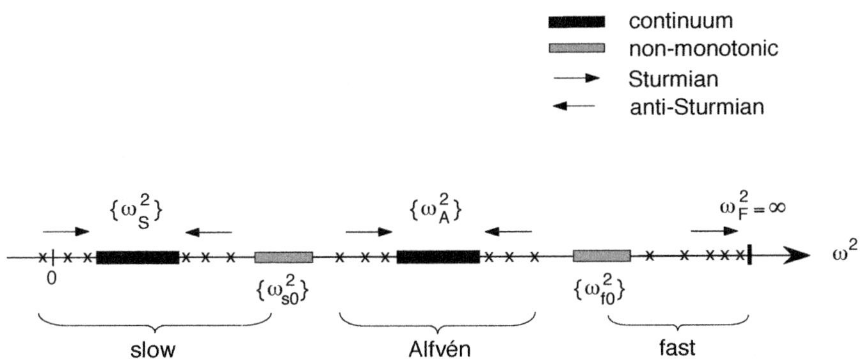

Fig. 7.14 Schematic structure and monotonicity of the spectrum of a slightly inhomogeneous plasma layer with gravity: genuine continuum singularities (black) and apparent ones (grey). Without gravity, the Sturmian slow modes and all discrete Alfvén modes will be missing. (From Goedbloed [215].)

In conclusion, compared to a uniform plasma, the spectrum of an inhomogeneous plasma slab exhibits the following features.

– The infinite degeneracy of the Alfvén point spectrum is lifted and replaced by a continuum of improper modes; in addition, a finite or infinite number of Sturmian, as well as anti-Sturmian, discrete Alfvén modes may occur.

– The accumulation point of the slow magneto-sonic point eigenvalues is spread out into a continuum of improper slow magneto-sonic modes; in addition, a finite or infinite number of Sturmian discrete slow modes may occur.

– The fast magneto-sonic point spectrum still accumulates at infinity.

– The cutoff frequencies $\omega_{\rm s0}^2$ and $\omega_{\rm f0}^2$ are spread out into ranges, which are *not* part of the spectrum, where the discrete spectrum may be non-monotonic there.

The clear separation of the sub-spectra shown in Fig. 7.14 is only obtained if the inhomogeneity is not too strong. For strong inhomogeneity the different parts of the spectrum, and their separators, in general fold over each other so that very complicated structures may result. To analyse those, one could consider the inhomogeneous plasma as consisting of a sequence of thin layers, each of which is only weakly inhomogeneous. For each of those layers the local modes (continua and cluster spectra) are well ordered according to the inequalities (7.148), but the non-local discrete modes may appear at apparently arbitrary positions in the spectrum since they may be due to composite structures displaying Alfvénic properties in one part of the plasma and magneto-sonic properties in another part.

Fortunately, to compute stability, in particular the growth rates of instabilities ($\omega^2 < 0$), the relevant part of the spectral structure is extremely simple. There are no continuous spectra for $\omega^2 < 0$ and the discrete modes are Sturmian there. This is exploited in the shooting method, discussed in Section 7.5.1(b).

7.4.4 Cluster spectra★

To show that all four branches of the Alfvén and slow discrete sub-spectra actually may occur, we now derive the conditions for clustering at the edges of the Alfvén and slow continua. To that end, we continue with the analysis of case (b) of Section 7.4.1, where ω^2 is chosen to correspond with the maximum or minimum of the Alfvén or slow continuum; see Fig. 7.10(b). The expression for the singular Alfvén factor at such an extremum is given in Eq. (7.115), and a similar expression holds for the singular slow factor. This yields a Frobenius expansion with indices $\nu_{1,2}$ given by Eq. (7.125). These indices become complex when the factor under the square root sign is negative, i.e. for

$$q_0 \equiv (s^2 Q/P)_0 < -\tfrac{1}{4}\,, \tag{7.158}$$

where $s \equiv x - x_0$ is the distance to the singular point. We will show in Section 7.5.3, for the similar case (c) of a cluster point at marginal stability, that complex indices imply that ξ oscillates infinitely rapidly when the singular point is approached. This implies, from the oscillation theorem, that an infinity of discrete modes is found accumulating at the chosen values of ω^2. We will work out the explicit forms of this inequality for the extrema of the Alfvén and slow continua.

Recall that $P \equiv N/D$. For the Alfvén continuum, one then finds from the expressions (7.80) and (7.81) that

$$\begin{aligned} &(N/s^2)_0 = -\tfrac{1}{2}k_\parallel^2 b^4 \rho(\omega_A^2)''\,, \quad D_0 = -k_\parallel^2 k_\perp^2 b^4\,, \quad Q_0 = -\rho' g\,, \\ &\Rightarrow \quad q_0 = -k_\perp^2 \rho' g/[\tfrac{1}{2}\rho(\omega_A^2)'']\,. \end{aligned} \tag{7.159}$$

This implies that, in order to have a *Sturmian sequence of discrete Alfvén modes clustering at the lower edge of the Alfvén continuum*, i.e. at the minimum of ω_A^2 (where $\omega_A^{2\,\prime\prime} > 0$), the following condition should be satisfied:

$$k_\perp^2 \rho' g > \tfrac{1}{8}\rho(\omega_A^2)'' > 0\,; \tag{7.160}$$

vice versa, for an *anti-Sturmian sequence at the upper edge of the Alfvén continuum*, i.e. at the maximum of ω_A^2 (where $\omega_A^{2\,\prime\prime} < 0$):

$$k_\perp^2 \rho' g \;<\; \tfrac{1}{8}\rho(\omega_A^2)'' \;<\; 0\,. \tag{7.161}$$

We have encountered an example of the latter in Fig. 7.9, with the cluster spectrum given by Eq. (7.109)(b), for the exponential atmosphere (where $\omega_A^{2\,\prime\prime} = 0$). The counterpart of a Sturmian Alfvén sequence was missing there because that requires an inverted density profile (which is very well possible for more general equilibria).

Similarly, for extrema of the slow continua, one finds the following expressions:

$$(N/s^2)_0 = \tfrac{1}{2}k_\parallel^2 b^4 \rho(\omega_S^2)''\,, \quad D_0 = -\omega_S^4\,,$$
$$Q_0 = \big[\rho b^2/(c^2\omega_S^2)\big]\,(\omega_S^4 - k_0^2 g^2) - \big[(b^2+c^2)/c^2)\,\rho g\big]'\,. \tag{7.162}$$

Hence, in order to have a *Sturmian sequence of discrete slow modes clustering at the lower edge of the slow continuum*, the following condition should be satisfied:

$$\frac{1}{b^2+c^2}\Big[-\rho(\omega_S^4 - k_0^2 g^2) + \frac{c^2}{b^2}\,\omega_S^2\Big(\frac{b^2+c^2}{c^2}\,\rho g\Big)'\Big] \;>\; \tfrac{1}{8}\,\rho(\omega_S^2)'' \;>\; 0\,; \tag{7.163}$$

vice versa, for an *anti-Sturmian sequence at the upper edge of the slow continuum*:

$$\frac{1}{b^2+c^2}\Big[-\rho(\omega_S^4 - k_0^2 g^2) + \frac{c^2}{b^2}\,\omega_S^2\Big(\frac{b^2+c^2}{c^2}\,\rho g\Big)'\Big] \;<\; \tfrac{1}{8}\,\rho(\omega_S^2)'' \;<\; 0\,. \tag{7.164}$$

We have encountered an example of the latter behaviour for homogeneous plasmas (where the gravity terms are missing) and of the former in Fig. 7.9(b), with the cluster spectrum given by Eq. (7.109)(c), for the exponential atmosphere (where $\omega_S^{2\,\prime\prime} = 0$). Also recall that transition from Sturmian to anti-Sturmian behaviour occurs for a value of $\bar{k}_0^2$ far to the right of the plotted region in Fig. 7.9(b).

Consequently, for general equilibria, the five discrete cluster spectra indicated in Fig. 7.14 do actually occur. In the absence of complex indices, the Alfvén and slow cluster points plus the high n modes will be missing, but global (lower n) modes may still be present. To compute the global modes requires numerical solution of the MHD wave equation (7.80).

7.5 Gravitational instabilities of a magnetized plasma slab

In this section, we apply the stability theory by means of the *energy principle*, as presented in Chapter 6, to the inhomogeneous gravitating plasma slab. This provides another opportunity to illustrate the equivalence of the variational approach with that of the differential equations, as developed in Sections 7.2–7.4.

Gravity does not influence the continua, but it does influence the appearance or disappearance of discrete cluster spectra (Section 7.4.4). This is clear from the fact that gravity does not appear in the expressions (7.84) and (7.85) for the genuine and apparent singularities ω_A^2, ω_S^2, ω_{s0}^2, ω_{f0}^2 of the basic differential equation (7.80), but it does appear in the other coefficients of the ODE. Hence, in the logical exposition of the theory, we first had to pay attention to the more difficult issue of the singularities (Section 7.4) whereas the simpler problems of local (related to cluster points) and global stability could be delayed until the present section.

7.5.1 Energy principle for a gravitating plasma slab

Recall that stability can be determined by just studying the sign of the quadratic form $W[\boldsymbol{\xi}]$ for the potential energy for all possible perturbations $\boldsymbol{\xi}$. The general expression of $W[\boldsymbol{\xi}]$ is given in Eq. (6.84). We now reduce this expression for the special case of a gravitating plasma slab. This will be the quadratic forms counterpart of the differential equation analysis of Sections 7.3 and 7.4. Of course, the starting point is the same, so that the equilibrium equations (7.62)–(7.65), the 2D Fourier expressions (7.69) for the perturbations, and the projections (7.70)–(7.73) still hold. Again, we study separate harmonics, denoted by $\boldsymbol{\xi}(x)\ \exp \mathrm{i}(k_y y + k_z z)$.

(a) Reduction to a one-dimensional variational problem Consider the different contributions to the integrand of W. Recall that we defined the three components ξ, η and ζ such that they can be assumed to be real. Hence, the expression for the compressibility, $\nabla\cdot\boldsymbol{\xi} = \xi' + k_\perp\eta + k_\parallel\zeta$ is real. The Cartesian components of the magnetic field perturbation $\mathbf{Q}$ were worked out in Eq. (7.77), from which Q_x is imaginary, whereas Q_y and Q_z are real. A typical term of the integrand,

$$\begin{aligned}\mathbf{j}\cdot\boldsymbol{\xi}^*\times\mathbf{Q} &= j_y(\xi_z^* Q_x - \xi_x^* Q_z) + j_z(\xi_x^* Q_y - \xi_y^* Q_x)\\ &= -B_z{}'\Big[-\frac{\mathrm{i}}{B}(B_y\eta - B_z\zeta)\,Q_x - \xi\,Q_z\Big] + B_y{}'\Big[\xi\,Q_y - \frac{\mathrm{i}}{B}(B_z\eta + B_y\zeta)\,Q_x\Big],\end{aligned}$$

is now easily seen to be real when the expressions (7.77) are inserted. The remaining contributions to the integrand are also real:

$$(\boldsymbol{\xi}^*\cdot\nabla p)\,\nabla\cdot\boldsymbol{\xi} = p'\xi\,\nabla\cdot\boldsymbol{\xi}\,, \qquad (\boldsymbol{\xi}^*\cdot\nabla\Phi_{\rm gr})\,\nabla\cdot(\rho\boldsymbol{\xi}) = -g\xi\,(\rho\nabla\cdot\boldsymbol{\xi} + \rho'\xi)\,.$$

Putting everything together, and cancelling terms by exploiting the equilibrium relation (7.65), the following expression for W is obtained:

$$\begin{aligned}W = \tfrac{1}{2}\int\Big[&\frac{k_\parallel^2 B^2}{k_0^2}\xi'^2 + \Big(k_\parallel^2 B^2 - \rho' g - \frac{\rho^2 g^2}{\gamma p}\Big)\xi^2\\ &+ B^2\Big(k_0\eta + \frac{k_\perp}{k_0}\xi'\Big)^2 + \gamma p\Big(\nabla\cdot\boldsymbol{\xi} - \frac{\rho g}{\gamma p}\xi\Big)^2\Big]\,dV\,.\end{aligned} \qquad (7.165)$$

This quadratic form only depends on ξ, ξ', η and ζ so that, upon minimization, one may expect a differential equation for ξ, and algebraic relations for η and ζ.

In fact, minimization with respect to the variables η and ζ is algebraic:

$$\eta = -\frac{k_\perp}{k_0^2}\,\xi'\,, \quad \text{and} \quad \nabla\cdot\boldsymbol{\xi} = \frac{\rho g}{\gamma p}\,\xi \quad \Rightarrow\ \zeta = -\frac{k_\parallel}{k_0^2}\,\xi' + \frac{\rho g}{\gamma p k_\parallel}\,\xi\,. \qquad (7.166)$$

Consequently, only the first two terms remain in the expression (7.165):

$$W = \tfrac{1}{2}\int_{x_1}^{x_2}\Big[\frac{k_\parallel^2 B^2}{k_0^2}\xi'^2 + \Big(k_\parallel^2 B^2 - \rho' g - \frac{\rho^2 g^2}{\gamma p}\Big)\xi^2\Big]\,dx\,, \qquad (7.167)$$

where we have suppressed again the infinite area A of the horizontal plane by renormalizing ($\overline{W} \equiv W/A$) and dropping the bar. One immediately notices that the plasma slab is trivially stable in the absence of gravity ($g = 0$) since $W \geq 0$ then. An inequality like this illustrates a typical use of the energy principle which has no simple counterpart in the differential equations approach.

Systematic minimization of the remaining 1D problem (7.167) is a standard variational problem, as explained in Section 6.6.4. For a general quadratic form

$$W = \tfrac{1}{2} \int_{x_1}^{x_2} \Big(P_0 \xi'^2 + Q_0 \xi^2 \Big)\, dx \,, \tag{7.168}$$

the minimizing solution is a solution of the *Euler–Lagrange equation*

$$\frac{d}{dx}\Big(P_0 \, \frac{d\xi}{dx} \Big) - Q_0 \, \xi = 0 \,, \tag{7.169}$$

subject to the boundary conditions

$$\xi(x_1) = \xi(x_2) = 0 \,. \tag{7.170}$$

For the present problem, the Euler–Lagrange equation corresponding to the quadratic form (7.167) reads

$$\frac{d}{dx}\Big(\frac{F^2}{k_0^2} \frac{d\xi}{dx} \Big) - \Big(F^2 - \rho' g - \frac{\rho^2 g^2}{\gamma p} \Big)\, \xi = 0 \,, \tag{7.171}$$

where we now exploit the notation $F \equiv -\mathrm{i}\mathbf{B} \cdot \nabla \equiv k_y B_y + k_z B_z$ to stress the central importance of the parallel gradient operator in stability theory.

The Euler–Lagrange equation (7.171) is just the *marginal equation of motion*, obtained from the general MHD wave equation (7.80) by substituting $\omega^2 = 0$:

$$P_0 \equiv P(\omega^2 \!=\! 0) \,, \qquad Q_0 \equiv Q(\omega^2 \!=\! 0) \,. \tag{7.172}$$

(In contrast to Section 7.4, the index $_0$ on P_0 and Q_0 now denotes this substitution.) This connection nearly closes the circle of our presentation. We just need to interpret the physical meaning of the 'solutions' of Eq. (7.171) and of the singularities that occur when $F = \mathbf{k}_0 \cdot \mathbf{B} = 0$ somewhere.

(b) Variational procedure (absence of singularities) We put quotation marks on 'solutions' since, in general, it is impossible to solve Eq. (7.171) if the boundary conditions (7.170) are imposed. This follows from the simple fact that *there is no eigenvalue parameter* in that differential equation. Hence, all one can do is, for example, start from the left by satisfying the left boundary condition and then integrate to the right and check whether or not more zeros are encountered on the interval (x_1, x_2). In general, the right boundary condition will not be satisfied for that 'solution'. By means of the calculus of variations, as carried out systematically by Newcomb [461] for the analogous case of a cylindrical plasma, one may relate the oscillatory behaviour of ξ to stability as follows:

Insert the 'solution' ξ of the Euler–Lagrange equation (7.169) in Eq. (7.168) for the energy W, and integrate by parts:

$$\begin{aligned} W &= \tfrac{1}{2} \int_{x_1}^{x_2} (P_0 \, \xi'^2 + Q_0 \, \xi^2)\, dx \\ &= \tfrac{1}{2} \int_{x_1}^{x_2} \Big[P_0 \, \xi'^2 + \xi \, (P_0 \xi')' \Big]\, dx = \tfrac{1}{2} \Big[P_0 \, \xi \, \xi' \Big]_{x_1}^{x_2} . \end{aligned} \tag{7.173}$$

If the interval (x_1, x_2) is larger than the distance between two consecutive zeros of a solution to the Euler–Lagrange equation, one may split it into two sub-intervals (x_1, x_0) and (x_0, x_2) such that a solution ξ_a which vanishes at $x = x_1$ does not vanish again on (x_1, x_0) and a solution ξ_b which vanishes at $x = x_2$ does not vanish a second time on (x_0, x_2). At $x = x_0$ the amplitudes of the two solutions may be chosen equal (Fig. 7.15(a)). By applying Eq. (7.173) to a solution composed of ξ_a on (x_1, x_0) and ξ_b on (x_0, x_2), one then obtains

$$W = \tfrac{1}{2} P_0 \Big(\xi_a \xi_a' - \xi_b \xi_b' \Big) \Big|_{x=x_0} = \tfrac{1}{2} \Big(P_0\, \xi (\xi_a' - \xi_b') \Big) \Big|_{x=x_0} < 0 \,, \tag{7.174}$$

so that the contribution to the energy is negative. In conclusion:

(1) if the 'solution' ξ_0 of the Euler–Lagrange equation (7.171) (i.e. the marginal equation of motion) that satisfies the left boundary condition $\xi_0(x_1) = 0$ has another zero on the interval (x_1, x_2), then a trial function ξ_1 can be constructed (the composite function of Fig. 7.15(a)) that satisfies both boundary conditions and for which the energy $W(\xi_1) < 0$: the system is unstable;

(2) if ξ_0 has no other zeros on the interval, that construction fails and $W(\xi_1) \geq 0$ for all trial functions: the system is stable.

This is the simple part of Newcomb's variational procedure, valid in the absence of singularities ($F \neq 0$ on the entire interval). The complicated part is the proper handling of those singularities. This will be addressed in Section 7.5.3.

(c) 'Shooting' method By means of the oscillation theorem of Section 7.4.3, the variational procedure can be replaced by a more transparent one related to the calculation of the *growth rates of the instabilities*. This approach is just a special case of solving for the discrete modes for values of ω^2 outside the ranges of $\omega_A^2(x)$, $\omega_S^2(x)$, $\omega_{s0}^2(x)$, $\omega_{f0}^2(x)$ on the interval $x_1 \leq x \leq x_2$ depicted in Fig. 7.14. In that case, the full MHD wave equation (7.80) is non-singular and a straightforward method to find discrete eigenvalues is to apply the so-called *shooting method*. This amounts to a procedure in which the second order differential equation (7.80) is first solved as an initial value problem where a value of ω^2 is guessed, say $\omega^2 = \omega^{2(i)}$, so that the functions appearing in Eq. (7.80) are determined, and the differential equation is solved numerically starting from the left 'initial' data $\xi(x_1) = 0$, $\xi'(x_1) = \text{const.}$ By some numerical integration scheme the solution $\xi(x; \omega^{2(i)})$ is found. Since the RHS boundary condition (7.82) will not be satisfied in general, $\xi(x_2; \omega^{2(i)}) \neq 0$, one needs to shoot again until this boundary condition is satisfied as well (see Fig. 7.15(b)). To stay with the metaphor: this is not done by changing the inclination of the gun (i.e., by changing the value of $\xi'(x_1)$, which would just change the amplitude of the solution and not the location of the zeros) but by changing the amount of gun powder (i.e., by changing the value of ω^2). Since, according to the oscillation theorem, the eigenvalue ω^2 is monotonic in the distance between the zeros of ξ, one knows precisely in which direction this change has to go: for Sturmian frequency ranges, one should decrease ω^2 to move out the next zero to the position $x = x_2$ to get a genuine eigenfunction, for anti-Sturmian frequency ranges one should increase ω^2.

Unstable discrete modes are always Sturmian because $N/D > 0$ for $\omega^2 < 0$. Since the Euler–Lagrange equation (7.171) is just the marginal form ($\omega^2 = 0$) of the general MHD wave equation (7.80), the connection with the variational procedure for stability is obvious. Referring to Fig. 7.15(b): if one shoots with $\omega^2 = 0$ and one finds a zero (the curve labelled $\omega^{2(0)}$), there will be a genuine eigenvalue $\omega^{2(2)} < 0$ so that the system is unstable. On the other hand, if one shoots with $\omega^2 = 0$ and one finds no zero (the curve labelled $\omega^{2(1)}$), the eigenvalue $\omega^{2(2)} > 0$ so that the system is stable. Hence, the shooting method not only supplies an answer to the stability problem, but

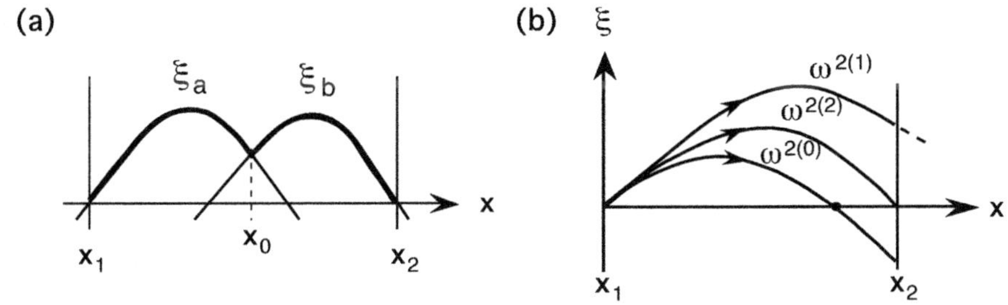

Fig. 7.15 (a) Composite trial function. (b) 'Shooting' for discrete eigenvalues.

also to the 'danger' of the instabilities, i.e. a value for their growth rate. In this manner, the global stability problem appears to have been solved completely for one-dimensional inhomogeneities.

The singularities $F = 0$ of the marginal stability equation are absent in the shooting method as long as $\omega^2 < 0$. From a practical point of view, this is sufficient to solve any stability problem because one can always exploit the σ-stability concept (Section 6.5.3) to stay away from these singularities (any small σ^2 suffices). However, singularities also offer the advantage of increased analytical insight in stability problems, as will be shown in Section 7.5.3.

7.5.2 Interchanges in shearless magnetic fields

A special case occurs when the magnetic field has *no shear*, i.e. it is unidirectional. Then, perturbations may be found with $F \equiv 0$ over the whole domain, so that the horizontal wave vector $\mathbf{k}_0$ is perpendicular to the magnetic field everywhere. In the following Section 7.5.3 on interchanges in sheared magnetic fields, the general stability condition (7.191) is derived. Naively, one would expect the no-shear limit ($\varphi' = 0$) of that condition to be the worst case, with stability determined by the familiar condition $N_{\rm B}^2 \geq 0$. However, if $F \equiv k_\parallel B = 0$, the variable ζ disappears from the expression given in Section 7.5.1 for $\nabla \cdot \boldsymbol{\xi}$ and, thereby, also from the expression (7.165) for the energy. Hence, the two last terms of Eq. (7.165) cannot be considered as independent, so that the minimization is different from the case with shear. For $k_\parallel = 0$, $k_\perp = k_0$, the expression for the energy reduces to

$$W = \tfrac{1}{2} \int \Big[-\Big(\rho' g + \frac{\rho^2 g^2}{\gamma p}\Big)\xi^2 + B^2\Big(k_0\eta + \xi'\Big)^2 + \gamma p\Big(\xi' + k_0\eta - \frac{\rho g}{\gamma p}\xi\Big)^2\Big]\, dV$$

$$= \tfrac{1}{2} \int \Big\{ -\Big(\rho' g + \frac{\rho^2 g^2}{B^2 + \gamma p}\Big)\xi^2 + k_0^2(B^2 + \gamma p)\Big[\eta + \frac{(B^2 + \gamma p)\xi' - \rho g \xi}{k_0(B^2 + \gamma p)}\Big]^2\Big\}\, dV\,. \quad (7.175)$$

By minimization with respect to η, the last term disappears and we get

$$W = -\tfrac{1}{2} \int \Big(\rho' g + \frac{\rho^2 g^2}{B^2 + \gamma p}\Big)\, \xi^2\, dx\,, \quad (7.176)$$

so that the stability criterion becomes

$$\rho' g + \frac{\rho^2 g^2}{B^2 + \gamma p} \quad \Big(\equiv -\rho N_{\rm m}^2\Big) \;\leq 0\,. \quad (7.177)$$

Consequently, the stability criterion obtained for stability with respect to *pure interchanges* is less severe than the condition (7.191) in the no-shear limit. The expression is familiar: $N_{\rm m}^2 \geq 0$, where $N_{\rm m}^2$ is the square of the magnetically modified Brunt–Väisäläa frequency defined in Eq. (7.102) of Section 7.3.3.

The criterion $\mathrm{N}_{\rm m}^2 \geq 0$ is automatically satisfied for the exponential atmosphere analysed in that section, consistent with the stability of the solutions (7.101) for perpendicular propagation. However, we also found out there, in the intriguing equation (7.112) where both $N_{\rm B}^2$ and $N_{\rm m}^2$ appeared, that *the worst instabilities occur for near-perpendicular propagation*, not for purely perpendicular propagation. This subtlety in the stability of general (not restricted to exponential profiles) shearless magnetic fields was already pointed out in 1961 by Newcomb [462]. We summarize the results of that paper since it throws light on the relationship between stability and spectral analysis.

As is clear from Section 7.3.3, the relevant modes for instability are the low-frequency Alfvén and slow modes. Hence, it is expedient to get rid of the high-frequency fast modes by means of an ordering. This we do by introducing a scale length L of the equilibrium variations,

$$L^{-1} \equiv \rho g/(p + \tfrac{1}{2}B^2) \tag{7.178}$$

(the parameter α of Section 7.3.3), and imposing the following ordering on the frequencies and wave numbers:

$$\rho\omega^2 \sim L^{-1}\rho g \sim k_\parallel^2 b^2 \ll k_0^2 b^2 \,, \tag{7.179}$$

assuming near perpendicular propagation of the modes ($|k_\parallel| \ll |k_\perp|$). The MHD wave equation (7.80) then reduces to

$$\frac{d}{dx}\,\rho(\omega^2 - k_\parallel^2 b^2)\,\frac{d\xi}{dx} - k_0^2\Big[\rho(\omega^2 - k_\parallel^2 b^2) + \rho' g + \rho g^2\,\frac{\omega^2 - k_\parallel^2 b^2}{(b^2 + c^2)\omega^2 - k_\parallel^2 b^2 c^2}\Big]\,\xi = 0\,. \tag{7.180}$$

This equation still contains the exact marginal equation (7.171) for $\omega^2 \to 0$. We solve it for local modes, varying rapidly over the length scale L:

$$\xi \sim \mathrm{e}^{\mathrm{i}qx}\,, \qquad qL \gg 1\,. \tag{7.181}$$

This yields the following dispersion equation:

$$\omega^4 - \Big(\frac{b^2 + 2c^2}{b^2 + c^2}\,k_\parallel^2 b^2 + \frac{k_0^2}{k_0^2 + q^2}\,\mathrm{N}_{\rm m}^2\Big) + \frac{c^2}{b^2 + c^2}\,k_\parallel^2 b^2\Big(k_\parallel^2 b^2 + \frac{k_0^2}{k_0^2 + q^2}\,\mathrm{N}_{\rm B}^2\Big) = 0\,, \tag{7.182}$$

having two qualitatively different solutions. For $F^2/\rho \equiv k_\parallel^2 b^2 \ll |N_{\rm B}^2| \sim |N_{\rm m}^2|$, we recover the solutions (7.112) found in Section 7.3.3:

$$\omega_1^2 \approx \frac{k_0^2}{k_0^2 + q^2}\,N_{\rm m}^2\,, \qquad \omega_2^2 \approx \frac{\gamma p}{\gamma p + B^2}\,\frac{N_{\rm B}^2}{N_{\rm m}^2}\,F^2/\rho\,. \tag{7.183}$$

The first mode is a *pure interchange* ($F = 0$) which becomes unstable when $N_{\rm m}^2 < 0$. The second mode is called a *quasi-interchange* ($F \neq 0$) since only finite segments of the field lines can be involved in interchanging plasma and magnetic field. They become unstable when $N_{\rm B}^2 < 0$. Since this is the more severe criterion, it should be considered as the overall boundary for stability of gravitational instabilities in shearless magnetic fields. With the consideration of the quasi-interchanges, the discrepancy in the stability criteria between sheared magnetic fields in the limit $\varphi \to 0$ and shearless magnetic fields disappears.

To establish the danger of instabilities, it is not sufficient to derive stability criteria, one should also calculate the maximum growth rate to find out which mode dominates. This involves maximizing the growth rate obtained from Eq. (7.182) with respect to the mode numbers $k_\parallel$, $k_\perp$ and q. The reader may wish to check this using the following notation:

$$\Gamma \equiv -\frac{\rho'}{\rho}\, g\,, \quad \Gamma_{\rm B} \equiv \frac{\rho g^2}{\gamma p}\,, \quad \Gamma_{\rm m} \equiv \frac{\rho g^2}{\gamma p + B^2}\,, \quad \Gamma_0 \equiv \frac{\gamma p\, \rho g^2}{(\gamma p + B^2)^2}\,, \tag{7.184}$$

where $\Gamma_{\rm B} \geq \Gamma_m \geq \Gamma_0$. The result of the optimization can then be summarized as follows:

(1) if $\Gamma \geq \Gamma_{\rm B}$, the plasma is stable;

(2) if $\Gamma_0 \leq \Gamma < \Gamma_{\rm B}$, the fastest growing mode is a quasi-interchange ($k_\parallel \neq 0$) :

$$\omega^2 = -(\rho g^2/B^2)\big[1 - (\Gamma/\Gamma_{\rm B})^{1/2}\big]^2\,; \tag{7.185}$$

(3) if $\Gamma \leq \Gamma_0$, the fastest growing mode is a pure interchange ($k_\parallel = 0$):

$$\omega^2 = \Gamma - \Gamma_{\rm m} \equiv N_{\rm m}^2\,. \tag{7.186}$$

Hence, calculating the eigenfrequencies of the modes not only removes apparent discrepancies from stability theory, it also introduces new transitions ($\Gamma = \Gamma_0$) that are relevant for the description of the dynamics. Spectral theory is not only a mathematical beauty but also a physical necessity for understanding the dynamics of inhomogeneous plasmas.

7.5.3 Interchange instabilities in sheared magnetic fields

The remaining problem in the stability theory of gravitating plasmas is the proper interpretation of the *singularities* $F \equiv \mathbf{k}_0 \cdot \mathbf{B} = 0$ of the marginal equation (7.171). This is again related to the continuous spectrum since the expressions (7.84) for the Alfvén and slow continuum frequencies both degenerate into the marginal frequency $\omega^2 = 0$ when $F = 0$. Hence, if there is such a point on the interval $[x_1, x_2]$, the Alfvén and slow continua both extend to the origin $\omega^2 = 0$. The physical significance of these points is that the perturbations do not disturb the magnetic field since the horizontal wave vector $\mathbf{k}_0$ is perpendicular to $\mathbf{B}$. The magnetic part of the potential energy of the Alfvén wave perturbations vanishes there because the field lines are not bent. Consequently, these are the positions where the driving forces of instability are minimally counterbalanced by magnetic tensions so that instabilities are predominantly localized there.

On the other hand, by means of *magnetic shear* ($F' \neq 0$), the region of minimal field line bending can be minimized so that stability can be restored, as we will see. We have already encountered this effect in Section 6.6.4 when discussing the stabilization of the Rayleigh–Taylor instability of interface plasmas by means of a magnetic field with different directions in the plasma and the vacuum (Fig. 6.16). In that particular case, the magnetic shear was entirely localized to the infinitely narrow surface layer separating plasma and magnetic field. In the present section, we deal with diffuse plasmas where shear is present everywhere.

To establish the significance of the singularities $F(x_0) = 0$, we expand all functions in the neighbourhood of that singularity. Hence, we continue the analysis of case (c) of Section 7.4.1, with the expansion (7.116) of the Alfvén factor, illustrated in Fig. 7.10(c). Introducing the variable angle $\varphi(x) \equiv \arccos(B_z/B)$ between the magnetic field and the z-axis (Fig. 7.8(b)), and the constant angle $\theta \equiv \arccos(k_z/k_0)$ between the horizontal wave vector and the z-axis, we have

$$F \equiv \mathbf{k}_0 \cdot \mathbf{B} = k_0 B(x) \cos[\,\varphi(x) - \theta\,]\,. \tag{7.187}$$

At the singularity, $\varphi(x_0) = \theta \pm \pi/2$, we have

$$F'(x_0) = \mp k_0 (B\varphi')_0 \quad \Rightarrow \quad \rho\omega_{\rm A}^2 \approx (F'^2)_0\, s^2 = k_0^2 (B^2 \varphi'^2)_0\, s^2\,, \tag{7.188}$$

where $s \equiv x - x_0$. The quantity φ' represents *the shear of the magnetic field* caused by the parallel component of the current: $\varphi' = (B_y B_z' - B_y' B_z)/B^2 = -\mathbf{j}\cdot\mathbf{B}/B^2$, according to Eq. (7.63).

Close to the singularity, the marginal equation (7.171) is approximated by

$$\frac{d}{ds}\left[s^2\,(1 + \cdots)\,\frac{d\xi}{ds}\right] - q_0 (1 + \cdots)\,\xi = 0\,, \qquad q_0 \equiv -\left(\frac{\rho' g + (\rho^2 g^2/\gamma p)}{B^2 \varphi'^2} \right)_0\,, \tag{7.189}$$

which has the form of the standard differential equation (7.119). Consequently, the Frobenius expansion (7.121) may be exploited. To leading order, the solutions of equation (7.189) behave as s^{ν_1} and s^{ν_2}, where ν_1 and ν_2 are the roots of the indicial equation $\nu(\nu+1) - q_0 = 0$ given by Eq. (7.125). Depending on whether $1 + 4q_0$ is positive or negative, the indices are real or complex.

(a) Local stability (complex indices) The most interesting case is obtained for $1 + 4q_0 < 0$, *when the indices are complex* so that the solutions of the marginal stability equation are oscillatory:

$$\begin{aligned} \xi_1 &= s^{-1/2+{\rm i}w} + s^{-1/2-{\rm i}w} &&= 2 s^{-1/2} \cos(w \ln s)\,, \\ \xi_2 &= {\rm i}(s^{-1/2+{\rm i}w} - s^{-1/2-{\rm i}w}) &&= -2 s^{-1/2} \sin(w \ln s)\,, \end{aligned} \tag{7.190}$$

where $w \equiv \frac{1}{2}\sqrt{-(1+4q_0)}$. The kind of oscillatory behaviour obtained is quite extreme, since the solutions not only oscillate infinitely rapidly but their amplitude also blows up when $s \to 0$, as schematically illustrated in Fig. 7.16(a). According to the oscillation theorem of Section 7.4.3(b), such marginal 'solutions' imply instability, since the zeros of the solutions of the full MHD spectral equation are peeled off one by one as the value of ω^2 is decreased from 0 to the actual growth rate. Consequently, one obtains an infinity of unstable point eigenvalues, clustering at $\omega^2 = 0$ for $n \to \infty$, where n is the number of nodes of the corresponding eigenfunction: *when the indices are complex, the marginal point is a cluster point of the unstable modes of the discrete spectrum* (Fig. 7.17). This should not distract attention from the dominant physics, viz. that the fastest growing instability is obtained for $n = 1$. Although the condition for instability is local, the instability need not be local at all (see Fig. 7.16(b)), where the width of this mode is determined by the width of the region over which the local stability criterion is violated.

To avoid these instabilities, one should demand that $1 + 4q_0 > 0$, so that the indices are real. This leads to the *stability criterion for interchange modes:*

$$\rho' g + \frac{\rho^2 g^2}{\gamma p} \quad \left(\equiv -\rho N_{\rm B}^2 \right) \;\leq \tfrac{1}{4} B^2 \varphi'^2\,. \tag{7.191}$$

The three terms represent the driving force of the *gravitational* or *Rayleigh–Taylor instability* (heavy fluid on top of a lighter one), *modified by adiabatic effects* (term with γ), and *stabilized by magnetic*

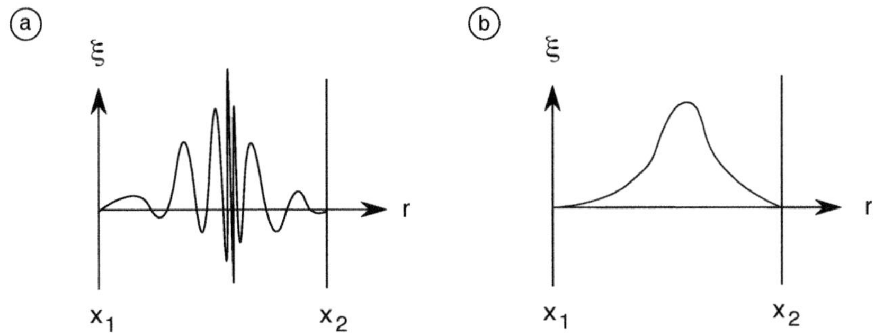

Fig. 7.16 Interchange criterion violated: (a) marginal mode ($\omega^2 = 0\,, n = \infty$), (b) the associated most global mode ($\omega^2 < 0\,, n = 1$).

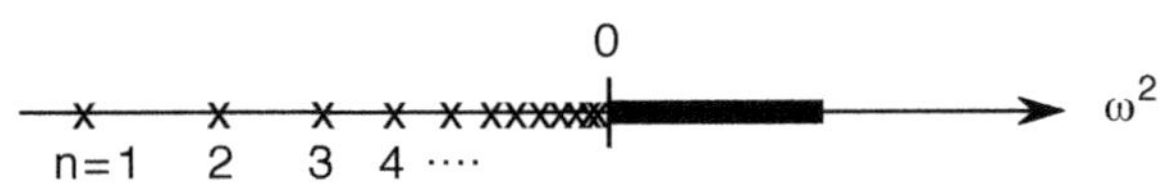

Fig. 7.17 Interchange criterion violated: appearance of a cluster point at $\omega^2 = 0$.

shear (term on the right hand side). To return to our introductory discussion of Section 6.1.1: a glass of magnetized plasma may be turned upside down without the contents dropping out, if the magnetic shear is large enough! Note that the criterion reduces to the HD criterion (7.35) of Section 7.2.2 for convective stability, $N_{\rm B}^2 > 0$, when $\varphi' = 0$. Here, $N_{\rm B}^2$ is the square of the Brunt–Väisäläa frequency, which also turned up in the condition (7.98) of Section 7.3.3 for the stability of gravito-MHD modes.

The analysis has been a *local* one, so that it may be repeated for every point on the interval $[x_1, x_2]$. Hence, for overall stability, one should at least demand that the inequality (7.191) is satisfied everywhere: the criterion is a necessary one for stability. A local criterion of this kind, known as *Suydam's criterion*, was first derived by Suydam (1958) [581] for a diffuse cylindrical plasma column (pinch) where the driving force of the instability is the pressure gradient in combination with curvature of the magnetic field lines. The derivation is completely analogous to the one given here. We will discuss it more extensively in Chapter 9.

The terminology 'interchange mode' dates from the early days of fusion research when there was hope of confining extremely high-β plasmas, where plasma and magnetic field are nearly completely separated in space. Such plasmas are virtually always unstable with respect to 'interchange' of plasma and magnetic field.

(b) Global stability (real indices) For $1 + 4q_0 > 0$, when *the indices are real*, the two solutions at the singularity behave as

$$\xi_{\rm s} \sim s^{\nu_{\rm s}}\,, \quad \nu_{\rm s} = -\tfrac{1}{2} + \tfrac{1}{2}\sqrt{1 + 4q_0} \;>\; -\tfrac{1}{2} \qquad (\text{'small' solution})\,,$$

$$\xi_l \sim s^{\nu_\ell}\,, \quad \nu_\ell = -\tfrac{1}{2} - \tfrac{1}{2}\sqrt{1 + 4q_0} \;<\; -\tfrac{1}{2} \qquad (\text{large solution})\,. \tag{7.192}$$

Hence, the large solution ξ_ℓ always blows up at $s = 0$, whereas the 'small' solution may or may not blow up depending on whether the square root is smaller or larger than 1. Similar to the singu-

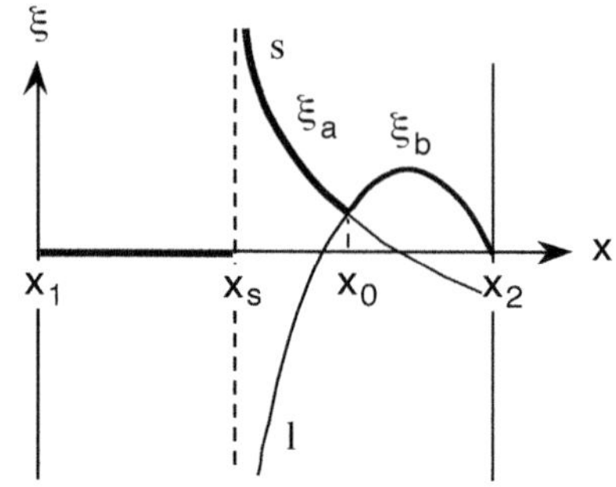

Fig. 7.18 Composite (right) trial function in the presence of a singularity; the label s indicates the 'small' solution and l indicates the large solution.

larities for $\omega^2 > 0$, discussed in Section 7.4.2, the present ones (at $\omega^2 = 0$) also split the interval (x_1, x_2) into independent sub-intervals (x_1, x_0) and (x_0, x_2), in the following sense. According to Eq. (7.173), the contribution to the energy of the sub-interval on the left of the singularity is:

$$W(x_1, x_0) = \tfrac{1}{2}\Big[P_0 \xi \xi' \Big]_{x_1}^{x_0} . \tag{7.193}$$

At the singularity, this expression behaves as

$$W(x_0) = \tfrac{1}{2}\Big(P_0 \xi \xi' \Big)(x_0) \sim s^{2\nu+1} \rightarrow \begin{cases} 0 & \text{for } \nu = \nu_s , \\ \infty & \text{for } \nu = \nu_\ell , \end{cases} \tag{7.194}$$

so that the energy contribution of the large solution diverges, but that of the 'small' one vanishes (hence: 'small', also when $-\frac{1}{2} < \nu_s < 0$). Consequently, testing for stability while keeping the energy of the perturbations finite implies that we have to exclude the large solution. This means that a kind of internal boundary condition is to be imposed at the singularity, viz. that ξ should be 'small' there. This can be done since jumps in the 'small' solution are allowed by an argument similar to that of Section 7.4.2. Such jumps do not contribute to the energy:

$$P_0 \xi_s H(s) \Big[\xi_s H(s) \Big]' \sim s^{2\nu+1} \Big[\nu H(s) + s\delta(s) \Big] H(s) \rightarrow 0 . \tag{7.195}$$

Therefore, the intervals (x_1, x_0) and (x_0, x_2) may be tested separately with respect to stability (i.e. the sign of W) by means of trial functions that are 'small' at $x = x_0$ and vanish identically either to the right or to the left of $x = x_0$.

The stability test is carried out with the following modification of the non-singular case (described in Section 7.5.1(b), Fig. 7.15(a)), where we now exploit the symbol x_s for the singularity and x_0 for an ordinary interior point. Consider a solution ξ_a of the Euler–Lagrange equation (7.171) which vanishes on the left interval (x_1, x_s), is 'small' to the right of the singularity $x = x_\mathrm{s}$, and vanishes once in the interval (x_s, x_2). Such a solution may be joined at a point x_0 in between the singularity x_s and the zero point of ξ_a to another solution ξ_b which vanishes at the right end point $x = x_2$, but does not vanish in the open interval (x_0, x_2) (Fig. 7.18). The energy of the Euler–Lagrange solution consisting of $\xi = 0$ on $(0, x_\mathrm{s})$, $\xi = \xi_\mathrm{a}$ on (x_s, x_0), and $\xi = \xi_\mathrm{b}$ on (x_0, x_2) may be shown to be negative by a completely analogous argument to that used in the derivation of Eq. (7.174). Hence, on independent sub-intervals the 'smallness' of a solution should be counted as a zero, so that for stability a solution that is 'small' at the singularity should not vanish somewhere in the interval. Thus, using Newcomb's wording [461] on the analogous case of a cylindrical 'diffuse linear pinch', we obtain the following *Stability theorem of a plane gravitating plasma* for the case that the interval (x_1, x_2) contains one singularity $F \equiv k_y B_y + k_z B_z = 0$ at $x = x_s$.

For specified values of k_y and k_z such that F vanishes at some point $x = x_s$ of the interval (x_1, x_2), the gravitating plasma slab is stable if, and only if, (1) the interchange criterion (7.191) is satisfied at $x = x_s$; (2) the non-trivial solution $\xi_{\rm left}$ of the Euler–Lagrange equation (7.171) that is 'small' to the left of $x = x_s$ does not vanish in the open interval (x_1, x_s); (3) the non- trivial solution $\xi_{\rm right}$ that is 'small' to the right of $x = x_s$ does not vanish in the open interval (x_s, x_2).

If there is more than one singularity, there will be more than two independent sub-intervals that have to be tested for oscillatory behaviour.

In conclusion: Through a careful analysis of the interchange singularities, as pioneered by Newcomb [461], the implications of the approach to the Alfvén and slow singularities of the continuous spectrum have been shown to be directly related to the MHD spectral oscillation theorem [240] discussed in Section 7.4.3.

7.6 Literature and exercises

Notes on literature

Hydrodynamics of the solar interior and corona

- Priest, *Solar Magnetohydrodynamics of the Sun* [510], Chapter 4 on waves.
- Stix, *The Sun* [569], Chapter 2 on internal structure of the Sun, Chapter 5 on oscillations.
- Christensen-Dalsgaard, *Stellar Oscillations* [123], Chapter 2 on the analysis of oscillation data, Chapter 5 on properties of solar and stellar oscillations.

Spectral theory of gravitating plasma slab

- Goedbloed [214], in a series of papers on 'Stabilization of magnetohydrodynamic instabilities by force-free magnetic fields', also derives the MHD wave equation for a gravitating plasma slab and quasi-interchanges in shearless magnetic fields.
- Goedbloed and Sakanaka [240], in a 'New approach to magnetohydrodynamic stability', prove the oscillation theorem for the MHD wave equation and introduce the concept of σ-stability.
- Lifschitz, *Magnetohydrodynamics and Spectral Theory* [410], Chapter 7 on MHD oscillations of a gravitating plasma slab.

Singular differential equations

- Ince, *Ordinary Differential Equations* [325], Chapter 18 on the solution of linear differential equations.
- Bender and Orszag, *Advanced Mathematical Methods for Scientists and Engineers* [50], Chapter 3 on approximate solutions of linear differential equations.

Gravitational instabilities

- Newcomb [462], on quasi-interchanges in 'Convective instability induced by gravity in a plasma with a frozen-in magnetic field'.
- Parker [482], on the Parker instability in 'The dynamical state of the interstellar gas and field'.
- Mouschovias [452], on 'The Parker instability in the interstellar medium'.

Exercises

[7.1] *The Sun – radiative transport*

The transport of radiation through a medium can be regarded as a random walk process. On average, a photon will travel one mean free path $\lambda_{\rm mfp}$ and collide with a particle, which redistributes it into a random direction. Due to the gradient in particle density, the net flux of radiation will be pointing outward.

- The mean free path is defined by $\lambda_{\rm mfp} \equiv (\kappa\rho)^{-1}$. Choosing for the opacity κ (also called the Rosseland absorption coefficient) the value $\kappa = 0.12\,{\rm m}^2\ {\rm kg}^{-1}$, calculate the mean free path of a photon, using the solar radius and mass.
- In n steps, the photon travels a distance $d = \sqrt{n}\,\lambda_{\rm mfp}$. Estimate the number of steps needed for a photon to travel one solar radius and the time elapsed.

[7.2] *The Sun – convection*

At some radius, the energy produced no longer escapes as radiation, but is transported by convection. Derive the Schwarzschild criterion for the onset of convection.

- Write down the equation of state for an ideal gas in terms of ρ and T, and derive the temperature gradient from this.
- The bulk motion of gas is only stable if the specific entropy, $S \equiv p\rho^{-\gamma}$, is constant in space. Derive the density gradient and the isentropic temperature gradient from this.
- By comparing these temperature gradients, the criterion for instability is obtained. Convert this criterion by means of the gravitational equilibrium equation to another form, relating to an apparently different instability. Which one is it?

[7.3] *Waves in a gravitating hydrodynamic slab*

Derive the HD wave equation for a plane parallel atmosphere from the nonlinear equations for the density, pressure and velocity, ignoring non-adiabatic effects.

- Using the Cowling approximation (what does that mean?), derive the differential equations in terms of the displacement field $\boldsymbol{\xi}$ for the horizontal Fourier modes.
- Rotate the coordinate system such that one of the axes is in the direction of the horizontal wave vector (why is that expedient?), and then derive the second order differential equation for the vertical displacement.

[7.4] *HD waves in an exponentially stratified atmosphere*

We continue with the HD wave equation obtained in the previous exercise.

- What is the physical significance of choosing exponential dependence on height, $\exp(-\alpha x)$, for both the pressure and the density? What does it imply for the sound speed? Using the equilibrium condition, express α in the gravitational parameter g.
- Derive the second order differential equation for ξ_x in terms of the Brunt–Väisäläa frequency $N_{\rm B}^2$. Obtain the dispersion equation from it and solve it.

[7.5] *MHD waves in an exponentially stratified atmosphere*

Look up the MHD wave equation for general one-dimensional inhomogeneity.

- In addition to the exponential pressure and density profiles of the previous exercise, introduce an appropriate magnetic field as a function of the vertical coordinate to get an atmosphere with constant Alfvén speed. Calculate the new expression for α and use it to simplify the MHD wave equation and resulting dispersion equation.
- For these magnetic field, pressure and density profiles, calculate the different MHD singularities. What do they represent?
- Give the solutions for parallel and purely perpendicular propagation of the waves, exploiting the magnetically modified Brunt–Väisäläa frequency, $N_{\rm m}^2 = \alpha g - g^2/(b^2 + c^2)$, and $N_{\rm B}^2$. Which wave may become unstable? Why is the other one stable?

[7.6]* *Dispersion equation for gravito-MHD waves*

Write a numerical program, exploiting the explicit solutions given in the text, to solve for the three gravito-MHD waves in an exponential atmosphere of magnetized plasma.

- Plot the three branches $\bar{\omega}^2(\bar{k}_0^2)$, similar to Fig. 7.9, for $\bar{\alpha} = 20$, $\beta = 1$, $\vartheta = \pi/2 - 0.1$ with $\bar{q} = \pi, 2\pi, \ldots$. How wide is the unstable region? What kind of modes are they?
- Estimate the value of $\bar{k}_0^2$ where the slow modes transform from Sturmian to anti-Sturmian. Continue the scan to that value to check the answer. What do you learn?

[7.7] *Spectrum of an incompressible gravitating slab*

Derive the MHD wave equation for an incompressible ($\gamma \to \infty$) gravitating plasma slab from the general wave equation.

- Show that the Alfvén and slow singularities coincide so that the two continua are degenerate. How do the associated 'improper' modes differ?
- Neglecting gravity, assuming $\rho = \text{const}$, and a linear dependence of $\omega_A^2(x)$ about the singularity at $x = x_0$, show that the wave equation reduces to the zero order modified Bessel equation, $\xi'' + z^{-1}\xi' - \xi = 0$, $z \equiv k_0(x - x_0)$, with solutions

$$I_0 = 1 + \tfrac{1}{4}z^2 + \tfrac{1}{32}z^4 + \cdots, \qquad K_0 = -[\ln(\tfrac{1}{2}z) + \gamma_0]\, I_0(z) + \tfrac{1}{4}z^2 + \tfrac{3}{128}z^4 + \cdots,$$

  where $\gamma_0 \approx 0.5772$ is Euler's constant. Are there any 'solutions' other than the ones associated with the continua?
- What changes when you allow for gravity?

[7.8] *Interchange instabilities*

Unstable modes have frequencies that are automatically outside the ranges of the continua, at least when their growth rate is finite. At $\omega^2 = 0$, one may exploit the potential energy of the perturbations to investigate stability, but one has to worry about a particular singularity.

- Assume a plane gravitating plasma with a sheared magnetic field. Write down the expression for the potential energy, W, and reduce it to a one-dimensional form.
- Explain why, usually, instabilities are localized at positions $x = x_0$ where $F(x_0) \equiv \mathbf{k}_0 \cdot \mathbf{B} = 0$. Expand F around $x = x_0$ exploiting the angle $\varphi(x) \equiv \arccos(B_z/B)$.
- Insert this expansion in the Euler–Lagrange equation, and solve it by means of a Frobenius expansion. Show that this yields a leading order equation of the form $(s^2\xi')' - q_0\xi = 0$ with indicial equation $\nu(\nu+1) - q_0 = 0$. Compute q_0.
- How do you obtain the stability criterion for interchange modes from that equation? Explain the meaning of the three terms.

8

Magnetic structures and dynamics of the solar system

8.1 Plasma dynamics in laboratory and nature

In this chapter we will make an excursion to the vast territory of magnetic structures and dynamics of the different plasmas encountered in the solar system, in particular the Sun and the planetary magnetospheres. While laboratory plasma confinement for the eventual goal of energy production also provides diverse magnetic structures, their topology and dynamics is always constrained by the presence of a fixed set of coils with programmed currents that should control the spatial and temporal behaviour of the magnetic fields. The reason is clear: for the success of thermonuclear energy production, plasma dynamics and complexity are not really desired. The best thing would be to extract energy from a plasma that just sits quietly inside a toroidal vessel and the engineering approach to plasma confinement is to try to approach this ideal as closely as possible. The history of thermonuclear fusion research demonstrates impressive progress along this line but also the obstacles, due to complex plasma dynamics, that have to be overcome. In astrophysical plasmas, no such human engineering constraints exist: plasmas and their associated magnetic structures appear to be almost free to exhibit the bewildering variety of different dynamics that are observed on virtually all length and time scales.

Space missions in the second part of the twentieth century have played an important role in demonstrating the magnetic structures and dynamics of plasmas in the solar system. The Skylab missions of 1973 revealed new solar structures in X-rays (like coronal holes) due to magnetic fields, the Voyager missions of the 1980s and 1990s provided completely new facts on the magnetic fields and magnetospheres of the planets, while the SOHO satellite, launched in 1995, provided visualizations of the dynamics of the solar corona that have become website hits.

With the tremendous progress in satellite and ground-based observations of solar system plasma dynamics, the need for a theoretical framework to describe it all has become pressing. At this point, it is probably wise not to hide the embarrassment about the absence of such an all-encompassing vision. Theory is lagging far behind observations at the present moment. Frequently, cartoons representing a particular plasma phenomenon take the place of genuine theoretical analysis. Whereas cartoons may be a useful way of communicating ideas, they cannot replace genuine physical understanding based on mathematical analysis. Computational MHD, stimulated by recent programmes on *space weather*, is presently filling in some gaps. The field is open for lots of new ideas.

Hence, the purpose of this chapter is not to provide a detailed description of the observational facts (excellent textbooks exist: Parker [483], Priest [510], Friedman [194], Zirin [669], Stix [569], Foukal [181], Mestel [438], Schrijver and Zwaan [544], from which we have freely borrowed in this chapter), but just to remind the reader that all theory eventually has to be confronted with empirical reality. This should lead to an attempt to answer the following questions: ‘Is the MHD model developed in Chapters 4–7 an adequate starting point for the description of observed plasma dynamics?’, ‘Are important theoretical pieces missing in this approach?’, and ‘What should be the main goals to be pursued in the following chapters?’ The phenomenology of magnetic structures and associated dynamics presented in the present chapter is used in later chapters to provide some physical reality to the model problems presented in Section 4.6.

8.2 Solar magnetism

We start with the central object of the solar system, the Sun, and ask the central question: *Where does its magnetism come from?* To answer that question, we recall some basic facts about the solar structure and then discuss some of the observational evidence leading to a model of solar magnetism.

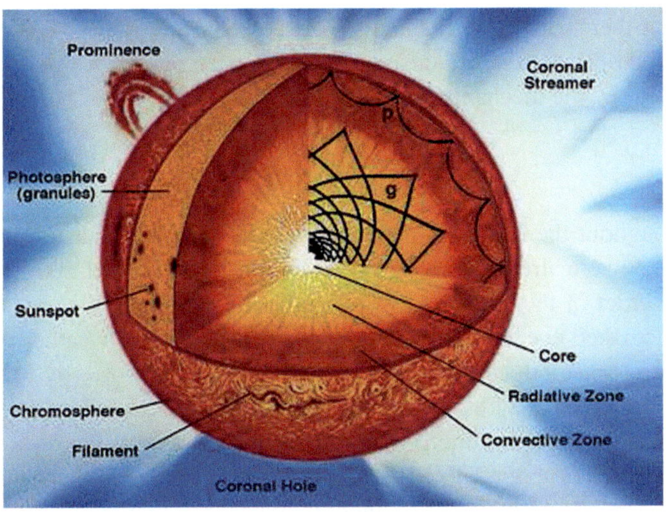

Fig. 8.1 Global model of the Sun showing gravity waves (g-modes) and sound waves (p-modes) propagating in the interior, and photospheric, chromospheric and coronal magnetic features in the exterior. (Courtesy SOHO, ESA–NASA.)

The standard solar model, already encountered in Chapter 7, is depicted in Fig. 8.1. The *interior* of the Sun cannot be observed directly and our knowledge of its structure is based on theoretical models and helioseismology. On the basis of differences in physical properties, the dense interior of the Sun can be divided into three layers, viz. the core, the radiative zone and the convection zone. The *core* of the Sun ($r \leq 0.25\,R_\odot$) is the region where the solar energy is generated by means of thermonuclear conversion of hydrogen into helium (see Section 1.2.1). This region is surrounded by the *radiative zone* ($0.25\,R_\odot \leq r \leq 0.71\,R_\odot$) where the produced energy is radiatively transported outward. Finally, in the relatively shallow *convection zone* ($0.71\,R_\odot \leq r \leq R_\odot$) the temperature gradient is so steep that the plasma becomes convectively unstable. This region is considered to be the seat of the solar dynamo (see Section 8.2.1).

Recent results of helioseismology have led to the idea that the shear in the differential rotation, which is the main driving force of the solar dynamo, is concentrated at the bottom of the convection zone in a relatively thin (just 13 000 km thick) and narrow (25°–30° latitude) layer. Estimates of the azimuthal field strength in that shear layer yield fields up to 5–10 T. However, there is no explanation yet for such high fields at the bottom of the convection zone.

The visible *atmosphere* of the Sun also consists of three layers. The *photosphere* is the region where the visible light of the Sun escapes. It is only 500 km thick. At the bottom of the photosphere the temperature is about 6600 K and it decreases to about 4300 K at the top. The next layer is the *chromosphere*, with a thickness of about 2500 km, where the temperature starts to increase again (from 4300 K to about 10^6 K) connecting smoothly onto the very hot (millions of degrees) and very tenuous *corona*, which stretches out into the whole *heliosphere*. The mechanism for this dramatic temperature rise is one of the major, as yet not really resolved, issues of solar MHD but it is generally agreed that the magnetic field in the corona is the essential carrier of the energy transport.

According to our present knowledge all of the mentioned regions, except for the core and the radiative zone, form the scene of spectacular magnetic activity of different kinds which are characteristic for each of them. Since the magnetism of the solar atmosphere is a result of that of the solar interior it is appropriate to consider the latter first, even though it presents the hardest of all theoretical questions, viz. the mechanism of the solar dynamo.

8.2.1 The solar cycle

(a) Sunspots Throughout the centuries people have been observing dark spots on the Sun: the *sunspots* (in fact, they were already mentioned in Chinese chronicles of 800 BC). Since they have typical sizes in the order of the diameter of the Earth and sometimes as large as 40 000 km, they can be seen with the naked eye, e.g. at sunset or sunrise during hazy weather. When the telescope was invented in the seventeenth century it was immediately put to use (by Fabricius, Galilei and Scheiner) to observe these structures. It was found that they can last from 2 to 30 days, i.e. long enough to observe their systematic motion from west to east across the disc, demonstrating that the Sun is actually rotating about a fixed axis. Moreover, these early astronomers observed that the sunspot motion depends on the solar latitude: they occur in two bands around the equator and they move faster at the solar equator (with a rotation period of 25 days) than at higher latitudes (with periods of 27 days at 40° and 30 days at 70°). Hence, already the first observations – with very primitive telescopes according to present day standards – revealed the basic fact behind solar magnetism: the *Sun rotates differentially*.

The *association of sunspots with magnetic fields* only came in the twentieth century, notably through the spectroscopic work of George Ellery Hale at the Mt Wilson Observatory (1908) and culminating in the development of the magnetograph by Harold Babcock (1953). Whereas the magnetic fields inside sunspots are relatively easy to measure because they are so intense (1000–4000 gauss), the magnetograph enables one to measure the Zeeman splitting by the much weaker magnetic fields outside the sunspots. Thus, the spatial variation of the magnetic field across sunspots may be determined. It has been firmly established that sunspots represent the most intense magnetic flux concentrations on the surface of the Sun. The average field strength increases with the area of the spot.

From the numbers quoted above one easily calculates that the speed of sunspots (i.e., the solar rotation rate) at the equator is about 2 $\mathrm{km\,s^{-1}}$, which corresponds to $10''\,\mathrm{h^{-1}}$. This simple dynamics is embedded in the much more complicated one of the surrounding photosphere which exhibits *a cellular convection pattern* due to the thermal instability of the underlying convection zone. High-resolution photographs of the photosphere by means of telescopes on board balloons (early 1960s) and dedicated solar telescopes (like the Dutch Open Telescope, see Fig. 8.8) reveal the different scales of the *granules* with a typical size of 1000 km, horizontal flow velocities of $1.5\ \mathrm{km\,s^{-1}}$, and a turn-over time ($\approx$ a lifetime) of the eddies of the order of 5 min and of the *supergranules* with characteristic sizes of 30 000 km, flow velocities of $0.5\ \mathrm{km\,s^{-1}}$, and turn-over times of the order of 20 hours. The latter structures are the result of an outflowing velocity field pushing the upwelling magnetic field to the boundaries of the eddies, which become visible as the *photospheric network*. Note that these convective motions take place at widely separated space and time scales.

Returning to the sunspots, the most important aspect of solar magnetism is yet to be mentioned, viz. the long time scale periodicity of the large-scale magnetic field and the associated magnetic activity: the *solar cycle*. This periodicity was discovered by Heinrich Schwabe, who systematically recorded the occurrence of sunspots during the period 1826–1851 in the hope of detecting a planet inside the orbit of Mercury. Instead, he found the more lasting result that the *number of sunspots varies periodically in time* with a period of about 11 years. Important for our subject is the observation that maxima and minima of sunspot numbers coincide with increased and decreased magnetic activity of the Sun as a whole so that they are appropriately called *solar maxima* and *solar minima*. By means of historical records one has been able to reconstruct the solar cycle back to the time of Galilei (Fig. 8.2). The periodicity is not precise, shorter and longer periods do occur (from 7 to 17 years). Also, the amplitudes vary considerably. A particularly quiet time occurred in the second part of the seventeenth century, the Maunder minimum, which coincided with the cold period of the little ice age on Earth.

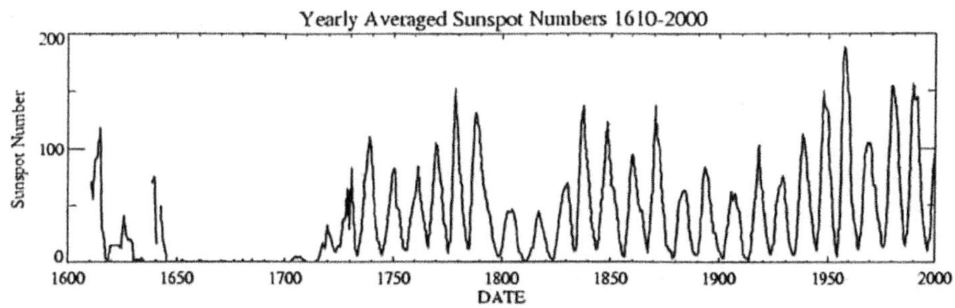

Fig. 8.2 The solar cycle: number of sunspots versus time. Note the virtual absence of sunspots during the little ice age. (Courtesy David H. Hathaway, NASA, Huntsville.)

Most sunspots appear in two belts between the equator and the latitudes $\pm 35°$. However, in 1859 Christopher Carrington discovered that the average latitude of occurrence of the sunspots depends on the phase of the solar cycle. During the 11 year period the sunspots gradually drift from latitudes $25°-30°$, where they first appear, to the equator, where they disappear again at the end of the cycle. This drift towards the equator (called Spörer's law) yields the *butterfly diagram* (Fig. 8.3).

Again, these phenomena would only be associated with magnetic fields in the twentieth century. In particular, only when the polarity of the fields was taken into account, it was realized that the physically relevant period of the solar cycle is actually 22 years with a reversal of the overall

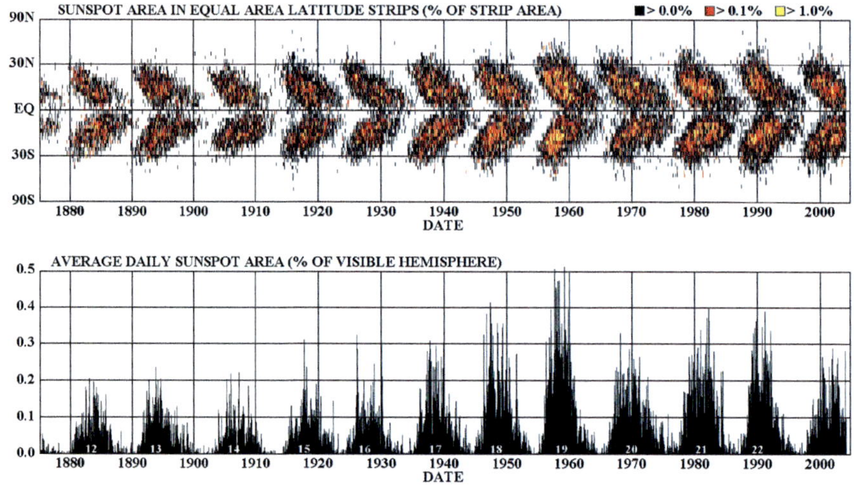

Fig. 8.3 Butterfly diagram: daily sunspot area averaged over solar rotations. Detailed observations of sunspots have been obtained by the Royal Greenwich Observatory since 1874. These data show that sunspots are concentrated in two latitude bands which first form at mid-latitudes, widen, and then move toward the equator as each cycle progresses. (Courtesy David H. Hathaway, NASA, Huntsville.)

magnetic field direction every 11 years. In general, sunspots occur in pairs with a leading spot of a certain polarity and a following one of opposite polarity at a slightly higher latitude. Moreover, it was found by Hale and Nicholson (1925) that the leading sunspots on the northern hemisphere all have the same polarity whereas those on the southern hemisphere all have the opposite polarity during the first 11 years of the cycle, and the roles are reversed during the second period of 11 years. Hence, adding the sign of the sunspot magnetic field to the picture given in Fig. 8.3 reveals that the *solar cycle is really a magnetic oscillation*.

(b) The solar dynamo By way of introduction, consider a particular example of the cartoon approach to dynamo action provided by the Babcock model of the solar cycle (Fig. 8.4, extensively discussed by Foukal [181]). Due to differential rotation and the 'frozen in' condition of magnetic fields in highly conductive plasmas, an initially poloidal magnetic field (like the usual dipole field created by toroidal currents in the convection zone) is stretched and wound up at the equator. After many periods of the solar rotation, the toroidal magnetic field component, which was negligible at first, now has become the dominant one. This field is expelled from the convection zone and breaks through the photosphere at isolated places where it forms sunspots with preceding (p) and following (f) polarities of fixed sign on the northern hemisphere, and of the opposite sign on the southern hemisphere. The outer portions of the flux loops expand into the corona and reconnect with the original poloidal field. This results in reversal of the direction of the latter field: the second half of the solar cycle has started. Note that reconnection, i.e. breaking and rejoining of field lines due to some anomalous resistive process in the corona, is necessary here.

We have seen that the Sun rotates differentially, that the photospheric surface exhibits cellular convection patterns characteristic of thermal convection, and that the solar magnetic field is periodic in time. These are the main facts behind the *solar dynamo*, which results in the conversion of

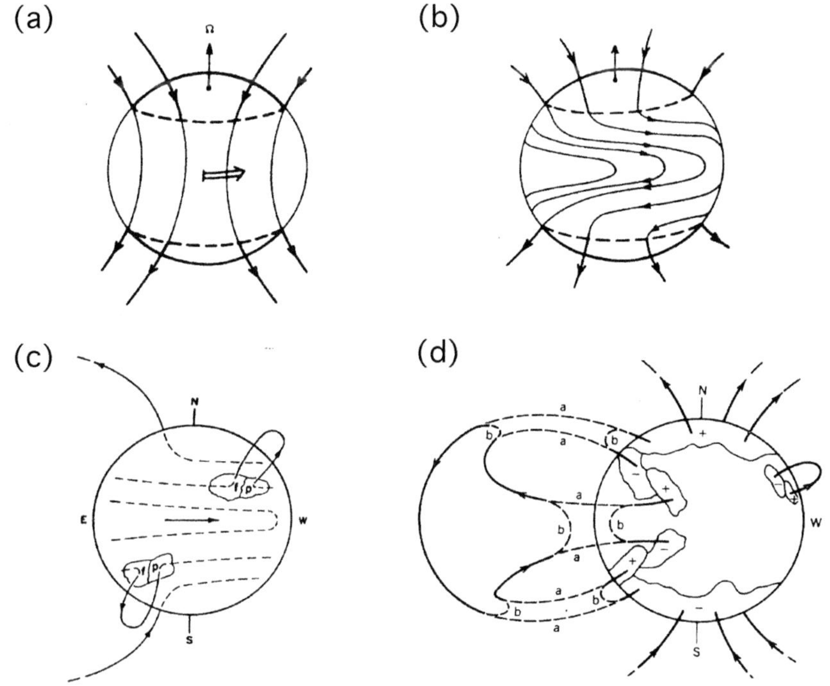

Fig. 8.4 Babcock model for solar magnetism. (a) An initially poloidal magnetic field is (b) stretched and wound up by differential rotation, producing a toroidal magnetic field. (c) This field is expelled from the photosphere, creating sunspots and (d) reconnecting to the original poloidal field, after which the process starts all over again with reversed polarities. (From P. Foukal [181], after H. Babcock, *Astrophys. J.* **133**, 572 (1961).)

mechanical energy into magnetic energy. We now come to one of the main questions of this chapter, viz. can all this be explained by the MHD equations? Recall the (near) conservation laws of Section 4.1 and 4.4:

$$\frac{\partial \rho}{\partial t} = -\nabla \cdot (\rho \mathbf{v}) \qquad (mass), \tag{8.1}$$

$$\rho \frac{\mathrm{D}\mathbf{v}}{\mathrm{D}t} = -\nabla p - \frac{1}{\mu_0}\mathbf{B} \times (\nabla \times \mathbf{B}) + \rho \mathbf{g} \qquad (momentum), \tag{8.2}$$

$$\frac{\mathrm{D}p}{\mathrm{D}t} = -\gamma p \nabla \cdot \mathbf{v} + \Big[(\gamma - 1)(H - \nabla \cdot \mathbf{h})\Big] \qquad (entropy), \tag{8.3}$$

$$\frac{\partial \mathbf{B}}{\partial t} = \nabla \times (\mathbf{v} \times \mathbf{B}) - \nabla \times \left[\frac{\eta}{\mu_0}(\nabla \times \mathbf{B})\right], \qquad \nabla \cdot \mathbf{B} = 0 \qquad (flux). \tag{8.4}$$

Here, the terms in square brackets are the ones which spoil the conservation of entropy and magnetic flux. In particular, the pressure evolution equation (8.3), which expresses (near) entropy conservation, and which also can be expressed in terms of the internal energy $e \equiv p/[(\gamma - 1)\rho]$ or the temperature $T \equiv e/C_V$, contains the generated heat per unit volume, H, and the heat flow $\mathbf{h}$. For

Ohmic dissipation, $H = \eta j^2$, with $\mathbf{j} = \mu_0^{-1}\nabla \times \mathbf{B}$, and $\mathbf{h} = -\lambda \nabla T$, where λ is the coefficient of thermal conductivity with dimension $[\lambda] = \mathrm{W\,K^{-1}\,m^{-1}}$. (Recall that we replaced the symbol κ, introduced in Section 4.4.2 for the thermal conductivity, by λ in Section 7.1.1.) The resistive term in the (near) flux conservation equation (8.4) results in magnetic field diffusion. Our use of the words 'near conservation' implies that the transport coefficients λ and η are small.

One part of the dynamo problem involves the combined effects of differential rotation, gravitational contraction, and the presence of a highly conducting medium. This part is well described by the combined equations of gas dynamics and electrodynamics, i.e. by the *partial differential equations* (8.1)–(8.4) of magnetohydrodynamics. The other part, which is just as important, involves the coupling to the external world through the *boundary conditions*. Together this constitutes a complicated nonlinear problem in space and time which is essentially four-dimensional and, hence, necessarily numerical.

Let us approach this problem in three steps. First, the crudest approximation would be to neglect the velocity $\mathbf{v}$. Equation (8.2) then results in the magneto-static equilibrium equation where gravitational forces are to balance pressure gradients and magnetic expansion forces (so-called magnetic buoyancy by which flux tubes are expelled from the solar interior). The equations (8.3) and (8.4), with $\mathbf{v} = 0$, are diffusion equations for the pressure (or temperature) and the magnetic field. In particular, assuming constant resistivity for simplicity, Eq. (8.4) becomes

$$\frac{\partial \mathbf{B}}{\partial t} = -\nabla \times \left[\frac{\eta}{\mu_0}(\nabla \times \mathbf{B}) \right] = -\frac{\eta}{\mu_0}\nabla \times \nabla \times \mathbf{B} \overset{(A.5)}{=} \frac{\eta}{\mu_0}\nabla^2 \mathbf{B} \equiv \bar{\eta}\,\nabla^2 \mathbf{B}\,, \tag{8.5}$$

where the reference to Eq. (A.5) above the last equal sign refers to a vector identity of Appendix A. Hence, the *inhomogeneity of the magnetic field* (created by currents) *will decay* on a time scale τ_d determined by the resistivity η and the length scale $l_0 \sim \nabla^{-1}$ of the inhomogeneity:

$$\tau_\mathrm{d} = \mu_0 l_0^2/\eta = l_0^2/\bar{\eta}\,. \tag{8.6}$$

We have introduced a new quantity $\bar{\eta} \equiv \eta/\mu_0 = \eta/(4\pi \times 10^{-7})$ which just absorbs the awkward factor μ_0. It is called the *magnetic diffusivity* and it has the convenient dimension of $\mathrm{m^2\,s^{-1}}$ (since $[\eta] = \Omega\,\mathrm{m}$ and $[\mu_0] = \mathrm{H\,m^{-1}} = \Omega\,\mathrm{s\,m^{-1}}$). In the astrophysical literature the bar on $\bar{\eta}$ is usually omitted and the conductivity $\sigma \equiv 1/\eta$ rather than the resistivity η is exploited. Here, we stick to the plasma physics convention by considering the resistivity η as the basic parameter from which the magnetic diffusivity $\bar{\eta}$ is derived.

The very first question to be answered is: what value to take for the resistivity? Spitzer and Härm (1953) [562] derived an expression for the resistivity of plasmas due to collisions between electrons and ions which has become known as the *Spitzer, or classical, resistivity* for reasons that soon will become apparent:

$$\eta_\parallel = \frac{e^2\sqrt{m_\mathrm{e}}}{6\epsilon_0^2(2\pi k)^{3/2}}\, Z \ln\Lambda\, T_\mathrm{e}^{-3/2} \approx 65\, Z \ln\Lambda\, T_\mathrm{e}^{-3/2}\,. \tag{8.7}$$

Here, $[\eta] = \Omega\,\mathrm{m}$, $[T_\mathrm{e}] = \mathrm{K}$, Z is the charge number of the ions and $\ln\Lambda$ is the Coulomb logarithm, which only weakly depends on the electron temperature T_e and the particle density n. It has a value between 10 and 20 for plasmas of interest.

For the solar corona, with $T_\mathrm{e} \approx 10^6\,\mathrm{K}$, $n_\mathrm{e} \approx 10^{16}\,\mathrm{m^{-3}}$ (see Table B.4), $\ln\Lambda \approx 18$ and the resistivity $\eta \approx 10^{-6}\,\Omega\,\mathrm{m}$ (i.e. $\bar{\eta} \approx 0.8\,\mathrm{m^2\,s^{-1}}$), so that $\tau_\mathrm{d} \approx 10^{14}\,\mathrm{s} \approx 3 \times 10^6\,\mathrm{y}$ for a loop with a transverse length scale $l_0 \approx 10\,000\,\mathrm{km}$. Yet, coronal loops may disrupt in a solar flare on a time scale of minutes: this certainly cannot be described by classical resistive diffusion.

For the solar interior, in particular the bottom of the convection zone (at present considered to be the origin of the solar dynamo), where $T_e = 1.9 \times 10^6\,\mathrm{K}$ and $n = 1.9 \times 10^{29}\,\mathrm{m}^{-3}$, the value of the resistivity $\eta \approx 1.7 \times 10^{-7}\,\Omega\,\mathrm{m}$ (i.e. $\bar{\eta} \approx 0.14\,\mathrm{m}^2\,\mathrm{s}^{-1}$) and the length scale of the inhomogeneity $l \approx R_\odot = 700\,000\,\mathrm{km}$, so that $\tau_d \approx 3.6 \times 10^{18}\,\mathrm{s} \approx 2 \times 10^{11}\,\mathrm{y}$! It is clear that the factor l_0^2 in the expression (8.6) for τ_d beats everything. In astrophysical plasmas, effects of classical resistivity are extremely small, so that ideal MHD (i.e. $\eta = 0$) is an excellent approximation for many purposes. However, for the present purpose of explaining why the solar dynamo has a 22 year period, it is clear that classical resistive diffusion cannot be the controlling factor since it is completely negligible. We would be left with a static equilibrium at this point: certainly not a dynamo.

The second step is to add flow. One would have hoped that the apparent rotational symmetry of the differential rotation of the Sun would get rid of at least one coordinate so that we could assume this flow to be axi-symmetric. However, it is clear from the behaviour of the polarity of the magnetic field that this symmetry is not respected by the solar cycle. In fact, it cannot be respected since *axi-symmetric motion does not lead to a dynamo*. The latter statement is the content of *Cowling's theorem*, which we will not prove here. Thus, *non-axi-symmetry of the flow* is the necessary third step to get dynamo action. Such a flow is created automatically by the convective motions of the convection zone. Here, the main complication is the fact that these motions are *turbulent*, involving the interaction of many small-scale vortices. It is generally assumed that these turbulent processes are also responsible for an anomalous increase of the magnetic diffusivity and associated decay of the magnetic field so as to correspond to the time scales of the solar cycle. This implies an increase from the classical value computed above, $\bar{\eta}_{\mathrm{class}} \sim 0.1\,\mathrm{m}^2\,\mathrm{s}^{-1}$, to $\bar{\eta}_{\mathrm{turb}} \sim 10^9\,\mathrm{m}^2\,\mathrm{s}^{-1}$, i.e. an increase by a factor of 10^{10}!

Although the latter fact leads us far away from our present theme (which is to present a simple survey of solar phenomena which are, in principle, tractable by means of the equations of magnetohydrodynamics), we will just indicate the kind of arguments involved in the analysis of a dynamo based on *turbulent eddy magnetic diffusivity*. This theory is based on the idea that in a turbulent fluid the mean values of the variables can be distinguished from the fluctuating, turbulent, ones. Thus,

$$\mathbf{v} = \langle \mathbf{v} \rangle + \mathbf{v}' , \qquad \mathbf{B} = \langle \mathbf{B} \rangle + \mathbf{B}' , \tag{8.8}$$

where angular brackets denote average values and primes denote fluctuating parts. Averaging is to be understood in a statistical sense but, in the solar case, it could be interpreted as averaging over a solar rotation period ($\sim$ 27 days). Inserting the expressions (8.8) into Eq. (8.4), one obtains the following averaged form of the induction equation:

$$\frac{\partial \langle \mathbf{B} \rangle}{\partial t} = \nabla \times \Big(\langle \mathbf{v} \rangle \times \langle \mathbf{B} \rangle \Big) + \nabla \times \langle \mathbf{v}' \times \mathbf{B}' \rangle - \nabla \times \Big(\bar{\eta} \nabla \times \langle \mathbf{B} \rangle \Big) . \tag{8.9}$$

Obviously, all complications of the theory are hidden in the second term on the right hand side which involves the average cross-product of the fluctuating parts of the velocity and the magnetic field. These have to come from the counterparts of Eq. (8.9) describing the evolution of the turbulent variables $\mathbf{v}'$ and $\mathbf{B}'$. By means of a number of drastic assumptions, this cross-product is written as

$$\langle \mathbf{v}' \times \mathbf{B}' \rangle \approx \alpha \, \langle \mathbf{B} \rangle - \beta \, \nabla \times \langle \mathbf{B} \rangle + \cdots , \tag{8.10}$$

where α and β are correlation functions of the turbulent velocity $\mathbf{v}'$. This transforms Eq. (8.9) into

$$\frac{\partial\langle\mathbf{B}\rangle}{\partial t} = \nabla\times\Big(\langle\mathbf{v}\rangle\times\langle\mathbf{B}\rangle\Big) + \nabla\times\Big(\alpha\,\langle\mathbf{B}\rangle\Big) - \nabla\times\Big[(\bar{\eta}+\beta)\,\nabla\times\langle\mathbf{B}\rangle\Big]. \tag{8.11}$$

This equation describes both magnification of the magnetic field, i.e. possible dynamo action, through the term with α, and decay due to turbulent magnetic diffusivity $\bar{\eta}_{\rm turb} = \bar{\eta}+\beta \approx \beta$, through the term with β. Everything is hidden now in the derivation of expressions for the coefficients α and β. Order of magnitude estimates based on the length scale $l \sim 1000\,{\rm km}$ and the time scale $\tau \sim 1000\,{\rm s}$ of the turbulent velocity fluctuations $v' \approx l/\tau \sim 1\,{\rm km\,s^{-1}}$ are $\alpha \sim l\,\Omega \sim 3\,{\rm m\,s^{-1}}$ (where $\Omega \sim 2.6\times10^{-6}\,{\rm rad\,s^{-1}}$ is the angular velocity corresponding to a solar rotation period of 27 days) and $\bar{\eta}_{\rm turb} \approx \beta \sim v'\,l \sim 10^9\,{\rm m^2\,s^{-1}}$. The miracle has happened: the required enhancement factor of 10^{10} for $\bar{\eta}$ has been obtained!

For the demonstration of an α-effect and a turbulent diffusivity, the back-reaction of the magnetic field on the plasma flow has been ignored. This reduced problem is called the *kinematic dynamo problem*. Even if taken for granted, one still has to solve the full nonlinear *MHD dynamo problem*, i.e. one needs to show that plasma motion and a quasi-oscillatory magnetic field can maintain each other. This is a 'hot' item in computational plasma-astrophysics.

8.2.2 Magnetic structures in the solar atmosphere

We now describe the enormous variety of the structures and dynamics in the solar atmosphere that are caused by the magnetic fields.

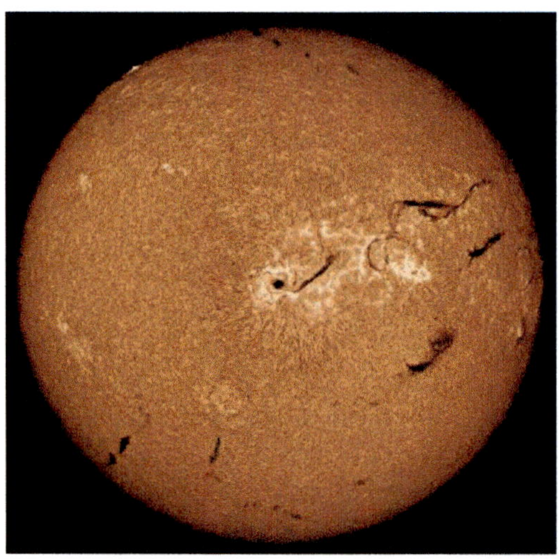

Fig. 8.5 Image of the Sun taken through a filter centred on a spectral line of hydrogen (Hα, wavelength = 6563 Å). This line forms above the surface of the Sun, but large *sunspots* are still visible and *active regions* and *plages* show up brighter than their surroundings. Also visible are condensations of cooler gas high up in the solar atmosphere which show up as *filaments*, dark string-like structures visible on the disc, and *prominences*, bright structures extending outward over the limb. (Courtesy National Solar Observatory/Sacramento Peak.)

(a) Photosphere and chromosphere The energy radiated by the Sun comes from the photosphere. The Sun emits a continuous spectrum but in the spectrum received on Earth dark absorption lines occur, the Fraunhofer lines. They are dark because the lower gas transparency, due to the absorption at specific wavelengths by specific particles such as iron atoms, causes radiation escape at greater height where the temperature is lower. Most of the absorption lines are formed in the photosphere. In the optical solar spectrum there are about 20 000 of these Fraunhofer spectral lines. From the comparison with laboratory experiments these absorption lines allow us not only to study the abundance of the elements of the Sun (all 92 elements are present although most of them are very rare), but they also allow us to measure density, temperature, velocity and field vectors.

The *intensity* of the absorption lines in the solar spectrum provides us information on the *temperature*, their *Zeeman splitting* provides information on the *strength of the magnetic field*, and their *Doppler shifts* provide information on the *velocity* along the line of sight. Different absorption lines are formed at different heights so that one can look at different levels in the solar atmosphere by using different filters. For instance, selecting the Hα Balmer line one can look at the middle chromosphere. By observing in the Ca^+ K line one sees the low chromosphere, and in soft X-rays one gets a direct view of the corona. Figure 8.5 is an Hα-picture of the lower chromosphere. The bright areas are active regions (above sunspots). The elongated dark filaments are prominences which, in the line of sight, appear to be vertically flat structures. Their protruding shape is visible though in observations of the limb (see below).

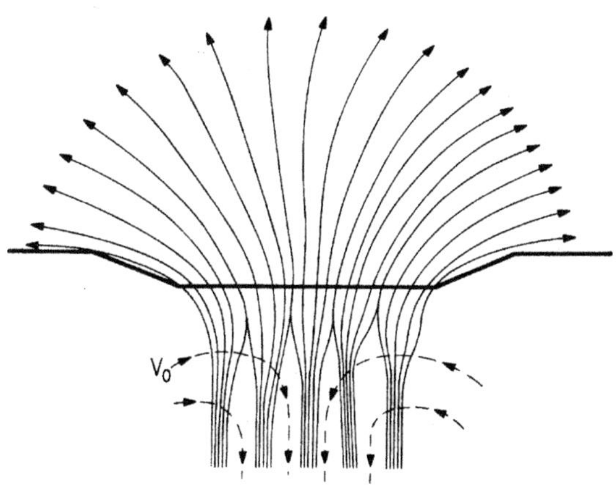

Fig. 8.6 Parker's sunspot model: bundles of magnetic flux tubes confined by the granulation vortices. (Courtesy Eugene N. Parker, *Astrophys. J.* **230**, 905 (1979).)

Standard magnetohydrostatic models of the equilibrium configuration of sunspots picture them as a single 'monolithic' flux tube with almost vertical magnetic field lines under the photosphere. Above the photosphere the external pressure decreases exponentially with height and, as a result, the magnetic field lines fan out in the chromosphere. Such monolithic flux tube configurations are not compatible with the observed external pressure stratification and the non-axi-symmetric fine structure of the penumbra. In 1979 Parker suggested an alternative model for sunspots, depicted in Fig. 8.6. In this alternative model, the single flux tube splits up in a large number of narrow flux

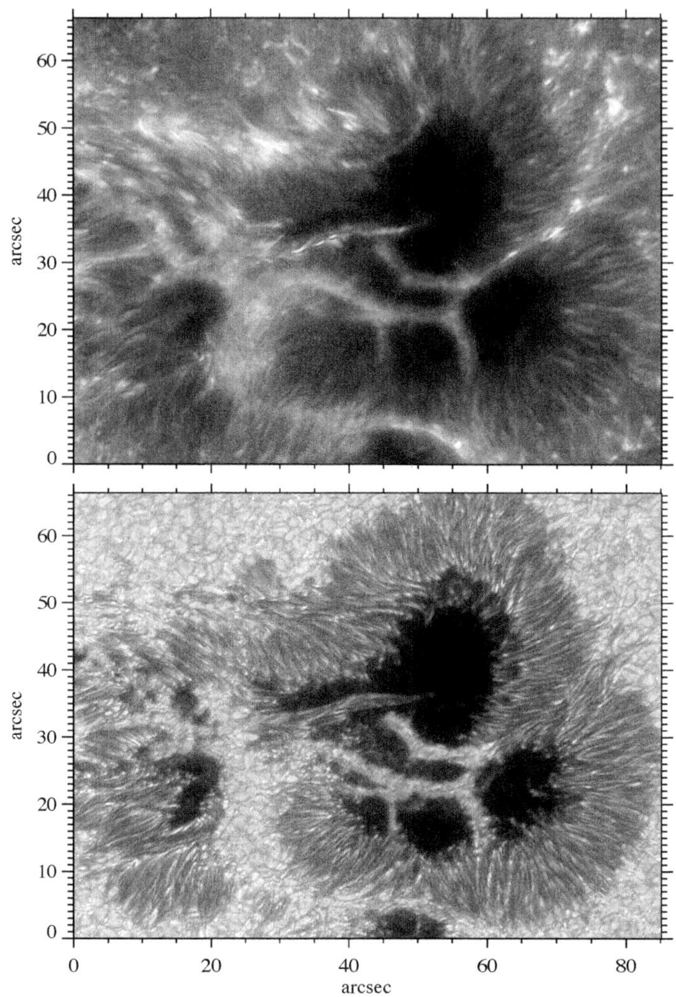

Fig. 8.7 Mature sunspot, observed with the Dutch Open Telescope on La Palma, 1 May 2003. The axes are in arcsec (1 arcsec = 725 km on the Sun). Upper frame: Ca II H (3968 Å) sampling the low chromosphere. Lower frame: G band around 4305 Å sampling the deep photosphere. At the photospheric level the solar magnetic fields are very finely structured; they spread and become more diffuse in the chromosphere. (Courtesy Rob Rutten, Utrecht University.)

tubes at a depth of about 1000 km under the photosphere. This bundle of magnetic flux tubes is held together by the convective motions at the granulation boundaries, indicated by the dashed arrows.

In Section 8.2.1 sunspots were just considered as 'tracers' of the magnetic processes going on in the interior of the Sun. However, sunspots can also be studied as entities by themselves. They are also of interest as emerging magnetic flux. Most sunspots disappear again after a couple of days but the larger ones can last much longer, even up to several months. These large sunspots have diameters between 40 000 km and 60 000 km. Such a large sunspot is displayed in Fig. 8.7. The central dark area is called the *umbra*. It has a diameter of 10 000–20 000 km (about 40% of

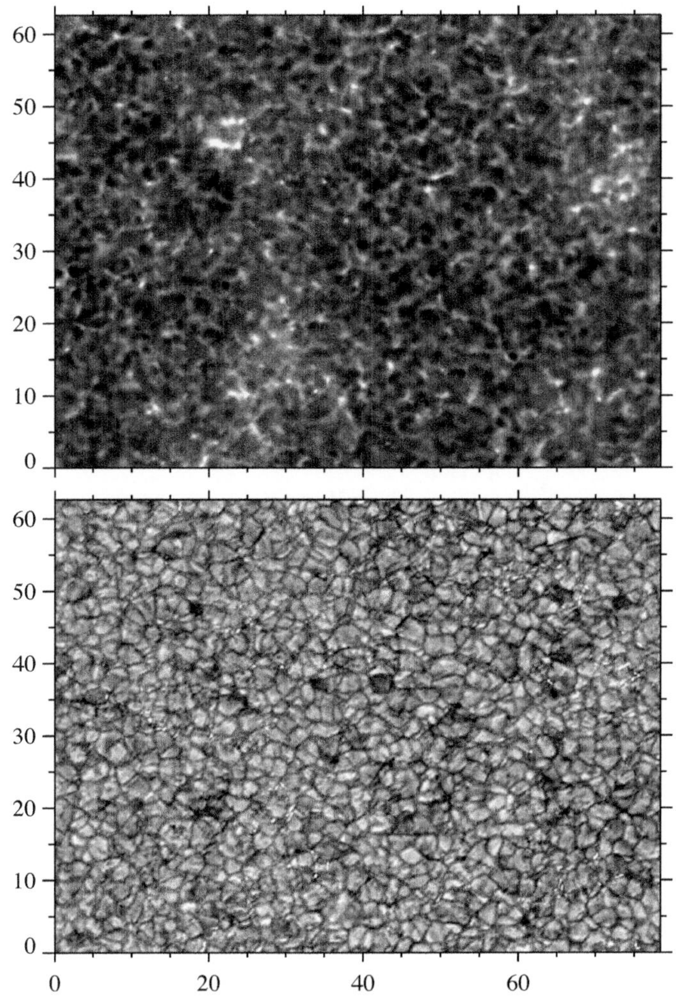

Fig. 8.8 Granulation with network of the quiet Sun, observed with the Dutch Open Telescope, La Palma, 3 July 2002. The axes are in arcsec (1 arcsec = 725 km on the Sun). Upper frame: Ca II H (3968 Å) sampling the low chromosphere. Lower frame: G band around 4305 Å sampling the deep photosphere. (Courtesy Rob Rutten, Utrecht University.)

the total diameter of the spot) and the magnetic field and the temperature are almost uniform in the umbra. The magnetic field is typically 2000–3000 gauss but can be as high as 4000 gauss. The temperature is about 3700 K, which is lower than the temperature of the surrounding photosphere. This is the reason why the umbra appears dark: the brighter surroundings dictate short exposure times. By itself, against a non-emitting background, the umbra would be bright. Longer exposure times reveal that the umbra is not uniformly dark. Bright dots with diameters of 150–200 km and life times of typically 20 minutes occur in the umbra and may be indications of Parker's 'spaghetti model' (Fig. 8.6). The umbra is surrounded by the *penumbra* which consists of light and dark radial filaments. These filaments are 5000–7000 km long and 300–400 km wide and live from half an hour up to six hours. The magnetic field strength decreases in the penumbra and at the interface between the penumbra and the photosphere it is about 1000–1500 gauss.

The solar photosphere is covered by 3 to 4 million granules at any time. A close-up of the granulation pattern is shown in Fig. 8.8: clear evidence of the nonlinear dynamics of the underlying convection zone. The granules are the tops of convection cells overshooting the convection zone. Their centres appear brighter because of the hot, rising plasma there. The diameter of these cells is about 1 to 2″, corresponding to 700–1500 km. They exist typically for 5 minutes, i.e. shorter than the turnover time. The upper frame of Fig. 8.8 shows that the magnetic flux tubes become relatively bright in the low chromosphere due to a heating mechanism that is not understood. The granulation is replaced by a brightness pattern due to convective overshoot, acoustic waves and gravity waves. The lower frame of Fig. 8.8 shows the solar surface granulation caused by the abrupt transition from convective to radiative energy transport. The tiny bright points in some dark intergranular lanes mark magnetic flux tubes with field strengths around 1400 gauss.

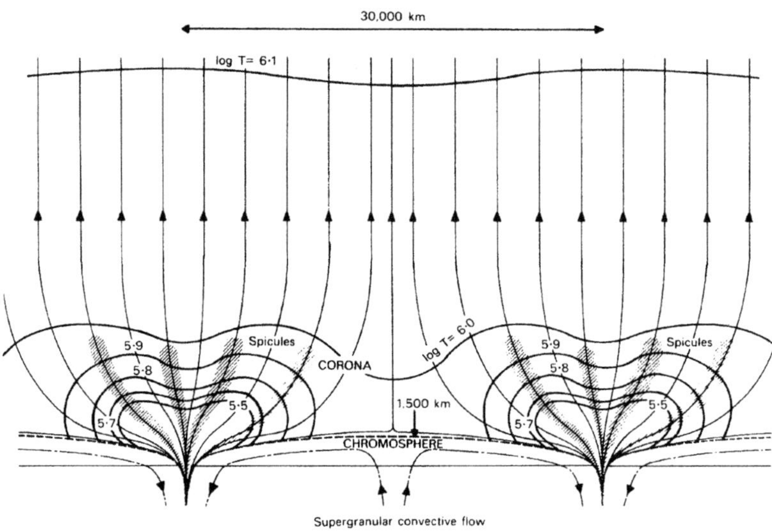

Fig. 8.9 Canopy fields: model of the overall magnetic field structure of the solar atmosphere. (From A. H. Gabriel, *Phil. Trans. Roy. Soc. London* **A281**, 339 (1976).)

In the quiet photosphere, the magnetic field is concentrated in intense flux tubes in the intergranular lanes, i.e. the downdrafts of the granulation. They are driven to the boundaries of the supergranulation cells by horizontal outflows, as indicated in Fig. 8.9. Supergranulation cells were discovered by studying the vertical motions of the photosphere by means of the Doppler shifts of the absorption lines in the spectrum. Horizontal outflows and large-scale velocity patterns were identified as the tops of large convection cells. The diameter of these supergranules varies from 20 000 to 54 000 km and is about 30 000 km on average. In the centre, hot plasma rises and then flows out horizontally at 1.3–1.5 $\mathrm{km\,s^{-1}}$. The typical lifetime of the supergranules is 1 to 2 days and the turnover time is of the same order of magnitude. Their boundaries are very prominent in the chromosphere in which the magnetic field lines fan out due to the decrease of the external plasma pressure. As a result, the magnetic field becomes more uniform in the upper chromosphere and the corona, as illustrated in Fig. 8.9, which also displays the temperature contours in the atmosphere above a supergranule cell.

(b) Corona Spectacular evidence for magnetic fields in the corona comes from the shapes of *prominences* and *loops* protruding into the corona. In eclipse or coronagraph pictures they appear bright at the limb. Prominences are cool and dense structures. Their temperature is about 100 times lower than that of the surrounding corona and their density is 100 to 1000 times higher than coronal values. There are two types of prominences: quiescent and active. The active prominences are located in active regions and exhibit violent motions that may give rise to *solar flares*. An example of an erupting prominence is shown in Fig. 8.10. Active prominences exist for minutes or hours, quiescent prominences last much longer.

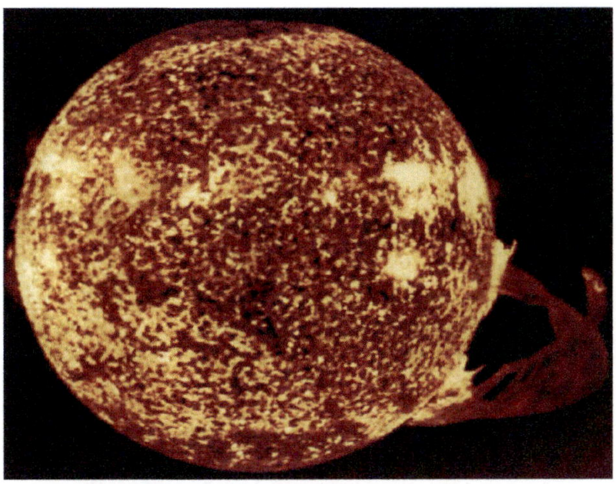

Fig. 8.10 Erupting prominence as seen from the NASA Skylab Space Station on 10 December 1973. (From www.solarviews.com.)

An old example (from 1871!) is shown in Fig. 8.11(a), where coronal loops and chromospheric spicules (shooting jets) are *drawn* from spectrohelioscope observations. A modern counterpart is shown in Fig. 8.11(b). The loops extend to heights of 500 000 km and contain hot plasma that is heated and rises along the magnetic field and then cools again and falls back with speeds of about 100 km per second. We now know that we are looking at *magnetic* structures with properties that are extremely well known. However, even though we now have these high-resolution observations, a satisfactory theory describing the dynamics and the heating of these loops, and predicting when they develop into flares, does not exist yet.

Solar flares are amongst the most impressive explosive phenomena due to the *release of huge amounts of magnetic energy*. Frequently, the irrelevant comparison is made with nuclear explosions (with energy releases of 10^{13}–10^{15} J). Large solar flares of 10^{24} J release the energy equivalent to a billion hydrogen bombs. Of course, the relevant comparison is with the solar luminosity $L_\odot = 4\times10^{26}\,\mathrm{J\,s^{-1}}$, which is hardly affected by a large flare on a time scale of minutes. However, recently discovered superflares of 10^{26}–10^{31} J on ordinary main sequence stars, including those that were

Fig. 8.11 (a) Coronal loops drawn by A. Secchi in 1871. (From C. Young, *The Sun* (London, 1882).) (b) Coronal loops observed in unprecedented detail with NASA's Transition Region and Coronal Explorer (TRACE) spacecraft. Coronal heating is deduced to be located at the bases of these loops. (From vestige.lmsal.com/TRACE.)

discovered to have planets, have evoked discussion on the consequences for possible life on those planets [539]. For our subject, the effects on our own magnetosphere of *coronal mass ejections* (CMEs), often concomitant with solar flares, are more relevant for human enterprises like radio communication and power transmission. See Section 8.3 on space weather.

The solar corona emits thermally in soft X-rays. This means that it can be observed directly in this frequency range since the contribution of the much colder lower atmosphere is negligible. Soft X-ray pictures of the solar corona, such as routinely taken since the Skylab missions in 1973, show that the corona is highly inhomogeneous and structured. Based on the topology of the magnetic field, the corona can be divided into two types of regions, viz. open and closed regions. These regions are associated with magnetic field lines that either are fanning out into interplanetary space or return to the photosphere, forming closed magnetic loops. The 'open' regions are colder and, hence, appear dark on soft X-ray pictures of the whole Sun. They are called *coronal holes* since the solar wind escapes here along the 'open' field lines and is accelerated to enormous speeds (about 400 km s^{-1} at 1 AU). The *closed* or *active* regions appear brighter because they are hotter. They consist of bundles of hot magnetic loops with temperatures of 2 to 3 million Kelvin. They are typically 200 000 km long and have a lifetime of the order of 1 day, although a system of loops may last for months.

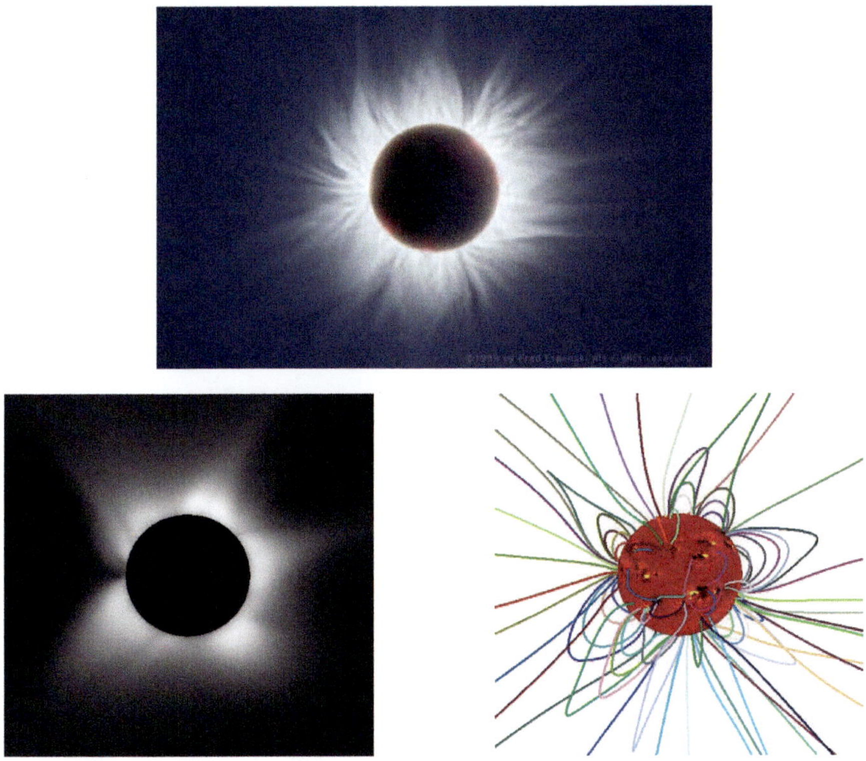

Fig. 8.12 Comparison of a photograph of the total solar eclipse of 11 August 1999 with 3D MHD model predictions performed on 28 July 1999. The MHD model used photospheric magnetic field data from Carrington rotation 1951 corresponding to 24 June–July 1999. Top frame: eclipse image constructed by adding 22 separate photographs taken in Turkey at different exposures to compensate for the rapid radial fall-off of the brightness in the corona and digitally processed to enhance fine details of the corona. (From F. Espenak.) Bottom left: predicted polarization brightness for 11 August 1999 at the time of totality in Eastern Turkey. Bottom right: magnetic field line traces in the 3D MHD model prediction. (Courtesy Zoran Mikić, haven.saic.com/corona.)

In white light the corona can only be seen during an eclipse, or by means of a coronagraph (a telescope with a small disc in it which creates an artificial eclipse), because the photospheric emission completely overwhelms the coronal emission. White-light eclipse photographs of the corona show radial structures, stretching out to 1–10 $R_\odot$, called coronal streamers. Special ones are the 'helmet streamers', that exhibit a cusp-like structure and usually appear above prominences. Near the poles one sees 'polar plumes', but otherwise these regions are dark and associated with coronal holes. In the bottom frames of Fig. 8.12 a predicted polarization brightness and computed magnetic field line traces are shown for a total solar eclipse. This illustrates the correspondence between the open or closed magnetic field configurations and the coronal structures, and convincingly shows that these structures are due to these two types of magnetic fields. The simulation was performed before the eclipse to predict what the solar corona would look like during it.

The closed and open magnetic field configurations have been the subject of many investigations. As remarked by B. C. Low [418], the helmet streamer structure plays an important role in the dynamics of coronal plasmas. It both represents the large-scale relaxed state of the quiescent corona and acts as an agent for coronal reconfiguration by flows and CMEs. Concerning the underlying filament-prominence configuration, it has been noted by Martens and Zwaan [431] that they are usually found at the location of inversion of the magnetic polarity in the active regions of the photosphere. At those locations, emergent magnetic flux produces loop-like filaments with helical magnetic field lines that are stable through line-tying to the photosphere. Through some form of reconnection, several such filaments with the same sense of helicity of the magnetic field join to form long structures that cannot be stabilized any more by line-tying so that they erupt.

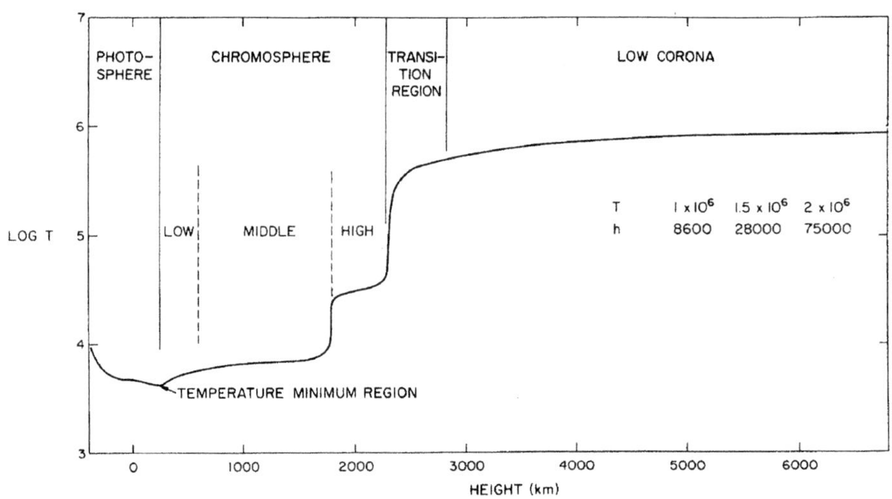

Fig. 8.13 Coronal temperature variation with height. (From R. G. Athay, *The Solar Chromosphere and Corona: Quiet Sun* (Reidel, Dordrecht, 1976).)

At the surface of the Sun, the temperature is about 6600 K and it decreases further in the photosphere to about 4300 K at the top of the photosphere. Above the photosphere, however, the temperature starts to increase again, as shown in Fig. 8.13 computed by Athay. (Note the logarithmic scaling of the vertical axis.) In the lower and middle chromosphere the temperature first increases relatively slowly, but in the higher chromosphere and the transition region to the corona the final increase to $T \sim 2\text{–}3 \times 10^6$ K is so counter-intuitive from a thermodynamic point of view that it has become known as the *coronal heating problem*. We have already mentioned that the mechanism behind this increasing temperature with increasing distance from the Sun is still one of the major problems in solar MHD research. It will be discussed more extensively in Chapter 11.

8.2.3 Inspiration from solar magnetism

Magnetic fields also play a key role in the dynamics (rotation, convection and mass-loss) of other stars and, thus, in stellar evolution. New observational techniques and numerical modelling, often starting from solar models, has enabled substantial progress of insight into the source and role of magnetic fields of stars in general. Schrijver and Zwaan [544] describe the similarities and differences of solar and stellar magnetism in a systematic manner. This includes dynamo processes in the interior, flux emergence from the convection zone, magnetic field evolution in the coronae, diffusion, annihilation and transport by stellar winds. Not all solar-like stars show a magnetic cycle such as our Sun. The various evolution time scales of starspots appear to be crucial for models of stellar differential rotation and magnetic activity cycles [51]. Observations of 'astrospheres', defined as the interaction regions created by collisions between stellar winds and the interstellar medium, also contribute to our understanding of those winds [655].

Different plasma laboratory experiments also take inspiration from the magnetic phenomena on the Sun. The apparently very stable magnetic structures of the quiescent solar loops shown in Fig. 8.11 are sometimes taken as an example of how curved magnetic fields can be kept in stable equilibrium for a long time. As occasionally stated, "more stable than obtained after many years of concentrated efforts in thermonuclear fusion devices devoted to the goal of eventually producing energy". However, this is misleading. Actually, the coronal loops shown in the figure are so stable due to the fact that the magnetic fields are 'stuck' in the photosphere by what is called 'line-tying', which prevents free perpendicular motion. That mechanism is not available in thermonuclear experiments. More important for stability is the pertinent time scale. For a magnetic structure to 'know' that it is unstable, a magnetic perturbation, that is an Alfvén wave, has to travel at least once along the length of the loop. In Table B.4, the number of Alfvén transits during the life times of typical tokamaks and coronal loops are compared. The conclusive evidence is that the life time of present tokamaks, *measured in terms of the relevant Alfvén transit time*, exceeds that of typical coronal loops by a very large factor!

8.2.4 Solar wind and heliosphere

The density of the corona is many orders of magnitude smaller than that of the photosphere and the temperature increases by a factor 100 to 1000. Typical numbers are:

$$\begin{aligned} n_{\text{photosphere}} &\sim 10^{23}\,\text{m}^{-3}\,, \qquad & n_{\text{corona}} &\sim 10^{14}\,\text{m}^{-3} \quad \text{(at 3000 km)}\,, \\ T_{\text{photosphere}} &\sim 6000\,\text{K}\,, & T_{\text{corona}} &\sim 10^{6}\,\text{K}\,. \end{aligned}$$

Whatever the mechanism behind this remarkable phenomenon, these numbers imply that the solar corona cannot be in static equilibrium but suffers a continuous outflow of mass: *the solar wind*. It was predicted by E. N. Parker in 1958 [481] and observed by satellites in 1959. This plasma escapes along the open magnetic field lines, mainly originating in the coronal holes. The question then arises as to what kind of stationary equilibria (i.e. equilibria with a time-independent outflow velocity) such a corona permits. To sketch the answer to that question, we will consider the simplest possible, gas dynamic, model of the corona.

Following Hundhausen [314], the principal arguments can be obtained from a model in which the energy equation is ignored, i.e. an *isothermal model of the corona* ($T \approx T_e \approx T_i$) with a hydrogen plasma ($n \equiv n_\mathrm{e} \approx n_\mathrm{i}$). This implies the following equations:

$$\frac{\partial \rho}{\partial t} + \nabla \cdot (\rho \mathbf{v}) = 0 \qquad \textit{(mass conservation)}, \tag{8.12}$$

$$\rho \frac{\partial \mathbf{v}}{\partial t} + \rho \mathbf{v} \cdot \nabla \mathbf{v} + \nabla p - \rho \mathbf{g} = 0 \qquad \textit{(momentum conservation)}, \tag{8.13}$$

$$p \equiv n_\mathrm{e} k T_\mathrm{e} + n_\mathrm{i} k T_\mathrm{i} \approx 2nkT \qquad \textit{(equation of state with given } T\textit{)}, \tag{8.14}$$

$$\rho \equiv n_\mathrm{e} m_\mathrm{e} + n_\mathrm{i} m_\mathrm{i} \approx nm \qquad \textit{(definition of the density)}, \tag{8.15}$$

where m is the sum of the proton and electron masses. Hence, p and ρ are related through the isothermal sound speed,

$$p/\rho \approx 2kT/m \equiv v_\mathrm{th}^2 \,, \tag{8.16}$$

which is assumed constant in this model.

A spherically symmetric *static* ($\mathbf{v} = 0$) equilibrium would imply

$$\nabla p = \rho \mathbf{g} \quad \Rightarrow \quad \frac{dp}{dr} = -\rho \frac{GM_\odot}{r^2} = -\alpha R_\odot \frac{p}{r^2} \,, \qquad \alpha \equiv \frac{GM_\odot}{R_\odot v_\mathrm{th}^2} \,, \tag{8.17}$$

so that

$$p = p_0 \exp[-\alpha(1 - R_\odot/r)] \,. \tag{8.18}$$

With the typical parameters $T = 10^6\,\mathrm{K}$ and $n = 1.5 \times 10^{15}\,\mathrm{m}^{-3}$ for the base of the corona, we should have $p_0 = 4 \times 10^{-3}\,\mathrm{N\,m^{-2}}$ and $\alpha = 11.53$. Hence, far away from the Sun, in interstellar space, we would have $p_\infty = \mathrm{e}^{-\alpha} p_0 = 4\mathrm{e}^{-11.53} \times 10^{-3} = 4 \times 10^{-8}\,\mathrm{N\,m^{-2}}$, i.e. many orders of magnitude too big as compared to the actual values there, viz. $p_\infty \approx 3 \times 10^{-14}\,\mathrm{N\,m^{-2}}$. Consequently, a hot corona (10^6 K) cannot be in static equilibrium with the interstellar pressure. (For a cold atmosphere at photospheric temperature, $T = 6000\,\mathrm{K}$, such a problem does not arise: $\alpha = 1920$, and p_∞ would be completely negligible.) This is the reason for the solar wind solutions, i.e. stationary ($\mathbf{v} \neq 0$) equilibria.

Hence, consider a spherically symmetric *stationary state* with $v \equiv v_r$. The mass conservation equation (8.12) then gives

$$\frac{d}{dr}(r^2 \rho v) = 0 \quad \Rightarrow \quad r^2 \rho v = \mathrm{const} \,, \tag{8.19}$$

where the constant is related to the total mass loss rate of the Sun,

$$\dot{M}_\odot \equiv \frac{d}{dt} \int \rho \, dV = - \oint \rho \mathbf{v} \cdot \mathbf{n} \, dS = -4\pi r^2 \rho v \,. \tag{8.20}$$

Note that, in contrast to Eq. (4.84) of Section 4.3.2, we now have an open system where mass is lost. Obviously, the representation by a stationary state is restricted to time scales $\tau \ll M_\odot / \dot{M}_\odot$. The momentum equation (8.13) with the relation (8.16) then yields

$$\rho v \frac{dv}{dr} + v_\mathrm{th}^2 \frac{d\rho}{dr} + GM_\odot \frac{\rho}{r^2} = 0 \,. \tag{8.21}$$

Introducing dimensionless variables $\bar{v} \equiv v/v_\mathrm{th}$ (the Mach number of the flow) and $\bar{r} \equiv r/R_\odot$, the latter equation becomes

$$\bar{v}\frac{d\bar{v}}{d\bar{r}} + \frac{1}{\rho}\frac{d\rho}{d\bar{r}} + \frac{\alpha}{\bar{r}^2} = 0\,, \tag{8.22}$$

which yields a one-parameter ordinary differential equation after elimination of ρ by means of Eq. (8.19):

$$\left(\bar{v} - \frac{1}{\bar{v}}\right)\frac{d\bar{v}}{d\bar{r}} - \frac{2}{\bar{r}} + \frac{\alpha}{\bar{r}^2} = 0\,. \tag{8.23}$$

The solutions are implicitly obtained from the first integral:

$$F(\bar{v},\bar{r}) \equiv \tfrac{1}{2}\bar{v}^2 - \ln\bar{v} - 2\,\ln\left(\frac{\bar{r}}{\bar{r}_c}\right) - 2\,\frac{\bar{r}_c}{\bar{r}} + \tfrac{3}{2} = C\,, \qquad \bar{r}_c \equiv \tfrac{1}{2}\alpha\,, \tag{8.24}$$

where the constant C labels the solutions corresponding to the different boundary data at $\bar{r} = 1$ (the solar surface or, rather, the base of the corona). The normalization has been chosen such that solutions *going through the critical point*, where $\partial F/\partial\bar{v} = 0$, $\partial F/\partial\bar{r} = 0$, and located at $\bar{v} = \bar{v}_c = 1$, $\bar{r} = \bar{r}_c = \frac{1}{2}\alpha$, are labelled by $C = 0$. At this point the flow speed crosses the sound speed (hence, it is also called the *sonic point*) and the position is determined by the parameter α, i.e. by the temperature of the corona (assumed to be constant for the present purpose). The solutions are schematically represented in Fig. 8.14.

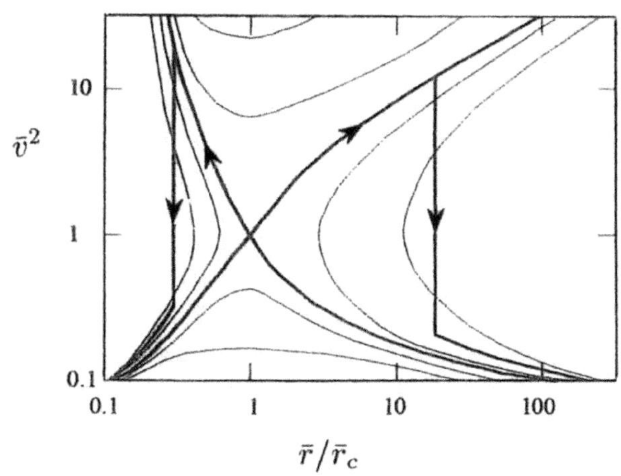

Fig. 8.14 Continuous solar wind solutions (thick lines). The lines with arrows indicate the possibility of shocked wind outflow and accretion flow. (Adapted from Holzer and Leer [309].)

Solutions of interest are those which connect the solar surface with interplanetary space. They are of two types, viz.

(a) *Solar breeze*, i.e. subsonic, solutions which decelerate for $\bar{r} \to \infty$:

$$-\ln\bar{v} \approx 2\ln\bar{r} + \text{const} \quad \Rightarrow \quad \bar{v} \sim \bar{r}^{-2}\,.$$

For these solutions, $\rho \sim \text{const}$, $p \sim \text{const}$ as $\bar{r} \to \infty$. Obviously, the static solution (8.18) is one of them so that the pressure is again too high to be balanced by the interstellar pressure. However, this behaviour at ∞ could be cured by the consideration of a decreasing temperature profile $v_{\text{th}}^2(r)$.

(b) *Solar wind*, i.e. transonic, solutions through the critical point and accelerating for $\bar{r} \to \infty$:

$$\bar{v}^2 \sim 4 \ln \bar{r} \quad \Rightarrow \quad \bar{v} \sim (\ln \bar{r})^{1/2} .$$

For these solutions, $p \sim \bar{r}^{-2}(\ln \bar{r})^{-1/2}$ as $\bar{r} \to \infty$, which is the acceptable behaviour at ∞. The flow is *subsonic* at the Sun, and *supersonic* at 1 AU.

As an aside, the two critical solutions shown in Fig. 8.14 may be modified to illustrate an important aspect of transonic flow, viz. the possibility of the formation of shocks (indicated by the lines with arrows). Since the direction of the flow may be reversed in Fig. 8.14, the accelerating solar wind solution represents a typical example of stellar outflow, whereas the reversed solar breeze critical solution could represent a transonic accretion flow. In the first case, the supersonic flow with ever decreasing pressure eventually meets the small but finite pressure of the interstellar medium and ends there with a shock (the termination shock). Similarly, supersonic accretion (e.g. onto a compact object) is stopped by a shock situated at some location inside the sonic point. The solutions shown are just the simplest examples of transonic flows, considered from a gas dynamic point of view. However, the solar wind is a tenuous plasma convecting the *interplanetary magnetic field*. Hence, a highly complex magnetic structure arises which engulfs all of the planets: *the heliosphere*. Obviously, for the description of the heliosphere and the interaction of the solar wind with the magnetospheres of the planets, the magnetic field cannot be neglected and the much richer variety of MHD shocks is needed. This will be the subject of Chapter 21.

As we have seen, the solar wind escapes from the solar coronal holes along the open field lines. These field lines again form a giant magnetic structure. Looking down at the Sun from a position above one of the poles, one would see that the wind rotates with the Sun and that the magnetic field lines form so-called Archimedean spirals. Since the magnetic field has a different polarity in the two hemispheres, a thin magnetically *neutral current sheet* separates these opposite polarities. Because the flux in the solar wind is not nicely up–down symmetric, this neutral current sheet is warped. Magnetometers on board satellites in the 1960s showed that the rotating solar wind had four magnetic sectors at that time. During the Skylab mission in the summer of 1973, this warped neutral current sheet looked schematically as shown in Fig. 8.15. This large-scale rotating magnetic structure already provides a very complex system by itself, but at the planetary magnetospheres it produces solar wind magnetic field orientations that systematically change direction with respect to the dipolar fields of the planets.

The solar wind that one finds is bimodal: it has two quite different components. The 'fast' solar wind is a tenuous, almost uniform stream with typical velocities of 750 km/s and higher. It escapes along the open magnetic field lines originating in the coronal holes. The 'slow' solar wind, on the other hand, is a more dense and turbulent flow of plasma that escapes from the tips and edges of helmet streamers at typically 300–400 km/s. The chemical composition of the two types of wind also differs. Superposed epoch plots [208] show that in the high speed streams from the polar coronal holes, the O^{7+}/O^{6+} and Mg/O ratios are almost identical and low, while they both increase very steeply to much higher values in the surrounding slow wind. This shows that there is a close relationship between the coronal temperature (represented by the O^{7+}/O^{6+} ratio) and chromospheric conditions (represented by the first ionization potential controlled by the Mg/O ratio). In other words, the abundances and temperatures of the slow and fast solar winds give away their different origins.

As mentioned, Parker [481] showed that the increase of the temperature with increasing height

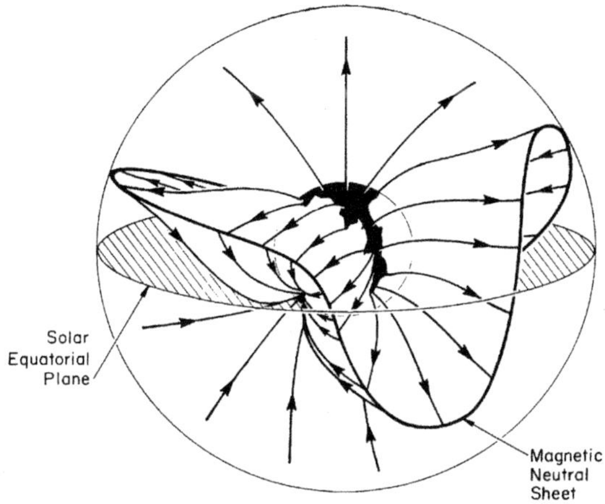

Fig. 8.15 Neutral current sheet. (From A. J. Hundhausen, in *Encyclopedia of Science and Technology* (McGraw-Hill, 1981).)

through the solar corona results in a pressure-driven plasma outflow. This pushes the magnetic field out into the heliosphere and after a few solar radii, the magnetic pressure is already comparable to the flow momentum. Parker, assuming that the magnetic field lines are somehow fixed in the rotating photosphere and 'frozen into' the outflowing plasma, predicted that the heliospheric magnetic field forms a spiral in the equatorial plane. This Parker spiral model provides a very good description of the interplanetary magnetic field (IMF). Near the Earth, the average solar wind velocity is about 400 km/s and its number density is reduced to an average of $10\,\mathrm{cm}^{-3}$. Hence, the collisionality in the solar wind is very low and a kinetic plasma description seems to be required. However, Verscharen et al. [633] recently studied slow modes in the solar wind and compared results from kinetic theory, two-fluid theory and MHD to each other and to in-situ observations of the solar-wind turbulence. They conclude that the solar wind plasma behaves more like a fluid, and that the MHD formulation is appropriate.

8.3 Space weather

8.3.1 Technological and economic implications

Life on Earth is determined and affected in surprisingly many different ways by the Sun. The whole set of complex effects of the radiation and plasma streams from the Sun on the Earth, her magnetosphere, our technological systems, the climate and the people determines the so-called *space weather.* From a socio-economic perspective, among the most important manifestations of solar activity are the solar eruptions that occur frequently on the Sun, and especially the magnetic plasma clouds – the coronal mass ejections – associated to them, due to their role as most important solar drivers of the space weather. The detectable effects on Earth appear in a broad spectrum of time and length scales and have various harmful effects for human health and for our technologies, on which we are ever more dependent. Polar light (the aurora) is one of the nicest and least harmful space weather effects but, alas, space weather can also have less pleasant effects on Earth.

Bad space weather conditions can hinder or damage satellite operations and communication and navigation systems and even cause power grid outages leading to a variety of tremendous socio-economic losses. *Extreme space weather events*, such as the one which occurred in September 1859 (the so-called Carrington event) are expected to occur only once in 100 to 200 years. 'Normal' space weather events accumulate to an estimated economic loss of several billion euros or dollars per year. Hence, the considerable efforts of space agencies to model space weather events and to develop tools to reliably predict its effects on Earth so that protective measures can be taken.

We have introduced the mechanism of the solar wind generation in Section 8.2.4, continue to discuss the propagation of CMEs in the solar wind in Section 8.3.2, then review various numerical modelling efforts of space weather in Section 8.3.3, and finally turn to the interaction of the solar wind and CMEs with the planetary magnetospheres in Section 8.3.4.

8.3.2 Coronal mass ejections

Coronal mass ejections (CMEs) are huge plasma bubbles threaded with magnetic field lines that are ejected from the Sun over the course of several hours. CMEs are spectacular manifestations of solar activity and the most energetic phenomena observed in the solar system. In large CMEs, such as the one depicted in Fig. 8.16, up to 10^{13} kg of coronal material is ejected, but the average value is closer to 10^{12} kg. The average speed of a CME is about 300–400 $\mathrm{km\,s^{-1}}$, but it can be as low as 50 $\mathrm{km\,s^{-1}}$ while fast CMEs have speeds up to 3000 $\mathrm{km\,s^{-1}}$. The energy associated with CMEs amounts to 10^{24}–10^{25} Joule. In spite of this, the existence of CMEs was not realized until the space age and the earliest evidence for these dynamical events came from coronagraph observations made by the 7th Orbiting Solar Observatory (OSO 7) from 1971 to 1973; see Brueckner [97]. In contrast, ground based coronagraphs only make the innermost corona visible. The term 'CME' comes from Burlaga et al. [99].

Close to the Sun, many CMEs have a three-part structure consisting of a bright core (the eruptive prominence), a dark cavity and a bright loop. This structure is lost in interplanetary space. CMEs are often associated with solar flares, or prominence eruptions, but they can also occur independently. The frequency of CMEs varies with the sunspot cycle from about one CME per week at solar minimum to an average of 2 to 3 CMEs per day near solar maximum.

Coronal Mass Ejections constitute transient deviations from the Parker spiral IMF as they also expel large magnetic clouds into interplanetary space at very high velocities. When sampled in situ by a spacecraft in the interplanetary medium, they are termed Interplanetary CMEs (ICMEs). The space weather effects from CMEs are major and twofold. Firstly, upon arrival at Earth, CMEs interact directly with the Earth's magnetosphere, driving significant changes in the geomagnetic field and causing strong variations in the ionospheric current systems. Secondly, CME-driven shock waves accelerate protons to extremely high energies causing radiation storms at the Earth. Even a single high-energy charged particle may seriously damage delicate satellite electronics. While it takes on average a few days for a CME to reach the Earth, Solar Energetic Particle (SEPs) arrive very shortly after the eruption on the Sun. Therefore, the modelling of CME onset and their interplanetary propagation up to the impact on the Earth's magnetosphere (affecting the ionosphere, thermosphere, radiation belts, etc.) is a key issue for reliable space weather forecasts.

Despite the plethora of CME models and observations, the exact trigger mechanism of the related instability remains unknown. The power source involved is undoubtedly the energy that is stored in

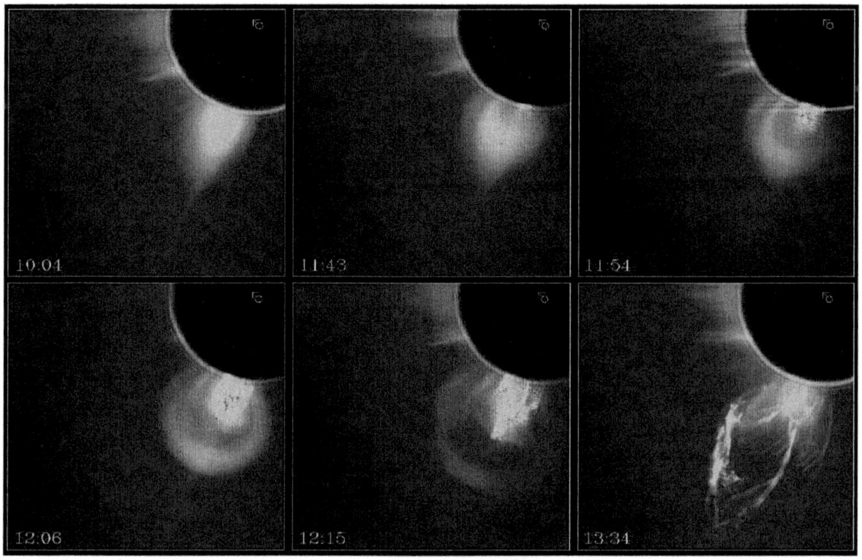

Fig. 8.16 Six snapshots showing the evolution of a *coronal mass ejection* on 18 August 1980. The dark disc in the upper right corner is not the solar disc, but the occulting disc (radius 60% larger than the solar disc) of the Solar Maximum Mission (SMM) coronagraph, used to take these images. (From www.hao.ucar.edu.)

the electric currents in the corona. This 'free energy' can be estimated by constructing the coronal magnetic field and computing its excess magnetic energy over the energy of a potential magnetic field that is reconstructed with the same boundary conditions. The photospheric (and sometimes the chromospheric) magnetic field is known from vector-magnetograms. The coronal magnetic field is 'extrapolated' from these by means of a model. These extrapolation models usually make some assumptions regarding the boundary conditions and the magnetic field itself, e.g. assuming that the magnetic field is force free, i.e. the current density is parallel to $\mathbf{B}$ so that $\mathbf{J} \times \mathbf{B} = \mathbf{0}$. However, there are also complete MHD models of the solar corona. This is justified because the particles are almost all ionized due to the high temperatures in the solar corona so that the global dynamics of the plasma is governed by the magnetic field. The CME onset mechanism is presumably also provided through the magnetic field by shearing motions (e.g. sunspot rotations) and magnetic flux emergence or cancellation. These pump energy into a pre-existing magnetic structure up to the point where some threshold is exceeded and the structure becomes unstable and is ejected into the atmosphere. Hence, closed magnetic structures that can temporarily 'store' excess energy play a key role in CME initiation. All CME initiation models adopt an initial magnetic topology with either a pre-existing flux rope or an arcade that is sheared, resulting in the formation of a flux rope that becomes unstable at some point [371]. An example is the so-called 'breakout' model [15, 16].

Such a breakout model embedded in a background solar wind model was studied by van der Holst et al. [619]. Their 2.5D axi-symmetric simulation model started from a steady-state wind with a triple arcade structure with alternating magnetic polarities, which is kept in place by the magnetic tension in an overlying helmet streamer. This stationary equilibrium is then perturbed by shearing the inner arcade magnetic foot points by applying an azimuthal velocity profile as a time-

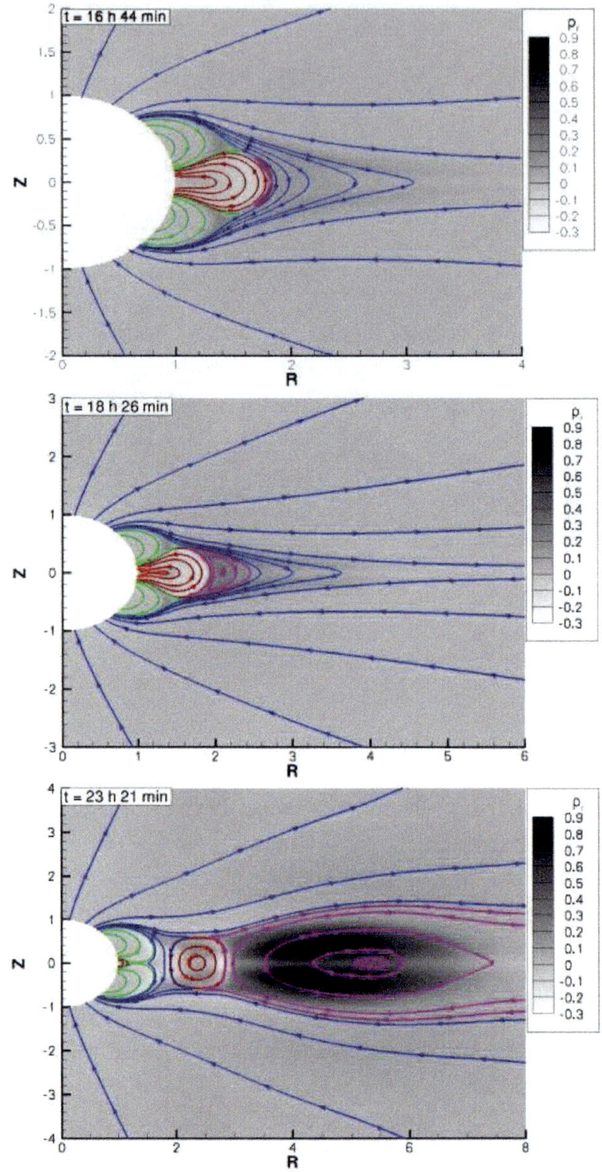

Fig. 8.17 Snapshots of the formation of the breakout CME from a triple arcade embedded in a helmet streamer in the solar wind, showing the relative mass density (grey scale) and some projected field lines (blue). (From van der Holst et al. [619].)

dependent boundary condition. This results in a time evolution that is characterized by an initial phase that closely follows the original breakout scenario [15, 16] with a flattened X-point above the expanding arcade (Fig. 8.17, top panel). In a subsequent phase, however, the breakout reconnection at the leading edge stops and two new reconnection locations form on the flanks of the erupting central arcade. Ultimately, these two side reconnections disconnect the top of the overlying helmet streamer from the Sun (Fig. 8.17, central panel), which then results in a slow but massive CME (bottom panel of Fig. 8.17).

Some models use magnetic flux emergence or cancellation as triggering mechanism for the CME eruption, or a combination of shearing motions and flux emergence/cancellation, as in Zuccarello et al. [670]. In these models, the magnetic flux rope is formed by the shearing motions and flux emergence. Other models use a combination of converging and shearing flows to create a flux rope, see e.g. the simulation by Xia, Keppens and Guo [659], or start from an analytic flux rope model. An example of the latter is discussed in detail in Chapter 20 (see the figures in Section 20.3.2).

Most models agree that the majority of CMEs occur because coronal magnetic fields suffer a sudden loss of equilibrium. Although the vast majority of models are relatively simple, they agree in describing the general properties of CMEs. The most advanced models are 'data-driven' and use observational data on sheared velocity patterns and flux emergence in the photosphere as boundary conditions driving the CME onset. Some of these models have been applied to case studies of particular events. The self-consistent model of Roussev et al. [531] demonstrated that the solar interior regions, down to the convection zone, can be linked to the solar atmosphere using magnetic flux emergence as a driving mechanism for CMEs. The injection of magnetic flux is shown to result in a catastrophic evolution of the corona as the newly emerged flux system interacts with the global surrounding magnetic field. The use of flux emergence simulation data yields realistic electric fields at the photospheric boundary.

8.3.3 Numerical modelling of space weather

Continuous monitoring of the solar corona using space and ground based solar telescopes has dramatically increased our awareness and understanding of CMEs. To a fair degree of accuracy, instantaneous solar coronal data gives us telltale signatures for pre-eruptive events. Ultimately, one may envision a space weather forecasting scenario where both past and instantaneous pre-eruption conditions from observations are used to initialize a 3D computational model of the solar corona and the inner heliosphere. In order to have relevance for forecasting, the simulation must yield a prediction of the eventual local heliospheric conditions at Earth orbit, prior to the actual arrival of the CME itself. The latter translates to being able to compute 3D MHD dynamics on an enormous range of length scales (resolving the coronal field structure in the corona, covering a distance of 1 AU, and having sufficiently detailed resolution near the Earth's orbit). This is currently heavily pursued using massively parallel, grid-adaptive MHD codes.

A pioneering example is shown in Fig. 8.18, from Manchester et al. [427]. The authors used the BATS-R-US code, acronym for block adaptive tree solar wind Roe upwind scheme, which features an octree block based Adaptive Mesh Refinement (AMR) method in a parallel (MPI-based) implementation (see Chapter 20 on *Computational nonlinear MHD* for the numerical terminology). The code solves the ideal MHD equations, augmented with the inclusion of the solar gravitational field and a parameterized volumetric heating term in the energy equation. The latter is designed to give a good agreement with the latitudinal variation of the solar wind measured by the Ulysses spacecraft, for representative solar minimum conditions. The code uses the Powell source term strategy for handling magnetic monopole errors and employs an approximate Riemann solver of Roe-type. The local $t = 0$ field configuration is shown in the top panel of Fig. 8.18: the simulation starts by inserting an analytic model for a helically magnetized flux rope (red field lines) into steady-state

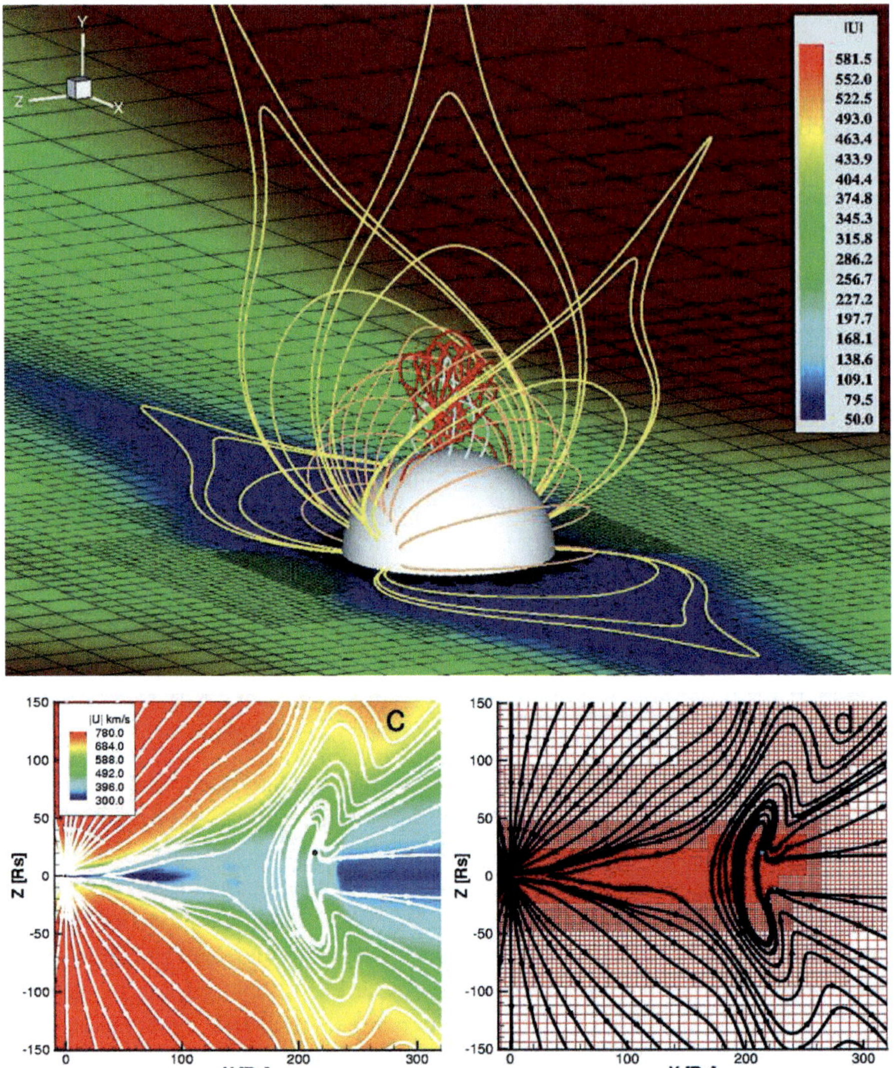

Fig. 8.18 Top: close-up view of an initial coronal magnetic field configuration. Orange and yellow lines correspond to the steady-state equatorial streamer belt. The inserted flux rope is shown with red and light blue field lines. Bottom: global view of the simulation when the CME arrives at Earth (little blue dot at right). Shown are the velocity magnitude, field lines and the grid in a planar cross-section. (From Manchester et al. [427].)

solar wind conditions. The background wind is for a predominantly dipolar solar magnetic field, and this yields an equatorial streamer belt characterized by a helmet streamer configuration (shown in the yellow field lines), together with open field lines starting from both polar regions. The fast wind then originates from these coronal holes. This initial configuration is not in force balance and the flux rope expands and gives rise to a CME with initial speeds exceeding 1000 km/s. The bottom frames of Fig. 8.18 show the global view of the end result, when the CME arrives at the Earth's

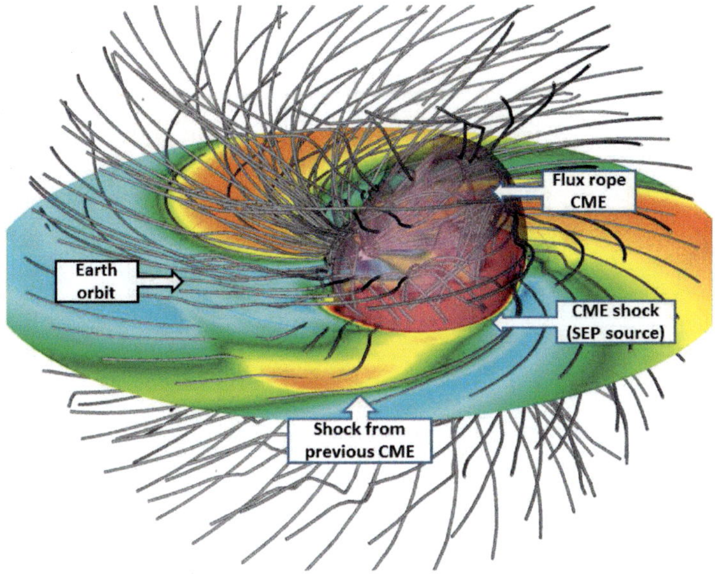

Fig. 8.19 Snapshot of a flux-rope CME simulation with EUHFORIA, showing 3D magnetic field lines and the radial velocity (colour coded) in the equatorial plane. (From Pomoell and Poedts [503].)

orbit. It also shows the grid in a cross-section through the 3D domain, highlighting the need for grid refinement to obtain conditions at or even within the Earth's magnetosphere.

Naturally, the model is continuously improved with more realistic initial conditions, and interfaced with codes that better represent the Earth's magnetosphere, all the way down to ionospheric layers. This kind of activity is pursued in very sophisticated frameworks such as the Space Weather Modeling Framework [608], where different codes are run concurrently to model the dynamics from Sun to Earth. At the core of the different framework components, one finds MHD solvers employing shock-capturing algorithms.

Current operational CME propagation models, including the most known one, viz. ENLIL [470] made available to the research community via the NASA Community Coordinated Modeling Center (CCMC, http://ccmc.gsfc.nasa.gov), all use a steady-state background solar wind model. They do not evolve during the superimposed CME evolution, they use simplified CME models, and they describe the CME early propagation only in a simplistic way or not at all (e.g. when introduced only at 0.1 AU). More recent CME models take into account the crucial internal magnetic structure of the magnetic cloud, as illustrated in Figs. 8.19 and 8.20. The space weather forecasting-targeted inner heliosphere model EUHFORIA ("European Heliospheric Forecasting Information Asset") consists of a coronal model and a heliospheric model with superposed coronal mass ejections [503]. The coronal model just provides data-driven solar wind plasma parameters at 0.1 AU, employing empirical relations to determine the plasma state such as the solar wind speed and mass density, while the magnetic field information is obtained from a reconstruction of the coronal field. These are then used as boundary conditions driving a three-dimensional time-dependent magnetohydrodynamics model of the solar wind up to 2 AU. CMEs are injected into the ambient solar wind. These are usually modelled using the cone model, consisting of a simple density and pressure enhancement. The

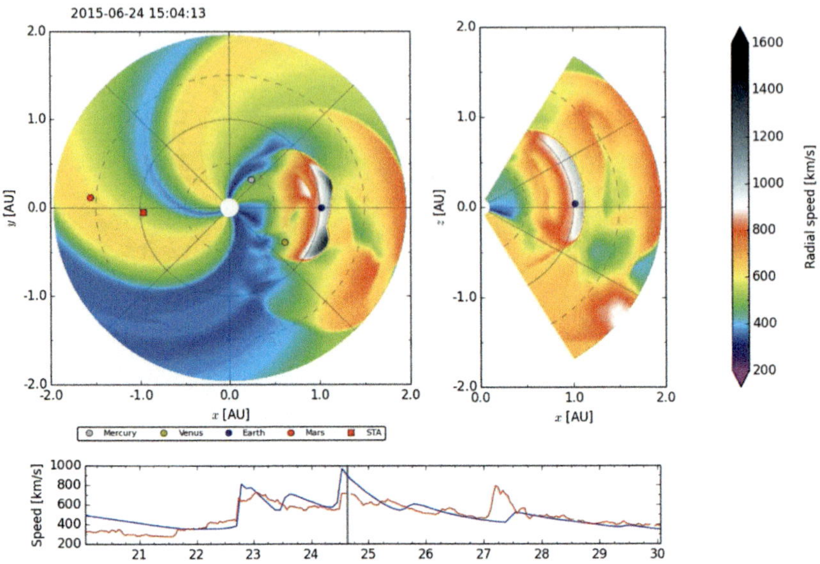

Fig. 8.20 Snapshot of a forecast simulation with EUHFORIA, showing the radial velocity in the equatorial plane (top left, viewed from above) and in the meridional plane through the position of Earth (top right, side view). Bottom: comparison of the radial velocity at L1 simulated with EUHFORIA (in blue) and measured with ACE (in red). (From Pomoell and Poedts [503].)

snapshot illustrated in Figs. 8.19 and 8.20, however, contains a flux-rope CME. Such flux-rope CME models enable more reliable CME evolution simulations, taking into account the effects of *erosion* (through magnetic reconnection of the internal magnetic field with the magnetic field of the ambient solar wind) and *deformation* (due to the interaction with the ambient solar wind), and predictions of the *geo-effectiveness* of an event (depending on the sign and magnitude of the component of the magnetic field in the ICME cloud perpendicular to the equatorial plane).

8.3.4 Solar wind and planetary magnetospheres

Concerning the magnetic fields of the planets and their associated magnetospheres: from the numbers in Table B.6 it is clear that the largest magnetic fields are found for the *Earth* and the *Jovian (or giant) planets*. Of the latter, Jupiter has by far the largest magnetic dipole moment and, consequently, the most extended magnetospheric system. The question of why these planets have large magnetic fields, whereas the fields of the remaining terrestrial planets are very weak, is a difficult one that has not been answered satisfactorily yet. The answer would require a detailed knowledge of both the *internal structure of the planets* and of the theoretical solutions of the *nonlinear dynamo equations*. In a sense, these problems are *mathematically ill-posed:* solutions depend extremely sensitively on the boundary conditions imposed on the outside, which is the only place where magnetic field or seismological measurements can be made. It is beyond the scope of this book to analyse these problems in detail.

Quite some progress has been made on the numerical solution of the dynamo problem for the

Earth by Glatzmaier and Roberts [212]. In particular, their numerical simulation through several 100 000 years of geo-dynamo activity exhibited, for the first time, a clear reversal event with a change of magnetic field orientation from north-south to south-north, as required by geophysical observations. See Fig. 19.9 of Chapter 19, where the nonlinear numerics involved is discussed.

One of the interesting features of the solar system is the variety of the physical properties of the planets. It is clear that the existence and the strength of a planetary field will depend on the *size*, the *internal structure and composition*, and the *rotation* of the planet. The relative importance of these three factors is still far from obvious at present. From this point of view it is nice that there are more planets than just one so that we are prevented from drawing oversimplified and erroneous conclusions. For example, the fast rotation of the *giant planets* correlates well with their sizeable magnetic fields, whereas the slow rotation of *Venus* also correlates with the virtual absence of an internal magnetic field (see Table B.6). However, the similarity in the rotation periods of *Mars* and *Earth* does not give a clue as to why the magnetic field of Mars is three orders of magnitude smaller. Also notice that the smaller planet *Mercury*, which also rotates much slower than Mars, still has a magnetic field larger than that of Mars.

These magnetospheric structures may be disturbed by sudden changes of the solar wind variables (e.g. by CMEs, see above), giving rise to *MHD waves* like ultra-low frequency (ULF, $< 10\,\mathrm{Hz}$) waves, or higher frequency plasma waves, or may be disrupted by *reconnection* of field lines causing flux transfer events (FTEs) on the day-side as well as on the night-side of the magnetosphere, *Kelvin–Helmholtz instabilities* driven by shear flow at the magnetopause, etc. All of these topics will return as genuine plasma dynamics in later chapters. The geo-effectiveness of coronal mass ejections, i.e. their impact at Earth, often causing severe magnetic storms, is determined by a complex sequence of different processes. The severe magnetic storms they often cause are quantified by geomagnetic indices, including the so-called AA* and Ap* indices (referring to the major magnetic storms), the Auroral Electrojet (AE) index (high resolution and hourly data), the Disturbance storm time (Dst) Index, and the Kp and Ap indices.

Obviously, the magnetospheres of the other planets also interact with the variable solar wind. This interaction is often studied via 3D global MHD simulations. Chané et al. [114] showed that density variations in the solar wind strongly affect the Jovian magnetosphere. Density increases in the solar wind cause the magnetosphere to decrease and the magnetopause moves inward, enhancing dawn–dusk asymmetries, and compressing the magnetic field on the dayside while elongating them at the nightside. A denser wind also brightens the main oval on the nightside as well as on the dayside, after first darkening there for a few hours.

8.4 Perspective

In spite of the current 3D time-dependent data-driven solar wind and CME initiation models, there is still a lot of missing or neglected physics. Most importantly, the detailed theory of magnetic reconnection in 3D is still lacking. Almost all CME onset models invoke magnetic reconnection, but it is often caused by numerical diffusion. Also the interaction of the neutral wind with the solar plasma is poorly understood, but it may play a role in twisting and shearing of the magnetic field. There is a need for better MHD modelling of the solar atmosphere and the self-consistent evolution of CMEs, from their onset in the low solar corona all the way through interplanetary space. Such models can clarify how they interact with, and affect, the background solar wind and satellites, magnetospheres of planets and some moons, comets, etc. This requires chains of consistently coupled models to capture the many complex physical processes involved.

8.5 Literature and exercises

Notes on literature

Solar MHD

- Priest [510], *Magnetohydrodynamics of the Sun*, is the basic text on this subject.

Solar physics

- Stix [569], *The Sun* (2nd edition), Chapter 8 on solar magnetism treats the subjects of flux tubes, sunspots and the solar cycle, and has a discussion of mean field electrodynamics and the solar dynamo.
- Foukal [181], *Solar Astrophysics*, Chapter 11 on dynamics of the solar magnetic field reviews the concepts of solar flux tubes, sunspots, butterfly diagram and the Babcock model for the solar dynamo.
- Schrijver and Zwaan [544], *Solar and Stellar Magnetic Activity*, is a comprehensive review putting solar magnetism in the wider context of magnetic fields in stars that have a convective envelope immediately below their photosphere.
- Aschwanden [28], *Physics of the Solar Corona*, provides a comprehensive plasma physics introduction into the basic phenomena in the solar corona. Based on the results of recent solar space missions, coronal instabilities and heating, magnetic reconnection, particle acceleration and radiation processes, flares and CMEs are discussed.

Magnetospheric physics

- Saunders [538], 'The Earth's magnetosphere', contains an MHD description of the magnetosphere, with the magnetic field and electric current structures coupled to the solar wind, and the dynamics of MHD waves and shocks, reconnection, FTEs and substorms.

Solar and stellar winds, space weather

- Hundhausen [314], in 'The solar wind', reviews the properties of the solar wind, fluid theories on formation in the corona, the large-scale magnetic structure of the heliosphere and the termination shock.
- Lamers and Cassinelli [390], *Introduction to Stellar Winds*, review the different mechanisms proposed to explain the stellar winds of different types of stars, starting with the Sun in Chapter 5 on coronal winds.
- Goossens [248], *An Introduction to Plasma Astrophysics and Magnetohydrodynamics*, presents in Chapter 6 on the solar wind a careful overview of rotating thermally driven winds from a hydrodynamics as well as a magnetohydrodynamics view point.
- Strangeway, Luhmann, and Russell [575], *Space Physics: An Introduction*, contains an overview of current understanding of the physics behind space weather based on lectures of the authors on the subject, including chapters on the solar wind, magnetospheres, waves, shocks, etc.

Exercises

[8.1] *The solar cycle*

The solar cycle is a periodic phenomenon which can be related to the number of sunspots on the surface of the Sun. They are present in a narrow strip around the equator only.

- Sketch the so-called 'butterfly diagram' and explain what it shows.
- Discuss the cartoons illustrating the Babcock model of the solar cycle. Does it explain the narrow strip around the equator?
- Write down the evolution equation for the magnetic field, including resistivity, but assuming a static

configuration. Express the typical time scale for the decay of the magnetic field inhomogeneity using the Spitzer resistivity $\eta \approx 64Z\ln\Lambda T_e{}^{-3/2}$, where $\ln\Lambda \approx 15$. Find the required quantities at the bottom of the convection zone, where the source of the solar dynamo is located, and estimate the time scale. Does it approach the required time scale of the solar cycle? Comment on the implications.

[8.2] *Turbulent magnetic diffusivity*

Consider the evolution of the solar magnetic field by turbulent flow. Split the magnetic field and the velocity field into an average part, indicated by $\langle \ldots \rangle$, and a perturbation, indicated by a prime. Assume $\langle \mathbf{v}' \times \mathbf{B}' \rangle \approx \alpha \langle \mathbf{B} \rangle - \beta \nabla \times \langle \mathbf{B} \rangle$, where β models the turbulent magnetic diffusivity.

- Comment on the interpretation of the coefficients α and β, and give estimates of their order of magnitude from the solar parameters you can find. How did the situation improve with respect to the previous exercise?

[8.3] *Alfvén waves in the solar corona and in the magnetosphere*

Consider an MHD wave propagating in a thin flux tube in the dipolar magnetic field of the Sun or the magnetosphere of the Earth. The magnitude of the magnetic field varies both along and across the flux tube. However, in this problem, we ignore the latter dependence. Effectively, the problem is then one-dimensional, with variations of the background quantities in the z-direction only. We assume that the flux tube is pressureless and carries no current. Choose the perpendicular coordinates x and y such that the Alfvén wave is represented by ξ_x and one of the magneto-sonic waves by ξ_y. The Alfvén waves, for the magnetic field $\mathbf{B} = B(z)\mathbf{e}_z$ and the density $\rho = \rho(z)$, obey the wave equation

$$\partial^2 \xi_x / \partial t^2 = v_A^2 \, \partial^2 \xi_x / \partial z^2 \, .$$

Now insert the actual magnitude of a dipolar field given by $\mathbf{B}(\mathbf{r}) = \mu_0 (4\pi r^3)^{-1} (3\mathbf{m} \cdot \mathbf{e}_r \, \mathbf{e}_r - \mathbf{m})$, where $\mathbf{m}$ is the magnetic dipole moment vector of a current loop, $\mathbf{e}_r$ is the unit vector in the direction of the position vector $\mathbf{r}$ and $r \equiv |\mathbf{r}|$.

- Find an expression for the effective coordinate z in terms of the spherical coordinates $r(\theta), \theta$ along a field line, where θ runs from $\theta = \theta_0$ to $\pi - \theta_0$. This will give an expression for $B(z)$, or $B(\theta)$.
- Numerically solve the wave equation for the Alfvén waves with the magnetic field just found. Assume constant density and periodic boundary conditions on ξ_x. This may represent standing Alfvén waves in a solar or magnetospheric loop. What is the influence of the variation of $B(z)$?
- Now specify to a solar coronal loop, where the photospheric boundary conditions are simulated by a large density increase at the ends of the loop. Use, for example, the density profile exploited by Beliën et al. [46]:

$$\rho(z) = 1 + (\rho^{\mathrm{p}} - 1) \exp\big[- (0.5/\sigma^2) \sin^2(\pi z/L) \big] \, ,$$

where ρ^{p} is the density at the photospheric boundaries $z = 0$ and $z = L$ and σ is the density scale length. For large σ, how are the waves modified compared to the case of periodic boundary conditions?

[8.4]* *Solar wind interaction with the magnetosphere*

The interaction of the solar wind with the magnetosphere of the Earth is a complicated three-dimensional MHD problem giving rise to many kinds of time-dependent disturbances that influence the magnetic structure of the magnetosphere as a whole. For example, when the solar wind carries a magnetic field pointing in the same direction as the dipole field of the Earth, the magnetosphere is compressed and the structure remains closed. However, when the solar wind magnetic field has opposite direction, a flux transfer event may occur where the day-side magnetosphere is suddenly opened up and an entirely different magnetic configuration is formed. In this problem, we will model the two types of magnetospheric structures (closed and open) by means of a very crude (static and two-dimensional) model where the one-sided solar wind with embedded magnetic field is replaced by a rotationally symmetric vertical magnetic field that exerts pressure on the dipole field of the Earth from all sides. We will exploit spherical coordinates (see Appendix A.2).

- Show that $\nabla \cdot \mathbf{B} = 0$ is solved in these coordinates by

$$B_r = -\frac{1}{Rr}\frac{\partial \psi}{\partial \theta} \, , \quad B_\theta = \frac{1}{R}\frac{\partial \psi}{\partial r} \, , \qquad \text{i.e. } \mathbf{B} = -\frac{1}{R}\mathbf{e}_\phi \times \nabla \psi \, ,$$

where ψ is the poloidal flux (evaluated through a circle R in the ecliptic plane).

- Show that $(\nabla \times \mathbf{B})_\phi = 0$ yields a second order partial differential equation for ψ,

$$\frac{\partial^2 \psi}{\partial r^2} + \frac{\sin\theta}{r^2} \frac{\partial}{\partial \theta} \frac{1}{\sin\theta} \frac{\partial \psi}{\partial \theta} = 0\,,$$

  which is the Grad–Shafranov equation for this problem. (In plasmas with pressure and current, like in tokamaks, this equation has a non-vanishing RHS.)
- Show that the dipole field $\mathbf{B}_\mathrm{d} = r^{-3}(3\mathbf{M}\cdot\mathbf{e}_r\,\mathbf{e}_r - \mathbf{M})$ is obtained from the particular solution

$$\psi_\mathrm{d} = C + (M/r)\sin^2\theta\,,$$

  where we exploit the value $M = -8.1 \times 10^{15}\,\mathrm{T\,m^3}$ for the dipole moment of the Earth. Note that $\mathbf{M} \equiv \mu_0(4\pi)^{-1}\mathbf{m}$, so that the values of $\mathbf{m}$ in $\mathrm{A\,m^2}$ of Table B.6 are to be multiplied with 10^{-7} to get the values of $\mathbf{M}$ in $\mathrm{T\,m^3}$.
- Now add a vertical field $B_\mathrm{v} = B_0\mathbf{e}_z = B_0(\cos\theta\,\mathbf{e}_r - \sin\theta\,\mathbf{e}_\theta)$, derivable from the flux function

$$\psi_\mathrm{v} = \tfrac{1}{2}B_0 R^2\,,$$

  representing the magnetic pressure of the solar wind.
- Find out what the relevant dimensionless parameters are for this problem. Make contour plots of $\psi \equiv \psi_\mathrm{d} + \psi_\mathrm{v}$ and study the magnetic structures you obtain for $B_0 > 0$ and $B_0 < 0$.
- Having obtained these qualitatively different solutions, determine the special points (x-points and stagnation points) analytically. Insert numbers and make estimates for solar wind parameters. Comment on the results obtained.

9
Cylindrical plasmas

9.1 Equilibrium of cylindrical plasmas

We have considered the effects of plasma inhomogeneity on MHD waves and instabilities in Chapter 7 for the model of a plane gravitating plasma slab where inhomogeneity is restricted to the vertical direction. For the description of laboratory and astrophysical plasma dynamics, the concept of *magnetic flux tube* is quite central, as we have seen in Chapter 8. This automatically leads to the consideration of cylindrical plasmas where the inhomogeneities are operating in the radial direction. Whereas the model remains one-dimensional, so that most of the analytical techniques developed in Chapter 7 remain valid, the introduction of *curvature of the magnetic field* brings in qualitatively different physical effects that significantly influence the dynamics of flux tubes. We will now neglect gravity since it plays no role in laboratory plasmas and, for astrophysical plasmas, it is more adequately incorporated in an axi-symmetric model with a central gravitating object (the subject of Chapter 18). We will see though that curvature of the magnetic field in a cylindrical plasma enters the equations in a very similar way to gravity in the plasma slab that was treated in Chapter 7.

9.1.1 Diffuse plasmas

For the study of confined plasmas, the diffuse cylindrical plasma column (called 'diffuse linear pinch' in the older plasma literature) is one of the most useful models. It is probably the most widely studied model in plasma stability theory. Since we have obtained a basic understanding of the spectrum of inhomogeneous one-dimensional systems, the analysis of the diffuse linear pinch can now be undertaken more fruitfully than was possible in the early days of fusion research when this configuration was first investigated. Also, we will consider this configuration as a first approximation to toroidal systems, where the addition of a second direction of inhomogeneity leads to partial differential equations and, therefore, to substantial complications of the analysis. For those systems, the construction of a coherent picture of the spectrum of waves and instabilities is a very demanding task, which is still far from completion.

Consider an infinite diffuse cylindrical plasma of radius a (Fig. 9.1). The equilibrium equations,

$$\mathbf{j} \times \mathbf{B} = \nabla p\,, \qquad \mathbf{j} = \nabla \times \mathbf{B}\,, \qquad \nabla \cdot \mathbf{B} = 0\,, \tag{9.1}$$

in cylindrical r, θ, z-coordinates, i.e. rotational symmetry in θ and translational symmetry in z, yield

$$p' = j_\theta B_z - j_z B_\theta\,, \qquad j_\theta = -B_z'\,, \qquad j_z = \frac{1}{r}(rB_\theta)'\,, \tag{9.2}$$

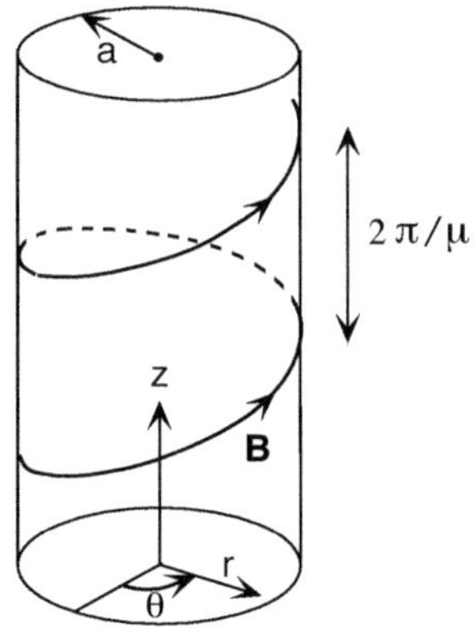

Fig. 9.1 Diffuse cylindrical plasma column with a helical magnetic field $\mathbf{B}$ of inverse pitch $\mu = \mu(r)$, drawn here at the wall radius $r = a$.

where the primes denote derivatives with respect to r. Eliminating j_θ and j_z, the equilibrium turns out to be characterized by the pressure profile $p(r)$ and the magnetic field profiles $B_\theta(r)$ and $B_z(r)$, subject to just one differential equation:

$$\left[p(r) + \tfrac{1}{2}B^2(r)\right]' + \frac{B_\theta^2(r)}{r} = 0 \,. \tag{9.3}$$

Hence, we may choose two of these three profiles arbitrarily, whereas the density profile $\rho(r)$ may be chosen arbitrarily as well since it does not appear in the equilibrium equations when gravity is neglected. Special cases of such diffuse cylindrical equilibria are the z-pinch and θ-pinch configurations introduced in Section 1.2.3 (Fig. 1.4) and Sections 2.4.3 (Fig. 2.10).

We here assume a model I plasma (Section 4.6.1), where the magnetic field is not considered beyond the radius $r = a$. For the equilibrium, this implies the presence of a rigid wall absorbing the mechanical forces (laboratory plasmas) or anything else that justifies the assumption of a radially confined plasma. For the perturbations, the flows have to be tangential at that radius. It is expedient to introduce a function that describes the radial variation of the helicity of the magnetic field, which is the *inverse pitch of the magnetic field lines:*

$$\mu(r) \equiv \frac{B_\theta(r)}{rB_z(r)} \,. \tag{9.4}$$

Also useful is the *kinetic pressure contained versus the magnetic pressure on axis:*

$$\beta(r) \equiv \frac{2p(r)}{B_0^2} \,. \tag{9.5}$$

The diffuse cylindrical equilibrium is now completely determined by, for example, prescribing the functions $p(r)$ (or $\beta(r)$) and $B_\theta(r)$ (or $\mu(r)$), whereas $B_z(r)$ follows from the solution of the equilibrium equation (9.3) and $\rho(r)$ is arbitrary.

▷ **Dimensionless scaling of the equilibrium** It is important (e.g. for numerical applications) to construct the *smallest set of parameters* that characterize these equilibria. To that end, we exploit scale independence (Section 4.1.2) to eliminate three *trivial* parameters,

$$a \,, \quad B_0 \; (\equiv B_{z0}) \,, \quad \rho_0 \,, \tag{9.6}$$

by normalizing everything with respect to those scales of lengths, field strengths and densities (i.e. Alfvén speeds). Here, we exploit the subscript o to indicate function values on axis ($r = 0$). On the other hand, the two *essential* parameters

$$\beta_0 \equiv \beta(r=0) = \frac{2p_0}{B_0^2}\,, \qquad \bar{\mu}_0 \equiv a\mu(r=0) = \frac{aj_{z0}}{2B_0}\,, \tag{9.7}$$

fix the magnitude of the plasma pressure and the inverse pitch of the magnetic field lines ($\sim$ the current density) on axis. Having fixed the amplitudes, all physical functions are distributed on the unit plasma interval $0 \le \bar{r} \equiv r/a \le 1$ according to *shape functions:*

$$\bar{p}(\bar{r}) \equiv \frac{1}{p_0}\,p(r)\,, \quad \bar{\mu}(\bar{r}) \equiv a\,\mu(r)\,, \quad \bar{\rho}(\bar{r}) \equiv \frac{1}{\rho_0}\,\rho(r)$$

$$\stackrel{(9.3),(9.4)}{\Longrightarrow} \quad \bar{B}_z(\bar{r}) \equiv \frac{1}{B_0}\,B_z(r)\,, \quad \bar{B}_\theta(\bar{r}) \equiv \bar{\mu}(\bar{r})\,\bar{r}\bar{B}_z(\bar{r})\,. \tag{9.8}$$

In conclusion: the equilibrium is determined by choosing the two parameter values β_0 and $\bar{\mu}_0$, and the (infinitely many parameters of the) shape functions $\bar{p}(\bar{r})$ and $\bar{\mu}(\bar{r})$. In the dynamics of the perturbations, the additional shape function $\bar{\rho}(\bar{r})$ enters. ◁

(a) Force-free magnetic fields An interesting class of cylindrical equilibria is obtained for very low β when pressure gradients can be neglected so that the magnetic field becomes 'force-free', satisfying the condition

$$\mathbf{j} = \alpha\,\mathbf{B}\,, \tag{9.9}$$

where the function $\alpha(r)$ is completely free. From this expression and the components of $\mathbf{j} = \nabla\times\mathbf{B}$, the relationship between α and μ is obtained:

$$\left.\begin{aligned} j_\theta &= -B_z' = \alpha B_\theta \\ j_z &= \tfrac{1}{r}(rB_\theta)' = \alpha B_z \end{aligned}\right\} \quad \Rightarrow \quad \alpha = \frac{2\mu + \mu' r}{1+\mu^2 r^2}\,. \tag{9.10}$$

For the choice of constant pitch, $\mu' = 0$, one obtains from these first order differential equations the following explicit solutions:

$$B_z(r) = \frac{B_0}{1+\mu^2 r^2}\,, \qquad B_\theta(r) = \frac{B_0 \mu r}{1+\mu^2 r^2}\,. \tag{9.11}$$

For the special choice of constant α, another one-parameter family of equilibrium solutions is obtained:

$$B_z(r) = B_0 J_0(\alpha r)\,, \qquad B_\theta(r) = B_0 J_1(\alpha r)\,, \tag{9.12}$$

where J_0 and J_1 are the zeroth and first order Bessel functions. These force-free magnetic field solutions were already discussed in Section 4.3.4 (Fig. 4.8). Because of their apparent simplicity, they have been the subject of numerous investigations with respect to stability and slow dissipative dynamics. On axis, the inverse pitch μ is simply related to α through $\mu_0 = \frac{1}{2}\alpha$, but away from the axis, μ varies from $+\infty$ to $-\infty$ when αr progresses through the various zeros of the Bessel function J_0: an entirely non-trivial class of current-carrying equilibria.

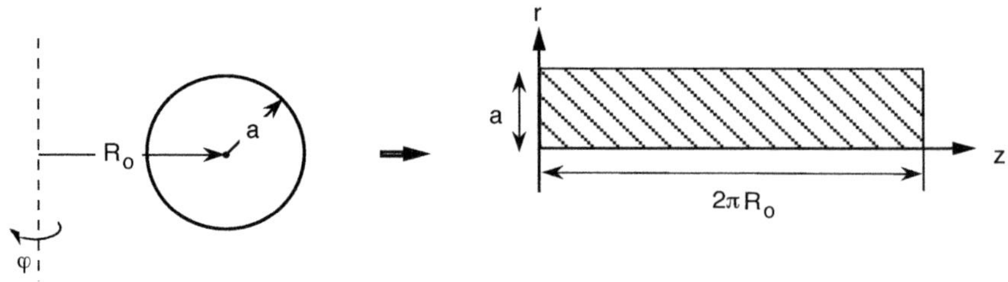

Fig. 9.2 Slender torus with inverse aspect ratio $\epsilon \equiv a/R_0 \ll 1$ represented as a periodic cylinder with length $2\pi R_0$.

(b) 'Straight tokamak' limit A slender torus, with small *inverse aspect ratio* $\epsilon \equiv a/R_0 \ll 1$ (Fig. 9.2, left frame), may be approximated by a straight cylinder of finite length $L = 2\pi R_0$ (Fig. 9.2, right frame). This becomes a mathematical torus when the ends are identified. This gives a first approximation of toroidal equilibrium. Some of the toroidal effects are well represented in this manner (e.g. that the wave number has to be quantized in the toroidal direction). By consistently developing all physical quantities to the relevant order in ϵ, the dynamics of these equilibria may be computed to leading order. A meaningful choice of the order of magnitude of the two essential parameters β_0 and μ_0 then becomes crucial.

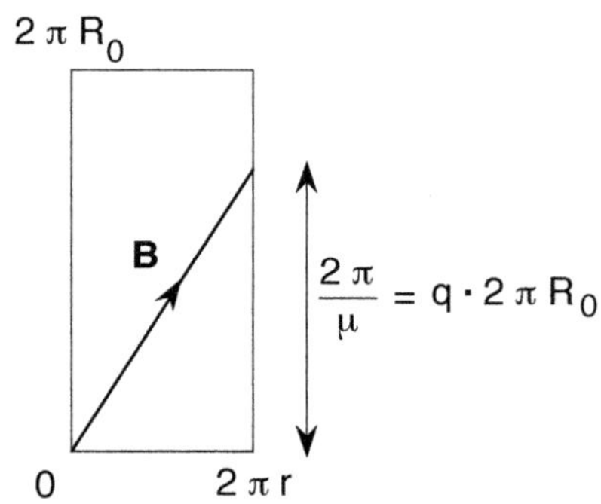

Fig. 9.3 Inverse pitch μ and safety factor q of the magnetic field lines in a 'straight tokamak' periodic cylinder model of a toroidal plasma.

In the periodic cylinder representation of tokamaks, called the 'straight tokamak' limit, the variable μ is replaced by a variable that measures the pitch of the field lines relative to the circumference of the torus (Fig. 9.3), i.e. the *'safety factor'* (already introduced in Section 2.4.3, Fig. 2.12),

$$q(r) \equiv \frac{rB_z(r)}{R_0 B_\theta(r)} \equiv \frac{1}{\mu(r) R_0} \quad \left[\equiv \frac{\epsilon}{\mu(r) a} \right] . \tag{9.13}$$

In the so-called *low-beta tokamak regime*, the order of magnitude of the two essential parameters is chosen as

$$\beta_0 \sim \epsilon^2 \ll 1 , \qquad q_0 \sim 1 . \tag{9.14}$$

In the cylindrical 'straight tokamak' approximation, the normalized q-profile,

$$\bar{q}(\bar{r}) \equiv q(r)/q_0 \,, \tag{9.15}$$

then becomes the only shape function entering the leading order expressions of the equilibrium and the perturbations. The reason is that, because β is small, pressure effects, leading to an outward shift of the magnetic axis in a torus (the Shafranov shift), only enter in a higher order where poloidal θ-variations of the equilibrium through toroidal curvature are permitted. The proper treatment of the latter toroidal effects requires the two-dimensional theory developed in Chapters 16–18.

9.1.2 Interface plasmas

In laboratory fusion research, the plasma is usually isolated from the wall by means of a region of rather cold plasma with low pressure and small current density. Such a plasma configuration can be idealized by means of either one of the two interface models II or II* introduced in Section 4.6.1. Here, we will assume that the outer plasma is pressureless and carries no current, so that there is no difference with respect to the equilibrium between model II and model II*: the outer region is characterized by a *vacuum magnetic field* configuration. With respect to the perturbations, model II and model II* interface plasmas may behave very differently, though, even with identical equilibrium fields, since a plasma (model II*) allows for induction of perturbed currents whereas a vacuum (model II) does not. We will see in Section 9.2.2 that this leads to major differences in the stability properties.

We consider cylindrical model II and model II* configurations consisting of a diffuse plasma in an infinite cylinder of radius a, surrounded by a vacuum magnetic field $\hat{\mathbf{B}}$, and enclosed by a perfectly conducting wall at $r = b$. In the plasma region $0 \le r \le a$ the equilibrium is described by the functions $p(r), B_\theta(r), B_z(r)$ (and $\rho(r)$), that satisfy the equilibrium equation (9.3). At the plasma surface $r = a$, surface currents produce jumps in the variables p, B_θ and B_z which are restricted to satisfy pressure balance:

$$p_1 + \tfrac{1}{2}(B_{\theta 1}^2 + B_{z1}^2) = \tfrac{1}{2}(\hat{B}_{\theta 1}^2 + \hat{B}_{z1}^2) \,, \tag{9.16}$$

where the subscript 1 indicates equilibrium values at the plasma surface. The surface currents are given by

$$j_\theta^\star = -[\![B_z]\!] \equiv -\hat{B}_{z1} + B_{z1} \,, \qquad j_z^\star = [\![B_\theta]\!] \equiv \hat{B}_{\theta 1} - B_{\theta 1} \,. \tag{9.17}$$

The outer vacuum magnetic field in the region $a < r \le b$ is given by

$$\hat{B}_z(r) = \hat{B}_{z1} \,, \qquad \hat{B}_\theta(r) = \frac{\hat{\mu}_1 a^2}{r} \hat{B}_{z1} \,. \tag{9.18}$$

Assuming the interior plasma to be prescribed as in the previous section, the pressure jump condition (9.16) fixes the magnitude of the vacuum magnetic field $\hat{\mathbf{B}}$ but leaves its direction free. Hence, the interface model with an outer vacuum magnetic field adds two parameters to the problem: the pitch of the outer magnetic field, $\hat{\mu}_1 \neq \mu_1$, and the relative wall position b/a.

A special case is the early sharp-boundary model of pinch discharges, investigated by Kruskal and Schwarzschild [381], Kruskal and Tuck [382], Rosenbluth [528] and Tayler [589]. The inner plasma was taken to be homogeneous, with $B_\theta = 0$, $B_z = B_0$, $p = p_0 = p_1$, so that the only relevant parameter for the inner plasma is $\beta \equiv 2p/B_0^2$, which was taken to be high since dreams of confining plasma by separating it completely from the magnetic field were still alive. The current is then exclusively confined to the surface $r = a$, and the vacuum magnetic field there is determined from Eq. (9.16) by the two parameters β and $\hat{\mu}_1$:

$$\hat{B}_{\theta 1} = \hat{\mu}_1 a \, \hat{B}_{z1} \,, \qquad 1 + \beta = (1 + \hat{\mu}_1^2 a^2) \, \hat{B}_{z1}^2 / B_0^2 \,. \tag{9.19}$$

This model successfully described the threat posed by external kink modes to high-β confinement (see Section 9.3.2).

9.2 MHD wave equation for cylindrical plasmas

9.2.1 Derivation of the MHD wave equation for a cylinder

We derive the equation of motion for cylindrical equilibria along the same lines as for the gravitating slab (Section 7.3.2). Our starting point is the equation of motion

$$\mathbf{F}(\boldsymbol{\xi}) \equiv -\nabla \pi - \mathbf{B} \times (\nabla \times \mathbf{Q}) + (\nabla \times \mathbf{B}) \times \mathbf{Q} = \rho \frac{\partial^2 \boldsymbol{\xi}}{\partial t^2} \,, \tag{9.20}$$

with the usual abbreviations

$$\pi \equiv -\gamma p \nabla \cdot \boldsymbol{\xi} - \boldsymbol{\xi} \cdot \nabla p \,, \qquad \mathbf{Q} \equiv \nabla \times (\boldsymbol{\xi} \times \mathbf{B}) \,. \tag{9.21}$$

Because of the symmetry, we may study normal mode solutions of the form

$$\boldsymbol{\xi}(r, \theta, z, t) = \Big(\xi_{r,mk}(r), \xi_{\theta,mk}(r), \xi_{z,mk}(r) \Big) \, \mathrm{e}^{\mathrm{i}(m\theta + kz - \omega t)} \,, \tag{9.22}$$

where the subscripts m and k will again be dropped in the following. For these separate modes the equation of motion may be reduced to an ordinary second order differential equation in terms of the component $\xi_r(r)$.

As in the analysis of the plasma slab, we exploit a projection based on the magnetic field lines with unit vectors

$$\mathbf{e}_r \,, \quad \mathbf{e}_\perp \equiv (0, B_z, -B_\theta)/B \,, \quad \mathbf{e}_\parallel \equiv (0, B_\theta, B_z)/B \,. \tag{9.23}$$

In this projection, the result of the gradient operator applied to a perturbed quantity as given in Eq. (9.22) may be written as

$$\nabla = \mathbf{e}_r \, \partial_r + \mathrm{i}\mathbf{e}_\perp \, k_\perp + \mathrm{i}\mathbf{e}_\parallel \, k_\parallel \quad (+ \text{ possible terms from } \partial_\theta \mathbf{e}_r \text{ and } \partial_\theta \mathbf{e}_\theta) \,, \tag{9.24}$$

where the perpendicular and parallel gradient operators become algebraic:

$$k_\perp \equiv (m B_z / r - k B_\theta)/B = G/B \,, \qquad k_\parallel \equiv (m B_\theta / r + k B_z)/B = F/B \,. \tag{9.25}$$

Beyond Eq. (9.28), we will use the symbols G and F instead of $k_\perp$ and $k_\parallel$.

▷ **Pitfalls** (1) One should not denote the vector in large brackets on the RHS of Eq. (9.22) as $\boldsymbol{\xi}_{mk}(r)$. On a circle, this would indicate a vector of constant amplitude and direction (as shown in Fig. 9.4(a)), which is not meant here. The correct notation is $\boldsymbol{\xi}_{mk}(r, \theta)$ (Fig. 9.4(b)). This respects the θ-dependence of the unit vectors: $\partial \boldsymbol{\xi}_{mk}(r, \theta)/\partial \theta \neq 0$.

(2) The representation (9.24) for the gradient operator should not be considered as a recipe to be applied blindly, but just as a kind of short-hand notation for the expressions obtained after conversion to cylindrical coordinates by means of Appendix A.2.2. In the analogous projection (7.71) for the plane slab with shear, this representation could be used also for computing divergences and curls by properly accounting for the

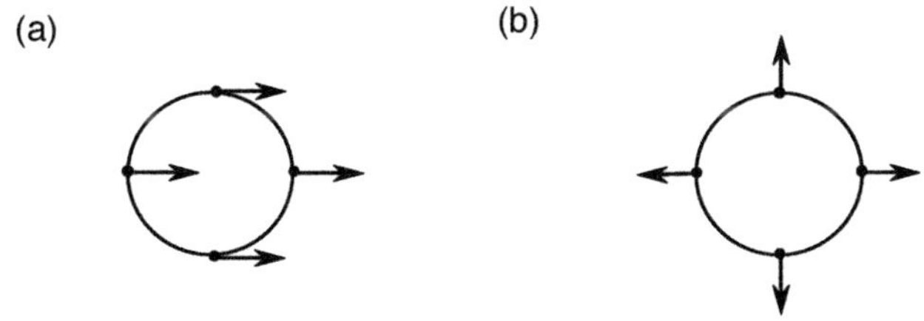

Fig. 9.4 Different vectors: (a) $\boldsymbol{\xi}_{mk}(r)$ and (b) $\boldsymbol{\xi}_{mk}(r,\theta)$.

dependence of the unit vectors $\mathbf{e}_\perp$ and $\mathbf{e}_\parallel$ on the normal coordinate x. Here, the situation is fundamentally different since the cylindrical coordinate system has a scale factor h_2 $(= r)$, and the unit vectors $\mathbf{e}_r$ and $\mathbf{e}_\theta$ depend also on the ignorable coordinate θ: $\partial\mathbf{e}_r/\partial\theta = \mathbf{e}_\theta\,,\ \partial\mathbf{e}_\theta/\partial\theta = -\mathbf{e}_r\,.$ ◁

The projection of the displacement vector on the field line triad is denoted by

$$\xi \equiv \mathbf{e}_r\cdot\boldsymbol{\xi} = \xi_r\,, \quad \eta \equiv \mathrm{i}\mathbf{e}_\perp\cdot\boldsymbol{\xi} = \mathrm{i}(B_z\xi_\theta - B_\theta\xi_z)/B\,, \quad \zeta \equiv \mathrm{i}\mathbf{e}_\parallel\cdot\boldsymbol{\xi} = \mathrm{i}(B_\theta\xi_\theta + B_z\xi_z)/B\,. \quad (9.26)$$

In terms of these variables we obtain

$$\begin{aligned} \mathbf{Q} &= \mathrm{i}k_\parallel B\xi\,\mathbf{e}_r - [\,(B_\theta\xi)' - kB\eta\,]\,\mathbf{e}_\theta - \frac{1}{r}[\,(rB_z\xi)' + mB\eta\,]\,\mathbf{e}_z\,, \\ \pi &= -p'\xi - \gamma p\nabla\cdot\boldsymbol{\xi}\,, \qquad \nabla\cdot\boldsymbol{\xi} = \frac{1}{r}(r\xi)' + k_\perp\eta + k_\parallel\zeta\,, \end{aligned} \quad (9.27)$$

where the factor $\exp[\mathrm{i}(m\theta + kz - \omega t)]$ is dropped for simplicity. By means of these expressions, and the equilibrium relation (9.3), the equation of motion (9.20) can be evaluated in the same manner as for the plane gravitating slab in Section 7.3.2. This yields the following matrix representation of the spectral equation:

$$\mathsf{F}_0\cdot\mathbf{X} = -\rho\omega^2\mathbf{X}\,, \qquad \mathbf{X} \equiv (\xi,\eta,\zeta)^{\mathrm{T}}\,, \quad (9.28)$$

with the force operator

$$\mathsf{F}_0 \equiv \begin{pmatrix} \dfrac{d}{dr}\dfrac{\gamma p + B^2}{r}\dfrac{d}{dr}r - k_\parallel^2B^2 - r\left(\dfrac{B_\theta^2}{r^2}\right)' & \dfrac{d}{dr}k_\perp(\gamma p + B^2) - \dfrac{2kB_\theta B}{r} & \dfrac{d}{dr}k_\parallel\gamma p \\ -\dfrac{k_\perp(\gamma p + B^2)}{r}\dfrac{d}{dr}r - \dfrac{2kB_\theta B}{r} & -k_\perp^2(\gamma p + B^2) - k_\parallel^2B^2 & -k_\parallel k_\perp\gamma p \\ -\dfrac{k_\parallel\gamma p}{r}\dfrac{d}{dr}r & -k_\parallel k_\perp\gamma p & -k_\parallel^2\gamma p \end{pmatrix}.$$

The subscript 0 is introduced here for later use (in Section 13.3.1). This formulation is symmetric, apart from the occurrence of some factors r. Those may be eliminated by exploiting the radial variable $\chi \equiv r\xi$ instead of ξ.

Note the similarity of the matrices (9.28) for the cylindrical plasma, and (7.78) for the gravitating slab, where now off-diagonal curvature terms (with B_θ) appear instead of gravitational ones. Since these terms do not involve derivatives of $\boldsymbol{\xi}$, they do not affect the essential spectrum (i.e. the part of the spectrum associated with the singularities). Of course, the discrete spectrum and, hence, stability is quite significantly affected by these terms.

(a) Hain–Lüst equation The structure of Eq. (9.28), with lower order differential equations for the tangential components η and ζ, allows us to reduce the system to a single second order differential equation by expressing the tangential components in terms of the radial variable $\chi \equiv r\xi$:

$$
\begin{aligned}
\eta &= \frac{G\Big[(\gamma p + B^2)\rho\omega^2 - \gamma p F^2\Big]\, r\chi' + 2kB_\theta(B^2\rho\omega^2 - \gamma p F^2)\,\chi}{r^2 BD}\,, \\
\zeta &= \frac{\gamma p F\Big[(\rho\omega^2 - F^2)\, r\chi' + 2kB_\theta G\,\chi\Big]}{r^2 BD}\,.
\end{aligned}
\tag{9.29}
$$

Substituting these expressions into the first component of Eq. (9.28) yields the slightly generalized *Hain–Lüst equation*:

$$
\frac{d}{dr}\left[\frac{N}{D}\frac{d\chi}{dr}\right] + \left[A + \frac{B}{D} + \left(\frac{C}{D}\right)'\right]\chi = 0\,. \tag{9.30}
$$

(The derivation by Hain and Lüst [276] was restricted to isothermal perturbations, $\gamma = 1$; it was generalized to adiabatic perturbations by Goedbloed [214](b).) Here,

$$
N \equiv \frac{1}{r}(\rho\omega^2 - F^2)\Big[(\gamma p + B^2)\rho\omega^2 - \gamma p F^2\Big]\,, \tag{9.31}
$$

$$
D \equiv \rho^2\omega^4 - (m^2/r^2 + k^2)\Big[(\gamma p + B^2)\rho\omega^2 + \gamma p F^2\Big]\,, \tag{9.32}
$$

and the three other coefficients are defined by

$$
\begin{aligned}
A &\equiv \frac{1}{r}(\rho\omega^2 - F^2) + \left(\frac{B_\theta^2}{r^2}\right)'\,, \qquad B \equiv -\frac{4}{r^3}k^2 B_\theta^2 (B^2\rho\omega^2 - \gamma p F^2)\,, \\
C &\equiv -\frac{2B_\theta^2}{r^2}D + \frac{2kB_\theta G}{r^2}\Big[(\gamma p + B^2)\rho\omega^2 - \gamma p F^2\Big] \\
&\equiv -\frac{2B_\theta^2}{r^2}\rho^2\omega^4 + \frac{2mB_\theta F}{r^3}\Big[(\gamma p + B^2)\rho\omega^2 - \gamma p F^2\Big]\,.
\end{aligned}
\tag{9.33}
$$

Note that the term $(B_\theta^2/r^2)'$ of the coefficient A is not combined with the similar term $(C/D)'$ from the first of the equivalent expressions for C. This is dictated by the physical significance of the three coefficients A, B and C, as will become clear when discussing the equivalent first order system (9.43) and the conditions (9.45) and (9.46) for the absence of spurious singularities.

Comparing Eq. (9.30) with the corresponding equation (7.80) for the plane slab, it is clear that the terms caused by the curvature of the poloidal field B_θ play a similar role as the gravitational terms $\rho\hat{g}$. These terms disappear when $B_\theta = 0$ (θ-pinch) and we obtain a problem almost equal to a plane slab without gravity. Since the latter is stable, it follows that the linear θ-pinch is also stable.

With the wall at the plasma (model I), the appropriate boundary conditions are:

$$
\chi(0) = \chi(a) = 0\,. \tag{9.34}
$$

In the presence of an external vacuum (model II) or of a pressureless plasma (model II*), the boundary condition at $r = a$ becomes a rather complicated expression. It is derived in Section 9.2.2.

The boundary condition at $r = 0$ deserves extra care since the wave equation has a *singularity due to the cylindrical geometry* there. The Frobenius expansion (see Section 7.4.1) about the origin yields an indicial equation $\nu^2 - m^2 = 0$, implying a small solution $\chi_s \sim r^{|m|}$ and a singular solution with a logarithm that is excluded by the boundary condition (9.34). For $m = 0$, the indicial equation is $\nu(\nu - 2) = 0$, so that the small solution behaves like $\chi_s \sim r^2$ and the solution with the logarithm is again excluded by the boundary condition. Although the boundary condition (9.34) on χ conveniently combines the different possibilities, it is useful to realize the different meaning of it for the different modes when expressed in terms of the physical variable ξ. Using ξ, one has to distinguish between the $|m| = 1$ modes, which have a finite value of ξ on axis (corresponding to a finite displacement on axis by these modes) and all other modes which have $\xi(0) = 0$. Consequently, when using ξ, the appropriate boundary conditions become:

$$\xi'(0) = 0 \quad \text{for } |m| = 1\,, \qquad \xi(0) = 0 \quad \text{for } |m| \neq 1\,. \tag{9.35}$$

Clearly, the geometrical singularity at $r = 0$ is of a completely different nature to the physical singularities associated with the continuous spectra.

(b) Singularities For the purpose of the analysis, we abbreviate the differential equation (9.30) as

$$\Big[P(r;\omega^2)\,\chi' \Big]' - Q(r;\omega^2)\,\chi = 0\,, \tag{9.36}$$

where

$$P(r;\omega^2) \equiv \frac{N}{D}\,, \qquad \begin{aligned} N(r;\omega^2) &\equiv \frac{\rho^2}{r}(\gamma p + B^2)\,[\omega^2 - \omega_A^2(r)]\,[\omega^2 - \omega_S^2(r)]\,, \\ D(r;\omega^2) &\equiv \rho^2\,[\omega^2 - \omega_{s0}^2(r)]\,[\omega^2 - \omega_{f0}^2(r)]\,, \end{aligned} \tag{9.37}$$

and

$$Q(r;\omega^2) = -A - \frac{B}{D} - \left(\frac{C}{D}\right)'. \tag{9.38}$$

The expressions for the singular frequencies ω_A^2, ω_S^2, ω_{s0}^2 and ω_{f0}^2 are given by

$$\begin{aligned} &\omega_A^2(r) \equiv F^2/\rho\,, \qquad \omega_S^2(r) \equiv \frac{\gamma p}{\gamma p + B^2}\,F^2/\rho\,, \\ &\omega_{s0,f0}^2(r) \equiv \tfrac{1}{2}\omega_m^2\left[1 \pm \sqrt{1 - 4\omega_S^2/\omega_m^2}\,\right], \quad \omega_m^2 \equiv (m^2/r^2 + k^2)(\gamma p + B^2)/\rho\,. \end{aligned} \tag{9.39}$$

They are almost identical to the expressions (7.83)–(7.85) for the plane gravitating plasma slab (except that the square of the horizontal wave number, $k_0^2 \equiv k_y^2 + k_z^2$, is replaced by the expression $m^2/r^2 + k^2$, which is not constant and suffers from the geometrical singularity at $r = 0$ discussed above). Consequently, we may refer to the analysis of Section 7.4 and conclude that the cylindrical plasma also has *two continuous spectra*, the Alfvén continuum $\{\omega_A^2(r)\}$ and the slow continuum $\{\omega_S^2(r)\}$, and that the apparently singular sets $\{\omega_{s0}^2(r)\}$ and $\{\omega_{f0}^2(r)\}$ are ranges of *turning point frequencies*.

For every radius r, these genuinely and apparently singular frequencies are well ordered according to the scheme

$$0 \le \omega_S^2 \le \omega_{s0}^2 \le \omega_A^2 \le \omega_{f0}^2 \le \omega_F^2 \equiv \infty\,. \tag{9.40}$$

The collections of these frequencies for the whole interval $(0, 1)$ may be represented by a diagram similar to Fig. 7.18, with one important difference: no matter how small the inhomogeneity, the geometrical singularity at $r = 0$ causes overlap between the slow turning point frequencies and the slow continua and between the fast turning point frequencies and the (formal) fast continuum because $\omega_{s0}^2(r\to 0) \to \omega_S^2(0)$ and $\omega_{f0}^2(r\to 0) \to \omega_F^2 \equiv \infty$.

(c) Equivalent system of first order differential equations As in Section 7.3.2(d), we transform the second order differential equation (9.30) into a system of two first order equations. This turns out to be quite illuminating. Rather than just rewriting the equation in terms of the variables χ and χ', we use a variable with physical significance, viz. the perturbation Π of the total pressure $p + \frac{1}{2}B^2$:

$$\Pi = \pi + \mathbf{B}\cdot\mathbf{Q}\,. \tag{9.41}$$

(This is the Eulerian pressure perturbation, Π_E, related to the Lagrangian pressure perturbation, Π_L, by $\Pi_E = \Pi_L + B_\theta^2\,\chi/r^2$.) Inserting the expressions (9.27) for $\mathbf{Q}$ and π, and reducing them to expressions in terms of χ' and χ yields

$$\Pi = -\frac{1}{D}\,(N\chi' + C\chi)\,. \tag{9.42}$$

Notice that all terms with radial derivatives in the Hain–Lüst equation (9.30), except for the term $(B_\theta^2/r^2)'$, appear undifferentiated in the expression for Π.

By straightforward algebra, the Hain–Lüst equation is then transformed into the following pair of first order differential equations:

$$N\begin{pmatrix}\chi\\ \Pi\end{pmatrix}' + \begin{pmatrix} C & D\\ E & -C\end{pmatrix}\begin{pmatrix}\chi\\ \Pi\end{pmatrix} = 0\,, \tag{9.43}$$

where

$$E \equiv -\Big(A + \frac{B}{D}\Big)N - \frac{C^2}{D} \equiv -N\left[\frac{1}{r}(\rho\omega^2 - F^2) + \Big(\frac{B_\theta^2}{r^2}\Big)'\right] \\ - \frac{4B_\theta^2}{r^4}\left[B_\theta^2\rho^2\omega^4 - F^2\Big((\gamma p + B^2)\rho\omega^2 - \gamma p F^2\Big)\right], \tag{9.44}$$

and N and D were defined in Eqs. (9.31) and (9.32). The determinant of the matrix,

$$DE + C^2 = -(AD + B)N \to 0\,, \quad \text{when } N \to 0\,, \tag{9.45}$$

exhibits the proportionality with N required to cancel out one of the two factors, that was introduced by multiplying both derivatives by N.

This formulation, which is due to Appert, Gruber and Vaclavik [18], again shows that the slow and Alfvén continua $\{\omega_A^2\}$ and $\{\omega_S^2\}$ originate from the zeros of the factor N in front of the derivatives. The virtue of this formulation, over that in terms of the second order differential equation, is that the singularities $D = 0$ are immediately seen to be apparent ones since nothing singular shows up there. In the numerical problem of solving Eq. (9.43) by means of a shooting method, one multiplies the equations by $1/N$ and proceeds to calculate the derivatives. Given initial data χ_0 and Π_0 at a certain point $r = r_1$, one then calculates $\chi_0{}'$ and $\Pi_0{}'$, from which one obtains new initial data χ_1 and Π_1 at a point $r = r_1$, and so forth. The only difficulty which may arise is the occurrence of $N = 0$ singularities. For $D = 0$ no problem turns up. This not evident in the Hain–Lüst formulation.

▷ **Apparent singularities** In the formulation in terms of the second order differential equation, one has to prove that expansion about the $D = 0$ locations again yields the condition (7.130) of Section 7.4.1. To that end, one notes that the coefficients of the functions P and Q of Eqs. (9.37) and (9.38) and the coefficient E defined in Eq. (9.44) are related by

$$C^2 + NB = -(AN + E)D \to 0 \quad \text{when } D \to 0\,, \tag{9.46}$$

which is the required condition; QED. ◁

(d) Limiting forms of the Hain–Lüst equation For reference purposes we list four significant limits of the Hain–Lüst equation.

(1) *Low-frequency limit* $(|\rho\omega^2| \ll (m^2/r^2 + k^2)(\gamma p + B^2))$:

$$\frac{d}{dr}\left[\frac{\rho\omega^2 - F^2}{m^2 + k^2r^2}\, r\, \frac{d\chi}{dr}\right] - \frac{1}{r}\Bigg[\rho\omega^2 - F^2 - \frac{4k^2B_\theta^2}{m^2 + k^2r^2}\,\frac{\gamma p\rho\omega^2}{(\gamma p + B^2)\rho\omega^2 - \gamma pF^2} - \frac{2B_\theta^2}{rB^2}\,p'$$
$$+ \frac{4mB_\theta F}{(m^2 + k^2r^2)B^2}\left(p' - \frac{rB_zG}{2m}\frac{\mu'}{\mu} + \frac{k^2rB^2}{m^2 + k^2r^2} + \frac{B_\theta F}{m}\right)\Bigg]\chi = 0\,. \tag{9.47}$$

This equation was obtained by Goedbloed and Hagebeuk [234] by assuming small frequencies (or growth rates) and expressing all derivatives B_θ' and B_z' in terms of p' and μ' through the equilibrium relation (9.3) and the definition $\mu \equiv B_\theta/(rB_z)$. The number of terms is increased in this way, but the important ones involved in instabilities (p', μ' and F) are now clearly distinguished so that different orderings may be designed to optimize their effect. This equation still contains the exact limits of the incompressible and marginal equations (see below).

(2) *Incompressible limit* $(\gamma \to \infty)$:

$$\frac{d}{dr}\left[\frac{\rho\omega^2 - F^2}{m^2 + k^2r^2}\, r\, \frac{d\chi}{dr}\right] - \Bigg[\frac{1}{r}(\rho\omega^2 - F^2) + \left(\frac{B_\theta^2}{r^2}\right)'$$
$$- \frac{4k^2B_\theta^2F^2}{r(m^2 + k^2r^2)(\rho\omega^2 - F^2)} - \left(\frac{2mB_\theta F/r}{m^2 + k^2r^2}\right)'\Bigg]\chi = 0\,. \tag{9.48}$$

This equation was derived by Freidberg [184]. It contains the exact marginal stability equation, listed below, in the limit $\omega^2 \to 0$. This shows that stability is not affected by compressibility, although growth rates are. (Note that the singularity $(\rho\omega^2 - F^2)^{-1}$ in the second term in square brackets is an apparent one since it originates from the factor $\omega^2 - \omega_{\rm s0}^2$ of D.)

(3) *Marginal equation of motion* $(\rho\omega^2 = 0)$:

$$\frac{d}{dr}\left[\frac{rF^2}{m^2 + k^2r^2}\frac{d\chi}{dr}\right] - \left[\frac{1}{r}F^2 - \left(\frac{B_\theta^2}{r^2}\right)' - \frac{4k^2B_\theta^2}{r(m^2 + k^2r^2)} + \left(\frac{2mB_\theta F/r}{m^2 + k^2r^2}\right)'\right]\chi = 0\,. \tag{9.49}$$

This is one form of Newcomb's [461] Euler–Lagrange equation describing the stability of cylindrical plasmas (for an alternative form, see Eqs. (9.103) and (9.104) of Section 9.4.1). With respect to the singularity $F = 0$, the equation is analogous (but physically not the same of course) to the marginal equation of motion (7.171) for the gravitating slab, so that singular behaviour can be discussed in complete analogy with Section 7.5.

(4) *'Straight' low-β tokamak* ($\gamma p \ll B^2$, $|\rho\omega^2| \ll (m^2/r^2 + k^2)B^2$):

$$\frac{d}{dr}\left[\frac{\rho\omega^2 - F^2}{m^2}\, r\, \frac{d\chi}{dr}\right] - \left[\frac{1}{r}(\rho\omega^2 - F^2) - \frac{2F}{m}\left(\frac{B_\theta}{r}\right)'\right]\chi = 0\,. \qquad (9.50)$$

Here, a periodic cylinder with length $2\pi R \gg a$ is considered (Fig. 9.2), so that $k = n/R$, and only leading order terms are kept. In terms of the safety factor $q \equiv (B_z/R)(r/B_\theta)$, this implies that $F = (B_z/R)(n + m/q)$, where $B_z/R \approx$ const.

9.2.2 Boundary conditions for cylindrical interfaces

If there is an external vacuum (model II), or pressureless plasma (model II*), surrounding the central plasma column, the right (model I) boundary condition, $\chi(a) = 0$, should be replaced by conditions determining the amplitude and the normal derivative of $\chi(a)$ describing the freely moving plasma surface. This problem turns up in the investigation of free-boundary modes (e.g. external kinks). For the cylindrical interface equilibrium of Section 9.1.2, the calculation of ξ (or χ) in the interior region should then be complemented with the appropriate 'extension' of the perturbation in the outer region. For model II, this involves the calculation of the perturbation $\hat{\mathbf{Q}}$ of the vacuum magnetic field plus the boundary conditions connecting $\hat{\mathbf{Q}}$ to $\boldsymbol{\xi}$ at the plasma–vacuum interface at $r = a$.

(a) Plasma–vacuum interface (model II) The appropriate boundary conditions were derived in Section 6.6.1: the first interface condition (6.131) and the second interface condition (6.134). The first boundary condition is easily transformed to

$$\mathrm{i}\hat{F}\chi = r\hat{Q}_r \quad (\text{at } r = a), \quad \text{where} \quad \hat{F} = m\hat{B}_{\theta 1}/a + k\hat{B}_{z1}\,. \qquad (9.51)$$

The LHS of the second boundary condition is the Lagrangian perturbation of the total pressure, so that this condition may be transformed by means of Eq. (9.41) to:

$$\Pi - (B_\theta^2/r^2)\,\chi = \hat{B}_\theta\hat{Q}_\theta + \hat{B}_z\hat{Q}_z - (\hat{B}_\theta^2/r^2)\,\chi \qquad (\text{at } r = a)\,, \qquad (9.52)$$

where Π is given by Eq. (9.42). The equations (9.51) and (9.52) determine the plasma variables Π (or χ') and χ at the plasma surface completely if the vacuum solutions are known. This part of the problem can be carried out explicitly since the solutions in the vacuum are Bessel functions.

From the vacuum equations

$$\nabla \times \hat{\mathbf{Q}} = 0\,, \qquad \nabla \cdot \hat{\mathbf{Q}} = 0\,, \qquad (9.53)$$

we obtain the tangential components of $\hat{\mathbf{Q}}$ in terms of the radial component:

$$\hat{Q}_\theta = \mathrm{i}\,\frac{m}{m^2 + k^2 r^2}\,(r\hat{Q}_r)'\,, \qquad \hat{Q}_z = \mathrm{i}\,\frac{kr}{m^2 + k^2 r^2}\,(r\hat{Q}_r)'\,, \qquad (9.54)$$

$$\Rightarrow \quad \hat{\mathbf{B}} \cdot \hat{\mathbf{Q}} = \mathrm{i}\,\frac{r\hat{F}}{m^2 + k^2 r^2}\,(r\hat{Q}_r)'\,. \qquad (9.55)$$

The radial component satisfies the second order differential equation

$$\left[\frac{r}{m^2 + k^2 r^2}\,(r\hat{Q}_r)'\right]' - \hat{Q}_r = 0\,, \qquad (9.56)$$

which has derivatives of the modified Bessel functions as solutions:

$$\hat{Q}_r = C_1 I'_m(kr) + C_2 K'_m(kr) . \tag{9.57}$$

One of the constants is determined by the boundary condition (6.122) at the conducting wall,

$$\hat{Q}_r(b) = 0 , \tag{9.58}$$

so that the final solution for $\hat{Q}_r$ on the vacuum interval (a, b) becomes

$$\hat{Q}_r = C[\, I'_m(kb)K'_m(kr) - K'_m(kb)I'_m(kr)\,] . \tag{9.59}$$

The constant C is eliminated by inserting this solution in Eq. (9.52) and dividing that equation by Eq. (9.51), which leads to a *single boundary condition*:

$$\frac{\Pi}{\chi}\bigg|_{r=a} = \frac{B_\theta^2(a) - \hat{B}_\theta^2(a)}{a^2} - \frac{\hat{F}^2(a)}{ka}\,\frac{I_m(ka)K'_m(kb) - K_m(ka)I'_m(kb)}{I'_m(ka)K'_m(kb) - K'_m(ka)I'_m(kb)} . \tag{9.60}$$

In this boundary condition, Π is to be expressed in terms of χ' and χ by means of Eq. (9.42), and χ' and χ, in turn, are found by solving the Hain–Lüst equation (9.30). This determines the free-boundary modes of model II.

The replacement of the two boundary conditions (9.51) and (9.52) by the single boundary condition (9.60) is possible because the choice of the amplitude of the eigenfunction does not influence the eigenvalue. Equation (9.60) corresponds to normalizing the eigenfunctions with $\chi(a) = 1$. If $\chi(a)$ happens to vanish, i.e. when there is already an eigensolution with the wall at the plasma, one should exploit a different normalization.

For the numerical solution of the Hain–Lüst equation (9.30), one may exploit a shooting method, as described in Section 7.5.1(c). One chooses a value of ω^2 and integrates in the outward direction, starting from $\chi = 0$ at $r = 0$. One then keeps changing ω^2 until $(\Pi/\chi)_{r=a}$ reaches the value prescribed by the RHS of Eq. (9.60). Then, ω^2 has become an eigenvalue. For this procedure to be practical, a guiding principle should exist on how to change the parameter ω^2 in such a way that the solution for the next try is closer to satisfying the boundary condition at $r = a$ than it was in the previous run. This is provided by the oscillation theorem, proved in Section 7.4.3 for the gravitating plasma slab, but equally valid for the cylindrical plasma (see Section 9.3.1).

(b) Plasma–'ghost' plasma interface (model II)* We replace the vacuum by a pressureless plasma ($p_0 = 0$), carrying no current ($\hat{\mathbf{j}}_0 = 0$), with negligible density ($\hat{\rho}_0 = 0$), and having the same distribution of the magnetic field $\hat{\mathbf{B}}_0$ as the vacuum, given by Eq. (9.18). Let us call such a plasma a 'ghost' plasma. With respect to the equilibrium, there is evidently no difference with a vacuum. With respect to the perturbations, from Eqs. (6.19) and (6.21), the perturbed pressure and density also vanish: $\hat{p}_1 \equiv \hat{\pi} = 0$, $\hat{\rho}_1 = 0$. However, since the medium is assumed to be perfectly conducting, the magnetic field perturbation will be associated with the displacement $\hat{\boldsymbol{\xi}}$ of the 'ghost' plasma by $\hat{\mathbf{Q}} = \nabla \times (\hat{\boldsymbol{\xi}} \times \hat{\mathbf{B}})$, and there is nothing to prevent the development of a perturbed current density $\hat{\mathbf{j}}_1 = \nabla \times \hat{\mathbf{Q}}$. The question is: do those currents actually develop, and is a 'ghost' plasma any different from a vacuum with respect to stability?

Let us solve the pertinent limiting form of the Hain–Lüst equation for the 'ghost' plasma. Since we have assumed $\hat\rho_0 = 0$, that equation turns out to be identical to Newcomb's marginal equation of motion (9.49). To facilitate the solution, we transform this equation to $\hat Q_r$. From Eq. (9.27),

$$r\hat Q_r = \mathrm{i}\hat F(r\hat\xi_r) \quad\Rightarrow\quad \hat\chi = -\mathrm{i}\hat F^{-1}(r\hat Q_r)\,, \tag{9.61}$$

where $\hat F$ is the parallel gradient operator for a vacuum magnetic field distribution:

$$\hat F = (k + \hat\mu m)\hat B_{z1}\,, \qquad \hat\mu = \hat\mu_1(a^2/r^2)\,. \tag{9.62}$$

Now, it is just a matter of diligent algebra to show that the resulting equation for $\hat Q_r$ is just Eq. (9.56) and, hence, that the expression (9.55) for $\hat{\mathbf B}\cdot\hat{\mathbf Q}$ entering the second interface condition is unchanged! Apparently, no current perturbation is induced and the boundary condition (9.60), derived for the vacuum, also applies for the 'ghost' plasma. With respect to the perturbations, there appears to be no difference between a 'ghost' plasma and a vacuum.

That answer is **wrong** though. It is true that the vacuum field equation (9.56) for $\hat Q_r$ correctly describes the dynamics of the 'ghost' plasma, so that no perturbed currents develop, but only *when the inversion* (9.61) *can be carried out*, i.e. for all points where $k + \hat\mu m \neq 0$. For mode number k and m such that there is a point $r = r_s$ in the interval (a, b) where $k + \hat\mu(r_s)m = 0$, the inversion has to be reconsidered. Such points are not 'seen' by the magnetic field equation (9.56), so that $\hat Q_r$ is finite, in general, which implies that $\hat\chi$ blows up. The Frobenius expansion of the marginal equation (9.49) for $\hat\chi$ around such points is a special case of Newcomb's expansion around interchange points, described in Section 7.5.3 for a plane gravitating plasma. For the cylinder, the parameter q_0 of the indicial equation becomes proportional to p' (see Section 9.4.1) which vanishes for the 'ghost' plasma so that the indices, given by Eq. (7.192), become $\nu_1 = 0$ and $\nu_2 = -1$. Hence, there is a small solution with a finite amplitude at the singularity and a large solution with a logarithmic term. To be physically acceptable, the large solution is to be excluded because its contribution to the energy W blows up, according to Eq. (7.194), so that the physically acceptable small solution (with finite energy) becomes

$$\hat\chi_s = c_1 + c_2(r - r_s) + \cdots\,. \tag{9.63}$$

From Eq. (9.61), this implies that $\hat Q_r$ vanishes at $r = r_s$. In other words: the differential equation (9.56) for the magnetic field perturbation $\hat Q_r$ can be exploited, but it should be subjected to the boundary condition

$$\hat Q_r(r_s) = 0\,, \quad \text{when } k + \hat\mu(r_s)m = 0\,. \tag{9.64}$$

However, since the solution (9.59) for $\hat Q_r$ is not oscillatory, this would be in conflict with also satisfying the boundary condition (9.58) at the wall. Similarly, when describing the problem in terms of $\hat\chi$, there would be a conflict with the model II* boundary condition (6.123), viz. $\hat\chi(b) = 0$. This is reconciled by a property of the 'small' solution, viz. that it may jump at the singularity (see Section 7.5.3). Hence, the physically acceptable solution may look like that depicted in Fig. 9.5.

On (a, r_s) the perturbation is finite, it jumps to zero at the singularity, and on (r_s, b) it vanishes identically. Hence, *the boundary condition for a plasma–'ghost' plasma interface is given by the boundary condition (9.60) for a plasma–vacuum interface with the modification* $b \to b^*$, where

$$b^* \equiv \begin{cases} b & \text{if } k + \hat\mu(r)m \neq 0 \text{ for all } r \text{ on } (a,b)\,, \\ r_s = a\sqrt{-\hat\mu_1 m/k} & \text{if } k + \hat\mu(r_s)m = 0 \text{ for } r_s \text{ on } (a,b)\,. \end{cases} \tag{9.65}$$

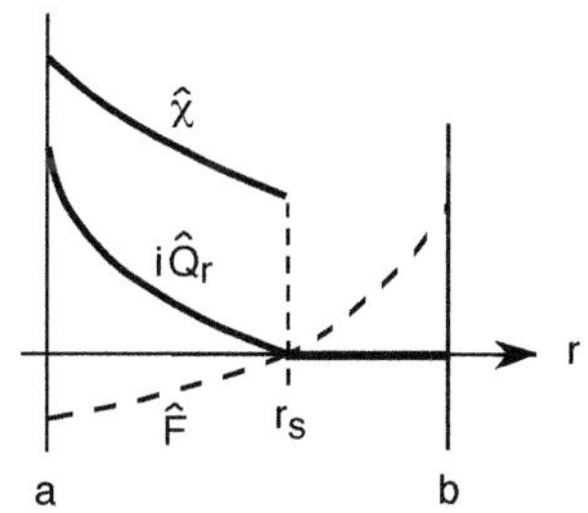

Fig. 9.5 Perturbations at an interchange singularity for a 'ghost' plasma.

The 'ghost' plasma has a big effect on stability: *the wall is effectively placed at the singularity*. This is due to the perturbed current density $\hat{\mathbf{j}}_1 = \nabla \times \hat{\mathbf{Q}}$, which vanishes nearly everywhere in the exterior region, but exhibits a surface current layer (Section 4.5.2) at $r = r_s$ with an associated jump of $\hat{\mathbf{Q}}$:

$$\hat{\mathbf{j}}_1^\star = \mathbf{n} \times [\![\hat{\mathbf{Q}}]\!] . \tag{9.66}$$

In ideal MHD (infinite conductivity), this induced surface current has the same effect as a solid wall put at the position of the singularity!

We now have the complete machinery available to describe the waves and instabilities of cylindrical plasmas, to be applied in the following sections. The present subsection should have illustrated, once more, that singularities are nearly always present. They are not a mathematical frivolity, but they determine the dominant dynamics of the plasma. Bender and Orszag's motto in Section 7.4 stays with us.

9.3 Spectral structure

9.3.1 One-dimensional inhomogeneity

At this point, the reader may well wonder why the algebra of the spectral analysis is so complicated and how one can be sure of results when so many factors contribute and subtle cancellations are a rule rather than an exception. We here point out: (a) the intrinsic reason for these complications, and (b) several alleviating factors helping to restore confidence in the spectral enterprise.

(a) Corresponding problems in quantum mechanics In Chapters 6 and 7, we frequently stressed the analogy between MHD and quantum mechanical spectral theory. Since the Hain–Lüst equation has direct relevance for realistic plasma confinement problems, it is instructive to compare it with a corresponding concrete problem in quantum mechanics. To that end, we contrast the normal mode equation $\mathbf{F}(\boldsymbol{\xi}) = -\rho\omega^2\boldsymbol{\xi}$ with the Schrödinger equation $H\Psi = E\Psi$ for an electron moving in the potential field $V(\mathbf{r})$ of a nucleus with mass M,

$$\left[-\frac{\hbar^2}{2M}\Delta + V(\mathbf{r}) \right] \Psi(\mathbf{r}) = E\Psi(\mathbf{r}) . \tag{9.67}$$

One-dimensional problems are obtained for a potential that is spherically symmetric, like the H-atom where $V = V(r)$. In that case, one writes the wave function as a superposition of spherical harmonics which may be studied separately,

$$\Psi(r,\theta,\phi) = R(r)\, Y_\ell^m(\theta,\phi)\,, \tag{9.68}$$

in much the same way as the separate Fourier components (9.22) for a cylindrical plasma. Inserting the expression (9.68) in Eq. (9.67) leads to a second order differential equation for the radial wave function:

$$\frac{1}{r}\frac{d^2}{dr^2}(rR) - \left[\frac{\ell(\ell+1)}{r^2} + \frac{2M}{\hbar^2}\Big(V(r) - E\Big)\right] R = 0\,. \tag{9.69}$$

This equation will be compared with the Hain–Lüst equation.

It is clear that the spectral problem of calculating the waves and instabilities of a cylindrical plasma is much more complicated than the determination of the energy levels of the hydrogen atom, or even the general quantum mechanical problem of scattering of particles in an *arbitrary* one-dimensional potential field. In the latter, the only controlling function is $V(r)$, whereas four functions, $\rho(r)$, $p(r)$, $B_\theta(r)$, $B_z(r)$, enter the MHD equation. Moreover, the reduction from a vector equation implies a complicated dependence of the coefficients P and Q of Eq. (9.36) on the eigenvalue ω^2. Consequently, whereas the radial wave equation (9.69) is a classical Sturm–Liouville differential equation, with monotonic dependence of the eigenvalue E on the number of nodes of the radial eigenfunction $R(r)$, the Hain–Lüst equation (9.30) is not of such a classical type, so that the dependence of ω^2 on the number of nodes of $\chi(r)$ is much more complicated.

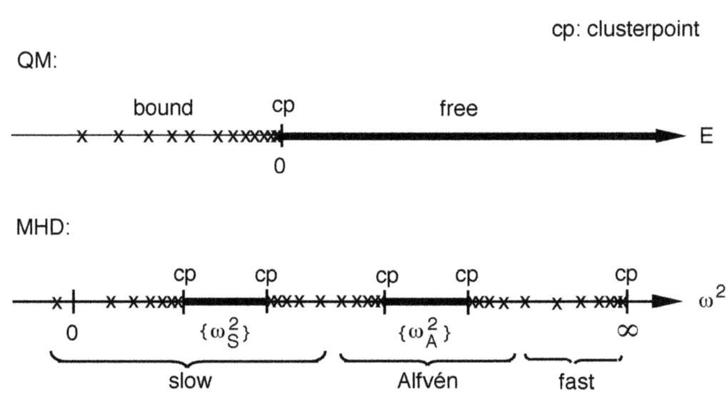

Fig. 9.6 Comparison of schematic spectra in quantum mechanics (for fixed quantum numbers m, l) and in MHD (for fixed wave numbers m, k).

The vector character of ideal MHD is reflected in the occurrence of three sub-spectra. However, the general structure of each of these sub-spectra is similar to the complete spectrum of quantum mechanical systems (Fig. 9.6). If one fixes the quantum numbers m and ℓ for the H-atom one finds a discrete spectrum of bound states for $E < 0$ clustering at $E = 0$, which is the edge of a continuum of free states for $E > 0$. Likewise, for the cylindrical plasma the Alfvén and slow sub-spectra consist of discrete modes that may cluster at the edge of the continua $\{\omega_{\rm A}^2\}$ and $\{\omega_{\rm S}^2\}$, whereas the fast sub-spectrum accumulates at $\omega^2 = \infty$.

Hence, the two problems have much more in common than suggested by the evident differences: (1) both concern the determination of the spectrum of a *self-adjoint linear operator in Hilbert space*;

(2) the operators have a *discrete spectrum* as well as a *continuous spectrum*, with different but intrinsically physical reasons for the distinction between them; (3) for one-dimensional inhomogeneity, *the discrete spectrum (or sub-spectra for vector problems) is asymptotically monotonic in the number of nodes of the eigenfunctions*. The latter property connects the discrete spectrum, or sub-spectra, to the essential spectrum, consisting of the continuous spectrum of free states in quantum mechanics, and of the Alfvén and slow continua and the fast cluster point in MHD.

(b) Common properties of inhomogeneous plasmas With the extensive preparation of Chapter 7 on the plane gravitating plasma slab, for the analogous problem of the cylindrical plasma there is no need to repeat (a) the proof of the existence of continuous Alfvén and slow continua (Section 7.4.2); (b) the demonstration of apparent singularities (Section 7.3.2 and 7.4.1); (c) the proof of the oscillation theorem (Section 7.4.3); (d) the variational procedures for stability (Sections 7.5.1 and 7.5.3). All this immediately carries over to the cylindrical case so that we can concentrate on the surprisingly many different effects of the curvature term associated with the transverse magnetic field component B_θ. The fact that all these features carry over from one inhomogeneous problem to another illustrates that there is good reason for trust in the results of spectral problems in MHD.

One should be aware though that these chapters focus on *one-dimensional* inhomogeneous plasmas described by *ideal* MHD. When toroidal curvature is introduced, separability fails and ODEs are replaced by PDEs. Naive expectations about similar monotonicity properties of the eigenvalues in the toroidal case, with 2D nodal lines taking the place of 1D nodal points, are quickly shattered when one realizes what could happen (see the example given by Courant and Hilbert [132](a), p. 455, on the peculiar behaviour of nodal lines of the Helmholtz equation on a square). When dissipation is admitted, self-adjointness is lost and spectral problems become much more complicated. Dissipative and toroidal spectral problems will be discussed extensively in Chapters 14 and 16–18.

9.3.2 Cylindrical model problems

In cylindrical geometry, analytically solvable models inevitably involve Bessel functions. We illustrate this with three examples that are frequently used in cylindrical stability problems.

(a) Waves in a homogeneous $\boldsymbol{\theta}$*-pinch* We start with the simplest model, a linear θ-pinch with a homogeneous magnetic field, pressure and density,

$$B_\theta = 0\,, \quad B_z = B_0\,, \quad \rho = \rho_0\,, \quad p = p_0\,, \tag{9.70}$$

so that the Alfvén speed and sound speed are constant, related to each other by the parameter β:

$$b \equiv \sqrt{B_0^2/\rho_0}\,, \quad c \equiv \sqrt{\gamma p_0/\rho_0} \quad \Rightarrow \quad \tfrac{1}{2}\gamma\beta \equiv c^2/b^2\,. \tag{9.71}$$

The Hain–Lüst equation (9.30) then simplifies to

$$(\omega^2 - k^2 b^2)\left[\left(\frac{r}{m^2 + k^{*2} r^2}\,\chi'\right)' - \frac{1}{r}\,\chi\right] = 0\,, \tag{9.72}$$

where a kind of modified longitudinal wave number appears:

$$k^* \equiv \left[\frac{(k^2 - \omega^2/b^2)(k^2 - \omega^2/c^2)}{k^2 - \omega^2/b^2 - \omega^2/c^2}\right]^{1/2}. \tag{9.73}$$

For internal modes (model I), the differential equation (9.72) is to be solved subject to the boundary conditions

$$\chi(0) = \chi(a) = 0\,. \tag{9.74}$$

Equation (9.72) yields, first of all, *an infinitely degenerate spectrum of Alfvén waves* with frequency $\omega^2 = \omega_A^2 = k^2 b^2$, and with arbitrary radial dependence of the eigenfunction χ. They propagate along the z-axis with the Alfvén speed b.

For $\omega^2 \neq \omega_A^2$, the solutions of Eq. (9.72) are Bessel functions, where it depends on the sign of k^{*2} whether they are of the modified, exponential, kind (giving evanescence) or of the ordinary, oscillatory, kind (giving propagation):

$$\begin{aligned} \chi &= C r I_m'(k^* r)\,, \quad \text{if } k^{*2} > 0 \quad (\text{for } \omega^2 < \omega_S^2 \text{ and } k^2c^2 < \omega^2 < \omega_A^2)\,, \\ \chi &= C r J_m'(\mathrm{i}k^* r)\,, \quad \text{if } k^{*2} < 0 \quad (\text{for } \omega_A^2 < \omega^2 \text{ and } \omega_S^2 < \omega^2 < k^2c^2)\,, \end{aligned} \tag{9.75}$$

where $c^2 < b^2$ has been assumed in the inequalities. The first expression does not permit satisfaction of the boundary condition for internal modes. It will be used below in the expression for the external kink modes. From the second expression, the boundary condition $\chi(a) = 0$ implicitly fixes the eigenfrequencies through

$$\mathrm{i}\, k^* a = j_{mn}'\,, \tag{9.76}$$

where j_{mn}' is the nth zero of the Bessel function $J_m'(x)$. This yields the dispersion equation for *the slow and fast magneto-acoustic waves in a homogeneous θ-pinch*:

$$\omega^4 - (k^2 + j_{mn}'^2/a^2)(b^2 + c^2)\,\omega^2 + k^2(k^2 + j_{mn}'^2/a^2)\, b^2 c^2 = 0\,. \tag{9.77}$$

This equation is fully analogous to the magneto-sonic factor of the dispersion equation (5.54) for homogeneous plasmas, where the parallel wave number is now indicated by k and the effective total wave number by $\sqrt{k^2 + j_{mn}'^2/a^2}$.

(b) Free-boundary modes of interface plasmas For modes of a plasma–vacuum (model II) or a plasma–'ghost' plasma (model II*) system, we need to solve the Hain–Lüst equation (9.30) subject to the boundary conditions discussed in Section 9.2.2. As in model I, the perturbation on axis is restricted by the regularity condition $\chi(0) = 0$, whereas the boundary condition at the interface involves the perturbation of the total pressure, $(\Pi/\chi)_{r=a}$ as given by Eq. (9.60). For a 'ghost' plasma, the substitution (9.65) should be made in the latter expression.

We wish to study this problem for a sharp-boundary plasma where the current is confined to the plasma surface $r = a$ (*skin-current model*, Fig. 9.7). Then, the equilibrium quantities for the interior of the plasma column are those of a homogeneous θ-pinch given by Eq. (9.70), whereas the external magnetic field is given by Eqs. (9.18) and (9.19). This model provides a very useful first approximation to the study of *external kink modes*, which are the most dangerous instabilities occurring in a cylindrical plasma column. Here, most dangerous is meant in the sense of affecting the bulk of the plasma and having large growth rates. For typical densities of high-β pinches they exponentiate on the microsecond time scale.

With a free boundary, exploiting the oscillatory Bessel functions (9.75)(b), the frequencies of the modes will be slightly shifted from those discussed under (a). However, the really interesting

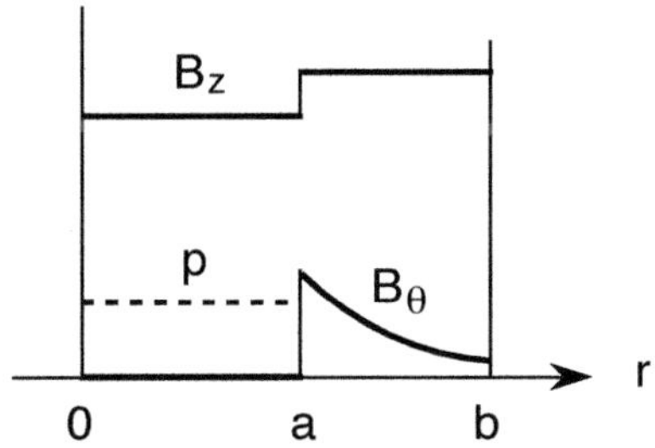

Fig. 9.7 Skin-current model of plasma confinement.

new feature of free-boundary plasmas is the appearance of a *surface mode*, like the f-mode in the gravito-acoustic spectrum (Section 7.2.3). This mode is just the lowest one in the evanescent region, which may even become unstable ($\omega^2 < 0$), so that one should exploit the exponential Bessel functions (9.75)(a).

To obtain the dispersion equation of the modes, we first compute the perturbation Π of the total pressure for the first solution (9.75) for χ,

$$\Pi = -\frac{N}{D}\chi' = -\frac{\omega^2 - k^2 b^2}{m^2 + k^{*2} r^2}\, r\chi' = -\frac{\omega^2 - k^2 b^2}{k^*} C I_m(k^* r)\,, \tag{9.78}$$

and then insert this expression and χ in the boundary condition (9.60). This yields *the dispersion equation for free-boundary modes*:

$$\begin{aligned}\omega^2 &= \frac{k^2 B_0^2}{\rho_0} \\ &\quad - \frac{k^* a\, I_m'(k^* a)}{\rho_0 I_m(k^* a)}\left[\frac{\hat{B}_\theta^2}{a^2} + \frac{(m\hat{B}_\theta/a + k\hat{B}_z)^2}{ka}\,\frac{I_m(ka)K_m'(kb) - K_m(ka)I_m'(kb)}{I_m'(ka)K_m'(kb) - K_m'(ka)I_m'(kb)}\right].\end{aligned} \tag{9.79}$$

At this point, the dispersion equation is still a highly transcendental equation in the eigenvalue because of the dependence of k^* on ω^2. (Note that the symbol b now indicates the wall position.)

Many different limits may be studied for this equation, but the most interesting one is obtained for *the tokamak approximation* where we again consider a cylindrical plasma of length $2\pi R_0$ as a first approximation to a torus of major radius R_0 (Fig. 9.2). In that case, *the wave number k is quantized with integer toroidal mode number n* (not to be confused with the radial node number of the Bessel functions):

$$k = n/R_0\,, \quad \text{so that} \quad ka = \epsilon n \quad \Big[\ll 1 \ \text{ for } n \sim 1 \Big]. \tag{9.80}$$

The approximation in square brackets is the long-wavelength approximation for the longitudinal mode number, which is crucial here. Furthermore, the magnetic field components are ordered as

$$\hat{B}_\theta \sim \epsilon \hat{B}_z\,, \quad \text{so that} \quad \hat{q} = \epsilon \hat{B}_z/\hat{B}_\theta \sim 1\,. \tag{9.81}$$

We assume (and easily justify this from the result obtained below) that the eigenvalues of the modes (or the growth rates of the modes) are much smaller than the Alfvén frequency:

$$|\omega^2| \ll \omega_A^2 \equiv k^2 b^2 \quad \Rightarrow \quad k^* \approx k\,. \tag{9.82}$$

In view of Eq. (9.80), the arguments of all occurring Bessel functions are small, so that we may use the following approximations for $m \neq 0$:

$$\begin{aligned} k^*a\, I'_m(k^*a)/I_m(k^*a) \;&\approx\; |m|\,, \\ \frac{I_m(ka)K'_m(kb) - K_m(ka)I'_m(kb)}{I'_m(ka)K'_m(kb) - K'_m(ka)I'_m(kb)} \;&\approx\; -\frac{\epsilon n}{|m|}\,\frac{(b/a)^{|m|} + (b/a)^{-|m|}}{(b/a)^{|m|} - (b/a)^{-|m|}}\,. \end{aligned} \tag{9.83}$$

Inserting these approximations in Eq. (9.79) leads to the following approximate form of the dispersion equation:

$$\omega^2 \approx \frac{\epsilon^2 B_0^2}{a^2\rho_0}\left[n^2 - \frac{\hat{B}_\theta^2}{\epsilon^2 B_0^2}\left(|m| - (m + n\hat{q})^2\,\frac{(b/a)^{|m|} + (b/a)^{-|m|}}{(b/a)^{|m|} - (b/a)^{-|m|}}\right)\right]. \tag{9.84}$$

From the equilibrium expressions (9.19), and Eq. (9.13) relating $\hat{\mu}$ to $\hat{q}$, one may convert the factor involving $\hat{B}_\theta$ in terms of β and $\hat{q}_1$:

$$\frac{\hat{B}_\theta^2}{\epsilon^2 B_0^2} = \frac{1+\beta}{\hat{q}_1^2 + \epsilon^2} \approx \frac{1}{\hat{q}_1^2}\,. \tag{9.85}$$

In the rightmost approximation, we have neglected small terms β and ϵ^2 in agreement with *the low-β tokamak ordering* (9.14). The dispersion equation for the 'straight tokamak' then becomes:

$$\omega^2 \approx \frac{\epsilon^2 B_0^2}{a^2\rho_0\hat{q}^2}\left[n^2\hat{q}^2 - |m| + (m + n\hat{q})^2\,\frac{(b/a)^{|m|} + (b/a)^{-|m|}}{(b/a)^{|m|} - (b/a)^{-|m|}}\right]. \tag{9.86}$$

This expression may be considered as an end product since it has the trivial dimensional factors B_0, a, ρ_0 in the appropriate way (as required by scale independence) to give the dimension of a growth rate squared, whereas the essential parameters $\hat{q}$, b/a and ϵ describing the equilibrium features, and the mode number m and n describing the perturbations, appear in a physically significant way.

Rearranging terms, Eq. (9.86) may be written as

$$\omega^2 \approx \frac{\epsilon^2 B_0^2}{a^2\rho_0\hat{q}^2}\left[\tfrac{1}{2}|m|(|m| - 2) + \tfrac{1}{2}(2n\hat{q} + m)^2 + \frac{2(n\hat{q} + m)^2}{(b/a)^{2|m|} - 1}\right]. \tag{9.87}$$

This reveals some of the physical mechanisms at work in this model. First, there is *the kink* term which is only negative when $|m| = 1$. Then, there is a stabilizing term representing *the average field line-bending* across the plasma boundary which disappears for modes that propagate perpendicular to the average direction of the field across the surface layer at $r = a$ (recall that $q = \infty$ for $r = a^-$ and $q = \hat{q}$ for $r = a^+$). The last term represents *the stabilizing influence of the wall*, ranging from infinitely stabilizing when $b/a = 1$ to no effect when $b/a \to \infty$.

Since only $|m| = 1$ is unstable, we may restrict the analysis to that mode:

$$\omega^2(m\!=\!-1) = \frac{2\epsilon^2 B_0^2}{a^2\rho_0\hat{q}^2}\,\frac{(n\hat{q} - 1)(n\hat{q} - a^2/b^2)}{1 - a^2/b^2}\,. \tag{9.88}$$

This growth rate is plotted in Fig. 9.8. Clearly, the external kink mode is always unstable for this model in the region

$$a^2/b^2 < n\hat{q} < 1\,. \tag{9.89}$$

This suggests a simple way of eliminating unstable external kink modes by prescribing the geometry

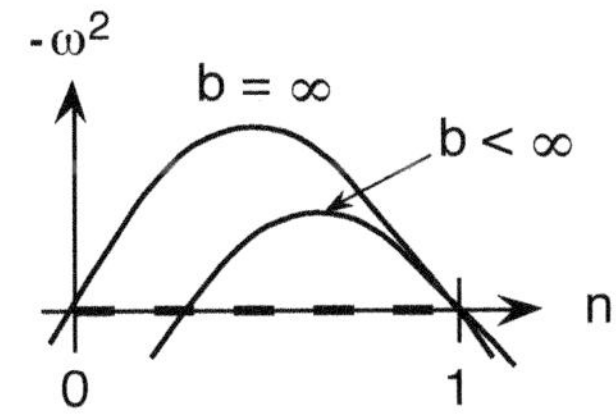

Fig. 9.8 Growth rate of the external kink mode for a skin-current plasma–vacuum model. A conducting wall reduces the growth rate, but replacing the vacuum by a 'ghost' plasma yields complete stabilization (thick dashed line).

of the torus and the total plasma current I_z such that

$$\hat{q} = 2\pi a^2 B_0/(R_0 I_z) > 1\,, \tag{9.90}$$

so that the unstable $n = 1$ modes (and, hence, all the $n > 1$ modes as well) simply do not fit into the torus. This condition is called *the Kruskal–Shafranov limit*. The limit imposed on the plasma currents by Eq. (9.90) is a quite important consideration in the operation of tokamaks. It is appropriate to repeat here the remark made in Section 2.4.3 that the fact that $\hat{q} = 1$ corresponds to a topology with closed magnetic field lines has nothing to do with the stability mechanism of the external kink mode. This is a purely accidental coincidence which disappears as soon as one introduces genuine toroidal effects in the theory (see Chapter 17).

Let us now consider the model II* version of the free-boundary external kink mode. According to Eq. (9.65), one should replace the actual wall position by a virtual wall position b^* because of stabilization by induced skin currents at the singular position given by $m+n\hat{q}(r_s) = 0$. The unstable region given by Eq. (9.89) indicates that $|m| = 1$ instability precisely occurs when there is such a singularity. Hence, in that regime, we should exploit the virtual wall position given by

$$b^*/a = \sqrt{-\frac{\hat{\mu}_1 m}{k}} = \sqrt{-\frac{m}{n\hat{q}_1}}\,. \tag{9.91}$$

Hence, in the expression (9.88) for the growth rate, the factor

$$n\hat{q} - a^2/b^2 \quad \rightarrow \quad n\hat{q} - (a/b^*)^2 = 0 \quad \text{for} \quad |m| = 1\,. \tag{9.92}$$

The external kink mode is completely stabilized by the singular currents in the 'ghost' plasma! This agrees with the theory of internal kink modes (Section 9.4.4), which are stable to leading order in the inverse aspect ratio ϵ. The external kink mode of the plasma–vacuum model II actually becomes an internal kink mode in the plasma–'ghost' plasma model II*.

Of course, for the operation of an actual fusion experiment, one would not rely on stabilization by an external 'ghost' plasma to push the current beyond the Kruskal–Shafranov limit. That stabilization depends on perfect conductivity, whereas the outermost plasma is also subject to resistive instabilities. Moreover, stability with respect to internal kink modes is lost when the next order of the toroidal effects is taken into account. Yet, it is true that an external 'ghost' plasma does slow down the growth rate of the kink mode as compared to a vacuum.

(c) Modes of an incompressible plasma in a shearless magnetic field In our next model, we move a bit closer to physical reality by admitting a distributed current in the plasma. To facilitate the analysis, we exploit Freidberg's simplified form (9.48) of the Hain–Lüst equation for an incompressible plasma. Moreover, we assume a constant longitudinal magnetic field component B_z and a linearly increasing transverse magnetic field component B_θ, so that the pitch μ of the magnetic field lines is constant and the current is evenly distributed over the plasma:

$$B_\theta = Ar \quad \Rightarrow \quad j_z = \frac{1}{r}(rB_\theta)' = 2A\,. \tag{9.93}$$

According to the equilibrium equation (9.3), the pressure distribution then becomes parabolic and may be chosen to vanish at the plasma–vacuum boundary:

$$p = p_0 - A^2r^2 = a^2A^2(1 - r^2/a^2)\,. \tag{9.94}$$

The parallel gradient operator $F \equiv mB_\theta/r + kB_z = B_z(k + \mu m) = \text{const}$, so that the Alfvén factor $\rho\omega^2 - F^2$ also becomes constant if we assume constant density. Then, the incompressible equation of motion (9.48) simplifies to

$$\frac{d}{dr}\left[\frac{r}{m^2 + k^2r^2}\frac{d\chi}{dr}\right] - \frac{1}{r}\left[1 + \left(\frac{2m\alpha}{m^2 + k^2r^2} - \alpha^2\right)\frac{k^2r^2}{m^2 + k^2r^2}\right]\chi = 0\,, \quad \alpha \equiv \frac{2AF}{\rho\omega^2 - F^2}\,. \tag{9.95}$$

The solution of this equation may again be expressed in terms of Bessel functions, in the following combination:

$$\chi = C\Big[\,k^*rI_m'(k^*r) - \alpha I_m(k^*r)\Big]\,, \quad k^* \equiv k\sqrt{1 - \alpha^2}\,. \tag{9.96}$$

How does one produce such a miraculous answer? By transforming to another variable! The reader may wish to check that, in this case, one obtains Bessel's equation itself for the total pressure perturbation, with the solution $\Pi \sim I_m(k^*r)$.

This model was probably considered first by Alfvén to study unstable loops as a mechanism for the generation of cosmic magnetic fields; see the second (1963) edition of Ref. [11] co-authored with C. Fälthammar. For internal modes, it is clear that the boundary condition $\chi(a) = 0$ can only be satisfied if the Bessel functions are oscillatory. This requires $\alpha^2 > 1$, or $4A^2F^2 > (\rho\omega^2 - F^2)^2$. For $F \to 0$, this implies $-\rho\omega^2 < 2|AF|$ and it follows that the growth rate tends to zero. On the other hand, for $\omega^2 = 0$, one always finds oscillatory solutions in the limit $F \to 0$, so that there will be infinitely many unstable branches according to the oscillation theorem. We have encountered this behaviour before, in Chapter 7 for the gravitational quasi-interchanges (Section 7.3.3).

Alfvén's model for the incompressible plasma cylinder with a constant pitch magnetic field is unstable with respect to *quasi-interchanges* (see Section 9.4.2), having maximum growth rate for $F \neq 0$. If one relaxes the incompressibility constraint, the plasma becomes unstable with respect to *pure interchanges*, having maximum growth rate at $F = 0$. This model has been used for the benchmarking of large-scale eigenvalue solvers exploiting finite elements. One particularly nice example, due to Chance et al. [110], is reproduced in Fig. 9.9 in the representation by Kerner [357]. It shows the full structure of the ideal MHD spectrum in the presence of an instability. The spectrum is resolved in a practical way (where continua are represented by as many densely spaced eigenvalues as there are grid points in the calculation) over the many orders of magnitude from high-frequency fast modes to unstable interchanges. Note that the interchange instabilities occur in the region where

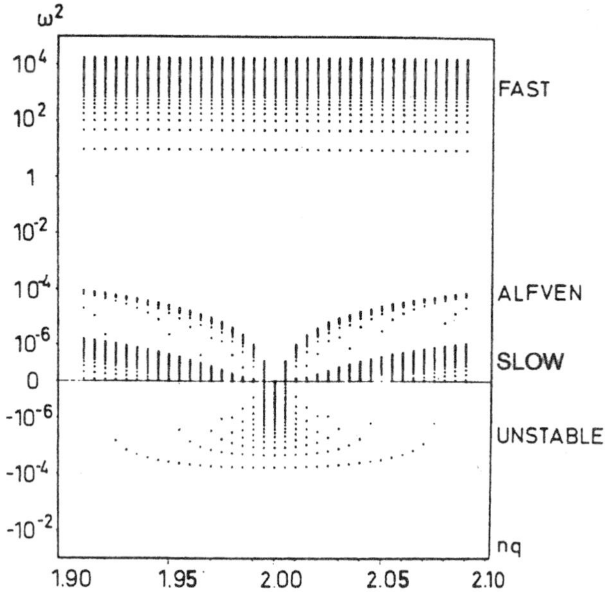

Fig. 9.9 Complete MHD spectrum for a compressible plasma with a shearless magnetic field, which is interchange unstable, for $m = -2$. (Adapted from Kerner [357].)

$F = 0$, so that the Alfvén and slow continua extend to the origin. In a sense, these instabilities belong to both the Alfvén and slow sub-spectra.

For external modes, exploiting again the low-β tokamak ordering to simplify the Bessel function expressions of the incompressible model, we obtain the following dispersion equation for the distributed current model:

$$\omega^2 = \frac{2\epsilon^2 B_0^2}{a^2 \rho_0 q^2} \frac{m + nq}{1 - (a/b)^{2|m|}} \Big[m + nq - sg(m)\Big(1 - (a/b)^{2|m|}\Big)\Big] . \tag{9.97}$$

Here, the constant value of q in the interior plasma has been chosen equal to the value $\hat{q}_1$ of the vacuum magnetic field at $r = a$. Note that, for $m = -1$, the expression reduces to that of Eq. (9.88) so that this mode is not sensitive for the current distribution. However, we now obtain external instabilities for all values of m, in regions to the left of the integer values of nq, similar to the $m = -1$ mode of the skin-current model shown in Fig. 9.8. The constant current distribution is violently unstable, with no easy stabilization in the manner of the Kruskal–Shafranov limit for the $|m| = 1$ external kink modes.

9.3.3 Cluster spectra*

A cluster point analysis analogous to that of Section 7.4.4 for a plane gravitating slab may also be carried out for a cylindrical plasma. The result is as follows.

(1) For a *Sturmian (or anti-Sturmian) sequence of discrete Alfvén modes to cluster at the lower (or upper) edge of the Alfvén continuum*, i.e. at the minimum (or maximum) of ω_A^2, the following

condition should be satisfied:

$$0 < \tfrac{1}{8}\rho(\omega_A^2)'' < Q_A \qquad \big(\text{or } 0 > \tfrac{1}{8}\rho(\text{or}\,\omega_A^2)'' > Q_A\big)\,,$$

$$Q_A \equiv \frac{rG^2}{B^2}\Big(\frac{B_\theta^2}{r^2} - \frac{2B_\theta B_z F}{r^2 G}\Big)' + \frac{4k^2 B_\theta^2 (B^2 - \gamma p)}{r^2 B^2}\,. \qquad (9.98)$$

These solutions correspond to a sequence of stable *Global Alfvén Eigenmodes* (GAEs), if the cluster point is not overlapped by other parts of the continua.

(2) For a *Sturmian (or anti-Sturmian) sequence of discrete slow modes to cluster at the lower (or upper) edge of the slow continuum*, i.e. at the minimum (or maximum) of ω_S^2, the following condition should be satisfied:

$$0 < \tfrac{1}{8}\rho(\omega_S^2)'' < Q_S \qquad \big(\text{or } 0 > \tfrac{1}{8}\rho(\omega_S^2)'' > Q_S\big)\,,$$

$$Q_S \equiv -\Big(\frac{\gamma p}{\gamma p + B^2}\Big)^2 \Big[\frac{F^4}{\gamma p + B^2} + \frac{rF^2}{B^2}\Big(\frac{B_\theta^2}{r^2}\Big)' - \frac{4k^2 B_\theta^2(\gamma p + B^2)}{r^2 B^2}\Big]. \qquad (9.99)$$

This results in stable *Global Slow Eigenmodes* (GSEs) if continua do not overlap.

The GAEs, and their prospect of low-frequency plasma heating, were described by Appert et al. [19]. The Alfvén and slow cluster conditions were first presented by Goedbloed [219]. Those conditions demonstrate that, in principle, all branches of the discrete spectrum may occur in a cylindrical plasma with $B_\theta \neq 0$.

9.4 Stability of cylindrical plasmas

9.4.1 Oscillation theorems for stability

Most of the classical stability theory for plasmas with one-dimensional inhomogeneity can be derived from the MHD oscillation theorem (Section 7.4.3), where the interchange singularities $F = 0$ is the main complicating factor. We will not repeat the analysis but only present the main steps in so far as they lead to different expressions, viz. magnetic curvature terms (B_θ) instead of the gravitational terms ($\rho'\hat{g}$) of Chapter 7. Recall that the eigenfrequencies are monotonic in the number of nodes of the radial component ξ of the eigenfunction outside the ranges of the continua and the turning point frequencies. Hence, the discrete sub-spectra are asymptotically either Sturmian or anti-Sturmian. Since the unstable range is always Sturmian, there is an immediate connection between the MHD oscillation theorem and Newcomb's stability theory.

(a) Newcomb's variational procedure and Suydam's criterion For the study of stability, we start from the energy principle (6.95). For the diffuse cylindrical plasma, the reduction of W proceeds analogously as for the plane gravitating slab (Section 7.5.1). Exploiting the expressions (9.27) for $\mathbf{Q}$ and $\nabla \cdot \boldsymbol{\xi}$, consistently expressing all variables in ξ (or $\chi \equiv r\xi$), η and ζ, and integrating by parts to get rid of a term $\xi'\xi$, one obtains the following form:

$$W = \pi L \int_0^a \Big\{ P_0[\chi']^2 + Q_0\chi^2 + (m^2 + k^2 r^2)\Big[\frac{B}{r}\eta + \frac{G\chi' + 2kB_\theta\chi/ri}{m^2 + k^2 r^2}\Big]^2 + \gamma p\Big[\frac{1}{r}\chi' + \frac{G\eta + F\zeta}{B}\Big]^2 \Big\}\, r\,dr\,, \qquad (9.100)$$

where the length $L \to \infty$ for the full cylinder. Minimization with respect to η and ζ is trivial (leading to the expressions (9.29) for $\omega^2 \to 0$), so that W reduces to

$$W = \pi L \int_0^a \left\{ P_0(\chi')^2 + Q_0\chi^2 \right\} r\, dr \,. \tag{9.101}$$

This expression is minimized by solutions to the Euler–Lagrange equation (9.49), which is the Hain–Lüst equation in the limit $\omega^2 \to 0$. The explicit form of P_0 and Q_0 may be read off from the marginal equation of motion (9.49) given in Section 9.2.1(d).

For many stability applications it is convenient to transform to the variable ξ again, where we now exploit the notation f_0 and g_0 of Newcomb:

$$W = \pi L \int_0^a \left(f_0 \xi'^2 + g_0 \xi^2 \right) dr \,, \tag{9.102}$$

leading to the Euler–Lagrange equation

$$(f_0\, \xi')' - g_0\, \xi = 0 \,, \tag{9.103}$$

where $f_0 \equiv r^2 P_0$ and $g_0 \equiv r^2 Q_0 - r P_0'$ are obtained from P_0 and Q_0 of Eq. (9.49) by a straightforward transformation involving the equilibrium relation (9.3):

$$\begin{aligned} f_0 &= \frac{r^3 F^2}{m^2 + k^2 r^2} \,, \\ g_0 &= \frac{2k^2 r^2}{m^2 + k^2 r^2}\, p' + \frac{m^2 + k^2 r^2 - 1}{m^2 + k^2 r^2}\, r F^2 - \frac{2k^2 r^3 (m B_\theta / r - k B_z)}{(m^2 + k^2 r^2)^2}\, F \,. \end{aligned} \tag{9.104}$$

Since Eq. (9.103) is equivalent to the Hain–Lüst equation for $\omega^2 = 0$, the MHD oscillation theorem directly yields Newcomb's [461] *Theorem for the stability of the diffuse cylindrical plasma in the absence of a singularity* $F = 0$:

For specified values of m *and* k *such that* $F \equiv m B_\theta / r + k B_z \neq 0$ *on* $(0, a)$, *the cylindrical plasma is stable if, and only if, the non-trivial solution* $\chi \equiv r\xi$ *of the marginal equation (9.49), corresponding to the Euler–Lagrange equation (9.103) that vanishes at* $r = 0$ *does not have a zero in the open interval* $(0, a)$.

Not all values of m need to be investigated since W has two convenient monotonicity properties.

(1) For $m = 0$, the energy integral becomes

$$W_{m=0} = W_0 + \pi L k^2 \int r B_z^2\, dr \,, \quad W_0 \equiv \pi L \int \left[r B_z^2 \xi'^2 + \Big(2p' + \frac{B_z^2}{r} \Big) \xi^2 \right] dr \,. \tag{9.105}$$

Hence, if W is positive for $m = 0$, $k \to 0$, it is positive for $m = 0$, all $k \neq 0$. The case $m = 0$, $k = 0$ is not contained since the derivation of Eq. (9.100) is invalid then (there is a division by 0). From the original expression for W one finds:

$$W_{m=0,k=0} = W_0 + \pi L \int \left\{ B_\theta^2 \left[r \Big(\frac{\xi}{r} \Big)' \right]^2 + \gamma p \left[\frac{1}{r} (r\xi)' \right]^2 \right\} r dr \,. \tag{9.106}$$

Hence, if W is positive for $m = 0$, $k \to 0$, it is also positive for $m = 0$, $k = 0$.

(2) For $m \neq 0$, a different trick is performed. Keeping the mode number m, but replacing k by the parameter $\lambda \equiv k/m$, one finds:

$$f_0 = f_0(r;\lambda)\,, \qquad g_0 = h_0(r;\lambda) + rm^2(B_\theta + \lambda B_z)^2\,. \tag{9.107}$$

Since the only term in m is positive definite, the worst case is $m = 1$, so that one may restrict the analysis to that mode.

In conclusion, *the cylindrical plasma is stable for all m and k, if it is stable for the cases $m = 0$, $k \to 0$ and $m = 1$, all k.*

As in Section 7.5, the main complication is the proper analysis of the singularities $F = 0$. These singularities are just the lower edges of the Alfvén and slow continua $\{\omega_{\rm A}^2\}$ and $\{\omega_{\rm S}^2\}$, which extend to $\omega^2 = 0$ if the interval $(0, a)$ contains a point where $F = 0$, i.e. $k + \mu m = 0$. For these values of the wave numbers m and k, the tangential wave vector is perpendicular to $\mathbf{B}$. In that case, the phase of the perturbation is constant along the field lines at the position $r = r_{\rm s}$ of the singularity. Expanding all quantities in terms of the variable $s \equiv r - r_{\rm s}$, so that

$$F \approx mB_z\mu' s\,, \qquad m^2 + k^2r^2 \approx m^2(1 + \mu^2 r^2)\,, \tag{9.108}$$

we obtain

$$f_0 \approx \frac{r^3 B_z^2 \mu'^2}{1 + \mu^2 r^2} s^2\,, \qquad g_0 \approx \frac{2\mu^2 r^2}{1 + \mu^2 r^2} p'\,. \tag{9.109}$$

Consequently, close to the singularity, Eq. (9.103) reduces to

$$(s^2\xi')' - \alpha\,\xi = 0\,, \quad \alpha \equiv 2\mu^2 p'/(rB_z^2\mu'^2)\,. \tag{9.110}$$

The solutions of the equation (9.110) are s^{ν_1} and s^{ν_2}, where ν_1 and ν_2 are the roots of the indicial equation $\nu(\nu + 1) - \alpha = 0$:

$$\nu_{1,2} = -\tfrac{1}{2} \pm \tfrac{1}{2}\sqrt{1 + 4\alpha}\,. \tag{9.111}$$

The discussion of the implications of real or complex indices is again identical to that given in Section 7.5.3.

The condition $1 + 4\alpha > 0$, which is necessary for the absence of oscillatory solutions, was derived by Suydam [581] and known as *Suydam's criterion:*

$$p' + \tfrac{1}{8} r B_z^2 \left(\frac{\mu'}{\mu}\right)^2 > 0\,. \tag{9.112}$$

Its violation implies the existence of highly localized instabilities close to a singular surface where $k + \mu m = 0$. These instabilities are so-called flute modes which interchange the magnetic field lines without appreciable bending. They are driven by the pressure gradient p' and stabilized by the magnetic shear, if the second term is large enough. One of the merits of Suydam's criterion is that it provides a simple condition that may be tested easily and that, at least for laboratory fusion research, suggests measures (increasing the shear or lowering the pressure gradient) to be taken to ensure its satisfaction. A considerably more complicated toroidal version of this condition is known as *Mercier's criterion* (1960) [437].

The real importance of the localized Suydam solutions resides in the implications from the MHD oscillation theorem. Violation of Suydam's criterion (9.112) is the condition that the marginal point

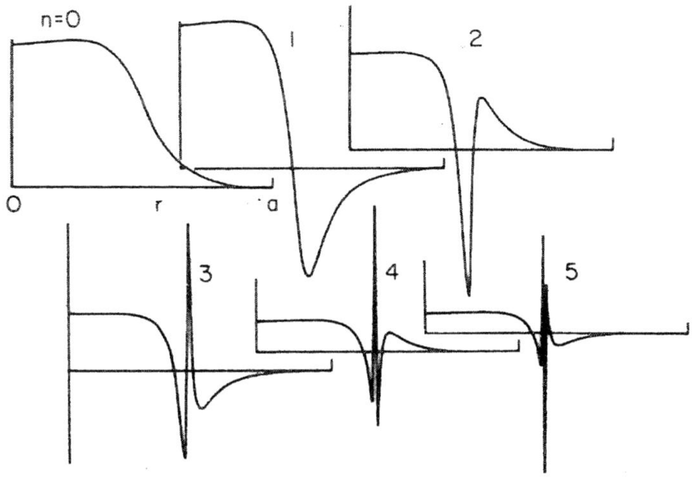

Fig. 9.10 Suydam unstable $m = 1$ modes with an increasing number of radial nodes of the eigenfunctions. (From Goedbloed and Sakanaka [240].)

$\omega^2 = 0$ is a *cluster point* of the unstable part of the discrete spectrum. The marginal equation then has solutions that oscillate infinitely rapidly and the oscillation theorem asserts that a global $n = 0$ solution to the full equation of motion exists for which the growth rate $-\omega^2$ is larger than that of all the higher n solutions. Hence, *violation of Suydam's criterion implies the existence of a global* $n = 0$ *instability* (Fig. 9.10). This instability may also be global in the azimuthal direction (e.g. $m = 1$) if the mode number k may be chosen such that $k + \mu m = 0$ somewhere on the interval $(0, a)$. Hence, Suydam's criterion provides a first test of stability which is quite significant.

Finally, Newcomb's stability test in the presence of a singular point $F = 0$ involves both complex and real indices. For the case of a gravitating slab, this yields the stability theorem of Section 7.5.3(b). For the cylinder, the interchange criterion (7.191) is then to be replaced by Suydam's criterion (9.112).

(b) ***σ****-stability* In Section 7.4.3 we have proved the oscillation theorem for the plane plasma slab. This proof carries over to cylindrical plasmas with proper modifications, exploiting the expressions (9.41) and (9.42) for the perturbation Π of the total pressure. Sturmian branches of the slow and Alfvén sub-spectra were foreseen in the proof of Chapter 7, when it still had to be shown that such branches actually exist. In the meantime, we have encountered plenty of examples demonstrating this. The most important one has just been discussed, viz. *instabilities* for values of m and k such that $F = 0$ at some point in the interval $(0, a)$. Since the continua $\{\omega_S^2\}$ and $\{\omega_A^2\}$ then stretch out to $\omega^2 = 0$, the mere existence of instabilities indicates that at least one of the Alfvén or slow branches of the discrete spectrum has become Sturmian. It is convenient that the function N/D never changes sign on the unstable side of the spectrum, so that *unstable modes are always Sturmian*. This is in agreement with our intuition that moving the wall inward does not increase the growth rate of an unstable mode.

Since the unstable side of the spectrum is non-singular, we immediately obtain a theorem for σ-stability of the diffuse cylindrical plasma. To that end, we notice that the *σ-marginal equation of* *motion* (6.111) for the diffuse cylindrical plasma is obtained from the Hain–Lüst equation (9.30) by just replacing ω^2 by $-\sigma^2$:

$$\Big[P(r;-\sigma^2)\,\chi'\Big]' - Q(r;-\sigma^2)\,\chi = 0\,, \tag{9.113}$$

with the boundary conditions $\chi(0) = \chi(a) = 0$. The corresponding *modified energy principle* reads:

$$W^\sigma[\chi] = \pi L \int_0^a \Big[P(r;-\sigma^2)\,\chi'^2 + Q(r;-\sigma^2)\,\chi^2\Big]\, r\, dr\,. \tag{9.114}$$

It could have been derived from Eq. (6.112) by a similar analysis to the one leading to the Hain–Lüst equation. Here, we have simply posed it directly as that functional which produces Eq. (9.113) as the σ-Euler equation.

In general, the σ-marginal equation (9.113) does not have solutions satisfying both boundary conditions. This problem is solved in the same way as in the ordinary stability theory (see Section 7.5.1). Suppose that we integrate Eq. (9.113) starting from the left end point $r = 0$ where we satisfy the boundary condition $\chi(0) = 0$. If the solution $\chi(r)$ thus obtained does not have a zero in the open interval $(0, a)$, our oscillation theorem asserts that a discrete eigenvalue $\omega^2 < -\sigma^2$ does not exist, so that the system is σ-stable. On the other hand, if the solution $\chi(r)$ vanishes somewhere on the open interval $(0, a)$, a discrete eigenvalue $\omega^2 < -\sigma^2$ does exist for which both boundary conditions are satisfied. We then have the following *Theorem for σ-stability of the cylindrical plasma*.

For specified values of m and k, the cylindrical plasma is σ-stable if, and only if, the non-trivial solution χ of the σ-marginal equation of motion (9.113) that vanishes at $r = 0$ does not have a zero in the open interval $(0, a)$.

The wording of this theorem is the same as that of the parallel theorem of Newcomb, discussed above, for the theory of marginal stability in the usual sense. However, since in the latter theory the singularities associated with the continua at $\omega^2 = 0$ have to be accounted for, the marginal theory in the usual sense is much more complicated than the corresponding theory for σ-stability.

On the basis of the σ-stability theorem it is possible to systematically search for σ-stable configurations while taking a reasonable choice for σ, e.g. one which corresponds to the msec time scale. From a large number of numerical runs the qualitative picture of Fig. 9.11 emerged at the time of the investigation, but still valid today as a leading order approximation of toroidal systems. (Conforming to present conventions, the parameter q is used here instead of μ, although these results strictly refer to infinite cylinder theory.) There are, broadly speaking, four categories of cylindrical configurations that are σ-stable with respect to internal modes. All four of them are characterized by a monotonically increasing or decreasing q-profile, representing shear of the field lines, which turns out to facilitate stability. The q and j_z profiles shown for these configurations are the most characteristic ones to distinguish the different configurations. As the current profile is broadened the maximum allowable β for stability in general increases from a few per cent for tokamaks to some 40% for the reversed field pinch. Except for the latter configuration all other configurations require $q > 1$, either on axis when the q-profile is increasing as in a tokamak, or at the wall when the q-profile is decreasing as in a screw pinch. In the latter, stability at high β is obtained due to force-free currents (rather than a 'vacuum') in the outer region of the plasma [214].

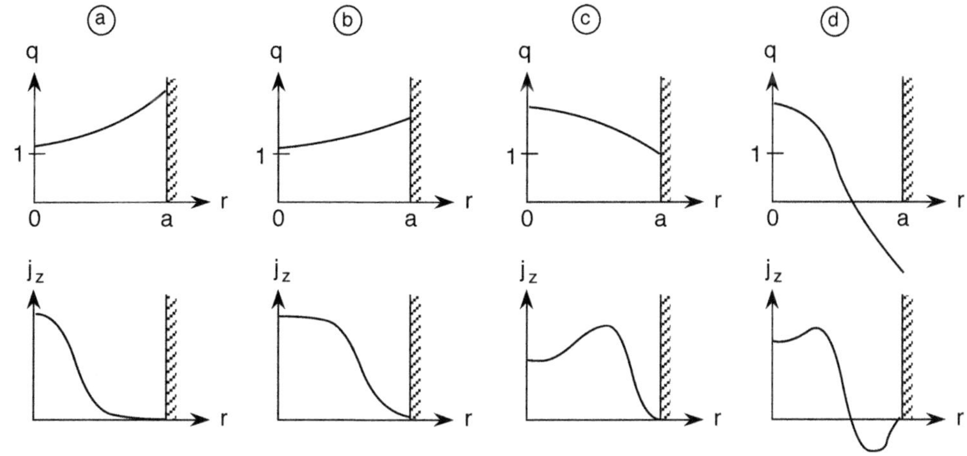

Fig. 9.11 Schematic overview of the different σ-stable cylindrical configurations: (a) tokamak; (b) flux-conserving tokamak; (c) screw pinch; (d) reversed field pinch. (After Sakanaka and Goedbloed [536].)

9.4.2 *Stability of plasmas with shearless magnetic fields*

(a) Instabilities of a z-pinch The most well-known MHD instabilities are the *sausage* ($m = 0$) *and kink* ($m = 1$) *instabilities of a z-pinch* ($B_z = 0,\, B = B_\theta$). Since the magnetic field is exclusively azimuthal, it has no shear so that Newcomb's theory does not apply to the $m = 0$ modes, which are pure interchanges (similar to the gravitational interchanges in the absence of shear in Section 7.5.2). For those modes, $F = 0$, so that the variable ζ disappears from the expression (9.100) for the energy and the last two terms have to be combined:

$$W = \pi L \int_0^a \Big\{ \Big[\frac{2p'}{r} + \frac{4\gamma p B^2}{r^2(\gamma p + B^2)} \Big] \xi^2 + k^2(\gamma p + B^2) \Big[\eta - \frac{1}{kr}\Big((r\xi)' - \frac{2B^2}{\gamma p + B^2}\, \xi \Big) \Big]^2 \Big\} \, r\, dr \,. \tag{9.115}$$

Upon minimization, the last term disappears and we obtain the following stability criterion for the $m = 0$ sausage modes:

$$-rp' < \frac{2\gamma p B^2}{\gamma p + B^2} \qquad \text{(everywhere)}\,. \tag{9.116}$$

This implies that there is a limit on each point of the radial pressure profile of a confined plasma in a z-pinch. Again, as in Section 7.5.2, the stability criterion for pure interchanges is less severe than the local condition (Suydam's criterion in this case) in the limit of no shear.

For the $m \neq 0$ modes, Newcomb's analysis applies. Here, it is interesting to consider the limit $k \to \infty$ since that approaches the interchange condition most closely. In that case,

$$f_0 \to 0\,, \qquad g_0 \approx 2p' + m^2 B^2/r\,, \tag{9.117}$$

so that the stability criterion for $m \neq 0$ modes becomes:

$$-rp' < \tfrac{1}{2}m^2 B^2 \qquad \text{(everywhere)}\,. \tag{9.118}$$

Notice that the limit of Suydam's criterion, $p' > 0$, would be obtained for $m \to 0$, just like the quasi-interchange stability criterion of Section 7.5.2. However, in a cylinder, the smooth approach of this limit is excluded because of azimuthal periodicity, and the limit $m = 0$ itself is governed by the stability criterion (9.116). Hence, $m = 1$ becomes the worst case.

Comparing the two stability criteria: for local values of $\gamma\beta \equiv c^2/b^2 < 1/3$, the $m = 0$ criterion (9.116) is the more restrictive condition. For $\gamma\beta > 1/3$, the criterion (9.118) with $m = 1$ is the more restrictive one; see Kadomtsev [338]. Hence, it is possible to construct stable pressure profiles for a z-pinch, but they 'live' in a sea of violently unstable profiles, so that experimental control would be extremely complicated and risky.

It is instructive to consider the special case of an incompressible z-pinch with the current and pressure distributions (9.93) and (9.94) of Section 9.3.2(c), but taking $B_z = 0$. For that equilibrium, $-rp' = 2A^2r^2 = 2B^2$, so that the $m = 0$ criterion is marginally satisfied, but the $m = 1$ criterion is violated. To compute the instability threshold $k = k_0$ for the $m = 1$ modes, we exploit the eigenfunction (9.96) with

$$B_z = 0 \quad \Rightarrow \quad \alpha = \frac{2A^2}{\rho\omega^2 - A^2}\,, \qquad k^{*2} = k^2(1 - \alpha^2)\,. \tag{9.119}$$

For marginal stability, $\alpha = -2$ so that $k^* = \mathrm{i}k\sqrt{3}$ and the solution becomes:

$$\chi_0 = Cx^{-1}[x^2 J_1(x)]'\,, \qquad \text{where } x \equiv kr\sqrt{3}\,. \tag{9.120}$$

Increasing the value of $|k|$, instability sets in when $\chi_0(a) = 0$ for the first time, i.e. for $|k_0|a = 1.58$. Hence, this z-pinch is $m = 1$ unstable for $|k|a > 1.58$.

An estimate of the growth rate of the $m = 1$ modes may be obtained as well from the eigenfunction (9.96). Since $1 - \alpha^2 < 0$ for $-A^2 < \rho\omega^2 < 0$, the Bessel function oscillates infinitely rapidly in that range when $k \to \infty$, unless $\alpha^2 \approx 1$. Hence, the growth rate of the $m = 1$ kink mode in the limit $k \to \infty$ is given by:

$$\rho\omega^2 \approx -A^2 = -\frac{B_\theta^2}{\mu_0 r^2} \quad \Rightarrow \quad -\mathrm{i}\omega \approx \frac{B_\theta}{r\sqrt{\mu_0\rho}} = \tfrac{1}{2}\sqrt{\frac{\mu_0}{\rho}}\, j_z\,. \tag{9.121}$$

This expression shows that the internal kink modes of a z-pinch exponentiate on the same time scale as the external kink modes, for which the growth rate was given by Eq. (2.162). Since growth rates of this order of magnitude cannot be tolerated in magnetic fusion devices, the z-pinch has been abandoned there long ago. However, in inertial confinement fusion and in discharges for laser wake field acceleration, where time scales are very much shorter, the z-pinch is a valuable plasma confinement scheme.

(b) Interchanges and quasi-interchanges in a shearless magnetic field The peculiar crossing of the stability criteria for the z-pinch, at the interchange point $F = 0$, is a general property of shearless magnetic fields. As we have seen, for sheared magnetic fields, the singular points $F = 0$ significantly influence the stability properties of the plasma. In shearless magnetic fields, either such singularities are absent or the whole interval is singular. This causes discontinuity of the stability

criteria. That discontinuity disappears in the expressions for the growth rates of the instabilities. We have already encountered this effect in Section 7.5.2 for the gravitational interchanges. Here, we present the cylindrical counterpart developed in the papers by Ware [639], Goedbloed [214](c), and Goedbloed and Hagebeuk [234].

For analytical calculations, it is expedient to exploit the simplified form (9.47) of the Hain–Lüst for low-frequency waves or instabilities. We here exploit the special case of a shearless magnetic field ($\mu' = 0$) so that the last term may be dropped. The resulting form of this equation clearly exhibits the terms driving the interchange instabilities ($\sim p'$) and the terms stabilizing them ($\sim F$). To further simplify it, we exploit the *low-β tokamak ordering* (introduced in Section 9.1.1) and an ordering of the wave numbers to focus on *the range of the interchanges* ($F = 0$) *and quasi-interchanges* ($F \neq 0$).

We demonstrate these orderings with a particular equilibrium that is obtained by generalizing the constant-pitch force-free field of Eq. (9.11):

$$B_z = \frac{B_0}{1+\delta^2 r^2} \approx B_0(1-\delta^2 r^2)\,, \qquad B_\theta = \frac{B_0 \mu r}{1+\delta^2 r^2} \approx B_0 \mu r\,,$$

$$p = \mathrm{const} + \frac{B_0^2(\mu^2-\delta^2)}{2\delta^2(1+\delta^2 r^2)^2} \approx B_0^2\left[\tfrac{1}{2}\beta - (\mu^2-\delta^2)r^2\right], \tag{9.122}$$

where the approximate expressions result from the low-β tokamak ordering,

$$\beta \sim \mu^2 a^2 \sim \delta^2 a^2 \sim \epsilon^2\,. \tag{9.123}$$

Since the pitch of the field lines is constant, the parallel wave number is approximately constant,

$$F \equiv k_\parallel B\,, \quad k_\parallel = \frac{k+\mu m}{\sqrt{1+\mu^2 r^2}} \approx k + \mu m\,, \tag{9.124}$$

and the wave numbers may be ordered as

$$k_\parallel a \ll ka \ll m^2 \sim 1 \quad \Rightarrow \quad m^2 + k^2 r^2 \approx m^2\,, \quad k^2/m^2 \approx \mu^2\,. \tag{9.125}$$

With $\rho = \mathrm{const}$ and $F = \mathrm{const}$, the Alfvén factor $\rho\omega^2 - F^2$ can be extracted, so that we obtain the following form of the eigenvalue problem:

$$\frac{d}{dr}\left(r\frac{d\chi}{dr}\right) - \frac{1}{r}\,(m^2 - \lambda^2 r^2)\,\chi = 0\,, \qquad \chi(0) = \chi(a) = 0\,, \tag{9.126}$$

where

$$\lambda \equiv \frac{m^2 B_\theta^2}{r^2(\rho\omega^2 - F^2)}\left[\frac{2p'}{rB^2} + 4\mu^2\,\frac{\gamma p\,\rho\omega^2}{(\gamma p + B^2)\rho\omega^2 - \gamma p F^2}\right] \approx \mathrm{const}\,. \tag{9.127}$$

The assumption $\lambda \approx \mathrm{const}$ is justified for the particular equilibrium chosen, but it may also be assumed for the radially localized modes of more general equilibria.

The solutions of Eq. (9.126) are Bessel functions:

$$\chi = C J_m(\sqrt{\lambda}\,r)\,, \quad \lambda = j_{mn}^2/a^2\,, \tag{9.128}$$

where j_{mn} indicates the zeros of J_m. This gives the dispersion equation

$$\rho^2\omega^4 - \left[\alpha(\Pi - \Pi_1) + (1+M_c^2)F^2\right]\rho\omega^2 + M_c^2 F^2(\alpha\Pi + F^2) = 0\,, \tag{9.129}$$

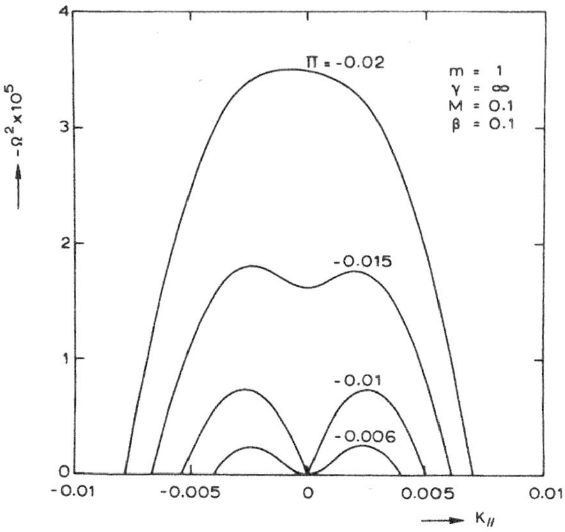

Fig. 9.12 Normalized growth rate, $-\Omega^2 \equiv -(\rho a^2/B^2)\,\omega^2$, of the interchange and quasi-interchange instabilities in a constant-pitch magnetic field as a function of $K_\parallel \equiv k_\parallel a$ for different values of the normalized pressure gradient Π; $M \equiv \mu a$ is the normalized pitch of the field lines. (From Goedbloed and Hagebeuk [234].)

with the abbreviations

$$\alpha \equiv \frac{4m^2 B_\theta^2}{j_{mn}^2 r^2}\,, \quad \Pi \equiv \frac{a^2 p'}{2rB^2}\,, \quad \Pi_1 \equiv -\mu^2 a^2 M_c^2\,, \quad M_c^2 \equiv \frac{\gamma p}{\gamma p + B^2}\,.$$

The unstable solution

$$\begin{aligned} \rho\omega^2 &= \tfrac{1}{2}\alpha(\Pi - \Pi_1) + \tfrac{1}{2}(1 + M_c^2)F^2 \\ &\quad - \sqrt{\left[\tfrac{1}{2}\alpha(\Pi - \Pi_1) + \tfrac{1}{2}(1 + M_c^2)F^2\right]^2 - M_c^2 F^2(\alpha\Pi + F^2)}\,, \end{aligned} \tag{9.130}$$

represents a quasi-interchange ($F \neq 0$) or a pure interchange ($F = 0$) depending on the values of the pressure gradient Π and the parallel wave factor F^2.

This implies that the following instability regimes are crossed when the negative pressure gradient is increased (as illustrated in Fig. 9.12 for an incompressible plasma, $\gamma \to \infty$ or $M_c^2 \to 1$):

(1) For $\Pi > 0$, the plasma is stable (i.e. under the approximations made, eliminating the quasi-kink instability, see below);

(2) For $\Pi_1 \leq \Pi \leq 0$, the plasma is marginally stable at $F^2 = 0$ and the most unstable mode is a quasi-interchange at

$$\begin{aligned} F^2 &= -\frac{\alpha\Pi_1}{1 - M_c^2}\left\{\frac{\Pi}{\Pi_1} - \frac{1 + M_c^2}{1 - M_c^2}\left[1 - \sqrt{1 - (1 - M_c^2)\frac{\Pi}{\Pi_1}}\,\right]\right\} \\ \Rightarrow \quad \left(-\rho\omega^2\right)_{\max} &= -\frac{\alpha M_c^2 \Pi_1}{(1 - M_c^2)^2}\left[1 - \sqrt{1 - (1 - M_c^2)\frac{\Pi}{\Pi_1}}\,\right]^2; \end{aligned} \tag{9.131}$$

(3) For $\Pi_2 \equiv (1 + M_c^2) \le \Pi < \Pi_1$, the most unstable mode is still a quasi-interchange, as given by Eq. (9.131), but the interchange instability also kicks in at

$$F^2 = 0 \quad \Rightarrow \quad \rho\omega^2 = \alpha(\Pi - \Pi_1)\,; \tag{9.132}$$

(4) for $\Pi < \Pi_2$, the quasi-interchanges are stable and the most unstable mode is now a pure interchange with the growth rate as given by Eq. (9.132).

Consequently, for an equilibrium with a constant-pitch magnetic field, Suydam's stability criterion (9.112) degenerates into the quasi-interchange stability condition $\Pi > 0$ (i.e. $p' > 0$) and not into the pure interchange condition $\Pi > \Pi_1$, as one might have expected. This example demonstrates that stability conditions, considered for fixed wave number, may exhibit discontinuities that disappear when the growth rates are considered. Figure 9.12 clearly shows the gradual change of the squared growth rate $-\omega^2(k_\parallel)$ when the negative pressure gradient is increased. Note that, if one insists on absolute stability, the quasi-interchange condition $\Pi > 0$ is not enough either, because other instabilities, viz. *quasi-kinks* are still unstable in the absence of pressure gradients. This requires a different ordering of the parameters; see Ref. [234] and Exercise 9.4.

9.4.3 Stability of force-free magnetic fields*

Stability analysis becomes much more complicated when the restriction of a constant pitch magnetic field is dropped, which usually means that solutions can only be obtained numerically (as in the final analysis of Section 9.4.1). In this section and the next, we discuss two important configurations, one relevant for astrophysical plasmas and one for laboratory fusion plasmas, where semi-analytical solutions can be constructed which, again, centre about the $F = 0$ singularities.

For many astrophysical plasmas, like magnetic flux loops in the solar corona, gravity and the kinetic pressure of the plasma are negligible compared to the magnetic pressure, so that $\nabla p \approx 0$ and the magnetic field is force-free:

$$\mathbf{j} \times \mathbf{B} = 0 \quad \Rightarrow \quad \mathbf{j} = \alpha(\mathbf{r})\mathbf{B}\,. \tag{9.133}$$

One of the simplest, non-trivial, examples is the Lundquist field, a cylindrical force-free magnetic field with a constant value of α. This field has been discussed extensively in Section 4.3.4, in the context of magnetic helicity, and in Section 9.1.1(a). The explicit Bessel function expressions for B_θ and B_z are given in Eq. (4.114), or Eq. (9.12), and illustrated in Fig. 4.8. Since the instabilities of the Lundquist field essentially occur for $B_\theta \sim B_z$, we cannot exploit the low-β tokamak approximation to simplify the Bessel function expressions.

(a) Solution of the marginal equation of motion Rather than exploiting Newcomb's equations, it is expedient to derive the marginal stability equations from the original expressions (6.29) and (6.30) for the force operator. The pressure terms disappear, since $\nabla p = 0$ and $\nabla \cdot \boldsymbol{\xi} = 0$ at marginal stability, so that the marginal equation of motion becomes

$$\mathbf{F}(\boldsymbol{\xi}) = -\mathbf{B} \times (\nabla \times \mathbf{Q}) + \mathbf{j} \times \mathbf{Q} = -\mathbf{B} \times (\nabla \times \mathbf{Q} - \alpha\mathbf{Q}) = 0\,. \tag{9.134}$$

Hence, the magnetic field perturbation $\mathbf{Q} = \nabla \times (\boldsymbol{\xi} \times \mathbf{B})$ satisfies the differential equation

$$\nabla \times \mathbf{Q} - \alpha\mathbf{Q} = \lambda\mathbf{B}\,, \tag{9.135}$$

where $\lambda(r)$ is a first order quantity (with suppressed phase factor $\exp \mathrm{i}(m\theta + kz)$). To determine λ, we take the divergence of Eq. (9.135), which only gives a contribution from the RHS,

$$\mathbf{B} \cdot \nabla\lambda = \mathrm{i}F\lambda = 0 \quad \Rightarrow \quad \lambda = u(r)\delta(r - r_\mathrm{s}) \,, \tag{9.136}$$

when there is a singularity $F = 0$ at $r = r_\mathrm{s}$. Clearly, the quantity λ represents a skin current induced by the perturbation at the singularity, where the strength $u(r)$ is to be determined yet.

We solve Eq. (9.135) by eliminating Q_r and Q_θ from the first two components,

$$\begin{aligned} Q_r &= \frac{\mathrm{i}}{(\alpha^2 - k^2)r} \Big[kr(Q_z' + B_\theta\lambda) + m\alpha Q_z \Big] \,, \\ Q_\theta &= \frac{-1}{(\alpha^2 - k^2)r} \Big[\alpha r(Q_z' + B_\theta\lambda) + kmQ_z \Big] \,, \end{aligned} \tag{9.137}$$

and substituting them in the third component:

$$\frac{1}{r}\frac{d}{dr}\Big(r\frac{dQ_z}{dr} \Big) + \Big(\alpha^2 - k^2 - \frac{m^2}{r^2} \Big) Q_z = -B_\theta\lambda' - 2\alpha B_z\lambda \,. \tag{9.138}$$

Here, we have used the force-free field condition $j_z = (1/r)(rB_\theta)' = \alpha B_z$ and cancelled a term $F\lambda = 0$. The solution of the homogeneous equation that is regular at $r = 0$ is a Bessel function of the first kind:

$$\frac{1}{r}\frac{d}{dr}\Big(r\frac{d\phi}{dr} \Big) + \Big(\alpha^2 - k^2 - \frac{m^2}{r^2} \Big)\phi = 0 \quad \Rightarrow \quad \phi = CJ_m(\sqrt{\alpha^2 - k^2}\, r) \,, \tag{9.139}$$

where we may restrict the analysis to $\alpha^2 > k^2$ since the boundary condition at $r = a$ (to be discussed below) requires oscillatory solutions. A solution of the inhomogeneous equation (9.138) can then be constructed that is only different from zero on the interval between the axis ($r = 0$) and the first singularity ($r = r_\mathrm{s}$):

$$Q_z = \phi(r)H(r_\mathrm{s} - r) = CJ_m(\sqrt{\alpha^2 - k^2}\, r)H(r_\mathrm{s} - r) \,. \tag{9.140}$$

Substituting this expressing back into Eq. (9.138), and eliminating the homogeneous contributions, yields *two* conditions from the factors multiplying δ' and δ:

$$\phi\delta' + \Big(2\phi' + \frac{\phi}{r} \Big)\delta = B_\theta u\delta' + (B_\theta u' + 2\alpha B_z u)\delta \quad \Rightarrow \quad u = \phi/B_\theta \,, \quad B_\theta\phi' - \alpha B_z\phi = 0 \,. \tag{9.141}$$

The first one is the relation between u and ϕ we were looking for. However, we obtain an additional condition, a kind of internal boundary condition, to be satisfied by ϕ in order for the discontinuous solution (9.140) to be acceptable.

Since these conditions fix everything, let us consider them in detail. From $\mathbf{j}_1 = \nabla \times \mathbf{Q} = \alpha\mathbf{Q} + \lambda\mathbf{B}$, the skin current at the singularity is now determined:

$$\mathbf{j}_1^\star = \int_{r_\mathrm{s}^-}^{r_\mathrm{s}^+} \mathbf{j}_1 \, dr = \int_{r_\mathrm{s}^-}^{r_\mathrm{s}^+} u\delta(r - r_\mathrm{s})\mathbf{B} \, dr = \Big[(\mathbf{B}/B_\theta)\phi \Big]_{r_\mathrm{s}} \,. \tag{9.142}$$

According to Eq. (4.165) of Section 4.5.2, this skin current also fixes the two tangential components of $\mathbf{Q}$ at r_s^-:

$$\mathbf{j}_1^\star = \mathbf{n} \times [\![\mathbf{Q}]\!] \quad \Rightarrow \quad \begin{cases} j_{1\theta}^\star = -[\![Q_z]\!] = Q_z(r_s^-) = \phi\Big|_{r_s} \\ j_{1z}^\star = [\![Q_\theta]\!] = -Q_\theta(r_s^-) = \Big[(B_z/B_\theta)\phi\Big]_{r_s} \end{cases} . \tag{9.143}$$

Evaluating everything at r_s^-, where $F \equiv mB_\theta/r + kB_z = 0$, this yields:

$$\begin{aligned} B_z j_{1\theta}^\star - B_\theta j_{1z}^\star &= B_z Q_z + B_\theta Q_\theta = \frac{rB_z}{m}\Big(\frac{m}{r}Q_z - kQ_\theta\Big) \overset{(9.135)}{=} -\mathrm{i}\frac{\alpha r B_z}{m} Q_r \\ &\overset{(9.137)}{=} \frac{\alpha B_z}{m(\alpha^2 - k^2)}(kr\phi' + m\alpha\phi) = -\frac{\alpha}{\alpha^2 - k^2}(B_\theta\phi' - \alpha B_z\phi) = 0. \end{aligned} \tag{9.144}$$

Hence, the internal boundary condition (9.141)(b) turns out to be equivalent to the requirement that the perturbed radial magnetic field Q_r *has to vanish at the singularity*, i.e. we have recovered the condition that the displacement $\xi_r = -\mathrm{i}Q_r/F$ has to be 'small' there: the circle is closed.

Inserting the expression (9.137)(a) for Q_r, with the solution (9.140) for Q_z, in the expression for ξ_r, and transforming the Bessel function derivatives, finally yields the explicit solution of the marginal equation of motion:

$$\begin{aligned} \xi_r &= \frac{C\alpha}{B_0(\alpha^2 - k^2)} \frac{m(1 - k/\alpha)J_m(\beta r) + (k/\alpha)\beta r J_{m-1}(\beta r)}{mJ_1(\alpha r) + krJ_0(\alpha r)} H(r_s - r) \\ &\equiv \frac{C\alpha}{B_0(\alpha^2 - k^2)} \frac{\widetilde{Q}(m, k/\alpha, \alpha r)}{\widetilde{F}(m, k/\alpha, \alpha r)} H(r_s - r), \quad \text{where} \quad \beta \equiv \alpha\sqrt{1 - (k/\alpha)^2}. \end{aligned} \tag{9.145}$$

The auxiliary quantities $\widetilde{Q}$ and $\widetilde{F}$, proportional to Q_r and F, are introduced here to discuss the oscillatory behaviour of these functions. It would have been extremely difficult to produce this solution directly from Newcomb's equations (9.103) and (9.104). More important, we now have the physical reason of how 'small' solutions come about: the skin currents effectively produce 'a virtual wall' at the singularity, as illustrated in Fig. 9.5 for the case of a current-free 'ghost' plasma.

(b) Stability analysis and calculation of the growth rates According to Newcomb's stability theorem, or the MHD oscillation theorem, the expression (9.145) is to be studied with respect to the zeros of both Q_r and F. Since both functions are oscillatory in the relevant parameter domain, this involves a rather subtle analysis, that was carried out by Voslamber and Callebaut [637]. Roughly speaking, for instability, the function $\widetilde{Q}$ in the numerator should oscillate faster than the function $\widetilde{F}$ in the denominator, since the zeros of the latter function delimit the independent sub-intervals. The first transition to instability is found when $F \sim k + \mu m = 0$ somewhere (e.g. at the position indicated by the first dotted line in the top panel of Fig. 9.14) and Q_r also vanishes there, so that it may be chosen to vanish identically to the right of that point. Hence, at that transition (where $\omega^2 = 0$), a kind of block-function displacement is obtained for ξ_r.

Recall from Section 9.4.1 that only the $m = 0\,, k \to 0$ and $|m| = 1$ modes have to be investigated to determine stability of a particular configuration. The $m = 0\,, k \to 0$ modes are stable since $\xi_r \sim J_1(\alpha r)/J_0(\alpha r)$ in that case, and the first zero of J_0 is smaller than the first zero of J_1. Hence, only the $|m| = 1$ modes need to be investigated. Stability of the Lundquist field turns out to depend on two parameters, αa and k/α, and the unstable region for the $m = 1$ modes falls

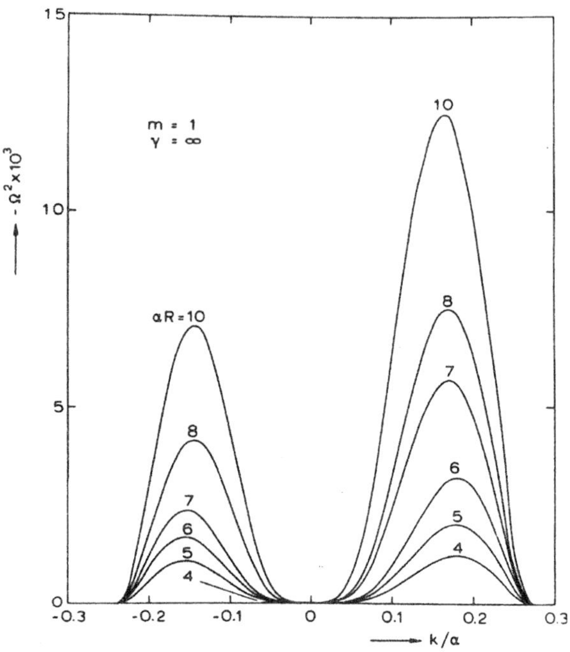

Fig. 9.13 Normalized growth rate, $-\Omega^2 \equiv -(\rho_0 a^2/B_0^2)\,\omega^2$, of the $m = 1$ kink mode in a force-free Lundquist field as a function of k/α for several values of the parameter αa. In this figure, the wall position is indicated by R instead of a. (From Goedbloed and Hagebeuk [234].)

within the strip ($\alpha a > 3.175\,,\ -0.237 < k/\alpha < 0.273$) of the $(\alpha a\,, k/\alpha)$-plane. Instability with respect to the $m = -1$ modes is governed by similar conditions with the sign of k reversed. The result is quite reasonable: $m = \pm 1$ kink modes are long-wavelength, current driven, instabilities which occur when the total current ($\sim \alpha a$) is large enough and when the longitudinal wavelength ($\sim k^{-1}$) is large compared to a typical length scale of the radial inhomogeneity ($\sim \alpha^{-1}$). Readers interested in the details of the stability analysis should consult the original paper by Voslamber and Callebaut [637], or the recent extension by Goedbloed [229](b), in particular Figs. 1 and 2.

The computation of the growth rates of the instabilities of the Lundquist field by means of the Hain–Lüst equation is significantly simpler than the stability analysis (granted that a computer program for the solution of this equation has been written), since the equation is non-singular for $\omega^2 < 0$ so that the shooting method (Section 7.5.1) can be applied. The results of such computations are shown in Figs. 9.13 and Fig. 9.14. In the first figure, the growth rate of the $m = 1$ kink mode is shown for the unstable range of the parameters αa and k/α, found by Voslamber and Callebaut. Note the typical asymmetry of the growth rate with respect to the wave number k. The top panel of Fig. 9.14 shows the radial dependence of the inverse pitch μ of the magnetic field lines for a highly unstable Lundquist field, and two singular positions $k + \mu m = 0$ for a particular choice of the wave numbers. The bottom panel of this figure shows the eigenfunction of the unstable kink mode for that case, consisting of two approximately rigid displacements of different amplitude in the two independent sub-intervals of the marginal stability analysis. Note that the jumps of the marginal solution (the trial function of the energy principle) have disappeared for the finite value

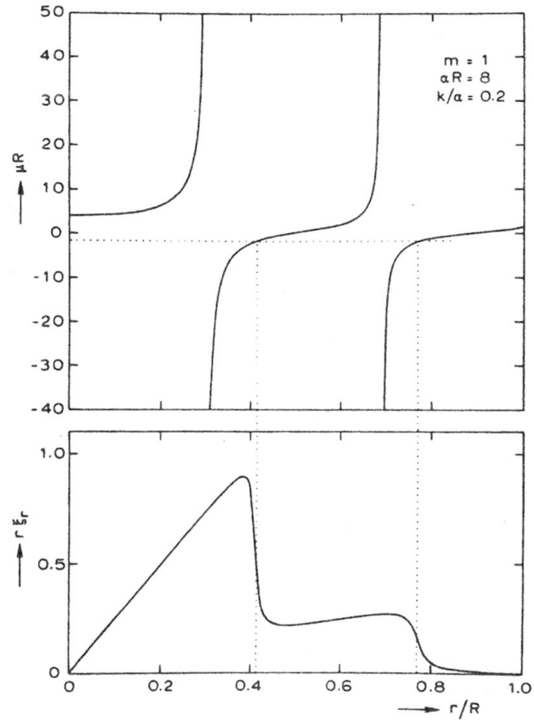

Fig. 9.14 Eigenfunction of the kink mode in a Lundquist field, $\alpha a = 8, k/\alpha = 0.2$. In this figure, the wall position is indicated by R instead of a. Top: inverse pitch of the field lines, indicating the radial positions of the singularity $k+\mu m = 0$ (dotted lines). Bottom: the $m = 1$ eigenfunction has smoothed out the jumps of the marginal mode at the singularities. (From Goedbloed and Hagebeuk [234].)

of $-\omega^2$ so that the actual eigenfunction is much smoother and, hence, more realistic than the trial function. With respect to the numerics, this implies that the computation of the growth rates requires significantly fewer grid points for the same accuracy of the result.

In conclusion, the stability of the force-free magnetic field configuration with respect to kink modes is determined by the competition between the destabilizing factor of the force-free equilibrium currents and the stabilizing factor of the skin current perturbations at the singularities. This produces a very delicate balance in the stability analysis, where the force-free current destabilization just dominates, but only in a very tiny part of parameter space. Of course, this does not imply that the instabilities are hard to realize (nature immediately finds the route to the lowest energy state), but it does imply that the growth rates of the instabilities are significantly lower than they would be without skin-current stabilization.

9.4.4 Stability of the 'straight tokamak'

In this section, we present the energy principle counterpart of the normal mode analysis of Section 9.2 for a cylindrical plasma surrounded by vacuum. We depart from the extended energy principle of Section 6.6.3 with $W = W^{\rm p} + W^{\rm s} + W^{\rm v}$, where the expressions for the plasma energy $W^{\rm p}$, the surface energy $W^{\rm s}$ and the vacuum energy $W^{\rm v}$ are given by Eqs. (6.146)–(6.148), respectively. The variables $\boldsymbol{\xi}$ and $\hat{\mathbf{Q}}$ are subject to the boundary conditions (6.150)–(6.152). This general formalism is applied to the straight tokamak, where a toroidal configuration is represented by a periodic cylinder of finite length $L = 2\pi R_0$, so that the volume element becomes $dV = 4\pi^2 R_0 dr$.

Whereas some concepts of the preceding section on force-free fields will return (like singular trial functions and skin-current stabilization), the tokamak configuration is essentially different in two ways: (1) the cylinder has a finite length, so that only integer values of the longitudinal wave number $n = k/R_0$ need to be considered; (2) the q-profile is mainly increasing with radius and has a magnitude of order unity, in contrast to the Lundquist field, where $q \equiv (\mu R_0)^{-1}$ is radially decreasing (even going through zero) and $q_0 = 2\epsilon(\alpha a)^{-1} \ll 1$. These two factors make the stability properties completely different.

(a) Energy expression for a general cylindrical plasma–vacuum configuration In the expressions of Newcomb's minimized form (9.102) of the plasma energy, a boundary term has been dropped (since only internal modes were considered in Section 9.4.1) that needs to be restored when we also permit external modes:

$$W^{\rm p} = 2\pi^2 R_0 \int_0^a \Big(f_0 \xi'^2 + g_0 \xi^2 \Big)\, dr - 2\pi^2 R_0 \left[\frac{m^2 B_\theta^2 - k^2 r^2 B_z^2}{m^2 + k^2 r^2}\, \xi^2 \right]_{r=a} . \tag{9.146}$$

As in Section 6.6.4, Eq. (6.181), inserting back the minimizing solution of the Euler–Lagrange equation (9.103) into $W^{\rm p}$ yields:

$$W^{\rm p} = 2\pi^2 R_0 \left[\frac{r^3 F^2}{m^2 + k^2 r^2} \frac{\xi'}{\xi} - \frac{m^2 B_\theta^2 - k^2 r^2 B_z^2}{m^2 + k^2 r^2} \right]_{r=a} \xi^2(a) , \tag{9.147}$$

where the logarithmic derivative $(\xi'/\xi)|_{r=a}$ contains now all the necessary information of the plasma interval. The expression for the surface energy becomes

$$W^{\rm s} = -2\pi^2 R_0 \Big[\hat{B}_\theta^2 - B_\theta^2 \Big]_{r=a} \xi^2(a) . \tag{9.148}$$

As in Section 6.6.4(b), the vacuum energy is obtained by eliminating the variable $\hat{Q}_z$ and rearranging the resulting expression to get one positive definite term involving $\hat{Q}_\theta$:

$$\begin{aligned} W^{\rm v} &= 2\pi^2 R_0 \int_a^b \Big[|\hat{Q}_r|^2 + |\hat{Q}_\theta|^2 + \frac{1}{k^2 r^2} \Big| (r\hat{Q}_r)' + {\rm i} m \hat{Q}_\theta \Big|^2 \Big]\, r\, dr \\ &= 2\pi^2 R_0 \int_a^b \Big[\frac{1}{m^2 + k^2 r^2} \Big| (r\hat{Q}_r)' \Big|^2 + |\hat{Q}_r|^2 \\ &\qquad + \frac{m^2 + k^2 r^2}{k^2 r^2} \Big| {\rm i}\hat{Q}_\theta + \frac{m}{m^2 + k^2 r^2} (r\hat{Q}_r)' \Big|^2 \Big]\, r\, dr . \end{aligned} \tag{9.149}$$

Upon minimization, the last term vanishes and the first terms yield the Euler–Lagrange equation (9.56) with the Bessel function derivative solution (9.59), derived in Section 9.2.2. Inserting this solution back into $W^{\rm v}$, and expressing $\hat{Q}_r$ in ξ_r through the boundary condition (9.51), gives

$$W^{\rm v} = -2\pi^2 R_0 \left[\frac{r}{m^2 + k^2 r^2} (r\hat{Q}_r)(r\hat{Q}_r)' \right]_{r=a} = -2\pi^2 R_0 \left[\frac{r^3 \hat{F}^2}{m^2 + k^2 r^2} \frac{(r\hat{Q}_r)'}{r\hat{Q}_r} \right]_{r=a} \xi^2(a) . \tag{9.150}$$

Collecting terms provides the general expression for the energy of a cylindrical plasma–vacuum configuration:

$$W = 2\pi^2 R_0 \left[\frac{r^3 F^2}{m^2 + k^2 r^2} \frac{\xi'}{\xi} - \frac{m^2 B_\theta^2 - k^2 r^2 B_z^2}{m^2 + k^2 r^2} + B_\theta^2 - \hat{B}_\theta^2 - \frac{r^3 \hat{F}^2}{m^2 + k^2 r^2} \frac{(r\hat{Q}_r)'}{r\hat{Q}_r} \right]_{r=a} \xi^2(a) \,, \tag{9.151}$$

where the logarithmic derivative of ξ is determined by solving the Euler–Lagrange equation (9.103), and the logarithmic derivative for $r\hat{Q}_r$ follows from Eq. (9.59):

$$\left[\frac{(r\hat{Q}_r)'}{\hat{Q}_r} \right]_{r=a} = \frac{m^2 + k^2 a^2}{ka} \frac{I_m(ka) K'_m(kb) - K_m(ka) I'_m(kb)}{I'_m(ka) K'_m(kb) - K'_m(ka) I'_m(kb)} \left(\approx -|m| \frac{1 + (a/b)^{2|m|}}{1 - (a/b)^{2|m|}} \right). \tag{9.152}$$

The approximation in large brackets holds for long-wavelength perturbations ($k^2 a^2 \sim k^2 b^2 \ll m^2 \sim 1$), appropriate for the tokamak problem considered below.

(b) Energy expression for the 'straight tokamak' We now construct the leading order expression for the energy in the low-β tokamak ordering. This ordering continues with genuine toroidal contributions in the next orders, see Chapter 17. In the leading order (straight cylindrical) contribution, only the safety factor q enters:

$$q \approx \frac{1}{\mu R_0} = \frac{r B_z}{R_0 B_\theta} \approx \frac{B_0}{R_0} \frac{r}{B_\theta} \sim 1 \,; \tag{9.153}$$

the longitudinal field $B_z \approx B_0$, and pressure effects are negligible:

$$\beta \equiv 2p_0 / B_0^2 \sim \epsilon^2 \ll 1 \,. \tag{9.154}$$

Hence, to leading order in the low-β tokamak ordering, we may forget about the cylindrical equilibrium equation (9.3), and just arbitrarily specify either one of the profiles for $q(r)$, $B_\theta(r)$, or the current density

$$j_z(r) = \frac{1}{r}(r B_\theta)' = \frac{B_0}{R_0} \frac{1}{r} \left(\frac{r^2}{q} \right)' . \tag{9.155}$$

Next, as in Eq. (9.80), we replace the longitudinal wave number k by the toroidal mode number n, $k \equiv n/R_0$, and assume both, toroidal and poloidal, mode numbers to be of order unity:

$$m \sim n \sim 1 \quad \Rightarrow \quad m^2 + k^2 r^2 \approx m^2 \,, \qquad F \approx \frac{B_0}{R_0} \left(n + \frac{m}{q} \right). \tag{9.156}$$

The expression (9.151) for W then simplifies to:

$$W = \frac{2\pi^2 a^2 B_0^2}{R_0} \left[\left(\frac{n}{m} + \frac{1}{q} \right)^2 \frac{r\xi'}{\xi} + \frac{n^2}{m^2} - \frac{1}{\hat{q}^2} - \left(\frac{n}{m} + \frac{1}{\hat{q}} \right)^2 \frac{(r\hat{Q}_r)'}{\hat{Q}_r} \right]_{r=a} \xi^2(a) \,. \tag{9.157}$$

To complete this expression, we need to solve the Euler–Lagrange equation (9.103) for ξ with the approximations (9.153)–(9.156):

$$\frac{d}{dr} \left[r^3 \left(\frac{n}{m} + \frac{1}{q} \right)^2 \frac{d\xi}{dr} \right] - (m^2 - 1) r \left(\frac{n}{m} + \frac{1}{q} \right)^2 \xi = 0 \,, \tag{9.158}$$

where $q(r)$ needs to be specified. The equation for $r\hat{Q}_r$ simplifies from the Bessel equation (9.56) to an elementary differential equation that can be solved directly:

$$\frac{d}{dr}\left[r\frac{d}{dr}(r\hat{Q}_r)\right] - m^2\hat{Q}_r = 0 \quad \Rightarrow \quad r\hat{Q}_r = C\big[(r/b)^{|m|} - (r/b)^{-|m|}\big]\,, \tag{9.159}$$

giving the approximate form of Eq. (9.152) for the logarithmic derivative. Restricting the analysis to configurations without surface currents now, so that $q_a = \hat{q}_a$, the expression for W becomes:

$$W = \frac{2\pi^2 a^2 B_0^2}{R_0}\bigg[\Big(\frac{n}{m}+\frac{1}{q}\Big)^2\frac{r\xi'}{\xi} + \frac{n^2}{m^2} - \frac{1}{q^2} + \Big(\frac{n}{m}+\frac{1}{q}\Big)^2 |m|\frac{1+(a/b)^{2|m|}}{1-(a/b)^{2|m|}}\bigg]_{r=a}\xi^2(a)\,. \tag{9.160}$$

Clearly, stability only depends on the profile $q(r)$, the relative wall position b/a and the mode numbers n and m.

For a flat current profile, $j_z = \text{const}$, also $q = \text{const}$ so that Eq. (9.158) reduces to an elementary equation that is easily solved:

$$(r^3\xi')' - (m^2-1)r\xi = 0 \quad \Rightarrow \quad \xi = r^{|m|-1} \quad \Rightarrow \quad \Big(\frac{r\xi'}{\xi}\Big)_{r=a} = |m|-1\,. \tag{9.161}$$

Then,

$$W = \frac{4\pi^2 a^2 B_0^2}{R_0}\,\frac{(nq+m)\,[\,nq+m-\text{sgn}(m)(1-(a/b)^{2|m|})\,]}{|m|q^2\,(1-(a/b)^{2|m|})}\,, \tag{9.162}$$

in agreement with the earlier derived expression (9.97) for the growth rates of external kink modes in a constant-pitch magnetic field. Considering negative values of m only (for $n > 0$, instability only occurs when $m < 0$, and vice versa for $n < 0$ and $m > 0$), the plasma becomes unstable in the ranges

$$\big[|m| - 1 + (a/b)^{2|m|}\big]/n < q < |m|/n\,, \tag{9.163}$$

that is everywhere when the wall is at infinity. With the wall at a finite distance, stable windows for q are found, but only for single values of m and n. When all values of m and n are considered, no stable window of finite size remains: a flat current profile is unacceptable in a tokamak.

(c) Internal and external kink modes of the 'straight tokamak' It remains to solve Eq. (9.158) for realistic choices of $q(r)$ and to show that stable operating windows in parameter space can be found. This problem has been solved by Shafranov [550] in a very satisfactory manner, as we will see. The current profile is now assumed to be peaked in the centre ($r = 0$) and to fall off to zero at the plasma boundary ($r = a$), so that the q-profile is monotonically increasing:

$$q_0 \le q(r) \le q_a \qquad (0 \le r \le a)\,. \tag{9.164}$$

We already know from Section 9.3.2(b), Eq. (9.90), that there is a simple (geometrical) cure for the external $|m| = 1$ kink instability (which is independent of the shape of the current profile), viz. the Kruskal–Shafranov limit:

$$q_a \equiv \frac{2\pi a^2 B_0}{R_0 I_z} > 1\,. \tag{9.165}$$

With this limitation of the total plasma current I_z, we have to determine whether a reasonable current density profile $j_z(r)$ can be found for which the other modes are stable as well. 'Reasonable' here means: a current profile that can be experimentally realized and maintained on the time scale needed for fusion (i.e. many orders of magnitude longer than the characteristic time scale of these ideal MHD instabilities!). This leads to further restrictions on the value of q in the centre, q_0, and on the shape of the q-profile.

First, consider the $n = 1$, $|m| = 1$ *internal kink mode*. If $q_0 < 1$ and $q_a > 1$, there is a singular point in the plasma at $r = r_s$, where $q(r_s) = 1$. According to Newcomb's theorem, at such a point, the 'small' solution may jump so that we get the following solution of the Euler–Lagrange equation (9.158):

$$\xi = CH(r_s - r)\,. \tag{9.166}$$

Inside the cylinder of radius $r = r_s$ the plasma is displaced rigidly with amplitude C, at $r = r_s$ the displacement jumps to zero (with a concomitant skin current), outside $r = r_s$ the displacement vanishes identically. Consequently, the full expression (9.160) for W vanishes: *the internal kink mode is marginally stable* in the 'straight tokamak' limit. At this point, one should realize that the expression (9.160) for W is actually only the leading order expression of the low-β-tokamak ordering, giving a growth rate $-\omega^2 \sim \epsilon^2$ when it is negative; see, e.g., Eq. (9.97). Exploiting the subscript 2 for this order, the result is a typical one for singular expansions, viz. that the leading order expression vanishes, $W_2 = 0$, so that the question immediately becomes: what is the sign of the next order, W_4? To answer that question, genuine toroidal contributions need to be calculated, which is beyond the present cylindrical analysis. However, the result of the toroidal calculation can be expressed simply as a condition on the q-profile: $W_4 > 0$ if $q_0 > 1$. Accidentally, the same condition is also obtained from a toroidal expansion of the Mercier criterion [437] for interchange modes in a low-β tokamak with confined pressure profile ($p' < 0$), typically involving mode numbers $|m|, |n| \gg 1$. Hence,

$$q_0 > 1 \tag{9.167}$$

is a condition for stability of the internal kink mode as well as the interchange modes in a low-β tokamak.

Large-amplitude sawtooth oscillations (oscillations of the central electron temperature with a period of 10–100 msec and leading to periodic loss of plasma confinement in the central region) occur in a tokamak when $q_0 < 1$. Hence, it was generally believed that the value of q_0 can never be smaller than 1 in a tokamak, until it was finally measured with the necessary precision by Soltwisch [556] and found to be in the range of 0.73–0.78 in a particular tokamak (TEXTOR) and since then in many other tokamaks as well! (Precision measurement of the current profile is still very much needed for laboratory plasmas, and even more so for astrophysical plasmas, in order for MHD spectroscopy (Section 7.2.4) to become a mature method to determine the internal characteristics of plasmas.) Clearly, internal $|m| = 1$ kink modes (strictly limiting the central value of q to $q_0 = 1$) should not be considered as a similar threat to tokamak confinement as the external $|m| = 1$ kink modes (limiting the boundary value of q to the Kruskal–Shafranov limit $q_a = 1$). However, since the present section is concerned with the leading order ideal MHD stability of the 'straight tokamak', we will maintain the condition (9.167) for consistency of the analysis. Satisfying it, for an increasing q-profile, the Kruskal–Shafranov limit (9.165) is automatically satisfied as well.

Next, consider the *external* $|m| \geq 2$ *kink modes*. To determine the marginal stability boundaries, one needs to solve the differential equation (9.158) for ξ and subject it to the boundary condition

$$\left(\frac{r\xi'}{\xi}\right)_{r=a} = S_0 \equiv \frac{m - nq}{m + nq} - |m|\frac{1 + (a/b)^{2|m|}}{1 - (a/b)^{2|m|}}, \tag{9.168}$$

where the expression S_0 is obtained from Eq. (9.160) by putting $W = 0$. This boundary condition brings in the destabilizing free-boundary motion of the plasma–vacuum interface, as well as the stabilizing reaction of image currents induced in the conducting wall by the perturbations. For a flat current distribution, this boundary value problem yields the unstable windows (9.163): a kind of worst case scenario. To avoid these instabilities, one needs to shape the q-profile, i.e. the current density profile.

For definiteness, we now assume current profile distributions and associated q-profiles, as investigated by Wesson [646]:

$$j_z = j_0(1 - r^2/a^2)^\nu \quad \Rightarrow \quad q = q_0\,\frac{(\nu + 1)\,r^2/a^2}{1 - (1 - r^2/a^2)^{\nu+1}}, \quad q_0 = \frac{2B_0}{R_0 j_0}. \tag{9.169}$$

This provides a minimum number of parameters to fix q_0 and the overall shape q_a/q_0 of the q-profile (shear of the magnetic field), expressed by the parameter ν:

$$q_a/q_0 = \nu + 1. \tag{9.170}$$

As illustrated in the rightmost part of Fig. 9.15, the parameter ν conveniently ranges from $\nu = 0$ (flat current), through $\nu = 1$ (parabolic current), to $\nu > 1$, corresponding to current distributions with an ever smaller gradient at the plasma boundary. The latter property guarantees stability with respect to the higher-$|m|$ kink modes. These current profiles also produce a smooth transition from the q-profile in the plasma to the $\hat{q}$-profile in the vacuum,

$$\hat{q}(r) = q_a(r/a)^2 \qquad (a \leq r \leq b), \tag{9.171}$$

following from Eq. (9.155) with $j_z = 0$.

The expressions (9.169) and (9.171) for the q-profiles permit one to precisely locate the position of the *rational magnetic surfaces* for each pair of mode numbers (m, n),

$$\text{in the plasma:} \qquad m + nq(r_\mathrm{s}) = 0 \qquad (0 \leq r_\mathrm{s} \leq a), \tag{9.172}$$

$$\text{in the vacuum:} \qquad m + n\hat{q}(\hat{r}_\mathrm{s}) = 0 \qquad (a < \hat{r}_\mathrm{s} \leq b). \tag{9.173}$$

Pairs (m, n) satisfying the condition (9.172), corresponding to a rational surface in the plasma, give rise to the interchange singularity that has been central to most of the stability analysis of the previous sections. In fact, the internal kink mode stability criterion (9.167) is just the condition for the absence of such a singularity for the $n = 1$, $m = -1$ internal kink mode in a plasma with an increasing q-profile. Because of the enormous simplification due to the low-β tokamak approximation, this was the only internal kink mode singularity we had to worry about. On the other hand, pairs (m, n) satisfying the condition (9.173), corresponding to a rational surface in the vacuum, do not give rise to a singularity at all: nothing prevents vacuum magnetic field lines from breaking and rejoining. (One would have to bring in the 'ghost' plasma of Section 9.2.2(b) to permit the induction of skin currents on the rational magnetic surfaces of the outer region (which would stabilize the external kink modes, as illustrated in Fig. 9.8).) It is a direct consequence of the

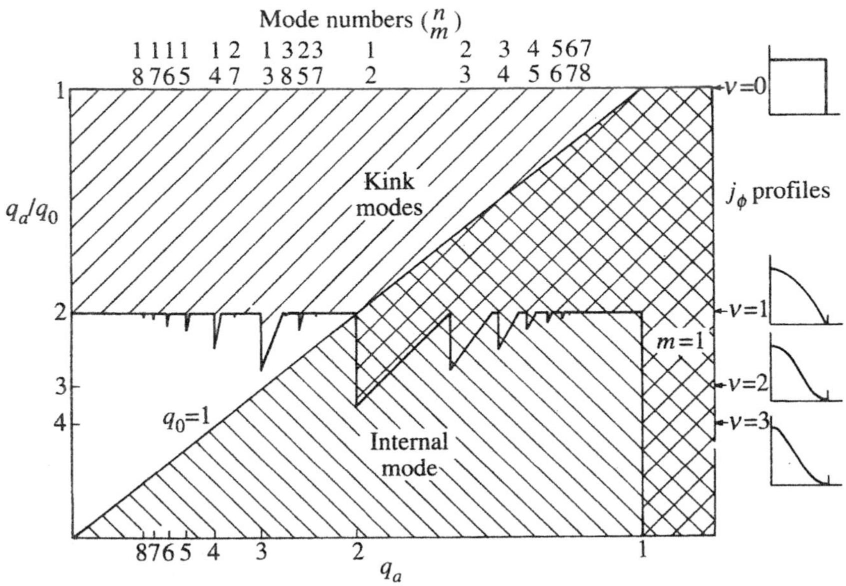

Fig. 9.15 Stability diagram for kink modes for the current distribution $j = j_0[1-(r/a)^2]^\nu$, without a conducting wall. The vertical axis measures the peaking of the current as given by q_a/q_0 $(= \nu + 1)$, and the horizontal coordinate is proportional to $1/q_a$ and therefore the total current. (In this figure, the definition of q is chosen such that the instabilities have positive m and n.) The $n = m = 1$ internal kink modes are unstable in the hatched area below the diagonal $q_0 = 1$. The $m = 1$ external kink modes are unstable for $q_a < 1$, and the $m \geq 2$ external kink modes are unstable in the upper hatched area. (From Wesson [646].)

stability criterion (9.163) (assuming that it is, in fact, the worst case scenario, which is confirmed by the numerical results) that potentially unstable pairs (m, n) with respect to external kink modes necessarily correspond to a rational surface in the vacuum:

$$q_a < |m|/n \equiv \hat{q}(\hat{r}_s)\,. \tag{9.174}$$

Hence, external kink modes are due to the absence of electrical conductivity of the outer region, called 'vacuum'. In a tokamak, such a region is produced by a limiter scraping off the outer plasma layers.

For the analytically minded, a trace of the influence of the interchange singularity $m + nq = 0$ is still to be found in the differential equation (9.158) and the first term of the expression (9.168) for S_0. The latter term is positive (destabilizing) and can be made large (not infinite) by choosing $m+nq$ small (but always $\neq 0$) so that there is a 'virtual singularity' located in the vacuum, but close to the plasma boundary. By means of analytic continuation of the plasma equations, one can make a Frobenius expansion around that point, compute the indices, and construct a solution where the large part, involving terms $\ln(r_s - r)$ and $(r_s - r)^{-1}$, still dominates at the plasma boundary when $r_s - a$ can be made small enough. Consequently, the logarithmic derivative $(r\xi'/\xi)_{r=a}$ is obtained as a local quantity, depending on derivatives of the current density or the q-profile at the plasma boundary only. For the equilibrium (9.169), the results of this analysis are as follows:

(1) if $j_z(a) \neq 0$, the plasma is kink unstable for all $|m|$ (this involves extension of the current density profile with a pedestal, showing that current peaking is not enough);

(2) if $\nu \leq 1$, the plasma is unstable for any wall position (since the local analysis already indicates instability);

(3) if $\nu > 1$, the plasma is only stable for $b < b_{\rm crit}$, where $b_{\rm crit}$ is to be determined by the global boundary value problem (in other words: the local analysis fails).

This analysis is tedious and requires severe testing of the limits of validity. It is clear that the borders of the kingdom where singularity reigns have been reached here: analysis is superior to numerics in the presence of singularities, in the absence of singularities the opposite holds.

In contrast, numerical solution of the boundary value problem (9.158), (9.168) for the external $|m| \geq 2$ kink modes is nearly trivial. The following steps are taken.

– Specify a monotonically increasing profile $q(r)$, a relative wall position b/a and a mode pair (m, n) in the unstable range (9.174);

– integrate the differential equation (9.158) for ξ from the magnetic axis, where $\xi(0) = 0$, to the plasma boundary, determine the logarithmic derivative $L \equiv (r\xi'/\xi)_{r=a}$, and compare it with the quantity S_0 defined in Eq. (9.168):

$$\text{if } L < S_0 \;\Rightarrow\; \text{unstable}\,, \quad \text{if } L = S_0 \;\Rightarrow\; \text{marginal}\,, \quad \text{if } L > S_0 \;\Rightarrow\; \text{stable}\,; \tag{9.175}$$

– to determine the marginal boundaries in parameter space, repeat this process for all pertinent mode pairs (m, n), iterating on the values of the parameters q_0 and q_a until $L = S_0$.

The differential equation for ξ is not only non-singular, but also admits no oscillatory solutions (the coefficients are positive definite). Hence, numerical integration is extremely fast and accurate. The only issue is the proper bookkeeping of the mode number pairs. Also, once the numerical scheme is established, extension with the computation of the growth rate of the instabilities is completely straightforward. This we leave for the Exercises [9.5] and [9.6].

The stability diagram of Fig. 9.15 is obtained by such a numerical procedure applied to an equilibrium with the $q(r)$-profile (9.169). Without the internal kink mode condition $q_0 > 1$, there is a fairly wide region of parameters where the $|m| \geq 2$ external kink modes are stable, roughly corresponding to a current profile with $q_a/q_0 \sim 3$ ($\nu \sim 2$) and $q_a > 2$. This confirms the earlier conclusion of Shafranov [550] that complete stability of the 'straight tokamak' may be obtained if the radial current density profile is sufficiently peaked on axis. Imposing the condition $q_0 > 1$ as well, the stable region becomes significantly smaller, although an operating region remains (the white, roughly triangular, region), which is, however, rather hard to control experimentally. It has already been indicated that relaxation of the internal kink mode condition may not be disastrous, whereas relaxation of the condition for the higher $|m|$ external kink modes (excluding the $|m| = 2$ and 3 modes) may just give rise to 'enhanced MHD activity'. Also, stabilization by the external wall has not been taken into account in Fig. 9.15. This will slow down the ideal MHD modes to the resistive skin time of the wall, where feedback stabilization techniques may be applied.

To assess the implications for tokamak operation, one should realize that a number of important effects have not been taken into account here: the presence of *plasma flow* in the edge region, the *finite conductivity* of that plasma, the influence of β, i.e. of the *toroidicity*. These are topics, necessarily involving advanced numerical solution techniques, addressed in Chapters 12–18.

9.5 Literature and exercises

Notes on literature

MHD spectral theory of cylindrical plasmas

- Hain and Lüst [276], in 'Zur Stabilität zylinder-symmetrischer Plasmakonfigurationen mit Volumenströmen', derive an ordinary differential equation to determine the growth rates of MHD instabilities in diffuse linear pinch configurations.
- Goedbloed [214](b), in 'Stabilization of magnetohydrodynamic instabilities by force-free magnetic fields – Linear pinch', rederives the (apparently forgotten) Hain–Lüst equation, generalizes it to $\gamma \neq 1$, and applies it to stabilization of external kink modes by a force-free magnetic field in the outer region.
- Grad [254], in 'Magnetofluid-dynamic spectrum and low shear stability', puts the subject in the context of spectral theory, points out the presence of four types of singularities (unfortunately associating a continuous spectrum with each of them), and applies the theory to demonstrate stability of a large class of low-shear systems.
- Appert, Gruber and Vaclavik [18], in 'Continuous spectra of a cylindrical magnetohydrodynamic equilibrium', derive the equivalent system of first order differential equations and demonstrate that the $D = 0$ singularities are apparent.

MHD stability of cylindrical plasmas

- Newcomb [461], in 'Hydromagnetic stability of a diffuse linear pinch', presents the classical treatise of the stability of cylindrical plasmas, with a careful exposition of the techniques needed from variational analysis.
- Voslamber and Callebaut [637], in 'Stability of force-free magnetic fields', present a beautiful example of the subtleties of MHD stability theory applied to the Lundquist field.
- Shafranov [550], in 'Hydrodynamic stability of a current-carrying pinch in a strong longitudinal field', demonstrates stability of low-β tokamaks with respect to all kink modes for realistic current distributions.
- Robinson [523], in 'High-β diffuse pinch configurations', analyses the different diffuse pinch configurations from the point of view of maximizing β, demonstrating the favourable properties of the shear profile (opposite to that of the tokamak) of the reversed field pinch.
- Goedbloed and Hagebeuk [234] numerically solve the Hain–Lüst equation to obtain the growth rates of instabilities of the Lundquist and constant pitch magnetic fields; Goedbloed and Sakanaka [240] and Sakanaka and Goedbloed [536] continue to construct the different classes of σ-stable diffuse linear pinch configurations this way.
- Freidberg, *Ideal MHD* [186], Chapter 11 on stability of one-dimensional configurations, presents the MHD stability theory of cylindrical plasmas, with applications to a variety of experimental fusion devices.
- Wesson, *Tokamaks* [647], Chapter 6 on MHD stability, contains the essential elements (with the simple diagrams that are Wesson's hallmark) of the review paper [646] on 'Magnetohydrodynamic stability of tokamaks'.

Exercises

[9.1] *Cylindrical force-free magnetic fields*

Force-free magnetic fields, $\mathbf{j} = \alpha\mathbf{B}$, with cylindrical symmetry can be fixed in several different ways, viz. by prescribing the function $\alpha(r)$, or $\mu(r)$, or $B^2(r)$. Each of these quantities has a physical meaning: α is the ratio of the current and the magnetic field, μ is the inverse pitch of the field lines and B^2 is the magnetic energy density (all disregarding constant factors). Derive the differential equations for B_z and B_θ for the three cases where either $\alpha(r)$, or $\mu(r)$, or $B^2(r)$ is prescribed. Comment on the conditions that have to be imposed for physical reality and on the advantages and disadvantages of the three prescriptions. Construct as many explicit solutions as you can for one of those cases. (They are useful as explicit equilibria when you want to check stability calculations.) Also, using the equilibrium relations, find relations between the three quantities.

[9.2] *Newcomb's stability equations*

Derive Newcomb's Euler–Lagrange equation $(f_0\xi')' - g_0\xi = 0$ from Hain–Lüst's equation $(P\chi')' - Q\chi = 0$ by inserting the value $\omega^2 = 0$ in the latter and transforming from the variable $\chi \equiv r\xi$ to ξ. Pay attention to how the derivative term is transformed (recall that it is associated with the perturbed total pressure).

- Is marginal stability analysis equivalent to the variational analysis associated with the energy principle? Comment on the role of singularities.
- Derive the one-dimensional form of the energy principle for a cylindrical plasma with length L and radius a from the Euler–Lagrange equation (i.e. the other way around with respect to the usual order) and get rid of the boundary terms.
- Expand about a point where $F = mB_\theta/r + kB_z = 0$ and derive Suydam's local stability criterion from the indicial equation.
- What type of solution is associated with the violation of Suydam's criterion? How is it related to a genuine eigenfunction?

[9.3] *WKB solution of the Hain–Lüst equation*

Consider a cylindrical plasma with weak inhomogeneity so that the Hain–Lüst equation, $(P\chi')' - Q\chi = 0$, can be solved by means of the WKB method. Writing

$$\chi(r) = p(r) \exp\left[\mathrm{i} \int q(r)\, dr \right],$$

the expressions $p(r)$ and $q(r)$ are determined by requiring that the solution be correct to leading order in the inhomogeneity. This yields

$$p \approx (-PQ)^{-1/4}, \qquad q \approx (-Q/P)^{1/2},$$

where we have to demand that $|qL| \gg 1$ in order for the WKB approximation to be valid. Here, L is the scale length for the inhomogeneities. This yields a local dispersion equation relating ω^2 and the local radial wave number q,

$$q^2 = -Q/P,$$

which is a quintic in ω^2. (No, this has nothing to do with the fact that the discrete spectrum consists of five sub-spectra!) This equation may be solved in the neighbourhood of the Alfvén and slow continua, when these are sufficiently far apart. Show that this gives cluster spectra of the form

$$\rho\omega^2 \approx F^2 - \frac{A}{q^2}, \qquad \rho\omega^2 \approx \left(\frac{\gamma p}{\gamma p + B^2} \right)^2 F^2 - \frac{S}{q^2}.$$

Determine the expressions for A and S. Compare them with the exact cluster conditions of Section 9.3.3 and comment. Discuss the validity of the WKB approximation.

[9.4] *Instabilities of shearless magnetic fields*

In Section 9.4.2(b) the pressure-driven instabilities of plasmas with shearless magnetic fields were analysed. Using the same techniques, investigate the residual instabilities that occur when the pressure gradient is much smaller, $\Pi \equiv a^2 p'/(2r^2B^2) \sim \epsilon^4$. The stability threshold for these modes can be found from the condition that there should be no real values of $k_\parallel$ where marginal stability occurs. Show that this gives the following local criterion for high-m modes:

$$p' > \frac{2B_\theta^4}{rm^2B^2}.$$

For $m \to \infty$, this criterion transforms into Suydam's criterion for shearless magnetic fields. For low-m modes, the criterion is much more stringent, in particular when B_θ is large. Derive the dispersion equation for these modes, called quasi-kink modes. Show that they are unstable for values of $k_\parallel$ on only one side of $k_\parallel = 0$. Derive the expression for the maximum growth rate and estimate the time scale.

[9.5]* *Marginal stability of external kink modes in a tokamak*

Write a computer program constructing the Wesson diagram of Fig. 9.15 from the boundary value problem

(9.158), (9.168) outlined in Section 9.4.4. Use the criterion $L = S_0$ of Eq. (9.175) to determine the stable regions. Figure out how to iterate on the parameters such that rapid convergence to marginal stability is obtained.

- When you have obtained agreement with the figure, extend the program with the finite wall positions $b/a = 1.1, 1.2, 1.5, 2.0$. Comment on the results.
- Run the program for $\nu = 0.5$ and try to stabilize by moving the wall in. What happens? Plot the solution of the function ξ.

[9.6]* *Growth rates of external kink modes in a tokamak*

Assuming $\rho = \rho_0 = $ const, and normalizing the eigenvalue $\bar{\omega}^2 \equiv (\rho_0 R_0^2/B_0^2)\omega^2$, derive the equation of motion for a 'straight tokamak' from the low-frequency limit (9.47) of the Hain–Lüst equation:

$$\frac{d}{dr}\left[r^3 \left\{ \bar{\omega}^2 - (n + m/q)^2 \right\} \frac{d\xi}{dr} \right] - (m^2 - 1) r \left\{ \bar{\omega}^2 - (n + m/q)^2 \right\} \xi = 0 \,.$$

Also show that this is equivalent to the more complete expression (9.50) when $\rho = $ const. Derive the associated plasma–vacuum boundary condition from Eq. (9.60):

$$\left. \frac{r\xi'}{\xi} \right|_{r=a} = S \equiv \frac{1}{(n + m/q)^2 - \bar{\omega}^2} \left[\bar{\omega}^2 - n^2 + m^2/q^2 - (n + m/q)^2 |m| \frac{1 + (a/b)^{2|m|}}{1 - (a/b)^{2|m|}} \right] .$$

- Noting that this boundary value problem only requires some additional terms $\bar{\omega}^2 < 0$, modify the marginal stability program of exercise [9.5] to calculate the growth rates of the external kink modes.
- Pick a reasonable cutoff value for $\sigma^2 \equiv -\bar{\omega}^2$ and construct the σ-stability contour diagram for that value. What do you conclude from the result?

10

Initial value problem and wave damping*

10.1 Implications of the continuous spectrum*

We now embark on the solution of the initial value problem (IVP) as outlined in Section 6.3.2. In general, the IVP in ideal MHD arises when one wishes to solve the equation of motion (6.23), which we now write as

$$\rho^{-1}\mathbf{F}(\boldsymbol{\xi}) - \frac{\partial^2 \boldsymbol{\xi}}{\partial t^2} = 0 \,, \tag{10.1}$$

for arbitrary initial values $\boldsymbol{\xi}_{\rm i}$,

$$\boldsymbol{\xi}(\mathbf{r}; t\!=\!0) = \boldsymbol{\xi}_{\rm i}(\mathbf{r}) \,. \tag{10.2}$$

Obviously, the spatial boundary value problem (BVP) must be solved simultaneously, but we have studied that part of the problem for inhomogeneous plasmas sufficiently in Chapters 7 and 9. In the present chapter, we wish to investigate how such plasmas evolve in time. What is the role of the intricate MHD spectra derived in the previous chapters? How do the discrete and continuous parts of those spectra enter the temporal description of an arbitrarily excited plasma?

The IVP for macroscopic plasmas is a natural analogue of the IVP for microscopic plasmas first analysed in 1946 by Landau [393], who derived the surprising result that electrostatic waves ('plasma oscillations'), in the absence of any dissipation mechanism, are damped (Section 2.3.3). The mechanism is called Landau damping and, since it is associated with a microscopic description in terms of the velocity space distribution function, it was thought that such dissipationless damping processes are restricted to the microscopic picture. However, in 1971 Sedláček [546] showed that a macroscopic description of plasma oscillations of a cold plasma also leads to dissipationless damping, due to inhomogeneity in ordinary space (whereas Landau damping is due to inhomogeneity in velocity space). Next, Tataronis and Grossmann [588] showed that ideal MHD waves actually also exhibit damping, due to spatial inhomogeneities of the plasma.

The bottom line appears to be that, *in conservative systems with a continuous spectrum* (whether it concerns a quantum mechanical system of scattering particles, microscopic plasma oscillations of a collection of charged particles, or macroscopic Alfvén waves of a confined plasma), *damping of the initial perturbations occurs through redistribution over the different improper continuum modes*. In plasmas described by MHD, the dissipationless development, on an ideal MHD time scale, of very localized structures builds up large spatial gradients of the macroscopic variables that will enormously enhance the genuine dissipation rate associated with the resistivity (Section 4.3.1). We will call the ideal MHD part of this evolution 'quasi-dissipation'. Its practical consequences for resonant wave absorption and heating, in particular for astrophysical plasmas, will be discussed in Chapter 11. The latter chapter deals with the implications of the continuous spectrum for systems that are actively excited (described by model III of Section 4.6.1). In the present chapter, we will deal with the implications of the continuous spectrum for IVP and wave damping for passive systems, left to themselves.

10.2 Initial value problem★

To solve the equation of motion (10.1) with the initial data (10.2), one may exploit the techniques of forward and inverse Laplace transformation that were already introduced in Section 6.3, and we will repeat the necessary equations for convenience of reading. The forward Laplace transformation to the complex ω-plane introduces the initial data in the equations, whereas the inverse Laplace transformation back to the time domain then has to deal with the special values of ω that belong to the spectrum. In this manner, the contributions of the different parts of the spectrum become manifest. The forward Laplace transformation

$$\hat{\boldsymbol{\xi}}(\mathbf{r};\omega) \equiv \int_0^\infty \boldsymbol{\xi}(\mathbf{r};t)\, \mathrm{e}^{\mathrm{i}\omega t}\, dt \tag{10.3}$$

transforms the homogeneous equation (10.1) into the inhomogeneous equation

$$(\rho^{-1}\mathbf{F} + \omega^2 \mathbf{I}) \cdot \hat{\boldsymbol{\xi}}(\mathbf{r};\omega) = \mathrm{i}\omega \boldsymbol{\xi}_{\mathrm{i}}(\mathbf{r}) - \dot{\boldsymbol{\xi}}_{\mathrm{i}}(\mathbf{r}) \equiv \mathrm{i}\omega \mathbf{X}(\mathbf{r})\,, \tag{10.4}$$

where the initial displacement $\boldsymbol{\xi}_{\mathrm{i}}$ and the initial velocity $\dot{\boldsymbol{\xi}}_{\mathrm{i}}$ are absorbed in the definition of the vector $\mathbf{X}$. The formal solution of the inhomogeneous equation (10.4) involves the inversion of the differential operator on the left hand side:

$$\hat{\boldsymbol{\xi}}(\mathbf{r};\omega) = (\rho^{-1}\mathbf{F} + \omega^2 \mathbf{I})^{-1} \cdot \mathrm{i}\omega \mathbf{X}(\mathbf{r};\omega)\,. \tag{10.5}$$

Clearly, the construction of the inverse operator $(\rho^{-1}\mathbf{F} + \omega^2\mathbf{I})^{-1}$ will be one of the major tasks of this section. This operator is *the resolvent operator* for this problem, which is the most compact expression of the different spectral alternatives, as we saw in Section 6.3.1. Not surprisingly, since the resolvent operator is the inverse of a differential operator, it will turn out to be an integral operator involving Green's functions. To obtain the actual solution of the IVP, the inverse Laplace transformation is applied to Eq. (10.5):

$$\boldsymbol{\xi}(\mathbf{r};t) = \frac{1}{2\pi} \int_{\mathrm{i}\nu_0 - \infty}^{\mathrm{i}\nu_0 + \infty} \hat{\boldsymbol{\xi}}(\mathbf{r};\omega)\, \mathrm{e}^{-\mathrm{i}\omega t}\, d\omega = \frac{1}{2\pi} \int_{\mathrm{C}} (\rho^{-1}\mathbf{F} + \omega^2)^{-1} \cdot \mathrm{i}\omega \mathbf{X}(\mathbf{r};\omega) \mathrm{e}^{-\mathrm{i}\omega t}\, d\omega\,, \tag{10.6}$$

as we already saw in Section 6.3.2. Here, judicious deformation of the contour C in the complex ω-plane (Fig. 6.8) will reveal the characteristic temporal behaviour of the different contributions of the spectrum.

10.2.1 Reduction to a one-dimensional representation★

Let us now specify the plasma to be a plane gravitating slab, as considered in Chapter 7. (We follow the highly delayed publication of Goedbloed [221], based on an unpublished memorandum of 1973, which contained the construction of the resolvent operator in the apparently singular ranges $D = 0$

for the analogous case of cylindrical geometry.) The direction of inhomogeneity is represented by the x-coordinate and the dependence on the symmetry coordinates y and z may be eliminated by considering Fourier modes $\hat{f}_{k_y k_z}(x;\omega)\exp \mathrm{i}(k_y y + k_z z)$, where we omit the subscripts k_y and k_z and suppress the exponential factor from now on. It is expedient again to exploit the field line projection of Section 7.3.2, with vector components in the directions $\mathbf{e}_x, \mathbf{e}_\perp, \mathbf{e}_\parallel$:

$$\hat{\boldsymbol{\xi}} \equiv (\hat{\xi}, \hat{\eta}, \hat{\zeta})^{\mathrm{T}}, \qquad \mathbf{X} \equiv (X, Y, Z)^{\mathrm{T}}. \tag{10.7}$$

Exploiting the same reductions as in the derivation of the wave equation (7.78), this leads to the following representation of the inhomogeneous problem (10.4):

$$(\mathbf{F} + \rho\omega^2 \mathbf{I}) \cdot \hat{\boldsymbol{\xi}}(x;\omega) = \mathrm{i}\rho\omega \mathbf{X}(x), \tag{10.8}$$

where the matrix $\mathbf{F}$ is the same as in Eq. (7.78). Our next task is the inversion of Eq. (10.8), producing the explicit solution of $\hat{\boldsymbol{\xi}}$ as a function of $\mathbf{X}$ as formally expressed by Eq. (10.5).

The elimination of the tangential components $\hat{\eta}$ and $\hat{\zeta}$ in terms of $\hat{\xi}$ and the initial data takes the following form:

$$\hat{\eta} = S\hat{\xi}' + K\hat{\xi} + \mathrm{i}\rho\omega(HY + IZ), \qquad \hat{\zeta} = A\hat{\xi}' + L\hat{\xi} + \mathrm{i}\rho\omega(IY + JZ), \tag{10.9}$$

where the expressions S, A, K, L correspond to the homogeneous Eqs. (7.79):

$$S \equiv \frac{k_\perp(b^2 + c^2)}{D}(\omega^2 - \omega_S^2), \qquad K \equiv -\frac{k_\perp g}{D}\omega^2,$$
$$A \equiv \frac{k_\parallel c^2}{D}(\omega^2 - \omega_{\mathrm{A}}^2), \qquad L \equiv -\frac{k_\parallel g}{D}(\omega^2 - k_0^2 b^2), \tag{10.10}$$

and new expressions H, I, J enter with the initial data:

$$H \equiv \frac{\omega^2 - k_\parallel^2 c^2}{\rho D}, \quad I \equiv \frac{k_\parallel k_\perp^2 c^2}{\rho D}, \quad J \equiv \frac{\omega^2 - k_\parallel^2 b^2 - k_\perp^2(b^2 + c^2)}{\rho D}. \tag{10.11}$$

Here, S and A introduce the slow and Alfvén continuum frequencies:

$$\omega_{\mathrm{S}}^2(x) \equiv k_\parallel^2 \frac{b^2 c^2}{b^2 + c^2}, \qquad \omega_{\mathrm{A}}^2(x) \equiv k_\parallel^2 b^2. \tag{10.12}$$

The elimination of the tangential components involves the determinant D of the four lower right corner elements of the matrix of Eq. (10.8):

$$D(x) \equiv \omega^4 - k_0^2(b^2 + c^2)\,\omega^2 + k_0^2 k_\parallel^2 b^2 c^2 = (\omega^2 - \omega_{\mathrm{s0}}^2)(\omega^2 - \omega_{\mathrm{f0}}^2), \tag{10.13}$$

where the notation of the local frequencies,

$$\omega_{\mathrm{s0,f0}}^2 \equiv \tfrac{1}{2}k_0^2(b^2 + c^2)\left[1 \pm \sqrt{1 - \frac{4k_\parallel^2 b^2 c^2}{k_0^2(b^2 + c^2)^2}}\right], \tag{10.14}$$

indicates a relationship to the slow and fast wave motion. This brings the issue of the nature of the apparent $D = 0$ singularities into focus. By now, we know that those frequency ranges do not constitute continuous spectra, in contrast to the genuine slow and Alfvén continua, $\{\omega_{\mathrm{S}}^2(x)\}$ and $\{\omega_{\mathrm{A}}^2(x)\}$. The latter continua manifest themselves in the numerators of the expressions (10.10) for S and A, multiplying the derivative $\hat{\xi}'$ in the expressions (10.9) for $\hat{\eta}$ and $\hat{\zeta}$. This is more decisive for the local dynamics than the apparent $D = 0$ singularities, which all turn out to cancel in the final analysis.

Substitution of $\hat{\eta}$ and $\hat{\zeta}$ into Eq. (10.8) leads to an inhomogeneous differential equation for $\hat{\xi}$:

$$(P\hat{\xi}')' - Q\hat{\xi} = R \equiv \mathrm{i}\omega\Big\{\rho X - \big[\rho(SY + AZ)\big]' + \rho(KY + LZ)\Big\}\,, \tag{10.15}$$

where

$$P \equiv \frac{N}{D}\,, \qquad N(x) \equiv \rho(b^2 + c^2)(\omega^2 - \omega_{\mathrm{A}}^2)(\omega^2 - \omega_{\mathrm{S}}^2)\,, \tag{10.16}$$

$$Q \equiv -\bigg[\rho(\omega^2 - k_\parallel^2 b^2) + \rho' g - k_0^2 \rho g^2\,\frac{\omega^2 - k_\parallel^2 b^2}{D} - \Big\{\rho g\,\frac{\omega^2(\omega^2 - k_\parallel^2 b^2)}{D}\Big\}'\bigg]\,.$$

Equation (10.15) is solved by the integral

$$\hat{\xi}(x) = \int_{x_1}^{x_2} G(x, x') R(x')\, dx'\,, \tag{10.17}$$

where $G(x, x')$ is the *Green's function* satisfying the equation

$$\frac{d}{dx}\bigg[P(x)\frac{dG(x, x')}{dx}\bigg] - Q(x)\, G(x, x') = \delta(x - x')\,, \tag{10.18}$$

subject to the boundary conditions

$$G(x{=}x_1, x') = G(x{=}x_2, x') = 0\,,$$
$$[\![G]\!]_{x=x'} = 0\,, \qquad \bigg[\!\!\bigg[P\frac{\partial G}{\partial x}\bigg]\!\!\bigg]_{x=x'} = 1\,. \tag{10.19}$$

The double brackets indicate jumps, $[\![f]\!] \equiv \lim_{\epsilon\to 0}[f(x' + \epsilon) - f(x' - \epsilon)]$. See Fig. 10.1(a) of Section 10.3.1 for a picture of such a Green's function.

For simplicity, we here consider a slab $[x_1, x_2]$, assuming weak inhomogeneity or a slab thin enough that the frequency ranges $\{\omega_{\mathrm{S}}^2\}$, $\{\omega_{\mathrm{A}}^2\}$, $\{\omega_{\mathrm{s0}}^2\}$, $\{\omega_{\mathrm{f0}}^2\}$ do not overlap. This is possible since the frequencies are well ordered when considered for fixed position x:

$$0 \le \omega_{\mathrm{S}}^2 \le \omega_{\mathrm{s0}}^2 \le \omega_{\mathrm{A}}^2 \le \omega_{\mathrm{f0}}^2 \le \omega_{\mathrm{F}}^2 = \infty\,. \tag{10.20}$$

Overlapping could hardly be avoided in a cylinder when the origin is included since $\omega_{\mathrm{s0}}^2 \to \omega_{\mathrm{S}}^2$ and $\omega_{\mathrm{f0}}^2 \to \infty$ there. This does not happen in a plane slab. Hence, $\{\omega_{\mathrm{f0}}^2\}$ is extremely dependent on the choice of coordinates (another clear indication that the $D = 0$ singularities cannot represent continuous spectra).

The Green's function is built from solutions of the homogeneous equations

$$(PU_i')' - QU_i = 0 \qquad (i = 1, 2)\,, \tag{10.21}$$

producing a left component $U_1(x)$ satisfying the left boundary condition $U_1(x_1) = 0$ and a right component $U_2(x)$ satisfying the right boundary condition $U_2(x_2) = 0$. The formal solution of Eq. (10.18) then reads:

$$G(x, x'; \omega^2) = \Gamma(x, x'; \omega^2)/\Delta(\omega^2)\,, \tag{10.22}$$

where

$$\begin{aligned}\Gamma(x,x';\omega^2) &\equiv U_1(x;\omega^2)U_2(x';\omega^2)H(x'-x)+U_1(x';\omega^2)U_2(x;\omega^2)H(x-x')\,,\\ \Delta(\omega^2) &\equiv P(x;\omega^2)\left[U_1(x;\omega^2)U_2'(x;\omega^2)-U_1'(x;\omega^2)U_2(x;\omega^2)\right].\end{aligned}\tag{10.23}$$

For the study of the initial value problem with respect to the response of the Alfvén and slow continua $N(\omega^2)=0$, the consideration of the logarithmic singularities of the components $U_{1,2}(x)$ is imperative. They lead to corresponding branch cuts of Γ and Δ when considered as a function of complex ω, due to contributions like $\ln(\omega^2-\omega_{\rm A}^2(x))$. The analysis of these singularities is fully analogous to that of cold plasma oscillations of an inhomogeneous plasma which has been given in great detail by Sedláček [546]. His analysis explicitly shows that the zeros of the *conjunct* $\Delta(\omega^2)$ constitute the discrete spectrum whereas the branch cuts of Γ and Δ constitute the continuous spectrum. The corresponding analysis for Alfvén waves has been given by Tataronis [587].

In Sections 10.3 and 10.4, we will consider explicit solutions of the Green's function for a simplified model. Here, we keep the analysis general to stay close to the mentioned Ref. [221] on which this section is based. The goal is to construct the full three-dimensional response for MHD waves in the frequency range $D(\omega^2)=0$ and to show that no singularities of G occur there other than possible poles $\Delta(\omega^2)=0$ corresponding to discrete modes. Expansion of the coefficients P and Q of Eq. (10.18) in terms of the distance $s\equiv x-x_0$ from an apparent singularity $D(x_0)=0$ satisfies the special property, due to J. M. Greene [258],

$$P=\frac{P_0}{s}+\cdots\,,\qquad Q=\frac{Q_0}{s^2}+\frac{Q_1}{s}+\cdots\quad\Rightarrow\quad Q_0^2-P_0Q_1=0\,,\tag{10.24}$$

encountered in Eq. (7.130). This guarantees absence of logarithmic singularities so that both independent solutions are regular close to the apparent singularity:

$$U_s=s^2+\cdots\,,\qquad U_\ell=1+\alpha s+\cdots\,,\qquad \alpha\equiv -Q_0/P_0=\left(\frac{g\omega^2}{(b^2+c^2)(\omega^2-\omega_S^2)}\right)_0.\tag{10.25}$$

Hence, in the mentioned frequency range, there are no logarithmic singularities, no branch cuts, but an almost forbidding number of occurrences of a vanishing denominator D in the coefficients (10.10)–(10.11). We will show that this is not just an apparent obstacle, but even necessary for the response to the genuine continua.

10.2.2 Restoring the three-dimensional picture*

Having solved for the normal response $\hat\xi(x)$, expressed by Eq. (10.17) with the Green's function $G(x,x')$ given by Eqs. (10.22)–(10.23), it remains to construct the tangential response $\hat\eta(x)$, $\hat\zeta(x)$ according to Eqs. (10.9). This part of the analysis is usually underexposed since it does not require the solution of additional differential equations but just involves the substitution of the normal solutions obtained. However, this part is really the most significant in this case. We first complete the response (10.17) for $\hat\xi(x)$ by partially integrating the derivative term on the RHS of Eq. (10.15):

$$\hat\xi={\rm i}\omega\int\Big\{GX+\Big[\Big(S\frac{\partial}{\partial x'}+K\Big)G\Big]Y+\Big[\Big(A\frac{\partial}{\partial x'}+L\Big)G\Big]Z\Big\}\rho(x')\,dx'\,.\tag{10.26}$$

Here, the functions in the integrand are considered as functions of x' and the square brackets indicate that the derivative $\partial/\partial x'$ is to be taken on $G(x,x')$ only. This expression reveals two special normal operators, already encountered in Eq. (10.9) for the tangential components, but now they act on the Green's function itself:

$$\tilde{S}(x) \equiv S(x)\frac{\partial}{\partial x} + K(x)\,, \qquad \tilde{A}(x) \equiv A(x)\frac{\partial}{\partial x} + L(x)\,. \tag{10.27}$$

Equation (10.26) provides the complete normal response $\hat{\xi}$ in terms of the initial data X, Y, Z. In turn, substitution of this expression in Eqs. (10.9) provides the response of the tangential components $\hat{\eta}$ and $\hat{\zeta}$ in terms of the initial data. The latter involves the operation of $\tilde{S}$ and $\tilde{A}$ on the Green's function twice in succession.

We now compose *the full three-dimensional response*:

$$\hat{\boldsymbol{\xi}}(\mathbf{r};\omega^2) = \mathrm{i}\omega \int dx'\rho(x')\,\mathbf{G}(\mathbf{r},\mathbf{r}';\omega^2)\cdot\mathbf{X}(\mathbf{r}';\omega)\,, \tag{10.28}$$

where we recall that we have suppressed the Fourier factors providing the dependence on the coordinates y and z and, consequently, also the corresponding factors in $\mathbf{G} \sim \delta(y-y')\delta(z-z')$ since integration over those coordinates is trivial. In terms of the field line projection, this response may be written as

$$\Big(\hat{\xi}(x),\hat{\eta}(x),\hat{\zeta}(x)\Big)^{\mathrm{T}} = \mathrm{i}\omega \int dx'\rho(x')\,\mathbf{G}(\mathbf{r},\mathbf{r}';\omega^2)\cdot\Big(X(x'),Y(x'),Z(x')\Big)^{\mathrm{T}}\,, \tag{10.29}$$

where *the Green's dyadic* $\mathbf{G}$ is given by

$$\mathbf{G} = \begin{pmatrix} 1 & \tilde{S}(x') & \tilde{A}(x') \\ \tilde{S}(x) & \tilde{S}(x)\tilde{S}(x') & \tilde{S}(x)\tilde{A}(x') \\ \tilde{A}(x) & \tilde{A}(x)\tilde{S}(x') & \tilde{A}(x)\tilde{A}(x') \end{pmatrix} G(x,x') + \begin{pmatrix} 0 & 0 & 0 \\ 0 & H(x) & I(x) \\ 0 & I(x) & J(x) \end{pmatrix} \delta(x-x')\,, \tag{10.30}$$

with the one-dimensional Green's function $G(x,x')$ given by Eqs. (10.22)–(10.23). This provides the three-dimensional response, but, since all constituent functions S, A, K, L, H, I, J are proportional to $1/D$, the representation is full of apparent singularities. However, we should not lose sight of the essential feature here, viz. that the operators $\tilde{S}$ and $\tilde{A}$ separately 'kill' the localized response associated with large normal derivatives ($\partial/\partial x \to \infty$), for the two tangential components, leaving the components $\hat{\eta}$ and $\hat{\zeta}$ as the dominant responses for the Alfvén frequencies ω_{A} and the slow frequencies ω_{S}, respectively.

With the representation (10.30), we have the key to open up the structure of the Green's dyadic $\mathbf{G}$ to finally produce the complete three-dimensional response in a form that is free of apparent singularities and also exhibits the genuine ones. To that end, we study $\mathbf{G}$ in the neighbourhood of the point $x = x_0$ (and $x' = x_0'$) where $D(x) = 0$, of course, exploiting the fact that the constituent functions $U_{1,2}(x)$ of the one-dimensional Green's function $G(x,x')$ are regular there, according to Eqs. (10.25). We now transform to a three-dimensional picture exploiting solution *vectors* $\mathbf{U}_{1,2}(x)$ with components $(U_{1,2}, V_{1,2}, W_{1,2})$, where the tangential components satisfy the homogeneous counterparts of Eqs. (10.9):

$$\mathrm{V}_{1,2}(x) \equiv \tilde{S}(x)U_{1,2}\,, \qquad W_{1,2}(x) \equiv \tilde{A}(x)U_{1,2}\,. \tag{10.31}$$

In the neighbourhood of $x = x_0$ we then have

$$\begin{aligned} U &= C_s U_s + C_\ell U_\ell = C_s(s^2 + \cdots) + C_\ell(1 + \alpha s + \cdots)\,, \\ V &\equiv \tilde{S} U = C_s(2Ss + Ks^2 + \cdots) + C_\ell(S\alpha + K + \cdots)\,, \\ W &\equiv \tilde{A} U = C_s(2As + Ls^2 + \cdots) + C_\ell(A\alpha + L + \cdots)\,, \end{aligned} \tag{10.32}$$

where the leading order terms are all finite due to the equalities

$$(K/S)_0 = (L/A)_0 = -\alpha\,. \tag{10.33}$$

Hence, the operators $\tilde{S}$ and $\tilde{A}$ produce the finite expressions (10.32) for $V_{1,2}$ and $W_{1,2}$, which may then be used to define the three-dimensional generalization of the function $\Gamma(x, x'; \omega^2)$ defined in Eq. (10.23):

$$\Gamma_{UV}(x, x'; \omega^2) \equiv U_1(x; \omega^2) V_2(x'; \omega^2) H(x' - x) + V_1(x'; \omega^2) U_2(x; \omega^2) H(x - x')\,, \tag{10.34}$$

and similarly for Γ_{UW}, Γ_{VU}, etc., whereas $\Gamma_{UU} \equiv \Gamma$ is defined in Eq. (10.23).

With the definitions (10.34) of the Γs, the different linear contributions to the Green's dyadic $\mathbf{G}$ given in Eq. (10.30) may be decomposed as follows:

$$\begin{aligned} \tilde{S}(x)\Gamma(x, x') &= \big[\tilde{S}U_1(x)\big] U_2(x') H(x' - x) + U_1(x')\big[\tilde{S}U_2(x)\big] H(x - x') \\ &\equiv \Gamma_{VU}(x, x')\,, \\ \tilde{S}(x')\Gamma(x, x') &\equiv \Gamma_{UV}(x, x')\,, \qquad \text{etc.} \end{aligned} \tag{10.35}$$

Here, the square brackets are put around the first factor only because operation of the normal derivatives on the Heaviside functions does not produce additional terms since G is continuous according to the boundary conditions (10.19)(a). However, the boundary conditions (10.19)(b) show that the derivative of G is not continuous, so that the quadratic contributions to the Green's dyadic produce decompositions with additional δ-functions:

$$\begin{aligned} \tilde{S}(x)\tilde{S}(x')\Gamma(x, x') &= \big[\tilde{S}U_1(x)\big]\big[\tilde{S}U_2(x')\big] H(x' - x) + \big[\tilde{S}U_1(x')\big]\big[\tilde{S}U_2(x)\big] H(x - x') \\ &\qquad - S^2\big(U_1' U_2 - U_1 U_2'\big)\delta(x - x') \\ &\equiv \Gamma_{VV}(x, x') - (S^2/P)\Delta\,\delta(x - x')\,, \\ \tilde{S}(x)\tilde{A}(x')\Gamma(x, x') &= \Gamma_{VW}(x, x') - (SA/P)\Delta\,\delta(x - x')\,, \\ \tilde{A}(x)\tilde{S}(x')\Gamma(x, x') &= \Gamma_{WV}(x, x') - (SA/P)\Delta\,\delta(x - x')\,, \\ \tilde{A}(x)\tilde{A}(x')\Gamma(x, x') &= \Gamma_{WW}(x, x') - (A^2/P)\Delta\,\delta(x - x')\,. \end{aligned} \tag{10.36}$$

The δ-function contributions combine with those already present in the second matrix of Eq. (10.30) according to the equalities

$$-\frac{S^2}{P} + H = \frac{1}{\rho(\omega^2 - \omega_{\rm A}^2)}\,, \qquad -\frac{SA}{P} + I = 0\,, \qquad -\frac{A^2}{P} + J = \frac{1}{\rho(\omega^2 - \omega_{\rm S}^2)}\,. \tag{10.37}$$

Now, insert the expressions (10.35)–(10.37) into Eq. (10.30) for $\mathbf{G}$ and watch the miracle happen:

$$\mathbf{G}(x,x';\omega^2) = \frac{1}{\Delta(\omega^2)}\begin{pmatrix} \Gamma_{UU} & \Gamma_{UV} & \Gamma_{UW} \\ \Gamma_{VU} & \Gamma_{VV} & \Gamma_{VW} \\ \Gamma_{WU} & \Gamma_{WV} & \Gamma_{WW} \end{pmatrix} + \begin{pmatrix} 0 & 0 & 0 \\ 0 & \dfrac{\delta(x-x')}{\rho(\omega^2-\omega_A^2)} & 0 \\ 0 & 0 & \dfrac{\delta(x-x')}{\rho(\omega^2-\omega_S^2)} \end{pmatrix}. \quad (10.38)$$

Since the Γs defined in Eq. (10.34) are all finite for $D=0$, this representation is manifestly free of apparent singularities. More importantly, in addition to the implicit presence in the branch cuts of the Γs and Δ, the genuine Alfvén and slow singularities $N=0$ are now distinctly present on the diagonal. Starting from this dyadic form of $\mathbf{G}$, the resolution of the identity, the expansion in eigenfunctions, and the question of completeness of the MHD spectrum can be established, in complete analogy with the analysis of Sedláček [546].

We have now completed the construction of the resolvent operator $(\mathbf{F}+\rho\omega^2\mathbf{I})^{-1}$ connecting the Laplace transformed variable $\hat{\boldsymbol{\xi}}$ with the initial data $\mathbf{X}$,

$$\hat{\boldsymbol{\xi}} = (\mathbf{F}+\rho\omega^2\mathbf{I})^{-1}\cdot \mathrm{i}\rho\omega\mathbf{X}\,, \quad (10.39)$$

explicitly given by Eq. (10.28) with the Green's dyadic $\mathbf{G}$ as given by Eq. (10.38). The solution of the initial value problem involves the inverse Laplace transform:

$$\begin{aligned} \boldsymbol{\xi}(\mathbf{r};t) &= \frac{1}{2\pi}\int_{\mathrm{i}\nu_0-\infty}^{\mathrm{i}\nu_0+\infty}\hat{\boldsymbol{\xi}}(\mathbf{r};\omega)\,\mathrm{e}^{-\mathrm{i}\omega t}\,d\omega \\ &= \frac{1}{2\pi}\int_C d\omega\,\mathrm{e}^{-\mathrm{i}\omega t}\int_{x_1}^{x_2} dx'\rho(x')\,\mathbf{G}(\mathbf{r},\mathbf{r}';\omega^2)\cdot\left[\mathrm{i}\omega\boldsymbol{\xi}_{\mathrm{i}}(x')-\dot{\boldsymbol{\xi}}_{\mathrm{i}}(x')\right]. \end{aligned} \quad (10.40)$$

The integration contour of Fig. 6.8 is to be placed above the largest point eigenvalue $\mathrm{i}\nu_{\max}$ (i.e. $\nu_0 > \nu_{\max}$) of $\mathbf{F}$. Deforming this contour, *branch cuts of the* Γ*s and* Δ corresponding to the Alfvén and slow *continua* $\{\pm\omega_A\}$, and $\{\pm\omega_S\}$ and *zeros of* Δ corresponding to the different *discrete modes* are encountered.

The explicit evaluation of the integrals in Eq. (10.40) requires the specification of an equilibrium and of initial data. For example, equilibria with a steep gradient in the Alfvén frequency exhibit damping of Alfvén waves [588] which may be described by deforming the integration contour off the principal Riemann sheet across the branch cuts to another Riemann sheet where poles corresponding to quasi-modes may be encountered (see Section 10.3). Thus the great example of Landau's prescription of handling the poles in plasma kinetic theory [393] can be applied to ideal MHD theory as well. Preparation of special initial data, singling out the δ-functions of Eq. (10.38), leads to improper Alfvén and slow modes where a single magnetic surface oscillates with the frequency ω_A or ω_S. This is the analogue of the van Kampen modes [628] for ideal MHD. Consequently, the phenomena of damping and singular oscillation are exclusively associated with the genuine $N=0$ singularities, corresponding to *local* perturbations (large 'wave numbers') which rapidly vary in the direction of inhomogeneity.

Once more: the $D=0$ singularities are apparent, not genuine, and correspond in an average sense to the *global* slow and fast discrete modes in the turning point frequency ranges $\{\omega_{s0}\}$ and $\{\omega_{f0}\}$. They just complicate the analysis of the discrete modes with respect to their monotonicity properties (expressed by the oscillation theorem of Section 7.4.4) since they usually overlap with the genuine $N = 0$ continua. However, there is no place for additional continua besides the slow and Alfvén continua, except for the cluster point $\omega_{\rm F}^2 = \infty$ which provides the asymptotic behaviour of the localized fast modes for large 'wave numbers'. In that sense, there are *three continua* in ideal MHD [215]. They correspond to the slow, Alfvén and fast degrees of freedom. This structure is already present in homogeneous plasmas. Cylindrical or toroidal inhomogeneity extends this structure by quite a number of additional features, but it does not change the fundamental number of degrees of freedom: *the threefold spectrum of ideal MHD is complete!*

10.3 Damping of Alfvén waves★

In principle, the initial value problem has now been solved. However, this solution consists of the simultaneous evolution of all the MHD modes. In order not to get lost by all formal generalities, let us now concentrate on the important features. To that end, we make some simplifying assumptions to the effect that the three sub-spectra become widely separated. We may then study the separate influence of one sub-spectrum, in this case the Alfvén continuum.

For the study of the Alfvén continuum by itself, we may ignore the gravitational terms. We also assume the density to be constant and the magnetic field $\mathbf{B}$ to be *unidirectional*, so that $k_\parallel$ and $k_\perp$ become constant wave numbers. Next, we consider a *low* β plasma ($\beta \equiv 2p/B^2 \ll 1$), so that

$$c^2 \ll b^2 \,. \tag{10.41}$$

This assumption separates the slow from the Alfvén modes:

$$\omega_{\rm S}^2 \approx k_\parallel^2 c^2 \approx \omega_{\rm s0}^2 \;\ll\; \omega_{\rm A}^2 = k_\parallel^2 b^2 \,. \tag{10.42}$$

In order to separate off the influence of the fast modes as well, we concentrate our study on *nearly perpendicular propagation*,

$$k_\parallel \ll k_\perp \approx k_0 \,, \tag{10.43}$$

so that

$$\omega_{\rm A}^2 = k_\parallel^2 b^2 \;\ll\; \omega_{\rm f0}^2 \approx k_0^2 b^2 \,. \tag{10.44}$$

Consequently. the three squared frequencies $\omega_{\rm s0}^2$, $\omega_{\rm A}^2$ and $\omega_{\rm f0}^2$ are wide apart.

Under the mentioned conditions, there is no parallel motion to leading order, so that Eq. (10.8) simplifies to an equation for transverse motion alone:

$$\begin{pmatrix} \dfrac{d}{dx} b^2 \dfrac{d}{dx} - k_\parallel^2 b^2 + \omega^2 & \dfrac{d}{dx} k_\perp b^2 \\ -k_\perp b^2 \dfrac{d}{dx} & -k_\perp^2 b^2 - k_\parallel^2 b^2 + \omega^2 \end{pmatrix} \begin{pmatrix} \hat\xi \\ \hat\eta \end{pmatrix} = {\rm i}\omega \begin{pmatrix} X \\ Y \end{pmatrix} . \tag{10.45}$$

In this equation we have kept terms of unequal order in $k_\parallel$ and $k_\perp$ because large terms cancel upon elimination of $\hat\eta$. After elimination we keep terms of comparable order only, resulting in the following equations:

$$-\frac{1}{k_0^2}\Big[(\omega^2 - \omega_{\rm A}^2)\,\hat\xi'\Big]' + (\omega^2 - \omega_{\rm A}^2)\,\hat\xi = {\rm i}\omega\Big(X + \frac{1}{k_0} Y'\Big) , \quad \hat\eta = -\frac{1}{k_0}\hat\xi' - {\rm i}\omega \frac{Y}{k_0^2 b^2} , \tag{10.46}$$

where all equilibrium variations are expressed by the Alfvén frequency,

$$\omega_A^2 = \omega_A^2(x) = k_\parallel^2 b^2(x)\,. \tag{10.47}$$

Introducing the short-hand notation

$$P(x;\omega^2) \equiv -\rho(\omega^2-\omega_A^2)/k_0^2\,,\quad Q(x;\omega^2) \equiv -\rho(\omega^2-\omega_A^2)\,,\quad R(x;\omega^2) \equiv \mathrm{i}\rho\omega\Big(X + Y'/k_0\Big)\,, \tag{10.48}$$

the inhomogeneous second order differential equation (10.46) may be written as

$$(P\hat\xi')' - Q\,\hat\xi = R\,. \tag{10.49}$$

Of course, the basic equation is of the same form as the general equation (10.15) so that the solution $\hat\xi(x)$ may be represented by the integral (10.17) involving a Green's function $G(x,x')$ which is a solution of the differential equation (10.18) subject to the boundary conditions (10.19).

10.3.1 Green's function★

The inhomogeneous equation (10.18) gives a unique solution of the Green's function (Fig. 10.1(a)) when the homogeneous equation does not have a non-trivial solution (*Fredholm alternative*). Proper and improper solutions of the homogeneous equation (Figs. 10.1(b) and (c)) occur for values of ω^2 inside the spectrum, which is confined to the real ω^2-axis, so that we certainly have a unique Green's function for complex values of ω on the Laplace contour. The procedure is then to construct the Green's function for complex values of ω^2 where existence is guaranteed and to deform the contour in such a way that the spectrum is approached.

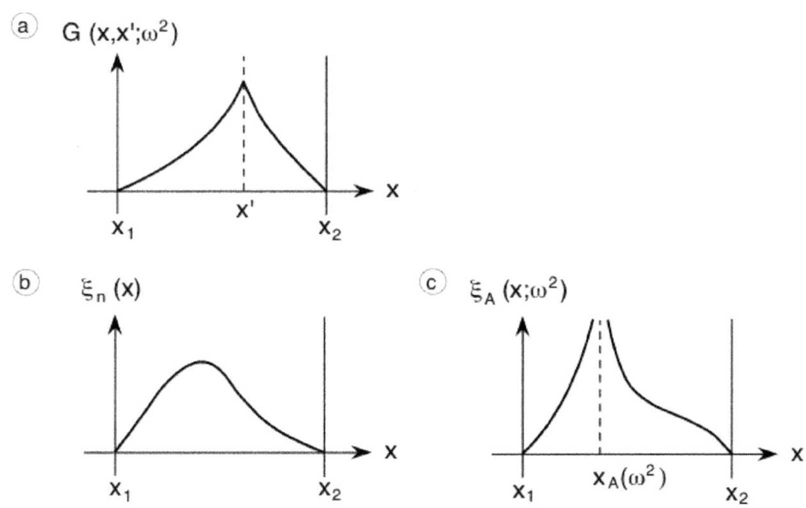

Fig. 10.1 (a) Green's function for ω^2 outside the spectrum of proper and improper eigenvalues; (b) proper eigenfunction for $\omega^2 = \omega_n^2$; (c) improper eigenfunction for $\omega^2 \in \{\omega_A^2(x)\}$.

As in Section 10.2.1, the symmetric expression for $G(x,x';\omega^2)$ is found in terms of the fundamental solutions $U_1(x;\omega^2)$ and $U_2(x;\omega^2)$ of the homogeneous equations Eqs. (10.21) and satisfying the left and right boundary conditions. In terms of these functions one constructs the expression (10.22) for the Green's function with the definitions (10.23) for Γ and Δ. The expression inside the square brackets in the definition of Δ is recognized as the Wronskian. By means of Eqs. (10.21) one proves

$$\begin{aligned}\frac{\partial\Delta}{\partial x} &= P'(U_1U_2' - U_1'U_2) + P\,(U_1U_2'' - U_1''U_2)\\ &= U_1(PU_2{}')' - U_2(PU_1{}')' = Q\,U_1U_2 - Q\,U_2U_1 = 0\,,\end{aligned} \tag{10.50}$$

so that $\Delta \neq \Delta(x)$. For eigenfunctions, the solution of the homogeneous equation satisfies both left and right boundary conditions, so that $U_1 = U_2$. In that case $\Delta(\omega^2) = 0$. For that reason, $\Delta(\omega^2)$ is called the *dispersion function*.

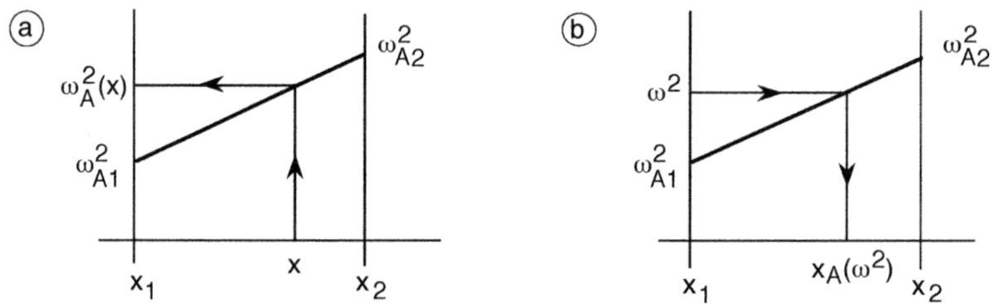

Fig. 10.2 Inversion of the function $\omega_A^2 = \omega_A^2(x)$ for linear profiles.

Let us again specify the profile $\omega_A^2 = \omega_A^2(x)$ to be monotonically increasing on the interval (x_1, x_2), as in Section 7.4, and construct the inverse profile $x_A = x_A(\omega^2)$. For a simple linear profile (Fig. 10.2) the explicit functions would read:

$$\begin{aligned}\omega_A^2(x) &= \omega_0^2(x) + \omega_A^2{}'(x - x_0)\,, &\quad \omega_0^2 &\equiv \tfrac{1}{2}(\omega_{A1}^2 + \omega_{A2}^2)\,,\\ x_A(\omega^2) &= x_0 + (\omega^2 - \omega_0^2)/\omega_A^2{}'\,, &\quad x_0 &\equiv \tfrac{1}{2}(x_1 + x_2)\,.\end{aligned} \tag{10.51}$$

In Section 7.4 we expanded around the singularity $x = x_A(\omega^2)$ by exploiting the variable $s = x - x_A(\omega^2)$. Here, ω^2 is complex so that the corresponding singularity of Eq. (10.21) occurs in the complex z-plane for $z = z_A(\omega^2)$ (see Fig. 10.3) where $z_A(\omega^2)$ is the analytic continuation of $x_A(\omega^2)$. For the linear profile the explicit expression for $z_A(\omega^2)$ would be

$$z_A(\omega^2) = x_0 + (\omega^2 - \omega_0^2)/\omega_A^2{}'\,. \tag{10.52}$$

Introducing a complex variable ζ replacing s,

$$\zeta = \zeta(x;\omega^2) \equiv x - z_A(\omega^2)\,, \tag{10.53}$$

the solutions U_1 and U_2 of the equations (10.21) may be expressed as a linear combination of the functions

$$u(\zeta)\,, \quad \text{and} \quad u(\zeta)\ln\zeta + v(\zeta)\,. \tag{10.54}$$

These are the analytic continuations of the functions $u(s)$ and $v(s)$ introduced in Eq. (7.127), which

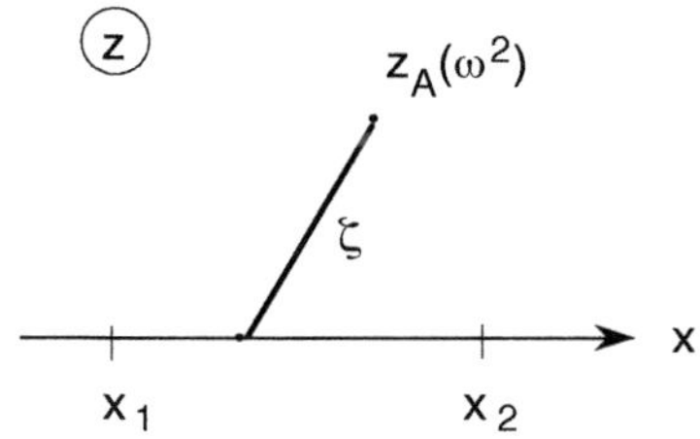

Fig. 10.3 Analytic continuation of the inverse Alfvén function $x_{\rm A}(\omega^2)$.

may be written as a power series in ζ, $u = a + b\zeta + \cdots$, and similarly for v. Hence,

$$
\begin{aligned}
U_1(\zeta) &= \left[\ln \frac{\zeta(x;\omega^2)}{\zeta_1(\omega^2)} - \frac{v_1(\omega^2)}{u_1(\omega^2)} \right] u(\zeta;\omega^2) + v(\zeta;\omega^2)\,, \\
U_2(\zeta) &= \left[\ln \frac{\zeta(x;\omega^2)}{\zeta_2(\omega^2)} - \frac{v_2(\omega^2)}{u_2(\omega^2)} \right] u(\zeta;\omega^2) + v(\zeta;\omega^2)\,.
\end{aligned} \tag{10.55}
$$

Substituting these expressions into Eq. (10.22) provides us with the formal solution of the Green's function:

$$
\begin{aligned}
G(x, x';\omega^2) &= \left\{ \left[\ln \frac{\omega^2 - \omega_{\rm A}^2(x_<)}{\omega^2 - \omega_{\rm A1}^2} - \frac{v_1(\omega^2)}{u_1(\omega^2)} \right] u(x_<;\omega^2) + v(x_<;\omega^2) \right\} \\
&\quad \times \left\{ \left[\ln \frac{\omega^2 - \omega_{\rm A}^2(x_>)}{\omega^2 - \omega_{\rm A2}^2} - \frac{v_2(\omega^2)}{u_2(\omega^2)} \right] u(x_>;\omega^2) + v(x_>;\omega^2) \right\} \\
&\quad \bigg/ \left[\ln \frac{\omega^2 - \omega_{\rm A2}^2}{\omega^2 - \omega_{\rm A1}^2} - \frac{v_1(\omega^2)}{u_1(\omega^2)} + \frac{v_2(\omega^2)}{u_2(\omega^2)} \right],
\end{aligned} \tag{10.56}
$$

where we have introduced the notation

$$
x_< \equiv \inf(x, x')\,, \qquad x_> \equiv \sup(x, x')\,. \tag{10.57}
$$

The logarithmic expression in terms of ζ has been converted into the more transparent form in terms of $\omega^2 - \omega_{\rm A}^2(x)$ by means of the relation

$$
\zeta = x - z_{\rm A}(\omega^2) = -(\omega^2 - \omega_{\rm A}^2(x))/\omega_{\rm A}^{2\,\prime}\,, \tag{10.58}
$$

which is, strictly speaking, only valid for the linear profile. However, for an arbitrary monotonically increasing profile, Eq. (10.56) is also valid if the functions u and v are redefined such that the expressions for the basic solutions are written as

$$
u(\omega^2 - \omega_{\rm A}^2(x))\,, \quad \text{and} \quad u(\omega^2 - \omega_{\rm A}^2(x))\, \ln(\omega^2 - \omega_{\rm A}^2(x)) + v(x;\omega^2) \tag{10.59}
$$

instead of Eq. (10.54). Clearly, for the derivation of the expression (10.56) of the Green's function no other property has been used than the fact that $\omega_{\rm A}^2(x)$ is a monotonic function, and that the slow continuum is far away so that we are dealing with only one singularity at a time.

10.3.2 Spectral cuts★

For the completion of the initial value problem we now need to study the behaviour of the Green's function when ω approaches the spectrum. We have already seen that the zeros of the denominator $\Delta(\omega^2)$ represent the discrete spectrum. The continuous spectrum arises as a result of the multi-valuedness of the logarithmic terms appearing in both $\Gamma(x, x'; \omega^2)$ and $\Delta(\omega^2)$. In order to make these logarithmic terms single-valued one needs to cut the complex ω-plane along branch cuts that precisely correspond to the continuous spectra $\pm\{\omega_{\rm A}(x)\}$, as we will see.

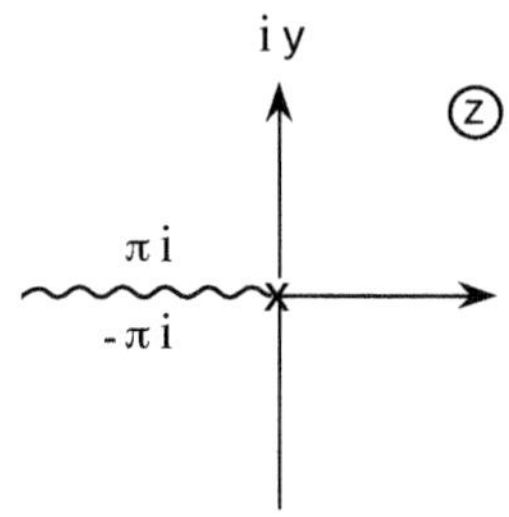

Fig. 10.4 Branch cut for ln z .

In order to make a logarithmic function $\ln z$ single-valued one may cut the z-plane along any curve starting at the branch point $z = 0$ and extending to ∞ . Let us choose the negative real axis as a branch cut (Fig. 10.4). Along this branch cut one may write

$$\lim_{y\to 0^\pm} \ln z = \ln|z| \pm \pi{\rm i} \tag{10.60}$$

on the principal ($n = 0$) Riemann sheet, where $+\pi{\rm i}$ is the value immediately above the branch cut and $-\pi{\rm i}$ immediately below. If one wishes to deform a contour across a branch cut one moves to another Riemann sheet of the logarithmic function. These sheets are labelled by n, and the logarithmic function increases by an amount $2\pi{\rm i}$ every time one encircles the branch point and moves to the next Riemann sheet. Therefore, the general expression for the logarithmic function when approaching the real axis may be written as

$$\lim_{y\to 0^\pm} \ln z = \ln|x| \pm \pi{\rm i}H(-x) + 2n\pi{\rm i}\,, \tag{10.61}$$

where the jump of the Heaviside function occurs at the branch point. Accordingly, for complex values of $\omega = {\rm Re}\,\omega + {\rm i}\nu$, one may write for a logarithmic expression $\ln[(\omega^2 - \omega_\beta^2)/(\omega^2 - \omega_\alpha^2)]$ when approaching the real axis

$$\lim_{\nu\to 0^\pm} \ln\frac{\omega^2 - \omega_\beta^2}{\omega^2 - \omega_\alpha^2} = \ln\left|\frac{\omega^2 - \omega_\beta^2}{\omega^2 - \omega_\alpha^2}\right| \pm {\rm i}\pi\Big[H(\omega - \omega_\alpha) - H(\omega - \omega_\beta) - H(\omega + \omega_\alpha) + H(\omega + \omega_\beta)\Big] + 2n\pi{\rm i}\,. \tag{10.62}$$

Hence, assuming $\omega_\beta^2 > \omega_\alpha^2$, the branch cuts and jumps are as indicated in Fig. 10.5. Here, we have indicated how one moves from the principal sheet to the $n = 1$ and $n = -1$ sheets when crossing the branch cuts.

On the basis of the second expression (10.59) we find that the function $\Gamma(x, x'; \omega^2)$ has branch

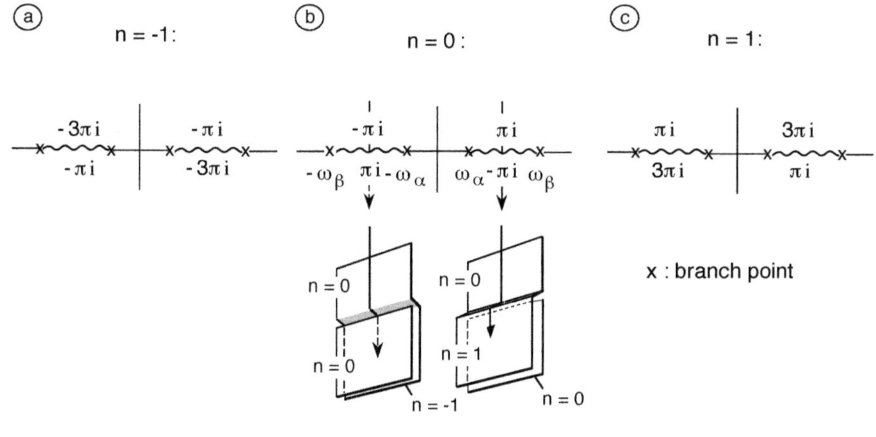

Fig. 10.5 The different Riemann sheets of the function $\ln[(\omega^2 - \omega_\beta^2)/(\omega^2 - \omega_\alpha^2)]$.

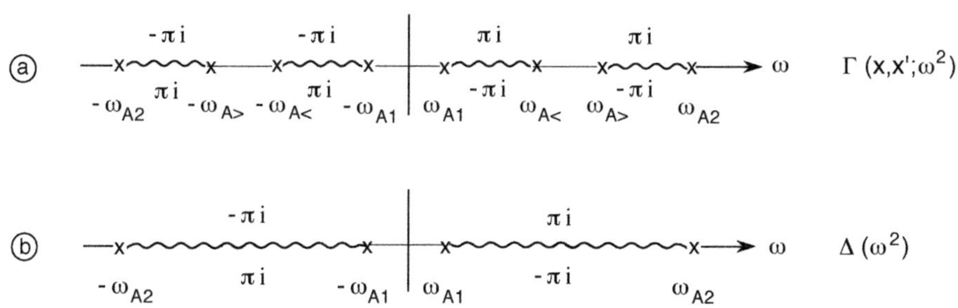

Fig. 10.6 Branch cuts for (a) the numerator Γ, and (b) the denominator Δ of the Green's function.

points $\omega_{A<}^2 \equiv \omega_{\rm A}^2(x_<)$, $\omega_{A>}^2 \equiv \omega_{\rm A}^2(x_>)$, $\omega_{\rm A1}^2$ and $\omega_{\rm A2}^2$, whereas the function $\Delta(\omega^2)$ only has branch points at $\omega_{\rm A1}^2$ and $\omega_{\rm A2}^2$. One may connect these branch points as indicated in Fig. 10.6. For the Green's function $G = \Gamma/\Delta$ these branch points should be joined. One may do this by choosing the branch cuts for Δ differently, so that the Laplace contour C may be deformed to a contour C′ as shown in Fig. 10.7 (see Sedláček [546]). This clearly shows that the contribution of the continuous spectrum is due to the jump in the logarithmic function along the branch cuts.

Let us now calculate the typical contributions of the spectral cuts to the solution of the initial value problem. Take special initial data: $\xi_{\rm i}(x) \neq 0$, $\dot\xi_{\rm i}(x) = \eta_{\rm i}(x) = \dot\eta_{\rm i}(x) = 0$. The solution of the initial value problem can then be written from the Eqs. (10.1), (10.17) and (10.46)(b) as:

$$\begin{aligned} \xi(x;t) &= \frac{1}{2\pi}\int_{\rm C} d\omega \frac{{\rm i}\omega}{\Delta(\omega^2)} {\rm e}^{-{\rm i}\omega t} \int_{x_1}^{x_2} dx'\, \Gamma(x,x';\omega^2)\xi_{\rm i}(x')\,, \\ \eta(x;t) &= -\frac{1}{k_0}\frac{\partial}{\partial x}\xi(x;t)\,. \end{aligned} \qquad (10.63)$$

From Eq. (10.62) one then finds as the typical contribution from a jump of the logarithmic function

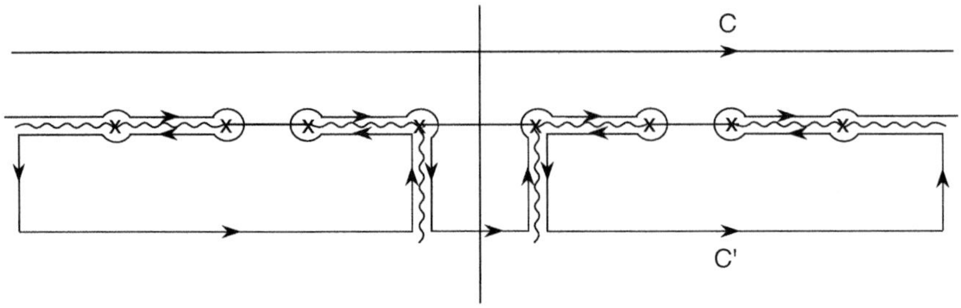

Fig. 10.7 Laplace contours for the Green's function $G(x,x';\omega^2) \equiv \Gamma(x,x';\omega^2)/\Delta(\omega^2)$.

at some real frequency ω_α :

$$\xi(t) \sim \int_C \mathrm{i}\omega \mathrm{e}^{-\mathrm{i}\omega t} H(\omega-\omega_\alpha)\, d\omega = -\int_C \omega \frac{\partial}{\partial\omega}\Big(\frac{\mathrm{e}^{-\mathrm{i}\omega t}}{t}\Big) H(\omega-\omega_\alpha)\, d\omega$$

$$= \int_C \frac{\mathrm{e}^{-\mathrm{i}\omega t}}{t} H(\omega-\omega_\alpha)\, d\omega + \int_C \omega \frac{\mathrm{e}^{-\mathrm{i}\omega t}}{t} \delta(\omega-\omega_\alpha)\, d\omega\,. \tag{10.64}$$

Asymptotically, the first integral may be neglected because the rapidly oscillating integrand kills this contribution for large t . Thus, as shown by Tataronis [587], the asymptotic behaviour in time of the Alfvén continuum modes is given by:

$$\xi(t) \sim \omega_\alpha t^{-1} \mathrm{e}^{-\mathrm{i}\omega_\alpha t}\,, \qquad \eta(t) \sim -\mathrm{i}\,(\omega_\alpha \omega'_\alpha / k_0)\, \mathrm{e}^{-\mathrm{i}\omega_\alpha t}\,. \tag{10.65}$$

Consequently, the continuous spectrum of Alfvén modes yields oscillatory normal components that are damped like t^{-1} and *undamped* oscillatory tangential components, perpendicular to the field lines, where each point oscillates with its own local Alfvén frequency. As time goes on, the factor $\exp(-\mathrm{i}\omega_\alpha t)$ gives rise to an ever more fluctuating spatial structure of the motion, finally resulting in completely uncoordinated oscillations, which is called *phase mixing*. In this way, large spatial gradients are built up so that, eventually, dissipative effects lead to dissipation of the energy of the continuum modes and heating of the plasma (see Chapter 11).

10.4 Quasi-modes★

In contrast to the situation just described, another kind of motion exists that displays coherent oscillations. To exhibit this, let us start with a profile $\omega_A^2(x)$ that has a step discontinuity at some value of x, say in the middle of the slab at $x = x_0 \equiv \frac{1}{2}(x_1 + x_2)$; see Fig. 10.8(a). The singularities of the continuous spectrum $\omega_{A1}^2 \le \omega^2 \le \omega_{A2}^2$ are now all concentrated in the point $x = x_0$. This gives rise to a special mode which is called *a surface mode*.

This mode may be found from the homogeneous counterpart of Eq. (10.46),

$$\frac{1}{k_0^2}\Big[(\omega^2-\omega_A^2)\xi'\Big]' - (\omega^2-\omega_A^2)\xi = 0\,, \tag{10.66}$$

where $\omega_A^2(x) = \omega_{A1}^2 H(x_0 - x) + \omega_{A2}^2 H(x - x_0)$. On the left and right intervals $x_1 \le x < x_0$

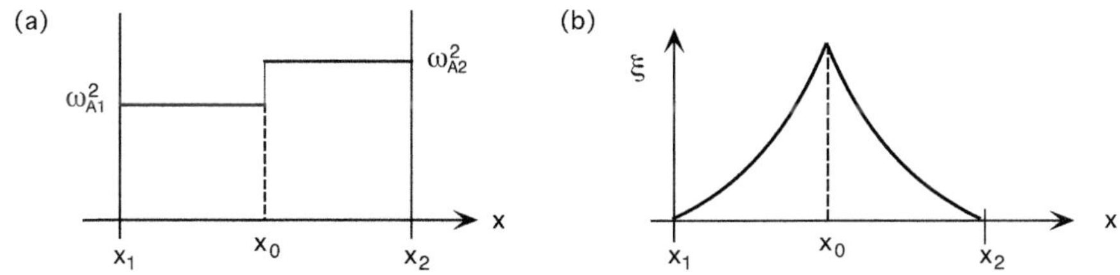

Fig. 10.8 (a) Step discontinuity of the Alfvén frequency and (b) surface mode for this configuration.

and $x_0 \leq x < x_2$ this equation reduces to

$$\xi'' - k_0^2 \xi = 0\,, \tag{10.67}$$

having the solutions $\exp(k_0 x)$ and $\exp(-k_0 x)$, when $\omega^2 \neq \omega_{\rm A1}^2$ and $\omega^2 \neq \omega_{\rm A2}^2$, respectively. The solution $\xi_1 = \sinh[k_0(x - x_1)]$ satisfying the left hand boundary condition may be combined with the solution $\xi_2 = \sinh[k_0(x_2 - x)]$ satisfying the right hand boundary condition to form a cusp-shaped perturbation which is an eigenfunction of the system; see Fig. 10.8(b). This may be seen by applying the proper boundary condition to join ξ_1 to ξ_2 by integrating Eq. (10.66) across the jump:

$$\int_{x_0^-}^{x_0^+} \Big\{ \Big[\omega^2 - \omega_{\rm A1}^2 H(x_0 - x) - \omega_{\rm A2}^2 H(x - x_0)\Big]\xi' \Big\}' dx = 0 \quad \Rightarrow \quad \Big[(\omega^2 - \omega_{\rm A}^2)\xi'\Big]_{x=x_0} = 0\,. \tag{10.68}$$

This condition is fulfilled for $\omega^2 = \omega_0^2 \equiv \frac{1}{2}(\omega_{\rm A1}^2 + \omega_{\rm A2}^2)$, which is the eigenfrequency of the cusped surface wave.

(a) Dispersion equation Let us now remove the degeneracy of the step and introduce a genuine continuum by smoothing out the discontinuity (Fig. 10.9(a)). This we do by replacing the step by a linearly increasing profile between $x = -a$ and $x = a$, where we have fixed $x_0 = 0$. For simplicity, we also take $x_1 \to -\infty$ and $x_2 \to +\infty$. The spectrum of the system then changes as shown in Fig. 10.9(b).

Notice that for the stepped and the continuous profile there are also infinitely many discrete Alfvén modes with eigenfrequencies $\omega = \pm\omega_{A1}$ and $\omega = \pm\omega_{A2}$. These modes are localized to the left of $x = -a$, or to the right of $x = a$, respectively, but otherwise have a completely arbitrary shape. That this is so may be seen from Eq. (10.66) by pulling out the factor $\omega^2 - \omega_{\rm A}^2$ which is constant on those intervals:

$$(\omega^2 - \omega_{\rm A}^2)(\xi'' - k_0^2 \xi) = 0\,. \tag{10.69}$$

Hence, for $\omega^2 = \omega_{\rm A1}^2$ on the left homogeneous interval ξ may be chosen arbitrarily. Each choice of this function is a proper Alfvén eigenfunction. Likewise, for $\omega^2 = \omega_{\rm A2}^2$ on the right interval. However, here we wish to concentrate on the influence of the inhomogeneity. In particular, we want to see what happened to the surface wave by the introduction of the linear profile. Does the appearance of a continuous spectrum imply that all of a sudden the coherent oscillations of the surface wave have disappeared to make place for the kind of chaotic response expressed by Eq. (10.65)? This is hard to believe.

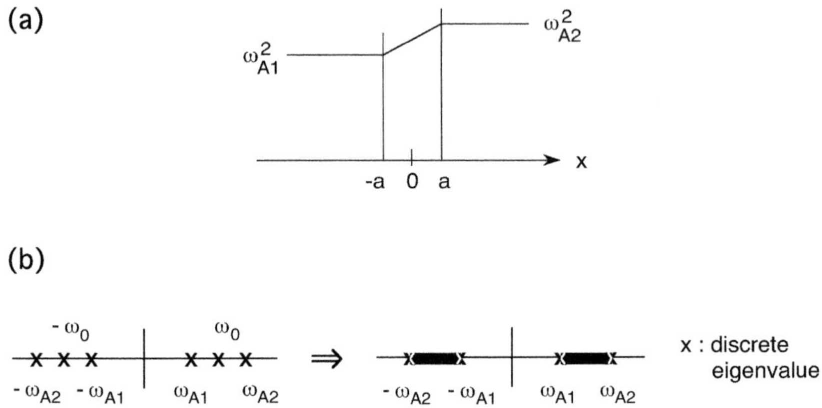

Fig. 10.9 (a) Smoothing the discontinuity and (b) change of the spectrum due to this smoothing.

We already noticed that the discrete spectrum comes about from the poles of the Green's function, i.e. the zeros of the dispersion function $\Delta(\omega^2)$. Let us, therefore, study the expression $\Delta(\omega^2)$ for the present case. To that end, we need the explicit solutions U_1 and U_2 to the homogeneous equations (10.21) on the three intervals $(-\infty,-a)$, $(-a,a)$ and (a,∞) . The virtue of the choice of a linear profile on $(-a,a)$ is that the homogeneous equation for this interval may be written as

$$\frac{d}{d\zeta}\,\zeta\,\frac{dU}{d\zeta} - k_0^2\zeta\,U = 0\,, \qquad \zeta \equiv -2a\,\frac{\omega^2-\omega_A^2(x)}{\omega_{A2}^2-\omega_{A1}^2}\,, \tag{10.70}$$

so that we obtain modified Bessel functions of complex argument as solutions:

$$I_0(k_0\zeta) = 1 + \tfrac{1}{4}(k_0\zeta)^2 + \cdots\,, \quad K_0(k_0\zeta) = -(\ln\tfrac{1}{2}k_0\zeta + \gamma)I_0(k_0\zeta) + \tfrac{1}{4}(k_0\zeta)^2 + \cdots\,, \tag{10.71}$$

where $\gamma \approx 0.577$ is Euler's constant. Consequently, the following solutions are obtained:

$$U_1 = \begin{cases} e^{k_0x} \\ A_1 I_0(k_0\zeta) + B_1 K_0(k_0\zeta)\,, \\ C_1 e^{k_0x} + D_1 e^{-k_0x} \end{cases} \qquad U_2 = \begin{cases} C_2 e^{k_0x} + D_2 e^{-k_0x} & (-\infty,-a) \\ A_2 I_0(k_0\zeta) + B_2 K_0(k_0\zeta) & (-a,a) \\ e^{-k_0x} & (a,\infty) \end{cases}. \tag{10.72}$$

The constants $A_{1,2}$, $B_{1,2}$, $C_{1,2}$ and $D_{1,2}$ are fixed by equating functions and first derivatives at the boundaries of the intervals. For the calculation of $\Delta(\omega^2)$ we actually only need to compute $A_{1,2}$ and $B_{1,2}$ because $\Delta(\omega^2)$ is independent of x, so that we may choose to evaluate it in the inhomogeneous layer. The solutions U_1 and U_2 on $(-a,a)$ read:

$$\begin{aligned} U_1 = &\ \ k_0\zeta_1 e^{-k_0a}\Big\{\Big[K_0(k_0\zeta_1) + K_1(k_0\zeta_1)\Big]I_0(k_0\zeta) - \Big[I_0(k_0\zeta_1) - I_1(k_0\zeta_1)\Big]K_0(k_0\zeta)\Big\}\,, \\ U_2 = &\ -k_0\zeta_2 e^{-k_0a}\Big\{\Big[K_0(k_0\zeta_2) - K_1(k_0\zeta_2)\Big]I_0(k_0\zeta) - \Big[I_0(k_0\zeta_2) + I_1(k_0\zeta_2)\Big]K_0(k_0\zeta)\Big\}\,, \end{aligned} \tag{10.73}$$

where

$$\zeta_{1,2} \equiv -2a\,(\omega^2 - \omega_{A1,2}^2)/(\omega_{A2}^2 - \omega_{A1}^2)\,. \tag{10.74}$$

Inserting these solutions into the dispersion function (10.23) yields

$$\Delta = C\zeta_1\zeta_2\Big\{\Big[I_0(k_0\zeta_1) - I_1(k_0\zeta_1)\Big]\Big[K_0(k_0\zeta_2) - K_1(k_0\zeta_2)\Big] - \Big[K_0(k_0\zeta_1) + K_1(k_0\zeta_1)\Big]\Big[I_0(k_0\zeta_2) + I_1(k_0\zeta_2)\Big]\Big\}, \tag{10.75}$$

where C is a constant that is not important for the present purpose. To obtain Eq. (10.75) we have used the property $z[I_0(z)K_1(z) + I_1(z)K_0(z)] = 1$.

The dispersion equation is given by

$$\Delta(\omega^2) = 0. \tag{10.76}$$

Note that the two trivial solutions $\zeta_1 = 0$ and $\zeta_2 = 0$ on the homogeneous intervals, corresponding to the two discrete eigenvalues $\omega^2 = \omega_{A1}^2$ and $\omega^2 = \omega_{A2}^2$ discussed above, are *not* contained in this dispersion equation because the factors ζ_1 and ζ_2 cancel out against the singularities of the exponential Bessel functions.

(b) Exponential damping Let us now investigate whether some more solutions exist, hopefully corresponding to the surface wave solution of the step function model. To that end we study a situation where the continuous profile is close to the step function model, i.e., a is considered to be small. Since the other intervals are infinite, the only scale to compare a with is the perpendicular wavelength k_0^{-1}. Hence, we assume $k_0 a \ll 1$ and expand Eq. (10.76) in orders of $k_0 a$. By means of the expansions (10.71) of the Bessel functions we find to leading order

$$\ln\frac{\zeta_2}{\zeta_1} + \frac{1}{k_0}\left(\frac{1}{\zeta_1} + \frac{1}{\zeta_2}\right) = 0 \quad\Rightarrow\quad \ln\frac{\omega^2 - \omega_{A2}^2}{\omega^2 - \omega_{A1}^2} - \frac{\omega_{A2}^2 - \omega_{A1}^2}{2k_0 a}\left[\frac{1}{\omega^2 - \omega_{A1}^2} + \frac{1}{\omega^2 - \omega_{A2}^2}\right] = 0. \tag{10.77}$$

We study this expression in the neighbourhood of the real axis so that $\nu \ll \omega$. We then have from Eq. (10.62) for ω in the range of the continua

$$\ln\frac{\omega^2 - \omega_{A2}^2}{\omega^2 - \omega_{A1}^2} \approx \ln\left|\frac{\omega^2 - \omega_{A2}^2}{\omega^2 - \omega_{A1}^2}\right| + \mathrm{sgn}(\omega)\mathrm{sgn}(\nu)\,\pi\mathrm{i} + 2n\pi\mathrm{i} + 2\mathrm{i}\nu\omega\frac{\omega_{A2}^2 - \omega_{A1}^2}{(\omega^2 - \omega_{A1}^2)(\omega^2 - \omega_{A2}^2)}, \tag{10.78}$$

where the last term may be dropped again as it is small compared to the other imaginary contributions. This gives

$$\ln\left|\frac{\omega^2 - \omega_{A2}^2}{\omega^2 - \omega_{A1}^2}\right| - \frac{\omega_{A2}^2 - \omega_{A1}^2}{k_0 a}\,\frac{\omega^2 - \frac{1}{2}(\omega_{A1}^2 + \omega_{A2}^2)}{(\omega^2 - \omega_{A1}^2)(\omega^2 - \omega_{A2}^2)} + \mathrm{sgn}(\omega)\mathrm{sgn}(\nu)\,\pi\mathrm{i} + 2n\pi\mathrm{i} + \mathrm{i}\,\frac{\nu\omega(\omega_{A2}^2 - \omega_{A1}^2)}{k_0 a}\,\frac{(\omega^2 - \omega_{A1}^2)^2 + (\omega^2 - \omega_{A2}^2)^2}{(\omega^2 - \omega_{A1}^2)^2(\omega^2 - \omega_{A2}^2)^2} = 0. \tag{10.79}$$

The real and imaginary parts of this dispersion equation give the roots we are looking for:

$$\begin{aligned} \mathrm{Re}\,\omega &= \pm\,\omega_0 \equiv \pm\sqrt{\tfrac{1}{2}(\omega_{A1}^2 + \omega_{A2}^2)}\,, \\ \nu \equiv \mathrm{Im}\,\omega &= \nu_0 \equiv \tfrac{1}{8}\pi k_0 a\,\Big[\mathrm{sgn}(\nu)\mathrm{sgn}(\omega) + 2n\Big]\,\frac{\omega_{A2}^2 - \omega_{A1}^2}{\mathrm{sgn}(\omega)\omega_0}\,. \end{aligned} \tag{10.80}$$

This seems to give a satisfactory generalization of the surface mode as it reduces to $\omega = \omega_0$ for $a = 0$. If $a \neq 0$, a 'mode' is obtained which has a small imaginary part to the 'eigenfrequency'. We have put quotation marks here because we have proved already that in ideal MHD normal modes cannot have complex eigenvalues. On the other hand, we have obtained a genuine pole of the Green's function, which certainly will influence the response to the initial data.

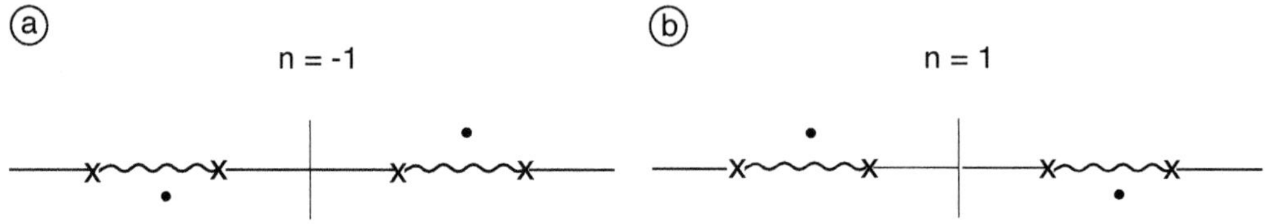

Fig. 10.10 Poles of the Green's function for (a) the $n = -1$ and (b) the $n = 1$ Riemann sheet.

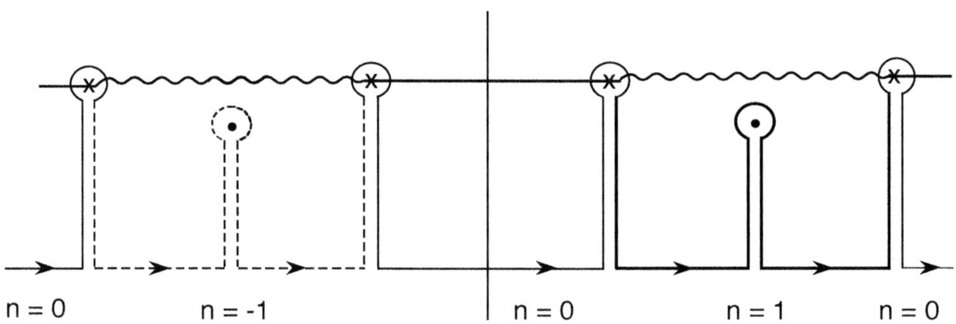

Fig. 10.11 Deformation of the Laplace contour across the branch cuts and about the poles.

For $n = 0$ the expression for ν_0 in Eq. (10.80) gives a contradiction, so that no solutions are found on the principal Riemann sheet, corresponding to the fact that complex eigenvalues do not exist in ideal MHD. For $n = 1$ and $n = -1$, however, we find two poles (see Fig. 10.10) with

$$|\nu_0| = \frac{1}{8}\pi k_0 a\,(\omega_{A2}^2 - \omega_{A1}^2)/\omega_0\,. \tag{10.81}$$

We may now deform the Laplace contour across the branch cuts so that the contributions of the complex poles on the neighbouring Riemann sheets are picked up (Fig. 10.11). Ignoring the contributions of the branch cuts corresponding to the continuous spectrum (and also the contribution of the branch points, which are simultaneously poles from the degenerate Alfvén modes), we find asymptotically for large t for the contributions of these poles from Cauchy's theorem

$$\xi(t) \sim \frac{1}{2\pi}\oint d\omega\,\frac{\mathrm{i}\omega}{\Delta(\omega^2)}\,\mathrm{e}^{-\mathrm{i}\omega t} \sim \frac{1}{2\pi}\oint d\omega\,\frac{\mathrm{i}\omega}{\omega-\omega_0}\,\mathrm{e}^{-\mathrm{i}\omega t} = \omega_0 \mathrm{e}^{-\mathrm{i}\omega_0 t} \sim \mathrm{e}^{-|\nu_0|t}\mathrm{e}^{-\mathrm{i}(\mathrm{Re}\,\omega_0)t}\,,$$

$$\text{and similarly}\quad \eta(t) \sim \mathrm{e}^{-|\nu_0|t}\mathrm{e}^{-\mathrm{i}(\mathrm{Re}\,\omega_0)t}\,. \tag{10.82}$$

Hence, we have found a 'mode' that is exponentially damped. Since the pole is not on the principal branch of the Green's function, there is no contradiction with the general proof that complex eigenvalues do not occur for self-adjoint linear operators. On the other hand, it is clear that the present 'mode' of the plasma is of physical interest as it represents a coherent oscillation of the inhomogeneous system. In contrast to the chaotic response produced by the branch cuts of the continuous spectrum this 'mode' constitutes a very orderly motion. The plasma as a whole oscillates with a definite frequency that cannot be distinguished from a true eigenmode during times $\tau \ll \nu_0^{-1}$. 'Modes' like these occur in many branches of physics and, accordingly, they have received many different names, like *quasi-modes, collective modes, virtual eigenmodes, resonances*, etc. The damping is completely analogous to the well-known phenomenon of Landau damping in the Vlasov description of plasmas. The physical difference is that Landau damping is due to inhomogeneity of the equilibrium in velocity space, whereas damping of Alfvén waves is due to inhomogeneity of the equilibrium in ordinary space.

The expression (10.80) for the frequency of a quasi-mode in a plasma–plasma interface configuration has been derived under the assumption that the density is constant (so that $\omega_{\rm A}^2$ variations are due to the magnetic field). Permitting a jump in the density, the expression for the real part of the quasi-mode becomes a weighted average of the Alfvén frequencies on both sides,

$$\mathrm{Re}\,\omega = \pm\sqrt{\frac{\rho_1\omega_{\rm A1}^2 + \rho_2\omega_{\rm A2}^2}{\rho_1 + \rho_2}}. \tag{10.83}$$

This expression also describes the quasi-modes of a plasma–vacuum interface configuration (in the limit $\rho_2 \to 0$), which play an important role in resonant absorption processes (Section 11.1).

(c) Different kinds of quasi-modes The quasi-modes derived above originate from surface waves propagating along a thin transition region between two homogeneous plasmas. However, quasi-modes can originate from many different kinds of waves and the term is used for any 'discrete' mode with an oscillatory part of the frequency in the range of the continuous spectrum, so that it couples to the continuum modes resulting in damping. In MHD, quasi-modes can be due to fast and slow magneto-sonic waves, discrete Alfvén waves, and different kinds of 'gap' modes in two-dimensional configurations. Even external kink modes in a plasma–vacuum configuration with a wall may turn into quasi-modes when the wall is moved in and the kink mode moves into the Alfvén continuum, see Chance et al. [110].

To demonstrate how easily fast magneto-sonic waves can turn into quasi-modes, we consider a pressureless plasma slab with a uniform magnetic field $B_0 = B_{0z} = 1$. Assume that the slab has a finite width a in the x-direction, L in the z-direction, and is infinite in the y-direction. The wave number in the z-direction is then quantized, $k_z = n\pi/L$, and the eigenfrequencies of the first three fast magneto-sonic eigenmodes $(0, 1$ and 2 nodes in the x-direction) are shown in Fig. 10.12 as functions of k_z. The different gray lines connect fast modes with the same number of nodes in the x-direction. The figure also shows the upper and lower limits of the Alfvén continuum as functions of k_z. The density for this case is chosen as $\rho_0(x) = 0.6 + 0.4\cos(\pi/a\,x)$. Keeping k_y fixed while increasing k_z then results in ever more fast magneto-sonic modes 'swallowed' by the continuum. Since $k_y \neq 0$, the fast eigenmodes with an eigenfrequency within the range of the Alfvén continuum couple to the shear Alfvén continuum modes and become quasi-modes. For more realistic (larger) values of L, many more quasi-modes are present, see De Groof et al. [144].

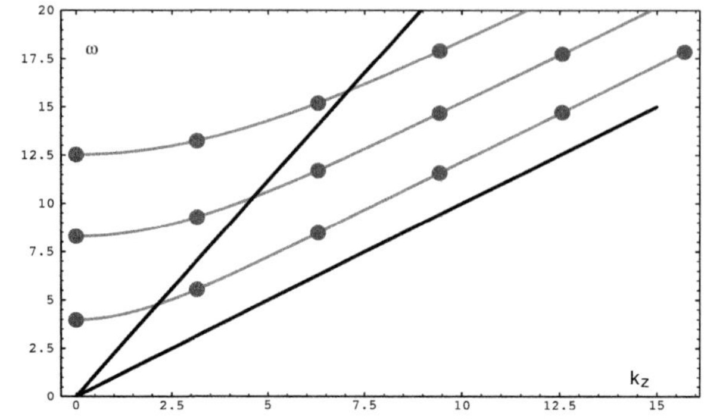

Fig. 10.12 Eigenfrequencies of the first three families of fast eigenmodes (grey dots) with upper and lower bound of the Alfvén continuum (black lines) as a function of k_z, for $L = a = 1$ and $k_y = 0$. (From De Groof et al. [144].)

10.5 Leaky modes★

Most of the plasmas considered so far were isolated from their surroundings by either a perfectly conducting wall or a vacuum, or both. In solar astrophysics, one frequently considers magnetic structures, such as loops and arcades, embedded in another plasma that may, or may not, be magnetized. The possible wave solutions in the magnetic structures are then classified on the basis of their spatial and temporal character inside and outside the magnetic structure (see Fig. 10.13). A wave that is propagating along the boundary of the magnetic structure, and shows exponentially damped behaviour both in the central and in the surrounding plasma, is called a *surface wave*. An oscillating wave solution that shows evanescent behaviour in the surrounding plasma is called a *body mode*. Both surface and body modes thus have a non-propagating character in the surrounding plasma and are therefore called *non-leaky modes*, as they do not leak out energy from the magnetic structure to the environment. But waves that do have an outward propagating behaviour in the surrounding plasma are called *leaky waves*.

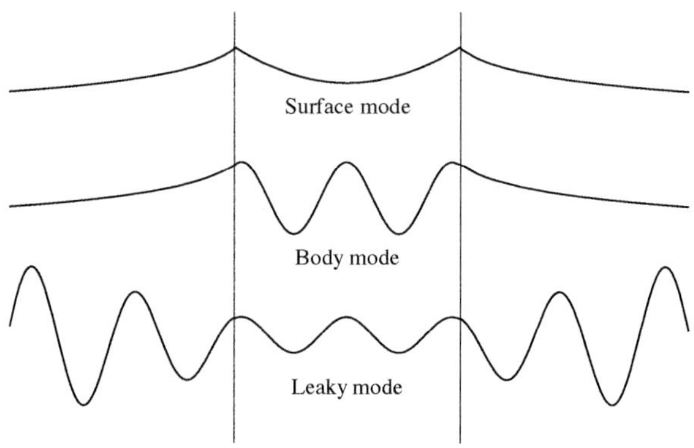

Fig. 10.13 Typical structures of surface waves, body waves and leaky waves inside a plasma, within another plasma with boundaries indicated by the two vertical lines. The leaky waves are defined by their outward propagating external behaviour. (Courtesy of A. De Groof.)

In this section, we will derive leaky wave solutions in a simple configuration so that analytical methods can be exploited and the derivation is relatively straightforward. First, we discuss the equations to be solved and the boundary conditions to be imposed. Next, we discuss the normal-mode analysis and solve the eigenvalue problem. Finally, the initial value problem is solved by means of the Laplace transform. The wave equation is Laplace transformed with respect to time to obtain a second order ODE with a RHS given by the initial conditions. Just like in Section 10.2, the Green's function is constructed by joining two linearly independent solutions. The conjunct of these solutions is a function depending only on the complex frequency. The Laplace transform of the solution is then expressed by means of this Green's function. In this initial value problem approach, the leaky modes correspond to the poles of the Green's function (the zeros of the conjunct). These poles are independent of the spatial coordinates and correspond to the eigenvalues found, demonstrating that the two approaches of the problem are equivalent.

(a) Model equations and boundary conditions Consider a uniform, pressureless, plasma slab confined in the x-direction between $x = \pm a$ and infinite in the y- and z-directions. In the x-direction, the plasma is surrounded on both sides by another uniform plasma that supports waves and extends up to $\pm\infty$. The Alfvén velocity inside the plasma slab is indicated by b. It is lower than the Alfvén velocity in the external plasma, indicated by b_e. The profiles of the Alfvén frequency and the cutoff frequency (ω_{f0}) are illustrated in Fig. 10.14, indicating the frequencies of the surface mode and the first three 'fast' modes. The lowest two of the latter modes are situated below the cutoff frequency in the external plasma so that these two modes will have an exponentially decreasing behaviour in the external plasma. Therefore, as will be shown, these modes correspond to 'body' modes and have real frequencies. The third mode, however, lies above the external cut off frequency and thus has a propagating character in the external region. This is a 'leaky' mode with a complex frequency.

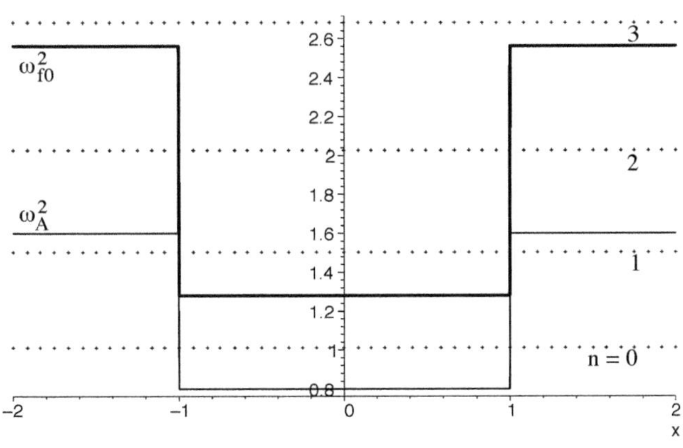

Fig. 10.14 Profiles of the Alfvén frequency (thin solid line) and the cutoff frequency ω_{f0} (thick solid line); $a = 1, b = 0.5, b_e = 1, k_y = 2, k_z = 1.6$. The horizontal dotted lines indicate the real parts of the frequencies of the surface mode (lowest) and the first three 'fast' modes.

Consider normal-mode solutions of the form

$$\boldsymbol{\xi}(\mathbf{r}, t) = \hat{\boldsymbol{\xi}}(x)\, e^{i(k_y y + k_z z - \omega t)} . \tag{10.84}$$

In the pressureless, uniform, plasma slab considered here, Eq. (7.80) for $\xi \equiv \hat{\xi}_x$ reduces to a simple Helmholtz equation:

$$\xi'' + k_x^2\,\xi = 0\,, \quad \text{with} \quad k_x^2 \equiv \frac{\omega^2}{b^2} - k_0^2, \tag{10.85}$$

where $k_0^2 \equiv k_y^2 + k_z^2$. This equation also applies to the displacement in the external plasma, ξ_e, with k_x^2 replaced by $k_{xe}^2 \equiv \omega^2/b_e^2 - k_0^2$. The general solution of these equations can be written as

$$\xi = \begin{cases} \alpha_L e^{ik_{xe}x} + \beta_L e^{-ik_{xe}x} & \text{for } x < -a, \\ \alpha e^{ik_x x} + \beta e^{-ik_x x} & \text{for } -a < x < a, \\ \alpha_R e^{ik_{xe}x} + \beta_R e^{-ik_{xe}x} & \text{for } a < x, \end{cases} \tag{10.86}$$

where the coefficients are determined by applying the boundary conditions.

Note that the results of Chapter 7 are recovered when perfectly conducting walls are put at $x = \pm a$. The boundary conditions $\xi(-a) = \xi(a) = 0$ then yield the dispersion relation $\sin(2ak_x) = 0$, so that

$$\omega_n^2 = (k_x^2 + k_0^2)\, b^2\,, \quad k_x = \frac{n\pi}{2a} \quad (n = 0, \pm 1, \pm 2, \dots)\,, \tag{10.87}$$

for the fast magneto-sonic modes of a finite homogeneous slab.

In the plasma–plasma case considered here, however, there is an external plasma and $k_{xe} \equiv (\omega^2/b_e^2 - k_0^2)^{1/2}$ can be real or imaginary, corresponding to oscillatory or evanescent behaviour in the external region. The boundary conditions to be considered in this case are

$$[\![\xi]\!]_{x=-a} = 0\,, \quad [\![\xi]\!]_{x=a} = 0\,, \quad \text{and} \quad [\![\Pi]\!]_{x=-a} = 0\,, \quad [\![\Pi]\!]_{x=a} = 0\,, \tag{10.88}$$

where Π denotes the total perturbed pressure. These boundary conditions can be written in terms of the mechanical impedance, the ratio of the alternating force to the alternating velocity:

$$\left[\!\!\left[\frac{\Pi}{-i\omega\xi}\right]\!\!\right]_{x=-a} = 0\,, \qquad \left[\!\!\left[\frac{\Pi}{-i\omega\xi}\right]\!\!\right]_{x=a} = 0\,. \tag{10.89}$$

Imposing these equivalent boundary conditions is known as 'impedance matching'. The internal solution gives rise to the 'transmitted' or 'internal mechanical' impedance $Z_{\rm trans} \equiv \Pi/(-i\omega\xi)$. The external solution can be split into outgoing and incoming waves (see Keppens [343]), each with a corresponding impedance:

$$Z_{\rm in} \equiv \frac{\Pi_{\rm in}}{-i\omega\xi_{\rm in}}, \qquad \text{and} \qquad Z_{\rm out} \equiv \frac{\Pi_{\rm out}}{-i\omega\xi_{\rm out}}. \tag{10.90}$$

The boundary conditions can be written in terms of these impedances as follows:

$$\frac{\xi_{\rm out}}{\xi_{\rm in}} = \frac{Z_{\rm in} - Z_{\rm trans}}{Z_{\rm trans} - Z_{\rm out}}. \tag{10.91}$$

To determine the possibly complex eigenfrequencies of the leaky modes, we have to consider the case of no incoming waves. The impedance criterion to be satisfied is then $Z_{\rm trans} = Z_{\rm out}$. (In the next chapter, we have to impose the impedance criterion $Z_{\rm trans} = Z_{\rm in}$ to find the frequencies that yield 100% absorption of the corresponding incoming wave, i.e. no outgoing waves.)

A similar treatment for the boundary conditions was carried out by Stenuit et al. [568] in the case of a cylindrical flux tube. The cylindrical geometry complicates the analysis considerably and equation (10.85) takes the form of a Bessel equation. The in- and outgoing wave solutions can then be expressed in terms of Hankel functions, where the boundary conditions becomes non-trivial. In order to determine what boundary conditions have to be imposed, one has to check the asymptotic behaviour of the Hankel functions and their contribution to the radial energy flux. In the slab geometry considered here this is trivial. The energy flux averaged over a period is defined as $S_x \equiv \frac{1}{2}\mathrm{Re}(-\Pi^* \mathrm{i}\omega\xi)$, where the asterisk denotes the complex conjugate. It is clear that, in the RHS plasma for instance, the solution $\alpha_\mathrm{R}\exp(\mathrm{i}k_{xe}x)$ yields a positive outward energy flux in the limit of infinitely large x, i.e. $\lim_{x\to\infty} S_x^{\mathrm{Out}} > 0$. On the other hand, the solution $\beta_\mathrm{R}\exp(-\mathrm{i}k_{xe}x)$ yields $\lim_{x\to\infty} S_x^{\mathrm{in}} < 0$. Hence, to find the leaky modes, the incoming wave has to be rejected and one has to set $\beta_\mathrm{R} = 0$. For non-leaky or body modes, the outgoing wave has to be rejected, i.e. $\alpha_\mathrm{R} = 0$. Similarly, for the LHS plasma.

(b) Normal-mode analysis To get the leaky modes, we set $\alpha_\mathrm{L} = \beta_\mathrm{R} = 0$ and define

$$F \equiv -\rho b \frac{\omega^2 - \omega_\mathrm{A}^2}{\omega^2 - \omega_\mathrm{f0}^2}\mathrm{i}k_x\,, \qquad F_\mathrm{e} \equiv -\rho_\mathrm{e} b_\mathrm{e} \frac{\omega^2 - \omega_\mathrm{A}^2}{\omega^2 - \omega_\mathrm{f0}^2}\mathrm{i}k_{xe}\,. \tag{10.92}$$

Impedance matching at $s = \pm a$ discussed in the previous section then yields a homogeneous system for the constants α and β:

$$\frac{F\big(\alpha\mathrm{e}^{\mathrm{i}k_x a} - \beta\mathrm{e}^{-\mathrm{i}k_x a}\big)}{\alpha\mathrm{e}^{\mathrm{i}k_x a} + \beta\mathrm{e}^{-\mathrm{i}k_x a}} = F_\mathrm{e}\,, \qquad \frac{F\big(\alpha\mathrm{e}^{-\mathrm{i}k_x a} - \beta\mathrm{e}^{\mathrm{i}k_x a}\big)}{\alpha\mathrm{e}^{-\mathrm{i}k_x a} + \beta\mathrm{e}^{\mathrm{i}k_x a}} = -F_\mathrm{e}\,. \tag{10.93}$$

This system has non-trivial solutions if

$$\frac{(F + F_\mathrm{e})^2}{(F - F_\mathrm{e})^2} = \mathrm{e}^{4\mathrm{i}k_x a}. \tag{10.94}$$

This is the dispersion relation for the leaky modes we were looking for.

Let us consider the simple case with $k_y = k_z = 0$, so that $\omega_\mathrm{A} = \omega_\mathrm{f0} = 0$ and $k_x^2 = \omega^2/b^2$, while $k_x\mathrm{e}^2 = \omega^2/b_\mathrm{e}^2$. The dispersion relation can then be solved analytically:

$$\omega = -\frac{\pi b}{2a}n - \frac{ib}{2a}\ln\left(\frac{\rho b + \rho_\mathrm{e} b_\mathrm{e}}{\rho b - \rho_\mathrm{e} b_\mathrm{e}}\right), \qquad (n = 0, \pm 1, \pm 2, \dots)\,. \tag{10.95}$$

For $b = 0.5, b_\mathrm{e} = 1$ and $a = 1$, this then yields the following leaky modes:

$$\omega = -\tfrac{1}{4}(n\pi + \mathrm{i}\ln 3) \qquad (n = 0, \pm 1, \pm 2, \dots)\,. \tag{10.96}$$

A typical eigenfunction is shown in Fig. 10.15.

Let us also consider the non-leaky modes. As the slab plasma is uniform, no coupling with continuum modes is possible. For body modes, the external solution is non-propagating so that there is no damping mechanism and the corresponding eigenfrequencies must be purely real. Thus, $k_{xe}^2 = \omega^2/b_\mathrm{e}^2 - k_0^2 \le 0$. A similar analysis as above, but now ignoring the outgoing wave solutions in the external plasma, gives rise to the dispersion relation

$$1 = \mathrm{e}^{4\mathrm{i}k_x a}. \tag{10.97}$$

For $b = 0.5, b_\mathrm{e} = 1$ and $a = 1$, this then yields the following body modes:

$$\omega = -\tfrac{1}{4}n\pi \quad (n = 0, \pm 1, \pm 2, \dots)\,. \tag{10.98}$$

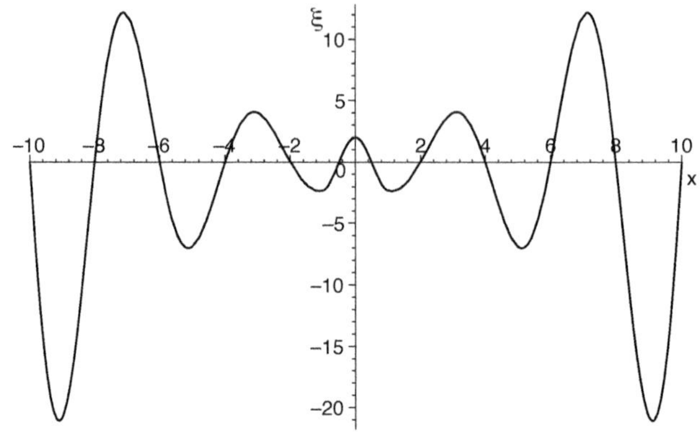

Fig. 10.15 Leaky mode eigenfunction ξ for $n = 2\,, b = 0.5\,, b_{\rm e} = a = 1\,, k_y = k_z = 0\,.$

Notice that in the limit $\rho_{\rm e} \to 0$, i.e. $\omega_{\rm f0e} \to \infty$ and $b_{\rm e} \to \infty$, no propagation is possible in the external region so that only body modes are found. Applying the limit to the dispersion relation found above indeed yields the dispersion relation for the body modes. In the opposite limit, $\rho_{\rm e} \to \infty$ and $b_{\rm e} \to 0$, we also only find body modes so that no outward propagation is possible.

(c) Initial value problem approach In order to show how the modes found appear in the response of the slab to an initial perturbation, we now formulate the corresponding initial value problem which we will solve by means of a Laplace transform (10.3). Applying this transform to the wave equation (10.1) and reducing it to the one-dimensional representation in terms of ξ alone (see Section 10.2.1), yields

$$\hat{\xi}''(x;\omega) + k_x^2\,\hat{\xi}(x;\omega) = \frac{{\rm i}\omega}{b^2}\,\xi(x;0) - \frac{1}{b^2}\,\dot{\xi}(x;0), \tag{10.99}$$

which is a simplified form of Eq. (10.4) applied to the uniform, pressureless, slab plasma considered here. The solution of this ODE for $\hat{\xi}(x;\omega)$ can be obtained by means of a Green's function, expressed in terms of two linearly independent solutions $\hat{\xi}_1$ and $\hat{\xi}_2$ of the homogeneous equation, where the first satisfies the boundary condition at $x = -\infty$, and the second satisfies the boundary condition at $x = +\infty$. The two solutions read:

$$\hat{\xi}_1 = \begin{cases} {\rm e}^{-{\rm i}k_{x{\rm e}}x} \\ \alpha_1{\rm e}^{{\rm i}k_x x} + \beta_1{\rm e}^{-{\rm i}k_x x} \\ \alpha_{\rm R}{\rm e}^{{\rm i}k_{x{\rm e}}x} + \beta_{\rm R}{\rm e}^{-{\rm i}k_{x{\rm e}}x} \end{cases}, \quad \hat{\xi}_2 = \begin{cases} \alpha_{\rm L}{\rm e}^{{\rm i}k_{x{\rm e}}x} + \beta_{\rm L}{\rm e}^{-{\rm i}k_{x{\rm e}}x} & (\text{for } x < -a) \\ \alpha_2{\rm e}^{{\rm i}k_x x} + \beta_2{\rm e}^{-{\rm i}k_x x} & (\text{for } -a < x < a) \\ {\rm e}^{{\rm i}k_{x{\rm e}}x} & (\text{for } a < x) \end{cases}. \tag{10.100}$$

According to the general theory, the conjunct $\hat{\xi}_2'\hat{\xi}_1 - \hat{\xi}_1'\hat{\xi}_2$ is independent of x. Hence, we determine the two solutions only in the region $-a < x < a$, i.e. we determine the constants $\alpha_{1,2}$ and $\beta_{1,2}$ by applying the boundary conditions specified above. The zeros of this conjunct give rise to poles of the Green's function and determine the discrete spectrum of the problem. These poles are the solutions of the same dispersion relation (10.94) already found with the normal mode analysis. Since the solutions are also the same, the normal-mode and initial value approach yield the same discrete spectrum, as they should.

10.6 Literature and exercises★

Notes on literature

Alfvén resonance: mode conversion

- An alternative treatment of waves in inhomogeneous plasmas can be found in 'Waves in plasmas' by Stix [570]. Chapter 13 considers wave propagation through an inhomogeneous plasma. Starting from a WKB approach, the Alfvén resonance is discussed in terms of the singular-turning-point theory and mode conversion.

Continuous spectrum in cold plasma oscillations

- Barston [36] solves the problem of cold plasma oscillations by means of a normal-mode analysis.
- Sedláček [546] analyses 'Electrostatic oscillations in cold inhomogeneous plasmas' and shows that inhomogeneities of the plasma lead to dissipationless damping. He applies both normal-mode analysis and the initial value problem approach and shows that the two are equivalent. Since the normal-mode analysis may overlook collective modes, he concludes that it cannot fully replace the Laplace transform technique.

Continuous spectrum and damping in MHD

- Tataronis and Grossmann [588] show that the continuous spectrum of ideal MHD, due to spatial inhomogeneities of the plasma, leads to dissipationless damping as a result of phase mixing (see Section 11.1).
- Chen and Hasegawa [118] extend the analysis of Tataronis & Grossmann [588] to compressible plasmas with magnetic shear (see further in Section 11.1).
- Goedbloed [221] constructs the resolvent operator, involved in the general solution of the initial value problem, for one-dimensional MHD problems (as presented in Section 10.2).

Leaky modes

- Wilson [650] studies the eigenspectrum of a flux tube of finite width embedded in a non-magnetized medium and considers complex frequencies and leakage of energy into the surroundings. This leakage of wave energy is further developed by Spruit [563] in the thin flux tube approximation, and by Cally [103] for a tube with arbitrary radius giving a classification of seven types of non-leaky modes.

Exercises

[10.1] *Derivation of Eq. (10.83)*

Generalize the derivation in Section 10.4 by considering a density profile that is constant but different in each of the two plasmas to derive Eq. (10.83).

[10.2] *Continuum damping in cylindrical plasmas*

In Chapter 9 we studied the continuous MHD spectrum in cylindrical plasmas. Consider a uniform cylindrical plasma with radius 1 and Alfvén frequency ω_{Ai} separated from another uniform surrounding plasma, characterized by ω_{Ae}, by a thin transition region $1-a \leq r \leq 1+a$ in which the square of the local Alfvén frequency varies linearly from ω_{Ai}^2 to ω_{Ae}^2. Repeat the derivation of the quasi-mode in Section 10.4 in this set-up.

[10.3] *Quasi-modes in real plasmas*

Use the tables in Appendix B to get a rough estimate of the oscillatory frequency and the damping time scales of the quasi-mode resulting from a surface wave in laboratory plasmas and in solar coronal loops (cf. Eq. (10.80)).

[10.4] *Damping of quasi-modes*

The damping rate of ideal quasi-modes does not correspond to a heating rate of the plasma. Why not? What is the meaning of this damping rate then? Explain.

[10.5] *Understanding Fig. 10.12*

Consider a finite ($-a < x < +a$), homogeneous, magnetized slab plasma surrounded by another homogeneous plasma, also with uniform magnetic field (in the z-direction). Write a Maple or Mathematica work sheet to plot the internal and external Alfvén frequency, ω_{Ai} and ω_{Ae}, versus k_z (choose parameter values so that these two frequencies are different). Also plot the frequency ω_{f0}, both for the internal and external plasma. Now add plots of the frequencies of the first three fast magneto-sonic modes (notice that k_x is quantized). Play around with the parameters (k_y, internal and external magnetic field and density) and observe that the frequencies of the fast magneto-sonic modes lie between ω_{Ai} and ω_{Ae} for large k_z. What is the fundamental difference of the figure you get from Fig. 10.12? How do you get Fig. 10.12 from here?

[10.6] *Leaky waves in cylindrical geometry*

Consider a uniform, cylindrical magnetic flux tube embedded in a uniform, wave carrying plasma and repeat the derivation of the leaky waves in Section 10.5 for this configuration.

[10.7] *Resonant leaky modes*

Combine the derivations in Sections 10.4 and 10.5 and derive the dispersion relation for resonant leaky modes in a non-uniform plasma. (Consider a plane slab.)

11

Resonant absorption and wave heating

In Chapters 7 and 9, ideal MHD spectral analysis of a plasma with inhomogeneities in one spatial direction led to a singular second order differential equation for the plasma displacement in the direction of inhomogeneity: Eqs. (7.80) and (9.30). The two singularities of these equations give rise to the Alfvén and slow magneto-sonic continua. The corresponding eigenfunctions possess non-square integrable tangential components leading to extreme anisotropic behaviour. This has a dramatic effect on the dynamics of inhomogeneous plasmas. In this chapter, we will discuss the consequences of the continuous spectra for the dynamical response of an inhomogeneous plasma to periodic, multi-periodic or random external drivers. This leads to the concepts of *resonant 'absorption'* (or 'dissipation') of the waves and *phase mixing* of the magnetic field lines.

Since all plasmas occurring in nature are inhomogeneous and since waves can be excited easily in plasmas, these phenomena frequently occur. They have been studied in the context of wave damping and heating for controlled thermonuclear fusion experiments and for solar and astrophysical plasmas, whereas resonant mode conversion is also studied in magnetospheric physics [429]. In the solar context, resonant absorption and phase mixing play a dominant role in 'sunspot seismology' [78] and 'coronal seismology' [458] where the interactions of MHD waves with magnetic structures are exploited as diagnostic tools, as in helioseismology and 'MHD spectroscopy' (see Section 7.2.4).

The physical mechanisms of resonant absorption and phase mixing are explained in detail for a simple slab configuration in Section 11.1. In Section 11.2 the applications of this mechanism to wave damping and heating are then illustrated for a generic cylindrical plasma. In Section 11.3, alternative configurations with different boundary conditions are presented together with their consequences for applications to solar and magnetospheric physics.

11.1 Ideal MHD theory of resonant absorption

11.1.1 Analytical solution of a simple model problem

Consider a semi-infinite plasma that occupies the half space $x < 0$ next to a vacuum in the half space $x > 0$. This is the plane slab version of model problem III introduced in Section 4.6.1. Let us assume, for simplicity, that there is a uni-directional magnetic field in both vacuum and plasma and choose the z-axis in the direction of the magnetic field. We also assume that all equilibrium quantities are constant for $x < x_1$ ($\equiv -a$) and only depend on x in a region $x_1 \leq x \leq x_2$ ($\equiv 0$).

In the vacuum ($x > 0$), we assume the presence of a sheet current $\mathbf{j}_c^\star$ in a coil which represents the external driving source or 'antenna' at $x = c$ (see Fig. 11.1, where the plasma inhomogeneity indicated by the quantities ϵ_1 and ϵ_2 is still to be defined).

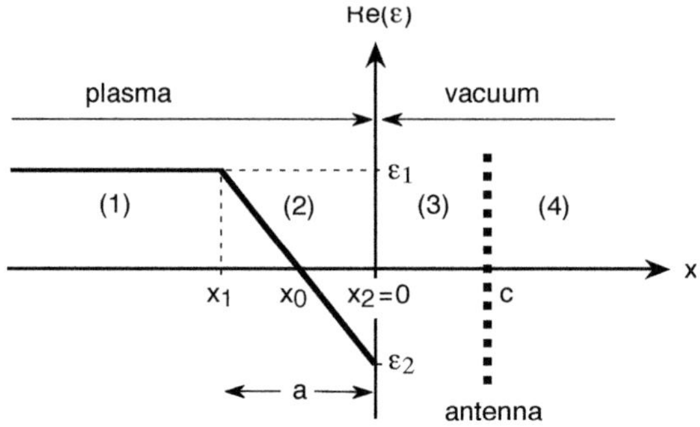

Fig. 11.1 Sketch of the configuration in which the resonant absorption of Alfvén wave energy is studied. The power source consists of an external 'antenna' with surface current $\mathbf{j}_c^\star$ situated in the vacuum region next to the plasma. Here $\epsilon(x) \equiv \rho(x)[\omega^2 - \omega_A^2(x)]$. (Adapted from Chen and Hasegawa [118].)

This configuration is a simplified version of the set-up studied in one of the first papers in which the mechanism of resonant 'absorption' was investigated in the framework of thermonuclear fusion [118] as a possible supplementary heating mechanism to bring tokamak plasmas into the ignition regime. (In the original set-up by Chen and Hasegawa [118] the magnetic field is sheared. They generalized earlier results of Tataronis and Grossmann [588] by also including compressibility, so that the important role of fast magneto-sonic waves could be considered.) As a matter of fact, the singular behaviour of the excited waves results in the small length scales required for efficient dissipation of wave energy in highly conductive plasmas. The solution derived by Chen and Hasegawa will be discussed for the special case of a uni-directional magnetic field.

Since the equilibrium quantities are constant in time and do not depend on y and z, we assume linear perturbations of the form

$$\boldsymbol{\xi}(x, y, z; t) = \boldsymbol{\xi}(x)\, \mathrm{e}^{\mathrm{i}(k_y y + k_z z - \omega t)}\,, \tag{11.1}$$

where $\omega = \omega_d + \mathrm{i}\nu$, with ω_d the frequency of the external driver and ν a small positive constant ($0 \le \nu \ll \omega_d$) to guarantee causality. This is further elaborated in Section 11.1.2. We concentrate on Alfvén wave absorption and separate off the slow magneto-sonic waves by considering the low-β approximation ($p \ll \frac{1}{2}B^2$) of Eq. (7.78) of Section 7.3.2. Since the magnetic field is uni-directional, the parallel and perpendicular 'wave numbers' $k_\parallel$ and $k_\perp$ are constant. For vanishing gravity, we then get a system of two differential equations:

$$\begin{pmatrix} \rho\omega^2 + \dfrac{d}{dx}B^2\dfrac{d}{dx} - k_\parallel^2 B^2 & \dfrac{d}{dx}k_\perp B^2 \\ -k_\perp B^2 \dfrac{d}{dx} & \rho\omega^2 - k_0^2 B^2 \end{pmatrix} \begin{pmatrix} \xi_x \\ \xi_\perp \end{pmatrix} = 0\,, \tag{11.2}$$

where $k_0^2 \equiv k_\perp^2 + k_\parallel^2$. This system describes Alfvén modes and fast magneto-sonic modes. (Note that we now exploit $\xi_x \equiv \xi$ and $\mathrm{i}\xi_\perp \equiv \eta$ as variables since we will need the symbol η for the plasma resistivity in Section 11.1.3.) In general, i.e. for $k_\perp \neq 0$, these two modes are coupled since the equations (11.2) are coupled. This *coupling is of vital importance* for the mechanism of resonant absorption. We will see that the mechanism *does not work for* $k_\perp = 0$, at least not in this set-up with sideways excitation (see Section 11.2). The second equation is algebraic in $\xi_\perp$ so that it allows us to express $\xi_\perp$ in terms of ξ_x:

$$\mathrm{i}\xi_\perp = \frac{k_\perp B^2}{\rho\omega^2 - k_0^2 B^2}\,\frac{d\xi_x}{dx}\,. \tag{11.3}$$

Upon substitution of this expression in the first line of Eq. (11.2) one obtains a second order ODE for the variable ξ_x:

$$\frac{d}{dx}\left(\frac{\rho\omega^2 - k_\parallel^2 B^2}{\rho\omega^2 - k_0^2 B^2} B^2 \frac{d\xi_x}{dx}\right) + (\rho\omega^2 - k_\parallel^2 B^2)\,\xi_x = 0\,, \tag{11.4}$$

which is the simplified form of Eq. (7.80) under the assumptions made.

Note that the Alfvén frequency,

$$\omega_\mathrm{A} \equiv k_\parallel v_\mathrm{A}\,, \quad v_\mathrm{A} \equiv B(x)\big/\sqrt{\rho(x)}\,, \tag{11.5}$$

is a function of x in the layer $-a \le x \le 0$ through the inhomogeneity of the magnetic field or the density, or both. These inhomogeneities may be considered to be arbitrary, but the equilibrium condition $(p + \frac{1}{2}B^2)' = 0$ requires the gradient of the magnetic pressure to be balanced by $-p'$, which should be small because of the low-β assumption. For definiteness, we will assume that the magnetic field is constant and that inhomogeneities are exclusively due to the density profile $\rho(x)$.

We now focus on nearly perpendicular propagation, i.e. $k_\parallel \ll k_\perp \approx k_0$, and very strong coupling (although the fast wave sub-spectrum is well separated from the Alfvén wave sub-spectrum in this case: $\omega_\mathrm{A}^2 = k_\parallel^2 v_\mathrm{A}^2 \ll \omega_\mathrm{f0}^2 \approx k_0^2 v_\mathrm{A}^2$). We also assume that the oscillatory part ω_d of the frequency ω lies in the range of the Alfvén continuum. Equation (11.4) then reduces to

$$\frac{d}{dx}\left(\rho\left[\omega^2 - \omega_\mathrm{A}^2(x)\right]\frac{d\xi_x}{dx}\right) - k_\perp^2 \rho\left[\omega^2 - \omega_\mathrm{A}^2(x)\right]\xi_x = 0\,, \tag{11.6}$$

which corresponds to Eq. (10.49) of Section 10.3 for $k_0^2 \approx k_\perp^2$ and vanishing right hand side (since we here consider the boundary value problem, instead of the initial value problem that was considered in Section 11.3). Defining a complex quantity $\epsilon(x) \equiv \rho(x)\left[\omega^2 - \omega_\mathrm{A}^2(x)\right]$, we again get an equation that is singular at $\epsilon = 0$:

$$\frac{d^2\xi_x}{dx^2} + \frac{1}{\epsilon}\,\frac{d\epsilon}{dx}\,\frac{d\xi_x}{dx} - k_\perp^2\xi_x = 0\,. \tag{11.7}$$

Hence, the solutions will have the general form (10.54), with one solution exhibiting a logarithmic singularity. We now consider the simple case where $\mathrm{Re}[\epsilon(x)]$ depends linearly on x in the inhomogeneous layer:

$$\mathrm{Re}[\epsilon(x)] = \begin{cases} \epsilon_1 & (x < x_1) \\ \epsilon_1 + \dfrac{\epsilon_2 - \epsilon_1}{x_2 - x_1}(x - x_1) & (x_1 \le x \le x_2) \end{cases}\,, \tag{11.8}$$

as displayed in Fig. 11.1. The zero of the function $\mathrm{Re}[\epsilon(x)]$ corresponds to the resonant position

$x = x_0 \equiv (\epsilon_1 x_2 - \epsilon_2 x_1)/(\epsilon_1 - \epsilon_2)$, where the driving frequency matches the local Alfvén frequency: $|\omega_{\rm d}| = |\omega_{\rm A}(x_0)|$. For this simple configuration, Eq. (11.7) can be solved analytically, just as Eq. (10.70) of Section 10.4.

We can distinguish four regions in the set-up as sketched in Fig. 11.1. The plasma occupies regions (1) and (2), while the vacuum is situated in regions (3) and (4). The equations for the vacuum regions will be derived below. They can also be solved analytically. The main issue is then to connect the solutions across those regions. This 'simple' problem actually involves a mixture of all the laboratory plasma models introduced in Section 4.6.1, and elaborated for the linearized interface problems in Section 6.6.1. Going from left to right, subsequently, the following boundary conditions (BCs) are to be imposed:

– At $x = x_{\rm L}$ ($\rightarrow -\infty$): the plasma–rigid wall (model I) BCs (4.167)–(4.168) for the plasma displacement at a conducting wall which, since it is actually absent in the present case, is equivalent to imposing a boundary condition at $-\infty$:

$$\xi_x \rightarrow 0 \qquad (\text{for } x \rightarrow -\infty)\,; \tag{11.9}$$

– At $x = x_1$ ($\equiv -a$): the plasma–plasma (model II*) BCs (4.162)–(4.164) which, upon linearization about the *perturbed* boundary, yield the first and second interface conditions (6.135) and (6.137) derived in Section 6.6.1:

$$[\![\xi_x]\!] = 0 \qquad (\text{at } x = x_1)\,, \tag{11.10}$$

$$[\![\xi_x{}']\!] = 0 \qquad (\text{at } x = x_1)\,; \tag{11.11}$$

– At $x = x_2$ ($\equiv 0$): the plasma–vacuum (model II) BCs (4.173)–(4.174) which, upon linearization about the *perturbed* boundary, yield the first and second interface conditions (6.131) and (6.134) derived in Section 6.6.1:

$$\mathrm{i}k_\parallel \hat{B}\xi_x = \hat{Q}_x \qquad (\text{at } x = x_2)\,, \tag{11.12}$$

$$Q_z = -B(\xi_x{}' + \mathrm{i}k_\perp \xi_y) = \hat{Q}_z \qquad (\text{at } x = x_2)\,; \tag{11.13}$$

– At $x = c$: the antenna (model III) BCs (4.177)–(4.178) for the jump of the perturbed magnetic field are produced by the currents in the antenna:

$$[\![\hat{Q}_x]\!] = 0 \qquad (\text{at } x = c)\,, \tag{11.14}$$

$$\mathbf{n} \times [\![\hat{\mathbf{Q}}]\!] = \mathbf{j}_{\rm c}^\star \qquad (\text{at } x = c)\,; \tag{11.15}$$

– At $x = x_{\rm R}$ ($\rightarrow \infty$): the vacuum–rigid wall (model II) BC (4.171) for the perturbed magnetic field at a conducting wall which, since it is again absent in the present case, is equivalent to imposing a BC at ∞:

$$\hat{Q}_x \rightarrow 0 \qquad (\text{for } x \rightarrow \infty)\,. \tag{11.16}$$

Now count: Eqs. (11.9)–(11.16) provide eight BCs to determine the eight arbitrary amplitudes of the independent solutions of the second order differential equations for the plasma variable ξ_x and the vacuum variable $\hat{Q}_z$ in the four regions (1)–(4). Also note the difference between the present forced oscillation problem, where arbitrary values of $\omega_{\rm d}$ and the amplitude of the antenna current $\mathbf{j}_{\rm c}^\star$ determine everything else, and the problem of free oscillations, where $\mathbf{j}_{\rm c}^\star = 0$ and ω becomes an eigenvalue (because eight BCs is one too much in that case, since the amplitude of a linear eigenoscillation is arbitrary then).

(a) Plasma solution In the left plasma region (1), the coefficient ϵ is constant so that Eq. (11.7) reduces to a simple Helmholtz equation with the general solution

$$\xi_x^{(1)} = A_1\, \mathrm{e}^{k_\perp (x+a)} + B_1\, \mathrm{e}^{-k_\perp (x+a)} , \tag{11.17}$$

where $B_1 = 0$ because of the BC (11.9), and A_1 is determined by one of the BCs (11.10), (11.11), which are simple to implement for this specific solution.

In the right plasma region (2), the coefficient function ϵ varies linearly with x. In terms of the normalized variable

$$X \equiv k_\perp \epsilon / \epsilon' , \quad \text{so that } \ \mathrm{Re}(X) = k_\perp (x - x_0) , \tag{11.18}$$

where $\epsilon' \equiv d\epsilon/dx$, Eq. (11.7) reduces to the zeroth order modified Bessel equation:

$$\frac{d^2 \xi_x}{dX^2} + \frac{1}{X}\frac{d\xi_x}{dX} - \xi_x = 0 . \tag{11.19}$$

Hence, the solution for ξ_x in region (2) can be written as a linear combination of the modified Bessel functions of the first and second kind [3]:

$$\xi_x^{(2)} = A_1 \left[A_2 I_0(X) + B_2 K_0(X) \right] , \tag{11.20}$$

where multiplication with A_1 simplifies implementation of the BCs at $x = x_1$.

▷ **Bessel function relations** The following identities are needed here:

$$\begin{aligned} &I_0{}' = I_1 , && I_1{}' = I_0 - X^{-1} I_1 , \\ &K_0{}' = -K_1 , && K_1{}' = -K_0 - X^{-1} K_1 , \\ &I_0 K_1 + K_0 I_1 = X^{-1} , && \end{aligned} \tag{11.21}$$

where the prime denotes differentiation with respect to X. For small X, the expansions, given by Eqs. (10.71) with $k_0 \zeta \to X$, exhibit a logarithmic singularity for $K_0(X)$. ◁

The boundary conditions (11.10) and (11.11) at $x = x_1$ then yield

$$\begin{aligned} A_2 I_0(X_1) + B_2 K_0(X_1) &= 1 , \\ A_2 I_1(X_1) - B_2 K_1(X_1) &= 1 , \end{aligned}$$

where the Bessel functions are evaluated at the left boundary $X_1 \equiv X(x = x_1)$ of interval (2). From this system, the constants A_2 and B_2 become

$$A_2 = X_1 \Big[K_0(X_1) + K_1(X_1) \Big] , \qquad B_2 = -X_1 \Big[I_0(X_1) - I_1(X_1) \Big] , \tag{11.22}$$

so that ξ_x is determined in the whole plasma region up to a scale factor A_1. This scale factor depends on the current density in the external antenna and follows from linking the plasma solution to the vacuum solution, which will be determined now.

(b) Vacuum solution The magnetic field perturbations in the vacuum regions (3) and (4) satisfy the equations $\nabla \times \hat{\mathbf{Q}} = 0$ and $\nabla \cdot \hat{\mathbf{Q}} = 0$. The first one gives

$$\hat{Q}_y = \frac{k_\perp}{k_\parallel}\hat{Q}_z\,, \qquad \hat{Q}_x = -\frac{\mathrm{i}}{k_\parallel}\frac{d\hat{Q}_z}{dx}\,, \tag{11.23}$$

which upon substitution in the second one yields a Helmholtz equation for $\hat{Q}_z$:

$$\frac{d^2\hat{Q}_z}{dx^2} - k_0^2\hat{Q}_z = 0\,. \tag{11.24}$$

Hence, the solutions in regions (3) and (4) can be written in the convenient form

$$\hat{Q}_z^{(3)} = A_3 \mathrm{e}^{-k_0(x-c)} + B_3 \mathrm{e}^{k_0(x-c)}\,, \qquad \hat{Q}_z^{(4)} = A_4 \mathrm{e}^{-k_0(x-c)} + B_4 \mathrm{e}^{k_0(x-c)}\,, \tag{11.25}$$

where the BC (11.16) implies that $B_4 = 0$.

Now comes an important part. The spatial and temporal dependence of the dynamics is forced onto the system by means of the antenna surface current

$$\mathbf{j}_\mathrm{c}^\star = (0, j_y^\star, j_z^\star)\, \mathrm{e}^{\mathrm{i}(k_y y + k_z z - \omega t)}\,, \tag{11.26}$$

which has the dimension of current per unit length. Since $\nabla \cdot \mathbf{j}_\mathrm{c}^\star = 0$, the wave vector $\mathbf{k}_0 \equiv (0, k_\perp, k_\parallel)$ will be perpendicular to the antenna current: $\mathbf{k}_0 \cdot \mathbf{j}_\mathrm{c}^\star = 0$. Hence, four of the five constants $j_y^\star, j_z^\star, k_\perp, k_\parallel, \omega_\mathrm{d}$ can be freely chosen, corresponding to a possible construction of the antenna as a series of wires with a current per unit length of amplitude $|\mathbf{j}_\mathrm{c}^\star|$ that varies harmonically from wire to wire, with the same temporal constant ω_d, and with a spatial constant $\mathbf{k}_0$ with an inclination with respect to the z-axis determined by $k_\parallel / k_\perp = -j_y^\star / j_z^\star$. The experimentalist appears to have complete control over everything! Not quite though: in addition to the continuum modes, the external antenna also excites the 'quasi-mode' in the plasma that is related to the plasma–vacuum surface mode (see Section 10.4). This collective mode has a frequency $\omega = \omega_0 + \mathrm{i}\nu_0$, where the parameter ν_0 (which is negative) contains the information of the continuous spectrum of Alfvén frequencies (11.5) in the inhomogeneous layer. Such easy-to-excite quasi-modes will turn out to determine the efficiency of the resonant absorption mechanism.

The jump of the vacuum magnetic field components at $x = c$ is determined by the BCs (11.14) and (11.15) giving two independent relations,

$$[\![\hat{Q}_x]\!] = \hat{Q}_x^{(4)}\Big|_{x=c} - \hat{Q}_x^{(3)}\Big|_{x=c} = \mathrm{i}(k_0/k_\parallel)\,(A_4 - A_3 + B_3) = 0\,, \tag{11.27}$$

$$[\![\hat{Q}_z]\!] = \hat{Q}_z^{(4)}\Big|_{x=c} - \hat{Q}_z^{(3)}\Big|_{x=c} = A_4 - A_3 - B_3 = -j_y^\star\,, \tag{11.28}$$

whereas the third relation is not independent, $[\![\hat{Q}_y]\!] = (k_\perp/k_\parallel)[\![\hat{Q}_z]\!]$, from Eq. (11.23)(a). With these two conditions, the constants A_3 and B_3 can be expressed in terms of A_4 and the current in the antenna coil:

$$A_3 = \tfrac{1}{2} j_y^\star + A_4\,, \qquad B_3 = \tfrac{1}{2} j_y^\star\,, \tag{11.29}$$

so that the magnetic field perturbation in regions (3) and (4) becomes

$$\hat{Q}_z^{(3)} = j_y^\star \cosh[k_0(x-c)] + A_4 \mathrm{e}^{-k_0(x-c)}\,, \qquad \hat{Q}_z^{(4)} = A_4 \mathrm{e}^{-k_0(x-c)}\,. \tag{11.30}$$

Hence, the vacuum solution consists of a *driven part*, confined to region (3), that depends on the current in the external coil, and an *induced part* with scale factor A_4 that depends on the plasma response, which is still unknown at this point.

(c) Linking the vacuum solution to the plasma solution We applied six of the eight boundary conditions determining the plasma and vacuum solutions up to the scale factors A_1 and A_4. Since we are dealing with a driven problem, there is a *unique solution*. In order to determine it, we only have to find expressions for these scale factors. They follow from linking the vacuum to the plasma solution by means of the two BCs (11.12) and (11.13) at the plasma–vacuum interface:

$$k_\parallel B \xi_x^{(2)}\Big|_{x=x_2} = -\frac{1}{k_\parallel}\frac{d}{dx}\hat{Q}_z^{(3)}\Big|_{x=x_2}, \tag{11.31}$$

$$-(k_\parallel^2/k_\perp^2) B \frac{d}{dx}\xi_x^{(2)}\Big|_{x=x_2} = \hat{Q}_z^{(3)}\Big|_{x=x_2}. \tag{11.32}$$

By substituting the derived expressions $\xi_x^{(2)}$ and $\hat{Q}_z^{(3)}$ in these equations, one obtains two equations for the determination of the amplitudes A_1 and A_4,

$$\alpha A_1 - \mathrm{e}^{k_0 c} A_4 = j_y^\star \sinh(k_0 c)\,, \qquad -\beta A_1 - \mathrm{e}^{k_0 c} A_4 = j_y^\star \cosh(k_0 c)\,,$$

where α and β are abbreviations involving the constants A_2 and B_2 determined in Eqs. (11.22) and the Bessel functions evaluated at x_2:

$$\alpha \equiv (k_\parallel^2 B/k_\perp)\Big[A_2 I_0(X_2) + B_2 K_0(X_2)\Big]\,, \qquad \beta \equiv (k_\parallel^2 B/k_\perp)\Big[A_2 I_1(X_2) - B_2 K_1(X_2)\Big]\,. \tag{11.33}$$

The final solution is

$$A_1 = -j_y^\star \frac{\mathrm{e}^{-k_0 c}}{\alpha+\beta}\,, \qquad A_4 = -\tfrac{1}{2} j_y^\star \Big(1 + \frac{\alpha-\beta}{\alpha+\beta}\mathrm{e}^{-2k_0 c}\Big)\,, \tag{11.34}$$

so that all constants turn out to be proportional to $j_y^\star$, as they should.

11.1.2 Role of the singularity

The solution appears to be determined completely now, in the plasma as well as in the vacuum: the eight constants A_i, B_i $(i = 1, \dots 4)$ are determined by eight boundary conditions. Clearly, the whole solution is proportional to the current in the external coil, $\mathbf{j}_c = (0, 1, -k_\perp/j_\parallel) j_y^\star$, and vanishes when there is no external driving current. Concerning the dependence on the driving frequency ω_d: when ω_d^2 is outside the range of the Alfvén continuum $\{\omega_A^2(x) | x_1 \le x \le x_2\}$, we could assume $\nu = 0$ and all constants to be real so that the obtained solution would be adequate. On the other hand, when ω_d^2 is in the continuum range, the singularity at x_0 where $\omega_d = \omega_A(x_0)$ may be avoided by considering a *complex* value $\omega = \omega_d + i\nu$ with $\nu = 0^+$. As mentioned in Section 11.1.1, this guarantees causality since the response of the wave amplitude (11.1) to the external driver vanishes as $t \to -\infty$ (see, e.g., Clemmov and Dougherty [124], p. 32).

In the solution of the initial value problem for plasma oscillations by means of the Laplace transform, the corresponding *Ansatz* is the Landau prescription for the integration contour in the complex ω-plane to be taken parallel to the real axis with a small positive imaginary part (Fig. 2.7(a)). To determine the asymptotic behaviour for $t \to \infty$, this contour is to be deformed into the lower half of the ω-plane around the branch points and zeros of the dispersion equation. This yields damped solutions. All this is not specific for the kinetic description but carries over to MHD waves, as outlined in Sections 6.3.2 (Fig. 6.8), 10.3 and 10.4.

An alternative way of treating the Alfvén singularity in the driven system is to consider a complex driving frequency $\omega = \omega_\mathrm{d} - \mathrm{i}\delta$ with an 'artificial' damping coefficient $\delta > 0$ (see Appert et al. [20]). The role of this 'artificial' damping is to mimic genuine resistive (or viscous) dissipation while keeping the analysis tractable. Below, it will be shown that the actual energy 'absorption' thus obtained *corresponds to real damping when dissipation is taken into account*. This may appear strange since in *ideal* MHD there can be no dissipation or heating. Hence, the early ideal MHD studies introduced the term resonant 'absorption' rather than resonant 'heating' or 'dissipation'. The point is, however, that *the actual energy absorption (or dissipation) rate does not depend on the value δ (nor on the resistivity or viscosity)*. It is the logarithmic singularity that causes the wave phase mixing and, hence, the dissipation of the energy of the excited wave. This process is reversible in ideal MHD, but genuine dissipation (no matter how small) destroys the reversibility and converts the dissipated energy into thermal energy.

Before we discuss the numerical results obtained with the artificial damping coefficient δ, let us complete the analysis of Section 11.1.1 and, next, derive the energy absorption rate by means of the first approach (with infinitesimal $\nu > 0$). The consequences of the logarithmic singularity of the solution of Eq. (11.19) need to be investigated first. We encountered this singularity many times before in Chapters 7, 9 and 10. It was shown there, notably in Eqs. (7.135) and (10.54) that, close to the Alfvén singularity, the general solution ξ_x consists of a 'small' and a logarithmic 'large' part, where the latter should be continuous but the former may jump. The associated tangential component $\xi_y \sim \xi_x'$ then exhibits a δ-function dependence. Close enough to the singularity, where $|\epsilon| \ll 1$ and $|X| \ll 1$, the expansion (10.71) of the modified Bessel functions yields $I_0(X) \approx 1$ and $K_0(X) \approx -\ln(X)$ so that the solution (11.20) may be approximated there by

$$\xi_x \approx C_0 \ln(X)\,, \qquad C_0 \equiv -A_1 B_2\,. \tag{11.35}$$

To evaluate the complex function $\ln(X)$ on the appropriate side of the branch cut emanating from $X = 0$ (see Fig. 11.1), we need to relate the complex function $X(x;\omega) \equiv k_\perp \epsilon/\epsilon'$ to the complex function $\epsilon(x;\omega) \equiv \rho(\omega^2 - \omega_\mathrm{A}^2)$ in the neighbourhood of that point while accounting for the signs of the different quantities:

$$\epsilon \equiv \rho\big[(\omega_\mathrm{d} + \mathrm{i}\nu)^2 - \omega_\mathrm{A}^2\big] \approx \rho(\omega_\mathrm{d}^2 - \omega_\mathrm{A}^2 + 2\mathrm{i}\omega_\mathrm{d}\nu) = \epsilon_r'(x - x_0) + \mathrm{i}\epsilon_i\,, \tag{11.36}$$

where $\epsilon_r \equiv \mathrm{Re}(\epsilon)$ and $\epsilon_i \equiv \mathrm{Im}(\epsilon) = 2\rho\omega_\mathrm{d}\nu$, so that

$$X \approx k_\perp \epsilon(x)/\epsilon_r'(x_0) \approx k_\perp(x - x_0) + 2\mathrm{i}k_\perp\rho(\omega_\mathrm{d}/\epsilon_r')\nu\,. \tag{11.37}$$

Hence, the dominant contributions for ξ_x and ξ_y in the range of the Alfvén continuum, i.e. for $x_1 < x < x_2$, become

$$\begin{aligned} \xi_x &= \lim_{\nu \downarrow 0} C_0 \ln(X) = C_0\Big[\ln|k_\perp(x - x_0)| + \mathrm{i}\pi\,\mathrm{sgn}(\omega_\mathrm{d}/\epsilon_r')H(x_0 - x)\Big]\,, \\ \xi_y &\approx (\mathrm{i}/k_\perp)\xi_x'\,, \qquad \xi_x' = C_0\Big[(x - x_0)^{-1} - \mathrm{i}\pi\,\mathrm{sgn}(\omega_\mathrm{d}/\epsilon_r')\delta(x - x_0)\Big]\,. \end{aligned} \tag{11.38}$$

For definiteness, we have assumed here that $k_\perp > 0$, so that the points where $H(x_0 - x) \neq 0$ are to the left of the singularity $x = x_0$.

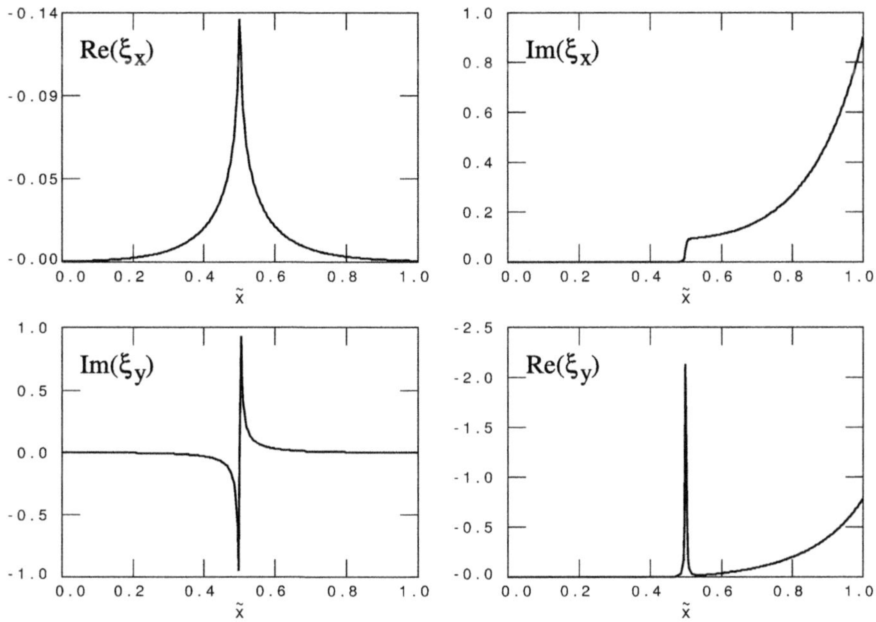

Fig. 11.2 Solutions ξ_x and ξ_y in region (2) as a function of the shifted coordinate $\tilde{x} \equiv x/a + 1$ for $a = B = 1$, $\omega_\mathrm{d} = k_\parallel = j_y^* = 1$, and a density such that $\mathrm{Re}(\epsilon_1) = 1$ and $\mathrm{Re}(\epsilon_2) = -1$ so that $\tilde{x}_0 = 0.5$; $\delta = 0.001$, $k_\perp = 5$.

In Fig. 11.2, the numerical solution in region (2) has been plotted for a choice of parameters that yields a singularity at $x = 0.5$, whereas a small value of the artificial damping coefficient δ prevents unlimited amplitudes but still permits nearly singular behaviour. The ideal MHD solution (11.38) is clearly recognizable, with the logarithmic behaviour near $x = x_0$ in the real part of ξ_x, the jump in the imaginary part of ξ_x, and the corresponding $(x - x_0)^{-1}$ and $\delta(x - x_0)$ behaviour of the tangential perpendicular component ξ_y. This jump determines the energy absorption rate, as will be shown below.

The different panels of Fig. 11.2 combine with the physical time dependence of the perturbations in an intricate manner. Note that we have exploited complex notation throughout this book. However, physical quantities are *real*. They are obtained from the complex mathematical ones by taking the real part at any point in the calculation: $\mathrm{Re}(A\mathrm{e}^{-\mathrm{i}\omega t}) = A_r \cos\omega t + A_i \sin\omega t$. Hence, the complex quantity $A\mathrm{e}^{-\mathrm{i}\omega t}$ represents a harmonic with a component A_r in phase with $\cos\omega t$ and a component A_i in phase with $\sin\omega t$. Accordingly, the temporal behaviour of the solution of Fig. 11.2 will show the real parts (left for ξ_x, right for ξ_y) at the beginning of each oscillation period and the imaginary parts half a period later. At any other moment, a combination of the two will be displayed.

(a) Energy 'absorption rate' With the solution (11.38), the energy 'absorption rate' resulting from the resonant plasma dynamics can be determined explicitly for the simple configuration considered. At this point, instead of following Chen and Hasegawa [118], who exploited the approximated expression for the energy flow in terms of the Poynting vector (neglecting the effects of the pressure

and the kinetic energy), we here exploit the exact expression derived by Canobbio [105]. In this way, the present example will provide a beautiful illustration of the implications of the general conservation concepts developed in Chapters 4 and 6.

Recall the nonlinear energy conservation equation derived in Section 4.3.2 for a plasma surrounded by a rigid wall and extended in Section 4.6.2 for a plasma–vacuum system with an antenna, injecting electromagnetic energy into the plasma. In the linear version, the energy conservation law is governed by the property of self-adjointness of the force operator $\mathbf{F}$, proved in Section 6.2.3 for a plasma surrounded by a rigid wall and in Section 6.6.2 for a plasma–vacuum system. The latter two sections contain all crucial relations needed to derive the energy conservation law for the combined plasma–vacuum–antenna system in terms of the kinetic energy K defined in Eq. (6.40) and the potential energy W defined in Eq. (6.83). The derivation is done by forming the inner product of the equation of motion (6.23) with $\partial\boldsymbol{\xi}^*/\partial t$, adding the complex conjugate expression, integrating over the plasma volume, and splitting off the plasma, surface and vacuum energies $W^{\rm p}$, $W^{\rm s}$ and $W^{\rm v}$ defined in Eqs. (6.146)–(6.148) by consistent integration by parts. This is effected by exploiting the self-adjointness properties (6.56) and (6.142), where the jumps at the resonance ($x = x_0$) and at the antenna ($x = c$) give rise to separate surface integrals. With these hints, the serious reader should be able to test her/his mastery of the MHD analysis expanded in this book by performing this feat by her/himself. Here is the desired result:

$$\frac{d}{dt}(W^{\rm p} + W^{\rm s} + W^{\rm v} + K) + {\rm Re}(P_{\rm res}) = {\rm Re}(P_{\rm ant})\,. \tag{11.39}$$

For generality, the surface integral $W^{\rm s}$ is included for equilibria that have a surface current on the plasma–vacuum interface. In our model problem, $W^{\rm s} = 0$. All volume integrals consist of two contributions, one from the region to the left and one from the region to the right of the pertinent jump. The power ${\rm Re}(P_{\rm res})$ absorbed in the resonant layer is expressed in terms of a jump across the layer, defined by $[\![f]\!]_{x_0} \equiv f(x_0^+) - f(x_0^-)$, which involves the total pressure perturbation $\Pi \equiv -\gamma p\nabla\cdot\boldsymbol{\xi} - \boldsymbol{\xi}\cdot\nabla p + \mathbf{B}\cdot\mathbf{Q}$ and the normal displacement $\xi \equiv \mathbf{n}\cdot\boldsymbol{\xi}$:

$$P_{\rm res} \equiv -\int \Big[\!\!\Big[\frac{\partial\xi^*}{\partial t}\,\Pi(\boldsymbol{\xi})\Big]\!\!\Big]_{x_0} dS_{x_0}\,. \tag{11.40}$$

The power ${\rm Re}(P_{\rm ant})$ emitted by the antenna involves the electric and magnetic field perturbations of the vacuum, derived from a vector potential, $\hat{\mathbf{E}} = -\partial\hat{\mathbf{A}}/\partial t$ and $\hat{\mathbf{Q}} = \nabla\times\hat{\mathbf{A}}$, with a jump in the tangential components of the magnetic field caused by the surface current density $\mathbf{j}_{\rm c}^\star = [\![\mathbf{n}\times\hat{\mathbf{Q}}]\!]_{\rm c}$ in the antenna:

$$P_{\rm ant} \equiv \int \Big[\!\!\Big[\frac{\partial\hat{\mathbf{A}}^*}{\partial t}\cdot\mathbf{n}\times(\nabla\times\hat{\mathbf{A}})\Big]\!\!\Big]_{\rm c} dS_{\rm c} = -\int \mathbf{E}_{\rm c}^*\cdot\mathbf{j}_{\rm c}^\star\, dS_{\rm c}\,. \tag{11.41}$$

(Note the different symbols * for complex conjugation and * for surface current.) Thus, after tedious analysis, the usual reward of a physically meaningful result is obtained: *In the stationary state, the power emitted by the antenna is precisely balanced by the power absorbed in the resonant layer.*

Substituting the solution (11.38) into the expression for Π, given in Eq. (7.88) for a plane plasma layer, and applying the low-β approximation, yields

$$\Pi = -\frac{N}{D}\xi' \approx \frac{\rho(\omega^2 - \omega_{\rm A}^2)}{k_\perp^2}\,\xi' = \frac{\epsilon}{k_\perp^2}\,\xi' = C_0\,\frac{\epsilon_{\rm r}'}{k_\perp^2}\,. \tag{11.42}$$

Hence, Π is both finite (the singularities are cancelled by the multiplication with ϵ) and continuous, so that the discontinuity exclusively resides in the jump of ξ:

$$[\![\xi^*]\!] = \mathrm{i}C_0^*\pi\,\mathrm{sgn}(\omega_\mathrm{d}/\epsilon_\mathrm{r}') \,. \tag{11.43}$$

Consequently,

$$\mathrm{Re}(P_\mathrm{res}) = L_y L_z \mathrm{Im}\Big(\omega^* [\![\xi^*]\!]\Pi\Big) = \frac{L_y L_z}{k_\perp^2}|C_0|^2\pi\,|\omega_\mathrm{d}|\,|\epsilon_\mathrm{r}'| \,. \tag{11.44}$$

Note that this correct result (positive absorption) is only obtained because extreme care was taken in accounting for the signs of the different quantities. Also note that Chen and Hasegawa [118] (and other authors) find a two times smaller absorption rate, because they consider the time average of the absorption rate for modes that vary like $\cos(\omega t)$ or $\sin(\omega t)$, whereas we consider those two simultaneously with modes of the form $\exp(-\mathrm{i}\omega t)$. Of course, in an actual experiment, the antenna is only excited with one of the two real modes, but there is no need to account for this in the energy balance (11.39) since all quadratic forms scale with the same factor. Hence, we will not burden the analysis with these additional factors of $1/2$.

Due to the jump in ξ, energy is absorbed collisionlessly by the plasma. As time progresses, the energy supplied by the driver accumulates in an ever diminishing plasma layer around $x = x_0$, where the frequency of the driver matches the local Alfvén frequency. In this singular point the fields grow unbounded as time proceeds and the external driver continues to pump energy into the plasma.

From Eq. (11.44) it is clear that the average energy absorption rate depends on the driving frequency. The novel idea of Chen and Hasegawa was to exploit the frequency of the surface mode, located at the plasma–vacuum surface. In an inhomogeneous plasma, this mode becomes a weakly damped quasi-mode (see Section 10.4). The oscillatory part of the frequency of this mode is given by

$$\omega_0 = \sqrt{2}\,\omega_{\mathrm{A}1}, \tag{11.45}$$

as follows from Eq. (10.83) for a uniform magnetic field and $\rho_2 = 0$ (vacuum).

▷ **Explicit expression of the absorption rate** The increase of the power absorption in the range of the quasi-mode frequency may be seen from the expression (11.44) by completing the calculation of C_0 defined in Eq. (11.35) for long-wavelength modes, $k_\perp a \ll 1$, so that the Bessel functions may be approximated by means of the relations (11.87) and (11.21). Inserting the expressions (11.34) for A_1 and (11.22) for B_2 in C_0 then gives

$$C_0 = -A_1 B_2 = -\dot{j}_y^\star\,\frac{k_\perp \mathrm{e}^{-k_0 c}}{k_\parallel^2 B}\,\frac{I_0(X_1) - I_1(X_1)}{\bar{\Delta}} \approx -\dot{j}_y^\star\,\frac{k_\perp \mathrm{e}^{-k_0 c}}{k_\parallel^2 B\,\bar{\Delta}} \,. \tag{11.46}$$

Here, the function $\bar{\Delta}$, involving the coefficients α and β defined in Eq. (11.33), turns out to be a rescaled version of the *dispersion function* (10.75) of Section 10.4:

$$\begin{aligned}
\bar{\Delta} \;&\equiv\; \frac{k_\perp}{k_\parallel^2 B}\,\frac{\alpha+\beta}{X_1} \equiv \big[K_0(X_1) + K_1(X_1)\big]\big[I_0(X_2) + I_1(X_2)\big] \\
&\qquad\qquad - \big[I_0(X_1) - I_1(X_1)\big]\big[K_0(X_2) - K_1(X_2)\big] \\
&\approx\; \ln\frac{X_2}{X_1} + \frac{1}{X_1} + \frac{1}{X_2} \approx \ln\left|\frac{\epsilon_2}{\epsilon_1}\right| + \frac{\epsilon_r'(\epsilon_1+\epsilon_2)}{k_\perp\epsilon_1\epsilon_2} + \mathrm{i}\pi\,\mathrm{sgn}(\omega) \\
&\approx\; \frac{\bar{\omega}^2(\bar{\omega}^2-2)}{k_\perp a\,(\bar{\omega}^2-1)} + \mathrm{i}\pi\,\mathrm{sgn}(\omega) \,, \quad \text{where} \quad \bar{\omega}^2 \equiv \omega^2/\omega_{\mathrm{A}1}^2 \,.
\end{aligned} \tag{11.47}$$

Substitution into Eq. (11.44) yields the final expression for the absorption rate:

$$\mathrm{Re}(P_{\mathrm{res}}) \approx aL_y L_z |j_y^\star|^2 \,|\omega|\, \frac{\pi\, \mathrm{e}^{-2k_0 c}\,(\omega^2/\omega_{\mathrm{A1}}^2)}{k_\parallel^2 a^2\, |\bar\Delta|^2}\,. \tag{11.48}$$

Clearly, at the quasi-mode frequency $\omega = \sqrt{2}\,\omega_{\mathrm{A1}}$, the absolute value of the dispersion function attains the minimum value $|\bar\Delta|_{\mathrm{min}} = |\bar\Delta_{\mathrm{i}}| = \pi$, and the expression (11.48) attains a maximum, meaning maximum absorption. To avoid misunderstanding: this value of $|\bar\Delta_{\mathrm{i}}|$ is to be taken according to our prescription of choosing $\lim \nu \downarrow 0$ along the proper branch cut of the logarithmic function, and *not* at the imaginary part ν_0 of the complex quasi-mode frequency (10.80). Yet, the resonant absorption process appears to 'feel' the proximity of the quasi-mode localized close to the surface. This indicates that it is easy to produce heating at the plasma edge, but much harder in the interior of the plasma. ◁

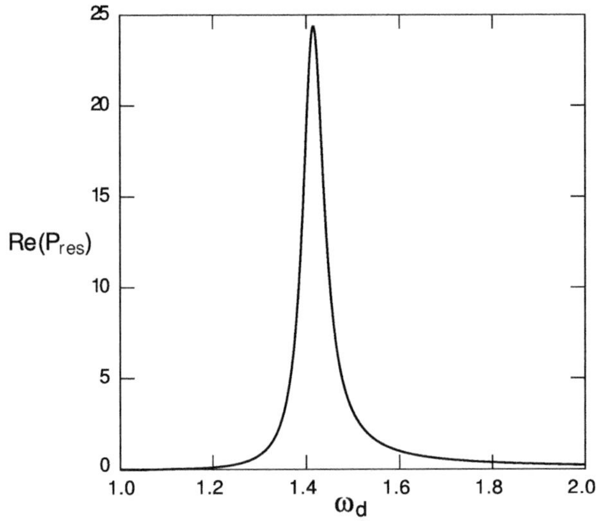

Fig. 11.3 Average energy absorption rate $\mathrm{Re}(P_{\mathrm{res}})$ in the steady state versus the driving frequency for $1 \le \omega_{\mathrm{d}} \le 2$; $a = 0.005$, $c = 0.1$, $k_\parallel = 1$, $k_\perp = 10$, $\rho_1 = 1$, $\rho_2 = 0$, $j_y^\star = 1$, $L_y = L_z = 1$.

In Fig. 11.3, the average energy absorption rate is plotted versus the normalized driving frequency. Here, the normalized component of the antenna current perpendicular to the magnetic field was fixed to 1. The peak in the power absorption profile is due to the presence of the quasi-mode. As expected, driving at the real part of its frequency yields a much larger power absorption than driving at other continuum frequencies.

Due to the coupling of the quasi-mode to the singular shear Alfvén continuum modes, the energy pumped into the quasi-mode is effectively absorbed (and, next, dissipated by non-ideal MHD effects). On the other hand, since the quasi- mode is the natural oscillation mode of the system, it is easy to excite: it guarantees good plasma–driver coupling. This brings us to another important issue, viz. the efficiency of the resonant absorption or heating mechanism.

(b) Efficiency and role of the 'quasi-modes' In the literature, the efficiency of the resonant absorption mechanism is expressed in terms of the coupling factor C, the fractional absorption f_{a}, and the antenna impedance. The definitions of these quantities are closely related to the energetics of

the driven system. In accordance with Eq. (11.41), the power emitted by the antenna is given by $P_{\rm ant}(\omega_{\rm d})$. In the steady state, the real part of this power corresponds to the energy absorbed by the plasma, i.e. the real part of $P_{\rm res}$ defined in Eq. (11.40). The imaginary part, however, corresponds to the energy fluctuating in the system, i.e. the kinetic and potential energy of the plasma and the potential energy of the vacuum.

To prove the latter statement, we need to derive the imaginary counterpart of the energy conservation equation (11.39). To that end, we exploit again the self-adjointness relation (6.144) of the force operator for plasma–vacuum systems, extended with contributions of jumps at the resonance and the antenna. This yields:

$$W = W^{\rm p} + W^{\rm s} + W^{\rm v} - \tfrac{1}{2}\int \Big[\!\!\Big[\xi^* \Pi(\boldsymbol{\xi}) \Big]\!\!\Big]_{x_0} dS_{x_0} - \tfrac{1}{2}\int \Big[\!\!\Big[\hat{\mathbf{A}}^* \cdot \mathbf{n}\times(\nabla\times\hat{\mathbf{A}}) \Big]\!\!\Big]_{\rm c} dS_{\rm c} \,. \tag{11.49}$$

The extension with the latter two terms implies that the force operator $\mathbf{F}$ is no longer self-adjoint, so that W becomes complex, whereas the symmetric expressions $W^{\rm p}$, $W^{\rm s}$ and $W^{\rm v}$, defined in (6.146)–(6.148), remain real. For a time dependence $e^{-{\rm i}\omega t}$ of the perturbations, the complex expression W is related to the real kinetic energy K through $W = (\omega/\omega^*)K$, as follows from the equation of motion (6.23), whereas the last two terms of Eq. (11.49) are related to $P_{\rm res}$ and $P_{\rm ant}$, defined in Eqs. (11.40) and (11.41), by $-\frac{1}{2}({\rm i}/\omega^*)P_{\rm res}$ and $\frac{1}{2}({\rm i}/\omega^*)P_{\rm ant}$, respectively. Thus, the energies and powers satisfy the following complex relation:

$$2{\rm i}\Big[\omega^*(W^{\rm p} + W^{\rm s} + W^{\rm v}) - \omega K\Big] + P_{\rm res} = P_{\rm ant} \,. \tag{11.50}$$

For the time dependence specified, the real part of this relation complies with the energy conservation equation (11.39), since $d/dt \to 2{\rm Im}(\omega)$ for quadratic forms, and the imaginary part yields the counterpart we were looking for:

$$2{\rm Re}(\omega)\Big[W^{\rm p} + W^{\rm s} + W^{\rm v} - K\Big] + {\rm Im}(P_{\rm res}) = {\rm Im}(P_{\rm ant}) \,. \tag{11.51}$$

The term ${\rm Im}(P_{\rm res})$ is negligible, since Eqs. (11.40), (11.42) and (11.43) give

$${\rm Im}(P_{\rm res}) = -L_y L_z {\rm Re}\Big(\omega^* [\![\xi^*]\!]\Pi\Big) = -\frac{L_y L_z}{k_\perp^2}|C_0|^2 \pi\nu\,{\rm sgn}(\omega_{\rm d})|\epsilon_{\rm r}'| \to 0 \,, \tag{11.52}$$

which proves the assertion. Hence, whereas the total energy disappears from the energy conservation equation (11.39) for the stationary state that is eventually reached, since $d/dt \to 0$, the difference of the potential and the kinetic energy remains and is determined by the imaginary counterpart (11.51) with ${\rm Re}(\omega) = \omega_{\rm d}$ and ${\rm Im}(P_{\rm res}) = 0$.

The *coupling factor* is defined as the ratio of the imaginary and real parts of the antenna power. Hence, in the steady state,

$$C(\omega_{\rm d}) \equiv \frac{{\rm Im}(P_{\rm ant})}{{\rm Re}(P_{\rm ant})} \approx \frac{2\omega_{\rm d}\big(W^{\rm p} + W^{\rm s} + W^{\rm v} - K\big)}{{\rm Re}(P_{\rm res})} \tag{11.53}$$

expresses the amount of energy fluctuating in the system compared to that eventually (in a dissipative system) converted into heat. Thus, C is closely related to the impedance of the system. As we will see, it depends sensitively on the driving frequency (see also [34], but note that the coupling factor is denoted in that reference by Q).

For the description of resonant absorption, the same information is more effectively represented by the *fractional absorption*, defined [263] as

$$f_{\rm a}(\omega_{\rm d}) \equiv \frac{{\rm Re}(P_{\rm ant})}{|P_{\rm ant}|} \equiv \frac{1}{\sqrt{1+C^2}} . \tag{11.54}$$

It compares the absorbed power to the total power of the disturbance of the system in the steady state. Hence, *perfect* coupling, i.e. $C = 0$, or rather $f_{\rm a} = 1$ (100%), means that all energy that enters the plasma remains inside the plasma and is eventually converted into heat. In that case, the Poynting vector at the plasma surface is pointing inwards all the time. On the other hand, no coupling at all corresponds to $C \to \infty$, or rather $f_{\rm a} = 0$. (Clearly, calling C the *de-coupling* factor would have been a more adequate terminology.)

To evaluate the functions $C(\omega_{\rm d})$ and $f_{\rm a}(\omega_{\rm d})$ for the explicit solution of the model problem of Section 11.1.1, a similar derivation is needed as for the resonant absorption rate ${\rm Re}(P_{\rm res})$. This involves, first, converting the expression (11.41) for $P_{\rm ant}$ in terms of the vector potential $\hat{\mathbf{A}}$ into the magnetic field perturbation $\hat{\mathbf{Q}}$, and, next, transforming the normal component $\hat{Q}_x^*$ by means of Eqs. (11.23) and (11.25), i.e. $\hat{Q}_x^* = -{\rm i}(k_0/k_\parallel)A_4^*$, with A_4 given by the expression (11.34), and transforming the jump $[\![\hat{Q}_z]\!]$ by means of Eq. (11.15), i.e. $[\![\hat{Q}_z]\!] = -j_y^\star$. This yields:

$$\begin{aligned} P_{\rm ant} &= -L_y L_z(\omega^*/k_\parallel)\hat{Q}_x^*[\![\hat{Q}_z]\!] \\ &= \tfrac{1}{2}{\rm i}L_y L_z\omega^*(k_0/k_\parallel^2)|j_y^\star|^2\Big[1+\Big(\frac{\alpha-\beta}{\alpha+\beta}\Big)^*{\rm e}^{-2k_0c}\Big] , \end{aligned} \tag{11.55}$$

so that the coupling factor is just determined by the coefficients α and β and the position c of the antenna:

$$C = \frac{1+{\rm Re}\Big(\dfrac{\alpha-\beta}{\alpha+\beta}\Big){\rm e}^{-2k_0c}}{{\rm Im}\Big(\dfrac{\alpha-\beta}{\alpha+\beta}\Big){\rm e}^{-2k_0c}} . \tag{11.56}$$

▷ **Explicit expression of the coupling factor** To complete the analysis, we need to introduce a similar expression $\bar{\Lambda}$ for $\alpha-\beta$ as the expression (11.47) for $\bar{\Delta}$ in terms of $\alpha+\beta$, and expand the Bessel functions:

$$\begin{aligned} \bar{\Lambda} &\equiv \frac{k_\perp}{k_\parallel^2 B}\frac{\alpha-\beta}{X_1} \equiv \big[K_0(X_1)+K_1(X_1)\big]\big[I_0(X_2)-I_1(X_2)\big] \\ &\qquad\qquad\qquad - \big[I_0(X_1)-I_1(X_1)\big]\big[K_0(X_2)+K_1(X_2)\big] \\ &\approx \ln\frac{X_2}{X_1}+\frac{1}{X_1}-\frac{1}{X_2} \approx -\frac{\bar{\omega}^4}{k_\perp a(\bar{\omega}^2-1)}+{\rm i}\pi\,{\rm sgn}(\omega) . \end{aligned} \tag{11.57}$$

Substitution of the relations (11.47) and (11.57) for $\bar{\Delta}$ and $\bar{\Lambda}$ yields the final expression for the coupling factor:

$$\begin{aligned} C &= \frac{1+{\rm Re}(\bar{\Lambda}/\bar{\Delta})\,{\rm e}^{-2k_0c}}{{\rm Im}(\bar{\Lambda}/\bar{\Delta})\,{\rm e}^{-2k_0c}} \\ &\approx \frac{{\rm sgn}(\omega)\,{\rm e}^{2k_0c}}{2\pi k_\perp a\bar{\omega}^2}\Big\{\frac{\bar{\omega}^4(\bar{\omega}^2-2)}{(\bar{\omega}^2-1)^2}\Big[\bar{\omega}^2\big(1-{\rm e}^{-2k_0c}\big)-2\Big]+(k_\perp a)^2\pi^2\big(1+{\rm e}^{-2k_0c}\big)\Big\} . \end{aligned} \tag{11.58}$$

This compound expression shows that C is small, though not zero, for $\bar{\omega} = \sqrt{2}$ and for $\bar{\omega} = \sqrt{2/(1-{\rm e}^{-2k_0c})}$, i.e. at the frequency (11.45) of the quasi-mode and at the frequency of another 'mode', to be discussed below. Actually, the dependencies of C and $f_{\rm a}$ on the driving frequency $\omega_{\rm d}$, shown in Figs. 11.4(a) and 11.4(b) for the parameters of Fig. 11.3, exhibit two zeros for C and a double peak for $f_{\rm a}$, one at $\bar{\omega} \approx 1.425$ and another one at $\bar{\omega} \approx 1.509$. These zeros correspond to the frequencies just mentioned, but shifted through the factor with π^2 originating from the imaginary contributions of the logarithm that appears in the expansions of $\bar{\Delta}$ and $\bar{\Lambda}$. ◁

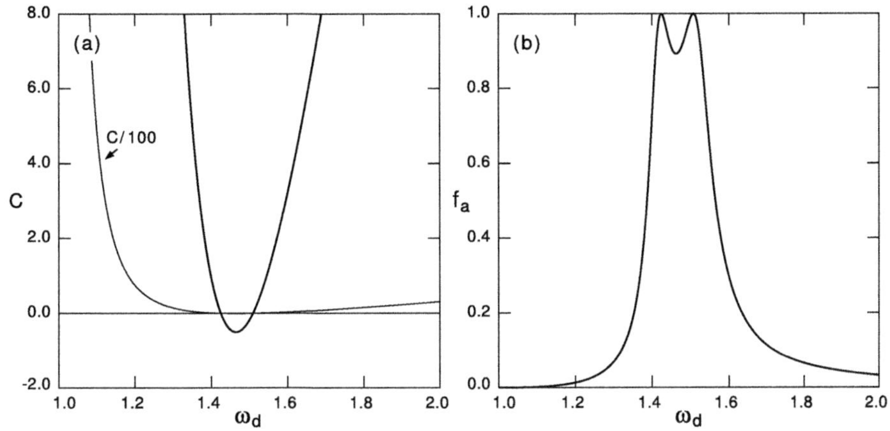

Fig. 11.4 (a) Coupling factor and (b) fractional absorption in the steady state as a function of the driving frequency for $1 \leq \omega_{\rm d} \leq 2$; same parameters as in Fig. 11.3.

As already mentioned, the presence of a quasi-mode in the Alfvén continuum is crucial for the efficiency of the resonant absorption mechanism since the global character of this mode guarantees good plasma–vacuum coupling, while its coupling with the singular Alfvén continuum modes guarantees efficient dissipation. The good plasma–vacuum coupling at the quasi-mode frequency $\omega_{\rm d} = \sqrt{2}$ is confirmed by Figure 11.4. However, this figure also reveals a second driving frequency that yields excellent plasma–vacuum coupling. That one is due to a 'mode' that perfectly fits in the plasma–vacuum–antenna cavity.

The existence of two modes with perfect coupling can easily be seen from the expression (11.55) for the power emitted by the antenna in the driven system: $P_{\rm ant} \sim \hat{Q}_x^* [\![\hat{Q}_z]\!]$. The imaginary part of that expression vanishes both for the quasi-mode, with $[\![\hat{Q}_z]\!] = -j_y^\star = 0$, and for the other 'mode' with $\mathrm{Im}(\hat{Q}_x^*) = 0$, which yields $A_3 = B_3 = -\frac{1}{2} j_y^\star$ $(\neq 0!)$ and $A_4 = 0$. The latter condition means that the induced solution vanishes in the region outside the antenna ($x \geq c$). For that case, linking the vacuum solution to the plasma solution yields an 'eigenvalue problem' with the 'eigenfrequency'

$$\omega = \frac{\sqrt{2}\omega_{\rm A1}}{\sqrt{1 - {\rm e}^{-2k_0 c}}} \,. \tag{11.59}$$

In contrast to the quasi-mode, this second 'mode' is not an eigenmode of the system without antenna (hence, the quotation marks) since it requires an antenna with a finite current. It would become a quasi-mode of that system though if a wall were placed at the position $x = c$. Accordingly, when that wall (or the antenna) is moved to ∞, the frequency (11.59) of the second 'mode' transforms into the frequency (11.45) of the quasi-mode.

The frequencies of the quasi-mode and the second 'mode' become complex by coupling to the continuous spectrum, whereas their real parts shift to the values that can be calculated from Eq. (11.58). As shown in Fig. 11.4, there is perfect absorption, i.e. $f_{\rm a} = 1$, for the frequencies $\bar{\omega} \approx 1.425$ and $\bar{\omega} \approx 1.509$. This implies that, in the steady state, all energy that enters the plasma stays there for those frequencies. For other driving frequencies the driver–plasma coupling is not perfect. However, one should not lose sight of the first part of our exposition, which is the most important one, viz. about the magnitude of the resonant energy absorption rate $\mathrm{Re}(P_{\rm res})$. According to Eq. (11.48) and Figure 11.3, this rate is reduced about eight times for the frequency of the second 'mode' compared to that for the frequency of the quasi-mode.

11.1.3 Resonant 'absorption' versus resonant 'dissipation'

Clearly, *dissipative effects* will prevent unlimited growth and will cause a real, dissipative, plasma to reach a *stationary state* after a finite time. In this stationary state all physical quantities oscillate harmonically with a constant amplitude and with the frequency imposed by the external source. The power supplied by the external source is exactly balanced by the energy dissipation rate in the plasma. In the above ideal MHD solution obtained in Section 11.1.1, this steady state was simulated in ideal MHD by considering artificial damping in the form of an imaginary part of the external driving frequency.

(a) Dissipative MHD solutions Solving the dissipative MHD equations is more complicated. The visco-resistive MHD equivalent of Eq. (11.2) reads

$$\begin{pmatrix} \rho\omega^2 + \dfrac{d}{dx}B^2\dfrac{d}{dx} - k_\parallel^2 B^2 - \mathrm{i}\omega\rho(\eta+\nu)\dfrac{d^2}{dx^2} & \dfrac{d}{dx}k_\perp B^2 \\ - k_\perp B^2 \dfrac{d}{dx} & \rho\omega^2 - k_0^2 B^2 - \mathrm{i}\omega\rho(\eta+\nu)\dfrac{d^2}{dx^2} \end{pmatrix} \begin{pmatrix} \xi_x \\ \mathrm{i}\xi_\perp \end{pmatrix} = 0\,, \qquad (11.60)$$

where the scalar viscosity ν and the electrical resistivity η are only retained in terms with derivatives in the x-direction. The reason for doing so is that these dissipative effects are extremely small in hot plasmas (e.g. in the solar corona or in tokamaks) and the dissipative terms only contribute in combination with x-derivatives which become extremely large in the neighbourhood of ideal singularities, as we have seen in the above ideal solution. The visco-resistive terms have the same effect as the artificial damping term in the previous sections, viz. to remove the singularity from the equations and, hence, to keep the solution finite. The fact that the dissipative effects are extremely small in the hot plasmas of interest results in nearly singular behaviour in the neighbourhood of the ideal MHD resonance positions in a very similar way to that in the above solution. We will see examples of this below when some numerical solutions of the dissipative MHD equations are discussed.

Kappraff and Tataronis [339] considered a configuration of a planar sheet pinch, similar to that of Fig. 11.1, but with sheet currents on both sides of the plasma and with finite plasma resistivity. They demonstrated that the high absorption rate obtained in ideal MHD indeed corresponds to an effective heating rate in highly conductive tokamak plasmas. Their analysis involves asymptotic matching of the solution in the resistive layer with the ideal MHD solution outside the layer and inverse Laplace transformation of the resistive layer solution. This reveals two time periods which characterize the energy absorption. Initially the energy accumulates at the singular layer, just like in ideal MHD, up to a critical time $t_{\rm h}$ which scales as $\eta^{-1/3}$. In our notation their solution reads

$$\xi_\perp(t) \approx \left\{ \begin{array}{ll} N(\omega_{\rm d})\, t\, {\rm e}^{-{\rm i}\omega_{\rm d} t} & (\text{for } t < t_{\rm h}) \\ N(\omega_{\rm d})\, t_{\rm h}\, {\rm e}^{-{\rm i}\omega_{\rm d} t} & (\text{for } t > t_{\rm h}) \end{array} \right. , \tag{11.61}$$

with $N(\omega_{\rm d})$ a multiplicative constant, that depends on the nature of the equilibrium and the source but is independent of η, and with

$$t_{\rm h} = \left(\frac{24\mu_0}{\omega_{\rm d}^2 \eta} \right)^{1/3} \left(\frac{2B_1'(x_0)}{B} - \frac{\rho'(x_0)}{\rho} \right)^{-2/3} . \tag{11.62}$$

Hence, for $t > t_{\rm h}$ the growing solution in the resistive layer has reached saturation and the absorbed energy is transferred to heat through Ohmic dissipation.

Sakurai et al. [537] and Goossens et al. [251] followed a similar approach for cylindrical geometry. These authors used asymptotic expansions of the solutions of the dissipative MHD equations to derive connection formulas that are used to integrate the equations through the near-singularities at the resonance points. This method was used in several papers on resonant absorption later on.

(b) Resistive energetics In resistive MHD, the plasma heating may be treated consistently. This makes it possible to get a more detailed picture of the energetics in the driven system derived from the linearized MHD equations. Below, we will denote the perturbed current density by $\mathbf{j}_1$ (here we need the subscript 1 because there is also $\mathbf{j}_0$). The perturbed electric field is still denoted by $\mathbf{E}$. From the combined laws of Ampère and Faraday one gets

$$-\nabla \cdot \mathbf{S} \equiv -\nabla \cdot (\mathbf{E}^* \times \mathbf{Q}) = \mathbf{j}_1 \cdot \mathbf{E}^* + \mathbf{Q} \cdot \frac{\partial \mathbf{Q}^*}{\partial t}, \tag{11.63}$$

which means that the inflow of electromagnetic energy (represented by minus the divergence of the Poynting vector) produces electrical energy in the plasma and gives an increase of the magnetic energy. In turn, the electrical energy given to the plasma by the disturbance of the electromagnetic field may be rewritten:

$$\mathbf{j}_1 \cdot \mathbf{E}^* = \eta\, |\mathbf{j}_1|^2 + \mathbf{v}^* \cdot \mathbf{j}_1 \times \mathbf{B}_0 . \tag{11.64}$$

Thus, the electrical energy appears as heat by Ohmic dissipation plus work done by the $\mathbf{j}_1 \times \mathbf{B}_0$-force. Furthermore, the scalar product of $\mathbf{v}^*$ with the equation of motion yields the mechanical energy equation:

$$\rho_0\, \mathbf{v}^* \cdot \frac{\partial \mathbf{v}}{\partial t} = -\mathbf{v}^* \cdot \nabla p_1 + \mathbf{v}^* \cdot \mathbf{j}_0 \times \mathbf{Q} + \mathbf{v}^* \cdot \mathbf{j}_1 \times \mathbf{B}_0 . \tag{11.65}$$

Hence,

$$-\nabla \cdot \mathbf{S} = \rho_0\, \mathbf{v}^* \cdot \frac{\partial \mathbf{v}}{\partial t} + \mathbf{v}^* \cdot (\nabla p_1 - \mathbf{j}_0 \times \mathbf{Q}) + \mathbf{Q} \cdot \frac{\partial \mathbf{Q}^*}{\partial t} + \eta\, |\mathbf{j}_1|^2 . \tag{11.66}$$

These expressions are complex and actually involve energy densities.

Extracting the real part by integrating half of the sum of Eq. (11.66) and its complex conjugate over the plasma volume V gives the *resistive energy balance:*:

$$
\begin{aligned}
-\,\mathrm{Re}\int \nabla\cdot\mathbf{S}^*\,dV = \underbrace{\mathrm{Re}\int \rho_0\,\mathbf{v}^*\cdot\frac{\partial\mathbf{v}}{\partial t}\,dV}_{\dot{K}} \\
+\underbrace{\mathrm{Re}\int\Big[\mathbf{v}^*\cdot(\nabla p_1-\mathbf{j}_0\times\mathbf{Q})+\mathbf{Q}\cdot\frac{\partial\mathbf{Q}^*}{\partial t}\Big]\,dV}_{\dot{W}^{\mathrm{p}}}+\underbrace{\int\eta|\mathbf{j}_1|^2\,dV}_{D}\,.
\end{aligned}
\qquad (11.67)
$$

This means that the inflow of electromagnetic energy (LHS term) into the plasma region yields a rate of change of the kinetic energy of the plasma ($\dot{K}$), a rate of change of the potential energy of the plasma ($\dot{W}^{\mathrm{p}}$) and a heating rate due to Ohmic dissipation (D). The LHS term in the above equation is related to the power emitted by the antenna. For $\eta \neq 0$ the tangential components of the magnetic field perturbation are continuous at the plasma–vacuum boundary. With the help of Gauss' theorem, this means that the LHS term can be written as

$$
\begin{aligned}
-\,\mathrm{Re}\int \mathbf{S}^*\cdot\mathbf{n}\,dS_{\mathrm{p}} &= -\,\mathrm{Re}\int \hat{\mathbf{S}}^*\cdot\mathbf{n}\,dS_{\mathrm{p}} = \mathrm{Re}\int\nabla\cdot(\hat{\mathbf{E}}^*\times\hat{\mathbf{Q}})\,d\hat{V} \\
&= \underbrace{-\mathrm{Re}\int \mathbf{j}_{\mathrm{c}}^\star\cdot\mathbf{E}_{\mathrm{c}}^*\,dS_{\mathrm{c}}}_{\mathrm{Re}(P_{\mathrm{ant}})}\;\underbrace{-\;\mathrm{Re}\int\hat{\mathbf{Q}}\cdot\frac{\partial\hat{\mathbf{Q}}^*}{\partial t}\,d\hat{V}}_{\dot{W}^{\mathrm{v}}}\,,
\end{aligned}
\qquad (11.68)
$$

i.e. the inflow of electromagnetic energy into the plasma equals the outflow of electromagnetic energy out of the vacuum region, which is the power emitted by the antenna ($\mathrm{Re}(P_{\mathrm{ant}})$) minus the rate of change of the vacuum magnetic energy ($\dot{W}^{\mathrm{v}}$). Combining Eq. (11.67) and Eq. (11.68) yields

$$
\mathrm{Re}(P_{\mathrm{ant}}) = \dot{K}+\dot{W}^{\mathrm{p}}+\dot{W}^{\mathrm{v}}+D\,, \qquad (11.69)
$$

i.e., the power emitted by the antenna produces a change of the kinetic and potential energy of the plasma and the vacuum plus heat by Ohmic dissipation. In the steady state, $\mathrm{Re}(P_{\mathrm{ant}}) = D$, which means that the energy supply by the external harmonic source is exactly balanced by the dissipation rate in the plasma.

The reader may wish to check the following details of the above derivations:

(a) Identification of the second RHS term of Eq. (11.67) with the time derivative of the plasma energy W^{p}, as defined in Eq. (6.84), actually involves an integration by parts producing a surface integral, which vanishes by virtue of the second interface BC (6.134) if $p_1 = -\mathbf{B}\cdot\mathbf{Q}+\hat{\mathbf{B}}\cdot\hat{\mathbf{Q}} = 0$ at the plasma–vacuum boundary (which is the case in resistive MHD);

(b) Identification of the first RHS term of Eq. (11.68) with the power emitted by the antenna follows by converting an integral over the 'vacuum' region $\hat{\mathrm{V}}$ (with an embedded antenna of infinitesimal thickness and infinite current density) into an integral over the surface $\mathrm{S_c}$ of the antenna only involving the finite surface current $\mathbf{j}_{\mathrm{c}}^\star$, in agreement with the expression defined in Eq. (11.41).

For η sufficiently small, the Ohmic heating rate is very well approximated by the energy accumulation rate obtained in ideal MHD. Hence, Kappraff and Tataronis obtained the following *resonant dissipation rate* (adapted to our notation):

$$
D_0 \equiv \lim_{\eta\to 0}\int \eta|j_1|^2\,dV = \mathrm{Re}(P_{\mathrm{res}})\Big|_{\mathrm{ideal\ MHD}}\,. \qquad (11.70)
$$

Substitution into Eq. (11.69) reproduces the energy conservation equation (11.39).

The coupling factor and the fractional absorption express the efficiency of the resonant absorption mechanism in the steady state by comparing quantities averaged over a driving period. Hence, they do not contain any information on the time scales of the mechanism, which are nonetheless important. As a matter of fact, efficient absorption in the steady state means nothing if the steady state is only reached in an asymptotic way. The latter remark is related to the quality factor Q of the resonance which is defined in Ref. [492] as

$$Q = \frac{|\dot{K}| + |\dot{W}^{\rm p} + \dot{W}^{\rm v}|}{2\pi D}, \tag{11.71}$$

i.e. as the ratio of the total energy change of the system and the Ohmic heating per driving cycle in the stationary state. Notice that the norms of the components of the total energy have been taken separately in the definition of Q. In perfect coupling cases these terms can be quite large, but $\dot{K}$ has the opposite sign of $\dot{W}^{\rm p} + \dot{W}^{\rm v}$ and, therefore, these terms cancel and $\mathrm{Im}(P_{\rm ant})$ vanishes, so that the fractional absorption becomes 1. In the quality factor, however, these terms do not cancel and Q compares the amount of dissipation in one driving period to the amount of energy that has to be pumped into the system before the steady state is reached. It thus tells us something about the time scales of the process. Good quality resonances have little losses and thus high Q-values. This means that a lot of energy needs to be put into the system to reach the steady state. Hence, for efficient heating on short time scales, low Q-values are required. Good plasma–driver coupling does not guarantee a low-Q resonance, and vice versa.

The quality of LRC-circuits or resonant cavities is sometimes defined as

$$\widetilde{Q} = \frac{\omega_0}{\Delta\omega}, \tag{11.72}$$

where ω_0 is the resonance frequency and $\Delta\omega$ denotes the width at half-maximum of the resonance peak obtained by plotting the power absorbed by the oscillator as a function of the driving frequency (see, e.g., Ref. [303](b)). The tilde is added to distinguish the two definitions. The definitions are equivalent in the simple case of a resonant circuit or cavity, at least when the damping is weak so that the resonance is sharp. In general, the two definitions are not equivalent [492].

11.2 Heating and wave damping in tokamaks and coronal loops

As mentioned, resonant absorption of Alfvén waves was first studied in the context of controlled thermonuclear fusion research to provide an additional heating mechanism for tokamak plasmas. In 1978, Ionson [326] proposed the same heating mechanism for solar coronal loops to explain the high temperature of the solar corona. Since then, the efficiency of resonant absorption or 'Alfvén wave heating' has been studied extensively in this context. In this section we briefly discuss these applications, illustrating the main results for a cylindrical plasma.

11.2.1 Tokamaks

(a) Heating of tokamak plasmas In large tokamaks, Ohmic heating due to the plasma current of a few MA yields ion temperatures of a few keV, typically 30–40% of the required $10^8\,$K for fusion of deuterium and tritium (see Chapter 1). Hence, a considerable amount of supplementary

heating is required.[1] In the early 1970s it was suggested that the Alfvén continuum resonances be exploited for this purpose [588]. Due to the low frequencies involved (a few MHz), it was relatively easy and cheap to build antennas that generate the required power at the required frequencies. Experiments, such as the 'Tokamak Chauffage Alfvén' (TCA) and its successor the 'Tokamak à Configuration Variable' (TCV) in Lausanne [57], were done to check the theoretical predictions and it was demonstrated that the antennas can indeed couple electromagnetic energy to the plasma efficiently and that resonant absorption is an efficient heating mechanism. However, it turned out that most of the resonances occur close to the plasma edge so that the experiments yielded edge heating instead of heating of the inner core [34]. The TCA tokamak is living its second life in Brazil as the TCA-BR [13].

Nowadays, similar heating mechanisms are used to deposit energy in the inner plasma layers of tokamaks but involving higher frequencies, such as ion cyclotron resonance heating (ICRH), with frequencies Ω_i in the range 20–120 MHz, lower hybrid resonance heating (LHRH) with frequencies between Ω_i and Ω_e (1–8 GHz) and electron cyclotron resonance heating (ECRH), with frequencies Ω_e in the range 50–200 GHz. These methods are known under the common name radio frequency (RF) heating.

(b) Damping of global Alfvén waves Additional plasma heating by neutral beam injection or ICRH at frequencies of the order of the MHD continuum frequencies generates energetic He nuclei flying through the plasma at supra-thermal speeds. These α-particles can then destabilize the global toroidicity induced Alfvén eigenmodes (so-called TAEs) by a particle–wave interaction which leads to the loss of them before their energy can be thermalized. Since confinement of α-particles is essential for ignition, this interaction has been studied in detail. TAEs are observed routinely in large tokamaks, such as TFTR, DIII-D, JET, JT-60 [613]. Experiments have shown that TAEs can indeed be destabilized by the energetic particles, but the threshold for instability turned out to be much higher than expected from theoretical estimates based on collisionless Landau damping (see Section 2.3.3). Cylindrical or slab geometry simplifications cannot be applied here since TAEs owe their existence to the toroidicity of the plasma, which results in 'gaps' in the continuous spectrum due to coupling of the poloidal wave numbers. Hence, the models are necessarily two-dimensional (to be discussed in detail in the later Chapters 16–18). Numerical investigations showed that the global Alfvén modes also experience resonant damping, which might explain the increased instability threshold [498].

11.2.2 Coronal loops and arcades

The temperature increase above the photosphere of the Sun (Fig. 8.14) requires work which is, most probably, done by motions of the convection zone. These motions produce much more energy than required to heat the upper solar atmosphere, which has a very low heat capacity due to its extremely low density. Hence, the only problems left are to explain (1) how this energy gets to the upper layers of the solar atmosphere, and (2) how it is transformed into heat there. There is an additional problem associated with the inhomogeneity of the upper atmosphere which consists of

[1] An alternative would be to increase the plasma current, which requires a tokamak with a very large magnetic field. Such an experiment has been suggested in the IGNITOR project, but it has not been built so far for various economic and political reasons [86].

many magnetic structures. All of these structures, with a variety of length scales, have to be heated. Many solar physicists think it is not likely that one single heating mechanism is responsible for the heating of all these magnetic structures but that, instead, several heating mechanisms operate. These may be more or less effective in coronal holes, short coronal loops, long coronal loops, etc. However, it is almost certain that the heating mechanisms responsible for heating of the upper layers of the solar atmosphere are magnetic in nature.

The magnetic heating mechanisms that have been proposed can be divided into two types [386] differing with respect to the time scales of the driving source. The first one is the *wave heating mechanism*. Here, problem (1) is 'solved' by MHD waves generated by upwelling convective motions in the photosphere and propagating through the photosphere and the chromosphere into the corona. For efficient dissipation of this wave energy in the upper atmospheric layers (problem (2)), small length scales have to be created in order to make the time scale for magnetic diffusion ($\tau_{\rm d} \sim \mu_0 l^2/\eta$) reasonably short. Such small length scales could be produced by transformation of the waves into shocks or by 'resonant absorption'. Clearly, the time scale of the driving source must be short compared to the time a perturbation needs to cross the loop in order to generate waves in the loop. The second kind of magnetic heating mechanism involves *Joule dissipation of electric current sheets* along the coronal magnetic field lines and involves much longer time scales. These currents could be generated by slow twisting of the flux tubes by photospheric convective motions (problem (1)) and release of energy could take place in narrow current layers (problem (2)). We will concentrate on resonant heating.

11.2.3 Numerical analysis of resonant absorption

The efficiency of the resonant absorption mechanism has been investigated quantitatively in tokamak plasmas, coronal plasmas and magnetospheric plasmas. The coronal loops and arcades are usually modelled by a straight cylinder or even a slab, so that the curvature of the loops along their length is not taken into account (see Fig. 11.5). As we have seen, in the early studies the geometry of tokamak plasmas was simplified in the same way. Later, however, numerical studies quantifying resonant absorption took into account the full geometry of the plasma, including the toroidal curvature and the non-circularity of the poloidal cross-section as, e.g., in Ref. [498]. Here, the main numerical results for cylindrical plasmas are briefly reviewed. We consider a generic cylindrical equilibrium in which the profile of the local Alfvén frequency $\omega_{\rm A}(r) = (mB_\theta/r + kB_z)/\sqrt{\rho}$ is parabolic (Fig. 11.6). For $m = 2$ and $k_z = 0.05$, the ideal Alfvén continuum then ranges from $\omega_{\rm A}(1) = 0.15$ to $\omega_{\rm A}(0) = 0.25$ in dimensionless units (where $a = B_z = \rho = 1$).

(a) Temporal evolution This equilibrium is now excited periodically at a frequency $\omega_{\rm d} = 0.205$ located in the range of the continuum. The ideal singular layer that corresponds to this oscillation frequency is located at $r = 0.671$, i.e. $\omega_{\rm d} = \omega_{\rm A}(0.671)$. In Fig. 11.7, snapshots of the radial profile of the velocity perturbation tangential to the magnetic surfaces and perpendicular to the magnetic field lines, $v_{1\perp} = (v_{1\theta}B_z - v_{1z}B_\theta)/B$, are plotted versus dimensionless time. The time step between the snapshots is five driving periods and the magnetic Reynolds number $R_{\rm m} = 10^6$. Initially, the relatively small plasma response is global and phase-mixing only occurs as time progresses. The plasma response localizes in a gradually diminishing plasma layer around $r = r_0 = 0.671$, where large gradients build up. This localization and the growth of the oscillation amplitude in the resonant

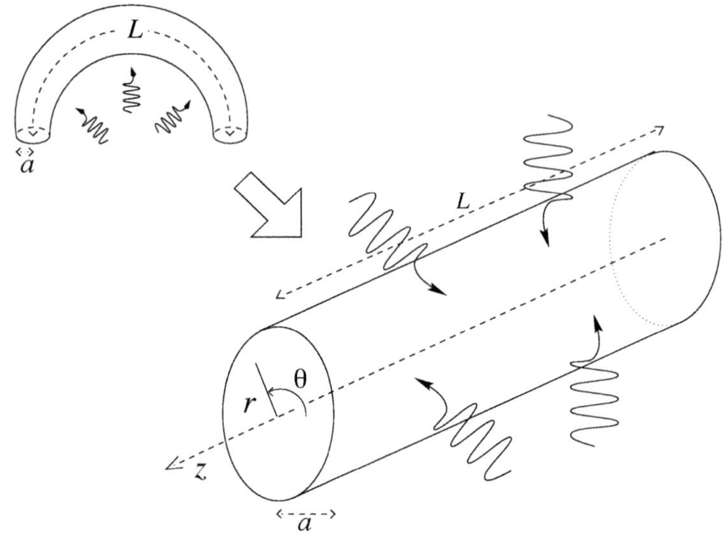

Fig. 11.5 Straightening a coronal loop with sideways wave excitation into a cylindrical model.

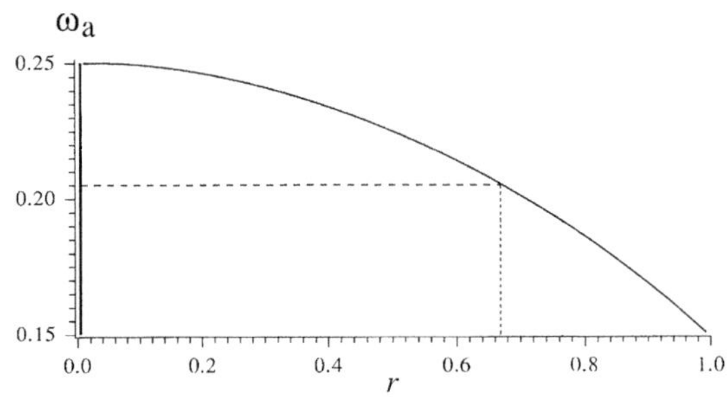

Fig. 11.6 Profile of the local Alfvén frequency resulting from a parabolic current density profile $j_z(r)$, constant magnetic field B_z, plasma density ρ, and wave numbers $m = 2$ and $k_z = 0.05$.

layer end after about 50 driving periods. From then on, the plasma oscillates purely harmonically, i.e. all the magnetic surfaces oscillate with constant amplitude and with the frequency $\omega_{\rm d}$ of the external source. In other words, the system has reached a *stationary state*. The delta function $\delta(r - r_0)$ and singularity $(r - r_0)^{-1}$, characteristic for the perpendicular tangential component of the ideal Alfvén mode (see Eq. (11.38)), are still recognizable in this resistive solution.

(b) Energetics The time dependence of the energetics of the externally driven resistive plasma column is shown in Fig. 11.8. The mean power P emitted by the external antenna, the mean rate of change of the kinetic energy $\dot{K}$ and of the potential energy $\dot{W}^{\rm p}$ of the plasma, the mean Ohmic dissipation rate D and the mean rate of change of the vacuum magnetic energy $\dot{W}^{\rm v}$ are plotted versus the number of driving periods. These quantities are defined in Eqs. (11.67)–(11.69). The initial, oscillatory, behaviour of these quantities is due to 'beats' resulting from the initial excitation

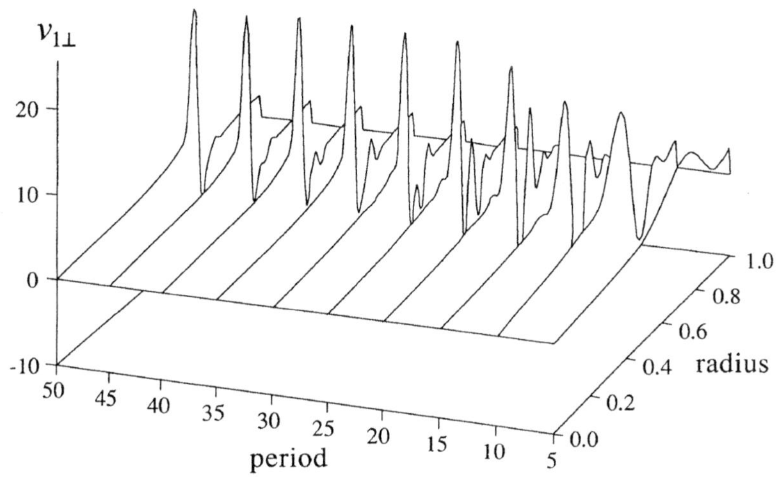

Fig. 11.7 Snapshots of $v_{1\perp}(r)$ for $\omega_d = 0.205$, $\eta = 10^{-6}$ (matching Fig. 11.6).

of the ideal quasi-mode. (The beat frequency that results from $\sin(\omega_d t) + \sin(\omega_{qm} t)$ is given by $\frac{1}{2}(\omega_d - \omega_{qm})$.)

Initially, in the first 10 to 15 driving periods, the power supplied by the external source produces mainly a change of the kinetic and potential energy of the plasma and also, to a lesser degree, of the vacuum magnetic energy. In this phase, the Ohmic dissipation rate is very low. As time progresses, however, the rate of change of the kinetic and potential plasma energy and of the vacuum magnetic energy gradually decrease and the Ohmic dissipation rate increases. Then, after about 55 driving periods the system attains a stationary state: the kinetic and potential plasma energy and the vacuum magnetic energy do not change any more ($\dot{K} = \dot{W}^{\rm p} = \dot{W}^{\rm v} = 0$) and the power supplied by the external source is exactly balanced by the Ohmic dissipation rate in the plasma ($P = D$).

The time interval needed to reach the stationary state, τ_{ss}, depends on the plasma resistivity. We computed this dependence by means of numerical simulations keeping all parameters fixed except the plasma resistivity η. The time needed to reach the steady state is proportional to the cubic root of the magnetic Reynolds number. Hence, in terms of η, we get $\tau_{ss} \sim \eta^{-1/3}$ (see Fig. 11.9), in agreement with the analytical result of Kappraff and Tataronis [339] (see also Eq. (11.62)). However, the additional excitation of an ideal quasi-mode in the initial phase yields a different η-scaling, viz. $\tau_{ss} \sim \eta^{-1/5}$ (upper curve of Fig. 11.9).

(c) Discussion We have seen that the ideal quasi-modes play an important role in resonant absorption and affect both the efficiency and the time scales of this heating mechanism considerably. Excitation of a plasma at the frequency of such a weakly damped quasi-mode yields 100% absorption, which means that all the energy that is supplied by the external source is absorbed by the plasma and, in the stationary state, converted into heat by Ohmic dissipation. Driving at an arbitrary frequency in the range of the ideal Alfvén continuum yields a less efficient plasma–driver coupling. This is a result of the fact that Alfvén waves *cannot be driven directly by a sideways external driver*

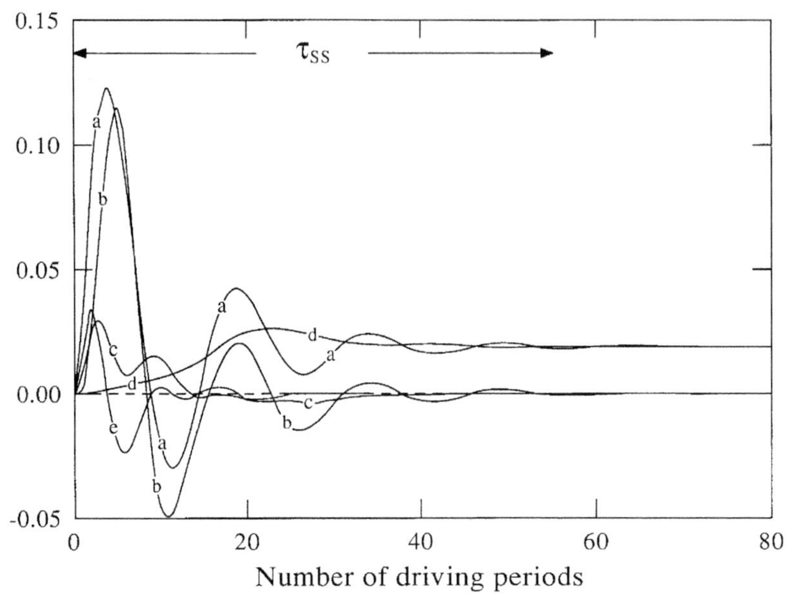

Fig. 11.8 (a) Time-averaged power, (b) change of kinetic energy, (c) change of potential plasma energy, (d) Ohmic dissipation rate and (e) change of vacuum energy versus number of driving periods for $\omega_d = 0.205$ and $\eta = 10^{-6}$ (matching Fig. 11.6).

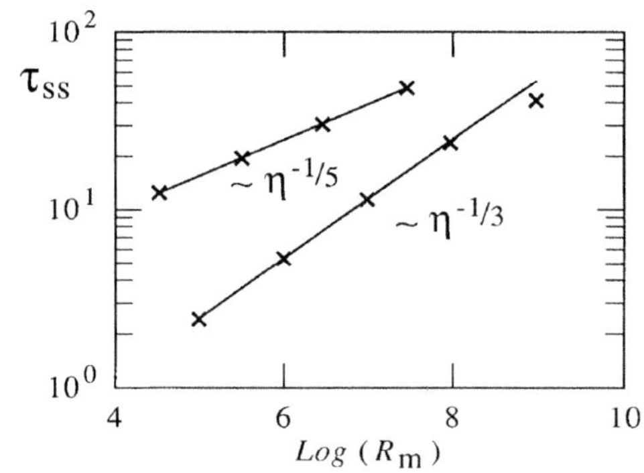

Fig. 11.9 Scaling of the time needed to reach the stationary state with the magnetic Reynolds number for typical coronal loop parameter values.

as they propagate the energy mainly along the magnetic field lines. The contribution of the global mode is that it can transport the energy from the external driver *across the magnetic surfaces* to the inner plasma. The global mode has a frequency inside the continuous spectrum and thus couples to an Alfvén wave, and it is this (nearly singular) wave energy that is finally dissipated. However, for 'ordinary' continuum frequencies the fractional absorption can still be very high, depending on the equilibrium profiles. For typical coronal loop parameter values, driving at frequencies in the continuous spectrum often yields more than 90% 'absorption', i.e. more than 90% of the energy supplied by the external driver is actually converted into heat. Driving at the foot points turns out to be even more efficient. This is due to the fact that in this way the Alfvén waves can be driven *directly* and there is *no need for a global mode to transport the energy from the external driver to the inner plasma* (see Section 11.3.1).

For *sideways driving*, it has been confirmed now that the basic time scale of resonant absorption depends on the proximity of the driving frequency to the quasi-mode frequency: for driving at the quasi-mode frequency, $\tau_{ss} \sim \eta^{-1/5}$, while for driving frequencies in the range of the ideal continuum but not close to the quasi-mode frequency, $\tau_{ss} \sim \eta^{-1/3}$. In active region loops on the Sun, the Alfvén continuum frequencies $\omega \approx 0.1$–1 typically correspond to 0.005–$2\,\mathrm{Hz}$. For realistic loop equilibria, τ_{ss} varies from a few minutes to a few hours, i.e. much shorter than the typical life time of coronal loops (1 day), even for realistic values of η. Hence, resonant absorption is very efficient for typical coronal loop values and a viable mechanism for heating of solar coronal loops. In tokamaks, the time scales for resonant absorption are much shorter. For a fully ionized tokamak plasma with an ion number density of $10^{20}\,\mathrm{m}^{-3}$, a small radius of $0.2\,\mathrm{m}$, and a toroidal magnetic field of $2\,\mathrm{T}$, the driving frequency $\omega = 0.205$ corresponds to a frequency of $1\,\mathrm{MHz}$. Hence, for typical parameter values for small tokamaks, the time scale to reach the steady state is very short: for $R_{\mathrm{m}} = 10^8$ the steady state is reached after about 150 driving periods in the simulation discussed, i.e. after about 150μs.

The cylindrical model considered here in the framework of linearized MHD is only a first approximation. For a final conclusion on the role of resonant absorption in coronal heating, more realistic simulations including effects of line-tying, curvature, foot-point excitation and nonlinearity need to be taken into account. Such simulations are being conducted at present. Line-tying has a drastic influence on the waves that appear in the loops and on the continuous spectrum. Also, for realistic magnetic Reynolds numbers the dynamics of the shear flow in the resonant layers turns out to be very nonlinear [471, 494, 499]. Nonlinear effects include background flows (leading to Doppler shifts of the continuum frequencies), nonlinear mode coupling and Kelvin–Helmholtz instability of the resonant layers [471, 499]. Last but not least, a final conclusion also requires observational data on the power spectrum of the waves that are incident on the coronal loops.

11.3 Alternative excitation mechanisms

In the previous section we considered the simplest possible model to describe the mechanism of resonant absorption in a laboratory situation where the plasma is excited by a current in an external coil. In that case, the fast magneto-sonic wave mode plays the *crucial role of energy carrier*. As a matter of fact, shear Alfvén waves are unable to transfer the energy emitted by the external antenna from the antenna to the resonant layer since Alfvén waves can only transport energy *along* the magnetic field lines. Their group velocity is directed along the magnetic field (see Section 5.3.2). This is also clear from the system of equations (11.2) and the solution derived in the previous section: for $k_\perp = 0$ the Alfvén mode decouples completely from the fast magneto-sonic mode and resonant absorption *simply does not work*. Indeed, the constants A_3 and B_3 are zero in that case (since $X = 0$ when $k_\perp = 0$) and the plasma does not respond to the external driver.

As mentioned in the introduction, the resonant absorption mechanism is intensively studied for

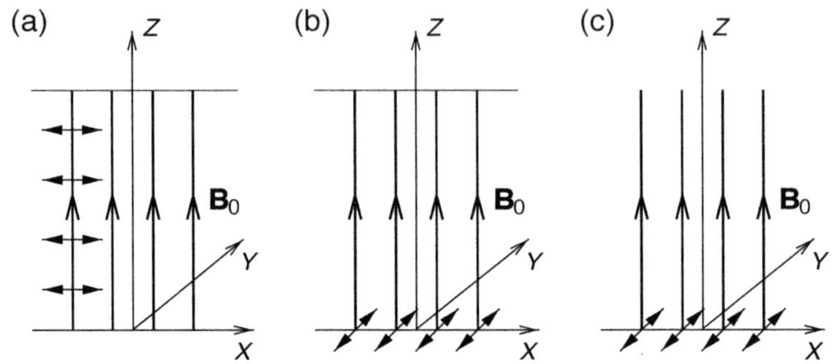

Fig. 11.10 Three different configurations corresponding to model problems III, IV and V of Section 4.6: (a) a sideways driven finite ('closed') loop, (b) the same system but driven at the 'foot points' and (c) an 'open' system without resonances but with phase mixing.

many different situations. From a topological point of view these different physical set-ups can be classified into three configurations which are schematically displayed in Fig. 11.10. The first configuration, displayed in Fig. 11.10(a), is the plane slab version of the model problem III, discussed in Section 4.6. It corresponds to the sideways driven plasma discussed in the previous section. This configuration is used to study resonant absorption in tokamak devices, sideways driven coronal loops, and the Earth's magnetopause (Section 11.3.3). The second configuration (Fig. 11.10(b)) refers to the model problem IV and is used to model the heating of solar coronal loops by foot point driving. From the physical point of view, this set-up is simpler than the previous one in the sense that the Alfvén waves can be excited directly, i.e. without the need of fast magneto-sonic wave modes, since the location of the energy source does not require the transfer of energy *through* magnetic flux surfaces here. Hence, this situation is mathematically tractable, even when dissipation is taken into account. It will be discussed in further detail below in Section 11.3.1. The first and second configurations have a finite length in the direction parallel to the magnetic field. The configuration can be periodic in this direction (as in a tokamak), or the ends of the field lines may be fixed (as in a coronal loop). The third configuration (Fig. 11.10(c)) corresponds to the plane slab version of model problem V of Section 4.6. It differs from the two previous ones by the fact that the field lines are 'open', i.e. very much longer than the wavelength in the parallel direction. It is representative for extremely long loops (as compared to the excited wave lengths) or coronal holes. In this situation, no resonances can occur but small length scales can still be created by phase-mixing of the field lines. This process also occurs in resonant absorption and will be discussed in further detail in Section 11.3.2.

11.3.1 Foot point driving

Consider a simple slab model with a uniform magnetic field B in the z-direction and a density that is stratified in the x-direction only. Let us neglect the effects of plasma pressure and gravity and suppose the slab is bounded in the z-direction by two boundaries at $z = 0$ and $z = L$, as illustrated in Fig. 11.10(b). This could be a model for a solar coronal loop, ignoring curvature effects of the field lines and magnetic shear effects but taking into account 'line-tying', i.e. 'anchoring' of the 'foot points' of the magnetic field lines in the dense photosphere. With $L = 10^8\,\mathrm{m}$ and an average Alfvén velocity $\overline{v}_{\mathrm{A}} = 2 \times 10^6\,\mathrm{m\,s^{-1}}$, the characteristic time scale of such a loop would be $\tau_{\mathrm{A}} = L/\overline{v}_{\mathrm{A}} = 50\,\mathrm{s}$. As a result of the non-uniformity of the density, the Alfvén velocity and, hence, the local Alfvén frequency are functions of x. Thus, the ideal MHD spectrum of this simple model plasma contains an Alfvén continuum.

Let us assume that the above system is excited by a mono-periodic driver at one of the foot points of the loop, e.g. at $z = 0$, *with a frequency* ω_{d} *in the range of the continuous spectrum*. Then, as in the situation discussed in the previous section, a resonance will occur at $x = x_0$, where $\omega_{\mathrm{d}} = \omega_{\mathrm{A}}(x_0) = k_z B/[\rho(x_0)]^{1/2}$. Hence, in the stationary state a solution of the form

$$\xi_y(x,z,t) = A(x)\,\mathrm{e}^{\mathrm{i}(k_z z - \omega_{\mathrm{d}} t)}\,, \tag{11.73}$$

is obtained. (We here ignore the fact that, in general, a Fourier series is required in the z-direction in order to satisfy the line-tying boundary conditions at $z = 0$ and $z = L$ [235, 277], according to Eq. (4.186) for model IV of Section 4.6.3.) Two important remarks are to be made at this point. First, the driving frequency is *real* now. This is because we include viscosity and resistivity so that there is no need to include artificial damping to circumvent the singularity in the equations. Second, we simplified the model by considering perturbations which are polarized in the y-direction. The divergence-free condition for the perturbed magnetic field then reduces to $k_y \hat{Q}_y = 0$ so that $k_\perp = k_y = 0$ and the physical quantities do not depend on the y-coordinate. (Notice that, as a result of this assumption, the perturbations become automatically incompressible, $\nabla \cdot \mathbf{v}_1 = \mathrm{i}k_y v_{1y} = 0$, so that the assumption $k_\parallel \ll k_\perp$ must be dropped. This gives an additional dissipative term $\sim k_z^2$ in Eq. (11.74) below.) We can then take dissipative effects into account while keeping the analysis tractable. As a matter of fact, the equations in the system (11.60) then decouple and the dynamic equation for the Alfvén waves can be studied separately. This decoupling of the Alfvén and fast wave modes means that in the present configuration the mechanism discussed in the previous section would not work. With the foot point driver, however, the fast wave modes are not required since the energy does not have to be transferred across the magnetic surfaces and the Alfvén waves are excited *directly* by foot point motions that are polarized in the tangential perpendicular (y) direction. As we will see below, considering $k_\perp = k_y \neq 0$, i.e. coupling between Alfvén and fast waves, affects the efficiency substantially though, especially when random driving is considered.

The visco-resistive dynamic equation for the Alfvén wave then reads:

$$\frac{\partial^2 \xi_y}{\partial t^2} - (\eta + \nu)\left(\frac{\partial^2}{\partial x^2} - k_z^2\right)\frac{\partial \xi_y}{\partial t} + \omega_{\mathrm{A}}^2(x)\,\xi_y = F(x)\,\mathrm{e}^{\mathrm{i}(k_z z - \omega_{\mathrm{d}} t)}, \tag{11.74}$$

where the effect of an external harmonic driving term F with frequency ω_{d} is taken into account. The explicit form of $F(x)$ depends on details of the driver, i.e. the velocity profiles given on the boundaries at $z = 0$ and $z = L$. Equation (11.74) is the equation of a continuum of coupled forced oscillators which are *damped*.

▷ **Amplitude and phase at resonance** The resonant behaviour described by Eq. (11.74) is similar to the well-known resonance of a driven oscillator described in many textbooks; see e.g. Ref. [90]. We can trace this comparison further by making a local analysis around a point $x = x_0$, i.e. a specific field line. This yields the classic equation for a forced and damped oscillator:

$$\frac{\partial^2 \xi_y}{\partial t^2} + (\eta + \nu)(k_x^2 + k_z^2)\frac{\partial \xi_y}{\partial t} + \omega_0^2\,\xi_y = F_0\,\mathrm{e}^{-\mathrm{i}\omega_{\mathrm{d}} t}, \tag{11.75}$$

where $F_0 \equiv F(x_0)$ and $\omega_0 \equiv \omega_A(x_0)$. Substitution of a solution $\xi_y(x, z; t) = A_0 \exp\{i(k_x x + k_z z - \omega_d t)\}$ yields

$$A_0 = \frac{F_0}{(\omega_0^2 - \omega_d^2) - i\omega_d(\eta + \nu)(k_x^2 + k_z^2)} . \tag{11.76}$$

For $\eta = \nu = 0$, the amplitude A_0 becomes real (the plasma response is in phase or in opposite phase with the external driving force) and infinite when the driver frequency ω_d matches the 'natural' frequency ω_0 of the system. This is called 'resonance'. The behaviour of the field line resonator changes considerably if we include dissipation ($\eta \neq 0$ and/or $\nu \neq 0$). First of all, the absolute value of A_0 is reduced for every value of ω_d and remains finite for $\omega_d = \omega_0$. Next, the phase angle between the motion of the field line and the external driving force depends on the frequency,

$$\tan^{-1} \frac{\mathrm{Im}(A_0)}{\mathrm{Re}(A_0)} = \tan^{-1} \frac{\omega_d(\eta + \nu)(k_x^2 + k_z^2)}{(\omega_0^2 - \omega_d^2)} , \tag{11.77}$$

going through $\pi/2$ at the resonance. ◁

The characteristic time and length scales of the resonant absorption process can be estimated fairly easily. Let us assume that the system has been driven at a single frequency, viz. ω_d, for a long time so that it has reached a steady state. In this steady state, all physical quantities have a time behaviour of the form $e^{-i\omega_d t}$. Making the resonance condition explicit by using the linear Taylor expansion of $\omega_d^2 - \omega_A^2$ around the point x_0 where $\omega_d = \omega_A(x_0)$,

$$\omega_d^2 - \omega_A^2 = -(x - x_0) \left. \frac{d\omega_A^2}{dx} \right|_{x=x_0} , \tag{11.78}$$

a simplified version of Eq. (11.74) is obtained which is valid only in the vicinity of the resonance position:

$$(x - x_0)\, 2\,\omega_A(x_0)\, \frac{d\omega_A}{dx}(x_0)\, \xi_y + i\omega_d(\eta + \nu) \frac{\partial^2 \xi_y}{\partial x^2} = F(x, z)\, e^{-i\omega_d t} . \tag{11.79}$$

Here, we also exploited the fact that $|k_z| = |\partial/\partial z| \ll |\partial/\partial x|$ in the steady state to drop the terms with k_z^2. The dissipation is only significant when the absolute value of the second, dissipative, term in the above equation becomes comparable with the first term. This comparison leads to an estimate of the characteristic length scale l_0 of the resonant absorption process:

$$l_0 \sim \left(\frac{\eta + \nu}{2\, |\omega_A{}'|_{x=x_0}} \right)^{1/3} , \tag{11.80}$$

where the prime denotes the derivative with respect to x. Upon substitution of this characteristic dissipation length scale in the expressions for the time scales of resistive diffusion, $\tau_{res} = l_0^2/\eta$, and viscous diffusion, $\tau_{vis} = l_0^2/\nu$, we find that the characteristic time scale for resonant absorption scales as

$$\tau_{ra} \sim (\nu + \eta)^{-1/3}\, |\omega_A{}'|^{-2/3} . \tag{11.81}$$

Notice that, if the profile of the local Alfvén frequency is not monotonic, there may exist two or more dissipation layers, even if the driver is mono-periodic. Also, a multi-periodic driver will create many dissipation layers in the plasma so that the whole plasma volume can be heated by resonant absorption.

11.3.2 Phase mixing

Let us again consider the 'closed' configurations discussed in the previous subsection. So far we assumed a mono-periodic driver. In a more realistic model for the solar corona, the driver has a broad spectrum and the time evolution of the wave amplitude is given by a superposition of the nearly singular solutions discussed above. When each field line picks up its own characteristic frequency from the broad spectrum of the driver, we get a solution of the form

$$\xi_y(x, z, t) = A(x)\, \mathrm{e}^{\mathrm{i}[k_z z - \omega_\mathrm{A}(x) t]}, \tag{11.82}$$

which means that the field lines get more and more *out of phase* as time evolves, i.e. we get *phase mixing*. This yields an effective wave number

$$k_{x,\mathrm{eff}} = \omega_\mathrm{A}{}' \, t\,, \tag{11.83}$$

which is proportional to t, meaning that the effective wavelength becomes smaller and smaller as time proceeds (see Fig. 11.11). This 'cascade' of the energy to ever smaller length scales continues until the effective wavelength becomes of the order of the length scale l_0 of Eq. (11.80) when the dissipative terms become important. Hence, we can define a phase mixing time τ_mix as the time at which $k_{x,\mathrm{eff}} = 1/l_0$, i.e.

$$\tau_\mathrm{mix} = \frac{1}{l_0 |\omega_\mathrm{A}{}'|} \sim (\eta + \nu)^{-1/3} \, |\omega_\mathrm{A}{}'|^{-2/3}. \tag{11.84}$$

This phenomenon of phase mixing also occurs in the resonant absorption process discussed in the previous section. There too the field lines are initially in phase, but in the steady state shown in Fig. 11.2 the phases of the field lines are clearly not the same any more. In fact, the process of bringing neighbouring field lines out of phase is essential to get short length scales. Notice that we do not really need a resonance condition to get phase mixing. This process can also occur in coronal holes in which running waves along the 'open' field lines get phase mixed as time evolves. As a matter of fact, considering now the configuration displayed in Fig. 11.10(c) and assuming the foot points of the field lines are excited by a mono-periodic driver, we obtain *running wave* solutions of the form

$$\xi_y(x, z, t) \sim \mathrm{e}^{\mathrm{i}[k_z(x) z - \omega_\mathrm{d} t]}\,, \tag{11.85}$$

where now, instead of the frequency, the *wave number depends on* x: $k_z(x) = \omega_\mathrm{d}/v_\mathrm{A}(x)$. Since the phase speed of the waves depends on x, the running waves phase mix as they propagate along the field lines. This results in an effective wave number in the x-direction which scales with z:

$$k_{x,\mathrm{eff}} \sim \frac{1}{\xi_y}\frac{\partial \xi_y}{\partial x} \sim k_z' \, z. \tag{11.86}$$

Hence, large wave gradients are built up in the wave front as the waves propagate and z increases. These large gradients will appear at lower heights when the plasma is more inhomogeneous, i.e. when $k_z'(x)$ is larger. This means that we can define a phase mixing height z_mix as the height where $k_{x,\mathrm{eff}} = 1/l_0$:

$$z_\mathrm{mix} = 1/(l_0 k_z')\,. \tag{11.87}$$

Phase mixing of running Alfvén waves has been proposed as a possible means for explaining the heating of coronal holes and the acceleration of the solar wind [471]. The energy dissipation is now

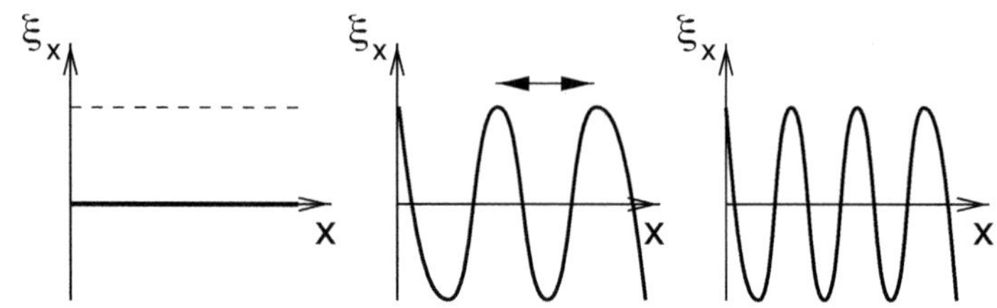

Fig. 11.11 'Cascade' of energy to small length scales due to phase mixing: (a) in the initial state ($t = 0$ and/or $z = 0$) all field lines are in phase; (b) at $t = t_1 > 0$ phase mixing sets in; (c) at later times small length scales develop.

spread over the whole plasma volume even for a mono-periodic driver, in contrast to the resonant absorption case described in the previous section where it was limited to a narrow dissipative layer around the ideal resonance position in that case.

11.3.3 Applications to solar and magnetospheric plasmas

(a) Heating of line-tied loops and arcades As mentioned above, early loop heating models 'copied' the fusion set-up and considered a one-dimensional (periodic in two directions) flux tube excited *sideways* by incident fast magneto-sonic waves. Soon, however, it was realized that the foot points of the coronal loops are effectively 'anchored' ('line-tied') in the photosphere due to the high conductivity and the relatively high mass density of the latter. Moreover, the loops are also, perhaps mainly, excited at their foot points as a result of the 'anchoring' in the turbulent sub-photospheric plasma, and also due to the magnetic reconnection events that take place at the ends of the loops.

In Ref. [577] resonant heating or dissipation was simulated in an incompressible cylindrical line-tied plasma excited at its foot points. It was remarked that the locations of the resonances differ from those in a periodic plasma. These conclusions were confirmed analytically [235, 277]: line-tying couples the Alfvén and fast magneto-sonic waves and modifies the continuous spectrum such that, for a cylindrical plasma with a sheared magnetic field, it is given by

$$\omega_{\rm A}(r) = \frac{n\pi}{L} \frac{B_z(r)}{\sqrt{\rho(r)}} , \tag{11.88}$$

i.e. *independent of the poloidal magnetic field component B_θ and the poloidal wave number*. For realistic coronal parameter values there are many quasi-modes in this line-tied continuum which positively affects the plasma driver coupling [597], although they are, strictly speaking, not necessary here since the shear Alfvén waves can be driven *directly*.

Wave heating of line-tied loops has been studied extensively with linear and nonlinear methods, numerically and analytically, computing the steady state, the eigenvalues and the time evolution (see Refs. [495, 471], and literature quoted there). Poedts and Boynton [493] showed with nonlinear MHD simulations that the heating can be very efficient (even without quasi-modes), and that it can easily compensate the radiative and conductive losses in a loop. Some studies also take into account the variation of the density [277] and magnetic field strength *along* the loops [47]. For realistic input power spectra the supplied energy is stored in the quasi-modes, which couple to Alfvén waves

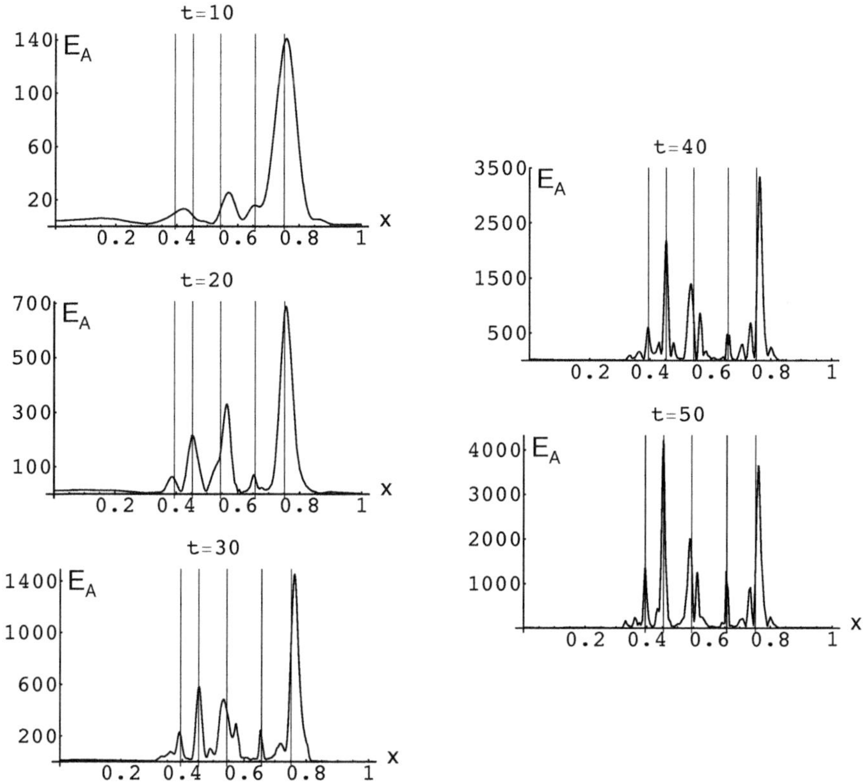

Fig. 11.12 Temporal evolution of the spatial distribution of the energy stored in Alfvén modes for a short loop with five quasi-modes. (From De Groof et al. [145].)

resulting in global heating of the loop on realistic time scales [597, 144, 145]. The outcome of all these studies is that *resonant dissipation is a viable heating mechanism*, even when the heating rate drops due to 3D 'Kelvin–Helmholtz-like vortices' at the resonance layers [471].

However, it is important to include the coupling of the coronal parts of the loops with the transition region and chromosphere in order to take into account the effect of leakage [54] and tuning/detuning effects due to variation of the lengths of the loops. In a first attempt, Ofman et al. [472] tried this by using scaling laws in a 1D model to update the density due to the expected chromospheric evaporation. This yielded efficient heating concentrated in multiple resonant layers even for mono-periodic driving. Beliën et al. [45] even take into account the thermal structuring of the transition region and the higher chromosphere in a nonlinear numerical study. They note that only about 30% of the energy supplied by the driver takes the form of Alfvén waves. The bulk of this energy goes to slow magneto-sonic waves which are resonantly *dissipated in the foot points* and never reach the coronal part of the loops. This is actually compatible with the observed uniform temperature along the loops, which requires a more efficient heating of the foot points in order to compensate the larger radiation there due to the higher density. These results for mono-periodic driving should be extended with the consideration of more realistic drivers.

Such radial and azimuthal drivers have been considered in Refs. [144, 145], and reviewed in

Ref. [249]. These linear MHD studies confirmed that for $k_y \neq 0$ the fast magneto-sonic waves and the Alfvén waves are coupled. The fast magneto-sonic body-modes yield efficient coupling of the loops to the random (broadband) driver, and the coupling to the Alfvén waves yields efficient dissipation which is spread over the entire loop volume (Fig. 11.12).

(b) Absorption of acoustic waves by sunspots Since 1989, the resonant absorption mechanism has also been studied in the context of p-mode absorption by sunspots. Observations by Braun, Duvall and Labonte [93] of high degree p-mode oscillations in regions around sunspots have revealed that sunspots act as strong absorbers of p-mode wave energy. Adopting a cylindrical coordinate system centred on the sunspot, the amplitudes of the waves travelling inward and outward from the sunspot were determined and it was found that as much as 50% of the acoustic wave power can be lost. In a subsequent investigation [94], they explored the horizontal spatial distribution of high degree p-mode absorption in solar active regions. They found that the absorption reaches a maximum in the visible sunspot, but that it is not limited to the location of the visible spot and it is also associated with magnetic fields in the surrounding plage. Larger sunspots are observed to absorb even more p-mode energy. The discovery that sunspots are strong absorbers of acoustic wave energy opens up the new avenue of *sunspot seismology* [78], or *active region seismology* [594]: the effect of active regions on solar oscillations can be directly observed. The aim is to use observations of p-mode oscillations outside sunspots to derive information on the conditions inside the spots. This requires a basic theoretical model that describes the observed properties of p-mode oscillations, in particular the strong absorption.

As a model for a sunspot we adopt an ideal static cylindrically symmetric magnetic flux tube in which the equilibrium quantities are functions of the radial distance to the axis of symmetry only, similar to a model exploited by Lou [417]. At the boundary $r = a$ of the sunspot the total (plasma and magnetic) pressure has to be continuous. In the idealized equilibrium state considered the magnetic field in the sunspot gradually drops to zero with increasing radial distance r, and the non-uniform sunspot is surrounded by a non-magnetic and uniform plasma, so that for $r > a$ the equilibrium density, pressure and temperature are constant.

Denoting the amplitude of the incoming wave by $A^{\rm i}$ and the amplitude of the outgoing wave by $A^{\rm o}$, the quantity we are interested in is the absorption coefficient

$$\alpha \equiv 1 - (A^{\rm o}/A^{\rm i})^2 \,. \tag{11.89}$$

Numerical simulations of the resonant absorption of incident sound waves for this simple sunspot model [250] reveal that the absorption coefficient α can indeed be of the order of 50% for typical sunspot parameters and for typical solar p-modes. This is illustrated in Fig. 11.13(a) where the absorption coefficient is plotted versus the wave number k of the incident sound waves for a fixed sunspot radius. In Fig. 11.13(b), α is plotted versus the sunspot radius for a fixed wave number. These results are for a spot with $B_z(0) = 0.2\,\mathrm{T}$ ($= 2000\,$gauss) and total photospheric pressure $p = 2.4 \times 10^2\,\mathrm{N\,m^{-2}}$ ($= 2.4 \times 10^4\,$Pa), as taken by Lou in a *viscous MHD model* [417] so that comparison is straightforward. In each figure, results are displayed for oscillations with $m = 1, 2, 3$ and 5 to see how the absorption coefficients vary as a function of azimuthal wave number and radius. Since the magnetic field is straight, the absorption is independent of the sign of m so that it suffices to consider positive values only. Figure 11.13 applies to p-mode oscillations with a frequency $\omega = 0.02\,\mathrm{rad\,s^{-1}}$. For comparison, Lou's results for oscillations with $m = 1$ are also displayed.

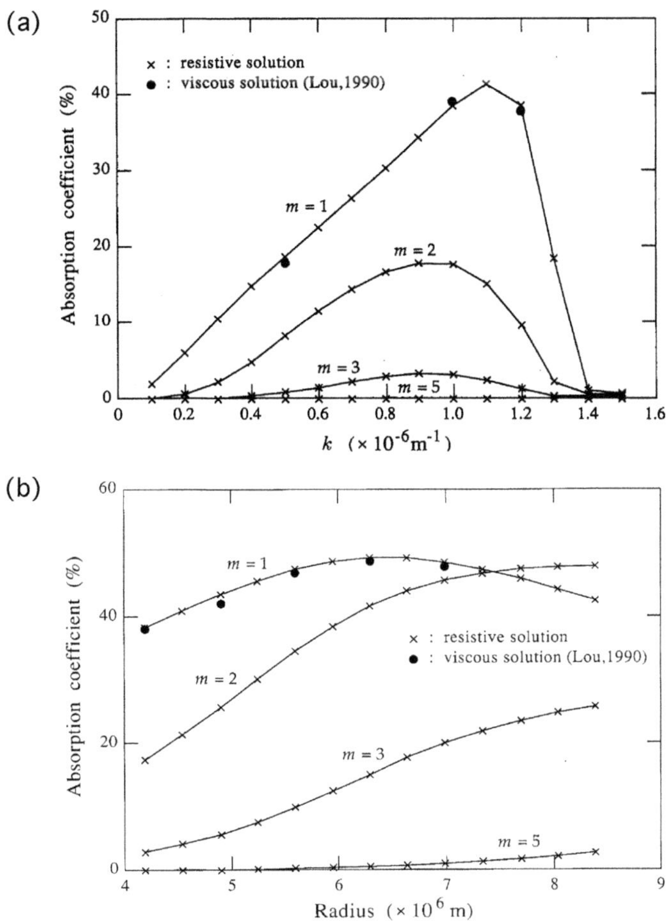

Fig. 11.13 Absorption coefficient α for the acoustic oscillations of a sunspot with a straight magnetic field, (a) plotted as a function of the wave number, $R = 4.2 \times 10^6$ m, (b) plotted as a function of the radius of the sunspot. $B_z(0) = 0.2\,\mathrm{T}$, $\omega = 0.02\,\mathrm{rad\,s^{-1}}$, $k = 1 \times 10^{-6}\,\mathrm{m^{-1}}$ and $m = 1, 2, 3, 5$. Crosses refer to resistive MHD, dots to viscous MHD [417].

From the parameter studies discussed, the following conclusions can be drawn:

– Results obtained in viscous MHD [417] are recovered in resistive MHD [250], so that *the energy absorption rate is independent of the actual dissipation mechanism*;

– Observed absorption of 50% and higher can be explained by resonant absorption;

– The variation of the absorption coefficient with the azimuthal wave number depends in a complicated way on the sunspot equilibrium and the characteristics of the incident acoustic mode (i.e. wavelength and angle of incidence);

– Resonant absorption of p-modes is more efficient in larger sunspots, and also in sunspots with twisted magnetic fields for higher azimuthal wave numbers.

The results discussed above are obtained for rather simple equilibrium models for sunspots.

Clearly, for final conclusions numerical simulations are required based on more realistic sunspot models. These should take into account the flaring out of the field lines in the upper photospheric part (due to the decrease of the external pressure) and the stratification of the density with height (due to gravity). Also, there are other possible explanations for the p-mode absorption by sunspots which deserve further attention such as p-mode to s-mode conversion. In that mechanism, the incident p-modes couple to the slow magneto-sonic (s-)modes in the sunspot, in effect redirecting the energy flux downwards (or upwards) in the sunspot. Another explanation can be found in fibril models for sunspots. In such bundles of flux tubes multiple scattering resonances occur which can drastically damp the waves [344].

(c) Phase mixing of coronal loops and holes Phase mixing does not really need resonances to be effective. In 'open field' configurations, such as coronal holes, the mechanism also works. This was studied by many authors. It was realized that flaring out of the magnetic field lines at the base of the coronal holes or long loops affects the slope of the local Alfvén frequency substantially [499, 532, 149]. In the last reference, a realistic two-dimensional configuration was considered, taking flow along the field lines into account. Strong damping was found in the layer where the velocity shear is concentrated. The heat deposited in the coronal hole seems not to be affected by the vertical stratification, but the efficiency of the heating mechanism does depend on the geometry and the wave amplitudes.

(d) Magnetospheric applications In 1861(!), Stewart observed ultra-low frequency (ULF) oscillations of magnetic fields on the surface of the Earth. These oscillations turned out to be caused by waves with periods ranging from seconds to minutes. They correspond to standing Alfvén waves with fixed ends in the ionosphere. In the 'box' model, the magnetic field lines are straightened and the natural frequencies are determined only by the length of the dipole magnetic flux tubes, the strength of the magnetic field, and the plasma density. Hence, for a given (known) magnetic geometry, the plasma density can be derived from the observed Alfvén wave frequencies [368]. This resonant field line model with the solar wind as sideways external driver is very successful in explaining many important properties of the ULF waves, e.g. the variation of the amplitude with latitude [368, 369].

11.4 Literature and exercises

Notes on literature

Resonant absorption in tokamaks

- Tataronis and Grossmann [588] and Chen and Hasegawa [118] propose resonant absorption as supplementary heating for tokamaks. Tataronis and Grossmann consider an incompressible plasma and show that MHD waves decay due to phase mixing. Chen and Hasegawa consider a compressible plasma slab with shear, using the surface wave for coupling to avoid 'undesirable effects' of the fast magneto-sonic wave.
- Kappraff and Tataronis [339] analyse the time evolution of resonant absorption in linearized resistive MHD for a plane slab geometry. They confirm earlier ideal MHD results on time scales, efficiency and energetics of the heating process.
- Balet, Appert and Vaclavik [34] review results of the TCA experiment in Lausanne where the theory

of resonant absorption is tested, reporting detection of a 'collective' mode. Vaclavik and Appert [617] present the theory of magnetic pumping and Alfvén resonance heating.

- Poedts and Kerner [497] compute the resistive MHD spectrum and show that the ideal quasi-modes (important for excitation) correspond to resistive global eigenmodes. Poedts et al. [498] also analyse another application of resonant absorption, viz. damping of toroidicity-induced Alfvén eigenmodes ('gap' modes).

Resonant absorption and phase mixing in coronal heating

- Ionson [326] suggests resonant absorption as a heating mechanism for coronal loops, applying the model of Chen and Hasegawa [118] with solar corona parameter values.
- Heyvaerts and Priest [301] suggest a simple model for phase mixing as a mechanism to heat coronal loops as well as coronal holes.
- Kuperus, Ionson and Spicer [386] compare the different heating mechanisms proposed for the solar corona up to 1981, Goossens [247] reviews the various resonant absorption models up to 1991, whereas Aschwanden [27] gives an evaluation for active regions based on observations by Yohkoh, SOHO and TRACE up to 2001.
- Grossmann and Smith [263] investigate the efficiency of resonant absorption for coronal heating with the power spectrum of driving photospheric oscillations as input.
- Hollweg and Yang [304], in 'Resonance absorption of compressible MHD waves at thin surfaces', calculate the damping of quasi-modes in finite loops.
- Most numerical results on resonant absorption are obtained with ideal MHD codes. Poedts, Goossens and Kerner [495, 491] solve the linearized resistive MHD equations with a finite-element numerical code to obtain the resonant dissipation rate. An alternative, integrating the ideal MHD ODEs and replacing the steep gradients in the dissipative layer by jump conditions, is given by Sakurai et al. [537, 251].

Resonant absorption in sunspots

- Thomas, Cram and Nye [594] propose the concept of sunspot seismology. They suggest that the response of spots to forcing by the 5-minute oscillations from the surrounding photosphere "may be used as a probe of sunspot structure below the surface".
- Braun, Duvall and LaBonte [93, 94] describe the interaction of p-modes with sunspots by comparing the amplitudes of inward and outward propagating waves in an annular region around the sunspot or active region. They show that the loss, or absorption, of power of the incoming acoustic waves is robust enough to be measured.
- Hollweg [303](a) suggests that the observed p-mode absorption by sunspots is due to resonant absorption. His restrictions of plane slab and thin transition region are lifted by Lou [417], who solves the viscous MHD equations and obtains absorption coefficients of 40–50%. This is confirmed by Goossens and Poedts [250] in resistive MHD, demonstrating that the actual dissipation mechanism is of secondary importance.

Resonant absorption in magnetospheres

- Kivelson and Southwood [369] discuss the role of global quasi-modes coupling to field line resonances in magnetospheric plasmas.
- Kivelson [367] exploits a box model of the magnetosphere to describe how waves at the magnetopause pump energy into the magnetospheric cavity and deposit it near magnetic shells where the conditions for transverse resonances are satisfied.

Exercises

[11.1] *Time average*

In Section 11.1.2 it was stated that the *time average* of two harmonic quantities, which are the real parts of $A\mathrm{e}^{\mathrm{i}\omega t}$ and $B\mathrm{e}^{\mathrm{i}\omega t}$, is half the product of A and the complex conjugate of B, i.e. $\frac{1}{2}AB^*$. Check this statement.

[11.2] *Radial excitation*

In Section 11.3.1 we assumed that the perturbations were polarized in the tangential perpendicular (y) direction. Under which condition(s) can the Alfvén wave heating mechanism work with *radially* polarized foot point oscillations? Explain why.

[11.3] *Resonant absorption*

Write a Maple or Mathematica work sheet to reconstruct Fig. 11.4(b) and verify how the parameters a, c, $k_\parallel$ and $k_\perp$ affect the fractional absorption.

[11.4] *Time and length scales resonant absorption*

Use the tables in Appendix B to get a rough estimate of the resonant heating time scales and the width of the resonant layers both in laboratory plasmas and in solar coronal loops.

[11.5] *Basic phase mixing result*

Consider Eq. (11.74) without the external driving term and assume that the gradients in the x-direction are much larger than the gradients in the z-direction ($\sim k_z$):

$$\frac{\partial^2 \xi_y}{\partial t^2} = v_{\rm A}^2(x)\,\frac{\partial^2 \xi_y}{\partial z^2} + (\eta + \nu)\frac{\partial^2}{\partial x^2}\frac{\partial \xi_y}{\partial t}\,.$$

Heyvaerts and Priest [301] assume in their analysis a solution to this equation of the form

$$\xi_y \sim \xi_y(x,z)\ \exp i[\Omega t - k(x)z].$$

Derive the solution for $\xi_y(x,z)$ obtained by Heyvaerts and Priest, viz.

$$\xi_y(x,z) = \xi_y(x,0)\ \exp\Big[-\frac{\big(k(x)z\big)^3}{6R_T}\Big]\,, \qquad \text{where } R_T = \frac{\Omega}{\eta+\nu}\Big[\frac{d}{dx}\log k(x)\Big]^{-2}\,,$$

which is valid under the condition of weak damping and strong phase mixing:

$$\frac{1}{k}\frac{\partial}{\partial z} \ll 1, \quad \text{and} \quad \frac{z}{k}\frac{\partial k}{\partial x} \gg 1,$$

and thus only for large z.

[11.6] *Phase mixing time and length scales*

Use the tables in Appendix B to get a rough estimate of the phase mixing time scales in solar coronal loops and the phase mixing height $z_{\rm mix}$ in coronal holes.

Part IV

Flow and Dissipation

12

Waves and instabilities of stationary plasmas

In this chapter and the next, the most urgent extension of the theory presented so far will be developed. This concerns the waves and instabilities in plasmas with stationary background flows, which is a theme of common interest for laboratory fusion and astrophysical plasma research. The old problem of how to find the complex eigenvalues of stationary plasmas is solved by means of a new method of constructing paths in the complex plane that constitute what has been called *the Spectral Web* [229]. This method is applied to the classical topics of shear flow in plane plasma slabs, and to rotation in cylindrical plasmas (Chapter 13). In the later chapters of Part IV, we include dissipation, in particular the resistive spectrum and reconnection (Chapter 14), and we discuss in detail the enormous variety of numerical solution techniques needed for the spectral calculations of ideal and dissipative plasmas (Chapter 15).

Only after these basic techniques have been elaborated, another urgent extension will be presented in Part V on 'Toroidal plasmas', viz. the spectral theory of toroidal plasmas. Chapters 16 and 17 are devoted to the description of toroidal equilibrium and stability for the standard picture of static plasmas, whereas the substantial extension for toroidal plasmas with equilibrium flow is elaborated in Chapter 18. The latter preludes some aspects of transonic MHD flows, only analysed in the later Chapter 21.

Finally, in Part VI on 'Nonlinear dynamics', the analysis of moving plasmas will be extended to the nonlinear domain where separation of equilibrium and stability is no longer feasible. After an exposition of the subject of turbulence in magneto-fluids (Chapter 19) and, again, of the enormous variety of numerical solution techniques needed to analyse the nonlinear dynamics (Chapter 20), the subject of moving plasmas is resumed with the analysis of transonic flow and shocks (Chapter 21), and ultimately of the extreme speeds of relativistic plasma flows encountered in the different collisions and explosions that occur in the final phases of stellar and galactic evolution (Chapter 22).

12.1 Laboratory and astrophysical plasmas

12.1.1 Grand vision: magnetized plasma on all scales!

In Chapter 1 we pointed out that, since more than 90% of visible matter in the Universe is plasma, plasmas and the associated magnetic fields are important constituents of nature. In Chapter 4, we then showed that *the equations of magnetohydrodynamics (MHD) are scale-independent:* the scales of length, density and magnetic field strength of a magnetized plasma may be transformed away. This simple fact has the amazing consequence that the macroscopic dynamics of plasmas in both laboratory fusion devices (tokamaks, stellarators, etc.) and astrophysical objects (stellar coronae, accretion disks, galaxies, etc.) may be described by the same equations, viz. the equations of MHD. We encountered several examples of this before. Here, we will continue the investigation of this common field of research by means of the new 'wide-angle MHD telescope'.

Figure 12.1 shows two representative, but very different, examples from science and technology, viz. the design drawing of the international tokamak experimental reactor ITER, presently under construction, and an image of the Whirlpool Galaxy M51 made by the Hubble Space Telescope (HST), overlaid with radio intensity contours and magnetic field vectors from the radio telescopes Effelsberg and VLA. The consequence of scale-independence is that the most obvious difference of the two configurations, their length scale indicated next to the figure, is actually irrelevant for the description of macroscopic plasma dynamics!

▷ **Scale-dependent models** To avoid misunderstanding: small-scale kinetic or two-fluid effects like electron inertia [32], described by the scale-dependent model of Hall-MHD, can have macroscopic consequences like reconnection and waves (see Section 14.4), which may even be detectable by spacecrafts flying through the bow shock of the magnetosphere; see Stasiewicz [566]. Likewise, in the description of hot plasmas in thermonuclear confinement experiments, kinetic effects exhibit a bewildering range of dynamical phenomena on many spatial and temporal scales presenting a challenge to the computational modelling of these plasmas by different, scale-dependent, fluid closures; see Schnack et al. [543]. ◁

For our purpose, the usual Hubble Space Telescope pictures, without the radio emission contours, are somewhat misleading since they only show the stars and dust. Roughly an equal amount of plasma is present in the plasma component of galaxies (not counting the plasma interiors of the stars). According to the review by Fukugita [200], for the Universe as a whole the balance is shifted significantly towards plasma: ten times more mass is present in plasmas than in stars (again, not counting that stars themselves are mostly plasma). However, much more mass should be present in the dark matter component. Since we have no clue about the physics of dark matter, it might be advisable to first investigate the plasma component with all techniques that are presently available. Recalling our critical discussion of the standard view of nature (Section 1.3.4), which does not articulate the distinction between neutral gas and plasma, as schematically represented in Figure 1.8, one would expect on the contrary that the abundance of plasma ($\equiv$ abundance of magnetic fields $\equiv$ global anisotropic dynamics) should play a much more prominent role in the description of the Universe than it has up till now.

In fact, there are many signs that astrophysics is beginning to fill in this gap. For example, when Land and Magueijo [391] established that there is a small statistically significant anisotropy, with a preferred axis, in the cosmic background radiation observed with the WMAP satellite, the far-going implications for cosmology were immediately realized. Also, from alignment of quasar polarizations with large-scale structures found by Hutsemékers et al. [315], some researchers started to speculate that a large-scale cosmic magnetic field might be involved.

As another example, Kaastra et al. [337], and several others (see Peterson and Fabian [485]), pointed out that magnetic fields may play an important role in the dynamics of clusters of galaxies. From X-ray spectra obtained by the XMM-Newton satellite, they conclude that magnetized plasmas in huge magnetic loops, of similar spatial structure to those in stellar coronae, may be responsible for the temperature decrement observed for cooling plasma flows in those clusters.

Likewise, filaments of warm hot intergalactic matter (WHIM) connecting clusters of galaxies have been detected by Werner et al. [644] by means of X-ray images obtained from the same satellite. This discovery appears to agree with simulations that ascribe this 'cosmic web' mainly to dark

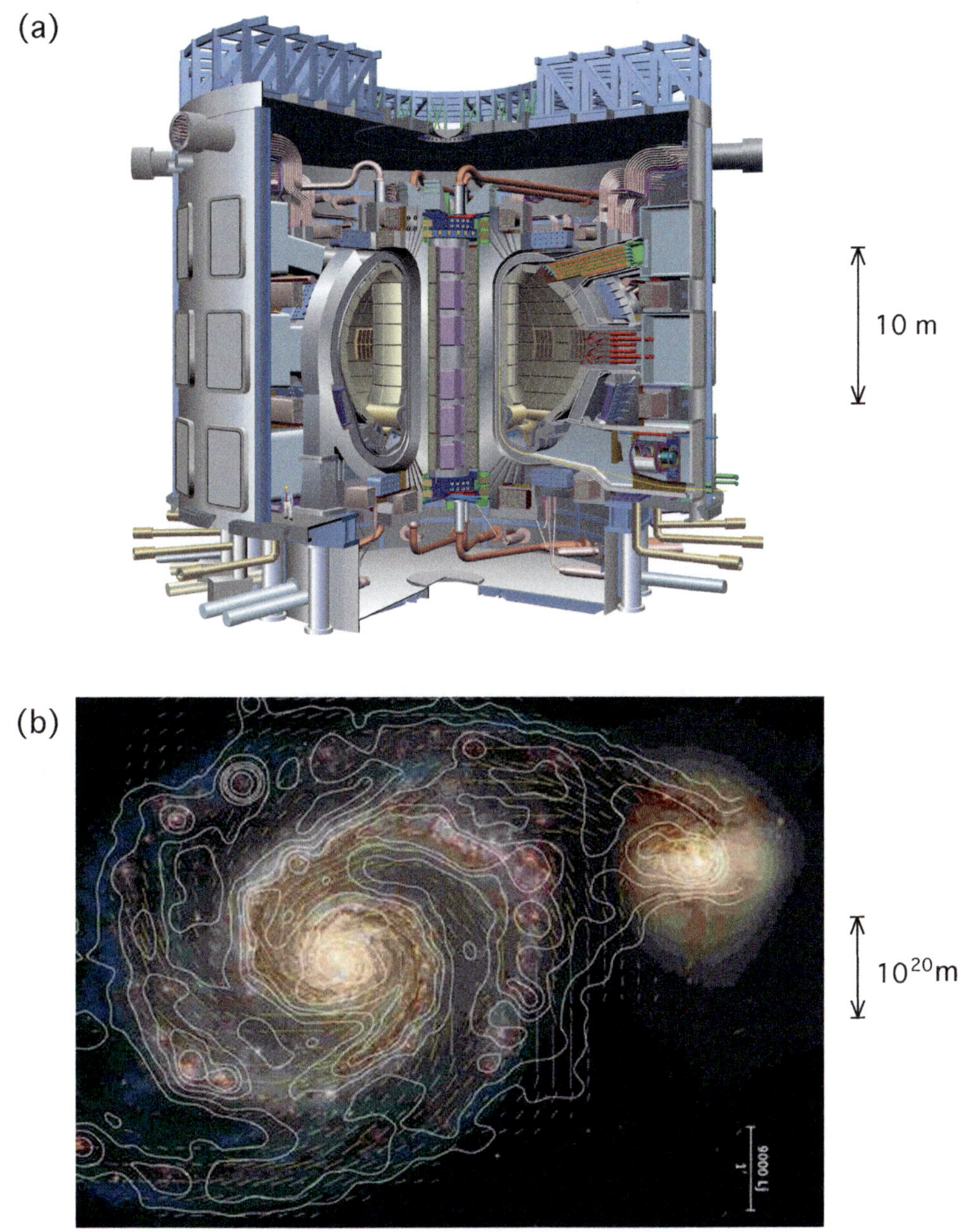

Fig. 12.1 Magnetized plasmas in the laboratory and in the Universe: (a) the international tokamak experimental reactor ITER; (b) the Whirlpool spiral Galaxy M51.

matter, but the filamentary structure could be associated with a magnetized plasma component just as well. Also, it was pointed out by Gu et al. [268] that the observed 3.5 eV emission lines in clusters of galaxies, that are speculatively attributed to decay of sterile neutrinos, can adequately be explained by charge transfer between neutral gas and hot plasma. The bookkeeping of the gravitational effects ascribed to dark matter might well change in the direction of a larger contribution of plasma and the associated magnetic field!

Whatever the final outcome of these debates will be, it is probable that plasma and, hence, magnetic fields will become much more central for our understanding of the dynamics of the Universe at large than presently accounted for.

12.1.2 Laboratory and astrophysical plasmas

Although scale-independence of the MHD equations permits analysis of global plasma dynamics in laboratory and astrophysical plasmas by the same techniques, the important differences of the parameters that govern overall force balance should not be lost sight of. For example, the parameter $\beta \equiv 2\mu_0 p/B^2$ is small for tokamak plasmas and usually large for astrophysical plasmas, so that plasma dynamics in tokamaks is always dominated by magnetic fields whereas this may not be the case for astrophysical plasmas.

Roughly speaking, one could distinguish the two kinds of plasma configurations on the basis of the following global equilibrium characteristics.

(a) Tokamaks are *magneto*-hydrodynamic plasmas, with a magnetic field that is approximately a force-free field (FFF),

$$\begin{aligned} \mathbf{j}\times\mathbf{B} &\approx 0 \qquad \textit{(FFF to leading order)} \\ &= \nabla p \sim \beta \ll 1 \qquad \textit{(important correction)}. \end{aligned} \tag{12.1}$$

The equilibrium is nearly exclusively determined by the magnetic field geometry, but the pressure corrections are essential since they determine the power output of a future fusion reactor.

(b) Most astrophysical objects are *hydro*-magnetic plasmas, with sizeable flows, and the gravitational acceleration is usually not negligible,

$$\begin{aligned} \rho\mathbf{v}\cdot\nabla\mathbf{v} + \nabla p + \rho\nabla\Phi_{\rm gr} &\approx 0 \qquad \textit{(Keplerian flow to leading order)} \\ &= \mathbf{j}\times\mathbf{B} \sim \beta^{-1} \ll 1 \qquad \textit{(important correction)}. \end{aligned} \tag{12.2}$$

Gravity and rotation usually dominate over the magnetic terms, but the latter may be crucial for the growth of instabilities (such as for the Parker instability, discussed below, and the magneto-rotational instability which even operates when the magnetic field is infinitesimal, see Section 13.5).

It is well known that a force-free magnetic field cannot be extended indefinitely, as follows from the virial theorem (see Shafranov [549], p. 106). Eventually, the magnetic pressure has to be balanced by something. In tokamaks, equilibrium is due to balancing of the Lorentz forces on the plasma by mechanical forces on the induction coils, which have to be firmly fixed to the laboratory by 'nuts and bolts'. (Without those, the configuration would simply fly apart: a magnetic field of 5 T exerts a pressure of $B^2/(2\mu_0) \approx 10^7\,\mathrm{N\,m^{-2}} \approx 100\,\mathrm{atm}$!) The mechanical counterpart for accretion disks or galaxies is balancing of the centrifugal acceleration by the gravitational pull of the central objects, which may include a black hole. The implications of this difference for stability are much more wide-ranging than generally realized, as will be illustrated by contrasting 'intuition' developed on tokamak stability to some major instabilities operating in astrophysical plasmas.

12.1.3 Interchanges and the Parker instability

To appreciate the issue, let us pronounce some general features of tokamak stability theory, based on the results from the quasi-cylindrical approximation presented in Section 9.4 and anticipating the exact toroidal representation to be developed in Chapter 17. For the present purpose, the difference between the cylindrical approximation in terms of r, θ and z and the toroidal representation in terms of ψ (the poloidal magnetic flux, the 'radial' coordinate), ϑ (the poloidal angle) and φ (the toroidal angle) may be ignored. Without exaggeration, it may then be said that the wide variety of MHD instabilities operating in tokamaks, represented by normal modes of the form

$$f(\psi, \vartheta, \varphi, t) = \sum_m \tilde{f}_m(\psi)\, e^{i(m\vartheta + n\varphi - \omega t)}\,, \tag{12.3}$$

is unstable only for (approximately) perpendicular wave vectors,

$$\mathbf{k}_0 \perp \mathbf{B} \quad \Rightarrow \quad -i\mathbf{B}\cdot\nabla \sim m + nq \approx 0\,. \tag{12.4}$$

The reason is the enormous field line bending energy of the Alfvén waves,

$$W_A \approx \tfrac{1}{2}\int \big[(\mathbf{k}_0\cdot\mathbf{B})^2\, |\mathbf{n}\cdot\boldsymbol{\xi}|^2 + \cdots \big]\, dV \gg 0\,, \tag{12.5}$$

so that field line localization ($k_\parallel \ll k_\perp$) is necessary to eliminate this term and to get instability from the different higher order terms due to, e.g. pressure gradients and currents. The Ansatz (12.4) is made in virtually all tokamak stability calculations, like in the derivation of the Mercier criterion [437] involving *interchanges* on rational magnetic surfaces, of *ballooning modes* [125](b) involving localization about rational magnetic field lines, of *internal kink modes*, of *neo-classical tearing modes*, of *external kink modes*, etc.; see Sections 17.2 and 17.3. All involve localization about rational magnetic surfaces, either inside the plasma or in an outer vacuum. Hence, it became a kind of 'intuition' in tokamak physics to assume that this field line localization is a general truth about plasma instabilities.

In contrast, some major instabilities in astrophysical plasmas turn out to operate under precisely the opposite conditions:

$$\mathbf{k}_0 \parallel \mathbf{B} \quad \Rightarrow \quad -i\mathbf{B}\cdot\nabla \sim m + nq \sim 1\,. \tag{12.6}$$

These include the *Jeans instability* discussed in Section 5.3.4, the *Parker instability* operating in spiral arms of galaxies [482] and the *magneto-rotational instability* [631, 111, 29] which is held responsible for the turbulent dissipation in accretion disks about a compact object. All have their largest growth rates when the wave vector $\mathbf{k}_0$ is about parallel to the magnetic field, and certainly not perpendicular! (It is most peculiar that this apparent contradiction with stability of laboratory plasmas went unnoticed so far.) How is the above argument about the dominance of field line bending energy circumvented for astrophysical plasmas?

In order to answer that question, let us compare how the two entirely different pairs of equilibrium conditions (12.1) and (12.2), and the associated pairs of instability conditions (12.4) and (12.6), appear in the analysis of the gravitational interchange [225]. This instability has played an important role in modelling both the stability of laboratory plasmas (where gravity is used as just a way to simulate magnetic field line curvature) and the Parker instability [482] which is concerned with instability due to genuine gravity in spiral arms of galaxies.

To that end, we recapitulate the major conclusions on the gravitational interchange from Sections 7.5.2 and 7.5.3, Eqs. (7.177), (7.183) and (7.191). The stability criterion for gravitational interchanges of a plane plasma slab reads:

$$-\rho N_{\rm B}^2 \equiv \rho' g + \frac{\rho^2 g^2}{\gamma p} \leq \tfrac{1}{4} B^2 \varphi'^2 \,, \tag{12.7}$$

where $N_{\rm B}$ is the Brunt–Väisäläa frequency and φ' is the magnetic shear. Without magnetic shear, stability just appears to depend on the square of the Brunt–Väisäläa frequency: $N_{\rm B}^2 \geq 0$, which amounts to the Schwarzschild criterion for convective stability when expressed in terms of the equilibrium temperature gradient. This criterion is obtained from the marginal equation of motion ($\omega^2 = 0$) in the limit of small parallel wave number ($k_\parallel \to 0$). However, when these two limits are interchanged ($k_\parallel = 0$ and $\omega^2 \to 0$), an entirely different criterion is obtained:

$$-\rho N_{\rm M}^2 \equiv \rho' g + \frac{\rho^2 g^2}{\gamma p + B^2} \leq 0 \,, \tag{12.8}$$

where $N_{\rm M}$ is the magnetically modified Brunt–Väisäläa frequency. The apparent discrepancy between these stability criteria was resolved by Newcomb [462] who noted that there is a cross-over of two branches of the local dispersion equation with the solutions

$$\omega_1^2 \;=\; (k_0^2/k_{\rm eff}^2)\, N_{\rm M}^2 \qquad \textit{(pure interchanges)}, \tag{12.9}$$

$$\omega_2^2 \;=\; \frac{N_{\rm B}^2}{N_{\rm M}^2}\, \frac{\gamma p}{\gamma p + B^2}\, \frac{1}{\rho}\, (\mathbf{k}_0 \cdot \mathbf{B})^2 \qquad \textit{(quasi-interchanges)}, \tag{12.10}$$

where the last mode is the first to become unstable when the density gradient is increased. The first expression holds for $k_\parallel = 0$, where the factor $k_{\rm eff}^2 \equiv k_0^2 + n^2\pi^2/a^2$ indicates clustering of the modes, $\omega_1^2 \to 0$, when the vertical mode number becomes large, $n \to \infty$ (this n should not be confused with the toroidal mode number n introduced above), and the second expression is only valid for $k_\parallel \ll k_\perp$. Hence, *field line bending is small in both cases.*

In cylindrical and toroidal plasmas, magnetic field line curvature is unavoidable and interchange instabilities then arise when the negative pressure gradient (associated with confinement) exceeds the shear of the magnetic field lines, analogous to the gravitational interchange criterion (12.7). This is expressed by the criteria of Suydam [581], Eq. (9.112), and Mercier [437], Eq. (17.97). For cylindrical plasmas without magnetic shear, expressions were derived for the growth rates of interchanges and quasi-interchanges [639, 214, 234], analogous to Eqs. (12.9) and (12.10) with the following replacements:

$$N_{\rm M}^2 \to \frac{2B_\theta^2}{\rho r B^2}\Big(p' + \frac{\gamma p}{\gamma p + B^2}\frac{2B_\theta^2}{r}\Big), \qquad N_{\rm B}^2 \to \frac{2B_\theta^2}{\rho r B^2}\, p' \,. \tag{12.11}$$

As illustrated in Figs. 9.12 and 9.9, when p' becomes negative (violation of the shearless limit of Suydam's criterion) first the quasi-interchanges become unstable and the pure interchanges become unstable when $p' \leq -\gamma p\, (\gamma p + B^2)^{-1}\, (2B_\theta^2)/r$, in agreement with the expression for the z-pinch derived by Kadomtsev [338].

It would appear that the analogy between plasmas with curved magnetic fields and gravitational plasmas is perfect: instability only occurs at the interchange value $k_\parallel = 0$ or close to it. However (we now complete the analysis of the gravito-MHD waves started in Section 7.3.3), the *Parker instability* operates under precisely opposite conditions ($k_\perp \approx 0$). For an exponential atmosphere, its growth rate is given by expanding the expression (7.100):

$$\omega^2 \approx \Big(1 + \frac{\rho N_B^2}{k_{\rm eff}^2 B^2}\Big) \frac{\gamma p}{\gamma p + B^2} \frac{1}{\rho} (k_0 B)^2 . \tag{12.12}$$

This looks similar to the expression (12.10) for the quasi-interchanges, which gives the growth rate at $k_\parallel \approx 0$ for *localized modes* ($n \to \infty$), but it is actually completely different since the expression (12.12) for the Parker instability requires $k_\perp \approx 0$ and only yields instability for *global modes* ($n \approx 1$). This is so because the criterion for the Parker instability, $k_{\rm eff}^2 B^2 + \rho N_B^2 < 0$, cannot be satisfied for $n \to \infty$, since $k_{\rm eff} \to \infty$ then. In other words, it is very well possible to have a global instability when the field line bending energy (12.5) is not small at all! This is also the case for the magneto-rotational instability (see Section 13.5).

Hence, MHD instabilities occur in astrophysical plasmas under conditions that do not allow instability in laboratory plasmas. The reason is the stabilizing 'backbone' of a large toroidal magnetic field in the latter. Estimating orders of magnitudes for an equilibrium with inhomogeneity length scale L, the Parker instability requires

$$N_B^2 \sim -\frac{B^2}{\rho L^2} , \quad \text{with } \beta \sim 1 . \tag{12.13}$$

In contrast, the corresponding term of Eq. (12.11)(b) for curvature–pressure gradient driven interchanges in quasi-cylindrical/toroidal equilibria requires

$$\frac{2B_\theta^2}{\rho r B^2} p' \sim -\epsilon^2 \beta \frac{B^2}{\rho L^2} , \quad \text{with } \epsilon \equiv a/R_0 \ll 1 , \ \ \beta \sim \epsilon^2 . \tag{12.14}$$

In the first case, the driving force of the instability can compete with the field line bending contributions (12.5). In the latter case, because of the small factor $\epsilon^2\beta \sim \epsilon^4$, this is impossible so that pressure-driven interchanges in cylindrical and toroidal plasmas never occur for $\mathbf{k}_0 \parallel \mathbf{B}$. Consequently, tokamak 'intuition' focusing on rational magnetic surfaces and field lines as exclusively determining stability may be very misleading for astrophysical plasmas [228].

The two different viewpoints can be reconciled as follows. Whereas the condition (12.4) for tokamak instability automatically leads to study of the degeneracy and couplings of the Alfvén and slow continua close to marginal stability ($\omega \approx 0$), an entirely different path to avoid the stabilizing contribution (12.5) of the Alfvén waves is exploited by the Parker instabilities. These are actually modified slow magneto-acoustic waves avoiding the coupling to the Alfvén waves by remaining orthogonal to them: the polarization (expressed by the eigenvector $\boldsymbol{\xi}$) of the Parker (slow) modes is parallel to $\mathbf{B}$ (flow along the magnetic field is essential), whereas the polarization of the Alfvén waves is mainly perpendicular to $\mathbf{B}$. This orthogonality is clearly exhibited by Fig. 12.2, which shows the complete low-frequency part of the spectrum of modes for a gravitating plasma slab with exponential dependence on height of the density, magnetic field and pressure, for different values of the angle ϑ between the horizontal wave vector $\mathbf{k}_0$ and the magnetic field.

The exponentially stratified equilibrium was analysed in Section 7.3.2, resulting in the dispersion equation (7.104) with solutions shown in Fig. 7.9 for fixed angle ϑ. These solutions are now shown in Fig. 12.2(a) for all directions of $\mathbf{k}_0$. At $\vartheta = 0$ ($\mathbf{k}_0 \parallel \mathbf{B}_0$), the Parker instability has its largest growth rate, whereas around $\vartheta = \frac{1}{2}\pi$ ($\mathbf{k}_0 \perp \mathbf{B}_0$), the interchanges and quasi-interchanges operate.

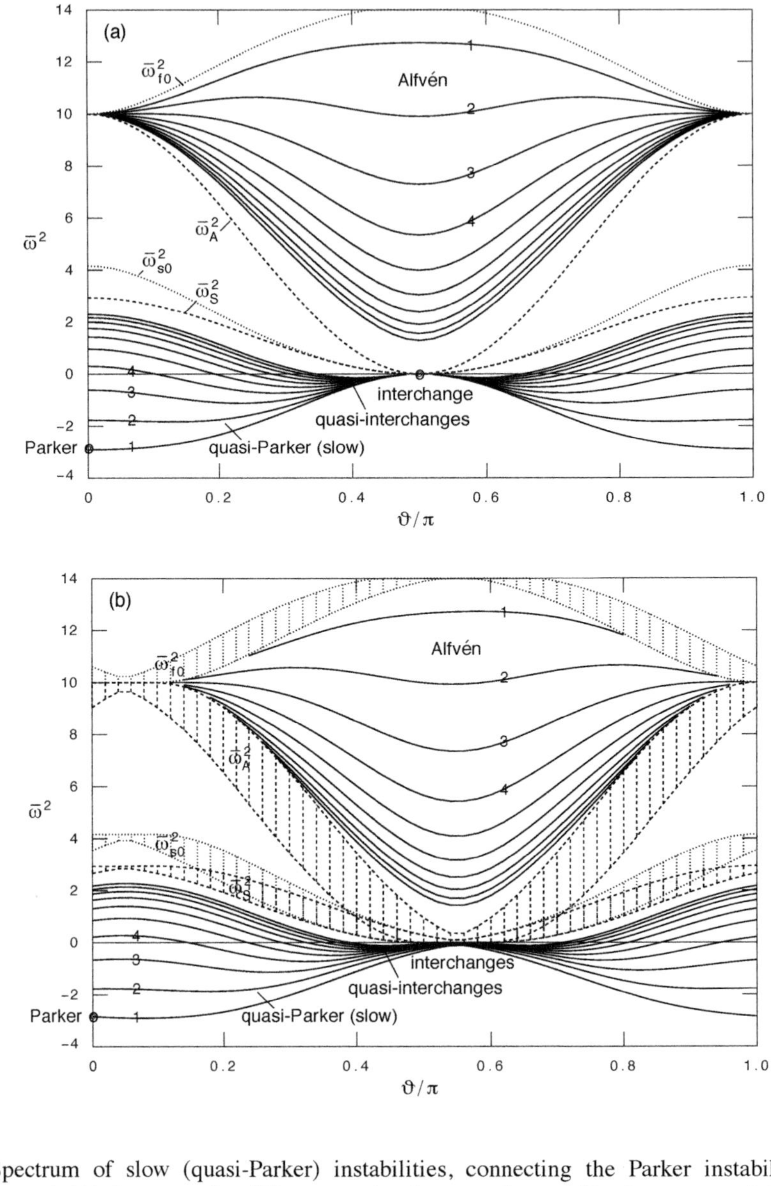

Fig. 12.2 Spectrum of slow (quasi-Parker) instabilities, connecting the Parker instability to the quasi-interchanges, and Alfvén waves for different angles between $\mathbf{k}_0$ and $\mathbf{B}_0$ for exponential atmosphere with (a) a uni-directional field ($\bar{\lambda} = 0$), (b) magnetic shear ($\bar{\lambda} = 0.3$) and genuine continua $\bar{\omega}_A^2$ and $\bar{\omega}_S^2$; $\bar{\alpha} = 20$, $\beta = 0.5$, $\bar{k}_0^2 = 10$, $\bar{q} = n\pi$ $(n = 1, 2, \ldots, 10)$.

These two ranges correspond to two different instability mechanisms: in the $k_\parallel \approx 0$ range, coupling of *local* (high n) slow and Alfvén modes causes interchange or quasi-interchange instability, whereas in the range $k_\perp \approx 0$, *global* (low n) instability of the slow magneto-sonic branch, viz. the Parker instability, occurs. In the intermediate range, there is a smooth transformation from the Parker instability to interchanges via modes that we have termed *quasi-Parker instabilities*.

The local interchange and quasi-interchange instabilities are modified substantially by the introduction of magnetic shear,

$$\mathbf{B} = B_0 \mathrm{e}^{-\frac{1}{2}\alpha x} \left[\sin(\lambda x)\mathbf{e}_y + \cos(\lambda x)\mathbf{e}_z \right] . \tag{12.15}$$

Except for modifying the stability properties, it also leads to the bands of continuous spectra ω_{A}^2 and ω_{S}^2 separating the Alfvén waves from the gravitational instabilities, as shown in Fig. 12.2(b), obtained by solving the full implicit eigenvalue problem (7.80), (7.82) with the 'shooting' method. Clearly, the consideration of magnetic shear is essential for the analysis of local stability criteria, as usual in tokamak stability studies, but not so important for the Parker instability.

In conclusion, a full spectral analysis of gravitational instabilities exhibits the existence of a large class of instabilities, called *quasi-Parker instabilities*, that smoothly connect the Parker instability (operating at $\mathbf{k} \parallel \mathbf{B}$) of astrophysical plasmas to the quasi-interchanges and interchanges of laboratory plasmas (operating at $\mathbf{k} \perp \mathbf{B}$). Eventually (stretching the imagination now), full spectral studies of the gravitational waves in galaxies could lead to *MHD spectroscopy of galactic plasmas* [345], i.e. determination of the internal characteristics of the galactic plasma by computing and observing the spatial structure of the modes (the frequencies are unobservable on a human time scale).

12.2 Spectral theory of stationary plasmas

12.2.1 Plasmas with background flow

So far, we have developed the theory of equilibrium, waves and instabilities mainly from the idealized picture of a plasma at rest in *static equilibrium*. This view on plasmas from laboratory fusion research has led to many fruitful insights, but it fails to account for the dynamics of the vast majority of astrophysical plasmas where this assumption is simply wrong. Also, in fusion research, the ultimate goal of an energy producing machine requires the presence of substantial plasma flows caused by the injection of neutral beams to heat the plasma fuel and by divertors to absorb exhaust and impurities. Hence, the study of the *influence of background flows on equilibrium, waves and instabilities* is of major importance for both laboratory and astrophysical plasmas. This implies that the hydrodynamics part of MHD has to be taken more seriously. Hence, this chapter is devoted to the urgent modifications of MHD spectral theory by the consequences of equilibrium plasma flows. This involves accounting for the effects of the Doppler shift on the spectra, eliminating misconceptions about supposed non-self-adjointness of the operators, and construction of a method to compute eigenvalues in the complex plane.

To describe the dynamics of plasmas with stationary flow, we start from the nonlinear ideal MHD equations (see Chapter 4) for the density ρ, the velocity $\mathbf{v}$, the pressure p and the magnetic field $\mathbf{B}$:

$$\frac{\partial \rho}{\partial t} + \nabla \cdot (\rho \mathbf{v}) = 0 , \tag{12.16}$$

$$\rho\Big(\frac{\partial \mathbf{v}}{\partial t} + \mathbf{v} \cdot \nabla \mathbf{v}\Big) + \nabla p - \mathbf{j} \times \mathbf{B} - \rho \mathbf{g} = 0 , \quad \mathbf{j} = \nabla \times \mathbf{B} , \tag{12.17}$$

$$\frac{\partial p}{\partial t} + \mathbf{v} \cdot \nabla p + \gamma p \nabla \cdot \mathbf{v} = 0 , \tag{12.18}$$

$$\frac{\partial \mathbf{B}}{\partial t} - \nabla \times (\mathbf{v} \times \mathbf{B}) = 0 , \qquad \nabla \cdot \mathbf{B} = 0 . \tag{12.19}$$

This set is to be complemented with appropriate initial and boundary conditions, e.g. the BCs for model I (plasma confined inside a rigid wall):

$$\mathbf{n}\cdot\mathbf{v} = 0\,, \qquad \mathbf{n}\cdot\mathbf{B} = 0 \qquad \text{(at the wall)}\,, \tag{12.20}$$

as discussed in Section 4.6.1. The gravitational acceleration $\mathbf{g} = -\nabla\Phi_{\rm gr}$ is here considered to be caused by a fixed external gravitational field $\Phi_{\rm gr}$. We will consistently exploit scale independence (see Section 4.1.2) and drop factors μ_0.

Most of the previous chapters were concerned with *static equilibria* ($\mathbf{v} = 0$),

$$\nabla p = \mathbf{j}\times\mathbf{B} + \rho\mathbf{g}\,, \qquad \mathbf{j} = \nabla\times\mathbf{B}\,, \qquad \nabla\cdot\mathbf{B} = 0\,, \tag{12.21}$$

and perturbations of these equilibria described by the equation of motion with the force operator $\mathbf{F}$ acting on the plasma displacement vector $\boldsymbol{\xi}$:

$$\mathbf{F}(\boldsymbol{\xi}) = \rho\frac{\partial^2\boldsymbol{\xi}}{\partial t^2}\,. \tag{12.22}$$

The spectral problem for normal mode solutions $\hat{\boldsymbol{\xi}}(\mathbf{r})\exp(-\mathrm{i}\omega t)$,

$$\mathbf{F}(\hat{\boldsymbol{\xi}}) = -\rho\omega^2\hat{\boldsymbol{\xi}}\,, \tag{12.23}$$

was particularly effective since the linear operator $\mathbf{F}$ is self-adjoint, so that the eigenvalues ω^2 are real.

All this changes when *stationary equilibria* ($\mathbf{v} \neq 0$) are considered. In the first place, none of the MHD equations (12.16)–(12.19) is trivially satisfied now, so that the description of stationary equilibria requires four differential equations:

$$\nabla\cdot(\rho\mathbf{v}) = 0\,, \tag{12.24}$$

$$\rho\mathbf{v}\cdot\nabla\mathbf{v} + \nabla p - \mathbf{j}\times\mathbf{B} - \rho\mathbf{g} = 0\,, \quad \mathbf{j} = \nabla\times\mathbf{B}\,, \tag{12.25}$$

$$\mathbf{v}\cdot\nabla p + \gamma p\nabla\cdot\mathbf{v} = 0\,, \tag{12.26}$$

$$\nabla\times(\mathbf{v}\times\mathbf{B}) = 0\,, \qquad \nabla\cdot\mathbf{B} = 0\,. \tag{12.27}$$

Note that the presence of a background flow not only enlarges the set of equilibrium solutions with an additional function $\mathbf{v}(\mathbf{r})$, but it also enlarges the freedom of choice for the functions $\rho(\mathbf{r})$, $p(\mathbf{r})$ and $\mathbf{B}(\mathbf{r})$. Paradoxically, though more equations are to be satisfied, more solutions are permitted. To convince yourself of this basic fact, consider a variant of Fig. 6.1 with two balls on opposite sides of the top of a hill: not an equilibrium. However, if the two balls are constrained by a wire connecting them, all of a sudden infinitely many equilibrium positions are obtained. In other words, constraints produce a more intricate energy landscape.

Before we turn to the implications for the waves and instabilities, let us consider the effects of flow on the two generic classes of equilibria, of plane plasma slabs and cylindrical flux tubes, that were studied in Chapters 7 and 9. For the stationary equilibrium of a plane slab with gravity, $\mathbf{g} = -g\mathbf{e}_x$, and the other inhomogeneities also in the vertical direction, one easily convinces oneself that the equilibrium equations (12.24), (12.26), (12.27) are trivially satisfied whereas the momentum equation (12.25) is unchanged with respect to the static case:

$$(p + \tfrac{1}{2}B^2)' = -\rho g \qquad \left(' \equiv \frac{d}{dx}\right). \tag{12.28}$$

Hence, the functions $v_y(x)$ and $v_z(x)$ are entirely free, whereas $\rho(x)$, $p(x)$, $B_y(x)$ and $B_z(x)$ are constrained by Eq. (12.28). This is nice since we can now study the effects of plasma flow on the *same* equilibria as studied for the static case.

For cylindrical equilibria, Eqs. (12.24), (12.26) and (12.27) are also trivially satisfied but the momentum equation has the important additional contribution $\rho\mathbf{v}\cdot\nabla\mathbf{v}$ compared to the static case. In the evaluation of this expression for cylindrical coordinates (using Appendix A), one should account for the non-vanishing derivatives of the unit vectors, $\partial\mathbf{e}_r/\partial\theta = \mathbf{e}_\theta$ and $\partial\mathbf{e}_\theta/\partial\theta = -\mathbf{e}_r$, resulting in the nasty vector expression (A.50). This yields the *centrifugal acceleration*

$$-\mathbf{v}\cdot\nabla\mathbf{v} = (v_\theta^2/r)\mathbf{e}_r\,, \tag{12.29}$$

so that the equilibrium equation becomes

$$(p + \tfrac{1}{2}B^2)' = (\rho v_\theta^2 - B_\theta^2)/r - \rho\Phi_{\rm gr}' \qquad \Big(\,' \equiv \frac{d}{dr}\Big)\,. \tag{12.30}$$

For the present discussion, the gravitational potential is better ignored (cylindrical gravity fields are not very physical) but it is included here since we will need it later, in Section 13.5, when discussing the cylindrical limit of accretion disks. Hence, only the translational component $v_z(x)$ can be chosen arbitrarily, the remaining quantities $\rho(x)$, $p(x)$, $v_\theta(x)$, $B_\theta(x)$ and $B_z(x)$ are constrained by the equilibrium equation (12.30). The presence of the rotational component v_θ essentially changes the cylindrical equilibrium with respect to the static case. Moreover, the computation of waves and instabilities becomes much more complicated than for the plane slab: translation and rotation are physically quite different phenomena.

Finally, the Eulerian perturbations ρ_1, $\mathbf{v}_1$, p_1, $\mathbf{B}_1$ of stationary equilibria are described by the following linear differential equations:

$$\Big(\frac{\partial}{\partial t} + \mathbf{v}\cdot\nabla\Big)\rho_1 + \rho_1\nabla\cdot\mathbf{v} + \nabla\cdot(\rho\mathbf{v}_1) = 0\,, \tag{12.31}$$

$$\begin{aligned}\Big(\frac{\partial}{\partial t} + \mathbf{v}\cdot\nabla\Big)\mathbf{v}_1 &+ \rho\mathbf{v}_1\cdot\nabla\mathbf{v} + \rho_1(\mathbf{v}\cdot\nabla\mathbf{v} + \nabla\Phi_{\rm gr})\\ &+ \nabla p_1 + \mathbf{B}\times(\nabla\times\mathbf{B}_1) - (\nabla\times\mathbf{B})\times\mathbf{B}_1 = 0\,,\end{aligned} \tag{12.32}$$

$$\Big(\frac{\partial}{\partial t} + \mathbf{v}\cdot\nabla\Big)p_1 + \gamma p_1\nabla\cdot\mathbf{v} + \mathbf{v}_1\cdot\nabla p + \gamma p\nabla\cdot\mathbf{v}_1 = 0\,, \tag{12.33}$$

$$\begin{aligned}\Big(\frac{\partial}{\partial t} + \mathbf{v}\cdot\nabla\Big)\mathbf{B}_1 &- \mathbf{B}_1\cdot\nabla\mathbf{v} + \mathbf{B}_1\nabla\cdot\mathbf{v}\\ &- \mathbf{B}\cdot\nabla\mathbf{v}_1 + \mathbf{B}\,\nabla\cdot\mathbf{v}_1 + \mathbf{v}_1\cdot\nabla\mathbf{B} = 0\,, \qquad \nabla\cdot\mathbf{B}_1 = 0\,,\end{aligned} \tag{12.34}$$

where the variables $\rho, \mathbf{v}, p, \mathbf{B}$ without a subscript should satisfy the equilibrium conditions (12.24)–(12.27).

In this chapter, instead of the seven Eulerian perturbations, we will exploit a Lagrangian representation in terms of the three components of a displacement vector $\boldsymbol{\xi}$. This reduction will be elaborated in the following section. The Lagrangian representation implies a significant simplification for the analysis of ideal plasmas, but it has three implicit shortcomings: (a) it cannot be generalized to dissipative plasmas; (b) it involves second order instead of first order derivatives; (c) it ignores the *entropy modes*. Because of the first two items, we will exploit the Eulerian representation in Chapters 14 and 15.

Concerning the third item, recall the discussion of the similar apparent loss of a degree of freedom for static plasmas in Section 5.1. This may be elucidated by the entropy evolution equation, obtained from Eqs. (12.16) and (12.18):

$$\frac{\mathrm{D}S}{\mathrm{D}t} = 0 \quad \Rightarrow \quad \Big(\frac{\partial}{\partial t} + \mathbf{v}\cdot\nabla\Big) S_1 + \mathbf{v}_1 \cdot \nabla S = 0\,, \tag{12.35}$$

where $\mathrm{D}/\mathrm{D}t$ denotes the Lagrangian time derivative. In the absence of a plasma displacement $\boldsymbol{\xi}$, i.e. a velocity perturbation, the Lagrangian representation will automatically deduce from the first expression that $S_{1\mathrm{L}} = 0$ and, hence, also $S_{1\mathrm{E}} = 0$ (since the two are connected by $\boldsymbol{\xi}$, see Eq. (12.54) below). On the other hand, from the second expression, the Eulerian representation will permit $S_{1\mathrm{E}} \neq 0$, even in the absence of a plasma displacement (when $\mathbf{v}_1 = 0$ as well). This corresponds to so-called non-holonomic initial data, i.e. the freedom to choose initial perturbations that do not arise from perturbations satisfying the ideal MHD constraints. In Section 6.1, we have called this "cheating on the initial data" since it also applies to the other variables, like the magnetic field (see Fig. 6.5). In other words: although the Lagrangian representation misses out on this class of perturbations, one should consider this an advantage since those modes are actually inconsistent with the ideal MHD model. In particular, associated with the possible vanishing of the operator $\partial/\partial t + \mathbf{v}\cdot\nabla$ acting on S_1 in Eq. (12.35), a continuous spectrum of non-holonomic *Eulerian entropy continuum* modes is obtained, which is rightly ignored in the Lagrangian representation. It is important to distinguish these rather flimsy modes from the physically significant Lagrangian *flow continuum modes* of hydrodynamics (introduced by Case [108] in 1960) that are related to the vanishing of the same operator, but for variables that are expressible in terms of $\boldsymbol{\xi}$.

12.2.2 Frieman–Rotenberg formulation

Recall the discussion of MHD spectral theory in Chapter 6. That discussion was based on the reduction from a representation of the perturbations in terms of the primitive, Eulerian, variables ρ_1, $\mathbf{v}_1$, p_1 and $\mathbf{B}_1$ to a representation in terms of the plasma displacement vector $\boldsymbol{\xi}$ alone. This led to the powerful force operator formalism, with spectral representations in complete analogy with the mathematics of quantum mechanics and, hence, the possible transfer of methods and insights from that field. Can this be generalized for plasmas with background flow?

Clearly, the crucial part is to construct a displacement vector $\boldsymbol{\xi}$ that connects the perturbed flow with the unperturbed flow (Fig. 12.3) and to specify coordinates that exploit the fact that the stationary equilibrium is actually independent of time. This would not be the case if the perturbed flow were described, in a truly Lagrangian fashion, by the initial positions $\mathbf{r}_0$ of the fluid elements. Instead, a quasi-Lagrangian representation will be exploited, where the position vector $\mathbf{r}$ of a fluid element of the perturbed flow is connected to the position vector $\mathbf{r}^0$ of that same element on the unperturbed flow:

$$\mathbf{r} = \mathbf{r}(\mathbf{r}^0, t) = \mathbf{r}^0 + \boldsymbol{\xi}(\mathbf{r}^0, t)\,. \tag{12.36}$$

In the coordinates $(\mathbf{r}^0, t)$, the equilibrium variables are *time-independent,*

$$\rho^0 = \rho^0(\mathbf{r}^0)\,, \quad \mathbf{v}^0 \equiv \mathbf{v}^0(\mathbf{r}^0)\,, \quad p^0 = p^0(\mathbf{r}^0)\,, \quad \mathbf{B}^0 \equiv \mathbf{B}^0(\mathbf{r}^0)\,, \tag{12.37}$$

by definition satisfying the equilibrium equations (12.24)–(12.27), with a superscript ^{0} on all variables and $\nabla \to \nabla^0 \equiv \partial/\partial\mathbf{r}^0$. In the end, we will drop these superscripts again, but for now we keep

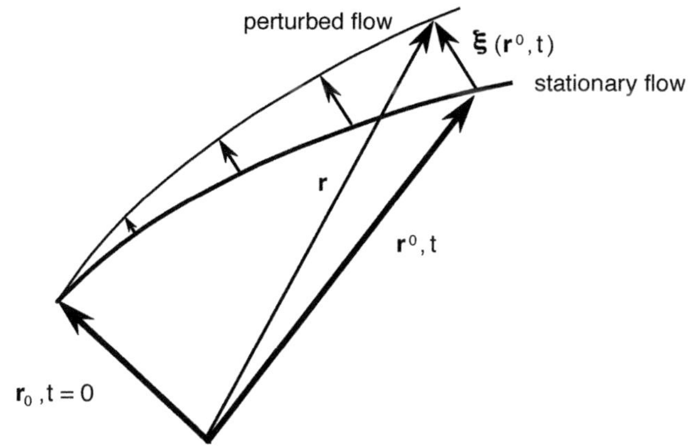

Fig. 12.3 Displacement vector field for plasma with stationary background flow. (Adapted from Frieman and Rotenberg [197].)

them in order to distinguish perturbed and unperturbed quantities. It has to be demonstrated that the perturbations are expressible in terms of $\boldsymbol{\xi}(\mathbf{r}^0, t)$ alone. Needless to say, $\boldsymbol{\xi}$ is always considered to be small compared to the length scales of the pertinent magnetic geometry. A nonlinear generalization, without that restriction, was made by Newcomb [463, 464]. Here, we will exploit the linear counterparts of his expressions.

To systematically exploit the coordinates $(\mathbf{r}^0, t)$, we need to express the gradient operator at the perturbed position in terms of that at the unperturbed position:

$$\nabla \equiv \frac{\partial}{\partial \mathbf{r}} = \frac{\partial \mathbf{r}^0}{\partial \mathbf{r}} \cdot \frac{\partial}{\partial \mathbf{r}^0} \equiv (\nabla \mathbf{r}^0) \cdot \nabla^0 = (\nabla(\mathbf{r} - \boldsymbol{\xi})) \cdot \nabla^0 = (\mathbf{I} - \nabla \boldsymbol{\xi}) \cdot \nabla^0 .$$

Newcomb's exact expression for the gradient operator is obtained by iteration,

$$\nabla \boldsymbol{\xi} = \nabla^0 \boldsymbol{\xi} - (\nabla^0 \boldsymbol{\xi}) \cdot \nabla^0 \boldsymbol{\xi} + (\nabla^0 \boldsymbol{\xi}) \cdot (\nabla^0 \boldsymbol{\xi}) \cdot \nabla^0 \boldsymbol{\xi} - \cdots = (\mathbf{I} + \nabla^0 \boldsymbol{\xi})^{-1} \cdot \nabla^0 \boldsymbol{\xi}$$

$$\Rightarrow \quad \nabla = (\mathbf{I} + \nabla^0 \boldsymbol{\xi})^{-1} \cdot \nabla^0 \approx \nabla^0 - (\nabla^0 \boldsymbol{\xi}) \cdot \nabla^0 , \tag{12.38}$$

where the latter approximation holds for $\boldsymbol{\xi}$ small. The Lagrangian time derivative is expressed by

$$\frac{\mathrm{D}}{\mathrm{D}t} \equiv \left. \frac{\partial}{\partial t} \right|_{\mathbf{r}^0} + \mathbf{v}^0 \cdot \nabla^0 , \tag{12.39}$$

so that the velocity at the perturbed position becomes

$$\mathbf{v}(\mathbf{r}^0 + \boldsymbol{\xi}, t) \equiv \frac{\mathrm{D}\mathbf{r}}{\mathrm{D}t} = \frac{\mathrm{D}\mathbf{r}^0}{\mathrm{D}t} + \frac{\mathrm{D}\boldsymbol{\xi}}{\mathrm{D}t} = \mathbf{v}^0 + \mathbf{v}^0 \cdot \nabla^0 \boldsymbol{\xi} + \frac{\partial \boldsymbol{\xi}}{\partial t} , \tag{12.40}$$

where $\mathbf{v}^0 = \mathbf{v}^0(\mathbf{r}^0)$ is the equilibrium velocity. As compared to the expression (6.18) for static equilibria, the velocity perturbation now contains the important additional contribution $\mathbf{v}^0 \cdot \nabla^0 \boldsymbol{\xi}$ due to the *gradient parallel to the velocity field*. This gives rise to a Doppler shift and possibly, depending on the geometry, curvature contributions.

Similar to the procedure for static equilibria (Section 6.1), we will integrate the equations (12.16), (12.18) and (12.19) for ρ, p and $\mathbf{B}$ to get the exact perturbed quantities, that we will expand to first order in $\boldsymbol{\xi}$, and then substitute the result in Eq. (12.17) for $\mathbf{v}$ to get the linearized equation of motion in terms of $\boldsymbol{\xi}$. This could be done by direct substitution of the expressions (12.38)–(12.40), but it is more instructive to exploit the local conservation laws derived in Section 4.3.3, which were based on the kinematic expressions (4.88)–(4.90) for the time evolution of the line, surface and volume elements $d\mathbf{r}$, $d\boldsymbol{\sigma}$ and $d\tau$.

▷ **Kinematic transformation** of the line, surface and volume elements $d\mathbf{r}$, $d\boldsymbol{\sigma}$ and $d\tau$ (at position $\mathbf{r}$) to $d\mathbf{r}^0$, $d\boldsymbol{\sigma}^0$ and $d\tau^0$ (at position $\mathbf{r}^0$) involves the coordinate transformation (12.36), which gives rise to the transpose $\mathbf{R}^{\mathrm{T}}$ of the connection matrix $\mathbf{R}$ with elements

$$r_{ij} \equiv \frac{\partial r_i}{\partial r_j^0} = \delta_{ij} + \frac{\partial \xi_i}{\partial r_j^0} \quad \Rightarrow \quad \mathbf{R}^{\mathrm{T}} \equiv \nabla^0 \mathbf{r} = \mathbf{I} + \nabla^0 \boldsymbol{\xi}\,, \tag{12.41}$$

and the associated Jacobian with Levi-Civita symbols ϵ_{jmn} defined in Appendix A.3,

$$J \equiv \frac{\partial(r_1, r_2, r_3)}{\partial(r_1^0, r_2^0, r_3^0)} \equiv \det(r_{ij}) = \tfrac{1}{6}\, \epsilon_{ikl}\, \epsilon_{jmn} r_{ij} r_{km} r_{ln} \equiv \det\left(\mathbf{I} + \nabla^0 \boldsymbol{\xi}\right) \approx 1 + \nabla^0 \cdot \boldsymbol{\xi}\,. \tag{12.42}$$

These expressions immediately provide the transformations of line and volume element, Eqs. (12.45) and (12.47) below. As always, transformation of the surface element $d\boldsymbol{\sigma}$ ($\equiv d\mathbf{r}_\alpha \times d\mathbf{r}_\beta$) is a bit more complicated since it involves the inverse $\mathbf{R}^{-1}$. This requires evaluation of the matrix $\mathbf{C}$ of cofactors c_{ij}, defined as the determinants of the minors of $\mathbf{R}$ (obtained by taking out the ith row and jth column) multiplied by $(-1)^{i+j}$, e.g.,

$$c_{23} \equiv - \begin{vmatrix} r_{11} & r_{12} \\ r_{31} & r_{32} \end{vmatrix} = -r_{11}\, r_{32} + r_{12}\, r_{31}\,.$$

The inverse $\mathbf{R}^{-1}$ is then related to the transpose $\mathbf{C}^{\mathrm{T}}$ through

$$\mathbf{R}^{-1} = (1/J)\, \mathbf{C}^{\mathrm{T}} \quad \Rightarrow \quad c_{ki} r_{kj} = J \delta_{ij}\,, \tag{12.43}$$

whereas the cofactors also have the following properties:

$$c_{ij} = \tfrac{1}{2} \epsilon_{ikl}\, \epsilon_{jmn} r_{km} r_{ln} = \frac{\partial J}{\partial r_{ij}}\,, \qquad \frac{\partial c_{ij}}{\partial r_j^0} = 0\,. \tag{12.44}$$

This provides the *transformations of line, surface and volume elements:*

$$d\mathbf{r} = \mathbf{R} \cdot d\mathbf{r}^0 = d\mathbf{r}^0 \cdot (\mathbf{I} + \nabla^0 \boldsymbol{\xi})\,, \tag{12.45}$$

$$d\boldsymbol{\sigma} = \mathbf{C} \cdot d\boldsymbol{\sigma}^0\,, \tag{12.46}$$

$$d\tau = J d\tau^0 = \det(\mathbf{I} + \nabla^0 \boldsymbol{\xi}) d\tau^0 \approx (1 + \nabla^0 \cdot \boldsymbol{\xi})\, d\tau^0\,, \tag{12.47}$$

as one may check by substitution and use of the properties (12.44). ◁

Integration of the local mass, entropy and magnetic flux conservation equations,

$$\mathrm{D}(dM)/\mathrm{D}t \equiv \mathrm{D}(\rho d\tau)/\mathrm{D}t = 0 \qquad (\textit{mass})\,, \tag{12.48}$$

$$\mathrm{D}(S)/\mathrm{D}t \equiv \mathrm{D}(p\rho^{-\gamma})/\mathrm{D}t = 0 \qquad (\textit{entropy})\,, \tag{12.49}$$

$$\mathrm{D}(d\psi)/\mathrm{D}t \equiv \mathrm{D}(\mathbf{B} \cdot d\boldsymbol{\sigma})/\mathrm{D}t = 0 \qquad (\textit{magnetic flux})\,, \tag{12.50}$$

is now straightforward:

$$\rho d\tau = \rho^0 d\tau^0 \quad \Rightarrow \quad \rho = J^{-1} \rho^0 \approx \rho^0 - \rho^0 \nabla^0 \cdot \boldsymbol{\xi}\,, \tag{12.51}$$

$$p\rho^{-\gamma} = p^0 (\rho^0)^{-\gamma} \quad \Rightarrow \quad p = J^{-\gamma} p^0 \approx p^0 - \gamma p^0 \nabla^0 \cdot \boldsymbol{\xi}\,, \tag{12.52}$$

$$B_i d\sigma_i = B_i^0 d\sigma_i^0 \quad \Rightarrow \quad \mathbf{B} = J^{-1} \mathbf{R} \cdot \mathbf{B}^0 \approx (1 - \nabla^0 \cdot \boldsymbol{\xi})\, \mathbf{B}^0 \cdot (\mathbf{I} + \nabla^0 \boldsymbol{\xi}) \approx \mathbf{B}^0 + \mathbf{B}^0 \cdot \nabla^0 \boldsymbol{\xi} - \mathbf{B}^0 \nabla^0 \cdot \boldsymbol{\xi}\,. \tag{12.53}$$

This completes the first half of the programme, viz. construction of the Lagrangian representation of density, pressure and magnetic field perturbations. Since

$$f_{1\mathrm{L}} = f_{1\mathrm{E}} + \boldsymbol{\xi}\cdot\nabla^0 f^0\,, \tag{12.54}$$

the Lagrangian expressions (12.51)–(12.53) are fully compatible with the earlier obtained Eulerian counterparts (6.19)–(6.21) for static equilibria,

$$\begin{aligned}
\rho_{1\mathrm{E}} &= -\nabla^0\cdot(\rho^0\boldsymbol{\xi})\,, &(12.55)\\
\pi \equiv p_{1\mathrm{E}} &= -\gamma p^0\nabla^0\cdot\boldsymbol{\xi} - \boldsymbol{\xi}\cdot\nabla^0 p^0\,, &(12.56)\\
\mathbf{Q} \equiv \mathbf{B}_{1\mathrm{E}} &= \nabla^0\times(\boldsymbol{\xi}\times\mathbf{B}^0) = \mathbf{B}^0\cdot\nabla^0\boldsymbol{\xi} - \mathbf{B}^0\,\nabla^0\cdot\boldsymbol{\xi} - \boldsymbol{\xi}\cdot\nabla^0\mathbf{B}^0\,. &(12.57)
\end{aligned}$$

The abbreviations π and $\mathbf{Q}$ will be used in the reductions below.

It remains to substitute the expressions obtained in the equation of motion,

$$\rho\,\frac{\mathrm{D}\mathbf{v}}{\mathrm{D}t} + \nabla p - (\nabla\times\mathbf{B})\times\mathbf{B} - \rho\mathbf{g} = 0 \qquad \textit{(momentum)}. \tag{12.58}$$

This involves a number of steps that are put in small print below.

▷ **Reduction of the equation of motion** involves the following contributions, with the full perturbed expressions on the LHS, while in the first order expansions on the RHS the superscripts ^{0} on the unperturbed quantities are now consistently dropped:

$$\begin{aligned}
\rho\,\frac{\mathrm{D}\mathbf{v}}{\mathrm{D}t} &\approx (\rho - \rho\nabla\cdot\boldsymbol{\xi})\Big(\frac{\partial}{\partial t} + \mathbf{v}\cdot\nabla\Big)\Big(\mathbf{v} + \mathbf{v}\cdot\nabla\boldsymbol{\xi} + \frac{\partial\boldsymbol{\xi}}{\partial t}\Big)\\
&\approx \rho\mathbf{v}\cdot\nabla\mathbf{v} - (\nabla\cdot\boldsymbol{\xi})\rho\mathbf{v}\cdot\nabla\mathbf{v} + \rho(\mathbf{v}\cdot\nabla)^2\boldsymbol{\xi} + 2\rho\mathbf{v}\cdot\nabla\,\frac{\partial\boldsymbol{\xi}}{\partial t} + \rho\,\frac{\partial^2\boldsymbol{\xi}}{\partial t^2}\\
&= \rho\mathbf{v}\cdot\nabla\mathbf{v} + \boldsymbol{\xi}\cdot\nabla(\rho\mathbf{v}\cdot\nabla\mathbf{v})\\
&\quad - \nabla\cdot(\boldsymbol{\xi}\rho\mathbf{v}\cdot\nabla\mathbf{v} - \rho\mathbf{v}\mathbf{v}\cdot\nabla\boldsymbol{\xi}) + 2\rho\mathbf{v}\cdot\nabla\,\frac{\partial\boldsymbol{\xi}}{\partial t} + \rho\,\frac{\partial^2\boldsymbol{\xi}}{\partial t^2}\,, &(12.59)\\
\nabla p &\approx [\nabla - (\nabla\boldsymbol{\xi})\cdot\nabla]\,(p + \pi + \boldsymbol{\xi}\cdot\nabla p)\\
&\approx \nabla p - (\nabla\boldsymbol{\xi})\cdot\nabla p + \nabla(\boldsymbol{\xi}\cdot\nabla p) + \nabla\pi = \nabla p + \boldsymbol{\xi}\cdot\nabla\nabla p + \nabla\pi\,, &(12.60)\\
-(\nabla\times\mathbf{B})\times\mathbf{B} &\approx -\big\{[\nabla - (\nabla\boldsymbol{\xi})\cdot\nabla]\times(\mathbf{B} + \mathbf{Q} + \boldsymbol{\xi}\cdot\nabla\mathbf{B})\big\}\times(\mathbf{B} + \mathbf{Q} + \boldsymbol{\xi}\cdot\nabla\mathbf{B})\\
&\approx -(\nabla\times\mathbf{B})\times\mathbf{B} - (\nabla\times\mathbf{B})\times(\boldsymbol{\xi}\cdot\nabla\mathbf{B}) - (\nabla\times(\boldsymbol{\xi}\cdot\nabla\mathbf{B}))\times\mathbf{B}\\
&\quad + \big[((\nabla\boldsymbol{\xi})\cdot\nabla)\times\mathbf{B}\big]\times\mathbf{B} + \mathbf{B}\times(\nabla\times\mathbf{Q}) - (\nabla\times\mathbf{B})\times\mathbf{Q}\\
&= -(\nabla\times\mathbf{B})\times\mathbf{B} - \boldsymbol{\xi}\cdot\nabla\big[(\nabla\times\mathbf{B})\times\mathbf{B}\big]\\
&\quad + \mathbf{B}\times(\nabla\times\mathbf{Q}) - (\nabla\times\mathbf{B})\times\mathbf{Q}\,, &(12.61)\\
\rho\nabla\Phi_{\mathrm{gr}} &\approx (\rho - \rho\nabla\cdot\boldsymbol{\xi})[\nabla - (\nabla\boldsymbol{\xi})\cdot\nabla](\Phi_{\mathrm{gr}} + \boldsymbol{\xi}\cdot\nabla\Phi_{\mathrm{gr}})\\
&\approx \rho\nabla\Phi_{\mathrm{gr}} - \rho(\nabla\cdot\boldsymbol{\xi})\nabla\Phi_{\mathrm{gr}} - \rho(\nabla\boldsymbol{\xi})\cdot\nabla\Phi_{\mathrm{gr}} + \rho\nabla(\boldsymbol{\xi}\cdot\nabla\Phi_{\mathrm{gr}})\\
&= \rho\nabla\Phi_{\mathrm{gr}} + \boldsymbol{\xi}\cdot\nabla(\rho\nabla\Phi_{\mathrm{gr}}) - (\nabla\Phi_{\mathrm{gr}})\nabla\cdot(\rho\boldsymbol{\xi})\,. &(12.62)
\end{aligned}$$

The derivation of the last line of Eq. (12.59) involves the equilibrium condition (12.24). Adding the expressions (12.59)–(12.62), the four first terms on the RHSs cancel because of the equilibrium condition (12.25) and the four second terms cancel as well because they represent the Lagrangian perturbation $\boldsymbol{\xi}\cdot\nabla$ of the same equilibrium condition. The remaining terms constitute the desired perturbation of the equation of motion. ◁

This finally yields the linearized equation of motion for perturbations of stationary equilibria that was first derived by Frieman and Rotenberg [197]:

$$\mathbf{G}(\boldsymbol{\xi}) - 2\rho\mathbf{v}\cdot\nabla\frac{\partial\boldsymbol{\xi}}{\partial t} - \rho\,\frac{\partial^2\boldsymbol{\xi}}{\partial t^2} = 0\,, \tag{12.63}$$

where $\mathbf{G}$ is the *generalized force operator,*

$$\mathbf{G}(\boldsymbol{\xi}) \equiv \mathbf{F}(\boldsymbol{\xi}) + \nabla\cdot\big(\boldsymbol{\xi}\rho\mathbf{v}\cdot\nabla\mathbf{v} - \rho\mathbf{v}\mathbf{v}\cdot\nabla\boldsymbol{\xi}\big)\,, \tag{12.64}$$

involving the standard force operator expression (6.23) of static equilibria,

$$\mathbf{F}(\boldsymbol{\xi}) \equiv \nabla(\gamma p\nabla\cdot\boldsymbol{\xi}) - \mathbf{B}\times(\nabla\times\mathbf{Q}) + \nabla(\boldsymbol{\xi}\cdot\nabla p) + \mathbf{j}\times\mathbf{Q} + (\nabla\Phi_{\rm gr})\,\nabla\cdot(\rho\boldsymbol{\xi})\,. \tag{12.65}$$

We have introduced the new notation $\mathbf{G}$ for the generalized force operator (Frieman and Rotenberg indicate that quantity also by $\mathbf{F}$), so that we can use the operator $\mathbf{F}$ as a convenient abbreviation for the velocity-independent terms.

Because of the equilibrium, the last term of $\mathbf{G}$ may be transformed to

$$-\nabla\cdot\big(\rho\mathbf{v}\mathbf{v}\cdot\nabla\boldsymbol{\xi}\big) \stackrel{(12.24)}{=} -\rho(\mathbf{v}\cdot\nabla)^2\boldsymbol{\xi}\,, \tag{12.66}$$

which yields an equivalent representation for the linearized equation of motion:

$$\mathbf{F}(\boldsymbol{\xi}) + \nabla\cdot\big(\boldsymbol{\xi}\rho\mathbf{v}\cdot\nabla\mathbf{v}\big) - \rho\Big(\frac{\partial}{\partial t} + \mathbf{v}\cdot\nabla\Big)^2\boldsymbol{\xi} = 0\,. \tag{12.67}$$

For plane shear flow, where $\mathbf{v}\cdot\nabla\mathbf{v} = 0$, the only change with respect to the static problem is the appearance of the operator $\mathbf{v}\cdot\nabla$ associated with the *Doppler shift*. However, since this term varies from place to place, the relationship between the waves and instabilities of plane static and those of plane stationary plasmas is not as straightforward as it might appear. For more general flow fields, like rotations in a cylinder, the equation of motion becomes much more involved since $\mathbf{v}\cdot\nabla\mathbf{v}$ yields the centrifugal acceleration (12.29) and the operator $\mathbf{v}\cdot\nabla$ gives Coriolis contributions to the frequency in addition to the Doppler shift (see Chapter 13).

For normal modes $\hat{\boldsymbol{\xi}}(\mathbf{r})\exp(-\mathrm{i}\omega t)$, the associated *spectral equation* reads:

$$\mathbf{G}(\hat{\boldsymbol{\xi}}) + 2\mathrm{i}\rho\omega\mathbf{v}\cdot\nabla\hat{\boldsymbol{\xi}} + \rho\omega^2\hat{\boldsymbol{\xi}} = 0\,, \tag{12.68}$$

or, equivalently, from Eq. (12.67),

$$\mathbf{F}(\hat{\boldsymbol{\xi}}) + \nabla\cdot\big(\hat{\boldsymbol{\xi}}\rho\mathbf{v}\cdot\nabla\mathbf{v}\big) + \rho\big(\omega + \mathrm{i}\mathbf{v}\cdot\nabla\big)^2\hat{\boldsymbol{\xi}} = 0\,. \tag{12.69}$$

This becomes an eigenvalue problem by supplementing appropriate BCs, e.g.

$$\mathbf{n}\cdot\hat{\boldsymbol{\xi}} = 0 \qquad \text{(at the wall)} \tag{12.70}$$

for model I (plasma confined inside a rigid wall). This eigenvalue problem is now a *quadratic* one (involving both ω and ω^2), in contrast to the static spectral problem (12.23) which is linear in the eigenvalue ω^2. This implies that the eigenvalues ω are no longer restricted to the real and imaginary axes but may be genuinely complex, so that *overstable modes* occur. This represents a major complication in the theory of waves and instabilities of plasmas with background flow, as will be extensively illustrated in the following sections. From now on, we will drop the hat on $\hat{\boldsymbol{\xi}}$.

The far reaching implications of the Frieman–Rotenberg spectral equation also for the nonlinear plasma dynamics has been pointed out recently by Keppens and Demaerel [346].

12.2.3 Self-adjointness of the generalized force operator★

In order to put the spectral theory of stationary plasmas on a firm mathematical basis, we need to study the adjointness properties of the basic spectral equation (12.68). As a first step, we will prove that the generalized force operator $\mathbf{G}$ itself is actually self-adjoint. The proof will be analogous to that of self-adjointness of the operator $\mathbf{F}$ for static equilibria of Section 6.2.3, but it cannot be copied blindly since the static equilibrium relations were used there, so that we need to retrace our steps and use the stationary equilibrium relations (12.24)–(12.27) instead.

▷ **Transformation of the inhomogeneity terms of the generalized force operator** The four terms with the gradients ∇p, $\mathbf{j}$, $\nabla\Phi_{\rm gr}$, $\mathbf{v}\cdot\nabla\mathbf{v}$ may be transformed as follows:

$$\begin{aligned}
\nabla(\boldsymbol{\xi}\cdot\nabla p) &= (\nabla p\times\nabla)\times\boldsymbol{\xi} + (\nabla p)\,\nabla\cdot\boldsymbol{\xi} + \boldsymbol{\xi}\cdot\nabla\nabla p \\
&\overset{(12.25)}{=} (\mathbf{B}\,\mathbf{j}\cdot\nabla - \mathbf{j}\,\mathbf{B}\cdot\nabla)\times\boldsymbol{\xi} - \rho(\nabla\Phi_{\rm gr}\times\nabla)\times\boldsymbol{\xi} \\
&\qquad - [\rho(\mathbf{v}\cdot\nabla\mathbf{v})\times\nabla]\times\boldsymbol{\xi} + (\nabla p)\,\nabla\cdot\boldsymbol{\xi} + \boldsymbol{\xi}\cdot\nabla\nabla p \\
&= \mathbf{B}\times(\mathbf{j}\cdot\nabla\boldsymbol{\xi}) - \mathbf{j}\times(\mathbf{B}\cdot\nabla\boldsymbol{\xi}) - \rho(\nabla\boldsymbol{\xi})\cdot\nabla\Phi_{\rm gr} + \rho(\nabla\Phi_{\rm gr})\,\nabla\cdot\boldsymbol{\xi} \\
&\qquad - \rho(\nabla\boldsymbol{\xi})\cdot(\mathbf{v}\cdot\nabla\mathbf{v}) + \rho(\mathbf{v}\cdot\nabla\mathbf{v})\,\nabla\cdot\boldsymbol{\xi} + (\nabla p)\,\nabla\cdot\boldsymbol{\xi} + \boldsymbol{\xi}\cdot\nabla\nabla p\,, \\
\mathbf{j}\times\mathbf{Q} &= \mathbf{j}\times(\mathbf{B}\cdot\nabla\boldsymbol{\xi}) - \mathbf{j}\times\mathbf{B}\,\nabla\cdot\boldsymbol{\xi} - \boldsymbol{\xi}\cdot\nabla(\mathbf{j}\times\mathbf{B}) - \mathbf{B}\times(\boldsymbol{\xi}\cdot\nabla\mathbf{j})\,, \\
(\nabla\Phi_{\rm gr})\,\nabla\cdot(\rho\boldsymbol{\xi}) &= \rho(\nabla\Phi_{\rm gr})\,\nabla\cdot\boldsymbol{\xi} + (\nabla\Phi_{\rm gr})\,(\nabla\rho)\cdot\boldsymbol{\xi}\,, \\
\nabla\cdot(\rho\boldsymbol{\xi}\mathbf{v}\cdot\nabla\mathbf{v}) &= \boldsymbol{\xi}\cdot\nabla\,(\rho\mathbf{v}\cdot\nabla\mathbf{v}) + \rho(\mathbf{v}\cdot\nabla\mathbf{v})\,\nabla\cdot\boldsymbol{\xi}\,.
\end{aligned}$$

Hence, the sum of the four inhomogeneity terms becomes

$$\begin{aligned}
&\nabla(\boldsymbol{\xi}\cdot\nabla p) + \mathbf{j}\times\mathbf{Q} + (\nabla\Phi_{\rm gr})\,\nabla\cdot(\rho\boldsymbol{\xi}) + \nabla\cdot(\rho\boldsymbol{\xi}\mathbf{v}\cdot\nabla\mathbf{v}) \\
&\quad= -\mathbf{B}\times(\nabla\times(\mathbf{j}\times\boldsymbol{\xi})) + (\nabla p - 2\mathbf{j}\times\mathbf{B} + 2\rho\nabla\Phi_{\rm gr} + 2\rho\mathbf{v}\cdot\nabla\mathbf{v})\,\nabla\cdot\boldsymbol{\xi} \\
&\qquad + \boldsymbol{\xi}\cdot\nabla\,(\nabla p - \mathbf{j}\times\mathbf{B} + \rho\mathbf{v}\cdot\nabla\mathbf{v}) - \rho(\nabla\boldsymbol{\xi})\cdot\nabla\Phi_{\rm gr} + (\nabla\Phi_{\rm gr})(\nabla\rho)\cdot\boldsymbol{\xi} - \rho(\nabla\boldsymbol{\xi})\cdot(\mathbf{v}\cdot\nabla\mathbf{v}) \\
&\quad\overset{(12.25)}{=} -\mathbf{B}\times(\nabla\times(\mathbf{j}\times\boldsymbol{\xi})) - (\nabla p)\,\nabla\cdot\boldsymbol{\xi} - \rho\nabla(\boldsymbol{\xi}\cdot\nabla\Phi_{\rm gr}) - \rho(\nabla\boldsymbol{\xi})\cdot(\mathbf{v}\cdot\nabla\mathbf{v}), \qquad (12.71)
\end{aligned}$$

which provides the revised expression for $\mathbf{G}$. ◁

The above transformation yields an *equivalent form of the generalized force operator,*

$$\begin{aligned}
\mathbf{G}(\boldsymbol{\xi}) &= \nabla(\gamma p\nabla\cdot\boldsymbol{\xi}) - \mathbf{B}\times[\,\nabla\times\mathbf{Q} + \nabla\times(\mathbf{j}\times\boldsymbol{\xi})\,] - (\nabla p)\,\nabla\cdot\boldsymbol{\xi} \\
&\qquad - \rho\nabla(\boldsymbol{\xi}\cdot\nabla\Phi_{\rm gr}) - \rho(\nabla\boldsymbol{\xi})\cdot(\mathbf{v}\cdot\nabla\mathbf{v}) - \nabla\cdot(\rho\mathbf{v}\mathbf{v}\cdot\nabla\boldsymbol{\xi})\,, \qquad (12.72)
\end{aligned}$$

more clearly exhibiting the different physical effects at work. The first two terms correspond to the magneto-sonic and Alfvén wave contributions of homogeneous plasmas, the next four terms correspond to the current, pressure gradient, gravitation and velocity gradient driven instabilities of inhomogeneous plasmas (interchanges, kinks, gravitational and rotational instabilities, and all their combinations), whereas the last term gives the contribution (12.66).

Self-adjointness of the generalized force operator is expressed by

$$\int \boldsymbol{\eta}\cdot\mathbf{G}(\boldsymbol{\xi})\,dV = \int \boldsymbol{\xi}\cdot\mathbf{G}(\boldsymbol{\eta})\,dV\,, \qquad (12.73)$$

where we again (as in Chapter 6) exploit an inner product with *real* vectors $\boldsymbol{\xi}$ and $\boldsymbol{\eta}$, satisfying the BCs $\mathbf{n}\cdot\boldsymbol{\xi} = 0$ and $\mathbf{n}\cdot\boldsymbol{\eta} = 0$ at a wall, with associated Eulerian magnetic perturbations $\mathbf{Q}$ and $\mathbf{R}$.

The integrand is split into a simple part with the homogeneity terms $\mathbf{G}_1$ and a complicated part with the inhomogeneity terms $\mathbf{G}_2$: $\boldsymbol{\eta}\cdot\mathbf{G} = \boldsymbol{\eta}\cdot\mathbf{G}_1 + \boldsymbol{\eta}\cdot\mathbf{G}_2$. The required transformation to a symmetric expression plus a divergence is straightforward for the first part:

$$\begin{aligned}\boldsymbol{\eta}\cdot\mathbf{G}_1 &\equiv \boldsymbol{\eta}\cdot\big[\nabla(\gamma p\nabla\cdot\boldsymbol{\xi}) - \mathbf{B}\times(\nabla\times\mathbf{Q})\big]\\ &= -\gamma p(\nabla\cdot\boldsymbol{\xi})(\nabla\cdot\boldsymbol{\eta}) - \mathbf{Q}\cdot\mathbf{R} + \nabla\cdot\big[\boldsymbol{\eta}\gamma p\nabla\cdot\boldsymbol{\xi} + (\mathbf{B}\boldsymbol{\eta} - \boldsymbol{\eta}\mathbf{B})\cdot\mathbf{Q}\big]\,. \quad (12.74)\end{aligned}$$

The transformation of the second part starts from the original definition (12.64) of the operator $\mathbf{G}$ and then proceeds by inserting the transformed form (12.72) of $\mathbf{G}$:

$$\begin{aligned}\boldsymbol{\eta}\cdot\mathbf{G}_2 &\equiv \boldsymbol{\eta}\cdot\big[\nabla(\boldsymbol{\xi}\cdot\nabla p) + \mathbf{j}\times\mathbf{Q} + (\nabla\Phi_{\rm gr})\nabla\cdot(\rho\boldsymbol{\xi})\\ &\qquad + \nabla\cdot\big(\boldsymbol{\xi}\rho\mathbf{v}\cdot\nabla\mathbf{v} - \rho\mathbf{v}\mathbf{v}\cdot\nabla\boldsymbol{\xi}\big)\big] \quad (12.75)\\ &= \boldsymbol{\eta}\cdot\big[-\mathbf{B}\times(\nabla\times(\mathbf{j}\times\boldsymbol{\xi})) - (\nabla p)\nabla\cdot\boldsymbol{\xi} - \rho\nabla(\boldsymbol{\xi}\cdot\nabla\Phi_{\rm gr})\\ &\qquad - \rho(\nabla\boldsymbol{\xi})\cdot(\mathbf{v}\cdot\nabla\mathbf{v}) - \nabla\cdot(\rho\mathbf{v}\mathbf{v}\cdot\nabla\boldsymbol{\xi})\big]\,. \quad (12.76)\end{aligned}$$

The substantial algebra involved in reworking the expression (12.76) into a mirror image of the expression (12.75) plus a divergence is put in small print below.

▷ **Transformation of the inhomogeneity terms of the integrand** The first term is transformed similar to the corresponding analysis of Section 6.2.3 for the static case. Next, the following steps are taken:

$$\begin{aligned}\boldsymbol{\eta}\cdot\mathbf{G}_2 \;&\equiv\; \boldsymbol{\xi}\cdot\mathbf{j}\times\mathbf{R} + \nabla\cdot[\mathbf{j}\mathbf{B}\cdot(\boldsymbol{\xi}\times\boldsymbol{\eta})] - \boldsymbol{\eta}\cdot(\nabla p - \mathbf{j}\times\mathbf{B})\nabla\cdot\boldsymbol{\xi} + \boldsymbol{\xi}\cdot\nabla(\boldsymbol{\eta}\cdot\mathbf{j}\times\mathbf{B})\\ &\quad - \rho\boldsymbol{\eta}\cdot\nabla(\boldsymbol{\xi}\cdot\nabla\Phi_{\rm gr}) - \rho(\boldsymbol{\eta}\cdot\nabla\boldsymbol{\xi})\cdot(\mathbf{v}\cdot\nabla\mathbf{v}) - \big[\nabla\cdot(\rho\mathbf{v}\mathbf{v}\cdot\nabla\boldsymbol{\xi})\big]\cdot\boldsymbol{\eta}\\ &=\; \boldsymbol{\xi}\cdot\big[\nabla(\boldsymbol{\eta}\cdot\nabla p) + \mathbf{j}\times\mathbf{R} + (\nabla\Phi_{\rm gr})\nabla\cdot(\rho\boldsymbol{\eta})\big]\\ &\qquad - \rho(\boldsymbol{\eta}\cdot\nabla\boldsymbol{\xi})\cdot(\mathbf{v}\cdot\nabla\mathbf{v}) - \big[\nabla\cdot(\rho\mathbf{v}\mathbf{v}\cdot\nabla\boldsymbol{\xi})\big]\cdot\boldsymbol{\eta}\\ &\quad + \nabla\cdot\big[\mathbf{j}\mathbf{B}\cdot(\boldsymbol{\xi}\times\boldsymbol{\eta}) - (\nabla p - \mathbf{j}\times\mathbf{B})\cdot\boldsymbol{\eta}\boldsymbol{\xi} - \rho\boldsymbol{\eta}\boldsymbol{\xi}\cdot\nabla\Phi_{\rm gr}\big]\\ &\overset{(12.25)}{=}\; \boldsymbol{\xi}\cdot\big[\nabla(\boldsymbol{\eta}\cdot\nabla p) + \mathbf{j}\times\mathbf{R} + (\nabla\Phi_{\rm gr})\nabla\cdot(\rho\boldsymbol{\eta})\big]\\ &\qquad - \rho(\boldsymbol{\eta}\cdot\nabla\boldsymbol{\xi})\cdot(\mathbf{v}\cdot\nabla\mathbf{v}) - \big[\nabla\cdot(\rho\mathbf{v}\mathbf{v}\cdot\nabla\boldsymbol{\xi})\big]\cdot\boldsymbol{\eta}\\ &\quad + \nabla\cdot\big[\mathbf{j}\mathbf{B}\cdot(\boldsymbol{\xi}\times\boldsymbol{\eta}) + (\nabla p - \mathbf{j}\times\mathbf{B})\cdot(\boldsymbol{\xi}\boldsymbol{\eta} - \boldsymbol{\eta}\boldsymbol{\xi}) + \rho(\mathbf{v}\cdot\nabla\mathbf{v})\cdot\boldsymbol{\xi}\boldsymbol{\eta}\big]\\ &=\; \boldsymbol{\xi}\cdot\big[\nabla(\boldsymbol{\eta}\cdot\nabla p) + \mathbf{j}\times\mathbf{R} + (\nabla\Phi_{\rm gr})\nabla\cdot(\rho\boldsymbol{\eta}) + \nabla\cdot\big(\boldsymbol{\eta}\rho\mathbf{v}\cdot\nabla\mathbf{v} - \rho\mathbf{v}\mathbf{v}\cdot\nabla\boldsymbol{\eta}\big)\big]\\ &\quad + \nabla\cdot\big[\mathbf{j}\mathbf{B}\cdot(\boldsymbol{\xi}\times\boldsymbol{\eta}) + (\nabla p - \mathbf{j}\times\mathbf{B})\cdot(\boldsymbol{\xi}\boldsymbol{\eta} - \boldsymbol{\eta}\boldsymbol{\xi})\\ &\qquad - \rho\mathbf{v}\big((\mathbf{v}\cdot\nabla\boldsymbol{\xi})\cdot\boldsymbol{\eta} - (\mathbf{v}\cdot\nabla\boldsymbol{\eta})\cdot\boldsymbol{\xi}\big)\big]\,. \quad (12.77)\end{aligned}$$

The latter expression has the required form of a mirror image of Eq. (12.75) plus a divergence. A symmetric expression is obtained by just averaging Eqs. (12.75) + (12.77):

$$\begin{aligned}\boldsymbol{\eta}\cdot\mathbf{G}_2 &\equiv \tfrac{1}{2}\boldsymbol{\eta}\cdot\big[\nabla(\boldsymbol{\xi}\cdot\nabla p) + \mathbf{j}\times\mathbf{Q} + (\nabla\Phi_{\rm gr})\nabla\cdot(\rho\boldsymbol{\xi}) + \nabla\cdot\big(\boldsymbol{\xi}\rho\mathbf{v}\cdot\nabla\mathbf{v} - \rho\mathbf{v}\mathbf{v}\cdot\nabla\boldsymbol{\xi}\big)\big]\\ &\quad + \tfrac{1}{2}\boldsymbol{\xi}\cdot\big[\nabla(\boldsymbol{\eta}\cdot\nabla p) + \mathbf{j}\times\mathbf{R} + (\nabla\Phi_{\rm gr})\nabla\cdot(\rho\boldsymbol{\eta}) + \nabla\cdot\big(\boldsymbol{\eta}\rho\mathbf{v}\cdot\nabla\mathbf{v} - \rho\mathbf{v}\mathbf{v}\cdot\nabla\boldsymbol{\eta}\big)\big]\\ &\quad + \tfrac{1}{2}\nabla\cdot\big[\mathbf{j}\mathbf{B}\cdot(\boldsymbol{\xi}\times\boldsymbol{\eta}) + (\nabla p - \mathbf{j}\times\mathbf{B})\cdot(\boldsymbol{\xi}\boldsymbol{\eta} - \boldsymbol{\eta}\boldsymbol{\xi}) - \rho\mathbf{v}\big((\mathbf{v}\cdot\nabla\boldsymbol{\xi})\cdot\boldsymbol{\eta} - (\mathbf{v}\cdot\nabla\boldsymbol{\eta})\cdot\boldsymbol{\xi}\big)\big]\\ &= -\tfrac{1}{2}(\nabla p)\cdot(\boldsymbol{\xi}\nabla\cdot\boldsymbol{\eta} + \boldsymbol{\eta}\nabla\cdot\boldsymbol{\xi}) - \tfrac{1}{2}\mathbf{j}\cdot(\boldsymbol{\eta}\times\mathbf{Q} + \boldsymbol{\xi}\times\mathbf{R}) + \tfrac{1}{2}(\nabla\Phi_{\rm gr})\cdot[\boldsymbol{\eta}\nabla\cdot(\rho\boldsymbol{\xi}) + \boldsymbol{\xi}\nabla\cdot(\rho\boldsymbol{\eta})]\\ &\quad + \tfrac{1}{2}\boldsymbol{\eta}\cdot[\nabla\cdot(\boldsymbol{\xi}\rho\mathbf{v}\cdot\nabla\mathbf{v})] + \tfrac{1}{2}\boldsymbol{\xi}\cdot[\nabla\cdot(\boldsymbol{\eta}\rho\mathbf{v}\cdot\nabla\mathbf{v})] + \rho(\mathbf{v}\cdot\nabla\boldsymbol{\xi})\cdot(\mathbf{v}\cdot\nabla\boldsymbol{\eta})\\ &\quad + \nabla\cdot\big[\boldsymbol{\eta}\boldsymbol{\xi}\cdot\nabla p + \tfrac{1}{2}\mathbf{j}\mathbf{B}\cdot(\boldsymbol{\xi}\times\boldsymbol{\eta}) - \tfrac{1}{2}(\mathbf{j}\times\mathbf{B})\cdot(\boldsymbol{\xi}\boldsymbol{\eta} - \boldsymbol{\eta}\boldsymbol{\xi}) - \rho\mathbf{v}(\mathbf{v}\cdot\nabla\boldsymbol{\xi})\cdot\boldsymbol{\eta}\big]\,. \quad (12.78)\end{aligned}$$

This finally has the requisite form of a symmetric expression plus a divergence. ◁

Adding Eqs. (12.74) and (12.78) yields the required form for the integrand $\boldsymbol{\eta}\cdot\mathbf{G}$. By using Gauss' theorem, the integral of this expression may be transformed into the sum of a symmetric volume integral and a surface integral, from which self-adjointness follows directly by applying the appropriate BCs. Before we complete the proof though, we simplify the integrand still further by exploiting the equality

$$(\mathbf{j}\times\mathbf{B})\cdot(\boldsymbol{\xi}\eta_n-\boldsymbol{\eta}\xi_n)=j_n\,\mathbf{B}\cdot(\boldsymbol{\xi}\times\boldsymbol{\eta})\,,$$

and introducing the Eulerian perturbation of the total pressure,

$$\Pi\equiv\left(p+\tfrac{1}{2}B^2\right)_{1\mathrm{E}}\equiv\pi+\mathbf{B}\cdot\mathbf{Q}=-\gamma p\nabla\cdot\boldsymbol{\xi}-\boldsymbol{\xi}\cdot\nabla p+\mathbf{B}\cdot\mathbf{Q}\,. \qquad (12.79)$$

These additional transformations yield an integral that applies to all the different model boundary condition problems that were listed in Section 4.6:

$$\begin{aligned}\int\boldsymbol{\eta}\cdot\mathbf{G}(\boldsymbol{\xi})\,dV&\\ =-\int\Big\{&\gamma p\,(\nabla\cdot\boldsymbol{\xi})\,\nabla\cdot\boldsymbol{\eta}+\mathbf{Q}\cdot\mathbf{R}+\tfrac{1}{2}(\nabla p)\cdot(\boldsymbol{\xi}\,\nabla\cdot\boldsymbol{\eta}+\boldsymbol{\eta}\,\nabla\cdot\boldsymbol{\xi})\\ &+\tfrac{1}{2}\mathbf{j}\cdot(\boldsymbol{\eta}\times\mathbf{Q}+\boldsymbol{\xi}\times\mathbf{R})-\tfrac{1}{2}(\nabla\Phi_{\mathrm{gr}})\cdot[\,\boldsymbol{\eta}\,\nabla\cdot(\rho\boldsymbol{\xi})+\boldsymbol{\xi}\,\nabla\cdot(\rho\boldsymbol{\eta})\,]\\ &-\tfrac{1}{2}\boldsymbol{\eta}\cdot[\nabla\cdot(\boldsymbol{\xi}\rho\mathbf{v}\cdot\nabla\mathbf{v})]-\tfrac{1}{2}\boldsymbol{\xi}\cdot[\nabla\cdot(\boldsymbol{\eta}\rho\mathbf{v}\cdot\nabla\mathbf{v})]-\rho(\mathbf{v}\cdot\nabla\boldsymbol{\xi})\cdot(\mathbf{v}\cdot\nabla\boldsymbol{\eta})\Big\}\,dV\\ -\int\Big[&\eta_n\Pi(\boldsymbol{\xi})-B_n\,\boldsymbol{\eta}\cdot\mathbf{Q}-j_n\,\mathbf{B}\cdot(\boldsymbol{\xi}\times\boldsymbol{\eta})+\rho v_n(\mathbf{v}\cdot\nabla\boldsymbol{\xi})\cdot\boldsymbol{\eta}\Big]\,dS\,. \qquad (12.80)\end{aligned}$$

For model I (plasma inside a rigid wall), as well as for models II and II* (configurations with a plasma–plasma or plasma–vacuum interface), the last three terms of the surface integral vanish because the equilibrium ensures that $B_n=0,\,j_n=0,\,v_n=0$ not only at a wall (model I), but also at a *stationary* interface (models II and II*). Leaving the outer boundary still open, this expression yields the important *pre-self-adjointness relation*,

$$\int\boldsymbol{\eta}\cdot\mathbf{G}(\boldsymbol{\xi})\,dV=\int\big[\text{symm. in }\boldsymbol{\xi}\text{ and }\boldsymbol{\eta}\big]\,dV-\int\eta_n\Pi(\boldsymbol{\xi})\,dS \qquad \textit{(open)}, \qquad (12.81)$$

that will be exploited extensively in the following sections.

For model I perturbations, the difference expression of this integral vanishes because of the BCs, $\xi_n=\eta_n=0$, at the walls:

$$\int\Big[\boldsymbol{\eta}\cdot\mathbf{G}(\boldsymbol{\xi})-\boldsymbol{\xi}\cdot\mathbf{G}(\boldsymbol{\eta})\Big]\,dV=0 \qquad \textit{(outer BCs satisfied)}. \qquad (12.82)$$

Hence, the generalized force operator $\mathbf{G}$ is self-adjoint for model I, QED.

Next, consider two plasmas divided by an interface S with a tangential discontinuity (model II*). We distinguish the quantities of the second plasma by a hat, indicate interface discontinuities by the notation $[\![f]\!]\equiv\hat{f}-f$, and let the normal $\mathbf{n}$ point into the second plasma. The equilibrium pressure balance BC then reads

$$[\![p+\tfrac{1}{2}B^2]\!]=0 \qquad (\text{on S})\,. \qquad (12.83)$$

The surface integral of Eq. (12.80) now needs to be converted into a symmetric expression by means of the BCs on the perturbations at the perturbed interface. Those are the same as given in Eqs. (6.135) and (6.137) for static plasmas:

$$[\![\, \xi_n]\!] = 0 \qquad \text{(on S)}\,, \tag{12.84}$$

$$[\![\, \Pi + \xi_n \mathbf{n}\cdot\nabla(p + \tfrac{1}{2}B^2)]\!] = 0 \qquad \text{(on S)}\,. \tag{12.85}$$

Recall that the latter BC is obtained by evaluating pressure balance at $\mathbf{r}^0 + \xi_n \mathbf{n}$ (rather than at $\mathbf{r}^0 + \boldsymbol{\xi}$), since the tangential component of $\boldsymbol{\xi}$ is not continuous in general (see Fig. 6.14). This gives the following generalization of Eq. (12.80) for the integral over the combined volume $\mathrm{V}_{\mathrm{all}} \equiv \mathrm{V} + \hat{\mathrm{V}}$ of the two plasmas:

$$\begin{aligned}
&\int \boldsymbol{\eta}\cdot\mathbf{G}(\boldsymbol{\xi})\, dV_{\mathrm{all}} \\
&\quad = -\int \Big\{ \cdots \text{as in (12.80)} \cdots \Big\}\, dV - \int \eta_n\, \xi_n\, \mathbf{n}\cdot [\![\nabla(p + \tfrac{1}{2}B^2)]\!]\, dS \\
&\qquad - \int \Big[\gamma\hat{p}\,(\nabla\cdot\hat{\boldsymbol{\xi}})\,\nabla\cdot\hat{\boldsymbol{\eta}} + \hat{\mathbf{Q}}\cdot\hat{\mathbf{R}} + \tfrac{1}{2}(\nabla\hat{p})\cdot(\hat{\boldsymbol{\xi}}\,\nabla\cdot\hat{\boldsymbol{\eta}} + \hat{\boldsymbol{\eta}}\,\nabla\cdot\hat{\boldsymbol{\xi}}) \\
&\qquad\quad + \tfrac{1}{2}\hat{\mathbf{j}}\cdot(\hat{\boldsymbol{\eta}}\times\hat{\mathbf{Q}} + \hat{\boldsymbol{\xi}}\times\hat{\mathbf{R}}) - \tfrac{1}{2}(\nabla\hat{\Phi}_{\mathrm{gr}})\cdot[\,\hat{\boldsymbol{\eta}}\,\nabla\cdot(\hat{\rho}\hat{\boldsymbol{\xi}}) + \hat{\boldsymbol{\xi}}\,\nabla\cdot(\hat{\rho}\hat{\boldsymbol{\eta}})\,] \\
&\qquad\quad - \tfrac{1}{2}\hat{\boldsymbol{\eta}}\cdot[\nabla\cdot(\hat{\boldsymbol{\xi}}\,\hat{\rho}\hat{\mathbf{v}}\cdot\nabla\hat{\mathbf{v}})] - \tfrac{1}{2}\hat{\boldsymbol{\xi}}\cdot[\nabla\cdot(\hat{\boldsymbol{\eta}}\,\hat{\rho}\hat{\mathbf{v}}\cdot\nabla\hat{\mathbf{v}})] \\
&\qquad\qquad - \hat{\rho}(\hat{\mathbf{v}}\cdot\nabla\hat{\boldsymbol{\xi}})\cdot(\hat{\mathbf{v}}\cdot\nabla\hat{\boldsymbol{\eta}}) \Big]\, d\hat{V}\,.
\end{aligned} \tag{12.86}$$

Since the surface integral is now symmetric as well, this again leads to the equality (12.82), with V replaced by V_{all}. Hence, the generalized force operator $\mathbf{G}$ is also self-adjoint for model II*, QED.

The proof of self-adjointness of the operator $\mathbf{G}$ for model II (plasma–vacuum configurations) is left as an exercise for the reader. The resulting quadratic forms are the same as for model II* in the limit $\hat{\rho} \to 0$, $\hat{p} \to 0$, $\hat{\mathbf{j}} \to 0$ and $\hat{\mathbf{v}} \to 0$. This limit also applies to BC (12.85), but the BC (12.84) needs to be replaced by

$$\mathbf{n}\cdot\nabla\times(\boldsymbol{\xi}\times\hat{\mathbf{B}}) = \hat{\mathbf{B}}\cdot\nabla\xi_n - \mathbf{n}\cdot(\nabla\hat{\mathbf{B}})\cdot\mathbf{n}\,\xi_n = \hat{Q}_n \qquad \text{(on S)}\,, \tag{12.87}$$

as already demonstrated in Section 6.6.1, Eqs. (6.131) and (6.132).

▷ **Exercise** Complete the proof of self-adjointness of the operator $\mathbf{G}$ for model II equilibria. Exploit the vector potential $\hat{\mathbf{A}}$ and the plasma–interface BCs of Section 6.6.1. Leaving the outer boundary open, you may also derive the pre-self-adjointness relation for model III. ◁

12.2.4 Energy conservation and stability

On the basis of Section 12.2.3, we will demonstrate that the stationary plasma is actually *conservative:* the total energy of the stable waves and instabilities is conserved. Of course, this, and the fact that the operator $\mathbf{G}$ is self-adjoint, does not mean that spectral theory of MHD waves and instabilities for plasmas with background flow is now at the same level as the spectral theory for static plasmas. The crucial difference is the appearance of *the gradient operator parallel to the velocity*, $\mathbf{v}\cdot\nabla$, in the second term of the equation of motion (12.63). Since

$$\nabla\cdot(\rho\mathbf{v}\boldsymbol{\xi}\cdot\boldsymbol{\eta}) \overset{(12.24)}{=} \rho\mathbf{v}\cdot\nabla(\boldsymbol{\xi}\cdot\boldsymbol{\eta}) = \rho(\mathbf{v}\cdot\nabla\boldsymbol{\xi})\cdot\boldsymbol{\eta} + \rho(\mathbf{v}\cdot\nabla\boldsymbol{\eta})\cdot\boldsymbol{\xi}\,,$$

that operator *is anti-symmetric:*

$$\int \Big[\boldsymbol{\eta} \cdot (\rho \mathbf{v} \cdot \nabla \boldsymbol{\xi}) + \boldsymbol{\xi} \cdot (\rho \mathbf{v} \cdot \nabla \boldsymbol{\eta}) \Big] \, dV = \int \rho v_n \, \boldsymbol{\xi} \cdot \boldsymbol{\eta} \, dS = 0 \,. \qquad (12.88)$$

Here the surface integral vanishes because $v_n = 0$, both at a wall (model I) and at stationary interfaces (models II and II*). The energy conservation equation is obtained from the equation of motion (12.63):

$$- \int \frac{\partial \boldsymbol{\xi}}{\partial t} \cdot \mathbf{G}(\boldsymbol{\xi}) \, dV + 2 \int \frac{\partial \boldsymbol{\xi}}{\partial t} \cdot \Big(\rho \mathbf{v} \cdot \nabla \frac{\partial \boldsymbol{\xi}}{\partial t} \Big) \, dV + \int \rho \, \frac{\partial \boldsymbol{\xi}}{\partial t} \cdot \frac{\partial^2 \boldsymbol{\xi}}{\partial t^2} \, dV = 0 \,,$$

where the second integral vanishes by virtue of the anti-symmetry (12.88). Also exploiting the self-adjointness of the operator $\mathbf{G}$ then gives

$$\frac{d}{dt} \Big[- \tfrac{1}{2} \int \boldsymbol{\xi} \cdot \mathbf{G}(\boldsymbol{\xi}) \, dV + \tfrac{1}{2} \int \rho \Big| \frac{\partial \boldsymbol{\xi}}{\partial t} \Big|^2 dV \Big] = 0$$

$$\Rightarrow \quad H = \int \mathcal{H} \, dV = \text{const} \,, \quad \mathcal{H} \equiv \tfrac{1}{2} \rho \Big| \frac{\partial \boldsymbol{\xi}}{\partial t} \Big|^2 - \tfrac{1}{2} \boldsymbol{\xi} \cdot \mathbf{G}(\boldsymbol{\xi}) \,, \qquad (12.89)$$

so that the total energy, or Hamiltonian $\mathcal{H}$, of the perturbations is conserved, QED. It should be stated immediately: this does not imply the existence of an energy principle with the possibility of inserting trial functions to monitor stability, as for static plasmas. The appearance of the operator $\mathbf{v} \cdot \nabla$ forbids that. However, we need to address the widespread misunderstanding that this implies non-self-adjointness of the system a whole. We will return to this after the intermezzo.

▷ **Intermezzo: Hamiltonian formulations** The formulation of a Lagrangian from which the equations of motion can be derived by means of Hamilton's principle is the most concise description of a dynamical system; see Chapter 12 of Goldstein [243]. One may consider a branch of physics to have become part of the classical curriculum if one succeeds in constructing the appropriate Lagrangian. For nonlinear ideal MHD, this was accomplished by Cotsaftis [129] and Newcomb [463] in 1962. For linear ideal MHD, Frieman and Rotenberg [197] showed that the equation of motion (12.63) can be derived from Hamilton's principle (see Section 6.4.2), involving the volume integral of the Lagrangian density $\mathcal{L}$:

$$\delta \int_{t_1}^{t_2} \Big[\int \mathcal{L} \, dV \Big] \, dt = 0 \,, \quad \mathcal{L} \equiv \tfrac{1}{2} \rho \Big| \frac{\partial \boldsymbol{\xi}}{\partial t} \Big|^2 - \rho \Big(\mathbf{v} \cdot \nabla \frac{\partial \boldsymbol{\xi}}{\partial t} \Big) \cdot \boldsymbol{\xi} + \tfrac{1}{2} \boldsymbol{\xi} \cdot \mathbf{G}(\boldsymbol{\xi}) \,. \qquad (12.90)$$

Defining the canonical momentum $\boldsymbol{\pi}$, the corresponding Hamiltonian density is obtained,

$$\boldsymbol{\pi} \equiv \rho \Big(\frac{\partial \boldsymbol{\xi}}{\partial t} + \mathbf{v} \cdot \nabla \boldsymbol{\xi} \Big) \quad \Rightarrow \quad \mathcal{H} = \frac{1}{2\rho} \big(\boldsymbol{\pi} - \rho \mathbf{v} \cdot \nabla \boldsymbol{\xi} \big)^2 - \tfrac{1}{2} \boldsymbol{\xi} \cdot \mathbf{G}(\boldsymbol{\xi}) \,, \qquad (12.91)$$

from which Hamilton's equations emerge:

$$\begin{aligned} \frac{\partial \xi_i}{\partial t} &= \frac{\partial \mathcal{H}}{\partial \pi_i} & \Rightarrow \quad & \frac{\partial \boldsymbol{\xi}}{\partial t} = \frac{1}{\rho} (\boldsymbol{\pi} - \rho \mathbf{v} \cdot \nabla \boldsymbol{\xi}) \,, \\ \frac{\partial \pi_i}{\partial t} &= \frac{\partial}{\partial r_j} \Big(\frac{\partial \mathcal{H}}{\partial (\partial \xi_i / \partial r_j)} \Big) - \frac{\partial \mathcal{H}}{\partial \xi_i} & \Rightarrow \quad & \frac{\partial \boldsymbol{\pi}}{\partial t} = \mathbf{G}(\boldsymbol{\xi}) - \rho \mathbf{v} \cdot \nabla \big(\boldsymbol{\pi}/\rho - \mathbf{v} \cdot \nabla \boldsymbol{\xi} \big) \,. \end{aligned} \qquad (12.92)$$

The first equation just rephrases the definition of the canonical momentum, and the latter equation reproduces the equation of motion (12.63), QED.

Next, Newcomb [463] integrated the nonlinear MHD equations, and constructed the associated Lagrangian, by replacing the above quasi-Lagrangian description by a truly Lagrangian description in terms of the displacement vector field $\boldsymbol{\xi}$ with respect to the initial positions ($\mathbf{r}_0$ in Fig. 12.3). Later, this approach was put in the framework of modern Hamiltonian methods [24] and elaborated by a growing number of authors: Holm, Marsden, Ratiu and Weinstein [305], Morrison and Greene [449], Hameiri [284, 285], etc. ◁

So far, we have exploited real vectors $\boldsymbol{\xi}$ and a real inner product. Evidently, for normal modes $\boldsymbol{\xi}(\mathbf{r})\exp(-\mathrm{i}\omega t)$, we should revert to a complex inner product. Hence, we now assume a Hilbert space with an inner product and a finite norm,

$$\langle \boldsymbol{\xi}, \boldsymbol{\eta} \rangle \equiv \tfrac{1}{2} \int \rho\, \boldsymbol{\xi}^* \cdot \boldsymbol{\eta}\, dV \,, \qquad I[\boldsymbol{\xi}] \equiv \|\boldsymbol{\xi}\|^2 \equiv \langle \boldsymbol{\xi}, \boldsymbol{\xi} \rangle < \infty \,. \tag{12.93}$$

With this inner product, the example of the momentum operator in quantum mechanics immediately suggests to absorb a factor $-\mathrm{i}$ in *a revised definition* of the gradient operator parallel to the velocity, that we will call *the Doppler–Coriolis operator:*

$$U \equiv -\mathrm{i}\mathbf{v}\cdot\nabla \,. \tag{12.94}$$

The equation of motion (12.63) and the spectral equation (12.68) then become

$$\mathbf{G}(\boldsymbol{\xi}) - 2\mathrm{i}\rho U \frac{\partial \boldsymbol{\xi}}{\partial t} - \rho \frac{\partial^2 \boldsymbol{\xi}}{\partial t^2} = 0 \quad \Rightarrow \quad \mathbf{G}(\boldsymbol{\xi}) - 2\rho\omega U \boldsymbol{\xi} + \rho\omega^2 \boldsymbol{\xi} = 0 \,, \tag{12.95}$$

involving two operators, $\rho^{-1}\mathbf{G}$ and U, that are *both self-adjoint (or Hermitian):*

$$\begin{aligned} \langle \boldsymbol{\eta}, \rho^{-1}\mathbf{G}(\boldsymbol{\xi}) \rangle &\equiv \tfrac{1}{2} \int \boldsymbol{\eta}^* \cdot \mathbf{G}(\boldsymbol{\xi})\, dV = \tfrac{1}{2} \int \boldsymbol{\xi} \cdot \mathbf{G}(\boldsymbol{\eta}^*)\, dV \;\equiv \langle \rho^{-1}\mathbf{G}(\boldsymbol{\eta}), \boldsymbol{\xi} \rangle \,, \\ \langle \boldsymbol{\eta}, U\boldsymbol{\xi} \rangle &\equiv \tfrac{1}{2} \int \rho \boldsymbol{\eta}^* \cdot U\boldsymbol{\xi}\, dV \;= \tfrac{1}{2} \int \rho \boldsymbol{\xi} \cdot (U\boldsymbol{\eta})^*\, dV \;\equiv \langle U\boldsymbol{\eta}, \boldsymbol{\xi} \rangle \,. \end{aligned} \tag{12.96}$$

Clearly, the latter equality does not require the kind of laborious derivations as involved in the proof of self-adjointness of $\mathbf{G}$ but it follows directly from the anti-symmetry relation (12.88). Notice that it does not even require satisfaction of boundary conditions of the functions $\boldsymbol{\xi}$ and $\boldsymbol{\eta}$. In that sense, it is appropriate to call the operator U *super self-adjoint!* We conclude that the complexity of the spectral problem in plasmas with stationary flow is not due to non-self-adjointness (as frequently stated in the literature), but to the fact that *the eigenvalue problem is nonlinear* (quadratic), governed by the two physical operators $\mathbf{G}$ and U, so that the eigenvalues ω are essentially complex.

▷ **Non-orthogonality of the eigenfunctions** One of the implications of the nonlinearity of the Frieman–Rotenberg spectral equation (12.95)(b) is that the eigenfunctions are non-orthogonal in general. This follows by taking the inner product of the operators $\mathbf{G}$ and U applied to eigenfunctions $\boldsymbol{\xi}_\alpha$ and $\boldsymbol{\xi}^*_\beta$ corresponding to eigenvalues ω_α and ω^*_β:

$$\langle \boldsymbol{\xi}_\beta, \rho^{-1}\mathbf{G}(\boldsymbol{\xi}_\alpha) \rangle = \omega^*_\beta\, \omega_\alpha \langle \boldsymbol{\xi}_\beta, \boldsymbol{\xi}_\alpha \rangle \,, \tag{12.97}$$

$$\langle \boldsymbol{\xi}_\beta, U\boldsymbol{\xi}_\alpha \rangle = \tfrac{1}{2}(\omega^*_\beta + \omega_\alpha) \langle \boldsymbol{\xi}_\beta, \boldsymbol{\xi}_\alpha \rangle \,, \tag{12.98}$$

where $\omega^*_\beta \neq \omega_\alpha$ since a common factor $\omega^*_\beta - \omega_\alpha$ has been divided out in both expressions. These relations, from Barston [37], exhibit a fundamental difference between the static and stationary spectral problems. For static plasmas, $U \equiv 0$, so that $\langle \boldsymbol{\xi}_\beta, \boldsymbol{\xi}_\alpha \rangle = 0$ for the eigenfunctions. This is no longer the case for stationary plasmas, so that *the eigenfunctions are not orthogonal*, in general. However, the new method of the Spectral Web, introduced in Section 12.3 and extensively exploited in Chapter 13, does not require orthogonality. ◁

The Hamiltonian (12.89) may be represented by $H \equiv K + W$ associating the following expression for the *kinetic energy* K with the dynamical variable $\boldsymbol{\xi}$:

$$K \equiv \tfrac{1}{2} \int \rho \frac{\partial \boldsymbol{\xi}^*}{\partial t} \cdot \frac{\partial \boldsymbol{\xi}}{\partial t}\, dV_{\rm all} \,, \tag{12.99}$$

so that boundedness of I is related to the condition of finite kinetic energy,

$$K[\boldsymbol{\xi}] \equiv \|\partial\boldsymbol{\xi}/\partial t\|^2 = |\omega|^2 I < \infty\,. \tag{12.100}$$

The corresponding expression for the *potential energy* W may be obtained from the manifestly symmetric quadratic form (12.86) by substituting $\boldsymbol{\eta} \equiv \boldsymbol{\xi}^*$. A more compact expression is obtained by summing Eqs. (12.74) and (12.75) and integrating by parts to eliminate the second derivatives of the equilibrium quantities:

$$W \equiv -\tfrac{1}{2}\int \boldsymbol{\xi}^* \cdot \mathbf{G}(\boldsymbol{\xi})\, dV_{\rm all} = W^{\rm p}[\boldsymbol{\xi}] + W^{\rm s}[\xi_n] + W^{\hat{\rm p}}[\hat{\boldsymbol{\xi}}]\,, \tag{12.101}$$

where

$$\begin{aligned} W^{\rm p} \;&\equiv\; \tfrac{1}{2}\int \Big[\gamma p|\nabla\cdot\boldsymbol{\xi}|^2 + |\mathbf{Q}|^2 + (\nabla p)\cdot\boldsymbol{\xi}\,\nabla\cdot\boldsymbol{\xi}^* + \mathbf{j}\cdot(\boldsymbol{\xi}^*\times\mathbf{Q}) \\ &\qquad - (\nabla\Phi_{\rm gr})\cdot\boldsymbol{\xi}^*\,\nabla\cdot(\rho\boldsymbol{\xi}) - \boldsymbol{\xi}^*\cdot[\nabla\cdot(\boldsymbol{\xi}\,\rho\mathbf{v}\cdot\nabla\mathbf{v})] - \rho|\mathbf{v}\cdot\nabla\boldsymbol{\xi}|^2\Big]\, dV, \end{aligned} \tag{12.102}$$

$$W^{\rm s} \;\equiv\; \tfrac{1}{2}\int |\xi_n|^2\, \mathbf{n}\cdot\big[\!\big[\nabla(p+\tfrac{1}{2}B^2)\big]\!\big]\, dS\,, \tag{12.103}$$

$$W^{\hat{\rm p}} \;\equiv\; \begin{cases} \tfrac{1}{2}\int |\hat{\mathbf{Q}}|^2\, d\hat{V} \;\; \big(\equiv W^{\rm v}\big) & \textit{(model II)} \\ \tfrac{1}{2}\int \Big[\gamma\hat{p}|\nabla\cdot\hat{\boldsymbol{\xi}}|^2 + \cdots \text{as (12.102)} \cdots\Big]\, d\hat{V} & \textit{(model II*)} \end{cases}\,. \tag{12.104}$$

For model I (plasma–wall) equilibria, $W^{\rm s} = 0$ because $\xi_n = 0$ at the wall, and the term $W^{\hat{\rm p}}$ is evidently absent. The energy expressions for model II (plasma–vacuum) equilibria may be obtained from the ones for model II* (plasma–plasma) equilibria by taking the limits $\hat{\rho} \to 0\,, \hat{p} \to 0\,, \hat{\mathbf{j}} \to 0$ and $\hat{\mathbf{v}} \to 0$. As noticed at the end of Section 12.2.3, this also applies to the BC (12.85) on the perturbed pressure, but the BC (12.84) on $\hat{\xi}_n$ needs to be replaced by the condition (12.87) on $\hat{Q}_n$. Compared to static equilibria, the background velocity $\mathbf{v}$ adds two terms to the volume integrals $W^{\rm p}$ and $W^{\hat{\rm p}}$, one due to the centrifugal acceleration and another one due to the square of the operator U (last two terms of Eq. (12.102)).

For clarity of notation, we now drop the external contributions and restrict the presentation to model I configurations. With the explicit expressions for the potential energy $W[\boldsymbol{\xi}]$ given above, including them is straightforward. The eigenvalue problem for waves and instabilities is then given by the spectral differential equation (12.95)(b), subject to appropriate BCs. To complete the discussion of the implications of the potential energy on stability of stationary plasmas, we also need to define the quadratic form corresponding to the Doppler–Coriolis operator U:

$$V \equiv \tfrac{1}{2}\int \rho\boldsymbol{\xi}^*\cdot U\boldsymbol{\xi}\, dV\,. \tag{12.105}$$

With the three quadratic forms W, V and I, corresponding to the operators $\mathbf{G}$, U and 1, we can convert the spectral equation (12.95) into a quadratic equation for the eigenvalues by dotting it with $\boldsymbol{\xi}^*$ and integrating over the volume:

$$\omega^2 - 2\overline{V}\omega - \overline{W} = 0\,. \tag{12.106}$$

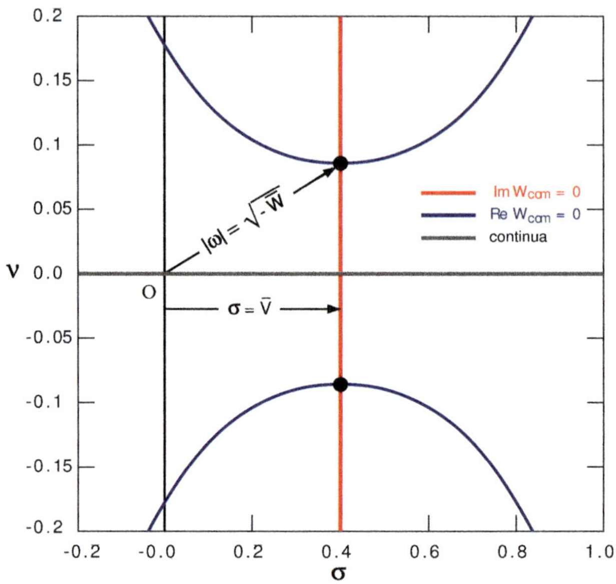

Fig. 12.4 Elementary Spectral Web, consisting of solution path (red) and conjugate path (blue), with the location of modes (black dots) in the complex ω-plane for a stationary equilibrium. This example refers to the $m = 1$, $k = 1.28$ internal kink instability of a cylindrical force-free magnetic field of constant $\bar{\alpha} = 8.0$ (see Figs. 9.13 and 9.14) , subjected to a constant Doppler shift, $\sigma = kv_z = 0.4$.

Here, the quadratic forms V and W have been normalized to correspond, in the terminology of quantum mechanics, to *the expectation values* (or averages) of the operators U and $-\rho^{-1}\mathbf{G}$,

$$\overline{V} \equiv V/I \equiv \langle U \rangle\,, \qquad \overline{W} \equiv W/I \equiv \langle -\rho^{-1}\mathbf{G} \rangle\,. \tag{12.107}$$

These quantities are *real* because of the proved self-adjointness of the operators U and $\rho^{-1}\mathbf{G}$. The solution of the quadratic equation (12.106) is straightforward,

$$\omega = \overline{V} \pm \big(\overline{V}^2 + \overline{W}\big)^{1/2}\,. \tag{12.108}$$

Of course, it does not solve the eigenvalue problem, since it involves the unknown solution $\boldsymbol{\xi}$, but it does provide very significant physical relationships.

Recall that, for static plasmas, the relation $\omega^2 = \overline{W}$ could be transformed into the Rayleigh–Ritz variational principle $\delta\Lambda = 0$, where the eigenfunctions $\boldsymbol{\xi}$ yield the stationary values ω^2 of the Rayleigh quotient $\Lambda \equiv W[\boldsymbol{\xi}]/I[\boldsymbol{\xi}]$. This provided the connection with the energy principle $W[\boldsymbol{\xi}] > 0$ for stability (Sections 6.4.3 and 6.4.4). This connection is lost for stationary plasmas.

Splitting the eigenvalue parameter into real and imaginary parts, by writing $\omega = \sigma + \mathrm{i}\nu$, the solutions (12.108) divide into waves ($\nu = 0$) and instabilities ($\nu \neq 0$):

$$\begin{cases} \sigma = \overline{V} \pm \big(\overline{V}^2 + \overline{W}\big)^{1/2}\,, \qquad \nu = 0 & \textit{(stable waves)}\,, \\ \sigma = \overline{V}\,, \quad \nu = \pm\big(-\,\overline{V}^2 - \overline{W}\big)^{1/2} \;\Rightarrow\; |\omega| = (-\overline{W})^{1/2} & \textit{(instabilities)}\,. \end{cases} \tag{12.109}$$

Hence, the marginal state is no longer at the origin of the ω-plane (as for static plasmas), but shifted to the point $\sigma = \overline{V}$, $\nu = 0$, where stable waves divide into faster and slower ones, whereas ω becomes complex for instabilities. This point is no longer at a fixed position in the ω-plane, but it depends on the solution $\boldsymbol{\xi}$ itself. Figure 12.4 exhibits the effect of flow on the modes of a stationary equilibrium obtained from a static one by a trivial, homogeneous, flow along the z-axis producing a Doppler shift $\sigma = \overline{V} = kv_z$ (exploiting the Spectral Web method [229] described in the following section). According to Eq. (12.109)(b), the unstable modes become located at a horizontal distance $\overline{V}$ to the imaginary axis and at a distance $\sqrt{-\overline{W}}$ to the origin. For every complex eigenvalue ω, a corresponding eigenvalue ω^* can be found. Since stationary equilibria are invariant under reflection of the background velocity ($\overline{V} \to -\overline{V}$), the pairs $-\omega$ and $-\omega^*$ will also be eigenvalues.

In conclusion, the most important effect of the operator U is the *Doppler shift* of eigenfrequencies. For plane waves $\boldsymbol{\xi} \sim \exp(\mathrm{i}\mathbf{k}\cdot\mathbf{r})$ in homogeneous plasmas, that relationship is straightforward: $\overline{V} = \mathbf{k}\cdot\mathbf{v}$. For inhomogeneous plasmas with curved velocity fields, the differential operator U also acts on the unit vectors so that $\overline{V}$ will contain contributions from the Coriolis acceleration. Accordingly, the quantity $\overline{V}$ will be called the *solution-averaged Doppler–Coriolis shift*.

With respect to stability, the standard criterion $\overline{W}[\boldsymbol{\xi}] > 0$ of static plasmas is still sufficient for the stability of stationary plasmas, but the above expressions provide much sharper criteria involving the square of the average Doppler–Coriolis shift:

A necessary and sufficient condition for stability of stationary plasmas is that

$$\overline{W}[\boldsymbol{\xi}] > -\overline{V}^2[\boldsymbol{\xi}] \tag{12.110}$$

for all $\boldsymbol{\xi}$ that are bound in norm and satisfy the appropriate boundary conditions. Vice versa, *a sufficient condition for instability* is that

$$\overline{W}[\boldsymbol{\xi}] < -\overline{V}^2[\boldsymbol{\xi}] \tag{12.111}$$

for a particular choice of $\boldsymbol{\xi}$. The proviso 'bound in norm' is crucial here since it has been shown that stationary rotating plasmas may have unstable continuous spectra [299, 286, 465, 230]. Hence, these criteria refer to discrete modes only.

The stability criterion (12.110) does not imply that the effect of the average Doppler–Coriolis shift $\overline{V}$ is stabilizing. For plane shear flows, the opposite is true since the centrifugal term $\sim \rho\mathbf{v}\cdot\nabla\mathbf{v}$ may be neglected and the only velocity-dependent term in the expression (12.102) for W^{p} is the negative term $-\rho|\mathbf{v}\cdot\nabla\boldsymbol{\xi}|^2$. Splitting this term off, the stability criterion (12.110) becomes

$$\overline{W} + \overline{V}^2 \equiv \overline{W}_0 + \overline{V}^2 - \overline{\Delta}_v > 0\,, \qquad \Delta_v \equiv \tfrac{1}{2}\int \rho|\mathbf{v}\cdot\nabla\boldsymbol{\xi}|^2\, dV\,, \tag{12.112}$$

where W_0 is the potential energy of a static plasma. From Schwarz' inequality,

$$\Big[\int \mathbf{f}^*\cdot\mathbf{g}\, dV\Big]^2 \le \int |\mathbf{f}|^2\, dV \int |\mathbf{g}|^2\, dV\,, \quad \text{with} \quad \mathbf{f} \equiv \sqrt{\rho}\,\boldsymbol{\xi}\,, \quad \mathbf{g} \equiv -\mathrm{i}\sqrt{\rho}\,\mathbf{v}\cdot\nabla\boldsymbol{\xi}\,,$$

$$\Rightarrow \quad \overline{V}^2 - \overline{\Delta}_v \equiv \langle -\mathrm{i}\mathbf{v}\cdot\nabla\rangle^2 - \langle(-\mathrm{i}\mathbf{v}\cdot\nabla)^2\rangle \le 0\,, \tag{12.113}$$

the net result is zero or negative: in general, *plane shear flows are destabilizing*. That is why such flows may be Kelvin–Helmholtz unstable (Section 13.1).

Let us now leave the esoteric realm of spectral theory and descend into the low lands of explicit calculations. To really get a grip on the waves and instabilities of specific stationary equilibria with plane shear flow or rotation, we need a general solution method to compute the complex eigenvalues.

This will be introduced in the following section. Applications to shear flow in plane geometry and rotation in cylindrical plasmas will be presented in Chapter 13.

12.3 The Spectral Web

12.3.1 Opening up the boundaries

Recall the 'simplicity' of spectral theory for static plasmas, based on the fact that the eigenvalue problem is linear in ω^2, so that ω itself is either real or imaginary. With flow, because of the Doppler–Coriolis shift operator U, the eigenvalue problem is nonlinear so that the eigenvalues are no longer confined to the real and imaginary axes but become essentially complex. Fortunately, we do not have to search randomly for eigenvalues in the complex plane. We have the powerful expressions (12.109) at our disposal telling us that the eigenvalues are to be found in particular locations. To find those locations, we will systematically exploit the self-adjointness of the operators U and $\mathbf{G}$, associated with the Doppler–Coriolis shift of moving plasmas and the conservation of energy.

We start from the basic spectral equation (12.95), repeated here for emphasis:

$$\mathbf{G}(\boldsymbol{\xi}) - 2\rho\omega\, U\boldsymbol{\xi} + \rho\omega^2\boldsymbol{\xi} = 0\,. \tag{12.114}$$

In order to solve this equation for a particular configuration, we need to specify the actual magnetic geometry of the confined plasma and the associated BCs. The essential aspects of the BCs used for different geometries are well represented by considering the model I BCs (introduced in Section 4.6.1) for plasma confined between two 'walls', one at $x = x_\ell$ and another one at $x = x_\mathrm{r}$:

$$\xi(x_\ell) \;=\; 0 \qquad \text{(inner wall, 'left' BC)}\,, \tag{12.115}$$

$$\xi(x_\mathrm{r}) \;=\; 0 \qquad \text{(outer wall, 'right' BC)}\,. \tag{12.116}$$

For simplicity, we drop the subscript on ξ_n and write *(non-bold!)* $\xi \equiv \boldsymbol{\xi}\cdot\mathbf{n}$. The coordinate x may represent the vertical coordinate of a plane gravitating slab (Section 13.1), or the radial coordinate r of a cylindrical plasma (Section 13.3), or the poloidal magnetic flux ψ of a toroidal plasma (Chapters 16–18), replacing the inner BC by regularity at the magnetic axis ($x_\ell = 0$) in the latter two cases.

Let us assume that we have an effective solver at our disposal for the solution of the set of differential equations (12.114). Imposing the left BC (12.115), but dropping the right BC (12.116), these equations can then be solved for *arbitrary complex values* of the parameter ω. This provides solutions $\boldsymbol{\xi}^\ell(x;\omega)$, which we will call *the left solutions*. (Recall that x labels the flux surfaces and the other two coordinates are suppressed.) Obviously, integrating from left to right, the right BC will not be satisfied in general. The question is: what is the physical significance of this defect? To answer it, we turn to the proof of self-adjointness of the operator $\mathbf{G}$ expressed by the equality (12.73). That proof consists of demonstrating that both the left and the right hand side of that equality can be converted into a symmetric expression and a divergence, which is then transformed into two surface integrals over the bounding surfaces S_left and $\mathrm{S}_\mathrm{right}$. Those integrals would vanish if both BCs (12.115) and (12.116) were imposed, so that only a symmetric expression would remain, which completes the proof. With an open right boundary this is no longer the case, and the pre-self-adjointness property (12.81) then yields

$$\int \boldsymbol{\eta}\cdot\mathbf{G}(\boldsymbol{\xi})\,dV = \int \big[\text{symm. in } \boldsymbol{\xi} \text{ and } \boldsymbol{\eta}\big]\,dV - \int \eta\Pi(\boldsymbol{\xi})\,dS_\mathrm{right}\,, \tag{12.117}$$

so that the operator $\mathbf{G}$ becomes essentially non-self-adjoint. Here, $\boldsymbol{\xi}$ and $\boldsymbol{\eta}$ are understood to be left solutions (the superscript ℓ will usually be suppressed). Substituting $\boldsymbol{\eta} = \boldsymbol{\xi}^*$ into Eq. (12.117) and multiplying with $-\frac{1}{2}$ yields an expression for the energy. The non-self-adjointness of $\mathbf{G}$ then turns out to be associated with the fact that the energy of an open system is complex:

$$W \equiv -\tfrac{1}{2} \int \boldsymbol{\xi}^* \cdot \mathbf{G}(\boldsymbol{\xi})\, dV = W^{\mathrm{p}} + W_{\mathrm{com}} \,. \tag{12.118}$$

Here, the symmetric part is the usual volume integral for the plasma energy W^{p}, which is *real*, whereas the non-symmetric *complex* surface integral

$$W^{\ell}_{\mathrm{com}}\big[\boldsymbol{\xi}^{\ell}(x_{\mathrm{r}};\omega)\big] \equiv \tfrac{1}{2} \int \xi^{\ell *}\Pi(\boldsymbol{\xi}^{\ell})\, dS_{\mathrm{right}} \tag{12.119}$$

will be called the *complementary energy*. This is the amount of energy, to be provided in addition to the plasma energy, to maintain the time dependence $\exp(-\mathrm{i}\omega t)$ of the perturbations of the system. This expression is related to the real (in-phase) and imaginary (out-of-phase) parts of the complementary power.

The expression (12.119) for the complementary energy will be used for the actual solution of the closed boundary value problem (BVP) (12.114)–(12.116),including the right BC. We claim that obtaining the eigenvalues,i.e. also satisfying the latter BC, is equivalent to concurrently solving two algebraic equations:

$$W_{\mathrm{com},2} \;=\; \tfrac{1}{2} \int \big(\xi_1\Pi_2 - \xi_2\Pi_1\big)\, dS_{\mathrm{right}} = 0 \qquad \Rightarrow \;\; \textit{solution path}\,, \tag{12.120}$$

$$W_{\mathrm{com},1} \;=\; \tfrac{1}{2} \int \big(\xi_1\Pi_1 + \xi_2\Pi_2\big)\, dS_{\mathrm{right}} = 0 \qquad \Rightarrow \;\; \textit{conjugate path}\,. \tag{12.121}$$

Here, $f_1 \equiv \mathrm{Re}(f)$ and $f_2 \equiv \mathrm{Im}(f)$ indicate the real and imaginary parts of the complex quantity f. The two indicated paths consist of *all* solutions of the respective equations (12.120) and (12.121), but they need to be obtained only *in a strip of the ω-plane*, as will be demonstrated below. The eigenvalues are situated on the intersections of the two paths, so that the complex spectrum of stationary plasmas is found by determining the common zeros of the functions $W_{\mathrm{com},2}\{\boldsymbol{\xi}^{\ell}(x_{\mathrm{r}};\omega)\}$ and $W_{\mathrm{com},1}\{\boldsymbol{\xi}^{\ell}(x_{\mathrm{r}};\omega)\}$. These equations are highly implicit, since they involve the left solutions of the open BVP (12.114)–(12.115) for the arbitrarily chosen values of the parameter ω, but their solution is a straightforward numerical task. Evaluation of the two surface integrals just involves substitution of the boundary values of the left solutions, which are available in any numerical solution scheme.

The construction of the two paths in the complex ω-plane, and their intersections, produces *the Spectral Web:* the complete spectrum of complex eigenvalues of a stationary plasma *together with the connecting structure* of the solution path and the conjugate path. Concerning the terminology: the term 'solution path' was introduced in Ref. [227], the concept of 'conjugate path' introduced here is meant in the sense of conjugate to the solution path (somewhat like the momentum conjugate to a generalized coordinate in classical mechanics). In Fig. 12.4 we already showed an elementary example for the internal kink mode of a cylindrical force-free equilibrium. The power of the method is illustrated in Fig. 12.5 which shows the Spectral Web for the same equilibrium, but subjected to a genuine flow profile that is maximum on axis and vanishes at the wall. As might be expected, the solution path is not straight anymore but slightly curved, since the Doppler shift is not constant, but quite unexpectedly a second branch of the solution path appears with a new instability on it.

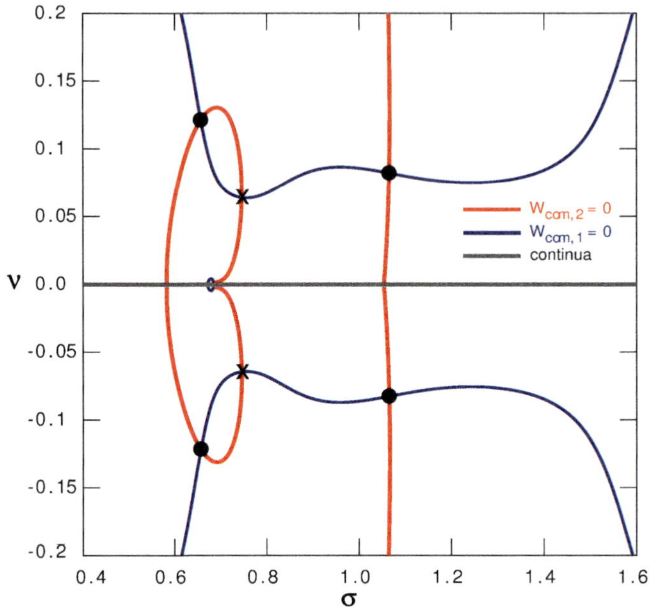

Fig. 12.5 Spectral Web for the same mode numbers and the same cylindrical force-free equilibrium as in Fig. 12.4, but subjected to a shear velocity profile $v_z = 1 - x^2$. Genuine eigenvalues are indicated by a dot, false ones by a cross.

(False solutions, indicated by the crosses, appear since $W_{\rm com}$ also vanishes when $\Pi = 0$ at the right boundary; those solutions are easily discarded by imposing the subsidiary condition $\Pi(\boldsymbol{\xi}^\ell(x_{\rm r})) \neq 0$. The alternation of genuine and false eigenvalues is actually one of the characteristic features of the spectra of stationary equilibria, as we will see in the next section.) Also, an additional small-scale loop (hardly visible) appears close to the real axis with infinitely many modes on it! The study of these interesting spectra will be continued in the next chapter. For now, it is to be noted that *the Spectral Web is a powerful new tool to study the complex spectra of stationary plasma equilibria.*

The order of the two equations (12.120) and (12.121) is chosen to stress that the solution path is the more important one. To underline this, consider how the quadratic (12.106) and its solutions (12.109) for the eigenvalues are modified for arbitrary values of ω by the expression (12.118) for the energy of the open system:

$$\omega^2 - 2\overline{V}\omega - \overline{W}^p - \overline{W}_{\rm com} = 0\,. \tag{12.122}$$

Along the solution path, the solutions of this quadratic become

$$\overline{W}_{\rm com,2} = 0 \quad \Rightarrow \quad \begin{cases} \sigma = \overline{V} \pm \big(\overline{V}^2 + \overline{W}^p + \overline{W}_{\rm com,1}\big)^{1/2}\,, & \nu = 0\,, \\ \sigma = \overline{V}\,, \qquad \nu = \big(-\overline{V}^2 - \overline{W}^p - \overline{W}_{\rm com,1}\big)^{1/2}\,. \end{cases} \tag{12.123}$$

Comparison with the expressions (12.109) shows that the real component of the complementary energy determines the distance to the eigenvalues. Moreover, the expressions show that the solution path consists of the real axis for stable waves and of the curve $\sigma = \overline{V}$ for instabilities. The latter provides an alternative expression for the solution path of the instabilities:

$$\sigma - \overline{V}\big[\boldsymbol{\xi}^\ell(x;\omega)\big] \equiv \sigma - \frac{\displaystyle\int \rho\boldsymbol{\xi}^{\ell*}\cdot U\boldsymbol{\xi}^\ell\,dV}{\displaystyle\int \rho|\boldsymbol{\xi}^\ell|^2\,dV} = 0 \qquad \Rightarrow \; \textit{solution path}\,. \tag{12.124}$$

This shows the physical significance of the solution path: on it, the solution averaged Doppler–Coriolis shifted frequency vanishes, as required for instabilities. In a sense, the solution path is the counterpart of the imaginary axis on which the instabilities of static plasmas are restricted to lie. Of course, this does not imply that it remains a simple, or even a single, curve for arbitrary flows (as demonstrated in Fig. 12.5). However, it does suggest that the solution path and, hence, all unstable eigenvalues, are restricted to lie in a strip of the complex ω-plane,

$$\sigma_{\rm min} \le \sigma \le \sigma_{\rm max}\,, \tag{12.125}$$

where $\sigma_{\rm min}$ and $\sigma_{\rm max}$ are (loosely speaking) related to the minima and maxima of $\mathbf{k}.\mathbf{v}$ and (exactly) determined by the solutions of Eq. (12.124). For the actual computation of the solution path, the equivalent but simpler expression (12.120) is to be preferred since it is a surface integral rather than a volume integral.

Mixed representations So far, we have restricted the analysis to *internal modes,* integrating from left to right as required for left solutions. However, for some annular plasma models (e.g. as exploited for accretion disks; see Section 13.5), it is frequently advantageous to integrate from right to left. This is the case when the solutions are localized near the inner boundary so that the magnitude of the left solution $\xi^\ell(x_{\rm r})$ at the outer boundary may be negligible. The required expression for the complementary energy $W^{\rm r}_{\rm com}$ of the right solutions is completely analogous to the expression (12.119) for $W^\ell_{\rm com}$ of the left solutions:

$$W^{\rm r}_{\rm com}\big[\boldsymbol{\xi}^{\rm r}(x_\ell;\omega)\big] \;\equiv\; -\tfrac{1}{2}\int \xi^{\rm r*}\Pi(\boldsymbol{\xi}^{\rm r})\,dS_{\rm left}\,. \tag{12.126}$$

A minus sign appears here since the normal is pointing to the left.

In general, internal modes have a large amplitude somewhere in the middle of the domain. It is then expedient to exploit a mixed representation of left and right solutions and to join them at some location $x = x_{\rm mix}$ by renormalizing them:

$$\xi^{\rm mix} \equiv \xi^\ell(x_{\rm mix}) = \xi^{\rm r}(x_{\rm mix})\,. \tag{12.127}$$

The resulting expression for the *mixed* complementary energy then becomes

$$\begin{aligned} &W^{\rm mix}_{\rm com}\big[\boldsymbol{\xi}^\ell(x_{\rm mix};\omega);\boldsymbol{\xi}^{\rm r}(x_{\rm mix};\omega)\big] \\ &\quad\equiv W^\ell_{\rm com}(x_{\rm mix}) + W^{\rm r}_{\rm com}(x_{\rm mix}) = -\tfrac{1}{2}\int \xi^{\rm mix*}\,[\![\Pi]\!]\,dS_{\rm mix}\,, \end{aligned} \tag{12.128}$$

where the jump is denoted by $[\![f]\!] \equiv f^{\rm r}(x^+_{\rm mix}) - f^\ell(x^-_{\rm mix})$. The system is now open at the surface $x = x_{\rm mix}$, where energy of the amount $W^{\rm mix}_{\rm com}$ is to be provided in order for the system to exhibit the time dependence $\exp(-{\rm i}\omega t)$. The eigenvalues are again obtained by imposing the condition that the complementary energy vanishes, i.e. that the jump in the total pressure perturbation vanishes so that the final solution becomes continuous. The arbitrary value of $x_{\rm mix}$ may be chosen on the basis of accuracy considerations. Note that this choice will give different spectral webs, but the spectrum itself (i.e. the intersections of the solution and conjugate paths) of course will be identical.

In the mixed representation, spurious solutions may occur when one of the solutions ξ^ℓ or $\xi^{\rm r}$ is oscillatory, so that it may accidentally vanish at $x = x_{\rm mix}$. Those solutions may be eliminated by applying the subsidiary condition $|\xi(x_{\rm mix})| \neq 0$. The mixed representation will be exploited extensively since it is the most flexible one (e.g. for Figures 12.4 and 12.5 the value $x_{\rm mix} = 0.3$ was chosen). In the following sections, we will restrict the exposition to the left representation. Generalization to the mixed representation is straightforward.

The expression (12.128) for the mixed internal solutions also suggests how the eigenvalues of *external modes* in a plasma–vacuum system may be obtained. In that case, the plasma is bounded by a vacuum where the magnetic field perturbation $\hat{\mathbf{Q}}$ may be expressed in terms of the scalar potential $\hat{\Phi}$, satisfying Laplace's equation:

$$\hat{\mathbf{Q}} = -\mathrm{i}\nabla\hat{\Phi}\,, \qquad \nabla^2\hat{\Phi} = 0 \qquad (x_{\rm a} \leq x < x_{\rm b})\,. \tag{12.129}$$

The latter differential equation is equivalent to two first order ones for $\hat{\Phi}$ and its normal derivative $\partial\hat{\Phi}/\partial n \equiv \mathbf{n}\cdot\nabla\hat{\Phi}$. Here, the coordinate x is continued into the vacuum, extending from the plasma–vacuum boundary at $x = x_{\rm a}$ to an outer wall at $x = x_{\rm b}$. The solution $\hat{Q} \equiv \mathbf{n}\cdot\hat{\mathbf{Q}}$ should satisfy the boundary condition $\hat{Q} = -\mathrm{i}\partial\hat{\Phi}/\partial n = 0$ at the wall.

To arrive at an expression for the complementary energy, the (left) plasma solution $\boldsymbol{\xi}^\ell$ is to be connected to the (right) vacuum solution $\hat{\mathbf{Q}}^{\rm r}$ by satisfying the first interface BC (6.131) at the plasma–vacuum boundary. This amounts to a mere renormalization of the plasma or vacuum variables, analogous to the renormalization (12.127) for the internal modes. The second interface BC (6.134) relating the total pressure perturbation $\Pi(\boldsymbol{\xi})$ in the plasma to the magnetic pressure perturbation $\hat{\Pi}(\hat{\mathbf{Q}}) \equiv \hat{\mathbf{B}}\cdot\hat{\mathbf{Q}}$ in the vacuum is *not* to be satisfied in the solution path method since it will naturally emerge in the final solution for the EVs. Without giving the detailed analysis (see Ref. [229](a) for that), we then obtain the following expression for the complementary energy of the system:

$$W^{\rm pv}_{\rm com}\big[\boldsymbol{\xi}^\ell(x_{\rm a};\omega);\hat{\mathbf{Q}}^{\rm r}(x_{\rm a};\omega)\big] \equiv -\tfrac{1}{2}\int \xi^*\,[\![\Pi + \xi\,\partial_n(p+\tfrac{1}{2}B^2)]\!]\,dS_{\rm a}\,. \tag{12.130}$$

Clearly, the eventual solution $W^{\rm pv}_{\rm com} = 0$ corresponds to satisfaction of the second interface condition and, thus, yields the required eigenvalues.

12.3.2 Oscillation theorems in the complex plane

For the computation of the eigenfrequencies of waves and instabilities, monotonicity properties are extremely important. For example, for static plasmas they establish the connection of the MHD spectrum with the stability of the plasma. The prototype monotonicity is that of the Sturm–Liouville equation, connecting the eigenvalues with the number of nodes of the eigenfunctions. For static equilibria, a Sturm–Liouville kind of oscillation theorem could be proved [240], see Section 7.4.3, because the spectra are actually real in terms of ω^2. They are also real for the stable part of the spectrum of stationary equilibria [231]. It was generally assumed that generalization to the complex part of the spectrum is not possible since there is no unique way of counting nodes for complex functions. However, *the Spectral Web method changes that,* because the solution and conjugate paths effectively guide the solutions to the eigenvalues *along curves* in the ω-plane. The analysis systematically exploits the self-adjointness of $\mathbf{G}$ and U.

(a) Sequencing integrals Consider two neighbouring points ω_α and ω_β, either on the solution path or on the conjugate path, with the corresponding left solutions $\boldsymbol{\xi}_\alpha$ and $\boldsymbol{\xi}_\beta$ of the spectral equation (12.114), symmetrically positioned about the point $\omega \equiv \sigma + \mathrm{i}\nu$ corresponding to the central left solution $\boldsymbol{\xi}$:

$$\begin{aligned} \omega_\alpha &\equiv \omega + \lambda\,, \qquad \boldsymbol{\xi}_\alpha \equiv \boldsymbol{\xi} + \lambda\boldsymbol{\eta}\,, \quad \Pi_\alpha \equiv \Pi(\boldsymbol{\xi}) + \lambda\Pi(\boldsymbol{\eta})\,, \\ \omega_\beta &\equiv \omega - \lambda\,, \qquad \boldsymbol{\xi}_\beta \equiv \boldsymbol{\xi} - \lambda\boldsymbol{\eta}\,, \quad \Pi_\beta \equiv \Pi(\boldsymbol{\xi}) - \lambda\Pi(\boldsymbol{\eta})\,. \end{aligned} \tag{12.131}$$

Here, $|\lambda| \ll |\omega|$, where the frequencies are made dimensionless in the usual way, and the perturbation $\lambda\boldsymbol{\eta}$ is small in norm compared to $\boldsymbol{\xi}$. From the homogeneous Frieman–Rotenberg equations (12.114) for $\boldsymbol{\xi}_\alpha$ and $\boldsymbol{\xi}_\beta$, one easily shows that $\boldsymbol{\eta}$ has to satisfy an inhomogeneous Frieman–Rotenberg equation,

$$\mathbf{G}(\boldsymbol{\eta}) - 2\rho\omega U\boldsymbol{\eta} + \rho\omega^2\boldsymbol{\eta} = 2\rho(U\boldsymbol{\xi} - \omega\boldsymbol{\xi})\,, \tag{12.132}$$

so that $\boldsymbol{\eta}$ is of the same order as, and determined by, the function $\boldsymbol{\xi}$.

From the mentioned equations for $\mathbf{G}(\boldsymbol{\xi})$ and $\mathbf{G}(\boldsymbol{\eta}^*)$, two quadratic forms may be constructed involving two surface integrals which are perturbations of the complementary energy. They result from the pre-self-adjointness relation (12.81), which holds for *any* two left vector fields $\boldsymbol{\xi}$ and $\boldsymbol{\eta}^*$:

$$\tfrac{1}{2}\int \eta^*\,\Pi(\boldsymbol{\xi})\,dS_{\mathrm{right}} \;\equiv\; -\,Q(\boldsymbol{\xi},\boldsymbol{\eta}^*) - \tfrac{1}{2}\int \boldsymbol{\eta}^*\cdot\mathbf{G}(\boldsymbol{\xi})\,dV\,, \tag{12.133}$$

$$\tfrac{1}{2}\int \xi\,\Pi(\boldsymbol{\eta}^*)\,dS_{\mathrm{right}} \;\equiv\; -\,Q(\boldsymbol{\xi},\boldsymbol{\eta}^*) - \tfrac{1}{2}\int \boldsymbol{\xi}\cdot\mathbf{G}(\boldsymbol{\eta}^*)\,dV\,. \tag{12.134}$$

The explicit form of the quadratic expression $Q(\boldsymbol{\xi},\boldsymbol{\eta}^*)$ is not needed here, but only its symmetry. [That expression may be obtained from Eq. (12.80); notice the relation $W^{\mathrm{p}}[\boldsymbol{\xi}] \equiv Q(\boldsymbol{\xi},\boldsymbol{\xi}^*)$.] Subtracting and adding the two equations, multiplied with λ^*, yields *two integrals that control the sequencing of the eigenvalues:*

$$\begin{aligned} \Gamma &\equiv \tfrac{1}{2}\lambda^*\int \big[\eta^*\,\Pi(\boldsymbol{\xi}) - \xi\,\Pi(\boldsymbol{\eta}^*)\big]\,dS_{\mathrm{right}} \\ &= -\,\tfrac{1}{2}\lambda^*\int \big[\boldsymbol{\eta}^*\cdot\mathbf{G}(\boldsymbol{\xi}) - \boldsymbol{\xi}\cdot\mathbf{G}(\boldsymbol{\eta}^*)\big]\,dV = 2\lambda^*\big[\mathrm{i}\nu X + (\overline{V} - \sigma)I\big], \end{aligned} \tag{12.135}$$

$$\begin{aligned} \Delta &\equiv \tfrac{1}{2}\lambda^*\int \big[\eta^*\,\Pi(\boldsymbol{\xi}) + \xi\,\Pi(\boldsymbol{\eta}^*)\big]\,dS_{\mathrm{right}} \\ &= -\,\tfrac{1}{2}\lambda^*\int \big[\boldsymbol{\eta}^*\cdot\mathbf{G}(\boldsymbol{\xi}) + \boldsymbol{\xi}\cdot\mathbf{G}(\boldsymbol{\eta}^*)\big]\,dV - 2\lambda^* Q(\boldsymbol{\xi},\boldsymbol{\eta}^*) = -2\lambda^* Y, \end{aligned} \tag{12.136}$$

where two auxiliary integrals of order unity have been introduced:

$$X \equiv I - \int \rho\boldsymbol{\eta}^*\cdot(U\boldsymbol{\xi} - \sigma\boldsymbol{\xi})\,dV\,, \tag{12.137}$$

$$Y \equiv (\overline{V} + \mathrm{i}\nu)I - \sigma X + \tfrac{1}{2}|\omega|^2\int \rho\boldsymbol{\eta}^*\cdot\boldsymbol{\xi}\,dV + Q(\boldsymbol{\xi},\boldsymbol{\eta}^*)\,. \tag{12.138}$$

Surprisingly little information on these integrals is needed for our purpose.

At this point, we need to distinguish between real and complex frequencies, and also split $\lambda \equiv \frac{1}{2}(\omega_\alpha - \omega_\beta) \equiv \delta + \mathrm{i}\epsilon$ to distinguish between horizontal (δ) and vertical (ϵ) components of arc length along the different paths. Also, we will restrict the analysis to one-dimensional systems, although this is not really necessary (see Section IID of Ref. [229] for generalization to toroidal systems).

(b) Real Oscillation Theorem For real frequencies, transforming back to the original vectors $\boldsymbol{\xi}_\alpha$ and $\boldsymbol{\xi}_\beta$, we obtain from Eq. (12.135) the following expression for the sequencing integral along the solution path (the real axis):

$$\Gamma \equiv \tfrac{1}{4} \int (\xi_\alpha \Pi_\beta - \xi_\beta \Pi_\alpha)\, dS_{\rm right} = 2\delta(\overline{V} - \sigma) I \approx (\sigma_\alpha - \sigma_\beta)[\overline{V}(\sigma_\alpha) - \sigma_\alpha] I\,. \qquad (12.139)$$

Consequently, along the real axis, but outside the Doppler–Coriolis indefinite range (12.125), where $\sigma = \{\overline{V}(x)\}$, and outside the continua and apparent singularity ranges (so that Γ and $\overline{V}$ are well defined), *the sequencing integral* Γ *is a (positive or negative) definite function of the frequency* σ. This is as far as we can get without restricting the equilibrium configuration.

For one-dimensional equilibria (plane slab or cylinder), the expression for Γ for a finite surface $S = \int dS_{\rm right}$ may be further reduced by exploiting the familiar relation $\Pi = -(N/D)\xi' - (C/D)\xi$, as in Sections 7.3.2 and 9.2.1. The modified expressions of the coefficients N, D, and C for stationary equilibria will be derived in Chapter 13. Here, we only need to know that the zeros of the quartics N and D represent the Doppler-shifted continuous spectra and apparent singularities, respectively. Equation (12.139) for a left solution ξ_α that also vanishes at the right end point then yields the generalization of Eq. (7.156) for stationary equilibria:

$$\begin{aligned} \Gamma &\equiv \tfrac{1}{4} S(\xi_\alpha \Pi_\beta - \xi_\beta \Pi_\alpha) && \textit{(real axis)} \\ &= \tfrac{1}{4} S\Big[\xi_\beta \frac{N(\sigma_\alpha)}{D(\sigma_\alpha)} \xi'_\alpha\Big]_{\rm right} \approx (\sigma_\beta - \sigma_\alpha)\big[\sigma_\alpha - \overline{V}(\sigma_\alpha)\big] I\,. && (12.140) \end{aligned}$$

Applying the Sturm–Liouville kind of reasoning to the left solution ξ_β gives the required result: Outside the Doppler–Coriolis indefinite range $\{\overline{V}(x) = \sigma\}$, the continua $\{N(x) = 0\}$ and the apparent singularities $\{D(x) = 0\}$, the expression

$$\frac{N(\sigma)}{[\sigma - \overline{V}(\sigma)]\, D(\sigma)} \begin{cases} > 1 & \textit{(Sturmian behaviour)} \\ < 1 & \textit{(anti-Sturmian behaviour)} \end{cases} \qquad (12.141)$$

determines whether ξ_β oscillates faster, or slower, than ξ_α when $\sigma_\beta > \sigma_\alpha$; QED. This kind of reasoning is essentially restricted to real frequencies.

A much more general approach, also applicable to complex frequencies, is obtained by exploiting the ratio of the two basic functions ξ and Π at the boundary. For reasons that will become clear, we call this expression *the alternator:*

$$R \equiv (\xi/\Pi)_{\rm right}\,. \qquad (12.142)$$

The eigenvalues are determined by $R = 0$, whereas for the spurious solutions $\Pi_{\rm right} = 0$ the alternator blows up and reverses sign to start a new stretch of monotonicity. Hence, the alternator is a powerful bookkeeper of the alternating zeros of the oscillating functions ξ and Π. We will prove that R is monotonic on the pieces of the solution path delimited by the zeros of $\Pi_{\rm right}$ so that the eigenvalues (where $\xi_{\rm right} = 0$) are guaranteed to be situated in between.

Substitution of the alternator into the sequencing integral (12.139) yields

$$\begin{aligned}\Gamma &\equiv \tfrac{1}{4}S(\xi_\alpha\Pi_\beta - \xi_\beta\Pi_\alpha) && \textit{(real axis)}\\ &\approx \tfrac{1}{4}S(R_\beta - R_\alpha)|\Pi_\alpha|^2 \approx (\sigma_\beta - \sigma_\alpha)\big[\sigma_\alpha - \overline{V}(\sigma_\alpha)\big]I\,. && (12.143)\end{aligned}$$

Hence, excluding again the mentioned frequency ranges (where R is not defined), we arrive at the final formulation of the Real Oscillation Theorem:

Along the real axis, outside the Doppler–Coriolis indefinite range, the continua and the apparent singularities, the alternator R is a monotonic function of the frequency σ in between the zeros of $\Pi_{\rm right}$; QED.

(c) Complex Oscillation Theorem For complex frequencies, also transforming back to the original vectors $\boldsymbol{\xi}_\alpha$ and $\boldsymbol{\xi}_\beta$, Eqs. (12.135) and (12.136) yield

$$\Gamma \equiv \tfrac{1}{4}\mathrm{Re}\int \big(\xi_\alpha^*\Pi_\beta - \xi_\beta^*\Pi_\alpha\big)\,dS_{\rm right} \quad \Big\{+\tfrac{1}{2}\mathrm{i}\,\mathrm{Im}\big(W_{{\rm com},\alpha} - W_{{\rm com},\beta}\big)\Big\}, \qquad (12.144)$$

$$\Delta \equiv \tfrac{1}{4}\mathrm{i}\,\mathrm{Im}\int \big(\xi_\alpha^*\Pi_\beta - \xi_\beta^*\Pi_\alpha\big)\,dS_{\rm right} \quad \Big\{+\tfrac{1}{2}\mathrm{Re}\big(W_{{\rm com},\alpha} - W_{{\rm com},\beta}\big)\Big\}, \qquad (12.145)$$

where the expressions in curly brackets vanish on the pertinent path in the ω-plane. Hence, the integral Γ is real along the solution path, so that

$$\Gamma = 2\nu\,\mathrm{sgn}(x)\,|\lambda|\,|X| = \nu\,\mathrm{sgn}(x)\,|X|\,|\omega_\alpha - \omega_\beta|\,, \quad x \equiv X_1/\epsilon = -X_2/\delta\,, \qquad (12.146)$$

and the integral Δ is purely imaginary along the conjugate path, so that

$$\mathrm{Im}\,\Delta = 2\,\mathrm{sgn}(y)\,|\lambda|\,|Y| = \mathrm{sgn}(y)\,|Y|\,|\omega_\alpha - \omega_\beta|\,, \quad y \equiv Y_1/\epsilon = -Y_2/\delta\,. \qquad (12.147)$$

In the first expression, $|X|$ is a leading order quantity (depending on σ and ν only) and the value of $\mathrm{sgn}(x)$ is constant (either $+1$ or -1) along the solution path. This is true as long as $X \neq 0$, so that X_1 changes sign at horizontal parts ($\epsilon = 0$) and X_2 changes sign at vertical parts ($\delta = 0$) of the solution path. However, if $X = 0$ somewhere on the solution path, $\mathrm{sgn}(x)$ and hence Γ would fail to be definite. Explicit examples in Chapter 13 will show that the solution path actually breaks up to avoid its occurrence. Consequently, excepting the case $X = 0$, *the sequencing integral Γ is a real and (positive or negative) definite function of arc length along the solution path*. An analogous reasoning holds for the implications of the expression for $\mathrm{Im}\,\Delta$ in terms of $|Y|$. Hence, excepting the case $Y = 0$, *the sequencing integral Δ is purely imaginary and its imaginary part is a (positive or negative) definite function of arc length along the conjugate path.*

For one-dimensional equilibria, the expression for Γ on the solution path may be reduced by means of the complex generalization of the definition (12.142) for the alternator, $R \equiv (\xi/\Pi)_{\rm right}$. However, the crucial consequence of the path expression (12.120) is that *the alternator is real for complex frequencies on the solution path:*

$$R = R_1 = \xi_1/\Pi_1 = \xi_2/\Pi_2 \qquad (\text{at } x = x_{\rm right})\,. \qquad (12.148)$$

Substitution into Eqs. (12.144) and (12.146) yields

$$\begin{aligned}\Gamma &= \tfrac{1}{4}S(R_\alpha - R_\beta)(\Pi_\alpha^*\Pi_\beta + \Pi_\alpha\Pi_\beta^*) && \textit{(solution path)}\\ &\approx \tfrac{1}{4}S(R_\alpha - R_\beta)|\Pi_\alpha|^2 = \nu_\alpha\,\mathrm{sgn}(x)\,|X|\,|\omega_\alpha - \omega_\beta|\,. && (12.149)\end{aligned}$$

As in the real case, the eigenvalues are determined by $R_{\rm right} = 0$, whereas for the spurious solutions $\Pi_{\rm right} = 0$ the alternator blows up and reverses sign to start a new stretch of monotonicity of the solution path.

Similarly, the expression for Δ on the conjugate path may also be reduced by means of the complex alternator R. The path expression (12.121) then implies that *the alternator is imaginary for complex frequencies on the conjugate path:*

$$R = \mathrm{i}\, R_2 = \mathrm{i}\, \xi_2/\Pi_1 = -\mathrm{i}\, \xi_1/\Pi_2 \qquad (\text{at } x = x_{\rm right})\,. \tag{12.150}$$

Substitution into Eqs. (12.145) and (12.147) then yields

$$\begin{aligned} \mathrm{Im}\,\Delta &= -\tfrac{1}{4} S(R_\alpha - R_\beta)(\Pi_\alpha^* \Pi_\beta + \Pi_\alpha \Pi_\beta^*) \qquad \textit{(conjugate path)} \\ &\approx -\tfrac{1}{4} S(R_{\alpha,2} - R_{\beta,2})|\Pi_\alpha|^2 = \mathrm{sgn}(y)\,|Y|\,|\omega_\alpha - \omega_\beta|\,. \end{aligned} \tag{12.151}$$

The eigenvalues are again determined by $R_{\rm right} = 0$, whereas for the spurious solutions $\Pi_{\rm right} = 0$ the alternator blows up and reverses sign to start a new stretch of monotonicity of the conjugate path. Consequently, we arrive at the formulation of Complex Oscillation Theorem:

Along the solution path, in between the zeros of Π*, the alternator* R *is a real and monotonic function of arc length; along the conjugate path, in between the zeros of* Π*, the alternator* R *is purely imaginary and its imaginary part* R_2 *is a monotonic function of arc length;* QED.

Thus, the eigenvalues on either one of the paths are consistently connected by counting the number of consecutive branches of the alternator. On the real sections of the solution path, this number coincides with the number of nodes of the eigenfunction, so that a genuine generalization of the Sturm–Liouville counting procedure has been obtained.

In conclusion Since $\Gamma(\boldsymbol{\xi}_\alpha, \boldsymbol{\xi}_\beta)$ and $\mathrm{Im}\Delta(\boldsymbol{\xi}_\alpha, \boldsymbol{\xi}_\beta)$ are (positive or negative) definite functions on their respective paths, the progression from ω_α to ω_β, from ω_β to ω_γ, etc., is essentially a monotonic one for one-dimensional equilibria. This is one of the powerful consequences of the self-adjointness of the operators $\mathbf{G}$ and U, which guarantees fast convergence to the eigenvalues for iterative numerical procedures, in particular along the solution path. However, a genuine complication is due to the nonlinearity of the eigenvalue problem of stationary equilibria: the solution path and the conjugate path are usually not single curves but fragmented into many branches (hence the terminology 'Spectral Web'), even containing closed loops. This appears to be how occurrence of $X = 0$ or $Y = 0$, which would imply non-monotonicity, is avoided. An important (and very convenient!) consequence of the monotonicity properties is that closed loops of the solution or of the conjugate path are guaranteed to contain at least one eigenvalue. Chapter 13 will give many illustrations of these fundamental properties.

We are now fully prepared to complete this study with the explicit determination of the spectra of stationary gravitating plane plasma slabs (Sections 13.1 and 13.2) and cylindrical plasmas (Sections 13.3–13.5). The method developed will be exploited in two ways: (a) determining the overall structure of the Spectral Web by blindly contour plotting the curves $\mathrm{Im}\, W_{\rm com} = 0$ and $\mathrm{Re}\, W_{\rm com} = 0$ in relevant parts of the complex ω-plane (a trivially parallel procedure); (b) exploiting the Oscillation Theorems to accurately determine the eigenvalues and the associated eigenfunctions by iterating along pieces of the solution or conjugate path.

12.4 Literature and exercises

Notes on literature

Stability of stationary equilibria

- Frieman and Rotenberg, 'On the hydromagnetic stability of stationary equilibria' [197], outline the stationary counterpart to the widely used MHD stability theory of static plasmas [275, 55], extensively discussed in Chapter 6, when flows are admitted. Surprisingly (certainly for plasma-astrophysical applications where incorporation of flow is imperative), this more general theory remains underdeveloped.
- Chandrasekhar, *Hydrodynamic and Hydromagnetic Stability* [112], is one of the first systematic presentations of the subject with numerous carefully worked out explicit examples: it is a real pleasure to see the master at work!

Spectral Web

- Goedbloed, 'Spectral Web of stationary plasma equilibria: I. General theory', and the sequel 'II. Internal modes' [229], develops the new tool of the Spectral Web to compute the complex spectrum of MHD waves and instabilities of moving plasmas discussed in this chapter and the next. Exploiting only self-adjoint operators, it is an attempt to mend the above mentioned underdevelopment of MHD spectral theory and applications for stationary equilibria.

Exercises

[12.1] *Tokamak physics versus astrophysics*

Laboratory and astrophysical plasmas appear to be very different forms of matter. In this book, the common features are stressed.

- Why and when can a plasma in a fusion device, like a tokamak, be described with the same theoretical model as a plasma in the Universe?
- Using the same theory, there are important differences between the two cases though, in particular with respect to plasma equilibrium. What are those differences?
- Also with respect to the waves and instabilities, what are the important differences?

[12.2] *Displacement field*

Newcomb has derived an expression which relates the gradient operator at the perturbed position to that at the unperturbed position. Explain why this expression needs special attention in nonlinear MHD, in particular in numerical nonlinear MHD.

[12.3] *Incompressible plasma*

In the exercises of the following chapters, the incompressible limit will be exploited frequently. Here, you will investigate what kind of implications this limit has. First, obtain the equations for an incompressible plasma from the ones for a compressible plasma by taking the limit $\gamma \to \infty$.

- Show that $\nabla \cdot \mathbf{v} = 0$ in the incompressible limit.
- Also show that $\nabla^0 \cdot \mathbf{v}^0 = 0$.
- Show that the displacement field satisfies $\nabla^0 \cdot \boldsymbol{\xi} = 0$.
- What does this limit imply for the generalized force operator $\mathbf{G}(\boldsymbol{\xi})$? What are the advantages of taking the incompressible limit?

[12.4] *Flow and self-adjointness*

Show that the operator $U \equiv -i\mathbf{v} \cdot \nabla$ is self-adjoint. Motivate for every step what kind of argument you have used. Start from the inner product $\langle \boldsymbol{\eta}, U\boldsymbol{\xi} \rangle$.

- Indicate the differences in self-adjointness between the operators U and $\mathbf{G}$. Which one is the stronger? (Hint: consider the BCs to be satisfied by $\boldsymbol{\xi}$ in both cases.)

[12.5] *Determination of the solution path*

In this exercise you will make a summary and flow diagram of how to compute the solution paths for a one-dimensional system. This system can be described either in Cartesian or in cylindrical coordinates.

- Suppose you have selected a growth rate ν_u. It looks like there are still infinitely many possibilities to choose a value for the real part σ_u of the frequency. This is not so, σ_u should be selected from a certain range. Determine that range.
- Once you have selected a ν_u and a σ_u, you need to solve a certain differential equation. Which differential equation is that and what are the boundary conditions?
- Which condition has to be satisfied for the selected ν_u and σ_u to be part of the solution path and, therefore, to be an appropriate guess for the eigenvalue ω. Why are the ν_u and σ_u selected just a guess and not already the eigenvalue ω?
- Make a flow diagram of the computation of the solution path.

13
Shear flow and rotation

The theory of the Spectral Web [229], described in Chapter 12, is applied to construct the basic structure of the spectrum of waves and instabilities for representative stationary 1D equilibria, viz. plane slabs in Sections 13.1 and 13.2 and cylindrical plasmas in Sections 13.3–13.5. These two classes of problems are the first-principle ones in any explicit analysis of the waves and instabilities of fluids and plasmas, as demonstrated for static plasmas in Chapters 7 and 9. With respect to MHD spectral theory, study of these 1D problems has the same function as study of the spectrum of the hydrogen atom in quantum mechanics. It reveals the basic complexity of the system, to be understood before the more complicated multi-dimensional systems can be studied fruitfully. In spectral theory of stationary plasmas, this is even more urgent than in quantum mechanics since the basic spectral equation is now a quadratic one governed by two operators, instead of just one.

13.1 Spectral theory of plane plasmas with shear flow

13.1.1 Gravito-MHD wave equation for plane plasma flow

Consider the model of a plane magnetized and gravitating plasma slab, introduced in Section 7.3.2 but now extended with a *plane shear flow* field (Fig. 13.1(a)):

$$\begin{aligned} &\mathbf{B} = B_y(x)\,\mathbf{e}_y + B_z(x)\,\mathbf{e}_z\,, \quad \rho = \rho(x)\,, \quad p = p(x)\,, \\ &\mathbf{v} = v_y(x)\mathbf{e}_y + v_z(x)\mathbf{e}_z\,, \quad \mathbf{g} = -g\,\mathbf{e}_x\,. \end{aligned} \tag{13.1}$$

The functions $B_y(x)$, $B_z(x)$, $\rho(x)$ and $p(x)$ should satisfy the equilibrium equation $(p + \frac{1}{2}B^2)' = -\rho g$, but $v_y(x)$ and $v_z(x)$ are completely arbitrary. The slab is confined between two solid boundaries at $x = x_1$ and $x = x_2$.

For applications, it is quite important that the magnetic field has *magnetic shear*, so that one can assume that the magnitude of $\mathbf{B}(x)$ is fixed but its direction should change with the vertical position x. On the other hand, *velocity shear* effects can be obtained by just assuming variation of the amplitude of a uni-directional flow field $\mathbf{v}(x)$. Without loss of generality, one can also choose the direction of the horizontal wave vector $\mathbf{k}_0$ to be along the z axis, making an angle θ with $\mathbf{v}$ and an angle $\varphi(x)$ with $\mathbf{B}$ (see Fig. 13.1(b)). For the time being, we will not make these simplifying assumptions and just derive completely general expressions for the spectral equations of these equilibria.

In the reduction of the spectral equation for these equilibria, one of the two velocity-dependent terms in the expression (12.64) for the generalized force operator $\mathbf{G}$ vanishes since there are no

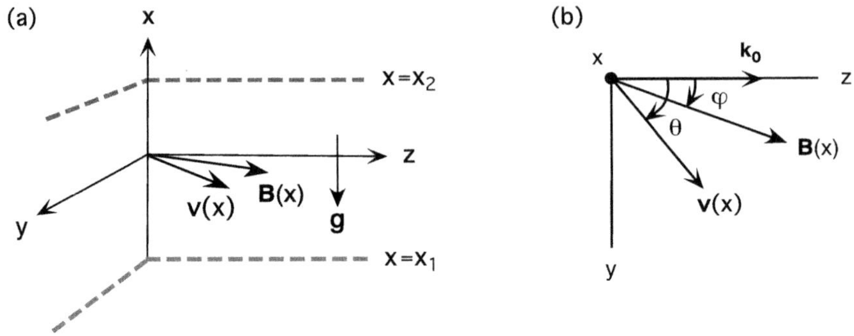

Fig. 13.1 (a) Gravitating magnetized slab with impenetrable walls at x_1 and x_2; (b) Directions of the horizontal wave vector, the magnetic field and the velocity.

centrifugal forces present, so that $\mathbf{v}\cdot\nabla\mathbf{v} = 0$, and the other one yields $-\rho(\mathbf{v}\cdot\nabla)^2\boldsymbol{\xi}$, so that the eigenvalue problem (12.68), or rather (12.69), 'nearly' simplifies to the old static one, with the same operator $\mathbf{F}$ but ω replaced by the Doppler shifted expression:

$$\mathbf{F}(\boldsymbol{\xi}) = -\rho(\omega - U)^2\,\boldsymbol{\xi} \equiv -\rho(\omega + \mathrm{i}\mathbf{v}\cdot\nabla)^2\,\boldsymbol{\xi}\,. \tag{13.2}$$

From the discussion in the previous chapter, it is evident that all the new physics associated with stationary plasma flows is hidden in this replacement.

As in Section 7.3.2, we assume normal modes with plane wave dependence in the ignorable (y and z) directions,

$$\boldsymbol{\xi}(x,y,z,t) = \boldsymbol{\xi}(x;k_y,k_z)\,\mathrm{e}^{\mathrm{i}(k_y y + k_z z - \omega t)}\,, \tag{13.3}$$

where we recall that a distinctive hat for the Fourier amplitude was dropped for simplicity of the notation. For these Fourier normal modes (indicated by the label 'Fnm' above the equal sign), the operator U in Eq. (13.2) becomes a multiplication:

$$\omega + \mathrm{i}\mathbf{v}\cdot\nabla \overset{\text{Fnm}}{=} \widetilde{\omega}(x) \equiv \omega - \Omega_0(x)\,, \qquad \Omega_0 \equiv \mathbf{k}_0\cdot\mathbf{v}(x)\,, \tag{13.4}$$

with $\mathbf{k}_0 \equiv k_y\mathbf{e}_y + k_z\mathbf{e}_z$ the horizontal wave vector, $\Omega_0(x)$ the *local Doppler shift* and $\widetilde{\omega}(x)$ the *local Doppler shifted frequency* observed in a frame *co-moving* with the plasma layer at the position x. Hence, the eigenvalue problem becomes

$$\mathbf{F}(\boldsymbol{\xi}) = -\rho\widetilde{\omega}^2\,\boldsymbol{\xi}\,, \tag{13.5}$$

subject to model I BCs on $\xi \equiv \xi_x$ at the walls:

$$\xi(x_1) = 0 \quad (\text{left})\,, \qquad \xi(x_2) = 0 \quad (\text{right})\,. \tag{13.6}$$

Since the equilibrium condition (12.28) is the same as for static plasmas, this implies that the equations derived in Section 7.3.2 for the static plane slab remain valid for the stationary slab if just the replacement $\omega \to \widetilde{\omega}(x)$ is made. Of course, the x-dependence of the local Doppler shift $\Omega_0(x)$ is the real complicating factor. It involves the possibility of new flow-driven (Kelvin–Helmholtz) instabilities.

Since $\widetilde{\omega}$ depends on x through $\Omega_0(x)$, every discrete eigenvalue will be subject to a different

Doppler shift involving the solution $\boldsymbol{\xi}$ of this eigenvalue problem across the layer. This is *the solution-averaged Doppler–Coriolis shift* $\overline{V}$ defined in Eq. (12.107), which simplifies to

$$\overline{V} \equiv \langle \Omega_0(x) \rangle = \int \rho\, \mathbf{k}_0 \cdot \mathbf{v}\, |\boldsymbol{\xi}|^2\, dx \Big/ \int \rho |\boldsymbol{\xi}|^2\, dx\,, \tag{13.7}$$

since there is no Coriolis contribution for plane flows. Inserting the left solution $\boldsymbol{\xi}^\ell$ of the one-sided BVP (13.5), (13.6)(a) into the expression (12.124) yields

$$\sigma = \overline{V}\big[\boldsymbol{\xi}^\ell(x; \sigma + \mathrm{i}\nu)\big]\,, \tag{13.8}$$

which yields *the left solution path of unstable modes*. Recall though that exploiting the complementary energy expression (12.120) is the better way of computing it.

When an equilibrium parameter is varied, two stable modes may approach each other along the real σ-axis, coalesce, and then split into a complex conjugate pair of modes, one of which is unstable. (Note that the path along which the modes move when an equilibrium parameter is varied need not be the solution path.) In the static case, the transition to instability is always through the marginal point $\omega = 0$. In the stationary case, transitions occur at solutions of the marginal equation

$$\sigma_0 = \overline{V}[\boldsymbol{\xi}(x; \sigma_0)]\,, \tag{13.9}$$

where, in contrast to Eq. (13.8), $\boldsymbol{\xi}$ should be a solution of the *complete* BVP (13.5), (13.6). Those solutions are usually away from the origin so that *overstable* modes appear. Also. the marginal equation may allow for multiple solutions so that the stationary stability problem is much more complicated than the static one.

The spectral equation (13.5) is reduced by exploiting the *field line projection,* as introduced in Section 7.3.2 in the derivation of the gravito-MHD spectral equation for the analogous case of static equilibria. We exploit the *spatially dependent* wave numbers $k_\parallel(x)$ and $k_\perp(x)$:

$$\begin{aligned} \mathbf{e}_x &\equiv \nabla x\,, & \mathbf{e}_\perp &\equiv (\mathbf{B}/B) \times \mathbf{e}_x\,, & \mathbf{e}_\parallel &\equiv (\mathbf{B}/B)\,, \\ \partial_x &\equiv \mathbf{e}_x \cdot \nabla\,, & k_\perp &\equiv -\mathrm{i}\mathbf{e}_\perp \cdot \nabla\,, & k_\parallel &\equiv -\mathrm{i}\mathbf{e}_\parallel \cdot \nabla\,, \\ \xi &\equiv \xi_x\,, & \eta &\equiv \mathrm{i}\mathbf{e}_\perp \cdot \boldsymbol{\xi}\,, & \zeta &\equiv \mathrm{i}\mathbf{e}_\parallel \cdot \boldsymbol{\xi}\,. \end{aligned} \tag{13.10}$$

In contrast to the static case, *the variables ξ, η and ζ are no longer purely real* now since ω is genuinely complex (with both $\sigma \neq 0$ and $\nu \neq 0$) for instabilities. The resulting vectorial form of the *gravito-MHD wave equation for plane plasma flow* [620] is just Eq. (7.78) of Section 7.3.2, and eliminating η and ζ yields the scalar ordinary differential equation (ODE) form (7.80) of the gravito-MHD wave equation with the replacement $\omega \to \widetilde{\omega}(x)$:

$$\frac{d}{dx}\left(\frac{N}{D}\frac{d\xi}{dx}\right) + \left[A + \frac{B}{D} + \left(\frac{C}{D}\right)'\right]\xi = 0\,. \tag{13.11}$$

This involves three regular coefficients with different powers of the acceleration of gravity (including the driving term $\rho' g$ of the Rayleigh–Taylor instability),

$$\begin{aligned} A(x;\widetilde{\omega}^2) &\equiv \rho(\widetilde{\omega}^2 - \omega_{\mathrm{A}}^2) + \rho' g\,, \qquad B(x;\widetilde{\omega}^2) \equiv -k_0^2 \rho g^2\, (\widetilde{\omega}^2 - \omega_{\mathrm{A}}^2)\,, \\ C(x;\widetilde{\omega}^2) &\equiv -\,\rho g\, \widetilde{\omega}^2(\widetilde{\omega}^2 - \omega_{\mathrm{A}}^2)\,, \end{aligned} \tag{13.12}$$

the numerator N determining the (Alfvén and slow) genuine singularities,

$$\begin{aligned} N(x;\widetilde{\omega}^2) &\equiv (\gamma p + B^2)(\widetilde{\omega}^2 - \omega_{\rm A}^2)(\widetilde{\omega}^2 - \omega_{\rm S}^2)\,, \\ \omega_{\rm A}^2(x) &\equiv k_\parallel^2 B^2/\rho\,, \qquad \omega_{\rm S}^2(x) \equiv [\gamma p/(\gamma p + B^2)]\omega_{\rm A}^2\,, \end{aligned} \tag{13.13}$$

and the denominator D determining the (fast and slow) apparent singularities,

$$\begin{aligned} D(x;\widetilde{\omega}^2) &\equiv (\widetilde{\omega}^2 - \omega_{\rm f0}^2)(\widetilde{\omega}^2 - \omega_{\rm s0}^2) = \widetilde{\omega}^4 - k_0^2(\gamma p + B^2)(\widetilde{\omega}^2 - \omega_{\rm S}^2)/\rho\,, \\ \omega_{\rm s0,f0}^2(x) &\equiv \tfrac{1}{2}\omega_{\rm m}^2\Big[1 \pm \sqrt{1 - 4\omega_{\rm S}^2/\omega_{\rm m}^2}\Big]\,, \quad \omega_{\rm m}^2 \equiv k_0^2(\gamma p + B^2)/\rho\,. \end{aligned} \tag{13.14}$$

The ODE (13.11), with the BCs (13.6), uniquely determines the eigenvalues and eigenfunctions. The tangential components then follow by algebraic relations:

$$\begin{aligned} \eta &= k_\perp\big[(\gamma p + B^2)(\widetilde{\omega}^2 - \omega_{\rm S}^2)\,\xi' - \rho g\,\widetilde{\omega}^2\,\xi\big]/(\rho D)\,, \\ \zeta &= k_\parallel\big[\gamma p(\widetilde{\omega}^2 - \omega_{\rm A}^2)\,\xi' - g\,(\rho\widetilde{\omega}^2 - k_0^2 B^2)\,\xi\big]/(\rho D)\,. \end{aligned} \tag{13.15}$$

The only difference with the equations for the static case is the replacement of the constant ω^2 by the function $\widetilde{\omega}^2(x;\omega)$, but with far reaching consequences!

For numerical integration, exploiting iterative solution of the two-sided BVP by means of a succession of one-sided (left or right) BVPs, the second order ODE is suitably converted into an equivalent pair of first order ODEs for ξ and the Eulerian total pressure perturbation Π, defined in Eq. (12.79),

$$\Pi = -(N\xi' + C\xi)/D = -(\gamma p + B^2)(\xi' + k_\perp\eta) - k_\parallel\gamma p\,\zeta + \rho g\xi\,. \tag{13.16}$$

This yields

$$N\frac{d}{dx}\begin{pmatrix} \xi \\ \Pi \end{pmatrix} + \begin{pmatrix} C & D \\ E & -C \end{pmatrix}\begin{pmatrix} \xi \\ \Pi \end{pmatrix} = 0\,, \tag{13.17}$$

involving a new function

$$E(x;\widetilde{\omega}^2) \equiv -N\big(A + B/D\big) - C^2/D = -\big[\rho(\widetilde{\omega}^2 - \omega_{\rm A}^2) + \rho' g\big]N - \rho^2 g^2(\widetilde{\omega}^2 - \omega_{\rm A}^2)^2 \tag{13.18}$$

connected with the functions of the second order formulation as indicated.

▷ **Apparent and spurious singularities** Recall from the analysis of the analogous problems for static plasma slabs (Sections 7.3.2 and 7.4.1) and cylinders (Section 9.2.1) that the formulation in terms of the second order ODE contains the apparent singularities $D = 0$, whereas the formulation in terms of first ODEs contains spurious singularities $N = 0$. The apparent singularities gave rise to considerable confusion, as they were originally thought to be associated with additional continua by Grad [254], which was shown by Appert, Gruber and Vaclavik [18] to be incorrect by means of the formulation in terms of the first order system where these singularities simply did not turn up. The absence of $D = 0$ continua was further substantiated by Goedbloed [221] from the solution of the initial value problem (Section 10.2). In the second order formulation, the condition for the singularities $D = 0$ to be apparent, found by Greene [258], is

$$NB + C^2 \sim D\,. \tag{13.19}$$

On the other hand, the first order formulation also has a defect, viz. the occurrence of a set of spurious singularities $N = 0$ on top of the genuine ones (since N multiplies both ξ' and Π' in Eq. (13.17): one factor too much). It was pointed out by Bondeson, Iacono and Bhattacharjee [79], for the analogous cylindrical problem with flow (see Section 13.3.1), that it is essential for the first order formulation that the determinant of the matrix in Eq. (13.17) is proportional to N,

$$DE + C^2 \sim N\,. \tag{13.20}$$

This guarantees that the spurious $N = 0$ singularities can be eliminated (leaving the genuine ones). It now emerges that *the apparent and spurious singularities are complementary evils*: the condition (13.19) for the absence of $D = 0$ singularities and the condition (13.20) for the absence of spurious $N = 0$ singularities are just rearrangements of the same terms of the relation (13.18) connecting the two formulations. ◁

So far, we have stressed the formal similarities of the static and the stationary problems. We now need to concentrate on the differences. It is evident that the genuine ($N = 0$) and apparent ($D = 0$) singularities of the eigenvalue equation (13.11) have to be considered in detail in order to understand their role in the spectrum of stationary plasmas. Factoring these polynomials, e.g. the Alfvén component

$$\widetilde{\omega}^2 - \omega_A^2 \equiv (\omega - \Omega_0)^2 - \omega_A^2 = (\omega - \Omega_A^+)(\omega - \Omega_A^-)\,, \quad \text{where } \Omega_0 \equiv \mathbf{k}_0 \cdot \mathbf{v}\,,$$

and also the slow one, yields the *singular frequencies for equilibria with flow:*

$$\begin{gathered} N(x;\omega) = (\gamma p + B^2)(\omega - \Omega_A^+)(\omega - \Omega_A^-)(\omega - \Omega_S^+)(\omega - \Omega_S^-)\,, \\ \Omega_A^\pm(x) \equiv \Omega_0(x) \pm \omega_A(x)\,, \qquad \Omega_S^\pm(x) \equiv \Omega_0(x) \pm \omega_S(x)\,, \end{gathered} \tag{13.21}$$

and the new *apparent singularities with flow:*

$$\begin{gathered} D(x;\omega) = (\omega - \Omega_{f0}^+)(\omega - \Omega_{f0}^-)(\omega - \Omega_{s0}^+)(\omega - \Omega_{s0}^-)\,, \\ \Omega_{f0}^\pm(x) \equiv \Omega_0(x) \pm \omega_{f0}(x)\,, \qquad \Omega_{s0}^\pm(x) \equiv \Omega_0(x) \pm \omega_{s0}(x)\,. \end{gathered} \tag{13.22}$$

Here, the frequencies $\Omega_A^\pm$ and $\Omega_S^\pm$ indicate the *forward*$^{(+)}$ or *backward*$^{(-)}$ local Doppler-shifted Alfvén and slow frequencies in the *laboratory frame*. (Hence, the shift is $+\Omega_0$ here, whereas it is $-\Omega_0$ in Eq. (13.4) since that equation refers to the co-moving frame.) Since the analysis of the continuous spectra of static plasmas can be applied in completely the same way to stationary plasmas, the only change is that the static Alfvén continuum $\{\omega_A^2(x)\}$ is split into forward and backward Alfvén continua $\{\Omega_A^+(x)\}$ and $\{\Omega_A^-(x)\}$, and the static slow continuum $\{\omega_S^2(x)\}$ is split into forward and backward slow continua $\{\Omega_S^+(x)\}$ and $\{\Omega_S^-(x)\}$. Similarly, the collection of apparent singularities $\{\omega_{f0,s0}^2(x)\}$ is split into the sets of forward and backward turning point frequencies $\{\Omega_{f0,s0}^+(x)\}$ and $\{\Omega_{f0,s0}^-(x)\}$.

As in the static case, to complete the essential spectrum (consisting of continua *and* cluster points), one also needs to include the cluster points of the forward and backward fast magnetosonic waves:

$$\Omega_F^\pm \equiv \pm\infty\,. \tag{13.23}$$

One easily proves from the definitions that the following ordering of the local frequencies (which are all real!) holds at each point $x_1 \le x \le x_2$ of the plasma:

$$\Omega_F^- \le \Omega_{f0}^- \le \Omega_A^- \le \Omega_{s0}^- \le \Omega_S^- \le \Omega_0 \le \Omega_S^+ \le \Omega_{s0}^+ \le \Omega_A^+ \le \Omega_{f0}^+ \le \Omega_F^+\,. \tag{13.24}$$

Computation of the collections of these frequencies over the interval $[x_1, x_2]$ should precede any computation of the spectrum since they determine its overall structure. One might object that, to determine global stability of e.g. gravitational instabilities, one is not really interested in these 'trouble makers'. However, even in that case, one needs to know where these frequencies are located in order to properly deal with the marginal stability transitions, i.e. the approach to the real ω-axis.

13.1.2 Kelvin–Helmholtz instabilities in interface plasmas

As a non-trivial first example, consider the following extension of the interface problem discussed at the end of Chapter 6, illustrated in Fig. 6.15. We now replace the bottom vacuum layer by a second plasma with different flow and magnetic field parameters. In the resulting configuration, two homogeneous plane plasma layers with embedded constant magnetic fields $\mathbf{B}$ and $\hat{\mathbf{B}}$ and constant velocity fields $\mathbf{v}$ and $\hat{\mathbf{v}}$ are superposed, with a surface current $\mathbf{j}^*$ and a surface vorticity $\boldsymbol{\omega}^*$ creating jumps in the magnitudes and directions of the magnetic fields and velocities at the interface. The jump of the velocity causes the plasma to be Kelvin–Helmholtz unstable. Since we also keep the vertical gravity field $\mathbf{g} = -g\mathbf{e}_x$, Rayleigh–Taylor instabilities will also be present. The change of direction of the magnetic field at the interface will cause stabilization by magnetic shear.

We summarize the assumptions on the various equilibrium quantities and recall that the equilibrium is essentially the same as for the static case.

– Upper layer ($0 < x \le a$):

$$\rho\,,\mathbf{v}\,,\mathbf{B}\ \text{const}\,,\qquad p' = -\rho g \ \Rightarrow\ p = p_0 - \rho g x \quad (\text{with } p_0 \ge \rho g a)\,. \tag{13.25}$$

– Lower layer ($-b \le x < 0$):

$$\hat\rho\,,\hat{\mathbf{v}}\,,\hat{\mathbf{B}}\ \text{const}\,,\qquad \hat p' = -\hat\rho g \ \Rightarrow\ \hat p = \hat p_0 - \hat\rho g x \quad (\text{with } \hat p_0 \ge 0)\,. \tag{13.26}$$

– Jumps at the interface ($x = 0$):

$$p_0 + \tfrac12 B_0^2 = \hat p_0 + \tfrac12 \hat B_0^2 \qquad \textit{(pressure balance)}\,, \tag{13.27}$$

$$\mathbf{j}^\star = \mathbf{n}\times [\![\mathbf{B}]\!] = \mathbf{e}_x \times (\mathbf{B} - \hat{\mathbf{B}}) \qquad \textit{(surface current)}\,, \tag{13.28}$$

$$\boldsymbol{\omega}^\star = \mathbf{n}\times [\![\mathbf{v}]\!] = \mathbf{e}_x \times (\mathbf{v} - \hat{\mathbf{v}}) \qquad \textit{(surface vorticity)}\,. \tag{13.29}$$

Recall from Section 4.5.2 that the pressure balance equation (13.27) is a genuine boundary condition whereas the latter two equations are just implications of the jumps permitted by this equilibrium.

We perform a normal mode analysis of this configuration, with Fourier harmonics in the ignorable directions, as in Eq. (13.3). Above, the example was called 'simple' since the plasmas are taken incompressible and homogeneous so that the differential equations become trivial. Inhomogeneities and associated singularities will be considered in the following sections. We derive the general expression for the incompressible form of the wave equation (13.11) by taking the limit $\gamma \to \infty$. This yields the *wave equation for incompressible plasmas:*

$$\frac{d}{dx}\Big[\rho(\widetilde\omega^2 - \omega_A^2)\,\frac{d\xi}{dx}\Big] - k_0^2\Big[\rho(\widetilde\omega^2 - \omega_A^2) + \rho' g\Big]\,\xi = 0\,, \tag{13.30}$$

where $\widetilde\omega \equiv \omega - \Omega_0 = \omega - \mathbf{k}_0\cdot\mathbf{v}$ is the Doppler shifted frequency and $\omega_A \equiv \mathbf{k}_0\cdot\mathbf{B}/\sqrt{\rho_0}$ is the Alfvén frequency. For completeness, we also present the incompressible limit of the expressions (13.15) and (13.16) for the tangential variables η and ζ and the total pressure perturbation Π:

$$\eta = -(k_\perp/k_0^2)\xi' , \qquad \zeta = -(k_\parallel/k_0^2)\xi' , \qquad \Pi = (\rho/k_0^2)(\widetilde{\omega}^2 - \omega_A^2)\xi' . \tag{13.31}$$

The incompressible counterpart of the system of first order ODEs (13.17) for ξ and Π can easily be constructed from these expressions.

In the present case, all equilibrium quantities are constant in the respective layers so that the ODEs simplify to equations with constant coefficients that can easily be solved. This yields the following solutions, for the variable ξ in the top layer:

$$\xi'' - k_0^2\xi = 0 , \quad \text{with BC } \xi(a) = 0 \qquad \Rightarrow \quad \xi = C\,\frac{\sinh[k_0(a-x)]}{\sinh(k_0 a)} , \tag{13.32}$$

and for the variable $\hat{\xi}$ in the bottom layer:

$$\hat{\xi}'' - k_0^2\hat{\xi} = 0 , \quad \text{with BC } \hat{\xi}(-b) = 0 \qquad \Rightarrow \quad \hat{\xi} = \hat{C}\,\frac{\sinh[k_0(x+b)]}{\sinh(k_0 b)} . \tag{13.33}$$

These eigenfunctions have the usual cusp-shaped form of *surface modes*.

Since the bottom layer is a plasma rather than a vacuum, the pertinent BCs connecting ξ and $\hat{\xi}$ at $x = 0$ are the model II* interface conditions (6.135) and (6.137):

– *First interface condition* (continuity of the normal velocity perturbation):

$$[\![\mathbf{n}\cdot\boldsymbol{\xi}]\!] = 0 \quad \Rightarrow \quad \xi(0) = \hat{\xi}(0)\ (\neq 0) \quad \Rightarrow \quad C = \hat{C} . \tag{13.34}$$

– *Second interface condition* (pressure balance perturbation),

$$[\![\,\Pi + \mathbf{n}\cdot\boldsymbol{\xi}\,\mathbf{n}\cdot\nabla(p + \tfrac{1}{2}B^2)\,]\!] = 0 , \quad \Pi \equiv -\gamma p \nabla\cdot\boldsymbol{\xi} - \xi\cdot\nabla p + \mathbf{B}\cdot\mathbf{Q} . \tag{13.35}$$

The latter equation needs to be reworked since $\gamma p\nabla\cdot\boldsymbol{\xi}$ is actually an undetermined quantity for incompressible plasmas ($\gamma \to \infty , \nabla\cdot\boldsymbol{\xi} \to 0$). We determine it by exploiting the incompressible expression (13.31) for Π. Inserting this expression into the second interface condition, exploiting the equilibrium conditions (13.25) and (13.26), and dividing by the first interface condition then yields

$$\left[\!\!\left[\, \frac{\rho}{k_0^2}(\widetilde{\omega}^2 - \omega_A^2)\frac{\xi'}{\xi} - \rho g \,\right]\!\!\right] = 0 . \tag{13.36}$$

Substituting the solutions (13.32) and (13.33) gives *the dispersion equation,*

$$-\rho\big[(\omega-\Omega_0)^2 - \omega_A^2\big]\coth(k_0 a) - k_0\rho g = \hat{\rho}\big[(\omega-\hat{\Omega}_0)^2 - \hat{\omega}_A^2\big]\coth(k_0 b) - k_0\hat{\rho} g , \tag{13.37}$$

which clearly demonstrates the different physical mechanisms at work.

In the limit of small wavelength perturbations, $\coth(k_0 a) \approx \coth(k_0 b) \approx 1$ (walls effectively at ∞ and $-\infty$), the solutions of this dispersion equation become:

$$\omega = \frac{\rho\Omega_0 + \hat{\rho}\hat{\Omega}_0}{\rho+\hat{\rho}} \pm \left[-\frac{\rho\hat{\rho}(\Omega_0 - \hat{\Omega}_0)^2}{(\rho+\hat{\rho})^2} + \frac{\rho\omega_A^2 + \hat{\rho}\hat{\omega}_A^2}{\rho+\hat{\rho}} - \frac{k_0(\rho-\hat{\rho})g}{\rho+\hat{\rho}} \right]^{1/2} . \tag{13.38}$$

These solutions represent either two waves with real frequencies or two modes with complex conjugate frequencies (an overstable and a damped wave), depending on the sign of the expression under the square root. If it is negative, an instability is obtained driven by a destabilizing *Kelvin–Helmholtz* contribution when $\Omega_0 \neq \hat{\Omega}_0$, a stabilizing *magnetic field line bending* contribution when $\omega_A^2 \neq 0$ and/or $\hat{\omega}_A^2 \neq 0$, and a destabilizing *Rayleigh–Taylor* contribution when $\rho > \hat{\rho}$ (or stabilizing when $\rho < \hat{\rho}$). Magnetic stabilization of both the Kelvin–Helmholtz and the Rayleigh–Taylor instability is obtained when

$$(\mathbf{k}_0 \cdot \mathbf{B})^2 + (\mathbf{k}_0 \cdot \hat{\mathbf{B}})^2 > \big[\rho\hat{\rho}/(\rho + \hat{\rho})\big]\,\big[\mathbf{k}_0 \cdot (\mathbf{v} - \hat{\mathbf{v}})\big]^2 + k_0(\rho - \hat{\rho})g\,. \tag{13.39}$$

Note that *magnetic shear* (different directions of $\mathbf{B}$ and $\hat{\mathbf{B}}$) is effective because it prevents vanishing of the magnetic terms for directions of the wave vector $\mathbf{k}_0 \perp \mathbf{B}$. The expression for arbitrary wavelengths is easily derived from Eq. (13.37) by noting that every ρ occurs together with the term $\coth(k_0 a)$ and every $\hat{\rho}$ together with $\coth(k_0 b)$, except for the Rayleigh–Taylor term $-k_0(\rho - \hat{\rho})g$. For long-wavelength perturbations, $\coth(k_0 a) \approx (k_0 a)^{-1} \gg 1$ and $\coth(k_0 b) \approx (k_0 b)^{-1} \gg 1$ (walls effectively close), this leads to genuine competition between the three terms (all $\sim k_0$), so that stability depends on the precise choice of all those parameters.

Summarizing, this sub-section demonstrates how equilibrium background flow may create new (Kelvin–Helmholtz) instabilities and affect existing (Rayleigh–Taylor) instabilities in magnetized plasmas. The eigenvalues of these instabilities are located on the (mixed) solution path $\sigma = \langle\Omega_0\rangle = (\rho\Omega_0 + \hat{\rho}\hat{\Omega}_0)/(\rho + \hat{\rho})$, which is a vertical straight line in this case. Because of the simplifying assumption of homogeneous plasma layers, the present example hardly reveals the complexity of the general problem for stationary diffuse plasmas though.

13.1.3 Continua and the real oscillation theorem

To appreciate the complexity of the spectral problem when background flow is involved, consider the hydrodynamic (HD) case of a plane incompressible and inviscid inhomogeneous fluid without gravity, but with a horizontal flow velocity,

$$\mathbf{v} = v_y(x)\mathbf{e}_y + v_z(x)\mathbf{e}_z\,. \tag{13.40}$$

The Lagrangian time derivative of any one of the Eulerian perturbations $\rho_1, \mathbf{v}_1, p_1$ occurring in the HD counterpart of the MHD equations (12.31)–(12.34) again involves the Doppler shifted frequency $\widetilde{\omega}(x)$, defined in Eq. (13.4):

$$\Big(\frac{Df}{Dt}\Big)_1 \equiv \Big(\frac{\partial f}{\partial t} + \mathbf{v}\cdot\nabla f\Big)_1 \overset{\rm Fnm}{=} -\mathrm{i}\,\widetilde{\omega} f_1 + f_0{}' v_{1x}\,, \qquad \widetilde{\omega} \equiv \omega - \Omega_0(x)\,. \tag{13.41}$$

In the HD case, the gradient operator parallel to the background velocity,

$$U \equiv -\mathrm{i}\mathbf{v}\cdot\nabla \overset{\rm Fnm}{=} \mathbf{k}_0 \cdot \mathbf{v} \equiv \Omega_0(x)\,, \tag{13.42}$$

gives rise to two continuous spectra, the *flow continua*

$$\omega \in \{\Omega_{\mathrm{H}i} \equiv \Omega_0(x) | x_1 \leq x \leq x_2\} \qquad (i = 1, 2)\,, \tag{13.43}$$

consisting of all local Doppler shifts, for which the Eulerian equations are singular. The multiplicity index i will be explained below.

Prior to the discovery of the flow continua by Case [108] in 1960 (see also Drazin and Reid [164], p. 149), the HD literature on the inviscid limit of the Navier–Stokes equations [412] was confused because the absence of discrete modes for the simplest flow profiles appeared to imply absence of stable oscillations. This paradox follows from the Eulerian representation in terms of the velocity perturbation $\mathbf{v}_1$, related to both the stream function χ, which is usually exploited for incompressible HD problems, and the fluid displacement $\boldsymbol{\xi}$, according to Eq. (12.40):

$$\mathbf{v}_1 = \mathbf{e}_z \times \nabla\chi = \Big(\mathbf{v}_0 \cdot \nabla + \frac{\partial}{\partial t}\Big)\boldsymbol{\xi} \overset{\rm Fnm}{=} -\mathrm{i}\widetilde{\omega}\boldsymbol{\xi}$$
$$\Rightarrow \quad v_{1x} = -\mathrm{i}k_0\chi = -\mathrm{i}\widetilde{\omega}\xi\,, \quad v_{1y} = \chi' = -\widetilde{\omega}\eta\,. \tag{13.44}$$

For constant density ρ, these variables should satisfy the ODEs

$$\frac{d^2\chi}{dx^2} - k_0^2\,\chi - (\widetilde{\omega}''/\widetilde{\omega})\chi = 0\,, \qquad \text{or} \qquad \frac{d}{dx}\Big(\widetilde{\omega}^2\,\frac{d\xi}{dx}\Big) - k_0^2\widetilde{\omega}^2\,\xi = 0\,, \tag{13.45}$$

and be subject to the BCs $\chi(x_1) = \chi(x_2) = 0$, respectively $\xi(x_1) = \xi(x_2) = 0$. Assuming $\widetilde{\omega} \neq 0$ and a linear velocity profile, $v_0 = a + bx$, so that $\widetilde{\omega}'' = 0$, these equations only possess solutions $\sim \exp(\pm k_0 x)$ which cannot satisfy the BCs. Clearly, the assumption $\widetilde{\omega} \neq 0$ should be dropped and one should consider singular modes and solve the initial value problem à la Landau [393] (see Section 2.3.3) in order to construct the stable response to an initial perturbation, which amounts to an integral representation over the continuous spectra (13.43). This is what Case [108] did, demonstrating that the initial value problem is well-posed for HD and that the continuum contributions damp out as t^{-1}.

For MHD, Hameiri [280] introduced the distinction between the Eulerian and Lagrangian descriptions, noting that the former admits solutions that are absent in the latter. As pointed out at the end of Section 12.2.1, these additional solutions correspond to the *non-holonomic Eulerian entropy continua*

$$\omega \in \{\Omega_{\rm E} \equiv \Omega_0(x)\}\,, \tag{13.46}$$

for which the entropy perturbations $S_{\rm E1}$ cannot be expressed in terms of $\boldsymbol{\xi}$. Unfortunately, since the frequencies of these continua coincide with those of the flow continua (13.43) (which, on the contrary, are expressible in terms of $\boldsymbol{\xi}$!), the misunderstanding could arise that these continua are the same as the flow continua and, hence, that the flow continua also exist in MHD [559, 299, 60]. This confusion was eliminated by Goedbloed et al. [231] by means of an initial value approach similar to that of Case, demonstrating that the plasma response due to the Alfvén and slow continua in MHD is completely analogous to the fluid response due to the flow continua $\Omega_{{\rm H}i}$ in HD. That there is no place in MHD for continua in addition to the *Alfvén and slow MHD continua* (13.21),

$$\omega \in \{\Omega_{\rm A}^{\pm} \equiv \Omega_0(x) \pm \omega_{\rm A}(x)\}\,, \qquad \omega \in \{\Omega_{\rm S}^{\pm} \equiv \Omega_0(x) \pm \omega_{\rm S}(x)\}\,, \tag{13.47}$$

is obvious since these expressions transform into the flow continua (13.43) in the limit of vanishing magnetic field. Hence, the HD flow continua (13.43) and the MHD Alfvén and slow continua (13.47) are 'robust' (describing the dynamics in terms of holonomic variables), whereas the Eulerian entropy continua (13.46) are 'flimsy' (originating from initial data incompatible with the ideal MHD constraints). There is no coupling between the Eulerian entropy continuum modes and the HD flow or MHD Alfvén and slow continua, so that it is legitimate to restrict the analysis to the holonomic Lagrangian description in terms of $\boldsymbol{\xi}$.

Extending the discussion to compressible fluids *and* plasmas, the continua and the associated dominant physical variables are indicated in Table 13.1. The degeneracy of the Alfvén and slow

Table 13.1 *Essential spectra, with dominant variables, in HD and MHD.*

HD				MHD			
entropy:	$\Omega_E \equiv \{\Omega_0\}$	S_{1E}	$(1\times)$	entropy:	$\Omega_E \equiv \{\Omega_0\}$	S_{1E}	$(1\times)$
flow 1:	$\Omega_{H1} \equiv \{\Omega_0\}$	v_{1y}	$(1\times)$	Alfvén:	$\Omega_A^\pm \equiv \{\Omega_0 \pm \omega_A\}$	η	$(2\times)$
flow 2:	$\Omega_{H2} \equiv \{\Omega_0\}$	v_{1z}	$(1\times)$	slow:	$\Omega_S^\pm \equiv \{\Omega_0 \pm \omega_S\}$	ζ	$(2\times)$
sound:	$\Omega_P^\pm \equiv \pm\infty$	v_{1x}	$(2\times)$	fast:	$\Omega_F^\pm \equiv \pm\infty$	ξ	$(2\times)$

continua is then resolved and cluster spectra of fast waves with limiting frequencies $\Omega_F^\pm \equiv \pm\infty$ and normal polarization ($\boldsymbol{\xi} \to \xi \mathbf{e}_x$) appear in MHD, whereas analogous cluster spectra of sound waves with limiting frequencies $\Omega_P^\pm \equiv \pm\infty$ and normal polarization ($\mathbf{v} \to v_{1x}\mathbf{e}_x$) appear in HD. Like in MHD where the two tangential degrees of freedom asymptotically give rise to the Alfvén and slow continua, in HD they lead to the two flow continua Ω_{H1} and Ω_{H2}. Thus, the Lagrangian description involves four modes in HD and six in MHD, in the Eulerian description extended with the Eulerian entropy modes. This is how the continuous spectra on the real ω-axis asymptotically account for the different degrees of freedom of the linearized dynamics of stationary plasmas.

Let us consider how the continua control the *full* spectral structure of waves and instabilities for stationary plasmas. This involves the study of the spectral equation (13.11) for MHD, *with gravity*. The analogous analysis of static plasmas in Section 7.4 suggests to try to involve an oscillation theorem linking the sequences of discrete modes to the extrema of the continua by means of a monotonicity property. The required theorems for stationary plasmas were introduced in Section 12.3.2, both for real and for complex frequencies. For real frequencies, where $\widetilde{\omega}$ and ξ remain real, the Sturm–Liouville type expression (12.140), with the quotient (12.141) determining Sturmian or anti-Sturmian behaviour, leads to the **Real Oscillation Theorem for stable waves** [231]:
The eigenvalues of stable oscillations of a stationary plasma slab [or cylinder] are monotonic (Sturmian or anti-Sturmian) in the number of nodes of the eigenfunction ξ [or χ] for real values of $\omega = \sigma$ outside the continua $\{\Omega_A^\pm(x)\}$ and $\{\Omega_S^\pm(x)\}$, the apparent singularity ranges $\{\Omega_{s0}^\pm(x)\}$ and $\{\Omega_{f0}^\pm(x)\}$, and the Doppler–Coriolis indefinite range $\{\Omega_0(x)\}$.

The alternative formulation exploiting the expression (12.143) in terms of the alternator, involves the viewpoint of the Spectral Web and establishes the link with the oscillation theorem for complex frequencies of the following section.

The collection of Doppler shifts $\{\Omega_0(x)\}$, though in general not part of the continuous spectrum in MHD, significantly alters the spectrum of stable waves. In particular, a marginal stability threshold σ_0, following from Eq. (13.9), is necessarily situated inside that range. In the expression (12.141), the denominator $(\sigma - \overline{V})D(\sigma)$ may then be replaced by $(\sigma - \sigma_0)D(\sigma)$. When outside any of the continua, such a frequency marks the transition from Sturmian to anti-Sturmian behaviour of the stable discrete modes and they may mark the creation of a pair of overstable and damped waves. Also, the new continua $\{\Omega_A^\pm(x)\}$ and $\{\Omega_S^\pm(x)\}$ become asymmetric with respect to

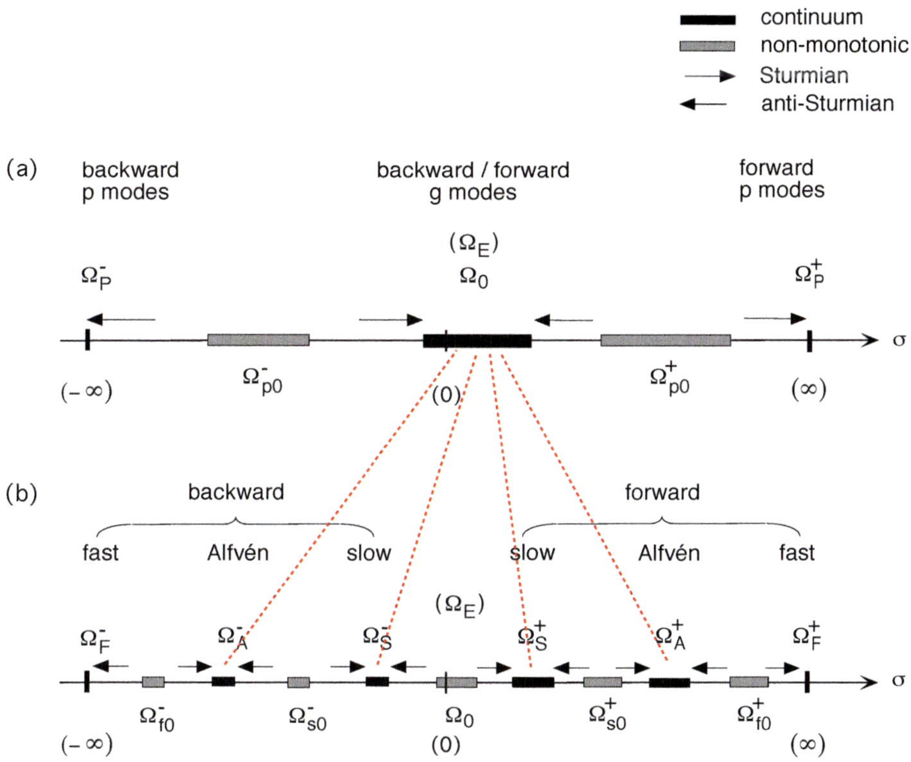

Fig. 13.2 Schematic spectrum of the stable oscillations of a stationary plane flow (a) in hydrodynamics and (b) in magnetohydrodynamics. Continua are labelled above, apparent singularities below the axis. Marginal transitions σ_0 in $\{\Omega_0\}$ are not indicated; the location $\sigma = 0$ inside $\{\Omega_0\}$ is accidental. The red dashes indicate how the HD flow continua degeneracy is lifted in the MHD continua by $B \neq 0$.

the origin $\omega = 0$ through the dependence on $\Omega_0(x)$. Schematically, this yields the spectral structures for the real eigenvalues in HD and MHD as illustrated in Fig. 13.2.

In compressible HD, the Doppler shifts create the flow continua $\{\Omega_{\mathrm{H}i} \equiv \Omega_0\}$, with possible cluster spectra of forward and backward gravity-driven g modes at the ends [108](b), [168], and they also split apart forward and backward pressure-driven p modes (sound waves). See Section 7.2 for the corresponding static case, where the spectral equation for the stationary case may be obtained from Eq. (7.29) by replacing ω by $\widetilde{\omega}$. This also provides the expression for the apparent HD singularities $\Omega_{\mathrm{p}0}^{\pm} \equiv \Omega_0 \pm k_0 c$, where c is the sound speed, separating the p and the g modes. In MHD (neglecting the Eulerian entropy modes), the Doppler shifts do not create continua but they split apart the forward and backward parts of the spectrum. Thus, forward and backward continua $\Omega_{\mathrm{S}}^{\pm}$, $\Omega_{\mathrm{A}}^{\pm}$, $\Omega_{\mathrm{F}}^{\pm}$ are separated by regions Ω_0, $\Omega_{\mathrm{s}0}^{\pm}$, $\Omega_{\mathrm{f}0}^{\pm}$ of non-monotonicity of the discrete spectrum which, otherwise, is either Sturmian ($\rightarrow$) or anti-Sturmian ($\leftarrow$) along the real ω-axis [620]. The sub-spectra shown in Fig. 13.2 just indicate *possible* structures for small inhomogeneity, as obtained for a thin slice of plasma across the normal direction. For realistic equilibria, the continua may fold over themselves, they may overlap, and they usually 'swallow' large parts of the discrete sub-spectra.

13.1.4 Spectral Web and the complex oscillation theorem

For instabilities of plasmas with background equilibrium flow, we have to address the fact that, due to the variable Doppler shift $\Omega_0(x)$, the eigenvalues $\omega = \sigma + \mathrm{i}\nu$ and the eigenfunctions $\boldsymbol{\xi} = \boldsymbol{\xi}_1 + \mathrm{i}\boldsymbol{\xi}_2$ are irreducibly complex. Notice though that the x-dependence of the crucial factor $\widetilde{\omega}(x)$ only resides in the real part:

$$\widetilde{\omega}(x) = \widetilde{\sigma}(x) + \mathrm{i}\nu\,, \quad \widetilde{\sigma}(x) \equiv \sigma - \Omega_0(x)\,. \tag{13.48}$$

For numerical solution, we will exploit the first order eigenvalue system (13.17), where we split all functions into real and imaginary parts:

$$\begin{pmatrix} \xi_1' \\ \xi_2' \\ \Pi_1' \\ \Pi_2' \end{pmatrix} + \begin{pmatrix} \hat{C}_1 & -\hat{C}_2 & \hat{D}_1 & -\hat{D}_2 \\ \hat{C}_2 & \hat{C}_1 & \hat{D}_2 & \hat{D}_1 \\ \hat{E}_1 & -\hat{E}_2 & -\hat{C}_1 & \hat{C}_2 \\ \hat{E}_2 & \hat{E}_1 & -\hat{C}_2 & -\hat{C}_1 \end{pmatrix} \begin{pmatrix} \xi_1 \\ \xi_2 \\ \Pi_1 \\ \Pi_2 \end{pmatrix} = 0\,. \tag{13.49}$$

The explicit expressions of the real and imaginary parts of the normalized coefficients $\hat{C} \equiv C/N$, $\hat{D} \equiv D/N$ and $\hat{E} \equiv E/N$ are put in small print below.

▷ **Explicit expressions of the matrix elements** The elements of the first order system of ODEs (13.49) are given by the expressions

$$\begin{aligned} \hat{C}_1 &\equiv (N_1 C_1 + N_2 C_2)/|N|^2\,, & \hat{C}_2 &\equiv (N_1 C_2 - N_2 C_1)/|N|^2\,, \\ \hat{D}_1 &\equiv (N_1 D_1 + N_2 D_2)/|N|^2\,, & \hat{D}_2 &\equiv (N_1 D_2 - N_2 D_1)/|N|^2\,, \\ \hat{E}_1 &\equiv (N_1 E_1 + N_2 E_2)/|N|^2\,, & \hat{E}_2 &\equiv (N_1 E_2 - N_2 E_1)/|N|^2\,, \end{aligned} \tag{13.50}$$

with the factors

$$\begin{aligned} N_1 &= (\gamma p + B^2)\left[(R - \omega_A^2)(R - \omega_S^2) - I^2\right], \quad N_2 = I(\gamma p + B^2)\,(2R - \omega_A^2 - \omega_S^2)\,, \\ D_1 &= R^2 - I^2 - k_0^2\left[(\gamma p + B^2)/\rho\right](R - \omega_S^2)\,, \quad D_2 = I\left[2R - k_0^2(\gamma p + B^2)/\rho\right], \\ C_1 &= -\rho g\left[R(R - \omega_A^2) - I^2\right], \qquad C_2 = -I\rho g\,(2R - \omega_A^2)\,, \\ E_1 &= -\rho(\gamma p + B^2)\Big\{(R - \omega_A^2 + \rho' g/\rho)\left[(R - \omega_A^2)(R - \omega_S^2) - I^2\right] \\ &\quad - I^2(2R - \omega_A^2 - \omega_S^2)\Big\} - \rho^2 g^2\left[(R - \omega_A^2)^2 - I^2\right], \\ E_2 &= -I\Big\{\rho(\gamma p + B^2)\Big[(R - \omega_A^2)(R - \omega_S^2) - I^2 \\ &\quad + (R - \omega_A^2 + \rho' g/\rho)(2R - \omega_A^2 - \omega_S^2)\Big] + 2\rho^2 g^2 (R - \omega_A^2)\Big\}, \end{aligned} \tag{13.51}$$

where σ and ν enter through $R \equiv \mathrm{Re}(\widetilde{\omega}^2) = \widetilde{\sigma}^2 - \nu^2$, and $I \equiv \mathrm{Im}(\widetilde{\omega}^2) = 2\widetilde{\sigma}\nu$. For *incompressible* plasmas, $\hat{C} \to 0$, $\hat{D} \to -k_0^2/[\rho(\widetilde{\omega}^2) - \omega_A^2]$, and $\hat{E} \to -\rho(\widetilde{\omega}^2 - \omega_A^2) - \rho' g$, so that only the off-diagonal elements $\hat{D}_{1,2}$ and $\hat{E}_{1,2}$ survive. ◁

The eigenvalues are determined by solving the ODEs (13.49) subject to the BCs (13.6), which are also split into real and imaginary parts:

$$\xi_1(x_1) = \xi_2(x_1) = 0 \quad \text{(left)}\,, \qquad \xi_1(x_2) = \xi_2(x_2) = 0 \quad \text{(right)}\,. \tag{13.52}$$

(Note the different meanings of the subscripts 1 and 2.) Numerical solution of the ODEs for a given value of ω is straightforward if this two-sided BVP is turned into a one-sided BVP by 'shooting', either from the left or the right. (Accurate numerical routines exist to solve one-sided BVPs for a system of first order ODEs.) Hence, the problem may be turned into a left one-sided BVP with solutions ξ^ℓ,

$$\text{left BVP:} \qquad \xi_1^\ell(x_1) = \xi_2^\ell(x_1) = 0\,, \quad \Pi_1^\ell(x_1) = 1\,, \quad \Pi_2^\ell(x_1) = 0\,, \tag{13.53}$$

or a right one-sided BVP with solutions $\xi^{\rm r}$,

$$\text{right BVP:} \qquad \xi_1^{\rm r}(x_2) = \xi_2^{\rm r}(x_2) = 0\,, \quad \Pi_1^{\rm r}(x_2) = 1\,, \quad \Pi_2^{\rm r}(x_2) = 0\,. \tag{13.54}$$

These solutions determine the solution and conjugate paths from the respective complementary energy expressions (12.119) for $W_{\rm com}^\ell$, (12.126) for $W_{\rm com}^{\rm r}$, and (12.128) for $W_{\rm com}^{\rm mix}$. The resulting solution paths are then determined from the expression (12.120), and the conjugate paths from the expression (12.121). The alternators associated with the three pairs of paths are defined by

$$R^\ell \equiv (\xi/\Pi)^\ell_{x=x_2}\,, \quad R^{\rm r} \equiv (\xi/\Pi)^{\rm r}_{x=x_1}\,, \quad R^{\rm mix} \equiv ([\![\Pi]\!]/\xi)^{\rm mix}_{x=x_{\rm mix}}\,. \tag{13.55}$$

All these quantities are complex, but only their real parts (or only their imaginary parts) enters into our final construction of the different Spectral Webs and oscillation theorems. The choice of unit boundary values $\Pi_1^\ell(x_1)$, $\Pi_1^{\rm r}(x_2)$ in the BCs (13.53), (13.54) is arbitrary, it just fixes a complex phase factor $\exp({\rm i}\phi)$ which drops out in the complementary energy and alternator expressions.

We can now reformulate the complex theorem of Section 12.3.2 in the following **Complex Oscillation Theorem for instabilities**:

The complex eigenvalues of a stationary plasma slab [or cylinder] are situated on the intersections of the solution and conjugate paths where the complementary energy $W_{\rm com}^\ell \equiv \frac{1}{2}\int \xi^{\ell *}\Pi^\ell\, dS_{\rm right}$ *vanishes, while* $\Pi^\ell(x_2)$ *should not vanish. The alternator* $R^\ell \equiv \xi^\ell(x_2)/\Pi^\ell(x_2)$ *is a real and monotonic function of arc length along the solution path, and a purely imaginary and monotonic function along the conjugate path, in between the zeros of* $\Pi^\ell(x_2)$ *separating the eigenvalues.*

The theorem is phrased here for the left solutions, the formulation for the right and mixed solutions is completely analogous. The different solution paths may also be obtained from the alternative expression (12.124) involving the solution averaged Doppler–Coriolis shifted frequency.

We now have all the necessary tools at our disposal to investigate the spectra of the various waves and instabilities of stationary one-dimensional equilibria. Before we do that, it is important to notice that the concept of Spectral Web applies equally well to toroidal equilibria, like tokamaks. Hence, the first part of the Complex Oscillation Theorem, referring to the solution path and the conjugate path, is also valid for toroidal systems. It involves the computation of *solutions* of the Frieman–Rotenberg partial differential equation, satisfying only the left boundary condition, for arbitrary values of the spectral parameter ω. (Italics to stress that the solutions involved in the Spectral Web method are not trial functions as exploited in variational principles.) The solutions are completely determined when the value of ω is specified. Different solution and conjugate paths will be obtained for the different modes, which e.g. could be considered as resulting from the toroidal mode coupling of cylindrical Fourier modes. Those modes are usually characterized by a dominant cylindrical mode number with side bands, which could be preferentially excited by specifying the amplitude of just that particular Fourier mode $\Pi_{1,m}^\ell(x_1)$ in the BC (13.53). The amplitudes of the sidebands will be determined by the tangential derivatives of the Frieman–Rotenberg equation. Since those coupled modes have different solution averaged Doppler–Coriolis shifts, the corresponding solution paths will be located in different parts of the complex ω-plane.

The numerical implementation of the Spectral Web method involves two stages, which may be executed independently:

(a) *Contour plotting* of the full Spectral Web by solving the solution and conjugate path equations (12.120) and (12.121) in the strip of the complex ω-plane where the unstable eigenvalues are restricted to lie according to the solution averaged Doppler-Coriolis expression (12.124). Since those solutions for different values of ω are completely independent of each other, this stage of computing is trivially parallel.

(b) *Separate eigenvalue search* by, first, constructing the solution path from Eq. (12.120) and, next, finding the eigenvalue(s) and associated eigenfunctions on that path by determining the zeros of the alternator. This search is facilitated by the monotonicity of the alternator. It may be restricted to a rectangle $\{\sigma_1 \le \sigma \le \sigma_2\,,\ \nu_1 \le \nu \le \nu_2\}$ of the ω-plane where one expects the eigenvalue(s) to lie from the contour plot.

With respect to the contour plotting, the iterative solutions of the paths may be obtained by a very robust root finding routine [509], provided the pertinent quantities ($W_{\rm com,2}$ and $W_{\rm com,1}$) have opposite signs on the left and right borders of the range in σ chosen (as is the case if it is as large as the Doppler range itself). With respect to the eigenvalue search, the same holds for the iteration on the zero of R where the path employed should intersect the bottom and top borders of the rectangle, which should be small enough to admit not more than one solution.

In this manner, the physically meaningful structure of the Spectral Web is obtained connecting all eigenvalues in the complex ω-plane. The implications for the spectral theory of stationary equilibria will be demonstrated by the explicit examples of the following sections.

13.2 Analysis of flow-driven instabilities in plane plasmas

We now possess the appropriate tools to explore the spectra of some representative stationary equilibria with respect to the effects of *gravity*, due to a non-constant density $\rho(x)$, of *velocity shear*, due to the variation of the magnitude of a uni-directional flow field $\mathbf{v}(x)$, and of *magnetic shear*, due to the varying direction of a constant amplitude magnetic field $\mathbf{B}(x)$; see Fig. 13.1. We will choose parameters such that gravity-driven Rayleigh–Taylor instabilities and velocity-driven Kelvin–Helmholtz instabilities may occur simultaneously. For the time being, we restrict the analysis to incompressible perturbations of the following equilibria:

$$\begin{aligned}
&\mathbf{k}_0 = k_0\mathbf{e}_z\,, \qquad \rho(x) = 1-\delta x\,, \qquad \mathbf{g} = -g\mathbf{e}_x \qquad\qquad (0\le x\le 1)\,,\\
&\mathbf{v}(x) = \big[v_0 + v_1(x-\tfrac12) + v_2\sin\tau(x-\tfrac12)\big]\,(\sin\theta\,\mathbf{e}_y + \cos\theta\,\mathbf{e}_z)\,,\\
&\mathbf{B}(x) = \sin\varphi(x)\mathbf{e}_y + \cos\varphi(x)\mathbf{e}_z\,, \qquad \varphi(x) = \varphi_0 + \alpha(x-\tfrac12)\,.
\end{aligned} \tag{13.56}$$

The magnetic field is force-free, $\mathbf{j} = \nabla\times\mathbf{B} = \alpha\mathbf{B}$, where α yields the magnetic shear, $\varphi' = \alpha$, restricted to be stabilizing only. (Current-driven instabilities will be investigated in Chapter 14.) As always, we exploit dimensionless variables, based on unit scales of length $a \equiv x_2 - x_1$, density ρ_0 and magnetic field strength B_0, leaving ten free parameters: $k_0, \delta, g, \theta, v_0, v_1, v_2, \tau, \varphi_0, \alpha$.

The equilibrium pressure may be computed, $p(x) = p_0 - (x - \frac12\delta x^2)g$, but it does not enter in

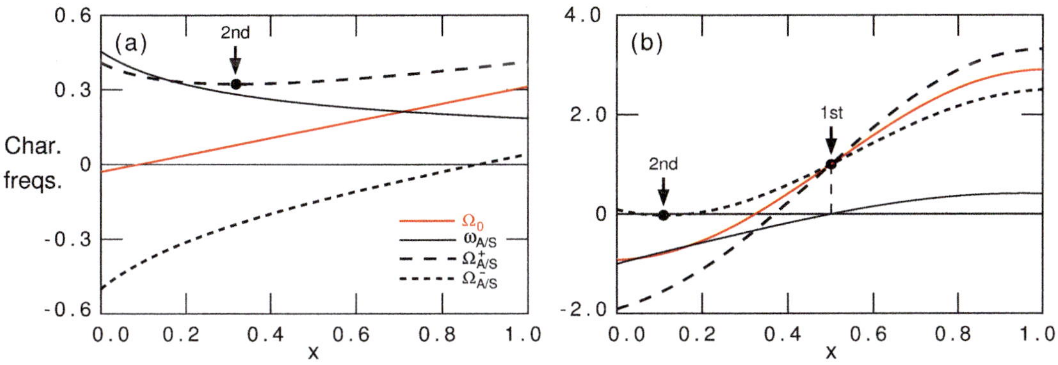

Fig. 13.3 Characteristic frequencies of two equilibria: (a) magnetized gravitating plasma without magnetic shear and with linear density and flow profiles (RTI), (b) idem with magnetic shear and with sinusoidal flow profile, $\tau = 4$ (RTI+KHI) [arrows indicate potential clustering from criteria with '2nd' or '1st' derivatives].

the spectral equations for incompressible perturbations. The equilibria are then characterized by the spatial dependence of the frequencies of the degenerate Alfvén/slow continua (in MHD), or of the flow continua (in HD):

$$\Omega_{A/S}(x) = \Omega_0(x) \pm k_0 \cos\varphi(x)/\sqrt{\rho}(x)\,, \qquad \Omega_0(x) = k_0 \cos\theta\, v(x)\,. \tag{13.57}$$

These frequencies are plotted in Fig. 13.3 for two representative MHD examples. We here present the list of cases investigated in this section, together with their parameter values, in the order in which they will be treated:

(a) *RTI in MHD:* Rayleigh–Taylor instabilities of a stationary plasma with uni-directional magnetic field and linear velocity profile (Section 13.2.1, Figs. 13.4–13.5),

$$k_0 = 1\,,\quad \delta = -5\,,\quad g = 15\,,\quad \varphi_0 = -0.35\pi\,,\quad \alpha = 0\,,$$

$$\theta = 0.35\pi\,,\quad v_0 = 0.2\,,\quad v_1 = 0.6\,,\quad v_2 = 0\,,\quad \tau = 0\,;$$

(b) *KHI in HD:* Kelvin–Helmholtz instabilities of a stationary fluid for three cases of a sinusoidal velocity profile (Section 13.2.2, Figs. 13.6–13.8),

$$k_0 = 1\,,\quad \delta = 0\,,\quad g = 0\,,\quad \theta = 0\,,\quad v_0 = v_1 = 0\,,\quad v_2 = 1\,,\quad \tau = 5/8/11\,;$$

(c) *RTI + KHI in MHD:* Rayleigh–Taylor and Kelvin–Helmholtz instabilities of a stationary plasma with magnetic shear and sinusoidal velocity (Section 13.2.3, Figs. 13.9–13.10),

$$k_0 = 1\,,\quad \delta = -5\,,\quad g = 100\,,\quad \varphi_0 = 0.5\pi\,,\quad \alpha = -\pi\,,$$

$$\theta = 0\,,\quad v_0 = 1\,,\quad v_1 = 2\,,\quad v_2 = 1\,,\quad \tau = 4\,.$$

We will refer to the different cases by the stated acronyms. In the MHD examples, we have freely chosen $\rho' > 0$ in order to study unstable 'inverted atmospheres' (cf. Case [108](c) for a corresponding HD example), without addressing the question of what would cause such a density distribution (obviously, not gravity).

13.2.1 Rayleigh–Taylor instabilities of magnetized plasmas

The first example (RTI in MHD) concerns the pure Rayleigh–Taylor instability mechanism. The velocity field $\mathbf{v}$ and the magnetic field $\mathbf{B}$ are both taken uni-directional, at angles of $+0.35\pi$ and -0.35π with the horizontal wave vector. The resulting forward and backward continua are well separated from each other and from the Doppler range. This creates a gap between the continua accommodating an infinite sequence of gravitational (g) modes clustering at the minimum $x_0 \approx 0.318$ of the degenerate forward continua, where $\big(\Omega^+_{\rm A/S}\big)_{\rm min} \approx 0.323$, indicated by '2nd' in Fig. 13.3(a). The condition for clustering is derived from the spectral equation (13.30) for incompressible plasmas, analogous to the derivation for static compressible plasmas of Section 7.4.4. This admits clustering of the gravitational modes at the *real* frequency $\omega = \Omega^{\pm}_{\rm A/S}(x_0)$, where $\Omega^{\pm'}_{\rm A/S}(x_0) = 0$, if

$$k_0^2 \rho' g < \pm\tfrac{1}{4}\rho\,\omega_{\rm A,S}\Omega^{\pm''}_{\rm A/S} < 0\,, \qquad \text{or} \qquad k_0^2 \rho' g > \pm\tfrac{1}{4}\rho\,\omega_{\rm A,S}\Omega^{\pm''}_{\rm A/S} > 0\,. \tag{13.58}$$

The second condition is satisfied in the present example so that clustering will occur at the lower edge of the forward continua $\Omega^+_{\rm A/S}$. The cluster conditions for *compressible* plasmas (with separate Alfvén and slow continua) have been derived by van der Holst et al. [620]. One does *not* obtain the above criteria in the limit $\gamma \to \infty$ from those conditions (different factors $k_\perp^2$ and $k_\parallel^2$ appear) because the above ones result from the confluence of six of the eight $N = 0$ and $D = 0$ singularities of the spectral equation (13.11) in the incompressible limit.

The Spectral Web for this configuration is shown in Fig. 13.4. Six unstable Rayleigh–Taylor modes (labelled $n = 1, 2, \ldots 6$) are located on both the solution path and the conjugate path, and above the Doppler range $[-0.0454, 0.2270]$, as required by the theory. The wiggling solution path is continuous, but the conjugate path is split into separate branches for reasons that are yet to be discovered. Recall that the geometry of the two paths is different for the different choices (left, right or mixed) of one-sided boundary value problems, but note that those problems are more than just a mathematical trick to solve the eigenvalue problem. They actually correspond to the physical problem of excitation at the pertinent boundary (right, left or middle of the plasma). The location of the eigenvalues themselves is independent of that choice, and the Spectral Web is (to our knowledge) the first method to connect them in a physically meaningful manner.

The solution path meets the real axis at $\sigma \approx 0.129$, where the sequence of modes is split into stable backward and forward Alfvén/slow modes that are situated on the real axis. At this point, the labelling of the modes (by the number of branches of the alternator) gets the additional meaning of referring to the number of nodes of the eigenfunction ξ, which is then real. Only the backward mode labelled $n = 7'$ and the forward mode $n = 7$ appear in the frame shown, the rest of the cluster sequence, $n = 8, 9, \ldots$, including the cluster point, falls outside the frame.

Notice that all genuine eigenvalues, which occur at intersections of the two paths where $|\xi_{\rm e}| = 0$ (the subscript e referring to the end point, $x = x_2$ in this case), alternate with 'false' eigenvalues, which occur at intersections where $|\Pi_{\rm e}| = 0$. This alternation is actually continued along the real axis (which is also part of the solution path), where the stable eigenvalue $n = 7'$ lies on the left leg of a tiny loop of the conjugate path that is not visible on the scale of the figure.

The value of the alternator for the unstable modes along the solution path is plotted in Fig. 13.5(a), illustrating the monotonicity property involved in the complex oscillation theorem. For simplicity, R is plotted versus the growth rate ν, rather than arc length λ along the solution path, as would be required by the theorem. For the present oscillatory shape of the solution path, this makes no

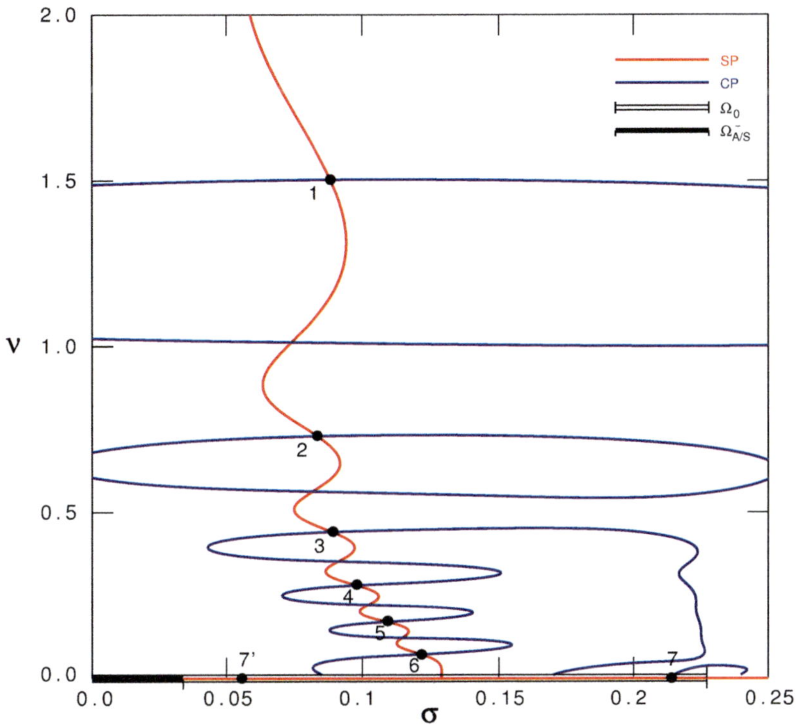

Fig. 13.4 Spectral Web consisting of solution path (in red) and conjugate path (in blue) for the left solutions for the Rayleigh–Taylor instabilities of the equilibrium of case RTI, with the characteristic frequencies shown in Fig. 13.3(a). The six unstable and two of the stable modes are located at the intersections of the two paths.

difference though since λ is still a monotonic function of ν. The amplitudes $|\xi_e|$ and $|\Pi_e|$ for the solution at the other end of the plasma interval are plotted in Fig. 13.5(b). This demonstrates that, in fact, the zeros of $|\xi_e|$ and $|\Pi_e|$ nicely alternate. Hence the appropriate name 'alternator' for R. It turns out to be a perfect counter, where every new zero $|\Pi_e| = 0$ initiates a new branch of R, monotonically running from $+\infty$ to $-\infty$, so that the zeros $|\xi_e| = 0$ may be computed with very high precision. In this way, we have constructed a counter for complex eigenfunctions that does not require counting of the nodes of either the real or the imaginary part of the eigenfunctions, and obtained a generalization for complex eigenvalues of the known real oscillation theorem of one-dimensional equilibria [240].

13.2.2 Kelvin–Helmholtz instabilities of ordinary fluids

The Kelvin–Helmholtz instability in the simplest configurations gives rise to surprisingly complicated Spectral Webs. We will first study them in HD and then in MHD. We have encountered the instability in Section 13.1.2 when discussing instabilities of interface plasmas consisting of two layers with different densities, velocities and magnetic fields. The resulting expression (13.37) for the growth rate of instabilities appears to be quite 'general', but it could falsely be taken to suggest that

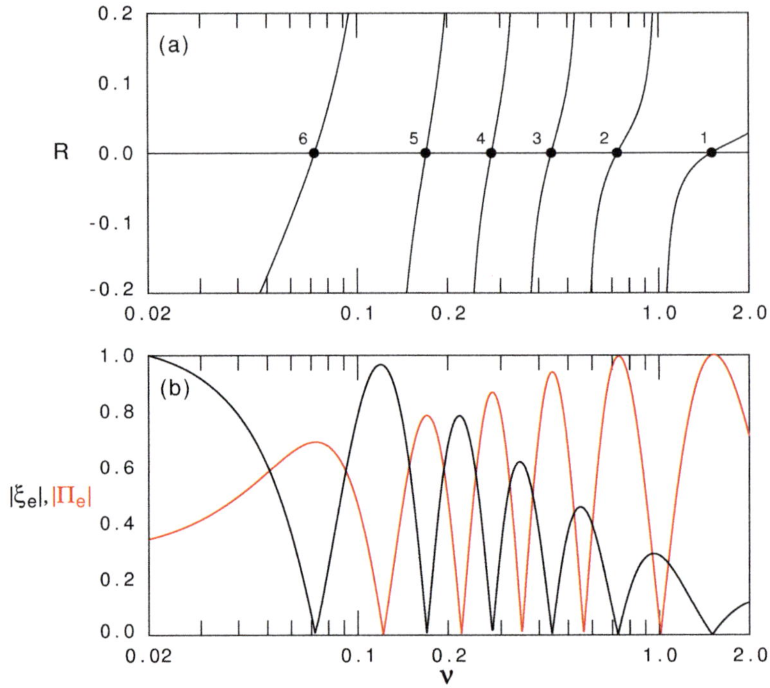

Fig. 13.5 (a) Alternator $R(\nu)$ and (b) absolute values $|\xi_e(\nu)|$ and $|\Pi_e(\nu)|$ along the left solution path for the Rayleigh–Taylor unstable modes of the equilibrium of case RTI shown in Fig. 13.3(a). The eigenvalues occur for $R = 0$, where $|\xi_e| = 0$.

the Kelvin–Helmholtz instability just requires a velocity gradient to drive it. In a streaming diffuse fluid or plasma, the conditions for instability are much more subtle than that, as could have been expected from the fact that there is no discrete spectrum of stable modes that could be studied perturbatively to find out about transitions to instability. In a sense, the Kelvin–Helmholtz instability is a perturbation of the flow continua Ω_{Hi}, expressed by Eqs. (13.43).

In our second example (KHI in HD), we return to an incompressible fluid layer with a unidirectional velocity of varying amplitude, $\mathbf{v} = v(x)\mathbf{e}_z$. The velocity profile is chosen sinusoidal with the zero in the middle, so that the flow has opposite directions in the lower ($0 \leq x < 0.5$) and upper ($0.5 < x \leq 1$) parts of the fluid:

$$v(x) = \sin \tau(x - \tfrac{1}{2}) \,. \tag{13.59}$$

Quite significant for the stability analysis, discussed below, is the fact that the zero in the middle is also an inflexion point of the flow, $v''(x = 0.5) = 0$. The Kelvin–Helmholtz instability depends critically on the value of the parameter τ, which is a measure of the velocity gradient at $x = 0.5$, but it also introduces additional counter flows in the configuration. This has important effects on the different kinds of eigenmodes of the instability, as illustrated by Fig. 13.6 where the Spectral Web is depicted for (a) $\tau = 5$, (b) $\tau = 8$ and (c) $\tau = 11$. Including the complex conjugate mirror images (not shown of course), Fig 13.6(a) demonstrates the existence of two modes (a purely exponential instability and its damped counterpart), Fig 13.6(b) the existence of four modes (two

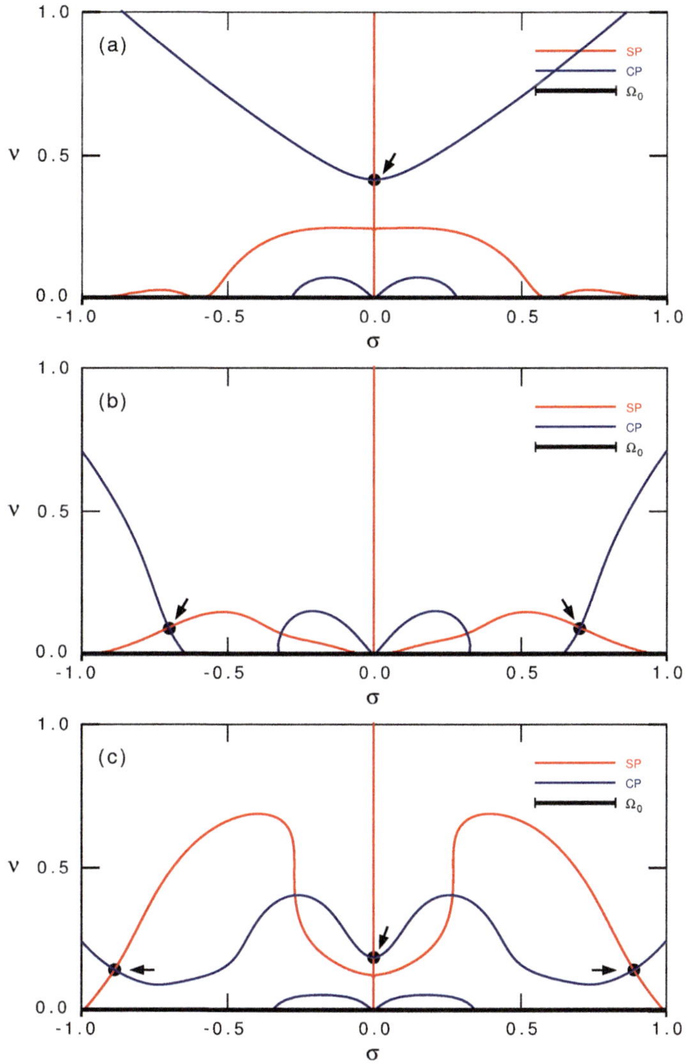

Fig. 13.6 Spectral Webs for the Kelvin–Helmholtz instability of an ordinary fluid with sinusoidal flow profile for (a) $\tau = 5$, (b) $\tau = 8$ and (c) $\tau = 11$. Mixed solution and conjugate paths are for $x_{\rm mix} = 0.5$; eigenvalues are indicated by arrows.

overstable modes with opposite values of σ and their damped counterparts), and Fig 13.6(c) the existence of six modes (one purely exponential instability and two overstable modes, and their damped counterparts). The values of the corresponding eigenvalues are: (a) $\nu = 0.4169$, (b) $\nu = 0.09519$, $\sigma = 0.7024$, (c) $\nu = 0.1880$ and $\nu = 0.1445$, $\sigma = 0.8869$. The eigenfunctions are shown in Fig. 13.7, omitting the overstable ones for $\sigma < 0$ (since they have a simple left–right relationship to the ones for $\sigma > 0$).

To appreciate the subtleties of the Kelvin–Helmholtz instability for the flow profile (13.59), note that in either one of the pertinent spectral equations (13.45), the only expression that occurs (in addition to the factor k_0) is precisely the one that determines the flow continua according to Eq. (13.4). For a linearly increasing velocity profile, the spectral equation for χ immediately shows that the fluid is stable since $\widetilde{\omega}'' = -\Omega_0'' = -k_0 v'' = 0$ so that only the flow continua $\widetilde{\omega} = 0$ remain. Hence, it is necessary for instability to have a velocity profile with $v'' \neq 0$ over some region. This appears to imply that only local profile conditions matter, but above we made the parenthetical remark "in addition to the factor k_0". That factor cannot be scaled out of the problem since the length scale of the inhomogeneities is fixed by the thickness $a \equiv x_2 - x_1$ of the fluid layer. This observation is crucial because the solutions of the spectral equation (13.45) become exponentially dominant for large k_0 (small wavelength), so that it is impossible to satisfy the boundary conditions then. Hence, in addition to the local criteria that are discussed below, the Kelvin–Helmholtz instability requires k_0 to be small enough, so that it is a *global, long-wavelength, instability*.

The local conditions on the velocity profile are obtained following Drazin and Reid [164], Section 22. Depart from the spectral equation (13.45) for χ, multiply with χ^*, integrate over the plasma volume, transform the second derivative of χ by integration by parts, and cancel the boundary terms by satisfying the BCs:

$$\begin{aligned} -\mathrm{Re} \int (\widetilde{\omega}''/\widetilde{\omega})\, |\chi|^2 \, dx &= \int \frac{(\sigma - \Omega_0)\Omega_0''}{(\sigma - \Omega_0)^2 + \nu^2}\, dx = \int \left(|\chi'|^2 + k_0^2 |\chi|^2 \right) dx \,, \\ -\mathrm{Im} \int (\widetilde{\omega}''/\widetilde{\omega})\, |\chi|^2 \, dx &= -\,\nu \int \frac{\Omega_0''}{(\sigma - \Omega_0)^2 + \nu^2}\, dx = 0 \,. \end{aligned} \tag{13.60}$$

From the imaginary part it follows that it is necessary for instability ($\nu \neq 0$) that the velocity profile should have an inflexion point, $v''(x_i) = 0$, somewhere in the flow (*Rayleigh's inflexion point theorem*). A sharper criterion follows by adding the two expressions, after multiplying the second one with $[\sigma - \Omega(x_i)]/\nu$:

$$-\int \frac{(\Omega_0 - \Omega_0(x_i))\Omega_0''}{(\sigma - \Omega_0)^2 + \nu^2}\, dx = \int \left(|\chi'|^2 + k_0^2 |\chi|^2 \right) dx > 0 \,. \tag{13.61}$$

This implies that it is necessary for instability that the velocity profile should satisfy $[v - v(x_i)]v'' < 0$ somewhere in the flow (*Fjørtoft's theorem*).

For the velocity profile (13.59), Fjørtoft's theorem is satisfied (nearly) everywhere. This does not imply that the fluid is unstable for every value of τ since we still have to consider the global boundary value problem, involving k_0. This implies numerical analysis, but we can obtain the explicit stability thresholds for this example because of the symmetry of the equilibrium with a sinusoidal flow profile. As a result, these thresholds occur for $\omega = 0$, when $\widetilde{\omega}''/\widetilde{\omega} \to v''/v = -\tau^2$, so that marginal solutions are obtained for $\tau = \tau_n$ ($n = 1, 2, \cdots$):

$$\chi'' - (k_0^2 - \tau^2)\chi = 0 \qquad \Rightarrow \; \chi = A \sin n\pi x \,, \quad \tau_n \equiv (k_0^2 + n^2\pi^2)^{1/2}. \tag{13.62}$$

In our example $k_0 = 1$, so that instability sets in for $\tau > \tau_1 \approx 3.297$, whereas for $\tau > \tau_2 \approx 6.362$ and for $\tau > \tau_3 \approx 9.478$, etc., new unstable windows appear in the stability diagram, depicted in Fig. 13.8. It should be noted that all eigenvalues shown correspond to a different Spectral Web. The most striking feature of the stability diagram is the fact that, in general, the growth rate ν is not an increasing function of the parameter τ, and even vanishes for $\tau = \tau_2$.

For these HD examples, the Doppler range has turned into continuous spectra (the flow continua $-1.0 \leq \Omega_0 \leq 1.0$), from which the solution and conjugate paths (and the eigenvalues on them)

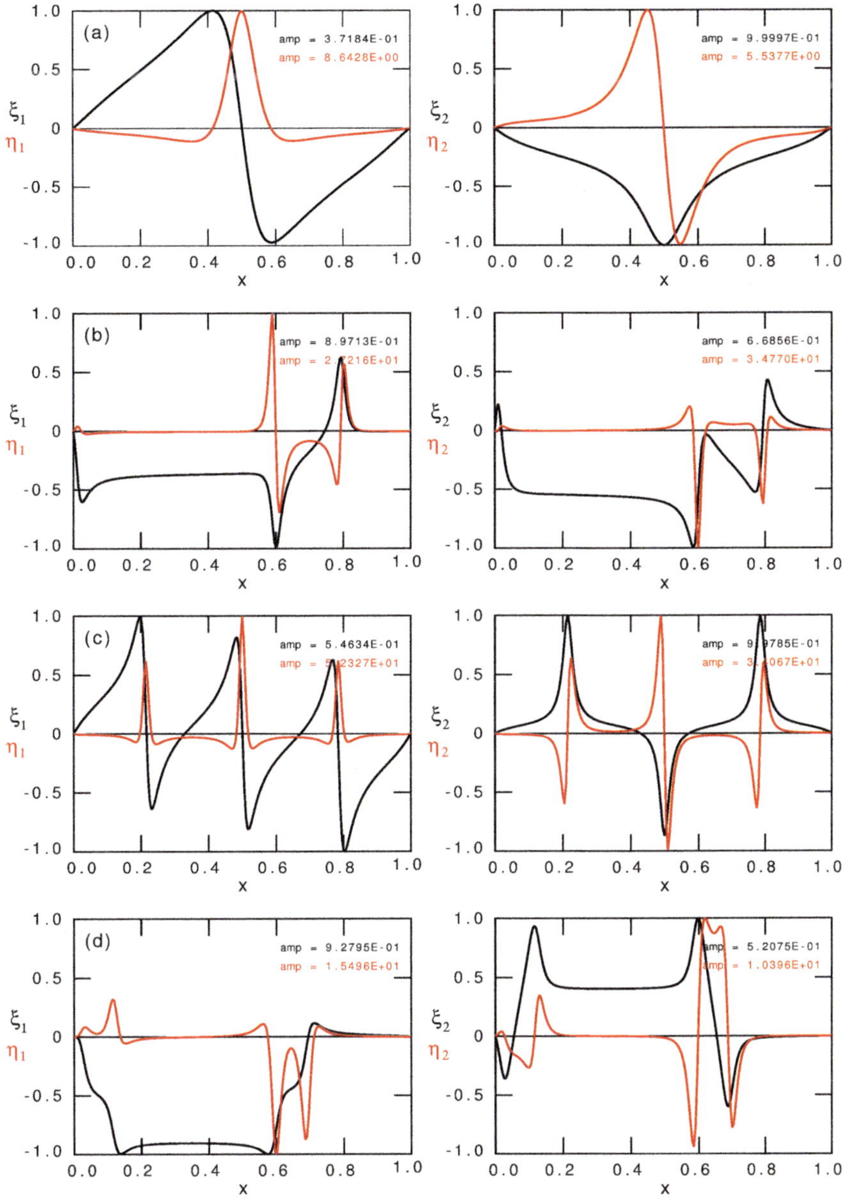

Fig. 13.7 Eigenfunctions of the Kelvin–Helmholtz instability (KHI) for sinusoidal flow for (a) $\tau = 5$ (purely exponential mode), (b) $\tau = 8$ (overstability), (c) $\tau = 11$ (purely exponential mode) and (d) $\tau = 11$ (overstability).

protrude into the complex ω-plane. In this case, since we have chosen the mixed representation with $x_{\rm mix}$ precisely in the middle of the plasma, the Spectral Webs are symmetrical with respect to $\sigma = 0$, appropriate for the symmetry of the problem. The eigenfunctions of Fig. 13.7 are represented in terms of the perpendicular displacement η (instead of the pressure perturbation Π) because this is

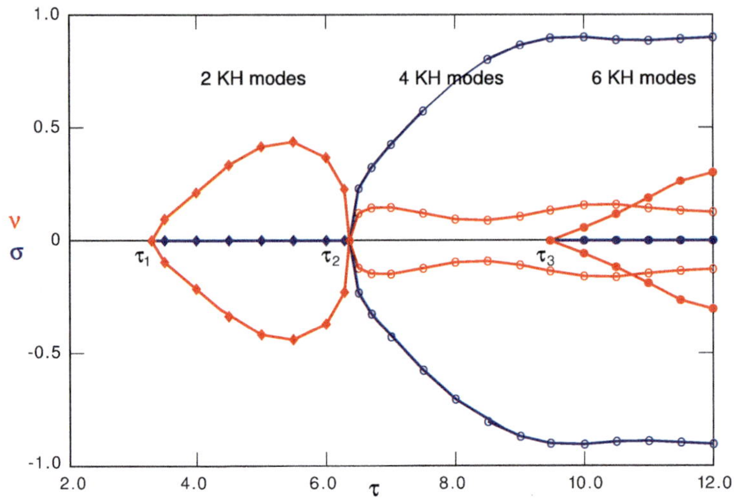

Fig. 13.8 Real (σ) and imaginary (ν) part of the eigenfrequency of the Kelvin–Helmholtz instabilities as function of the parameter τ measuring the period of the sinusoidal flow profile. Every time τ increases beyond a critical $\tau_n \equiv [k_0^2 + (n\pi)^2]^{1/2}$, two more complex eigenvalues emerge from the origin of the ω-plane.

most suitable for incompressible modes. In that case, the expressions (13.15) for η and ζ transform into $\eta = -(k_\perp/k_0^2)\xi'$ and $\zeta = -(k_\parallel/k_0^2)\xi'$, so that $\nabla \cdot \boldsymbol{\xi} \equiv \xi' + k_\perp \eta + k_\parallel \zeta = 0$.

The eigenfunctions clearly demonstrate the two aspects of the Kelvin–Helmholtz instability: they are global, but they do require the proximity of the flow continua to develop large, but rather local, variations of the amplitude of the displacement ξ. The corresponding plots of the perturbed velocity $v_1 \sim \widetilde{\omega}\xi$ (not shown) do not exhibit such large variations because of the smallness of the factor $\widetilde{\omega}$ close to the continua. Note the qualitative difference between the symmetric/anti-symmetric eigenfunctions depicted in Figs. 13.7(a) and (c), for $\tau = 5$ and $\tau = 11$, and the asymmetric eigenfunctions depicted in Figs. 13.7(b) and (d), for $\tau = 8$ and $\tau = 11$. The symmetric eigenfunctions correspond to purely exponentially growing modes, with 'smoothed jumps' precisely at the positions where $\sigma = 0$ and $\Omega_0(x) = 0$, and the asymmetric eigenfunctions correspond to overstable modes, where 'smoothed jumps' occur at the asymmetric positions $\sigma - \Omega_0(x) = 0$.

13.2.3 Combined instabilities of magnetized plasmas

In our third example (RTI+KHI in MHD), we superpose Rayleigh–Taylor and Kelvin–Helmholtz mechanisms in a plasma with a sheared magnetic field. In the first example (RTI, see Fig. 13.3(a)), a minimum of the forward Alfvén/slow continua gave rise to Rayleigh–Taylor instabilities connected with a *stable* cluster sequence concentrated at the lower edge of those continua. We now choose the parameters of the equilibria (13.56) such that a minimum of the *backward* continua $\Omega^-_{A/S}$ occurs inside the Doppler range (RTI+KHI, see Fig. 13.3(b)), so that the full cluster sequence of Rayleigh–Taylor modes becomes unstable. Since the backward continua are also embedded in the much larger range of the forward continua, a very intricate spectral structure arises.

As in the HD examples, the flow profile is chosen sinusoidal so that Kelvin–Helmholtz instabilities appear as well. The minimum $(\Omega^{-}_{\rm A/S})_{\rm min} \approx -0.0242$ occurs at $x_0 \approx 0.112$ (indicated by '2nd' in Fig. 13.3(b)), so that the Rayleigh–Taylor modes are expected to be localized on the left part of the plasma interval, whereas Fjørtoft's theorem (13.61) suggests that the Kelvin–Helmholtz modes should be concentrated on the right. In fact, the Spectral Web and eigenfunctions for the unstable modes shown in Figs. 13.9 and 13.10 below support these expectations. However, we first have to address some features that are less intuitive.

First, for an ordinary fluid in the presence of gravity and density stratification, the criteria of Rayleigh and Fjørtoft are superseded by criteria that involve the *Richardson number* J. This is a measure of the gravitational potential energy associated with the differences of the density with respect to the kinetic energy residing in the relative motions of the different fluid layers. By an estimate of the energy change due to the displacement of a fluid element (Chandrasekhar [112], p. 491), or by the construction of a special quadratic form (Drazin and Reid [164], p. 328), one may derive *a sufficient criterion for HD stability*, due to Howard [312],

$$J \equiv -\frac{\rho' g}{\rho v'^2} \geq \tfrac{1}{4}\,, \quad \text{or} \quad \rho' g + \tfrac{1}{4}\rho v'^2 \leq 0 \quad \text{(everywhere)}\,, \tag{13.63}$$

and a necessary condition for instability is that $J < \frac{1}{4}$ somewhere in the fluid. Hence, a fluid with a gravitationally stable density stratification ($\rho' \leq 0$) may still be unstable if the shear flow is large enough. For the present example, the criterion (13.63) is violated everywhere. This does not guarantee instability though since Howard's criterion is not valid for magnetized plasmas: magnetic shear may overcome both the gravitational and the shear flow instability drives.

Second, concerning the stability of a magnetized plasma, except for the mentioned extremum $\Omega^{-}_{\rm A/S}$, one also notices an additional special point in Fig. 13.3(b) (indicated by '1st') that lies precisely in the middle of the plasma ($x = 0.5$), where $\omega_{\rm A/S} = 0$ and, consequently, the two continuum frequencies $\Omega^{+}_{\rm A/S}$ and $\Omega^{-}_{\rm A/S}$ degenerate into the Doppler frequency Ω_0. For this frequency, the spectral equation (13.30) develops a quadratic singularity so that, again, a cluster sequence could appear there, but governed by an entirely different cluster criterion than the previous (13.58), viz. the criterion (13.66) below involving first derivatives. The detour on its derivation and discussion of implications for HD and MHD stability really interrupts the flow of the argument. Therefore, it is put in small print.

▷ **Cluster criteria and local gravitational interchanges** If a resonant surface $\mathbf{k}_0 \cdot \mathbf{B} = 0$ occurs at some point $x_{\rm res}$ in the plasma, i.e. $\omega_{\rm A}(x_{\rm res}) = 0$, the frequencies of the forward and backward continua coincide with the Doppler frequency. Hence, at the *real* frequency $\sigma = \Omega^{+}_{\rm A/S} = \Omega^{-}_{\rm A/S} = \Omega_0$, i.e. in the limit $\nu \longrightarrow 0$, the singular factor of the spectral equation (13.30) may be expanded as

$$\widetilde{\omega}^2 - \omega_{\rm A}^2 \equiv (\sigma - \Omega_0)^2 - \omega_{\rm A}^2 \approx (\Omega_0'^2 - \omega_{\rm A}'^2)\, s^2\,, \qquad s \equiv x - x_{\rm res}\,, \tag{13.64}$$

giving an equation similar to Eq. (7.189) governing local gravitational interchanges:

$$\frac{d}{ds}\left[s^2\,(1+\cdots)\,\frac{d\xi}{ds} \right] - q_0\,(1+\cdots)\,\xi = 0\,, \qquad q_0 \equiv \left(\frac{k_0^2 \rho' g}{\rho(\Omega_0'^2 - \omega_{\rm A}'^2)} \right)_{x_{\rm res}}\,. \tag{13.65}$$

The indices $\mu_{1,2} = -\frac{1}{2} \pm \sqrt{1+4q_0}$ are complex, associated with infinitely oscillatory behaviour of the solutions at $x = x_{\rm res}$, when $1 + 4q_0 < 0$, so that clustering occurs if

$$\text{either } k_0^2 \rho' g < -\tfrac{1}{4}\rho(\Omega_0'^2 - \omega_{\rm A}'^2) < 0\,, \qquad \text{or } k_0^2 \rho' g > \tfrac{1}{4}\rho(\omega_{\rm A}'^2 - \Omega_0'^2) > 0\,. \tag{13.66}$$

The first condition applies to gravitationally stable plasmas ($\rho' < 0$) when shear flow dominates over magnetic shear, and the second one applies to gravitationally unstable plasmas ($\rho' > 0$) when magnetic shear dominates over shear flow. In the present example, the latter applies but the right part of the cluster condition is not satisfied, as is evident from the continuum profiles of Fig. 13.3(b) at the resonant point. In the limits of HD ($\omega_A \to 0$) and static MHD ($\Omega_0 \to 0$), the two mutually exclusive conditions (13.66) for clustering reduce to a single possibility for each case, with precisely the opposite relation to stability. Considering the limit to HD first, the cluster condition (13.66)(a) reduces to the inequality (13.63) (without the equality part of the signs) so that it refers to clustering of stable modes. This condition was studied by Case in his second 1960 paper [108](b) for an exponentially decreasing density and a linear velocity profile, based on the analysis of the complex zeros of the solutions of the corresponding spectral equation by Dyson [168]. When the condition (13.63) is satisfied, in addition to the flow continua, the spectrum consists of a cluster sequence of stable g modes clustering towards the edge of the flow continua. It was also shown that this particular class of equilibria is stable for all values of J, demonstrating that Howard's criterion (13.63) is only sufficient, not necessary for stability. For completeness: a third paper in 1960 by Case [108](c) concerns the spectrum of static HD, where an unstable continuum is found, but this is due to the consideration of a gravitationally unstable inverted atmosphere *of infinite extent*.

In the limit to static MHD, the negation of the cluster condition (13.66)(b) yields the Schwarzschild–'Suydam' stability criterion (7.191) (in the limit $\gamma \to \infty$):

$$\rho' g - \tfrac{1}{4}B^2\varphi'^2 \leq 0 \qquad \text{(necessary for stability in static MHD)}\,. \tag{13.67}$$

When satisfied, it predicts stability because the stabilizing magnetic shear is larger than the destabilizing gravitational drive. It appears logical to combine this stability criterion for static MHD with Howard's HD criterion (13.63), but there is no justification for that (there exists no MHD counterpart of Howard's lucky juggling with quadratic forms). All one gets from the cluster condition (13.66)(b) is a sufficient criterion for instability involving two conditions that both have to be satisfied:

$$\rho' g > \tfrac{1}{4}B^2\varphi'^2 - \tfrac{1}{4}\rho v'^2 > 0 \qquad \text{(sufficient for \textit{instability} in stationary MHD)}\,. \tag{13.68}$$

Since this criterion is not satisfied in our RTI+KHI example, this implies that there are no local instabilities clustering at the resonant frequency $\sigma = 1.0$ in the middle of the plasma. Of course, this does not exclude the occurrence of global instabilities or of local instabilities clustering elsewhere. Finally, the generalization of the cluster criterion (13.66)(b) for compressible plasmas yields two separate criteria for Alfvén and slow resonances, which exhibit subtle cross-over limits that were studied for pressure-driven interchanges in cylindrical geometry by Hameiri [280] and Bondeson et al. [79]. The counterpart for gravitational interchanges in plane geometry is due to van der Holst et al. [620]. ◁

The Spectral Web for this example, shown in Fig. 13.9, exhibits all the stabilizing and destabilizing mechanisms mentioned. For large growth rates ($\nu > 3$), the solution path is tuned towards the most global gravitational interchange (labelled '1') associated with the resonant frequency $\sigma = \Omega^+_{A/S} = \Omega^-_{A/S} = \Omega_0 = 1.0$ at $x = 0.5$. The eigenfunction (not shown in Fig. 13.10) is completely global there. Thus, although local resonant instabilities do not occur in this example since the right inequality of the conditions (13.68) is not satisfied, the three terms of the left inequality do indicate the three mechanisms that produce global instability.

For smaller values of the growth rate ν, with the approach of the continua, the solution path breaks up in an infinite sequence of closed loops, each with an eigenvalue on it, and a number of branches fanning out from the point $\sigma = 1.0, \nu = 0$, but without actual eigenvalues there since the first derivative criterion (13.68) is not satisfied. However, the infinite series of eigenvalues on the closed loops of both the solution path and the conjugate path (shown in more detail in the inset (b) of Fig 13.9) is the result of violation of the second derivative condition (13.58)(b) at the extremum of the backward continua. This permits an infinite sequence of local interchange instabilities clustering towards the edge of the backward continua, which are embedded in the forward continua. One of the corresponding eigenfunctions is shown in the top frame of Fig. 13.10. Notice, again, that the

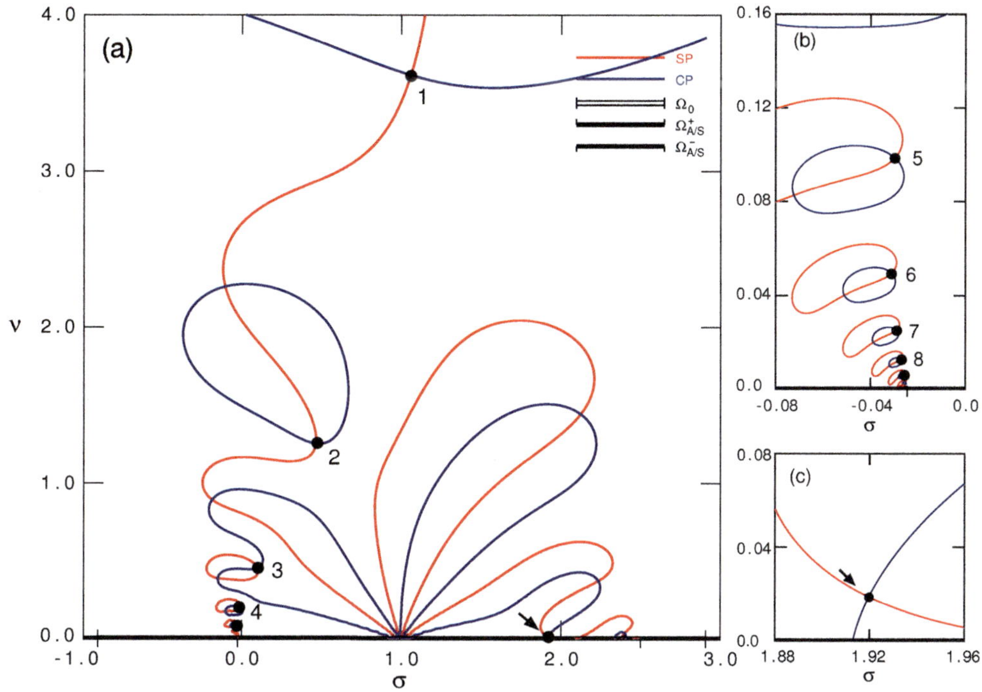

Fig. 13.9 Spectral Web for a gravitating plasma with magnetic shear and sinusoidal flow (RTI+KHI). (a) The mixed solution path ($x_{\rm mix} = 0.5$) at first connects the cluster sequence of Rayleigh–Taylor instabilities by a continuous curve, and then transforms into (b) a series of ever smaller loops about the eigenvalues; (c) an isolated Kelvin–Helmholtz instability is indicated by the arrow.

three terms of the instability conditions (13.58)(b) indicate the three mechanisms at work for this local instability which, although described by second instead of first derivatives of the equilibrium profiles, are the same as for the global interchanges.

The curves of the Spectral Web clearly guide the eye towards the significant regions in the ω-plane where eigenvalues are to be found. Thus, by zooming in onto the region about $\sigma \approx 1.920$, $\nu \approx 0.01867$ (inset (c) of Fig. 13.9) one finds a Kelvin–Helmholtz instability emerging from the continuous spectra below. This instability has an eigenfunction that is localized on the right part of the plasma (Fig. 13.10(b)). From the two eigenfunctions presented in Fig. 13.10 one cannot easily draw the above conclusions, viz. that the left one (shown on top) belongs to a gravitational interchange cluster sequence and that the right one (shown at the bottom) is actually a gravitationally modified Kelvin–Helmholtz instability. That conclusion was reached by some additional 'numerical experiments'. If one moves in the right boundary (e.g. by just considering the slice $0 \leq x \leq 0.8$), the growth rates of the most global modes decrease, but the cluster sequence of local modes remains concentrated at the same cluster point. However, if one moves in the left boundary (i.e. one removes a sizeable fraction of the gravitationally unstable part), the growth rate of the Kelvin–Helmholtz instability increases! Most convincingly, if one just switches off gravity (which one can do for these equilibria without affecting the continuum frequencies), the growth rate increases enormously (the eigenvalue moves from $\sigma \approx 1.92$, $\nu \approx 0.0187$ to $\sigma \approx 1.72$, $\nu \approx 0.168$) and the eigenfunc-

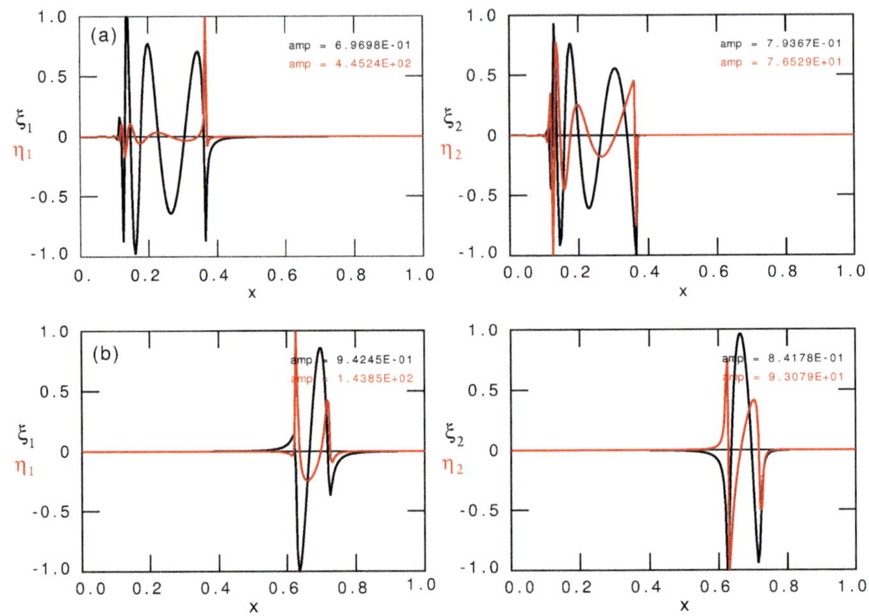

Fig. 13.10 Eigenfunctions for a gravitating plasma with magnetic shear and sinusoidal flow profile corresponding to the spectrum shown in Fig. 13.9 (RTI+KHI): (a) one of the Rayleigh–Taylor modes ($n = 12$) of the cluster spectrum, (b) an isolated Kelvin–Helmholtz instability.

tion then precisely shows the kind of dependence of the Kelvin–Helmholtz instability exhibited in Fig. 13.7(a). In other words: the Kelvin–Helmholtz instability is suppressed by the gravitational instability in these equilibria.

13.3 Spectral theory of rotating plasmas

13.3.1 MHD wave equation for cylindrical flow in 3D

Recall the basic spectral equation (12.95) for stationary equilibria:

$$\mathbf{G}(\boldsymbol{\xi}) - 2\rho\omega U\boldsymbol{\xi} + \rho\omega^2\boldsymbol{\xi} = 0\,. \tag{13.69}$$

For static ($U \equiv 0$) cylindrical equilibria, the reduction to a single second order ODE (the Hain–Lüst equation) was presented in Chapter 9. The substantial modification of this equation for *cylindrical plasmas with flow,* in particular *rotation,* was first obtained by Hameiri [280, 282], and later studied by Bondeson et al. [79]. We here follow the extension by Keppens et al. [345] with the effects of a gravitational field of cylindrical symmetry, which was applied to the thin slice approximation of the Magneto-Rotational Instability (MRI) of accretion discs.

The one-dimensional cylindrical equilibrium, described by the density $\rho(r)$, the pressure $p(r)$, the magnetic field components $B_\theta(r)$ and $B_z(r)$, and the velocity components $v_\theta(r)$ and $v_z(r)$, is restrained by the single differential equation (12.30) in r. For now, the gravitational potential $\Phi_{\rm gr}(r)$ is considered to be due to a singular mass distribution on axis. In Section 13.5.1, we will consider the more appropriate limit for application to a genuine gravitational field, viz. that of a compact central mass surrounded by an accretion disc.

As in Chapter 9, we consider Fourier normal modes of the form (9.22). The operators $\mathbf{G}$ and U now have contributions that prevent casting the spectral equation (13.69) into the simple form (13.5). The generalized force reads:

$$\mathbf{G}(\boldsymbol{\xi}) \equiv \hat{\mathbf{F}}_0(\boldsymbol{\xi}) + (\nabla\Phi_{\rm gr})\,\nabla\cdot(\rho\boldsymbol{\xi}) + \nabla\cdot\big(\boldsymbol{\xi}\,\rho\mathbf{v}\cdot\nabla\mathbf{v}\big) - \rho(\mathbf{v}\cdot\nabla)^2\boldsymbol{\xi}\,, \qquad (13.70)$$

where $\hat{\mathbf{F}}_0$ is the force operator of static equilibria without the gravitational contribution (indicated by the subscript 0; the hat indicates a subtlety discussed below). In cylindrical geometry, the three additional terms become

$$\begin{aligned}
(\nabla\Phi_{\rm gr})\,\nabla\cdot(\rho\boldsymbol{\xi}) &= \Phi'_{\rm gr}(\rho\nabla\cdot\boldsymbol{\xi} + \rho'\xi_r)\mathbf{e}_r\,,\\
\nabla\cdot(\boldsymbol{\xi}\rho\mathbf{v}\cdot\nabla\mathbf{v}) &= -\Big[(\rho v_\theta^2/r)\nabla\cdot\boldsymbol{\xi} + (\rho v_\theta^2/r)'\,\xi_r\Big]\mathbf{e}_r - (\rho v_\theta^2/r^2)\xi_\theta\,\mathbf{e}_\theta\,,\\
-\rho(\mathbf{v}\cdot\nabla)^2\boldsymbol{\xi} &= \rho(v_\theta^2/r^2)(\xi_r\mathbf{e}_r + \xi_\theta\mathbf{e}_\theta) + 2\rho\Omega_0\mathbf{C}(\boldsymbol{\xi}) + \rho\Omega_0^2\boldsymbol{\xi}\,. \qquad (13.71)
\end{aligned}$$

Since the gradient operator not only acts on the vector components of $\boldsymbol{\xi}$ but also on the unit vectors, the operator $U \equiv -\mathrm{i}\mathbf{v}\cdot\nabla$ not only produces a local Doppler shift $\Omega_0(r)$, but also an additional local *Coriolis shift* proportional to the vector product of the rotation vector $(v_\theta/r)\mathbf{e}_z$ and $\boldsymbol{\xi}$:

$$U\boldsymbol{\xi} \equiv -\mathrm{i}\mathbf{v}\cdot\nabla\boldsymbol{\xi} = \Omega_0\boldsymbol{\xi} + \mathbf{C}(\boldsymbol{\xi})\,, \quad \Omega_0 \equiv mv_\theta/r + kv_z\,, \quad \mathbf{C}(\boldsymbol{\xi}) \equiv -\mathrm{i}(v_\theta/r)\mathbf{e}_z\times\boldsymbol{\xi}\,. \quad (13.72)$$

For plane flow along the z-axis, the simple 'quasi-static' form (13.5) of the spectral equation still applies, with the eigenvalue only appearing as the square of the Doppler-shifted frequency

$$\widetilde{\omega}(r) \equiv \omega - \Omega_0(r)\,, \qquad (13.73)$$

not involving the Coriolis shift. With rotation, the full quadratic eigenvalue problem (13.69) applies, bringing in both the linear term in ω and the Coriolis terms.

We exploit the field line projection (9.23)–(9.26), involving the dimensionless 'wave numbers' $k_\perp(r) \equiv G/B$ and $k_\parallel(r) \equiv F/B$, where $G \equiv mB_z/r - kB_\theta$ and $F \equiv mB_\theta/r + kB_z$, the dimensionless field magnitudes $b_\theta \equiv B_\theta/B$ and $b_z \equiv B_z/B$, and the displacement $\boldsymbol{\xi}$ with components $\xi \equiv \xi_r$, $\eta \equiv \mathrm{i}(B_z\xi_\theta - B_\theta\xi_z)/B$, $\zeta \equiv \mathrm{i}(B_\theta\xi_\theta + B_z\xi_z)/B$. To construct the counterpart of the static spectral problem (9.28) recall that a term $\mathbf{T}(\boldsymbol{\xi})$ could be dropped there due to the static equilibrium. That term now contributes:

$$\hat{\mathbf{F}}_0(\boldsymbol{\xi}) = \mathbf{F}_0(\boldsymbol{\xi}) + \mathbf{T}(\boldsymbol{\xi})\,, \qquad \mathbf{T}(\boldsymbol{\xi}) \equiv \nabla(r\Lambda\xi)\,, \qquad (13.74)$$

where the function

$$\Lambda(r) \equiv \rho(v_\theta^2 - r\Phi'_{\rm gr})/r^2 = \big[r(p + \tfrac{1}{2}B^2)' + B_\theta^2\big]/r^2 \qquad (13.75)$$

represents the deviation from static MHD equilibrium caused by rotation and gravity, or, in the context of accretion flows, the deviation from pure Keplerian HD flow caused by pressure gradients and Lorentz forces.

The terms (13.71) and (13.74) combine into a symmetric expression,

$$\mathbf{R}(\boldsymbol{\xi}) \equiv -\big[\rho r(\Phi'_{\rm gr}/r)'\xi + r\Lambda(k_\perp\eta + k_\parallel\zeta)\big]\,\mathbf{e}_r + \mathrm{i}r\Lambda\xi(k_\perp\mathbf{e}_\perp + k_\parallel\mathbf{e}_\parallel)\,, \qquad (13.76)$$

plus two terms involving the Doppler shift Ω_0. The matrix representation of the generalized force operator then becomes

$$\mathbf{G} = \mathbf{F}_0 + \mathbf{R} + 2\rho\Omega_0\mathbf{C} + \rho\Omega_0^2\mathbf{I}\,, \qquad (13.77)$$

so that the spectral equation (13.69) transforms into

$$(\mathbf{F}_0 + \mathbf{R} - 2\rho\widetilde{\omega}\mathbf{C} + \rho\widetilde{\omega}^2\mathbf{I})\cdot\mathbf{X} = 0\,, \qquad \mathbf{X} \equiv (\xi,\eta,\zeta)^{\mathrm{T}}\,. \qquad (13.78)$$

Here, $\mathbf{F}_0$ is the previous matrix of the static case without gravity and flow, defined in Eq. (9.28), $\mathbf{R}$ represents the effects of the gravity and partially of the rotation, whereas $\mathbf{C}$ represents the Coriolis effects of the rotation:

$$\mathbf{R} \equiv -\begin{pmatrix} \rho r(\Phi'_{\mathrm{gr}}/r)' & rk_\perp\Lambda & rk_\parallel\Lambda \\ rk_\perp\Lambda & 0 & 0 \\ rk_\parallel\Lambda & 0 & 0 \end{pmatrix}, \qquad \mathbf{C} \equiv \frac{v_\theta}{r}\begin{pmatrix} 0 & b_z & b_\theta \\ b_z & 0 & 0 \\ b_\theta & 0 & 0 \end{pmatrix}. \qquad (13.79)$$

The self-adjointness of the operators **G** and U is now expressed by the symmetry of the matrices $\mathbf{F}_0$, $\mathbf{R}$ and $\mathbf{C}$. This completes the three-dimensional representation.

13.3.2 Reduction to a second order differential equation

Expressing the tangential components η and ζ in terms of the radial component $\chi \equiv r\xi$ of the plasma displacement through the tangential components of Eq. (13.78), see Eqs. (13.93) and (13.94) below, and substitution into the radial component, results in the following one-dimensional form of the MHD wave equation:

$$\frac{d}{dr}\left(\frac{N}{D}\frac{d\chi}{dr}\right) + \left[A + \frac{B}{D} + \left(\frac{C}{D}\right)'\right]\chi = 0\,, \qquad (13.80)$$

with polynomial coefficients defined below in Eqs. (13.84), (13.85), (13.89)–(13.91). An eigenvalue problem is obtained by subjecting the solutions to the BCs

$$\chi(0) = 0\,, \qquad \chi(a) = 0\,. \qquad (13.81)$$

Recall that ξ has been replaced by the variable χ just to be able to express the BCs on the origin in this way, rather than to have to distinguish between the finite $|m| = 1$ Fourier components of $\xi(0)$ and the vanishing ones (see Section 9.2.1).

The corresponding formulation of two coupled first order differential equations for χ and the Eulerian total pressure variation $\Pi \equiv -(1/D)(N\chi' + C\chi)$ reads:

$$N\frac{d}{dr}\begin{pmatrix} \chi \\ \Pi \end{pmatrix} + \begin{pmatrix} C & D \\ E & -C \end{pmatrix}\begin{pmatrix} \chi \\ \Pi \end{pmatrix} = 0\,, \qquad (13.82)$$

where the new function

$$E \equiv -N\big(A + B/D\big) - C^2/D \qquad (13.83)$$

may again be transformed in a polynomial in ω, see Eq. (13.92) below.

With the addition of stationary flows to the equilibrium, the expressions with the eigenvalue ω transform into expressions involving the local Doppler shifted frequency $\widetilde{\omega}$. Together with the parallel gradient operator $F \equiv mB_\theta/r + kB_z$, they determine the *Alfvén and slow continuum singularities* $N = 0$, where

$$N \equiv \frac{1}{r}\widetilde{A}\widetilde{S} \equiv \frac{\rho^2(\gamma p + B^2)}{r}(\omega - \Omega_\mathrm{A}^+)(\omega - \Omega_\mathrm{A}^-)(\omega - \Omega_\mathrm{S}^+)(\omega - \Omega_\mathrm{S}^-),$$
$$\widetilde{A} \equiv \rho(\widetilde{\omega}^2 - \omega_\mathrm{A}^2), \qquad \widetilde{S} \equiv \rho(\gamma p + B^2)(\widetilde{\omega}^2 - \omega_\mathrm{S}^2), \tag{13.84}$$

and the *apparent (fast and slow turning point) singularities* $D = 0$, where

$$D \equiv \rho^2\widetilde{\omega}^4 - h^2\widetilde{S} \equiv \rho^2(\omega - \Omega_\mathrm{f0}^+)(\omega - \Omega_\mathrm{f0}^-)(\omega - \Omega_\mathrm{s0}^+)(\omega - \Omega_\mathrm{s0}^-),$$
$$\text{with} \quad h^2 \equiv k_\perp^2 + k_\parallel^2 \equiv m^2/r^2 + k^2. \tag{13.85}$$

The associated definitions of the *forward and backward Alfvén and slow continua,*

$$\Omega_\mathrm{A}^\pm(r) \equiv \Omega_0(r) \pm \omega_\mathrm{A}(r), \qquad \Omega_\mathrm{S}^\pm(r) \equiv \Omega_0(r) \pm \omega_\mathrm{S}(r),$$
$$\omega_\mathrm{A}^2(r) \equiv F^2/\rho, \qquad \omega_\mathrm{S}^2(r) \equiv \frac{\gamma p}{\gamma p + B^2}F^2/\rho, \tag{13.86}$$

and of the *forward and backward apparent fast and slow singularities,*

$$\Omega_\mathrm{f0}^\pm(r) \equiv \Omega_0(r) \pm \omega_\mathrm{f0}(r), \qquad \Omega_\mathrm{s0}^\pm(r) \equiv \Omega_0(r) \pm \omega_\mathrm{s0}(r), \tag{13.87}$$
$$\omega_\mathrm{f0,s0}^2(x) \equiv \tfrac{1}{2}\omega_\mathrm{m}^2\Big[1 \pm \sqrt{1 - 4\omega_\mathrm{S}^2/\omega_\mathrm{m}^2}\,\Big], \quad \omega_\mathrm{m}^2 \equiv h^2(\gamma p + B^2)/\rho,$$

only differ from the plane slab definitions in that they contain an 'azimuthal wave number' contribution m/r in the effective wave number h, which replaces k_0. With these definitions, and also supplementing the cluster frequencies $\Omega_\mathrm{F}^\pm \equiv \pm\infty$ of the fast magneto-sonic modes, the ordering (13.24) of the local frequencies of the genuine and apparent singularities again applies. Notice though that $\Omega_\mathrm{f0}^\pm \to \pm\infty$ for $r \to 0$ (another sign that this singularity is not genuine).

With the above definitions for N and D, and those for A, B, C and E given in Eqs. (13.89)–(13.92) below, the formulations in terms of the second order ODE (13.80) for χ and the system of first order ODEs (13.82) for χ and Π are formally identical to the formulations (13.11) and (13.17) for the stationary plane slab. Consequently, the fundamental relationship (13.83) between the two formulations is identical to the relation (13.18). Thus, the expression for the coefficient E can again be rewritten in the two forms given by the relations (13.19) and (13.20), which, successively, guarantee that the $D = 0$ singularities of the second order ODE are apparent ones, and that no spurious $N = 0$ singularities are introduced in the system of first order ODEs. Hence, the conclusion of Section 13.1.1 applies again: *The apparent singularities of the second order representation and the redundant singularities of the first order representation are complementary evils*. The controversies on the different formulations of the spectral problem [254], [18], [221], related to the elimination of apparent singularities, are thus finally resolved by observing that one representation is not superior to the other and that the relationship (13.83) is essential to remove the spurious singularities of both.

▷ **Explicit expressions of the coefficients of the differential equations** For numerical purposes, it is expedient to renormalize the variables with r by redefining $\bar{h}^2 \equiv r^2h^2$, $\bar{B}_\theta \equiv B_\theta/r$ and $\bar{v}_\theta \equiv v_\theta/r$, and to introduce abbreviations for two equilibrium functions, $\Delta(r)$ and $\Lambda(r)$, and three perturbation functions, $P(r;\widetilde{\omega})$, $Q(r;\widetilde{\omega})$ and $R(r;\widetilde{\omega})$:

$$\Delta \equiv r\big(\bar{B}_\theta^2 - \rho\bar{v}_\theta^2\big)' + \rho'\Phi_{\rm gr}' , \qquad \Lambda \equiv \rho(\bar{v}_\theta^2 - \Phi_{\rm gr}'/r) , \tag{13.88}$$

$$P \equiv \bar{B}_\theta F + \rho\bar{v}_\theta\widetilde{\omega} , \qquad Q \equiv \bar{B}_\theta F P + \bar{B}_\theta^2 \widetilde{A} , \qquad R \equiv \bar{B}_\theta^2\big[2mP - \bar{h}^2(\bar{B}_\theta^2 + \rho\bar{v}_\theta^2)\big] .$$

The coefficients A, B, C and E of the ODEs (13.80) and (13.82) are then defined by

$$A \equiv \frac{1}{r}\big(\widetilde{A} + \Delta\big) , \tag{13.89}$$

$$B \equiv -\frac{4}{r}\Big\{(\rho\widetilde{\omega}^2 - k^2\gamma p)P^2 - \rho\widetilde{\omega}^2 R + \big(m\rho\widetilde{\omega}^2 P - \bar{h}^2 Q\big)\Lambda + \tfrac{1}{4}\bar{h}^2\widetilde{A}\Lambda^2\Big\} , \tag{13.90}$$

$$C \equiv \frac{2}{r^2}\Big\{m\widetilde{S}P - r^2\rho\widetilde{\omega}^2\big(Q - \tfrac{1}{2}\widetilde{A}\Lambda\big)\Big\} , \tag{13.91}$$

$$E \equiv -\frac{1}{r^2}\Big\{\widetilde{A}\widetilde{S}\big(\widetilde{A} + \Delta\big) - 4\widetilde{S}P^2 + 4r^2\big(Q - \tfrac{1}{2}\widetilde{A}\Lambda\big)^2\Big\} . \tag{13.92}$$

Hence, cylindrical flow does not change the formal form of the singular expressions N and D, but the coefficients A, B, C and E are modified substantially. The function $A(r;\omega)$ is a quadratic polynomial in the parameter ω, whereas the functions N, D, B, C are quartic, and E is a sixth order polynomial in ω.

The perpendicular and parallel components of $\boldsymbol{\xi}$ follow directly from χ:

$$\eta = \frac{G\widetilde{S}\chi' - 2\big[k\gamma pFP - r(\bar{B}_z P - G\bar{B}_\theta^2 + \tfrac{1}{2}G\Lambda)\rho\widetilde{\omega}^2\big]\chi}{rBD} , \tag{13.93}$$

$$\zeta = \frac{F\gamma p\widetilde{A}\chi' + 2\big[k\gamma pGP + r(\bar{B}_\theta P - F\bar{B}_\theta^2 + \tfrac{1}{2}F\Lambda)(\rho\widetilde{\omega}^2 - h^2B^2)\big]\chi}{rBD} . \tag{13.94}$$

The expressions $\widetilde{A}$, $\widetilde{S}$, D and h^2 are defined in Eqs. (13.84) and (13.85).

For numerical integration of the ODEs (13.82) for complex values of ω, all coefficients are split in real and imaginary parts, resulting in four coupled ODEs for the components χ_1. χ_2, Π_1 and Π_2, as in Eq. (13.49) for the plane case, where the coefficients $\hat{C}_{1,2}$, $\hat{D}_{1,2}$ and $\hat{E}_{1,2}$ are normalized as in Eq. (13.50). Their explicit form is obtained by straightforward expansion of Eqs. (13.89)–(13.92). ◁

The differential equation (13.80) is formally identical to the spectral wave equation (13.11) for the stationary plane slab, but also to the spectral equations for the static case presented in Chapter 7 for the plane slab and Chapter 9 for the cylindrical plasma. In particular, the expressions A, B, C transform into the corresponding ones (9.33) of the generalized Hain–Lüst equation in the limit of vanishing flow and gravity ($\widetilde{\omega} \to \omega$, $v_\theta \to 0$, $\Phi_{\rm gr}' \to 0$). Of course, this does not imply that the spectral problem for the stationary cylinder is just as simple as that for the static cylinder, but it does imply that the analytical tools of *Spectral Web* and *oscillation theorems* governing the distribution of discrete eigenvalues on them, developed in Sections 12.3, 13.1 and 13.2, and applied there to shear flow of plane plasmas, can be applied to cylindrical plasmas as well. This we have already indicated in the formulations of the two oscillation theorems of Section 13.1.3 and 13.1.4.

13.3.3 Singular expansions★

The singularities $\widetilde{\omega}^2 = \omega_{\rm A}^2$ and $\widetilde{\omega}^2 = \omega_{\rm S}^2$ of the differential equation (13.80) permit expansions leading to local stability criteria and cluster criteria about special points in the continuous spectra. Their derivation is completely analogous to that of Suydam's criterion in Section 9.4.1 and of the cluster criteria in Section 9.3.3 for static equilibria. The fundamental new feature for stationary equilibria is the appearance of the *local shear Alfvén Mach number*, together with its slow critical value:

$$\widehat{M} \equiv \frac{\Omega_0 - \Omega_0(r_s)}{\omega_A - \omega_A(r_s)} \approx \frac{\Omega_0{}'}{\omega_A{}'}\bigg|_{r_s}, \qquad M_c \equiv \sqrt{\frac{\gamma p}{\gamma p + B^2}}\bigg|_{r_s}, \tag{13.95}$$

where $\widehat{M}/M_c \approx (\Omega_0{}'/\omega_S')_{r_s}$. As before, the distance to a singular point is indicated by $s \equiv r - r_s$, where $r = r_s$ is the singular point.

(a) Suydam rational surface expansion Recall from the analysis of Section 9.4.1 that Suydam's criterion is obtained by expanding the marginal equation of motion ($\omega = 0$) about a singular point where $F \equiv k_\parallel B$ vanishes. The dominant magnetic field perturbations are then minimized and interchanges, driven by the usually much smaller pressure gradient effects, can develop. At such points, four genuine and two apparent singularities of the spectral differential equation coalesce ($\omega^2 = \omega_A^2 = \omega_S^2 = \omega_{s0}^2 = 0$). For the analogous theory of stationary plasmas, we need to consider resonant modes ($F = 0$) that are marginal in the co-moving Doppler-shifted frame ($\widetilde{\omega} = 0$):

$$F = 0\,, \quad \widetilde{\omega} = 0 \qquad \Rightarrow \qquad \omega = \Omega_0 = \Omega_A^{\pm} = \Omega_S^{\pm} = \Omega_{s0}^{\pm} \qquad (\text{at } r = r_s)\,, \tag{13.96}$$

so that, again, confluence of six singularities occurs. Expansion of the coefficients of the spectral differential equation (13.80) about the singular point $r = r_s$ yields

$$\widetilde{\omega} \approx -\Omega_0' s\,, \quad F \approx -kB_z(\mu'/\mu)\, s\,, \quad \text{where } \mu \equiv B_\theta/(rB_z)\,, \tag{13.97}$$

so that the functions $\widetilde{A}$ and $\widetilde{S}$ defined in Eqs. (13.84) become

$$\widetilde{A} \approx -(1 - \widehat{M}^2)F^2\,, \qquad \widetilde{S} \approx -(\gamma p + B^2)(M_c^2 - \widehat{M}^2)F^2\,. \tag{13.98}$$

The rest of the analysis proceeds analogous to that of Section 9.4.1.

In the expansion of N/D, the familiar magnetic shear term $(\mu'/\mu)^2$ is now multiplied by the trans-Alfvénic factor,

$$\frac{N}{D} \approx \alpha s^2\,, \qquad \alpha \equiv \frac{B_\theta^2 B_z^2}{rB^2}(1 - \widehat{M}^2)\left(\frac{\mu'}{\mu}\right)^2, \tag{13.99}$$

and the spectral differential equation for $\omega \approx \Omega_0(r_s)$ reduces to

$$\alpha \frac{d}{ds}\left(s^2 \frac{d\chi}{ds}\right) - \beta\chi = 0\,, \qquad \beta \equiv -\left[A + \frac{B}{D} + \left(\frac{C}{D}\right)'\right]_{r_s}. \tag{13.100}$$

Close to the singularity $s = 0$, the solutions behave as

$$\chi = as^{\nu_1} + bs^{\nu_2}\,, \qquad \nu_{1,2} = -\tfrac{1}{2} \pm \tfrac{1}{2}\sqrt{1 + 4\beta/\alpha}\,, \tag{13.101}$$

which gives rise to solutions with infinitely many oscillations as $s \to 0$ if the indices are complex. Hence, local stability demands the opposite:

$$1 + 4\beta/\alpha > 0\,. \tag{13.102}$$

Substitution of the coefficients α and β in this inequality produces a formidable expression, the generalization of Suydam's criterion of a cylindrical plasma with background flow. It was first obtained by Hameiri [280, 282] and investigated numerically by Bondeson et al. [79]. The implications of violation of this criterion, and the resulting eigenfunctions of the modes, were investigated by means of a boundary layer analysis; see the references cited.

Without rotation, shear flow and gravity, Suydam's criterion [581] is recovered. Without rotation but *with shear flow* $v_z(r)$ *in the longitudinal direction,* the pressure gradient of the generalized Suydam's criterion is divided by the trans-Alfvénic factor and a compressional term appears that is divided by both factors:

$$\frac{1}{1-\widehat{M}^2}p' - \frac{2B_\theta^2 M_c^2 \widehat{M}^2}{r(M_c^2-\widehat{M}^2)(1-\widehat{M}^2)} + \tfrac{1}{8} r B_z^2 \Big(\frac{\mu'}{\mu}\Big)^2 > 0\,. \tag{13.103}$$

Clearly, the simple picture of instability driven by a pressure gradient and stabilized by magnetic shear is no longer valid in the presence of shear flow. New terms occur and their influence on stability or instability changes depending on the magnitude of $\widehat{M}^2$, giving rise to the following shear flow regimes:

$$\begin{cases} 0 \le \widehat{M}^2 < M_c^2 & \textit{(sub-slow)}, \\ M_c^2 \le \widehat{M}^2 < 1 & \textit{(slow/sub-Alfvénic)}, \\ 1 \le \widehat{M}^2 & \textit{(super-Alfvénic)}. \end{cases} \tag{13.104}$$

The flow regimes will return in a different guise in Chapter 18 (unstable toroidal Alfvén–slow continua) and Chapter 21 (transonic MHD flows and shocks). The Suydam analysis presented is evidently invalid for $\widehat{M}^2 \to 1$ and $\widehat{M}^2 \to M_c^2$.

(b) Alfvén and slow cluster point expansions At an extremum of one of the Alfvén continua, where $\Omega_A^{\pm\prime}(x_s) = 0$ so that $\widehat{M} = \pm 1$, the relevant expansion becomes: $\sigma - \Omega_A^{\pm} \approx -\frac{1}{2}\Omega_A^{\pm\prime\prime} s^2$. Again, the exponent n of the dominant solution is complex, and a cluster point of infinitely rapidly oscillating solutions results, if

$$\begin{aligned} &\text{either} \quad 0 < \pm\tfrac{1}{4}\omega_A \Omega_A^{\pm\prime\prime} < Q_A\,, \qquad \text{or} \quad 0 > \pm\tfrac{1}{4}\omega_A \Omega_A^{\pm\prime\prime} > Q_A\,, \\ &\text{where} \quad Q_A \equiv \frac{rG^2}{B^2}\Big(\frac{B_\theta^2}{r^2} - \frac{2B_\theta B_z F}{r^2 G}\Big)' + \frac{4k^2 B_\theta^2}{r^2}\frac{1-2M_c^2}{1-M_c^2}\,, \end{aligned} \tag{13.105}$$

corresponding to an infinite sequence of stable *Global Alfvén Eigenmodes* (GAEs). Similarly, at an extremum of one of the slow continua, where $\Omega_S^{\pm\prime}(x_s) = 0$ so that $M = \pm M_c$, the expansion yields the conditions for a slow cluster point:

$$\begin{aligned} &\text{either} \quad 0 < \pm\tfrac{1}{4}\rho\omega_S \Omega_S^{\pm\prime\prime} < Q_S\,, \qquad \text{or} \quad 0 > \pm\tfrac{1}{4}\rho\omega_S \Omega_S^{\pm\prime\prime} > Q_S\,, \\ &\text{where} \quad Q_S \equiv -\frac{M_c^4}{B^2}\Big[F^4(1-M_c^2) + rF^2\Big(\frac{B_\theta^2}{r^2}\Big)' - \frac{4k^2 B_\theta^2 B^2}{r^2(1-M_c^2)}\Big]\,, \end{aligned} \tag{13.106}$$

corresponding to an infinite sequence of stable *Global Slow Eigenmodes* (GSEs). These conditions transform into the ones presented in Section 9.3.3 for the static case (where $\pm\Omega_{A,S}^{\pm\prime\prime} \to \omega_{A,S}''$). Clearly, for incompressible modes ($M_c^2 = 1$), the expansion breaks down and a revision is needed; see Ref. [229].

13.3.4 Doppler–Coriolis shift and solution path

We have presented all the basic equations for the analysis of the spectrum of stationary plasma flow in cylindrical geometry. The most important difference with the analysis of the plane slab in the preceding sections is the contribution of rotation. This clearly shows up in the expression for the *solution averaged Doppler–Coriolis shift* in the equation for the solution path of unstable modes:

$$\sigma = \overline{V} = \int \rho \Big[r\Omega_0 |\boldsymbol{\xi}|^2 + \mathrm{i}(v_\theta/r)(\chi^* \xi_\theta - \chi \xi_\theta^*) \Big] \, dr \Big/ \!\! \Big/ \int \rho |\boldsymbol{\xi}|^2 r \, dr \,. \tag{13.107}$$

The term with Ω_0 represents the Doppler part, as in Eq. (13.7) for the plane slab, and the term with v_θ/r represents the Coriolis part of the shift. Both shifts are real. To evaluate the volume integrals, one needs to substitute ξ_θ and ξ_z in terms of the solutions χ and Π of the ODEs (13.82).

In the Spectral Web method, instead, the iteration on the eigenvalues is performed by determining the zeros of the two components of the complementary energy according to the scheme (12.120)–(12.121). The explicit expressions for $W_{\rm com}$ are given by Eqs. (12.119), (12.126) and (12.128) of Chapter 12 for the left, right and mixed internal solutions, respectively. For the left solutions of a cylinder with finite length L, the complementary energy reduces to

$$W^\ell_{\rm com} = \pi L \big(\chi^* \Pi\big)_{r=a} \quad \Rightarrow \left\{ \begin{array}{ll} \pi L \big(\chi_1 \Pi_2 - \chi_2 \Pi_1\big)_{r=a} = 0 & \textit{(solution path)} \\[2ex] \pi L \big(\chi_1 \Pi_1 + \chi_2 \Pi_2\big)_{r=a} = 0 & \textit{(conjugate path)} \end{array} \right. . \tag{13.108}$$

Similar simple expressions are easily derived for $W^{\rm r}_{\rm com}$ and $W^{\rm mix}_{\rm com}$. Whereas the volume integral (13.107) for the solution path is physically quite significant, the equivalent relation (13.108)(a) is clearly to be preferred since it just requires evaluation of an expression involving the known boundary data of χ and Π.

Recall that the complex oscillation theorem involves the complementary energy through the *alternator*:

$$W^\ell_{\rm com} = \pi L \big|\Pi^\ell\big|^2_{r=a} R^{\ell *} \,, \qquad R^\ell \equiv \chi^\ell(a)/\Pi^\ell(a) \,, \tag{13.109}$$

and similarly for $R^{\rm r}$ and $R^{\rm mix}$ as defined in Eqs. (13.55) replacing ξ by χ. Along the solution path, where $W_{\rm com}$ is real, the alternator is also real, it vanishes at the eigenvalues, and it is monotonic in between the zeros of the boundary values of the other variable (where the alternator blows up). Similarly, along the conjugate path, where $W_{\rm com}$ is purely imaginary, the alternator vanishes at the eigenvalues, and its imaginary part is monotonic in between the zeros of the boundary values of the other variable. The behaviour of the alternator as a tangent (see Fig. 13.5) facilitates extremely rapid convergence to the eigenvalues.

The examples of the following sections exploit the mixed representation with the general spectral equations (13.82) for compressible perturbations, where the sole restriction will be satisfaction of the equilibrium equation (12.30). The solutions are obtained by 'shooting' from the left ($r = 0$) and from the right ($r = a$), and joining the two fragments at some internal point ($r = r_{\rm mix}$) by equating χ there. The Spectral Web is then obtained by contour plotting the real and imaginary parts of $W^{\rm mix}_{\rm com} = 0$ and the associated eigenvalues by iterating along the solution path until $R^{\rm mix}$ is smaller than some specified small number. This has been applied to some of the classical MHD instabilities in cylinder geometry. A first application to the internal kink modes of a force-free magnetic field (a continuation of Figs 12.4 and 12.5 of Section 12.4) is skipped here because it 'only' involves uni-directional shear flow. The reader should consult Section III of the original Spectral Web paper [229] for the complexities this problem entails though. Two other examples of instabilities from the cited paper, involving rotating equilibria with a significant contribution of the Coriolis effects, are presented here, viz. the Rayleigh–Taylor instability of a rotating theta-pinch (Section 13.4) and the Magneto-Rotational Instability (MRI) of an accretion disc (Section 13.5).

13.4 Rayleigh–Taylor instabilities in rotating theta-pinches

Our first example concerns the $|m| = 1$ modes of rigidly rotating equilibria of theta pinch-like plasma configurations in the straight cylinder approximation, as exemplified by the ordinary theta pinch [190, 189] and the Field Reversed Configuration (FRC) [567]. Although of historical interest only, we will treat the rather disastrous instabilities of the rotating theta pinch more extensively since they provide an excellent illustration of the Spectral Web method which also applies to the more subtle instabilities of modern devices.

The stability of the rotating theta pinch was used in the original literature on the stabilization of MHD instabilities by the finiteness of the ion gyro radius [590], which is least effective for the $|m| = 1$ modes. This literature contains some confusing results that have not been clarified before. In particular, Taylor [590] makes the erroneous remark that the growth rate of the instability for vanishing longitudinal mode number ($k = 0$), i.e. the imaginary part of his expression for the complex frequency,

$$\omega = \Omega\big[m - \mathrm{sgn}(m) \pm \mathrm{i}\sqrt{|m| - 1}\big], \tag{13.110}$$

is "quite independent of the plasma profile". In Eq. (13.110), the Coriolis shift $(\Delta\sigma)_{\mathrm{C}} = -\mathrm{sgn}(m)\,\Omega$ precisely cancels the Doppler shift $(\Delta\sigma)_{\mathrm{D}} = m\Omega$ so that the real part of the complex frequency also vanishes. This would imply marginal stability of the $|m| = 1$ modes for these configurations in general. However, this expression is only valid for the special case he investigated, viz. a theta pinch with a step function density profile without a wall. Effectively, for the incompressible $|m| = 1$, $k = 0$ modes of that equilibrium, there is no perturbation at all since that 'perturbation' simply corresponds to a translation to a new equilibrium state. A detailed numerical calculation by Freidberg and Wesson [190] appeared to confirm Taylor's result, by searching in the complex ω-plane for solutions in the neighbourhood of the imaginary axis as suggested by Eq. (13.110). A later study by Freidberg and Pearlstein [189] actually found substantial deviations from Eq. (13.110) for an additional class of internal $|m| = 1$ modes. The Spectral Web method will highlight one of the intricacies involved in the stability of rotating stationary equilibria, viz. that the more local internal $|m| = 1$ modes actually grow faster than the most global one!

Rigidly rotating theta pinch equilibria are described by satisfying Eq. (12.30) with constant angular frequency $\Omega \equiv v_\theta / r$ and the following choice of the profiles:

$$\begin{aligned} \rho(r) &= \rho_0\,\mathrm{sech}^2 f(x)\,, \qquad p(r) = p_0\,\mathrm{sech}^2 f(x)\,, \\ B_z(r) &= B_\infty\big[\delta + (1-\delta)\tanh f(x)\big]\,, \quad f(x) \equiv \alpha^2(x^2 - x_0^2)\,, \end{aligned} \tag{13.111}$$

where $x \equiv r/a$, whereas $p_0 = \frac{1}{2}(1-\delta)^2 B_\infty^2$ and $\Omega = \alpha\sqrt{2\delta(1-\delta)}\,B_\infty/(a\sqrt{\rho_0})$. The plasma radius a, the maximum density ρ_0, and the asymptotic value B_∞ of the magnetic field provide parameters to make the basic quantities dimensionless, like the unit plasma interval $0 \le x \equiv r/a \le 1$, the longitudinal wave number $\bar{k} \equiv k\,a$ and the complex frequency $\overline{\omega} \equiv \omega\, a\sqrt{\rho_0}/B_\infty$. Consequently, only three genuinely free dimensionless parameters remain, viz. the normalized radius x_0 at maximum density, the stretching parameter α, and the magnetic field deviation parameter δ. The parameter x_0 (describing the FRC equilibria) vanishes for the theta pinch equilibrium. The parameters α and δ are more conveniently represented by the dimensionless pressure, $\beta \equiv 2\bar{p}_0 = (1-\delta)^2$, and the dimensionless angular rotation frequency, $\bar{\Omega} = \alpha\sqrt{2\delta(1-\delta)}$. We will exploit these parameters, dropping the bars. A representative equilibrium is depicted in Fig. 13.11.

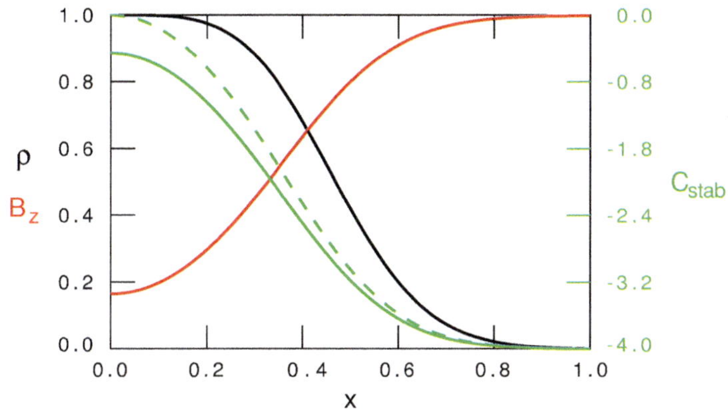

Fig. 13.11 Equilibrium density (black) and magnetic field (red) of a rotating theta pinch for $\beta = 0.6944$ and $\Omega = 1.0542$ (corresponding to $x_0 = 0$, $\alpha = 2$, $\delta = 0.1667$). The local HD stability criterion $C_{\rm stab}$ is indicated in green for compressible (drawn) and incompressible (dashed) plasmas.

The rotating theta pinch equilibrium of Fig. 13.11 has decreasing density and pressure profiles over the whole plasma domain, and a uni-directional increasing magnetic field profile. Since the stabilizing magnetic field terms $k^2B_z^2$ vanish for the $k=0$ modes, the instabilities are essentially hydrodynamic, Rayleigh–Taylor, instabilities driven by the negative density gradient and the centrifugal acceleration [281]. Magnetic stabilization occurs for the $k \neq 0$ modes, but a large range $0 \le |k| \le |k_0|$ remains unstable. Similar conclusions hold for the straight cylinder approximation of the FRC equilibrium, but the instability is less severe because the density increases in the region $0 \le x \le x_0$, so that the inner region is stable, whereas the magnetic field reversal, and associated singularity $\mathbf{k}\cdot\mathbf{B} = 0$ close to $x = x_0$, significantly modifies the modes at marginal stability.

13.4.1 Hydrodynamic modes ($k = 0$)

The Spectral Web for the compressible $m=1, k=0$ modes of the rotating theta pinch equilibrium is shown in Fig. 13.12. One notices the interlacing structure of the solution path and the conjugate path, with genuine and false eigenvalues (EVs) alternating at the intersections. Furthermore, the solution path is not a single curve, but it is split into a separate open curve with the most global $n=1$ mode on it and a closed loop with the $n=2$ mode and all the rest of infinitely many clustering modes. The latter have a Doppler shifted frequency $\widetilde{\sigma} \equiv \sigma - m\Omega$ which tends to zero, whereas the most global modes have a large Coriolis shift, opposite to the Doppler shift, such that the combined Doppler–Coriolis shift $\overline{V}$, and hence σ, becomes very small. The most striking feature

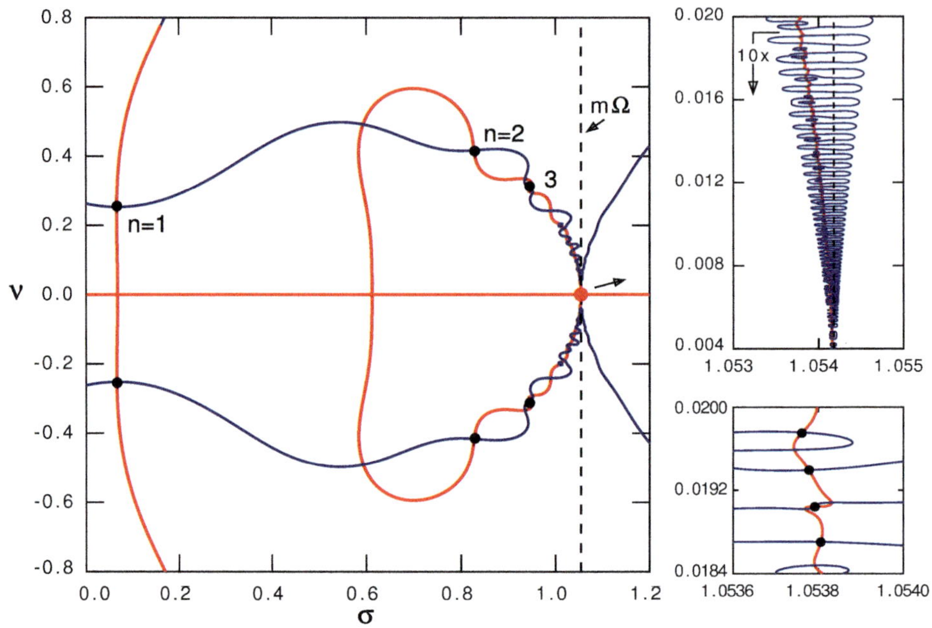

Fig. 13.12 Spectral Web consisting of solution path (red) and conjugate path (blue), with some eigenvalues highlighted by black dots, for the $m = 1$, $k = 0$ (HD) modes of the rotating theta pinch equilibrium of Fig. 13.11. The side frame shows the cluster sequence toward the HD flow continuum (red dot in the main frame), with a zoom of the $n = 53, \ldots 56$ eigenvalues in the bottom frame.

is that the $n = 2$ mode has a larger growth rate than the $n = 1$ mode. Such behaviour never occurs for static equilibria. This calls for justification of the labels $n = 1, n = 2$, etc.

For the $n \geq 2$ modes, the justification of the labelling is provided by the monotonicity of the alternator along the solution path, which yields a well-ordered sequence toward the cluster point $\omega = m\Omega$. This sequence does not include the $n = 1$ EVs since they are on a separate sub-path. At this point, it is important to note that the Spectral Webs of Fig. 13.12 and 13.14 were obtained for the specific parameter choice $x_{\rm mix} = 0.7$, which is optimal for this problem. Whereas the eigenvalues are independent of this choice (as they should be), the geometries of the solution and conjugate paths are not. In particular, their split into disconnected sub-paths, as in Fig. 13.12, depends on the value of $x_{\rm mix}$. Frequently, disconnected sub-paths may be merged by just changing that value. However, in this case, the attempt of connecting the sub-paths by a different choice of $x_{\rm mix}$ fails: for $x_{\rm mix} = 0.7493$, the topology of the solution path is still as shown in Fig. 13.12, but for $x_{\rm mix} = 0.7494$, it has abruptly changed to one where now $n = 1$ is on a closed loop and the $n \geq 2$ EVs are on an open curve. Apparently, a possible merger (with $|X| = 0$ at the reconnection point) is simply avoided. However, the conjugate path does connect the two EVs and, with the proof of monotonicity of the imaginary part of the alternator along that path, this provides clinching evidence for the validity of the collective labelling $n = 1, n = 2$, etc.

To understand the physics of these instabilities, consider the *local* (high n) modes. For $k = 0$, the four MHD continua $\{\Omega_{\rm A}^{\pm}(r)\}$ and $\{\Omega_{\rm S}^{\pm}(r)\}$ collapse into the HD flow continuum $\{\Omega_0(r)\}$, which is infinitely degenerate here since $\Omega_0 = m\Omega$ is constant. Approaching this continuum along the vertical line $\widetilde{\sigma} \equiv \sigma - m\Omega = 0$, the approximate differential equation obtained from Eq. (13.80),

$$\widetilde{\omega}^2 \frac{d}{dr}\Big(\rho r \frac{d\chi}{dr}\Big) + m^2\Omega^2\Big(\rho' - \frac{r\rho^2\Omega^2}{\gamma p + B^2}\Big)\chi = 0\,, \qquad (13.112)$$

gives rise to an infinite sequence of ever more rapidly oscillating solutions. This is associated with a cluster sequence of unstable discrete modes when the well-known local stability criterion for convective, Rayleigh–Taylor, instability,

$$C_{\rm stab} \equiv \frac{1}{\alpha^2}\Big(\frac{\rho'}{\rho r} - \frac{\rho\Omega^2}{\gamma p + B^2}\Big) > 0\,, \qquad (13.113)$$

is violated. This is the case for the equilibrium of Fig. 13.11 over the whole plasma (as shown by the green curves). The second term, representing destabilization due to compressibility, becomes more important at higher densities, faster rotations, and smaller magnetic field strengths. For the present equilibrium, the first term dominates over most of the plasma (except close to the magnetic axis), so that the assumption of incompressibility appears to be reasonable for most modes (except for the higher n and the marginal $k \neq 0$ modes discussed below).

To understand the overall geometry of the Spectral Web of Fig. 13.12, it is useful to temporarily make the approximation of *incompressibility* for the $k = 0$ modes. The spectral differential equation (13.14) then reduces to

$$r\frac{d}{dr}\Big(\rho r \frac{d\chi}{dr}\Big) - \big(m^2\rho + \lambda r\rho'\big)\chi = 0\,, \quad \lambda \equiv -m\Omega(2\widetilde{\omega} + m\Omega)/\widetilde{\omega}^2\,, \qquad (13.114)$$

where λ is a convenient eigenvalue replacing ω. This equation does not have the compressibility term of Eq. (13.113), but it is no longer restricted to the neighbourhood of the cluster point $\widetilde{\omega} = 0$. For this special case, as noted by Spies [559], the complex eigenvalue problem has degenerated into a Sturm–Liouville problem with *real* eigenvalues λ and *real* eigenfunctions χ, where the actual complex eigenvalues ω follow from the definition of λ. One may construct the following quadratic equation for the eigenvalue $\widetilde{\omega}$ from Eq. (13.114):

$$\widetilde{\omega}^2 + 2Qm\Omega\,\widetilde{\omega} + Qm^2\Omega^2 = 0\,, \qquad (13.115)$$

where $Q \equiv \lambda^{-1}$ turns out to be a quotient of two integrals,

$$Q = -(1/m^2)\int \rho'|\chi|^2\,dr \Big/ \int \rho\big[(1/r)|\chi|^2 + (r/m^2)|\chi'|^2\big]\,dr\,. \qquad (13.116)$$

Hence, the eigenvalues of the instabilities are given by

$$\omega = m\Omega\Big[1 - Q \pm \mathrm{i}\sqrt{Q(1-Q)}\Big]\,. \qquad (13.117)$$

Eliminating Q, one finds that the locus of eigenvalues is a circle $|\omega - \frac{1}{2}m\Omega| = \frac{1}{2}|m|\Omega$ through the origin (where $Q = 1$) and the cluster point (where $Q = 0$) is approached by EVs associated with eigenfunctions that have an increasing number of oscillations. It is clear that instability occurs for $\rho' < 0$, when the eigenvalue Q is in the range $0 < Q < 1$. However, the value of Q and, hence, *the values of σ and ν cannot be evaluated a priori:* they depend on the non-trivial solution χ of the boundary value problem (13.114), (13.81). Since the effects of compressibility are small for most of the $k = 0$ modes of the present equilibrium, the eigenvalues shown in Fig. 13.12 are rather close to the mentioned circle.

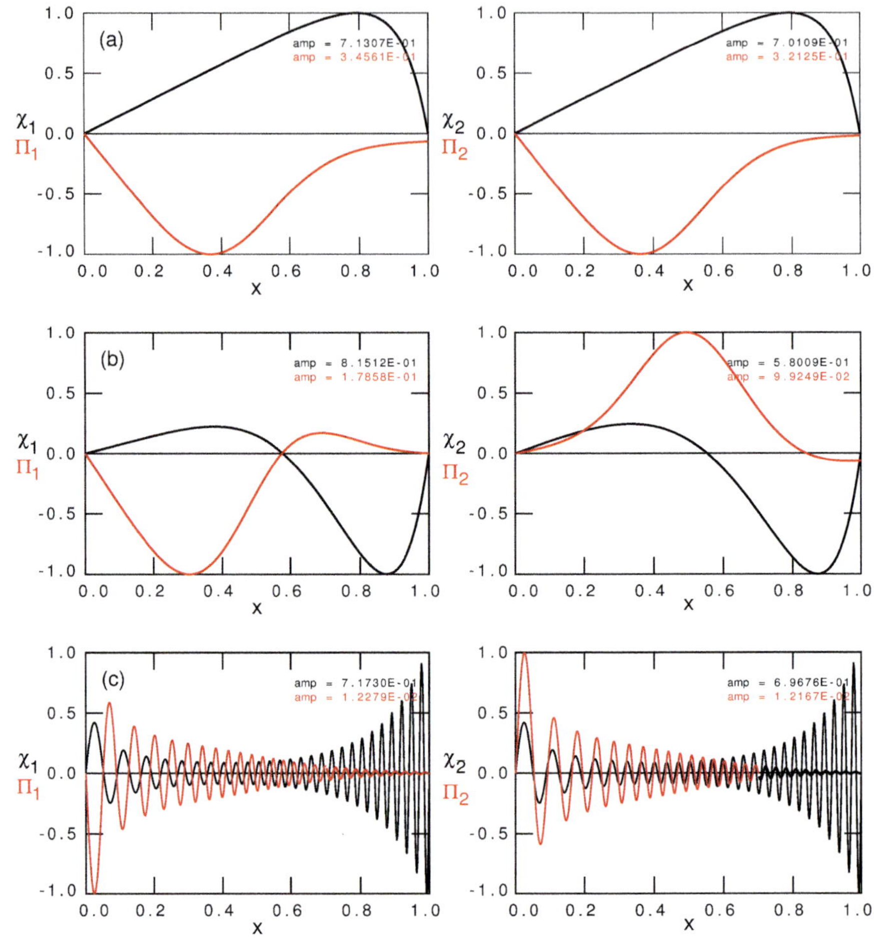

Fig. 13.13 The first two global eigenfunctions, and a local one, for the rotating theta pinch equilibrium of Fig. 13.11: (a) $n = 1$ ($\sigma = 0.06547$, $\nu = 0.25248$), (b) $n = 2$ ($\sigma = 0.82729$, $\nu = 0.41698$), (c) $n = 54$ ($\sigma = 1.0538$, $\nu = 0.01939$).

Comparison of the expression (13.117) with the general expression (12.109)(b) for the eigenvalues yields the following intriguing relations for the solution averages of the Doppler–Coriolis shift and the potential energy for this case:

$$\overline{V} = (1 - Q)m\Omega\,, \qquad \overline{W} = -(1 - Q)m^2\Omega^2\,. \tag{13.118}$$

They clearly demonstrate that, for the investigation of the stability of stationary equilibria, the two quadratic forms $\overline{W}$ and $\overline{V}$ play equally important roles.

We return to the actual numerical results for the *compressible* $k = 0$ modes. The $n = 1$ and $n = 2$ eigenfunctions shown in Figs. 13.13(a) and (b), display the usual features of low n modes, where $n = 1$ is the most global one. If the plasma column were more confined (by increasing the value of the stretching parameter α), the configuration would tend to the one discussed by Taylor [590] and that eigenvalue would approach the origin (where $Q = 1$ in the incompressible analysis). Hence, the $n = 1$ mode is intrinsically more stable than the $n = 2$ mode, which is located farthest away from both the origin and the cluster point (where $Q = 0$).

The side frame of Fig. 13.12 shows a zoom of the cluster spectrum, with a further zoom in the bottom frame highlighting the $n = 53, \ldots 56$ modes. One notices the characteristic alternation of genuine and false eigenvalues, with monotonicity of the alternator $[\![\Pi]\!]_{\rm mix}/\chi_{\rm mix}$ in between the false ones. This is in perfect agreement with the oscillation theorem for complex eigenvalues permitting counting of modes in the complex ω plane. The adjacent location of the $n = 55$ and 56 modes on the same loop of the conjugate path, which would conflict with this theorem, is only apparent. Small loops of the conjugate path may become entangled with bigger ones, leading to a convoluted structure at very small scales which only becomes visible by further zooming in (as shown for the $n = 55$ mode in the bottom side frame of Fig. 13.12). These higher n compressible modes localize both at the maximum $C_{\rm stab}$ (on the outside) and at the origin (where compressibility is dominant), as shown in Fig. 13.13(c) for the compressible $n = 54$ mode. For comparable growth rate of the incompressible modes, the number of oscillations about the origin decreases, indicative of the next significant difference with static equilibria: *for stationary equilibria, compressibility destabilizes*.

13.4.2 Magnetohydrodynamic modifications ($k \neq 0$)

The magnetic field completely changes the topology of the Spectral Web and, with it, the location of the unstable eigenvalues, in particular of the more global modes. This is shown in Fig. 13.14 by the Spectral Web for the $m = 1, k = 0.1$ modes. For this value of k, the two separate branches of the solution path with the $n = 1$ and $n = 2$ modes on them have merged, but those two modes are still wide apart in the ω-plane, with very different values of σ. It is convenient to express this part of the EV in terms of the Coriolis shift, which may be equated with the Doppler shifted frequency $\widetilde{\sigma}$, since, *along the solution path,*

$$\sigma = \overline{V} = (\Delta\sigma)_{\rm D} + (\Delta\sigma)_{\rm C} \quad \Rightarrow \quad \widetilde{\sigma} \equiv \sigma - m\Omega = (\Delta\sigma)_{\rm C} \,. \tag{13.119}$$

As for $k = 0$, the growth rate ν of the $n = 2$ mode is still much larger than the growth rate for the $n = 1$ mode, whereas the Coriolis shift $\widetilde{\sigma}$ is much smaller.

Concerning the higher n modes, the Spectral Web of Fig. 13.14 illustrates how the majority of unstable modes of the hydrodynamic cluster spectrum of Fig. 13.12 is now stabilized by the magnetic field. It also illustrates how the degeneracy of the HD flow continuum is lifted and how the approach of the marginal stability point is now governed by the resulting spread of the MHD continuum frequencies. In particular, the backward continua are well separated from the forward continua, leaving a narrow gap for marginal modes $\omega = \sigma_{\rm marg}$ to pass:

$$\begin{aligned} \{\Omega_{\rm A}^-\} &= [-1.6751, 1.0375]\,, \qquad & \{\Omega_{\rm A}^+\} &= [1.0706, 3.7835]\,, \\ \{\Omega_{\rm S}^-\} &= [\ \ 0.9781, 1.0379]\,, \qquad & \{\Omega_{\rm S}^+\} &= [1.0705, 1.1302]\,, \end{aligned}$$

so that $\{\Omega_{\rm S}^-, \Omega_{\rm A}^-\}_{\rm max} < \sigma_{\rm marg} < \{\Omega_{\rm S}^+, \Omega_{\rm A}^+\}_{\rm min}$. One can clearly see the transition from the pairs of complex modes ($n = 1 - 4$ are highlighted in the main frame, $n = 5 - 13$ in the side frame) to the pairs of real stable modes ($n \geq 14$). The numbering $n = 1, 2, \ldots 13$ of the complex eigenvalues (obtained by counting the number of branches of the alternator) smoothly joins that of the stable eigenvalues on the real axis (where n counts the number of nodes of the real eigenfunctions). In

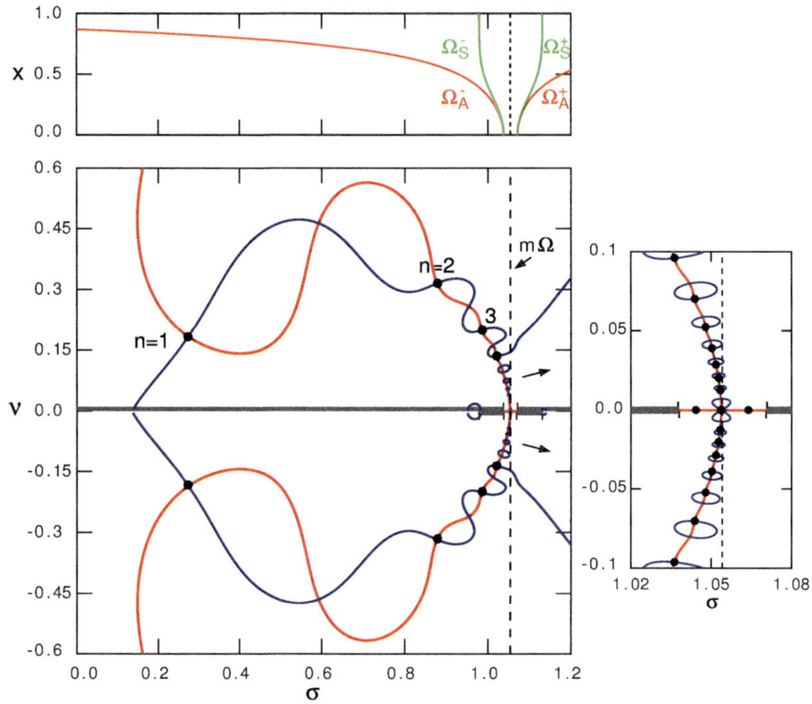

Fig. 13.14 Spectral Web consisting of solution path (red) and conjugate path (blue), with some eigenvalues highlighted by black dots, for the $m = 1$, $k = 0.1$ (MHD) modes of the rotating theta pinch equilibrium of Fig. 13.11. The range of the continua is indicated by dark strips above (Alfvén) and below (slow) the horizontal axis, their spatial dependence is shown in the top frame. The side frame shows a zoom of the unstable $n = 5, \ldots 13$ eigenvalues and the stable $n \geq 14$ eigenvalues (only the $n = 14$ ones are highlighted).

order not to clutter the picture, only the two $n = 14$ modes are shown since there is a large number of stable modes close to the continua, whereas the highest ones have been 'swallowed' by them.

We finally apply the Spectral Web method to compute the complete curve of the growth rate $\nu(k)$ and of the Coriolis shift $\widetilde{\sigma}(k)$ for all values of k, from the most unstable modes at $k = 0$ to the marginally stable ones at $k = k_0$. Of course, all the underlying Spectral Webs will not be shown. One of the advantages of this method is that one may restrict the investigation to small pieces of the ω-plane if, in a parameter scan, one already roughly knows where the next eigenvalue will be located. Because of the increasing magnetic field stabilization, the curves shown in Fig. 13.15 are overall decreasing. At larger k ($k > 0.254$), the dominance of the instability of the $n = 2$ modes (the blue and green curves in Fig. 13.15) disappears and the $n = 1$ modes (the red and black curves) become more unstable, so that eventually marginal stability is determined by the $n = 1$ mode.

For larger k, the values of ν and $\widetilde{\sigma}$ are also quite different for incompressible and compressible modes (note the deviation of the dashed curves from the drawn ones in Fig. 13.15). It is clear from the equilibrium profiles shown in Fig. 13.11 that magnetic field line bending, expressed by the terms $k^2 B_z^2$ in the spectral equation, is strongest on the outside so that, eventually, the marginal stability transition at $k = k_0$ is determined by the internal region, where compressibility is dominant. This

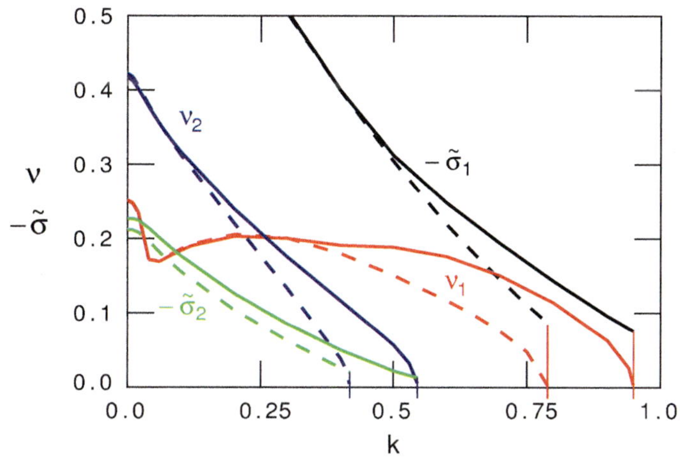

Fig. 13.15 Growth rate $\nu \equiv \text{Im}(\omega)$ (in red) and Coriolis shift $\widetilde{\sigma} \equiv \text{Re}(\omega) - m\Omega$ (in black) as a function of k for the compressible (drawn) and incompressible (dashed) $m = 1, n = 1$ modes of the rotating theta pinch equilibrium of Fig. 13.11. The blue and green curves refer to ν and $\widetilde{\sigma}$ for the $n = 2$ modes. The vertical red and blue dashes indicate the marginal stability transitions.

explains the large differences at marginal stability: $k_0 = 0.948$ for the compressible $n = 1$ modes ($k_0 = 0.548$ for $n = 2$), versus $k_0 = 0.792$ for the incompressible $n = 1$ modes ($k_0 = 0.419$ for $n = 2$). Recall that such differences never occur for static equilibria, since the marginal modes are incompressible for such equilibria.

In conclusion: the Spectral Web analysis of the rotating theta pinch demonstrates that three 'intuitive' assumptions originating from numerous investigations of the stability of static equilibria, viz. (1) that stability is determined by the sign of the potential energy W, (2) that the most global modes have the largest growth rate, (3) that it is sufficient for stability to investigate the incompressible modes only, are false for stationary equilibria.

13.5 Magneto-rotational instability in accretion discs

Our second example of instabilities in rotating plasmas concerns the magneto-rotational instability (MRI), independently found by Velikhov [631] and Chandrasekhar [111] and later applied by Balbus and Hawley [29] to clarify the mechanism of accretion onto compact objects. The compact object may be a young stellar object (mass $M_* \sim M_\odot$) or an active galactic nucleus (massive black hole with $M_* \sim 10^9 M_\odot$). For this incredible range of objects, the same fundamental problem arises, viz. how can accretion onto such objects occur at all on a reasonable time scale? Without dissipation this would be impossible, because a disc (that would already be there for whatever reason) would conserve angular momentum and, hence, rotate forever without change. Some form of viscosity is needed to facilitate the transfer of angular momentum to larger distances. But with viscosity there is a problem since the ordinary molecular viscosity coefficient is much too small to produce accretion of the correct order of magnitude. A turbulent increase of this coefficient might give the right answer. For such a process, small-scale instabilities are needed. However, no hydrodynamic instabilities are known for these discs. It is generally assumed that the solution of this problem involves the magneto-rotational instability. We present a derivation of the instability conditions to demonstrate that the analysis of Section 13.3 is actually applicable to genuine astrophysical objects.

13.5.1 Analytical preliminaries

The schematic axi-symmetric (2D) representation of an accretion disc in Fig. 4.1 is even further simplified by, first, estimating the scale height H due to the vertical pressure gradient in the quasi-two-dimensional model due to Shakura and Sunyaev [551] depicted in Fig. 13.16(a), discussed in Section 13.5.2, and, next, also *neglecting the vertical equilibrium variations* so that the annular cylindrical (1D) thin slice model of Fig. 13.16(b) is obtained. The gravitational potential due to the compact object of mass M_* at the origin is given by

$$\Phi_{\rm gr} = -\frac{GM_*}{\sqrt{r^2+z^2}} \approx -\frac{GM_*}{r}\,, \tag{13.120}$$

where the latter, cylindrical, approximation is appropriate for *short wavelengths fitting the disc in the vertical direction:*

$$k\,\Delta z \gg 1\,. \tag{13.121}$$

The magneto-rotational instabilities occur in the limit of *small magnetic fields:*

$$\beta \equiv 2p/B^2 \gg 1\,, \tag{13.122}$$

which justifies exploiting the *incompressible limit* of the spectral equation (13.80):

$$r\frac{d}{dr}\Big[\frac{\rho(\widetilde{\omega}^2-\omega_{\rm A}^2)}{h^2 r}\frac{d\chi}{dr}\Big] - \Big[\rho(\widetilde{\omega}^2-\omega_{\rm A}^2) + \Delta - \frac{4k^2(B_\theta F + \rho r\Omega\widetilde{\omega})^2}{h^2r^2\rho(\omega^2-\omega_{\rm A}^2)} - 2mr\Big(\frac{B_\theta F + \rho r\Omega\widetilde{\omega}}{h^2r^3}\Big)'\Big]\chi = 0\,, \tag{13.123}$$

where $\Omega(r) \equiv v_\theta/r$ is the angular rotation frequency, not to be confused with Ω_0, and Δ, defined in Eq. (13.88), becomes

$$\Delta = -\rho'(r\Omega^2 - GM_*/r^2) - r\rho\Omega^{2'} + r(B_\theta^2/r^2)'\,. \tag{13.124}$$

This at once demonstrates the crucial difference with the rotating theta-pinch discussed in Section 13.4: for *Keplerian rotation,* the centrifugal term is exactly cancelled by the gravitational term, so that the first term, with the Rayleigh- –Taylor instability drive, trivially vanishes. The last term may be neglected as well since $\beta \gg 1$.

Next, we simplify the magnetic configuration to the most simple form consistent with a rotating magnetized disc by choosing a *purely vertical and constant magnetic field*, a *constant density* and a *purely azimuthal velocity:*

$$B_\theta = 0\,,\quad v_z = 0 \qquad \Rightarrow\ \omega_{\rm A} = kB_z/\sqrt{\rho} = \text{const}\,,\quad \Omega_0 = m\Omega\,. \tag{13.125}$$

Furthermore, we restrict the analysis to *vertical wave numbers* k only:

$$m = 0 \qquad \Rightarrow\ \Omega_0 = 0\,,\ \text{ so that } \widetilde{\omega} = \omega\,. \tag{13.126}$$

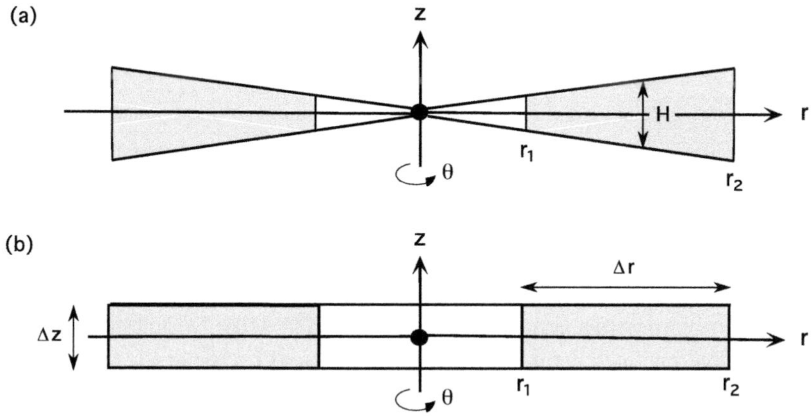

Fig. 13.16 Schematic geometries for the rotating accretion disc about a compact object (black dot): (a) quasi two-dimensional model; (b) cylindrical slice model.

With these approximations, nearly everything is constant, except for the angular rotation frequency $\Omega(r)$. A derived quantity is the *squared epicyclic frequency*

$$\kappa_e^2(r) \equiv r^{-3}(r^4\Omega^2)' = 2r\Omega\Omega' + 4\Omega^2 \,, \tag{13.127}$$

which can be negative in general, and which is a measure of how much the specific angular momentum $L \equiv rv_\theta \equiv r^2\Omega$ of the disc deviates from a constant distribution: $L' = 0$ if $\kappa_e^2 = 0$. The incompressible spectral problem then reads:

$$r\,\frac{d}{dr}\Big[\frac{\omega^2-\omega_A^2}{r}\,\frac{d\chi}{dr}\Big] - k^2\Big[\omega^2-\omega_A^2-\kappa_e^2-\frac{4\omega_A^2\Omega^2}{\omega^2-\omega_A^2}\Big]\,\chi = 0\,, \tag{13.128}$$

subject to the BCs $\chi(r_1) = \chi(r_2) = 0$. This equation exposes the differences between the stability of a hydrodynamic disc ($\omega_A^2 = 0$) and a magnetohydrodynamic disc ($\omega_A^2 \neq 0$), even when the magnetic field is arbitrarily small.

In the hydrodynamic limit, a direct result is *Rayleigh's circulation criterion*, stating that the fluid is stable to axi-symmetric disturbances ($m = 0$) when

$$\kappa_e^2(r) \geq 0 \quad \text{everywhere}\,, \tag{13.129}$$

which is why κ_e^2 is called Rayleigh's discriminant in hydrodynamics. This criterion is satisfied for Keplerian rotation, $\kappa_{e,K}^2 = GM_*/r^3 > 0$, so that such equilibria are hydrodynamically stable under rather wide assumptions. This explains why interest shifted to magnetohydrodynamic instabilities to explain the turbulent increase of the dissipation processes in accretion discs.

To obtain a stability criterion for the magnetohydrodynamic case, we first construct the Doppler–Coriolis shift from Eq. (13.107). Under the assumptions made, the Doppler shift $\Omega_0 = 0$ and the expressions (13.93) and (13.94) reduce to

$$\eta = \mathrm{i}\xi_\theta = 2\omega\Omega\big[r(\omega^2-\omega_A^2)\big]^{-1}\chi\,, \qquad \zeta = \mathrm{i}\xi_z = (kr)^{-1}\chi'\,, \tag{13.130}$$

so that the remaining, Coriolis, part of the shift becomes

$$\overline{V} = \frac{4\omega}{\omega^2 - \omega_A^2} \int \Omega^2 \, |\xi|^2 r \, dr \Big/ \int (|\xi|^2 + |\eta|^2 + |\zeta|^2) \, r \, dr \,. \tag{13.131}$$

Since we know that the Doppler–Coriolis shift is real, the extracted constant factor also has to be real. Hence, a trivial solution (the imaginary ω axis) of the solution path equation (13.107) for instabilities is found immediately:

$$\sigma = \overline{V}(\omega = \mathrm{i}\nu) = 0 \,. \tag{13.132}$$

Consequently, in a problem with a genuine rotation profile $\Omega(r)$, the Doppler–Coriolis shift vanishes altogether and we have produced solution paths (the real and imaginary ω axes) that are identical to those of a static equilibrium! Needless to say that this lucky circumstance will get lost for more general perturbations and equilibria with $B_\theta \neq 0$ (see Section 13.5.2).

Next, from the spectral equation (13.128), a quadratic form is constructed,

$$\int \frac{\rho}{r} \Big\{ (\omega^2 - \omega_A^2) |\chi'|^2 + k^2 \Big[\omega^2 - \omega_A^2 - \kappa_e^2 - \frac{4\omega_A^2 \Omega^2}{\omega^2 - \omega_A^2} \Big] |\chi|^2 \Big\} \, dr = 0 \,, \tag{13.133}$$

which immediately yields a *local instability criterion for MRIs:*

$$\omega_A^2 (\omega_A^2 + \kappa_e^2 - 4\Omega^2) < 0 \,, \quad \text{or} \quad 0 < k^2 B_z^2 < -\rho(\kappa_e^2 - 4\Omega^2) \,, \tag{13.134}$$

showing that the magnetic field is necessary for the instability but eventually stabilizes it again. Inserting the definition (13.127) for κ_e^2, this criterion is satisfied for Keplerian discs *when the magnetic field is sufficiently small* ($\omega_A^2 \to 0$) whereas

$$\kappa_e^2 - 4\Omega^2 \equiv r\,(\Omega^2)' = -3\Omega^2 < 0 \quad \text{(Keplerian rotation)} \,. \tag{13.135}$$

Hence, in contrast to hydrodynamic discs, discs with a small magnetic field are always unstable with respect to the magneto-rotational instability [29]. This is the reason for the popularity of the topic of MRIs in accretion discs.

The apparent contradiction between the hydrodynamic criterion (13.129) (obtained for $\omega_A^2 = 0$) and the hydrodynamic limit ($\omega_A^2 \to 0$) of the MHD criterion (13.134) is resolved when the growth rate of instabilities is taken into account. This is best illustrated by considering modes that are sufficiently *localized radially* to exploit a WKB solution of the form $\chi(r) = p(r) \exp\big[\mathrm{i} \int q(r)\, dr\big]$ with

$$q\,\widetilde{\Delta r} \gg 1 \,, \qquad \widetilde{\Delta r} \sim r_1 \ll \Delta r \equiv r_2 - r_1 \,. \tag{13.136}$$

The restriction to $\widetilde{\Delta r}$ instead of Δr is necessary here since the equilibrium, i.e. $\Phi_{gr}(r)$, has an $\mathcal{O}(1)$ variation over a much narrower range than the width of the disc. With the coefficients of the approximate spectral equation (13.128), the *local dispersion equation* reduces to

$$(k^2 + q^2)(\omega^2 - \omega_A^2)^2 - k^2 \kappa_e^2 (\omega^2 - \omega_A^2) - 4k^2 \omega_A^2 \Omega^2 = 0 \,, \tag{13.137}$$

giving two solutions,

$$\omega^2 \approx \omega_A^2 + (1/E) \Big[\tfrac{1}{2}\kappa_e^2 \pm \tfrac{1}{2} \big(\kappa_e^4 + 16E\Omega^2\omega_A^2 \big)^{1/2} \Big] \,, \quad E \equiv 1 + q^2/k^2 \,. \tag{13.138}$$

This yields the following approximations under various assumptions.

(a) The *most local modes* ($q^2 \gg k^2 \gg 1$) yield cluster spectra immediately above and below the degenerate Alfvén/slow continua $\Omega^{\pm}_{A/S} = \pm\omega_A$:

$$\omega^2 \approx \omega_A^2 \pm 2(k/q)\omega_A\Omega \qquad \text{(continua and cluster spectra)}. \tag{13.139}$$

(b) The *most global modes* ($q^2 \ll k^2$) yield the instability criteria:

$$\omega^2 \approx \omega_A^2 + \tfrac{1}{2}\kappa_e^2 \pm \tfrac{1}{2}\sqrt{\kappa_e^4 + 16\omega_A^2\Omega^2} \;\Rightarrow \begin{cases} \kappa_e^2 < 0 & \text{(HD)} \\ \kappa_e^2 - 4\Omega^2 < -\omega_A^2 & \text{(MHD)} \end{cases}. \tag{13.140}$$

(c) *Small magnetic fields* ($\omega_A^2 \ll \kappa_e^2 \sim \Omega^2$) yield the two relevant kinds of modes:

$$\omega^2 \approx \begin{cases} (1/E)\,\kappa_e^2 + (4\Omega^2/\kappa_e^2 + 1)\,\omega_A^2 & \text{(epicyclic modes)} \\ -(4\Omega^2/\kappa_e^2 - 1)\,\omega_A^2 & \text{(MRIs)} \end{cases}. \tag{13.141}$$

In the limit of vanishing magnetic field ($\omega_A^2 \to 0$), the upper solution is unstable if Rayleigh's stability criterion (13.129) is violated, i.e. for rotation profiles $\Omega(r)$ that fall off faster than r^{-2}. For those profiles, the lower solution is stable. However, for rotation falling off with a lower power, like Keplerian rotation $\Omega \sim r^{-3/2}$, the upper solution is stable and the lower solution (corresponding to MRI) is unstable, but with a vanishing growth rate. For those rotation profiles, the MRI dominates as long as ω_A^2 is finite, but sufficiently small.

The maximum growth rate for MRIs is obtained from the bottom expression (13.141)(b), which is valid for the most global modes ($q^2 \ll k^2$). For increasing values of q, a sequence of MRIs is obtained, crossing the marginal point $\omega = 0$ and bifurcating there, and joining the $-$ pair of cluster spectra (13.139) for $q \to \infty$. The epicyclic modes (13.141)(a) then transform into the $+$ pair.

With respect to the order of magnitude of the physical parameters for which the MRI will be operating in the disc, we first utilize scale-independence by setting three parameters equal to 1. This time, since we want to study the limits of small and vanishing magnetic field strength, it is not expedient to exploit B_z for that purpose. Instead, we normalize frequencies with respect to the gravitational potential at $r = r_1$ and eliminate dimensions by fixing the following three parameters:

$$r_1 \equiv 1\,, \quad \rho(r_1) \equiv 1\,, \quad -\Phi_{gr}(r_1) = GM_* \equiv 1 \qquad \textit{(scale independence)}. \tag{13.142}$$

The geometry of a thin and wide accretion disc is then represented by a small parameter ϵ and a large parameter Δ:

$$\epsilon \equiv \Delta z \ll 1 \ll \Delta \equiv \Delta r \equiv r_2 - r_1 \qquad \textit{(thin and wide disc)}. \tag{13.143}$$

The longitudinal wave number k and the radial 'wave number' q should be chosen such that the conditions (13.121) and (13.136) are satisfied, e.g.

$$k \sim \epsilon^{-2} \gg \epsilon^{-1} \sim q \gg 1 \qquad \textit{(local disturbances)}. \tag{13.144}$$

We can now estimate the order of magnitude of the magnetic field B_z, i.e. of the Alfvén frequency ω_A, required to give significant growth of the MRI during one rotation period (the condition (13.134) excludes growth faster than that):

$$\nu_{MRI} \sim \omega_A \equiv kB_z \sim \Omega(1) \approx 1 \quad \Rightarrow \quad B_z \sim k^{-1} \ll \epsilon \ll 1\,. \tag{13.145}$$

Assuming $k \sim \epsilon^{-2}$, this implies a very small order of magnitude of the magnetic field, $B_z \sim \epsilon^2 \Omega(1)$ (in dimensionless units), *above which the MRI switches off*. Even with the successes of modelling turbulent processes in accretion discs by means of the magneto-rotational instability (see Balbus and Hawley [30], Stone et al. [572]), this still begs the question whether nature actually makes use of these processes. Eventually (in a distant future), this question can only be answered by means of polarimetry and high-resolution observations, unequivocally establishing the magnetic signature of the modes involved.

13.5.2 Numerical Spectral Web solutions

Since the window in parameter space of the MRI is so narrow, it is logical that intensive research was conducted on extended models of accretion discs where the mentioned restrictive conditions were dropped one by one. We here discuss one particular model, by Keppens et al. [345], since it illustrates the complexity of the spectrum of waves and instabilities of rotating equilibria with 'numerically exact' solutions of the full spectral equation (13.80) for a cylinder with a relevant choice of the radial distribution of the equilibrium variables for accretion discs. This brings in dependence on *compressibility* (i.e. lifting of the degeneracy of Alfvén and slow sub-spectra and appearance of the fast sub-spectrum) and a *toroidal magnetic field* ($B_\theta \neq 0$). (Note that 'toroidal' here refers to the azimuthal component, in contrast to straight cylinder models for tokamaks where that angle is called 'poloidal'.) The results presented were obtained with the numerical Spectral Web code ROC described in Ref. [229].

The cylindrical equilibrium equation (12.30) is satisfied for discs where the variables have a power dependence on the radial variable r that is determined by the density $\rho(r) = r^e$. In the present context of an *ideal* stationary equilibrium, the exponent e is completely arbitrary. It will be fixed to the value $-3/2$, corresponding to a particular class of self-similar *dissipative* stationary disc solutions, given by Spruit et al. [564], with radial accretion velocity v_r ($\ll v_\theta$) and associated outward angular momentum transport due to turbulent viscosity, modelled with the parameter α introduced by Shakura and Sunyaev [551]. For the quasi-2D model of Fig. 13.16(a) the pressure is estimated from the two-dimensional equilibrium equation by balancing pressure gradient and gravitational acceleration in the vertical direction (see Frank, King and Raine [183], p. 87):

$$\rho^{-1}\partial p/\partial z = -\partial\Phi_{\rm gr}/\partial z \approx -z/r^3 \qquad \Rightarrow \quad p \approx p_1 r^{-5/2}(1 - \tfrac{1}{2}z^2/H^2)\,. \tag{13.146}$$

Here, $H \approx \sqrt{p_1}\,r$ is the scale height of the disc, assumed to be determined by thermal effects only. This very crude estimate of the two-dimensional structure of the equilibrium permits us to relate the pressure to the inverse aspect ratio of the disc, $\sqrt{p_1} = \epsilon \equiv H/r$. These considerations on dissipation and geometry only fix the radial powers of the *ideal one-dimensional* equilibrium that will be investigated with respect to the MRI.

On the fast time scale of the pertinent MHD processes, where dissipation may be neglected, one then obtains the following thin disc cylindrical equilibrium satisfying (12.30), while incorporating the arguments on self-similarity of the radial profiles and the viscosity needed for accretion:

$$\begin{aligned} &\rho = r^{-3/2}\,, \quad v_\theta = v_1 r^{-1/2}\,, \quad p = p_1 r^{-5/2}\,, \quad B_\theta = B_{\theta 1} r^{-5/4}\,, \quad B_z = B_{z1} r^{-5/4}\,, \\ &\text{where} \quad \Omega_1 \equiv v_1 - \Big[1 - \tfrac{5}{2}p_1 \quad \tfrac{1}{4}B_{\theta 1}^2 - \tfrac{5}{4}B_{z1}^2\Big]^{1/2}. \end{aligned} \tag{13.147}$$

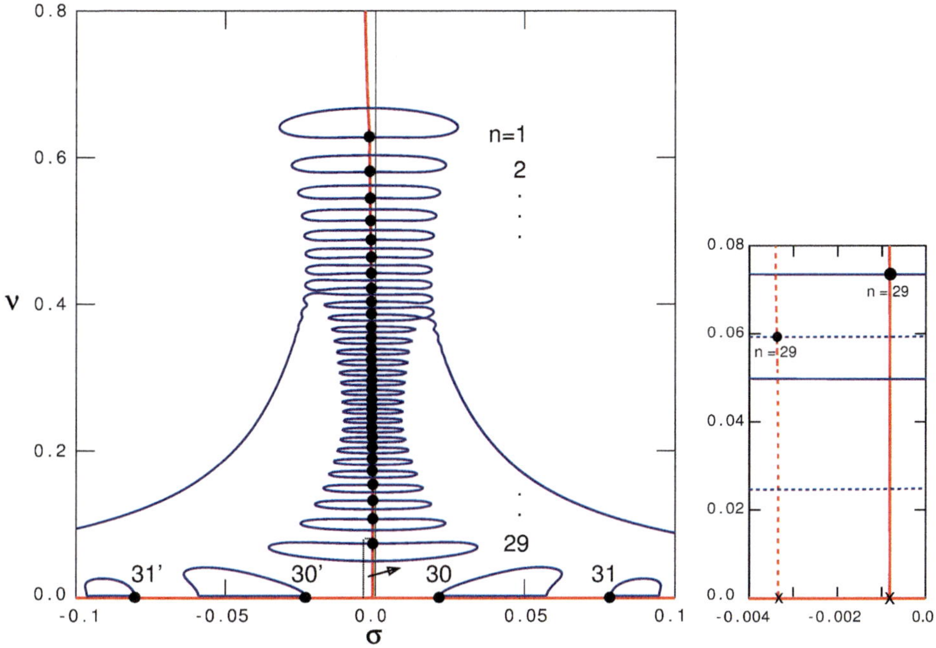

Fig. 13.17 Spectral Web of MRIs for an accretion disc equilibrium with $\epsilon = 0.1, \beta = 100, \mu_1 = 1$, $\delta = 2$, and mode numbers $m = 0, k = 70$. The eigenvalues of the $n = 1$–29 unstable modes and of the $n = 30, 31$ and $n = 30', 31'$ stable modes are highlighted by black dots. The side frame shows a zoom of the Spectral Web close to the origin with the marginal Coriolis shift σ_0 indicated by the cross; the dashed curves refer to incompressible modes.

Instead of the three parameters p_1, $B_{\theta 1}$, B_{z1}, it is convenient to exploit their relative magnitudes and a parameter measuring the radial size of the disc:

$$\epsilon \equiv \sqrt{p_1} \ll 1\,, \quad \beta \equiv 2p_1/B_1^2 \gg 1\,, \quad \mu_1 \equiv B_{\theta 1}/B_{z1} \sim 1\,, \quad \delta \equiv r_2/r_1 \gg 1\,. \qquad (13.148)$$

This ordering has been chosen such that the angular rotation parameter is of the order unity, $\Omega_1 \approx 1 - (5/4)\epsilon^2$, implying that the rotation is close to Keplerian, whereas the pressure is small enough to warrant the thin disc approximation and the magnetic fields are extremely small, yet important enough to yield the MRIs.

Figures 13.17–13.19 show the numerically calculated Spectral Web and corresponding eigenfunctions for the MRIs, where none of the approximations of Section 13.5.1 were exploited. In fact, the numerical results exhibit significant differences with respect to compressibility and the Coriolis shift of the eigenvalues. They were obtained by solving the full compressible ($\gamma = 5/3$) equations (13.82) for complex values of ω, substituting the explicit equilibrium profiles (13.147). The right solutions $\boldsymbol{\xi}^r$ were exploited, i.e. the complementary energy $W^{\rm r}_{\rm com}$ was evaluated at the inner edge of the disc. The values of ϵ, β and μ_1 were chosen according to the ordering (13.148). However, the radial width δ was chosen of order unity, rather than much larger than 1, in order to be able to distinguish the different modes on the solution path. We will see below that extrapolation to larger radial widths, $\delta \gg 1$, can easily be inferred from the results obtained for $\delta \sim 1$.

The Spectral Web shown in Fig. 13.17 has a very clear structure. All modes are located on the solution path. For the unstable modes, that path slightly deviates from the imaginary axis due to the Coriolis shift caused by the non-vanishing 'toroidal' field component B_θ, in agreement with earlier MRI calculations [167, 75]. Due to the choice of a moderate value of the radial width δ of the disc, the conjugate path consists of a series of separate 'pancakes', with precisely one genuine (and one false) EV on each pancake, thus determining the sequence number n ($= 1, \ldots 29$) of the unstable modes with growth rates that diminish monotonically as n increases, in agreement with the Complex Oscillation Theorem. This counting is exactly continued by the counting of nodes of the real eigenfunctions of the stable modes ($n \geq 30$ and $n' \geq 30$) along the real parts of the solution path, where the Real Oscillation Theorem dictates monotonicity of the eigenvalue ω as a function of n outside the genuine and apparent singularity regions. The latter include the collection of Doppler shifts $\omega = \Omega_0 \equiv m v_\theta / r + k v_z$ (here degenerate at $\omega = 0$), at the same frequencies as the HD flow continuum singularities (but not a continuum in ideal MHD! [231]). At $\nu = 0$ (but $\sigma = \sigma_0 \neq 0$), the stable modes split into two sequences, an anti-Sturmian one with modes approaching the backward continua and a Sturmian one approaching the forward continua.

Three representative MRI eigenfunctions are shown in Fig. 13.18. The first one ($n = 1$), with the maximum growth rate, is clearly localized at the position where the stability criterion (13.134) is violated most, i.e. at the inner edge of the disc. As n increases, the growth rate ν decreases and the eigenfunction spreads over the whole disc. For the last unstable mode ($n = 29$), the amplitude of the imaginary part of the eigenfunction has decreased, whereas the eigenfunctions of the stable modes ($n \geq 30$) become real. For the purpose of reference, we report the precise eigenvalues of the lowest n stable modes highlighted in Fig. 13.17:

$$
\begin{aligned}
n = 30' : \quad \sigma &= -2.2985 \times 10^{-2} , \qquad n = 30 : \quad \sigma = 2.1366 \times 10^{-2} , \\
n = 31' : \quad \sigma &= -8.0106 \times 10^{-2} , \qquad n = 31 : \quad \sigma = 7.8519 \times 10^{-2} .
\end{aligned}
$$

The Coriolis shift causes these values to be asymmetric with respect to $\sigma = \sigma_0$.

For larger values of n, the eigenfunctions have an increasing number of oscillations in the region corresponding to the boundaries of the approached continua. This approach is shown for $n = 100$ and $n' = 100$ in the zoomed out Spectral Web of Fig. 13.19(a), with a plot of the radial dependence of the continua in the upper frame. The eigenfunction for $n = 100$ shown in Fig. 13.19(b) illustrates that, eventually, an infinite number of oscillations occurs at the radial location corresponding to the boundary of the pertinent (forward or backward) slow magneto-sonic continuum, i.e. at the outer edge of the disc. Hence, those boundaries of the continua are cluster points of the discrete spectrum of stable "MRIs" (quotation marks to indicate the contradiction in terminology).

It is now evident how the Spectral Web of Figs. 13.17 and 13.19 will be affected by choosing $\delta \gg 1$ (i.e. $r_2 \to \infty$). The continua $\Omega^{-}_{\rm A,S}$ and $\Omega^{-}_{\rm A,S}$ will then extend to $\omega = 0$, pushing the infinity of cluster modes onto the complex branch of the solution path. The Coriolis shift will vanish at $\omega = 0$, which then becomes the branching point of the solution path and a cluster point of infinitely many unstable MRIs. Hence, the solution path in between the most unstable $n = 1$ mode and the $n \to \infty$ modes at the origin will be densely packed (not 'dense' in the mathematical sense though) with discrete modes. However, the fastest growing one will not be affected much since it only depends on the conditions of where it is localized, viz. at the inner edge of the disc. Thus, in peculiar contrast with instabilities of laboratory plasmas (e.g. tokamaks), the lowest n modes with

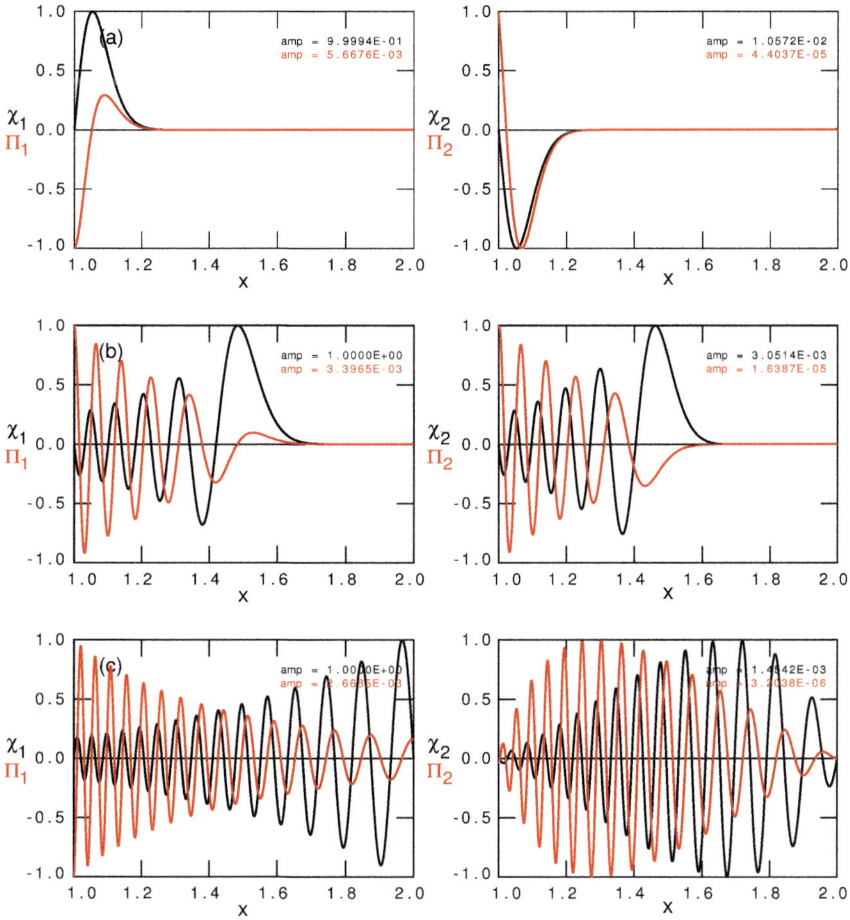

Fig. 13.18 Eigenfunctions of MRIs corresponding to some of the eigenvalues of the Spectral Web shown in Fig. 13.17: (a) $n = 1$, (b) $n = 10$, (c) $n = 29$.

the highest growth rates are here the 'local' ones, whereas the higher n modes with much reduced growth rates are the 'global' ones spreading over the whole disc.

For finite δ, division of unstable MRIs and stable modes occurs at the marginal ($\nu = 0$) value of the solution-averaged Doppler–Coriolis shift $\sigma_0 = \overline{V}[\chi(\sigma_0)]$, i.e. at the frequency where the solution path intersects the real axis (indicated by the crosses in the side frame of Fig. 13.17). An explicit expression for σ_0 may be obtained from the general expression (13.107) by, first, noting that the Doppler part vanishes for the present case (since $m = 0$), and, next, exploiting the approximations (13.148) and assuming σ to be small to simplify the expressions for the two tangential components of $\boldsymbol{\xi}$ resulting from Eqs. (13.93) and (13.94):

$$\mathrm{i}\xi_\theta \approx -\frac{2B_\theta}{kr^2 B_z}\Big[1 + \frac{r^2\Lambda}{2\gamma p} + \frac{r\rho\Omega\sigma}{kB_\theta B_z}\Big]\chi\,, \qquad \mathrm{i}\xi_z \approx -(kr)^{-1}\chi'\,. \tag{13.149}$$

Here, the second term in square brackets involves the deviation Λ from Keplerian rotation, which is

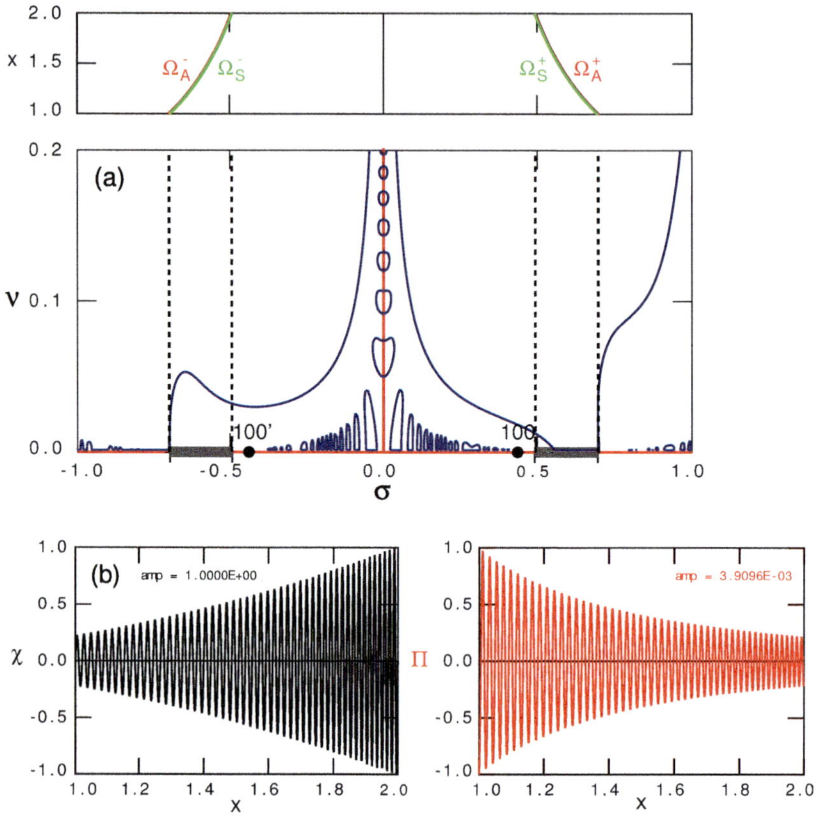

Fig. 13.19 (a) Zoomed out lower part of the Spectral Web of Fig. 13.17, highlighting the continua and the stable $n = 100'$ ($\sigma = -0.43682$) and $n = 100$ ($\sigma = 0.43632$) modes. The range of the continua is indicated by dark strips above (Alfvén) and below (slow) the horizontal axis, their spatial dependence is shown in the top frame. (b) Eigenfunction of the $n = 100$ mode.

small but not at all negligible for the present purpose:

$$\Lambda \equiv \rho(\Omega^2 - GM_*/r^3) \approx -5p/(2r^2) \sim \epsilon^2 \quad \Rightarrow \quad \frac{r^2\Lambda}{2\gamma p} \approx -\frac{5}{4\gamma}\,. \tag{13.150}$$

The last term in square brackets corresponds to the last but one term of Eq. (13.123) that was neglected in the analysis of Eqs. (13.133)–(13.138). Since this term involves the unknown σ, the corresponding integral needs to be moved to the LHS of Eq. (13.107). This yields the following expression for the Coriolis shift:

$$\sigma_0 \approx -\frac{(4\mu_1/k)\big[1 - 5/(4\gamma)\big] \int \rho(\Omega/r^2)|\chi|^2\, dr}{\int (\rho/r)\big[|\chi|^2 + |\chi'/k|^2 + 4(\Omega^2/\omega_{\rm A}^2)|\chi|^2\big]\, dr}\,. \tag{13.151}$$

This is still an implicit equation for the determination of σ_0 since the integrals involve the solution $\chi(\sigma_0)$ of the spectral equation. However, they may be estimated for the explicit equilibrium (13.147) by straightforward approximations. The values of σ_0 are in excellent agreement with the results of the numerical code.

The factor $1 - 5/(4\gamma)$, which equals $1/4$ for $\gamma = 5/3$ (but equals 1 for $\gamma = \infty$), produces a large effect at marginal stability (as illustrated in the side frame of Fig. 13.17): $\sigma_0 = -8.262 \times 10^{-4}$ for the compressible modes, whereas $\sigma_0 = -3.373 \times 10^{-3}$ for the incompressible ones. A similar large factor of about 4 occurs in the deviation of the solution path from the imaginary axis (illustrated in the main frame of Fig. 13.17 for the compressible modes). For example, the EVs corresponding to the eigenfunctions of Fig. 13.19 are given by:

$$
\begin{aligned}
n = 1: &\quad \sigma = -2.0312\,(-8.2883) \times 10^{-3}\,, \quad \nu = 6.2772\,(6.2077) \times 10^{-1}\,,\\
n = 10: &\quad \sigma = -1.2866\,(-5.2491) \times 10^{-3}\,, \quad \nu = 3.8605\,(3.8104) \times 10^{-1}\,,\\
n = 29: &\quad \sigma = -8.2615\,(-33.728) \times 10^{-4}\,, \quad \nu = 7.3537\,(5.9292) \times 10^{-2}\,,
\end{aligned}
$$

where the values in brackets are for incompressibility. The latter approximation gives reasonable estimates for the growth rates, but the real parts of the eigenvalues are completely off.

For finite size discs, the split in unstable and stable modes is of considerable physical significance, since the unstable MRIs are relevant for turbulence studies of the accretion process, whereas the frequencies and eigenfunctions of the stable modes contain important information about the morphology of the disc because of the strong radial dependence of the equilibrium profiles. The latter, stable, part of the spectrum may become relevant in some distant future when the subject of magneto-seismology (correlating spectral studies with observations) of accretion discs [345] has come of age. As for the present, it appears that we have succeeded in incorporating the MRI into the general MHD spectral theory of stationary equilibria by means of the method of the Spectral Web.

13.6 Literature and exercises

Notes on literature

Hydrodynamic and magnetohydrodynamic stability

- Chandrasekhar, *Hydrodynamic and Hydromagnetic Stability* [112], discusses a great variety of topics in fluid dynamic stability, like thermal instability, effects of shear flow, rotation, gravity and stability of jets and cylinders. Magnetic field effects enter at many places, e.g. in Chapter IX in the discussion of the magneto-rotational instability, discovered by Velikhov [631] and independently by Chandrasekhar [111].
- Drazin and Reid, *Hydrodynamic Stability* [164], is the classical compendium on the subject of linear hydrodynamic stability, treating both inviscid and viscous flows with respect to thermal convection, shear flow and rotation, and containing a short introduction of some topics in nonlinear stability.
- Hameiri, 'Spectral estimates, stability conditions, and the rotating screw-pinch' [282], presents the first systematic study of the spectrum of stationary cylindrical plasmas.
- Bondeson, Iacono and Bhattacharjee, 'Local magnetohydrodynamic instabilities of cylindrical plasma with sheared equilibrium flows' [79], extend Hameiri's results with a numerical analysis of local instabilities at resonant magnetic surfaces.

Astrophysical flows

- Balbus and Hawley, 'Instability, turbulence, and enhanced transport in accretion discs' [30], point out the significance of the magneto-rotational instability as a possible mechanism for turbulent enhancement of angular momentum transport in accretion discs about compact objects.
- Pringle and King, *Astrophysical Flows* [512], is a basic textbook on the fluid dynamical processes relevant to astrophysics, covering wave propagation, shocks, spherical flows, stellar oscillations and instabilities driven by magnetic fields, thermal conduction, gravity, shear flow and rotation.

Exercises

[13.1] *Convection in a plasma slab*

In this exercise we investigate the convective instability of a plasma slab. Assume that the slab has a constant flow and magnetic field.

- Derive the spectral equation in the incompressible limit.
- Derive the local dispersion equation using a solution of the form $\xi \sim e^{iqx}$, assuming that $q\Delta x \gg 1$.
- What are the solutions of this dispersion equation? What is the role of the magnetic field?

[13.2] *Gravito-acoustic waves in a stationary fluid*

Similarly to Section 7.2.3, we are going to investigate gravito-acoustic waves, i.e. neglect the magnetic field, but we do take flow into account.

- Derive an equation for the pressure assuming that the density is exponentially decaying, $\rho = \rho_0 e^{-\alpha x}$, and that the gravity is constant. Show that the sound speed c and the decay parameter α are constant.
- Writing $(\tilde{\omega}^2 - k_0^2 c^2)[\widetilde{\omega}^2/(\tilde{\omega}^2 - k_0^2 c^2)]' = -k_0^2 \lambda$, find the expression for $\lambda(x, \omega)$.
- Derive from the general spectral equation a second order differential equation for the displacement ξ. Write this equation in the form $\xi'' + f(x,\omega)\xi' + g(x,\omega)\xi = 0$.
- Looking at this differential equation, one might think to try solutions of the form $\xi \sim C \exp[(a \pm iq)x]$. Explain, why this kind of solutions cannot be used in general.
- Assume that the velocity is constant. Explain why the form mentioned in the previous question can now be used. Show that both $f(x,\omega)$ and $g(x,\omega)$ are real.
- Derive the dispersion equation from the spectral equation. Determine the eigenfrequencies.

[13.3] *Magneto-rotational instability and convection*

In Section 13.5.1, the magneto-rotational instability is discussed assuming constant density. Here, we drop this assumption, which allows us to study convective instabilities.

- Derive the spectral equation in the incompressible limit. Assume a constant magnetic field in the vertical direction, and a purely azimuthal velocity field. Consider only axi-symmetric perturbations. (Hint: use the function $\Lambda(r)$ defined in Eq. (13.75).)
- Derive the local dispersion equation from the spectral equation, assuming oscillatory solutions of the form $\chi \sim \exp(iqr)$ with $q\Delta r \gg 1$.
- Determine the solutions of the dispersion equation.
- Derive a stability criterion from them and explain the role of the magnetic field.

[13.4] *Magneto-rotational instability and non-zero azimuthal magnetic field*

In the previous exercise, you have investigated the magneto-rotational instability in combination with convection in the absence of an azimuthal magnetic field. Here, you will look at the case of a non-vanishing azimuthal magnetic field.

- Convince yourself that the azimuthal magnetic field cannot be constant.
- Repeat the first two questions of the previous exercise, including an azimuthal magnetic field. Find a local dispersion equation of the form $\omega^4 + a_2\omega^2 + a_1\omega + a_0 = 0$, and determine the coefficients.
- The solutions of this dispersion equation cannot be obtained analytically. You have to find them numerically. To do so, an equilibrium has to be specified. Work out the coefficients for the equilibrium specified in Eq. (13.147) for an accretion disc.
- Compute the solutions of the dispersion equation making use of Laguerre's method (see *Numerical Recipes* [509]; in IDL use FZ_ROOTS, in Matlab use ROOTS1). Use the following parameters for the equilibrium: $\epsilon = 0.1$ (thin disc), $\mu_1 = 1$ (inclusion of azimuthal magnetic field), $\beta = 1000$ (weakly magnetized plasma), and $GM_* = 1$. As a starting value for Laguerre's method use the complex number $\widehat{\omega} = i$. What do you notice? What happens if you include an azimuthal magnetic field?
- Make a plot of the growth rate and oscillation frequency as a function of the vertical wave number k. What is the minimum value for the wave number k? What do you conclude from this plot?

14

Resistive plasma dynamics

14.1 Plasmas with dissipation

14.1.1 Conservative versus dissipative dynamical systems

We have already come across the enormous difference between conservative (ideal) MHD and dissipative (resistive, viscous, etc.) MHD in Chapter 4. This difference runs through all of classical dynamics of discrete and continuous media. It involves quite different physical assumptions and corresponding different mathematical solution techniques. An instructive example is spectral theory (Chapter 6) which is classical, consistent and misleadingly beautiful for ideal MHD, but full of unresolved problems in resistive MHD. The classical part concerns self-adjoint linear operators in Hilbert space, analogous to quantum mechanics, and stability analysis by means of an energy principle. When dissipation is important, precisely these two 'sledgehammers' are missing in the dynamical systems workshop. Even the definition of what is an important, i.e. physically dominant, contribution to the dynamics deserves extreme care. This is best illustrated by the general description of the dynamics of ordinary fluids which is fundamentally different for ideal fluids, characterized by an infinite Reynolds number, and viscous fluids, characterized by a finite Reynolds number. This is even so for extremely large Reynolds numbers, in a certain sense irrespective of how large this number is. Viscous boundary layers always arise in real fluids. This qualitative difference between ideal and dissipative dynamics, with the occurrence of boundary layers, also applies to MHD when resistivity is introduced. This gives rise to internal resistive boundary layers, facilitating new modes of instability, as we will see in Section 14.2. The physical cause of these instabilities is the loss of conservation of magnetic flux, leading to reconnection of magnetic field lines (Section 14.4). The implications for the structure of the resistive spectrum are only partly understood (Section 14.3). Resistive MHD is only one of a number of extensions of the ideal MHD model which come under the name of 'extended MHD' (Section 14.4.4).

14.1.2 Stability of force-free magnetic fields: a trap

As a preliminary to the study of resistive instabilities, to be undertaken in Section 14.2, let us first investigate the ideal MHD stability of about the simplest system that deserves analysis, viz. a plane current-carrying plasma slab. The reader is warned in advance that an educational trap has been set in this section.

The current will be chosen such that the magnetic field has a constant magnitude but its direction varies. The simplest case to treat is a so-called force-free magnetic field with a constant ratio α

between the current and the magnetic field:

$$\mathbf{j} = \nabla \times \mathbf{B} = \alpha \mathbf{B} , \qquad \alpha = \text{const} , \tag{14.1}$$

or, in components,

$$j_y = -B_z' = \alpha B_y , \qquad j_z = B_y' = \alpha B_z .$$

This equation can easily be integrated:

$$\mathbf{B} = B_0 [\, \sin \varphi(x) \mathbf{e}_y + \cos \varphi(x) \mathbf{e}_z] , \quad \varphi(x) = \alpha x , \quad B_0 = \text{const} , \tag{14.2}$$

representing a field with a uniformly varying direction (Fig. 14.1). The plasma is considered to be confined between two perfectly conducting plates at $x = x_1$ and $x = x_2$. Hence, the parameter αa , where $a \equiv x_2 - x_1$, is a measure for the total current through the plasma. In the absence of gravity, the equilibrium would still permit a finite, but constant, pressure. However, we will neglect pressure altogether by considering a zero-β plasma. We wish to investigate the stability of this configuration.

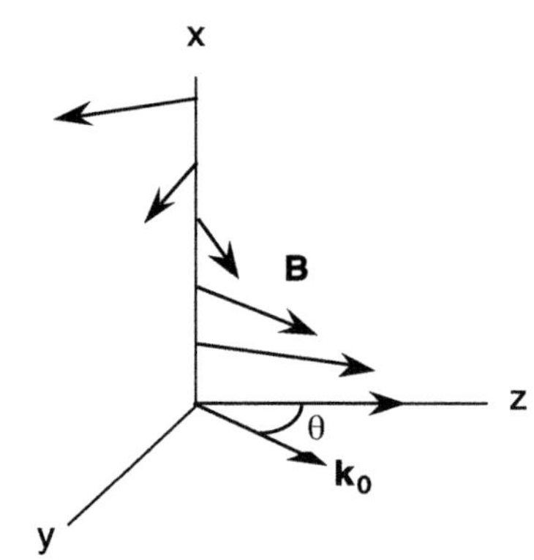

Fig. 14.1 Plane force-free magnetic field: changing direction caused by current $\|$ $\mathbf{B}$.

(a) An interesting stability result . . .

Stability may be investigated by means of the energy principle, i.e. the study of the sign of the energy $W[\boldsymbol{\xi}]$ of the perturbations $\boldsymbol{\xi}$, exploiting the explicit expression (6.84) for the fluid energy. As in Section 7.3.2, we decompose the displacement vector $\boldsymbol{\xi}(\mathbf{r})$ in Fourier components, according to Eq. (7.69), and we study the stability of the separate modes. For the present problem, the expression for the fluid energy W simplifies to

$$W = \tfrac{1}{2} \int \big(|\mathbf{Q}|^2 + \alpha \mathbf{B} \cdot \boldsymbol{\xi}^* \times \mathbf{Q} \big) \, dx , \tag{14.3}$$

where we have normalized W with respect to the area in the $y - z$ plane. Following the textbook of Schmidt [542], p. 141, we may simplify the algebra by using the vector potential $\mathbf{A}$,

$$\mathbf{Q} = \nabla \times \mathbf{A} , \qquad \mathbf{A} \equiv \boldsymbol{\xi} \times \mathbf{B} , \tag{14.4}$$

so that

$$W = \tfrac{1}{2} \int \Big[|\nabla \times \mathbf{A}|^2 - \alpha \mathbf{A}^* \cdot \nabla \times \mathbf{A} \Big] \, dx \,. \tag{14.5}$$

According to Section 6.4.4, we may now minimize W subject to some convenient normalization, for which we choose the helicity (see Section 4.3.4) of the perturbations:

$$K \equiv \tfrac{1}{2} \int \mathbf{A}^* \cdot \nabla \times \mathbf{A} \, dx = \text{const} \,. \tag{14.6}$$

The proper way to minimize W subject to such a constraint is to minimize another quadratic form, viz.

$$\widetilde{W} \equiv W + \lambda K = \tfrac{1}{2} \int \Big[|\nabla \times \mathbf{A}|^2 - (\alpha - \lambda) \mathbf{A}^* \cdot \nabla \times \mathbf{A} \Big] \, dx \,, \tag{14.7}$$

where the constraint is absorbed by means of an undetermined Lagrange multiplier λ that is to be determined together with $\mathbf{A}$. Since

$$\nabla \cdot \Big[\mathbf{A}^* \times (\nabla \times \mathbf{A}) \Big] \stackrel{(A.12)}{=} \nabla \times \mathbf{A}^* \cdot \nabla \times \mathbf{A} - \mathbf{A}^* \cdot \nabla \times \nabla \times \mathbf{A} \,, \tag{14.8}$$

we may integrate the expression for $\widetilde{W}$ by parts:

$$\widetilde{W} = \tfrac{1}{2} \Big[\mathbf{A}^* \times (\nabla \times \mathbf{A}) \cdot \mathbf{n} \Big]_{x_1}^{x_2} + \tfrac{1}{2} \int \mathbf{A}^* \cdot \Big[\nabla \times \nabla \times \mathbf{A} - (\alpha - \lambda) \nabla \times \mathbf{A} \Big] \, dx \,. \tag{14.9}$$

The boundary term vanishes by virtue of the boundary conditions $\mathbf{B} \cdot \mathbf{n} = 0$ and $\boldsymbol{\xi}^* \cdot \mathbf{n} = 0$ applied to $\mathbf{A} \equiv \boldsymbol{\xi} \times \mathbf{B}$. Consequently, for arbitrary $\mathbf{A}^*$, the quadratic form $\widetilde{W}$ is minimized by solutions of the Euler–Lagrange equation

$$\nabla \times \nabla \times \mathbf{A} - (\alpha - \lambda) \nabla \times \mathbf{A} = 0 \,. \tag{14.10}$$

This may be written as another force-free field equation for the perturbations:

$$\nabla \times \mathbf{Q} = \tilde{\alpha} \, \mathbf{Q} \,, \quad \tilde{\alpha} \equiv \alpha - \lambda \,. \tag{14.11}$$

Equation (14.11) is an eigenvalue equation, where $\tilde{\alpha}$ (and, hence, λ) is determined by imposing the boundary condition $\mathbf{n} \cdot \mathbf{Q} = 0$ at $x = x_1$ and $x = x_2$. Inserting such a solution into the expression (14.9) for $\widetilde{W}$ gives $\widetilde{W} = 0$, so that

$$\begin{aligned} W &= \widetilde{W} - \lambda K = (\tilde{\alpha} - \alpha) \, \tfrac{1}{2} \int \mathbf{A}^* \cdot \nabla \times \mathbf{A} \, dx \\ &\stackrel{(14.10)}{=} \frac{\tilde{\alpha} - \alpha}{\tilde{\alpha}} \, \tfrac{1}{2} \int \mathbf{A}^* \cdot \nabla \times \nabla \times \mathbf{A} \, dx \stackrel{(14.8)}{=} \frac{\tilde{\alpha} - \alpha}{\tilde{\alpha}} \, \tfrac{1}{2} \int |\nabla \times \mathbf{A}|^2 \, dx \\ &\stackrel{(14.4)}{=} \frac{\tilde{\alpha} - \alpha}{\tilde{\alpha}} \, \tfrac{1}{2} \int |\mathbf{Q}|^2 \, dx \,. \end{aligned} \tag{14.12}$$

Hence, $W < 0$ and the system appears to be unstable if the equation (14.11) has an eigenvalue $\tilde{\alpha}$ such that

$$0 < \tilde{\alpha} < \alpha \,. \tag{14.13}$$

It remains to determine the eigenvalue $\tilde{\alpha}$.

We now have to study the Euler equation (14.11) in detail to find out whether the condition (14.13) can be satisfied for the slab model. In this model, the Euler equation may be reduced to an ordinary second order differential equation for the normal component of $\mathbf{Q}$, which may be solved analytically so that we find an explicit stability criterion. To that end, write Eq. (14.11) in components,

$$\mathrm{i}k_y Q_z - \mathrm{i}k_z Q_y = \tilde{\alpha} Q_x \,, \qquad \mathrm{i}k_z Q_x - Q_z{}' = \tilde{\alpha} Q_y \,, \qquad Q_y{}' - \mathrm{i}k_y Q_x = \tilde{\alpha} Q_z \,. \tag{14.14}$$

Note that only two of these three equations are independent since $\mathbf{Q}$ is a magnetic field perturbation, so that

$$\nabla \cdot \mathbf{Q} = Q_x' + \mathrm{i}k_y Q_y + \mathrm{i}k_z Q_z = 0 \,. \tag{14.15}$$

Hence, two variables suffice, for which we choose the normal components of the magnetic field and the current:

$$Q \equiv -\mathrm{i}Q_x \,, \qquad R \equiv \mathrm{i}(\nabla \times \mathbf{Q})_x = k_z Q_y - k_y Q_z \,. \tag{14.16}$$

Substituting the reverse expressions,

$$Q_x = \mathrm{i}Q \,, \qquad Q_y = -(k_y Q' - k_z R)/k_0^2 \,, \qquad Q_z = -(k_z Q' + k_y R)/k_0^2 \,, \tag{14.17}$$

into Eq. (14.14)(a) produces a simple relationship between R and Q,

$$R = -\tilde{\alpha} Q \,, \tag{14.18}$$

whereas the two other components (14.14)(b),(c) just reduce to the same, second order, differential equation for Q:

$$Q'' + (\tilde{\alpha}^2 - k_0^2) Q = 0 \,, \quad k_0^2 \equiv k_y^2 + k_z^2 \,. \tag{14.19}$$

This is the Euler equation we were looking for.

The solution of Eq. (14.19) which vanishes for $x = x_1$ and $x = x_2$ reads:

$$Q = C \sin \sqrt{\tilde{\alpha}^2 - k_0^2}\, x \,, \quad \text{where } \sqrt{\tilde{\alpha}^2 - k_0^2} = n\pi/a \,, \quad a \equiv x_2 - x_1 \,. \tag{14.20}$$

Hence, the instability criterion (14.13) is fulfilled for

$$\tilde{\alpha}^2 = k_0^2 + \frac{n^2 \pi^2}{a^2} < \alpha^2 \,, \qquad \text{or} \quad (k_0/\alpha)^2 + (n\pi/(\alpha a))^2 < 1 \,. \tag{14.21}$$

This gives an unstable region in the $k_0/\alpha - \alpha a$ plane, as sketched in Fig. 14.2(a). Moving to the right in the shaded area subsequently $n = 1\,, n = 2\,, \ldots$ become unstable. The marginal modes (for which $\tilde{\alpha} = \alpha$) are distinguished by the number of nodes, $n - 1$, of Q on the interval (x_1, x_2) (Fig. 14.2(b)). Notice that in the long-wavelength limit, $k_0 = 0$, every time αa increases with πa, i.e. every time the magnetic field has changed its direction by $180°$, a mode with one more node becomes unstable. This appears to be a perfectly reasonable result: the analysis indicates the existence of *long-wavelength instabilities that are driven by the current,* which has to surpass a certain critical value given by $\alpha a = \pi$.

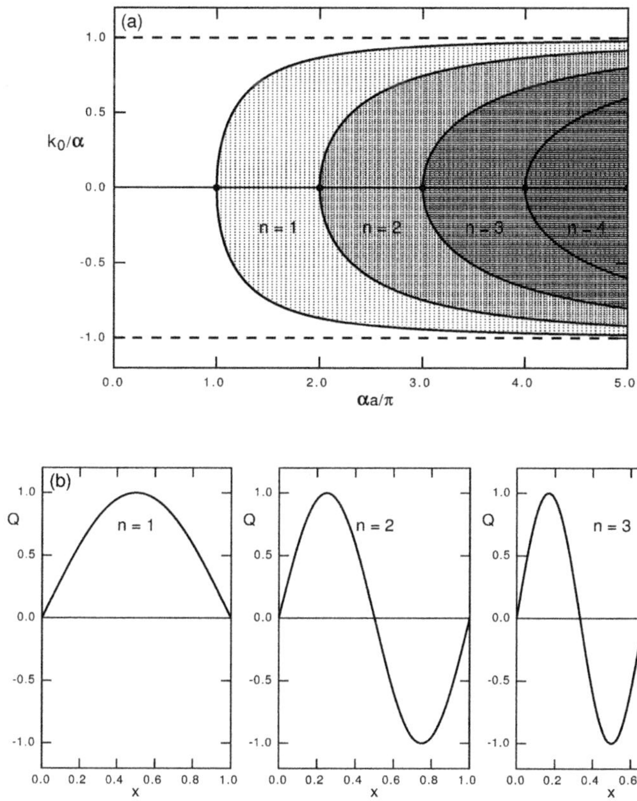

Fig. 14.2 Plane force-free magnetic field: (a) 'stability' diagram; (b) marginal modes.

(b) ... which, however, does not make sense ...

Let us double-check the result obtained by rederiving it from a formulation in terms of the displacement $\boldsymbol{\xi}$ rather than the magnetic field perturbation $\mathbf{Q}$. It is then expedient to exploit *the field line projection* for $\boldsymbol{\xi}$, which was also introduced in Section 7.3.2:

$$\boldsymbol{\xi} = \xi \mathbf{e}_x - \mathrm{i}\eta \mathbf{e}_\perp - \mathrm{i}\zeta \mathbf{e}_\parallel \,, \tag{14.22}$$

where $\mathbf{e}_\perp$ and $\mathbf{e}_\parallel$ were defined in Eq. (7.70) and the relationship with the Cartesian components was given in Eq. (7.73). Using this projection, when computing derivatives, recall that one should take care of the fact that these unit vectors are x-dependent: $\partial_x \mathbf{e}_\perp = -\alpha \mathbf{e}_\parallel$, $\partial_x \mathbf{e}_\parallel = -\alpha \mathbf{e}_\perp$. (This, incidentally, shows that α is proportional to the variation with height of the directional angle φ of the magnetic field shown in Fig. 7.8: $\alpha = \varphi'$.) Next, express the components of $\mathbf{Q} = \nabla \times (\boldsymbol{\xi} \times \mathbf{B})$ in the components ξ, η and ζ of the displacement vector, where we notice that the last component does not appear:

$$\begin{aligned} Q_x = \mathrm{i}F\xi \,, \qquad Q_y &= -(B_y\xi)' + k_z B\eta = -B_y\xi' - \alpha B_z \xi + k_z B\eta \,, \\ Q_z &= -(B_z\xi)' - k_y B\eta = -B_z\xi' + \alpha B_y \xi - k_y B\eta \,, \end{aligned} \tag{14.23}$$

so that

$$Q \equiv -\mathrm{i}Q_x = F\xi\,, \qquad R \equiv k_z Q_y - k_y Q_z = G\xi' - \alpha F\xi + k_0^2 B\eta\,, \tag{14.24}$$

where

$$F \equiv k_y B_y + k_z B_z\,, \quad \text{and} \quad G \equiv k_y B_z - k_z B_y \tag{14.25}$$

are the two projections of the horizontal wave vector having the properties

$$F' = \alpha G \quad \text{and} \quad G' = -\alpha F\,. \tag{14.26}$$

Using Eqs. (14.23), the potential energy (14.3) is expressed in terms of ξ and η:

$$\begin{aligned} W &= \tfrac{1}{2}\int_{x_1}^{x_2} \Big[|Q_x|^2 + |Q_y|^2 + |Q_z|^2 - \alpha\xi_x^*(B_y Q_z - B_z Q_y) - \alpha(B_z\xi_y^* - B_y\xi_z^*)\,Q_x\Big]\,dx \\ &= \tfrac{1}{2}\int_{x_1}^{x_2} \Big[F^2\xi^2 + (\alpha B\xi - F\eta)^2 + (B\xi' + G\eta)^2 - \alpha^2 B^2\xi^2 + 2\alpha BF\xi\eta\Big]\,dx \\ &= \tfrac{1}{2}\int_{x_1}^{x_2} \Big[F^2(\xi^2 + \eta^2) + (B\xi' + G\eta)^2\Big]\,dx \,>\, 0\,. \end{aligned} \tag{14.27}$$

Hence, the potential energy of the perturbations is positive definite so that we conclude that *the slab is trivially stable!* See Ref. [232].

We may obtain the minimizing perturbations by rearranging terms:

$$W = \tfrac{1}{2}\int_{x_1}^{x_2} \Big[F^2(\xi'^2/k_0^2 + \xi^2) + (k_0 B\eta + G\xi'/k_0)^2\Big]\,dx$$

so that W is minimized for perturbations that satisfy

$$k_0 B\eta + G\xi'/k_0 = 0 \tag{14.28}$$

and

$$(F^2\xi')' - k_0^2 F^2\xi = 0\,. \tag{14.29}$$

One easily checks that the latter equation corresponds to Eq. (14.19) with $Q = F\xi$ for $\tilde{\alpha} = \alpha$: the minimizing equations are equivalent. There is no mistake in the algebra!

(c) ... and why

To see what went wrong let us plot the eigenfunctions ξ corresponding to the eigenfunctions Q shown in Fig. 14.2(b). Writing $F = k_0 B_0 \cos(\alpha x - \theta)$, with θ as defined in Fig. 14.1, we find:

$$\xi = \frac{Q}{F} = \frac{1}{k_0 B_0}\,\frac{\sin(n\pi x/a)}{\cos(\alpha x - \theta)}\,, \tag{14.30}$$

as plotted in Fig. 14.3(a). Hence, if a solution Q exists such that W as given in Eq. (14.12) is negative, $\alpha a > \pi$ and ξ then develops a singularity (see Fig. 14.3(b)). For every zero that is added in Q, at least one zero is added to the function F because F oscillates faster than or at least as fast as Q. It is clear that these singularities are of such a nature that the norm

$$\|\boldsymbol{\xi}\|^2 = \int (\xi^2 + \eta^2 + \zeta^2)\,\rho\,dx = \int \Big[\xi^2 + G^2\xi'^2/(k_0^4 B^2)\Big]\,\rho\,dx \to \infty\,,$$

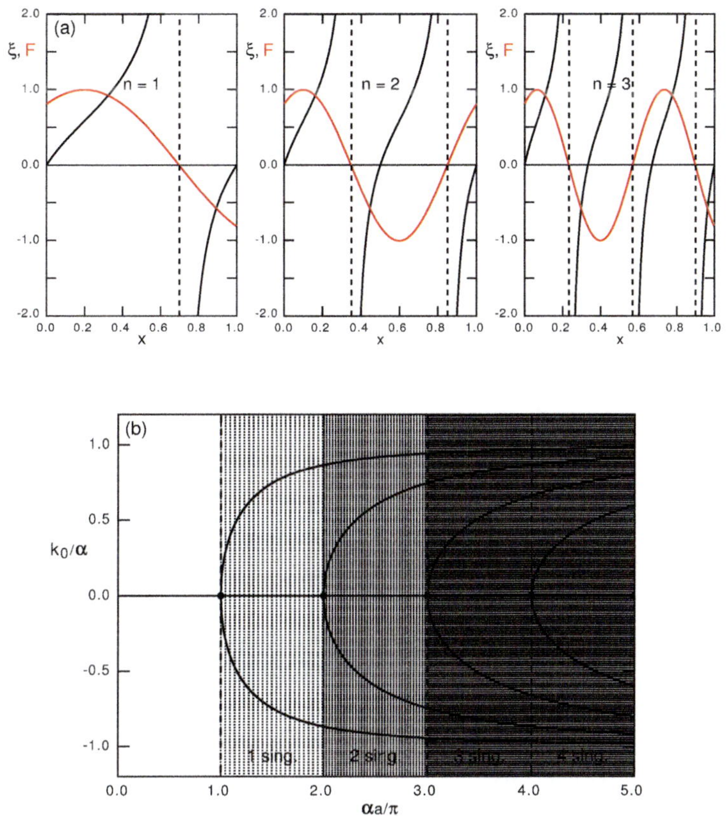

Fig. 14.3 Plane force-free magnetic field: (a) marginal modes in terms of ξ ; (b) singularities $F = 0$ of ξ in the shaded areas $\alpha a \geq n\pi$ $(n = 1, 2, \ldots)$.

where η from Eq. (14.28) and $\zeta = 0$ have been substituted. Hence, *the trial functions Q used in the derivation of the stability criterion (14.21) do not correspond to physically permissible displacements $\boldsymbol{\xi}$.*

Fortunately, there is still a use for the nice stability diagram 14.2. Observe that apparently a reservoir of energy is available that could drive instabilities if the associated displacement $\boldsymbol{\xi}$ only were realizable. Such is the case if we allow a small amount of resistivity in the system so that the relation $Q = F\xi$ of ideal MHD has to be replaced by one that has extra terms proportional to the resistivity. These terms limit the amplitude of the displacement $\boldsymbol{\xi}$ at the singularity (and, therefore, also the perturbed current that is flowing there). As a result, the unstable energy reservoir is tapped and resistive instabilities develop. Such instabilities are called *tearing modes*. We will investigate these modes in detail in the next section and prove that the stability diagram 14.2(a) applies to those modes.

In cylindrical geometry, the picture becomes more complicated yet. Then, ideal MHD instabilities of the force-free magnetic fields do develop. This is a subtle effect, due to the additional magnetic curvature, where the magnetic field variable Q may just oscillate a little faster than the function F in certain regions of the $k_0/\alpha - \alpha a$ parameter plane. This was shown by Voslamber and Callebaut

[637] by a careful analysis taking proper care of the singularities. The corresponding calculation of the growth rates and eigenfunctions was carried out by Goedbloed and Hagebeuk [234] (see Chapter 9, Figs. 9.13 and 9.14).

14.2 Resistive instabilities

14.2.1 Basic equations

We now present the *resistive normal mode analysis* of the plane slab. The starting point is the nonlinear resistive MHD equations as given in Chapter 2, Eqs. (2.129)–(2.132), which we here summarize for the convenience of the reader:

$$\frac{\partial \rho}{\partial t} = -\nabla \cdot (\rho \mathbf{v}), \tag{14.31}$$

$$\rho\Big(\frac{\partial \mathbf{v}}{\partial t} + \mathbf{v}\cdot\nabla\mathbf{v}\Big) = -\nabla p + \rho\mathbf{g} + \mathbf{j}\times\mathbf{B}, \quad \mathbf{j} = \nabla\times\mathbf{B}, \tag{14.32}$$

$$\frac{\partial p}{\partial t} = -\mathbf{v}\cdot\nabla p - \gamma p \nabla\cdot\mathbf{v} + (\gamma-1)\,\eta|\mathbf{j}|^2, \tag{14.33}$$

$$\frac{\partial \mathbf{B}}{\partial t} = -\nabla\times\mathbf{E} = \nabla\times(\mathbf{v}\times\mathbf{B}) - \nabla\times(\eta\mathbf{j}). \tag{14.34}$$

Note that resistivity enters through the Ohmic dissipation term in the pressure equation and through the resistive diffusion in the flux equation. In particular, the latter effect is responsible for substantial modifications of the stability analysis.

We will linearize Eqs. (14.31)–(14.34) for small amplitude perturbations about a static equilibrium. Strictly speaking, the assumption of static equilibrium is not justified since resistivity causes the magnetic field to decay. However, since *the magnetic Reynolds number $R_{\rm m}$ is very large* for situations of interest, this is a very slow process operating on a time scale of the order of $R_{\rm m}\,\tau_{\rm A}$, where $\tau_{\rm A}$ is the characteristic Alfvén time for ideal MHD phenomena. The resistive modes considered in this section will turn out to exponentiate on a much faster time scale, proportional to a broken power of the magnetic Reynolds number, so that the background equilibrium may be considered static. (We here follow standard practice and write 'magnetic Reynolds number', although 'Lundquist number' would be the more appropriate terminology since the background flow velocity is neglected so that the equilibrium is properly characterized by the value of the Alfvén velocity; see Section 4.4.1.)

Considering a plasma slab with background equilibrium quantities depending on the transverse coordinate x only, we make the usual Ansatz

$$f(\mathbf{r},t) = f_0(x) + f_1(x)\mathrm{e}^{\mathrm{i}(k_y y + k_z z - \omega t)}, \tag{14.35}$$

where $f_0(x)$ refers to equilibrium quantities and $f_1(x)$ to perturbations. The equilibrium is described by the variables ρ_0, p_0 and $\mathbf{B}_0$, where we will suppress the subscript 0 for convenience, and the perturbations are described by variables that are renamed as $\delta \equiv \rho_1$, $\mathbf{v} \equiv \mathbf{v}_1$, $\pi \equiv p_1$ and $\mathbf{Q} \equiv \mathbf{B}_1$, so that the subscript 1 may be dropped (which is a recommendable practice for lengthy calculations with many variables).

Assuming a constant resistivity η, linearization of the basic equations (14.31)–(14.34) results in the following set of linear evolution equations:

$$\begin{aligned} \frac{\partial \delta}{\partial t} &= -\nabla \cdot (\rho \mathbf{v}) \,, & (14.36)\\ \rho \frac{\partial \mathbf{v}}{\partial t} &= -\nabla \pi + \delta \mathbf{g} - \mathbf{B} \times (\nabla \times \mathbf{Q}) + (\nabla \times \mathbf{B}) \times \mathbf{Q} \,, & (14.37)\\ \frac{\partial \pi}{\partial t} &= -\mathbf{v} \cdot \nabla p - \gamma p \nabla \cdot \mathbf{v} + 2(\gamma - 1)\eta \nabla \times \mathbf{B} \cdot \nabla \times \mathbf{Q} \,, & (14.38)\\ \frac{\partial \mathbf{Q}}{\partial t} &= \nabla \times (\mathbf{v} \times \mathbf{B}) + \eta \nabla^2 \mathbf{Q} \,. & (14.39) \end{aligned}$$

The resistive terms spoil the possibility of integrating the equations for δ, π and $\mathbf{Q}$ to get expressions in terms of the displacement vector $\boldsymbol{\xi}$ alone, as in ideal MHD. We can still exploit the latter variable, but it will not be possible to eliminate the magnetic field perturbation $\mathbf{Q}$. Thus, the new feature of resistive MHD is the distinction between fluid flow, described by $\boldsymbol{\xi}$, and magnetic field evolution, described by $\mathbf{Q}$, since *magnetic field and fluid do not necessarily move together anymore*.

We now introduce *a projection* based on *the direction of inhomogeneity* (x) and *the two directions in the horizontal plane* defined with respect to the horizontal wave vector $\mathbf{k}_0 = (0, k_y, k_z)$:

$$\begin{aligned} &u \equiv v_x \,, \quad v \equiv (\nabla \times \mathbf{v})_x = -\mathrm{i}(k_z v_y - k_y v_z) \,, \quad w \equiv \nabla \cdot \mathbf{v} - v_x{}' = \mathrm{i}(k_y v_y + k_z v_z) \,, \\ &Q \equiv -\mathrm{i} Q_x \,, \quad R \equiv \mathrm{i}(\nabla \times \mathbf{Q})_x = \mathrm{i}\, j_{1x} = k_z Q_y - k_y Q_z \,. \qquad (14.40) \end{aligned}$$

Here, u is the normal velocity, v is the normal vorticity, w is the horizontal part of the compressibility, Q is the normal magnetic field perturbation and R is the perturbed normal current. In terms of these variables the eigenvalue problem becomes

$$\begin{aligned} -\mathrm{i}\omega\, \delta &= -(\rho u)' - \rho w \,, \\ -\mathrm{i}\rho\omega\, u &= -\pi' - g\delta + k_0^{-2}(FQ' + GR)' - FQ \,, \\ -\mathrm{i}\rho\omega\, v &= -G'Q + FR \,, \\ -\mathrm{i}\rho\omega\, w &= k_0^2 \pi - F'Q - GR \,, \\ -\mathrm{i}\omega\, \pi &= -p'u - \gamma p(u' + w) - 2(\gamma - 1)\eta\, k_0^{-2} \left[F'(Q'' - k_0^2 Q) + G'R' \right] , \\ -\mathrm{i}\omega\, Q &= Fu + \eta(Q'' - k_0^2 Q) \,, \\ -\mathrm{i}\omega\, R &= (Gu)' - Fv + Gw + \eta(R'' - k_0^2 R) \,, \qquad (14.41) \end{aligned}$$

where $k_0^2 \equiv k_y^2 + k_z^2$, and F and G are the projections of the horizontal wave vector $\mathbf{k}_0$ onto the magnetic field:

$$G \equiv \mathbf{e}_x \cdot (\mathbf{k}_0 \times \mathbf{B}) = k_y B_z - k_z B_y \,, \qquad F \equiv \mathbf{k}_0 \cdot \mathbf{B} = k_y B_y + k_z B_z \,. \qquad (14.42)$$

Recall that the equilibrium is inhomogeneous through the quantities $\rho(x)$, $p(x)$, $B_y(x)$, $B_z(x)$, so that F and G also depend on x. As usual, the prime denotes differentiation with respect to x. The system (14.41) is suitable for numerical solution but the main difficulty is the presence of a complex eigenvalue spectrum.

Alternatively, one could introduce the components of the displacement vector $\boldsymbol{\xi}$,

$$u \equiv -\mathrm{i}\omega\xi \,, \qquad v \equiv -\mathrm{i}\omega\sigma \,, \qquad w \equiv -\mathrm{i}\omega\tau \,, \qquad (14.43)$$

and eliminate the variables δ, σ, τ and π to obtain a sixth order system of three coupled second order differential equations for ξ, Q and R. The ideal MHD second order differential equation (7.80) for ξ would be contained as the limiting case $\eta \to 0$. We will not pursue this line here since it would involve too many terms in an exposition that is already complicated enough if the mere essentials are presented. This we intend to do.

An important simplification results from *the assumption of incompressibility*. This is justified for the kind of resistive modes we will study, as can be checked after the solutions have been obtained. The incompressible limit is formally obtained from Eqs. (14.41) by taking the limits $\gamma \to \infty$ and $\nabla \cdot \mathbf{v} \to 0$ simultaneously in such a way that the product $\gamma p \nabla \cdot \mathbf{v}$ and, hence, π remains finite but undetermined. Consequently, Eq. (14.41)(e) for π should be dropped (so that the complicated Ohmic dissipation term in square brackets also disappears from the problem) and replaced by the constraint of incompressibility, $\nabla \cdot \mathbf{v} = 0$. The latter implies that $w = -u'$, so that the variable w is known in terms of u, and Eq. (14.41)(d) for w can then be used to determine π. Furthermore, the variables δ, v and w may be expressed in terms of $\xi \equiv u/(-\mathrm{i}\omega)$, Q and R so that we obtain the following sixth order system:

$$\eta \Big[(\rho\omega^2 \xi')' - k_0^2(\rho\omega^2 + \rho' g)\, \xi + F'' Q \Big] + \mathrm{i}\omega F\, (Q - F\xi) = 0\,,$$

$$\eta\, (Q'' - k_0^2 Q) + \mathrm{i}\omega\, (Q - F\xi) = 0\,,$$

$$\eta\, (R'' - k_0^2 R) + \mathrm{i}\omega\, (R - G'\xi) - \frac{\mathrm{i}F}{\rho\omega}\, (FR - G'Q) = 0\,. \qquad (14.44)$$

A pleasant surprise is that the variable R does not occur in the first pair of equations, so that we may drop the last equation and restrict the study to the fourth order system for the variables ξ and Q alone.

Obtaining the incompressible ideal MHD equations from these equations by taking the limit $\eta \to 0$ is tricky. First, we have to expand Eq. (14.44)(b) to first order,

$$Q = F\xi + \frac{\mathrm{i}\eta}{\omega}(Q'' - k_0^2 Q) \approx F\xi + \frac{\mathrm{i}\eta}{\omega}\big[\, (F\xi)'' - k_0^2 F\xi \,\big]\,, \qquad (14.45)$$

and then insert the result in Eq. (14.44)(a):

$$\Big[(\rho\omega^2 - F^2)\, \xi' \Big]' - k_0^2(\rho\omega^2 - F^2 + \rho' g)\xi = 0\,, \qquad Q = F\xi\,. \qquad (14.46)$$

This agrees with the ideal MHD Eq. (7.80), in the incompressible limit $\gamma p \to \infty$, where field and fluid move together again. Our task now is to analyse what happens when these equations are replaced by Eqs. (14.44) for the resistive evolution and what is the role of the ideal MHD limiting case represented by Eq. (14.46).

14.2.2 Tearing modes

We will now study the so-called *tearing modes*, which result in breaking and rejoining of the magnetic field lines. Starting from the incompressible resistive MHD equations (14.44), the following assumptions are appropriate:

– the analysis is restricted to eigenvalues corresponding to purely exponential instability so that a real and positive eigenvalue parameter can be defined,

$$\lambda \equiv -\mathrm{i}\omega > 0\,; \tag{14.47}$$

– it is assumed that the density $\rho = \text{const}$, which eliminates ideal MHD gravitational instabilities, since $\rho' g = 0$, so that the slab is stable in the ideal MHD description.

The modes are then described by the resistive MHD equations in the form

$$\begin{aligned} &\eta\,\big[\,\lambda^2(\xi'' - k_0^2\xi) - (F''/\rho)\,Q\,\big] + \lambda(F/\rho)\,(Q - F\xi) = 0\,,\\ &\eta\,(Q'' - k_0^2 Q) - \lambda\,(Q - F\xi) = 0\,. \end{aligned} \tag{14.48}$$

Note that all terms in these equations are real now.

In a problem like this it is imperative to enumerate the degrees of freedom by defining *dimensionless parameters*. One can then make various assumptions on the smallness of those parameters to exploit them in *asymptotic expansions*. In the present problem, the thickness of the slab a, the density ρ and the magnitude of magnetic field at the mid plane B_0 are taken as units of length, mass and time, exploiting the definition of the Alfvén velocity, $v_{\rm A} \equiv B_0/\sqrt{\rho}$, so that $\tau_{\rm A} \equiv a/v_{\rm A}$. This implies that a, ρ and B_0 should not be counted as free parameters since they simply fix the dimensions and then disappear from the problem, only to return in the end when actual dimensional numbers need to be computed for comparison with observed quantities.

Our first assumption on parameters is that the wavelength in the horizontal plane is comparable to the transverse size a of the plasma:

$$k_0 a \sim 1\,. \tag{14.49}$$

This expresses the fact that tearing modes should be considered as *large-scale macroscopic MHD modes* which do involve a small-scale resistive effect operating in the normal (x) direction, as we will see, but it does not require localization or small wavelengths in the transverse (y, z) directions. Next, as already mentioned, we will *exploit the magnetic Reynolds number*[1] *as an ordering parameter*:

$$(R_{\rm m})^{-1} \equiv \eta/(\mu_0 a v_{\rm A}) \ll 1\,. \tag{14.50}$$

The equilibrium decays on a diffusion time scale $\tau_{\rm d}$ that is much longer than the characteristic Alfvén time $\tau_{\rm A}$ for ideal MHD. We wish to study resistive modes that exponentiate much faster than the resistive diffusion time, but much slower than the ideal MHD times:

$$(\tau_{\rm d})^{-1} \equiv (R_{\rm m})^{-1} v_{\rm A}/a \;\ll\; \lambda \;\ll\; v_{\rm A}/a \equiv (\tau_{\rm A})^{-1}\,. \tag{14.51}$$

This is possible if we can find modes with a growth rate λ that scales, e.g., as a broken power of the magnetic Reynolds number: $\lambda \sim (R_{\rm m})^{-\nu}\, v_{\rm A}/a$, where $0 < \nu < 1$. This will turn out to be the case. Since $R_{\rm m}$ is huge, this provides enough parameter space for our asymptotic analysis.

With this ordering, for small η, the resistive equations (14.48) automatically would lead to the ideal MHD equations,

$$\big[\,(\lambda^2 + F^2/\rho)\,\xi'\,\big]' - k_0^2(\lambda^2 + F^2/\rho)\,\xi = 0\,, \qquad Q = F\xi\,, \tag{14.52}$$

[1] For once, we have explicitly written the constant μ_0 in the definition of $R_{\rm m}$ to refresh our memory, although it is suppressed again in most of the rest of this chapter.

since the expansion (14.45) yields

$$Q \approx F\xi + (\eta/\lambda)\,\big[\,(F\xi)'' - k_0^2 F\xi\,\big] \approx \big[\,1 + \mathcal{O}(\lambda\tau_{\rm d})^{-1}\,\big]\,F\xi \approx F\xi\,. \tag{14.53}$$

Here, the resistive correction of order $(\lambda\tau_{\rm d})^{-1} \equiv \eta/(\lambda a^2)$ is negligible according to the left part of the approximation (14.51) *if ξ is assumed to have $\mathcal{O}(1)$ variations only*. However, we know from our previous analysis of Section 14.1 that the assumption of finite variation on ξ is not justified when ideal MHD singularities $F = 0$ occur. Then (choosing the origin of the x-coordinate to coincide with the singularity), $\xi_{\rm ideal} \sim 1/x \to \infty$ while the magnetic field variable Q remains finite. Hence, the resistive terms in Eq. (14.53) cannot be neglected in a small layer surrounding the ideal MHD singularity where they limit the amplitude of ξ and the related current density perturbation. Outside this layer, ideal MHD is appropriate. Consequently, the problem may be analysed by distinguishing three regions, viz. *two outer ideal MHD regions* where F is not small, and *one inner resistive layer* surrounding the point $F = 0$. The solutions of the three regions have to be matched to each other so that there should exist overlap regions where the resistive as well as the ideal solutions are valid.

The singularity $F = 0$ can occur anywhere on the interval $x_1 \le x \le x_2$, but we will position it at $x = x_0 = 0$ for simplicity. This involves both a simplification of notation (expansions in x rather than $x - x_0$) and of some part of the analysis (which distinguishes between modes that are even or odd with respect to $x = 0$). Of course, these restrictions can easily be relaxed. Our choice of the zero point does not limit the physics of the problem in any way, since we are free to shift the x-axis to simplify the algebra.

(a) Outer ideal MHD regions The outer ideal MHD regions cover most of the real axis from $x = x_1 = -\frac{1}{2}a$ to $x = x_2 = \frac{1}{2}a$, where $F \neq 0$, except for a small region around $x = 0$ where F is small and resistive effects dominate. Obviously, the size of the resistive region is determined by the resistivity, but in a way that has to be determined yet. In the outer ideal MHD regions we may simplify the equations (14.52) even further by noting that

$$\frac{\lambda^2}{F^2/\rho} \sim (\lambda a/v_{\rm A})^2 \ll 1\,,$$

according to the estimates (14.49) and the right part of the inequalities (14.51). Hence, in the outer layer the *marginal* ($\lambda = 0$) ideal MHD equations apply:

$$(F^2\xi')' - k_0^2 F^2\,\xi = 0\,, \qquad Q = F\xi\,. \tag{14.54}$$

This implies that the evolution of tearing modes is so slow that inertia is negligible on the ideal MHD time scale. The associated boundary conditions are

$$\xi(x_1) = \xi(x_2) = 0\,. \tag{14.55}$$

In terms of Q, the basic differential equation (14.54) transforms into

$$Q'' - (k_0^2 + F''/F)\,Q = 0\,. \tag{14.56}$$

If there is no ideal MHD singularity ($F \neq 0$ everywhere) Eq. (14.54) has no solutions satisfying the boundary conditions (14.55), as we have seen in Section 14.1, and hence there are no resistive instabilities in that case. To get resistive instabilities we need a point $F = 0$ so that the ideal MHD solution ξ blows up and Q' is discontinuous at that point (dashed lines in Fig. 14.4).

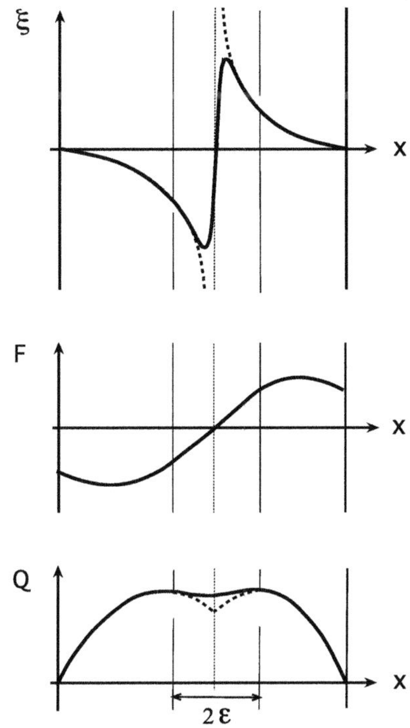

Fig. 14.4 Sketch of the resistive (drawn) and the ideal (dashed) MHD solutions at the ideal MHD singularity $F = 0$.

Let us formally solve Eq. (14.54) in the two outer regions $(x_1, -\epsilon)$ and (ϵ, x_2), where matching to the resistive layer solutions should take place at $x = \pm\epsilon$. The main idea is that the singularity $x = 0$ is eliminated, but that the ideal MHD solutions are valid close enough to this point to permit a singular expansion. Since $F \approx F_0' x$ for $|x| \sim \epsilon$, two solutions are obtained, a large one, ξ_ℓ, and a small one, ξ_s, which may be continued up to the outer boundaries $x = x_{1,2}$. The solutions in the two outer regions are distinguished by a superscript $+$ for $x > \epsilon$ and $-$ for $x < -\epsilon$. Consequently, the general solutions in the two outer regions may be written as

$$\xi^\pm(x) = A^\pm \xi_\ell^\pm(x) + B^\pm \xi_s^\pm(x) , \tag{14.57}$$

where

$$\xi_\ell = \frac{1}{x}\left[1 + \tfrac{1}{2}(k_0 x)^2 + \cdots\right], \qquad \xi_s = 1 - \tfrac{1}{4}(k_0 x)^2 + \cdots . \tag{14.58}$$

The two ratios B^+/A^+ and B^-/A^- between the small and large contributions are determined by the boundary conditions (14.55). Whereas close to the singularity the large contributions dominate, for matching to the resistive inner layer solutions the small contributions are essential as well. Matching will require a prescribed mix of large and small solutions, as we will see. This is expressed most clearly by the logarithmic derivatives of the magnetic field perturbation which are completely determined by the solutions in the ideal regions:

$$\frac{1}{Q}\frac{dQ}{dx}\bigg|^{\rm outer}_{x=\epsilon\downarrow 0} \approx \frac{B^+}{A^+} = -\frac{\xi_\ell^+(x_2)}{\xi_{\rm s}^+(x_2)}\,, \qquad \frac{1}{Q}\frac{dQ}{dx}\bigg|^{\rm outer}_{x=-\epsilon\uparrow 0} \approx \frac{B^-}{A^-} = -\frac{\xi_\ell^-(x_1)}{\xi_{\rm s}^-(x_1)}\,. \tag{14.59}$$

These quantities are all that the outer ideal MHD regions provide. It will turn out to be sufficient.

Notice that the two logarithmic derivatives (14.59) are not equal in the limit $\epsilon \longrightarrow 0$, so that a jump is obtained. This jump is permitted in ideal MHD and it corresponds to the occurrence of a *surface current perturbation*, which is extremely stabilizing. Neglecting this contribution is the physical cause of the erroneous instability results obtained in Section 14.1. In resistive MHD, the jump is smoothed through the resistive layer solutions, which will be computed next, and the instability results expressed by Fig. 14.2 will turn out to become applicable then.

(b) Inner resistive layer The resistive layer surrounds the ideal MHD singularity $F = 0$, which we have positioned at $x = 0$. In this region, the function F may be approximated as

$$F \equiv \mathbf{k}_0 \cdot \mathbf{B} \approx F_0'\,x\,. \tag{14.60}$$

We will call the width of the resistive layer δ. Obviously,

$$\delta \ll a\,, \tag{14.61}$$

because of the small value of the resistivity, and it is logical to expect it to scale with a negative power of the magnetic Reynolds number. One of our aims is to determine this power. Hence, we are in the peculiar situation of having to prescribe boundary conditions at an unknown location. This problem is solved by following the example of *boundary layer analysis* of fluid mechanics.

Note that matching to the outer ideal MHD solutions is considered to take place at $x = \pm\epsilon$ so that we formally need to impose

$$\delta \ll \epsilon \ll a\,. \tag{14.62}$$

This can easily be arranged again by means of some broken power of the magnetic Reynolds number, which we could choose after the scaling of δ with $R_{\rm m}$ has been determined. The precise magnitude of ϵ should not matter, though, as long as the approximation (14.60) is valid there.

In the resistive layer, the differential equations (14.48) hold. They may be further simplified on the basis of the approximations (14.49), (14.51) and (14.61). To that end, the orders of magnitude of the different terms are compared:

$$\left|\frac{k_0^2\,\xi}{\xi''}\right| \sim \delta^2/a^2 \ll 1\,, \qquad \left|\frac{\eta(F''/\rho)\,Q}{\lambda(F/\rho)\,Q}\right| \sim \left|\frac{\eta k_0^2\,Q}{\lambda\,Q}\right| \sim \frac{(R_{\rm m})^{-1}v_{\rm A}/a}{\lambda} \ll 1\,.$$

Consequently, the resistive layer equations may be simplified to

$$\begin{aligned} \eta\,\lambda\,\xi'' - (F^2/\rho)\,\xi &= -\,(F/\rho)\,Q\,, \\ \eta\,Q'' - \lambda\,Q &= -\,\lambda F\,\xi\,. \end{aligned} \tag{14.63}$$

It is to be noted that this significant simplification now only holds in the resistive layer since the approximate ideal MHD equations (14.54) are no longer contained as a limiting case (which was the case for the original pairs of equations (14.48) and (14.52)). Consequently, we have lost the option of (numerically) integrating the resistive equations all the way to the plasma boundaries at $x = x_{1,2}$, i.e. we are now forced to consider the matching problem.

(c) Scaling and matching Before we do this, we finally perform the promised scaling of the resistive layer equations (14.63) exploiting three of the four dimensionless parameters that occur in this problem, viz.

$$\lambda\, a/v_{\rm A}\,, \quad R_{\rm m} \equiv a v_{\rm A}/\eta\,, \quad H \equiv F_0' a/(k_0 B_0)\,, \quad K \equiv k_0 a\,. \tag{14.64}$$

Note that K appears as a separate parameter in the outer ideal MHD equations (14.54). By means of these parameters all variables may be rescaled:

$$\begin{aligned} s &\equiv \Big[R_{\rm m} K H\Big]^{1/3} (x/a)\,, & \bar\lambda &\equiv \Big[R_{\rm m}/(K^2H^2)\Big]^{1/3} (\lambda a/v_{\rm A})\,,\\ \Phi &\equiv \Big[R_{\rm m}/(K^2H^2)\Big]^{-1/3} (\xi/a)\,, & \Psi &\equiv Q/B_0\,. \end{aligned} \tag{14.65}$$

The explicit dependence on the parameters $R_{\rm m}$, K and H then disappears and only the scaled eigenvalue $\bar\lambda$ appears in the resistive equations:

$$\begin{aligned} \bar\lambda \frac{d^2\Phi}{ds^2} - s^2\,\Phi &= -\,s\,\Psi\,,\\ \frac{d^2\Psi}{ds^2} - \bar\lambda\,\Psi &= -\,\bar\lambda\, s\,\Phi\,, \qquad \text{or} \qquad \frac{d^2\Psi}{ds^2} = -\bar\lambda^2 \frac{1}{s}\frac{d^2\Phi}{ds^2}\,. \end{aligned} \tag{14.66}$$

The problem has now been reduced to the solution of a fourth order system of ODEs with one parameter ($\bar\lambda$). This has to be solved on an interval of the stretched coordinate s corresponding to a very narrow region in physical space. Notice that the tendency of the original variable ξ to blow up in the limit of small resistivity has been absorbed in the normalization of the scaled variable Φ.

▷ **Analysis of the fourth order differential equation** The equations (14.66) may be written explicitly as a fourth order equation in terms of Φ alone:

$$\bar\lambda s^2 \frac{d^4\Phi}{ds^4} - 2\bar\lambda s \frac{d^3\Phi}{ds^3} + (2\bar\lambda - \bar\lambda^2 s^2 - s^4)\frac{d^2\Phi}{ds^2} - 2s^3\frac{d\Phi}{ds} = 0\,. \tag{14.67}$$

The lowest order term with Φ is missing so that one of the four solutions is a constant. Since this is not essential (for other resistive modes it does not occur, Section 14.2.3), we will not base the analysis on it. Eq. (14.67) may be investigated with respect to singular behaviour by means of the usual series expansion methods [325], [50]. One finds out that the point $s = 0$ has become a regular point (which was the whole point of replacing the ideal by the resistive MHD equations) and the points $s = \pm\infty$ are irregular singular points. The latter permit a singular expansion, to be matched asymptotically to the outer ideal MHD solutions. It is useful to consider the three formal expansions in the resistive regions (about $s = 0, \infty, -\infty$) to clarify the behaviour of the solutions and the kind of BCs needed. Once this is clear, Eq. (14.67) can be solved by whatever method is convenient.

By straightforward substitution, we find the following expansions about $s = 0$:

$$\begin{aligned} \Phi &= \Phi^0(s) = \sum_{i=1}^{4} C_i^0 \phi_i^0\,, & \Psi &= \Psi^0(s) = \sum_{i=1}^{4} C_i^0 \psi_i^0\,,\\ \phi_1^0 &= 1\,, & \psi_1^0 &= s\,,\\ \phi_2^0 &= s + \frac{1}{20\bar\lambda}s^5 + \cdots\,, & \psi_2^0 &= -\frac{1}{12}\bar\lambda s^4 + \cdots\,,\\ \phi_3^0 &= s^3 + \frac{3\bar\lambda}{20}s^5 + \cdots\,, & \psi_3^0 &= -6\bar\lambda + \cdots\,,\\ \phi_4^0 &= s^4 + \frac{\bar\lambda}{15}s^6 + \cdots\,, & \psi_4^0 &= -12\bar\lambda\, s + \cdots\,. \end{aligned} \tag{14.68}$$

As expected, all solutions are regular near $s = 0$, and they are either odd or even, consistent with the reflection symmetry about $s = 0$ of the resistive equations (14.66) or (14.67).

The behaviour at the outer edges is considerably more complicated, as shown by the following expansions for $s \to \infty$ and $s \to -\infty$ (distinguished by the superscripts + and −):

$$\begin{aligned}
\Phi &= \Phi^{\pm}(s) = \sum_{i=1}^{4} C_i^{\pm}\phi_i^{\pm}\,, & \Psi &= \Psi^{\pm}(s) = \sum_{i=1}^{4} C_i^{\pm}\psi_i^{\pm}\,,\\
\phi_1^{\pm} &= 1\,, & \psi_1^{\pm} &= s\,\phi_1^{\pm}\,,\\
\phi_2^{\pm} &= \frac{1}{s}\Big(1 - \frac{\bar\lambda^2}{3s^2} + \cdots\Big)\,, & \psi_2^{\pm} &\approx s\,\phi_2^{\pm}\,,\\
\phi_3^{\pm} &= \mathrm{e}^{s^2/(2\sqrt{\bar\lambda})}\,\frac{1}{s}\,|s|^{\frac{1}{2}+\frac{1}{2}\bar\lambda\sqrt{\bar\lambda}}\Big(1 + \frac{\alpha^{\pm}}{s^2} + \cdots\Big)\,, & \psi_3^{\pm} &\approx -\frac{\bar\lambda^2}{s}\,\phi_3^{\pm}\,,\\
\phi_4^{\pm} &= \mathrm{e}^{-s^2/(2\sqrt{\bar\lambda})}\,|s|^{-\frac{1}{2}-\frac{1}{2}\bar\lambda\sqrt{\bar\lambda}}\Big(1 + \frac{\beta^{\pm}}{s^2} + \cdots\Big)\,, & \psi_4^{\pm} &\approx -\frac{\bar\lambda^2}{s}\,\phi_4^{\pm}\,.
\end{aligned} \tag{14.69}$$

The coefficients $\alpha^{\pm}$ and $\beta^{\pm}$ are known, in principle, but they are not important. Clearly, at the outer edges of the resistive layer, there are two pairs of solutions, $\phi_1^{\pm}$ and $\phi_2^{\pm}$, which exhibit the ideal MHD relationship $\Psi \approx s\Phi$, and two pairs, $\phi_3^{\pm}$ and $\phi_4^{\pm}$, which do not. Consequently, the first two solutions can be used to match to the outer ideal MHD solutions, and the latter two should not play a role. For the solutions $\phi_4^{\pm}$ this is automatically the case since they are exponentially small there. The two solutions $\phi_3^{\pm}$, though, are exponentially unbounded and should be excluded by imposing the boundary conditions of *regularity*, so that $C_3^- = C_3^+ = 0$. What remains is the other boundary conditions of *matching of the logarithmic derivatives of the magnetic field perturbation* to the outer ideal MHD expressions given in Eq. (14.59). One easily checks that these expressions just involve the two solutions $\phi_1^{\pm}$ and $\phi_2^{\pm}$: $(\Psi'/\Psi)^- = C_1^-/C_2^-$ and $(\Psi'/\Psi)^+ = C_1^+/C_2^+$. Notice that the two exponentially small solutions $\phi_4^{\pm}$ have not been excluded; they just do not contribute to the boundary conditions directly. Hence, everything now counts correctly: we have two regularity conditions and two conditions on the logarithmic derivatives, i.e. four boundary conditions on a fourth order ODE.

Of course, the three expansions (14.68) and (14.69) all refer to one and the same solution of the fourth order ODE (14.67), i.e. we may write

$$\Phi = \sum_{i=1}^{4} C_i^-\phi_i^- = \sum_{i=1}^{4} C_i^0\phi_i^0 = \sum_{i=1}^{4} C_i^+\phi_i^+\,. \tag{14.70}$$

Therefore, the three sets of fundamental solutions $\{\phi_i^-\}$, $\{\phi_i^0\}$ and $\{\phi_i^+\}$ possess a linear relationship to each other which could be found explicitly, if desired, by analytic continuation or by straightforward numerical integration of Eq. (14.67). Consequently, the three sets of constants $\{C_i^-\}$, $\{C_i^0\}$ and $\{C_i^+\}$ are linearly dependent as well:

$$C_i^- = \sum_{j=1}^{4} \gamma_{i,j}\,C_j^0\,, \qquad C_i^0 = \sum_{j=1}^{4} \delta_{i,j}\,C_j^+\,, \tag{14.71}$$

where the coefficients $\gamma_{i,j}$ and $\delta_{i,j}$ are then known, in principle. The point is that one set of four constants is sufficient to fix the solution, but these four can not coincide with any one of the three separate sets since boundary conditions at different locations are involved. For example, from the present analysis, it should be clear that these four constants could be C_4^-, C_1^-/C_2^-, C_4^+ and C_1^+/C_2^+. ◁

In conclusion, appropriate boundary conditions to be imposed on the resistive layer equations (14.66) are:

$$\begin{aligned}
&\Phi\Big|_{\substack{\text{inner}\\ s\to-\infty}}\ \text{regular}\,, && \frac{1}{\Psi}\frac{d\Psi}{ds}\Big|_{\substack{\text{inner}\\ s\to-\infty}} \approx \Big(\frac{x}{s}\Big)\frac{1}{Q}\frac{dQ}{dx}\Big|_{\substack{\text{outer}\\ x\uparrow 0}}\,,\\
&\Phi\Big|_{\substack{\text{inner}\\ s\to\infty}}\ \text{regular}\,, && \frac{1}{\Psi}\frac{d\Psi}{ds}\Big|_{\substack{\text{inner}\\ s\to\infty}} \approx \Big(\frac{x}{s}\Big)\frac{1}{Q}\frac{dQ}{dx}\Big|_{\substack{\text{outer}\\ x\downarrow 0}}\,,
\end{aligned} \tag{14.72}$$

which translate into the following conditions on the constants:

$$C_3^- = 0\,, \qquad C_1^-/C_2^- = B^-/A^-\,,$$
$$C_3^+ = 0\,, \qquad C_1^+/C_2^+ = B^+/A^+\,. \tag{14.73}$$

Alternatively, since we have arranged the singularity to be located in the middle of the interval, we can restrict the analysis to even and odd modes about the mid-plane. In particular, we can study modes that are odd in Φ and even in Ψ since they are the first ones to become unstable. The appropriate boundary conditions for this system read:

$$\Phi(0) = 0\,, \qquad \Psi'(0) = 0\,,$$
$$\Phi\,\Big|_{\substack{\text{inner}\\ s\to\infty}} \text{regular}\,, \qquad \frac{1}{\Psi}\frac{d\Psi}{ds}\Big|_{\substack{\text{inner}\\ s\to\infty}} \approx \left(\frac{x}{s}\right)\frac{1}{Q}\frac{dQ}{dx}\Big|_{\substack{\text{outer}\\ x\downarrow 0}}\,, \tag{14.74}$$

which translates into the following conditions on the constants:

$$C_1^0 = 0\,, \qquad C_4^0 = 0\,,$$
$$C_3^+ = 0\,, \qquad C_1^+/C_2^+ = B^+/A^+\,. \tag{14.75}$$

Notice that the factor (x/s) just represents the different normalizations of the inner and outer regions according to Eq. (14.65).

We could now solve Eqs. (14.66). This would have to be done numerically, since the system is still not elementary enough to allow for analytical expressions. Therefore, one additional step will be taken to obtain closed answers justifying the scaling assumptions.

(d) Approximate solution From a study of the numerical results, Furth, Killeen and Rosenbluth in their classical paper on the resistive instabilities [202] proposed the so-called *'constant Ψ' approximation* to obtain explicit solutions. The idea is that the function Φ is the one which misbehaves in ideal MHD and, consequently, exhibits large variations in the limit of small resistivity, but the magnetic field variable Ψ only exhibits moderate variations. Its behaviour only counts to produce the correct numerical magnitude of the derivative to connect to the outer solutions through the boundary conditions (14.72). If Φ were known, that part of the information could be obtained with high accuracy by just integrating the second form of Eq. (14.66)(b). On the other hand, the correct behaviour of Φ can be obtained from Eq. (14.66)(a) without much influence of the magnetic field contribution on the right hand side: just putting $\Psi \approx \Psi_0 = \text{const}$ will do. This approximation turns out to produce results that are quite close to the numerical values. Once it has been made, the rest of the analysis is straightforward.

In the 'constant Ψ' approximation, the variables can be scaled once more:

$$\bar{s} \equiv \bar{\lambda}^{-1/4}\, s\,, \qquad \bar{\Phi} \equiv \bar{\lambda}^{1/4}\, \Phi/\Psi_0\,, \qquad \bar{\Psi} \equiv \Psi/\Psi_0\,, \tag{14.76}$$

so that the eigenvalue $\bar{\lambda}$ is also eliminated from the ODE for Φ, and the equations (14.66) transform into

$$\frac{d^2\bar{\Phi}}{d\bar{s}^2} - \bar{s}^2\,\bar{\Phi} = -\,\bar{s}\,\bar{\Psi} \approx -\bar{s}\,,$$
$$\frac{d^2\bar{\Psi}}{d\bar{s}^2} = -\,\bar{\lambda}^{3/2}\,\frac{1}{\bar{s}}\frac{d^2\bar{\Phi}}{d\bar{s}^2}\,. \tag{14.77}$$

In other words: we obtain an inhomogeneous second order ODE for $\bar{\Phi}$, whereas the logarithmic derivative of $\bar{\Psi}$, i.e. $\bar{\Psi}'$, may be obtained by integrating Eq. (14.77)(b) once. We will now exploit the symmetry of the problem, i.e. we assume $\bar{\Phi}$ to be odd, by using the boundary conditions (14.74) which, for the present variables, become

$$\begin{aligned} &\bar{\Phi}(0) = 0\,, && \bar{\Psi}'(0) = 0\,, \\ &\bar{\Phi}\Big|_{\bar{s}\to\infty}\ \text{regular}\,, && \bar{\Psi}'\Big|_{\bar{s}\to\infty} \approx \text{const}\ \Big[= (x/\bar{s})\, Q'/Q\Big|_{x\downarrow 0}\Big]\,, \end{aligned} \tag{14.78}$$

where the constant is just the logarithmic derivative of the magnetic field of the outer ideal MHD solutions.

The solution of the ODE (14.77)(a) for $\bar{\Phi}$, subject to the boundary conditions (14.78), can be represented in terms of a definite integral over an auxiliary variable u:

$$\bar{\Phi} = \tfrac{1}{2}\bar{s} \int_0^1 (1-u^2)^{-1/4}\, \mathrm{e}^{-\frac{1}{2}\bar{s}^2 u}\, du\,. \tag{14.79}$$

One easily checks the correctness of this expression by substitution into the differential equation, whereas the behaviour for small and large $\bar{s}$ demonstrates satisfaction of the boundary conditions:

$$\begin{aligned} \bar{\Phi}(\bar{s}) &\approx \frac{C}{2\sqrt{\pi}}\,\bar{s}\,, && \bar{\Psi}'(\bar{s}) \approx \bar{\lambda}^{3/2}\bar{s} && (\bar{s} \ll 1)\,, \\ \bar{\Phi}(\bar{s}) &\approx \frac{1}{\bar{s}}\,, && \bar{\Psi}'(\bar{s}) \approx \tfrac{1}{2} C\bar{\lambda}^{3/2} && (\bar{s} \gg 1)\,, \end{aligned} \tag{14.80}$$

where the constant

$$C = \sqrt{2\pi} \int_0^1 \sqrt{u}\,(1-u^2)^{-1/4}\, du = 2\pi\Gamma(\tfrac{3}{4})/\Gamma(\tfrac{1}{4}) = 2.1236\ldots \tag{14.81}$$

results from an integration over the auxiliary variable.

The actual identification of $\bar{\Psi}'\big|_{\bar{s}\to\infty}$ with the logarithmic derivative of the outer ideal MHD solutions, according to the boundary condition (14.78)(b), provides the dispersion equation for the computation of the growth rate of the tearing modes. We will rewrite this condition in terms of the contributions of both outer boundaries:

$$\frac{d\bar{\Psi}}{d\bar{s}}\Big|^{\text{inner}}_{\bar{s}\to\infty} - \frac{d\bar{\Psi}}{d\bar{s}}\Big|^{\text{inner}}_{\bar{s}\to-\infty} = -\,\bar{\lambda}^{3/2}\int_{-\infty}^{\infty} d\,\bar{s}\,\frac{1}{\bar{s}}\frac{d^2\bar{\Phi}}{d\bar{s}^2} = C\,\bar{\lambda}^{3/2} = \Big(\frac{x/a}{\bar{s}}\Big)\,\Delta'\,, \tag{14.82}$$

where Δ' is the jump of the logarithmic derivative of the magnetic field perturbation of the outside ideal MHD solution,

$$\Delta' \equiv \frac{a}{Q}\left(\frac{dQ}{dx}\Big|^{\text{outer}}_{x\downarrow 0} - \frac{dQ}{dx}\Big|^{\text{outer}}_{x\uparrow 0} \right), \tag{14.83}$$

which is determined by the solutions of Eq. (14.56).

Substituting the scaling factors from the definitions (14.65) and (14.76) back into Eq. (14.82) then results in the explicit expression for *the growth rate of the tearing mode*:

$$\lambda = R_{\mathrm{m}}^{-3/5}(KH)^{2/5}\Big(\frac{\Delta'}{C}\Big)^{4/5} v_{\mathrm{A}}/a\,, \tag{14.84}$$

which justifies our original assumption of broken powers of the magnetic Reynolds number. An estimate of *the resistive layer width* is obtained from the relation $s \sim 1$, which gives:

$$\delta \sim R_{\rm m}^{-2/5}(KH)^{-2/5}\left(\frac{\Delta'}{C}\right)^{1/5} a\,, \tag{14.85}$$

which also conforms to our assumptions.

The tearing mode requires Δ' to be positive so that the outer ideal MHD solutions determine its stability. It should be noted that the growth rate λ of Eq. (14.84) depends on the dimensionless parameters K and H both through the explicit power $(KH)^{2/5}$ and through the implicit dependence $\Delta' = \Delta'(H, K)$. This may be illustrated by the force free magnetic field of Section 14.1, where the value of Δ' can be computed easily. In that case (taking $\theta = \pi/2$, so that the singularity occurs at $x = 0$ and $F = k_0 B_0 \sin \alpha x$), $Q^{\pm} = \mp D \sin \sqrt{\alpha^2 - k_0^2}\,(x \mp \frac{1}{2}a)$, so that

$$\Delta' = -2a\sqrt{\alpha^2 - k_0^2}\,\cot\left(\tfrac{1}{2}a\sqrt{\alpha^2 - k_0^2}\right) = -2\sqrt{H^2 - K^2}\,\cot\left(\tfrac{1}{2}\sqrt{H^2 - K^2}\right). \tag{14.86}$$

Its value is positive, as required for tearing instability, when $(\alpha a)^2 - (k_0 a)^2 \equiv H^2 - K^2 > (n\pi)^2$. This agrees with the stability diagrams of Section 14.1 so that this analysis has finally received its proper re-interpretation now in the context of resistive MHD: *the plane force-free field is unstable with respect to long-wavelength tearing instabilities*. The expressions (14.84) and (14.86) clearly show that tearing modes are long-wavelength instabilities ($K \sim 1$ and $|K| < |H|$) *driven by the current*, as expressed by the parameter H.

14.2.3 Resistive interchange modes

A completely different kind of resistive mode is the *resistive gravitational interchange mode*. The driving force of this instability may be expressed by the dimensionless parameter

$$G \equiv \rho' g a^2 / B_0^2\,, \tag{14.87}$$

which was neglected when transforming Eq. (14.44) to Eq. (14.48). We will not enter the detailed analysis of this mode, but just give some of the final results in order to illustrate the wide range of possible outcomes of a boundary layer analysis.

First of all, it should be remarked that the stability criterion for the corresponding ideal MHD gravitational interchanges is given by Eq. (7.191) in the incompressible limit ($\gamma p \to \infty$):

$$\rho' g \le \tfrac{1}{4} B^2 \varphi'^2\,, \quad \text{i.e.} \quad G \le \tfrac{1}{4} H^2\,. \tag{14.88}$$

This criterion is completely analogous to the well-known Suydam criterion where a pressure gradient driven instability is balanced by the shear of the magnetic field lines. Similarly, the gravitational interchanges are driven by *a density gradient* (heavy fluid on top of a lighter fluid), but possibly stabilized by *the shear* term H. Hence, the same parameter H enters as in the tearing mode analysis, but with an entirely different result: tearing instabilities are driven by the current, interchanges are stabilized by shear which is also created by the current.

However, *stabilisation by shear of the magnetic field lines is another ideal MHD property that is lost when resistivity is introduced*. The stability criterion for resistive interchange modes simply becomes

$$G \leq 0\,, \tag{14.89}$$

i.e. only lighter fluid on top of heavier fluid allowed! By a similar boundary layer analysis as for the tearing modes, the growth rate is found to be given by

$$\lambda \sim R_{\rm m}^{-1/3} \left(\frac{KG}{H} \right)^{2/3} v_{\rm A}/a\,, \tag{14.90}$$

whereas the resistive layer width, in this case, is given by

$$\delta \sim R_{\rm m}^{-1/3} \left(\frac{G}{K^2 H^4} \right)^{1/6} a\,. \tag{14.91}$$

Hence, resistive interchanges are no longer stabilized by magnetic shear (the parameter H), but their growth rate is strongly diminished with the $-1/3$ power of the magnetic Reynolds number according to Eq. (14.90).

14.3 Resistive spectrum

14.3.1 Resistive wall mode

The basic mechanism of resistive instability due to magnetic flux reconnection is also operating in an entirely different location than inside the plasma, viz. in the conducting wall that is supposed to stabilize the external MHD instabilities. In particular, if operating parameters of fusion experiments are pushed up to obtain higher values of beta, current-driven external instabilities frequently interfere, and they are usually stabilized by the presence of a conducting wall. Apart from the fact that a conducting wall close to a plasma is not very desirable in a fusion reactor, the conductivity itself turns out to be a problem.

Consider an external kink mode which is stabilized by a conducting wall, but which would be unstable without that wall. For simplicity, we analyse this mode for a flat current distribution in the 'straight tokamak' approximation, appropriate for low-beta plasmas. Normalizing the wall position to the plasma radius, $w \equiv b/a$, the expression for the growth rate is given by Eq. (9.97):

$$\bar{\omega}^2 \equiv \tau_{\rm A}^2 \omega^2 = \frac{2(n + m/q)}{1 - w^{-2|m|}} \Big[n + m/q - (\mathrm{sgn}(m)/q)(1 - w^{-2|m|}) \Big]\,. \tag{14.92}$$

This expression is valid for a slender torus represented by a periodic straight cylinder of length $2\pi R_0$ and plasma radius a (so that the inverse aspect ratio is small, $\epsilon \equiv a/R_0 \ll 1$), whereas $\tau_{\rm A}$ is the characteristic Alfvén time along that torus:

$$\tau_{\rm A} \equiv \frac{R_0}{v_{\rm A}} = \frac{R_0 \sqrt{\mu_0 \rho_0}}{B_0}\,. \tag{14.93}$$

For the external $n = -1, m = 2$ (or $n = 1, m = -2$) kink mode with the wall close to the plasma, only a narrow instability range $1 + w^{-4} < q < 2$ remains and the mode is stable in the wide range $1 < q < 1 + w^{-4}$, but violently unstable when the wall is taken away. Let us consider what happens in that range when the wall is not taken away but the finite resistivity of it is taken into account.

The expression (14.92) for the ideal MHD growth rate is found by solving both the ODE for the radial displacement ξ, obtained from the Hain–Lüst equation in the low-beta 'straight tokamak' approximation, Eq. (9.50), with $\bar{\rho} \equiv \rho(r)/\rho_0$,

$$\frac{d}{dr}\left\{\frac{r}{m^2}\left[\bar{\rho}\bar{\omega}^2 - (n+m/q)^2\right]\frac{d}{dr}(r\xi)\right\} - \left[\bar{\rho}\bar{\omega}^2 - (n+m/q)^2 - \frac{2r}{m}(n+m/q)(1/q)'\right]\xi = 0 \qquad (0 \le r \le 1), \qquad (14.94)$$

and the ODE for the radial perturbation $\hat{Q}_{\rm i}$ of the magnetic field in the vacuum,

$$\frac{d}{dr}\left[\frac{r}{m^2}\frac{d}{dr}(r\hat{Q}_{\rm i})\right] - \hat{Q}_{\rm i} = 0 \qquad (1 \le r \le w), \qquad (14.95)$$

joining them by applying the BC at the plasma–vacuum interface,

$$\frac{\bar{\omega}^2 - (n+m/q)^2}{m^2(n+m/q)}\frac{(r\xi)'}{\xi} + \frac{2}{mq} = -\frac{n+m/q}{m^2}\frac{(r\hat{Q}_{\rm i})'}{\hat{Q}_{\rm i}} \qquad (r=1), \qquad (14.96)$$

and applying either one of the following BCs on the outside:

$$\hat{Q}_{\rm i} = 0, \quad \text{for } r = \begin{cases} w & \textit{(infinitely conducting wall)} \\ \infty & \textit{(no wall)} \end{cases}. \qquad (14.97)$$

The BC (14.96) is obtained by dividing the two plasma–vacuum interface conditions (6.131) and (6.134) of Section 6.6.1, and applying the low-β tokamak approximation. The usual problem with a vanishing denominator for $\xi(1) = 0$ does not occur here since $\xi(1) \neq 0$ by definition of the external modes that we are concerned about.

For constant plasma density, $\bar{\rho} = 1$, and constant current density, $q = \text{const}$, the factors in square brackets in the ODE (14.94) become constant and equal, so that we obtain the following simple solutions for ξ and $r\hat{Q}_{\rm i}$:

$$r\xi = A_1 r^{|m|}, \qquad r\hat{Q}_{\rm i} = A_2(r^{|m|} - Cr^{-|m|}). \qquad (14.98)$$

Substituting these solutions into the plasma–vacuum BC (14.96), the constants A_1 and A_2 drop out and the remaining constant C is determined:

$$C = \frac{\bar{\omega}^2 + 2(\text{sgn}(m)/q)(n+m/q)}{\bar{\omega}^2 + 2(\text{sgn}(m)/q)(n+m/q) - 2(n+m/q)^2}. \qquad (14.99)$$

The perfectly conducting wall BC (14.97) also fixes this constant:

$$C = w^{2|m|}. \qquad (14.100)$$

Equating these two expressions for C yields the dispersion equation (14.92). If desired, one could assume a more realistic current profile and solve the ODE (14.94), but the equation determining the growth rate would be implicit. This would change the magnitude of the growth rate, also of the resistive wall mode, but not the formal scaling of the expression that we will derive.

The problem of the loss of wall stabilization of external kink modes when the resistivity of the wall is taken into account was first discussed by Pfirsch and Tasso [488]. They proved that an MHD unstable configuration cannot be stabilized by the introduction of resistive walls, i.e. with respect to stability it does not make a difference whether or not a resistive wall is present. Of course, the crucial question then becomes whether the growth rate of the ensuing *resistive wall mode* (RWM) is compatible with plasma confinement on the characteristic time scale needed for nuclear fusion. This problem was addressed by Goedbloed et al. [239], who solved the dispersion equation of resistive wall modes for a high-beta screw pinch (a toroidal device differing from a tokamak by the presence of a stabilizing layer of force-free currents in a tenuous outer plasma replacing the 'vacuum'). The outcome was positive for the high-beta device, considered promising at the time, but rather marginal for tokamaks. In this section, we will follow that paper but replace the Bessel functions, needed to represent perturbations of high-beta pinches, by the much simpler representation in terms of powers of r, appropriate for the low-beta tokamak approximation. This changes the numerical magnitude of the growth rates, but not the qualitative spectral picture presented in Fig. 14.5.

Obviously, the difference between a perfectly conducting and a resistive wall resides in the BCs applied there. Hence, instead of applying the infinitely conducting wall BC (14.97)(a) to $\mathbf{Q}_\mathrm{i}$, which implies that no flux diffuses through the wall, we now have to apply resistive BCs, connecting the internal vacuum field perturbation $\mathbf{Q}_\mathrm{i}$ to the external perturbation $\mathbf{Q}_\mathrm{e}$ through a discontinuity. This is obtained by integrating the equations for the perturbations across the resistive wall, similar to the procedure described in Section 4.5 to obtain jump conditions for plasma discontinuities. This implies considering a thin wall of thickness δ and resistivity η, with a current density perturbation $\mathbf{j}_1$, and taking the limits $\delta \to 0, \eta \to 0$ and $\mathbf{j}_1 \to \infty$, such that the surface resistivity $\eta^* \equiv \eta/\delta$ and the surface current density $\mathbf{j}_1^* \equiv \mathbf{j}_1 \delta$ remain finite. In that limit, integration of the combined laws of Faraday, Ohm and Ampère gives

$$\frac{\partial \hat{\mathbf{Q}}}{\partial t} = -\nabla \times \mathbf{E}_1 = -\nabla \times (\eta \mathbf{j}_1) = -\nabla \times (\eta^* \mathbf{j}_1^*) = -\frac{\eta^*}{\mu_0} \nabla \times \big(\mathbf{n} \times [\![\hat{\mathbf{Q}}]\!]\big) ,$$

where, as usual, jumps are denoted by $[\![f]\!] \equiv f_\mathrm{ex} - f_\mathrm{in}$. Since the normal component of $\hat{\mathbf{Q}}$ should be continuous, so that $\mathbf{n} \cdot \nabla \times (\mathbf{n} \times [\![\hat{\mathbf{Q}}]\!]) = \mathbf{n} \cdot [\![\nabla \hat{\mathbf{Q}}]\!] \cdot \mathbf{n}$, this yields the resistive wall BCs:

$$\left.\begin{aligned} & [\![\mathbf{n} \cdot \hat{\mathbf{Q}}]\!] = 0 \\ & \omega \mathbf{n} \cdot \hat{\mathbf{Q}} - \mathrm{i}\frac{\eta^*}{\mu_0} \mathbf{n} \cdot [\![\nabla \hat{\mathbf{Q}}]\!] \cdot \mathbf{n} = 0 \end{aligned}\right\} \qquad (\text{at } r = w) , \tag{14.101}$$

whereas the external field should satisfy

$$\mathbf{n} \cdot \hat{\mathbf{Q}} \to 0 \qquad (\text{for } r \to \infty) . \tag{14.102}$$

Note that these boundary conditions are generally valid, independent of assumptions on the geometry of the configuration.

For the present case of a cylindrical plasma, it is expedient to introduce an abbreviation for the characteristic time scale associated with resistive diffusion of the magnetic field perturbation through the wall, and an associated small parameter comparing it with the Alfvén time scale:

$$\tau_\mathrm{d} \equiv \mu_0 w / \eta^* , \qquad \bar{\eta}^* \equiv \tau_\mathrm{A} / \tau_\mathrm{d} \ll 1 . \tag{14.103}$$

The combined resistive wall BC, replacing the combined ideal wall BC (14.96), is obtained by dividing the two expressions (14.101) and inserting τ_d:

$$\frac{(r\hat{Q}_\mathrm{i})'}{\hat{Q}_\mathrm{i}} = \frac{(r\hat{Q}_\mathrm{e})'}{\hat{Q}_\mathrm{e}} + \mathrm{i}\tau_\mathrm{d}\omega \qquad (\text{at } r = w) . \tag{14.104}$$

The exterior solution $\hat{Q}_e$ should obey the BC (14.102), i.e.

$$\hat{Q}_e \to 0 \qquad (r \to \infty)\,, \tag{14.105}$$

so that

$$r\hat{Q}_e = A_3 r^{-|m|}\,. \tag{14.106}$$

The full eigenvalue problem now becomes to solve the ODEs (14.94) and (14.95) for $\xi(r)$ and $\hat{Q}_i$, connect them by applying the plasma–vacuum BC (14.96), solve the ODE analogous to (14.95) for $\hat{Q}_e$, subject it to the BC (14.105) at infinity, and finally connect $\hat{Q}_i$ and $\hat{Q}_e$ by applying the resistive wall BC (14.104).

Substitution of the expressions (14.98) for ξ and $\hat{Q}_i$ into the plasma–vacuum BCs (14.96) yields, of course, the same expression (14.99) for the constant C, with the difference that C is now complex since $\bar{\omega}^2$ is complex. It remains to substitute the expression (14.106) for $\hat{Q}_e$ into the resistive wall BC (14.104). This gives another expression for the complex constant C:

$$C = w^{2|m|}\Big(1 + 2\mathrm{i}|m|\frac{\bar{\eta}^*}{\bar{\omega}}\Big)\,. \tag{14.107}$$

This one replaces the perfectly conducting wall expression (14.100). Equating the two expressions (14.99) and (14.107) for C yields the *dispersion equation for resistive wall modes*, which is a cubic in $\bar{\omega}$:

$$\bar{\omega}\big(\bar{\omega}^2 - S\big) + \mathrm{i}\bar{\eta}^* E\big(\bar{\omega}^2 - T\big) = 0\,, \tag{14.108}$$

where

$$\begin{aligned} S &\equiv \frac{2(n+m/q)^2}{1-w^{-2|m|}} - 2(\mathrm{sgn}(m)/q)(n+m/q)\,, \quad E \equiv \frac{2|m|}{1-w^{-2|m|}}\,, \\ T &\equiv 2(n+m/q)^2 - 2(\mathrm{sgn}(m)/q)(n+m/q)\,. \end{aligned} \tag{14.109}$$

From this expression, when the resistivity of the wall is increased, from $\bar{\eta}^* = 0$ to $\bar{\eta}^* = \infty$, the spectrum of eigenvalues of the external waves and instabilities qualitatively changes as indicated in Fig. 14.5:

(a) for vanishing resistivity of the wall, $\bar{\eta}^* = 0$, from the first term, we obtain the two stable kink oscillations described by Eq. (14.92):

$$\bar{\omega}^2 = S \equiv \bar{\omega}_{\mathrm{w}}^2 > 0\,; \tag{14.110}$$

(b) for small, but non-vanishing, resistivity, the two stable kink modes become resistively damped, and a new mode, the resistive wall mode, pops out of the origin:

$$\bar{\omega}_{1,2} \approx \pm\sqrt{S} - \tfrac{1}{2}\mathrm{i}\bar{\eta}^* E(1 - T/S)\,, \qquad \bar{\omega}_3 \approx -\mathrm{i}\bar{\eta}^* ET/S\,; \tag{14.111}$$

(c) increasing the resistivity still further, the real part of the stable oscillations vanishes for a certain critical value $\bar{\eta}^* = \bar{\eta}^*_{\mathrm{crit}}$, so that a degenerate stable oscillation together with a rapidly growing resistive wall mode is obtained;

(d) for very large $\bar{\eta}^*$, the two degenerate modes split apart, one moves off to $-\mathrm{i}\infty$ and the other one becomes the complex conjugate partner of the unstable kink mode in the absence of a wall;

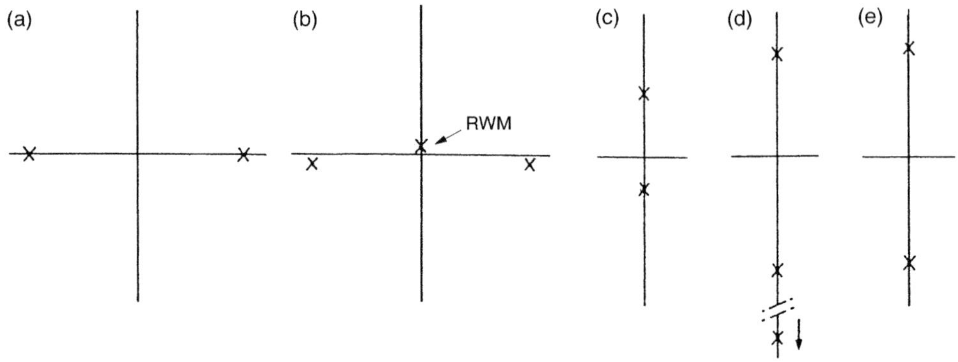

Fig. 14.5 Transition of wall-stabilized external kink modes to unstable and damped kink modes by increasing the resistivity of the wall: (a) for $\bar{\eta}^* = 0$, the two kink modes are wall stabilized, (b) for small $\bar{\eta}^*$, the kink modes become damped and a purely growing resistive wall mode pops out of the origin, (c) for $\bar{\eta}^* = \bar{\eta}^*_{\rm crit}$, the damped kink modes coalesce and become purely imaginary, (d) for $\bar{\eta}^* > \bar{\eta}^*_{\rm crit}$, one of the damped modes moves off to $-\mathrm{i}\infty$, (e) for $\bar{\eta}^* = \infty$, the unstable kink mode and its complex conjugate is obtained. (From Goedbloed et al. [239].)

(e) finally, for $\bar{\eta}^* \to \infty$, from the second term, we obtain the unstable kink mode and its conjugate partner described by Eq. (14.92) in the limit $w \to \infty$:

$$\bar{\omega}^2 = T \equiv \bar{\omega}_\infty^2 < 0 \,. \tag{14.112}$$

Clearly, the crucial expression above is the imaginary $\bar{\omega}_3$ given by Eq. (14.111), which is the *growth rate of the resistive wall mode* for small resistivity:

$$\bar{\omega}_{\rm rwm} \approx -\mathrm{i}\bar{\eta}^* ET/S = -\mathrm{i}\bar{\eta}^* \frac{2|m|}{1 - w^{-2|m|}} \frac{\bar{\omega}_\infty^2}{\bar{\omega}_{\rm w}^2} \,. \tag{14.113}$$

Because of our starting assumption, the two factors $\bar{\omega}_\infty^2$ and $\bar{\omega}_{\rm w}^2$ have opposite signs so that the mode is unstable. Hence, with respect to stability, a resistive wall cannot change a mode from unstable to stable, but it does decrease the growth rate in proportion to the conductivity of the wall.

It is now just of matter of inserting numbers to find out whether the growth rate can be made small enough to be of no concern for fusion. Of course, for that purpose, realistic current profiles and a proper description of the toroidal geometry need to be taken into account. Several ways of eliminating the resistive wall mode for future fusion devices have been investigated. We just mention the consideration of RF stabilization [233], plasma rotation [80, 59, 267] and active feedback control [179, 177] to slow down the modes. The fact that there is a lot of ongoing research on this topic shows that the answer to the question is far from comforting.

14.3.2 Spectrum of homogeneous plasma

We will now consider the influence of resistivity of the plasma itself, starting with the stable part of the spectrum, in particular of the Alfvén waves. As we will see, similar spectral structures as in the resistive wall mode occur. However, in this case, there is no formal justification of the neglect of the resistive decay of the background equilibrium since the frequencies of the waves will be modified by a damping that may operate on a time scale comparable to that of resistive diffusion. Nevertheless, we will follow standard practice and neglect this effect, pending an overall theory that incorporates both resistive diffusion of the background and resistive decay of the waves.

First, let us study the case when the background equilibrium is completely homogeneous. In that case, Eqs. (14.44) for incompressible plasmas transform into

$$\eta\, \rho\omega^2(\xi'' - k_0^2\xi) - \mathrm{i}\omega F\,(Q - F\xi) = 0\,,$$
$$\eta\,(Q'' - k_0^2 Q) + \mathrm{i}\omega\,(Q - F\xi) = 0\,. \qquad (14.114)$$

In ideal MHD, this reduces to

$$(\omega^2 - \omega_A^2)(\xi'' - k_0^2\xi) = 0\,, \qquad (14.115)$$

so that $\pm\omega_A$ represents a continuum of two infinitely degenerate eigenvalues.

Now, we may also assume harmonic dependence in the x-direction, so that $d/dx = \mathrm{i}k_x$. Introducing the resistive parameter

$$\widetilde{\eta} \equiv \frac{\eta k^2}{2\omega_A}\,, \qquad k^2 \equiv k_x^2 + k_y^2 + k_z^2\,. \qquad (14.116)$$

the dispersion equation may be written as

$$\omega^2 + 2\mathrm{i}\widetilde{\eta}\,\omega_A\omega - \omega_A^2 = 0\,, \qquad (14.117)$$

with the solution

$$\omega = \omega_A\Big(\pm\sqrt{1-\widetilde{\eta}^2} - \mathrm{i}\widetilde{\eta}\Big)\,. \qquad (14.118)$$

For varying values of $\widetilde{\eta}$, the modes are lying on a semi-circle in the lower half plane, starting from the real axis at $\omega = \pm\omega_A$ ($\widetilde{\eta} = 0$), intersecting the negative imaginary axis at $\omega = -\mathrm{i}\omega_A$ ($\widetilde{\eta} = 1$), and then moving along the negative imaginary axis to $\omega = 0$ and $\omega = -\mathrm{i}\infty$ ($\widetilde{\eta} \to \infty$).

The analysis above centred on the Alfvén modes for a homogeneous, incompressible plasma, and showed analytically that resistivity introduces specific curves (semi- circles) in the damped frequency half-plane on which the modes reside. The spacing of the resistive Alfvén eigenmodes on these curves is influenced by the resistivity in accord with Eq. (14.116), but the radius of the semi-circle is independent of the resistivity η.

In Fig. 14.6, we show the effect of finite resistivity on the entire MHD spectrum consisting of fast, Alfvén and slow modes of a homogeneous, compressible plasma. Using numerical techniques discussed more extensively in Chapter 15, an overview of all modes with mode numbers $k_y = 0$, $k_z = 1$, for a uniform layer of density $\rho = 1$, magnetic field $\mathbf{B} = (0,0,1)$ and $\beta = 0.25$ (with $\gamma = 5/3$) is computed easily. The figure combines the spectrum obtained under ideal MHD $\eta = 0$ conditions, together with its modification when the resistivity parameter has a constant value $\eta = 0.001$ or $\eta = 0.005$. Note that, in ideal MHD, the homogeneous plasma slab has an infinite Sturmian sequence of fast modes accumulating to infinity, a degenerate Alfvén frequency ω_A and an anti-Sturmian sequence of slow modes accumulating to the (constant) slow frequency ω_S. The latter is shown in more detail in the top insert of the figure. This behaviour was explained previously in Chapter 7 and also shown in Fig. 7.7. When resistivity is included, the bottom part of Fig. 14.6 shows that the fast mode sequence shifts from purely oscillatory to damped, where the resistivity *does influence the precise location of the damped fast mode sequence* in the complex plane. As

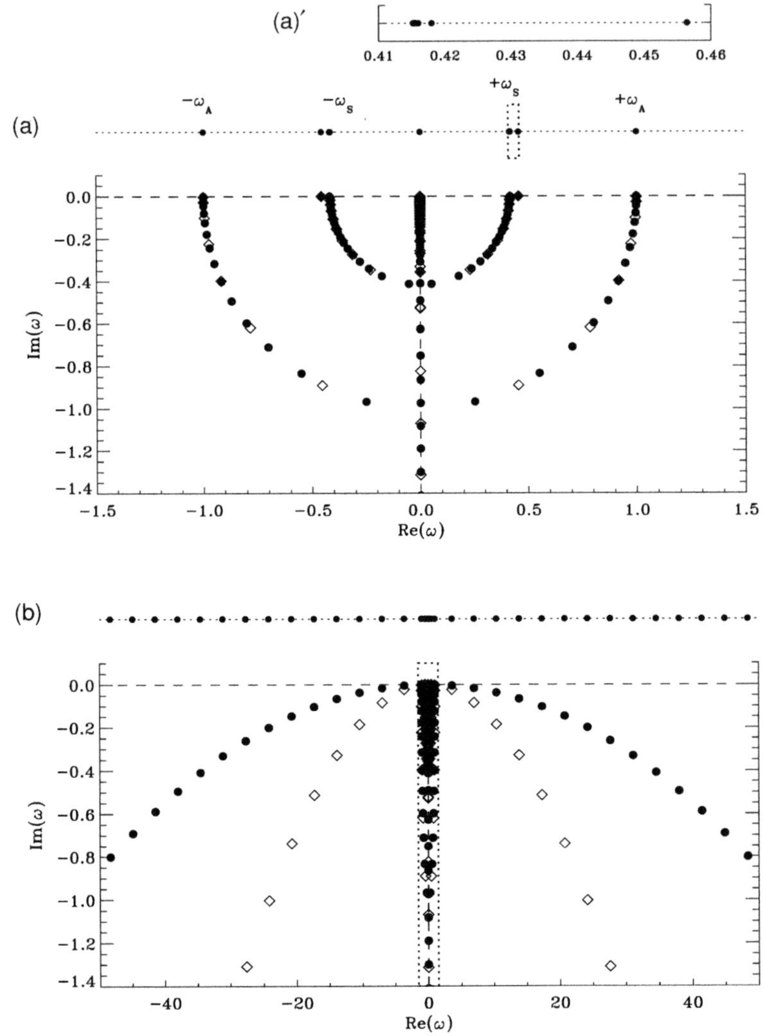

Fig. 14.6 Linear ideal versus resistive MHD spectra for a uniform, compressible plasma. The ideal (purely stable) results are indicated above the two panels, which combine results for $\eta = 0.001$ (filled circles) and $\eta = 0.005$ (diamonds). Note the difference in the horizontal axis (oscillation frequency): the bottom panel (b) shows most clearly the fast mode sequence, while the top panel (a) zooms in on Alfvén and slow modes. Dashed boxes indicate the range of a zoomed view above. The insert (a)′ illustrates the anti-Sturmian sequence of ideal MHD slow modes.

expected, the higher the mode frequency in the Sturmian sequence, the more it is damped. The top panel demonstrates clearly that both the Alfvén and the slow modes relocate to semi-circles in the damped half-plane (and the negative imaginary axis), and these semi-circle locations remain uninfluenced by resistivity. However, the spacing of the individual resistive Alfvén or slow modes on these curves is affected by resistivity changes. This complements and agrees with the above analytical results for the resistive Alfvén modes in the incompressible case.

14.3.3 Spectrum of inhomogeneous plasma

For inhomogeneous plasmas, the resistive spectrum should be some kind of modification of the continuous spectrum of ideal MHD. Resistivity changes the order of the system so that the singularities due to the vanishing of the coefficients in front of the highest derivatives disappear. Hence, one should expect the continua to split up into discrete modes. This is indeed what happens. However, it happens in a rather unexpected fashion, as already suggested by the case of the homogeneous plasma of the previous section. There, the infinitely degenerate ideal MHD continuum frequencies relocated to sequences of discrete resistive eigenmodes on semi-circles and the negative imaginary axis in the stable half-plane. As a mathematical analysis of the resistive spectrum of an inhomogeneous plasma becomes rather formidable, we here restrict the discussion to exemplary spectra computed numerically. This has the advantage that no simplifying assumptions on, e.g., incompressibility, or non-overlapping continua have to be made.

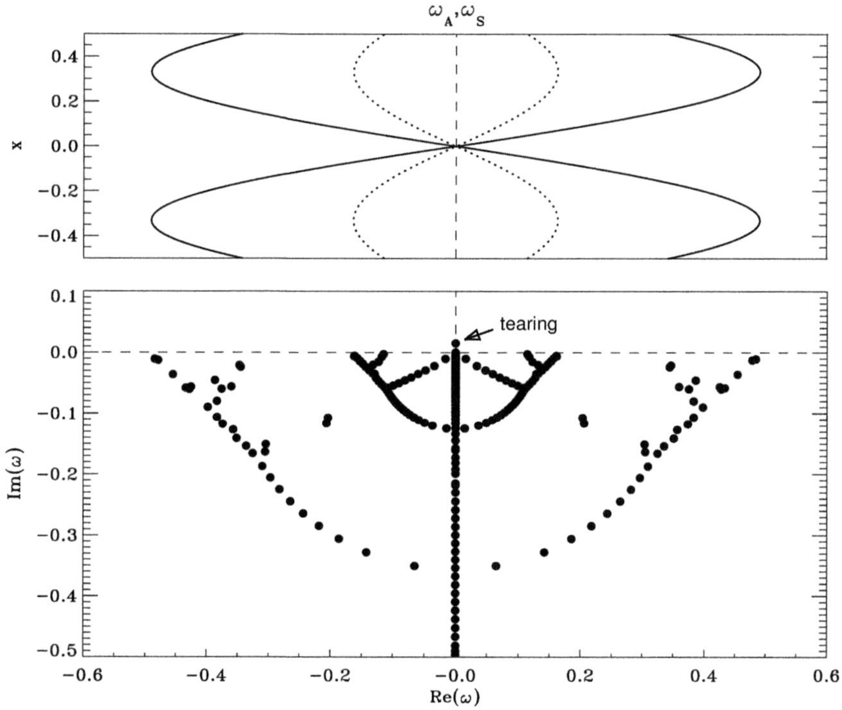

Fig. 14.7 Typical Alfvén and slow magneto-sonic parts of the resistive MHD spectrum for a force-free slab. Note the isolated unstable tearing mode. The ideal continuous spectra indicated in the top frame (x-location in the slab versus oscillation frequency) show the connection between their end points ($x = \pm 0.5$) and the extremal values with the curves on which resistive modes are found.

A relevant example revisits the force-free equilibrium from Eq. (14.2) analysed in Section 14.1. While most of the analysis presented there used incompressible conditions, the main conclusions on stability against resistive tearing modes as governed by criteria in Eq. (14.21) or Eq. (14.86) can be expected to hold in the compressible case as well. When considering a plasma of constant density $\rho = 1, \beta = 0.15$ and magnetic field $\mathbf{B} = (0, \sin(\alpha\, x), \cos(\alpha\, x))$ on $x \in [-0.5, 0.5]$ with specific

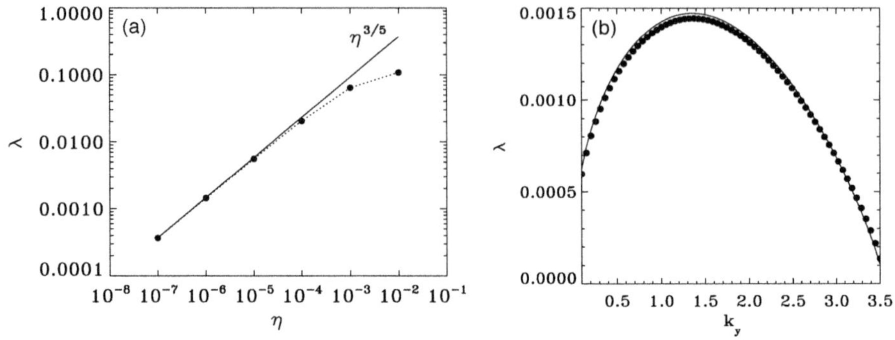

Fig. 14.8 Linear resistive MHD results for the tearing mode in a force-free slab: (a) growth rate as a function of η for $k_y = 1.5$ and all other parameters as in Fig. 14.7; (b) predicted versus computed growth rate variation at fixed $\eta = 10^{-6}$.

heat ratio $\gamma = 5/3$, the tearing mode stability criterion is violated for the choice $k_y = 0.49, k_z = 0$, when $\alpha = 4.73884$. For this combination of wave numbers and equilibrium parameters, the Alfvén and slow continua all go through zero at $x = 0$, and contain internal extrema. These overlapping continuum ranges are shown in the top panel of Fig. 14.7. The bottom panel shows the resistive MHD spectrum for a resistivity of $\eta = 0.0001$. Only the Alfvén and slow modes are seen in this frequency range, and one recognizes the following.

– The ideal, stable continuum ranges are replaced by sequences of discrete resistive eigenmodes which lie on specific curves in the stable half-plane. Repeating the computation for differing values of η influences the spacing of the modes on these curves, but not the curves themselves.

– For the case of the overlapping continua shown in the figure, the curves still vaguely resemble semi-circles, as for the homogeneous case in Fig. 14.6, but complicated by branches which split off in almost triangular patterns. The tips of the triangles approach the stable real axis at oscillation frequencies that correspond to an internal extremum, an end value (at $x = \pm 0.5$) or a zero value of the ideal continua.

– While the ideal continua are thus dramatically transformed into an intricate structure of discrete modes in the stable half-plane, the unstable tearing mode is still prominently present with its growth rate in agreement with the scaling given by Eq. (14.84) for varying resistivity. This latter point is demonstrated in Fig. 14.8, where the theoretical predictions are compared with numerically obtained growth rates.

The latter conclusion underlines the fact that instabilities form the most robust part of the MHD spectrum (when going from ideal to resistive conditions), and that the concepts of σ-stability (Chapter 6) can be expected to remain of practical use.

In addition to the unstable tearing mode and the resistive discrete eigenmodes on curves that are the counterpart of the ideal continua, Fig. 14.7 also shows some more isolated modes within the triangular sections of the curves. A better example is discussed in Chapter 15 (Fig. 15.11), where half of the resistive sub-spectrum of Alfvén and slow modes is shown (rotated over 90 degrees) for a cylindrical equilibrium and mode numbers with non-overlapping, well-separated ideal Alfvén and slow continuum ranges. The isolated mode in the triangular section of the Alfvén sub-spectrum

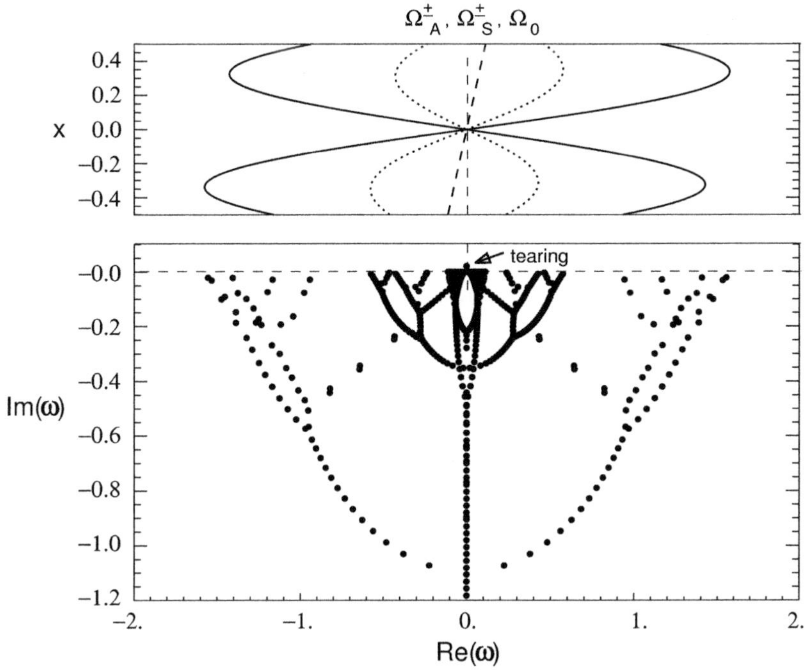

Fig. 14.9 Typical Alfvén and slow magneto-sonic parts of the resistive MHD spectrum for a force-free slab with a linear flow profile. Note the (isolated) unstable tearing mode. The Doppler-shifted ideal continuous spectra are indicated above together with the local Doppler shift, and show the distinct connection between their end point ($x = \pm 0.5$) and internal extremal values with the curves on which resistive modes are found.

in Fig. 15.11 has been identified as an ideal quasi-mode, and these locally damped global modes form the most robust part of the *stable* frequency spectrum. They represent the natural oscillation modes of the system and explain how perfect coupling between a driver and its excited plasma loops can be achieved. As mentioned in Chapter 11, these quasi-modes have a damping rate which becomes *independent of resistivity* for large Reynolds numbers, distinguishing them from the resistive modes on the curves. In the study [627] their intricate connection with the resistive eigenmodes was investigated.

Finally, we may wonder how the combined ingredients of resistivity *and equilibrium flow* manifest themselves in this tearing-unstable force-free slab. For the static equilibrium of Figs. 14.7–14.8, maximal growth occurs for modes with $k_y = 1.5$. To investigate the effect of equilibrium flow on this mode, we add a linear flow profile $v_y(x) = 0.15\,x$, leading to Doppler shifted frequency ranges $\{\Omega^{\pm}_{\rm A,S}\}$. With this flow profile, the spectrum still contains the tearing instability, and the non-trivial effect of flow on the linear MHD spectrum is illustrated in Fig. 14.9. The discrete modes seen in the damped half-plane are again found on curves, which still show a clear link with the extremal edge values of the Doppler shifted continua. The computational result also shows the purely oscillatory Eulerian entropy continuum modes, as well as evidence for more curves appearing in the frequency range $\{\Omega_0(x)\}$ of the local Doppler shift. It should be emphasized that the combined effects of equilibrium flow, magnetic shear and finite resistivity on the full MHD spectrum are not fully understood yet. Future MHD spectroscopy studies for these configurations are called for. By varying the equilibrium flow profile, one can then investigate subtle effects on global unstable modes: the tearing mode in this force-free field configuration can either be suppressed or rendered more unstable, depending on the precise flow strength and profile.

Extensive research has been devoted in hydrodynamics to the similar problem of the effect of viscosity on the spectrum of waves and instabilities. In local WKB solutions of the Orr–Sommerfeld equation, an important role is played by the Stokes lines in the complex ω plane (see Drazin and Reid [164], Section 27.3). Eigenvalues associated with wildly oscillating eigenfunctions are found on particular curves, like those of Figs. 14.6, 14.7, 14.9. These techniques have been transferred to magnetohydrodynamics (see Pao and Kerner [478], Riedel [519] and Lifschitz [410], Section 7.15). However, in contrast to the resistive instabilities, the role of the stable damped part of the spectrum remains rather unclear. Some research ties this to the concept of *pseudo-spectrum* (see Borba et al. [83]), developed by Trefethen and Embree [609] for hydrodynamics.

14.4 Reconnection

We thus far concentrated on how resistivity modifies the linear MHD spectrum, pointing out the existence of various unstable tearing-type modes, involving reconnection. Reconnection plays an important role in nearly all dynamic phenomena in space and laboratory plasmas, where it manifests itself in intrinsically nonlinear evolutions. Magnetic reconnection rightfully deserves fundamental study on its own, a fact reflected in excellent modern textbooks on the subject (e.g. Priest and Forbes [511] and Biskamp [66]). In what follows, we restrict ourselves to discussing its role in the temporal evolution of a very simple, planar configuration, the so-called Harris sheet. This configuration has played a central role in several collaborative modelling 'challenges', such as the 'Geospace Environmental Modeling (GEM) Magnetic Reconnection Challenge' [62] and the 'Newton Challenge' [63]. In both challenges, the nonlinear evolution of the 2D system was computed and compared with a large variety of codes: ranging from conventional resistive MHD, over various extended MHD models, to fully particle based, kinetic treatments. In Section 14.4.4, we briefly summarize the important findings of this multi-code approach, but we first discuss what can be concluded from standard resistive MHD modelling alone.

14.4.1 Reconnection in a 2D Harris sheet

(a) Linear stability properties The Harris sheet configuration is a planar MHD equilibrium with a horizontal magnetic field given by

$$B_x(y) = B_0 \tanh(y/\lambda_B)\,. \tag{14.119}$$

The corresponding current layer thus has a thickness $2\lambda_B$. One further assumes a uniform temperature $T = T_0$ and a density profile given by

$$\rho(y) = \rho_0 \cosh^{-2}(y/\lambda_B) + \rho_\infty\,. \tag{14.120}$$

Exploiting a scaling where $B_0 = 1$, $\rho_0 = 1$ and $2\lambda_B = 1$, all velocities become normalized to the Alfvén speed and time is measured in Alfvén crossing times of the current sheet width. From pressure balance, one finds $T_0 = 0.5$, and the reference equilibrium configuration takes $\rho_\infty/\rho_0 = 0.2$ and a ratio of specific heats $\gamma = 5/3$. The problem is fully specified when supplemented with the domain size and the employed boundary conditions. The standard case takes $-\frac{1}{2}L_x < x < \frac{1}{2}L_x$ and $-\frac{1}{2}L_y < y < \frac{1}{2}L_y$ with $L_x = 25.6$ and $L_y = 12.8$. The simulations use periodic boundary conditions horizontally, while perfectly conducting impenetrable walls are assumed at $y = \pm\frac{1}{2}L_y$. In the GEM challenge, this initial equilibrium is perturbed by an additional magnetic perturbation

$$\begin{aligned} B_{1x}(x,y) &= -\psi_1(\pi/L_y)\cos(2\pi x/L_x)\sin(\pi y/L_y)\,, \\ B_{1y}(x,y) &= +\psi_1(2\pi/L_x)\sin(2\pi x/L_x)\cos(\pi y/L_y)\,. \end{aligned} \tag{14.121}$$

In the original study, one adopts an amplitude $\psi_1 = 0.1$, which is deliberately taken so large that nonlinearity dominates instantly. When we restrict the analysis to uniform resistivity, the only parameter is the value of the magnetic Reynolds number, quantified by its dimensionless inverse η. In what follows, we will first point out the linear stability properties for this choice of parameters, to make a link with our previous discussion of tearing modes. Then, we will discuss its nonlinear evolution: not unexpectedly, the decisive factor will turn out to be the value of the resistivity η, as it implies radically different reconnection scenarios.

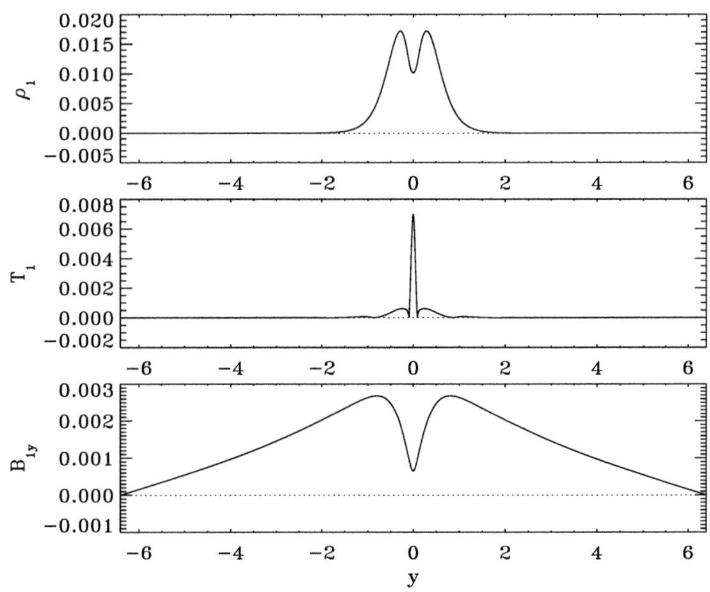

Fig. 14.10 Linear resistive MHD eigenfunctions for the Harris sheet equilibrium exploited in the GEM challenge.

We first present in Fig. 14.10 the numerically determined eigenfunctions for the most unstable linear tearing mode of this configuration, taking $\eta = 10^{-4}$. The linear mode is assumed to be purely two-dimensional, and adopts a Fourier dependence $\exp(\mathrm{i}k_x x)$ with $k_x = 2\pi/L_x$, so that its horizontal wavelength exactly fits the chosen box size. The growth rate λ of the exponential growth $\exp(\lambda t)$ for this mode has the same dependence on the resistivity η and on the wave number k_x, as shown by Fig. 14.8. (In contrast to the previous sections, the direction of inhomogeneity is now taken in the y-direction.) The tearing mode shows the characteristic magnetic field perturbations discussed earlier, with the component $B_{1y}(y)$ to be compared with Q as sketched in Fig. 14.4. The density eigenfunction clearly shows that compressibility effects can no longer be neglected in this case, while the tearing mode also manifests a distinct temperature variation across the resistive inner layer. Varying the growth rate versus wave number at constant $\eta = 10^{-4}$ shows that this mode attains its maximal growth (like the tearing mode of Fig. 14.8) at a somewhat shorter wavelength than that preferred by the horizontal box size $L_x = 25.6$. Finally, the variation of the growth rate with resistivity confirms the analytic scaling $\lambda \sim \eta^{3/5}$, although convincingly only for values of the corresponding magnetic Reynolds number beyond 10^4. It must be stressed that linear MHD codes actually do possess the required accuracy to compute eigenmodes at realistic Reynolds numbers of order 10^9 or more. This requires the use of a sufficiently accurate numerical representation, if needed combined with grid accumulation, and aptly chosen generalized eigenvalue solvers (as explained in Chapter 15). In contrast, for the nonlinear simulations discussed next, direct numerical simulations that achieve magnetic Reynolds numbers of 10^4 are still at the limit of many of the high resolution methods exploited to date (such as the shock-capturing finite volume methods described in detail in Chapter 20).

(b) Nonlinear evolution Although the reference Harris sheet equilibrium exploited for the nonlinear simulations is tearing-unstable, the large amplitude perturbation used in Eq. (14.121) imposes an initial magnetic island perturbation with a width comparable to the current layer. This is mainly motivated by the GEM challenge's aim to compare nonlinear reconnection rates found from both fluid and kinetic models. For purely uniform resistivity, the original resistive MHD simulations from Birn and Hesse [64] already demonstrated pronounced differences in the evolution when varying η from 0.005 down to 0.001. Their fairly low resolution simulations followed the current sheet dynamics for several hundreds of Alfvén crossing times. At their highest resistivity value $\eta = 0.005$, the initial current sheet collapses centrally to form an elongated, narrow dissipative layer connecting the two halves of the magnetic island located at the periodic sides. This island gradually grows in size, at a slow rate set by the near-steady reconnection occurring through the Y-shaped end points of the elongated current layer. This rate can be computed analytically from a consideration of the stationary ($\partial/\partial t = 0$) resistive MHD equations across a diffusion region of macroscopic length $2L$ and width $2l$, e.g. observing that mass conservation demands $Lv_{\rm i} = lv_{\rm o}$, where $v_{\rm i}$ and $v_{\rm o}$ denote the (vertical) inflow and (lateral) outflow velocities, respectively. This was originally done by Sweet and further analysed by Parker [480], and the resulting so-called Sweet–Parker reconnection layer is characterized by a very slow inflow speed $v_{\rm i} \sim \sqrt{\eta}v_{\rm A}$, an Alfvénic outflow velocity $v_{\rm o} \sim v_{\rm A}$ and a current sheet width related to the length L and resistivity by $l \approx \sqrt{\eta}L$. Figure 14.11 shows the evolution of the perturbed Harris sheet when taking $\eta = 0.005$. The simulation here employs automated grid-adaptivity to achieve an effective resolution of 1920×1920, and essentially recovers the original results from [64]. During the entire simulation, a single dominant island grows in size, as mediated by the near-steady reconnection occurring across the central current layer.

When the magnetic Reynolds number is increased, the same Harris sheet configuration can demonstrate pronouncedly different evolutions. Although the original results by Birn and Hesse [64] already hinted at this possibility by including a lower $\eta = 0.001$ case as well, only more recent work by Lapenta [396] has convincingly demonstrated the complexity attainable by visco-resistive MHD evolutions at the larger, physically more realistic, magnetic Reynolds numbers. Figure 14.12

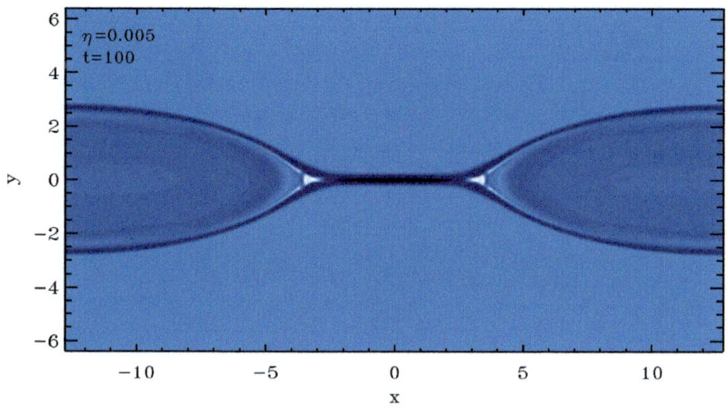

Fig. 14.11 Nonlinear evolution of a Harris sheet for uniform resistivity $\eta = 0.005$. Shown is the out-of-plane current distribution at $t = 100$.

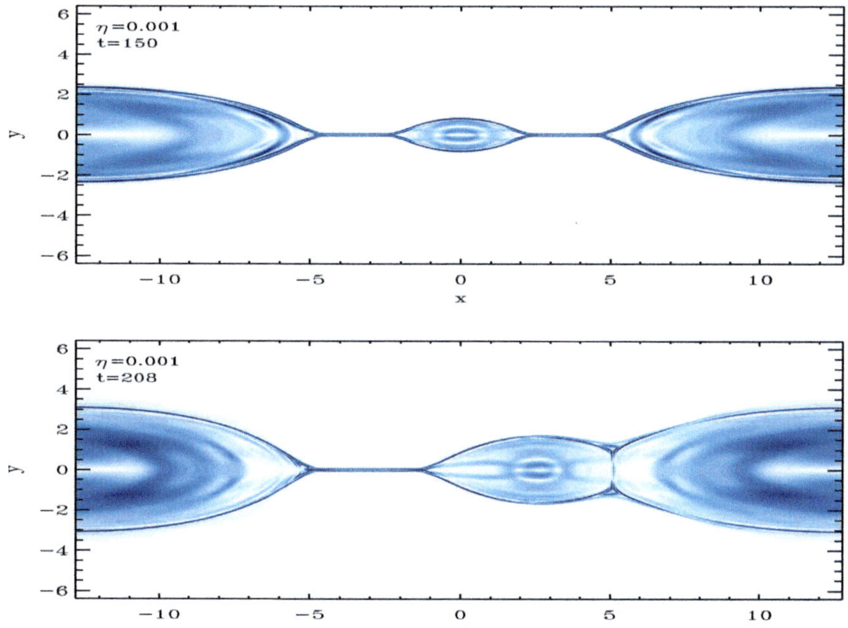

Fig. 14.12 Nonlinear evolution of a Harris sheet equilibrium for uniform resistivity $\eta = 0.001$. Shown are Schlieren plots of the density at consecutive times.

presents the evolution of the identical configuration as simulated in Fig. 14.11, for $\eta = 0.001$. Now, the central current sheet still collapses to initiate the slow reconnection process, but in addition forms a central secondary island which grows to macroscopic dimensions. This was already found in [64], and interpreted as due to growing linearly unstable eigenmodes of smaller wavelength. This secondary island eventually undergoes a sudden merger with the larger island structure, an effect attributable to the so-called coalescence instability mediated by the mutual attraction of parallel cur-

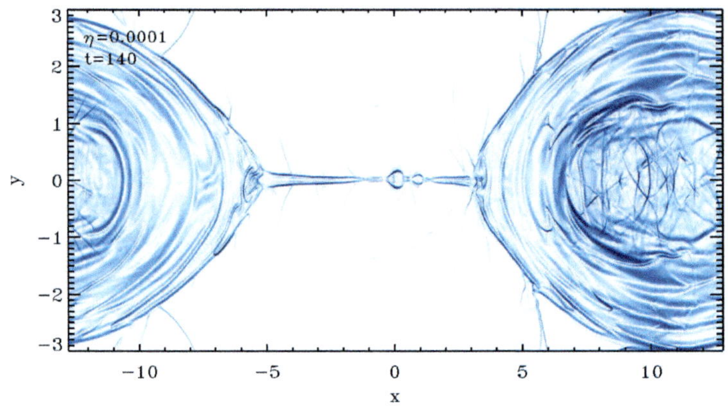

Fig. 14.13 Nonlinear evolution of a Harris sheet equilibrium for uniform resistivity $\eta = 0.0001$. Shown is a Schlieren plot of the density at $t = 140$.

rent filaments. Continuing the simulation further shows the renewed appearance of a smaller island structure in the central current layer, which later on is again seen to coalesce. This process already indicates the transition to a new, highly non-steady, reconnection regime that is characterized by sudden tearing-type disruptions of the current layer.

As first pointed out by Lapenta [396], even lower resistivity values where $\eta = 10^{-4}$ show a dramatic changeover from the original Sweet–Parker reconnection regime to one where the collapsed current layer spontaneously disrupts chaotically in repeated island chains. The resulting reconnection rate in the self-feeding, turbulent reconnection phase is dramatically increased beyond the slow Sweet–Parker rate. Figure 14.13 again shows a representative snapshot of the evolution, this time characterized by almost randomly appearing, small island chains, which again show sudden mergers with the larger islands due to the coalescence instability. He attributes the onset of the resulting fast reconnection regime to the formation of closed circulation patterns centred on the multiple reconnection sites, self-feeding at X-type points. In Fig. 14.13, the sites where this fast reconnection occurs demonstrate a clear resemblance to that obtained in the other well-known stationary Petschek reconnection model, involving slow MHD shocks and co-spatial current sheets found in an X-type configuration. In the next subsection, this stationary Petschek reconnection process is briefly discussed. The self-feeding, turbulent reconnection found at sufficiently low values of η is highly non-stationary, but physically very relevant since it occurs on Alfvénic time scales, and by nature distributes the reconnection over a macroscopic region.

14.4.2 Petschek reconnection

The slow reconnection rate obtained in the stationary Sweet–Parker regime is in stark contrast with that found in violent solar flares, or the ones realized in several laboratory reconnection experiments. To overcome this shortcoming of the modelling based on stationary resistive MHD, Petschek [487] proposed an alternative involving the formation of slow shock fronts across which most of the energy conversion takes places. The central diffusion region is in this model reduced in size as compared to the elongated current layer of the Sweet–Parker regime, and four standing shock waves emerge from it in an overall X-type configuration.

Tóth et al. [606] applied their resistive MHD Versatile Advection Code (see Section 20.3.2) to study Petschek-type reconnection of magnetic field lines. Initially, there is a Harris sheet equilibrium with $v_x = v_y = B_x = 0$ and $B_y = \tanh(x/L)$ with L the width of the initial current layer. This initial state spontaneously transits to a configuration containing a localized dissipative layer by imposing a spatially non-uniform anomalous resistivity centred about the origin. A pair of slow mode shocks propagate away from the reconnection layer, eventually forming a standing X configuration. The material crossing those shocks accelerates to Alfvénic velocities and gets heated in the process. The configuration eventually evolves to a steady-state Petschek reconnection regime. While most computations assume non-uniform resistivity, it has been shown that at nearly uniform resistivity a stationary configuration can be obtained numerically [41]. At the same time, this classical stationary Petschek regime is just one realization of the various fast reconnection mechanisms known to date. For a discussion of these, we refer the interested reader to Priest and Forbes [511], and to Biskamp [66].

14.4.3 Kelvin–Helmholtz induced tearing instabilities

The Harris sheet configuration discussed thus far clearly demonstrates the surprising complexity for planar resistive MHD evolutions. With various tearing-type instabilities already present for static equilibria, one can anticipate even more complex scenarios for stationary equilibria (of the preceding Chapters 12 and 13) containing initial current sheets. We end our discussion of resistive MHD reconnection with an example of tearing-type disruptions induced by pure Kelvin–Helmholtz instabilities. In Keppens et al. [356], a numerical survey of a planar shear flow $v_x = v_0 \tanh(y/a)$, embedded in a uniform plasma, was performed. At the high initial plasma beta $\beta = 120.2$, the configuration is known to be Kelvin–Helmholtz unstable for subsonic flow conditions $v_0 < c \equiv \sqrt{\gamma p_0/\rho_0}$, with a weak stabilizing influence on the growth rate by the tension of the flow-aligned magnetic field. The linear Kelvin–Helmholtz growth rates, as well as the nonlinear evolution of the system, was compared for a uni-directional as well as a 'reversed' magnetic field at $t = 0$. The reversed case discontinuously changes the direction of the magnetic field at the $y = 0$ mid-plane, and represents the limit of a co-spatial magnetic Harris sheet $B_x = B_0 \tanh(y/b)$ for the limit $b \to 0$. At sufficiently high resolution, combined with low values of the uniform resistivity, the reversed case demonstrated the distinct possibility of sudden tearing-type events. Small-scale magnetic islands appear suddenly and are amplified by the Kelvin–Helmholtz mode development. The vortical flow accompanying the Kelvin–Helmholtz mode warps the initial field in spiral patterns at the vortex periphery. A case which contrasts the evolution of a uniform versus a reversed field configuration is shown in Fig. 14.14. The density 'islands' coincide with magnetic islands. Their sudden appearance allows a rapid transition to magneto-turbulent flow conditions.

This example shows that the nonlinear dynamics of flowing, current-carrying, plasmas is a topic that deserves further study within a purely resistive MHD context. Even their linear stability properties have not been charted to the amount of detail obtained for static equilibria. Moreover, reconnection in more complicated 3D configurations is even more intricate, with a staggering increase in topological possibilities for reconnecting anti-parallel field lines. Chapter 8 of Priest and Forbes [511] provides a readable account of its possible manifestations in 3D.

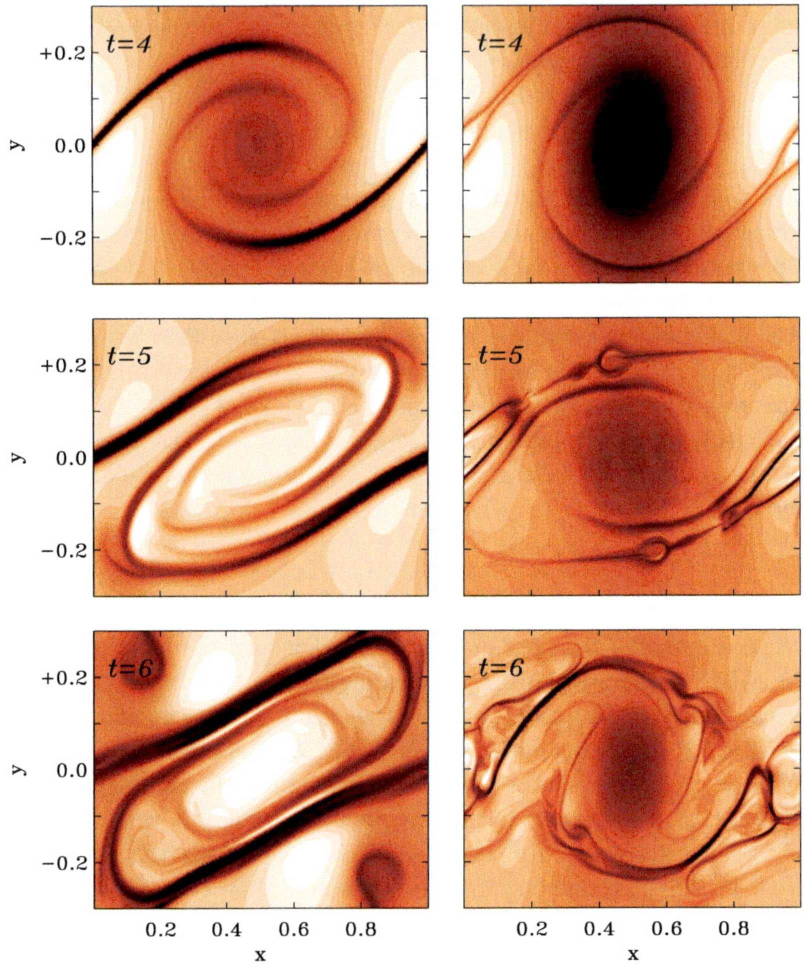

Fig. 14.14 Evolution of the density for a Kelvin–Helmholtz unstable shear flow for an initially uniform (left) versus a reversed (right) flow-aligned magnetic field. With an initial narrow current layer (right panels), the Kelvin–Helmholtz mode triggers tearing-type instabilities at the vortex periphery. (From Keppens et al. [356]).

14.4.4 Extended MHD and reconnection

The 2D Harris sheet configuration has been used as a benchmark configuration in both the GEM and the Newton challenge projects to identify the essential physics required to properly model collisionless magnetic reconnection. Whereas the GEM challenge adds the fairly large perturbation given by Eq. (14.121) to bring the system to a nonlinear reconnection regime, the Newton challenge differs in the way this nonlinear regime gets accessed. Rather than imposing a perturbation in the current sheet at $t = 0$, it gradually moves the field lines at $y = \pm L_y/2$ inwards at a prescribed velocity. This inflow is at most 10% of the Alfvén velocity and decays beyond about 60 Alfvén crossings, so it gently forces the central reconnection. Both challenges used a large variety of models to simulate the evolution, with codes ranging from resistive MHD to full kinetic treatments. The findings

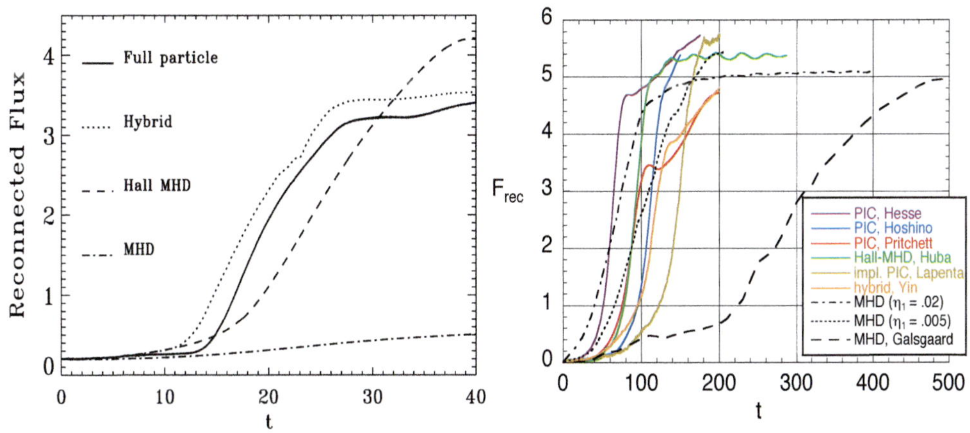

Fig. 14.15 Reconnection rates as obtained for the GEM challenge [62] (left panel) and Newton challenge [63] (right panel) simulations.

of both efforts were similar. The reconnection rates obtained from different models are shown in Fig. 14.15. We summarize their results as follows.

– The reconnection rate found in conventional resistive MHD simulations was significantly smaller than that in any simulation which allows a minimal decoupling of the electron and ion dynamics. However, the resistivity values adopted for uniform resistivity models were typically higher than $\eta = 0.001$, so that the eventual transition to the chaotic, fast reconnection regime as shown in Fig. 14.13 was not appreciated fully.

– All treatments that consider at least the effect of the Hall term in a generalized Ohm's law show that reconnection occurs fast (with Alfvénic inflow rates). This reconnection rate is surprisingly similar between Hall-MHD, hybrid, or full particle models. This could indicate that reconnection is insensitive to details of the electron dynamics and the dissipation mechanism.

– To raise the reconnection rate from resistive MHD to the faster rate found from the more advanced models, one can adopt an anomalously raised local resistivity.

These conclusions from both 2D magnetic reconnection studies then call for suitable extensions of the standard MHD model. The simplest of such models is that of Hall-MHD, which can be regarded as a straightforward generalization of the single fluid MHD description. It includes the effect of the Hall current in the generalized Ohm's law, and brings in whistler type waves, which have a faster phase speed at shorter wavelengths or higher frequencies. The Hall-MHD model thereby enables a faster reconnection than obtained in pure resistive MHD. Extended MHD models of increasing physical complexity can be rigorously derived along the lines given in Chapter 3. In so doing, one can gradually relax the inherent assumptions of the single fluid MHD description, allowing a descent in the length and time scale hierarchy. Recalling these hierarchies from Eq. (3.143), we have

$$\begin{aligned} \lambda_{\rm MHD} &\equiv |\nabla|^{-1} \sim a \gg R_{\rm i} \Big[\gg \delta_{\rm e} \gg R_{\rm e} \gg \lambda_{\rm D} \Big] , \\ \tau_{\rm MHD} &\equiv |\partial/\partial t|^{-1} \sim a/v_{\rm A} \gg \Omega_{\rm i}^{-1} \Big[\gg \omega_{\rm p}^{-1} \sim \Omega_{\rm e}^{-1} \Big] , \end{aligned} \tag{14.122}$$

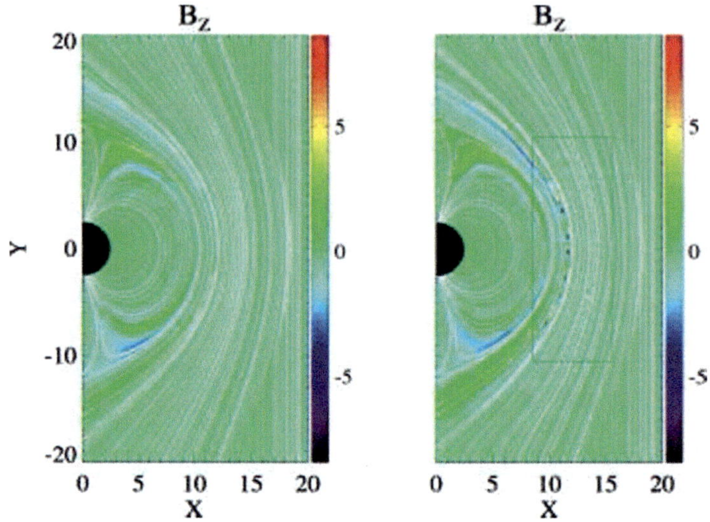

Fig. 14.16 Global 2D Hall-MHD simulation of a magnetosphere without (left) and with (right) an embedded PIC region. Where magnetic field lines (in white) allow for dayside reconnection, true kinetic effects are accounted for, with two-way feedback between the global and local models. (From Daldorff et al. [139].)

where a and v_A are the typical MHD length scale and Alfvén velocity, R_i and R_e denote the ion and electron gyro-radii, δ_e is the electron skin depth, λ_D the Debye length, Ω_i and Ω_e denote the ion and electron gyro-frequencies, while the plasma frequency is ω_p. One must then similarly invoke a kind of maximal ordering (Section 3.4.1) to derive various extended MHD descriptions, valid up to a pre-chosen length and time scale. Obviously, each extended model must adopt a suitable closure strategy and somehow parameterize the effect of the higher order moments of the distribution function, i.e. enter the realm of modern transport theory.

Since the computational techniques differ significantly from one extended MHD model to the next (with particle in cell or PIC treatments usually prevailing at the kinetic levels, while finite volume or finite difference methods reign at fluid-like levels), this is the domain of many advanced numerical efforts. Ultimately, one would like to develop adaptive physics solvers, where not only the employed (grid or particle) resolution changes dynamically, but also the used model differs on different parts of the computational domain. An important step in this direction was realized in Ref. [139], demonstrating a two-way coupled global Hall-MHD model with local implicit PIC regions. This MHD-EPIC approach (MHD with Embedded PIC regions) confines the kinetic modelling to a priori known regions where true kinetic effects are accounted for. It has since been used for multi-dimensional simulations of magnetosphere dynamics, as shown in Fig. 14.16.

▷ **Generalized Ohm's law and Hall-MHD** As an example of an extended MHD model, we here present the Hall-MHD model. Our starting point will be the intermediate set of four single fluid equations, Eqs. (3.135), (3.148)–(3.150), describing the time evolution of the macroscopic variables ρ, $\mathbf{v}$, $\mathbf{j}$ and p, derived from kinetic theory. The first equation, (3.135), represents total mass conservation where $\rho = n_e m_e + n_i m_i$, the second one, Eq. (3.148), is the total momentum equation where $\rho\mathbf{v} = n_e m_e \mathbf{u}_e + n_i m_i \mathbf{u}_i$, obtained as a mass-weighted combination of both electron and ion momentum equations, the third equation, (3.149), is the generalized Ohm's law, while Eq. (3.150) is the heat balance equation for the total pressure $p = p_e + p_i$. In the latter equation, we neglected pressure anisotropies (no ion/electron viscosities) so that we are left with a scalar pressure. We also implicitly assumed an interest in time scales beyond the temperature equilibration time scale, adopted quasi-neutrality such that $n_e = Zn_i$, while the ion–electron momentum transfer due to collisions has been quantified by means of a scalar resistivity η. The ratio $\mu = Zm_e/m_i$ of mass over charge appears as a small parameter. Note that electron and ion pressures are to a good approximation given by $p_e = Zp/(1+Z)$ and $p_i = p/(1+Z)$. The generalized Ohm's law (3.149) was obtained as a charge-weighted average of the ion and electron momentum equations and it can be used to consistently extend the ideal or resistive MHD model. For the standard MHD descriptions, the electric field is calculated from the simple form $\mathbf{E}+\mathbf{v}\times\mathbf{B} = \eta\mathbf{j}$. This is then used in combination with the pre-Maxwell equations

$$\mathbf{j} = \mu_0^{-1}\nabla\times\mathbf{B}\,, \qquad \frac{\partial\mathbf{B}}{\partial t} + \nabla\times\mathbf{E} = 0\,, \qquad \nabla\cdot\mathbf{B} = 0\,. \tag{14.123}$$

When we exploit the smallness of μ, we can alternatively write

$$\begin{aligned} \mathbf{E} &= -\left(\mathbf{v} - \frac{\mathbf{j}}{en_e}\right)\times\mathbf{B} + \eta\mathbf{j} - \frac{\nabla p_e}{en_e} + \frac{m_e}{e^2 n_e}\left[\frac{\partial\mathbf{j}}{\partial t} + \nabla\cdot(\mathbf{jv}+\mathbf{vj})\right] \\ &= -\,\mathbf{u}_e\times\mathbf{B} + \eta\mathbf{j} - \frac{\nabla p_e}{en_e} + \frac{m_e}{e^2 n_e}\left[\frac{\partial\mathbf{j}}{\partial t} + \nabla\cdot(\mathbf{jv}+\mathbf{vj})\right]. \end{aligned} \tag{14.124}$$

The Hall current $(en_e)^{-1}\mathbf{j}\times\mathbf{B}$ contribution to the electric field breaks the frozen-in condition of the ideal MHD induction equation, in a collisionless manner. Hence, even when only the first term in Eq. (14.124) is taken along, field lines are no longer forced to follow the flow due to decoupled ion–electron dynamics, in a way which is separate from the ion–electron collisional effects encoded in the scalar resistivity. One can incorporate various effects beyond the resistive MHD model, by computing the electric field from this generalized Ohm's law and inserting the result in Eq. (14.123)(b). The extra terms in Eq. (14.124) can be omitted when electron pressure is not important ($p_e = 0$) and electron inertia can be neglected. In practice, one may also find Hall-MHD models where anisotropic pressure effects are incorporated, or one partially accounts for electron inertia, noting that this generalized Ohm's law is to be seen as a form of the electron momentum equation. ◁

14.5 Excursion: Hall-MHD wave diagrams

As noted above, extensions beyond ideal MHD are paramount to properly describe the complex physics of magnetic reconnection, where the multi-scale (from fluid to kinetic) character of plasma physics has to be accounted for. At the same time, the simplest Hall-MHD extension of the ideal one-fluid MHD model already gives a satisfactory description of magnetic reconnection, as it allows for electron-ion decoupling at the scale of the ion gyro-radius. Hall-MHD in essence describes a two-fluid model with massless electrons, whereas the inertia of the ions remains finite, so that the magnetic field is 'frozen into' the electron fluid only. In Section 3.3.4, the electron cyclotron and the electron plasma frequencies appeared along with their ion equivalents, when the general dispersion equation for waves in a two-fluid plasma was derived. A typical dispersion diagram for a homogeneous plasma configuration, such as Fig. 3.1, then clearly shows how the two-fluid model extends the global, low-frequency MHD limit with its three waves (slow, Alfvén, fast). In this section, we will discuss the modifications to these waves induced by the Hall-MHD model, and generalize the familiar phase and group speed diagrams.

The most basic Hall-MHD model merely takes the first term in Eq. (14.124) along, and then notes that the remaining equations contain essentially ion dynamics information where $\rho = n_i m_i$ while $n_e = Zn_i$ (charge neutrality holds at all times). The massless electron limit allows us to write $\mathbf{v} = \mathbf{u}_i$ and $\mathbf{u}_e = \mathbf{v} - (en_e)^{-1}\mathbf{j}$, while the model does allow for a finite electron and ion pressure. In what follows, we assume negligible resistivity, $\eta = 0$, to focus on how the ideal MHD waves get modified and to make a direct link with the ideal two-fluid case from Chapter 3.

Following Hameiri et. al. [287], we derive the Hall-MHD dispersion relation from the following starting equations, exploiting a generalization of the $(S, \mathbf{v}, p, \mathbf{B})$ representation of Section 5.2.2:

$$\frac{\partial S}{\partial t} + \mathbf{v} \cdot \nabla S = 0\,,$$

$$\rho \frac{\partial \mathbf{v}}{\partial t} + \rho \mathbf{v} \cdot \nabla \mathbf{v} + \nabla p = \mathbf{j} \times \mathbf{B}\,,$$

$$\frac{\partial p}{\partial t} + \mathbf{v} \cdot \nabla p + \gamma p \nabla \cdot \mathbf{v} = 0\,,$$

$$\frac{\partial \mathbf{B}}{\partial t} - \nabla \times \Big[\Big(\mathbf{v} - \frac{m_{\rm i}}{Ze\rho} \mathbf{j} \Big) \times \mathbf{B} \Big] = 0\,, \qquad \nabla \cdot \mathbf{B} = 0\,. \qquad (14.125)$$

The waves of the system are found by using a geometric optics Ansatz for the perturbations as $\tilde{\mathbf{B}} = \exp(\mathrm{i}\phi/\epsilon)\,[\tilde{\mathbf{B}}^0 + \epsilon \tilde{\mathbf{B}}^1 + \epsilon^2 \tilde{\mathbf{B}}^2 + \ldots]$, and then working out the linearized equations at order $\mathcal{O}(1/\epsilon)$. The phase function $\phi(\mathbf{x}, t)$ is shared for all perturbed quantities. The Hall-modified induction equation involves *the ion inertial length* $\delta_{\rm i} \equiv c/\omega_{\rm pi} \approx m_{\rm i}/(Ze\sqrt{\mu_0 \rho})$, with c the speed of light, and the Ansatz is valid for waves with local wave number $2\pi\epsilon/|\nabla\phi|$ much smaller than typical equilibrium scales. If the equilibrium varies on scales much greater than $\delta_{\rm i}$, one can choose $\epsilon \sim \mathcal{O}(\delta_{\rm i})$ and find that the dispersion relation directly generalizes the ideal MHD case where for any real wave number k (for wave vector $\mathbf{k} \equiv \nabla\phi/\epsilon$), we obtain eight real values for the local frequencies $\omega \equiv -(\partial\phi/\partial t + \mathbf{v} \cdot \nabla\phi)/\epsilon$. These contain two zero frequency solutions $\omega = 0$ for the trivial divergence and entropy waves, identical to ideal MHD. Note that this zero frequency corresponds to a frame moving with the (ion) equilibrium flow $\mathbf{v}$, which accounts for the Doppler shift. Furthermore, this derivation is valid for general, moving, equilibria that vary on scales much beyond $\delta_{\rm i}$, of which the homogeneous plasma at rest is the simplest example.

The six remaining eigenvalues are found to obey the following generalization of the ideal MHD dispersion equation (5.54) of Chapter 5:

$$(\omega^2 - k_\parallel^2 b^2) \left[\omega^4 - k^2 (b^2 + c^2) \omega^2 + k_\parallel^2 k^2 b^2 c^2 \right] - \lambda_{\rm H} \omega^2 k_\parallel^2 b^2 \left(\omega^2 - k^2 c^2 \right) = 0\,. \qquad (14.126)$$

Denoting $\lambda_{\rm H} \equiv (k\delta_{\rm i})^2$ as the Hall parameter, one can use the dispersion relation to show how the waves modify from ideal MHD ($\lambda_{\rm H} = 0$) to cases with a very strong Hall effect. This is best shown by displaying the phase diagrams, where the phase speed ω/k is quantified for all angles ϑ between wave vector and equilibrium magnetic field. These diagrams dramatically deviate from the ideal MHD regime as $\lambda_{\rm H}$ becomes large. Examples are shown in Fig. 14.17 for the case with ratio of squared sound and Alfvén speeds $c^2/b^2 = 0.5$, whereas the Hall parameter is $\lambda_{\rm H} = 0.4$ (top left panel) and $\lambda_{\rm H} = 10$ (top right). A careful study of the dispersion relation, along with the corresponding eigenfunctions, shows that the Alfvén wave ultimately transforms into a compressible ion acoustic wave, tending to propagate spherically at the sound speed c, except in near-perpendicular directions (the right panel shows how the pair of red curves tends to become a single circle). The fast wave transitions at large $\lambda_{\rm H}$ (short wavelengths) to a strongly incompressible whistler wave, satisfying the dispersion relation $\omega_{\rm f} \approx \pm k^2 b^2 \cos\vartheta/\Omega_{\rm i}$ (or $\omega_{\rm f}/\Omega_{\rm i} \approx \pm\lambda_{\rm H} \cos\vartheta$ since $b^2 = \delta_{\rm i}^2 \Omega_{\rm i}^2$) involving the ion cyclotron frequency $\Omega_{\rm i}$. Hence, the shortest wavelengths travel fastest (which means trouble for computational approaches), or the highest frequencies arrive first and the wave has a

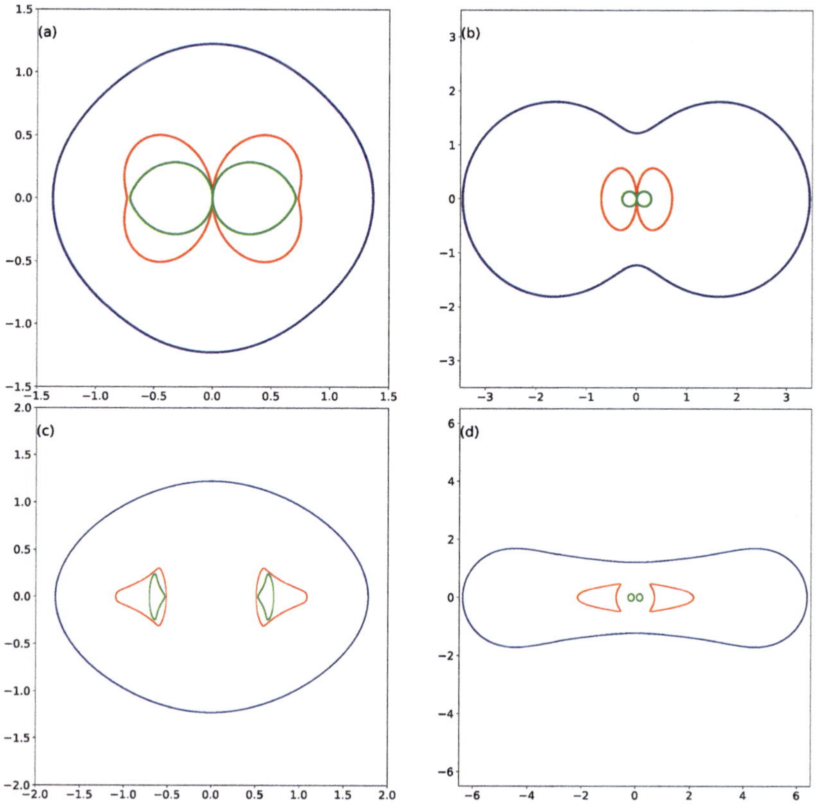

Fig. 14.17 Phase (top) and group (bottom) speed diagrams for a Hall-MHD plasma with $c^2/b^2 = 0.5$. Curves are coloured blue for the fast, red for the Alfvén and green for the slow waves. (a) and (c): Hall parameter $\lambda_H = 0.4$. (b) and (d): $\lambda_H = 10$, showing the transition to short wavelength behaviour.

descending pitch on arrival, hence the name 'whistler'. For the slow wave, one finds for $\lambda_H \to \infty$ the relation $\omega_s \approx \pm\Omega_i \cos\vartheta$.

The dispersion relation Eq. (14.126) can also be used to compute and visualize the Hall-MHD analogue of the group, or Friedrichs, diagrams. Although in contrast to the MHD case, the Hall-MHD system displays dispersive waves, one can still use the dispersion relation to quantify the shape of a propagating wave front resulting from a localized disturbance, which yields the ray surfaces. Formally, one uses implicit differentiation on the relation (14.126) to obtain the group velocity $\partial\omega/\partial\mathbf{k}$. The result is given in explicit formulas in Ref. [287], and can be used to plot corresponding ray surfaces, such as depicted in Fig. 14.17 (bottom panels). Note that values of λ_H exist for which the relative ordering of the three ray surfaces is not the familiar slow, Alfvén, fast. However, this ordering does persist at large Hall parameters for the ray surfaces and phase diagrams.

Finally, to show how waves in Hall-MHD connect to the ideal two-fluid picture, we use the dispersion relation (14.126) to quantify the variation of frequency ω/Ω_i with dimensionless wave number $\sqrt{\lambda_H} = k\,\delta_i$. This is shown in Fig. 14.18 for three values of the angle ϑ between wave vector and equilibrium magnetic field. This can be compared to Fig. 3.1 where a representative ideal two-

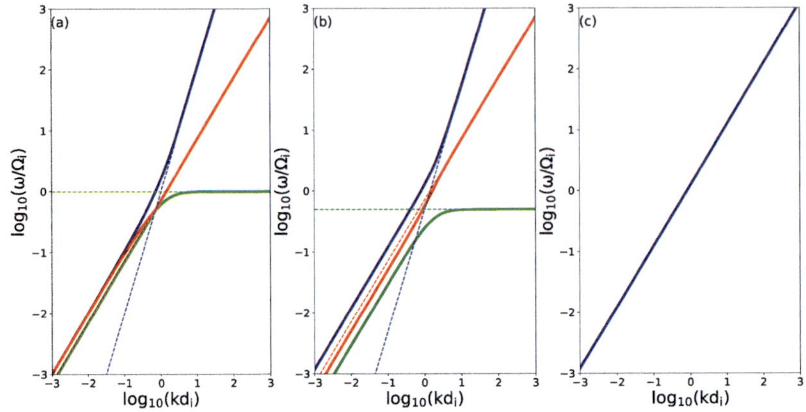

Fig. 14.18 Dispersion diagrams for a Hall-MHD plasma with ratio $c^2/b^2 = 0.5$. (a) Parallel propagation, (b) oblique propagation ($\vartheta = \pi/3$), (c) perpendicular propagation, where only the fast wave (in blue) survives. Dashed lines indicate the short wavelength limits.

fluid dispersion diagram was shown. Then we note that the Hall-MHD dispersion diagram correctly describes the limit to the ion cyclotron resonance for the slow magneto-sonic waves. It also formally (re)moves the three additional wave pairs of the ideal two-fluid picture to infinite frequencies, due to the assumption of massless electrons.

14.6 Literature and exercises

Notes on literature

Resistive instabilities

- The threat of resistive instabilities to plasma confinement for fusion was realized very early. Two seminal papers on the subject already appeared in 1963: 'Finite resistivity instabilities of a sheet pinch' by Furth, Killeen and Rosenbluth [202] and 'On the stability of hydromagnetic systems with dissipation' by Coppi [126].
- Most textbooks on plasma physics have chapters on resistive instabilities, e.g. *Theory of Toroidally Confined Plasmas* (Chapter 5) by White [649]; *Plasma Physics* (Chapter 17) by Sturrock [580]; *Introduction to Plasma Physics* (Chapter 20) by Goldston and Rutherford [244]; *Fundamentals of Plasma Physics* (Chapter 12) by Bellan [49].

Resistive spectrum

- The subject of MHD spectral theory of dissipative plasmas has not developed yet to the textbook level. A start can be found in Section 7.15 of *Magnetohydrodynamics and Spectral Theory* by Lifschitz [410].
- The subject of resistive MHD instabilities has been investigated by too many authors to even attempt to give a representative sample. However, fundamental analytical studies by Pao and Kerner [478], by Riedel [519] and by Borba et al. [83], and systematic numerical studies by Kerner et al. on cylindrical plasmas [362], quasi-modes [497], and large-scale computing for tokamaks [360], and similar studies by McMillan et al. [435] come close to the viewpoint of the present book.

Reconnection

- The subject of reconnection is central in the exposition of the sequence of the three books *Nonlinear*

Magnetohydrodynamics [65], *Magnetic Reconnection in Plasmas* [66], and *Magnetohydrodynamic Turbulence* [67] by Biskamp.

- *Magnetic Reconnection; MHD Theory and Applications* by Priest and Forbes [511] provides the fundamentals of the subject, including reconnection in 3D, with many applications to solar and astrophysical plasmas.

Exercises

[14.1] *Ideal MHD versus resistive MHD*

The difference between ideal and resistive MHD is that one takes resistivity into account (where η is usually assumed to be small).

- What is the main consequence of this?
- Why can we not use the same strategy for analysing the stability of resistive plasmas as for ideal plasmas?

[14.2] *Tearing modes*

In this chapter, a detailed analysis of tearing modes is presented.

- What essential condition is needed to make sure that tearing modes may appear?
- The analysis of tearing modes is carried out in three regions. Specify for each region what kind of assumptions has been used and which equations have to be solved.

[14.3] *Resistive spectrum of a homogeneous incompressible plasma*

The resistive spectrum of a homogeneous incompressible plasma is discussed in Section 14.3. In this exercise, you derive the equations and you make a plot of the resistive spectrum. Consider a plasma slab between two perfect conducting plates at a distance L.

- Start from the resistive equations (14.44) and reduce them for a homogeneous plasma.
- In a homogeneous plasma, the displacement field component ξ can be written as follows:
  $$\xi(x,y,z;t) = \hat{\xi}(k_x,k_y,k_z;\omega)\exp[\mathrm{i}(k_x x + k_y y + k_z z - \omega t)]\,.$$
  Explain why you can make this assumption. What are the boundary conditions on the walls?
- Derive the resistive equations using the assumption of the previous question.
- Write the derived equations in matrix form and derive the dispersion equation.
- Solve this dispersion equation.
- Look at the solutions of the dispersion equation and note that there is one special solution. Which one is it and why is it special?
- Plot both solutions of the dispersion equation in the complex ω-plane using your favourite plotting program. What kind of scaling can be used to plot the solutions?

[14.4] *Resistive spectrum of a homogeneous compressible plasma*

Figure 14.6 shows the resistive spectrum of a homogeneous compressible plasma. This spectrum has been created with one of the numerical methods discussed in Chapter 15. The dispersion equation can also be derived as in the previous exercise.

- Start from the resistive equations (14.41) and reduce them for a homogeneous plasma representing perturbations $\xi(x,y,z;t)$ as in Exercise [14.3].
- Write the equations in matrix form.
- Show that the resulting dispersion equation can be written as
  $$\omega\big(\omega^2 - k_\parallel^2 b^2 + \mathrm{i}\eta k^2\omega\big)\big[\omega^4 - k^2(b^2+c^2)\omega^2 + k^2k_\parallel^2 b^2c^2 + \mathrm{i}\eta k^2\omega(\omega^2 - k^2c^2)\big] = 0\,,$$
  where $b^2 \equiv B^2/\rho$ and $c^2 \equiv \gamma p/\rho$ are the squared Alfvén and sound speeds.
- Show that this dispersion equation reduces to Eq. (14.117) in the incompressible limit.
- Take the limit of vanishing resistivity and discuss the resulting solutions.
- Reproduce the plots of Fig. 14.6 using your favourite plotting program. Make sure that you can use

Laguerre's method (see Numerical recipes [509], in IDL use FZ_ROOTS, in Matlab use ROOTS1) to compute the roots of the fourth order polynomial. What kind of boundary conditions should be used?

[14.5] *Hall MHD waves*

Following Ref. [287], derive the dispersion relation Eq. (14.126) for the ideal Hall-MHD description of a two-fluid plasma. Also derive the expression for the group speeds, and make plots for both representative phase and group speed diagrams. Starting from the dispersion relation, study the limit of a cold Hall-MHD model, where the sound speed $c = 0$. What happens to the Alfvén wave at short wavelengths in this cold Hall-MHD model? (Answer: it obeys $\omega_{\rm A} = \Omega_{\rm i}$, except for perpendicular propagation where it vanishes.)

15

Computational linear MHD

Computational magnetohydrodynamics is a very active research field due to the increasing demand for quantitative results for realistic magnetic configurations on the one hand and the availability of ever more computer power on the other [490]. Many MHD phenomena can not be described by analytical methods in all of their complexity although simplified analytical models have led to indispensable insight into the fundamental physics of various magnetohydrodynamic processes. The intricate geometry of present tokamaks, for instance, forces theory to resort to computer simulations as the mathematics is not fully tractable anymore.

While the governing ideal MHD equations form a set of *nonlinear, hyperbolic, partial differential equations*, we have already encountered many magneto-fluid phenomena which are adequately modelled by means of the *linearized* MHD equations. In this chapter, we concentrate on computational approaches for linear MHD problems, and introduce several basic numerical concepts and techniques. We give an overview of frequently encountered spatial discretizations to translate any problem expressed as a set of differential equations into a discrete linear algebraic problem, where one solves generalized eigenvalue problems. Representative applications cover MHD spectroscopy computations for diagnosing eigenoscillations and stability, as well as steady-state and time-dependent solutions to externally driven configurations, both in ideal and in non-ideal MHD.

In Section 15.1, the different spatial discretization techniques that are most common in linear MHD are discussed by means of a generic *model steady-state problem*. The application of these techniques to solve linear MHD boundary value problems is illustrated in Section 15.2. Section 15.3 solves some linear MHD *initial value problems*. We conclude, in Section 15.4, with final remarks on the different numerical methods for solving linear MHD problems.

15.1 Spatial discretization techniques

Even after linearization, the MHD equations remain a fairly complicated set of time-dependent partial differential equations. To introduce the basic concepts connected to spatial discretization techniques, we consider a generic one-dimensional *model problem*. This is related to the *Sturm–Liouville equation*, which is a second order linear differential equation of the form

$$-\frac{d}{dx}\left[p(x)\frac{du(x)}{dx}\right] + q(x)u(x) = \lambda\, w(x)u(x)\,, \tag{15.1}$$

where the coefficients $p(x)$ and $q(x)$ and the weight function $w(x)$ are given and real, and λ is one of the eigenvalues to be computed together with the associated unknown solution $u(x)$. The boundary conditions originate from specific physical problems, and they are frequently chosen to guarantee real eigenvalues. The self-adjoint form of the equation is related to the quadratic forms that may be associated with it. A central topic is the construction of suitable spaces of orthogonal eigenfunctions. One can then try to find a *Sturmian monotonicity property* of the eigenvalues λ_i in terms of the number of nodes of those eigenfunctions $u_i(x)$.

This classical problem occurs in many parts of physics, in particular it has been instrumental in unravelling the spectra of elementary quantum mechanical systems. Also, in the study of the eigenvalues of the MHD spectrum of one-dimensional equilibria, like a gravitating slab or cylindrical plasma, a second order differential equation occurs, as we have seen in Chapters 7 and 9, which is not in Sturm–Liouville form, but, nevertheless exhibits similar monotonicity properties, like Sturmian or anti-Sturmian behaviour of the eigenvalues. Although the coefficients of the differential equation have a much more involved dependence on the eigenvalue parameter, similarly powerful oscillation theorems could be formulated and proven. They could be extended to stationary configurations, with complex eigenvalues, as discussed in Chapters 12 and 13. This shows that the basic theory of the Sturm–Liouville equation, and its generalizations, have wide applicability.

In the present chapter, the different discretization methods will be illustrated for the related *inhomogeneous* problem

$$-\frac{d}{dx}\left[p(x)\frac{du(x)}{dx}\right] + q(x)u(x) = f(x)\,, \tag{15.2}$$

on the finite closed interval $[0,1]$, again with the functions $p(x)$, $q(x)$ and $f(x)$ given and real, but without an eigenvalue parameter for the time being. The solution $u(x)$ will be subjected to the following boundary conditions. At $x=0$ the function value is assumed to be known,

$$u(0) = 0\,, \tag{15.3}$$

which is a boundary condition of the 'Dirichlet' type. At $x=1$ we impose

$$\alpha u(1) + \beta\frac{du}{dx}(1) = \gamma\,, \tag{15.4}$$

which is a more general, inhomogeneous, boundary condition involving both u and du/dx, where α, β and γ are given constants.

Physically, the above problem might correspond to the displacement of a string with variable density that is fixed at one end ($x=0$) and neither fixed nor completely free at the other end (e.g. connected to a spring), or to the distribution of the temperature of a gas or a plasma which is kept at a fixed temperature at one end ($x=0$) and heated at the other end ($x=1$). Note that Eq. (15.2) and the boundary conditions (15.3) and (15.4) represent a *steady-state problem*, and not an initial value problem. Initial value problems will be discussed in Section 15.3.

The differential form of Eq. (15.2) inherently assumes that first and second order derivatives of $u(x)$ exist on the domain $[0,1]$. Associated with the differential form, the function $u(x)$ must obey the *integral form*

$$\int_0^1 \left[-\frac{d}{dx}\left(p(x)\frac{du(x)}{dx}\right) + q(x)u(x) - f(x)\right] dx = 0\,. \tag{15.5}$$

For our model problem, this reduces to

$$\left[-p(x)\frac{du(x)}{dx}\right]_0^1 + \int_0^1 [q(x)u(x) - f(x)]\, dx = 0\,, \tag{15.6}$$

requiring less regularity of the sought function $u(x)$.

15.1.1 Basic concepts for discrete representations

To solve the two-point boundary-value problem (15.2) with the boundary conditions (15.3) and (15.4), we need to make a choice for the numerical representation of the unknown function $u(x)$. When the functions $p'(x)/p(x)$ and $q(x)/p(x)$ are analytic in all $x \in [0,1]$, the exact solution $u(x)$ is mathematically well-behaved and defined on the entire continuous interval $[0,1]$. However, a computational approach always needs to represent $u(x)$ *discretely*, i.e. involving a *finite* number of unknowns. We will indicate the discrete representation for $u(x)$ involving N unknowns with $\hat{u}^N(x)$. As will be explained in following sections, depending on the chosen representation, these unknowns may directly relate to function values $u(x_i)$ at a finite set of particular pre-chosen points $x_i \in [0,1]$, with $i = 1,\ldots,N$ (note that in $x_0 = 0$ we have $u(0) = 0$ due to boundary condition (15.3)), or to the average values of $u(x)$ in N sub-intervals of $[0,1]$, or even to more general expansion coefficients used in finite function series representations. Either way, the discretization function $\hat{u}^N(x)$ is an approximation of $u(x)$.

In order to quantify how good this approximation is, we need to define a norm so that we can measure the 'distance' between the two functions. Several choices are possible here. For example, for quadratically integrable functions on a domain $[a,b]$, the L^2-norm quantifying the distance between functions f and g is

$$\|f-g\|_2 \equiv \left[\int_a^b |f(x)-g(x)|^2\, dx \Big/ (b-a)\right]^{1/2}. \tag{15.7}$$

This L^2-norm is especially useful for linear problems because Fourier analysis is often used for these problems, and Parseval's relation says that a function and its Fourier transform have the same L^2-norm. Frequently encountered norms, written for a finite-dimensional vector $\mathbf{u} = \{u_i\}$, with $i = 1,\ldots,N$, are the L^p-norm and the L^∞-norm defined as

$$\|\mathbf{u}\|_p \equiv \left[\frac{1}{N}\sum_{i=1}^{N} |u_i|^p\right]^{1/p} \quad \text{and} \quad \|\mathbf{u}\|_\infty \equiv \max_i |u_i|\,, \tag{15.8}$$

respectively, where $p=1$ and $p=2$ are the most popular L^1-norm and L^2-norm in their discrete form. The concept of 'convergence' of a numerical solution is then always connected to a certain norm. The series $\{\hat{u}^N(x)\}_{N\to\infty}$ *converges* to the function $u(x)$ *in the L^2-norm* when

$$\lim_{N\to\infty} \|u(x) - \hat{u}^N(x)\|_2 = 0\,. \tag{15.9}$$

Note that this involves the *global truncation error*, i.e. the difference between the exact solution $u(x)$ and the approximation $\hat{u}^N(x)$ over the entire domain. Although the global truncation error is important, it will typically be easier to quantify the so-called *local truncation error* E_i between a local function value (or derivative) such as $u_i \equiv u(x_i)$ (or $u'_i \equiv du/dx(x_i)$) and the local approximation $\hat{u}^N(x_i)$, i.e. $E_i = u_i - \hat{u}^N(x_i)$. Knowledge of the local truncation error will often enable an estimate of the maximal global error.

Another important cause of errors in numerical solutions is the fact that computers calculate with a finite number of decimals while real numbers usually have infinitely many decimals. This yields a *round-off error* R_i:

$$R_i = \hat{u}^N(x_i) - U_i \,, \tag{15.10}$$

where U_i denotes the actually calculated value with the given method. The absolute value of the *total error* made in calculating u_i is given by

$$\begin{aligned} |u_i - U_i| &= |u_i - \hat{u}^N(x_i) + \hat{u}^N(x_i) - U_i| \\ &\le |u_i - \hat{u}^N(x_i)| + |\hat{u}^N(x_i) - U_i| \equiv |E_i| + |R_i| \,. \end{aligned} \tag{15.11}$$

The total error on the function value u_i is thus limited by the sum of the absolute values of the local truncation error and the round-off error. In the context of spatio-temporal problems, it will therefore be important to use a method where truncation errors do not grow unbounded, as discussed further in this chapter.

A frequently encountered measure of the quality of the approximation is the *residual*, which is the equation to be solved as evaluated for the approximation. For our model Eq. (15.2) this means

$$\hat{r}^N(x) \equiv -\frac{d}{dx}\left[p(x)\frac{d\hat{u}^N}{dx}\right] + q(x)\hat{u}^N - f(x) \,. \tag{15.12}$$

This residual $\hat{r}^N(x)$ can be quantified even when the exact solution is not available. A numerical scheme is said to be *consistent* when this residual vanishes in the limit $N \to \infty$ for all x, i.e. when the approximation $\hat{u}^N(x)$ converges for all x to the exact solution $u(x)$.

Other useful criteria to evaluate numerical schemes are *accuracy*, *stability* and *efficiency*,. The first two express the requirement that the deviation of the computed values from the exact solution of the differential equation is small. *Accuracy* is concerned with the global and local truncation errors and the round-off errors discussed above. Usually, round-off errors may be ignored in comparison with the truncation errors if the scheme is *stable*. *Numerical stability* is concerned with error propagation. Even if truncation and round-off errors are small, a numerical scheme will be of little use for time-dependent problems when small errors grow rapidly in time. In Section 15.3 we will provide a quantitative measure for both accuracy and stability of some of the numerical schemes discussed in this chapter. Finally, *efficiency* is concerned with optimizing the costs in terms of CPU time and computer memory of the simulation model.

15.1.2 Finite difference methods

The finite difference approximation is the most widely used computational method. In the finite difference method (FDM), the continuous domain of the independent variable x is replaced by a finite number of discrete points, the *grid points* or *mesh points*. The grid points can be equidistant, as in Fig. 15.1, where we subdivided the domain $[0, 1]$ into N intervals of the same width (with $N+1$ grid points), but they can also be accumulated in places where a higher resolution is required. All dependent functions $u(x)$ on the continuous domain are then approximated by their local values $\{u_i\}$ on the mesh, $u_i \equiv u(x_i)$ with $i = 0, \ldots, N$.

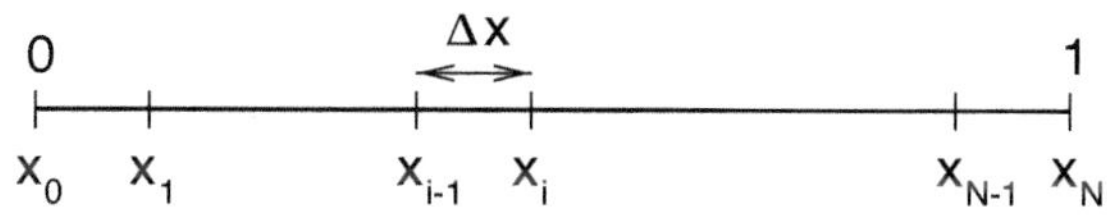

Fig. 15.1 Discrete representation of the domain $[0, 1]$ with $N + 1$ equidistant grid points, where $x_i = i\Delta x$ with $\Delta x = 1/N$.

The representation of the local derivatives is based on truncated Taylor series expansions. For example, when u_i denotes the value of the variable u in grid point x_i, the value of u in grid point x_{i+1} is given by

$$u_{i+1} = u_i + u'_i \Delta x + \tfrac{1}{2} u''_i (\Delta x)^2 + \mathcal{O}(\Delta x)^3 . \tag{15.13}$$

Hence, up to *first order accuracy* in $\Delta x \equiv x_{i+1} - x_i$, we have

$$u_{i+1} = u_i + u'_i \Delta x . \tag{15.14}$$

Solving Eq. (15.14) for the first order derivative u'_i in x_i yields

$$u'_i = (u_{i+1} - u_i)/\Delta x + \mathcal{O}(\Delta x) , \tag{15.15}$$

which is a *first order forward difference* expression for the derivative. Similarly, a Taylor series expansion for u_{i-1} about u_i yields a *first order backward difference* expression. By subtracting the two Taylor series expansions one gets

$$u'_i = \tfrac{1}{2}(u_{i+1} - u_{i-1})/\Delta x + \mathcal{O}(\Delta x)^2 , \tag{15.16}$$

which is a *second order central difference* expression for the first derivative.

▷ **Exercise** Show that the expression

$$u'_i = \left[\tfrac{1}{2}\alpha\,(u_{i+1} - u_{i-1}) + \tfrac{1}{4}(1-\alpha)\,(u_{i+2} - u_{i-2}) \right] / \Delta x$$

yields a fourth order approximation when $\alpha = 4/3$. ◁

Upon substitution of Eq. (15.16) into Eq. (15.13) one obtains a *second order* accurate expression for the second derivative u''_i of u in x_i, as the third order term in the Taylor series in Eq. (15.13) cancels out:

$$u''_i = (u_{i+1} - 2u_i + u_{i-1})/(\Delta x)^2 + \mathcal{O}(\Delta x)^2 . \tag{15.17}$$

Expressions for higher derivatives of u can be found in a similar way. Hence, finite difference approximations are simple to derive and easy to code on regular meshes.

▷ **Exercise** Show that the formula given by Eq. (15.17) can be interpreted as the forward difference of the first derivative, when backward differences are used to discretely evaluate these first derivatives. ◁

The FDM applied to Eq. (15.2) replaces the differential equation by its finite difference representation in all mesh points. For all interior mesh points x_i with $i \neq 0$ or $i \neq N$, use of the central difference formulas (15.16)–(15.17) leads obviously to a representation with a second order local truncation error. For this model problem, to guarantee the same second order accuracy for the full solution on the grid, a corresponding second order treatment of the boundary points $x_0 = 0$ and $x_N = 1$ is used. The boundary condition (15.3) is easily imposed: it corresponds directly to $u_0 = 0$.

However, condition (15.4) involves the first order derivative in $x_N = 1$, so we need to derive a second order backward difference expression for use in this right boundary condition.

While it is possible to do this by suitably combining Taylor expansions for u_i, u_{i-1} and u_{i-2} as above, an alternative and equivalent method to derive such expressions is based on local polynomial fits. In the case at hand, where we seek a second order backward difference expression for u'_i, we envision a locally linear dependence for u' which corresponds to a quadratic formula for u. Hence, for grid point $x_N = 1$, we write locally $x = x_N + y$, and assume that $u(x)$ can be expressed by a second order polynomial, namely:

$$u(x_N + y) = a_0 + a_1 y + a_2 y^2 \quad \Rightarrow \quad u'(x_N + y) = a_1 + 2a_2 y \,. \tag{15.18}$$

Applying this expression to the last three grid points x_N, x_{N-1} ($= x_N - \Delta x$) and x_{N-2} ($= x_N - 2\Delta x$) yields:

$$\begin{aligned} u_N &= a_0 \,, \\ u_{N-1} &= a_0 - a_1 \Delta x + a_2 (\Delta x)^2 \,, \\ u_{N-2} &= a_0 - a_1 (2\Delta x) + a_2 (2\Delta x)^2 \,. \end{aligned} \tag{15.19}$$

Solving these three equations for the wanted $u'(x_N) = a_1$ gives the expression:

$$u'_N = a_1 = \tfrac{1}{2}(3u_N - 4u_{N-1} + u_{N-2})/\Delta x + \mathcal{O}(\Delta x)^2 \,. \tag{15.20}$$

This formula can be used to impose the boundary condition (15.4) in a second order treatment of the model problem. Note that the second order central difference discretization of our model problem described by Eq. (15.2) leads to an algebraic system with a tridiagonal coefficient matrix. Once solved, the representation $\{u_i\}$ is clearly an incomplete description of $u(x)$ for $x \in [0, 1]$, but the function $u(x)$ can be approximated at any point of the interval by, e.g., using a second order, linear interpolation between adjacent grid points. Thus, when x lies in the interval $[x_i, x_{i+1}]$ we have:

$$u(x) \approx \hat{u}(x) = \left(\frac{x - x_i}{x_{i+1} - x_i} \right) u_{i+1} + \left(\frac{x_{i+1} - x}{x_{i+1} - x_i} \right) u_i \,. \tag{15.21}$$

▷ **Exercise** Derive a second order *forward* difference expression for u'_N using the polynomial fit method to find

$$u'_N = \tfrac{1}{2}(-u_{N+2} + 4u_{N+1} - 3u_N)/\Delta x + \mathcal{O}(\Delta x)^2 \,. \tag{15.22}$$

Also, derive a second order one-sided difference expression for the second derivative u''_i, which requires a polynomial for $u(x_i + y)$ which is cubic in y, and find

$$u''_i = (-u_{i+3} + 4u_{i+2} - 5u_{i+1} + 2u_i)/(\Delta x)^2 + \mathcal{O}(\Delta x)^2 \,. \tag{15.23}$$

Note that this second order formula involves four grid points. The number of grid points needed to evaluate derivatives discretely is referred to as the 'stencil' of the method. The stencil of formula (15.23) thus includes grid point x_i and three grid points to its right. ◁

For use in multi-dimensional problems (as in most MHD applications), finite difference expressions for partial derivatives in any spatial direction are given by similar expressions. Expressions for mixed derivatives $\partial^2/(\partial x \partial y)$ are easily derived by writing the x-derivative as a central difference of the y-derivatives, according to Eq. (15.16), and also using this expression for the y-derivatives.

This yields a second order accurate expression for the mixed derivative:

$$\left(\frac{\partial^2 u}{\partial x \partial y}\right)_{i,j} = \frac{u_{i+1,j+1} + u_{i-1,j-1} - u_{i-1,j+1} - u_{i+1,j-1}}{4\Delta x \Delta y} + \mathcal{O}\big((\Delta x)^2, (\Delta y)^2\big).$$

Generalizations to higher order Achieving higher than second order accuracy with the FDM is conceptually simple, but normally comes at the price of handling wider stencils. This also means that boundary treatments need particular attention to achieve the same overall order of accuracy. Fourth order central finite difference formulas for the first and second derivative involve up to five grid points, so that adjacent to boundaries both one-sided and semi-one-sided formulas are needed. These formulas can again most directly be obtained by means of the local polynomial fit method. Representative $\mathcal{O}(\Delta x)^4$ formulas are

$$\begin{aligned}
u_i' &= \tfrac{1}{12}(-u_{i+2} + 8u_{i+1} - 8u_{i-1} + u_{i-2})/\Delta x\,, \\
u_i' &= \tfrac{1}{12}(u_{i+3} - 6u_{i+2} + 18u_{i+1} - 10u_i - 3u_{i-1})/\Delta x\,, \\
u_i' &= \tfrac{1}{12}(-3u_{i+4} + 16u_{i+3} - 36u_{i+2} + 48u_{i+1} - 25u_i)/\Delta x\,, \\
u_i'' &= \tfrac{1}{12}(-u_{i+2} + 16u_{i+1} - 30u_i + 16u_{i-1} - u_{i-2})/(\Delta x)^2\,, \\
u_i'' &= \tfrac{1}{12}(u_{i+4} - 6u_{i+3} + 14u_{i+2} - 4u_{i+1} - 15u_i + 10u_{i-1})/(\Delta x)^2\,, \\
u_i'' &= \tfrac{1}{12}(-10u_{i+5} + 61u_{i+4} - 156u_{i+3} + 214u_{i+2} - 154u_{i+1} + 45u_i)/(\Delta x)^2\,.
\end{aligned} \tag{15.24}$$

▷ **Exercise** Derive these formulas and also derive the equivalent formulas at the other boundary of the computational domain, i.e. at $x = 1$. Your result should be consistent with replacing indices $i \pm * \leftrightarrow i \mp *$ and $\Delta x \leftrightarrow -\Delta x$. ◁

An alternative means to achieve higher order accuracy within the finite difference framework, while maintaining a compact stencil, is to exploit so-called compact or implicit FD schemes. The basic idea is to exploit the local values for the derivatives u_i' and u_i'' as additional unknowns, and complement the equations to solve with implicit formulas linking these additional unknowns to the local function values in discrete expressions with the desired accuracy, as e.g. in

$$\tfrac{1}{2}(u_i' + u_{i+1}') = (u_{i+1} - u_i)/\Delta x + \mathcal{O}(\Delta x)^2, \tag{15.25}$$

which is easily found from combining the central second order formula (15.16) for u_i' with the backwards formula (15.20) for u_{i+1}'. Fourth order expressions are

$$\begin{aligned}
\tfrac{1}{6}(u_{i+1}' + 4u_i' + u_{i-1}') &= \tfrac{1}{2}(u_{i+1} - u_{i-1})/\Delta x + \mathcal{O}(\Delta x)^4, \\
\tfrac{1}{12}(u_{i+1}'' + 10u_i'' + u_{i-1}'') &= (u_{i+1} - 2u_i + u_{i-1})/(\Delta x)^2 + \mathcal{O}(\Delta x)^4.
\end{aligned} \tag{15.26}$$

The above examples yield, for the second order relation (15.25) and for the fourth order relations (15.26), a bidiagonal and two tridiagonal systems in the derivatives, respectively, to be solved together with the discretized equation. Since these systems effectively couple all grid points, no explicit relation can be written for u_i' in terms of neighbouring function values u_i alone, hence their implicit nature.

A detailed description of FDM concepts can be found in Hirsch [302], Chapter 4. The most attractive feature of the FDM is that it is easy to implement. A disadvantage is the poor approximation between the grid points. This is addressed in the finite element approach, which focuses more on the approximation to the solution of the differential equation than on the equation itself.

15.1.3 Finite element method

In the *finite element method* (FEM), the dependent variables are approximated by a finite set of local piecewise polynomials. Consider again the problem posed by Eq. (15.2) and the boundary conditions (15.3) and (15.4). The domain $D = [0,1]$ can be divided into a finite number of equally (as in Fig. 15.1) or unequally sized sub-domains $[x_{i-1}, x_i]$ for $i = 1, \dots N$, which are now called the 'elements' connecting the 'nodes' x_i. The solution $u(x)$ is then approximated by a linear combination of *basis functions*, which are local piecewise polynomials on the sub-intervals $[x_{i-1}, x_{i+1}]$ and taken to vanish outside these finite intervals. Hence the name 'finite' elements. The basis functions are in turn constructed from *shape functions* of a local coordinate ξ, defined on the interval of the local coordinate $\xi \in [-1, 1]$. There are two linear finite element shape functions, corresponding to the two degrees of freedom in a linear profile, fixed by requiring a unit value at one node and a zero value at the other node:

$$N_1(\xi) \equiv \tfrac{1}{2}(1-\xi) \quad \text{and} \quad N_2(\xi) \equiv \tfrac{1}{2}(1+\xi)\,. \tag{15.27}$$

These shape functions are combined into the 'tent' functions $h_i(x)$ depicted in Fig. 15.2, which form the linear basis functions to approximate:

$$u(x) \approx \hat{u}(x) \equiv \sum_{i=0}^{N} u_i h_i(x)\,. \tag{15.28}$$

The approximate solution $\hat{u}$ can be interpreted as belonging to the function space spanned by the linear finite element basis functions h_i given by

$$h_i(x) \equiv \begin{cases} \dfrac{x - x_{i-1}}{x_i - x_{i-1}} & \text{for } x_{i-1} \le x \le x_i\,, \\ \dfrac{x_{i+1} - x}{x_{i+1} - x_i} & \text{for } x_i \le x \le x_{i+1}\,, \\ 0 & \text{elsewhere.} \end{cases} \tag{15.29}$$

Approximations to derivatives are obtained by differentiating Eq. (15.28). Note that these linear elements are such that the coefficients in the expansion (15.28) directly relate to local grid point values u_i. Obviously, other interpolation schemes are possible, based on other shape functions. For instance, the approximation $\hat{u}$ can be obtained by considering piecewise constant shape functions. In this case, the FDM can be considered as a special case of the FEM approach. Other choices include piecewise quadratic and/or cubic shape functions, to be discussed below.

A linear system of algebraic equations for the coefficients $\{u_i\}$ is obtained by requiring that the $N+1$ weighted integrals (15.12) of the *residual* vanish for the approximation $\hat{u}$. For the model problem defined on $[0,1]$, this implies

$$\int_0^1 w_l \left[\Big(-\frac{d}{dx} p(x) \frac{d}{dx} + q(x) \Big) \hat{u}(x) - f(x) \right] dx = 0\,, \tag{15.30}$$

where the *weight functions* w_l are still unspecified. Here, and in the expressions (15.31)–(15.34)

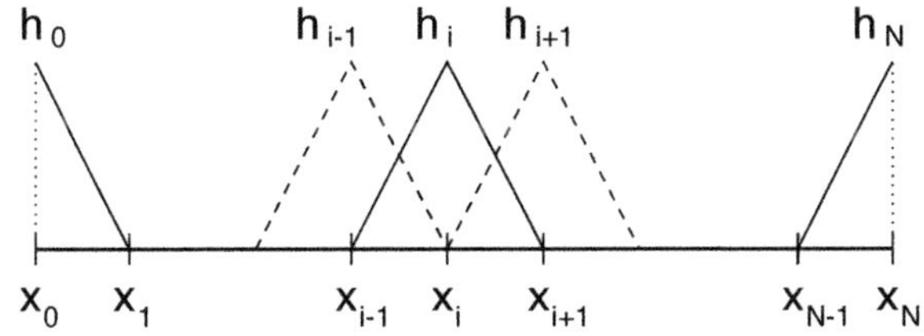

Fig. 15.2 Linear finite elements h_i $(i = 0, 1, \dots N)$.

below, the index $l = 0, 1, \dots N$. In general, the function space consisting of all linear combinations of the basis functions $\{h_i;\ i = 0, 1, \dots N\}$ is finite. Hence, the function $\hat{u}(x)$ defined by Eq. (15.28) may not be equal to the exact solution $u(x)$ at all points of the interval $[0, 1]$. Convergence requires completeness of the function space of the basis functions *and* of the space of the weight functions, i.e. as N increases, the approximation (15.28) must become better and the residual must vanish in the limit $N \to \infty$. The form (15.30) is called the *weighted residual formulation* and, with a certain choice for the weight functions, this formulation yields an approximate solution for each N.

In the finite element method, the weight functions are often chosen to be equal to the basis functions themselves, i.e. $w_l = h_l$, such that the residual $r(x)$ is made orthogonal, in the sense of an inner product defined by the integral $\int_0^1 h_l(x)\, r(x)\, dx$, to the function space of the basis functions. The method is then called the *Galerkin method*. For the model problem (15.2) this yields

$$\int_0^1 h_l \left[\Big(-\frac{d}{dx} p(x) \frac{d}{dx} + q(x) \Big) \hat{u}(x) - f(x) \right] dx = 0\,, \tag{15.31}$$

where $\hat{u}$ is given by Eq. (15.28). Hence, we have a linear algebraic system of $N+1$ equations for the $N+1$ unknowns $\{u_i\}_{i=0,1,\dots N}$.

The integrals occurring in the Galerkin method can often be simplified by performing integrations by parts on the integrands with the highest order derivatives. For problem (15.2) we could proceed with the Galerkin method (15.31) as follows:

$$-\left[h_l\, p(x) \frac{d\hat{u}}{dx} \right]_0^1 + \int_0^1 \frac{dh_l}{dx}\, p(x) \frac{d\hat{u}}{dx}\, dx + \int_0^1 h_l(x) [q(x)\hat{u} - f(x)]\, dx = 0\,. \tag{15.32}$$

The integration by parts results in a formulation which allows solutions that are less 'smooth', in the sense that they are continuously differentiable to a lower degree. Therefore, this formulation is called the *weak formulation* of the problem. For the weak form (15.32), this means that functions with discontinuous first derivative are allowed, whereas in the original differential form (15.2) the first order derivatives have to be continuous and differentiable. In this sense, the weak formulation is closer to the integral form of the equations.

The boundary terms generated by the integrations by parts can be used to impose the boundary conditions, which are called *natural boundary conditions*. For example, for our model problem (15.2) this yields for the boundary condition (15.4), with linear elements:

$$-\left[h_l p(x) \frac{d\hat{u}}{dx} \right]_0^1 = \delta_{lN} h_N(1) p(1) \Big(\frac{\alpha}{\beta} h_N(1) u_N - \frac{\gamma}{\beta} \Big)\,, \tag{15.33}$$

since h_N is the only basis function different from zero in $x = x_N = 1$ and h_0 has to be left out to

Table 15.1 *Weights and evaluation points in double precision accuracy for n-point Gaussian quadrature. (From Abramowitz and Stegun [3].)*

n	weights w_i	evaluation points ξ_i
1	2	0
2	1	$\pm 1/\sqrt{3}$
3	0.888 888 888 888 889	0
	0.555 555 555 555 556	±0.774 596 669 241 483
4	0.652 145 154 862 546	±0.339 981 043 584 856
	0.347 854 845 137 454	±0.861 136 311 594 053

satisfy the boundary condition (15.3), as we will see below. The first term in the RHS of Eq. (15.33) yields a contribution to the coefficient matrix of the system, called the 'stiffness matrix' in FEM terminology, while the second term defines a 'source' term. The other boundary conditions, such as condition (15.3) of our model problem, have to be imposed explicitly and are called *essential boundary conditions*. This is done by limiting the space of basis functions to those basis functions that satisfy these boundary conditions. For boundary condition (15.3) of our model problem this means that all basis functions that are non-zero in $x = 0$ have to be left out, i.e. the basis function h_0 has to be left out in the case of linear elements.

▷ **Exercise** Consider model problem (15.2) with $p(x) \equiv 1$ and $f(x) \equiv 0$ and show that the Galerkin method in combination with linear finite elements reduces to the standard second order finite difference scheme when the trapezoidal rule $\int_a^b f(x)dx \approx \frac{1}{2}(b-a)[f(a)+f(b)]$ is used and partial integration is performed on the term with the second derivative. ◁

Construction of the system matrix For the model problem (15.2), the system of equations for the coefficients $\{u_i\}_{i=0}^N$ is obtained upon substitution of the expansion (15.28) for $\hat{u}$ in the 'weak' form (15.32),

$$-h_l \delta_{lN}\, p \frac{d\hat{u}}{dx}\bigg|_1 + \sum_{i=0}^{N} \left[\int_0^1 \Big(\frac{dh_l}{dx}\frac{dh_i}{dx} p + h_l q h_i \Big) dx \right] u_i = \sum_{i=0}^{N} \left[\int_0^1 h_l f dx \right], \tag{15.34}$$

where the boundary condition (15.4) still has to be substituted in the first (surface) term. For simple functions $p(x)$, $q(x)$ and $f(x)$, the integrals can be calculated by hand, and many are actually vanishing due to the chosen *finite* elements. In general, however, the coefficient functions of the differential equation(s) can be complicated functions of x and the finite elements themselves can be taken as higher order polynomials too. The integrals are then calculated by numerical integration or 'quadrature' formulas, which themselves need to reach a particular order of accuracy to guarantee the overall accuracy of the solution. Gaussian quadrature formulas for integration on the standard interval $\xi \in [-1,1]$ involve the judicious choice of n weights w_i and the locations of n integration

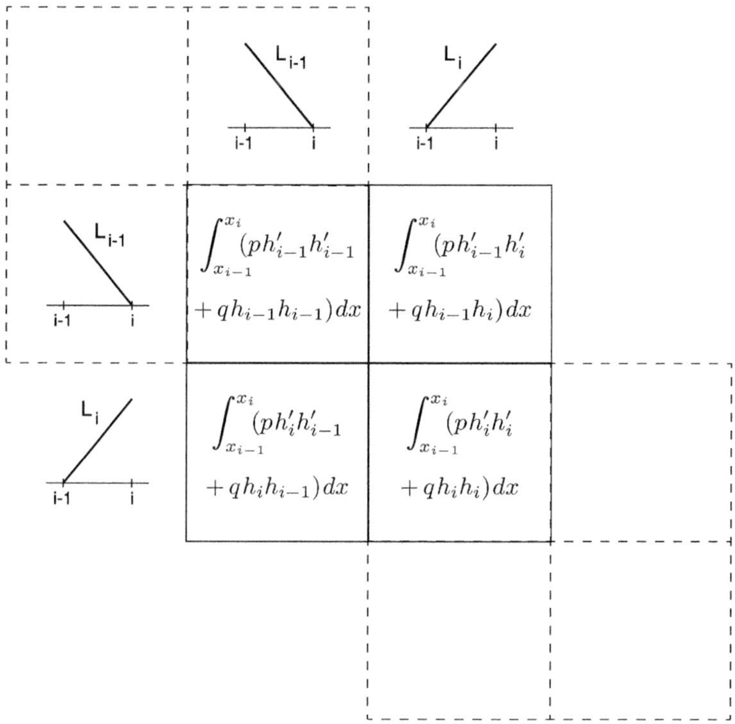

Fig. 15.3 The sub-matrix resulting from integrations over the element $[x_{i-1}, x_i]$ with linear finite elements. The position of the sub-matrices resulting from the neighbouring elements $[x_{i-2}, x_{i-1}]$ and $[x_i, x_{i+1}]$ are indicated with dashed boxes. Notice that each diagonal entry gets contributions from two neighbouring elements.

points ξ_i such that all polynomials of degree $2n-1$ are evaluated exactly by the discrete formula

$$\int_{-1}^{1} f(\xi)d\xi = \sum_{i=1}^{n} w_i f(\xi_i). \tag{15.35}$$

Table 15.1 gives the positions of the evaluation points ξ_i and the weights w_i for this Gaussian integration up to $n = 4$.

The most practical way to implement the coefficient matrix is by performing a loop over the elements $[x_{i-1}, x_i]$ instead of the intervals $[x_{i-1}, x_{i+1}]$. This is the easiest way because the integration can be done in *exactly the same way* for all linear element shape functions. In the interval $[x_{i-1}, x_i]$ *only* the linear basis functions h_{i-1} and h_i are non-zero. This yields four combinations ($h_{i-1}\,h_{i-1}$, $h_{i-1}\,h_i$, $h_i\,h_{i-1}$, $h_i\,h_i$) that are all computed in the same step (Fig. 15.3). The contribution of the neighbouring elements is added in the next step of the loop. Each row of the coefficient matrix, except for the first and the last one, is then computed in two steps.

Generalizations to higher order So far we have only discussed linear finite elements. With linear elements, the FEM is equivalent to the FDM so that there seem to be no advantages that justify the use of this more complex method. However, a local polynomial fit of a function $u(x)$ becomes more accurate when higher order polynomials are used. Hence, one may exploit higher order finite elements like quadratic ones. For example, one can add the midpoint $x_{i-\frac{1}{2}}$ (or $\xi = 0$ on the standard interval $\xi \in [-1, 1]$) as an extra node internal to the element. Three quadratic element shape functions can be defined to attain a unit value in one node and zero at both other nodes, namely

$$N_1(\xi) = -\tfrac{1}{2}\xi(1-\xi)\,, \quad N_2(\xi) = 1-\xi^2\,, \quad N_3(\xi) = \tfrac{1}{2}\xi(1+\xi)\,. \tag{15.36}$$

These can be combined into two basis functions for the quadratic finite elements:

$$Q_i^1(x) \equiv \begin{cases} \dfrac{4(x-x_{i-1})(x_i-x)}{(x_i-x_{i-1})^2} & \text{for } x_{i-1} \le x \le x_i\,, \\ 0 & \text{elsewhere,} \end{cases} \tag{15.37}$$

and

$$Q_i^2(x) \equiv \begin{cases} \dfrac{(2x-x_i-x_{i-1})\,(x-x_{i-1})}{(x_i-x_{i-1})^2} & \text{for } x_{i-1} \le x \le x_i\,, \\ \dfrac{(2x-x_{i+1}-x_i)\,(x-x_{i+1})}{(x_{i+1}-x_i)^2} & \text{for } x_i \le x \le x_{i+1}\,, \\ 0 & \text{elsewhere.} \end{cases} \tag{15.38}$$

The expansion used to approximate $u(x)$ with quadratic elements is written as

$$u(x) \approx \hat{u}(x) = \sum_{i=0}^{N} \Big[Q_i^1(x) u_{i1} + Q_i^2(x) u_{i2} \Big]\,. \tag{15.39}$$

Note that the expansion coefficients now correspond to local function values $u_i = u_{i2}$ and some non-trivial relation between u_{i1}, u_{i2} and the local approximations to derivatives u_i' (in fact, u_{i1} is the approximated function value at the midpoint $(x_{i-1}+x_i)/2$). The derivatives of these quadratic elements, however, are not continuous at the nodes (see the schematic representation of Fig. 15.4).

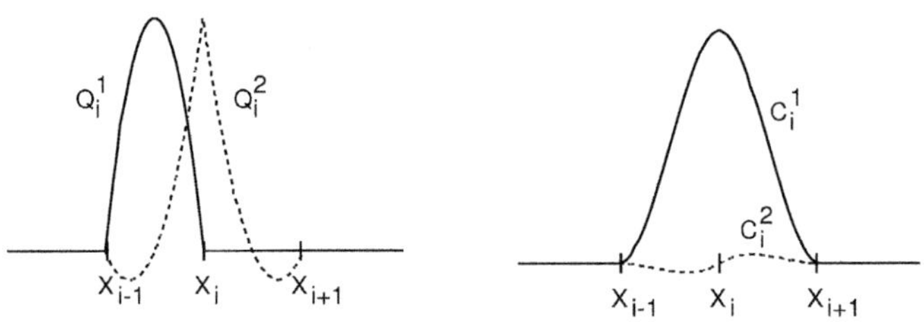

Fig. 15.4 Quadratic (left) and cubic 'Hermite' (right) finite elements.

With two cubic elements per interval one can also make the derivatives continuous at the nodes and approximate both the original differential equation and its derivative. In this way, a higher accuracy can be achieved (fourth order for the cubic elements [574]) and derivatives of higher order can be approximated *without abandoning the local nature of the elements, so that the system matrix remains compact*. This should be compared with the higher order expressions for the FD method, which typically require wider stencils and, hence, more computer storage. The approximation of

$u(x)$ with cubic Hermite elements is written as

$$u(x) \approx \hat{u}(x) = \sum_{i=0}^{N} \left[C_i^1(x) u_{i1} + C_i^2(x) u_{i2} \right] . \tag{15.40}$$

The basis functions for cubic Hermite elements $C_i^1(x)$ and $C_i^2(x)$ are defined by

$$C_i^1(x) \equiv \begin{cases} \left(\dfrac{x - x_{i-1}}{x_i - x_{i-1}}\right)^2 \left(3 - 2\dfrac{x - x_{i-1}}{x_i - x_{i-1}}\right) & \text{for } x_{i-1} \le x \le x_i, \\ \left(\dfrac{x_{i+1} - x}{x_{i+1} - x_i}\right)^2 \left(3 - 2\dfrac{x_{i+1} - x}{x_{i+1} - x_i}\right) & \text{for } x_i \le x \le x_{i+1}, \\ 0 & \text{elsewhere,} \end{cases} \tag{15.41}$$

and

$$C_i^2(x) \equiv \begin{cases} (x - x_i)\left(\dfrac{x - x_{i-1}}{x_i - x_{i-1}}\right)^2 & \text{for } x_{i-1} \le x \le x_i, \\ (x - x_i)\left(\dfrac{x_{i+1} - x}{x_{i+1} - x_i}\right)^2 & \text{for } x_i \le x \le x_{i+1}, \\ 0 & \text{elsewhere.} \end{cases} \tag{15.42}$$

Now, the expansion coefficients in (15.40) directly correspond to local function values $u_i = u_{i1}$ and derivatives $u_i' = u_{i2}$. These Hermite cubic elements are built up from the four shape functions

$$\begin{aligned} N_1(\xi) &= \tfrac{1}{4}(1-\xi)^2(2+\xi), & N_2(\xi) &= \tfrac{1}{4}(1-\xi)^2(1+\xi), \\ N_3(\xi) &= \tfrac{1}{4}(1+\xi)^2(2-\xi), & N_4(\xi) &= -\tfrac{1}{4}(1+\xi)^2(1-\xi). \end{aligned} \tag{15.43}$$

The four degrees of freedom in a cubic formulation have then been fixed by requiring a double zero at one node, and either a unit value and zero derivative *or* a zero value and unit derivative at the other node. The four shape functions combine to the two basis functions C_i^1 and C_i^2 per node and yield 4×4 combinations in each element $[x_{i-1}, x_i]$ so that the 'stiffness' matrix gets a block–tridiagonal structure with sub-blocks of size 2×2 (see Fig. 15.5). Thus, with this better approximation for our 1D model problem, we obtain a system of $2(N+1)$ equations with a 'sparse' banded coefficient matrix: only the diagonal and the first three upper and lower diagonals contain non-zero entries.

Multi-dimensional techniques and MHD applications The finite element technique can be generalized to more than one spatial dimension. In two dimensions, linear elements become 'pyramids' (generalized 'tent' functions) e.g. exploited on a triangular grid. The 'elements' can have curved sides which yields a more accurate representation of domains with curved boundaries (cf. Section 16.3.3 where two-dimensional equilibria determined with isoparametric mapping are discussed). Therefore, the finite element method is very popular as it combines high accuracy with high flexibility. The superiority of the FEM with respect to the FDM becomes more apparent the more irregular the grid is and the more curved the boundaries are. In MHD, with systems of equations in both scalar and vector quantities, it becomes possible to use a different kind of element (linear, quadratic, ...) for each unknown function. We will see in Section 15.1.5 that this hybrid choice is necessary to avoid 'spectral pollution' in the computations of linear eigenmodes [17].

Finite elements are typically used for discretization in the direction normal to the flux surfaces in

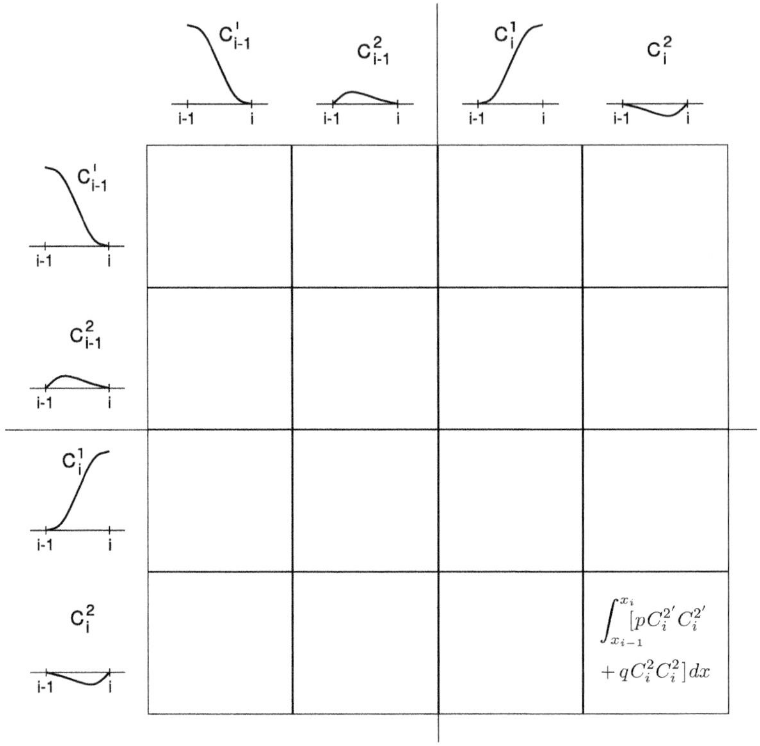

Fig. 15.5 The sub-matrix resulting from the integrations over the subinterval $[x_{i-1}, x_i]$ with cubic 'Hermite' finite elements.

linear and nonlinear MHD codes. For 1D slab or cylindrical equilibrium configurations, the spatial dependence of the linear perturbations about the equilibrium state in the symmetry directions is handled trivially by selecting one Fourier mode pair at a time. The non-trivial normal direction is treated with the FE method in 1D MHD spectral codes such as LEDA (large-scale eigenvalue solver for the dissipative Alfvén spectrum) [362], or its extension LEDAFLOW to stationary equilibria with external gravity, suitable for MHD spectroscopic studies of stratified atmospheres, astrophysical jets or accretion disC models [466]. Also, the same technique returns for diagnosing the linear dynamics of 2D MHD equilibria, with translational or axi-symmetric invariance. Linear MHD codes in this category that exploit FEs for treating the direction normal to the flux surfaces are, e.g. POLLUX (line-tied coronal loop configurations) [278], MARS (magnetohydrodynamic resistive spectrum) [414], NOVA (non-variational code) [121] and CASTOR (complex Alfvén spectrum of toroidal plasmas) [360]. Examples of earlier tokamak spectral codes based on the ideal MHD variational formulation are ERATO [266] and PEST [261]. The FE method is also used in nonlinear MHD computations, in particular to compute the 2D MHD equilibria themselves. Examples include tokamak equilibrium codes like CHEASE (cubic Hermite element axi-symmetric static equilibrium) [422, 423] and HELENA [319], or more general axi-symmetric stationary MHD equilibrium solvers such as FINESSE (finite element solver for stationary equilibria) [44]. In these MHD equilibrium solvers, one uses 2D finite elements for both the normal and the poloidal direction.

15.1.4 Spectral methods

Since the FDM essentially works with local function values and the FEM with local polynomial representations, both the FDM and the FEM are very useful for approaching very localized solutions. This is of great practical value for MHD applications, and one can optimally benefit from this in combination with irregular grids, using accumulation of the mesh points in the region(s) where the solution is expected to have large gradients. Often, however, the solution is not localized at all in one or more spatial directions. Also, the solution may be periodic in one or more spatial directions, e.g. in the toroidal and poloidal directions in a tokamak. In such cases the *spectral method* provides a valuable alternative. The spectral method is in some sense very similar to the finite element method. The important difference is that the shape functions used to approximate the solution are now 'global' functions. These are functions that are non-zero on the whole domain, i.e. the same domain as the differential operator itself. There are several possibilities, and the most convenient choices involve 'orthonormal' functions, such as Legendre and Chebyshev polynomials on $[-1, +1]$ and $\{\sin nx\}_{n=1}^{\infty}$ on the interval $[0, \pi]$.

Fourier harmonics are the most well-known choice for the global expansion functions. They are used for linear MHD problems with periodic boundary conditions. Every truncation of the Fourier series of the solution yields an approximation. Hence, for a function $u(x)$ on a finite domain $[0, L]$, we may write

$$u(x) \approx \hat{u}(x) = \sum_{k=0}^{N-1} \hat{u}_k \mathrm{e}^{\mathrm{i}\,k2\pi x/L} \,. \tag{15.44}$$

Since every term used in the RHS expansion is periodic, $\hat{u}(x=0) = \hat{u}(x=L)$. In the case when the exact solution does not obey this relation (e.g. in model problem (15.2) when $\beta \neq 0$ or $\gamma \neq 0$ in the RHS boundary condition (15.4)), other global functions should be used, or, when the equation is linear, one can split the problem and first solve the homogeneous equation with inhomogeneous BCs and then the inhomogeneous equation with homogeneous BCs. The same periodicity is then typically implied for any function appearing in the equation to be solved, like the functions $p(x)$ and $q(x)$ in Eq. (15.2). The procedure to obtain a linear system for the N complex-valued expansion coefficients $\hat{u}_k$ closely follows the procedure explained for the FEs. In the Fourier representation, derivatives with respect to x turn into multiplications with factors $\mathrm{i}k2\pi/L$, as a result of differentiating the basis functions $\exp(\mathrm{i}k2\pi x/L)$ and replacing the equations by their projections onto the basis functions (as in the Galerkin method and the weak form). In the complex notation of (15.44), one has to multiply with the complex conjugates of the shape functions, i.e. $\exp(-\mathrm{i}l2\pi x/L)$, and then use the orthogonality relation

$$\frac{1}{L}\int_0^L \mathrm{e}^{\mathrm{i}\,(k-l)2\pi x/L}\, dx = \delta_{kl} \tag{15.45}$$

to obtain the weak Galerkin form of the equations.

For our model problem (15.2), we write $q(x) = \sum_{m=0}^{N-1} \hat{q}_m \exp(\mathrm{i}m2\pi x/L)$,, and likewise for p and f, This yields contributions of the corresponding terms to the kth row of the coefficient matrix, e.g. the second LHS term in Eq. (15.2):

$$\text{row } k \quad \Rightarrow \quad \frac{1}{L}\int_0^L \mathrm{e}^{-\mathrm{i}\,k2\pi x/L} \sum_{m=0}^{N-1}\sum_{l=0}^{N-1} \hat{q}_m \hat{u}_l \mathrm{e}^{\mathrm{i}\,(m+l)2\pi x/L}\, dx = \sum_{l=0}^{N-1} \hat{q}_{k-l}\hat{u}_l \,, \tag{15.46}$$

due to the orthogonality relation (15.45) of the Fourier modes, making the integral vanish except when $m = k - l$. Notice that, due to the x-dependence of the coefficient function $q(x)$, there still is a summation in Eq. (15.46), which means that all $\hat{u}_l$-components couple. This summation is the discrete convolution of both Fourier series. Its presence results in a *full* coefficient matrix for the weak formulation, in contrast to the sparse coefficient matrices in the FDM and FEM.

For periodic functions with all derivatives periodic as well, the truncated Fourier series can be shown to yield exponential convergence, faster than any algebraic power (N^{-p}). This so-called spectral or infinite-order accuracy makes the spectral approach the method of choice for periodic problems. Its advantage lies in its tendency to make minimal or no phase errors. For nonlinear incompressible flow simulations of transitions to turbulent flows it is unrivalled in accuracy.

Fourier expansion coefficients and local function values The expansion (15.44) appears to have N complex valued unknown Fourier coefficients, while in the discussion of the FDM and the FEM we always considered only N real-valued degrees of freedom. At the same time, the expansion is meant to approximate the real function $u(x)$ on the finite domain $[0, L]$. The relation between the Fourier coefficients and local function values u_j on an equidistant grid $x_j = j\Delta x$ for $j = 0, \ldots, N$ with fixed grid spacing $\Delta x = L/N$ (as shown in Fig. 15.1 for $L = 1$) is then as follows. First, the periodicity implies $u_0 = u_N$, so that we have only N possibly different function values u_j to determine. Introducing the Nth root of unity $w = \exp(\mathrm{i}\, 2\pi/N)$, local function values u_j form the discrete Fourier transform (DFT) of the Fourier coefficients $u_j = \sum_{k=0}^{N-1} \hat{u}_k w^{j\,k}$. Reversely, the expansion coefficients themselves are the inverse DFT of the local function values:

$$\hat{u}_k = \frac{1}{N} \sum_{j=0}^{N-1} u_j \left(w^{-1}\right)^{j\,k} . \tag{15.47}$$

Since we also have $w^m = w^{m+n\,N}$ for all integer values m, n, several symmetry properties between the N complex Fourier components can be obtained from the knowledge that the function $u(x)$ we are approximating is real, so that $u_j = u_j^*$. Indeed, from the orthogonality relation (15.45) one deduces that

$$\hat{u}_0 = \hat{u}_0^* = \frac{1}{N} \sum_{j=0}^{N-1} u_j \,, \tag{15.48}$$

relating the first (real) Fourier coefficient to the arithmetic mean of the local function values. Similarly, one finds a symmetry about $N/2$ as

$$\hat{u}_k = \hat{u}_{N-k}^* \,, \tag{15.49}$$

meaning that indeed only N real numbers are to be determined, consistent with the equivalent number of real local function values u_j, with $j = 0, \ldots N-1$. Finally, the same observation is true between positive and negative indices, namely

$$\hat{u}_k = \hat{u}_{-k}^* \,, \tag{15.50}$$

which is the reason for restricting the sum to positive indices only in (15.44). Note that these relations also play a role for evaluating the discrete convolution term in (15.46), which we can write as

$$\sum_{l=0}^{N-1} \hat{f}_{k-l}\hat{u}_l \;\Rightarrow\; \sum_{l=0}^{k} \hat{f}_{k-l}\hat{u}_l + \sum_{l=k+1}^{N-1} \hat{f}^*_{l-k}\hat{u}_l\,. \tag{15.51}$$

Finally, once the Fourier coefficients are determined as solutions of the resulting linear system, the local function values u_j are most efficiently computed using a fast Fourier transform algorithm, which computes the N DFTs in order $N\log_2 N$ operations. This relies on the following split in even and odd terms of the DFT formula $u_j = \sum_{k=0}^{N-1} \hat{u}_k \exp(\mathrm{i}\,2\pi\, jk/N)$ for the special case where $N = 2^s$:

$$u_j = \sum_{m=0}^{N/2-1} \hat{u}_{2m}\mathrm{e}^{\mathrm{i}\,2\pi\, jm/(N/2)} + \mathrm{e}^{\mathrm{i}\,2\pi\, j/N} \sum_{m=0}^{N/2-1} \hat{u}_{2m+1}\mathrm{e}^{\mathrm{i}\,2\pi\, jm/(N/2)}\,. \tag{15.52}$$

Each occurring sum is itself recognized as a DFT, yielding a recursive algorithm of order $N\log_2 N$.

The periodicity of the Fourier modes can be a disadvantage when the solution is *not periodic*. The obtained approximation is then poor at the boundaries, where 'Gibbs' phenomena occur: the Fourier series of a piecewise continuously differentiable periodic function behaves peculiarly as a result of the fact that the nth partial sum of the Fourier series has large oscillations near the jump, and often the maximum of the partial sum is higher than that of the function itself (cf. Section 20.1.3). For non-periodic problems on a finite domain $[-1,1]$, one can take the orthonormal *Legendre polynomials*

$$P_n(x) = \frac{1}{2^n}\sum_{l=0}^{n/2}(-1)^l \frac{(2n-2l)!}{l!\,(n-l)!\,(n-2l)!}x^{n-2l} \qquad (n = 0, 1, 2, \ldots) \tag{15.53}$$

as global expansion functions. These functions obey the orthogonality relation

$$\int_{-1}^{1} P_m(x)\, P_n(x)\, dx = \frac{2}{2n+1}\delta_{mn}\,, \tag{15.54}$$

and any function $u(x)$ on the domain $[-1,1]$ can be approximated by truncating its Legendre series given by

$$u(x) = \sum_{k=0}^{\infty} (k+\tfrac{1}{2})\, P_k(x) \int_{-1}^{1} u(x')\, P_k(x')\, dx' \equiv \sum_{k=0}^{\infty} \tilde{u}_k P_k(x)\,. \tag{15.55}$$

For any arbitrary continuous function $u(x)$ (or even quadratically integrable u, under additional assumptions), the completeness of the set of Legendre polynomials guarantees that this series converges to $u(x)$ for any point $x \in [-1,1]$. As for the Fourier series, equations for the expansion coefficients $\tilde{u}_k$ for any finite approximation $\hat{u}(x) = \sum_{k=0}^{N-1} \tilde{u}_k P_k(x)$ are derived from the weak form, using Eq. (15.54).

In a completely analogous fashion, one could use *Chebyshev polynomials* instead, which are defined as

$$\left\{ \begin{array}{lcl} T_0(x) & = & 1\,, \\ T_n(x) & = & \frac{1}{2}n \sum_{l=0}^{n/2}(-1)^l \dfrac{(n-l-1)!}{l!\,(n-2l)!}(2x)^{n-2l} \qquad (n = 1, 2, \ldots)\,, \end{array} \right. \tag{15.56}$$

with corresponding orthogonality relations (involving a weight function):

$$\int_{-1}^{1} T_m(x)\, T_n(x)\, \frac{1}{\sqrt{1-x^2}}\, dx = \begin{cases} 0 & n \neq m\,, \\ \pi/2 & n = m > 0\,, \\ \pi & n = m = 0\,. \end{cases} \tag{15.57}$$

These Chebyshev polynomials obey the recursion formula $T_{n+1}(x) = 2xT_n(x) - T_{n-1}(x)$ for $n \geq 1$, with $T_0(x) = 1$ and $T_1(x) = x$. These polynomials are, e.g., used in the S (spectral compressible linear stability) code for linear HD and MHD stability computations. This Chebyshev collocation code was first developed for compressible hydrodynamic linear eigenvalue problems for one-dimensional stationary planar shear flows by Macaraeg et al. [424], and subsequently extended to the compressible MHD case by Dahlburg and Einaudi [137].

Non-Galerkin spectral approaches In the description of the spectral methods, we only mentioned the Galerkin approach where the weight functions in the weighted residual (15.30) are taken identical to the global expansion functions. In this spectral Galerkin approach, essential boundary conditions exclude those expansion functions not obeying the boundary conditions from the set of expansion functions used, similar to what we discussed for dealing with essential boundary conditions in the FE Galerkin method. There are two other popular variants of the spectral methods, namely the *collocation approach* and the *tau approach*. The spectral collocation technique uses as weight functions in (15.30) the delta function $\delta(x - x_i)$ in a suitably chosen set of collocation points x_i. Hence, the collocation approach uses again local function values, and the global expansion functions are used only to evaluate local derivatives. In the spectral tau method, one can handle non-periodic boundary conditions too by allowing also weight functions which do not satisfy the boundary conditions. In essence, the boundary conditions themselves then need to be expressed in a weighted residual form as well. A thorough discussion of spectral methods is the textbook by Canuto et al. [106].

15.1.5 Mixed representations

Different discretization techniques are often combined in MHD calculations, as we need to deal with up to three spatial dimensions and eight scalar fields. If we consider linear perturbations about an MHD equilibrium, the equilibrium introduces preferred directions inside and normal to its magnetic flux surfaces. The direction normal to the magnetic flux surfaces is then usually discretized by a finite difference or a finite element method because the solutions are often localized due to singular or nearly singular behaviour in this direction. In addition, these two methods have the advantage that they allow mesh accumulation in boundary layers or other regions where the solutions vary rapidly, e.g. the resonant layers created by resonant heating (cf. Chapter 11). On the other hand, for the two periodic coordinates in a tokamak, spectral methods are widely used. Below, we make some additional observations about certain discretization combinations for linear MHD computations. We discuss some exemplary combinations of FD and FE discretizations with (pseudo-)spectral methods and their consequences for the discrete spatial representations used for the three components of the occurring vector quantities (velocity and magnetic field).

Assuming for simplicity a cylindrical coordinate system for a generic periodic tubular flux configuration, Fourier representations in the directions about and along the 'loop' or 'flux rope' axis

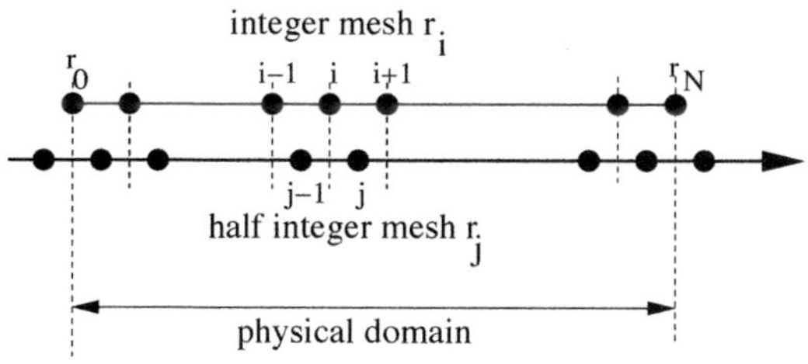

Fig. 15.6 Staggered mesh with an equidistant integer mesh r_i and a half-integer mesh $r_j = r_{i+\frac{1}{2}}$.

can be exploited for all quantities. The radial dependence may then be handled by a finite difference approach. As explained in what follows, solenoidal magnetic fields can be ensured on the numerical level by exploiting a *staggered mesh*, essentially using different grid point locations for different vector components. In the simplest case of an equidistant base grid $r_i = i\Delta r$ for $i = 0, 1, \ldots, N$, where N intervals divide the loop radius $r_N = N\Delta r$, one uses as an additional radial grid the half-integer locations $r_j \equiv r_{i+\frac{1}{2}} = (r_i + r_{i+1})/2$, as shown in Fig. 15.6. The main advantage is that the first order forward difference formula for u'_i given by Eq. (15.16) can be interpreted as a second order central difference formula for $u'_j = u'_{i+\frac{1}{2}}$. This second order formula now involves only u_i and u_{i+1}. To handle boundary conditions on this half-integer mesh, one extends this mesh with *ghost cells* appearing *outside* the domain $[0, r_N]$ (corresponding to $r_{-\frac{1}{2}}$ and $r_{N+\frac{1}{2}}$). For the three components of the magnetic field vector, one can then choose to represent the radial field component B_r on the integer mesh, while the B_θ and B_z components are linked to the half-integer mesh. In combination with the Fourier representation used in their (θ, z) dependence, the following discrete formula for $\nabla \cdot \mathbf{B} = 0$ is obtained for each Fourier mode pair (m, n) from $\exp[\mathrm{i}\,(m\theta + 2\pi\, n/L)]$ (where L is the loop length) on the half-integer mesh:

$$\left(\nabla \cdot \hat{\mathbf{B}}\right)_j = \frac{1}{r_j \Delta r}\left(r_{i+1}\hat{B}_{r,i+1} - r_i \hat{B}_{r,i}\right) + \frac{\mathrm{i}m}{r_j}\hat{B}_{\theta,j} + \frac{\mathrm{i}2\pi n}{L}\hat{B}_{z,j} = 0\,. \tag{15.58}$$

This can be seen as a bi-diagonal system for the Fourier coefficients $\hat{B}_{r,i}$, once $\hat{B}_{\theta,j}$ and $\hat{B}_{z,j}$ are known. This procedure to determine the radial magnetic field component can then replace the discretized radial component of the induction equation. Additionally, one can guarantee the identity $\nabla \cdot (\nabla \times \mathbf{B}) = 0$ in its discrete equivalent by taking

$$\begin{aligned}
\left(\nabla \times \hat{\mathbf{B}}\right)_{r,i} &= \frac{\mathrm{i}m}{2r_i}\left(\hat{B}_{z,j} + \hat{B}_{z,j-1}\right) - \frac{\mathrm{i}\pi n}{L}\left(\hat{B}_{\theta,j} + \hat{B}_{\theta,j-1}\right), \\
\left(\nabla \times \hat{\mathbf{B}}\right)_{\theta,j} &= \frac{\mathrm{i}\pi n}{L}\left(\hat{B}_{r,i+1} + \hat{B}_{r,i}\right) - \frac{1}{2\Delta r}\left(\hat{B}_{z,j+1} - \hat{B}_{z,j-1}\right), \\
\left(\nabla \times \hat{\mathbf{B}}\right)_{z,j} &= \frac{1}{2r_j \Delta r}\Big[r_{i+1}\left(\hat{B}_{\theta,j+1} + \hat{B}_{\theta,j}\right) - r_i\left(\hat{B}_{\theta,j} + \hat{B}_{\theta,j-1}\right)\Big] - \frac{\mathrm{i}m}{2r_j}\left(\hat{B}_{r,i+1} + \hat{B}_{r,i}\right).
\end{aligned} \tag{15.59}$$

This is obviously desirable physically as well. It boils down to using simple second order linear interpolation between the two staggered grids.

A related issue is relevant for all MHD eigenvalue computations. For numerical diagnosis of the stability of a particular MHD equilibrium, it is desirable that the numerical eigenvalues converge to the correct physical eigenvalues when the number of grid points is increased. However, both finite difference and finite element discretizations of eigenvalue problems may exhibit 'spectral pollution', where unphysical eigenvalues belong to strongly distorted or spuriously introduced branches of the discrete equivalent of the dispersion equation. Avoiding spectral pollution implies that the chosen discretization for the eigenfunctions should be able to satisfy constraints like

$$\nabla \cdot \mathbf{v} = 0\,, \qquad \nabla \cdot \mathbf{B} = 0 \tag{15.60}$$

at every point, i.e. also in the entire interval between subsequent mesh points. In the case of a FE representation, it is easily seen that this necessitates the use of elements of a different order for the components of the velocity and the magnetic field. Suppose a cylindrical coordinate system is used and the radial variation is represented by FEs while Fourier modes handle the periodic directions. The radial derivative on the radial velocity in the constraint (15.60)(a) forces the use of finite elements for this component of one order higher than for the other velocity components. This combination is termed a 'conforming' FE discretization. Of course, the MHD equations do not impose these constraints, but some eigenmodes 'choose' to satisfy them. Hence, to get uniform convergence of the entire computed spectrum, i.e. for every eigenvalue, one should be able to satisfy constraints like Eqs. (15.60). An analysis of 'spectral pollution' occurring in simple model eigenvalue problems for both FD and FE discretizations can be found in Ref. [415], where remedies involve the use of staggered representations for FD schemes, or the use of a 'finite hybrid element method' [264]. The latter involves the choice of different order FE representations for vector components and their derivatives as occurring in the corresponding weak forms. The precise mixture of FE orders must be such that each term in the weak form can have the same functional dependence, e.g. mixing piecewise linear and piecewise constant elements to get an overall piecewise constant dependence. More on 'spectral pollution' can be found in the papers [17, 514], and in the book [265], treating finite element methods as used in linear MHD spectral solvers.

For linear MHD eigenvalue computations, ideal spectral solvers can exploit the variational formulation in terms of the Lagrangian displacement vector $\boldsymbol{\xi}$. Examples of such codes for tokamak spectroscopic studies are PEST and ERATO. The PEST code Fourier analyses the poloidal angle variation, and uses a conforming linear and piecewise constant FE mixture in the direction normal to the flux surfaces. ERATO uses 2D hybrid FE discretizations in the poloidal cross-section, and exploits the finite hybrid element approach mixing constant and linear finite shape functions. Also, in cases where the three components of the linearized velocity field and magnetic field vector are handled with a FE representation, the use of mixed FE representations has become a standard practice. For example, the MARS code uses Fourier modes in the poloidal direction and mixes constant and linear finite elements in the radial flux coordinate. CASTOR again handles the poloidal angle in Fourier representation, and uses conforming quadratic and cubic FE representations in the flux coordinate direction.

15.2 Linear MHD: boundary value problems

We now specify the discussion to the linearized MHD equations. We will explain how a choice of spatial discretization for all occurring variables, together with an assumed time-dependence, turns the mathematical problem into a boundary value problem. First, steady-state calculations of externally driven dissipative plasmas are discussed. Next, the corresponding eigenvalue problems are introduced. In both cases, a linear system of equations containing a large number of unknowns is obtained. The determination of the full time-accurate solution of the initial value problem, treated in Section 15.3, also involves the discretization of time.

15.2.1 Linearized MHD equations

Linearized MHD studies the dynamic response of a plasma, initially in equilibrium, to 'small' perturbations. The equations are linearized around the equilibrium, which is usually assumed to be static. This latter assumption is not essential for the numerical methods discussed. For the static case, when we also include the effect of finite resistivity on the linear response, the linearized resistive MHD equations can be written in the following dimensionless form, where the indices 0 refer to equilibrium quantities and the indices 1 refer to perturbed quantities:

$$\frac{\partial \rho_1}{\partial t} = -\nabla \cdot (\rho_0 \mathbf{v}_1)\,, \tag{15.61}$$

$$\rho_0 \frac{\partial \mathbf{v}_1}{\partial t} = -\nabla(\rho_0 T_1 + \rho_1 T_0) + (\nabla \times \mathbf{B}_0) \times (\nabla \times \mathbf{A}_1) - \mathbf{B}_0 \times (\nabla \times \nabla \times \mathbf{A}_1)\,, \tag{15.62}$$

$$\rho_0 \frac{\partial T_1}{\partial t} = -\rho_0 \mathbf{v}_1 \cdot \nabla T_0 - (\gamma - 1)\big[\rho_0 T_0 \nabla \cdot \mathbf{v}_1 + 2\eta(\nabla \times \mathbf{B}_0) \cdot (\nabla \times \nabla \times \mathbf{A}_1)\big]\,, \tag{15.63}$$

$$\frac{\partial \mathbf{A}_1}{\partial t} = -\mathbf{B}_0 \times \mathbf{v}_1 - \eta \nabla \times \nabla \times \mathbf{A}_1\,. \tag{15.64}$$

The temperature T is used instead of the pressure $p = \rho T$, and the vector potential $\mathbf{A}_1$ is used instead of $\mathbf{B}_1$ itself, $\mathbf{B}_1 = \nabla \times \mathbf{A}_1$. As a result, the magnetic field is guaranteed to be 'divergence-free', i.e. $\nabla \cdot \mathbf{B}_1$ is always zero. (While the identity $\nabla \cdot \nabla \times \mathbf{A}_1 = 0$ is obviously true analytically, in a numerical approach it requires that also the discrete equivalent of the divergence and curl operators behave in this fashion. Also note that the vector potential $\mathbf{A}_1$ is only determined up to an arbitrary gauge $\nabla\Phi$. This freedom can be used to set the potential Φ or a specific component of $\mathbf{A}_1$ equal to zero.)

Gravity is ignored for now, as it does not influence the numerics significantly (at least for external gravitational fields). Dissipative effects, however, do affect the choice of discretization method for both space and time dependencies. Electrical resistivity is taken into account to illustrate how. In one-dimensional configurations, i.e. configurations in which the equilibrium quantities depend on one spatial coordinate only, the equilibrium force balance $\nabla p_0 = \mathbf{j}_0 \times \mathbf{B}_0$ corresponds to an ordinary differential equation which can be solved easily. We have seen examples of this in Chapters 7 and 9 and here we will assume that the equilibrium is known.

In three spatial dimensions the system (15.61)–(15.64) consists of eight partial differential equations for eight unknowns. Interchanging LHS and RHS, the system (15.61)–(15.64) can be written symbolically in the form

$$\mathsf{L} \cdot \mathbf{u} = \mathsf{R} \cdot \frac{\partial \mathbf{u}}{\partial t}\,, \tag{15.65}$$

when we introduce the state vector $\mathbf{u}$. In cylindrical coordinates, this vector can be taken as

$$\mathbf{u}^{\mathrm{T}} = (\rho_1,\, v_{1r},\, v_{1\theta},\, v_{1z},\, T_1,\, A_{1r},\, A_{1\theta},\, A_{1z})\,. \tag{15.66}$$

However, other possibilities for $\mathbf{u}$ may exploit e.g. projections of the velocity perturbation on the three orthogonal directions of the local field line triad. In any case, the operators **L** and **R** contain equilibrium quantities and spatial differential operators. With the state vector as in Eq. (15.66), the operator **R** is diagonal and can be read off from the LHS of the system (15.61)–(15.64). The derivative with respect to time is written explicitly in the form (15.65) because it plays a *key role* in the distinction between *three types of problem*. This relates to the three possibilities of having (1) a prescribed time-dependence (as e.g. in *steady, externally driven* problems); (2) an exponential or oscillatory time-dependence, with growth rates or frequencies to be computed as physically realizable eigenfrequencies for the given equilibrium (i.e. *eigenvalue* problems); or (3) an unknown time-dependence, to be determined along with the spatial dependence of the solution. In many cases, these three possibilities correspond to *three different numerical approaches to the same problem* which yield complementary information (see also Section 6.3.1, discussed in different order). Since the cases (1) and (2) can be interpreted as boundary value problems, where in essence only the spatial dependence of linear perturbations is left to determine, we will discuss them first in the following sections.

15.2.2 Steady solutions to linearly driven problems

Many dynamical systems evolve to a steady state and one is often not interested in the temporal evolution but just in the eventual steady state itself. For parameter studies of the efficiency of the plasma–driver coupling, time-dependent computer simulations are much too CPU intensive. The steady-state approach yields the stationary state at once ('in one time step' so to speak). This makes very extensive parameter studies possible, at the price of skipping almost all information on the time scales of the heating process. We already encountered an analytically tractable example in Section 11.1 of a periodically driven dissipative system where the steady-state was of particular interest. There we analysed a semi-infinite plasma slab adjacent to a vacuum in which an external antenna (surface current) induces periodic perturbations which are resonantly absorbed. We assumed that the system had evolved to a steady state in which all physical quantities oscillate harmonically with the frequency $\omega_{\rm d}$ imposed by the external source. Representative numerical results of the steady-state approach have been encountered previously as well: steady-state quantifications of solar p-mode absorption by sunspots taken from [250] were given in Fig. 11.13. The parameter scans shown in those figures were performed numerically, using the generic approach explained below. An example is the parametric study of the efficiency of the plasma–driver coupling in MHD wave heating schemes for solar coronal loops by means of resonant absorption. In the same spirit, impedance scans of the response of a tokamak plasma to an external 'antenna' current can be performed in a systematic fashion. In the context of MHD spectroscopy for laboratory plasmas, such computer simulated impedance scans must be compared to measured ones, in turn yielding information on the internal profiles of the equilibrium quantities.

In computational linear MHD, the steady state of a problem involving a driving frequency $\omega_{\rm d}$ can easily be determined by assuming that it has been reached already. In practice this is done by imposing the time behaviour $\exp(-{\rm i}\omega_{\rm d}t)$ in the equations, i.e. by *replacing the time derivative by a multiplication with* $-{\rm i}\omega_{\rm d}$. After discretization of all spatial dependencies of all eight components of the state vector $\mathbf{u}$ in Eq. (15.66), using any of the discrete representations introduced in Section 15.1, we need to apply the appropriate method to translate the set of linear PDEs (15.65) into a linear system for all the expansion coefficients. Indicating the corresponding vector of unknowns with $\mathbf{x}$, the system (15.65) then reduces to

$$(\mathsf{A} + \mathrm{i}\omega_{\mathrm{d}}\mathsf{B}) \cdot \mathbf{x} = \mathbf{f}, \tag{15.67}$$

where A and B are now algebraic matrices. The vector $\mathbf{f}$ results from imposing the boundary conditions related to the external driving source (see below).

As a concrete example, we consider a combination of model configurations II and III of Section 4.6: a static cylindrical equilibrium plasma in which the equilibrium quantities only depend on the radial coordinate r, surrounded by a vacuum and a perfectly conducting wall at $r = r_{\mathrm{w}}$. The plasma is perturbed by a periodic (AC) current in a helical coil in the vacuum region (at $r = r_{\mathrm{a}}$, with $r_{\mathrm{p}} < r_{\mathrm{a}} < r_{\mathrm{w}}$). This is the cylindrical equivalent of the slab configuration studied analytically in Chapter 11 (apart from the conducting wall). The linear dynamics of this system is described by Eqs. (15.61)–(15.64) with appropriate boundary conditions. With a mono-periodic driving current, the dissipative system will reach a steady state after a finite time. In order to determine this steady state, we assume all perturbed quantities to have a time behaviour of the form $\exp(-\mathrm{i}\omega_{\mathrm{d}}t)$, with ω_{d} the frequency of the 'antenna'. We will describe the system in the common cylindrical coordinates. Due to the one-dimensional nature of the equilibrium, the θ and z-dependence of the perturbed quantities is handled trivially in a Fourier fashion, with in this case only one Fourier mode to consider for each of these coordinates, hence $\exp[\mathrm{i}\,(m\theta + k\,z)]$. In the radial direction, two kinds of finite element are used to avoid spectral pollution. This is accomplished by choosing finite elements of one order higher for the (suitably scaled) components v_{1r}, $A_{1\theta}$ and A_{1z} than for the other ones since these components appear differentiated in the r-direction in discrete expressions for $\nabla \cdot \mathbf{v}_1$ and $\nabla \times \mathbf{A}_1$. For their scaled counterparts $\bar{v}_1$, $\bar{A}_2$ and $\bar{A}_3$ cubic Hermite finite elements are used, while $\bar{\rho}$, $\bar{v}_2$, $\bar{v}_3$, $\bar{T}$ and $\bar{A}_1$ are approximated by quadratic finite elements. As explained in Section 15.1.3, we then typically use two finite elements per grid point in the expansion, making the total number of unknowns $2 \times 8 \times N$, with N the number of radial grid points. In the Galerkin procedure, the discretization (15.39) or (15.40) is inserted in the system (15.61)–(15.64), and the weak form of this system is obtained by multiplying it with each of the $2 \times 8 \times N$ finite elements and integrating over the plasma volume [574]. The application of the Galerkin procedure then yields the algebraic system (15.67). The matrices appearing in (15.67) are block-tridiagonal matrices with sub-blocks of dimension 16×16, where A is a non-Hermitian matrix and B is a positive-definite matrix.[1]

In order to obtain a well-defined problem with a unique solution, boundary conditions need to be specified. As in the similar slab problem studied in Chapter 11, the vacuum solution can be determined analytically. In cylindrical geometry, the vacuum solution is obtained in terms of modified Bessel functions. The system (15.61)–(15.64) is of order six in r (due to the dissipative terms) while the modified Bessel equation describing the vacuum solutions in each of the regions $[r_{\mathrm{p}}, r_{\mathrm{a}}]$ and $[r_{\mathrm{a}}, r_{\mathrm{w}}]$ is of order two. Hence, ten boundary conditions are needed for a unique solution. These boundary conditions are summarized in Fig. 15.7. The two regularity conditions at the magnetic axis are imposed by dropping the finite elements that do not satisfy them in the space of shape functions. Hence, these boundary conditions are treated as *essential* boundary conditions (cf. Section 15.1.3). The boundary conditions at the plasma–vacuum interface become *natural* [320] in a very similar way to that in the model problem considered in Section 15.1: they are imposed by exploiting the

[1] A complex square matrix $\mathsf{A} = (a_{ij})$ is called Hermitian when it is equal to its conjugate transpose, $\mathsf{A}^{\mathrm{H}} = (a^*_{ji}) = \mathsf{A}$. For a real matrix, this reduces to a symmetric matrix. The complex matrix A is positive definite when, for all non-zero complex column vectors $\mathbf{x}$, we have $\mathbf{x}^{\mathrm{H}}\mathsf{A}\mathbf{x} > 0$.

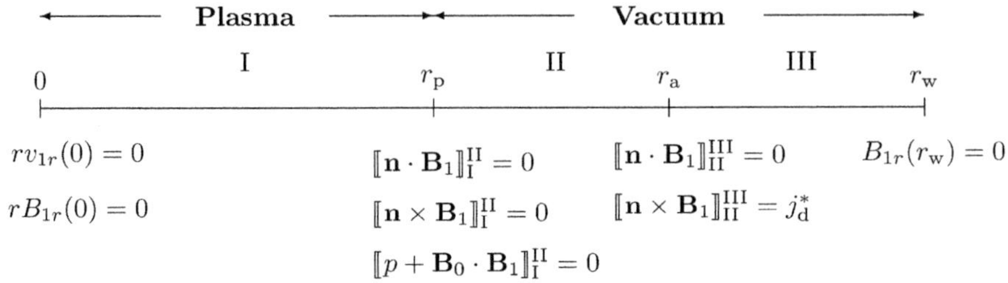

Fig. 15.7 Boundary conditions needed to obtain a unique solution of the steady state of a periodic cylindrical plasma that is driven externally by a current $\mathbf{j}_{\rm d}^*$ in an external helical coil at $r = r_{\rm a}$.

surface terms that arise upon integrating by parts, i.e. by formulating the problem in the weak form. This eventually yields a linear system of the form

$$(\mathbf{A}' + \mathrm{i}\omega_{\rm d}\,\mathbf{B}) \cdot \mathbf{x} = \mathbf{f}', \tag{15.68}$$

with $\mathbf{A}'$ and $\mathbf{f}'$ a slight modification of $\mathbf{A}$ and $\mathbf{f}$, respectively, both resulting from imposing the natural boundary conditions. As explained above, the coefficient matrix $\mathbf{A}' + \mathrm{i}\omega_{\rm d}\,\mathbf{B}$ is of order $16N$. While this can become quite large, it is a sparse block tridiagonal matrix, which can be stored in band storage mode.

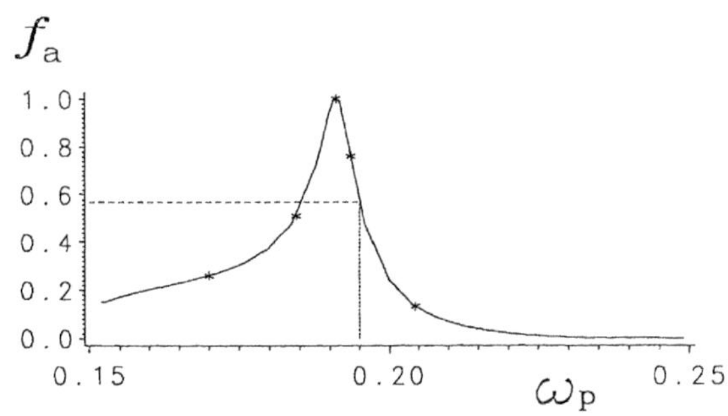

Fig. 15.8 Fractional absorption $f_{\rm a}$ versus driving frequency $\omega_{\rm p}$ of the driven plasma—vacuum–antenna–wall system of Fig. 15.7. This steady-state computation identified the possibility for perfect coupling (at driving frequency $\omega_{\rm p} = 0.1911$), where all energy supplied is converted into heat. (From Poedts et al. [495].)

Figure 15.8 shows the result of a numerically obtained parameter scan for the 1D driven plasma configuration from Fig. 15.7. As a function of the driving frequency $\omega_{\rm d}$ (in the figure indicated as $\omega_{\rm p}$), the fractional absorption $f_{\rm a}$ measures the ratio of the Ohmically dissipated energy to the total energy emitted by the antenna. The equilibrium in the plasma region was characterized by a parabolic axial current density, a constant axial magnetic field component and uniform density. For a magnetic Reynolds number of $R_{\rm m} = 10^8$, and mode numbers of the helical antenna surface current $(m, n) = (2, 1)$ in an $\exp(im\theta + inkz)$ Fourier dependence, the scan revealed the existence of an optimal driving frequency with perfect coupling, i.e. $f_{\rm a} = 1$ or 100%. Through the systematic exploration of the effects of the equilibrium, of the perturbation parameters (driving frequency and mode numbers), and of the model parameters (resistivity), it has been proved that the resistively damped collective 'quasi'-modes, located in the frequency range of the Alfvén continuum, play a prominent role in the energetics of driven magnetic loop configurations.

15.2.3 MHD eigenvalue problems

In the normal mode approach, more information on the linear dynamics is obtained by computing the spectrum of all eigenoscillations. Imposing an a priori unknown time dependence of the form $\exp(\lambda t)$ with $\lambda \equiv -i\omega$ results, in a similar fashion as explained for the driven case (Eq. (15.67)), in the discrete equivalent of Eq. (15.65) for the general eigenvalue problem:

$$(\mathsf{A} - \lambda \mathsf{B}) \cdot \mathbf{x} = 0. \tag{15.69}$$

The possibly complex eigenvalue λ now needs to be determined as part of the solution procedure, and $\mathbf{x}$ corresponds to the discrete representation of the eigenvector. Quite a variety of algorithms exists for the solution of such large-scale eigenvalue problems [357].

This eigenvalue formulation is an efficient approach to determine growth rates of instabilities. In addition, it allows one to explore the stable part of the spectrum and to study e.g. continuum damping of quasi-modes such as encountered in Chapter 10. An example for a cylindrical plasma showing all three (slow, Alfvén and fast) mode types and the possibility of unstable interchanges is depicted in Fig. 9.10. Note that when we use the same spatial discretization technique as for the corresponding driven problem, e.g. a FE and its weak Galerkin formulation of the resistive MHD eigenvalue problem, this leads to a complex non-Hermitian matrix eigenvalue problem (15.69) with the matrices A and B of course *exactly the same* as considered before in the driven problem. It is of interest to note that in the non-ideal MHD eigenvalue solvers based on these FE discretizations, realistically low values of the resistivity coefficient η can be handled accurately, confirming predictions of growth rate dependence as fractional powers of η for analytically tractable unstable modes.

In Fig. 15.9 part of the resistive MHD spectrum (including a so-called ideal MHD quasi-mode) is shown for $\eta = 5 \times 10^{-5}$, as calculated by Poedts and Kerner [497]. These authors determined the full resistive MHD spectrum of a periodic cylindrical plasma column, using 51 grid points in the radial direction. The static equilibrium considered in this calculation consisted of a finite length 'periodic cylinder' with aspect ratio 10, limited by a perfectly conducting wall and with a constant plasma density and a parabolic current density profile. The weakly damped discrete resistive eigenmodes lie on typical curves in the complex plane. The resistive Alfvén eigenmodes lie on the upper λ-shaped curve with a bifurcation point where two branches approach the ends of the ideal Alfvén continuum. The ideal Alfvén continuum, ranging from $\lambda = (0, 0.15)$ to $\lambda = (0, 0.25)$ in this case, is also indicated on the plot. Decreasing the value of the plasma resistivity results in an upward shift of the resistive eigenmodes along the same curves so that the density of the eigenmodes on these curves increases. The weakly damped eigenmode with frequency situated in the triangle formed by the ideal Alfvén continuum and the two legs of the λ-shaped curve with resistive Alfvén eigenmodes in the complex plane, corresponds to an 'ideal quasi-mode'. The oscillatory part of the frequency of this ideal quasi-mode is precisely 0.191, consistent with the result in Fig. 15.8. This weakly damped global mode manifests itself as the natural oscillation of the plasma and explains the temporal evolution of the driven system, as discussed in Chapter 11. The complex frequencies in the lower part of Fig. 15.9, near the origin, belong to the slow magneto-sonic sub-spectrum which does

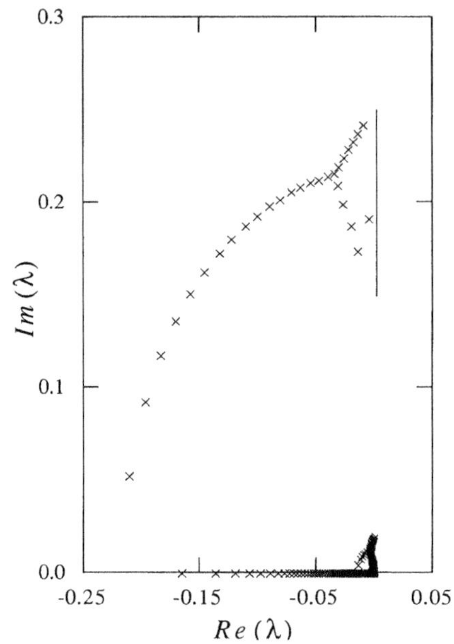

Fig. 15.9 Typical Alfvén and slow magneto-sonic parts of the resistive MHD spectrum, with a quasi-mode in the triangle; $\lambda \equiv -i\omega$. (From Poedts and Kerner [497].)

not couple to the Alfvén sub-spectrum in first order. Notice the discrepancy in time scales between the Alfvén modes and the slow magneto-sonic modes. The fast magneto-sonic modes have much higher frequencies and are not shown in the plot.

15.2.4 Extended MHD examples

The significant advantage of a numerical approach over an analytical one for computing the linear eigenmodes of an ideal MHD equilibrium configuration is the relative ease of including more physical effects in the description of the linear dynamics. To illustrate this, we discuss three examples of eigenvalue computations of essentially 1D MHD equilibria which use a different spatial discretization method for the non-trivial direction.

(a) Magneto-thermal instabilities for solar coronal arcades Our first example is an MHD eigenvalue computation for a solar coronal 'arcade' configuration, taken from Van der Linden, Goossens and Hood [624]. The 1D equilibrium taken for the 'arcade' is rather rudimentary and is actually a pure z-pinch field with $\mathbf{B}_0 = [r/(1+r^2)]B_c\mathbf{e}_\theta$ and with a constant temperature throughout (isothermal). The 'arcade' is then the top half of the cylinder when it is oriented such that its symmetry axis is the magnetic neutral line of the overarching arcade field lines. The eigenvalue computation includes various non-adiabatic effects of importance in the solar coronal environment, in particular optically thin radiative losses and a parameterized coronal heating prescription. This implies that both a source and sink of internal energy is included, appearing as a term $-(\gamma-1)\rho\mathcal{L}$ on the RHS

of the governing nonlinear equation for the temperature evolution:

$$\rho \frac{\partial T}{\partial t} = -\rho \mathbf{v} \cdot \nabla T - (\gamma - 1) \big[p \nabla \cdot \mathbf{v} + \rho \mathcal{L} - \nabla \cdot (\boldsymbol{\kappa} \cdot \nabla T) \big] . \qquad (15.70)$$

The function $\mathcal{L}(\rho, T)$ represents the energy gain–loss per unit mass, due to non-adiabatic processes *other than those due to thermal conduction*. The thermal conduction itself will be anisotropic in strongly magnetized plasmas; hence the appearance of the tensorial conductivity coefficient $\boldsymbol{\kappa}$.

For the internal energy losses due to optically thin radiation, a dependency of the form $\rho^2 T^\alpha$ is appropriate, with the precise proportionality coefficient as well as the fractional temperature dependence α varying for different temperature ranges. Writing $\rho\mathcal{L} = \chi\rho^2 T^\alpha - \rho\, h$, in which h denotes the unknown 'coronal heating' per unit mass, the equilibrium itself is then assumed to obey $\rho_0 \mathcal{L}(\rho_0, T_0) = \nabla \cdot (\boldsymbol{\kappa}_0 \cdot \nabla T_0)$. In Ref. [624], the z-pinch equilibrium was used to compare the analytic predictions obtained for unstable magnetic and *thermal condensation modes* using WKB techniques to those found by means of a numerical eigenvalue solution. The numerical analysis used the same (mixed order) FE discretization discussed in Section 15.2.2 for the radial variation of the eigenfunctions, with a fixed Fourier dependence $\exp \mathrm{i}(m\theta + kz)$. Boundary conditions impose a regularity condition at the axis, while a fixed perfectly conducting wall was adopted at the cylinder radius for simplicity.

For the axi-symmetric $m = 0$ eigenmodes, $\mathbf{k} \cdot \mathbf{B}_0 = 0$ and the influence of purely field-aligned thermal conduction can be dropped. Both the slow and Alfvén continuum collapse onto the origin, but an infinite sequence of unstable discrete $m = 0$ magnetic pinching modes with real growth rates λ can accumulate to the marginal frequency $\lambda = 0$. In the presence of non-adiabatic effects, another continuous range of purely exponentially growing (or damped) modes exists known as the *thermal continuum*. This thermal continuum range can be derived analytically for 1D planar or cylindrical equilibria [623](b). It represents an additional continuous range in the eigenfrequency plane corresponding to purely exponentially growing or damped perturbations, in contrast to the wave-like Alfvén and slow continua. In the presence of anisotropic thermal conduction, the neglect of cross-field thermal conduction is needed to maintain a continuous range, as any finite $\kappa_\perp$ will replace the continuum with a dense set of discrete modes [623](a). This highlights that this thermal continuum is due to the fact that each field line can cool or heat independently of the adjoining ones when perpendicular thermal conduction is absent. Incorporating a heat-loss function and only field-aligned thermal conduction (and no resistivity) introduces the thermal continuum, leaves the familiar Alfvén continuum unmodified from the ideal case, and renders the slow continuum into a continuous range of complex overstable or damped wave modes influenced by non-adiabatic effects. In addition to the continuum range for the thermal condensation modes, an additional sequence of discrete thermal modes may appear at its edges.

Figure 15.10 shows the computed unstable part of the MHD spectrum for axi-symmetric modes with axial wave number $k = 1$. (In this figure, the eigenvalue parameter λ is indicated by s.) For this equilibrium, (1) the entire thermal continuum is situated in the unstable half-plane (seen as solid line segments on the $\mathrm{Im}(s) = 0$ axis); (2) the infinite sequence of discrete unstable magnetic pinching modes is seen to start as indicated by the mode labelled with I_{+1}; (3) discrete thermal modes with fundamental mode labelled with T_1 exist beyond the thermal continuum range. The numerical analysis shows that from a certain overtone onwards, the two discrete sequences merge in coalesced overstable magneto-thermal mode pairs (with complex conjugate eigenvalues λ and λ^*).

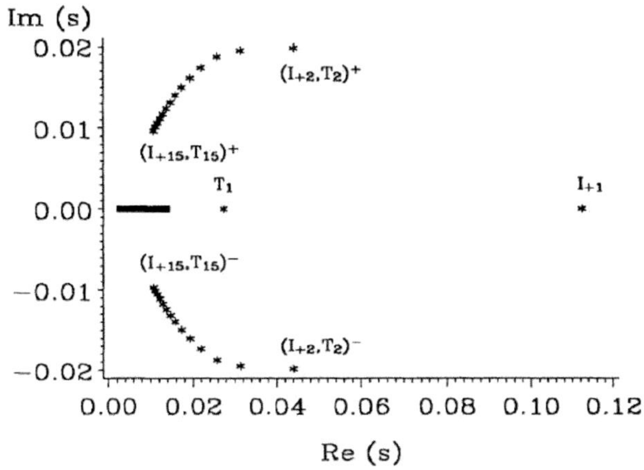

Fig. 15.10 Unstable part of the MHD spectrum for $m = 0$ and $k = 1$ perturbations of a z-pinch 'solar arcade model'. Discrete thermal instabilities labelled with T_n (where n is the number of nodes of the radial eigenfunction) coalesce from a certain n onwards with unstable pinching modes I_n to form overstable magneto-thermal mode pairs; $s \equiv -i\omega$. (From Van der Linden et al. [624].)

This is a particular example of a computational MHD eigenmode analysis where an analytic treatment alone of these intricate mode couplings is challenging but almost necessarily incomplete. In further computational studies [623](a) of more realistic helical equilibrium fields of coronal loops, the effects of anisotropic thermal conduction as well as finite resistivity [327] have been included. The inclusion of a finite (but small) cross-field conduction replaces the thermal continuum with a dense set of discrete thermal eigenmodes, and introduces very localized rapidly varying density eigenfunctions for those modes with eigenvalues located in the original continuum range. These could be responsible for the formation of fine structure during the thermal condensation process leading to solar prominences. In tokamak plasmas, the thermal instability is believed to cause the multifaceted asymmetric radiation from the edge (MARFE) phenomenon, where low charge state impurity radiation gives rise to a condensation instability that is primarily located at the high field side of the tokamak plasma edge (hence it is asymmetric in the poloidal cross-section, but toroidally symmetric.

▷ **Thermal condensation modes** Thermal instability can be encountered as a result of the precise thermodynamic dependence of the energy gain–loss function in Eq. (15.70). The linearized version of this equation will contain the partial derivatives $\partial\mathcal{L}/\partial\rho|_T$ taken at constant temperature and $\partial\mathcal{L}/\partial T|_\rho$ taken at constant density. Non-gravitational unstable thermal condensation modes can be predicted to form even in a static unmagnetized medium of constant density and temperature, if the isobaric criterion

$$\left.\frac{\partial\mathcal{L}}{\partial T}\right|_p = \left.\frac{\partial\mathcal{L}}{\partial T}\right|_\rho - \frac{\rho}{T}\left.\frac{\partial\mathcal{L}}{\partial\rho}\right|_T < 0 \qquad (15.71)$$

is encountered. The ideal gas law $p \sim \rho T$ has been assumed. A runaway condensation can result when a local cooling causes a net increase of the radiative energy losses, resulting in a further cooling triggering yet more radiative losses and hence an unstable situation. This condensation mode is an exponentially growing mode. It is the marginal entropy mode discussed in Section 5.2.2 driven unstable due to non-adiabatic influences, as already analysed in detail in a seminal paper by Field [175], following on an earlier suggestive work by Parker [479]. In the same purely hydrodynamic case, the sound waves, which are purely oscillatory wave modes in adiabatic situations, can be driven overstable if the isentropic instability criterion

$$\left.\frac{\partial \mathcal{L}}{\partial T}\right|_S = \left.\frac{\partial \mathcal{L}}{\partial T}\right|_\rho + \frac{\rho^2}{p(\gamma-1)}\left.\frac{\partial \mathcal{L}}{\partial \rho}\right|_T < 0 \tag{15.72}$$

is met. In the presence of isotropic thermal conduction, these criteria are modified with a stabilizing influence of thermal conduction on both the purely exponential condensation mode and the overstable sound waves for short wavelengths. ◁

(b) Shear flow instabilities in current sheet configurations A second example of MHD eigenvalue computations concerns the linear eigenmodes of a 1D current-vortex sheet, taken from Dahlburg and Einaudi [137], see Fig. 15.11. The planar current-vortex sheet is a *stationary* equilibrium configuration in ideal MHD, characterized by a uniform density $\rho_0 = 1$ and pressure $p_0 = 1/(\gamma M^2)$, but with a shear flow profile given by $\mathbf{v}_0 = \tanh y\, \mathbf{e}_x$ and a co-spatial current sheet with $\mathbf{B}_0 = A \tanh y\, \mathbf{e}_x + A\, \mathrm{sech}\, y\, \mathbf{e}_z$. The dimensionless parameters M and A denote the sonic Mach number and the Alfvén Mach number, respectively. The magnetic field configuration is a force-free field where the magnetic field rotates about the y-axis in an out-of-plane fashion near $y = 0$ to become aligned with the anti-parallel planar flow in both half spaces at $y = \pm\infty$.

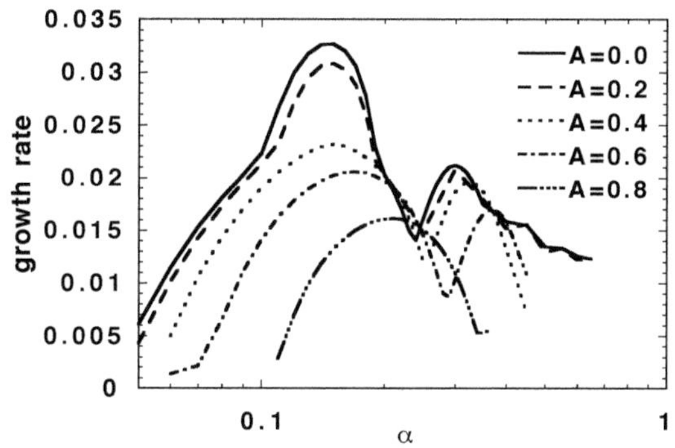

Fig. 15.11 Growth rates of overstable, oscillatory and weakly evanescent, eigenmodes for varying wave number α of a planar current-vortex sheet. The stationary equilibrium is characterized by a sonic Mach number $M = 1$ and Alfvén Mach number A. (From Dahlburg and Einaudi [137].)

The linearized MHD equations now contain many additional terms due to the equilibrium velocity profile, and in Ref. [137] the effects due to (isotropic) thermal conduction, finite magnetic resistivity and viscous stresses are also taken along. Both the streamwise x-direction and the spanwise z-direction are handled in a single Fourier mode pair for the perturbations $\exp[\mathrm{i}(\alpha x + \beta z)]$, while the spatial variation of the eigenfunctions in the cross stream y-direction is discretized using Chebyshev polynomials in a spectral collocation formalism. The SPECLS code, already mentioned in Section 15.1.4, was exploited. The boundary conditions in the cross stream directions represent free slip conditions: setting the gradients of all fluctuating quantities to zero at fixed distances $y = \pm L$. This allows one to compute eigenmodes which are only weakly evanescent and oscillatory in their asymptotic $y \to \pm\infty$ behaviour.

The numerical approach then investigated parametrically the unstable part of the MHD spectrum for flow and magnetically dominated regimes. Depending on their relative strength, surface type ordinary Kelvin–Helmholtz modes can be found (for $M < 1$ and small A), but also a class of overstable, oscillatory, weakly evanescent modes occur when $M \approx 1$ and $A < 1$. The growth rate of the latter modes is shown in Fig. 15.11 for $M = 1$ and varying A, over a range of streamwise wave numbers α. Extensive numerical parameter studies are an essential means to categorize the many intricate relationships between shear-flow driven Kelvin–Helmholtz, resistive tearing modes, and all other ideal and non-ideal instabilities for MHD equilibrium configurations with co-spatial velocity and magnetic shear.

(c) Line-tied MHD spectrum Our examples of MHD eigenvalue computations so far only considered 1D equilibrium configurations (infinite slab or cylinder). The spatial dependence of the eigenfunctions in the ignorable directions is then handled by a single Fourier mode pair. However, virtually all astrophysical and laboratory MHD equilibrium states are at least two-dimensional in nature. For example, coronal loops are anchored (line-tied) at both ends in the dense photospheric environment and can show considerable curvature depending on their loop length to radius ratio. Even when neglecting their curvature and treating only their internal variation with cylinder radius, the line-tying boundary conditions render the MHD stability and eigenoscillation problem intrinsically 2D. For axi-symmetric tokamak plasmas, only the toroidal direction for the linear perturbations can be treated with a single Fourier mode at a time. In both cases, one needs to choose appropriate discretization combinations for the two non-trivial spatial directions in the corresponding MHD spectroscopic analysis.

In the case of line-tied loops, several authors [632, 142, 625] considered the 2D eigenvalue problem for axi-symmetric cylindrical models. The static MHD equilibrium is then still only constrained by the 1D radial equilibrium condition

$$\frac{d}{dr}\left(p_0 + \tfrac{1}{2}B_0^2\right) = -B_{\theta 0}^2/r\,. \tag{15.73}$$

However, the eigenvalue problem takes account of a finite loop length and considers eigenoscillations which obey line-tying boundary conditions. Line-tying typically imposes rigid wall conditions for the velocity perturbations

$$\mathbf{v}_1 = 0 \qquad (\text{at } z = \pm L)\,, \tag{15.74}$$

for a loop of length $2L$. The poloidal θ-direction can be handled trivially as $\exp(\mathrm{i}m\theta)$, but the line-tying will introduce coupling between all axial Fourier modes $\exp(\mathrm{i}n\pi z/L)$. In the case when Eqs. (15.61)–(15.64) are considered, all linear quantities are written as

$$f_1(r,\theta,z) \rightarrow \sum_{n=-\infty}^{n=+\infty} (-1)^n \hat{f}_{nm}(r)\, \mathrm{e}^{\mathrm{i}(n\pi z/L + m\theta)}\,, \tag{15.75}$$

and all θ-derivatives are, as usual, replaced by multiplications with factors $\mathrm{i}m$. However, both first and second order derivatives with respect to z have to be handled with care: not all perturbed quantities or their z-derivatives have equal values at $z = \pm L$ and hence the periodic continuation of them will not be continuous at $z = \pm L$. Partial integration of the (multiplied) equations by $\int_{-L}^{+L} dz\, \exp(-\mathrm{i}k\pi z/L)$ introduces unknown 'surface terms', with jumps across $z = \pm L$, in addition to the usual Fourier substitutions of first and second derivatives. One then uses a truncation of the series representation (15.75) up to N terms. The $8N$ equations for the different Fourier modes $\exp(\mathrm{i}n\pi z/L)$ following from (15.61)–(15.64) are then coupled through these unknown surface terms. The procedure to remove these unknown surface terms subtracts the Nth equation for each linear variable from the $N-1$ previous equations, and replaces those equations of the Nth set containing them by the rigid wall line-tying conditions, which read in the form (15.75) as

$$\sum_{n=-\infty}^{n=+\infty} \hat{v}_{r1;nm}(r) = 0\,. \tag{15.76}$$

Truncating this and similar relations for the other velocity components to N terms yields a finite set of coupled ordinary differential equations for the radial variation of the eigenfunctions, supplemented with boundary conditions on axis and at the loop radius. Velli et al. [632, 142] combined this axial representation of the eigenfunctions using a truncated Fourier series with an FD treatment of the radial direction, while Van der Linden et al. [625] combined it with the FE representation discussed above. Both used it to quantify the stabilizing influence of line-tying and finite loop lengths on selected kink $m = \pm 1$ modes. In such cases, it suffices to take a reasonably small number (up to 10) of axial Fourier modes to get a converged eigenvalue, when these axial modes are chosen to correspond in wavelength to the most unstable modes for the equivalent infinite length cylinder.

15.3 Linear MHD: initial value problems

The computation of the steady solution of a linearly excited, driven, MHD configuration, discussed in Section 15.2.2, or the computation of its spectrum of eigenoscillations and instabilities presented in Sections 15.2.3 and 15.2.4, involved the solution of a boundary value problem and a suitable choice for the spatial discretization. In the efficient calculation of the steady state of a driven dissipative configuration, we directly assumed the asymptotically expected temporal periodicity of the driver. We then lost all information on the time scales to reach this steady state and on the transient phenomena occurring in the initial driving phase. The study of this transient phase requires accurate determination of the temporal evolution of the system, i.e. integration of the MHD equations in time. We first introduce several basic concepts associated with spatio-temporal discretizations, and then discuss specific strategies for and examples of initial value problems in linear MHD.

15.3.1 Temporal discretizations: explicit methods

Similar to our presentation of the various spatial discretization methods, we introduce some basic terminology concerning temporal discretization techniques by first considering a suitable model problem. It should be clear how these can subsequently be applied to the more complicated linearized MHD equations. Following e.g. LeVeque [404], it is instructive to focus on a seemingly trivial problem at first, namely the numerical solution to the linear advection equation. This equation considers a scalar variable u depending on one spatial coordinate x and on time t which obeys the following prototype hyperbolic equation:

$$\frac{\partial u}{\partial t} = -v\,\frac{\partial u}{\partial x}\,, \tag{15.77}$$

with v a given constant. The above equation is the simplest hyperbolic equation containing an 'advection' term in the RHS. The equation just states that u is carried along (advected) without change of form in its arbitrary initial shape $u_0(x, t = 0)$ and the solution is a wave propagating in the positive x-direction (when $v > 0$):

$$u(x,t) = u_0(x - vt)\,. \tag{15.78}$$

Since we know the exact analytic solution at any time, the intricacies of numerical methods can easily be quantified. Moreover, Eq. (15.77) can be written in the form of a conservation law, viz.

$$\frac{\partial u}{\partial t} + \frac{\partial F}{\partial x} = 0\,, \qquad F \equiv vu\,, \tag{15.79}$$

showing that u is a conserved quantity with F its associated flux.

The numerical integration of Eq. (15.77) involves a discretization in both space and time. When we discretize the time coordinate in discrete time levels $t^n = n\Delta t$, the temporal treatment can be explicit, implicit or 'semi-implicit'. In *explicit time integration*, values of quantities at a new time level t^{n+1} are computed from explicitly available information on time level t^n. Implicit or semi-implicit schemes will also involve the evaluation of the unknowns at the new time level, and will be discussed in Section 15.3.3. An explicit method for Eq. (15.77) could replace the spatial derivative with the second order central difference (15.16) evaluated at t^n and the time derivative with the first order accurate expression (15.15) to get

$$\frac{u_i^{n+1} - u_i^n}{\Delta t} = -v\,\frac{u_{i+1}^n - u_{i-1}^n}{2\,\Delta x}\,, \tag{15.80}$$

or, rearranging the terms,

$$u_i^{n+1} = u_i^n - v\,\frac{\Delta t}{\Delta x}\,\frac{u_{i+1}^n - u_{i-1}^n}{2}\,. \tag{15.81}$$

As before, the subscript refers to the grid point and the superscript to the time step, such that $u_i^n = u(i\Delta x, n\Delta t)$. The above scheme is called the *forward FTCS scheme* (forward time central space), but we will shortly explain why this method is utterly useless. Formally, it is second order accurate in space and first order accurate in time. Its 'stencil' in the x–t-plane is shown in Fig. 15.12(a). The difference equation (15.81) is a *consistent* representation of the model PDE (15.77): the local truncation errors vanish in the limit $\Delta x \to 0$ and $\Delta t \to 0$.

In order to be of any practical value, a spatio-temporal discretization should not only be consistent, but also obey numerical stability constraints. Computers always round off numbers and a scheme solving the differential equation (15.77) is *stable* only if these round-off errors shrink or at least do not grow (accumulate) during the time progression. The growth or decay of these round-off errors in a numerical method can usually be evaluated by the *von Neumann method*. This is a *local* stability analysis, i.e. it is assumed that the coefficients of the difference equations vary slowly so that they can be considered constant in space and time. The round-off error $\epsilon(x,t)$ can then be Fourier analysed in space and time, and each Fourier term

$$\epsilon_k(x,t) = \hat{\epsilon}_k \mathrm{e}^{\lambda t} \mathrm{e}^{ikx} \tag{15.82}$$

can be studied separately. The condition for a numerically stable scheme then reduces to

$$\left|\frac{\epsilon_k^{n+1}}{\epsilon_k^n}\right| = \left|\mathrm{e}^{\lambda \Delta t}\right| \le 1 \qquad \text{(for every } k)\,. \tag{15.83}$$

When we analyse the numerical stability of the difference equation (15.81), we must admit that the

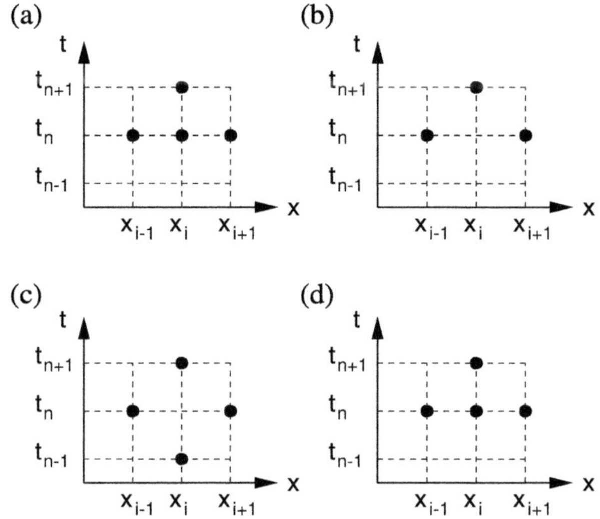

Fig. 15.12 Stencils in the x–t plane for (a) Euler's FTCS scheme, (b) Lax–Friedrichs scheme, (c) leapfrog scheme, (d) one-step Lax–Wendroff scheme.

numerical solution is at best the exact solution, say E, plus the round-off error ϵ. Since the exact solution E satisfies the equation by definition, we need to insert Eq. (15.82) into Eq. (15.81) and we get, after dividing by ϵ_k^n,

$$\mathrm{e}^{\lambda \Delta t} = 1 - \frac{v\Delta t}{\Delta x}\,\tfrac{1}{2}(\mathrm{e}^{\mathrm{i}k\Delta x} - \mathrm{e}^{-\mathrm{i}k\Delta x}) = 1 - \mathrm{i}\,\frac{v\Delta t}{\Delta x}\,\sin(k\Delta x)\,. \tag{15.84}$$

We are thus led to conclude that this scheme is *unconditionally unstable* since

$$\left|\frac{\epsilon_k^{n+1}}{\epsilon_k^n}\right| > 1 \qquad \text{(for every } k)\,, \tag{15.85}$$

which makes the Euler FTCS scheme useless. Note that if v were a function of x and t (or even u, which would bring in nonlinearity), it would still be treated as constant in the von Neumann method since this performs a *local* analysis. Nevertheless, this method is easy to apply and generally yields correct answers.

▷ **Exercise** Analyse a scheme similar to Eq. (15.81), viz. Euler's FTFS method (forward time forward space). How does the discretization of our model PDE look like in this scheme? Show that this method is also unconditionally unstable for $v < 0$. Use a physical argument to explain why the scheme is unstable then. ◁

One may alter the discrete formula (15.81) in several ways to resolve this numerical stability problem. Generally speaking, there are three types of solution. First, one can restore stability by the addition of 'numerical diffusion' to damp the numerical (non-physical) instability. A second 'cure' involves the use of a discretization with the same space-time symmetry as the original PDE. A third way out is the use of an *implicit* scheme. We start with examples of the first two strategies, while the third one is discussed in Section 15.3.3. Upon replacing u_i^n in Eq. (15.81) by an average value of u^n between x_{i-1} and x_{i+1}, we get

$$u_i^{n+1} = \tfrac{1}{2}(u_{i+1}^n + u_{i-1}^n) - v\,\frac{\Delta t}{\Delta x}\,\tfrac{1}{2}(u_{i+1}^n - u_{i-1}^n)\,, \tag{15.86}$$

which is called the *Lax–Friedrichs scheme* (or Lax scheme). The scheme (15.86) is also a consistent, explicit scheme with a simple stencil shown in Fig. 15.12(b). Rearranging the terms in Eq. (15.86) yields

$$\frac{u_i^{n+1} - u_i^n}{\Delta t} = -v\,\frac{u_{i+1}^n - u_{i-1}^n}{2\Delta x} + \frac{(\Delta x)^2}{2\Delta t}\,\frac{u_{i+1}^n - 2u_i^n + u_{i-1}^n}{(\Delta x)^2}\,, \tag{15.87}$$

which can be recognized as a discretization of

$$\frac{\partial u}{\partial t} = -v\,\frac{\partial u}{\partial x} + \frac{(\Delta x)^2}{2\Delta t}\,\frac{\partial^2 u}{\partial x^2}\,. \tag{15.88}$$

In other words, we have added a diffusion term to the original equation which introduces 'numerical dissipation' or 'numerical viscosity'. Hence, this is an example of the first cure to the stability problem. As a matter of fact, the von Neumann stability analysis for the Lax–Friedrichs scheme (15.86) yields

$$e^{\lambda \Delta t} = \cos(k\Delta x) - i\,\frac{v\Delta t}{\Delta x}\,\sin(k\Delta x)\,. \tag{15.89}$$

The resulting condition for stability reads:

$$C \equiv \frac{|v|\Delta t}{\Delta x} \le 1\,, \tag{15.90}$$

which is a limitation of the time step Δt for a given resolution Δx. The condition (15.90) is called the *Courant–Friedrichs–Lewy* condition (CFL) after Courant, Friedrichs and Lewy who derived it in one of the first papers on finite difference methods in 1928 [131], and C is called the *Courant number*. The CFL condition is only a necessary condition for stability, not sufficient. Its physical meaning is that the explicit time step has to be smaller than the time required for the fastest wave in the system to propagate from one grid point to the next. Stated differently, "the domain of dependence of the differential equation should be contained in the domain of dependence of the discretized equations" [302, 643]. If the time step is too big, the physical domain of dependency is larger than the stencil dependency provided by the scheme and this lack of information leads to an instability. The physical domain of dependence is bounded by the physical characteristics in the x–t plane corresponding to the fastest accessible signal speed. For our model problem the physical characteristics are given by $dx/dt = v$, i.e. they are straight parallel lines (see Fig. 15.13).

Let us now use the 'flux notation', i.e. the model equation (15.79) considered in 'conservative form'. For our model equation the 'flux' is $F = vu$. This conservative form makes generalizations to the conservation laws of ideal MHD more obvious and ensures that the numerical schemes discussed below are directly applicable to the finite volume method often used for hyperbolic nonlinear problems (discussed in Section 20.1.4). Above, the unstable forward Euler scheme has been 'stabilized' by adding numerical dissipation. Now, we discuss an example of the second 'cure', viz. mimicking the symmetry of the PDE in the discretized equation. Using a central discretization for x and t, one gets the 'leapfrog' scheme:

$$u_i^{n+1} = u_i^{n-1} - \frac{\Delta t}{\Delta x}\,(F_{i+1}^n - F_{i-1}^n)\,, \tag{15.91}$$

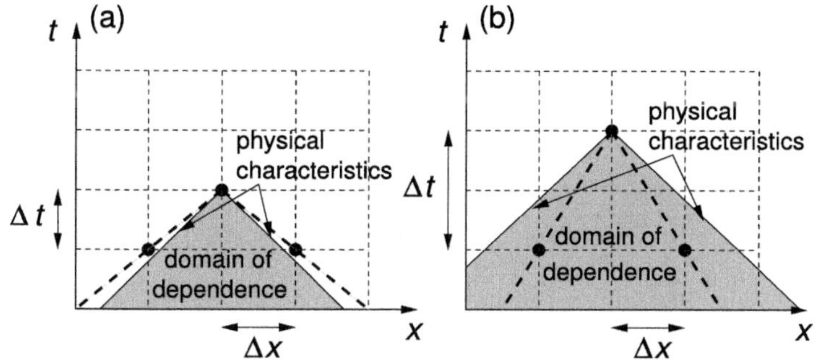

Fig. 15.13 Physical meaning of the CFL condition. (a) If Δt is small enough, the 'physical' domain of dependence lies within the 'numerical' domain of dependence. (b) If not, the scheme can not provide all information needed to determine the solution in the next time step and becomes unstable. In the model problem of the text there is only one characteristic through each point, so that the situation sketched here applies to a second order wave equation with two characteristics $dx/dt = \pm v$.

with the 'numerical' flux function $F_i^n \equiv v u_i^n$. This scheme is *second* order in both time and space. The time levels in the t-derivative 'leapfrog' over the time levels in the x-derivative, as illustrated in Fig. 15.12(c). This scheme requires storage of u^n *and* u^{n-1} to determine u^{n+1}.

Another method which is also second order in time and very extensively used for MHD and CFD calculations, is the *Lax–Wendroff scheme*. The Lax–Wendroff scheme is based on a truncated Taylor series expansion around $u(x,t)$:

$$u(x, t + \Delta t) \approx u(x,t) + \Delta t\, \frac{\partial u}{\partial t}(x,t) + \tfrac{1}{2}(\Delta t)^2\, \frac{\partial^2 u}{\partial t^2}(x,t)\,. \tag{15.92}$$

Using Eq. (15.77), the time-derivatives can be replaced by derivatives with respect to x. This yields the second order accurate Lax–Wendroff scheme:

$$u_i^{n+1} = u_i^n - \tfrac{1}{2}\frac{\Delta t}{\Delta x}\, v\,(u_{i+1}^n - u_{i-1}^n) + \tfrac{1}{2}\frac{(\Delta t)^2}{(\Delta x)^2}\, v^2\,(u_{i+1}^n - 2u_i^n + u_{i-1}^n)\,. \tag{15.93}$$

Clearly, this is another example of the first cure to the numerical instability problem: the LHS term and the first two RHS terms are identical, as in the forward Euler scheme, while the third term in the RHS adds numerical dissipation. The scheme is explicit and, hence, only conditionally stable. Its stencil is the same as for the Euler FTCS scheme and displayed in Fig. 15.12(d).

▷ **Exercise** Show that for our model problem the leapfrog scheme (15.91) requires the CFL condition to be satisfied for numerical stability. Notice that you obtain a quadratic equation in $\exp(\lambda\Delta t)$ now due to the occurrence of three time levels in the scheme. Further, show that the Lax–Wendroff scheme (15.93) again requires the CFL condition to be satisfied for numerical stability. ◁

Example MHD application: p-mode interactions with sunspots An example linear MHD application where a Lax–Wendroff type scheme is used for the temporal integration of the linearized MHD equations is taken from Cally and Bogdan [104]. These authors simulated f- and p-mode interactions with a stratified 'sunspot' in a purely planar approximation. The spot was represented by a mere slab of vertical field of sunspot strength, but took into account the realistically strong gravitational stratification of the plasma. Both internal 'spot' plasma and the external unmagnetized

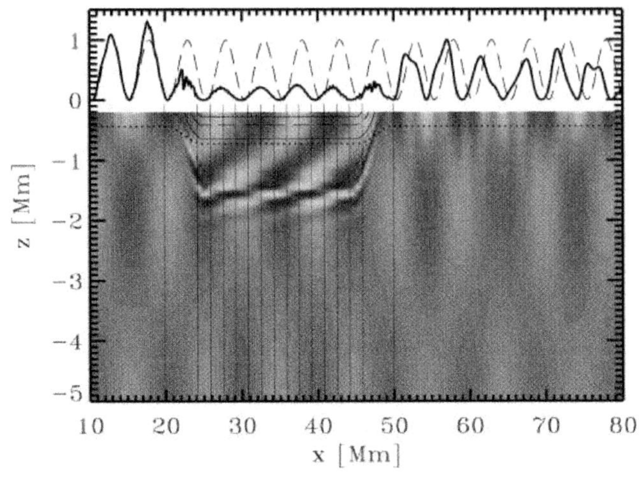

Fig. 15.14 A snapshot of the temporal evolution of a p-mode impinging on a stratified sunspot. The grey scale of the horizontal velocity shows that part of the incoming acoustic power (driver is at left) gets converted into downward propagating magneto-acoustic slow modes in the magnetized slab representing the sunspot. The top solid line indicates the square of the vertical velocity along a temperature iso-therm marked by the dotted line. If the spot is removed, this velocity profile would follow the dashed line shown. (From Cally and Bogdan [104].)

medium had a polytropic stratification (taking account of the offsets required to ensure horizontal pressure balance) and the background density resulted from vertical hydrostatic equilibrium. Using a finite difference spatial discretization of the linearized MHD equations in conservation form, the evolution of the driven problem was simulated. The driver consisted of prescribing the horizontal velocity field at one side of the slab from the exact eigenfunction of the non-magnetic polytrope fitting the horizontal domain size. A staggered grid was employed to ensure numerical conservation. These simulations demonstrated convincingly that impinging f- or p-modes are partially converted to slow magneto-atmospheric gravity waves within the stratified magnetic slab. In turn, this leads to a clear deficit in the amplitude of the emerging, non-magnetic modes exiting the 'spot' on the far side of the driver. As seen in Fig. 15.14, showing the horizontal velocity field when an incident p-mode is driving the slab from the left, the incoming acoustic mode gets coupled at the fore boundary of the magnetic slab (between $x = 20$ Mm and $x = 50$ Mm), at depths where $\beta \approx 1$, to downward propagating magneto-acoustic slow or s-modes. These propagate away from this conversion layer along the vertical field lines, and form a very natural explanation for the observed p-mode power absorption in sunspot and plage regions. Indeed, at the far side, the exciting p-mode re-emerges but is clearly reduced in amplitude and also mixed with an f-mode of the same frequency but shorter horizontal wavelength. In more realistically stratified sunspot models, the additional effects of resonant absorption could act as an extra sink of acoustic power, see Chapter 11.

Semi-discretization (I) So far, all schemes are presented as if space and time are discretized simultaneously. In practice, one may use the 'method of lines' also known as *semi-discretization* for treating a system of PDEs numerically. This implies that one first only discretizes in space, turning the problem into a set of ODEs in time. The latter can be solved by one of the numerous methods available for integrating ODEs, e.g. a second or fourth order Runge–Kutta scheme or a predictor–corrector approach. Obviously, semi-discretization is very useful when higher-order (> 2) accuracy is needed in time. The approach is also powerful because any spatial discretization method of any accuracy can then be coupled to the ODE solver for the time discretization. For explicit schemes such as those discussed before, the CFL stability constraint on the time step will typically require the use of the same order of accuracy for treating the then separate temporal and spatial discretizations. This is no longer the case when dealing with implicit treatments, where the stability constraint will be lifted.

Runge–Kutta methods The application of the semi-discretization method on the model problem given by Eq. (15.77) yields an initial value problem determined by an ODE of the form

$$\frac{du}{dt} = f(t, u)\,, \tag{15.94}$$

and an initial condition

$$u(t^0) = u^0\,. \tag{15.95}$$

When f and $\partial f/\partial u$ are continuous on an open rectangle in the (t, u)-plane containing the point (t^0, u^0) it can be proven that there exists a unique solution $u = \phi(t)$ of this problem in an interval around t^0. When Eq. (15.94) is nonlinear, the solution of this problem is not trivial and often needs to be solved numerically, e.g. by the forward Euler scheme discussed in Section 15.3.1 or the backward Euler scheme discussed in Section 15.3.3. Both schemes, however, are only first order accurate in time. There exists a class of schemes, called the *Runge–Kutta methods,* that enables much faster convergence in time. The Runge–Kutta formulas contain weighted averages of the value of $f(t, u)$ in different points of the time interval $[t^n, t^{n+1}]$. The second order formula uses a trial step at the mid-point of the interval to cancel out lower order error terms. It is given by

$$u^{n+1} = u^n + \Delta t\, k^{n2} + \mathcal{O}(\Delta t)^3\,, \tag{15.96}$$

with

$$k^{n1} = f(t^n, u^n), \qquad k^{n2} = f(t^n + \tfrac{1}{2}\Delta t, u^n + \tfrac{1}{2}\Delta t\, k^{n1}). \tag{15.97}$$

This two-step method is also known as the *predictor–corrector method*, where k^{n2} corresponds to evaluating the function f at time $t^{n+\frac{1}{2}}$. The classic fourth order four-step Runge–Kutta method is given by

$$u^{n+1} = u^n + \Delta t\, \tfrac{1}{6}(k^{n1} + 2k^{n2} + 2k^{n3} + k^{n4}) + \mathcal{O}(\Delta t)^5\,, \tag{15.98}$$

where

$$\begin{aligned} k^{n1} &= f(t^n, u^n)\,, & k^{n2} &= f(t^n + \tfrac{1}{2}\Delta t, u^n + \tfrac{1}{2}\Delta t\, k^{n1})\,, \\ k^{n3} &= f(t^n + \tfrac{1}{2}\Delta t, u^n + \tfrac{1}{2}\Delta t\, k^{n2})\,, & k^{n4} &= f(t^n + \Delta t, u^n + \Delta t\, k^{n3})\,. \end{aligned} \tag{15.99}$$

It is not difficult to show that the scheme (15.98) differs from the Taylor expansion of the exact solution by terms that are proportional to $(\Delta t)^5$, though the derivation is rather lengthy. On a finite time interval, the global truncation error is smaller than a constant times $(\Delta t)^4$. The scheme is thus *fourth order accurate* for the time derivative in the step size Δt. In other words, the order of consistency of this method is equal to four, and there are four intermediate steps necessary, viz. the determination of $k^{n1}, \ldots, k^{n4}$. The Runge–Kutta method thus converges three orders of magnitude faster than the Euler method with respect to the time step. Yet, it is relatively easy to implement and sufficiently accurate to tackle most problems efficiently. This is also true for so-called *adaptive* Runge–Kutta methods which take variable step sizes Δt where needed.

15.3.2 Disparateness of MHD time scales

In MHD, the CFL time step limitation for explicit codes is very severe due to the disparate time scales associated with the various MHD wave types. In practice, this makes explicit schemes useless for the computation of resistive instabilities, as it would involve too many discrete time steps and CPU time. In particular, the fast wave sub-spectrum is responsible for this problem since it accumulates at infinity. Let us analyse the consequences of the disparate time scales for simulating the linear (or nonlinear) dynamical behaviour of a tokamak plasma. In tokamak geometry, the time scale associated with the compressional fast magneto-sonic wave is measured by the transit time of the fast magneto-sonic wave over the small plasma radius a, $\tau_{\rm fast} \equiv a/v_{\rm f}$, with $v_{\rm f}$ the phase velocity of the fast magneto-sonic wave. For the low-β plasmas in tokamaks, the fast magneto-sonic speed is nearly equal to the Alfvén speed, $v_{\rm f} \approx v_{\rm A}$, and, hence, $\tau_{\rm fast} \approx a/v_{\rm A}$. Shear Alfvén waves, on the other hand, propagate mainly along the magnetic field lines. Hence, the time scale related to these waves is measured by the transit time of the shear Alfvén wave over the *length* $2\pi R_0$ of the torus: $\tau_{\rm Alfv} \equiv 2\pi R_0/v_{\rm A}$, where R_0 is the major radius of the tokamak. In large aspect ratio $R_0/a \gg 1$ tokamaks (the argument also holds for large aspect ratio solar coronal loops), the time scales $\tau_{\rm fast}$ and $\tau_{\rm Alfv}$ differ substantially.

When resistive instabilities and Ohmic dissipation are studied, there is a third time scale of interest, viz. the time scale of resistive diffusion of magnetic fields, $\tau_{\rm diff} \equiv \mu_0 a^2/\eta$. The ratio of the compressional time scale to the time scale of resistive diffusion is equal to the magnetic Reynolds number $R_{\rm m}$. In tokamak plasmas, $R_{\rm m}$ is typically 10^6–10^8 and in the large aspect-ratio solar coronal loops $R_{\rm m}$ may reach values up to 10^{10}. Clearly, the resistive diffusion time scale is much longer than both $\tau_{\rm fast}$ and $\tau_{\rm Alfv}$ so that we have

$$\tau_{\rm fast} \ll \tau_{\rm Alfv} \ll \tau_{\rm diff} \,. \qquad (15.100)$$

Consequently, the study of wave problems in tokamaks or solar coronal loops where dissipation is important leads inevitably to 'stiff' equations in which the dependent variables change on two or more very different scales. Stiff problems make explicit methods inefficient to use, or may even render them unstable, and the cure to solve this problem is the use of implicit methods.

15.3.3 Temporal discretizations: implicit methods

Consider again Euler's FTCS scheme. By evaluating the spatial derivative in this scheme in the $(n+1)$th time step instead of the nth, one obtains the 'BTCS Euler scheme' (backward in time now):

$$u_i^{n+1} = u_i^n \quad v \frac{\Delta t}{\Delta x} \tfrac{1}{2}(u_{i+1}^{n+1} - u_{i-1}^{n+1}) \,. \qquad (15.101)$$

Notice that now u_i^{n+1} cannot be expressed in terms of known values at time index n anymore. It also depends on unknown values at time index $n+1$, viz. u_{i+1}^{n+1} and u_{i-1}^{n+1}. The scheme is therefore called *implicit*. For (15.101), this results in a tridiagonal system for $u_i^{n+1}, i = 1, \dots N$. A von Neumann stability analysis of the scheme (15.101) yields

$$e^{-\lambda \Delta t} = 1 + iv \frac{\Delta t}{\Delta x} \sin(k\Delta x) , \tag{15.102}$$

so that

$$\left| e^{\lambda \Delta t} \right| = \frac{1}{|1 + iC \sin(k\Delta x)|} < 1 \qquad \text{(for all } k) . \tag{15.103}$$

Hence, this scheme is *unconditionally stable*, and any large time step Δt is in principle allowed. However, for wave problems, the time step must be much smaller than the wave period of interest to avoid numerical damping. This is so since the implicit treatment improves the stability of the scheme, but *not* the accuracy. Larger time steps introduce larger truncation errors. The stencil of scheme (15.101) is given in Fig. 15.15(a) where the three dots at time level $n + 1$ imply the tridiagonal system solve. This introduces more computational work per time step, so the use of an implicit scheme only pays off when large enough time steps can be taken to make the total amount of work to reach a fixed future time decrease. For wave problems, this may often not be the case.

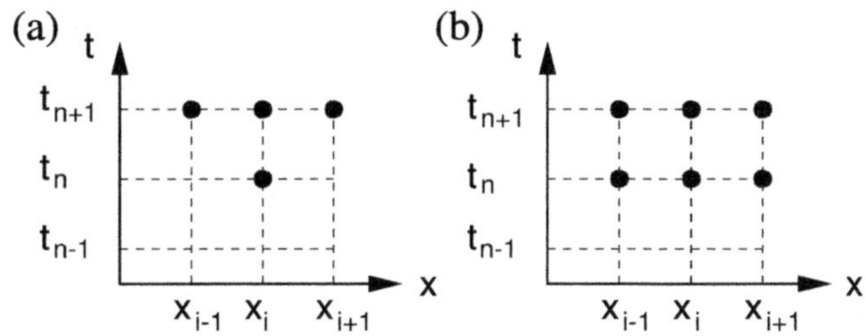

Fig. 15.15 Stencils of (a) BTCS Euler scheme and (b) Crank–Nicolson method.

The implicit backward Euler scheme (15.101) is only first order in time. Upon writing the spatial differences in the right hand side of the explicit unstable forward Euler scheme (15.81) in terms of averages between the nth and the $(n + 1)$th time step, one gets

$$u_i^{n+1} = u_i^n - \tfrac{1}{4} v \frac{\Delta t}{\Delta x} \left(u_{i+1}^{n+1} + u_{i+1}^n - u_{i-1}^{n+1} - u_{i-1}^n \right) , \tag{15.104}$$

which is called the *Crank–Nicolson method*. This scheme is second order accurate in both time and space and again requires a tridiagonal system to be solved at each time step. Its stencil is shown in Fig. 15.15(b). The Crank–Nicolson method is widely used in CFD and computational MHD.

▷ **Exercise** Show that the scheme (15.104) is unconditionally stable. ◁

▷ **Semi-implicit schemes** The term 'semi-implicit' for temporal discretization techniques is used for several schemes which are neither explicit nor fully implicit. Different options can be, e.g., to treat some variables explicitly and others implicitly, or to update only specific terms in the equations implicitly, or to relax the CFL condition somewhat by treating only the fastest sub-spectrum of waves implicitly. The latter alternative has been applied in particular for nonlinear MHD with success [288]. When the fast magneto-sonic waves are handled in an implicit way, the time step is limited by the CFL condition on the Alfvén waves. This is much less restrictive in tokamaks because of Eq. (15.100). ◁

Semi-discretization (II) Returning to the semi-discretization method, we can consider the quite general model problem involving a vector $\mathbf{u}$ of unknowns. These unknowns are introduced by the spatial discretization (FDM, FEM, (pseudo-)spectral, etc.) and one obtains an ordinary differential equation for $\mathbf{u}$:

$$\frac{d\mathbf{u}}{dt} = \mathbf{f}(\mathbf{u}), \tag{15.105}$$

where $\mathbf{f}$ could even be a nonlinear function of $\mathbf{u}$. Discretize this ODE in time by the following scheme:

$$\mathbf{u}^{n+1} = \mathbf{u}^n + \Delta t \left[\alpha \mathbf{f}(\mathbf{u}^{n+1}) + (1-\alpha)\mathbf{f}(\mathbf{u}^n)\right], \tag{15.106}$$

with parameter α. For $\alpha = 1/2$ the above implicit scheme is called the *trapezoidal method*, while for $\alpha = 1$ the scheme reduces to the implicit 'backward Euler' scheme. For $\alpha = 0$ we get the explicit forward Euler method. The scheme is only second order accurate for $\alpha = 1/2$. As a matter of fact, with a central formula for the flux the trapezoidal method is exactly the same as the Crank–Nicolson method discussed above for the model problem (15.77). For a nonlinear $\mathbf{f}$, the above scheme involves the (iterative) solution of a nonlinear system at each time step. The system can be simplified by linearizing $\mathbf{f}(\mathbf{u}^{n+1})$, i.e. by replacing this term by

$$\mathbf{f}(\mathbf{u}^{n+1}) \approx \mathbf{f}(\mathbf{u}^n) + \frac{\partial \mathbf{f}^n}{\partial \mathbf{u}} \cdot (\mathbf{u}^{n+1} - \mathbf{u}^n), \tag{15.107}$$

where the matrix $\dfrac{\partial \mathbf{f}^n}{\partial \mathbf{u}}$ is called the *'Jacobian matrix'* of $\mathbf{f}$. With this linearization (15.107), the scheme (15.106) reduces to

$$\begin{aligned} \mathbf{u}^{n+1} &= \mathbf{u}^n + \Delta t \alpha \frac{\partial \mathbf{f}^n}{\partial \mathbf{u}} \cdot (\mathbf{u}^{n+1} - \mathbf{u}^n) + \Delta t \mathbf{f}(\mathbf{u}^n) \\ \Rightarrow &\quad \left[\mathsf{I} - \Delta t \alpha \frac{\partial \mathbf{f}^n}{\partial \mathbf{u}}\right] \cdot \delta \mathbf{u} = \Delta t \mathbf{f}(\mathbf{u}^n), \end{aligned} \tag{15.108}$$

which is a linear system for the unknowns $\delta \mathbf{u} \equiv \mathbf{u}^{n+1} - \mathbf{u}^n$. Notice that this equation is equivalent to the first step of a Newton iteration on Eq. (15.106), which may need to be expanded to a full Newton iteration in the case of nonlinear problems.

15.3.4 Applications: linear MHD evolutions

If we follow the semi-discretization approach for the linearized MHD equations (15.61)–(15.64), we obtain a system of ODEs in time. In a manner analogous to the discussion of the driven problem in Section 15.2.2, but now keeping the unknown temporal behaviour, we can write

$$\mathsf{A} \cdot \mathbf{x} = \mathsf{B} \cdot \frac{d\mathbf{x}}{dt} + \mathbf{f}, \tag{15.109}$$

where $\mathbf{f}$ again results from a possible external driving source or initial perturbation. For linear (and even for nonlinear) stability calculations, the solution of this system as a function of time will automatically yield the expected growth at early times of the most unstable mode out of a random initial vector $\mathbf{f}$. Moreover, this initial value approach also enables us to simulate the nonlinear evolution of instabilities. We discuss two example linear MHD applications where a different implicit time discretization was used to overcome the CFL time step constraint, thus enabling us to obtain quantitative insight into long-term, resistively controlled, phenomena.

(a) Determining growth rates of linear resistive instabilities Specifying the discussion to cylindrical geometry and linear MHD, we can combine, as in the steady-state approach, the finite-element discretization for the radial direction and the spectral discretization for the other two spatial directions. The Galerkin method then yields system (15.109) (without $\mathbf{f}$ for a non-driven case) with the matrices A and B exactly the same as before. If we perform a forward time discretization, we get the integration formula

$$\mathbf{x}^{n+1} = (\Delta t\, \mathsf{B}^{-1} \cdot \mathsf{A} + \mathsf{I}) \cdot \mathbf{x}^n . \tag{15.110}$$

Numerical stability limits the time step to

$$\Delta t \leq 2/\lambda_{\max} , \tag{15.111}$$

where $\lambda_{\max}$ is the largest eigenvalue of the system $\mathsf{A}\mathbf{x} = \lambda\, \mathsf{B}\mathbf{x}$. This CFL condition is often not acceptable in MHD because $\lambda_{\max}$ tends to infinity in the limit of infinite resolution in the direction normal to the magnetic flux surfaces. The generalized trapezoidal method (15.106) then gives the following algorithm for the time advance:

$$[\, -\mathsf{B} + \alpha\, \Delta t\, \mathsf{A}\,] \cdot \mathbf{x}^{n+1} = -[\, \mathsf{B} + (1-\alpha)\, \Delta t\, \mathsf{A}\,] \cdot \mathbf{x}^n , \tag{15.112}$$

which is unconditionally stable for $\alpha \geq 0.5$.

Kerner, Jakoby and Lerbinger [361] applied this technique to analyse resistive waves (which requires an appropriate initialization), and current- and pressure-driven resistive instabilities for cylindrical equilibrium configurations. They typically used $\alpha = 0.52$ in order to introduce some numerical damping to damp out the fast modes. An example is shown in Fig. 15.16, where the growth rate of a resistive instability is quantified for such a cylindrical 'tokamak-like' equilibrium configuration with a peaked current density and constant toroidal field. They considered the class of profiles

$$j_z = j_0 \big(1 - r^2/a^2\big)^\nu , \qquad B_z = 1, \qquad \rho = 1,$$

with a the radius of the cylinder. For this class, the ratio of the safety factor on the surface to that on the axis is given by $q(a)/q(0) = \nu + 1$. They set $\nu = 1$ and adjusted the constant j_0 to vary $q(0)$, since j_0 is connected with $q(0)$ by $j_0 = 2k/q(0)$, where $k = 2\pi/L$ defines the periodicity length of the cylinder (simulating a tokamak with large aspect ratio). The wall is placed directly at the plasma surface (cf. model I in Section 4.6.1). This equilibrium is known to be unstable against the $m = 1$ tearing mode (see Section 14.2.2) if the $q = 1$ surface is located inside the plasma. This instability is avoided by making sure that $q > 1$ over the whole plasma radius. The $m = 2$ tearing mode is then the most dangerous instability. The growth rate of the most unstable mode is plotted versus $q(a)$ in Fig. 15.16. When the wall is placed directly at the surface, the $m = 2$ tearing mode is unstable for $2.20 \leq q(a) \leq 4.00$. The growth rates were obtained by studying the time evolution of an initial random starting vector with a time step $\Delta t = 400$ for the strong instabilities near $nq(a) = 3$, and $\Delta t = 1000$ for the weaker instabilities around $nq(a) = 2.1$. For $nq(a) \leq 2.3$ the instability changes from a current-driven into a pressure-driven mode.

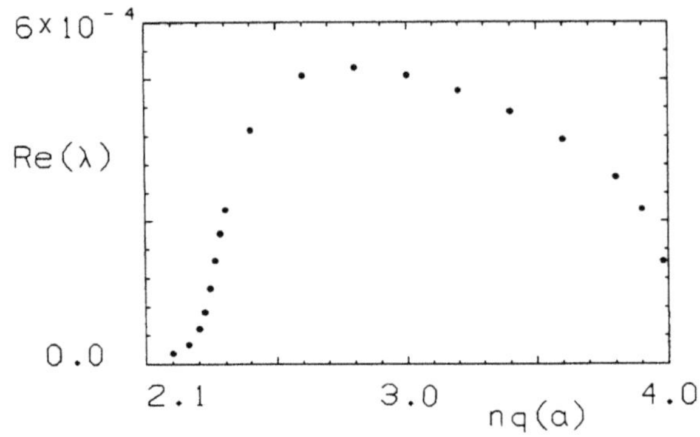

Fig. 15.16 Growth rates of resistive instabilities versus total current, parameterized by $q(a)$, showing the transition (at $nq = 2.3$) from pressure driven to current driven instabilities. (From Kerner et al. [361].)

When an external driving source excites the plasma, the implementation of the boundary conditions through the boundary terms in the weak form of the equations yields a 'force' term $\mathbf{f}$ and a small change to the matrix $\mathbf{A} \to \mathbf{A}'$, as discussed in Section 15.2.2. Hence, we now retain $\mathbf{f}$ in (15.109) and the algorithm for the time advance is modified to

$$\underbrace{\left[-\mathbf{B} + \alpha\,\Delta t\,\mathbf{A}'\right]}_{\equiv \hat{\mathbf{A}}} \cdot \mathbf{x}^{n+1} = \underbrace{-\left[\mathbf{B} + (1-\alpha)\,\Delta t\,\mathbf{A}'\right]}_{\equiv \hat{\mathbf{B}}} \cdot \mathbf{x}^{n} + \underbrace{\Delta t\left[(1-\alpha)\mathbf{f}^{n} + \alpha \mathbf{f}^{n+1}\right]}_{\equiv \hat{\mathbf{f}}(t)} . \quad (15.113)$$

This equation for $\mathbf{x}^{n+1}$ is of the same form as Eq. (15.112) with the exception that the RHS is now *time dependent* through the driving term $\hat{\mathbf{f}}(t)$. However, the matrix operations are straightforward and preserve the block-tridiagonal structure. Hence, in each time step the above system (15.113) can be solved in the same way as the steady-state problem (Section 15.2.2). Notice that Δt is constant so that the matrices $\hat{\mathbf{A}}$ and $\hat{\mathbf{B}}$ are constant in time. This means that the CPU time consuming LU factorization needs to be carried out only once. The solution process itself, however, must be carried out several 10^4 times since the time step Δt is limited by the period of the external driving source. As concrete examples of this approach, see the results on time scales and transient phenomena involving resonant absorption and wave heating effects discussed in Chapter 11. For example, Figs. 11.7 and 11.8 have been obtained with the above semi-discretization method.

(b) Damped coronal loop oscillations A representative example of a linear MHD time evolution concentrates on solar coronal loop dynamics. The advent of modern high-resolution, high cadence, views on the highly structured coronal plasma by space missions such as the transition region and coronal explorer (TRACE, see e.g. the image in Fig. 8.11), has identified many MHD wave modes in coronal loops. Routinely, transverse oscillations are detected in these loops, e.g. in connection with a solar flare disruption of a neighboring arcade. Detailed information on the oscillation amplitudes, periods and damping times has been collected. Although the observed loop displacements are large, linear MHD theory can still be applicable since the associated velocities are small with respect to the Alfvén speed in the low-β coronal conditions. Terradas et al. [593] solved the linearized resistive MHD equations for a cylindrical coronal loop model exploiting a zero-β limit.

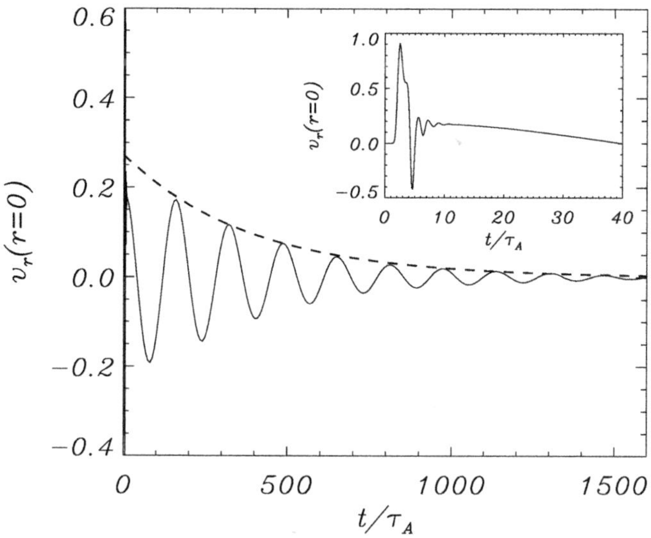

Fig. 15.17 The temporal evolution of the radial velocity at the loop axis, for a suddenly perturbed coronal loop. The inset shows the initial leaky transient, due to exitation of fast modes, while the main plot shows the longer-period attenuated oscillation identified with a resonantly damped kink mode. (From Terradas et al. [593].)

In this limit, slow modes are eliminated and the governing equations need only retain radial and azimuthal velocity perturbations together with the linear field $\mathbf{B}_1$. The set of equations was semi-discretized, and the spatial discretization employed a finite element treatment of the radial direction, and a Fourier handling in the ignorable directions of the equilibrium. A single Fourier mode pair $(m, k_z) = (1, \pi/L)$ was chosen to isolate kink displacements of a line-tied loop of length L. The equilibrium taken considered a uniform axial magnetic field $\mathbf{B}_0 = B_0\mathbf{e}_z$, and a smoothly varying density profile connecting the higher internal loop density ρ_i to the external coronal value ρ_e. The temporal discretization used the second order Crank–Nicolson approach, and simulated the long-term evolution of the response of the loop to a sudden radial velocity perturbation impinging on it. In Fig. 15.17, the typical temporal behaviour of the radial velocity at the loop axis is shown: several phases characterized by totally different time scales are evident. The authors identified the first series of strongly attenuated short-period oscillations seen in the inset of Fig. 15.17 with a leaky eigenmode of the loop configuration. As the perturbation was selected to deposit a considerable amount of energy in the loop, this first phase corresponds to the resulting excitation of fast waves carrying energy radially off to infinity. The damped longer period oscillation eventually dominating the loop dynamics was then found to be in exact agreement with the presence of a resonantly damped quasi-mode. This global kink eigenoscillation is coupled to torsional Alfvén motions in a thin resonant layer due to the density variation. The strongly localized resistive dissipation quickly leads to the observed amplitude decay. By also computing the eigenmodes of the system, a one-to-one correspondence with the damped quasi-mode was established. This time-dependent analysis of loop oscillations convincingly showed the combined interpretative power of both eigenvalue and initial value approaches.

15.4 Concluding remarks

Computational MHD is essential for present-day modelling of fundamental plasma processes. We briefly discussed basic concepts of spatial and temporal discretization techniques, and gave various applications in computational linear MHD. Applications considered steady state as well as normal mode codes, together with time evolution codes for simulating linear wave dynamics. After discretization, all computational linear MHD applications for determining steady solutions to driven problems eventually lead to a large linear algebraic system. Similarly, determining the MHD spectrum of a particular equilibrium configuration mathematically boils down to the numerical solution of a generalized eigenvalue problem. We then rely on linear algebraic methods for their solution, where known properties of the coefficient matrix (symmetry or Hermitian property, sparseness pattern, etc.) will influence the optimal algorithm choice for solving the linear system or eigenvalue problem. Both direct and iterative methods are in use. A good insight into the construction of iterative methods for the solution of linear systems with a large number of unknowns is provided by H. A. van der Vorst in *Iterative Krylov Methods for Large Linear Systems* [626]. Linear algebra library routines are usually available on all computing platforms, and are highly optimized. As a result, numerical MHD simulations become more realistic as increasing computing power allows the consideration of higher Reynolds numbers (introducing smaller length scales and longer time scales), more complex geometries (curvature, stratification, X-points), and more 'physics' (background flow, thermal conduction, etc.). In Chapter 20, we continue the discussion of numerical methods, with particular emphasis on nonlinear MHD computations.

15.5 Literature and exercises

Notes on literature

Finite differences and finite elements

- A good introduction to the basic theory and the practical implementation of the finite element method is given by Strang and Fix in *An Analysis of the Finite Element Method* [574]. Some sections are dated but the book is well-written, easy reading and contains many ideas on the behaviour of finite elements which are interesting from both a mathematical and a practical point of view.
- A good introduction to the theory of finite difference and finite element methods is given by Larsson and Thomée in *Partial Differential Equations with Numerical Methods* [397], Chapter 2, Sections 4.1 and 5.1.
- O'Leary's *Scientific Computing with Case Studies* [473] is a practical guide to the numerical solution of linear and nonlinear differential equations, optimization problems and eigenvalue problems.

Exercises

[15.1] *Finite difference method*

We consider the simple problem

$$-[p(x)u'(x)]' + q(x)u(x) = f(x), \quad \text{for } x \in [0,1],$$

where the functions $p(x), q(x)$ and $f(x)$ are given and $u(0) = u(1) = 0$. We will assume that $0 < p(x) \le p_0$ and $q(x) \le 0$ for $x \in [0,1]$.

- First rewrite the derivative as $-p(x)u''(x) - p'(x)u'(x)$. Then, use the first order backward difference method (the equivalent of Eq. (15.15) of Section 15.1.2 for $u'(x)$ and the formula (15.17) for the second order derivative $u''(x)$. Note that $p'(x)$ will be computed analytically so that no approximation to it is needed.

- Choose mesh points $x_j = jh$, where $h = 1/(M-1)$ for some integer M, and solve for $u_j \approx u(x_j)$ for $j = 1, \ldots, M-2$ to obtain an equation for each unknown by substituting the finite difference approximations for u'' and u' in the differential equation and then evaluating the equation at $x = x_j$.
- Let $M = 6$ and write the four finite difference equations at $x = 0.2, 0.4, 0.6$ and 0.8.

Notice that the matrix you constructed has non-zeros on only three bands around the main diagonal; all other elements are zero. The full matrix requires $(M-2)^2$ storage locations, but, if we are careful, we can instead store all of the data in $\mathcal{O}(M)$ locations by agreeing to store only the non-zero elements, along with their row and column indices. This is a standard technique for storing sparse matrices, where most elements vanish.

[15.2] *Finite difference method: improvement*

In the previous exercise, there is a mismatch between our approximation to u'', which is second order in h, and our approximation to u', which is only first order. We can compute a better solution, for the same cost, by using a second order (central difference) approximation to u'. Revisit exercise [15.1] by making this change.

[15.3] *Finite element method: Galerkin approach*

For the finite element method, we keep the differential equation of Exercise [15.1].

- Apply the Galerkin approach and use integration by parts to obtain

$$\int_0^1 \left[p(x)u_h'(x)\,\phi_h'(x) + q(x)u_h(x)\,\phi_h(x)\right]\,dx = \int_0^1 f(x)\,\phi_h(x)\,dx\,,$$

for all functions $\phi_h \in S_h$, a subspace of H_0^1 (all functions that satisfy the boundary conditions and have a first derivative), with the integral of $(\phi_h'(x))^2$ on $[0,1]$ finite. Then choose for S_h the set of piecewise linear elements that are continuous and linear on each interval $[jh, (j+1)h]$, $j = 0, \ldots, M-2$, where $h = 1/(M-1)$. Use the set of hat functions ϕ_j, $j = 1.\ldots, M-2$ defined by Eq. (15.29) as basis of S_h. These functions are designed to satisfy $\phi_j(x_j) = 1$ and $\phi_j(x_k) = 0$ if $j \neq k$.

- Put the unknowns u_j in a vector $\mathbf{u}$ and write the resulting system of equations as $\mathbf{Au} = \mathbf{g}$, where the (j,k) entry in $\mathbf{A}$ is $a(\phi_j, \phi_k)$ and the jth entry in $\mathbf{g}$ is (f, ϕ_j).

[15.4] *Comparison of finite element methods versus finite difference methods*

Use your favourite software package or programming language to implement the three approaches from the above exercises. To compare them, define a set of functions for experimentation:

$$u_1(x) = x(1-x)\,e^x, \qquad u_2(x) = \begin{cases} u_1(x) \\ x(1-x)\,e^{2/3} \end{cases}, \qquad u_3(x) = \begin{cases} u_1(x) & (x \le 2/3) \\ x(1-x) & (x > 2/3) \end{cases},$$

$$p_1(x) = 1, \qquad p_2(x) = 1 + x^2, \qquad p_3(x) = \begin{cases} p_2(x) & (x \le 1/3) \\ (x - 1/3) + 10/9 & (x > 1/3) \end{cases},$$

$$q_1(x) = 0, \qquad q_2(x) = 2, \qquad q_3(x) = 2x\,.$$

The function $f(x)$ is the one obtained under the specified true solutions mentioned below. Use your four algorithms to solve seven problems:

- p_1 with $q_j (j = 1, 2, 3)$ and true solution u_1.
- $p_j (j = 2, 3)$ with q_1 and true solution u_1.
- p_1 with q_1 with true solution $u_j (j = 2, 3)$.

Compute three approximations for each algorithm and each problem, choose for the number of unknowns 9, 99 and 999. For each approximation, print $||\mathbf{u}_{\text{computed}} - \mathbf{u}_{\text{true}}||$ where $\mathbf{u}_{\text{true}}$ is the vector of true values at the $M-1$ mesh points. Discuss the results.

- How easy is it to programme each of the methods? Estimate how much work it is to form and solve the linear systems.
- For each problem, note the observed convergence rate r: if the error drops by a factor of 10^r when M is increased by a factor of 10, the observed convergence rate is r.

– Explain any deviations from the theoretical convergence rate: $r = 1$ and $r = 2$ for the two finite difference implementations, $r = 2$ for the (linear) finite element one.

[15.5] *Wave modes in unmagnetized, uniform, radiating gases*

Compute the dispersion equation of linear perturbations about a static (non-magnetized) homogeneous gas of constant density ρ_0, pressure p_0 and temperature T_0, which can lose internal energy by means of radiative losses.

– Assuming a radiative loss function $\mathcal{L}(\rho, T)$, which vanishes for the equilibrium conditions $\mathcal{L}(\rho_0, T_0) = 0$, obtain this dispersion equation by linearizing the set of (dimensionless) equations given by

$$\frac{\partial \rho}{\partial t} + \nabla \cdot (\rho \mathbf{v}) = 0, \qquad \rho\Big(\frac{\partial}{\partial t} + \mathbf{v} \cdot \nabla\Big)\mathbf{v} + \nabla p = 0,$$
$$\Big(\frac{\partial}{\partial t} + \mathbf{v} \cdot \nabla\Big)p + \gamma p \nabla \cdot \mathbf{v} + (\gamma - 1)\rho \mathcal{L}(\rho, T) = 0, \qquad p = \rho T.$$

– Assume for the linear perturbations $\rho_1, \mathbf{v}_1, p_1$ and T_1 the usual $\exp \mathrm{i}(\mathbf{k} \cdot \mathbf{r} - \omega t)$ dependence, with wave vector $\mathbf{k}$ and eigenfrequency ω. (Note: since the precise dependence of $\mathcal{L}$ on thermodynamic variables ρ and T is further unspecified, linearization will introduce partial derivatives $\mathcal{L}_\rho \equiv \partial \mathcal{L}/\partial \rho|_T$ and $\mathcal{L}_T \equiv \partial \mathcal{L}/\partial T|_\rho$.)

– Renaming $\lambda \equiv -\mathrm{i}\omega$, discuss the essential cubic dispersion equation in λ in terms of its physical implications: what changes from the case when radiative losses are absent?

– Assuming that the radiative terms are small and the cubic in λ has then one real root and a pair of complex conjugate roots, can you derive an instability criterion for the real root from the constant term in the cubic?

– Discuss the basic physical mechanism to form condensations in a uniform radiating gas in the absence of gravity through this ‘thermal instability’.

Part V

Toroidal Geometry

16

Static equilibrium of toroidal plasmas

16.1 Axi-symmetric equilibrium

16.1.1 Equilibrium in tokamaks

The aim of the theory of plasma equilibrium in any configuration is to *determine the global magnetic confinement topology* and the physical characteristics of the underlying basic equilibrium state. For most fusion applications, to first approximation, this state is assumed to be static, i.e. the background plasma velocity and the time derivative of the other variables vanish, $\mathbf{v} = 0$ and $\partial/\partial t\{\rho, p, \mathbf{B}\} = 0$. A superficial impression might be that this must correspond to the most boring example of plasma behaviour, viz. total absence of dynamics: a corresponding fluid dynamics problem hardly exists. Of course, the reason for our interest in this state is the prospect of obtaining clean, abundant and cheap energy from controlled thermonuclear fusion reactions. At present, the most promising candidate to reach this goal is the tokamak configuration, in which the assumption of static equilibrium is satisfied to a rather high degree of precision.

The MHD equations for static equilibrium are about the best satisfied plasma equations we know. If a plasma is sitting at rest, it is hard to imagine it satisfying any other conditions than

$$\mathbf{j} \times \mathbf{B} = \nabla p \qquad \textit{(pressure balance)}, \tag{16.1}$$

$$\mathbf{j} = \nabla \times \mathbf{B} \qquad \textit{(Ampère's law)}, \tag{16.2}$$

$$\nabla \cdot \mathbf{B} = 0 \qquad \textit{(basic law of magnetic flux)}. \tag{16.3}$$

In fact, if the equation of pressure balance is not satisfied, the plasma is immediately accelerated to huge velocities and there is no way to prevent it from being smashed into the wall and causing severe damage to the equipment.

In Section 1.2.3, we have schematically summarized the history of magnetic plasma confinement experiments aimed at the eventual construction of a thermonuclear reactor. After the declassification of the subject in 1958, some ten years of rather unsuccessful research of high-beta plasmas in z-pinch and θ-pinch configurations followed, exhibiting dramatic demonstration of lack of equilibrium, instabilities and end losses, all operating on the typical MHD time scale of microseconds. It was entirely unclear how time scales of these fusion experiments could ever be extended to seconds or minutes, let alone to steady-state operation. All this changed by the announcement of progress by Soviet scientists with an entirely different confinement scheme, called *tokamak*, at the IAEA Novosibirsk conference of 1968 [25]. (See the review paper on tokamaks by Artsimovich [26] and Section 5.2 of Braams and Stott [86] on the history of this period.) Soon after this, tokamaks were constructed in many countries, and the next thirty years witnessed steady increase of the triple product $n\tau_{\rm E}\tilde{T}$, of density, energy confinement time and temperature in these devices, with a factor of about 10^6 nearing the required value for ignition of $5 \times 10^{21}\ {\rm m}^{-3}\,{\rm s}\,{\rm keV}$ by the end of the twentieth century (see Fig. 1.1.1 of Wesson [647]). Based on this successful upgrading, the International Tokamak Experimental Reactor (ITER) is presently being constructed in Cadarache, France.

Whereas the simple schemes of θ- and z-pinch easily produced the temperatures needed for thermonuclear ignition by shock heating, they fell short by a factor of at least a million with respect to the required confinement times. Crudely speaking, the tokamak configuration cures the main problems of the z-pinch (its instability due to the curvature of the magnetic field B_θ) and of the θ-pinch (its end losses) by combining them into a single configuration (see Fig. 1.4). With respect to the θ-pinch end losses, those are simply eliminated by closing the plasma column onto itself by means of a 'toroidal' confinement chamber and the current-driven instabilities of the z-pinch are eliminated by keeping the toroidal plasma current I_φ below the Kruskal–Shafranov limit (2.163) so that the resulting 'poloidal' magnetic field $B_{\rm p}$ (corresponding to B_θ in cylinder geometry) remains much smaller than the dominant toroidal magnetic field B_φ (corresponding to B_z). Hence, the magnetic field structure is now helical and the confinement geometry is toroidal, rather than cylindrical. Most important, the plasma β ($\equiv 2p/B^2$) is significantly decreased and the induction of currents is slow, so that shock heating has to be abandoned and replaced by other heating methods (Ohmic, neutral beams, ion and electron cyclotron resonance heating and, eventually, heating by the fusion produced α particles themselves.). Although the tokamak configuration thus eliminates the mentioned insurmountable problems of θ-pinches and z-pinches, quite a variety of lesser equilibrium and stability problems have to be addressed still. These will be the subject of the present and the next chapter.

The very first remark to be made about equilibrium in tokamaks described by the misleadingly simple looking equations (16.1)–(16.3), is that these equations are actually *nonlinear* partial differential equations, as follows immediately by substituting Ampère's law (16.2) into the pressure balance equation (16.1). We have encountered the central importance of the basic law (16.3) for magnetic flux many times. Hence, the solution of the combined equations

$$(\nabla \times \mathbf{B}) \times \mathbf{B} = \nabla p\,, \qquad \text{or} \qquad -\nabla(p + \tfrac{1}{2}B^2) + \mathbf{B}\cdot\nabla\mathbf{B} = 0\,, \tag{16.4}$$

$$\text{and} \qquad \nabla\cdot\mathbf{B} = 0\,, \tag{16.5}$$

with the single boundary condition

$$\mathbf{n}\cdot\mathbf{B} = 0 \qquad \text{(at the plasma boundary)}\,, \tag{16.6}$$

for plasmas without circular cylinder symmetry (2D: tokamaks, 3D: stellarators) constitutes a nonlinear problem with many intriguing features that turns out not to be a boring problem at all!

From the original equations (16.1) and (16.2), it follows that the magnetic field and the current density are orthogonal to the pressure gradient:

$$\mathbf{B}\cdot\nabla p = 0\,, \qquad \mathbf{j}\cdot\nabla p = 0\,. \tag{16.7}$$

Hence, the *magnetic surfaces* spanned by the magnetic field lines and the current density lines are also constant pressure surfaces. If none of the quantities ∇p, $\mathbf{B}$ and $\mathbf{j}$ vanishes anywhere in the plasma volume, these surfaces form an infinite continuous sequence of nested magnetic toroids around a single closed curve, called the *magnetic axis* [380]. Consequently, the magnetic surfaces may be labelled by a single variable Ψ, so that $p = p(\Psi)$. We will soon see that the most effective choice for that variable is the poloidal magnetic flux.

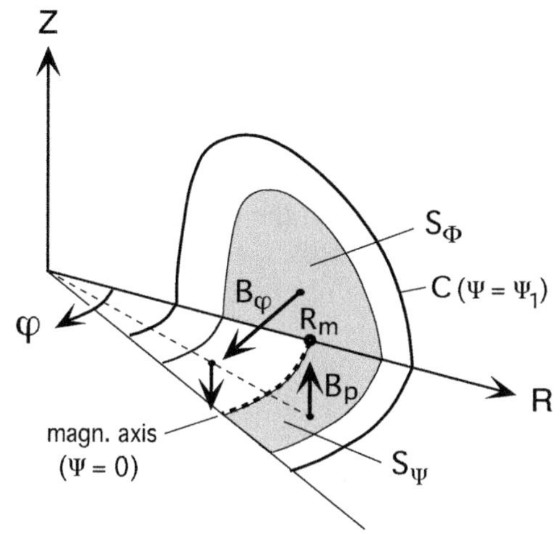

Fig. 16.1 Toroidal and poloidal magnetic field components B_φ and $B_{\rm p}$ in a tokamak and cross-sections S_Φ and S_Ψ for the corresponding magnetic fluxes Φ and Ψ within a magnetic surface lying between the magnetic axis and the plasma boundary, indicated by its cross-section (the curve C) with the poloidal plane $\varphi = 0$. Of course, the surface S_Φ also includes the reflected part $Z \leq 0$ of the shaded area and it can be taken at any angle φ. Similarly, the surface S_Ψ covers the full range $0 \leq \varphi \leq 2\pi$ and it can be taken between the magnetic axis and any circle $R = \mathrm{const}$, $Z = \mathrm{const}$ lying in the magnetic surface.

For tokamaks, the equilibria still have the additional property of *axi-symmetry* ($\partial/\partial\varphi = 0$), which is of great help in explicit calculations. Figure 16.1 shows a schematic representation of the magnetic configuration. Because of the axi-symmetry of the configuration, it is expedient to exploit special cylindrical coordinates R, Z, φ, where R is the distance from the symmetry axis, Z is the vertical coordinate and φ is the (ignorable) toroidal angle (see Appendix A.2.4). Although the toroidal magnetic field component is much larger than the poloidal component in tokamaks, the latter is usually considered as the primary quantity since it effectively describes the magnetic geometry of the tokamak. This is done by defining the toroidal and poloidal magnetic fluxes through the respective surfaces S_Φ and S_Ψ (indicated by the shaded areas in Fig. 16.1):

$$\Phi \equiv \frac{1}{2\pi}\int B_\varphi \, dS_\Phi \,, \qquad \Psi \equiv \frac{1}{2\pi}\int B_{\rm p} \, dS_\Psi \,, \tag{16.8}$$

where the normalization factor $1/(2\pi)$ is introduced to simplify the relation between the poloidal flux and the poloidal magnetic field; see Eq. (16.84) below. In Fig. 16.1, the poloidal flux surface S_Ψ is chosen in the horizontal plane $Z = 0$, containing the magnetic axis (the dashed circle $R = R_{\rm m}$), so that the poloidal field is pointing up for $R > R_{\rm m}$ and down for $R < R_{\rm m}$. From the flux tube concept (see Section 4.2.1), it is obvious though that the outer boundary of S_Ψ may be any horizontal circle lying in the magnetic surface. Hence, the infinitely many nested magnetic flux surfaces are most effectively labelled with the value of Ψ, which thus plays the role of a 'radial' coordinate, running from $\Psi = 0$ at the magnetic axis to $\Psi = \Psi_1$ at the plasma boundary, where Ψ_1 is the total poloidal flux confined within the plasma. The toroidal magnetic flux is then considered as a function of the poloidal flux, $\Phi = \Phi(\Psi)$.

Once this is established, the helicity of the magnetic field lines is expressed as the derivative of the toroidal flux with respect to the poloidal flux. This quantity is known as the *safety factor*, or *inverse rotational transform*, of the field lines:

$$q(\Psi) \equiv \frac{d\Phi}{d\Psi} \,. \tag{16.9}$$

This function, which runs from $q_0 \equiv q(0)$ on the magnetic axis to $q_1 \equiv q(\Psi_1)$ on the plasma boundary, is the most important physical variable in the stability analysis of tokamaks. An alternative expression for the safety factor may be obtained from Eq. (16.9) by considering the poloidal flux $d\Psi = RB_{\rm p}\,dx$ through an infinitesimal annulus between two flux surfaces separated by a local distance dx and the corresponding toroidal flux $d\Phi = (2\pi)^{-1}\oint(B_\varphi dx)\,d\ell$, where $\oint d\ell$ is the line integral along the closed boundary curve of the toroidal flux surface S_Φ (see Wesson [647], Section 3.4). This yields

$$q(\Psi) = \frac{1}{2\pi}\oint_\Psi \frac{B_\varphi}{RB_{\rm p}}\,d\ell \,, \tag{16.10}$$

which now expresses the safety factor as some average of the tangent $B_\varphi/B_{\rm p}$ of the field line.

Before we discuss the implications of the details of the Ψ-dependence of the safety factor, it is useful to stress the importance of the global properties of the equilibrium, which are mainly determined by two physical parameters and one geometrical parameter. Concerning the latter, the geometry is mainly controlled by *the inverse aspect ratio of the torus,*

$$\epsilon \equiv a/R_0 \,, \tag{16.11}$$

where a is the half width of the plasma column and R_0 is the distance of the centre of that column to the symmetry axis. For simplicity, we will usually (except in Sections 16.1.4 and 16.2.3) assume that the plasma boundary coincides with a 'wall' surrounding it so that a and R_0 are just geometrical properties of that wall. Moreover, we will assume up–down symmetry of the equilibrium with respect to the horizontal plane $Z = 0$. None of these assumptions is essential for the argument, they just simplify the analysis. (In particular, divertors for impurity control usually destroy up–down symmetry.) The physical parameters are the value of the *safety factor at the edge of the plasma,*

$$q_1 \equiv q(\Psi_1) \,, \tag{16.12}$$

and the value of the *average kinetic pressure compared to the magnetic pressure,*

$$\beta \equiv \frac{2\langle p\rangle}{\langle B_\varphi^2\rangle} \,, \qquad \langle f\rangle \equiv \int f\,dV \,. \tag{16.13}$$

We will return to these determining parameters in Section 16.1.4 when we discuss global confinement in tokamaks, i.e. equilibrium 'in the large'. Of course, all the local dependence of the plasma variables has to be determined as well. This can only be done if the magnetic geometry is known in detail, i.e. if the nonlinear partial differential equations (16.1)–(16.3) have been solved. This will concern us in Section 16.2, but we will first define the relevant properties of the magnetic field lines and coordinates describing them.

16.1.2 Magnetic field geometry

In order to appreciate the meaning of the average in the definition (16.10) of the safety factor, we need to consider the magnetic field lines themselves. Introducing the infinitesimal tangent vector $d\mathbf{s}_{\rm fl}$ to them, the equation for the field lines may be written as

$$\mathbf{B} \times d\mathbf{s}_{\rm fl} = 0\,. \tag{16.14}$$

To work out the consequences of this equation, it is expedient to exploit coordinates based on the poloidal flux Ψ and the toroidal angle φ, rather than the cylindrical coordinates (R, Z, φ), which do not 'see' the geometry of the magnetic field. The third coordinate must be a poloidal angle indicating the position on the flux surface for given φ. That coordinate, θ, could be a polar angle but, since any function of θ that increases monotonically from 0 to 2π is again an acceptable poloidal coordinate, it is convenient to exploit the arbitrariness to construct coordinates with simplifying properties. Two such coordinate systems are frequently exploited, viz. *orthogonal flux coordinates*, where the gradient of the poloidal angle is chosen orthogonal to $\nabla\Psi$ and $\nabla\varphi$, and *straight-field-line coordinates*, where the poloidal angle is chosen such that the field lines are straightened out in the representation of the tangential plane. The details of these coordinates are put in small print since they are a detour of the present exposition, but the reader who is not familiar with these constructions is advised *not* to skip this part since it is central to understanding of the geometry of the equilibrium needed in this chapter and the next.

▷ **Flux coordinates** The construction of these coordinates involves a task that recurs in equilibrium calculations, viz. the determination of the connection between the ordinary geometrical coordinates and the coordinates based on the flux:

$$\big[(X\,,Y\,,Z) \;\Rightarrow\big] \quad (R\,,Z\,,\varphi) \quad \Rightarrow \quad (\Psi\,,\theta\,,\varphi)\,. \tag{16.15}$$

This is actually an entirely non-trivial problem since it involves, first, the solution of the Grad–Shafranov equation (see Section 16.2), providing $\Psi = \Psi(R,Z)$, and, next, the construction of $\theta = \theta(R,Z)$ from the condition that describes the property desired of the poloidal angle. For now, we simply assume that these problems have been solved (we will return to them in Sections 16.2 and 16.3), so that these two functions are known.

For *orthogonal flux coordinates* (Ψ, χ, φ), for which $\nabla\chi\cdot\nabla\Psi = 0$, we can then construct the line element,

$$(ds)^2 = h_1^2\,(d\Psi)^2 + h_2^2\,(d\chi)^2 + h_3^2\,(d\varphi)^2\,, \tag{16.16}$$

the volume element and the Jacobian,

$$dV = J_{\rm o} d\Psi d\chi d\varphi\,, \qquad J_{\rm o} \equiv (\nabla\Psi\times\nabla\chi\cdot\nabla\varphi)^{-1} = h_1 h_2 h_3\,, \tag{16.17}$$

the gradient operator,

$$\nabla = \mathbf{e}_\Psi \frac{1}{h_1}\frac{\partial}{\partial\Psi} + \mathbf{e}_\chi \frac{1}{h_2}\frac{\partial}{\partial\chi} + \mathbf{e}_\varphi \frac{1}{h_3}\frac{\partial}{\partial\varphi}\,, \tag{16.18}$$

and all other vector expressions (see Appendix A.2.7), which involve the three scale factors

$$h_1 \equiv \frac{1}{|\nabla\Psi|} = \frac{1}{RB_{\rm p}}\,, \quad h_3 \equiv \frac{1}{|\nabla\varphi|} = R \quad \Rightarrow \quad h_2 \equiv \frac{1}{|\nabla\chi|} = \frac{J_0}{h_1 h_3} = J_{\rm o}B_{\rm p}\,. \tag{16.19}$$

This provides all expressions needed to account for the geometry of magnetic field lines in orthogonal flux coordinates (assuming the coordinate connections (16.15) are known). These coordinates are frequently used in analytical calculations, but not in numerical ones since they provide poor angular resolution in plasmas with elongated cross-sections.

For *straight-field-line coordinates* $(\Psi, \vartheta, \varphi)$, we determine the function $\vartheta(\theta)$ by straightening out the field lines in the tangential $(\vartheta\,,\varphi)$ plane (see Fig. 16.2) for each flux surface $\Psi = \text{const}$. This coordinate system has

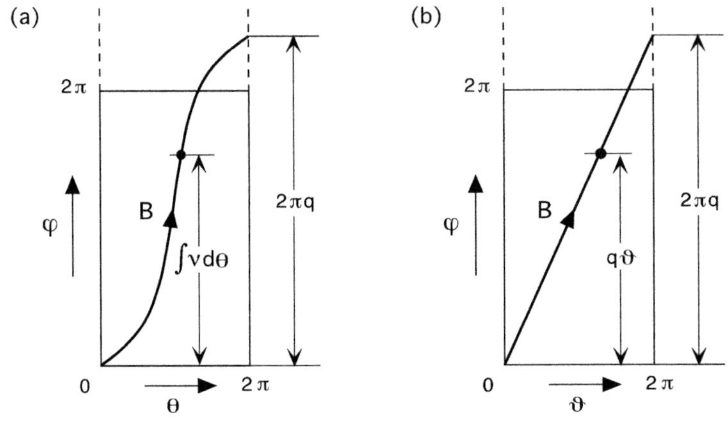

Fig. 16.2 Field line and safety factor q in plane tangential to a magnetic surface for (a) arbitrary flux coordinates (Ψ, θ, φ); (b) straight-field-line coordinates $(\Psi, \vartheta, \varphi)$.

the obvious advantage of a simple representation of the field lines, which is extremely important in stability calculations. However, it is non-orthogonal so that its use requires the knowledge of the four elements of the metric tensor (see Appendix A.3.2):

$$g_{11} \equiv \left|\frac{\partial \mathbf{r}}{\partial \Psi}\right|^2, \quad g_{12} \equiv \frac{\partial \mathbf{r}}{\partial \Psi} \cdot \frac{\partial \mathbf{r}}{\partial \vartheta}, \quad g_{22} \equiv \left|\frac{\partial \mathbf{r}}{\partial \vartheta}\right|^2, \quad g_{33} \equiv \left|\frac{\partial \mathbf{r}}{\partial \varphi}\right|^2 = R^2 . \tag{16.20}$$

We just provide the expressions for the volume element and the Jacobian,

$$dV = \mathcal{J} d\Psi d\vartheta d\varphi , \qquad \mathcal{J} \equiv (\nabla\Psi \times \nabla\vartheta \cdot \nabla\varphi)^{-1} = \sqrt{[g_{11}g_{22} - (g_{12})^2]\, g_{33}} , \tag{16.21}$$

and leave explicit determination of the metric elements for later. ◁

In the orthogonal flux coordinate system (Ψ, χ, φ), the magnetic field and the infinitesimal tangent vector are expressed as

$$\mathbf{B} = (0, B_{\rm p}, B_\varphi) , \qquad d\mathbf{s}_{\rm fl} = (0, J_{\rm o} B_{\rm p} d\chi, R d\varphi) , \tag{16.22}$$

so that Eq. (16.14) yields the local direction ν of the field lines in the representation of the plane tangential to the magnetic surfaces:

$$B_{\rm p}(R d\varphi - J_{\rm o} B_\varphi d\chi) = 0 \qquad \Rightarrow \quad \nu \equiv \left.\frac{d\varphi}{d\chi}\right|_{\rm fl} = \frac{J_{\rm o} B_\varphi}{R} . \tag{16.23}$$

Hence, since $d\ell = J_{\rm o} B_{\rm p} d\chi$, the safety factor q just represents the *progression of the field line over the toroidal angle after one full poloidal revolution:*

$$q(\Psi) = \frac{1}{2\pi} \oint_\Psi \frac{B_\varphi}{R B_{\rm p}}\, d\ell = \frac{1}{2\pi} \oint_\Psi \nu(\Psi, \chi)\, d\chi . \tag{16.24}$$

This is illustrated in Fig. 16.2(a) for arbitrary flux coordinates (Ψ, θ, φ) on a magnetic surface with $q(\Psi) > 1$.

Actually, inside a magnetic surface, orthogonality of $\nabla\Psi$ and $\nabla\chi$ is not of much help. It is much more expedient to exploit the arbitrariness of the poloidal coordinate to construct a coordinate ϑ for which the field lines are straight, as shown in Fig. 16.2(b). In the definition of q, the line element along the poloidal circumference of a flux surface differs for the different coordinate systems since their Jacobians are different. In particular,

$$d\ell = J B_{\rm p}\, d\theta = J_{\rm o} B_{\rm p}\, d\chi = \mathcal{J} B_{\rm p}\, d\vartheta \tag{16.25}$$

for arbitrary flux coordinates (Ψ, θ, φ), with Jacobian J; orthogonal flux coordinates (Ψ, χ, φ), with Jacobian $J_{\rm o}$; and straight-field-line coordinates, with Jacobian $\mathcal{J}$, respectively. Obviously, the last expression is the best choice since the local direction of the field lines then coincides with the global direction, as expressed by the relationship between the safety factor and the Jacobian that does not depend on the position on the flux surface:

$$q(\Psi) = \frac{\mathcal{J} B_\varphi}{R} \qquad \textit{(straight-field-line coordinates)}. \tag{16.26}$$

Running ahead of our presentation (see Chapter 17), these coordinates are most appropriate to describe local stability since that basically involves Alfvén waves travelling along the field lines in a curved magnetic geometry.

An important concept in this context is that of *rational magnetic field lines*, situated on *a rational magnetic surface*. The latter is a surface where the field lines close upon themselves after M revolutions the short way (poloidally) and N revolutions the long way (toroidally) around the torus (see Fig. 2.12, for $M = 2$ and $N = 3$), so that the safety factor is a rational number there:

$$q_{\rm rat} = \frac{N}{M} \quad \left(= \frac{\text{number of toroidal revolutions}}{\text{number of poloidal revolutions}} \right). \tag{16.27}$$

If q is irrational, the field line does not close onto itself but covers the magnetic surface ergodically. The importance of rational field lines can be seen from the expression of magnetic perturbations on a rational magnetic surface. Expressing the doubly periodic perturbations in the form of an expansion in Fourier components of the plasma displacement normal to the magnetic surfaces,

$$\xi(\Psi, \vartheta, \varphi) = \sum_m \sum_n \xi_{mn}(\Psi)\, {\rm e}^{{\rm i}(m\vartheta + n\varphi)}, \tag{16.28}$$

where m is the poloidal mode number and n is the toroidal mode number, the main field line bending perturbation is roughly (neglecting curvature contributions) proportional to the parallel gradient operator acting on ξ:

$$\mathbf{B} \cdot \nabla \xi \sim \mathbf{k}_0 \cdot \mathbf{B}\, \xi \sim (m + nq)\, \xi\,. \tag{16.29}$$

Since field line bending is associated with a large increase of the potential energy, this term must vanish, or be small, for almost all instabilities that occur in tokamaks (not for astrophysical plasmas, recall the discussion of Section 12.1.3). The expression vanishes when the 'wave vector' $\mathbf{k}_0$ is perpendicular to the magnetic field, i.e. when the wave fronts of the perturbation on the magnetic surface are parallel to the magnetic field lines so that they are minimally bent. According to Eq. (16.29), this happens when the ratio of mode numbers is rational, which only happens on a rational surface where *the field lines are resonant with the perturbation*:

$$q_{\rm rat} = -\frac{m}{n} \quad \left(= -\frac{\text{poloidal mode number}}{\text{toroidal mode number}} \right). \tag{16.30}$$

(Note that the adjectives 'poloidal' and 'toroidal' have switched position here.) Consequently, most local MHD stability analyses (Suydam's criterion (9.138), the Mercier criterion and ballooning mode theory, see Chapter 17) centres about the resonant surfaces and field lines. (Therefore, if such resonances are forbidden because of different longitudinal boundary conditions, like line-tying of solar coronal flux tubes at the photospheric boundary, stability theory is completely changed and configurations are encountered which are much more stable; see Goedbloed and Halberstadt [235].)

16.1.3 Cylindrical limits

With the introduction of straight-field-line coordinates, the magnetic geometry has been reduced to a representation resembling that of a cylindrical plasma with circular cross-section. This is helpful because much intuition on equilibrium and stability comes from circular cylinder theory. Hence, let us consider the limit of a slender torus to a 'periodic' straight cylinder, the so-called *straight tokamak* approximation, already introduced in Sections 9.1.1 and 9.4.4. We will use this limit to discuss the connection of the rather subtle effects of the safety factor and flux function distributions to the robust ones of the total current flowing in the plasma and global pressure balance requirements.

Recall from Section 9.1.1 that in a straight plasma cylinder with a circular cross-section, because of the symmetries $\partial/\partial\theta = 0$ and $\partial/\partial z = 0$, the solution of the equilibrium equations (16.1)–(16.3) is a function of the radius r alone. Moreover, $B_r = 0$ and $j_r = 0$, whereas the density profile $\rho(r)$ is completely arbitrary in static equilibria. The remaining equilibrium functions $p(r), B_\theta(r), B_z(r)$ have to satisfy just one differential equation, viz.

$$\frac{d}{dr}\Big(p + \tfrac{1}{2}B^2\Big) = -\frac{B_\theta^2}{r}\,, \tag{16.31}$$

so that two of those three functions can be chosen arbitrarily. The components of the current density are then determined by Ampère's law,

$$j_\theta = -\frac{dB_z}{dr}\,, \qquad j_z = \frac{1}{r}\frac{d}{dr}(rB_\theta)\,, \tag{16.32}$$

which completes the description of cylindrical equilibria. In conclusion, in addition to $\rho(r)$, two of the three functions $p(r), B_\theta(r), B_z(r)$ can be freely chosen. This corresponds to the experimental freedom to create different magnetic confinement configurations, like those of a θ-pinch or a z-pinch (Figs. 16.3(a) and (b)). The same freedom is present in toroidal equilibria, where it needs to be represented very carefully in order to enable comparison of experimental data with theoretical assumptions on the equilibrium (see Section 16.3).

As discussed in Section 16.1.1 on the history of the different approaches to plasma confinement for fusion, the tokamak configuration may be considered as the combination of a θ-pinch with a much smaller pressure (and hence a smaller dip in the longitudinal field) and a z-pinch with a smaller plasma current (Fig.16.3(c)). This gives rise to a finite safety factor with a radially increasing profile, typically (but not necessarily) increasing monotonically from $q_0 \sim 0.8$ to $q_1 \sim 2.5$.

To define the safety factor q in straight circular cylinder geometry, one first needs to consider the original toroidal problem for a slender torus with *major radius* R_0 and *minor radius* $r = a$ (Fig. 16.4), so that the *inverse aspect ratio*, defined in Eq. (16.11), is small:

$$\epsilon \equiv a/R_0 \ll 1\,. \tag{16.33}$$

In the limit $\epsilon \to 0$, this torus is represented as a 'periodic cylinder' of length L_z and radius a, where

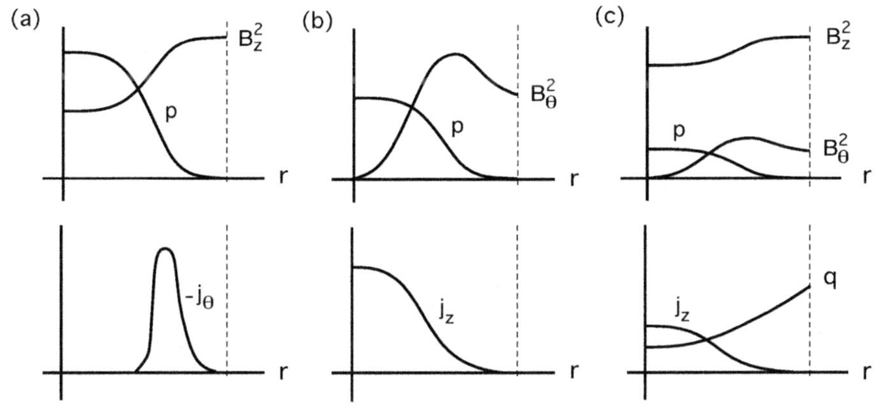

Fig. 16.3 Schematic cylindrical equilibrium profiles for (a) θ-pinch, (b) z-pinch, (c) 'straight tokamak' with periodicity over $L_z \equiv 2\pi R_0$.

the cylindrical coordinate z is related to the toroidal angle φ and the periodicity length L_z according to

$$z = [\varphi/(2\pi)]\, L_z \,, \qquad L_z \equiv 2\pi R_0 \,. \tag{16.34}$$

The nested magnetic surfaces become nested periodic cylinders of radius $r \leq a$. On each of those cylinders, the safety factor is easily visualized by unrolling that cylinder, which results in the straight-field-line representation of Fig. 16.2(b) and Fig. 2.12. In the limit $\epsilon \to 0$, the two magnetic fluxes defined in Eq. (16.8) become

$$\Phi_{\rm cyl} = \int_0^r B_z\, r\, dr \,, \qquad \Psi_{\rm cyl} = R_0 \int_0^r B_\theta\, dr \,, \tag{16.35}$$

so that any of the definitions (16.9), (16.10) or (16.24) of the safety factor yields

$$q_{\rm cyl}(r) = \frac{d\Phi}{d\Psi} = \frac{1}{2\pi} \oint \frac{d\varphi}{d\theta}\bigg|_{\rm fl} d\theta = \frac{r B_z}{R_0 B_\theta} \,. \tag{16.36}$$

The value of the safety factor at the edge of the plasma, defined in Eq. (16.12) is then a *direct measure for the total toroidal current flowing in the plasma:*

$$q_{\rm cyl}(a) = \frac{a B_z(a)}{R_0 B_\theta(a)} = \frac{2\pi a^2 B_z(a)}{R_0 I_z} \equiv (2\pi\epsilon a B_0)/I_z \,, \tag{16.37}$$

where the constant value $B_0 \equiv B_z(a)$ in the last expression anticipates the dimensional scaling that will be made in Section 16.3.1.

The importance of extracting the magnitude of the safety factor at the plasma boundary, $q_{\rm cyl}(a)$, from the detailed radial distribution $q_{\rm cyl}(r)$ is to stress the very different physical consequences of the toroidal current and the field line geometry. In particular, as already noted in Section 2.4.3, when this difference is not appreciated, it leads to confusion on the cause of the $m = 1$ external kink mode instability. In a periodic cylinder with circular cross-section, the condition for stability of the external kink mode is the celebrated *Kruskal–Shafranov limit:*

$$q_{\rm cyl}(a) > 1 \,, \quad \text{or} \quad I_z < 2\pi\epsilon a B_0 \,, \tag{16.38}$$

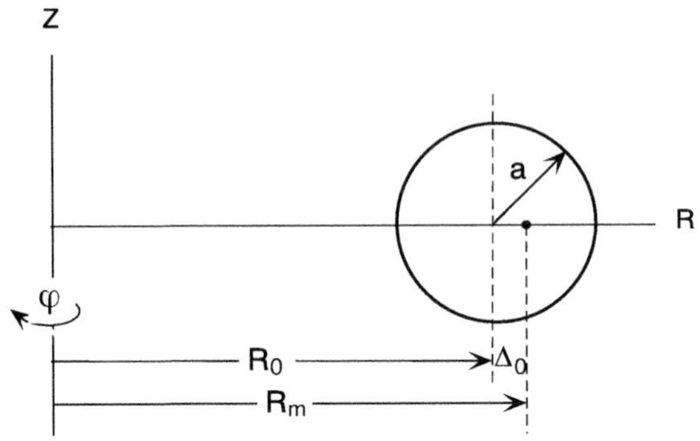

Fig. 16.4 Large aspect ratio torus with Shafranov shift Δ_0 of the magnetic axis.

already encountered in Eqs. (2.163) and (9.90). This condition just depends on the value of $q_{\rm cyl}(a)$, not on the details of $q_{\rm cyl}(r)$, i.e. the current distribution, as clearly illustrated in Wesson's stability diagram [646], reproduced in Fig. 9.15. If this condition is violated, the result is just as bad as lack of equilibrium: the plasma is smashed onto the wall on a time scale of microseconds. Hence, the Kruskal–Shafranov limit is an essential limit on the parameter regime of tokamaks (the very reason of the necessity to decrease the toroidal current of a z-pinch to that of a tokamak, illustrated in Figs. 16.3(b) and(c)). The confusion arises because Eq. (16.38) appears to suggest, as frequently stated in the literature, that external kink mode instability occurs because the field lines close onto themselves after one revolution the short way and one revolution the long way around the torus. However, this is just an unfortunate consequence of degeneracy of the circular cross-section cylinder, which disappears for genuine toroidal equilibria with non-circular cross-section. (To avoid further confusion: field line topology does play a dominant role in the stability of *internal* $m = 1$ kink modes, requiring $q_{\rm cyl}(0) > 1$. However, this is a much softer condition, as demonstrated by the fact that tokamaks usually operate in a regime where this condition is violated.)

The other global parameter describing the overall equilibrium of tokamaks is the *ratio of the average plasma pressure confined to the average pressure of the confining toroidal magnetic field*, defined in Eq. (16.13):

$$\beta \equiv \frac{2\int p\,dV}{\int B_\varphi^2\,dV} \approx \frac{2\int p\,dS}{\int B_z^2\,dS} \approx \frac{2}{B_0^2}\,\langle p\rangle\,, \qquad \langle p\rangle \equiv \frac{\int p\,dS}{\int dS}\,. \tag{16.39}$$

Here, the volume integral is over the total plasma volume and the surface integral is over the poloidal cross-section of it. The last two approximations come from the straight cylinder approximation, whereas the first one is due to the general orders of magnitude in a tokamak:

$$p \sim B_{\rm p}^2\ (\approx B_\theta^2) \ll B_\varphi^2\ (\approx B_z^2) \quad \Rightarrow \quad \beta \sim \epsilon^2 \ll 1\,, \tag{16.40}$$

as illustrated in Fig. 16.3(c). The importance of β is that it is a measure of the thermonuclear power obtained for a given magnetic field strength (Wesson [647], Section 3.5; Freidberg [185], Section 5.5.6). Because of the estimate (16.40), the total pressure is dominated by the contribution of the magnetic field:

$$P = p + \tfrac{1}{2}B^2 \approx \tfrac{1}{2}B^2\,. \tag{16.41}$$

One gets an impression of the magnitude of this pressure by just inserting typical values of tokamaks, e.g. for $B = 5\,\mathrm{T}$, we already found in Section 12.1.1 that

$$P \approx 10^7\,\mathrm{N/\,m^2} = 1000\,\text{metric tons} = 100\,\mathrm{atm}\,! \tag{16.42}$$

Clearly, magnetic pressures are huge and need to be balanced carefully. It is to be noted though that the dominant contribution of the above pressure is exerted on the external coils of the tokamak which, therefore, need a very strong supporting structure. The internal pressures, even though smaller by an order of magnitude, are not negligible either (e.g., for $\beta = 5\%$ the plasma pressure $p \approx 5\,\mathrm{atm}$). Moreover, since β is a figure of merit for future fusion reactors, there is an urgent need to try to increase it. This implies that some part of the huge magnetic pressure estimated above should ultimately be shifted towards the plasma interior, that would then resemble more the θ-pinch configuration of Fig. 16.3(a). However, we will see in Section 16.1.4 that, within the tokamak confinement scheme, the best one can do eventually is to obtain values of $\beta \sim \epsilon$.

This upper estimate of the order of magnitude of β implies that the toroidal geometry is essential for the proper description of overall pressure balance, so that the cylindrical approximation fails. This is most clearly illustrated by one of the effects of toroidal pressure balance, viz. the outward shift of the magnetic surfaces relative to those of a circular cross-section cylinder: the *Shafranov shift* (Fig. 16.4). In the next section, we will see that the order of magnitude of this shift depends on the order of magnitude of β, as expressed by the following two expansion schemes:

$$q_1 \sim 1\,, \quad \beta \sim \epsilon^2 \ll 1 \quad \Rightarrow \quad \Delta_0 \sim \epsilon \qquad \textit{(low } \beta \textit{ tokamak)}\,, \tag{16.43}$$

$$q_1 \sim 1\,, \quad \beta \sim \epsilon \ll 1 \quad \Rightarrow \quad \Delta_0 \sim 1 \qquad \textit{(high } \beta \textit{ tokamak)}\,. \tag{16.44}$$

In the first expansion scheme, the zeroth order is the straight circular cylinder and toroidal effects enter as subsequent higher order corrections. In the second one, the cylindrical approximation fails since the equilibrium is essentially two-dimensional. Hence, we now have to turn to a proper toroidal description of tokamak equilibrium.

To summarize: except for the freedom of choice of the equilibrium profiles, two other properties of toroidal configurations can be anticipated in the context of straight cylinder theory. These have to do with the total toroidal current flowing in the plasma, i.e. the global magnetic field line geometry expressed by the edge safety factor, and with some aspects of the bulk forces associated with the magnetic pressure. However, a proper description of the magnetic pressure and of the kinetic pressure effects in the plasma requires a genuine toroidal theory.

16.1.4 Global confinement and parameters

Why can one not obtain the desirable high values of β by just bending the cylindrical θ-pinch column of Fig. 16.3(a) into a torus, i.e. by closing the configuration onto itself? To answer that question, we need to consider the complete equilibrium problem, both of *internal pressure balance* inside the plasma and of the position control of the plasma column as a whole by means of magnetic fields produced by currents in *external coils* (Shafranov [549], p. 124; Miyamoto [442](a), Section 7.7, and [442](b), Section 6.3; Wesson [647], Section 3.1; Freidberg [185], Section 11.7). In a tokamak, the large magnetic field B_z of the θ-pinch becomes the main toroidal magnetic field component B_φ. That component is primarily produced by poloidal currents $I_{\mathrm{p}}^{\mathrm{ex}}$ in the external toroidal field coils, whereas the internal poloidal plasma current density $\mathbf{j}_p$ produces a relatively small deviation from that externally produced 'vacuum magnetic field' distribution. Neglecting the poloidal plasma currents for the time being, the toroidal magnetic field is obtained by integrating Ampère's law (16.2) along a circle of radius R in the mid-plane ($Z = 0$), enclosing a surface S_R, and applying Gauss' law (A.14):

$$I_p^{ex} \approx \int \mathbf{j} \cdot \mathbf{n} \, dS_R = \oint \mathbf{B} \cdot d\boldsymbol{l}_R = 2\pi R B_\varphi \quad \Rightarrow B_\varphi \approx R_0 B_0 / R \,. \qquad (16.45)$$

Here, B_0 is the value of the external toroidal magnetic field on the plasma boundary at $R = R_0$, see Fig. 16.5. (For the scaling to be made later, in Section 16.3.1, it is important to note that this parameter is exact, independent of the approximation made.) The $1/R$ dependence of the toroidal magnetic field implies that the plasma column cannot be in equilibrium with this field since it produces a magnetic pressure that is much larger on the inside, the high field side ($R = R_0 - a$), than on the outside, the low field side ($R = R_0 + a$). The result is an outward force in the R-direction, which is larger than the inward force due to magnetic tension, as we will see below. To ensure equilibrium, counter measures have to be taken.

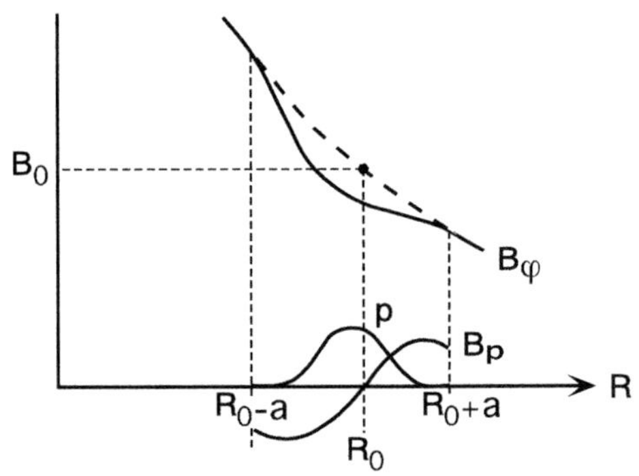

Fig. 16.5 Schematic tokamak equilibrium at the mid plane ($Z = 0$).

The counter measures taken are, first, induction of a (secondary) *toroidal plasma current* I_φ by means of coupling, due to the change in time of the poloidal magnetic flux through the central hole of the torus, to the toroidal current I_φ^{ex} in a set of (primary) windings on the outer legs of a transformer (see Fig. 1.5). This contributes the z-pinch part (Fig. 16.3(b)) to the tokamak confinement scheme, with the advantage of enhanced flexibility due to the additional parameter I_φ^{ex}. An obvious disadvantage is that tokamak operation now becomes limited to the time scale of resistive decay of the plasma current. This is the main reason for continued interest in the stellarator approach to plasma confinement, where such currents are not needed. (Eventual steady-state operation of tokamaks has become feasible, though, through the kinetic effects of current drive by radiofrequency heating and the possible production of a toroidal bootstrap current by pressure gradients; see the review papers by Fisch [176] and Boozer [82].)

The induction of the toroidal plasma current appears to have an adverse effect on the equilibrium, viz. the production of an additional outward force, called the *hoop force*. This follows from the basic fact of electrodynamics that a current-carrying ring tends to increase its size in order to reduce the magnetic field strength for given magnetic flux trapped inside the ring. The hoop force can be obtained from the expression for the magnetic energy of the poloidal field of a thin current-carrying ring ($a \ll R$) with circular cross-section (Shafranov [549], p. 122):

$$W_{\rm p} = \tfrac{1}{2} \int B_{\rm p}^2 \, dV = \tfrac{1}{2} L_\varphi I_\varphi^2 \,, \quad L_\varphi = R[\,\ln(8R/a) + \tfrac{1}{2}\ell_{\rm i} - 2\,]\,, \tag{16.46}$$

where L_φ is the self-inductance of the ring. The internal contribution to the self-inductance, $\ell_{\rm i}$, defined below, is a positive quantity of order unity which depends on the distribution of the current (Landau and Lifschitz [395], p. 124; recall that the factor 4π of the Gaussian system of units is to be replaced by the factor μ_0 of the mks system of units, which is consistently dropped here). The hoop force is then given by:

$$F_{\rm h} = -\frac{\partial W_{\rm p}}{\partial R}\bigg|_{L_\varphi I_\varphi} = \tfrac{1}{2} I_\varphi^2 \frac{\partial L_\varphi}{\partial R} \approx \tfrac{1}{2} I_\varphi^2 \,[\,\ln(8R_0/a) + \tfrac{1}{2}\ell_{\rm i} - 1\,] > 0\,, \tag{16.47}$$

i.e. it points outward, QED.

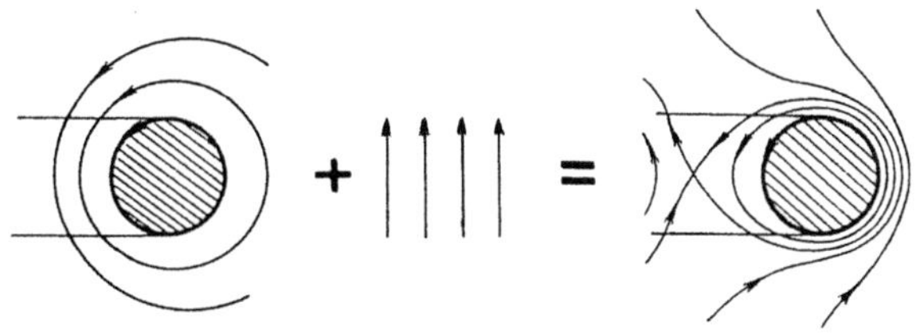

Fig. 16.6 Adding an external vertical field to ensure equilibrium in a tokamak. (From Mukhovatov and Shafranov [453].)

The second, crucial, counter measure is the creation of a homogeneous *vertical magnetic field* (in the Z-direction) by means of a toroidal current $I^{\rm ex}_{\varphi,{\rm vf}}$ in a set of external vertical field coils (Fig. 16.6). Of course, this counter measure is to be taken together with the first one, since a homogeneous vertical magnetic field has no effect on the toroidal θ-pinch (non-)equilibrium part of the tokamak, but it does have an effect on the toroidal z-pinch part by interacting with the toroidal plasma current. This is easily seen from the direction of the poloidal magnetic field vectors of Fig. 16.1: the external vertical field will increase the magnitude of the poloidal field on the outside, but decrease it on the inside. The resulting magnetic pressure will be inward. This inward force can be estimated from the Lorentz force of the vertical field on a wire carrying the toroidal plasma current I_φ:

$$F_{\rm vf} \approx -2\pi R_0 I_\varphi B_{\rm v}\,. \tag{16.48}$$

It is now simply a matter of turning the knob on the current $I^{\rm ex}_{\varphi,{\rm vf}}$ in the vertical field coils to keep the plasma column at the desired equilibrium position.

By means of the three mentioned external current parameters $I^{\rm ex}_{\rm p}$ (controlling the toroidal magnetic field B_φ), $I^{\rm ex}_\varphi$ (controlling the poloidal magnetic field $B_{\rm p}$) and $I^{\rm ex}_{\varphi,{\rm vf}}$ (controlling the external vertical magnetic field $B_{\rm v}$), the tokamak configuration obtains the necessary flexibility which has produced the impressive increase in performance towards controlled fusion described in Section 16.1.1. It remains to be shown how the value of β is determined by these parameters. This requires the consideration of the volume averaged effects of the gradients of the plasma pressure and of the toroidal field pressure, giving rise to forces F_p and F_{B_φ}.

Following Freidberg [185], Section 11.7.7, these forces are approximated by computing the different weighting due to the larger volume on the outside than on the inside. Exploiting polar coordinates r, θ, with the origin in the centre of the plasma, $R = R_0, Z = 0$, i.e. neglecting the Shafranov shift, so that r and θ effectively become 'cylindrical' coordinates, this only involves toroidal corrections of the volume element,

$$dV = 2\pi R r\, dr\, d\theta \approx 2\pi R_0 [1 + (r/R_0)\cos\theta] r\, dr\, d\theta\,, \tag{16.49}$$

coupled to the $\cos\theta$ variations of the outward unit vector $\mathbf{e}_R$,

$$\mathbf{e}_R = \cos\theta\, \mathbf{e}_r - \sin\theta\, \mathbf{e}_\theta\,, \tag{16.50}$$

so that the volume integrals just involve averaging of the $\cos^2\theta$ contributions:

$$\begin{aligned} -\int \mathbf{e}_R \cdot \nabla f\, dV &\approx -2\pi R_0 \int_0^a \int_0^{2\pi} \big[1 + (r/R_0)\cos\theta\big] \cos\theta\, f' r\, dr\, d\theta \\ &= -2\pi^2 \int_0^a f' r^2\, dr = 2\pi^2 \Big[2\int_0^a f r\, dr - a^2 f(a)\Big] \equiv 2\pi^2 a^2 \big[\langle f\rangle - f(a)\big]\,. \end{aligned} \tag{16.51}$$

Hence, the final volume averages just reduce to integration of the leading order, 'cylindrical', contributions of the variables over the radius r:

$$\langle f\rangle \equiv \frac{1}{V}\int f\, dV \approx \frac{2}{a^2}\int_0^a f r\, dr\,. \tag{16.52}$$

All this will be justified rigorously in Section 16.2.2 from a large aspect ratio expansion of the Grad–Shafranov equation in the 'shifted circle' approximation.

In effect, the leading order outward force due to pressure gradients becomes

$$F_p \equiv -\int \mathbf{e}_R \cdot \nabla p\, dV \approx 2\pi^2 a^2 \langle p\rangle\,. \tag{16.53}$$

For the calculation of the outward force due to the toroidal magnetic field, the $1/R$ dependence of B_φ, indicated by Eq. (16.45), should be accounted for:

$$B_\varphi(r,\theta) \approx (R_0/R) B_{\varphi 0}(r) \approx \big[1 - (r/R_0)\cos\theta\big] B_{\varphi 0}(r)\,, \tag{16.54}$$

where we now also include internal structure due to poloidal currents, indicated by the 'cylindrical' factor $B_{\varphi 0}(r)$. As a result of the $1/R^2$ dependence of the toroidal magnetic field pressure, the integrand of Eq. (16.51) flips sign, so that the associated outward force becomes

$$\begin{aligned} F_{B_\varphi} &\equiv -\int \mathbf{e}_R \cdot \Big[\nabla(\tfrac{1}{2}B_\varphi^2) + B_\varphi^2/R\Big]\, dV \approx -\int \mathbf{e}_R \cdot \nabla(\tfrac{1}{2}B_\varphi^2)\, dV \\ &\approx -\int \frac{R_0^2}{R^2}\, \mathbf{e}_R \cdot \nabla(\tfrac{1}{2}B_{\varphi 0}^2)\, dV \approx -2\pi^2 a^2 \Big[\langle \tfrac{1}{2}B_{\varphi 0}^2\rangle - \tfrac{1}{2}B_{\varphi 0}^2(a)\Big]. \end{aligned} \tag{16.55}$$

The poloidal field also has a $1/R$ dependence, $B_{\rm p}(r,\theta) \approx (R_0/R) B_{\rm p0}(r)$, producing an outward force that has already been accounted for by the hoop force (16.47). Of course, the three functions $B_{\rm p0}(r)$, $p(r)$ and $B_{\varphi 0}(r)$ have to satisfy the cylindrical equilibrium equation (16.31), which becomes

$$(p + \tfrac{1}{2}B_{\varphi 0}^2)' = -\frac{B_{\mathrm{p}0}}{r}\,(rB_{\mathrm{p}0})'\,. \qquad (16.56)$$

It is now expedient to convert the toroidal field force (16.55) to an expression in terms of the plasma pressure and the poloidal field. This is done by applying the operator $\int dr\, r^2$ to Eq. (16.56) and integrating by parts, giving

$$-\langle \tfrac{1}{2}B_{\varphi 0}^2\rangle + \tfrac{1}{2}B_{\varphi 0}^2(a) = \langle p\rangle - \tfrac{1}{2}B_{\mathrm{p}0}^2(a)\,. \qquad (16.57)$$

Hence

$$F_p + F_{B_\varphi} = 2\pi^2 a^2\left[2\langle p\rangle - \tfrac{1}{2}B_{\mathrm{p}0}^2(a)\right] = \tfrac{1}{2}I_\varphi^2(\beta_{\mathrm{p}} - \tfrac{1}{2})\,, \qquad (16.58)$$

where the new parameter

$$\beta_{\mathrm{p}} \equiv \frac{2\langle p\rangle}{B_{\mathrm{p}0}^2(a)} = \frac{8\pi^2 a^2}{I_\varphi^2}\,\langle p\rangle \qquad (16.59)$$

represents the average plasma pressure compared to the magnetic pressure of the poloidal field at the plasma boundary, i.e. measured in terms of the square of the total toroidal current flowing in the plasma.

Adding up the four forces yields global equilibrium,

$$F_{\mathrm{tot}} \equiv F_p + F_{B_\varphi} + F_{\mathrm{h}} + F_{\mathrm{vf}} \qquad (16.60)$$

$$= \tfrac{1}{2}I_\varphi^2\left[\beta_{\mathrm{p}} + \ln(8R_0/a) + \tfrac{1}{2}\ell_{\mathrm{i}} - \tfrac{3}{2} - \frac{4\pi R_0}{I_\varphi}B_{\mathrm{v}}\right] = 0\,, \qquad (16.61)$$

provided the vertical field has the proper magnitude

$$B_{\mathrm{v}} = \frac{I_\varphi}{4\pi R_0}\left[\beta_{\mathrm{p}} + \ln(8R_0/a) + \tfrac{1}{2}\ell_{\mathrm{i}} - \tfrac{3}{2}\right]. \qquad (16.62)$$

Here, the parameter

$$\ell_{\mathrm{i}} \equiv \frac{\langle B_{\mathrm{p}0}^2\rangle}{B_{\mathrm{p}0}^2(a)} \qquad (16.63)$$

represents the average poloidal magnetic field pressure, which is a function of the radial distribution of the toroidal current density.

We have now encountered the *main parameters governing global equilibrium* in a tokamak with circular plasma cross-section. Taking the external 'engineering' parameters $I_{\mathrm{p}}^{\mathrm{ex}}$, I_φ^{ex} and $I_{\varphi,\mathrm{vf}}^{\mathrm{ex}}$, respectively producing the toroidal, the poloidal and the vertical magnetic field, for granted, and zooming in onto the resultant equilibrium characteristics of the plasma column itself, we may distinguish

(a) the *trivial scaling parameters* a (plasma radius) and B_0 (external magnetic field at the centre of the plasma, $R = R_0$) that fix the size of the plasma and the overall magnetic field strength: just to be used to create dimensionless quantities according to our usual scale-independence argument;

(b) the *geometry parameter* $\epsilon \equiv a/R_0$ that fixes the aspect radius of the plasma torus, and possible elongation parameters of the plasma cross-section;

(c) the *global confinement parameters* β_{p} and $B_{\mathrm{p}0}(a) \sim I_\varphi$, fixing the average amount of plasma pressure confined and the total toroidal current flowing in the plasma;

(d) *distribution parameters* describing the details of the pressure $p(r)$ and the toroidal current density $j_{\varphi 0}(r)$, or the poloidal field $B_{\rm p0}(r)$, for given global parameters. (The equilibrium condition (16.56), later to be replaced by the exact conditions from the Grad–Shafranov equation, then fixes the third profile, that of the toroidal field B_φ.)

We will consider these scaling considerations in more detail in Section 16.3.1, but item (3) needs to be elaborated here since it still involves the dimensional poloidal magnetic field and toroidal current variables that appear to be rather different from the safety factor q, that might be expected at this point.

In fact, an obvious way to create a dimensionless parameter representing the poloidal field at the plasma edge, $B_{\rm p0}(a)$, or the toroidal plasma current, I_φ, is to relate them to the safety factor (16.10), evaluated at the plasma edge:

$$\left.\begin{aligned} q_1 &= \frac{1}{2\pi}\oint_{\Psi_1}\frac{B_\varphi}{RB_{\rm p}}\,d\ell \\ I_\varphi &\equiv \iint \nabla\times\mathbf{B}\cdot\mathbf{e}_\varphi\,dS = \oint B_{\rm p}\,d\ell \end{aligned}\right\} \Rightarrow\; q_{1,\rm cyl} = \frac{aB_0}{R_0B_{\rm p0}(a)} = \frac{2\pi a^2B_0}{R_0I_\varphi}, \tag{16.64}$$

as already suggested in Eq. (16.37). Since elongating the plasma cross-section has become an important method of increasing the admissible value of β in tokamaks, with respect to both equilibrium and stability, let us consider the corresponding expression for a straight cylinder with elliptical cross-section [399]:

$$q_{1,\rm cyl} = \frac{\pi(a^2+b^2)B_0}{R_0I_\varphi}, \tag{16.65}$$

where a and b define the plasma cross-section through the points $R = R_0\pm a$ and $Z = \pm b$. It would appear that $\beta_{\rm p}$ and $q_{1,\rm cyl}$ become the most appropriate global dimensionless parameters to describe equilibrium in a tokamak.

A serious objection needs to be raised, though, against the tacit identification of the robust effects of the toroidal current, expressed by the cylindrical parameter $q_{1,\rm cyl}$, and the subtle toroidal effects of field line geometry, expressed by the parameter $q(\Psi_1)$. For example, in tokamaks with a divertor, the line integral (16.64)(a) for the edge safety factor q_1 blows up due to the x-point at the plasma boundary, where $B_{\rm p} = 0$, whereas the line integral (16.64)(b) for I_φ just stays finite. Similarly, the ultimate limit of tokamak equilibrium by means of a vertical field, just discussed, implies that the x-point depicted in Fig. 16.6 intrudes into the plasma, again implying that $q_1 \to \infty$, whereas I_φ stays finite. Hence, we introduce a parameter that is formally identical to $q_{1,\rm cyl}$, but actually only measures the magnitude of the toroidal current, viz. the *modified safety factor* q^*:

$$q^* \equiv \frac{L^2B_0}{2\pi R_0I_\varphi}, \qquad L \equiv \oint d\ell \approx \pi\sqrt{2(a^2+b^2)}\,. \tag{16.66}$$

This definition arose in stability studies of skin current high-β tokamaks [159], but turns out to be most adequate for our present general discussion of tokamak equilibrium. (We here drop the later modification [205, 236] of the definition of q^* and return to the earlier one since it conserves the mentioned relation with $q_{1,\rm cyl}$ for non-circular cross-sections.)

It is to be noted that the definition (16.66) is an exact one for toroidal geometry: it only involves the trivial item (1) scaling parameters a and B_0 and the item (2) geometry parameters b and R_0, that may be considered to be *external parameters*, i.e. they can be accurately prescribed (e.g. by the size of the limiters) and they do not require determination through plasma diagnostics. In contrast, the safety factor, both on edge and at the magnetic axis, though extremely important for local plasma dynamics, is only indirectly known by means of various diagnostics (usually delivering no better than 10% accuracy, up till now).

We now also need to replace the quasi-cylindrical definition (16.59) for $\beta_{\rm p}$ by an exact toroidal definition, for which we choose [205]

$$\beta_{\rm p} \equiv \frac{8\pi S \langle p \rangle}{I_\varphi^2}, \qquad S \equiv \iint dx\, dy \approx \pi a b\,, \tag{16.67}$$

where, again, an accurate external geometry parameter appears, viz. the surface area S of the plasma cross-section. Finally, we redefine the basic confinement parameter β by comparing the average plasma pressure with the magnetic pressure of the external toroidal field, i.e. again in terms of an external parameter:

$$\beta \equiv \frac{2\langle p \rangle}{B_0^2}\,. \tag{16.68}$$

With the global equilibrium parameters $\beta_{\rm p}$ and q^* thus defined, the parameter β, though a figure of merit for fusion reactors, just becomes a secondary quantity:

$$\frac{\beta}{\epsilon} = C\,\frac{\epsilon\beta_{\rm p}}{q^{*2}}\,, \qquad C \equiv \frac{L^4}{16\pi^3 a^2 S} \approx \begin{cases} 1 & \textit{(circle)} \\ \dfrac{(1 + b^2/a^2)^2}{4b/a} & \textit{(ellipse)} \end{cases}\,. \tag{16.69}$$

Here, the constant C just depends on the cross-sectional shape of the plasma, i.e. it may be considered as an external parameter that is accurately known. With this parameterization, the two orderings introduced in Eqs. (16.43) and (16.44) of Section 16.1.3 can be expressed more appropriately as

$$q^* \sim 1\,, \qquad \beta/\epsilon \sim \epsilon\beta_{\rm p} \sim \begin{cases} \epsilon & \textit{(low-beta tokamak)} \\ 1 & \textit{(high-beta tokamak)} \end{cases}\,, \tag{16.70}$$

where the high-beta tokamak regime is the one where the limiting values of the parameters are found.

We have purposely replaced the usual definition of β, involving the average of the toroidal magnetic field pressure $\langle \frac{1}{2} B_\varphi^2 \rangle$, given in Eq. (16.39), by the above one which just involves the external magnetic field pressure $\frac{1}{2} B_0^2$. Also, the volume averages of plasma variables, defined in Eq. (16.52), have been replaced with cross-sectional averages. This way, the influence of poorly known internal distributions of plasma variables is restricted to the necessary minimum, viz. to the average pressure $\langle p \rangle$. Of course, once an accurate equilibrium solver is constructed, all desired derived quantities can be computed 'exactly', *necessarily making bold assumptions on those poorly known distributions though*, but this is just a numerical detail, not be confused with the basic scaling of plasma equilibrium variables.

The relation (16.69) summarizes the main constituents in the optimization procedure of tokamaks with respect to equilibrium and stability. Roughly speaking, as illustrated in Fig. 16.7, the value of the parameter $\epsilon\beta_{\rm p}$ is limited from above by equilibrium considerations, and the value of q^* is limited from below by stability conditions. With respect to equilibrium, we have already indicated that there

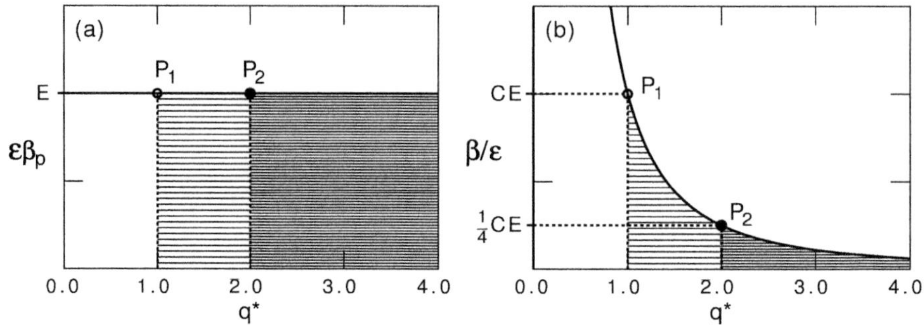

Fig. 16.7 Schematic equilibrium–stability diagram. (a) An x point intruding into the plasma imposes a limit $\epsilon\beta_{\rm p} \le E$ to the equilibrium, external $m = 1$ and $m = 2$ instabilities impose the limits $q^* \ge 1$, respectively $q^* \ge 2$, to the stability. (b) This translates into the limits $\beta/\epsilon \le CE$ (point P_1), resp. $\beta/\epsilon \le \frac{1}{4}CE$ (point P_2).

is a limit due to the fact that the vertical magnetic field produces an x-point that will hit the plasma, and thus destroy the coherence of the nesting of the magnetic surfaces, if $\epsilon\beta_{\rm p}$ is increased beyond a certain value. Let us call that value E. For circular plasma cross-section, this number is easily estimated from the expression for the poloidal magnetic field $B_{\rm p}$, which involves a parameter Λ that describes the poloidal variation (see Eq. (16.109) of the following section):

$$E \equiv \epsilon\beta_{\rm p,max} = \epsilon\Lambda_{\rm max} \approx 1\,. \tag{16.71}$$

From the relation (16.69), it is clear that to obtain a maximum value of β, one wishes to push the toroidal current to a maximum, i.e. to choose the value of q^* as small as possible. External kink mode stability will ultimately limit that value to $q^* = 1$, which is the Kruskal–Shafranov limit. However (see Wesson's stability diagram, reproduced in Fig. 9.15), the $m = 2$ external kink mode presents a much severer condition since it increases the limit to $q^* = 2$. Consequently, a rough estimate of the combined equilibrium–stability limit on β is given by

$$\beta_{\rm max} = C\,\frac{\epsilon^2\beta_{\rm p,max}}{q^{*2}} = \tfrac{1}{4}CE\,\epsilon\,. \tag{16.72}$$

For a circular plasma cross-section, this implies that β should not exceed $\frac{1}{4}\epsilon$, i.e. less than 0.1 for usual aspect ratios. The increase of this value by cross–sectional shaping is most desirable, but one should also heed ordinary engineering wisdom to stay away from ultimate operating boundaries with a rather wide margin.

In principle, these severe equilibrium and stability limits can easily be overcome by enclosing the plasma with a conducting wall. With perfect conductivity, the value of $\beta_{\rm p}$ can even be increased without a limit since the poloidal flux trapped between the plasma and the wall will always prevent the plasma from actually hitting the wall. Similarly, a conducting wall that is close enough will stabilize any external kink mode with $m \ge 2$. However, proximity of a conducting wall is prohibited in a fusion reactor. More fundamentally, even when close, a conducting wall is no longer a cure for improving either equilibrium or stability limitations since the perfect conductivity required is simply no longer there on fusion time scales: the poloidal flux required for equilibrium leaks away and the resistive wall mode grows unimpeded on those time scales (see Section 14.3.1). Although Fig. 16.7 gives a good impression of the order of magnitude of the limits on β, it is clearly an over-simplification: toroidal effects deform the straight lines of external kink stability limits at higher β. Most important, in toroidal geometry, new pressure-driven instabilities appear that restrict the operating windows even further.

16.2 Grad–Shafranov equation

16.2.1 Derivation of the Grad–Shafranov equation

The previous section presented the qualitative features and global parameterization of plasma equilibrium in a torus. To proceed further, to the construction of general axi-symmetric equilibria, which is presently possible virtually without any real limitations of the numerical precision, we need to derive the central partial differential equation (PDE) that describes the spatial dependence of the poloidal magnetic flux. This equation is generally known as the *Grad–Shafranov equation*, after the authors that published it in the 1950s [255, 548], although Lüst and Schlüter [421] also published it independently at the same time.

We will exploit again the special cylindrical coordinates R, Z, φ, where φ is the toroidal angle, R is the distance to the symmetry axis and Z is the vertical coordinate, so that R and Z are just Cartesian coordinates in the poloidal plane. Note that the order of these cylindrical coordinates is not the usual one. It is chosen to provide the most logical connection with the different toroidal coordinates (Appendices A.2.4–A.2.7). In particular, notice that the replacements (A.62) to convert vector operators from the usual r, θ, z coordinates to the present R, Z, φ coordinates involves a minus sign: $d\theta = -d\varphi$.

The derivation of the Grad–Shafranov equation involves the following steps.

(a) From the divergence equation (16.3) and Ampère's law (16.2), the poloidal field and the poloidal current are derivable from *stream functions* $\Psi(R, Z)$ and $I(R, Z)$:

$$\nabla\cdot\mathbf{B} = \frac{1}{R}\frac{\partial(RB_R)}{\partial R} + \frac{\partial B_Z}{\partial Z} = 0 \quad \Rightarrow \; B_R = -\frac{1}{R}\frac{\partial\Psi}{\partial Z}, \qquad B_Z = \frac{1}{R}\frac{\partial\Psi}{\partial R}, \tag{16.73}$$

$$\nabla\cdot\mathbf{j} = \frac{1}{R}\frac{\partial(Rj_R)}{\partial R} + \frac{\partial j_Z}{\partial Z} = 0 \quad \Rightarrow \; j_R = \frac{1}{R}\frac{\partial I}{\partial Z}, \qquad j_Z = -\frac{1}{R}\frac{\partial I}{\partial R}. \tag{16.74}$$

Of course, Ψ is the poloidal magnetic flux (normalized by dividing by a factor 2π). Alternatively, solving $\nabla\cdot\mathbf{B} = 0$ by means of $\mathbf{B} = \nabla\times\mathbf{A}$, the poloidal flux turns out to be related to the toroidal component of the vector potential:

$$\Psi = -RA_\varphi\,. \tag{16.75}$$

(b) The toroidal and poloidal components of Ampère's law (16.2) provide associated expressions for the toroidal current density j_φ and the poloidal current stream function I:

$$Rj_\varphi = R\Big(\frac{\partial B_Z}{\partial R} - \frac{\partial B_R}{\partial Z}\Big) = R\frac{\partial}{\partial R}\Big(\frac{1}{R}\frac{\partial\Psi}{\partial R}\Big) + \frac{\partial^2\Psi}{\partial Z^2} \equiv \Delta^*\Psi\,, \tag{16.76}$$

$$j_R = \frac{\partial B_\varphi}{\partial Z}, \quad j_Z = -\frac{1}{R}\frac{\partial(RB_\varphi)}{\partial R} \quad \Rightarrow \; I \equiv RB_\varphi\,. \tag{16.77}$$

Here, the special symbol Δ^* indicates a Laplacian-like operator where the order of the factors R

and $1/R$ is reversed with respect to the ordinary Laplacian.

(c) Finally, the toroidal and poloidal components of the pressure balance equation (16.1) imply that the stream function I of the poloidal current and the pressure p are *flux functions* (i.e. functions of the flux Ψ) which are related to the toroidal current j_φ:

$$\frac{\partial p}{\partial \varphi} = j_R B_Z - j_Z B_R = \frac{1}{R^2}\left(\frac{\partial I}{\partial Z}\frac{\partial \Psi}{\partial R} - \frac{\partial I}{\partial R}\frac{\partial \Psi}{\partial Z}\right) = 0 \qquad \Rightarrow \quad I \equiv I(\Psi)\,, \tag{16.78}$$

$$\left.\begin{aligned} \frac{\partial p}{\partial R} &= j_Z B_\varphi - j_\varphi B_Z = \left(-\frac{II'}{R^2} - \frac{j_\varphi}{R}\right)\frac{\partial \Psi}{\partial R} \\ \frac{\partial p}{\partial Z} &= j_\varphi B_R - j_R B_\varphi = \left(-\frac{II'}{R^2} - \frac{j_\varphi}{R}\right)\frac{\partial \Psi}{\partial Z} \end{aligned}\right\} \quad \Rightarrow \quad p = p(\Psi)\,, \tag{16.79}$$

$$p' = -\frac{II'}{R^2} - \frac{j_\varphi}{R}\,. \tag{16.80}$$

Here and in the following, the prime indicates differentiation with respect to Ψ.

Summarizing, from Eqs. (16.76) and (16.80) it follows that the equilibrium is described by an elliptic nonlinear PDE, *the Grad–Shafranov equation,* for the poloidal flux $\Psi = \Psi(R, Z)$:

$$\left[\Delta^*\Psi \equiv R^2\nabla\cdot\left(\frac{1}{R^2}\nabla\Psi\right) \equiv \Delta\Psi - \frac{2}{R}\frac{\partial\Psi}{\partial R} \equiv\right]$$

$$R\frac{\partial}{\partial R}\left(\frac{1}{R}\frac{\partial\Psi}{\partial R}\right) + \frac{\partial^2\Psi}{\partial Z^2} = -II' - R^2 p' \quad \left[= R j_\varphi\right], \tag{16.81}$$

which has to satisfy the boundary condition

$$\Psi = \Psi_1 = \text{const} \quad \text{(on the plasma cross-section)}\,. \tag{16.82}$$

A considerable complication is that this plasma cross-section, in general, is also unknown since it represents the interface between the plasma and the external vacuum region which is determined by another nonlinear problem, viz. the *external free-boundary problem* with given currents in external coils. Here, we will assume that this problem is solved separately so that the cross-sectional shape is known. (Alternatively, one could assume that the plasma is surrounded by a closely fitting external wall of the desired shape, or, as we will do in the next sections, one could prescribe a desired shape of the plasma cross-section and compute what external field would produce it: an ill-posed, though physically very relevant, problem.) From Eqs. (16.77), (16.78), (16.79) it follows that the RHS of the Grad–Shafranov equation (16.81) contains *two completely arbitrary flux functions*,

$$I \equiv RB_\varphi = I(\Psi)\,, \quad \text{and} \quad p = p(\Psi)\,. \tag{16.83}$$

This arbitrariness is the toroidal counterpart of the freedom to specify two variables in cylindrical equilibria, that we frequently encountered. Specifying the functions $I = I(\Psi)$ and $p = p(\Psi)$, everything else will be determined from the solution of the Grad–Shafranov equation. For example, *the poloidal field and current* are determined from Eqs. (16.73) and (16.74):

$$\mathbf{B}_\mathrm{p} = \frac{1}{R}\mathbf{e}_\varphi \times \nabla\Psi\,, \qquad \mathbf{j}_\mathrm{p} = -I'\,\mathbf{B}_\mathrm{p}\,. \tag{16.84}$$

This completes the derivation of the Grad–Shafranov equation.

16.2.2 Large aspect ratio expansion: internal solution

As a first application that requires the solution of the Grad–Shafranov equation, we revisit the equilibrium problem of Section 16.1.4. We now justify all the rather ad hoc approximations that were made there and derive precise expressions for all quantities, both inside the plasma (this section) and in the outer vacuum region that produces the necessary vertical field (next section).

The poloidal flux Ψ of the internal plasma region is determined by the full Grad–Shafranov equation (16.81). To solve it, we anticipate that, for a circular outer cross-section of the plasma, the cross-sections of the magnetic surfaces inside are approximately shifted circles, where the shift Δ varies from Δ_0 at the magnetic axis to 0 at the plasma surface. We assume this shift to be small with respect to the plasma radius: $\Delta/a \sim \epsilon \ll 1$. Of course, these assumptions are to be justified by the solutions that will be obtained. Our main task is to determine the magnitude of Δ for the different magnetic surfaces. For the solution of this problem, we exploit the *non-orthogonal shifted circle coordinate system* $\hat{r}$, $\hat{\theta}$, φ (Fig. 16.8) based on the 'Shafranov' shift Δ:

$$R = R_0 + \hat{r}\cos\hat{\theta} + \Delta(\hat{r})\,, \qquad Z = \hat{r}\sin\hat{\theta}\,. \tag{16.85}$$

Note that the three terms of R are of the order ϵ^{-1}, 1 and ϵ, respectively. Actually, our definition of $\Delta(\hat{r})$ differs from the shift $\Delta_{\rm Shafr}(\hat{r}) \equiv \Delta_0 - \Delta(\hat{r})$ defined by Shafranov in that we measure it with respect to the centre of the plasma ($R = R_0$), whereas Shafranov defined it with respect to the magnetic axis ($R = R_{\rm m}$). This modification is made to implement the strategy discussed in Section 16.1.4 to define parameters in terms of precisely known external quantities.

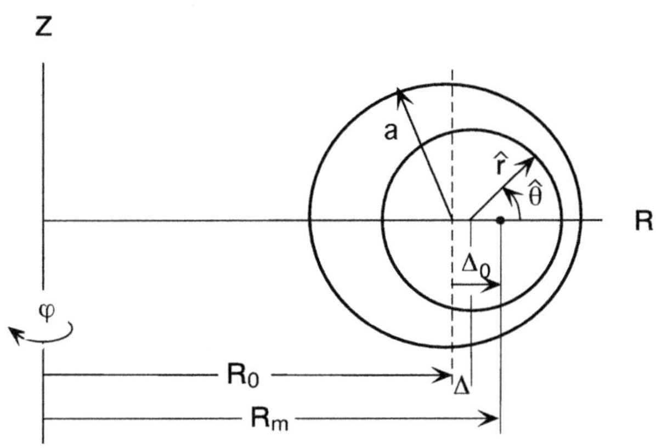

Fig. 16.8 'Shafranov' shifted circle approximation of the flux surfaces: non-orthogonal coordinates $\hat{r}, \hat{\theta}$ are defined with respect to the centres of the flux surfaces, which are shifted by an amount $\Delta(\hat{r})$, to be computed, with respect to the centre of the plasma ($R = R_0$); the magnetic axis ($R = R_{\rm m}$) is indicated by a dot.

The covariant metric elements of the shifted circle coordinates are given by

$$\begin{aligned} &\hat{g}_{11} = 1 + 2\Delta'\cos\hat{\theta} + \Delta'^2 \approx 1 + 2\Delta'\cos\hat{\theta}\,, \qquad \hat{g}_{12} = -\hat{r}\Delta'\sin\hat{\theta}\,, \\ &\hat{g}_{22} = \hat{r}^2\,, \qquad \hat{g}_{33} = R^2 \approx R_0^2\,[1 + 2(\hat{r}/R_0)\cos\hat{\theta}]\,, \\ &\hat{J} = \sqrt{\det(\hat{g}_{ij})} = R\hat{r}(1 + \Delta'\cos\hat{\theta}) \approx R_0\hat{r}[1 + (\hat{r}/R_0 + \Delta')\cos\hat{\theta}], \end{aligned} \tag{16.86}$$

where each first expression is exact and each second expression is correct to first order in ϵ. We also need the contravariant components of the metric tensor:

$$\hat{g}^{11} = (R^2/J^2)\,\hat{g}_{22}\,, \qquad \hat{g}^{12} = -(R^2/J^2)\,\hat{g}_{12}\,,$$
$$\hat{g}^{22} = (R^2/J^2)\,\hat{g}_{11}\,, \qquad \hat{g}^{33} = 1/R^2. \tag{16.87}$$

To first order in the inverse aspect ratio, the Grad–Shafranov equation becomes

$$\begin{aligned}\Delta^*\Psi &= \frac{R^2}{\hat{J}}\,\partial_i\Big(\frac{\hat{J}}{R^2}\,\hat{g}^{ij}\,\partial_j\Psi\Big)\\ &\approx (1-2\Delta'\cos\hat\theta)\,\frac{1}{\hat r}\frac{\partial}{\partial\hat r}\Big(\hat r\,\frac{\partial\Psi}{\partial\hat r}\Big) - (1/R_0-\Delta'/\hat r+\Delta'')\cos\hat\theta\,\frac{\partial\Psi}{\partial\hat r} + \frac{1}{\hat r^2}\frac{\partial^2\Psi}{\partial\hat\theta^2}\\ &= -\,I\frac{dI}{d\Psi} - R^2\frac{dp}{d\Psi} \approx -\frac{d}{d\Psi}\Big(\tfrac12 I^2 + R_0^2 p\Big) - 2R_0\hat r\,\frac{dp}{d\Psi}\cos\hat\theta\,,\end{aligned} \tag{16.88}$$

where the arbitrary functions $I(\Psi)$ and $p(\Psi)$ are yet to be specified. Here, the terms with $\partial\Psi/\partial\hat\theta$ and $\partial^2\Psi/\partial\hat r\partial\hat\theta$ have been neglected since first order multipliers turn them into second order quantities.

We now apply the low-β tokamak approximation introduced in Section 16.1.3. To get dimensionless quantities that one can compare, we exploit a characteristic length scale and magnetic field strength, for which we choose the plasma radius a and the 'vacuum' magnetic field strength B_0 at $R = R_0$, i.e. the trivial scaling parameters of Section 16.1.4. This yields the following orders of magnitude for the main quantities:

$$\hat r/a \sim 1\,, \quad B_\varphi/B_0 \sim 1\,, \quad \beta \sim p/B_0^2 \sim B_{\rm p}^2/B_0^2 \sim \epsilon^2\,. \tag{16.89}$$

The merit of the shifted circle approximation is that, in the usual expansion

$$\Psi(\hat r,\hat\theta) = \Psi_0(\hat r) + \Psi_1(\hat r)\cos\hat\theta + \cdots\,, \tag{16.90}$$

the first order flux Ψ_1 vanishes identically since the angular distortion of the magnetic surfaces it represents is already accounted for by the coordinates. Hence, satisfaction of the BC (16.82) is automatic in this order.

Before we continue with the solution of the Grad–Shafranov equation, it is useful to digress on certain technicalities of the use of non-orthogonal coordinates. The details are given in Appendix A.3, but we here illustrate some of the subtleties by working out the expression (16.84) for the poloidal field. Denoting the basis vectors of the non-orthogonal coordinates $\hat r, \hat\theta, \varphi$ as

$$\mathbf{a}^1 \equiv \nabla\hat r\,, \qquad \mathbf{a}^2 \equiv \nabla\hat\theta\,, \qquad \mathbf{a}^3 \equiv \nabla\varphi\,, \tag{16.91}$$

the gradient of the poloidal flux is written as

$$\nabla\Psi = (\partial_i\Psi)\,\mathbf{a}^i = \frac{\partial\Psi}{\partial\hat r}\,\mathbf{a}^1 + \frac{\partial\Psi}{\partial\hat\theta}\,\mathbf{a}^2\,, \tag{16.92}$$

so that the poloidal field becomes

$$\begin{aligned}\mathbf{B}_{\rm p} &= \frac{1}{R}\mathbf{e}_\varphi\times\nabla\Psi = \nabla\varphi\times\nabla\Psi = \mathbf{a}^3\times\Big(\frac{\partial\Psi}{\partial\hat r}\,\mathbf{a}^1 + \frac{\partial\Psi}{\partial\hat\theta}\,\mathbf{a}^2\Big)\\ &\overset{(A.69)}{=} \frac{1}{\hat J}\frac{\partial\Psi}{\partial\hat r}\,\mathbf{a}_2 - \frac{1}{\hat J}\frac{\partial\Psi}{\partial\hat\theta}\,\mathbf{a}_1 \approx \frac{1}{\hat J}\frac{\partial\Psi}{\partial\hat r}\,\mathbf{a}_2\,.\end{aligned} \tag{16.93}$$

This yields the angular dependence of the poloidal magnetic field amplitude,

$$B_{\rm p} = \frac{\sqrt{g_{22}}}{\hat{J}}\frac{\partial\Psi}{\partial\hat{r}} \approx \frac{1}{R_0}\Big[1 - (\hat{r}/R_0 + \Delta')\cos\hat{\theta}\Big]\frac{\partial\Psi}{\partial\hat{r}}\,, \tag{16.94}$$

which plays a central role in these sections.

We now work out the expanded Grad–Shafranov equation (16.88) in the low-β tokamak ordering. This yields the following *leading order contribution:*

$$\frac{1}{\hat{r}}\frac{d}{d\hat{r}}\Big(\hat{r}\frac{d\Psi_0}{d\hat{r}}\Big) = -\frac{d}{d\Psi_0}\Big(\tfrac{1}{2}I^2 + R_0^2 p\Big)\,. \tag{16.95}$$

From the leading order part of the poloidal field expression (16.94), derivatives with respect to Ψ_0 may now be converted into derivatives with respect to $\hat{r}$:

$$B_{\rm p0} = \frac{1}{R_0}\frac{d\Psi_0}{d\hat{r}} \qquad \Rightarrow \qquad \frac{d}{d\Psi_0} = \frac{1}{R_0 B_{\rm p0}}\frac{d}{d\hat{r}}\,. \tag{16.96}$$

Hence, to significant order, the poloidal field expression (16.94) becomes

$$B_{\rm p}(\hat{r},\hat{\theta}) \approx [1 - (\hat{r}/R_0 + \Delta')\cos\hat{\theta}]B_{\rm p0}(\hat{r})\,, \tag{16.97}$$

where $B_{\rm p0}(\hat{r})$ is free so far. Moreover, to get a balance between the LHS and RHS of Eq. (16.95), the function I^2 must be constant to leading and first order, so that

$$\begin{aligned} I \equiv RB_\varphi &\approx (R_0 + \hat{r}\cos\hat{\theta} + \Delta)(B_\varphi^{(0)} + B_\varphi^{(1)} + B_\varphi^{(2)}) \\ &\approx R_0 B_\varphi^{(0)} + R_0 B_\varphi^{(1)} + \hat{r}\cos\hat{\theta}\,B_\varphi^{(0)} + R_0 B_\varphi^{(2)} + \hat{r}\cos\hat{\theta}\,B_\varphi^{(1)} + \Delta B_\varphi^{(0)} \\ &\Rightarrow \;\; B_\varphi^{(0)} = \text{const}\,, \qquad B_\varphi^{(1)} = -(\hat{r}/R_0)\cos\hat{\theta}\,B_\varphi^{(0)}\,. \end{aligned} \tag{16.98}$$

Hence, to significant order, the expression for the toroidal field becomes

$$\begin{aligned} B_\varphi &\approx [1 - (\hat{r}/R_0)\cos\hat{\theta}]B_\varphi^{(0)} + B_\varphi^{(2)} \approx [1 - (\hat{r}/R_0)\cos\hat{\theta}](B_\varphi^{(0)} + B_\varphi^{(2)}) \\ &\Rightarrow \;\; B_\varphi(\hat{r},\hat{\theta}) \approx [1 - (\hat{r}/R_0)\cos\hat{\theta}]B_{\varphi 0}(\hat{r})\,, \end{aligned} \tag{16.99}$$

where we have lumped together the zeroth and second order into the 'cylindrical' function $B_{\varphi 0}(\hat{r})$, which is also free so far. Consequently, the leading order equilibrium relation (16.95) reduces to the cylindrical equilibrium relation (16.56),

$$\frac{d}{d\hat{r}}(p + \tfrac{1}{2}B_{\varphi 0}^2) + \frac{B_{\rm p0}}{\hat{r}}\frac{d}{d\hat{r}}(\hat{r}B_{\rm p0}) = 0\,, \tag{16.100}$$

where the variables p, $B_{\rm p0}$ and $B_{\varphi 0}$ should be considered as functions of $\hat{r}$, *two of which may be chosen freely*, as in the analogous cylindrical equilibrium problem. An expedient choice is $p(\hat{r})$ and $B_{\rm p0}(\hat{r})$, or the related current density $j_{\varphi 0}(\hat{r})$.

The next, and final, order of the expanded Grad–Shafranov equation (16.88) is the *first order contribution:*

$$-2\Delta'\frac{d^2\Psi_0}{d\hat{r}^2} - (\Delta'' + \Delta'/\hat{r} + 1/R_0)\frac{d\Psi_0}{d\hat{r}} = -2R_0\hat{r}\frac{dp}{d\Psi_0}\,. \tag{16.101}$$

Converting $d\Psi_0/dr$ into $R_0 B_{\rm p0}$ in this relation yields the crucial inhomogeneous differential equation for the determination of $\Delta(\hat{r})$:

$$\frac{d}{d\hat{r}}\Big(\hat{r}B_{\rm p0}^2\frac{d}{d\hat{r}}\Delta\Big) = (\hat{r}/R_0)(2\hat{r}p' - B_{\rm p0}^2)\,. \tag{16.102}$$

For given $p = p(\hat{r})$ and $B_{\rm p0}^2(\hat{r})$, the two integrations required are straightforward, so that the problem may be considered solved. We first construct the first integral of Eq. (16.102) and transform it by integration by parts:

$$\begin{aligned}\Delta'(\hat{r}) &= (1/R_0)\cdot\frac{1}{\hat{r}B_{\rm p0}^2(\hat{r})}\int_0^{\hat{r}}(2\hat{r}p' - B_{\rm p0}^2)\,\hat{r}d\hat{r}\\ &= (1/R_0)\cdot\frac{1}{\hat{r}B_{\rm p0}^2(\hat{r})}\Big[2\hat{r}^2p(\hat{r}) - \int_0^{\hat{r}}(4p + B_{\rm p0}^2)\,\hat{r}d\hat{r}\Big]\,,\end{aligned} \tag{16.103}$$

and then integrate the result to produce the shift itself, satisfying the BC at the plasma boundary, viz. $\Delta(a) = 0$:

$$\Delta(\hat{r}) = -\int_{\hat{r}}^a \Delta'(\hat{r})\,d\hat{r} \quad\Rightarrow\quad \Delta_0 \equiv \Delta(0) = -\int_0^a \Delta'(\hat{r})\,d\hat{r}\,. \tag{16.104}$$

This finally gives the unknown displacement Δ_0 of the magnetic axis. Incidentally, notice that, only now, after obtaining the solution characterized by $\Psi_0 = \Psi_0(\hat{r})$ and $\Delta = \Delta(\hat{r})$, has the coordinate system become explicitly known. This is the general a posteriori feature of exploiting magnetic flux based coordinates.

It remains to extract the global physical characteristics from these solutions. To that end, we average over the toroidal plasma volume,

$$dV = \hat{J}\,d\hat{r}\,d\hat{\theta}\,d\varphi \approx 4\pi^2 R_0\,\hat{r}\,d\hat{r} \quad\Rightarrow\quad V \approx 2\pi^2a^2R_0\,, \tag{16.105}$$

so that the rough approximation (16.52) for the average of leading order quantities gets a precise meaning in terms of the shifted circle coordinate $\hat{r}$:

$$\langle f\rangle \equiv \frac{1}{V}\int f(\hat{r},\hat{\theta})\,dV \approx \frac{2}{a^2}\int_0^a f_0(\hat{r})\,\hat{r}\,d\hat{r}\,. \tag{16.106}$$

With this understanding, the definitions (16.59) and (16.63) of Section 16.1.4 for the *poloidal beta*, $\beta_{\rm p}$, and of the *internal inductance*, $\ell_{\rm i}$, of the plasma, now also get a precise meaning. Their normalization with respect to the average poloidal field at the plasma boundary demonstrates the central importance of the *total toroidal current flowing in the plasma,*

$$B_{\rm p0}(a) = \frac{I_\varphi}{2\pi a}\,, \tag{16.107}$$

and, hence, of the modified safety factor q^* defined in Eq. (16.66).

With the mentioned definitions of the two global parameters $\beta_{\rm p}$ and $\ell_{\rm i}$, the solution (16.103) for Δ' at the plasma boundary becomes

$$\Delta'(a) = -\epsilon(\beta_{\rm p} + \tfrac{1}{2}\ell_{\rm i})\,, \tag{16.108}$$

where the edge pressure is assumed to vanish, $p(a) = 0$. From Eq. (16.97), this yields the required expression for the poloidal field at the plasma boundary:

$$B_{\rm p}(a,\theta) = \frac{I_\varphi}{2\pi a}\Big(1 + \epsilon\Lambda\cos\theta\Big)\,, \qquad \Lambda \equiv -1 - \Delta'(a)/\epsilon = \beta_{\rm p} + \tfrac{1}{2}\ell_{\rm i} - 1\,. \qquad (16.109)$$

We have replaced $\hat\theta$ by the ordinary polar angle θ because $\Delta(a) = 0$ implies that the Shafranov coordinates coincide with the ordinary polar coordinates at the plasma boundary: $\hat r = r = a \;\Rightarrow\; \hat\theta = \theta$. The poloidal dependence of $B_{\rm p}(a,\theta)$, expressed through Λ, summarizes the main physical properties of the internal solution Ψ, which will be *applied as a BC on the external solution* $\Psi^{\rm ex}$ in the next section.

Extrapolating the expression (16.109) to the high-beta tokamak regime, which is strictly speaking invalid here since we have assumed the low-beta tokamak ordering, but which nevertheless catches the essential physics, a limit on the equilibrium appears when the vertical field has to be increased so much that the poloidal field vanishes on the inside plasma boundary, $B_{\rm p}(a,\pi) = 0$:

$$\epsilon\Lambda \approx \epsilon\beta_{\rm p} = 1\,. \qquad (16.110)$$

This yields the limiting equilibrium value E of Eq. (16.71).

Finally, it is of interest to notice that $\beta_{\rm p}$ also determines the *overall* radial dependence of the 'cylindrical' part $B^2_{\varphi 0}(\hat r)$ of the toroidal magnetic field pressure [647]. This follows by applying a similar reasoning as led to Eq. (16.57):

$$\int_0^a \frac{dB^2_{\varphi 0}}{d\hat r}\,\hat r^2\,d\hat r = -2\int_0^a j_{\rm p0}B_{\varphi 0}\hat r^2\,d\hat r = a^2\Big(B_0^2 - \langle B^2_{\varphi 0}\rangle\Big) = \frac{I_\varphi^2}{4\pi^2}\,(\beta_{\rm p} - 1)\,. \qquad (16.111)$$

Clearly, the value $\beta_{\rm p} = 1$ separates *paramagnetic equilibria* with the 'wrong' direction of the poloidal current (viz. increasing the toroidal field in the plasma with respect to that in the vacuum) from *diamagnetic equilibria* where the poloidal current is in the proper direction (viz. θ-pinch like) to facilitate higher values of the plasma pressure confined:

$$\begin{cases} j_{\rm p0}B_{\varphi 0} > 0 \quad \textit{(outward Lorentz force)}: & \beta_{\rm p} < 1\,, \\ j_{\rm p0}B_{\varphi 0} < 0 \quad \textit{(inward Lorentz force)}: & \beta_{\rm p} > 1\,. \end{cases} \qquad (16.112)$$

For the latter case, on average, the plasma pressure 'digs a hole' in the toroidal magnetic field pressure, as shown in Fig. 16.5.

▷ **Explicit solutions for Wesson profiles** It is instructive to substitute in the expressions derived the simple model distributions for the plasma pressure and the toroidal current density introduced by Wesson [647], Section 3.7, viz.

$$p = p_0(1-\bar r^2)\,, \quad j_{\varphi 0} = j_0(1-\bar r^2)^\nu\,, \quad \text{where} \quad \bar r \equiv \hat r/a\,. \qquad (16.113)$$

Recall that this class of current profiles was extensively exploited to study the stability of 'straight tokamaks' (see Section 9.4.4 and Fig. 9.15). In particular, recall that the overall magnetic shear, expressed by the ratio of the cylindrical safety factors at the plasma surface and the magnetic axis, is $q_1/q_0 = \nu + 1$. For $\nu = 1$ (parabolic pressure and current profile), the explicit expressions of the 'cylindrical' quantities become

$$\begin{aligned} B_{\rm p0}(\bar r) &= \frac{I_\varphi}{2\pi a}\cdot \bar r(2-\bar r^2)\,, \qquad I_\varphi = \tfrac{1}{2}\pi a^2 j_0\,, \\ \beta_{\rm p} &= \Big(\frac{2\pi a}{I_\varphi}\Big)^2 p_0\,, \qquad \ell_{\rm i} = \tfrac{11}{12} \approx 0.917\,, \end{aligned} \qquad (16.114)$$

and the final integrations (16.103) and (16.104) of the Shafranov shift yield

$$\Delta'(\hat{r}) \equiv \frac{1}{a}\frac{d}{d\bar{r}}\Delta(\bar{r}) = -\epsilon \cdot \frac{\bar{r}(\beta_{\rm p} + 1 - \frac{2}{3}\bar{r}^2 + \frac{1}{8}\bar{r}^4)}{(2-\bar{r}^2)^2} \quad \Rightarrow \quad \Delta'(a) = -\epsilon(\beta_{\rm p} + \tfrac{11}{24})\,, \tag{16.115}$$

$$\Delta(\bar{r}) = \tfrac{1}{4}\epsilon a \cdot \left[\frac{(2\beta_{\rm p} + \frac{5}{6} - \frac{1}{4}\bar{r}^2)(1-\bar{r}^2)}{2-\bar{r}^2} + \tfrac{1}{3}\ln(2-\bar{r}^2)\right]$$
$$\Rightarrow \quad \Delta_0 \equiv \Delta(0) = \tfrac{1}{4}\epsilon a\,(\beta_{\rm p} + \tfrac{5}{12} + \tfrac{1}{3}\ln 2)\,. \tag{16.116}$$

The first relation reproduces Eq. (16.108) for the derivative of the shift at the plasma surface. The second one yields the shift of the magnetic axis, in agreement with Fig. 3.7.2 of [647], where graphs of Δ_0 for varying ν (i.e. $\ell_{\rm i}$) are given. Explicit calculation of the toroidal field is left as an exercise for the reader. ◁

16.2.3 Large aspect ratio expansion: external solution

In this section, we determine the vertical magnetic field needed for equilibrium, extending the procedure first presented by Shafranov [548](c), applied by Greene et al. [259] to a tokamak compression experiment, and extensively discussed by Miyamoto [442]. This requires solution of the external 'vacuum' magnetic field equations

$$\nabla \times \mathbf{B}^{\rm ex} = 0\,, \qquad \nabla \cdot \mathbf{B}^{\rm ex} = 0 \tag{16.117}$$

$$\Rightarrow \quad \mathbf{B}_{\rm p}^{\rm ex} = \frac{1}{R}\mathbf{e}_\varphi \times \nabla\Psi^{\rm ex}\,, \qquad \Delta^*\Psi^{\rm ex} = 0\,, \tag{16.118}$$

so that the poloidal flux $\Psi^{\rm ex}$ of the external region is determined by the Grad–Shafranov equation (16.81) with vanishing RHS, i.e. vanishing toroidal current. The source of the external field, viz. the distribution of the toroidal current $I^{\rm ex}_{\varphi,\rm vf}$ in the vertical field coil(s) (Section 16.1.4), is not considered here but, instead, a BC is imposed on $\Psi^{\rm ex}$ that yields a homogeneous vertical field $\mathbf{B}_{\rm v}$ at large distances from the plasma. This is sufficient to determine the amplitude $B_{\rm v}$. Of course, for the design of external coils for plasma control or the interpretation of magnetic diagnostics, the actual current distribution can not be ignored [12, 387, 23].

In the absence of skin currents on the plasma boundary, the pertinent BCs for this problem follow from prescribing the dependence (16.109) of $B_{\rm p}$ at the plasma boundary and the dependence of $\Psi^{\rm ex}$ far away (at 'infinity') from the plasma:

$$\left.\begin{aligned} \Psi^{\rm ex}(a,\theta) &= \text{const} \\ B_{\rm p}^{\rm ex}(a,\theta) &= \frac{I_\varphi}{2\pi a}(1 + \epsilon\Lambda\cos\theta) \end{aligned}\right\} \quad \textit{(at the plasma boundary)}\,, \tag{16.119}$$

$$\Psi^{\rm ex} = \Psi^{\rm ex}_{\rm pl} + \Psi^{\rm ex}_{\rm vf} \;\to\; C + \Psi^{\rm ex}_{\rm pl,\infty} + \tfrac{1}{2}B_{\rm v}R^2 \qquad \textit{(at 'infinity')}\,, \tag{16.120}$$

where $\Lambda \equiv \beta_{\rm p} + \frac{1}{2}\ell_{\rm i} - 1$, the flux $\Psi^{\rm ex}_{\rm pl,\infty}$ corresponds to the far field due to the plasma current and $\Psi^{\rm ex}_{\rm vf}$ is the flux of the vertical field.

It should be noticed that the boundary value problem (BVP) of solving the PDE (16.118) with BCs (16.119)–(16.120), suffers from two defects.

(a) Disregarding the BC (16.120), the remaining BVP is already sufficient to completely determine the solution $\Psi^{\rm ex}$. However, it is *ill-posed*: instead of solving the elliptic problem starting from

Dirichlet conditions on the two boundaries, the problem is solved by posing Cauchy conditions on the internal boundary and integrating outward. This is not forbidding by itself (many problems in science and engineering are ill-posed and effective numerical procedures exist to solve them), but it does have peculiar consequences, as we will see.

(b) The additional BC (16.120) really makes the problem *over-determined*, which implies that we will have to relax it in some way. We will discuss how this may be done when we have obtained the solution of the restricted BVP (16.118)–(16.119).

To solve the 'Grad–Shafranov' equation (16.118) with the BC (16.119), we exploit *orthogonal toroidal coordinates* μ, η, φ, using the notation of Morse and Feshbach [450](b), p. 1301. They have the following relationship to the cylindrical coordinates R, Z, φ and resultant scale factors of the differential operators:

$$R = R_\mathrm{c}\frac{\sinh\mu}{D}\,, \qquad Z = R_\mathrm{c}\frac{\sin\eta}{D}\,, \qquad D \equiv \cosh\mu - \cos\eta\,,$$
$$h_1 = h_2 = \frac{R_\mathrm{c}}{D}\,, \qquad h_3 = R\,, \qquad J = h_1 h_2 h_3\,. \tag{16.121}$$

The full range of these coordinates is indicated by the following scheme:

$$\mu = \begin{cases} 0\ : & R = 0\,, \text{ or } Z = \pm\infty\,, \text{ or } R = \infty & \textit{(infinity)} \\ \mu_1 : & R = R_0 + a\cos\theta\,, \quad Z = a\sin\theta & \textit{(plasma boundary)} \\ \infty : & R = R_\mathrm{c}\,, \quad Z = 0 & \textit{(concentration point)} \end{cases}\,. \tag{16.122}$$

The 'radial' coordinate curves $\mu =$ const of this system are circles with centres $R = R_\mathrm{c}\coth\mu$ and radii $r = R_\mathrm{c}/\sinh\mu$, fitted to the circular plasma boundary that is indicated by the value $\mu = \mu_1$, so that

$$R_0 = R_\mathrm{c}\coth\mu_1\,, \quad a = R_\mathrm{c}/\sinh\mu_1 \tag{16.123}$$
$$\Rightarrow\ \epsilon^{-1} \equiv R_0/a \equiv \cosh\mu_1 \approx \tfrac{1}{2}\mathrm{e}^{\mu_1} \gg 1\,, \qquad R_\mathrm{c} = R_0\sqrt{1-\epsilon^2}\,.$$

These coordinates are not very practical inside the plasma since the concentration point $(R_\mathrm{c}, 0)$ is shifted inward with respect to the plasma centre $(R_0, 0)$, whereas the magnetic axis is shifted outward. We exploit them only in the outer range,

$$0 \le \mu \le \mu_1 \equiv \cosh^{-1}(R_0/a) \approx \ln(2R_0/a)\,, \tag{16.124}$$

where μ_1 need not be very large for $\mathrm{e}^{-\mu_1} \approx \frac{1}{2}\epsilon$ to be very small.

In terms of these coordinates, the vacuum Grad–Shafranov equation becomes

$$\Delta^*\Psi^\mathrm{ex} \equiv \frac{R^2}{J}\left[\frac{\partial}{\partial\mu}\left(\frac{1}{R}\frac{\partial\Psi^\mathrm{ex}}{\partial\mu}\right) + \frac{\partial}{\partial\eta}\left(\frac{1}{R}\frac{\partial\Psi^\mathrm{ex}}{\partial\eta}\right)\right] = 0\,. \tag{16.125}$$

This equation may be solved by the transformation

$$\Psi^\mathrm{ex}(\mu,\eta) = \frac{G(\mu,\eta)}{\sqrt{D(\mu,\eta)}} = \frac{1}{\sqrt{D(\mu,\eta)}}\sum_{m=-\infty}^{\infty}\hat{G}_m(\mu)\cos m\eta\,, \tag{16.126}$$

where the Fourier harmonics $\hat{G}_m$ of the function $G(\mu,\eta)$ satisfy the ODEs

$$\sinh\mu\,\frac{d}{d\mu}\left(\frac{1}{\sinh\mu}\,\frac{d\hat{G}_m}{d\mu}\right) - (m^2 - \tfrac{1}{4})\hat{G}_m = 0\,, \tag{16.127}$$

having derivatives of the two kinds of Legendre functions as solutions. Hence, the general up–down symmetric solution of Eq. (16.125) may be written as

$$\Psi^{\rm ex} = \frac{1}{\sqrt{D(\mu,\eta)}}\sum_{m=0}^{\infty}(2-\delta_{m0})\Big[a_m S_m(\mu) + b_m T_m(\mu)\Big]\cos m\eta\,,$$

$$\begin{cases} S_m(\mu) \equiv \sinh\mu\,\dfrac{d}{d\mu}P_{m-\frac{1}{2}}(\cosh\mu) \equiv \sinh\mu\,P^1_{m-\frac{1}{2}}(\cosh\mu)\,, \\[2ex] T_m(\mu) \equiv \sinh\mu\,\dfrac{d}{d\mu}Q_{m-\frac{1}{2}}(\cosh\mu) \equiv \sinh\mu\,Q^1_{m-\frac{1}{2}}(\cosh\mu)\,, \end{cases} \tag{16.128}$$

where $P_{m-\frac{1}{2}}$ and $Q_{m-\frac{1}{2}}$ are the zero order toroidal harmonics (or Legendre functions of the first and second kind), and $P^1_{m-\frac{1}{2}}$ and $Q^1_{m-\frac{1}{2}}$ are the *first order toroidal harmonics*. (The higher order toroidal harmonics may be exploited to calculate 3D perturbations with toroidal dependence $\exp(in\varphi)$, like external kink modes.) The functions S_m and T_m are related to the Fock functions f_m and g_m, exploited by Shafranov [548](b),(c), through $S_m \equiv (m^2-\frac{1}{4})f_m$ and $T_m \equiv (m^2-\frac{1}{4})\pi g_m$. They satisfy the Wronskian identity $S_m T'_m - T_m S'_m = (m^2-\frac{1}{4})\sinh\mu$.

▷ **Scalar potential** Just for completeness: the vacuum field equations (16.117) may also be solved by means of the scalar potential, $\mathbf{B}^{\rm ex} = \nabla\Phi^{\rm ex}$, satisfying $\Delta\Phi^{\rm ex} = 0$; see Biermann et al. [61]. The latter is solved by the transformation $\Phi = \sqrt{D}F$, where the Fourier harmonics $\hat{F}_m$ of $F(\mu,\eta)$ satisfy Legendre's equation proper, so that the solution

$$\Phi^{\rm ex}(\mu,\eta) = \sqrt{D(\mu,\eta)}\sum_{m=-\infty}^{\infty}\Big[c_m P_{m-\frac{1}{2}}(\cosh\mu) + d_m Q_{m-\frac{1}{2}}(\cosh\mu)\Big]\cos m\eta \tag{16.129}$$

involves the Legendre functions themselves; see [450](b), p. 1303. ◁

Close to the plasma boundary (large μ), the exact solution (16.128) for $\Psi^{\rm ex}(\mu,\eta)$ may be approximated by the leading terms in an expansion in powers of $e^{-\mu}$:

$$[D(\mu,\eta)]^{-1/2} \approx \sqrt{2}e^{-\frac{1}{2}\mu}(1+e^{-\mu}\cos\eta)\,, \tag{16.130}$$

$$\begin{cases} S_0(\mu) \approx -\dfrac{1}{2\pi}(\mu + \ln 4 - 2)e^{\frac{1}{2}\mu}\,, \quad S_m(\mu) \approx \dfrac{(m-1)!}{2\Gamma(m-\frac{1}{2})\sqrt{\pi}}\,e^{(m+\frac{1}{2})\mu} \quad (m\ge 1)\,, \\[2ex] T_m(\mu) \approx -\dfrac{\Gamma(m+\frac{3}{2})\sqrt{\pi}}{2m!}\,e^{-(m-\frac{1}{2})\mu} \qquad (\text{all } m)\,. \end{cases} \tag{16.131}$$

For the full hypergeometric expressions, see Abramowitz and Stegun [3], p. 332, or Morse and Feshbach [450](b), pp. 1302 and 1329. To assist the reader who wishes to reproduce the algebra, in the rest of this section, we indicate by ≪...≫ the different stumbling blocks from errors and confusing notations encountered in the literature.

≪ *In the latter reference, the functions P and Q are defined with extra powers of* i *and* -1 *respectively, the expression for Q on p. 1302 is correct, but the one on p. 1329 is incorrect, and the Fourier coefficients of the expansion for* $1/\sqrt{D}$ *should be a factor 2 larger for* $m \neq 0$ *than those given on pp. 1304 and 1330.* ≫

Since the low-β tokamak approximation of the internal solution contains only two Fourier harmonics, to solve the restricted BVP (16.118)–(16.119) with these approximate expressions, we only need to substitute the boundary values of the $m = 0$ and $m = 1$ harmonics into the external solutions $\Psi^{\rm ex}$ and $B_{\rm p}^{\rm ex}$:

$$\begin{aligned}\Psi^{\rm ex}(\mu_1,\eta) &\approx -\frac{1}{\pi\sqrt{2}}\Big\{(\ln(8/\epsilon)-2)a_0+\tfrac{1}{2}\pi^2 b_0 \\ &\quad +\tfrac{1}{2}\epsilon\Big[(\ln(8/\epsilon)-2)a_0-8\epsilon^{-2}a_1+\tfrac{1}{2}\pi^2(b_0+3b_1)\Big]\cos\eta\Big\} = \text{const}\,, \quad (16.132)\end{aligned}$$

$$\begin{aligned}B_{\rm p}^{\rm ex}(\mu_1,\eta) &= -\left(\frac{1}{h_1 R}\frac{\partial\Psi^{\rm ex}}{\partial\mu}\right)_{\mu_1} \approx \frac{\epsilon^{-1}}{\pi\sqrt{2}R_0^2}\Big\{a_0-\tfrac{1}{2}\epsilon\Big[(\ln(8/\epsilon)+1)a_0 \\ &\quad +8\epsilon^{-2}a_1+\tfrac{1}{2}\pi^2(b_0+3b_1)\Big]\cos\eta\Big\} = \frac{I_\varphi}{2\pi a}(1+\epsilon\Lambda\cos\eta)\,. \quad (16.133)\end{aligned}$$

In the last equality we used $\theta \approx \eta$, obtained from the coordinate relationship $R(\mu,\eta) = R_0 + r\cos\theta$, $Z(\mu,\eta) = r\sin\theta$. To the order required, this yields

$$\cos\eta \approx \cos\theta\,, \qquad \mu-\mu_1 \approx \ln(a/r) - \tfrac{1}{2}\epsilon(a/r - r/a)\cos\theta\,. \quad (16.134)$$

The transformed BCs (16.132) and (16.133) determine three of the four constants,

$$a_0 = \frac{1}{\sqrt{2}}R_0 I_\varphi\,, \quad a_1 = -\tfrac{1}{8}\epsilon^2(\Lambda+\tfrac{3}{2})a_0\,, \quad b_0+3b_1 = -\frac{2}{\pi^2}[\Lambda+\ln(8/\epsilon)-\tfrac{1}{2}]a_0\,, \quad (16.135)$$

so that the solution close to the plasma boundary is represented by

$$\begin{aligned}\Psi^{\rm ex}(\mu,\eta) &\approx -\frac{R_0 I_\varphi}{2\pi}\Big\{\mu-\mu_1+\ln(8/\epsilon)-2+\tfrac{1}{2}\pi^2 b_0/a_0 \\ &\quad +\Big[\mu-\mu_1-(\Lambda+\tfrac{3}{2})(1-{\rm e}^{2(\mu-\mu_1)})\Big]{\rm e}^{-\mu}\cos\theta\Big\}\,, \quad (16.136)\end{aligned}$$

or

$$\begin{aligned}\Psi^{\rm ex}(r,\theta) &\approx -\frac{R_0 I_\varphi}{2\pi}\Big\{\ln(8R_0/r)-2+\tfrac{1}{2}\pi^2 b_0/a_0 \\ &\quad -\tfrac{1}{2}\Big[\ln(r/a)+(\Lambda+\tfrac{1}{2})(1-a^2/r^2)\Big](r/R_0)\cos\theta\Big\}\,. \quad (16.137)\end{aligned}$$

Apart from the arbitrary constant b_0/a_0, the solution is now completely determined in terms of a truncated Fourier expansion around the plasma boundary.

We now have to consider whether it is possible at all to satisfy the additional 'BC' (16.120) for the vertical field and, if so, to determine its amplitude $B_{\rm v}$. To that end, we first extract the plasma self field $\Psi^{\rm ex}_{{\rm pl},\infty}$ from the solution obtained and then evaluate whether the remainder conforms to the vertical flux $\Psi^{\rm ex}_{\rm pl,v} = \frac{1}{2}B_{\rm v}R^2$. (For simplicity, we exploit the arbitrariness of b_0/a_0 to set $C = 0$.)

Far away from the plasma, the actual current distribution in the plasma can be ignored so that the plasma self field is related to the exact expression of the vector potential for a ring current I_φ at $R = R_0$ (see Jackson [331], Section 5.5):

$$\begin{aligned}\Psi^{\rm ex}_{{\rm pl},\infty} &= -RA_\varphi = -R_0 I_\varphi \cdot \sqrt{R/R_0}\,\frac{(2-k^2)K(k^2)-2E(k^2)}{2\pi k}\,, \\ k^2 &\equiv \frac{4R/R_0}{(1+R/R_0)^2+(Z/R_0)^2}\,. \quad (16.138)\end{aligned}$$

Converting these expressions into μ, η coordinates, the square root $\sqrt{R/R_0} \approx \sqrt{\sinh\mu/D}$, the parameter k becomes a function of μ alone, $k^2 \equiv 1 - \mathrm{e}^{-2\mu}$ and the quotient with the elliptic integrals brings in the function $-S_0(\mu)/\sqrt{2\sinh\mu}$, where $S_0(\mu) = -\frac{1}{2}\cosh\mu\, P_{-1/2}(\cosh\mu) + \frac{1}{2}P_{1/2}(\cosh\mu)$. Hence,

$$\Psi^{\rm ex}_{\rm pl,\infty} = R_0 I_\varphi \frac{S_0}{\sqrt{2D}} = a_0 \frac{S_0(\mu)}{\sqrt{D(\mu,\eta)}} \,. \tag{16.139}$$

This expression for the plasma far field not only justifies the neglect of $a_1 S_1(\mu)$, since $a_1 \sim \epsilon^2$ according to Eq. (16.135), but of all terms $a_m S_m(\mu)$ for $m \geq 1$.

≪ *In the derivation of these expressions, a serious source of confusion had to be identified: in contrast to Jackson [331] and others [256], we considered it more logical to indicate the argument of the elliptic integrals as it appears in the integrals, viz. as k^2 and not as k, i.e. to follow the convention of Abramowitz and Stegun [3], Chapter 17. However, in their expressions for the Legendre functions in terms of elliptic integrals [3], Section 8.13, they inconsistently exploit the wrong argument k.* [1] ≫

It remains to express the vertical field in μ, η coordinates and to convert it into a series of the functions $T_m(\mu)$. To that end, we exploit a general identity (see, e.g., B. Braams [85]) relating those functions to the Fourier coefficients of $\sqrt{2D}$:

$$\pi\sqrt{2D(\mu,\eta)} = \sum_{m=0}^{\infty} \frac{2-\delta_{m0}}{m^2 - \frac{1}{4}}\, T_m(\mu)\cos m\eta \tag{16.140}$$

$$\Rightarrow\quad T_m(\mu) = (m^2 - \tfrac{1}{4}) \int_0^\pi \sqrt{2D(\mu,\eta')}\cos m\eta'\, d\eta' \,. \tag{16.141}$$

By differentiating this identity twice, and exploiting the ODE (16.127) for $T_m(\mu)$, we obtain the following representation of the vertical flux:

$$\begin{aligned}\Psi^{\rm ex}_{\rm vf} &= \tfrac{1}{2}B_{\rm v}R^2 \approx \frac{1}{\sqrt{D}}\,\tfrac{1}{2}B_{\rm v}R_0^2\,\frac{\sinh^2\mu}{D^{3/2}}\\ &= -\frac{\sqrt{2}}{\pi\sqrt{D}}\sum_{m=0}^{\infty}(2-\delta_{m0})B_{\rm v}R_0^2\,T_m(\mu)\cos m\eta = \Psi^{\rm ex}(a_m = 0)\end{aligned} \tag{16.142}$$

$$\Rightarrow\quad b_m = -\frac{\sqrt{2}}{\pi}B_{\rm v}R_0^2 \quad (\text{all } m)\,. \tag{16.143}$$

Hence, all coefficients b_m should be equal! However, from the relation (16.135) it follows that only b_0 and b_1 are needed to determine $B_{\rm v}$. Substituting them, the required relation for the vertical field results:

$$B_{\rm v} \approx -\frac{\pi}{4\sqrt{2}R_0^2}(b_0 + 3b_1) = \frac{I_\varphi}{4\pi R_0}\Big[\beta_{\rm p} + \ln(8R_0/a) + \tfrac{1}{2}\ell_{\rm i} - \tfrac{3}{2}\Big]\,, \tag{16.144}$$

justifying the heuristic expression (16.62) derived in Section 16.1.4; QED.

This illustrates the *intermediate asymptotic* character of the ill-posed equilibrium problem. Since the exact expression (16.142) for the vertical field involves all coefficients b_m, but only two of them are determined from the present low-β tokamak expansion, the sum $\Psi^{\rm ex} = \Psi^{\rm ex}_{\rm pl,\infty} + \Psi^{\rm ex}_{\rm vf}$ of the

[1] This has been corrected in the new version, *NIST Handbook of Mathematical Functions* [3], p. 360 and 486. Unfortunately, the change in this new edition is not in the expressions for the Legendre functions, but in the arguments of the elliptic integrals: $K(k)$ and $E(k)$, whereas we write $K(k^2)$ and $E(k^2)$ for these expressions.

expressions (16.138) and (16.142) represents the solution 'far away' from the plasma, *but not too far away* (in particular not at $\mu \ll 1$, see small text below) so that the functions $S_0(\mu)$, $T_0(\mu)$ and $T_1(\mu)$ suffice to approximate the solution. Once this is established, one can also extract the contribution of the vertical field from the solution (16.137) at the plasma boundary, as Shafranov does [548](c):

$$\begin{aligned} \Psi^{\rm ex}_{\rm vf} &\equiv \Psi^{\rm ex}(a_m = 0) \approx -\frac{\pi}{2\sqrt{2}}\Big[b_0 + \tfrac{1}{2}(b_0 + 3b_1)\frac{r}{R_0}\cos\theta \Big] \\ &= \tfrac{1}{2}B_{\rm v}R^2 \approx \tfrac{1}{2}B_{\rm v}R_0^2\Big(1 + 2\frac{r}{R_0}\cos\theta\Big)\,. \end{aligned} \tag{16.145}$$

This again yields the expression (16.144) for $B_{\rm v}$.

Since the external solution (16.128) is exact, the method may be generalized to arbitrary aspect ratio. The coefficients a_m and b_m are then determined by the plasma current distribution; see Zakharov and Shafranov [664], Section 3.3.

▷ **Supplementary expressions** From Eq. (16.141), one may obtain exact relations of the functions $T_m(\mu)$ in terms of elliptic integrals, alternative to the hypergeometric series:

$$\begin{aligned} T_0(\mu) &= -\tfrac{1}{4}\int_0^\pi \sqrt{2D(\mu,\eta')}\,d\eta' = -\frac{1}{\lambda}E(\lambda^2)\,, \qquad \lambda^2 \equiv [\,\cosh(\mu/2)]^{-2}\,, \\ T_1(\mu) &= \tfrac{3}{4}\int_0^\pi \sqrt{2D(\mu,\eta')}\cos\eta'\,d\eta' = \frac{(2-2\lambda^2)K(\lambda^2) - (2-\lambda^2)E(\lambda^2)}{\lambda^3}\,. \end{aligned} \tag{16.146}$$

These provide checks on the expansions (16.131), and (16.147) below.

The approximations of S_m and T_m for $\mu \ll 1$ may be derived from the functions P of Morse and Feshbach [450](b), p. 1329, and the functions Q of Bateman [39], p. 149:

$$\left\{\begin{aligned} &S_m(\mu) \approx \tfrac{1}{2}(m^2 - \tfrac{1}{4})\mu^2\,, \\ &T_m(\mu) \approx -1 + \tfrac{1}{2}(m^2 - \tfrac{1}{4})\Big[\mu^2\ln(8/\mu) + \Big(\tfrac{1}{2} - \sum_{n=1}^{m}\frac{2}{2n-1}\Big)\mu^2\Big]\,. \end{aligned}\right. \tag{16.147}$$

The logarithmic terms demonstrate the ill-posedness from another angle: to give finite results for the flux $\Psi^{\rm ex}$ and its derivatives in the limit $\mu \to 0$ (e.g. on the axis of symmetry), the constants b_m should be such that $\sum b_m T_m$ cancels, which is a hopeless task to accomplish when the solution is obtained by integrating from the inside to the outside. ◁

16.3 Exact equilibrium solutions

16.3.1 Poloidal flux scaling

Recall the analysis of the scaling properties of the MHD equations in Section 4.1.2. We found that in any MHD problem three trivial quantities appear that have no other function than to provide magnitude and dimensions for the occurring lengths, magnetic field strengths and time scales. Hence, they can, and should, be scaled out by dividing all occurring variables and parameters by appropriate powers of them, so that the resulting problem becomes dimensionless. Since equilibrium problems are time-independent, only two such trivial scaling quantities occur, viz. a length and a magnetic field strength. After these have been taken out, we can concentrate on the quantities of real physical interest, e.g. the field line geometry, represented by the parameter q_1, the amount of plasma confined, represented by the parameter β/ϵ and the 'fatness' of the torus, represented by the parameter ϵ. In

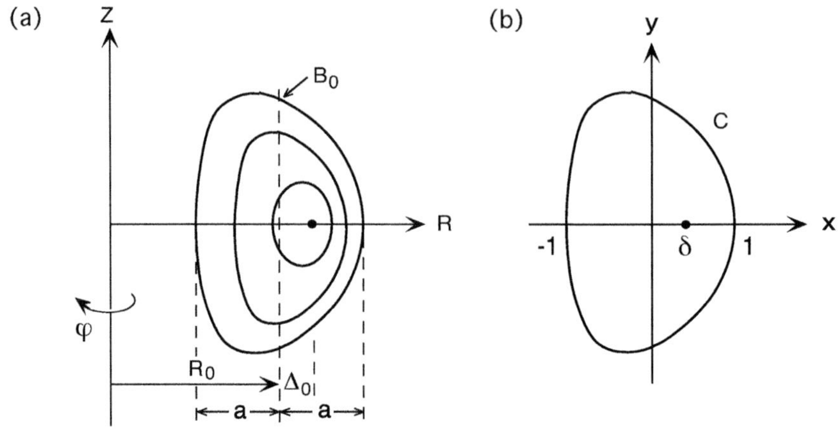

Fig. 16.9 Cross-sectional geometry of flux surfaces: (a) in R, Z, φ coordinates; (b) in the dimensionless poloidal x-y plane.

Section 16.1.4, we have already demonstrated that a more effective parameterization is obtained by replacing the parameter pair $\beta/\epsilon\,, q_1$ by the parameter pair $\epsilon\beta_{\rm p}\,, q^*$. We will now complete that analysis by systematically subjecting the Grad–Shafranov equation to this scaling, where also the degrees of freedom residing in the free flux function profiles will be accounted for.

This procedure involves the following steps.

(a) Construct *a minimum number of dimensionless quantities* of order unity by dividing lengths by the half-width a of the plasma, and magnetic fields by the strength B_0 of the *external* vacuum magnetic field at $R = R_0$, i.e. at the middle of the plasma, but on the outside (indicated by B_0 in Fig. 16.9(a)). This yields dimensionless poloidal coordinates (Fig. 16.9(b)),

$$x \equiv (R - R_0)/a\,, \qquad y \equiv Z/a\,, \tag{16.148}$$

the *inverse aspect ratio,*

$$\epsilon \equiv a/R_0 \quad (\,\ll 1 \ \text{ in asymptotic expansions}\,)\,, \tag{16.149}$$

the *dimensionless poloidal flux,*

$$\overline{\Psi} \equiv \Psi/(a^2 B_0)\,, \tag{16.150}$$

a separate parameter for the *inverse of the total poloidal flux through the plasma*, which is proportional to the modified safety factor q^*,

$$\alpha \equiv a^2 B_0/\Psi_1 \equiv 1/\overline{\Psi}_1 \ \ (\,\sim q^*)\,, \tag{16.151}$$

and a dimensionless 'radial' *flux coordinate of unit range,*

$$\psi \equiv \Psi/\Psi_1 \equiv \alpha\overline{\Psi} \quad \Rightarrow \quad 0 \leq \psi \leq 1\,. \tag{16.152}$$

From now on, we will leave this construction understood and drop the bars.

(b) Introduce *scaled dimensionless profiles* for the pressure and the 'diamagnetism',

$$P(\psi) \equiv \frac{\alpha^2}{\epsilon B_0^2}\,p(\Psi)\,, \qquad Q(\psi) \equiv -\frac{\epsilon\alpha^2}{a^2 B_0^2}\,\tfrac{1}{2}\big[I^2(\Psi) - R_0^2 B_0^2\big]\,, \tag{16.153}$$

so that the Grad–Shafranov equation transforms into

$$\psi_{xx} + \psi_{yy} - \frac{\epsilon}{1+\epsilon x}\,\psi_x = \frac{1}{\epsilon}\Big[Q' - (1+\epsilon x)^2 P'\Big]\,. \tag{16.154}$$

To get a RHS of order unity, we introduce yet another profile that expresses the mentioned fact that tokamak equilibria at 'high' β ($\sim \epsilon$) become θ-pinch like so that the pressure and diamagnetism profiles must approximately balance:

$$G(\psi) \equiv -\frac{1}{\epsilon}\Big[Q(\psi) - P(\psi)\Big]\,. \tag{16.155}$$

The Grad–Shafranov equation then transforms into the desired form where the order of magnitude of each term is manifest:

$$\psi_{xx} + \psi_{yy} - \frac{\epsilon}{1+\epsilon x}\,\psi_x = -G' - 2x(1 + \tfrac{1}{2}\epsilon x)P'\,. \tag{16.156}$$

Notice that, even in the limit $\epsilon \to 0$, there is still poloidal asymmetry resulting in an outward shift of the flux surfaces due to the term $-2xP'$. This is the essence of the *high β tomakak ordering*, but with $\epsilon \neq 0$ the equation is exact.

(c) We are not content yet: the arbitrary profiles $G'(\psi)$ and $P'(\psi)$ represent an infinite amount of freedom which makes it hard to compare experimental with theoretical stability results obtained for different equilibria. We should distinguish between the amplitudes of these functions (related to the global parameters $\epsilon\beta_\mathrm{p}$ and q^*) and their shapes. Therefore, we introduce *unit profiles* $\Gamma(\psi)$ and $\Pi(\psi)$, or rather $\gamma(\psi)$ and $\pi(\psi)$, and amplitudes A and B:

$$G' \equiv -A\Gamma(\psi)\,, \qquad \Gamma(\psi) = \Gamma_1 + (1-\Gamma_1)\gamma(\psi)\,, \tag{16.157}$$

$$\text{with boundary values} \quad \Gamma(0) = \gamma(0) = 1\,, \quad \Gamma(1) = \Gamma_1\,, \;\; \gamma(1) = 0\,,$$

$$P' \equiv -\tfrac{1}{2}B\Pi(\psi)\,, \qquad \Pi(\psi) = \Pi_1 + (1-\Pi_1)\pi(\psi)\,, \tag{16.158}$$

$$\text{with boundary values} \quad \Pi(0) = \pi(0) = 1\,, \quad \Pi(1) = \Pi_1\,, \;\; \pi(1) = 0\,,$$

where the parameters Γ_1 and Π_1 roughly represent a possible non-vanishing current density and pressure gradient at the plasma edge. When G' and P' are integrated with respect to ψ, two additional arbitrary integration constants appear that represent possible non-vanishing surface currents; they will be neglected here. Hence,

$$G(\psi) = A\int_\psi^1 \Gamma(\psi')\,d\psi'\,, \quad P(\psi) = \tfrac{1}{2}B\int_\psi^1 \Pi(\psi')\,d\psi'\,, \quad Q(\psi) = P(\psi) - \epsilon G(\psi)\,. \tag{16.159}$$

The *scaled Grad–Shafranov equation* then becomes:

$$\Delta^*\psi \equiv \psi_{xx} + \psi_{yy} - \frac{\epsilon}{1+\epsilon x}\,\psi_x = A\Gamma(\psi) + Bx(1 + \tfrac{1}{2}\epsilon x)\Pi(\psi)\,, \tag{16.160}$$

where $\Gamma(\psi)$ and $\Pi(\psi)$ approximately represent the shapes of the toroidal current density and the pressure gradient profiles. This appears to be a final, definite, form of the Grad–Shafranov equation,

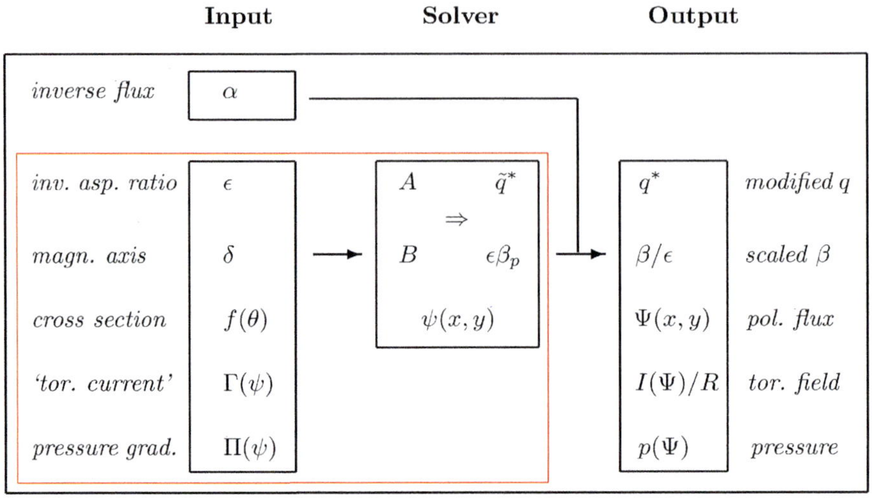

Fig. 16.10 Independent parameters and profiles in a numerical equilibrium solver. Inside the red box: core part of the solver, which is independent of the inverse flux parameter $\alpha \equiv 1/\bar{\Psi}_1$ or the modified safety factor q^*.

with no arbitrariness in the normalizations.

(d) Finally, we transform the original boundary value problem (16.81)–(16.82) into a *non-standard boundary value problem* by considering not only the cross-sectional shape C, but also the dimensionless position $\delta \equiv \Delta_0/a$ of the magnetic axis (Fig. 16.9(b)) *as given*, so that the boundary conditions become

$$\psi = 1 \qquad \text{(at the plasma boundary C, } r \equiv \sqrt{x^2+y^2} = f(\theta)\text{)}, \tag{16.161}$$

$$\psi = \psi_x = \psi_y = 0 \qquad \text{(at the magnetic axis, } x = \delta\,,\ y = 0\text{)}. \tag{16.162}$$

The boundary value problem (16.160)–(16.162) is then over-determined, so that the global parameters A and B become *eigenvalues* that should be determined together with the solution ψ. Effective methods exist to do this (see Section 16.3.3).

Now count! Since the Grad–Shafranov equation is a nonlinear equation, it is impossible to distinguish between cause and effect. This translates into arbitrariness of the choice of input and output variables of a numerical equilibrium code that solves this problem. It is essential, though, that the number of independent global and shape parameters is carefully counted in order to be able to compare results from different equilibrium solvers and to avoid spurious parameter scans. This is true in general, but more pertinent in MHD equilibrium because of the twofold infinite freedom of the equilibrium profiles. Therefore, in Section 16.1.4, we have introduced a distinction between four kinds of parameter stressing the difference between precisely known external parameters and parameters that require knowledge of the plasma profiles which is only very indirectly known from diagnostics.

The mentioned distinction between the four kinds of parameters has now been completed by the associated construction for the Grad–Shafranov equation:

(1) we have eliminated the *trivial scaling parameters* a and B_0;

(2) we have defined the *geometry parameters* ϵ and the shape $r = f(\theta)$ of the plasma cross-sectional boundary curve C;

(3) we have (temporarily) replaced the *global confinement parameters* $\epsilon\beta_{\rm p}$ and q^* by the shift δ of the magnetic axis and the inverse poloidal flux α;

(4) we have defined the *distribution functions* $\Gamma(\psi)$ and $\Pi(\psi)$ for the toroidal plasma current and the pressure profile.

The peculiar construction of item (3) requires some explanation. Whereas $\epsilon\beta_{\rm p}$ and q^* remain the global equilibrium parameters of interest, in a numerical solution procedure it is expedient to consider δ and α as the primary, input, parameters. Specifying δ then fixes one of the major geometric features of the solution, viz. the position of the magnetic axis, and specifying α fixes the other major geometric feature of the magnetic geometry, viz. the total poloidal flux.

There is another, more fundamental, reason for the procedure put forward: the *poloidal flux scaling*, described by the parameter α, permits the solution of the core of the equilibrium problem, described by the scaled Grad–Shafranov equation (16.160), without actually specifying α! This is evident from the fact that the latter equation does not contain α. The eigenvalue parameters A and B of this equation will be completely determined, and hence the solution $\psi(x, y)$ as well, when all item (1), (2) and (4) parameters are given, but from the item (3) parameters only δ, which then uniquely determines the parameter $\epsilon\beta_{\rm p}$. However, the parameter q^* is not determined by the scaled equations, but only its relative magnitude with respect to α,

$$\widetilde{q}^* \equiv q^*/\alpha\,. \tag{16.163}$$

Hence, *the poloidal flux scaling* described by the parameter α, *and the toroidal current scaling* introduced in Section 16.1.4 and described by the parameter q^*, *are one and the same*. This is illustrated in Fig. 16.10 by the red box, which contains the α or q^* independent part of the equilibrium solver. In tokamak stability studies, when constructing stability diagrams as schematically indicated in Fig. 16.7 usually q_1, or rather q^*, is varied while the rest of the equilibrium is kept fixed. With the poloidal flux scaling, this computer time saving procedure is now uniquely defined.

After the core problem for the flux ψ and the eigenvalues A and B has been solved, the explicit spatial dependence of the physical variables is found by substituting the value of α in the following expressions:

$$\begin{aligned}
&\psi = \psi(x,y) \quad\Rightarrow\quad (\alpha/\epsilon)\mathbf{B}_{\rm p} = (1+\epsilon x)^{-1}\mathbf{e}_\varphi \times \nabla\psi\,,\\
&(\alpha^2/\epsilon)p = P(\psi)\,, \qquad B_\varphi = (1+\epsilon x)^{-1}\Big[1 - 2(\epsilon/\alpha^2)Q(\psi)\Big]^{1/2}\,,\\
&\alpha\mathbf{j}_{\rm p} = Q'(\psi)\Big[1 - 2(\epsilon/\alpha^2)Q(\psi)\Big]^{-1/2}\mathbf{B}_{\rm p}\,, \qquad (\alpha/\epsilon)j_\varphi = (1+\epsilon x)^{-1}\Delta^*\psi\,.
\end{aligned} \tag{16.164}$$

The global parameters defined in Eqs. (16.66) [(16.163)] and (16.67),

$$\widetilde{q}^* = \frac{L^2}{2\pi}\Big[\iint_S (1+\epsilon x)^{-1}\Delta^*\psi\, dx\,dy\Big]^{-1}\,, \tag{16.165}$$

$$\epsilon\beta_{\rm p} = 8\pi\iint_S P(\psi)\,dx\,dy\,\Big[\iint_S (1+\epsilon x)^{-1}\Delta^*\psi\,dx\,dy\Big]^{-2}\,, \tag{16.166}$$

do not involve α, but just the eigenvalues A and B, according to Eqs. (16.159) and (16.160), so that they may be determined inside the red box of Fig. 16.10. At this level, the scaled counterpart of Eq. (16.69) may also be computed:

$$\widetilde{\beta}/\epsilon \equiv \alpha^2\beta/\epsilon = 2S^{-1}\iint_S P(\psi)\,dx\,dy = C\,\epsilon\beta_{\rm p}/\widetilde{q}^{*2}\,. \tag{16.167}$$

However, the scaled safety factor profile depends on α and, hence, does not belong to the core solver:

$$\widetilde{q}(\psi) \equiv q(\psi)/\alpha = \frac{1}{2\pi}\oint_\psi \frac{\sqrt{1-2(\epsilon/\alpha^2)Q(\psi)}}{(1+\epsilon x)|\nabla\psi|}\,d\ell\,. \tag{16.168}$$

Notice that it requires accurate calculation over a flux surface: another reason not to exploit the boundary value q_1 as a global parameter (nor the arbitrary parameter q_{95}, that is sometimes used to indicate the value of q at the flux surface containing 95% of the flux in order to avoid the infinity when a separatrix occurs). To avoid misunderstanding: this does not imply that the safety factor profile $q(\psi)$ loses its central importance, in particular not because it is crucial in the construction of straight-field-line coordinates for stability analysis.

Summarizing, we have enumerated the freedom in the choice of MHD equilibrium parameters and profiles by prescribing three geometric quantities, viz. the inverse aspect ratio ϵ, the shift δ of the magnetic axis, the plasma cross-sectional shape C, and two arbitrary unit profiles $\Gamma(\psi)$ and $\Pi(\psi)$, corresponding to the toroidal current and the pressure gradient. Solving the Grad–Shafranov equation with these independent input parameters turns A and B into eigenvalues which should be computed together with the solution $\psi(x,y)$. These parameters are directly related to the scaled physical parameters $\widetilde{q}^*$ and $\epsilon\beta_{\rm p}$. The unscaled parameters q^* and β/ϵ, and all other parameters and functions of interest, are obtained by means of a simple scaling in terms of the parameter $\alpha \equiv \Psi_1^{-1}$. As a bonus, in the *'high'-beta tokamak ordering*, where $\epsilon \ll 1$ while $\epsilon\beta_{\rm p} \sim 1$ and $q^* \sim 1$ are kept finite, the scaling becomes trivial: all RHSs of Eqs. (16.164)–(16.168) become independent of both α and ϵ, whereas the LHSs scale as simple powers of those parameters.

16.3.2 Soloviev equilibrium

A useful special solution of the Grad–Shafranov equation (16.81) was obtained by Soloviev [554] by assuming linear profiles for $I^2(\Psi)$ and $p(\Psi)$,

$$\begin{aligned} &\tfrac{1}{2}I^2(\Psi) = \tfrac{1}{2}I_0^2 - E\Psi\,, \quad I_0 \equiv R_0B_0 \qquad \Rightarrow\ \tfrac{1}{2}I^{2\prime} = -E\,,\\ &p(\Psi) = p_0 - F\Psi\,, \qquad p_0 \equiv p(0) \qquad \Rightarrow\quad p' = -F\,, \end{aligned} \tag{16.169}$$

so that the Grad–Shafranov equation becomes a linear, inhomogeneous, PDE:

$$R\frac{\partial}{\partial R}\left(\frac{1}{R}\frac{\partial\Psi}{\partial R}\right) + \frac{\partial^2\Psi}{\partial Z^2} = E + FR^2\,. \tag{16.170}$$

If one ignores the elliptic character of this equation, by replacing the BC (16.82) by the condition that Ψ should vanish at the magnetic axis, which is accepted to be located at some position determined by

$$\Psi = \Psi_R = \Psi_Z = 0 \qquad \text{(at the magnetic axis } R = R_{\rm m}\,,\ Z = 0)\,, \tag{16.171}$$

one easily checks (just substitute) that the Grad–Shafranov equation is solved by the following polynomial expression:

$$\Psi(R,Z) = (C - DR^2)^2 + \tfrac{1}{2}\Big[E + (F - 8D^2)R^2\Big] Z^2\,, \qquad R_\mathrm{m} = \sqrt{C/D}\,. \tag{16.172}$$

This is a four-parameter family of solutions of the original equilibrium problem (16.81)–(16.82), where the position of the magnetic axis (at $R = R_\mathrm{m}$) is controlled by the parameters C and D, but the cross-section of the outer wall can only be specified a posteriori by cutting out some suitable part of the solution where the flux surfaces are still nested around the magnetic axis. Obviously, this is not what one needs in the control room of a tokamak experiment, but the Soloviev solution is, nevertheless, important to test equilibrium solvers and the accuracy of the associated construction of flux coordinates for stability analysis.

In order to develop intuition about the parameterization introduced in the previous section, let us now consider the scaled counterpart of the Soloviev construction. The flux functions $\Gamma(\psi)$ and $\Pi(\psi)$ corresponding to the linear profiles (16.169) become constants, $\Gamma_1 = \Pi_1 = 1$, and the scaled Grad–Shafranov equation (16.160) becomes

$$\psi_{xx} + \psi_{yy} - \frac{\epsilon}{1+\epsilon x}\psi_x = A + Bx(1 + \tfrac{1}{2}\epsilon x)\,. \tag{16.173}$$

The scaled Soloviev solution now involves only three arbitrary parameters, viz. the inverse aspect ratio ϵ, the elongation $\sigma = b/a$ of the outer flux surface $\psi = 1$ at $x = 0$ and the triangularity τ of the outer flux surface. It may be written as

$$\psi(x,y) = \Big[x - \tfrac{1}{2}\epsilon(1 - x^2)\Big]^2 + (1 - \tfrac{1}{4}\epsilon^2)\Big[1 + \epsilon\tau x(2 + \epsilon x)\Big]\Big(\frac{y}{\sigma}\Big)^2\,, \tag{16.174}$$

where the position of the magnetic axis at $x = \delta$, $y = 0$, found a posteriori from the 'boundary condition' $\psi = \psi_x = \psi_y = 0$, is directly related to ϵ:

$$\delta = \frac{1}{\epsilon}\Big[\sqrt{1+\epsilon^2} - 1\Big] \quad \Big(\approx \tfrac{1}{2}\epsilon\,, \text{ if } \epsilon \ll 1\Big)\,. \tag{16.175}$$

The eigenvalue parameters A and B of the scaled Grad–Shafranov equation are found by substitution of Eq. (16.174) into Eq. (16.173):

$$A = 2\Big[1 + \frac{1 - \frac{1}{4}\epsilon^2}{\sigma^2}\Big]\,, \qquad B = 4\epsilon\Big[1 + \frac{1 - \frac{1}{4}\epsilon^2}{\sigma^2}\tau\Big]\,. \tag{16.176}$$

They are independent of α, as expected, and they determine the scaled parameters $\tilde{q}^*$ and $\epsilon\beta_\mathrm{p}$ by the integrals (16.165) and (16.166). Also specifying the parameter α, the three parameters ϵ, σ and τ then finally completely determine the global parameters q^* and β/ϵ, and the safety factor profile according to Eq. (16.168).

According to our normalization, the plasma boundary curve C of the Soloviev equilibria is given by

$$\psi(x,y) = 1\,, \tag{16.177}$$

determined by the expression (16.174) in terms of the parameters ϵ, σ and τ. Resulting flux contour plots are shown in Fig. 16.11 for some representative parameter values. The qualitative geometry

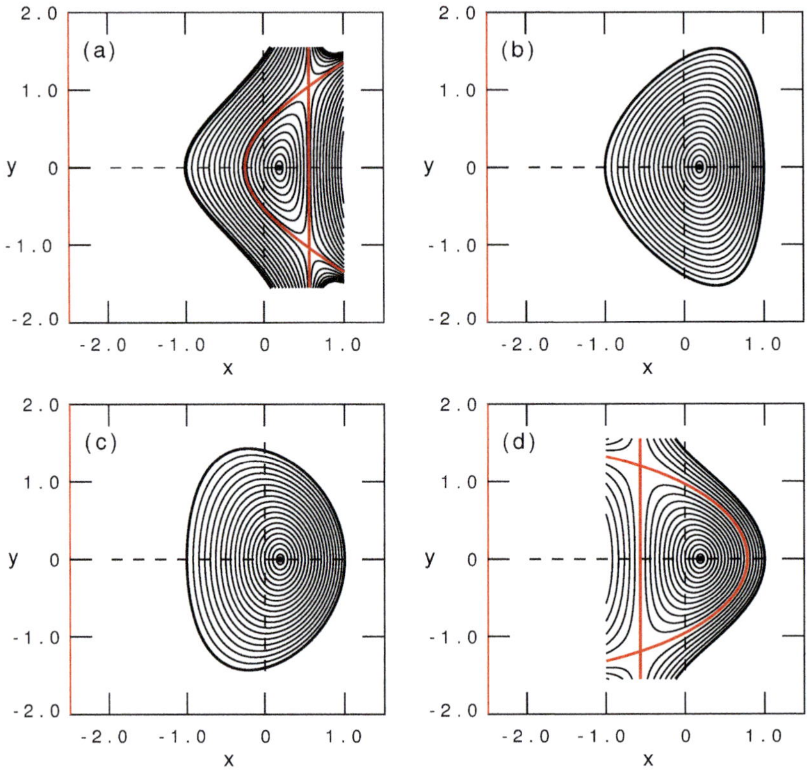

Fig. 16.11 Flux contours $0 \le \sqrt{\psi} \le 1$ for Soloviev equilibria, $\epsilon = 0.4$, $\sigma = 1.4$, (a) open configuration with separatrix on the right, $\tau = -2.0$; (b) closed configuration with negative triangularity, $\tau = -0.5$; (c) closed configuration with positive triangularity, $\tau = 1.0$; (d) open configuration with separatrix on the left, $\tau = 2.5$. The plasma boundary ($\psi = 1$) is indicated in thick black, the axis of symmetry and the separatrices in red.

of the flux contours is determined by the extrema of the flux, $\psi_x = \psi_y = 0$, which are elliptic or hyperbolic points depending on the value of the discriminant,

$$D \equiv \psi_{xx}\psi_{yy} - 4\psi_{xy}^2 \begin{cases} > 0 & \text{(elliptic point)} \\ < 0 & \text{(hyperbolic point)} \end{cases} . \tag{16.178}$$

This yields the following characteristic features of the Soloviev equilibria:

(a) the central point on the symmetry axis, $x = -1/\epsilon, y = 0$, is elliptic for $\tau > 1$ and hyperbolic for $\tau < 1$;

(b) the magnetic axis at $x = \delta, y = 0$, given by Eq. (16.175), is elliptic if $\tau > -1/\epsilon^2$;

(c) the separatrices through the hyperbolic points

$$x_s = -(1/\epsilon)\big[1 - \sqrt{1 - 1/\tau}\,\big], \quad y_s = \pm[\sigma/(\epsilon\tau)]\sqrt{\tfrac{1}{2}(1 + \epsilon^2\tau)/(1 - \tfrac{1}{4}\epsilon^2)} \tag{16.179}$$

enter the plasma domain from the right ($x_s < 1$) if $\tau < -1/\big[\epsilon(2 + \epsilon)\big]$, and they enter from the left if $\tau > 1/[\epsilon(2-\epsilon)]$. A system of closed flux contours requires the value of τ to be restricted within these limits. These features are manifest in Fig. 16.11.

These restrictions of the Soloviev equilibria should not be taken to be illustrative of the ultimate limit $\epsilon\beta_{\rm p,max} \sim 1$ of tokamaks at high-beta, defined in Eq. (16.71) of Section 16.1.4. This is evident from the orders of magnitude $\delta \sim \epsilon$ and $B \sim \epsilon$, which imply that Soloviev equilibria are essentially low-β equilibria with artificial limits $\epsilon\beta_{\rm p,max} \sim \epsilon$ due to the way in which they were constructed. If one chooses $\delta = \mathcal{O}(1)$ and solves Eq. (16.173) with the same boundary (16.177), one obtains valid high-beta equilibria, but they will not be of the Soloviev polynomial type.

To analyse the waves and instabilities of a toroidal equilibrium, we need to be able to construct the poloidal flux coordinates corresponding to the magnetic surfaces. One of the merits of the Soloviev equilibrium is that this can be done explicitly by means of a Pythagorean decomposition of ψ:

$$\begin{aligned} f &\equiv \sqrt{x - \tfrac{1}{2}\epsilon(1-x^2)} \\ g &\equiv \sqrt{(1-\tfrac{1}{4}\epsilon^2)[1+\epsilon\tau x(2+\epsilon x)]}\,\frac{y}{\sigma} \end{aligned} \quad \Rightarrow \quad \left\{ \begin{aligned} \psi(x,y) &= f^2(x) + g^2(x,y) \\ \theta(x,y) &\equiv \arctan(g/f) \end{aligned} \right. , \tag{16.180}$$

with the inverse:

$$\left\{ \begin{aligned} x(\psi,\theta) &= \epsilon^{-1}\Big[\sqrt{1+2\epsilon\sqrt{\psi}\cos\theta+\epsilon^2} - 1\Big] \\ y(\psi,\theta) &= \frac{\sigma\sqrt{\psi}\sin\theta}{\sqrt{(1-\tfrac{1}{4}\epsilon^2)(1+2\epsilon\tau\sqrt{\psi}\cos\theta+\epsilon^2\tau)}} \end{aligned} \right. . \tag{16.181}$$

An example of the coordinates is shown in Fig. A.9 of Appendix A.3.2. In the limit $\epsilon \to 0$, they transform into genuine straight-field-line coordinates ψ, ϑ, with inverse $x = \sqrt{\psi}\cos\vartheta$, $y = \sigma\sqrt{\psi}\sin\vartheta$, describing a 'straight low-beta tokamak' with elliptical cross-section.

16.3.3 Numerical equilibria*

In this section, we will discuss two numerical methods for solving the Grad–Shafranov equation, where the plasma boundary shape is fixed and given a priori as a bounding flux contour. The emphasis is on computing the nested flux contours interior to this boundary to a high degree of accuracy, ensuring among other things a continuous representation of ψ and $\nabla\psi$. The essential form of the Grad–Shafranov equation can be written as

$$\Delta^*\psi = \widetilde{F}(\psi)\,, \tag{16.182}$$

where the nonlinear ψ dependence is found in the right hand side. It is then convenient to use Picard iteration to converge onto the solution such that, from a starting guess $\psi^{(0)}(x,y)$, the iterate solves the linearized problem

$$\Delta^*\psi^{(n+1)} = \widetilde{F}(\psi^{(n)})\,, \tag{16.183}$$

which is stopped when e.g. $\max|\psi^{(n+1)} - \psi^{(n)}| \leq \epsilon_{\rm tol}$. The max-norm runs over all grid points and tolerances $\epsilon_{\rm tol}$ can be taken as low as, or smaller than, 10^{-8}, depending on machine precision.

(a) Conformal mapping One of the earlier numerical approaches to tokamak equilibrium exploited a conformal mapping to transform the circular plasma cross-section of a high-beta tokamak onto itself while shifting the image of the magnetic axis to the centre of the coordinates [187, 216]. The Moebius transformation $z \equiv x + \mathrm{i}y \to w$, and its inverse, effecting this are given by

$$w(z) = \frac{z - \delta}{1 - \delta z} \qquad \Longleftrightarrow \qquad z(w) = \frac{w + \delta}{1 + \delta w} . \tag{16.184}$$

It is then expedient to exploit polar coordinates in the mapped plane, $w \equiv s \exp(\mathrm{i}t)$, so that the two-dimensional Laplacian part of $\Delta^* \psi$ simply transforms through the explicitly known scale factor $h(s,t)$:

$$\psi_{xx} + \psi_{yy} = \frac{1}{h^2}\Big(\frac{1}{s}\frac{\partial}{\partial s}\, s\, \frac{\partial \psi}{\partial s} + \frac{1}{s^2}\frac{\partial^2 \psi}{\partial t^2}\Big) , \qquad h(s,t) \equiv |dz/dw| . \tag{16.185}$$

To fully exploit the power of analytic functions, it is essential to modify the Picard iterate of the scaled Grad–Shafranov equation (16.160) such that the two-dimensional Laplacian appears on the LHS,

$$\begin{aligned} \Big(\frac{1}{s}\frac{\partial}{\partial s}\, s\, \frac{\partial}{\partial s} + \frac{1}{s^2}\frac{\partial^2}{\partial t^2}\Big) \psi^{(n+1)} &= F^{(n)}(s,t) \\ &\equiv h^2 \Big[A^{(n)} \Gamma(\psi^{(n)}) + B^{(n)} x (1 + \tfrac{1}{2}\epsilon x) \Pi(\psi^{(n)}) + \frac{\epsilon}{1 + \epsilon x}\, \psi_x^{(n)} \Big] , \end{aligned} \tag{16.186}$$

whereas all the terms on the RHS are explicitly known on the (s,t) grid from the previous step, including the eigenvalues $A^{(n)}$ and $B^{(n)}$, as we will see. (A further simplification, which we will not exploit here, is to assume the high-beta tokamak approximation, so that the two terms with ϵ on the RHS vanish.)

Although the flux surfaces in the mapped plane are not circular (except for the outermost one), it is clear that mapping the magnetic axis onto the origin is the crucial step to facilitate Fourier analysis of the solutions with only few harmonics. This is very much like the Shafranov shifted circle representation, except that the present one is not restricted to small shifts. Exploiting a truncated Fourier series,

$$\psi^{(n+1)}(s,t) = \sum_{m=0}^{M} (1 - \tfrac{1}{2}\delta_{m0}) \hat{\psi}_m^{(n+1)}(s) \cos mt , \tag{16.187}$$

and similarly for $F^{(n)}(s,t)$, the Fourier components of the Picard iterate become simple ODEs,

$$\frac{1}{s}\frac{d}{ds}\, s\, \frac{d}{ds} \hat{\psi}_m^{(n+1)} - \frac{m^2}{s^2} \hat{\psi}_m^{(n+1)} = \hat{F}_m^{(n)} , \tag{16.188}$$

whereas the two boundary conditions (16.161), prescribing the value $\psi = 1$ at the plasma boundary, and (16.162), prescribing $\psi = \psi_x = \psi_y = 0$ at the position $x = \delta$, $y = 0$ of the magnetic axis, become conditions on the harmonics:

$$\hat{\psi}_m^{(n+1)}(1) = 2\delta_{m0} , \qquad \hat{\psi}_m^{(n+1)}(0) = \hat{\psi}_m^{(n+1)\,\prime}(0) = 0 . \tag{16.189}$$

The solutions of the ODEs (16.188) satisfying all these BCs, except the first one for $m = 0$ and the second one for $m = 1$, are simple integrals:

$$\hat{\psi}_0^{(n+1)}(s) = \int_0^s ds'\, s'^{-1} \int_0^{s'} ds'' s'' \hat{F}_0^{(n)}(s'') ,$$

$$\hat{\psi}_m^{(n+1)}(s) = s^m \int_1^s ds'\, s'^{-2m-1} \int_0^{s'} ds'' s''^{m+1} \hat{F}_m^{(n)}(s'') \quad (m \neq 0) . \tag{16.190}$$

These expressions still contain the unknown values of A and B. They are determined by the BC on $m = 0$,

$$\hat{\psi}_0^{(n+1)}(1) = 2 \quad \Rightarrow \quad A , \tag{16.191}$$

and the BC on $m = 1$,

$$\hat{\psi}_1^{(n+1)\,\prime}(0) = -\tfrac{1}{2} \int_0^1 (1 - s^2) \hat{F}_1^{(n)}(s)\, ds = 0 \quad \Rightarrow \quad B . \tag{16.192}$$

Thus, the full solution $\psi^{(n+1)}(s,t)$ at the $(n+1)$th step is found, requiring no additional iterations to determine the eigenvalues A and B.

This completes the solution of the Grad–Shafranov equation with conformal mapping. For reasonable profiles $\Gamma(\psi)$ and $\Pi(\psi)$, the convergence is fast and accurate with few harmonics. No further operations are needed than fast Fourier transforms in the angular variable and integrations over the radial variable. Important for stability analysis is that the first and second derivatives (magnetic field and currents) are obtained with the same accuracy as the flux function $\psi(s,t)$ itself.

To generalize the conformal mapping technique to arbitrary cross-sections, pre-mappings could be applied that are known from the design of wings for airplanes and biplanes; see Fig. A.6 of Appendix B. These map an arbitrary simply connected domain onto a circle, or of a doubly connected domain onto an annulus, by means of the nonlinear integral equations of Theodorsen or Garrick, respectively. These integral equations were converted by Henrici [300] into effective numerical algorithms introducing the fast Hilbert transform (FHT), a powerful counterpart of the fast Fourier transform (FFT), for conjugate periodic functions. This results in beautiful analysis of the equilibrium and stability of high-beta tokamaks, in particular when the plasma current is confined to the surface so that both the plasma region and the vacuum are described by two-dimensional Laplace equations [217]. In that case, complex analysis provides the full solution in terms of a one-dimensional variational principle involving a double angular integral over the plasma boundary. For diffuse current distributions, the radial dependence of the perturbations could be effectively described by means of a new set of polynomials [216](c) that, together with FFTs, provide the general solution of incompressible fluid flow in arbitrary domains. Stability of tokamaks at high beta was extensively analysed with the resulting program HBT [205, 236].

One complication of the conformal mapping of an arbitrary domain onto a circular disc is the fact that the distribution of angular points is not free. For example, mapping of an elliptical domain onto a circle results in a distribution which is sparse on the curved parts and crowded on the flat parts of the ellipse (just like the orthogonal Ψ, χ, φ coordinates described in Section 16.1.2), precisely opposite to what is needed in stability analysis. One could remedy this by angular grid accumulation, but this spoils most of the beauty of obtaining explicit results from complex analysis. This is one of the reasons to consider the more flexible technique of the next sub-section.

(b) Two-dimensional finite elements Later numerical approaches of solving the Grad–Shafranov equation still exploit the Picard iteration scheme (16.183), but they are more tuned to general solution strategies, exploiting two-dimensional finite elements and Galerkin methods with the essential elements of proven implementations, already encountered in Section 15.1.3, as found in [319, 47, 44, 422].

To specify the newer solution algorithms, we need to explain the choice of the grid on which the solution is computed, the way in which the solution is approximated numerically on this grid and the manner in which the PDE (16.182) is turned into a discrete (linear algebraic) problem. A proven strategy which allows for a continuous $\nabla\psi$ is to use a third order finite element method (FEM). On a unit square $[-1,1]^2$, we can define up to sixteen bicubic Hermite polynomials, four per corner $(x_0, y_0) = (\pm 1, \pm 1)$, namely

$$\begin{aligned}
H_{00}(x,y) &= \tfrac{1}{16}(x+x_0)^2(xx_0-2)(y+y_0)^2(yy_0-2)\,,\\
H_{10}(x,y) &= -\tfrac{1}{16}x_0(x+x_0)^2(xx_0-1)(y+y_0)^2(yy_0-2)\,,\\
H_{01}(x,y) &= -\tfrac{1}{16}(x+x_0)^2(xx_0-2)y_0(y+y_0)^2(yy_0-1)\,,\\
H_{11}(x,y) &= \tfrac{1}{16}x_0(x+x_0)^2(xx_0-1)y_0(y+y_0)^2(yy_0-1)\,.
\end{aligned} \tag{16.193}$$

This allows us to approximate any function $f(x,y)$ on $[-1,1]^2$ by the expansion

$$\begin{aligned}
f(x,y) = \sum_{x_0,y_0} \Big[& H_{00}(x,y) f(x_0,y_0) + H_{10}(x,y)\frac{\partial f}{\partial x}(x_0,y_0)\\
& + H_{01}(x,y)\frac{\partial f}{\partial y}(x_0,y_0) + H_{11}(x,y)\frac{\partial^2 f}{\partial y \partial x}(x_0,y_0)\Big]\,.
\end{aligned} \tag{16.194}$$

The function then has a prescribed functional dependence on the square as encoded in the bicubic expansion polynomials, and the sixteen expansion coefficients are the local corner values of the function f and its derivatives.

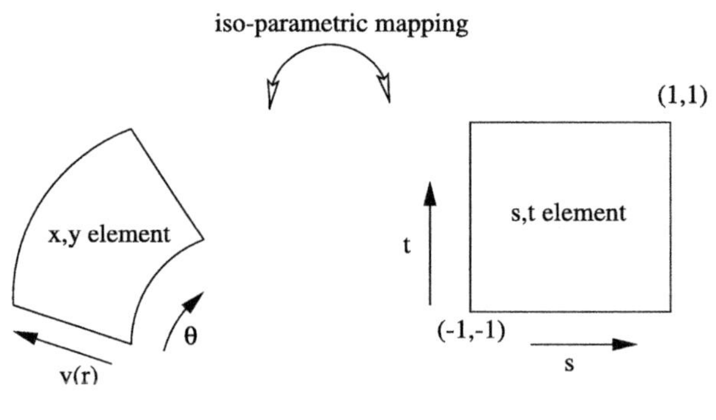

Fig. 16.12 Iso-parametric coordinate mapping of a curved quadrilateral to the local s, t coordinates on the square $[-1,1]^2$.

For a given cross-sectional shape, the grid should allow alignment with the usually curved boundary. In the (x, y) plane, polar coordinates (r, θ) centred at $(x, y) = (0, 0)$ can be used to represent a given boundary curve in a (suitably truncated) Fourier series $r = f_b(\theta) = \sum_m a_m \exp(im\theta)$. We

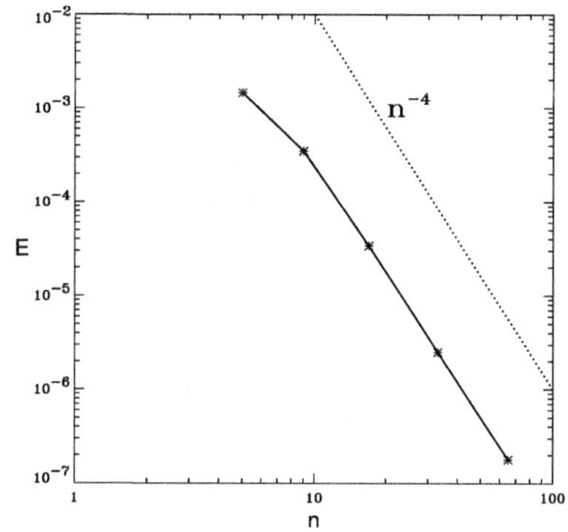

Fig. 16.13 Fourth order convergence in the max-norm of the error $E \equiv \max|\psi - \psi_{\rm exact}|$ in a computed Soloviev equilibrium for increasing grid resolution $n \equiv n_r = n_\theta$.

can then use an arbitrary 'radial' function $v(r)$ from 0 (centre) to 1 (boundary) in global coordinates following the boundary shape, where

$$x = v(r) f_{\rm b}(\theta) \cos\theta \,, \qquad y = v(r) f_{\rm b}(\theta) \sin\theta \,. \tag{16.195}$$

Our grid is then formed by taking a discrete number of n_r radial points r_i, and of n_θ angular points θ_j. The grid consists of curved quadrilateral elements, while we discussed the FE representation of a function on a unit element $[-1,1]^2$. The numerical representation of the solution on the boundary-fitted grid is complete by making use of an iso-parametric mapping, where each curved quadrilateral is mapped onto $[-1,1]^2$ by changing from (x,y) to local (s,t) coordinates (see Fig. 16.12). In fact, this mapping consists of representing both the solution and the coordinates x, y in the same FE representation, i.e.

$$\begin{aligned} \psi(x,y) &= \sum \Big[H_{00}(s,t)\psi(s_0,t_0) + H_{10}(s,t)\frac{\partial\psi}{\partial s}(s_0,t_0) \\ &\qquad + H_{01}(s,t)\frac{\partial\psi}{\partial t}(s_0,t_0) + H_{11}(s,t)\frac{\partial^2\psi}{\partial s\partial t}(s_0,t_0)\Big] \,, \\ x(s,t) &= \sum \Big[H_{00}(s,t)x(s_0,t_0) + H_{10}(s,t)\frac{\partial x}{\partial s}(s_0,t_0) + \cdots \Big] \,, \\ y(s,t) &= \sum \Big[H_{00}(s,t)y(s_0,t_0) + H_{10}(s,t)\frac{\partial y}{\partial s}(s_0,t_0) + \cdots \Big] \,. \end{aligned} \tag{16.196}$$

Identifying $v(r) \equiv s$ and $\theta \equiv t$, it is possible to compute the consistent FE representation of the coordinates (as one can calculate x, $\partial x/\partial s$ etc. at all (s_0,t_0)). One can use this flexibility (after each Picard iterate or after Picard convergence) to take the local s coordinate to become some function of ψ, and align the grid with ψ flux contours.

Within each Picard step, one solves the discrete equivalent of Eq. (16.183). In FE methodology, we formulate the problem in its weak form, where we select a space of test-functions χ and look for

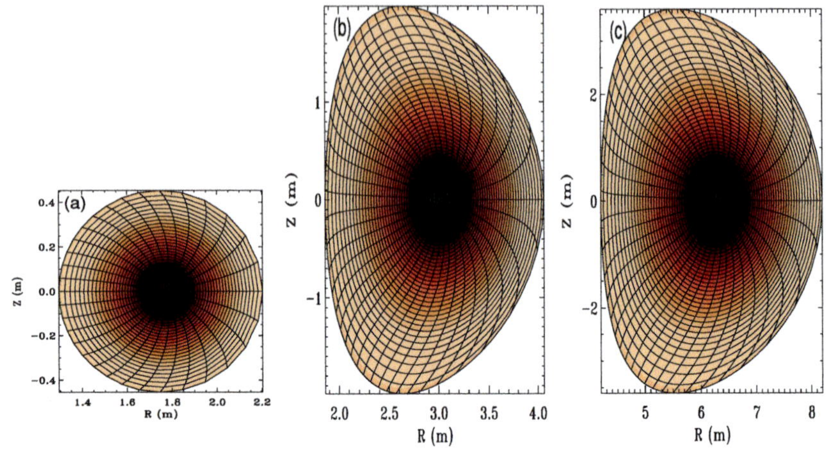

Fig. 16.14 The equilibrium flux surfaces of (a) TEXTOR, (b) JET and (c) ITER, normalized to the same width and calculated for the same normalized pressure and current distribution.

the solution ψ such that for all test functions we have

$$\int_V \chi \nabla \cdot (R^{-2} \nabla \psi)\, dV = \int_V \chi\, (R^{-2} \widetilde{F})\, dV \,, \tag{16.197}$$

where V is the plasma volume. In the Galerkin method, the test functions are simply the finite elements (H_{00}, etc.) which are already used in the representation for ψ. This reduces the problem to a linear system $\mathsf{K} \cdot \mathbf{x} = \mathbf{b}$, where the vector $\mathbf{x}$ represents the unknown coefficients in the FE representation, while the K matrix and $\mathbf{b}$ vector elements are integrals of known functions (involving the FEs and their derivatives). These integrals can e.g. be evaluated numerically by Gaussian quadrature. Figure 16.13 shows the expected quartic convergence reached when a Soloviev solution is determined numerically for increasing grid resolution $n \equiv n_r = n_\theta$ going from 5 up to 65. As the exact solution is known, the error can be quantified precisely and the fourth order convergence in ψ shows that very accurate solutions are already obtained for rather coarse grids.

Figure 16.14 shows the flux surface distributions for three ‘generations’ of tokamaks, calculated with the program HELENA [319]/FINESSE [44] described in this section. The three cases assumed the same flux profiles and correspond to low-beta configurations that differ in their geometric parameters: TEXTOR has a circular cross-section and $a = 0.45\,\mathrm{m}$, $R_0 = 1.75\,\mathrm{m}$, JET has a D-shaped cross-section (finite ellipticity and triangularity) with $a = 1.1\,\mathrm{m}$, $R_0 = 2.96\,\mathrm{m}$, while ITER will scale up to $a = 2\,\mathrm{m}$, $R_0 = 6.2\,\mathrm{m}$. Shown are computed pressure distributions and the nested flux surfaces, where the angular coordinate lines shown represent the straight-field-line coordinates.

16.4 Extensions

16.4.1 Toroidal rotation

The purely static, axi-symmetric, equilibrium can be generalized in several ways. Avoiding the significant complications due to poloidal flow (see Chapter 18), a first non-trivial extension of the Grad–Shafranov equilibrium includes the effect of toroidal rotation and the associated centrifugal force. The toroidal component of the stationary induction equation $\nabla \times (\mathbf{v} \times \mathbf{B}) = 0$ then dictates that each flux surface rotates at fixed angular velocity $\Omega(\Psi) = v_\varphi/R$.

The azimuthal component of the force balance equation still prescribes the current stream function I to be a flux function $I(\Psi) = RB_\varphi$. However, the poloidal part now requires two equations to be satisfied simultaneously. In the $\nabla\Psi$ direction, a Grad–Shafranov-like equation is obtained, namely

$$R\frac{\partial}{\partial R}\left(\frac{1}{R}\frac{\partial \Psi}{\partial R}\right) + \frac{\partial^2 \Psi}{\partial Z^2} = -I\frac{dI}{d\Psi} - \frac{\partial p}{\partial \Psi}R^2 = Rj_\varphi\,, \tag{16.198}$$

where the pressure is no longer a flux function. The dependence of the pressure $p = p(\Psi, R)$ is such that along the poloidal flux contours the force balance is ensured [618](b) by

$$\left.\frac{\partial p}{\partial R}\right|_\Psi = \rho R \Omega^2(\Psi)\,. \tag{16.199}$$

The latter equation can, e.g., be solved analytically under the additional assumption that the entropy is a flux function, $S \equiv p\rho^{-\gamma} = S(\Psi)$, a result which generalizes to all stationary MHD equilibria since $\mathbf{v}\cdot\nabla S = 0$ (with poloidal flows as well). The pressure can then be written as

$$p(\Psi, R) = \left[R^2\frac{\gamma-1}{\gamma}\frac{\Omega^2}{2S^{1/\gamma}} + f\right]^{\gamma/(\gamma-1)}\,, \tag{16.200}$$

so that for toroidally rotating equilibria four flux functions can be freely chosen: besides $I(\Psi)$ and $\Omega(\Psi)$ we get, e.g., $S(\Psi)$ and $f(\Psi)$.

For the purpose of equilibrium reconstruction for toroidally rotating tokamak plasmas where diagnostic information on density and temperature profiles is available, a convenient parameterization uses the corresponding static pressure $p_{\rm st}(\Psi)$ and density $\rho_{\rm st}(\Psi)$ profiles with the same entropy variation $S(\Psi)$, as follows:

$$p(\Psi, R) = p_{\rm st}(\Psi)\left[(R^2 - R_0^2)\frac{\gamma-1}{\gamma}\frac{\Omega^2\rho_{\rm st}}{2p_{\rm st}} + 1\right]^{\gamma/(\gamma-1)}\,. \tag{16.201}$$

This is useful as long as the toroidal rotation is low and its small influence on the equilibrium properties is to be quantified in comparison with a similar static reconstruction. The pressure can be shown to have its maximum shifted radially outward with respect to the magnetic axis, which in turn is shifted outward due to toroidal rotation.

As an example, consider the influence of a relatively large toroidal rotation on the pressure profile, as shown in Fig. 16.15 for a circular tokamak cross-section of inverse aspect ratio $\epsilon = 0.26$, and parameterized as follows:

$$\begin{aligned} I^2 &= A\left[1 - 0.01\psi + 0.005\psi^2\right]\,, & \Omega &= 0.075\left[1 - 0.999\psi\right]\,, \\ p_{\rm st} &= 0.0125\,A\left[1 - 2\psi + 1.001\psi^2\right]\,, & \rho_{\rm st} &= A\left[1 - \psi + 0.6\psi^2 - 0.5\psi^4\right]\,. \end{aligned} \tag{16.202}$$

The last two profiles only enter in the computation when rotation is considered. The eigenvalue A determined numerically is 155 for the rotating equilibrium, versus 164 in the static case. The rotation would correspond to a maximal sonic Mach number $M_{\rm s} = v_\varphi/\sqrt{\gamma p/\rho} = 1.06$, which is rather high but illustrates well the influence of toroidal rotation.

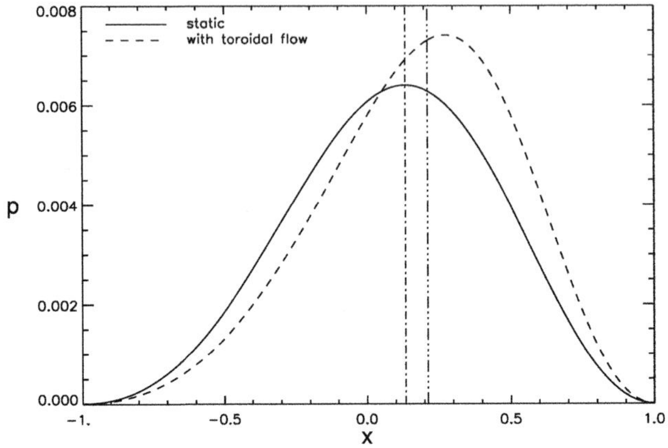

Fig. 16.15 The radial pressure profile for a toroidally rotating equilibrium compared to that of a static one. Note the outward shift of the magnetic axis (vertical dashed lines) and the separation of the pressure maximum from the magnetic axis when a large rotation is present.

16.4.2 Gravitating plasma equilibria*

In astrophysical contexts, the influence of external or self-gravity can become very important in the overall force balance. In accretion discs, for example, the pure Keplerian disc balances a central gravitational field with centrifugal forces. In the solar corona, quiescent prominences can be observed suspended above the solar limb, characterized by much denser and much cooler (typical factors 10–100) plasma than the coronal environment. There, the main force balance is again determined by Lorentz forces, now opposing gravity and pressure gradients. An analytical solution for quiescent prominences can be found in [419]. A similar analysis of the static force balance, now assuming translational invariance in the z-direction, finds for force balance in the $\nabla\Psi$ direction

$$\frac{\partial^2 \Psi}{\partial x^2} + \frac{\partial^2 \Psi}{\partial y^2} = -I\frac{dI}{d\Psi} - \frac{\partial p}{\partial \Psi} = j_z \,. \tag{16.203}$$

Again, $I(\Psi) = B_z$ is a free flux function and $p = p(\Psi, y)$. Along a poloidal flux,

$$\left.\frac{\partial p}{\partial y}\right|_{\Psi} = -\rho g \tag{16.204}$$

should be satisfied when considering a constant external gravitational acceleration $\mathbf{g} = -g\mathbf{e}_y$. Assuming the temperature to be a flux function $T(\Psi)$, three free flux functions can be chosen as the pressure is then $p(\Psi, y) = k(\Psi)\exp[-g\,y/T(\Psi)]$. Using a scaling with respect to half the horizontal diameter a, the outermost flux value Ψ_1 and a reference field strength B_0, a suitable scaling writes the governing Grad–Shafranov-like equation as

$$\frac{\partial^2 \psi}{\partial x^2} + \frac{\partial^2 \psi}{\partial y^2} = -\alpha^2\left[I\frac{dI}{d\psi} + \left(\frac{d\ell}{d\psi} + \frac{\ell\, y}{H^2}\frac{dH}{d\psi}\right)\mathrm{e}^{-y/H}\right], \tag{16.205}$$

where $\alpha = a\,B_0/\Psi_1$, and the pressure $p = \ell(\psi)\mathrm{e}^{-y/H}$ uses the scale height $H = T(\psi)/(ag)$. All quantities appearing are again dimensionless, e.g., $(x/a, (y - y_0)/a) \rightarrow (x, y)$ with y_0 a reference height in the prominence.

Numerical magneto-hydrostatic solutions describing the gravitationally stratified equilibrium of cool prominence plasma embedded in a near-potential coronal field are described by Petrie et al. [486]. The solutions are calculated using the FINESSE equilibrium solver. They describe the morphologies of the magnetic field distributions in and around prominences and the cool prominence plasma that these fields support. It reproduces the three-part structure encountered in observations: a cool dense prominence within a cavity/flux rope embedded in a hot corona.

16.4.3 Challenges

To be able to describe perturbations with minimum field line bending, it is necessary to represent the parallel gradient operator (16.29) as accurately as possible. An effective method is to replace the poloidal angle θ, which could be any angle as long as it increases by 2π after one revolution the short way around the torus, by a poloidal angle ϑ such that the field lines become straight lines in the φ–ϑ plane (Fig. 16.2). This construction has to be carried out on each flux surface labelled by ψ. To exploit the constructed flux variable ψ and the new angle ϑ as coordinates for the description of instabilities, inversion of the coordinates is needed,

$$\left.\begin{aligned} \psi &= \psi(x_i, y_j) \\ \vartheta &= \vartheta(x_i, y_j) \end{aligned}\right\} \quad \Rightarrow \quad \left\{\begin{aligned} x &= x(\psi_i, \vartheta_j) \\ y &= y(\psi_i, \vartheta_j) \end{aligned}\right. , \tag{16.206}$$

as exemplified by Eq. (16.181) for the Soloviev equilibrium. All quantities occurring in the stability analysis are then to be transformed to ψ, ϑ, φ coordinates, e.g. the normal field line curvature κ_n, the geodesic curvature κ_g, the toroidal current j_φ, etc. (see Section 17.1.2 of the next chapter). These quantities involve second derivatives with respect to ψ, so that the *equilibrium solutions need to be surprisingly accurate* for a reliable stability analysis.

In this chapter, we have presented a theoretical description of toroidal equilibrium, which involves the required accurate solution of the Grad–Shafranov equation and a careful choice of the parameters. The experimental counterpart is the production of *accurate equilibrium data* by means of all available diagnostics. To get agreement between these two is a long-term iterative process. At the present time, the situation is far from satisfactory: important parameters and profiles are only known to about 10%. Consequently, tokamak diagnostics need to be improved if we ever are to arrive at the agreement between theory and experiment that is standard in physics.

A systematic way of approaching this problem has been called *MHD spectroscopy* [237]; see Section 7.2.4. Here, the frequency spectrum of MHD waves is used to obtain equilibrium information (the inverse spectral problem). This spectrum is calculated by means of a spectral code with input of an equilibrium obtained by fitting experimental data to get an approximation of the free profiles. However, this involves considerable *uncertainty* as to how to handle deviations from nested flux surfaces, see e.g. the rather crude method of fitting diagnostic data with a symmetric flux function profile presented in Fig. 16.16.

We have extensively discussed the problem of *freedom* in the choice of equilibrium profiles. This should be clearly distinguished from the just mentioned inaccuracy and uncertainty. In principle, the latter freedom can be eliminated by experimental data. More substantial are the problems of *three-dimensionality* of the equilibrium, associated with deviations from axi-symmetry due to field errors, saturated nonlinear instabilities, possibly leading to non axi-symmetric equilibrium states,

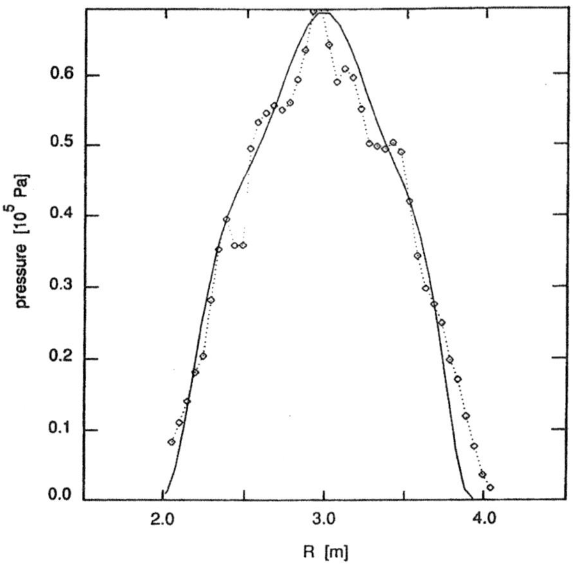

Fig. 16.16 Uncertainty: electron pressure profile obtained by LIDAR diagnostic at JET (dashed curve with diamonds) and reconstructed pressure profile by means of an equilibrium solver (drawn curve). (From Huysmans et al. [320].)

stochastic field lines, etc. Another area of study is the effect of *stationary flow* ($\mathbf{v}_0 \neq 0$) on equilibria, i.e. departure from static equilibrium in connection with the effects of neutral beams, pumped divertors, etc. (see Chapter 18). Here, three more free equilibrium profiles appear (the velocity and the density). The problem of infinite freedom remains with us!

16.5 Literature and exercises

Notes on literature

Axi-symmetric equilibrium

- The original papers on tokamak equilibrium by the 'founding father' Shafranov [548] remain beautiful examples of down-to-earth theoretical analysis with inescapable experimental consequences. Over the years, his approach of general calculations interspersed with explicit applications has widely expanded, as presented in subsequent volumes of *Reviews of Plasma Physics* by himself [549], and with collaborators Soloviev [555], and Zakharov [664].
- Classical treatises on the structure of magnetic fields by Morozov and Soloviev [447], in *Reviews of Plasma Physics*, Volume 2, and on closed magnetic configurations by Soloviev [554], in Volume 6, contain a wealth of information on stability of magnetic fields, natural coordinates, equilibria with shaped cross-sections, etc.

Basic equilibrium concepts

- The recently revised classical textbook *Ideal MHD* by Freidberg [186], Chapters 4–7, presents the basic equilibrium considerations and extends this to cylindrical, 2D (tokamaks) and 3D equilibria (stellarators). His textbook on *Plasma Physics and Fusion Energy* [185], in particular Chapter 11, continues this practical presentation, always keeping eventual application to fusion energy in mind.

- Axi-symmetric equilibrium is extensively discussed in the textbooks by Miyamoto, *Plasma Physics for Nuclear Fusion* [442](a), Chapter 7, and *Plasma Physics and Controlled Fusion* [442](b), Chapter 6. The latter also discusses the different experimental approaches of tokamak, reversed field pinch, stellarator and inertial confinement to fusion.
- Chapter 3 of the 'manual' on tokamaks by Wesson and collaborators [647] contains the essential concepts of flux functions, safety factor, beta, and many simple one-dimensional model calculations (à la Shafranov) on tokamak equilibrium problems.
- Summerschool proceedings [310, 162] present equilibrium theory together with many experimental contributions, demonstrating the lively research field of controlled thermonuclear fusion.

Exercises

[16.1] *Thin plasma slab*

In this exercise you will take a look at a thin plasma slab about the mid-plane. In this slab, it is reasonable to assume that all physical quantities depend only on the radius R. Show that the Grad–Shafranov equation reduces to

$$\frac{d}{dR}\left(p + \tfrac{1}{2}B^2\right) + \frac{B_\varphi^2}{R} = 0\,,$$

which is the force balance equation of cylindrical plasma as discussed in Section 12.2.1.

[16.2] *Shafranov shift*

In this exercise you are going to derive an analytical expression for the Shafranov shift. This is the difference between the centre of each flux surface and the geometric axis. Assume that the inverse aspect ratio ϵ is small ($\ll 1$) and that the outer plasma boundary is circular to first order. In this case the magnetic flux function Ψ can be represented as $\Psi(\hat{r},\hat{\theta}) = \Psi(\hat{r})$, where $(\hat{r},\hat{\theta})$ are non-orthogonal polar coordinates defined with respect to the centre of the flux surface. The connection with the cylindrical coordinates is given by

$$R = R_0 + r\cos(\hat{\theta}) + \Delta(\hat{r})\,, \qquad Z = \hat{r}\sin(\hat{\theta})\,,$$

where $\Delta(\hat{r})$ is the Shafranov shift. This shift has the boundary conditions $\Delta(0) = \Delta_0$ and $\Delta(1) = 0$, where Δ_0 is the shift of the magnetic axis with respect to $R = R_0$.

- Derive expressions for $R_{\hat{r}}$, $R_{\hat{\theta}}$, $Z_{\hat{r}}$ and $Z_{\hat{\theta}}$.
- Derive the co- and contravariant metric tensors in these new coordinates $(\hat{r},\hat{\theta})$.
- Derive an expression for the Jacobian J.
- Derive an expression for the covariant magnetic field components.
- Show that

$$R^2\nabla\cdot\left(\frac{1}{R^2}\nabla\Psi\right) \approx \frac{\partial^2\Psi}{\partial\hat{r}^2} + \frac{1}{\hat{r}}\frac{\partial\Psi}{\partial\hat{r}} - \cos\hat{\theta}\left[2\Delta'\frac{\partial^2\Psi}{\partial\hat{r}^2} - \frac{1}{\hat{r}}\left(\epsilon\hat{r} + \Delta' + \Delta''\hat{r}\right)\frac{\partial\Psi}{\partial\hat{r}}\right].$$

- Show that

$$-\tfrac{1}{2}\frac{dI^2}{d\Psi} \approx -\frac{1}{2R_0B_{\mathrm{p}0}}\frac{dI^2}{d\hat{r}}\,, \qquad -R^2\frac{dp}{d\Psi} \approx -\frac{R_0}{B_{\mathrm{p}0}}\left(1 + 2\frac{\hat{r}}{R_0}\cos\hat{\theta}\right)\frac{dp}{d\hat{r}}\,.$$

- Show that the zeroth order reduces to

$$\frac{d}{d\hat{r}}\left(p + \tfrac{1}{2}B_{\varphi 0}^2 + \tfrac{1}{2}B_{\mathrm{p}0}^2\right) + \frac{B_{\mathrm{p}0}^2}{\hat{r}} = 0\,.$$

Compare this with the equation found in Exercise [16.1] and discuss the differences.

– Show that the first order yields the equation for the Shafranov shift,

$$\frac{d\Delta}{d\hat{r}} = -\frac{1}{\hat{r}R_0 B_{\mathrm{p}0}^2}\int_0^{\hat{r}} \left(\hat{r}B_{\mathrm{p}0}^2 - 2\hat{r}^2\frac{dp}{d\hat{r}}\right)d\hat{r}\,.$$

– Derive the expression for the Shafranov shift defined by Shafranov himself.

[16.3] *Shafranov shift and Wesson profiles*

In this exercise, you will derive an analytical expression for the Shafranov shift using the profiles defined by Wesson [647] for the pressure and toroidal current density,

$$p = p_0(1-\overline{r}^2)\,, \qquad j_{\varphi 0} = j_0(1-\overline{r}^2)^{\nu}\,.$$

Here, $\overline{r} \equiv \hat{r}/a$, where a is the plasma radius. For this exercise we set $\nu = 1$, which means that both the pressure and the toroidal current density are parabolic functions.

– Derive the expressions for the poloidal and the toroidal magnetic field, B_{p} and B_{φ}.

– From the toroidal magnetic field, derive the expression for the safety factor q.

– Find the expression for the poloidal beta, β_{p}.

– Finally, derive the expression for the derivative of the shift, Δ'.

[16.4] *Equilibria with toroidal flow and gravity*

In this exercise, you will extend the equilibrium with toroidal flow and gravity. The extension with flow is relevant in present tokamak experiments because the plasma rotates due to neutral beam injection. On the other hand, toroidal flow and gravity are essential ingredients of accretion discs. The additional inclusion of poloidal flow will be discussed in Chapter 18. The equilibrium has to satisfy Eqs. (12.24)–(12.27).

– Show that $\Omega = v_{\varphi}/R$ is a flux function, where v_{φ} is the toroidal velocity.

– Show that the projection of the momentum equation parallel to the poloidal magnetic field leads to the following two equations:

$$\left.\frac{\partial p}{\partial R}\right|_{\Psi=\mathrm{const}} = \rho\left(R\Omega^2 - \frac{\partial\Phi_{\mathrm{gr}}}{\partial R}\right), \qquad \left.\frac{\partial p}{\partial Z}\right|_{\Psi=\mathrm{const}} = -\rho\frac{\partial\Phi_{\mathrm{gr}}}{\partial Z}\,.$$

– Show that the projection perpendicular to the poloidal magnetic field lines results in the extended Grad–Shafranov equation (note the partial derivative on p!)

$$R^2\nabla\cdot\left(\frac{1}{R^2}\nabla\Psi\right) = -I\frac{dI}{d\Psi} - R^2\frac{\partial p}{\partial\Psi}\,.$$

– This extended Grad–Shafranov equation can be further specialized by assuming that the temperature, the density or the entropy is a flux function. Derive the equation for the pressure for all three cases.

– Derive the specialized extended Grad–Shafranov equation for all three choices.

[16.5] *Shafranov shift and toroidal flow*

In Exercise [16.2] you have derived the Shafranov shift for static toroidal plasmas. Repeat this exercise, including toroidal flow now. Show that the three specialized extended Grad–Shafranov equations just derived result in the same equation for the 'Shafranov' shift, i.e.

$$\frac{d\Delta}{d\hat{r}} = -\frac{1}{\hat{r}R_0 B_{\mathrm{p}0}^2}\int_0^{\hat{r}} \left[\hat{r}B_{\mathrm{p}0}^2 - \hat{r}^2\frac{d}{d\hat{r}}\left(2p + \rho v_{\varphi 0}^2\right)\right]d\hat{r}\,.$$

Comment on this result.

17

Linear dynamics of static toroidal plasmas

17.1 "Ad more geometrico"

17.1.1 Alfvén wave dynamics in toroidal geometry

It was shown in Chapters 6–12 that spectral theory of MHD waves and instabilities essentially concerns the *dynamics of Alfvén waves* in the environment of magnetized plasmas. Since Alfvén waves travel along the magnetic field lines, and the field lines in turn are constrained to the nested magnetic surfaces in an axi-symmetric toroidal plasma, this implies that *the geometry of the magnetic field lines and of the constraining magnetic surfaces becomes the all-determining factor for MHD spectral theory of toroidal plasmas*. Recalling the 'grand vision' of Section 12.1.1 on magnetized plasmas occurring everywhere in the Universe, it is appropriate at this point to call upon the great examples of general relativity, where light waves propagate along geodesics, and upon the dream of philosophers (Spinoza) to construct the theoretical understanding of the world "ad more geometrico" (in the geometrical manner). As we will see, even when fusion applications are the main concern, it pays off to exploit the ready-made concepts of geometry expanded by the great scientists of the past.

Recall that Alfvén wave dynamics is dominated by the gradient operator parallel to the magnetic field lines, $\mathbf{B} \cdot \nabla$. In toroidal geometry, this leads to such intricate dynamics that a very accurate description is needed of the geometry of the field lines and of the magnetic surfaces, i.e. of the equilibrium, if one wishes to study stability. Hence, we will start this chapter by describing the mapping to straight-field-line coordinates (Section 17.1.2), which is a rather compelling representation for the (numerical) analysis of the stability of toroidal plasmas. We then list the different characteristics of the equilibrium, which turn out to be really geometrical properties, like the curvatures of the field lines and of the magnetic surfaces (Section 17.1.3). These play a central role in the waves and instabilities of toroidal plasmas described in the Sections 17.2–17.3. Broadly speaking, for fusion applications, the MHD waves of the system are desirable, as they provide information on the equilibrium distribution through MHD spectroscopy, and the MHD instabilities are to be avoided, as they lead to premature loss of the plasma.

17.1.2 Coordinates and mapping

To enable accurate description of Alfvén wave dynamics in toroidal geometry, the following steps should be taken in the analysis (see Fig. 17.1).

(a) *Solve the Grad–Shafranov equation* (16.81), or the scaled version (16.160) derived in Section 16.3.1. For the present purpose, we will exploit the unscaled poloidal flux Ψ (rather than the scaled flux $\psi \equiv \alpha\Psi$), and we will assume that the required solution

$$\Psi = \Psi(x, y) \tag{17.1}$$

is known in the relevant portion of the poloidal plane. In practice, this implies that a numerical equilibrium program has to be available, such as the finite element code described in Section 16.3.3.

(b) Rather than exploiting an arbitrary poloidal angle θ, or the angle χ belonging to the orthogonal flux coordinate triad, *construct the straight-field-line coordinates* Ψ, ϑ, φ, introduced in Section 16.1.2 (Fig. 16.2). Again, for the present purpose, we will assume that the solution of this problem, i.e. the distribution of the poloidal angle coordinate

$$\vartheta = \vartheta(x, y)\,, \tag{17.2}$$

is explicitly known on a grid in the x–y plane. For these coordinates, all tangential information on the field lines has been lumped into the definition of the coordinates, while the normal behaviour is described by the safety factor $q(\Psi)$. As shown below, for the harmonic dependence $\sim \exp \mathrm{i}(m\vartheta + n\varphi)$ of perturbations, the parallel gradient operator then becomes a multiplier,

$$F \equiv -\mathrm{i}\,\mathbf{B}\cdot\nabla \qquad \Rightarrow \qquad F\,\mathrm{e}^{\mathrm{i}(m\vartheta+n\varphi)} = \mathcal{J}^{-1}(m + nq)\;\mathrm{e}^{\mathrm{i}(m\vartheta+n\varphi)}\,, \tag{17.3}$$

so that it can be made to vanish exactly for rational field lines and surfaces (see Section 16.1.2). Consequently, straight-field-line coordinates are the optimal representation of the geometry of the field lines for the description of Alfvén wave dynamics. This establishes the Ψ, ϑ, φ coordinates as an important analytical tool, but it is not enough to exploit them for the explicit numerical construction of waves and instabilities. For that purpose one more step is needed.

(c) *Invert the poloidal coordinates* (17.1)–(17.2), as already announced in Eq. (16.206),

$$\left.\begin{aligned} \Psi &= \Psi(x, y) \\ \vartheta &= \vartheta(x, y) \end{aligned}\right\} \quad \Rightarrow \quad \left\{\begin{aligned} x &= x(\Psi, \vartheta) \\ y &= y(\Psi, \vartheta) \end{aligned}\right.\,, \tag{17.4}$$

so that we can represent all geometrical and physical quantities on a computational grid $\{\Psi_i, \vartheta_j\}$ in the poloidal plane. Once this has been established by construction (i.e., again having written a numerical program to do this), we can go ahead and study the waves and instabilities of the system.

17.1.3 Geometrical–physical characteristics

After the *straight-field-line* (SFL) *coordinates* have been constructed, there really is no distinction anymore between physical and geometrical quantities. Moreover, the basic nonlinearity of the equilibrium implies that these quantities do not have the simple relationship of cause and effect to each other. Therefore, in somewhat arbitrary order, we now list the different geometrical–physical characteristics that will be needed in the analysis of MHD waves and instabilities in axi-symmetric toroidal systems.

(a) It is expedient to exploit the general tensor machinery for *curvilinear coordinates* (see Appendix A.3) by denoting the three coordinates as

$$x^1 \equiv \Psi\,, \quad x^2 \equiv \vartheta\,, \quad x^3 \equiv \varphi\,, \tag{17.5}$$

EQUIL

Grad–Shafranov $\Rightarrow \begin{cases} \Psi = \Psi(x,y) \\ \vartheta = \vartheta(x,y) \end{cases}$

$+$ equil. variables $V_0(x,y)$

$\Downarrow$

MAP

Coord. inversion $\Rightarrow \begin{cases} x = x(\Psi,\vartheta) \\ y = y(\Psi,\vartheta) \end{cases}$

$+$ equil. variables $V_0(\Psi,\vartheta)$

$\Downarrow$

STAB

Spectral equation $\Rightarrow$ Eigenvalues $\{\omega_i\}$

$+$ perturbations $V_1(\Psi,\vartheta)\mathrm{e}^{in\varphi}$

Fig. 17.1 The three parts of a numerical stability analysis of toroidal plasmas.

so that the two sets of basis vectors $\{\mathbf{a}^i\}$ and $\{\mathbf{a}_i\}$ determine the *contravariant and covariant elements of the metric tensor*,

$$\mathbf{a}^i \equiv \nabla x^i\,, \quad \mathbf{a}_i \equiv \frac{\partial \mathbf{r}}{\partial x^i} \qquad \Rightarrow \quad g^{ij} = \mathbf{a}^i \cdot \mathbf{a}^j\,, \quad g_{ij} = \mathbf{a}_i \cdot \mathbf{a}_j\,, \tag{17.6}$$

where the latter are directly available in the SFL coordinates:

$$\begin{pmatrix} g_{11} & g_{12} & 0 \\ g_{12} & g_{22} & 0 \\ 0 & 0 & g_{33} \end{pmatrix} = \begin{pmatrix} x_\Psi^2 + y_\Psi^2 & x_\Psi x_\vartheta + y_\Psi y_\vartheta & 0 \\ x_\Psi x_\vartheta + y_\Psi y_\vartheta & x_\vartheta^2 + y_\vartheta^2 & 0 \\ 0 & 0 & R^2 \end{pmatrix}, \quad R = \epsilon^{-1}(1+\epsilon x)\,. \tag{17.7}$$

The characteristic difference in length scales of the poloidal and toroidal coordinates is manifest from the expression for the *complete Jacobian* $\mathcal{J}$ in terms of the product of the distance R from the axis of symmetry and the Jacobian $\mathcal{D}$ of the poloidal plane,

$$\mathcal{J} \equiv (\nabla\Psi \times \nabla\vartheta \cdot \nabla\varphi)^{-1} = R\mathcal{D}\,, \quad \mathcal{D} \equiv \frac{\partial(x,y)}{\partial(\Psi,\vartheta)} = x_\Psi y_\vartheta - x_\vartheta y_\Psi = \sqrt{g_{11}g_{22} - g_{12}^2}\,, \tag{17.8}$$

from which the contravariant elements of the metric tensor may be obtained:

$$g^{11} = g_{22}/\mathcal{D}^2\,, \quad g^{12} = -g_{12}/\mathcal{D}^2\,, \quad g^{22} = g_{11}/\mathcal{D}^2\,, \quad g^{33} = 1/R^2\,. \tag{17.9}$$

We can now compute all geometric–physical quantities of interest.

(b) From Eq. (16.84) of Section 16.2.1, the *magnetic field* is expressed by

$$\mathbf{B} = \nabla\varphi \times \nabla\Psi + I\nabla\varphi = \mathbf{a}^3 \times \mathbf{a}^1 + I\mathbf{a}^3 = \mathcal{J}^{-1}\mathbf{a}_2 + I\mathbf{a}^3\,, \tag{17.10}$$

so that the physical components are given by

$$B_\mathrm{p} = |\nabla\Psi|/R = \sqrt{g^{11}}/R = \mathcal{J}^{-1}\sqrt{g_{22}}\,, \qquad B_\varphi = I/R\,, \tag{17.11}$$

whereas the contravariant and covariant components are related to them by

$$B^1 = 0\,, \quad B^2 = \mathcal{J}^{-1} = B_\mathrm{p}/\sqrt{g_{22}}\,, \quad B^3 = I/R^2 = B_\varphi/R\,,$$
$$B_1 = \mathcal{J}^{-1}g_{12} = (g_{12}/\sqrt{g_{22}})B_\mathrm{p}\,, \quad B_2 = \mathcal{J}^{-1}g_{22} = \sqrt{g_{22}}B_\mathrm{p}\,, \quad B_3 = I = RB_\varphi\,. \tag{17.12}$$

The field line equation (16.14) of Section 16.1.2 then provides the *safety factor in SFL coordinates,*

$$q(\Psi) \equiv \frac{d\varphi}{d\vartheta}\bigg|_\mathrm{fl} = \frac{B^3}{B^2} = \frac{\sqrt{g_{22}}B_\varphi}{RB_\mathrm{p}} = \frac{\mathcal{J}B_\varphi}{R} = \frac{\mathcal{J}I}{R^2}\,, \tag{17.13}$$

so that the expression for the parallel gradient operator becomes

$$F \equiv -\mathrm{i}\mathbf{B}\cdot\nabla = -\mathrm{i}(B^2\partial_\vartheta + B^3\partial_\varphi) = -\mathrm{i}\mathcal{J}^{-1}(\partial_\vartheta + q\partial_\varphi)\,. \tag{17.14}$$

This produces the desired effect on harmonic perturbations presented in Eq. (17.3).

(c) From Eqs. (16.84) and (16.81) of Section 16.2.1, the *current density* is expressed by

$$\mathbf{j} = -I'\nabla\varphi\times\nabla\Psi + Rj_\varphi\nabla\varphi\,, \tag{17.15}$$

from which the contravariant and covariant components may be constructed, analogous to the above derivation of the components of $\mathbf{B}$. The toroidal component j_φ is related to the profiles $p(\Psi)$ and $I(\Psi)$ in the RHS of the original Grad–Shafranov equation (16.81). That equation is supposed to have been solved at this stage, so that the *transformed Grad–Shafranov equation in SFL coordinates* becomes a first order differential relation between the different geometrical–physical quantities that is satisfied exactly (of course, to the extent that the equilibrium is solved exactly):

$$\Delta^*\Psi(\Psi,\vartheta) = \frac{R^2}{\mathcal{J}}\Big(\partial_\psi - \partial_\vartheta\frac{g_{12}}{g_{22}}\Big)\mathcal{J}B_\mathrm{p}^2 = -II' - R^2p' = Rj_\varphi\,. \tag{17.16}$$

Recall that the functions $p = p(\Psi)$ and $I = I(\Psi)$ should be considered arbitrary.

(d) For the propagation of waves and instabilities, the geometry of the magnetic field lines and magnetic surfaces is important, in particular the *different curvatures* that appear. This involves projections onto three different triads of unit vectors, that need to be distinguished carefully. They refer to the field lines by themselves, the magnetic surfaces by themselves and the field lines with respect to the magnetic surfaces, respectively:

(1) The traditional *Serret–Frenet triad for the field lines* $\{\mathbf{b},\boldsymbol{\nu},\boldsymbol{\beta}\}$, consisting of the tangent, the normal and the binormal, is defined by

$$\mathbf{b} \equiv \mathbf{B}/|\mathbf{B}|\,, \qquad \boldsymbol{\nu} \equiv \mathbf{b}\cdot\nabla\mathbf{b}/|\mathbf{b}\cdot\nabla\mathbf{b}|\,, \qquad \boldsymbol{\beta} \equiv \mathbf{b}\times\boldsymbol{\nu}\,. \tag{17.17}$$

Only the first of the three Serret–Frenet formulas for the tangential derivatives is needed. This defines the *curvature vector* $\boldsymbol{\kappa}$ *of the field line* as the tangential derivative of the tangent,

$$\boldsymbol{\kappa} \equiv \kappa\boldsymbol{\nu} \equiv \frac{d\mathbf{b}}{ds} \equiv \mathbf{b}\cdot\nabla\mathbf{b}\,, \tag{17.18}$$

which, by definition, is pointing in the direction of the normal to the field line.

(2) The *magnetic surface symmetry triad* $\{\mathbf{n}, \mathbf{t}, \mathbf{e}_\varphi\}$, consisting of the triad of the normal to the magnetic surface, the tangent to the magnetic surface in the poloidal plane and the unit vector in the toroidal direction, is defined by

$$\mathbf{n} \equiv \frac{\nabla\Psi}{|\nabla\Psi|} = \frac{\mathbf{a}^1}{RB_\mathrm{p}}\,, \qquad \mathbf{t} \equiv \mathbf{e}_\varphi \times \mathbf{n} = \frac{\mathbf{a}_2}{\mathcal{J}B_\mathrm{p}}\,, \qquad \mathbf{e}_\varphi \equiv \frac{\nabla\varphi}{|\nabla\varphi|} = R\mathbf{a}^3\,. \tag{17.19}$$

This determines the principal curvatures, viz. the *poloidal and toroidal curvature of the magnetic surface:*

$$\kappa_\mathrm{p} \;\equiv\; \mathbf{t}\cdot(\nabla\mathbf{n})\cdot\mathbf{t} \quad = -\mathbf{t}\cdot(\nabla\mathbf{t})\cdot\mathbf{n} \quad = RD^\dagger B_\mathrm{p}\,, \tag{17.20}$$

$$\kappa_\varphi \;\equiv\; \mathbf{e}_\varphi\cdot(\nabla\mathbf{n})\cdot\mathbf{e}_\varphi = -\mathbf{e}_\varphi\cdot(\nabla\mathbf{e}_\varphi)\cdot\mathbf{n} = B_\mathrm{p}DR\,, \tag{17.21}$$

where D and $D^\dagger$ indicate the normal derivative and its adjoint:

$$D \equiv \frac{\mathbf{n}}{RB_\mathrm{p}}\cdot\nabla = \partial_\Psi - \frac{g_{12}}{g_{22}}\partial_\vartheta\,, \quad D^\dagger \equiv \nabla\cdot\frac{\mathbf{n}}{RB_\mathrm{p}} = \frac{1}{\mathcal{J}}\Big(\partial_\Psi - \partial_\vartheta\frac{g_{12}}{g_{22}}\Big)\mathcal{J}\,. \tag{17.22}$$

The equalities for κ_p and κ_φ are proved below, under Eqs. (17.29)–(17.32).

(3) The *mixed field line/magnetic surface triad* $\{\mathbf{b}, \mathbf{n}, \boldsymbol{\pi}\}$, consisting of the tangent to the field line, the normal to the magnetic surface, and the tangent to the magnetic surface normal to the field line, is defined by

$$\begin{aligned} \mathbf{b} &\equiv \frac{\mathbf{B}}{|\mathbf{B}|} = \frac{1}{\mathcal{J}B}(\mathbf{a}_2 + q\mathbf{a}_3)\,, \qquad \mathbf{n} \equiv \frac{\nabla\Psi}{|\nabla\Psi|} = \frac{\mathbf{a}^1}{RB_\mathrm{p}}\,,\\ \boldsymbol{\pi} &\equiv \mathbf{b}\times\mathbf{n} = \frac{1}{\mathcal{J}B}\Big(\frac{B_\varphi}{B_\mathrm{p}}\mathbf{a}_2 - q\frac{B_\mathrm{p}}{B_\varphi}\mathbf{a}_3\Big)\,. \end{aligned} \tag{17.23}$$

We have exploited this projection frequently in our spectral studies, and we will continue to use it (in the cyclically permuted order $\mathbf{n}$, $\boldsymbol{\pi} \equiv \mathbf{e}_\perp$, $\mathbf{b} \equiv \mathbf{e}_\parallel$) to distinguish the different components of perturbations. It determines the two essential projections of the field line curvature vector with respect to the magnetic surface, viz. the *normal and the geodesic curvature of the magnetic field line:*

$$\kappa_n \;\equiv\; \boldsymbol{\kappa}\cdot\mathbf{n} \equiv \mathbf{b}\cdot(\nabla\mathbf{b})\cdot\mathbf{n} = -(B_\mathrm{p}^2/B^2)\kappa_\mathrm{p} - (B_\varphi^2/B^2)\kappa_\varphi\,, \tag{17.24}$$

$$\kappa_\mathrm{g} \;\equiv\; \boldsymbol{\kappa}\cdot\boldsymbol{\pi} \equiv \mathbf{b}\cdot(\nabla\mathbf{b})\cdot\boldsymbol{\pi} = \frac{B_\varphi}{\mathcal{J}B_\mathrm{p}B^2}\frac{\partial B}{\partial\vartheta}\,. \tag{17.25}$$

These equalities are proved below, under Eqs. (17.36)–(17.37). It should be noted that the magnetic surface curvatures κ_p and κ_φ and the field line curvatures κ_n and κ_g, though exploiting the same stem symbol κ, refer to entirely different geometrical objects. What is worse, to stick to the conventions of the literature, the sign of the magnetic surface curvatures κ_p and κ_φ is chosen positive when the surface is concave with respect to the plasma, whereas the sign of the normal field line curvature κ_n is chosen negative in that case (see Fig. 17.2). The reader be aware!

▷ **Christoffel symbols and curvatures** To obtain the expressions for the curvatures of the magnetic surfaces and of the field lines, we need to calculate the derivatives of the basis vectors, i.e. the Christoffel symbols (see Appendix A.3):

$$\Gamma^k_{ij} \equiv \mathbf{a}^k\cdot\partial_i\mathbf{a}_j = \tfrac{1}{2}g^{kl}\big(\partial_i g_{lj} + \partial_j g_{il} - \partial_l g_{ij}\big)\,. \tag{17.26}$$

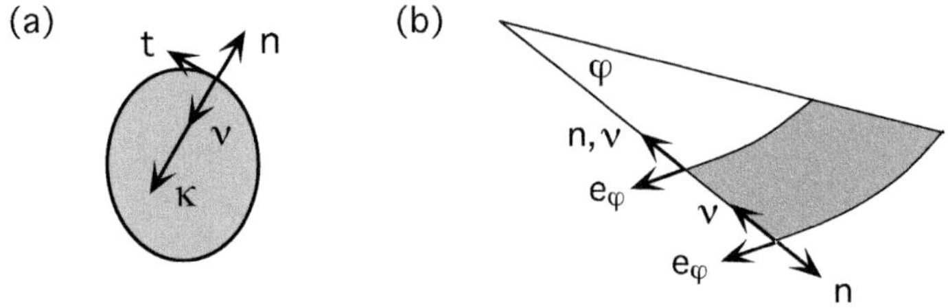

Fig. 17.2 Relation of the magnetic surface curvatures $\kappa_{\rm p}$ and κ_φ to the normal field line curvature κ_n: (a) for the poloidal field component, $\kappa_{\rm p} = -\kappa_n > 0$; (b) for the toroidal field component, $\kappa_\varphi = -\kappa_n < 0$ (inside) and $\kappa_\varphi = -\kappa_n > 0$ (outside).

There are ten non-vanishing Christoffel symbols, viz. $\Gamma^1_{11}, \Gamma^1_{12}, \Gamma^1_{22}, \Gamma^1_{33}, \Gamma^2_{11}, \Gamma^2_{12}, \Gamma^2_{22}, \Gamma^2_{33}, \Gamma^3_{13}$ and Γ^3_{23}, which may be computed from the coordinate definitions (17.4), e.g.

$$\Gamma^1_{11} = (1/\mathcal{D})\,(y_\vartheta x_{\Psi\Psi} - x_\vartheta y_{\Psi\Psi})\,, \qquad \Gamma^1_{33} = -(R/\mathcal{D})\,y_\vartheta\,, \quad \text{etc.} \tag{17.27}$$

Consequently, all necessary information about the coordinates $\{x(\Psi,\vartheta), y(\Psi,\vartheta), \varphi\}$, i.e. the first and second derivatives of x and y, is contained in the four metric elements and the six Christoffel symbols not containing the index 3:

$$\begin{aligned} &x\,,\ x_\Psi\,,\ x_\vartheta\,,\ y_\Psi\,,\ y_\vartheta && \Rightarrow\quad g_{11}\,,\ g_{12}\,,\ g_{22}\,,\ g_{33}\,,\\ &x_{\Psi\Psi}\,, x_{\Psi\vartheta}\,, x_{\vartheta\vartheta}\,, y_{\Psi\Psi}\,, y_{\Psi\vartheta}\,, y_{\vartheta\vartheta} && \Rightarrow\quad \Gamma^1_{11}\,,\ \Gamma^1_{12}\,,\ \Gamma^1_{22}\,,\ \Gamma^2_{11}\,,\ \Gamma^2_{12}\,,\ \Gamma^2_{22}\,. \end{aligned} \tag{17.28}$$

Thus, the expressions for the poloidal and toroidal magnetic surfacles reduce to

$$\kappa_{\rm p} \equiv \mathbf{t}\cdot(\nabla\mathbf{n})\cdot\mathbf{t} = \frac{1}{\mathcal{J}^2RB_{\rm p}^3}\frac{\partial\mathbf{a}^1}{\partial\vartheta}\cdot\mathbf{a}_2 = -\frac{1}{\mathcal{J}^2RB_{\rm p}^3}\mathbf{a}^1\cdot\frac{\partial\mathbf{a}_2}{\partial\vartheta} \equiv -\frac{1}{\mathcal{J}^2RB_{\rm p}^3}\Gamma^1_{22}, \tag{17.29}$$

$$\kappa_\varphi \equiv \mathbf{e}_\varphi\cdot(\nabla\mathbf{n})\cdot\mathbf{e}_\varphi = \frac{1}{R^3B_{\rm p}}\frac{\partial\mathbf{a}^1}{\partial\varphi}\cdot\mathbf{a}_3 = -\frac{1}{R^3B_{\rm p}}\mathbf{a}^1\cdot\frac{\partial\mathbf{a}_3}{\partial\varphi} \equiv -\frac{1}{R^3B_{\rm p}}\Gamma^1_{33}. \tag{17.30}$$

The explicit expressions for the latter two Christoffel symbols read:

$$\begin{aligned} \Gamma^1_{22} &= \tfrac12 g^{1\ell}(\partial_2 g_{\ell 2} + \partial_2 g_{2\ell} - \partial_\ell g_{22}) = R^2B_{\rm p}^2\Big(\partial_2 g_{12} - \tfrac12\partial_1 g_{22} - \tfrac12\frac{g_{12}}{g_{22}}\,\partial_1 g_{22}\Big)\\ &= -\,\mathcal{J}R^2B_{\rm p}^3\Big(\partial_\Psi - \partial_\vartheta\frac{g_{12}}{g_{22}}\Big)\mathcal{J}B_{\rm p} \qquad\Rightarrow\quad \kappa_{\rm p} = RD^\dagger B_{\rm p}\,, \end{aligned} \tag{17.31}$$

$$\begin{aligned} \Gamma^1_{33} &= \tfrac12 g^{1\ell}(\partial_3 g_{\ell 3} + \partial_3 g_{3\ell} - \partial_\ell g_{33}) = -\tfrac12 R^3B_{\rm p}^2\Big(\partial_1 R^2 - \frac{g_{12}}{g_{22}}\,\partial_2 R^2\Big)\\ &= -\,R^3B_{\rm p}^2\Big(\partial_\Psi - \frac{g_{12}}{g_{22}}\,\partial_\vartheta\Big)R \qquad\Rightarrow\quad \kappa_\varphi = B_{\rm p}DR\,, \end{aligned} \tag{17.32}$$

which proves the assertions (17.20) and (17.21).

The curvature of the magnetic field lines may be written in the alternative form

$$\boldsymbol{\kappa} \equiv \mathbf{b}\cdot\nabla\mathbf{b} \overset{(A.10)}{=} -\mathbf{b}\times(\nabla\times\mathbf{b}) \qquad\Rightarrow\quad \boldsymbol{\kappa}\cdot\mathbf{b} = 0\,, \tag{17.33}$$

so that we need to express the curls of the tangent vector,

$$(\nabla\times\mathbf{b})^1 = \mathcal{J}^{-1}\partial_2 b_3\,,\quad (\nabla\times\mathbf{b})^2 = \mathcal{J}^{-1}\partial_1 b_3\,,\quad (\nabla\times\mathbf{b})^3 = \mathcal{J}^{-1}(\partial_1 b_2 - \partial_2 b_1)\,, \tag{17.34}$$

in terms of the co- or contravariant components of the unit vectors (17.23):

$$n_1 = 1/(RB_\mathrm{p})\,, \qquad n_2 = n_3 = 0\,,$$
$$b^1 = 0\,, \qquad b^2 = 1/(\mathcal{J}B)\,, \qquad b^3 = q/(\mathcal{J}B)\,,$$
$$\pi^1 = 0\,, \qquad \pi^2 = B_\varphi/(\mathcal{J}B_\mathrm{p}B)\,, \qquad \pi^3 = -B_\mathrm{p}/(RB)\,. \tag{17.35}$$

Substitution of these equalities leads to the two desired expressions for the normal and the geodesic curvature of the field lines:

$$\kappa_n = \boldsymbol{\kappa}\cdot\mathbf{n} = -(\mathbf{b}\times(\nabla\times\mathbf{b}))\cdot\mathbf{n} = (\mathbf{b}\times\mathbf{n})\cdot(\nabla\times\mathbf{b}) = \boldsymbol{\pi}\cdot(\nabla\times\mathbf{b})$$
$$= \mathcal{J}^{-1}[\pi_1\partial_2 b_3 - \pi_2\partial_1 b_3 + \pi_3(\partial_1 b_2 - \partial_2 b_1)] = -(B_\mathrm{p}^2/B^2)\kappa_\mathrm{p} - (B_\varphi^2/B^2)\kappa_\varphi\,, \tag{17.36}$$

$$\kappa_\mathrm{g} = \boldsymbol{\kappa}\cdot\boldsymbol{\pi} = -(\mathbf{b}\times(\nabla\times\mathbf{b}))\cdot\boldsymbol{\pi} = -(\boldsymbol{\pi}\times\mathbf{b})\cdot(\nabla\times\mathbf{b}) = -\mathbf{n}\cdot(\nabla\times\mathbf{b})$$
$$= -n_1(\nabla\times\mathbf{b})^1 = -\frac{1}{\mathcal{J}RB_\mathrm{p}}\frac{\partial}{\partial\vartheta}\left(\frac{qR^2}{\mathcal{J}B}\right) = \frac{B_\varphi}{\mathcal{J}B_\mathrm{p}B^2}\frac{\partial B}{\partial\vartheta}\,, \tag{17.37}$$

where, in the last step, we used the fact that the functions $I(\Psi)$ and $q(\Psi)$ are independent of ϑ. This proves the assertions (17.24) and (17.25). ◁

As compared to cylindrical plasmas (Chapter 9), two additional curvatures have now entered the picture, viz. the geodesic curvature κ_g and the toroidal curvature κ_φ. They both play an important role in spectral theory of toroidal plasmas. The geodesic curvature of the field lines indicates that field lines are no longer geodesics in toroidal geometry[1], due to the fact that, in general, the magnitude of the magnetic field is not constant in the magnetic surfaces. Hence, as first noticed in [215], one may expect that Alfvén waves will have to decide whether to follow the shortest path (the geodesic) or the magnetic field line. As we will see in Section 17.2.3, they 'make the best of it' doing both through coupling to the slow magneto-sonic modes. This complex feature is related to the *geodesic acoustic modes* (GAMs) [652], which have become a subject of intense research recently [329] (see Section 17.2.3).

The distinguishing feature of the toroidal curvature is its change in sign on the inside ('high field side') of the torus compared to that on the outside. As we have stressed many times, interchange modes driven by pressure gradients in toroidal plasmas are only possible if the overwhelming stabilizing field line bending energy contribution of the Alfvén waves is minimized by modes for which the parallel gradient operator vanishes. This requires localization to rational magnetic surfaces. Interchange modes in toroidal geometry, satisfying this requirement, will become unstable if a counterpart to Suydam's cylindrical stability criterion is violated, viz. Mercier's toroidal stability criterion (Section 17.2.5). However, a much more subtle localization of the modes, called ballooning, was theoretically proposed [125] (and, of course, utilized by the plasma) that couples a pressure gradient to the negative toroidal curvature on the inside. This leads to much more severe stability limitations of pressure driven toroidal modes since they include *ballooning instabilities*, which are absent in a cylinder (Section 17.2.5).

[1] To avoid possible misunderstanding: that the field lines are straight in the SFL representation is a convenience that has nothing to do with the geodesics (which are independent of the coordinates). This is manifest from the occurrence of the coordinate independent length element $d\ell = \mathcal{J}\,d\vartheta$ in the expressions.

17.2 Analysis of waves and instabilities in toroidal geometry

17.2.1 Spectral wave equation

We will now present the main stages of the analysis of wave propagation and stability of toroidal plasmas, in close analogy with the analysis of cylindrical plasmas (Chapter 9). Since toroidal equilibria are two-dimensional, we will obtain partial differential equations, instead of the ordinary differential equations, like the Hain–Lüst equation, that were obtained for one-dimensional equilibria.

As always, we start from the linearized equation of motion

$$\mathbf{F}(\boldsymbol{\xi}) = \rho \frac{\partial^2 \boldsymbol{\xi}}{\partial t^2} \,, \tag{17.38}$$

with the static MHD force operator defined in Eqs. (6.23)–(6.25), and assume *Fourier-normal mode solutions* of the form

$$\boldsymbol{\xi}(\mathbf{r}, t) = \hat{\boldsymbol{\xi}}(\Psi, \vartheta)\, \mathrm{e}^{\mathrm{i}(n\varphi - \omega t)} \,, \tag{17.39}$$

where $\hat{\boldsymbol{\xi}}$ is the amplitude of the normal modes and n is the toroidal mode number. As in Chapters 7 and 9, we exploit the *field line/magnetic surface projection*, expressed by the field line triad $\{\mathbf{n}, \boldsymbol{\pi}, \mathbf{b}\}$ defined in Eq. (17.23). In this projection, the gradient operators become

$$\begin{aligned}
D &\equiv \frac{1}{RB_\mathrm{p}} \mathbf{n} \cdot \nabla &&= \partial_\Psi - (g_{12}/g_{22})\, \partial_\vartheta \,, \\
G &\equiv -\mathrm{i} RB_\mathrm{p} B \boldsymbol{\pi} \cdot \nabla &&= -\frac{\mathrm{i} RB_\varphi}{\mathcal{J}} \Big(\partial_\vartheta - q(B_\mathrm{p}^2/B_\varphi^2) \partial_\varphi \Big) \,, \\
F &\equiv -\mathrm{i} B \mathbf{b} \cdot \nabla &&= -\frac{\mathrm{i}}{\mathcal{J}} (\partial_\vartheta + q \partial_\varphi) \,.
\end{aligned} \tag{17.40}$$

(Watch out when converting these expressions for vectors $\boldsymbol{\xi}$: the partial derivatives ∂_ϑ and ∂_φ act not only on the vector components but also on the basis vectors and the unit vectors!) The projections of the displacement vector will be denoted as

$$\begin{aligned}
X &\equiv RB_\mathrm{p}\, \boldsymbol{\xi} \cdot \mathbf{n} = \xi^1 \,, \\
Y &\equiv \frac{\mathrm{i} B}{RB_\mathrm{p}} \boldsymbol{\xi} \cdot \boldsymbol{\pi} = \frac{\mathrm{i} B_\varphi}{\mathcal{J} RB_\mathrm{p}^2} \Big[\xi_2 - q(B_\mathrm{p}^2/B_\varphi^2) \xi_3 \Big] \,, \\
Z &\equiv \frac{\mathrm{i}}{B} \boldsymbol{\xi} \cdot \mathbf{b} = \frac{\mathrm{i}}{\mathcal{J} B^2} (\xi_2 + q \xi_3) \,,
\end{aligned} \tag{17.41}$$

where the factors i in the components Y and Z are introduced to obtain real expressions in the final Fourier analysed form of the spectral equation, whereas the factors RB_p and B, found by trial and error, just simplify that equation.

At this stage in our exposition, we will not give the complete derivations anymore, but just indicate the steps and give the final result. (The student wishing to enter this field is best advised anyway to spend some time deriving these equations by her- or himself.) The important steps are to first Fourier analyse the spectral equation (17.38) with respect to φ, i.e. to replace ∂_φ by $\mathrm{i}n$, then to project the equation onto the orthogonal triad $\{\mathbf{n}, \boldsymbol{\pi}, \mathbf{b}\}$, and finally to eliminate all second order derivatives of equilibrium quantities by exploiting the properties listed in Section 17.1.3. For example, the pressure terms entering **F** and the magnetic field perturbation **Q** become

$$\boldsymbol{\xi}\cdot\nabla p = p'X\,,\qquad \nabla\cdot\boldsymbol{\xi} = D^\dagger X + GB^{-2}Y + FZ\,,$$
$$Q_n = \frac{\mathrm{i}}{RB_\mathrm{p}}X\,,\qquad Q_\perp = \frac{RB_\mathrm{p}}{B}\Big\{\Big[D^\dagger\Big(\frac{B_\varphi}{R}\Big)\Big]X + FY\Big\}\,,$$
$$Q_\parallel = -BDX - \frac{RB_\varphi}{B}\Big[D^\dagger\Big(\frac{B_\varphi}{R}\Big)\Big]X - \frac{1}{B}GY\,. \tag{17.42}$$

After some tedious, but straightforward, algebra one then finds the following formulation of *the spectral problem*:

$$\begin{pmatrix} \mathcal{A}_{11} & \mathcal{A}_{12} & \mathcal{A}_{13} \\ \mathcal{A}_{21} & \mathcal{A}_{22} & \mathcal{A}_{23} \\ \mathcal{A}_{31} & \mathcal{A}_{32} & \mathcal{A}_{33} \end{pmatrix}\begin{pmatrix} X \\ Y \\ Z \end{pmatrix} = -\rho\omega^2 \begin{pmatrix} \mathcal{B}_{11}\,X \\ \mathcal{B}_{22}\,Y \\ \mathcal{B}_{33}\,Z \end{pmatrix}. \tag{17.43}$$

This equation was first derived by Goedbloed [215] exploiting orthogonal Ψ, χ, φ coordinates. (Note that the erroneous expression (9) of that paper for the geodesic curvature should be multiplied by B_p/B to get the correct expression (17.25).)

▷ **Matrix elements** The explicit expressions for the matrix elements read:

$$\mathcal{A}_{11} \equiv D(\gamma p + B^2)D^\dagger - F\frac{1}{R^2B_\mathrm{p}^2}F - E\,,\qquad E \equiv 2\Big[D\Big(\frac{B_\mathrm{p}}{R}\kappa_\mathrm{p}\Big) + \frac{B_\varphi}{R}{}^\dagger D\Big(\frac{B_\varphi}{B_\mathrm{p}}\kappa_\varphi\Big)\Big]\,,$$
$$\mathcal{A}_{21} \equiv -\frac{\gamma p + B^2}{B^2}GD^\dagger - 2\Big(\frac{\mathrm{i}}{\mathcal{J}}\partial_\vartheta\frac{B_\varphi\kappa_\varphi}{B_\mathrm{p}} + n\frac{B_\mathrm{p}\kappa_\mathrm{p}}{R}\Big)\,,$$
$$\mathcal{A}_{31} \equiv -F\gamma pD^\dagger\,,$$
$$\mathcal{A}_{12} \equiv DG\frac{\gamma p + B^2}{B^2} - 2\Big(\frac{B_\varphi\kappa_\varphi}{B_\mathrm{p}}\frac{\mathrm{i}}{\mathcal{J}}\partial_\vartheta + \frac{B_\mathrm{p}\kappa_\mathrm{p}}{R}n\Big)\,,$$
$$\mathcal{A}_{22} \equiv -\frac{1}{B^2}G\gamma pG\frac{1}{B^2} - G\frac{1}{B^2}G - F\frac{R^2B_\mathrm{p}^2}{B^2}F\,,$$
$$\mathcal{A}_{32} \equiv -F\gamma pG\frac{1}{B^2}\,,$$
$$\mathcal{A}_{13} \equiv D\gamma pF\,,\qquad \mathcal{B}_{11} \equiv \frac{1}{R^2B_\mathrm{p}^2}\,,$$
$$\mathcal{A}_{23} \equiv -\frac{1}{B^2}G\gamma pF\,,\qquad \mathcal{B}_{22} \equiv \frac{R^2B_\mathrm{p}^2}{B^2}\,,$$
$$\mathcal{A}_{33} \equiv -F\gamma pF\,,\qquad \mathcal{B}_{33} \equiv B^2\,, \tag{17.44}$$

with $D^\dagger$ and (for lack of a better symbol) ${}^\dagger D$ defined by

$$D^\dagger \equiv \frac{1}{\mathcal{J}}D\mathcal{J} = \frac{1}{\mathcal{J}}\Big(\partial_\Psi - \partial_\vartheta\frac{g_{12}}{g_{22}}\Big)\mathcal{J}\,,\qquad {}^\dagger D \equiv \mathcal{J}\Big(\partial_\Psi + \partial_\vartheta\frac{g_{12}}{g_{22}}\Big)\frac{1}{\mathcal{J}}\,. \tag{17.45}$$

The square brackets in the definition of E indicate that the action of the differential operator is to be restricted to the terms inside those brackets. ◁

The two-dimensional spectral equation (17.43) may be considered as the generalization of the cylindrical MHD wave equation (9.28). The latter equation is recovered by exploiting the leading order relationship (16.96) between Ψ and $B_{\rm p}$ of the Shafranov shifted circle approximation. This yields the following translation recipe to the cylindrical wave equation:

$$\begin{aligned}
&\Psi \to R_0 \int B_\theta\, dr\,, \quad \vartheta \to \theta\,, \quad \varphi \to z/R_0\,, \quad \mathcal{J} \to r/B_\theta\,, \\
&R \to R_0\,, \quad n \to kR_0\,, \quad \kappa_{\rm p} \to 1/r\,, \quad \kappa_\varphi \to 0\,, \quad g_{12} \to 0\,, \\
&D \to \frac{1}{R_0 B_\theta}\frac{d}{dr}\,, \quad D^\dagger \to \frac{1}{R_0 r}\frac{d}{dr}\frac{r}{B_\theta}\,, \quad G \to R_0 B_\theta G_{\rm cyl}\,, \quad F \to F_{\rm cyl}\,, \\
&X \to R_0 B_\theta \xi_{\rm cyl}\,, \quad Y \to (B/R_0 B_\theta)\eta_{\rm cyl}\,, \quad X \to (1/B)\zeta_{\rm cyl}\,.
\end{aligned} \qquad (17.46)$$

However, the next step, to the construction of a scalar wave equation for the normal component of $\boldsymbol{\xi}$, i.e. the analogue of the Hain–Lüst equation, is not taken now: the elimination of the tangential components is no longer algebraic but involves PDEs. If one wishes to continue along this line, one usually exploits an ordering in a small parameter, like the low-β tokamak ordering. This involves the counterpart of the equilibrium expansion described in Section 16.2.2, i.e. expanding the vector wave equation (17.43) to the first non-trivial order (considering the cylindrical solution as the trivial leading order; after all, one only goes to toroidal corrections if the cylindrical basis is fully understood). This generally leads to a system of coupled ODEs describing the Fourier harmonics of the vector $\boldsymbol{\xi}$ in the angle ϑ. An example along this line may be found in Section 18.3.3 of the next chapter.

The equation of motion may be used as a starting point for further analysis. For numerical work, the corresponding quadratic forms of the next section are to be preferred. In the following Sections 17.2.3 and 17.2.5, we will then use the representation (17.43) to derive the analytical 'core structure' for toroidal systems, consisting of the ODEs describing the continuous spectra and ballooning stability, and the explicit Mercier criterion.

17.2.2 Spectral variational principle

Recall from Section 6.1.1 that there are two equivalent approaches to spectral theory, one based on differential equations and the other one on quadratic forms (as in the Schrödinger and Heisenberg pictures of quantum mechanics). The quadratic forms approach for MHD was formulated in Eq. (6.89) as a variational principle for the eigenvalues and eigenfunctions in terms of the Rayleigh quotient,

$$\delta\Lambda = 0\,, \qquad \Lambda[\boldsymbol{\xi}] \equiv \frac{W[\boldsymbol{\xi}]}{K[\boldsymbol{\xi}]}\,, \qquad (17.47)$$

where W is the potential energy and K is the norm of the perturbations. (The notation for the norm is changed here to K since I is used for another purpose.) The eigenvalues ω^2 are the stationary values of Λ. Exploiting the same techniques as in the derivation of the spectral wave equation (17.43), the diligent student will be able to derive the following explicit expressions for the quadratic forms of the potential energy and of the norm of the perturbations in the field line projection:

$$W = \pi \iint \Big[\frac{1}{R^2 B_{\rm p}^2}\big|FX\big|^2 + B^2 \Big| D^\dagger X + \frac{2\kappa_n}{RB_{\rm p}} X + \frac{1}{B^2} GY \Big|^2$$

$$+\frac{R^2B_{\rm p}^2}{B^2}\Big|FY+\frac{2B_\varphi(\kappa_{\rm p}-\kappa_\varphi)}{R^2B_{\rm p}}X\Big|^2+\gamma p\Big|D^\dagger X+G\frac{1}{B^2}Y+FZ\Big|^2$$
$$+\frac{2}{RB_{\rm p}}\Big\{p'\kappa_{\rm p}+\frac{II'}{R^2}(\kappa_{\rm p}-\kappa_\varphi)\Big\}|X|^2\Big]\mathcal{J}d\Psi d\vartheta\,, \tag{17.48}$$

$$K=\pi\iint\rho\Big[\frac{1}{R^2B_{\rm p}^2}|X|^2+\frac{R^2B_{\rm p}^2}{B^2}|Y|^2+B^2|Z|^2\Big]\mathcal{J}d\Psi d\vartheta\,. \tag{17.49}$$

As discussed in Chapter 6, these expressions are completely equivalent to the spectral wave equation when used with the variational principle (17.47). The cylindrical expression (9.100) for W may be obtained from the toroidal expression (17.48) by again exploiting the translation recipe (17.46) (and an integration by parts to combine the second and third term).

Since W has been the starting point for a substantial number of investigations in ideal MHD stability of tokamaks, it is of some interest to consider different alternative expressions for the term in curly brackets, which potentially gives rise to instabilities:

$$U\equiv\frac{2}{RB_{\rm p}}\Big\{p'\kappa_{\rm p}+\frac{II'}{R^2}(\kappa_{\rm p}-\kappa_\varphi)\Big\}=-\frac{2}{R^2B_{\rm p}}\Big\{j_\varphi(\kappa_{\rm p}-\kappa_\varphi)-Rp'\kappa_\varphi\Big\}$$
$$=-\frac{2}{R^2B_{\rm p}}\Big\{j_\parallel\frac{B_\varphi}{B}(\kappa_{\rm p}-\kappa_\varphi)+Rp'\kappa_n\Big\}\,. \tag{17.50}$$

For the reduction, the first order differential form (17.16) of the Grad–Shafranov equation has been exploited. Clearly, all the ingredients for instability are there, e.g. the parallel current $j_\parallel$, driving *kink modes*, and the pressure gradient–curvature term $p'\kappa_n$ ($\approx -p'\kappa_\varphi$), driving *ballooning modes*, but it is not so clear in what form they should be presented. There is also a connection with the curvature terms of the equation of motion (17.43), in particular the expression E, but it is not instructive enough to reproduce it here. One may wonder what all these expressions are good for. Actually, the sole purpose of this paragraph is to exhibit the ambiguity of certain ingrained terminology in tokamak literature. The terms 'kink' and 'ballooning' do not have a unique meaning. There are many ways of transforming the terms using the equilibrium conditions and which presentation is the most meaningful depends entirely on the approximations made to study a particular case.

The expression for the energy has been used extensively in analytic studies of MHD stability of internal modes, e.g. the *internal kink mode*, involving a singular perturbation at a rational $q=1$ surface, which requires delicate balancing of terms; see e.g. [101], [141]. This has its counterpart in extreme conditions on numerical accuracy of the equilibrium, its inversion and the spectral representation needed to describe these modes; see e.g. [360].

17.2.3 Alfvén and slow continuum modes

Recall that the continuous spectra in cylindrical plasmas are obtained by considering perturbations that are *localized to a particular magnetic surface*. This construction can be generalized to toroidal geometry, as shown by Goedbloed [215] and, independently, by Pao [477]. Since the equation of motion (17.43) is written in the field line projection, which already incorporates one of the essential properties of the singular Alfvén and slow continuum modes, the construction is relatively straightforward. The first step is to consider the limit $D\to\infty$ so as to obtain modes localized to a single magnetic surface. To leading order, the first component of Eq. (17.43) then becomes a derivative with respect to Ψ, which can be integrated once to give

$$D^\dagger X \approx -\frac{1}{\gamma p + B^2} G \frac{\gamma p + B^2}{B^2} Y - \frac{\gamma p}{\gamma p + B^2} F Z \,. \tag{17.51}$$

Substitution of this expression into the second and third component of Eq. (17.43), and the use of commutation relations like

$$G B^{-1} - B^{-1} G = \mathrm{i} R B_{\mathrm{p}} \kappa_{\mathrm{g}} \,, \tag{17.52}$$

leads to a system of two ordinary differential equations for Y and Z where the normal derivatives of the equilibrium (i.e. the curvatures κ_{p} and κ_φ of the magnetic surfaces) no longer appear. As a result, a system of equations is obtained that is *intrinsic to every magnetic surface:* a two-dimensional creature living on such a surface will not notice the three-dimensional embedding. Consequently, we have effectively obtained modes that are localized about a single magnetic surface, i.e. of the form

$$\boldsymbol{\xi}(\Psi, \vartheta, \varphi) \approx -\mathrm{i}\delta(\Psi - \Psi_0)\,[\,\eta(\vartheta)\boldsymbol{\pi} + \zeta(\vartheta)\mathbf{b}\,]\,\mathrm{e}^{\mathrm{i}n\varphi} \,, \tag{17.53}$$

where the variables

$$\eta \equiv \mathrm{i}\boldsymbol{\xi} \cdot \boldsymbol{\pi} \equiv (R B_{\mathrm{p}}/B) Y \,, \quad \text{and} \quad \zeta \equiv \mathrm{i}\boldsymbol{\xi} \cdot \mathbf{b} \equiv B Z$$

are more convenient than Y and Z for the present purpose. These variables describe the *Alfvén and slow magnetic surface resonances*, satisfying a system of two coupled ODEs [215]:

$$\begin{pmatrix} \alpha_{11} & \alpha_{12} \\ \alpha_{21} & \alpha_{22} \end{pmatrix} \begin{pmatrix} \eta \\ \zeta \end{pmatrix} = \rho\omega^2 \begin{pmatrix} \eta \\ \zeta \end{pmatrix} , \tag{17.54}$$

where

$$\alpha_{11} \equiv \frac{B}{R B_{\mathrm{p}}} F \frac{R^2 B_{\mathrm{p}}^2}{B^2} F \frac{B}{R B_{\mathrm{p}}} + \frac{4\gamma p B^2}{\gamma p + B^2} \kappa_{\mathrm{g}}^2 \,, \qquad \alpha_{12} \equiv -\frac{2\mathrm{i}\gamma p B^2}{\gamma p + B^2} \kappa_{\mathrm{g}} F \frac{1}{B} \,,$$

$$\alpha_{21} \equiv \frac{\mathrm{i}}{B} F \frac{2\gamma p B^2}{\gamma p + B^2} \kappa_{\mathrm{g}} \,, \qquad \alpha_{22} \equiv \frac{1}{B} F \frac{\gamma p B^2}{\gamma p + B^2} F \frac{1}{B} \,. \tag{17.55}$$

The appropriate boundary conditions to impose on these equations are poloidal periodicity of the variables η and ζ and of their first derivatives.

Notice that the only curvature that has survived in this representation is the geodesic curvature κ_{g} of the field lines inside a particular magnetic surface. The two-dimensional creature introduced above will be able to draw the correct geometrical and, hence, physical conclusions about the dynamics of the waves on the magnetic surface, in particular their anisotropy with respect to the field lines. From the representation (17.54)–(17.55) it is clear that the MHD continuum modes approximately behave like such two-dimensional creatures. It is also clear, from the occurrence of the off-diagonal matrix elements $\sim \gamma p \kappa_{\mathrm{g}}$ in the ODEs (17.54), that the Alfvén waves are no longer polarized purely perpendicular to the field lines (as they are in the one-dimensional plane slab and cylindrical equilibria), but that they get a parallel component by coupling to the slow magneto-acoustic modes. Vice versa, the polarization of the slow waves will no longer be purely parallel to the field lines due to geodesic coupling to the Alfvén waves. Consequently, in toroidal configurations, the Alfvén and slow magneto-acoustic modes can no longer be distinguished on the basis of their polarization.

Since the determination of the continuous spectra has now been turned into the construction of the discrete, doubly infinite, set of eigenvalues $\{\omega^2_{\mathrm{A/S},i}(\Psi_0)\}$ on each magnetic surface, we may also obtain those eigenvalues from a variational principle in terms of the two-vector $\mathbf{v} \equiv (\eta, \zeta)^{\mathrm{T}}$:

$$\delta\hat{\Lambda} = 0\,, \quad \hat{\Lambda}[\mathbf{v}] \equiv \frac{\displaystyle\int \Big\{ \Big|\frac{RB_{\mathrm{p}}}{B} F \frac{B}{RB_{\mathrm{p}}}\eta\Big|^2 + \frac{\gamma p}{\gamma p + B^2}\Big| BF\frac{1}{B}\zeta + 2\mathrm{i}\kappa_{\mathrm{g}} B\eta\Big|^2 \Big\} \mathcal{J}\, d\vartheta}{\displaystyle\int \rho(\eta^2 + \zeta^2)\mathcal{J}\, d\vartheta}\,. \tag{17.56}$$

This expression shows that the continuous spectra of static toroidal plasmas are exclusively stable, $\omega^2_{\mathrm{A/S},i} \geq 0$: geodesic curvature does not create instability. An additional contribution from the normal curvature is needed to create instability, but this involves another kind of localization (see Section 17.2.5). On the other hand, in plasmas with background flow, the continua may become unstable due to a generalization of the geodesic curvature term (see Section 18.3).

▷ **Geodesic acoustic modes** A special example of the Alfvén–slow wave coupling due to geodesic curvature is the geodesic acoustic mode (GAM), first described by Winsor, Johnson and Dawson [652]. This is an electrostatic wave in low-beta plasmas ($\gamma p \ll B^2$), where the perpendicular displacement $[B/(RB_{\mathrm{p}})]\eta \approx$ const in the magnetic surfaces so that the magnetic (Alfvén wave) perturbations, indicated by the first term of the variational principle (17.56), are negligible. Note that this does not imply that η is negligible: the GAM is a slow wave having both perpendicular and parallel components, constrained by the electrostatic condition. Therefore, it is expedient to convert η and ζ to the density perturbations $\widetilde{\rho}$ that are relevant for these modes:

$$\begin{aligned} \widetilde{\rho} &= -\nabla\cdot(\rho\boldsymbol{\xi}) = -\rho(D^{\dagger}X + GB^{-2}Y + FZ) - (RB_{\mathrm{p}})^{-2}(\nabla\rho\cdot\nabla\Psi)X \\ &\approx -\rho\frac{B^2}{\gamma p + B^2}(2\mathrm{i}\kappa_{\mathrm{g}}\eta + FB^{-1}\zeta) \approx -\rho(2\mathrm{i}\kappa_{\mathrm{g}}\eta + FB^{-1}\zeta)\,, \end{aligned} \tag{17.57}$$

where the localization assumption (17.51) and the commutation relation (17.52) were used for the reductions to the last line. The ODEs (17.54) then transform into

$$\begin{aligned} 4\gamma p\kappa_{\mathrm{g}}^2\eta - 2\mathrm{i}\gamma p\kappa_{\mathrm{g}} FB^{-1}\zeta &= 2\mathrm{i}\gamma p\kappa_{\mathrm{g}}(\widetilde{\rho}/\rho) = \rho\omega^2\eta\,, \\ 2\mathrm{i}\gamma pB^{-1}F\kappa_{\mathrm{g}}\eta + B^{-1}F\gamma pFB^{-1}\zeta &= -\gamma pB^{-1}F(\widetilde{\rho}/\rho) = \rho\omega^2\zeta\,. \end{aligned} \tag{17.58}$$

Inserting these expressions into the variational principle (17.56) yields the expression for the GAM frequencies derived by Winsor et al. [652], except for a different normalization factor:

$$\omega^2 = \frac{\gamma p}{\rho}\int \Big[|B^{-1}F\widetilde{\rho}|^2 + |2\kappa_{\mathrm{g}}\widetilde{\rho}|^2\Big]\mathcal{J}d\vartheta \Big/ \int |\widetilde{\rho}|^2\mathcal{J}d\vartheta\,. \tag{17.59}$$

As a companion to MHD spectroscopy, Itoh et al. proposed *GAM spectroscopy* [329], exploiting these modes to determine the velocity profiles of the different ion species. ◁

In the equivalent approach by Pao [477], the continua in toroidal geometry are obtained by starting from the six first order ODEs for the primitive variables $\mathbf{v}_1$, p_1 and $\mathbf{B}_1$. In producing a set of equations for the normal component of $\mathbf{v}_1$ and the total pressure perturbation (a generalization of the one-dimensional Appert–Gruber–Vaclavik [18] representation in terms of ξ and Π), a four by four matrix operator acting on the tangential components of $\mathbf{v}_1$ and $\mathbf{B}_1$ has to be inverted. The continuous spectra are obtained for those frequencies where this inverse does not exist; see also Kieras and Tataronis [363].

17.2.4 Poloidal mode coupling

In the cylindrical limit, all equilibrium quantities are constant on a magnetic surface, so that the tangential dependence of the perturbations reduces to just the Fourier amplitude dependence $\exp(im\vartheta)$ with the familiar result of uncoupled Alfvén and slow continua:

$$\begin{aligned} \omega_{\rm A} &= \pm \frac{(n + m/q)B_{\varphi 0}}{R_0\sqrt{\rho}}\,, & \eta_{\rm A} \neq 0\,, \quad & \zeta_{\rm A} = 0 \quad \textit{(Alfvén)}\,, \\ \omega_{\rm S} &= \pm \sqrt{\frac{\gamma p}{\gamma p + B^2}}\,\frac{(n + m/q)B_{\varphi 0}}{R_0\sqrt{\rho}}\,, & \eta_{\rm S} = 0\,, \quad & \zeta_{\rm S} \neq 0 \quad \textit{(slow)}\,. \end{aligned} \tag{17.60}$$

In toroidal geometry, as we have noted above, the Alfvén and slow continua are coupled through the combination of finite compressibility and finite beta (γp terms) and geodesic curvature ($\kappa_{\rm g}$). In low-beta plasmas, this coupling is weak. However, an entirely different, usually much stronger, coupling arises from the fact that all continua are degenerate at the cylindrical cross-over points $r = r_{\rm cross}$, where $q_{\rm cross}$ has a rational value determined by

$$n + m/q = -n - m'/q \qquad \Rightarrow \qquad q_{\rm cross} = -\frac{m + m'}{2n}\,. \tag{17.61}$$

At those values, two Alfvén wave branches (and two slow wave branches, at much lower frequency) cross, creating the possibility of lifting of the degeneracy by mode coupling due to poloidal variation of the equilibrium. In particular, the poloidal modulation of the equilibrium with the large radius R creates coupling between modes separated by $\Delta m \equiv |m - m'| = 1$, the ellipticity of the plasma cross-section causes $\Delta m = 2$, the triangularity causes $\Delta m = 3$, etc. This coupling produces *gaps in the continua* (see Fig. 17.3), very analogous to the band structure of electrons in a crystal lattice [160].

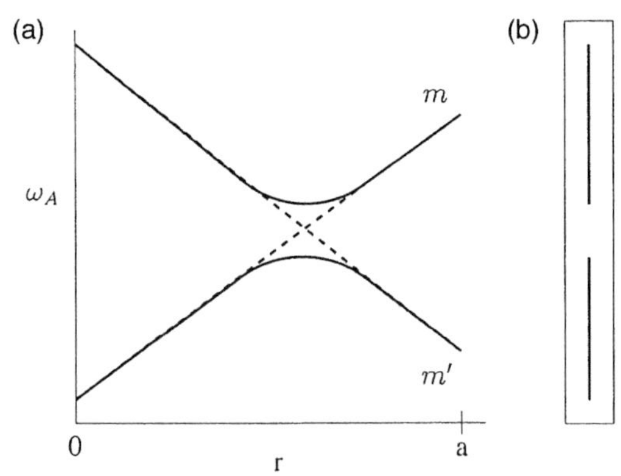

Fig. 17.3 (a) The poloidal mode coupling of two cylindrical modes, labelled with m and m', creates (b) a 'gap' in the continuous spectrum of a toroidal plasma.

An illuminating example of poloidal mode coupling is shown in Fig. 17.4, which presents the results of an analytic calculation of the spectrum of free-boundary kink modes (a particular kind of Alfvén waves) of a skin-current low-beta tokamak, where the toroidal current flows exclusively

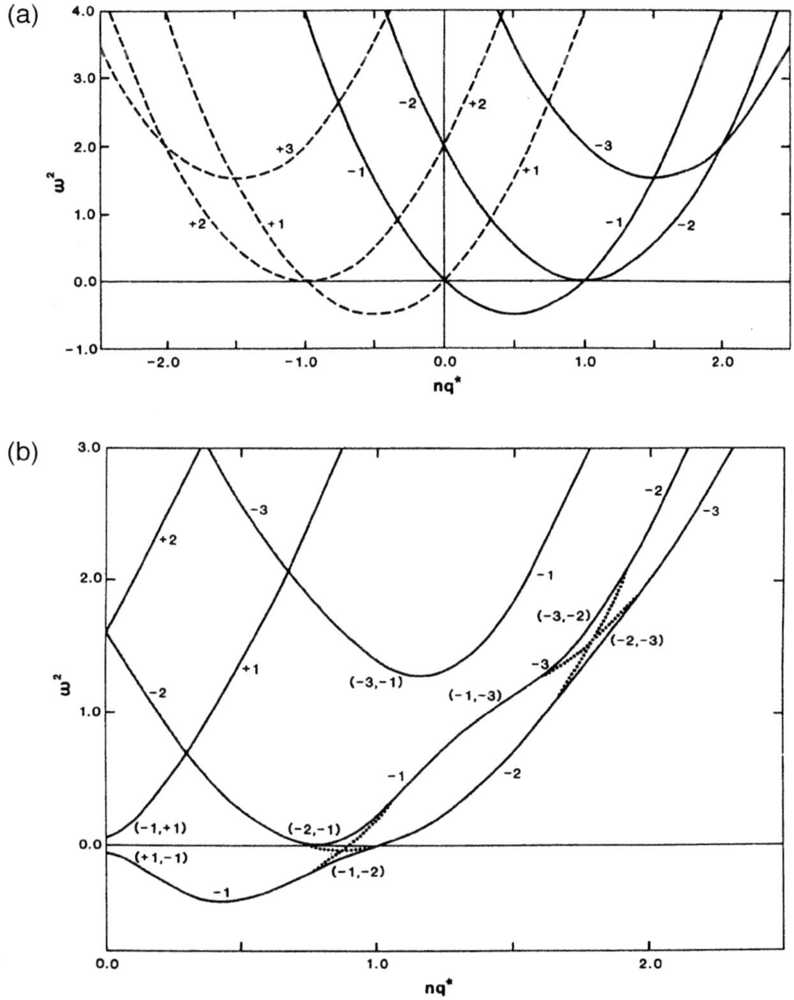

Fig. 17.4 Mode coupling of free-boundary kink modes in a skin-current model of a low-beta tokamak: (a) The spectrum of uncoupled modes for a circular cylinder, $\epsilon\beta_{\rm p} = 0$; (b) the spectrum of coupled modes for a low-beta tokamak with an elliptical cross-section, $\epsilon\beta_{\rm p} = 0.1, b/a = 1.2$. (From D'Ippolito and Goedbloed [160](a).)

on the plasma–vacuum surface. The top part of the figure shows the spectrum for the 'zero-order' cylinder, where the different modes (labelled by the poloidal mode number m) are uncoupled:

$$\bar{\omega}_m^2 = (nq^*)^2 - |m| + (nq^* + m)^2 \,. \tag{17.62}$$

The bottom part shows the results of standard first order perturbation theory of the coupling between those modes due to the toroidicity or, what amounts to the same in the low-beta tokamak ordering, due to beta (causing $\Delta m = 1$ splitting) and due to the ellipticity of the plasma cross-section (causing

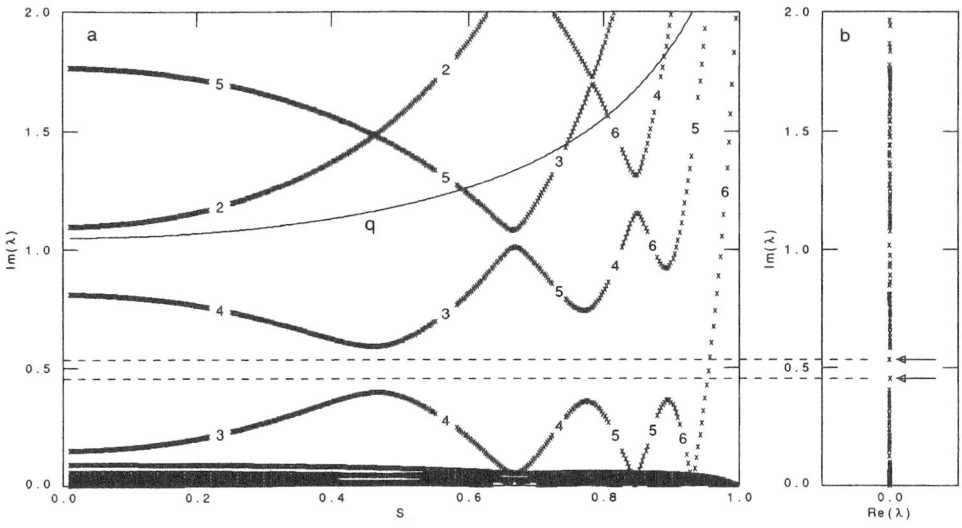

Fig. 17.5 Continuous spectrum and gap modes in a model tokamak equilibrium. (a) The continua, with frequency parameter $\lambda \equiv -\mathrm{i}\omega$, are plotted as a function of $s \equiv \sqrt{\psi}$; the labels refer to the poloidal mode number m; slow continua crowd at the bottom of the diagram. (b) Discrete representation of the complete spectrum with two 'gap modes' (indicated by the arrows). (From Poedts et al. [498].)

$\Delta m = 2$ splitting). These couplings are due to a surface energy integral (unstable when $S > 0$) with an integrand of the form

$$\begin{aligned} S(\theta) &\equiv [b_{\rm p}^2(\theta)\kappa_{\rm p}(\theta) + \epsilon\beta_{\rm p}\kappa_{\varphi}(\theta)]/h(\theta) \\ &\approx \tfrac{1}{2}(3 - b^2/a^2) + 3\epsilon\beta_{\rm p}\cos\theta - (b^2/a^2 - 1)\cos 2\theta\,, \end{aligned} \tag{17.63}$$

where the three terms correspond to *kink, ballooning and ellipticity*, respectively. Due to these splittings, there is a complete reordering of the structure of the spectrum, with important effects on the $|m| = 1$ kink mode (recall that the Kruskal–Shafranov limit is at $q^* = 1$) and elliptical splitting for the axi-symmetric modes (at $q^* = 0$). Mode coupling spectra for diffuse high-beta tokamak equilibria, exploiting the techniques of conformal mapping and polynomials described in Section 16.3.3, and exhibiting even more striking similarity with the electron band structures, may be found in [370].

Since the continuous spectra really come about from the diffuse inhomogeneity of the equilibrium, let us now consider the full spectrum of a diffuse low-beta tokamak. Here, 'full' is meant is the sense of collecting all values of the continua over the full range $0 \le \psi \le 1$ (in plots, usually ψ is replaced by the quasi-radial variable $s \equiv \sqrt{\psi}$). In the calculation of the continuum gaps, important global modes, named *toroidal Alfvén eigenmodes* (TAEs), were found by Cheng and Chance [120]. These modes are not continuum modes, but discrete global modes with frequencies located inside the gaps, again very much like the discrete modes that occur in the forbidden bands of solid state physics. The TAEs should not be confused with the global Alfvén eigenmodes (GAEs), which are part of possible cluster sequences that occur at the extrema of the continua, as extensively discussed in Chapters 7 and 9. Whereas the GAEs already occur in 1D cylindrical equilibria, the TAEs require poloidal mode coupling to produce the continuum gaps. The different kinds of TAE caused by cross-sectional shaping, finite beta, and toroidal rotation are called ellipticity induced Alfvén eigenmodes (EAEs) [58], beta induced Alfvén eigenmodes (BAEs) [612] and toroidal flow induced Alfvén eigenmodes (TFAEs) [618]. Their importance resides in the fact that they may be driven unstable by fusion-produced alpha particles [199, 119] and, thus, pose a threat to alpha particle confinement in future fusion reactors. Their calculation requires large-scale computations of the low mode number gaps ($n, m \sim 1$) and of the global TAE modes. An example is given in Fig. 17.5, taken from work by Poedts et al. [498] (exploiting a simple technique to compute the continua with a finite element spectral code [500]), which shows both the gaps in the continua and the discrete 'gap modes' located in those gaps.

17.2.5 Alfvén and slow ballooning modes

Alfvén and slow continuum modes are constrained to 'live' on the magnetic surfaces, where they necessarily remain stable, as is manifest from the expression for the potential energy in the numerator of the Rayleigh quotient (17.56). However, this does not exhaust the ways in which MHD waves can follow the magnetic field lines. In particular, notice that vanishing of the parallel gradient operator expression, $F \sim m + nq \approx 0$, leads to completely degenerate continua in the origin $\omega^2 = 0$. In the study of MHD stability, this is always the sign that higher order contributions should be considered. The obvious constraint to be dropped is the restriction of the displacement vector $\boldsymbol{\xi}$ to be tangential to the magnetic surfaces. This leads to a different kind of localized modes, called ballooning modes, which may be considered as *Alfvén and slow field line resonances*.

The ballooning mode analysis in tokamaks centres about the description of *instabilities at finite beta, driven by a combination of the pressure gradient p' and the curvature of the field lines* which produces a negative potential energy on the outside of the torus. Hence the name 'ballooning'. To properly describe this requires a subtle localization about single magnetic field lines, developed by Connor, Hastie and Taylor [125], and others (see, e.g. Coppi et al. [127, 128]), where the main difficulty to be resolved is the basic incompatibility of field line localization with poloidal and toroidal periodicity. This is resolved by the so-called ballooning transformation to an extended domain $-\infty < \vartheta < \infty$ of the poloidal angle or, equivalently, by the consideration of a covering space; see Dewar and Glasser [157]. In this section, we will follow the analysis of the latter paper.

To describe modes that are localized about field lines, we exploit a representation of the magnetic field in terms of the two Clebsch potentials α and Ψ,

$$\mathbf{B} = \nabla\alpha \times \nabla\Psi\,, \qquad \alpha \equiv \varphi - q\vartheta\,, \tag{17.64}$$

where α has been chosen such that it varies linearly perpendicular to the straight field lines in the ϑ–φ plane, so that α and Ψ may be considered as field line labels (see footnote on Section 7.3.2). Substitution of α in the expression for $\mathbf{B}$ reproduces the basic magnetic field representation (17.10) for axi-symmetric equilibria:

$$\mathbf{B} = \nabla(\varphi - q\vartheta) \times \nabla\Psi = \nabla\varphi \times \nabla\Psi + I(\Psi)\nabla\varphi\,. \tag{17.65}$$

With this representation, the angles φ and ϑ can be continued indefinitely: the *covering space* [157] corresponding to a magnetic surface. The ballooning equations will turn out to be a system of two second order differential equations in terms of the extended angular variable ϑ, for each field line indicated by Ψ_0 and α_0.

Field line localization is effected by the assumption of large mode numbers, in particular the toroidal wave number, $n \gg 1$. We need to expand our equations in powers of the small parameter $n^{-1} \ll 1$. Such expansions are usually more conveniently carried out starting from the quadratic forms than from the differential equations, since they typically require expanding one order higher for the latter than is needed with the quadratic forms. Hence, we will work out the consequences of field line localization from the two quadratic forms (17.48) and (17.49) for W and K.

We now consider WKB solutions of the plasma displacement of the form

$$X(\Psi,\vartheta,\varphi) = \widetilde{X}(\Psi,\vartheta)\,\mathrm{e}^{\mathrm{i}nS(\Psi,\vartheta,\varphi)}\,, \tag{17.66}$$

and similarly for the components Y and Z, with a split in a slowly varying amplitude $\widetilde{X}$ and a rapidly varying phase nS, where S is known as the *eikonal* [243]. An obvious choice for the eikonal is

$$S(\Psi,\alpha) = \alpha + q(\Psi)\vartheta_0 \equiv \varphi - q(\Psi)(\vartheta - \vartheta_0) = S(\Psi,\vartheta,\varphi)\,, \tag{17.67}$$

where ϑ_0 is an arbitrary constant. Associated with this eikonal is a local wave vector $\mathbf{k} = n\nabla S$, which is normal to the field lines since $\mathbf{b}\cdot\nabla S = 0$ and, hence, normal to the eikonal wave fronts, that we will project onto the field line/magnetic surface triad $\{\mathbf{n},\boldsymbol{\pi},\mathbf{b}\}$. The possible confusion of the distinction between the normal vector $\mathbf{n}$ and the wave number n we eliminate right away by renormalizing $\mathbf{k}$ onto a wave vector $\widetilde{\mathbf{k}}$ of unit order of magnitude:

$$\mathbf{k} = n\nabla S \quad\Rightarrow\quad \widetilde{\mathbf{k}} \equiv n^{-1}\mathbf{k} = \nabla S = \widetilde{k}_n\mathbf{n} + \widetilde{k}_\pi\boldsymbol{\pi}\,, \tag{17.68}$$

$$\begin{aligned} \widetilde{k}_n &= \mathbf{n}\cdot\nabla S = \mathbf{n}\cdot[-q'(\vartheta-\vartheta_0)\nabla\Psi - q\nabla\vartheta] = -RB_\mathrm{p}[q'(\vartheta-\vartheta_0) - (g_{12}/g_{22})q]\,,\\ \widetilde{k}_\pi &= \boldsymbol{\pi}\cdot\nabla S = \boldsymbol{\pi}\cdot[-q\nabla\vartheta + \nabla\varphi] = -B/(RB_\mathrm{p})\,. \end{aligned} \tag{17.69}$$

These wave numbers are local wave numbers in the WKB sense, i.e. they depend on the coordinates, $\widetilde{k}_n = \widetilde{k}_n(\Psi,\vartheta)$ and $\widetilde{k}_\pi = \widetilde{k}_\pi(\Psi,\vartheta)$, and, hence, they become capricious quantities twisting with the field lines.

▷ **Meaning of the constant ϑ_0** One should well distinguish between the two coordinate dependencies $S(\Psi,\alpha)$ and $S(\Psi,\vartheta,\varphi)$ indicated in the equalities (17.67). From the former, we obtain the following, equally meaningful, expressions for the components of $\widetilde{\mathbf{k}}$:

$$\widetilde{\mathbf{k}} = \frac{\partial S}{\partial\Psi}\nabla\Psi + \frac{\partial S}{\partial\alpha}\nabla\alpha = q'\vartheta_0\nabla\Psi + \nabla\alpha = \vartheta_0\nabla q + \nabla\alpha \qquad\Rightarrow\quad \widetilde{k}_{\hat q} = \vartheta_0\,,\quad \widetilde{k}_{\hat\alpha} = 1\,, \tag{17.70}$$

where $\widetilde{k}_{\hat q}$ and $\widetilde{k}_{\hat\alpha}$ are the covariant components of $\widetilde{\mathbf{k}}$ with respect to the coordinates q and α. Here, the coordinate Ψ has been replaced by the equivalent coordinate $q(\Psi)$. Hence, ϑ_0 is the covariant component of $\widetilde{\mathbf{k}}$ with respect to q, considered as a radial coordinate. ◁

For the analysis of ballooning modes it is expedient to exploit a projection based on a new orthogonal triad $\mathbf{d}, \mathbf{e}, \mathbf{b}$, where $\mathbf{d} \equiv \mathbf{k}/|\mathbf{k}|$ and $\mathbf{e} \equiv \mathbf{b}\times\mathbf{d}$, with corresponding components $\widetilde{\xi}_d, \widetilde{\xi}_c, \widetilde{\xi}_b$ of the vector $\widetilde{\boldsymbol{\xi}}$. This is illustrated in Fig. 17.6, where the result of the analysis below ($\widetilde{\xi}_d \approx 0$) is already indicated.

We now work out the effects of the operators D, G, F, defined in Eqs. (17.40), acting on the

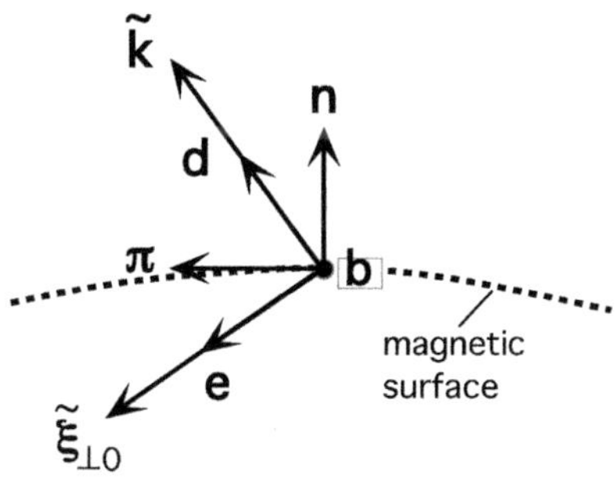

Fig. 17.6 Ballooning triad $\{\mathbf{b}, \mathbf{d}, \mathbf{e}\}$, based on the local wave vector $\widetilde{\mathbf{k}} \equiv n^{-1}\mathbf{k}$, with dominant perpendicular component $\widetilde{\boldsymbol{\xi}}_{\perp 0}$ of the ballooning perturbation.

perturbations X, Y, Z, defined in Eqs. (17.41):

$$\begin{aligned} DX &= (RB_\mathrm{p})^{-1}\mathbf{n}\cdot\nabla(\widetilde{X}\mathrm{e}^{\mathrm{i}nS}) = \Big[\Big(D + \mathrm{i}n\widetilde{k}_n/(RB_\mathrm{p})\Big)\widetilde{X}\Big]\mathrm{e}^{\mathrm{i}nS} \\ &\equiv [\widetilde{D}\widetilde{X}]\mathrm{e}^{\mathrm{i}nS} \qquad \Rightarrow \qquad \widetilde{D} \equiv D + \mathrm{i}n\widetilde{k}_n/(RB_\mathrm{p}), \end{aligned} \tag{17.71}$$

$$\begin{aligned} GX &= -\mathrm{i}RB_\mathrm{p}B\boldsymbol{\pi}\cdot\nabla(\widetilde{X}\mathrm{e}^{\mathrm{i}nS}) = -\mathrm{i}RB_\mathrm{p}B\Big[\boldsymbol{\pi}\cdot\nabla\widetilde{X} + \mathrm{i}n\widetilde{k}_\pi\widetilde{X}\Big]\mathrm{e}^{\mathrm{i}nS} \\ &\equiv [\widetilde{G}\widetilde{X}]\mathrm{e}^{\mathrm{i}nS} \qquad \Rightarrow \qquad \widetilde{G} \equiv -\mathrm{i}RB_\varphi\mathcal{J}^{-1}\partial_\vartheta - nB^2, \end{aligned} \tag{17.72}$$

$$\begin{aligned} FX &= -\mathrm{i}B\mathbf{b}\cdot\nabla(\widetilde{X}\mathrm{e}^{\mathrm{i}nS}) = -\mathrm{i}B\Big[(\mathcal{J}B)^{-1}\partial_\vartheta\widetilde{X}\Big]\mathrm{e}^{\mathrm{i}nS} \\ &\equiv [\widetilde{F}\widetilde{X}]\mathrm{e}^{\mathrm{i}nS} \qquad \Rightarrow \qquad \widetilde{F} \equiv -\mathrm{i}\mathcal{J}^{-1}\partial_\vartheta, \end{aligned} \tag{17.73}$$

and, of course, similarly for Y and Z. The new operators $\widetilde{D}$, $\widetilde{G}$ and $\widetilde{F}$ act on the slowly varying amplitude functions $\widetilde{X}$, $\widetilde{Y}$ and $\widetilde{Z}$ only, which is indicated by enclosing them by square brackets. Expanding in the small parameter n^{-1}, it is obvious that the leading order expressions will be obtained from the terms proportional to n in the operators $\widetilde{D}$ and $\widetilde{G}$.

We now substitute these expressions into the quadratic forms (17.48) and (17.49) for W and K, and reduce them order by order according to the following scheme:

$$\begin{aligned} &\widetilde{X} = \widetilde{X}_0 + n^{-1}\widetilde{X}_1 + \cdots, \quad \text{and similarly for } \widetilde{Y} \text{ and } \widetilde{Z}, \\ &\Rightarrow \quad W = n^2 W_0 + W_2 + \cdots, \qquad K = K_0 + \cdots. \end{aligned} \tag{17.74}$$

This yields the following lowest order expressions:

$$W_0 = \pi\iint(\gamma p + B^2)\Big|\mathrm{i}[\widetilde{k}_n/(RB_\mathrm{p})]\widetilde{X}_0 - \widetilde{Y}_0\Big|^2\mathcal{J}\,d\Psi\,d\vartheta, \tag{17.75}$$

$$K_0 = \pi\iint\rho\Big[\frac{1}{R^2B_\mathrm{p}^2}|\widetilde{X}_0|^2 + \frac{R^2B_\mathrm{p}^2}{B^2}|\widetilde{Y}_0|^2 + B^2|\widetilde{Z}_0|^2\Big]\mathcal{J}d\Psi d\vartheta, \tag{17.76}$$

with the obvious minimizing relation

$$\mathrm{i}\widetilde{\mathbf{k}}\cdot\widetilde{\boldsymbol{\xi}}_0 \equiv \mathrm{i}(\widetilde{k}_n\widetilde{\xi}_{n0} + \widetilde{k}_\pi\widetilde{\xi}_{\pi 0}) = \mathrm{i}[\widetilde{k}_n/(RB_\mathrm{p})]\widetilde{X}_0 - \widetilde{Y}_0 = 0 \quad \Rightarrow \quad W_0 = 0. \tag{17.77}$$

Hence, to leading order, the ballooning perturbation $\widetilde{\xi}_{d0} = 0$ (see Fig. 17.6). The next order, W_2, contains the zero order variables $\widetilde{X}_0$, $\widetilde{Y}_0$ and $\widetilde{Z}_0$, as well as the first order variables $\widetilde{X}_1$ and $\widetilde{Y}_1$, in the combination $\widetilde{\xi}_{d1}$. The latter variable is eliminated by minimizing W_2 with respect to it. This involves first completing the squares of the second and fourth term (involving $\widetilde{D}^\dagger \widetilde{X}_0$) of the expression (17.48) for W, then combining them into a new square, and finally minimizing the resulting expression for W_2 with respect to $\widetilde{\xi}_{d1}$, which now remains finite:

$$\begin{aligned} \mathrm{i}\widetilde{\mathbf{k}} \cdot \widetilde{\boldsymbol{\xi}}_1 &\equiv \mathrm{i}[\widetilde{k}_n/(RB_\mathrm{p})]\widetilde{X}_1 - \widetilde{Y}_1 \\ &= -\,\widetilde{D}^\dagger \widetilde{X}_0 - \frac{B^2}{\gamma p + B^2}\Big(\frac{2\kappa_n}{RB_\mathrm{p}}\widetilde{X}_0 + \frac{1}{B^2} I\widetilde{F}\widetilde{Y}_0\Big) - \frac{\gamma p}{\gamma p + B^2}\Big(I\widetilde{F}\frac{1}{B^2}\widetilde{Y}_0 + \widetilde{F}\widetilde{Z}_0\Big). \end{aligned} \tag{17.78}$$

Substitution of this expression back into W_2 gives the final result:

$$\begin{aligned} W_2 = \pi \iint \Big\{ &\frac{1}{R^2 B_\mathrm{p}^2}|\widetilde{F}\widetilde{X}_0|^2 + \frac{R^2 B_\mathrm{p}^2}{B^2}\Big|\widetilde{F}\widetilde{Y}_0 + \frac{2B_\varphi(\kappa_\mathrm{p} - \kappa_\varphi)}{R^2 B_\mathrm{p}}\widetilde{X}_0\Big|^2 \\ &+ \frac{\gamma p B^2}{\gamma p + B^2}\Big|\widetilde{F}\widetilde{Z}_0 - \frac{2\kappa_n}{RB_\mathrm{p}}\widetilde{X}_0 + 2\mathrm{i}\frac{RB_\mathrm{p}}{B}\kappa_\mathrm{g}\widetilde{Y}_0\Big|^2 + U|\widetilde{X}_0|^2\Big\} \mathcal{J}\, d\Psi\, d\vartheta \\ = \pi \iint \Big\{ &\frac{1}{R^2 B_\mathrm{p}^2}(1 + \widetilde{k}_n^2/\widetilde{k}_\pi^2)|\widetilde{F}\widetilde{X}_0|^2 - \frac{2}{RB_\mathrm{p}}[\kappa_n - (\widetilde{k}_n/\widetilde{k}_\pi)\kappa_\mathrm{g}]p'|\widetilde{X}_0|^2 \\ &+ \frac{\gamma p B^2}{\gamma p + B^2}\Big|\widetilde{F}\widetilde{Z}_0 - \frac{2}{RB_\mathrm{p}}[\kappa_n - (\widetilde{k}_n/\widetilde{k}_\pi)\kappa_\mathrm{g}]\widetilde{X}_0\Big|^2\Big\} \mathcal{J}\, d\Psi\, d\vartheta\,, \end{aligned} \tag{17.79}$$

where U is defined in Eq. (17.50), and the last expression follows by substituting $\widetilde{Y}_0$ from (17.77), integration by parts and use of the relations of Section 17.1.3.

In conclusion, the variational principle (17.47) has been reduced to

$$\delta\Lambda_2 = 0\,, \qquad \Lambda_2[\widetilde{\boldsymbol{\xi}}_0] \equiv \frac{W_2[\widetilde{\boldsymbol{\xi}}_0]}{K_0[\widetilde{\boldsymbol{\xi}}_0]}\,, \tag{17.80}$$

with W_2 and K_0, given by the expressions (17.79) and (17.76), converted into expressions in terms of the leading order displacement vector $\widetilde{\boldsymbol{\xi}}_0$ of the ballooning perturbations. Since this vector is directed perpendicular to $\widetilde{\mathbf{k}}$, it is appropriate to convert $(\widetilde{u}, \widetilde{v}, \widetilde{\zeta})$ to a two-component representation in terms of $\widetilde{v}$ and $\widetilde{\zeta}$:

$$\begin{aligned} &\widetilde{\boldsymbol{\xi}}_0(\Psi, \vartheta) = -\mathrm{i}\,[\widetilde{v}(\Psi, \vartheta)\mathbf{e} + \widetilde{\zeta}(\Psi, \vartheta)\mathbf{b}] \\ &\Rightarrow\quad \widetilde{u} \equiv \mathrm{i}\widetilde{\boldsymbol{\xi}}_0 \cdot \mathbf{d} = (1/\widetilde{k})[\mathrm{i}(\widetilde{k}_n/(RB_\mathrm{p}))\widetilde{X}_0 - \widetilde{Y}_0] = 0\,, \\ &\qquad\quad \widetilde{v} \equiv \mathrm{i}\widetilde{\boldsymbol{\xi}}_0 \cdot \mathbf{e} = (1/\widetilde{k})[(B/(R^2 B_\mathrm{p}^2))\widetilde{X}_0 - \mathrm{i}(RB_\mathrm{p}/B)\widetilde{Y}_0] = \widetilde{k}\widetilde{X}_0\,, \\ &\qquad\quad \widetilde{\zeta} \equiv \mathrm{i}\widetilde{\boldsymbol{\xi}}_0 \cdot \mathbf{b} = B\widetilde{Z}_0\,. \end{aligned} \tag{17.81}$$

This yields the most symmetric representation of the Euler equations associated with the variational principle (17.80) in terms of two coupled second order ordinary differential equations in the extended variable ϑ, with parametric dependence of the coefficients on Ψ, as derived by Dewar and

Glasser [157] (see also [410]):

$$\begin{pmatrix} \widetilde{\alpha}_{11} & \widetilde{\alpha}_{12} \\ \widetilde{\alpha}_{21} & \widetilde{\alpha}_{22} \end{pmatrix} \begin{pmatrix} \widetilde{v} \\ \widetilde{\zeta} \end{pmatrix} = \rho\omega^2 \begin{pmatrix} \widetilde{v} \\ \widetilde{\zeta} \end{pmatrix}, \tag{17.82}$$

where

$$\widetilde{\alpha}_{11} \equiv \frac{B}{\widetilde{k}}\widetilde{F}\frac{\widetilde{k}^2}{B^2}\widetilde{F}\frac{B}{\widetilde{k}} + \frac{4\gamma p B^2}{\gamma p + B^2}\kappa_e^2 - 2\frac{B}{\widetilde{k}}\kappa_e p', \qquad \widetilde{\alpha}_{12} \equiv -\frac{2\mathrm{i}\gamma p B^2}{\gamma p + B^2}\kappa_e \widetilde{F}\frac{1}{B},$$

$$\widetilde{\alpha}_{21} \equiv \frac{\mathrm{i}}{B}\widetilde{F}\frac{2\gamma p B^2}{\gamma p + B^2}\kappa_e, \qquad \widetilde{\alpha}_{22} \equiv \frac{1}{B}\widetilde{F}\frac{\gamma p B^2}{\gamma p + B^2}\widetilde{F}\frac{1}{B}. \tag{17.83}$$

Here, $\kappa_e \equiv \mathbf{e}\cdot\boldsymbol{\kappa} = -(\widetilde{k}_\pi/\widetilde{k})\kappa_n + (\widetilde{k}_n/\widetilde{k})\kappa_g$ is the component of the field line curvature perpendicular to $\widetilde{\mathbf{k}}$ and the operator

$$\widetilde{F} \equiv -\mathrm{i}\mathbf{B}\cdot\widetilde{\nabla} = -(\mathrm{i}/\mathcal{J})\,\partial_\vartheta \tag{17.84}$$

is just the part of the parallel gradient operator which acts on the slowly varying amplitude functions. Since the functions $\widetilde{v}$ and $\widetilde{\zeta}$ are defined on an infinite domain for periodic configurations, the appropriate boundary conditions at $\vartheta \to \pm\infty$ are that they should decay rapidly enough to remain square integrable.

As compared to the magnetic-surface-localized Alfvén and slow continua of Eq. (17.55), due to the additional freedom of field line localization, there is now a new, potentially negative term, proportional to $-\kappa_e p'$, in the upper left diagonal element $\widetilde{\alpha}_{11}$ of the matrix (17.82). This is the driving term of high-n ballooning modes, corresponding to the low-n ballooning term $\epsilon\beta_p\kappa_\varphi$ encountered in the coupling term (17.63) of the high-beta tokamak model of the previous section.

Finally, the ballooning equation of Connor, Hastie and Taylor [125] is obtained from the two ODEs (17.82) by neglecting the compressibility terms ($\gamma p \ll B^2$, appropriate for low-beta) so that $\widetilde{\zeta} = 0$ (or $\widetilde{Z}_0 = 0$). This gives:

$$-\frac{B^2}{\mathcal{J}\widetilde{k}^2}\frac{\partial}{\partial\vartheta}\left(\frac{\widetilde{k}^2}{\mathcal{J}B^2}\frac{\partial\widetilde{X}_0}{\partial\vartheta}\right) - \frac{2B}{\widetilde{k}}\kappa_e p'\widetilde{X}_0 = \rho\omega^2\widetilde{X}_0. \tag{17.85}$$

This appears to yield a very simple prescription for the study of the stability of tokamaks with respect to high-n ballooning modes: investigate the Sturmian properties of the eigenfunctions $\widetilde{X}_0$ on the infinite domain $-\infty < \vartheta < \infty$ and determine the lowest eigenvalues $\omega^2(\Psi, \vartheta_0)$. If they are positive, the configuration is locally stable, if there is a negative one it is unstable.

One aspect of the ballooning theory presented so far is entirely unclear. How does this analysis reconcile poloidal periodicity with field line localization? Following Dewar and Glasser [157] again (see also [156], [297], [484] for the associated small-scale radial periodicity), this question is answered by observing that the eigenvalue problem (17.82), or (17.85), is infinitely degenerate with respect to poloidal periodicity. In particular, notice that the Rayleigh quotient Λ_2 is invariant under a change of the poloidal angle ϑ over a period 2π, provided that the value of the parameter ϑ_0 is replaced by $\vartheta_0 + 2\pi$ as well. Exploiting the coordinates Ψ and α to indicate the field lines, and considering the variable X only (dropping the subscript 0), this implies the following:

$$\begin{aligned} \omega^2(\Psi, \alpha - 2\pi q, \vartheta_0 + 2\pi) &= \omega^2(\Psi, \alpha, \vartheta_0) \qquad \text{(for eigenvalues)}, \\ X(\vartheta + 2\pi; \Psi, \alpha - 2\pi q, \vartheta_0 + 2\pi) &= X(\vartheta; \Psi, \alpha, \vartheta_0) \quad \text{(for eigenfunctions)}. \end{aligned} \tag{17.86}$$

Hence, instead of the single non-periodic mode (17.66), the superposition

$$X(\Psi,\vartheta,\varphi) = \sum_{\ell=-\infty}^{\infty} \widetilde{X}(\Psi,\vartheta+2\pi\ell)\,\mathrm{e}^{\mathrm{i}n[\varphi-q(\vartheta+2\pi\ell-\vartheta_0)]} \tag{17.87}$$

is a periodic solution of the ballooning equations, with the eigenvalue ω^2, provided that the amplitude $\widetilde{X}$ falls off rapidly enough at infinity to leave the quadratic forms finite. This behaviour is dictated by the indicial equation, which turns out to be associated with the Mercier criterion at marginal stability (see below).

▷ **Analogies between continuum and ballooning equations** The analogy between the coupled ODEs (17.54), describing the Alfvén and slow continua, and (17.82), describing the Alfvén and slow ballooning modes, is striking but misleading. Formally, one may obtain Eq. (17.54) from Eq. (17.82), as noted by Lifschitz [410], by taking the limit $\vartheta_0 \to \infty$ so that $\widetilde{k} \approx \widetilde{k}_n \approx q'\vartheta_0 \cdot RB_\mathrm{p} \to \infty$, $\kappa_e \to \kappa_\mathrm{g}$, $\widetilde{v} \to \widetilde{\eta}$ and $\kappa_e p'/\widetilde{k} \to 0$. However, we then get an equation in terms of $\widetilde{F}$ acting on $\widetilde{\eta}$, rather than F acting on η, whereas the relationship between η and $\widetilde{\eta}$, viz.

$$\eta(\vartheta) = \mathrm{e}^{\mathrm{i}nq\vartheta_0}\,\widetilde{\eta}(\vartheta)\,\mathrm{e}^{-\mathrm{i}nq\vartheta}\,, \tag{17.88}$$

becomes a tricky one in view of the wildly oscillating factor involving ϑ_0. In order to permit the consideration of $n \sim 1$ modes one would have to re-interpret a posteriori the ballooning approximation, which assumed $n \gg 1$. Physically, this is related to the absence in Eq. (17.54) of the most important term in the ballooning equation, viz. the one that drives instabilities: $\kappa_e p'$. In order to derive it, it was necessary to relax on the constraint of displacements lying in the magnetic surfaces (necessary for Eq. (17.54)), on the one hand, but to restrict the displacements to be localized around the field lines (implicit in the Ansatz (17.66)), on the other. The latter constraint is necessary to minimize field line bending, which is quite appropriate in stability studies but completely spurious for the study of Alfvén wave heating, where the global exciting wave just enforces the field line bendings that correspond to their scale lengths. For example, in the case of Alfvén wave heating of the solar corona, it is a safe assumption that these scale lengths are arbitrary, in the sense of being dictated by the photospheric convection patterns and not by the field line structure of the coronal loops. We conclude that each of the formalisms associated with the two types of equation has its own domain of application, where the typical ballooning term $\kappa_e p'$ is crucial for stability studies but a minor correction in heating scenarios and, vice versa, the low mode number structure of the continua (with gaps in the case of tokamaks) is crucial for heating but strictly lost in the ballooning formalism. ◁

Mercier criterion The Mercier criterion was originally derived by Mercier [437] as a generalization of Suydam's criterion. With the introduction of the ballooning formalism, it became clear that it can also be interpreted as a condition on the solutions of the ballooning equation for $\vartheta \to \infty$. This leads again to a cluster point analysis, but now in the stretched angular variable ϑ, rather than the radial flux variable Ψ. The derivation is put in small print.

▷ **Derivation of the Mercier criterion** The Mercier criterion may be obtained from the marginal ($\omega^2 = 0$) version of the ballooning equation (17.85),

$$\frac{1}{\mathcal{J}}\frac{d}{d\vartheta}\left[\frac{1}{\mathcal{J}R^2B_\mathrm{p}^2}\left(1+\frac{\widetilde{k}_n^2}{\widetilde{k}_\pi^2}\right)\frac{d\widetilde{X}_0}{d\vartheta}\right] + \frac{2}{RB_\mathrm{p}}\left(\kappa_n - \frac{\widetilde{k}_n}{\widetilde{k}_\pi}\kappa_\mathrm{g}\right)p'\widetilde{X}_0 = 0\,, \tag{17.89}$$

where the equilibrium functions are periodic, but $\widetilde{k}_n/\widetilde{k}_\pi$ has a secular dependence on ϑ:

$$\widetilde{k}_n/\widetilde{k}_\pi = (R^2B_\mathrm{p}^2/B^2)\,z\,, \qquad z \equiv q'(\vartheta-\vartheta_0) - (g_{12}/g_{22})\,q\,. \tag{17.90}$$

Splitting periodic and secular dependencies, the ballooning equation is written as

$$\frac{d}{d\vartheta}\left[(az^2+b)\,\frac{d\widetilde{X}_0}{d\vartheta}\right] + (\dot{c}z+d)\,\widetilde{X}_0 = 0\,, \tag{17.91}$$

$$a \equiv \frac{R^2 B_\mathrm{p}^2}{\mathcal{J} B^2}\,, \qquad b \equiv \frac{1}{\mathcal{J} R^2 B_\mathrm{p}^2}\,, \qquad c \equiv \frac{I p'}{B^2}\,, \qquad d \equiv \frac{2\mathcal{J} \kappa_n p'}{R B_\mathrm{p}}\,,$$

where the dot indicates differentiation with respect to ϑ. We try solutions of the form

$$\widetilde{X}_0 = z^\nu \Big[f_0(\vartheta) + \frac{f_1(\vartheta)}{z} + \frac{f_2(\vartheta)}{z^2} + \cdots \Big]\,, \tag{17.92}$$

where the f_is are periodic functions of ϑ. The task is to determine the index ν (similar to the derivation of Suydam's criterion in Section 9.4.1, but now on the extended angular domain). This is done by substituting (17.92) into (17.91) and balancing the powers of z:

$$z^{\nu+2}: \quad \frac{d}{d\vartheta}\Big(a \frac{df_0}{d\vartheta} \Big) = 0 \qquad \Rightarrow \quad f_0 = C_1\,, \tag{17.93}$$

$$z^{\nu+1}: \quad \frac{d}{d\vartheta}\Big(a \frac{df_1}{d\vartheta} \Big) = -\Big[\nu (a\dot{z})^{\cdot} + \dot{c} \Big] f_0$$

$$\Rightarrow \quad f_1 = -C_1 \Big[\nu z + \int (c/a)\, d\vartheta \Big] + C_2 \int (1/a)\, d\vartheta + C_3\,,$$

$$- C_1 \Big[2\pi \nu q' + \oint (c/a)\, d\vartheta \Big] + C_2 \oint (1/a)\, d\vartheta = 0\,, \tag{17.94}$$

$$z^{\nu} \quad : \quad \frac{d}{d\vartheta}\Big(a \frac{df_2}{d\vartheta} \Big) = -2\nu a \dot{z} \frac{df_1}{d\vartheta} - \Big[(\nu - 1)(a\dot{z})^{\cdot} + \dot{c} \Big] f_1 - \Big[\nu(\nu+1) a \dot{z}^2 + d \Big] f_0\,,$$

$$- C_1 \oint \Big[c\dot{z} - (c^2/a) - d \Big]\, d\vartheta + C_2 \Big[(\nu+1) 2\pi q' - \oint (c/a)\, d\vartheta \Big] = 0\,. \tag{17.95}$$

Here, periodicity of f_0 is manifest, periodicity of f_1 requires the relationship (17.94) between the constants C_1 and C_2, whereas periodicity of f_2 requires the relationship (17.95). Compatibility of the latter two conditions yields the *indicial equation:*

$$\nu(\nu+1) + D = 0\,,$$

$$D \equiv \frac{1}{4\pi^2 q'^2} \Big\{ \oint (c/a)\, d\vartheta \Big[2\pi q' - \oint (c/a)\, d\vartheta \Big] - \oint (1/a)\, d\vartheta \oint \Big[c\dot{z} - (c^2/a) - d \Big]\, d\vartheta \Big\}\,. \tag{17.96}$$

This implies infinitely oscillatory solutions (local instability with respect to interchanges) when the indices are complex ($D < 1/4$) and stability when the indices are real. ◁

By exploiting the Grad–Shafranov relation (17.16) and the other relations of Section 17.1.3, many equivalent forms of the Mercier criterion $D > 1/4$ can be derived. The following one, due to Pao [476], is probably the most appealing one:

$$\tfrac{1}{2}\pi^2 q'^2 + p' \Big[\oint \frac{B^2}{R^2 B_\mathrm{p}^2} \mathcal{J}\, d\vartheta \oint \frac{\kappa_\mathrm{p}}{R B_\mathrm{p}} \mathcal{J}\, d\vartheta + \oint \frac{1}{R^2 B_\mathrm{p}^2} \mathcal{J}\, d\vartheta \oint \frac{(-\kappa_\mathrm{p} + \kappa_\varphi) B_\varphi^2}{R B_\mathrm{p}} \mathcal{J}\, d\vartheta \Big] > 0\,. \tag{17.97}$$

In this form, it clearly exhibits the competing contributions of the magnetic shear (first term), which is stabilizing, and the pressure gradient, which drives interchange instabilities if the term in square brackets is positive. In the cylindrical limit, $\kappa_\varphi = 0$ and $B^2 = \text{const}$, that term is positive definite and Suydam's criterion is recovered. In the toroidal case, the term may change sign through the contribution of the toroidal curvature. In general, numerical analysis of specific equilibria is needed to determine stability with respect to interchange modes.

17.3 Computation of waves and instabilities in tokamaks

We now turn to numerical theory to provide us with the procedures that produce the explicit answers needed in fusion research. Once large-scale computing is embraced, some of the analytical methods lose their attraction and new possibilities present themselves. Thus, we will only discuss ideal MHD in short, since it is contained in resistive MHD as far as the numerics is concerned, and its numerical implementation appears to be too restrictive at the present time.

17.3.1 Ideal MHD versus resistive MHD in computations

There is a long history on the subject of ideal MHD stability of tokamaks. There appears to be no point trying to summarize this in a few lines. Excellent texts on this subject exist [186], [647]. The reader is advised to consult those. The present chapter is more concerned with the structures of the different theoretical methods than with specific results. We will merely use the results to illustrate the methods.

Troyon beta limit There is also a long history on the development and subsequent use of toroidal ideal MHD stability programs, the most well known of them ERATO [266, 265] and PEST [260]. Again, we will not attempt to summarize the results reached. There is one famous result, though, that illustrates one of the points of this section. This is the Troyon beta limit [610] for stability of tokamaks with respect to ideal MHD kink modes. It summarizes the results of a great number of numerical optimization studies with respect to high-β stability of tokamaks. The result is a quite simple scaling law of the maximum value of β with respect to kink mode stability as a function of the toroidal current I_φ, the vacuum magnetic field B_0 and the plasma radius a (with the units indicated in brackets):

$$\beta\,(\%) < g_{\rm T}\,\frac{I_\varphi\,(\mathrm{MA})}{a\,(\mathrm{m})\,B_0\,(\mathrm{T})}\,, \tag{17.98}$$

where $g_{\rm T}$ is a factor obtained from the numerical studies, originally put at $g_{\rm T} \approx 3$. How is it that all the complicated Alfvén wave dynamics in the curved environment of toroidal confinement systems just leads to such a simple answer? Is it a genuine scaling law or just a very stimulating way of getting experimentalists to improve tokamak performance? (Like Moore's law in computer chips: not a law of nature, but demonstration of the possibility of obtaining desirable scaling by intense technological efforts.) Figure 17.7 appears to point in the latter direction, with the swarm of experimental points continually moving up, in JET reaching $\beta \approx 6\%$ in 1990, but soon afterwards (1995) put into the shade by the record values of $\beta \approx 12.6\%$ obtained in the DIII-D tokamak [592], with some points far above the curve. Not surprisingly, the slope of the curve also turned out to depend on details of the current distribution, parameterized by the internal inductance $\ell_{\rm i}$. Thus, ideal MHD computations have served not only as a theoretical framework for the interpretation of tokamak stability results, but also as a stimulus for experimental progress. (In this respect, the comparison with Moore's law should be taken as positive evaluation of the Troyon scaling law.)

▷ **High-beta tokamak scaling** The high-beta scaling of Section 16.1.4 actually yields a quadratic dependence on the toroidal current, following from Eqs. (16.69) and (16.66):

$$\beta = \frac{0.04\pi a^2\,\beta_{\rm p}}{S}\left(\frac{10^{-6} I_\varphi}{aB_0}\right)^2 \qquad \text{(all in mks units)}\,. \tag{17.99}$$

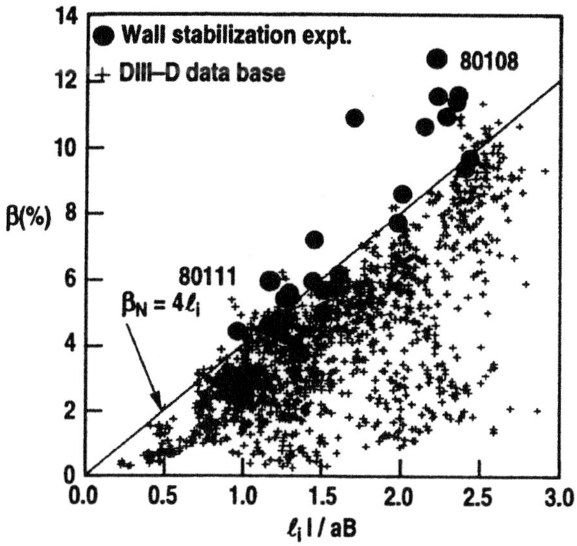

Fig. 17.7 Toroidal beta as a function of normalized current in DIII-D discharges. The drawn line is a modified form of the Troyon limit. (From Taylor et al. [592]).

For limiting values $\beta_p \sim \epsilon^{-1}$, roughly independent of the toroidal current, this expression appears not to conflict with the general tendency of the points shown in Fig. 17.7. ◁

Turning to resistive MHD calculations Progress in computer equipment and computational methods not only served to narrow down the gap between theory of highly simplified analytical models and numerical analysis of more realistic models of the magnetic confinement geometry of tokamaks, but it also revealed a false dilemma between ideal (conservative) and resistive (dissipative) MHD. Historically, the vast majority of tokamak stability investigations have been done by means of the energy principle of ideal MHD [275, 55], and its spectral generalizations, extensively discussed in this book. The reason is obvious: that theory is the simplest and also the most appealing one with respect to mathematical and physical properties (the power of spectral theory combined with the beauty of conservation laws). Also, with the velocity representation (or the displacement vector $\boldsymbol{\xi}$), reduction to the smallest number of unknowns was effected, a very desirable feature in the early days of computing with very restricted central memory sizes. However, it falsely created the impression that computing for the more extended MHD models, including resistive MHD, would necessarily be less accurate and also much slower than for ideal MHD. In order to appreciate this misunderstanding, we first need to introduce the newer numerical methods used in resistive MHD calculations. At the end of this section, we will then return to an explanation of why and how the superiority of the conservative model over the dissipative model turned out to be a false dilemma.

Moving to dissipative MHD, we immediately realize that the basic variable $\boldsymbol{\xi}$ is no longer available since its definition is strictly based on flux conservation. Therefore, let us return to the basic equations of resistive MHD, Eqs. (4.120)–(4.123), already exploited in Section 14.2, but now replacing the pressure by the temperature using the relation $p = (\gamma - 1)C_v \rho T$:

$$\frac{\partial \rho}{\partial t} = -\nabla \cdot (\rho \mathbf{v}), \tag{17.100}$$

$$\rho \frac{d\mathbf{v}}{dt} = -(\gamma - 1)C_v \nabla(\rho T) + \mu_0^{-1}(\nabla \times \mathbf{B}) \times \mathbf{B}, \tag{17.101}$$

$$C_v \rho \frac{dT}{dt} = -(\gamma - 1)C_v \rho T \nabla \cdot \mathbf{v} + \mu_0^{-2}\eta(\nabla \times \mathbf{B})^2, \tag{17.102}$$

$$\frac{\partial \mathbf{B}}{\partial t} = \nabla \times (\mathbf{v} \times \mathbf{B}) - \mu_o^{-1}\nabla \times (\eta \nabla \times \mathbf{B}), \tag{17.103}$$

$$\nabla \cdot \mathbf{B} = 0. \tag{17.104}$$

We will eliminate the factors C_v and μ_0 by defining $\bar{T} \equiv (\gamma - 1)C_v T$, $\bar{\eta} \equiv \eta/\mu_0$ and $\bar{\mathbf{B}} \equiv \mathbf{B}/\sqrt{\mu_0}$, but we will drop the bars immediately. The basic equations (17.100)–(17.103) govern the temporal evolution of the density ρ, the velocity $\mathbf{v}$, the temperature T and the magnetic field $\mathbf{B}$, whereas Eq. (17.104) is a constraint on $\mathbf{B}$ that should be satisfied with the same accuracy as the other equations. (See Tóth [603], and Section 20.3.1, for a discussion of the numerical aspects.)

Let us now linearize these equations according to the scheme

$$f(\mathbf{r}, t) \approx f_0(\psi, \vartheta) + f_1(\psi, \vartheta)e^{i(n\varphi - \omega t)}, \tag{17.105}$$

where the f_0s describe the equilibrium quantities, the f_1s are the amplitudes of the normal mode solutions and n is the toroidal mode number. Recall that for ideal MHD, dropping the terms with η and writing $\mathbf{v}_1 \equiv \partial\boldsymbol{\xi}/\partial t$, three of the four equations can be integrated directly and the spectral problem

$$\mathbf{F}(\boldsymbol{\xi}) = -\rho\omega^2\boldsymbol{\xi} \tag{17.106}$$

is obtained, where $\mathbf{F}$ is the force operator, which is self-adjoint, and the eigenvalues ω^2 are real. For resistive MHD, this scheme does not work and we are forced to consider the eight components ρ_1, $\mathbf{v}_1$, T_1, $\mathbf{B}_1$ describing the perturbed state. In general, the eigenvalues ω will then have both real and imaginary parts. Instead of ω, it is customary to exploit the complex eigenvalue parameter $\lambda \equiv -i\omega$, the real part of which measures the exponential growth rate of instabilities.

We will simply remove the initial condition $\nabla \cdot \mathbf{B}_1 = 0$ from the eigenvalue problem by exploiting the vector potential $\mathbf{A}_1$ as a variable, $\mathbf{B}_1 = \nabla \times \mathbf{A}_1$, and choosing the gauge condition $\Phi_1 = 0$ on the scalar potential. This elimination of the initial condition will introduce spurious eigenvalues $\lambda = 0$, which have to be removed later on. However, this is a relatively easy numerical task.

The equations of the resulting eigenvalue problem, already presented as Eqs. (15.61)–(15.64) of Section 15.2.1, are repeated here for convenience:

$$\lambda\rho_1 = -\nabla \cdot (\rho \mathbf{v}_1), \tag{17.107}$$

$$\lambda\rho\mathbf{v}_1 = -\nabla\Big(\rho T_1 + (p/\rho)\,\rho_1\Big) + (\nabla \times \mathbf{B}) \times (\nabla \times \mathbf{A}_1) - \mathbf{B} \times (\nabla \times \nabla \times \mathbf{A}_1), \tag{17.108}$$

$$\lambda\rho T_1 = -\rho\mathbf{v}_1 \cdot \nabla(p/\rho) - p\nabla \cdot \mathbf{v}_1 + 2\eta\,(\gamma - 1)\,(\nabla \times \mathbf{B}) \cdot (\nabla \times \nabla \times \mathbf{A}_1), \tag{17.109}$$

$$\lambda\mathbf{A}_1 = -\mathbf{B} \times \mathbf{v}_1 - \eta\,\nabla \times \nabla \times \mathbf{A}_1. \tag{17.110}$$

Here, explicit knowledge of the equilibrium, characterized by the functions $\rho(\Psi, \vartheta)$, $p(\Psi, \vartheta)$ and

$\mathbf{B}(\Psi, \vartheta)$, is presupposed. Hence, the equilibrium part of the calculation (Section 16.3.3), the inversion of the coordinates (Section 17.1.2) and the equilibrium properties (Section 17.1.3) remain in effect; we just discuss a different implementation of the box labelled 'STAB' in Fig. 17.1. Introducing the basic state eight-vector

$$\mathbf{u} \equiv (\,\rho_1, \mathbf{v}_1, T_1, \mathbf{A}_1\,)^{\mathrm{T}}\,, \tag{17.111}$$

the system of equations (17.107)–(17.110) may be written in matrix form as

$$\mathbf{L} \cdot \mathbf{u} = \lambda\, \mathbf{R} \cdot \mathbf{u}\,, \tag{17.112}$$

exploiting the same operators $\mathbf{L}$ and $\mathbf{R}$ as in Eq. (15.65). (Note again the interchange of LHS and RHS!) This is the basic eigenvalue problem of resistive MHD, to be compared with Eq. (17.106) of ideal MHD.

Before we discuss the numerical implementation of this analysis, we observe some simple facts.

(a) The resistive eigenvalue problem (17.107)–(17.110) in terms of state vectors $\mathbf{u}$ with eight rather than three components, implies *a substantial increase of necessary memory size:* a distinct disadvantage of resistive MHD compared to ideal MHD. This is one of the reasons that dissipative spectral computations have become feasible only recently, with the advent of cheap memory in large-scale computing.

(b) Since no special tricks have been used in the derivation of the resistive equations, *the extension with other dissipation mechanisms like viscosity, heat conduction, etc. is easy:* a distinct advantage of the resistive MHD computations.

(c) The eigenvalue λ appears linearly in the resistive eigenvalue problem (17.112), whereas the ideal MHD problem (17.106) has a quadratic eigenvalue ω^2. For a given accuracy of the computed equilibrium and of the eigenvalue solvers exploited, the number of accurate digits obtained in λ with resistive calculations is of the same order as that in ω^2 with ideal calculations. Hence, for a given time scale of interest, it appears that *the resistive calculations are trivially more accurate than the ideal ones!* (A sobering thought after twenty years of intensive numerical research in ideal MHD.)

(d) Likewise, the spatial derivatives of the equilibrium quantities appearing in the dissipative system are typically first order, versus second order in the ideal case: a distinct disadvantage for ideal MHD computations. Again, *the resistive calculations are trivially more accurate than the ideal ones!*

To sum it up: we appear to lose a powerful tool, but we get a much more powerful numerical tool in return.

Let us now see how it works in practice; we here summarize some of the numerical implementations of these ideas by Kerner [357, 358]. The eigenvalue problem (17.107)–(17.110) is solved by the Galerkin method. *A weak form* of Eq. (17.112) is constructed by multiplying with an arbitrary test function $\mathbf{v}$ and integrating over the domain of interest, i.e. the plasma interior (restricting the analysis to internal modes for the time being):

$$\int \mathbf{v}^{\mathrm{T}} \cdot \mathbf{L} \cdot \mathbf{u}\, dV = \lambda \int \mathbf{v}^{\mathrm{T}} \cdot \mathbf{R} \cdot \mathbf{u}\, dV\,. \tag{17.113}$$

$\mathbf{v}$ is a solution of Eq. (17.112) in the weak sense if Eq. (17.113) is satisfied for every test function $\mathbf{v}$ of the appropriate space, i.e. the space of functions that satisfy the pertinent boundary conditions.

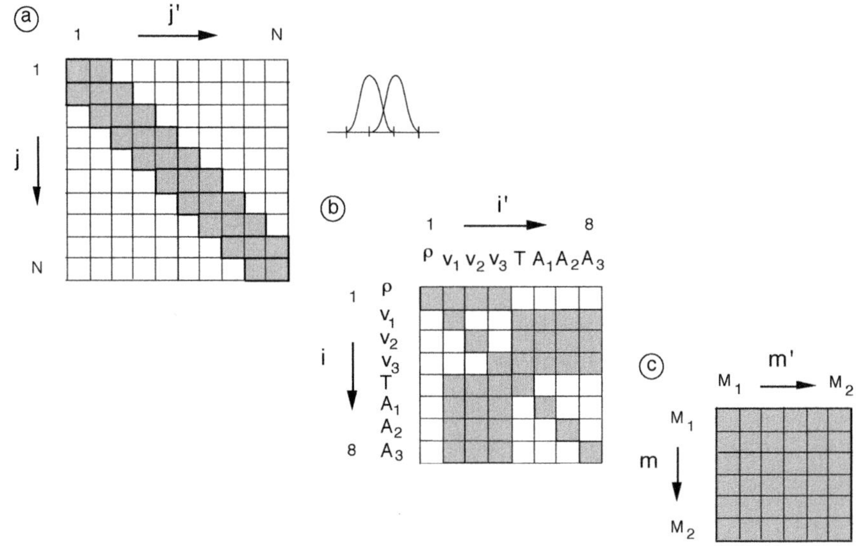

Fig. 17.8 Block structure of the CASTOR eigenvalue matrices: (a) tridiagonal structure due to finite element radial discretization; the finite elements can be accumulated at positions where the modes tend to localize (e.g. at the locations of the ideal MHD singularities); (b) block structure due to cross-products of the physical variables; (c) block structure due to poloidal mode coupling. The blocks of (b) and (c) are essentially full, sparseness is solely due to the finite element discretization in the radial direction. This discretization reflects the difference of wave propagation inside and across magnetic surfaces. On each grid point two kinds of finite element are exploited so that there are $(2 \times 8 \times (M_2 - M_1 + 1))^2 \times 3 \times N$ non-zero complex matrix elements, where N is the number of radial points, and M_1 and M_2 are the lower and upper limit of the range of the poloidal mode number m.

The set of test functions $\mathbf{v}$ in the weak form (17.113) is chosen to be the same as used in the discretization of the physical variables $\mathbf{u}$. For both of them, we exploit a finite element representation, extensively discussed in Section 15.1.3, with pairs of quadratic and cubic finite elements on each grid point labelled j. For simplicity, we first consider the 1D cylindrical case with radial variations only. Hence, if the ith component of $\mathbf{u}$ is approximated by

$$u^i \simeq \widetilde{u}^i = \sum_{j=1}^{2N} x^i_j\, h^i_j(r) \qquad (i = 1, \ldots, 8)\,, \tag{17.114}$$

where N is the number of pairs of finite elements employed (which is the same as the number of radial grid points), the following set of equations is obtained for the coefficients x^i_j:

$$\sum_{i'}\sum_{j'}\Big(\int h^i_j\, L^{ii'} h^{i'}_{j'}\, dV\Big)\, x^{i'}_{j'} = \lambda \sum_{i'}\sum_{j'}\Big(\int h^i_j\, R^{ii'} h^{i'}_{j'}\, dV\Big)\, x^{i'}_{j'}\,. \tag{17.115}$$

These equations lead to the non-symmetric eigenvalue problem

$$\mathbf{A}\cdot\mathbf{x} = \lambda\,\mathbf{B}\cdot\mathbf{x}\,, \tag{17.116}$$

where $\mathbf{A}$ and $\mathbf{B}$ are large *non-Hermitian matrices* and $\lambda \equiv -i\omega$ is the complex eigenvalue. The ideal MHD spectral problem is contained as a special case ($\eta = 0$), with Hermitian matrices and λ purely real or purely imaginary.

In the toroidal case, except for the radial finite elements, one also needs a discretization of the poloidal variation. Because of poloidal periodicity, it is most effective to exploit fast Fourier transforms for this purpose, so that the volume integrations of Eq. (17.115) also contain angular integrations over products of Fourier harmonics involving the poloidal variation of the equilibrium quantities. The resulting block-tridiagonal structure of the matrix **A** is illustrated in Fig. 17.8 for the particular implementation used in the pair of codes HELENA-CASTOR [319, 360] for the calculation of static axi-symmetric equilibria in tokamaks and their ideal or resistive spectra of waves and instabilities. For the computation of external kink modes, extension with an external vacuum region surrounded by a conducting wall, or a boundary of a prescribed shape, has been constructed using the same numerical discretization methods [320]. Those methods have also been used in the succeeding pair of codes FINESSE-PHOENIX [44, 74] for toroidally and poloidally rotating plasmas, that are exploited in the next chapter.

In conclusion, numerical computation with the resistive set of equations (17.107)–(17.110) is much more flexible, and even more accurate, than with the corresponding ideal MHD spectral equation (17.106). There is also no need to develop separate codes for ideal MHD since it is contained by means of the simple switch $\eta = 0$. This does not imply that ideal MHD stability is now superseded. Actually, quite the opposite: as we will see in the following sections, all ideal MHD instabilities remain present, and even much more pronounced, since accurately computable.

17.3.2 Internal modes

Tearing modes Plasmas with a fixed boundary display *internal modes,* like the tearing modes and resistive interchanges discussed in Sections 14.2.2 and 14.2.3. These modes can be stabilized by the combined effect of good average curvature of the field lines and finite pressure of the plasma [211]. In toroidal geometry, as shown by Glasser et al. in [211], the dependence of the growth rate on the resistivity is much more complicated than suggested by the simple tearing scaling (14.84) or the interchange scaling (14.90) of Section 14.2. In particular, the eigenvalues of the resistive modes become complex and exhibit an intriguing pattern of coalescence and splitting in the complex λ plane, as shown in Fig. 17.9(a). For $\eta = 10^{-6}$, the frequency of one stable mode (imaginary $\lambda \equiv$ real ω) decreases as η decreases and the frequency of a second stable mode increases as η decreases. At $\eta = 2.3 \times 10^{-7}$, the two modes coalesce and split into two overstable modes (complex conjugate in terms of ω) which become purely exponential for $\eta = 4 \times 10^{-8}$. Finally, if the resistivity is still further decreased, the two modes approach the origin with the resistive interchange power $\eta^{1/3}$ of Eq. (14.90).

Figure 17.9(b) shows the influence of the pressure (parameterized by $\beta_{\rm p}$) on the growth rate of the internal $m = 2, n = 1$ tearing mode. As is evident from this figure, the pressure has a stabilizing influence on the internal tearing modes, so that an increase of $\beta_{\rm p}$ results in a decrease of the instability window in the plot of the growth rate versus the safety factor at the edge. In contrast, the pressure has almost no effect on the external, free boundary, tearing modes, which means that those modes are essentially incompressible. At zero pressure the instability window of the external tearing modes is much larger than for the internal modes, and at $\beta_{\rm p} = 0.26$ the growth rates are four times larger than for the internal, fixed boundary, modes. Thus, the stabilizing influence on the

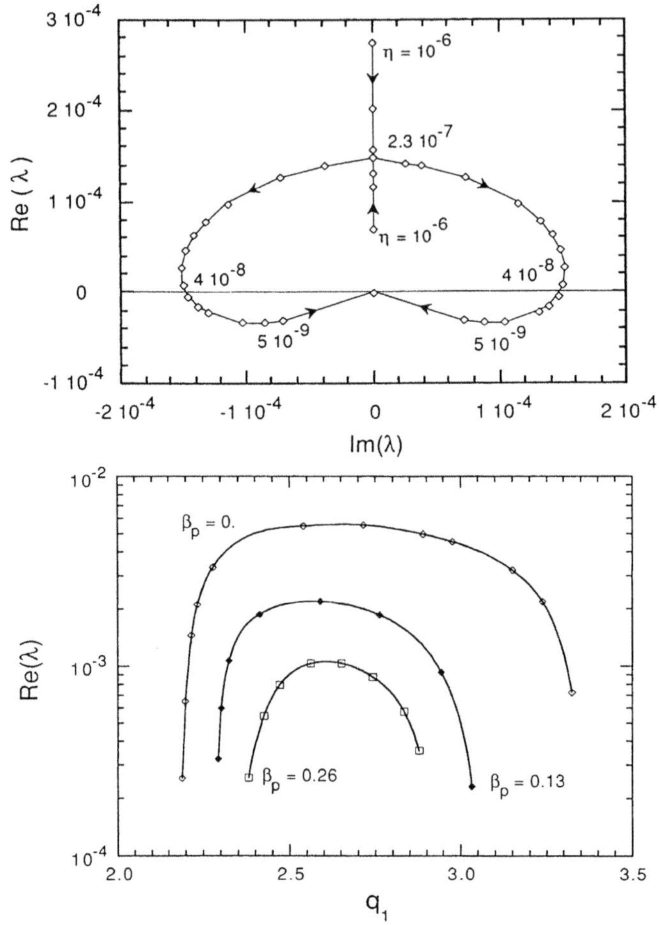

Fig. 17.9 Spectral behaviour of the internal, $m = 2, n = 1$, resistive tearing mode. Top: Locus of the eigenvalue λ ($\equiv -i\omega$) in the complex plane as a function of resistivity; $q_1 = 2.65$, $\beta_p = 0.26$. Bottom: Growth rate versus total current for fixed resistivity $\eta = 10^{-6}$. (From Huysmans et al. [320].)

tearing mode, which can be very effective for modes in the plasma centre, is not effective for tearing modes at the plasma edge [320].

Infernal modes The long-term prospect of steady-state operation in tokamaks, facilitated by bootstrap currents (see Wesson [647], Section 4.9), has led to intense research of the advanced Tokamak (AT) scenario (see e.g. Freidberg [185], Section 13.7.4). This scenario requires unusual (hollow) current and q-profiles, i.e. negative shear in the plasma centre, and, thus, reopened many questions of MHD stability of tokamaks with respect to these new profiles. We discuss one example.

The standard ballooning theory, discussed in Section 17.2.5, leads to the one-dimensional problem of solving an ODE for the ballooning perturbation, which may be investigated for each magnetic surface separately. This approach is justified for toroidal mode numbers $n \gg 1$, and may be extended to lower values of n, provided that the magnetic shear is large enough (the standard tokamak scenario). In that case, growth rates of ballooning instabilities decrease with decreasing n, so that establishing high-n ballooning stability is sufficient. However, for the low shear region that is unavoidable in the AT scenarios, stability with respect to intermediate values of n is completely unrelated to the standard ballooning conditions. For those low-shear q-profiles, even when the ballooning stability criterion is satisfied, a new class of unstable pressure-driven modes, called *infernal modes* (a play on words), was found by Manickam et al. [428]. For those modes, the growth rate is a wildly oscillating function of n with instability in low-n regions sensitively dependent on the value of q.

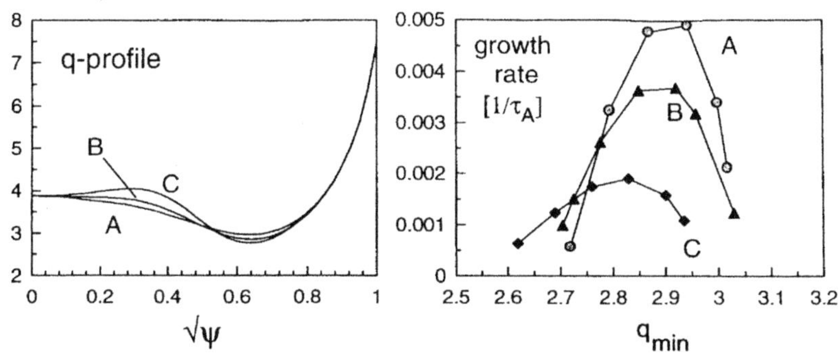

Fig. 17.10 Safety factor and growth rates of the infernal mode. (From Holties et al. [306, 308].)

The infernal modes were investigated for advanced tokamak regimes in JET obtained with pellet injection [306, 308]. The improved confinement turned out to be transient and ended in a collapse due to an MHD instability. The *infernal* mode was considered a likely candidate for this MHD instability since it is driven by the large pressure gradient in the region of negative shear. Increasing the shear around $q_{\min}$ reduces the growth rates of the infernal mode but enlarges the instability window as the infernal mode becomes more localized and gives a less deteriorating effect on the plasma (Fig. 17.10). Broadening the pressure profile, i.e. moving the pressure gradient away from $q_{\min}$, has a stabilizing effect. It is possible to completely stabilize the infernal mode by making the pressure gradient equal to zero in a large enough region around $q_{\min}$. One might hope that the effect of the instability is precisely this switching off (therefore called 'self healing').

17.3.3 Edge localized modes

Edge-localized modes (ELMs) typically occur in the *H-mode confinement regime* in tokamaks. This operating regime, discovered in ASDEX [638], is characterized by improved energy confinement with a factor of two as compared to the L-mode regime, due to the suppression of plasma edge turbulence and large edge gradients of the temperature and the density (and also in the shear flow and the associated radial electric field) [262]. The H-mode may be degraded by ELMs of large magnitude, destroying the desired transport barrier. On the other hand, ELMs of small magnitude may have the effect of providing a control mechanism of the impurity content, so that quasi-stationary state H-mode operation is facilitated.

The traditional, over-simplified, approach to MHD stability of tokamaks is to separately consider

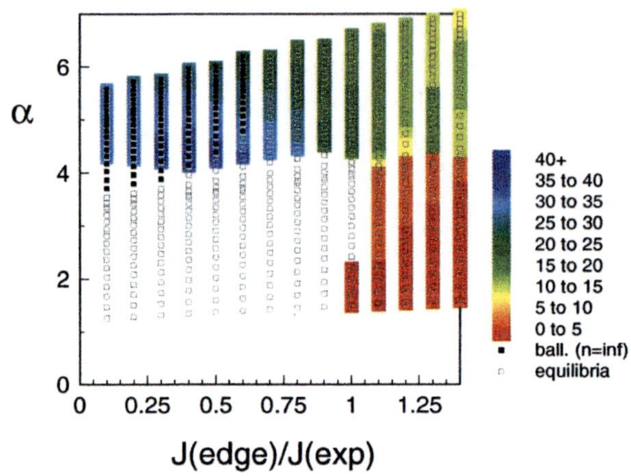

Fig. 17.11 MHD stability limits for toroidal mode number $n = 1$–40 as a function of the edge pressure gradient α and the edge current density, relative to the experimental value, for a low triangularity JET H-mode discharge. Black squares indicate $n = \infty$ ballooning instability, coloured ones peeling–ballooning instability with the most unstable n indicated by the colour. (From Huysmans [316](b).)

the global external kink modes, with toroidal mode number $n = 1$, and the highly localized ballooning modes, with $n = \infty$. However, to properly describe ELMs in the H-mode regime, the full variety of combined modes should be considered. In particular, the large gradients of the pressure and of the current density at the plasma edge give rise to mixed *peeling–ballooning modes with a wide range of intermediate values of* n; see Huysmans [316](b). The peeling mode proper is an external kink mode, with extreme localization at the edge, that is driven by a finite current density, or its derivatives, at the edge of the plasma; see Frieman et al. [196]. In the intermediate n range, these modes couple to pressure gradient driven ballooning modes, so that the distinction between the two becomes meaningless; hence, the terminology 'peeling–ballooning'.

The stability diagram for these modes is shown in Fig. 17.11. Whereas $n = \infty$ ballooning modes become unstable if the pressure gradient parameter $\alpha > 3.7$ (black squares) and become stable again when the edge current density becomes larger than 65% of the experimental value, the intermediate n peeling–ballooning modes clearly are unstable for parameters that are experimentally relevant for the onset of ELMs (coloured squares). It should be noted, though, that, whereas the maximum pressure gradient observed before the occurrence of ELMs is a rather well-known quantity, the edge current density is really dependent on details of the equilibrium reconstruction that are much less known. Furthermore, the shape of the plasma boundary, in particular its triangularity, may significantly enhance stability with respect to these modes.

Obviously, this does not exhaust the dynamics of different MHD modes in the plasma edge. For example, *low n external kink modes* have been observed during an initially ELM free period. Because these modes are driven by the edge current density, or its gradient, they should be classified as peeling modes as well. Since their instability hinges on the proximity of a rational surface in the vacuum, the plasma boundary shape strongly influences the growth rate of these instabilities; see Fig. 17.12(a). For example, the presence of the X point of a divertor close to the plasma boundary stabilizes the ideal $n = 1$ peeling modes (Fig. 17.12(b)), whereas it has little influence on resistive

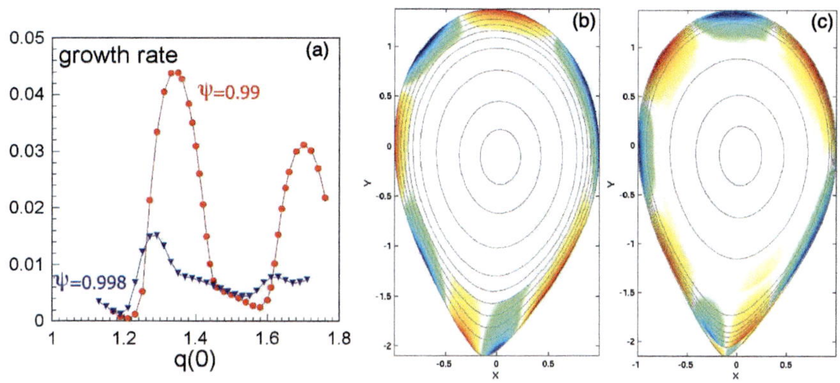

Fig. 17.12 (a) Growth rate of the $n = 1$ peeling–tearing instability for two shapes of the plasma boundary: $\psi_b = 0.99$ (modest triangularity, red dots) and $\psi_b = 0.998$ (high triangularity, blue triangles), resistivity $\eta = 2 \times 10^{-8}$; (b) ideal peeling mode for $\psi_b = 0.99$, $q_0 = 1.45$, $q_b = 4.06$; (c) resistive peeling–tearing mode for $\psi_b = 0.998$, $q_0 = 1.54$, $q_b = 4.4$. (From Huysmans [316](a).)

$n = 1$ peeling–tearing modes (Fig. 17.12(c)). Hence, *peeling–tearing instabilities* may be considered as a possible mechanism for the more benign, small magnitude, ELMs.

The parameterization of Fig. 17.12 requires explanation. In stability codes exploiting straight-field-line coordinates, like CASTOR, the separatrix region itself cannot be studied because the Jacobian diverges. Instead, the geometry close to an X point is mimicked by considering boundaries parameterized by the flux ψ_b as a fraction of the flux $\psi = 1$ at the X point. Also, because q_b is a rather poor measure for kink modes (q^* would have been better; see Section 16.1.4), the total current has been parameterized by the safety factor on axis, q_0. One then finds instability of ideal peeling modes in the range $1.2 < q_0 < 1.45$ (red peak in Fig. 17.12(a)) and of resistive peeling modes in the range $1.45 < q_0 < 1.54$ (blue curve, indicating even increased growth rate for higher triangularity). As illustrated in Fig. 17.12(b) and (c), these two types of mode are similar in the bulk of the plasma, but exhibit a distinct parity difference in the immediate neighbourhood of the X point.

The eventual effect of ELMs on plasma confinement has to be studied for mode amplitudes that have grown into the nonlinear regime. This requires a nonlinear evolution code, possibly with input of initial data from an equilibrium-spectral code such as HELENA-CASTOR. A popular model to reduce the complexity of the nonlinear calculation is Strauss' *reduced MHD model* [576], further developed by many authors, see e.g. [330], [95]. A direct extension of the above analysis into the nonlinear regime has been initiated by the development and application of the nonlinear evolution code JOREK by Huysmans and Czarny [318], based on such a reduced resistive MHD model. The equations solved are evolution equations for the poloidal flux, the poloidal vorticity, the density and the temperature, with a number of transport coefficients, such as resistivity, viscosity and heat conductivity. Significant is the fact that most of the discretization methods, like two-dimensional finite elements for the poloidal plane, Fourier transformation for the toroidal direction, grid refinement techniques, etc. are developments of the methods used in the spectral code. In this respect, an important extension is the use of bicubic Bezier finite elements [135]. The time evolution is treated fully implicitly (see Sections 15.4 and 20.4). Most important is the incorporation of the X point geometry for divertor tokamaks, like JET and ITER, since ELMs usually consist of an external kink

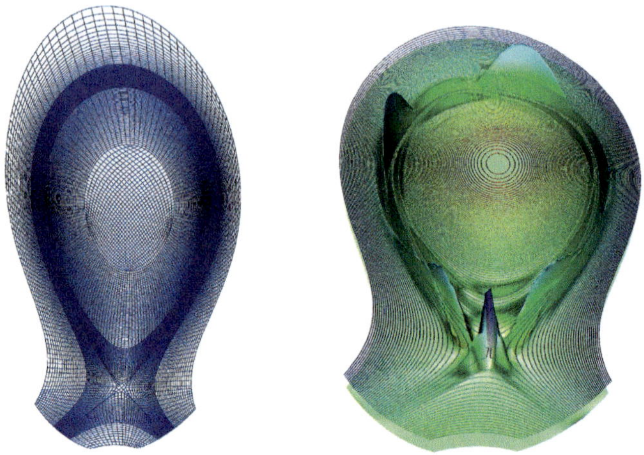

Fig. 17.13 Separatrix geometry and a peeling mode current perturbation computed with the nonlinear evolution code JOREK. (From Huysmans and Czarny [318].)

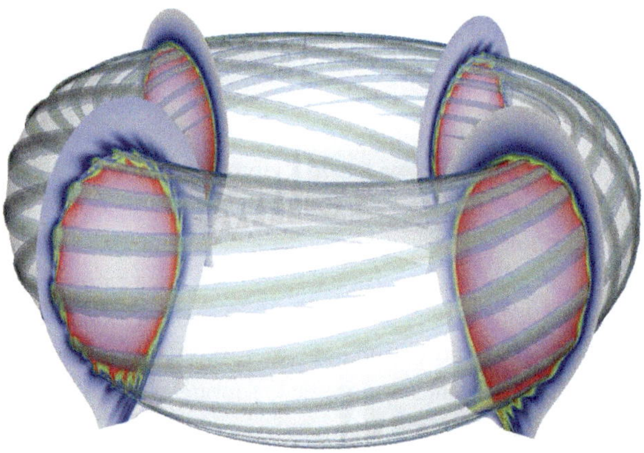

Fig. 17.14 JOREK2 simulation of an ELM crash: The 3D contours show the ejected filaments (density outside the plasma); the poloidal cross sections show the density (left) and the temperature (right). (Courtesy Guido Huysmans.)

mode component of the peeling mode kind, as discussed above. Consequently, these modes are very sensitively dependent on the separatrix geometry. Figure 17.13 shows the flux aligned finite element grid used in JOREK and the localized current perturbation of a peeling mode that is ideally stable, but resistively still unstable, in agreement with the linear results [316](a). An example of a peeling–ballooning mode computed with a recent version of this code is shown in Fig. 17.14.

A complete nonlinear evolution study of medium-n ideal ballooning instabilities was performed by Huysmans and Czarny [318] by means of the JOREK code. The initial equilibrium has a large pressure gradient (the edge pedestal) inside the separatrix. However, as time proceeds in the computation, the modes more and more stretch out into the 'vacuum', so that the distinction between internal and external modes becomes meaningless. In effect, the medium-n ballooning modes develop into 'blobs' or density filaments, very similar to the experimentally observed ELMs. Further work along this line was performed with a new version of the code, JOREK2, where the mentioned Bezier finite elements were implemented. The result of a simulation of an ELM crash with this code is shown in Fig. 17.14. Numerical work on the dynamics and control of ELMs in ITER is intensively investigated, see the review paper by Huijsmans et al. [317], and the paper by Becoulet et al. on ELM mitigation by resonant magnetic perturbations [43].

17.3.4 Toroidal Alfvén eigenmodes and MHD spectroscopy

(a) External excitation of Toroidal Alfvén Eigenmodes We encountered the TAE modes in connection with the poloidal mode coupling of the Alfvén and slow continua in toroidal systems (Section 17.2.4, Fig. 17.3). Due to this coupling, gaps appear in the continua in which the TAEs may be found (see Fig. 17.5). They are naturally excited by energetic particles (neutral beam injection, fusion α-particles) and, in turn, these destabilized TAEs may cause severe losses of α-particles in future ignited plasmas. Hence, a lot of research is devoted to possible damping mechanisms [613].

TAE modes not only pose a threat to plasma confinement for future fusion machines, but they may also be used for a positive purpose, viz. to diagnose the plasma by means of measurements of the magnetic signals they produce. This activity, proposed by Goedbloed et al. [220, 237], has been called *MHD spectroscopy*. Recall from our discussion of the example of *helioseismology* in Section 7.2.4 how agreement between observed Doppler shifts of spectral lines, due to solar oscillations [123], could be brought into agreement with the calculated ones for a standard solar model to within 0.1%! This impressive agreement may serve as an example for what is possible in a purely classical, fluid dynamical, kind of spectroscopy and, hence, in the MHD spectroscopy of plasmas.

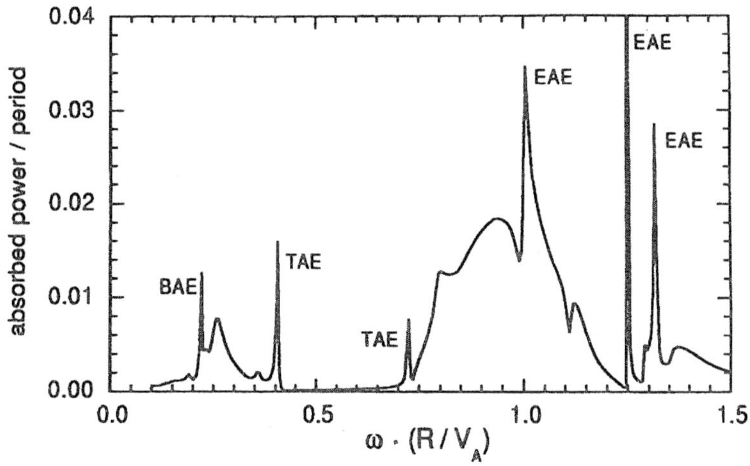

Fig. 17.15 Power absorbed by the plasma as a function of the normalized antenna frequency ($n = 1$, $\eta = 10^{-7}$, $V_A/R = 332$ kHz). (From Huysmans et al. [321].)

TAEs could be artificially excited at JET by means of an external antenna. The saddle coils for disruption control were adapted to permit scanning of the driving frequency in the Alfvén frequency range 30–500 kHz. The frequencies of the TAE modes showed up as resonances of the power absorbed by the plasma, as measured by the antenna impedance.

The theoretical counterpart of this activity was undertaken by means of a modification of CASTOR. Replacing the spectral problem (17.116) by a representation for external driving,

$$\big(\mathbf{A} + \mathrm{i}\omega_{\mathrm{d}}\mathbf{B}\big) \cdot \mathbf{x} = \mathbf{f}\,, \tag{17.117}$$

where ω_{d} is the driving frequency and $\mathbf{f}$ is the driving term, the response of the continuous spectra and other modes could be calculated. The response of the plasma due to an $n = 1$ magnetic field perturbation which would be induced by the saddle coils is shown in Fig. 17.15. On top of the broad continuum shoulders several sharp peaks are found labelled with TAE (toroidicity induced: $\Delta m = 1$ couplings dominate in the eigenfunctions), EAE (ellipticity induced: $\Delta m = 2$ couplings) and BAE (β induced, actually compressibility induced: γp). As may be deduced from the corresponding eigenfunctions, these peaks in the antenna impedance are indicative of the background plasma equilibrium profiles [322, 359]. Observing such peaks and correlating them with the background profiles would be an example of *active MHD spectroscopy*. We will see below how a more effective, passive, way of MHD spectroscopy was advanced later at JET.

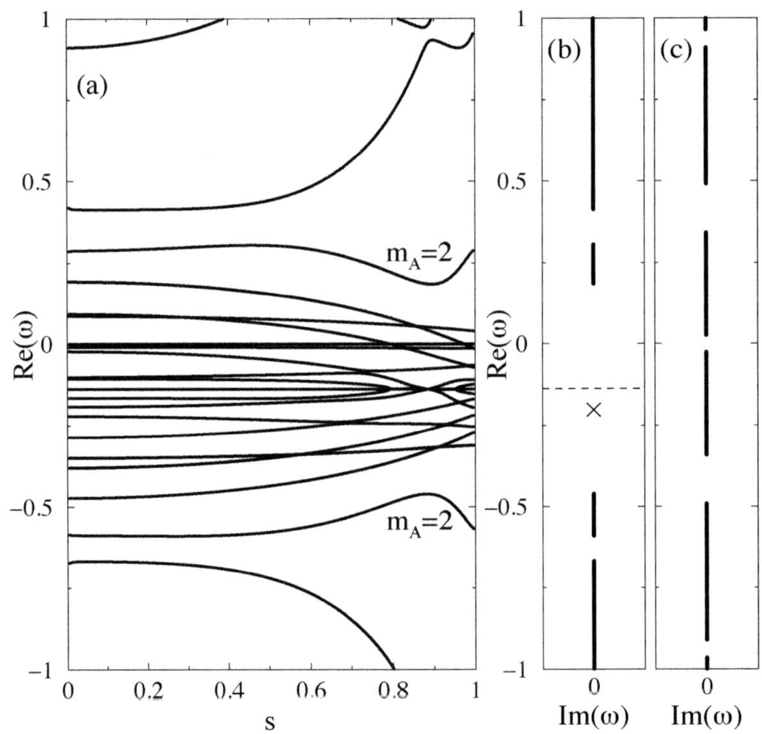

Fig. 17.16 Toroidal flow induced Alfvén eigenmode (TFAE): (a) Alfvén and slow continua plotted against $s = \sqrt{\psi}$; (b) large gap in the Alfvén spectrum with TFAE (indicated by the cross); (c) for the corresponding static equilibrium, the gap is much narrower and no TFAE occurs. (From van der Holst et al. [618].)

(b) Toroidal flow induced Alfvén eigenmodes One particular kind of TAE mode, which might become important in MHD spectroscopy because it occurs at very low frequencies (in the slow magneto-sonic range), is the *toroidal flow induced Alfvén eigenmode* (TFAE), found by van der Holst et al. [618]. It is illustrated in Fig. 17.16. The mode occurs in the $\Delta m = 0$ gap caused by the coupling of the Alfvén and slow continua, similar to the geodesic coupling described in Section 17.2.3. However, this gap is not caused by geodesic coupling but by centrifugal and Coriolis forces, and it is significantly wider than the static $\Delta m = 0$ gap. Because of the flow, the continua are Doppler shifted and the symmetry with respect to the Doppler shifted frequency $n\Omega$ is broken through the Coriolis effect. In Fig. 17.16(b), the Alfvén part of the spectrum is highlighted by omitting the slow continua. Inside this gap, the TFAE is found with a frequency $\mathrm{Re}(\omega) = -0.202$. For comparison, the gaps for the corresponding static case are shown in Fig. 17.16(c) (omitting two symmetric BAEs in the narrow $\Delta m = 0$ gap). It is evident that toroidal flow changes the low-frequency part of the continuous spectra completely. The gap is determined by a three mode interaction involving a central Alfvén mode and two sideband slow modes.

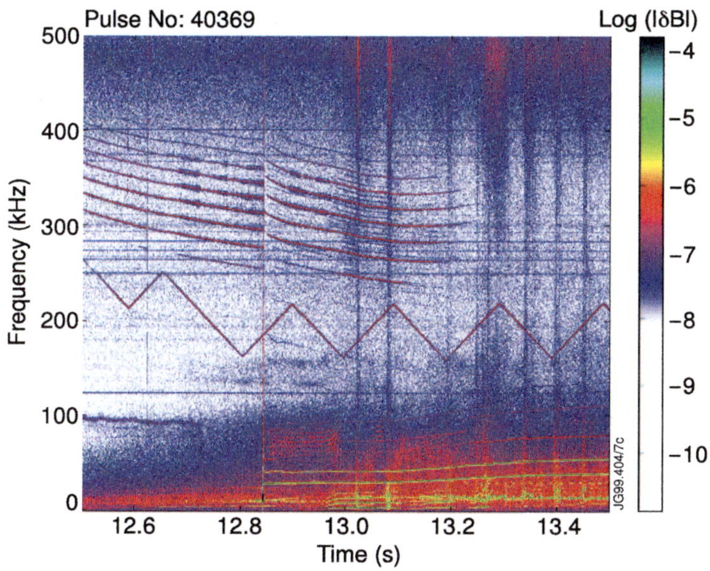

Fig. 17.17 MHD spectroscopy: magnetic perturbations measured at the vessel wall for determination of the toroidal mode numbers of TAEs in JET discharge #40369. (From Sharapov et al. [552].)

(c) MHD spectroscopy in rotating plasmas In principle, measurement of the frequencies of TAEs yields information about the safety factor profile $q(\Psi)$ since the TAE frequency is determined by plasma parameters in a narrow range of width associated with the magnetic cross-over surface (Fig. 17.3), as shown in many studies, see e.g. Holties et al. [307]. Neutral beam injection in tokamaks causes the plasma to rotate in the toroidal direction and, thus, the TAE frequencies are Doppler shifted with a factor that depends on the toroidal mode number n times the rotation frequency. Information about the radial profile of the rotation frequency may be obtained from independent (charge-exchange) measurements. From these combined data, the q profile may be obtained, as shown by

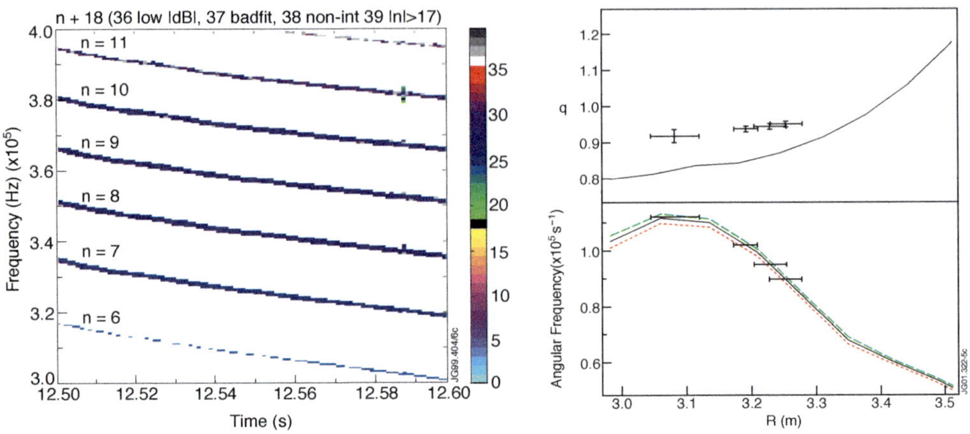

Fig. 17.18 MHD spectroscopy: TAEs observed with magnetic pick-up coils (left) and resulting safety factor and rotation profiles (right) for JET discharge #40369. (From Sharapov et al. [552].)

Sharapov et al. [552]. Figure 17.17 shows the measured magnetic field perturbations at the vessel wall in JET for different values of n. The TAEs occur in the frequency range 300–400 kHz, whereas low-frequency MHD instabilities occur in the range 0–80 kHz. The profiles of the safety factor and the rotation profile that were obtained by inversion of these data are shown in Fig. 17.18.

One may also use the unstable Alfvén waves in the lower frequency range, excited by energetic ions, to determine the q profile [552]. In tokamaks with a non-monotonic profile of the safety factor, these occur in the form of *Alfvén wave cascades* where the frequency changes upward or downward when the minimum value of the safety factor decreases in time during the discharge. This recently discovered class of modes offers an additional way of identifying the plasma parameters from the dependence on the Alfvén wave spectrum, another example of MHD spectroscopy [552].

At the end of Chapter 16, we noted the lack of accurate information on the profile of the safety factor q, a crucial parameter for the operation of future fusion reactors. This deficiency is presently being addressed by a rapidly growing body of research devoted to the mentioned Alfvén wave cascades, also called *chirping modes*, see e.g. Edlund et al. [169], which provide information on the precise location of the $q = 1$ surface in reversed shear profiles (relevant for advanced tokamak scenarios). Clearly, MHD spectroscopy is well underway in tokamaks.

17.4 Literature and exercises

Notes on literature

Basic papers on MHD stability

- 'Zur Stabilität eines Plasmas' by Hain, Lüst and Schlüter [275] and 'An energy principle for hydromagnetic stability problems' by Bernstein, Frieman, Kruskal and Kulsrud [55] are the first (and very unevenly cited) papers on the subject.
- 'Hydromagnetic stability of a plasma' in Volume 2 of *Reviews of Plasma Physics* by Kadomtsev [338] is one of the first overviews of the subject.
- 'Kink instabilities in a high-β tokamak' by Freidberg and Haas [188].

- 'Study of the MHD spectrum of an elliptic plasma column' by Chance, Greene, Grimm and Johnson [110].

Textbook chapters on MHD stability of toroidal plasmas

- *Ideal MHD* (Chapter 12) by Freidberg [186], and *Plasma Physics and Fusion Energy* (Chapter 12 and Section 13.7) by Freidberg [185].
- *Theory of Toroidally Confined Plasmas* (Chapter 4) by White [649].
- *Magnetohydrodynamics and Spectral Theory* (Chapter 9) by Lifschitz [410].
- *Plasma Confinement* (Chapter 7) by Hazeltine and Meiss [296].
- *Tokamaks* (Chapters 6 and 7) by Wesson [647].

Ballooning modes

- 'Ballooning effects and plasma stability in tokamaks' in Volume 11 of *Reviews of Plasma Physics* by Pogutse and Yurchenko [501] is a very broad overview of ballooning effects on tokamak stability with many references up to 1980.
- Clear expositions of the ballooning representation can be found in *Ideal Magnetohydrodynamics* (Section 12.6) by Freidberg [186] and *Plasma Confinement* (Sections 7.10–12) by Hazeltine and Meiss [296].

Exercises

[17.1] *Mapping for stability*★

According to Section 17.1.2, stability analysis preferably should be done in straight-field-line coordinates. In this exercise, we will investigate how one can invert the coordinates.

- Derive the co- and contravariant metric tensor by making use of the transformation

$$g_{ij} = \frac{\partial \tilde{x}^\mu}{\partial x^i} \frac{\partial \tilde{x}^\nu}{\partial x^j} \tilde{g}_{\mu\nu} ,$$

  where $\tilde{x}^\mu$ and $\tilde{g}_{\mu\nu}$ are the coordinates and the metric tensor of the old coordinate system (R, Z, φ), while x^i are the coordinates of the new system $(\Psi, \vartheta, \varphi)$.
- Express the covariant elements of the magnetic field in the new coordinate system in terms of the old one, by means of

$$B_i = \frac{\partial \tilde{x}^\mu}{\partial x^i} \tilde{B}_\mu ,$$

  and also the other way around.
- Derive expressions for R_Ψ, R_ϑ, Z_Ψ, Z_ϑ and for the co- and contravariant elements of the metric tensor.
- Derive an expression for the Jacobian $\mathcal{J}$. Is it defined everywhere?
- Now we have derived the basic geometric elements for stability analysis in straight-field-line coordinates. As an extra, also derive the expressions for the co- and contravariant magnetic field components in terms of the metric elements.
- For toroidal plasmas, the safety factor $q(\Psi)$ is defined as

$$q(\Psi) = \frac{\mathbf{B} \cdot \nabla\varphi}{\mathbf{B} \cdot \nabla\vartheta} .$$

  Show how that this can be written in terms of the equilibrium quantities derived.

[17.2] *The magnetic axis*

In this chapter, the powerful spectral equation (17.43) has been derived using straight field line coordinates. Special care is required at the magnetic axis. Here, you will see why.

- What kind of properties does the magnetic field have on the magnetic axis?
- What consequences does this have for the metric elements g_{ij} and the Jacobian $\mathcal{J}$?

- What does this all mean for the spectral equation (17.43)?
- How would you solve this problem?

[17.3] *Projections of the displacement field*

In this exercise, you will derive expressions for the projections X, Y, Z of the displacement vector in straight-field-line coordinates. This is the first step in the derivation of the spectral equation (17.43).

- Derive the expressions for the co- and contravariant components of the vector $\mathbf{n}$. Show that the toroidal contravariant component is equal to zero.
- Do the same for the vectors $\boldsymbol{\pi}$ and $\mathbf{b}$. Show that one contravariant component of each vector vanishes.
- Convert the unit vector pair $\{\boldsymbol{\pi}, \mathbf{b}\}$ into $\{\mathbf{n}, \mathbf{e}_\varphi\}$. Show that the projections X, Y, Z can be written in terms of the physical components $\xi_n, \xi_\mathrm{p}, \xi_\varphi$ as

$$X = RB_\mathrm{p}\,\xi_n\,, \quad Y = \frac{\mathrm{i}}{RB_\mathrm{p}}(B_\varphi \xi_p - B_p \xi_\varphi)\,, \quad Z = \frac{\mathrm{i}}{B^2}(B_p \xi_p + B_\varphi \xi_\varphi)\,.$$

  Also write the inverse relations.

[17.4] *Thin plasma slab*

Similar to Exercise [17.1], you are going to consider a thin plasma slab about the mid-plane. You will show that the spectral equation (17.43) reduces to the spectral equation (9.28). It is reasonable to assume that all physical quantities only depend on the radius R in this approximation. Furthermore, $R \approx R_0 + r\,,\ Z \approx a\theta$, where a is small.

- Derive the expression for R_Ψ, R_θ, Z_Ψ and Z_θ.
- Derive the expressions for the elements of the metric tensor g_{ij}.
- Show that the Jacobian reduces to $J = a/B_Z$.
- Show that the poloidal and toroidal curvature are zero and $1/R$, respectively. Explain why the poloidal curvature is zero.
- Reduce the projections X, Y, Z of the displacement field for this case.
- Show that the matrix elements $\mathcal{A}_{ij}$ and $\mathcal{B}_{ij}$ reduce to the ones of Eq. (9.28), up to a factor. Make use of $k = m/a$.

[17.5] *Small inverse aspect ratio*

In the previous exercise, you have investigated a thin plasma slab about the mid-plane. Now, you do the same for a plasma with small inverse aspect ratio, $\epsilon \ll 1$. As stated in Exercise [17.2], assume that the outer plasma boundary is circular up to first order. Under these two conditions, the magnetic flux function Ψ can be represented as $\Psi(\hat{r}, \hat{\theta}) = \Psi(\hat{r})$. Here, $(\hat{r}, \hat{\theta})$ are non-orthogonal polar coordinates defined with respect to the centre of the flux surfaces. The relation with the cylindrical coordinates is the following:

$$R = R_0 + \hat{r}\cos\hat{\theta}, \quad Z = \hat{r}\sin\hat{\theta}\,.$$

Furthermore, the relation between the angle $\hat{\theta}$ and the straight field line angle ϑ is $\hat{\theta} = \vartheta$.

- Derive the expressions for $R_\Psi, R_\vartheta, Z_\Psi, Z_\vartheta$ and the elements of the metric tensor $\hat{g}_{ij}$.
- Show that the Jacobian reduces to $J = \hat{r}/B_p$.
- Show that the poloidal and toroidal curvature are $1/\hat{r}$ and zero, respectively. Explain.
- Derive the expressions for the projections X, Y, Z in these coordinates.
- Derive the matrix elements $\mathcal{A}_{ij}$ and $\mathcal{B}_{ij}$ in these coordinates. When it is not really necessary DO NOT replace R with the expression above.
- Show that these matrix elements are the same as the ones of Eq. (9.28) using that $R_0 \gg \hat{r}$ and defining $k \equiv n/R_0$.

18

Linear dynamics of toroidal plasmas with flow*

18.1 Transonic toroidal plasmas

Waves and instabilities of transonically rotating axi-symmetric plasmas is a highly complex problem that is of interest for the two unrelated fields of laboratory plasma confinement, aimed at eventual thermonuclear energy production, and the dynamics of a vast number of astrophysical plasmas rotating about compact objects, broadly indicated as accretion discs. The complexity comes from the *transonic transitions of the poloidal flow* which causes the character of the rotating equilibrium states to change dramatically, from elliptic to hyperbolic or vice versa, when the poloidal velocity surpasses certain critical speeds. Associated with these transitions the different types of magnetohydrodynamic shocks may appear (see Chapter 21). Obviously, at such transitions the possible waves and instabilities of the system also change dramatically. We here describe these changes for the two mentioned classes of physical systems, starting from the point of view that the continuous spectrum of ideal MHD presents the best organizing principle for the structure of the complete spectrum of waves and instabilities since it is the most robust part of it. It provides the simplest approach to local waves and instabilities of the system and, possibly, to the onset of MHD turbulence.

The equilibrium problem of translation symmetric and axi-symmetric plasmas was formulated by Goedbloed and Lifschitz [238] in terms of three generic functions (see Section 18.2) that permit analysis of the different singularities and the resolution of the concomitant discontinuities that occur in transonic MHD flows. In the present chapter, we exploit that equilibrium formulation and summarize whatever expressions are needed to perform a spectral analysis of the waves and instabilities of these plasmas.

A cartoon of the physical system is sketched in Fig. 18.1, which shows the compact central object generating the gravitational acceleration $g = -\nabla\Phi_{gr}$ (absent for the tokamak case), and a toroidal plasma with poloidal (indicated by the subscript p) and toroidal (subscript φ) velocity and magnetic field components. For the purpose of the present study, treating rotating laboratory and astrophysical plasmas on an equal footing, two important restrictions have to be made.

(a) We neglect the dynamics external to the toroidal plasma and assume the outer plasma boundary to be fixed in space. For tokamaks, this condition is evidently justified by the presence of a conducting wall and current-carrying coils fixed to the laboratory which absorb the mechanical forces. For accretion discs, this may be justified by considering the toroidal plasma to be embedded in a gas which toroidally rotates at approximately the Keplerian velocity. This assumption amounts to the neglect of the accretion flow itself, i.e. we assume that the external accretion process proceeds on a much slower time scale than the internal plasma rotations. In this manner, we concentrate our

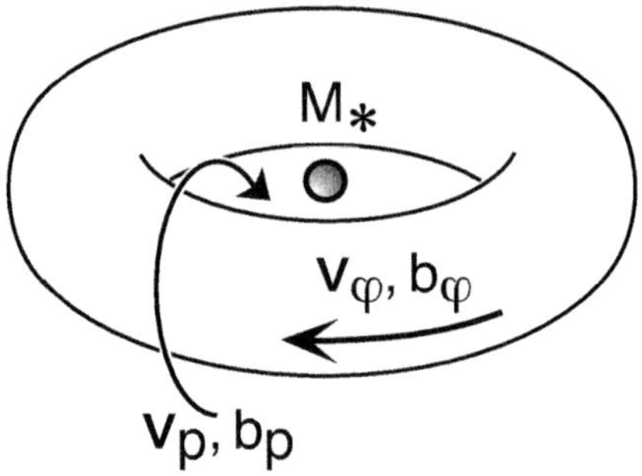

Fig. 18.1 Magnetized plasma rotating about a compact object (forming a thick accretion disc) or inside a tokamak (where $M_* = 0$). Lower case **b** indicates the Alfvén speed in the direction of the magnetic field.

study on the *internal* plasma dynamics as influenced by the toroidal and poloidal rotations and the gravitational field.

(b) We assume the PDEs describing the background flow to be *elliptic*. To our knowledge, this assumption has always been made, explicitly or implicitly, in wave and stability studies of axi-symmetric systems. In fact, this appears to be basic to the classical paradigm of splitting the study of the dynamics of a system into a time-independent equilibrium and the linear perturbations of it [222]. Certainly, all numerical programs for studying linear stability of tokamaks start from an equilibrium with nested flux surfaces, where prescription of the shape of the outer boundary is sufficient to determine the solutions. In this chapter, using the numerical programs FINESSE [44] and PHOENIX [230, 74] which exploit algorithmic techniques explained in Chapters 15–17, we will stay with this paradigm. Entering the hyperbolic flow regime appears to require a full nonlinear evolution study, without making the split into equilibrium and perturbations. Such studies, exploiting the Versatile Advection Code (VAC) [602], are also carried out at present [347, 109]. Here we push the spectral approach as far as possible by studying the linear dynamics in the elliptic flow regions while approaching the boundaries of transition to hyperbolic flow. This yields convincing evidence of qualitatively new linear dynamics caused by the transonic transitions.

In current astrophysical terminology, configurations like that of Fig. 18.1 are called 'thick accretion discs' or 'accretion tori' [183]. However, whereas the cited authors are concerned with accretion discs that are thick because of pressure effects, so that the thermal speed approaches the toroidal rotation velocity ($v_{\rm th} \sim v_\varphi$), we are here concerned with *magnetically dominated accretion tori*, where the magnetic pressure causes the accretion disc to be thick ($v_{\rm th} \sim \sqrt{\beta}\, v_{\rm A} \ll v_\varphi \ll v_{\rm A}$). As opposed to the extreme high-β regime where the magneto-rotational instability operates [29, 30] (see Section 13.5), in this regime, where magnetic fields dominate ($\beta \ll 1$), anomalous dissipation mechanisms due to local MHD instabilities have not been investigated. Such dissipation processes are needed to set up turbulence that would cause the co-moving condition of plasma and magnetic field to be broken so that jets could emerge from the disc (discussed in more detail in Sections 21.4 and 22.3). We analyse the high-β regime from a fundamental point of view, i.e. we formulate the equations for tokamaks and astrophysical objects on the same theoretical footing and investigate their similarities and differences with respect to local modes of the continuous spectrum.

Whereas transonic poloidal flow substantially upsets 'intuition' obtained from static tokamak equilibrium, one extremely important feature remains intact: the poloidal velocity and magnetic field are parallel so that *flow surfaces and magnetic surfaces coincide!* This feature alone justifies the study of modes of the continuous spectrum since these modes are preferentially localized on these surfaces.

On first reading, the reader may want to jump to Section 18.4 with the main results.

18.2 Axi-symmetric equilibrium of transonic stationary states★

18.2.1 Equilibrium flux functions★

We start from the MHD equations for stationary equilibrium, Eqs. (12.24)–(12.27) of Chapter 12, and exploit a right-handed system of cylinder coordinates R, Z, φ, where R is the distance to the symmetry axis (see Appendix A.2.4). For stationary axi-symmetric equilibrium flows, these equilibrium equations permit the introduction of five flux functions [654, 283], as follows. From Eqs. (12.27)(b) and (12.24), the poloidal components of the magnetic field and of the velocity may be expressed in terms of *the poloidal magnetic flux function* $\Psi = \Psi(R, Z)$, and *the poloidal velocity stream function* $\chi = \chi(R, Z)$:

$$\nabla \cdot \mathbf{B} = 0 \qquad \Rightarrow \quad \mathbf{B} = R^{-1}\,\mathbf{e}_\varphi \times \nabla\Psi + B_\varphi \mathbf{e}_\varphi \,, \tag{18.1}$$

$$\nabla \cdot (\rho\mathbf{v}) = 0 \quad \Rightarrow \quad \mathbf{v} = (\rho R)^{-1}\,\mathbf{e}_\varphi \times \nabla\chi + v_\varphi \mathbf{e}_\varphi \,. \tag{18.2}$$

We introduce the Poisson bracket notation for functions of the poloidal coordinates,

$$\{F, G\} \equiv \mathbf{e}_\varphi \cdot (\nabla F \times \nabla G) = F_R G_Z - F_Z G_R \,, \tag{18.3}$$

so that the derivative along a magnetic field line of a function F may be written as $\mathbf{B} \cdot \nabla F \equiv R^{-1}\{F, \Psi\}$. If this expression vanishes, F is a function of Ψ. Using this notation, the magnetic field equation (12.27)(a) may be written as

$$\nabla \times (\mathbf{v} \times \mathbf{B}) = -\mathbf{e}_\varphi \times \nabla\Big[\frac{1}{\rho R^2}\{\chi, \Psi\}\Big] + \mathbf{e}_\varphi\Big[\Big\{\frac{B_\varphi}{\rho R}, \chi\Big\} - \Big\{\frac{v_\varphi}{R}, \Psi\Big\}\Big] = 0 \,. \tag{18.4}$$

The poloidal component of this equation gives $\{\chi, \Psi\} = C\rho R$, where the constant C has to vanish because the toroidal electric field has to vanish [283]:

$$E_\varphi = -(\mathbf{v} \times \mathbf{B})_\varphi = -(\rho R)^{-1}\{\chi, \Psi\} = -C = 0 \quad \Rightarrow \quad \chi = \chi(\Psi) \,. \tag{18.5}$$

Hence, *the poloidal stream function* χ *is a flux function*. The toroidal component of Eq. (18.4) then yields another flux function:

$$\widehat{\Omega} \equiv R^{-1}[v_\varphi - (\chi'/\rho)B_\varphi] = \widehat{\Omega}(\Psi) \,, \tag{18.6}$$

where the prime denotes the derivative with respect to Ψ. The flux function $\widehat{\Omega}$ is related to the poloidal electric field; it is *the derivative of the electric potential:* $\widehat{\Omega} = -\Phi'_{\rm el}$. (The hat should remind us of the difference from the toroidal angular velocity $\Omega \equiv v_\varphi/R$.) From Eqs. (18.1), (18.2) and (18.5), the velocity may then be written in terms of a component parallel to $\mathbf{B}$ and a toroidal component $\sim \widehat{\Omega}$:

$$\mathbf{v}_{\rm p} = (\chi'/\rho)\mathbf{B}_{\rm p} \quad \Rightarrow \quad \mathbf{v} = (\chi'/\rho)\mathbf{B} + R\widehat{\Omega}\mathbf{e}_\varphi \,. \tag{18.7}$$

Converting the stationary equilibrium equation (12.26) from the pressure into the entropy, $S \equiv \rho^{-\gamma} p$, yields the result that *the entropy S is a flux function:*

$$\mathbf{v} \cdot \nabla S = \left[\frac{\chi'}{\rho R} \right] \mathbf{e}_\varphi \times \nabla \Psi \cdot \nabla S \equiv - \left[\frac{\chi'}{\rho R} \right] \{S, \Psi\} = 0 \quad \Rightarrow \quad S = S(\Psi) \,. \tag{18.8}$$

We have now derived three of the five flux functions, two more to go!

It remains to work out the consequences of the momentum equation (12.25). It may be written in the following, rather convoluted, form:

$$\rho \nabla \Big(\tfrac{1}{2} v^2 + \frac{\gamma}{\gamma - 1} \rho^{\gamma - 1} S + \Phi_{\rm gr} \Big) - \frac{\rho^\gamma}{\gamma - 1} \nabla S - \rho \mathbf{v} \times \mathbf{w} + \mathbf{B} \times \mathbf{j} = 0 \,, \tag{18.9}$$

where $\mathbf{w} \equiv \nabla \times \mathbf{v}$ is the vorticity and $\Phi_{\rm gr} = -GM_* (R^2 + Z^2)^{-1/2}$ is the gravitational potential due to the mass M_* of the central object. Diligently working out all the vector expressions yields yet another convoluted expression, but much closer to where we want to get:

$$\begin{aligned} &\Big[\nabla \cdot \Big(\frac{1}{R^2} \nabla \Psi \Big) - \chi' \nabla \cdot \Big(\frac{\chi'}{\rho R^2} \nabla \Psi \Big) - \frac{\rho^\gamma}{\gamma - 1} S' \Big] \nabla \Psi \\ &\quad + \rho \nabla \Big(\tfrac{1}{2} v^2 + \frac{\gamma}{\gamma - 1} \rho^{\gamma - 1} S + \Phi_{\rm gr} \Big) - \frac{\rho v_\varphi}{R} \nabla (R v_\varphi) + \frac{B_\varphi}{R} \nabla (R B_\varphi) \\ &\quad - \mathbf{e}_\varphi \frac{1}{R^2} \mathbf{e}_\varphi \cdot \big[\chi' R v_\varphi - \nabla (R B_\varphi) \times \nabla \Psi \big] = 0 \,. \end{aligned} \tag{18.10}$$

The toroidal component on the last line yields our fourth flux function:

$$-\{\chi' R v_\varphi - R B_\varphi, \Psi\} = 0 \quad \Rightarrow \quad K \equiv R v_\varphi - (1/\chi') R B_\varphi = K(\Psi) \,, \tag{18.11}$$

which will be commented on below. From the definitions (18.6) and (18.11) of the flux functions $\widehat{\Omega}$ and K, we may express v_φ and B_φ as follows:

$$v_\varphi = \frac{(\chi'^2/\rho) K - R^2 \widehat{\Omega}}{\chi'^2/\rho - 1} \,, \qquad R B_\varphi = \frac{\chi' (K - R^2 \widehat{\Omega})}{\chi'^2/\rho - 1} \,. \tag{18.12}$$

By means of these relations, the two poloidal components of the momentum equation (18.10) may be further transformed to produce the final relations, viz. the last flux function and the partial differential equation to determine Ψ. After some algebra, the component parallel to $\mathbf{B}_{\rm p}$ yields the fifth flux function:

$$H \equiv \tfrac{1}{2} \frac{\chi'^2}{\rho^2} \Big(\frac{|\nabla \Psi|^2}{R^2} + B_\varphi^2 \Big) - \tfrac{1}{2} R^2 \widehat{\Omega}^2 + \frac{\gamma}{\gamma - 1} \rho^{\gamma - 1} S + \Phi_{\rm gr} = H(\Psi) \,, \tag{18.13}$$

which is called *the Bernoulli function*. Here, the expression (18.12)(b) is to substituted for B_φ, and $|\nabla \Psi|/R = B_{\rm p}$ can only be obtained after the PDE below for Ψ is solved. Finally, exploiting all these relations, the component of the momentum equation (18.10) perpendicular to $\mathbf{B}_{\rm p}$ may be converted into *the central partial differential equation of poloidally and toroidally rotating equilibra:*

$$\begin{aligned} &\nabla \cdot \Big(\frac{1 - \chi'^2/\rho}{R^2} \nabla \Psi \Big) + \frac{\rho}{\chi'^2} \Big[\chi'^2 \big(H + \tfrac{1}{2} R^2 \widehat{\Omega}^2 - \Phi_{\rm gr} \big) \Big]' \\ &\quad - \frac{1}{\gamma - 1} \Big(\frac{\rho}{\chi'^2} \Big)^\gamma \Big[(\chi'^2)^\gamma S \Big]' + \frac{\chi' (R^2 \widehat{\Omega} - K) \big[\chi' (R^2 \widehat{\Omega} - K) \big]'}{R^2 (1 - \chi'^2/\rho)} = 0 \,. \end{aligned} \tag{18.14}$$

This PDE is to be solved for any choice of the five flux functions that one considers to be a fair representation of some stationary equilibrium: much more complicated than solving the Grad–Shafranov equation with only two flux functions for a static tokamak equilbrium (Chapter 16), where a lot of diagnostic information is available on the distribution of the pressure and the q-profile!

The six physical variables ρ, p, $B_{\rm p}$, B_φ, $v_{\rm p}$ and v_φ, which are functions of R and Z, are now determined by the solution $\Psi(R,Z)$ of the PDE (18.14) and by the choice of the five flux functions $\chi'(\Psi)$, $H(\Psi)$, $S(\Psi)$, $K(\Psi)$ and $\widehat{\Omega}(\Psi)$. For the final determination of the physical variables, the constant GM_* can also to be considered as an additional (trivial) flux function.

We abandon the misnomer 'specific angular momentum flux' that has been used to indicate the flux function K by quite a number of authors [640, 71, 510] and many others, until recently including ourselves [347]. The point is that the specific angular momentum density vector $\mathbf{L} \equiv \mathbf{r} \times \mathbf{v}$ (with the toroidal component $L_\varphi = Rv_\varphi$) has no magnetic contributions. Such contributions enter the general expression for the angular momentum of relativistic matter and field, but they are negligible in non-relativistic MHD; see Jackson [331], p. 288 (2nd edition, p. 264). Also, for static equilibrium (see Chapter 16), the flux function $I \equiv RB_\varphi$, corresponding to K in the limit of vanishing toroidal flow, plays an analogous role in the equilibrium. It is obvious that angular momentum does not have anything to do with this flux function. However, the analogy does reveal the essential relation with the poloidal current density and vorticity, which K expresses. The poloidal components of the vorticity $\mathbf{w} \equiv \nabla \times \mathbf{v}$ and the current density $\mathbf{j} \equiv \nabla \times \mathbf{B}$ may be derived from stream functions V and I,

$$\mathbf{w}_{\rm p} = -\frac{1}{R}\,\mathbf{e}_\varphi \times \nabla V\,, \quad V \equiv Rv_\varphi\,, \qquad \mathbf{j}_{\rm p} = -\frac{1}{R}\,\mathbf{e}_\varphi \times \nabla I\,, \quad I \equiv RB_\varphi\,, \tag{18.15}$$

which, separately, are *not* flux functions, but the combination

$$K \equiv V - I/\chi'\,, \quad \text{so that} \quad K\nabla\chi = V\nabla\chi - I\nabla\Psi\,, \tag{18.16}$$

is a flux function. It exhibits the proper limits to HD, where $I = 0$, and to static MHD, where $V = 0$ and angular momentum is not a relevant concept. We will indicate K as *the poloidal vorticity–current density stream function*.

The transonic poloidal flow field is best characterized by the value of *the squared poloidal Alfvén Mach number:*

$$M^2 \equiv \frac{\rho v_{\rm p}^2}{B_{\rm p}^2} \equiv \frac{v_{\rm p}^2}{v_{\rm Ap}^2} = \frac{\chi'^2}{\rho}\,, \tag{18.17}$$

which determines the density $\rho(R,Z)$ for an assumed (known) choice for the streamfunction $\chi(\psi)$. Exploiting this quantity, the relationship between the plasma velocity and the vectorial Alfvén velocity $\mathbf{b} \equiv \mathbf{B}/\sqrt{\rho}$ may be written as

$$\mathbf{v} = M\mathbf{b} + R\widehat{\Omega}\mathbf{e}_\varphi\,, \tag{18.18}$$

so that $\mathbf{v}_{\rm p} = M\mathbf{b}_{\rm p}$, but the toroidal velocity consists of a Mach number component Mb_φ and a supplementary component $R\widehat{\Omega}$ (Fig. 18.2). Hence, the physical interpretation of the quantity $R\widehat{\Omega}$ is quite different here from both HD and MHD in the absence of poloidal rotation (where it represents the complete toroidal rotation velocity): $R\widehat{\Omega}$ is the toroidal velocity of the rotating frame in which the plasma velocity becomes parallel to the Alfvén speed.

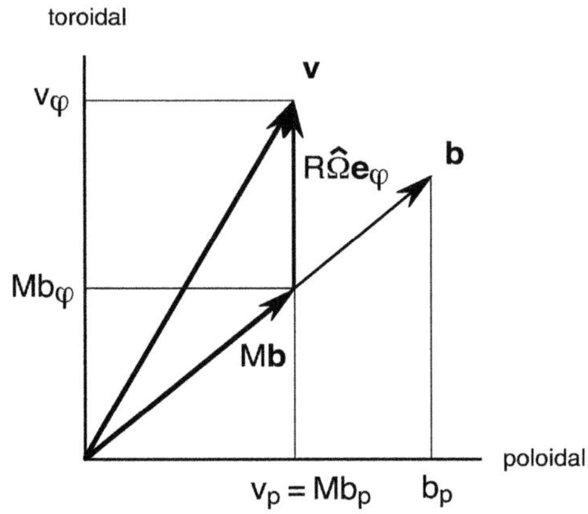

Fig. 18.2 Representation of the plasma velocity relative to the vectorial Alfvén speed $\mathbf{b} \equiv \mathbf{B}/\sqrt{\rho}$. The poloidal velocity is measured in terms of the poloidal Alfvén Mach number: $v_\mathrm{p} = M b_\mathrm{p}$, but the toroidal velocity has an additional contribution: $v_\varphi = M b_\varphi + R\widehat{\Omega}$.

18.2.2 Equilibrium variational principle and rescaling★

From the way in which the representations (18.13) and (18.14) of the two components of the poloidal momentum equation have been arranged, it is evident that it makes sense to compose the flux functions $H, S, K, \widehat{\Omega}$ and GM_* with χ' to obtain the following five *scaled flux functions* $\Lambda_i(\Psi)$:

$$\Lambda_1 \equiv \chi'^2 H\,, \qquad \Lambda_2 \equiv \frac{\gamma}{\gamma - 1}\,(\chi'^2)^\gamma S\,,$$
$$\Lambda_3 \equiv \frac{1}{\sqrt{2}}\,\chi' K\,, \qquad \Lambda_4 \equiv \frac{1}{\sqrt{2}}\,\chi'\widehat{\Omega}\,, \qquad \Lambda_5 \equiv GM_*\chi'^2\,. \tag{18.19}$$

Note that only the first four are needed to describe the equilibrium of laboratory plasmas. Prescribing the (completely arbitrary) functions $\Lambda_i(\Psi)$, the only remaining constraints on the physical variables then come from the mentioned two poloidal components of the momentum equation. They may be written as a nonlinear PDE for the poloidal flux $\Psi(R, Z)$:

$$\nabla \cdot \Big(\frac{1 - M^2}{R^2}\nabla\Psi\Big) + \frac{1}{M^2}\frac{\partial \Pi_1}{\partial \Psi} - \frac{1}{\gamma M^{2\gamma}}\frac{\partial \Pi_2}{\partial \Psi} - \frac{1}{M^2 - 1}\frac{\partial \Pi_3}{\partial \Psi} = 0\,, \tag{18.20}$$

and an algebraic Bernoulli equation for the squared poloidal Alfvén Mach number $M^2(R, Z)$:

$$\frac{1}{2R^2}|\nabla\Psi|^2 - \frac{\Pi_1}{M^4} + \frac{\Pi_2}{M^{2(\gamma+1)}} + \frac{\Pi_3}{(M^2 - 1)^2} = 0\,. \tag{18.21}$$

The problem may be cast in the form of a variational principle,

$$\delta \int \Big[\frac{1}{2R^2}(1 - M^2)\,|\nabla\Psi|^2 - \frac{\Pi_1}{M^2} + \frac{\Pi_2}{\gamma M^{2\gamma}} + \frac{\Pi_3}{M^2 - 1}\Big]\,dV = 0\,, \tag{18.22}$$

giving as Euler–Lagrange equations the above core equations (18.20) and (18.21). Here, three generic functions Π_j appear that depend on the five scaled flux functions as well as on the coordinates:

$$
\begin{aligned}
\Pi_1(\Psi; R, Z) &\equiv \Lambda_1 + R^2\Lambda_4^2 + \frac{\Lambda_5}{\sqrt{R^2+Z^2}} &&= M^2\left(\tfrac{1}{2}M^2B^2 + \frac{\gamma}{\gamma-1}p\right),\\
\Pi_2(\Psi) &\equiv \Lambda_2 &&= \frac{\gamma}{\gamma-1}M^{2\gamma}\,p\,,\\
\Pi_3(\Psi; R) &\equiv (R^{-1}\Lambda_3 - R\Lambda_4)^2 &&= \tfrac{1}{2}(M^2-1)^2\,B_\varphi^2\,. \qquad (18.23)
\end{aligned}
$$

The utmost right expressions show the relationship with the physical variables. The equations (18.20) and (18.21) determine the spatial dependence of $B_{\rm p}(R, Z)$, and of the five remaining physical variables through the five flux functions. Hence, solution of these two equations constitutes the core of the equilibrium problem. The rest is trivial algebra, except for the choice of meaningful magnitudes for the free functions, which is our next topic.

▷ **Toroidal rescaling** We introduce a reference point $R = R_0\,, Z = 0$ somewhere in the plasma, and exploit the plasma half-width a and the magnetic field B_0 at the reference point as units of length and magnetic field strength. Dimensionless coordinates with respect to the reference point are then given by

$$x \equiv (R - R_0)/a\,, \quad y \equiv Z/a\,, \qquad (18.24)$$

$$\Rightarrow \quad R = R_0(1+\epsilon x)\,, \quad Z = ay\,, \quad \epsilon \equiv a/R_0\,, \qquad (18.25)$$

where ϵ may be used for small inverse aspect ratio expansions ($\epsilon \ll 1$). For the time being, we will not make this assumption. However, it is clear that the dependence on this geometric factor has a big influence on what the magnitudes of the five flux functions physically represent. Since there is hardly any observational information on those functions, it is expedient to separate their influence as much as possible from that of the geometric factors that occur. This is done similar to the poloidal flux scaling of Section 16.3.1.

(1) All quantities are made dimensionless by putting $a = 1$ and $B_0 = 1$:

$$a\nabla \;\to\; \nabla \equiv \mathbf{e}_x\partial_x + \mathbf{e}_y\partial_y\,, \quad (a^2B_0)^{-1}\psi \;\to\; \psi\,, \quad B_0^{-2}\Pi_j \;\to\; \Pi_j \;\; (j=1,2,3)\,, \qquad (18.26)$$

i.e. we indicate dimensionless variables by the same symbol as the original dimensional variables. This operation is clearly to be distinguished from the non-trivial rescalings with ϵ and the total dimensionless flux Ψ_1, to be discussed. Now (in contrast to Chapter 16), we fix the reference point to be the (unknown) location $x = \delta, y = 0$ of the magnetic axis, where δ is the shift relative to the geometric centre of the plasma, and we assume the plasma shape to be up–down symmetric. Since $\Psi = \Psi_0 = 0$ at this reference point, we obtain a dimensionless flux variable of unit range by normalizing with respect to the flux Ψ_1 at the plasma boundary:

$$\psi \equiv \Psi/\Psi_1\,. \qquad (18.27)$$

Next, we rescale the generic functions and the flux functions with ϵ and Ψ_1:

$$
\begin{aligned}
&\bar\Pi_j \equiv (\epsilon\Psi_1)^{-2}\Pi_j \;\; (j=1,2,3)\,, \qquad \bar\Lambda_{1,2} \equiv (\epsilon\Psi_1)^{-2}\Lambda_{1,2}\,,\\
&\bar\Lambda_3 \equiv \Psi_1^{-1}\Lambda_3\,, \quad \bar\Lambda_4 \equiv (\epsilon^2\Psi_1)^{-1}\Lambda_4\,, \quad \bar\Lambda_5 \equiv (\epsilon\Psi_1^2)^{-1}\Lambda_5\,. \qquad (18.28)
\end{aligned}
$$

In effect, Eqs. (18.20) and (18.21) transform to the following dimensionless representation of the core equations for $\psi(x, y)$ and $M^2(x, y)$:

$$\nabla\cdot\left[\frac{1-M^2}{(1+\epsilon x)^2}\nabla\psi\right] + \frac{1}{M^2}\frac{\partial\bar\Pi_1}{\partial\psi} - \frac{1}{\gamma M^{2\gamma}}\frac{\partial\bar\Pi_2}{\partial\psi} - \frac{1}{M^2-1}\frac{\partial\bar\Pi_3}{\partial\psi} = 0\,, \qquad (18.29)$$

$$\frac{1}{2(1+\epsilon x)^2}|\nabla\psi|^2 - \frac{\bar\Pi_1}{M^4} + \frac{\bar\Pi_2}{M^{2(\gamma+1)}} + \frac{\bar\Pi_3}{(M^2-1)^2} = 0\,, \qquad (18.30)$$

where the relationship of the rescaled generic functions and flux functions will be given below. In this manner, the arbitrary shapes of the five rescaled flux functions $\bar\Lambda_i(\psi)$ become well-defined, the parameter ϵ only describes the (usually small) effects of toroidicity and the value of Ψ_1 *appears* to have vanished from the equilibrium core equations.

(2) The problem (18.29)–(18.30) still suffers from cancellations of large factors associated with the gravitational potential and the toroidal rotation. This may be cured by extracting the separate horizontal (x) and vertical (y) dependencies of those factors to facilitate maximal orderings for *thin accretion discs* (both effects appearing), *thick accretion discs* (vertical gravity negligible), and *tokamaks* (horizontal gravity absent as well). This is effected by redefining the first flux function,

$$\bar{\Lambda}_1^* \; \Big(\equiv (\epsilon\Psi_1)^{-2}\Lambda_1^* \equiv (\epsilon\Psi_1)^{-2}\chi'^2 H^* \Big) \equiv \bar{\Lambda}_1 + \bar{\Lambda}_4^2 + \bar{\Lambda}_5 \,, \tag{18.31}$$

i.e. by introducing a *modified Bernoulli function* H^* that incorporates the large leading order rotational and gravitational flux function contributions, and that may be assumed to be positive definite:

$$\begin{aligned} H^* &\equiv H + \tfrac{1}{2}R_0^2\widehat{\Omega}^2 + R_0^{-1}GM_* \\ &\equiv \tfrac{1}{2}v_p^2\frac{B^2}{B_p^2} + \frac{\gamma}{\gamma-1}\frac{p}{\rho} + \tfrac{1}{2}(R_0^2 - R^2)\widehat{\Omega}^2 + \Big(\frac{1}{R_0} - \frac{1}{\sqrt{R^2+Z^2}}\Big)GM_* \,. \end{aligned} \tag{18.32}$$

The rescaled generic functions then become

$$\begin{aligned} \bar{\Pi}_1(\psi; x, y) &\equiv \bar{\Lambda}_1^* + \epsilon\, e\big(2\bar{\Lambda}_4^2 - \bar{\Lambda}_5\big) + \epsilon^2 f \bar{\Lambda}_5 \,, \\ \bar{\Pi}_2(\psi) &\equiv \bar{\Lambda}_2 \,, \\ \bar{\Pi}_3(\psi; x) &\equiv \big[\bar{\Lambda}_3 - \bar{\Lambda}_4 - \epsilon\big(g\bar{\Lambda}_3 + h\bar{\Lambda}_4\big)\big]^2 \,, \end{aligned} \tag{18.33}$$

where four geometry functions appear,

$$\begin{aligned} e(x) &\equiv x(1 + \tfrac{1}{2}\epsilon x) && \approx x \,, \\ f(x,y) &\equiv \epsilon^{-2}\Big\{\big[(1+\epsilon x)^2 + \epsilon^2 y^2\big]^{-1/2} - 1 + \epsilon x + \tfrac{1}{2}\epsilon^2 x^2\Big\} && \approx \tfrac{1}{2}(3x^2 - y^2) \,, \\ g(x) &\equiv x(1+\epsilon x)^{-1} && \approx x \,, \\ h(x) &\equiv x \,, \end{aligned} \tag{18.34}$$

that have order unit magnitude as shown by the approximations (valid for $\epsilon \ll 1$) on the right. In the limit of vanishing toroidicity ($\epsilon \to 0$), the three generic functions become flux functions and the previous translation symmetric equations of Goedbloed and Lifschitz [238] are recovered. This completes the separation of flux functions and geometry dependencies as far as possible without solving the differential equations.

We will use the rescaled flux functions $\bar{\Lambda}_i$ in the rest of the analysis, where we exploit ϵ as a small parameter to obtain analytical results but we keep ϵ arbitrary in the numerical analysis. Since $e \sim f \sim 1$, assuming $\bar{\Lambda}_i \sim 1$ implies that horizontal gravity and toroidal rotation effects (factor e) will enter to first order, but vertical gravity effects (factor f) only to second order. To get horizontal gravity and toroidal rotation to appear in leading order (for tokamaks and thick accretion discs), we may keep $\bar{\Lambda}_1^* \sim \bar{\Lambda}_2 \sim 1$ but we need to push the orders of magnitude of the other flux functions:

$$\bar{\Lambda}_3 \approx \bar{\Lambda}_4 \sim \epsilon^{-1/2} \,, \quad \bar{\Lambda}_5 \sim \epsilon^{-1} \,, \quad \text{such that} \quad \bar{\Lambda}_3 - \bar{\Lambda}_4 \sim 1 \,. \tag{18.35}$$

To get vertical gravity to appear in the leading order as well (for thin accretion discs), we need to push the orders of magnitude of those flux functions even further:

$$\bar{\Lambda}_3 \approx \bar{\Lambda}_4 \sim \epsilon^{-1} \,, \quad 2\bar{\Lambda}_4^2 \approx \bar{\Lambda}_5 \sim \epsilon^{-2} \quad \Rightarrow \quad \bar{\Lambda}_3 - \bar{\Lambda}_4 \sim 1 \,, \quad 2\bar{\Lambda}_4^2 - \bar{\Lambda}_5 \sim \epsilon^{-1} \,. \tag{18.36}$$

In this manner, we guarantee that explicit results from $\epsilon \ll 1$ analysis may be generalized immediately to 'exact' results by means of the numerical codes.

Note that the physical meaning of the different geometrical terms is not straightforward because of the flux function constraints on the variables. In particular, the horizontal force balance, which requires approximate Keplerian toroidal rotation in HD, is determined by the flux function combination $2\bar{\Lambda}_4^2 - \bar{\Lambda}_5$ in MHD:

$$\begin{aligned}\big(\epsilon\Psi_1/\chi'\big)^2\,\big(2\bar{\Lambda}_4^2 - \bar{\Lambda}_5\big) &\equiv R_0^2\widehat{\Omega}^2 - R_0^{-1}GM_* \\ &\to v_\varphi^2 - R_0^{-1}GM_* \quad \text{when either } M = 0 \text{ or } B_\varphi = 0\,. \end{aligned} \tag{18.37}$$

Only in the absence of poloidal flow or toroidal magnetic field, one obtains the standard Keplerian expression on the second line of Eq. (18.37). In general, horizontal force balance is expressed by the first line of Eq. (18.37), i.e. instead of v_φ only the toroidal deviation from the Alfvén speed (Fig. 18.2) counts.

The transonic equilibrium problem (18.29), (18.30), (18.33), (18.34) is solved by first choosing the functional dependencies of the five rescaled flux functions $\bar{\Lambda}_i(\psi)$ (based on whatever observational evidence is available) and then iteratively solving the nonlinear PDE (18.29) for the flux $\psi(x,y)$ in an elliptic flow domain, subject to the BCs

$$\psi = 0 \ \text{(on the magnetic axis)}\,, \qquad \psi = 1 \ \text{(on the plasma boundary)}\,, \tag{18.38}$$

while inserting the algebraic solutions $M^2(\psi(x,y);x,y)$ of the Bernoulli equation (18.30) for each step of the iteration. This internal equilibrium problem is strictly determined by the shape of the boundary curve, which should be determined by the external dynamics. The latter part we take for granted and we prescribe it to be a circle or an ellipse (to study the influence of external flattening of the disc). Thus, the external flattening dominates the internal one described by the geometry function f and the flux function $\bar{\Lambda}_5$. ◁

18.2.3 Elliptic and hyperbolic flow regimes★

As stated in Section 18.1, we will have to restrict the analysis to flow regimes described by *elliptic partial differential equations*. This is a complication since the condition for ellipticity or hyperbolicity is determined by the solutions ψ and M^2 of Eqs. (18.29) and (18.30) themselves, according to the following expression involving the Alfvén Mach numbers [238]:

$$\begin{aligned}\Delta_{\rm char} &= (M^2-1)\,\frac{\bar{\Pi}_1 - \frac{1}{2}\big[\gamma+1-(\gamma-1)M^2\big]M^{-2(\gamma-1)}\bar{\Pi}_2}{\bar{\Pi}_1 - \frac{1}{2}(\gamma+1)M^{-2(\gamma-1)}\bar{\Pi}_2 - M^6(M^2-1)^{-3}\bar{\Pi}_3} \\ &= \frac{\gamma p + B^2}{B_{\rm p}^2}\,\frac{(M^2-1)^2(M^2-\widetilde{M}_{\rm c}^2)}{(M^2-\widetilde{M}_{\rm s}^2)(M^2-\widetilde{M}_{\rm f}^2)}\ \left\{\begin{array}{l} <0: \text{elliptic} \\ =0: \text{parabolic} \\ >0: \text{hyperbolic}\end{array}\right.\,,\end{aligned} \tag{18.39}$$

$$\widetilde{M}_{\rm c}^2 \equiv \frac{\gamma p}{\gamma p + B^2}\,, \qquad \widetilde{M}_{\rm f,s}^2 \equiv \frac{\gamma p + B^2}{2B_{\rm p}^2}\left[1 \pm \sqrt{1 - \frac{4\gamma p B_{\rm p}^2}{(\gamma p + B^2)^2}}\,\right]. \tag{18.40}$$

The quantity $\Delta_{\rm char}$ appears as a square root in the directional derivatives of the characteristics, giving two real characteristics in the hyperbolic flow regimes and no characteristics at all in the elliptic flow regimes. Hence, the transitions from ellipticity to hyperbolicity [665, 283, 7] occur at positions on the magnetic/flow surfaces where either the numerator or the denominator of $\Delta_{\rm char}$ vanishes. The tildes on the quantities indicate that the explicit expressions for the transition values of M^2 are only obtained after solving an implicit relation, e.g.

$$M^2 - \widetilde{M}_{\rm c}^2(M^2,\psi;x,y) = 0 \quad \Rightarrow \quad M^2 = M_{\rm c}^2(\psi;x,y)\,, \tag{18.41}$$

from the numerator expression on the first line of Eq. (18.39), and, similarly, $M^2 = M_{\rm f,s}^2(\psi;x,y)$ are obtained from the denominator expression. The numerical aspects of this problem are discussed in [44] and implemented in FINESSE.

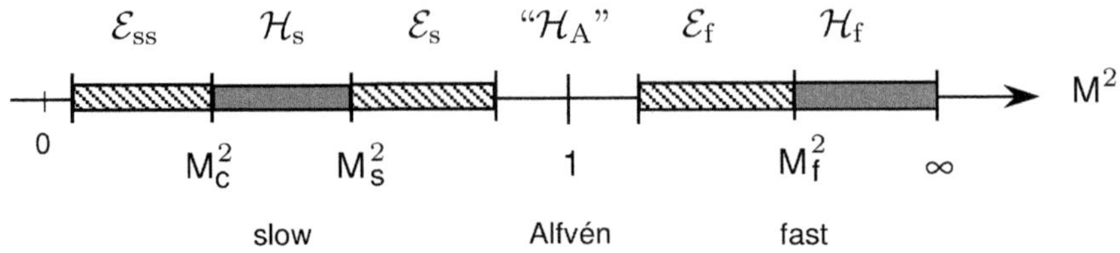

Fig. 18.3 Elliptic (hatched) and hyperbolic (dark) flow regimes corresponding to the value of the poloidal Alfvén Mach number M. Flow should be in the first elliptic ($\mathcal{E}_{ss}$: sub-slow), the second elliptic ($\mathcal{E}_s$: slow), or the third elliptic ($\mathcal{E}_f$: fast) flow regime for the spectral results of this chapter to be valid.

As follows from the definitions (18.40), the transition values for M^2 are well-ordered in magnitude giving five different flow regimes (Fig. 18.3). We just need to specify the flow to be either in the first (sub-slow) elliptic flow regime ($\mathcal{E}_{ss}$, where $M^2 < M_c^2$), or the second (slow) elliptic flow regime ($\mathcal{E}_s$, where $M_s^2 < M^2 < 1$), or the third (fast) elliptic flow regime ($\mathcal{E}_f$, where $1 < M^2 < M_f^2$). Since the effects of transition through a hyperbolic flow regime should be first noticeable in the regime $\mathcal{E}_s$, we will pay special attention to equilibria in that regime.

In Ref. [238], a similar problem for stationary magnetohydrodynamic flows was solved for a translation symmetric case, exploiting self-similar solutions. The occurring flow regimes were analogous to the ones discussed here. It was found that limiting line and Alfvén singularities leads to a division into four distinct velocity regimes which cannot be connected by continuous flows. Instead the three types of MHD shocks appear, enabling the system to pass from the high to the low speed regimes.

18.2.4 Expansion of the equilibrium in small toroidicity★

(a) Small inverse aspect ratio expansion We assume the inverse aspect ratio to be small, $\epsilon \ll 1$, and impose the outer plasma boundary to be circular, and try to find solutions of the equilibrium equations (18.29) and (18.30) in the form of the *Shafranov shifted circle approximation*, where the poloidal magnetic flux/flow surfaces are assumed to have circular cross-sections that are shifted by an amount $\Delta(\hat{r})$ with respect to the centre of the plasma (see Section 16.2.2, Fig. 16.8). To first order, these solutions may then be represented as

$$\psi(\hat{r},\hat{\theta}) \approx \psi_0(\hat{r}), \tag{18.42}$$

$$M^2(\hat{r},\hat{\theta}) \approx M_0^2(\hat{r}) + \mu(\hat{r})\cos\theta, \tag{18.43}$$

where $\hat{r}$, $\hat{\theta}$ are non-orthogonal polar coordinates defined with respect to the centres of the flux surfaces. Note that $\psi_1(\hat{r}) \equiv 0$ since the perturbation of the flux surfaces is absorbed in the definition of the coordinates, i.e. it is represented by the shift $\Delta(\hat{r})$. The latter quantity satisfies the boundary conditions $\Delta(0) = \delta$ and $\Delta(1) = 0$, where δ is the shift of the magnetic axis. The perturbation of the squared poloidal Alfvén Mach number is denoted by $\mu = M_1^2(\hat{r})$. Due to the shifted circle approximation (18.42), the flux functions will now be determined by the radial dependence $\psi_0(\hat{r})$:

$$\bar{\Lambda}_i(\psi) \approx \bar{\Lambda}_i(\psi_0(\hat{r})) \equiv \bar{\Lambda}_{i,0}(\hat{r}). \tag{18.44}$$

Since we have uniquely settled the notation, we will drop the subscript 0 on ψ_0 and M_0^2 from now on. (This is done because we will need the subscripts 0 and 1 later on to indicate equilibrium values on the magnetic axis and at the boundary.) The expansion procedure will then consist of a zeroth order part determining the radial dependencies of $\psi(\hat{r})$ and $M^2(\hat{r})$ and a first order part determining the shift $\Delta(\hat{r}) \sim \epsilon$ (and, hence, δ) and the Alfvén Mach number perturbation $\mu(\hat{r}) \sim \epsilon$.

As in Section 16.2.2, the connection with the Cartesian coordinates is given by

$$\begin{aligned} x = x(\hat{r}, \hat{\theta}) &= \hat{r}\cos\theta + \Delta(\hat{r}) \,, \\ y = y(\hat{r}, \hat{\theta}) &= \hat{r}\sin\theta \,. \end{aligned} \tag{18.45}$$

To convert the equations to the $\hat{r}, \hat{\theta}, \varphi$ coordinates, we insert these into Eq. (18.25):

$$\begin{aligned} R = R(\hat{r}, \hat{\theta}) &= \epsilon^{-1} + \hat{r}\cos\theta + \Delta \,, \\ Z = Z(\hat{r}, \hat{\theta}) &= \hat{r}\sin\theta \,, \end{aligned} \tag{18.46}$$

from which the metric tensor, the poloidal field and the Jacobian follow, to first order:

$$\hat{g}^{ij} \approx \begin{pmatrix} 1 - 2\Delta'\cos\theta & (\Delta'/\hat{r})\sin\hat{\theta} & 0 \\ (\Delta'/\hat{r})\sin\theta & 1/\hat{r}^2 & 0 \\ 0 & 0 & \epsilon^2(1 - 2\epsilon\hat{r}\cos\theta) \end{pmatrix} , \tag{18.47}$$

$$B_{\rm p} \approx \hat{r}\Psi_1\psi'/\hat{J} \,, \qquad \hat{J} \equiv (\nabla\hat{r} \times \nabla\hat{\theta} \cdot \nabla\varphi)^{-1} \approx \epsilon^{-1}\hat{r}\big[1 + (\epsilon\hat{r} - \Delta')\cos\hat{\theta}\big] \,. \tag{18.48}$$

The hats distinguish these metric coefficients from the ones for the straight-field-line coordinates, that we will eventually exploit in Section 18.3.

(b) Zeroth and first order solutions Choosing $\bar{\Lambda}_1^* \sim \bar{\Lambda}_2 \sim \cdots \bar{\Lambda}_5 \sim 1$, the expansions of the generic functions, accurate to first order, are given by

$$\begin{aligned} \bar{\Pi}_1(\hat{r}, \hat{\theta}) &\approx \bar{\Lambda}_{1,0}^* + \epsilon x (2\bar{\Lambda}_{4,0}^2 - \bar{\Lambda}_{5,0}) \,, \\ \bar{\Pi}_2(\hat{r}) &\approx \bar{\Lambda}_{2,0} \,, \\ \bar{\Pi}_3(\hat{r}, \hat{\theta}) &\approx \big(\bar{\Lambda}_{3,0} - \bar{\Lambda}_{4,0}\big)^2 - 2\epsilon x (\bar{\Lambda}_{3,0}^2 - \bar{\Lambda}_{4,0}^2) \,, \end{aligned} \tag{18.49}$$

where three flux function combinations appear in zeroth order:

$$\begin{aligned} (\epsilon\Psi_1)^2 \bar{\Lambda}_{1,0}^* &= \tfrac{1}{2} M^4 B^2 + \frac{\gamma}{\gamma - 1} M^2 p \,, \\ (\epsilon\Psi_1)^2 \bar{\Lambda}_{2,0} &= \frac{\gamma}{\gamma - 1} M^{2\gamma} p \,, \\ (\epsilon\Psi_1)^2 \big(\bar{\Lambda}_{3,0} - \bar{\Lambda}_{4,0}\big)^2 &= \tfrac{1}{2}(M^2 - 1)^2 B_\varphi^2 \,, \end{aligned} \tag{18.50}$$

and two in first order:

$$\begin{aligned} (\epsilon\Psi_1)^2 (2\bar{\Lambda}_{4,0}^2 - \bar{\Lambda}_{5,0}) &= M^2\big[(\sqrt{\rho}v_\varphi - MB_\varphi)^2 - \rho R_0^{-1} G M_*\big] \,, \\ (\epsilon\Psi_1)^2 (\bar{\Lambda}_{3,0}^2 - \bar{\Lambda}_{4,0}^2) &= -\tfrac{1}{2}(M^2 - 1) B_\varphi\big[(1 + M^2) B_\varphi - 2M\sqrt{\rho}v_\varphi\big] . \end{aligned} \tag{18.51}$$

We have again omitted the subscript 0 on the original physical variables.

As compared to the original 2D problem, where the flux surfaces $\psi(\hat{r},\hat{\theta}) = \text{const}$ and the Bernoulli surfaces $M^2(\hat{r},\hat{\theta}) = \text{const}$ intersect, in the present 1D problem $\psi(\hat{r})$ and $M^2(\hat{r})$ trivially label the same surfaces. This implies that there is a much closer relationship now between the free flux functions and the physical variables (both being functions of $\hat{r}$ alone). In the zeroth order, the choice of the three functions (18.50) determines three of the four physical variables p, B_p, B_φ and M^2 (leaving room for only one extra condition: see below). In the first order, the choice of the two functions (18.51) determines the two remaining physical variables ρ and v_φ (except for the contribution of GM_*).

Inserting the metric coefficients (18.47)–(18.48) and the expressions (18.49)–(18.51) into the equilibrium (18.29) and (18.30), we obtain two simple equations for ψ and M^2 in the zeroth order and two more complicated ones for Δ and μ in the first order. The *zeroth order flux equation* is a second order ODE for $\psi(\hat{r})$:

$$\frac{\psi'}{\hat{r}}\big[(1-M^2)\hat{r}\psi'\big]' + \frac{\bar{\Lambda}^{*\prime}_{1,0}}{M^2} - \frac{\bar{\Lambda}'_{2,0}}{\gamma M^{2\gamma}} - \frac{(\bar{\Lambda}_{3,0}-\bar{\Lambda}_{4,0})^{2\prime}}{M^2-1} = 0\,, \tag{18.52}$$

where the primes now denote differentiation with respect to $\hat{r}$. Since $B_\mathrm{p} \approx (\epsilon\Psi_1)\psi'$ to leading order, this boils down to the cylindrical equilibrium relation (cf. Eq. (12.30)):

$$\frac{d}{d\hat{r}}(p+\tfrac{1}{2}B^2) - (M^2-1)\frac{B_\mathrm{p}^2}{\hat{r}} = 0\,. \tag{18.53}$$

However, the zeroth order Bernoulli equation reduces to an identity so that there is no longer an equilibrium restriction on the choice of $M^2(\hat{r})$. Hence, Eq. (18.53) is the only condition now on the six physical equilibrium functions (of $\hat{r}$).

After some lengthy, but straightforward, reductions the *first order magnetic flux and Bernoulli equations* yield two coupled differential equations for Δ and μ:

$$\big[(1-M^2)\hat{r}B_\mathrm{p}^2\Delta'\big]' + \big(\hat{r}B_\mathrm{p}^2\mu\big)' = -\epsilon\hat{r}E\,, \tag{18.54}$$

$$M^2(M^2-1)B_\mathrm{p}^2\Delta' + \frac{B_\mathrm{p}^2}{M^2}(M^2-M_\mathrm{s}^2)(M^2-M_\mathrm{f}^2)\mu = -\epsilon\hat{r}F\,, \tag{18.55}$$

where the two RHS expressions are defined by

$$E \equiv -(M^2-1)B_\mathrm{p}^2 - \hat{r}\big[2p+\rho(v^2-R_0^{-1}GM_*)\big]'\,, \tag{18.56}$$

$$F \equiv (M^2-1)\rho(v^2-R_0^{-1}GM_*) + 2MB_\varphi(\sqrt{\rho}v_\varphi - MB_\varphi)\,. \tag{18.57}$$

The appearance in Eq. (18.55) of the transition values of M^2 defined in Eqs. (18.40) is no accident, as will be explained below. (Because $M^2 = M^2(\hat{r})$ to leading order, according to Eq. (18.43), there is no need to solve an implicit equation in this case so the tildes have been dropped.) Rather than integrating Eqs. (18.54) and (18.55) directly, it is more instructive to first decouple them to a separate second order ODE for $\Delta(\hat{r})$ and a first order ODE for $\mu(\hat{r})$:

$$\frac{d}{d\hat{r}}\left[\frac{(\gamma p+B^2)(M^2-1)(M^2-M_\mathrm{c}^2)}{(M^2-M_\mathrm{s}^2)(M^2-M_\mathrm{f}^2)}\,\hat{r}\,\frac{d\Delta}{d\hat{r}}\right] = \epsilon\left\{\hat{r}E + \left[\frac{M^2\hat{r}^2F}{(M^2-M_\mathrm{s}^2)(M^2-M_\mathrm{f}^2)}\right]'\right\}, \tag{18.58}$$

$$\frac{d}{d\hat{r}}\left[\frac{(\gamma p+B^2)(M^2-M_\mathrm{c}^2)}{M^4}\,\hat{r}\mu\right] = -\epsilon\left\{\hat{r}E + \left(\frac{r^2F}{M^2}\right)'\right\}. \tag{18.59}$$

Integration yields

$$\Delta' \;=\; -\,\epsilon\hat{r}\,\frac{(M^2-M_{\rm s}^2)(M^2-M_{\rm f}^2)I + M^2 F}{(\gamma p + B^2)(M^2-1)(M^2-M_{\rm c}^2)}\,, \tag{18.60}$$

$$\frac{\mu}{M^2} \;=\; -\,\frac{\rho_1}{\rho} = -\epsilon\hat{r}\,\frac{M^2 I + F}{(\gamma p + B^2)(M^2-M_{\rm c}^2)}\,, \qquad I \equiv \frac{1}{\hat{r}^2}\int_0^{\hat{r}} E\hat{r}\,d\hat{r}\,. \tag{18.61}$$

In the spectral analysis of the following sections, only Δ' and μ appear, so that no further integration is needed. Of course, for the limiting case of static tokamak equilibria, the well-known expression for Δ' is recovered [647].

At this point, a truly amazing confluence of apparently completely unrelated topics may be noted. The coefficient in front of the highest derivative of the differential equation (18.58) for Δ is of exactly the same form as the analogous coefficient appearing in the spectral Hain–Lüst equation for cylindrical plasmas [276, 214], where the eigenvalue ω^2 of the latter is now replaced by the squared poloidal Alfvén Mach number, M^2. As we have seen in Section 18.2.3, the transonic transitions from elliptic to hyperbolic flows occur at the values $M^2 = 1$ (recall that this 1 just corresponds to the normalization with respect to the poloidal Alfvén speed), $M^2 = M_{\rm c}^2$, $M^2 = M_{\rm s}^2$, and $M^2 = M_{\rm f}^2$, defined in Eq. (18.40). On the other hand, in the Hain–Lüst equation, the spectrum of MHD waves is concentrated about continua [614, 254] at the Alfvén and slow (or cusp) frequencies $\omega^2 = \omega_{\rm A}^2$ and $\omega^2 = \omega_{\rm S}^2$, whereas the special values $\omega^2 = \omega_{\rm s}^2$ and $\omega^2 = \omega_{\rm f}^2$ do not correspond to continua [18] but to monotonicity transitions in the spectrum [240]. Apparently, there is a deep correspondence between the linear waves and the nonlinear stationary states. This is another example of the recent insight that equilibrium and perturbations are not really separate issues in transonic magnetohydrodynamics [222]. We will see more of this in the following sections.

Obviously, for the shifted circle approximation to be valid, i.e. $\Delta' \ll 1$, the singularities $M^2 = 1$ and $M^2 = M_{\rm c}^2$ must be avoided. However, as in the analogous spectral case, the denominator zeros of Eq. (18.58) just constitute apparent singularities of the differential equation so that approaching $M^2 \downarrow M_{\rm s}^2$ or $M^2 \uparrow M_{\rm f}^2$ from within the elliptic flow regimes poses no fundamental problem. We will use this simplifying feature in the following sections where we will investigate the effect of the transonic transition from sub-slow to slow flow on changes of the continuous spectra in the second elliptic flow regime.

(c) Approximate solutions in the second elliptic flow regime We will simplify the solutions obtained by means of approximations that are valid in the second elliptic flow regime by extending the usual low-β tokamak approximation ($B_\varphi \sim 1$, $B_{\rm p} \sim \epsilon$, $\beta \equiv p/(2B^2) \sim \epsilon^2$) with poloidal rotation. So far, we have made no assumption on M^2. We will now push M^2 in the region $M_{\rm s}^2 \le M^2 < 1$ as much as possible within the low-β approximation. Since $M_{\rm s}^2 \approx M_{\rm c}^2(1+\mathcal{O}(\epsilon^4)) \sim \epsilon^2$ and $M_{\rm f}^2 \sim \epsilon^{-2}$, the validity of the expressions (18.60) and (18.61) for Δ' and μ/M^2 demands that M^2 is larger than $\mathcal{O}(\epsilon^2)$ but smaller than $\mathcal{O}(\epsilon)$:

$$M^2 \sim \epsilon^{2-\nu} \quad (0 \le \nu < 1)\,. \tag{18.62}$$

Hence, we may choose ν such that the poloidal Alfvén Mach number squared, M^2, represents flows much faster than the slow but much slower than the Alfvén speed:

$$\tfrac{1}{2}\gamma\beta \approx M_{\rm c}^2 \ll M^2 \ll 1\,. \tag{18.63}$$

As we will see, this ordering is quite effective to study a particular type of violent instability caused by the coupling of slow and Alfvén modes and driven by poloidal flows in the second elliptic flow domain. Let us call it the *trans-slow poloidal flow ordering* (avoiding the intuitively contradictory adjective 'super-slow').

From the equilibrium equations (18.29) and (18.30) and the definitions (18.33) of the generic functions, it is obvious that we must then choose

$$\bar{\Lambda}_1^{*\prime} \sim M^2(\bar{\Lambda}_3 - \bar{\Lambda}_4)^{2\prime}\,, \qquad \bar{\Lambda}_1^{*} \sim M^4(\bar{\Lambda}_3 - \bar{\Lambda}_4)^{2}\,, \tag{18.64}$$

in order to balance the magnetic terms. This dictates how the orders of magnitude of the different parameters appearing in the flux functions $\bar{\Lambda}_i$ have to be chosen to be consistent with the trans-slow ordering.

For our present purpose, we will consider a particular class of equilibria with negligible pressure ($p \approx 0$) and negligible non-parallel velocity ($\widehat{\Omega} \approx 0$). These additional assumptions are not necessary for the trans-slow ordering, but they are expedient since they simplify the equilibrium expressions considerably:

$$\Delta' \approx \frac{\epsilon}{\hat{r}B_{\rm p}^2}\int_0^{\hat{r}} \hat{r}^2\left[M^2B^2(1-\Gamma)\right]' d\hat{r}\,, \tag{18.65}$$

$$\frac{\mu}{M^2} = -\frac{\rho_1}{\rho} \approx \epsilon\hat{r}(1-\Gamma)\,. \tag{18.66}$$

Here, the flux function $\Gamma(\psi(\hat{r}))$ measures the relative strength of the gravitational interaction:

$$\Gamma(\psi) \equiv \frac{\bar{\Lambda}_5(\psi)}{2\bar{\Lambda}_1^*(\psi)} \approx \frac{\rho G M_*}{R_0 M^2 B^2}\,, \tag{18.67}$$

where we have restored the dimensional factors in the last expression.

Since we have assumed approximately parallel flow ($\widehat{\Omega} \approx 0$), M^2 now represents both the poloidal and toroidal Alfvén Mach number, $M^2 \equiv \rho v_{\rm p}^2/B_{\rm p}^2 \approx \rho v_\varphi^2/B_\varphi^2$, so that $\Gamma \approx GM_*/(Rv_\varphi^2)$ becomes a measure for the deviation from Keplerian flow (where $\Gamma = 1$). However, there is no need to stick to Keplerian flow in the presence of poloidal rotation, so that we obtain a class of equilibria that may be changed continuously from tokamak ($\Gamma = 0$) to accretion disc ($\Gamma \neq 0$). From Eqs. (18.65) and (18.66) it is clear that weak gravitational 'tokamak-like' equilibria ($\Gamma < 1$) have a density that is larger on the inside than on the outside of the torus ($\rho_1 < 0$), as is well known for poloidal flows in the second elliptic regime [667]. For strong gravitational interaction ($\Gamma > 1$), the opposite holds: the density is larger on the outside, i.e. in the region of smaller gravity in order to balance the stronger gravitational pull of the inner region. Note that the trans-slow ordering (18.63) permits the study of the effects of very massive central objects, as long as the expressions (18.65) and (18.66) for Δ' and μ remain small, e.g. by considering $\Gamma \sim \epsilon^{-1/2} \gg 1$.

The mentioned features are confirmed by the solutions shown in Fig. 18.4, obtained using the transonic equilibrium program FINESSE [44]. In the code, no small ϵ approximations were made, but the parameters of the flux functions were chosen to agree with the trans-slow ordering. (Explicit parameters may be found in [230].) As a result, there is perfect agreement between the numerical results and the present analysis.

The solutions obtained are representative examples of the following transonic equilibria, which will be investigated with respect to stability of continuum modes:

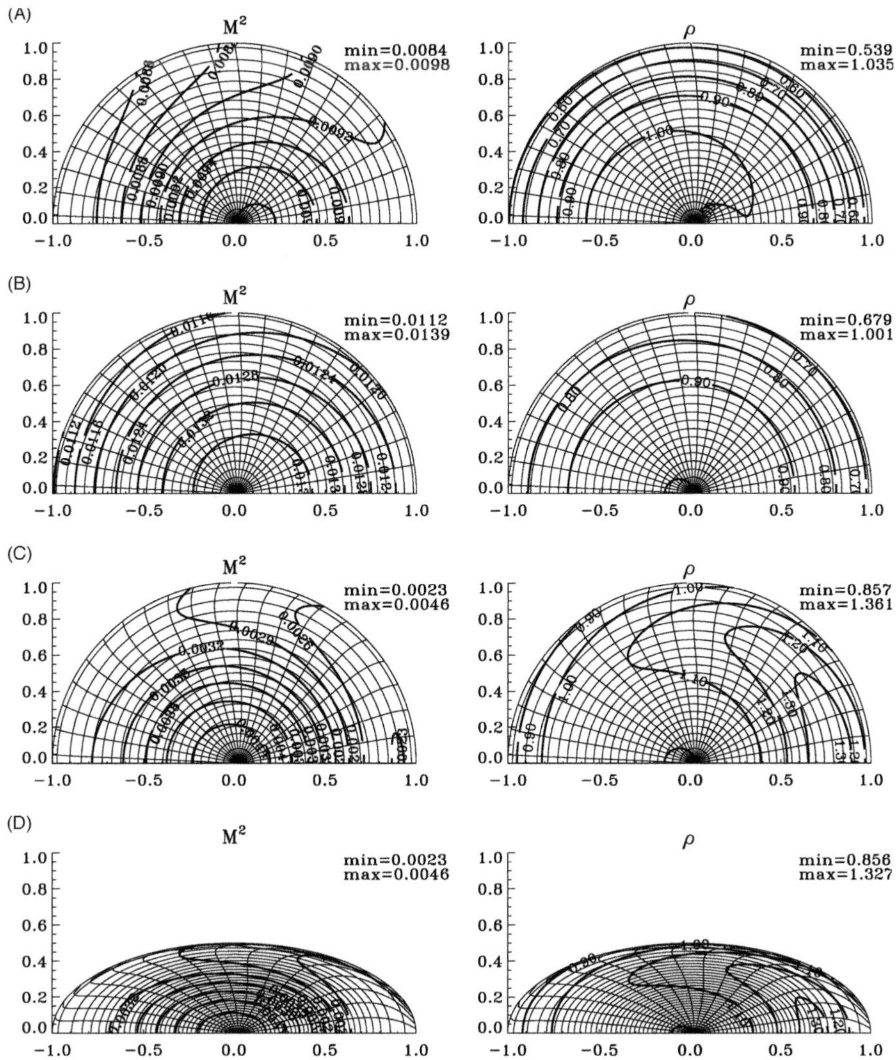

Fig. 18.4 Contours of the squared poloidal Alfvén Mach number M^2 and of the density ρ: (A) tokamak ($\Gamma_0 = 0$); (B) thick accretion disc, small central mass ($\Gamma_0 = 0.25$); (C) thick accretion disc, large central mass ($\Gamma_0 = 2$); (D) flat accretion disc, large central mass ($\Gamma_0 = 2$).

(A) *Tokamak* ($\Gamma_0 = 0$);

(B) *Thick accretion disc with a small central mass* ($\Gamma_0 = 0.25$);

(C) *Thick accretion disc with a massive central object* ($\Gamma_0 = 2$);

(D) *Flat accretion disc with a massive central object* ($b/a = 0.5, \Gamma_0 = 2$).

The stability of these equilibria will be discussed in Sections 18.3 and 18.4, after we have developed the theory of continuous spectra for transonic equilibria.

18.3 Equations for the continuous spectrum★

18.3.1 Reduction for straight-field-line coordinates★

The perturbations of the stationary equilibria of Section 18.2 will be described by the force-operator formalism in terms of the displacement $\boldsymbol{\xi}$ developed by Frieman and Rotenberg [197], as discussed in Chapter 12. Recall from Section 12.2.2 that, for normal modes $\sim \mathrm{e}^{-\mathrm{i}\omega t}$, the spectral equation takes the form

$$\mathbf{F}(\boldsymbol{\xi}) + \nabla\cdot\big(\rho\boldsymbol{\xi}\mathbf{v}\cdot\nabla\mathbf{v} - \rho\mathbf{v}\mathbf{v}\cdot\nabla\boldsymbol{\xi}\big) + 2\mathrm{i}\rho\omega\mathbf{v}\cdot\nabla\boldsymbol{\xi} + \rho\omega^2\boldsymbol{\xi} = 0\,, \tag{18.68}$$

where

$$\mathbf{F}(\boldsymbol{\xi}) \equiv \nabla\big(\gamma p\nabla\cdot\boldsymbol{\xi} + \boldsymbol{\xi}\cdot\nabla p\big) - \mathbf{B}\times(\nabla\times\mathbf{Q}) + (\nabla\times\mathbf{B})\times\mathbf{Q} + (\nabla\Phi_{\mathrm{gr}})\,\nabla\cdot(\rho\boldsymbol{\xi})\,, \tag{18.69}$$

with $\mathbf{Q} \equiv \nabla\times(\boldsymbol{\xi}\times\mathbf{B})$, is the usual force-operator for static equilibria [55]. Note that the eigenvalue problem (18.68) is *quadratic* (involving both ω and ω^2), in contrast to the static spectral problem, which is linear in the eigenvalue ω^2.

The Frieman–Rotenberg equation (18.68) will be expressed in the straight-field-line coordinates Ψ, ϑ, φ, with Jacobian $\mathcal{J} \equiv (\nabla\Psi\times\nabla\vartheta\cdot\nabla\varphi)^{-1}$, where the poloidal angle ϑ is constructed such that the magnetic field lines are represented by straight lines in the ϑ–φ plane so that the inverse rotational transform of the field lines (the 'safety factor') is constant on the magnetic/flow surfaces:

$$q(\Psi) \equiv \frac{d\varphi}{d\vartheta}\bigg|_{\text{field line}} = \frac{\mathcal{J}B_\varphi}{R} = \frac{\mathcal{J}}{R^2}\,\frac{\sqrt{\rho}M(K - R^2\widehat{\Omega})}{M^2 - 1}\,. \tag{18.70}$$

As compared to static plasmas or plasmas with toroidal flow only, where $q = (\mathcal{J}/R^2)I$ with the constant factor $I \equiv RB_\varphi$, the corresponding complicated quotient is not constant on the magnetic/flow surfaces. In the straight-field-line coordinates, the flow lines are not represented by straight lines in general (except when $\widehat{\Omega} = 0$, see Fig. 18.2) so that their inverse rotational transform is not constant on the magnetic/flow surfaces:

$$u(\Psi,\vartheta) \equiv \frac{d\varphi}{d\vartheta}\bigg|_{\text{flow line}} = \frac{\mathcal{J}\sqrt{\rho}v_\varphi}{MR} = \frac{K - R^2\widehat{\Omega}/M^2}{K - R^2\widehat{\Omega}}\,q(\psi)\,. \tag{18.71}$$

Obviously, one could also construct a poloidal angle such that the flow lines become straight. In that case, the roles of q and u would be interchanged. However, the field lines turn out to play the more fundamental role in transonic MHD, as will become clear from the spectral analysis in the next section.

We now project the spectral equation (18.68) onto the three preferred physical directions expressed by the magnetic surface/field line triad

$$\mathbf{n} \equiv (RB_{\mathrm{p}})^{-1}\nabla\Psi\,, \qquad \boldsymbol{\pi} \equiv \mathbf{b}\times\mathbf{n}\,, \qquad \mathbf{b} \equiv \mathbf{B}/B\,. \tag{18.72}$$

The displacement vector $\boldsymbol{\xi}$ will be represented by the three components

$$X \equiv RB_{\mathrm{p}}\,\boldsymbol{\xi}\cdot\mathbf{n}\,, \quad Y \equiv \mathrm{i}B(RB_{\mathrm{p}})^{-1}\,\boldsymbol{\xi}\cdot\boldsymbol{\pi}\,, \quad Z \equiv \mathrm{i}B^{-1}\,\boldsymbol{\xi}\cdot\mathbf{b}\,, \tag{18.73}$$

and the gradient operator, for harmonics $\sim \mathrm{e}^{-\mathrm{i}n\varphi}$, will be represented by the three scalar operators

$$D \;\equiv\; (RB_{\mathrm{p}})^{-1}\mathbf{n}\cdot\nabla \;=\; \frac{\partial}{\partial\psi} - \frac{g_{12}}{g_{22}}\frac{\partial}{\partial\vartheta}\,,$$

$$G \equiv -\mathrm{i}RB_{\rm p}B\boldsymbol{\pi}\cdot\nabla = -\mathrm{i}\frac{RB_\varphi}{\mathcal{J}}\frac{\partial}{\partial\vartheta} - nB_{\rm p}^2\,,$$

$$F \equiv -\mathrm{i}B\mathbf{b}\cdot\nabla = \frac{1}{\mathcal{J}}\Big(-\mathrm{i}\frac{\partial}{\partial\vartheta} + nq\Big)\,, \qquad (18.74)$$

where g_{ij} is the covariant metric tensor of the Ψ, ϑ, φ coordinate system. Recall that the main reason for exploiting straight-field-line coordinates is the fact that the operator F has the simple representation $\mathcal{J}^{-1}(m+nq)$ for single poloidal harmonics $\mathrm{e}^{\mathrm{i}m\vartheta}$ so that resonant modes are well represented.

Associated with the magnetic field and flow lines, two parallel gradient operators enter the formalism, viz. the *parallel field operator* $F \equiv -\mathrm{i}\mathbf{B}\cdot\nabla$, just defined, and the *parallel flow operator*

$$U \equiv -\mathrm{i}\mathbf{v}\cdot\nabla = -\frac{\mathrm{i}M}{\mathcal{J}\sqrt{\rho}}\frac{\partial}{\partial\vartheta} + \frac{nv_\varphi}{R} = \frac{M}{\mathcal{J}\sqrt{\rho}}\Big(-\mathrm{i}\frac{\partial}{\partial\vartheta} + nu\Big) = (M/\sqrt{\rho})F + n\widehat{\Omega}\,. \qquad (18.75)$$

Here, the very last expression represents a bit of luck since $n\widehat{\Omega}$ is constant on magnetic surfaces. Whereas the parallel gradient operator F enters as usual when converting the expression for the static force operator part $\mathbf{F}$, the qualitatively new features of rotating plasmas originate from the inertial term $\mathbf{v}\cdot\nabla\mathbf{v}$ and the Doppler shift operator $\omega+\mathrm{i}\mathbf{v}\cdot\nabla$ (second and third term of the first line of Eq. (18.68)). The latter expression involves the parallel flow operator U, but it may be transformed in terms of F by means of Eq. (18.75):

$$\omega + \mathrm{i}\mathbf{v}\cdot\nabla = \omega - U = \widetilde{\omega} - (M/\sqrt{\rho})F\,, \qquad (18.76)$$

where we have introduced a frequency $\widetilde{\omega}$ that represents the Doppler shifted frequency for a frame rotating with the angular frequency $\widehat{\Omega}$:

$$\widetilde{\omega} \equiv \omega - n\widehat{\Omega}\,. \qquad (18.77)$$

Note that this frequency is not the one corresponding to the full toroidal rotation velocity v_φ, but only the part that represents the deviation from the Alfvén speed (see Fig. 18.2). Consequently, the optimal representation of the parallel field operator F for straight-field-line coordinates is also fully exploited for the parallel flow operator. Moreover, for the present calculation of continuous spectra, where modes are localized on a single magnetic surface, by using the Doppler shifted frequency $\widetilde{\omega}$ we may eliminate most of the terms with $\widehat{\Omega}$ so that the characteristic dependence on the poloidal Alfvén Mach number M becomes manifest.

By means of the projection (18.72), the eigenvalue problem (18.68) becomes a 3×3 matrix equation, involving the gradient operators D, G and F, acting on the three-vector $\mathbf{X} \equiv (X, Y, Z)^{\rm T}$:

$$\mathsf{A}\cdot\mathbf{X} = \mathsf{B}\cdot\mathbf{X}\,, \qquad (18.78)$$

where the matrix B contains all terms involving $\omega - U$. To get the equations for the continuous spectrum, we exploit the same technique as used in Section 17.2.3 for the similar problem in the static toroidal case [215] (where the alternative method of exploiting the primitive variables [477] leads to the same result). We consider modes that are localized around a particular magnetic surface $\Psi = \Psi_0$ by taking the limit $\partial/\partial\Psi \to \infty$. For finite $\widetilde{\omega}$, the normal component of the spectral equation (18.78) then reduces to a derivative with respect to Ψ that may be integrated to give the following leading order relation, similar to Eq. (17.51) of the static case:

$$D^\dagger X \approx -\frac{1}{B^2} GY + \mathrm{i}\frac{\gamma p}{\gamma p + B^2}\frac{1}{\rho}\Big[\partial\Big(\frac{\rho R B\varphi}{B^2}\Big)\Big] Y - \frac{\gamma p}{\gamma p + B^2}\frac{1}{\rho} F \rho Z\,. \tag{18.79}$$

Here, we exploit a short-hand notation for the invariant tangential derivative:

$$[\partial f] \equiv [\mathbf{B}_\mathrm{p}\cdot\nabla f] = \frac{1}{\mathcal{J}}\frac{\partial f}{\partial\vartheta}\,, \tag{18.80}$$

where the square brackets are used to indicate that the derivative only acts on the quantity inside. Equation (18.79) implies that $\partial X/\partial\Psi$, Y and Z are of the same order, so that X itself is small compared to Y and Z. Hence, these modes are dominantly polarized tangential to the flux surfaces.

Substituting Eq. (18.79) into the tangential components of the spectral equation (18.78) then leads to a 2×2 matrix eigenvalue problem in terms of the two-vector $\mathbf{V} \equiv (Y, Z)^\mathrm{T}$ that only involves the parallel operators F and U, i.e. it no longer contains derivatives of $\mathbf{V}$ with respect to Ψ. Due to this property, the reduced 2×2 problem is non-singular. Therefore, it is advantageous to separate the improper Ψ-dependence from the proper tangential dependence:

$$\begin{pmatrix} Y(\Psi,\vartheta) \\ Z(\Psi,\vartheta)\end{pmatrix} \mathrm{e}^{in\varphi} \approx \delta(\Psi-\Psi_0)\begin{pmatrix} \hat{Y}(\vartheta) \\ \hat{Z}(\vartheta)\end{pmatrix}\mathrm{e}^{in\varphi}\,. \tag{18.81}$$

In this manner, a proper eigenvalue problem is obtained *for each magnetic/flow surface*. The collection of discrete eigenvalues obtained from this reduced one-dimensional problem in terms of $\hat{Y}(\vartheta)$ and $\hat{Z}(\vartheta)$ will map out the continuous spectra of the original two-dimensional problem in terms of $Y(\psi,\vartheta)$ and $Z(\psi,\vartheta)$.

We will omit the details of the formidable amount of manipulations needed to reduce the expressions to the compact form of the continuum equations that will be presented in the following subsection. The reductions basically consist of commuting the parallel gradient operators with equilibrium quantities while consistently exploiting the invariance properties of the magnetic/flow surfaces based on the five basic equilibrium flux functions of Section 18.2.1. Some of the crucial geometrical properties needed are presented in the paragraph in small print below. Here, the basic handicap in the derivation has been the bias of the previously derived expressions for the case of toroidal flow in the absence of poloidal flow [618]. Only when it was realized that the case of poloidal flow is fundamentally different (since it has entropy conservation built into the equilibrium equations so that the previous possibility of entropy instabilities is absent now), the necessary freedom of expression was obtained to produce the beautiful expressions (18.88)–(18.90) where poloidal flow completely changes the picture obtained previously for static, toroidally rotating, and gravitating equilibria [618].

▷ **Intrinsic properties of the magnetic/flow surfaces** A two-dimensional creature (cf. Section 17.2.3) living on a magnetic/flow surface would be able to test the validity of:

(a) relationships between the physical variables due to the flux functions of Section 18.2.1:

$$\partial M^2 = -\big(M^2/\rho\big)\,\partial\rho\,, \qquad \partial\big(\tfrac{1}{2}M^2B^2/\rho - \tfrac{1}{2}\widehat{\Omega}^2R^2 + \Phi_\mathrm{gr}\big) = -(1/\rho)\,\partial p\,,$$

$$\partial p = \big(\gamma p/\rho\big)\,\partial\rho\,, \qquad (M^2-1)\,\partial(RB_\varphi) = -RB_\varphi\,\partial M^2 - \sqrt{\rho}M\widehat{\Omega}\,\partial R^2\,, \qquad \partial\widehat{\Omega} = 0\,; \tag{18.82}$$

(b) the presence of a normal vorticity and a normal current density (i.e. non-divergence-free vorticity and current density in the two dimensions of the surface):

$$w_{\rm n} = \frac{1}{RB_{\rm p}}\,\partial(Rv_\varphi)\,; \qquad j_{\rm n} = \frac{1}{RB_{\rm p}}\,\partial(RB_\varphi) = \sqrt{\rho}Mw_{\rm n}\,; \tag{18.83}$$

(c) a relationship between the coupling factor of the matrix $\hat{\mathbf{A}}$ and the Coriolis coupling factor of the matrix $\hat{\mathbf{B}}$ occurring in the eigenvalue problem (18.88)–(18.90) below:

$$\frac{M^2B^2}{\rho}\,\partial\Big(\frac{\rho RB_\varphi}{B^2}\Big) = M\alpha - \partial(rB_\varphi)\,; \tag{18.84}$$

(d) the geodesic curvature of the magnetic field lines:

$$\kappa_{\rm g} \equiv \mathbf{b}\cdot(\nabla\mathbf{b})\cdot\boldsymbol{\pi} = -\frac{1}{RB_{\rm p}}\,\partial\Big(\frac{RB_\varphi}{B}\Big)\,; \tag{18.85}$$

(e) and, finally, the most surprising intrinsic, Gaussian, curvature of the magnetic/flow surface:

$$K \equiv \kappa_{\rm p}\kappa_\varphi = -\frac{1}{RB_{\rm p}}\,\partial\Big(\frac{1}{B_{\rm p}}\,\partial R\Big)\,, \tag{18.86}$$

where the two principal curvatures of the surface,

$$\kappa_{\rm p} \equiv R\,D^\dagger B_{\rm p} = \frac{R}{J}\Big(\frac{\partial}{\partial\Psi} - \frac{\partial}{\partial\vartheta}\frac{g_{12}}{g_{22}}\Big)\mathcal{J}B_{\rm p}\,, \qquad \kappa_\varphi \equiv B_{\rm p}DR = B_{\rm p}\Big(\frac{\partial}{\partial\Psi} - \frac{g_{12}}{g_{22}}\frac{\partial}{\partial\vartheta}\Big)R\,, \tag{18.87}$$

would be beyond his understanding, but he would highly appreciate the last part of Eq. (18.86). And so did we, when we tried to derive the equations for the transonic continua and got stuck with products of normal derivatives, that should not be there in tangential equations, until we realized the beauty of this equation.

With these geometric counterparts of Section 17.1.3, we are now prepared to analyse the continuous spectra of stationary transonic equilibra. ◁

18.3.2 Continua of poloidally and toroidally rotating plasmas★

By means of the magnetic surface localization technique described in the previous sub-section, we arrive at the following formulation. The continuous spectra of axi-symmetric plasmas, transonically rotating in both the toroidal and the poloidal direction and gravitating about a massive object, are obtained by solving the following set of ordinary differential equations on each magnetic/flow surface:

$$\hat{\mathbf{A}}\cdot\hat{\mathbf{V}} = \hat{\mathbf{B}}\cdot\hat{\mathbf{V}}\,, \qquad \hat{\mathbf{V}} \equiv \begin{pmatrix} \hat{Y} \\ \hat{Z} \end{pmatrix}, \tag{18.88}$$

where the matrix operator $\hat{\mathbf{A}}$ represents the magnetic contributions,

$$\hat{\mathbf{A}} \equiv \begin{pmatrix} F\dfrac{R^2B_{\rm p}^2}{B^2}F - (M^2-M_{\rm c}^2)B^2P^2 & -{\rm i}(M^2-M_{\rm c}^2)\dfrac{B^2}{\rho}PF\rho \\ {\rm i}\rho F(M^2-M_{\rm c}^2)\dfrac{B^2}{\rho}P & FM_{\rm c}^2B^2F + Q \end{pmatrix},$$

$$P \equiv \frac{1}{\rho}\Big[\partial\Big(\frac{\rho RB_\varphi}{B^2}\Big)\Big]\,, \qquad Q \equiv \rho\Big[\partial\Big((M^2-M_{\rm c}^2)\frac{B^2}{\rho^2}\partial\rho\Big)\Big]\,, \tag{18.89}$$

and the matrix operator $\hat{\mathbf{B}}$ represents the hydrodynamic contributions,

$$\hat{\mathbf{B}} \equiv \begin{pmatrix} (\sqrt{\rho}\,\widetilde{\omega} - FM)\dfrac{R^2 B_{\mathrm{p}}^2}{B^2}(\sqrt{\rho}\,\widetilde{\omega} - MF) & -\,\mathrm{i}\alpha\sqrt{\rho}\,\widetilde{\omega} \\ \mathrm{i}\alpha\sqrt{\rho}\,\widetilde{\omega} & (\sqrt{\rho}\,\widetilde{\omega} - FM)B^2(\sqrt{\rho}\,\widetilde{\omega} - MF) \end{pmatrix}. \quad (18.90)$$

The off-diagonal elements of the latter matrix contain the *Coriolis coupling factor*

$$\alpha \equiv 2\sqrt{\rho}\Big[R v_{\mathrm{p}} \partial\big(\arctan(B_\varphi/B_{\mathrm{p}})\big) + v_\varphi \partial R \Big] = M \frac{B^2}{R B_\varphi}\Big[\partial\Big(\frac{R^2 B_\varphi^2}{B^2}\Big)\Big] + \sqrt{\rho}\,\widehat{\Omega}\big[\partial R^2\big], \quad (18.91)$$

where we have transformed the expression involving v_{p} and v_φ to an equivalent form in terms of M and $\widehat{\Omega}$ (as we have consistently done in all of the analysis; see the above paragraph in small print for further intrinsic properties of the magnetic/flow surfaces). For economy of representation, the parallel field operator F, rather than the parallel flow operator U, has been exploited in the matrix $\hat{\mathbf{B}}$ so that the Doppler shifted frequency $\widetilde{\omega} \equiv \omega - n\widehat{\Omega}$ appears instead of ω itself. Transforming the diagonal elements of $\hat{\mathbf{B}}$ back to the original form,

$$\big(\sqrt{\rho}\,\widetilde{\omega} - FM\big)\cdots\big(\sqrt{\rho}\,\widetilde{\omega} - MF\big) = \rho(\omega - U)\cdots(\omega - U)\,, \quad (18.92)$$

clearly demonstrates that these contributions are, in fact, the hydrodynamic ones: the eigenvalue parameter ω is Doppler shifted by an amount determined by the operator U. Recall the extensive analysis of Chapter 12 on this topic.

In the matrix $\hat{\mathbf{A}}$, the critical 'cusp' value of the square of the poloidal Alfvén Mach number,

$$M_{\mathrm{c}}^2 \equiv \frac{\gamma p}{\gamma p + B^2} \quad (18.93)$$

(omitting the tilde for simplicity), has been introduced to highlight the main issue of this chapter. When $M^2 > M_{\mathrm{c}}^2$, there is a big negative definite contribution in the upper diagonal element, i.e. precisely in the location where the driving terms of all usual MHD instabilities (including ballooning modes) are to be found. Considering field-aligned modes, giving small contributions of the F operator terms, appears to be sufficient to bring out the instability mechanism: no subtle ballooning-type localizations are necessary. It is not quite that simple, though, because the Coriolis coupling factor α in the matrix $\hat{\mathbf{B}}$ not only strongly couples the two possible polarizations $\hat{Y}$ and $\hat{Z}$ of the modes (as do the off-diagonal elements of the matrix $\hat{\mathbf{A}}$), but also makes the system *non-Hermitian*. We will investigate the peculiar consequences of this fact in the explicit analytical and numerical examples of Sections 18.3.3 and 18.4.

Where does the gravitational potential reside in this formulation? It has been eliminated by making use of the invariance of the Bernoulli function, $\partial H = 0$ (see paragraph in small print in Section 18.3.1), so that it is mainly hidden in the resultant density variations over the magnetic/flow surfaces. However, those cause very significant differences in the stability properties of tokamaks and accretion discs, as we will see.

The previously derived expressions for the continuous spectra of static toroidal equilibria [215] (Section 17.2.3) are recovered in the limit of vanishing poloidal and toroidal velocities ($M \to 0$, $\widehat{\Omega} \to 0$). In that case, the functions ρ and RB_φ become constant on the magnetic surfaces and the remaining poloidal variation ∂B^2 becomes proportional to the geodesic curvature (17.37), which becomes the sole cause of coupling between the two polarizations $\hat{Y}$ and $\hat{Z}$. In the presence of poloidal flow, the expression $\partial(\rho R B_\varphi/B^2)$ no longer represents the geodesic curvature (see paragraph in small print in Section 18.3.1) and the density variations become the more important effect. The other relevant limits of static toroidal equilibria with gravity [496], toroidally rotating θ-pinches [299] and tokamaks [618] are not fully recovered from the present formulation since those cases admit equilibria with $\partial S \neq 0$, which may drive a kind of Brunt–Väisäläa instability of the continuum modes. As in the convection zone of the Sun, one might expect these instabilities to switch off when they have generated a constant entropy distribution on the magnetic/flow surfaces. Such instabilities are absent from the present analysis since the poloidal flow itself imposes the constraint $\partial S = 0$. The mentioned cases are recovered for equilibria that obey this additional constraint.

The present analysis is actually closer to the one by Hameiri and Hammer [286] for incompressible plasmas. They showed that a poloidally rotating straight θ-pinch with a non-circular cross-section has an unstable continuous spectrum for super-Alfvénic speeds, $M^2 > 1$, which is caused by parametric instability due to coupling of the continuum modes with the plasma rotation. The generalization to toroidal plasmas, and to general cylindrical plasmas with a poloidal magnetic field component [465], may be obtained from Eqs. (18.88)–(18.90) by taking the limit $\gamma \to \infty$, so that $M_c^2 \to 1$. In the incompressible limit, the hyperbolic flow regimes disappear and the elliptic flow regimes coalesce, with $M^2 = 1$ in the middle. Hence, our present instability threshold $M^2 = M_c^2$ moves to $M^2 = 1$ and the poloidal variation of the poloidal magnetic field, $\partial B_p \neq 0$ due to the non-circular cross-section, becomes the driving force of the instability.

Two coupled differential equations describing the unstable continuum modes in toroidally and poloidally rotating plasmas, like the ones described above but in a completely different representation, were obtained later by Lakhin and Ilgisonis [389]. They also describe some solutions for large aspect ratio tokamaks.

(a) Cylindrical limits of the continua With respect to the meaning of the dominant diagonal terms in the spectral equations (18.88)–(18.90), recall the asymptotic spectral structure of one-dimensional configurations in the limit of large normal 'wave numbers' (rapid variations in the direction of inhomogeneity). For static plasmas, the structure of the spectrum then proves to centre about the slow continua ω_S^2, the Alfvén continua ω_A^2, and the fast continua $\omega_F^2 \equiv \infty$, where the three MHD waves become purely polarized in the three orthogonal directions associated with the magnetic surfaces and the field lines. Considering the limit of an infinitely slender torus ($\epsilon \to 0$), or a periodic cylinder with longitudinal coordinate $\varphi = z/R_0$, the tangential dependence of the modes becomes $e^{i(m\vartheta + n\varphi)}$, where m and n do not couple, so that

$$\omega_S \equiv M_c \omega_A , \qquad \omega_A \equiv (\sqrt{\rho})^{-1} F , \quad F \equiv \mathcal{J}^{-1}(m + nq) = \epsilon(m/q + n) , \tag{18.94}$$

where F is the algebraic multiplication factor resulting from the parallel field operator F. For stationary plasmas, the left–right symmetry with respect to the origin $\omega = 0$ is lifted by the Doppler shift Ω_0, and the continuum frequencies become (see Chapter 13):

$$\Omega_S^\pm \equiv \Omega_0 \pm \omega_S , \qquad \Omega_A^\pm \equiv \Omega_0 \pm \omega_A , \qquad \Omega_F^\pm \equiv \pm\infty . \tag{18.95}$$

In plane slab geometry, $\Omega_0 \equiv \mathbf{k}\cdot\mathbf{v}$, where $\mathbf{k}$ is the horizontal wave number. In the periodic cylinder, according to Eq. (18.75), Ω_0 is the result of the parallel flow operator U:

$$\Omega_0 \equiv U \equiv M(\mathcal{J}\sqrt{\rho})^{-1}(m + nu) = n\widehat{\Omega} + M(\sqrt{\rho})^{-1} F = n\widehat{\Omega} + M\omega_A . \tag{18.96}$$

For fixed m and n, the spectrum becomes schematically as indicated in Fig. 13.3, with essential asymmetry between the continuum modes propagating in the forward (+) and in the backward (−) directions with respect to the background flow. We stress that Fig. 13.3 is essentially restricted to small inhomogeneity in 1D systems (as obtained in a thin slab) so that the different parts of the spectrum do not overlap and the continua remain purely real. Under these restrictions, monotonicity of the discrete spectra on the real axis outside the continua and separator frequencies was proved by a modification of the proof for static 1D systems [240]; see Section 13.1.3.

In the limit of infinitely large aspect ratio ($\epsilon \to 0$), all poloidal variations disappear from the matrices $\hat{\mathbf{A}}$ and $\hat{\mathbf{B}}$. This directly yields the two continua associated with singular perturbations tangential to the magnetic/flow surfaces:

(1) the *Alfvén continua* $\mathrm{A}_m^{\pm}$, with polarizations $\hat{Y} \neq 0, \hat{Z} = 0$ and frequencies

$$\big(\sqrt{\rho}\,\widetilde{\omega} - MF\big)^2 - F^2 = 0 \quad \Rightarrow \quad \omega = \Omega_{\mathrm{A}}^{\pm} \equiv \Omega_0 \pm \omega_{\mathrm{A}} = n\widehat{\Omega} + (M \pm 1)\,\omega_{\mathrm{A}}\,; \tag{18.97}$$

(2) the *slow continua* $\mathrm{S}_m^{\pm}$, with polarizations $\hat{Y} = 0, \hat{Z} \neq 0$ and frequencies

$$\big(\sqrt{\rho}\,\widetilde{\omega} - MF\big)^2 - M_{\mathrm{c}}^2 F^2 = 0 \quad \Rightarrow \quad \omega = \Omega_{\mathrm{S}}^{\pm} \equiv \Omega_0 \pm \omega_{\mathrm{S}} = n\widehat{\Omega} + (M \pm M_{\mathrm{c}})\,\omega_{\mathrm{A}}\,. \tag{18.98}$$

Note that, in the limit of vanishing magnetic field ($\omega_{\mathrm{A}} \to 0$), the matrix $\hat{\mathbf{A}}$ disappears and the matrix $\hat{\mathbf{B}}$ yields the usual hydrodynamic flow continua [108] at Ω_0, which may be considered as degeneracies of the Alfvén and slow continua in that limit. In other words: contrary to statements in the literature [559, 299, 60], *there are no separate, additional, flow continua in MHD, but they are absorbed by the Alfvén and slow continua when a magnetic field is present.* As already discussed in Section 13.1.3, this is consistent with the fact that the Alfvén and slow polarizations represent all possible degrees of freedom tangential to the magnetic/flow surfaces. Accordingly, in MHD, the frequencies Ω_0 are not continua but just act as separator frequencies, similar to the frequency ranges $\Omega_{\mathrm{s0}}^{\pm}$ and $\Omega_{\mathrm{f0}}^{\pm}$ indicated by the grey boxes with labels below the axis in Fig. 13.3.

One more fundamental issue (also discussed in Section 13.1.3) remains to be addressed. Describing plasma perturbations by means of the primitive Eulerian variables $\rho_1, \mathbf{v}_1, p_1, \mathbf{B}_1$, instead of in terms of the Lagrangian variable $\boldsymbol{\xi}$, an additional class of continua is obtained, viz. continua with entropy perturbations $S_1 \sim p_1 - (\gamma p_0/\rho_0)\rho_1 \neq 0$. This class of continua is excluded in the Frieman–Rotenberg formalism because they are not representable in terms of $\boldsymbol{\xi}$ and they just represent initial entropy perturbations that are passively carried by the plasma flow, without influencing the rest of the dynamics ($\mathbf{v}_1 = 0, \mathbf{B}_1 = 0$). Hence, these continua do not represent important physics. Nevertheless, they are extremely useful for bookkeeping purposes, as we will see below when we compare the results of our Lagrangian analysis with the numerical results from the PHOENIX code, which exploits an Eulerian representation. Hence, we also list this third class of continua:

(3) the *Eulerian entropy continua* E_m, with $\delta S \neq 0, \hat{Y} = \hat{Z} = 0$, and frequencies

$$\omega = \Omega_{\mathrm{E}} \equiv \Omega_0 = n\widehat{\Omega} + M\omega_{\mathrm{A}}\,. \tag{18.99}$$

Although their frequencies coincide with the mentioned frequencies of the hydrodynamic flow continua, they should not be confused with the latter. The flow continua are transformed into the Alfvén and slow continua when $B \neq 0$, but the Eulerian entropy continua are unaffected by the magnetic field. (Actually, the latter were even absent in the original analysis of the HD flow continua by Case [108] because he assumed incompressibility.)

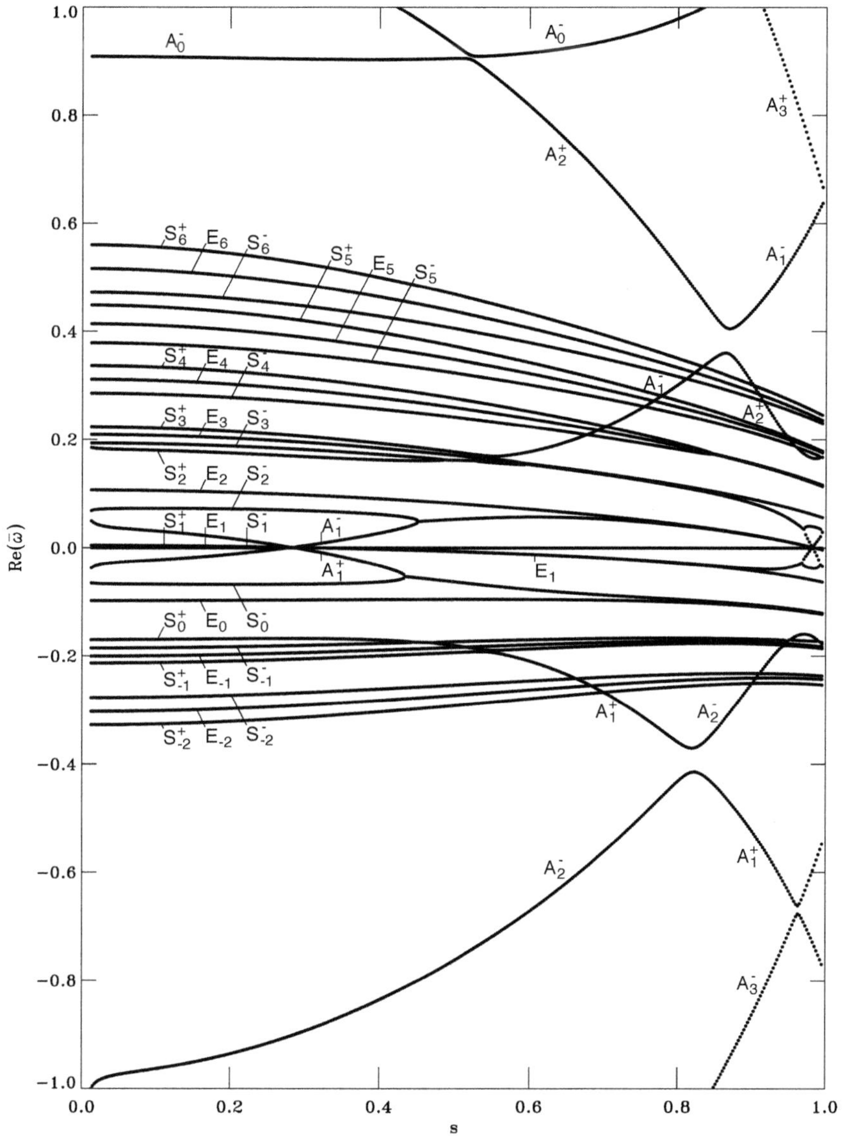

Fig. 18.5 Poloidal mode labelling and coupling of the real parts of the continuum frequencies $\bar{\omega} \equiv \epsilon^{-1}\omega$ as a function of the radial flux coordinate $s \equiv \sqrt{\psi}$ for equilibrium (A) of Fig. 18.4; toroidal mode number $n = -1$, poloidal mode numbers $-2 \leq m \leq 6$.

(b) Poloidal mode couplings We now have the necessary labels $\mathrm{A}_m^{\pm}$, $\mathrm{S}_m^{\pm}$ and E_m at our disposal to discuss the different mode couplings that occur in the MHD spectra of stationary toroidal plasmas. These couplings are caused by the poloidal variation of the quantities inside the square brackets in Eqs. (18.89) and (18.91), which will be explicitly calculated in Sections 18.3.3 and 18.4. Antici-

pating those calculations, we just present one result of a full numerical calculation with PHOENIX (or, rather, CSPHOENIX, see Section 18.4) of the continuous spectra of a tokamak equilibrium in the second elliptic flow regime, viz. equilibrium (A) of Fig. 18.4. Figure 18.5 shows the real part of the normalized continuum frequencies $\bar{\omega} \equiv \epsilon^{-1}\omega$, together with the labels of the modes wherever this makes sense (i.e. away from mode crossings). Every point in the plot (the appearance of curves is just the result of many runs of the code with small increments of s) corresponds to one constituent frequency of the continuous spectrum corresponding to an improper mode that becomes singular at the indicated position s. Note that the curves for the entropy continua E_m do not interact with the other ones, as expected, but they do provide a very convenient indication of the radial variation of Ω_0 (mainly due to $\rho(s)$ and $q(s)$), which is the central frequency for both the Alfvén and the slow modes, according to Eqs. (18.97) and (18.98). For all the numerical examples of the equilibria A–D (corresponding to Fig. 18.4), parameters have been chosen such as to agree with the trans-slow ordering (18.63), whereas the contribution $\widehat{\Omega}$ of the non-parallel toroidal flow has been chosen to be negligible so that

$$\omega_\mathrm{S}/\omega_\mathrm{A} \equiv M_\mathrm{c} \; (\approx 0.008) \;\ll\; \Omega_0/\omega_\mathrm{A} \approx M \; (\approx 0.1) \;\ll\; 1 \,. \tag{18.100}$$

As a result, the spacing $\Delta\Omega_\mathrm{E} \sim M$ of the entropy curves E_m is just intermediate between the narrow spacing $\Delta\Omega_\mathrm{S} \sim M_\mathrm{c}$ of the slow curves $\mathrm{S}_m^\pm$ and the very wide spacing $\Delta\Omega_\mathrm{A} \sim 1$ of the Alfvén curves $\mathrm{A}_m^\pm$: triplets $\mathrm{S}_m^+, \mathrm{E}_m, \mathrm{S}_m^-$ are easily distinguished, but the complete quintuplets A_m^-, $\mathrm{S}_m^+, \mathrm{E}_m, \mathrm{S}_m^-, \mathrm{A}_m^+$ are hardly visible on the scale of Fig. 18.5.

Of course, the main point of Fig. 18.5 is the illustration of how poloidal mode coupling completely upsets the near cylindrical radial variation of the triplets $\mathrm{S}_m^+, \mathrm{E}_m, \mathrm{S}_m^-$ by the crossing of the lower m Alfvén branches at the rational surfaces $q = 1$ (at $s \approx 0.3$) and $q = 2$ (at $s \approx 0.98$), where ω_A vanishes for the $m = 1$ and $m = 2$ harmonics, respectively. In a wide neighbourhood of the $q = 1$ rational surface, the slow modes $\mathrm{S}_2^\mp$ and $\mathrm{S}_0^\mp$ are pushed far apart by the $\mathrm{A}_1^\mp$ harmonics and, similarly, strong mode interaction around the $q = 2$ surface pushes the $\mathrm{S}_3^\mp$ and $\mathrm{S}_1^\mp$ modes apart by the $\mathrm{A}_2^\mp$ modes. More spectacular yet is the merging of the S_2^- and A_1^- branches (at $s = 0.46$) and of the A_1^+ and S_0^- branches (at $s = 0.43$), where the frequency $\bar{\omega}$ becomes complex and the instabilities originate that are the main subject of this chapter (the reader may want to glance at Fig. 18.9). *Through the six mode interaction* $\mathrm{S}_{m+1}^\pm$, $\mathrm{A}_m^\pm$, $\mathrm{S}_{m-1}^\pm$ *at the rational surfaces, the continuous spectra of transonically rotating plasmas become unstable.* (Fortunately, for the analysis, the slow modes $\mathrm{S}_m^\pm$, which are also degenerate at the rational surfaces, $\omega_\mathrm{S} = 0$, appear not to interact with the six modes.) We will investigate these six-mode interactions analytically in Section 18.3.3 and numerically in Section 18.4.

Another interesting feature of the continuous spectra shown in Fig. 18.5 is the avoided crossings of the A_m^+ and A_{m-1}^- and of the A_m^- and A_{m-1}^+ Alfvén modes at

$$q = -\frac{1}{n}\Big[m - \tfrac{1}{2}(1 \mp M)\Big] \,, \tag{18.101}$$

resp., i.e., for $m = 1$: $q = 1.55$ (at $s = 0.86$) and $q = 1.45$ (at $s = 0.82$). The resulting $|\Delta m| = 1$ gaps in the continuous Alfvén spectra are well-known for static tokamak equilibria. Equation (18.101) shows that poloidal rotation removes the forward/backward degeneracy of the gap frequencies with important consequences for the discrete gap modes that may be found there.

18.3.3 Analysis of trans-slow continua for small toroidicity★

(a) Fourier representation (exact) We exploit the equilibria of Section 18.2.4 to study the transonic continuous spectra by means of the formalism of Section 18.3.2. To that end, we write the matrix eigenvalue problem (18.88) as

$$\begin{pmatrix} \mathcal{K} & \mathcal{M} \\ \mathcal{N} & \mathcal{L} \end{pmatrix} \begin{pmatrix} \hat{Y} \\ \hat{Z} \end{pmatrix} = 0 \,, \tag{18.102}$$

where the elements

$$\begin{aligned} \mathcal{K} &\equiv \mathcal{J}^{-1}\big[(\widetilde{\omega} - \mathcal{J}F f_2)\, f_1\, (\widetilde{\omega} - f_2\mathcal{J}F) - \mathcal{J}F f_3\, \mathcal{J}F + f_4\big]\,, \\ \mathcal{M} &\equiv \mathcal{J}^{-1}\big[-\mathrm{i}f_5\,\widetilde{\omega} + \mathrm{i}f_6\,\mathcal{J}F + f_7\big]\,, \\ \mathcal{N} &\equiv \mathcal{J}^{-1}\big[\mathrm{i}f_5\,\widetilde{\omega} - \mathrm{i}\mathcal{J}F f_6 + f_7\big]\,, \\ \mathcal{L} &\equiv \mathcal{J}^{-1}\big[(\widetilde{\omega} - \mathcal{J}F f_2)\, f_8\, (\widetilde{\omega} - f_2\mathcal{J}F) - \mathcal{J}F f_9\, \mathcal{J}F - f_{10}\big] \end{aligned} \tag{18.103}$$

need evaluation of the following functions of ϑ:

$$\begin{aligned} &f_1 \equiv \frac{\mathcal{J}\rho R^2 B_{\rm p}^2}{B^2}\,, \quad f_2 \equiv \frac{M}{\mathcal{J}\sqrt{\rho}}\,, \quad f_3 \equiv \frac{R^2 B_{\rm p}^2}{\mathcal{J}B^2}\,, \quad f_4 \equiv \mathcal{J}(M^2 - M_{\rm c}^2)B^2U^2, \\ &f_5 \equiv \mathcal{J}\alpha\sqrt{\rho}\,, \quad f_6 \equiv (M^2 - M_{\rm c}^2)B^2U\,, \quad f_7 \equiv \mathcal{J}(M^2 - M_{\rm c}^2)\frac{B^2}{\rho}\,U\,[\partial\rho], \\ &f_8 \equiv \mathcal{J}\rho B^2\,, \quad f_9 \equiv \frac{M_{\rm c}^2 B^2}{\mathcal{J}}\,, \quad f_{10} \equiv \mathcal{J}V\,. \end{aligned} \tag{18.104}$$

Fourier analysis of Eq. (18.102) yields an infinite matrix eigenvalue problem,

$$\sum_{m'} \begin{pmatrix} K_{mm'} & M_{mm'} \\ N_{mm'} & L_{mm'} \end{pmatrix} \begin{pmatrix} Y_{m'} \\ Z_{m'} \end{pmatrix} = 0 \,, \tag{18.105}$$

where $Y_{m'}$ and $Z_{m'}$ are the Fourier components of $\hat{Y}(\vartheta)$ and $\hat{Z}(\vartheta)$. The matrix elements

$$K_{mm'} \equiv \frac{1}{2\pi}\oint d\vartheta\, \mathrm{e}^{-\mathrm{i}m\vartheta}\mathcal{J}\mathcal{K}\mathrm{e}^{\mathrm{i}m'\vartheta} \equiv \frac{1}{2\pi}\oint K(m,m';\vartheta)\mathrm{e}^{-\mathrm{i}(m-m')\vartheta}\, d\vartheta\,, \ \text{ etc.}\,, \tag{18.106}$$

involve integrands in terms of the operators $\mathcal{K}$, etc., which are converted into algebraic expressions $K(m,m';\vartheta)$, etc., by means of integration by parts:

$$\begin{aligned} K &\equiv f_1\big[\widetilde{\omega} - f_2(m+nq)\big]\big[\widetilde{\omega} - f_2(m'+nq)\big] - f_3(m+nq)(m'+nq) + f_4\,, \\ M &\equiv -\mathrm{i}f_5\,\widetilde{\omega} + \mathrm{i}f_6(m'+nq) + f_7\,, \\ N &\equiv \mathrm{i}f_5\,\widetilde{\omega} - \mathrm{i}f_6(m+nq) + f_7\,, \\ L &\equiv f_8\big[\widetilde{\omega} - f_2(m+nq)\big]\big[\widetilde{\omega} - f_2(m'+nq)\big] - f_9(m+nq)(m'+nq) - f_{10}\,. \end{aligned} \tag{18.107}$$

Obviously, M should not be confused here with the Alfvén Mach number.

Before we can continue with the evaluation of the matrix elements for small toroidicity, we need to consider the way in which the density enters the problem. Recall that ρ did not enter the equilibrium core equations (18.29) and (18.30). However, the spectral equation (18.88) manifestly depends on ρ, both through the weight factor $\sqrt{\rho}$ in front of the eigenvalue parameter $\widetilde{\omega}$ and through the angular derivative $\partial\rho$, that is crucial for the instabilities analysed here. At this point, we introduce a third dimensional unit (after length a and magnetic field strength B_0), viz. the density ρ_0 on the magnetic axis. Similar to the transformations (18.26), we will exploit ρ_0 to make all density-dependent quantities dimensionless by putting $\rho_0 = 1$ without indicating this by further subscripts:

$$\rho/\rho_0 \rightarrow \rho\,, \qquad \big(a\sqrt{\rho}/B_0\big)\,\omega \rightarrow \omega\,. \tag{18.108}$$

The radial and angular dependencies of the dimensionless density are determined by M^2 and χ'^2, according to Eq. (18.17). Whereas M^2 should be a solution of the Bernoulli equation (18.30), $\chi'^2(\psi)$ is an arbitrary function, except that its amplitude is given by the value of M^2 on the magnetic axis: $\chi'^2(0) \equiv M_0^2$. Hence, it is expedient to introduce a sixth flux function, with unit amplitude:

$$\bar{\Lambda}_0(\psi) \equiv M_0^{-2}\chi'^2(\psi)\,\Big[\equiv \bar{\Lambda}_5(\psi)/\bar{\Lambda}_5(0)\Big]\,, \qquad \bar{\Lambda}_0(0) = 1\,, \tag{18.109}$$

where we have indicated in brackets that this flux function is not really an additional free function, for accretion discs, since it has to have the same shape as $\bar{\Lambda}_5(\psi)$ according to Eq. (18.19). The density is then determined by

$$\rho = \bar{\Lambda}_0(\psi)\,(M^2/M_0^2)^{-1}\,, \tag{18.110}$$

so that the radial variation is due to both $\bar{\Lambda}_0$ and M^2, but the more important angular variation is solely due to M^2.

Since we are concerned with continuum modes in the Alfvénic range of frequencies, according to Eq. (18.94) we need to rescale ω with ϵ to get $\mathcal{O}(1)$ expressions:

$$\bar{\omega} \equiv \epsilon^{-1}\omega\ \ \big[\equiv (R_0\sqrt{\rho}/B_0)\,\omega\big]\,. \tag{18.111}$$

We have not used the smallness of ϵ yet, so that the expressions so far are exact.

(b) Expansion in small toroidicity We now apply the trans-slow poloidal flow ordering (18.63) to the expressions f_i of Eqs. (18.104). This requires conversion of the metric of Section 18.2.4 to the straight-field-line coordinates Ψ, ϑ, φ:

$$\vartheta \approx \hat{\theta} - (\epsilon\hat{r} - \Delta' - \sigma)\sin\theta\,. \tag{18.112}$$

Here, σ is the angular variation of RB_φ,

$$RB_\varphi \sim 1 + \sigma\cos\vartheta\,, \quad \sigma \approx -2\epsilon\hat{r}\,\frac{\bar{\Lambda}_{4,0}}{\bar{\Lambda}_{3,0} - \bar{\Lambda}_{4,0}}\,, \tag{18.113}$$

which we may neglect in the present case ($\bar{\Lambda}_4 \ll \bar{\Lambda}_3$). The metric tensor and the Jacobian then become

$$\begin{aligned}
&\hat{g}^{11} \equiv \psi'^2(1 - 2\Delta'\cos\theta)\,, \qquad \hat{g}^{12} \equiv -\psi'^2[\epsilon - \hat{r}^{-1}(\hat{r}\Delta')' - \sigma']\sin\vartheta\,,\\
&\hat{g}^{22} \equiv \hat{r}^{-2}[1 - 2(\epsilon\hat{r} - \Delta' - \sigma)\cos\vartheta]\,, \qquad \hat{g}^{33} \equiv \epsilon^2(1 - 2\epsilon\hat{r}\cos\theta)\,,\\
&J \equiv (\nabla\psi\times\nabla\vartheta\cdot\nabla\varphi)^{-1} \approx (\epsilon\Psi')^{-1}\hat{r}\big[1 + (2\epsilon\hat{r} - \sigma)\cos\vartheta\big]\,.
\end{aligned} \tag{18.114}$$

In these coordinates, we may remove all factors ψ' in favour of the safety factor q:

$$q \equiv JB_\varphi/R \approx \hat{r}/\Psi'\,, \qquad B_{\rm p} \approx \epsilon r q^{-1}\big[1 - (\epsilon\hat{r} + \Delta')\cos\vartheta\big]\,. \tag{18.115}$$

Finally, the most crucial quantity is the perturbed density,

$$\rho(\hat{r})[1 - (\mu/M^2)\cos\vartheta] = \rho(\hat{r})[1 - \epsilon(1-\Gamma)\hat{r}\cos\vartheta]\,, \tag{18.116}$$

where now, in contrast to the previous subsection, $\rho(\hat{r})$ only indicates the zeroth order radial density profile. The gravitational interaction coefficient Γ, defined in Eq. (18.67), determines the first order angular variation of the density.

We evaluate the factors f_i of Eqs. (18.104) to first order in terms of Δ' and Γ:

$$\begin{aligned}
f_1 &\approx (\epsilon q)^{-1}\rho\hat{r}^2\,(1+\epsilon a_1\hat{r}\cos\vartheta)\,, && a_1 \equiv 3+\Gamma-2(\epsilon r)^{-1}\Delta'\,,\\
f_2 &\approx \epsilon M(q\sqrt{\rho})^{-1}\,(1+\epsilon a_2\hat{r}\cos\vartheta)\,, && a_2 \equiv -1-\Gamma\,,\\
f_3 &\approx \epsilon\hat{r}^2 q^{-3}\,(1+\epsilon a_3\hat{r}\cos\vartheta)\,, && a_3 \equiv -2(\epsilon\hat{r})^{-1}\Delta'\,,\\
f_4 &\approx \epsilon M^2 q^{-1}\,a_4^2\hat{r}^2\sin^2\vartheta\,, && a_4 \equiv -1-\Gamma\,,\\
f_5 &\approx -\,(\sqrt{\rho}M+\epsilon^{-1}\rho\widehat{\Omega})\,a_5\hat{r}\sin\vartheta\,, && a_5 \equiv 2\,,\\
f_6 &\approx \epsilon M^2 q^{-1}\,a_6\hat{r}\sin\vartheta\,, && a_6 \equiv -1-\Gamma\,,\\
f_7 &\approx \epsilon^2 M^2 q^{-1}\,a_6 a_7\hat{r}^2\sin^2\vartheta\,, && a_7 \equiv 1-\Gamma\,,\\
f_8 &\approx \epsilon^{-1}\rho q\,(1+\epsilon a_8\hat{r}\cos\vartheta)\,, && a_8 \equiv -1+\Gamma\,,\\
f_9 &\approx \epsilon q^{-1}\,M_c^2\,, &&\\
f_{10} &\approx \epsilon^2 M^2 q^{-1}\,a_{10}\hat{r}\cos\vartheta\,, && a_{10} \equiv 1-\Gamma\,.
\end{aligned} \tag{18.117}$$

Scaling the matrix elements and the eigenvector as

$$\begin{aligned}
&\bar{K} \equiv (\epsilon\hat{r}^2)^{-1}q\,K\,,\quad \bar{M} \equiv (\epsilon\hat{r})^{-1}M\,,\quad \bar{N} \equiv (\epsilon\hat{r})^{-1}N\,,\quad \bar{L} \equiv (\epsilon q)^{-1}L\,,\\
&\bar{Y} \equiv \hat{r}q^{-1}Y\,,\quad \bar{Z} \equiv Z\,,
\end{aligned} \tag{18.118}$$

the transformed matrix eigenvalue problem (18.105) will contain the following matrix elements (accurate to first order):

$$\begin{aligned}
\bar{K}_{mm'} &\approx \Big\{\big[\sqrt{\rho}\bar{\omega} - M(m/q+n)\big]^2 - (m/q+n)^2\Big\}\delta_{mm'}\\
&\qquad\qquad + \tfrac{1}{2}M^2 a_4^2(\delta_{mm'} - \delta_{m,m'-2} - \delta_{m,m'+2})\,,\\
\bar{M}_{mm'} &\approx -\tfrac{1}{2}\Big\{Ma_5\sqrt{\rho}\,\bar{\omega} + M^2 a_6(m'/q+n)\Big\}(\delta_{m,m'-1} - \delta_{m,m'+1})\,,\\
\bar{N}_{mm'} &\approx \tfrac{1}{2}\Big\{Ma_5\sqrt{\rho}\,\bar{\omega} + M^2 a_6(m/q+n)\Big\}(\delta_{m,m'-1} - \delta_{m,m'+1})\,,\\
\bar{L}_{mm'} &\approx \Big\{\big[\sqrt{\rho}\bar{\omega} - M(m/q+n)\big]^2 - M_c^2(m/q+n)^2\Big\}\delta_{mm'}\,.
\end{aligned} \tag{18.119}$$

Since Ω is negligible in that order for the equilibria we are considering, $\widetilde{\omega} \approx \omega$, so that we do not need to distinguish the two in the normalized eigenvalue $\bar{\omega}$.

(c) 'Dispersion equation' for trans-slow Alfvén continuum modes The matrix elements (18.119) give rise to two separate coupling schemes, with either the Alfvén waves or the slow waves as the

central perturbation. Since the most important instabilities result from the first scheme, we will concentrate on that. The Alfvén wave coupling scheme

$$\begin{pmatrix} \bar{L}_{m-1,m-1} & \bar{N}_{m-1,m} & 0 \\ \bar{M}_{m,m-1} & \bar{K}_{m,m} & \bar{M}_{m,m+1} \\ 0 & \bar{N}_{m+1,m} & \bar{L}_{m+1,m+1} \end{pmatrix} \begin{pmatrix} \bar{Z}_{m-1} \\ \bar{Y}_m \\ \bar{Z}_{m+1} \end{pmatrix} = 0\,, \tag{18.120}$$

precisely provides us with the means to analyse the interaction of the six modes $\mathrm{A}_m^\pm$, $\mathrm{S}_{m-1}^\pm$ and $\mathrm{S}_{m+1}^\pm$ that we encountered in Fig. 18.5 of Section 18.3.2. In the neighbourhood of the Alfvén resonance $m/q + n \sim \epsilon$ this gives a 'dispersion equation' in terms of a_4, a_5, a_6. Inserting the expressions E160, and defining a rescaled eigenvalue $\hat{\omega} \equiv (\sqrt{\rho}q/M)\,\bar{\omega}$ and a measure $Q \equiv (n/M)(m/n+q)$ for the 'distance' to the rational value of q, the 'dispersion equation' becomes

$$\begin{vmatrix} (\hat{\omega}+1)^2 & q\big[\hat{\omega}+\tfrac{1}{2}(1+\Gamma)\big] & 0 \\ q\big[\hat{\omega}+\tfrac{1}{2}(1+\Gamma)\big] & \hat{\omega}^2 - Q^2 + \tfrac{1}{2}q^2(1+\Gamma)^2 & -\,q\big[\hat{\omega}-\tfrac{1}{2}(1+\Gamma)\big] \\ 0 & -\,q\big[\hat{\omega}-\tfrac{1}{2}(1+\Gamma)\big] & (\hat{\omega}-1)^2 \end{vmatrix} = 0\,, \tag{18.121}$$

which turns out to be a cubic equation in $\hat{\omega}^2$,

$$\hat{\omega}^6 - \big[2+Q^2+\tfrac{1}{2}q^2(1-\Gamma)(3+\Gamma)\big]\hat{\omega}^4 + \big[1+2Q^2+\tfrac{1}{2}q^2(1-\Gamma)(1+3\Gamma)\big]\hat{\omega}^2 - Q^2 = 0\,, \tag{18.122}$$

that may be solved by means of the Cardano expressions.

At a rational surface, $Q^2 = 0$ and we get two marginal modes $\hat{\omega}^2 = 0$ and four solutions

$$\hat{\omega}^2 = 1 + \tfrac{1}{4}q^2(1-\Gamma)(3+\Gamma) \pm q|1-\Gamma|\sqrt{1+\tfrac{1}{16}q^2(3+\Gamma)^2}\,. \tag{18.123}$$

These correspond to four stable waves ($\hat{\omega}^2 > 0$) when $\Gamma < 1$, but two solutions may become exponentially unstable ($\hat{\omega}^2 < 0$) when $\Gamma > 1$. Clearly, $\Gamma = 1$ marks a qualitative change in the continuous spectrum, corresponding to the qualitative change in the equilibria given by Eq. (18.66) and illustrated in Fig. 18.4.

The solutions $\bar{\omega} \equiv M(\sqrt{\rho}q)^{-1}\hat{\omega}$ of Eq. (18.122) are plotted as a function of q in Figs. 18.6 and 18.7 for four representative values of Γ, roughly corresponding to the equilibria (A) and (C) of Fig. 18.4, for which the 'exact' numerical continuous spectra will be discussed in Section 18.4 (Figs. 18.8–18.11). Starting with $\Gamma = 0$ (tokamak, i.e. no gravitating central object, Fig. 18.6(a)), the six-mode interaction $\mathrm{S}_2^\pm$, $\mathrm{A}_1^\pm$, $\mathrm{S}_0^\pm$ describes the real part of the 'exact' spectrum shown in Fig. 18.5 quite well around the $q = 1$ resonance (where $\omega_{\mathrm{A}} = 0$ for $n = -1$, $m = 1$). Away from the resonance (increasing Q^2), the curves $\mathrm{Re}\,(\bar{\omega})$ of the pair S_2^-, A_1^- and of the pair A_1^+, S_0^- coalesce (at the same value of q: the asymmetry present in the 'exact' solution of Fig. 18.5 is due to higher order contributions that are neglected here) and an imaginary contribution $\mathrm{Im}\,(\bar{\omega})$ emerges (indicated in red): *the trans-slow continuous spectrum becomes overstable*. The growth rate of the instability is roughly M times the real part of the Alfvén frequency: quite sizeable! The spectrum for an accretion disc with moderate central mass (roughly corresponding to equilibrium (B) with $\Gamma = 0.25$, not shown) is similar to that of the tokamak, but has smaller imaginary parts (corresponding to the reduction of the density variation expressed by Eq. (18.66)). This tendency continues until $\Gamma = 1$, which is completely stable (Fig. 18.6(b)): Eq. (18.122) degenerates into two

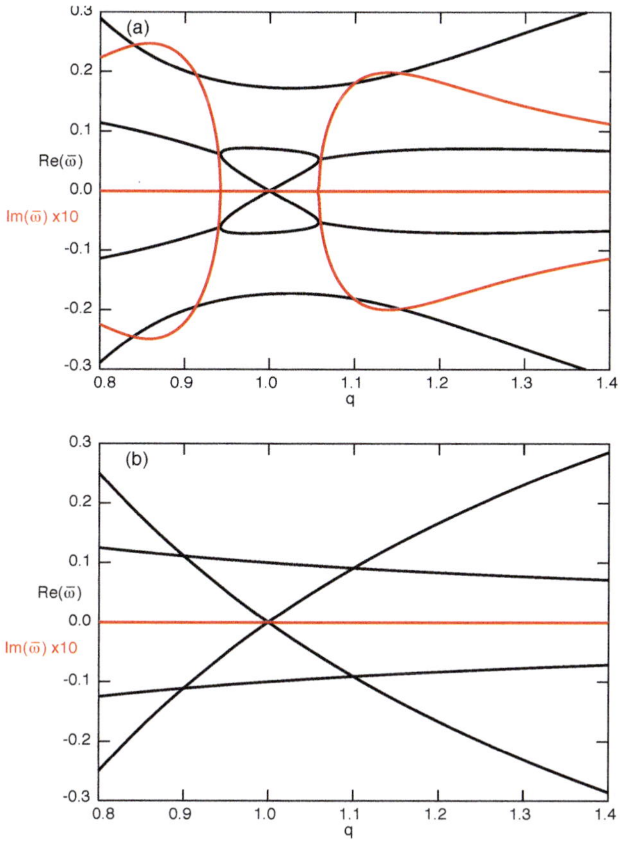

Fig. 18.6 Real and imaginary part of the roots of the 'dispersion equation' (18.122) for the trans-slow Alfvénic continuous spectrum of (a) tokamak ($\Gamma = 0$) and (b) Keplerian accretion disc ($\Gamma = 1$) for $n = -1, m = 1, M = 0.1, \rho = 1$.

solutions $\bar{\omega} = \pm(\sqrt{\rho})^{-1}(m/q + n)$ corresponding to the Alfvén modes $A_1^{\pm}$ and four solutions $\bar{\omega} = \pm(\sqrt{\rho}q)^{-1}M$ corresponding to the degenerate slow modes $S_2^{\pm}$, $S_0^{\pm}$ (since the contribution of M_c has been neglected). Beyond $\Gamma = 1$, an entirely different spectral structure emerges. For $\Gamma = 2$ (roughly corresponding to equilibrium (C), Fig. 18.7(a)), far away from the resonance, two real and four complex roots (i.e. four overstable modes) are found. In the neighbourhood of the resonance $q = 1$, the complex roots coalesce two by two to produce two purely imaginary roots, i.e. instabilities with maximum growth rate at the resonance. These growth rates are enormous, of the same order of magnitude as the Alfvén frequency!

Note that the solution $\hat{\omega}^2 = \hat{\omega}^2(Q, q, \Gamma)$ of the 'dispersion equation' (18.122) involves the equilibrium parameters q, M, Γ, but that Q also brings in the mode numbers n and m in a physically significant manner. This is shown by taking the limits

$$|n| \to \infty\,, \quad \frac{m}{n} + q \to 0\,, \quad \text{for fixed} \quad Q \equiv \frac{n}{M}\Big(\frac{m}{n} + q\Big)\,, \tag{18.124}$$

which implies that the instabilities of the plots shown, that are obtained for certain values of the

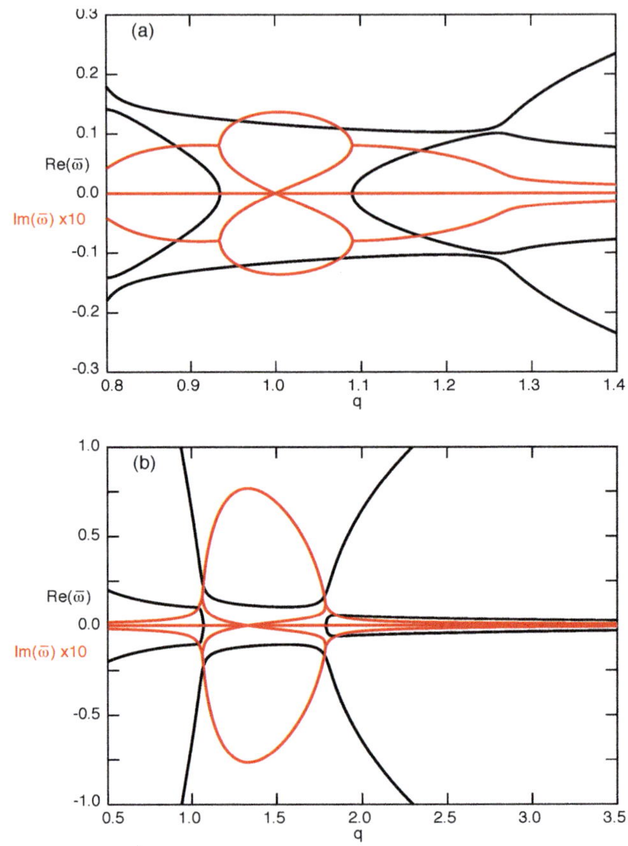

Fig. 18.7 Real and imaginary part of the roots of the 'dispersion equation' (18.122) for the trans-slow Alfvénic continuous spectrum of accretion discs with massive central object: (a) $\Gamma = 2$ for $n = -1$, $m = 1$ and (b) $\Gamma = 10$ for $n = -3, m = 4$; $M = 0.1, \rho = 1$.

parameter Q, will be obtained for any sextuple $\mathrm{A}^{\pm}_{m}, \mathrm{S}^{\pm}_{m-1}, \mathrm{S}^{\pm}_{m+1}$: *all toroidal mode numbers n are unstable close to or at the rational surfaces $q = -m/n$ for trans-slow ($M^2 > M_c^2$) Alfvénic continuum modes*. Their growth rates are just determined by the values of the equilibrium parameters, i.e. the density, ρ, the inverse rotational transform of the magnetic field lines, q, the poloidal Alfvén Mach number of the flow, M, and the strength of the gravitational interaction, Γ, at that particular location.

To demonstrate this, we have plotted the roots of the 'dispersion equation' for mode numbers corresponding to a rational surface, at $q = 4/3$, for $\Gamma = 10$ (Fig. 18.7(b)). In that case, the very massive central object drives the growth of the instability to values far in excess of the Alfvén frequency. We obtain the following estimate of the asymptotic growth rate for $\Gamma \gg 1$ from Eq. (18.122):

$$\bar{\omega} \approx \pm \mathrm{i}\, M\Gamma/\sqrt{2\rho}\,. \tag{18.125}$$

In this limit, the modes become dominantly Alfvénic ($|Y_m| \gg |Z_{m-1}|, |Z_{m+1}|$) and the growth rates are so large that nonlinear effects soon should take over. Hence, these modes might produce the levels of MHD turbulence in thick magnetically dominated accretion discs with trans-slow poloidal velocities that are needed to break the co-moving constraint of the poloidal flow and magnetic field in order to produce accretion and jets.

18.4 Trans-slow continua in tokamaks and accretion discs★

In this section, we turn to a full numerical study of the trans-slow Alfvénic continuous spectrum for the equilibria (A)–(D) of Fig. 18.4, where the gravitational strength parameter $\Gamma_0 \equiv \Gamma(0)$ is increased from 0 (tokamak) to 2 (thick accretion disc with massive central object). The effect of flattening of the cross-section of the disc was also studied, but it is not shown here (see [230]).

We exploit the numerical programs FINESSE [44] and PHOENIX [230, 74], developed in the context of MHD spectroscopy for laboratory and astrophysical plasmas [237], relying on the techniques discussed in Chapters 15–17. The first code computes the equilibrium of an axi-symmetric plasma with transonic rotations in both the toroidal and poloidal direction in the presence of the gravitational field of a central object. The discretization techniques for the flux equation are the same as exploited in the HELENA code [319] for static equilibria. However, the additional implementation of the Bernoulli equation required a careful monitoring of the parameters of the new flux functions $\bar{\Lambda}_i(\psi)$ to remain in a particular elliptic flow regime. The second code, PHOENIX, computes the complete ideal or resistive MHD spectrum of this equilibrium, exploiting an Eulerian representation so that the Eulerian entropy continua are computed as well, as we have seen in Section 18.3.2. This spectral code is a further development of the CASTOR code [360] for static plasmas, with the addition of the rotational and gravitational contributions and implementation of the powerful Jacobi–Davidson algorithm [553] for scanning the spectrum of non-Hermitian operators in the complex ω-plane. Here, the resistivity was switched off. For the computation of the continuous spectrum, the auxiliary program CSPHOENIX was exploited. Through a special limiting procedure [500] (also exploited to construct Fig. 17.5), this program uses the same discretized matrix elements as PHOENIX, but evaluated on a single magnetic/flow surface only. In this manner, we obtain the continuous spectrum of transonic equilibria, characterized by the flux functions $\bar{\Lambda}_i(\psi)$, in the form of plots of $\mathrm{Re}(\bar{\omega})$ and $\mathrm{Im}(\bar{\omega})$ as a function of the radial flux coordinate $s \equiv \sqrt{\psi}$.

These plots are shown in two ways, viz. (1) as separate plots of $\mathrm{Re}(\bar{\omega})$ and $\mathrm{Im}(\bar{\omega})$ versus s or q (Figs. 18.8 and 18.10); (2) as plots in the complex $\bar{\omega}$ plane, with s as parameter, in colour coding (Figs. 18.9 and 18.11). We have restricted the analysis to one toroidal harmonic ($n = -1$), with a band of nine coupling poloidal harmonics ($-2 \leq m \leq 6$), and a flux range that usually contains only one rational surface $q = 1$ (except for the tokamak case of Fig. 18.8, which also contains the $q = 2$ rational surface). The numerical results clearly exhibit the central six-mode interaction analysed in Section 18.3.3, but without neglect of the higher order terms in M^2 and M_c^2. Hence, asymmetries arise in the continuous spectra and many more unstable roots become visible. When comparing the numerical values of $\bar{\omega}$ shown in Fig. 18.8 with the analytical solutions of the 'dispersion equation' (18.122) shown in Fig. 18.5, one should realize that, in the latter plots, only q varies with the radial position whereas the equilibrium quantities ρ, M^2 and Γ have been kept constant. Of course, in the numerical solutions of the present section, the full 2D dependencies of the equilibrium quantities have been used. When this is taken into account, the quantitative agreement between the two approaches is excellent.

18.4.1 Tokamaks and magnetically dominated accretion discs★

The continuous spectra for the tokamak equilibrium (A), where $\Gamma_0 = 0$, are shown in Fig. 18.8. Figure 18.8(a) represents the central interaction region of the larger plot already shown in Fig. 18.5 with the mode labels. The imaginary parts of the continuum frequencies in Fig. 18.8(b) show the sudden onset of instability, with complex values of $\bar{\omega}$ (i.e. overstability), at $s \approx 0.43$–0.45 that we have already interpreted as due to the six-mode interaction $\mathrm{S}_2^{\pm}, \mathrm{A}_1^{\pm}, \mathrm{S}_0^{\pm}$.

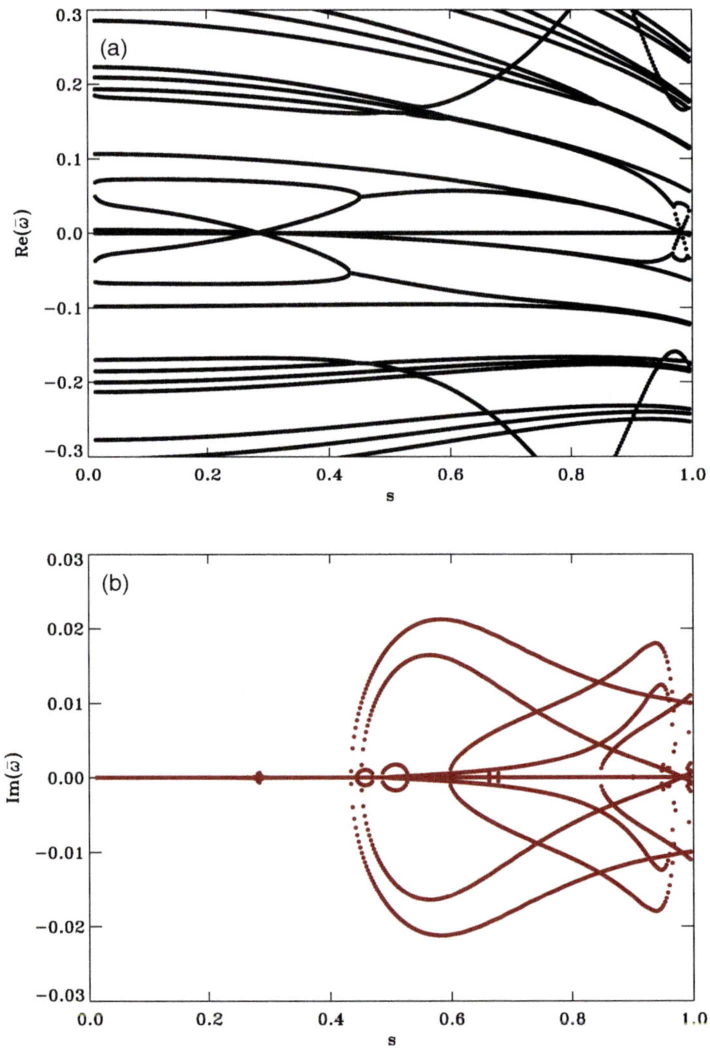

Fig. 18.8 (a) Real and (b) imaginary parts of the frequencies $\bar{\omega}$ of the trans-slow continuous spectrum as a function of the radial flux coordinate $s \equiv \sqrt{\psi}$ for a tokamak with circular cross-section, equilibrium (A); $n = -1$.

Some additional structure is to be seen in Fig. 18.8(b). Going from $s = 0.28$ ($q = 1$) to the right

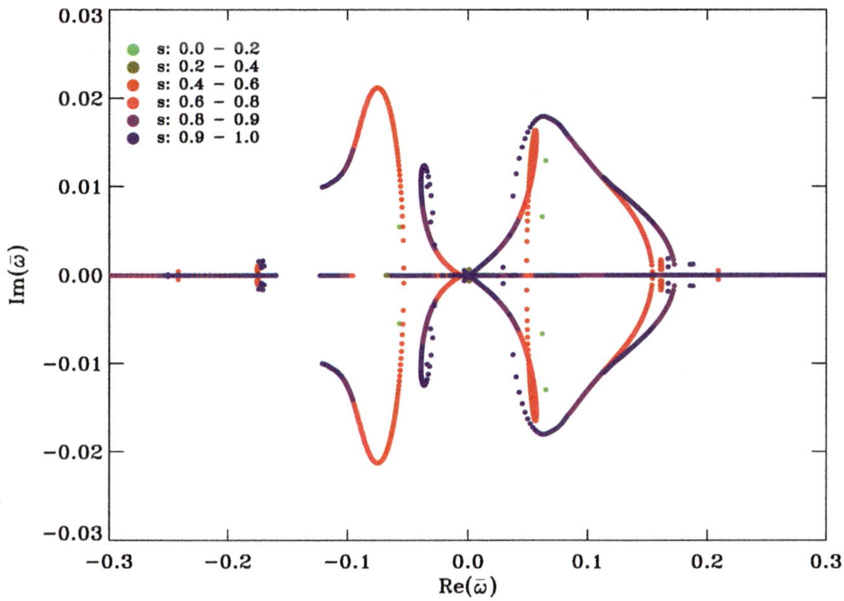

Fig. 18.9 Complex spectrum corresponding to Fig. 18.8, equilibrium (A).

in this figure, we subsequently encounter the following features.

(a) The degenerate six-mode interaction analysed in Section 18.3.3 splits apart into the three-mode interaction A_1^+, $S_0^\pm$, which becomes unstable at $s = 0.43$ and remains unstable in the rest of the interval, and the three-mode interaction A_1^+, $S_2^\pm$, which becomes unstable at $s = 0.45$ and becomes stable again at $s = 0.98$ ($q = 2$). Due to this lifting of the degeneracy, the first triple is more unstable than the second one;

(b) Just to the right of the onset of these instabilities, Re($\bar{\omega}$) of the slow constituent S_0^+ of the first triple coalesces with that of S_{-1}^+ to produce a secondary instability with Im($\bar{\omega}$) $\sim 10^{-3}$ (small circle starting at $s = 0.44$). Similarly, a secondary instability is produced by the S_2^+ constituent of the second triple, which gets entangled in a higher order slow mode interaction with S_3^- (bigger circle starting at $s = 0.48$);

(c) Moving further to the right, we encounter the two triples of the six modes associated with the $q = 2$ rational surface, viz. the triple A_2^+, $S_3^\pm$, which becomes unstable at $s = 0.60$, has its maximum growth rate Im($\bar{\omega}$) ≈ 0.018 at $s = 0.94$, and becomes stable again just left of $s = 0.98$ ($q = 2$), and the triple A_2^+, $S_1^\pm$ with the smaller growth rate, which becomes unstable earlier and also becomes stable again to the left of $q = 2$;

(d) Finally, the Re($\bar{\omega}$)'s of the slow modes $S_4^\pm$ coalesce at $s = 0.85$ and the first unstable triple A_3^+, $S_4^\pm$ of the $q = 3$ rational surface is encountered. Even though that rational surface falls far outside the domain, the effects of the associated six-mode interaction $S_4^\pm$, $A_3^\pm$, $S_2^\pm$ are clearly noticed.

In Fig. 18.9, the composite continuous spectrum is shown in the complex $\bar{\omega}$-plane with the radial position s in colour coding. The plot clearly exhibits the cause of the lifting of the degeneracy of the six-mode interaction, viz. asymmetry with respect to left and right propagation of the continuum modes. Here, left and right are to be interpreted as follows. For $m > 0$ and $n < 0$, lines of constant phase $\Phi \equiv m\vartheta + n\varphi - \omega t$ in the φ–ϑ plane (i.e. the magnetic/flow surface) move to the left when $\mathrm{Re}(\bar\omega) < 0$ and to the right when $\mathrm{Re}(\bar\omega) > 0$. Hence, the triples A_1^+, $\mathrm{S}_0^\pm$ (L) and A_2^-, $\mathrm{S}_1^\pm$ (L) correspond to continuum modes that rotate leftward, and the triples A_2^+, $\mathrm{S}_3^\pm$ (R) and A_1^-, $\mathrm{S}_2^\pm$ (R) correspond to continuum modes that rotate rightward.

Whereas the frequencies of the continuous spectrum can be calculated one-by-one for each magnetic/flow surface separately, without the radial dependencies of $\mathrm{Re}(\bar\omega)$ and $\mathrm{Im}(\bar\omega)$ shown in the plots, it would have been extremely difficult to make sense of all these interactions. In the remaining plots, for the equilibrium (C), we use $q(\psi)$ instead of s as the radial flux parameter.

18.4.2 Gravity dominated accretion discs*

We have seen in Section 18.3.3 that the structure of the continuous spectra changes dramatically at $\Gamma = 1$. In order to avoid misunderstanding about this transition, note that (1) the function $\Gamma(\psi)$ will not be constant in realistic equilibria, so that a condition $\Gamma = 1$ can, in general, be satisfied only at one particular magnetic/flow surface, leaving the other surfaces unstable to either $\Gamma < 1$ or $\Gamma > 1$ types of instabilities; (2) the condition $\Gamma = 1$, though associated with Keplerian rotation, does not represent a privileged equilibrium state in the case of transonic poloidal flows. It just implies that the poloidal density variation changes from tokamak-like (with density maxima on the inside of the torus) to gravity-dominated (with density maxima on the outside), as indicated by Eq. (18.66) and illustrated in Fig. 18.4. Rather than unduly complicating the discussion by presenting the instabilities of the continuous spectrum for equilibria having a local $\Gamma = 1$ condition, we move on to gravity-dominated equilibria with $\Gamma > 1$ everywhere. We skip presentation of the spectra of equilibria (B) and (D) which can be found in Ref. [230].

The continuous spectra of the gravity-dominated accretion disc equilibrium (C), where $\Gamma_0 = 2$, are shown in Fig. 18.10. Comparison with the solutions of the dispersion equation of Section 18.3.3 shows that the spectrum does, in fact, exhibit the six-mode instability structure characteristic of $\Gamma > 1$ equilibria, with much increased growth rates: $\mathrm{Re}(\bar\omega) \sim \mathrm{Im}(\bar\omega)$ now. Again, the left–right symmetry of the rotating modes is broken in the numerical equilibrium, as clearly demonstrated by the asymmetric (with respect to $\mathrm{Re}(\bar\omega) = 0$) spectrum of Fig. 18.11. New for this class of equilibria is that, moving to the rational surface $q = 1$, the real part of the frequencies of the leftward and rightward rotating continuum modes decreases until there is *mode locking* at $q = 1$ when $\bar\omega$ becomes imaginary: a purely exponentially growing instability is found with a huge growth rate, in agreement with the estimate given in Eq. (18.125). At this point, the calculation might be repeated for $n = -2$, $n = -3$, etc., but the analysis of Section 18.3.3 already indicates that all of these modes will be unstable with approximately the same growth rate and localized at the respective rational surfaces $q = -m/n$.

Finally, the continuous spectra of the gravity-dominated accretion disc equilibrium (D), with $\Gamma_0 = 2$ and elliptical cross-section ($b/a = 0.5$), were also computed (not shown here, see [230]). The mode-locked instabilities found in equilibrium (C) are now completely swamped by other instabilities that grow an order of magnitude faster. The instabilities remain to be analysed in detail, but it is already clear that flattening of the disc makes the trans-slow Alfvén continuum modes even more unstable. From these results it appears likely that the instabilities in flat accretion discs involve localization at the most curved sections of the magnetic/flow surfaces, i.e. on the inside and on the outside of the torus. Together with the finding of Section 18.3.3 that all higher toroidal harmonics n with their associated poloidal harmonics m are unstable, as expressed by Eqs. (18.124) and (18.125), this shows that the instabilities of the trans-slow Alfvén continua do not require that the magnetic/flow surfaces are closed, but just that they exist. In general accretion flows on open magnetic/flow surfaces, and the same instabilities will operate as the ones found here for closed surfaces, in particular on the highly curved inside where jet formation is assumed to take place.

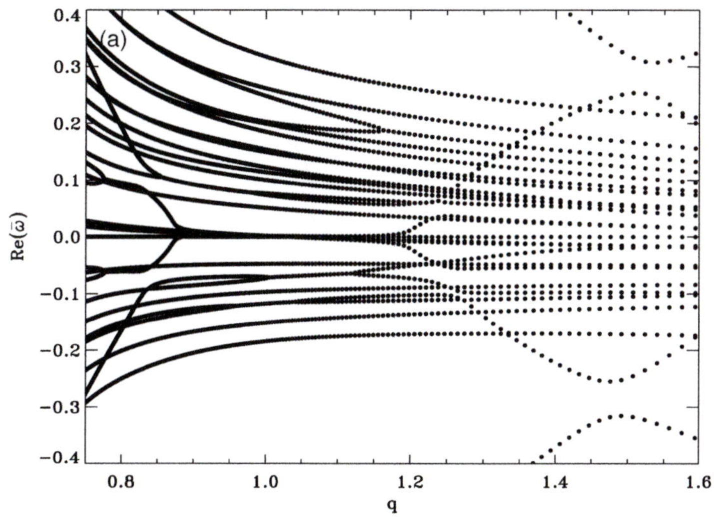

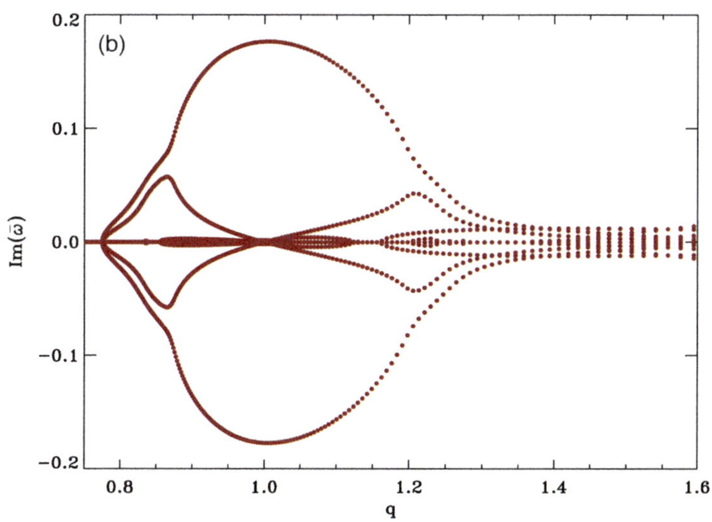

Fig. 18.10 (a) Real and (b) imaginary parts of the frequencies $\bar{\omega}$ of the trans-slow continuous spectrum as a function of $q(\psi)$ for an accretion disc with circular cross-section, $\Gamma_0 = 2$, equilibrium (C); $n = -1$.

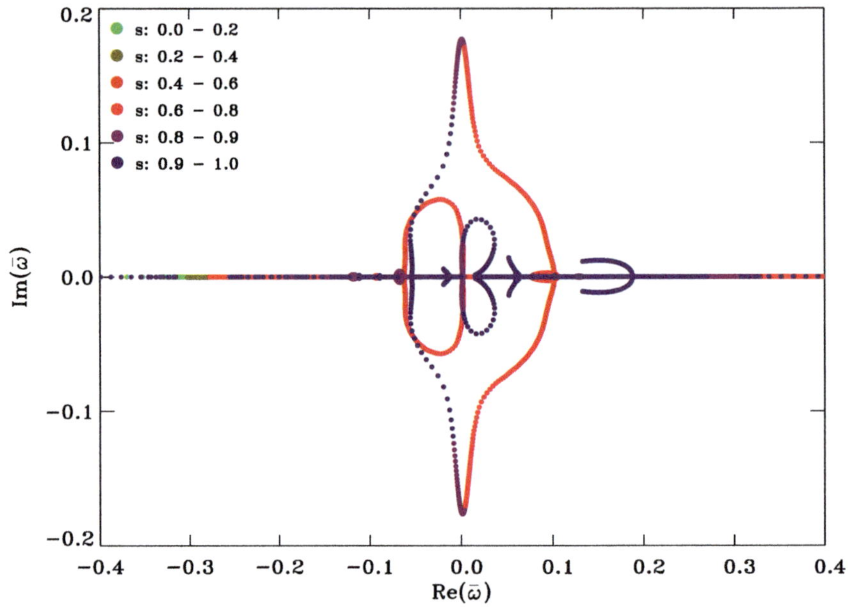

Fig. 18.11 Complex spectrum corresponding to Fig. 18.10, equilibrium (C).

18.4.3 Trans-slow Alfvén continuum instabilities

In summary, we have studied 'transonic' axi-symmetric MHD equilibria and stability of toroidal plasmas without (tokamak) and with (accretion disc) the gravitational field of a massive central object (Fig. 18.1). 'Transonic' here refers to the poloidal flow speed, measured by the value M^2 of the poloidal Alfvén Mach number squared defined in Eq. (18.17), when it surpasses the characteristic slow magneto-sonic values M_c^2 and M_s^2, the Alfvén value 1 or the fast magneto-sonic value M_f^2. For the validity of the analysis, the hyperbolic flow regimes $M_c^2 < M^2 < M_s^2$ and $M^2 > M_f^2$ of Fig. 18.3 should be excluded. This leaves three elliptic flow regimes, called slow ($0 < M^2 < M_c^2$), trans-slow ($M_s^2 < M^2 < 1$), and fast ($1 < M^2 < M_f^2$). We have studied the trans-slow regime in detail since it is the first 'transonic' one encountered when increasing the poloidal flow speed.

The equilibria are characterized by five flux functions, where we have eliminated misunderstanding in the literature about the nature of the poloidal vorticity–current density stream function $K(\psi)$, defined in Eqs. (18.8) and (18.16), previously associated with the specific angular momentum. These functions are rescaled in Eqs. (18.19) and (18.28) to facilitate the analysis of the core equations (18.29) and (18.30) for ψ and M^2 when the centrifugal and gravitational forces approximately balance. The analysis is done both by means of an extension of the low-β tokamak ordering, called the trans-slow ordering defined in Eq. (18.63), and by numerical means, exploiting the FINESSE code [44]. The former yields the second order ODE (18.58) for the Shafranov shift $\Delta(\hat{r})$ and the first order ODE (18.59) for the toroidal perturbation μ of M^2. The structure of the ODE for Δ (in terms of M^2) turns out to be completely analogous to the structure of the Hain–Lüst equation [276] (in terms of the eigenfrequency parameter ω^2), in agreement with recent insight on transonic MHD [222]. The trans-slow equilibria computed with FINESSE (Fig. 18.4) are characterized by the relative strength $\Gamma(\psi)$ of the gravitational interaction, defined in Eq. (18.67), where $\Gamma \ll 1$ refers to tokamak-like equilibria and $\Gamma \gg 1$ to accretion discs with a very massive central object.

Next, we derived the equations (18.88)–(18.90) governing the radial asymptotics of the slow and Alfvén continuous spectra associated with perturbations of a single magnetic/flow surface. They clearly exhibit a major change of the coupling between Alfvén and slow modes when the critical value $M^2 = M_c^2$ is surpassed. Again, the analysis of the continuous spectra is performed both by means of an expansion in small inverse aspect ratio of the equations (18.88)–(18.90) for the Lagrangian displacement vector and by means of the full spectral code PHOENIX [230, 74] exploiting the primitive Eulerian variables. This enabled labelling of all spectral curves obtained (Fig. 18.5) with perfect agreement between the Lagrangian analysis and the Eulerian numerics for the coupling between the Alfvén and slow continuum modes, defined in Eqs. (18.97) and (18.98), whereas the PHOENIX code adds the Eulerian entropy continuum modes, defined in Eq. (18.99), that do not couple to the other modes.

The small inverse aspect ratio expansion yields the dispersion equation (18.121) for the coupling of six modes, labelled $\mathrm{A}^{\pm}_{m}$, $\mathrm{S}^{\pm}_{m-1}$, and $\mathrm{S}^{\pm}_{m+1}$, which yield unstable Alfvénic continuum modes in the neighbourhood of rational surfaces when $M^2 > M_c^2$. This is demonstrated in Figs. 18.6 and 18.7 for different values of Γ. For accretion discs with a very massive central object ($\Gamma \gg 1$), the instability growth rates become huge (larger than the Alfvén frequency) when the toroidal and poloidal mode number n and m are large. These results are confirmed by the PHOENIX code where no assumption on the aspect ratio needs to be made. Figures 18.8–18.11 show the complex frequencies of the trans-slow Alfvén continuum modes for the equilibria depicted in Fig. 18.4. Increasing the gravitational parameter Γ from 0 (tokamak) to 2 (accretion disc) and changing the cross-sectional shape from circular to elliptical ($b/a = 0.5$), the growth rate increases from fractions of the Alfvén frequency to exceeding it significantly.

In conclusion, one of the important issues in accreting flows about massive central objects is to explain how there can be accretion of magnetized plasma, and even ejection of jets, at all. If one takes the achievements of tokamak research seriously, plasma flow and magnetic field should remain confined near the toroidal magnetic/flow surfaces. Some violent dissipative mechanism appears to be needed to break this constraint. Obviously, for tokamaks, such a mechanism is not desired and, if it were found to operate under fusion conditions, one would certainly need to find ways to eliminate it. Fortunately, two important features of stationary flows in toroidal plasmas are absent in tokamaks but dominant in accretion discs, viz. the existence of large, 'transonic', poloidal flows and the presence of a strong gravitational field. Since the MHD description of plasmas is independent of the size of the magnetic configuration, it appeared to be attractive to find a class of equilibria common to tokamaks and thick accretion discs, but distinctly different with respect to their stability properties. We have found such a class, as described in this chapter.

The instabilities found are due to the continuous spectrum, which is stable for the usually sub-slow poloidal flow speeds in tokamaks, but which becomes unstable when the poloidal flow surpasses the critical threshold associated with the slow magneto-sonic cusp speed. The instability becomes explosive when the gravitational field is increased to the values encountered in accretion discs about massive black holes.

For magnetically dominated thick accretion discs, the trans-slow poloidal flow ordering turns out to be quite effective to describe the instabilities. It assumes low β and sub-Alfvénic poloidal flow of the following orders of magnitude:

$$\beta \ll M^2 \equiv \frac{\rho v_{\rm p}^2}{B_{\rm p}^2} \left(\sim \frac{\rho v_\varphi^2}{B_\varphi^2} \sim \frac{\rho G M_*}{R B_\varphi^2} \right) \ll 1 \tag{18.126}$$

(where the estimates in brackets are for flow parallel to the magnetic field, $\widehat{\Omega} = 0$, and for Keplerian toroidal rotation, $\Gamma = 1$, resp.). Under these conditions, the continuous spectrum becomes unstable over most of the plasma volume. The instability is switched on whenever the poloidal Alfvén Mach number exceeds the critical value $M_{\rm c}^2 \approx \frac{1}{2}\gamma\beta$. Hence, it appears to be unavoidable in accretion flows onto a compact object. At this point, it may be objected that we have not treated an actual accretion flow, but just a poloidal rotation. However, the discussion of the dispersion equation' in Section 18.3.3 shows that the instabilities operate for *all* mode number pairs n, m, including the higher ones. The latter represent localization on the magnetic/flow surfaces independent of whether or not these surfaces are closed or open, or whether the flow is directed towards or away from the gravitational centre. Whereas the use of state-of-the-art numerical tools for the computation of equilibrium and perturbations required us to stay away from the hyperbolic flow regimes, it appears not too farfetched now to assume that transition of the critical value $M^2 = M_{\rm c}^2$ in accretion flows will turn on *the trans-slow Alfvén continuum instabilities* precisely at the location of that transition. That would be the location where the flow becomes detached from the magnetic field due to a high level of MHD turbulence. Obviously, this conjecture needs further investigation by means of nonlinear numerical codes, along the lines of recent efforts by various authors (see [294], [324], [109](b) and references quoted there).

We have found a very wide class of local instabilities of the magnetic/flow surfaces of transonic axi-symmetric plasmas, which may be called *trans-slow Alfvén continuum (TSAC) modes*. These instabilities would exclude operation of tokamaks in the trans-slow poloidal flow regimes (recall the caveat of Section 18.1 though), except possibly for a thin outer layer in divertor-operated fusion machines. For magnetically dominated accretion tori, these instabilities may provide a route to MHD turbulence and associated anomalous dissipation, breaking the co-moving condition of plasma and magnetic field, completely independent of the magneto-rotational instability.

18.5 Literature and exercises*

Notes on literature

Equilibrium of rotating axi-symmetric plasmas

- Zehrfeld and Green, 'Stationary toroidal equilibria at finite beta' [665].
- Morozov and Soloviev, 'Steady-state plasma flow in a magnetic field', in *Reviews of Plasma Physics*, Vol. 8 [448].
- Hameiri, 'The equilibrium and stability of rotating plasmas' [283].

Transonic MHD equilibria and instabilities

- This chapter is based on the paper 'Unstable continuous spectra of transonic axi-symmetric plasmas' by Goedbloed et al. [230], where more examples and the numerical data of the equilibria can be found.

Transonic two-fluid equilibria

- A generalization to the equations for transonic two-fluid equilibria was initiated by McClements and

Thyagaraja [434]. The equations were cast into the form of a general variational principle with a Lagrangian for seven fields (electron and ion densities and stream functions, magnetic flux, electric potential and gravitational potential) and the 'Bernoulli nightmare' for these equations was pointed out by Goedbloed [223]; see also the comment [596] and the response [224].

Exercises

[18.1] *Thin plasma slab: equilibrium*

In Exercises [16.1] and [17.4] you have exploited the thin plasma slab approximation for the equilibrium as well as the perturbations. Here, you will apply this approximation to the generalized Grad–Shafranov equation (18.20). This equation describes the axi-symmetric equilibrium of a tokamak or accretion disc.

- Explain why you cannot use the utmost right expressions of Eq. (18.23).
- Show that the generalized Grad–Shafranov equation (18.20) in the thin slab approximation reduces to

$$\frac{d}{dR}\Big(p + \tfrac{1}{2}B^2\Big) = \rho\Big(\frac{v_\varphi^2}{R} - \frac{GM_*}{R^2}\Big).$$

What does this equation describe?

[18.2] *Thin plasma slab: continuous spectrum*

In the previous exercise, you have looked at the equilibrium in the thin plasma slab approximation. Now, you do the same for equations (18.88) describing the continuous spectra for axi-symmetric plasmas with both toroidal and poloidal flow. Show that these equations decouple in this approximation and that the eigenvalues are the Doppler shifted Alfvén and slow frequencies.

[18.3] *Connection with a static tokamak plasma: equilibrium*

In this exercise, you will show that the generalized Grad–Shafranov equation (18.20) reduces to the ordinary Grad–Shafranov equation (16.81) in the limit of no gravity and no toroidal and poloidal flow. As you will see, these limits have to be taken with special care.

- Show that the limit of no poloidal flow is the same as $M^2 \to 0$.
- Show that

$$\lim_{M^2\to 0} \frac{\Pi_1}{M^2} = \lim_{M^2\to 0} \frac{\Pi_2}{M^{2\gamma}}.$$

- Now take a look at the Bernoulli equation (18.21). Show that this equation reduces to a trivial equation in the no poloidal flow limit.
- Before further reducing the generalized Grad–Shafranov equation (18.20), take a closer look at Π_1 and Π_3. Show that Π_1 reveals that the temperature becomes a flux function in the no flow and no gravity limit.
- Again, using the no flow and no gravity limit, show that Π_3 can be written as

$$\Pi_3 = \tfrac{1}{2}{\chi'}^2 (K/R)^2.$$

- Notice that the limits of no gravity and no toroidal flow can be taken without any problem. However, the no poloidal flow limit is a very special one. In this limit, the poloidal Alfvén Mach number also vanishes, which upsets an assumption of the generalized Grad–Shafranov equation (18.20). Show that

$$\lim_{M^2\to 0} \frac{1}{{\chi'}^2}\frac{d{\chi'}^2}{d\Psi} = 0.$$

- Show that the generalized Grad–Shafranov equation (18.20) reduces to the ordinary Grad–Shafranov equation (16.81). What extra assumption do you have to use?

[18.4] *Connection with a static tokamak plasma: continuous spectrum*

In this exercise, you will find the connection between the equation (18.88) for the continua in rotating plasmas and the corresponding equation (17.54) for the static case. Reduce the first equation to the latter one by setting the flow to zero. What extra assumption do you have to use to get this result? (Hint: What are the flux functions for a static axi-symmetric equilibrium?)

[18.5] *Dispersion equation*

Reproduce the plots shown in Figs. 18.6 and 18.7. This can be done using IDL, Matlab or a similar program. For the computation of the roots, you can use the Cardano expressions or Laguerre's method. Make the program such that you can easily use both of them.

- Make a function which computes the coefficients of the dispersion equation (18.122) for given poloidal mode number m, toroidal mode number n, poloidal Alfvén Mach number M, gravitational interaction Γ and safety factor q.
- Make a function which computes the roots using the Cardano expressions. The input of this function should be the coefficients of a third order polynomial.
- Do the same for Laguerre's method. If you use IDL or Matlab, this method is part of their standard library, so you can skip this task.
- Make a function which finds the roots of the dispersion equation (18.122). The input for this should be the poloidal mode number m, the toroidal mode number n, the poloidal Alfvén Mach number M, the density ρ, the gravitational interaction Γ, the minimum and maximum value of the safety factor q and the method to be used. The function should return the roots found.
- Reproduce the plots shown in Fig. 18.6 and 18.7.
- Explain why the first plot of Fig. 18.7 does not look the same as the plots shown in Fig. 18.10.

Part VI

Nonlinear Dynamics

19

Turbulence in incompressible magneto-fluids

In this chapter, we focus on incompressible MHD evolutions, where the governing nonlinear equations dictate solenoidal ($\equiv$ divergence-free) flow and magnetic field topologies. As a preliminary to the full nonlinear magnetohydrodynamic system, we first revisit the hydrodynamic (HD) equations, reduced to the incompressible case. Various formulations differ in their representation of the solenoidal velocity field. We intentionally discuss a wave-like view on the inviscid Euler system, even though the absence of (sound) waves distinguishes incompressible from compressible gas dynamics. The energy conservation equation forms our starting point to discuss the Kolmogorov scaling law, introducing the concept of an inertial range across which energy cascades. Pure planar hydrodynamic evolutions have additional constraints, making 2D versus 3D incompressible turbulence in fluids radically different. We illustrate this with early and contemporary numerical examples, going from turbulence evolutions in fully periodic settings to hints of singular structure formation in ideal Euler flow simulations.

In incompressible MHD, we touch on differences between ideal versus visco-resistive evolutions. We point out kinematic and full MHD insights into magnetic field amplification. The Elsässer formulation of the governing equations allows us to generalize insights from linear to nonlinear wave package behaviour, which prominently appear in MHD turbulence theories. We restrict ourselves to dimensional reasoning to introduce the various scaling laws predicted for the energy spectra. High resolution (pseudo-)spectral simulations have given important clues to the anisotropic nature of MHD turbulence. We also discuss numerical evidence for singular structure growth in incompressible MHD and for small-scale dynamo action by summarizing selected simulation-based studies. We end the chapter with an introduction to compressible MHD, where the linear wave picture is richer and allows wave steepening, paving the way to shock-dominated plasma behaviour.

19.1 Incompressible hydrodynamics preliminaries

19.1.1 The incompressible hydro model

The incompressible HD equations provide a continuum description of the macroscopic dynamics of a fluid, with density ρ, velocity vector $\mathbf{v}$ and pressure p as variables entering the PDE system. The incompressibility constraint leads directly to $\nabla \cdot \mathbf{v} = 0$, which is a topological constraint on the possible morphologies of streamlines. They can form closed loops, cover surfaces or volumes ergodically, or simply start and end at infinity. This constraint impacts the formulation of mass conservation, which normally allows the local density value to change in two distinct ways: as a result of local compressions or decompressions, or through flow-advected density gradients, since

$$\frac{\partial \rho}{\partial t} = -\rho \nabla \cdot \mathbf{v} - \mathbf{v} \cdot \nabla \rho \,. \tag{19.1}$$

Assuming incompressibility implies that one only retains advection, making

$$\frac{\mathrm{D}\rho}{\mathrm{D}t} \equiv \frac{\partial \rho}{\partial t} + \mathbf{v} \cdot \nabla \rho = 0 \,, \tag{19.2}$$

where $\mathrm{D}/\mathrm{D}t$ is the Lagrangian derivative. This means that the density is advected as a passive scalar quantity, although different fluid elements could still assume different density values. In many incompressible hydro studies, this is usually simplified even further, by adopting a spatio-temporally constant $\rho(\mathbf{x}; t) = \rho_0$. In what follows, we will adopt this restriction to constant density conditions, leaving the constant ρ_0 in all formulas. The governing equations then reduce to a set of nonlinear PDEs for the velocity $\mathbf{v}(\mathbf{x}; t)$, where the pressure $p(\mathbf{x}; t)$ also appears:

$$\frac{\partial \mathbf{v}}{\partial t} + \mathbf{v} \cdot \nabla \mathbf{v} = -\rho_0^{-1} \nabla p + \nu \nabla^2 \mathbf{v} + \mathbf{F}_{\mathrm{driv}} \,, \qquad \nabla \cdot \mathbf{v} = 0 \,. \tag{19.3}$$

We have introduced the viscous diffusion term $\nu \nabla^2 \mathbf{v}$, which is omitted in the inviscid Euler case. This term employs the kinematic viscosity coefficient ν, which has the dimension $[\ell^2 t^{-1}]$ (where ℓ stands for length and t for time). Although in reality this coefficient is dependent on the local fluid properties, we will assume a constant ν throughout. The momentum equation as formulated above contains the combined effects of fluid inertia (with the nonlinear term $\mathbf{v} \cdot \nabla \mathbf{v}$) and pressure gradients, while additionally accounting for an imposed driver $\mathbf{F}_{\mathrm{driv}}$ which may have certain statistical properties.

Although the pressure p appears in the governing incompressible hydro system, one should note that it plays a mathematically peculiar role. Obviously, pressure must be positive, $p \geq 0$, but in addition we can operate with $\nabla \cdot$ on the momentum equation to find that it obeys

$$\rho_0^{-1} \nabla^2 p = -\nabla \cdot (\mathbf{v} \cdot \nabla \mathbf{v}) + \nabla \cdot \left(\nu \nabla^2 \mathbf{v} + \mathbf{F}_{\mathrm{driv}} \right) . \tag{19.4}$$

Hence, the scalar pressure distribution is instantaneously determined from the velocity field by the solution of the above Poisson equation. Therefore, the three components of the velocity, constrained by solenoidality, are the main unknowns to solve for incompressible HD scenarios. This can also be illustrated by the following change of variables. Start from taking the curl of the momentum equation (19.3) and introduce the vorticity vector $\boldsymbol{\omega} \equiv \nabla \times \mathbf{v}$. Noting that $\nabla \cdot \boldsymbol{\omega} = 0$, and using the vector identity $\mathbf{v} \cdot \nabla \mathbf{v} = \nabla \frac{1}{2} v^2 - \mathbf{v} \times (\nabla \times \mathbf{v})$, one then obtains the vorticity evolution equation in the form

$$\frac{\partial \boldsymbol{\omega}}{\partial t} - \nabla \times (\mathbf{v} \times \boldsymbol{\omega}) = \nu \nabla \times \nabla^2 \mathbf{v} + \nabla \times \mathbf{F}_{\mathrm{driv}} \,. \tag{19.5}$$

Due to the assumed incompressibility, $\nabla \times (\mathbf{v} \times \boldsymbol{\omega}) = \boldsymbol{\omega} \cdot \nabla \mathbf{v} - \mathbf{v} \cdot \nabla \boldsymbol{\omega}$. and exploiting the vector identity $\nabla^2 \mathbf{v} = -\nabla \times (\nabla \times \mathbf{v})$, since $\nabla \cdot \mathbf{v} = 0$, we can rewrite the dissipative term as $\nu \nabla \times \nabla^2 \mathbf{v} = \nu \nabla^2 \boldsymbol{\omega}$. Gathering these formulas, the governing equations become:

$$\frac{\partial \boldsymbol{\omega}}{\partial t} - \boldsymbol{\omega} \cdot \nabla \mathbf{v} + \mathbf{v} \cdot \nabla \boldsymbol{\omega} = \nu \nabla^2 \boldsymbol{\omega} + \nabla \times \mathbf{F}_{\mathrm{driv}} \,,$$

$$\nabla \cdot \mathbf{v} = 0 \quad \Rightarrow \quad \boldsymbol{\omega} = \nabla \times \mathbf{v} \,. \tag{19.6}$$

Note, in particular, the absence of the pressure p and of the density ρ_0.

19.1.2 Two-dimensional formulations

Further equivalent formulations can be worked out, restricting the fluid flow to a purely 2D, planar, flow field $\mathbf{v} = v_x(x, y; t)\mathbf{e}_x + v_y(x, y; t)\mathbf{e}_y$, obeying

$$\frac{\partial v_x}{\partial x} = -\frac{\partial v_y}{\partial y} . \tag{19.7}$$

Then, the vector vorticity only involves a scalar function ω:

$$\boldsymbol{\omega} = \omega(x, y; t)\mathbf{e}_z . \tag{19.8}$$

Since this implies $\boldsymbol{\omega} \cdot \nabla \equiv \omega \partial/\partial z = 0$, whereas $\mathbf{v} \cdot \nabla\omega = v_x \partial\omega/\partial x + v_y \partial\omega/\partial y$, one can write the 2D incompressible Navier–Stokes equations in the so-called vorticity–velocity formulation (where we now drop the driver contribution):

$$\begin{aligned} &\frac{\partial \omega}{\partial t} + v_x \frac{\partial \omega}{\partial x} + v_y \frac{\partial \omega}{\partial y} = \nu \nabla^2 \omega , \\ &\nabla^2 v_x = -\frac{\partial \omega}{\partial y} , \qquad \nabla^2 v_y = \frac{\partial \omega}{\partial x} . \end{aligned} \tag{19.9}$$

Equivalently, a general means to ensure incompressibility can be written as

$$\mathbf{v} = \nabla \times \boldsymbol{\psi} . \tag{19.10}$$

Again, the simplification to 2D reduces the flow vector potential $\boldsymbol{\psi} = \psi(x, y; t)\mathbf{e}_z$ to a scalar streamfunction ψ, so that $v_x = \partial\psi/\partial y$ and $v_y = -\partial\psi/\partial x$. The contours of constant ψ represent streamlines. Without the driver term, the 2D incompressible Navier–Stokes equations in vorticity–streamfunction formulation can be written as

$$\frac{\partial \omega}{\partial t} + v_x \frac{\partial \omega}{\partial x} + v_y \frac{\partial \omega}{\partial y} = \nu \nabla^2 \omega , \quad \omega = -\nabla^2 \psi . \tag{19.11}$$

These 2D incompressible hydro equations have been widely used in the literature, adopting various spatio-temporal discretization strategies for numerical solutions. Such approaches benefit from the knowledge of conserved or invariant quantities, to quantify possible discretization errors. For the 2D incompressible case, the inviscid Euler flows must obey

$$\frac{\mathrm{D}\omega}{\mathrm{D}t} = 0 , \tag{19.12}$$

stating that the vorticity of a fluid element is conserved during its motion.

19.1.3 'Wave' analysis for incompressible Euler

We now perform a rather unusual analysis on the incompressible fluid equations. Incompressibility means that there are no sound waves, as we can formally obtain the incompressible HD system as the peculiar limit of the compressible HD system where we allow the sound speed to become infinite. We therefore anticipate that no waves exist, but we will linearize the incompressible Euler equations (without the driver contribution). This starts from

$$\frac{\partial \mathbf{v}}{\partial t} + \mathbf{v} \cdot \nabla \mathbf{v} = -\rho_0^{-1} \nabla p , \qquad \nabla \cdot \mathbf{v} = 0 , \tag{19.13}$$

and adopts a pressure split up as $p = p_0 + p_1(\mathbf{x}, t)$ and a velocity field split into $\mathbf{v} = \mathbf{v}_0 + \mathbf{v}_1(\mathbf{x}, t)$, where we take a uniform p_0 and homogeneous uni-directional flow $\mathbf{v}_0$. As usual, linearization considers only small perturbations $p_1 \ll p_0$, along with a 'small' flow field $\mathbf{v}_1$ (although flows have no natural reference strength anymore, which in the compressible gas case was identifiable with the sound speed). One finds that the zeroth order is trivial, whereas the equations in first order yield

$$\frac{\partial \mathbf{v}_1}{\partial t} + \mathbf{v}_0 \cdot \nabla \mathbf{v}_1 = -\rho_0^{-1} \nabla p_1 \,, \qquad \nabla \cdot \mathbf{v}_1 = 0 \,. \tag{19.14}$$

Inserting a plane wave decomposition, with wave vector $\mathbf{k}$ and frequency ω (not to be confused with the earlier scalar vorticity!), the perturbed variables can be written as

$$\mathbf{v}_1 = \sum_{\mathbf{k}} \hat{\mathbf{v}}_{\mathbf{k}} \, e^{i(\mathbf{k}\cdot\mathbf{x} - \omega t)} \,, \qquad p_1 = \sum_{\mathbf{k}} \hat{p}_{\mathbf{k}} \, e^{i(\mathbf{k}\cdot\mathbf{x} - \omega t)} \,. \tag{19.15}$$

Inserting into the linearized equations, the individual Fourier modes decouple so that it is convenient to go to Fourier space, where $\partial/\partial t \to -i\omega$ and $\nabla \to i\mathbf{k}$:

$$(\omega - \mathbf{k} \cdot \mathbf{v}_0) \, \hat{\mathbf{v}}_{\mathbf{k}} = \mathbf{k} \hat{p}_{\mathbf{k}} / \rho_0 \,, \qquad \mathbf{k} \cdot \hat{\mathbf{v}}_{\mathbf{k}} = 0 \,. \tag{19.16}$$

The last equation expresses orthogonality of the wave vector $\mathbf{k}$ and the Fourier amplitude $\hat{\mathbf{v}}_{\mathbf{k}}$ of the perturbation. Projecting the first relation onto $\hat{\mathbf{v}}_{\mathbf{k}}$, one finds

$$(\omega - \mathbf{k} \cdot \mathbf{v}_0) \, \hat{v}_{\mathbf{k}}^2 = 0 \,. \tag{19.17}$$

The only solution with a non-trivial velocity perturbation $\hat{v}_{\mathbf{k}} \neq 0$ is found at

$$\omega = \mathbf{k} \cdot \mathbf{v}_0 \,, \tag{19.18}$$

i.e. a single frequency at the constant Doppler shift. Without background flow, $\mathbf{v}_0 = 0$, we see that only the marginal (zero) frequency is allowed: no waves exist in incompressible hydrodynamics.

However, we may look at it differently, and try to picture the result using familiar plane wave concepts. Plane waves have a phase velocity $\mathbf{v}_{\mathrm{ph}} \equiv (\omega/k)(\mathbf{k}/k)$, and denoting the angle between $\mathbf{v}_0$ and $\mathbf{k}$ as α, this becomes

$$\mathbf{v}_{\mathrm{ph}} = v_0 \cos\alpha \, \mathbf{k}/k \,. \tag{19.19}$$

In this artificial 'wave' view of our trivial result (19.19), we may envision how a point perturbation (i.e. a δ-function localized incompressible 'wave package') launches 'plane waves' in all directions, and obtain the phase diagram as the locus of end points of the phase vector $\mathbf{v}_{\mathrm{ph}}$ for all angles α. This becomes a circle of diameter v_0, see Fig 19.1(a).

The group speed of a wave-package is found from $\mathbf{v}_{\mathrm{gr}} \equiv \partial\omega/\partial\mathbf{k}$, so that

$$\mathbf{v}_{\mathrm{gr}} = \mathbf{v}_0 \,. \tag{19.20}$$

Carrying through this 'wave' view to the extreme, we may use the Huygens' construction to find the loci of constructive interference of all wave fronts, which is the group diagram. Of course, our trivial case renders this merely into a displaced point perturbation, as indeed we are over-interpreting the translation due to the constant flow $\mathbf{v}_0$. In this 'wave' view on advected local perturbations, we should point out that actually two 'wave' solutions exist, both at the constant Doppler shift $\omega = \mathbf{k} \cdot \mathbf{v}_0$. This distinction comes from recognizing that there is still a freedom in the polarization of the velocity perturbation $\hat{\mathbf{v}}_{\mathbf{k}}$, which must obey $\mathbf{k} \cdot \hat{\mathbf{v}}_{\mathbf{k}} = 0$. Graphically this is shown in

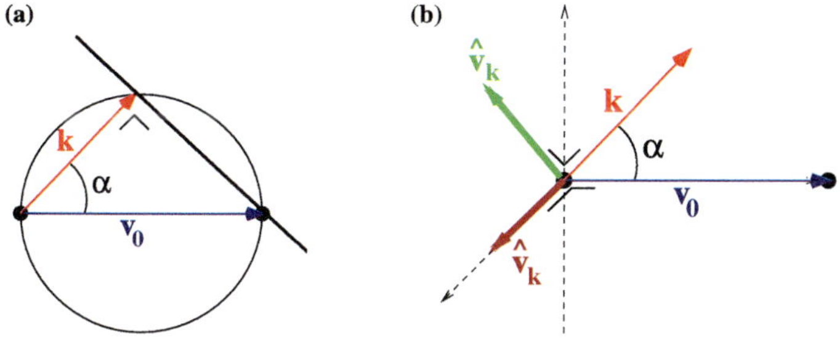

Fig. 19.1 (a) Wave diagram view on how a point perturbation is advected with constant speed $\mathbf{v}_0$; (b) Freedom in polarization for incompressible perturbations orthogonal to the wave vector $\mathbf{k}$.

Fig 19.1(b), where we envision two 'shear' waves, either in or perpendicular to the wave vector-flow plane. As a final note on this peculiar excursion into 'wave' theory, we point out that the number of solutions obtained counts right: we appeared to start from the $3+1$ equations (19.13) but, since the last equation is a constraint, this actually counts for 2 solutions. They represent incompressible transverse translations (encountered earlier in Chapter 5, Section 5.1) and they remind us how things may change for more general fluids, like magneto-fluids described by MHD.

19.1.4 Energy equation and Kolmogorov scaling

To make the link to the most prominent effect of nonlinearity in the governing equations and its role in the transition to turbulent flow, we consider again the full momentum equation (19.3). We can reduce this vector relation by projecting it onto $\mathbf{v}$ and manipulating it to the scalar equation

$$\frac{\partial}{\partial t}\left(\tfrac{1}{2}v^2\right) + \mathbf{v}\cdot\nabla\left(\tfrac{1}{2}v^2 + \rho_0^{-1}p\right) = \nu\mathbf{v}\cdot\left(\nabla^2\mathbf{v}\right) + \mathbf{v}\cdot\mathbf{F}_{\rm driv}\,. \tag{19.21}$$

In the inviscid Euler case without forcing, the right hand side vanishes and the relation yields the kinetic energy conservation law,

$$\frac{\partial}{\partial t}\left(\tfrac{1}{2}v^2\right) + \nabla\cdot\left[\left(\tfrac{1}{2}v^2 + \rho_0^{-1}p\right)\mathbf{v}\right] = 0\,, \tag{19.22}$$

from which the energy flux can be read off. We can then use dimensional arguments on both this energy conservation law and the momentum equation to introduce the concept of energy transfer across an inertial length scale range.

To do so, consider a typical velocity magnitude v_0 and a length scale l_0 as a typical length over which the flow varies. By comparing terms in the momentum equation, in particular the nonlinear term $\mathbf{v}\cdot\nabla\mathbf{v}$ with the dissipative term $\nu\nabla^2\mathbf{v}$, we can introduce the dimensionless ratio of terms known as the Reynolds number,

$$R_{\rm e} = v_0 l_0/\nu\,. \tag{19.23}$$

Further, we just demonstrated that the energy flux expression goes like $\nabla\cdot\left(\tfrac{1}{2}v^2\mathbf{v}\right)$. If we now consider this flux at a length scale l_k (or wavelength $l_k = 2\pi/k$ corresponding to wave number k), where the scale-specific velocity is v_k, we may estimate this expression by

$$\nabla\cdot\left(\tfrac{1}{2}v^2\mathbf{v}\right) \sim (v_k/l_k)\,v_k^2\,. \tag{19.24}$$

The basic idea in the seminal Kolmogorov turbulence theory [374] is the following hypothesis. Suppose that a continuous energy input (i.e. a continuous driving $\mathbf{F}_{\rm driv}$) happens at a macroscopic scale $l_0 = 2\pi/k_0$, then energy transfer will be possible across a range of length scales l_k, such that the energy flux becomes independent of the length scale, up to a small length scale l_ν where ultimately dissipation matters. This simple argument implies the existence of an intermediate length scale/wave number region, the 'inertial range' where neither driving nor dissipation is relevant and only a kind of self-similar energy transfer is operative. This constancy of energy flux is written as

$$\epsilon \sim (v_k/l_k)\, v_k^2 \,, \tag{19.25}$$

where (v_k/l_k) represents the inverse of eddy-turnover time at scale l_k, v_k^2 represents the energy in eddies at scale l_k and ϵ represents the energy flux. For constant energy flux ϵ, we get $v_k \sim \epsilon^{1/3} k^{-1/3}$ so that the kinetic energy itself obeys

$$v_k^2 \sim \epsilon^{2/3} k^{-2/3} \,. \tag{19.26}$$

Now we must realize that this energy is expressed in Fourier space, so that it relates to the actual energy $E(k)$ in physical space through

$$v_k^2 \sim \epsilon^{2/3}\, k^{-2/3} \sim \int_k^{k+\Delta k} E(k')\, dk' \,. \tag{19.27}$$

Differentiating with respect to k, while ϵ is constant, yields

$$E(k) \sim \epsilon^{2/3}\, k^{-5/3} \,, \tag{19.28}$$

which is the famous $-5/3$ scaling in the inertial range due to the Russian mathematician Andrey Kolmogorov.

This inertial range lies intermediate between the large scale l_0 (at which continuous driving feeds energy into the system) and the small scale l_ν where dissipation matters. The latter scale l_ν can be quantified as follows: l_ν is found where the nonlinear term balances diffusion, so that $l_\nu v_\nu \sim \nu$. Since $v_k \sim \epsilon^{1/3} k^{-1/3}$, we find

$$l_\nu \sim \nu^{3/4}\, \epsilon^{-1/4} \,. \tag{19.29}$$

Schematically, turbulent self-similar energy transfer then gives an energy spectrum which is as sketched in Fig. 19.2 on a log–log scale.

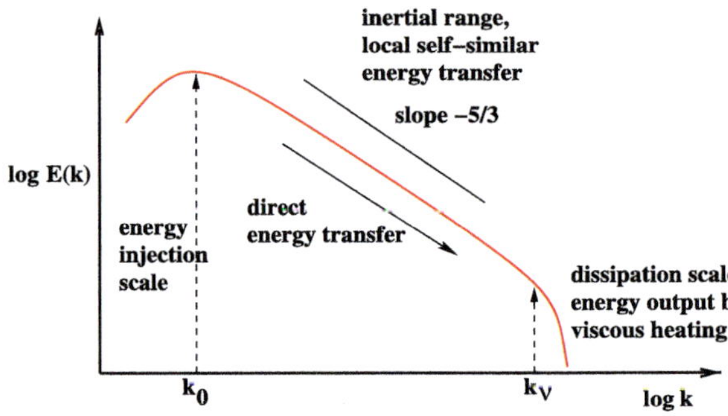

Fig. 19.2 Energy spectrum with typical Kolmogorov inertial range behaviour.

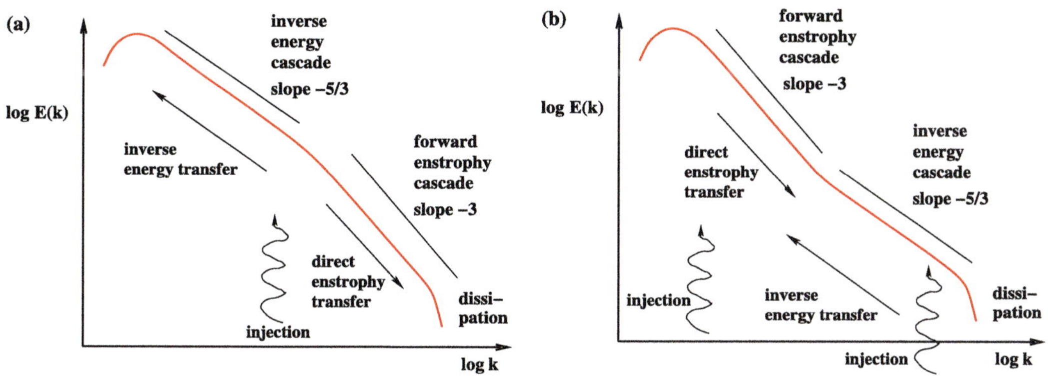

Fig. 19.3 Variations of a typical incompressible fluid energy spectrum, where (a) single to (b) multiple energy injection length scales can lead to both inverse and forward cascade ranges.

While Kolmogorov's result is seminal in turbulence theories, the direct energy transfer from large to small length scales, or 'direct cascade', is actually not always applicable. In fact, incompressible fluid dynamics behaves rather differently when 2D, versus 3D, turbulence is at play. While fluid turbulence always invokes a 'cascading' of energy across an inertial length scale range, the direction of this transfer can be from large length scales to small ones (as in the usual direct cascade assumed thus far), or it can manifest itself as an 'inverse cascade' where energy is transferred from smaller to larger length scales. The latter possibility occurs in practice for purely 2D fluid turbulence. This difference between 3D and 2D turbulent flows is intricately related to the vorticity equation which, for the general inviscid 3D case, can be written as

$$\frac{\mathrm{D}\boldsymbol{\omega}}{\mathrm{D}t} = \boldsymbol{\omega} \cdot \nabla \mathbf{v} \,. \tag{19.30}$$

We indicated earlier that the 2D case further simplifies due to the vanishing of the right hand side, to express preservation of the fluid element vorticity: $\mathrm{D}\omega/\mathrm{D}t = 0$. In contrast, the 3D case allows a fluid element to increase its vorticity. Indeed, suppose in the above equation that the vectorial vorticity $\boldsymbol{\omega}$ is oriented along z. If a fluid element then undergoes an instantaneous vertical velocity increase with z (i.e. $\partial v_z/\partial z > 0$), corresponding to a vertical stretching, the fluid element vorticity will grow. This effect of vortex stretching is only possible in 3D, and it is the main reason for the distinction of 3D versus 2D turbulence.

Another argument to distinguish 3D versus 2D turbulence stresses the fact that the 2D case has another invariant besides the energy conservation in (19.22), with the latter shown previously for the inviscid non-driven case. Vorticity conservation together with incompressibility yields the following conservation law for incompressible 2D flows:

$$\frac{\partial}{\partial t}\left(\tfrac{1}{2}\omega^2\right) + \nabla \cdot \left(\tfrac{1}{2}\omega^2 \mathbf{v}\right) = 0 \,. \tag{19.31}$$

This law is the conservation of enstrophy $\Omega = \iint \omega^2 \, dx \, dy$ (holding exactly for inviscid non-driven cases on doubly periodic domains), which is additional to the conservation of total kinetic energy $\frac{1}{2} \iint v^2 \, dx \, dy$.

▷ **Exercise** Use dimensional reasoning invoking a constant enstrophy flux χ, to derive that this leads to an inertial range scaling where $E(k) \sim \chi^{2/3} k^{-3}$. ◁

Therefore, while the energy spectrum is still following $E(k) \sim k^{-5/3}$, 2D turbulence is found to have an inverse energy cascade (from small to large) and a direct enstrophy cascade (from large to small) [377](b), as schematically shown in Fig. 19.3. Depending on the presence of single or multiple energy injection length scales, and their position with respect to the large (system size) and the small (dissipative) scales, different parts of the spectrum may exhibit different slopes.

This peculiar behaviour of planar fluid flows has meanwhile been confirmed both experimentally and numerically. 2D flow patterns with turbulent small-scale features (e.g. introduced by stirring) will tend to self-organize themselves into larger scale patterns, where the pertinent nonlinearity involved is mediated by vortex interactions and mergers on a variety of temporal and spatial scales.

19.1.5 Selected numerical examples

To illustrate some aspects of incompressible HD evolutions, later to be contrasted with MHD counterparts, we discuss selected computational results. On fully periodic domains, Fourier based spectral discretizations are usually exploited. An early example of a 2D evolution, made by Fox and Orszag [182], compared a fully spectral Galerkin technique with a pseudo-spectral collocation method to simulate the decay of an initially isotropic turbulent flow. Using the 2D incompressible HD equations written in vorticity–streamfunction form (19.11), they simulated the evolution from an initial turbulent state. With resolutions of up to 128×128 Fourier mode pairs, the authors varied the kinematic viscosities ($\nu = 0.0025, 0.0018, 0.001$) such that the corresponding Reynolds number varied by a factor of about two. The main aim of this work was to demonstrate that the pseudo-spectral approach (where the convolutions in Fourier space are avoided by using FFTs to evaluate the nonlinear terms in real space) can rival a fully spectral treatment. Figure 19.4 compares the obtained vorticity field at time $t = 1$ for $\nu = 0.001$, for a spectral (left) and a pseudo-spectral (right) computation, and visually confirms this statement. The final vorticity map is characterized by larger scale structures than in its initial distribution, a typical finding for freely decaying 2D turbulence. Modern higher resolution simulations confirm how random initial 2D flow fields ultimately self-organize to the largest vortical structures that fit the computational domain and eventually, on the very long viscous time scale, decay to uniform static conditions.

A Direct Numerical Simulation (DNS) result for 3D turbulence evolutions is taken from the work of Yokokawa et al. [661]. With increasing computer power, the higher resolution needed at lower Reynolds numbers can be afforded. The authors exploited resolutions with up to 4096^3 Fourier modes, and reached corresponding kinematic viscosities $\nu \sim \mathcal{O}(10^{-4}\text{–}10^{-5})$. Unlike the just mentioned freely decaying 2D turbulence case without a driver contribution, this 3D study adopted a continuous driving force, and was targeted to resolve the self-similar nature of turbulence throughout the inertial range of length scales. The DNS simulation used a pseudo-spectral approach involving 3D-FFTs on a triple periodic box. The temporal advance of the incompressible Navier–Stokes equations used Runge–Kutta methods. Figure 19.5 shows iso-surfaces of the magnitude of the vorticity, in consecutive zooms into the indicated boxed areas, clearly illustrating the aspect of invariance across ever-smaller scales. The role of stretching of vortex tubes explained earlier returns into the clearly elongated character of the iso-surfaces.

Fig. 19.4 Snapshots at time $t = 1$ of the vorticity, comparing a spectral (left) with a pseudo-spectral (right) discretization approach. (From Fox and Orszag [182].)

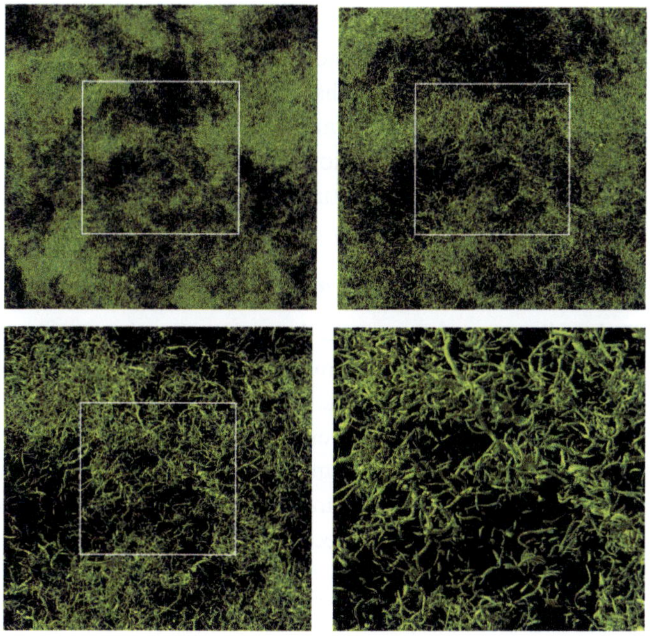

Fig. 19.5 Vorticity iso-surfaces found in a 3D incompressible hydro simulation by Yokokawa et al. [661]. Consecutive zooms demonstrate the scale-invariance.

As a final example of a DNS study in incompressible HD, we mention a study of Brachet et al. [87] focusing on 3D Euler scenarios and the distinct possibility of so-called 'singular' structure formation. The question addressed by these inviscid Euler ($\nu = 0$, without forcing) simulations is whether 3D vortex stretching can lead to singular structures, infinitely stretched in a finite amount of time. To ensure that pure Euler evolutions are numerically well-represented, the 3D pseudo-spectral simulations in a triple periodic box follow the evolutions at varying resolutions, from 256^3 to 864^3. The authors used a second order leapfrog or an Adams–Bashforth finite difference scheme.

The initial conditions employ a deterministic large-scale flow pattern, known as the Taylor–Green vortex, given by the smooth and large-scale initial flow field

$$v_x = \sin x \cos y \cos z \,, \quad v_y = -\cos x \sin y \cos z \,, \quad v_z = 0 \,. \tag{19.32}$$

At the employed resolutions, numerical convergence to pure Euler flow evolutions could be proven, e.g. by plotting the total enstrophy evolution for this Taylor–Green vortex for increasing resolutions. The authors argue that their simulations give evidence for continuously shrinking singular structures, such as pancake-like structures corresponding to vorticity layers whose thickness shrinks in time without distortion. This result demonstrates that the processes leading to small-scale structures in turbulent flows can well be of a purely inviscid nature. Similar small-scale structure formation is at play in incompressible MHD studies as well. We now continue to discuss MHD counterparts.

19.2 Incompressible magnetohydrodynamics

We start with formulating the governing equations for visco-resistive, incompressible MHD, where the non-ideal effects are parameterized by the ordinary and magnetic Reynolds numbers, as well as their ratio known as the Prandtl number. These incompressible MHD equations can be manipulated to the symmetric Elsässer formulation, used to discuss wave aspects. We first illustrate some basic concepts following from Ampère's law, concentrating on kinematic MHD evolutions in ideal versus resistive regimes. In 3D kinematic MHD evolutions, several insights into dynamo action emerged, nowadays also realized in full 3D geo-dynamo studies.

19.2.1 Governing equations

Incompressible MHD describes the macroscopic dynamics of a plasma, in a single fluid fashion, exploiting density ρ, velocity $\mathbf{v}$, pressure p and magnetic field $\mathbf{B}$. Incompressibility yields a topological constraint on the streamlines just like the divergence-free condition of the magnetic field expresses a topological constraint on the field lines. In contrast to HD, deviations from ideal evolution now come in two flavours. Viscous effects impact the momentum evolution. Resistive effects, causing Ohmic heating and dissipation, enter the induction equation. The governing equations translate into a set nonlinear PDEs for $\mathbf{v}(\mathbf{x};t)$ and $\mathbf{B}(\mathbf{x};t)$, with pressure $p(\mathbf{x};t)$:

$$\begin{aligned} \frac{\partial \mathbf{v}}{\partial t} + \mathbf{v}\cdot\nabla\mathbf{v} &= -\frac{1}{\rho_0}\nabla p + \frac{1}{\mu_0\rho_0}(\nabla\times\mathbf{B})\times\mathbf{B} + \nu\nabla^2\mathbf{v} + \mathbf{F}_{\mathrm{driv}}\,, \qquad \nabla\cdot\mathbf{v} = 0\,, \\ \frac{\partial \mathbf{B}}{\partial t} &= \nabla\times(\mathbf{v}\times\mathbf{B}) + \tilde{\eta}\nabla^2\mathbf{B}\,, \qquad \nabla\cdot\mathbf{B} = 0\,. \end{aligned} \tag{19.33}$$

The viscous diffusion term $\nu\nabla^2\mathbf{v}$ contains the kinematic viscosity coefficient ν, with dimension $[\ell^2 t^{-1}]$. In the original induction equation, the resistivity η appears, with dimension $[m\,\ell^3\,t^{-1}q^{-2}]$ (with m for mass and q for charge). We here exploit the more convenient parameter $\tilde{\eta} \equiv \eta/\mu_0$, with dimension $[\ell^2 t^{-1}]$ putting ν and $\tilde{\eta}$ on an equal footing. To simplify our discussion, and as adopted in many incompressible MHD studies, we assume ν and $\tilde{\eta}$ to be constant throughout.

The non-ideal effects of the momentum equation are quantified by the ratio of the nonlinear term $\mathbf{v}\cdot\nabla\mathbf{v}$ to the diffusive term $\nu\nabla^2\mathbf{v}$, i.e. by the Reynolds number

$$R_{\mathrm{e}} = v_0\,l_0/\nu\,. \tag{19.34}$$

Here, l_0 is a typical length scale for the velocity variation and v_0 is a reference speed. Analogously, the non-ideal effects of the induction equation are quantified by the ratio $\nabla \times (\mathbf{v} \times \mathbf{B})$ to $\tilde{\eta}\nabla^2\mathbf{B}$, i.e. by the magnetic Reynolds number

$$R_{\rm m} = v_0 \, l_0/\tilde{\eta} \,. \tag{19.35}$$

The relative importance of viscous to resistive diffusion is quantified by the ratio

$$P_{\rm r} = R_{\rm m}/R_{\rm e} = \nu/\tilde{\eta} \,, \tag{19.36}$$

which is known as the Prandtl number.

19.2.2 Elsässer formulation

To obtain the symmetric Elsässer formulation [170] for incompressible MHD, we note that the Lorentz force can be split:

$$\mathbf{j} \times \mathbf{B} = -\mu_0^{-1}\nabla\big(\tfrac{1}{2}B^2\big) + \mu_0^{-1}\mathbf{B}\cdot\nabla\mathbf{B} \,. \tag{19.37}$$

This exhibits the magnetic influences through an isotropic magnetic pressure and a field line tension part. Using a typical length scale l_0, field strength B_0 and density ρ_0, a speed dimension arises from the familiar Alfvén speed, $v_0 = B_0/\sqrt{\mu_0\rho_0}$, as curved field lines have magnetic tension as their restoring force; see Fig. 2.9.

If we start from the incompressible visco-resistive MHD equations (19.33), we can rewrite the momentum equation with the Lorentz force split into pressure and tension contributions, and introduce a total pressure $P \equiv p + B^2/(2\mu_0)$. When we also rewrite $\nabla \times (\mathbf{v} \times \mathbf{B}) = \mathbf{B}\cdot\nabla\mathbf{v} - \mathbf{v}\cdot\nabla\mathbf{B}$, we get

$$\begin{aligned} \frac{\partial \mathbf{v}}{\partial t} &= -\,\mathbf{v}\cdot\nabla\mathbf{v} - \frac{1}{\rho_0}\nabla P + \frac{1}{\mu_0\rho_0}\mathbf{B}\cdot\nabla\mathbf{B} + \nu\nabla^2\mathbf{v} + \mathbf{F}_{\rm driv} \,, \qquad \nabla\cdot\mathbf{v} = 0 \,, \\ \frac{\partial \mathbf{B}}{\partial t} &= \mathbf{B}\cdot\nabla\mathbf{v} - \mathbf{v}\cdot\nabla\mathbf{B} + \tilde{\eta}\nabla^2\mathbf{B} \,, \qquad \nabla\cdot\mathbf{B} = 0 \,. \end{aligned} \tag{19.38}$$

As early as 1950, Elsässer realized that one can make linear combinations from these equations to get the temporal evolution for the now-called Elsässer fields

$$\mathbf{z}^{\pm} \equiv \mathbf{v} \pm \mathbf{B}/\sqrt{\mu_0\rho_0} \,. \tag{19.39}$$

These are seen to represent forward and backward Alfvén related signals. The governing equations for these Elsässer variables $\mathbf{z}^{\pm}$ are

$$\begin{aligned} \frac{\partial \mathbf{z}^+}{\partial t} &= -\,\mathbf{z}^-\cdot\nabla\mathbf{z}^+ - \rho_0^{-1}\nabla P + \mathbf{F}_{\rm driv} + \tfrac{1}{2}\nabla^2\left[(\nu+\tilde{\eta})\,\mathbf{z}^+ + (\nu-\tilde{\eta})\,\mathbf{z}^-\right] , \\ \frac{\partial \mathbf{z}^-}{\partial t} &= -\,\mathbf{z}^+\cdot\nabla\mathbf{z}^- - \rho_0^{-1}\nabla P + \mathbf{F}_{\rm driv} + \tfrac{1}{2}\nabla^2\left[(\nu-\tilde{\eta})\,\mathbf{z}^+ + (\nu+\tilde{\eta})\,\mathbf{z}^-\right] , \end{aligned} \tag{19.40}$$

while still $\nabla\cdot\mathbf{z}^{\pm} = 0$. We here exploited the rescaled $\tilde{\eta} \equiv \eta/\mu_0$ introduced earlier. Thus, a nice symmetric formulation of the incompressible MHD equations is obtained, where the limit to HD is especially clear, as then $\mathbf{z}^{\pm} = \mathbf{v}$.

▷ **Exercise** Derive the governing equations in this symmetric form. ◁

19.2.3 Kinematic MHD modelling

In so-called kinematic MHD studies, the induction equation is used by itself, with the constraint $\nabla \cdot \mathbf{B} = 0$, to solve for the evolution of the magnetic field with a prescribed velocity field. Kinematic modelling may invoke a typically time independent velocity field, obeying $\nabla \cdot \mathbf{v} = 0$. By ignoring the momentum equation, this kinematic approach can only be justified when no back reaction of the magnetic field on the flow is expected, since the Lorentz force is ignored. This can be argued when the magnetic energy is far less than the kinetic energy, $\frac{1}{2}B^2/\mu_0 \ll \frac{1}{2}\rho_0 v^2$. This approach suffices to demonstrate field amplification, and makes the link to true dynamo activity where kinetic energy is feeding magnetic energy.

The first term on the RHS of the induction equation, $\nabla \times (\mathbf{v} \times \mathbf{B})$, represents conversion of mechanical motion into electromagnetic induction, happening when a conductor moves with velocity $\mathbf{v}$ in a magnetic field $\mathbf{B}$. This creates an electric field $-(\mathbf{v} \times \mathbf{B})$. The second RHS term yields, by dimensional reasoning, a time scale for resistive diffusion:

$$\tau_{\mathrm{R}} = l_0^2/\tilde{\eta} \equiv R_{\mathrm{m}} l_0/v_0 \,. \tag{19.41}$$

Its role can be appreciated from early kinematic models, which solved for the magnetic field evolution from the induction equation with resistivity, employing the vector potential $\mathbf{A}$ ensuring solenoidal fields:

$$\mathbf{B} = \nabla \times \mathbf{A} \quad \Rightarrow \quad \nabla \cdot (\nabla \times \mathbf{A}) = 0 \,. \tag{19.42}$$

Reformulated for the spatio-temporal evolution of the vector potential, kinematic modelling solves the following equation:

$$\frac{\partial \mathbf{A}}{\partial t} = \mathbf{v} \times \mathbf{B} - \tilde{\eta}\, \nabla \times \mathbf{B} \,. \tag{19.43}$$

One may note that the last term involves second order derivatives of the components of $\mathbf{A}$, and that the current is also related to $\mathbf{A}$ by second order derivatives, translating into accuracy requirements when solving this PDE numerically.

Seminal work by Weiss [642] performed 2D numerical simulations, to study the effect of magnetic flux expulsion by convective eddies. The restriction to 2D implies that only the z-component of the vector potential $A_z \equiv A(x, y; t)$ is non-vanishing, related to the magnetic field through

$$\mathbf{B} = (\partial A/\partial y, -\partial A/\partial x, 0) \,. \tag{19.44}$$

The $A = \mathrm{const}$ contours are field lines, and the induction equation can now be rewritten as

$$\frac{\partial A}{\partial t} = -\mathbf{v} \cdot \nabla A + \tilde{\eta}\, \nabla^2 A \,. \tag{19.45}$$

This advection-diffusion equation was studied by Weiss for a variety of flow fields and resistivities. As a concrete case, we considered the 2D time-independent incompressible velocity field $\mathbf{v} = [\sin(2\pi x)\sin(2\pi y - \frac{1}{2}\pi), \sin(2\pi x + \frac{1}{2}\pi)\sin(2\pi y)]$ modelling four convection cells on the unit square $[0,1]^2$. When one solves the 2D induction equation numerically, for cases where $\tilde{\eta} = 0.1$, $0.01, 0.001$, an initially uniform vertical field $\mathbf{B} = (0, 1)$ will evolve due to the assumed fixed flow pattern. By going to lower values of $\tilde{\eta}$, one can approach the limit of ideal MHD, but reaching the ideal $\tilde{\eta} = 0$ case itself is impossible (if only due to numerical diffusion). In Fig. 19.6, we performed simulations of resolution 30^2 with a second order accurate numerical scheme, and show for the

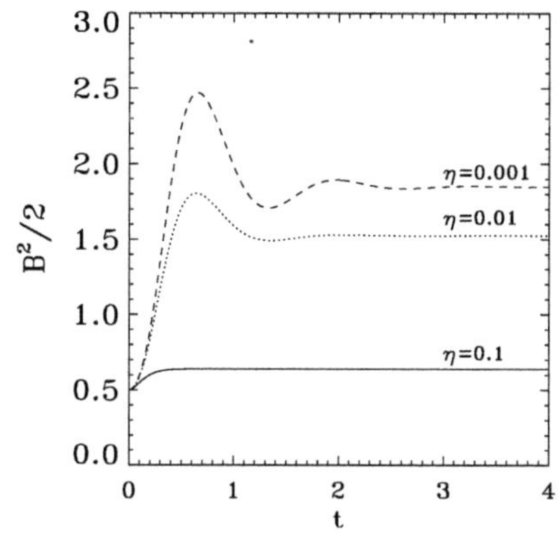

Fig. 19.6 The magnetic energy evolution in three kinematic MHD simulations with varying resistivity $\tilde{\eta}$, towards the steady-state field patterns shown in Fig. 19.7.

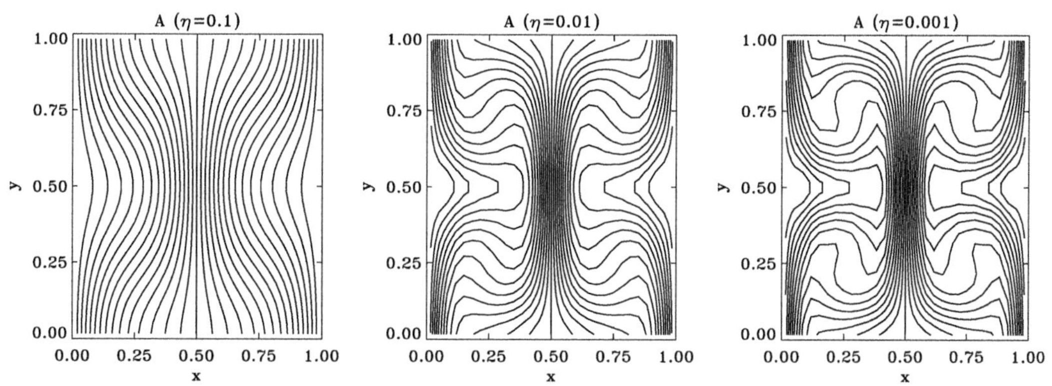

Fig. 19.7 Steady magnetic field configurations obtained for the 2D kinematic MHD evolutions from Fig. 19.6; $\tilde{\eta}$ decreases from left to right.

three cases the evolution of the total box-integrated magnetic energy. (The values of η indicated in Figs. 19.6 and 19.7 actually refer to $\tilde{\eta}$.) All cases exhibit three distinct phases, namely an initial phase of field amplification, a turnover phase when the convective term becomes comparable to the resistive term and ultimately a settling into a steady-state configuration. Visualizing the field configuration in their steady-state regime in Fig. 19.7, there is a clear trend to magnetic flux being expelled from the centre of the eddies as $\tilde{\eta} \to 0$. Hence, this study nicely shows how magnetic flux concentrates at the edges of convective cells. This aspect returns in full 2D and 3D magneto-convection models of solar and stellar convection zones.

19.2.4 Dynamo aspects

The kinematic MHD approach also allows us to study dynamo action aspects, as was done by Galloway and Proctor [204]. We already noted that the time scale for magnetic field diffusion from the (linear) kinematic induction equation scales with the magnetic Reynolds number R_m according to

Eq. (19.41). In all astrophysically relevant cases, the magnetic Reynolds number is very large (up to order 10^{10}). However, the kinematic induction equation with resistivity included may allow for exponentially growing linear eigen-solutions, with a temporal variation $\exp(st)$ with $\mathrm{Re}(s) > 0$. We then distinguish slow versus fast dynamos when the assumed flows that give growth of the magnetic field behave as follows. A flow pattern gives slow dynamo action when its growth rate $\mathrm{Re}(s) \to 0$ for $R_\mathrm{m} \to \infty$. A pattern yields fast dynamo action on the other hand, when the growth rate $\mathrm{Re}(s)$ becomes independent of R_m for $R_\mathrm{m} \to \infty$. Even though the governing equation is a linear equation for the field evolution, the flow pattern itself can render the problem analytically difficult or even impossible to solve, as we are encountering a singular limit. Galloway and Proctor studied this linear kinematic dynamo problem numerically. For the flow field, they studied several variants that all represent smooth flow fields which possess chaotic regions (i.e. individual streamlines traverse spatial regions chaotically). The 'ABC' flow field takes a time-independent 3D incompressible $\mathbf{v}(x, y, z)$:

$$\mathbf{v}_{ABC} = \big(A \sin z + C \cos y,\ B \sin x + A \cos z,\ C \sin y + B \cos x\big) \qquad (19.46)$$

in a $[0, 2\pi]^3$ box. The authors also analysed a 2D time-dependent 'circularly polarized' (CP) prescription (independent on x), given by

$$\mathbf{v}_\mathrm{CP} = \big(A \sin(z + \sin \omega t) + C \cos(y + \cos \omega t), A \cos(z + \sin \omega t), C \sin(y + \cos \omega t)\big) . \quad (19.47)$$

Finally, a third scenario considered a 2D time-dependent 'linearly polarized' (LP) prescription as

$$\mathbf{v}_\mathrm{LP} = \big(A \sin(z + \cos \omega t) + C \cos(y + \cos \omega t), A \cos(z + \cos \omega t), C \sin(y + \cos \omega t)\big) . \quad (19.48)$$

Using a triple Fourier description on a fully periodic domain, the growth of linear eigen-modes was studied numerically for all three cases. It can be noted that the circularly or linearly polarized cases are 2D (or 2.5D) problems since the modes $\exp(ikx)$ decouple. This allowed the mode number k to be fixed and a 2D code to be used instead, which immediately allows for higher values of R_m to be reached. It was found that when taking $A = C = \sqrt{3/2}$ and $\omega = 1$, the CP case (a) achieved a highest growth for $k = 0.57$, while the LP case (b) attained highest growth for $k = 0.62$. As the full ABC case (c) is truly 3D, it was done at more modest resolutions of 96^3 (translating into $R_\mathrm{m} \approx 800$), where the parameters were fixed at $A = B = C = 1$. The three sets of simulation results for varying R_m are collected in Fig. 19.8. As especially the CP case (a) is seen to have a clear R_m-independent limit, these results gave convincing evidence for fast dynamo action.

These kinematic demonstrations of fast dynamo action gained further attention in a study by Tobias and Cattaneo [599]. Indeed, although the original 2.5D study showed fast dynamo action (and even categorized as a 'quick dynamo', reaching its asymptotic growth rate even at moderate R_m), the self-amplified magnetic structure obtained was all at small-scales. In contrast, the solar cycle (discussed in Section 8.2) confronts us through the butterfly diagram (Fig. 8.3) with a clear large-scale coherence, evolving over time scales that are long compared to the convective turnover times. Revisiting the kinematic dynamo problem, Tobias and Cattaneo demonstrated that the addition of a large-scale shear flow, varying on the scale of the simulation domain, together with imposed small-scale cellular flow patterns, can act to generate organized fields at high R_m, and thus lead to propagating dynamo waves very reminiscent of the solar cycle. The key observation of their study was that the added background large-scale flow actually acts to suppress the small-scale dynamo action, such that the large-scale dynamo action survives.

Of course, true dynamo behaviour must consider the Lorentz force back reaction and, hence,

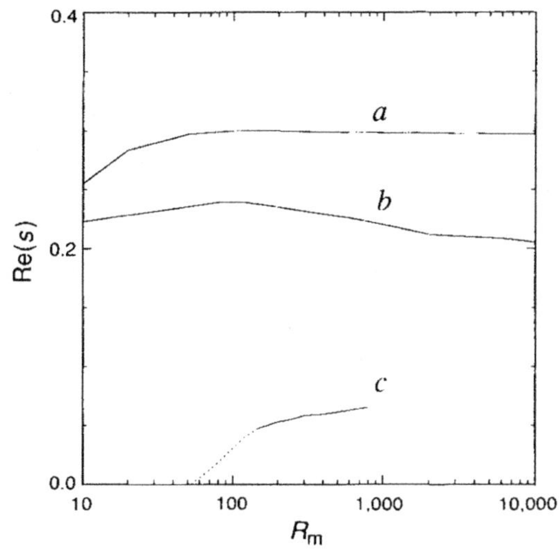

Fig. 19.8 Variation of the linear growth rate with magnetic Reynolds number from the numerical kinematic dynamo runs described by Galloway and Proctor [204]. Cases a and b are from 2D runs with a specific time-dependent imposed flow, case c uses a 3D ABC flow.

the full set of incompressible MHD equations in three space dimensions. For the case of the Earth dynamo, one should account for the spherical geometry, include thermal convection (and gravity) aspects for self-consistency, and involve rotation. The latter usually leads to computations in the rotating frame, with centrifugal and Coriolis effects where the rotation rate enters. One can then distinguish between the so-called 'weak-field' regime where the magnetic influence can be treated as a linear perturbation, or the 'strong-field' regime where the Lorentz force is estimated to be as large as the Coriolis force. For the Earth dynamo, it turns out that nature operates in the strong field regime, where only numerics helps out.

Such full 3D geo-dynamo simulations have been performed extensively, and provided crucial insight in the way the geo-dynamo operates. We know from geophysical surveys that the Earth's magnetic field is currently mostly dipolar, but there is evidence from magnetized rocks that its orientation reverses every few 100 000 years, taking a few 1000 years for a full reversal. The Earth structure is believed to consist of an inner solid core, a liquid outer core, a mantle, and a crust. The liquid iron outer core, situated at $1300 < R < 3400$ km must somehow maintain the magnetic field. This involves rotation and convection in a moving conducting fluid described by Ohm's law

$$\mathbf{E} + \mathbf{v} \times \mathbf{B} = \eta \mathbf{j} \,. \tag{19.49}$$

Dynamo action must take place, since the inhomogeneity of the magnetic field decays in a resistive time τ_{R}, involving resistivity η and the inhomogeneity length scale $l_0 \sim \nabla^{-1}$. For values typical of the inner Earth conditions, one estimates

$$\tau_{\mathrm{R}} = \mu_0 l_0^2 / \eta \sim 5 \times 10^{11}\mathrm{s} = 16\,000\,\text{years} \,. \tag{19.50}$$

Contrasted with the fact that the dipole survives for several 100 000 years, this introduces the need for sustained $\mathbf{B}$ generation by the molten iron motion in the outer core. There, convection is driven

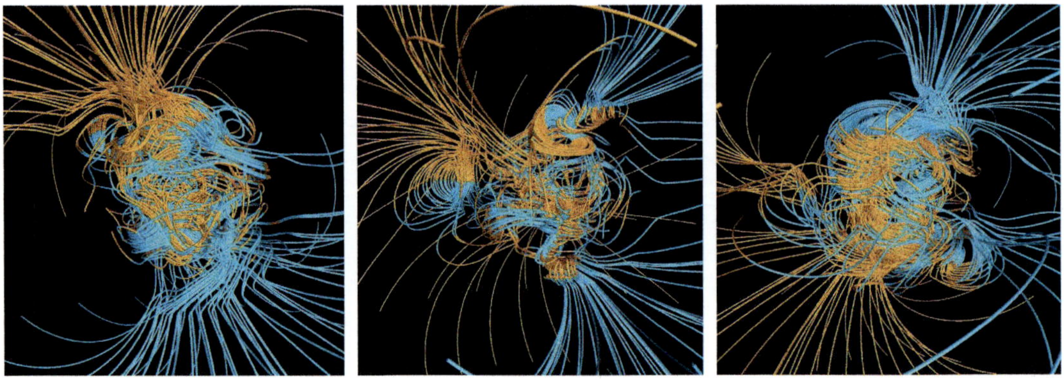

Fig. 19.9 Three snapshots of the 3D magnetic field structure simulated with the Glatzmaier–Roberts geodynamo model, showing how the magnetic field orientation changes from north–south to south–north through the reversal event. The rotation axis of the model Earth is vertical and through the centre. A transition occurs at the core-mantle boundary from the intense complicated field structure in the fluid core, where the field is generated, to the smooth potential field structure outside the core. (Courtesy of Gary A. Glatzmaier (UCSC) and Paul H. Roberts (UCLA).)

by heat from radioactive decay in the inner core, continuously feeding energy into the system. The challenge is to show that this system, confined between the solid inner core and the mantle, where both may be considered as electric insulators but which do exchange heat with the fluid, gives rise to solutions exhibiting the highly non-stationary behaviour of the geomagnetic field.

In a full 3D incompressible MHD simulation, Glatzmaier and Roberts [212] were able to simulate several 100 000 years of geo-dynamo activity. In this simulated time frame, they captured a full dipole reversal, with the dipole orientation changing in about 1000 years. The study showed that the inner core mediates reversals by the associated field anchoring and that the **B** change itself happens on the diffusion time. Some snapshots of their simulation are shown in Fig. 19.9, showing the field orientation changing from north–south to south–north in snapshots of the field structure during the reversal. This study made some necessary simplifications, such as to consider incompressible flows, and to neglect inertial terms in the momentum equation. Their work included (external) gravity and the Coriolis force, as they adopted a rotating reference frame. An extra equation for the temperature evolution was incorporated, to correctly model the effect of buoyancy and heat flux.

▷ **Numerical aspects of the Glatzmaier and Roberts study** Numerically, the Glatzmaier and Roberts dynamo simulations employed a spectral method (as in Section 15.1.4), with all variables expanded in global functions. In particular, the authors used Chebyshev polynomials for the radial variation and spherical harmonics to represent both angular variations. By ignoring inertia in the momentum equation, the CFL time step constraint for explicit integration was alleviated, as normally the Alfvén velocity $B/\sqrt{\mu_0 \rho_0}$ would severely limit the computational resources. Dropping inertia effectively replaced the Alfvén related CFL limit by one involving convective velocity alone. The authors further treated the linear terms in the equations implicitly with a Crank–Nicolson approach, while all nonlinear terms used an explicit Adams–Bashforth scheme. ◁

19.3 Waves in incompressible MHD

In this section, we highlight the role of Alfvén waves, both by linear as well as nonlinear analysis of the governing equations. The ideal incompressible case has conservation laws, such as those for total energy, magnetic helicity and cross-helicity, which all have influenced MHD turbulence theories. The latter theories are discussed using dimensional arguments, where prevailing ideas are contrasted with some views obtained by high resolution numerical simulations.

19.3.1 Linear wave analysis

In the ideal limit, the governing nonlinear PDEs for the velocity $\mathbf{v}(\mathbf{x};t)$ and the magnetic field $\mathbf{B}(\mathbf{x};t)$ are obtained from either Eq. (19.33), or Eq. (19.38), dropping the driving force and the dissipative terms. (Analogous to the HD case of Section 19.1.3, these $6+2$ PDEs actually mean $6-2$ solutions.) We will linearize these equations about a uniform plasma, by adopting a constant p_0, $\mathbf{v}_0$ and $\mathbf{B}_0$. As usual, we split $p = p_0 + p_1$, etc., where the variables with subscript 1 indicate perturbations, and retain only linear terms:

$$\begin{aligned} \frac{\partial \mathbf{v}_1}{\partial t} + \mathbf{v}_0 \cdot \nabla \mathbf{v}_1 &= -\frac{1}{\rho_0}\nabla p_1 + \frac{1}{\mu_0 \rho_0}(\nabla \times \mathbf{B}_1) \times \mathbf{B}_0 \,, \qquad \nabla \cdot \mathbf{v}_1 = 0\,, \\ \frac{\partial \mathbf{B}_1}{\partial t} &= (\mathbf{B}_0 \cdot \nabla)\,\mathbf{v}_1 - (\mathbf{v}_0 \cdot \nabla)\,\mathbf{B}_1\,, \qquad \nabla \cdot \mathbf{B}_1 = 0\,. \end{aligned} \tag{19.51}$$

We will analyse plane wave solutions, $\exp[\mathrm{i}(\mathbf{k}\cdot\mathbf{x} - \omega t)]$. As in the incompressible HD case, the linear wave modes decouple, and the Fourier amplitudes obey:

$$\begin{aligned} (\omega - \mathbf{k}\cdot\mathbf{v}_0)\,\hat{\mathbf{v}}_\mathbf{k} &= \frac{\hat{p}_\mathbf{k}}{\rho_0}\mathbf{k} + \frac{\hat{\mathbf{B}}_\mathbf{k}\cdot\mathbf{B}_0}{\mu_0\rho_0}\mathbf{k} - \frac{\mathbf{k}\cdot\mathbf{B}_0}{\mu_0\rho_0}\hat{\mathbf{B}}_\mathbf{k}\,, \qquad \mathbf{k}\cdot\hat{\mathbf{v}}_\mathbf{k} = 0\,, \\ (\omega - \mathbf{k}\cdot\mathbf{v}_0)\,\hat{\mathbf{B}}_\mathbf{k} &= -(\mathbf{k}\cdot\mathbf{B}_0)\,\hat{\mathbf{v}}_\mathbf{k}\,, \qquad \mathbf{k}\cdot\hat{\mathbf{B}}_\mathbf{k} = 0\,. \end{aligned} \tag{19.52}$$

We thus find that the wave vector $\mathbf{k}$ is orthogonal to both $\hat{\mathbf{v}}_\mathbf{k}$ and $\hat{\mathbf{B}}_\mathbf{k}$, and that the Doppler shifted frequency $\widetilde{\omega} \equiv \omega - \mathbf{k}\cdot\mathbf{v}_0$ appears. By projecting the first equation onto $\hat{\mathbf{v}}_\mathbf{k}$, and substituting $\hat{\mathbf{B}}_\mathbf{k}$ from the third equation, one then finds that

$$\left[\widetilde{\omega}^2 - \frac{(\mathbf{k}\cdot\mathbf{B}_0)^2}{\mu_0\rho_0}\right]\hat{v}_\mathbf{k}^2 = 0\,, \qquad \hat{\mathbf{B}}_\mathbf{k} = -\frac{\mathbf{k}\cdot\mathbf{B}_0}{\widetilde{\omega}}\hat{\mathbf{v}}_\mathbf{k}\,, \qquad \mathbf{k}\cdot\hat{\mathbf{v}}_\mathbf{k} = 0\,, \qquad \mathbf{k}\cdot\hat{\mathbf{B}}_\mathbf{k} = 0\,. \tag{19.53}$$

Hence, non-trivial ($\hat{v}_\mathbf{k} \neq 0$) solutions should satisfy the dispersion equation,

$$\widetilde{\omega}^2 = (\mathbf{k}\cdot\mathbf{B}_0)^2/(\mu_0\rho_0) = k^2\cos^2\vartheta\, v_{\mathrm{A}0}^2 \equiv k_\parallel^2 v_{\mathrm{A}0}^2 \equiv \omega_\mathrm{A}^2\,, \tag{19.54}$$

where ϑ is the angle between $\mathbf{k}$ and $\mathbf{B}_0$, and $v_{\mathrm{A}0} \equiv B_0/\sqrt{\mu_0\rho_0}$ is the Alfvén speed. Introducing $\Omega_0 = \mathbf{k}\cdot\mathbf{v}_0$ (which is constant here), the solutions can be written as

$$\widetilde{\omega} = \pm\omega_\mathrm{A} \;\Rightarrow\; \omega = \Omega_\mathrm{A}^\pm \equiv \Omega_0 \pm \omega_A \equiv \mathbf{k}\cdot\left(\mathbf{v}_0 \pm \mathbf{B}_0/\sqrt{\mu_0\rho_0}\right). \tag{19.55}$$

We hence find both forward and backward Doppler shifted Alfvén frequencies.

Since the solenoidality conditions on $\mathbf{v}_1$ and $\mathbf{B}_1$ are actually constraints, we should expect 4 solutions. This is precisely what we get: 2 forward (+) or backward (−) propagating Alfvén waves and 2 similar slow magneto-sonic waves, pairwise degenerate at the Doppler shifted Alfvén frequencies. These two types of waves are also present without background flow ($\mathbf{v}_0 = 0$); the wave perturbations and other directions involved were shown in Figs. 5.1(a) and 5.1(b).

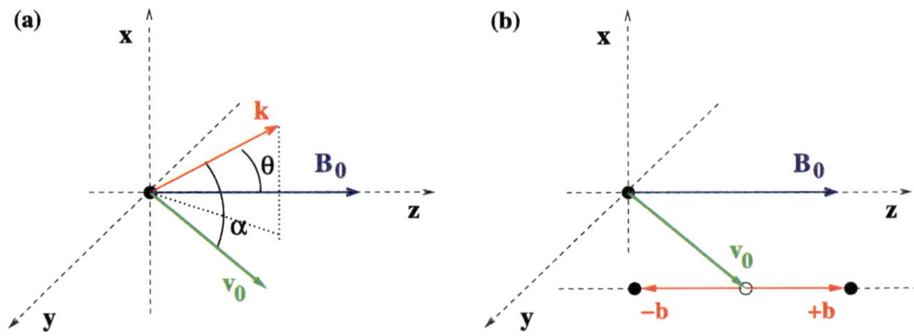

Fig. 19.10 (a) Angles between the magnetic field $\mathbf{B}_0$, the flow $\mathbf{v}_0$ and the wave vector $\mathbf{k}$. (b) Resulting group diagram for point disturbances at the origin.

Exploiting the angles α and ϑ as shown in Fig. 19.10(a), we can construct the phase speed from the expression (19.55) for ω:

$$\mathbf{v}_{\rm ph} = (\omega/k)\,\mathbf{k}/k = \big(v_0 \cos\alpha \pm v_{\rm A0} \cos\vartheta\big)\,\mathbf{k}/k\,. \tag{19.56}$$

In the co-moving frame ($\mathbf{v}_0 = 0$), the end point of the $\mathbf{k}$ vector as ϑ varies yields two spheres, left and right of the origin, which is the phase diagram. One can construct the group diagram by a Huygens construction, as done in Fig. 5.3 for a case without background flow. This gives two point perturbations, from a single point disturbance, travelling with speed $\mathbf{b} \equiv \mathbf{B}_0/\sqrt{\mu_0 \rho_0}$ along the field line. In the laboratory frame ($\mathbf{v}_0 \neq 0$), the group speed is given by

$$\mathbf{v}_{\rm gr} = \partial\omega/\partial\mathbf{k} = \mathbf{v}_0 \pm \mathbf{b}\,. \tag{19.57}$$

The corresponding group diagram is shown in Fig. 19.10(b).

All this has to do with the fact that ideal MHD ensures 'frozen-in' conditions. Hence, while a point perturbation on a given field line will split into forward and backward Alfvén pulses, the field line itself gets displaced according to the flow component orthogonal to the original field line. On the field line, the pulses spread away from the Doppler shifted location (involving the parallel velocity only) at the Alfvén speed. Note the extreme anisotropy: this is purely along the perturbed field line. The Alfvén and slow waves differ with respect to their polarization. Recall that the slow wave is slower than the Alfvén wave in compressible MHD. We exploit the Elsässer variable prescription to show that nonlinear wave packages are exact solutions too!

19.3.2 Nonlinear wave solutions and conservation laws

To extend the wave analysis to nonlinear wave packages, we start from the incompressible nonlinear equations (19.38), without driver and dissipation. Exploiting the Elsässer variables (19.39) then yields a simplified form of Eqs. (19.40):

$$\begin{aligned} \frac{\partial \mathbf{z}^+}{\partial t} + \mathbf{z}^- \cdot \nabla \mathbf{z}^+ &= -\,\rho_0^{-1} \nabla P\,, \\ \frac{\partial \mathbf{z}^-}{\partial t} + \mathbf{z}^+ \cdot \nabla \mathbf{z}^- &= \rho_0^{-1} \nabla P\,, \qquad \nabla \cdot \mathbf{z}^\pm = 0\,. \end{aligned} \tag{19.58}$$

Suppose there are magnetic field fluctuations (e.g. in the presence of fully developed turbulence), but that one can identify a uniform background field $\mathbf{B}_0$, constant in space and time, by averaging out the fluctuations, then

$$\mathbf{B} = \mathbf{B}_0 + \Delta\mathbf{B} . \tag{19.59}$$

The only difference with the earlier split of the linear wave analysis is that we now allow nonlinear deviations $\Delta\mathbf{B}$ of any magnitude. We choose the z-direction to be along the background field, $\mathbf{B}_0 = B_0\mathbf{e}_z$, so that

$$\mathbf{z}^{\pm} = \mathbf{v} \pm v_{A0}\mathbf{e}_z \pm \Delta\mathbf{B}/\sqrt{\mu_0\rho_0} . \tag{19.60}$$

We define new Elsässer-type variables, involving the fluctuating field only:

$$\begin{aligned} \mathbf{w}_{\downarrow} &\equiv \mathbf{z}^{+} - v_{A0}\mathbf{e}_z = \mathbf{v} + \Delta\mathbf{B}/\sqrt{\mu_0\rho_0} , \\ \mathbf{w}_{\uparrow} &\equiv \mathbf{z}^{-} + v_{A0}\mathbf{e}_z = \mathbf{v} - \Delta\mathbf{B}/\sqrt{\mu_0\rho_0} . \end{aligned} \tag{19.61}$$

In these new variables, the governing equations become

$$\begin{aligned} \frac{\partial\mathbf{w}_{\downarrow}}{\partial t} - v_{A0}\frac{\partial\mathbf{w}_{\downarrow}}{\partial z} &= -\left(\mathbf{w}_{\uparrow}\cdot\nabla\right)\mathbf{w}_{\downarrow} - \rho_0^{-1}\nabla P , \\ \frac{\partial\mathbf{w}_{\uparrow}}{\partial t} + v_{A0}\frac{\partial\mathbf{w}_{\uparrow}}{\partial z} &= -\left(\mathbf{w}_{\downarrow}\cdot\nabla\right)\mathbf{w}_{\uparrow} - \rho_0^{-1}\nabla P , \qquad \nabla\cdot\mathbf{w}_{\uparrow\downarrow} = 0 . \end{aligned} \tag{19.62}$$

▷ **Exercise** Derive these equations. ◁

We now consider the physical significance of the variables $\mathbf{w}_{\uparrow\downarrow}$. They represent up/down propagating nonlinear wave packages, which can be travelling along the background field at the Alfvén speed v_{A0} when the RHSs of the above first two equations vanish. This happens, in particular, when only uni-directional (either up or down) packages are present, e.g. when $\mathbf{w}_{\uparrow} = 0$. Then there is no nonlinear cross-term $(\mathbf{w}_{\uparrow}\cdot\nabla)\,\mathbf{w}_{\downarrow}$. In that case, the remaining equations,

$$\frac{\partial\mathbf{w}_{\downarrow}}{\partial t} - v_{A0}\frac{\partial\mathbf{w}_{\downarrow}}{\partial z} = 0 , \qquad \nabla P = 0 , \tag{19.63}$$

allow for an exact solution

$$\mathbf{w}_{\downarrow} = \mathbf{w}\left(z + v_{A0}t\right) \tag{19.64}$$

representing an arbitrary, uni-directionally propagating, nonlinear wave package. Similarly, when $\mathbf{w}_{\downarrow} = 0$, we find an exact solution $\mathbf{w}_{\uparrow} = \mathbf{w}\left(z - v_{A0}t\right)$, representing an upwardly propagating wave package.

With this reformulation one can show that: (1) nonlinear wave package distortions only occur when oppositely directed wave packages interact nonlinearly; (2) when no viscous or resistive effects are at play, energy exchange does not occur between two oppositely travelling interacting wave packages during the collision. These collisions are elastic (their form can change but not their energy content) since the energy of an isolated up/down propagating wave package is proportional to $\int w_{\uparrow\downarrow}^2\, dV$, which can be shown to be time-invariant.

We finish this section by noting that several conservation laws are respected by incompressible ideal MHD. First, by adding $[\rho_0\mathbf{v}\cdot$ (momentum equation)$]$ and $[\mathbf{B}/\mu_0\cdot$ (induction equation)$]$

in Eqs. (19.38) without driver and dissipation, we obtain a conservation law for the energy density $\mathcal{H} \equiv \frac{1}{2}\rho_0 v^2 + \frac{1}{2}B^2/\mu_0$:

$$\frac{\partial}{\partial t}\Big[\tfrac{1}{2}\rho_0 v^2 + \tfrac{1}{2}B^2/\mu_0\Big] + \nabla \cdot \Big[\mathbf{v}\big(\tfrac{1}{2}\rho_0 v^2 + p\big) + \mu_0^{-1}\big(\mathbf{v}B^2 - \mathbf{B}\mathbf{v}\cdot\mathbf{B}\big)\Big] = 0\,. \tag{19.65}$$

The flux of energy corresponds to the advected part, which is already present in HD and is now augmented with the electromagnetic energy flux. The latter corresponds to the Poynting flux, where $\mathbf{E} = -\mathbf{v} \times \mathbf{B}$ for a perfectly conducting plasma. In fact, integrating the conservation law (19.65) over the total plasma volume yields conservation for the total energy, $\int \mathcal{H}\, dV = \text{const}$.

A second important conserved quantity is the cross-helicity, implying that the volume integral $\int \mathbf{v}\cdot\mathbf{B}\, dV = \text{const}$. The underlying conservation law is found by adding [$\mathbf{B}\cdot$ (momentum equation)] and [$\mathbf{v}\cdot$ (induction equation)]:

$$\frac{\partial}{\partial t}(\mathbf{v}\cdot\mathbf{B}) + \nabla \cdot \Big[\mathbf{v}(\mathbf{v}\cdot\mathbf{B}) + \mathbf{B}\big(\rho^{-1}p - \tfrac{1}{2}v^2\big)\Big] = 0\,. \tag{19.66}$$

Furthermore, one can exploit the vector potential in $\mathbf{B} = \nabla \times \mathbf{A}$ to define the magnetic helicity, which is also conserved, $\int \mathbf{A}\cdot\mathbf{B}\, dV = \text{const}$, where integration is over the volume of the flux tube. We already mentioned the usual magnetic flux conservation law of ideal MHD, key to the concept of frozen-in field lines. All these conservation laws, together with the aforementioned insights on nonlinear wave package interactions, have inspired incompressible MHD turbulence theories, to which we now turn our attention.

19.3.3 MHD turbulence scaling laws

We now discuss incompressible MHD turbulence (see e.g. Biskamp [67]) for an extensive monograph on the subject), using a Kolmogorov-type reasoning. We will contrast early theories of Iroshnikov [328] and Kraichnan [377](a) with more recent ideas by Goldreich and Sridhar [565, 242] and hints from modern simulations.

The Iroshnikov–Kraichnan (IK) theory reasons that only oppositely directed waves interact in incompressible MHD, that turbulence is still isotropic, and that a direct energy cascade from long to short wavelengths exists. More implicit is that the dominant wave package interactions must involve three-wave coupling, since two-wave interactions are elastic, as stated before. The IK dimensional reasoning follows from the Elsässer terms $\partial\mathbf{z}^+/\partial t$ and $\mathbf{z}^- \cdot \nabla\mathbf{z}^+$. Their dimensions involve the background Alfvén speed $v_{\rm A}$ and typical velocity, time and length scales v_k, t_k and l_k ($= 2\pi/k$) for an eddy. This translates into a time scale for wave package interactions at length scale l_k:

$$t_k \sim l_k v_{\rm A}/v_k^2\,. \tag{19.67}$$

Invoking the usual argument of a constant energy flux $\epsilon = v_k^2/t_k$, we obtain an estimate for the scaling of the kinetic energy density:

$$v_k^2 \sim (l_k v_{\rm A}\epsilon)^{1/2} = \sqrt{\epsilon v_{\rm A}}\, k^{-1/2}\,. \tag{19.68}$$

The Fourier analysed energy power spectrum $\int_k^{k+\Delta k} E(k')\, dk' \sim v_k^2$ leads to a one-dimensional energy power spectrum with

$$E(k) \sim \sqrt{\epsilon v_{\rm A}}\, k^{-3/2}\,. \tag{19.69}$$

Clearly, the IK scaling law has an inertial range with a power law index which is different from Kolmogorov's $-5/3$ power law.

▷ **Variants on IK scalings** Related IK scalings can be found in the literature. For example, suppose that a steady excitation at amplitude $v_L \ll v_A$ occurs at an intermediate length scale L. The above arguments remain valid, so that the scaling (19.67) for the time scale $t_{\mathbf{k}}$ of wave package interactions, (19.68) for the kinetic energy density v_k^2 and (19.69) for the power spectrum still hold, with $v_A \rightarrow v_L$. Instead of relating the estimate of v_k^2 to a one-dimensional energy power spectrum $E(k)$, it is sometimes useful to exploit the three-dimensional energy power spectrum. Carrying on the analysis for steady excitation, where the associated energy injection sets $\epsilon = v_L^3/L$, gives

$$v_k^2 \sim \iiint E^*(\mathbf{k}')\, dk'^3 \sim \sqrt{\epsilon v_L}\, k^{-1/2} = v_L^2\,(kL)^{-1/2} . \tag{19.70}$$

Formally differentiating left and right, we find

$$E^*(\mathbf{k}) \sim v_L^2\, L^{-1/2} k^{-7/2} . \tag{19.71}$$

Recall: this variant of the IK scaling would hold when energy injection is at length scale L with associated velocity $v_L \ll v_A$. ◁

As all scalings presented thus far merely involve dimensional arguments, it should come as no surprise that different combinations of the relevant quantities give different scaling laws for inertial range behaviour. In 1994–95, Goldreich and Sridhar (GS) proposed alternative scalings, arguing that it is not justified to assume isotropy for the MHD cascade. They point out that for the IK theory to work, one needs three-wave interactions, which are inconsistent with the assumption of isotropy. With respect to the directions parallel ($\|$) and perpendicular ($\perp$) to the mean field $\mathbf{B}_0$, three-wave interactions will not cascade energy in the parallel direction. Therefore, one must distinguish between the correlation lengths $l_{k\|}$ and $l_{k\perp}$ (or $k_\|$ and $k_\perp$) in the presence of such a background field. A scaling analysis for GS MHD turbulence behaviour replaces the time scale (19.67) by

$$t_{k\perp} \sim \frac{v_A}{l_{k\|}} \left(\frac{l_{k\perp}}{v_{k\perp}} \right)^2 . \tag{19.72}$$

Keeping $l_{k\|}$ constant, since no energy transfer occurs along the mean field, the energy flux now reads $\epsilon = v_{k\perp}^2/t_{k\perp}$, so that the typical velocity perpendicular to the field yields the energy dependence on the perpendicular length scales:

$$v_{k\perp}^2 \sim \int E(k'_\perp)\, dk'_\perp \sim \left(\epsilon v_A l_{k\perp}^2 / l_{k\|}\right)^{1/2} = \sqrt{\epsilon v_A k_\|}\, k_\perp^{-1} . \tag{19.73}$$

The final anisotropic GS model for intermediate MHD turbulence thus gives a one-dimensional perpendicular energy power spectrum

$$E(k_\perp) \sim \sqrt{\epsilon v_A k_\|}\, k_\perp^{-2} . \tag{19.74}$$

Variations of this dimensional scaling theme can argue for yet different power law dependences when going to strong turbulence, where perturbations become of the same order as the background field. One then considers the additional argument (known as 'critical balance') that the cascade time for wave package distortions must become of the order of the wave period. This gives the scaling

$$E(k_\perp) \sim v_A^2\, L^{-2/3} k_\perp^{-5/3} , \tag{19.75}$$

i.e. an anisotropic version of the Kolmogorov scaling pertaining to the perpendicular direction.

With this abundance of theoretically predicted scalings, the need for high resolution numerical MHD turbulence studies is evident, if only to confirm or to falsify the underlying physical arguments. As an example, we summarize: (1) 3D studies by Müller and Biskamp [454], focusing on freely decaying MHD turbulence in a 512^3 simulation using pseudo-spectral methods, and (2) a parametric study by Maron and Goldreich [430], with resolutions up to $256 \times 256 \times 512$ for both driven and decaying turbulence.

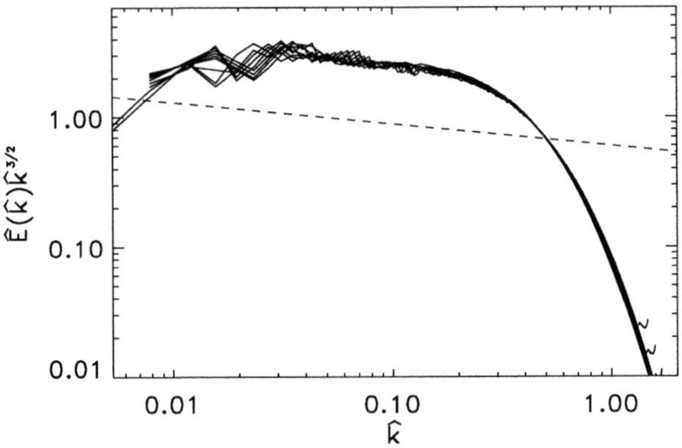

Fig. 19.11 One-dimensional energy spectrum at various times in the simulations by Müller and Biskamp [454]. The dashed line corresponds to a $-3/2$ scaling.

Müller and Biskamp considered a 3D triple periodic box, where an initial turbulent field was introduced with random phases in the incompressible flow/field modes and an overall finite value of the magnetic helicity. The code used pseudo-spectral (Fourier) modes with resolutions 128^3 to 512^3, and cases with fixed Prandtl number ($= 1$) explored Reynolds numbers up to 3000. A recurring aspect of turbulence studies is the use of so-called hyper-diffusion, where the diffusive operator ∇^2 is replaced by a higher order variant such as ∇^4. In their study, both cases with normal and with hyper-diffusion were investigated. Letting the turbulence develop fully, the resulting energy spectrum was steeper than $k^{-3/2}$ and rather close to $k^{-5/3}$. Figure 19.11 shows the $k^{3/2}$ corrected one-dimensional energy spectrum at various moments in the turbulent phase, where the dashed line would indicate the $-5/3$ scaling. Hence, Müller and Biskamp find no IK scaling, and concentrate on 'isotropic' turbulence. In a follow-up paper, Biskamp and Müller [68] analysed a total of 14 simulations, and emphasized the role played by small-scale structures. Compared to HD runs at the same Reynolds number (as in Fig. 19.5 of the Yokokawa et al. [661] study), the occurring structures in MHD realizations are clearly larger and behave anisotropically, especially when comparing cases without and with a guide field. The guide field case finds current sheets preferably elongated along the background field.

Soon after that study, Maron and Goldreich [430] presented an analysis of 14 parametric runs, with cases of both forced and decaying turbulence. They used a 3D pseudo-spectral code with a standard 2/3-1/3 de-aliasing rule, a second order Runge–Kutta time advance, and handled the diffusive term with an integrating factor. They took a unit guide field along z and adopted a box of size $[0.002, 0.002, 1]$, maximizing the resolution in the perpendicular plane. They went up to $256 \times 256 \times 512$ mode triples, used hyper-diffusion and took a unit Prandtl number throughout. Figure 19.12 shows the dependence $E(k_\perp)$ of the energy spectrum on the perpendicular wave number, for a forced case. They found a slope of $k_\perp^{-3/2}$, and could split up their wave spectrum into the four modes at play: up and down, Alfvén and slow waves. In further analysis, they quantified the occurring energy losses, due to the hyper-viscous dissipation and errors/dissipation due to de-aliasing. None of those affected the inertial range behaviour.

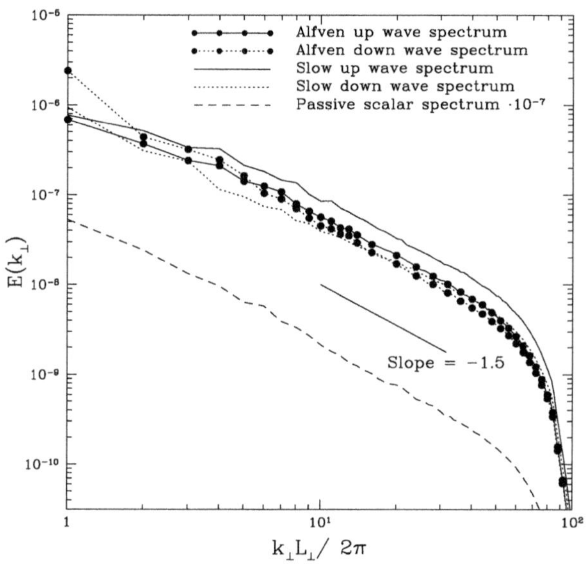

Fig. 19.12 Dependence of the energy spectrum on the perpendicular wave number as found in a forced turbulence simulation. Contributions from up and downward Alfvén and slow waves are identified. (From Maron and Goldreich [430].)

Summarizing the Maron and Goldreich (2001) results, they find 1D power spectra exhibiting close to $-3/2$-scaling, which is different from the one predicted by Goldreich and Sridhar (with a slope of $-5/3$ for strong anisotropic turbulence). They confirm the anisotropic behaviour of turbulence and clearly show how the various wave families are involved. Since higher resolutions may yield further clues, this is obviously an area of ongoing research, encompassing weak to strong turbulence studies, the influence of finite helicity, elucidating the role of slow versus Alfvén waves and the role played by the formation of 'singular' structures.

19.4 Incompressible MHD simulations

We close this chapter by sampling computational results throughout the literature, in a manner which touches on the behaviour of 2D MHD turbulence and the role of singular structure formation in 2D and 3D evolutions. We then continue to emphasize dynamo related aspects, by revisiting the Taylor–Green vortex evolution, and summarize modern small-scale dynamo studies at varying Prandtl numbers. We end with an excursion to compressible MHD, where more waves introduce correspondingly richer physical evolutions.

19.4.1 Structure formation in incompressible MHD studies

Recall the visco-resistive governing PDEs (19.33) for the velocity $\mathbf{v}(\mathbf{x};t)$ and the magnetic field $\mathbf{B}(\mathbf{x};t)$, where again we drop $\mathbf{F}_{\text{driv}}$. As in the hydrodynamic case, alternative formulations exploiting other variables exist, in particular for purely planar (x-y) 2D cases. Then, the vorticity vector $\boldsymbol{\omega} = \nabla \times \mathbf{v}$ simplifies to $\boldsymbol{\omega} = \omega(x,y;t)\,\mathbf{e}_z$ and the vector potential from $\mathbf{B} = \nabla \times \mathbf{A}$ has similarly $\mathbf{A} = A(x,y;t)\,\mathbf{e}_z$. Velocity components relate to scalar vorticity through Eqs. (19.9)(b), whereas the current density is found from

$$\mathbf{j} = j(x,y;t)\mathbf{e}_z = \mu_0^{-1}\nabla \times \mathbf{B} = -\mu_0^{-1}\nabla^2 A\,\mathbf{e}_z\,. \tag{19.76}$$

The governing equations stated in vorticity–vector potential form can then be written as

$$\begin{aligned}\frac{\partial \omega}{\partial t} + (\mathbf{v}\cdot\nabla)\omega &= \rho_0^{-1}(\mathbf{B}\cdot\nabla)j + \nu\nabla^2\omega\,,\\ \frac{\partial A}{\partial t} + (\mathbf{v}\cdot\nabla)A &= \tilde{\eta}\,\nabla^2 A\,.\end{aligned} \tag{19.77}$$

Note that the first equation is the modified form of the HD expression (19.9)(a), and the second one (with $\tilde{\eta} \equiv \eta/\mu_0$) has been presented before in Eq. (19.45).

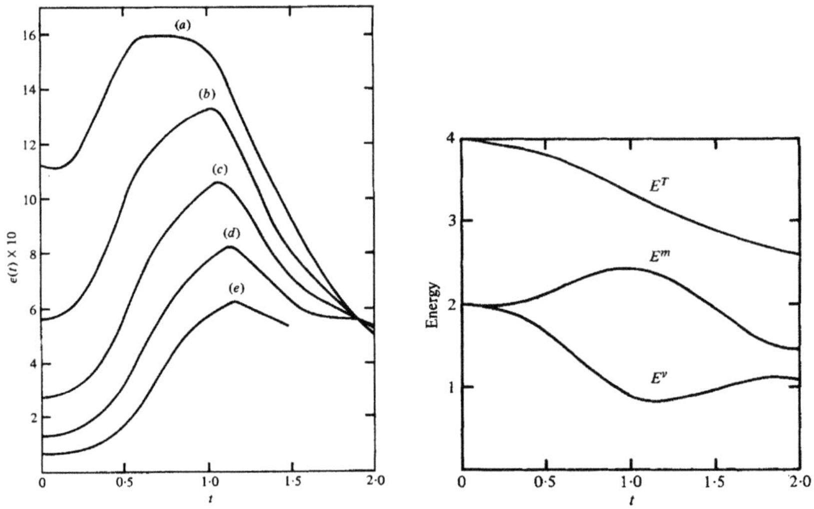

Fig. 19.13 Left: Total energy dissipation as a function of time for varying visco-resistive evolutions. Right: Evolution of the total energy and of the magnetic and kinetic contributions for case (c). (From Orszag and Tang [474].)

This formulation was exploited in a study by Orszag and Tang [474], where a 2D pseudo-spectral code was used with up to 256×256 modes. They took a double-periodic box $[0, 2\pi]^2$, a unit Prandtl number, i.e. $\nu = \tilde{\eta}$ with both coefficients varying from 0.08 to 0.005, and investigated the evolution from the initial condition

$$\mathbf{v}(x,y;t{=}0) = -\sin y\,\mathbf{e}_x + \sin x\,\mathbf{e}_y\,, \qquad A(x,y;t{=}0) = \cos y + \tfrac{1}{2}\cos(2x)\,. \tag{19.78}$$

The configuration invokes a vortical flow field with a superposed double magnetic island structure.

From this initial condition, one can study the ensuing nonlinear evolution governed by the incompressible visco-resistive MHD equations and monitor the total energy dissipation

$$\epsilon(t) = \frac{\rho_0 \nu}{4\pi^2} \iint_0^{2\pi} \omega^2 \, dx \, dy + \frac{\eta}{4\pi^2} \iint_0^{2\pi} j^2 \, dx \, dy \,. \tag{19.79}$$

(We have here re-inserted the constant factor ρ_0, that was previously suppressed, to get the correct relative dimensions of the two contributions.) The authors ran several cases, shown in Fig. 19.13, comparing the evolutions for different values of ν, viz. (a) 0.08, (b) 0.04, (c) 0.02, (d) 0.01, (e) 0.005. This MHD case stands in clear contrast with 2D HD scenarios, where the dissipated energy must decrease with time, $\epsilon(t) \le \epsilon(0)$, since enstrophy is conserved in ideal HD. Figure 19.13 (right panel) shows how energy dissipation has a pronounced maximum at intermediate times. When quantifying the total (magnetic and kinetic) energy evolution for $\nu = 0.02$, they demonstrate a clear energy transfer between kinetic and magnetic components, responsible for the non-monotonically decreasing total energy dissipation. This energy is dissipated in current sheets that form during the energy transfer phase.

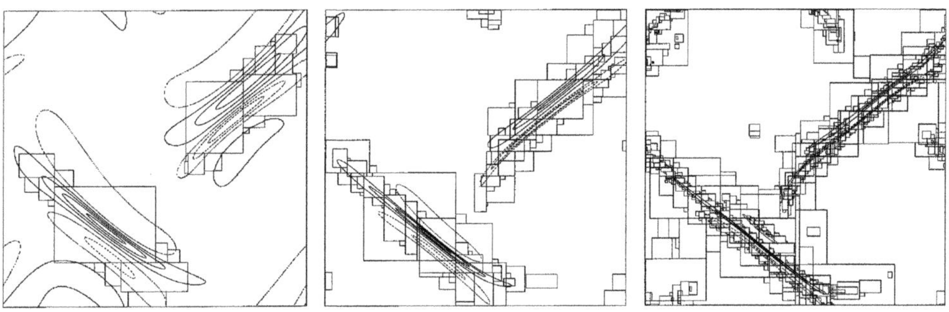

Fig. 19.14 Adaptive Mesh Refinement used in the incompressible MHD simulation of an Orszag–Tang setup. (From Friedel et al. [192].)

While eventually all structure must decay to zero due to dissipation, intermittent flow/current structures of arbitrarily small sizes may be realized. It is of interest to investigate whether such small-scale structures can persist in the limit $\nu \to 0$. This requires very high resolution. One means to achieve this at affordable CPU and memory cost is through Adaptive Mesh Refinement (AMR), an automated procedure to dynamically create and destroy hierarchically nested grids. This technique was used for incompressible MHD by Friedel et al. [192], who revisited the Orszag–Tang evolution, modified to the less symmetric initial condition

$$\begin{aligned} v_x &= -\sin(y+2) \,, \qquad v_y = \sin(x+1.4) \,, \\ B_x &= -\tfrac{1}{3}\sin(y+6.2) \,, \quad B_y = \tfrac{2}{3}\sin(2x+2.3) \,. \end{aligned} \tag{19.80}$$

Figure 19.14 shows the current density and grid structure at the consecutive times $t = 1.6, 2.2, 2.7$. The simulations could quantify the associated amplification of the current density and, by plotting the instantaneous maximal value together with the exponential fit $\frac{1}{2}\exp(2.115\,t)$, a numerical 'proof' for the possibility of singular structure formation in a finite time was given. Their numerical method used a second order 'upwind' scheme, with a projection approach to handle the solenoidality

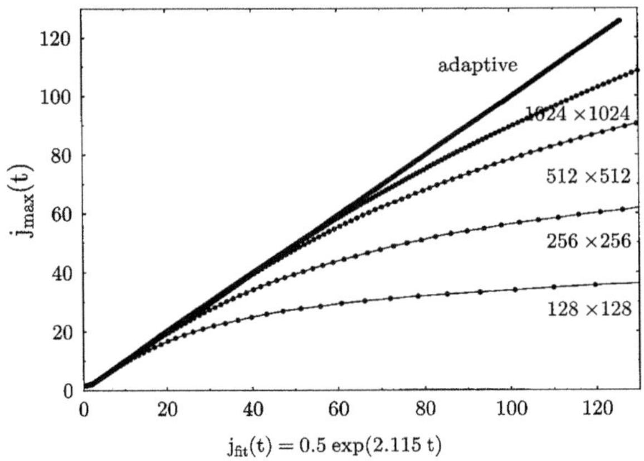

Fig. 19.15 Only by exploiting adaptive mesh refinement does one capture the exponential growth of the localized current: a hint for singular structure formation in a finite time. (From Friedel et al. [192].)

across the AMR grid structure. The code solved the ideal incompressible MHD system (no visco-resistive terms), verified by demonstrating that energy conservation held up to within 1%. The need for extreme resolution (effectively 2048 × 2048) was made clear by showing the numerical results for the maximal current evolution for increasing effective resolutions (Fig. 19.15).

While the results discussed so far were all 2D, the AMR approach was adopted to 3D inviscid incompressible MHD as well. Graueand and Marliani [257] studied several 3D scenarios, among them a 3D Orszag–Tang kind of velocity/field structure. They concluded that this 3D evolution resembles the 2D one, and settles into a similar exponential growth. The same paper also studied related 3D configurations, such as parallel flux tubes. Reaching impressive effective resolutions of up to 4096^3, those again gave hints of exponential growth of the current density, with almost 2D pancake-like current sheets forming locally, where the two flux tubes meet and where ultimately reconnection should occur.

19.4.2 Dynamo aspects continued

We already discussed dynamo related results from kinematic and full Earth dynamo targeted studies in Section 19.2.3. Numerous numerical studies of 3D incompressible MHD evolutions have yielded further insights. Following the 3D hydrodynamic result of Brachet et al. [87], an MHD counterpart by Nore et al. [468] revisited the magnetized Taylor–Green vortex evolution. They assumed an external driver such that the (k_0, k_0, k_0) mode remains fixed at

$$\mathbf{v}_{\mathrm{TG}} = \big(\sin(k_0\, x)\, \cos(k_0\, y),\, -\cos(k_0\, x)\, \sin(k_0\, y),\, 0\big)\, \cos(k_0\, z)\,. \tag{19.81}$$

Taking $k_0 = 1$ in a triple periodic box $[0, 2\pi]^3$, they employed resolutions up to 200^3 to study the behaviour in the presence of a small seed magnetic field breaking the initial symmetry. A pseudo-spectral code was applied with 2/3–1/3 de-aliasing. The seed field was such that the initial ratio of mean magnetic to mean kinetic energy was about $\mathcal{O}(10^{-12})$. At Prandtl number $P_{\mathrm{r}} \simeq 4$ (set by the two Reynolds numbers $R_{\mathrm{e}} = 10.3$ and $R_{\mathrm{m}} = 41$), they found that magnetic energy significantly grew in distinct phases. When the symmetry of the flow was clearly broken, at time $t \sim 220$, quadrupling of the magnetic energy followed in a short time span. Figure 19.16 shows the evolution of the magnetic energy (long dashes), the kinetic energy (solid), the current density (dash-dot), and the squared vorticity (dashed). The rapid increase is followed by an oscillation and nonlinear saturation where the magnetic energy levels remain roughly constant. This study showed that numerical dynamo action is feasible when the symmetry can break. Since it was performed at Prandtl number 4, viscosity dominated over resistivity.

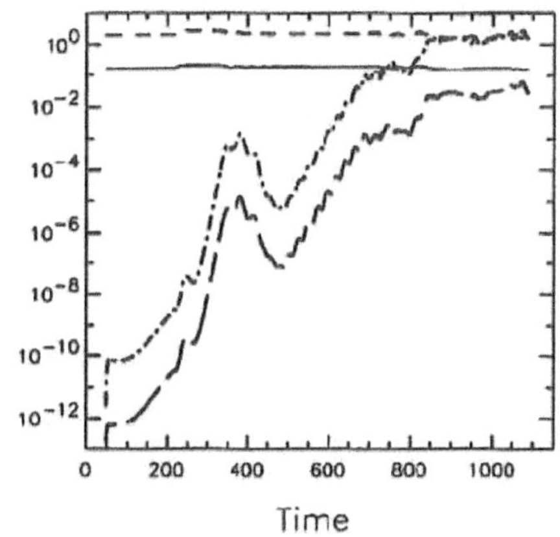

Fig. 19.16 Visco-resistive MHD evolution of the Taylor–Green vortex shows dynamo action by rapid magnetic energy increase after symmetry-breaking. (From Nore et al. [468].)

Modern variants of the above result are found by varying the Prandtl number. This again requires very high resolutions. One such study by Schekochihin et al. [540] targets small-scale dynamo action at large Prandtl numbers. The previous result already indicated that, when viscosity dominates over resistivity, one can have a small-scale dynamo in which the magnetic energy first grows and subsequently decays. In fact, it would be feasible to have the energy spectrum behave schematically as indicated in Fig. 19.17 (with forcing included). This schematic view translates into severe numerical challenges, as it would ask first to fully resolve the inertial range behaviour up to the viscous scale, to then resolve the small-scale dynamo range where magnetic energy grows, and finally the decay due to resistive dissipation at the smallest scales. This kind of scale separation in numerical studies needs sacrifices, and Schekochihin et al. did so by making the forcing and viscous scales similar (hence giving up resolving the hydro-like inertial range). Even with this simplification, they could barely reach $P_r \sim 2500$ at resolutions of 256^3. The setup considered the standard triple periodic box in a pseudo-spectral approach (following the Maron and Goldreich 2001 prescriptions). The simulations indeed demonstrate growth and saturation of the magnetic energy (Fig. 19.18, left panel) for Prandtl numbers beyond 50. Links with analogous kinematic studies were made, in the nonlinearly saturated states found, and the magnetic energy spectra in this saturated state were quantified (Fig. 19.18, right panel).

These quantifications showed that, without imposed mean field, all small-scale turbulent fields are dynamo generated by random stretching and folding of field lines. This was demonstrated in the limit of large P_r, which is only realizable numerically by setting the Reynolds number $R_e \sim 1$

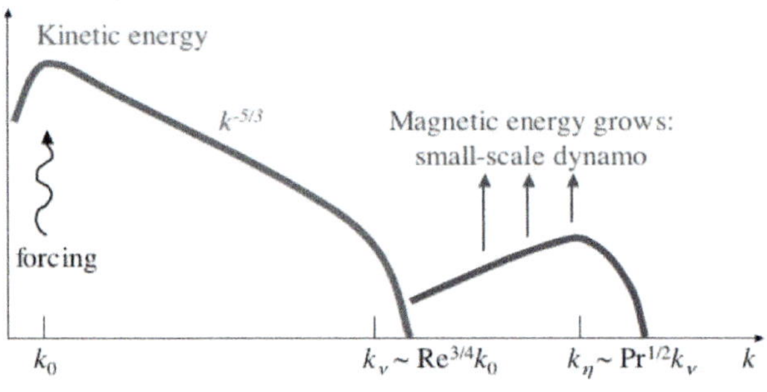

Fig. 19.17 Schematic view of the energy spectrum expected at large Prandtl numbers, with a classical inertial range and small-scale dynamo action. (From Schekochihin et al. [540].)

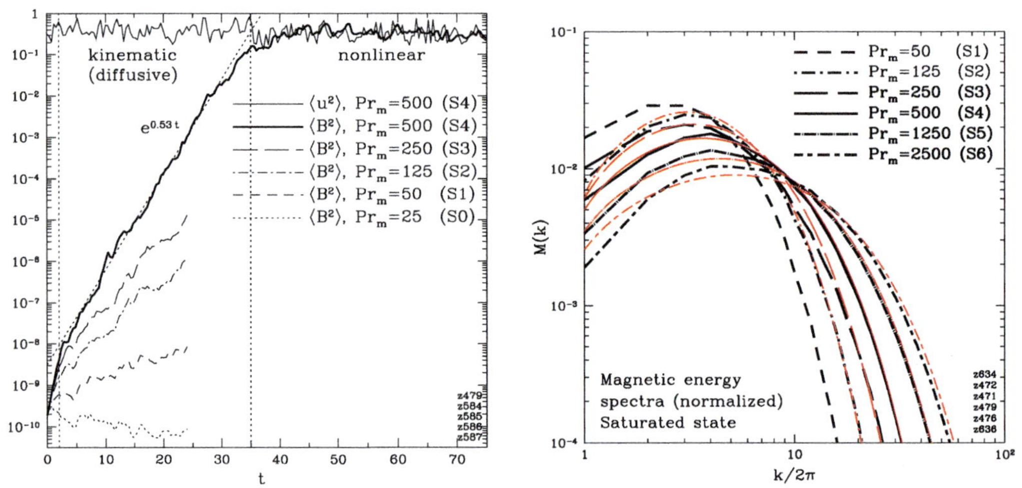

Fig. 19.18 Small-scale dynamo mitigated growth of magnetic energy (left) and magnetic energy spectra (right). (From Schekochihin et al. [540].)

and then follow cases such that P_{r} went up to about 2500. A clear role for highly intermittent and structured fields was evident, which was at the same time organized at the (larger) flow scale. Similarly modern studies have gone to explore the opposite, small P_{r}, regime [541]. This resistivity dominated case is relevant for the various ongoing liquid metal laboratory dynamo experiments. Using 3D pseudo-spectral incompressible MHD simulations, where the flow is randomly driven in the box by a non-helical body force, Schekochihin et al. [541] demonstrated dynamo action above a critical magnetic Reynolds number of about 200, for Reynolds numbers beyond 6000.

19.5 Extension to compressible MHD and concluding remarks

We finish this chapter by contrasting the incompressible, assumed thus far, with the full MHD description. The latter expresses conservation of mass, momentum, energy and magnetic flux, by a set of 8 non-linear PDEs for $\rho, \mathbf{v}, p$ and $\mathbf{B}$, with $\nabla \cdot \mathbf{B} = 0$ as an additional constraint. Mathematically, ideal MHD is a hyperbolic system, where 7 waves appear, labelled as the entropy, forward/backward slow, Alfvén and fast families, which are all anisotropic. According to Section 4.3, this system can be written as conservation laws for the density ρ, momentum density $\boldsymbol{\pi} = \rho\mathbf{v}$, total energy density $\mathcal{H}$ and magnetic field $\mathbf{B}$. As will be discussed in the following chapter, effective numerical strategies exist to solve these compressible ideal MHD equations numerically. Doing so, we can dynamically trigger the emergence of the group diagram by locally perturbing a homogeneous magnetized plasma at rest. For $\gamma = 5/3$, and initial values $\rho = 1, p = 0.6$ and $\mathbf{B} = 0.9\mathbf{e}_x$, we obtain a unit sound speed and an Alfvén speed $b = 0.9$. Simulating the triggering of linear waves by perturbing this uniform plasma in a δ-type fashion, one can compare the outcome with the predictions of a linear wave analysis.

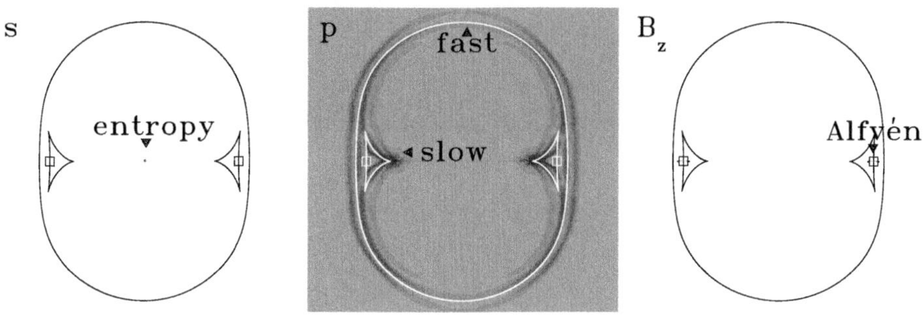

Fig. 19.19 Numerically obtained group diagram compared with the theoretical curves. At left the entropy signal remains at the origin. Middle total pressure view shows fast and slow wave fronts. In the magnetic field component perpendicular to the plane (right), Alfvén waves travel purely along the horizontal background field.

In Fig. 19.19 this comparison is done for a simulation on the square $[-0.5, 0.5]^2$ in a 2.5D scenario which includes constant out-of-plane components v_z and B_z. The state was perturbed at the origin with a density variation $\Delta\rho = 0.1$, velocity perturbation $\Delta v_z = 0.01$ and pressure change $\Delta p = 0.06$. The outcome is a self-similarly travelling set of wave fronts: the MHD group diagram, known from the analytic theory of Chapter 5 (overplotted on the numerical result in Fig. 19.19). The left panel shows the entropy field, where only the point perturbation at the origin (the entropy mode) is visible. The middle panel shows the total pressure, where both the fast (outer) and slow (triangularly shaped) wave fronts are seen. The right panel shows the variation of B_z, showing how Alfvénic modes travel along the perturbed field line. In the incompressible limit, the fast waves move off to infinity, the slow and Alfvén waves degenerate into point-like perturbations travelling at the Alfvén speed, whereas the entropy mode is absent. The compressible MHD system is thus far richer, and even more troubled by anisotropic wave behaviour. In fact, fast and slow waves can steepen into shocks, while the Alfvén waves can carry rotational discontinuities (Chapter 21). This may result in complicated shock-governed evolutions, for steady as well as unsteady magnetized flows.

This aspect of compressible MHD evolutions is best visualized by re-computing the Orszag–Tang vortex for flow conditions where the rotation speed goes up to the local sound speed, as was done by Picone and Dahlburg [489]. The setup remains simple initially, with a double periodic 2D domain $[0, 2\pi]^2$ filled at $t = 0$ with uniform $\rho = 25/9, p = 5/3$, and a velocity circulation $(-\sin y, \sin x)$. Magnetic islands with unequal horizontal wavelength prescribed by $\mathbf{B} = (-\sin y, \sin 2x)$ are used, together with an ideal gas law with $\gamma = 5/3$. The initial state is then found to evolve to compressible

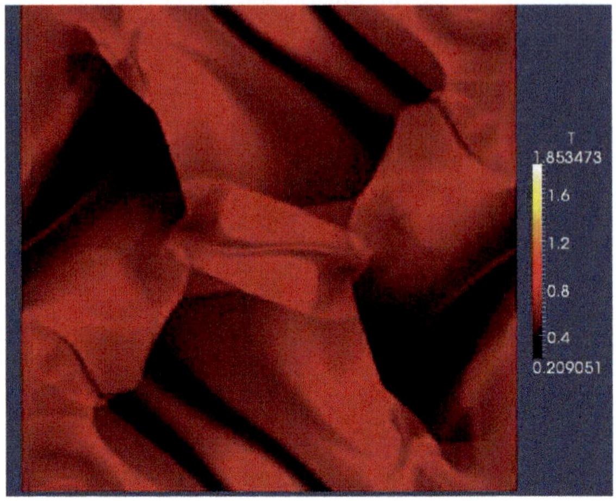

Fig. 19.20 Temperature distribution for the compressible Orszag–Tang setup. (From Picone and Dahlburg [489].)

(supersonic) MHD turbulence. A view of the temperature structure at $t \sim 3.1$ is given in Fig. 19.20. This evolution shows the formation of thin current sheets, where compression occurs, and it is found that shock fronts separate off from the compressed current sheet locations, travelling against the original vortical flow patterns. Their interaction causes intricate smaller-scale dynamics and, when repeated at high resolution and continued beyond the time shown above, modern shock-capturing schemes demonstrate that the thinning current sheets become amenable to tearing, with localized reconnections forming islands. This is most affordably realized by AMR runs.

In conclusion, incompressible MHD has traditionally focused on topics like turbulence, dynamo aspects and singular structure formation. All these translate into high performance computing challenges for realistic Reynolds numbers R_{e} and R_{m} (or P_{r}). For fully periodic settings, it is customary to use spectral or pseudo-spectral methods, effectively giving the highest numerical accuracy possible. Although many issues remain to be resolved in incompressible MHD, attention has turned to *compressible* MHD turbulence, an active research field inspired by the numerous violent astrophysical events. There, one must face the fact that more waves enter the picture, with possible shock formation due to wave steepening. For compressible MHD simulations, advanced shock-capturing schemes as well as AMR techniques are in use, as discussed in the following chapter.

19.6 Literature and exercises

Notes on literature

Nonlinearity and turbulence

– The two excellent monographs by Biskamp on *Nonlinear Magnetohydrodynamics* [65] and on *Magnetohydrodynamic Turbulence* [67] are recommended additional reading for this chapter.

Exercises

[19.1] *Kinematic magnetohydrodynamic evolutions*

In this assignment, we revisit results originally obtained in a seminal paper by Nigel Weiss [642]. The paper discusses how an initially uniform magnetic field becomes distorted and amplified in total energy content, when embedded in a time-independent, prescribed incompressible flow field consisting of several convective eddies. The problem is purely 2D, and in fact linear in the unknowns. We specify to the following setup:

- Take a $[0, 1]^2$ square domain, and use a 2D incompressible flow field
$$\mathbf{v} = \big(\sin(2\pi x)\sin(2\pi y - \pi/2), \sin(2\pi x + \pi/2)\sin(2\pi y)\big)\,;$$
- solve the induction equation for the evolution of the magnetic field for varying resistivity parameter $\tilde{\eta} \equiv \eta/\mu_0$, from 0.1 to 0.001 ;
- use a (pseudo-)spectral approach ;
- monitor the total energy $\iint_0^1 \frac{1}{2}B^2/\mu_0\, dx\, dy$ as a function of time.

As a matter of fact, two formulations can be exploited; pick whichever you prefer.

- Formulation A: Exploit the vector potential $A(x, y; t)$, and solve
$$\frac{\partial A}{\partial t} = -v_x\frac{\partial A}{\partial x} - v_y\frac{\partial A}{\partial y} + \tilde{\eta}\left[\frac{\partial^2 A}{\partial x^2} + \frac{\partial^2 A}{\partial y^2}\right].$$
Due to the incompressibility, one can also write this as a conservation-type equation,
$$\frac{\partial A}{\partial t} = -\nabla\cdot\big(A\mathbf{v} + \tilde{\eta}\nabla A\big)\,.$$
Use periodic top–bottom boundary conditions, but the initial constant $\mathbf{B} = (0, 1)$ means an initially linear $A(x, y; t{=}0) = 1 - x$ (which is not periodic!). Implement a fixed boundary given by $A(x{=}0, y; t) = 1$ and $A(x{=}1, y; t) = 0$.
- Formulation B: Exploit vector components $B_x(x, y; t)$, $B_y(x, y; t)$ to solve
$$\begin{aligned}\frac{\partial B_x}{\partial t} &= -v_y\frac{\partial B_x}{\partial y} - B_x\frac{\partial v_y}{\partial y} + v_x\frac{\partial B_y}{\partial y} + B_y\frac{\partial v_x}{\partial y} + \tilde{\eta}\left[\frac{\partial^2 B_x}{\partial x^2} + \frac{\partial^2 B_x}{\partial y^2}\right],\\ \frac{\partial B_y}{\partial t} &= +v_y\frac{\partial B_x}{\partial x} + B_x\frac{\partial v_y}{\partial x} - v_x\frac{\partial B_y}{\partial x} - B_y\frac{\partial v_x}{\partial x} + \tilde{\eta}\left[\frac{\partial^2 B_y}{\partial x^2} + \frac{\partial^2 B_y}{\partial y^2}\right].\end{aligned}$$
This has the advantage of having a periodic top–bottom, and a periodic left–right boundary. However, we need to ensure $\nabla\cdot\mathbf{B} = 0$ at all times. One may use a projection approach where, after each time step, yielding a field $\mathbf{B}^*$, we correct this possibly non-solenoidal field to $\mathbf{B} = \mathbf{B}^* - \nabla\phi$. The field $\mathbf{B}$ should be solenoidal,
$$\nabla\cdot\mathbf{B} = 0 \quad\Rightarrow\quad \nabla^2\phi = \nabla\cdot\mathbf{B}^*\,,$$
so that we need to solve a Poisson equation for ϕ.

[19.2] *Orszag–Tang incompressible MHD study*

In this assignment, the aim is to use/develop a modern incompressible MHD code to revisit the original results by Orszag and Tang [474]. We wish to solve a 2D incompressible MHD evolution on a double-periodic $[0, 2\pi]^2$ domain. The initial velocity and magnetic field are given by Eqs. (19.78). This mimics a large vortical flow field (repeated due to periodicity), with a double magnetic island structure superimposed (plot the contours for A, coincident with magnetic field lines). The idea is to study the temporal evolution from this initial condition, as governed by the 2D visco-resistive incompressible MHD equations. Follow the original paper by taking a unit Prandtl number, varying $\nu = \tilde{\eta}$ from 0.08 to 0.005. Monitor the total energy dissipation (19.79).

One may try to develop a full 2D incompressible MHD solver, using (pseudo-)spectral discretizations. A possibility is to use an existing code, with a related description of its numerical approach; see Goméz, Mininni and Dmitruk [246].

20

Computational nonlinear MHD

In Chapter 15, we introduced basic concepts for solving linear MHD problems computationally. In principle, the various options discussed there for discretizing a set of partial differential equations in space and time can all be adopted for simulating nonlinear MHD phenomena. Many suitable combinations of spatial (finite difference, finite element, spectral) and temporal (explicit or implicit) discretizations have successfully been applied in problem-specific contexts. However, the conservation law nature of the ideal nonlinear MHD equations poses additional challenges when discontinuous, shock-dominated evolutions are computed. In this chapter, we pay particular attention to 'shock-capturing' schemes, which have proven to be of general use for nonlinear hyperbolic equations. We explain fundamental strategies and some of the difficulties in applying these schemes to the MHD system. This is illustrated with selected examples including simulations of flux rope eruptions and prominence formation in the solar atmosphere. Some of the many alternative algorithmic approaches are discussed as well, complemented with representative applications.

For simulations involving a hierarchy of temporal scales, one must again handle the time scale problem by some (semi-)implicit time integration method. We limit their discussion to some exemplary treatments in computational nonlinear MHD, concluded with an impression of state-of-the-art global simulations of laboratory tokamak plasmas, where extended MHD models are applied.

20.1 General considerations for nonlinear conservation laws

As discussed extensively in Chapter 4, the ideal MHD equations represent a set of *nonlinear* conservation laws. In the nonlinear dynamics, shocks can form spontaneously out of continuous initial data by wave steepening, and contact as well as tangential discontinuities may be encountered. All these discontinuities constitute moving internal interfaces across which the Rankine–Hugoniot conditions must hold, requiring an appropriate numerical representation. In what follows, we introduce various general concepts for conservative systems, which will be useful to design and validate proper numerical treatments.

20.1.1 Conservative versus primitive variable formulations

In general, a system of conservation laws can be expressed in conservation form as

$$\frac{\partial \mathbf{U}}{\partial t} + \nabla \cdot \mathbf{F}(\mathbf{U}) = 0\,. \tag{20.1}$$

This form exploits the set of conservative variables $\mathbf{U}$, and specifies that any local Eulerian change in a conservative variable is due to spatially diverging (or converging) fluxes $\mathsf{F}(\mathbf{U})$. In Section 4.3, we derived the conservation form (4.71)–(4.74) for the ideal MHD equations, and denoted the conservative variables as $\mathbf{U} \equiv (\rho, \boldsymbol{\pi}, \mathcal{H}, \mathbf{B})^{\mathrm{T}}$. In turn, these conservative variables and their fluxes were specific functions of the density ρ, velocity $\mathbf{v}$, pressure p and magnetic field $\mathbf{B}$, or of any other set involving $\mathbf{v}$, $\mathbf{B}$ and two thermodynamic quantities (such as the internal energy e, temperature T, specific entropy s, or the related $S \equiv p\rho^{-\gamma} = f(s), \ln(\rho)$, etc.). Writing the ideal MHD equations in terms of any of the latter set of primitive variables is mathematically and physically equivalent, but the equations then deviate from the conservation form (20.1). It was pointed out in Section 4.5 that only from the conservation form, a simple substitution recipe could be followed to obtain jump conditions, relating changes in the dynamical variables across moving interfaces, i.e. across shock fronts and co-moving plasma–plasma interfaces. When solving the equations numerically, one must obey the conservation laws also at the discrete level, or at least be consistent with their conservation properties within discretization errors.

(a) Quasi-linear forms, flux Jacobians and characteristic speeds Simplifying the discussion to 1D, a set of conservation laws can be written as

$$\frac{\partial \mathbf{U}}{\partial t} + \frac{\partial \mathbf{F}(\mathbf{U})}{\partial x} = 0\,, \tag{20.2}$$

when expressed in terms of the conservative variables $\mathbf{U}(x,t)$. Many equivalent formulations may exist in terms of other primitive variables $\mathbf{V}(x,t)$, which will have a quasi-linear form,

$$\frac{\partial \mathbf{V}}{\partial t} + \mathsf{W} \cdot \frac{\partial \mathbf{V}}{\partial x} = 0\,, \tag{20.3}$$

involving a square matrix $\mathsf{W}(\mathbf{V})$. If the number of (conservative or primitive) variables is indicated by n, W will be an $n \times n$ matrix. The change in variables from $\mathbf{U}$ to $\mathbf{V}$ is quantified by a transformation matrix $\mathsf{U}_\mathbf{V}$ found from $d\mathbf{U} = \mathsf{U}_\mathbf{V} \cdot d\mathbf{V}$. This is also an $n \times n$ square matrix, and we will assume that this transformation is well-defined and thus invertible, $d\mathbf{V} = \mathsf{U}_\mathbf{V}^{-1} \cdot d\mathbf{U}$. The conservation law can similarly be written in quasi-linear form by exploiting the flux Jacobian matrix F_U, such that

$$\frac{\partial \mathbf{U}}{\partial t} + \mathsf{F}_\mathrm{U} \cdot \frac{\partial \mathbf{U}}{\partial x} = 0\,. \tag{20.4}$$

For any equivalent set of primitive variables $\mathbf{V}$ with their coefficient matrix $\mathsf{W}(\mathbf{V})$, we then necessarily find

$$\mathsf{F}_\mathrm{U} = \mathsf{U}_\mathbf{V} \cdot \mathsf{W}(\mathbf{V}) \cdot \mathsf{U}_\mathbf{V}^{-1}\,. \tag{20.5}$$

This similarity relation between matrices F_U and $\mathsf{W}(\mathbf{V})$ guarantees that they have identical eigenvalues. These can thus be computed from $|\mathsf{F}_\mathrm{U} - \lambda_p \mathsf{I}| = 0$ or from $|\mathsf{W} - \lambda_p \mathsf{I}| = 0$. We deduce from dimensional analysis of Eq. (20.4) that these eigenvalues indicate velocities, called the characteristic speeds. For hyperbolic equations, like the ideal MHD equations, these n eigenvalues λ_p for $p = 1, 2, \ldots n$ are real and correspond to sets of right $\mathbf{r}^p$ as well as left eigenvectors $\mathbf{l}^p$ from

$$\mathsf{F}_\mathrm{U} \cdot \mathbf{r}^p = \lambda_p \mathbf{r}^p\,, \qquad \mathbf{l}^p \cdot \mathsf{F}_\mathrm{U} = \mathbf{l}^p \lambda_p\,. \tag{20.6}$$

As always, eigenvectors are determined up to a multiplicative factor only, so that we are free to use any appropriate scaling for these eigenvectors. We can write the set of right eigenvectors $\mathbf{r}^p$ as columns of a matrix $\mathbf{R}$, and the first set of the relations (20.6) then can be written as

$$\mathbf{F_U} \cdot \mathbf{R} = \mathbf{R} \cdot \mathbf{\Lambda}\,, \tag{20.7}$$

where we introduced the diagonal matrix $\mathbf{\Lambda} \equiv \mathrm{diag}(\lambda_1, \ldots, \lambda_n)$ containing the eigenvalues. If all these eigenvalues are always distinct, the system of equations is said to be strictly hyperbolic. We then usually assume the ordering $\lambda_1 < \lambda_2 < \ldots < \lambda_n$. When identical eigenvalues occur, their degeneracy can be accounted for, e.g. by reducing the matrix rank and regrouping variables with the same characteristic velocities together. For the ideal MHD equations, we will have to face such additional complications, since the occurring eigenvalues may coincide and the system is not strictly hyperbolic. In the strictly hyperbolic case, we are sure to have linearly independent right eigenvectors in $\mathbf{R}$. This means that we can compute the left eigenvectors $\mathbf{l}^p$ by inverting the matrix $\mathbf{R}$, and identifying the rows of $\mathbf{R}^{-1}$ with the left eigenvectors, since $\mathbf{R}^{-1} \cdot \mathbf{F_U} = \mathbf{\Lambda} \cdot \mathbf{R}^{-1}$. This makes left and right eigenvectors orthonormal, i.e. $\mathbf{l}^q \cdot \mathbf{r}^p = \delta_{qp}$. With each primitive variable formulation, different (albeit related) sets of right and left eigenvectors are found. Denoting $\widehat{\mathbf{R}}$ as the matrix with the right eigenvectors $\hat{\mathbf{r}}^p$ for $\mathbf{W}(\mathbf{V})$ in its columns, we find from (20.5) that $\widehat{\mathbf{R}} = \mathbf{U_V}^{-1} \cdot \mathbf{R}$. Its left eigenvectors form the rows of $\widehat{\mathbf{R}}^{-1} = \mathbf{R}^{-1} \cdot \mathbf{U_V}$.

Finally, we may try to find a set of variables $\widetilde{\mathbf{R}}$ for which the governing equation (20.2) becomes

$$\frac{\partial \widetilde{\mathbf{R}}}{\partial t} + \mathbf{\Lambda} \cdot \frac{\partial \widetilde{\mathbf{R}}}{\partial x} = 0\,. \tag{20.8}$$

Such so-called Riemann invariants would remain constant on curves $dx = \lambda_p dt$ in the (x, t)-plane. These curves are the characteristics of the hyperbolic partial differential equations at hand, in the same sense as discussed for the ideal MHD case in Section 5.4. When dealing with a system of $n = 2$ unknowns, one can always find both Riemann invariants, one for each characteristic $p = 1, 2$, and write them collectively as in Eq. (20.8). They can be used to solve the initial value problem by drawing the pair of characteristics (with slopes locally determined by λ_p) from each point on the x-axis at $t = 0$ into the (x, t) plane, and using the fact that they 'propagate' a constant value of their respective Riemann invariant. Using this method of characteristics, the value of $\mathbf{U} = (U_1(x,t), U_2(x,t))^{\mathrm{T}}$ at any later time and location can in principle be determined by back tracing the local values of the Riemann invariants along their corresponding characteristics to $t = 0$. This is schematically illustrated in Fig. 20.1. The pair of Riemann invariants $\widetilde{\mathbf{R}} = (\tilde{R}_1(U_1, U_2), \tilde{R}_2(U_1, U_2))^{\mathrm{T}}$ provides a mapping from (U_1, U_2) space to the (x, t) plane of the characteristic curves. An extended region in (x, t) space in which one of the two Riemann invariants remains constant is then called a simple wave region, and through this mapping we then have $U_2(U_1)$ in this region.

When there are $n > 2$ conserved quantities, one may not always be able to find a full set of Riemann invariants obeying Eq. (20.8). A full set of Riemann invariants should obey the proportionalities

$$\frac{\partial \widetilde{\mathbf{R}}}{\partial t} \sim \mathbf{R}^{-1} \cdot \frac{\partial \mathbf{U}}{\partial t} = \widehat{\mathbf{R}}^{-1} \cdot \frac{\partial \mathbf{V}}{\partial t}\,, \qquad \frac{\partial \widetilde{\mathbf{R}}}{\partial x} \sim \mathbf{R}^{-1} \cdot \frac{\partial \mathbf{U}}{\partial x} = \widehat{\mathbf{R}}^{-1} \cdot \frac{\partial \mathbf{V}}{\partial x}\,. \tag{20.9}$$

The related variables found by writing $\mathbf{R}^{-1} \cdot \mathbf{U}$ or equivalently $\widehat{\mathbf{R}}^{-1} \cdot \mathbf{V}$ are termed characteristic variables. In order to generalize the simple wave concept mentioned above for the case of two unknowns to a system of $n > 2$ variables $\mathbf{U} = (U_1(x,t), \ldots, U_n(x,t))^{\mathrm{T}}$, we can again ask whether an extended region of (x, t) space can correspond to an essentially one-parameter variation where

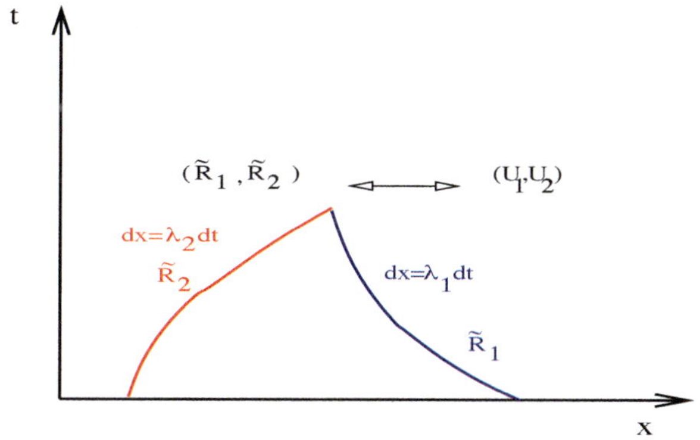

Fig. 20.1 The method of characteristics, schematically illustrated for the case of two variables $\big(U_1(x,t), U_2(x,t)\big)$.

$U_i(U_1)$ for all $i = 1, 2, \ldots n$. One can appreciate that there will be n such simple wave constructions possible, one for each eigenvalue λ_p. In fact, in the (x, t) plane, a pth simple wave region will have straight lines for the pth characteristic curve $dx = \lambda_p dt$ (of varying slope $\lambda_p(U_1)$), along which the solution $\mathbf{U}$ is constant (since $U_i(U_1)$). For the simple wave associated with the pth characteristic field, one arrives at a set of $n - 1$ *generalized* Riemann invariants, defined

$$\text{from either} \quad \frac{dU_i}{r_i^p} = \frac{dU_j}{r_j^p}, \quad \text{or from} \quad \frac{dV_i}{\hat{r}_i^p} = \frac{dV_j}{\hat{r}_j^p} \qquad (\text{for} \quad i \neq j \in \{1, 2, \ldots n\}). \tag{20.10}$$

When integrated, these $n - 1$ relations yield $n - 1$ functions $J^p(\mathbf{U})$ which obey $\nabla_{\mathbf{U}} J^p \cdot \mathbf{r}^p = 0$, or equivalently $\nabla_{\mathbf{V}} J^p \cdot \hat{\mathbf{r}}^p = 0$. Simple wave regions can be proven to be essential ingredients in nonlinear wave problems. In particular, a region in (x, t) bordering a constant state $\mathbf{U} = \mathbf{U}_0$ region will be a simple wave region (or another constant state, which can be regarded as a trivial simple wave).

In what follows, we will encounter specific examples for the various matrices, their eigenvalue and eigenvector pairs, and the (generalized) Riemann invariants. They usually play a prominent role in the design of shock-capturing numerical algorithms for integrating nonlinear hyperbolic systems. They are also useful for thoroughly checking numerically obtained solutions.

(b) Iso-thermal MHD in 1.5D and Riemann invariants The simplest MHD problem where we can directly illustrate the above concepts is posed by the 1.5D iso-thermal MHD equations. The assumption of iso-thermal conditions can be justified in various astrophysical contexts, when dealing with the dynamics of a gas whose cooling time due to radiative processes is very short with respect to all other dynamical time scales. These conditions can be met in dilute environments which are constantly fed energy from a central source (e.g. irradiation from a star). To keep the gas at a constant temperature, one must thus relax energy conservation. Instead of explicitly handling source and sink terms in an equation governing the energy evolution, the iso-thermal magnetohydro system avoids the solution of this equation altogether, and assumes that the net effect is to maintain a constant temperature. The $\nabla \cdot \mathbf{B} = 0$ condition in a 1D configuration with quantities varying along x turns B_x into a constant parameter. Therefore, the first non-trivial MHD counterpart of a 1D iso-

thermal gas problem needs to account for a non-vanishing component $B_y(x)$ (and hence also v_y) in a translationally invariant y-direction. This is typically referred to as a 1.5D problem.

The governing equations can be written in the conservation form (20.2) as

$$\begin{pmatrix} \rho \\ m_x \\ m_y \\ B_y \end{pmatrix}_t + \begin{pmatrix} m_x \\ m_x^2/\rho - B_x^2 + c_\mathrm{i}^2 \rho + \frac{1}{2}(B_x^2 + B_y^2) \\ m_x m_y/\rho - B_x B_y \\ B_y m_x/\rho - B_x m_y/\rho \end{pmatrix}_x = 0\,. \tag{20.11}$$

This introduces the four conserved variables consisting of density, both momentum components and the translational magnetic field component. In the x-momentum equation, the gas pressure is written as $p = c_\mathrm{i}^2 \rho$, with c_i^2 denoting the squared iso-thermal sound speed, a constant proportional to the fixed temperature. The flux Jacobian matrix is then found to be

$$\mathbf{F_U} = \begin{pmatrix} 0 & 1 & 0 & 0 \\ c_\mathrm{i}^2 - m_x^2/\rho^2 & 2m_x/\rho & 0 & B_y \\ -m_x m_y/\rho^2 & m_y/\rho & m_x/\rho & -B_x \\ (-m_x B_y + B_x m_y)/\rho^2 & B_y/\rho & -B_x/\rho & m_x/\rho \end{pmatrix}. \tag{20.12}$$

A primitive variable formulation (20.3) exploits the velocity components:

$$\begin{pmatrix} \rho \\ v_x \\ v_y \\ B_y \end{pmatrix}_t + \begin{pmatrix} v_x & \rho & 0 & 0 \\ c_\mathrm{i}^2/\rho & v_x & 0 & B_y/\rho \\ 0 & 0 & v_x & -B_x/\rho \\ 0 & B_y & -B_x & v_x \end{pmatrix} \begin{pmatrix} \rho \\ v_x \\ v_y \\ B_y \end{pmatrix}_x = 0\,. \tag{20.13}$$

The eigenvalues from both the flux Jacobian $\mathbf{F_U}$ and the matrix $\mathbf{W}(\mathbf{V})$ are

$$\lambda_1 = v_x - c_\mathrm{f}\,, \quad \lambda_2 = v_x - c_\mathrm{s}\,, \quad \lambda_3 = v_x + c_\mathrm{s}\,, \quad \lambda_4 = v_x + c_\mathrm{f}\,, \tag{20.14}$$

where the familiar slow and fast magneto-acoustic speeds are computed from

$$c_{\mathrm{f,s}}^2 = \tfrac{1}{2}\left[c_\mathrm{i}^2 + (B_x^2 + B_y^2)/\rho\right] \pm \tfrac{1}{2}\sqrt{\left[c_\mathrm{i}^2 + (B_x^2 + B_y^2)/\rho\right]^2 - 4c_\mathrm{i}^2 B_x^2/\rho}\,. \tag{20.15}$$

As already mentioned, some of the four eigenvalues coincide in certain limits, such that the MHD system is not strictly hyperbolic.

▷ **Eigenvectors** The right eigenvectors for the Jacobian matrix are the columns of

$$\mathbf{R} = \begin{pmatrix} 1 & 1 & 1 & 1 \\ v_x - c_\mathrm{f} & v_x - c_\mathrm{s} & v_x + c_\mathrm{s} & v_x + c_\mathrm{f} \\ v_y + \dfrac{c_\mathrm{f} B_x B_y}{\rho c_\mathrm{f}^2 - B_x^2} & v_y + \dfrac{c_\mathrm{s} B_x B_y}{\rho c_\mathrm{s}^2 - B_x^2} & v_y - \dfrac{c_\mathrm{s} B_x B_y}{\rho c_\mathrm{s}^2 - B_x^2} & v_y - \dfrac{c_\mathrm{f} B_x B_y}{\rho c_\mathrm{f}^2 - B_x^2} \\ \dfrac{c_\mathrm{f}^2 B_y}{\rho c_\mathrm{f}^2 - B_x^2} & \dfrac{c_\mathrm{s}^2 B_y}{\rho c_\mathrm{s}^2 - B_x^2} & \dfrac{c_\mathrm{s}^2 B_y}{\rho c_\mathrm{s}^2 - B_x^2} & \dfrac{c_\mathrm{f}^2 B_y}{\rho c_\mathrm{f}^2 - B_x^2} \end{pmatrix}. \tag{20.16}$$

Note that we are free to adopt another scaling for these eigenvectors. This will be needed for proper handling of the indeterminacies where $\rho c_{\mathrm{s,f}}^2 = B_x^2$. The left eigenvectors of $\mathbf{F_U}$ can be found in the rows of

$$
\mathbf{R}^{-1} = \begin{pmatrix}
-\frac{v_y c_\mathrm{f}}{2c_\mathrm{i}^2}\frac{B_y}{B_x}\frac{c_\mathrm{s}^2}{c_\mathrm{f}^2 - c_\mathrm{s}^2} + \frac{v_x c_\mathrm{f} + c_\mathrm{i}^2}{2c_\mathrm{i}^2}\alpha_\mathrm{f}^2 & -\alpha_\mathrm{f}^2\frac{c_\mathrm{f}}{2c_\mathrm{i}^2} & \frac{B_y}{B_x}\frac{2c_\mathrm{s}^2 c_\mathrm{f}}{(c_\mathrm{f}^2 - c_\mathrm{s}^2)c_\mathrm{i}^2} & \frac{B_y}{2(c_\mathrm{f}^2 - c_\mathrm{s}^2)} \\
\frac{v_y c_\mathrm{s}}{2c_\mathrm{i}^2}\frac{B_y}{B_x}\frac{c_\mathrm{f}^2}{c_\mathrm{f}^2 - c_\mathrm{s}^2} + \frac{v_x c_\mathrm{s} + c_\mathrm{i}^2}{2c_\mathrm{i}^2}\alpha_\mathrm{s}^2 & -\alpha_\mathrm{s}^2\frac{c_\mathrm{s}}{2c_\mathrm{i}^2} & -\frac{B_y}{B_x}\frac{2c_\mathrm{f}^2 c_\mathrm{s}}{(c_\mathrm{f}^2 - c_\mathrm{s}^2)c_\mathrm{i}^2} & \frac{-B_y}{2(c_\mathrm{f}^2 - c_\mathrm{s}^2)} \\
-\frac{v_y c_\mathrm{s}}{2c_\mathrm{i}^2}\frac{B_y}{B_x}\frac{c_\mathrm{f}^2}{c_\mathrm{f}^2 - c_\mathrm{s}^2} - \frac{v_x c_\mathrm{s} - c_\mathrm{i}^2}{2c_\mathrm{i}^2}\alpha_\mathrm{s}^2 & \alpha_\mathrm{s}^2\frac{c_\mathrm{s}}{2c_\mathrm{i}^2} & \frac{B_y}{B_x}\frac{2c_\mathrm{f}^2 c_\mathrm{s}}{(c_\mathrm{f}^2 - c_\mathrm{s}^2)c_\mathrm{i}^2} & \frac{-B_y}{2(c_\mathrm{f}^2 - c_\mathrm{s}^2)} \\
\frac{v_y c_\mathrm{f}}{2c_\mathrm{i}^2}\frac{B_y}{B_x}\frac{c_\mathrm{s}^2}{c_\mathrm{f}^2 - c_\mathrm{s}^2} - \frac{v_x c_\mathrm{f} - c_\mathrm{i}^2}{2c_\mathrm{i}^2}\alpha_\mathrm{f}^2 & \alpha_\mathrm{f}^2\frac{c_\mathrm{f}}{2c_\mathrm{i}^2} & -\frac{B_y}{B_x}\frac{2c_\mathrm{s}^2 c_\mathrm{f}}{(c_\mathrm{f}^2 - c_\mathrm{s}^2)c_\mathrm{i}^2} & \frac{B_y}{2(c_\mathrm{f}^2 - c_\mathrm{s}^2)}
\end{pmatrix}. \tag{20.17}
$$

In these expressions, we introduced the positive parameters

$$
\alpha_\mathrm{f}^2 = \frac{c_\mathrm{i}^2 - c_\mathrm{s}^2}{c_\mathrm{f}^2 - c_\mathrm{s}^2}, \qquad \alpha_\mathrm{s}^2 = \frac{c_\mathrm{f}^2 - c_\mathrm{i}^2}{c_\mathrm{f}^2 - c_\mathrm{s}^2}. \tag{20.18}
$$

For the primitive variable formulation, the right eigenvectors are found to be

$$
\widehat{\mathbf{R}} = \begin{pmatrix}
1 & 1 & 1 & 1 \\
-c_\mathrm{f}/\rho & -c_\mathrm{s}/\rho & c_\mathrm{s}/\rho & c_\mathrm{f}/\rho \\
\frac{c_\mathrm{f}}{\rho}\frac{B_x B_y}{\rho c_\mathrm{f}^2 - B_x^2} & \frac{c_\mathrm{s}}{\rho}\frac{B_x B_y}{\rho c_\mathrm{s}^2 - B_x^2} & -\frac{c_\mathrm{s}}{\rho}\frac{B_x B_y}{\rho c_\mathrm{s}^2 - B_x^2} & -\frac{c_\mathrm{f}}{\rho}\frac{B_x B_y}{\rho c_\mathrm{f}^2 - B_x^2} \\
\frac{c_\mathrm{f}^2 B_y}{\rho c_\mathrm{f}^2 - B_x^2} & \frac{c_\mathrm{s}^2 B_y}{\rho c_\mathrm{s}^2 - B_x^2} & \frac{c_\mathrm{s}^2 B_y}{\rho c_\mathrm{s}^2 - B_x^2} & \frac{c_\mathrm{f}^2 B_y}{\rho c_\mathrm{f}^2 - B_x^2}
\end{pmatrix}. \tag{20.19}
$$

Finally, the left eigenvectors for $\mathbf{W}(\mathbf{V})$ are the rows of

$$
\widehat{\mathbf{R}}^{-1} = \begin{pmatrix}
\frac{1}{2}\alpha_\mathrm{f}^2 & -\frac{1}{2}\alpha_\mathrm{f}^2\frac{\rho c_\mathrm{f}}{c_\mathrm{i}^2} & \frac{1}{2}\frac{B_x B_y}{c_\mathrm{f}(c_\mathrm{f}^2 - c_\mathrm{s}^2)} & \frac{B_y}{2(c_\mathrm{f}^2 - c_\mathrm{s}^2)} \\
\frac{1}{2}\alpha_\mathrm{s}^2 & -\frac{1}{2}\alpha_\mathrm{s}^2\frac{\rho c_\mathrm{s}}{c_\mathrm{i}^2} & -\frac{1}{2}\frac{B_x B_y}{c_\mathrm{s}(c_\mathrm{f}^2 - c_\mathrm{s}^2)} & \frac{-B_y}{2(c_\mathrm{f}^2 - c_\mathrm{s}^2)} \\
\frac{1}{2}\alpha_\mathrm{s}^2 & \frac{1}{2}\alpha_\mathrm{s}^2\frac{\rho c_\mathrm{s}}{c_\mathrm{i}^2} & \frac{1}{2}\frac{B_x B_y}{c_\mathrm{s}(c_\mathrm{f}^2 - c_\mathrm{s}^2)} & \frac{-B_y}{2(c_\mathrm{f}^2 - c_\mathrm{s}^2)} \\
\frac{1}{2}\alpha_\mathrm{f}^2 & \frac{1}{2}\alpha_\mathrm{f}^2\frac{\rho c_\mathrm{f}}{c_\mathrm{i}^2} & -\frac{1}{2}\frac{B_x B_y}{c_\mathrm{f}(c_\mathrm{f}^2 - c_\mathrm{s}^2)} & \frac{B_y}{2(c_\mathrm{f}^2 - c_\mathrm{s}^2)}
\end{pmatrix}. \tag{20.20}
$$

Using the columns in $\widehat{\mathbf{R}}$ of Eq. (20.19), we can derive generalized Riemann invariants. ◁

Following the definition from Eq. (20.10), we get for the first wave family associated with $\lambda_1 = v_x - c_\mathrm{f}$ the system

$$
c_\mathrm{f}\, d\rho + \rho\, dv_x = 0\,, \quad B_x(c_\mathrm{f}^2 - c_\mathrm{i}^2)\, dv_x + B_y c_\mathrm{f}^2\, dv_y = 0\,, \quad \rho c_\mathrm{f}\, dv_y - B_x\, dB_y = 0\,. \tag{20.21}
$$

Changing to dimensionless variables $r = \rho c_\mathrm{i}^2 / B_x^2$ and $q = c_\mathrm{f}^2 / c_\mathrm{i}^2$, we find

$$
dv_x = -c_\mathrm{i}\frac{\sqrt{q}}{r}\, dr\,, \quad dv_y = c_\mathrm{i}\frac{\sqrt{(q-1)(qr-1)}}{r(qr-1)}\,\mathrm{sgn}(B_x B_y)\, dr\,, \quad dq = \frac{q^2(q-1)}{q^2 r - 1}\, dr\,. \tag{20.22}
$$

The function $J = r/(q-1) + \int dq\, q^{-2}(q-1)^{-2}$ satisfies $dJ = \partial J/\partial r\, dr + \partial J/\partial q\, dq = 0$. This essentially integrates the third equation in the set (20.22). Since this generalized Riemann invariant does not depend on the velocity components, it will be referred to as the magneto-acoustic Riemann invariant. The two other generalized Riemann invariants for simple waves associated with λ_1 are found from integrating the first two equations in (20.22). Note the requirement $q > 1$ and $qr > 1$, restricting (r, q) state space, so that better parameterizations may exist [457]. Similar expressions can be found for the three generalized Riemann invariants associated with each simple wave family for $\lambda_{2,3,4}$. For the iso-thermal system considered, we can explicitly evaluate the integral appearing in the fast and slow magneto-acoustic (generalized) Riemann invariants, to get the analytic forms

$$J_{\mathrm{s,f}} = \frac{-c_{\mathrm{s,f}}^2 + \rho c_{\mathrm{i}}^4 / B_x^2}{c_{\mathrm{s,f}}^2 - c_{\mathrm{i}}^2} + \frac{c_{\mathrm{s,f}}^2 - c_{\mathrm{i}}^2}{c_{\mathrm{s,f}}^2} - 2 \ln \left| \frac{c_{\mathrm{s,f}}^2 - c_{\mathrm{i}}^2}{c_{\mathrm{s,f}}^2} \right| . \tag{20.23}$$

These can be used to verify numerical solutions to Riemann problems, as will be explained and demonstrated further on.

(c) Linear hyperbolic PDE system It is instructive to briefly reconsider the linear system case, where the flux is given by $\mathbf{F} = \mathsf{A} \cdot \mathbf{U}$ and the Jacobian is a constant $n \times n$ matrix $\mathsf{A} = \mathsf{F}_{\mathbf{U}}$. Again assuming strict hyperbolicity, such that A has real and distinct eigenvalues, we then find immediately from Eq. (20.9) that Riemann invariants and characteristic variables coincide, $\widetilde{\mathbf{R}} = \mathsf{R}^{-1} \cdot \mathbf{U}$, since the matrix R^{-1} is then also constant. In terms of these characteristic variables, the system of equations becomes a set of n decoupled advection equations

$$\frac{\partial \tilde{R}_p}{\partial t} + \lambda_p \frac{\partial \tilde{R}_p}{\partial x} = 0 \qquad (p = 1, \ldots n) . \tag{20.24}$$

Each of these equations has a trivial, exact solution in terms of its initial condition,

$$\tilde{R}_p(x,t) = \tilde{R}_p(x - \lambda_p t, 0) . \tag{20.25}$$

The exact solution to the linear hyperbolic PDE system in terms of the conservative variables is then found from $\mathbf{U} = \mathsf{R} \cdot \widetilde{\mathbf{R}}$, or

$$\mathbf{U}(x,t) = \sum_{p=1}^{n} \tilde{R}_p(x - \lambda_p t, 0) \, \mathbf{r}^p . \tag{20.26}$$

For every point x, the solution at any time t thus depends only on the value of the initial data at n discrete points. In particular, these points are at the intersection of the $t = 0$ axis with all the p-characteristics through (x,t). The p-characteristics are in this linear case all straight lines $x = x_0 + \lambda_p t$. Since we know the exact solution when the initial condition is translated to the characteristic variables, the linear hyperbolic system can be used to validate numerical approaches for handling, in particular, discontinuous data.

20.1.2 Scalar conservation law and the Riemann problem

The consequences of nonlinearity in a system of conservation laws can already be illustrated in the case of a single scalar conservation law. A general nonlinear scalar conservation law for the quantity $u(x,t)$ is written as

$$u_t + \big(f(u)\big)_x = 0 , \tag{20.27}$$

where the flux function $f(u)$ depends nonlinearly on u. The inviscid Burgers' equation is obtained for $f(u) = \frac{1}{2}u^2$. Assuming differentiability, Burgers' equation can be written as

$$u_t + u \, u_x = 0 , \tag{20.28}$$

so that we can speak of a local 'advection' speed u, which is also the characteristic speed quantified by the Jacobian, i.e. derivative, $f_u = u \equiv f'(u)$. Identifying u as a density, the local density determines the local advection speed, with denser regions travelling faster than more rarified ones. Note further that the spatial integral $\int_{x_1}^{x_2} u\, dx$ determining the total 'mass' over the interval $[x_1, x_2]$ will be the same for all times if the fluxes $f(u)$ at the edges x_1 and x_2 vanish at all times. This is the conservation property of Eq. (20.27). From these considerations, it is now easily understood how wave steepening and shock formation can occur. Envision a triangular pulse as initial data, given by

$$u(x,0) = \begin{cases} u_0 & (x \le -x_0)\,, \\ u_0 + h_0(x_0 + x)/x_0 & (-x_0 < x \le 0)\,, \\ u_0 + h_0(x_0 - x)/x_0 & (0 < x \le x_0)\,, \\ u_0 & (x > x_0)\,. \end{cases} \tag{20.29}$$

Since the characteristic speed is given by the local value of u, the tip of the triangle experiences the fastest rightward advection (we assume $u_0 > 0$ and $h_0 > 0$). In conserving the total area underneath the triangle, the front edge steepens. At the space-time point when the tip of the triangle has caught up with the rightmost point of the front edge, a discontinuity appears in the solution. This happens at the time when their characteristics meet, namely at time $t_s = x_0/h_0$. From this time of shock formation onwards, conservation now demands the discrete equivalent of the conservation law to hold across the discontinuity. This means, in terms of the adjacent left u_ℓ and right u_r values, that

$$f(u_\ell) - f(u_r) = s\,(u_\ell - u_r)\ . \tag{20.30}$$

Here, the discontinuity is assumed to travel with shock speed s, and for the inviscid Burgers' case, we find that $s = \frac{1}{2}(u_\ell + u_r)$. Precisely at time t_s, we then have $s(t_s) = u_0 + \frac{1}{2}h_0$. From this time on, the base of the triangle will widen due to the speed difference between the left edge travelling with u_0 and the shocked right edge travelling at speed $s(t)$. In accord with conservation, the height of the triangle must therefore decrease in time. The full solution for times $t > t_s = x_0/h_0$ thus works out to be

$$u(x,t) = \begin{cases} u_0 & (x \le -x_0 + u_0 t), \\ u_0 + 2h_0 \dfrac{x + x_0 - u_0 t}{\left(\sqrt{h_0 t} + \sqrt{x_0}\right)^2} & (-x_0 + u_0 t < x \le u_0 t + \sqrt{x_0 h_0 t}), \\ u_0 & (x > u_0 t + \sqrt{x_0 h_0 t}). \end{cases} \tag{20.31}$$

The shock speed for the right edge is then $s(t) = u_0 + h_0\sqrt{x_0}/\left(\sqrt{x_0} + \sqrt{h_0 t}\right)$. The evolution is schematically indicated in Fig. 20.2.

Fig. 20.2 Evolution of an initially triangular pulse in the inviscid Burgers' equation.

The relation (20.30) is the jump condition or Rankine–Hugoniot relation for the simple scalar conservation law. Note that it is symmetric in its two arguments representing a left and right state

u_ℓ, u_r. Obviously, though, for Burgers' equation we appreciate immediately that a different solution must emerge when we interchange the two states in the particular discontinuous initial condition where

$$u = u_\ell \quad (\text{for } x \leq 0), \qquad u = u_r \quad (\text{for } x > 0). \tag{20.32}$$

This represents the Riemann problem where two constant states in contact are left to evolve. When $u_\ell > u_r$, we can expect a pure shock solution, with shock speed found from (20.30). Conversely, when $u_\ell < u_r$, we expect the right state to 'run away' from the left one, turning the initial discontinuity into a continuously increasing profile. Therefore, the jump relation (20.30) is not sufficient to discriminate between allowed (weak) discontinuous solutions. A physically admissible shock must obey another relation as well, known as the *Lax entropy condition*. This Lax entropy condition states that a shock with speed s satisfying the Rankine–Hugoniot relations must additionally obey

$$f'(u_\ell) > s > f'(u_r). \tag{20.33}$$

We identified $f'(u)$ from the scalar conservation law as the characteristic speed, so this expresses that the shock speed must lie in-between the characteristic speeds of the two adjacent states. This condition is asymmetric in left versus right state. For Burgers' equation, $u_\ell > \frac{1}{2}(u_\ell + u_r) > u_r$, as expected. The space-time characteristics for Burgers' equation on either side of this admissible shock have constant slopes $u_\ell > u_r$, so that the characteristics 'go into the shock' [404].

▷ **Generalization to nonlinear systems** We can readily generalize the Rankine–Hugoniot relations for shocks to nonlinear systems. From the conservation form (20.2), we obtain the equivalent of Eq. (20.30), expressing conservation across a moving discontinuity,

$$\mathbf{F}(\mathbf{U}) - \mathbf{F}(\mathbf{U}^*) = s(\mathbf{U} - \mathbf{U}^*). \tag{20.34}$$

For a fixed state $\mathbf{U}^*$, relation (20.34) defines its Hugoniot locus, consisting of all states $\mathbf{U}$ which can be connected to $\mathbf{U}^*$ via a discontinuous shock moving at (scalar) speed s. For a system of n equations, we will then find n one-parameter families. When writing $\mathbf{U}(\zeta, \mathbf{U}^*)$ and $s(\zeta, \mathbf{U}^*)$, where ζ is chosen such that $\mathbf{U}(0, \mathbf{U}^*) = \mathbf{U}^*$, we can look in particular at weak shocks. Then $\mathbf{F}(\mathbf{U}(\zeta, \mathbf{U}^*)) \approx \mathbf{F}(\mathbf{U}(0, \mathbf{U}^*)) + \mathsf{F}_\mathbf{U}(\mathbf{U}(0, \mathbf{U}^*)) \cdot (\mathbf{U}(\zeta, \mathbf{U}^*) - \mathbf{U}^*)$. We then differentiate relation (20.34) with respect to ζ, writing $d\mathbf{U}/d\zeta = \mathbf{U}'$, and evaluate it at $\zeta = 0$:

$$\mathsf{F}_\mathbf{U}(\mathbf{U}(0, \mathbf{U}^*)) \cdot \mathbf{U}'(0, \mathbf{U}^*) = s(0, \mathbf{U}^*)\, \mathbf{U}'(0, \mathbf{U}^*). \tag{20.35}$$

Since this expresses that $\mathbf{U}'(0, \mathbf{U}^*)$ must be an eigenvector of the flux Jacobian evaluated in $\mathbf{U}^*$, with eigenvalue $s(0, \mathbf{U}^*)$, the n parameter families for shocks can be associated with the n eigenvalues of $\mathsf{F}_\mathbf{U}$. For weak shocks, we find $s_p = \lambda_p(\mathbf{U}^*)$ so that shock speeds are given by the characteristic speeds. However, this correspondence for weak shocks does not hold for strong shocks, where the nonlinear character of Eq. (20.34) can cause significant deviations. Furthermore, we again will need selection criteria to distinguish which part of the Hugoniot locus represents physically admissible shocks. One such criterion is the generalization of the Lax entropy condition. This states that a jump in the pth wave family obeying Rankine–Hugoniot which travels with speed s is allowed when

$$\lambda_p(\mathbf{U}_\ell) > s > \lambda_p(\mathbf{U}_r). \tag{20.36}$$

The p-characteristics then enter the p-shock from both sides. From every point on the shock, one can then travel along characteristics backward in time. This indicates how information reaches the shock from the past, not from the future, as required by 'causality'. This time-irreversibility argument leads to the nomenclature of 'entropy' condition for (20.36). However, additional arguments may be needed to select physically admissible shocks, since for a system of n nonlinear equations, a discontinuity travelling at speed s can have more than one of the n characteristic families on each side converge into the shock. This situation represents so-called overcompressive shocks. ◁

(a) Rarefaction waves Returning to Burgers' equation, we can also analyse the case $u_\ell < u_\mathrm{r}$, and a continuously increasing profile is expected. In fact, there are infinitely many weak solutions in this case, including a pure shock one with shock speed $s = \frac{1}{2}(u_\ell + u_\mathrm{r})$. But now the characteristics go out of the shock, which makes this solution unstable to perturbations. A small change in initial data can yield a completely different solution in this 'entropy-violating' shock [404]. The physically correct continuous solution to the Riemann problem will obey an x/t self-similarity in the (x,t) plane. Writing $u(x,t) = u(x/t) \equiv u(\xi)$, we find that the conservation law (20.27) translates into

$$f'(u)\,\frac{du}{d\xi} = \xi\,\frac{du}{d\xi}\,, \tag{20.37}$$

or we must have $\xi = f'(u)$. For Burgers' equation, this means that the solution to the Riemann problem, for which $u_\ell < u_\mathrm{r}$ with the discontinuity initially at $x = 0$, is given by

$$u(x,t) = u(x/t) = \begin{cases} u_\ell & (x < u_\ell t) \\ x/t & (u_\ell t < x < u_\mathrm{r} t) \\ u_\mathrm{r} & (x > u_\mathrm{r} t) \end{cases}\,. \tag{20.38}$$

This is a so-called rarefaction wave, as u decreases (in terms of a density: the medium gets rarefied) when the signal passes. Hence, the solution to the Riemann problem for the inviscid Burgers' equation is either a shock wave (case $u_\ell > u_\mathrm{r}$) or a rarefaction wave (case $u_\ell < u_\mathrm{r}$).

▷ **Generalization to nonlinear systems** The self-similar rarefaction wave solution for the nonlinear system starts with the quasi-linear form (20.4), as it analyses the possibility for continuously varying solutions where $\mathbf{U}(x,t) = \mathbf{U}(x/t) \equiv \mathbf{U}(\xi)$. Note that this is the simple wave construction mentioned earlier, with the dependence on x/t sometimes referred to as a centred simple wave. We deduce that, for such self-similar solutions, $\partial\mathbf{U}/\partial t = -x\mathbf{U}'/t^2$ and $\partial\mathbf{U}/\partial x = \mathbf{U}'/t$, leading to

$$\mathsf{F}_\mathbf{U} \cdot \mathbf{U}' = \xi\mathbf{U}'\,. \tag{20.39}$$

This expression means that ξ must be an eigenvalue λ_p of the flux Jacobian $\mathsf{F}_\mathbf{U}$, and that $\mathbf{U}'$ (with the prime indicating the derivative with respect to ξ) must be proportional to the corresponding right eigenvector $\mathbf{r}^p$. Note that we already found this relation for the *local* Hugoniot locus (weak shock case), but here it is true along the entire parameterized $\xi = x/t$ range. Curves in state space $\mathbf{U}$ along which the tangent always coincides with an eigenvector are termed integral curves. Hence, from Eq. (20.39), we can write

$$\xi = \lambda_p(\mathbf{U}(\xi)) \quad \text{and} \quad \mathbf{U}' = \alpha(\xi)\mathbf{r}^p(\mathbf{U}(\xi))\,. \tag{20.40}$$

Differentiating the first expression with respect to ξ we find

$$1 = (\nabla_\mathbf{U}\lambda_p) \cdot \mathbf{U}'\,. \tag{20.41}$$

Using the correspondence between $\mathbf{U}'$ and $\mathbf{r}^p$, the proportionality constant is found from

$$\alpha(\xi) = \frac{1}{(\nabla_\mathbf{U}\lambda_p) \cdot \mathbf{r}^p}\,. \tag{20.42}$$

Obviously, the construction fails for wave families where the denominator in Eq. (20.42), the so-called structure coefficient $s_p \equiv (\nabla_\mathbf{U}\lambda_p) \cdot \mathbf{r}^p$, vanishes identically. In such a case, the pth wave field is termed linearly degenerate. When it is always strictly positive (or strictly negative), the field is genuinely nonlinear. In MHD, we will see that slow and fast magneto-sonic wave families are neither linearly degenerate nor genuinely nonlinear. This is in contrast with the Euler system for gas dynamics, where all fields are either linearly degenerate or genuinely nonlinear, which is the characterizing property of a convex system of conservation laws. Note that we can also compute the structure coefficients from the primitive eigenvectors through $s_p = (\nabla_\mathbf{V}\lambda_p) \cdot \hat{\mathbf{r}}^p$. These coefficients relate to the tendency of a wave family to steepen or spread, and when $s_p = 0$ identically, the pth wave mode propagates by means of finite discontinuities. Finally, when a rarefaction wave solution connects a left constant state $\mathbf{U}_\ell$ with a right constant state $\mathbf{U}_\mathrm{r}$, we need $\xi = \lambda_p$ to increase monotonically from its value $\xi_\ell = \lambda_p(\mathbf{U}_\ell) < \xi = x/t < \xi_r = \lambda_p(\mathbf{U}_r)$, which also acts as a selection criterion for admissible rarefaction waves. Note the reversal in the order as compared to the Lax entropy condition (20.36). Through a p- rarefaction, all $n-1$ generalized Riemann invariants for the pth wave family remain constant. ◁

(b) Compound waves: a scalar example The flux $f(u)$ in the nonlinear conservation law (20.27) is said to be convex when the second derivative $f''(u)$ has the same sign everywhere, expressing that the first derivative $f'(u)$ varies monotonically. Burgers' equation with flux function $f(u) = \frac{1}{2}u^2$ has this property, and the Riemann problem gave rise to either a rarefaction solution or a shock. When we take as nonlinear flux function $f(u) \equiv u^3$, the flux is non-convex, as $f''(u) = 6u$ changes sign at $u = 0$. The characteristic speed is now locally $3u^2$. In the solution of the Riemann problem for a scalar equation with a non-convex flux function, a new possible outcome arises. The familiar cases return as long as we stay on the monotonic side by e.g. restricting $u > 0$. Then, we conclude by analogy with Burgers' equation that for $f(u) \equiv u^3$ the rarefaction wave will occur when $0 < u_\ell < u_r$ with solution

$$u(x,t) = u(x/t) = \begin{cases} u_\ell & (x < 3\,u_\ell^2 t) \\ \sqrt{x/(3t)} & (3\,u_\ell^2 t < x < 3\,u_r^2 t) \\ u_r & (x > 3\,u_r^2 t) \end{cases} . \qquad (20.43)$$

In contrast, when $0 < u_r < u_\ell$, the discontinuity is maintained and travels at the shock speed $s = u_\ell^2 + u_\ell\, u_r + u_r^2$.

When we allow for negative states u, we can find a 'compound' solution as follows. Taking $u_\ell > 0 > u_r$, we can seek an intermediate state $u_m < 0$ which connects to u_ℓ discontinuously, hence at shock speed $s = u_\ell^2 + u_\ell\, u_m + u_m^2$, while at the same time u_m is connecting to the right state u_r by a rarefaction solution. This compound solution then has

$$u(x,t) = u(x/t) = \begin{cases} u_\ell & (x < st = 3\,u_m^2 t) \\ -\sqrt{x/(3t)} & (3\,u_m^2 t < x < 3\,u_r^2 t) \\ u_r & (x > 3\,u_r^2 t) \end{cases} . \qquad (20.44)$$

For the case considered here, $u_m = -\frac{1}{2}u_\ell$, and thus this compound solution will only emerge when $u_\ell > 0 > -\frac{1}{2}u_\ell > u_r$. We will see that the MHD system allows for such compound solutions in both fast and slow wave families.

20.1.3 Numerical discretizations for scalar conservation

Turning to numerical treatments for conservation laws, we would like the employed discretization to allow for a proper treatment of discontinuous solutions as well, and only give rise to physically admissible shocks. To stress the importance of using a discretization which is able to handle discontinuities properly, suppose we were to apply the following first order scheme to the inviscid Burgers' equation, directly discretizing its quasi-linear form (20.28) as

$$u_i^{n+1} - u_i^n + \frac{\Delta t}{\Delta x} u_i^n \left(u_i^n - u_{i-1}^n\right) = 0, \qquad (20.45)$$

where we assumed $u \geq 0$. The particular initial data $[1,1,1,0,0,0]$ is seen [404] to remain a discrete solution that does not change from time t^n to $t^{n+1} = t^n + \Delta t$. However, we know from the discussion above that the correct solution should have the discontinuous jump travelling rightward, at the shock speed $s = 0.5$. The reason why this discretization fails for discontinuous solutions is that it cannot be written in a conservative form. A conservative scheme for a set of conservation laws with conserved variables $\mathbf{U}$ is of the form

$$\mathbf{U}_i^{n+1} = \mathbf{U}_i^n - \frac{\Delta t}{\Delta x}\left[\mathbf{F}_{i+1/2} - \mathbf{F}_{i-1/2}\right], \tag{20.46}$$

where the numerical fluxes $\mathbf{F}_{i+1/2}$ are interpreted as time-average fluxes over cell edges $x_i + \frac{1}{2}\Delta x \equiv x_{i+\frac{1}{2}}$. Note that the non-conservative scheme (20.45) would do fine for smooth solutions to the inviscid Burgers' equation, but because it is not in conservation form, the discrete equivalent of the Rankine–Hugoniot relation (20.30) is not guaranteed. In fact, a rigorous mathematical analysis by Hou and Le Floch [311] demonstrates that using any finite difference scheme in non-conservative form to numerically solve a scalar conservation law will lead to significant errors which grow in time. A demonstration of how bad a non-conservative MHD code may perform on discontinuity dominated test problems can be found in [172]. The Lax–Wendroff theorem ensures that a convergent conservative scheme does converge to a weak solution of the conservation law (see, e.g. [404] and references therein).

(a) Lax–Friedrichs and Lax–Wendroff schemes revisited In the case of a single, scalar conservation law as expressed by (20.27), we now revisit and readily generalize to the nonlinear case some of the conditionally stable explicit discretizations encountered in Chapter 15, which are conservative. The first order Lax–Friedrichs scheme (15.86) is written as

$$u_i^{n+1} = \tfrac{1}{2}(u_{i+1}^n + u_{i-1}^n) - \tfrac{1}{2}\frac{\Delta t}{\Delta x}\left(f_{i+1}^n - f_{i-1}^n\right). \tag{20.47}$$

This scheme is stable when the CFL condition

$$\left|\frac{\Delta t}{\Delta x} f'(u_i)\right| \le 1, \tag{20.48}$$

is satisfied at all grid points. This Lax–Friedrichs scheme is conservative, since we can write it in the generic form (20.46) when we identify the numerical flux as

$$F_{i+1/2}^{\mathrm{LF}} = \tfrac{1}{2}\left[f_{i+1} + f_i - \frac{\Delta x}{\Delta t}(u_{i+1} - u_i)\right]. \tag{20.49}$$

To generalize the second order Lax–Wendroff scheme (15.93) to a nonlinear solver, the most popular formulation is in the form of a predictor-corrector type method as proposed by Richtmyer. The Richtmyer two-step version of the Lax–Wendroff method reads:

$$\begin{aligned} \text{`predictor'}: \quad u_{i+\frac{1}{2}}^{n+\frac{1}{2}} &= \tfrac{1}{2}(u_{i+1}^n + u_i^n) - \tfrac{1}{2}\frac{\Delta t}{\Delta x}(f_{i+1}^n - f_i^n), \\ \text{`corrector'}: \quad u_i^{n+1} &= u_i^n - \frac{\Delta t}{\Delta x}\left(f_{i+\frac{1}{2}}^{n+\frac{1}{2}} - f_{i-\frac{1}{2}}^{n+\frac{1}{2}}\right). \end{aligned} \tag{20.50}$$

Hence, in the predictor step the *intermediate* values at half time step and at half mesh points, $u_{i+1/2}^{n+1/2}$, are calculated by the Lax–Friedrichs scheme and these intermediate values are used in the corrector step. Clearly, the scheme is conservative as it uses the fluxes $f_{i\pm\frac{1}{2}}^{n+\frac{1}{2}}$ and determines the values u_i^{n+1}

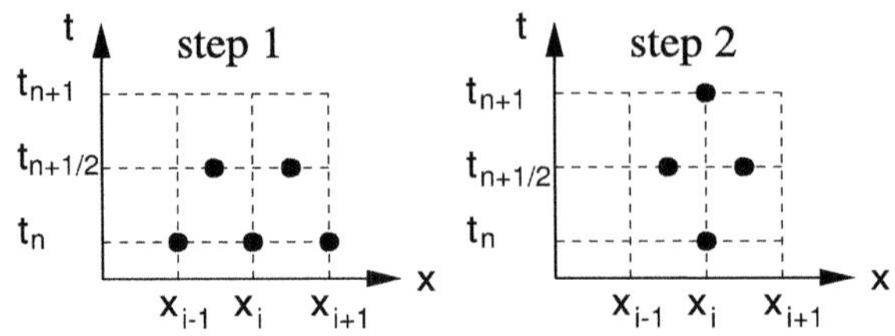

Fig. 20.3 Stencils of the predictor (left) and corrector (right) steps in the Lax–Wendroff method.

in the next time step with the leapfrog scheme (15.91). The stencils of the two steps are illustrated in Fig. 20.3.

It is this two-step version of the Lax–Wendroff scheme that has been used extensively in MHD simulations. As a concrete example, Ofman and Davila [471] used it to address the nonlinear evolution of both standing and travelling Alfvén waves in 3D slab models of driven coronal 'loops'. The loop equilibria were approximated as channels with a density depletion in uniformly magnetized zero-β slabs. Their nonlinear resistive MHD simulations demonstrated that nonlinear effects play a crucial role in the resonant absorption of the wave energy (discussed in Chapter 11 for the linear regime). In the vicinity of the resonant dissipation layers, highly sheared flows were formed, and Kelvin–Helmholtz instabilities developed.

The Lax–Wendroff scheme (20.50) is a better approximation to the PDE (20.27) than the Lax–Friedrichs scheme (20.47), since the former scheme is second order accurate. Second order accuracy in both space and time translates into local truncation errors given by $\mathcal{O}(C^3)$ or $\mathcal{O}(C\delta^2)$ in the dimensionless parameters $\delta = \big|(\Delta x/u)\, \partial u/\partial x\big|$ and the Courant number $C = f'(u)\, \Delta t/\Delta x$. Second order schemes yield satisfactory results as long as the solution $u(x)$ and the coefficient function $f'(u(x))$ are smooth. However, in the neighbourhood of shocks and steep gradients $\delta \sim \mathcal{O}(1)$ and in regions where the characteristic speed $f'(u)$ is large, we can have $C \sim \mathcal{O}(1)$. As a result, in such regions the truncation errors can become as large as the solution itself, yielding 'numerical pollution' in the form of short-wavelength oscillations. These can already be demonstrated for a linear advection problem where $f(u) = vu$ with constant v, by numerically solving its Riemann problem (20.32) using the Lax–Wendroff scheme. This is illustrated in Fig. 20.4, where we show the effect in combination with increasing the number of grid points. The result of solving a Riemann problem with $u_\ell = 1$ and $u_\mathrm{r} = 0.1$, initially at $x = 0.5$ with $v = 1$, is shown at time $t = 0.4$. The initial discontinuity is captured numerically by a steep negative gradient region, immediately trailed by an oscillatory part. This oscillation does not disappear when more grid points are used, but covers roughly the same number of grid points throughout. This so-called *Gibbs phenomenon* is an indication of the *dispersive* properties of this scheme, where numerical wave signals of different wavelengths travel at different speeds. Note that the Riemann problem has an initially monotone (although discontinuous) variation with x, i.e. there is no internal extremum of $u(x, t = 0)$. The Lax–Wendroff method can introduce internal extrema from initially monotone data, and is therefore 'non-monotonicity preserving'. In fact, for a scalar nonlinear conservation law, the exact solution should maintain monotonicity if the initial data is monotone (even including discontinuities), so it would be nice to guarantee 'monotonicity preserving' schemes. We return to this issue in the next paragraph on total variation diminishing schemes.

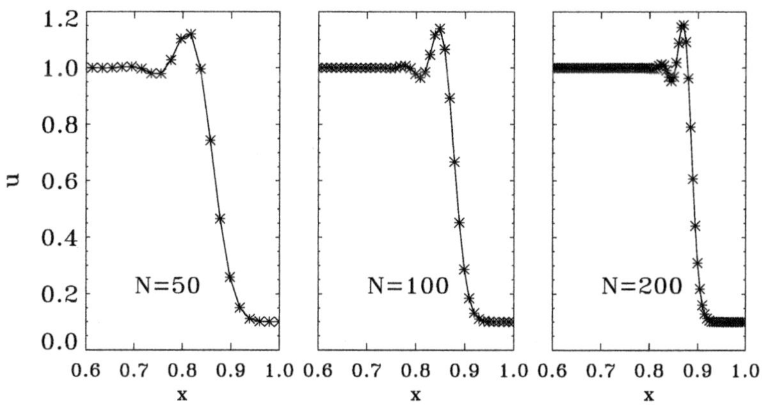

Fig. 20.4 Gibbs phenomenon, as illustrated for the linear advection of a single discontinuity, as obtained with the Lax–Wendroff method for different resolutions. The initial discontinuity was at $x = 0.5$, and the solution corresponds to time $t = 0.4$. The exact solution should jump at $x = 0.9$.

(b) Total variation diminishing schemes One way to ensure a monotonicity preserving scheme for a scalar nonlinear conservation law is to require a more stringent property, namely that the scheme is total variation diminishing, or TVD. The total variation of a function $u(x)$ on its domain, e.g. $[0, 1]$, is defined as

$$TV(u) \equiv \int_0^1 \left| \frac{du}{dx} \right| \mathrm{d}x \,. \tag{20.51}$$

In case of discontinuous profiles $u(x)$, the derivative is to be interpreted in the sense of distribution functions. In analogy with this definition, the total variation of the numerical approximation of u is then

$$TV(u^n) = \sum_{i=0}^{N-1} \left| u_{i+1}^n - u_i^n \right| \,. \tag{20.52}$$

A scheme is said to be *'total variation diminishing'* (TVD) or actually has a non-increasing total variation in time if and only if, for every discrete time level n,

$$TV(u^{n+1}) \leq TV(u^n) \,. \tag{20.53}$$

Again, it can be shown that the true solution of a scalar conservation law has this TVD property, i.e. $TV(u(x, t_2)) \leq TV(u(x, t_1))$, for all $t_2 > t_1$. When a scheme is TVD as defined by (20.53), it is rather obvious that it will also be monotonicity preserving: if a local extremum in the discrete solution appeared from an initially monotone sequence u_i with $i = 0, \ldots N$, it would naturally raise the total variation too. Hence, we may build in the TVD property (20.53) into the scheme, and we are then sure that no spurious oscillations will be introduced when numerically solving the Riemann problem. A result due to Harten [290] is thereby very useful. It states that any scheme which can be written in the general form

$$u_i^{n+1} = u_i^n + A_{i+1/2} \underbrace{\left(u_{i+1}^n - u_i^n\right)}_{\Delta u_{i+1/2}^n} - B_{i-1/2} \underbrace{\left(u_i^n - u_{i-1}^n\right)}_{\Delta u_{i-1/2}^n} \tag{20.54}$$

is TVD when the scheme dependent coefficients $A_{i+1/2}$ and $B_{i-1/2}$ obey

$$A_{i+1/2} \geq 0\,, \quad B_{i-1/2} \geq 0\,, \quad 0 \leq A_{i+1/2} + B_{i+1/2} \leq 1 \tag{20.55}$$

at all grid indices i. As a particular example, the first order Lax–Friedrichs scheme (20.47) has this property, as we may formally manipulate the discrete formula to

$$u_i^{n+1} = u_i^n + \tfrac{1}{2}\left(1 - \frac{\Delta t}{\Delta x}\frac{f_{i+1}^n - f_i^n}{\Delta u_{i+\frac{1}{2}}}\right)\Delta u_{i+\frac{1}{2}} - \tfrac{1}{2}\left(1 + \frac{\Delta t}{\Delta x}\frac{f_i^n - f_{i-1}^n}{\Delta u_{i-\frac{1}{2}}}\right)\Delta u_{i-\frac{1}{2}}\,. \tag{20.56}$$

The requirements (20.55) translate into the CFL condition

$$\left|\frac{\Delta t}{\Delta x}\frac{f_{i+1}^n - f_i^n}{u_{i+1}^n - u_i^n}\right| \leq 1\,, \tag{20.57}$$

as stated in (20.48).

We can improve the rather diffusive nature of the first order Lax–Friedrichs scheme by changing the numerical flux (20.49) to the following:

$$F_{i+1/2}^{\mathrm{LLF}} = \tfrac{1}{2}\left\{f_{i+1} + f_i - |\alpha_{i+\frac{1}{2}}|\,[u_{i+1} - u_i]\right\}\,, \tag{20.58}$$

in which the coefficient $\alpha_{i+\frac{1}{2}}$ is found from

$$\alpha_{i+1/2} \equiv \begin{cases} \Delta x/\Delta t & (u_{i+1}^n - u_i^n = 0)\,, \\ \dfrac{f_{i+1}^n - f_i^n}{u_{i+1}^n - u_i^n} & (u_{i+1}^n - u_i^n \neq 0)\,. \end{cases} \tag{20.59}$$

This definition ensures that the derivatives are taken in the 'upwind' direction, meaning that we effectively switch between one-sided left or right derivative evaluation depending on the sign of $\alpha_{i+\frac{1}{2}}$. Indeed, we get

$$u_i^{n+1} = u_i^n - \frac{\Delta t}{\Delta x}\begin{cases} (f_{i+1}^n - f_i^n) & (\text{for } \alpha_{i+1/2}, \alpha_{i-1/2} < 0)\,, \\ (f_i^n - f_{i-1}^n) & (\text{for } \alpha_{i+1/2}, \alpha_{i-1/2} > 0)\,. \end{cases} \tag{20.60}$$

In the linear advection problem, the sign of the constant advection speed v determines which of these discretizations represents the upwind scheme. The full scheme with numerical flux (20.58) can be rewritten in different forms, e.g.

$$\begin{aligned} u_i^{n+1} \;=\; u_i^n - \tfrac{1}{2}\frac{\Delta t}{\Delta x}\Big[&\alpha_{i+1/2}\Delta u_{i+1/2}^n - |\alpha_{i+1/2}|\Delta u_{i+1/2}^n \\ &+ \alpha_{i-1/2}\Delta u_{i-1/2}^n + |\alpha_{i-1/2}|\Delta u_{i-1/2}^n\Big]\,. \end{aligned} \tag{20.61}$$

This is of the general form (20.54) with

$$\begin{cases} A_{i+1/2} = \tfrac{1}{2}\dfrac{\Delta t}{\Delta x}\left(|\alpha_{i+1/2}| - \alpha_{i+1/2}\right) \;\geq 0\,, \\ B_{i-1/2} = \tfrac{1}{2}\dfrac{\Delta t}{\Delta x}\left(|\alpha_{i-1/2}| + \alpha_{i-1/2}\right) \;\geq 0\,. \end{cases} \tag{20.62}$$

Hence, two of the three sufficient conditions (20.55) for the scheme to have the TVD property are satisfied. The third condition reads

$$0 \leq \left|\alpha_{i+1/2}\right| \frac{\Delta t}{\Delta x} \leq 1\,, \tag{20.63}$$

which is a CFL-type condition on the time step for a given spatial resolution. When it is satisfied, the upwind scheme has the TVD property.

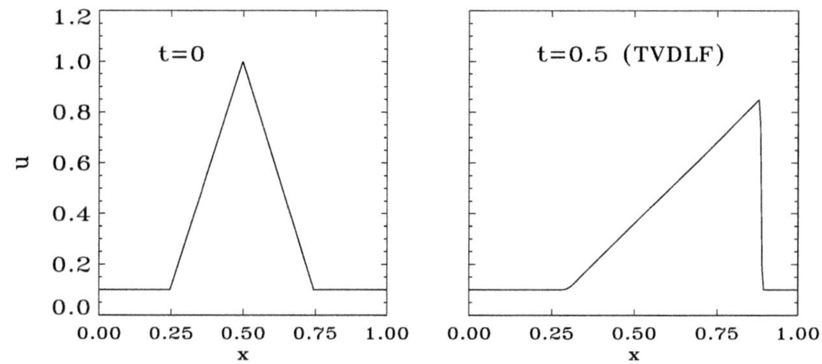

Fig. 20.5 Initial triangular pulse (left) and shocked numerical solution of Burgers' equation as obtained with the second order TVDLF method.

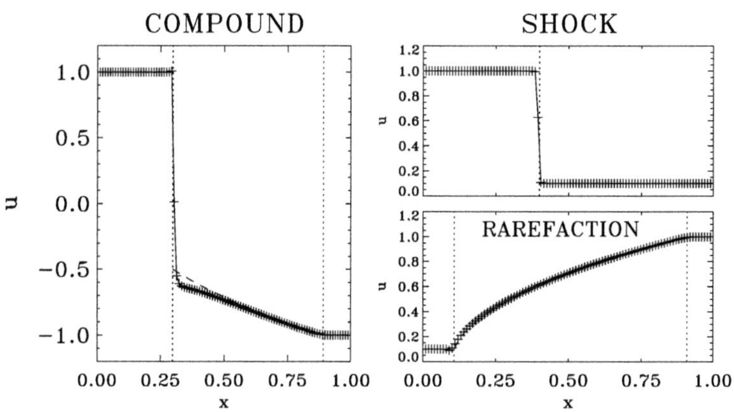

Fig. 20.6 Riemann problems as solved with the TVDLF scheme for a non-convex scalar equation. The discontinuity was placed at $x = 0.1$ at $t = 0$, and the frames shown are at time $t = 0.27$. The vertical dotted lines in all frames correspond to the analytic locations of shocks or rarefaction edges. In the left panel, the dashed line indicates the exact solution for the compound wave structure.

So far, we only gave examples of first order accurate TVD schemes, which lead to oscillation-free solutions. In what follows, we will present second order TVD schemes. However, the scheme then needs a truly nonlinear dependence on the discrete data, even for the case of a linear advection equation. This is a direct consequence of the Godunov theorem, which states that a linear monotonicity preserving scheme is at most first order accurate. Second order TVD schemes can be developed by introducing nonlinear flux limiters. Such schemes are second order only in 'smooth' regions and first order at extrema and shocks. Still, in comparison with fully first order schemes, the shock smearing will be reduced. An example second order TVD type solution for the triangular pulse in the Burgers problem is illustrated in Fig. 20.5. The same second order TVDLF method (discussed fully below) is used in Fig. 20.6, where we demonstrate the three possible outcomes of the Riemann problem for the non-convex equation $u_t + (u^3)_x = 0$ discussed earlier. The numerical solutions agree in all cases with the known analytic solutions.

Higher order TVD schemes are thus a type of 'hybrid' scheme, combining a higher order scheme in smooth regions with a first order scheme in regions where high gradients occur. In what follows, we will discuss such *hybrid* schemes, also called 'high-resolution methods', that circumvent the Godunov theorem. TVD methods are widely used in MHD computations. Other modern schemes somewhat relax the requirement of monotone schemes, and achieve higher than first order at extrema. Popular categories are known as (weighted) essentially non-oscillatory, or (W)ENO, schemes and discontinuous Galerkin methods. All these schemes are frequently exploited in combination with so-called finite volume treatments, the ideas of which are briefly discussed next.

20.1.4 Finite volume treatments

In many 'high resolution shock-capturing' schemes, finite volume methods are used for the spatial discretization. This discretization technique refers directly to the integral form of the conservation laws. For a single scalar conservation law as given by Eq. (20.27), the frequently used differential form actually follows from the integral form *only when the solution is smooth or regular enough* (as it must e.g. be continuously differentiable). This integral form states that a conserved scalar quantity $u(x,t)$ changes in the space-time volume $[x_1, x_2] \times [t_1, t_2]$ in accord with

$$\int_{x_1}^{x_2} u(x,t_2)\,dx = \int_{x_1}^{x_2} u(x,t_1)\,dx + \int_{t_1}^{t_2} f(x_1,t)\,dt - \int_{t_1}^{t_2} f(x_2,t)\,dt. \qquad (20.64)$$

The instantaneous volume average changes only through the temporally varying fluxes across the domain boundaries. This is also true in general, when we deal with a multi-dimensional set of conservation laws as in Eq. (20.1),

$$\frac{\partial \mathbf{U}}{\partial t} + \nabla \cdot \mathsf{F} = 0\,, \qquad (20.65)$$

where $\mathbf{U}$ is the state vector of n conserved variables and F is the flux. For the MHD system, where $n = 8$, recall from the analysis of Section 4.3.2 the explicit expressions for $\mathbf{U} = (\rho, \boldsymbol{\pi}, \mathcal{H}, \mathbf{B})^{\mathrm{T}}$ and $\mathsf{F} = (\boldsymbol{\pi}, \mathsf{T}, \widetilde{\mathbf{U}}, \mathbf{Y})^{\mathrm{T}}$ of Eqs (4.66)–(4.70). Note that in 3D, the term $\nabla \cdot \mathsf{F}$ in Eq. (20.65) can be expressed in terms of fluxes along three Cartesian coordinate axes, by writing

$$\nabla \cdot \mathsf{F} = \frac{\partial \mathbf{F}_x}{\partial x} + \frac{\partial \mathbf{F}_y}{\partial y} + \frac{\partial \mathbf{F}_z}{\partial z}\,. \qquad (20.66)$$

When we discretize space in control volumes V_i with bounding surfaces $\partial \mathrm{V}_i$, with outward unit normal $\mathbf{n} = (n_x, n_y, n_z)$, we find from Gauss' theorem that

$$\frac{d}{dt}\int_{\mathrm{V}_i} \mathbf{U}(\mathbf{x},t)\,d\mathbf{x} = -\int_{\partial \mathrm{V}_i} \mathsf{F}\cdot \mathbf{n}\,dS = -\int_{\partial \mathrm{V}_i} (\mathbf{F}_x n_x + \mathbf{F}_y n_y + \mathbf{F}_z n_z)\,dS\,. \qquad (20.67)$$

This formula can easily be generalized to account for source and sink terms as well, if they exist for certain components of $\mathbf{U}$. We now introduce the matrix $\mathsf{T}(\mathbf{n})$ which rotates all occurring vector quantities to a local orthogonal coordinate system formed by $\mathbf{n}, \mathbf{t}, \mathbf{s}$ $(\equiv \mathbf{n} \times \mathbf{t})$, where the latter are tangential unit vectors within the bounding surface $\partial \mathrm{V}_i$. For the MHD system as above,

$$\mathsf{T} = \begin{pmatrix} 1 & 0 & 0 & 0 & 0 & 0 & 0 & 0 \\ 0 & \sin\theta\cos\phi & \sin\theta\sin\phi & \cos\theta & 0 & 0 & 0 & 0 \\ 0 & \cos\theta\cos\phi & \cos\theta\sin\phi & -\sin\theta & 0 & 0 & 0 & 0 \\ 0 & -\sin\phi & \cos\phi & 0 & 0 & 0 & 0 & 0 \\ 0 & 0 & 0 & 0 & 1 & 0 & 0 & 0 \\ 0 & 0 & 0 & 0 & 0 & \sin\theta\cos\phi & \sin\theta\sin\phi & \cos\theta \\ 0 & 0 & 0 & 0 & 0 & \cos\theta\cos\phi & \cos\theta\sin\phi & -\sin\theta \\ 0 & 0 & 0 & 0 & 0 & -\sin\phi & \cos\phi & 0 \end{pmatrix} . \tag{20.68}$$

In the latter expression, we introduced the spherical coordinate angles (θ, ϕ) fully determining the direction of the normal $\mathbf{n}$ and the tangent vectors. One can then verify, or understand from the fact that the MHD equations must be unchanged under rotation, that

$$\mathbf{F}_x n_x + \mathbf{F}_y n_y + \mathbf{F}_z n_z = \mathsf{T}^{-1}(\mathbf{n}) \cdot \mathbf{F}_x \left(\mathsf{T}(\mathbf{n}) \cdot \mathbf{U}\right) . \tag{20.69}$$

Ultimately, this means that we obtain an essentially 1D problem in the direction normal to the control volume boundary, since Eq. (20.67) now becomes

$$\frac{d}{dt} \int_{\mathrm{V}_i} \mathbf{U}(\mathbf{x}, t)\, d\mathbf{x} = - \int_{\partial \mathrm{V}_i} \mathsf{T}^{-1}(\mathbf{n}) \cdot \mathbf{F}_x \left(\mathsf{T}(\mathbf{n}) \cdot \mathbf{U}\right)\, dS . \tag{20.70}$$

We then usually employ control volumes with multiple flat surface segments, turning the integral over the boundary into a discrete sum over its sides. This means that, in practice, the only information of the grid needed in the process consists of the volumes V_i of the grid cells and the geometry of the cells such as the number of bounding surface segments, their surface area and their normal directions.

20.2 Upwind-like finite volume treatments for one-dimensional MHD

The finite volume method turned the solution of the conservation laws into, in essence, 1D flux updates across the segmented boundaries of control volumes. We will therefore now pay further attention to the 1D case in particular. Recapitulating, in a 1D configuration, a shock-capturing finite volume scheme thus interprets $\mathbf{U}_i$ as the average value of the solution $\mathbf{U}(x, t)$ in the interval $[x_{i-1/2}, x_{i+1/2}]$:

$$\mathbf{U}_i(t) \equiv \frac{1}{\Delta x_i} \int_{x_{i-1/2}}^{x_{i+1/2}} \mathbf{U}(x, t)\, \mathrm{d}x , \tag{20.71}$$

where the domain is now subdivided into a set of equal-sized 'cells' covering the whole domain. The update of these volume averages becomes simply

$$\frac{d\mathbf{U}_i}{dt} + \frac{1}{\Delta x_i} \left(\mathbf{F}_{i+1/2} - \mathbf{F}_{i-1/2}\right) = 0 . \tag{20.72}$$

The advantage of the volume average interpretation is that the discretized equation itself can be seen as an integral law, rather than a differential law, and that its weak solutions will obey conservation by construction. In 1D MHD, where we can allow for translational invariance in ignorable y and z directions, the normal B_x component remains constant so that we have at most seven components for $\mathbf{U}$.

20.2.1 The Godunov method

A method originally proposed by Godunov was to consider the cell averaged values $\mathbf{U}_i^n$ for time $t = t_n$ as piecewise constant data throughout the cells, and concentrate on the discontinuous Riemann problems that then arise at the cell interfaces. The solution of each local Riemann problem is in essence self-similar in x/t, so one can properly restrict the time step such that no wave interaction occurs in one cell of size Δx within Δt^{n+1}. This is achieved by imposing $\Delta t^{n+1} < \Delta x/(2\max|\lambda_p^n|)$ on the discrete (but variable) time step, with the maximum taken over all eigenvalues of the flux Jacobian $\mathsf{F}_{\mathbf{U}}$. For 1D MHD, this is set by the fast eigenvalues and is $|v_x| \pm c_{\mathrm{f}}$. The Godunov method consists of using the exact nonlinear solution of the Riemann problems at the cell interfaces in the numerical flux. Denoting the exact Riemann problem solution for state $\mathbf{U}_i^n$ and $\mathbf{U}_{i+1}^n$ as $\hat{\mathbf{U}}\big((x - x_{i+1/2})/t, \mathbf{U}_i^n, \mathbf{U}_{i+1}^n\big)$, the Godunov scheme then uses as numerical flux

$$\mathbf{F}_{i+1/2}(\mathbf{U}_i, \mathbf{U}_{i+1}) = \mathbf{F}\big(\hat{\mathbf{U}}(0, \mathbf{U}_i^n, \mathbf{U}_{i+1}^n)\big)\,. \tag{20.73}$$

Due to its piecewise constant representation of cell values, this original Godunov scheme is at best first order accurate. In any case, it requires an exact solution to the nonlinear Riemann problem. For 1D MHD, where seven wave families are involved, this is a far from trivial exercise involving adequate procedures to handle the nonlinearities, whereby one must face existence and uniqueness issues as well. We take in what follows a more pragmatic approach, discussing various upwind-like methods which avoid the full solution of the nonlinear Riemann problem.

(a) The linearized MHD system and the Riemann problem A full solution to the Riemann problem can be given for any linear hyperbolic system, as was explained before. Specifying how this works for 1D MHD, we can, e.g., look at the primitive variable formulation $\mathbf{V}_t + \mathsf{W}\cdot\mathbf{V}_x = 0$, where

$$\begin{pmatrix} \rho \\ v_x \\ v_y \\ v_z \\ p \\ B_y \\ B_z \end{pmatrix}_t + \begin{pmatrix} v_x & \rho & 0 & 0 & 0 & 0 & 0 \\ 0 & v_x & 0 & 0 & 1/\rho & B_y/\rho & B_z/\rho \\ 0 & 0 & v_x & 0 & 0 & -B_x/\rho & 0 \\ 0 & 0 & 0 & v_x & 0 & 0 & -B_x/\rho \\ 0 & \rho c^2 & 0 & 0 & v_x & 0 & 0 \\ 0 & B_y & -B_x & 0 & 0 & v_x & 0 \\ 0 & B_z & 0 & -B_x & 0 & 0 & v_x \end{pmatrix} \begin{pmatrix} \rho \\ v_x \\ v_y \\ v_z \\ p \\ B_y \\ B_z \end{pmatrix}_x = 0\,. \tag{20.74}$$

To turn this into a linear hyperbolic PDE, we will evaluate W in a fixed state $\mathbf{V}^*$, and assume that the eigenvalues for the matrix $\mathsf{W}(\mathbf{V}^*)$ are distinct. They are the familiar ordered set

$$v_x^* - c_{\mathrm{f}}^*\,,\quad v_x^* - b_x^*\,,\quad v_x^* - c_{\mathrm{s}}^*\,,\quad v_x^*\,,\quad v_x^* + c_{\mathrm{s}}^*\,,\quad v_x^* + b_x^*\,,\quad v_x^* + c_{\mathrm{f}}^*\,, \tag{20.75}$$

where the Alfvén speed $b_x^* = |B_x^*|/\sqrt{\rho^*}$, and the slow and fast magneto-acoustic speeds are as in

Eq. (20.15), with the iso-thermal sound speed squared c_i^2 replaced by the sound speed squared c^{*2}. We wish to solve the Riemann problem for this linear system exactly, with given initial constant left and right states $\mathbf{V}_\ell$ and $\mathbf{V}_r$. We will need the left $\hat{\mathbf{l}}^p$ and right $\hat{\mathbf{r}}^p$ eigenvectors of the matrix W in the process, and as explained before, these are collected in the matrices $\widehat{\mathsf{R}}^{-1}$ and $\widehat{\mathsf{R}}$, respectively. Introducing dimensionless parameters

$$\alpha_s^2 \equiv \frac{c_f^2 - c^2}{c_f^2 - c_s^2}, \quad \alpha_f^2 \equiv \frac{c^2 - c_s^2}{c_f^2 - c_s^2}, \quad \beta_x \equiv \frac{B_x}{|B_x|}, \quad \beta_y \equiv \frac{B_y}{B_\perp}, \quad \beta_z \equiv \frac{B_z}{B_\perp}, \tag{20.76}$$

where $B_\perp = \sqrt{B_y^2 + B_z^2}$, the matrix $\widehat{\mathsf{R}}$, with columns $\hat{\mathbf{r}}^p$, is given by

$$\widehat{\mathsf{R}} = \begin{pmatrix} \alpha_f \rho & 0 & \alpha_s \rho & 1 & \alpha_s \rho & 0 & \alpha_f \rho \\ -\alpha_f c_f & 0 & -\alpha_s c_s & 0 & \alpha_s c_s & 0 & \alpha_f c_f \\ \alpha_s c_s \beta_y \beta_x & -\beta_z & -\alpha_f c_f \beta_y \beta_x & 0 & \alpha_f c_f \beta_y \beta_x & \beta_z & -\alpha_s c_s \beta_y \beta_x \\ \alpha_s c_s \beta_z \beta_x & \beta_y & -\alpha_f c_f \beta_z \beta_x & 0 & \alpha_f c_f \beta_z \beta_x & -\beta_y & -\alpha_s c_s \beta_z \beta_x \\ \alpha_f \rho c^2 & 0 & \alpha_s \rho c^2 & 0 & \alpha_s \rho c^2 & 0 & \alpha_f \rho c^2 \\ \alpha_s c \beta_y \sqrt{\rho} & -\beta_z \sqrt{\rho} \beta_x & -\alpha_f c \beta_y \sqrt{\rho} & 0 & -\alpha_f c \beta_y \sqrt{\rho} & -\beta_z \sqrt{\rho} \beta_x & \alpha_s c \beta_y \sqrt{\rho} \\ \alpha_s c \beta_z \sqrt{\rho} & \beta_y \sqrt{\rho} \beta_x & -\alpha_f c \beta_z \sqrt{\rho} & 0 & -\alpha_f c \beta_z \sqrt{\rho} & \beta_y \sqrt{\rho} \beta_x & \alpha_s c \beta_z \sqrt{\rho} \end{pmatrix}, \tag{20.77}$$

while its inverse $\widehat{\mathsf{R}}^{-1}$, with $\hat{\mathbf{l}}^p$ in the rows, is given by

$$\widehat{\mathsf{R}}^{-1} \begin{pmatrix} 0 & -\frac{c_f \alpha_f}{2c^2} & \frac{\alpha_s c_s}{2c^2} \beta_y \beta_x & \frac{\alpha_s c_s}{2c^2} \beta_z \beta_x & \frac{\alpha_f}{2c^2 \rho} & \frac{\alpha_s}{2c\sqrt{\rho}} \beta_y & \frac{\alpha_s}{2c\sqrt{\rho}} \beta_z \\ 0 & 0 & -\frac{1}{2}\beta_z & \frac{1}{2}\beta_y & 0 & -\beta_z \frac{\beta_x}{2\sqrt{\rho}} & \beta_y \frac{\beta_x}{2\sqrt{\rho}} \\ 0 & -\frac{c_s \alpha_s}{2c^2} & -\frac{\alpha_f c_f}{2c^2} \beta_y \beta_x & -\frac{\alpha_f c_f}{2c^2} \beta_z \beta_x & \frac{\alpha_s}{2c^2 \rho} & -\frac{\alpha_f}{2c\sqrt{\rho}} \beta_y & -\frac{\alpha_f}{2c\sqrt{\rho}} \beta_z \\ 1 & 0 & 0 & 0 & -\frac{1}{c^2} & 0 & 0 \\ 0 & \frac{c_s \alpha_s}{2c^2} & \frac{\alpha_f c_f}{2c^2} \beta_y \beta_x & \frac{\alpha_f c_f}{2c^2} \beta_z \beta_x & \frac{\alpha_s}{2c^2 \rho} & -\frac{\alpha_f}{2c\sqrt{\rho}} \beta_y & -\frac{\alpha_f}{2c\sqrt{\rho}} \beta_z \\ 0 & 0 & \frac{1}{2}\beta_z & -\frac{1}{2}\beta_y & 0 & -\beta_z \frac{\beta_x}{2\sqrt{\rho}} & \beta_y \frac{\beta_x}{2\sqrt{\rho}} \\ 0 & \frac{c_f \alpha_f}{2c^2} & -\frac{\alpha_s c_s}{2c^2} \beta_y \beta_x & -\frac{\alpha_s c_s}{2c^2} \beta_z \beta_x & \frac{\alpha_f}{2c^2 \rho} & \frac{\alpha_s}{2c\sqrt{\rho}} \beta_y & \frac{\alpha_s}{2c\sqrt{\rho}} \beta_z \end{pmatrix}. \tag{20.78}$$

In the form given in (20.77), the right eigenvectors have the same dimension as the primitive variables $\mathbf{V}$. Also, the fast eigenvectors can be artificially decomposed in an acoustic ($\sim \alpha_f$) and a magnetic ($\sim \alpha_s$) contribution, and a similar argument is true for the slow eigenvectors. To solve the linear Riemann problem posed above, we write the constant left and right states $\mathbf{V}_{l,r}$ as linear combinations of the right eigenvectors of $\mathsf{W}(\mathbf{V}^*)$. The orthonormal set $\hat{\mathbf{r}}^{*p}$ and $\hat{\mathbf{l}}^{*p}$ allows us to write

$$\mathbf{V}_\ell = \sum_{p=1}^{7} (\hat{\mathbf{l}}^{*p} \cdot \mathbf{V}_\ell) \hat{\mathbf{r}}^{*p} = \sum_{p=1}^{7} \beta_p \hat{\mathbf{r}}^{*p}, \qquad \mathbf{V}_r = \sum_{p=1}^{7} (\hat{\mathbf{l}}^{*p} \cdot \mathbf{V}_r) \hat{\mathbf{r}}^{*p} = \sum_{p=1}^{7} \gamma_p \hat{\mathbf{r}}^{*p}. \tag{20.79}$$

The solution of the Riemann problem can then use the general expression given in Eq. (20.26), which due to our initial data becomes

$$\mathbf{V}(x,t) = \sum_{x/t<\lambda_p^*} (\hat{\mathbf{l}}^{*p} \cdot \mathbf{V}_\ell)\hat{\mathbf{r}}^{*p} + \sum_{x/t>\lambda_p^*} (\hat{\mathbf{l}}^{*p} \cdot \mathbf{V}_\mathrm{r})\hat{\mathbf{r}}^{*p} . \tag{20.80}$$

In the (x,t) plane, we can then graphically represent this solution as shown by the seven characteristics in Fig. 5.10 of Chapter 5: backward and forward travelling fast, Alfvén and slow discontinuities, with speeds $v \pm c_\mathrm{f}$, $v \pm b_x$, $v \pm c_\mathrm{s}$ (indicated by the symbols $\mathrm{f}^\pm$, $\mathrm{A}^\pm$, $\mathrm{s}^\pm$ in the figure) are separated by a contact discontinuity travelling at speed v (indicated by the symbol E).

At $t = 0$, the two constant states $\mathbf{V}_\ell$ and $\mathbf{V}_\mathrm{r}$ are denoted by their seven coefficients from the expansion (20.79), $(\beta_1, \beta_2, \ldots, \beta_7)$ and $(\gamma_1, \gamma_2, \ldots, \gamma_7)$, respectively. Assuming distinct and ordered eigenvalues $\lambda_1^* < \lambda_2^* < \cdots < \lambda_7^*$, the seven characteristic curves $x = \lambda_p(\mathbf{V}^*)t$ through the origin $(x = 0, t = 0)$ divide the half plane $t > 0$ into eight regions. In each of those, the solution of the linear Riemann problem is a constant state. Indeed, according to the formula (20.80), the coefficients used in writing $\mathbf{V}(x,t)$ as a linear combination of the (constant) right eigenvectors $\hat{\mathbf{r}}^p$ can be found by drawing the seven p-characteristics backwards in time from the point (x,t). Their intersection with the initial data determines which of the two coefficients β_p or γ_p prevails. Hence, out of one discontinuity separating two constant MHD states for a system of seven linear equations, up to a maximum of eight constant states (six on top of the given left and right state) separated by seven discontinuities emerge. The discontinuities travel at the characteristic speeds given by the eigenvalues of the coefficient matrix, while the jumps across the seven discontinuities are proportional to the right eigenvectors of the matrix.

▷ **Exercise** Note that the scaling of the eigenvectors given here for the full MHD equations is different from the one used in the expressions (20.19)–(20.20) for the 1.5D iso-thermal MHD system. It is a matter of algebra to show how a similarly scaled system can be used there, and this is left as an exercise. This rescaling will be needed for a convenient numerical handling of the degeneracies in the eigenvalue expressions. Historically, eigenvector expressions scaled as in Eq. (20.19) were given in [336], but in this form the expressions contain indefinite limits when strict hyperbolicity is lost. Better behaved scaled versions were introduced by [96], and the slightly different scaling used here follows [525]. ◁

(b) Degeneracies of the MHD characteristic speeds The seven wave speeds (20.75) (where we will drop the $*$-superscript) are degenerate if one of the following situations arises:

– When both tangential field components vanish, i.e. $B_y = B_z = 0$, and $b_x^2 \neq c^2$, either of both fast $(b_x > c)$ or both slow $(b_x < c)$ characteristic speeds coincide with the Alfvén signals at $v_x \pm b_x$. This is termed a *double umbilic point*.

– When the tangential field components vanish as well as $b_x^2 = c^2$, the slow, Alfvén and fast wave speeds coincide, making $v_x \pm b_x$ *triple umbilic points*.

– When $B_x = 0$, the Alfvén and slow pairs collapse to a *quintuple umbilic point* v_x.

Therefore, the MHD equations are not strictly hyperbolic and this complicates the strategy to design a true nonlinear MHD Riemann solver. Similar degeneracies are already present in the 1.5D iso-thermal MHD system discussed earlier, where only four wave speeds are at play. For this simplified system, the outcome of the true nonlinear Riemann problem can be analysed to potentially lead to 289 distinct (mathematically allowed) outcomes, in which the trivial case of equal left and right states is included too [457]. In fact, both fast and slow wave families can arise as shocks, rarefaction waves or compound waves. Therefore, $4^4 = 256$ cases allow four 'waves' with each wave either absent or present in one of its three manifestations. The other 33 possibilities account for overcompressive shock situations, with (1) both left-going waves merged and 16 possibilities arising out of the two right-going waves, (2) both right-going waves merged, yielding 16 possibilities for the left pair of waves and (3) both left and right going wave pairs merged into overcompressive shocks. Since the full MHD system additionally includes the entropy and Alfvén wave families, allowing for contact and rotational discontinuities, it is practically impossible to appropriately handle that many distinct possibilities. As a result, the nonlinear Riemann solvers available in the literature necessarily handle the MHD Riemann problem in a more tractable simplified manner. One way to do so is to allow only discontinuous shock transitions in all wave families. This was done by Dai and Woodward [138] who, in an all-shock nonlinear Riemann solver, exploited the nonlinear Rankine–Hugoniot relations at up to seven transitions, conveniently written in a Lagrangian mass coordinate. A more elaborate nonlinear 1D MHD Riemann solver was presented in [535], where, in addition to shocks, the solver allowed for either slow or fast rarefaction waves. To do so, the differential equations governing the various Riemann invariants were used through the rarefactions. Still, these nonlinear Riemann solvers which only use regular (shock or rarefaction) waves are fairly computationally involved, and a tendency emerged to avoid using them as the basic building block for a shock-capturing method. Instead, approximate Riemann solvers gained in popularity. These will be discussed later on.

▷ **Eigenvectors of the flux Jacobian** Further on in the discussion, we will also need the eigenvectors of the flux Jacobian. The right eigenvectors can be found directly from the relation $\mathsf{R} = \mathsf{U_V} \cdot \widehat{\mathsf{R}}$, giving column vectors $\mathbf{r}^p$. The gives the following expressions for the fast eigenvectors $\mathbf{r}_\mathrm{f}^{1,7}$, the slow eigenvectors $\mathbf{r}_\mathrm{s}^{3,5}$, the Alfvén eigenvectors $\mathbf{r}_\mathrm{a}^{2,6}$ and the entropy eigenvectors $\mathbf{r}_\mathrm{e}^{4}$, respectively:

$$\begin{pmatrix} \alpha_\mathrm{f}\rho \\ \alpha_\mathrm{f}\rho(v_x \mp c_\mathrm{f}) \\ \alpha_\mathrm{f}\rho v_y \pm \alpha_\mathrm{s}\rho c_\mathrm{s}\beta_y\beta_x \\ \alpha_\mathrm{f}\rho v_z \pm \alpha_\mathrm{s}\rho c_\mathrm{s}\beta_z\beta_x \\ g_\mathrm{f}^{\mp} \\ \alpha_\mathrm{s} c\beta_y\sqrt{\rho} \\ \alpha_\mathrm{s} c\beta_z\sqrt{\rho} \end{pmatrix}, \quad \begin{pmatrix} \alpha_\mathrm{s}\rho \\ \alpha_\mathrm{s}\rho(v_x \mp c_\mathrm{s}) \\ \alpha_\mathrm{s}\rho v_y \mp \alpha_\mathrm{f}\rho c_\mathrm{f}\beta_y\beta_x \\ \alpha_\mathrm{s}\rho v_z \mp \alpha_\mathrm{f}\rho c_\mathrm{f}\beta_z\beta_x \\ g_\mathrm{s}^{\mp} \\ -\alpha_\mathrm{f} c\beta_y\sqrt{\rho} \\ -\alpha_\mathrm{f} c\beta_z\sqrt{\rho} \end{pmatrix}, \quad \begin{pmatrix} 0 \\ 0 \\ \mp\rho\beta_z \\ \pm\rho\beta_y \\ g_\mathrm{a}^{\mp} \\ -\beta_z\sqrt{\rho}\beta_x \\ \beta_y\sqrt{\rho}\beta_x \end{pmatrix}, \quad \begin{pmatrix} 1 \\ v_x \\ v_y \\ v_z \\ \frac{1}{2}v^2 \\ 0 \\ 0 \end{pmatrix}, \tag{20.81}$$

where we introduced the symbols

$$\begin{aligned} g_\mathrm{f}^{\mp} &\equiv \alpha_\mathrm{f}\rho\left(\tfrac{1}{2}v^2 + \tfrac{1}{\gamma-1}c^2 \mp c_\mathrm{f}v_x\right) \pm \alpha_\mathrm{s}\rho c_\mathrm{s}\beta_x\left(v_y\beta_y + v_z\beta_z\right) + \alpha_\mathrm{s} c\sqrt{\rho}B_\perp\,, \\ g_\mathrm{a}^{\mp} &\equiv \mp\rho(v_y\beta_z - v_z\beta_y)\,, \\ g_\mathrm{s}^{\mp} &\equiv \alpha_\mathrm{s}\rho\left(\tfrac{1}{2}v^2 + \tfrac{1}{\gamma-1}c^2 \mp c_\mathrm{s}v_x\right) \mp \alpha_\mathrm{f}\rho c_\mathrm{f}\beta_x\left(v_y\beta_y + v_z\beta_z\right) - \alpha_\mathrm{f} c\sqrt{\rho}B_\perp\,. \end{aligned} \tag{20.82}$$

One can then verify that the structure coefficients

$$\left(\frac{\partial\lambda_p}{\partial\rho}, \frac{\partial\lambda_p}{\partial m_x}, \frac{\partial\lambda_p}{\partial m_y}, \frac{\partial\lambda_p}{\partial m_z}, \frac{\partial\lambda_p}{\partial\mathcal{H}}, \frac{\partial\lambda_p}{\partial B_y}, \frac{\partial\lambda_p}{\partial B_z}\right) \cdot \mathbf{r}^p = 0\,, \tag{20.83}$$

for the entropy $p = 4$ and the Alfvén wave $p = 2, 6$ families. They are thus linearly degenerate and do not allow for rarefaction wave constructions using Eq. (20.42). They do allow for discontinuous shock-type solutions, and the generalized Riemann invariants for these characteristic fields then directly demonstrate

$$\begin{aligned} &\text{entropy:} \quad \lambda_4 = m_x/\rho\,, \quad dv_x = dv_y = dv_z = dp = dB_y = dB_z = 0\,, \\ &\text{Alfvén:} \quad \lambda_{2,6} = m_x/\rho \mp \beta_x B_x/\sqrt{\rho}\,, \quad d\rho = dv_x = dp = dB_\perp = 0\,, \quad dv_{y,z} = \pm(\beta_x/\sqrt{\rho})dB_{y,z}. \end{aligned} \tag{20.84}$$

The entropy family thus leads to contact discontinuities, where only a jump in density (or entropy, temperature or total energy $\mathcal{H}$) occurs. For all wave families other than the entropy one, one true Riemann invariant can directly be identified with the entropy $S = p\rho^{-\gamma}$, since we know that in ideal MHD this variable obeys Eq. (4.64). From Eq. (20.84), we see that the Alfvén families yield rotational discontinuities, where the total field magnitude is constant and the tangential field $\mathbf{B}_\perp$ is rotated. For the fast and slow wave families, one computes the structure coefficient from $(\nabla_{\mathbf{V}}\lambda_p)\cdot\hat{\mathbf{r}}^p$ of the primitive right eigenvector set. To compute partial derivatives with respect to density, pressure and tangential magnetic field components, one can use

$$c_{\rm s,f}^4 - (c^2 + b_x^2 + b_\perp^2)c_{\rm s,f}^2 + c^2 b_x^2 = 0\,, \quad b_\perp = B_\perp/\sqrt{\rho}\,, \tag{20.85}$$

which one differentiates to obtain expressions for e.g. $\partial c_{\rm s,f}/\partial\rho$. This yields

$$\begin{aligned} s_1 &= -\,\alpha_{\rm f}c_{\rm f}\big(\tfrac{1}{2}\gamma\alpha_{\rm f}^2 + \alpha_{\rm s}^2 + \tfrac{1}{2}\big)\,, \quad s_3 = -\alpha_{\rm s}c_{\rm s}\big(\tfrac{1}{2}\gamma\alpha_{\rm s}^2 + \alpha_{\rm f}^2 + \tfrac{1}{2}\big)\,,\\ s_5 &= \alpha_{\rm s}c_{\rm s}\big(\tfrac{1}{2}\gamma\alpha_{\rm s}^2 + \alpha_{\rm f}^2 + \tfrac{1}{2}\big)\,, \qquad s_7 = \alpha_{\rm f}c_{\rm f}\left(\tfrac{1}{2}\gamma\alpha_{\rm f}^2 + \alpha_{\rm s}^2 + \tfrac{1}{2}\right)\,. \end{aligned} \tag{20.86}$$

These were also presented in [525], and they reduce to the gas dynamic expression when either a fast or a slow wave becomes a purely acoustic wave (e.g. when $\alpha_{\rm f} = 1, c_{\rm f} = c$ and $\alpha_{\rm s} = 0$). The non-convex nature of these waves is now hidden in the parameters $\alpha_{\rm f}^2$ and $\alpha_{\rm s}^2$, which always obey $\alpha_{\rm f}^2 + \alpha_{\rm s}^2 = 1$. With these expressions, we can write out the slow and fast rarefaction wave (centred simple wave) integral curves, as given by Eq. (20.40). There, the integral curves are formulated as curves in state space $\mathbf{U}$ everywhere tangent to the eigenvectors $\mathbf{r}^p$. Similarly, we can use the primitive state space $\mathbf{V}$ and write integral curves from $\hat{\mathbf{r}}^p$. Using a dimensionless parameter θ along the integral curve instead of the dimensional $\xi = x/t$, we find for the fast integral curve for the left going wave ($p = 1$) the governing differential equations

$$\begin{aligned} \rho' &= -\rho\,, \qquad p' = -\gamma p\,, \qquad v_x' = c_{\rm f}\,,\\ \mathbf{v}_{\rm t}' &= \frac{c_{\rm f}}{1 - c_{\rm f}^2/b_x^2}\frac{\mathbf{B}_{\rm t}}{B_x}\,, \qquad \mathbf{B}_t' = \frac{c_{\rm f}^2/b_x^2}{1 - c_{\rm f}^2/b_x^2}\mathbf{B}_{\rm t}\,. \end{aligned} \tag{20.87}$$

The prime stands for differentiation with respect to θ, and we denote the transverse vector components with the subscript t. Expressions for all six generalized Riemann invariants associated with either fast and slow wave families can be obtained in a way similar to the one outlined for the 1.5D iso-thermal MHD system [336]. Apart from the entropy, not many can be given in easily verified closed form expressions, with the exception of fast and slow magneto-acoustic Riemann invariants for an ideal gas with the fixed value of $\gamma = 5/3$. Since

$$\frac{dv_y}{dv_z} = \frac{dB_y}{dB_z} = \frac{B_y}{B_z}\,, \tag{20.88}$$

the generalized Riemann invariants show that, through fast or slow simple waves, the transverse field and flow vary in magnitude but do not rotate, as opposed to Alfvén waves. ◁

20.2.2 A robust shock-capturing method: TVDLF

Before presenting numerical schemes which, to varying degree, exploit the detailed knowledge of the characteristic fields in sets of conservation laws, we can easily generalize the TVD upwind method presented in Eq. (20.58) from a scalar conservation law to any system of conservation laws. We know that we need to impose a conservative scheme, of the form given in Eq. (20.46). We recognize that the coefficient $\alpha_{1+\frac{1}{2}}$ in the numerical flux function given by Eq. (20.58) is a (numerical) proxy for the characteristic speed $f'(u)$. A straightforward generalization to the case of a nonlinear system then adopts the numerical flux

$$\mathbf{F}_{i+\frac{1}{2}} = \tfrac{1}{2}\Big[\mathbf{F}(\mathbf{U}^{\rm L}_{i+\frac{1}{2}}) + \mathbf{F}(\mathbf{U}^{\rm R}_{i+\frac{1}{2}}) - |c^{\rm max}_{i+1/2}|\big(\mathbf{U}^{\rm R}_{i+\frac{1}{2}} - \mathbf{U}^{\rm L}_{i+\frac{1}{2}}\big)\Big]\,. \tag{20.89}$$

In this expression, $\mathbf{U}^{\rm L}$ and $\mathbf{U}^{\rm R}$ denote the state on the cell interface $i + \frac{1}{2}$, to the left or right, respectively. The scalar coefficient $c^{\rm max}$ then denotes the maximal physical propagation speed for the system under consideration. Obviously, for 1D MHD we get $c^{\max} = |v_x| + c_{\mathrm{f}}$, the maximal speed set for fast magneto-acoustic disturbances. This scheme has the advantage that we may even exploit it for a zero-beta plasma.

Minor variations of formula (20.89) can be found, depending on how one computes $c^{\max}$ at the cell interface. One can use the maximal physical propagation speed of the average $\frac{1}{2}(\mathbf{U}^{\mathrm{L}} + \mathbf{U}^{\mathrm{R}})$ (which could also average the primitive variables), or the maximum of each state separately, i.e.

$$\max\left(|c^{\max}(\mathbf{U}^{\mathrm{L}})|, |c^{\max}(\mathbf{U}^{\mathrm{R}})|\right). \tag{20.90}$$

The TVDLF method is then also referred to as local Lax–Friedrichs method, since it uses a local value for the maximal propagation speed. A first order variant was exploited as early as in 1961 by Rusanov [534] to solve several shock-dominated 2D gas dynamic problems. A first order scheme indeed results when we take the states $\mathbf{U}^{\mathrm{L}} = \mathbf{U}_i$ and $\mathbf{U}^{\mathrm{R}} = \mathbf{U}_{i+1}$, combined with the single step temporal advance written in Eq. (20.46). A temporally second order variant is obtained by using a two-step predictor-corrector approach, and use the time-centred (i.e. at $n + \frac{1}{2}$) fluxes (20.89) in the corrector step. For full second order accuracy, one also needs to raise the spatial order. To that effect, a linear reconstruction from cell-centre to cell edge quantities $\mathbf{U}^{\mathrm{L,R}}$ needs to be performed as well, which ideally remains in agreement with the TVD concept. This method was introduced and compared in [607] on a variety of HD and MHD problems.

(a) Linear reconstruction and slope limiters In the finite volume discretization, the evaluation of fluxes at the cell edges implies that some extrapolation within the cell i is needed from the volume averaged value $\mathbf{U}_i$. Simply using a constant extrapolation where $\mathbf{U}(x \in [x_{i-1/2}, x_{i+1/2}]) = \mathbf{U}_i$ is of course consistent with $\mathbf{U}_i$ as the cell average, but leads to first order accuracy where averaged edge fluxes $\mathbf{F}_{i+1/2} = \frac{1}{2}\big(\mathbf{F}(\mathbf{U}_i) + \mathbf{F}(\mathbf{U}_{i+1})\big)$ are used. Better, and still consistent with $\mathbf{U}_i$ as a volume average, is to use a linear extrapolation within the cell with slope $\boldsymbol{\sigma}_i$, hence

$$\mathbf{U}(x \in [x_{i-1/2}, x_{i+1/2}]) = \mathbf{U}_i + \boldsymbol{\sigma}_i (x - x_i)/\Delta x_i\,. \tag{20.91}$$

The slope is then the difference

$$\boldsymbol{\sigma}_i = \mathbf{U}_{i+1/2} - \mathbf{U}_{i-1/2} \equiv \Delta\mathbf{U}_i\,. \tag{20.92}$$

This linear reconstruction is exploited to get a *left* and *right* edge centred state,

$$\mathbf{U}^{\mathrm{L}}_{i+1/2} = \mathbf{U}_i + \Delta\mathbf{U}_i/2 \quad \text{and} \quad \mathbf{U}^{\mathrm{R}}_{i+1/2} = \mathbf{U}_{i+1} - \tfrac{1}{2}\Delta\mathbf{U}_{i+1}\,. \tag{20.93}$$

The flux at the cell edge then takes the average from $\mathbf{F}_{i+1/2} = \frac{1}{2}\big(\mathbf{F}(\mathbf{U}^{\mathrm{L}}_{i+1/2}) + \mathbf{F}(\mathbf{U}^{\mathrm{R}}_{i+1/2})\big)$. This process of linear reconstruction raises the spatial order of accuracy to second order. In practice, one must limit the slopes used in the linear reconstruction in order to avoid the introduction of spurious oscillations. This means that we rather write, denoting this slope limiting process by an overbar,

$$\mathbf{U}^{\mathrm{L}}_{i+\frac{1}{2}} = \mathbf{U}_i^{n+1/2} + \tfrac{1}{2}\Delta\bar{\mathbf{U}}_i^{n+1/2}, \qquad \mathbf{U}^{\mathrm{R}}_{i+\frac{1}{2}} = \mathbf{U}_{i+1}^{n+1/2} - \tfrac{1}{2}\Delta\bar{\mathbf{U}}_{i+1}^{n+1/2}\,. \tag{20.94}$$

On comparing the slopes as obtained from using neighbouring left or right cell values, one typically needs to take the least steep slope of the two, and to fall back on constant extrapolation within a cell when these slopes conflict in sign. Schematically, this is shown in Fig. 20.7. Different slope limiters exist which ensure the total variation diminishing (TVD) concept, thereby avoiding the creation of spurious oscillations in the numerical solution to the Riemann problem. However, these statements

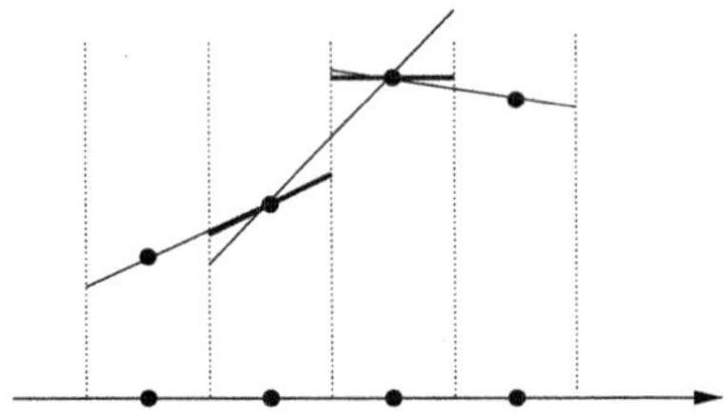

Fig. 20.7 Slope limited linear reconstruction process.

are strictly speaking only true for a single scalar nonlinear equation. A particularly robust (albeit diffusive) slope limiter is the 'minmod' limiter, where

$$\Delta \bar{\mathbf{U}}_i = \mathrm{sgn}(\mathbf{U}_i - \mathbf{U}_{i-1}) \max\left[0, \min\{|\mathbf{U}_i - \mathbf{U}_{i-1}|, (\mathbf{U}_{i+1} - \mathbf{U}_i)\,\mathrm{sgn}(\mathbf{U}_i - \mathbf{U}_{i-1})\}\right]. \quad (20.95)$$

By generalizing the minmod idea where left and right-sided slopes are compared, the monotonized central-difference (MC) limiter compares three slopes, and is in practice a better choice. These and various other slope limiter flavours can be found in [405]. The stencil of the resulting second order TVDLF scheme is five cells wide, through the slope limited linear reconstruction. In the second order variant, slight improvements (in robustness) may occur when performing the linear reconstruction on the primitive variables $\mathbf{V}$ instead of the conservative variables $\mathbf{U}$.

(b) Example Riemann problems for 1.5D iso-thermal MHD Exploiting the second order TVDLF method, we can now generate numerical solutions for 1D MHD problems. In particular, we solve the 1.5D iso-thermal MHD equations, and generate numerically the solution for the following initial conditions. In the examples, we actually use an excessive amount of grid points (order 1000 to 50 000), with an emphasis on the fully converged end state.

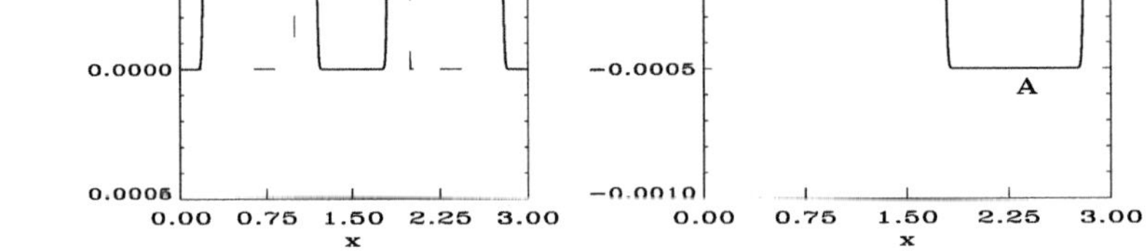

Fig. 20.8 An Alfvén pulse simulated in 1.5D zero-beta MHD.

A first test considers linear shear Alfvén waves, and takes $B_x = 1$, in a zero-beta plasma, i.e. $p = 0$. This zero-beta limit will pose difficulties for characteristic based schemes, since the slow magneto-sonic speed vanishes and the matrices collecting the eigenvectors become singular. The density $\rho = 1$, and we set up a pulse in transverse velocity $v_y = 0.001$ at $x \in [1, 2]$, on a total domain extending from $x \in [0, 3]$. The pulse will split into two equal sized Alfvén signals, shown at $t = 0.8$ in Fig. 20.8. The figure shows both the initial (dotted) and final numerical solution, for the transverse velocity and magnetic field. The propagation speed and polarization are in exact accord with the expected outcome, where both pulses travel at the Alfvén speed along the x-directed field. Note that we actually set up a double Riemann problem.

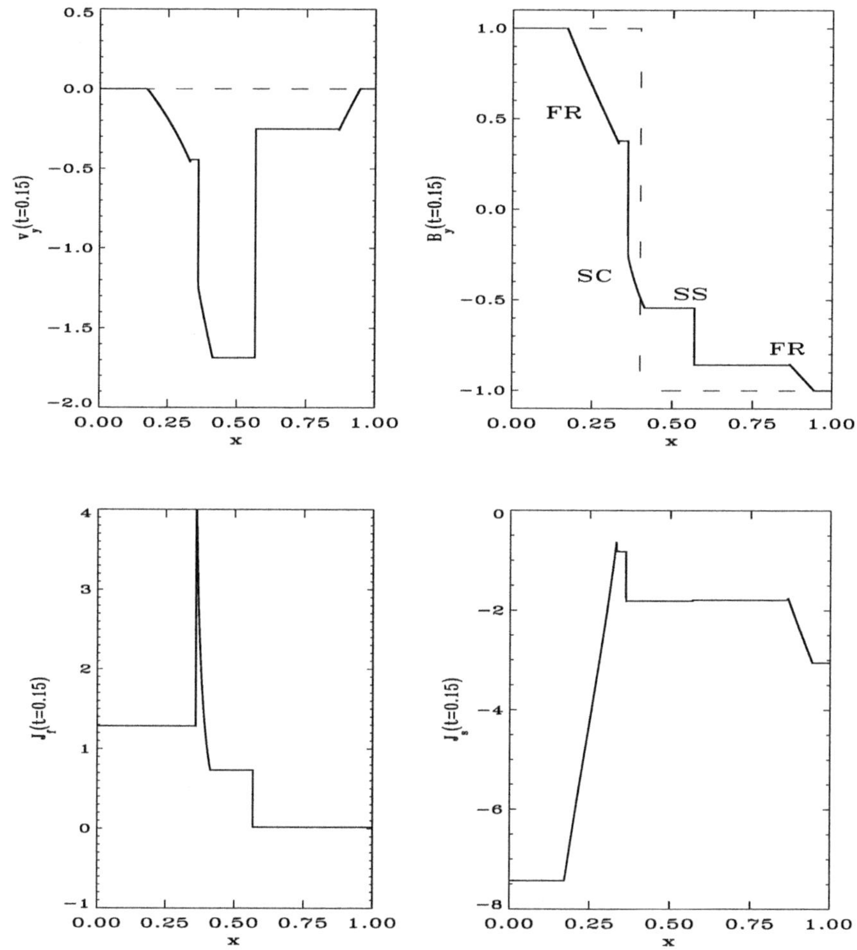

Fig. 20.9 Top: A Riemann solution in 1.5D iso-thermal MHD with a left going fast rarefaction, a slow compound wave, a slow shock and a right going fast rarefaction. Bottom: The fast and slow magneto-acoustic Riemann invariants.

The next two tests consider Riemann problems, chosen to demonstrate all flavours of the nonlinear wave patterns one may encounter for this system. On a unit domain, a discontinuity is placed at $x = 0.4$ separating $\mathbf{V}_\ell = (1, 0, 0, 1)$ from $\mathbf{V}_\mathrm{r} = (0.125, 0, 0, -1)$. The constant parameters are set to $B_x = 0.75$ and $c_\mathrm{i}^2 = 1$. This test is analogous to that introduced by Brio and Wu [96] for the 1.5D MHD system. It leads to a left going fast rarefaction, a slow compound wave (shock with rarefaction wave attached to it), a right going slow shock and fast rarefaction. These can be seen in Fig. 20.9(a) at $t = 0.15$, again showing the transverse vector components. Figure 20.9(b) plots the fast and slow magneto-acoustic Riemann invariants from Eq. (20.23). Through the rarefactions, we indeed find constant values as evaluated from the numerical solutions. They change discontinuously at shocks, while the fast invariant stays constant through fast rarefactions, and the slow invariant is constant through the rarefaction part of the slow compound wave.

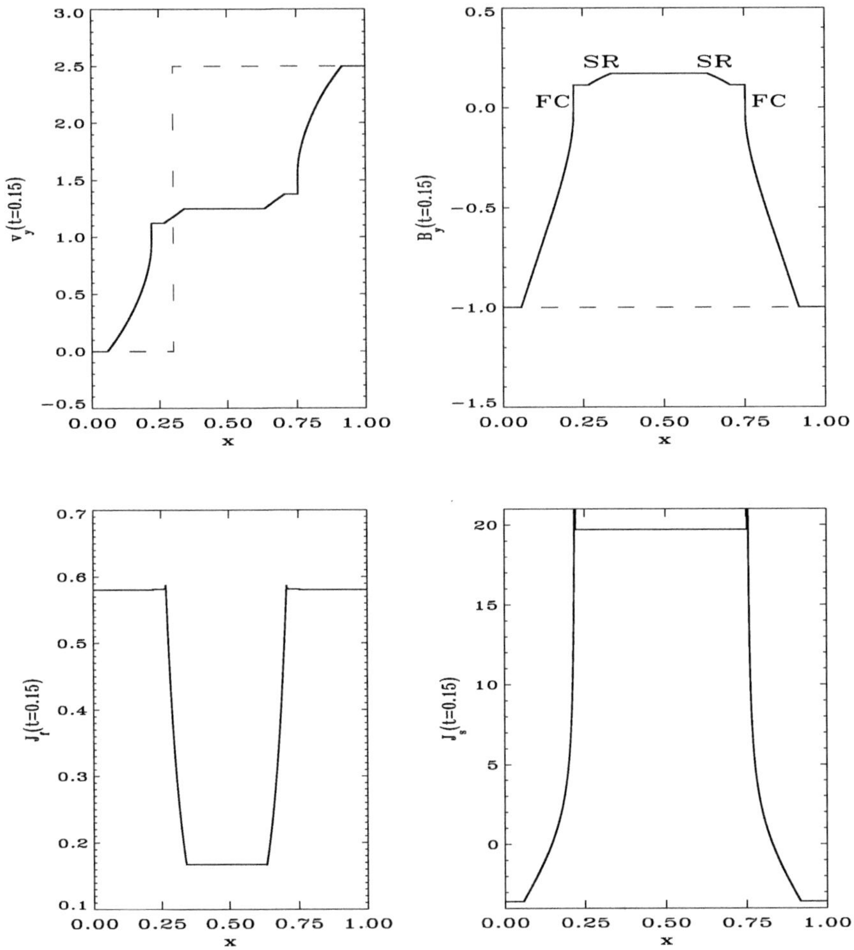

Fig. 20.10 Top: A Riemann solution in 1.5D iso-thermal MHD with a left going fast compound wave, two slow rarefactions, and a right going fast compound wave. Bottom: The fast and slow magneto-acoustic Riemann invariants.

A final Riemann problem, analogous to a test in [457], leads to fast compound waves. The discontinuity, initially at $x = 0.3$, brings the states $\mathbf{V}_\ell = (1, 0, 0, -1)$ and $\mathbf{V}_\mathrm{r} = (1, 2.5, 2.5, -1)$ into contact. The parameters fix $B_x = 1$ and $c_\mathrm{i}^2 = 1$, and we generate a solution up to $t = 0.15$. We find a fast compound wave, with a rarefaction part running ahead of the fast shock attached to it, two slow rarefaction waves, and a similar fast compound wave at right. Figure 20.10(a) plots the solutions obtained together with initial conditions, while the evaluation of the magneto-acoustic Riemann invariants is shown in Fig. 20.10(b). Slight overshoots in the solutions occur, despite the excessive resolution used. However, the main point demonstrated here is the feasibility of using a simple TVDLF scheme for generating numerical solutions in excellent agreement with theory.

20.2.3 Approximate Riemann solver schemes

When discussing the Godunov method, we noted that the basic ingredient is the (exact) solution of the nonlinear Riemann problem encountered at cell edges. It was soon realized that one might just as well solve these local Riemann problems in an approximate fashion. *Approximate Riemann solver* based schemes use a linearization of the nonlinear problem, and capitalize on the fact that the exact analytic solution for a linear hyperbolic system is known. In hydrodynamics, a method initiated by Roe has proven to be very powerful, and a useful review on characteristic based schemes for the Euler equations is given in [524]. We present its extension to MHD, which has been initiated by the seminal work of Brio and Wu [96].

(a) Roe-type approximate Riemann solver As a general procedure to numerically solve the system $\mathbf{U}_t + (\mathbf{F}(\mathbf{U}))_x = 0$, one considers again the local Riemann problem obtained from a left and right cell interface value $\mathbf{U}_\ell$ and $\mathbf{U}_\mathrm{r}$. Instead of using the exact nonlinear solution, Roe suggested solving a linear Riemann problem instead, namely

$$\mathbf{U}_t + \big(\mathbf{G}(\mathbf{U})\big)_x = 0\,, \tag{20.96}$$

where $\mathbf{G}(\mathbf{U}) = \mathbf{F}(\mathbf{U}_\mathrm{r}) + \mathbf{A}\cdot(\mathbf{U}-\mathbf{U}_\mathrm{r})$ includes a constant matrix $\mathbf{A} = \mathbf{A}\,(\mathbf{U}_\ell, \mathbf{U}_\mathrm{r})$. This matrix must satisfy the conditions

- $\mathbf{F}(\mathbf{U}_\ell) - \mathbf{F}(\mathbf{U}_\mathrm{r}) = \mathbf{A}(\mathbf{U}_\ell, \mathbf{U}_\mathrm{r})\cdot(\mathbf{U}_\ell - \mathbf{U}_\mathrm{r})$,
- $\mathbf{A}(\mathbf{U}_\ell, \mathbf{U}_\mathrm{r}) \rightarrow \mathbf{F}_\mathbf{U}(\mathbf{U}_\mathrm{r})$ as $\mathbf{U}_\ell \rightarrow \mathbf{U}_\mathrm{r}$,
- $\mathbf{A}(\mathbf{U}_\ell, \mathbf{U}_\mathrm{r})$ has only real eigenvalues,
- $\mathbf{A}(\mathbf{U}_\ell, \mathbf{U}_\mathrm{r})$ has a complete system of eigenvectors.

These conditions ensure that the exact Riemann problem solution is obtained when the initial states obey the Rankine–Hugoniot relations, i.e. are separated by a single shock transition or contact discontinuity. Indeed, when $(\mathbf{U}_\ell, \mathbf{U}_\mathrm{r})$ are such that $\mathbf{F}(\mathbf{U}_\ell) - \mathbf{F}(\mathbf{U}_\mathrm{r}) = s\,(\mathbf{U}_\ell - \mathbf{U}_\mathrm{r})$, the first condition ensures that $\mathbf{U}_\ell - \mathbf{U}_\mathrm{r}$ is an eigenvector of $\mathbf{A}$ with eigenvalue equal to the shock speed s. The solution of the linear Riemann problem merely maintains this jump at the physically correct shock speed. The second condition ensures consistency with the original nonlinear equations. The last two guarantee solvability of the linear Riemann problem.

Provided such a Roe matrix $\mathbf{A}$ can be found, the Roe scheme uses the numerical Roe flux expression at the interface between cells $i, i+1$ given by

$$\mathbf{F}_{i+1/2}\,(\mathbf{U}_i, \mathbf{U}_{i+1}) = \mathbf{F}(\mathbf{U}_i) + \mathbf{A}_{i+1/2}\cdot\big(\hat{\mathbf{U}} - \mathbf{U}_i\big)\,. \tag{20.97}$$

In this expression, $\hat{\mathbf{U}} = \hat{\mathbf{U}}(0, \mathbf{U}_i, \mathbf{U}_{i+1})$ indicates the exact solution of the linear Riemann problem and the matrix depends on $\mathbf{A}_{i+1/2}(\mathbf{U}_i, \mathbf{U}_{i+1})$. Let us write the eigenvalues λ_p and right eigenvectors $\mathbf{r}^p$ of the local Roe matrix $\mathbf{A}_{i+1/2}(\mathbf{U}_i, \mathbf{U}_{i+1})$, hence $\mathbf{A}_{i+1/2}\cdot\mathbf{r}^p = \lambda_p\mathbf{r}^p$. More explicit expressions for the Roe flux can be obtained as follows. As encountered in the discussion of the linear hyperbolic system, one decomposes both states in terms of the eigenvectors $\mathbf{r}^p$, writing $\mathbf{U}_i = \sum \beta_p \mathbf{r}^p$ and $\mathbf{U}_{i+1} = \sum \gamma_p \mathbf{r}^p$. The same decomposition of the difference $\mathbf{U}_{i+1} - \mathbf{U}_i$ then introduces coefficients α_p from

$$\mathbf{U}_{i+1} - \mathbf{U}_i = \sum (\gamma_p - \beta_p)\, \mathbf{r}^p \equiv \sum \alpha_p \mathbf{r}^p \,. \tag{20.98}$$

As found from the analytic solution for the Riemann problem for the linear hyperbolic system, the solution along the (x,t) ray $(x - x_{i+\frac{1}{2}})/t = 0$ is thus

$$\hat{\mathbf{U}} = \sum_{0<\lambda_p} \beta_p \mathbf{r}^p + \sum_{0>\lambda_p} \gamma_p \mathbf{r}^p \,. \tag{20.99}$$

This can be written as

$$\hat{\mathbf{U}} = \tfrac{1}{2}(\mathbf{U}_i + \mathbf{U}_{i+1}) + \tfrac{1}{2}\left[\sum_{\lambda_p<0} - \sum_{\lambda_p>0} \right] \alpha_p \mathbf{r}^p \,. \tag{20.100}$$

The Roe flux (20.97) then becomes, as a result of the first Roe condition imposed on the matrix A,

$$\mathbf{F}_{i+1/2} = \tfrac{1}{2}\big(\mathbf{F}(\mathbf{U}_i) + \mathbf{F}(\mathbf{U}_{i+1})\big) - \tfrac{1}{2} \sum \mid \lambda_p \mid \alpha_p \mathbf{r}^p \,. \tag{20.101}$$

The solver is then determined once the matrix $\mathsf{A}_{i+1/2}$ that satisfies the Roe conditions is constructed. For the Euler system of compressible gas dynamics, the matrix satisfying all conditions is the flux Jacobian of the system $\mathsf{F}_{\mathrm{U}}(\bar{\mathbf{U}})$, evaluated in a special, so-called Roe-average, state computed from the edge values $\bar{\mathbf{U}}(\mathbf{U}_\ell, \mathbf{U}_\mathrm{r})$. When generalized to the MHD system, Brio and Wu [96] identified a similar Roe-average for the special case of $\gamma = 2$ only. A general procedure to construct Roe matrices for systems of conservation laws, including the gas dynamic and ideal MHD systems, was presented in [107]. There, an MHD Roe solver without hypothesis on the value of γ is presented. For the MHD case, Brio and Wu already noted that one may use the flux Jacobian, merely evaluated in a state obtained from a simple averaging (e.g. the arithmetic average of both edge states). This then no longer captures stationary discontinuities as exact steady numerical solutions, but instead resolves these jumps with few grid points. In practice, one thus uses expression (20.101) for computing the flux, which evaluates the right eigenvectors of the flux Jacobian in the chosen average state $\mathbf{r}^p(\bar{\mathbf{U}})$, evaluates the eigenvalues for this state $\lambda_p(\bar{\mathbf{U}})$, and computes the wave strength coefficients $\alpha_p = \mathbf{l}^p \cdot (\mathbf{U}_{i+1} - \mathbf{U}_i)$, using the left eigenvectors of the Jacobian $\mathbf{l}^p(\bar{\mathbf{U}})$. The latter can also be computed from the jumps in (any choice of) the associated primitive variables $\mathbf{V}$, where we write $\alpha_p = \hat{\mathbf{l}}^p \cdot (\mathbf{V}_{i+1} - \mathbf{V}_i)$. Earlier, we gave the needed expressions for the right eigenvectors'$\mathbf{r}^p$ in Eqs. (20.81)–(20.82), for the eigenvalues in Eq. (20.75), and for the left eigenvectors (of the primitive formulation) in Eq. (20.78), so that all ingredients to the above flux formula (20.101) are known. This Roe-type approximate Riemann solver is similarly extended to second order by using a limited linear reconstruction from cell centre to cell edge values, and replacing $\mathbf{U}_{i+1}$ with $\mathbf{U}^{\mathrm{R}}_{i+1/2}$ and $\mathbf{U}_i$ with $\mathbf{U}^{\mathrm{L}}_{i+1/2}$. The limiting can actually be done on the characteristic wave contributions separately, too, and again exists in many flavours.

▷ **Roe average, entropy fixes and further details** The Roe average state, as presented for $\gamma = 2$ by Brio and Wu [96], has been used for the general case as well, and reduces to the gas dynamical Roe average for vanishing magnetic field. For left and right states $\mathbf{U}^{\mathrm{L}}$ and $\mathbf{U}^{\mathrm{R}}$, all average variables are computed from

$$\bar{\rho} \equiv \sqrt{\rho^{\mathrm{L}}\rho^{\mathrm{R}}}, \qquad \bar{\mathbf{v}} \equiv \frac{\sqrt{\rho^{\mathrm{L}}}\mathbf{v}^{\mathrm{L}} + \sqrt{\rho^{\mathrm{R}}}\mathbf{v}^{\mathrm{R}}}{\sqrt{\rho^{\mathrm{L}}} + \sqrt{\rho^{\mathrm{R}}}},$$
$$\bar{\mathbf{B}} \equiv \frac{\sqrt{\rho^{\mathrm{R}}}\mathbf{B}^{\mathrm{L}} + \sqrt{\rho^{\mathrm{L}}}\mathbf{B}^{\mathrm{R}}}{\sqrt{\rho^{\mathrm{L}}} + \sqrt{\rho^{\mathrm{R}}}}, \qquad \bar{h} \equiv \frac{\sqrt{\rho^{\mathrm{L}}}h^{\mathrm{L}} + \sqrt{\rho^{\mathrm{R}}}h^{\mathrm{R}}}{\sqrt{\rho^{\mathrm{L}}} + \sqrt{\rho^{\mathrm{R}}}}. \tag{20.102}$$

The latter quantity is the specific enthalpy $h \equiv (\mathcal{H} + p + B^2)/\rho$, and can be used to obtain the average squared sound speed $\bar{c}^2 = (\gamma - 1)(\bar{h} - \frac{1}{2}\bar{v}^2 - \bar{B}^2/\bar{\rho})$. Roe and Balsara [525] analysed the sensitivity of the approximate Riemann solver to the precise choice of averaging near the triple umbilic point, and concluded that any reasonable averaging would perform well. Therefore, less computationally elaborate (e.g. arithmetic) averagings have found widespread use. As was quickly noted upon application of the Roe solver to the Euler gas dynamic system, the solver may fail in cases where a transonic rarefaction is part of the solution, in a sense that it would rather detect an entropy violating weak shock solution. For this purpose, entropy fixes are used, which merely replace the near vanishing eigenvalue by some small value. For the wave speeds appearing in (20.101) for the MHD system, in particular for the fast and slow families $p = 1, 3, 5, 7$, we thus write

$$\lambda_p \longrightarrow \tfrac{1}{2}\left(\lambda_p^2/\epsilon_{\mathrm{e}} + \epsilon_{\mathrm{e}}\right), \tag{20.103}$$

if $|\lambda_p| < \epsilon_{\mathrm{e}}$. This is thus only effective when transonic solutions $|v_x| \approx c_{\mathrm{s,f}}$ occur, and introduces a small parameter ϵ_{e}. In the numerical evaluation of the various terms in the eigenvector expressions, we still encounter difficulties in certain degenerate cases. In particular, the eigenvectors contain the ratios $\beta_y = B_y/B_\perp$ and β_z, which become ill-defined when the transverse field vanishes completely. This is remedied by replacing them by

$$\beta_y = \frac{B_y + \epsilon}{\sqrt{B_y^2 + B_z^2 + 2\epsilon^2}}, \qquad \beta_z = \frac{B_z + \epsilon}{\sqrt{B_y^2 + B_z^2 + 2\epsilon^2}}. \tag{20.104}$$

With ϵ a very small number above machine precision, we then in effect replace the indeterminacies by $1/\sqrt{2}$ constants. Similarly, when $c_{\mathrm{s}}^2 = c_{\mathrm{f}}^2$, one uses $\alpha_{\mathrm{f}} = \alpha_{\mathrm{s}} = 1/\sqrt{2}$ which still leaves $\alpha_{\mathrm{f}}^2 + \alpha_{\mathrm{s}}^2 = 1$. ◁

(b) HLL and HLLC or HLLD discretizations The approximate Riemann solver just discussed uses a linearization of the Riemann problem encountered at cell interfaces, which is at the heart of all Godunov-type schemes. The Roe-type solver clearly incorporates much knowledge of the underlying physical problem, and in effect uses the proper characteristic decomposition to 'weight' the information propagating in different directions. It approximates the time-averaged flux appropriate for that state of the Riemann fan sketched schematically in Fig. 5.10 which is found along the vertical t-axis. This 'upwind' concept is a key issue for generating numerically stable conservative schemes. Characteristic based schemes are by far the most accurate numerical methods for handling problems with discontinuities, and several variants have improved on the work initiated in MHD by Brio and Wu [96]. Zachary and Collela [663] constructed an MHD Godunov method in which the approximation to the upwind flux was built up in the primitive variable space, organizing the states in the Riemann fan about some mean entropy wave speed. A TVD Roe-type upwind scheme was presented by Ryu and Jones [535], benchmarking against the predictions of a nonlinear Riemann solver. A large number of 1D MHD Riemann problem tests are documented there.

The methods mentioned all incorporate the full 7-wave structure which is a natural outcome of the MHD Riemann problem. Contrasting the flux expression (20.101) with the TVDLF method with flux (20.89), we may re-interpret the latter as one where the 7-wave structure is essentially replaced by an approximation where only the maximum wave speed enters, replacing all $|\lambda_p|$ by $|c^{\max}|$.

This retains a crude form of upwinding, but introduces more numerical diffusion. The latter can in fact be advantageous to deal with some of the robustness issues that plague the approximate Riemann solvers in particular. For the Euler system, a catalogue of failings on shock-dominated problems has been presented by Quirk [513]. Some of these flaws can be cured by appropriate fixes, but a fairly common strategy is to add numerical dissipation in some form or other. A trend away from using the full 7-wave structure towards the other extreme of using only the fastest wave (TVDLF) is noticeable in the extensive literature on the subject. Versions of solvers originated in gas dynamics have been adapted to MHD, which in effect replace the Riemann fan by a 2-wave or 3-wave structure. Appropriate recipes for computing meaningful intermediate states then ensure desirable properties like positivity (ensuring always positive pressures and densities), the ability to capture isolated discontinuities, etc. A 2-wave approximation is used in the HLL solver (due to Harten, Lax and van Leer [291]), and this involves an intermediate state $\mathbf{U}^*$ which generalizes the Rankine–Hugoniot relation (20.34) to ensure

$$\mathbf{F}(\mathbf{U}_\ell) - \mathbf{F}(\mathbf{U}_\mathrm{r}) = \lambda^-(\mathbf{U}_\ell - \mathbf{U}^*) + \lambda^+(\mathbf{U}^* - \mathbf{U}_\mathrm{r}), \tag{20.105}$$

where λ^- and λ^+ are both fastest wave speeds still retained in the Riemann fan approximation. This method then switches the flux evaluation between $\mathbf{F}(\mathbf{U}_\ell)$ when $\lambda^- \geq 0$, $\mathbf{F}(\mathbf{U}_\mathrm{r})$ when $\lambda^+ \leq 0$, or a weighted flux expression

$$\mathbf{F}^* = \frac{\lambda^+\mathbf{F}(\mathbf{U}_\ell) - \lambda^-\mathbf{F}(\mathbf{U}_\mathrm{r}) + \lambda^-\lambda^+(\mathbf{U}_\mathrm{r} - \mathbf{U}_\ell)}{\lambda^+ - \lambda^-} \tag{20.106}$$

for the case where the intermediate state should apply. Note that $\mathbf{F}^* \neq \mathbf{F}(\mathbf{U}^*)$. In practical application, this HLL method has proven to work very similarly to the TVDLF method, demonstrating minor improvements over the latter. For contact discontinuities, both methods introduce fairly large diffusion, but at the same time the schemes are very robust and computationally simple.

In an attempt to remedy the smearing of contact discontinuities, the HLLC approach "restores the contact surface in the HLL-solver" [600], hence the added C for 'contact'. Its adaptation to MHD has been undertaken by various authors, e.g. by [413] and [407]. One can readily generalize the relation (20.105) to introduce a third central wave λ^0 separating a left $\mathbf{U}^*_\ell$ and right $\mathbf{U}^*_\mathrm{r}$ middle state, and choose a well-motivated construction to uniquely compute these two additional states. Four corresponding flux expressions are in use, depending on the orientation of the three waves in the Riemann fan with respect to the time axis in the (x, t) plane. Schematically then, the Riemann fan for HLL and HLLC type methods is shown in Fig. 20.11. Other variants of the HLL family of solvers exist, and they all share a relative straightforward application to any system of conservation laws (with more than two unknowns) without requiring full decompositions in the many $(n > 2)$ wave families. In particular, HLLC has been adapted to relativistic MHD computations as well, a subject which will return in Chapter 22.

A final noteworthy solver for ideal MHD is the HLLD (where the D stands for discontinuity) variant by Miyoshi and Kusano [443]. Depending on whether B_x vanishes or not, HLLD switches between the use of two intermediate states (like HLLC, to which it reduces for vanishing magnetic field) versus four intermediate states. This solver then exactly resolves isolated discontinuities and shocks, and tests demonstrate that it is more robust than using a full approximate Riemann solver. The HLLD solver can also be combined with a splitting strategy introduced by Tanaka [586] (dis-

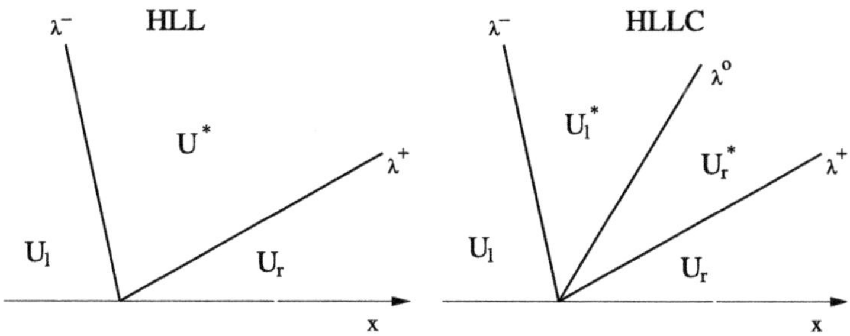

Fig. 20.11 The approximate representation of the Riemann fan employed in the HLL (left) and HLLC (right) methodology.

cussed further in Section 20.3.1), where the magnetic field one solves for is written as a deviation from a static potential field. This is then useful for handling low plasma beta conditions [271].

20.2.4 Simulating 1D MHD Riemann problems

A vast number of 1D MHD Riemann problem tests can be found throughout the literature, with the paper by Ryu and Jones [535] collecting a good number of distinct possibilities. One distinguishes the discontinuous transitions according to their type, as will be discussed in Chapter 21. We can have the already mentioned *contact and rotational discontinuities*, but also *tangential discontinuities* as B_x tends to zero, which can be regarded as degeneracies where a contact, two slow, and two rotational discontinuities coincide. *Fast and slow shocks*, as well as *rarefaction waves*, can be of special type, called *switch-off* (or *switch-on*) when they reduce (increase) the tangential magnetic field to (from) zero, and a fast shock is of pure *magneto-sonic* kind when B_x vanishes. Slow and fast signals can become of *compound* nature, where they become of *'intermediate'* type, i.e. cross the unit Alfvén Mach number and include a $180°$ degree rotation of the transverse field. Whether the latter are physically realizable transitions has received a lot of attention since they were first discovered by the Brio and Wu numerical solutions. In the coplanar 1.5D MHD case considered there, one can hardly avoid them in numerical solutions, such as those generated for the iso-thermal case above. In principle, they may then be replaced by a π-rotational wave, combined with a slow or fast shock, so that they represent cases where the Riemann problem has a non-unique solution. A clear discussion of the subject is found in Torrilhon [601], where it was shown that any non-planar Riemann problem will have a unique solution containing only regular waves if the initial jump in transverse velocity vanishes. Conversely, there exist also non-planar Riemann problems, which can have both a regular and a non-regular solution, similar to some coplanar problems. A thorough discussion, where on the basis of evolutionary arguments (i.e. stability against small perturbations) the admissibility of intermediate shocks is disputed, is found in [173].

We give now two examples of Riemann problems, using an approximate Riemann solver based scheme to generate the solution. The first example shown in Fig. 20.12 takes initial conditions also presented in Torrilhon [601], namely a left state $\mathbf{V}_\ell = (3,0,0,0,3,1,0)$ adjacent to a right state $\mathbf{V}_r = (1,0,0,0,1,\cos(1.5),\sin(1.5))$ where $\gamma = 5/3$ and $B_x = 3/2$. Our numerical solution is shown in the same format as Fig. 2 from [601], except for the bottom left panel where we show

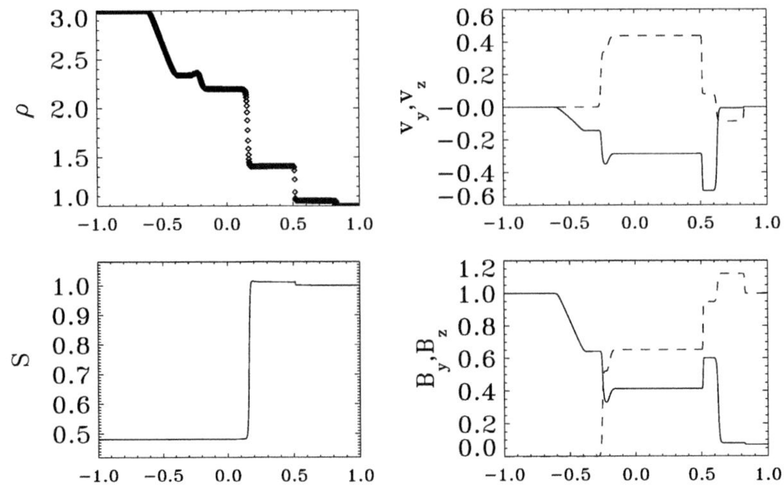

Fig. 20.12 A Riemann problem in 1D MHD, numerically solved with a TVD scheme using a Roe-type approximate Riemann solver. From left to right: a fast rarefaction, a rotational discontinuity, a slow rarefaction, a contact discontinuity, a slow shock, a rotational discontinuity and a fast shock.

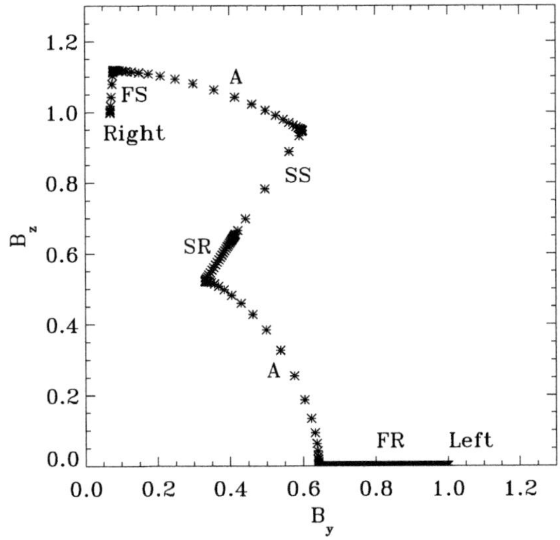

Fig. 20.13 The same numerical solution from Fig. 20.12, plotted in the (B_y, B_z) state plane. Fast and slow signals change the magnitude of the transverse field, while Alfvén signals rotate the field. In reality, the parts corresponding to shocks and rotational discontinuities should contain no points; the arcs and line segments seen relate to the numerical representation.

entropy $S = p\rho^{-\gamma}$ instead of pressure. This latter shows its invariance through fast and slow rarefactions, and Alfvén discontinuities. Further, we plotted the solution in the state space spanned by the transverse magnetic field components. The latter can be compared with Fig. 3 in [601]. One can detect from left to right in Fig. 20.12: a fast rarefaction, a rotational discontinuity, a slow rarefaction, a contact discontinuity, a slow shock, a rotational discontinuity and a fast shock. The representation of the numerical solution in the state space (Fig. 20.13) shows the expected outcome: only Alfvén signals rotate the transverse field, while fast and slow shocks and rarefactions alter the magnitude of the transverse field. Obviously, the solution is not perfect, in the sense that several grid points are needed to represent the discontinuities, and one may detect overshoots in density and a diffused contact discontinuity. Nevertheless, a very satisfactory agreement with the exact solution is obtained, and this is generally true for all modern upwind-type algorithms.

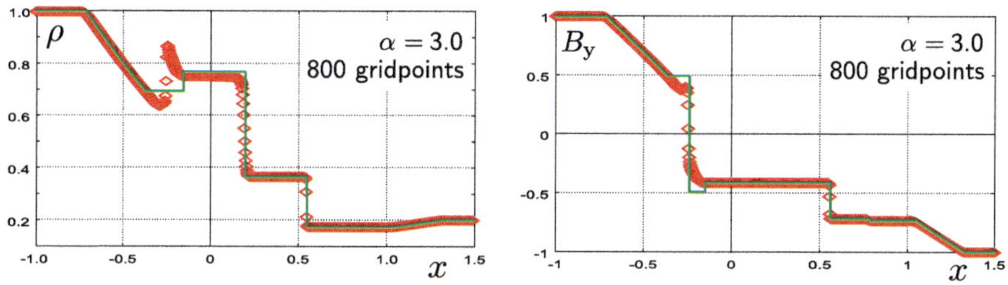

Fig. 20.14 A Riemann problem in 1D MHD, with a unique regular solution (solid line), and a typical 'high' resolution numerical result already exploiting 800 grid points: a wrong compound structure plagues the scheme to pseudo-converge. (From Torrilhon [601](b).)

A second and final Riemann problem shown is taken from Torrilhon [601](b), where a non-coplanar Riemann problem which has a unique, regular, solution was solved numerically using a variety of modern upwind algorithms. The initial conditions are taken as $\mathbf{V}_\ell = (1, 0, 0, 0, 1, 1, 0)$ and $\mathbf{V}_r = (0.2, 0, 0, 0, 0.2, \cos(3), \sin(3))$ with $B_x = 1$ and $\gamma = 5/3$. This is an almost coplanar configuration (exact coplanar when 3 is replaced by π) without jump in transverse velocity, where the unique exact solution has a fast rarefaction, a rotational discontinuity, a slow shock, a contact discontinuity, a slow shock, a rotational discontinuity and a fast rarefaction. It was demonstrated convincingly how all modern methods pseudo-converge to a solution containing a compound wave structure for up to several 1000 grid points. Figure 20.14 shows how the numerical solution significantly errs at $x \approx -0.25$. Only after heavy refinement, the pseudo-convergence changes to true convergence to the exact regular solution of the Riemann problem. This is a serious shortcoming of all modern shock-capturing schemes, which can only be remedied thus far by combining these methods with some form of grid-adaptivity.

20.3 Multi-dimensional MHD computations

Our description of numerical methods thus far concentrated on 1D problems. Conceptually, multi-dimensional simulations can be reduced to a succession of 1D problems in the various coordinate directions. As explained earlier, even finite volume methods on general (unstructured) grids can also be reduced to successive 1D problems only, in directions normal to the grid cell faces. A common strategy is to use a Strang type dimensional splitting, which for a 2D case in a Cartesian (x, y) setting can be written as

$$\mathbf{U}^{n+1} = L^x_{\Delta t/2} L^y_{\Delta t} L^x_{\Delta t/2} \mathbf{U}^n \,, \tag{20.107}$$

where the operator $L^x_{\Delta t/2}$ indicates an update with a half time step $\Delta t/2$ taking account of only the fluxes in the x-direction. Alternatively, we can alternate coordinate directions using the so-called

Godunov splitting

$$\mathbf{U}^{n+2} = L^x_{\Delta t} L^y_{\Delta t} L^y_{\Delta t} L^x_{\Delta t} \mathbf{U}^n \,. \tag{20.108}$$

These dimensional splitting strategies retain the second order accuracy of the individual dimensional sweeps.

The relative ease with which multi-dimensional codes are developed from 1D solvers has led to a rapid surge in multi-dimensional MHD modelling, once shock-capturing algorithms for the MHD system emerged in earnest. Dai and Woodward [138](b) used their own approximate nonlinear all-shock Riemann solver as a basic ingredient in a multi-dimensional ideal MHD solver. They used Strang splitting and exploited an essentially third order scheme, using parabolic instead of linear reconstructions (the piecewise parabolic method, or PPM). In a variety of shock-dominated 2D tests, they demonstrated robustness of their scheme, despite the fact that no additional measures for controlling numerical monopole errors were employed. A multi-dimensional TVD scheme, also using dimensional splitting, was described in Ryu et al. [535](b). A serious comparison between different shock-capturing schemes (among them TVD and TVDLF) for 1D and 2D hydro- and magnetohydrodynamic problems can be found in [607]. Both these latter works in essence use second order accurate, shock-capturing schemes.

It should be emphasized that the assumption of directional splitting strategies may not always be justified: characteristic based schemes then implicitly assume that waves travel orthogonal to grid cell interfaces, which is typically not the case. Multi-dimensional upwind strategies may need to be incorporated. Other modern efforts, such as [416], introduce higher order interpolations with a fully consistent strategy for Maxwell's law $\nabla \cdot \mathbf{B} = 0$ incorporated in the scheme.

20.3.1 $\nabla \cdot \mathbf{B} = 0$ *condition for shock-capturing schemes*

For multi-dimensional MHD, one also needs to handle the non-trivial $\nabla \cdot \mathbf{B} = 0$ constraint. Even if satisfied exactly at $t = 0$, one can still numerically generate $\nabla \cdot \mathbf{B} \neq 0$ due to the nonlinearities of the various shock-capturing methods. Besides the fact that this is clearly undesirable physically, it became clear that when no corrective action is taken at all, the accumulated errors may cause fatal numerical instabilities. However, exact solenoidal fields may not be needed in numerical simulations, as one always faces discretization and machine precision errors.

In practice, it turns out to be very difficult, although possible [604], to insist on (1) a conservative form which is needed for correctly handling shocks, (2) ensuring solenoidal $\mathbf{B}$ in some discrete sense, *and* (3) having the discretized Lorentz force orthogonal to the magnetic field in the cell centres. The conservative form of the MHD equations uses the divergence of the Maxwell stress tensor, which is equal to the Lorentz force provided $\nabla \cdot \mathbf{B} = 0$ since

$$\nabla \cdot \left(\tfrac{1}{2}B^2 \mathsf{I} - \mathbf{BB}\right) = -(\nabla \times \mathbf{B}) \times \mathbf{B} - \mathbf{B}\,(\nabla \cdot \mathbf{B}) \,. \tag{20.109}$$

In the context of shock-capturing schemes discussed thus far, many strategies to cope with the solenoidal constraint have been developed. A thorough comparison of seven different strategies on a series of nine 2D MHD tests can be found in [603], where the same high resolution base scheme (a TVD method) was used in combination with varying strategies for numerical monopoles. In what follows, we briefly discuss some of the most frequently used methods to handle the $\nabla \cdot \mathbf{B}$ problem in state-of-the-art MHD codes.

(a) Vector potential One way of handling $\nabla \cdot \mathbf{B}$ is by replacing the magnetic field variable by a corresponding vector potential defined from

$$\mathbf{B} = \nabla \times \mathbf{A} \,. \tag{20.110}$$

Analytically, this guarantees a solenoidal magnetic field, but when evaluating $\mathbf{B}$ in some discrete manner from $\mathbf{A}$, the operators should still be chosen to satisfy $\nabla \cdot (\nabla \times = 0$. Using $\mathbf{A}$ unavoidably increases the order of the occurring spatial derivatives to evaluate the Lorentz force. The need to express boundary conditions on $\mathbf{A}$ instead of $\mathbf{B}$ when using a ghost cell prescription for boundaries is also a non-trivial complication. This can be avoided by directly prescribing fluxes at the boundary instead. From the discussion of the characteristic based solvers, it is clear that they 'conflict' with using the vector potential as a basic variable. In many solvers, though, the vector potential can successfully be exploited.

(b) Projection method for $\nabla \cdot \mathbf{B}$ Originally mentioned by Brackbill and Barnes [89], one can combine any multi-dimensional method with a projection scheme strategy, which controls the numerical value of $\nabla \cdot \mathbf{B}$ in a particular discretization to a given accuracy. The basic idea [89] is to correct the $\mathbf{B}^*$ computed by a scheme with $\nabla \cdot \mathbf{B}^* \neq 0$, by projecting it on the sub-space of zero divergence solutions. Hence, we modify $\mathbf{B}^*$ by subtracting the gradient of a scalar field ϕ, to be computed from the Poisson equation

$$\nabla^2 \phi = \nabla \cdot \mathbf{B}^* \,. \tag{20.111}$$

By construction, this yields a solenoidal $\mathbf{B} = \mathbf{B}^* - \nabla\phi$ which is then used in the next time step. This process can be repeated after each time step or only after a certain number of time steps. It is important to note that the accuracy up to which the Poisson problem (20.111) is solved need not be machine precision, and that this approach keeps the order of accuracy of the base scheme, while not violating its conservation properties [603]. Note that the projection does not change the current density $\mathbf{j} = \nabla \times \mathbf{B}$. This 'elliptic divergence cleaning' indirectly affects the local balance of thermal to kinetic contributions to the total energy. It is possible to rather keep thermal energy unmodified, at the cost of giving up exact conservation of total energy.

The projection scheme in essence uses numerical concepts from incompressible fluid flow (where $\nabla \cdot \mathbf{v} = 0$). On uniform Cartesian grids, it was shown [603] to make the smallest possible correction to remove the divergence created by the base scheme: it corresponds to minimizing $|\mathbf{B} - \mathbf{B}^*|$ under the solenoidal constraint. For cases where the initial field is not known analytically, it may even be useful to eliminate finite numerical divergence errors of the discrete initial $\mathbf{B}$.

(c) Powell's source terms and divergence wave Another popular means of handling monopole errors was introduced by Powell [507]. In essence, the method uses additional source terms proportional to $\sim \nabla \cdot \mathbf{B}$ in the original set of ideal MHD equations in conservation form (20.65). These terms are

$$-\nabla \cdot \mathbf{B} \left(0\,, \mathbf{B}\,, \mathbf{v} \cdot \mathbf{B}\,, \mathbf{v}\right)^{\mathrm{T}} . \tag{20.112}$$

The source term for the momentum equation, $-\nabla \cdot \mathbf{B}\mathbf{B}$, directly follows from the relation (20.109). One of the reasons to add these terms is that the resulting modified set of PDEs restores Galilean invariance, as originally already noted by Godunov [213]. In Section 5.2, we also noted that the unmodified system has eight eigenvalues, one of which is zero while all others lie symmetrically about the fluid velocity v. It is the spurious zero eigenvalue which conflicts with the $\nabla \cdot \mathbf{B} = 0$ constraint, as it carries a jump in the normal component of the field. When adding the source terms (20.112), the zero eigenvalue gets replaced by an additional v. In a characteristic based scheme, it involves an additional 'eighth wave' contribution. The effect of this operation can be clearly illustrated on the (ideal) induction equation. The original form of this equation can be written as

$$\frac{\partial \mathbf{B}}{\partial t} = \nabla \times (\mathbf{v} \times \mathbf{B}), \quad \text{so that} \quad \frac{\partial}{\partial t} \nabla \cdot \mathbf{B} = 0 \tag{20.113}$$

and $\nabla \cdot \mathbf{B}$ errors remain undamped in time. With the source term included, however, we have

$$\frac{\partial \mathbf{B}}{\partial t} - \nabla \times (\mathbf{v} \times \mathbf{B}) + (\nabla \cdot \mathbf{B})\mathbf{v} = 0, \tag{20.114}$$

so that numerical errors in $\nabla \cdot \mathbf{B}$ become compensated, noting the equivalence of $\nabla \times (\mathbf{v} \times \mathbf{B}) = \mathbf{B} \cdot \nabla \mathbf{v} + \mathbf{v} \nabla \cdot \mathbf{B} - \mathbf{v} \cdot \nabla \mathbf{B} - \mathbf{B} \nabla \cdot \mathbf{v}$. We can further deduce from the divergence of (20.114) and mass conservation that

$$\frac{\partial}{\partial t}\big((\nabla \cdot \mathbf{B})/\rho\big) + \mathbf{v} \cdot \nabla\big((\nabla \cdot \mathbf{B})/\rho\big) = 0. \tag{20.115}$$

The latter equation is an advection equation for $\nabla \cdot \mathbf{B}/\rho$ and it ensures that $\nabla \cdot \mathbf{B}$ errors are advected away with the flow. (However, it fails to do anything in stagnation points.) The use of these source terms was exploited in combination with different base schemes in [607], with a clearly stabilizing influence overall.

However, just like any method which does not obey the strict conservation form of the equations, it can potentially introduce incorrect jumps across discontinuities, and an example was given in [603]. One can, of course, easily restore conservation of energy and momentum, by taking only the source term for the induction equation along, and additional arguments for doing so were given by [332] and [147]. Any of these source term strategies carry over easily to grid-adaptive computations, such as demonstrated in Powell et al. [508].

(d) Constrained transport Another commonly used strategy is to insist on *machine precision accuracy* in maintaining $\nabla \cdot \mathbf{B}$ *in one particular discretization*. The original concept for this so-called constrained transport is due to Evans and Hawley [171], and has since been used in many MHD codes. One needs to ensure an initial magnetic field representation that fully complies with zero divergence in the chosen discretization, and make sure that boundary conditions remain compatible with this constraint. Originally, typical constrained transport approaches used a staggering with thermodynamic quantities at cell centres, magnetic fields at cell face centres (at the middle of the edges in 2D) and electric field vectors (i.e. $\mathbf{E} = -\mathbf{v} \times \mathbf{B}$) at cell vertices (corners in 2D). By updating the magnetic field using a discrete evaluation of Stokes' law, the initial divergence of the magnetic field is maintained. In Fig. 20.15(a), this staggering is shown for a 2D Cartesian grid, where Ω denotes the variable E_z. The update for $\mathbf{B} = (B^x, B^y)$ from the induction equation can be written as

$$\begin{aligned}
\big(B^{x,n+1}_{j+\frac{1}{2},k} - B^{x,n}_{j+\frac{1}{2},k}\big)/\Delta t &= -\big(\Omega_{j+\frac{1}{2},k+\frac{1}{2}} - \Omega_{j+\frac{1}{2},k-\frac{1}{2}}\big)/\Delta y\,, \\
\big(B^{y,n+1}_{j,k+\frac{1}{2}} - B^{y,n}_{j,k+\frac{1}{2}}\big)/\Delta t &= +\big(\Omega_{j+\frac{1}{2},k+\frac{1}{2}} - \Omega_{j-\frac{1}{2},k+\frac{1}{2}}\big)/\Delta x\,.
\end{aligned} \tag{20.116}$$

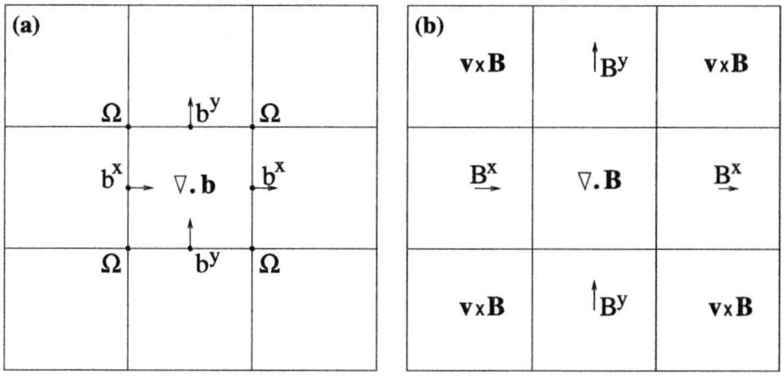

Fig. 20.15 (a) The staggering typically employed in constrained transport codes. (b) An alternative, fully cell-centred oriented scheme was introduced by Tóth: the field interpolated central difference scheme. (From Tóth [603].)

One can verify that, with this update, the particular discretized divergence found from

$$(\nabla \cdot \mathbf{B})_{j,k} = \left(B^x_{j+\frac{1}{2},k} - B^x_{j-\frac{1}{2},k}\right)/\Delta x + \left(B^y_{j,k+\frac{1}{2}} - B^y_{j,k-\frac{1}{2}}\right)/\Delta y \tag{20.117}$$

is maintained at its original value, i.e. $\nabla \cdot \mathbf{B}^{n+1} = \nabla \cdot \mathbf{B}^n$. Note that truncation errors still persist in any other discretization (certainly at shock locations or at other discontinuities). Tóth [603] realized that this constrained transport concept can be recast fully in a finite volume sense, i.e. in terms of (average) fluxes across grid cell edges for *cell centred magnetic field* evaluations involving spatio-temporal averaging of the original staggered variables. This could be reversed, so that one no longer needs to stagger the variables in actual implementations. Once this realization was made, a fair variety of variants on the CT idea could be introduced. One of these in essence 'zooms out', to employ the cell-centred variables at neighbouring grid points, as shown in Fig. 20.15(b). This eliminates the need for spatial interpolations needed in CT schemes (as magnetic fields and velocities need to be evaluated at corners and cell centres), and the corresponding field interpolated central difference scheme [603] only uses temporal interpolations.

(e) Parabolic and hyperbolic cleaning A last strategy mentioned here is a parabolic and hyperbolic variant of the initial elliptic strategy to use the numerical truncation errors in $\nabla \cdot \mathbf{B}$ as source terms. One can add diffusion type source terms to both the energy equation and the induction equation, or to the induction equation alone, which diffuse $\nabla \cdot \mathbf{B}$ at a maximal rate. This maximal rate is one that does not further constrain the discrete time step beyond the usual CFL condition. For a purely parabolic equation with a diffusion coefficient $\eta_{\rm D}$ as in

$$\frac{\partial \mathbf{B}}{\partial t} = \eta_{\rm D} \nabla^2 \mathbf{B} \,, \tag{20.118}$$

one gets an additional constraint for $\Delta t < \Delta x^2/\eta_{\rm D}$. Taking $\eta_{\rm D} \sim \Delta x^2$ then suggests the use of parabolic source terms for (20.65) given by

$$\left(0\,,0\,,\mathbf{B}\cdot\nabla C_{\rm d}\Delta x^2(\nabla\cdot\mathbf{B})\,,\nabla C_{\rm d}\Delta x^2(\nabla\cdot\mathbf{B})\right)^{\rm T} . \tag{20.119}$$

The coefficient C_d is a numerical factor of order unity (depending on the scheme to treat source additions and to evaluate the gradients). This strategy can be easily adapted to curvilinear hierarchically nested grids, as shown in [621]. Obviously, like any source term strategy, it can potentially cause violations of the discrete Rankine–Hugoniot relations across shock fronts, so one best uses only the term for the induction equation. In grid-adaptive computations, this strategy compared favourably with all other source term approaches on a variety of shock dominated MHD problems [352].

While this parabolic treatment tries to diffuse numerical monopole errors, and the elliptic approach with Powell's source terms passively advects $\nabla \cdot \mathbf{B}/\rho$, a new *hyperbolic cleaning* variant was proposed by Dedner et al. [143]. The idea is to combine the two ideas, namely to transport $\nabla \cdot \mathbf{B}$ at a maximal admissible speed and damp the error simultaneously. This comes at the price of one additional scalar quantity, whose spatio-temporal evolution is coupled with $\nabla \cdot \mathbf{B}$. Various variants can be found in [143].

▷ **Splitting strategy for strong potential background fields** In a seminal paper by Tanaka [586], one of the first finite volume TVD schemes for 3D MHD simulations was presented and applied to model the Earth's magnetosphere in the supersonic solar wind. Figure 20.16 shows a 3D impression of the field lines obtained in the steady-state solution, with the sunward direction at right. A key feature of the method was to account

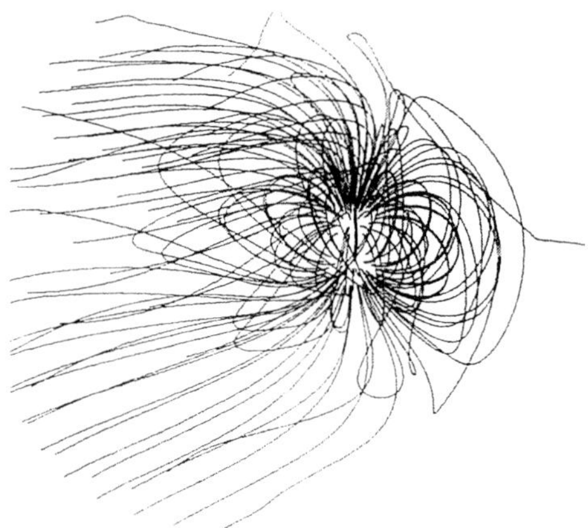

Fig. 20.16 A 3D model of the earth's magnetosphere embedded in supersonic solar wind flow. The Sun direction is towards the right. (From Tanaka [586].)

for the fact that the strong dipole field of the Earth will cause serious difficulties when the total magnetic field is taken as dependent variable, as both high- and low-β regions need to be computed accurately in the shock-dominated interaction. An elegant solution to this problem is to split the magnetic field into a static potential background field (e.g. only the dipole) and solve for the deviation from this field. Hence, we write

$$\mathbf{B} = \mathbf{B}_0 + \mathbf{B}_1 , \quad \text{where} \quad \frac{\partial \mathbf{B}_0}{\partial t} = 0, \quad \nabla \cdot \mathbf{B}_0 = 0, \quad \nabla \times \mathbf{B}_0 = \mathbf{0} , \tag{20.120}$$

and take $\mathbf{B}_1$ as dependent variable. Also, the equation for total energy conservation needs replacing by an equation governing the partial energy,

$$\mathcal{H}_1 = \tfrac{1}{2}\rho v^2 + \frac{p}{\gamma - 1} + \tfrac{1}{2}B_1^2 . \tag{20.121}$$

In combination with Powell's strategy for handling monopole errors, the method replaces the basic conservation law (20.65), now written for $\mathbf{U} = (\rho, \rho\mathbf{v}, \mathcal{H}_1, \mathbf{B}_1)^{\mathrm{T}}$, with

$$\frac{\partial \mathbf{U}}{\partial t} + \nabla \cdot \mathsf{F} + \nabla \cdot \mathsf{G} = \mathbf{S}\,, \tag{20.122}$$

where the fluxes F have *exactly the same form as for the full system under replacing* $\mathcal{H} \leftrightarrow \mathcal{H}_1$ *and* $\mathbf{B} \leftrightarrow \mathbf{B}_1$, but with additional fluxes and source terms given by

$$\mathcal{G} = \begin{pmatrix} 0 \\ \mathbf{B}_1 \cdot \mathbf{B}_0 \mathsf{I} - (\mathbf{B}_0\mathbf{B}_1 + \mathbf{B}_1\mathbf{B}_0) \\ \mathbf{B}_1 \cdot \mathbf{B}_0 \mathbf{v} - \mathbf{B}_1 \cdot \mathbf{v}\mathbf{B}_0 \\ \mathbf{v}\mathbf{B}_0 - \mathbf{B}_0\mathbf{v} \end{pmatrix}, \qquad \mathbf{S} = \begin{pmatrix} 0 \\ -(\mathbf{B}_0 + \mathbf{B}_1)\nabla \cdot \mathbf{B}_1 \\ -\mathbf{v} \cdot \mathbf{B}_1 \nabla \cdot \mathbf{B}_1 \\ -\mathbf{v}\nabla \cdot \mathbf{B}_1 \end{pmatrix}. \tag{20.123}$$

Note that the method involves just a few extra flux and source terms to be implemented. With e.g. a basic TVDLF solver, this is easily incorporated. The CFL limit still needs to account for the full magnetic field influence. In the original paper, the magnetic monopole treatment was actually done by a projection scheme, and the solver was a full characteristic based scheme. The linearized Riemann solver was suitably modified for the changed choice of variables, and it was implemented for unstructured grids. The corresponding set of left and right eigenvectors for the modified system is given in Powell et al. [508]. ◁

20.3.2 Example nonlinear MHD scenarios

The aforementioned shock-capturing algorithms for nonlinear MHD have been successfully applied in many studies of magnetized plasmas. In this section, we briefly discuss two example applications taken from a solar physics context. Nowadays, many mostly freely distributed codes exist, with more advanced ones offering a fair variety of choice to the user with respect to the employed spatial and/or temporal discretizations, as well as the option to select additional physics modules that go beyond ideal MHD. With an early example set by the Versatile Advection Code (VAC) [602, 607, 603, 604, 606, 347, 353, 352], the incomplete and continuously growing code list includes MPI-AMRVAC [351, 506], FCTMHD3D [155], BATS-R-US [245, 508], PLUTO [441], ATHENA [207], FLASH [102], CO5BOLD [641], NIRVANA [668], Clawpack [405], RAMSES [198], ASTRO-BEAR [134], Racoon [165], etc. Many of these codes even allow for automated grid adaptivity, where finer grids are created and eliminated dynamically during the computation, in direct correspondence with the evolving flow features. This adaptive mesh refinement (AMR) approach [53] provides an efficient means to perform very high-resolution simulations at affordable computational cost.

(a) Solar flux rope eruptions An example simulation taken from Mei et al. [436] considers the evolution of a solar magnetic flux rope triggered by the kink instability. Using AMR with up to 6 refinement levels, an effective resolution of 1728^3 is reached which allows us to resolve details of the reconnection occurring during the eruption. The simulation solves the isothermal MHD equations, representing a coronal volume at finite plasma beta (ranging from 0.03 to 0.37, with lower beta conditions internal to the flux rope). The initial condition adopts a widely used analytic equilibrium configuration for a twisted flux rope in magnetized surroundings, introduced by Titov and Démoulin [598], where part of a current-carrying torus above the 'photospheric' plane captures the basic aspects of a line-tied, curved flux rope structure. In [436], the flux rope parameters render it unstable to external kink development, despite its line-tied nature. Figure 20.17 presents several

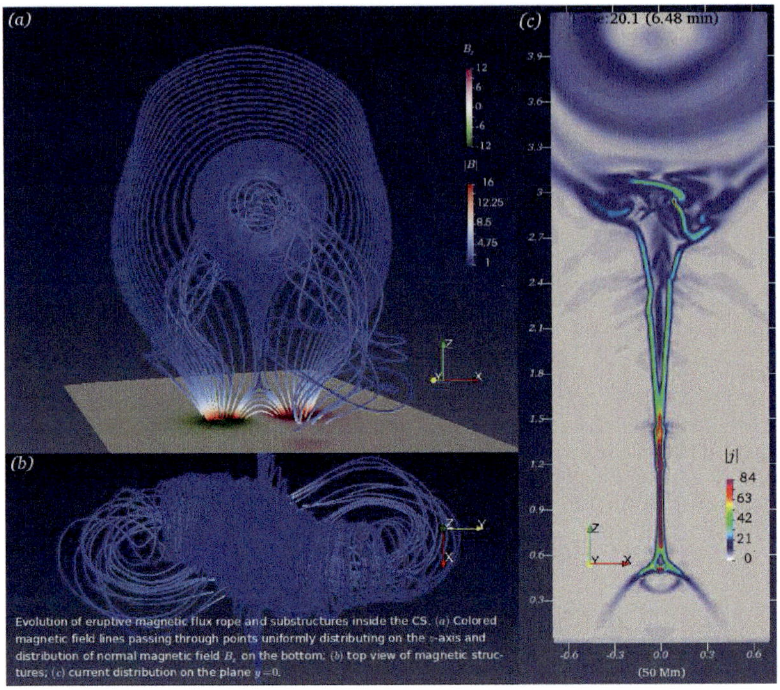

Fig. 20.17 A 3D isothermal MHD simulation of an erupting flux rope, where the fine structure due to the reconnection processes in the current sheet underneath the expanding flux rope is captured. (From Mei et al. [436].)

different views on the eruption that follows, showing representative field lines in a side (panel (a)) and top (panel (b)) view, where it is seen how the overarching magnetic field can not withstand the upward eruptive process resulting from the kink instability. The simulation especially focused on the 3D morphology of the current sheet developing underneath the expanding flux rope. The right panel (c) shows in a zoomed cross-sectional view of the current distribution, that one can detect the typical bifurcating slow shock structures reminiscent of fast Petschek type reconnection (as discussed in Section 14.4.2), with internal island substructure (see also Section 14.4.1) due to tearing. Reconnection jets impact on the tip of the arcade structure that forms at low altitudes, as well as all along the bottom of the erupting flux rope, where a more turbulent region develops. These details are essential ingredients for studies concerned with charged particle acceleration due to solar flares. Similar parameterized flux rope eruption models are exploited in modern space weather simulations, as mentioned in Section 8.3.3.

(b) Prominence formation Our second example considers the intricate physics of solar prominences, which involves the thermal instability mechanism encountered in Section 15.2.4. Xia and Keppens [660] simulated the various stages involved in the life cycle of a quiescent solar prominence, representing a 100-fold cooler and denser plasma structure, magnetically suspended in the solar corona (see Chapter 8 and Fig. 8.5). A first stage involves creating a stable and stratified magnetic flux rope structure, which can still be done in isothermal MHD, when external gravity

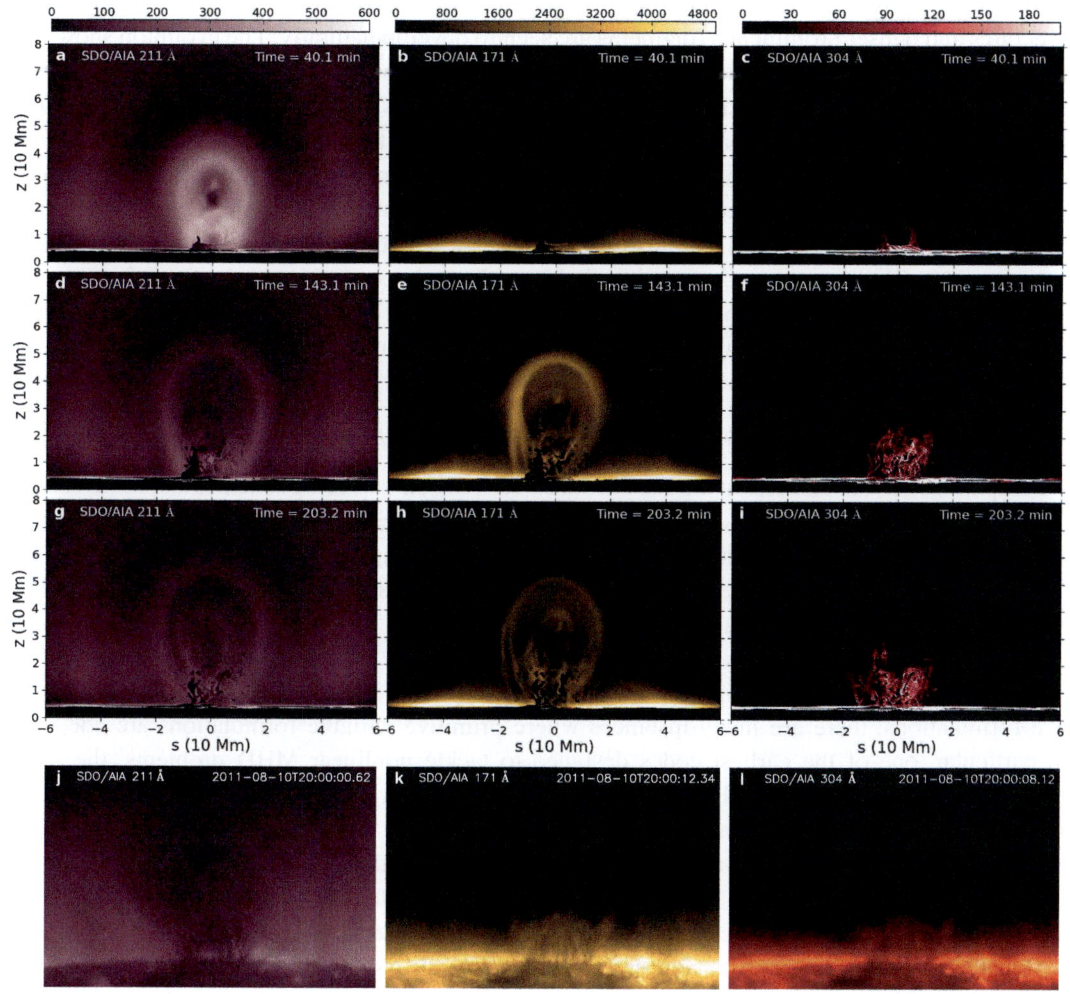

Fig. 20.18 Simulating the formation of a prominence, where the 3D MHD evolution is translated to synthetic views at EUV wavelengths. The bottom panels confront the model with an actual prominence observation. (From Xia and Keppens [660].)

is included. Starting from a bipolar arcade, a helical magnetic flux rope can form due to converging and shearing flow patterns imposed as bottom boundary conditions. A second stage relaxes the flux rope and its surrounding bipolar arcade in a full thermodynamic evolution, where the energy equation is included to deal with anisotropic field-aligned heat conduction, together with optically thin radiative losses and parameterized heating to maintain a hot corona. A third stage then concentrates on the gradual evolution of this realistically stratified and thermodynamically evolving flux rope, when it is subject to a systematic foot point heating. This causes chromospheric matter to slowly evaporate and fill up the coronal flux rope volume, until the local density–temperature conditions allow for thermal condensations to occur. Figure 20.18 presents synthetic views in the extreme ultraviolet wavebands accessible to the Solar Dynamics Observatory, where both heated loop structures and denser prominence fragments are seen to develop naturally. The evolution shows the typical coronal cavity hosting the prominence matter on the upwardly curved, low-lying field lines, and captures well the internal dynamics of observed filaments. The bottom panels of Fig. 20.18 are taken from an actual observation. The simulation followed the evolution for several hours real time, and could demonstrate how the fragmented and dynamic nature of prominences establishes a stable chromosphere–corona mass cycle, where in-situ evaporating and falling blobs get compensated by continuous heating and chromospheric evaporation, in turn driving new thread-like plasma condensations. Since prominences are also vital ingredients in coronal mass ejections, future space weather models will need to combine prominence formation and evolution models, with the ultimate flux rope destabilization and eruption scenarios from the previous paragraph.

20.3.3 Alternative numerical methods

We emphasized in our discussion the use of conservative shock-capturing schemes and finite volume treatments for nonlinear MHD scenarios, but it must be clear that virtually all other (combinations of) discretizations mentioned in Chapter 15 have found widespread use in the astrophysical and plasma physical literature. In the following, we point out various alternative approaches, along with brief summaries of typical example applications.

(a) MHD solvers using primitive variable formulations While we stressed the need for conservative formulations, there are many instances where primitive variable formulations are adequate. In particular, one of the earliest codes designed to tackle nonlinear MHD problems, the ZEUS code [573], is directly solving the MHD equations in a primitive variable formulation. In essence, ZEUS solves the gas dynamic part of the MHD equations using upwinding for the advection terms only (the terms of the form $\nabla \cdot (\rho \mathbf{v})$), and then augments this with finite difference treatments to add Lorentz force and gas pressure 'source' contributions. To handle shocks, artificial viscosity terms are then needed which automatically take effect in compression zones only, both for stability and for physical reasons (restoring the correct shock propagation speeds since the conservative form is not obeyed discretely). These involve usually problem-tailored parameter settings, and it was demonstrated in [172] that erroneous shocks can be easily produced, a statement which holds generally for all non-conservative treatments. In ZEUS, the induction equation is treated separately, in its Stokes' type form, using a constrained transport approach. The precise form of the latter approach pioneered in ZEUS is referred to as MOC-CT (method of characteristics–constrained transport) [573], which specifically takes the Alfvén wave space-time characteristics into account to evaluate the electromotive force (i.e. the cell edge located electric field $-\mathbf{v} \times \mathbf{B}$). This allowed for many nonlinear MHD computations, and one of the first test suites, still used for numerical MHD simulations, was presented in [571].

An early, impressive application of the MOC-CT scheme concentrated on the nonlinear evolution of the magneto-rotational instability (MRI) encountered in Section 13.4. In the radially sheared near Keplerian flows in accretion discs, the MRI provides a linear MHD route to instability, which requires the presence of a small seed magnetic field. Only by numerical simulations could one assess the nonlinear evolution of the MRI, and demonstrate that it would lead to sustained MHD turbulence, accompanied by enhanced outward angular momentum transport. The latter could explain

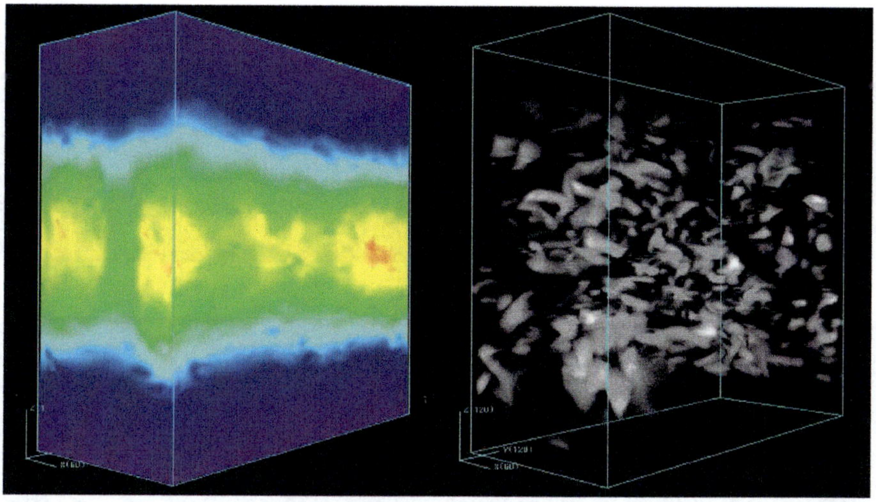

Fig. 20.19 3D rendering of the turbulent local dynamics in a magneto-rotationally unstable gravitationally stratified and magnetized accretion disc. Left panel: density structure, right panel: magnetic pressure iso-surfaces after about 20 full orbital revolutions. (From Stone et al. [572].)

the observationally inferred high accretion rates, which translate into 'anomalously high' effective viscosities needed in simplified Navier–Stokes accretion disc models (the so-called α-disc prescriptions). Realizing that it is practically impossible to perform global simulations while assessing turbulent dynamics, Hawley et al. [295] introduced the local 'shearing box' model in which the 3D dynamics is simulated in a patch located at a global disc radius R_0, and the equation of motion is then formulated in a co-rotating frame where Coriolis, centrifugal and gravitational forces are accounted for. The local nature justifies an expansion of the effective potential due to centrifugal and gravitational effects, and the radial velocity shear becomes a linear profile. The 'shearing box' is thought to be surrounded by similar local slabs which, due to the velocity profile, slide past each other in the 'toroidal' direction in direct relation to the linear velocity variation. This means that a kind of shifted periodic boundary condition can be adopted in the radial direction (artificially connecting front and back planes of the shearing box), while other boundaries can be taken as standard periodic pairs. In [295], the vertical component of gravity was neglected, and it was found that a saturated, turbulent state indeed forms, with increased outward angular momentum. The angular momentum flux in the saturated state was dominated by magnetic stresses. The code used an MOC-CT scheme, and was rather similar to the ZEUS implementation. A follow-up study with ZEUS [572] extended this work to include vertical gravitational stratification. Figure 20.19 shows a snapshot of the nonlinear end state from that study, as seen in density (left) and magnetic pressure iso-surfaces (right). Anisotropic MHD turbulence has developed, as a result of the MRI. The simulation shown in Fig. 20.19 started with a weak vertical magnetic seed field, and shows the local shearing box dynamics after about twenty orbital revolution times.

(b) Spectral and pseudo-spectral MHD simulations Spectral methods are extensively used in MHD turbulence studies, specifically for 3D incompressible simulations. An example of a compressible

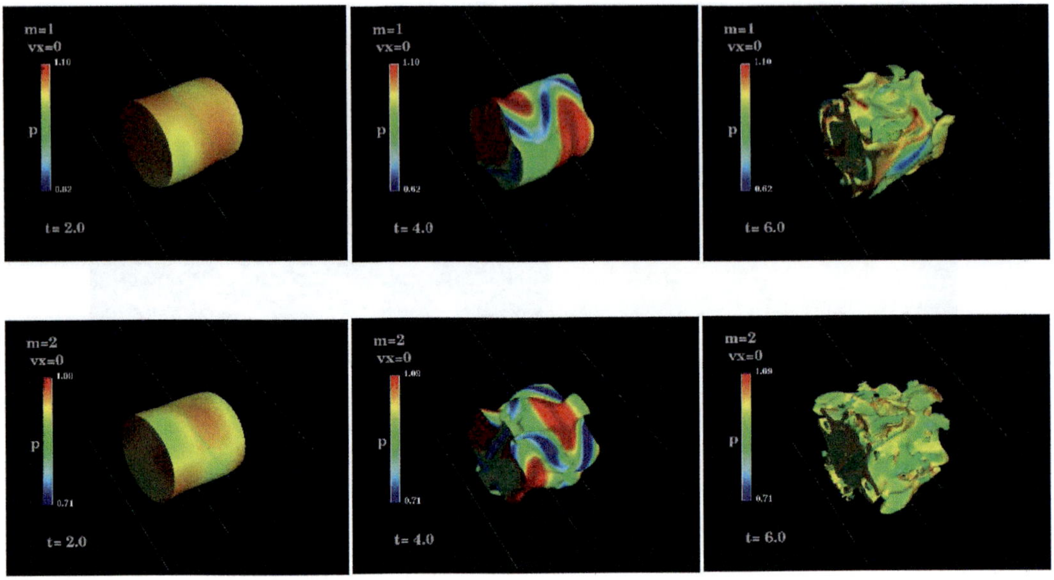

Fig. 20.20 3D jets deform and disrupt due to Kelvin–Helmholtz instabilities at the jet surface. Top: the jet iso-surface is coloured by thermal pressure at consecutive times during the evolution for an initial $m = \pm 1$ perturbation. Bottom: the same for $m = \pm 2$. (From Keppens and Tóth [354].)

MHD code based on Fourier collocation methods is the CRUNCH3D code, which addressed MHD turbulence and magnetic reconnection in Cartesian triple periodic domains (demonstrating the distinct possibility for magnetic flux tube tunnelling, see [136]). In cylindrical flux-tube like configurations, Fourier representations along and about the loop can be used in conjunction with finite differences for the radial direction. This mixed discretization was employed in a parallel *pseudo-spectral* code, which was used in [354] to study Kelvin–Helmholtz instabilities in 3D magnetized jet flows. Figure 20.20 shows a visualization of the disruption of such jets, for different initial perturbations selecting linearly unstable modes of prescribed axial k_z and poloidal m mode numbers. The linear growth rates were found to correspond exactly with linear analysis, while the knowledge on the evolution of all individual Fourier mode pairs (k_z, m) provided by the pseudo-spectral code allowed us to identify and distinguish the most dominant nonlinear interactions.

▷ **Pseudo-spectral method** For numerically solving nonlinear PDEs, the spectral method employing Fourier representations becomes very expensive in terms of CPU time. This is a consequence of the convolutions required to evaluate the nonlinear terms. Suppose, e.g., that two functions $f(x)$ and $g(x)$ are both approximated by means of Fourier harmonics. Their product $f(x)g(x)$ then requires the convolution of the two Fourier sums,

$$p(x) \equiv f(x)g(x) \approx \hat{f}(x)\hat{g}(x) = \sum_{k=-N}^{N} f_k \mathrm{e}^{\mathrm{i}kx} \times \sum_{k'=-N}^{N} g_{k'} \mathrm{e}^{\mathrm{i}k'x} = \sum_{k=-N}^{N} \sum_{k'=-N+k}^{N} f_{k'} g_{k-k'} \mathrm{e}^{\mathrm{i}kx} , \tag{20.124}$$

which involves $\sim N^2$ calculations. Higher order terms become even more CPU time consuming, so that the spectral method is often no longer practical in nonlinear problems.

For nonlinear problems, the pseudo-spectral method often provides an attractive combination of accuracy and computational speed. This method makes handy use of the fact that the Fourier transform of a convolution product is equal to $\sqrt{2\pi}$ times the product of the Fourier transforms. In the pseudo-spectral method, the terms in the equations are computed in real space but the unknowns are updated in Fourier space (Fig. 20.21). This involves FFTs from Fourier space to real space, and back, which is much faster for the determination of the nonlinear terms. For example, for quadratic terms, the CPU time for convolution of the two truncated Fourier series is proportional to N^2 in the spectral method and to $N \log N$ for the pseudo-spectral method. Hence, for large N, the pseudo-spectral method is much more efficient than the spectral method. ◁

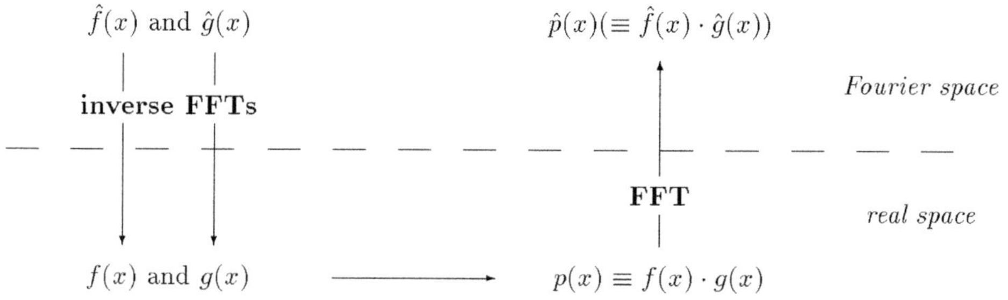

Fig. 20.21 Schematic representation of the determination of a quadratic term in the pseudo-spectral method. All quantities are updated in Fourier space but the products are calculated in real space.

(c) Lagrangian versus Eulerian approaches In essentially all methods mentioned thus far, we implicitly assumed a static grid on which we wish to follow the temporal variations of the basic MHD variables. Hence, an Eulerian viewpoint was adopted all along. In Chapter 4, both Eulerian and Lagrangian formalisms were discussed, and it is also possible to directly solve the Lagrangian equations and thus obtain the evolving MHD dynamics in the co-moving frame. In practice, an initial grid on which one expresses the initial conditions then moves and deforms, and volume changes directly yield density changes from mass conservation. However, numerical errors may then eventually lead to grid entangling in more than 1D. An effective way to avoid this problem, known as arbitrary Lagrangian–Eulerian (or ALE), is to use a hybrid approach where one or more Lagrangian steps are taken, consecutively remapped to a less deformed grid, and then the next cycle of Lagrangian updates is started. An early approach by Brackbill [88] used particle-in-cell type methods for MHD simulations. A modern example of a 3D Lagrangian–Eulerian remap code, Lare3D [22], applies this to nonlinear shock-dominated MHD. In Lare3D, a single Lagrangian step is followed by a remap to a fixed Eulerian grid, and the latter remapping is fully conservative. The Lagrangian update is done at second order accuracy with the use of artificial viscosity (and resistivity) to cope with shocks. Lare3D employs a staggered prescription with magnetic field components on cell faces, scalars in cell volumes and velocities at cell vertices. The $\nabla \cdot \mathbf{B} = 0$ constraint is then maintained using constrained transport, applied to the advection, as well as to the remap stage. Although the implementation does not guarantee perfect conservation of total energy, it was shown by Arber et al. [22] that it can give results as accurate as approximate Riemann solver based methods for shock dominated problems. Another advantage is that it maintains a positive pressure, so that one can handle very low values of beta.

(d) Solar physics applications with higher order centred differences As a final example of an alternative to the Riemann-solver-type shock-capturing schemes discussed in this chapter, we now turn

to a modern MHD computation of solar photospheric magneto-convection taken from Vögler et al. [635]. The employed code MURaM (acronym for Max-Planck-Institute for Aeronomy/University of Chicago Radiation Magnetohydrodynamics) is tailored to solve the non-ideal 3D MHD equations in a Cartesian box, and in essence employs straightforward fourth order centred difference formulas, in combination with an explicit fourth order Runge–Kutta time stepping. These, and similar higher than second order centred difference (CD) approaches, have found widespread use in many solar and astrophysics applications. This is presumably because the current emphasis is more on coping with more realistic (i.e. beyond ideal MHD) physics than on handling very strong shocks accurately. MURaM includes detailed non-local radiative transfer effects, and even accounts fully for partial ionization effects. The latter introduces the need for interpolations through pre-computed tabulated equations of state, to determine pressure and temperature from density and internal energy values, and in MURaM this allows for the eleven most abundant elements in the solar photosphere. This is in sharp contrast to the simple ideal equation of state assumed in most of this book, and it would clearly complicate the use of approximate Riemann solvers. The fourth order CD in MURaM operates on the usual set of conservative variables, and is in fact equivalent to a conservative finite volume scheme. However, the price to pay for the CD approach is the need to incorporate artificial diffusion terms, even for the mass conservation equation. These terms are not totally undesirable, since one needs them to dissipate energy at the unresolved sub-grid scales anyway. These artificial diffusion coefficients then have a shock-resolving contribution that is effective in converging flow regions alone, and a contribution for so-called 'hyper-diffusion', which merely serves to detect and dissipate small-scale fluctuations (in essence depending on numerical approximations of third to first order derivative ratios). For modest Mach number flows, such as typical in the solar photosphere, this approach has proven its worth many times.

With the aim of studying the upper solar convection zone as well as the directly observable photospheric dynamics, present state-of-the-art simulations spend most computational effort on handling the radiative transfer as accurately as possible. As an example simulation result from [635], Figure 20.22 shows a 3D view of a spontaneously formed thin magnetic sheet of kilo Gauss field strength. This represents a face-on view (in the direction as indicated on the right panel of field intensity at the indicated photospheric 'plane') of the sub- to upper photospheric field configuration. As a general tendency, magnetic field gets expelled from up-welling adiabatically expanding convection flows and concentrates along the narrow downdraft lanes bordering the cellular convection patterns. These high resolution radiative MHD simulations (employing $100 \times 288 \times 288$ resolution) extend from 800 km down into optical depth unity (i.e. the photosphere) to 600 km above, and cover an area of 6000^2 km. The simulations actually followed an initially uniform 200 Gauss magnetic field, and gave unprecedented insights into evolving solar plage dynamics over many hours. The snapshot shown in Figure 20.22 is representative for the quasi-two-dimensional flux sheets occasionally found in the simulated time interval, and shows that they are in fact rather shallow phenomena, as the magnetic configuration is clearly disrupted in the deeper convective layers. This is fully consistent with the vertical variation of the plasma beta parameter in the strong field regions, found to vary from values above 100 at the bottom, to about 0.1 near the top boundary. Besides the sheet-like structures accumulating at the convection cell edges, the simulations also witnessed the formation of a more coherent 'micropore' of kilo Gauss field strength at the vertices where several down flow lanes merge. This micropore appeared dark in intensity since temporarily the convective transport got almost completely suppressed. This structure only survived for about two convective turnover

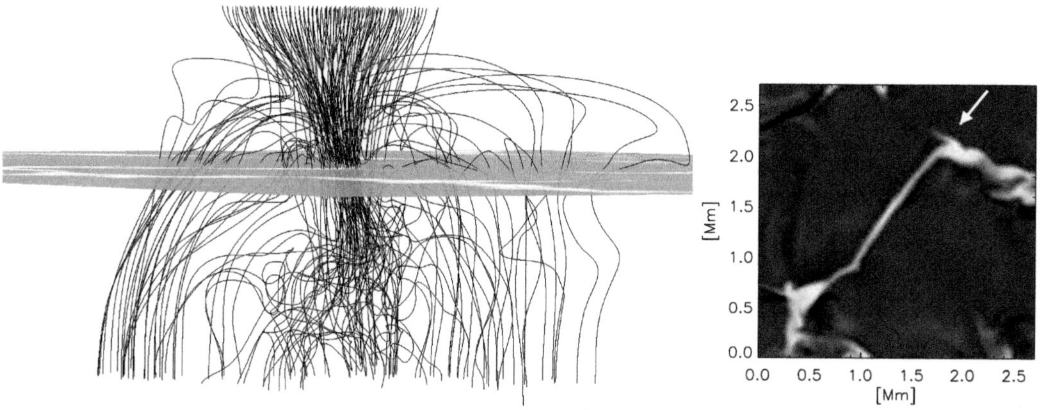

Fig. 20.22 3D view of the magnetic field in a kG magnetic sheet, in a radiative MHD simulation spanning a vertical scale of 800 km down into optical depth unity (grey iso-surface) up to 600 km above the solar photosphere. The view angle is along the arrow as indicated in the photospheric field intensity plot at right. (From Vögler et al. [635].)

times, and showed remarkable internal structure on shorter time scales. This represents a current frontier of MHD modelling in which predictions for future generation solar telescopes can be made, as it gives insights into details below current spatial and temporal resolution limits. The MURaM radiative MHD code has subsequently been used to e.g. simulate full sunspots [516], flux emergence, small-scale dynamo action in magneto-convection [515](a), etc. usually spanning sub- photosphere to low chromosphere/coronal regions where the treatment of the detailed (de)coupling between radiation and matter fields is key (as the photosphere represents the layer where the so-called optical depth is unity). The code has more recently been extended [515](b) to handle solar coronal physics conditions as well, where optically thin radiative losses, anisotropic thermal conduction and a more sophisticated equation of state is employed. This then paves the way for fully self-consistent numerical studies of flux emergence, through the convection zone up into the solar corona, to e.g. assess the physics of solar flares. This type of realistic solar modelling is also undertaken with the Bifrost code [270], which similarly combines radiation approximations going from optically thin to more complete radiative transfer treatments, as well as non-local thermal equilibrium aspects in their hydrogen ionization module.

20.4 Implicit approaches for extended MHD simulations

All examples and methods discussed thus far in this chapter use explicit time integration strategies, generalizing the basic explicit schemes mentioned in Section 15.3.1. In Section 15.3.2, we pointed out that explicit schemes may face insurmountable difficulties, due to the disparate time scales inherent in the MHD wave dynamics, especially when it is necessary to study resistive phenomena occurring on the longest diffusion time scale. To that end, the use of implicit time integration methods was discussed in Section 15.3.3, and linear MHD computations employing implicit time integration strategies were covered in Section 15.3.4. Also, for the full nonlinear MHD system, the use of implicit (or semi-implicit) time integration methods is needed to alleviate the stringent CFL limit on the time step. Examples of the typical use of implicit methods in MHD context for both steady and unsteady problems can be found in [606], and an extended description of the numerical algorithms is given in [355]. A key observation in the use of fully implicit schemes is that (1) the nonlinearity of the system will call for some suitable (Newton) iteration strategy translating the problem into repeated solves of linear systems which involve the flux Jacobian; and (2) the solution of such huge algebraic linear systems. In 1D problems, the latter is feasible with the use of direct methods since we then deal with typically block-tridiagonal systems. However, in multi-dimensional problems, we face the need to solve many block penta-diagonal (in 2D) or block hepta-diagonal (3D) systems, which necessitate the use of iterative linear solvers and adequate preconditioners to accelerate their convergence. The computation of the flux Jacobian can be avoided altogether in an approach where the Newton iteration is performed in combination with iterative Krylov sub-space methods using a matrix-free approach. (In the iterative scheme, only matrix-vector products are required.) This is an area of active research, with a modern example found in Tóth et al. [605]. In what follows, we discuss the so-called semi-implicit methods, introduced in MHD by Harned and Kerner [288], as employed in state-of-the-art simulations of laboratory tokamak plasmas.

20.4.1 Semi-implicit methods

As already mentioned in Section 15.3.3, the term 'semi-implicit' is used for several alternative schemes which are neither explicit nor fully implicit. Different options can be, e.g., to treat some dependent variables explicitly and other ones implicitly, or to update only specific terms in the equations implicitly. In the following discussion, semi-implicit refers to a means to relax the CFL condition somewhat by treating only the fastest sub-spectrum of waves implicitly. This has been applied in nonlinear MHD with success. In hot elongated plasmas, the fast magneto-sonic waves have extremely short time scales and, hence, cause a severe limit on the time step for explicit schemes due to the CFL condition for numerical stability. When these fast magneto-sonic waves are handled in an implicit way, the time step is limited by the CFL condition on the Alfvén waves, which is much less restrictive because of the time scale discrepancy between fast and Alfvén waves. The implicit treatment of only the sub-spectrum of fast magneto-sonic waves can be achieved in a simple and elegant way. The procedure involves three steps: (1) identification of the mode that produces numerical instability when the time step is raised; (2) subtraction of a simple approximation of the term producing this mode from both sides of the equation; (3) evaluation of the new approximate term on the RHS explicitly and on the LHS implicitly. We will demonstrate this procedure on the simple linear advection model problem considered in Chapter 15 and then apply it to MHD.

(a) Demonstration of the procedure and physical interpretation Consider again the model problem

$$\frac{\partial u}{\partial t} = -v\,\frac{\partial u}{\partial x}\,. \tag{20.125}$$

This simple linear equation describes only one type of mode, but nevertheless we can demonstrate the idea of the semi-implicit method on it. The von Neuman stability analysis of the explicit Lax–Friedrichs scheme from Eq. (15.86), namely

$$u_i^{n+1} = \tfrac{1}{2}(u_{i+1}^n + u_{i-1}^n) - \tfrac{1}{2}v\,\frac{\Delta t}{\Delta x}\,(u_{i+1}^n - u_{i-1}^n)\,, \tag{20.126}$$

yielded the CFL stability condition. This gave a restriction of the time step for a given spatial discretization, viz. $|v\Delta t/\Delta x| \leq 1$. In an implicit scheme, the spatial derivative is evaluated at the next time level:

$$u_i^{n+1} = u_i^n - v\Delta t \, \frac{\partial u}{\partial x}_i^{n+1} , \tag{20.127}$$

which yields an unconditionally stable numerical scheme. For this model problem (which easily generalizes to nonlinear cases, with v a function of u or of (x, y, z), or both), the three steps of the semi-implicit procedure become very simple. The first step can be skipped since there is only one wave mode. The term that generates this mode is also obvious, since there is only one RHS term in Eq. (20.125). A semi-implicit method for this model problem is

$$\begin{aligned} u_i^{n+1} &+ \tfrac{1}{2}v_0 \, \frac{\Delta t}{\Delta x} \, (u_{i+1}^{n+1} - u_{i-1}^{n+1}) \\ &= \tfrac{1}{2}(u_{i+1}^n + u_{i-1}^n) - \tfrac{1}{2}v \, \frac{\Delta t}{\Delta x} \, (u_{i+1}^n - u_{i-1}^n) + \tfrac{1}{2}v_0 \, \frac{\Delta t}{\Delta x} \, (u_{i+1}^n - u_{i-1}^n) , \end{aligned} \tag{20.128}$$

where the last term on the LHS and RHS are additional to the Lax–Friedrichs scheme (20.126). The coefficient v_0 is a constant which is to be chosen from numerical stability considerations. By consequence, the LHS operator to be inverted is linear and the obtained tridiagonal system is easy to solve.

▷ **Exercise** Show that the semi-implicit scheme (20.128) is unconditionally stable for $v_0 = \frac{1}{2}v$ when v is constant. Apply the von Neumann stability analysis of Chapter 15. ◁

Notice that the terms added to the LHS and to the RHS in Eq. (20.128) are identical, except for the fact that the LHS term is evaluated at time level $n + 1$ while the RHS term is evaluated at time level n. This comes down to adding some numerical dissipation which slows down the high-k (short wavelength) modes. As a matter of fact, the dispersion relation of the original model equation is $\omega/k = v$, so that there is no dispersion for a constant v. The semi-implicit equation (20.128), however, is a discretization of

$$\frac{\partial u}{\partial t} = -v \, \frac{\partial u}{\partial x} - v_0 \Delta t \, \frac{\partial}{\partial x} \frac{\partial u}{\partial t} , \tag{20.129}$$

with the dispersion relation $\omega = kv - \mathrm{i}v_0\omega k\Delta t$, or $\omega/k = v/(1 + \mathrm{i}v_0 k\Delta t)$. Hence,

$$\left|\frac{\omega}{k}\right| = \frac{v}{\sqrt{1 + (kv_0\Delta t)^2}} , \tag{20.130}$$

so that the high-k modes are indeed slowed down. In the MHD application discussed below, the semi-implicit term is chosen such that compressible modes propagating perpendicular to the magnetic field are slowed down in a similar way.

(b) Application to nonlinear MHD The semi-implicit method is most powerful when used in conjunction with a spectral discretization, although it can also be used in finite difference schemes. In MHD applications for tokamak and solar coronal loop simulations, it has been successfully used in 3D, often applying a finite difference method in the direction normal to the initial magnetic flux surfaces and a (pseudo-)spectral method in the two spatial directions spanning these surfaces. Below, we describe the three-step procedure mentioned, in the way it was originally introduced to MHD by Harned and Kerner [288].

The first step, identification of the trouble mode, is obvious: the fast magneto-sonic mode is the one that causes the severe time step limitation in explicit schemes. This wave mode propagates in the direction normal to the flux surfaces and this is precisely the direction in which the mesh size has to be very small in order to resolve the singular layers that develop in e.g. kink instabilities or resonantly excited continuum modes. The simple approximation of the term that generates these modes follows from Eq. (5.52) of Section 5.2.3, where we do not assume normal modes, i.e. the linearized ideal MHD equations in terms of the amplitude $\hat{\mathbf{v}}$ for plane waves,

$$\Big\{\Big[-\frac{\partial^2}{\partial t^2}-(\mathbf{k}\cdot\mathbf{b})^2\Big]\mathsf{I}-(b^2+c^2)\mathbf{k}\mathbf{k}+\mathbf{k}\cdot\mathbf{b}(\mathbf{k}\mathbf{b}+\mathbf{b}\mathbf{k})\Big\}\cdot\hat{\mathbf{v}}=0\,. \qquad (20.131)$$

Retaining only the 'fast compressional modes' ($\mathbf{k}\perp\mathbf{b}$), one obtains a representative equation for the fast evolution:

$$\frac{\partial^2\hat{\mathbf{v}}}{\partial t^2}=-(b^2+c^2)\mathbf{k}\mathbf{k}\cdot\hat{\mathbf{v}}\,. \qquad (20.132)$$

For a general (inhomogeneous) equilibrium, remembering that the ∇ operator was replaced by $i\mathbf{k}$, this equation becomes

$$\frac{\partial^2\hat{\mathbf{v}}}{\partial t^2}=\Big(\frac{\gamma p_0+B_0^2}{\rho_0}\Big)\nabla\nabla\cdot\hat{\mathbf{v}}\,. \qquad (20.133)$$

The semi-implicit method then approximates this simply by replacing the coefficient by a constant factor A_0^2, introducing the linear operator

$$S(\hat{\mathbf{v}}_\perp)\equiv A_0^2\,\nabla\nabla\cdot\hat{\mathbf{v}}_\perp\,, \qquad (20.134)$$

where only the perpendicular component of $\hat{\mathbf{v}}$ is included since the severest time step limitations occur when $k_\parallel \ll k_\perp$. In that case, the parallel component of the fast magneto-sonic wave is negligible. The constant A_0^2 is to be determined from stability considerations.

▷ **Predictor-corrector scheme for MHD** In this scheme used to evolve the MHD equations, the semi-implicit term is applied in the corrector step only, as follows.

– *predictor step:*

$$\begin{aligned}
\mathbf{v}^* &= \mathbf{v}^n+(\alpha\Delta t/\rho^n)\,\mathbf{F}(\mathbf{v}^n,\mathbf{B}^n,p^n)\,,\\
\mathbf{B}^* &= \mathbf{B}^n+\alpha\Delta t\,\nabla\times(\mathbf{v}^n\times\mathbf{B}^n)\,,\\
\rho^* &= \rho^n-\alpha\Delta t\,\nabla\cdot(\rho^n\mathbf{v}^n)\,,\\
p^* &= p^n-\alpha\Delta t\,(\mathbf{v}^n\cdot\nabla p^n+\gamma p^n\nabla\cdot\mathbf{v}^n)\,;
\end{aligned}$$

– *corrector step:*

$$\begin{aligned}
v_z^{n+1} &= v_z^n+(\Delta t/\rho^*)\,F_z(\mathbf{v}^*,\mathbf{B}^*,p^*)\,,\\
(1-S)\,\mathbf{v}_\perp^{n+1} &= \mathbf{v}_\perp^n+(\Delta t/\rho^*)\,\mathbf{F}_\perp(\mathbf{v}^*,\mathbf{B}^*,p^*)-S(\mathbf{v}_\perp^n)\,,\\
\mathbf{B}^{**} &= \mathbf{B}^n+\tfrac12\Delta t\,\nabla\times\big[(\mathbf{v}^{n+1}+\mathbf{v}^n)\times\mathbf{B}^*\big]\,,\\
\rho^{n+1} &= \rho^n-\tfrac12\Delta t\,\nabla\cdot\big[\rho^*(\mathbf{v}^{n+1}+\mathbf{v}^n)\big]\,,\\
p^{n+1} &= p^n-\tfrac12\Delta t\,\big[\big(\mathbf{v}^{n+1}+\mathbf{v}^n\big)\cdot\nabla p^*+\gamma p^*\nabla\cdot\big(\mathbf{v}^{n+1}+\mathbf{v}^n\big)\big]\,;
\end{aligned}$$

– *diffusion step:*

$$(1-\eta_0\Delta t\nabla^2)\,\mathbf{B}^{n+1} = \mathbf{B}^{**}+\Delta t\,\nabla\times(\eta\nabla\times\mathbf{B}^{**})-\eta_0\Delta t\,\nabla^2\mathbf{B}^{**}\,.$$

In the above scheme, $\mathbf{F}$ denotes the force operator in the momentum equation. Notice that the semi-implicit term has been added to the corrector step in the momentum equation, *implicitly* in the LHS and *explicitly* in the RHS. The z-component of the momentum equation is treated separately because the semi-implicit term only involves the r- and θ-component of the velocity. The magnetic field is always advanced after the velocity field and requires no special treatment. The parameter α determines the time step size. For $\alpha = 0.5$ the ideal MHD part of the scheme (apart from the fast magneto-sonic waves) is second order accurate in time. In practice, one chooses α equal to 0.51 or 0.52 in order to add a small amount of numerical damping to make the scheme more stable. Also notice the Crank–Nicolson type treatment (see Section 15.3.3) of the velocity in the corrector step: all information on $\mathbf{v}$ is used as soon as it becomes available. ◁

The stability of the semi-implicit algorithm was analysed in [288], where it was found that the following constraints must be satisfied throughout the plasma for linear stability:

$$A_0^2 > \frac{1}{16\rho}(B_z^2 + B_\theta^2 + \gamma p)(1 + 2\alpha)^2 \,, \tag{20.135}$$

$$\frac{1}{\rho}\Big(\frac{m}{r}B_\theta + \frac{2\pi n}{L}B_z\Big)^2 (\Delta t)^2 < \frac{16}{(1 + 2\alpha)^2} \,. \tag{20.136}$$

The constraint (20.135) determines the constant A_0 and follows from the requirement of unconditional stability with respect to the fast magneto-sonic modes. The second constraint is a CFL-like condition on the time step following from the explicit treatment of the shear Alfvén term $\omega_{\rm A}^2 = (\mathbf{k} \cdot \mathbf{v}_{\rm A})^2$ on the LHS.

The semi-implicit scheme was applied [288] in combination with a finite difference discretization for the direction normal to the flux surfaces and a spectral discretization in the other two spatial directions. The constant A_0 in the semi-implicit term then avoids the CPU time consuming convolutions that would complicate a fully implicit time advance. In fact, the only 'overhead' the semi-implicit scheme produces involves the solution of $N_{\rm F}$ tri-diagonal systems in every time step, with $N_{\rm F}$ the number of Fourier modes. However, there are very efficient solvers for such simple systems. Kerner [357](b) derived the following expression for the maximal time step of the semi-implicit scheme, $\Delta t_{\rm SI}$, compared to the CFL limited time step of an explicit scheme, $\Delta t_{\rm CFL}$. From numerical simulations in a cylindrical plasma of finite length $2\pi R_0$ and radius a (i.e. a 'straight tokamak'),

$$\frac{\Delta t_{\rm SI}}{\Delta t_{\rm CFL}} = 15\,\frac{R_0}{a}\,\frac{N}{100} \,. \tag{20.137}$$

This means that the ratio is proportional to the aspect ratio of the cylinder and to the number of radial grid points used in the discretization. Indeed, the explicit time step size depends on this number because the fast magneto-sonic wave propagates in the radial direction, but the semi-implicit scheme treats these waves implicitly so that the gain is larger for higher radial resolutions. A high resolution run with $N = 1000$ yields a 1500 times larger time step for the semi-implicit scheme for an aspect ratio of 10.

The semi-implicit method is widely used in MHD and has been generalized in various ways. Harned and Schnack [289] adjusted the semi-implicit term in order to treat both the fast magneto-sonic waves and the Alfvén waves implicitly. Their scheme is more complicated but allows even larger time steps as the shear Alfvén time step constraint is eliminated as well. Lerbinger and Luciani [403] went even further and used the full linearized MHD force operator in the semi-implicit term. As a result, the time step in their scheme is limited only by nonlinear physical phenomena and they show results on nonlinear tearing modes obtained with time steps as large as 10 Alfvén crossing times in the linear phase and 2 Alfvén transit times in the nonlinear saturated phase. Poedts and Goedbloed [494] applied the method to study driven coronal loop systems, where resonant layers form as discussed in Chapter 11, to study nonlinear effects of the shear flows that develop near the dissipative layer.

20.4.2 Simulating ideal and resistive instabilities

The semi-implicit method is often used in MHD to determine the linear and nonlinear growth rates and to study the nonlinear behaviour and saturation of ideal and resistive instabilities. The method has been applied to the study of so-called 'relaxation' instabilities, which occur periodically and each time transfer energy and plasma particles from one part of the plasma to another or even out of the plasma. There are two known relaxation instabilities: 'sawtooth' instabilities, which affect only the plasma centre and 'edge localized modes' (ELMs) which only occur at the plasma surface. The latter have been discussed in Chapter 15 in the context of linear MHD. Sawtooth instabilities have been simulated [69, 100] using a semi-implicit scheme very similar to the original one of Harned and Kerner [288].

An exemplary challenging calculation involves the nonlinear saturation of the ideal internal kink mode, predicted analytically by Rosenbluth et al. [529]. The $m = 1$ internal kink mode develops nonlinearly into a helical equilibrium state with a singular current sheet on the $q = 1$ surface. This saturated nonlinear ideal equilibrium state turns out to be unstable when finite resistivity is taken into account. The singular current sheet, in which the magnetic field changes sign, is unstable in a resistive plasma and the magnetic field lines reconnect in this layer. Kadomtsev derived the time scale for this reconnection process and showed that it is proportional to $\eta^{-1/2}$. Numerical simulations [357](b) confirmed this property of Kadomtsev's model for sawtooth instabilities by semi-implicitly evolving the nonlinear ideal equilibrium in resistive MHD (Fig. 20.23).

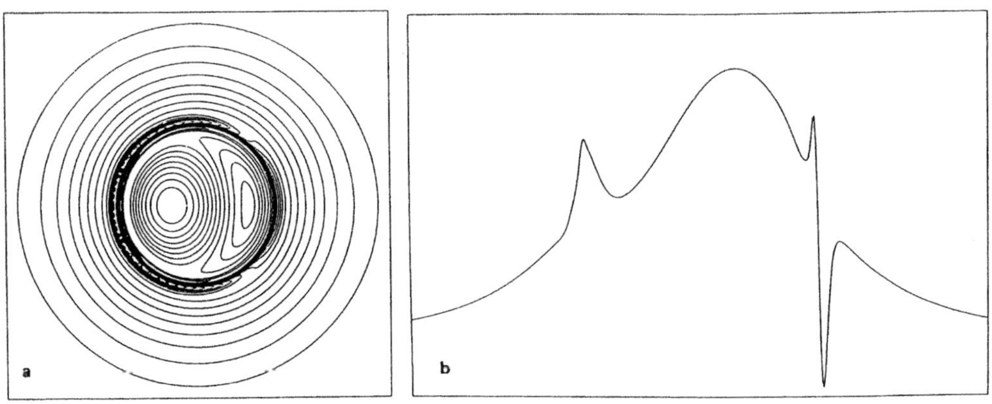

Fig. 20.23 (a) Contour plot of the magnetic flux and (b) current density as a function of radius through the symmetry plane for a nonlinearly saturated state of the $m = 1$ internal kink mode in a 'straight tokamak'. There is a current sheet at the rational surface $q = 1$. (From Kerner [357](b).)

20.4.3 Global simulations for tokamak plasmas

As a final example where nonlinear MHD simulations, using semi-implicit algorithms, are extensively used, we discuss a state-of-the-art study from laboratory fusion context. A US-based consortium, the NIMROD team (non-ideal MHD with rotation, open discussion), has progressed to simulating long-term plasma dynamics in global tokamak geometry. In NIMROD [557], the 3D toroidal geometry is discretized using a 2D FEM representation in the poloidal cross-section (as was found optimal for handling also the equilibrium problem for axi-symmetric tokamak plasmas, see Chapter 16), with a Fourier handling of the toroidal direction. The Galerkin method is used, whereby nonlinear terms which depend on φ are handled pseudo-spectrally in the periodic toroidal direction, using FFTs. The major challenge is to account for the wide range in time scales of interest (as already discussed earlier in this chapter), as well as for the extreme and anisotropic properties of high temperature, magnetized plasmas. The latter implies that one wishes to compute plasma evolution over resistive time scales while facing magnetic Reynolds numbers of order 10^9 or higher, in conjunction with thermal conduction processes qualified by ratios $\kappa_\parallel/\kappa_\perp \approx 10^9$. In order to handle these or more reasonable numerical values, the temporal discretization must use implicit strategies. A modification akin to the semi-implicit strategies described earlier is made to the momentum equation, involving the full self-adjoint linear ideal MHD force operator about a steady state. In fact, the nonlinear equations are also written in a manner which accounts for this strong background equilibrium field, reminiscent of the $\mathbf{B}_0 + \mathbf{B}_1$ strategy introduced by Tanaka [586] for shock-capturing methods. The $\nabla \cdot \mathbf{B} = 0$ constraint is handled by introducing a diffusive source term to the induction equation, recognized as the parabolic cleaning strategy mentioned in Section 20.3.1. The semi-implicit operator is modified in a manner which nearly guarantees no numerical dissipation for linear stable modes, while unstable modes will grow at their physically correct growth rates, if the employed time step is suitably restricted. In addition to this, physical dissipation terms (resistivity, anisotropic thermal conduction, viscosity) are handled fully implicitly.

The main strength of the NIMROD strategy is its demonstrated ability to solve for long-term non-ideal evolution in global tokamak plasmas. The order of the FEM used in the poloidal discretization must be higher than 2, but provided such second or higher order elements are used, several stringent tests for tearing modes, and extreme anisotropic heat conduction conditions were demonstrated successfully in [557]. Figure 20.24 shows a snapshot of a simulation where a disruption (shot # 87009) in the DIII-D tokamak was simulated from beginning to end. The initial MHD equilibrium was found unstable to a helical ($2/1$ in poloidal/toroidal mode numbers) deformation with a prominent magnetic island structure. The instability grows and leads to a global disruption of the tokamak plasma. Figure 20.24 shows an impression of the plasma at the time of maximal heat load to the tokamak walls. The field lines are coloured by local temperature and iso-surfaces of temperature show significant deformation from their initial nested tori configuration. The heat flux on the wall is also indicated: the reddish features at top and bottom indicate where most heat is deposited at this instant. In an animation of the disruption, one can follow the sudden loss of equilibrium, and see how eventually a single field line connects top and bottom areas of maximal heat load on the tokamak walls. The plasma as a whole cools rapidly: one obviously wants to avoid this kind of global disruption in actual fusion plasmas. Global tokamak modelling will thereby serve a crucial role: it offers the only way to quantify macroscopic plasma properties throughout nonlinear evolutions in the entire tokamak geometry. Combined with detailed knowledge and feedback from MHD

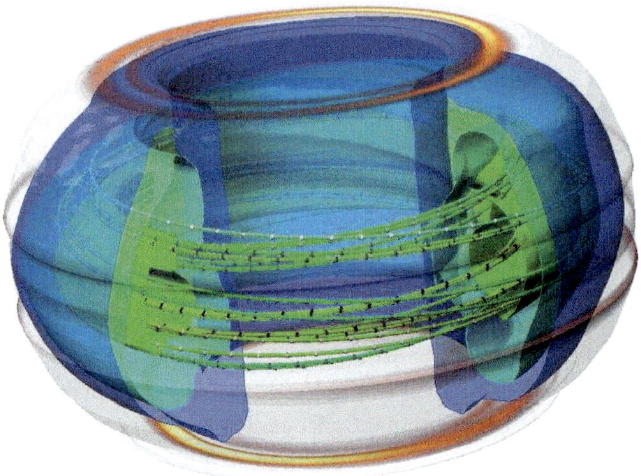

Fig. 20.24 Visualizing temperature iso-surfaces, various magnetic field lines coloured by the local temperature, and the heat load on the wall (orange). This 3D MHD simulation models a disruption in shot #87009 of the DIII-D tokamak. (An animation is available at `nimrodteam.org`.)

equilibrium reconstructions and linear MHD spectral codes, one may hope to devise strategies for controlling self-burning fusion plasmas. Eventually, fluid models of tokamak plasmas can be expected to be replaced by more elaborate fluid-like treatments which gradually incorporate more and more kinetic effects. This requires detailed knowledge of the temporal and spatial scales inherent in the more elaborate closures, such that suitable numerical strategies can be tailored to the challenge. An up-to-date review of these issues for a hierarchy of ever more sophisticated fluid models can be found in [543], where various 'extended MHD' models are discussed.

20.5 Literature and exercises

Notes on literature

Numerical methods

- Many excellent books cover computational hydrodynamics, e.g. Wesseling, *Principles of Computational Fluid Dynamics* [645]. Riemann solver based methods for gas dynamics are thoroughly discussed in Toro, *Riemann Solvers and Numerical Methods for Fluid Dynamics* [600]. Their application to MHD was pioneered in the paper by Brio and Wu, 'An upwind differencing scheme for the equations of ideal magnetohydrodynamics' [96].
- The lecture notes for the 1997 Saas Fee Advanced Course on *Computational Methods for Astrophysical Fluid Flow* [406] include material on finite volume methods for conservation laws, with a section on MHD, by LeVeque. The monograph by Murawski, *Analytical and Numerical Methods for Wave Propagation in Fluid Media* [455], discusses aspects of modern shock-capturing schemes for the MHD system.
- Introductions to the broad topics of computational plasma physics and computational astrophysics, with sections on numerical MHD, are the books by Tajima, *Computational Plasma Physics* [585] and Bodenheimer et al., *Numerical Methods in Astrophysics* [77]. A contemporary monograph on techniques to simulate magnetically confined plasmas is *Computational Methods in Plasma Physics* [333] by Jardin.

Exercises

[20.1] *Conservative versus primitive variable formulations for iso-thermal hydrodynamiecs*

Consider the 1D iso-thermal hydrodynamic equations given by

$$\frac{\partial \rho}{\partial t} + \frac{\partial (\rho v)}{\partial x} = 0 \,, \qquad \frac{\partial m}{\partial t} + \frac{\partial (mv + p)}{\partial x} = 0 \,.$$

These express conservation of mass in terms of the density ρ, and momentum $m = \rho v$ conservation with velocity v. For an iso-thermal gas of uniform temperature, the gas pressure is directly proportional to the density, i.e. $p = c_{\rm i}^2 \rho$, where $c_{\rm i}$ denotes the constant iso-thermal sound speed. We can alternatively write the governing equations in the form (20.3) in terms of primitive variables $\mathbf{V} = (\rho, v)^{\rm T}$.

– Compute the flux Jacobian $\mathsf{F}_{\mathbf{U}}$ and verify that

$$\mathsf{F}_{\mathbf{U}} = \begin{pmatrix} 0 & 1 \\ c_{\rm i}^2 - m^2/\rho^2 & 2m/\rho \end{pmatrix} \,, \quad \mathsf{W}(\mathbf{V}) = \begin{pmatrix} v & \rho \\ c_{\rm i}^2/\rho & v \end{pmatrix} \quad \begin{matrix} \lambda_1 = v - c_{\rm i} \\ \lambda_2 = v + c_{\rm i} \end{matrix} \,,$$

$$\mathsf{R} = \begin{pmatrix} 1 & 1 \\ v - c_{\rm i} & v + c_{\rm i} \end{pmatrix} \,, \quad \mathsf{R}^{-1} = \begin{pmatrix} (v + c_{\rm i})/(2c_{\rm i}) & -1/(2c_{\rm i}) \\ -(v - c_{\rm i})/(2c_{\rm i}) & 1/(2c_{\rm i}) \end{pmatrix} \,, \quad \widehat{\mathsf{R}} = \begin{pmatrix} 1 & 1 \\ -c_{\rm i}/\rho & c_{\rm i}/\rho \end{pmatrix} \,,$$

$$\widehat{\mathsf{R}}^{-1} = \begin{pmatrix} 1/2 & -\rho/(2c_{\rm i}) \\ 1/2 & \rho/(2c_{\rm i}) \end{pmatrix} \,, \quad \widetilde{\mathsf{R}} = \begin{pmatrix} v - c_{\rm i} \ln \rho \\ v + c_{\rm i} \ln \rho \end{pmatrix} \,.$$

– Note that the equations for the Riemann invariants (generally found from Eq.(20.8)) can be manipulated to the following characteristic equations, as also found from the generalized Riemann invariants construction (20.10),

$$dv \pm (c_{\rm i}/\rho) d\rho = 0 \,, \ \text{along curves} \ dx/dt = v \mp c_{\rm i} \,.$$

In these equations, $dv = (\partial v/\partial t)dt + (\partial v/\partial x)dx$ denotes the total velocity change.

[20.2] *Linear hyperbolic systems*

Consider the 2×2 linear system

$$\begin{cases} u_t + 0.5u_x + w_x &= 0 \,, \\ w_t - 1.25u_x + 3.5w_x &= 0 \,. \end{cases}$$

Show that this system is strictly hyperbolic. Compute its eigenvalues and the left and right eigenvectors. Calculate the characteristic variables and verify that their evolution is governed by constant coefficient linear advection equations. Compute the analytic solution for the particular initial data $(u, w) = (4, 1)$ for $x < 0$ and $(u, w) = (2, 100)$ for $x > 0$.

[20.3] *Shallow water magnetohydrodynamics*

An interesting set of nonlinear conservation laws, known as the 'shallow water' magnetohydrodynamic equations, is given by

$$\frac{\partial}{\partial t} \begin{pmatrix} h \\ h\mathbf{v} \\ h\mathbf{B} \end{pmatrix} + \nabla \cdot \begin{pmatrix} h\mathbf{v} \\ h\mathbf{v}\mathbf{v} - h\mathbf{B}\mathbf{B} + \frac{1}{2} g h^2 \mathbf{I} \\ h\mathbf{v}\mathbf{B} - h\mathbf{B}\mathbf{v} \end{pmatrix} = 0 \,.$$

These equations govern the two-dimensional motion of a narrow layer of local height $h(x, y)$ of conducting fluid, with purely planar $\mathbf{v} = (v_x, v_y)$ and $\mathbf{B} = (B_x, B_y)$. The magnetic field divergence constraint complements this with $\nabla \cdot (h\mathbf{B}) = 0$. The parameter g represents a constant gravitational acceleration, perpendicular to the (x, y) plane.

– Compute the flux Jacobian of this five component system. Derive an equivalent quasi-linear form in terms of primitive variables $\mathbf{V} = (h, v_x, v_y, B_x, B_y)^{\rm T}$.

- Compute all the characteristic speeds, and the left and right eigenvectors for both primitive and conservative variable formulations.
- Try to derive generalized Riemann invariants, and compute the structure coefficients for the five wave modes. Discuss the Rankine–Hugoniot relations and the equations governing (centred) simple waves.

You can verify your results and find physical explanations for the occurring wave modes in this system in [152].

[20.4] *A Roe scheme for iso-thermal MHD*

Construct a Roe-type Riemann solver based scheme for the 1.5D iso-thermal MHD system. This can then be compared to the algorithm presented in [364], where a TVD scheme of Roe-type was exploited for multi-dimensional iso-thermal MHD. Expressions for suitably scaled eigenvectors are given there as well, but note the difference between our coefficients α_f^2, α_s^2 and theirs.

21
Transonic MHD flows and shocks

21.1 Transonic flows

We studied the effects of background flow on the waves and instabilities of plasmas in Chapter 12. We also considered the modifications of the equilibrium caused by the flow. These are rather trivial for plane shear flows, but considerable for rotating plasmas due to the centrifugal acceleration. However, except for the forebodings in Chapter 18, the most substantial effects of flows have not been faced yet. The adjective 'substantial' obviously should refer to some standard of what is a sizeable velocity. For ordinary gas dynamics, it is clear that it is the sound speed. For the macroscopic description of plasmas, the three MHD speeds (slow, Alfvén and fast) collectively take over the role of the sound speed. This implies that 'transonic' MHD flows will be characterized by different flow regimes depending on the magnitude of the flow velocity relative to those three speeds. In addition, the direction of the background magnetic field $\mathbf{B}_0$ introduces an anisotropy in plasma dynamics that is not present in ordinary gas dynamics.

Because of the important implications, the theory of MHD flows is extensively investigated at present, both in laboratory and in astrophysical contexts. In the linear analysis, the usual approach of a split into background equilibrium and perturbations by waves and instabilities leads to substantial difficulties since these two topics become intermingled for transonic flows [223, 225], as described in Section 18.2.3. For the computation of transonic equilibria, the usual assumption of ellipticity breaks down, with concomitant complexity of bookkeeping of solutions, whereas the appearance of new overstable modes due to transonic transitions (Section 18.4.3) signals that a vast field of research on the interaction of transonic flows and waves is opened up. This may imply that, eventually, nonlinear analysis of the complete dynamics (without making the split into equilibrium and perturbations) is the more appropriate approach. This necessarily involves large-scale computing using the computational tools spelled out in Chapter 20. At present, the deep relationship between linear instabilities in transonic MHD flows and the resulting nonlinear 3D dynamics has hardly been investigated, let alone properly understood. As a first step, we now turn to the most basic part of magnetohydrodynamics, leaving the linearization techniques of the equilibrium–stability split and the resulting differential equations, and we just consider the algebraic effects of the nonlinearity at *a single point* of the flow. We will find out that, once again, a rich structure is waiting there, with quite some common features with linear dynamics that have not been investigated, even today. MHD never disappoints!

21.1.1 Characteristics and shocks

Recall the discussion of the ideal MHD characteristics in Section 5.4.3, with the seven characteristic speeds given by Eqs. (5.120)–(5.123). That discussion centred on the waves and instabilities of static equilibria, illustrated by the Friedrichs diagrams of Fig 5.9 for the propagation of plane waves or point disturbances in plasmas without background flow. The generalization to *plasmas with background flow* (stationary equilibria), as given in Chapter 12, yields the *seven characteristic speeds* of the flow illustrated in Fig. 5.10:

$$u_{\rm E} = v_{\rm n} \equiv \mathbf{n}\cdot\mathbf{v}\,, \quad u_{\rm s}^{\pm} = v_{\rm n} \pm v_{\rm sn}\,, \quad u_{\rm A}^{\pm} = v_{\rm n} \pm v_{\rm An}\,, \quad u_{\rm f}^{\pm} = v_{\rm n} \pm v_{\rm fn}\,, \tag{21.1}$$

involving the familiar three phase velocities,

$$v_{\rm An} \equiv \frac{B}{\sqrt{\rho}}\cos\vartheta\,, \quad v_{\rm s,fn} \equiv \Big[\frac{1}{2\rho}\Big(\gamma p + B^2 \mp \sqrt{(\gamma p + B^2)^2 - 4\gamma p B^2 \cos\vartheta^2}\Big)\Big]^{1/2}. \tag{21.2}$$

Here, ϑ is the angle between the magnetic field and the wave normal $\mathbf{n}$, the minus sign is for the slow waves and the plus sign for the fast waves. By varying the direction of the normal $\mathbf{n}$, the expressions (21.1) yield the space-time manifolds along which perturbations propagate.

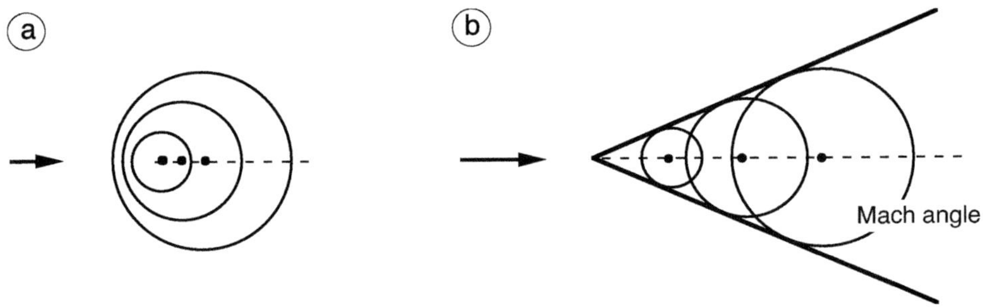

Fig. 21.1 Sound characteristics in (a) subsonic and (b) supersonic gas flow about a point source.

Permitting two spatial dimensions, the temporal snapshots of the three MHD perturbations become the phase and group diagrams of Fig. 5.9. These figures may produce an interesting new feature, depending on the magnitude of the background flow. This is illustrated in Fig. 21.1 for the case of sound waves in ordinary fluids: when the flow velocity becomes supersonic, the spatial part of the characteristics forms envelopes where information accumulates so that discontinuous solutions, *or shocks*, are formed. Whereas in subsonic flows the solutions propagate everywhere, in supersonic flows these discontinuities separate *hyperbolic regions*, where the solutions propagate, from *elliptic regions*, where they do not.

One of the deep problems in transonic flows is that the transitions from ellipticity to hyperbolicity occur at locations that are not known a priori. Those locations are part of the solution of the problem. In MHD, we will also use the term 'transonic' to indicate flows that surpass one of the three characteristic speeds $v_{\rm A}$, $v_{\rm s}$, $v_{\rm f}$ defined in Eqs. (21.2). The theory of MHD shocks, associated with those transonic transitions, is discussed in Sections 21.2 and 21.3. We will present the subject from a new angle, time reversal duality (Section 21.3.2), and show how this connects with the classification of MHD shocks in terms of converging characteristics, where a central role is played by the intermediate shocks [656].

Another fundamental problem associated with the transonic dynamics is the construction of the stationary equilibrium itself. So far, the tacit assumption for the basic equilibrium state, which is the space in which the waves and instabilities 'live' (Section 17.1), has been that the governing nonlinear partial differential equations are *elliptic* (like the Grad–Shafranov equation of Chapter 16). In fact, all numerical methods in use for MHD spectral analysis assume that the equilibria are described by elliptic equations and the perturbations by hyperbolic ones. However, when the poloidal flow velocity increases beyond certain critical values, the stationary equilibrium equations become hyperbolic (see Section 18.2) and the classical paradigm of a split into equilibrium and perturbations breaks down. As a result, the standard equilibrium solvers diverge and we need to rethink the problem completely.

The breakdown of the classical split into equilibrium and perturbations is associated with the Lagrangian time derivative $D/Dt \equiv \partial/\partial t + \mathbf{v}\cdot\nabla$ in the MHD equations. Whereas the Eulerian time derivative $\partial/\partial t$ produces the eigenfrequencies ω of the waves, the spatial derivative $\mathbf{v}\cdot\nabla$ produces not only the Doppler shifts of the perturbations but also the possibility of spatial discontinuities of the equilibria. The two pieces of the Lagrangian time derivative really belong together so that *the waves and the stationary equilibria, with transitions from ellipticity to hyperbolicity, are not separate issues*. This is the reason for the difficulties mentioned in the study of stability of transonic MHD flows.

The most striking phenomenon in transonic flows is the occurrence of shocks, where virtually all physical variables are discontinuous. We have encountered discontinuities at plasma–plasma and plasma–vacuum interfaces for static and subsonic stationary flows in Chapter 4 when discussing the different model problems of laboratory and astrophysical plasmas. In contrast, in the theory of transonic flows, one of the essential problems is the occurrence of *internal surfaces of discontinuity* where the same jump conditions provide boundary conditions that are to be imposed at the positions of the shock fronts, which are a priori unknown and to be determined together with the solution. In the present chapter, we develop the subject of shock conditions just to the point where they can be implemented in one of the general Riemann solvers discussed in Chapter 20. Although straightforward in principle, this involves a surprising amount of algebra.

The jump conditions for the density, the velocity, the pressure and the magnetic field were derived from the MHD conservation equations in Section 4.5, Eqs. (4.149)–(4.154). The basic 'trick' consisted of integrating the evolution equations across a thin layer, of thickness δ, in which nonlinear and dissipative effects smooth the discontinuities to large, but finite, variations in the normal direction, and then taking the limit $\delta \to 0$. This turns the partial differential equation into algebraic relations between the variables on the two sides of the shock. This procedure translates the basic nonlinearities of the MHD equations into algebraic relations at a single point. Recall that the jumps are denoted by the notation

$$[\![f]\!] \equiv f_1 - f_2 \,, \tag{21.3}$$

where f_1 indicates the value of a physical variable in the undisturbed, *upstream,* part of the fluid in front of the shock, and f_2 indicates the value in the shocked, *downstream,* part behind the shock.

For convenience, we repeat this general form of the *jump conditions in the shock frame*, in which the shock is stationary and the fluid velocities $\mathbf{v}'$ are evaluated with respect to that frame. We change the order of the equations to facilitate the reductions to the final shock relations of Section 21.2:

$$[\![\rho v_{\rm n}']\!] = 0 \qquad \textit{(mass)}, \tag{21.4}$$

$$[\![B_{\rm n}]\!] = 0 \qquad \textit{(normal flux)}, \tag{21.5}$$

$$\rho v_{\rm n}' [\![\mathbf{v}_{\rm t}']\!] = B_{\rm n} [\![\mathbf{B}_{\rm t}]\!] \qquad \textit{(tangential momentum)}, \tag{21.6}$$

$$\rho v_{\rm n}' [\![\mathbf{B}_{\rm t}/\rho]\!] = B_{\rm n} [\![\mathbf{v}_{\rm t}']\!] \qquad \textit{(tangential magnetic flux)}, \tag{21.7}$$

$$[\![\rho v_{\rm n}'^{\,2} + p + \tfrac{1}{2} B_{\rm t}^2]\!] = 0 \qquad \textit{(normal momentum)}, \tag{21.8}$$

$$\rho v_{\rm n}' [\![\tfrac{1}{2}(v_{\rm n}'^{\,2} + v_{\rm t}'^{\,2}) + e + p/\rho + B_{\rm t}^2/\rho]\!] = B_{\rm n} [\![\mathbf{v}_{\rm t}' \cdot \mathbf{B}_{\rm t}]\!] \qquad \textit{(energy)}, \tag{21.9}$$

where the internal energy $e = (\gamma - 1)^{-1} p/\rho$ for the ideal plasmas that we focus on. Recall that the momentum and magnetic flux equations have been projected normal (subscript n) and tangential (subscript t) to the shock front. As stated in Eq. (4.148), to properly discuss shocks, we need to restrict the possible jumps by demanding that the entropy increases across the shock front:

$$[\![S]\!] \equiv [\![\rho^{-\gamma} p]\!] \le 0 \qquad \textit{(entropy)}. \tag{21.10}$$

This additional condition models the effects of a thin dissipative layer of thickness δ, where the large gradients of the physical variables become jumps in the limit $\delta \to 0$ and entropy production is the only dissipative effect that remains.

21.1.2 Gas dynamic shocks

As a preliminary to the analysis of the MHD shocks on the basis of these equations, we will consider the special case of ordinary gas dynamic shocks. But first, starting with an even more restricted case, viz. the *absence of mass flow across the surface of discontinuity* ($\rho v_{\rm n}' = 0$), note that the general jump conditions (21.4)–(21.10) for $\mathbf{B} = 0$ admit the solutions

$$m \equiv \rho v_{\rm n}' = 0 \quad \Rightarrow \quad [\![p]\!] = 0\,, \quad \text{but} \quad [\![\rho]\!] \neq 0\,, \quad [\![\mathbf{v}_{\rm t}]\!] \neq 0\,. \tag{21.11}$$

These imply that the fluid density and the tangential velocity display jumps of arbitrary magnitude, as is usual at the *interface between two fluids*.

Next, turning to the study of genuine shocks by *admitting mass flow across the surface of discontinuity* ($\rho v_{\rm n}' \neq 0$), the jump conditions reduce to:

$$[\![\rho v_{\rm n}']\!] = 0\,, \qquad [\![\mathbf{v}_{\rm t}']\!] = 0\,, \tag{21.12}$$

$$[\![\rho v_{\rm n}'^{\,2} + p]\!] = 0\,, \tag{21.13}$$

$$[\![\tfrac{1}{2} v_{\rm n}'^{\,2} + e + p/\rho]\!] = 0\,, \qquad e = \frac{p}{(\gamma - 1)\rho}\,. \tag{21.14}$$

Since $[\![\mathbf{v}_{\rm t}']\!] = 0$ now, we may transform to a coordinate system moving with the tangential flow so

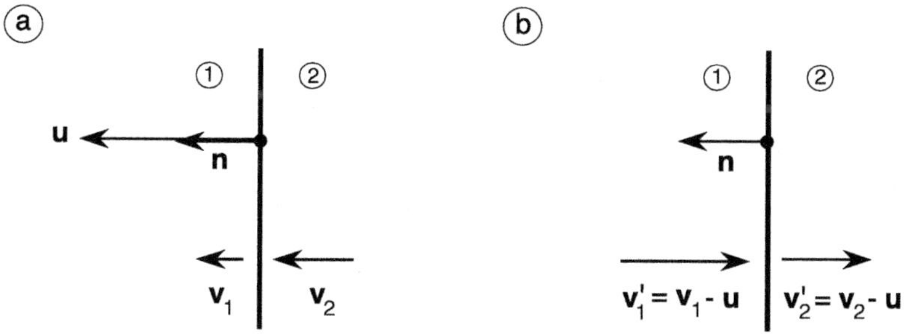

Fig. 21.2 Gas dynamic shock: (a) laboratory frame; (b) shock frame.

that $\mathbf{v}'_t = 0$. The velocities in the laboratory and shock frames then have the relative magnitudes as illustrated in Fig. 21.2.

Dropping the primes for simplicity of the notation, the shock conditions become

$$\rho_1 v_1 = \rho_2 v_2 \,, \tag{21.15}$$

$$\rho_1 v_1^2 + p_1 = \rho_2 v_2^2 + p_2 \,, \tag{21.16}$$

$$\tfrac{1}{2} v_1^2 + e_1 + p_1/\rho_1 = \tfrac{1}{2} v_2^2 + e_2 + p_2/\rho_2 \,, \quad e_{1,2} = \frac{p_{1,2}}{(\gamma - 1)\rho_{1,2}} \,. \tag{21.17}$$

They determine the values of the three downstream parameters ρ_2, v_2, p_2 in terms of the three upstream parameters ρ_1, v_1, p_1. They may be reduced to more compact expressions involving only one parameter since *gas dynamic shocks are essentially a one-parameter family.* We will show this by two methods, one following the traditional exposition, as in Landau and Lifshitz, *Fluid Mechanics* [394], and another one that anticipates the sequence of steps involved in the analogous, but more complex, reduction for shocks in magnetohydrodynamics.

In the first approach, the shock conditions are considered from the thermodynamic point of view, not necessarily restricted to ideal gases. Defining the *mass flow through the shock*, $m \equiv \rho v_{\rm n}$, *the specific volume*, $V \equiv \rho^{-1}$, *the internal energy*, e, and *the enthalpy* (or heat function), $w \equiv e + pV$, the jump conditions (21.12)–(21.14) transform into

$$[\![m]\!] = 0 \,, \tag{21.18}$$

$$m^2 = -[\![p]\!]/[\![V]\!] \,, \tag{21.19}$$

$$[\![\tfrac{1}{2} m^2 V^2 + w]\!] = 0 \,, \quad \text{or} \quad e_1 - e_2 + \tfrac{1}{2}(p_1 + p_2)(V_1 - V_2) = 0 \,. \tag{21.20}$$

Since $m^2 > 0$, the jump condition (21.19) permits solutions $p_2 > p_1$, $V_2 < V_1$, but also solutions $p_2 < p_1$, $V_2 > V_1$. The latter ones are eliminated by the entropy condition (21.10), as will be shown. The expression (21.20)(b) is called the Hugoniot or shock 'adiabatic', not to be confused with the ordinary Poisson adiabatics relating p and V for genuine adiabatic (constant entropy) processes.

As shown in Fig. 21.3 for the case of ideal gases, the Hugoniot adiabatic can also be represented by a curve in the p–V diagram, viz. the curve which connects the representative upstream point (p_1, V_1) to all possible downstream points (p_2, V_2) for the different values of the parameter m. According to Eq. (21.19), the slope of the chord connecting these points is $-m^2$. The explicit

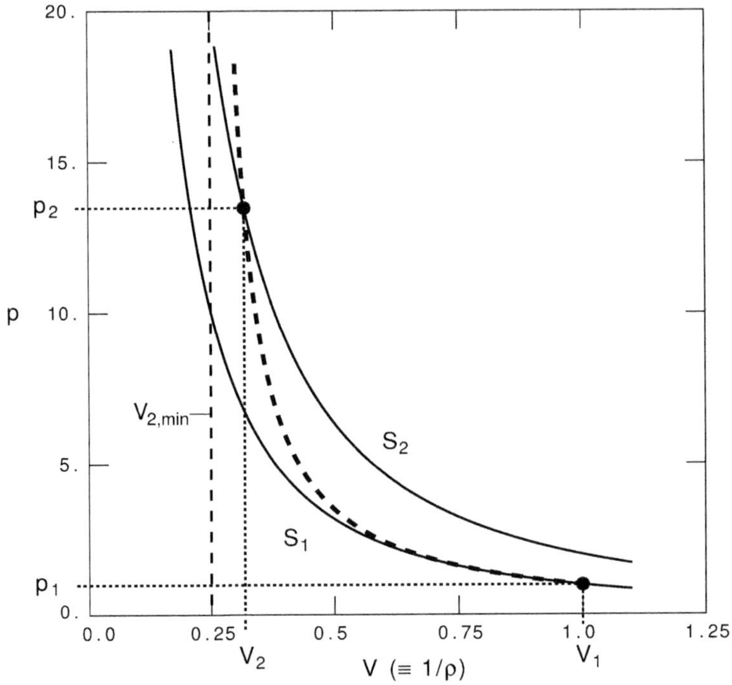

Fig. 21.3 Hugoniot, or shock, 'adiabatic' (thick dashed curve) connecting the upstream and downstream states on the Poisson adiabatics (S_1 and S_2). Maximum compression is indicated by the vertical dashed asymptote labelled $V_{2,\rm min}$.

expression of the Hugoniot adiabatic for ideal gases is obtained from the expression (21.20) by substituting the equation of state, $e = pV/(\gamma - 1)$:

$$\frac{p_2}{p_1} = \frac{\gamma + 1 - (\gamma - 1)\, V_2/V_1}{(\gamma + 1)\, V_2/V_1 - \gamma + 1} \,. \tag{21.21}$$

This relation is shown in Fig. 21.3 (dashed curve) together with the two Poisson adiabatics labelled S_1 and S_2 for the upstream and downstream states. Notice the distinction between the Hugoniot and the Poisson adiabatics: the Hugoniot adiabatic 'nestles' against the upstream Poisson adiabatic S_1 at (p_1, V_1), where the values of the two functions and their first and second derivatives are equal, but it intersects the downstream Poisson adiabatic S_2 at (p_2, V_2). This illustrates the nonlinearity of shocks: they are not additive. If a point on a Hugoniot adiabatic is taken as an upstream state for a second shock, the corresponding Hugoniot adiabatic will be different from the first one and the new shock will overtake the old one.

Introducing the *shock strength*, $\sigma \equiv 1 - V_2/V_1$, which ranges from 0 for weak shocks ($V_2 \approx V_1$) to 1 for infinite compression ($V_2 = 0$), it is noted from Eq. (21.21) that there is actually a limit to the amount of compression, viz. $(V_2/V_1)_{\rm min} = (\gamma - 1)/(\gamma + 1)$ $\big[= 1/4$ for $\gamma = 5/3\big]$ where $p_2 \to \infty$. This is indicated by the vertical asymptote to the Hugoniot adiabatic in Fig. 21.3 which corresponds to maximum shock strength, $\sigma \to \sigma_{\rm max} \equiv 1 - (V_2/V_1)_{\rm min} \equiv 2/(\gamma + 1)$ $\big[= 3/4$ for $\gamma = 5/3\big]$.

In our second approach, we stress the central role of sound waves in the formation of gas dynamic shocks. This is most conveniently expressed by means of the squared upstream and downstream *Mach numbers*, relating the flow velocities $v_{1,2}$ to the sound speeds $c_{s1,2}$ in the two regions:

$$M_1^2 \equiv \frac{v_1^2}{c_{s1}^2} \equiv \frac{\rho_1 v_1^2}{\gamma p_1}, \qquad M_2^2 \equiv \frac{v_2^2}{c_{s2}^2} \equiv \frac{\rho_2 v_2^2}{\gamma p_2}. \tag{21.22}$$

The mass and momentum conservation relations (21.15) and (21.16) then yield the following ratios of the primitive variables on the two sides of the shock:

$$\frac{\rho_2}{\rho_1} = \frac{v_1}{v_2} = \frac{M_1^2(\gamma M_2^2 + 1)}{M_2^2(\gamma M_1^2 + 1)} \quad \left[= \frac{(\gamma+1)M_1^2}{(\gamma-1)M_1^2 + 2} \right], \tag{21.23}$$

$$\frac{p_2}{p_1} = \frac{\gamma M_1^2 + 1}{\gamma M_2^2 + 1} \quad \left[= \frac{2\gamma M_1^2 - \gamma + 1}{\gamma + 1} \right]. \tag{21.24}$$

Except for the trivial solution without jumps ($M_2^2 = M_1^2$), the energy conservation relation (21.17) yields the *distilled energy jump condition*, relating the downstream to the upstream squared Mach number:

$$M_2^2 = \frac{(\gamma-1)M_1^2 + 2}{2\gamma M_1^2 - \gamma + 1}. \tag{21.25}$$

By substituting this condition, Eqs. (21.23) and (21.24) reduce to the expressions in square brackets.

We now exploit the *scale-independence* of the MHD equations, introduced in Section 4.1.2, which also applies to the equations of hydrodynamics (HD). This implies that in any ideal MHD, or HD, problem three parameters can be found that should not be considered as free since they just fix the scales of length, time and mass, which can be eliminated by defining appropriate dimensionless variables. Since the present problem is essentially time-independent, only two such scale parameters occur, viz. ρ_1 and p_1. Two parameters have already been eliminated by moving to a frame with vanishing tangential velocity. Hence, of the five arbitrary upstream parameters $\rho_1, \mathbf{v}_1, p_1$, only one should be counted as free. This is expressed by the upstream Mach number M_1. The downstream variables are then expressed by the dimensionless parameters ρ_2/ρ_1 and p_2/p_1, determined by the conditions (21.23) and (21.24), and the Mach number M_2 following from the distilled jump condition (21.25). This proves that M_1 is the only control parameter for gas dynamic shocks.

The distilled energy condition (21.25) is shown in Fig. 21.4 as a curve (the jump curve) in the M_1^2–M_2^2 plane, consisting of forbidden (dashed) and permitted (drawn) parts. It is defined for all $M_{\rm min}^2 \le M_1^2 < \infty$ and $M_{\rm min}^2 \le M_2^2 < \infty$, where $M_{\rm min}^2 \equiv \frac{1}{2}(1 - 1/\gamma)$ $[= 1/5$ for $\gamma = 5/3]$. Here, $M_1^2 = M_{\rm min}^2$ marks the boundary of an unphysical domain (indicated in grey) where $p_2 < 0$, and $M_2^2 = M_{\rm min}^2$ marks the boundary of another unphysical domain (also grey) beyond the maximum compression, $(\rho_2/\rho_1)_{\max} \equiv (\gamma+1)/(\gamma-1)$ $[= 4$ for $\gamma = 5/3]$. Finally, the distinguishing entropy condition (21.10) yields the *distilled entropy condition:*

$$\frac{S_2}{S_1} \equiv \frac{p_2}{p_1}\left(\frac{\rho_1}{\rho_2}\right)^\gamma = \left(\frac{M_2^2}{M_1^2}\right)^\gamma \left(\frac{\gamma M_1^2 + 1}{\gamma M_2^2 + 1}\right)^{\gamma+1} \ge 1 \tag{21.26}$$

$$\left[\Rightarrow \; \frac{2\gamma M_1^2 - \gamma + 1}{\gamma + 1} \left(\frac{(\gamma-1)M_1^2 + 2}{(\gamma+1)M_1^2}\right)^\gamma \ge 1 \right].$$

This condition is only satisfied in the dotted areas of the M_1^2–M_2^2 plane of Fig. 21.4. The jump curve $M_2^2 = M_2^2(M_1^2)$ enters one of these for $M_1^2 \ge 1$, where $M_2^2 \le 1$, $v_1/v_2 = \rho_2/\rho_1 \ge 1$ and

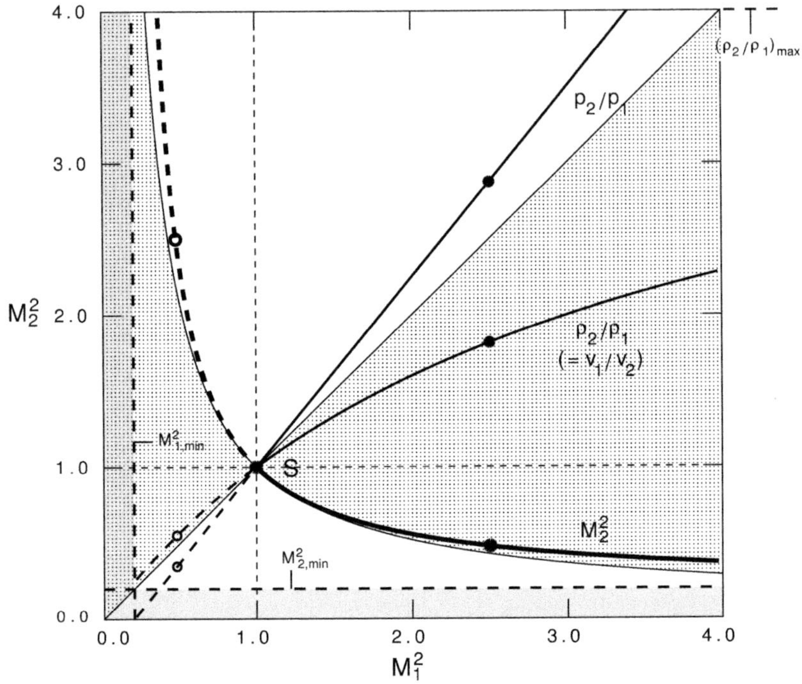

Fig. 21.4 Duality between entropy-forbidden jumps (thick dashed curve) and entropy-permitted shocks (thick solid curve) in gas dynamics. An arbitrary point on the permitted part of the jump curve (dot) and its forbidden dual (open circle) are indicated together with corresponding points on the curves for p_2/p_1 and ρ_2/ρ_1. The point S indicates sound waves (HD shocks of infinitesimal strength).

$p_2/p_1 \geq 1$. Hence, entropy-permitted shocks (thick solid curve) are only obtained for *supersonic upstream flow, corresponding to subsonic downstream flow*. Vice versa, entropy-forbidden jumps (dashed part of the jump curve) are obtained for $M_1^2 < 1$, where $M_2^2 > 1$. Note that this part of the jump curve lies just outside the other dotted area of positive entropy jump, but it touches it at the central sonic point S, where $M_1^2 = M_2^2 = 1$. This point corresponds to sound waves, i.e. shocks of infinitesimal strength ($\sigma \to 0$). It marks the transition from forbidden jumps to permitted shocks. The central position of the sonic point and the symmetry of the jump curve with respect to it has important consequences, as we will show now.

The procedure of replacing entropy conservation of ideal HD or MHD by the dissipative concept of entropy increase severely restricts the permitted dynamics. However, since entropy increase may be considered as *the arrow of time,* the discarded entropy-forbidden jumps may be turned into physically acceptable solutions by just reversing the direction of the flow. This does not affect the jump conditions since they do not involve the Mach numbers themselves, only their squares, but the roles of upstream and downstream states are interchanged and what previously was a forbidden jump becomes a permitted shock for the reversed flow, and vice versa. Exploiting the symmetry of the jump curve by defining parameters that measure the 'distance' to the sonic point S,

$$\Delta_1 \equiv M_1^2 - 1\,, \qquad \Delta_2 \equiv M_2^2 - 1\,, \tag{21.27}$$

the distilled jump condition (21.25) is transformed into the more concise form

$$\Delta_2 = -\frac{\Delta_1}{1-\Delta_1/\Delta_{\rm min}}\,, \qquad \Delta_{\rm min} \equiv -\tfrac{1}{2}(1+1/\gamma)\,, \tag{21.28}$$

where $\Delta_{\rm min}$ $\big[= -4/5$ for $\gamma = 5/3\big]$ corresponds to $M^2_{\rm min}$ defined above. We now introduce the principle of *time reversal duality between entropy-forbidden jumps and entropy-permitted shocks*, which is expressed by the correspondence

$$\Delta_2 = \Delta_2^{\pm}(\Delta_1) \quad \Longleftrightarrow \quad \Delta_1 = \Delta_1^{\mp}(\Delta_2) \tag{21.29}$$

with associated parameter ranges

$$\left\{\begin{array}{ll} \Delta_2^+ : & 0 \le \Delta_1 < \infty \\ \Delta_2^- : & \Delta_{\rm min} < \Delta_1 \le 0 \end{array}\right. \quad \Longleftrightarrow \quad \left\{\begin{array}{ll} \Delta_1^- : & \Delta_{\rm min} < \Delta_2 \le 0 \\ \Delta_1^+ : & 0 \le \Delta_2 < \infty \end{array}\right. .$$

Except that reflected shocks frequently occur in nature, this duality also may be exploited in problems without reflection to express shock conditions in terms of downstream instead of upstream control parameters. In HD this involves the simple operation of inverting the distilled jump relation (21.25), but in MHD this extremely useful operation is no longer simple at all, as we will see.

A final, parenthetical, remark: the subject of gas dynamic shocks is a beautiful topic; too bad it is stained by the activities of war. Just note the dates of the original contributions to the field as cited in, e.g., Refs. [130], [269], [394] and [547].

21.1.3 Misnomers

It should be stressed: in this chapter, *the sound Mach number M refers to velocities measured with respect to the sound speed in ordinary gases,* as defined in Eqs. (21.22) of Section 21.1.2, and *the Alfvén Mach number M refers to velocities measured with respect to the Alfvén speed in magnetized plasmas*, as defined in Eq. (21.43) of Section 21.2.2 below (where the subscript A will be dropped immediately). Hence, we purposely exploit the same symbol for these two different quantities since they play the same role in 'transonic' phenomena in gases and plasmas, whereas we never have to consider the two simultaneously: A fluid is either an isotropic ordinary gas (dominated by binary collisions between neutral particles) or a very anisotropic plasma (dominated by collective effects and magnetic fields). The relevant speeds to compare fluid velocities with are clearly, in the first case, the sound speed $c_{\rm s} \equiv \sqrt{\gamma p/\rho}$ (earlier indicated by the symbol c, for emphasis here subscripted with the letter s, clearly to be distinguished from the slow magneto-sonic velocity $v_{\rm s}$, and obviously also from the velocity of light c exploited in the next chapter) and, in the second case, the Alfvén speed $v_{\rm A} \equiv B/\sqrt{\rho}$ (earlier indicated by the symbol b). That also brings in the essential difference between gases and plasmas, i.e. the magnetic field and, with it, the two additional parameters that are needed to describe MHD discontinuities and shocks, viz. the angle ϑ between the normal to the front and the magnetic field, and the ratio β of kinetic and magnetic field pressures [see Eq. (21.57) of Section 21.2.2].

This nomenclature is specifically addressed against the misunderstanding that abounds in the plasma physics literature on the supposed role of the sound Mach number to describe 'transonic' MHD phenomena: there are no sound waves in a magnetized plasma! Instead, there are three kinds of waves, viz. the central Alfvén wave, and the 'satellite' slow and fast magneto-sonic waves with the characteristic speeds given by Eqs. (21.2). The misconceptions originate, on one hand, from the tokamak community, where *slow* magneto-sonic waves are frequently called 'sound waves' since, for $\beta \ll 1$,

$$v_{\rm sn} \approx (\sqrt{\gamma p}/B)\, v_{\rm An} \equiv (B_n/B)\, c_{\rm s}\,, \qquad v_{\rm fn} \approx (B/B_n)\, v_{\rm An}\,, \tag{21.30}$$

and, hence, the misnomer 'sound Mach number' is used to express velocities when really comparison with the slow magneto-sonic speed is meant. This clearly misses the crucial aspect of anisotropy with respect to the direction of the magnetic field. On the other hand, in the plasma-astrophysics community concerned with high β phenomena, e.g. accretion flows about compact objects, the opposite misconception abounds, viz. also using the misnomer 'sound Mach number' but now when comparison with the *fast* magneto-sonic speed is involved since, for $\beta \gg 1$,

$$v_{\rm sn} \approx \sqrt{1-\alpha}\, v_{\rm An} \approx v_{\rm An}\,, \qquad v_{\rm fn} \approx \sqrt{1+\alpha}\, c_{\rm s} \approx c_{\rm s}\,, \tag{21.31}$$

where $\alpha \equiv (B^2 - B_n^2)/(\gamma p) \sim \beta^{-1} \ll 1$. Even worse, sometimes, both the sound and the Alfvén Mach number are provided to describe MHD shocks: a very roundabout way of providing the value of β! Clearly, use of these misnomers makes no sense at all for intermediate values of β. It should be avoided at all cost since it confuses the basic issues of 'transonic' MHD.

21.2 MHD shock conditions

21.2.1 MHD discontinuities without mass flow

We now switch to the proper MHD discontinuities. For HD, we saw in Section 21.1.2 that, in the absence of mass flow across the discontinuity, the jump conditions admit jumps across that interface. In MHD, this gives rise to the contact and tangential discontinuities already introduced in Section 4.5.2. With the presence of a magnetic field, it is clear from the tangential jump conditions (21.6) and (21.7), both involving $\mathbf{v}'_{\rm t}$ and $\mathbf{B}_{\rm t}$, that a transformation to $\mathbf{v}'_{\rm t} = 0$, as illustrated in Fig. 21.2, is not possible in general. The tangential velocity is an essential constituent of MHD discontinuities, both for the present discontinuities without mass flow and for the ones with mass flow discussed in the next section. Recall that the jump conditions (21.4)–(21.10) refer to the shock frame so that the condition for absence of mass flow should be written with a prime: $v'_{\rm n1,2} \equiv v_{\rm n1,2} - u = 0$ [Fig. 21.2(b)], whereas $v_{\rm n1} = v_{\rm n2} = u$ in the laboratory frame [Fig. 21.2(a)]. This implies that these discontinuities are just carried with the fluid flow. As in Section 21.1.2, we will consistently exploit the shock frame but drop the primes.

Two very different kinds of discontinuity without mass flow are obtained depending on whether the magnetic field $\mathbf{B}$ has a component normal to the discontinuity, or not. In the former case ($B_{\rm n} \neq 0$), discontinuity of the tangential velocity is not possible and only the density may jump, so that *contact discontinuities* (paragraph *(a)* below) are more restrictive in plasmas than in ordinary fluids. In the latter case ($B_{\rm n} = 0$), both ρ and $\mathbf{v}_{\rm t}$ may jump, as in the gas dynamic case, but continuity of p is replaced by continuity of $p + \frac{1}{2}B^2$ so that *tangential discontinuities* (paragraph *(b)* below) display a much wider variety in plasmas than in ordinary fluids. We now discuss the two discontinuities without mass flow separately.

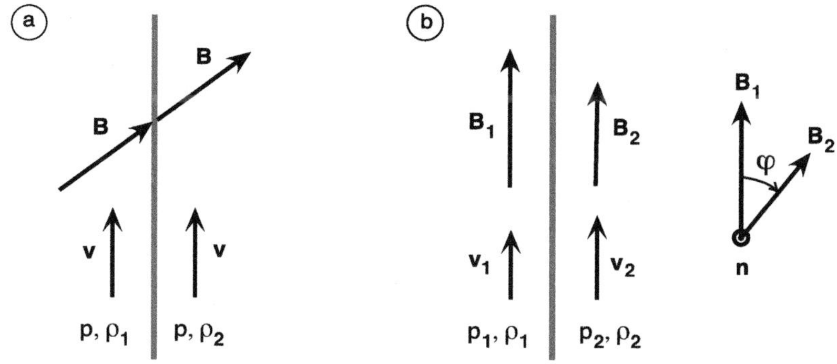

Fig. 21.5 (a) Contact discontinuities; (b) tangential discontinuities.

(a) Contact discontinuities If the mass flow vanishes ($\rho v_{\mathrm{n}} = 0$) and the normal magnetic field does not vanish ($B_{\mathrm{n}} \neq 0$) at the discontinuity, then

$$[\![\mathbf{v}_{\mathrm{t}}]\!] = 0\,, \qquad [\![p]\!] = 0\,, \qquad [\![\mathbf{B}_{\mathrm{t}}]\!] = 0\,, \qquad [\![B_{\mathrm{n}}]\!] = 0\,, \tag{21.32}$$

$$[\![\rho]\!] \neq 0\,. \tag{21.33}$$

In contrast to the gas dynamic contact discontinuity of Eq. (21.11), jumps in the tangential flow are now inhibited by the normal magnetic field. Hence, all quantities, except the density, are continuous across the discontinuity (Fig. 21.5(a)).

Since the density jumps, the entropy also jumps, $[\![S]\!] \neq 0$. Such a jump is just carried with the fluid without interaction with the other quantities. For small amplitudes, when linear theory applies, the perturbations just transform into the *entropy waves* discussed in Section 5.2.2, Eqs. (5.43). We encountered these jump conditions in Sections 4.5 and 4.6 as the boundary conditions for plasmas with a jump in the density (the astrophysical plasma models IV–VI).

(b) Tangential discontinuities If both the mass flow and the normal magnetic field vanish at the discontinuity ($\rho v_{\mathrm{n}} = 0$ and $B_{\mathrm{n}} = 0$), then

$$[\![p + \tfrac{1}{2}B_{\mathrm{t}}^2]\!] = 0\,, \tag{21.34}$$

$$[\![\rho]\!] \neq 0\,, \qquad [\![\mathbf{v}_{\mathrm{t}}]\!] \neq 0\,, \qquad [\![p]\!] \neq 0\,, \qquad [\![\mathbf{B}_{\mathrm{t}}]\!] \neq 0\,. \tag{21.35}$$

Hence, nearly arbitrary jumps of $\rho, \mathbf{v}_{\mathrm{t}}, p, \mathbf{B}_{\mathrm{t}}$ are possible and the only restriction is that the total pressure should be constant across the discontinuity (Fig. 21.5(b)). Such discontinuities are of considerable interest since they provide a tool for modelling the dynamics of plasma–plasma interfaces separating plasmas with different properties. We have encountered this type of discontinuity when discussing the boundary conditions for the laboratory plasma models II and II* in Section 4.6.

The jumps in the tangential velocity and magnetic field are due to singularities of the vorticity $\boldsymbol{\omega} \equiv \nabla \times \mathbf{v}$ and of the current density $\mathbf{j} = \nabla \times \mathbf{B}$, i.e. vortex and current sheets of strength

$$\boldsymbol{\omega}^* = \mathbf{n} \times [\![\mathbf{v}_{\mathrm{t}}]\!] \neq 0\,, \qquad \mathbf{j}^* = \mathbf{n} \times [\![\mathbf{B}_{\mathrm{t}}]\!] \neq 0\,. \tag{21.36}$$

They produce changes not only of the magnitudes, but also of the directions of $\mathbf{v}_\mathrm{t}$ and $\mathbf{B}_\mathrm{t}$, as indicated in the right part of Fig. 21.5(b) by the angle φ for $\mathbf{B}_\mathrm{t}$.

The jumps in the pressure and the density admit solutions both with and without entropy jump. This indicates that these discontinuities represent some kind of degeneracy. In particular, for infinitesimal amplitudes, they transform into a superposition of *zero-frequency Alfvén and magneto-sonic waves*. This degeneracy is lifted by contact discontinuities as well as by the discontinuities of the next section.

21.2.2 MHD discontinuities with mass flow

The generalization of the gas dynamic shocks of Section 21.1.2 to plasmas with a magnetic field requires *mass flow across the surface of discontinuity*. It leads to the two major classes of MHD discontinuities that operate in transonic plasmas, viz. rotational (or Alfvén) discontinuities and magneto-acoustic shocks. Their analysis is quite different, but we will keep it on the same footing as long as possible.

In this case, the full potential of the jump conditions (21.4)–(21.10) is realized, so that it is expedient to recast them into dimensionless form. This can be effected by means of the normal mass flow ρv_n and the normal magnetic field B_n, since they are constant across the discontinuity according to the first two conditions. We first reformulate the remaining jump conditions (21.6)–(21.10):

$$\rho v_\mathrm{n} [\![\mathbf{v}_\mathrm{t}]\!] = B_\mathrm{n} [\![\mathbf{B}_\mathrm{t}]\!] \qquad \textit{(tangential momentum)}, \tag{21.37}$$

$$B_\mathrm{n}^2 [\![\mathbf{B}_\mathrm{t}]\!] = \rho^2 v_\mathrm{n}^2 [\![\mathbf{B}_\mathrm{t}/\rho]\!] \qquad \textit{(tangential magnetic flux)}, \tag{21.38}$$

$$[\![p + \tfrac{1}{2}B_\mathrm{t}^2]\!] + \rho^2 v_\mathrm{n}^2 [\![1/\rho]\!] = 0 \qquad \textit{(normal momentum)}, \tag{21.39}$$

$$\Big[\!\!\Big[e + p/\rho \Big]\!\!\Big] + \tfrac{1}{2}(\rho v_\mathrm{n}/B_\mathrm{n})^2 \Big[\!\!\Big[B^2/\rho^2 \Big]\!\!\Big] = 0 \qquad \textit{(energy)}, \tag{21.40}$$

$$[\![\rho^{-\gamma} p]\!] \le 0 \qquad \textit{(entropy)}. \tag{21.41}$$

Here, the jump condition (21.38) is obtained by combining the tangential conditions (21.6) and (21.7), whereas the derivation of the energy condition (21.40) from Eq. (21.9) involves a number of steps that are given in small print below.

▷ **Reduction of the energy condition** The energy conservation relation (21.9) is usually transformed (see Landau and Lifshitz [395]) by generalizing the shock adiabatic (21.20) with magnetic contributions,

$$[\![e]\!] + \left\{ \tfrac{1}{2}(p_1 + p_2) + \tfrac{1}{4}(B_{\mathrm{t}1} - B_{\mathrm{t}2})^2 \right\} [\![1/\rho]\!] = 0\,. \tag{21.42}$$

However, for the purpose of reducing the jump conditions to a minimum number of free parameters, it is more expedient to transform it into

$$\left[\!\!\left[e + \frac{p}{\rho} + \tfrac{1}{2}v_\mathrm{n}^2 + \tfrac{1}{2}\left| \mathbf{v}_\mathrm{t} - \frac{B_\mathrm{n}}{\rho v_\mathrm{n}}\mathbf{B}_\mathrm{t} \right|^2 + \frac{B_\mathrm{t}^2}{\rho}\left(1 - \frac{B_\mathrm{n}^2}{2\rho v_\mathrm{n}^2}\right) \right]\!\!\right] = 0\,,$$

where the fourth term vanishes due to Eq. (21.6), and the third and fifth term combine to

$$\left[\!\!\left[\tfrac{1}{2}v_\mathrm{n}^2 + \frac{B_\mathrm{t}^2}{\rho}\left(1 - \frac{B_\mathrm{n}^2}{2\rho v_\mathrm{n}^2}\right) \right]\!\!\right] = \left[\!\!\left[\frac{v_\mathrm{n}^2 B^2}{2B_\mathrm{n}^2} \right]\!\!\right] - \frac{B_\mathrm{n}^2}{2\rho^2 v_\mathrm{n}^2}\left[\!\!\left[(\rho v_\mathrm{n}^2/B_\mathrm{n}^2 - 1)^2 B_\mathrm{t}^2 \right]\!\!\right] = \left[\!\!\left[\frac{v_\mathrm{n}^2 B^2}{2B_\mathrm{n}^2} \right]\!\!\right],$$

due to Eq. (21.38). This yields the energy conservation condition (21.40). ◁

We now exploit $\rho v_{\rm n}$ and $B_{\rm n}$ to create dimensionless variables that characterize the states on the two sides of the shock. To that end, we first introduce the square of the *normal Alfvén Mach number*,

$$M_{\rm An}^2 \equiv \frac{v_{\rm n}^2}{v_{\rm An}^2} \equiv \frac{\rho v_{\rm n}^2}{B_{\rm n}^2} \quad \left(= \frac{\rho v_{\rm n}}{B_{\rm n}^2} \cdot v_{\rm n} = \frac{\rho^2 v_{\rm n}^2}{B_{\rm n}^2} \cdot \frac{1}{\rho} \right), \tag{21.43}$$

which is the MHD counterpart of the ordinary Mach number of hydrodynamics. Since the latter does not occur in MHD, there is no confusion if we suppress the subscripts and indicate the square of the normal Alfvén Mach number also by M^2. According to the equalities (21.43) in brackets, this quantity may be considered as the dimensionless normal speed (apart from the sign, see below), or as the dimensionless inverse density (or specific volume). Hence, the ratio of the normal speeds and the inverse ratio of the densities are proportional to the ratio of the squared normal Alfvén Mach numbers across the discontinuity,

$$\frac{v_{\rm n2}}{v_{\rm n1}} = \frac{\rho_1}{\rho_2} = \frac{M_2^2}{M_1^2} \,. \tag{21.44}$$

The jump conditions (21.37)–(21.41) then simplify considerably by introducing the following dimensionless variables:

$$\begin{aligned} &\bar{v}_{{\rm n}i} \equiv \frac{\rho |v_{\rm n}|}{B_{\rm n}^2}\, v_{{\rm n}i} = -M_i^2 \,, \qquad \bar{B}_{\rm n} \equiv \frac{B_n}{|B_{\rm n}|} = -1 \,, \\ &\bar{\mathbf{v}}_{{\rm t}i} \equiv \frac{\rho |v_{\rm n}|}{B_{\rm n}^2}\, \mathbf{v}_{{\rm t}i} \,, \qquad \bar{\mathbf{B}}_{{\rm t}i} \equiv \frac{\mathbf{B}_{{\rm t}i}}{|B_{\rm n}|} \,, \qquad \bar{p}_i \equiv \frac{p_i}{B_{\rm n}^2} \qquad (i = 1, 2) \,, \end{aligned} \tag{21.45}$$

where $\bar{B}_{{\rm t}i}$ is related to the angle ϑ_i between the magnetic field and the normal, and $\bar{p}_i$ is related to the ratio β_i of the plasma and magnetic field pressures:

$$\tan \vartheta_i \equiv B_{{\rm t}i}/|B_{\rm n}| \equiv \bar{B}_{{\rm t}i} \,, \qquad \beta_i \equiv 2p_i/B_i^2 \equiv 2\bar{p}_i \cos^2 \vartheta_i \,. \tag{21.46}$$

Since the direction of the flow is opposite to the normal and the sign of the magnetic field is not relevant for the shock conditions, in Eq. (21.45) we have assumed both $\rho v_{\rm n} < 0$ and $B_{\rm n} < 0$ (as illustrated in Fig. 21.6). Since $B_n = 0$ for $\vartheta_i = \pi/2$, the singularities $\tan \vartheta_i \to \infty$ that will appear in the shock relations of Section 21.3 are artificial, just a result of having chosen B_n rather than B for the normalization.

With this normalization, the jump conditions (21.37)–(21.41) transform into:

$$[\![\bar{\mathbf{v}}_{\rm t}]\!] = [\![\bar{\mathbf{B}}_{\rm t}]\!] \qquad \left(\Rightarrow \ \bar{\boldsymbol{\omega}}^* = \mathbf{n} \times [\![\bar{\mathbf{v}}_{\rm t}]\!] = \bar{\mathbf{j}}^* = \mathbf{n} \times [\![\bar{\mathbf{B}}_{\rm t}]\!] \right), \tag{21.47}$$

$$[\![(M^2 - 1)\bar{\mathbf{B}}_{\rm t}]\!] = 0 \,, \tag{21.48}$$

$$[\![M^2 + \bar{p} + \tfrac{1}{2}\bar{B}_{\rm t}^2]\!] = 0 \,, \tag{21.49}$$

$$\left[\!\!\left[\frac{\gamma}{\gamma - 1}\bar{p} M^2 + \tfrac{1}{2}(1 + \bar{B}_{\rm t}^2) M^4\right]\!\!\right] = 0 \,, \tag{21.50}$$

$$[\![\bar{p} M^{2\gamma}]\!] \le 0 \,. \tag{21.51}$$

Apart from the degree of freedom of the rotational discontinuities, see below, the system (21.47)–(21.50) completely determines the jumps across the discontinuity for given values of the upstream parameters, whereas the entropy inequality (21.51) determines whether such a jump qualifies as a shock or not.

Again two possibilities arise according to whether the thermodynamic variables jump (genuine shock) or not across the discontinuity. The latter leads to discontinuities without a gas dynamic analogue, viz. *rotational discontinuities* (paragraph *(c)* below), which may be considered as finite amplitude Alfvén waves. The former leads to *magneto-acoustic shocks* (paragraph *(d)* below), which come in the three kinds of *slow, intermediate and fast* (Section 21.2.3). They are associated with the magneto-acoustic waves, whereas the Alfvén speed plays a central role. We now discuss the two types of discontinuity with mass flow separately.

(c) Rotational (or Alfvén) discontinuities If the density is continuous across the discontinuity, $[\![\rho]\!] = 0$, the normal velocity is also continuous, $[\![v_{\rm n}]\!] = 0$, so that

$$M_1^2 = M_2^2 = 1\,, \qquad [\![\bar{p}]\!] = 0\,, \qquad [\![\bar{B}_{\rm t}^2]\!] = 0\,, \tag{21.52}$$

$$[\![\bar{\mathbf{v}}_{\rm t}]\!] = [\![\bar{\mathbf{B}}_{\rm t}]\!] \neq 0\,. \tag{21.53}$$

Hence, all thermodynamic variables (p, ρ, e, S) and the magnitude of $\mathbf{B}$ are continuous, but the tangential component $\mathbf{B}_{\rm t}$ turns through an angle φ about the normal $\mathbf{n}$ (proportional to the vortex and current sheets defined by Eq. (21.47)). Moreover, the normal velocity and the jump in the tangential velocity are equal to their respective Alfvén velocities. This discontinuity is properly called a *rotational, or Alfvén, discontinuity* since a rotational jump of the tangential magnetic field about the normal propagates with the jump of the tangential Alfvén speed.

Rotational discontinuities are essentially determined by prescribing $M^2 = 1$ and choosing values for the three free parameters,

$$\varphi \equiv 2\arcsin\left(\tfrac{1}{2}\left|[\![\bar{\mathbf{B}}_{\rm t}]\!]\right|/\bar{B}_{\rm t}\right), \quad \vartheta \equiv \arctan \bar{B}_{\rm t}\,, \quad \beta \equiv 2\bar{p}\cos^2\vartheta\,, \tag{21.54}$$

which describe the change of the direction of the tangential magnetic field, the fixed angle of the magnetic field with the normal and the fixed value of the pressure. Note that, whereas genuine shocks are (nearly) completely determined by the values of the upstream parameters, the free parameter φ of rotational discontinuities is not determined at all by the values of the upstream parameters.

(d) Magneto-acoustic shocks If the density is discontinuous across the discontinuity, $[\![\rho]\!] \neq 0$, the normal velocity is also discontinuous, $[\![v_{\rm n}]\!] \neq 0$, so that the full system of jump conditions (21.47)–(21.51) is needed to fix the parameters. The essential features are given by

$$M_1^2 \neq M_2^2\,, \qquad [\![\bar{p}]\!] \neq 0\,, \qquad [\![\bar{B}_{\rm t}^2]\!] \neq 0\,, \tag{21.55}$$

$$[\![\bar{\mathbf{v}}_{\rm t}]\!] = [\![\bar{\mathbf{B}}_{\rm t}]\!] \parallel \bar{\mathbf{B}}_{\rm t1} \parallel \bar{\mathbf{B}}_{\rm t2}\,, \qquad (M_1^2 - 1)\bar{B}_{\rm t1} = (M_2^2 - 1)\bar{B}_{\rm t2}\,. \tag{21.56}$$

Hence, all thermodynamic variables (p, ρ, e and S) and the magnitude of $\bar{\mathbf{B}}$ are discontinuous, but the vectors $\bar{\mathbf{B}}_{\rm t1}$, $\bar{\mathbf{B}}_{\rm t2}$, $\mathbf{n}$ and $[\![\bar{\mathbf{v}}_{\rm t}]\!]$ all lie in the same plane. This discontinuity is called a *magnetohydrodynamic shock*. It is a genuine generalization of the gas dynamic shocks for magnetized plasmas.

MHD shocks are essentially determined by fixing $\varphi = 0$ and choosing values for the three free upstream parameters

$$M_1^2 \equiv -\bar{v}_{n1} \equiv 1/\bar{\rho}_1\,, \quad \vartheta_1 \equiv \arctan \bar{B}_{t1}\,, \quad \beta_1 \equiv 2\bar{p}_1 \cos^2 \vartheta_1\,, \tag{21.57}$$

which determine the values of the downstream parameters M_2^2, ϑ_2 and β_2 according to the jump conditions (21.47)–(21.50). The substantial algebra involved in the reduction of these conditions will be completed in Section 21.3.

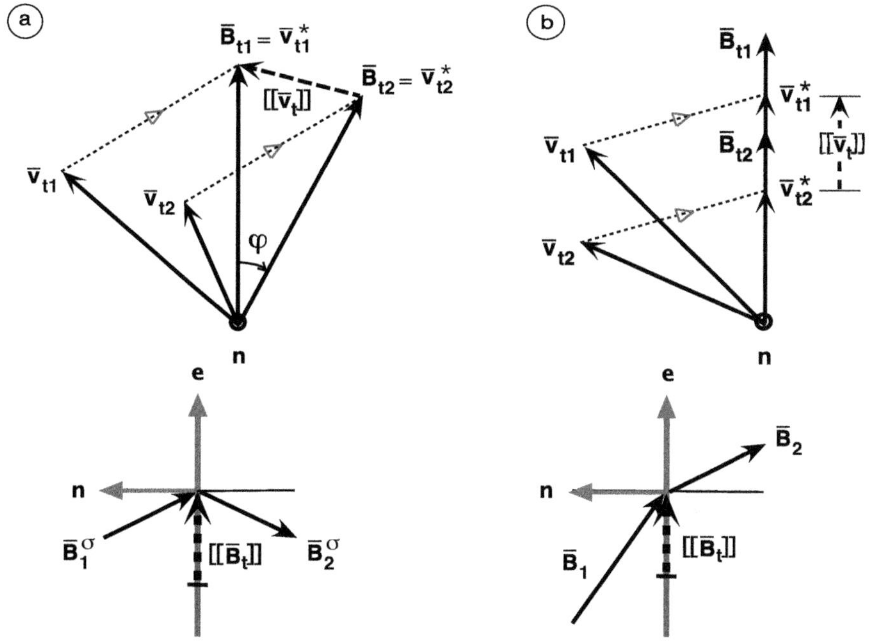

Fig. 21.6 Top: tangential magnetic fields and velocities in the de Hoffmann–Teller frame for (a) rotational discontinuities and (b) shocks; $\bar{\mathbf{v}}_{ti}^* = M_i^2 \bar{\mathbf{B}}_{ti}$ $(i = 1, 2)$. Bottom: projection (on a different scale) in the $\mathbf{e}$–$\mathbf{n}$ plane, with $\mathbf{e} \parallel [\![\bar{\mathbf{v}}_t]\!] = [\![\bar{\mathbf{B}}_t]\!]$. For (a), the constant part $\bar{\mathbf{B}}^\tau$ of $\bar{\mathbf{B}}_{ti} \perp \mathbf{e}$ does not show up in this projection, where $\bar{\mathbf{B}}_i = \bar{\mathbf{B}}_i^\sigma + \bar{\mathbf{B}}^\tau$, $\bar{\mathbf{B}}_i^\sigma = \pm \bar{B}_t \sin\frac{1}{2}\varphi\, \mathbf{e} + \bar{B}_n \mathbf{n}$, $\bar{B}_n = -1$, $\bar{\mathbf{B}}^\tau = \bar{B}_t \cos\frac{1}{2}\varphi\, \mathbf{e} \times \mathbf{n}$.

The geometric meaning of the jump conditions for the MHD discontinuities becomes much clearer when the tangential velocities and magnetic fields are aligned by means of a transformation to a tangentially moving frame in which the normal electric field vanishes. Since there are infinitely many such tangential transformations, one can find a particular one involving the normal velocities as well such that *the complete electric field vanishes,* $\mathbf{E}_i^* = \mathbf{v}_i^* \times \mathbf{B}_i = 0$, both in front and behind the shock. Since, by definition, $\bar{v}_{ni} \equiv -M_i^2$, this alignment of the transformed velocities $\mathbf{v}_i^*$ with the magnetic field $\mathbf{B}_i$ is obtained by transformation to the *de Hoffmann–Teller frame* [146],

$$\bar{\mathbf{v}}_i \rightarrow \bar{\mathbf{v}}_i^* \equiv M_i^2 \bar{\mathbf{B}}_i \qquad (i = 1, 2)\,. \tag{21.58}$$

The boost $\bar{\mathbf{w}} \equiv \bar{\mathbf{v}}^* - \bar{\mathbf{v}}$ effecting this is the same in front and behind the shock, as follows from the jump conditions:

$$[\![\bar{\mathbf{w}}_{\mathrm{t}}]\!] \equiv [\![\bar{\mathbf{v}}_{\mathrm{t}}^{*} - \bar{\mathbf{v}}_{\mathrm{t}}]\!] \equiv [\![M^2\bar{\mathbf{B}}_{\mathrm{t}} - \bar{\mathbf{v}}_{\mathrm{t}}]\!] \stackrel{(21.47)}{=} [\![(M^2-1)\bar{\mathbf{B}}_{\mathrm{t}}]\!] \stackrel{(21.48)}{=} 0\,,$$

$$\bar{w}_{\mathrm{n}i} \equiv \bar{v}_{\mathrm{n}i}^{*} - \bar{v}_{\mathrm{n}i} \equiv -M_i^2 - \bar{v}_{\mathrm{n}i} \equiv 0 \qquad \Rightarrow \quad [\![\bar{w}_{\mathrm{n}}]\!] = 0\,. \tag{21.59}$$

The first part of the last line shows that the boost is actually only tangential. The de Hoffmann–Teller transformation also holds for relativistic shocks. As noted by the authors in Ref. [146], it fails for perpendicular shocks ($\vartheta = \pi/2$) since then $B_{\mathrm{n}} = 0$ but $v_{\mathrm{n}} \neq 0$. In that case, M_i^2 and the normalizations (21.46) become undefined. This defect will be cured in Section 21.3.3 by our final representation (21.93) for the velocities in the de Hoffmann–Teller frame.

In conclusion, in the de Hoffmann–Teller frame, *the constant amplitude tangential vectors* $\bar{\mathbf{v}}_{\mathrm{t}}^{*}$ *and* $\bar{\mathbf{B}}_{\mathrm{t}}$ *are rotated over the same angle* φ *for rotational (Alfvén) discontinuities* [Fig. 21.6(a)], whereas *they have changing amplitudes but constant direction for shocks* [Fig. 21.6(b)]. The directional changes of $\bar{\mathbf{B}}$ and $\bar{\mathbf{v}}^{*}$ in the plane through the jumps and the normal (bottom panels of Fig. 21.6) alone do not permit us to distinguish between rotational discontinuities and shocks because intermediate shocks show similar behaviour to rotational discontinuities in that plane [see Fig. 21.7(b) of the next sub-section]. Those two kinds of discontinuity are distinguished by the presence or absence of a constant contribution of $\bar{\mathbf{B}}_{\mathrm{t}i}$ in the direction orthogonal to that plane.

21.2.3 Slow, intermediate and fast shocks

Restricting the discussion to shocks proper, it may appear surprising that they come in three, rather than two, flavours. Anticipating the analysis of Section 21.3, where we will show that the entropy condition only permits jumps with $M_1^2 \geq M_2^2$, those three arise due to the relationship (21.56)(b) between $\bar{B}_{\mathrm{t}1}$ and $\bar{B}_{\mathrm{t}2}$:

$$\begin{array}{llll} M_2^2 \leq M_1^2 \leq 1 & \Rightarrow & |\bar{B}_{\mathrm{t}1}| \geq |\bar{B}_{\mathrm{t}2}| & \textit{(slow shocks)}, \\ M_2^2 \leq 1 \leq M_1^2 & \Rightarrow & \bar{B}_{\mathrm{t}1}/\bar{B}_{\mathrm{t}2} < 0 & \textit{(intermediate shocks)}, \\ 1 \leq M_2^2 \leq M_1^2 & \Rightarrow & |\bar{B}_{\mathrm{t}1}| \leq |\bar{B}_{\mathrm{t}2}| & \textit{(fast shocks)}. \end{array} \tag{21.60}$$

The magnetic fields $\mathbf{B}_{\mathrm{t}1}$and $\mathbf{B}_{\mathrm{t}2}$, and hence the velocities $\mathbf{v}_{\mathrm{t}1}^{*}$ and $\mathbf{v}_{\mathrm{t}2}^{*}$, have the same directions for slow and fast shocks, with breaking towards the normal for slow shocks (Fig 21.7(a)) and away from it for fast shocks (Fig 21.7(c)), but they have opposite directions for intermediate shocks (Fig 21.7(b)).

When the angles ϑ_i of the directions of the magnetic fields are varied, several significant *limiting cases* are encountered.

- ***Perpendicular shocks*** ($\vartheta_1 = \vartheta_2 = \pi/2$) In this limit, the normal Alfvén Mach number is not defined, so that the normalizations (21.43) and (21.45) can no longer be exploited and one should go back to the original jump conditions (21.37)–(21.40) with $B_{\mathrm{n}} = 0$ and $\rho v_{\mathrm{n}} \neq 0$. Equation (21.38) then yields $[\![\mathbf{B}_{\mathrm{t}}/\rho]\!] = 0$, so that $\varphi = 0$ and a rotational counterpart of the tangential discontinuity illustrated in Fig. 21.5(b) becomes impossible. However, perpendicular shocks are perfectly valid generalizations of tangential discontinuities with $\varphi = 0$ and jumps in p and $\bar{B}_{\mathrm{t}}$ such that $[\![p + \frac{1}{2}\bar{B}_{\mathrm{t}}^2]\!] = 0$. The shock relations for this case are obtained from the general relations derived by replacing the variables M_1^2 and M_2^2 by $v_{\mathrm{n}1}$ and $v_{\mathrm{n}2}$, and properly taking the limit $B_{\mathrm{n}} \to 0$.

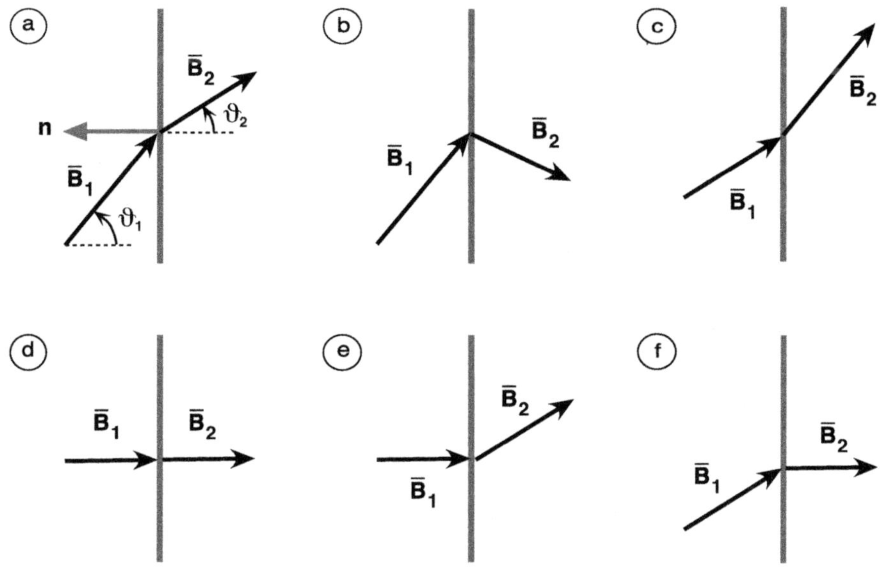

Fig. 21.7 MHD shocks: (a) slow shock, (b) intermediate shock, (c) fast shock. Limiting cases: (d) parallel ('HD') shock, (e) switch-on shock, (f) switch-off shock. The normalization is given by $\bar{B}_{\mathrm{t}i} \equiv B_{\mathrm{t}i}/|B_\mathrm{n}| = \tan\vartheta_i$, $\bar{B}_n \equiv B_\mathrm{n}/|B_\mathrm{n}| = -1$.

• ***Parallel shocks*** ($\vartheta_1 = \vartheta_2 = 0$) When the magnetic field is parallel to the normal, both upstream and downstream (Fig 21.7(d)), and hence also the velocities $\mathbf{v}_i^*$, the relation (21.56) is trivially satisfied: $\bar{B}_{\mathrm{t}1} = \bar{B}_{\mathrm{t}2} = 0$. In that case, the jump conditions (21.47)–(21.50) reduce to the hydrodynamic jump conditions (21.15)–(21.17). This follows from the relationship between the hydrodynamical (sound) Mach number and the magnetohydrodynamical (normal Alfvén) Mach number,

$$\rho_i v_{\mathrm{n}i}^2 \equiv \gamma p_i M_i^2\big|_{\mathrm{HD}} \equiv B_\mathrm{n}^2 M_i^2\big|_{\mathrm{MHD}} \,. \tag{21.61}$$

These shocks are sometimes called *hydrodynamic shocks*. Yet, the normal Alfvén Mach number is defined so that one can still distinguish fast and slow parallel shocks.

• ***Switch-on shocks*** ($\vartheta_1 = 0, \vartheta_2 \neq 0$) When the upstream magnetic field is parallel to the normal ($\bar{B}_{\mathrm{t}1} = 0$), the downstream magnetic field may have a different direction $\big[$it is switched on, $\bar{B}_{\mathrm{t}2} \neq 0$: Fig 21.7(e)$\big]$. From Eq. (21.56), this happens when $M_2^2 = 1$ so that $M_1^2 > 1$: switch-on shocks should be termed intermediate or fast.

• ***Switch-off shocks*** ($\vartheta_1 \neq 0, \vartheta_2 = 0$) The relation (21.56) between $\bar{B}_{\mathrm{t}1}$ and $\bar{B}_{\mathrm{t}2}$ is also satisfied in the opposite case ($\bar{B}_{\mathrm{t}1} \neq 0\,, \bar{B}_{\mathrm{t}2} = 0$), when the magnetic field is switched off $\big[$Fig 21.7(f)$\big]$. This happens for $M_1^2 = 1$ so that $M_2^2 < 1$: switch-off shocks should be termed intermediate or slow.

This appears to complete the classification of the different kinds of MHD shocks. We still have to demonstrate though that the *complete* set of jump conditions, including the ones associated with energy conservation and entropy production, can be satisfied for all those cases. This is the subject of the next section.

21.3 Advanced classification of MHD shocks

21.3.1 Distilled shock conditions

We will now complete the reduction of the MHD shock conditions (21.47)–(21.51) to obtain explicit expressions for $\bar{v}_{t2}, \bar{B}_{t2}, \bar{p}_2, M_2^2$ in terms of $\bar{v}_{t1}, \bar{B}_{t1}, \bar{p}_1, M_1^2$. The first condition (21.47) is actually superseded by the de Hoffman–Teller transformation (21.58) since it expresses all relationships between the downstream and upstream velocities in that preferred frame in terms of the M_i's and ϑ_i's:

$$\left\{ \begin{array}{ll} \bar{v}^*_{n1} = M_1^2 \bar{B}_n & = -M_1^2 \\ \bar{v}^*_{t1} = M_1^2 \bar{B}_{t1} & = M_1^2 \tan\vartheta_1 \end{array} \right. \Rightarrow \left\{ \begin{array}{ll} \bar{v}^*_{n2} = M_2^2 \bar{B}_n & = -M_2^2 \\ \bar{v}^*_{t2} = M_2^2 \bar{B}_{t2} & = M_2^2 \tan\vartheta_2 \end{array} \right. . \tag{21.62}$$

The next two relations (21.48) and (21.49) determine $\bar{B}_{t2}$ and $\bar{p}_2$:

$$\bar{B}_{t2} = \frac{M_1^2 - 1}{M_2^2 - 1} \bar{B}_{t1} , \tag{21.63}$$

$$\bar{p}_2 = \bar{p}_1 + (M_1^2 - M_2^2)\Big[1 - \tfrac{1}{2}\bar{B}_{t1}^2 \frac{M_1^2 + M_2^2 - 2}{(M_2^2 - 1)^2}\Big] , \tag{21.64}$$

where $\bar{B}_{ti} = \tan\vartheta_i$ and $\bar{p}_i = \frac{1}{2}\beta_i(1 + \tan^2\vartheta_i)$, $i = 1, 2$, according to the normalizations (21.46), so that these relations only involve the triples $M_i^2, \vartheta_i, \beta_i$.

Substituting these expressions into the remaining energy jump condition (21.50), and dividing out a factor $M_2^2 - M_1^2$ to eliminate trivial solutions without jumps, the *distilled energy jump condition* is obtained:

$$\begin{aligned} f \equiv\ & \tfrac{1}{2}(M_2^2 - 1)^2 \Big\{(\gamma + 1)M_2^2 - (\gamma - 1)M_1^2 - 2\gamma\bar{p}_1\Big\} \\ & + \tfrac{1}{2}\bar{B}_{t1}^2 \Big\{(\gamma - 1)(M_2^2 - 1)(M_1^2 - M_2^2) - M_2^2(M_1^2 + M_2^2 - 2)\Big\} \\ =\ & f(M_2^2; M_1^2, \bar{B}_{t1}^2, \bar{p}_1) = \tilde{f}(M_2^2; M_1^2, \vartheta_1, \beta_1) = 0 . \end{aligned} \tag{21.65}$$

Similarly, the inequality (21.51) for entropy increase across the shock provides the *distilled entropy jump condition*:

$$\begin{aligned} g \equiv\ & (M_2^2 - 1)^2 \Big\{M_1^2 - M_2^2 - \bar{p}_1\Big[(M_1^2/M_2^2)^\gamma - 1\Big]\Big\} \\ & - \tfrac{1}{2}\bar{B}_{t1}^2 (M_1^2 - M_2^2)(M_1^2 + M_2^2 - 2) \\ =\ & g(M_2^2, M_1^2, \bar{B}_{t1}^2, \bar{p}_1) = \tilde{g}(M_2^2, M_1^2, \vartheta_1, \beta_1) \geq 0 . \end{aligned} \tag{21.66}$$

All solutions $M_2^2 = M_2^2(M_1^2, \vartheta_1, \beta_1)$ of the distilled energy jump condition are permitted jumps. They only qualify as shocks when they also satisfy the distilled entropy inequality.

Analogous to our discussion of the central role of the sonic point S in gas dynamic shocks (Fig. 21.4 of Section 21.1.2), we now exploit the similar role of the Alfvén point A (where $M_1^2 = M_2^2 = 1$), and of the intermediate shocks about that point, in the theory of MHD shocks. To that end, we again define parameters

$$\Delta_i \equiv M_i^2 - 1 \qquad (-1 \leq \Delta_i \leq \infty) , \tag{21.67}$$

measuring the 'distances' to A, and again replace $\bar{B}_{ti}$ and $\bar{p}_i$ by ϑ_i and β_i, according to the definitions (21.46). By means of these new parameter sets $\Delta_i, \vartheta_i, \beta_i$, the jump conditions (21.63)–(21.66) acquire their most compact expression. The distilled energy jump condition becomes a cubic:

$$
\begin{aligned}
&(\gamma+1)\Delta_2^3 + p\,\Delta_2^2 + q\,\Delta_2 + r = 0 \qquad \Rightarrow \quad \Delta_2 = \Delta_2(\Delta_1, \vartheta_1, \beta_1)\,, \\
&\quad p \equiv 2 - (\gamma-1)\Delta_1 - \gamma\beta_1 - \gamma(1+\beta_1)\tan^2\vartheta_1\,, \\
&\quad q \equiv -[\,1 + (2-\gamma)\Delta_1]\tan^2\vartheta_1\,, \qquad r \equiv -\Delta_1\tan^2\vartheta_1\,.
\end{aligned} \tag{21.68}
$$

Once this equation is solved, the parameters $\vartheta_2 = \vartheta_2(\Delta_1, \vartheta_1, \beta_1)$ and $\beta_2 = \beta_2(\Delta_1, \vartheta_1, \beta_1)$, are found by substitution,

$$
\vartheta_2 = \arctan\Big(\frac{\Delta_1}{\Delta_2}\tan\vartheta_1\Big)\,, \tag{21.69}
$$

$$
\beta_2 = \frac{\beta_1 + 2(\Delta_1 - \Delta_2) + (1 + \beta_1 - \Delta_1^2/\Delta_2^2)\tan^2\vartheta_1}{1 + (\Delta_1^2/\Delta_2^2)\tan^2\vartheta_1}\,. \tag{21.70}
$$

The equations (21.68)–(21.70) constitute the *distilled ideal MHD jump problem*. With the addition of the distilled entropy condition (21.66), written in terms of the variables $\Delta_2, \Delta_1, \vartheta_1, \beta_1$,

$$
\Delta_2^2\Big\{\Delta_1 - \Delta_2 - \tfrac{1}{2}\beta_1(1+\tan^2\vartheta_1)\Big[\big((1+\Delta_1)/(1+\Delta_2)\big)^\gamma - 1\Big]\Big\} - \tfrac{1}{2}\tan^2\vartheta_1(\Delta_1^2 - \Delta_2^2) \geq 0\,, \tag{21.71}
$$

our formulation of the *distilled MHD shock problem* is complete.

By means of the de Hoffmann–Teller transformation and the scale-independence of the MHD equations, the six downstream parameters $\rho_1, p_1, v_{n1}, v_{t1}, B_{n1}, B_{t1}$ of the problem have now been reduced to three basic parameters, Δ_1, ϑ_1, β_1. Hence, *the MHD shocks are essentially a three-parameter family of solutions*. After this maximal reduction, the MHD shock problem is still much more complex than the gas dynamic one, as will be illustrated by Figs. 21.8–21.10, where the conditions (21.68) and (21.71) are plotted for a typical choice of the parameter values. The solutions presented demonstrate the wide variety of MHD shocks, however before the analysis involved actually has been completed. That will be done in Section 21.3.2 where the full meaning of all the symbols along the curves and of the parameter regions I, II and III is explained.

In Fig. 21.8, the complete jump curve $M_2^2 = M_2^2(M_1^2)$ is shown without suppressing the entropy-forbidden parts, like in Fig. 21.4 for gas dynamics. This exposes the time-reversal duality, to be discussed in Section 21.3.2, but it also reveals the character of the intermediate jumps/shocks as a *continuous transition in parameter space between the slow and fast jumps/shocks*. The S-shaped jump curve intersects the diagonal $M_1^2 = M_2^2$ at the points F, A and S, where the entropy-permitted (dotted) and the entropy-forbidden (white) areas interchange position. Precisely at those points, the jump curve leaves or enters the dotted areas, distinguishing entropy-forbidden jumps from entropy-permitted shocks. The shocks only occur for values of M_1^2 and M_2^2 below the diagonal

$$
M_2^2 \leq M_1^2 \quad \Rightarrow \quad \Delta_2 \leq \Delta_1\,. \tag{21.72}
$$

According to Eq. (21.44), this implies that *MHD shocks are compressive*.

Going downward, the overall trend of the jump curve is *prograde*, i.e. decreasing with respect to both M_1^2 and M_2^2, up to the point B and past the point C, but it is *retrograde*, i.e. increasing with respect to M_1^2 and decreasing with respect to M_2^2 in-between B and C. The implications for the intermediate jumps and shock are analysed in depth in Section 21.3.2. As a mnemonic, the labels B–A–C have been chosen to indicate the *backward* (retrograde) part of the jump curve and the labels D–E to *demarcate* the boundaries of the intermediate jumps.

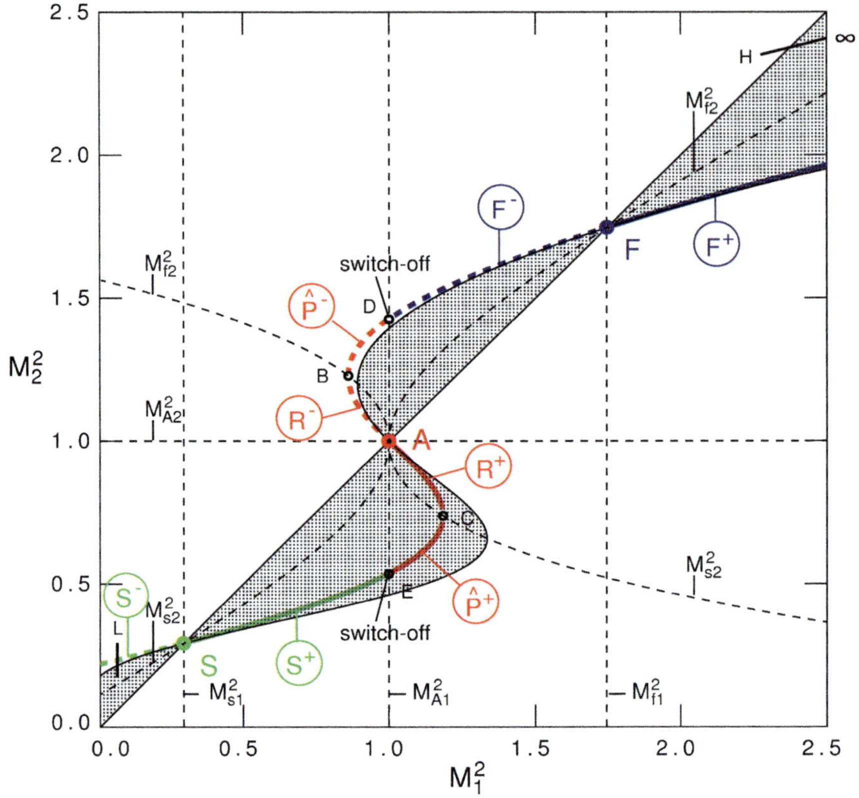

Fig. 21.8 Jump curve for $\vartheta_1 = 0.2\pi$, $\beta_1 = 0.4$ (parameter region II) showing the slow (green), intermediate (red) and fast (blue) entropy-forbidden jumps (dashed) as well as entropy-permitted shocks (drawn). The latter occur when the jump curve intersects the dotted areas, where the entropy increases downstream. The points S, A and F indicate slow, Alfvén and fast waves ($\equiv$ weak discontinuities).

The corresponding solutions $\vartheta_2 = \vartheta_2(M_1^2)$ and $\beta_2 = \beta_2(M_1^2)$ of Eqs. (21.69) and (21.70) are shown in Fig. 21.9 for the same values of ϑ_1 and β_1 as in Fig. 21.8. The first plot clearly shows the distinct directional differences between the different kinds of shock described by Eq. (21.60): $\vartheta_2 \le \vartheta_1$ for slow shocks, $\vartheta_2 \le 0$ for intermediate shocks and $\vartheta_2 \ge \vartheta_1$ for fast shocks. The second plot shows the enormous variation of the downstream pressure when the upstream Alfvén Mach number is varied. Similar to the situation for gas dynamic jumps and shocks, the downstream value of the pressure $\bar{p}_2$ (or β_2) from Eq. (21.64) cannot be guaranteed to be positive for arbitrary values of the upstream parameters. Of course, the unphysical parts of the jump curve, where $\beta_2 < 0$, have been suppressed. For the parameters of Figs. 21.8 and 21.9, this occurs for $0 \le M_1^2 < L$.

At the points F, A and S, where the jump curve of Fig. 21.8 crosses the diagonal, both upstream and downstream squared Alfvén Mach numbers take on the values

$$M_A^2 \equiv 1\,, \qquad M_{f,s}^2 \equiv \frac{1}{2\cos^2\vartheta}\Big[1 + \tfrac{1}{2}\gamma\beta \pm \sqrt{(1+\tfrac{1}{2}\gamma\beta)^2 - 2\gamma\beta\cos^2\vartheta}\,\Big]\,, \qquad (21.73)$$

corresponding to the characteristic speeds of the Alfvén, fast and slow waves of Eqs. (21.2). The

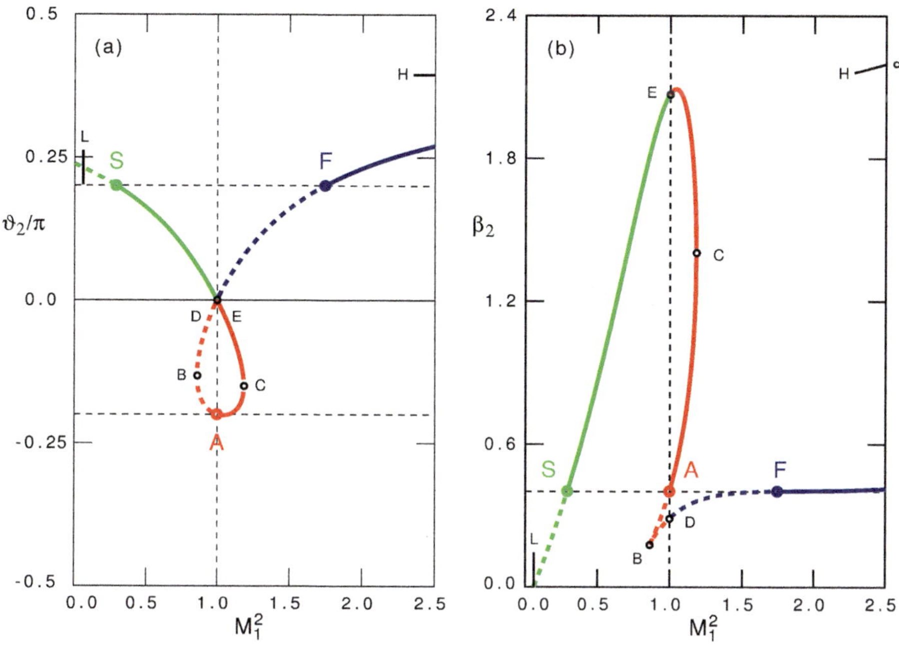

Fig. 21.9 (a) Downstream angle $\vartheta_2(M_1^2)$ and (b) relative pressure $\beta_2(M_1^2)$ of the MHD jumps/shocks for the parameter values of Fig. 21.8 ($\vartheta_1 = 0.2\pi$, $\beta_1 = 0.4$). For $M_1^2 < L$, the jump curve would be unphysical because $\beta_2 < 0$. The symbol H indicates the asymptote of the curves for $M_1^2 \to \infty$.

shocks may be classified according to the ranges of M_1^2 and M_2^2 with respect to these transition values. This is related to the *shock strength*, $\sigma \equiv 1 - M_2^2/M_1^2$. Expressing the distilled energy jump condition (21.65) in terms of σ instead of M_2^2, and then taking the limit $\sigma \to 0$, we obtain

$$\lim_{\sigma\to 0} \tilde{f}(M_2^2; M_1^2, \vartheta_1, \beta_1) = (M^2 - M_S^2)(M^2 - 1)(M^2 - M_F^2) = 0\,, \tag{21.74}$$

where M_S^2 indicates the intersection of the curves $M_1^2 = M_{s1}^2$ and $M_2^2 = M_{s2}^2$, defined below, and similarly for M_F^2. Hence, the slow, Alfvén and fast waves of linear MHD may be considered as shocks of infinitesimal strength. In this way, the threefold dynamics of linear MHD returns in the classification of the nonlinear shocks. The entropy condition (21.66) eliminates the possibility of a finite amplitude analogue of the non-propagating entropy wave of Fig. 5.10.

The upstream states of the shocks are divided into four groups by means of the three vertical lines $M_1^2 = M_{s1}^2 \equiv M_s^2(\vartheta_1, \beta_1) \le 1$, $M_1^2 = M_{A1}^2 \equiv 1$ and $M_1^2 = M_{f1}^2 \equiv M_f^2(\vartheta_1, \beta_1) \ge 1$. The downstream transitions are more involved (except for the Alfvén line $M_2^2 = M_{A2}^2 \equiv 1$), as they need to solve $M_2^2 = M_{s,f2}^2 \equiv M_{s,f}^2(\vartheta_2, \beta_2)$, with ϑ_2, β_2 in terms of $M_1^2, \vartheta_1, \beta_1$ from Eqs. (21.63), (21.64), (21.46). This yields a quartic for the *downstream magneto-sonic transition values:*

$$\begin{aligned} h \;\equiv\; & (M_2^2 - 1)^3 \Big\{ \gamma M_1^2 - (\gamma+1) M_2^2 + \gamma \bar{p}_1 \Big\} \\ & + \bar{B}_{t1}^2 \Big\{ M_2^2 (M_1^2 - 1)^2 - \tfrac{1}{2}\gamma (M_2^2 - 1)(M_1^2 - M_2^2)(M_1^2 + M_2^2 - 2) \Big\} \\ =\; & h(M_2^2; M_1^2, \bar{B}_{t1}^2, \bar{p}_1) = \tilde{h}(M_2^2; M_1^2, \vartheta_1, \beta_1) = 0\,, \end{aligned} \tag{21.75}$$

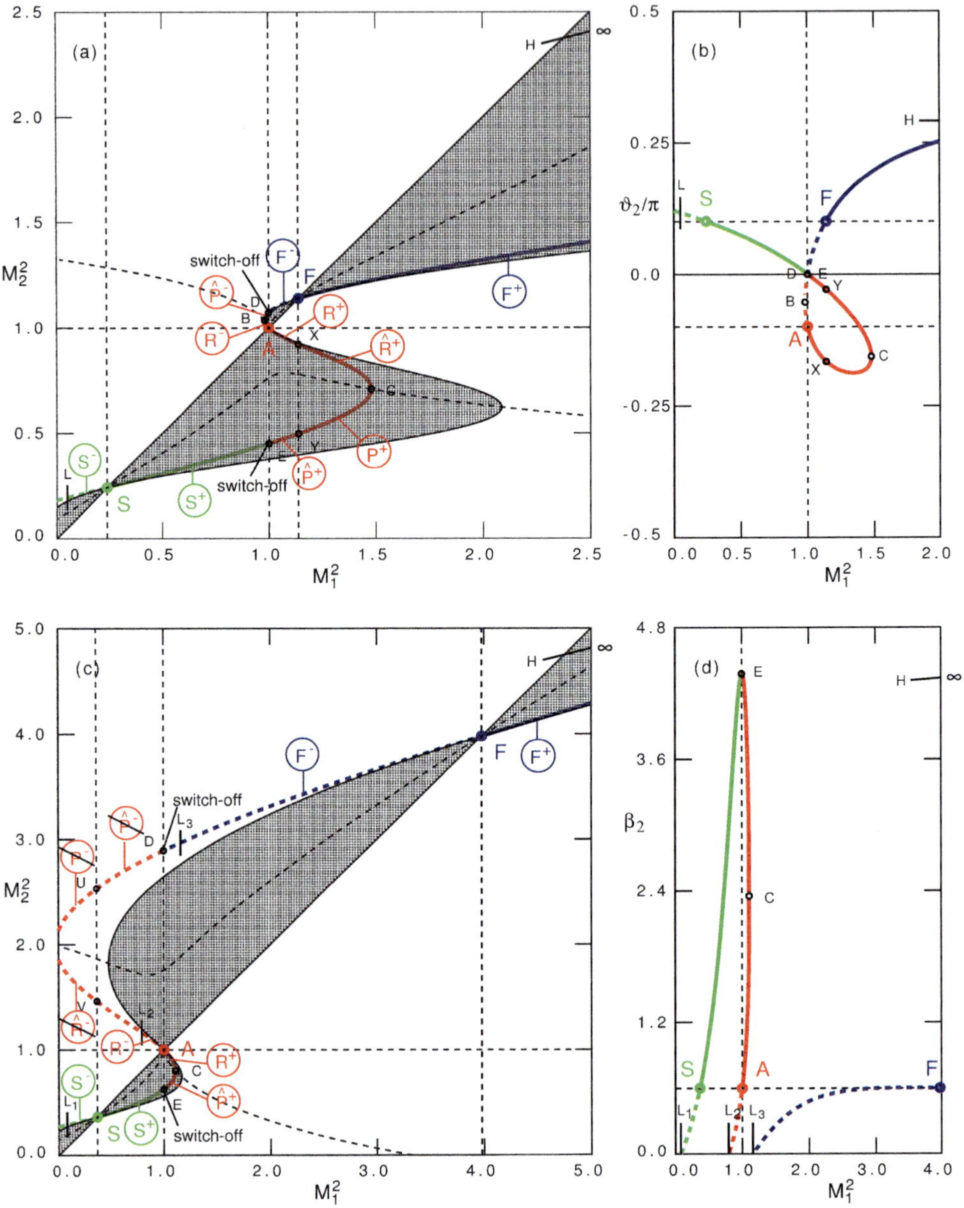

Fig. 21.10 (a) Jump curve and (b) downstream angle for $\vartheta_1 = 0.1\pi$, $\beta_1 = 0.3$ (parameter region I); jumps are unphysical to the left of point L on the jump curve. (c) Jump curve and (d) pressure for $\vartheta_1 = 0.3\pi$, $\beta_1 = 0.6$ (parameter region III); jumps are unphysical to the left of L_1 and between L_2 and L_3 on the jump curve.

determining the slow and fast transition curves labelled M_{s2}^2 and M_{f2}^2 in Fig. 21.8. One of the two curves touches the central Alfvén point $M_1^2 = M_2^2 = 1$ (viz. the fast curve for the parameters of Fig. 21.8) whereas the other one stays at a finite distance from it (viz. the slow curve, which leaves a tiny gap that is hardly visible). Together with the Alfvén line $M_2^2 = M_{A2}^2 \equiv 1$, this again yields three (dashed) curves separating the four kinds of downstream states.

The dashed transition curves divide the M_1^2–M_2^2 plane into pairs of upstream and downstream flow regimes, traditionally labelled from 1 to 4:

1 – *super-fast*, for $M^2 > M_f^2$ (i.e. $|v_n| > |v_{fn}|$);

2 – *sub-fast, super-Alfvénic*, for $M_A^2 < M^2 < M_f^2$ (i.e. $|v_{An}| < |v_n| < |v_{fn}|$);

3 – *super-slow, sub-Alfvénic*, for $M_s^2 < M^2 < M_A^2$ (i.e. $|v_{sn}| < |v_n| < |v_{An}|$);

4 – *sub-slow*, for $M^2 < M_s^2$ (i.e. $|v_n| < |v_{sn}|$).

Accordingly, the fast shocks shown in Fig. 21.8 are 1–2 (super-fast to sub-fast) transitions, the intermediate shocks are 2–3 (super-Alfvénic to super-slow) and 2–4 (super-Alfvénic to sub-slow) transitions, whereas the slow shocks are 3–4 (super-slow to sub-slow) transitions. For different parameter choices, the intermediate part of the shock curve (labelled A–X–C–Y–E now) protrudes into the super-fast regime 1, so that also 1–3 and 1–4 intermediate shocks appear (Fig. 21.10(a), parameter region I). Similarly, the entropy-forbidden part of the jump curve (labelled D–U–B–V–A, not all inside the frame) protrudes into the sub-slow regime 4, admitting 4–1 and 4–2 intermediate jumps (Fig. 21.10(b), region III).

All of these solutions were obtained from the distilled jump condition (21.65), which is a cubic equation in M_2^2. Extensive literature exists on the MHD jump conditions based on quite different representations with a cubic equation, see e.g. Refs. [408], [341] and [648]. One can show these to be equivalent, but there is surprisingly little overlap with the results presented here.

In conclusion, anticipating the terminology of the next section, we encounter the following succession of entropy-forbidden jumps ($-$) and permitted shocks ($+$) along the jump curves, separated by the indicated transition points:

$$\mathrm{F}_{12}^{+}, \mathrm{F}_{21}^{-}, \widehat{\mathrm{P}}_{31}^{-}, \left[\mathrm{P}_{41}^{-}, \widehat{\mathrm{R}}_{42}^{-},\right] \mathrm{R}_{32}^{-}, \mathrm{R}_{23}^{+}, \left\{\widehat{\mathrm{R}}_{13}^{+}, \mathrm{P}_{14}^{+},\right\} \widehat{\mathrm{P}}_{24}^{+}, \mathrm{S}_{34}^{+}, \mathrm{S}_{43}^{-}. \tag{21.76}$$
$$\begin{matrix}\mathrm{H} & \mathrm{F} & \mathrm{D} & \mathrm{U} & \mathrm{B} & \mathrm{V} & \mathrm{A} & \mathrm{X} & \mathrm{C} & \mathrm{Y} & \mathrm{E} & \mathrm{S} & \mathrm{L}\end{matrix}$$

The discontinuities in square brackets are missing in parameter regions I and II, and the ones in curly brackets are missing in parameter regions II and III.

21.3.2 Time reversal duality

The principle of time reversal duality has been formulated for gas dynamic shocks in Section 21.1.2. To apply it to MHD shocks, we need to be able to properly distinguish the different sub-classes of solutions, beyond what has been presented so far in the literature. To that end, we need to introduce a number of further sub-divisions by exploiting the different algebraic properties of the distilled jump relations.

First, since the distilled jump condition (21.68) is cubic in Δ_2, but linear in Δ_1, it is more easily solved from the inverse jump relation in terms of Δ_1:

$$\Delta_1 = \frac{(\gamma+1)\Delta_2^2 + \left[2 - \gamma\beta_1 - \gamma(1+\beta_1)\tan^2\vartheta_1\right]\Delta_2 - \tan^2\vartheta_1}{(\gamma-1)\Delta_2^2 + (2-\gamma)\tan^2\vartheta_1\Delta_2 + \tan^2\vartheta_1}\,\Delta_2\,. \tag{21.77}$$

The limits of this relation reveal the following peculiarities of the jump curve:

– Around the Alfvén point $M_1^2 = M_2^2 = 1$ (or $\Delta_1 = \Delta_2 = 0$), the jump curve is *retrograde*, i.e. M_1^2 increases when M_2^2 decreases:

$$|\Delta_1| \sim |\Delta_2| \ll 1 \quad \Rightarrow \quad \Delta_1 \approx -\Delta_2 \,. \tag{21.78}$$

– For large M_1^2 and M_2^2, the jump curve has a *constant slope* ($= 4$ for $\gamma = 5/3$):

$$\Delta_1 \sim \Delta_2 \gg 1 \quad \Rightarrow \quad \Delta_1 \approx \frac{\gamma+1}{\gamma-1}\,\Delta_2 \,. \tag{21.79}$$

– For $\vartheta_1 = 0$, when $\vartheta_2 = 0$ as well, the jump curve degenerates into a straight line:

$$(\gamma-1)\Delta_1 - (\gamma+1)\Delta_2 = 2 - \gamma\beta_1 \,, \tag{21.80}$$

corresponding to the *limit of parallel shocks* (Figs. 21.7(d)). This expression is transformed into the HD expression (21.25) by converting the Alfvén Mach number into the HD Mach number through Eq. (21.61). This implies that the striking difference of decreasing HD shock curves (Fig. 21.4) and increasing MHD shock curves (Fig. 21.8) is not deep, but just result from the different definitions of the Mach numbers.

– The expressions for *perpendicular shocks* may also be obtained from Eq. (21.77) in the limit $\vartheta_1 \to \pm\pi/2$. This is left as an exercise for the reader.

– The denominator of the expression (21.77) never vanishes for physical values of the parameters (since $\Delta_2 \geq -1$), except when both $\vartheta_1 = 0$ and $\Delta_2 = 0$. The numerator then also vanishes, so that the limits $\vartheta_1 \to 0$ and $\Delta_2 \to 0$ may be taken such that $\vartheta_2 \neq 0$ according to Eq. (21.69). This yields *switch-on shocks* (Figs. 21.7(e)).

Similarly, the quartic (21.75) for the magneto-sonic transition values $M_{\mathrm{f,s2}}^2(M_1^2)$ is more easily solved from the inverse relation $\Delta_{\mathrm{f,s1}}(\Delta_2)$, which is a quadratic:

$$\Delta_{\mathrm{f,s1}} = -s \pm \sqrt{s^2 - t}\,, \qquad s \equiv \frac{\gamma\Delta_2^3}{[\,2 + (2-\gamma)\Delta_2\,]\tan^2\vartheta_1}\,, \tag{21.81}$$
$$t \equiv -\frac{[\,2 - \gamma\beta_1 - \gamma(1+\beta_1)\tan^2\vartheta_1 + 2(\gamma+1)\Delta_2\,]\,\Delta_2^3}{[\,2 + (2-\gamma)\Delta_2\,]\tan^2\vartheta_1}\,.$$

The slow and fast transition curves, labelled M_{s2}^2 and M_{f2}^2 in Fig. 21.8, are obtained by inverting these solutions again (a trivial plotting operation), where the two roots labelled $\pm$ correspond to the $\Delta_1 \geq 0$ and $\Delta_2 \leq 0$ pieces of each curve.

We can now study the consequences of the S-shape of the distilled jump curve, associated with the possibility of multiple solutions of the cubic $\Delta_2 = \Delta_2(\Delta_1)$, where intermediate jumps/shocks occur together with slow or fast jumps/shocks for the same values of the upstream parameters $\Delta_1, \vartheta_1, \beta_1$. The indeterminacy of the downstream state in the range of intermediate shocks has frequently been associated with lack of evolutionarity of these shocks and, consequently, abandoned. We will come back to this in Section 21.3.4, but for now we note that discarding these solutions, unfortunately, has seriously hampered the study of the structure of the MHD shock relations as presented here. This aspect of evolutionarity is extensively discussed by Kennel, Blandford and Coppi [341].

Since the original cubic (21.68) is a multi-valued function of Δ_1, but the inverse jump relation (21.77) is a single-valued function of Δ_2, the jump curve from the point H to L depicted in

Fig. 21.8 is monotonically decreasing in Δ_2, but either decreasing (along H–F–D–B and C–E–S–L) or increasing (along B–A–C) in Δ_1. We call the corresponding intermediate jumps in these ranges *prograde* (indicated by P) when Δ_1 and Δ_2 decrease or increase together, and *retrograde* (indicated by R) when they have opposite monotonicity. The distinctions of *quasi-prograde* (indicated by $\widehat{\mathrm{P}}$) and *quasi-retrograde* (indicated by $\widehat{\mathrm{R}}$) encountered in Figs. 21.8–21.10 are explained below, when the time reversed problem is considered.

To complete the classification, we introduce ***eleven transition points***, delimiting the twelve different kinds of MHD jumps and shocks that (may) occur on the jump curve. They are all determined by the following simple algebraic equations, only involving the parameters ϑ_1 and β_1:

– *Alfvén, fast and slow points* A, F *and* S,

$$\Delta_{\rm A} = 0\,, \qquad \Delta_{\rm F,S} = -a \pm \sqrt{a^2 - b}\,, \tag{21.82}$$
$$a \equiv \tfrac{1}{2}(1 - \tfrac{1}{2}\gamma\beta_1) - \tfrac{1}{2}(1 + \tfrac{1}{2}\gamma\beta_1)\tan^2\vartheta_1\,, \quad b \equiv -\tan^2\vartheta_1\,.$$

– *Intermediate demarcation points* D *and* E,

$$\Delta_{\rm 1D,E} = 0\,, \qquad \Delta_{\rm 2D,E} = -c \pm \sqrt{c^2 - d}\,, \tag{21.83}$$
$$c \equiv \frac{1 - \frac{1}{2}\gamma\beta_1 - \frac{1}{2}\gamma(1+\beta_1)\tan^2\vartheta_1}{\gamma + 1}\,, \quad d \equiv -\frac{\tan^2\vartheta_1}{\gamma+1}\,.$$

– *Intermediate turning points* B *and* C, *determined by a quartic for* $\Delta_{\rm 2B,C}$,

$$\Delta_2^4 + e\Delta_2^3 + f\Delta_2^2 + g\Delta_2 + h = 0\,, \quad e \equiv \frac{2(2-\gamma)}{\gamma - 1}\tan^2\vartheta_1\,, \tag{21.84}$$
$$f \equiv \frac{4\gamma + 2 + (2-\gamma)\,[\,2 - \gamma\beta_1 - \gamma(1+\beta_1)\tan^2\vartheta_1]}{(\gamma+1)(\gamma-1)}\tan^2\vartheta_1\,,$$
$$g \equiv \frac{2\,[\,2 - \gamma\beta_1 - \gamma(1+\beta_1)\tan^2\vartheta_1]}{(\gamma+1)(\gamma-1)}\tan^2\vartheta_1\,, \quad h \equiv -\frac{\tan^4\vartheta_1}{(\gamma+1)(\gamma-1)}\,,$$

with corresponding $\Delta_{\rm 1B,C}$ determined by the inverse jump condition (21.77).

– *Region* I *transition points* X *and* Y, *and region* III *transition points* U *and* V,

$$\Delta_{\rm 1X,Y} = \Delta_{\rm F}\,, \qquad \Delta_{\rm 2X,Y} = -u \pm \sqrt{u^2 - w}\,, \tag{21.85}$$
$$\Delta_{\rm 1U,V} = \Delta_{\rm S}\,, \qquad \Delta_{\rm 2U,V} = -v \pm \sqrt{v^2 - w}\,, \tag{21.86}$$
$$u \equiv \frac{\Delta_{\rm F} + 1 - \frac{1}{2}\gamma\beta_1 - \frac{1}{2}(1+\beta_1)\tan^2\vartheta_1}{\gamma+1}\,,$$
$$v \equiv \frac{\Delta_{\rm S} + 1 - \frac{1}{2}\gamma\beta_1 - \frac{1}{2}(1+\beta_1)\tan^2\vartheta_1}{\gamma+1}\,, \quad w \equiv \frac{\tan^2\vartheta_1}{\gamma+1}\,.$$

This enables the construction of all the points of the sequence (21.76) that appear in our new terminology for the different MHD jumps and shocks.

We have introduced the parameter regions I, II and III, illustrated in Fig. 21.11, to describe the different dependences of the intermediate discontinuities on the upstream parameters ϑ_1 and β_1. They can be determined before the shock problem is solved. The defining criteria for the parameter

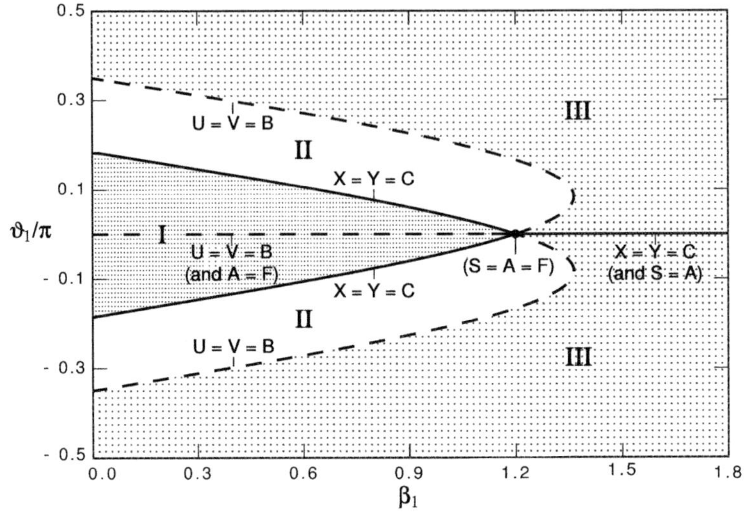

Fig. 21.11 Regions I, II and III of the β_1–ϑ_1 parameter plane determining the shapes of the jump curve associated with the possible intermediate discontinuities. Parameter regions I and II apply for $\Delta_1 \geq 0$, regions II and III apply for $\Delta_1 \leq 0$.

regions I and III follow from the discriminants of the above expressions (21.85) and (21.86):

$$D_{\rm I} \equiv u^2 - w \geq 0\,, \qquad D_{\rm III} \equiv v^2 - w \geq 0\,, \tag{21.87}$$

whereas parameter region II occurs when neither of these criteria is satisfied. The parameter region I refers to the presence ($D_{\rm I} \geq 0$) or absence ($D_{\rm I} < 0$) of the transition points C, X and Y of the intermediate shocks to the right ($\Delta_1 \geq 0$) of the Alfvén point. For $D_{\rm I} = 0$, the points C, X and Y coalesce, so that for $D_{\rm I} < 0$, the intermediate shocks $\widehat{\rm R}^+$ and $\rm P^+$ (shown in Fig. 21.10(a)) disappear. Parameter region III refers to the presence ($D_{\rm III} \geq 0$) or absence ($D_{\rm III} < 0$) of the transition points B, U and V of the intermediate jumps to the left ($\Delta_1 \leq 0$) of the Alfvén point. For $D_{\rm III} = 0$, the points B, U and V coalesce, so that for $D_{\rm III} < 0$, the intermediate jumps $\widehat{\rm R}^-$ and $\rm P^-$ (shown in Fig. 21.10(c), although unphysical for the particular parameter values chosen) disappear. Consequently, in parameter region II (Fig. 21.8) none of the labels X, Y and U, V appear, and only the intermediate shocks $\rm R^+$ and $\widehat{\rm P}^+$ and jumps $\rm R^-$ and $\widehat{\rm P}^-$ remain.

A further explanatory note on the symbols occurring in Figs. 21.8–21.10: The points H and L refer to the high and low extremes of the jump curve. The label H indicates either the limiting value of the asymptote (horizontal dash) or its slope (inclined dash with ∞ on the right) for $\Delta_1 \to \infty$, exploiting Eq. (21.79). The low side is either limited by $\Delta_1 = -1$ or by L, corresponding to vanishing numerator of the expression (21.70) for β_2. This may also happen in the middle of the jump curve (e.g. in-between $\rm L_2$ and $\rm L_3$ in Fig. 21.10(c), where three of the intermediate jumps have been struck through), so that the parts corresponding to $\beta_2 < 0$ have to be eliminated. Unfortunately, the criterion $\beta_2 \geq 0$ can only be applied after the shock problem has been solved, not before.

We now consider the *time reversed* problem in MHD, analogous to that of gas dynamic jumps and shocks (Section 21.1.2). Reversing the direction of time, i.e. reversing the direction of the flow, does not alter the distilled ideal MHD jump problem (21.68)–(21.70) since the Alfvén Mach number

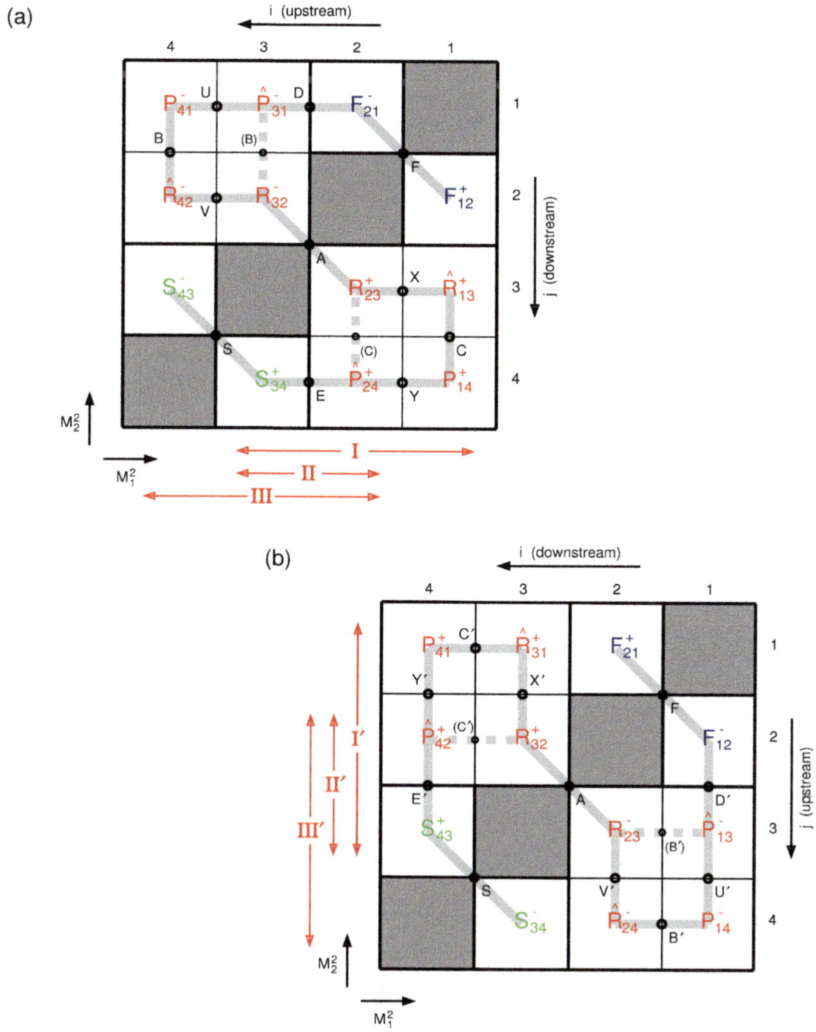

Fig. 21.12 Time reversal duality and classification of discontinuities in MHD: (a) entropy-permitted shocks and forbidden jumps joined by the forward jump curve, (b) entropy-permitted shocks and forbidden jumps joined by the inverse jump curve. Jump curves (grey) are schematic, dashed short cuts may exclude jumps.

only occurs squared. This is why jumps and shocks could be treated on an equal footing. Jumps and shocks are only discriminated when the additional entropy condition (21.71) is considered, i.e. when downstream entropy increase is used to fix the arrow of time for the enlarged dissipative system. For the complete distilled MHD shock problem (21.68)–(21.71), the shocks survive and the jumps should be discarded.

An alternative point of view is to interpret the time reversal operation as just interchanging the roles of upstream and downstream states. The different jumps and shocks obtained in Figs. 21.8–21.10 then exchange position. Instead of jump curves $M_2^2 = M_2^2(M_1^2; \vartheta_1, \beta_1)$, one obtains jump curves $M_1^2 = M_1^2(M_2^2; \vartheta_2, \beta_2)$. These remain S-shaped, but they connect the different jumps and shocks in a completely different order, as illustrated by Fig. 21.12. For example, the intermediate shock R_{23}^+ of Fig. 21.12(a) becomes the intermediate jump R_{23}^- of Fig. 21.12(b). These may be considered as the solutions of two different physical problems, but they may also be considered as solutions of the same problem described by two different representations. This permits us to classify the MHD shocks in a new way by noting that some of the intermediate (IM) jumps/shocks remain retrograde ($\mathrm{R}^\pm \to \mathrm{R}^\mp$) and prograde ($\mathrm{P}^\pm \to \mathrm{P}^\mp$) with respect to M_1^2–M_2^2 dependences, but others get the opposite sense of monotonicity that we will call *quasi-retrograde* ($\widehat{\mathrm{P}}^\pm \to \widehat{\mathrm{R}}^\mp$) and *quasi-prograde* ($\widehat{\mathrm{R}}^\pm \to \widehat{\mathrm{P}}^\mp$).

Time reversal duality between entropy-forbidden jumps and entropy-permitted shocks in MHD is now expressed by the following correspondence:

$$\left\{\begin{array}{rcl} \Delta_2 &=& \Delta_2^{\alpha\pm}(\Delta_1, \vartheta_1, \beta_1) \\ \vartheta_2 &=& \vartheta_2^{\alpha\pm}(\Delta_1, \vartheta_1, \beta_1) \\ \beta_2 &=& \beta_2^{\alpha\pm}(\Delta_1, \vartheta_1, \beta_1) \end{array}\right. \iff \left\{\begin{array}{rcl} \Delta_1 &=& \Delta_1^{\alpha\mp}(\Delta_2, \vartheta_2, \beta_2) \\ \vartheta_1 &=& \vartheta_1^{\alpha\mp}(\Delta_2, \vartheta_2, \beta_2) \\ \beta_1 &=& \beta_1^{\alpha\mp}(\Delta_2, \vartheta_2, \beta_2) \end{array}\right. , \tag{21.88}$$

$$\alpha^+ \equiv (\mathrm{F}^+, \widehat{\mathrm{R}}^+, \mathrm{R}^+, \mathrm{P}^+, \widehat{\mathrm{P}}^+, \mathrm{S}^+)\,, \quad \alpha^- \equiv (\mathrm{F}^-, \widehat{\mathrm{P}}^-, \mathrm{R}^-, \mathrm{P}^-, \widehat{\mathrm{R}}^-, \mathrm{S}^-)\,.$$

With the expressions (21.68)–(21.70) of the distilled jump conditions, this scheme not only yields a prescription for the construction of shocks at each point of a transonic plasma, but it also permits us to uniquely classify the MHD jumps and shocks by means of the monotonicity properties of the jump curves.

The following unique characterization of the *six dual pairs* of entropy-allowed MHD shocks (bold italic,$^+$) and entropy-forbidden jumps (italic,$^-$) emerges:

- ***Fast shocks*** F_{12}^+: $\Delta_\mathrm{F} \le \Delta_1 < \infty\,, \quad \Delta_\mathrm{F} \le \Delta_2 < \infty\,;$
- *Fast jumps* F_{21}^-: $0 \le \Delta_1 \le \Delta_\mathrm{F}\,, \quad \Delta_{2\mathrm{D}} \le \Delta_2 \le \Delta_\mathrm{F}\,;$
- ***Quasi-retrograde IM shocks*** $\widehat{\mathrm{R}}_{13}^+$: $\Delta_{1\mathrm{C}} \ge \Delta_1 > \Delta_{1\mathrm{X/C}}\,, \quad \Delta_{2\mathrm{C}} \le \Delta_2 < \Delta_{2\mathrm{X/C}}\,;$
- *Quasi-prograde IM jumps* $\widehat{\mathrm{P}}_{31}^-$: $\Delta_{1\mathrm{U/B}} \le \Delta_1 \le 0\,, \quad \Delta_{2\mathrm{U/B}} \le \Delta_2 \le \Delta_{2\mathrm{D}}\,;$
- ***Retrograde IM shocks*** R_{23}^+: $\Delta_{1\mathrm{X/C}} \ge \Delta_1 \ge 0\,, \quad \Delta_{2\mathrm{X/C}} \le \Delta_2 \le 0\,;$
- *Retrograde IM jumps* R_{32}^-: $0 \ge \Delta_1 \ge \Delta_{1\mathrm{V/B}}\,, \quad 0 \le \Delta_2 \le \Delta_{2\mathrm{V/B}}\,;$
- ***Prograde IM shocks*** P_{14}^+: $\Delta_{1\mathrm{Y/C}} < \Delta_1 \le \Delta_{1\mathrm{C}}\,, \quad \Delta_{2\mathrm{Y/C}} < \Delta_2 \le \Delta_{2\mathrm{C}}\,;$
- *Prograde IM jumps* P_{41}^-: $\Delta_{1\mathrm{B}} \le \Delta_1 < \Delta_{1\mathrm{U/B}}\,, \quad \Delta_{2\mathrm{B}} \le \Delta_2 < \Delta_{2\mathrm{U/B}}\,;$
- ***Quasi-prograde IM shocks*** $\widehat{\mathrm{P}}_{24}^+$: $0 \le \Delta_1 \le \Delta_{1\mathrm{Y/C}}\,, \quad \Delta_{2\mathrm{E}} \le \Delta_2 \le \Delta_{2\mathrm{Y/C}}\,;$
- *Quasi-retrograde IM jumps* $\widehat{\mathrm{R}}_{42}^-$: $\Delta_{1\mathrm{V/B}} > \Delta_1 \ge \Delta_{1\mathrm{B}}\,, \quad \Delta_{2\mathrm{V/B}} < \Delta_2 \le \Delta_{2\mathrm{B}}\,;$
- ***Slow shocks*** S_{34}^+: $\Delta_\mathrm{S} \le \Delta_1 \le 0\,, \quad \Delta_\mathrm{S} \le \Delta_2 \le \Delta_{2\mathrm{E}}\,;$
- *Slow jumps* S_{43}^-: $-1 \le \Delta_1 \le \Delta_\mathrm{S}\,, \quad -1 \le \Delta_2 \le \Delta_\mathrm{S}\,.$ (21.89)

For the sake of comparison with the standard (but incomplete) terminology, the notation in terms of the $\alpha^\pm$'s is extended with the redundant indices ij. Due to the peculiar properties of the intermediate jumps and shocks, the jump curve folds over in the Alfvénic range. As a result, the indicated parameter ranges from the relations (21.82)–(21.86) overlap in Δ_1 but not in Δ_2, so that any shock or jump is uniquely described by the value of M_2^2. Also note a sub-structure: the 'core' IM discontinuities $\mathrm{R}^\pm$, $\mathrm{P}^\pm$ (on the anti-diagonals of Figs. 21.12) are separated from the fast discontinuities $\mathrm{F}^\pm$ by the quasi-retrograde or prograde IM discontinuities $\widehat{\mathrm{R}}^+$, $\widehat{\mathrm{P}}^-$, and from the slow ones $\mathrm{S}^\pm$ by the discontinuities $\widehat{\mathrm{P}}^+$, $\widehat{\mathrm{R}}^-$.

21.3.3 Angular dependence of MHD shocks*

The problems of parameter reduction by the distilled jump conditions and of the related classification of MHD shocks are completely solved now. It remains to show how the downstream velocities of the different discontinuities depend on the angle of incidence ϑ_1 of the upstream flow velocity. This will complement the pictorial representations of Fig. 21.7 with a quantitative description. Since the three values M_S, M_A ($\equiv 1$) and M_F of the normal Alfvén Mach number corresponding to the three linear MHD waves play a central role in the classification of shocks, a generalization of the Friedrichs diagram (Figs. 5.4 and 5.9) would appear to give the most logical angular representation. In the present context, this becomes a superposition of the three pairs of equal upstream and downstream velocities at $M = M_\mathrm{S}$, M_A and M_F for all angles $|\vartheta_1| = |\vartheta_2| \leq \pi/2$, as represented in Fig. 21.13. Two caveats are in order here.

First, the shock strength, defined as $\sigma \equiv 1 - M_2^2/M_1^2$, and leading to the expression (21.74) for 'weak' shocks, is actually misleading with respect to the intermediate shocks and their relationship with Alfvén waves. In the limit $\sigma \to 0$, when both $M_1^2 \to 1$ and $M_2^2 \to 1$ along the jump curve, but from opposite sides of the Alfvén point according to Eq. (21.78), the retrograde intermediate jumps R^- and shocks R^+ coalesce, but they do not become weak (hence the quotation marks above). In general (except for $\vartheta_1 \to 0$), a finite jump of the directions of the magnetic field and velocity remains, since $\vartheta_1 = -\vartheta_2$. Hence, *intermediate shocks almost never become weak*, and, in the limit $\sigma \to 0$, *they do not become Alfvén waves*. Instead, Alfvén waves proper emerge from the rotational discontinuities (appropriately called Alfvén discontinuities) in the limit $\varphi \to 0$ or π. Whereas the central Alfvén point $M = M_\mathrm{A} = 1$ and the intermediate jumps and shocks around it clearly produce the shape of the jump curve, intermediate discontinuities remain magneto-acoustic phenomena, with jumps restricted to the plane through the magnetic field and the normal, whereas Alfvén discontinuities have jumps in a plane perpendicular to that. The two kinds of discontinuity may be superposed at a particular point in time, but in the subsequent nonlinear dynamics the orientations of those planes will change and a complicated, non-constant, mix of Alfvén-type and magneto-sonic type of discontinuity will emerge.

Second, the Friedrichs diagram is actually not an adequate representation for the angular dependence of MHD shocks since it presents the phase speeds of the different MHD waves in the direction of the normal $\mathbf{n}$. According to the expressions given in Eq. (21.2), this yields the well-known feature that the Alfvén and slow waves do not propagate in the perpendicular directions $\vartheta = \pm\pi/2$. Exploiting such a representation for strong shocks is undesirable since it requires separate treatment for the perpendicular shocks. In fact, the representation of Fig. 21.13 is a superposition of three plots, corresponding to the three special values M_A^2 and $M_\mathrm{f,s}^2$ of the squared Alfvén Mach number given in (21.73). The latter two cover a range of values when ϑ is varied. This is not what we want. To study the effect of the angle of incidence of a shock, we wish to keep the value of the Alfvén Mach number M_1 (and also β_1) fixed, and just vary ϑ_1. Note that, in contrast to the phase speeds (21.2),

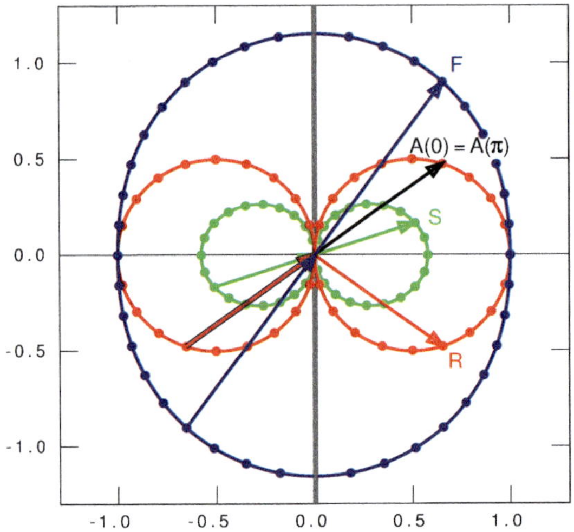

Fig. 21.13 Friedrichs-type diagram: 'weak' shocks for $\beta_1 = 0.4$. In the weak shock limit, slow and fast shocks become magneto-acoustic waves, but intermediate shocks almost never become weak since $\vartheta_2 = -\vartheta_1$. Instead, Alfvén waves proper emerge from the rotational discontinuity in the limit $\varphi \to 0$ or π. Dots correspond to a discrete set of angles $\vartheta_{1k} = \pm(k/20)\,\pi$, with $k = 0, 1, \ldots 10$. For distinction, the velocity vectors are shown at $k = 2$ for slow shocks, at $k = 4$ for intermediate shocks and rotational discontinuities, and at $k = 6$ for fast shocks.

the values for the Alfvén Mach numbers (21.73) do not vanish for $\vartheta = \pm\pi/2$. Clearly, dividing by $\cos\vartheta$ here makes a crucial difference.

Hence, we return to the expressions (21.62) for the velocities $\bar{\mathbf{v}}_1^*$ and $\bar{\mathbf{v}}_2^*$ in the de Hoffmann–Teller frame. For fixed M_1^2, but varying ϑ_1, the expression for $\bar{\mathbf{v}}_1^*$ yields a straight line in a polar velocity plot so that perpendicular shocks are properly described for $\vartheta_1 \to \pm\pi/2$. However, in that limit, $|\bar{\mathbf{v}}_1^*| \to \infty$. This is remedied by exploiting a powerful property of the transformation that simplifies the expressions for the velocities even further. Since $\mathbf{v}_i^*$ and $\mathbf{B}_i$ are parallel, the squared upstream and downstream normal Alfvén Mach numbers may be written as

$$M_i^2 \equiv \rho_i v_{\mathrm{n}i}^2 / B_{\mathrm{n}}^2 = \rho_i v_i^{*2} / B_i^2 \qquad (i = 1, 2)\,. \tag{21.90}$$

In other words: *for the velocities in de Hoffmann–Teller frame, the normal Alfvén Mach number coincides with the Alfvén Mach number itself!* This means that the Alfvén Mach number M can be exploited as a measure for the total velocity, not just for the normal component as implied by the expressions (21.62).

Recall that the bars on the velocities $\bar{\mathbf{v}}_1^*$ and $\bar{\mathbf{v}}_2^*$ come from the normalization (21.45) that we exploited so far, where the constancy of the factors ρv_{n} and B_{n} across the surface of discontinuity was utilized to symmetrize the jump conditions as far as possible with respect to upstream and downstream values of the parameters. This having been accomplished, we may now return to the real world, so to speak, and normalize the velocities in a more practical way, based on the upstream value $v_{\mathrm{A}1} = B_1/\sqrt{\rho_1}$ of the Alfvén speed:

$$\bar{\mathbf{v}}_1^* \equiv \frac{\rho|v_{\mathrm{n}}|}{B_{\mathrm{n}}^2}\,\mathbf{v}_1^* = \frac{M_1}{\cos\vartheta_1}\,\widetilde{\mathbf{v}}_1^*\,, \qquad \bar{\mathbf{v}}_2^* \equiv \frac{\rho|v_{\mathrm{n}}|}{B_{\mathrm{n}}^2}\,\mathbf{v}_2^* = \frac{M_2}{\lambda\cos\vartheta_2}\widetilde{\mathbf{v}}_2^*\,, \tag{21.91}$$

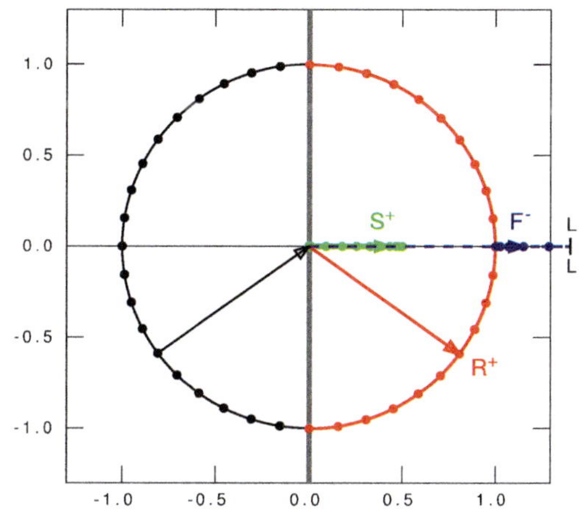

Fig. 21.14 Polar plot of the velocities $\mathbf{v}^*_{1,2}(\vartheta_1)$ in the de Hoffmann–Teller frame for $M_1 = 1$ and $\beta_1 = 0.4$. Vectors for the retrograde IM shock, the switch-off slow shock and the switch-off fast jump are shown at $\vartheta_1 = 0.2\pi$ ($k = 4$). The fast jump becomes unphysical ($\beta_2 < 0$) beyond L, corresponding to $\vartheta_1 > 0.279\pi$.

where $\widetilde{\mathbf{v}}^*_i \equiv \mathbf{v}^*_i / v_{\rm A1}$ and the factor λ converts the downstream velocity into the upstream normalization:

$$\lambda \equiv \frac{v_{\rm A2}}{v_{\rm A1}} \equiv \frac{B_2}{B_1}\sqrt{\frac{\rho_1}{\rho_2}} = \frac{\cos\vartheta_1}{\cos\vartheta_2}\,\frac{M_2}{M_1}\,. \tag{21.92}$$

Dropping the tildes, as usual, this yields the following asymmetric expressions for the upstream and downstream velocities in the de Hoffmann–Teller frame:

$$\left\{ \begin{array}{l} v^*_{\rm n1} = -M_1 \cos\vartheta_1 \\ v^*_{\rm t1} = M_1 \sin\vartheta_1 \end{array} \right. \quad \Rightarrow \quad \left\{ \begin{array}{l} v^*_{\rm n2} = -\lambda M_2 \cos\vartheta_2 \\ v^*_{\rm t2} = \lambda M_2 \sin\vartheta_2 \end{array} \right. \,. \tag{21.93}$$

Now, the upstream Alfvén Mach number M_1 and the effective downstream Alfvén Mach number λM_2 are just the normalized *total* upstream and downstream velocities. For fixed M_1 and β_1, the locus of the end points of the velocity $\mathbf{v}^*_1(\vartheta_1)$ is a half-circle and the corresponding locus for $\mathbf{v}^*_2(\vartheta_1)$ may be computed from the obtained explicit solutions of the distilled jump problem (21.68)–(21.70). Hence, with this normalization, a much more effective representation of the polar plots is obtained than that resulting from either the Friedrichs representation or the one given by the original expressions (21.62). This is illustrated by Fig. 21.14, which shows the velocities of the three jumps/shocks that occur at $M_1 = 1$.

Since the Alfvén point A of the jump curve of Fig. 21.8 is a fixed point of the transformation $M_1 \to M_2$ ($M_1 = M_2$, irrespective of the values of ϑ_1 and β_1), the absolute magnitude of the velocity of the retrograde IM shocks R^+ ($M_1 \downarrow 1$) is the same upstream and downstream. Hence, the polar plot for R^+ is a circle in Fig. 21.14. With respect to the angular dependence $\vartheta_2 = \vartheta_2(\vartheta_1)$ for the retrograde IM shocks, close to the point A, it follows from Eqs. (21.78) and Eq. (21.69) that

$$\vartheta_2 \approx -\vartheta_1\,, \tag{21.94}$$

thus producing a second reason for the terminology *retrograde*. At A, this becomes $\vartheta_2 = -\vartheta_1$

so that the directions of the upstream and downstream velocities for the $M = 1$ retrograde IM shocks on the circle of Fig. 21.14 continuously change from $\vartheta_1 = \vartheta_2 = 0$ (parallel shock) to $\vartheta_1 = -\vartheta_2 = \pi/2$ (perpendicular shock). This shows, once more, the central role of the intermediate shocks for the description of MHD discontinuities. In contrast, for the points D and E of Fig. 21.8, which demarcate the transitions from intermediate to fast and slow jumps or shocks, the downstream Mach number M_2 depends on ϑ_1, so that the magnitudes of the downstream velocities vary but, since $M_1 = 1$, the downstream angle stays $\vartheta_2 = 0$. Hence, these points correspond to *switch-off* fast–IM jumps F^- ($M_1 \uparrow 1$) and *switch-off* slow–IM shocks S^+ ($M_1 \downarrow 1$) for all values of the upstream angle ϑ_1. By the principle of time reversal duality, corresponding *switch-on* fast–IM shocks F^+ and *switch-on* slow–IM jumps S^- may be obtained from these solutions by just interchanging the upstream and downstream values of the parameters.

We now complete the description of the angular dependence of MHD shocks by discussing the complex behaviour for sub- and super-Alfvénic values of M_1, where we have selected just four representative cases in Fig. 21.15. These pictures should be considered as mere illustrations of the different kinds of shocks since the actual solution of the distilled MHD shock problem has already been obtained in Section 21.3.2. Also note that, for proper interpretation of the polar plots of Figs. 21.14 and 21.15, one would need the corresponding dependence of $\beta_2(\vartheta_1)$, which is not shown, with possible transitions to negative values. For example, the switch-off fast jumps F^- of Fig. 21.14 only exist for incident angles $0 \leq \vartheta_1 \leq 0.279\pi$. For larger angles they become unphysical since $\beta_2 < 0$: there is simply no discontinuity possible for those values of the upstream parameters.

In Figs. 21.15(a) and (b), the polar velocity plots are shown for *sub-Alfvénic* values of M_1. For those values, parameter region II (jump curves as in Fig. 21.8) applies for smaller ϑ_1 and parameter region III (jump curves as in Fig. 21.10(c)) applies for larger angles. The resulting polar plot of slow jumps and shocks for $M_1 = 0.55$ is shown in Fig. 21.15(a). Since M_1 is far below the Alfvénic value 1, the IM jumps (which would have had much larger values of M_2 than the slow discontinuities) are eliminated because they are all unphysical. The remaining slow discontinuities are S^- jumps for $\vartheta_1 < 0.161\pi$ and S^+ shocks for $\vartheta_1 \geq 0.161\pi$. Hence, in contrast to the Friedrichs diagram, weak slow shocks now only occur for the single angle $\vartheta_1 = \vartheta_2 = 0.161\pi$, separating slow jumps from slow shocks. The polar plot for a near-Alfvénic value, $M_1 = 0.95$, is shown in Fig. 21.15(b). The slow jumps S^- have disappeared and the slow shocks S^+ are focused close to the parallel direction (anticipating the switch-off behaviour of Fig. 21.14). Since M_1 is close to 1, the IM jumps have returned: retrograde IM jumps R^- and quasi-prograde IM jumps $\widehat{P}^-$ appear for $\vartheta_1 \geq 0.178\pi$. Starting from the turning point B (where $\vartheta_1 = 0.178$ and $\vartheta_2 = -0.111\pi$), for increasing upstream angle ϑ_1, the downstream angle ϑ_2 monotonically decreases towards $\vartheta_2 = -\pi/2$ (perpendicular jump) for R^-, whereas ϑ_1 monotonically increases towards $\vartheta_2 = -0.037\pi$ (point L, at $\vartheta_1 = 0.263\pi$) for P^-, where this jump becomes unphysical.

In Figs. 21.15(c) and (d), the polar plots are shown for *super-Alfvénic* values of M_1. Now, parameter region I (jump curves as in Fig. 21.10(a)) applies for smaller ϑ_1 and parameter region II applies for larger angles. In the polar plot for the near-Alfvénic value $M_1 = 1.05$ shown in Fig. 21.15(c), the IM discontinuities have become genuine shocks. For increasing upstream angle ϑ_1, retrograde R^+ shocks switch-on at $\vartheta_1 = 0$ (where $\vartheta_2 = -0.110\pi$), while ϑ_2 monotonically decreases towards $-\pi/2$ (perpendicular R^+ shock). Quasi-prograde $\widehat{P}^+$ shocks also appear for all upstream angles, but their value of ϑ_2 monotonically decreases from $\vartheta_2 = 0$ towards $-\pi/2$ (perpendicu-

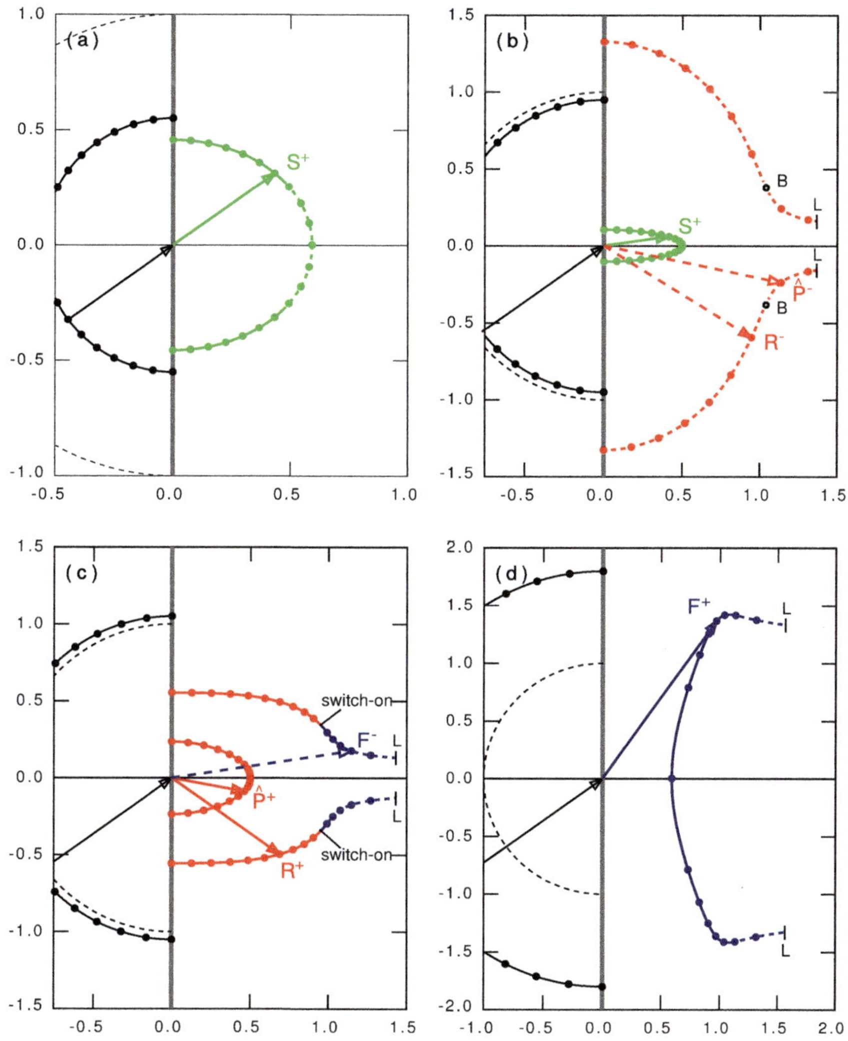

Fig. 21.15 Polar plot of the velocities $\mathbf{v}^*_{1,2}(\vartheta_1)$ in the de Hoffmann–Teller frame for $\beta_1 = 0.4$ and increasing Alfvén Mach number: (a) $M_1 = 0.55$, (b) $M_1 = 0.95$, (c) $M_1 = 1.05$, (d) $M_1 = 1.8$. Velocity vectors are shown at $\vartheta_1 = 0.2\pi$ ($k = 4$). The dashed circle on the upstream side indicates the Alfvén speed, $M_1 = 1$.

lar $\widehat{\mathrm{P}}^+$ shock). For this near-Alfvénic value of M_1, fast F^+ shocks only occur for a narrow range of ϑ_1: they switch-on at $\vartheta_1 = 0$ (where $\vartheta_2 = 0.110\pi$), their value of ϑ_2 monotonically decreases to $\vartheta_1 = \vartheta_2 = 0.081\pi$, where they turn into fast F^- jumps, which become unphysical at $\vartheta_1 = 0.292\pi$ ($\vartheta_2 = 0.028\pi$). For the large super-Alfvénic value $M_1 = 1.8$ illustrated in Fig. 21.15(d), only fast discontinuities survive. Fast F^+ shocks occur for the rather wide range $0 \le \vartheta_1 \le 0.289\pi$ and they turn into fast F^- jumps at $\vartheta_1 = 0.289\pi$ (where $\vartheta_2 = 0.288\pi$), which become unphysical beyond $\vartheta_1 = 0.386\pi$ (where $\vartheta_2 = 0.224\pi$).

A final cautionary remark on the angular dependence $\vartheta_2(\vartheta_1)$ of the polar plots is in order.

For fixed Δ_1 and β_1, this dependence follows from the misleadingly simple relationship (21.69), $\Delta_2 \tan\vartheta_2 = \Delta_1 \tan\vartheta_1$, involving the solution $\Delta_2(\vartheta_1)$ of the distilled jump condition (21.68). To avoid possible misunderstanding with respect to our terminology, which is based on the monotonicity of the jump condition with respect to the Alfvén Mach numbers: in general, because of this dependence $\Delta_2(\vartheta_1)$, the angular dependence $\vartheta_2(\vartheta_1)$ is *not monotonic*. For example, for the fast shocks and jumps of Fig. 21.15(d), ϑ_2 rapidly increases for small ϑ_1 (a remnant of the switch-on behaviour at $M_1 = 1$), becomes rather flat and reaches a maximum ($\vartheta_2 = 0.304\pi$ at $\vartheta_1 = 0.188\pi$), and then decreases again.

21.3.4 Observational considerations of MHD shocks

We have analysed the implications of the distilled jump conditions (21.68)–(21.70) from a wide variety of viewpoints, with the time reversal duality (21.88) and the associated classification (21.89) of shocks as the most prominent ones. They imply that a transonic plasma is uniquely described by assigning values to the three upstream parameters Δ_1, ϑ_1, β_1 at each point, at shock fronts complemented with the values of the three downstream parameters Δ_2, ϑ_2, β_2. As we have seen, the relationship between these two parameter spaces is a very intricate one, where the downstream parameters are not uniquely determined by the upstream ones since the distilled energy jump condition (21.68) is a cubic equation, having one, two or three solutions at each point of the shock front, dependent on the value of Δ_1. Moreover, some of the values for the upstream parameters have to be excluded a posteriori since they correspond to negative values of the downstream pressure. At this point of our exposition, the reader may wonder how such a rich structure, with far-reaching physical implications, could have emerged from just the set of algebraic relations (21.4)–(21.10), not even involving differential equations. The reason is that these relations express, at each point of the transonic plasma, the laws of conservation of mass, momentum, energy and magnetic flux of the ideal MHD model, supplemented with a prescription for the entropy change. In a certain sense, these equations bring together all the physical properties of the model of nonlinear magnetohydrodynamics, discussed so far, at a single point! It thus focuses on the nonlinearity of these equations, which, by definition, does not relate to the differential equations but to the algebraic aspects of the model.

Which of the multifarious MHD shocks may be realized in actual stationary or time-dependent transonic MHD flows will be determined by the dissipative and nonlinear processes acting in such an inhomogeneous configuration. The ones that survive are called *evolutionary.* Since intermediate shocks precisely occur when the distilled shock problem is multi-valued in terms of the upstream Alfvén Mach number M_1, such shocks typically have bifurcations and may split up into other types of discontinuity. For dissipative models in simple geometries in the limit of vanishing dissipation, the intermediate shocks were shown not to be stable against certain perturbations. Traditionally (Akhiezer et al. [9, 8], Germain [209], Jeffrey and Taniuti [336]), this failure of the test of evolutionarity has been used to rule out the physical reality of intermediate shocks. However, in 1988, Wu [656] showed intermediate shocks arising in a numerical solution of the dissipative MHD equations through nonlinear steepening from a continuous wave. This claim was further substantiated by calculations by Kennel et al. [342] for a dissipative model problem and by De Sterck et al. [153] on transonic flows in complex geometries like those encountered at solar wind–magnetosphere boundaries.

After Wu's papers, a flurry of numerical and observational papers appeared on the formation of intermediate shocks [191, 657, 658, 274], on their astrophysical implications [163, 115], on their relationship with magnetic reconnection [615, 388], on their significance for the Riemann problem [456, 457, 601], on their occurrence in bow-shock flows [154] and on their possible breakup [648]. On the other hand, Falle and Komissarov [173] presented arguments to stick to the traditional viewpoint, subjecting numerical calculations to strict rules on what to call a shock and what a transient feature. However, as noted by Kennel et al. [341], and reiterated in Section 21.3.2, this viewpoint has also hampered the development of the theory of MHD shock relations, as demonstrated by the fact that the solutions of this section were obtained only recently [226]. For the purpose of understanding how the MHD jump conditions permit a continuous description in parameter space from slow to fast shocks through the intermediate ones, the problem of the evolutionarity of intermediate shocks can be deferred to a future stage, when conclusive observational and numerical evidence has been collected. Note that such a strategy is customary (because effective) in experimental fusion research when slow dissipative tokamak evolution is analysed by means of the two-fold arbitrary family of equilibrium solutions of the Grad–Shafranov equation, and the decision on which of these equilibria is actually chosen by the experiment is left to evidence obtained from the different diagnostics. A glance at the list of plasma transport coefficients of Section 3.3.2, and the realization that these just present the classical picture, which nearly always needs to be modified by anomalous corrections of even the orders of magnitude, suffices to appreciate the practical impossibility of predictive transport calculations of the dynamics of magnetized plasmas. Similarly, in the related subjects of helioseismology and MHD spectroscopy [237], discussed in Section 7.2.4, and extended to accretion discs [345, 73] and general transonic flows [230, 223, 74], the full variety of MHD waves for all possible equilibrium configurations are simply catalogued and the final decision on which of those is realized in nature is left to data from observations by earth-based or space-born telescopes (in astrophysics) or from the various diagnostics (in the laboratory). In physics, the difference between theoretical truth and prejudice is decided by empirical evidence: nature has all the answers.

21.4 Example astrophysical transonic flows

Magnetohydrodynamic shocks play a prominent role in the wide variety of transonic laboratory and astrophysical plasmas. Concrete examples range from the early pinch implosion experiments [372], to more recent inertial confinement fusion experiments [133], over bow shocks encountered at planetary magnetospheres or at the heliosphere–interstellar medium boundary, up to shocks found in supersonic accretion flows onto neutron stars [527] and black holes [201]. As mentioned in Chapter 14, reconnection processes [511] may also create standing magnetohydrodynamic shock fronts. In turn, steady, transient or recurring shock fronts aid in the acceleration of particles. In many astrophysical outflows, like in extragalactic jets [21, 349], shock-accelerated particles can easily reach relativistic speeds [365]. Knowledge of the shock relations, extended to relativistic MHD as introduced in Chapter 22, is thus vital to our understanding of transonic flows and the physical processes involved in particle acceleration.

Explicit classification of all the different MHD shocks is also important for the numerical solution of the nonlinear evolution problem by means of characteristics. In two-dimensional equilibria with flow, the characteristics exhibit both spatial and temporal features such that equilibrium and waves appear to be no longer separate issues. Some of these features of transonic MHD flow have been analysed by Goedbloed and Lifschitz [238, 411] for a special class of ('self-similar') solutions that permits in-depth analysis of the transitions and shocks of the flow, by means of a system of coupled PDEs and the algebraic Bernoulli equation. For non-planar flows, in particular axi-symmetric stationary equilibria [238], a further division of the flow regimes discussed in Section 21.2.3 occurs due to the occurrence of *limiting line characteristics* in the hyperbolic magneto-sonic regions (as earlier shown for gas dynamics by Courant and Friedrichs [130]) and of flows that are disconnected at the Alfvén speed by a *forbidden flow regime due to the constraint of constant Bernoulli function*.

In the remainder of this chapter, we discuss some examples of continuously varying transonic flows. The solar wind is perhaps the best known example of such a 'smooth' transonic flow, on top of which the complex shock patterns associated with coronal mass ejections (CMEs) develop. The solar wind is just one manifestation of a smooth transonic flow whose acceleration is mainly thermally driven, while magnetized stellar winds or outflows emanating from accretion discs can also reach high speeds by magneto-centrifugal mechanisms. In young star environments, with circumstellar accretion discs, strong stellar magnetic fields even deviate the accreting matter to form transonic 'funnel' flows.

(a) The solar wind–magnetosphere boundary The existence of intermediate shocks in numerical solutions by De Sterck, Low and Poedts [153] of the dissipative MHD equations for transonic flows of the type encountered at the solar wind–magnetosphere boundary nicely illustrates the complexity

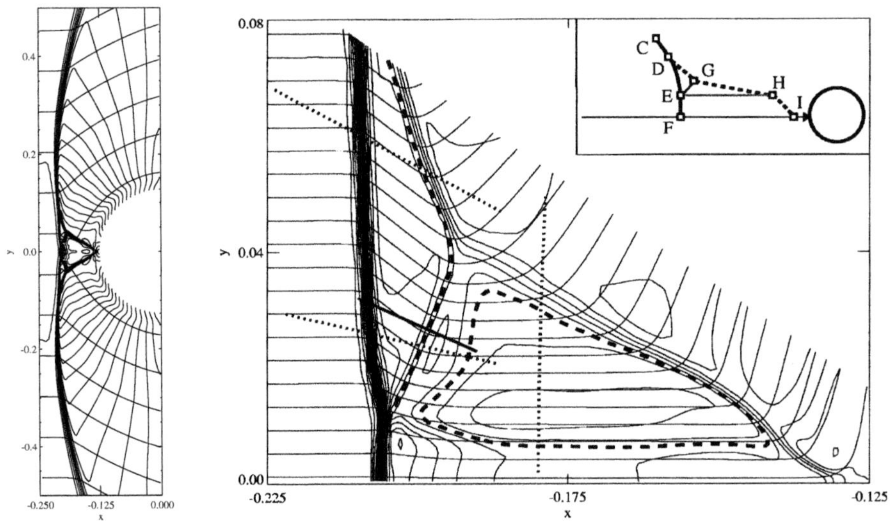

Fig. 21.16 Interacting MHD shocks in a homogeneous flow around a perfectly conducting cylinder. Global view (left) and detail of the flow in the half plane above the stagnation streamline (right). (From De Sterck et al. [153].)

of ideal MHD shock-dominated flow patterns. To model the interaction of the solar wind with a 'magnetosphere', or to study the shock fronts associated with CMEs traversing the solar wind, the authors considered the model of a uniform magnetized flow around perfectly conducting obstacles. When the upstream plasma β is smaller than $2/\gamma$, while the inflow is super-Alfvénic such that

$1 \leq M_1 \leq \sqrt{(\gamma+1-\gamma\beta_1)/(\gamma-1)}$, in agreement with Eq. (21.72) in the limit $\vartheta_1 = 0, \Delta_2 = 0$, fast switch-on shocks can occur and the shock structure in the flow around the obstacle can become quite complicated.

The steady shock structures shown in Fig 21.16 were obtained numerically using the shock-capturing schemes discussed in Chapter 20. A global view and a detail of the converged bow shock solution are given for an initially uniform field-aligned flow around a cylinder in the switch-on regime, where $\beta_1 = 0.4$ and $M_1 = 1.5$ The steady shock front was found to consist of multiple shock segments of various types. The shock D–E is a fast shock, almost a switch-on shock since $B_{y,1} \approx 0$ upstream. The leading shock front part E–F is a HD shock and E–G, as well as D–G–H–I, is an IM shock. The latter one is almost a slow switch-off shock. The part indicated with E–H is a tangential discontinuity and other tangential discontinuities stretch out from points D, G and H along the streamlines to infinity. Notice that, in this inflow regime, the shock front shape is not parabolic (as would be the case in a hydrodynamic supersonic inflow about the cylinder), but it shows a clear dimple in the leading shock front. This simulation may thus provide an explanation for CME related shocks, which are observed to have such a dimple.

(b) Jets from magnetized accretion discs Transonic magnetized plasma flows, collimated over very large distances from their source region, occur throughout our observable Universe. Astrophysical jets on light year scales are associated with young stellar objects (YSO) and the fossils of dead stars: neutron stars and stellar mass black holes in X-ray binary systems (XRB). Jets up to several millions of light years long occur in association with active galactic nuclei (AGN), containing a supermassive black hole. Many of these jets likely involve equipartition magnetic field strengths where the thermal and the magnetic energy contents are comparable. Observational and theoretical arguments favour that such strong magnetic fields will play a role in jet launching and propagation, as well as in termination of the jet through interaction with the surrounding medium. To explain their remarkable collimation, magnetic hoop stresses are needed to counteract the tendency of the outflow to widen by centrifugal and pressure effects. Common to the YSO, XRB and AGN type systems is also the presence of an accretion disc. Observational links have been established between accretion disc luminosity and jet emission, highly suggestive of a unifying jet launch scenario. The most promising scenario to explain the ubiquitous jet phenomenon relies on the interaction of a large-scale magnetic field with the accretion disc in order to give birth to bipolar self-collimated jets.

A magnetic field threading an accretion disc can brake rotating matter, allowing accretion, and can act to transfer part of the angular momentum into a jet. A seminal MHD model by Blandford and Payne [71], with transonic flows in a cold plasma, has paved the way for many investigations of magnetized jet dynamics. In equipartition thin accretion discs ($\beta \approx 1$), it is possible to realize sufficiently bent magnetic field configurations in the inner disc regions for magneto-centrifugal acceleration of jet material. To reach a stationary configuration in the simulations, one must model the disc internal regions in a resistive MHD framework, as material should be allowed to accrete without dragging in magnetic field lines. The anomalous resistivity mimics the effect of the magneto-turbulent nature of the inner disc plasma (with the turbulence presumably originating from magneto-rotational instabilities, or any of the transonic MHD instabilities discussed in Chapters 13 and 18). The jet regions and the surrounding medium are adequately modelled in ideal MHD. Most simulations of

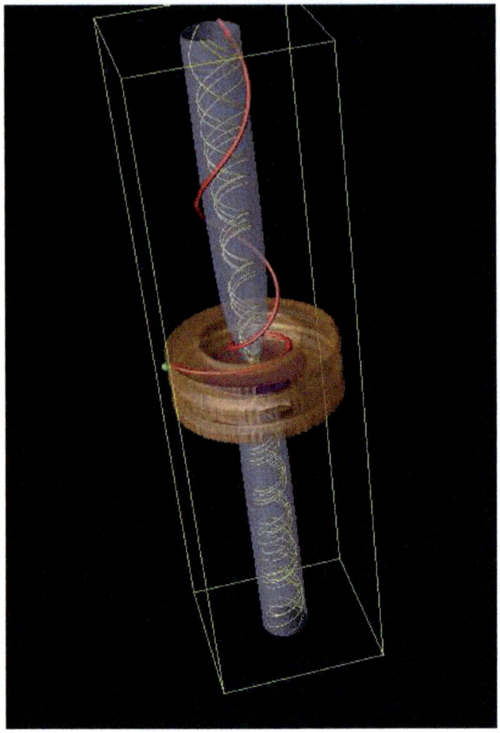

Fig. 21.17 A 3D impression of a transonic jet launched from an accretion disc. A high density iso-surface (brown) indicates the disc location, while selected magnetic field lines are shown in yellow. A fraction of the accreting disc matter gets lifted out of the disc and forms magneto-centrifugally accelerated jet matter: an indicative particle streamline (red) shows how spiralling fall-in motion gets diverted into a collimated jet. (From Casse and Keppens [109].)

astrophysical jets focus on acceleration and collimation and treat the dynamics of the disc as a mere boundary condition.

More recent work has numerically demonstrated the continuous launching of trans-magneto-sonic collimated jets from resistive accretion discs threaded by open large-scale magnetic fields. In particular, Casse and Keppens [109] performed MHD simulations where the disc launches a non-transient ideal MHD jet accelerated to super-fast magneto-sonic velocities. In [109] the authors assumed a simple polytropic relation between pressure and density profiles, and were able to evolve the magnetized disc over many rotational time scales. Figure 21.17 gives a 3D impression of the numerical end result, showing that a bipolar pair of self-collimated cool jets forms and is persistently ejected. The dominant part of disc matter is effectively accreted. A constant fraction reaches the inner disc surface, where the pressure gradient lifts the matter to be propelled in the jet. Jet material is then magneto-centrifugally accelerated to reach super-fast magneto-sonic speeds. In the follow-up paper [109](b), the energetics of the flow was computed, and the simulations produced hot jets with equipartition magnetic fields. In a self-consistent manner, the numerically obtained magnetized accretion-ejection structures (MAES), accounted for the observation that many systems display bright collimated jets together with under-luminous accretion discs.

(c) Funnel accretion flows Our final example of a continuous transonic flow configuration in astrophysics zooms in on the central region of the YSO-accretion disc system. While the previous example showed that jets can be launched from the inner equipartition disc regions, the fate of the major accreting fraction of the disc material was treated in a rudimentary fashion: a 'sink' region allowed matter to escape the simulation box near the origin. For a YSO system, this inner region would have a typical size of about 0.1 AU, while the disc and jet dynamics were covering a region extending to several AU. The jet launch simulations thus illustrated how self-collimated, hollow jets can form within 2–3 AU. We now shift our view to the star–disc dynamics happening within a range of a few stellar radii, accounting for the observational fact that young solar-type stars can harbour dynamically important global magnetic fields of 100 G to a few kG. In reality, the central accretion dynamics can be very complex, as the young star will rotate as it has gone through a contraction phase, and its main dipole moment does not need to be aligned with the rotation axis. We will avoid these complications, and discuss a computational example of magnetized star–disc interaction assuming axi- symmetry, with aligned magnetic dipole moment and rotation axis.

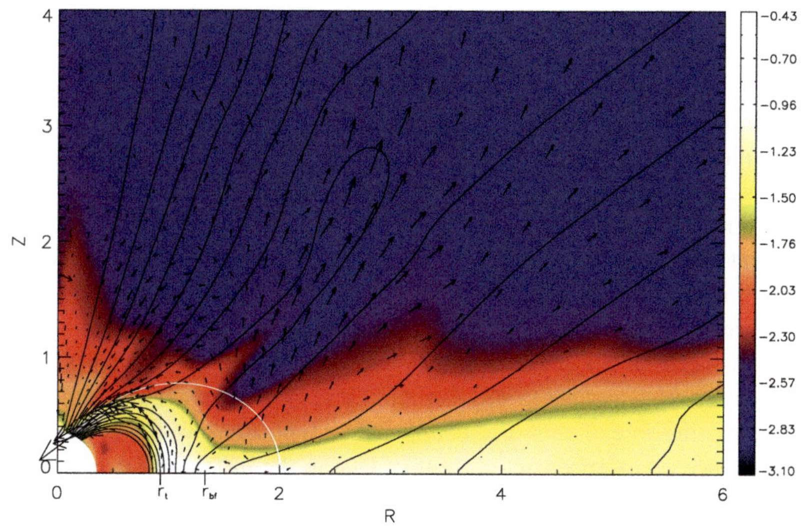

Fig. 21.18 In an axi-symmetric computation of accretion onto a young magnetized star, the accretion disc (high density matter in yellow) is magnetically halted within a disc truncation radius r_t. The infalling matter gets diverted into a transonic funnel flow along the dipolar magnetic field lines (black lines) connecting star and disc. The base of the funnel is located at r_{bf}. (From Bessolaz et al. [56].)

In the presence of a stellar magnetic field, the accretion flow onto the star can be significantly different from purely equatorial accretion. The stellar field can truncate the disc at an inner truncation radius r_t and, as pointed out by Romanova et al. [526] and refined by Bessolaz et al. [56], dynamical arguments can be used to estimate this truncation radius for a given accretion rate, stellar field strength and other parameters like its radius and mass. In [56], the estimate was derived for equipartition fields ($\beta \sim 1$) from the requirement that the accretion ram pressure ρv_r^2 is balanced by the poloidal magnetic pressure. Under typical T-Tauri star parameters the disc is truncated within a few stellar radii. The numerical study of [526] convincingly demonstrated the formation of accretion 'funnels' near the star, as material gets lifted out of the disc due to a vertical plasma pressure gradient and loaded onto the stellar field lines. While the original results were obtained for kG stellar fields, Fig. 21.18 shows the later result [56] which extended the findings to lower stellar field strengths of about 140 G, at even lower accretion rates of order $10^{-9}\ M_{\odot}/\mathrm{yr}$. The disc region was treated resistively, analogous to the jet launch studies mentioned above. The accretion funnels develop self-consistently from initial conditions where only a truncated accretion disc is embedded in a dipolar stellar magnetosphere. The end state shown in Fig. 21.18 is a stationary transonic flow configuration: it crosses the slow magneto-sonic point but remains sub-Alfvénic. The main forces acting within the funnel are the thermal pressure gradient at the base, as matter accumulates there, while eventually gravity overwhelms near the stellar radius: matter reaches the star with approximately free-fall velocities. While these funnel flows seem to be common to accreting magnetized stars, the stellar parameters may bring the YSO system to completely different transonic flow regimes, such as the 'propeller' flow predicted in [527].

The examples clearly illustrate that transonic MHD flows are ubiquitous in astrophysics, but the calculations concentrate on highly simplified configurations to isolate one aspect of what happens in reality. Continued progress is made on each aspect separately, e.g. by tackling 3D accretion flows onto young stars with unaligned rotation and magnetic axes, and considering higher order multi-pole fields. While the basic funnel flow physics is well understood from the axi-symmetric simulations, the bewildering variety of multi-column accretion funnels, and the possibility of intermittent accretion tongues, forming as a result of Rayleigh–Taylor instabilities developing at the inner truncation radius [384], illustrates that the observational task of inverting from spectroscopic data is simply daunting. Furthermore, the supersonic accretion columns themselves must ultimately end on the stellar surface, and thus involve a standing shock front, from which much of the radiative losses occur. This aspect needs to be studied within the global disc dynamics simulations. Finally, the angular momentum regulation in true YSO systems is still posing many puzzles: a large fraction of classical T-Tauri stars exhibit circumstellar discs rotating much slower than expected from pre-main sequence evolutionary models. Due to the fact that these stars contract and hence should spin up significantly, a regulating role of star–disc coupling was typically invoked to explain the observed slow rotation rates (see e.g. Keppens et al. [348]). Due to the different ranges in spatial and temporal scales, the full 3D demonstration of stellar outflows, shocked multicolumn accretion flows and disc jet/wind launching is still extremely challenging for current generation grid-adaptive, massively parallel software tools. A related challenge is the study of jet launch, propagation, stability and shock-dominated termination by interaction with the interstellar medium for systems ranging from the YSO to the AGN category. We will give a modern example of the latter in the next chapter, devoted to relativistically flowing magnetized plasmas.

21.5 Literature and exercises

Notes on literature

Classic treatments of gas dynamic and MHD shocks

– A classic treatment of gas dynamic shocks is to be found in *Supersonic Flow and Shock Waves* by Courant and Friedrichs [130]. Equally classic are the chapters on gas dynamic shocks and MHD shocks in *Course of Theoretical Physics* (Volumes 6 and 8) by Landau and Lifshitz [394, 395]. An in-depth treatment of MHD characteristics may be found in *Magnetohydrodynamics* by Kulikovskij and Lyubimov [383].

Textbook chapters on MHD shocks

- A number of textbooks on plasma physics have useful chapters on MHD shocks, e.g. *Magnetohydrodynamics* by Jeffrey [335]; *Magnetohydrodynamics of the Sun* by Priest [510]; *The Physics of Plasmas* by Boyd and Sanderson [84]; *Introduction to Plasma Physics* by Gurnett and Bhattacharjee [272].

MHD shock relations

- The material on shock relations presented in this chapter is based on the paper 'Time reversal duality of magnetohydrodynamical shocks' by Goedbloed [226]. It adds the viewpoint of scale-independence of the MHD equations to the powerful tangential transformation that can be found in the classical, and still very readable, 1950(!) paper 'Magneto-hydrodynamic shocks' by de Hoffmann and Teller [146].

Exercises

[21.1] *Entropy*

In this chapter, it is stated that the entropy has to increase across a shock front.

- Define entropy. Explain why it has to increase across a shock front.

[21.2] *Essential steps in the derivation of the distilled shock conditions*

What are the essential steps in deriving the distilled shock conditions from the original ones? Explain for each step why it is important.

[21.3] *Plotting the distilled energy jump condition*

In this exercise, you are going to reproduce the S-shape curve of Figs. 21.8 and 21.9. We assume that you use IDL, Matlab, Maple or any other similar program. Whatever program you use, make sure that Laguerre's method [509] is implemented.

- Write $\bar{p}_1$ and $\bar{B}_{t1}$ in terms of the angle ϑ_1 and the plasma beta β_1.
- Now, write an IDL program which reproduces the S-curve shown in Fig. 21.8. Create a file called `quantities.pro` in which you put two functions, one for the upstream pressure and one computing the tangential magnetic field. Both functions use the angle ϑ_1 and the plasma β_1 as input parameters.
- Write the distilled energy jump condition (21.65) as a polynomial in the squared Alfvén Mach number M_2^2. How many solutions are there and what are the expressions for each of the coefficients?
- Next, create a function computing the coefficients of the energy jump condition for given ratio of specific heats γ, pressure $\bar{p}_1$, tangential magnetic field $\bar{B}_{t1}$ and squared Alfvén Mach number M_1^2. Put this function in the file `shockconditions.pro`.
- Next, write a generic function computing the roots of a polynomial using Laguerre's method, which is part of the IDL library. Create the file `solvers.pro` which creates the function `rootfinder` with input parameters γ, $\bar{p}_1$, $\bar{B}_{t1}$, 'points' and $\max(M_1^2)$. By 'points' the number of points to be calculated between 0 and $\max(M_1^2)$ is meant. At each point, you should compute the coefficients of the energy jump condition needed to compute the roots with Laguerre's method. Store only the appropriate solutions. What kind of test can you design to check if a solution is appropriate or not? How do you sort the solutions? The generic function should return the appropriate solutions as a two-dimensional array with the values for M_1^2 and M_2^2.
- Make a plot of the roots found for the given parameters of Fig. 21.8.

[21.4] *Plotting the downstream magneto-sonic transition function*

The purpose of this exercise is to make a plot of the downstream magneto-sonic transition function (21.75) as a function of the upstream Alfvén Mach number. Similar to the previous exercise, you may use IDL.

- Write the downstream magneto-sonic transition equation as a polynomial in M_2^2. Find its coefficients.
- Add a function to the file `shockconditions.pro` which computes these coefficients. If you did not do the previous exercise, then create this file.

- Extend the function `rootfinder` so that it can also compute the roots of the downstream magneto-sonic transition function. Again, if you did not do the previous exercise, make a file `solvers.pro` creating the function `rootfinder`. What this function should do is specified in the last but one question of the previous exercise.
- Make a plot of the curves $M^2_{s2,A2,f2}$. If you did the previous exercise, you can just add the curves to the previously created plot.
- Finally, also add the curve $M_1^2 = M^2_{s1,A1,f1}$ to the plot.

[21.5] *Plotting the distilled entropy jump condition*

In this exercise, you are going to make a plot of the distilled entropy condition (21.66). We assume again that you use IDL, but you may use a similar program. Make sure you are able to use the van Wijngaarden–Dekker–Brent (WDB) method [509]. This is an advanced method to compute the zero of a monotonic function on a given domain.

- Why cannot you plot the distilled entropy condition using Laguerre's method?
- For a numerical reason, you have to multiply the entropy condition by $M_2^{2\gamma}$. What is that reason precisely?
- The entropy condition will be plotted using the WDB method. Add a function computing it for given $M_2^2, \gamma, \bar{p}_1, \bar{B}_{t1}, M_1^2$ to the file `shockconditions.pro`. Make sure that the first input parameter is the squared Alfvén Mach number M_2^2. If you did not do the two previous exercises, then create this file.
- Download the file `zbrent.pro` from the Internet. This is a part of the IDL Astronomy User's Library (idlastro.gsfc.nasa.gov) which contains the WDB method. Modify the function `zbrent` such that it accepts four additional parameters.
- Make a plot of the entropy condition for $\vartheta_1, \beta_1 = 0.4$, M_1^2 and $M_2^2 = [0, 2.5]$. What do you observe and what does it mean for the use of the WDB method?
- What method can be used to solve the problem of the previous question?
- Add a function to the file `solvers.pro` (create it if it does not exist), which computes where the entropy condition is zero for given $\gamma, \bar{p}_1, \bar{B}_{t1}$, 'points' and $\max(M_1^2)$. The zeros should be computed using the WDB method. By 'points' the number of points between zero and the maximum of M_1^2 is meant. The function should return a 2D array containing the values M_1^2 and M_2^2 for which the entropy jump is zero.
- Make a plot of the solutions found. If you did one of the two previous exercises, you can just add the solutions to the existing plot.

[21.6] *Beyond the presented results*

If you do not want to spend your time on the previous exercises, but wish to investigate the huge parameter space of MHD shocks beyond what has been presented in this chapter, you may request the Fortran program `Shock`, by which Figs. 21.8–21.10, and 21.13–21.15 were produced, from the author (goedbloed@differ.nl).

22
Ideal MHD in special relativity

We have seen that the MHD description for the macroscopic dynamics of plasmas offers a powerful, unifying, viewpoint on both laboratory and astrophysical plasmas. The applicability of the MHD viewpoint was discussed extensively in previous chapters, along with the various approximations made to arrive at the MHD equations from first principles. For most laboratory plasmas, the single fluid ideal or resistive MHD model eventually needs to be extended towards a multi-fluid model and by including important kinetic effects, since its continuum approach to plasma modelling neglected, e.g., Landau damping as well as many other velocity-space dependent physical phenomena. For many astrophysical plasmas, we face yet another shortcoming of the MHD model used thus far, namely that we restricted all attention to non-relativistic plasma velocities. This is perfectly adequate for most of the plasma found in our own solar system. However, astronomical observations indicate that, e.g., the extragalactic jets associated with Active Galactic Nuclei clearly harbour dynamically important magnetic fields and relativistically flowing plasmas. In order to model these plasmas in a continuum model, the restriction on the plasma velocities must be alleviated, by revisiting the ideal MHD equations in a frame-invariant relativistic formulation within four-dimensional space-time. In this chapter, we present such a formulation, restricting our attention to special relativity where we still have a 'flat' geometry. Modern computational techniques such as those discussed in Chapter 20 have been adapted and used in this more demanding relativistic MHD regime, and we summarize the numeric algorithmic challenges posed by the ideal MHD model in special relativity. We mention in this chapter selected example applications, covering relativistic jets from X-ray binaries, to the modelling of pulsar wind nebulae. For the Crab nebula in particular, we discuss insights gained from recent computations of magnetized, relativistic plasma dynamics, and show that they start to resemble actual observations.

22.1 Four-dimensional space-time: special relativistic concepts

Relativity implies that physical laws do not depend on the chosen reference frame. In special relativity, the speed of light c is explicitly recognized as the maximal speed with which information can travel between different spatial locations. Since this maximal speed c is the same for all observers in uniform relative motion with respect to each other, we must abandon the familiar notion of Galilean invariance. Indeed, under Galilean invariance between two such inertial reference frames, speeds expressed in the two frames transform by adding the speed of the relative motion. In order to account for an upper limit c in propagation speed, which is the same for two inertial systems moving with relative velocity $\mathbf{v}$, one must relax their notion of simultaneity. What appears simultaneous in one frame will occur at different times in any other frame, and as a result, the temporal duration of a physical event will differ from frame to frame. This leads to a viewpoint where time is an extra coordinate, augmenting the three spatial coordinate directions, to describe physical events in a four-dimensional space-time continuum. The four coordinates associated with different inertial space-time reference frames are related by the Lorentz transformation. In the following, we introduce the basic concepts leading to a special relativistic formulation of nonlinear compressible gas dynamics. Section 22.2 extends this with the inclusion of electromagnetic fields, and discusses how one successively obtains the ideal MHD model in special relativity. The material in this introductory treatment towards special relativistic MHD benefited from material presented in the more broadly oriented textbook by Thorne and Blandford [595], while much more complete, but also more technical, treatments can be found in earlier monographs by Anile [14] and Lichnerowicz [409].

22.1.1 Space-time coordinates and Lorentz transformations

In Section 2.2.2, we presented *the Lorentz transformation* of two inertial frames moving with relative velocity $\mathbf{v}$, given by

$$\begin{cases} \mathbf{x}' = \mathbf{x} + \dfrac{\Gamma - 1}{v^2}\,\mathbf{v}\,\mathbf{v}\cdot\mathbf{x} - \Gamma\mathbf{v}\,t\,, \\[2ex] t' = \Gamma\left(t - \dfrac{1}{c^2}\mathbf{v}\cdot\mathbf{x}\right), \qquad \Gamma \equiv \dfrac{1}{\sqrt{1 - v^2/c^2}}\,. \end{cases} \tag{22.1}$$

The Lorentz factor Γ depends nonlinearly on the relative velocity $\mathbf{v}$, which is still a vector with three components (v_1, v_2, v_3) along the three spatial (orthogonal) coordinates $\mathbf{x} \equiv (x_1, x_2, x_3)^{\mathrm{T}}$. We will write a Latin index $i = 1, 2, 3$ when a spatial coordinate direction is meant, e.g. in x_i or v_i. However, the Lorentz transformation (22.1) must rather be interpreted as a transformation between the four-dimensional space-time coordinates

$$\mathbf{X} \equiv \begin{pmatrix} ct \\ \mathbf{x} \end{pmatrix}, \qquad \mathbf{X}' \equiv \begin{pmatrix} ct' \\ \mathbf{x}' \end{pmatrix}, \tag{22.2}$$

when we write it as

$$\mathbf{X}' = \begin{pmatrix} \Gamma & -\Gamma v_1/c & -\Gamma v_2/c & -\Gamma v_3/c \\ -\Gamma v_1/c & 1 + (\Gamma - 1)v_1^2/v^2 & (\Gamma - 1)v_1 v_2/v^2 & (\Gamma - 1)v_1 v_3/v^2 \\ -\Gamma v_2/c & (\Gamma - 1)v_1 v_2/v^2 & 1 + (\Gamma - 1)v_2^2/v^2 & (\Gamma - 1)v_2 v_3/v^2 \\ -\Gamma v_3/c & (\Gamma - 1)v_1 v_3/v^2 & (\Gamma - 1)v_2 v_3/v^2 & 1 + (\Gamma - 1)v_3^2/v^2 \end{pmatrix} \mathbf{X}\,. \tag{22.3}$$

The space-time coordinates $\mathbf{X}$ are specific to a pre-chosen inertial Lorentzian reference frame. If we indicate the 4×4 matrix occurring in (22.3) with $L_\alpha^{\alpha'}$, and denote the four coordinate entries as X^α and $X^{\alpha'}$, with Greek index $\alpha = 0, 1, 2, 3$, we have the compact notation

$$X^{\alpha'} = L_\alpha^{\alpha'} X^\alpha\,, \tag{22.4}$$

omitting the summation sign over the common indices α (the Einstein summation convention). This formula implicitly assumes a choice of two Lorentzian reference frames, L′ with space-time coordinates $\mathbf{X}'$, and L with coordinates $\mathbf{X}$. The inertial frame L′ is moving at speed $\mathbf{v}$ with respect to frame L, and Eq. (22.4) merely states how the coordinates for a unique point in four-dimensional space-time, also referred to as an 'event', relate between the two selected frames. Obviously, all symbols appearing in Eq. (22.4) are frame-dependent. The inverse transformation,

$$X^{\alpha} = (L^{-1})^{\alpha}_{\alpha'} X^{\alpha'}, \tag{22.5}$$

can be computed by noting that the matrix $(L^{-1})^{\alpha}_{\alpha'} = (L^{\alpha'}_{\alpha})^{-1}$. As expected on physical grounds, the matrix $(L^{-1})^{\alpha}_{\alpha'}$ turns out to be found from $L^{\alpha'}_{\alpha}$ by merely replacing $\mathbf{v}$ by $-\mathbf{v}$. Note also that the Lorentz transformation matrix $L^{\alpha'}_{\alpha}$ (and thus its inverse) is symmetric.

Pure Lorentz boost, length contraction and time dilation We can specify the Lorentz transformation to a case where frame L$'$ is moving with a velocity $\mathbf{v}$ directed along the x_1 coordinate axis of frame L. Then, we speak of a pure Lorentz boost with direct transformation given by

$$\begin{pmatrix} ct' \\ x_1' \\ x_2' \\ x_3' \end{pmatrix} = \begin{pmatrix} \Gamma & -\Gamma v_1/c & 0 & 0 \\ -\Gamma v_1/c & \Gamma & 0 & 0 \\ 0 & 0 & 1 & 0 \\ 0 & 0 & 0 & 1 \end{pmatrix} \begin{pmatrix} ct \\ x_1 \\ x_2 \\ x_3 \end{pmatrix}. \tag{22.6}$$

The inverse transformation replaces $\mathbf{X}$ by $\mathbf{X}'$ and v_1 by $-v_1$. A length interval Δx_1 along the x_1 axis of frame L is measured in the L$'$ system at a fixed time t'. The inverse transformation yields immediately that, for the moving observer in frame L$'$, the length appears to contract since

$$\Delta x_1' = \Delta x_1 / \Gamma. \tag{22.7}$$

Since $\Gamma \geq 1$, lengths in the inertial frame L along the direction of motion of a moving observer in L$'$ appear shorter to him, which is the length contraction effect. Note that in the two directions perpendicular to the motion, no difference occurs, so volume measures will only differ by the single factor Γ^{-1}. For a comparison of time intervals, we use the direct transformation to compare time intervals as determined at fixed position ($\Delta x_1 = 0$) in frame L. We find similarly that

$$\Delta t' = \Gamma \Delta t, \tag{22.8}$$

telling that time in the inertial frame L appears to run slow to the moving observer in L$'$. Of course, when we change viewpoint from L$'$ to L, the same conclusions will be reached: L is also moving with velocity $-\mathbf{v}$ along the x_1' coordinate axis of frame L$'$. We will reach the (apparently paradoxical) conclusions that then Eqs. (22.7)–(22.8) will hold in reverse, i.e. when replacing t by t' and x_1' by x_1. This is just what relativity is all about: both observers come to the same conclusion. The paradoxical nature is a mere consequence of the fact that simultaneity has obtained a different meaning in both systems.

For the pure boost (22.6), we can graphically illustrate the different parameterization of space-time associated with the (x_1, t) and (x_1', t') systems. When frame L$'$ moves at speed v in frame L along x_1, and we draw the axes of L$'$ at the moment when origins of L and L$'$ coincide, the t' axis is given by the straight line $ct = (c/v)x_1$. Similarly, the x_1'-axis given by $t' = 0$ points is the line $ct = (v/c)x_1$. This is illustrated in Fig. 22.1. It is then clearly seen that simultaneous events in frame L are no longer simultaneous to frame L$'$. Such space-time diagrams can also be used to

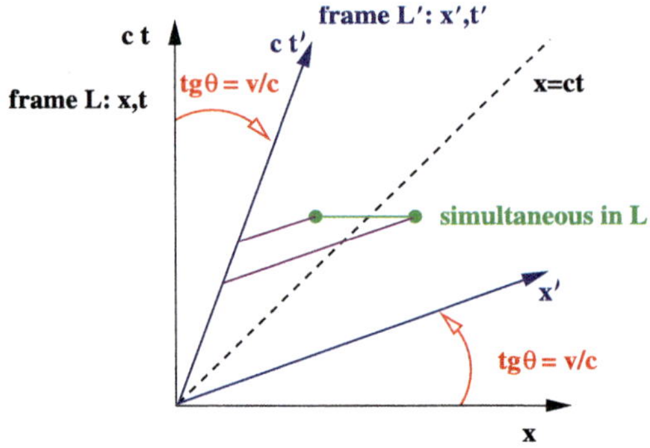

Fig. 22.1 Pure Lorentz boost and the loss of simultaneity.

prove in a more graphical sense the mentioned effects of length contraction and time dilation. Note finally that from Eq. (22.6) (as well as from Eq. (22.3)), we find the relation

$$-c^2t^2 + x_1^2 + x_2^2 + x_3^2 = -c^2t'^2 + {x'_1}^2 + {x'_2}^2 + {x'_3}^2 \,, \tag{22.9}$$

which will return in the discussion of the metric associated with flat space-time.

22.1.2 Four-vectors in flat space-time and invariants

We now consider the path traced in space-time by a single particle, which we will call the world line of this particle. Time progression as experienced by this particle is referred to as its 'proper time'. This proper time τ is measured by an ideal clock in the particle's local rest frame, i.e. in a reference frame moving along with the particle. Introducing a local Lorentzian reference frame with orthogonal coordinates X^α, the particle four-velocity is then given by

$$U^\alpha \equiv \frac{dX^\alpha}{d\tau} \,, \tag{22.10}$$

where $dX^\alpha \equiv (c\,dt, dx_1, dx_2, dx_3)^{\mathrm{T}}$ measures the 'distance' along each coordinate direction traversed in the proper time interval $d\tau$. These concepts are graphically shown in Fig. 22.2, depicting space-time as an (x, t) plane where two of the spatial directions are conveniently omitted.

The four components U^α in Eq. (22.10), as well as the coordinates X^α, are in fact contravariant components of physically meaningful four-vectors in four-dimensional space-time. The position four-vector $\mathbf{X} \equiv X^\alpha \mathbf{e}_\alpha$ connects the origin of our reference frame with the space-time point with coordinates X^α, whereby the basis vectors $\mathbf{e}_\alpha$ point along the four coordinate axes. The transformations given by Eqs. (22.4) and (22.5) can thus be understood as follows: two reference frames L and L′ each introduce their set of basis vectors $\mathbf{e}_\alpha$ and $\mathbf{e}_{\alpha'}$, respectively. The matrix $L_\alpha^{\alpha'}$ quantifies how each basis vector $\mathbf{e}_\alpha$ can be written as a linear combination of the other set of basis vectors, namely $\mathbf{e}_\alpha = L_\alpha^{\alpha'} \mathbf{e}_{\alpha'}$, exploiting the Einstein summation convention over the double index α'. The frame-independent position four-vector $\mathbf{X}$ is then clearly

$$\mathbf{X} = X^\alpha \mathbf{e}_\alpha = X^\alpha L_\alpha^{\alpha'} \mathbf{e}_{\alpha'} = X^{\alpha'} \mathbf{e}_{\alpha'} \,, \tag{22.11}$$

which directly leads to the transformation as given in Eq. (22.4).

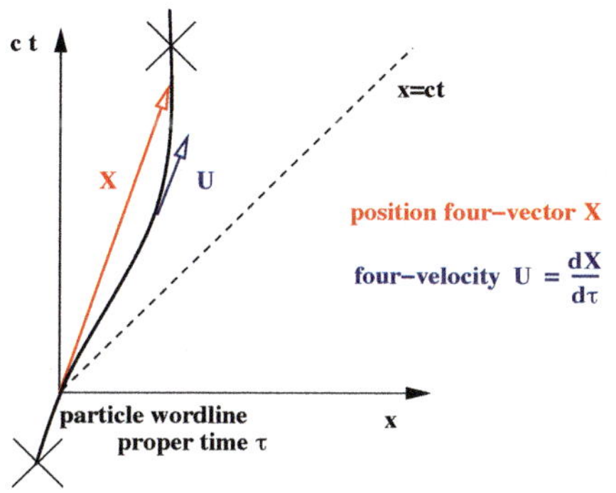

Fig. 22.2 The world line of a particle, its position four-vector $\mathbf{X}$ and four-velocity $\mathbf{U}$.

Once a set of basis vectors $\mathbf{e}_\alpha$ is given, we can in general introduce a set of dual or reciprocal basis vectors which obey $\mathbf{e}_\alpha \cdot \mathbf{e}^\beta = \delta^\beta_\alpha$, where δ^β_α is the Kronecker delta. Covariant vector components appear when working with this dual basis, e.g. for the position four-vector $\mathbf{X} = X_\alpha \mathbf{e}^\alpha$. These dual basis vectors $\mathbf{e}^\beta$ can then be used to express the two sets of basis vectors in terms of each other, namely

$$\mathbf{e}_\alpha = \mathbf{e}_\alpha \cdot \mathbf{e}_\beta \mathbf{e}^\beta \equiv g_{\alpha\beta}\mathbf{e}^\beta \,, \quad \text{or} \quad \mathbf{e}^\alpha = \mathbf{e}^\alpha \cdot \mathbf{e}^\beta \mathbf{e}_\beta \equiv g^{\alpha\beta}\mathbf{e}_\beta \,. \tag{22.12}$$

This introduces the 4×4 components of the metric tensor $g_{\alpha\beta}$ and its inverse $g^{\alpha\beta}$. They allow one to transform from contravariant components X^α to covariant components X_α by $X_\alpha = g_{\alpha\beta}X^\beta$. The metric tensor components associated with flat space-time, i.e. in a Lorentzian reference frame equipped with space-time coordinates, are given by the Minkowski metric

$$g_{\alpha\beta} = g^{\alpha\beta} = \begin{pmatrix} -1 & 0 & 0 & 0 \\ 0 & 1 & 0 & 0 \\ 0 & 0 & 1 & 0 \\ 0 & 0 & 0 & 1 \end{pmatrix} . \tag{22.13}$$

Note that the four basis vectors for this flat space-time are orthogonal to each other. Also, the three space-like directions form an orthonormal set with unit lengths since $\mathbf{e}_i \cdot \mathbf{e}_i = 1$. The squared length of the temporal basis vector $\mathbf{e}_0 \cdot \mathbf{e}_0$ has the negative value -1. For flat space-time, we then conclude that the dual basis vector $\mathbf{e}^0 = -\mathbf{e}_0$, while all spatial directions have basis vectors identical to their dual vector $\mathbf{e}^i = \mathbf{e}_i$. This simplifies the tensor calculus considerably, and one only needs to remember the sign reversal for any temporal tensor component.

In general, the squared length of a four-vector represents a frame-independent scalar invariant upon which different observers will agree. It can be evaluated by means of the metric since for a four-vector $\mathbf{X}$ we find

$$\mathbf{X} \cdot \mathbf{X} = (X^\alpha \mathbf{e}_\alpha) \cdot (X^\beta \mathbf{e}_\beta) = g_{\alpha\beta} X^\alpha X^\beta = X^\alpha X_\alpha \,. \tag{22.14}$$

In flat space-time, and for a differential vector $d\mathbf{X}$, this then is simply

$$(ds)^2 \equiv (d\mathbf{X}) \cdot (d\mathbf{X}) = -(cdt)^2 + (dx_1)^2 + (dx_2)^2 + (dx_3)^2 \,, \tag{22.15}$$

where we introduced the square of the line element ds. A four-vector with negative squared length is said to separate time-like events in space-time, while a four-vector with a positive squared length separates space-like events. A zero squared length is a light-like separation. World lines of particles will always have time-like tangent four-vectors in space-time, while light (a photon) travels along a path with light-like tangents.

From (22.15), we can deduce several useful relations connected to the proper time τ and a particle's four-velocity as introduced in Eq. (22.10). It is clear that, for a stationary particle, the proper time τ will be equal to the time coordinate t of an inertial frame attached to the particle. We then find from (22.15) that $(d\tau)^2 = -(ds)^2/c^2$. Since ds is an invariant quantity, this relation between proper time intervals and the (time-like) line element tangent to the world line of a particle will always be true. One then uses this equality to find the general relation

$$-c^2 = \left[-(cdt)^2 + (dx_1)^2 + (dx_2)^2 + (dx_3)^2\right]/(d\tau)^2 . \tag{22.16}$$

Noting further that

$$\frac{dx_i}{d\tau} = \frac{dx_i}{dt}\frac{dt}{d\tau} , \qquad \frac{dx_i}{dt} = v_i , \tag{22.17}$$

where v_i is a component of the spatial three-velocity $\mathbf{v}$, one finds that

$$\frac{dt}{d\tau} = \Gamma . \tag{22.18}$$

This is again the time-dilation effect, expressing the fact that the proper time associated with a moving particle will always appear to run slow from a stationary viewpoint, since $\Gamma \geq 1$. As a result of Eq. (22.18), the contravariant components of the four-velocity can be written as

$$U^\alpha = (c\Gamma, \Gamma\mathbf{v})^{\mathrm{T}} . \tag{22.19}$$

Using the general relation (22.14), we can then verify the invariance

$$U^\alpha U_\alpha = -c^2 . \tag{22.20}$$

This confirms the time-like nature of the four-velocity, which is the tangent vector to the particle world line. For a particle with rest mass m_0, the four-momentum is then the four-vector

$$P^\alpha = m_0 U^\alpha . \tag{22.21}$$

In Chapter 2, we used the notation $E = \Gamma m_0 c^2$ and $\mathbf{p} = \Gamma m_0 \mathbf{v}$ to indicate the relativistic energy and (three-)momentum, respectively. This allows us to write

$$P^\alpha = (E/c, \mathbf{p})^{\mathrm{T}} , \qquad E^2 = m_0^2 c^4 + p^2 c^2 , \tag{22.22}$$

where the latter relation comes from evaluating the invariant $P^\alpha P_\alpha$. Hence, energy includes the rest mass contribution $m_0 c^2$, and momentum and energy need to be treated on the same footing.

Particle dynamics and forces in space-time The dynamics of a particle is governed by forces exerted on it, and this requires the generalization of the force concept from a three-vector $\mathbf{F}$ to its four-vector equivalent. We can expect these two concepts to coincide in the rest frame of the particle, such that we write the contravariant components of the four-vector in this frame by $F^{\alpha'} = (0, \mathbf{F})^{\mathrm{T}}$.

When we then transform to the reference frame which observes the particle moving with velocity $\mathbf{v}$, we use the inverse Lorentz transform to compute

$$F^{\alpha} = (L^{-1})^{\alpha}_{\alpha'} F^{\alpha'} = (\Gamma \mathbf{v} \cdot \mathbf{F}/c, \Gamma \mathbf{F})^{\mathrm{T}} , \tag{22.23}$$

where we used the fact that the velocity $\mathbf{v}$ will typically be aligned with $\mathbf{F}$. The equation of motion is then

$$\frac{dP^{\alpha}}{d\tau} = F^{\alpha} , \tag{22.24}$$

which can be decomposed into its temporal and spatial parts as

$$\frac{dE}{dt} = \mathbf{v} \cdot \mathbf{F} , \qquad \frac{d\mathbf{p}}{dt} = \mathbf{F} . \tag{22.25}$$

We find familiar results: work done by a force changes the energy, and the equation of motion, expressed in three-vectors. Note that the latter contains the relativistic three-momentum $\mathbf{p} = \Gamma m_0 \mathbf{v}$. This is sometimes written as $\mathbf{p} = m(v)\mathbf{v}$, where the particle 'mass' $m(v) = \Gamma m_0$ increases from its rest mass value m_0 when the particle velocity approaches the speed of light.

Three-velocity addition law Since we need to use the Lorentz transformation to transform between inertial reference frames, three-velocities will no longer merely add up. We can find the special relativistic rule for the addition of three-velocities from the following consideration. Suppose that a particle is observed in a frame L to move with three-velocity $\mathbf{v}$. In the frame L′ co-moving with this particle, a signal propagates away from the particle at the speed $\mathbf{w}$ with respect to L′. When we denote the Lorentz factor evaluated for this three-velocity as $\Gamma_w = \left(1 - w^2/c^2\right)^{-1/2}$, frame L′ will ascribe a four-velocity $U^{\alpha'} = (c\Gamma_w, \Gamma_w \mathbf{w})^{\mathrm{T}}$ to that signal. We can then use the inverse Lorentz transform $U^{\alpha} = (L^{-1})^{\alpha}_{\alpha'} U^{\alpha'}$, to obtain the corresponding four-velocity in frame L as

$$\begin{pmatrix} c\Gamma_u \\ \Gamma_u \mathbf{u} \end{pmatrix} = \begin{pmatrix} \Gamma & \Gamma \mathbf{v}/c \\ \Gamma \mathbf{v}/c & \mathsf{I} + (\Gamma - 1)\mathbf{v}\mathbf{v}/v^2 \end{pmatrix} \begin{pmatrix} c\Gamma_w \\ \Gamma_w \mathbf{w} \end{pmatrix} , \tag{22.26}$$

where I is the 3×3 identity matrix. It is then found easily that

$$\mathbf{u} = \frac{\mathbf{v}\left[\Gamma + (\Gamma - 1)\mathbf{v} \cdot \mathbf{w}/v^2\right] + \mathbf{w}}{\Gamma\left(1 + \mathbf{v} \cdot \mathbf{w}/c^2\right)} . \tag{22.27}$$

One can show that $\Gamma_u = \Gamma \Gamma_w \left(1 + \mathbf{v} \cdot \mathbf{w}/c^2\right)$ is the Lorentz factor for the velocity given by this equation. One can reorganize this expression to one giving the three-speed $\mathbf{w}$ in frame L′,

$$\mathbf{w} = \frac{\mathbf{u} - \mathbf{v}\left[\Gamma - (\mathbf{u} \cdot \mathbf{v}/c^2)\Gamma^2/(\Gamma + 1)\right]}{\Gamma\left(1 - \mathbf{u} \cdot \mathbf{v}/c^2\right)} , \tag{22.28}$$

obtained by adding the velocities $\mathbf{u}$ and $\mathbf{v}$ known in L.

22.1.3 Relativistic gas dynamics and stress-energy tensor

We can now formulate the governing conservation laws in four-dimensional space-time. We begin with particle conservation. The proper density ρ is the mass per unit volume as seen in the rest frame of the gas. When we indicate the number density per unit volume in this frame by n_0, we have

$$\rho = m_0 n_0 \,. \tag{22.29}$$

In the inertial Lorentzian frame where the gas is seen to move with velocity $\mathbf{v}$, volumes differ due to the length contraction effect. As a result, the number density will be $n = \Gamma n_0$. A convenient variable to quantify the 'density' in this 'laboratory' frame is $D \equiv \Gamma m_0 n_0 = m_0 n = \Gamma \rho$. It should be noted that the actual lab frame density is ΓD, due to the mass increase $m(v) = \Gamma m_0$, but we will loosely refer to D as the density. Particle number conservation is then generally expressed by the vanishing divergence of the four-vector ρU^α, namely

$$\partial_\alpha (\rho U^\alpha) = 0 \,. \tag{22.30}$$

Written out in terms of the coordinates $(ct, \mathbf{x})$, we then find

$$\frac{\partial D}{\partial t} + \nabla \cdot (D\mathbf{v}) = 0 \,. \tag{22.31}$$

It is easily seen how the classical limit $\Gamma \to 1$ indeed reduces to the familiar mass conservation equation, since $D \to \rho$.

Because energy and momentum need to be treated as a single physical entity, the classical three-momentum and energy conservation laws will be unified in a conservation law in four dimensions involving the vanishing divergence of a four-tensor. This four-tensor is the stress-energy tensor which we will denote in contravariant components with $T^{\alpha\beta}$. Its components contain

$$\begin{pmatrix} T^{00} & T^{0i} \\ T^{i0} & T^{ij} \end{pmatrix} = \begin{pmatrix} \text{energy density} & \text{energy flux} \\ \text{momentum flux} & \text{stresses} \end{pmatrix} . \tag{22.32}$$

In the rest frame of the fluid, explicit expressions for these components are easily given since fluxes vanish there. The energy density $T^{0'0'}$ contains both a rest mass and an internal energy contribution. Writing the specific internal energy in the fluid rest frame as ϵ, the expression for the energy density is

$$T^{0'0'} = \rho c^2 + \rho\epsilon \,. \tag{22.33}$$

When the pressure in the fluid rest frame is denoted by p, isotropic pressure corresponds to stresses

$$T^{i'j'} = p\,\mathsf{I} \,. \tag{22.34}$$

When we now transform back to the laboratory frame, where the gas moves with velocity $\mathbf{v}$, we compute

$$T^{\alpha\beta} = (L^{-1})^{\alpha}_{\alpha'} (L^{-1})^{\beta}_{\beta'} T^{\alpha'\beta'} \,, \tag{22.35}$$

which yields the expression

$$T^{\alpha\beta} = \left(\rho c^2 + \rho\epsilon + p\right) \frac{U^\alpha U^\beta}{c^2} + p g^{\alpha\beta} \,. \tag{22.36}$$

The tensor is symmetric, $T^{\alpha\beta} = T^{\beta\alpha}$, and scalar invariants computed from this stress-energy tensor, upon which all observers will agree, are its trace defined as

$$T^{\alpha}_{\ \alpha} = T^{\alpha\beta} g_{\alpha\beta} = 3p - \rho c^2 - \rho\epsilon \,, \tag{22.37}$$

and

$$T^{\alpha\beta}T_{\alpha\beta} = \left(\rho c^2 + \rho\epsilon\right)^2 + 3p^2 \,. \tag{22.38}$$

We can write this tensor in the form of Eq. (22.32), and introduce some convenient notation as

$$\begin{pmatrix} T^{00} & T^{0i} \\ T^{i0} & T^{ij} \end{pmatrix} = \begin{pmatrix} \left(\rho c^2 + \rho\epsilon + p\right)\Gamma^2 - p & \left(\rho c^2 + \rho\epsilon + p\right)\Gamma^2 \mathbf{v}/c \\ \left(\rho c^2 + \rho\epsilon + p\right)\Gamma^2 \mathbf{v}/c & \rho\Gamma^2\mathbf{vv} + p\mathbf{I} + \left(\rho\epsilon + p\right)\Gamma^2\mathbf{vv}/c^2 \end{pmatrix}$$
$$\equiv \begin{pmatrix} \tau_{\rm g} + Dc^2 & \mathbf{S}_{\rm g}/c \\ \mathbf{S}_{\rm g}/c & \mathbf{S}_{\rm g}\mathbf{v}/c^2 + p\mathbf{I} \end{pmatrix} . \tag{22.39}$$

In terms of the variables $\tau_{\rm g}$ (the total energy density in the lab frame minus the rest mass contribution), and the three-vector $\mathbf{S}_{\rm g}$ (the relativistic energy flux), the divergence of the stress-energy tensor is then written as follows. The temporal part, i.e. the equation $\partial_\alpha T^{0\alpha} = 0$, will yield

$$\frac{\partial}{\partial t}\left(\tau_{\rm g} + Dc^2\right) + \nabla\cdot\mathbf{S}_{\rm g} = 0 \,. \tag{22.40}$$

We can combine this relation with particle conservation expressed by Eq. (22.31) to obtain

$$\frac{\partial \tau_{\rm g}}{\partial t} + \nabla\cdot\left[(\tau_{\rm g} + p)\mathbf{v}\right] = 0 \,. \tag{22.41}$$

The spatial part $\partial_\alpha T^{i\alpha} = 0$ works out to be

$$\frac{\partial \mathbf{S}_{\rm g}}{\partial t} + \nabla\cdot\left(\mathbf{S}_{\rm g}\mathbf{v} + pc^2\mathbf{I}\right) = 0 \,. \tag{22.42}$$

Note that in these variables, the classical Newtonian limits are directly obtained, since in that limit we find

$$\mathbf{S}_{\rm g} \overset{\Gamma\to 1}{\to} c^2\rho\mathbf{v} \,, \qquad \tau_{\rm g} \overset{\Gamma\to 1}{\to} \tfrac{1}{2}\rho v^2 + \rho\epsilon \,. \tag{22.43}$$

This set of equations still needs a specification of an equation of state, relating specific energy ϵ to the gas pressure p and the proper density ρ.

Thermodynamics and special relativity The combination of mass energy, internal energy and pressure appearing in both the energy density $\tau_{\rm g}$ and energy flux vector $\mathbf{S}_{\rm g}$ is known as the relativistic enthalpy. We can introduce the specific enthalpy h by

$$\rho h \equiv \rho c^2 + \rho\epsilon + p \equiv \rho(c^2 + h_{\rm g}) \,, \tag{22.44}$$

where $h_{\rm g} = \epsilon + p/\rho$ is the specific enthalpy of the gas as used in non-relativistic treatments. With these definitions, we can now work out the following scalar identity

$$U_\alpha \partial_\beta T^{\alpha\beta} = 0 \,, \tag{22.45}$$

which must hold since we already had vanishing $\partial_\beta T^{\alpha\beta}$ for each $\alpha = 0, 1, 2, 3$. In doing so, we can use relation (22.20) to find that for all $\beta = 0, 1, 2, 3$ we have $U^\alpha \partial_\beta U_\alpha = 0$. The temporal relation $\beta = 0$ gives an alternative expression for $\partial\Gamma/\partial t$, while the spatial ones yield expressions for $\partial\Gamma/\partial x_i$. When we also use the particle conservation law (22.31), we can rework Eq. (22.45) to

$$\rho\Gamma\Big(\frac{\partial h_{\rm g}}{\partial t} + \mathbf{v}\cdot\nabla h_{\rm g}\Big) - \Gamma\Big(\frac{\partial p}{\partial t} + \mathbf{v}\cdot\nabla p\Big) = 0\,. \qquad (22.46)$$

Note that $\Gamma(\partial/\partial t + \mathbf{v}\cdot\nabla) \equiv U^\alpha\partial_\alpha$ is the derivative along the world line.

When we now specify the discussion to an ideal gas law equation of state, we make a commonly encountered simplification with constant polytropic index γ appearing as

$$\rho\epsilon = p/(\gamma-1)\,. \qquad (22.47)$$

It is then a matter of algebra to manipulate Eq. (22.46) to one in terms of the entropy related quantity $S = p\rho^{-\gamma}$, written as

$$\Gamma\big(\partial/\partial t + \mathbf{v}\cdot\nabla\big)S = 0\,. \qquad (22.48)$$

Hence, as in the non-relativistic case, entropy is advected with the fluid. Furthermore, it is possible to show (see Section 22.1.4) that the relativistic expression for the sound speed $c_{\rm g}$ in the polytropic case becomes

$$c_{\rm g}/c = \sqrt{\gamma p/(\rho h)}\,. \qquad (22.49)$$

The non-relativistic limit clearly reduces to $c_{\rm g} = \sqrt{\gamma p/\rho}$, as expected. Furthermore, this relation also shows that $c_{\rm g}^2 \leq (\gamma-1)c^2$, so that there is an upper limit to sound propagation speeds.

▷ **The Synge gas** The constant γ in Eq. (22.47) is in practice always taken in the range $4/3 \leq \gamma \leq 5/3$. In fact, for a perfect gas law where $p = k_{\rm B}\rho T/m_0$, the lower value $\gamma = 4/3$ is applicable at relativistic internal energies, i.e. $k_{\rm B}T \geq m_0c^2$, while the non-relativistic limit for a monatomic gas is known to be $5/3$. A proper generalization of the perfect gas law to relativistic regimes needs to start from a relativistic kinetic plasma theory and the relativistic counterpart of the Boltzmann equation. For an equilibrium distribution function known as the Juttner distribution function (the relativistic counterpart of the Maxwell–Boltzmann distribution), one can again deduce the perfect gas law relation $p = k_{\rm B}\rho T/m_0$. Moreover, it is then found that

$$\rho c^2 + \rho\epsilon + p = \rho c^2 G(z)\,, \quad G(z) \equiv \frac{K_3(z)}{K_2(z)}\,, \qquad (22.50)$$

where K_2 and K_3 denote the modified Bessel functions of the second kind, with argument $z \equiv m_0c^2/k_{\rm B}T = \rho c^2/p$. If we then write in analogy with the polytropic case

$$\hat\gamma \equiv 1 + p/(\rho\epsilon)\,, \qquad (22.51)$$

we find the relation $1 + \hat\gamma/[z(\hat\gamma-1)] = G(z)$. To find the effective polytropic index $\hat\gamma(z)$ at low temperature ($z \to \infty$) and high temperature $z \to 0$, we can use the asymptotic expansions at small and large argument,

$$G(z) \overset{z\to\infty}{\approx} 1 + 5/(2z)\,, \qquad G(z) \overset{z\to 0}{\approx} 4/z\,. \qquad (22.52)$$

The actual polytropic index thus increases smoothly between $4/3$ at ultra-relativistic internal energies and its non-relativistic value $5/3$, when plotted as a function of z. This relativistically correct perfect gas description is known as the Synge gas [583]. The sound speed is then given by

$$\frac{c_{\rm g}^2}{c^2} = \frac{G'}{G(zG' + 1/z)}\,, \qquad (22.53)$$

where the derivative $G' = G^2 - 1 - 5G/z$, due to recurrence relations. For computational approaches, one can use an approximation due to Mathews [433], which avoids the (costly) Bessel function evaluations and sets

$$
\begin{aligned}
1+\epsilon/c^2 &= \frac{p}{(\gamma-1)\rho c^2}+\left[\left(\frac{p}{(\gamma-1)\rho c^2}\right)^2+1\right]^{1/2}, \\
h &= \tfrac{1}{2}c^2\left[(\gamma+1)(1+\epsilon/c^2)-(\gamma-1)/(1+\epsilon/c^2)\right]. \qquad (22.54)
\end{aligned}
$$

In this approximation, the sound speed is given by

$$
c_{\mathrm{g}}^2/c^2 = \left[p/(2\rho h)\right]\left[\gamma+1+(\gamma-1)/(1+\epsilon/c^2)^2\right], \qquad (22.55)
$$

while the locally effective polytropic index is then given by

$$
\hat{\gamma} = \gamma - \tfrac{1}{2}(\gamma-1)\left[1-1/(1+\epsilon/c^2)^2\right]. \qquad (22.56)
$$

In these expressions, setting the parameter $\gamma = 5/3$ yields an excellent approximation to the Synge gas variation, while non-adiabatic effects can be approximated using different γ values. ◁

22.1.4 Sound waves and shock relations in relativistic gases

To obtain the relativistic expression for the sound wave speed, one must perform the usual linearization of the relativistic hydrodynamic equations about an equilibrium configuration. We will do this for a homogeneous gas, and adopt a strategy which rewrites the governing covariant equations in four-dimensional space-time to equivalent expressions in a 3 + 1 formalism in terms of a convenient set of primitive variables. The 3 + 1 formalism splits temporal and spatial derivatives in a fixed Lorentzian reference frame, and shows clearly how the equations become more involved with respect to their Galilean versions used in all previous chapters. Also, the 3 + 1 formalism is the one adopted in modern numerical approaches, where one fixes a laboratory frame and uses shock-capturing schemes to solve the governing conservation laws. We choose as primitive variables the entropy S, the rest frame proper density ρ and the velocity $\mathbf{v}$ in our laboratory frame. For simplicity, we restrict the discussion here to a constant polytropic equation of state, where $S = p\rho^{-\gamma}$ and note that we already have Eq. (22.48) for the entropy. In a similar fashion, one can (after some algebra) obtain the following set of equations:

$$
\begin{aligned}
&\frac{\partial S}{\partial t}+\mathbf{v}\cdot\nabla S = 0, \\
&\frac{\partial \rho}{\partial t}+\mathbf{v}\cdot\nabla\rho+\frac{\rho h}{u}\nabla\cdot\mathbf{v}-\frac{1}{u\Gamma^2}\mathbf{v}\cdot\nabla\left(S\rho^{\gamma}\right) = 0, \\
&\frac{\partial \mathbf{v}}{\partial t}+(\mathbf{v}\cdot\nabla)\,\mathbf{v}+\frac{c^2}{\rho h\Gamma^2}\nabla\left(S\rho^{\gamma}\right)-\mathbf{v}\left(\nabla\cdot\mathbf{v}\right)\left(1-\frac{yc^2}{u}\right)-\mathbf{v}\frac{yc^2}{u\rho h\Gamma^2}\mathbf{v}\cdot\nabla\left(S\rho^{\gamma}\right) = 0. \qquad (22.57)
\end{aligned}
$$

In these equations, we have used the abbreviations

$$
\begin{aligned}
u &\equiv h-\frac{v^2}{c^2}\gamma S\rho^{\gamma-1}\quad(\rightarrow c^2), \qquad y \equiv \frac{2h}{c^2}-1-\frac{\gamma^2 S\rho^{\gamma-1}}{c^2(\gamma-1)}\quad(\rightarrow 1), \\
h &\equiv c^2+\frac{\gamma S\rho^{\gamma-1}}{\gamma-1}\quad(\rightarrow c^2), \qquad (22.58)
\end{aligned}
$$

where the expressions between brackets denote the Galilean limits when $\Gamma \rightarrow 1$. In this form, no approximations have been made yet, and we can clearly identify all terms denoting relativistic corrections. We will next use this form as a starting point to obtain linear wave speeds in two complementary approaches.

(a) Waves in a static homogeneous gas The equations (22.57) can easily be linearized about a static ($\mathbf{v}_0 = 0$) uniform gas with constant entropy and density S_0, ρ_0. As before, one conveniently assumes a plane wave variation $\exp(-\mathrm{i}\omega t + \mathrm{i}\mathbf{k}\cdot\mathbf{x})$ of all linear quantities $S_1, \rho_1, \mathbf{v}_1$ to arrive at

$$\omega S_1 = 0\,, \qquad \omega\rho_1 = \rho_0 \mathbf{k}\cdot\mathbf{v}_1\,, \qquad \omega\mathbf{v}_1 = \frac{c^2}{\rho_0 h_0}\mathbf{k}\left(S_0\gamma\rho_0^{\gamma-1}\rho_1 + \rho_0^{\gamma}S_1\right). \tag{22.59}$$

As already known from the non-relativistic case, this system admits five solutions, where three wave modes are at marginal frequency $\omega = 0$. These are the entropy wave with arbitrary S_1 but without density or velocity perturbation, together with the two transverse translations (shear waves), already encountered in Section 5.2.2. The physically more interesting modes are compressible perturbations with $\mathbf{k}\cdot\mathbf{v}_1 \neq 0$ that satisfy the dispersion equation

$$\frac{\omega^2}{k^2c^2} = \frac{\gamma S_0\rho^{\gamma-1}}{h_0} = \frac{\gamma p_0}{\rho_0 h_0}\,. \tag{22.60}$$

The sound speed c_{g} for the polytropic case is thus in accord with Eq. (22.49).

(b) Characteristic speeds A second approach to using the equations (22.57) follows the technique discussed in Section 20.1.1, noting that this form allows us to read off the components of the 5×5 coefficient matrix $\mathbf{W}$ in the quasi-linear form from Eq. (20.3), namely

$$\frac{\partial\mathbf{V}}{\partial t} + \mathbf{W}\cdot\frac{\partial\mathbf{V}}{\partial x} = 0\,, \tag{22.61}$$

where the primitive variables are $\mathbf{V} = (S, \rho, v_x, v_y, v_z)^{\mathrm{T}}$. When we compute the five eigenvalues λ of the $\mathbf{W}$ matrix, we obtain the characteristic equation

$$(\lambda - v_x)^3\left[\lambda^2 - 2\lambda v_x\frac{1-c_{\mathrm{g}}2/c^2}{1-v^2c_{\mathrm{g}}^2/c^4} + \frac{v_x^2\left(1-c_{\mathrm{g}}^2/c^2\right) - c_{\mathrm{g}}^2\left(1-v^2/c^2\right)}{1-v^2c_{\mathrm{g}}^2/c^4}\right] = 0\,. \tag{22.62}$$

Hence, the characteristic speeds either take the value $\lambda = v_x$ (which obviously correspond to the entropy and shear waves from above), or the value for the sound waves this time found from a more complicated quadratic expression. Naturally, both approaches must agree. The key observation is that by computing the characteristic speeds from the $\mathbf{W}$ matrix, we in fact linearized the equations about a moving plasma, and that we need to consider how plane waves in the gas rest frame transform relativistically to a moving reference frame. We therefore address how plane waves behave under the Lorentz transformation.

▷ **Exercise** Consider the purely 1D case where $v = v_x$, and show that the quadratic expression in Eq. (22.62) corresponds to relativistic speed addition between v_x and c_{g}. ◁

(c) Phase and group diagrams in special relativity As usual, we will consider frame L′ with coordinates $(ct', \mathbf{x}')$ to move with respect to frame L with a velocity $\mathbf{v}$. Equation (22.1) then relates the time and space coordinates, and we momentarily assume that in L′ we have a plane wave with variation $\exp[-\mathrm{i}(\omega' t' - \mathbf{k}'\cdot\mathbf{x}')]$. Anticipating Doppler shifts as well as a change in the wave vector, we indicate the frequency and wave vector to be specific to L′. One then finds directly that frame L

will still observe a plane wave with variation $\exp[-\mathrm{i}(\omega t - \mathbf{k}\cdot\mathbf{x})]$ with frequency and wave vector given by

$$\omega = \Gamma\left(\omega' + \mathbf{k}'\cdot\mathbf{v}\right), \qquad \mathbf{k} = \mathbf{k}' + \mathbf{v}\left[\frac{\omega'\Gamma}{c^2} + (\mathbf{k}'\cdot\mathbf{v})\frac{\Gamma - 1}{v^2}\right]. \tag{22.63}$$

These expressions quantify the relativistic Doppler effect (i.e. the change in frequency) and show that the wave vector changes direction when viewed from a moving vantage point. The latter effect is known as relativistic wave aberration. The inverse formulas (which are identical with $\omega' \leftrightarrow \omega$, $\mathbf{k}' \leftrightarrow \mathbf{k}$ and a sign change for $\mathbf{v} \leftrightarrow -\mathbf{v}$) then allow us to find the phase speed for frame L given by $v_{\mathrm{ph}} = \omega/k$ from the formula

$$\frac{v'^2_{\mathrm{ph}}}{c^2} = \frac{\Gamma^2\left(v_{\mathrm{ph}} - \mathbf{n}\cdot\mathbf{v}\right)^2}{c^2 + \Gamma^2\left(v_{\mathrm{ph}} - \mathbf{n}\cdot\mathbf{v}\right)^2 - v^2_{\mathrm{ph}}}. \tag{22.64}$$

We introduced, in analogy with the notation of Section 5.4.3, the unit vector $\mathbf{n} = \mathbf{k}/k$. When we invert this formula to find the phase speed in frame L, we get

$$v^2_{\mathrm{ph}} - 2v_{\mathrm{ph}}(\mathbf{n}\cdot\mathbf{v})\frac{1 - v'^2_{\mathrm{ph}}/c^2}{1 - v^2 v'^2_{\mathrm{ph}}/c^4} + \frac{(\mathbf{n}\cdot\mathbf{v})^2\left(1 - v'^2_{\mathrm{ph}}/c^2\right) - v'^2_{\mathrm{ph}}\left(1 - v^2/c^2\right)}{1 - v^2 v'^2_{\mathrm{ph}}/c^4} = 0. \tag{22.65}$$

This agrees with expression (22.62) for the sound waves in the moving reference frame.

To complete this discussion on plane wave propagation in uniform gases for special relativity, we can draw the phase and group speed diagrams in different reference frames. This is done in Fig. 22.3, where the $z-x$ plane is drawn. We show at top left the phase $v_{\mathrm{ph}}\mathbf{n}/c$ and group speed $\mathbf{v}_{\mathrm{gr}} = \partial\omega/\partial\mathbf{k}$ in the gas rest frame, where we find isotropic sound wave propagation in all directions, and where phase and group speed diagrams overlap. The sound wave diagram will always be interior to the inner dashed circle which corresponds to the upper limit $\sqrt{1/3}c$, while the outer circle indicates the light limit. In this diagram, we actually adopted the Mathews prescription described above, and took $\rho_0 = 1, p_0 = 1$ (a relativistically hot gas, making the sound speed very close to the upper limit) and scaled with $c = 1$. In the right panel, the phase speed diagram is plotted as seen in a reference frame where the plane wave emitter is moving at velocity $\mathbf{v} = 0.9c\mathbf{e}_z$ along the horizontal z-axis. The wave aberration effects deformed the single circle to a kind of double loop form. In the bottom panels, we indicated how a Huygens construction then yields the corresponding group speed diagram in the same reference frame. As before, the group diagram is what evolves from a point perturbation in a finite time. The final panel just shows this group speed diagram in that frame, and demonstrates how the wave front gets 'beamed' into an anti-symmetric (about the position of the point source) oval shape.

▷ **Exercise** Obtain the analytic expressions for the phase and group speed diagrams for sound waves in both reference frames, and show that the group speed diagram clearly follows the speed addition rule from the expression in Exercise [22.2]). ◁

(d) Gas dynamic shock relations To discuss gas dynamic shocks in special relativity, we need to go back to the actual conservation laws in four-dimensional space-time, and consider the limits where discontinuities occur across moving manifolds in space-time. The shock front is generally a surface in space-time, which is given by an equation $\phi(ct, \mathbf{x}) = \text{const}$. The normal to the shock

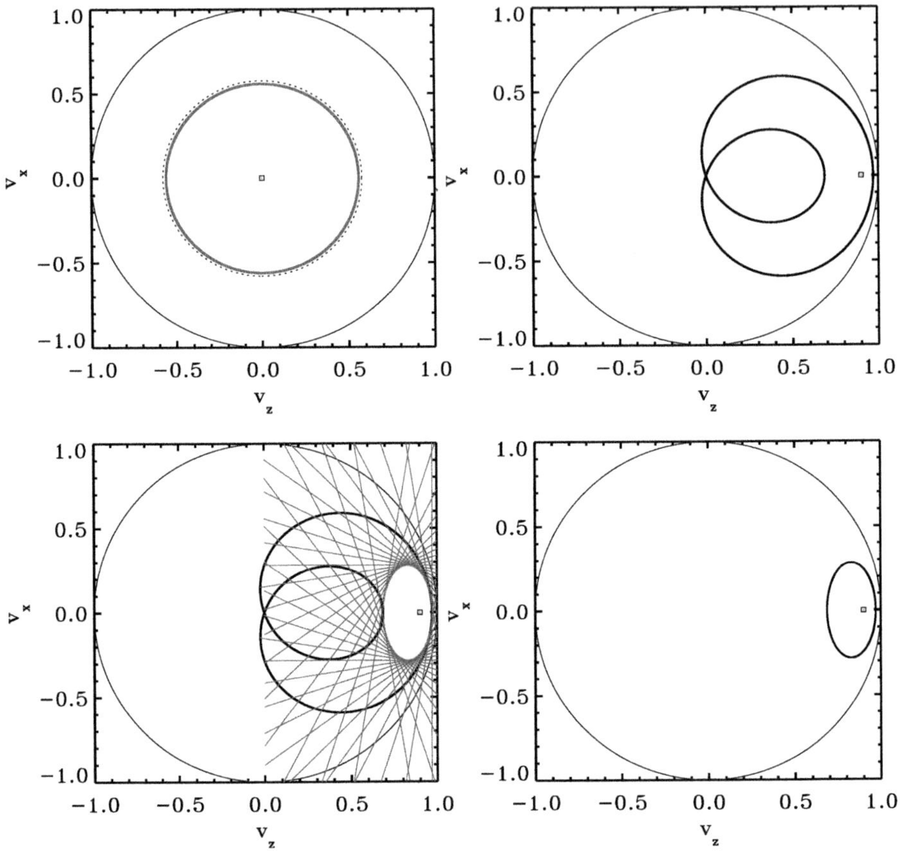

Fig. 22.3 Phase and group speed diagram in the gas rest frame (top left), compared to the phase speed in a frame where the source moves at $0.9c$ along the z-axis (top right). The bottom left panel shows the Huygens construction, which can be used to obtain the group speed diagram (bottom right) in the same lab frame.

front is a space-like four-vector $\mathbf{l}$ (i.e. one whose invariant $l^\alpha l_\alpha > 0$), whose components are given by $l_\alpha = \partial_\alpha \phi$. We are free to normalize this such that $l^\alpha l_\alpha = 1$. The Rankine–Hugoniot conditions always follow from the conservation laws across the manifold, expressed generally as

$$[\![\rho U^\alpha]\!] l_\alpha = 0\,, \tag{22.66}$$

$$[\![T^{\alpha\beta}]\!] l_\alpha = 0\,. \tag{22.67}$$

These expressions can be written out in various reference frames, and can be used to quantify shock compression ratios, or be manipulated to identify various shock invariants. In analogy with Chapter 21, we will write out the expressions in the shock rest frame (SRF), where by definition the four-velocity of the shock is $\mathbf{U}^{\mathrm{s}} = (c, \mathbf{0})$. Without loss of generality, we can assume the shock normal to be oriented along the x-axis, such that $\mathbf{l} = (0, \mathbf{e}_x) \equiv (0, \mathbf{n})$, where the space part is then indicated by a three-vector $\mathbf{n}$. The four-velocities at right and left of the shock front are denoted in

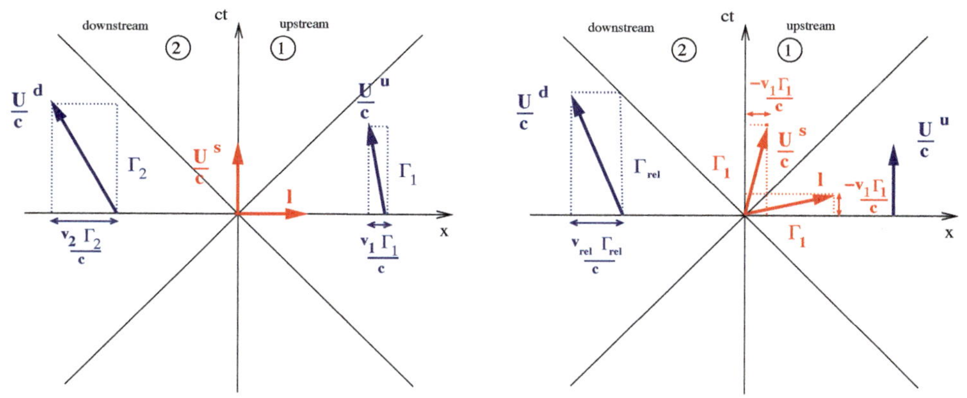

Fig. 22.4 The space-time diagrams appropriate for analysis of the Rankine–Hugoniot shock relations in the shock rest frame (SRF, left panel), and in the upstream rest frame (URF, right panel).

this SRF as

$$\mathbf{U}^{\mathrm{u}} = (c\Gamma_1, \Gamma_1 \mathbf{v}_1) \,, \qquad \mathbf{U}^{\mathrm{d}} = (c\Gamma_2, \Gamma_2 \mathbf{v}_2) \,, \tag{22.68}$$

where the index 1 is for the upstream state, while 2 denotes the downstream state.

A graphical representation of the various four-vectors in the SRF is given in the left panel of Fig. 22.4 (the right panel shows the same quantities in the upstream rest frame). Besides the three-velocities $\mathbf{v}_1$ and $\mathbf{v}_2$, which indicate the velocity of the upstream (downstream) gas with respect to the shock, a third relative velocity can be identified: the relative velocity of the upstream gas with respect to the downstream gas $\mathbf{v}_{\mathrm{rel}}$ (its explicit formula can be obtained from velocity addition rules). From the jump relations (22.66)–(22.67) in the SRF we find that

$$\begin{aligned} \rho_1 \Gamma_1 v_{1n} &= \rho_2 \Gamma_2 v_{2n} \,, \\ \rho_1 h_1 \Gamma_1^2 v_{1n} \mathbf{v}_1 + c^2 p_1 \mathbf{n} &= \rho_2 h_2 \Gamma_2^2 v_{2n} \mathbf{v}_2 + c^2 p_2 \mathbf{n} \,, \\ h_1 \Gamma_1 &= h_2 \Gamma_2 \,. \end{aligned} \tag{22.69}$$

From the second equation it follows that the tangential velocities do not jump, i.e. $[\![\mathbf{v}_{\mathrm{t}}]\!] = 0$, where $\mathbf{t} \perp \mathbf{n}$ is a unit three-vector perpendicular to $\mathbf{n}$ in the Euclidean sense. We can thus Lorentz transform to a frame which is moving with this constant, purely tangential, three-velocity. Denoting this reference frame by tangential reference frame or TRF, the gas dynamic shock relations are then reduced to 1D relations as all tangential velocities vanish, and we find

$$\rho_1 \Gamma_1 v_1 = \rho_2 \Gamma_2 v_2 \,, \qquad \rho_1 h_1 \Gamma_1^2 v_1^2 + c^2 p_1 = \rho_2 h_2 \Gamma_2^2 v_2^2 + c^2 p_2 \,, \qquad h_1 \Gamma_1 = h_2 \Gamma_2 \,. \tag{22.70}$$

Note that although we used the same symbols to denote the velocities up and downstream, these in Eq. (22.70) are in the TRF, while in Eq. (22.69) we have velocities as observed in the SRF, with the two related by a Lorentz transformation involving the tangential velocity seen in the SRF. Further discussion of these shock relations is then possible for various physically interesting limiting cases, and we refer to the topical review by Kirk and Duffy [366] to find some quantitative expressions. As a final note on gas dynamic shocks, one can manipulate the expressions to the following generalization of the Hugoniot adiabat from Eq. (21.20), which is known as the Taub adiabat:

$$h_1^2 - h_2^2 + (p_2 - p_1)\left(\frac{h_1}{\rho_1} + \frac{h_2}{\rho_2}\right) = 0\,. \tag{22.71}$$

This is an invariant across the shock, and thus holds in any reference frame.

▷ **Exercise** Verify that the coefficient of the term proportionate to c^2 in expression (22.71) reduces to the Hugoniot adiabat from Eq. (21.20). ◁

Example application: The relativistic precessing SS433 jet Three-dimensional treatments are a prerequisite to model the sub-parsec scale behaviour observed in the archetypical X-ray binary SS433. In that well observed binary system, a hard X-ray source is categorized as a 9 $M_\odot$ compact object (black hole), accreting matter from a supergiant star at a rate of up to $10^{-4} M_\odot$ per year. Mildly relativistic jets with speeds $0.26c$ are persistently ejected from the central accreting region, and these jets (and the underlying disc) precess with a period of 162 days, with an estimated inclination angle between disc and orbital plane of 20° [76]. Together, these values determine a purely kinematic model for the jet path. To go beyond this essentially 1D model, we need 3D relativistic hydro simulations to ultimately confront the detailed Very Large Array (VLA) radio observations of the jets below and up to the parsec scale, with SS433 itself being located at an inferred 5.5 kpc.

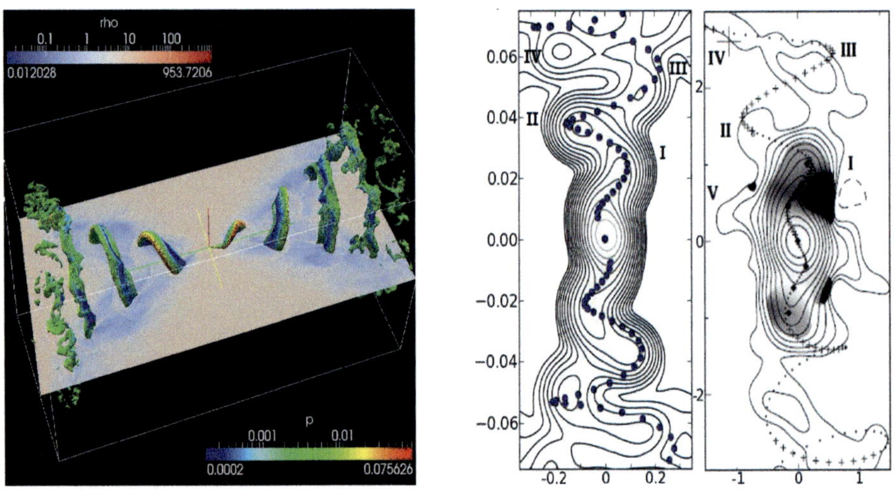

Fig. 22.5 Left: A 3D simulation for SS433 [445], where the isosurface identifies the jet location, as quantified by an advected tracer, while the pressure variation is used to colour this tracer isosurface. The density variation is shown in a cross-sectional plane. Right: a synthetic radio map produced from the simulation matches well with the observational VLA radio map [521]. Several of the bright features labelled as I–IV are reproduced. Both panels overlay a purely kinematic model as a spiral curve. Axes units are in parsec versus arcsecond, respectively.

To do this, Monceau-Baroux et al. [445] used relativistic hydro simulations, taking the observed energy flux of order $10^{39}\,\mathrm{erg\,s^{-1}}$, as well as the known dynamically evolving geometry, into account. To handle the precessional motion, the jet is treated through an internal boundary prescription which fills a precessing cone at the given 20° inclination angle. The jet radius at the end of these cones is 0.0007 pc, and the simulation covers a box of size $0.2 \times 0.1 \times 0.1$ pc^3, within which the jet propagation is followed over several years to obtain the jet morphology throughout the domain. Models were constructed that contrasted non-precessing with precessing jets, with initially overdense and slow ($0.26c$ or $\Gamma \approx 1.036$) to mildly underdense and fast ($\Gamma \approx 1.86$) cases. The findings highlighted that the overdense precessing case can reproduce many of the observed features at these scales, and especially the overall helical shape of the twin jet system. The precessing jets enlarge the volume affected by jet propagation significantly, while an underdense jet prescription is sweeping up so much of the surrounding material that it gets hampered in its propagation, despite its initial higher beam Lorentz factor. Using the instantaneous thermodynamical values, as well as energy quantifications of a passively treated electron distribution, one can produce synthetic radio maps which rival modern VLA analogs for SS433. Precessing motions clearly enrich earlier jet propagation findings based on axi-symmetric studies or 1D analytic momentum balance estimates, since the motion gets decelerated as it distributes the ram pressure of the jet over a geometrically involved time-evolving 3D area. An impression of the helical jet best matching SS433 observations is shown in Fig. 22.5.

22.2 Electromagnetism and special relativistic MHD

When dealing with plasmas, with charged particle constituents, electric and magnetic fields are generated by static as well as moving charged particles. The electric and magnetic fields themselves are governed by the Maxwell equations. In Section 2.2.2, we wrote these Maxwell equations in terms of three-vector electric and magnetic fields. However, static charges which act as sources for electric fields in one Lorentzian reference frame will be moving charges, i.e. currents, leading to magnetic fields in another frame in relative motion. The combined six components of both three-vectors will in fact form the six independent tensor components of an asymmetric four-tensor of rank two. In what follows, we revisit the Maxwell equations in the appropriate tensorial formulation applicable in space-time. We then generalize the stress-energy tensor to include electromagnetic field contributions. Finally, we specify the governing equations for relativistic plasma dynamics for perfectly conducting plasmas, where the electric field in the co-moving frame vanishes. This yields the ideal MHD equations in special relativity.

22.2.1 Electromagnetic field tensor and Maxwell's equations

In analogy with Eqs. (22.25), the relativistically correct equation of motion for a particle of charge q, written in three-vector notation, can be expected to be

$$\frac{d\mathbf{p}}{dt} = q\left(\mathbf{E} + \mathbf{v} \times \mathbf{B}\right) . \tag{22.72}$$

Since the Lorentz force is orthogonal to the spatial three-velocity, the change in relativistic energy $E = \Gamma m_0 c^2$ is given by

$$\frac{dE}{dt} = \mathbf{v} \cdot (q\mathbf{E}) . \tag{22.73}$$

As the Lorentz force involves the fluid velocity, we need to generalize Eq. (22.24) to a four-momentum formulation with a four-force involving the four-velocity in

$$\frac{dP^\alpha}{d\tau} = qF^{\alpha\beta}U_\beta \,. \tag{22.74}$$

The tensor components of the occurring electromagnetic field tensor $F^{\alpha\beta}$ can in fact be computed from requiring the spatial and temporal components of this expression (22.74) to reduce to Eq. (22.72) and Eq. (22.73), respectively. The end result is the anti-symmetric tensor with contravariant components given by

$$F^{\alpha\beta} = \begin{pmatrix} 0 & E_1/c & E_2/c & E_3/c \\ -E_1/c & 0 & B_3 & -B_2 \\ -E_2/c & -B_3 & 0 & B_1 \\ -E_3/c & B_2 & -B_1 & 0 \end{pmatrix} . \tag{22.75}$$

The components of the electric field $\mathbf{E}$ and magnetic field $\mathbf{B}$ as observed in this particular reference frame thus form the six independent entries of this tensor. Associated with this tensor, we can compute the invariant scalar quantity

$$F^{\alpha\beta}F_{\alpha\beta} = 2\left(\mathbf{B}\cdot\mathbf{B} - \mathbf{E}\cdot\mathbf{E}/c^2\right) . \tag{22.76}$$

All observers will thus agree on the relative magnitudes of electric and magnetic field vectors, $|\mathbf{E}|$ and $|\mathbf{B}|$. In particular, for electromagnetic (plane) waves in vacuum where $|\mathbf{E}| = c|\mathbf{B}|$, all observers will find a zero value for the invariant (22.76).

When we change from one Lorentzian frame to another by means of the Lorentz transformation (22.3), the components of the electromagnetic field tensor transform in the usual manner. When changing from frame L to frame L′, moving with velocity $\mathbf{v}$ with respect to L, we thus find from

$$F^{\alpha'\beta'} = \begin{pmatrix} 0 & E_1'/c & E_2'/c & E_3'/c \\ -E_1'/c & 0 & B_3' & -B_2' \\ -E_2'/c & -B_3' & 0 & B_1' \\ -E_3'/c & B_2' & -B_1' & 0 \end{pmatrix} = L^{\alpha'}_{\alpha} L^{\beta'}_{\beta} F^{\alpha\beta} \tag{22.77}$$

that terms of the three-vectors for electric and magnetic fields is as given by Eqs. (2.23) of Section 2.2.2.

We can introduce a four-vector equivalent of the electric field $\mathbf{E}$, when we write the governing equation of motion (22.74) as

$$\frac{dP^\alpha}{d\tau} = qF^{\alpha\beta}U_\beta = qe^\alpha \,. \tag{22.78}$$

The contravariant components e^α then quantify the electric field measured in the local rest frame of the moving plasma (analogous to $q\mathbf{E}$ for electrostatic acceleration of a static charge q in three-vector form). We can write this four-vector as

$$e^\alpha = \left[\Gamma\mathbf{v}\cdot\mathbf{E}/c, \Gamma(\mathbf{E} + \mathbf{v}\times\mathbf{B})\right]^{\mathrm{T}} . \tag{22.79}$$

Note that this four-vector vanishes when $\mathbf{E} = -\mathbf{v}\times\mathbf{B}$, and this observation will lead us in Section 22.2.2 to the ideal MHD limit in special relativity. Generally, though, the electric four-vector is orthogonal (in a space-time sense) to the four-velocity, since it is easily shown that

$$e^{\alpha} U_{\alpha} = 0 \,. \tag{22.80}$$

An invariant quantity is then also

$$e^{\alpha} e_{\alpha} = \Gamma^2 \Big[(\mathbf{E} + \mathbf{v} \times \mathbf{B}) \cdot (\mathbf{E} + \mathbf{v} \times \mathbf{B}) - (\mathbf{v} \cdot \mathbf{E}/c)^2 \Big] \,. \tag{22.81}$$

To arrive at Maxwell's equations in four-dimensional space-time notation, we will need to handle the source terms appearing in the non-homogeneous equations

$$\nabla \cdot \mathbf{E} \;= c^2 \sigma \,, \qquad \nabla \times \mathbf{B} = \mathbf{j} + \frac{1}{c^2} \frac{\partial \mathbf{E}}{\partial t} \,. \tag{22.82}$$

The charge density is given by $\sigma = qn = q\Gamma n_0$, where the charged particle number density is n (and n_0 in the local rest frame). The current density three-vector is $\mathbf{j} = qn\mathbf{v}$. Note that we used mks units, setting $\mu_0 = 1$ for convenience (and equivalently setting the permittivity $\epsilon_0 = 1/c^2$). In analogy with four-velocity or four-momentum, we introduce a four-current $J^{\alpha} = qn_0 U^{\alpha}$ such that

$$J^{\alpha} = (c\sigma, \mathbf{j})^{\mathrm{T}} \,. \tag{22.83}$$

Its invariant is $J^{\alpha} J_{\alpha} = -c^2 q^2 n_0^2$. Charge conservation is expressed by $\partial_{\alpha} J^{\alpha} = 0$, or

$$\frac{\partial \sigma}{\partial t} + \nabla \cdot \mathbf{j} = 0 \,. \tag{22.84}$$

The non-homogeneous Maxwell equations (22.82) are then unified in the following law in terms of the electromagnetic field tensor:

$$\partial_{\beta} F^{\alpha\beta} = J^{\alpha} \,. \tag{22.85}$$

Also, with the aid of this four-current J^{α}, we can combine the Lorentz force, and its contribution to the energy, for an ensemble of particles with number density n. The equivalent of the single particle expression Eq. (22.74) can be written as

$$F^{\alpha\beta} J_{\beta} = F^{\alpha}_{\;\beta} J^{\beta} = \begin{pmatrix} nq\mathbf{v} \cdot \mathbf{E}/c \\ nq(\mathbf{E} + \mathbf{v} \times \mathbf{B}) \end{pmatrix} \,. \tag{22.86}$$

As a consequence, the divergence of the stress-energy tensor (22.39) for a plasma now equals

$$\partial_{\beta} T^{\alpha\beta}_{\mathrm{pl}} = F^{\alpha}_{\;\beta} J^{\beta} \,. \tag{22.87}$$

▷ **Exercise** Alternatively, we can arrive at the expression for the four-current (22.83) by noting that in the local rest frame it must boil down to $(cqn_0, \mathbf{0})^{\mathrm{T}}$, and by performing the inverse Lorentz transformation to the frame where the plasma moves at velocity $\mathbf{v}$. ◁

For any anti-symmetric second rank tensor in space-time, such as the electromagnetic field tensor $F^{\alpha\beta}$, we can define a related anti-symmetric tensor by making use of the Levi-Civita symbol. In four-dimensional space-time, this symbol is defined as

$$\epsilon_{\mu\nu\alpha\beta} = \begin{cases} +1 & \text{for any even permutation of } 0,1,2,3 \\ -1 & \text{for uneven permutations of } 0,1,2,3 \\ \;\;0 & \text{for any case with repeated indices} \end{cases} \,. \tag{22.88}$$

▷ **Parity of permutations and Levi-Civita symbols** By definition, the parity of a permutation of $0, 1, 2, 3$ is given by the parity of the amount of numbers that appear in a non-increasing order, when going from left to right. As an example, the permutation $P = \{0, 1, 2, 3\}$ has zero numbers appearing in non-increasing order, hence its parity is even. However, the cyclic permutation of P given by $Q = \{3, 0, 1, 2\}$ has three numbers appear 'in the wrong order', namely the sequence $0, 1, 2$ following 3. Therefore, its parity is odd. Permuting Q in cyclic fashion once more, we get $R = \{2, 3, 0, 1\}$ where four numbers are non-increasing: $0, 1$ follow both 2 and 3. Hence R is even again. This means that the Levi-Civita symbol in four-dimensional form changes sign under cyclic permutations of its indices, a result which is different from the perhaps more familiar 3D case. In flat space-time, we also find easily that

$$\epsilon_{\mu\nu\alpha\beta} = -\epsilon^{\mu\nu\alpha\beta} . \tag{22.89}$$

Finally, we note the following equalities (e.g. see [584]):

$$\begin{aligned}
\epsilon^{\alpha\beta\mu\nu}\epsilon_{\alpha\gamma\lambda\kappa} &= -\delta^{\beta}{}_{\gamma}\delta^{\mu}{}_{\lambda}\delta^{\nu}{}_{\kappa} + \delta^{\beta}{}_{\gamma}\delta^{\mu}{}_{\kappa}\delta^{\nu}{}_{\lambda} + \delta^{\beta}{}_{\lambda}\delta^{\mu}{}_{\gamma}\delta^{\nu}{}_{\kappa} - \delta^{\beta}{}_{\lambda}\delta^{\mu}{}_{\kappa}\delta^{\nu}{}_{\gamma} \\
&\quad - \delta^{\beta}{}_{\kappa}\delta^{\mu}{}_{\gamma}\delta^{\nu}{}_{\lambda} + \delta^{\beta}{}_{\kappa}\delta^{\mu}{}_{\lambda}\delta^{\nu}{}_{\gamma} = -\delta^{\beta\mu\nu}_{\gamma\lambda\kappa} = -\begin{vmatrix} \delta^{\beta}_{\gamma} & \delta^{\beta}_{\lambda} & \delta^{\beta}_{\kappa} \\ \delta^{\mu}_{\gamma} & \delta^{\mu}_{\lambda} & \delta^{\mu}_{\kappa} \\ \delta^{\nu}_{\gamma} & \delta^{\nu}_{\lambda} & \delta^{\nu}_{\kappa} \end{vmatrix} , \\
\epsilon^{\alpha\beta\mu\nu}\epsilon_{\alpha\beta\lambda\kappa} &= -2\delta^{\mu\nu}_{\lambda\kappa} , \\
\epsilon^{\alpha\beta\mu\nu}\epsilon_{\alpha\beta\mu\kappa} &= -6\delta^{\nu}_{\kappa} ,
\end{aligned} \tag{22.90}$$

where we introduced the generalized Kronecker delta symbols, such as $\delta^{\beta\mu\nu}_{\gamma\lambda\kappa}$ which has components that vanish unless $\beta\mu\nu$ are mutually distinct *and* $\gamma\lambda\kappa$ is a permutation of $\beta\mu\nu$. When $\gamma\kappa\lambda$ is an even permutation of $\beta\mu\nu$, its value is $+1$, while it is -1 for an uneven permutation. ◁

With the aid of the Levi-Civita symbol, we define the dual electromagnetic tensor to be

$$F^{*}_{\mu\nu} = -[1/(2c)]\,\epsilon_{\mu\nu\alpha\beta}F^{\alpha\beta} . \tag{22.91}$$

Written out in components and transforming to contravariant notation, we get

$$F^{*\alpha\beta} = g^{\alpha\mu}g^{\beta\nu}F^{*}_{\mu\nu} = \begin{pmatrix} 0 & B_1/c & B_2/c & B_3/c \\ -B_1/c & 0 & -E_3/c^2 & E_2/c^2 \\ -B_2/c & E_3/c^2 & 0 & -E_1/c^2 \\ -B_3/c & -E_2/c^2 & E_1/c^2 & 0 \end{pmatrix} . \tag{22.92}$$

We can now compute another invariant from the combination of

$$F^{*\alpha\beta}F_{\alpha\beta} = -(4/c^2)\,\mathbf{E}\cdot\mathbf{B} . \tag{22.93}$$

Hence, when the three-vector electric and magnetic fields are orthogonal in one frame, they will be orthogonal in all other inertial frames. The homogeneous Maxwell equations $\nabla\cdot\mathbf{B} = 0$ and $\nabla\times\mathbf{E} = -\partial\mathbf{B}/\partial t$ are then the temporal and spatial components of the four-law

$$\partial_\beta F^{*\alpha\beta} = 0 . \tag{22.94}$$

▷ **Exercise** It is of course also possible to write the homogeneous Maxwell equations directly in terms of the tensor $F^{\alpha\beta}$. Show that alternative forms are

$$\epsilon^{\alpha\beta\lambda\kappa}\partial_\beta F_{\lambda\kappa} = 0 . \tag{22.95}$$

This can be rewritten, using the anti-symmetry of the tensor $F_{\lambda\kappa}$, as

$$\partial_\beta F_{\lambda\kappa} + \partial_\lambda F_{\kappa\beta} + \partial_\kappa F_{\beta\lambda} = 0 . \tag{22.96}$$

Note the cyclic permutation of the indices appearing in Eq. (22.96). ◁

In complete analogy with the four-electric field, we can also consider the four-vector given by

$$b^\alpha = F^{*\alpha\beta} U_\beta \,. \qquad (22.97)$$

Splitting into temporal and spatial components, we get

$$b^\alpha = \Big[\Gamma \mathbf{v} \cdot \mathbf{B}/c \,, \Gamma(\mathbf{B} - \mathbf{v} \times \mathbf{E}/c^2)\Big]^{\mathrm{T}} . \qquad (22.98)$$

This four-vector is again found to be orthogonal in space-time to the local four-velocity since $b^\alpha U_\alpha = 0$, and is then identified as the four-magnetic field measured in the local rest frame. The invariant associated with this field is

$$b^\alpha b_\alpha = \Gamma^2 \Big[(\mathbf{B} - \mathbf{v} \times \mathbf{E}/c^2) \cdot (\mathbf{B} - \mathbf{v} \times \mathbf{E}/c^2) - (\mathbf{v} \cdot \mathbf{B}/c)^2\Big] . \qquad (22.99)$$

▷ **Alternative expressions for the electromagnetic field tensor** The four-vectors for the electric field e^α and magnetic field b^β can also be used to express the electromagnetic field tensor as follows. First note that

$$b_\alpha = F^*_{\alpha\beta} U^\beta \,. \qquad (22.100)$$

Using this allows an alternative form for the components of $\epsilon^{\alpha\beta\mu\nu} b_\alpha U_\beta$, and manipulating the expression in terms of $F^{\lambda\kappa}$ with the aid of Eqs. (22.89)–(22.90), one finds

$$\begin{aligned} F^{\alpha\beta} &= -\frac{1}{c}\epsilon^{\mu\nu\alpha\beta} b_\mu U_\nu - \frac{1}{c^2}\left(e^\alpha U^\beta - e^\beta U^\alpha\right) , \\ F^{*\alpha\beta} &= \frac{1}{c^2}\left(U^\alpha b^\beta - U^\beta b^\alpha\right) + \frac{1}{2c^3}\epsilon^{\alpha\beta\mu\nu}\left(e_\mu U_\nu - e_\nu U_\mu\right) . \end{aligned} \qquad (22.101)$$

These expressions simplify considerably for vanishing e^α, as will be the case in ideal relativistic MHD. ◁

Stress-energy tensor for electromagnetic fields The electromagnetic field also has an energy density $\frac{1}{2}(B^2 + E^2/c^2)$. The energy flux associated with the electromagnetic field is given by the Poynting flux three-vector

$$\mathbf{S}_{\mathrm{em}} = \mathbf{E} \times \mathbf{B} \,. \qquad (22.102)$$

The full Maxwell stress tensor involving both electric and magnetic field contributions is further given in three-vector notation by

$$\tfrac{1}{2}\left(B^2 + E^2/c^2\right) \mathsf{I} - \mathbf{E}\mathbf{E}/c^2 - \mathbf{B}\mathbf{B} \,. \qquad (22.103)$$

Hence, the stress-energy tensor for the electromagnetic field, in analogy with the gas dynamic case from Eq. (22.39), can be written as

$$T^{\alpha\beta}_{\mathrm{em}} = \begin{pmatrix} \frac{1}{2}(B^2 + E^2/c^2) & \mathbf{S}_{\mathrm{em}}/c \\ \mathbf{S}_{\mathrm{em}}/c & \frac{1}{2}(B^2 + E^2/c^2)\mathsf{I} - \mathbf{E}\mathbf{E}/c^2 - \mathbf{B}\mathbf{B} \end{pmatrix} . \qquad (22.104)$$

This is actually the expression for electromagnetic fields in vacuum, but this restriction will be sufficient for our purposes. Using the four-tensor for the electromagnetic field introduced in the previous section, it is possible to write it as

$$T^{\alpha\beta}_{\mathrm{em}} = F^\alpha_{\ \gamma} F^{\beta\gamma} - \tfrac{1}{4} g^{\alpha\beta} F^{\gamma\delta} F_{\gamma\delta} \,. \qquad (22.105)$$

This stress-energy tensor for the electromagnetic field is symmetric and traceless, i.e. $T^{\alpha}_{\alpha} = 0$. When we compute the divergence of this electromagnetic stress-energy tensor from the above expression, we find after some algebra

$$\partial_\beta T^{\alpha\beta}_{\mathrm{em}} = -F^{\alpha}_{\ \gamma} J^{\gamma} \,. \tag{22.106}$$

In the derivation, one uses the inhomogeneous Maxwell equation (22.85), the anti-symmetry of $F^{\beta\mu}$ to write $2F^{\beta\mu}\partial_\beta F_{\nu\mu} = F^{\beta\mu}(\partial_\beta F_{\nu\mu} - \partial_\mu F_{\nu\beta})$ and the homogeneous Maxwell equation in the form (22.96). In combination with the stress-energy of the plasma given by Eq. (22.87), the governing conservation equation then considers the combination of plasma stress-energy and electromagnetic field stress-energy, such that

$$\partial_\beta \left(T^{\alpha\beta}_{\mathrm{pl}} + T^{\alpha\beta}_{\mathrm{em}} \right) = 0 \,. \tag{22.107}$$

▷ **Entropy conservation in a relativistic plasma** Due to the conservation law (22.107), we again can write

$$U_\alpha \partial_\beta \left(T^{\alpha\beta}_{\mathrm{pl}} + T^{\alpha\beta}_{\mathrm{em}} \right) = 0 \,. \tag{22.108}$$

However, the electromagnetic part in this double summation also vanishes separately, since

$$U_\alpha \partial_\beta T^{\alpha\beta}_{\mathrm{em}} = -U_\alpha F^{\alpha}_{\ \gamma} J^{\gamma} = q n_0 e^{\beta} U_\beta = 0 \,. \tag{22.109}$$

Therefore, the derivation for the entropy equation (22.48) given for a relativistic gas also applies in the plasma case. The entropy $S = p\rho^{-\gamma}$ thus obeys

$$U^\alpha \partial_\alpha S = 0 \,, \tag{22.110}$$

which, combined with particle number conservation (22.31), yields a conservation law for $DS = \Gamma \rho S$ since

$$\frac{\partial (DS)}{\partial t} + \nabla \cdot (DS\mathbf{v}) = 0 \,. \tag{22.111}$$

This equation can then be used instead of the energy conservation law. ◁

22.2.2 Ideal MHD in special relativity

To arrive at the ideal MHD limit, we now only need to consider the consequences of all the above, when the electric field in the co-moving frame vanishes. We already mentioned that this occurs when in the lab frame $\mathbf{E} = -\mathbf{v} \times \mathbf{B}$, or equivalently $e^\alpha = 0$ with e^α expressed by Eq. (22.79). Under this assumption, several expressions given earlier simplify. For example, the invariant associated with the magnetic field four-vector is then

$$b^\alpha b_\alpha = \frac{\mathbf{B} \cdot \mathbf{B}}{\Gamma^2} + \frac{(\mathbf{v} \cdot \mathbf{B})^2}{c^2} \,. \tag{22.112}$$

This is proportional to the magnetic pressure and we write $b^\alpha b_\alpha \equiv 2p_{\mathrm{mag}}$.

▷ **Exercise** Verify that we get the same result from the transformation laws for electric and magnetic field three-vectors given by Eqs. (2.23). Note that $\mathbf{E} = -\mathbf{v} \times \mathbf{B}$ will yield $\mathbf{E}' = 0$, and that Eq. (22.112) is obtained from $\mathbf{B}' \cdot \mathbf{B}'$. ◁

Revisiting the general stress-energy tensor for electromagnetic fields, we can now write for the case where co-moving electric fields vanish (i.e. $e^\alpha = 0$)

$$T^{\alpha\beta}_{\mathrm{em}} = 2p_{\mathrm{mag}} \frac{U^\alpha U^\beta}{c^2} + p_{\mathrm{mag}} g^{\alpha\beta} - b^\alpha b^\beta \,. \tag{22.113}$$

In a similar form to the general case given by (22.104), we get for vanishing co-moving electric fields the following expression for $T^{\alpha\beta}_{\rm em}$:

$$\begin{pmatrix} \frac{1}{2}B^2 + \dfrac{1}{2c^2}[B^2v^2 - (\mathbf{v}\cdot\mathbf{B})^2] & \dfrac{\mathbf{S}_{\rm em}}{c} \\ \dfrac{\mathbf{S}_{\rm em}}{c} & \dfrac{\mathbf{S}_{\rm em}\mathbf{v}}{c^2} + p_{\rm mag}\mathbf{I} - \dfrac{\mathbf{B}\mathbf{B}}{\Gamma^2} - \dfrac{(\mathbf{v}\cdot\mathbf{B})\mathbf{v}\mathbf{B}}{c^2} \end{pmatrix}. \tag{22.114}$$

In this expression, the Poynting flux three-vector now can be written as

$$\mathbf{S}_{\rm em} = B^2\mathbf{v} - (\mathbf{v}\cdot\mathbf{B})\mathbf{B}. \tag{22.115}$$

We can now write the total energy density for the gas plus electromagnetic field in the lab frame as $\tau_{\mathcal{H}} + Dc^2$, which separates off the rest mass contribution. Hence, we have the relations

$$\tau_{\mathcal{H}} = \tau_{\rm g} + \tfrac{1}{2}B^2 + \tfrac{1}{2}[B^2v^2 - (\mathbf{v}\cdot\mathbf{B})^2]/c^2\,, \tag{22.116}$$

where the two additions to $\tau_{\rm g}$ represent magnetic and electric field energy densities, respectively. It is then seen from Eqs. (22.114) and (22.39) that the temporal component of the conservation law (22.107) is written as

$$\frac{\partial}{\partial t}\left(\tau_{\mathcal{H}} + Dc^2\right) + \nabla\cdot(\mathbf{S}_{\rm g} + \mathbf{S}_{\rm em}) = 0\,. \tag{22.117}$$

When we write the total energy flux $\mathbf{S}_{\rm tot} = \mathbf{S}_{\rm g} + \mathbf{S}_{\rm em}$ as the added relativistic energy flux of the gas and the Poynting flux, this can be combined with particle conservation $\partial D/\partial t + \nabla\cdot(D\mathbf{v}) = 0$ to obtain

$$\frac{\partial \tau_{\mathcal{H}}}{\partial t} + \nabla\cdot\left[(\tau_{\mathcal{H}} + p_{\rm tot})\mathbf{v} - (\mathbf{v}\cdot\mathbf{B})\mathbf{B}\right] = 0\,. \tag{22.118}$$

In so writing, the total pressure is introduced as

$$p_{\rm tot} = p + p_{\rm mag} = p + \tfrac{1}{2}\left[\frac{\mathbf{B}\cdot\mathbf{B}}{\Gamma^2} + \frac{(\mathbf{v}\cdot\mathbf{B})^2}{c^2}\right]. \tag{22.119}$$

The equation governing the momentum/energy flux evolution is obtained from the spatial part of (22.107), and then becomes

$$\frac{\partial \mathbf{S}_{\rm tot}}{\partial t} + \nabla\cdot\left[\mathbf{S}_{\rm tot}\mathbf{v} + p_{\rm tot}c^2\mathbf{I} - c^2\frac{\mathbf{B}\mathbf{B}}{\Gamma^2} - (\mathbf{v}\cdot\mathbf{B})\mathbf{v}\mathbf{B}\right] = 0\,. \tag{22.120}$$

The Newtonian limits again reduce to the familiar non-relativistic ideal MHD equations, since

$$\mathbf{S}_{\rm tot} \overset{\Gamma\to 1}{\rightarrow} c^2\rho\mathbf{v}\,, \quad \tau_{\mathcal{H}} \overset{\Gamma\to 1}{\rightarrow} \tfrac{1}{2}\rho v^2 + \rho e + \tfrac{1}{2}\mathbf{B}\cdot\mathbf{B}\,, \quad p_{\rm tot} \overset{\Gamma\to 2}{\rightarrow} p + \tfrac{1}{2}\mathbf{B}\cdot\mathbf{B}\,. \tag{22.121}$$

▷ **Exercise** Verify that the expressions for the spatial flux three-vectors in Eq. (22.118) and (22.120) indeed reduce to their non-relativistic counterparts for $\Gamma \to 1$. ◁

To get a closed set of equations governing special relativistic MHD, we now need to combine the full set of Maxwell equations with particle conservation as in Eq. (22.31), and energy-momentum conservation. The latter splits in the fixed Lorentzian reference frame into Eq. (22.118) and Eq. (22.120). As for the gas dynamic case, we additionally need to provide an equation of state, such as the polytropic one given in Eq. (22.47). However, due to the vanishing electric field in the co-moving frame, it was possible to write energy-momentum conservation solely in terms of the magnetic field three-vector $\mathbf{B}$ and three-velocity $\mathbf{v}$. Hence, just as in the non-relativistic ideal MHD limit, we can close the system (mathematically speaking) by the homogeneous Maxwell equations alone, since Eq. (22.94) can use the identity

$$F^{*\alpha\beta} = (U^\alpha b^\beta - b^\alpha U^\beta)/c^2 . \tag{22.122}$$

Written out in spatial and temporal parts, Eq. (22.94) turns into the familiar set

$$\nabla \cdot \mathbf{B} = 0 , \qquad \frac{\partial \mathbf{B}}{\partial t} - \nabla \times (\mathbf{v} \times \mathbf{B}) = 0 . \tag{22.123}$$

In contrast to the classical ideal MHD formulation, the full non-homogeneous Maxwell equations (22.82) need to be used now for computing the lab frame charge density σ and current density three-vector $\mathbf{j}$. In particular, this latter three-vector has a contribution from the displacement current $c^{-2}\partial\mathbf{E}/\partial t$, which was appropriately neglected in the non-relativistic regime.

22.2.3 Wave dynamics in a homogeneous plasma

To obtain the propagation speeds for linear waves in relativistic MHD, the governing conservation laws in tensorial form can be linearized in space-time. The algebra involved can be substantial, even in the case of linearizing about a stationary homogeneous plasma. This is partly because of the wave aberration effects, which we mentioned already for the relativistic gas dynamic case. The analysis is tractable for the special case of the plasma rest frame, and very elegant relativistic variants of the slow, Alfvén and fast wave speeds can be found in the appendix of Komissarov [375].

We note that it is possible to write down the equivalent set of equations in a 3+1 formalism for the primitive variables $(S, \rho, \mathbf{v}, \mathbf{B})$, analogous to the equations (22.57) for gas dynamics. The equation for the entropy (22.48) was already shown to be identical, while Eqs. (22.123) for the magnetic field are familiar from the non-relativistic case. The continuity as well as the momentum equation become fairly cumbersome expressions, and we only mention what results from them, after linearizing with plane waves $\exp[-\mathrm{i}(\omega t - \mathbf{k}\cdot\mathbf{x})]$ about the homogeneous plasma rest frame. Indicating as usual the background quantities by $S_0, \rho_0, \mathbf{B}_0$, and the linear variables by $S_1, \rho_1, \mathbf{B}_1, \mathbf{v}_1$, we get

$$\begin{aligned}
\omega S_1 &= 0 , \\
\omega \rho_1 &= \rho_0 \mathbf{k}\cdot\mathbf{v}_1 , \\
\omega \mathbf{B}_1 &= \mathbf{B}_0(\mathbf{k}\cdot\mathbf{v}_1) - \mathbf{v}_1(\mathbf{k}\cdot\mathbf{B}_0) , \qquad \mathbf{k}\cdot\mathbf{B}_1 = 0 , \\
\omega \mathbf{v}_1 &= \frac{c^2}{w_0}\Big\{ \Big(\mathbf{k} + \frac{\mathbf{k}\cdot\mathbf{B}_0}{\rho_0 h_0}\mathbf{B}_0\Big)\big(S_0\gamma\rho_0^{\gamma-1}\rho_1 + \rho_0^\gamma S_1\big) + \mathbf{k}(\mathbf{B}_0\cdot\mathbf{B}_1) - \mathbf{B}_1(\mathbf{k}\cdot\mathbf{B}_0)\Big\} .
\end{aligned} \tag{22.124}$$

We here adopted a polytropic equation of state, where $h_0 = c^2 + [\gamma/(\gamma-1)]S_0\rho_0^{\gamma-1}$, and introduced the quantity

$$w_0 \equiv \rho_0 h_0 + B_0^2 . \tag{22.125}$$

One can directly compare these expressions with the non-relativistic expressions given in Section 5.2, and note that only the momentum equation yields an extra term (the last term is purely relativistic, and the coefficients for the other terms are changed to involve w_0). Not surprisingly then, the seven wave solutions return in slightly modified form. The marginal entropy mode is identical, being the solution at $\omega = 0$ for which $S_1 \neq 0$ only. The Alfvén waves return in a virtually unmodified form: they represent solutions with $\mathbf{v}_1 \neq 0$ and $\mathbf{B}_1 \neq 0$ while

$$\rho_1 = S_1 = \mathbf{k} \cdot \mathbf{v}_1 = \mathbf{k} \cdot \mathbf{B}_1 = \mathbf{B}_0 \cdot \mathbf{B}_1 = \mathbf{B}_0 \cdot \mathbf{v}_1 = 0\,, \tag{22.126}$$

this time given by the dispersion equation

$$\omega^2 = c^2 (\mathbf{k} \cdot \mathbf{B}_0)^2 / w_0\,. \tag{22.127}$$

They retain their field-sampling property familiar from non-relativistic MHD, and we can express their phase $\mathbf{v}_{\rm ph}$ and group velocity $\mathbf{v}_{\rm gr}$ as follows. For that purpose, assuming $\mathbf{n} = \mathbf{k}/k$ and denoting by ϑ the angle between $\mathbf{n}$ and $\mathbf{B}_0$, we find

$$\mathbf{v}_{\rm ph}/c = (B_0 \cos\vartheta / \sqrt{w_0})\,\mathbf{n}\,, \qquad \mathbf{v}_{\rm gr}/c = \mathbf{B}_0/\sqrt{w_0}\,. \tag{22.128}$$

The compressible modes are obtained from Eqs. (22.124) by straightforward algebraic manipulations to yield the dispersion equation

$$\omega^4 - \omega^2 \Big[k^2 c^2 \left(\rho_0 h_0 c_{\rm g}^2/c^2 + B_0^2\right) + c_{\rm g}^2 (\mathbf{k} \cdot \mathbf{B}_0)^2\Big] / w_0 + k^2 c^2 c_{\rm g}^2 (\mathbf{k} \cdot \mathbf{B}_0)^2 / w_0 = 0\,. \tag{22.129}$$

Here we purposely wrote this again in terms of the squared sound speed $c_{\rm g}^2$, while we can now also introduce the squared Alfvén speed $v_{\rm A}^2 = B_0^2 c^2 / w_0$. The expressions are then generally valid, with the expressions for specific enthalpy h_0 and sound speed $c_{\rm g}$ depending on the equation of state. It is then left as an exercise to the reader that their phase speeds are found from

$$\mathbf{v}_{\rm ph}/c = (v_{\rm ph}/c)\,\mathbf{n} = \mathbf{n}\sqrt{\tfrac{1}{2}[(\rho_0 h_0/w_0) c_{\rm g}^2 + v_{\rm A}^2]}\,\sqrt{1 + \delta\cos^2\vartheta \pm a}\,/c\,. \tag{22.130}$$

Here, the symbols δ and σ (no longer the charge density!) express the following dimensionless ratios

$$\delta \equiv \frac{c_{\rm g}^2 v_{\rm A}^2}{\left[(\rho_0 h_0/w_0) c_{\rm g}^2 + v_{\rm A}^2\right] c^2}\,, \qquad \sigma \equiv \frac{4 c_{\rm g}^2 v_{\rm A}^2}{\left[(\rho_0 h_0/w_0) c_{\rm g}^2 + v_{\rm A}^2\right]^2}\,, \tag{22.131}$$

and the symbol a follows from

$$a^2 \equiv (1 + \delta\cos^2\vartheta)^2 - \sigma\cos^2\vartheta\,. \tag{22.132}$$

Noting that $\rho_0 h_0 / w_0 = 1 - v_{\rm A}^2/c^2$, the phase speed for purely parallel propagation reduces to the same expression found in non-relativistic MHD, where we have

$$v_{{\rm ph},\parallel}/c = \left\{\tfrac{1}{2}\left(c_{\rm g}^2 + v_{\rm A}^2\right)\left[1 \pm \sqrt{1 - 4 c_{\rm g}^2 v_{\rm A}^2/(c_{\rm g}^2 + v_{\rm A}^2)^2}\right]\right\}^{1/2}/c\,. \tag{22.133}$$

The group speed, written in terms of the orthogonal directions $\mathbf{n} = \mathbf{k}/k$ and $\mathbf{t} = [(\mathbf{B}_0/B_0) \times \mathbf{n}] \times \mathbf{n}$, becomes

$$\mathbf{v}_{\rm gr}/c = (v_{\rm ph}/c)\left[\mathbf{n} \pm \mathbf{t}\,\frac{\left[\sigma \mp 2\delta\left(a \pm (1 + \delta\cos^2\vartheta)\right)\right]\sin\vartheta\cos\vartheta}{2\,(1 + \delta\cos^2\vartheta \pm a)\,a}\right]. \tag{22.134}$$

These can be compared directly with the non-relativistic expressions, and we can note that all relativistic effects are due to the parameter δ, together with the fact that both sound and Alfvén speeds get relativistic corrections.

As representative examples, we show in Fig. 22.6 the phase and group diagrams for three cases with (again in units making $c = 1$):

(a) $\rho_0 = 1, p_0 = 0.1, B_0 = 0.3$ $(c_g = 0.354,\ v_A = 0.258)$, Fig. 22.6 (top);

(b) $\rho_0 = 1, p_0 = 0.1, B_0 = 0.5$ $(c_g = 0.354,\ v_A = 0.406)$, Fig. 22.6 (middle);

(c) $\rho_0 = 0.01, p_0 = 0.001, B_0 = 1$ $(c_g = 0.354,\ v_A = 0.99)$, Fig. 22.6 (bottom).

These assume a Mathews equation of state. In Fig. 22.7, we also show the group diagram for the first and the third case, plotted in a frame which saw the point perturbation pass by along the field lines at a velocity $0.9c$. (Also see Ref. [350]).

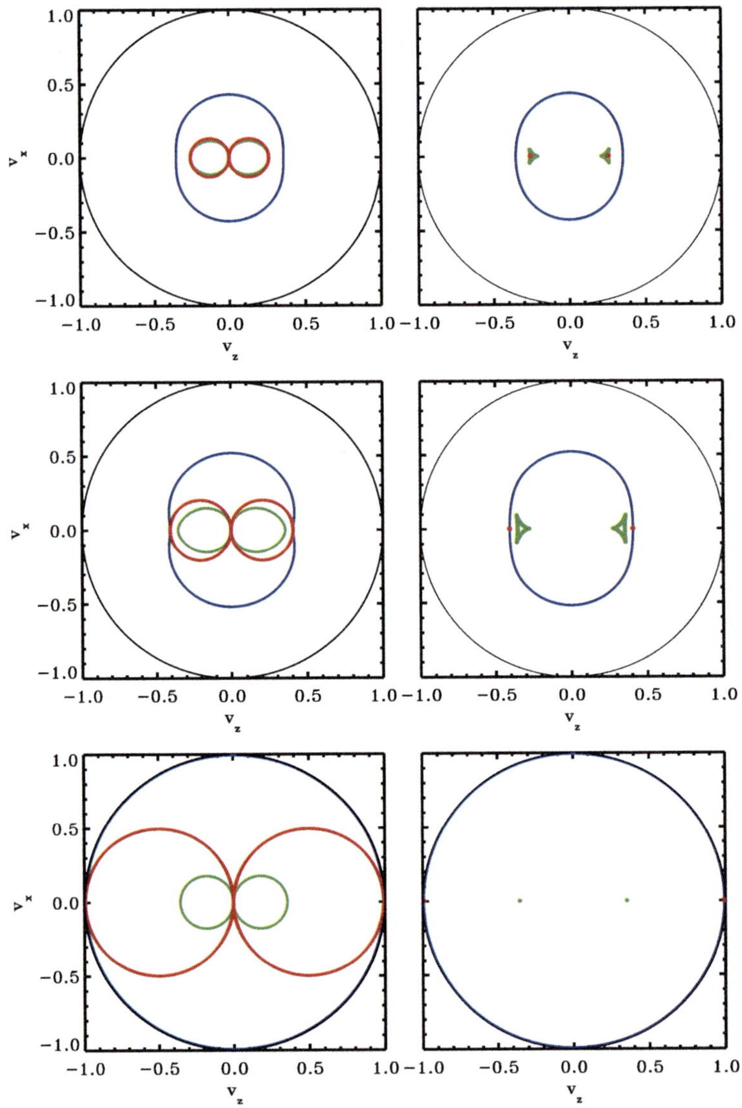

Fig. 22.6 Phase (left) and group (right) diagrams in the gas rest frame for three representative cases with uniform magnetic field. Thermodynamic quantities range as indicated in the text.

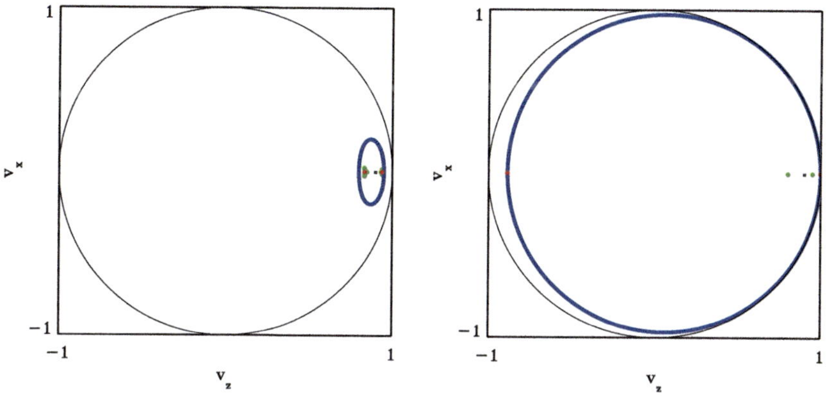

Fig. 22.7 Group diagrams as seen from a frame where the source moves at $0.9c$ along the z-axis aligned with the magnetic field for case (a) of Fig. 22.6 (left) and case (c) of Fig. 22.6 (right).

Characteristic speeds for relativistic MHD For the relativistic ideal MHD equations, expressions for the characteristic speeds can be derived which correspond to the phase speeds as seen from the laboratory frame (as opposed to the plasma rest frame discussed thus far). There will again be seven signal speeds, where one is related to the entropy equation (22.111). Entropy is passively advected with the characteristic speed $\lambda_{\rm E}$ in the i-coordinate direction given by

$$\lambda_{\rm E}/c = v_i/c\,. \tag{22.135}$$

Similar to the non-relativistic case (cf. Eqs. (5.124), (7.148) and (13.24)), the characteristic speeds are again ordered according to

$$-c \le \lambda_{\rm F}^- \le \lambda_{\rm A}^- \le \lambda_{\rm S}^- \le \lambda_{\rm E} \le \lambda_{\rm S}^+ \le \lambda_{\rm A}^+ \le \lambda_{\rm F}^+ \le c\,. \tag{22.136}$$

In contrast to the non-relativistic case, the forward and backward wave speeds of a given family (fast, Alfvén, slow) are no longer necessarily symmetric about $\lambda_{\rm E} = v_i$. The Alfvén wave speeds are found from

$$\frac{\lambda_{\rm A}^{\pm}}{c} = \frac{v_i}{c} \pm \frac{1}{\Gamma^2}\frac{B_i}{\sqrt{\rho h_{\rm tot}} \pm \mathbf{B}\cdot\mathbf{v}/c}\,. \tag{22.137}$$

We thereby introduced the total specific enthalpy, where

$$h_{\rm tot} = h + 2p_{\rm mag}/\rho\,. \tag{22.138}$$

The fast and slow characteristic speeds can most conveniently be found from the following quartic polynomial:

$$\rho h(c^2 - c_{\rm g}^2)\Gamma^4\big[(\lambda - v_i)/c\big]^4 - (1-\lambda^2/c^2)\Big\{\Gamma^2(\rho h c_{\rm g}^2 + 2p_{\rm mag}c^2)\big[(\lambda - v_i)/c\big]^2 - c_{\rm g}^2\big[\Gamma(\mathbf{v}\cdot\mathbf{B})(\lambda - v_i)/c^2 - B_i/\Gamma\big]^2\Big\} = 0\,. \tag{22.139}$$

As before, these expressions can in fact be obtained from the rest frame expressions given earlier. They become more involved due to relativistic aberration, and will involve the Lorentz transformation for the magnetic field as well, as the magnetic field component normal to the wave front will look different from differing inertial reference frames.

22.2.4 Shock conditions in relativistic MHD

The conservation laws of relativistic MHD can again be studied in the case of discontinuities across surfaces in space-time. Completely analogous to the relativistic hydro case, we write

$$[\![\rho U^\alpha]\!] l_\alpha = 0\,, \tag{22.140}$$

$$[\![T^{\alpha\beta}]\!] l_\alpha = 0\,, \tag{22.141}$$

$$[\![U^\alpha b^\beta - b^\alpha U^\beta]\!] l_\alpha = 0\,, \tag{22.142}$$

where l^α is the space-like normal to the shock surface. The analysis of these jump relations can be rather complicated in general, and their precise algebraic form in a $3+1$ manner depends on selecting a specific Lorentzian reference frame. We already noted this for relativistic hydro, where we mentioned the shock rest frame (SRF), upstream rest frame (URF), downstream rest frame (DRF) and the tangential reference frame (TRF). We will mention the significance of the important de Hoffman–Teller frame (HTF) further on, which, like the TRF, involves a transformation with a certain tangential velocity. If we, for now, consider the relations (22.140)–(22.142) in the shock rest frame, where we denote the shock normal as $\mathbf{l} = (0, \mathbf{n})$ with $\mathbf{n}$ the normal three-vector, we find

$$[\![\rho \Gamma v_{\rm n}]\!] = 0\,, \tag{22.143}$$

$$[\![B_{\rm n}]\!] = 0\,, \tag{22.144}$$

$$\left[\!\!\left[\left(\rho h \Gamma^2 + B^2\right)\frac{\mathbf{v}_{\rm t} v_{\rm n}}{c^2} - \frac{(\mathbf{v}\cdot\mathbf{B})}{c^2}\left(\mathbf{B}_{\rm t} v_{\rm n} + \mathbf{v}_{\rm t} B_{\rm n}\right) - \frac{\mathbf{B}_{\rm t} B_{\rm n}}{\Gamma^2}\right]\!\!\right] = 0\,, \tag{22.145}$$

$$\rho \Gamma v_{\rm n} \left[\!\!\left[\frac{\mathbf{B}_{\rm t}}{\rho \Gamma}\right]\!\!\right] = B_{\rm n} [\![\mathbf{v}_{\rm t}]\!]\,, \tag{22.146}$$

$$\left[\!\!\left[\left(\rho h \Gamma^2 + B^2\right)\frac{v_{\rm n}^2}{c^2} + p + p_{\rm mag} - 2\frac{(\mathbf{v}\cdot\mathbf{B})}{c^2} v_{\rm n} B_{\rm n} - \frac{B_{\rm n}^2}{\Gamma^2}\right]\!\!\right] = 0\,, \tag{22.147}$$

$$\rho \Gamma v_{\rm n} \left[\!\!\left[h\Gamma + \frac{B^2}{\rho \Gamma}\right]\!\!\right] = B_{\rm n} [\![\mathbf{v}\cdot\mathbf{B}]\!]\,, \tag{22.148}$$

where we wrote the equations in the same order as the non-relativistic variants (21.4)–(21.9) of Section 21.2. The relativistic hydro limit (22.69) is obtained for vanishing magnetic field.

The occurrence of the Lorentz factor in many terms of these expressions makes a discussion of the general case beyond the scope of this textbook. These relations have been studied in some detail by de Hoffmann and Teller [146], Majorana and Anile [425], Appl and Camenzind [21], and the procedure to obtain the downstream from the upstream state variables is briefly recapitulated in the appendix to the review by Kirk and Duffy [366]. The books by Lichnerowicz [409] and Anile [14] contain a precise mathematical treatment. We here only mention the important limiting case of normal perpendicular shocks, and point out the means to obtain various shock invariants from the relations (22.140)–(22.142).

Normal perpendicular shocks This limiting case assumes that in the SRF, where Eqs. (22.143)–(22.148) hold, the normal magnetic fields vanish, $B_{\rm n} = 0$ (making the field purely tangential to the shock front, hence the name 'perpendicular'). Moreover, we assume purely normal velocities for which $\mathbf{v}_{\rm t} = 0$ as well (hence the name 'normal'). It is then easy to verify that we are left with

$$[\![\rho\Gamma v_{\rm n}]\!] = 0\,, \tag{22.149}$$

$$[\![\mathbf{B}_{\rm t} v_{\rm n}]\!] = 0\,, \qquad [\![\Gamma\mathbf{B}'_{\rm t} v_{\rm n}]\!] = 0\,, \tag{22.150}$$

$$\left[\!\!\left[\left(\rho h\Gamma^2 + B_{\rm t}^2\right)\frac{v_{\rm n}^2}{c^2} + p + \frac{B_{\rm t}^2}{2\Gamma^2}\right]\!\!\right] = 0\,, \qquad [\![\rho h_{\rm tot}\Gamma^2 v_{\rm n}^2 + p_{\rm tot}]\!] = 0\,, \tag{22.151}$$

$$[\![\rho h\Gamma^2 v_{\rm n} + B_{\rm t}^2 v_{\rm n}]\!] = 0\,, \qquad [\![\rho h_{\rm tot}\Gamma^2 v_{\rm n}]\!] = 0\,. \tag{22.152}$$

The relations on the right express these SRF relations in terms of the plasma rest frame magnetic fields at left and right of the shock front (those we have in the DRF and URF, respectively), which in this normal perpendicular case are $\mathbf{B}' = \mathbf{B}/\Gamma$. Using them, the total enthalpy and pressure are written as $\rho h_{\rm tot} = \rho h + B_{\rm t}'^2$ and $p_{\rm tot} = p + B_{\rm t}'^2$. In this way, a rather direct generalization of the pure gas dynamic case is obtained, replacing pressure and enthalpy with their total counterparts.

▷ **Relativistic MHD shock invariants.** For completeness, we now list various relativistic MHD shock invariants, as described and proven in Anile [14]. They are most conveniently found from the general expressions (22.140)–(22.142), while we will give expressions as obtained in the SRF for some of them.

(a) Normal particle flux

$$\mathcal{M} = \rho U^\alpha l_\alpha \;\; [= \rho\Gamma v_{\rm n}]\,. \tag{22.153}$$

(b) Introducing $V^\beta = \left(U^\alpha b^\beta - b^\alpha U^\beta\right) l_\alpha$, we have the invariant $\mathcal{H} = -V^\alpha V_\alpha/\mathcal{M}^2$ or

$$\mathcal{H} = \frac{c^2\,(b^\mu l_\mu)^2}{\mathcal{M}^2} - \frac{b^\beta b_\beta}{\rho^2} \left[= \frac{c^2}{(\rho\Gamma v_{\rm n})^2}\left(\frac{B_{\rm n}}{\Gamma} + \frac{\Gamma(\mathbf{v}\cdot\mathbf{B})v_{\rm n}}{c^2}\right)^2 - \frac{2p_{\rm mag}}{\rho^2}\right]. \tag{22.154}$$

(c) Writing $W^\beta = T^{\alpha\beta} l_\alpha$ we have

$$\mathcal{B} = \frac{V_\alpha W^\alpha}{\mathcal{M}} = h b^\alpha l_\alpha \left[= h\left(\frac{B_{\rm n}}{\Gamma} + \frac{\Gamma(\mathbf{v}\cdot\mathbf{B})v_{\rm n}}{c^2}\right)\right]. \tag{22.155}$$

(d) The quantity $\mathcal{E}$ found from

$$\mathcal{E} = W^\mu l_\mu \left[= p + p_{\rm mag} + (\rho\Gamma v_{\rm n})^2\left(\frac{h}{\rho c^2} - \frac{\mathcal{H}}{c^2}\right)\right]. \tag{22.156}$$

(e) Introducing $X^\beta = W^\beta - (W^\alpha l_\alpha)\, l^\beta$, we have the invariance of

$$\mathcal{K} = -\frac{X^\alpha X_\alpha}{\mathcal{M}^2} = \frac{h^2}{c^2} + \frac{\mathcal{M}^2 h^2}{\rho^2 c^4} + \left(\frac{2h}{\rho c^2} - \frac{\mathcal{H}}{c^2}\right)\left(2p_{\rm mag} - \frac{\mathcal{M}^2\mathcal{H}}{c^2}\right). \tag{22.157}$$

(f) Also combinations of the above, such as

$$\mathcal{L} = -\mathcal{K}\mathcal{H} + \frac{\mathcal{B}^2}{\mathcal{M}^2} = c^2\left(2p_{\rm mag} - \frac{\mathcal{M}^2\mathcal{H}}{c^2}\right)\left(\frac{h}{\rho c^2} - \frac{\mathcal{H}}{c^2}\right)^2. \tag{22.158}$$

With the knowledge of all the above invariants, one can obtain the relativistic MHD generalization of the Taub adiabat, called the Lichnerowicz adiabat and given by

$$\frac{h_2^2}{c^2} - \frac{h_1^2}{c^2} - \left(\frac{h_1}{\rho_1 c^2} + \frac{h_2}{\rho_2 c^2}\right)(p_2 - p_1) + \tfrac{1}{2}\left(\frac{h_2}{\rho_2 c^2} - \frac{h_1}{\rho_1 c^2}\right)(\chi_1 + \chi_2) + \alpha_2\chi_2 - \alpha_1\chi_1 = 0\,, \tag{22.159}$$

where $\chi \equiv 2p_{\rm mag} - \mathcal{M}^2\mathcal{H}/c^2$ and $\alpha \equiv h/(\rho c^2) - \mathcal{H}/c^2$. This invariant then can be shown to reduce to the various limit cases we encountered thus far, namely the non-relativistic hydrodynamic Hugoniot adiabat (21.20), the relativistic Taub adiabat (22.71) and the classical MHD relation (21.42). ◁

We conclude the discussion of the relativistic MHD shock relations with pointing out that, in analogy with Chapter 21, it is likely that a more insightful classification for relativistic MHD shocks can be found than the one currently found in the literature. Such a discussion will best be done in the de Hoffmann–Teller reference frame, which writes the equations in a frame where the total electric field at left and right of the shock vanishes. This involves a Lorentz transformation from the SRF to a frame moving with a certain tangential speed with respect to the SRF. The existence of this frame was pointed out by de Hoffmann and Teller [146], and we saw that the non-relativistic case then became most insightful when we expressed all relations using the Alfvén normal Mach number in this reference frame. This is still to be pursued in future research for the relativistic case. Note as well that we thus far did not mention the extra inequality of entropy increase across a shock front, which obviously still acts to select physically realizable shocks.

22.3 Computing relativistic magnetized plasma dynamics

We demonstrated that relativistic magneto-fluids obey, in a fixed 'laboratory' Lorentz frame, a set of conservation laws given by Eqs. (22.31), (22.118), (22.120) and the familiar set of equations for the magnetic field given by Eqs. (22.123). As an alternative, also Eq. (22.111) can be used instead of (22.31) or (22.118). In any case, it is clear that we end up with a nonlinear system of conservation laws, which can be integrated numerically using shock-capturing techniques such as those presented in Chapter 20. A discussion of numerical relativistic hydrodynamics, including extensions to general relativistic hydrodynamics, can be found in the book by Wilson and Mathews [651]. Relativistic MHD poses its own challenges, as can be expected from the algorithmic complexity already needed when going from non-relativistic hydro to MHD. As an early example, Dubal [166] adopted an FCT algorithm to special relativistic MHD, where 1D Riemann problems of modest Lorentz factors (up to $\Gamma \sim 3$) were adequately resolved: Riemann invariants through rarefactions remained constant to within 1%, and even 2D spherical blast waves in initially uniform magnetic fields could be computed. Another approach was followed by van Putten [630], where a pseudo-spectral method in combination with a leapfrog scheme was exploited, demonstrating that the compound waves found in coplanar non-relativistic MHD Riemann problems persist in (numerical) relativistic MHD. As mentioned further on, more recent efforts successfully exploit approximate Riemann solvers or TVDLF schemes in challenging applications involving high Lorentz factor plasma flows.

Conservative to primitive variable transformation The most obvious complication in relativistic ideal MHD is that we have truly non-trivial relations between the primitive variables $(\rho, \mathbf{v}, p, \mathbf{B})$ and the conserved variables, since

$$\begin{aligned} D &= \Gamma\rho\,, \qquad \mathbf{S}_{\rm tot} = \rho h\Gamma^2\mathbf{v} + B^2\mathbf{v} - (\mathbf{v}\cdot\mathbf{B})\mathbf{B}\,, \\ \tau_{\mathcal{H}} &= \rho h\Gamma^2 - p - \Gamma\rho c^2 + \tfrac{1}{2}B^2 + \tfrac{1}{2}\big(B^2v^2 - (\mathbf{v}\cdot\mathbf{B})^2\big)/c^2\,. \end{aligned} \tag{22.160}$$

Especially due to the occurrence of the Lorentz factor Γ, the conversion from conservative to primitive variables, which is needed to evaluate the flux expressions, is non-algebraic and must be handled numerically. Defining an auxiliary variable $\xi \equiv \rho h \Gamma^2$, so that $\mathbf{S}_{\rm tot} \cdot \mathbf{B} = \xi \mathbf{v} \cdot \mathbf{B}$ and thus

$$\mathbf{v} = \frac{\mathbf{S}_{\rm tot} + \mathbf{S}_{\rm tot} \cdot \mathbf{B}\mathbf{B}/\xi}{\xi + B^2} . \tag{22.161}$$

The Lorentz factor can then be computed from ξ and the conservative variables $\mathbf{S}_{\rm tot}$ and $\mathbf{B}$, since $\Gamma^{-2} = 1 - v^2/c^2$. This in turn means that the defining equation for the energy variable $\tau_{\mathcal{H}}$ in expressions (22.160) provides a nonlinear relation for ξ as being the zero of

$$\xi - p - Dc^2 + B^2 - \frac{1}{2}\left[\frac{B^2}{\Gamma^2} + \frac{(\mathbf{S}_{\rm tot} \cdot \mathbf{B})^2}{\xi^2 c^2}\right] - \tau_{\mathcal{H}} . \tag{22.162}$$

For a constant polytropic index, we find also

$$p = \frac{\gamma - 1}{\gamma} \frac{\xi - \Gamma D c^2}{\Gamma^2} , \tag{22.163}$$

so one can determine ξ from a given set of conserved variables $(D, \mathbf{S}_{\rm tot}, \tau_{\mathcal{H}}, \mathbf{B})$ using root finding algorithms on Eq. (22.162).

The Riemann problem and modern solver strategies Just as in non-relativistic MHD, knowledge of the solution of the 1D Riemann problem in special relativistic MHD is extremely useful to test the various flavours of shock-capturing schemes. It is surprising that the exact solution for the general case where all seven nonlinear waves are accounted for has only recently been obtained [210](a), while the development of shock-capturing solvers for relativistic MHD problems started in earnest in the late 1990s and beginning of the twenty-first century. A first Godunov-type method for relativistic MHD, exploiting an approximate Riemann solver, was presented by Komissarov [375], and his paper collects all details needed for its implementation, such as the eigenvalue and eigenvector pairs, the means to handle degenerate cases, and how to convert from primitive to conservative variables. Using a staggered representation where the magnetic field components are defined on cell interfaces, a constrained transport type approach was taken to handle the solenoidal field constraint. This allowed multi-dimensional simulations involving considerably higher Lorentz factors than those achieved in early works [166, 630]. Following this, improved, but purely 1D, approximate Riemann solver implementations are found in [35] and in [373], where the former is of TVD type. Multi-dimensional relativistic MHD schemes exploiting more central approximations to the Riemann fan (HLL and HLLC) emerged in [148] and [440]. A TVDLF type method which uses only the fastest propagation speed is exploited in [622]. The latter uses parabolic source term treatments for handling $\nabla \cdot \mathbf{B} = 0$ in a grid-adaptive framework. There is currently a quickly growing research community focused on numerical MHD solvers for relativistic regimes, already involving general relativistic MHD simulations in dynamically evolving space-times [210](b).

An example taken from [622] compares, in Fig. 22.8, the exact solution with a grid-adaptive numerical solution of a Riemann problem, which was first presented in [35] and subsequently analytically solved in [210](a). The left state has

$$(\rho, \mathbf{v}, p, \mathbf{B})^{\rm L} = (1.08, 0.4, 0.3, 0.2, 0.95, 2.0, 0.3, 0.3)$$

adjacent to $(1.0, -0.45, -0.2, 0.2, 1.0, 2.0, -0.7, 0.5)$ and constant polytropic index $\gamma = 5/3$. Note that the grid-adaptive result recovers the solution containing a left-going fast shock, a left-going Alfvén discontinuity, a slow rarefaction, a contact discontinuity, as well as the right-going slow

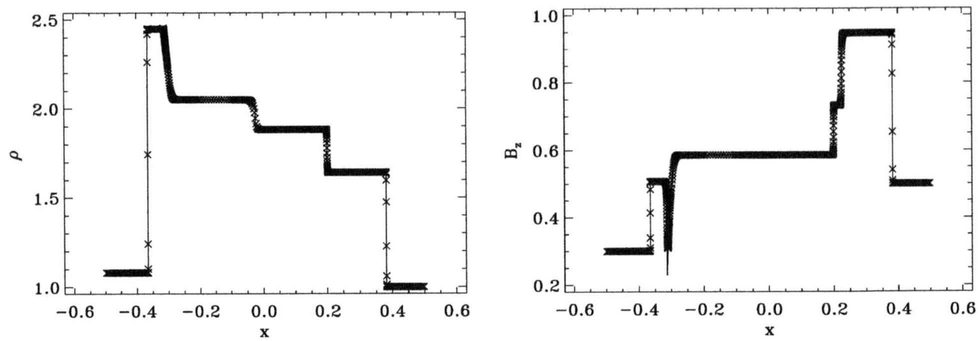

Fig. 22.8 Exact (solid line) versus numerical solution of a relativistic MHD Riemann problem. Shown is the proper density and a tangential field component B_z. (From van der Holst et al. [622].)

shock, Alfvén discontinuity and fast shock. However, close scrutiny reveals that the separation between left-going Alfvén signal and slow rarefaction is not quite resolved. The need for grid-adaptive simulations in relativistic MHD is already obvious for correctly solving even 1D Riemann problems.

22.3.1 Numerical challenges from relativistic MHD

In this paragraph, we point out various challenges associated with numerical, relativistic MHD. Some have already led to clever algorithmic approaches, while others may need new ideas for further progress. A fairly straightforward observation is that the equations of relativistic MHD contain in principle the physical constants, such as the speed of light c and the particle rest mass $m_{\rm p}$ (for protons). Obviously, in numerics one always exploits a proper scaling, which for relativistic MHD sets $c = 1$ and, together with a reference distance and number density, all other quantities are measured in this scaled unit system. This was already the case in non-relativistic MHD, where we noted scale-independence in Section 4.1.2 and stated that units are set by choosing a length, magnetic field strength and density at a reference position. One of these latter three can now no longer be independently fixed in space-time, since c is the reference speed.

In a previous paragraph, we pointed out that the conservative to primitive variable computation is a numerical problem by itself. It must be stressed that the accuracy with which this problem is solved numerically is a crucial element of modern solvers. Indeed, we have the obvious physical restrictions that $v < c$, which means $\Gamma \geq 1$, while $p > 0$, $\rho > 0$, and we also want the electric field density $E^2 > 0$, where $E^2 \sim B^2 v^2 - (\mathbf{v} \cdot \mathbf{B})^2$. The latter is obviously true analytically, but numerical precision is finite. All these constraints must be consistent with $\tau_{\mathcal{H}} > 0$ and $D \geq \rho$, and it helps to take explicit account of the numerical accuracy with which factors like $1 - v^2/c^2$ can actually be distinguished from zero. This can be done, by building in an appropriate 'upper limit' on velocities, up to which conversions are computationally feasible. Similar tricks can be employed to guarantee consistent conversions from primitive to conservative variables, while ensuring a lower limit on attainable densities and pressures. This is similar to classical MHD, where this conversion is algebraic and trivial, but still requires that all contributions to total energy $\mathcal{H}$ are positive separately, which may introduce numerical inaccuracies in regions of very low plasma beta. This problem returns in augmented form in relativistic settings, since rest mass, internal energy, kinetic energy, magnetic and electric field densities all may dominate in localized regions of the computational domain. A thorough comparison of several numerical strategies to obtain primitive from conservative variables, suitable for general relativistic MHD, has been presented by Noble et al. [467].

When using finite volume like treatments, we discussed in Chapter 20 how limited reconstructions from cell centre to cell edge are needed for raising the spatial order of accuracy, while avoiding spurious oscillations. This can still be used in relativistic MHD, but it then pays off to perform the limiting on quantities without physical bounds: one uses in practice the spatial part of the four-velocity U^α, i.e. $\Gamma \mathbf{v}$ to perform reconstructions, instead of the velocity itself $\mathbf{v}$, which must obey $v < c$. Similarly, when computing the roots of the quartic polynomial (22.139), which are needed for all shock-capturing methods exploiting knowledge of the characteristic speeds (even TVDLF which uses the maximal speed alone), one can better solve for $\Gamma(\lambda - v_i)$, or some other suitably scaled variable. In [622], a Laguerre method was then used to compute all four roots of this polynomial accurately. Also, this process is a non-trivial exercise in handling numerical accuracy, since its roots λ may come very close to each other and to 1 (the light speed in the scaled system).

Finally, just as in the non-relativistic case, the magnetic field must be solenoidal. For multi-dimensional computations, a suitable strategy must therefore be incorporated. It is noteworthy that the constrained transport idea [171] was from the beginning designed for (even general) relativistic MHD computations. On the other hand, since the induction equation appears identical to the non-relativistic case, also the more straightforward source terms treatments (Powell source, or parabolic cleaning, both limited to the induction equation) have been used successfully [622, 351]. In practice, the schemes in use today may still benefit from further algorithmic improvements, although robust relativistic MHD solvers exist in modern code developments [622], [441], [210](b), [351].

22.3.2 Pulsar Wind Nebulae modelling

The current suite of shock-capturing, high resolution, schemes for relativistic MHD has already been applied to a fair variety of astrophysical problems. The study of pulsar wind nebulae is one such problem, where computational relativistic MHD has led to novel insights. In the particular case of the Crab nebula, observations revealed a peculiar jet-torus structure at the centre. This precise shaping of the circum-pulsar environment has been shown to relate to the shock-governed interaction of an anisotropic, toroidally magnetized, relativistic pulsar wind, interacting with the slowly expanding ejecta of the supernova [376] which created the pulsar. Figure 22.9 superposes an infrared image from the Spitzer Space Telescope (red), an optical view as obtained with the Hubble Space Telescope (HST, the green to yellow filamentary structure), and an X-ray image from the Chandra X-ray observatory (light blue at centre). The latter shows clearly how the pulsar wind nebula (PWN) demonstrates an intricate central structure with a torus and jet like feature. The PWN itself is confined within freely expanding supernova remnant material, since the pulsar marks the strongly magnetized, spinning neutron star that remains after a violent core-collapse stellar explosion. Synchrotron radiation of the PWN photo-ionizes the ejecta seen with HST, and the optical view shows supernova ejecta reshaped into a highly filamentary pattern that interpenetrates the PWN.

Axi-symmetric 2.5D [376] and full 3D relativistic MHD simulations [504](b) have gradually revealed the role of MHD instabilities in obtaining these structures. The simulations typically start from an unshocked, highly relativistic pulsar wind zone, embedded in a radially expanding supernova shell, and letting this configuration adjust dynamically to one where the wind zone is bounded

Fig. 22.9 Observations of the Crab nebula in infrared (red), optical (green) and X-rays (light blue). (Credit: X-ray: NASA/CXC/ASU/J. Hester et al.; Optical: NASA/ESA/ASU/J. Hester and A. Loll; Infrared: NASA/JPL-Caltech/Univ. Minn./R. Gehrz.)

by a termination shock and gets surrounded by a gradually increasing PWN filled with shocked pulsar wind material. The artificial initial conditions cause the termination shock to first move in, and then rebound to reach a slow, asymptotic self-similar, expansion. A key aspect of the model is the purely azimuthal magnetic field prescription in the wind zone, setting the overall magnetization $\sigma \equiv B^2/\Gamma^2 \rho c^2$ of the inflow. This pulsar wind represents the far-out conditions in the pulsar magnetosphere, where the tilt between the rotation and magnetic axis ($\sim 45^\circ$ for the Crab pulsar) is expected to lead to a strongly warped current sheet appearance, as an extreme variant of the solar wind equatorial current sheet of Fig. 8.15. This gives rise to a striped wind zone with alternating azimuthal windings near the equator. Assuming that dissipation from reconnection in this striped wind zone already happens on the way from the pulsar to the PWN, the injected wind prescription introduces a latitudinal angle-dependent energy division over kinetic to magnetic contributions. This already lowers the magnetization σ on arrival at the termination shock for all polar angles $\pi/2 - \alpha \leq \theta \leq \pi - \alpha$, but not enough to solve the so-called sigma-problem. This riddle from pulsar wind theory asks to reconcile the expected high σ pulsar magnetospheres and high Lorentz factor pulsar wind zones (Γ values from 100 to 10^6 are often quoted) with the low σ conditions obtained in the PWN proper.

The answer to the sigma problem emerged with the first 3D relativistic MHD simulations of Crab-like PWN [504], and pointed to effective magnetic dissipation throughout the nebula, together with kink instabilities acting on the more toroidal field concentrations above the poles. Even though numerical limitations enforced the use of $\Gamma \sim 10$ pulsar wind outflows, the striped wind prescription does capture the correct asymptotic division of electromagnetic to kinetic luminosities for magnetizations with $\sigma > 1000$. An impression of the magnetic field structure obtained in the 3D simulations

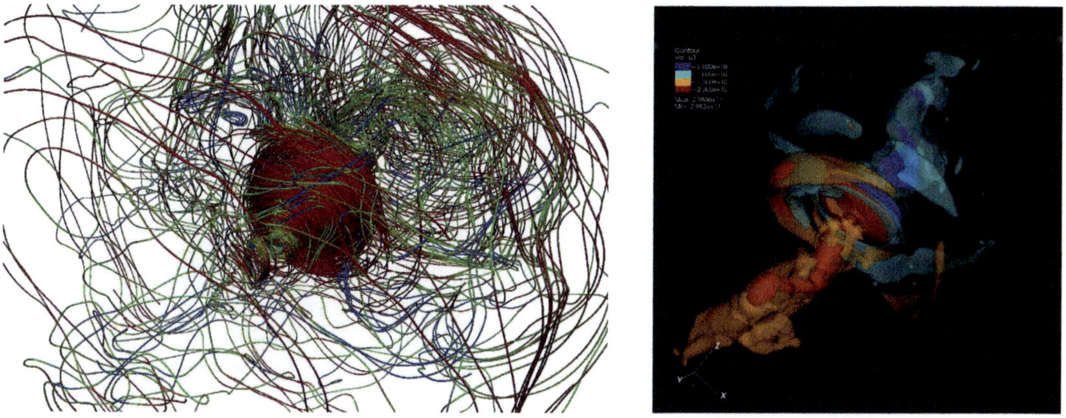

Fig. 22.10 Magnetic field (left) and velocity structure (right) in the central regions of the simulated Crab PWN. (From Porth et al. [504](b).)

is shown in Fig. 22.10, where the colour indicates toroidal (red) to poloidal (blue) field line orientations. The predominantly toroidal field zone in the centre marks the location of the pulsar wind termination shock, where plasma gets supplied to the PWN. It is seen how helical field distortions indicate kink instability development in the polar regions above and below this pulsar wind zone, while the entire PWN region has markedly randomized magnetic field. This randomization is influenced by the (numerical) magnetic dissipation happening throughout the downstream PWN volume, as turbulent flow conditions cause a rather violent meandering of the current sheet separating oppositely directed B_φ topologies. This effect is also there in 2.5D simulations which account for the B_φ sign change and do not enforce symmetry across the equator. The most striking difference between 2.5D and 3D setups is found in the polar flows or jets: axi-symmetric simulations find the flow that crosses the termination shock in the more polar regions to become focused by hoop stresses towards the axis to form a jet, but the resulting toothpaste effect is actually way too efficient and leads to pronounced polar jets not consistent with observations. The 3D results show a flow structure as visualized in the right panel of Fig. 22.10, showing iso-surfaces of axial velocities. The chosen snapshot shows clear helically distorted flow patterns that are seen to extend to a few termination shock radii. This 'plume' kinks and fragments to dissolve further out in the nebula, and corresponds well to the Crab jet feature.

Using a parameterized model for electrons accelerated at the termination shock of the striped wind only (excluding the stripe-free poles), the models can be turned into fairly realistic synthetic observations. An example optical synchrotron image is shown in Fig. 22.11, left panel, where several features known from actual observational campaigns are indicated on the plot. The pulsar location is at the central white cross, above which we find the 'knot' as a region of Doppler-boosted emission that samples the plasma just above and outside the central area of the termination shock. The 'sprite' feature is caused by streamlines from the striped wind that converge back towards the axis and thus transport emitting particles to the polar plume base. The jet itself is in the synthetic image too weak as compared to actual images, and asks for an additional recipe to light up this flow feature. The right panel of Fig. 22.11 takes a difference image of two optical images separated by 110 days, and the clear black to white striped arcs are known as moving wisps. The simulation finds these arc-like features to move out at apparent velocities of a third of the light speed, in nice correspondence with observations. These wisps are induced by termination shock oscillations, as quasi-periodic vortex shedding occurs near the tips of the Mach belt, that marks the end of the more equatorial pulsar wind. Also, the sprite feature is clearly highly variable.

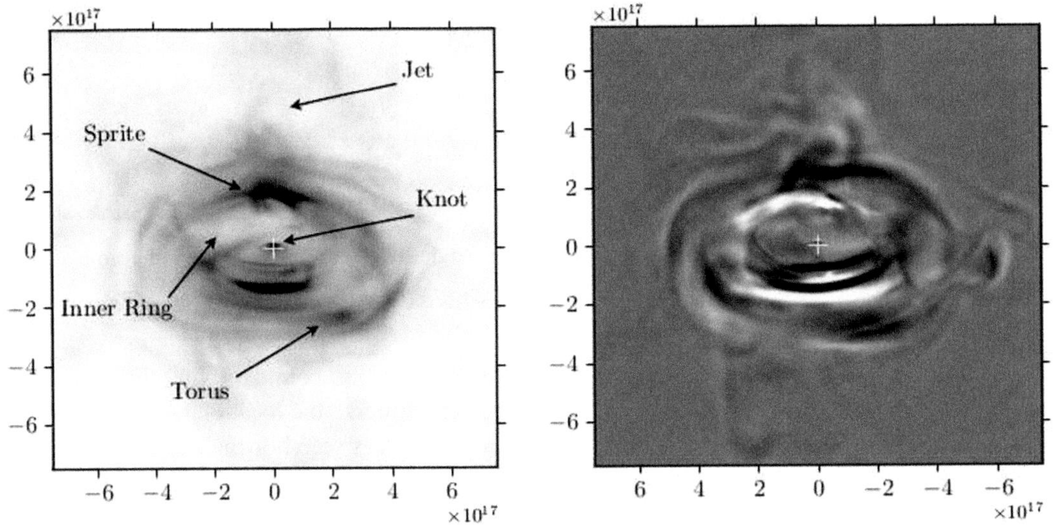

Fig. 22.11 Synthetic optical synchrotron image of the Crab nebula. At right, a difference image shows outward moving wisp features. (From Porth et al. [504](b).)

As a final note on pulsar wind nebulae simulations, the expanding PWN zone itself is expected to interact dynamically with the surrounding supernova remnant shell. Indeed, as the original ejecta are much heavier than the PWN bubble, we can expect Rayleigh–Taylor (RT) instability to develop when the contact interface between bubble and ejecta expands with increasing speed. To see whether the nonlinear RT development is in line with the filamentary structure seen in the HST images, simulations in 2014 [504](c) followed the expansion of the PWN for the entire 960 year lifetime of the Crab, after its observation by Chinese and Arab astronomers in 1054. Figure 22.12 shows the resulting structure from an axi-symmetric simulation (at relatively low σ) where RT fingers are seen to penetrate up to one quarter of the nebula radius. The central dark region marks the pulsar wind zone as bounded by the termination shock, and the white area is the PWN. The axi-symmetry adopted implies that the magnetic configuration can not bend and counteract the RT instabilities by tension. However, the 3D simulations discussed previously (which do not allow to go this far into the Crab lifetime history) indicate that the PWN is highly thermalized through effective dissipation. This keeps the magnetic field below equipartition near the interface, so that these 2D findings likely carry over to 3D settings as well.

The success of 2.5D and especially 3D relativistic PWN models in reproducing both optical and X-ray observations is striking, while remaining discrepancies point to physics beyond the relativistic ideal MHD prescription adopted. This is an aspect of current research, calling for a proper incorpo-

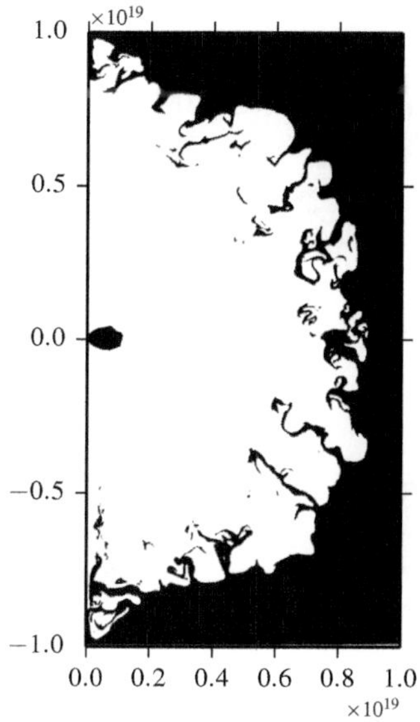

Fig. 22.12 Nonlinear Rayleigh–Taylor instabilities develop at the PWN interface with the ejecta, and fingers penetrate the surrounding supernova remnant shell. (From Porth et al. [504](c).)

ration of magnetic reconnection and particle acceleration physics, to e.g. explain the bright X-ray jet or the observed knotty appearance of the X-ray inner ring of the Crab nebula.

22.4 Outlook: General relativistic MHD simulations

Numerical simulations of relativistic magnetized plasmas have progressed to full general relativistic magnetohydrodynamics (GRMHD) studies, where the Einstein equations dictate the intricate interplay between matter and geometry. Codes like the Black Hole Accretion Code (BHAC) combine AMR possibilities in arbitrary stationary space-times with advanced shock-capturing methods. They can handle accretion flows onto compact objects and they have been validated against independent implementations [505]. When also the metric evolves, the instantaneous energy-momentum tensor appears as a source in the governing Einstein equations that determine this metric. Toolkits to solve for the metric tensor components are then exploited, in combination with numerical relativistic MHD codes. The latter then solve for the evolution of the matter-energy, as governed by the vanishing divergence of the energy-momentum tensor in the given (numerically computed) metric. This interplay between Einstein solvers and GRMHD codes is at the forefront of current research. An example is shown in Fig. 22.13, where the violent merger of a binary system of neutron stars is followed for tens of milliseconds [340]. Currently, these studies exploit an ideal MHD viewpoint on the occurring dynamics, and concentrate on how such systems lead to the launch of relativistic jets, powering the so-called short gamma-ray bursts, and how the merging translates into predicted gravitational wave signals. Topics of research include the details from the equation of state for the exotic neutron star systems, the possibility to get strong magnetic field amplification at the locations where the neutron stars collide and develop strong shear flows (i.e. favourable to Kelvin–Helmholtz

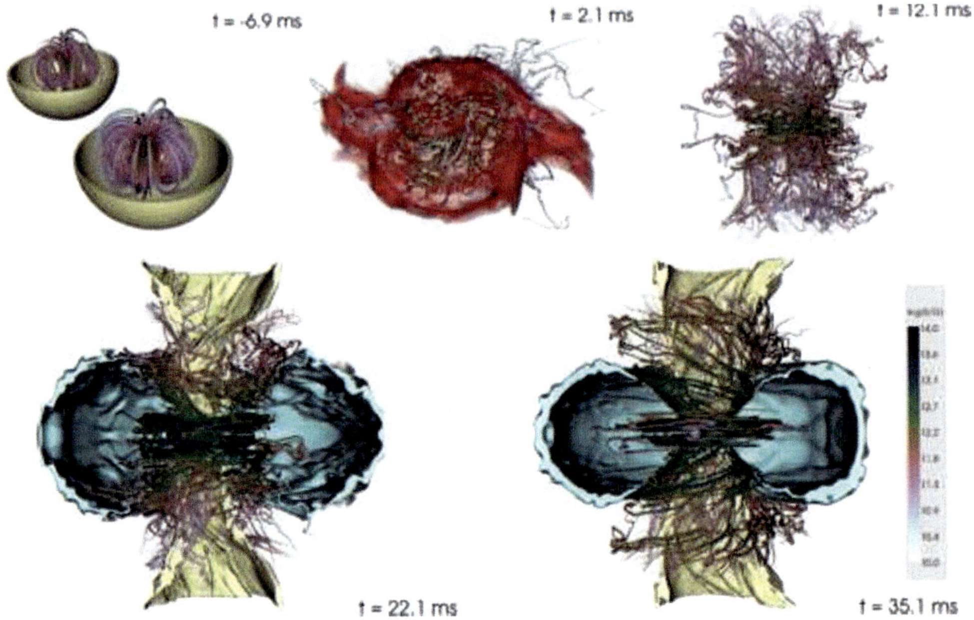

Fig. 22.13 Snapshots from GRMHD simulations of the merger of two binary neutron stars. These collapse to a spinning black hole surrounded by a magnetized accretion disc within tens of milliseconds. (From Kawamura et al. [340].)

instabilities), as well as properly reproducing the magneto-rotationally driven turbulence in the accretion disc about the final black hole. Suffice it to say that many aspects of MHD will continue to drive scientific research for decades to come, as one gradually gains realism in the fundamental multi-scale nature of plasma behaviour throughout our Universe.

22.5 Literature and exercises

Notes on literature

Relativistic MHD theory

- The monograph by Anile, *Relativistic fluids and magnetofluids* [14], is still the reference work on relativistic fluid treatments, with a clear mathematical orientation. This mathematical rigour is shared with the older work by Lichnerowicz, *Relativistic hydrodynamics and magnetohydrodynamics* [409].

Relativistic (M)HD and numerics

- The review article by Martí and Müller, 'Numerical hydrodynamics in special relativity' [432], discusses modern numerical methods and their applications for relativistic hydrodynamics. Relativistic hydrodynamics is also well covered in the book by Rezzolla and Zanotti [517]. Numerical relativistic magnetohydrodynamics making use of Riemann solver methodology emerged in earnest with the work by Komissarov, 'A Godunov-type scheme for relativistic magnetohydrodynamics' [375]. The review article by Font, 'Numerical hydrodynamics and magnetohydrodynamics in general relativity' [180] provides an overview of general-relativistic MHD formulations, giving numerical details as well as a summary of simulation based research efforts.

Exercises

[22.1] *Transformation laws for tensors*

In this exercise we derive the transformation laws for the components of first rank tensors (i.e. four-vectors) and second rank tensors within flat space-time.

- Find, in analogy with (22.11), but now exploiting *covariant* components and *dual* basis vectors, the transformation for *covariant* components of a four-vector in two sets of dual basis vectors $\mathbf{e}^{\alpha}$ and $\mathbf{e}^{\alpha'}$.
- Use the defining relations $\mathbf{e}_{\alpha} \cdot \mathbf{e}^{\beta} = \delta_{\alpha}^{\beta}$, as well as the symmetry in the Lorentz transformation matrix $L_{\alpha}^{\alpha'}$, to show that, when these two sets of dual basis vectors are associated with basis vectors related by $\mathbf{e}_{\alpha} = L_{\alpha}^{\alpha'} \mathbf{e}_{\alpha'}$, they necessarily obey $\mathbf{e}^{\alpha} = (L^{-1})^{\alpha}_{\alpha'} \mathbf{e}^{\alpha'}$.
- Find as well that the Lorentz transformation leaves the components of the Minkowski metric tensor unchanged, by working out the relations $g_{\alpha'\beta'} = (L^{-1})^{\alpha}_{\alpha'} (L^{-1})^{\beta}_{\beta'} g_{\alpha\beta}$ and $g^{\alpha'\beta'} = L_{\alpha}^{\alpha'} L_{\beta}^{\beta'} g^{\alpha\beta}$. These laws for component transformations apply generally to any second rank tensor.

[22.2] *Three-velocity addition law*

In this exercise, we revisit the three-velocity addition, written in various equivalent forms.

- Work out the velocity addition law given in Eq. (22.28), starting from the four-velocity formulation in Eq. (22.26). Prove that when only one spatial direction is involved, the addition law can be written as

$$u = \frac{v + w}{1 + v\,w/c^2} \,,$$

which clearly reduces to its Galilean $u = v + w$ counterpart for non-relativistic speeds.
- Also verify that we can write Eq. (22.28) as

$$w_{\|} = \frac{u_{\|} - v}{1 - u_{\|} v/c^2} \,, \quad \text{and } \mathbf{w}_{\perp} = \frac{\mathbf{u}_{\perp}}{\Gamma \left(1 - u_{\|} v/c^2\right)}$$

by decomposing $\mathbf{u} = \mathbf{u}_{\|} + \mathbf{u}_{\perp}$, where $\mathbf{u}_{\|} \parallel \mathbf{v}$.

[22.3] *Relativistic Doppler effect, aberration and beaming for light waves*

The transformation formulas from Eq. (22.63) are perhaps best known in their specific application to light waves in vacuum. Plane wave solutions to the Maxwell equations for empty space obey $\mathbf{k} \cdot \mathbf{B} = \mathbf{k} \cdot \mathbf{E} = \mathbf{E} \cdot \mathbf{B} = 0$ and the dispersion equation $\omega^2 = k^2 c^2$, whereas the electric and magnetic field (complex) wave amplitude vectors are related by $\mathbf{B} = \pm(\mathbf{n} \times \mathbf{E})/c$ and $\mathbf{E} = \mp c(\mathbf{n} \times \mathbf{B})$.

- Using the Lorentz transformation for electromagnetic fields given by Eqs. (2.23), prove that the amplitude of the light waves as observed in frames in relative motion to each other changes in accord with

$$\left(B'\right)^2 = \Gamma^2 B^2 \left(1 \mp \mathbf{n} \cdot \mathbf{v}/c\right)^2 \,.$$

This change in the intensity of the light is referred to as relativistic beaming.
- The relativistic Doppler effect can be written in terms of the frequency $\nu = \omega/2\pi$ for light waves as

$$\nu = \nu' \Gamma \left[1 \pm (v/c) \cos \phi'\right] \,,$$

where the angle between $\mathbf{k}'$ and $\mathbf{v}$ is denoted by ϕ'.
- Aberration is then also expressed for light waves as

$$\cos \phi = \frac{\cos \phi' \pm v/c}{1 \pm (v/c) \cos \phi'} \,.$$

Derive all these expressions yourself.

[22.4] *Relativistic hydro shock relations*

When discussing the gas dynamic shock relations, we wrote the jump relations (22.66)–(22.67) for the shock rest frame (SRF). Find the expressions for the four-vectors $\mathbf{U}^{\mathrm{u}}$, $\mathbf{U}^{\mathrm{d}}$, $\mathbf{U}^{\mathrm{s}}$, $\mathbf{l}$ in reference frames in which the upstream gas is at rest (upstream rest frame, or URF), and in which the downstream gas is at rest (downstream rest frame, or DRF). Make a graphical representation of the various reference frames using space-time diagrams as shown in Fig. 22.4. For simplicity, assume vanishing tangential velocities. Write down and analyse the shock relations in these reference frames.

[22.5] *Ohm's law in special relativity*

Note that we wrote the four-current as $J^\alpha = (c\sigma, \mathbf{j})^{\mathrm{T}} = (c\sigma, \sigma\mathbf{v})^{\mathrm{T}}$, where σ is the charge density.

- Verify the invariant $U_\alpha J^\alpha$.
- Find that, as a result,

$$J^\alpha + c^{-2}(U_\beta J^\beta)U^\alpha = 0\,.$$

  This is actually a simplified form of Ohm's law in covariant form, where more generally a right hand side $\sigma_{\mathrm{c}} e^\alpha$ should be written, with σ_{c} the conductivity parameter.
- Find that, instead of $\mathbf{j} = \sigma\mathbf{v}$, we must then write

$$\mathbf{j} = \sigma_{\mathrm{c}}\Gamma\left[\mathbf{E} + \mathbf{v}\times\mathbf{B} - \mathbf{v}\mathbf{v}\cdot\mathbf{E}/c^2\right] + \sigma\mathbf{v}\,.$$

  In the plasma rest frame, we have then indeed $\mathbf{j}' = \sigma_{\mathrm{c}}\mathbf{E}'$.

[22.6] *MHD waves dynamics*

In this exercise, we expand on the relations between the rest frame expressions for the characteristic speeds, and the expressions as given in the lab frame. For a more detailed discussion of the answers to the points below, we refer to [350].

- Show that the Alfvén wave expression (22.137) indeed corresponds with transforming from the rest frame expression, where the Alfvén phase speed is found from $v_{\mathrm{ph}}'^2 = (\mathbf{B}'\cdot\mathbf{k}')^2 c^2/(\rho h + B'^2)$.
- Verify that the expressions for the characteristic speeds from Eq. (22.139) reduce to their non-relativistic counterparts for $\Gamma \to 1$. Show that Eq. (22.139) yields the relativistic hydro result from Eq. (22.62) when the magnetic field vanishes. Furthermore, check the cold plasma limit where $c_{\mathrm{g}}^2 = 0$.
- Transform the rest frame expression (22.129) to the lab frame result (22.139). Note that you will need to combine the phase speed relation given by Eq. (22.64) with the Lorentz formulas (2.23) (in ideal relativistic MHD) for the magnetic field, together with the wave vector (and hence the wave front normal) relation from Eq. (22.63).

Appendix A

Vectors and coordinates

A.1 Vector identities

A list of the most frequently exploited identities:

$$\mathbf{a}\cdot(\mathbf{b}\times\mathbf{c}) = \mathbf{c}\cdot(\mathbf{a}\times\mathbf{b}) = \mathbf{b}\cdot(\mathbf{c}\times\mathbf{a})\,, \tag{A.1}$$

$$\mathbf{a}\times(\mathbf{b}\times\mathbf{c}) = \mathbf{a}\cdot\mathbf{c}\,\mathbf{b} - \mathbf{a}\cdot\mathbf{b}\,\mathbf{c}\,, \qquad (\mathbf{a}\times\mathbf{b})\times\mathbf{c} = \mathbf{a}\cdot\mathbf{c}\,\mathbf{b} - \mathbf{b}\cdot\mathbf{c}\,\mathbf{a}\,, \tag{A.2}$$

$$\nabla\times\nabla\Phi = 0\,, \tag{A.3}$$

$$\nabla\cdot(\nabla\times\mathbf{a}) = 0\,, \tag{A.4}$$

$$\nabla\times(\nabla\times\mathbf{a}) = \nabla\,\nabla\cdot\mathbf{a} - \nabla^2\mathbf{a}\,, \tag{A.5}$$

$$\nabla\cdot(\Phi\,\mathbf{a}) = \mathbf{a}\cdot\nabla\Phi + \Phi\,\nabla\cdot\mathbf{a}\,, \tag{A.6}$$

$$\nabla\times(\Phi\,\mathbf{a}) = \nabla\Phi\times\mathbf{a} + \Phi\,\nabla\times\mathbf{a}\,, \tag{A.7}$$

$$\mathbf{a}\times(\nabla\times\mathbf{b}) = (\nabla\mathbf{b})\cdot\mathbf{a} - \mathbf{a}\cdot\nabla\mathbf{b}\,, \tag{A.8}$$

$$(\mathbf{a}\times\nabla)\times\mathbf{b} = (\nabla\mathbf{b})\cdot\mathbf{a} - \mathbf{a}\,\nabla\cdot\mathbf{b}\,, \tag{A.9}$$

$$\begin{aligned}\nabla(\mathbf{a}\cdot\mathbf{b}) &= (\nabla\mathbf{a})\cdot\mathbf{b} + (\nabla\mathbf{b})\cdot\mathbf{a}\\ &= \mathbf{a}\cdot\nabla\mathbf{b} + \mathbf{b}\cdot\nabla\mathbf{a} + \mathbf{a}\times(\nabla\times\mathbf{b}) + \mathbf{b}\times(\nabla\times\mathbf{a})\,,\end{aligned} \tag{A.10}$$

$$\nabla\cdot(\mathbf{a}\,\mathbf{b}) = \mathbf{a}\cdot\nabla\mathbf{b} + \mathbf{b}\,\nabla\cdot\mathbf{a}\,, \tag{A.11}$$

$$\nabla\cdot(\mathbf{a}\times\mathbf{b}) = \mathbf{b}\cdot\nabla\times\mathbf{a} - \mathbf{a}\cdot\nabla\times\mathbf{b}\,, \tag{A.12}$$

$$\begin{aligned}\nabla\times(\mathbf{a}\times\mathbf{b}) &= \nabla\cdot(\mathbf{b}\,\mathbf{a} - \mathbf{a}\,\mathbf{b})\\ &= \mathbf{a}\,\nabla\cdot\mathbf{b} + \mathbf{b}\cdot\nabla\mathbf{a} - \mathbf{b}\,\nabla\cdot\mathbf{a} - \mathbf{a}\cdot\nabla\mathbf{b}\,,\end{aligned} \tag{A.13}$$

$$\iiint \nabla\cdot\mathbf{a}\,d\tau = \oiint \mathbf{a}\cdot\mathbf{n}\,d\sigma \qquad \textit{(Gauss)}, \tag{A.14}$$

$$\mathbf{a}\to\mathbf{a}\times\mathbf{c}(\mathrm{onst}) \quad\Rightarrow\quad \iiint \nabla\times\mathbf{a}\,d\tau = \oiint \mathbf{n}\times\mathbf{a}\,d\sigma\,, \tag{A.15}$$

$$\mathbf{a}\to\Phi\,\mathbf{c}(\mathrm{onst}) \quad\Rightarrow\quad \iiint \nabla\Phi\,d\tau = \oiint \Phi\,\mathbf{n}\,d\sigma\,, \tag{A.16}$$

$$\iiint (\Phi\nabla^2\Psi - \Psi\nabla^2\Phi)\,d\tau = \oiint (\Phi\nabla\Psi - \Psi\nabla\Phi)\cdot\mathbf{n}\,d\sigma \qquad \textit{(Green)}, \tag{A.17}$$

$$\iint (\nabla\times\mathbf{a})\cdot\mathbf{n}\,d\sigma = \oint \mathbf{a}\cdot d\boldsymbol{l} \qquad \textit{(Stokes)}, \tag{A.18}$$

$$\mathbf{a}\to\mathbf{a}\times\mathbf{c}(\mathrm{onst}) \quad\Rightarrow\quad \iint (\mathbf{n}\times\nabla)\times\mathbf{a}\,d\sigma = \oint d\boldsymbol{l}\times\mathbf{a}\,, \tag{A.19}$$

$$\mathbf{a}\to\Phi\,\mathbf{c}(\mathrm{onst}) \quad\Rightarrow\quad \iint \mathbf{n}\times\nabla\Phi\,d\sigma = \oint \Phi\,d\boldsymbol{l}\,. \tag{A.20}$$

A.2 Vector expressions in orthogonal coordinates

The position vector as a function of orthogonal coordinates x_i,

$$\mathbf{r} = \begin{cases} x(x_1,x_2,x_3) \\ y(x_1,x_2,x_3) \\ z(x_1,x_2,x_3) \end{cases}, \tag{A.21}$$

generates the following geometric quantities:

$$h_i \equiv \left|\partial\mathbf{r}/\partial x_i\right| \qquad \textit{(scale factors)}, \tag{A.22}$$

$$\mathbf{e}_i \equiv (1/h_i)\,\partial\mathbf{r}/\partial x_i\,, \quad \mathbf{e}_i\cdot\mathbf{e}_j = \delta_{ij} \qquad \textit{(dimensionless unit vectors)}, \tag{A.23}$$

$$d\ell = \sqrt{\textstyle\sum_i (h_i dx_i)^2} \qquad \textit{(line element)}, \tag{A.24}$$

$$d\tau = h_1 h_2 h_3\,dx_1 dx_2 dx_3 \qquad \textit{(volume element)}. \tag{A.25}$$

Vector representation:

$$\mathbf{V} = \textstyle\sum_i \hat{V}_i\,\mathbf{e}_i \quad \textit{(physical components } \hat{V}_i \textit{, the same dimension as } \mathbf{V}\textit{)}. \tag{A.26}$$

Products:

$$\mathbf{A}\cdot\mathbf{B} = \textstyle\sum_i \hat{A}_i\hat{B}_i \qquad \textit{(inner product)}, \tag{A.27}$$

$$\mathbf{A}\times\mathbf{B} = \textstyle\sum_i\sum_j\sum_k \epsilon_{ijk}\hat{A}_j\hat{B}_j\,\mathbf{e}_i \qquad \textit{(vector product)}, \tag{A.28}$$

$$\epsilon_{ijk} \equiv \begin{cases} 1 & \text{even permutation of 123} \\ -1 & \text{odd permutation of 123} \\ 0 & \text{otherwise} \end{cases} \qquad \textit{(permutation symbol)}, \tag{A.29}$$

$$\epsilon_{ijk}\epsilon_{ilm} = \delta_{jl}\delta_{km} - \delta_{jm}\delta_{kl}\,. \tag{A.30}$$

Differential operators:

$$\nabla\psi = \sum \frac{1}{h_i}\frac{\partial\psi}{\partial x_i}\,\mathbf{e}_i\,, \tag{A.31}$$

$$\nabla^2\psi = \frac{1}{h_1h_2h_3}\left[\frac{\partial}{\partial x_1}\left(\frac{h_2h_3}{h_1}\frac{\partial\psi}{\partial x_1}\right) + \frac{\partial}{\partial x_2}\left(\frac{h_1h_3}{h_2}\frac{\partial\psi}{\partial x_2}\right) + \frac{\partial}{\partial x_3}\left(\frac{h_1h_2}{h_3}\frac{\partial\psi}{\partial x_3}\right)\right], \tag{A.32}$$

$$\nabla\cdot\mathbf{A} = \frac{1}{h_1h_2h_3}\left[\frac{\partial}{\partial x_1}(h_2h_3\hat{A}_1) + \frac{\partial}{\partial x_2}(h_1h_3\hat{A}_2) + \frac{\partial}{\partial x_3}(h_1h_2\hat{A}_3)\right], \tag{A.33}$$

$$\begin{aligned}\nabla\times\mathbf{A} = {}& \frac{1}{h_2h_3}\left[\frac{\partial}{\partial x_2}(h_3\hat{A}_3) - \frac{\partial}{\partial x_3}(h_2\hat{A}_2)\right]\mathbf{e}_1 + \frac{1}{h_1h_3}\left[\frac{\partial}{\partial x_3}(h_1\hat{A}_1)\right.\\ & \left. - \frac{\partial}{\partial x_1}(h_3\hat{A}_3)\right]\mathbf{e}_2 + \frac{1}{h_1h_2}\left[\frac{\partial}{\partial x_1}(h_2\hat{A}_2) - \frac{\partial}{\partial x_2}(h_1\hat{A}_1)\right]\mathbf{e}_3\,.\end{aligned} \tag{A.34}$$

Derivatives of the unit vectors:

$$\begin{aligned}
&\frac{\partial\mathbf{e}_1}{\partial x_1} = -\frac{1}{h_2}\frac{\partial h_1}{\partial x_2}\mathbf{e}_2 - \frac{1}{h_3}\frac{\partial h_1}{\partial x_3}\mathbf{e}_3\,, \quad \frac{\partial\mathbf{e}_2}{\partial x_1} = \frac{1}{h_2}\frac{\partial h_1}{\partial x_2}\mathbf{e}_1\,, \quad \frac{\partial\mathbf{e}_3}{\partial x_1} = \frac{1}{h_3}\frac{\partial h_1}{\partial x_3}\mathbf{e}_1\,,\\
&\frac{\partial\mathbf{e}_1}{\partial x_2} = \frac{1}{h_1}\frac{\partial h_2}{\partial x_1}\mathbf{e}_2\,, \quad \frac{\partial\mathbf{e}_2}{\partial x_2} = -\frac{1}{h_1}\frac{\partial h_2}{\partial x_1}\mathbf{e}_1 - \frac{1}{h_3}\frac{\partial h_2}{\partial x_3}\mathbf{e}_3\,, \quad \frac{\partial\mathbf{e}_3}{\partial x_2} = \frac{1}{h_3}\frac{\partial h_2}{\partial x_3}\mathbf{e}_2\,,\\
&\frac{\partial\mathbf{e}_1}{\partial x_3} = \frac{1}{h_1}\frac{\partial h_3}{\partial x_1}\mathbf{e}_3\,, \quad \frac{\partial\mathbf{e}_2}{\partial x_3} = \frac{1}{h_2}\frac{\partial h_3}{\partial x_2}\mathbf{e}_3\,, \quad \frac{\partial\mathbf{e}_3}{\partial x_3} = -\frac{1}{h_1}\frac{\partial h_3}{\partial x_1}\mathbf{e}_1 - \frac{1}{h_2}\frac{\partial h_3}{\partial x_2}\mathbf{e}_2\,.
\end{aligned} \tag{A.35}$$

Hence,

$$\begin{aligned}\mathbf{A}\cdot\nabla\mathbf{B} = {}& \left[\frac{\hat{A}_1}{h_1}\left(\frac{\partial\hat{B}_1}{\partial x_1} + \frac{\partial h_1}{\partial x_2}\frac{\hat{B}_2}{h_2} + \frac{\partial h_1}{\partial x_3}\frac{\hat{B}_3}{h_3}\right) + \frac{\hat{A}_2}{h_2}\left(\frac{\partial\hat{B}_1}{\partial x_2} - \frac{\partial h_2}{\partial x_1}\frac{\hat{B}_2}{h_1}\right)\right.\\
& \left. + \frac{\hat{A}_3}{h_3}\left(\frac{\partial\hat{B}_1}{\partial x_3} - \frac{\partial h_3}{\partial x_1}\frac{\hat{B}_3}{h_1}\right)\right]\mathbf{e}_1 + \left[\frac{\hat{A}_1}{h_1}\left(\frac{\partial\hat{B}_2}{\partial x_1} - \frac{\partial h_1}{\partial x_2}\frac{\hat{B}_1}{h_2}\right)\right.\\
& \left. + \frac{\hat{A}_2}{h_2}\left(\frac{\partial\hat{B}_2}{\partial x_2} + \frac{\partial h_2}{\partial x_1}\frac{\hat{B}_1}{h_1} + \frac{\partial h_2}{\partial x_3}\frac{\hat{B}_3}{h_3}\right) + \frac{\hat{A}_3}{h_3}\left(\frac{\partial\hat{B}_2}{\partial x_3} - \frac{\partial h_3}{\partial x_2}\frac{\hat{B}_3}{h_2}\right)\right]\mathbf{e}_2\\
& + \left[\frac{\hat{A}_1}{h_1}\left(\frac{\partial\hat{B}_3}{\partial x_1} - \frac{\partial h_1}{\partial x_3}\frac{\hat{B}_1}{h_3}\right) + \frac{\hat{A}_2}{h_2}\left(\frac{\partial\hat{B}_3}{\partial x_2} - \frac{\partial h_2}{\partial x_3}\frac{\hat{B}_2}{h_3}\right)\right.\\
& \left. + \frac{\hat{A}_3}{h_3}\left(\frac{\partial\hat{B}_3}{\partial x_3} + \frac{\partial h_3}{\partial x_1}\frac{\hat{B}_1}{h_1} + \frac{\partial h_3}{\partial x_2}\frac{\hat{B}_2}{h_2}\right)\right]\mathbf{e}_3\,.\end{aligned} \tag{A.36}$$

▷ **Note**

The hat on the vector components is used here to avoid conflict with the covariant components of the non-orthogonal coordinate systems (Section A.3). It is dropped in the expressions for the coordinate systems below by writing A_{x_i} instead of $\hat{A}_i$. ◁

A.2.1 Cartesian coordinates (x, y, z)

For these coordinates, the scale factors reduce to unity,

$$\mathbf{r} = \begin{cases} x_1 \equiv x \\ x_2 \equiv y \\ x_3 \equiv z \end{cases}$$

$$\Rightarrow \quad h_1 = h_2 = h_3 = 1\,, \tag{A.37}$$

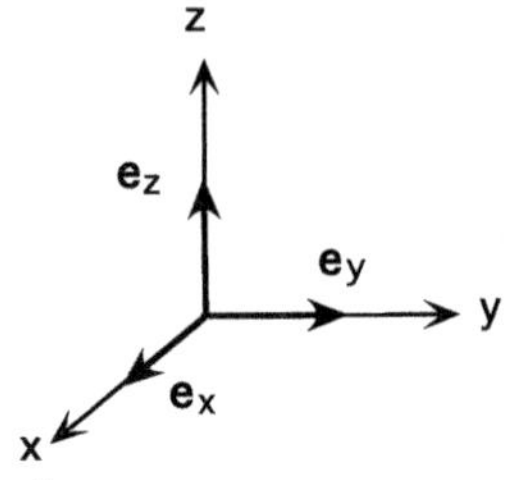

Fig. A.1 Cartesian.

so that the differential operators simplify to

$$\nabla\psi = \frac{\partial\psi}{\partial x}\mathbf{e}_x + \frac{\partial\psi}{\partial y}\mathbf{e}_y + \frac{\partial\psi}{\partial z}\mathbf{e}_z\,, \tag{A.38}$$

$$\nabla^2\psi = \frac{\partial^2\psi}{\partial x^2} + \frac{\partial^2\psi}{\partial y^2} + \frac{\partial^2\psi}{\partial z^2}\,, \tag{A.39}$$

$$\nabla\cdot\mathbf{A} = \frac{\partial A_x}{\partial x} + \frac{\partial A_y}{\partial y} + \frac{\partial A_z}{\partial z}\,, \tag{A.40}$$

$$\nabla\times\mathbf{A} = \left(\frac{\partial A_z}{\partial y} - \frac{\partial A_y}{\partial z}\right)\mathbf{e}_x + \left(\frac{\partial A_x}{\partial z} - \frac{\partial A_z}{\partial x}\right)\mathbf{e}_y + \left(\frac{\partial A_y}{\partial x} - \frac{\partial A_x}{\partial y}\right)\mathbf{e}_z\,. \tag{A.41}$$

▷ **Note**

The complicated vector identities of Section A.1, in particular the ones with the cross products, are most easily derived exploiting these coordinates, i.e. by substituting $h_i = 1$ into the expressions of Section A.2 (see Goldston and Rutherford [244], p. 481). ◁

A.2.2 Cylindrical coordinates (r, θ, z)

Position vector, scale factors and unit vector derivatives:

$$\mathbf{r} = \begin{cases} x = r\cos\theta \\ y = r\sin\theta \\ z = z \end{cases}$$

$$\Rightarrow \quad h_1 = 1\,, \quad h_2 = r\,, \quad h_3 = 1\,, \tag{A.42}$$

$$\frac{\partial\mathbf{e}_r}{\partial\theta} = \mathbf{e}_\theta\,, \quad \frac{\partial\mathbf{e}_\theta}{\partial\theta} = -\mathbf{e}_r \quad \text{(only ones} \neq 0)\,. \tag{A.43}$$

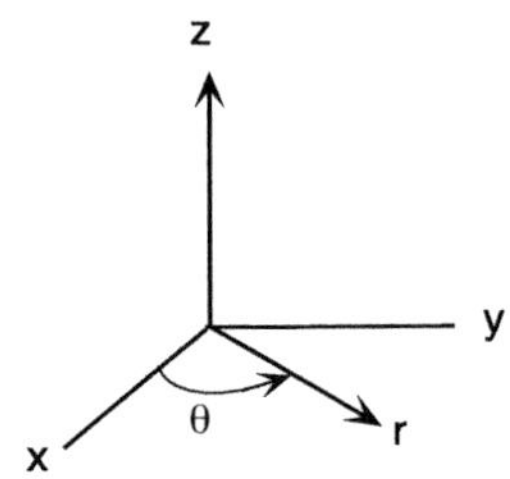

Fig. A.2 Cylindrical.

Differential operators:

$$\nabla\psi = \frac{\partial\psi}{\partial r}\mathbf{e}_r + \frac{1}{r}\frac{\partial\psi}{\partial\theta}\mathbf{e}_\theta + \frac{\partial\psi}{\partial z}\mathbf{e}_z\,, \tag{A.44}$$

$$\nabla^2\psi = \frac{1}{r}\frac{\partial}{\partial r}\left(r\frac{\partial\psi}{\partial r}\right) + \frac{1}{r^2}\frac{\partial^2\psi}{\partial\theta^2} + \frac{\partial^2\psi}{\partial z^2}\,, \tag{A.45}$$

$$\nabla \cdot \mathbf{A} = \frac{1}{r}\frac{\partial (rA_r)}{\partial r} + \frac{1}{r}\frac{\partial A_\theta}{\partial \theta} + \frac{\partial A_z}{\partial z}, \tag{A.46}$$

$$\nabla \times \mathbf{A} = \left(\frac{1}{r}\frac{\partial A_z}{\partial \theta} - \frac{\partial A_\theta}{\partial z}\right)\mathbf{e}_r + \left(\frac{\partial A_r}{\partial z} - \frac{\partial A_z}{\partial r}\right)\mathbf{e}_\theta + \left(\frac{1}{r}\frac{\partial (rA_\theta)}{\partial r} - \frac{1}{r}\frac{\partial A_r}{\partial \theta}\right)\mathbf{e}_z, \tag{A.47}$$

$$\nabla^2 \mathbf{A} = \left(\nabla^2 A_r - \frac{1}{r^2}A_r - \frac{2}{r^2}\frac{\partial A_\theta}{\partial \theta}\right)\mathbf{e}_r + \left(\nabla^2 A_\theta - \frac{1}{r^2}A_\theta + \frac{2}{r^2}\frac{\partial A_r}{\partial \theta}\right)\mathbf{e}_\theta + \nabla^2 A_z\,\mathbf{e}_z, \tag{A.48}$$

$$\begin{aligned}\nabla \times \nabla \times \mathbf{A} = &\left[-\frac{1}{r^2}\frac{\partial^2 A_r}{\partial \theta^2} - \frac{\partial^2 A_r}{\partial z^2} + \frac{1}{r^2}\frac{\partial^2 (rA_\theta)}{\partial \theta\,\partial r} + \frac{\partial^2 A_z}{\partial z \partial r}\right]\mathbf{e}_r \\ &+ \left[\frac{\partial}{\partial r}\left(\frac{1}{r}\frac{\partial A_r}{\partial \theta}\right) - \frac{\partial}{\partial r}\left(\frac{1}{r}\frac{\partial (rA_\theta)}{\partial r}\right) - \frac{\partial^2 A_\theta}{\partial z^2} + \frac{1}{r}\frac{\partial^2 A_z}{\partial z \partial \theta}\right]\mathbf{e}_\theta \\ &+ \left[\frac{1}{r}\frac{\partial}{\partial r}\left(r\frac{\partial A_r}{\partial z}\right) + \frac{1}{r}\frac{\partial^2 A_\theta}{\partial \theta \partial z} - \frac{1}{r}\frac{\partial}{\partial r}\left(r\frac{\partial A_z}{\partial r}\right) - \frac{1}{r^2}\frac{\partial^2 A_z}{\partial \theta^2}\right]\mathbf{e}_z,\end{aligned} \tag{A.49}$$

$$\begin{aligned}\mathbf{A} \cdot \nabla \mathbf{B} = &\left[A_r\frac{\partial B_r}{\partial r} + \frac{A_\theta}{r}\left(\frac{\partial B_r}{\partial \theta} - B_\theta\right) + A_z\frac{\partial B_r}{\partial z}\right]\mathbf{e}_r \\ &+ \left[A_r\frac{\partial B_\theta}{\partial r} + \frac{A_\theta}{r}\left(B_r + \frac{\partial B_\theta}{\partial \theta}\right) + A_z\frac{\partial B_\theta}{\partial z}\right]\mathbf{e}_\theta \\ &+ \left[A_r\frac{\partial B_z}{\partial r} + \frac{A_\theta}{r}\frac{\partial B_z}{\partial \theta} + A_z\frac{\partial B_z}{\partial z}\right]\mathbf{e}_z.\end{aligned} \tag{A.50}$$

A.2.3 Spherical coordinates (r, θ, ϕ)

Position vector, scale factors and unit vector derivatives:

$$\mathbf{r} = \begin{cases} x = R\cos\phi \\ y = R\sin\phi, \qquad R = R(r,\theta) \equiv r\sin\theta \\ z = r\cos\theta \end{cases}$$

$$\Rightarrow \quad h_1 = 1, \quad h_2 = r, \quad h_3 = r\sin\theta, \tag{A.51}$$

$$\frac{\partial \mathbf{e}_r}{\partial \theta} = \mathbf{e}_\theta, \qquad \frac{\partial \mathbf{e}_r}{\partial \phi} = \sin\theta\,\mathbf{e}_\phi,$$

$$\frac{\partial \mathbf{e}_\theta}{\partial \theta} = -\mathbf{e}_r, \qquad \frac{\partial \mathbf{e}_\theta}{\partial \phi} = \cos\theta\,\mathbf{e}_\phi,$$

$$\frac{\partial \mathbf{e}_\phi}{\partial \phi} = -\sin\theta\,\mathbf{e}_r - \cos\theta\,\mathbf{e}_\theta. \tag{A.52}$$

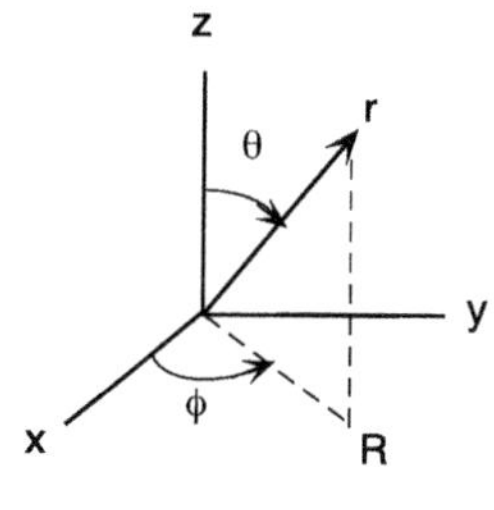

Fig. A.3 Spherical.

Differential operators:

$$\nabla\psi = \frac{\partial\psi}{\partial r}\,\mathbf{e}_r + \frac{1}{r}\frac{\partial\psi}{\partial\theta}\,\mathbf{e}_\theta + \frac{1}{r\sin\theta}\frac{\partial\psi}{\partial\phi}\,\mathbf{e}_\phi\,, \tag{A.53}$$

$$\nabla^2\psi = \frac{1}{r^2}\frac{\partial}{\partial r}\left(r^2\frac{\partial\psi}{\partial r}\right) + \frac{1}{r^2\sin\theta}\frac{\partial}{\partial\theta}\left(\sin\theta\frac{\partial\psi}{\partial\theta}\right) + \frac{1}{r^2\sin^2\theta}\frac{\partial^2\psi}{\partial\phi^2}\,, \tag{A.54}$$

$$\nabla\cdot\mathbf{A} = \frac{1}{r^2}\frac{\partial}{\partial r}(r^2A_r) + \frac{1}{r\sin\theta}\frac{\partial}{\partial\theta}(\sin\theta A_\theta) + \frac{1}{r\sin\theta}\frac{\partial A_\phi}{\partial\phi}\,, \tag{A.55}$$

$$\begin{aligned}\nabla\times\mathbf{A} = {} & \frac{1}{r\sin\theta}\left[\frac{\partial}{\partial\theta}(\sin\theta A_\phi) - \frac{\partial A_\theta}{\partial\phi}\right]\mathbf{e}_r \\ & + \frac{1}{r}\left[\frac{1}{\sin\theta}\frac{\partial A_r}{\partial\phi} - \frac{\partial}{\partial r}(rA_\phi)\right]\mathbf{e}_\theta + \frac{1}{r}\left[\frac{\partial}{\partial r}(rA_\theta) - \frac{\partial A_r}{\partial\theta}\right]\mathbf{e}_\phi\,,\end{aligned} \tag{A.56}$$

$$\begin{aligned}\nabla^2\mathbf{A} = {} & \left[(\nabla^2A_r) - \frac{2}{r^2}A_r - \frac{2}{r^2\sin\theta}\frac{\partial}{\partial\theta}(\sin\theta A_\theta) - \frac{2}{r^2\sin\theta}\frac{\partial A_\phi}{\partial\phi}\right]\mathbf{e}_r \\ & + \left[(\nabla^2A_\theta) + \frac{2}{r^2}\frac{\partial A_r}{\partial\theta} - \frac{1}{r^2\sin^2\theta}A_\theta - \frac{2\cos\theta}{r^2\sin^2\theta}\frac{\partial A_\phi}{\partial\phi}\right]\mathbf{e}_\theta \\ & + \left[(\nabla^2A_\phi) + \frac{2}{r^2\sin\theta}\frac{\partial A_r}{\partial\phi} + \frac{2\cos\theta}{r^2\sin^2\theta}\frac{\partial A_\theta}{\partial\phi} - \frac{1}{r^2\sin^2\theta}A_\phi\right]\mathbf{e}_\phi\,,\end{aligned} \tag{A.57}$$

$$\begin{aligned}\nabla\times\nabla\times\mathbf{A} = {} & \frac{1}{r^2\sin\theta}\left[-\frac{\partial}{\partial\theta}\left(\sin\theta\frac{\partial A_r}{\partial\theta}\right) - \frac{1}{\sin\theta}\frac{\partial^2A_r}{\partial\phi^2}\right. \\ & \left. + \frac{\partial}{\partial\theta}\left(\sin\theta\frac{\partial(rA_\theta)}{\partial r}\right) + \frac{\partial^2(rA_\phi)}{\partial\phi\,\partial r}\right]\mathbf{e}_r + \frac{1}{r^2}\left[r\frac{\partial^2A_r}{\partial r\partial\theta}\right. \\ & \left. - \frac{\partial}{\partial r}\left(r^2\frac{\partial A_\theta}{\partial r}\right) - \frac{1}{\sin^2\theta}\frac{\partial^2A_\theta}{\partial\phi^2} + \frac{1}{\sin^2\theta}\frac{\partial^2(\sin\theta A_\phi)}{\partial\phi\,\partial\theta}\right]\mathbf{e}_\theta \\ & + \frac{1}{r^2}\left[\frac{r}{\sin\theta}\frac{\partial^2A_r}{\partial r\partial\phi} + \frac{\partial}{\partial\theta}\left(\frac{1}{\sin\theta}\frac{\partial A_\theta}{\partial\phi}\right)\right. \\ & \left. - \frac{\partial}{\partial r}\left(r^2\frac{\partial A_\phi}{\partial r}\right) - \frac{\partial}{\partial\theta}\left(\frac{1}{\sin\theta}\frac{\partial(\sin\theta A_\phi)}{\partial\theta}\right)\right]\mathbf{e}_\phi\,,\end{aligned} \tag{A.58}$$

$$\begin{aligned}\mathbf{A}\cdot\nabla\mathbf{B} = {} & \left[A_r\frac{\partial B_r}{\partial r} + \frac{A_\theta}{r}\left(\frac{\partial B_r}{\partial\theta} - B_\theta\right) + \frac{A_\phi}{r}\left(\frac{1}{\sin\theta}\frac{\partial B_r}{\partial\phi} - B_\phi\right)\right]\mathbf{e}_r \\ & + \left[A_r\frac{\partial B_\theta}{\partial r} + \frac{A_\theta}{r}\left(B_r + \frac{\partial B_\theta}{\partial\theta}\right) + \frac{A_\phi}{r}\left(\frac{1}{\sin\theta}\frac{\partial B_\theta}{\partial\phi} - \cot\theta\,B_\phi\right)\right]\mathbf{e}_\theta \\ & + \left[A_r\frac{\partial B_\phi}{\partial r} + \frac{A_\theta}{r}\frac{\partial B_\phi}{\partial\theta} + \frac{A_\phi}{r}\left(B_r + \cot\theta\,B_\theta + \frac{1}{\sin\theta}\frac{\partial B_\phi}{\partial\phi}\right)\right]\mathbf{e}_\phi\,.\end{aligned} \tag{A.59}$$

A.2.4 Cylindrical-toroidal coordinates (R, Z, φ)

These coordinates are nearly equal to the cylindrical coordinates of Section A.2.2. They are an intermediate step to the toroidal coordinates (φ ignorable) of the following sections, where R and Z provide Cartesian coordinates in the poloidal plane. From now on, capital X, Y, Z are exploited for the Cartesian coordinates to enable later use of lower case x and y as scaled coordinates in the poloidal plane. Position vector, scale factors and unit vector derivatives:

$$\mathbf{r} = \begin{cases} x \;\to\; X = R\sin\varphi \,, \\ y \;\to\; Y = R\cos\varphi \,, \\ z \;\to\; Z \end{cases}$$

$$\Rightarrow \quad h_1 = 1 \,, \quad h_2 = 1 \,, \quad h_3 = R \,. \qquad \text{(A.60)}$$

$$\frac{\partial \mathbf{e}_R}{\partial \varphi} = \mathbf{e}_\varphi \,, \qquad \frac{\partial \mathbf{e}_\varphi}{\partial \varphi} = -\mathbf{e}_R \,. \qquad \text{(A.61)}$$

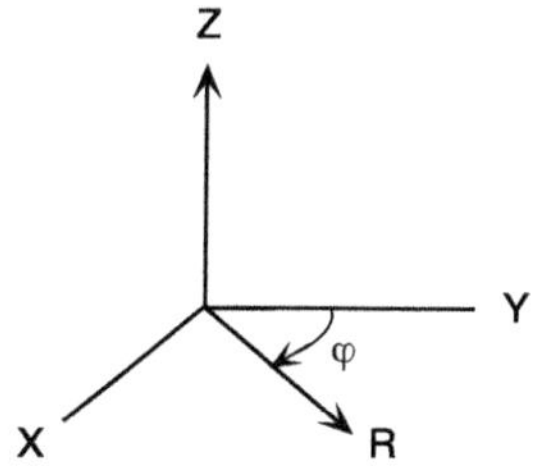

Fig. A.4 Cylindrical-toroidal.

The differential operators then follow from Eqs. (A.44)–(A.50) with the replacements

$$r \to R \,, \quad \theta \to \pi/2 - \varphi \,, \quad z \to Z \,; \qquad \partial/\partial\theta \to -\partial/\partial\varphi \,, \quad \mathbf{e}_\theta \to -\mathbf{e}_\varphi \,;$$
$$A_r \to A_R \,, \quad A_\theta \to -A_\varphi \,, \quad A_z \to A_Z \,. \qquad \text{(A.62)}$$

The only difference with the ordinary cylindrical coordinates is the order R, Z, φ.

A.2.5 Toroidal polar coordinates (r, θ, φ)

Position vector and scale factors:

$$\mathbf{r} = \begin{cases} X = R\sin\varphi \\ Y = R\cos\varphi \,, \qquad R(r,\theta) \equiv R_0 + r\cos\theta \\ Z = r\sin\theta \end{cases}$$

$$\Rightarrow \quad h_1 = 1 \,, \quad h_2 = r \,, \quad h_3 = R \,. \qquad \text{(A.63)}$$

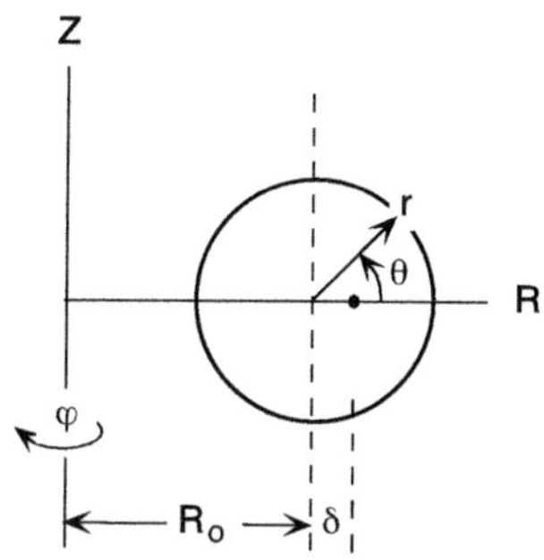

Fig. A.5 Toroidal polar.

The unit vector derivatives and the differential operators then follow by substitution into Eqs. (A.31)–(A.36) of Section A.2. These auxiliary coordinates may be used to describe the global toroidal geometrical features, like the vacuum fields outside the plasma. Since they do not incorporate the shift δ of the magnetic axis, they are not appropriate to describe toroidal equilibrium. For that purpose, the Shafranov shifted circle coordinates should be used; see Section 16.2.2.

A.2.6 Toroidal–conformal coordinates (s, t, φ)

Representing the poloidal plane by complex z-coordinates, $z \equiv x + \mathrm{i}y$, the Moebius transformation $z(w) = (w+\delta)/(1+\delta w)$, where $w \equiv s\exp(\mathrm{i}t)$, maps a circular plasma region with the magnetic axis at $x = \delta$ onto the unit disc $|w| \le 1$ with the image of the magnetic axis at the origin, $w = 0$. Position vector and scale factors then read:

$$\mathbf{r} = \begin{cases} X = R\sin\varphi \\ Y = R\cos\varphi\,, \quad R(r,\theta) \equiv R_0 + x(s,t) \\ Z = y(s,t)\,, \end{cases}$$

$$\Rightarrow \quad h_1 = h_2 = \frac{1-\delta^2}{1+2\delta s\cos t + \delta^2 s^2}\,, \qquad h_3 = R\,. \tag{A.64}$$

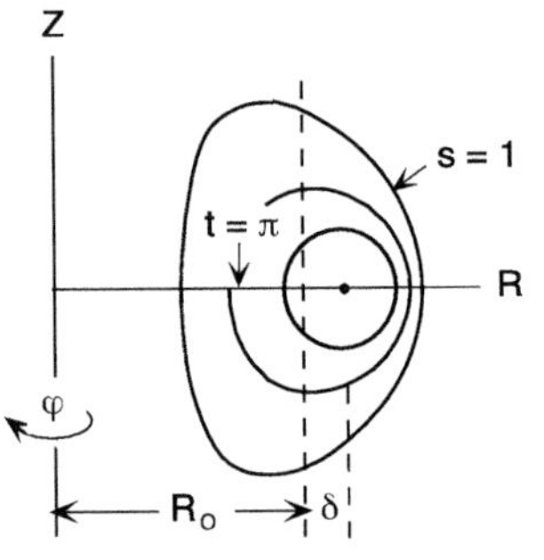

Fig. A.6 Toroidal-conformal.

For a non-circular boundary, the transformation is replaced by $z = z(\zeta(w))$, where $z(\zeta) = \sum_{m=1}^{\infty} \varphi_m \zeta^m$ is a numerical Henrici transformation [300] producing a circular boundary in the ζ-plane with the coefficients φ_m related to the FFTs of the boundary curve in the z-plane. These coordinates are *not flux coordinates* (except at $s = 0$ and $s = 1$), but they do respect the overall magnetic geometry; see Section 16.3.3. Together with special orthogonal polynomials [216], they are perfectly suited for the study of vortex dynamics on arbitrary two-dimensional domains.

A.2.7 Orthogonal flux coordinates (Ψ, χ, φ)

The use of these coordinates presupposes the explicit solution of the equilibrium equation for the poloidal flux Ψ and the construction of a poloidal angle χ such that $\nabla\Psi \cdot \nabla\chi = 0$, so that $\Psi(R,Z)$ and $\chi(R,Z)$ are known. Formal inversion yields the defining equations of the coordinates for the position vector and scale factors:

$$\mathbf{r} = \begin{cases} X = R(\Psi,\chi)\sin\varphi\,, \\ Y = R(\Psi,\chi)\cos\varphi\,, \\ Z = Z(\Psi,\chi) \end{cases}$$

$$\Rightarrow \quad h_1 = \frac{1}{RB_\mathrm{p}}\,, \qquad h_2 = J_\mathrm{o} B_\mathrm{p}\,, \qquad h_3 = R\,. \tag{A.65}$$

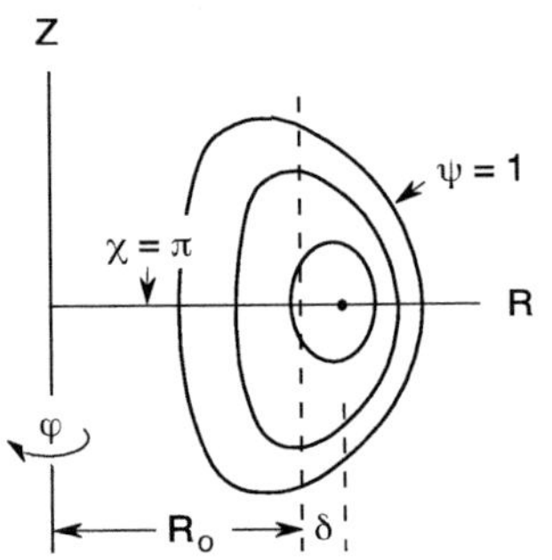

Fig. A.7 Orthogonal flux.

The subscript o on J_o distinguishes this Jacobian from the one of the non-orthogonal flux coordinates of Section A.3.1; see Eqs. (16.16)–(16.19). These coordinates are exclusively used for analytical calculations since the angular distribution of χ (not shown in Fig. A.7) suffers from logarithmic singularities: grid points evade the curved parts of the cross-section. For numerical calculations, the non-orthogonal flux coordinates of Section A.3.2 should be exploited.

A.3 Vector expressions in non-orthogonal coordinates

As mentioned above in Section A.2.7, the use of orthogonal coordinates may run into the problem of unacceptable spacing of grid lines in numerical calculations. In that case, use of non-orthogonal coordinates is imperative.

Considering the position vector as a function of curvilinear coordinates x^i,

$$\mathbf{r} = \begin{cases} x = x(x^1, x^2, x^3) \\ y = y(x^1, x^2, x^3) \ , \\ z = (x^1, x^2, x^3) \end{cases} \tag{A.66}$$

and denoting derivatives by $\partial_i \equiv \partial / \partial x^i$, the following geometric quantities are obtained:

$$\mathbf{a}_i \equiv \partial_i \mathbf{r} \,, \quad \mathbf{a}^i \equiv \nabla x^i \,, \quad \mathbf{a}_i \cdot \mathbf{a}^j = \delta_i{}^j \qquad \textit{(basis vectors)}\,, \tag{A.67}$$

[in general, these basis vectors are not dimensionless]

$$g_{ij} = \mathbf{a}_i \cdot \mathbf{a}_j \,, \quad g^{ij} = \mathbf{a}^i \cdot \mathbf{a}^j \,, \quad g_{ij}\, g^{jk} = \delta_i{}^k \qquad \textit{(metric tensor)}\,, \tag{A.68}$$

$$\mathbf{a}_1 \times \mathbf{a}_2 = J\,\mathbf{a}^3 \,, \qquad \mathbf{a}^1 \times \mathbf{a}^2 = \frac{1}{J}\,\mathbf{a}_3 \quad \text{(cyclic)}\,, \tag{A.69}$$

$$J \equiv \frac{\partial(x\ ,y\ ,z\ )}{\partial(x^1, x^2, x^3)} \equiv \begin{vmatrix} \partial_1 x & \partial_1 y & \partial_1 z \\ \partial_2 x & \partial_2 y & \partial_2 z \\ \partial_3 x & \partial_3 y & \partial_3 z \end{vmatrix} = \sqrt{\det(g_{ij})} \qquad \textit{(Jacobian)}\,, \tag{A.70}$$

[$J > 0$ for right-handed coordinate systems]

$$d\ell = \sqrt{g_{ij} dx^i dx^j} \qquad \textit{(line element)}\,, \tag{A.71}$$

[sum over repeated indices, unless stated otherwise]

$$d\tau = J\, dx^1 dx^2 dx^3 \qquad \textit{(volume element)}\,. \tag{A.72}$$

Vector representations (see Fig. A.8):

$$\mathbf{V} = V^i\, \mathbf{a}_i = V_i\, \mathbf{a}^i \qquad \textit{(contravariant } V^i\textit{, covariant } V_i\textit{)}\,, \tag{A.73}$$

[in general, these components have non-physical dimensions]

$$V^i = g^{ij} V_j \,, \qquad V_i = g_{ij} V^j \qquad \textit{(raising and lowering indices)}\,. \tag{A.74}$$

▷ **Note**

For orthogonal coordinates,

$$g_{ij} = (h_i)^2\, \delta_{ij} \,, \quad h_i = |\mathbf{a}_i| = |\mathbf{a}^i|^{-1} \quad \Rightarrow \quad \mathbf{e}_i = (h_i)^{-1}\, \mathbf{a}_i = h_i\, \mathbf{a}^i$$

$$\Rightarrow \quad \hat{V}_i \equiv \mathbf{V} \cdot \mathbf{e}_i = (h_i)^{-1}\, V_i = h_i\, V^i \quad \text{(not summing over } i\text{)}\,. \tag{A.75}$$

Hence, the hat on the physical components $\hat{V}_i$ in Section A.2. ◁

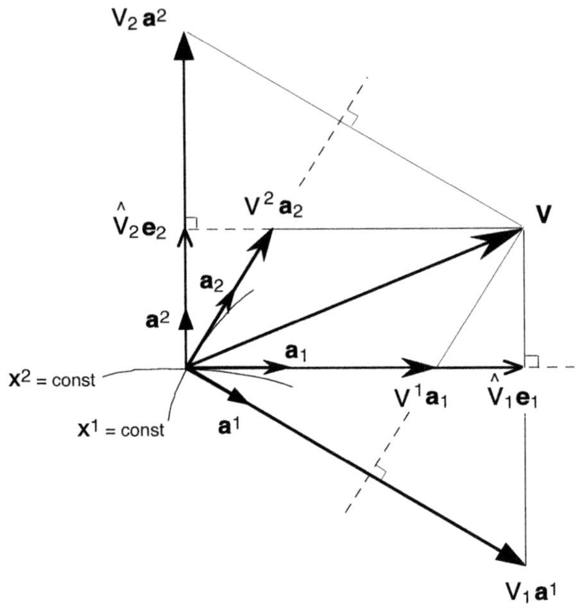

Fig. A.8 Covariant (V_i), contravariant (V^i) and physical ($\hat{V}_i$) vector components in two dimensions. The choice of the unit vector $\mathbf{e}_1$ to be parallel to $\mathbf{a}_1$ is arbitrary.

Products

$$\mathbf{A} \cdot \mathbf{B} = A_i B^i = g_{ij} A^i B^j = A^i B_i = g^{ij} A_i B_j \qquad \textit{(inner)}, \tag{A.76}$$

$$(\mathbf{A} \times \mathbf{B})_i = J \epsilon_{ijk} A^j B^k, \quad (\mathbf{A} \times \mathbf{B})^i = \frac{1}{J} \epsilon^{ijk} A_j B_k \qquad \textit{(vector)}, \tag{A.77}$$

involve the Levi-Civita symbols defined in Eq. (A.29), where now

$$\epsilon_{ijk} = \epsilon^{ijk}, \qquad \epsilon^{ijk} \epsilon_{ilm} = \delta^j_l \delta^k_m - \delta^j_m \delta^k_l . \tag{A.78}$$

▷ **Note**

(1) The Levi-Civita symbols ϵ_{ijk} and ϵ^{ijk} are tensor densities since it requires the Jacobian to turn them into the tensors $J\epsilon_{ijk}$ and $J^{-1}\epsilon^{ijk}$. Because of the dependence on J, the latter are further classified as pseudo-tensors since they change sign upon coordinate inversion ($J < 0$). Accordingly, the vector product is a pseudo-vector.

(2) Cyclic permutation of indices ($123 \to 312$, etc.) is an *even* operation in the 3D case, but in the 4D counterpart (22.88) of Section 22.2.2 cyclic permutation ($0123 \to 3012$, etc.) is *odd!* ◁

Differential operators:

$$\nabla \psi = (\partial_i \psi)\, \mathbf{a}^i, \qquad \nabla^2 \psi = \frac{1}{J} \partial_i \big(J g^{ij} \partial_j \psi \big), \tag{A.79}$$

$$\nabla \cdot \mathbf{A} = \frac{1}{J} \partial_i \big(J A^i \big), \qquad (\nabla \times \mathbf{A})^i = \frac{1}{J} \epsilon^{ijk} \partial_j A_k . \tag{A.80}$$

Derivatives of the basis vectors,

$$\partial_i \mathbf{a}_j = \Gamma^k_{ij}\, \mathbf{a}_k \,, \qquad \partial_i \mathbf{a}^k = -\Gamma^k_{ij}\, \mathbf{a}^j \,, \tag{A.81}$$

involve *the Christoffel symbols,*

$$\Gamma^k_{ij} = \Gamma^k_{ji} \equiv \mathbf{a}^k \cdot \partial_i \mathbf{a}_j = -\mathbf{a}_j \cdot \partial_i \mathbf{a}^k = \tfrac{1}{2} g^{kl} \big(\partial_i g_{lj} + \partial_j g_{il} - \partial_l g_{ij}\big) \,, \tag{A.82}$$

$$\partial_i g_{jk} = g_{kl}\Gamma^l_{ij} + g_{jl}\Gamma^l_{ik} \,, \quad \partial_i g^{jk} = -g^{jl}\Gamma^k_{il} - g^{kl}\Gamma^j_{il} \,, \quad \partial_i J = J\Gamma^j_{ij} \,, \tag{A.83}$$

so that

$$(\mathbf{A} \cdot \nabla \mathbf{B})^j = A^i \,(\partial_i B^j + \Gamma^j_{ik}\, B^k) \,. \tag{A.84}$$

Coordinate transformations,

$$x^j = x^j(\bar{x}^1, \bar{x}^2, \bar{x}^3) \quad \Rightarrow \quad dx^j = \alpha_i{}^j d\bar{x}^i \,, \quad \alpha_i{}^j \equiv \frac{\partial x^j}{\partial \bar{x}^i} \,, \quad \det(\alpha_i{}^j) \neq 0 \,, \tag{A.85}$$

involve the transformations

$$\bar{\mathbf{a}}_i \equiv \frac{\partial \mathbf{r}}{\partial \bar{x}^i} = \alpha_i{}^j\, \mathbf{a}_j \qquad \textit{(of the basis)}\,, \tag{A.86}$$

$$V^j = \alpha_i{}^j\, \bar{V}^i \qquad \textit{(of vector components, contragredient)}\,, \tag{A.87}$$

$$T^{kl} = \alpha_i{}^k \alpha_j{}^l\, \bar{T}^{ij} \,, \quad \text{etc.} \qquad \textit{(of tensor components)}\,. \tag{A.88}$$

A.3.1 General boundary-fitted coordinates (s, t, φ)

Boundary-fitted coordinates are quite generally employed in computational fluid dynamics. We encountered them in Section 5.4.2 when solving for the characteristics on a two-dimensional domain of arbitrary shape. Such coordinates only fit the outer boundary of the domain, not the whole set of flux surfaces enclosed, like the toroidal-conformal coordinates [216] of Section A.2.6. Another example is the set of isoparametric finite elements [319], discussed in Section 16.3.3. In the solution procedure of the nonlinear Grad–Shafranov equation by means of a Galerkin method, the iteration then starts with just boundary-fitted coordinates, but winds up with the solution in the non-orthogonal flux coordinates discussed below.

A.3.2 Non-orthogonal flux coordinates (Ψ, θ, φ)

As before, capitals X, Y, Z are exploited for the Cartesian coordinates and lower case x and y for the normalized coordinates in the poloidal plane:

$$\mathbf{r} = \begin{cases} X = R \sin\varphi \,, \\ Y = R\cos\varphi \,, \qquad R(\Psi, \theta) \equiv R_0 + x(\Psi, \theta) \,. \\ Z = y(\Psi, \theta) \end{cases} \tag{A.89}$$

Inversion of the equilibrium solution yields the poloidal coordinate dependence:

$$\text{Equilibrium} \begin{cases} \Psi = \Psi(x, y) \\ \theta = \theta(x, y) \end{cases} , \quad \text{Inversion} \;\Rightarrow\; \begin{cases} x = x(\Psi, \theta) \\ y = y(\Psi, \theta) \end{cases} , \tag{A.90}$$

giving the metric,

$$g_{ij} = \begin{pmatrix} g_{11} & g_{12} & 0 \\ g_{12} & g_{22} & 0 \\ 0 & 0 & g_{33} \end{pmatrix},$$

$$\begin{aligned} g_{11} &= (x_\Psi)^2 + (y_\Psi)^2 \,, \\ g_{12} &= x_\Psi \, x_\theta + y_\Psi \, y_\theta \,, \\ g_{22} &= (x_\theta)^2 + (y_\theta)^2 \,, \\ g_{33} &= R^2 = (R_0 + x)^2 \,, \end{aligned} \qquad \text{(A.91)}$$

and the Jacobian,

$$\begin{aligned} J &\equiv (\nabla\Psi \times \nabla\theta \cdot \nabla\varphi)^{-1} \\ &= R\sqrt{g_{11}g_{22} - (g_{12})^2} \\ &= R\,(x_\Psi \, y_\theta - x_\theta \, y_\Psi) \,. \end{aligned} \qquad \text{(A.92)}$$

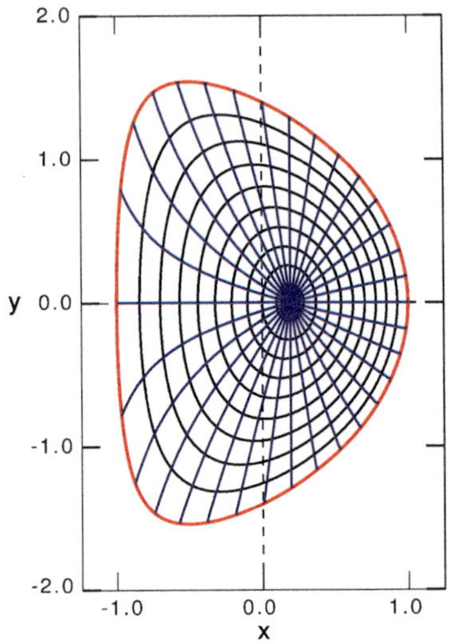

Fig. A.9 Non-orthogonal flux coordinates.

In Fig. A.9 these coordinates are illustrated for the Soloviev equilibrium analysed in Section 16.3.2. The coordinates lines are obtained from the inversion (16.181) with parameters $\epsilon = 0.4$. $\sigma = 1.4$, $\tau = 1.4$. The further specification to *straight-field-line coordinates* Ψ, ϑ, φ, with Jacobian $\mathcal{J} \equiv (\nabla\Psi \times \nabla\vartheta \cdot \nabla\varphi)^{-1}$, involves the transformation $\theta \to \vartheta$ illustrated in Fig. 16.2; see Section 16.1.2.

Notes on literature

Vector expressions and coordinate systems

- *NLR Plasma Formulary* by Book, and revised by Huba [313], has been in use by generations of plasma physicists to look up any of the standard formulas of vector analysis, systems of units, plasma parameters, transport coefficients, etc.
- *Flux Coordinates and Magnetic Field Structure* by D'haeseleer, Hitchon, Callen and Shohet [158] is a monograph entirely devoted to magnetic flux geometry, containing derivations of all coordinate systems in use for 2D and 3D geometries, like Hamada coordinates [279] and Boozer coordinates [81].
- *Plasma Physics and Fusion Energy* (Appendix C) by Freidberg [185] has an extensive derivation of Boozer coordinates and application to guiding centre orbit motion.

Appendix B
Tables of physical quantities

Table B.1

Physical constants (mks units)

Physical quantity	*Symbol*	*Value*
charge of the electron	e	$= 1.602 \times 10^{-19}$ C
mass of the electron	m_e	$= 9.109 \times 10^{-31}$ kg
mass of the proton	m_p	$= 1.673 \times 10^{-27}$ kg
velocity of light	c	$= 2.998 \times 10^{8}$ m s^{-1}
(vacuum permeability	μ_0	$\equiv 4\pi \times 10^{-7} = 1.257 \times 10^{-6}$ H m^{-1})
(vacuum permittivity	ϵ_0	$\equiv (\mu_0 c^2)^{-1} = 8.854 \times 10^{-12}$ F m^{-1})
Planck's constant	h	$= 6.626 \times 10^{-34}$ J s
Boltzmann's constant	k	$= 1.381 \times 10^{-23}$ J K^{-1}
gravitational constant	G	$= 6.673 \times 10^{-11}$ m^3 kg^{-1} s^{-2}

Table B.2

Plasma parameters

Physical quantity	*Definition*	*Expression* *
plasma frequency	$\omega_{\mathrm{p,e}} \equiv \sqrt{\dfrac{ne^2}{\epsilon_0 m_{\mathrm{e}}}}$	$= 56.6\sqrt{n}\ \mathrm{rad\,s^{-1}}$
Debye length	$\lambda_{\mathrm{D}} \equiv \sqrt{\dfrac{\epsilon_0 kT}{ne^2}}$	$= 69.0\sqrt{\dfrac{T}{n}}\ \mathrm{m}$
el. cyclotron frequency	$\Omega_{\mathrm{e}} \equiv \dfrac{eB}{m_{\mathrm{e}}}$	$= 1.76 \times 10^{11}\, B\ \mathrm{rad\,s^{-1}}$
el. cyclotron radius	$R_{\mathrm{e}} \equiv \dfrac{v_{\perp,\mathrm{e}}}{\Omega_{\mathrm{e}}}$	$= 3.13 \times 10^{-8}\, \dfrac{\sqrt{T}}{B}\ \mathrm{m}$
ion cyclotron frequency	$\Omega_{\mathrm{i}} \equiv \dfrac{ZeB}{m_{\mathrm{i}}}$	$= 9.58 \times 10^{7}\, \dfrac{Z}{A} B\ \mathrm{rad\,s^{-1}}$
ion cyclotron radius	$R_{\mathrm{i}} \equiv \dfrac{v_{\perp,\mathrm{i}}}{\Omega_{\mathrm{i}}}$	$= 1.34 \times 10^{-6}\, \dfrac{\sqrt{AT}}{ZB}\ \mathrm{m}$
el. thermal speed	$v_{\mathrm{th,e}} \equiv \sqrt{\dfrac{2kT}{m_{\mathrm{e}}}}$	$= 5.51 \times 10^{3}\sqrt{T}\ \mathrm{m\,s^{-1}}$
ion thermal speed	$v_{\mathrm{th,i}} \equiv \sqrt{\dfrac{2kT}{m_{\mathrm{i}}}}$	$= 1.28 \times 10^{2}\sqrt{\dfrac{T}{A}}\ \mathrm{m\,s^{-1}}$
sound speed	$v_{\mathrm{s}} \equiv \sqrt{\dfrac{\gamma p}{\rho}}$	$= 1.17 \times 10^{2}\sqrt{\dfrac{(1+Z)T}{A}}\ \mathrm{m\,s^{-1}}$
Alfvén speed	$v_{\mathrm{A}} \equiv \dfrac{B_0}{\sqrt{\mu_0 \rho}}$	$= 2.18 \times 10^{16}\sqrt{\dfrac{Z}{An}}\, B\ \mathrm{m\,s^{-1}}$

* in terms of $n\ (\equiv n_{\mathrm{e}} \approx Zn_{\mathrm{i}})$, $T\ (\equiv T_{\mathrm{e}} \approx T_{\mathrm{i}})$, B with $[n] = \mathrm{m^{-3}}$, $[T] = \mathrm{K}$, $[B] = \mathrm{T}$; A is ion mass number (multiples of m_{p}), Z is ion charge number (multiples of e). $1\,\mathrm{T} = 10^4\,\mathrm{G}$; $1\,\mathrm{eV} = 1.60 \times 10^{-19}\,\mathrm{J} \Leftrightarrow 1.16 \times 10^4\,\mathrm{K}$.

Table B.3

Orders of magnitude: kinetic theory

	Physical quantity	*Symbol*	**Tokamak**	**Coronal loop**
⇒	particle density	n	$10^{20}\,\mathrm{m}^{-3}$	$10^{16}\,\mathrm{m}^{-3}$
⇒	magnetic field	B	$3\,\mathrm{T}$ (= 30 kG)	$0.03\,\mathrm{T}$ (= 300 G)
⇒	temperature	T	$10^{8}\,\mathrm{K}\ (\approx 10\,\mathrm{keV})$	$10^{6}\,\mathrm{K}\ (\approx 100\,\mathrm{eV})$
	electron th. speed	$v_{\mathrm{th,e}}$	$5.5\times 10^{7}\,\mathrm{m\,s}^{-1}$	$5500\,\mathrm{km\,s}^{-1}$
	ion thermal speed*	$v_{\mathrm{th,i}}$	$1.3\times 10^{6}\,\mathrm{m\,s}^{-1}$	$130\,\mathrm{km\,s}^{-1}$
	electron gyro freq.	Ω_{e}	$\rightarrow 84\,\mathrm{GHz}$	$\rightarrow 0.84\,\mathrm{GHz}$
	electron gyro radius	R_{e}	$0.1\,\mathrm{mm}$	$1\,\mathrm{mm}$
	ion gyro freq.*	Ω_{i}	$\rightarrow 46\,\mathrm{MHz}$	$\rightarrow 0.46\,\mathrm{MHz}$
	ion gyro radius*	R_{i}	$4.5\,\mathrm{mm}$	$4.5\,\mathrm{cm}$
	plasma frequency	$\omega_{\mathrm{p,e}}$	$\rightarrow 90\,\mathrm{GHz}$	$\rightarrow 0.90\,\mathrm{GHz}$
	electron skin depth	δ_{e}	$0.53\,\mathrm{mm}$	$5.3\,\mathrm{cm}$
	Debye length	λ_{D}	$0.07\,\mathrm{mm}$	$0.7\,\mathrm{mm}$

⇒ independent parameters   → angular freq./(2π)

$*$ H plasma $(A = Z = 1)$

Table B.4

Orders of magnitude: MHD

	Physical quantity	*Symbol*	**Tokamak**	**Coronal loop**
⇒	half-width	a	$1\,\mathrm{m}$	$10\,000\,\mathrm{km}$
⇒	length	L	$20\,\mathrm{m}\ (= 2\pi R)$	$100\,000\,\mathrm{km}$
⇒	particle density	n	$10^{20}\,\mathrm{m}^{-3}$	$10^{16}\,\mathrm{m}^{-3}$
⇒	magnetic field	B	3 T (30 kG)	0.03 T (300 G)
⇒	temperature	T	$10^8\,\mathrm{K}\ (\approx 10\,\mathrm{keV})$	$10^6\,\mathrm{K}\ (\approx 100\,\mathrm{eV})$
	plasma density*	ρ	$1.7 \times 10^{-7}\,\mathrm{kg\,m^{-3}}$	$1.7 \times 10^{-11}\,\mathrm{kg\,m^{-3}}$
	pressure	p	$1.4 \times 10^5\,\mathrm{N\,m^{-2}}$	$0.14\,\mathrm{N/\,m^2}$
	beta	β	0.04	0.0004
	Alfvén speed*	v_A	$6500\,\mathrm{km\,s^{-1}}$	$6500\,\mathrm{km\,s^{-1}}$
	Alfvén transit time	τ_A	$3\,\mu\mathrm{s}$	$15\,\mathrm{s}$
	times of interest	τ	minute	week
	Alfvén transits	N	$20\,000\,000$	$40\,000$
	resist. decay time	τ_d	$16\,\mathrm{min}$	$3.2 \times 10^6\,\mathrm{y}$
	Spitzer resistivity*	$\eta_\parallel$	$1.3 \times 10^{-9}\,\Omega\,\mathrm{m}$	$1.2 \times 10^{-6}\,\Omega\,\mathrm{m}$
	mag. Reynolds nr.	R_m	6.3×10^9	6.8×10^{13}

⇒ independent parameters | $\rho \equiv (A/Z) n m_\mathrm{p}, \quad p \equiv nkT, \quad \beta \equiv 2\mu_0 p/B^2,$

* H plasma ($A = Z = 1$) | $v_\mathrm{A} \equiv B/\sqrt{\mu_0 \rho}, \quad \tau_\mathrm{A} \equiv L/v_\mathrm{A}, \quad N \equiv \tau/\tau_\mathrm{A},$

| $\tau_\mathrm{d} \equiv \mu_0 a^2/\eta, \quad \eta_\parallel = 65 Z \ln \Lambda T_\mathrm{e}^{-3/2}, \quad R_\mathrm{m} \equiv \mu_0 a v_A/\eta_\parallel .$

Table B.5

Orders of magnitude: MHD (cont'd)

		Solar wind	Magnetosphere	Your favourite plasma
⇒	a	1.5×10^6 km	6×10^3 km	....
⇒	L	1.5×10^8 km	4×10^4 km	....
⇒	n	$10^7\,\mathrm{m}^{-3}$	$10^{10}\,\mathrm{m}^{-3}$	....
⇒	B	6×10^{-9} T	3×10^{-5} T	....
⇒	T	10^5 K	10^4 K	....
	ρ	$1.7 \times 10^{-20}\,\mathrm{kg\,m}^{-3}$	$1.7 \times 10^{-17}\,\mathrm{kg\,m}^{-3}$	....
	p	$1.4 \times 10^{-11}\,\mathrm{N\,m}^{-2}$	$1.4 \times 10^{-9}\,\mathrm{N\,m}^{-2}$	....
	β	1	4×10^{-6}	....
	v_{A}	$41\,\mathrm{km\,s}^{-1}$	$6500\,\mathrm{km\,s}^{-1}$	....
	τ_{A}	42 days	6 s	....
	τ	month	hour	....
	N	< 1	600	....
	τ_{d}	1.7×10^9 y	1.2×10^3 y	....
	$\eta_{\parallel}$	$5.2 \times 10^{-5}\,\Omega\,\mathrm{m}$	$1.2 \times 10^{-3}\,\Omega\,\mathrm{m}$	....
	R_m	1.5×10^{12}	4.1×10^{10}	....

Table B.6

Physical data of Sun and planets and Planetary magnetic fields

	radius (km)	*mass* (kg) / *mag. mom.* (A m^2)	*rotation per.* (d) / B_{equator} $\times 10^{-4}$ T	*distance Sun* (AU) / *obliquity*	*orbital eccentr.* / *tilt mag. – rot. ax.*	*orbital per.* (y) / *day-side m.p.*
Sun	7.0×10^5	2.0×10^{30}	25–27	–	–	–
Mercury	2.4×10^3	3.3×10^{23}	58.6	0.39	0.206	0.24
		4.0×10^{19}	0.0032	0.0°	14.0°	1.4 R_M
Venus	6.1×10^3	4.9×10^{24}	243.0	0.72	0.007	0.615
		$< 1.0 \times 10^{18}$	< 0.0003	177.4°	–	–
Earth	6.4×10^3	6.0×10^{24}	1.0	$\equiv 1.0$	0.017	1.0
		8.1×10^{22}	0.31	23.5°	10.8°	10.4 R_E
Mars	3.4×10^3	6.4×10^{23}	1.03	1.52	0.093	1.88
		2.3×10^{19}	0.0006	25.2°	–	–
Jupiter	7.1×10^4	1.9×10^{27}	0.42	5.20	0.048	11.9
		1.5×10^{27}	4.28	3.1°	−9.6°	65 R_J
Saturn	6.0×10^4	5.7×10^{26}	0.43	9.54	0.056	29.5
		8.6×10^{25}	0.22	26.7°	−0°	20R_S
Uranus	2.6×10^4	8.7×10^{25}	0.72	19.2	0.047	84.0
		3.9×10^{24}	0.23	97.9°	−58.6°	18 R_U
Neptune	2.5×10^4	1.0×10^{26}	0.75	30.1	0.009	164.8
		2.0×10^{24}	0.14	29.6°	−46.8°	35R_N

$R_\odot = 6.96 \times 10^5$ km, 1 AU $= 215\, R_\odot = 1.496 \times 10^8$ km

1 ly $= 6.32 \times 10^4$ AU $= 9.46 \times 10^{12}$ km, 1 pc $= 3.26$ ly $= 3.09 \times 10^{13}$ km

References

[1] Previous Volume 1: J. P. Goedbloed and S. Poedts, *Principles of Magnetohydrodynamics* (Cambridge, Cambridge University Press, 2004). <xix>

[2] Previous Volume 2: J. P. Goedbloed, R. Keppens and S. Poedts, *Advanced Magnetohydrodynamics* (Cambridge, Cambridge University Press, 2010). <xix>

The starting pages of the sections, where the pertinent reference is cited, are indicated in angular brackets.

---

[3] (a) M. Abramowitz and I. A. Stegun, *Handbook of Mathematical Functions with Formulas, Graphs, and Mathematical Tables* (Washington, D.C., National Bureau of Standards, 1964);
(b) F. W. J. Olver, D. W. Lozier, R. F. Boisvert and C. W. Clarke, *NIST Handbook of Mathematical Functions* (National Institute of Standards and Technology and Cambridge University Press, 2010). <124, 399, 576, 642>

[4] A. Achterberg, 'Variational principle for relativistic magnetohydrodynamics', *Phys. Rev. A* **28** (1983), 2449–2458. <99>

[5] J. A. Adam, 'A nonlinear eigenvalue problem in astrophysical magnetohydrodynamics', *J. Math. Phys.* **30** (1989), 744–756. <258>

[6] C. Aerts, J. Christensen-Dalsgaard and D. W. Kurtz, *Asteroseismology* (New York, Springer Verlag, 2010). <239>

[7] Y. Z. Agim and J. A. Tataronis, 'General two-dimensional magnetohydrodynamic equilibria with mass flow', *J. Plasma Physics* **34** (1985), 337–360. <715>

[8] A. I. Akhiezer, I. A. Akhiezer, P. V. Polovin, A. G. Sitenko and K. N. Stepanov, *Plasma Electrodynamics, (a) Vol. 1, Linear Theory; (b) Vol. 2, Non-linear Theory and Fluctuations* (Oxford, Pergamon Press, 1975). <24, 101, 179, 870>

[9] A. I. Akhiezer, G. J. Lubarski and P. V. Polovin, 'The stability of shock waves in magnetohydrodynamics', *Soviet Phys.–JETP* **35** (1959), 507–511. <870>

[10] H. Alfvén, 'Existence of electromagnetic-hydrodynamic waves', *Nature* **3805** (1942), 405–405. <57, 64>

[11] H. Alfvén, *Cosmical Electrodynamics*, 1st edition (Oxford, Clarendon Press, 1950). <121, 341>

[12] F. Alladio and F. Crisanti, 'Analysis of MHD equilibria by toroidal multipolar expansions', *Nucl. Fusion* **26** (1986), 1143–1164. <642>

[13] G. Amarante-Segundo, A. G. Elfimov, D. W. Ross, R. M. O. Galvão and I. C. Nascimento, 'Calculations of wave excitation and dissipation in Tokamak Chauffage Alfvén wave heating experiment in Brazil', *Phys. Plasmas* **6** (1999), 2437–2442. <417>

[14] A.M. Anile, *Relativistic Fluids and Magneto-Fluids* (Cambridge, Cambridge University Press, 1989). <99, 879, 906, 916>

[15] S. K. Antiochos, 'The magnetic topology of solar eruptions', *Astrophys. J. (Lett.)* **502** (1998), L181–L184. <314>

[16] S. K. Antiochos, C. R. DeVore and J. A. Klimchuk, 'A model for solar coronal mass ejections', *Astrophys. J.* **510** (1999), 485–493. <314>

[17] K. Appert, D. Berger, R. Gruber and J. Rappaz, 'A new finite element approach to the normal mode

analysis in magnetohydrodynamics', *J. Comp. Physics* **18** (1975), 284–299. <576, 586>

[18] K. Appert, R. Gruber and J. Vaclavik, 'Continuous spectra of a cylindrical magnetohydrodynamic equilibrium', *Phys. Fluids* **17** (1974), 1471–1472. <252, 269, 330, 369, 473, 500, 677, 716>

[19] K. Appert, R. Gruber, F. Troyon and J. Vaclavik, 'Excitation of global eigenmodes of the Alfvén wave in tokamaks', *Plasma Physics* **24** (1982), 1147–1159. <347>

[20] K. Appert, B. Balet, R. Gruber, F. Troyon, T. Tsunematsu and J. Vaclavik, 'MHD computations for Alfvén wave heating in tokamaks', *Nucl. Fusion* **22** (1982), 903–919. <405>

[21] S. Appl and M. Camenzind, 'Shock conditions for relativistic MHD jets', *Astron. Astrophys.* **206** (1988), 258–268. <871, 906>

[22] T. D. Arber, A. W. Longbottom, C. L. Gerrard and A. M. Milne, 'A staggered grid, Lagrangian–Eulerian remap code for 3-D MHD simulations', *J. Comp. Physics* **171** (2001), 151–181. <822>

[23] M. Ariola and A. Pironti, *Magnetic Control of Tokamak Plasmas* (London, Springer-Verlag, 2008). <642>

[24] V. I. Arnold, *Mathematical Methods of Classical Mechanics* (New York, Springer-Verlag, 1980). <456>

[25] L. A. Artsimovich et al., 'Experiments in tokamak devices', in *Plasma Physics and Controlled Nuclear Fusion Research*, Proc. Third Intern. IAEA Conf., 1–7 August 1968, Novosibirsk, Vol. I (Vienna, IAEA, 1969), 157–173. [English translation: *Nucl. Fusion Suppl.* 1969, 17–24.] <617>

[26] L. A. Artsimovich, 'Tokamak devices', *Nucl. Fusion* **12** (1972), 215–252. <617>

[27] M. J. Aschwanden, 'An evaluation of coronal heating models for active regions based on Yohkoh, Soho, and TRACE observations', *Astrophys. J.* **560** (2001), 1035–1044. <432>

[28] M. J. Aschwanden, *Physics of the Solar Corona, an Introduction* (New York, Springer Verlag, 2004). <322>

[29] S. A. Balbus and J. F. Hawley, 'A powerful local shear instability in weakly magnetized disks. I. Linear analysis', *Astrophys. J.* **376** (1991), 214–222. <441, 513, 514, 707>

[30] W. A. Balbus and J. F. Hawley, 'Instability, turbulence, and enhanced transport in accretion disks', *Rev. Modern Physics* **70** (1998), 1–53. <105, 514, 523, 707>

[31] R. Balescu, 'Irreversible processes in ionised gases', *Phys. Fluids* **3** (1960), 52–63. <67>

[32] R. Balescu, (a) *Transport Processes in Plasmas, Vol. 1: Classical Transport Theory*; (b) *Vol. 2: Neoclassical Transport* (Amsterdam, North Holland, 1988). <33, 38, 64, 66, 67, 72, 83, 101, 437>

[33] R. Balescu, *Aspects of Anomalous Transport in Plasmas* (Bristol, Institute of Physics Publishing, 2005). <xix, 66, 101>

[34] B. Balet, K. Appert and J. Vaclavik, 'MHD studies on Alfvén wave heating of low-β plasmas', *Plasma Physics* **24** (1982), 1005–1023. <405, 417, 432>

[35] D. Balsara, 'Total variation diminishing scheme for relativistic magnetohydrodynamics', *Astrophys. J. Suppl. Ser.* **132** (2001), 83–101. <908>

[36] E. M. Barston, 'Electrostatic oscillations in inhomogeneous cold plasmas', *Ann. Phys. (New York)* **29** (1964), 282–303. <269, 397>

[37] E. M. Barston, 'Eigenvalue problem for Lagrangian systems', *J. Math. Phys.* **8** (1967), 523–532. <456>

[38] G. K. Batchelor, *An Introduction to Fluid Dynamics* (Cambridge, Cambridge University Press, 1967). <128>

[39] H. Bateman, *Higher Transcendental Functions, Volume I*, eds. A. Erdélyi, W. Magnus, F. Oberhettinger and F. G. Tricomi (New York, McGraw-Hill Book Company, 1953). <642>

[40] E. Battaner, *Astrophysical Fluid Dynamics* (Cambridge, Cambridge University Press, 1996). <24>

[41] H. Baty, E. R. Priest and T. G. Forbes, 'Effect of nonuniform resistivity in Petschek reconnection', *Phys. Plasmas* **13** (2006), 022312, 1–7. <558>

[42] (a) W. Baumjohann and R. A. Treumann, *Basic Space Plasma Physics* (London, Imperial College Press, 1996); (b) R. A. Treumann and W. Baumjohann, *Advanced Space Plasma Physics* (London, Imperial College Press, 1997). <24>

[43] M. Becoulet et al., 'Mechanism of edge localized mode mitigation by resonant magnetic perturbations', *Phys. Rev. Lett.* **113** (2014), 11500. <697>

[44] A. J. C. Beliën, M. A. Botchev, J. P. Goedbloed, B. van der Holst and R. Keppens, 'FINESSE: Axisymmetric MHD equilibria with flow', *J. Comp. Physics* **182** (2002), 91–117. <576, 655, 690, 707, 715, 716, 737, 742>

[45] A. J. C. Beliën, P. C. H. Martens and R. Keppens, 'Coronal heating by resonant absorption: The effects of chromospheric coupling', *Astrophys. J.* **526** (1999), 478–493. <428>

[46] A. J. C. Beliën, S. Poedts and J. P. Goedbloed, 'Magnetohydrodynamic continua and Stratification induced Alfvén Eigenmodes in coronal magnetic loops', *Phys. Rev. Lett.* **76** (1996), 567–570. <322>

[47] A. J. C. Beliën, S. Poedts and J. P. Goedbloed, 'Slow magnotosonic waves and instabilities in expanded flux tubes anchored in chromospheric/photospheric regions', in *The Corona and Solar Wind near Minimum Activity*, Proc. 5th SOHO Workshop, ESA SP-404 (1997), 193–197. <428, 655>

[48] R. B. Bellan, *Spheromaks* (London, Imperial College Press, 2000). <xix>

[49] P. M. Bellan, *Fundamentals of Plasma Physics* (Cambridge, Cambridge University Press, 2006). <566>

[50] C. M. Bender and S. Orszag, *Advanced Mathematical Methods for Scientists and Engineers* (Tokyo, McGraw-Hill International Book Company, 1978). <46, 265, 269, 289, 534>

[51] S. V. Berdyugina, 'Starspots: A key to the stellar dynamo', *Living Rev. Sol. Phys.* **2** (2005), 8. <309>

[52] M. A. Berger, 'Rigorous new limits on magnetic helicity dissipation in the solar corona', *Geophys. Astrophys. Fluid Dynamics* **30** (1984), 79–104. <124>

[53] M. J. Berger and P. Colella, 'Local adaptive mesh refinement for shock hydrodynamics', *J. Comp. Physics* **82** (1989), 64–84. <819>

[54] D. Berghmans and P. De Bruyne, 'Coronal loop oscillations driven by footpoint motions: analytical results for a model problem', *Astrophys. J.* **453** (1995), 495–504. <428>

[55] I. B. Bernstein, E. A. Frieman, M. D. Kruskal and R. M. Kulsrud, (a) 'An energy principle for hydromagnetic stability problems', *Proc. Roy. Soc. (London)* **A244** (1958), 17–40; (b) Appendix, Matterhorn Report NYO-7896 (Princeton, 1957). <183, 191, 204, 206, 213, 229, 471, 690, 704, 722>

[56] N. Bessolaz, C. Zanni, J. Ferreira, R. Keppens and J. Bouvier, 'Accretion funnels onto weakly magnetized young stars', *Astron. Astrophys.* **478** (2008), 155–162. <871>

[57] G. Besson, A. de Chambrier, G. A. Collins, B. Joye, A. Lietti, J. B. Lister, J. M. Moret, S. Nowak, C. Simm and H. Weisen, 'A review of Alfvén wave heating', *Plasma Phys. Control. F.* **28** (1986), 1291–1303. <417>

[58] R. Betti and J. P. Freidberg, 'Ellipticity induced Alfvén eigenmodes', *Phys. Fluids* **B3** (1991), 1865–1870. <680>

[59] R. Betti and J. P. Freidberg, 'Stability analysis of resistive wall kink modes in rotating plasmas', *Phys. Rev. Lett.* **74** (1995), 2949–2952. <544>

[60] S. N. Bhattacharyya and A. Bhattacharjee, 'Effect of the flow continuum on magnetohydrodynamic stablity', *Phys. Plasmas* **4** (1997), 3744–3748. <480, 725>

[61] L. Biermann, K. Hain, K. Jörgens and R. Lüst, 'Axialsymmetrische Lösungen der magnetohydrostatischen Gleichung mit Oberflächenströmen', *Z. Naturforsch.* **12a** (1957), 826–832. <642>

[62] J. Birn et al., 'Geospace Environmental Modeling (GEM) Magnetic Reconnection Challenge', *J. Geophys. Res.* **106** (2001), 3715–3719. <554, 560>

[63] J. Birn et al., 'Forced magnetic reconnection', *Geophys. Res. Lett.* **32** (2005) L06105. <554, 560>

[64] J. Birn and M. Hesse, 'Geospace Environmental Modeling (GEM) Magnetic Reconnection Challenge: resistive tearing, anisotropic pressure and Hall effects', *J. Geophys. Res.* **106** (2001), 3737–3750. <554>

[65] D. Biskamp, *Nonlinear Magnetohydrodynamics* (Cambridge, Cambridge University Press, 1993). <124, 128, 144, 566, 778>

[66] D. Biskamp, *Magnetic Reconnection in Plasmas* (Cambridge, Cambridge University Press, 2000). <554, 558, 566>

[67] D. Biskamp, *Magnetohydrodynamic Turbulence* (Cambridge, Cambridge University Press, 2003). <566, 768, 778>

[68] D. Biskamp and W. C. Müller, 'Scaling properties of three-dimensional isotropic magnetohydrodynamic turbulence', *Phys. Fluids* **7** (2000), 4889–4900. <768>

[69] D. Biskamp and H. Welter, 'Theory of sawtooth disruption', in *Plasma Physics and Controlled Fusion Research 1986*, Vol. 2 (Vienna, IAEA, 1987), 11–16. <832>

[70] J. A. Bittencourt, *Fundamentals of Plasma Physics* (New York, Pergamon Press, 1986); 2nd edition (Brazil, Foundation of the State of São Paulo for the Support of Research, 1995). <24, 38, 64, 67, 159, 179>

[71] R. D. Blandford and D. G. Payne, 'Hydromagnetic flows from accretion discs and the production of radio jets', *Monthly Not. Roy. Astron. Soc.* **199** (1982), 883–903. <709, 871>

[72] A. A. Blank, K.O. Friedrichs and H. Grad, *Notes on Magnetohydrodynamics, V. Theory of Maxwell's Equations without Displacement Current* (New York University, NYO-6486-V, 1957). <66, 99, 101>

[73] J. W. S. Blokland, R. Keppens and J. P. Goedbloed, 'Unstable magnetohydrodynamical continuous spectrum of accretion disks: a new route to magnetohydrodynamical turbulence in accretion disks', *Astron. Astrophys.* **467** (2007), 21–35. <870>

[74] J. W. S. Blokland, B. van der Holst, R. Keppens and J. P. Goedbloed, 'PHOENIX: MHD spectral code for rotating laboratory and gravitating astrophysical plasmas', *J. Comp. Physics* **226** (2007), 509–533. <690, 707, 737, 742, 870>

[75] J. W. S. Blokland, E. van der Swaluw, R. Keppens and J. P. Goedbloed, 'Magneto-rotational overstability in accretion disks', *Astron. Astrophys.* **444** (2005), 337–346. <518>

[76] K. M. Blundell and M. G. Bowler, 'Symmetry in the changing jets of SS 433 and its true distance from us', *Astrophys. J.* **616** (2004), L159–L162. <889>

[77] P. Bodenheimer, G. P. Laughlin, M. Rozyczka and H. W. Yorke, *Numerical Methods in Astrophysics* (New York, Taylor and Francis, 2007). <834>

[78] T. J. Bogdan and M. Knölker, 'Scattering of acoustic waves from a magnetic flux tube embedded in a radiating fluid', *Astrophys. J.* **369** (1991), 219–236. <399, 428>

[79] A. Bondeson, R. Iacono and A. H. Bhattacharjee, 'Local magnetohydrodynamic instabilities of cylindrical plasma with sheared equilibrium flows', *Phys. Fluids* **30** (1987), 2167–2180. <473, 494, 498, 502, 523>

[80] A. Bondeson and D. J. Ward, 'Stabilization of external modes in tokamaks by resistive walls and plasma rotation', *Phys. Rev. Lett.* **72** (1994), 2709–2712. <544>

[81] A. H. Boozer, 'Guiding center drift equations', *Phys. Fluids* **23** (1980), 904–908. <929>

[82] A. H. Boozer, 'Physics of magnetically confined plasmas', *Rev. Modern Physics* **76** (2004), 1071–1141. <627>

[83] A. Borba, K. S. Riedel, W. Kerner, G. T. A. Huysmans, M. Ottaviani and P. J. Schmid, 'The pseudospectrum of the resistive magnetohydrodynamic operator: resolving the resistive Alfvén paradox', *Phys. Plasmas* **1** (1994), 3151–3160. <551, 566>

[84] T. J. M. Boyd and J. J. Sanderson, (a) *Plasma Dynamics* (London, Nelson, 1969); (b) *The Physics of Plasmas* (Cambridge, Cambridge University Press, 2003). <24, 101, 876>

[85] B. J. Braams, 'The interpretation of tokamak magnetic diagnostics', *Plasma Phys. Control. F.* **33** (1991), 715–748. <642>

[86] C. M. Braams and P. E. Stott, *Nuclear Fusion: Half a Century of Magnetic Confinement Fusion research* (Bristol, Institute of Physics Publishing, 2002). <9, 24, 417, 617>

[87] M. E. Brachet, M. Meneguzzi, A. Vincent, H. Politano and P. L. Sulem, 'Numerical evidence of smooth self-similar dynamics and possibility of subsequent collapse for three-dimensional ideal flows', *Phys. Fluids* **A 4** (1992), 2845–2854. <756, 774>

[88] J. U. Brackbill, 'FLIP MHD – a particle-in-cell method for magnetohydrodynamics', *J. Comp. Physics* **96** (1991), 163–192. <822>

[89] J. U. Brackbill and D. C. Barnes, 'The effect of non-zero $\nabla \cdot \mathbf{B}$ on the numerical solution of the magnetohydrodynamic equations', *J. Comp. Physics* **35** (1980), 426–430. <152, 814>

[90] H. J. J. Braddick, *Vibrations, Waves, and Diffraction* (Maidenhead, McGraw-Hill Publishing Company Limited, 1965). <159, 424>

[91] S. I. Braginskii, 'On the modes of plasma oscillations in a magnetic field', *Sov. Phys. Dokl.* **2** (1957), 345–349. <86>

[92] S. I. Braginskii, 'Transport processes in a plasma', in *Reviews of Plasma Physics, Vol. 1*, ed. M. A. Leontovich (New York, Consultants Bureau, 1965), 205–311. <38, 66, 72, 79, 101>

[93] D. C. Braun, T. L. Duvall and B. J. LaBonte, (a) 'Acoustic absorption by sunspots', *Astrophys. J.* **319** (1987), L27–L31; (b) 'The absorption of high-degree p-mode oscillations in and around sunspots', *Astrophys. J.* **335** (1988), 1015–1025. <428, 432>

[94] D. C. Braun and T. L. Duvall, 'P-mode absorption in the giant active region of 10 March, 1989', *Solar Phys.* **129** (1990), 83–94. <428, 432>

[95] S. Briguglio, F. Zonca and G. Vlad, 'Hybrid magnetohydrodynamic–particle simulation of linear and nonlinear evolution of Alfvén modes in tokamaks', *Phys. Plasmas* **5** (1998), 3287–3301. <697>

[96] M. Brio and C. C. Wu, 'An upwind differencing scheme for the equations of ideal

magnetohydrodynamics', *J. Comp. Physics* **75** (1988), 400–422. <798, 802, 807, 834>

[97] G. E. Brueckner, 'Space observations of solar corona', *Astronaut. Aeronaut.* **10** (1972), 24. <314>

[98] D. Burgess, 'Collisionless shocks', in *Introduction to Space Physics*, eds. M. G. Kivelson and C. T. Russell (New York, Cambridge University Press, 1995), 129–163. <133>

[99] L. F. Burlaga, L. Klein, N. R. Sheeley Jr., D. J. Michels, R. A. Howard, M. J. Koomen, R. Schwenn and H. Rosenbauer, 'A magnetic cloud and a coronal mass ejection', *Solar Phys.* **9** (1982), 1317–1320. <314>

[100] M. N. Bussac and K. Lerbinger, 'Skin currents and internal kinks in tokamaks', *Phys. Rev. Lett.* **121** (1986), 327–341. <832>

[101] M. N. Bussac, R. Pellat, E. Edery and J. L. Soule, 'Internal kink modes in toroidal plasmas with circular cross sections', *Phys. Rev. Lett.* **35** (1975), 1638–1641. <676>

[102] A. C. Calder et al., 'On validating an astrophysical simulation code', *Astrophys. J. Suppl. Ser.* **143** (2002), 201–229. <819>

[103] P. S. Cally, 'Leaky and non-leaky oscillations in magnetic-flux tubes', *Solar Phys.* **103** (1986), 277–298. <397>

[104] P. S. Cally and T. J. Bogdan, 'Simulation of f- and p-mode interactions with a stratified magnetic field concentration', *Astrophys. J.* **486** (1997), L67–70. <599>

[105] E. Canobbio, 'Low-frequency heating revisited', *Nucl. Fusion* **21** (1981), 279–262. <405>

[106] C. Canuto, M. Y. Hussaini, A. Quarteroni and T. A. Zang, *Spectral Methods in Fluid Dynamics* (Berlin, Springer-Verlag, 1988). <583>

[107] P. Cargo and G. Gallice, 'Roe matrices for ideal MHD and systematic construction of Roe matrices for systems of conservation laws', *J. Comp. Physics* **136** (1997), 446–466. <807>

[108] K. M. Case, (a) 'Stability of inviscid plane Couette flow', *Phys. Fluids* **3** (1960), 143–148; (b) 'Stability of an idealized atmosphere. I. Discussion of results', *Phys. Fluids* **3** (1960), 149–154; (c) 'Taylor instability of an inverted atmosphere', *Phys. Fluids* **3** (1960), 366–368. <445, 480, 486, 494, 725>

[109] F. Casse and R. Keppens, (a) 'Magnetized accretion-ejection structures: 2.5 D MHD simulations of continuous ideal jet launching from resistive accretion disks', *Astrophys. J.* **581** (2002), 988–1001; (b) 'Radiatively inefficient magnetohydrodynamic accretion-ejection structures', *Astrophys. J.* **601** (2004), 90–103. <707, 742, 871>

[110] M. S. Chance, J. M. Greene, R. C. Grimm and J. L. Johnson, 'Study of the MHD spectrum of an elliptic plasma column', *Nucl. Fusion* **17** (1977), 65–83. <341, 386, 704>

[111] S. Chandrasekhar, 'The stability of non-dissipative Couette flow in hydromagnetics', *Proc. Nat. Acad. Sci. USA* **46** (1960), 253–257. <441, 513, 523>

[112] S. Chandrasekhar, *Hydrodynamic and Hydromagnetic Stability* (Oxford, Clarendon Press, 1961). <471, 494, 523>

[113] S. Chandrasekhar and E. Fermi, 'Problems of gravitational stability in the presence of a magnetic field', *Astrophys. J.* **118** (1953), 116–141. <166>

[114] E. Chané, J. Saur, R. Keppens and S. Poedts, 'How is the Jovian main auroral emission affected by the solar wind?', *J. Geophys. Res. Space Physics* **122** (2017), 1960–1978. <320>

[115] J. K. Chao, L. H. Lyu, B. H. Wu, A. J. Lazarus, T. S. Chang and R. P. Lepping, 'Observations of an intermediate shock in interplanetary space', *J. Geophys. Res.* **98** (1993), 17443–17450. <870>

[116] S. Chapman and T. G. Cowling, *The Mathematical Theory of Non-uniform Gases*, 3rd edition (Cambridge, Cambridge University Press, 1991). <66, 72, 101>

[117] F. F. Chen, *Introduction to Plasma Physics and Controlled Fusion*, 3rd edition (New York, Springer International Publisher, 2016). <19, 24, 43, 64, 86>

[118] L. Chen and A. Hasegawa, 'Plasma heating by spatial resonance of Alfvén waves', *Phys. Fluids* **17** (1974), 1399–1403. <269, 397, 399, 405, 432>

[119] C. Z. Cheng, 'Alpha particle destabilization of the toroidicity-induced Alfvén eigenmodes', *Phys. Fluids* **B3** (1991), 2463–2471. <680>

[120] C. Z. Cheng and M. S. Chance, 'Low-n shear Alfvén spectra in axisymmetric toroidal plasmas', *Phys. Fluids* **29** (1986), 3695–3701. <680>

[121] C. Z. Cheng and M. S. Chance, 'NOVA: a nonvariational code for solving the MHD stability of axisymmetric toroidal plasmas', *J. Comp. Physics* **71** (1987), 124–146. <576>

[122] A. R. Choudhuri, *The Physics of Fluids and Plasmas: An Introduction for Astrophysicists* (Cambridge, Cambridge University Press, 1998). <24>

[123] J. Christensen-Dalsgaard, *Stellar Oscillations* (Aarhus, Lecture Notes, Astronomisk Institut, Aarhus Universitet, 1989). <245, 289, 701>

[124] P. C. Clemmov and J. P. Dougherty, *Electrodynamics of Particles and Plasmas* (Reading, Addison-Wesley, 1969). <101, 405>

[125] J. W. Connor, R. J. Hastie and J. B. Taylor, (a) 'Shear, periodicity, and plasma ballooning modes', *Phys. Rev. Lett.* **40** (1978), 396–399; (b) 'High mode number stability of an axisymmetric toroidal plasma', *Proc. Roy. Soc. (London)* **A365** (1979), 1–17. <441, 668, 683>

[126] B. Coppi, 'On the stability of hydromagnetic systems with dissipation', in *Propagation and Instabilities in Plasmas*, ed. W. I. Futterman (Stanford, Stanford University Press, 1963), 70–86. <566>

[127] B. Coppi, 'Topology of ballooning modes', *Phys. Rev. Lett.* **39** (1977), 939–942. <683>

[128] B. Coppi, A. Ferreira, J. W.-K. Mark and J. J. Ramos, 'Ideal-MHD stability of finite-beta plasmas', *Nucl. Fusion* **19** (1979), 715–725. <683>

[129] M. Cotsaftis, 'Formulation Lagrangienne des équations de la magnétohydrodynamique appliquée à létude de la stabilité', *Nucl. Fusion*, 1962 Suppl., Part **2** (1962), 447–450. <456>

[130] R. Courant and K. O. Friedrichs, *Supersonic Flow and Shock Waves* (New York, Interscience Publishers, 1948). <133, 144, 840, 871, 876>

[131] R. Courant, K. O. Friedrichs and H. Lewy, 'Uber die Partiellen Differenzen-gleichungen der Mathematischen Physik', *Math. Ann.* **100** (1928), 32–74. [Trans.: 'On the Partial Difference Equations of Mathematical Physics', *IBM Journal* **11** (1967), 215–234.] <599>

[132] R. Courant and D. Hilbert, (a) *Methods of Mathematical Physics I* (New York, Interscience, 1953); (b) *Methods of Mathematical Physics II* (New York, Interscience, 1962). <171, 179, 339>

[133] C. A. Coverdale et al., 'Neutron production and implosion characteristics of a deuterium gas-puff Z pinch', *Phys. Plasmas* **14** (2007), 022706, 1–7. <871>

[134] A. J. Cunningham, A. Frank, A. C. Quillen and E. G. Blackman, 'Outflow-driven cavities: numerical simulations of intermediaries of protostellar turbulence', *Astrophys. J.* **653** (2006), 416–424. <819>

[135] O. Czarny and G. T. A. Huijsmans, 'Bézier surfaces and finite elements for MHD simulations', *J. Comp. Physics* **227** (2008), 7423–7445. <697>

[136] R. B. Dahlburg, S. K. Antiochos and D. Norton, 'Magnetic flux tube tunneling', *Phys. Rev. E* **56** (1997), 2094–2103. <822>

[137] R. B. Dahlburg and G. Einaudi, 'The compressible plane current-vortex sheet', *Phys. Plasmas* **7** (2000), 1356–1365. <583, 594>

[138] W. Dai and P. R. Woodward, (a) 'An approximate Riemann solver for ideal magnetohydrodynamics', *J. Comp. Physics* **111** (1994), 354–372; (b) 'Extension of the piecewise parabolic method to multidimensional ideal magnetohydrodynamics', *J. Comp. Physics* **115** (1994), 485–514. <798, 813>

[139] L. K. S. Daldorff, G. Tóth, T. I. Gombosi, G. Lapenta, J. Amaya, S. Markidis and J. U. Brackbill, 'Two-way coupling of a global Hall magnetohydrodynamics model with a local implicit particle-in-cell model', *J. Comp. Physics* **268** (2014), 236–254. <560>

[140] P. A. Davidson, *An Introduction to Magnetohydrodynamics*, 2nd edition (Cambridge, Cambridge University Press, 2016). <24>

[141] H. J. de Blank and T. J. Schep, 'Theory of the $m = 1$ kink mode in toroidal plasma', *Phys. Plasmas* **B3** (1991), 1136–1151. <676>

[142] P. de Bruyne, M. Velli and A. W. Hood, 'The ideal MHD stability of line-tied coronal loops: a truncated Fourier series approach', *Comput. Physics Commun.* **59** (1990), 55–59. <594>

[143] A. Dedner, F. Kemm, D. Kröner, C.-D. Munz, T. Schnitzer and M. Wesenberg, 'Hyperbolic divergence cleaning for the MHD equations', *J. Comp. Physics* **175** (2002), 645–673. <814>

[144] A. De Groof, W. J. Tirry and M. Goossens, (a) 'Random driven fast waves in coronal loops – I. Without coupling to Alfvén waves', *Astron. Astrophys.* **335** (1998), 329–340; (b) II. 'With coupling to Alfvén waves'*Astron. Astrophys.* **356** (2000), 724–734. <386, 428>

[145] A. De Groof and M. Goossens, (a) 'Fast and Alfvén waves driven by azimuthal footpoint motions - I. Periodic driver', *Astron. Astrophys.* **386** (2002), 681–690; (b) 'II. Random driver',*Astron. Astrophys.* **386** (2002), 691–698. <428>

[146] F. de Hoffmann and E. Teller, 'Magneto-hydrodynamic shocks', *Phys. Rev.* **80** (1950), 69–703. <848, 876, 906>

[147] P. J. Dellar, 'A note on magnetic monopoles and the one-dimensional MHD Riemann problem', *J. Comp. Physics* **172** (2001), 392–398. <152, 814>

[148] L. Del Zanna, N. Bucciantini and P. Londrillo, 'An efficient shock-capturing central-type scheme for multidimensional relativistic flows', *Astron. Astrophys.* **400** (2003), 397–413. <908>

[149] I. De Moortel, A. W. Hood, J. Ireland and T. D. Arber, 'Phase mixing of Alfvén waves in a stratified open atmosphere', *Astron. Astrophys.* **346** (1999), 641–651. <428>

[150] R. O. Dendy (ed.), *Plasma Physics: An Introductory Course* (Cambridge, Cambridge University Press, 1993). <24>

[151] J. F. Denisse and J. L. Delcroix, *Théorie des Ondes dans les Plasmas* (Paris, Dunod, 1961). [Transl.: *Plasma Waves* (New York, John Wiley, 1963).] <86>

[152] H. De Sterck, 'Hyperbolic theory of the "shallow water" magnetohydrodynamics equations', *Phys. Plasmas* **8** (2001), 3293–3305. <834>

[153] H. De Sterck, B. C. Low and S. Poedts, 'Complex MHD bow shock topology in field-aligned low-β flow around a perfectly conducting cylinder', *Phys. Plasmas* **5** (1998), 4015–4027. <870, 871>

[154] H. De Sterck and S. Poedts, (a) 'Intermediate shocks in three-dimensional magnetohydrodynamic bow-shock flows with multiple interacting shock fronts', *Phys. Rev. Lett.* **84** (2000), 5524–5527; (b) 'Disintegration and reformation of intermediate shock segments in three-dimensional MHD bow shock flows', *J. Geophys. Res.* **106** (2001), 30023–30037. <870>

[155] C. R. DeVore and S. K. Antiochos, 'Dynamical formation and stability of helical prominence magnetic fields', *Astrophys. J.* **539** (2000), 954–963. <819>

[156] R. L. Dewar, 'Spectrum of the ballooning Schrödinger equation', *Plasma Phys. Control. F.* **39** (1997), 453–470. <683>

[157] R. L. Dewar and A. H. Glasser, 'Ballooning mode spectrum in general toroidal systems', *Phys. Fluids* **26** (1983), 3038–3052. <683>

[158] W. D. D'haeseleer, W. N. G. Hitchon, J. D. Callen and J. L. Shohet, *Flux Coordinates and Magnetic Field Structure* (Berlin, Springer Verlag, 1991). <929>

[159] D. A. D'Ippolito, J. P. Freidberg, J. P. Goedbloed and J. Rem, 'High-beta tokamaks surrounded by force-free fields', *Phys. Fluids* **21** (1978), 1600–1616. <627>

[160] D. A. D'Ippolito and J. P. Goedbloed, 'Mode coupling in a toroidal sharp-boundary plasma, (a) I. Weak-coupling limit', *Plasma Physics* **22** (1980), 1091–1107; (b) II. Strong-coupling limit', *Plasma Physics* **25** (1983), 537–550. <680>

[161] P. A. M. Dirac, *The Principles of Quantum Mechanics*, 4th edition (London, Oxford University Press, 1958). <196, 210>

[162] A. J. H. Donné, A. L. Rogister, R. Koch and H. Soltwisch (eds.), *Proc. Second Carolus Magnus Summer School on Plasma Physics*, Aachen, 1995; *Trans. Fusion Technol.* **29** (1996), 1–432. <24, 664>

[163] B. T. Draine and C. F. McKee, 'Theory of interstellar shocks', *Ann. Rev. Astron. Astrophys.* **31** (1993), 373–432. <870>

[164] P. G. Drazin and W. H. Reid, *Hydrodynamic Stability*, 2nd edition (Cambridge, Cambridge University Press, 2004). <480, 489, 494, 523, 551>

[165] J. Dreher and R. Grauer, 'Racoon: a parallel mesh-adaptive framework for hyperbolic conservation laws', *Par. Comput.* **31** (2005), 913–932. <819>

[166] M. R. Dubal, 'Numerical simulations of special relativistic, magnetic gas flows', *Comput. Physics Commun.* **64** (1991), 221–234. <908>

[167] B. Dubrulle and E. Knobloch, 'On instabilities in magnetized accretion disks', *Astron. Astrophys.* **274** (1993), 667–674. <518>

[168] F. J. Dyson, 'Stability of an idealized atmosphere. II. Zeros of the confluent hypergeometric function', *Phys. Fluids* **3** (1960), 155–157. <480, 494>

[169] E. M. Edlund, M. Porkolab, G. J. Kramer, L. Lin, Y. Lin and S. J. Wukitch, 'Observation of reversed shear Alfvén eigenmodes between sawtooth crashes in the Alcator C-Mod tokamak', *Phys. Rev. Lett.* **102** (2009), 165003-1–4. <701>

[170] W. M. Elsässer, 'The hydromagnetic equations', *Phys. Rev.* **79** (1950), 183. <759>

[171] C. R. Evans and J. F. Hawley, 'Simulation of magnetohydrodynamic flows: a constrained transport method', *Astrophys. J.* **332** (1988), 659–677. <814, 910>

[172] S. A. E. G. Falle, 'Rarefaction shocks, shock errors and low order of accuracy in ZEUS', *Astrophys. J.* **577** (2002), L123–L126. <790, 822>

[173] S. A. E. G. Falle and S. S. Komissarov, 'On the inadmissibility of non-evolutionary shocks', *J. Plasma Physics* **65** (2001), 29–58. <811, 870>

[174] A. Ferriz-Mas and M. Núñez, *Advances in Nonlinear Dynamos; the Fluid Dynamics of Astrophysics and Geophysics* (New York, Springer, 2003). <xix>

[175] G. B. Field, 'Thermal instability', *Astrophys. J.* **142** (1965), 531–567. <594>

[176] N. J. Fisch, 'Theory of current drive in plasmas', *Rev. Modern Physics* **59** (1987), 175–234. <627>

[177] R. Fitzpatrick, 'A simple model of the resistive wall mode in tokamaks', *Phys. Plasmas* **6** (2002), 3459–3569. <544>

[178] R. Fitzpatrick, *Plasma Physics. An Introduction* (Boca Raton, CRC Press, Taylor and Francis Group, 2015). <24>

[179] R. Fitzpatrick and E. P. Yu, 'Feedback stabilization of resistive shell modes in a reversed field pinch', *Phys. Plasmas* **6** (1999), 3536–3547. <544>

[180] J. A. Font, 'Numerical hydrodynamics and magnetohydrodynamics in general relativity', *Living Rev. Relativity* **11** (2008) 7. <916>

[181] P. V. Foukal, *Solar Astrophysics* (New York, John Wiley and Sons, 1990). <24, 233, 234, 292, 294, 322>

[182] D. G. Fox and S. A. Orszag, 'Pseudospectral approximation to two-dimensional turbulence', *J. Comp. Physics* **11** (1973), 612–619. <756>

[183] J. Frank, A. King and D. Raine, *Accretion Power in Astrophysics*, 3rd edition (Cambridge, Cambridge University Press, 2002). <105, 518, 707>

[184] J. P. Freidberg, 'Magnetohydrodynamic stability of a diffuse screw pinch', *Phys. Fluids* **13** (1970), 1812–1818. <330>

[185] J. P. Freidberg, *Plasma Physics and Fusion Energy* (Cambridge, Cambridge University Press, 2007). <xix, 624, 627, 664, 695, 704, 929>

[186] J. P. Freidberg, *Ideal MHD* (Cambridge, Cambridge University Press, 2014). <24, 64, 75, 229, 369, 664, 690, 704>

[187] J. P. Freidberg and J. P. Goedbloed, 'Equilibrium and stability of a diffuse high-beta tokamak', in *Proc. Third Topical Conf. on Pulsed High Beta Plasmas*, 9–12 Sept. 1975, Culham, ed. D. E. Evans (Oxford, Pergamon Press) 117–121 (1976). <655>

[188] J. P. Freidberg and F. A. Haas, 'Kink instabilities in a high-β tokamak', *Phys. Fluids* **16** (1973), 1909–1916. <215, 704>

[189] J. P. Freidberg and L. H. Pearlstein, 'Rotational instabilities in a theta pinch', *Phys. Fluids* **21** (1978), 1207–1217. <506>

[190] J. P. Freidberg and J. A. Wesson, 'Instability of the m =1 mode in a rotating θ pinch', *Phys. Fluids* **13** (1970), 1117–1119. <506>

[191] H. Freistühler, 'Some remarks on the structure of intermediate magnetohydrodynamic shocks', *J. Geophys. Res.* **96** (1991), 3825–3827. <870>

[192] H. Friedel, R. Grauer and C. Marliani, 'Adaptive Mesh Refinement for singular current sheets in incompressible magnetohydrodynamic flows', *J. Comp. Physics* **134** (1997), 190–198. <772>

[193] B. Friedman, *Principles and Techniques of Applied Mathematics* (New York, Wiley and Sons., 1956). <196, 229>

[194] H. Friedman, *Sun and Earth* (New York, W. H. Freeman and Company, 1986). <292>

[195] K. O. Friedrichs and H. Kranzer, *Notes on Magnetohydrodynamics, VIII. Nonlinear wave motion* (New York University, NYO-4686-VIII, 1958). <161, 173, 179>

[196] E. A. Frieman, J. M. Greene, J. L. Johnson and K. E. Weimer, 'Toroidal effects on magnetohydrodynamic modes in tokamaks', *Phys. Fluids* **16** (1973), 1108–1125. <697>

[197] E. Frieman and M. Rotenberg, 'On the hydromagnetic stability of stationary equilibria', *Rev. Modern Physics* **32** (1960), 898–902. <448, 456, 471, 722>

[198] S. Fromang, P. Hennebelle and R. Teyssier, 'A high order Godunov scheme with constrained transport and adaptive mesh refinement for astrophysical magneto- hydrodynamics', *Astron. Astrophys.* **457** (2006), 371–384. <819>

[199] G. Y. Fu and J. W. Van Dam, 'Excitation of the toroidicity-induced shear Alfvén eigenmode by fusion alpha particles in an ignited tokamak', *Phys. Fluids* **B1** (1989), 1949–1952. <680>

[200] M. Fukugita, 'Cosmic matter distribution: Cosmic baryon budget revisited', in *International Astronomical Union Symposium* 220, eds. S. D. Ryder, D. J. Pisano, M. A. Walker and K. C. Freeman, 21–25 July, Sydney (San Fransico, 2003), 227–332. <437>

[201] K. Fukumura, M. Takahashi and S. Tsuruta, 'Magnetohydrodynamic shocks in nonequitorial plasma flows around a black hole', *Astrophys. J.* **657** (2007), 415–427. <871>

[202] H. P. Furth, J. Killeen and M. N. Rosenbluth, 'Finite resistivity instabilities of a sheet pinch', *Phys. Fluids* **6** (1963), 459–484. <534, 566>

[203] A. A. Galeev and R. Z. Sagdeev, 'Theory of neo-classical diffusion', in *Reviews of Plasma Physics, Vol. 7*, ed. M. A. Leontovich (New York, Consultants Bureau, 1979), 257–343. <72, 75, 101>

[204] D. J. Galloway and M. R. E. Proctor, 'Numerical calculations of fast dynamos in smooth velocity fields with realistic diffusion', *Nature* **356** (1992) 691–693. <761>

[205] R. M. O. Galvão, J. P. Goedbloed, J. Rem, P. H. Sakanaka, T. J. Schep and M. Venema, 'Global kink and ballooning modes in high-beta systems and stability of toroidal drift modes', *Proc. Ninth Intern. IAEA Conf. on Plasma Physics and Controlled Nuclear Fusion Research*, 1–8 Sept. 1982, Baltimore (IAEA, Vienna) Vol. III, 3–16 (1983). <627, 655>

[206] P. R. Garabedian, *Partial Differential Equations* (New York, John Wiley, 1964). <171, 179>

[207] T. A. Gardiner and J. M. Stone, 'An unsplit Godunov method for ideal MHD via constrained transport', *J. Comp. Physics* **205** (2005), 509–539. <819>

[208] J. Geiss, G. Gloeckner and R. Von Steiger, 'Origin of the solar wind from composition data', *Space Sc. Reviews* **72** (1995), 49–60. <309>

[209] P. Germain, 'Shock waves and shock-wave structure in magneto-fluid dynamics', *Rev. Modern Physics* **32** (1960), 951–958. <870>

[210] B. Giacomazzo and L. Rezzolla, (a) 'The exact solution of the Riemann problem in relativistic MHD', *J. Fl. Mech.* **562** (2006), 223–259; (b) 'WhiskyMHD: a new numerical code for general relativistic magnetohydrodynamics', *Classical Quant. Grav.* **24** (2007), S235–S258. <908, 910>

[211] A. H. Glasser, J. M. Greene and J. L. Johnson, 'Resistive instabilities in general toroidal plasma configurations', *Phys. Fluids* **18** (1975), 875–888. <695>

[212] G. A. Glatzmaier and P. H. Roberts, (a) 'A three-dimensional self-consistent computer simulation of a geomagnetic field reversal', *Nature* **377** (1995) 203–209; (b) 'A three-dimensional convective dynamo solution with rotating and finitely conducting inner core and mantle', *Phys. Earth and Planet. In.* **91** (1995), 63–75. <320, 761>

[213] S. K. Godunov, 'Symmetric form of the equations for magnetohydrodynamics', in *Numerical Methods for Mechanics of Continuous Media*, Vol. 1 (1972), 26–31, in Russian; translation by T. Linde (Report of the Computer Centre of the Siberian branch of the USSR Academy of Sciences, 1972). <152, 814>

[214] J. P. Goedbloed, 'Stabilization of magnetohydrodynamic instabilities by force-free magnetic fields – (a) I. Plane plasma layer', *Physica* **53** (1971), 412–444; (b) 'II. Linear pinch', *Physica* **53** (1971), 501–534; (c) 'III. Shearless magnetic fields', *Physica* **53** (1971), 535–570; (d) 'IV. Boundary conditions for plasma-plasma interface', *Physica* **100C** (1980), 273–275. <215, 229, 252, 289, 330, 348, 353, 369, 441, 716>

[215] J. P. Goedbloed, 'Spectrum of ideal magnetohydrodynamics of axisymmetric toroidal systems', *Phys. Fluids* **18** (1975), 1258–1268. <273, 376, 668, 674, 677, 722, 725>

[216] J. P. Goedbloed, (a) 'Conformal mapping methods in two-dimensional magnetohydrodynamics', *Comput. Physics Commun.* **24** (1981), 311–321; (b) 'Some remarks on computing axisymmetric equilibria', *Comput. Physics Commun.* **31** (1984), 123–135 [Erratum: **41** (1986), 196]; (c) 'Expansion functions for two-dimensional incompressible fluid flow in arbitrary domains', *J. Comp. Physics* **160** (2000), 283–297. <655, 926, 929>

[217] J. P. Goedbloed, 'Free-boundary high-beta tokamaks. (a) I. Free-boundary equilibrium', *Phys. Fluids* **25** (1982), 852–868; (b) 'II. Mathematical intermezzo: Hilbert transforms and conformal mapping', *Phys. Fluids* **25** (1982), 2062–2072; (c) 'III. Free-boundary stability', *Phys. Fluids* **25** (1982), 2073–2088. <655>

[218] J. P. Goedbloed, *Lecture Notes on Ideal Magnetohydrodynamics* (Universidade Estadual de Campinas, Brazil, 1978); Rijnhuizen Report 83-145 (FOM-Institute for Plasma Physics, Nieuwegein, the Netherlands, 1983). <144>

[219] J. P. Goedbloed, 'Plasma-vacuum interface problems in magnetohydrodynamics', *Physica* **12D** (1984), 107–132. <347>

[220] J. P. Goedbloed, 'MHD waves in thermonuclear and solar plasmas', in *Trends in Physics 1991*, Eighth General Conference EPS, Amsterdam, 4–8 Sept. 1990, ed. J. Kaczér, Vol. III (Prague, EPS,1991) 827–844. <701>

[221] J. P. Goedbloed, 'Once more: The continuous spectrum of ideal magnetohydrodynamics', *Phys. Plasmas* **5** (1998), 3143–3154. <269, 373, 397, 473, 500>

[222] J. P. Goedbloed, 'Transonic magnetohydrodynamic flows in laboratory and astrophysical plasmas',

Physica Scripta **T98** (2002), 43–47. <707, 716, 742>

[223] J. P. Goedbloed, 'Variational principles for stationary one- and two-fluid equilibria of axisymmetric laboratory and astrophysical plasmas', *Phys. Plasmas* **11** (2004), L81–L84. <744, 837, 870>

[224] J. P. Goedbloed, 'Response to "Comment on 'Variational principles for stationary one- and two-fluid equilibria of axisymmetric laboratory and astrophysical plasmas' " [*Phys. Plasmas* **12**, 064701 (2005)]', *Phys. Plasmas* **12** (2005), 064702–4. <744>

[225] J. P. Goedbloed, 'Alternatives and paradoxes in rotational and gravitational instabilities', in *Proc. International Workshop on Collective Phenomena in Macroscopic Systems*, 4–6 December 2006, Como, ed. G. Bertin, R. Pozzoli, and K.R. Sreenivasan (World Scientific, Singapore, 2007), 127–136. <441, 837>

[226] J. P. Goedbloed, 'Time reversal duality of magnetohydrodynamical shocks', *Phys. Plasmas* **15** (2008), 062101, 1–19. <870, 876>

[227] J. P. Goedbloed, 'New construction of the magnetohydrodynamic spectrum of stationary plasma flows. (a) I. Solution path and alternator', *Phys. Plasmas* **16** (2009), 122110, 1–14; (b) 'II. Rayleigh–Taylor and Kelvin–Helmholtz instability', *Phys. Plasmas* **16** (2009), 122111, 1–13. <462>

[228] J. P. Goedbloed, 'MHD instabilities in astrophysical plasmas: very different from MHD instabilities in tokamaks', *Plasma Phys. Control. F.* **60** (2017), 014001, 1–5. <441>

[229] J. P. Goedbloed, 'The Spectral Web of stationary plasma equilibria: (a) I. General theory', *Phys. Plasmas* **25** (2018), 032109, 1–13; (b) 'II. Internal modes ', *Phys. Plasmas* **25** (2018), 032110, 1–22. <357, 437, 456, 462, 466, 471, 473, 502, 505, 518>

[230] J. P. Goedbloed, A. J. C. Beliën, B. van der Holst and R. Keppens, 'Unstable continuous spectra of transonic axisymmetric plasmas', *Phys. Plasmas* **11** (2004), 28–54. <456, 707, 716, 737, 740, 742, 744, 870>

[231] J. P. Goedbloed, A. J. C. Beliën, B. van der Holst and R. Keppens, 'No additional flow continua in magnetohydrodynamics', *Phys. Plasmas* **11** (2004), 4332–4340. <466, 480, 518>

[232] J. P. Goedbloed and R. Y. Dagazian, 'Kinks and tearing modes in simple configurations', *Phys. Rev.* **A4** (1971), 1554–1560. <525>

[233] J. P. Goedbloed and D. A. D'Ippolito, 'RF stabilization of external kink modes in the presence of a resistive wall', *Phys. Fluids* **B2** (1990), 2366–2372. <544>

[234] J. P. Goedbloed and H. J. L. Hagebeuk, 'Growth rates of instabilities of a diffuse linear pinch', *Phys. Fluids* **15** (1972), 1090–1101. <330, 353, 357, 369, 441, 525>

[235] J. P. Goedbloed and G. Halberstadt, 'Magnetohydrodynamic waves in coronal flux tubes', *Astron. Astrophys.* **281** (1994), 265–301. <424, 428, 621>

[236] J. P. Goedbloed, G. M. D. Hogeweij, R. Kleibergen, J. Rem, R. M. O. Galvão and P. H. Sakanaka, 'Investigation of high-β tokamak stability with the program HBT', *Proc. 10th Intern. IAEA Conf. on Plasma Physics and Controlled Nuclear Fusion Research*, 12–19 Sept. 1984, London (IAEA, Vienna) II, 165–172 (1985). <627, 655>

[237] J. P. Goedbloed, G. T. A. Huysmans, H. A. Holties, W. Kerner and S. Poedts, 'MHD spectroscopy: free boundary modes (ELMs) and external excitation of TAE modes', *Plasma Phys. Control. F.* **35** (1993), B277–292. <245, 663, 701, 737, 870>

[238] J. P. Goedbloed and A. E. Lifschitz, 'Stationary symmetric magnetohydrodynamic flows', *Phys. Plasmas* **4** (1997), 3544–3564. <707, 712, 715, 871>

[239] J. P. Goedbloed, D. Pfirsch and H. Tasso, 'Instability of a pinch surrounded by a resistive wall', *Nucl. Fusion* **12** (1972), 649–657. <544>

[240] J. P. Goedbloed and P. H. Sakanaka, 'New approach to magnetohydrodynamic stability. I. A practical stability concept', *Phys. Fluids* **17** (1974), 908–918. <207, 229, 273, 285, 289, 348, 369, 466, 488, 716, 725>

[241] J. P. Goedbloed and J. W. A. Zwart, 'On the dynamics of the screw pinch', *Plasma Physics* **17** (1975), 45–67. <144>

[242] P. Goldreich and S. Sridhar, 'Toward a theory of interstellar turbulence. II. Strong Alfvénic turbulence', *Astrophys. J.* **438** (1995), 763–775. <768>

[243] H. Goldstein, *Classical Mechanics*, 2nd edition (Reading, Addison Wesley, 1980). <11, 33, 38, 64, 202, 456, 683>

[244] R. J. Goldston and P. H. Rutherford, *Introduction to Plasma Physics* (Bristol, Institute of Physics Publishing, 1995). <24, 33, 64, 67, 86, 566, 922>

[245] T. I. Gombosi, K. G. Powell and D. De Zeeuw, 'Axisymmetric modelling of cometary mass loading on

an adaptively refined grid, MHD results', *J. Geophys. Res.* **99** (1994), 525–539. <819>

[246] D. O. Gómez, P. D. Mininni and P. Dmitruk, 'MHD simulations and astrophysical applications', *Adv. Space Res.* **35** (2005), 899–907. <778>

[247] M. Goossens, 'MHD waves and wave heating in nonuniform plasmas', in *Advances in Solar System Magnetohydrodynamics*, eds. E. R. Priest and A. W. Hood (Cambridge, Cambridge University Press, 1991), 137–172. <432>

[248] M. Goossens, *An Introduction to Plasma Astrophysics and Magnetohydrodynamics* (Dordrecht, Springer Science + Business Media Dordrecht, 2003). <322>

[249] M. Goossens and A. De Groof, 'Resonant and phase-mixed magnetohydrodynamical waves in the solar atmosphere', *Plasma Physics* **8** (2001), 2371–2376. <428>

[250] M. Goossens and S. Poedts, 'Linear resistive MHD computations of resonant absorption of acoustic oscillations in sunspots', *Astrophys. J.* **384** (1982), 348–360. <428, 432, 590>

[251] M. Goossens, M. S. Ruderman and J. V. Hollweg, 'Dissipative MHD solutions for resonant Alfvén waves in 1-dimensional magnetic flux tubes', *Solar Phys.* **157** (1995), 75–102. <414, 432>

[252] D. Gough, 'Helioseismology: oscillations as a probe of the suns interior', *Nature* **304** (1983), 689–690. <245>

[253] H. Grad, *Notes on Magnetohydrodynamics, I. General Fluid Equations* (New York University, NYO-6486-I, 1956). <66, 101>

[254] H. Grad, 'Magnetofluid-dynamic spectrum and low shear stability', *Proc. Natl. Acad. Sci. USA* **70** (1973), 3377–3281. <207, 269, 369, 473, 500, 716>

[255] H. Grad and H. Rubin, 'Hydromagnetic equilibria and force-free fields', in *Proc. Second UN Intern. Conf. on Peaceful Uses of Atomic Energy* 31 (New York, Columbia University Press, 1959), 190–197. <635>

[256] I. S. Gradshsteyn and I. M. Ryzhik, *Table of Integrals, Series, and Products*, Corrected and enlarged edition prepared by A. Jeffrey (New York, Academic Press, 1980). <642>

[257] R. Grauer and C. Marliani, 'Current-sheet formation in 3D ideal incompressible magnetohydrodynamics', *Phys. Rev. Lett.* **84** (2000), 4850–4853. <772>

[258] J. M. Greene, private communication (unpublished, 1974). <265, 373, 473>

[259] J. M. Greene, J. L. Johnson and K. E. Weimer, 'Tokamak equilibrium', *Phys. Fluids* **49** (1971), 671–683. <642>

[260] R. C. Grimm, R. L. Dewar and J. Manickam, 'Ideal MHD stability calculations in axisymmetric toroidal coordinate systems', *J. Comp. Physics* **49** (1983), 94–117. <690>

[261] R. C. Grimm, J. M. Greene and J. L. Johnson, 'Computation of the MHD spectrum in axisymmetric toroidal confinement systems', in *Methods in Computational Physics*, ed. J. Killeen, Vol. 16 (New York, Academic Press, 1976), 253–280. <576>

[262] R. J. Groebner, 'An emerging understanding of H-mode discharges in tokamaks', *Phys. Fluids* **B5** (1993), 2343–2354. <697>

[263] W. Grossmann and R. A. Smith, 'Heating of solar coronal loops by resonant absorption of Alfvén waves', *Astrophys. J.* **332** (1988), 476–498. <405, 432>

[264] R. Gruber, 'Finite hybrid elements to compute the ideal MHD spectrum of an axisymmetric plasma, *J. Comp. Physics* **26** (1978), 379–389. <586>

[265] R. Gruber and J. Rappaz, *Finite Element Methods in Linear Ideal Magnetohydrodynamics* (New York, Springer, 1985). <586, 690>

[266] R. Gruber, F. Troyon, D. Berger, L. C. Bernard, S. Rousset, R. Schreiber, W. Kerner, W. Schneider and K. V. Roberts, 'ERATO stability code', *Comput. Physics Commun.* **21** (1981), 323–371. <576, 690>

[267] L. Guazzotto, J. P. Freidberg and R. Betti, 'A general formulation of magnetohydrodynamic stability including flow and a resistive wall, *Phys. Plasmas* **15** (2008), 072503, 1–9. <544>

[268] L. Gu, J. Kaastra, A. J. J. Raassen, P. D. Mullen, R. S. Cumbee, D. Lyons and P. C. Stancil, 'A novel scenario for the possible X-ray line feature at 3.5. keV. Charge exchange with bare sulfur ions', *Astron. Astrophys.* **584** (2015), L11-1–4. <437>

[269] K. G. Guderley, *The Theory of Transonic Flow* (Oxford, Pergamon Press, 1962; translated from the German edition of 1957 by J. R. Moszynski). <840>

[270] B. V. Gudiksen, M. Carlsson, V. H. Hansteen, W. Hayek, J. Leenaarts and J. Martínez-Sykora, 'The stellar atmosphere simulation code Bifrost. Code description and validation', *Astron. Astrophys.* **531** (2011), A154. <822>

[271] X. Guo, V. Florinski and C. Wang, 'The HLLD Riemann solver based on magnetic field

decomposition method for the numerical simulation of magnetohydrodynamics', *J. Comp. Physics* **327** (2016), 543–552. <807>

[272] D. A. Gurnett and A. Bhattacharjee, *Introduction to Plasma Physics* (Cambridge, Cambridge University Press, 2005). <24, 876>

[273] F. Haas, 'A magnetohydrodynamic model for quantum plasmas', *Phys. Plasmas* **12** (2005), 062117, 1–9. <30>

[274] T. Hada, 'Evolutionarity conditions in the dissipative MHD system: stability of intermediate MHD shock waves', *Geophys. Res. Lett.* **21** (1994), 2275–2278. <870>

[275] K. Hain, R. Lüst and A. Schlüter, 'Zur Stabilität eines Plasmas', *Z. Naturforsch.* **12a** (1957), 833–941. <204, 229, 471, 690, 704>

[276] K. Hain and R. Lüst, 'Zur Stabilität zylinder-symmetrischer Plasmakonfigurationen mit Volumenströmen', *Z. Naturforsch.* **13a** (1958), 936–940. <330, 369, 716, 742>

[277] G. Halberstadt and J. P. Goedbloed, 'The continuous Alfvén spectrum of line-tied coronal loops', *Astron. Astrophys.* **280** (1993), 647–660. <424, 428>

[278] G. Halberstadt and J. P. Goedbloed, 'Alfvén-wave heating of coronal loops - (a) Photospheric excitation', *Astron. Astrophys.* **301** (1995), 559-576; (b) 'Surface excitation revisited', *Astron. Astrophys.* **301** (1995), 577-592. <576>

[279] S. Hamada, 'Hydromagnetic equilibria and their proper coordinates', *Nucl. Fusion* **2** (1962), 23–37. <929>

[280] E. Hameiri, 'The stability of a particular MHD equilibrium with flow', PhD thesis New York University (Courant Institute of Mathematical Sciences Report MF-45, COO-3077-123, 1976). <480, 494, 498, 502>

[281] E. Hameiri, 'Shear stabilization of the Rayleigh–Taylor modes', *Phys. Fluids* **22** (1979), 89–98. <506>

[282] E. Hameiri, 'Spectral estimates, stability conditions, and the rotating screw-pinch', *J. Math. Phys.* **22** (1981), 2080–2088. <498, 502, 523>

[283] E. Hameiri, 'The equilibrium and stability of rotating plasmas', *Phys. Fluids* **26** (1983), 230–237. <709, 715, 744>

[284] E. Hameiri, 'Dynamically accessible perturbations and magnetohydrodynamic stability', *Phys. Plasmas* **10** (2003), 2643–2648. <456>

[285] E. Hameiri, 'The complete set of Casimir constants of the motion in magnetohydrodynamics', *Phys. Plasmas* **11** (2004), 3423–3431. <456>

[286] E. Hameiri and J. H. Hammer, 'Unstable continuous spectrum in magnetohydrodynamics', *Phys. Fluids* **22** (1979), 1700–1706. <456, 725>

[287] E. Hameiri, A. Ishizawa and A. Ishida, 'Waves in the Hall-magnetohydrodynamics model', *Phys. Plasmas* **12** (2005), 072109. <563, 566>

[288] D. S. Harned and W. Kerner, 'Semi-implicit method for three-dimensional compressible magnetohydrodynamics simulation', *J. Comp. Physics* **60** (1985), 62–75. <606, 827, 828, 832>

[289] D. S. Harned and D. D. Schnack, 'Semi-implicit method for long time scale magnetohydrodynamic computations in three dimensions', *J. Comp. Physics* **65** (1986), 57–70. <828>

[290] A. Harten, 'High resolution schemes for hyperbolic conservation laws', *J. Comp. Physics* **49** (1983), 357–393. <790>

[291] A. Harten, P. D. Lax and B. van Leer, 'On upstream differencing and Godunov-type schemes for hyperbolic conservation laws', *SIAM Rev.* **25** (1983), 35–61. <807>

[292] J. Harvey, 'Helioseismology', *Physics Today* **48** (1995), 32–38. <245>

[293] A. Hasegawa and T. Sato, *Space Plasma Physics, Vol. I. Stationary Processes* (New York, Springer Verlag, 1989). <24, 33, 64>

[294] J. F. Hawley and S. A. Balbus, 'The dynamical structure of nonradiative black hole accretion flows', *Astrophys. J.* **573** (2002), 738–748. <742>

[295] J. F. Hawley, C. F. Gammie and S. A. Balbus, 'Local three-dimensional simulations of accretion disks', *Astrophys. J.* **440** (1995), 742–763. <822>

[296] R. D. Hazeltine and J. D. Meiss, *Plasma Confinement* (Redwood City, Addison Wesley Publishing Company, 1992). <24, 101, 704>

[297] R. D. Hazeltine and W. A. Newcomb, 'Inversion of the ballooning transformation', *Phys. Fluids* **B2** (1990), 7–10. <683>

[298] R. D. Hazeltine and J. D. Waelbroeck, *The Framework of Plasma Physics* (Philadelphia, Westview

Press, 2004). <24>

[299] T. A. K. Hellsten and G. O. Spies, 'Continuous magnetohydrodynamic spectrum in axially symmetric rotating plasmas', *Phys. Fluids* **22** (1979), 743–746. <456, 480, 725>

[300] P. Henrici, (a) 'Fast Fourier methods in computational complex analysis', *SIAM Rev.* **21** (1979), 481–527; (b) *Applied and Computational Complex Analysis*, Volume 3: *Discrete Fourier Analysis, Cauchy Integrals, Construction of Conformal Maps, Univalent Functions* (New York, Wiley, 1993). <655, 926>

[301] J. Heyvaerts, E. R. Priest, 'Coronal heating by phase-mixed shear Alfvén waves', *Astron. Astrophys.* **117** (1983), 220–234. <432>

[302] C. Hirsch, *Numerical Computation of Internal and External Flow, 1: Fundamentals of Numerical Discretization* (Chichester, John Wiley and Sons, 1988). <572, 599>

[303] J. V. Hollweg, (a) 'Resonance-absorption of solar p-modes by sunspots', *Astrophys. J.* **335** (1988), 1005–1014; (b) 'Resonant decay of global MHD modes at thick interfaces', *J. Geophys. Res.* **95** (1990), 2319–2324. <414, 432>

[304] J. V. Hollweg and G. Yang, 'Resonance absorption of compressible MHD waves at thin surfaces', *J. Geophys. Res.* **93** (1988), 5423–5436. <432>

[305] D. D. Holm, J. E. Marsden, T. Ratiu and A. Weinstein, 'Nonlinear stability of fluid and plasma equilibria', *Phys. Rep.* **123** (1985), 1–116. <456>

[306] H. A. Holties, G. T. A. Huysmans, W. Kerner, J. P. Goedbloed, F. X. Söldner and V. V. Parail, 'MHD stability of advanced tokamak scenarios', *Proc. 21st Eur. Conf. on Controlled Fusion and Plasma Physics*, Vol. I (Montpellier, EPS, 1994), 234–237. <695>

[307] H. A. Holties, A. Fasoli, J. P. Goedbloed, G. T. A. Huysmans and W. Kerner, 'Determination of local tokamak parameters by magnetohydrodynamic spectroscopy', *Phys. Plasmas* **4** (1997), 709–719. <701>

[308] H. A. Holties, G. T. A. Huysmans, J. P. Goedbloed, W. Kerner, V. V. Parail and F. X. Söldner, 'Stability of infernal and ballooning modes in advanced tokamak scenarios', *Nucl. Fusion* **36** (1996), 973–986. <695>

[309] T. E. Holzer and E. Leer, 'Coronal hole structure and the high speed solar wind', in *The Corona and Solar Wind near Minimum Activity*, Proc. 5th SOHO Workshop, ESA SP-404 (1997), 65–74. <309>

[310] K. I. Hopcraft, 'Magnetohydrodynamcs', in *Plasma Physics: an Introductory Course*, ed. R. Dendy (Cambridge, Cambridge University Press, 1993), 77–101. <664>

[311] T. Y. Hou and P. G. Le Floch, 'Why nonconservative schemes converge to wrong solutions: error analysis', *Math. of Computation* **62** (1994), 497–530. <790>

[312] L. N. Howard, 'Note on a paper by John W. Miles', *J. Fluid Mech.* **10** (1961), 509–512. <494>

[313] J. D. Huba, (D. Book), *NLR Plasma Formulary* (Washington, Naval Research Laboratory, 2009). <929>

[314] A. J. Hundhausen, 'The solar wind', in *Introduction to Space Physics*, eds. M. G. Kivelson and C. T. Russell (New York, Cambridge University Press, 1995), 91–128. <309, 322>

[315] D. Hutsemékers, L. Braibant, V. Pelgrims and D. Sluse, 'Alignment of quasar polarizations with large-scale structures', *Astron. Astrophys.* **572** (2014), A18, 1–7. <437>

[316] G. T. A. Huysmans, (a) 'External kink (peeling) modes in x-point geometry', *Plasma Phys. Control. F.* **47** (2005), 2107–2121; (b) 'ELMs: MHD instabilities at the transport barrier', *Plasma Phys. Control. F.* **47** (2005), B165–B178. <697>

[317] G. T. A. Huijsmans et al., 'Modelling of edge localised modes and edge localised mode control', *Phys. Plasmas* **22** (2015), 021805, 1–19. <697>

[318] G. T. A. Huysmans and O. Czarny, 'MHD stability in X-point geometry: simulation of ELMs', *Nucl. Fusion* **47** (2007), 659–666. <697>

[319] G. T. A. Huysmans, J. P. Goedbloed and W. Kerner, 'Isoparametric bicubic Hermite elements for solution of the Grad–Shafranov equation', in *Proc. Second International Conference on Computational Physics*, Amsterdam, 10–14 September 1990, ed. A. Tenner (Singapore, World Scientific, 1991), 371–376. <576, 655, 690, 737, 929>

[320] G. T. A. Huysmans, J. P. Goedbloed and W. Kerner, 'Free boundary resistive modes in tokamaks', *Phys. Fluids* **B5** (1993), 1545–1558. <590, 663, 690, 695>

[321] G. T. A. Huysmans, W. Kerner, D. Borba, H. A. Holties and J. P. Goedbloed, 'Modeling the excitation of global Alfvén modes by an external antenna in the Joint European Tokamak', *Phys. Plasmas* **2** (1995), 1605–1613. <701>

[322] G. T. A. Huysmans, W. Kerner, J. P. Goedbloed, D. Borba and F. Porcelli, 'MHD spectroscopy: modelling the excitation of TAE modes by an external antenna', *Proc. 20th Eur. Conf. on Controlled Fusion and Plasma Physics*, Vol. I (Lisbon, EPS, 1993), 187–190. <701>

[323] S. Ichimaru, *Basic Principles of Plasma Physics; A Statistical Approach* (Reading, W. A. Benjamin Inc., 1973). <67>

[324] I. V. Igumenshchev, R. Narayan and M. A. Abramovicz, 'Three-dimensional MHD simulations of radiatively inefficient accretion flows', *Astrophys. J.* **592** (2003), 1042–1059. <742>

[325] E. L. Ince, *Ordinary Differential Equations* (New York, Dover Publications, 1956). <265, 273, 289, 534>

[326] J. A. Ionson, 'Resonant absorption of Alfvénic surface waves and the heating of solar coronal loops', *Astrophys. J.* **226** (1978), 650–673. <417, 432>

[327] R. C. Ireland, R. A. M. Van der Linden, A. W. Hood and M. Goossens, 'The thermal continuum in coronal loops: the influence of finite resistivity on the continuous spectrum', *Solar Phys.* **142** (1992), 265–289. <594>

[328] P. S. Iroshnikov, 'Turbulence of a conducting fluid in a strong magnetic field', *Astron. Zh.* **40** (1963) 742–750 [*Sov. Astron.* **1** (1964) 566–571]. <768>

[329] S-I. Itoh, K. Itoh, M. Sasaki, A. Fujisawa, T. Ido and Y. Nagashima, 'Geodesic acoustic mode spectroscopy', *Plasma Phys. Control. F.* **11** (2007), L7–L10. <668, 677>

[330] R. Izzo, D. A. Monticello, W. Park, J. Manickam, H. R. Strauss, R. Grimm and K. McGuire, 'Effects of toroidicity on resistive tearing modes', *Phys. Fluids* **26** (1983), 2240–2246. <697>

[331] J. D. Jackson, *Classical Electrodynamics*, 3rd edition (New York, John Wiley and Sons, 1999). <30, 33, 64, 105, 124, 141, 642, 709>

[332] P. Janhunen, 'A positive conservative method for magnetohydrodynamics based on HLL and Roe methods', *J. Comp. Physics* **160** (2000), 649–661. <152, 814>

[333] S. Jardin, *Computational Methods in Plasma Physics* (Boca Raton, CRC Press, Taylor and Francis Group, 2010). <834>

[334] J. H. Jeans, 'The stability of a spherical nebula', *Philos. T. Roy. Soc. Lon. A* **199**, (1902) 1–53. <166>

[335] A. Jeffrey, *Magnetohydrodynamics* (Edinburgh, Oliver and Boyd, 1966). <876>

[336] A. Jeffrey and T. Taniuti, *Non-linear Wave Propagation. With Applications to Physics and Magnetohydrodynamics* (New York, Academic Press, 1964). <798, 870>

[337] J. S. Kaastra et al., 'Spatially resolved X-ray spectroscopy of cooling clusters of galaxies', *Astron. Astrophys.* **413** (2004), 415–439. <437>

[338] B. B. Kadomtsev, 'Hydromagnetic stability of a plasma', in *Reviews of Plasma Physics, Vol. 2*, ed. M. A. Leontovich (New York, Consultants Bureau, 1966), 153–199. <213, 229, 353, 441, 704>

[339] J. M. Kappraff and J. A. Tataronis, 'Resistive effects on Alfvén wave heating', *J. Plasma Physics* **18** (1977), 209–226. <414, 419, 432>

[340] T. Kawamura, B. Giacomazzo, W. Kastaun, R. Ciolfi, A. Endrizzi, L. Baiotti and R. Perna, 'Binary neutron star mergers and short gamma-ray bursts: Effects of magnetic field orientation, equation of state, and mass ratio', *Phys. Rev. D* **94** (2016), 064012, 1–22. <915>

[341] C. F. Kennel, R. D. Blandford and P. Coppi, 'MHD intermediate shock discontinuities. Part 1. Rankine–Hugniot conditions', *J. Plasma Physics* **42** (1989), 299–319. <854, 859, 870>

[342] C. F. Kennel, R. D. Blandford and C. C. Wu, 'Structure and evolution of small-amplitude intermediate shock waves', *Phys. Fluids* **B2** (1990), 253–269. <870>

[343] R. Keppens, (a) 'Flux tubes with a thin transition layer – scattering and absorption properties', *Solar Phys.* **161** (1995), 25–267; (b) 'Hot magnetic fibrils: The slow continuum revisited', *Astrophys. J.* **468** (1996), 907–920. <392>

[344] R. Keppens, T. Bogdan and M. Goossens, 'Multiple-scattering and resonant absorption of p-modes by fibril sunspots', *Astrophys. J.* **436** (1994), 372–389. <428>

[345] R. Keppens, F. Casse and J. P. Goedbloed, 'Waves and instabilities in accretion disks: magnetohydrodynamic spectroscopic analysis', *Astrophys. J.* **569** (2002), L121–L126. <441, 498, 518, 870>

[346] R. Keppens and T. Demaerel, (a) 'Stability of ideal MHD configurations. I. Realizing the generality of the G operator', *Phys. Plasmas* **23** (2016), 122117; (b) Erratum: *Phys. Plasmas* **24** (2017), 029901. <166, 448>

[347] R. Keppens and J. P. Goedbloed, 'Stellar winds, dead zones, and coronal mass ejections', *Astrophys. J.* **530** (2000), 1036–1048. <707, 709, 819>

[348] R. Keppens, K. B. MacGregor and P. Charbonneau, 'On the evolution of rotational velocity distributions for solar-type stars', *Astron. Astrophys.* **294** (1995), 469–487. <871>

[349] R. Keppens, Z. Meliani, B. van der Holst and F. Casse, 'Extragalactic jets with helical magnetic fields: relativistic MHD simulations', *Astron. Astrophys.* **486** (2008), 663–678. <871>

[350] R. Keppens and Z. Meliani, 'Linear wave propagation in relativistic magnetohydrodynamics', *Phys. Plasmas* **15** (2008), 102103-1–11. <902, 916>

[351] R. Keppens, Z. Meliani, A. J. van Marle, P. Delmont, A. Vlasis and B. van der Holst, 'Parallel, grid-adaptive approaches for relativistic hydro and magnetohydrodynamics', *J. Comp. Physics* **231** (2012), 718–744. <819, 910>

[352] R. Keppens, M. Nool, G. Tóth and J. P. Goedbloed, 'Adaptive Mesh Refinement for conservative systems: multi-dimensional efficiency evaluation', *Comput. Physics Commun.* **153** (2003), 317–339. <814, 819>

[353] R. Keppens and G. Tóth, 'Simulating magnetized plasma with the Versatile Advection Code', in *Lecture Notes in Computer Sciences*, eds. J. M. L. M. Palma, J. Dongarra and V. Hernandez, Vol. 1573 (1999), 680–690. <819>

[354] R. Keppens and G. Tóth, 'Non-linear dynamics of Kelvin–Helmholtz unstable magnetized jets: three-dimensional effects', *Phys. Plasmas* **6** (1999), 1461–1469. <822>

[355] R. Keppens, G. Tóth, M. A. Botchev and A. van der Ploeg, 'Implicit and semi-implicit schemes: algorithms', *Intern. Journ. for Numer. Meth. in Fluids*, **30** (1999), 335–352. <827>

[356] R. Keppens, G. Tóth, R. H. J. Westermann and J. P. Goedbloed, 'Growth and saturation of the Kelvin–Helmholtz instability with parallel and antiparallel magnetic fields', *J. Plasma Physics* **61** (1999), 1–19. <559>

[357] W. Kerner, (a) 'Large-scale complex eigenvalue problems', *J. Comp. Physics* **85** (1989), 1–85; (b) 'Algorithms and software for linear and nonlinear MHD simulations', *Comput. Physics Commun.* **12** (1990), 135–175. <341, 593, 690, 828, 832>

[358] W. Kerner, 'Equilibrium and stability of tokamaks', *Int. J. of Numerical Methods in Fluids* **11** (1990), 791–809. <690>

[359] W. Kerner, D. Borba, G. T. A. Huysmans, F. Porcelli, S. Poedts, J. P. Goedbloed and R. Betti, 'Stability of global Alfvén waves (TAE, EAE) in JET tritium discharges', Proc. 20th Eur. Conf. on *Controlled Fusion and Plasma Physics*, Vol. I (Lisbon, EPS, 1993), 183–186. <701>

[360] W. Kerner, J. P. Goedbloed, G. T. A. Huysmans, S. Poedts and E. Schwarz, 'CASTOR: Normal-mode analysis of resistive MHD plasmas', *J. Comp. Physics* **142** (1998), 271-303. <566, 576, 676, 690, 737>

[361] W. Kerner, A. Jakoby and K. Lerbinger, 'Finite element semi discretization of linearized compressible resistive MHD', *J. Comp. Physics* **66** (1986), 332–355. <608>

[362] W. Kerner, K. Lerbinger, R. Gruber and T. Tsunematsu, 'Normal mode analysis for resistive cylindrical plasmas', *Comput. Physics Commun.* **36** (1985), 225–240. <566, 576>

[363] C. E. Kieras and J. A. Tataronis, 'The shear Alfvén continuous spectrum of axisymmetric toroidal equilibria in the large aspect ratio limit', *J. Plasma Physics* **28** (1982), 395–414. <677>

[364] J. Kim, D. Ryu, T. W. Jones and S. S. Hong, 'A multi-dimensional code for isothermal magnetohydrodynamic flows in astrophysics', *Astrophys. J.* **514** (1999), 506–519. <834>

[365] J. G. Kirk, 'Particle acceleration in relativistic current sheets', *Phys. Rev. Lett.* **92** (2004), 181101, 1–4. <871>

[366] J. G. Kirk and P. Duffy, 'Topical Review. Particle acceleration and relativistic shocks', *J. Phys. G* **25** (1999) R163–R194. <889, 906>

[367] M. G. Kivelson, 'Pulsations and magnetohydrodynamic waves', in *Introduction to Space Physics*, eds. M. G. Kivelson and C. T. Russell (New York, Cambridge University Press, 1995), 330–355. <432>

[368] M. G. Kivelson and C. T. Russell (eds.), *Introduction to Space Physics* (Cambridge, Cambridge University Press, 1995). <24, 428>

[369] M. G. Kivelson and D. J. Southwood, 'Coupling of global magnetospheric MHD eigenmodes to field line resonances', *J. Geophys. Res.* **91** (1986), 4345–4351. <428, 432>

[370] R. Kleibergen and J. P. Goedbloed, 'Alfvén wave spectrum of an analytic high-beta tokamak equilibrium', *Plasma Phys. Control. F.* **30** (1988), 1961–1987. <680>

[371] J. A. Klimchuk, 'Theory of coronal mass ejections', Chapter 15 in *Space Weather*, eds P. Song, H. J. Singer and G. L. Siscoe (Washington, D.C., American Geophysical Union, 2001). <314>

[372] A. C. Kolb, 'Magnetic compression of plasmas', *Rev. Modern Physics* **95** (1960) 748–757. <871>

[373] A. V. Koldoba, O. A. Kuznetsov and G. V. Ustyugova, 'An approximate Riemann solver for relativistic magnetohydrodynamics', *Monthly Not. Roy. Astron. Soc.* **333** (2002), 932–942. <908>

[374] A. N. Kolmogorov, 'The local structure of turbulence in incompressible viscous fluid for very large Reynolds numbers', *Dokl. Akad. Nauk SSSR* **30**, (1941) 299–303. [English transl.: *Proc. R. Soc. Lond.* **A 434** (1991), 9–13.] <753>

[375] S. S. Komissarov, 'A Godunov-type scheme for relativistic magnetohydrodynamics', *Monthly Not. Roy. Astron. Soc.* **303** (1999), 343–366. <902, 908, 916>

[376] S. S. Komissarov and Y. E. Lyubarsky, 'The origin of peculiar jet-torus structure in the Crab nebula', *Monthly Not. Roy. Astron. Soc.* **344** (2003), L93–L96. <911>

[377] R. H. Kraichnan, (a) 'Lagrangian-history closure approximation for turbulence', *Phys. Fluids* **8** (1965), 585–598; (b) 'Inertial ranges in two-dimensional turbulence', *Phys. Fluids* **10** (1967), 1417–1423. <753, 768>

[378] N. A. Krall and A. W. Trivelpiece, *Principles of Plasma Physics* (New York, McGraw-Hill, 1973). <101>

[379] M. D. Kruskal, 'Asymptotology', in *Plasma Physics*, eds. B. B. Kadomtsev, M. N. Rosenbluth and W. B. Thompson (Vienna, IAEA, 1965), 373–387. <95>

[380] M. D. Kruskal and R. M. Kulsrud, 'Equilibrium of a magnetically confined plasma in a toroid', *Phys. Fluids* **1** (1958), 265–274. <617>

[381] M. D. Kruskal and M. Schwarzschild, 'Some instabilities of a completely ionized plasma', *Proc. Roy. Soc. (London)* **A223** (1954), 348–360. <215, 221, 229, 329>

[382] M. D. Kruskal and J. L. Tuck, 'The instability of a pinched fluid with a longitudinal magnetic field', *Proc. Roy. Soc. (London)* **A245** (1958), 222–237. <215, 329>

[383] A. G. Kulikovskij and G. A. Lyubimov, *Magnetohydrodynamics* (Reading, Addison Wesley, 1965). <876>

[384] A. K. Kulkarni and M. M. Romanova, 'Accretion to magnetized stars through the Rayleigh–Taylor instability: global 3D simulations', *Monthly Not. Roy. Astron. Soc.* **386** (2008), 673–687. <871>

[385] R. M. Kulsrud, *Plasma Physics for Astrophysics* (Princeton, Princeton University Press, 2004). <24>

[386] M. Kuperus, J. A. Ionson and D. S. Spicer, 'On the theory of coronal heating mechanisms', *Ann. Rev. Astron. Astrophys.* **19** (1971), 7–40. <418, 432>

[387] K. Kurihara, 'Improvement of tokamak plasma shape identification with a Legendre–Fourier expansion of the vacuum poloidal flux function', *Fusion Technol.* **22** (1992), 334–349. <642>

[388] H. La Belle-Hamer, A. Otto and L. C. Lee, 'Magnetic reconnection in the presence of sheared plasma flow: intermediate shock formation', *Phys. Plasmas* **1** (1994), 706–713. <870>

[389] V. P. Lakhin and V. I. Ilgisonis, Continuum modes in rotating plasmas: General equations and continuous spectra for large aspect ratio tokamaks', *Phys. Plasmas* **18** (2011), 092103. <725>

[390] H. J. G. L. M. Lamers and J. P. Cassinelli, *Introduction to Stellar Winds* (Cambridge, Cambridge University Press, 1999). <322>

[391] K. Land and J. Magueijo, 'Examination of evidence for a preferred axis in the cosmic radiation anisotropy', *Phys. Rev. Lett.* **95** (2005), 071301, 1–4. <437>

[392] L. D. Landau, 'The transport equation in the case of Coulomb interactions', *J. Exp. Theor. Phys. USSR* **7** (1937), 203. [Transl.: *Phys. Z. Sowjet.* **10** (1936), 154; or *Collected Papers of L. D. Landau*, ed. D. ter Haar (Oxford, Pergamom Press, 1965), 163–170.] <38, 67>

[393] L. D. Landau, 'On the vibrations of the electronic plasma', *J. Phys. USSR* **10** (1946), 25. [Transl.: *JETP* **16** (1946), 574; or *Collected Papers of L.D. Landau*, ed. D. ter Haar (Oxford, Pergamom Press, 1965), 445–460.] <46, 269, 372, 376, 480>

[394] L. D. Landau and E. M. Lifshitz, *Course of Theoretical Physics, Vol. 6: Fluid Mechanics*, 2nd edition (Oxford, Pergamon Press, 1987). <133, 144, 840, 876>

[395] L. D. Landau and E. M. Lifshitz, *Course of Theoretical Physics, Vol. 8: Electrodynamics of Continuous Media*, 2nd edition (Oxford, Pergamon Press, 1984), Chap. VIII. <136, 144, 627, 848, 876>

[396] G. Lapenta, 'Self-feeding turbulent magnetic reconnection on macroscopic scales', *Phys. Rev. Lett.* **100** (2008), 235001, 1–4. <554>

[397] S. Larsson and V. Thomée, *Partial Differential Equations with Numerical Methods* (Berlin, Springer-Verlag, 2003). <612>

[398] G. Laval, C. Mercier and R. M. Pellat, 'Necessity of the energy principle for magnetostatic stability', *Nucl. Fusion* **5** (1965), 156–158. <207, 229>

[399] G. Laval, R. M. Pellat and J. S. Soule, 'Hydromagnetic stability of a current-carrying pinch with noncircular cross section', *Phys. Fluids* **17** (1974), 835–845. <627>

[400] J. D. Lawson, 'Some criteria for a power producing thermonuclear reactor', *Proc. Phys. Soc. (London)* **B70** (1975), 6–10. <6>

[401] A. Lenard, 'On Bogoliubov's kinetic equation for a spatially homogeneous plasma', *Ann. Phys. (New York)* **10** (1960), 390–400. <67>

[402] M. A. Leontovich (ed.), *Reviews of Plasma Physics, Vols. 1–5* (New York, Consultants Bureau, 1965–1970; translated from the Russian editions of 1963–1967). <24>

[403] K. Lerbinger and J. F. Luciani, 'A new semi-implicit method for MHD computations', *J. Comp. Physics* **97** (1991), 444–459. <828>

[404] R. J. LeVeque, *Numerical Methods for Conservation Laws* (Berlin, Birkhäuser Verlag, 1990). <133, 169, 179, 599, 786, 790>

[405] R. J. LeVeque, *Finite Volume Methods for Hyperbolic Problems* (Cambridge, Cambridge University Press, 2002). <802, 819>

[406] R. J. LeVeque, D. Mihalas, E. A. Dorfi and E. Müller, *Computational Methods for Astrophysical Fluid Flow* (Berlin, Springer Verlag, 1998). <169, 179, 834>

[407] S. Li, 'An HLLC Riemann solver for magneto-hydrodynamics', *J. Comp. Physics* **203** (2005), 344–357. <807>

[408] M. A. Liberman and A. L. Velikovich, *Physics of Shock Waves in Gases and Plasmas* (Berlin, Springer-Verlag, 1986). <854>

[409] A. Lichnerowicz, *Relativistic Hydrodynamics and Magnetohydrodynamics* (New York, W. A. Benjamin Inc., 1967). <99, 879, 906, 916>

[410] A. E. Lifschitz, *Magnetohydrodynamics and Spectral Theory* (Dordrecht, Kluwer Academic Publishers, 1989). <24, 190, 229, 289, 551, 566, 683, 704>

[411] A. Lifschitz and J. P. Goedbloed, 'Transonic magnetohydrodynamic flows', *J. Plasma Physics* **58** (1997), 61–99. <871>

[412] C. C. Lin, *The Theory of Hydrodynamic Stability* (Cambridge, Cambridge University Press, 1955). <480>

[413] T. J. Linde, 'A practical, general-purpose, two-state HLL Riemann solver for hyperbolic conservation laws', *Int. J. Numer. Meth. Fluids* **40** (2002), 391–402. <807>

[414] D. H. Liu and A. Bondeson, 'Improved poloidal convergence of the MARS code for MHD stability analysis', *Comput. Physics Commun.* **116** (1999), 55–64. <576>

[415] X. Llobet, K. Appert, A. Bondeson and J. Vaclavik, 'On spectral pollution', *Comput. Physics Commun.* **59** (1990), 199–216. <586>

[416] P. Londrillo and L. Del Zanna, 'High-order upwind schemes for multidimensional magnetohydrodynamics', *Astrophys. J.* **530** (2000), 508–524. <813>

[417] Y.-Q. Lou, 'Viscous MHD modes and p-mode absorption by sunspots', *Astrophys. J.* **350** (1990), 452–462. <428, 432>

[418] B. C. Low, 'Magnetohydrodynamic processes in the solar corona: Flares, coronal mass ejections, and magnetic helicity', *Phys. Plasmas* **1** (1994), 1684–1690. <300>

[419] B. C. Low and M. Zhang, 'Magnetostatic structures of the solar corona. III. Normal and inverse quiescent prominences', *Astrophys. J.* **609** (2004), 1098-1111. <662>

[420] S. Lundquist, 'Magneto-hydrostatic fields', *Arkiv för Fysik* **2** (1951), 361–365. <124>

[421] R. Lüst and A. Schlüter, 'Axialsymmetrische magnetohydrodynamische Gleichgewichtskonfigurationen', *Z. Naturforsch.* **12a** (1957), 850–854. <635>

[422] H. Lütjens, A. Bondeson and A. Roy, 'Axisymmetric MHD equilibrium solver with bicubic Hermite elements', *Comput. Physics Commun.* **69** (1992), 287–298. <576, 655>

[423] H. Lütjens, A. Bondeson and O. Sauter, 'The CHEASE code for toroidal MHD equilibria', *Comput. Physics Commun.* **97** (1996), 219–260. <576>

[424] M. G. Macaraeg, C. L. Streett and M. Y. Hussaini, 'A spectral collocation solution to the compressible stability eigenvalue problem', *NASA Technical Paper No. 2858* (NASA, Washington DC, 1988). <583>

[425] A. Majorana and A. M. Anile, 'Magnetoacoustic shock waves in a relativistic gas', *Phys. Fluids* **30** (1987), 3045–3049. <906>

[426] (a) J. H. Malmberg and C. B. Wharton, 'Dispersion of electron plasma waves', *Phys. Rev. Lett.* **17** (1966), 175–178;

(b) J. H. Malmberg, C. B. Wharton, R. W. Gould and T. M. O'Neil, 'Observation of plasma wave echos', *Phys. Fluids* **11** (1968), 1147–1153. <46>

[427] W. B. Manchester IV, T. I. Gombosi, I. Roussev, A. Ridley, D. L. De Zeeuw, I. V. Sokolov and K. G. Powell, 'Modeling a space weather event from the Sun to the Earth: CME generation and interplanetary propagation', *J. Geophys. Res.* **109** (2004), A02107, 1–15. <317>

[428] J. Manickam, N. Pomphrey and A. M. M. Todd, 'Ideal MHD stability properties of pressure driven modes in low shear tokamaks', *Nucl. Fusion* **27** (1987), 1461–1472. <695>

[429] I. R. Mann and A. N. Wright, 'Coupling of magnetospheric cavity modes to field line resonances – A study of resonance widths', *J. Geophys. Res.* **100** (1995), 19 441–19 456. <399>

[430] J. Maron and P. Goldreich, 'Simulations of incompressible magnetohydrodynamic turbulence', *Astrophys. J.* **554** (2001), 1175–1196. <768>

[431] P. Martens and C. Zwaan, 'Origin and evolution of filament-prominence systems', *Astrophys. J.* **558** (2001), 872–887. <300>

[432] J. M. Martí and E. Müller, 'Numerical hydrodynamics in special relativity', *Living Rev. Relativity* **6**, (2003), 1–100. <916>

[433] W. G. Mathews, 'The hydromagnetic free expansion of a relativistic gas', *Astrophys. J.* **165** (1971), 147–164. <885>

[434] K. G. McClements and A. Thyagaraja, 'Azimuthally symmetric magnetohydrodynamic and two-fluid equilibria with arbitrary flows', *Monthly Not. Roy. Astron. Soc.* **323** (2001), 733–742. <744>

[435] B. F. McMillan, R. L. Dewar and R. Storer, 'A comparison of incompressible limits for resistive plasmas', *Plasma Phys. Control. F.* **46** (2004), 1027–1038. <566>

[436] Z. X. Mei, R. Keppens, I. I. Roussev and J. Lin, 'Magnetic reconnection during eruptive magnetic flux ropes', *Astron. Astrophys.* **601** (2017), L7, 1–4. <819>

[437] C. Mercier, 'Un critère necessaire de stabilité hydromagnétique pour un plasma en symétrie de révolution', *Nucl. Fusion* **1** (1960), 47–53. <348, 361, 441, 683>

[438] L. Mestel, *Stellar Magnetism* (Oxford, Clarendon Press, 1999). <15, 24, 292>

[439] F. Meyer, 'Untersuchung der Stabilität eines gravitierenden Plasmas in gekreuzten Magnetfeldern', *Z. Naturforschg.* **13a** (1958), 1016–1020. <229>

[440] A. Mignone and G. Bodo, 'An HLLC Riemann solver for relativistic flows. II. Magnetohydrodynamics', *Monthly Not. Roy. Astron. Soc.* **368** (2006), 1040–1054. <908>

[441] A. Mignone, G. Bodo, S. Massaglia, T. Matsakos, O. Tesileanu, C. Zanni and A. Ferrari, 'PLUTO: a numerical code for computational astrophysics', *Astrophys. J. Suppl. Ser.* **170** (2007), 228–242. <819, 910>

[442] K. Miyamoto, (a) *Plasma Physics for Nuclear Fusion*, revised edition (Cambridge, Massachusetts Institute of Technology Press, 1989); (b) *Plasma Physics and Controlled Fusion* (Berlin, Springer, 2005). <4, 6, 627, 642, 664>

[443] T. Miyoshi and K. Kusano, 'A multi-state HLL approximate Riemann solver for ideal magnetohydrodynamics', *J. Comp. Physics* **208** (2005), 315–344. <807>

[444] H. K. Moffatt, *Magnetic Field Generation in Electrically Conducting Fluids* (Cambridge, Cambridge University Press, 1978). <xix, 124, 144>

[445] R. Monceau-Baroux, O. Porth, Z. Meliani and R. Keppens, 'Relativistic 3D precessing jet simulations for the X-ray binary SS433', *Astron. Astrophys.* **561** (2014), A30. <889>

[446] D. C. Montgomery and D. A. Tidman, *Plasma Kinetic Theory* (New York, McGraw-Hill Book Company, 1964). <101>

[447] A. I. Morozov and L .S. Soloviev, 'The structure of magnetic fields', in *Reviews of Plasma Physics, Vol. 2*, ed. M. A. Leontovich (New York, Consultants Bureau, 1966), 1–101. <664>

[448] A. I. Morozov and L .S. Soloviev, 'Steady-state plasma flow in a magnetic field', in *Reviews of Plasma Physics, Vol. 8*, ed. M. A. Leontovich (New York, Consultants Bureau, 1980; original Russian edition, 1974), 1–103. <744>

[449] P. J. Morrison and J. M. Greene, 'Noncanonical Hamiltonian density formulation of hydrodynamics and ideal magnetohydrodynamics', *Phys. Rev. Lett.* **45** (1980), 790–794 [Erratum: **48** (1982), 569]. <456>

[450] P. M. Morse and H. Feshbach, (a) *Methods of Theoretical Physics, Part I* (New York, McGraw-Hill, 1953); (b) *Methods of Theoretical Physics, Part II* (New York, McGraw-Hill, 1953). <171, 273, 642>

[451] C. Mouhat and C. Villani, 'On Landau damping', *Acta Math.* **207** (2011), 29–201. <46>

[452] T. C. Mouschovias, 'The Parker instability in the interstellar medium', in *Solar and Astrophysical*

Magnetohydrodynamic Flows, ed. K. C. Tsinganos (Dordrecht, Kluwer Academic Publishers, 1996), 475–504. <258, 289>

[453] V. S. Mukhovatov and V. D. Shafranov, 'Plasma equilibrium in a tokamak', *Nucl. Fusion* **11** (1971), 605–633. <627>

[454] W. C. Müller and D. Biskamp, 'Scaling properties of three-dimensional magnetohydrodynamic turbulence', *Phys. Rev. Lett.* **84** (2000), 475–478. <768>

[455] K. Murawski, *Analytical and Numerical Methods for Wave Propagation in Fluid Media* (Singapore, World Scientific, 2002). <834>

[456] R. S. Myong, 'Analytical results on MHD intermediate shocks', *Geophys. Res. Lett.* **24** (1997), 2929–2932. <870>

[457] R. S. Myong and P. L. Roe, 'Shock waves and rarefaction waves in magnetohydrodynamics; (a) 1. A model system ', *J. Plasma Physics* **58** (1997), 485–519; (b) 2. The MHD system', *J. Plasma Physics* **58** (1997), 521–552. <780, 798, 802, 870>

[458] V. M. Nakariakov, L. Ofman, E. E. DeLuca, B. Roberts and J. M. Davila, 'TRACE observation of damped coronal loop oscillations: implications for coronal heating', *Science* **285** (1999), 862–864. <399>

[459] Z. Nehari, *Conformal Mapping* (New York, Dover Publications, 1952). <46>

[460] W. A. Newcomb, 'Motion of magnetic lines of force', *Ann. Phys. (New York)* **3** (1958), 347–385. <113, 144>

[461] W. A. Newcomb, 'Hydromagnetic stability of a diffuse linear pinch', *Ann. Phys. (New York)* **10** (1960), 232–267. <280, 285, 330, 348, 369>

[462] W. A. Newcomb, 'Convective instability induced by gravity in a plasma with a frozen-in magnetic field', *Phys. Fluids* **4** (1961), 391–396. <258, 283, 289, 441>

[463] W. A. Newcomb, 'Lagrangian and Hamiltonian Methods in Magnetohydrodynamics', *Nucl. Fusion*, 1962 Suppl., Part **2** (1962), 451–463. <202, 448, 456>

[464] W. A. Newcomb, *Lecture Notes on Magnetohydrodynamics* (unpublished). <113, 448>

[465] R. J. Nijboer and J. P. Goedbloed, 'Mode coupling in two-dimensional magnetohydrodynamic flows', *J. Plasma Physics* **61** (1999), 241–262. <456, 725>

[466] R. J. Nijboer, B. van der Holst, S. Poedts and J. P. Goedbloed, 'Calculating magnetohydrodynamic flow spectra', *Comput. Physics Commun.* **106** (1997), 39–52. <576>

[467] S. C. Noble, C. F. Gammie, J. C. McKinney and L. Del Zanna, 'Primitive variable solvers for conservative general relativistic magnetohydrodynamics', *Astrophys. J.* **641** (2006), 626–637. <910>

[468] C. Nore, M. E. Brachet, H. Politano and A. Pouquet, 'Dynamo action in the Taylor–Green vortex near treshold', *Phys. Fluids* **4** (1997), 1–5. <774>

[469] T. G. Northrop, *The Adiabatic Motion of Charged Particles* (New York, Interscience Publishers, 1963). <33, 64>

[470] D. Odstrcil, 'Modeling 3-D solar wind structure', *Adv. Space Res.* **32 (4)** (2003), 497–506. <317>

[471] L. Ofman and J. M. Davila, (a) 'Nonlinear resonant absorption of Alfvén waves in three dimensions, scaling laws, and coronal heating', *J. Geophys. Res.* **100** (1995), 23 427–23 441; (b) 'Alfvén wave heating of coronal holes and the relation to the high-speed wind', *J. Geophys. Res.* **100** (1995), 23 413–23 425; (c) 'Nonlinear excitation of global modes and heating in randomly driven coronal loops', *Astrophys. J.* **456** (1996), L123–L126. <419, 427, 428, 790>

[472] L. Ofman, J.A. Klimchuk and J. M. Davila, 'A self-consistent model for the resonant heating of coronal loops: The effects of coupling with the chromosphere', *Astrophys. J.* **493** (1998), 474–479. <428>

[473] D. P. O'Leary, *Scientific Computing with Case Studies* (Philadelphia, SIAM Press, 2009). <612>

[474] S. A. Orszag and C.-M. Tang, 'Small-scale structure of two-dimensional magnetohydrodynamic turbulence', *J. Fluid Mech.* **90** (1979), 129–143. <772, 778>

[475] S. Ortolani and D. D Schnack, *Magnetohydrodynamics of Plasma Relaxation* (Singapore, World Scientific, 1993). <xix>

[476] Y. P. Pao, 'Instabilities in toroidal plasmas', *Phys. Fluids* **18** (1975), 1192–1196. <683>

[477] Y. P. Pao, 'The continuous MHD spectrum in toroidal geometries', *Nucl. Fusion* **15** (1975), 631–635. <677, 722>

[478] Y. P. Pao and W. Kerner, 'Analytic theory of stable resistive magnetohydrodynamics modes', *Phys. Fluids* **15** (1985), 287–293. <551, 566>

[479] E. N. Parker, 'Instability of thermal fields', *Astrophys. J.* **117** (1953), 431–436. <594>

[480] E. N. Parker, 'Sweet's mechanism for merging magnetic fields in conducting fluids', *J. Geophys. Res.* **62** (1957), 509–520. <554>

[481] E. N. Parker, 'Dynamics of the interplanetary gas and magnetic fields', *Astrophys. J.* **128** (1958), 664–676. <309>

[482] E. N. Parker, 'The dynamical state of the interstellar gas and field', *Astrophys. J.* **145** (1966), 811–833. <221, 258, 289, 441>

[483] E. N. Parker, *Cosmical Magnetic Fields* (Oxford, Clarendon Press, 1979). <292>

[484] F. Pegoraro and T. J. Schep, 'Low-frequency modes with high toroidal mode numbers: A general formulation', *Phys. Fluids* **24** (1981), 478–497. <683>

[485] J. R. Peterson and A. C. Fabian, 'X-ray spectroscopy of cooling clusters', *Phys. Rep.* **427** (2006), 1–39. <437>

[486] G. D. J. Petrie. J. W. S. Blokland and R. Keppens, 'Magnetohydrostatic solar prominences in near-potential coronal magnetic field', *Astrophys. J.* **865** (2007), 830–845. <662>

[487] H. E. Petschek, 'Magnetic field annihilation', in *Physics of Solar Flares*, ed. W. N. Hess (Washington, DC, NASA SP-50) (1964), 425–439. <558>

[488] D. Pfirsch and H. Tasso, 'A theorem on MHD instability of plasmas with resistive walls', *Nucl. Fusion* **11** (1971), 259–260. <544>

[489] J. M. Picone and R. B. Dahlburg, 'Evolution of the Orszag–Tang vortex system in a compressible medium. II. Supersonic flow', *Phys. Fluids* **B 3** (1991), 29–44. <776>

[490] S. Poedts, 'MHD instabilities: numerical methods and results', *Trans. Fusion Technol.* **29** (1996), 133–142. <569>

[491] S. Poedts, 'MHD waves and heating of the solar corona', in *Magnetic Coupling of the Solar Atmosphere*, Proc. Euroconference and IAU Colloquium 188, ESA SP-**505** (2002), 273–280. <432>

[492] S. Poedts, A. J. C. Beliën and J. P. Goedbloed, 'On the quality of resonant absorption as a coronal loop heating mechanism', *Solar Phys.* **151** (1994), 271–304. <414>

[493] S. Poedts and G. C. Boynton, 'Nonlinear MHD of footpoint driven coronal loops', *Astron. Astrophys.* **306** (1996), 610–620. <428>

[494] S. Poedts and J. P. Goedbloed, 'Nonlinear wave heating of solar coronal loops', *Astron. Astrophys.* **321** (1997), 935–944. <419, 828>

[495] S. Poedts, M. Goossens and W. Kerner, (a) 'Numerical simulation of coronal heating by resonant absorption of Alfvén waves', *Solar Phys.* **123** (1989), 83–115; (b) 'On the efficiency of coronal loop heating by resonant absorption', *Astrophys. J.* **360** (1990), 279–287. <428, 432, 590>

[496] S. Poedts, D. Hermans and M. Goossens, 'The continuous spectrum of an axisymmetric self-gravitating and static equilibrium with a mixed poloidal and toroidal magnetic field', *Astron. Astrophys.* **151** (1985), 16–27. <725>

[497] S. Poedts and W. Kerner, 'Ideal quasi-modes reviewed in resistive MHD', *Phys. Rev. Lett.* **66** (1991), 2871–2874. <432, 566, 593>

[498] S. Poedts, W. Kerner, J. P. Goedbloed, B. Keegan, G. T. A. Huysmans and E. Schwarz, 'Damping of global Alfvén waves in tokamaks due to resonant absorption', *Plasma Phys. Control. F.* **34** (1992), 1397–1422. <417, 419, 432, 680>

[499] S. Poedts, G. Tóth, A. J. C. Beliën and J. P. Goedbloed, 'Nonlinear MHD simulations of wave dissipation in flux tubes', *Solar Phys.* **172** (1997), 45–52. <419, 428>

[500] S. Poedts and E. Schwarz, 'Computation of the ideal MHD continuous spectrum in axisymmetric plasmas', *J. Comp. Physics* **105** (1993), 165–168. <680, 737>

[501] O. P. Pogutse and E. I. Yurchenko, 'Ballooning effects and plasma stability in tokamaks', in *Reviews of Plasma Physics, Vol. 11*, ed. M. A. Leontovich (New York, Consultants Bureau, 1982), 65–151. <704>

[502] R. V. Polovin and V. P. Demutskii, *Fundamentals of Magnetohydrodynamics* (New York, Consultants Bureau, 1990). <24>

[503] J. Pomoell and S. Poedts, 'EUHFORIA: EUropean Heliospheric FORecasting Information Asset', *J. Space Weather Space Clim.* **8** (2018), A35, 1–14. <317>

[504] O. Porth, S. S. Komissarov and R. Keppens, (a) 'Solution to the sigma problem of pulsar wind nebulae', *Monthly Not. Roy. Astron. Soc.* **431** (2013), L48–L52; (b) 'Three-dimensional magnetohydrodynamic simulations of the Crab nebula', *Monthly Not. Roy. Astron. Soc.* **438** (2014), 278–306; (c) 'Rayleigh–Taylor instability in magnetohydrodynamic simulations of the Crab nebula', *Monthly Not. Roy. Astron. Soc.* **443** (2014), 547–558. <911>

[505] O. Porth, H. Olivares, Y. Mizuno, Z. Younsi, L. Rezzolla, M. Moscibrodzka, H. Falcke and M. Kramer, 'The black hole accretion code', *Comput. Astrophys. Cosmology* **4** (2017), 1–42. <915>

[506] O. Porth, C. Xia, T. Hendrix, S.P. Moschou and R. Keppens, 'MPI-AMRVAC for solar and astrophysics, *Astrophys. J. Suppl. Ser.* **214** (2014), 4–29. <819>

[507] K. G. Powell, 'An approximate Riemann solver for magnetohydrodynamics (that works in more than one dimension)', *ICASE Report No 94-24* (Langley VA, 1994). <152, 814>

[508] K. G. Powell, P. L. Roe, T. J. Linde, T. I. Gombosi and D. L. De Zeeuw, 'A solution-adaptive upwind scheme for ideal magnetohydrodynamics', *J. Comp. Physics* **154** (1999), 284–309. <152, 814, 819>

[509] W. H. Press, B. P. Flannery, S. A. Teukolsky and W. T. Vettering, *Numerical Recipes* (Cambridge, Cambridge University Press, 1989). <484, 523, 566, 876>

[510] E. R. Priest, *Magnetohydrodynamics of the Sun* (Cambridge, Cambridge University Press, 2014). <xix, 4, 15, 24, 233, 258, 289, 292, 322, 709, 876>

[511] E. R. Priest and T. G. Forbes, *Magnetic Reconnection; MHD Theory and Applications* (Cambridge, Cambridge University Press, 2000). <124, 144, 554, 558, 559, 566, 871>

[512] J. Pringle and A. King, *Astrophysical Flows* (Cambridge, Cambridge University Press, 2007). <523>

[513] J. J. Quirk, 'A contribution to the great Riemann solver debate', *Int. J. Num. Meth. in Fluids* **18** (1994), 555–574. <807>

[514] J. Rappaz, 'Approximation of the spectrum of a noncompact operator given by the magnetohydrodynamic stability of plasma', *Num. Math.* **28** (1977), 15–24. <586>

[515] M. Rempel, (a) 'Numerical simulations of quiet sun magnetism: on the contribution from a small-scale dynamo', *Astrophys. J.* **789** (2014), 132–153; (b) 'Extension of the MURAM radiative MHD code for coronal simulations', *Astrophys. J.* **834** (2017), 10–32. <822>

[516] M. Rempel, M. Schüssler, R.H. Cameron and M. Knölker, 'Penumbral structure and outflows in simulated sunspots', Science **325** (2009), 171–174. <822>

[517] L. Rezzolla and O. Zanotti, *Relativistic Hydrodynamics* (Oxford, Oxford University Press, 2013). <916>

[518] A. Reiman, 'Minimum energy state of a toroidal discharge', *Phys. Fluids* **23** (1980), 230–231. <124>

[519] K. S. Riedel, 'The spectrum of resistive viscous magnetohydrodynamics', *Phys. Fluids* **29** (1986), 1093–1104. <551, 566>

[520] B. Roberts, 'Surface waves', in *Mechanisms of Chromospheric and Coronal Heating*, eds. P. Ulmschneider, E. R. Priest and R. Rosner (Berlin, Springer Verlag, 1991), 494–507. <242>

[521] D. H. Roberts, J. F. C. Wardle, S. L. Lipnick, P. L. Selesnick and S. Slutsky, 'Structure and magnetic fields in the precessing jet system SS 433. I. Multifrequency imaging from 1998', *Astrophys. J.* **676** (2008), 584–593. <889>

[522] P. H. Roberts, *An Introduction to Magnetohydrodynamics* (London, Longmans, Green and Co Ltd, 1967). <64, 131>

[523] D. C. Robinson, 'High-β diffuse pinch configurations', *Plasma Physics* **13** (1971), 439–462. <369>

[524] P. L. Roe, 'Characteristic-based schemes for the Euler equations', *Ann. Rev. Fluid Mech.* **18** (1986), 337–365. <807>

[525] P. L. Roe and D. S. Balsara, 'Notes on the eigensystem of magnetohydrodynamics', *SIAM J. Appl. Math.* **56** (1996), 57–67. <798, 807>

[526] M. M. Romanova, G. V. Ustyugova, A. V. Koldoba and R. V. E. Lovelace, 'Magnetohydrodynamic simulations of disk-magnetized star interactions in the quiescent regime: funnel flows and angular momentum transport', *Astrophys. J.* **578** (2002), 420–438. <871>

[527] M. M. Romanova, O. D. Toropina, Yu. M. Toropin and R. V. E. Lovelace, 'Magnetohydrodynamic simulations of accretion onto a star in the "propeller" regime', *Astrophys. J.* **588** (2003), 400–407. <871>

[528] M. N. Rosenbluth, 'Stability and heating in the pinch effect', in *Proc. Second UN Intern. Conf. on Peaceful Uses of Atomic Energy* 31 (New York, Columbia Univ. Press, 1959), 85–92. <329>

[529] M. N. Rosenbluth, R. Y. Dagazian and P. H. Rutherford, 'Nonlinear properties of the internal $m = 1$ instability in the cylindrical tokamak', *Phys. Fluids* **16** (1973), 1894–1902. <832>

[530] M. N. Rosenbluth, W. M. MacDonald and D. L. Judd, 'Fokker-Planck equation for an inverse-square force', *Phys. Rev.* **107** (1957), 1–6. <67, 101>

[531] I. I. Roussev, K. Galsgaard, C. Downs, N. Lugaz, I. V. Sokolov, E. Moise and J. Lin, 'Explaining fast ejections of plasma and exotic X-ray emission from the solar corona', *Nature Physics* **8** (2012), 845–849. <314>

[532] M. S. Ruderman, M. L. Goldstein, D. A. Roberts, A. Deane and L. Ofman, 'Alfvén wave phase mixing driven by velocity shear in two-dimensional open magnetic configurations', *J. Geophys. Res.* **104** (1999), 17 057-17 068. <428>

[533] G. Rüdiger and R. Hollerbach, *The Magnetic Universe; Geophysical and Astrophysical Dynamo Theory* (Weinheim, Wiley-VCH, 2004). <xix>

[534] V. V. Rusanov, 'The calculation of the interaction of non-stationary shock waves and obstacles', *USSR Comp. Math. and Math. Phys.* **1** (1961), 304–320. <802>

[535] D. Ryu and T. W. Jones, (a) 'Numerical magnetohydrodynamics in astrophysics: algorithm and tests for one-dimensional flow', *Astrophys. J.* **442** (1995), 228–258;
(b) D. Ryu, T. W. Jones and A. Frank,'Numerical magnetohydrodynamics in astrophysics: algorithm and tests for multidimensional flow', *Astrophys. J.* **452** (1995), 785–796. <798, 807, 811, 813>

[536] P. H. Sakanaka and J. P. Goedbloed, 'New approach to magnetohydrodynamic stability. II. Sigma-stable diffuse pinch configurations', *Phys. Fluids* **17** (1974), 919–929. <207, 348, 369>

[537] T. Sakurai, M. Goossens and J. V. Hollweg, 'Resonant behavior of MHD waves on magnetic-flux tubes. (a) I. Connection formulas at the resonant surfaces', *Solar Phys.* **133** (1991), 227–245;
(b) 'II. Absorption of sound waves by sunspots', *Solar Phys.* **133** (1991), 247–262. <414, 432>

[538] M. Saunders, 'The earth's magnetosphere', in *Advances in Solar System Magnetohydrodynamics*, eds. E. R. Priest and A. W. Hood (Cambridge, Cambridge University Press, 1991), 357–397. <322>

[539] B. E. Schaeffer, J. R. King and C. P. Deliyannis, 'Superflares on ordinary solar-type stars', *Astrophys. J.* **529** (2000), 1026–1030. <300>

[540] A. A. Schekochihin, S. C. Cowley, S. F. Taylor, J. L. Maron and J. C. McWilliams, 'Simulations of the small-scale turbulent dynamo', *Astrophys. J.* **612** (2004), 276–307. <774>

[541] A. A. Schekochihin, A. B. Iskakov, S. C. Cowley, J. C. McWilliams, M. R. E. Proctor and T. A. Yousef, 'Fluctuation dynamo and turbulent induction at low magnetic Prandtl numbers', *New J. Phys.* **9** (2007), 300–324. <774>

[542] G. Schmidt, *Physics of High Temperature Plasmas*, 2nd edition (New York, Academic Press, 1979). <101, 525>

[543] D. D. Schnack, D. C. Barnes, D. P. Brennan, C. C. Hegna, E. Held, C. C. Kim, S. E. Kruger, A. Y. Pankin and C. R. Sovinec, 'Computational modeling of fully ionized magnetized plasmas using the fluid approximation', *Phys. Plasmas* **13** (2006), 058103-1–058103-21. <437, 833>

[544] C. J. Schrijver and C. Zwaan, *Solar and Stellar Magnetic Activity* (Cambridge, Cambridge University Press, 2000). <15, 292, 309, 322>

[545] L. Schwartz, *Théorie des Distributions* (Paris, Hermann et cie, 1950). <196>

[546] Z. Sedláček, 'Electrostatic oscillations in cold inhomogeneous plasma', *J. Plasma Physics* **5** (1971), 239–263. <269, 372, 373, 376, 384, 397>

[547] L. I. Sedov, *Two-dimensional Problems in Hydrodynamica and Aerodynamics* (New York, Interscience Publishers, 1965; translated from the Russian edition of 1950 and edited by C. K. Chu, H. Cohen and B. Seckler). <840>

[548] V. D. Shafranov, (a) 'On magnetohydrodynamical equilibrium configurations', *Soviet Phys.–JETP* **33** (1957), 710–722; (b) 'Equilibrium of a plasma toroid in a magnetic field', *Soviet Phys.–JETP* **37** (1960), 775–779; (c) 'Equilibrium of a toroidal plasma in a magnetic field', *Journal of Nuclear Energy, Part C*, **5** (1963), 251–258. [*Atomnaya Energiya* **13** (1962), 521.] <635, 642, 664>

[549] V. D. Shafranov, 'Plasma equilibrium in a magnetic field', in *Reviews of Plasma Physics, Vol. 2*, ed. M. A. Leontovich (New York, Consultants Bureau, 1966), 103–151. <440, 627, 664>

[550] V. D. Shafranov, 'Hydrodynamic stability of a current-carrying pinch in a strong longitudinal field', *Sov. Phys.–Tech. Phys.* **15** (1970), 175–183. <361, 369>

[551] N. I. Shakura and R. A. Sunyaev, 'Black holes in binary systems. Observational appearance', *Astron. Astrophys.* **24** (1973), 337–355. <514, 518>

[552] S. E. Sharapov, D. Testa, B. Alper, D. N. Borba, A. Fasoli, N. C. Hawkes, R. F. Heeter, M. Mantsinen, M. G. Von Hellerman and contributors to the EFDA-JET work-programme, 'MHD spectroscopy through detecting toroidal Alfvén eigenmodes and Alfvén wave cascades', *Phys. Lett.* **A289** (2001) 127–134. <701>

[553] G. L. G. Sleijpen and H. A. van der Vorst, 'A Jacobi–Davidson iteration method for linear eigenvalue problems', *SIAM J. Matrix Anal. Appl.* **17** (1996), 401–425. <737>

[554] L .S. Soloviev, 'Hydromagnetic stability of closed plasma configurations', in *Reviews of Plasma Physics, Vol. 6*, ed. M. A. Leontovich (New York, Consultants Bureau, 1975; original Russian edition,

1972), 239–331. <652, 664>

[555] L .S. Soloviev and B. B. Shafranov, 'Plasma confinement in closed magnetic systems', in *Reviews of Plasma Physics, Vol. 5*, ed. M. A. Leontovich (New York, Consultants Bureau, 1970), 1–247. <664>

[556] H. Soltwisch, (a) 'Measurement of current-carrying changes during sawtooth activity in a tokamak by far-infrared polarimetry', *Rev. Sci. Instrum.* **59** (1992), 1599–1604; (b) 'Current density measurements in tokamak devices', *Plasma Phys. Control. F.* **34** (1992), 1669–1698. <361>

[557] C. R. Sovinec et al., 'Nonlinear magnetohydrodynamics simulation using high-order finite elements', *J. Comp. Physics* **195** (2004), 355–386. <833>

[558] G. O. Spies, 'Magnetohydrodynamic spectrum in a class of toroidal equilibria', *Phys. Fluids* **19** (1976), 427–437. <207>

[559] G. O. Spies, 'Magnetohydrodynamic spectrum of instabilities due to plasma rotation', *Phys. Fluids* **21** (1978), 580–587. <480, 507, 725>

[560] G. O. Spies and J. A. Tataronis, 'The accumulation continua in ideal magnetohydrodynamics', *Phys. Plasmas* **10** (2003), 413–418. <207>

[561] L. Spitzer, Jr, *Physics of Fully Ionized Gases* (New York, Interscience, 1962). <101>

[562] L. Spitzer and R. Härm, 'Transport phenomena in a completely ionized gas', *Phys. Rev.* **89** (1953), 977–981. <294>

[563] H. C. Spruit, 'Propagation speeds and acoustic damping of waves in magnetic flux tubes', *Solar Phys.* **75** (1982), 3–17. <397>

[564] H. C. Spruit, T. Matsuda, M. Inoue and K. Sawada, 'Spiral shocks and accretion in discs', *Monthly Not. Roy. Astron. Soc.* **229** (1987), 517–527. <518>

[565] S. Sridhar and P. Goldreich, 'Toward a theory of interstellar turbulence. I. Weak Alfvénic turbulence', *Astrophys. J.* **432** (1994), 612–621. <768>

[566] K. Stasiewicz, 'Nonlinear Alfvén, magnetosonic, sound, and electron inertial waves in fluid formalism', *J. Geophys. Res.* **110** (2005), A03220, 1–9. <437>

[567] L. C. Steinhauer,'Review of field-reversed configurations', *Phys. Plasmas* **18** (2011), 070501, 1–38. <506>

[568] H. Stenuit, R. Keppens and M. Goossens, 'Eigenfrequencies and optimal driving frequencies of 1D non-uniform magnetic flux tubes', *Astron. Astrophys.* **331** (1998), 392–404. <392>

[569] M. Stix, *The Sun, An Introduction*, 2nd edition (Berlin, Springer-Verlag, 2002). <24, 233, 234, 289, 292, 322>

[570] T. H. Stix, *Waves in Plasmas* (New York, American Institue of Physics, McGraw Hill, 1992). <46, 86, 397>

[571] J. M. Stone, J. F. Hawley, C. R. Evans and M. L. Norman, 'A test suite for magnetohydrodynamical simulations', *Astrophys. J.* **388** (1992), 415–437. <822>

[572] J. M. Stone, J. F. Hawley, C. F. Gammie and S. A. Balbus, 'Three-dimensional magnetohydrodynamical simulations of vertically stratified accretion disks', *Astrophys. J.* **463** (1996), 656–673. <514, 822>

[573] J. M. Stone and M. L. Norman, 'ZEUS-2D: a radiation magnetohydrodynamics code for astrophysical flows in two space dimensions. II. Magnetohydrodynamic algorithms and tests', *Astrophys. J. Suppl. Ser.* **80** (1992), 791–818. <822>

[574] G. Strang and G. J. Fix, *An Analysis of the Finite Element Method* (Englewood, Prentice Hall, 1973). <576, 590, 612>

[575] R. J. Strangeway, J. Luhmann and C. T. Russell, *Space Physics: An Introduction* (Cambridge, Cambridge University Press, 2016). <322>

[576] H. R. Strauss, 'Dynamics of high β tokamaks', *Phys. Fluids* **20** (1977), 1354–1360. <697>

[577] H. R. Strauss and W. S. Lawson, 'Computer-simulation of Alfvén resonance in a cylindrical, axially bounded flux tube', *Astrophys. J.* **346** (1989), 1035–1040. <428>

[578] T. E. Stringer, 'Low-frequency waves in an unbounded plasma', *Plasma Physics (J. Nucl. Energy Part C)* **5** (1963), 89–107. <86>

[579] P. A. Strittmatter, 'A note on gravitational instability in a magnetic medium', *Monthly Not. Roy. Astron. Soc.* **131** (1966), 491–498. <166>

[580] P. A. Sturrock, *Plasma Physics: An Introduction to the Theory of Astrophysical, Geophysical, and Laboratory Plasmas* (Cambridge, Cambridge University Press, 1994). <24, 64, 566>

[581] B. R. Suydam, 'Stability of a linear pinch', in *Proc. Second UN Intern. Conf. on Peaceful Uses of Atomic Energy* 31 (New York, Columbia Univ. Press, 1959), 157–159. <285, 348, 441, 502>

[582] D. G. Swanson, *Plasma Waves*, 2nd edition (Bristol, Institute of Physics Publishing, 2003). <46, 86>
[583] J. L. Synge, *The Relativistic Gas* (Amsterdam, North Holland, 1957). <885>
[584] J. L. Synge, *Relativity: the General Theory* (Amsterdam, North Holland, 1960). <895>
[585] T. Tajima, *Computational Plasma Physics, with Applications to Fusion and Astrophysics* (Redwood City, Addison-Wesley Publishing Company, 1989). <834>
[586] T. Tanaka, 'Finite volume TVD scheme on an unstructured grid system for three-dimensional MHD simulation of inhomogeneous systems including strong background potential fields', *J. Comp. Physics* **111** (1994), 381–389. <807, 814, 833>
[587] J. A. Tataronis, 'Energy absorption in the continuous spectrum of ideal MHD', *J. Plasma Physics* **13** (1975), 87–105. <373, 384>
[588] J. A. Tataronis and W. Grossmann, 'Decay of MHD waves by phase mixing', *Z. Physik* **261** (1973), 203–216. <269, 372, 376, 397, 399, 417, 432>
[589] R. J. Tayler, 'The influence of an axial magnetic field on the stability of a constricted gas discharge', *Proc. Roy. Soc. (London)* **B70** (1957), 1049–1063. <329>
[590] J. B. Taylor, 'Rotation and instability of plasma in fast Bz compression experiments', *Plasma Physics (J. Nucl. Energy Part C)* **4** (1962), 401–407. <506, 507>
[591] J. B. Taylor, 'Relaxation of toroidal plasma and generation of reversed magnetic fields', *Phys. Rev. Lett.* **33** (1974), 1139–1141. <124>
[592] T. S. Taylor et al., 'Wall stabilization of high beta plasmas in DIII-D', *Phys. Plasmas* **2** (1995), 2390–2396. <690>
[593] J. Terradas, R. Oliver and J. L. Ballester, 'Damped coronal loop oscillations: time-dependent results', *Astrophys. J.* **642** (2006), 533–540. <608>
[594] J. H. Thomas, L. E. Cram and A. H. Nye, 'Five-minute oscillations as a probe of a sunspot structure', *Nature* **297** (1982), 485–487. <428, 432>
[595] K. S. Thorne and R. D. Blandford, *Modern Classical Physics: Optics, Fluids, Plasmas, Elasticity, Relativity, and Statistical Physics* (Princeton, Princeton University Press, 2017). <879>
[596] A. Thyagaraja and K. G. McClements, 'Comment on "Variational principles for stationary one- and two-fluid equilibria of axisymmetric laboratory and astrophysical plasmas" [Phys. Plasmas 11, L81 (2004)]', *Phys. Plasmas* **12** (2005), 064701–2. <744>
[597] W. J. Tirry and D. Berghmans, 'Wave heating of coronal loops driven by azimuthally polarised footpoint motions, 2. The time-dependent behaviour in ideal MHD', *Astron. Astrophys.* **325** (1997), 329–340. <428>
[598] V. S. Titov and P. Démoulin, 'Basic topology of twisted magnetic configurations in solar flares', *Astron. Astrophys.* **351** (1999), 707–720. <819>
[599] S. M. Tobias and F. Cattaneo, 'Shear-driven dynamo waves at high magnetic Reynolds number', *Nature* **497** (2013), 463–465. <761>
[600] E. F. Toro, *Riemann Solvers and Numerical Methods for Fluid Dynamics* (Berlin, Springer-Verlag, 1997). <807, 834>
[601] M. Torrilhon, (a) 'Uniqueness conditions for Riemann problems of ideal magneto- hydrodynamics', *J. Plasma Physics* **69** (2003), 253–276; (b) 'Non-uniform convergence of finite volume schemes for Riemann problems of ideal magnetohydrodynamics', *J. Comp. Physics* **192** (2003), 73–94. <811, 870>
[602] G. Tóth, 'A general code for modeling MHD flows on parallel computers: Versatile Advection Code', *Astrophys. Lett. Commun.* **34** (1966), 245–250. <707, 819>
[603] G. Tóth, 'The $\nabla \cdot \mathbf{B} = 0$ constraint in shock-capturing magnetohydrodynamics codes', *J. Comp. Physics* **161** (2000), 605–652. <152, 690, 814, 819>
[604] G. Tóth, 'Conservative and orthogonal discretization for the Lorentz force', *J. Comp. Physics* **182** (2002), 346–354. <814, 819>
[605] G. Tóth, D. L. De Zeeuw, T. I. Gombosi and K. G. Powell, 'A parallel explicit/implicit time stepping scheme on block-adaptive grids', *J. Comp. Physics* **217** (2006), 722–758. <827>
[606] G. Tóth, R. Keppens and M. A. Botchev, 'Implicit and semi-implicit schemes in the Versatile Advection Code: numerical tests', *Astron. Astrophys.* **332** (1998), 1159–1170. <558, 819, 827>
[607] G. Tóth and D. Odstrčil, 'Comparison of some flux corrected transport and total variation diminishing numerical schemes for hydrodynamic and magnetohydrodynamic problems', *J. Comp. Physics* **128** (1996), 82–100. <802, 813, 814, 819>
[608] G. Tóth et al., 'Space weather modeling framework: a new tool for the space science community', *J.*

Geophys. Res. **110** (2005), A12226, 1–21. <317>

[609] L. N. Trefethen and M. Embree, *Spectra and Pseudospectra: The Behavior of Nonnormal Matrices and Operators* (Princeton, Princeton University Press, 2005). <551>

[610] F. Troyon, R. Gruber, H. Saurenmann, S. Semenzato and S. Succi, 'MHD limits to plasma confinement', *Plasma Phys. Control. F.* **26** (1984), 209–215. <690>

[611] B. A. Trubnikov, 'Particle interactions in a fully ionized plasma', in *Reviews of Plasma Physics, Vol. 1*, ed. M. A. Leontovich (New York, Consultants Bureau, 1965), 105–204. <38, 67, 101>

[612] A. D. Turnbull, M. S. Chu, M. S. Chance. J. M. Greene, L. L. Lao and E. J. Strait, 'Second-order toroidicity-induced Alfvén eigenmodes and mode splitting in a low-aspect-ratio tokamak', *Phys. Fluids* **B14** (1992), 3451–3453. <680>

[613] A. D. Turnbull, E. J. Strait, W. W. Heidbrink, M. S. Chu, H. H. Duong, J. M. Greene, L. L. Lao, T. S. Taylor and S. J. Thompson, 'Global Alfvén modes. Theory and experiment', *Phys. Fluids* **B5** (1993), 2546–2553. <417, 701>

[614] C. Uberoi, 'Alfvén waves in inhomogeneous magnetic fields', *Phys. Fluids* **15** (1972), 1673–1675. <269, 716>

[615] M. Ugai and T. Shimizu, 'Computer studies of noncoplanar slow and intermediate shocks associated with the sheared fast reconnection mechanism', *Phys. Plasmas* **1** (1994), 296–307. <870>

[616] W. Unno, Y. Osaki, H. Ando, H. Saio and H. Shibahashi, *Nonradial Oscillations of Stars*, 2nd edition (Tokyo, University of Tokyo Press, 1989). <239, 242>

[617] J. Vaclavik and K. Appert, 'Theory of plasma-heating by low-frequency waves – magnetic pumping and Alfvén resonance heating', *Nucl. Fusion* **31** (1991), 1945–1997. <432>

[618] B. van der Holst, A. J. C. Beliën and J. P. Goedbloed, (a) 'New Alfvén continuum gaps and global modes induced by toroidal flow', *Phys. Rev. Lett.* **84** (2000), 2865–2868; (b) 'Low frequency Alfvén waves induced by toroidal flows', *Phys. Plasmas* **7** (2000), 4208–4222. <660, 680, 701, 722, 725>

[619] B. van der Holst, C. Jacobs and S. Poedts, 'Simulation of a breakout coronal mass ejection in the solar wind', *Astrophys. J. (Lett.)* **671** (2007), L77–L80. <314>

[620] B. van der Holst, R. J. Nijboer and J. P. Goedbloed, 'Magnetohydrodynamic spectrum of gravitating plane plasmas with flow', *J. Plasma Physics* **61** (1999), 221–240. <473, 480, 488, 494>

[621] B. van der Holst and R. Keppens, 'Hybrid block-AMR in cartesian and curvilinear coordinates: MHD applications', *J. Comp. Physics* **226** (2007), 925–946. <814>

[622] B. van der Holst, R. Keppens and Z. Meliani, 'A multidimensional grid-adaptive relativistic magnetofluid code', *Comput. Physics Commun.* **179** (2008), 617–627. <908, 910>

[623] (a) R. A. M. Van der Linden and M. Goossens, 'The thermal continuum in coronal loops: instability criteria and the influence of perpendicular thermal conduction', *Solar Phys.* **134** (1991), 247–273; (b) R. A. M. Van der Linden, M. Goossens and J. P. Goedbloed, 'On the existence of a thermal continuum in non-adiabatic magnetohydrodynamic spectra', *Phys. Fluids* **B3** (1991), 866–868. <594>

[624] R. A. M. Van der Linden, M. Goossens and A. W. Hood, 'The relevance of the ballooning approximation for magnetic, thermal, and coalesced magnetothermal instabilities', *Solar Phys.* **140** (1992), 317–342. <594>

[625] R. A. M. Van der Linden, M. Goossens and W. Kerner, 'A combined finite element/Fourier series method for the numerical study of the stability of line-tied magnetic plasmas', *Comput. Physics Commun.* **59** (1990), 61–73. <594>

[626] H. A. van der Vorst, *Iterative Krylov Methods for Large Linear Systems* (Cambridge, Cambridge University Press, 2003). <612>

[627] T. Van Doorsselare and S. Poedts, 'Modifications to the resistive MHD spectrum due to changes in the equilibrium', *Plasma Phys. Control. F.* **49** (2007), 261–271. <551>

[628] N. G. van Kampen, 'On the theory of stationary waves in plasmas', *Physica* **21** (1955), 949–963. <46, 269, 376>

[629] N. G. van Kampen and B. U. Felderhof, *Theoretical Methods in Plasma Physics* (Amsterdam, North-Holland Publishing Company, 1967). <46, 67>

[630] M. H. P. M. van Putten, 'A numerical implementation of MHD in divergence form', *J. Comp. Physics* **105** (1993), 339–353. <908>

[631] E. P. Velikhov, 'Stability of an ideally conducting liquid flowing between cylinders rotating in a magnetic field', *Soviet Phys.–JETP Lett.* **36**, (1959) 995–998. <441, 513, 523>

[632] M. Velli, G. Einaudi and A. W. Hood, 'Ideal kink instabilities in line-tied coronal loops: growth rates

and geometrical properties', *Astrophys. J.* **350** (1990), 428–436. <594>

[633] D. Verscharen, Ch. H. K. Chen and R. T. Wicks, 'On kinetic slow modes, fluid slow modes, and pressure-balanced structures in the solar wind', *Astrophys. J.* **840** (2017), 106, 1–8. <309>

[634] A. A. Vlasov, (a) 'The oscillation properties of an electron gas', *Zhur. Eksp. Teor. Fiz.* **8** (1938), 291–318; (b) 'Vibrational poperties, crystal structure, non-dissipated counterdirected currents and spontaneous origin of these properties in a "gas" ', *J. Phys. USSR* **9** (1945), 25–40. <38, 46>

[635] A. Vögler, S. Shelyag, M. Schüssler, F. Cattaneo, T. Emonet and T. Linde, 'Simulations of magneto-convection in the solar photosphere. Equations, methods, and results of the MURaM code', *Astron. Astrophys.* **429** (2005), 335–351. <822>

[636] J. Von Neumann, *Mathematical Foundations of Quantum Mechanics* (Princeton, Princeton University Press, 1955). <196>

[637] D. Voslamber and D. K. Callebaut, 'Stability of force-free magnetic fields', *Phys. Rev.* **128** (1962), 2016–2021. <357, 369, 525>

[638] F. Wagner et al., 'Regime of improved confinement and high beta in neutral-beam heated divertor discharges of the ASDEX tokamak', *Phys. Rev. Lett.* **49** (1982), 1408–1412. <697>

[639] A. A. Ware, 'Role of compressibility in the magnetohydrodynamic stability of the diffuse pinch discharge', *Phys. Rev. Lett.* **12** (1964), 439–441. <353, 441>

[640] E. J. Weber and L. Davis Jr., 'The angular momentum of the solar wind', *Astrophys. J.* **148** (1967), 217–227. <709>

[641] S. Wedemeyer-Böhm, I. Kamp, J. Bruls and B. Freytag, 'Carbon monoxide in the solar atmosphere. I. Numerical method and two-dimensional models', *Astron. Astrophys.* **438** (2005), 1043–1057. <819>

[642] N. O. Weiss, 'The expulsion of magnetic flux by eddies', *Proc. Roy. Soc. London* **A 293** (1966), 310–328. <760, 778>

[643] J. F. Wendt (ed.), *Computational Fluid Dynamics* (Berlin, Springer-Verlag, 1992). <599>

[644] N. Werner, A. Finoguenov, J. S. Kaastra, A. Simionescu, J. P. Dietrich, J. Vink and H. Böhringer, 'Detection of hot gas in the filament connecting the clusters of galaxies Abell 222 and Abel 223', *Astron. Astrophys. Lett.* **482** (2008), L29–L33. <437>

[645] P. Wesseling, *Principles of Computational Fluid Dynamics* (Berlin, Springer-Verlag, 2000). <834>

[646] J. A. Wesson, 'Magnetohydrodynamic stability of tokamaks', *Nucl. Fusion* **18** (1978), 87–132. <361, 369, 624>

[647] J. Wesson, *Tokamaks*, 3rd edition (Oxford, Clarendon Press, 2004). <6, 24, 369, 617, 624, 627, 637, 664, 690, 695, 704, 716>

[648] V. Wheatley, D. I. Pullin and R. Samtaney, 'Regular shock refraction at an oblique planar density interface in magnetohydrodynamics', *J. Fluid Mech.* **522** (2005), 179–214. <854, 870>

[649] R. B. White, *The Theory of Toroidally Confined Plasmas*, 3rd edition (London, Imperial College Press, 2014). <xix, 24, 566, 704>

[650] P. R. Wilson, 'Free and forced oscillations of a flux tube', *Astrophys. J.* **251** (1981), 756–767. <397>

[651] J. R. Wilson and G. J. Mathews, *Relativistic Numerical Hydrodynamics*, (Cambridge, Cambridge University Press, 2003). <908>

[652] N. Winsor, J. L. Johnson and J. M. Dawson, 'Geodesic acoustic waves in hydro-magnetic systems', *Phys. Fluids* **11** (1968), 2448–2450. <668, 677>

[653] L. Woltjer, 'A theorem on force-free magnetic fields', *Proc. Natl. Acad. Sci. USA* **44** (1958), 489–491. <124>

[654] L. Woltjer, 'Hydromagnetic equilibrium. IV. Axisymmetric compressible media', *Astrophys. J.* **130** (1959), 405–413. <709>

[655] B. E. Wood, 'Astrospheres and solar-like stellar winds', *Living Rev. Sol. Phys.* **1** (2004), 1–52. <309>

[656] C. C. Wu, (a) 'The MHD intermediate shock interaction with an intermediate wave: are intermediate shocks physical?', *J. Geophys. Res.* **93** (1988), 987–990; (b) 'Formation, structure, and stability of MHD intermediate shocks', *J. Geophys. Res.* **95** (1990), 8149–8175; (c) 'New theory of MHD shock waves', in *Viscous Profiles and Numerical Methods for shock Waves*, Chapter 17, SIAM Proceedings Series, ed. M. Shearer (Philadelphia, SIAM, 1991), 209–236. <838, 870>

[657] C. C. Wu and T. Hada, 'Formations of intermediate shock in both two-fluid and hybrid models', *J. Geophys. Res.* **96** (1991), 3769–3778. <870>

[658] C. C. Wu and C. F. Kennel, 'Structure and evolution of time-dependent intermediate shocks', *Phys. Rev. Lett.* **68** (1992), 56–59. <870>

[659] C. Xia, R. Keppens and Y. Guo, 'Three-dimensional prominence-hosting magnetic configurations: creating a helical magnetic flux rope', *Astrophys. J.* **780** (2014), 130–141. <314>

[660] C. Xia and R. Keppens, 'Formation and plasma circulation of solar prominences', *Astrophys. J.* **823** (2016), 22-30. <819>

[661] M. Yokokawa, K. I. Itakura, A. Uno, T. Ishihara and Y. Kaneda, '16.4 Tflops direct numerical simulation of turbulence by a Fourier spectral method on the Earth Simulator', *Proc. ACM/IEEE Conf. on Supercomputing, 2002* (2006) 50–67. <756, 768>

[662] A. Yoshizawa, S.-I. Itoh and K. Itoh, *Plasma and Fluid Turbulence, Theory and Modelling* (Bristol, Institute of Physics, 2003). <xix, 66>

[663] A. L. Zachary and P. Colella, 'A higher-order Godunov method for the equations of ideal magnetohydrodynamics', *J. Comp. Physics* **99** (1992), 341–347. <807>

[664] L. E. Zakharov and V. D. Shafranov, 'Equilibrium of current-carrying plasmas in toroidal configurations', in *Reviews of Plasma Physics, Vol. 11*, ed. M. A. Leontovich (New York, Consultants Bureau, 1982), 153–302. <642, 664>

[665] H. P. Zehrfeld and B. J. Green, 'Stationary toroidal equilibria at finite beta', *Nucl. Fusion* **12** (1972), 569–575. <715, 744>

[666] M. Zeilik and E. v. P. Smith, *Introductory Astronomy and Astrophysics* (Philadelphia, Saunders College Publishing, 1987). <13>

[667] R. Zelazny, R. Stankiewicz, A. Galkowski and S. Potempski, 'Solutions to the flow equilibrium problem in elliptic regions', *Plasma Phys. Control. F.* **35** (1993), 1215–1227. <716>

[668] U. Ziegler, 'Self-gravitational adaptive mesh magnetohydrodynamics with the NIRVANA code', *Astron. Astrophys.* **435** (2005), 385–395. <819>

[669] E. R. Zirin, *Astrophysics of the Sun* (Cambridge, Cambridge University Press, 1988). <234, 292>

[670] F. P. Zuccarello, C. Jacobs, A. Soenen, S. Poedts, B. van der Holst and F. Zuccarello, 'On modelling the initiation of Coronal Mass Ejections: magnetic flux emergence versus shearing motions', *Astron. Astrophys.* **507** (2009), 441–452. <314>

Index

* Note: *italic page numbers* indicate a main section on the subject.